GENETIC Engineering

SMITA RASTOGI
Department of Biotechnology
Delhi Technological University
Delhi

NEELAM PATHAK
Department of Biotechnology
Integral University
Lucknow

OXFORD
UNIVERSITY PRESS

OXFORD
UNIVERSITY PRESS

Oxford University Press is a department of the University of Oxford. It furthers the University's objective of excellence in research, scholarship, and education by publishing worldwide. Oxford is a registered trademark of Oxford University Press in the UK and in certain other countries

Published in India by
Oxford University Press
YMCA Library Building, 1 Jai Singh Road, New Delhi 110001, India

First Edition published in 2009
Fifth impression 2013

ISBN-13: 978-0-19-569657-8
ISBN-10: 0-19-569657-3

Typeset in Times
by Anvi Composers, New Delhi 110063
Printed in India by Ram Printograph, Delhi 110051

Dedicated
to
our family members,
teachers,
and
students

Preface

Since its inception in the 1970s, genetic engineering has been the mainstay of many revolutionary developments in several areas of biology. It has contributed to the development of parallel fields such as biotechnology, bioinformatics, proteomics, genomics, etc. Major breakthroughs in biosciences like the identification of DNA being the hereditary material, discovery of the DNA double helix, and discovery of the recombinant DNA technology and restriction enzymes have led to the establishment of genetic engineering as a separate branch of modern biology.

As the name indicates, genetic engineering is a branch of science that deals with engineering or manipulating the genetic constitution of organisms. The genetic endowment of organisms can now be changed precisely in desired ways. Rapid developments in gene cloning methodologies fuelled the research in genetic engineering and made it a success. It is essentially the insertion of foreign DNA into a cell, through a suitable vector, in such a way that the inserted DNA replicates independently and is transferred to progenies during cell division. After proper selection and screening, the transformed cells containing the DNA can be used commercially for the production of useful compounds related to areas such as agriculture, medicine, environment, public health, and forensics. All these have direct applications to find suitable solutions to the major problems faced by mankind, such as gap in demand and supply of food, environmental pollution, and discovery of drugs and vaccines for the treatment of fatal diseases and syndromes.

About the Book

Designed primarily for students of biotechnology, *Genetic Engineering* will also serve as a ready reference for the students of life sciences, genetics, microbiology, biochemistry and other allied areas of biology. Genetic engineering is a prerequisite for specialization in advanced courses like genomics, proteomics, immunology, etc. Hence, the book will also be useful for students and research scholars who undertake research work in the areas of modern biology.

The book presents an enormous amount of information, collected from decades of research. Written in a lucid style, it will help students to gain a sound understanding of the underlying principles and potential applications of genetic engineering. It is the outcome of the authors' teaching and research experience, and knowledge of the subject. It provides an in-depth coverage of all major topics in genetic engineering, and includes numerous illustrations to supplement the text. Some pertinent interesting facts and anecdotes are also presented as exhibits in the chapters. A glossary at the end of the book provides a ready reference to some important terminology.

The additional key features of the text include the following:

- Discusses both the basic and applied aspects of genetic engineering
- Gives an overview of recombinant DNA biosafety guidelines and National Regulatory Mechanism for their implementation
- Provides elaborate tables for a quick view and easy grasp of the subject
- Includes numerous review questions for practice

Content and Structure

The entire content of the book is divided into four parts comprising a total of 21 chapters.

Part I (Fundamentals of Genetic Engineering) introduces the various steps of gene cloning experiments, and the enzymes involved in the process. Further, the strategies used in the first step of gene cloning, i.e., for obtaining DNA fragments are discussed in detail. Chapter 1 provides an overview of the steps used in gene cloning experiments, an elaborative account of DNA and RNA isolation procedures, and the properties of ideal cloning vectors. Chapter 2 deals with the properties and applications of various enzymes used in genetic engineering. Chapters 3 to 7 highlight the processes or techniques used to isolate gene fragments for cloning. In Chapter 3, types of restriction enzymes, their *in vivo* functions, nomenclature, structure, target sites, specificity of action, cognate methylases, and applications are discussed in detail. Chapter 4 talks about various methods used for solid-phase chemical synthesis of DNA. Besides, the chapter also presents various DNA chip fabrication techniques. In Chapter 5, various methods for cDNA synthesis are emphasized. mRNA purification procedures are also presented in the chapter.

Chapter 6 deals with the techniques used for visualization and fractionation of DNA fragments. In addition, details of techniques such as conventional agarose gel electrophoresis, pulsed field gel electrophoresis, polyacrylamide gel electrophoresis (PAGE), gel filtration chromatography, and density gradient centrifugation are described. In Chapter 7, the principle and working of polymerase chain reaction (PCR) are discussed. Further, variants of PCR and their applications in genetic engineering are also described.

Part II (Vectors used in Gene Cloning) discusses the vectors used in gene cloning. These are covered in Chapters 8 to 12. Chapter 8 discusses the biology of plasmids and their application as gene cloning vehicles. The vector maps of some commonly used plasmids and cloning strategies are also presented. Chapter 9 covers the biology of λ bacteriophage, types of vectors based on them, cloning strategies, and selection criteria used with some common λ phage-based vectors. Chapter 10 provides a detailed description on the biology of M13 bacteriophage, vectors based on wild-type phage, and their significance in DNA sequencing and site-directed mutagenesis. In Chapter 11, the relevance of yeast system for cloning of eukaryotic genes is discussed. Considerable emphasis is given on the significance of yeast artificial chromosome in genomic library construction. Chapter 12 deals with various chimeric vectors, *Agrobacterium* tumor-inducing plasmid-based binary and cointegrate vectors, viral vectors, artificial chromosomes, and vectors used for cloning in bacteria other than *E. coli*. Further, the significance of shuttle vectors, expression vectors, fusion vectors, special purpose vectors, and advanced gene tagging vectors in genetic engineering is also described.

Part III (Generation and Screening of Recombinants) is devoted to the second and third steps of gene cloning, *viz.* joining of insert DNA with vector, and introduction of recombinants into the host cells. Chapters 13 to 16 are included in this part. In Chapter 13, the roles of DNA ligase, linkers, adaptors, and homopolymer tailing in joining of DNA fragments are described. Chapter 14 gives an elaborate account of the methods used for introduction of recombinant DNA into a variety of host cells. Chapter 15 summarizes the advantages and procedures for constructing genomic and cDNA libraries. Chapter 16 gives an overview of nucleic acid or protein-based techniques used for screening of recombinants.

Part IV (Applied Genetic Engineering) comprises of Chapters 17 to 21. Chapter 17 deals with the procedures and applications of techniques, such as mutagenesis, recombination, antisense RNA technology, RNA interference, and cosuppression in gene manipulation. Chapter 18 highlights various types of molecular markers and their roles in map-based cloning and marker-assisted selection. Chapter 19 deals with the principles, functioning, and applications of various techniques used in genomics and proteomics. These include DNA and protein sequencing, DNA and protein microarray technology, matrix assisted laser desorption ionization-time of flight, 2-D electrophoresis, etc. Chapter 20 gives an elaborative account of applications of gene manipulation studies in both animals and plants. In Chapter 21, the recombinant DNA biosafety guidelines and National Regulatory Mechanism for their implementation are discussed.

A detailed glossary of important terminology is also given at the end of the book.

Acknowledgements

We wish to express our thanks to all who have helped us directly or indirectly in writing this book. We would like to extend our sincere gratitude to our revered Vice Chancellor, Prof. S. W. Akhtar and Pro-Vice Chancellor, Prof. S. M. Iqbal for their constant encouragement. Our sincere thanks are also due for our colleagues at the Department of Biotechnology for their kind cooperation.

We are also indebted to Prof. U. N. Dwivedi (Pro-Vice Chancellor and Head, Department of Biochemistry, Lucknow University, Lucknow) and Dr P. Nath (Scientist G, Plant Molecular Biology Division, NBRI, Lucknow) for showing confidence in our writing and technical skills and encouraging us to write this book. Prof. G. G. Sanwal (Department of Biochemistry, Lucknow University, Lucknow) and Prof. Theophenous Solomos (Department of NRLS, Plant Sciences, University of Maryland, USA) are also greatly acknowledged. We extend a special note of appreciation to our students at Integral University, Lucknow, who have been a constant source of motivation for us.

We thank each other for being unsparing critics of each other's work imparting tough comments accompanied with gentle inspirational support and ideas. We would also like to extend our sincere appreciation and thanks to the editorial team of Oxford University Press for their active cooperation, suggestions, and patience.

And finally, we would like to express our sincere gratitude to our parents and other family members for their unfailing emotional support, care, and cooperation during the process of writing this book.

We hope the reader finds the book both enjoyable and useful. All comments or suggestions for improvement in future editions would be gratefully received.

Smita Rastogi
Neelam Pathak

Contents

PART II : VECTORS USED IN GENE CLONING

PART III : GENERATION AND SCREENING OF RECOMBINANTS

PART IV: APPLIED GENETIC ENGINEERING

List of Abbreviations

λ phage *Bacteriophage lambda*
θ replication *Bidirectional replication*
σ replication *Sigma replication*
β-CE *β-cyanoethyl*
β-ME *β-mercaptoethanol*
γ-TMT *γ-Tocopherol methyltransferase*
2,5-DKG *2,5-Diketo-D-gluconic acid*
2DGE *Two-dimensional gel electrophoresis*
2-KLG *2-Keto-L-gulonic acid*
3-NBA *3-Nitrobenzyl alcohol*
5-FOA *5-Fluoro-orotic acid*
7ACA *7-Aminocephalosporanic acid*
A *Adenine*
A_{260} *Absorbance at 260 nm wavelength*
A_{280} *Absorbance at 280 nm wavelength*
AADP *ATP/ADP-transporter protein*
AAV *Adeno-associated virus*
ABA *Abscissic acid*
ABI *Applied Biosystems Instruments*
Ac/Ds *Activator/Dissociation*
ACC *1-Amino cyclopropane-1-carboxylic acid*
ACMV *African cassava mosaic virus*
***Ac*NPV** *Autographia californica nuclear polyhedrosis virus*
***Ac*MNPV** *Autographia californica multicapsid nucleopolyhedro virus*
AD *Activation domain*
ADP *Adenosine diphosphate*
AFLP *Amplified fragment length polymorphism*
AFP *Antifreeze protein*
AGPase *ADP-glucose pyrophosphorylase*
AHC *Acid hydrolyzable casein*
AIDS *Acquired immunodeficiency syndrome*
AIMV *Alfalfa mosaic alfamo virus*
AK *Aspartokinase*
AMP *Adenosine monophosphate*
AMP-PCR *Anchored microsatellite directed PCR*
AMPPD *Adamantly-1,2-dioxetane phosphate*
amp^r or ap^r *Ampicillin resistance gene*
amp^s or ap^s *Sensitivity to ampicillin*
AMV *Avian myeloblastosis virus*
AMV-RTase *Avian myeloblastosis virus reverse transcriptase*
AP site *Apurinic/apyrimidimic site*
AP-PCR *Allele specific PCR*
AP-PCR *Arbitrarily primed PCR*
APS *Adenosine phosphosulphate Ammonium persulphate*
ARS *Autonomously replicating sequence*
ASWY *Average step-wise yield*
Asx *Aspartate + asparagine*
ATP *Adenosine triphosphate*
ATPase *Adenosine triphosphatase*
***att* B** *Attachment site of bacteria*
***att* P** *Attachment site of phage*
ATZ *Anilinothiazolinone*
AUAP *Abridged universal amplification primer*
AuxREs *Auxin-response elements*
Avr *Avirulence factors*

BAC *Bacterial artificial chromosome*
BAP *Bacterial alkaline phosphatase*
BC *Biological containment*
BCIP *5-Bromo-4-chloro-3-indoyl phosphate*
BIBAC *Binary bacterial artificial chromosome*
BIRD *Blackbody infrared radiative dissociation*
Bis Tris-Cl 2-[*Bis*(2-*hydroxyethyl*)*iminotris*]-2-(*hydroxymethyl*)-1,3-*propanediol hydrochloride*
bla *β-Lactamase gene*
BLAST *Basic local alignment search tool*
***Bm*NPV** *Bombyx mori nuclear polyhedrosis virus*
BMV *Brome mosaic virus*
bom *Basis of mobility*
box *Conserved sequence*
bp *Base pair*
BPB *Bromophenol blue*
BSA *Bovine serum albumin*
BSA *Bulk segregant analysis*
bST *Bovine somatotropin*
Bt toxin *Bacillus thuringiensis toxin*
BV *Budded virus*

C *Cytosine*
cad *Cinnamyl aldehyde dehydrogenase*
CAD *Collisionally-activated dissociation*
CAM plasmid *Camphor degrading plasmid*
cAMP *Cyclic AMP; Cyclic adenosine monophosphate*
CaMV *Cauliflower mosaic virus*
CAP *Catabolite activating protein*
CAPS *Cleaved amplified polymorphic sequence*
CAT *Chloramphenicol acetyltransferase*
CBF *CRT/DRE element binding factor*
CCC DNA *Covalently closed circular DNA*
CCD camera *Charge-coupled device camera*
ccoaomt *Cinnamoyl CoA O-methyltransferase*
ccr *Cinnamoyl CoA reductase*
cDNA *Complementary DNA; copy DNA*
CDP *Cytidine diphosphate*
CE *Capillary electrophoresis*
CEN *Centromeric sequence*
CFGE *Crossed field gel electrophoresis*
cfu *Colony forming unit*
CGH array *Comparative genomic hybridization array*
CHEF *Contour clamped homogeneous electric fields*
***chi* site (χ site)** *Cross-over hotspot instigator*
CHO *Chinese hamster ovary*
CHS *Chalcone synthase*
chv *Chromosomal virulence genes*
CID *Collision induced dissociation*
CIP or CIAP *Calf intestinal phosphatase*
***cIts* or *cIts* 857** *Temperature sensitive mutation in λ or cI repressor*
CMP *Cytidine monophosphate*
cm^r *Chloramphenicol resistance gene*
CMV *Cytomegalovirus*
CNBr *Cyanogen bromide*
CNR *Cell nuclear replacement*
CoA *Coenzyme A*
COR *Cold responsive*
***cos* site** *Cohesive site*
CP *Coat protein*
CpDNA *Chloroplast DNA*
CPG *Controlled pore glass*
CpTI *Cowpea trypsin inhibitor*
Cre *Cyclic recombinase*
Cro *Control of repressor and other things*
Cry protein *Crystal protein (Bt toxin)*
CSO *Civil society organizations*
CsTFA *Cesium trifluoroacetate*
CTAB *Cetyl trimethyl ammonium bromide*
C-terminal *Carboxy terminal*
CTP *Cytidine triphosphate*
Cub *C-terminal fragment of ubiquitin*

dA *Deoxyadenosine*
DAF *DNA amplification fingerprinting*
Dam *Amber mutation in D gene*
Dam *DNA adenosine methylase or DNA adenosine methyltransferase*
dATP *Deoxyadenosine triphosphate*
DBD *DNA binding domain*
dC *Deoxycytidine*
DCA *Dichloroacetic acid*
DCC *Dicyclohexylcarbodiimide*
DCE *Dichloroethane*
DCGI *Drug Controller General of India*
DCM *Dichloromethane*
Dcm *DNA cytosine methylase or DNA cytosine methyltransferase*
dCTP *Deoxycytidine triphosphate*
DDRT-PCR *Differential display reverse transcriptase PCR*
DEAE *Diethyl aminoethyl*
DEPC *Diethylpyrocarbonate*
DET *Dithioerythritol; Cleland's reagent*
DFR *Dihydroflavonol reductase*
dG *Deoxyguanosine*
DGGE *Denaturing gradient gel electrophoresis*
DGMT *DNA-mediated gene transfer*
dGTP *Deoxyguanosine triphosphate*
DHDPS *Dihydrodipicolinic acid synthase*
DHFR *Dihydrofolate reductase*
dITP *Deoxy inosine triphosphate*
DLC *District Level Committee*
DMD *Digital micromirrors device*
DMS *Dimethyl sulfate*
DMSO *Dimethyl sulfoxide*
DMTr *Dimethoxytrityl*
DNA MTase *DNA methyltransferase or methylase*
DNA *Deoxyribonucleic acid*
DNase *Deoxyribonuclease*
dNMP *Deoxynucleoside monophosphate*
dNTP *Deoxynucleoside triphosphate; deoxyribonucleotide*
ds DNA *Double stranded DNA*
ds RNA *Double stranded RNA*
dT *Deoxythymidine*
DTAB *Dodecytrimethylammonium bromide*
DTT *Dithiothreitol*
dTTP *Deoxythymidine triphosphate*
dUTP *Deoxyuridine triphosphate*
dUTPase (DUT) *Deoxyuridine triphosphatase*

Eam *Amber mutation in E gene*
EBV *Epstein–Barr virus*
ECD *Electron capture dissociation*
EDTA *Ethylene diamine tetraacetic acid*
EEO *Electroendosmosis*
EGFP *Enhanced green fluorescent protein*
eIF-4E *Eukaryotic initiation factor 4E*
ELS *Evaporative light scattering*
EMBL *European Molecular Biology Laboratory*
EMSA *Electrophoretic mobility shift assay; Mobility shift or Band shift assay*
EPA *Environment Protection Act*
EPSPS *5-enol pyruvyl shikimate-3-phosphate synthase*
E-RCAT *Exponential rolling circle amplification technology*
Ert *Estrogen receptor*
ES cell *Embryonic stem cell*
ESI *Electrospray ionization*
ESTs *Expressed sequence tags*
ET *Energy transfer*
EtBr *Ethidium bromide*
ETD *Electron transfer dissociation*
EtOH *Ethanol*

F plasmid *Fertility plasmid*
F primer *Forward primer; Downstream primer*
f5h *Ferulate 5-hydroxylase*
FAB *Fast atom bombardment*
FAM *6-carboxyfluorescein*
FDA *Food and Drug Administration*
FDNB *1-Fluoro-2,4-dinitrobenzene; Sanger's reagent*
FICU 1-(2-deoxy-2-fluoro-β-*D*-*arabinofur-ano-syl*)-5-*iodouracil*
FIGE *Field inversion gel electrophoresis*
fin *Fertility inhibition*
Flp *Flp recombinase; Flippase*
FM primer *Forward mutagenic primer*
FRET *Fluorescence (or Förster) resonance energy transfer*
frt **site** *Flp (flip) recombination target site*
FSB *Frozen storage buffer*
FT-ICR *Fourier transform ion cyclotron resonance*

G *Guanine*
GAPDH *Glyceraldehyde-3-phosphate dehydrogenase*
GDEPT *Gene directed enzyme prodrug therapy*
GDP *Guanosine diphosphate*
GEAC *Genetic Engineering Approval Committee*
GEO *Genetically engineered organism*
GFP *Green fluorescent protein*
GGDP *Geranyl geranyl diphosphate*
GH *Growth hormone*
GL-7ACA 7-β-(4-*carboxy-butanamido*) *cephalosporanic acid*
Glc 1-P *Glucose 1-phosphate*
GLP *Good laboratory practice*
GLSP *Good large-scale practice*
Glx *Glutamate + Glutamine*
GM *Genetically modified*
GMO *Genetically modified organism*
GMP *Good manufacturing practice*
GMP *Guanosine monophosphate*
GPD **1** *Glycerol-3-phosphate dehydrogenase*
GRAS *Generally recognized as safe*
GSH *Glutathione (reduced)*
GSS syndrome *Gerstmann–Straussker–Scheinker syndrome*
GSSG *Glutathione (oxidized)*
GST *Glutathione S-transferase*
GSV *GFP selection vector*
GTP *Guanosine triphosphate*
gus *β-glucuronidase gene*
GV *Gemini virus*

HAC *Human artificial chromosome*
HAD *Helicase-dependent amplification*
HART *Hybrid arrest translation*
HAT medium *Hypoxanthine, aminopterin, and thymidine medium*
HBsAG *Hepatitis B surface antigen*
HDGS *Homology-dependent gene silencing*
HEPA *High efficiency particulate air*
HEPES *N-(2-hydroxyethyl)piperazine-N′(2-ethanesulfonic acid)*
hES *Human embryonic stem cell lines*
hfl *High frequency of lysogenization*
HFV *Human foamy virus*
hGH *Human growth hormone*
HGP *Human genome project*
HIV *Human immunodeficiency virus*
HNPP *2-Hydroxy-3-naphthoic acid 2′-phenylanillide phosphate*
HPLC *High performance liquid chromatography*
HPT *Homogentisic acid prenyltransferase*
HR *Hypersensitive response*
HRC *Herbicide resistant crops*
HPRT *Hypoxanthine-guanine phosphoribosyl transferase*
HRT *Hybrid release translation*
HSCT *Hematopoietic stem cell transplantation*
HSE *Heat-shock element*
HSF *Heat-shock transcription factor*
Hsp *Heat-shock protein*
hSS *Human somatic stem cells*
HSV *Herpes simplex virus*
HU *Histone-like proteins*

IAEC *Institutional Animal Ethics Committee*
IBSC *Institutional Biosafety Committee*
ICAR *Indian Council of Agricultural Research*
ice **site** *Inceptor site*
ICMR *Indian Council of Medical Research*
IC-SCRT *Institutional Committee for Stem Cell Research and Therapy*
IE promoter *Immediate early promoter*

IEC *Institutional Ethics Committee*
IEF *Isoelectric focusing*
IG region *Intergenic region*
IGF-I *Insulin-like growth factor I*
IgG *Immunoglobin G*
IGP dehydratase *Imidazole glycerol phosphate dehydratase*
IHF *Integration host factor*
IHGSC *International Human Genome Sequencing Consortium*
imm *Immunity gene*
IMP *Inosine monophosphate*
IMTECH *Institute of Microbial Technology, Chandigarh*
Inc *Incompatibility*
IOD *Integrated optical density*
IPG *Immobilized pH gradient*
IPM *Insect pest management*
IPTG *Isopropyl thiogalactoside*
IRE-PCR *Interspersed repeat element PCR*
IRESs *Internal ribosome entry sites*
IRMPD *Infrared multiphoton dissociation*
IS *Insertion sequence*
IS-PCR *In situ PCR*
ISA *Inter simple sequence repeats amplification*
ITRs *Inverted terminal repeats*
IVF *In vitro fertilization*
IVR *Inverted repeats*

keto-AD-7ACA 7-β-(*5-carboxy-5-oxopentanamido) cephalosporanic acid*
kn[r] **or *kan***[r] *Kanamycin resistance gene*

L arm *Left arm*
Lac Z β-*Galactosidase*
LAMP *Loop-mediated isothermal amplification*
LA-PCR *Ligation-anchored PCR*
LATE-PCR *Linear-after-the-exponential PCR*
LATs *Latency-associated transcripts*
LB medium *Luria Bertani medium*
LB *Left border*
LC *Liquid chromatography*
LC$_{50}$ *Lethal concentration* 50
LD$_{50}$ *Median lethal dose*
LEA *Late embryogenesis abundant*
***Le*GPAT** *Lycopersicon esculentum glycerol-3-phosphate acyltransferase*
LINEs *Long interspersed nuclear elements*
***lip* A** *Lipase gene*
LMP *Low melting point*
LMT *Low melting temperature*
LNA *Locked nucleic acid*
***lox* P** *Locus of X-over P*1
LSIMS *Liquid secondary ion mass spectrometry*
LTR *Long terminal repeat*

M in M13 *Munich*
M primer *Mutagenic primer*
MAC *Mammalian artificial chromosome*
MAHYCO *Maharashtra Hybrid Seed Company*
MALDI *Matrix-assisted laser desorption/ionization*
MAS *Marker assisted selection/Maskless array synthesizer*
***MAT*α** α-*mating factor* (*yeast*)
MBP *Myelin basic protein*
McAb *Monoclonal antibody*
MCS *Multiple cloning site; polylinker*
MDA *Multiple displacement amplification*
MDLC *Multi-dimensional liquid chromatography*
Me *Methyl*
MeNPOC α-*Methyl-6-nitropiperonyloxycarbonyl*
MEP pathway *Methylerythritol phosphate pathway*
MES 2-(*N-morpholino*) *ethanesulfonic acid*
MIC$_{50}$ *Median molt inhibitor concentration*
miRNA *Micro RNA*
MLT *Major late transcript*
MLVs *Murine viruses*
MMTr *Monomethoxytrityl*
MMTV *Mouse mammary tumor virus*
MMTVLTR *Mouse mammary tumor virus long terminal repeats*
MNCs *Multinational companies*
mob *Mobilization locus*
MoEF *Ministry of Environment and Forest*
Mo-MLV *Moloney murine leukemia virus*
MOPS 3-(*N-morpholino*) *propane sulphonic acid*
MOXp *Methanol oxidase promoter*
MOXt *Methanol oxidase transcription terminator*
mp *Max Planck*
MPBQ 2-*Methyl-6-phytylbenzoquinol*
MP-PCR *Microsatellite primed PCR*
Mr *Relative molecular weight*
MRC *Medical Research Council*
mRNA *Messenger RNA*
mRP *Mitochondrial retroplasmid*
MS *Mass spectrometry*
MSV *Maize streak virus*
MTase *Methyltransferase*
mtDNA *Mitochondrial DNA*
MUG 4-*methyl umbelliferyl*-β-*D-galactopyranoside*
MUG 4-*methyl umbelliferyl*-β-*D-glucuronide*

N *Any base* (*G/C/A/T*)
NAAT *Nicotianamine aminotransferase*
NAC-SCRT *National Apex Committee for Stem Cell Research and Therapy*
NAD *Nicotinamide adenine dinucleotide*
NAG *N-acetyl glucosamine*
NAH plasmid *Naphthalene degrading plasmid*
NAM *N-acetyl muramic acid*
NBPGR *National Bureau of Plant Genetic Resources*
NBT *Nitroblue tetrazolium*
NDP *Nucleoside diphosphate*
NGOs *Non governmental organizations*
NHEJ *Non-homologous end joining*
NHGRI *National Human Genome Research Institute*
***nic* site** *Nick site*
Nif *Nitrogen fixation proteins*
NIH *National Institutes of Health*
NIL *Nearly isogenic line*
nin N-*independent*
NLS *Nuclear localization signal*
NMN *Nicotinamide adenine mononucleotide*
NMP *Nucleoside monophosphate*
NMR *Nuclear magnetic resonance*
NOC *Nopaline catabolizing enzyme*
Nod *Nodulation proteins*
NOS *Nopaline synthase*
NP-40 *Nonidet P*-40
NPC *Nuclear pore complex*
NPE 2-*Nitrophenylethyl*
NPEOC 2-(4-*Nitrophenyl*) *ethoxycarbonyl*
NPPOC 2-(2-*Nitrophenyl*) *propoxycarbonyl*
***npt* I and II** *Neomycin phosphotransferase*
Nt *Nicking enzyme with top strand cleavage activity*
nt *Nucleotide*
N-terminal *Amino terminal*
NTP *Nucleoside triphosphate*
Nub *N-terminal fragment of ubiquitin*
Nus *N-utilization substance*
***nut* site** *N-utilization site*

O *Operator*
OC DNA *Open-circular DNA*
OCC *Octopine catabolizing enzyme*
OCS *Octopine synthase*
OCT plasmid *Octane degrading plasmid*
OD *Optical density*
ODAP *N-oxalyl-diamino propionic acid*
ODPFGE *One-dimensional pulsed field gel electrophoresis*
OFAGE *Orthogonal field alternation gel electrophoresis*
O_L *Left operator*
Omp *Outer membrane protein*
omt *O-methyltransferase*
ONPG *O-nitrophenyl*-β-*D-galactopyranoside*
opd *Organophosphorus dehydrogenase*
O_R *Right operator*
ORF *Open reading frame*
ori λ *Origin of replication* (λ *phage*)
***ori* C** *Origin of replication* (*E. coli*)

***ori* T** *Origin of transfer*
ori *Origin of replication*

P *Promoter*
PABP *Poly(A) tail binding protein*
PAC *P1-derived artificial chromosome*
PACE *Programmable autonomously controlled electrodes*
PAGE *Polyacrylamide gel electrophoresis*
PAP *Pokeweed antiviral protein*
P_{aQ} *Promoter for anti-Q*
par *Partition locus*
PAZ domain *Piwi-Argo-Zwiille domain; Pihead domain*
PC *Physical containment*
PC *Phytochelatin*
PCB *Polychlorinated biphenyl*
PCR *Polymerase chain reaction*
PDI *Protein disulfide isomerase*
PDMS *Polydimethylsiloxane*
PDR *Pathogen-derived resistance*
PEG *Polyethylene glycol*
PEPC *Phosphoenol pyruvate carboxylase*
PFGE *Pulsed field gel electrophoresis*
pfu *Plaque forming unit*
PG *Polygalacturonase*
PGA *Photo generated acid*
PGTS *Post-transcriptional gene silencing*
PHB *Poly (3-hydroxy butyric acid)*
PHOGE *Pulsed homogeneous orthogonal field gel electrophoresis*
Pi *Inorganic phosphate*
P_I *Promoter for integrase*
PIB *Polyhedral inclusion body*
PIG *Particle inflow gun*
pInt *Integrase*
PITC *Phenylisothiocyanate; Edman reagent*
P_L *Leftward promoter*
PLRV *Potato leaf roll luteovirus*
PMP *Paramagnetic particles*
PNK *Polynucleotide kinase*
PNPG *p-Nitrophenyl-β-D-glucuronide*
***poly* A** *Polyadenylation sequence*
POSaM *Piezoelectric oligonucleotide synthesizer and microarrayer*
pp2C *Serine/threonine protein phosphatase type 2C*
PPDK *Pyruvate orthophosphate dikinase*
PPi *Pyrophosphate*
PPL *Pharmaceuticals Proteins Ltd., Edinburgh, Scotland*
PPO *2,5-Diphenyloxazole*
PPO *Polyphenol oxidase*
PR *Pathogenesis-related proteins*
P_R *Rightward promoter*
P_{RE} *Promoter for repressor establishment*
P_{RM} *Promoter for repressor maintenance*
Pro *Proline*
PRSV *Papaya ringspot potyvirus*
PTC *Phenylthiocarbamoyl*
pTi *Tumor inducing plasmid (Agrobacte-rium)*
pUC *Plasmid University of California*
PUFA *Polyunsaturated fatty acid*
PVDF *Polyvinyl difluoride*
PVP *Plant varieties and farmers rights protection*
PVX *Potato X potexvirus*
PVY *Potato Y potyvirus*
pXis *Excisionase*

Q *Quencher*
QC *Quality control*
QTL *Quantitative trait locus*
qTOF *Quadrupole time of flight*
***qut* site** *Q-utilization site*

R arm *Right arm*
R plasmid *Resistance plasmid*
R primer *Reverse primer*
R *Purine*
RACE *Rapid amplification of cDNA ends*
RAGE *Recombinase-activated gene expression*
RAMP *Random amplified microsatellite polymorphisms*
RAPD *Randomly amplified polymorphic DNA*
RB *Right border*
RBS *Ribosome binding site*
RCAT *Rolling circle amplification technology*
RCGM *Review Committee on Genetic Manipulation*
RDAC *Recombinant DNA Advisory Committee*
rdl *Resistance to dieldrin*
rDNA *Recombinant DNA*
RE *Restriction enzyme*
Rec A *Recombination protein*
REV *Reversed phase evaporation*
RF *Radio frequency*
RF *Replicative form*
RFD *RNase-free DNase I*
RFLP *Restriction fragment length polymorphism*
RGE *Rotating gel electrophoresis*
RGEF *Ras guanine nucleotide exchange factor*
RI *Refractive index*
RIP *Ribosome inactivating protein*
RISCs *RNA-induced silencing complexes*
RM primer *Reverse mutagenic primer*
R-M system *Restriction and modification system*
RMCE *Recombinant-mediated cassette exchange*
RNA *Ribonucleic acid*
RNAi *RNA interference*
RNase H *Ribonuclease specific for degradation of RNA in RNA:DNA hybrid*
RNase *Ribonuclease*
rnc *RNase III gene*
rNTP *Ribonucleoside triphosphate; ribonucleotide*
ROFE *Rotating field electrophoresis*
Rom *RNA I modulator*
Rop *Repressor of primer*
ROS *Reactive oxygen species*
ROX *5-carboxy-X-rhodamine*
rRNA *Ribosomal RNA*
RSV *Rous sarcoma virus*
RT *Reverse transcription*
RTase *Reverse transcriptase; RNA-directed DNA polymerase*
RT-PCR *Reverse transcriptase-polymerase chain reaction*

S *Svedberg unit; sedimentation coefficient*
S1 nuclease *Single strand specific nuclease*
SA *Salicylic acid*
SAH *S-adenosyl homocysteine*
SAL plasmid *Salicylate degrading plasmid*
Sam *Amber mutation in S gene*
SAM; AdoMet *S-adenosyl methionine*
SAMPL *Selective amplification of microsatellite polymorphic loci*
SAP *Arctic shrimp alkaline phosphatase*
SBCC *State Biotechnology Coordination Committee*
SBE *Starch branching enzyme*
SCAR *Sequence characterized amplified regions*
scFV *Single chain Fv fragment*
SCNT *Somatic cell nuclear transfer*
scp plasmid *Saccharomyces cerevisiae plasmid*
SCP *Single cell protein*
SCRT guidelines *Stem cell research and therapy guidelines*
SDM *Site directed mutagenesis*
SDS *Sodium dodecyl sulfate*
SELEX *Systematic evolution of ligands by exponential enrichment*
SFV *Semliki forest virus*
ShRNA *Short hairpin RNA*
SiC *Silicon carbide whiskers or fibers*
SINEs *Small interspersed nuclear elements*
siRNA *Small interfering RNA*
SNP *Single nucleotide polymorphism*
SOD *Superoxide dismutase*
SOLiD *Sequencing by oligonucleotide ligation and detection*
***spa* A** *Surface protein antigen A*
SPAR *Single primer amplification reaction*
Spi *Sensitivity to P2 interference or inhibition*
SPS *Sucrose phosphate synthase*
SRS *SOS recruitment system*
Sry *Sex-determining region of the Y chromosome*
ss DNA *Single stranded DNA*
SSAP *Sequence specific amplification polymorphism*

SSB *Single strand binding protein*
SSCP *Single strand conformation polymorphism*
SSLP *Simple sequence length polymorphism*
SSR *Simple sequence repeats*
STCs *Sequence tagged connectors or BAC ends*
STR *Short tandem repeats; Microsatellites*
STSs *Sequence tagged sites*
sup$^+$ *Amber suppressor positive; Gene coding for suppressor tRNA that reads stop codon as an amino acid*
SuSy *Sucrose synthase*
SV 40 *Simian virus* 40

T *Thymine*
T4SS *Type-IV secretion system*
TAE *Tris-HCl, acetic acid, EDTA*
TAFE *Transverse alternating field electrophoresis*
TAMRA 5-*Carboxytetramethylrhodamine*
TAP-tagging *Tandem affinity purification-tagging*
TBE *Tris-borate, EDTA*
TBS *Tris-buffered saline*
TCA *Trichloroacetic acid*
T-DNA *Transferred-DNA*
TDP *Thymidine diphosphate*
Tdt *Terminal deoxynucleotidyl transferase; Terminal transferase*
TE *Tris-HCl, EDTA*
TEL *Telomeric sequence*
TEMED *N, N, N′, N′-Tetramethylene diamine*
TET *Tetrachlorofluorescein*
tet^r **or** ***tc***^{r} *Tetracycline resistance gene*
tet^s **or** ***tc***^{s} *Sensitivity to tetracycline*
TFA *Trifluoroacetic acid*
TGMV *Tomato golden virus*
TGS *Transcriptional gene silencing*
THF *Tetrahydrofolate/Tetrahydrofuran*
Thr *Threonine*
Tk *Thymidine kinase*
tk *Thymidine kinase gene*
TL *Tripartite leader*
T_m *Melting temperature*
TMP *Thymidine monophosphate*
TMV *Tobacco mosaic virus*
Tn *Transposon*
TNCs *Transnational companies*
TOF *Time of flight*
TOL plasmid *Toluene degrading plasmid*
TPE *Tris-phosphate, EDTA*
tra *Transfer locus*
tRNA *Transfer RNA*
tTA *Tetracycline transactivator*
TTP *Thymidine triphosphate*

U *Uracil*
UV *Ultraviolet*
UAP *Universal amplification primer*
UAS *Upstream activating sequences*
UBPs *Ubiquitin-specific proteases*
UDG *Uracil–DNA glycosylase*
UDP *Uridine diphosphate*
UMP *Uridine monophosphate*
UNG *Uracil N-glycosylase*
UTP *Uridine triphosphate*
UTRs *Untranslated regions*

v/v *Volume/volume*
VAD *Vitamin A deficiency*
V_e *Elution volume*
Vir proteins *Virulence proteins*
vir *Virulence gene*
VNTR *Variable number tandem repeats; Minisatellites*
V_o *Void volume*
V_t *Total volume*

W *A/T*
w/v *weight/volume*
WDV *Wheat dwarf virus*
WGA *Whole genome amplification*
WHO *World Health Organization*

XC FF *Xylene cyanol FF*
X-gal 5-*Bromo* 4-*chloro* 3-*indolyl* β-*D-galactoside*
X-gluc 5-*Bromo* 4-*chloro* 3-*indolyl* β-*D-glucuronide*
XYL plasmid *Xylene degrading plasmid*

Y *Pyrimidine*
YAC *Yeast artificial chromosome*
YCp *Yeast centromeric plasmid*
YEp *Yeast episomal plasmid*
YIp *Yeast integrative plasmid*
YLp *Yeast linear plasmid*
YRp *Yeast replicating plasmid*

ZIFE *Zero-integrated field electrophoresis*

Part I

Fundamentals of Genetic Engineering

- An Introduction to Gene Cloning
- Enzymes Used in Genetic Engineering
- Restriction Endonucleases
- Chemical Synthesis of Oligonucleotides
- Synthesis of Complementary DNA
- Techniques for Nucleic Acid Analysis and Size Fractionation
- Polymerase Chain Reaction

An Introduction to Gene Cloning

Key Concepts

In this chapter we will learn the following:

- Cloning vectors
- Steps involved in gene cloning
- Subcloning
- Advantages of gene or cDNA cloning

1.1 INTRODUCTION

Gene cloning, developed in the early 1970s, was a significant breakthrough in the field of molecular biology. The technique was pioneered by Paul Berg, Herbert Boyer, and Stanley Cohen and was the fruit of several decades of basic research on nucleic acids (Exhibits 1.1 and 1.2). Further contributing factors for the development of gene cloning technique were critical understanding of *E. coli* genetics, information about nucleic acid enzymology (i.e., having enzymes that cut, join, and replicate DNA or reverse transcribe RNA), availability of techniques for monitoring the cutting and joining reactions, and information about bacterial plasmids. The impetus for gene manipulation *in vitro* came with the discovery of natural gene transfer capability of viruses and *Agrobacterium tumefaciens* into hosts and development of techniques for genetic transformation of *E. coli*. Another area of research that laid the foundation of gene cloning was renaturation analysis that led to the discovery of base pairing characteristics of two complementary sequences (i.e., hybridization), indicating that complementary DNA or RNA probes can be used to detect specific nucleotide sequences.

The technique of gene cloning (also called molecular cloning) essentially involves the insertion of target DNA (also called foreign DNA/passenger DNA/exogenous DNA/insert DNA/DNA fragment/gene of interest) into a cell through a

Exhibit 1.1 Deoxyribonucleic acid (DNA)

DNA is a high molecular weight biopolymer comprising of mononucleotides as their repeating units, which are joined by 3′ → 5′ phosphodiester bonds. Each mononucleotide unit of DNA consists of purine and pyrimidine nitrogenous bases, phosphorus, and a pentose sugar 2′-deoxy-*D*-ribose (Figure A). Examples of purine bases are adenine (A) and guanine (G), and examples of pyrimidine bases are cytosine (C) and thymine (T) (Figure A).

The nitrogenous bases are planar due to their π-electron clouds, hydrophobic, and relatively insoluble in water at the near neutral pH of the cell. Purines exist in both *syn* or *anti* forms, while pyrimidines due to steric interference between the sugar and carbonyl oxygen at C2 position (carbon atom at position 2) of pyrimidine exist in *anti* form. Besides, the major nitrogenous bases, some minor bases called modified nitrogenous bases also occur in polynucleotide structures, for example, 5,6-dihydrouracil, pseudouracil, 4-thiouracil, 5-methylcytosine (5-MeC), and 5-hydroxymethylcytosine, etc. Phosphorus is present in the sugar-phosphate backbone of DNA as a constituent of phosphodiester bond linking the two sugar moieties. Sugars are always in closed ring β-furanose (ketose) form and hence are called furanose sugars. The base is linked to the sugar by β-*N*-glycosidic linkage. Five atoms in the sugar ring are denoted as 1′ to 5′, while six atoms in

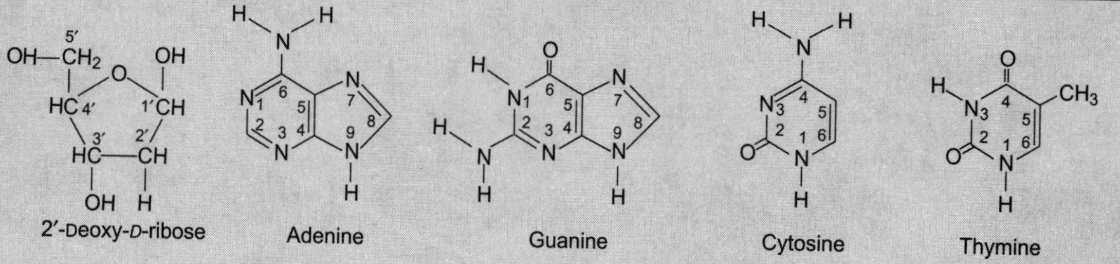

Figure A Structures of 2-deoxy-*D*-ribose sugar and four nitrogenous bases (A, G, C, and T)

pyrimidine ring (or nine in purine ring) are denoted as 1 to 6 (or 1 to 9) as shown in Figure A. The compounds in which the nitrogenous bases are conjugated to the pentose sugars by β-*N*-glycosidic linkages are called deoxyribonucleosides or deoxyribosides. These linkages are formed between the N1 atom of pyrimidine (or the N9 atom of purine) with the C1′ carbon atom of sugar. Thus, the purine deoxyribonucleosides are N-9 glycosides and the pyrimidine deoxyribonucleosides are N-1 glycosides. These are stable in alkali. The purine deoxyribonucleosides are readily hydrolyzed by acid, whereas pyrimidine deoxyribonucleosides are hydrolyzed only after prolonged treatment with concentrated acid. These deoxyribonucleosides are generally named for the particular purine and pyrimidine present. The nomenclature of deoxyribonucleosides differs from that of the bases. The trivial names of purine deoxyribonucleosides end with the suffix –sine and those of pyrimidine deoxyribonucleosides end with suffix –dine, for example, deoxyribonucleosides containing A, G, C, and T are called deoxyadenosine, deoxyguanosine, deoxycytidine, and deoxythymidine, respectively. The phosphate esters of deoxyribonucleosides are called deoxyribonucleotides or deoxyribonucleoside triphosphates. The phosphate is always esterified to the sugar moiety. Esterification can occur at any free hydroxyl group, but is most common at the 5′ and 3′ positions in sugars. The phosphate residue at α position is joined to the sugar ring at C5 by a phosphomonoester bond, while β- and γ-phosphate groups are joined in series by phosphoanhydride bonds. These deoxyribonucleotides occur either in the free form or as subunits in DNA linked by phosphodiester bond as shown in Figure B.

Chargaff (1950) formulated important generalizations about the structure of DNA, which form the Chargaff's equivalence rule. According to the rule, (i) The base composition of DNA varies from one species to another; (ii) DNA specimens isolated from different tissues of the same species have the same base composition; (iii) The base composition of DNA in a given species does not change with age, nutritional state, or changes in environment; (iv) Purines and pyrimidines are always equal such that amount of A is equal

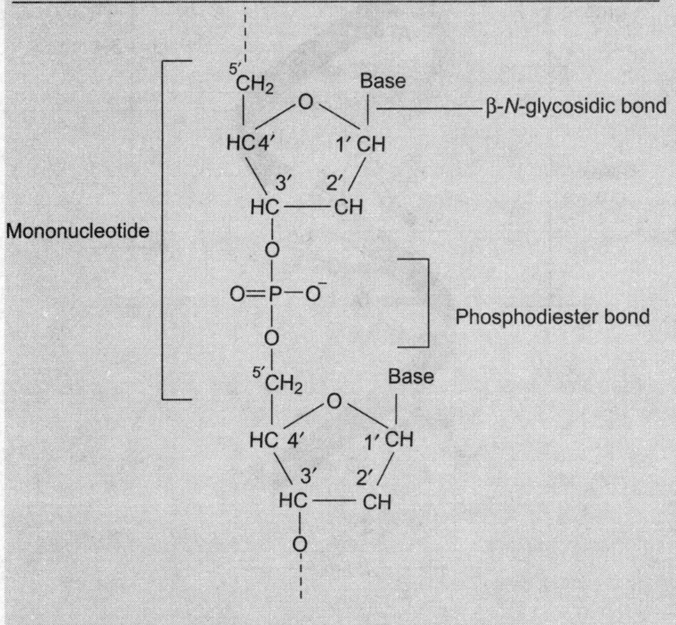

Figure B Structure of deoxyribonucleotide

to T, and the amount of G is always equal to C (molar equivalence of bases); (v) Base ratio A+T/G+C may vary from one species to other, but is constant for a given species, and this ratio is used to identify the source of DNA; and (vi) The deoxyribose sugar and phosphate components occur in equal proportions in the sugar–phosphate backbone. Based on the X-ray data of Franklin and Wilkins and the base equivalence observed by Chargaff, Watson and Crick (1953) proposed a model for the 3-D structure of DNA. This model accounted for many of the observations on the chemical and physical properties of DNA and also suggested a mechanism for the accurate replication of genetic information.

DNA contains two polynucleotide chains that are coiled in helical fashion around the same axis in right-handed or counterclockwise direction, thus forming a double helix. The two chains or strands are antiparallel, i.e., the 3′,5′-internucleotide phosphodiester linkages run in opposite directions on two strands. The antiparallel orientation is a stereochemical consequence of the way that A pairs with T and G pairs with C. All the phosphodiester linkages have the same orientation along the chain, giving each linear DNA strand a specific polarity, and distinct 5′ and 3′ ends. The backbone of helix consists of sugar and phosphate groups, while purine and pyrimidine bases are stacked inside the helix with their planes parallel to each other and perpendicular to the helix axis. Hydroxyl groups of sugar forms H-bonds with water. DNA is negatively charged due to its phosphate groups and negative charges are generally neutralized by ionic interactions with positive charges of protein, metals, and polyamines. Bases being hydrophobic, project inwards to the center, and hence shielded from water. Backbone is found on the periphery of the helix and is hydrophilic. It means that single stranded structure, in which the bases are exposed to aqueous environment, is unstable. Thus DNA is double helix, in which two strands are held together by H-bonding interactions between complementary base pairs and hydrophobic [(π–π) stacking interactions between adjacent bases] interactions. A perfect Watson–Crick base pair consists of a perfect match between hydrogen donor and acceptor sites on the two bases. Thus, a purine on one strand base pairs with a specific pyrimidine on the other strand; the resulting base pair exhibits proper spatial arrangement. Thus the base A is base paired to T by double H-bonds and G is bonded to C by triple H-bonds (Figure C). This forms the concept of specific base pairing.

The individual H-bond is weak in nature, but a large number of such bonds involved in the DNA molecule confer stability to it. However, the stability of DNA is primarily a consequence of van der Waals forces and hydrophobic (base stacking) interactions between the planes of stacked bases. On one hand, H-bonding is specific, which is responsible for complementarity of two strands, on the other hand, hydrophobic interactions are nonspecific and are responsible for the stability of the macromolecule. Note that the two DNA strands tend to stick together even in the absence of specific H-bonding interactions, although these specific interactions make the association stronger. The two helices are wound in such a way so as to produce two interchain spacings or grooves, a major or wide groove (width 12 Å, depth 8.5 Å), and a minor or narrow groove (width 6.0 Å, depth 7.5 Å). The major groove is slightly deeper than the minor one. The two grooves arise because the glycosidic bonds of a base pair are not diametrically opposite each other. The minor grove contains the O2 position of pyrimidine and the N3 positions of the purine of

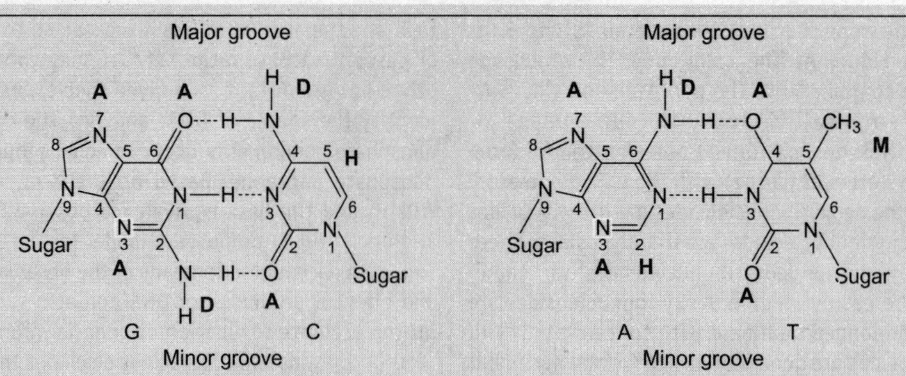

Figure C Watson–Crick base pairs: G:C and A:T

the base pair, while the major groove is on the opposite side of the pair. At each groove, potential H-bond donor and acceptor atoms are exposed to external environment, which serve as the interaction sites for the DNA-binding proteins. Thus the sequence of H-bonding donors and H-bonding acceptors exposed toward the major groove in a G:C base pair is AADH, while in an A:T base pair is ADAM (A is H-bond acceptor, D is H-bond donor, H is nonpolar hydrogen, and M is methyl group). Similarly, the sequence is MADA in T:A base pair, and HDAA in C:G base pair toward the major groove. The sequence of H-bonding donors and acceptors exposed toward the minor grove in G:C and C:G base pairs is ADA, and AHA in A:T and T:A base pairs (Figure C). Thus the major groove displays more information as compared to the minor groove. The double helices of DNA are plectonemic coils, i.e., coils that are interlocked about the same axis. These strands cannot be pulled apart, but can be separated by the unwinding process. The Watson and Crick structure of DNA is referred to as B-DNA (normal form). It is the biologically important one and exists under physiological conditions described below. The structural details of ds DNA as suggested by Watson and Crick (i.e., B-DNA) are shown in Figure D and Table A.

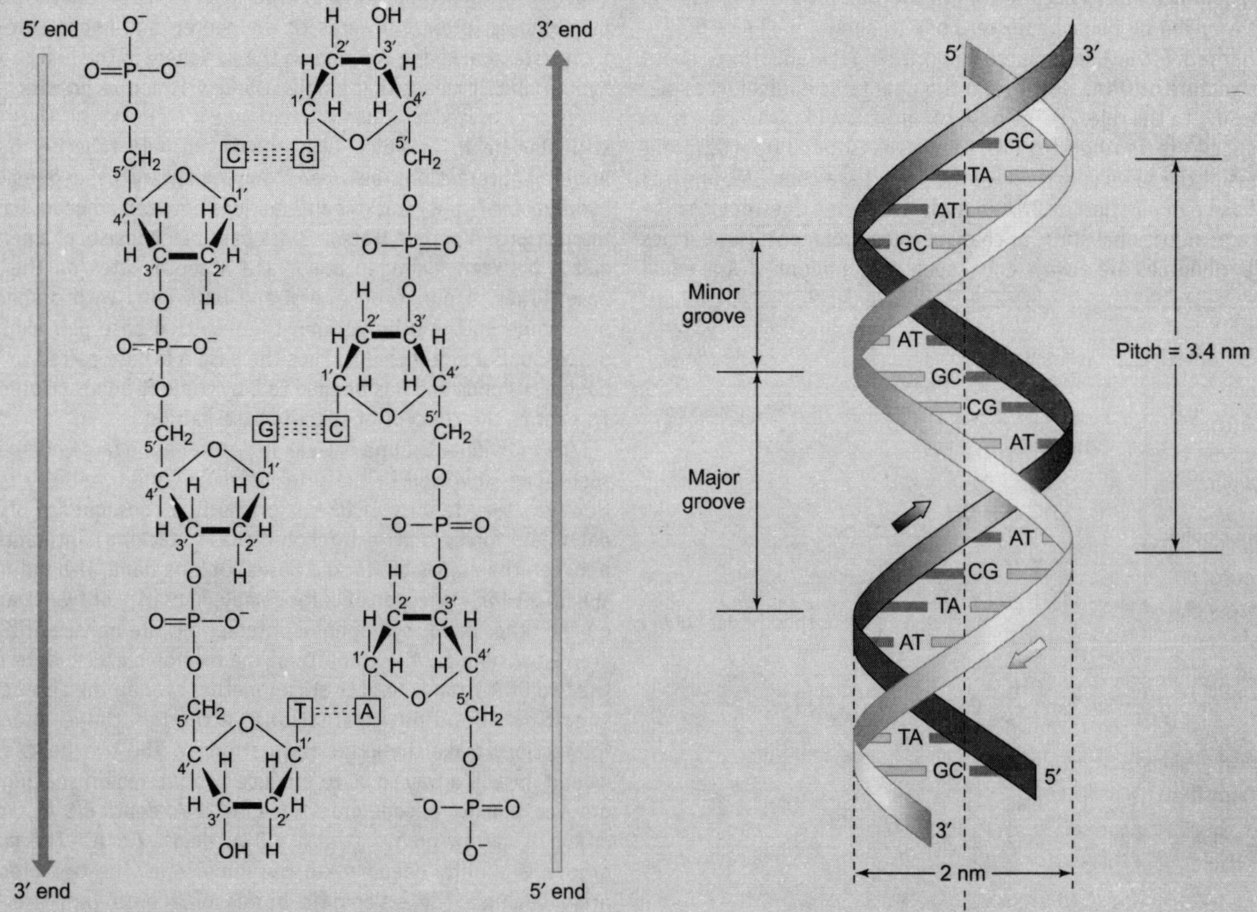

Figure D Watson–Crick double helical structure of DNA

Table A General Characteristics of Three Helical Conformers of DNA

Characteristics	Helical conformations of DNA		
	A	B	Z
Relative humidity	75%	92%	–
Ions required/salt concentration	Na^+, K^+, Cs^+ ions	Low ion strength	Very high salt concentration
Shape	Broadest	Intermediate	Narrowest
Helical state	Right	Right	Left
Pitch (base pairs per turn)	11	10.5	12 (=6 dimers)
Major groove	Deep, narrow	Wide	Flat
Minor groove	Broad, shallow	Narrow	Narrow and very deep
Helix diameter	~26 Å	~20 Å	~18 Å
Sugar pucker conformation	$C^{2'}$ endo	$C^{3'}$ endo	Alternating
Glycosidic bond angle	*anti*	*anti*	Alternating *anti/syn*
Displacement	–4.4	0.6	3.2
Twist	33	36	–49/–10
Helix rise per base pair	2.6 Å	3.4 Å	3.7 Å
Helix pitch	25.30 Å	35.36 Å	45.60 Å
Base tilt normal to helix axis	20°	6°	7°
Inclination	22	–2	–7
Rotation per base pair	+32.72°	+34.61°	–60° (per dimer)

DNA is strongly acidic molecule owing to the presence of phosphate groups. Double helical DNA is maximally stable between pH 4.0 and 11.0 (physiological range). Outside these physiological limits, DNA becomes unstable and unwinds. DNA is very flexible in nature, and due to thermal fluctuation, bending, stretching, and unpairing (melting) of strands can occur. However, when melted DNA is incubated at a temperature ~25°C below the denaturation temperature, the two separated strands reassociate (or reanneal or renature) to form a ds DNA molecule due to complementary sequences present in the two strands. DNA exhibits a strong positive rotation. Upon denaturation, optical rotation is highly decreased and becomes more negative. The solution of ds DNA possesses an absorption maximum at 260 nm. This characteristic absorption maximum is the property of its individual bases, and their corresponding deoxyribonucleotides. Upon denaturation of DNA, an increase in absorption of light (up to 40%) occurs even though the amount of DNA remains the same. This phenomenon is called hyperchromic effect. The temperature at the midpoint of melting curve is called melting temperature (T_m). Because of the rigidity of the double helix and the immense length of DNA in relation to its small diameter, the solution of DNA is highly viscous at pH 7.0 and room temperature (25°C). Mammalian DNA forms a band at position corresponding to 1.7 g/cm^3 buoyant density in a CsCl density gradient upon centrifugation.

Some structural variants of DNA may also arise due to difference in possible conformation of deoxyribose, rotation about phosphodiester bonds in the sugar–phosphate backbone, and free rotation about C1′-β-*N*-glycosidic bonds (*syn* or *anti*). Major helical conformers that are formed at different humidities are A and Z forms. These differ in the gross morphological features, bond angles, base inclination, displacement of the base pairs from the helical axis resulting from dehydration, helix parameters as base pairs per helical turn, helical twist, size, and shape of grooves. The variations in minor and major grooves potentially influence the nature of protein–DNA interactions and consequently affect the regulatory property of DNA. However, the key properties of DNA in different forms are not changed. A-DNA is observed when the relative humidity is reduced below 75% and is also favored in many solutions relatively devoid of water. Moreover, the double stranded regions of RNA (as in hairpins) and RNA:DNA hybrids also adopt a conformation similar to A-DNA. The structure of Z-DNA is characterized by alternating helical parameters and torsion angles with a two-base pair periodicity, causing the backbone of the helix to zig-zag and hence the name. The general characteristics of A-, B-, and Z-DNAs are tabulated in Table A.

Sources: Watson JD, Crick FHC (1953) *Nature* 171: 737–738; Dickerson RE (1983) *Sci. Am.* 249: 94–111; Watson JD, Baker TA, Bell SP, Gann A, Levine M, Losick R (2004) *Molecular Biology of the Gene*, Pearson Education (Singapore), pp. 97–128.

suitable vector in such a way that inserted DNA replicates independently and transferred to progenies during cell division, thereby generating genetically identical organisms. Note that the availability of different kinds of restriction endonucleases and DNA ligases made it feasible to treat sequences of DNA as modules, which can be moved at will from one DNA molecule to another. The artificially created DNA molecule formed by the insertion of foreign DNA into a vector is called recombinant DNA molecule or DNA chimera because of its analogy with the Chimera, a creature in mythology with the head of a lion, body of a goat, and tail of a serpent. The term cloning signifies that the technique leads to generation of a line of genetically identical organisms, all of which contain the recombinant DNA molecule that can be propagated

Exhibit 1.2 Ribonucleic acid (RNA)

RNA is a long, unbranched macromolecule consisting of nucleotides joined by 3′ → 5′ phosphodiester bonds. Chemically, RNA is very similar to DNA. It contains nitrogenous bases such as A, G, C, and uracil (U), phosphorus, and a *D*-ribose sugar (Figure A). The structural differences between the components of RNA and DNA include presence of U instead of T (its nucleoside is called uridine), and *D*-ribose sugar instead of 2′-deoxy-*D*-ribose sugar.

The linkages between sugar and phosphate in the sugar-phosphate backbone, and sugar and base (perpendicular to backbone) are the same as described in the structure of DNA in Exhibit 1.1. The structural difference in the sugars of RNA and DNA confers very different chemical and physical properties on RNA. RNA is much stiffer due to steric hindrance and more susceptible to hydrolysis in alkaline conditions (high pH). Unlike DNA, under physiological conditions, RNA is single stranded. However, the single strand can fold back on itself having potentially much greater structural diversity than DNA. These secondary structures arise due to intramolecular base pairing. If the two stretches of complementary sequences are near to each other, RNA may adopt stem loop structures in which the intervening RNA is looped out from the end of the double helical segment (e.g., secondary structures include hairpin, bulge, or simple loop). Similar to DNA, weak interactions, especially base stacking interactions, play a major role in stabilizing RNA structures. Where complementary sequences are present, the predominant double stranded structure is an A-form right handed double helix. The presence of 2′-OH on the sugar residue in the RNA backbone prevents RNA from adopting a B-form helix. A feature of RNA that adds to its propensity to form secondary structures is an additional non Watson–Crick G:U base pair. Note that G:U base pair contains two H-bonds, one between N3 position of U and carbonyl on C6 of G, and the other between the carbonyl on C2 of U and N1 of G, and due to these additional G:U base pairs, RNA chains have an enhanced capacity of self-complementarity.

On basis of size, function, and stability, there are three types of RNAs, *viz.*, ribosomal RNA (rRNA), messenger RNA (mRNA), and transfer RNA (tRNA). The rRNAs are the components of ribosomes (16S, 23S, and 5S rRNAs in prokaryotic 70S ribosomes; 18S, 28S, 5.8S, and 5S rRNAs in eukaryotic nuclear 80S ribosomes), which are the sites for protein synthesis. The mRNA carries genetic information from one or a few genes to a ribosome. The protein coding region(s) of each mRNA is composed of a contiguous, nonoverlapping string of codons called open reading frame (ORF), which is a sequence of DNA consisting of triplets that is translated into amino acids starting from initiation or start codon at 5′-end

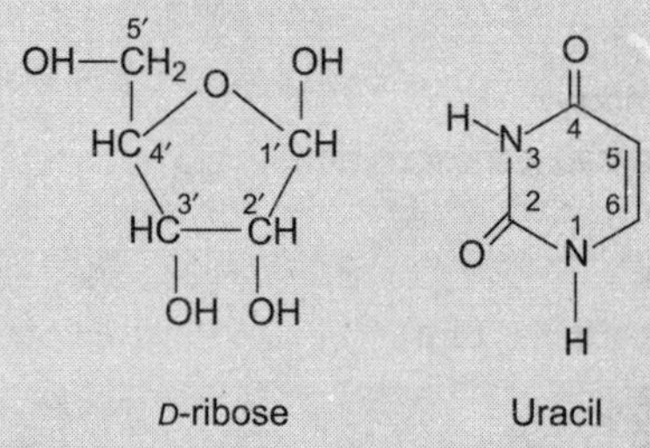

Figure A Structure of *D*-ribose and uracil

(e.g., AUG in most bacteria and eukaryotes or GUG or UUG in some bacteria), and ending with a termination or stop codon [e.g., UAG (amber) or UGA (opal) or UAA (ochre)] at 3′-end. Each ORF specifies a single polypeptide, and starts and ends at internal sites within the mRNA, i.e., the ends of an ORF are distinct from the ends of mRNA. Prokaryotic mRNA is polycistronic, i.e., a single mRNA molecule codes for two or more polypeptide chains or simply it contains multiple ORFs. The prokaryotic mRNA contains a ribosome-binding site (RBS; also referred to as Shine Dalgarno sequence). It is complementary to a sequence located near the 3′-end of 16S rRNA. RBS base pairs with 16S rRNA, thereby aligning the ribosome with the beginning of mRNA during the process of translation. Note that some mRNAs lack RBS and have translational coupling of termination with initiation due to the presence of sequence 5′-AUGA-3′ (in this overlapping sequence, UGA marks termination for previous transcript and AUG marks initiation for the next). In contrast, eukaryotic mRNA is monocistronic, i.e., a single mRNA codes for single polypeptide chain or simply it contains single ORF. There is no RBS, rather the eukaryotic mRNA is recognized by translation machinery by 5′ cap and the start codon (AUG) is reached by scanning. Note that most of the eukaryotic mRNAs contains a 5′ cap and a 3′ poly (A) tail. The tRNA serves as adapter molecule in translating the language of nucleic acids in mRNA into the language of proteins by serving as carriers of specific amino acids to specific sites on the ribosome. Thus anticodons of charged tRNAs (i.e., tRNAs covalently linked to an amino acid at 3′-end in a reaction catalyzed by amino acyl tRNA synthetase) pair with the codons of mRNA in such a way that amino acids are joined to form a polypeptide chain in a correct sequence.

Sources: Uhlenbeck OC, Pardi A, Feigon J (1997) *Cell* 90: 833–840; Watson JD, Baker TA, Bell SP, Gann A, Levine M, Losick R (2004) *Molecular Biology of the Gene*, Pearson Education (Singapore), pp. 97–128.

and grown in bulk hence amplifying the recombinant DNA molecule and any gene product whose synthesis it directs. The creation of this artificial recombinant DNA molecule by a variety of sophisticated techniques, and in many cases its subsequent introduction into living cells, is referred to as genetic engineering or gene manipulation because of the potential for creating novel genetic combination by biochemical means. By this technique, even the DNA sequences not usually found together in nature can be brought together. In the developed world, there is a precise legal definition of gene manipulation as a result of government legislation to control it. In the UK, for example, gene manipulation is defined as 'the formation of new combinations by the insertion of nucleic acid molecules, produced by whatever means outside the cell, into any virus, bacterial plasmid, or other vector system so as to allow their incorporation into a host organism in which these do not naturally occur but in which these are capable of continued propagation'. The resulting transformants are selected and screened for containing the DNA sequences of interest and later used either for research purpose or for commercial production of useful compounds related to the areas of medicine, agriculture, animal husbandry, environment, public health, forensics, etc. In this chapter, an overview of gene cloning procedure is presented; however, the details of each step are discussed in separate chapters in the book, a reference to which is given in each step. Here an emphasis is laid on the properties of ideal cloning vector and the procedures for nucleic acid isolation and purification.

1.2 CLONING VECTORS

A prime requisite for a gene cloning experiment is the selection of a suitable cloning vector (or cloning vehicle), i.e., a DNA molecule that acts as a vehicle for carrying a foreign DNA fragment when inserted into it and transports it into a host cell, which is usually a bacterium although other types of living cells can also be used.

A wide variety of natural replicons exhibit the properties that allow them to act as cloning vectors, however, vectors may also be designed to possess certain minimum qualifications to function as an efficient agent for transfer, maintenance, and amplification of target DNA (for details see Section 1.2.2). As will be evident from the list (Section 1.2.1), most of the naturally occurring or artificially constructed cloning vectors in use today are for *E. coli* as the host organism. This is not surprising in view of the central role that this bacterium has played in basic research over the last 50 years (Exhibit 1.3).

1.2.1 Examples of Cloning Vectors

The examples of naturally occurring or artificially constructed cloning vectors include vectors based on *E. coli* plasmids, bacteriophages (e.g., λ, M13, P1), viruses [animal viruses

Exhibit 1.3 *Escherichia coli (E. coli)*

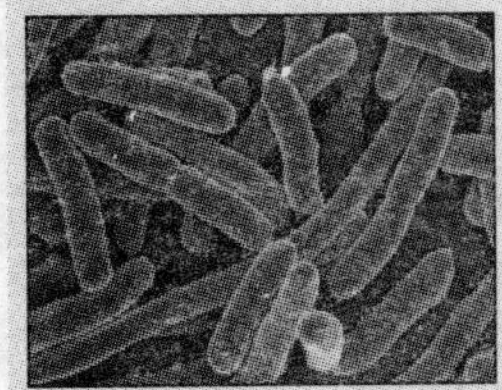

Theodor Escherich, a German pediatrician and bacteriologist, cultured '*Bacterium coli*' in 1885 from the feces of healthy individuals and concluded that this bacterium can be found almost universally in the large intestine or colon, and named it '*coli*'. It was renamed *Escherichia coli* (*E. coli*) in 1919 in a revision of bacteriological nomenclature to lend more specificity to this particular form of bacterium.

E. coli (Domain: Bacteria; Phylum: Proteobacteria; Class: Gammaproteobacteria; Order: Enterobacteriales; Family: Enterobacteriaceae; Genus: *Escherichia*; Species: *coli*) is a gram negative, unicellular, facultatively anaerobic chemoorganotroph capable of both respiratory and fermentative metabolism, mesophilic, nonphotosynthetic, and nonsporulating eubacteria. It is small (length ~2 μm, diameter 0.5 μm, and cell volume of 0.6–0.7 μm^3) and straight rod-shaped bacterium. The motile bacterium contains peritrichous flagella. It is one of the characteristic members of the normal intestinal (lower intestine) flora of warm-blooded animals, i.e., it is a coliform or enteric bacterium. However, it can also survive when released into the natural environment, allowing widespread dissemination to new hosts. The strains that are part of the normal flora of the gut are harmless and can benefit their hosts by producing Vitamin K2 or by preventing the establishment of pathogenic bacteria within the intestine. Most *E. coli* strains are harmless, but some pathogenic *E. coli* strains are responsible for infection of the enteric, urinary, pulmonary, and nervous systems, for example, serotype O157:H7 can cause serious food poisoning in humans.

The genome of *E. coli* is well studied. Sequence analysis of *E. coli* K-12 has revealed that the bacterium has a single circular chromosome of 46,39,221 bp and molecular weight of 2.7×10^9 Da. Its 4,288 protein-coding genes have been annotated, out of which 38% have no attributed function. The genome of *E. coli* K-12, like other *E. coli* genomes, has a 50.8% G+C content. Genes that code for proteins account for 87.8% of the genome, stable RNA-encoding genes make up 0.8%, noncoding repeats contribute to 0.7%, and about 11% is for regulatory and other functions. Comparison with five other sequenced microbes reveals ubiquitous as well as narrowly distributed gene families. Many families of similar genes within *E. coli* are also evident. The largest family of paralogous proteins contains 80 ABC transporters. The genome as a whole is strikingly organized with respect to the local direction of replication. The genome also contains insertion sequence (IS) elements, phage remnants, and many other patches of unusual composition indicating genome plasticity through horizontal transfer. *E. coli* and related bacteria possess the ability to transfer DNA via bacterial conjugation, transduction, or transformation, which allow genetic material to spread horizontally through an existing population.

E. coli is easy to cultivate and it can live on a wide variety of substrates. It can be easily cultivated on synthetic medium, with a minimal medium comprising of ingredients (g/l): 5 glucose, 6 Na_2HPO_4, 3 KH_2PO_4, 1 NH_4Cl, 0.5 NaCl, 0.12 $MgSO_4$, and 0.01 $CaCl_2$. It can also thrive well in complex medium, for example, Luria Bertani (LB) medium. The pH of the media is 7.0 and sterilized by autoclaving [15 psi (1.05 kg/cm^2), 121°C, 20 min]. On rich media, bacteria grow with a doubling time of 20 min, hence readily visible colonies can be seen overnight when plated on agar (Figure A). Specialized medium, like MacConkey's agar, was developed for the selective isolation and identification of *E. coli*, as this was used as a global indicator for the pollution of water supplies. Optimal growth of *E. coli* occurs at 37°C, but some laboratory strains can multiply at temperatures of up to 49°C.

From the beginning, although pathogenic strains were also found, *E. coli* was used as a representative, harmless bacterium that could be safely and easily cultivated, and it became one of the most studied bacterium in various fields of science. *E. coli* was an integral part of the first experiments to understand bacterial and phage genetics, process of replication, conjugation, gene regulation, the concept of operon, etc. Thus, it has served as prokaryotic model organism in various studies including molecular genetics, microbiology, and genetic engineering. First cloning experiments were also undertaken in *E. coli* and this organism became the primary cloning host. However, under some circumstances, it may be desirable to use a different host for a gene cloning experiment. This is especially true in biotechnology, where the aim may not be to study a gene, but to use cloning to control or improve synthesis of an important metabolic product (e.g., a hormone such as insulin) or to change the properties of the organism (e.g., to introduce herbicide resistance into a crop plant). Hence subsequently, cloning techniques were extended to a range of other microorganisms, such as *B. subtilis*, *Pseudomonas* spp., *A. tumefaciens*, yeasts, filamentous fungi, and higher eukaryotes. Despite these advances, *E. coli* remains the most widely used cloning host even today because gene manipulation in this bacterium is technically easier than in any other organism, and the widest variety of cloning vectors are available for this organism. As a result, it is unusual for researchers to clone DNA directly in other organisms. Rather DNA from the organism of choice is first manipulated in *E. coli* and subsequently transferred to the original host or another organism. Without the ability to clone and manipulate DNA in *E. coli*, the application of recombinant DNA technology to other organisms is greatly hindered. The dominant role that *E. coli* plays in recombinant DNA technology is due to construction of many well-characterized mutants, good understanding of gene regulation, establishment of growth medium, isolation of many plasmids, and establishment of genetic transformation procedure. *E. coli* is considered a very versatile host for the production of heterologous proteins and hence by introducing genes into these microbes, mass production of recombinant proteins in industrial fermentation processes is possible. One of the first useful applications of recombinant DNA technology is the manipulation of *E. coli* to produce human insulin. Other applications of modified *E. coli* include vaccine development, bioremediation, production of immobilized enzymes, etc.

Several *E. coli* strains find application in genetic engineering experiments. Few such examples include cultivated strain K-12 (isolated at Stanford University in 1922 from human feces), DH5α, C600, and XL-1 Red. The K-12 strain has universally been adopted for fundamental work in biochemistry, genetics, and physiology. *E. coli* K-12 serves as the precursor for almost all strains used by molecular biologists for propagating cloned DNA. This is because this strain is well-adapted to the laboratory environment, and unlike wild-type strains, it is nonpathogenic and has lost the ability to thrive in the intestine. Note that the K-12 strain harbors

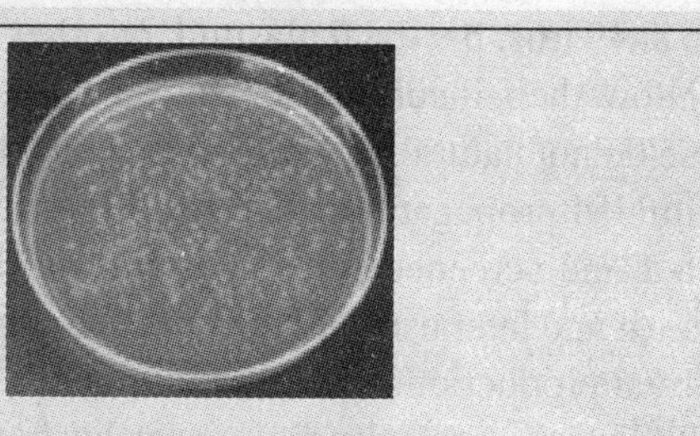

Figure A *E. coli* colonies on solid medium

a lysogenic λ bacteriophage and a number of plasmids. Many laboratory strains lose their ability to form biofilms. These features protect wild-type strains from antibodies and other chemical attacks, but require a large expenditure of energy and material resources. Improvements in *E. coli* K-12 for recombinant DNA experiments include: (i) Removal of *Eco* K restriction system (*hsd* R^-); (ii) Removal of *mcr* A/*mcr* B genes (*mcr* A^-/*mcr* B^-) that are responsible for degrading methylated foreign DNA, but not at the sites that *E. coli* K-12 recognize as its own (e.g., human and mouse DNA is CpG methylated); (iii) *rec* A^- mutation that suppresses homologous recombination, which makes it more sensitive to UV light; and (iv) *end* A^- mutation in the endonuclease A gene that greatly improves the quality of DNA isolated with biochemical techniques. Some derivatives of K-12 are XL1-Blue strain [*rec* A1 *end* A1 *gyr* A96 *thi*-1 *hsd* R17 *sup* E44 *rel* A1 *lac* {F′ *pro* AB *lac* I^q ZΔM15 *Tn* 10 (*tet*^{r})}], and XL1-Blue MR strain {Δ(*mcr* A) 183 Δ(*mcr* CB – *hsd* SMR – *mrr*) 173 *end* A1 *sup* E44 *thi*-1 *rec* A1 *gyr* A96 *rel* A1 *lac*}. Like many cloning strains, *E. coli* DH5α strain with a genotype [*fhu* A2 Δ(*arg* F - *lac* Z) *U* 169 *pho* A *gln* V44 ϕ80 (*lac* Z) ΔM15 *gyr* A96 *rec* A1 *rel* A1 *end* A1 *thi* -1 *hsd* R17] has several important features, which make it useful for recombinant DNA methods. These are as follows: (i) The strain transforms with high efficiency; (ii) The *end* A1 mutation inactivates an intracellular endonuclease that degrades plasmid DNA in many miniprep methods; (iii) The *hsd* R17 mutation eliminates the restriction endonuclease of the *Eco* KI restriction–modification system, so DNA lacking the *Eco* KI methylation will not be degraded. DNA prepared from *hsd* R strains that are wild-type for *hsd* M will be methylated and can be used to transform wild-type *E. coli* K-12 strains; (iv) *lac* ZΔM15 is the α-acceptor allele needed for blue-white screening with many *lac* Z-based vectors; (v) *rec* A eliminates homologous recombination. This reduces deletion formation and plasmid multimerization; and (vi) *sup* E44, with a systematic name *gln* V44, is an amber suppressor mutation. The chromosomal genotype of strain C600 is *sup* E44 *hsd* R *thi*-1 *leu* B6 *lac* Y1 *ton* A21 *hfl* 150 *chr* :: *Tn* 10 (*tet*^{r}). The genotype of XL-1 Red is *end* A1 *gyr* A96 *thi*-1 *hsd* R17 *sup* E44 *rel* A1 *lac mut* D5 *mut* S *mut* T. This strain is used for introducing random mutations into a gene of interest. Though this strain has the *Tn* 10 insertion with tetracycline, it should not be used for selection, as it is frequently lost.

Sources: Blattner FR, Plunkett IIIG, Bloch CA, Perna NT, Burland V, Riley M, Collado-Vides J, Glasner JD, Rode CK, Mayhew GF (1997) *Science* 277: 1453–1474; http://ecoliwiki.net/colipedia/index; Stanier RY, Lingraham JL, Wheelis ML, Painter PR (1986) *The Microbial World*, Prentice-Hall India, pp. 145–182.

(e.g., retrovirus, adenovirus, adeno-associated virus, Herpes simplex virus, Sindbis virus, Semliki forest virus, Lentivirus, *Vaccinia* virus, SV40, etc.), insect viruses (e.g., baculovirus), and plant viruses (e.g., cauliflower mosaic virus, gemini virus, tobacco mosaic virus, potato virus X, etc.)], *A. tumefaciens* tumor-inducing plasmid-based vectors, *A. rhizogenes* root-inducing plasmid-based vectors, chimeric plasmids (e.g., cosmid, phagemid, phasmid, fosmid), artificial chromosomes (e.g., yeast artificial chromosome, bacterial artificial chromosome, P1-derived artificial chromosome, mammalian artificial chromosome, and human artificial chromosome), and non-*E. coli* vectors (e.g., *Bacillus* and *Pseudomonas* vectors, etc.) (for details see Chapters 8 to 12).

1.2.2 Properties and Construction of a Vector DNA Molecule

A vector can be used for cloning to get DNA copies of the fragment inserted into it or as an expression vector to obtain expression of the cloned gene, i.e., to get either the RNA copies or protein corresponding to the cloned gene. To be ideal as a cloning vector, these should exhibit certain minimal properties; however, a few vectors should also possess certain specialist functions. These minimal or specialist properties are acquired either naturally or artificially (by retaining genes/loci for useful quality attributes and deleting nonessential ones from the potential vector DNA, or by gaining or combining useful genes/loci from different DNA molecules). The properties of ideal vectors to be used simply for multiplication of DNA or for expression of the cloned gene are as follows:

Capability of Autonomous Replication It is preferable to design a vector molecule that is capable of replication within the host cell so that numerous copies of recombinant DNA molecules are produced and passed to the daughter cells. Moreover, isolation of vector DNA should be possible independently of the host's genome. Thus, replicons containing an origin of replication, i.e., DNA capable of self-replication, are preferred as cloning vectors. Note that bacterial and viral genomes contain only one origin of replication, while eukaryotes contain multiple origins (i.e., autonomously replicating sequences (*ARS*).

On the contrary, some vectors are designed to lack the capability of autonomous replication (i.e., lack origin of replication). The purpose is to allow long-term expression of cloned genes to get several copies of DNA before expression or to regulate the expression of the genes toxic to the host cell. As these vectors lack origin of replication, the only way for their multiplication is by integration into the host genome. Such integrating vectors have low copy number and do not suffer from the problem of gene-dosage effect. Note that for integrating vectors, mapping of foreign DNA and the surrounding host sequences is required so as to facilitate integration by homologous recombination.

Small Size A cloning vehicle needs to be reasonably small in size (lower molecular weight) and manageable. Small molecules are easy to handle, isolate, and manipulate, as these are less susceptible to damage by shearing stress during purification. On the other hand, large molecules tend to break down or degrade due to shearing and are more difficult to handle and manipulate. Moreover, in small molecules, the chance of occurrence of unique sites for restriction enzyme(s) increases. In addition, the efficiency of gene transfer is also high with small vector molecules. Furthermore, small plasmids are usually high copy number plasmids and hence give higher yield of product (RNA or proteins) expressed from the cloned gene. The only disadvantage of high copy number plasmids is gene-dosage effect. Small wild-type plasmids are usually nonconjugative and hence can be biologically contained. Thus for cloning, a small wild-type vector may be used directly for cloning or a large vector DNA molecule may be subjected to size (or gene) diminution, i.e., reduction in size by deleting sites that do not affect its replicative ability, growth, and viability.

Presence of Selectable Marker Gene(s) Vector molecules should possess a selectable marker gene(s) (e.g., antibiotic resistance genes, *lac* Z, or resistance to toxin, etc.) that confer readily scorable phenotypic traits on host cells harboring recombinant DNA molecule, thereby allowing easy detection of the recombinants (for details see Chapter 16). If not present naturally, these selectable marker genes are artificially introduced into the vector molecule by different strategies including restriction digestion of source DNA followed by ligation of fragment containing selectable marker gene(s) into vector molecule, recombineering, or transposition. Note that some antibiotic resistance genes that are located on transposons are moved from one bacterial cell to another by transposition [e.g., *Tn* 1, *Tn* 5, *Tn* 9, and *Tn* 10 contain resistance genes for ampicillin (amp^r or ap^r), kanamycin (kn^r or kan^r), chloramphenicol (cm^r), and tetracycline (tet^r), respectively]. Some expression vectors also possess reporter gene(s) to determine the expression of the cloned gene and its level of expression (for details see Chapter 16).

Presence of Unique Restriction Enzyme Site(s) or Multiple Cloning Sites for Inserting the Target DNA For cloning the gene of interest, unique restriction enzyme site(s) should be present on the vector DNA molecule either individually or as cluster in the form of multiple cloning site (MCS; polylinker). The position of these restriction sites should be such that the insertion of a segment of DNA in any of these sites brings about a phenotypic change in the characteristics of vector DNA molecule, for example, loss of resistance to antibiotic or loss of expression of a gene whose product is an enzyme that normally carries out a reaction, leading to an easily recognizable trait (e.g., change in color). Note that MCS is chemically synthesized cluster of restriction enzyme recognition and cleavage sites. The clustering of restriction enzyme sites in the form of MCS allows a DNA fragment with two different ends to be cloned in particular orientation (i.e., directional cloning is possible provided the restriction enzyme sites in MCS are asymmetric) without resorting to additional manipulations such as linker/adaptor attachment. This also increases the flexibility in the use of restriction enzymes.

Ease of Purification Generating and purifying large amounts of vector DNA from the host cell should be easy and straightforward. This is possible when the size of the vector DNA molecule is small so that it is not susceptible to damage by shearing and the purification procedure is simple.

No Effect on the Replicative Ability of Vector due to Insertion of Target DNA The introduction of target DNA into vector DNA molecule to form recombinant DNA molecule followed by its introduction into host cell should not change the replicative ability of vector DNA molecule.

Ease of Reintroduction into Host Cell with High Efficiency The transformation/transfection protocols should be well-established and easy to perform. Moreover, the efficiency of these procedures should be high.

Biological Containment The self-catalyzed transfer of recombinant DNA from one host to another is not preferred. In simple words, vectors should be biologically contained with no possibility of gene escape. To achieve this, one way is to use nonconjugative and nonmobilizable plasmid vectors. The conjugative or mobilizable plasmids may be made nonconjugative or nonmobilizable by mutating or deleting responsible loci, for example, *tra*, *mob*, *nic*, or *bom*. Their deletion also leads to extensive reduction in plasmid size (for details see Chapter 8). For biological containment of λ bacteriophage, the strategy for biological containment is based on amber mutations in the lysis genes, for example, *S*, *R*, head, and tail genes, and their suppression using sup^+ host at a later stage when required (for details see Chapter 9).

Presence of Promoters and Ribosome Binding Site Through a vector, it is possible to give a cloned DNA fragment additional characteristics that may expand its use considerably. For example, expression vector containing promoter *in frame* with the cloned gene (transcriptional fusion vector) allows *in vitro* transcription of the cloned DNA fragment and that containing *in frame* promoter and ribosome binding site (RBS) (translational fusion vector) allows *in vitro* translation of the cloned DNA fragment (for details see Chapter 12). However, the transcription and/or translation of the cloned gene can also be achieved in a vector lacking promoter and RBS by including them *in frame* with the cloned gene itself. Note that if we are interested in simply getting the DNA copies of the cloned gene, then expression vectors need not be used.

Presence of Two Different Origins of Replication or Broad Host-range Origin of Replication Most of the routine manipulations involved in gene cloning experiments use *E. coli* as the host organism; consequently, most cloning vectors have origin of replication for multiplication in *E. coli*. *E. coli* is particularly used when the aim of the cloning experiment is to study gene structure and function. However, when the aim is not to study a gene, but is controlling or improving synthesis of an important metabolic product (e.g., a hormone such as insulin) or changing the properties of the organism (e.g., to introduce herbicide resistance into a crop plant), it is desirable to use a different host for a gene cloning experiment after gene manipulation studies in *E. coli*. Thus for such purposes, initial isolation and analysis of DNA fragments is almost always carried out using *E. coli* as the host organism and further manipulations are done in the second host. To easily carry out these steps, there should be some strategy that allows multiplication both in *E. coli* and other host. A solution to this is a shuttle vector, i.e., the vector containing two different origins of replication or a broad host-range origin of replication that allows multiplication in two different hosts, for example, *E. coli* and *A. tumefaciens* (for details see Chapter 12).

Specialist Vectors Expression vectors may be designed for some special purposes (for details see Chapter 12). For example, the expression vector design may be needed for the following: (i) Vector DNA molecules targeting the expressed protein corresponding to the gene of interest to extracellular medium for easy purification and reduced contamination with cellular extracts. This involves insertion of the target DNA *in frame* with signal sequence of secretory proteins; (ii) Vector DNA molecules allowing surface display of the expressed heterologous protein so that recombinants are easily recognized through immunological procedures. For this, the target DNA is cloned within the gene encoding outer membrane protein (in bacteria) or in the gene encoding a capsid protein (in phage); and (iii) Tagged vector DNA molecules that allow easy purification of the expressed heterologous protein through affinity purification or antigen–antibody reaction. This involves appending of a His tag or a marker peptide to the gene of interest.

1.3 STEPS INVOLVED IN GENE CLONING

As evident from the previous section, the prime requisite of a gene cloning experiment is the choice of a suitable vector.

Once this selection is done, a gene cloning experiment can be performed through a series of steps. The first step is the isolation of pure DNA from a particular source. This is followed by isolation of DNA fragments using different strategies, which are discussed in Section 1.3.3. Note that the gene of interest may become a part of a single or a few DNA fragments. DNA fragments containing the gene to be cloned (i.e., fragmented genomic DNA) are inserted into cloning vectors to produce composite molecules called the recombinant DNA molecules. Note that each DNA fragment becomes inserted into a different vector molecule to produce a family of recombinant DNA molecules, each carrying a different gene or part of a gene. The resulting mixture of recombinant DNA molecules thus represents the entire genetic complement of an organism, one (or more) of which carries the gene of interest. In the next step, recombinant DNA molecules are introduced into host cells, usually bacteria, or any other living cells. These recombinants multiply when plated on a suitable selection medium, producing colonies or plaques depending upon the vector used. The individuals in the colonies or plaques are genetically identical and are referred to as 'clones' and the collection of all the clones is called a 'genomic library'. Note that usually only one recombinant DNA molecule is transported into any single host cell, so that the final set of clones may contain multiple copies of just one molecule. The library is then screened to isolate the clone(s) of interest and the desired clone(s) is stored.

For cDNA library construction, total RNA is isolated from the source, followed by mRNA purification from total RNA and cDNA synthesis through reverse transcription (for details see Chapter 5). The end result is several cDNA molecules; one from each mRNA. Each cDNA is then cloned into vector DNA molecules and the resulting recombinant DNA molecules are introduced into host cells to get a cDNA library, which represents the entire complement of the expressed sequences within a given cell or tissue type at a particular period.

On the contrary, if the gene is small and its sequence is known from same or different source, it can be directly synthesized chemically taking account of codon bias (for details see Chapter 4) or amplified using gene-specific primers through polymerase chain reaction (PCR) (for details see Chapter 7). In either case, a single DNA fragment containing the gene of interest is obtained, which can be directly cloned into a vector DNA molecule resulting in the formation of a single recombinant DNA molecule, i.e., a library is not obtained.

A diagrammatic representation of the steps involved in a gene cloning experiment is presented in Figure 1.1, and in the following sections, these steps are discussed in detail.

1.3.1 Isolation and Purification of Total Cellular DNA

The first step in a gene cloning experiment is isolation of DNA from a particular source. There are several procedures for the isolation and purification of DNA. However, the fundamentals of all these procedures are same, and are discussed in this section. For easy understanding, the simplest type of DNA purification procedure from bacteria is discussed in detail, while only specific requirements are mentioned for the isolation and purification of DNA from other sources. The procedures for isolation and purification of vector DNA, for example, plasmids and phage (λ and M13) DNA is also based on the same principle as described below; however, the exact procedures are discussed in Chapters 8 to 10.

Isolation and Purification of Total DNA from Bacterial Cells

For all the procedures in molecular cloning experiments, the first step is the isolation of pure and good quality DNA. The procedure for total DNA preparation from a bacterial culture is divided into four stages: (i) Growing and harvesting a bacterial culture; (ii) Lysis of bacterial cells to release their contents and removal of cell debris (i.e., to get cell extract); (iii) Treatment of cell extract to remove all contaminating components (i.e., purification of DNA); and (iv) Precipitation and concentration of DNA (i.e., recovery of purified DNA). These steps are discussed in detail below and also summarized in Figure 1.2.

Growing and Harvesting a Bacterial Culture The first step in the isolation of DNA from bacterial cells is growing and harvesting a bacterial culture.

Growth medium Most bacteria are grown easily in a liquid medium (broth culture), which provides a balanced mixture of all the essential nutrients at concentrations that allow efficient growth and division of bacteria. Two typical bacterial growth media are LB and M9; however, there are a variety of other specific media. LB medium comprises of tryptone (10 g/l), yeast extract (5 g/l), NaCl (10 g/l), and pH adjusted to 7.0. Note that it is a complex or undefined medium, since the precise identity and quantity of its components are not known. This is because two of the ingredients, tryptone and yeast extract, are complicated mixtures of unknown chemical compounds. Tryptone supplies amino acids and small peptides, while yeast extract (a dried preparation of partially digested yeast cells) provides nitrogen along with sugars and inorganic and organic nutrients. This medium needs no further supplementation and supports the growth of a wide range of bacterial species. On the other hand, M9 is a defined medium in which all the components and their quantities are known. This medium contains a mixture of inorganic

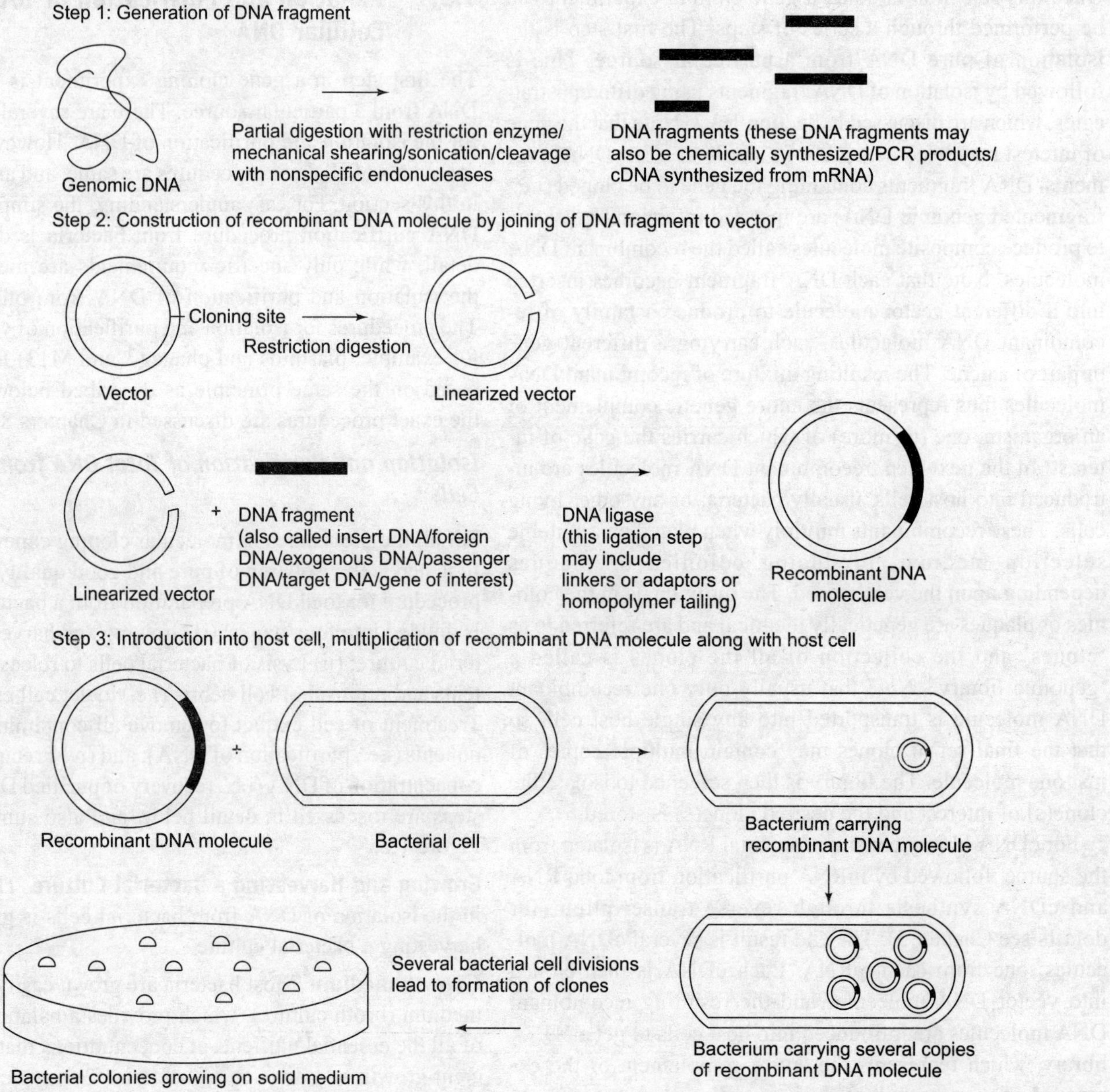

Figure 1.1 Basic steps in gene cloning [In this figure, the genomic DNA and vector are cut with blunt end cutter, however, a sticky end cutter can also be used depending upon the requirement.]

nutrients to provide essential elements such as nitrogen, magnesium, and calcium, as well as glucose to supply carbon and energy. The ingredients (per liter) of M9 minimal medium are 5 X M9 salts (200 ml; comprising 64.0 g/l $Na_2HPO_4 \cdot 7H_2O$, 15.0 g/l KH_2PO_4, 2.5 g/l NaCl, 5.0 g/l NH_4Cl), 1 M $MgSO_4$ (2 ml), 20% solution of appropriate carbon source (e.g., 20% glucose; 20 ml), 1 M $CaCl_2$ (0.1 ml), and rest water; pH adjusted to 7.0. Note that separately sterilized $MgSO_4$ and $CaCl_2$ solutions (by autoclaving) and glucose (by membrane sterilization through 0.22 μm filter) are added to autoclaved and diluted solution of 5 X M9 salts. In practice, additional growth factors such as amino acids, vitamins and trace elements are added to the M9 minimal medium to support bacterial growth, the selection of which depends on the species concerned. Defined media is used when the bacterial culture has to be grown under precisely controlled conditions, while complex medium is appropriate when the culture is being grown simply as a source of DNA.

Growth conditions As *E. coli* is mesophilic and facultative anaerobe, it is grown in LB medium at 37°C and aerated by shaking at 150–250 rpm on a rotary platform. Under these conditions, *E. coli* cells divide once in every 20 min approximately.

Harvesting by centrifugation after acquisition of appropriate cell density The density of a bacterial culture is measured by monitoring the optical density (OD) at 600

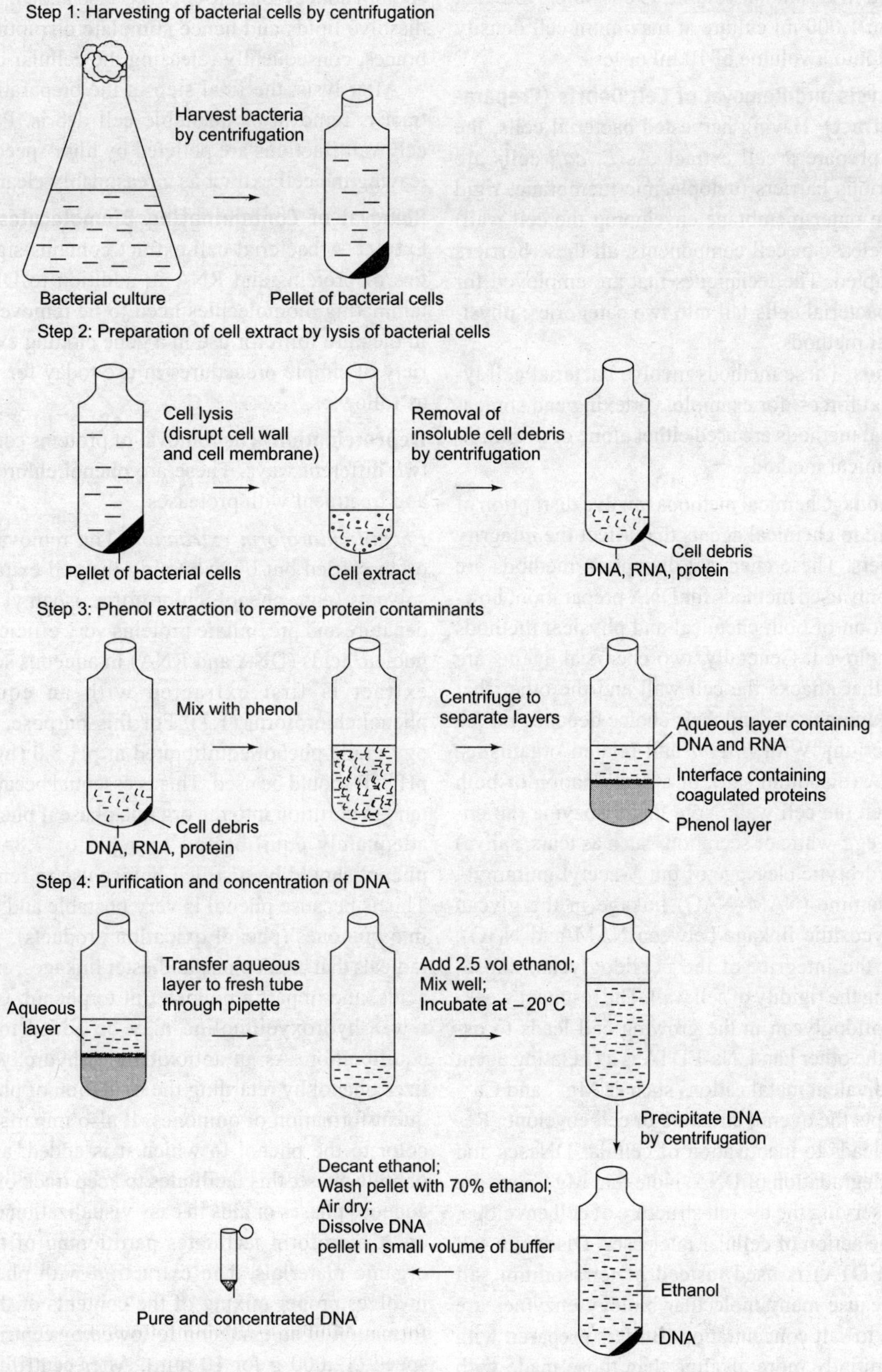

Figure 1.2 The basic steps in preparation of total cell DNA from a culture of bacteria

nm. At this wavelength, one OD corresponds to about 0.8 × 10^9 cells/ml. Once appropriate growth is obtained, the bacterial cells are harvested by spinning the culture in a centrifuge at fairly low centrifugation speeds for small time intervals (3,000–8,000 *g* for ≤5 min) to prevent the formation of compact pellets. After centrifugation, the bacterial cells pellet at the bottom of the centrifuge tube, allowing the culture medium to be decanted off. The harvested bacterial cells are then resuspended in as small a volume as possible. Note that the bacteria from 1,000 ml culture at maximum cell density are resuspended into a volume of 10 ml or less.

Bacterial Cell Lysis and Removal of Cell Debris (Preparation of Cell Extract) Having harvested bacterial cells, the next step is to prepare a cell extract. As *E. coli* cells are enclosed by various barriers (cytoplasmic membrane, rigid cell wall, and an outer membrane enveloping the cell wall) preventing the release of cell components, all these barriers have to be disrupted. The techniques that are employed for breaking open bacterial cells fall into two categories: physical and chemical methods.

Physical methods These methods involve bacterial cell lysis by mechanical forces, for example, vortexing and sonication. The physical methods are used either alone or in combination with chemical methods.

Chemical methods Chemical methods involve disruption of cells by exposure to chemical agents that affect the integrity of the cell barriers. These chemical disruption methods are the most commonly used methods for DNA preparation; however, a combination of both chemical and physical methods may also be employed. Generally, two chemical agents are employed, one that attacks the cell wall and the other that disrupts the cell membrane, and their choice depends on the species of bacterium. With *E. coli* and related organisms, lysozyme, EDTA (disodium salt), or a combination of both is used to weaken the cell wall. Note that lysozyme (an enzyme present in egg-white or secretions such as tears, saliva) catalyzes the hydrolytic cleavage of the *N*-acetyl muramyl–*N*-acetyl glucosamine (NAM–NAG) linkage in the glycan strands (i.e., glycosidic linkage between NAM and NAG), which destroys the integrity of the peptidoglycan, consequently disrupting the rigidity of cell wall. The resultant weakening of the peptidoglycan in the growing cell leads to osmotic lysis. On the other hand, Na_2EDTA is a chelating agent that sequesters divalent metal cations such as Mg^{2+} and Ca^{2+} and hence disrupts the overall structure of cell envelope. Removal of Mg^{2+} leads to inactivation of cellular DNases and hence prevents degradation of DNA. Note that Mg^{2+} ions are essential for preserving the overall structure of cell envelope, as well as for the action of cellular nucleases. Disodium salt of EDTA (Na_2EDTA) is used instead of tetrasodium salt (Na_4EDTA), because many molecular biology enzymes are highly sensitive to salt concentration. Buffers prepared with Na_4EDTA are initially more alkaline than those made with Na_2EDTA and the subsequent addition of HCl to establish pH 7.0–8.0 greatly increases the NaCl concentration in the final buffer, and likely have a negative influence on the outcome of an experiment. Under some conditions, weakening of the cell wall with lysozyme or Na_2EDTA is sufficient to cause bacterial cells to burst, but usually a detergent such as sodium dodecyl sulfate (SDS) is also added. These detergents dissolve lipids and hence stimulate disruption of cell membranes, consequently releasing the cellular components.

After lysis, the final step in the preparation of a cell extract is removal of insoluble cell debris. Partially digested cell wall fractions are pelleted by high-speed centrifugation, leaving the cell extract as a reasonably clear supernatant.

Removal of Contaminating Biomolecules from the Cell Extract A bacterial cell extract contains significant quantities of proteins and RNA in addition to DNA. These contaminating biomolecules need to be removed, leaving DNA in the pure form for use in a gene cloning experiment. A variety of simple procedures in use today for this purpose are as follows:

Deproteination The removal of proteins can be achieved by two different ways. These are phenol:chloroform extraction and treatment with proteases.

Phenol:chloroform extraction The removal of proteins is often carried out by extracting the cell extract with organic solvents (e.g., phenol, chloroform, isoamyl alcohol), which denature and precipitate proteins very efficiently, leaving the nucleic acids (DNA and RNA) in aqueous solution. The cell extract is first extracted with an equal volume of phenol:chloroform (1:1). For this purpose, molecular biology grade phenol equilibrated at pH 8.0 (by 1 M Tris-HCl, pH 8.0) should be used. This is essential because nucleic acids tend to partition into the organic phase if phenol has not been adequately equilibrated to a pH of 7.8–8.0. Moreover, phenol should be distilled before use to remove impurities. This is because phenol is very unstable and oxidizes rapidly into quinones (phenol oxidation products), which form free radicals that break phosphodiester linkages, cross-link nucleic acids, and impart a pinkish tint to phenol. Optionally, 0.1% w/v 8-hydroxyquinoline may be added to phenol during equilibration. As an antioxidant, 8-hydroxyquinoline stabilizes phenol by retarding the oxidation of phenol and consequent formation of quinones. It also imparts a bright yellow color to the phenol to which it is added, and hence to the organic phase; this facilitates to keep track of the organic and aqueous phases or aids in easy visualization of separated layers. Chloroform facilitates partitioning of the aqueous and organic materials. The extraction with phenol:chloroform involves proper mixing of the contents of the tube until the formation of an emulsion followed by centrifugation at high speed (12,000 *g* for 10 min). After centrifugation, aqueous phase forms the upper phase, while organic phase (yellow if 8-hydroxyquinoline is added) forms the lower phase. In case the aqueous phase contains >0.5 M salt or >10% sucrose, it becomes denser than organic phase and forms the lower phase. The precipitated protein molecules form a white coagulated mass at the interface between the aqueous and organic layers.

As nucleic acids (DNA and RNA) contain highly charged phosphate backbone, these are polar and hydrophilic (water soluble) in nature and hence get partitioned in the aqueous layer along with salts. This is explained by polar nature and high dielectric constant of water. [Note that water is also polar molecule with a partial negative charge near the oxygen atom due the unshared pairs of electrons and partial positive charges near the hydrogen atoms. It has a high dielectric constant (80.1 at 20°C), indicating that electrical force between any two charges in aqueous solutions is highly diminished compared with the force in vacuum or air. At an atomic level, this diminishing of force acting on charges results from water molecules forming hydration shells (solvation shells) around them. It makes water a very good solvent for charged compounds (nucleic acids or salts)]. The aqueous phase (containing nucleic acids and salts) is pipetted out and transferred to a fresh tube, while organic phase and interface (containing denatured proteins) are discarded. Some cell extracts contain such a high protein content that a single phenol extraction is not sufficient to completely purify nucleic acids. This problem can, however, be circumvented by performing several phenol:chloroform extractions one after the other until no protein is visible at the interface. Moreover, deproteination is more efficient when two or more different organic solvents are used instead of one. Hence, it is preferable to perform extraction with an equal volume of phenol:chloroform: isoamyl alcohol (25:24:1) after phenol:chloroform extraction, Isoamyl alcohol in the mixture reduces foaming generated by the mechanics of the extraction procedure. After organic extraction(s), the aqueous phase is extracted with an equal volume of chloroform. As phenol in extensively soluble in chloroform, any lingering traces of phenol are removed by extracting with an equal volume of chloroform. The removal of trace amounts of phenol from the nucleic acid preparation is essential so that quinones do not compromise the integrity of the nucleic acid sample. Note that phenol, chloroform, and their combinations are caustic and carcinogenic reagents that must be handled with extreme care. Phenol is highly corrosive. It anesthetizes the skin and can result in severe burns that scar on heating. It is therefore essential to work in a fume hood and wear a lab coat, gloves, and eye protection glasses. If any phenol comes into contact with skin, the affected area should be washed immediately with copious amounts of water. Note that ethanol (EtOH) should not be used for washing since this enhances phenol absorption and increases the severity of burn.

Treatment with proteolytic enzymes The removal of protein contaminants by several phenol:chloroform extractions is undesirable as each mixing and centrifugation step results in a certain amount of breakage of the DNA molecules, resulting in decrease in DNA yield. As an alternative, it is usual to remove most of the proteins before phenol extraction by digestion with proteolytic enzymes (proteases) such as pronase or proteinase K. These enzymes are active against a broad spectrum of native proteins and break down the polypeptides into smaller units that are easily removed by phenol. Proteinase K is proteolytic enzyme (a serine protease) that is purified from the mold *Tritirachium album*. In solution, it is stable over pH range 4–12.5 with an optimum of pH 8.0 and a temperature range 25–65°C. Although the enzyme has two binding sites for Ca^{2+}, in the absence of this divalent cation, some catalytic activity is retained for the degradation of proteins commonly found in nucleic acid preparations. Occasionaliy, proteinase K digestion is carried out in the presence of Na_2EDTA at 50°C to inhibit labile, Mg^{2+}-dependent nucleases. Proteinase K is commonly prepared as a 20 mg/ml stock solution in sterile water (stable for 1 year at –20°C) or in a solution of 50 mM Tris-HCl (pH 8.0) and 1 mM $CaCl_2$ (stable for months at 4°C). It is generally used at a working concentration of up to 50 μg/ml in any of a number of buffer formulations, one example of which is 10 mM Tris-HCl (pH 8.0), 1 mM Na_2EDTA, and 0.5% SDS. Maximum proteinase K activity is observed with the inclusion of 1 mM Ca^{2+} in the reaction buffer. Instead of proteinase K, pronase isolated from *Streptomyces griseus* can be used. In some cases, it may be necessary to perform a pronase self-digestion to eliminate contaminating DNase activity. If necessary, this is easily accomplished by incubation of the pronase stock [20 mg/ml in 10 mM Tris-HCl (pH 7.5) and 10 mM NaCl] at 37°C for 1 hour. This extra task is usually avoided by purchasing predigested pronase. In either case, suitable aliquots of pronase are stored at –20°C. Reaction conditions for pronase are identical to those for proteinase K except that the recommended working concentration for pronase is about 1 mg/ml.

Removal of RNA contamination The contaminating RNA molecules must also be removed from DNA before subjecting it to cloning experiments. Some RNA molecules, especially mRNAs, are removed by phenol extraction as described earlier. However, this is not a useful strategy for most RNA molecules, which remain with DNA in the aqueous layer even after phenol extraction. Traditionally used isopycnic ultracentrifugation (for details see Chapter 6) for partitioning grossly contaminated samples is also not the method of choice for removal of RNA. This is because the method is time-consuming, requires expensive and highly specialized equipment, and not preferable when the combined mass of DNA and RNA in the sample totals only a few micrograms. The strategies routinely employed for the removal of contaminating RNA molecules include the following:

Ribonuclease treatment After deproteination, the cell extract comprises of nucleic acids. Contaminating RNA molecules need to be removed from this nucleic acid preparation without compromising the integrity of DNA. The most effective way to remove RNA is to treat aqueous layer with the enzyme, ribonuclease (RNase). Two most commonly used RNases in DNA isolation procedure are RNase A and RNase T1 [stock solution of 10 mg/ml in water or TE buffer (10 mM Tris-HCl and 1 mM Na_2EDTA, pH 8.0)]. These enzymes rapidly degrade RNA into ribonucleotide subunits, which are then easily removed by phenol extraction. As crude preparations of RNases, if not specially treated, can harbor significant levels of DNase activity as well, RNase preparations should be purged of all intrinsic DNase activity (i.e., made DNase-free) in the laboratories. This is routinely done by heating RNase stock solutions to near boiling (90°C) for 10 min, followed by cooling on ice. This is because at 90°C, DNase activity is quickly eliminated without compromising the RNase activity of the reagent. The DNase-free RNase thus obtained is stored frozen in aliquots. DNase-free RNase that harbors no intrinsic DNase activity is also commercially available. Note that RNase is costly and should be used in minute amounts. Hence, RNase treatment of nucleic acid preparation to remove RNA contamination is done after concentration step.

Application of monophasic reagent The application of monophasic reagent containing acidified phenol, guanidinium or ammonium thiocyanate, and a phenol solubilizer as lysis buffer allows single-step simultaneous isolation of DNA, RNA, and proteins. This method involves lysis of cells with a monophasic solution of guanidine or ammonium isothiocyanate and phenol. Note that guanidinium salts are chaotropic and denaturing agents that function by destroying the 3-D structure of proteins and convert most proteins to a randomly coiled state through an unclear mechanism. Addition of chloroform generates a second (organic) phase into which DNA and proteins are extracted, leaving RNA in the aqueous supernatant. DNA is purified from the organic phase by precipitation with EtOH. DNA recovered from the organic phase is ~20 kbp in size and is a suitable template for PCR. The proteins, however, remain denatured as a consequence of their exposure to guanidine and are used chiefly for immunoblotting. As RNA is partitioned in aqueous phase, its contamination in DNA preparation is easily eliminated. Note that the same procedure is used for the isolation of pure RNA, where RNA is precipitated from aqueous phase with isopropanol and the contaminating DNA and proteins are eliminated in the organic phase (for details see Section 1.3.2). Contaminating biomolecules may also be removed by silica-based purification in combination with guanidinium thiocyanate. These methods also form the basis of some commercially available nucleic acid isolation kits.

Precipitation and Concentration of DNA The next step in a DNA isolation and purification procedure is recovery of DNA through easy and rapid processes, which are discussed below.

Precipitation with EtOH Coulomb's law dictates the electrostatic attractions between the positively charged ions in solution and the negatively charged phosphate ions, and this interaction is also affected by the dielectric constant of the solution. As water has a high dielectric constant, it becomes fairly difficult for the sodium and phosphate ions to come together. Consequently, even in the presence of positively charged ions in aqueous solution, the negatively charged phosphate groups on nucleic acid backbone forms relatively weak electric force, which prevents them from precipitating out of solution. This is in principle similar to weakening of electric force, which normally holds salt crystals together by way of ionic bonds in the presence of water, thereby separating ions from the crystal, and spread through solution. The addition of EtOH increases the stability of ionic interactions between positively charged ions and negative charged phosphate residues on nucleic acid backbone, thereby allowing nucleic acid precipitation. The principle behind EtOH precipitation is that EtOH being less polar than water with a much lower dielectric constant (24.3 at 25°C) disrupts the screening of charges by water (i.e., exposes negatively charged phosphate groups of nucleic acids by depleting the hydration shell from them), makes it much easier for positively charged ions to interact with phosphates, and makes the nucleic acid less hydrophilic, causing it to drop out of solution. In other words, the repulsive forces between the polynucleotide chains are reduced to an extent that a precipitate can form. Thus, if sufficient EtOH (~64% of the solution) is added, the electrical attraction between phosphate groups and any positively charged ions present in solution becomes strong enough to form stable ionic interactions.

Based on the principle described, it is best to recover nucleic acids from aqueous solutions by standard precipitation with 2–3 volumes of absolute EtOH (calculated after addition of salt) in the presence of monovalent cations. The most commonly used cations for this purpose are 0.3 M sodium acetate (pH 5.2), 0.2–0.5 M sodium chloride, and 2–2.5 M ammonium acetate (Table 1.1). These positively charged ions neutralize the negative charge on nucleic acid, making it far less hydrophilic and therefore much less soluble in water. Note that these ions should be removed at a later stage as these may interfere with the downstream applications of isolated DNA.

After EtOH precipitation, the next step is to recover the precipitated nucleic acids. This is done either by spooling or centrifugation at high speeds. Often a successful preparation of nucleic acids results in a very thick solution that does not need to be concentrated any further. With a thick solution of nucleic acid, EtOH is layered on top of the sample, causing

Table 1.1 Salt Solutions Used in Ethanol Precipitation of Nucleic Acids

Salt	Stock solution (M)	Final concentration (M)	Important remarks
Sodium acetate	3.0 (pH 5.2)	0.3	Stock solution (3.0 M) is added one-tenth the volume of aqueous phase to get a final concentration of 0.3 M; 0.3 M sodium acetate (pH 5.2) is used for most routine precipitations of DNA and RNA.
Ammonium acetate (NH_4OAc)	5.0	2.0–2.5	Stock solution (5.0 M) is added in a volume equal to that of aqueous phase to get a final concentration of 2.5 M; It is frequently used to reduce the coprecipitation of unwanted contaminants (e.g., dNTPs or oligosaccharides) with nucleic acids. For example, two sequential precipitations of DNA in the presence of 2.0 M ammonium acetate result in the removal of >99% of the dNTPs from preparations of DNA; It is also the best choice when nucleic acids are precipitated after digestion of agarose gels with agarase (for details see Chapter 6). This is because the use of ammonium ions reduces the possibility of coprecipitation of oligosaccharide digestion products; It is used frequently for the removal of unincorporated nucleotides following a DNA labeling reaction; It is not used when the precipitated nucleic acid is to be phosphorylated, as bacteriophage T4 polynucleotide kinase is inhibited by ammonium ions.
Sodium chloride (NaCl)	5.0	0.2	0.2 M should be used if the DNA sample contains SDS. The detergent remains soluble in 70% ethanol; It is not as soluble as NH_4OAc, NaOAc, or LiCl in EtOH-water, or isopropanol water.
Lithium chloride (LiCl)	8.0	0.8	Stock solution (8.0 M) is added one-tenth the volume of aqueous phase to get a final concentration of 0.8 M; It is frequently used when high concentrations of EtOH are required for precipitation (e.g., when precipitating RNA). LiCl is very soluble in ethanolic solutions and is not coprecipitated with the nucleic acids; Small RNAs (tRNAs and 5S rRNAs) are soluble in solutions of high ionic strength (without ethanol), whereas large RNAs are not. Because of this difference in solubility, precipitation in high concentrations of LiCl (0.8 M) can be used to purify large RNAs.

molecules to precipitate at the interface. A spectacular trick is to push a glass rod through the EtOH into the nucleic acid solution. When the rod is removed, nucleic acid molecules adhere to it and are pulled out (spooled out) of the solution in the form of a long fiber. In case the aqueous phase is in large volumes or the nucleic acid preparation is dilute, it becomes important to consider methods for increasing the concentration of nucleic acids. Thus, for precipitation of nucleic acids from dilute solutions, the organic (EtOH) and aqueous phases are mixed by vortexing when isolating small DNA molecules (<10 kbp) or by gentle shaking when isolating DNA molecules of moderate size (10–30 kbp). When isolating large DNA molecules (>30 kbp), precautions must be taken to avoid shearing, for example, using large bored pipette tips for transfer from one tube to another. If SDS is present, it can be readily removed in the supernatant by adding NaCl to a final concentration of only 0.2 M or by using LiCl (0.5 M) before adding the alcohol.

After centrifugation, the supernatant is discarded and the pellet (nucleic acid) is washed with 70% EtOH (its volume should be sufficient to at least cover the pellet and wet the sides of the tube) and air-dried (by leaving the tube open on the laboratory bench for a few minutes or in speed vac). Note as the pellet formed after EtOH precipitation contains both nucleic acids and salts and as nucleic acid molecules and salt form aggregates, demonstrating dramatically reduced solubility in alcohol, washing with 70% EtOH is important. Washing with 70% EtOH removes much of the salt present in the leftover supernatant or bound to DNA pellet and often loosens the precipitates from the wall of the tube, however, it is not sufficiently aqueous to redissolve DNA. DNA pellets should not be allowed to dry out completely, as it leads to denaturation of DNA and make it harder to resuspend. Moreover, it has been observed that slightly damp DNA pellets with EtOH dissolve more readily than completely dried ones. When dealing with small amounts (ng or pg) of DNA, it is prudent to save the ethanolic supernatant from each step (i.e., precipitation with absolute EtOH and washing with 70% EtOH) until the entire DNA has been recovered. The air-dried pellet (nucleic acids) is redissolved or rehydrated in an appropriate volume of 1 X TE buffer (pH 8.0) or sterile water. The aqueous solution of nucleic acids thus obtained is treated with RNase, extracted with phenol, and the DNA is precipitated with EtOH in the presence of salt. Besides the recovery of nucleic acids, EtOH precipitation is used for desalting and has the added advantage of leaving short-chain and monomeric nucleic acid components in solution. Hence ribonucleotides produced by RNase treatment are lost at this stage. The recovered pellet is washed with 70% EtOH, air-dried, and redissolved in TE buffer (pH 8.0). The resulting solution contains only DNA, which is stored at 4°C. Note that for long-term storage, DNA is stored as suspension in EtOH at –20°C.

The concentration of EtOH and salt added, time and temperature of incubation with EtOH, and the length and speed of centrifugation play a significant role in the process. The concentration of salt added should be appropriate, as too much of salt results in large amount of salt coprecipitating with nucleic acid and too little results in incomplete recovery of nucleic acid. EtOH precipitations are commonly performed at –70°C, –20°C, 0°C, or room temperature for periods of 5 minutes to overnight. The optimal incubation time depends on the length and concentration of DNA. For low DNA concentrations, higher final concentrations of EtOH, longer precipitations (for 1 hour to overnight), lower temperatures (–20°C to –70°C), and longer centrifugation times (up to 30 min) may be required for efficient recovery. This is because at lower temperatures, the viscosity of alcohol is greatly increased and centrifugation for longer times may be required to effectively pellet the precipitated DNA. The efficiency of precipitation for small concentrations of DNA may be increased by incubation at –70°C, but these reactions should be brought to 0°C before centrifugation. Due to increase in the viscosity of solution, some researchers are of the view that at lower temperatures precipitation efficiency is lowered and the highest precipitation efficiency is achieved at room temperature. However, when possible degradation at room temperature is considered, it is probably best to incubate nucleic acid preparation on wet ice or lower temperatures. Thus some researchers recommend that nucleic acids at concentrations as low as 20 ng/ml are precipitated at 0–4°C, so incubation for 15–30 min on ice is sufficient. Some researchers have also shown that this precipitation period is unnecessary and recommend direct centrifugation immediately after the addition of alcohol. During centrifugation, as the precipitated nucleic acids have to move through EtOH solution to the bottom of the tube, the time and speed of centrifugation have a significant effect on the recovery rates of nucleic acids. Smaller fragments and higher dilutions (lower nucleic acid concentrations) require longer and faster centrifugation as compared with larger fragments and lesser dilutions. Thus, for very small lengths and low concentrations overnight incubation is recommended. In such cases, the recovery of small quantities of nucleic acids is improved with the addition of carriers or coprecipitants, for example, yeast tRNA (10–20 μg/ml), glycogen (50 μg/ml), or linear polyacrylamide (10–20 μg/ml). Note that the carriers are inert substances that are insoluble in ethanolic solutions and form precipitates that trap the target nucleic acids. During centrifugation, carriers generate a visible pellet that facilitates handling of the target nucleic acids. Yeast tRNA is inexpensive, but it cannot be used for precipitating nucleic acids that are to be used as substrates in reactions catalyzed by polynucleotide kinase or terminal transferase. This is because the termini of yeast tRNA are excellent substrates for these enzymes and compete with the termini contributed by the target nucleic acid. Glycogen is used as a carrier when nucleic acids are precipitated with 0.5 M ammonium acetate and isopropanol. Glycogen does not compete with the target nucleic acids in subsequent enzymatic reactions. However, it interferes with the interactions between DNA and proteins. Linear polyacrylamide is an efficient neutral carrier for precipitating picogram amounts of nucleic acids with EtOH. Note that in the absence of a carrier, DNA concentrations as low as 20 ng/ml form a precipitate that can be quantitatively recovered by centrifugation in a microfuge, however, when lower concentrations of DNA or very small fragments (<100 nucleotides in length) are processed, more extensive centrifugation may be necessary to cause the pellet of nucleic acid to adhere tightly to the centrifuge tube. Centrifugation at 1,00,000 *g* for 20–30 min allows the recovery of picogram quantities of nucleic acid in the absence of carrier. In general, the length of time of centrifugation is more important for precipitating DNA than chilling the solution in –20°C or –70°C freezer.

Precipitation with isopropanol EtOH can also be replaced with isopropanol. This is advantageous because its precipitation efficiency is higher making 0.6–1.0 v/v (just half the volume of EtOH) enough for precipitation, although higher concentrations may be helpful when the DNA is at low concentration. Moreover, precipitation is done at room temperature, allows coprecipitation of fewer salts, and the pellet adheres less tightly to the tube. Isopropanol precipitation is more effective than EtOH in separating primers from PCR products. However, many salts are less soluble in isopropanol than in EtOH, hence a second 70% alcohol rinse of the pellet is recommended to more efficiently desalt the DNA pellet or else salts that are more soluble in isopropanol are preferred. Moreover, as isopropanol is less volatile than EtOH, more time is required for air-drying in the final step. Similar to EtOH precipitation procedure, subsequent desalting of the DNA pellet involves rinsing in 70% alcohol, recentrifugation, air-drying, and resuspension in appropriate buffer or water.

Extraction with butanol followed by EtOH precipitation DNA samples may also be concentrated by extraction with solvents such as secondary butyl alcohol (isobutanol) or *n*-butyl alcohol (*n*-butanol). This results in a reduction in the volume of DNA preparation to the point where the DNA can be recovered easily by precipitation with EtOH. This is because some of the water molecules are partitioned into the organic phase and by carrying out several cycles of extraction, the volume of a DNA solution can be reduced significantly. Thus slightly more than one volume of isobutanol (or *n*-butanol) is added and the solution is vortexed vigorosly and centrifuged to separate the two phases. The upper butanol phase is discarded and extracted once with water-saturated diethyl ether to remove residual isobutanol (or *n*-butanol). The nucleic acids are then precipitated with EtOH as described earlier.

Ultrafiltration Another alternative to alcohol precipitation for the concentration and desalting of DNA solutions is ultrafiltration. In this method, nucleic acid solution is forced under pressure or by centrifugal force through a semipermeable membranous disk. Aqueous medium and small solute molecules pass through the semipermeable membrane, while large molecules are retained, thus a more concentrated nucleic acid solution is left behind. The ultrafiltration membranes are available with different pore sizes and hence the technique can be used to separate different sized nucleic acids. By selecting ultrafiltration device (e.g., Microcon cartridge from Millipore) with a nucleic acid cut-off (the size of the smallest particle that cannot penetrate the membrane) equal to or smaller than the molecular size of the nucleic acid of interest, efficient desalting and concentration of nucleic acid samples can be achieved. This method does not require any phase change and is particularly useful for dealing with very low concentrations of nucleic acids. The Microcon cartridge is first inserted into a vial and to concentrate (without affecting salt concentration), 500 μl sample of DNA (or RNA) is pipetted into reservoir and centrifugation is performed for the recommended time, not exceeding the *g* force. To exchange salt, proper amount of appropriate diluent is added to bring the concentrated sample to 500 μl. The reservoir is removed from the vial, inverted into a new vial, centrifuged at 500–1,000 *g* for 2 min, and nucleic acids are recovered. Note that ultrafiltration does not change the buffer composition. The salt concentration in a sample concentrated by spinning in a Microcon is the same as that in the original sample. For desalting, the concentrated sample is diluted with water or buffer to its original volume and spun again. This is called discontinuous diafiltration and is used to remove salt by the concentration factor of the ultrafiltration. For example, if a 500 μl sample containing 100 mM salt is concentrated to 25 μl (20% concentration factor), 95% of the total salt in the sample is removed. Rediluting the sample to 500 μl in water brings the salt concentration to 5 mM. Concentrating to 25 μl once again removes 99% of the original total salt. The concentrated sample now contains only 0.25 mM salt. For more complete removal of salt, an additional redilution and spinning cycle removes 99.9% of the initial salt content.

Dialysis Another alternative to EtOH precipitation is dialysis, a process that separates molecules according to size through the use of semipermeable membrane (dialysis tubing or bag) containing pores of less than macromolecular dimensions. Small molecules (lower than cut-off value), such as salts, small biochemicals, and water diffuse across the membrane (through pores) driven by the concentration differential between the solutions on either side of the membrane but the passage of larger molecules is blocked. Dialysis allows desalting and concentration of DNA preparation. The technique is also used for the removal of CsCl after isopycnic ultracentrifugation and purification of DNA fragments from agarose gel (for details see Chapter 6). The most commonly used dialysis material is cellophane (cellulose acetate), although several other substances such as cellulose and colloidon are similarly employed. These are available in a wide variety of molecular weight cut-off values that range from 0.5 to 500 kDa. Besides the size of the molecule, the exact permeability of a solute is also dependent on the shape of the molecule, its degree of hydration, and its charge. These parameters are in turn influenced by the nature of the solvent, its pH, and its ionic strength. Before use, the dialysis tubing is pretreated in the given order: cutting into appropriately sized pieces (10–20 cm), washing for 10 min in a large volume of 2% w/v sodium bicarbonate and 1 mM Na_2EDTA (pH 8.0), thorough rinsing with distilled water, boiling for 10 min in 1 mM Na_2EDTA (pH 8.0) [or else autoclaving at 20 psi (i.e., 1.40 kg/cm^2) for 10 min on liquid cycle in a loosely capped jar filled with water], cooling, and storage at 4°C in submerged condition. After this treatment, the tubing is always handled with gloves, and before use, it is washed inside and outside with sterile distilled water. For concentrating DNA in the preparation, the DNA solution is transferred into a standard cellulose acetate dialysis bag with the help of a wide-bored pipette or cut tip. The dialysis bag is then placed on a bed of solid sucrose and additional sucrose is packed on top of the bag (this packing is best done at 4°C on a piece of aluminium foil spread on the bench in a cold room). Dialysis is allowed to proceed until the volume of the fluid in the dialysis bag is reduced by a factor of 5–10. The outside of the bag is rinsed with TE buffer (pH 8.0) to remove all of the adherent sucrose. The solution of DNA is gently massaged to one end of the bag and then the tubing is clamped just above the level of the fluid with a dialysis clip. The sample is dialyzed against large volume (4 l) of TE buffer (pH 8.0) for 16–24 hours with at least two changes of buffer. This method works more efficiently as compared to concentration in ultrafiltration devices, or in colloidon bags, and results in smaller losses of DNA.

Another variant called drop dialysis is used to remove low molecular weight contaminants from DNA in solution when high molecular weight DNA is to be used as a template in DNA sequencing reactions or if restriction enzymes fail to digest the DNA to completion. The method involves spotting of a drop (~50 μl) in the center of a Millipore Series V membrane (0.025 μm) floating shiny side up on 10 ml of sterile water in a petridish with a 10 cm diameter. DNA is then dialyzed for 10 minutes. The drop is then removed to a clean microfuge tube and aliquots of dialyzed DNA are used for restriction enzyme digestion and/or DNA sequencing experiments.

Isolation and Purification of Genomic DNA from Higher Eukaryotes

Total cell DNA from plants and animals is also required if the aim of the genetic engineering experiment is to clone genes from them. The basic steps in DNA purification are the same with all the organisms, however, some modifications need to be introduced to take account the special features of the cells being used. In this section, these special requirements are discussed.

Cell Lysis The major modifications from the basic approach discussed above are very likely to be needed at the cell breakage stage, as the chemicals used for disrupting bacterial cells (e.g., lysozyme) do not usually work with plant and animal cells. Like bacteria, a combination of physical and chemical disruption techniques is employed with plants and animals. Often physical techniques, such as grinding frozen source material with a mortar and pestle or homogenizer, are more efficient. For chemical disruption, specific plant wall degradative enzymes available for most cell wall types are used. On the other hand, most animal cells have no cell wall at all and can be lysed simply by treating with a detergent.

Removal of Contaminating Biomolecules and Recovery of DNA Another important consideration is the biochemical content of the cells from which DNA is being extracted. With most bacteria, the main biochemicals present in a cell extract are proteins, DNA, and RNA, so phenol extraction and/or protease treatment followed by removal of RNA with RNase or the application of monophasic reagents leaves a pure DNA sample. These treatments are, however, not sufficient to get pure DNA if the cells also contain significant quantities of other biochemicals, for example, polysaccharides or polyphenolics. Plant tissue is especially notorious for being a difficult source from which to isolate high-quality DNA with good yield. Problems encountered in the isolation and purification of high molecular weight DNA from plant species include degradation of DNA due to endonucleases, coisolation of highly viscous polysaccharides, and inhibitor compounds like polyphenols and other secondary metabolites released during cell disruption, which directly or indirectly interfere with the enzymatic reactions in the downstream reactions. Polysaccharides are not removed by phenol extraction, but these form complexes with nucleic acids during tissue extraction and coprecipitate during subsequent alcohol precipitation steps. Depending on the nature and the quantity of these contaminants, the resulting alcohol precipitates can be gelatinous and difficult to dissolve. Thus, DNA purification procedure from plant tissues should include a step for the removal of carbohydrates. The method used for removal of contaminating carbohydrates makes use of a detergent cetyl trimethyl ammonium bromide (CTAB), which forms an insoluble complex with nucleic acids. When CTAB is added to a plant cell extract, the nucleic acid–CTAB complex precipitates, leaving carbohydrate, protein, and other contaminants in the supernatant. The precipitate is then collected by centrifugation and resuspended in 1 M NaCl, which causes the complex to break down. The nucleic acids are then concentrated by EtOH precipitation and the RNA removed by RNase treatment. The CTAB method for purification of DNA from plant source is presented in Figure 1.3.

To eliminate browning effect of polyphenols and secondary metabolites leached out from plant tissues, polyvinyl pyrrolidone (PVP) may be included in the extraction buffer. PVP inhibits polyphenol oxidase and binds to the phenolic compounds, which are then eliminated by EtOH precipitation.

Other DNA Recovery Methods

The traditional method employed the isolation of genomic DNA bands from the cesium chloride (CsCl) or cesium trifluoroacetate (CsTFA) isopycnic gradient after ultracentrifugation (for details see Chapter 6). In the CsCl density gradient, protein and RNA contaminants are separated on the basis of differences in buoyant densities (for details see Chapter 6). For high yields of DNA, cells are lysed using a detergent and the lysate is alcohol precipitated. Resuspended DNA is mixed with CsCl and EtBr and centrifuged for several hours. The DNA band is collected from the centrifuge tube, extracted with isopropanol to remove EtBr, and then precipitated with EtOH to recover the DNA. This method allows the isolation

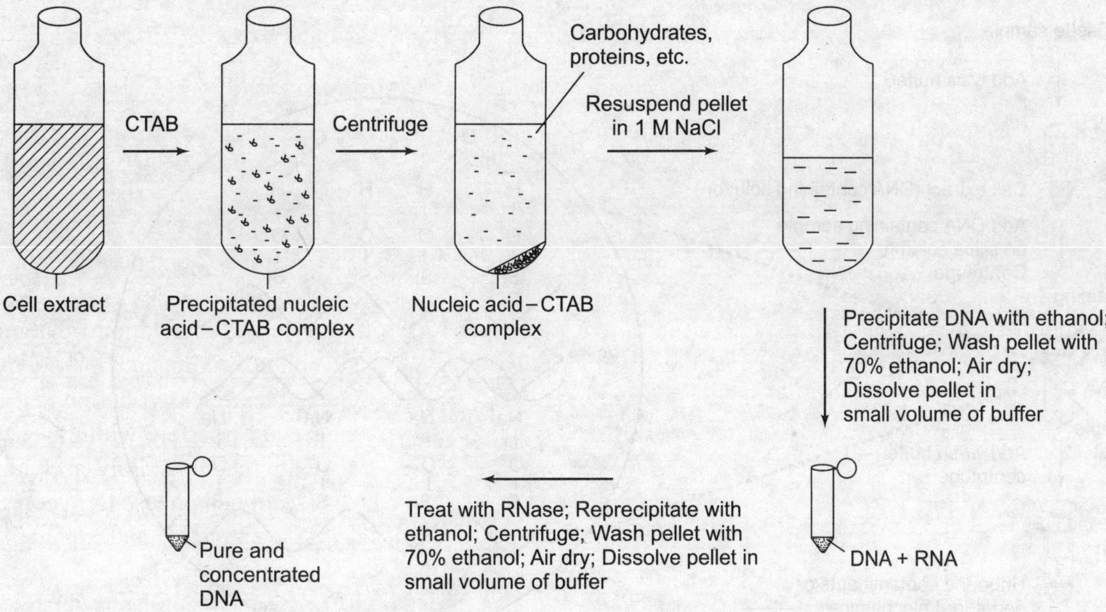

Figure 1.3 CTAB method for purification of plant DNA

of high-quality DNA, but is time-consuming, labor-intensive, and expensive (an ultracentrifuge is required), making it inappropriate for routine use. Moreover, this method uses toxic chemicals (e.g., CsCl) and is also impossible to automate. This traditional method has been replaced by conventional precipitation method involving EtOH or isopropanol as described above. The EtOH precipitation method is rapid, virtually foolproof, efficient, and even minute amounts of DNA can be quantitatively precipitated. For even faster and simpler recovery of DNA, certain variants or alternatives are presently available. These are summarized below.

Silica-based Spin Column Chromatography The spin column method involves selective adsorption of nucleic acids to silica in the presence of high concentrations of chaotropic salts. Earlier, nucleic acids were purified using glass powder or silica beads under alkaline conditions in the presence of chaotropic agents such as sodium iodide or sodium perchlorate. Note that silica structures are much more effective as packing material because these are etched into the microchannels during its fabrication by soft lithography. Moreover, these silica structures are easier to use in highly parallelized designs. This technique was later improved by using guanidinium salts (e.g., guanidinium thiocyanate or guanidinium hydrochloride) as chaotropic agent. The use of beads was later changed to minicolumns. Thus silica may be directly added to the cell extract or more conveniently packed in a chromatography column. In silica-based spin column chromatography, DNA binds tightly to silica particles in the presence of guanidinium thiocyanate, from where it is eluted at a later stage (Figure 1.4a). The principle behind the technique is that a chaotrope denatures biomolecules by disrupting the shell of hydration around them, thereby allowing positively charged ions to form salt bridges between the negatively charged silica and the negatively charged DNA backbone in high salt concentration (Figure 1.4b). Use of optimized buffers in the lysis procedure ensures that only DNA is adsorbed to silica while contaminating biochemicals such as cellular proteins and metabolites remain in solution, and pass through the column upon washing with high salt buffer and EtOH. DNA is ultimately eluted from the silica using a low-salt buffer or water, which destabilizes the interactions between DNA molecules and silica. Thus the method provides an easy way of recovering DNA from the denatured mix of biochemicals. In this technique, alcohol precipitation and resuspension of DNA (which is often difficult if the DNA has been over-dried) are not required. The technology is simpler and more effective than other methods where precipitation or extraction is required. The method is reliable, fast, inexpensive, and provides high-throughput isolation of high-quality DNA. Genomic DNA isolated using silica-based spin column technology is up to 50 kbp in size, with an average length of 20–30 kbp. Moreover, the isolated DNA is suitable for downstream applications such as Southern blotting, PCR, real-time PCR, random amplified polymorphic DNA (RAPD), restriction fragment length polymorphism (RFLP), and amplified fragment length polymorphism (AFLP) analyses (for details see Chapter 18). Currently, the main manufacturer of silica-based columns for purification of DNA is Qiagen and hence these columns are often referred to as Qiagen columns. For example, DNeasy tissue kits (Qiagen) are designed on the basis of silica-based column purification technique, which allows rapid isolation of pure total DNA

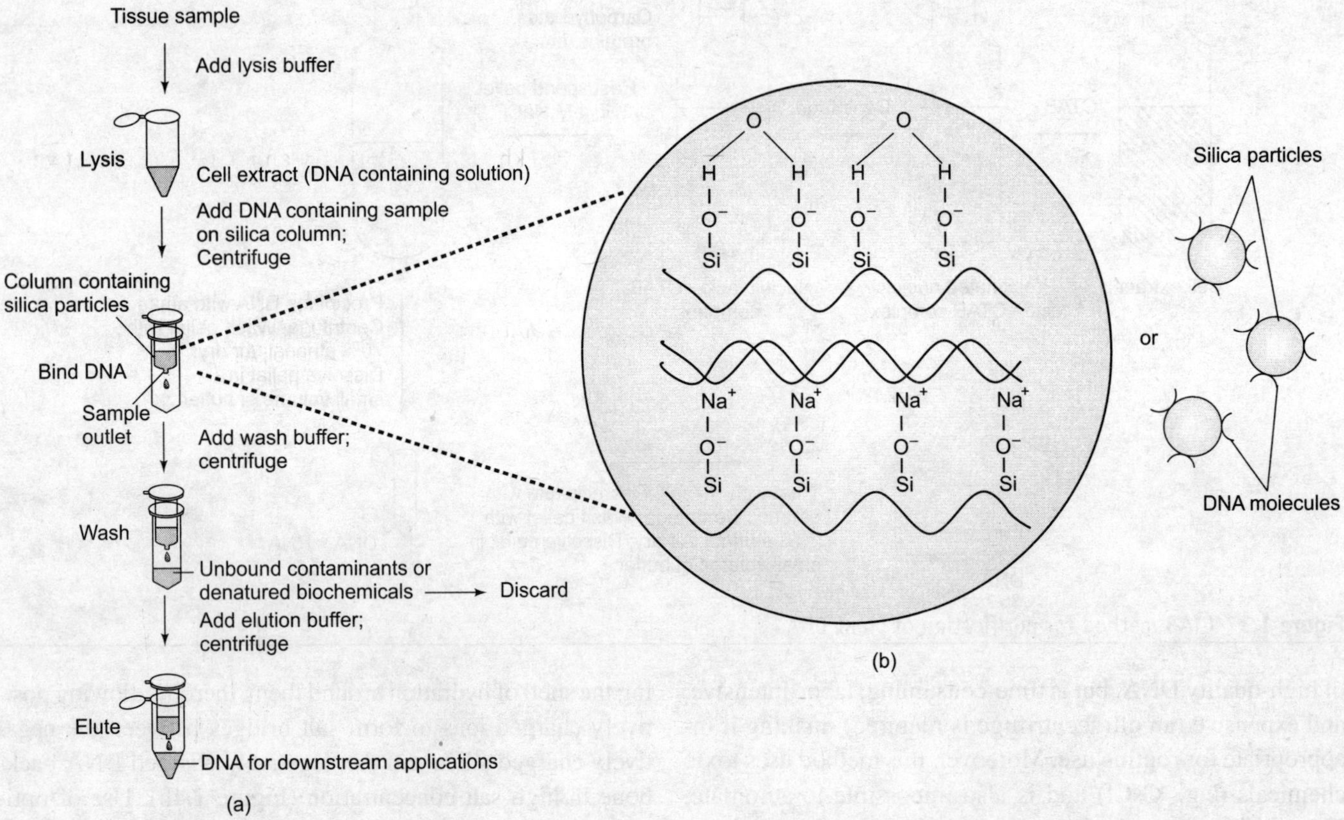

Figure 1.4 Silica-based spin column chromatography for purification of DNA; (a) Procedure for purification on silica-based spin column, (b) Magnified view showing binding of DNA to silica particles in the presence of guanidinium thiocyanate

(genomic/viral/mitochondrial) from a wide variety of sample sources, including fresh and frozen animal cells and tissues, yeasts, and blood. This method has, however, a few associated drawbacks. The beads and resins are highly variable and hence each loading of a microchannel can result in a different amount of packing, consequently changing the amount of DNA adsorbed to the channel. The technology is also not suitable for isolating genomic DNA >50 kbp in size.

Magnetic Bead Capture Technology This technology eliminates the conventional spin steps and employs magnetic bead capture for the isolation of nucleic acids. Using this technique, DNA is isolated with both high yield and purity. Moreover, the high magnetite content of the beads and the ease of handling make this technology highly adaptable to automation for isolation of DNA from different starting volumes as well as different sources. A general scheme for the isolation process is shown in Figure 1.5.

Anion Exchange Chromatography Solid phase anion exchange chromatography is based on the interaction between the negatively charged phosphates of the nucleic acid and positively charged surface molecules on the substrate (i.e., solid support). In this method, DNA is allowed to bind to the solid support under low salt conditions and the impurities such as RNA, cellular proteins, and metabolites are washed away using medium salt buffers. High-quality DNA is then eluted using a high salt buffer and DNA is recovered by alcohol precipitation. The advantage of anion exchange chromatography is that it completely avoids the use of toxic substances and can be used for different throughput requirements as well as for different scales of purification. The isolated DNA is sized up to 150 kbp, with an average length of 50–100 kbp. Moreover, the DNA preparation is suitable for all downstream applications.

DNA Isolation Kits Isolation and purification of DNA from different sources and different tissues can also be done using commercially available kits. These kits are based on one or the other methods described earlier and make the process of DNA isolation and purification rapid and simple.

Analysis of DNA Preparations

The DNA isolated by any of the above-mentioned techniques is then evaluated for quality (purity and integrity) and quantity. The techniques used for the purpose include the following:

Agarose Gel Electrophoresis and Visual Inspection Under UV Illumination The purity and integrity of DNA prepara-

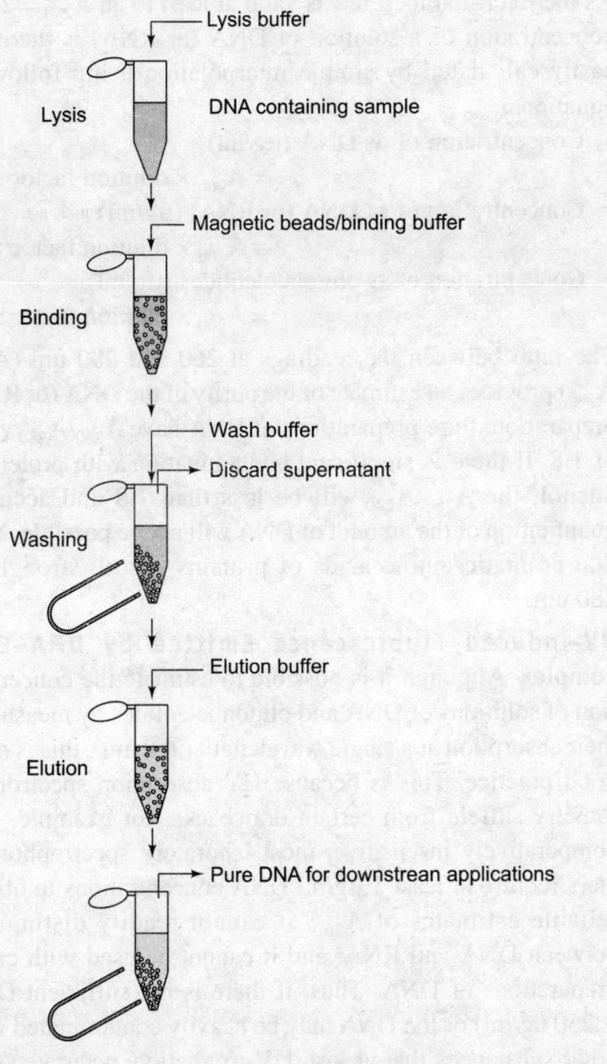

Figure 1.5 Magnetic bead capture technology for the purification of DNA

tion isolated from any source is assessed in the form of DNA–EtBr complex by agarose gel electrophoresis followed by UV illumination (for details see Chapter 6). Conventional agarose gel electrophoresis is used to determine the pattern of DNA fragments ranging in size from 50 bp to ~20 kbp, while pulse field gel electrophoresis (PFGE) is used for DNA molecules >20 kbp in size. Polyacrylamide gels are effective for visualizing and separating fragments of DNA ranging in size between 5 and 500 bp. Figure 1.6 shows the results of a conventional agarose gel electrophoresis of purified total DNA preparation. With this technique, the relative concentration of DNA solution with that of DNA size standard can also be estimated.

Agarose Gel Electrophoresis and Fluorescence Image Analysis For a gene cloning experiment, it is crucial to know the exact concentration of DNA solution. At present, fluorescence image analysis of fluorescent dye-stained DNA on gel is possible using image analysis software. This software affords an opportunity for quantifying nucleic acid concentration by digitizing a fluorescence image of a sample (after labeling with SYBR Green, Gel Star, or SYBR Gold) that has been run out on a gel and comparing it to a known mass standard (e.g., ϕX174 or λ DNA). Image analysis software counts the pixels that make up the image of all bands and smears in a lane and then compares the value for each unknown sample against a mass standard. This approach is also used to determine the quantity of synthesized cDNA. The method is rapid and usually quite accurate.

Visual Inspection of DNA Spots on Agarose Plate Under UV Illumination Another nonspectrophotometric method for determination of DNA concentration involves making an estimate by simple visual inspection. The method involves

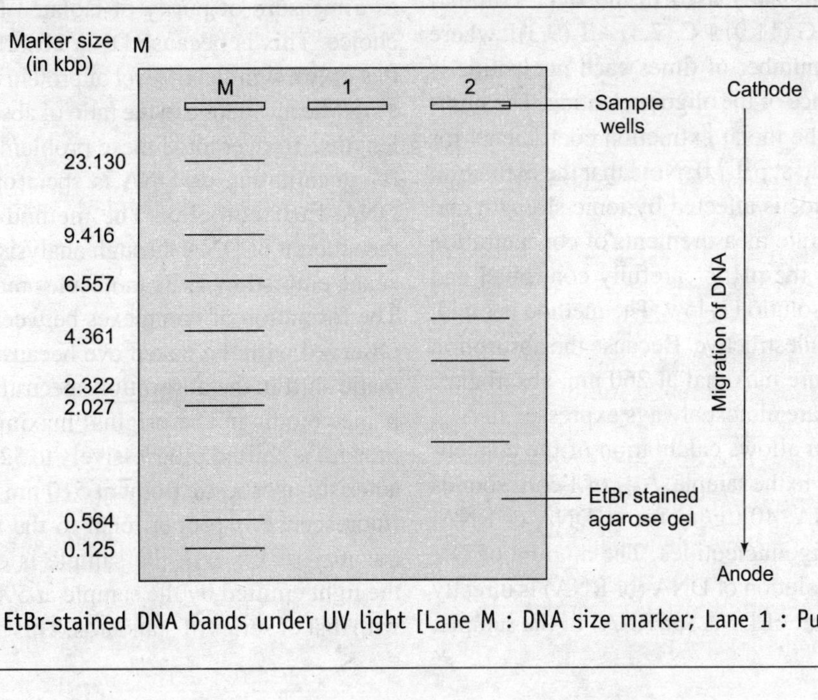

Figure 1.6 A pattern of EtBr-stained DNA bands under UV light [Lane M : DNA size marker; Lane 1 : Purified DNA; Lane 2 : DNA fragments.]

spotting dilutions of known amounts of standard DNA and samples of unknown concentrations on an agarose plate containing a fluorescent dye. On UV irradiation, the dye fluoresces in proportion to the amount of DNA present and comparing fluorescence intensity between the known and unknown samples gives a rough estimate of the amount of DNA in the unknown. Performing this somewhat imprecise assay requires the preparation of agarose plates containing 0.5 μg/ml EtBr or SYBR dyes. Note that the samples should be adsorbed into the agarose before assessing the fluorescence intensity of the dye. For ease of quantification, dilutions of the standards and the test samples are best applied in 5 μl (or less) aliquots, which avoids sample diffusion. These plates can be observed directly on the surface of a UV transilluminator or irradiated from above with a handheld UV monitor. Although this method is outdated, it may be a useful teaching tool when funds are limited.

UV Absorption Spectrophotometry For an accurate measurement of the concentration of DNA in pure solutions (i.e., without significant amounts of contaminants such as proteins, phenol, agarose, or RNA), the method of choice is UV absorbance spectrophotometry. The principle behind the technique is that the purines and pyrimidines in nucleic acids absorb UV light and the extinction coefficients of nucleic acids are the sum of the extinction coefficients of each of their constituent nucleotides. For large molecules, where it is both impractical and unnecessary to sum up the coefficients of all the nucleotides, an average extinction coefficient is used. For small molecules such as oligonucleotides, it is best to calculate an accurate extinction coefficient from the base composition. Because the concentrations of oligonucleotides are commonly reported as mmole/l, a millimolar extinction coefficient (ε) is conventionally used in the Beer–Lambert equation [$\varepsilon = A\,(15.3) + G\,(11.9) + C\,(7.4) + T\,(9.3)$], where A, G, C, and T are the number of times each nucleotide is represented in the sequence of the oligonucleotide. The numbers in parentheses are the molar extinction coefficients for the given deoxynucleotide at pH 7.0. Note that the extinction coefficient of nucleic acids is affected by ionic strength and pH of the solution. Accurate measurements of concentration can be made only when the pH is carefully controlled and the ionic strength of the solution is low. The method is rapid, simple, accurate, and nondestructive. Because the absorption spectra of nucleic acids are maximal at 260 nm, absorbance data for DNA (or RNA) are almost always expressed in A_{260} units. The OD at 260 nm allows calculation of the concentration of DNA (or RNA) in the sample. A_{260} of 1 corresponds to ~50 μg/ml for ds DNA, 40 μg/ml for ss DNA or RNA, and ~33 μg/ml for ss oligonucleotides. The amount of UV radiation absorbed by a solution of DNA (or RNA) is directly proportional to the amount of DNA (or RNA) in the sample. As the Beer–Lambert law is valid at least to an $A_{260} = 2$, the concentration of a solution of DNA (or RNA) is therefore easily calculated by simple interpolation using following equations:

Concentration of ds DNA (μg/ml)
$= A_{260} \times \text{dilution factor} \times 50$

Concentration of ss DNA (or RNA) (μg/ml)
$= A_{260} \times \text{dilution factor} \times 40$

Concentration of ss oligonucleotides (μg/ml)
$= A_{260} \times \text{dilution factor} \times 33$

The ratio between the readings at 260 and 280 nm (A_{260}/A_{280}) provides an estimate of the purity of the DNA (or RNA) preparation. Pure preparations of DNA have A_{260}/A_{280} value of 1.8. If there is significant contamination with protein or phenol, the A_{260}/A_{280} will be less than 1.8 and accurate quantitation of the amount of DNA will not be possible. Note that aromatic amino acids of proteins absorb strongly at 280 nm.

UV-induced Fluorescence Emitted by DNA–EtBr Complex Although it is possible to estimate the concentration of solutions of DNA and oligonucleotides by measuring their absorption at a single wavelength (260 nm), this is not a good practice. This is because UV absorption spectrophotometry suffers from certain drawbacks, for example, it is comparatively insensitive, most laboratory spectrophotometers require at least 1 μg/ml DNA concentrations to obtain reliable estimates of A_{260}, it cannot readily distinguish between DNA and RNA, and it cannot be used with crude preparations of DNA. Thus, if there is not sufficient DNA (<250 ng/ml) or the DNA may be heavily contaminated with other substances that absorb UV irradiation, accurate spectrophotometric analysis will be impeded. Moreover, A_{260}/A_{280} as a measure of purity of isolated DNA is also not a good choice. This is because DNA absorb so strongly at 260 nm that only a significant level of protein contamination will cause a significant change in the ratio of absorbance at the two wavelengths. To overcome these problems, an alternative method for quantitating ds DNA is spectrophotometric analysis of DNA–EtBr complex. The method involves estimation of the amount of DNA through analysis of UV-induced fluorescence emitted by EtBr molecules intercalated into the DNA. The formation of complexes between DNA and EtBr can be observed with the naked eye because of the large metachromatic shift in the absorption spectrum of the dye that accompanies binding. The original maximum at 480 nm (yellow-orange) is shifted progressively to 520 nm (pink) with a characteristic isosbestic point at 510 nm. Because the amount of fluorescence is proportional to the total mass of DNA, the quantity of DNA in the sample is estimated by comparing the light emitted by the sample at 590 nm (fluorescent yield) with that of series of standards. This provides a simple, faster, and more sensitive way to estimate the concentration of a sample of DNA, and as little as 1–5 ng of DNA is readily detected by this method.

It is suggested that the absorbance of the sample should be measured at several wavelengths since the ratio of absorbance at 260 nm to the absorbance at other wavelengths is a good indicator of the purity of the preparation. Note that significant absorption at 260 nm indicates contamination by phenolate ion, thiocyanates, and other organic compounds, whereas absorption at higher wavelengths (330 nm and higher) is usually caused by light scattering and indicates the presence of particulate matter.

Fluorometric Quantitation of DNA using Hoechst 33258 Hoechst 33258 is a class of *bis*-benzimidazole fluorescent dye that binds nonintercalatively and with high specificity into the minor groove of ds DNA. Unlike EtBr, Hoechst dyes are cell-permeant. Like many other nonintercalative dyes, Hoechst 33258 binds preferentially to A-T rich regions of the DNA helix. Hoechst 33258 in free solution has an excitation maximum at ~356 nm and an emission maximum at 492 nm. However, when bound to DNA, Hoechst 33258 absorbs maximally at 356 nm and emits maximally at 458 nm. After binding to ds DNA, the fluorescent yield increases from 0.01 to 0.6, with the $\log_{10}$ of the intensity of fluorescence increasing in proportion to the A+T content of DNA. The fluorescent yield of Hoechst 33258 is approximately three-fold lower with ss DNA. Owing to this property, Hoechst 33258 can be used for fluorometric detection and quantification of ds DNA in the solution. Hoechst 33258 is preferred to EtBr for this purpose because of its greater ability to differentiate ds DNA from RNA and ss DNA.

Measuring the concentration of DNA using fluorometry is simple and more sensitive than spectrophotometry and allows the detection of nanogram quantities of DNA. The concentration of DNA in the unknown sample is estimated from a standard curve constructed using a set of reference DNAs (10–250 ng/ml) whose base composition is the same as the unknown sample. In this assay, DNA preparations of known and unknown concentrations are incubated with Hoechst 33258 fluorochrome. Absorption values for the unknown sample are compared with those observed for the known series and the concentration of the unknown sample is estimated by interpolation. The assay can only be used to measure the concentration of DNAs whose sizes exceed ~1 kbp, as Hoechst 33258 binds poorly to smaller DNA fragments. Measurements should be carried out rapidly to minimize photobleaching and shifts in fluorescence emission due to changes in temperature. Either a fixed wavelength fluorometer (e.g., Hoefer model TKO 100) or a scanning fluorescence spectrometer (e.g., Hitachi Perkin–Elmer model MPF-2A) is used. Binding of Hoechst 33258 is adversely influenced by pH extremes, presence of detergents near or above their critical micelle concentrations, and salt concentrations above 3.0 M. Note that fluorometry assays with Hoechst 33258 do not work at extremes of pH and are affected by both detergents and salts. Assays are therefore usually carried out under standard conditions [0.2 M NaCl, 10 mM Na_2EDTA, pH 7.4]. However, two different salt concentrations are required to distinguish ds DNA from ss DNA and RNA. The concentration of Hoechst 33258 (M_r = 533.9) in the reaction should be kept low (5×10^{-7} M to 2.5×10^{-6} M), since quenching of fluorescence occurs when the ratio of dye to DNA is high. However, two concentrations of dye are sometimes used to extend the dynamic range of the assay. All DNAs and solutions should be free of EtBr, which quenches the fluorescence of Hoechst 33258. However, as Hoechst 33258 has little affinity for proteins or RNA, measurements can be carried using cell lysates or purified preparations of DNA.

Membrane Immobilization and Chromogenic Assay Another method for determining DNA concentration involves immobilization of a small aliquot of a DNA sample onto a membrane followed by a chromogenic (colorimetric) assay. Briefly, dilutions of DNA samples are spotted onto the membrane, followed by a series of washes and incubation in chromogenic substrate. DNA concentration is determined by comparing the resulting color intensity to standards. This technique permits efficient quantitation of very small quantities of ss DNA, ds DNA, RNA, and oligonucleotides, down to as little as 1 ng in aqueous buffer. Based on this technique, DNA Dipsticks kit (Invitrogen) does not register individual nucleotides and are thus quite handy for quantitating dilute template samples for PCR amplification and for monitoring the progress of the reaction.

Colorimetric Analysis using Diphenylamine (DPA) DPA indicator confirms the presence of DNA. The principle behind the procedure is that when a DNA solution is heated (at ≥95°C or simply in boiling water bath) with DPA under acid conditions, a blue compound is formed with a sharp absorption maximum at 595 nm. Note that DPA solution is prepared in a mixture of glacial acetic acid and concentrated sulfuric acid. In acid conditions, the straight chain form of 2-deoxypentose sugar of DNA is converted to the highly reactive β-hydroxylevulinyl aldehyde, which reacts with DPA to produce a blue-colored complex. DNA concentration is determined by measuring the intensity of absorbance of the blue-colored solution at 595 nm with a spectrophotometer, and comparing it with a standard curve of known DNA concentrations. In DNA, only the deoxyribose of the purine nucleotide reacts so that the value obtained represents half of the total deoxyribose present.

1.3.2 Isolation and Purification of Total RNA

RNA isolation is required for purification of mRNA, followed by cDNA synthesis to be used for the preparation of cDNA library. Besides cloning, other applications of RNA or mRNA include analysis of gene expression, poly (A) RNA selection, *in vitro* translation, reverse transcriptase-PCR (RT-PCR), northern hybridization, dot/slot blotting, RNase protection assay, S1 nuclease protection assay, RNase mapping, and primer extension experiments. Efficient methodologies have been empirically derived to accommodate the expedient isolation of RNA. In general, these methods yield cytoplasmic RNA, nuclear RNA, or mixtures of both, commonly known as cellular RNA. The conditions required for a successful isolation of RNA also differ significantly both between species and for the same species when grown under different environmental conditions. Similar to DNA isolation strategies, RNA isolation also requires same basic steps, i.e., cell lysis, removal of contaminating biomolecules, and recovery of RNA, however, a seemingly endless list of successful permutations on a few fundamental RNA extraction techniques exists. In the following sections, details of basic steps involved in RNA isolation strategies and the rationale behind each step is discussed. Note that throughout the RNA isolation and purification procedure, five Rs of molecular biology should be kept in mind. These are: rapid, representative, reproducible, RNase-free, and reliable.

Cell Lysis and Preparation of Cell Extract

The first step in RNA recovery process is cell disruption. The reagents included in the cell lysis buffer are chaotropic and denaturing agent [e.g., guanidinium salts such as guanidinium isothiocyanate (GIT) and guanidinium chloride], denaturant [e.g., β-mercaptoethanol (β-ME), dithiothreitol (DTT), and urea], deproteination agent [e.g., proteinase K, phenol, and chloroform], detergent [e.g., SDS, *N*-lauryl sarcosine (sarkosyl), Triton X-100, and sodium triisopropyl naphthalene sulfonate], chelating agent [e.g., sodium 4-amino salicylate and Na_2EDTA], and RNase inhibitor [e.g., heparin, iodoacetate, dextran sulfate, polyvinyl sulfate, macaloid, vanadyl ribonucleoside (VDR), and cationic surfactant].

Roles of Various Ingredients of Cell Lysis Buffer The inclusion of these reagents in the cell lysis buffer targets two aims, *viz.*, membrane solubilization and complete inhibition of RNase activity, and the selection of these reagents depends on the RNA population intended to be isolated, i.e., RNA population from the subcellular compartment(s) of interest.

Membrane solubilization The method of cell lysis determines the extent of subcellular disruption of the sample. For example, a lysis buffer that is used successfully with tissue-cultured cells may be entirely inappropriate for whole tissue samples. The method by which membrane solubilization is accomplished dictates whether additional steps are required to remove DNA from the RNA preparation and whether compartmentalized nuclear RNA and cytoplasmic RNA species can be purified independently of one another. Note that it is difficult, if not impossible, to determine the relative contribution of RNA from the nucleus and the cytoplasm once RNA from these two subcellular compartments have been copurified. A particular lysis procedure must likewise demonstrate compatibility with ensuing protocols once it has been recovered from the lysate.

Complete inhibition of RNase activity Setting up of conditions for the control of RNase activity is equally important. This includes speedy extraction and purification, purging RNase from reagents and equipments, and controlling RNase activity in a cell lysate. Some lysis reagents inhibit nuclease activity, while other lysis reagents require additional nuclease inhibitors for the elimination of RNase activity that are compatible with the lysis buffer. The methods used for prevention or inactivation of endogenous and exogenous RNase activities at different steps are summarized in Exhibit 1.4.

Exhibit 1.4 Ribonuclease (RNase) in RNA isolation and purification

Ribonucleases (RNases) are a family of enzymes that are small, very stable, and omnipresent (present in virtually all the living cells). These enzymes are present both in endogenous (intrinsic or internal) and exogenous (extrinsic or external) sources. These enzymes, if present, easily and quickly degrade RNA during extraction, purification, as well as storage. RNA exhibits short half-life due to its intrinsically labile nature and its degradation is further compounded by the ubiquity of these RNases. Further the resilient nature of RNases aggravates the problem, as these renature quickly following treatment with most denaturants even after boiling. This property is attributable to reformation and maintenance of their tertiary configurations by virtue of four disulfide bridges. RNases have minimal cofactor requirements and are active over a wide pH range.

To maintain the stability of the RNA before, during, and after its isolation from the cell (i.e., to isolate full-length RNA and store it for longer time periods), following precautions and preventive measures to overcome the problem of both endogenous and exogenous RNases should be taken into account. Extrinsic sources of potential RNase contamination must be identified and neutralized from the onset of the experiment. These include, but are not limited to, bottles and containers in which chemicals are packaged, RNase containing water, gel-boxes and combs, bacteria and molds present on airborne dust particles causing contamination of buffers, the hair or beards of investigator, and oil from user's fingertips, etc. These extrinsic sources of RNase lead to accidental contamination of an RNA preparation. Beyond the potential for accidental contamination of an RNA preparation with

RNases from the laboratory environment, one must be acutely aware of the fact that intracellular RNases, normally sequestered within the cell, are liberated on cellular lysis. The problems with exogenous RNase can be entirely avoided by vigilant use of prophylactic measures and the prudent application of common sense. Following precautions and preventive measures should be taken to eliminate the risk of contamination with both exogenous and endogenous RNases.

Wear Gloves Finger greases are notoriously rich in RNases and are generally accepted as the single greatest source of RNase contamination. Hence, rule number one for controlling extrinsic RNase activity is to wear rubber gloves throughout the isolation and purification process, i.e., during the preparation of reagents, handling of reagents and apparatus, and especially during the actual RNA extraction procedure. Furthermore, changing of gloves several times during the course of RNA related experiment is recommended because door-knobs, micropipettors, handles of refrigerator's door, and telephone receivers are also potential sources of RNase contamination.

Proper Handling and Maintenance of Cleanliness and Aseptic Conditions Proper microbiological sterile techniques should be observed for handling and preparation of reagents. Tubes should be kept closed when handling RNA samples. As reagents are used repeatedly, contamination must be prevented during opening and closing the reagent tubes. If the barrel or the metal ejector of the automatic pipettor comes in contact with the sides of tubes, it becomes a very efficient vector for the dissemination of RNase and hence one should be very careful in pipetting solutions.

Reserve Separate Glassware, Plasticware, Equipments, and Reagents for RNA Work Materials or stock solutions that have been used for purposes other than RNA isolation and purification in the laboratory should not be used for RNA work. Rather items of glassware, batches of plasticware, electrophoresis devices, and buffers that are to be reused should be reserved exclusively for RNA work and not be in general circulation in the laboratory. A special set of automatic pipettors for use when handling RNA should also be kept aside. Chemicals should be set aside for RNA work and should be handled with disposable spatulas or RNase-free spatula or more safely dispensed by tapping the bottle rather than using a spatula. Solutions/buffers should be stored in small aliquots of suitable volumes, rather than drawing repeatedly from the stock bottles, and each aliquot used once should be discarded. Although such actions may at first seem excessive, these may well preclude the accidental introduction of RNases and facilitate recovery of the highest possible quality RNA.

Use RNase-free and Disposable Plasticware Conical tubes, both polypropylene and polystyrene, are considered sterile if already capped and racked by the manufacturer. Individually wrapped serological pipettes are always preferred for RNA work because no special pretreatment is required. Sterile (marked as tissue culture sterile) disposable tips and microfuge tubes certified by a reputable manufacturer to be free of RNase should be used preferably. These materials are generally RNase-free and thus do not require pretreatment to inactivate RNase. Bulk packed polypropylene products are potential sources of RNase contamination, mainly due to handling and distribution from a single bag. This pertains to microfuge tubes and polypropylene micropipette tips because these can become contaminated and, in turn, contaminate stock solutions. Thus, any plastic product that comes into contact with an RNA sample at any time, either directly or indirectly, and that can withstand autoclaving should be autoclaved [sterilization using moist heat done under high temperature (121°C) and high pressure (15 psi)]. These microfuge tubes are then handled only with gloves and set aside exclusively for RNA work. To reduce the chances of contamination, it is best to use sterile forceps or gloved hands (use fresh gloves) for the distribution of small items of plasticware from original packages to laboratory racks or beakers. After distribution, these can be covered with aluminium foil and autoclaved. For the manipulation of organic extraction buffers, which typically contain mixtures of the organic solvents phenol and chloroform, individually wrapped borosilicate glass pipettes are strongly preferred.

Make Glassware, Plasticware, Equipments, and Reagents RNase-free Clearly, it is incumbent on the investigator to ensure that equipment, glassware, and plasticware are purged of RNases from the outset of the experiment. The temperature and pressure generated during the autoclaving cycle for the sterilization of solutions, laboratory plastics, and other apparatuses do not ensure complete elimination of RNase activity. Hence, some other strategies as mentioned below should be adopted.

- All reagents should also be maintained RNase-free at all times. Pre-made solutions that are certified as being RNase-free are widely available and it is worthwhile to invest in such solutions. Alternatively, stock solutions and buffers prepared in the laboratory can be treated directly or indirectly with diethyl pyrocarbonate (DEPC) (Figure A);

Moreover, all reagents should be prepared in high-purity biochemical quality water and in RNase-free glassware and plasticware. For indirect treatment with DEPC, solutions are made in DEPC–water (DEPC–water is prepared by treating water with 0.1% DEPC for at least 1 hour at 37°C and removed by autoclaving for 15 min at 15 psi on liquid cycle), followed by autoclaving of the prepared solution. Alternatively, DEPC is removed from large volumes of DEPC–water by boiling for 1 hour in a fume hood. On the other hand, for direct DEPC treatment, 0.1% DEPC is added to solutions and treatment is allowed for the required length of time, and later DEPC is removed by autoclaving. Thus wherever possible, solutions are treated with 0.1% DEPC for at least 1 hour at 37°C, or overnight at room temperature, and then autoclaved for 15 min at 15 psi on liquid cycle. For the preparation of solutions and buffers, DEPC is added to a final concentration of 0.1%, shaken for several hours on an orbital platform, or stirred vigorously with a magnetic stirrer for 20–30 min, and following

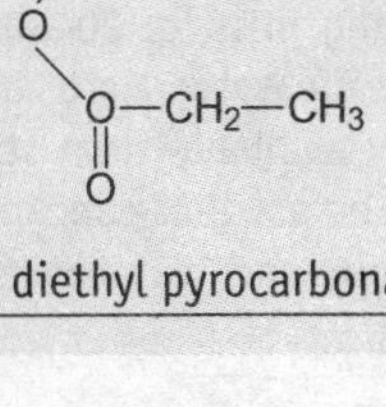

Figure A Structure of diethyl pyrocarbonate (DEPC)

this treatment, DEPC is destroyed completely by autoclaving. Complete removal of DEPC is further promoted by rapidly stirring the hot solutions with a nuclease-free magnetic stir bar. Frequently, if the autoclaving time is not adequate, the distinctive odour of the residual DEPC can be noticed. Alternatively, some solutions are maintained at 60°C overnight. Certain solutions such as those containing SDS, NP-40, or NaOH are not autoclaved and these nonautoclavable components are added to complete the solution formulation after the other components have been autoclaved as needed, or otherwise made RNase-free. DEPC is not added to any buffer containing mercaptans or primary amine groups, with which DEPC is reactive. Perhaps the most common buffers to which DEPC exposure is to be avoided are the Tris buffers [tris (hydroxymethyl) aminomethane]. This is because the hydrolysis of DEPC to CO_2 and ethanol is greatly accelerated by Tris and other amines, which themselves become consumed in the process. Note that in aqueous solution, DEPC is hydrolyzed rapidly to CO_2 and ethanol; half-life in phosphate buffer is ~20 min at pH 6.0 and 10 min at pH 7.0. Thus for RNase decontamination of Tris buffers, DEPC–water is used and the solution is autoclaved again. Those buffers consisting of chemicals that demonstrate or are known to have DEPC incompatibility are filtered twice through a nitrocellulose membrane to remove RNase activity, and other trace proteins. As DEPC is too difficult to remove from reagents and interferes with PCR and other reactions, a suitable alternative to DEPC treatment of water to render it RNase-free is to simply purchase RNase-free water;

• Nondisposable plasticware should also be treated before use to make them RNase-free. Plasticware is therefore filled with DEPC–water and allowed to stand for 1 hour at 37°C, or overnight at room temperature. The items are rinsed several times with DEPC-treated water and then autoclaved for 15 min at 15 psi on liquid cycle. Besides DEPC–water, commercially available products that inactivate RNase upon contact (e.g., RNaseZap from Ambion Inc.) can also be used to remove RNase contamination from pipettes or table-tops;

• Baking the glassware in a dry heat oven is a very effective method for purging glassware of RNase activity. Hence glassware should be cleaned scrupulously, rinsed with RNases-free water, and then baked for 3–4 hours or overnight at 300°C. It is important to note here that not all laboratory implements can withstand the heat generated in a dry heat oven. Such glassware (e.g., COREX® tubes) should be filled with DEPC–water and allowed to stand for 1 hour at 37°C or overnight at room temperature. The items are rinsed several times with DEPC-treated water and then autoclaved for 15 min. at 15 psi on liquid cycle; and

• The electrophoresis unit, including the comb, casting tray, and electrophoresis chambers, are treated to remove RNase contamination by rinsing and soaking in DEPC–water. For RNase decontamination, after cleaning of electrophoresis devices with detergent solution, these are rinsed in water, dried with ethanol, and then filled with a 3% solution of H_2O_2. H_2O_2 is an inexpensive and powerful oxidizing agent that can be used to render a surface nuclease-free by soaking in it for 20–30 min, then rinsing with copious amounts of sterile water. More concentrated form of H_2O_2 (e.g., 30%) commonly available from standard chemical supply companies should not be used, as concentrated solutions of H_2O_2 are extremely dangerous and may cause irreparable damage to acrylic gel-box components. After 10–20 min at room temperature, electrophoresis tank is washed thoroughly with sterile water. Note that the unit should never be exposed to DEPC, because acrylic is not resistant to DEPC.

Discard Old Chemicals By careless use of aseptic technique, buffers may become contaminated with bacteria or other microorganisms. As the growth of these microorganisms is not usually visible to the naked eye, solutions that are even suspected of being contaminated should be discarded. All stocks and working solutions in the laboratory should be labeled with a predetermined expiration date in addition to the date of preparation. A sad consequence of the use of out-of-date solutions in the laboratory is the unintentional introduction of RNase from microorganisms that may have taken up residence in the solution bottles.

Immediately Freeze Harvested Tissue The use of fresh tissue is preferred for RNA isolation. However, as it is easier to collect tissues in advance, the interest lies in preservation of the samples for days, weeks, or even months after tissue collection without sacrificing the integrity of the RNA. One way is to store tissue samples in RNAlater® tissue storage:RNA stabilization solution (Ambion). Dissected tissue or collected cells are simply dropped into the RNAlater® solution at room temperature. The solution permeates the cells, stabilizing the RNA. The samples are then stored at 4°C. Samples can be shipped on wet ice or even at room temperature if shipped overnight. Note that the use of RNAlater® for tissue storage is compatible with most RNA isolation procedures. Alternative to storage in RNAlater®, harvested tissues or cell pellets may be frozen immediately in liquid nitrogen and stored at –70°C to –80°C. When stored in this way, RNA can be purified up to a year later. Tissues stored in RNAlater® or in liquid nitrogen are simply removed and processed by homogenization or other mechanical apparatus in the lysis buffer specified in the RNA isolation procedure. Some investigators homogenize fresh tissue in guanidinium buffer on receipt and then freeze the homogenate at –80°C, continuing the RNA isolation procedure at a later date. Purified RNA is most stable when stored as an ethanol precipitate at –80°C. Under these conditions, the investigator can confidently store RNA for several months or even longer because the half-life of RNA is a direct function of the biological source.

Speedy Extraction Cellular RNases (endogenous RNases) are free to initiate degradation of the RNA that the investigator is attempting to isolate, unless these are inhibited without delay. Hence intrinsic RNases should be inactivated as quickly as possible at the very first stage in the RNA purification process (i.e., extraction). This demands speed as well as precautions while handling. Once the endogenous RNases have been destroyed, the immediate threat to the integrity of the RNA is greatly reduced and purification can proceed at a more graceful pace.

Inclusion of Chaotropic Agent and Reducing Agents in RNA Lysis Buffer RNA lysis buffer may combine the disruptive and protective properties of guanidine thiocyanate (GTC) and β-ME to inactivate RNases present in cell extracts. GTC is a chaotropic agent that disrupts nucleoprotein complexes, allowing RNA to be released into solution and isolated free of protein. β-ME is a reducing agent that breaks disulfide bonds in proteins, thereby inactivating RNases. The combination of GTC and β-ME makes the RNA lysis buffer a potent inhibitor of RNase activity.

8-Hydroxyquinoline, a partial RNase inhibitor, when included in lysis buffer, may optimize RNA purification efficiency. It chelates heavy metals that can cause RNA degradation when present with RNA for extended periods.

Application of RNase Inhibitors Creating an RNase-free environment or elimination of RNase activity is essential during the entire process. Because endogenous RNase activity varies tremendously from one biological source to the next, the degree to which action must be taken to inhibit RNase activity is a direct function of cell or tissue type. Knowledge of the extent of intrinsic RNase activity is derived from the literature and personal experience. Moreover, the method selected for controlling RNase activity must be compatible with the cell lysis procedure. It is also important to note that the method of nuclease inhibition must support the integrity of the RNA throughout the subsequent fractionation or purification steps, some of which can be quite time-consuming. In addition, the reagents used to inhibit RNase activity must be easily removed from purified RNA preparations so as not to interfere with subsequent manipulations. One such strategy is addition of RNase inhibitors in the lysis buffer, reaction buffers, and in the standard preparation of reagents intended for RNA work and storage. At the time of extraction, RNase inhibitors are most often added to relatively gentle lysis buffers when subcellular organelles (especially nuclei) have to be purified intact. Unfortunately, the inhibitors do not offer perfect protection because all of the available inhibitors inhibit only some part of the existing RNases. Consequently, these substances cannot be handled without working cleanly. RNase inhibitors are also used to protect mRNA in cDNA synthesis, and *in vitro* transcription/translation system. Note that RNase inhibitors are heat-sensitive and are stored at –20°C. The RNase inhibitors may be composed of mixtures of proteins that inhibit a more or less broad spectrum of RNases. There are two types of RNase inhibitors, *viz.*, specific and nonspecific inhibitors. The examples of these two types are discussed below:

- **Vanadyl ribonucleoside complexes** Amongst the category of specific RNase inhibitors are vanadyl ribonucleoside (VDR) complexes, protein inhibitors, and macaloid. VDR complexes were developed in the mid-1970s as a means of controlling RNase activity when using relatively gentle methods to support cellular lysis. VDR consists of complexes formed between the oxovanadium ion and any or all of the four ribonucleosides in which vanadium takes the place of phosphate. These complexes then function as transition state analogs that bind to many RNases and inhibit their activity almost completely. The four vanadyl ribonucleoside complexes are added to intact cells and used at a concentration of 10 mM during all stages of RNA extraction and purification. The resulting mRNA is isolated in a form that can be directly translated in frog oocytes and can be used as a template in some *in vitro* enzymatic reactions (e.g., reverse transcription of mRNA). However, vanadyl ribonucleoside complexes strongly inhibit translation of mRNA in cell-free systems and must be removed from the mRNA by multiple extractions with phenol (equilibrated with 0.01 M Tris–HCl, pH 7.8) containing 0.1% 8-hydroxyquinoline. Vanadyl ribonucleoside complexes are available from several commercial suppliers. In the absence of VDR, RNase-mediated cleavage of the phosphodiester backbone of RNA results in the transient formation of a dicyclic transition state intermediate, which is subsequently opened up by reaction with a water molecule. VDR is not widely used any longer due to associated drawbacks, *viz.*, (i) RNA isolation compounds that are more efficient inhibitors of RNase have been developed; (ii) VDR in trace quantities can inhibit *in vitro* translation of purified mRNA; and (iii) VDR inhibits reverse transcriptase and hence contraindicated for RT-PCR and the numerous permutations thereof. Thus, if purified RNA is to be subjected to either of these applications, the use of VDR is not recommended. 8-Hydroxyquinoline chelates heavy metals, making it very useful for removing VDR complexes from cell lysates. The color imparted by the 8-hydroxyquinoline changes from yellow to dark green upon binding VDR. When the phenol phase of the extraction buffer remains yellow, it indicates that all VDR has been removed .
- **Human placental inhibitor** Many RNases bind tightly ($K_i \approx 3 \times 10^{10}$) to a protein isolated from human placenta forming equimolar, noncovalent complexes that are enzymatically inactive. *In vivo*, the protein is probably an inhibitor of angiogenin, an angiogenic factor whose amino acid sequence and predicted tertiary structure are similar to that of pancreatic RNase. The inhibitor, which is sold by several manufacturers, is stored at –20°C in a solution containing 50% glycerol and 5 mM DTT. Preparations of the human placental inhibitor that have been frozen and thawed several times or stored under oxidizing conditions should not be used, as these treatments may denature the protein and release bound RNases. The inhibitor is therefore not used when denaturing agents are used to lyse mammalian cells in the initial stages of extraction of RNA. However, it should be included when more gentle methods of lysis are used and should be present at all stages during the subsequent purification of RNA. Fresh inhibitor should be added several times during the purification process, as it is removed by extraction with phenol. The inhibitor requires sulfydryl reagents for maximal activity and does not interfere with reverse transcription or cell-free translation.
- **Macaloid** Macaloid is clay that has been known for many years to adsorb RNase. The clay is prepared as slurry that is used at a final concentration of 0.015% w/v in buffers used to lyse cells. The clay, together with its adsorbed RNase, is removed by centrifugation at some stage during the purification of RNA (e.g., after extraction with phenol).
- **Diethyl pyrocarbonate** The most common example of nonspecific RNase inhibitor is DEPC. It is a highly reactive alkylating agent, which acts as a potent and efficient RNase inhibitor. It destroys the enzymatic activity of RNase chiefly by ethoxyformylation of histidyl groups. DEPC is commonly used to chemically inactivate trace amounts of RNases that may contaminate solutions, glassware, and plasticware to be used for the preparation of nuclear RNA. Note that endogenous RNases in cell suspensions or lysates are not inhibited by addition of DEPC. After treatment of glassware, plasticware, and solutions, DEPC has to be removed before use in RNA isolation procedure. This is because in addition to reacting with histidine residues in proteins, DEPC also carbethoxylates single stranded nucleic acids, and has strong affinity for adenosine nucleotides. It forms alkali-labile adduct with the imidazole ring N7 of unpaired purine, resulting in cleavage of the glycosidic bond and generation of an alkali-labile abasic site. Even trace amounts of residual DEPC result in chemical modification

of adenine, thereby changing the physical properties of RNA and compromising its utility for *in vitro* translation and other applications, including standard blot analysis and PCR. However, the ability of DEPC-treated RNA to form DNA:RNA hybrids is not seriously affected unless a large fraction of the purine residues have been modified. Note that due to ability of DEPC to modify unpaired adenine, DEPC is also carcinogenic and should be handled with extreme care. After treatment of glassware, plasticware, and solutions with DEPC for appropriate time period, it is removed readily by thermal degradation (autoclaving), leading to its decomposition into CO_2 and ethanol, both of which are quite volatile under the conditions of autoclaving. This is, however, problematic, because small amounts of these products lead to an increase in ionic strength, and lower the pH of unbuffered solutions. Samples of DEPC that are free of nucleophiles (e.g., water and ethanol) are perfectly stable, but even small amounts of these solvents can cause complete conversion of DEPC to diethylcarbonate. For this reason, DEPC should be protected against moisture. It should be stored under small aliquots in dry conditions and bottle should be allowed to reach temperature before being opened.

• **Other examples** Other examples of RNase inhibitors are heparin, iodoacetate, dextran sulfate, polyvinyl sulfate, and cationic surfactant, etc. Many RNase inhibitors are available commercially, for example, RNAsin from Promega, RNase Block from Stratagene, RNaseZap from Ambion Inc., Ribonuclease inhibitor from Clonetech, etc.

Sources: Farrell Jr RE (2005) *RNA Methodologies–A Laboratory Guide for Isolation and Characterization*, Elsevier Academic, New York, Appendix F; Tait RC (1997) *An Introduction to Molecular Biology*, Horizon Scientific Press; Sambrook J and Russel DW (2001) *Molecular Cloning: A Laboratory Manual*, 3rd ed. Cold Spring Harbor Laboratory Press, Cold Spring Harbor, NY, pp. 7.2–7.12.

Classification of Cell Lysis Buffers Based on differences in membrane solubilization and RNase inhibition properties, lysis buffers typically fall into two categories:

Lysis buffers containing plasma membrane and subcellular organelle membrane solubilizing agents This category of lysis buffers includes harsh chaotropic agents, denaturing agents, or surfactants, which rapidly disrupt the plasma membrane as well as subcellular organelle membranes of the cells, solubilize their components, and simultaneously inactivate RNase (i.e., addition of exogenous inhibitors of RNase is not required). Examples included in this category are GIT, guanidinium chloride, SDS, sarkosyl, urea, water-saturated phenol, or chloroform. Thus the extract obtained after direct cell lysis using such agents is a representative of final accumulation of RNA in the cell, as it includes the RNA from both the nucleus and cytoplasm. It is also known as steady state RNA.

Guanidinium salts function by destroying the 3-D structure of proteins and convert most proteins to a randomly coiled state. The chloride salt of guanidinium is a strong inhibitor of RNase, but it is not powerful enough as a denaturant to allow the extraction of intact RNA from tissues rich in RNase. On the other hand, GIT is a stronger chaotropic agent and contains potent cationic and anionic groups that form strong H-bonds. It is used in the presence of a reducing agent (e.g., β-ME) to break intramolecular protein disulfide bonds and in the presence of a detergent such as sarkosyl to disrupt hydrophobic interactions. Note that the efficiency of protein denaturation (including disruption of RNase) is enhanced by the inclusion of β-ME in the lysis buffer. Another common reducing reagent, DTT, is not used in combination with guanidinium salts because of their chemical reactivity. After cell lysis, partitioning of RNA, DNA, and protein in the resulting lysate is required. This is achieved in three ways: (i) Isopycnic ultracentrifugation using CsCl or CsTFA; (ii) Acid–phenol treatment; and (iii) Application of monophasic reagent containing acidified phenol, guanidinium or ammonium thiocyanate, and a phenol solubilizer (Figure 1.7). In the first and original method, guanidinium lysate is subjected to time-consuming isopycnic ultracentrifugation through CsCl/CsTFA. The original method has later been modified to eliminate ultracentrifugation step, for example, by inclusion of LiCl in the procedure or by the application of hot phenol (note that the hot phenol procedure results in high yields of RNA within hours). The second method involves guanidinium-based whole cell lysis followed by acid–phenol extraction. This method offers certain benefits, such as reduction in the length of RNA isolation step, collection of increased number of samples at a time, reduction in RNA loss, and purification of RNA from smaller sample sizes. Although such extraction buffers are easily prepared in the molecular biology laboratory by mixing water-saturated phenol with an acidic solution of sodium acetate, premixed monophasic formulations of phenol and guanidinium thiocyanate (e.g., TRIzol and TRI reagent) are readily available commercially. On phase separation, RNA is retained in the aqueous phase, while DNA and proteins partition into the interface and organic phase. Then RNA is recovered by precipitation with isopropanol and collected by centrifugation. In these procedures, RNA is efficiently isolated from as little as 1 mg of tissue or 10^6 cells usually in less than 1 hour. The single-step method yields the entire spectrum of RNA molecules, including small (4–5S) RNAs. The yield of RNA depends on the type of tissue used for isolation. Typically, 100–500 μg of total RNA is obtained from 100 mg of muscle and up to 800 μg of RNA from 100 mg of liver tissue. The yield of total RNA from cultured cells is in the range of 50–80 μg (fibroblasts, lymphocytes), or 100–120 μg (epithelial

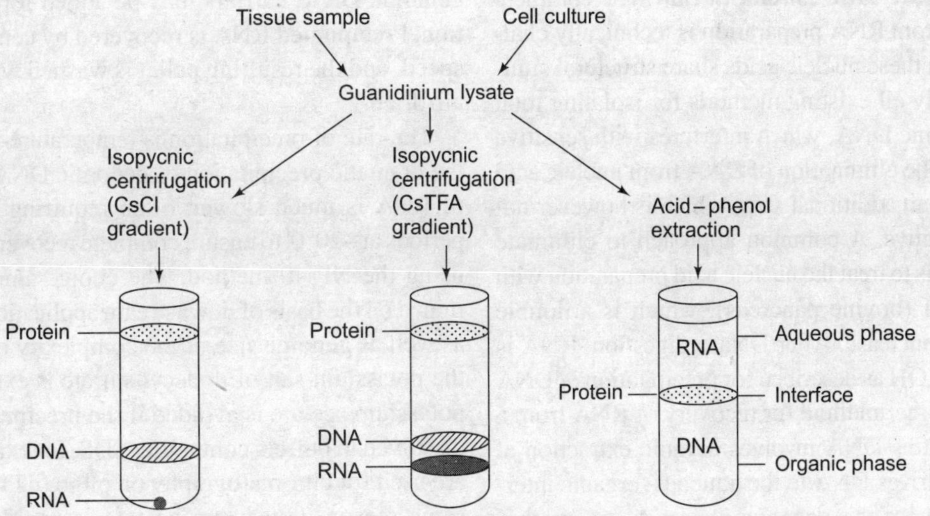

Figure 1.7 Methods for fractionation of total cellular RNA following cellular disruption with guanidinium buffer

cells) per 10^7 cells. The third method involves application of monophasic reagent to optimize the speed and extent of RNase inactivation (for details see Section 1.3.1). Several commercial RNA isolation kits are based on these monophasic lysis reagent, which are commercially available with a variety of trade names, for example, TRIzol reagent (Life Technologies), TRI reagent (Molecular Research Center), Isogen (Nippon Gene, Toyama, Japan), RNAzol and RNA-Stat-60 (Tel-Test Inc.) that allow partitioning of RNA in the aqueous phase from where it is precipitated with isopropanol and is further purified by chromatography on oligo (dT)-cellulose columns, and/or used for northern blot hybridization, reverse transcription, or RT-PCR. The yield of total RNA depends on the tissue or cell source, but it is generally 4–7 μg/mg starting tissue or 5–10 μg/10^6 cells.

Lysis buffers containing only plasma membrane solubilizing agents The abundance of specific steady-state RNA species may certainly be interpreted as an indication of how specific genes are modulated, but it does not furnish any information either about the rate at which these RNA molecules are transcribed or about their stability (or half-life) in the cell. This is the major disadvantage of evaluating the steady-state RNA alone. It is certainly clear that the cellular biochemistry responds to environmental change not only by modification of the rate of transcription but also through the processing efficiency of precursor RNA, the efficiency of nucleocytoplasmic transport, the stability of RNA in the cytoplasm, and the translatability of salient messages. In contrast to the protocols for the analysis of steady-state transcripts, transcription rate studies require the isolation of intact nuclei that are able to support elongation of, and label incorporation into, RNA molecules whose transcription was initiated at the time of cellular lysis. This family of techniques is known collectively as the nuclear runoff assay. Hence, another category of cellular lysis involves gentle solubilization of plasma membrane, while maintaining nuclear integrity using agents such as hypotonic Nonidet P-40 (NP-40) or Igepal CA-630 (Sigma) lysis buffers. Intact nuclei, other organelles, and cellular debris are then removed from the lysate by differential centrifugation. The reliability of this approach is often dependent on the inclusion of nuclease inhibitors in the lysis buffer and careful attention to the handling and storage of RNA so purified. Note that NP-40-based method is more appropriate for RNase-poor tissues or cells because the inactivation of the RNase is less efficient.

Removal of Contaminating Biomolecules

After successful cell lysis, cellular extract is obtained that contains biomolecules depending on the species under consideration. However, for various molecular biology studies, it is of paramount importance to isolate high-quality, high-purity, undegraded, or intact RNA. Hence, next step is the purification of RNA and the following strategies are adopted for removal of contaminating proteins, DNA, and carbohydrates.

Deproteination For complete removal of proteins from cellular lysate, RNA isolation procedure also includes a step for deproteination. Removal of proteins may be accomplished by any one or the combination of following procedures: (i) Digestion of the sample with the enzyme proteinase K; (ii) Repeated extraction with mixtures of organic solvents such as phenol and chloroform; (iii) Salting out of proteins; and (iv) Solubilization in guanidinium buffers (for details see Section 1.3.1). Note that although phenol denatures proteins efficiently, it does not completely inhibit RNase activity and it is a solvent for RNA molecules that contain long tracts of poly (A).

Prevention of Genomic DNA Contamination The complete separation of DNA from RNA preparation is technically challenging because both these nucleic acids share structural similarities and essentially all existing methods for isolating total RNA copurify genomic DNA, which interferes with sensitive detection methods. The elimination of DNA from nucleic acid preparation requires an additional step, which is however not always entirely effective. A common approach to eliminate DNA contamination is to treat the nucleic acid preparation with RNase-free DNase I (bovine pancreas), which is a double strand-specific endonuclease. After DNase digestion, RNA is reprecipitated with EtOH as described for precipitation of DNA in Section 1.3.1. Another method for recovery of RNA from a lysate that also contains DNA involves organic extraction at pH below 6.0. This drives DNA to the aqueous–organic interface, while RNA enters the aqueous phase. A yet another method is based on the application of guanidinium-based monophasic reagent. Besides, LiCl precipitation efficiently precipitates RNA, but not DNA, protein, or carbohydrate.

Elimination of Polysaccharides and Polyphenols Although osmotic cell lysis described above is one of the gentlest methods for cell disruption, it does not work well when carbohydrate-rich cell walls are present, as is the case with certain bacteria, fungi, and plant cells. These polysaccharides or polyphenolics form irreversible complexes with RNA during tissue extraction and coprecipitate during subsequent alcohol precipitation steps. Depending on the nature and quantity of these contaminants, the resulting alcohol precipitates can be viscous or gelatinous and difficult to dissolve. Moreover, an RNA solution contaminated with polysaccharides and/or polyphenols absorbs strongly at 230 nm and hence prevents an accurate quantitation of RNA preparation. Furthermore, RNA contaminated with polysaccharides or polyphenols is not suitable for cDNA synthesis, RT-PCR amplification, *in vitro* translation, or northern analysis. However, the problems associated with polysaccharides and polyphenolics can be rectified by including additional steps in the extraction and purification procedure, for example, treatment with hot phenol/SDS or hot borate.

Recovery of RNA

Recovery of RNA is the final step in RNA purification schemes. This is done by any of the following procedures:

Alcohol–Salt Precipitation Followed by Centrifugation The most versatile method for concentrating RNA is precipitation using various combinations of salt and alcohol. This is because nucleic acids and the salt that drives their precipitation form complexes with greatly reduced solubility in EtOH (or isopropanol). Thus, RNA is efficiently precipitated with 2.5–3.0 volumes of EtOH from solutions containing Na^+, K^+, Li^+, or NH_4^+ ions (Table 1.1). Similar to concentration of genomic DNA, carriers may be added for ease in visualization. Precipitated RNA is recovered by centrifugation at high speed, and the resulting pellet is washed with 70% EtOH and air-dried.

The rate of precipitation is temperature-dependent. Unlike the dramatic precipitation of genomic DNA, the precipitation of RNA is much slower, often requiring longer incubation periods at –20°C to ensure complete recovery, especially when using the NP-40 method. The choice among salts is determined on the basis of downstream application of isolated RNA as well as genome size or the complexity of organism. Since the potassium salt of dodecyl sulfate is extremely insoluble, potassium acetate is avoided if the precipitated RNA is to be dissolved in buffers containing SDS, for example, buffers that are used for chromatography on oligo (dT)-cellulose. For the same reason, potassium acetate is avoided if the RNA is already dissolved in a buffer containing SDS. LiCl is frequently used when high concentrations of EtOH are required for precipitation (i.e., for precipitating RNA). LiCl is very soluble in ethanolic solutions and is not coprecipitated with the nucleic acid. Small RNAs (tRNAs and 5S rRNAs) are soluble in solutions of high ionic strength (without EtOH), whereas large RNAs are not. Because of this difference in solubility, precipitation in high concentrations of LiCl can be used to purify large RNAs. LiCl precipitation offers major advantages over other RNA precipitation methods. It is the most appropriate method for the recovery of high molecular weight RNA to be used in cDNA library construction. Note that large RNAs (e.g., rRNAs and mRNAs) are insoluble in solutions of high ionic strength and can be removed by centrifugation, while small (<200 nucleotides) RNAs, for example, tRNA or 5S rRNA, are soluble or are not effectively precipitated. In other words, LiCl precipitation should be avoided if small RNAs are desired. It does not efficiently precipitate DNA, proteins, or carbohydrates. It is the method of choice for removing inhibitors of translation or cDNA synthesis from RNA preparations. It also provides a simple and rapid method for recovering RNA from *in vitro* transcription reactions. Precipitation of RNA with LiCl is also helpful in removing glycoproteins and yolky components from the preparations. Moreover, RNAs precipitated by this method give more accurate values when quantitated by UV spectrophotometry since LiCl is very effective at removing free nucleotides. However, LiCl is avoided when the RNA is to be used for cell-free translation or reverse transcription. This is because Li^+ ions inhibit initiation of protein synthesis in most cell-free systems and suppress the activity of RNA-dependent DNA polymerase.

Single-step Resin Binding Method An alternative to the alcohol precipitation is the single-step resin binding method, which employs RNA-binding resins [e.g., RNA Track^{TM} resin, (Biotecx), and RNA MATRIX^{TM} (BIO 101)]. The method leads to enhancement of the purification of total RNA, and avoids conventional lengthy precipitation methods of EtOH or isopropanol. RNA purified by this method is free from impurities such as traces of guanidinium salts, phenol, etc., which might interfere with subsequent RNA applications. This method requires no additional carrier protein in the case of low-yield RNA samples such as serum. This method also eliminates the requirement for density gradient centrifugation after guanidinium-based lysis. Although the yield of isolated RNA by this procedure is 10–20% lower than the conventional alcohol precipitation methods, the purity of RNA is improved.

Silica-based Column Purification Another modified single-step approach to quickly isolate milligram quantities of pure total RNA with undetectable levels of contaminating DNA is silica-based column purification. This solid-phase purification method relies on the fact that the nucleic acids bind (adsorb) to the solid phase (silica) depending on the pH and the salt content of the buffer. Similar to RNA-binding resins, the application of silica filters precludes the requirement for density gradient centrifugation. Moreover, high-quality total RNA is isolated with minimal effort from a wide variety of samples. One such example is the PureYield™ RNA midiprep system (Promega Corp.), which uses PureYield silica-membrane technology to isolate intact total RNA ranging in size from less than 100 bases to greater than 20 kb. Up to 300 mg of plant or animal tissue, 5×10^7 tissue culture cells, 1×10^{10} bacterial cells, 5×10^8 yeast cells, or 20 ml of blood can be processed per column. A disadvantage of this technology is that an RNA column is typically unsuitable for purification of short (<100 nucleotides) RNA, such as small interfering RNA (siRNA) and microRNA (miRNA).

Glass Fiber-based RNA Purification Glass fiber-based RNA purification step is done either as part of the RNA isolation strategy or as an additional clean-up step after RNA isolation. This purification technique increases the purity of RNA samples and dramatically reduces the amount of 5S rRNA and tRNA in samples. Moreover, RNA yields are higher than those obtained with simple organic extraction methods. This strategy is included in RiboPure™-bacteria kit (Ambion). This kit combines an efficient glass bead and RNAWIZ™-mediated disruption step followed by a glass filter-based RNA purification for high yields of exceptionally pure bacterial RNA. The kit also includes DNA-free™ reagents for quick and simple DNase treatment of samples without organic extraction, alcohol precipitation, or column purification.

Isopycnic Ultracentrifugation The application of guanidinium-based lysis mandates a procedure for partitioning of RNA, DNA, and protein in the resulting lysate. This is done by various methods, one of which is to subject guanidinium lysates to isopycnic ultracentrifugation using CsCl/CsTFA. In CsCl density gradient centrifugation, only ultrapure nuclease-free preparations are used because CsCl has only a limited ability to inhibit RNase activity. If necessary, impure, solid CsCl may be baked at 200°C for 6–8 hours to remove residual RNase activity prior to exposure to RNA. CsTFA is a highly soluble salt that solubilizes and dissociates proteins from nucleic acids without the use of detergents. CsTFA is an excellent inhibitor of RNase and its use precludes removal of proteins from a sample by more traditional methods (e.g., phenol:chloroform, proteinase K). As compared to CsCl, CsTFA is more chaotropic, inhibits RNase to a greater extent, and shows greater solubility in EtOH, which expedites its removal following isopycnic ultracentrifugation of RNA. RNA can be banded or pelleted in CsTFA as desired, because solution densities up to 2.6 g/ml (or 2.6 g/cm^3) are possible using this reagent. Because the buoyant density of RNA in CsCl (1.8 g/ml) is much greater than that of other cellular components, rRNAs and mRNAs migrate to the bottom of the tube during ultracentrifugation. As long as the step gradients are not overloaded, proteins remain in the guanidinium lysate, while DNA floats on the CsCl cushion. This method is useful when ultrapure RNA is required and the RNA extracted from cultured cells and most tissues can be directly used in further experiments without any further treatment. Depending on the cell type, typical RNA yields range from 50 to 75 μg/10^6 cells. The method is especially useful for kinetic studies where samples have to be taken at several different time points because different samples can be stored in guanidinium buffer for several days before ultracentrifugation. On the other hand, this method is cumbersome and involves extensive labor and expensive instrumentation (ultracentrifuge with a rotor), generally limiting the number of samples that can be easily isolated simultaneously. Furthermore, small RNAs (e.g., 5S rRNA and tRNAs), which do not sediment efficiently during centrifugation through CsCl, should not be prepared by this method. Moreover, running gradients is generally a time-consuming procedure (two days for RNA purification) that is usually not preferred for mainstream molecular biology applications. Currently, several methods have been developed where ultracentrifugation time with CsCl has been reduced from 12–18 hours to 3 hours utilizing a Beckman TL-100 ultracentrifuge and TLS-55 rotor.

RNA Isolation Kits A large number of kits are now commercially available from various manufacturers that allow rapid and easy isolation of high-quality RNA from different sources in good yields. These kits work by combining one or the other lysis, decontaminating, and recovery procedures described in Section 1.3.2.

Other Methods Alternatives to alcohol precipitation for concentration and desalting of RNA solutions are ultrafiltration and dialysis (for details see Section 1.3.1). Similar to DNA, magnetic bead technology and anion exchange chromatography can also be employed for the recovery of RNA (for details see Section 1.3.1).

Storage of RNA

The last step in every RNA isolation protocol, whether for total RNA or mRNA preparation, is to resuspend the purified RNA pellet. After painstakingly preparing an RNA sample, it is crucial that RNA be suspended and stored in a safe, RNase-free solution. Hence, for further use, RNA is stored in any of the following ways: (i) RNA precipitate is dissolved in deionized formamide and stored at –20°C. Note that formamide provides a chemically stable environment that protects RNA against degradation by RNases. Purified, salt-free RNA dissolves quickly in formamide up to a concentration of 4 mg/ml. At such concentrations, RNA samples are analyzed directly by gel electrophoresis and used for RT-PCR or RNase protection, saving time and avoiding potential degradation. When required, RNA is recovered by precipitation with four volumes of EtOH or by diluting the formamide four-fold with 0.2 M NaCl and then adding the conventional two volumes of EtOH. Note for long-term storage, RNA is dissolved in highly purified 100% formamide or a commercially available stabilized form of formamide known as FORMAzol (Molecular Research Center), in which RNA is stable for up to 2 years at –20°C. For northern blots, the RNA dissolved in formamide can be used directly although it must previously be precipitated using four volumes of EtOH for a cDNA synthesis reaction; (ii) RNA precipitate is dissolved in aqueous buffer, which minimizes hydrolysis of RNA (note that divalent cations catalyze the base hydrolysis of RNA) and is compatible with all of the common downstream applications of RNA. One such example is that of aqueous buffer containing 0.1–0.5% SDS in TE buffer (pH 7.6), followed by storage at –80°C. For downstream applications, SDS is removed by chloroform extraction and EtOH precipitation. Alternative storage solutions are TE buffer (pH 7.0)/RNA storage solution [Ambion; 1 mM sodium citrate (pH 6.4 ± 0.2); presence of low pH and sodium citrate as buffering and chelating agent minimizes base hydrolysis of RNA]/ RNAsecure™ resuspension solution [Ambion; RNA pellet is resuspended in the RNAsecure™ resuspension solution and heated to 60°C for 10 minutes to inactivate RNases. A unique feature of RNAsecure™ is that reheating after the initial treatment reactivates the RNase-destroying agent to eliminate any new contaminants]/DEPC–water containing 0.1 mM Na_2EDTA (pH 7.5); and (iii) RNA precipitate can be stored as a suspension at –20°C in EtOH.

Once total RNA is isolated from the given source, mRNA is purified from the preparation, which is then used for cDNA library construction (for details see Chapter 5) or other applications.

Analysis of RNA Preparation

Before proceeding for downstream applications, the RNA preparation is analyzed for its integrity, purity, and concentration. The following methods find application in analysis of RNA preparation.

Denaturing Agarose Gel Electrophoresis The quality of total RNA preparation (i.e., rRNAs, mRNAs, and tRNAs) is verified by electrophoresis on denaturing (formaldehyde or glyoxal containing) agarose gel (for details see Chapter 6). Running a gel is also the best diagnostic tool to assess the integrity of RNA preparation, and the probable utility of RNA sample in downstream applications. Because rRNA is the most prevalent RNA (80–85% of cellular RNA), an aliquot of total cellular or total cytoplasmic RNA should electrophoretically resolve into two very distinct, easily observable bands (28S and 18S rRNAs in eukaryotic RNA preparation, 23S and 16S rRNAs in prokaryotic RNA preparation) present within the smear of mRNAs. The appearance of these discrete bands is a convincing evidence of the integrity or intactness of the sample. Samples in which there has been any degree of degradation usually fail to manifest the characteristic formation of rRNA bands and light smearing of the mRNA. Note that in high-quality mammalian RNA, the ratio of 28S:18S eukaryotic rRNAs (judged by the intensity of bands through EtBr staining) should be ~2:1, however, this ratio approaches 1 as one moves down the evolutionary ladder. This ratio gives information about the integrity of preparation and a ratio of 2:1 indicates that no gross degradation of RNA has occurred. In RNA samples that have been degraded, this ratio is reversed, as the 28S rRNA is characteristically degraded to 18S-like species. In addition to the 28S and 18S rRNAs, an intact RNA sample manifests its mRNA component as a significantly lighter smear that appears slightly above, between, and below the rRNAs. This is the normal appearance of cellular mRNA because of its extremely heterogeneous nature and because the mRNA is usually less than 3% of the total mass of RNA in the cell. To achieve maximum resolution of larger molecular weight RNAs, electrophoresis is often allowed to continue to the extent that the small 5S and 5.8S rRNA and tRNA species run off the distal edge of the gel and into the running buffer. This is acceptable because the 5S and 5.8S rRNA species are too small to be useful as molecular weight markers for cellular RNA because these generally migrate through the gel along with the tRNA at the leading edge of the electrophoretic separation in the 300–400 base range. An agarose gel showing pure preparation of total cell RNA and mRNA from eukaryotes is depicted in Figure 1.8. The electrophoretic profile of plant RNA from green tissue is often different from that observed on electrophoresis of animal cells. In addition to the large and small rRNAs, an abundant number of other bands are visualized on staining. These are chloroplast transcripts, the presence of which is taken as an indicator of the integrity of the sample. Another bit of information that can be conveyed by examining an aliquot of RNA comes from the appearance of the well into which the sample is loaded. Fluorescence coming from within the well suggests that genomic DNA is present in the sample. Enormous fragments of chromosomal DNA, generated by shearing forces associated with the mechanics of RNA isolation, are unable to enter the gel during the course of the electrophoresis.

UV Spectrophotometry Similar to DNA, the concentration and purity of RNA preparations are easily determined by UV absorption spectrophotometry (for details see Section 1.3.1). As A_{260} of 1 corresponds to 40 µg/ml for RNA, the concentration of a RNA solution is easily calculated using following equation:

$$\text{Concentration of RNA } (\mu g/ml) = A_{260} \times \text{dilution factor} \times 40$$

For determination of the concentration of RNA, only OD values between 0.1 and 1 are significant because values between 0.02 and 0.1 may be associated with substantial errors. As UV spectrophotometry does not distinguish between pure samples of RNA and DNA-tainted samples, normalization based on A_{260} may be compromised because of the contribution to the total mass of the sample made by genomic DNA. Pure preparations of RNA have A_{260}/A_{280} value of 2.0. If there is significant contamination with protein or phenol, the A_{260}/A_{280} will be less than 2.0. The contamination of phenol may, however, be remedied by a final chloroform extraction. On the other hand, a low ratio may indicate incomplete solubilization of the RNA pellet.

Colorimetric Analysis by Orcinol Method Orcinol reaction confirms the presence of RNA in the sample. The principle behind the procedure is that when an RNA solution is heated in the presence of concentrated HCl, furfural is formed, which when reacted with orcinol in the presence of $FeCl_3$ as a catalyst forms a green-colored complex with a sharp absorption maximum at 665 nm. Note that only the purine nucleotides give any significant reaction. RNA concentration is determined by measuring the intensity of absorbance of the green-colored solution at 665 nm with a spectrophotometer and comparing with a standard curve of known RNA concentrations.

1.3.3 Construction of Recombinant DNA Molecule

After isolation and purification of DNA fragment containing the gene to be cloned and the vector DNA, the next requirement in a gene cloning experiment is construction of the recombinant DNA molecule. This involves cleavage of a vector as well as DNA to be cloned followed by their joining together in a controlled manner.

Cleavage of Vector DNA and Isolation of DNA Fragment from Genomic DNA

Each vector molecule is cut at a single position (e.g., in case of plasmids, λ insertional vectors, etc.) or at two different positions (e.g., in case of λ replacement vectors or artificial chromosomes, etc.) with restriction enzyme(s) and at exactly the same position(s) all the time. The large DNA molecule (genomic DNA) is cleaved or broken in precise and repro-

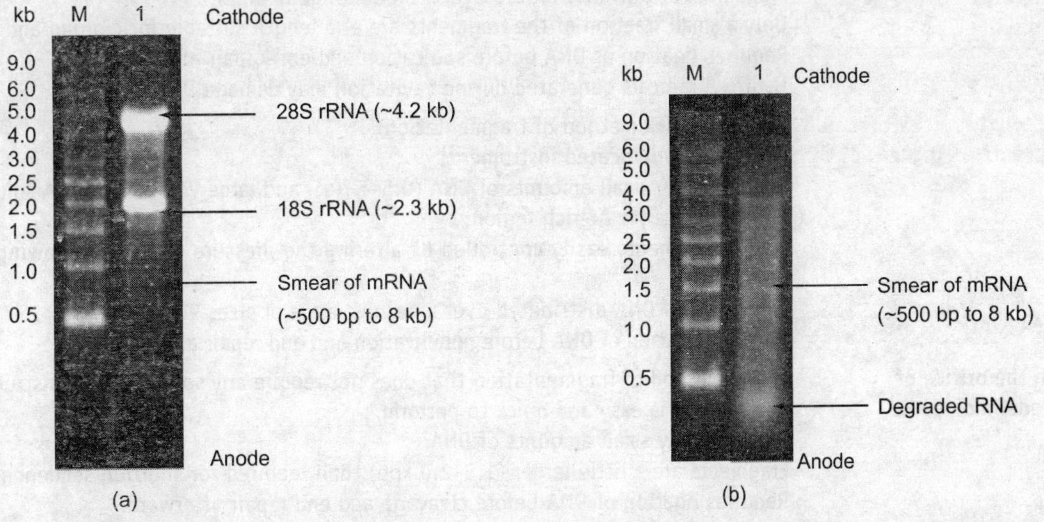

Figure 1.8 Electrophoresis of RNA on denaturing agarose gel; (a) Total RNA preparation [Lane M : RNA size marker; Lane 1 : Total cell RNA in eukaryotes; 18S and 28S rRNAs form clearly visible discrete bands at positions around 2.3 and 4.2 kb, respectively; Different mRNAs appear as smear between ~500 base and 8 kb, with the bulk mRNA lying between 1.5 kb and 2 kb. In prokaryotic total RNA preparation (not shown in figure) 16S and 23S rRNAs form two discrete bands at positions around 1.5 kb and 2.9 kb, respectively within the mRNA smear]. (b) Pure mRNA preparation [Lane M : RNA size marker; Lane 1 : Different mRNAs form smear between ~500 base and 8 kb; Degraded RNA appears as a lower molecular weight smear; Note that a pure mRNA preparation lacks two discrete bands of rRNAs.]

ducible manner. The purpose is either to isolate an intact single gene or to break down the large DNA into fragments small enough to be carried by the vector, which exhibits a preference for DNA fragments of a particular size range. Fragmentation is typically achieved by enzymatic or mechanical methods and the extent of treatment governs the average size of fragments produced. The most commonly used methods for fragmentation of genomic DNA are summarized below and also listed in Table 1.2.

Digestion with Restriction Enzymes Restriction digestion using site-specific restriction endonucleases is routinely used for obtaining DNA fragments in a precise and reproducible manner (for details see Chapter 3). The procedure is simple, reliable, and generates specific sized fragments.

Cleavage by Nonspecific Endonucleases Another enzymatic method for fragmentation of DNA is nonspecific endonuclease-catalyzed cleavage. This method gives nonuniform and random fragments.

Mechanical Shearing The long, thin duplex DNA molecules are sufficiently rigid to be very easily broken by shear forces in solution. By high-speed stirring in a blender, controlled shearing can be achieved. Typically, high molecular weight DNA is sheared to fragments with a mean size of ~8 kbp by stirring at 1,500 rev/min for 30 min. Similarly, passage through the orifice of a 28-gauge hypodermic needle leads to mechanical shearing of high molecular weight DNA (Table 1.2). The method is cheap, easy, and requires only small amounts of DNA. Both these procedures induce random breaks with respect to DNA sequence so that each time a DNA sample is treated, a different set of fragments are generated. Moreover, shearing leads to raggedness of ends. The termini consist of short, single stranded regions, which may have to be taken into account in subsequent joining procedures. These methods are thus useful for generating random, overlapping fragments of genomic DNA for use in chromosome walking experiments (for details see Chapter 19).

Automated Shearing Using High-Performance Liquid Chromatography (HPLC) During the last few years, a method for hydrodynamic shearing, initially based on the use of HPLC and called the 'point-sink' flow system, has become increasingly refined and finally automated. In the 'point-sink' flow system, an HPLC pump is used to apply pressure to the DNA sample, thereby forcing it through tubing of very small diameter. In the automated process known as HydroShear (commercially available from Gene Machines), a sample of DNA is repeatedly passed through a small hole until the

Table 1.2 Hydrodynamic Shearing Methods Used for DNA Fragmentation

Method	Advantages and disadvantages
Sonication	Easy and quick method of fragmentation; Requires sophisticated instrument; Requires relatively large amounts of DNA (10–100 μg); Fragments of DNA distributed over a broad range of sizes; Only a small fraction of the fragments are of a length suitable for cloning and sequencing; Requires ligation of DNA before sonication and end-repair afterward; Hydroxyl radicals generated during cavitation may damage DNA.
Nebulization	Easy and quick method of fragmentation; Requires sophisticated instrument; Requires only small amounts of DNA (0.5–5 μg), and large volumes of DNA solution; No preference for AT-rich region; Size of fragments easily controlled by altering the pressure of the gas blowing through the nebulizer; Fragments of DNA distributed over a narrow range of sizes 900–1,330 bp); Requires ligation of DNA before nebulization and end repair afterward.
Passage through the orifice of a 28-gauge hypodermic needle	Cheap method of fragmentation that does not require any sophisticated instrumentation; The method is easy and quick to perform; Requires only small amounts of DNA; Fragments are a little larger (1.5–2.0 kbp) than required for shotgun sequencing; Requires ligation of DNA before cleavage and end repair afterward.
Circulation through an HPLC pump	Requires expensive apparatus; Requires 1–100 μg of DNA; Fragments of DNA distributed over a narrow range of sizes that can be adjusted by changing the flow rate; Ligation of DNA required; End repair of fragments before cloning not necessary.

sample is fragmented to products of a certain size (Table 1.2). The final size distribution is determined by both the flow rate of the sample and the size of the opening. These parameters are controlled and monitored by the automated system. At any given setting, DNA fragments larger than a certain length are broken, whereas shorter fragments are unaffected by passage through the opening. The resulting sheared products therefore have a narrow size distribution. Typically 90% of the sheared DNA falls within a two-fold size range of the target length. Libraries constructed from these DNA fragments are likely to be of higher quality than those made using one of the old mechanical shearing methods. These certainly contain clones of more uniform size and possibly may be more comprehensive in their coverage of the genome.

Sonication In this method, isolated DNA is subjected to hydrodynamic shearing by exposure to brief pulses of sound waves (Table 1.2). A diagram showing functioning of sonicator is depicted in Figure 1.9. Most sonicators shear DNA to a size of 300–500 bp and sonication is continued until the entire population of DNA fragments has been reduced to this size. However, the yield of subclones is usually greater if sonication is stopped when the fragments of target DNA first reach a size of ~700 bp. Excessive sonication for long time

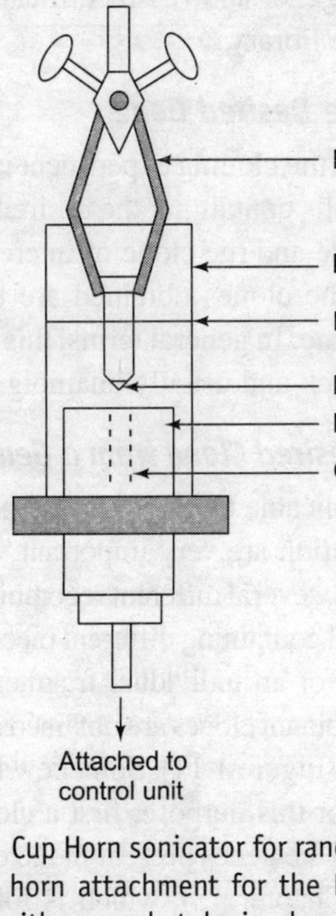

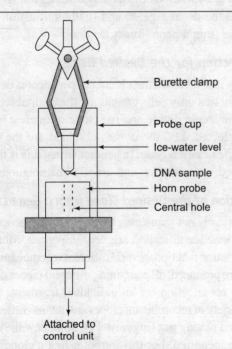

Figure 1.9 Cup Horn sonicator for random fragmentation of DNA [The cup horn attachment for the Heat Systems sonicator is depicted with a sample tube in place. The cup horn unit, which contains a large horn probe, is attached to the sonicator control unit and filled with ice water before the sample is sonicated. The sample tube is held in place from above by using a burette clamp and a ring stand. Alternatively, a tube holder can be fabricated from 1/4 inch plastic and used to hold up to eight tubes for simultaneous processing.]

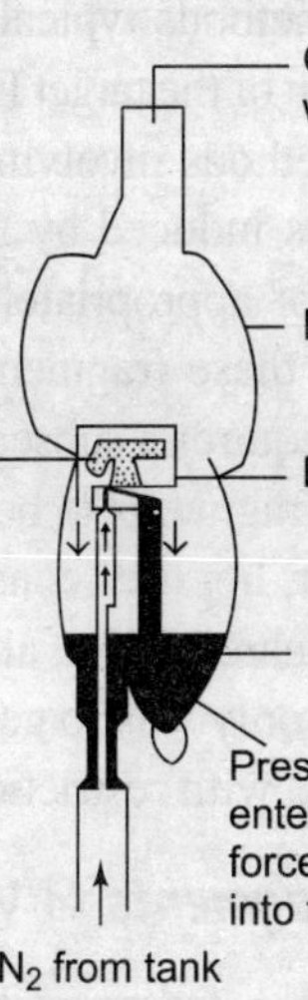

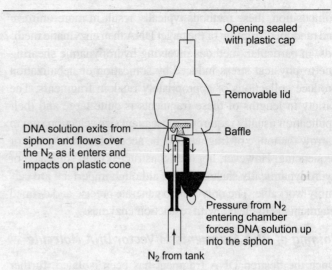

Figure 1.10 Functioning of nebulizer [A viscous DNA solution containing glycerol is placed in the nebulizer, which is attached to a nitrogen tank. Pressure from the nitrogen entering the chamber siphons the DNA solution from the bottom of the chamber to the top. The solution exits the siphon and impacts on a small plastic cone suspended near the top of the chamber, thus shearing the DNA.]

periods makes it extremely difficult to clone the sonicated DNA, perhaps due to damage caused by free radicals generated by cavitation.

Nebulization Nebulization, a form of hydrodynamic shearing, is another way to get DNA fragments from the large genomic DNA (Table 1.2). It is performed by collecting the fine mist created by forcing DNA in solution through a small hole in the nebulizer unit (e.g., CA-209 from CIS-US Inc.) (Figure 1.10). The speed of passage of DNA solution through the hole, the viscosity of the solution, and the temperature regulate the size of the fragments.

Other Methods of Obtaining DNA Fragment to be Cloned If the gene of interest is small and its sequences are known, it can be synthesized chemically for cloning in expression vector (for details see Chapter 4). Similarly, if the gene sequences from the same or related sources are known, DNA fragment may be obtained by PCR amplification using gene-specific primers and the resulting PCR product may be cloned directly into an expression vector (for details see Chapter 7). If the interest lies in gaining information about the expressed sequences or splice sites or developmentally regulated and tissue-specific sequences, mRNA molecules purified from the total cell RNA may be reverse transcribed to cDNA molecules, which when cloned into vector molecules form a cDNA library (for details see Chapters 5 and 15).

Although each approach is reasonably successful for generating a range of fragments from a large contiguous segment of DNA, each has its particular limitation. As physical methods used for shearing DNA are independent of sequence composition, these methods typically result in more uniform and random disruption of the target DNA than enzymatic methods. In particular, methods involving hydrodynamic shearing due to physical stress induced by sonication or nebulization produce collections of appropriately random fragments. The variety in lengths of these fragments is quite large and their application usually requires a subsequent size selection step to narrow the range of fragments to be acceptable for cloning or sequencing. However, libraries constructed from sonicated or hydrodynamically sheared DNA, although imperfect, are certainly workable. The only way to generate precise and defined fragments is to cleave with restriction enzymes.

Joining of DNA Fragments to Vector DNA Molecule

Once the desired DNA fragment has been isolated, further manipulations require them to be inserted into a cloning vector (for details see Chapter 13). This is because fragments of DNA are not replicons and in the absence of replication, these are diluted out of their host cells. It should be noted that even if a DNA molecule contains an origin of replication, this might not function in a foreign host cell. If fragments of DNA are not replicated, the obvious solution is to attach them to a suitable replicon. Such replicons are known as vectors (for details see Section 1.2). Thus isolated DNA fragments are joined with linearized or cleaved vector DNA molecules in a DNA ligase-catalyzed reaction, resulting in the formation of recombinant DNA molecules. The problems that may occur due to vector reconstruction or DNA fragment dimer formation are reduced at this stage by dephosphorylation (for details see Chapter 2) and the DNA fragments of appropriate size are obtained by size fractionation techniques (for details see Chapter 6). Short synthetic oligonucleotides, *viz.* linkers and adaptors, may be used to increase the versatility of ligation reaction (for details see Chapter 13). Other strategies for joining of DNA fragments with vector DNA molecule include homopolymer tailing and action of *Vaccinia* DNA topoisomerase (for details see Chapter 13).

1.3.4 Introduction of Recombinant DNA Molecule into Host Cell

The next step in a gene cloning experiment is to introduce the recombinant DNA molecule into living cells, usually bacterium, or any other living cell. Once introduced into host cell, with the multiplication of the host, a large number of recombinant DNA molecules may be produced from a limited amount of starting material. When the host cell divides, copies of the recombinant DNA molecules are passed to the progeny, and further vector replication takes place. After a number of cell divisions, a large number of genetically identical host cells (clones) are produced, which appear as a colony or plaque depending upon the vector molecules. Each cell in the clone contains one or more copies of a recombinant DNA molecule. This stage marks the 'cloning' of the gene. Different techniques used for introduction of recombinant DNA molecules into host cell include transformation, transfection, or *in vitro* packaging followed by natural infection by the assembled infectious phage (for details see Chapters 9 and 14).

1.3.5 Selection of Recombinants and Screening of Desired Clone

To isolate the clone of interest or to identify the clone containing the gene of interest, the next step in a gene cloning experiment is the selection and screening of recombinants or transformants. Screening of desired clone is almost invariably done at the plating out stage. Thus, by plating on selection medium (based on the product of selectable marker gene), recombinants are either distinguished from nonrecombinants by color reaction or are the only hosts capable of growth on that medium (for details see Chapter 16). Once it is established which colonies/plaques are recombinants, the next step is to search for the desired clone. Although there are many different procedures by which the desired clone is obtained, all are variations of two basic strategies. (i) Direct selection for the desired gene; and (ii) Identification of the desired clone from a gene library.

Direct Selection for the Desired Gene

This strategy is used when the cloning experiment is designed in such a way that only cells containing the desired recombinant DNA molecule divide and the clone of interest is automatically selected, i.e., the clones obtained are the clones containing the required gene. In general terms, this is the preferred method, as it is quick and usually unambiguous.

Identification of the Desired Clone from a Gene Library

Direct selection is not applicable to all genes and hence techniques for clone identification are very important. Moreover, during the ligation reaction, several different recombinant DNA molecules are produced, all containing different pieces of DNA and there is no selection for an individual fragment. Consequently, a variety of recombinant clones are obtained after transformation and plating out (Figure 1.11). Somehow the correct one must be identified. For this purpose, first a clone library representing all or most of the genes present in the cell is constructed (for details see Chapter 15), which is followed by analysis of the individual clones to identify the desired one from a mixture of lots of different clones. This strategy requires extensive screening because even the simplest organisms, such as *E. coli*, contain several thousand genes and a restriction digest of total cell DNA produces not only the fragment carrying the desired gene, but also many other fragments

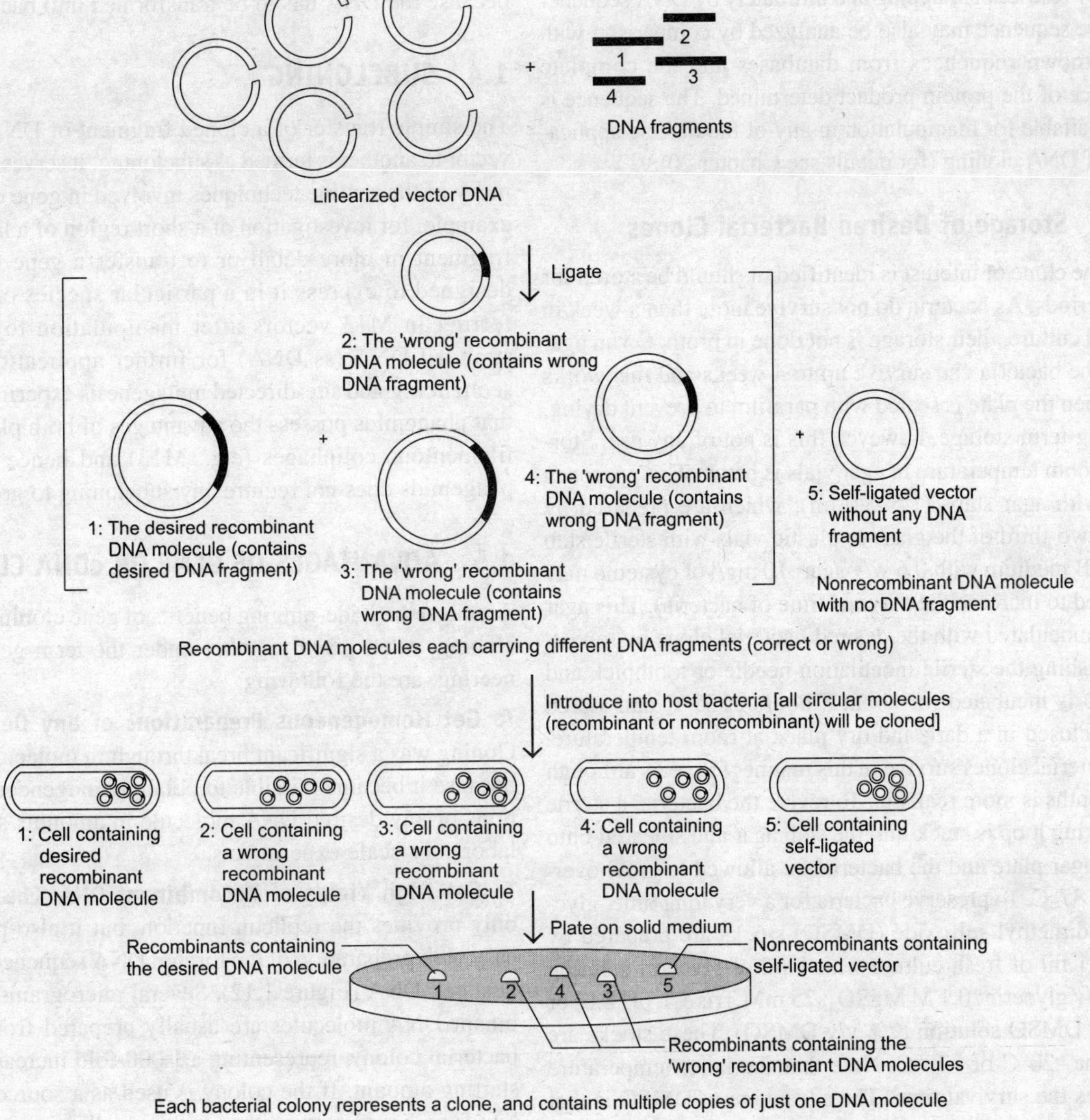

Figure 1.11 Cloning allows purification of individual fragments of DNA generated in the first step of cloning experiment [The DNA fragment to be cloned is one member of a mixture of many different fragments, each carrying a different gene or part of a gene. This mixture could indeed be the entire genetic complement of an organism, a human, for instance. Each of these fragments becomes inserted into a different vector molecule to produce a family of recombinant DNA molecules, one of which carries the gene of interest. Usually only one recombinant DNA molecule is transported into any single host cell so that the final set of clones contain multiple copies of just one DNA molecule. The gene is now separated away from all the other genes in the original mixture, and its specific features can be studied in detail. Cloning is analogous to purification. From a mixture of different molecules, clones containing copies of just one molecule can be obtained.]

carrying other genes. Moreover, there is no guarantee that the entire gene is present on a single clone. Thus, once a suitable library is prepared, a number of procedures based either on direct identification of the correct recombinant DNA molecule or detection of translation product of the cloned gene are employed to identify the desired clone (for details see Chapter 16). The DNA-based techniques are usually easier and involve hybridization probing, i.e., hybridization analysis by using either a radioactively or fluorescently labeled DNA probe complementary or partially complementary to a region of the gene sequence. On the other hand, other methods rely on the expression of coding sequences of the clones in the library and identification of the protein product by its activity or by antigen antibody reaction. Once identified, a cloned gene is analyzed by restriction mapping and ultimately by DNA sequencing. The sequence may also be analyzed by comparison with other known sequences from databases and the complete sequence of the protein product determined. The sequence is then available for manipulation in any of the diverse applications of DNA cloning (for details see Chapter 20).

1.3.6 Storage of Desired Bacterial Clones

Once the clone of interest is identified, it should be stored for long periods. As bacteria do not survive more than a week in a liquid culture, their storage is not done in broth. On an agar plate, the bacteria can survive up to 4 weeks and this works best when the plate is sealed with parafilm to prevent drying. For long-term storage, however, this is not of any use. Storage at room temperature in agar vials is better. This approach deals with agar stabs (1.5–5.0 ml), which are prepared by filling two-third of the glass or plastic vials with sterile stab agar (LB medium with 0.6 w/v agar; 10 mg/l of cysteine may be added to increase the survival time of bacteria). This agar stab is inoculated with the desired bacterial clone by repeatedly pushing the sterile inoculation needle or toothpick and the stab is incubated for about 8 hours at 37°C and stored tightly closed in a dark and dry place at room temperature. The bacterial clones survive in this manner for years although 3–6 months is more realistic. To revive the bacteria, a sterile inoculating loop is stuck into the stab agar and smeared onto an LB agar plate and the bacteria are allowed to grow overnight at 37°C. To preserve bacteria for a very long time, glycerin or dimethyl sulfoxide (DMSO) stocks are prepared by adding 1 ml of fresh culture with 1 ml of glycerin solution (65% v/v glycerin; 0.1 M $MgSO_4$; 25 mM Tris-Cl, pH 8.0) or 1 ml of DMSO solution (7% v/v DMSO). These stocks are stored at –20°C or –70°C. Note that the latter temperature increases the survival time. The advantage of DMSO is that it is easy to pipette as compared to glycerin. The bacteria are revived from these stocks by scratching the surfaces with a sterile toothpick or inoculation loop, inoculating an LB plate and growing overnight at 37°C. Another method for long-term storage of bacteria is freeze-drying or its low-cost version, vacuum drying. In this procedure, the bacteria are frozen, followed by removal of water by means of a vacuum, and collection of all the remnants in glass tubules in an anaerobic state. Depending on the mode of storage and the specific microorganisms, the survival rate can be up to 30 years. For revival of bacteria, the ampoule is broken carefully and the powdered bacteria are used for inoculation into a fresh LB medium. Besides storing bacteria, the most common and reliable method is to store bacterial DNA stocks at –20°C. As long as the DNA is free from DNases, it is unlikely to degrade. The concentration should be more than 0.1 μg/ml. Revival is somewhat laborious than with bacterial stocks because the DNA has to be transformed into bacteria.

1.4 SUBCLONING

The simple transfer of a cloned fragment of DNA from one vector to another is termed as subcloning. It serves to illustrate many of the routine techniques involved in gene cloning, for example, for investigation of a short region of a large cloned fragment in more detail or to transfer a gene to a vector designed to express it in a particular species or for transferring in M13 vectors after manipulation to get single stranded DNA (ss DNA) for further application in DNA sequencing and site-directed mutagenesis experiments. Note that phagemids possess the advantages of both plasmids and filamentous coliphages (e.g., M13) and hence cloning in phagemids does not require any subcloning to get ss DNA.

1.5 ADVANTAGES OF GENE OR cDNA CLONING

Amongst the wide-ranging benefits of gene cloning or cDNA cloning, often given together under the term genetic engineering, are the following:

To Get Homogeneous Preparations of any Desired DNA Cloning was a significant breakthrough in molecular biology because it became possible to obtain homogeneous preparations of any desired DNA molecule in amounts suitable for laboratory-scale experiments.

To Get High Yields of Recombinant DNA The vector not only provides the replicon function, but it also permits the easy bulk preparation of the foreign DNA sequence free from host cell DNA (Figure 1.12). Several micrograms of recombinant DNA molecules are usually prepared from a single bacterial colony, representing a 1,000-fold increase over the starting amount. If the colony is used as a source of inoculum for a liquid culture, the resulting cells may provide milligrams of DNA, a million-fold increase in yield.

To Obtain Pure Sample of a Gene Another important application of cloning is purification of an individual gene, separated from all the other genes in the cell. To understand exactly how cloning can provide a pure sample of a gene, consider the basic experiment from Figure 1.1, drawn in a slightly different way (Figure 1.12). The manipulation that results in a recombinant DNA molecule can rarely be controlled to the extent that no other DNA molecules are present at the end of the procedure. In addition to the desired recombinant DNA molecule, the ligation mixture may contain unligated vector molecules, unligated DNA fragments, DNA fragment dimers, recircularized or self-ligated vector molecules, and recombinant DNA molecules carrying the wrong DNA insert. Though

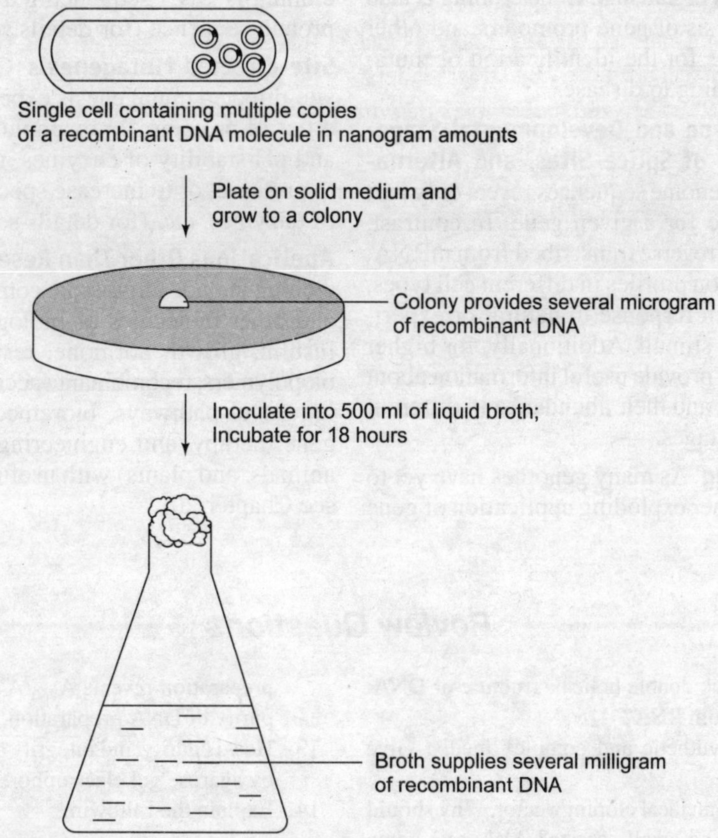

Figure 1.12 Cloning supplies large amounts of recombinant DNA

bacterial cells may take up unligated molecules or DNA fragment dimers, these rarely cause a problem because these either do not replicate or are subjected to degradation by the host enzymes (i.e., replicate only under exceptional circumstances). On the other hand, self-ligated vector molecules and incorrect recombinant vectors are replicated just as efficiently as the desired molecule. However, purification of the desired molecule can still be achieved through cloning because it is extremely unusual for any one cell to take up more than one DNA molecule. Each cell gives rise to a single colony, so each of the resulting clones consists of cells that contain the same molecule. Thus, different colonies contain different molecules, some contain the desired recombinant DNA molecule, some have wrong recombinant DNA molecule, and some contain self-ligated vector. If, somehow, the colonies containing the correct recombinant DNA molecule are identified, a pure sample of the gene of interest may be obtained.

Isolation and Manipulation of Fragments of an Organism's Genome DNA cloning facilitates the isolation and manipulation of fragments of an organism's genome by replicating them independently as part of an autonomous vector.

Isolation of Long Genes and Unknown Genes Gene cloning is the only way of isolating long genes with the application of 'chromosome walk' (for details see Chapter 19). Gene cloning is also the only procedure to isolate the genes that have never been studied before; such unknown genes cannot be isolated through PCR.

Elucidation of Gene Function, Promoter Analysis, and Identification of Mutations The task of functionally annotating genomes always lags way behind the structural annotation phase. Classically, detailed molecular analysis of proteins or other constituents of most organisms was rendered difficult or impossible by their scarcity and the consequent difficulty of their purification in large quantities. One approach is to isolate the expressed gene(s). Standard chemical or biochemical methods cannot be used to isolate a specific region of the genome for study, particularly as the required sequence of DNA is chemically identical to all the others. This is because every organism's genome is large and complex, and any sequence of interest usually occurs only once or twice per cell. Gene cloning strategies remain of value for the elucidation of gene function, for example, elucidation of expression profiles or biochemical functions of the protein encoded by the cloned gene. The technique is used for investigation of protein/enzyme/RNA function by large-scale production of normal and altered forms. Using this technique, new insights are emerging, for example, into the regulation of gene expression in cancer and development and the evolutionary history of proteins as well as organisms. Gene cloning is also used for isolation and analysis of gene promoters and other control sequences as well as for the identification of mutations, e.g., gene defects leading to disease.

Information About Cell Type and Developmental Stage-Specific Genes, Locations of Splice Sites, and Alternatively Spliced Genes The genome sequences reveal only part of the information available for a given gene. In contrast, cDNA sequences, which are reverse transcribed from mRNA, are cloned to reveal expression profiles in different cell types, developmental stages, and in response to natural or experimentally simulated external stimuli. Additionally, for higher organisms, cDNA sequences provide useful information about splice sites, splice isoforms, and their abundance in different tissues and developmental stages.

DNA or Genome Sequencing As many genomes have yet to be mapped or sequenced, other exploding application of gene cloning is DNA sequencing and consequently derivation of protein sequence (for details see Chapter 19).

Site-directed Mutagenesis Gene cloning is also helpful in site-directed mutagenesis experiments used to alter the properties of proteins, for example, to enhance thermal tolerance and pH stability of enzymes, increase resistance of proteins to protease, or to increase specificity and catalytic efficiency of enzymes, etc. (for details see Chapter 17).

Applications Other Than Research The applications of gene cloning include large-scale commercial production of proteins and other molecules of biological importance (e.g., human insulin, growth hormone, restriction enzymes, antibiotics, biopolymers, recombinant vaccines), edible vaccines, engineering novel pathways, bioremediation, reproductive cloning, gene therapy, and engineering organisms (microorganisms, animals, and plants) with useful quality attributes (for details see Chapter 20).

Review Questions

1. Discuss the Watson–Crick double helical structure of DNA. How does DNA differ from RNA?
2. Differentiate between synthetic and complex media. Give examples of each.
3. Discuss the properties of an ideal cloning vector. Why should an ideal cloning vector be small in size? Also give some examples of cloning vectors.
4. What is biological containment and what is its significance? How is it achieved?
5. What do you understand by the term 'gene cloning'? Give its experimental details. How is the process used for gene isolation?
6. Describe in detail the procedure for the preparation of total cell DNA from a bacterial cell. How does the process differ from that of isolation of total plant DNA?
7. Describe the principle and procedure of DNA purification using guanidinium thiocyanate and silica method.
8. Explain various strategies adopted to prevent contamination of DNA preparation by proteins, polysaccharides, and RNA.
9. Describe various strategies adopted for prevention of RNase contamination during RNA isolation procedure. What is the significance of wearing gloves in the process?
10. What are the differences between guanidinium containing lysis buffers and NP-40 containing lysis buffers used for RNA isolation?
11. Discuss the roles of following in nucleic acid isolation and purification procedures: lysozyme, β-ME, guanidinium thiocyanate, SDS, Na_2EDTA, monophasic reagents, proteinase K, phenol, chloroform, isoamyl alcohol, CTAB, DNase I, RNase, DEPC, sodium acetate, LiCl, 100% EtOH, silica columns, isopycnic ultracentrifugation, and washing with 70% EtOH.
12. How is purity of DNA and RNA preparations checked by UV absorbance measurement? If absorbance analysis of a DNA preparation reveals A_{260}/A_{280} ratio of 1.5, discuss about the purity of DNA preparation.
13. How is purity and integrity of total RNA preparation assessed by agarose gel electrophoresis?
14. Explain the following:
    (a) RNA is more reactive than DNA.
    (b) *E. coli* was the first bacteria to be used in cloning experiments.
    (c) Primary cloning is preferably done in *E. coli*.
    (d) PCR can be used as an alternative to cloning.
    (e) Gene cloning is the only way for isolating long genes or those that have never been studied before.
    (f) Nucleic acids are easily precipitated with EtOH in the presence of monovalent ions such as Na^+, NH_4^+, K^+, or Li^+.
    (g) One should wear gloves while isolating RNA.
    (h) 'Boiling inactivates DNases but not RNases' or 'RNases are sturdy molecules'.
    (i) After treatment of glassware, plasticware, and solutions with DEPC for appropriate time period, it is removed readily by autoclaving.
15. Why is fragmentation of genomic DNA required for cloning? Discuss different methods for obtaining DNA fragments for cloning experiments. Also discuss the advantages and disadvantages of each procedure.
16. What do you understand by the term 'subcloning'? Explain its significance.
17. Discuss the advantages of cDNA cloning. Also enumerate various applications of gene cloning.
18. The absorbance measurement at 260 nm with 2 μl of a 10 times diluted DNA sample reveals an OD of 0.8. Calculate the concentration of DNA (in μg) in a 100 μl DNA preparation.

Enzymes used in Genetic Engineering

Key Concepts

In this chapter we will learn the following:

- Restriction endonuclease
- DNA polymerase
- Reverse transcriptase
- RNA polymerase
- Alkaline phosphatase
- Polynucleotide kinase
- DNA ligase
- Deoxyribonuclease
- Ribonuclease
- Phosphodiesterase I
- β-Agarase
- Uracil–DNA glycosylase
- Proteinase K
- Lysozyme
- Topoisomerase

2.1 INTRODUCTION

Genetic engineering, i.e., recombinant DNA technology or simply gene cloning is a collective term that includes various experimental protocols resulting in the modification and transfer of DNA from one organism to another. This technology is absolutely dependent on different types of enzymes. Earlier, only procedures of classical genetics were available for analysis of the individual gene. During early 1970s the development of new techniques, stimulated by breakthrough in biochemical research, provided molecular biologists with enzymes that could be used to manipulate the DNA molecule *in vitro*. These enzymes are present naturally in the living cell and are essential for processes such as DNA replication, repair, and recombination, etc. These enzymes have been isolated and purified, and their properties have been analyzed *in vitro*. The purified enzymes are used as basic toolkit for manipulating DNA molecules in the desired manner, to make several copies of DNA; in cutting DNA to the desired size; to ligate the DNA in unique combinations; to modify the ends of DNA; and to express the DNA. These manipulations are the foundation of genetic engineering techniques. Various enzymes with different activities are used as indispensable gene manipulation tools in genetic engineering. The core concepts about the enzymes involved in gene manipulation are presented in the following topics.

2.2 RESTRICTION ENDONUCLEASE

The cutting of the DNA molecule is typically the first technique learned in the molecular biology laboratory and is fundamental to all recombinant DNA work. Such manipulations of DNA are conducted by molecular scissors known as restriction endonucleases or restriction enzymes. These are site-specific endonucleases that recognize and cut DNA molecules only at limited sites in a specific nucleotide sequence. The details are discussed in Chapter 3.

2.3 DNA POLYMERASE (DNA NUCLEOTIDYL-TRANSFERASE; EC 2.7.7.7)

A number of steps in molecular biology experiments require *in vitro* DNA synthesis, which is catalyzed by enzymes known as DNA-dependent DNA polymerases. DNA polymerases have a template-directed DNA synthesis activity (complementary to the template DNA), which allows it to extend from the free 3′-hydroxyl of a bound primer. DNA polymerases are widely used for many techniques in molecular biology. The following reaction is catalyzed by the enzyme:

$(dNMP)_n$	+ dNTP	$\rightarrow$	$(dNMP)_{n+1}$ +	PP_i
DNA	Incoming deoxyribonucleotide		Lengthened DNA	Pyrophosphate

where dNMP signifies deoxyribonucleoside monophosphate, n represents number of dNMPs, and dNTP signifies deoxyribonucleoside triphosphate.

Unit of Enzyme Activity For all DNA polymerases, one unit of enzyme activity is defined as the amount of enzyme that incorporates 10 nmoles of dNTPs into an acid precipitate material in 30 min at 37°C.

2.3.1 DNA Polymerase I

DNA polymerase I is a prokaryotic DNA polymerase belonging to the family 'A'. Under *in vivo* conditions it participates as a major repair enzyme catalyzing excision of primers and filling of gaps.

Enzyme Structure and Function

DNA polymerase I is generally isolated and purified from *E. coli.* It is a single-chain protein with a mass of about 109 kDa (928 amino acid residues) encoded by the *pol* A gene. The enzyme is readily cleaved by proteases into a small amino terminal fragment of 35 kDa and a large carboxy terminal fragment of 68 kDa. From N-terminal to C-terminal end of the enzyme, the three enzymatic activities present are: 5′ → 3′ exonuclease, 3′ → 5′ exonuclease, and 5′ → 3′ polymerase activities. The 5′ → 3′ polymerase activity (amino acids residues 521–928) helps in the addition of dNTP residues to the 3′-OH termini of DNA or RNA primers. These types of termini are generated by nicks or gaps present in double stranded DNA (ds DNA) and short segments of DNA or RNA that are base paired to a single stranded DNA (ss DNA) molecule. Mg^{2+} plays a significant role in the activity of the enzyme. Once a proper Watson–Crick base pair is formed (H-bonding between exposed base of the template and incoming dNTP), a catalytically competent complex that aids in phosphodiester bond formation is assembled at the enzyme's active site, involving two Mg^{2+} ions coordinately linked by three aspartic acid (Asp) residues. The 3′ → 5′ exonuclease activity is available on central domain (amino acids residues 324–517) and catalyzes the degradation of nucleotide residues from 3′-OH termini, thereby generating recessed 3′-ends. The 5′ → 3′ exonuclease activity of DNA Pol I (amino acids residues 1–323) catalyzes the cleavage of oligonucleotides from base paired 5′-ends. The enzyme with all the three activities is called holoenzyme (Figure 2.1). An intrinsic RNase H activity is also present, which is required for cell viability in *E. coli*, though not used in experiments. This enzyme first binds to short single stranded regions (nicks) in a ds DNA and then synthesizes a new strand complementary to the other strand, degrading the existing strand as it proceeds. As compared to other DNA polymerases, DNA Pol I possesses a unique feature, i.e., its capability of nick translation.

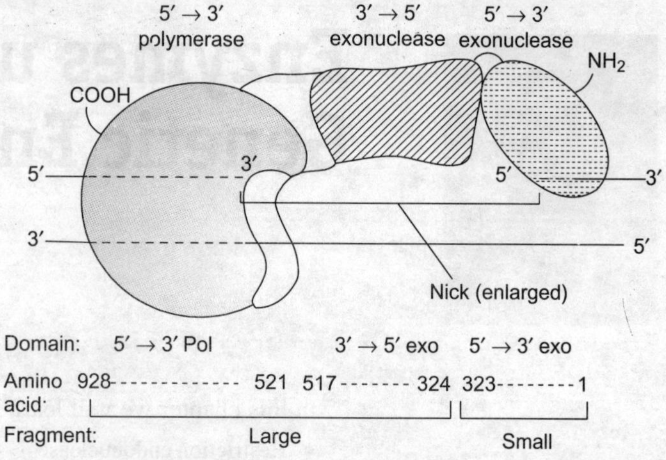

Figure 2.1 Domain structure of DNA Pol I indicating experimentally determined large and small fragments and the three enzymatic activities, their locations on the enzyme and on the nick in DNA

Applications

Because of its DNA synthesizing and nucleotide excision activities, the enzyme finds wide applications in genetic engineering experiments. The following are some applications of the enzyme in genetic engineering:

Synthesis of Second Strand of Complementary DNA (cDNA) The DNA Pol I holoenzyme is used for synthesis of second strand of cDNA during cDNA cloning. After first strand synthesis catalyzed by reverse transcriptase, an mRNA:cDNA hybrid is formed. RNase H cleaves the mRNA strand of the hybrid, producing gaps and nicks, and a series of RNA primers, which provide free 3′-OH termini for DNA polymerase I to initiate second strand cDNA synthesis. However, DNA polymerase I is no longer used for this purpose; instead, reverse transcriptase and Klenow fragment, which lack the 5′ → 3′ exonuclease activity are used. The details are discussed in Chapter 5.

Preparation of Radioactive Probes by End-labeling of DNA The use of DNA polymerase I for end-labeling of DNA molecules having 3′ protruding overhangs involves two steps. In the first step, the 3′ → 5′ exonuclease activity hydrolyzes the protruding 3′-end from the DNA and creates a 3′ recessed end. In the second stage, because of the presence of high concentrations of one radiolabeled dNTP (say d*CTP), the 3′ → 5′ exonuclease activity degrades ds DNA from the 3′-OH end until the base which is complementary to the added dNTP (i.e., dGTP) is exposed (Figure 2.2). After this, a series of synthesis and exchange reactions take place at this position. However, Klenow fragment and T4 DNA polymerase are better alternatives for end-labeling reactions.

DNA Labeling by Nick Translation Method DNA molecules can be easily radiolabeled at random positions by nick

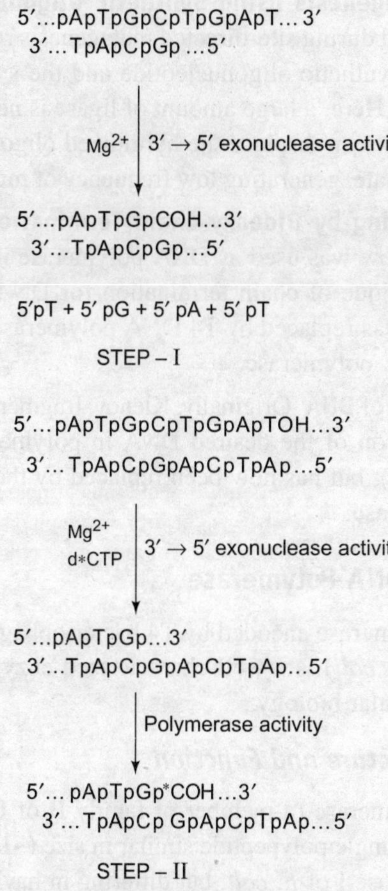

Figure 2.2 Application of DNA polymerase I in preparation of radioactive probe by end-labeling of DNA

translation method. Among all the known DNA polymerases, this reaction is performed only by DNA polymerase I. This reaction requires the addition of nucleotides, and if one of the dNTPs (say d*CTP) is radioactively labeled, the DNA molecule will automatically become labeled. The 5′ → 3′ exonuclease activity of this enzyme can degrade nucleotides from the DNA strand in front of the progressing enzyme. First, single stranded nicks are produced with the help of DNase I. Then, DNA polymerase I gets attached to the nicks and removes nucleotides from one strand of DNA, thus making a template for concomitant DNA synthesis. The joint action of 5′ → 3′ exonuclease and 5′ → 3′ polymerase activities leads to translation of the nick along the DNA molecule. During the repair of nick by DNA polymerase I, the addition of labeled dNTP (any one of the four) in the reaction mixture leads to its introduction into DNA molecule at complementary positions (Figure 2.3).

2.3.2 Klenow Fragment

The 5′ → 3′ exonuclease activity of *E. coli* DNA polymerase I makes it incompatible for many applications. However, this annoying enzymatic activity can be readily removed from

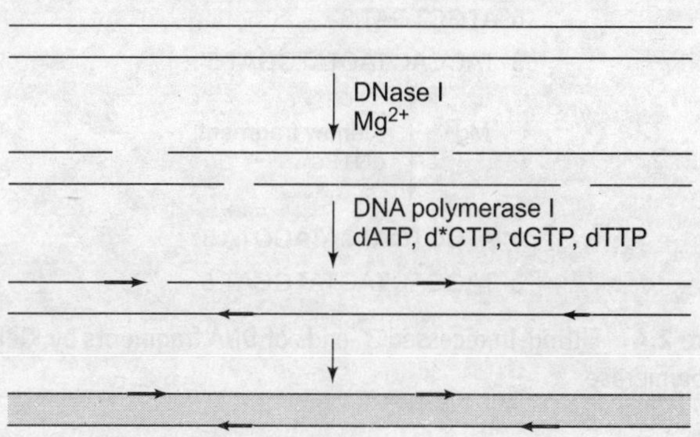

Figure 2.3 Application of DNA Polymerase I in DNA labeling by nick translation method

the holoenzyme. If DNA polymerase I is gently treated with the protease subtilisin, it can be cleaved into two fragments: the smaller N-terminal fragment (35 kDa) and the larger C-terminal fragment (68 kDa). The smaller fragment retains the 5′ → 3′ exonuclease activity, while the larger fragment retains the 5′ → 3′ polymerase and 3′ → 5′ exonuclease activities and is known as Klenow fragment, often referred to as the *E. coli* DNA polymerase I large fragment. This Klenow fragment thus lacks strand displacement (nick translation) activity and is widely used in molecular biology. It can also be expressed in bacteria from a truncated form of the DNA polymerase I gene. X-ray crystallography and mutational analysis of Klenow fragment have confirmed that the active sites of 5′ → 3′ polymerase and 3′ → 5′ exonuclease are present on different domains of Klenow fragment, lying 30–35 Å apart from each other. In some experiments the 3′ → 5′ exonuclease activity of Klenow fragment is either undesirable or not required. By introducing mutations in the gene that encodes the Klenow, forms of the enzyme that retain the polymerase activity but lack the 3′ → 5′ exonuclease activity can be expressed. These forms of the enzyme are usually called *exo*⁻ Klenow fragment.

Applications

Various applications of Klenow fragment in gene cloning experiments are as follows:

Synthesis of ds DNA from Single Stranded Templates During cDNA cloning, the second strand of cDNA can be synthesized with the help of Klenow fragment.

Filling in Recessed 3′-ends of DNA Fragments Klenow fragment is best-suited for filling-in 3′ recessed DNA ends. A filling-in reaction is used to generate blunt ends on fragments created by digestion with restriction enzymes that leave 5′-ends (Figure 2.4). The restriction and filling-in reactions can be performed by addition of dNTPs in single buffer, as Klenow fragment works well in buffer used for restriction enzymes.

5′ ATGCTGAT 3′
3′ TACGACTACTATCGAT 5′

↓ Mg^{2+} | Klenow fragment; dNTPs

5′ ATGCTGATGATAGCTA 3′
3′ TACGACTACTATCGAT 5′

Figure 2.4 Filling-in recessed 3′-ends of DNA fragments by Klenow polymerase

Digestion of Protruding 3′ Overhangs This is another method for producing blunt ends on DNA from 3′ overhangs generated by sticky end cutters. The protruding overhang is removed by 3′ → 5′ exonuclease activity of Klenow fragment. Moreover, digestion of nucleotides from the 3′-ends continues, but in the presence of dNTPs, the polymerase activity balances the exonuclease activity, creating blunt ends (Figure 2.5). However, T4 DNA polymerase conducts this reaction more efficiently (for details see Section 2.3.3).

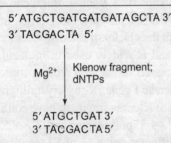

5′ ATGCTGATGATGATAGCTA 3′
3′ TACGACTA 5′

↓ Mg^{2+} | Klenow fragment; dNTPs

5′ ATGCTGAT 3′
3′ TACGACTA 5′

Figure 2.5 Digestion of protruding 3′ overhangs by Klenow fragment

Preparation of Radioactive Probes by End-labeling of DNA Klenow fragment is frequently used to prepare radiolabeled DNA. These labeling reactions are performed in the presence of three unlabeled dNTPs, each at a concentration in excess of the K_m, which lowers the potential of degradation of the 3′-ends of the template. The labeled dNTP is added at lower concentration, i.e., less than K_m. Labeling of the different ends is achieved in the following ways:

- The 3′ recessed ends can be easily labeled in the presence of radioactive (α-^{32}P) dNTP during the filling-in reaction.
- The labeling of protruding 3′-ends and blunt ended template is performed in two steps. The 3′ → 5′ exonuclease cleaves any protruding overhang and generates 3′ recessed ends. Radiolabeled dNTP is incorporated by an exchange reaction. T4 DNA polymerase then catalyzes end-labeling through its powerful 3′ → 5′ exonuclease activity.

Random Primer Labeling of DNA ss DNA or denatured ds DNA is easily labeled with the help of Klenow fragment in the presence of a random primer and four dNTPs, out of which one is labeled.

***In Vitro* Mutagenesis using Synthetic Oligonucleotides** Klenow is used during site-directed mutagenesis reaction for extension of synthetic oligonucleotide and the synthesis of circular DNA. Here, a large amount of ligase is needed; otherwise, Klenow can displace the hybridized oligonucleotide from the template, generating low frequency of mutagenesis.

DNA Sequencing by Dideoxy Chain Termination Method Initially, Klenow was used as DNA polymerase in Sanger's dideoxy technique of chain termination for DNA sequencing. Later, it was replaced by T4 DNA polymerase and thermostable DNA polymerase.

Amplification of DNA Originally, Klenow fragment was used for amplification of the desired DNA in polymerase chain reaction (PCR), but has now been replaced by thermostable DNA polymerase.

2.3.3 T4 DNA Polymerase

T4 DNA polymerase encoded by T4 bacteriophage (a bacteriophage of *E. coli*) is a well-characterized enzyme widely used in molecular biology.

Enzyme Structure and Function

T4 DNA polymerase (a member of family B of DNA polymerases) is a single polypeptide similar in size (~103 kDa) to DNA polymerase I of *E. coli*, but differing in having at least one essential sulphydryl group among its 15 cysteine residues. The sequence is surprisingly homologous to a family of polymerases that includes those of certain animal viruses, eukaryotic α-polymerases, and some phage polymerases, but does not include *E. coli* DNA polymerase I or III. The striking difference between T4 DNA polymerase and DNA polymerase I is in the template primer requirement. Unlike DNA polymerase I, T4 DNA polymerase, unaided by accessory proteins, cannot use a nicked duplex but requires a primed single stranded template. Such templates may have short or long gaps. In its ability to replicate DNA with extensive single stranded regions, the T4 enzyme differs from *E. coli* DNA polymerase I and DNA polymerase III (the core), which operate only on short gaps. The T4 enzyme displays the same requirements as DNA polymerase I for the primer terminus, template, and dNTPs in the assembly of a chain in the 5′ → 3′ direction. Thus, like all DNA polymerases, T4 DNA polymerase catalyzes a template directed DNA synthesis through extension of the free 3′-OH of a bound primer. The enzyme T4 DNA polymerase is somewhat similar to Klenow fragment as it contains 5′ → 3′ polymerase and 3′ → 5′ exonuclease activities and lacks 5′ → 3′ exonuclease activity. Its turnover number is about 400 nucleotides per second under optimal conditions. Moreover, because of the presence of a very active 3′ → 5′ exonuclease activity, the enzyme exhibits very high fidelity.

The kinetics of T4 DNA polymerase has been well-characterized for all the activities such as 3′ → 5′ exonuclease activity, polymerase activity, template dissociation, and misincorporation of dNTPs. The catalytic efficiency of the enzyme is defined both by its intrinsic polymerization rate and its processivity. The processivity of the enzyme varies according to the template tested, with natural templates exhibiting greater processivity than homopolymers. On a single stranded M13 (for details see Chapter 10) template, with a single primer, a processivity of less than 100 bases per binding event is observed. However, when T4 DNA polymerase is in excess, the apparent processivity is observed to be ~800 bases per binding event. This increase in apparent processivity may be due to nonspecific binding of the T4 DNA polymerase to the DNA template and the 2-D sliding of the polymerase to the elongation site. These studies suggest that enzyme: template ratios are important for achieving complete conversion to double stranded product. The processivity of the enzyme is greatly enhanced *in vivo* by the addition of accessory proteins such as gp44, gp62, gp45, and the single stranded binding protein gp32, which acts to lock the enzyme onto the DNA template.

Applications

Generally, T4 DNA polymerase is used for the same kind of reactions as Klenow fragment and/or DNA polymerase I. T4 DNA polymerase is preferred in a number of molecular biology applications because of the following reasons:

- The 3′ → 5′ exonuclease activity of T4 DNA polymerase is ~200 times more active than that of Klenow fragment.
- During polymerization, T4 DNA polymerase does not displace any downstream oligonucleotides.

T4 DNA polymerase is thus used in the following procedures:

Filling-in Recessed 3′-Ends of DNA Fragments The 3′ recessed ends are filled-in with the help of T4 DNA polymerase.

Preparation of Radioactive Probe by End-labeling of DNA

End-labeling of the recessed 3′-ends Labeling reaction of 3′ recessed ends is performed in the presence of high concentration of dNTPs so that polymerization activity exceeds the exonuclease activity.

End-labeling of the protruding 3′-ends This labeling reaction is carried out in two stages and has already been discussed in Sections 2.3.1 and 2.3.2.

Labeling DNA fragments for use as hybridization probes T4 DNA polymerase can be used to prepare strand-specific hybridization probes, which are better than probes prepared

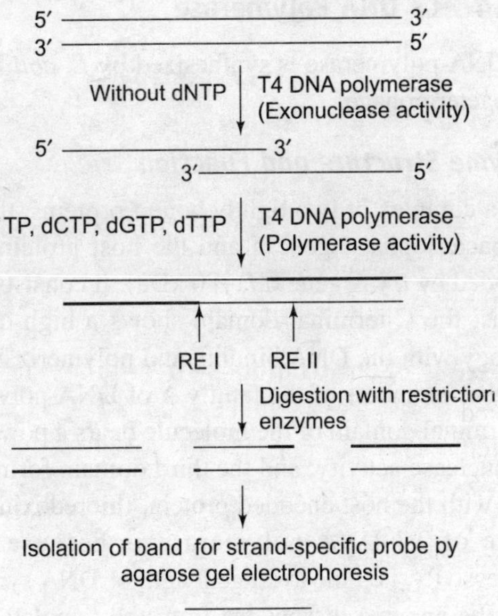

Figure 2.6 Role of T4 DNA polymerase in the preparation of radioactive probe by end-labeling of DNA

by nick translation, as these lack the hairpin structure. The specific strands are digested with two restriction enzymes and then treated with T4 DNA polymerase; the exonuclease activity of this enzyme creates recessed 3′-ends. These ends are filled with replacement reaction of polymerase in the presence of (^{32}P) dNTPs (Figure 2.6). Finally, the specific regions are digested with appropriate restriction enzymes and purified. The probes prepared by this method are fully labeled at their ends, but labeling decreases near the central regions.

Conversion of Cohesive Ends of Duplex DNA into Blunt End DNA Like Klenow reaction, T4 DNA polymerase can also repair 3′ overhangs and convert them into ds DNA. The powerful 3′ → 5′ exonuclease activity removes the protruding 3′-ends, and in the presence of high concentrations of dNTPs the degradation reaction is balanced by polymerization.

***In Vitro* Mutagenesis Using Synthetic Oligonucleotides** Unlike Klenow fragment, the T4 DNA polymerase does not have the intrinsic ability to extend from a nick. This property is important for applications such as site-directed mutagenesis where strand displacement by a polymerase results in displacement of the mutagenic oligonucleotide. Furthermore, T4 DNA polymerase lacks any detectable 5′ → 3′ exonuclease activity. This prevents digestion of annealed mutagenic oligonucleotides and reversion to the parental sequence. The high fidelity of the enzyme is also useful in reducing the number of undesirable secondary mutations, which can arise in strand synthesis.

2.3.4 T7 DNA Polymerase

T7 DNA polymerase is synthesized by *E. coli* infected with T7 bacteriophage.

Enzyme Structure and Function

It is a complex of two tightly bound proteins, the product of T7 bacteriophage gene *5* and the host protein thioredoxin encoded by *trx* A gene (M_r ~80 kDa). It consists of three domains: the C-terminal domain shows a high degree of homology with the DNA binding and polymerization domains of other members of the family A of DNA polymerases; the N-terminal domain of the molecule bears a powerful $3' \rightarrow 5'$ exonuclease activity; and the third domain forms a 1:1 complex with the host encoded protein, thioredoxin. The importance of T7 DNA polymerase is because of its high processivity, i.e., the average length of DNA synthesized before the enzyme dissociates from the template is considerably greater than for other thermolabile DNA polymerases. This enzyme is very similar in activity to Klenow fragment and T4 DNA polymerase. It has DNA polymerase and $3' \rightarrow 5'$ exonuclease activities, but lacks a $5' \rightarrow 3'$ exonuclease domain. The $3' \rightarrow 5'$ exonuclease activity is strong, like T4 DNA polymerase, i.e., ~1,000 times more than Klenow fragment. T7 DNA polymerase can be chemically treated or genetically engineered to remove its $3' \rightarrow 5'$ exonuclease activity. These forms of the enzyme are marketed under the name Sequenase 1.0 and Sequenase 2.0 and are widely used for DNA sequencing reactions (Exhibit 2.1).

Applications

T7 DNA polymerase is associated with the following applications:

End-labeling End-labeling of DNA molecule can be performed by either replacement or filling-in reactions (for details see Section 2.3.2).

Extension of Primer Primer extension reactions, which require the copying of long stretches of template DNA can be completed easily as this enzyme has high processivity.

DNA Sequencing T7 DNA polymerase is predominantly used in DNA sequencing by the chain termination technique because of its high processivity.

2.3.5 Thermostable DNA Polymerases

Thermostable DNA polymerases have been isolated from a number of organisms, mostly from thermophilic and hyperthermophilic eubacteria and thermophilic archea.

Enzyme Structure and Function

The enzymes from all sources are monomeric with molecular weights ranging from 60 kDa to 100 kDa. The first thermostable DNA polymerase was isolated and characterized from *Thermus aquaticus*.

Applications

These polymerases are a boon for *in vitro* DNA amplification reaction by PCR.

For more information on thermostable DNA polymerases, please refer to Section 7.3 and Table 7.1 of Chapter 7.

2.3.6 Terminal Deoxynucleotidyl Transferase (DNA Nucleotidylexotransferase; EC 2.7.7.31)

Terminal deoxynucleotidyl transferase is a specialized DNA polymerase expressed only in immature, pre-B, pre-T lymphoid cells, and acute lymphoblastic leukemia/lymphoma cells. It is also known as TdT and terminal transferase.

Enzyme Structure and Function

It is a single polypeptide with a molecular weight of 60 kDa. This enzyme adds a particular nucleotide to the 3′-end of DNA strands. Mg^{2+} and Co^{2+} ions are required as cofactors for the

Exhibit 2.1 Sequenase™ version 1.0 and 2.0

Sequenase™ Version 1.0 and 2.0, unlike the wild-type enzyme, has practically no $3' \rightarrow 5'$ exonuclease activity. In the first version of Sequenase, T7 polymerase was chemically modified and first marketed by USB (United States Biochemical) under the trade name Sequenase. The $3' \rightarrow 5'$ exonuclease activity of T7 DNA polymerase can be greatly reduced by incubating the enzyme for several days with a reducing agent, molecular oxygen, and Fe^{2+} ions. Here, the inactivation occurs due to site-specific modifications of the exonuclease domain by locally produced free radicals.

Sequenase 2.0, a genetically engineered form of T7 DNA polymerase, has superseded Sequenase 1.0. Sequenase Version 2.0 has two subunits, one is the *E. coli* protein thioredoxin (MW 12,000) and the other is a genetically engineered version of T7 bacteriophage gene *5* protein (MW 76,000). The genetic changes in this subunit (a deletion of 28 amino acids accomplished by *in vitro* mutagenesis) eliminates all measurable exonuclease activity without changing the DNA polymerase activity. It is highly processive, incorporates nucleotide analogs (deoxy inosine triphosphate (dITP), thio-dNTPs, dideoxy-NTPs, etc.), is not hindered by secondary structures, and can carry out strand displacement synthesis. It is an excellent enzyme for dideoxy sequencing and is useful in other applications, especially where the absence of associated exonuclease activity is desirable.

Source: www.usbweb.com

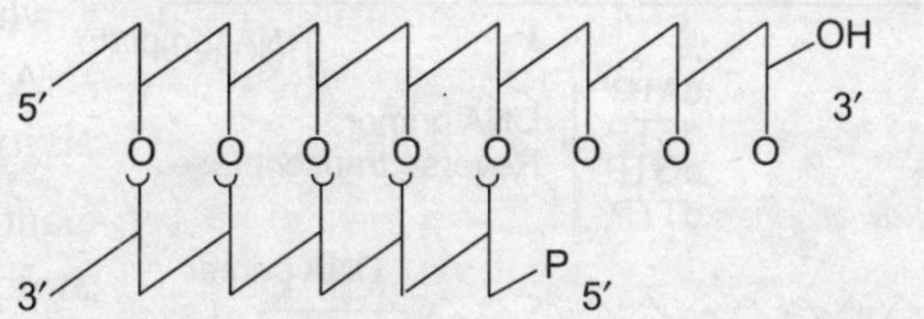

Figure 2.7 3′ overhang: a preferred substrate for terminal transferase

addition of purines and pyrimidines, respectively. Commercially available TdT is isolated and purified from recombinant *E. coli* cells expressing calf or rat or mouse thymus gene. Metal chelators and multivalent anions such as phosphate or pyrophosphate inhibit terminal transferase. Unlike most DNA polymerases it does not require a template. The preferred substrate of this enzyme is a protruding 3′ overhang (Figure 2.7), but it can also add nucleotides to the blunt or recessed 3′-ends less efficiently, in buffers of low ionic strength that contain Mg^{2+} or Co^{2+} ions. The minimum chain length of acceptor DNA is three nucleotides, and as many as thousands of dNTPs are added if the ratio of acceptor to nucleotide is adjusted correctly. A single nucleotide can be added at the 3′-end of DNA if dideoxy-NTP or cordycepin triposhphate are used as substrates. Homopolymers of rNTP can also be added at the 3′-end of the DNA molecule in the presence of a particular rNTP and Co^{2+}.

Applications

Terminal transferase is extremely useful for the following procedures:

Probe Preparation Radiolabeled probe can be prepared by labeling the 3′-ends of DNA molecule in the presence of TdT and single type of labeled dNTP (say dATP). Most commonly, the substrate for this reaction is a fragment of DNA generated by digestion with a restriction enzyme that leaves a 3′ overhang, but oligodeoxyribonucleotides can also be labeled. When such DNA is incubated with tagged oligodeoxyribonucleotides and terminal transferase, a string of the tagged nucleotides will be added to the 3′ overhang or to the 3′-end of the oligodeoxyribonucleotide (Figure 2.8).

Cloning by Homopolymer Tailing Complementary homopolymeric tail can be added to vector and cDNA (for details see Chapter 13).

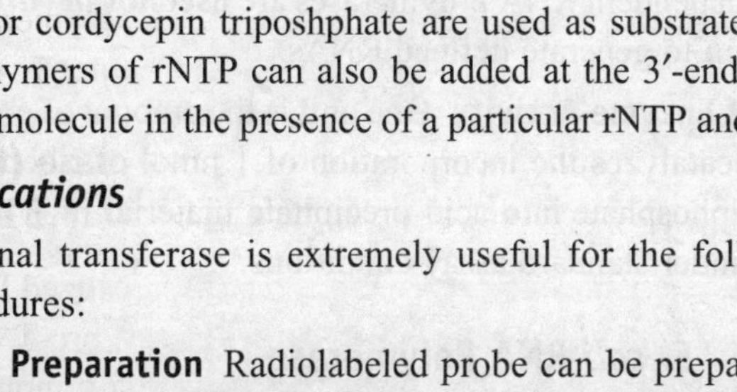

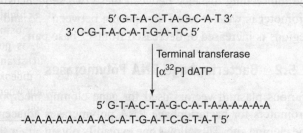

Figure 2.8 Probe preparation by using terminal transferase

Application of Homopolymer Tailing in Anchor PCR and 5′ Rapid Amplification of cDNA Ends (5′ RACE) Homopolymer tail can also be used in anchor PCR and 5′ RACE to add nucleotides which can then be used as a template for a primer in subsequent PCR (for details see Chapter 7).

Diagnosis of Acute Lymphoblastic Leukemia TdT is also used in the immunofluorescence assay for the diagnosis of acute lymphoblastic leukemia.

2.4 REVERSE TRANSCRIPTASE (RNA-DIRECTED DNA POLYMERASE, EC 2.7.7.49)

Reverse transcriptases are RNA-dependent DNA polymerases. These are encoded by retroviruses, where these copy the viral RNA genome into DNA prior to its integration into host cells.

Enzyme Structure and Function

Reverse transcriptases have two types of activities:

DNA Polymerase Activity In the retroviral life cycle, reverse transcriptase copies only RNA; but as used in the laboratory, it transcribes both ss RNA and ss DNA templates with fundamentally comparable efficiencies. In both the cases, RNA or DNA primer is required to initiate synthesis.

RNase H Activity RNase H is a ribonuclease that degrades the RNA in a RNA:DNA hybrid, which is formed during reverse transcription of an RNA template. This enzyme functions as both an endonuclease and an exonuclease in hydrolyzing its target.

All retroviruses have a reverse transcriptase, but the enzymes that are available commercially are derived from one of two retroviruses, either by purification from the virus or expression in *E. coli* (for details see Chapter 7). These retroviruses are:

- Moloney murine leukemia virus (Mo-MLV)
- Avian myeloblastosis virus (AMV)

Both enzymes have the same fundamental activities, but differ in a number of characteristics, including temperature and pH optima (for more details see Chapter 7). Most importantly, the murine leukemia virus enzyme has a very weak RNase H activity compared with the avian myeloblastosis enzyme, which makes it the clear choice when trying to synthesize cDNAs for long mRNAs.

Applications

Various applications of reverse transcriptase in genetic engineering are listed below:

***In Vitro* Reverse Transcription of mRNA** Reverse transcriptase enzyme is used to copy RNA into DNA, which is an integral task for cloning of cDNAs (for details see Chapter 5).

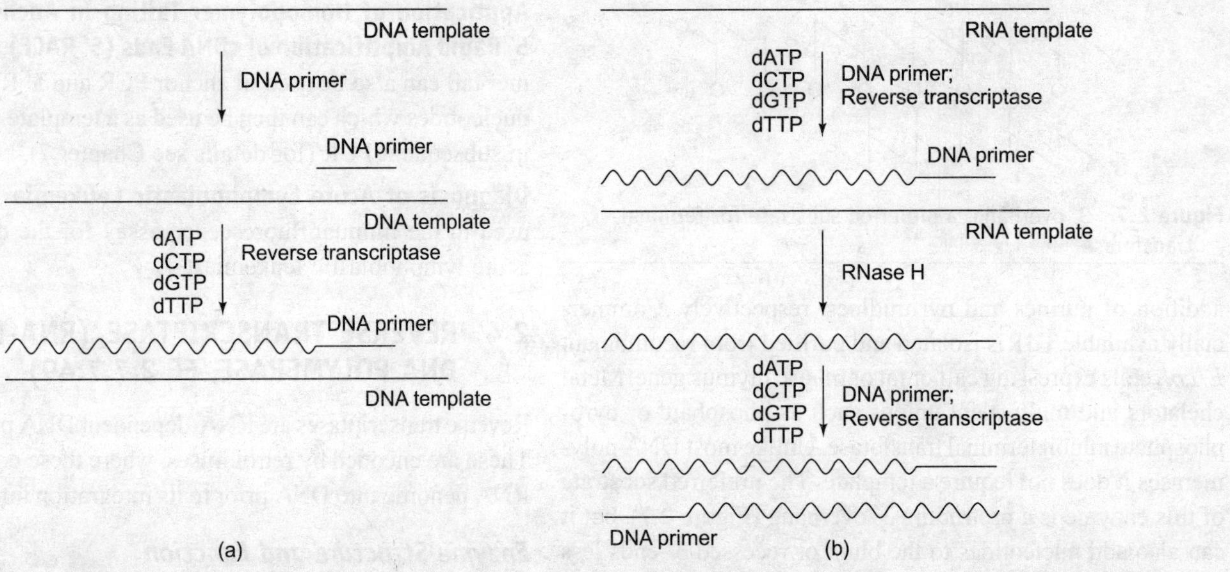

Figure 2.9 Reverse transcriptase-catalyzed labeling of DNA and RNA molecules. (a) Labeling of DNA, (b) labeling of RNA

Reverse Transcription PCR Reverse transcriptase is used to generate DNA copies of RNAs prior to amplifying that DNA by polymerase chain reaction, i.e., RT-PCR (for details see Chapter 7).

Labeling of DNA Molecule DNA or RNA molecule can be labeled by reverse transcriptase enzyme (Figure 2.9).

Sequencing of DNA Reverse transcriptase enzyme is also used for sequencing of DNA by Sanger's dideoxy method when other enzymes fail to perform efficiently.

2.5 RNA POLYMERASE (RNA NUCLEOTIDYL TRANSFERASE; EC 2.7.7.6)

RNA polymerase is an enzyme that makes a RNA copy of a DNA or RNA. Chemically, RNA polymerase is a nucleotidyl transferase that polymerizes ribonucleotides at the 3′-end of an RNA transcript. The reaction catalyzed is

$(NMP)_n$	+ NTP	$\rightarrow (NMP)_{n+1}$	+ PP_i
RNA	Incoming ribonucleotide	Lengthened RNA	Pyrophosphate

where NMP signifies ribonucleoside monophosphate; n represents number of NMPs.

RNA polymerase is used *in vitro* to produce large quantities of RNA complementary to one strand of gene of interest that has been placed immediately downstream from the promoter. Expression vectors are used if transcription of a cloned gene is required for the preparation of RNA probes or to purify large amounts of the gene product. Vectors carrying the promoter sequences are available, and by changing the orientation of the promoter with respect to the cloned DNA, it is possible to synthesize the RNA complementary to either strand. The promoter sequences can also be added to the target DNA by PCR primed with primers containing the sequences of specific promoters. Generally, *E. coli* and phage encoded DNA-dependent RNA polymerases are used for *in vitro* transcription to generate defined RNAs.

Unit of Enzyme Activity One unit is the amount of enzyme, which catalyzes the incorporation of 1 nmol of ribonucleoside triphosphate into acid precipitate material in 1 hour at 37°C under standard assay conditions.

2.5.1 *E. coli* RNA Polymerase

E. coli RNA polymerase is a multisubunit enzyme with a molecular weight of ~500 kDa. The core RNA polymerase of *E. coli* contains four types of subunits, $\alpha_2\beta\beta'\omega$. Another subunit called σ subunit binds only transiently to the core enzyme, forming a holoenzyme $\alpha_2\beta\beta'\omega\sigma$. The most common promoters are those that are recognized by the RNA polymerase with σ^{70}. In a strong promoter, the consensus sequences at –35 and –10 regions are 5′-TTGACA-3′ and 5′-TATAAT-3′, respectively. Thus, the rate of transcript production, i.e., the number of RNA copies synthesized per unit per enzyme molecule depends upon proximity of its sequence to the consensus. The strength of a promoter also depends on the intervening distance between these two sequences. The promoter is weaker when the spacing between –35 and –10 regions is increased or decreased from 17 base pairs.

2.5.2 Bacteriophage RNA Polymerases

Various plasmid vectors used for gene cloning incorporate promoters for bacteriophage RNA polymerases adjacent to the cloning site. This allows one to readily obtain either sense or antisense mRNA transcripts from the inserted DNA. The

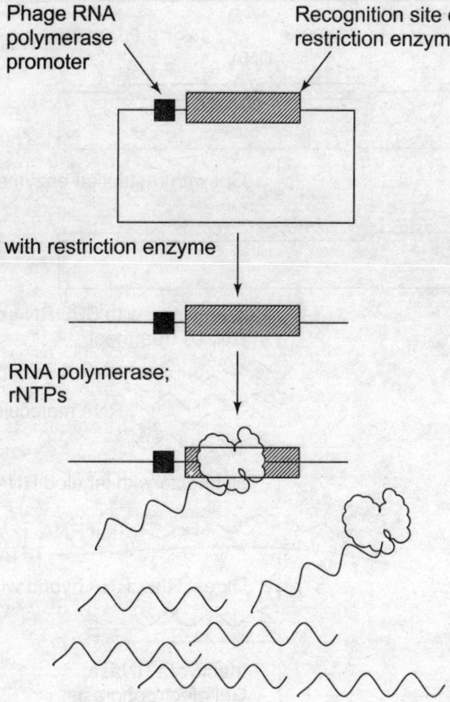

Figure 2.10 Run-off transcription by bacteriophage RNA polymerases

process is often called run-off transcription, because the plasmid is cut with a restriction enzyme downstream of the inserted DNA, which causes the polymerase to fall off the template when it reaches that position (Figure 2.10).

Enzyme Structure and Function

Several bacteriophage RNA polymerases are commercially available. These are named after the phage that encodes them and either purified from phage-infected bacteria or produced as recombinant proteins. Bacteriophage SP6, T3, and T7 RNA polymerases are DNA-dependent RNA polymerases with strict specificity for their respective double stranded promoters and also in their sequence preference for the first five coded bases (the start site). For example, only SP6 DNA or DNA cloned downstream from an SP6 promoter can serve as a template for SP6 RNA polymerase-directed RNA synthesis; SP6 RNA polymerase does not recognize T3 or T7 RNA polymerase promoters as a start site for transcription (Table 2.1). These RNA polymerases can catalyze the 5′ → 3′ synthesis of RNA using either ss DNA or ds DNA as template located downstream from their promoters in the presence of ribonucleoside triphosphates (rNTPs) and Mg^{2+}, which is required as a cofactor.

There are three reasons for using a phage promoter and its respective polymerase for *in vitro* transcription experiments.

- These promoters are strong, facilitating *in vitro* synthesis of a large amount of RNA.
- These promoters are not recognized by *E. coli* polymerases and so no transcription is performed within the cell.
- An active phage RNA polymerase consists of a single polypeptide, i.e., these are much simpler enzymes to work with than the *E. coli* enzyme, which is a multisubunit enzyme. *E. coli* RNA polymerase is not widely used in molecular biology laboratory.

Applications

RNA polymerases are indispensable for the following procedures:

Synthesis of ss RNA Transcripts ss RNA transcripts can be synthesized *in vitro* for use as hybridization probes, functional capped mRNAs transcripts, and for *in vitro* translation system. Generally, ss DNA can be used as a probe in hybridization reactions, while sometimes RNA probes are preferred as the rate of hybridization and the stability of RNA:DNA hybrids are greater than DNA:DNA hybrids. Bacteriophage RNA polymerase can be used to generate labeled RNA probes with high specific activity. RNA probes are made by cloning the gene under the control of phage-specific promoter followed by run-off transcription. SP6, T3, and T7 RNA polymerases incorporate ^{32}P, ^{35}S, and ^{3}H nucleotide phosphates.

Preparative quantities of defined length of RNA can be prepared by run-off transcription. Under proper reaction conditions, the enzyme produces at least 60 μg of RNA from 2 μg of DNA template. Reactions proceed quickly at moderate enzyme concentrations and can be scaled up easily. The high yields of RNA from SP6 RNA polymerase reactions are especially useful for biophysical applications where a large amount of material is required. This *in vitro* method of synthesizing RNAs has been applied by molecular biologists to produce small amounts of product as well as by structural biologists who need milligram quantities of RNAs. Since T7 RNA polymerase is closely related to SP6 and T3 RNA polymerases, it is to be expected that these polymerases also function in this synthetic system.

Table 2.1 Cognate Promoter Sequences of Bacteriophage RNA Polymerase

Polymerase name	Host of encoding phage	Molecular weight (Da)	Promoter sequence
SP6 RNA polymerase	*Salmonella typhimurium*	~ 98,500	5′-AATTTAGGTGACACTATAGAAGNG-3′
T3 RNA polymerase	*E. coli*	~ 99,804	5′-AATTAACCCTCACTAAAGGGAGA-3′
T7 RNA polymerase	*E. coli*	~ 98,800	5′-TAATACGACTCACTATAGGGAGA-3′

Expression of Cloned Gene into Bacteria Cloned DNA can be expressed easily in bacteria and also in yeast by utilizing the T7 bacteriophage system. Two types of T7 bacteriophage expression systems have been devised for *E. coli.*

First system Stable lysogens are introduced with λ bacteriophage carrying the phage T7 RNA polymerase gene under the control of the *E. coli lac* UV5 promoter. Plasmid containing the gene of interest under the control of the T7 promoter is then established into the lysogens containing the T7 RNA polymerase gene. The activation of T7 promoter is then attained by isopropyl thio-β galactoside (IPTG) induction of the *lac* UV5 promoter driving the T7 polymerase gene.

Second system The T7 promoter/plasmid carrying the gene of interest is introduced into the bacteria and the T7 promoter is activated by infecting the bacteria with λ bacteriophage containing the T7 RNA polymerase gene.

In yeast, T7 RNA polymerase gene is positioned under the control of a yeast promoter and stably introduced into the yeast cells on an autonomously replicating vector. Expression is achieved by introducing a second plasmid that contains the gene of interest under the control of T7 promoter into the yeast cells.

***In Vitro* Synthesis of Capped RNA Transcripts** T7 RNA polymerase is also used to generate capped mRNA for expression studies in oocytes and other cells.

RNase Protection Assays RNase protection assay is a technique which is used to quantify the abundance of RNA. This technique is also used to map the positions of 5′- and 3′-ends of mRNAs and positions of 5′ and 3′ splice sites. The target DNA is first cloned into a plasmid vector downstream from a bacteriophage SP6 or T7 promoter. The recombinant DNA is digested with a restriction enzyme that cleaves at the distal end of the gene of interest. Then the linear DNA is transcribed with the appropriate phage RNA polymerase and the template DNA is removed by digestion with RNase-free DNase. The unlabeled test RNAs are hybridized to complementary radiolabeled RNA probes. The unhybridized RNAs are finally digested with one or more single strand-specific ribonuclease, e.g., RNase A, RNase T2, and RNase I. At the end of the digestion, the ribonucleases are inactivated and the protected fragments of radiolabeled RNA are analyzed by polyacrylamide gel electrophoresis and autoradiography (Figure 2.11).

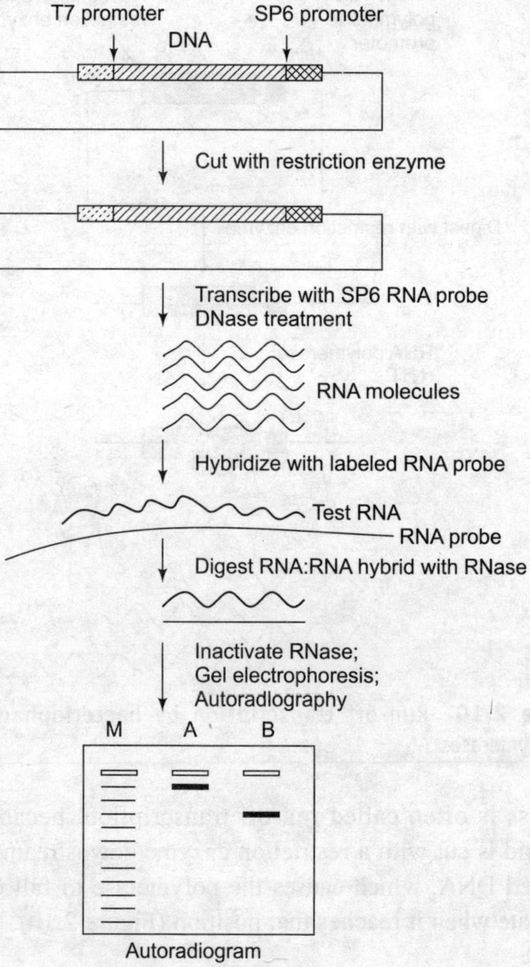

Figure 2.11 RNase protection assay [By using probes complementary to appropriate segments of the template DNA, it is possible to map the positions of the 5′- and 3′-termini of mRNAs and the positions of 5′ and 3′ splice sites. The samples loaded in different lanes are: Lane M (molecular weight standard); Lane A (radiolabeled RNA before digestion with RNase); Lane B (radiolabeled RNA after digestion with RNase).]

2.5.3 Poly (A) Polymerase (Polynucleotide Adenylyltransferase; EC 2.7.7.19)

Poly (A) polymerase is the enzyme which catalyzes the template independent polyadenylation at the 3′-terminus of a wide variety of RNAs. It is found in a wide variety of organisms, but for recombinant DNA technology it is isolated from *E. coli.* It is single polypeptide (~55 kDa) enzyme and catalyzes a linear polymerization reaction as follows:

$$\text{RNA} + n\text{ATP} \rightarrow \text{RNA-(pA)} + n\text{PP}_i$$

The poly (A) polymerase needs Mg^{2+}, Mn^{2+} or both, depending on several factors. The chain growth may be nonsynchronous; it is possible to obtain poly (A) products in defined size ranges by choosing proper reaction conditions.

Applications

Poly (A) polymerase is useful in the following RNA related techniques:

Production of Poly (A)-tailed RNA Poly (A) tails of RNAs have multiple uses *in vitro* in addition to the assumed roles of biological importance. The poly $(A)^-$ RNAs, e.g., rRNA, tRNA, and viral RNAs are converted into poly $(A)^+$ RNA. Poly $(A)^+$ RNA can be easily purified by using oligo (dT)-cellulose affinity chromatography. The polyadenylated RNAs can be cloned by conventional cDNA synthesis with oligo (dT) primers. Such cDNAs can also be used as hybridization probes.

3′ End-labeling of RNA RNA molecules can be radiolabeled at the 3′-end by the addition of short oligo (A) tails using [α-^{32}P] ATP or by the single addition of the chain terminating 3′-dA using [α-^{32}P] dATP. Such labeled RNAs are suitable substrates for RNA sequencing.

Determination of Poly (A)$^+$ RNA Content This enzyme catalyzes template-independent synthesis of poly (A)$^+$ tails on both poly (A)$^-$ and poly (A)$^+$ RNAs. In the presence of cordycepin 5′-triphosphate as the chain terminator and [α-^{32}P] ATP as the radiolabeled nucleotide, only oligo (A) tails having four residues or less in length are synthesized. Subsequently, these products are digested with RNase H in the presence of oligo (dT)$_{12\text{-}18}$. This reaction produces radioactive poly (A) tails only from the poly (A)$^+$ RNA, leaving the short oligo (A) tails on poly (A)$^-$ RNA intact. Radioactive tag is quantified which gives an estimate of the mole fraction of genuine poly (A)$^+$ RNA in mixed RNA sample.

2.6 ALKALINE PHOSPHATASE (ORTHOPHOSPHORIC-MONOESTER PHOSPHOHYDROLASE; EC 3.1.3.1)

Alkaline phosphatase catalyzes the removal of 5′-phosphate groups from DNA and RNA (Figure 2.12). It also removes phosphates from nucleotides (dNTPs and rNTPs) and proteins. The reaction catalyzed is

$$\text{5′-phospho-DNA (or RNA)} + H_2O \rightarrow \text{5′-dephospho-DNA (or RNA)} + \text{Pi}$$

These enzymes are most active at alkaline pH, hence their name. These are active on 5′ overhangs, recessed 5′-ends as well as blunt ends. Alkaline phosphatases are Zn^{2+} metalloenzymes that catalyze phosphate monoester hydrolysis through the formation of a phosphorylated serine intermediate.

Alkaline phosphatase are isolated and purified from numerous sources but the most commonly used for molecular biology experiments, are as follows:

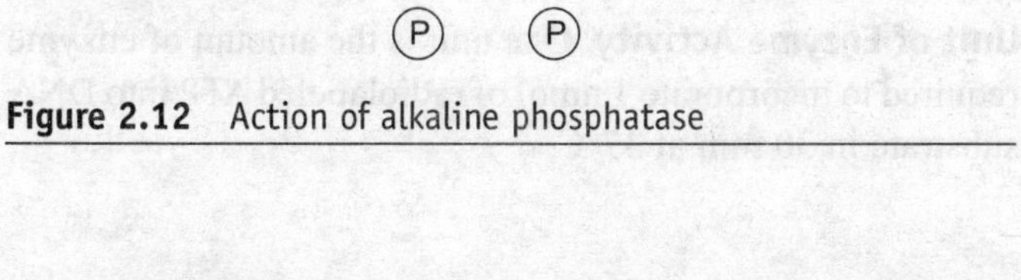

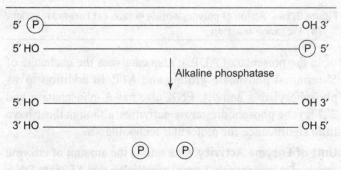

Figure 2.12 Action of alkaline phosphatase

Bacterial Alkaline Phosphatase (BAP) Bacterial alkaline phosphatase is isolated from *E. coli.* It is secreted in monomeric form (M_r ~ 47 kDa) into the periplasmic space of *E. coli* where it dimerizes and become catalytically active. At neutral or alkaline pH, dimeric BAP contains up to six Zn^{2+} ions, two of which are essential for enzymatic activity. One of the catalytic sites is active at low concentration of substrate, while both are activated at higher concentrations. It is the most active among the three enzymes and it is very difficult to destroy the enzyme at the end of the dephosphorylation reaction as it is resistant to heat and detergents.

Calf Intestinal Phosphatase (CIP or CIAP) CIP is isolated from calf intestinal mucosa. It is a dimeric glycoprotein (M_r ~ 110 kDa) that is bound to the plasma membrane by a phosphatidylinositol anchor and its optimal activity depends upon Mg^{2+} and Zn^{2+} concentration. Zn^{2+} is present at catalytic site and Mg^{2+} binds at different sites and act as allosteric activator. Normally CIP is used in routine experiments and it is either digested by treatment with proteinase K or inactivated by heating to 65°C for 1 hour in the presence of 5 mM EDTA. Finally, the dephosphorylated DNA can be purified by phenol:chloroform extraction before subsequent ligation reaction.

Arctic Shrimp Alkaline Phosphatase (SAP) SAP is isolated from *Pandalus borealis* (Arctic shrimp) and its enzymatic properties are quite similar to CIP. Native enzyme is a homodimer composed of 53 kDa subunits. It is unstable at higher temperature and completely and irreversibly inactivated by heating at 65°C for 15 min.

Unit of Enzyme Activity One unit is the amount of enzyme required to catalyze the hydrolysis of 1 μmol of *p*-nitrophenyl phosphate per minute at pH 9.6 (glycine/NaOH buffer) at 25°C.

Applications

Because of its dephosphorylating activity, this enzyme plays significant roles in genetic engineering experiments:

Prevention of Self-ligation of Vector Religation or recircularization of linearized vector DNA can be prevented by removing phosphate group from both termini (Figure 2.13).

Removal of 5′-Phosphate Group Before End-labeling 5′-phosphate group is removed from DNA or RNA prior to end-labeling with T4 polynucleotide kinase.

2.7 POLYNUCLEOTIDE KINASE (PNK) (ATP: 5′-DEPHOSPHOPOLYNUCLEOTIDE 5′-PHOSPHOTRANSFERASE; EC 2.7.1.78)

Polynucleotide 5′-hydroxyl kinase, commonly known as polynucleotide kinase (PNK) is homotetrameric protein with a molecular weight of ~140 kDa and encoded by T4 bacte-

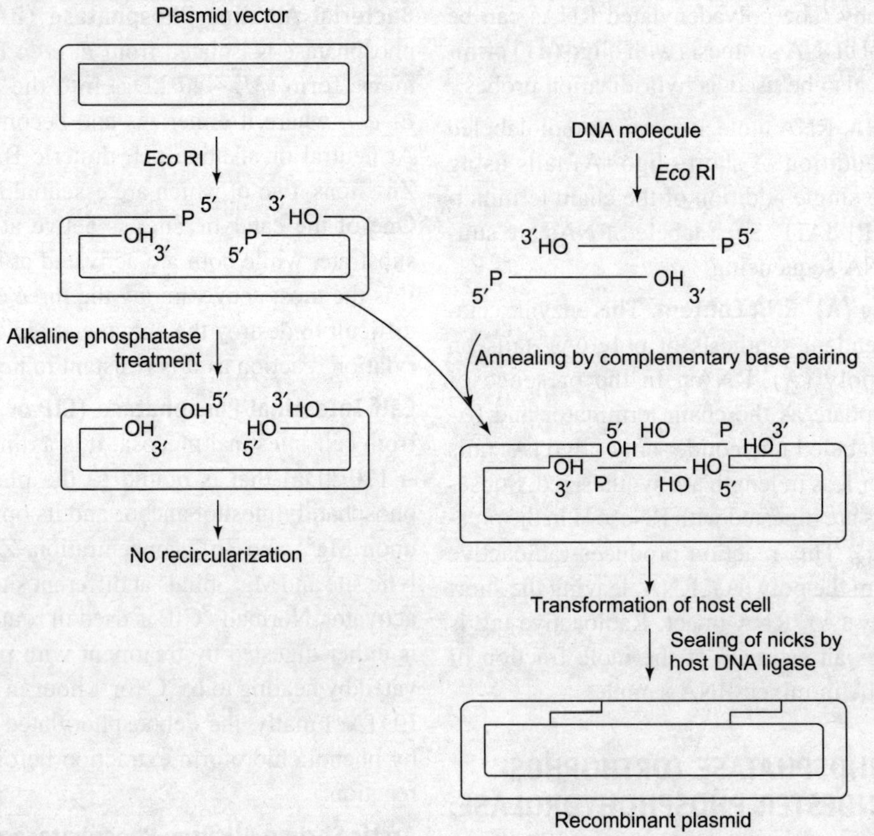

Figure 2.13 Prevention of self-ligation of vector by alkaline phosphatase treatment

riophage *pse* T gene. Its commercial preparations are usually products of the cloned phage gene expressed in *E. coli*. PNK catalyzes the transfer of a γ-phosphate from ATP to the 5′-hydroxyl termini of polynucleotides of either DNA or RNA in a following reaction:

ATP + 5′-dephospho-DNA (or RNA) → ADP
+ 5′-phospho-DNA (or RNA)

The enzymatic activity of PNK is utilized in two types of reactions:

Forward Reaction In the forward reaction, PNK transfers the γ-phosphate from ATP to the 5′-end of a polynucleotide (DNA or RNA). The target nucleotide is lacking a 5′-phosphate either because it has been dephosphorylated or has been synthesized chemically (Figure 2.14a).

Exchange Reaction In the exchange reaction, target DNA or RNA that has a 5′-phosphate is incubated with an excess of ADP. In this setting, PNK first transfers the phosphate from the nucleic acid onto an ADP, forming ATP and leaving a dephosphorylated target. The enzyme then performs a forward reaction and transfers a phosphate from ATP onto the target nucleic acid (Figure 2.14b). The efficiency of phosphorylation by the exchange reaction is considerably less than for the forward reaction.

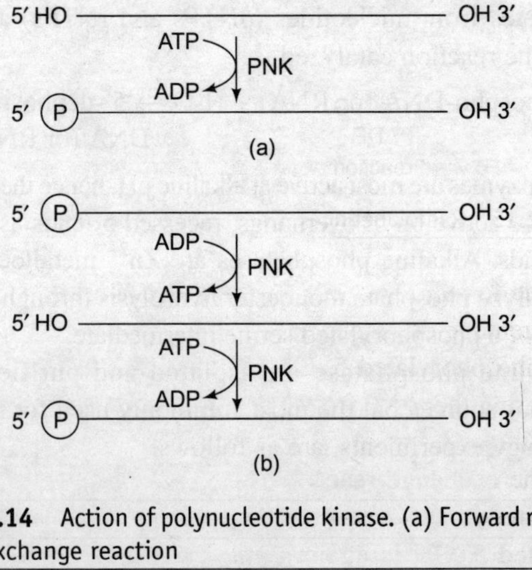

Figure 2.14 Action of polynucleotide kinase. (a) Forward reaction, (b) Exchange reaction

In the presence of ADP, it also catalyzes the exchange of 5′-terminal phosphate groups and ATP. In addition to its phosphorylating activity, PNK also has 3′-phosphatase and 2′,3′ cyclic phosphodiesterase activities, although these have little significance for molecular technologists.

Unit of Enzyme Activity One unit is the amount of enzyme required to incorporate 1 nmol of radiolabeled ATP into DNA substrate in 30 min at 37°C.

Applications

Owing to its phosphorylating ability, the enzyme plays significant roles in genetic engineering experiments:

Phosphorylation of Polynucleotide The enzyme phosphorylates linkers and adaptors, or fragments of DNA as a prelude to ligation, which requires a 5′-phosphate. This includes products of PCR, which are typically generated using nonphosphorylated primers.

Radiolabeling of 5′-Termini Hybridization probes are prepared by radiolabeling usually with γ-^{32}P. Radiolabeled 5′-termini are also used for sequencing by Maxam–Gilbert procedure and S1 nuclease analysis.

PNK facilitates transfer of γ-phosphate residue from (γ-^{32}P) ATP to the 5′-OH ends of dephosphorylated template. Labeling can be performed either with forward reaction or by exchange reaction (Figure 2.15).

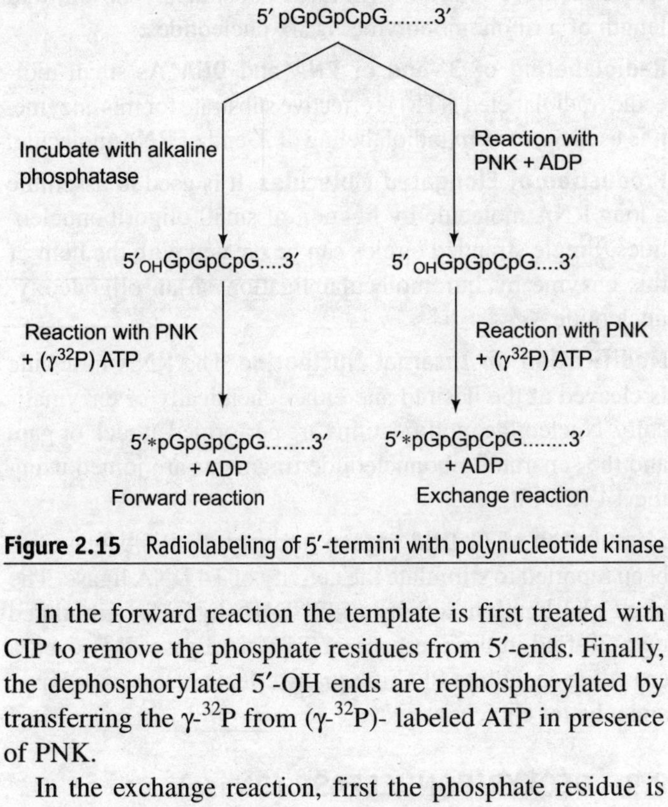

Figure 2.15 Radiolabeling of 5′-termini with polynucleotide kinase

In the forward reaction the template is first treated with CIP to remove the phosphate residues from 5′-ends. Finally, the dephosphorylated 5′-OH ends are rephosphorylated by transferring the γ-^{32}P from (γ-^{32}P)- labeled ATP in presence of PNK.

In the exchange reaction, first the phosphate residue is transferred from 5′-ends of template DNA to ADP. The newly generated 5′-OH is again rephosphorylated as in forward reaction.

Some important points associated with application of PNK are:

- Spermidine stimulates incorporation of (γ-^{32}P) ATP and inhibits the nuclease activity present in some preparations of T4 PNK.
- The minimum concentration of ATP must be 1 μM in the forward reaction and 2 μM in the exchange reaction.
- PNK is inhibited by small amounts of ammonium ions, so ammonium acetate should not be used to precipitate nucleic acids prior to phosphorylation.
- NaCl concentrations greater than about 50 μM also inhibit this enzyme.
- Low concentration of phosphate also inhibits T4 PNK. Imidazole buffer (pH 6.4) is, therefore, the buffer of choice for the exchange reaction and Tris buffer, for the forward reaction.

2.8 DNA LIGASE (POLYDEOXYRIBONUCLEOTIDE SYNTHETASE)

DNA ligase catalyzes the formation of a phosphodiester bond between the 5′-phosphate of one strand of DNA or RNA and the 3′-hydroxyl of another. This enzyme is used to covalently link or ligate fragments of DNA together. The DNA ligases used in molecular cloning differ in their abilities to ligate noncanonical substrates, such as blunt ended duplexes, DNA:RNA hybrids or ss DNAs. DNA ligases used in molecular cloning experiments are either of bacterial origin or encoded by bacteriophage, both of which are required to create new combinations of nucleic acids and to ligate them to vector molecule during cloning experiments. Depending on the source, the enzyme requires either ATP or NAD^+ as cofactors to form high-energy intermediates.

Unit of Enzyme Activity For all DNA ligases, one Weiss unit is the amount of enzyme required to convert 1 nmol of radiolabeled phosphate from pyrophosphate into absorbable material in 20 min at 37°C under standard assay conditions. One Weiss unit equals about 67 cohesive end ligation units.

The different types of DNA ligases are as follows:

2.8.1 Bacteriophage T4 DNA Ligase [Poly Deoxyribonucleotide Synthetase (ATP); EC 6.5.1.1]

The most widely used DNA ligase is derived from the T4 bacteriophage. It is a monomeric polypeptide of M_r ~68 kDa and is encoded by bacteriophage gene *30*. It catalyzes the formation of phosphodiester bonds between juxtaposed 5′-phosphate and 3′-OH ends in DNA. It has broader specificity and repairs single stranded nicks in duplex DNA, RNA or DNA:RNA hybrids. T4 DNA ligase requires ATP as a cofactor. Thus, T4 DNA ligase is the most commonly used enzyme in DNA ligations.

Applications

Various applications of T4 DNA ligase are summarized below:

Ligation of Cohesive Ends Intermolecular ligation is stimulated by low concentration of crowding agents, like PEG that promotes the efficient interaction of macromolecules in aqueous solution.

Ligation of Blunt Ended Termini This reaction is much slower than ligation of sticky ends. The rate of ligation is improved greatly by the addition of monovalent cations and low concentration of PEG.

Ligation of Synthetic Linkers or Adaptors T4 DNA ligase is also used for ligation of linkers and adaptors for increasing the versatility of restriction enzyme sites.

2.8.2 *E. coli* DNA Ligase (Polydeoxyribonucleotide Synthetase (NAD^+); EC 6.5.1.2)

E. coli DNA ligase is derived from *E. coli* cells and requires NAD^+ as cofactor. It is a monomeric enzyme of molecular weight ~74 kDa, which catalyzes the formation of the phosphodiester bond in duplex DNA containing cohesive ends (juxtaposed 5′-phosphoryl and 3′-hydroxyl termini). Also, in the presence of certain macromolecules such as PEG, i.e., macromolecular crowding, *E. coli* DNA ligase catalyzes blunt end ligation of DNA. This enzyme has narrower substrate specificities, making it a useful tool in specific applications.

Applications

E. coli DNA ligase is used for following applications:

Ligation of Cohesive Ends *E. coli* DNA ligase is used to catalyze sticky end ligation.

Cloning of Full-length cDNA *E. coli* DNA ligase has been employed in a procedure for high efficiency cloning of full-length cDNA.

T4 and *E. coli* DNA ligases are discussed in detail in Chapter 13.

2.8.3 *Taq* DNA Ligase (EC 6.5.1.2)

The genes encoding thermostable ligases have been identified from several thermophilic bacteria. Several of these ligases have been cloned and expressed to high levels in *E. coli*. *Taq* DNA ligase uses NAD^+ as cofactor and works at nicks of ds DNA. It can also catalyze blunt end ligation in the presence of crowding agents at elevated temperature.

Applications

Taq DNA ligase is useful in following the applications:

Detection of Mutation As thermostable DNA ligases retain their activities after exposure to higher temperature for multiple rounds, these are used in ligase amplification reaction to detect mutation in mammalian DNAs.

2.8.4 T4 RNA Ligase (Polyribonucleotide Synthetase (ATP); EC 6.5.1.3)

T4 RNA ligase is the only phage RNA ligase that has been extensively characterized and used in genetic engineering. This enzyme catalyzes the phosphodiester bond formation of RNA molecules with hydrolysis of ATP to AMP and PP_i. The enzyme can perform intermolecular as well as intramolecular ligation reactions to 3′-OH ends in ss DNA or RNA. This monomeric enzyme, with 373 deduced amino acid residues (M_r ~ 43 kDa), is a product of the T4 gene *63*.

Applications

Various applications of T4 RNA ligase include:

5′-Tailing of DNA with RNA The T4 RNA ligase is useful in joining defined oligoribonucleotides to the 5′-P end of DNA molecules. Various types of DNA donors up to 6 kbp long, single stranded or double stranded and with blunt or cohesive ends have similar affinity as substrates. The optimal length of a ribohomopolymer is six nucleotides.

Radiolabeling of 3′-end of RNA and DNA As small moleculc (radiolabeled NTP) is effective substrate for this enzyme, it is used for *in vitro* radiolabeling of 3′ end of RNA molecule.

Production of Elongated Molecules It is used to assemble a long RNA molecule by ligation of small oligoribonucleotides. Single stranded circles can be created with the help of this enzyme by intramolecular ligation on an oligodeoxynucleotide.

Modification on Internal Nucleotide The RNA molecule is cleaved at the desired site either chemically or enzymatically. Nucleotide modifications are performed at nick or gap, and the separated ribonucleotide fragments are joined using the T4 RNA ligase.

Stimulation of T4 DNA Ligase Activity T4 RNA ligase has been reported to stimulate the activity of T4 DNA ligase. The blunt end ligation activity of T4 DNA ligase is stimulated upto 20-fold in the presence of T4 RNA ligase. However, a crowding agent like PEG is equally effective, less expensive, and a better choice.

2.9 DEOXYRIBONUCLEASE (DNase)

Deoxyribonuclease is a type of nuclease enzyme, which catalyzes the hydrolytic cleavage of phosphodiester bonds of the DNA backbone. There are various types of DNases, which differ in their substrate specificities, chemical mechanisms and biological function. Some DNases remove nucleotide residues from the ends of DNA molecules (exodeoxyribonucleases). Alternatively, some deoxyribonu-leases are able to break internal phosphodiester bonds within a DNA molecule (endodeoxyribonucleases). Some DNases are sequence-specific like restriction enzyme (for details see Chapter 3), others degrade the DNA backbone randomly or nonspecifically as these do not discriminate between the DNA sequences. Some cleave only ss DNA, some are specific for double stranded molecules, and others are active towards both.

Various types of deoxyribonucleases are used in a variety of molecular biology applications. The commonly used deoxyribonucleases are discussed below.

2.9.1 DNase I (EC 3.1.21.1)

DNase I is an endonuclease that catalyzes the degradation of both ss DNA and ds DNA to yield di-, tri-, and oligonucleotides with 5′-phosphate and 3′-hydroxylated termini.

Enzyme Structure and Function

DNase I (molecular weight ~31 kDa) is a mixture of four glycoprotein components, DNase A (major), B, C, and D, of similar catalytic activity. DNase I acts on ss DNA and ds DNA, chromatin and RNA:DNA hybrids. The smallest substrate for DNase I is a trinucleotide. DNase I is stabilized by the addition of Ca^{2+} and requires a divalent metal ion, either Mg^{2+} or Mn^{2+}, for optimum activity. These ions function in synergistic manner, i.e., in the presence of both Ca^{2+} and Mg^{2+}, the rate of hydrolysis is greater than the sum of the rates for either cation alone. Since Ca^{2+} is known to bind tightly to DNase I and stabilize its active conformation, even micromolar levels of Ca^{2+} can act as a potent enzyme activator in the presence of Mg^{2+}. It randomly produces nicks independently into each strand of ds DNA in the presence of Mg^{2+} (Figure 2.16a). In the presence of Mn^{2+} or when high concentration of enzyme is used in the absence of monovalent cations, both strands of ds DNA are cleaved at approximately the same site (Figure 2.16b). Finally, the end product is either blunt end fragments or those with one or two protruding nucleotides. When DNase I digestion of heterogeneous ds DNA is performed it produces dinucleotides (60%), trinucleotides (25%), and oligonucleotides (15%). Though DNase I is usually supposed to cleave DNA nonspecifically, in practice it shows some sequence preference. For example, the enzyme is sensitive to the structure of the minor groove and shows preference for cleavage of purine–pyrimidine sequences. However, DNase I will cut at all four bases in heterogeneous ds DNA and the specificity of cleavage at a given base usually does not vary more than three-fold. The removal of DNase I is critical if the RNA is to be used to synthesize cDNA. It can be removed by phenol:chloroform extraction or by heat denaturation at 75°C for 10 min. EDTA must be added in excess relative to Mg^{2+} before heating to avoid chemical scission of RNA.

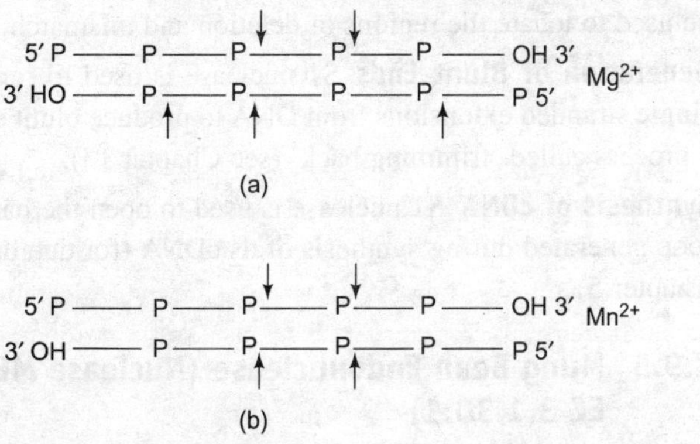

Figure 2.16 Activity of DNase I. (a) In the presence of Mg^{2+}, (b) In the presence of Mn^{2+}

Unit of Enzyme Activity For all DNases discussed below one unit of enzyme increases the absorbance at 260 nm by 0.001 per minute per ml at 25°C and pH 5.0 when acting on highly polymerized DNA in the presence of Ca^{2+} and Mg^{2+}.

Applications

DNase I is used for:

Removal of DNA Contamination The most common use of DNase I is to treat RNA preparations to degrade genomic DNA (up to 10 μg/ml) from trace to moderate amounts, which could otherwise result in false positive signals in subsequent studies. Normally, RNA preparations are contaminated with large amount of DNA, which must be removed before study by northern analysis, construction of cDNA library, and RT-PCR. DNase I does not cleave RNA, but crude preparations of the enzyme are contaminated with RNase A; RNase-free DNase I (RFD) is readily available. DNase is also used for the removal of DNA template after *in vitro* transcription.

Labeling of DNA By Nick Translation This enzyme is used in the generation of random single stranded nicks in ds DNA for nick translation reaction. The amount of DNase I should be standardized to prevent extensive destruction of DNA template. Please see Section 2.3.1 and Figure 2.3.

DNase I Footprinting DNA–protein interactions can be analyzed by DNase I footprinting. This technique is discussed in detail in Chapter 19.

2.9.2 Staphylococcal Nuclease (Micrococcal Nuclease, EC 3.1.31.1)

It is Ca^{2+}-dependent phosphodiesterase which cleaves both DNA and RNA to yield 3′-phosphomononucleotide and 3′-phosphooligonucleotide end products.

$$pN^1pN^2 \ldots\ldots\ldots pN^k \rightarrow pN^1p + (Np)^2 + (Np)^3 + \ldots + N^k$$

It is a single polypeptide having 149 amino acid residues (M_r 16.8 kDa). This enzyme is mainly used as a nonspecific endonuclease as is DNase I and can hydrolyze DNA and RNA with tailored site specificities.

Applications

The applications of this enzyme, which are quite similar to DNase I, are mentioned below:

Preparation of 3′-Phosphomononucleotides This enzyme can be used to produce 3′-phosphomononucleotides from DNA or RNA molecules. As a 3′-P generating endonuclease, it is useful in the analysis of nearest neighboring base in synthetic DNA and enzyme cleaved polynucleotide chains.

Removal of Nonspecific DNA and RNA It is frequently used to prepare eukaryotic cell-free extracts which support the *in vitro* translation of a wide variety of mRNAs. Brief exposure of endogenous mRNAs to this enzyme reduces the background of *in vitro* translation to a minimum, leaving the rRNAs and tRNAs intact.

DNase I Footprinting This enzyme is an excellent complement for DNase I in footprinting analysis. It recognizes a single strand of double helix DNA.

2.9.3 Shrimp DNase

This enzyme is isolated from *Pichia pastoris* strain containing DNase overproducing clone of *Pandalus borealis*. Like DNase I, shrimp DNase is an endonuclease that hydrolyzes phosphodiester bonds within the DNA molecule to yield di- and oligonucleotides with 5′-phosphate and 3′-hydroxyl termini. This DNase has a remarkably high specific activity for ds DNA, which is 5,000-fold higher for a ds DNA than a ss DNA and thus can be used selectively to degrade ds DNA, leaving ss DNA intact. Surprisingly, the sequence of this enzyme exhibits no similarity to that of DNase I, except for a 50-residue segment that encodes the nuclease active site. Shrimp DNase I enzyme exhibits intrinsic RNase activity in the presence of Mg^{2+} and Ca^{2+}, therefore, shrimp DNase is often referred to as shrimp nuclease. The activity of this enzyme depends on Mg^{2+} concentration and is stimulated by Ca^{2+}. However, the presence of Ca^{2+} should be avoided when RNA integrity is critical for the experiment. The DNase activity favors low ionic strength; activity decreases with increasing ionic strength. This recombinant enzyme can be heat inactivated by a moderate heat treatment without the use of EDTA. It is totally inactivated at 70°C after incubation for 25–30 min.

Applications

The Shrimp DNase finds various applications in: (i) Selective degradation of ds DNA leaving ss DNA and RNA intact; (ii) Removal of DNA from RNA prior to RT-PCR; (iii) Removal of DNA template after *in vitro* transcription; (iv) Nick translation with DNA polymerase I; (v) Footprint determination of DNA-binding protein; and (vi) Removal of carry-over contaminants in PCR.

2.9.4 S1 Nuclease (*Aspergillus* Nuclease S1, EC 3.1.21.31)

S1 nuclease is an extracellular enzyme secreted by the fungus *Aspergillus oryzae* that is encoded by the *nuc* O gene. This enzyme is heat stable and has molecular weight of ~29 kDa. The mature enzyme is glycosylated, having two disulfide bridges and a cluster of three Zn^{2+} ions that are required for enzymatic activity. The substrate depends on the amount of enzyme used and reaction conditions. In the presence of high ionic strength and 1 mM Zn^{2+}, low concentration of S1 nuclease digests ss DNAs or RNAs with high specificity. The rigorous exo- and endonucleolytic activities predominantly generate 5′-mononucleotides. The double stranded nucleic acids (DNA:DNA, DNA:RNA, and RNA:RNA) are degraded by high concentrations of enzyme. Moderate concentrations can be used to digest ds DNA at nicks or small gaps. Surprisingly, nuclease S1 can degrade ss DNA under very extreme conditions like 10% formamide, 25 mM glyoxal, and 30% sulfoxide. The enzyme activity is inhibited by phosphate ions, 5′ rNTPs, dNTPs, citrate, and EDTA. Nevertheless, it is stable in the presence of low concentrations of denaturing agents, e.g., SDS or urea.

$$5'\text{ (d or r) NpNpNpNpNpNpN } 3' \rightarrow 5'\text{ (d or r) NpN} + \text{pNpNpN} + \text{pN}$$

Unit of Enzyme Activity One unit of enzyme activity is usually defined as the amount of enzyme that liberates 1 µmol of acid-soluble nucleotides per minute under the standard assay conditions.

Applications

S1 nuclease is used in following processes:

S1 Nuclease Mapping S1 nuclease is used to analyze the structure of DNA:RNA hybrids. The process is discussed in detail in Exhibit 2.2.

Mutation Analysis Hybridization between mutant and wild-type DNAs and subsequent digestion with nuclease S1 can be used to locate the regions of deletion and mismatch.

Generation of Blunt Ends S1 nuclease is used to remove single stranded extensions from DNA to produce blunt ends, a process called 'trimming back' (see Chapter 13).

Synthesis of cDNA S1 nuclease is used to open the hairpin loop generated during synthesis of ds cDNA (for details see Chapter 5).

2.9.5 Mung Bean Endonuclease (Nuclease MB, EC 3.1.30.1)

Mung bean nuclease is a small glycoprotein having molecular weight ~39 kDa and made of two subunits linked by disulphide

Exhibit 2.2 S1 nuclease mapping

This technique was developed by Berk and Sharp[1]. In this technique RNA preparation containing the mRNA of interest is incubated with a complementary DNA or RNA probe under conditions that favor the formation of hybrids. The unhybridized single stranded RNA and DNA is degraded thereafter by nuclease enzyme. S1 nuclease is used in protection assays when the test RNA is hybridized to DNA. Ribonuclease is used when the test RNA is hybridized to RNA copy of template DNA. Exonuclease VII is used for more specialized purposes like for mapping of introns and to solve anomaly arising in nuclease protection assays.

In S1 nuclease mapping, the cloned segments of radiolabeled genomic DNA are used as probes. Hybrids formed between the transcribed strand of genomic DNA and mRNA form loops of single stranded DNA (introns). Digestion of these hybrids with S1 nuclease (100–1000 units/ml) at 20°C produces molecules whose RNA moieties are intact but DNAs contain gaps at the place on introns. At 20°C, S1 nuclease degrades loops of DNA but does not efficiently digest segments of RNA that bridge the loops of DNA. These molecules migrate as a single band on native PAGE, whereas in alkaline 5–6% PAGE/8M urea gels the RNA bridge is hydrolyzed and the individual fragments of DNA migrate according to their sizes. This property of S1 nuclease can be utilized for mapping of exon–intron boundary in fragment of genomic DNA. When the digestion with S1 nuclease (higher concentration) is performed at 45°C, both the RNA and DNA strands are cleaved and a series of smaller DNA–RNA hybrids are produced that can be separated by native as well on alkaline gels (Figure A).

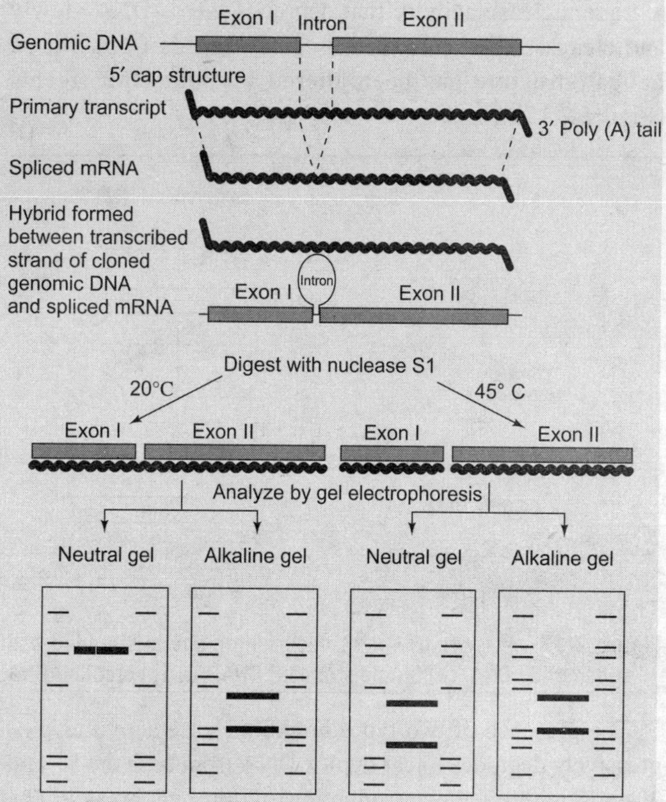

Figure A Mapping RNA by S1 Nuclease [First and third lanes in the gels are molecular weight markers]

[1] Berk AJ and Sharp PA (1977) Sizing and mapping of early aderovirus mRNAs by gel electrophoresis of endonuclease digested hybrids. *Cell* 12: 721–732.

Source: Sambrook J, Russel (2001) *Molecular Cloning: A Laboratory Manual*, 3rd edition, Cold Spring Harbor Laboratory Press, Cold Spring Harbor, New York.

bonds. Zn^{2+} is required for its activity and it shows optimum activity at low ionic strength and pH 5.0. Mung bean nuclease is a single strand-specific nuclease that degrades DNA and RNA to 5′-P mononucleotides and the ratio of its activity on single and double stranded substrate is ~30,000:1. ds DNA, ds RNA and DNA:RNA hybrids are somewhat resistant to this enzyme. Nevertheless, upon extensive exposure to high concentrations of enzyme, double stranded nucleic acids particularly rich in AT are totally degraded. Mung bean nuclease and S1 nuclease are quite similar to each other in their physical and catalytic properties. However, mung bean nuclease is less harsh in its activity as compared with S1 nuclease and it is easier to control. S1 nuclease can easily cleave DNA strand opposite to a nick in ds DNA, while mung bean nuclease works on nick after it has been enlarged to a gap of many nucleotides.

Unit of Enzyme Activity One unit is the amount of enzyme required to convert 1 μg of heat denatured DNA to acid-soluble form in 1 min at 37°C.

Applications

For most practical applications, mung bean nuclease can be used interchangeably with S1 nuclease. It is extremely useful in the following procedures: (i) High resolution mapping of termini in RNA transcripts; (ii) Conversion of 5′ protruding termini of DNA or RNA to blunt ends; (iii) Generating nested deletions, in conjunction with exonuclease III; (iv) Hairpin loop cleavage; (v) Transcriptional mapping; (vi) To create new restriction sites following overhang removal; and (vii) Investigation of DNA secondary structure.

2.9.6 *Bal* 31 Nuclease (EC 3.1.11)

Bal 31 is isolated from the marine bacteria, *Alteromonas espejiana.* It is Ca^{2+}-dependent nuclease and Mg^{2+} ions are also required for its full activation. *Bal* 31 consists of three kinds of activities:

- 3′→ 5′ exonuclease activity that eliminates mononucleotides from ds DNA

- $5' \rightarrow 3'$ exonuclease activity that works efficiently on ss DNA (Figure 2.17a).
- Endonuclease activity that degrades the ss DNA slowly and cleaves the supercoiled ds DNA and ds DNA whose helical structure has been altered by mutagenic agents (Figure 2.17b).

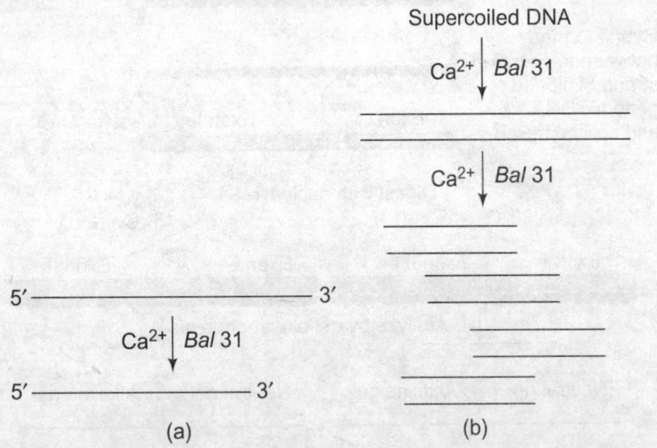

Figure 2.17 Activity of *Bal* 31 nuclease on single stranded and supercoiled DNA. (a) Single stranded DNA, (b) Supercoiled DNA

The presence of two types of exonuclease activities progressively degrades linear duplex DNA from both the 5′- and 3′-ends without the introduction of internal breaks. The mechanism of degradation involves fast exonucleolytic degradation followed by slow endonucleolytic degradation on the complementary strand. ds DNA with blunt or protruding 3′-OH ends are degraded to shorter double stranded molecules (Figure 2.18). The enzyme is also active at nicks, on ss DNA with 3′-OH termini, and on ds RNA molecules. Care must be taken while ligating the products of *Bal* 31 because the 3′ exonuclease activity of the enzyme works ~20-fold more efficiently than the endonuclease. At high concentration of enzyme, the average length of single stranded termini is ~5 nucleotides and 10–20% of the DNA molecules can be directly ligated to blunt ended DNA without any treatment, while at lower enzyme concentration the single stranded termini may be very long and the efficiency of blunt end ligation is very low. Filling-in with the help of T4 DNA polymerase or Klenow fragment is almost compulsory before cloning DNAs treated with *Bal* 31. The commercial preparations of this enzyme consist of two kinetically different forms of the enzyme, a fast (*Bal* 31F) and slow (*Bal* 31S). *Bal* 31F is a 109 kDa single polypeptide, whereas *Bal* 31S (85 kDa) is cleavage product of the F enzyme. Moreover, the pure fast form of enzyme is commercially available. However, it is very expensive and is exclusively used for removal of long, i.e., >100 bp fragments from the ends of ds DNA, ds RNA, and restriction sites mapping, while the slow form is used to remove short segment, i.e., 10–100 bp. *Bal* 31 exhibits preference for degradation of AT-rich fragments as compared to GC-rich regions. It is suggested to isolate the required size of DNA by agarose gel electrophoresis before cloning the DNA fragment which is generated after *Bal* 31 digestion and filling-in by T4 DNA polymerase (or Klenow fragment). *Bal* 31 should be stored at 4°C, as freezing causes loss in enzymatic activity.

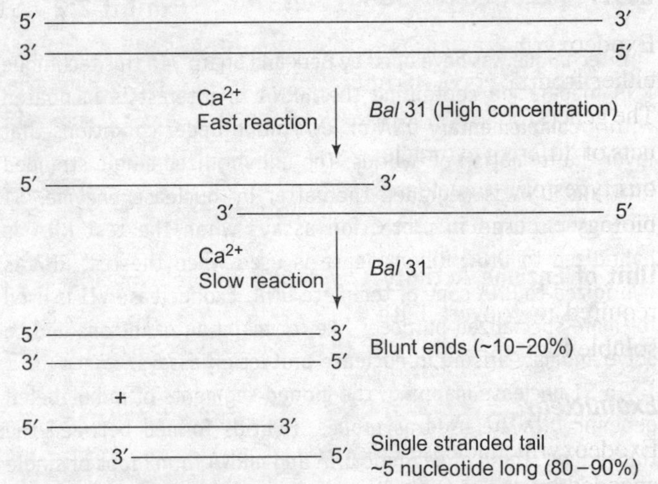

Figure 2.18 Degradation of linear duplex DNA into blunt end DNA molecule with *Bal* 31 nuclease

Unit of Enzyme Activity One unit is the amount of enzyme, which removes 200 bp from each end of linearized DNA in 10 min at 30°C at a DNA concentration of 50 μg/ml under standard assay conditions.

Applications

The enzyme finds wide applications in:

Creating Nested Deletions in DNA The nucleotides from the termini of ds DNA are removed in controlled manner. The shortened DNA can be used for number of purposes, for example, to produce deletion, to position desired sequence next to promoter or other controlling element, or to attach synthetic linkers at desired sites in the DNA.

Mapping Restriction Sites in DNA *Bal* 31 nuclease requires both Ca^{2+} and Mg^{2+} for activity, a property that is very useful in restriction fragment mapping, since *Bal* 31 nuclease digestions can be terminated with 20 mM EGTA, a specific chelator of Ca^{2+}, without affecting the Mg^{2+} concentration.

Mapping Secondary Structure in DNA The junctions between B-DNA and Z-DNA or sites of covalent or noncovalent modifications in ds DNA can be investigated with the *Bal* 31 degradation.

Removal of Nucleotides from ds RNA Nuclease also acts as a ribonuclease to catalyze the hydrolysis of rRNA and tRNA. It is also used in preparation of recombinant RNAs.

2.9.7 Exodeoxyribonucleases

Exodeoxyribonucleases catalyze the degradation of DNA either from 5′- or 3′-termini. These are a type of esterases. The gene(s) substrates, mode of action, and acid soluble products of different exonuclease are compared in Table 2.2. Various types of exonucleases that are widely used in molecular biology experiments are discussed below.

Unit of Enzyme Activity One unit is the amount of enzyme required to convert 1 μg of heat denatured DNA to acid-soluble form in 1 min at 37°C.

Exonuclease I (EC 3.1.11.1)

Exodeoxyribonuclease I, commonly called exonuclease I, is encoded by *sbc* B (*xon* A) gene of *E. coli* with a molecular weight of 55 kDa. The enzyme catalyzes the hydrolysis of ss DNA in the 3′ → 5′ direction, finally releasing 5′ mononucleotides and leaving the terminal 5′-dinucleotide intact. Hydrolysis is processive and cannot continue if the 3′-terminus is phosphorylated.

Applications The applications of exonuclease I are listed as following:

Elimination of residual ss DNA containing a 3′-terminus The enzyme is used for hydrolysis of single stranded DNA in the 3′ → 5′ direction.

Measuring endonucleolytic cleavage of covalently closed circular ss DNA Exonuclease I can be used to measure the endonucleolytic cleavage of covalently closed circular ss DNA.

Measuring DNA helicase activity DNA helicase activity can be measured with the help of exonuclease I.

Preparation of products of PCR This enzyme is predominantly useful in preparing the products of PCR for applications involving labeling and sequencing techniques. Moreover, the excess primers and any other irrelevant ss DNA present in PCR products interfere with subsequent enzymatic reactions involving DNA synthesis. The hydrolytic properties of exonuclease I degrade all ss DNA present in the PCR mixture allowing the product to be used more efficiently in further applications.

Exonuclease III (EC 3.1.11.2)

Exodeoxyribonuclease III, commonly called exonuclease III is a ~31 kDa globular monomeric protein encoded by *xth*, A gene of *E. coli*. It has multiple enzymatic activities like (Figure 2.19):

- 3′ → 5′ exonuclease activity that degrades ds DNA.
- 3′ phosphomonoesterase which removes a number of residues from 3′-termini of ds DNA.
- Nucleotidyl hydrolase that hydrolyzes 5′ residues to apurinic/apyrimidimic (AP) sites.
- Exonucleolytic ribonuclease H that degrades RNA strand of an RNA:DNA heteroduplex.

The different types of enzymatic activities are performed on a single active site. Exonuclease III does not hydrolyze ss DNA or ds DNA with a protruding 3′-terminus. It is nonprocessive enzyme. It produces a homogeneous population of DNA molecules that have been deleted to similar ex-

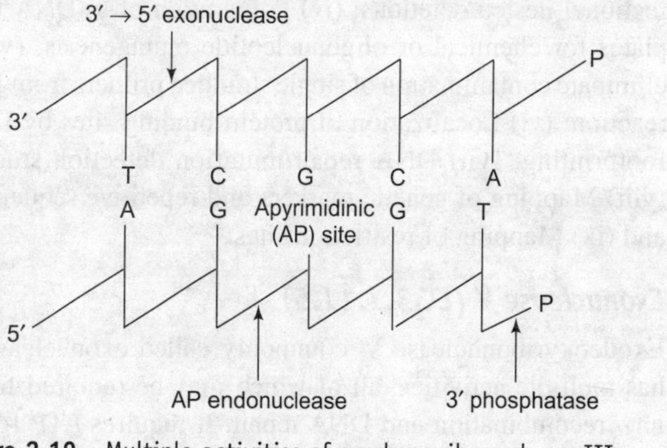

Figure 2.19 Multiple activities of exodeoxyribonuclease III

Table 2.2 Comparison of Various Exonucleases

Exonuclease	Gene(s)	DNA substrates	Mode of action	Acid soluble products
Exonuclease I	*sbc* B(*xon* A)	ss DNA	3′ → 5′	5′-dNMP; 5′-terminal pNpN
Exonuclease III	*xth* A	ds DNA	3′ → 5′	dNMP
Exonuclease V	*rec* B, *xse* C	ss and ds DNA	3′ → 5′ 5′ → 3′	Oligonucleotides
Exonuclease V, RecBCD	*rec* B, *rec* C, *rec* D	ss and ds DNA	3′ → 5′ 5′ → 3′	Oligonucleotides
Exonuclease VII	*xse* A, *xse* B	ss and ds DNA	3′ → 5′ 5′ → 3′	Oligonucleotides
λ exonuclease	λ *red* α	ds 5′ phosphorylated end	5′ → 3′	5′-dNMP
T7 gene *6* exonuclease	T7 *gene 6*	ds DNA	5′ → 3′	Oligonucleotides 5′-dNMP

Table 2.3 Exonucleolytic Activities of Exonuclease III

Reaction condition	Enzymatic activity
70 mM NaCl at 5°C	Removal of six nucleotides from the end of ds DNA and remaining bound as a stable complex
50 mM NaCl at 5°C	Initially removal of six nucleotides from the end of ds DNA followed by a slower rate of progressive hydrolysis
23–28°C	Exonucleolytic degradation is synchronous for ~250 nucleotides. At saturating enzyme concentrations, 5′ mononucleotides are removed at a rate of ~100 nucleotides/minute/3′-end
37°C	Exonucleolytic degradation is distributive and quasi-synchronous. At saturating enzyme concentrations, 5′ mononucleotides are removed at a rate of ~100 nucleotides/minute/3′-end

tents, and is helpful in isolating DNA molecules whose lengths have been reduced up to the desired extent. Exonuclease III contains bound divalent cations that are required for its activity. The purified enzyme does not need divalent cations unless it is exposed to EDTA. The mechanism by which exonuclease III degrades DNA exonucleolytically is affected by temperature, the concentration of monovalent cation and the concentration and structure of the 3′-termini (Table 2.3).

Applications Exonuclease III is used for the following purposes: (i) Preparation of strand-specific radioactive probes with a DNA polymerase; (ii) Preparation of ss DNA templates for dideoxy sequencing; (iii) Construction of unidirectional nested deletions; (iv) Preparation of ss DNA templates for chemical or oligonucleotide mutagenesis; (v) To eliminate contamination of single stranded primers from PCR reaction; (vi) Localization of protein binding sites by DNA footprinting; (vii) DNA repair/mutation detection studies; (viii) Mapping of genetic markers and repetitive sequences; and (ix) Mapping of restriction sites.

Exonuclease V (EC 3.1.11.5)

Exodeoxyribonuclease V, commonly called exonuclease V, has multiple activities, all of which may be required for *in vivo* recombination and DNA repair. It requires ATP for its activity and hydrolyzes nucleotides from both the 3′- and 5′-ends of linear ds and ss DNA. This enzyme acts preferably on ds DNA. Exonuclease V is encoded by *rec* B and *xse* C genes of *Micrococcus luteus.*

Applications Exonuclease V is used for: (i) Hydrolysis of nucleotides from 3′- and 5′-ends of linear ds DNA and ss DNA.

Exonuclease V (Rec B, C, D)

Exonuclease V (*Rec* B, C, D complex) can be isolated from *E. coli* strain containing an overproducing clone for the three subunits of *E. coli* Exonuclease V: *Rec* B, *Rec* C, and *Rec* D. The *Rec* BCD complex in *E. coli*, having molecular weight 330 kDa, has multiple activities. The enzyme has ATP-dependent DNase (DNA exonuclease) activity and is active on single stranded and double stranded DNA. This complex also possesses helicase activity. The hydrolysis in each case is bidirectional (from both the 3′- and 5′-ends) and processive, producing oligonucleotides. All exonuclease V activities exhibit divalent cation requirements, for example, Mg^{2+} is obligatory for the exonuclease activity; Ca^{2+} inhibits the exonuclease activity and allows ds DNA unwinding (helicase activity) without hydrolysis.

Applications Exonuclease V (*Rec* B, C, D complex) is used for: (i) Hydrolysis of nucleotides from 3′- and 5′-ends of linear ds DNA and ss DNA; (ii) Unwinding of ds DNA in the presence of Ca^{2+}; and (iii) D-loop cleavage.

λ Exonuclease (EC 3.1.11.3)

λ Exonuclease is an exodeoxyribonuclease, which is isolated from *E. coli* strain containing an overproducing clone of exonuclease from bacteriophage λ. This enzyme is encoded by λ red α gene and its molecular weight is ~25 kDa. Bacteriophage λ exonuclease catalyzes the stepwise and processive removal of 5′-mononucleotides from ds DNA liberating 5′-mononucleotides. This enzyme preferentially degrades the 5′ phosphorylated termini of ds DNA, i.e., λ exonuclease does not degrade 5′-hydroxyl termini. However, the enzyme also works on ss DNA, though the efficiency is ~100-fold less. Unlike exonuclease III, λ exonuclease does not act at a nick but it does remove a single stranded tail from a partially duplex DNA. A 5′-phosphate can be introduced into one strand of a PCR product by using one phosphorylated primer and one dephosphorylated primer during amplification. The phosphorylated strand is then removed by λ exonuclease, thus generating single stranded DNA for sequencing.

Applications The applications of λ exonuclease are as follows: (i) Generating ss DNA from ds DNA fragments; (ii) Preparation of PCR products for sequencing; and (iii) To modify the 5′-phosphate termini that are to be used by other enzymes.

T7 Gene 6 Exonuclease

T7 gene *6* exonuclease is purified from *E. coli* strain containing an overproducing clone of T7 gene *6* exonuclease. It has molecular weight ~22 kDa. This enzyme hydrolyzes ds DNA nonprocessively in the 5′ → 3′ direction from both 5′-phosphoryl or 5′-hydroxyl nucleotides by generating oligonucleotides as well as mononucleotides, until about 50% of the DNA is acid soluble. It also has capability to degrade nucleotides at the gaps and nicks of ds DNA from the 5′-termini. This enzyme also degrades RNA from RNA:DNA hybrids in the 5′ → 3′ direction but is unable to degrade either ds RNA or ss RNA. T7 gene *6* exonuclease is similar to λ exonuclease in that it catalyzes the stepwise hydrolysis of duplex DNA from the 5′-termini liberating 5′-mononucleotides. However, unlike λ exonuclease, the enzyme has low processivity and it removes both 5′-hydroxyl and 5′-phosphoryl termini.

Applications The T7 gene *6* exonuclease is used for following purposes: (i) Controlled stepwise digestion of ds DNA from the 5′-termini; (ii) Generation of ss DNA templates for sequencing via the chain termination method; and (iii) Generation of ss DNA templates for single nucleotide polymorphism (SNP) analysis.

2.10 RIBONUCLEASE (RNase)

Ribonuclease is a group of nucleases that catalyzes the hydrolysis of RNA into smaller components. Various types of RNases are being used in different experiments related to molecular biology. Some commonly used RNases are discussed below.

2.10.1 Ribonuclease A (EC 3.1.27.5)

Ribonuclease A is an endoribonuclease, having molecular weight of ~13.7 kDa. It is isolated from bovine pancreas and is one of the hardiest enzymes in common laboratory usage. This enzyme is sequence specific, hydrolyzes ss RNA at the 3′-end of pyrimidine (C and U) residues and the end products are pyrimidine 3′-phosphates and oligonucleotides having 3′-terminal pyrimidine 3′-phosphates.

5′ p-G-p-C-p↓A-p-U-p↓G-p-C-p↓G 3′ ⟶ pGpCp + A-p-U-p + G-p-C-p + G

One method of isolating it is to boil a crude cellular extract until all enzymes other than RNase A are denatured.

A variant of RNase A is now commercially available which is known as RNase I. RNase I is a preparation of 70% RNase A and the remaining 30% are other isozymes of RNase. RNase I catalyzes the hydrolysis of 3′ → 5′ phosphodiester bonds of RNA with the formation of oligoribonucleotides terminating in pyrimidine 2′ → 3′ cyclic phosphate.

Unit of Enzyme Activity One unit is the amount of enzyme required to catalyze the hydrolysis of RNA at a rate such that *K* (velocity constant) equals unity (Kunitz units) at pH 5.0 and 25°C.

Applications RNase A finds wide applications in genetic engineering experiments:

Removal of RNA contamination The enzyme is used for eliminating or reducing RNA contamination in preparations of plasmid DNA.

Removal of RNA in a hybrid The enzyme is used for removing unhybridized regions of RNA from DNA:RNA or RNA:RNA hybrids.

Mapping mutations in DNA or RNA by mismatch cleavage RNase cleaves the RNA in RNA:DNA hybrids at sites of single base mismatches and the cleavage products can be analyzed. A ^{32}P-labeled RNA probe complementary to wild type DNA or RNA is synthesized *in vitro*. The RNA probe is then annealed to test DNA or RNA containing a single base mutation. The resulting single base mutation is cleaved by RNase A and the site of mismatch is then determined by analyzing the sites of the cleavage products by gel electrophoresis. Approximately 50% single base mutations can be detected with the help of this method.

RNA sequencing RNase A cleaves RNA at pyrimidine bases, therefore partial digestion of an RNA sample results in fragments terminating with C or U. The fragments are then separated by PAGE to form a ladder, the sequence of which is determined by comparison with sequence ladders generated by other base-specific RNases such as RNase T1 (G specific), RNase U2 (A specific), and Phy M (U+A specific).

2.10.2 Ribonuclease H (EC 3.1.26.4)

Ribonuclease H (RNase H) is a monomer of ~18.4 kDa encoded by *rnh* gene of *E. coli.* RNase H specifically degrades only the RNA strand in RNA:DNA hybrids. It cleaves the 3′-O-P-bond of RNA in a RNA:DNA duplex to produce 3′-hydroxyl and 5′-phosphate terminated products of varying lengths.

-RpRpRpRpRpRpRpRpRp-
-DpDpDpDpDpDpDpDpDp-

RNA:DNA hybrid

↓ RNase H

-RpR + pRpRpRpR + pRpRpRp- + ss DNA

It was first recognized and isolated from calf thymus. Now it is known to be present in various mammalian tissues, yeasts, prokaryotes, and viruses. It does not hydrolyze the phosphodiester bonds within single stranded and double stranded DNA and RNA. RNase H is a nonspecific endonuclease and catalyzes the cleavage of RNA via hydrolytic mechanism, aided by an enzyme-bound divalent metal ion. Retroviral RNase H, a part of the viral reverse transcriptase enzyme, is an important pharmaceutical target, as it is absolutely necessary for the proliferation of retroviruses, such as HIV. Inhibitors of this enzyme could therefore provide new drugs against diseases like AIDS. As of 2004, there are no RNase H inhibitors in clinical trials, though some approaches employing DNA aptamers are in the preclinical stage.

Unit of Enzyme Activity One unit of the enzyme catalyzes the formation of 1 nmol of acid-soluble products in 20 min at 37°C.

Applications RNase H is used for following applications:

Second cDNA strand synthesis In a molecular biology laboratory, as RNase H specifically degrades the RNA in RNA:DNA hybrids and does not degrade DNA or unhybridized RNA, it is commonly used to destroy the RNA template after first cDNA strand synthesis by reverse transcription (for details see Chapter 5).

Removal of poly (A) tails of mRNA hybridized to poly (dT) RNase H can also be used to degrade specific RNA strands when the cDNA oligo is hybridized, such as the removal of the poly (A) tail from mRNA hybridized to oligo (dT), or the destruction of a chosen noncoding RNA inside or outside the living cell. To terminate the reaction, a chelator, such as EDTA, is often added to sequester the required metal ions in the reaction mixture.

Site-specific cleavage of RNA RNase H catalyzes site-specific cleavage of RNA.

Quantitation of poly (A) tailed mRNAs Poly (A) tailed mRNAs can be selectively measured from the total RNAs. The method is based on two enzymatic steps: (i) Poly (A) polymerase catalyzed labeling of the 3′-termini of all RNA species in the presence of ATP analog chain terminator, cordycepin triphosphate. This reaction adds one, or at the most, a few nucleotides from [α-^{32}] ATP; and (ii) RNase H digestion of the total sample in the presence of oligo $(dT)_{12-18}$, releasing ^{32}P-labeled poly (A) tails specifically from the poly (A) RNA molecules. The difference between the amounts of label that has been incorporated into the RNA and which remains TCA precipitated on glass filters represents the fraction of polyadenylated samples and thus the concentration of poly (A) tailed mRNA.

Detection of RNA-containing ds DNA structure The specificity of RNase H can be exploited to detect the presence of RNA in DNA structures. A covalently closed circular DNA that contains RNA can be converted to a relaxed form which, after denaturation, produces a single strand circle and a single strand linear molecule.

Role in nuclease protection assays RNase H specifically degrades the RNA in RNA:DNA hybrids and does not degrade DNA or unhybridized RNA, it is commonly used to destroy the RNA template in nuclease protection assays (Figure 2.11).

2.10.3 Ribonuclease T1 (EC 3.1.27.3)

Ribonuclease T1 (RNase T1) is a fungal endonuclease isolated from *Aspergillus oryzae*. The enzyme cleaves ss RNA after 3′-end of guanine residues, the final products are guanosine-3′-phosphates and oligonucleotides with guanosine-3′-phosphate terminal groups.

5′ApCpUpGpCpGpUpApGpGpCpU 3′

↓

ApCpUpGp + CpGp + UpApGp + Gp + CpU

RNase T1 is often used to digest denatured RNA prior to sequencing. Like other RNase, this enzyme is also used for folding studies. RNase T1 is monomeric enzyme composed of 104 amino acids (M_r ~11 kDa). Structurally, ribonuclease T1 is a small α+ β protein with a four-stranded, antiparallel β-sheet covering a long hydrophobic core of 4.5 turns α-helix. RNase T1 has two disulfide bonds, Cys2-Cys10 and Cys6-Cys103, of which the latter contributes more to its folding stability; complete reduction of both disulfides usually unfolds the protein, although its folding can be rescued with high salt concentrations.

Unit of Enzyme Activity One unit is the amount of enzyme that causes an absorbance increase of 1.0 at 260 nm under the standard assay conditions.

Applications In genetic engineering experiments, ribonuclease T1 is used for:

RNA sequencing As already described in Section 2.10.1, when RNA is partially digested with RNase T1 and separated by PAGE, the ladder yields information on the ordering of the fragments terminated with the G residues. When this reaction is performed with other base-specific RNases, RNA sequences can be read directly from the sequence ladder autoradiogram.

Site-directed cleavage of RNA This enzyme can site specifically cleave the target G site on the RNA strand of unhybridized regions of RNA from DNA:RNA and RNA:RNA hybrids.

RNase T1 fingerprinting The RNase T1 fingerprinting is the most frequently used technique for comparing and identifying the genomes of RNA viruses. Complete digestion of RNA molecules with RNase T1 yields oligonucleotides ending in G 3′-P. In high salt concentration which tends to promote a double-helical structure, RNase T1 selectively digest the ss RNA, while leaving ds RNA intact. As the stem region of hairpin structures is resistant to RNase T1, the digestion of ss RNA could be preceded by appropriate heat denaturation of the RNA at 60°C for 10 min, or 100°C for 1 min and rapid cooling on ice. The digested products of radiolabeled RNAs can be separated on 2D gel electrophoresis, producing a unique oligonucleotide pattern or RNA fingerprint.

2.11 PHOSPHODIESTERASE I (EC 3.1.4.41)

Phosphodiesterase I (venom exonuclease), isolated from *Crotalus adamanteus* venom, has a molecular weight of

~115 kDa and is widely used in studying nucleic acid structure and sequence. It hydrolyzes 5′-mononucleotides from 3′-hydroxy terminated DNA and RNA. Phosphodiesterase I cleaves ADP-ribosylated proteins at the pyrophosphate linkages to yield phosphoribosyl-AMP.

Unit of Enzyme Activity One unit of the enzyme hydrolyzes 1 μmol *p*-nitrophenyl thymidine-5-phosphate per minute at 25°C and pH 8.9.

Applications Phosphodiesterase I is used for: Base composition and nearest neighbor analysis.

2.12 β-AGARASE (EC 3.2.1.81)

β-Agarase specifically hydrolyzes the agarose polysaccharide core made up of repeating 1,3-linked β-D-galactopyranose and 1, 4-linked 3, 6-anhydro-β-L-galactopyranose into neoagaro-oligosaccharides. It specifically hydrolyzes the 1,3-β-*D*-galactosidic linkages. It is isolated from a strain of *E. coli* that carries a plasmid which encodes the β-*Agarase I* gene of *Pseudomonas atlantica*. β-Agarase I digests agarose, releasing trapped DNA and producing soluble carbohydrate molecules. β-Agarase I can be used to purify both large (>50 kbp) and small (<50 kbp) fragments of DNA from gels. The remaining carbohydrate molecules and β-Agarase I does not, in general, interfere with subsequent DNA manipulations such as restriction endonuclease digestion, ligation, and transformation. The recovered DNA may be directly used for sequencing, amplification, etc. The enzyme is quite active in various electrophoresis buffers, e.g., tris-acetate buffer (TAE) and tris-borate buffer (TBE).

Unit of Enzyme Activity One unit is defined as the amount of enzyme required for digestion of 200 μl of molten, low melting point or NuSieve agarose to nonprecipitable neoagaro-oligosaccharides in 1 hour at 42°C.

Applications β-Agarase is used for following purposes: (i) Gentle recovery of intact DNA and RNA; (ii) High yield recovery of DNA compared to other extraction methods; and (iii) Efficient recovery of long DNA fragments (>30 kbp).

2.13 URACIL–DNA GLYCOSYLASE (EC 3.2.2.2.3)

Uracil–DNA glycosylase (UDG) excises uracil from dU-containing DNA by cleaving the *N*-glycosidic bond between the uracil base and the sugar without disturbing the phosphodiester backbone. This cleavage generates alkali sensitive apyrimidinic sites that are blocked from replication by DNA polymerase or prevented from becoming a hybridization site. UDG works only on double and single stranded dU-containing DNA, whereas RNA and normal dT-containing DNA are resistant to UDG. UDG is isolated from *E. coli* strain containing an overproducing clone of *E. coli* UDG having a molecular weight of ~31 kDa.

Unit of Enzyme Activity One unit is defined as the amount of enzyme required to release 1 nmol of uracil from dU-containing DNA into acid-soluble form in 60 min at 37°C under standard assay conditions.

Applications UDG is used in following applications:

Study of DNA repair and mutation detection UDG is used in base excision repair and mutation detection.

Increase cloning efficiency of PCR products UDG can also be used to increase the cloning efficiency of PCR products having dU-containing primers incorporated into them. The use of dUTP in PCR reactions introduces uracil at the sites complementary to adenine. The ds DNA is subsequently treated with UDG, which generates single stranded ends that can be used in cloning of amplicons (Figure 2.20).

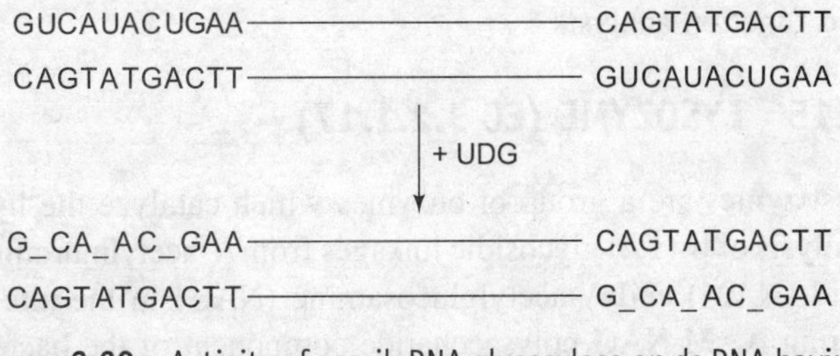

Figure 2.20 Activity of uracil–DNA glycosylase on ds DNA having U residues

Increase the efficiency of site-directed mutagenesis UDG is also used in increasing the efficiency of site-directed mutagenesis (for details see Chapter 17).

Protein–DNA interactions UDG is also used to study protein–DNA interaction.

2.14 PROTEINASE K (EC 3.4.21.64)

Proteinase K is a highly active and stable endopeptidase that is used in a wide range of applications. It is purified from the stationary culture of fungus *Tritirachium album* and has molecular weight of ~29 kDa. The letter K in the name of protease indicates that the total carbon and nitrogen requirement of the fungus can be fulfilled by the hydrolysis of keratin by the protease. The mature enzyme has two binding sites for Ca^{2+}, which are present at some distance from the catalytic sites. However, after the removal of Ca^{2+}, the catalytic site of the enzyme is lost to some extent, but the residual activity is sufficient to digest the protein. Digestion with proteinase *K* is generally performed in the presence of EDTA. Furthermore it is also active in the presence of SDS and urea, which are regularly used for lysis of cells. This enzyme catalyzes the hydrolysis of peptide bonds predominantly following the carboxyl group of N-substituted hydrophobic aliphatic and aromatic amino acids and is classified as a serine protease.

Proteinase K efficiently digests the RNases and DNases in cell lysates, which is useful in purifying high molecular weight DNA and RNA from cells and tissues. Commercially, proteinase K is available as lyophilized powder and should be dissolved in water at a concentration of 20 mg/ml.

Unit of Enzyme Activity One unit is the amount of enzyme that liberates folin positive amino acids and peptides corresponding to 1 μmol tyrosine in 1 min at 37°C using hemoglobin as substrate.

Applications Proteinase K is used in the following applications: (i) Isolation of native, high molecular weight DNA and RNA; (ii) To inactivate DNases, RNases, and to degrade proteins as a general protease; (iii) To specifically modify cell surface proteins and glycoproteins for analysis of membrane structures for protein localization; and (iv) To produce characteristic protein fragments used in enzyme/protein structure and function analysis.

2.15 LYSOZYME (EC 3.2.1.17)

Lysozymes are a group of enzymes which catalyze the hydrolysis of β-(1,4) glycosidic linkages from *N*-acetylmuramic acid (NAM) and *N*-acetylglucosamine (NAG) in the alternating NAM-NAG polysaccharide component of the bacterial cell wall peptidoglycans (Figure 2.21). Lysozymes are widely distributed in nature and are expressed wherever there is requirement to lyse bacterial cells. It is abundant in number of secretions such as tears, saliva, and mucous.

Unit of Enzyme Activity One unit is the amount of enzyme that causes a decrease in absorbance of 0.001 per minute at 450 nm, 25°C and pH 7.0 with *Micrococcus lysodeikticus* as a substrate.

Applications Lysozyme is used extensively for the following purposes:

Isolation of DNA Lysozyme (M_r ~14.4 kDa), isolated from egg white is used for cell wall lysis in cosmid and plasmid DNA isolation procedures. The reaction is performed at pH 8.0 in the presence of EDTA and detergents to release DNA from their bacterial hosts.

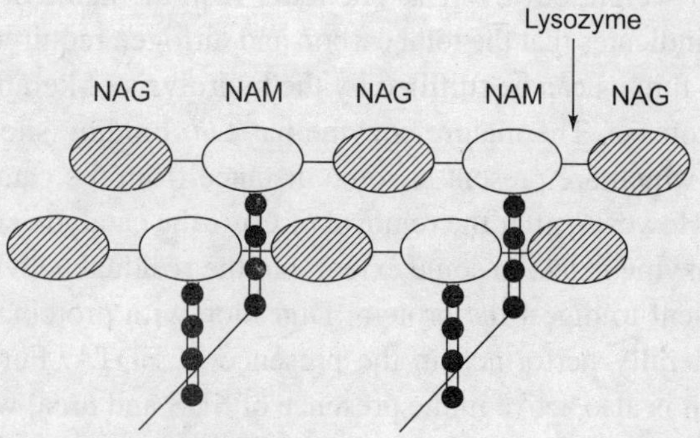

Figure 2.21 Hydrolysis of β (1,4) linkages between *N*-acetylglucosamine and *N*-acetylmuramic acid residues of peptidoglycan in bacterial cell walls by lysozyme

2.16 TOPOISOMERASE

Topoisomerase is a very important enzyme for DNA replication. There are two types of topoisomerases: type I and type II. Type I topoisomerase includes topoisomerase I and III, while type II includes topoisomerase II (DNA gyrase) and topoisomerase IV.

In general two types of topoisomerases are used in genetic engineering experiments, which are discussed below.

2.16.1 Topoisomerase I (EC 5.99.1.2)

Topoisomerase I, isolated from calf thymus, is a reversible nuclease that introduces transient single stranded breakage or nicks into one strand of DNA; it causes the other strand of DNA to untwist and then reseals the nick (i.e., rejoining of phosphodiester bonds). The enzyme thus helps in relaxation of both positively and negatively supercoiled DNA. During the course of reaction catalyzed by topoisomerase I, a transient covalent intermediate is formed between a specific tyrosine residue on the enzyme and one strand of DNA, thus conserving the energy of cleavage of phosphodiester bond in the form of phosphotyrosine linkage. This bond is broken by the activity of tyrosine–DNA phosphodiesterase. Topoisomerase I is active in the absence of divalent cations and can be used in solutions containing EDTA. It does not require ATP for its activity. In the absence of Mg^{2+}, topoisomerase I generates relaxed, covalently closed circles.

Unit of Enzyme Activity One unit is the amount of enzyme required to catalyze the conversion of 0.5 μg of supercoiled pBR322 DNA to a relaxed state in 30 min at 37°C.

Applications Topoisomerase I is used for following applications: (i) Analysis of DNA supercoiling and conformation; (ii) The enzyme is used to enhance the electrophoretic separation of plasmid DNAs. Thus, separation by gel electrophoresis of closed circular DNAs that have been relaxed by treatment with topoisomerase I leads to the resolution of molecules differing in length by a single base pair. BSA should be added to the reaction mixture to prevent sticking to glass or plastic material. The reaction mixture is prepared by adding, in order, reaction buffer, sterile water, template DNA, BSA, and the enzyme; white precipitate may form if 0.1% BSA is added directly to the reaction buffer (Figure 2.22). (iii) The enzyme produces knots in single stranded circular DNA; (iv) *In vitro* analysis of chromatin reconstitution; (v) DNA repair studies; and (vi) Studies on drug resistance, cell proliferation, and leukemia.

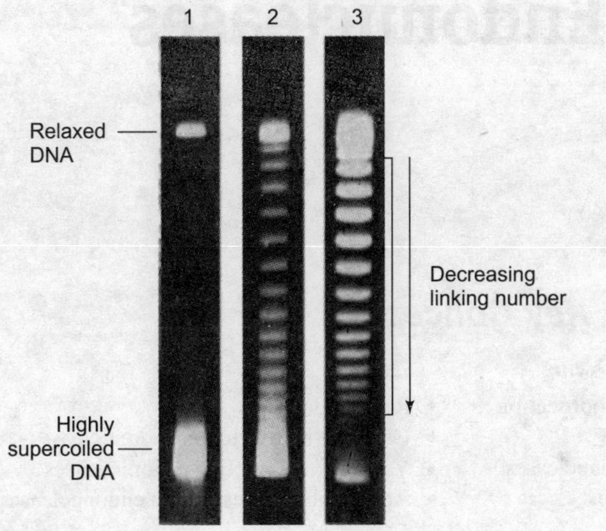

Figure 2.22 Enhancement of the electrophoretic separation of plasmid DNAs by topoisomerase I

2.16.2 Topoisomerase II Alpha (EC 5.99.1.3)

Topoisomerase II enzyme is isolated from *E. coli* containing a clone of the human *topoisomerase II* gene. Basically it is DNA relaxing enzyme. It has molecular weight of ~340 kDa. Topoisomerase II alters the topological state of nucleic acids by passing an intact DNA helix through a transient break, which generates a separate DNA helix. As a result of its ds DNA passage mechanism, the enzyme can relax negatively or positively supercoiled DNA as well as catenate/decatenate or knot/unknot DNA molecules. Topoisomerase II has an absolute requirement for divalent cation and ATP (or dATP).

Unit of Enzyme Activity One unit is the amount of enzyme required to fully relax 0.3 μg (5 nM) of negatively supercoiled pBR322 plasmid DNA in 15 min at 30°C under the standard assay conditions.

Applications Topoisomerase II is used for following applications: (i) Relaxing negatively or positively supercoiled DNA; (ii) Catenating or decatenating DNA; and (iii) Knotting or unknotting DNA.

Review Questions

1. Give an elaborative account of the important features of DNA polymerases and distinguish between the various DNA polymerases used in genetic engineering.
2. What are the two types of exonuclease activities that can be possessed by a DNA polymerase? Distinguish between them, and explain how these activities influence the potential applications of individual DNA polymerases in genetic engineering.
3. Illustrate the nick translation method of probe preparation by using DNA polymerases I with the help of suitable diagram.
4. How can DNA polymerase I be converted into Klenow fragment? Describe the major applications of Klenow fragment.
5. Explain why T4 DNA polymerase is better than Klenow fragment for site-directed mutagenesis.
6. Describe the reason why T7 DNA polymerase is best-suited for sequencing reaction. What is the modified version of this polymerase and how is this modification performed?
7. What do you understand by run-off transcription and why is this technique used in genetic engineering?
8. Give an account of bacteriophage T7 expression systems that have been developed for *E. coli*.
9. Describe the technique and application of RNase protection assays.
10. What are the most commonly used alkaline phosphatases for molecular biology experiments? For gene cloning experiments, why is the cleaved vector DNA often treated with alkaline phosphatase prior to the ligation step?
11. Describe briefly about poly (A) polymerase and also write the *in vitro* application of this enzyme.
12. Explain the two types of reaction mechanisms of polynucleotide kinase and 5′-end labeling of DNA in the presence of this enzyme.
13. Differentiate between T4 DNA ligase and *E. coli* DNA ligase.
14. Give a concise account of cation requirement of DNase I.
15. Although S1 nuclease is single strand-specific endonuclease but the substrate for this enzyme also depends on the amount of enzyme used and reaction conditions. Explain this statement.
16. Explain the technique for analysis of the structure of DNA:RNA hybrids using S1 nuclease enzyme.
17. Discuss the three types of enzymatic activities present in *Bal* 31 nuclease. What are the two kinetically different forms of this enzyme and also write the source of *Bal* 31 nuclease.
18. Differentiate between commonly used exonucleases on the basis of their gene, substrate, mode of action, and end products.
19. Give details of the multiple enzymatic activities of exonuclease III.
20. Briefly describe the technique of RNA sequencing with the help of different RNases.
21. Explain the mechanism of action and application of following enzymes in genetic engineering.
    (a) β-Agarase (b) Lysozyme
    (c) Uracil–DNA glycosylase (d) Topoisomerase I

Restriction Endonucleases

Key Concepts

In this chapter we will learn the following:

- Host controlled restriction and modification (R-M) system
- Nomenclature of restriction endonucleases
- Types of restriction endonucleases
- Recognition sites
- Cleavage by restriction endonucleases
- Variants of restriction endonucleases
- Applications of restriction endonucleases

3.1 INTRODUCTION

Restriction endonucleases, commonly known as restriction enzymes, are prokaryotic nucleases that recognize specific sequences in the DNA, and cleave the sugar–phosphate backbone between deoxyribose sugar and phosphate groups, either within or close to recognition site. The word 'restriction' indicates their ability to restrict the growth of bacteriophage. These enzymes make two incisions, one through each of the two strands of the double stranded DNA (ds DNA) without damaging the nitrogenous bases, and cleaves the DNA at a wide variety of locations along its length. The cleavage usually occurs in a predictable and reproducible way, giving rise to discrete fragments of defined length and sequence, and leaves a 3′-hydroxyl on one side of each cut and a 5′-phosphate on the other. Based on their function, restriction endonucleases are also called 'Molecular Scissors' or 'Nature's Exquisite Scalpels'. The discovery of restriction endonucleases that led to Nobel Prize in Physiology or Medicine in 1978 to Werner Arber jointly with Hamilton Othanel Smith and Daniel Nathans was one of the key breakthroughs in the development of genetic engineering and restriction mapping.

In the following sections, we shall study restriction endonucleases with respect to their roles *in vivo*, their nomenclature system, the different types, properties, mode of action, the variants, and their applications.

3.2 HOST CONTROLLED R-M SYSTEM

The host controlled restriction and modification (R-M) system comprises of a restriction enzyme (R), and its cognate methyltransferase (M or methylase or MTase). These two activities may be present as two separate enzyme(s) that act independently of each other, or the two activities may occur as separate subunits, or as separate domains of a larger multifunctional enzyme complex with combined restriction and modification activities (for details see Section 3.4). In this section, the physiological significance of R-M system, and its discovery is discussed.

3.2.1 Physiological Significance of R-M System

Restriction enzymes have apparently evolved as primitive immune system in bacteria. Their principal biological function is the protection of host from invasion by foreign DNA, such as virulent and temperate phages, and conjugative plasmids. Such defense system is prevalent in eubacteria; however, certain viruses and archaea have also been screened for their presence. The initial observation that led to the eventual discovery of restriction endonucleases was made in the early 1950s, when it was shown that some strains of bacteria are immune to bacteriophage infection. Further, it was observed that the phage particles that grow well and efficiently infect one strain of bacteria are often unable to grow well and infect other strains of the same bacterial species. Analysis indicated that the phage particles that efficiently grow and infect host cells have DNA with methylated (chemically and covalently modified) bases, while such methylation of DNA does not occur in poorly infecting phages. The cleavage and degradation of the unmethylated DNA in poorly infecting phages by the enzyme(s) of the host cell destroys the growth ability of the phage, a phenomenon referred to as 'host-controlled restriction'. The term 'restriction' comes from the fact that these enzymes within the cells degrade the foreign DNA before it has time to replicate and direct synthesis of new phage particles thereby restricting the infection. On the other hand, the methylated DNA is rendered insensitive to degradation. Even a single modified strand in the DNA duplex is

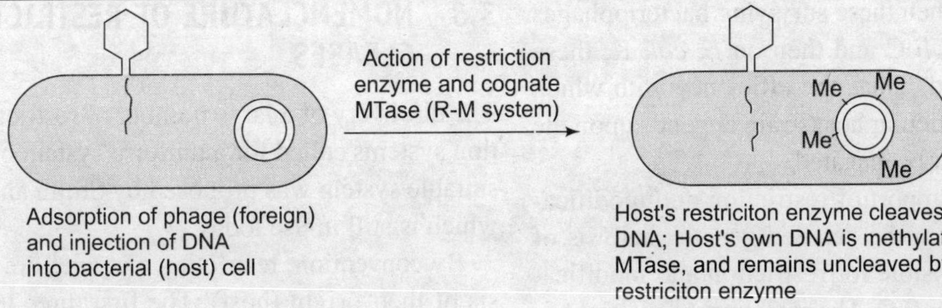

Figure. 3.1 Distinction of foreign and self DNA by R-M system [Incoming phage DNA is cleaved by host's restriction enzyme at the cognate recognition sites before the DNA has time to replicate, and direct synthesis of new phage particles. Host's own DNA is methylated within the recognition sequence by cognate methyltransferase (MTase), thereby preventing the cognate restriction enzyme from binding and cutting the DNA at the methylated site.]

sufficient to prevent restriction. The host DNA is modified, and hence remains protected from autorestriction, i.e., cleavage by its own enzyme(s) (for details see Section 3.6.1). This phenomenon is defined as 'host-controlled modification'. The two different enzymes responsible for phage DNA restriction and modification are 'restriction endonuclease,' that cleaves unmethylated DNA, and 'cognate methylase or methyltransferase' (MTase) that methylates the DNA, thereby preventing restriction. This combination of restriction endonuclease and its cognate methyltransferase forms the host-controlled R-M system of bacteria. The functions of restriction enzymes and cognate MTases in a bacterial cell is presented diagramatically in Figure 3.1. Gram negative bacteria have an added mode of protection that entails compartmentalization of restriction enzymes within the periplasmic space, and the MTases in the cell cytoplasm. This localization physically separates the restriction enzymes from the DNA while the chromosomal DNA is readily accessible to the modification enzymes. In addition to protecting the cell from the toxic

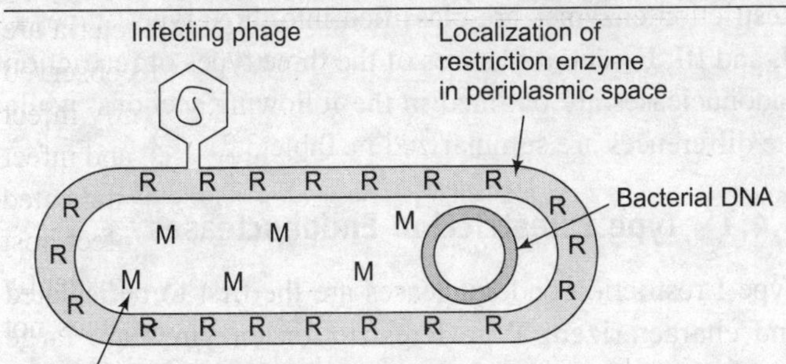

Figure 3.2 Cytoplasmic localization of modification or methyltransferase (MTase) enzyme (M), and periplasmic localization of restriction enzyme (R) in Gram negative bacteria. [The DNA of infecting phage when enters a bacterial cell is exposed first to restriction enzyme compartmentalized in the periplasmic space, and hence cleaved. On the other hand, host's own DNA is exposed to MTase first in the cytoplasm, and hence protected from cleavage by host's own restriction enzyme.]

effects of restriction enzymes, this segregation provides a cellular defense against attack by any foreign DNA, such as DNA from a phage that might enter the periplasm (Figure 3.2). Besides serving as defense system, restriction enzymes also play a role in preventing genetic exchange between groups of bacteria, while the modification system can promote recombination and aid in increasing diversity.

3.2.2 Discovery of R-M System

The phenomena of restriction and modification were well illustrated and studied by the behavior of λ phage on two *E. coli* host strains. Werner Arber and his colleagues demonstrated the two phenomena by studying the plating efficiency of λ phage (i.e., the number of plaques formed) onto two *E. coli* strains C and K (Figure 3.3). A stock preparation of λ phage, made by growth upon *E. coli* strain C, was later used to infect both *E. coli* C and *E. coli* K. The results indicated that the titres on these two strains differed by several orders of magnitude, the titre on *E. coli* K being the lower. This represented the restriction of growth of λ phage on *E. coli* K. However, a few bacteriophages escaped restriction, and the surviving bacteriophages were then replated on *E. coli* K, and no restric-

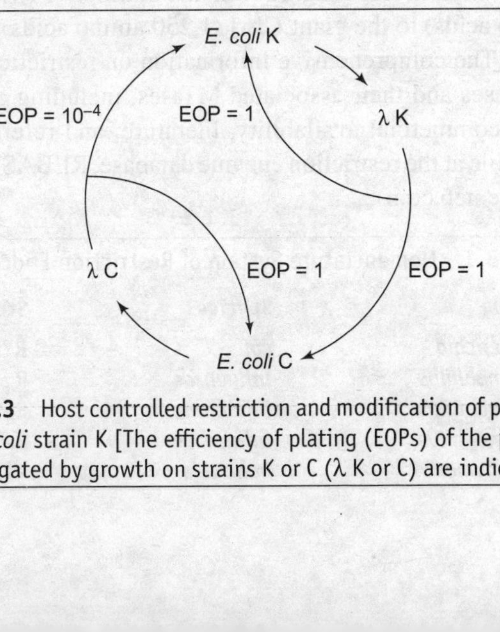

Figure 3.3 Host controlled restriction and modification of phage λ in *E. coli* strain K [The efficiency of plating (EOPs) of the phages propagated by growth on strains K or C (λ K or C) are indicated.]

tion was observed. But when these surviving bacteriophages were first plated on *E. coli* C and then on *E. coli* K, these were once again restricted. Thus, the efficiency with which phage λ plates upon a particular host strain depends upon the strain on which it was last propagated.

Another evidence in support of restriction and modification phenomena came from the detailed genetic analysis of the bacterial genes responsible for restriction and modification in *E. coli* K and *E. coli* C. The isolation of restriction-deficient and modification-deficient (R^-M^-), or restriction-deficient and modification-proficient (R^-M^+) bacterial mutants, but no restriction-proficient and modification-deficient (R^+M^-) mutants indicated that when DNA is modified, it is not susceptible to cleavage by restriciton enzyme, while the unmethylated DNA is cleaved by restriciton enzyme.

The biochemistry of restriction was advanced further with the isolation of the restriction endonuclease from *E. coli* K. It was evident that the restriction endonucleases from *E. coli* K and *E. coli* B were important examples of proteins that recognize specific structures in DNA, but the properties of these enzymes were complex. Although the recognition sites in the phage could be mapped genetically, determined efforts to define the DNA sequences cleaved were unsuccessful. Later Smith, Wilcox, and Kelley isolated and characterized the first restriction endonuclease, whose functioning depended on a specific DNA nucleotide sequence. Working with *Haemophilus influenzae* bacteria, this group isolated an enzyme, called *Hin*d II that always cuts phage T7 DNA at unmodified specific sequence of six base pairs (5′-GTY↓RAC-3′, where Y is any pyrimidine, and R is any purine). After the discovery of *Hin*d II, several type II restriction enzymes have been discovered, isolated, and purified. Presently, several restriction enzymes have been cloned and over-expressed to improve enzyme purity and yields. This not only simplifies purification procedure for commercialization but also aids in gene sequencing and X-ray crystallography of proteins. The studies indicate that the restriction enzymes are exceedingly varied in size ranging from the diminutive *Pvu* II (157 amino acids) to the giant *Cje* I (1,250 amino acids) and beyond. The comprehensive information on restriction endonucleases and their associated MTases, including cleavage sites, commercial availability, literature, and references is available at the restriction enzyme database, REBASE (http://rebase.neb.com).

3.3 NOMENCLATURE OF RESTRICTION ENDONUCLEASES

The discovery of a large number of restriction and modification systems called for a uniform system of nomenclature. A suitable system was proposed by Smith and Nathans (1973), which is still in use today.

By convention, restriction enzymes are named on the basis of their origin (host). The first three letters of the name are italicized because these abbreviate the genus and species names of the organism. The first letter of the name comes from the genus and the next two letters represent species of the prokaryotic cell from which these are discovered. The fourth letter (optional) designates the bacterial strain. Sometimes an enzyme is encoded by a plasmid, and the plasmid designation is used. When a particular host strain contains multiple restriction enzymes, the abbreviations for endonucleases are followed by roman numerals to identify them. These roman numerals typically depict the order in which these were discovered in a particular strain. A few examples are summarized in Table 3.1.

It is at times necessary to distinguish between the restriction and methylating activities. In such cases, the endonuclease and cognate methylase of a restriction–modification system are specified by the prefixes ‘R’ and ‘M’, respectively, e.g., R.*Bam* HI, M.*Bam* HI.

Besides restriciton endonucleases, there are certain homing endonucleases (Exhibit 3.1), which are named in a fashion similar to the type II restriction enzymes except that intron-encoded endonucleases are given the prefix ‘I-’ and intein-encoded endonucleases have the prefix ‘PI-’.

3.4 TYPES OF RESTRICTION ENDONUCLEASES

Restriction enzymes are classified into three types: types I, II, and III. Essential features of the three types of restriction endonucleases are outlined in the following sections, while the differences are summarized in Table 3.2.

3.4.1 Type I Restriction Endonucleases

Type I restriction endonucleases are the first to be isolated and characterized. Type I restriction enzymes are large multisubunit enzyme complexes of 300–400 kDa size. A single enzyme consists of specificity or recognition (S),

Table 3.1 Nomenclature System of Restriction Endonucleases

Genus	Species	Strain	Order of discovery	Nomenclature
Escherichia	*coli*	RY13	**First** from *E. coli*	*Eco* RI
Haemophilus	*influenzae*	R_d	**Second** from *H. influenzae* strain R_d	*Hind* II
Haemophilus	*influenzae*	R_d	**Third** from *H. influenzae* strain R_d	*Hind* III
Haemophilus	*influenzae*	R_f	**First** from *H. influenzae* strain R_f	*Hinf* I
Xanthomonas	*holicicola*	–	**First** from *X. holicicola*	*Xho* I

Exhibit 3.1 Homing endonucleases

'Homing' is the lateral transfer of an intervening genetic sequence, either an intron or an intein, to a cognate allele that lacks that element. Homing endonucleases are a large class of proteins (several hundreds of members) found in bacteria, archaebacteria, and eukaryotic nuclear as well as organellar genome. These enzymes are the products of selfish DNA elements, mostly present in many open reading frames (ORFs) in group I and II introns (I-homing endonucleases) or encoded *in frame* with a precursor protein as an intein (PI-homing endonucleases). The enzyme makes a highly site-specific ds DNA break at a homing site (extended recognition sequence). The intron containing the endonuclease gene is subsequently transferred to the cleaved recipient allele (an 'intron-less' allele) by a double strand break repair/gene conversion event. This double strand break is subsequently repaired by homologous recombination thereby integrating the homing endonuclease gene and completing the homing process. These endonucleases thus promote the lateral transfer of their encoding introns by a targeted transposition mechanism termed intron mobility or homing. The intein-encoded homing endonucleases combine two catalytic functions in one molecule, endonucleolytic activity and protein splicing activity, with which these liberate themselves from a precursor protein, leading to the generation of two functional and independent proteins.

Several features of these proteins make them attractive subjects for structural and functional studies. These endonucleases are unique, but may be contrasted with a variety of enzymes involved in nucleic acid strand breakage and rearrangement, particularly, restriction endonucleases. These exhibit extreme specificity and display a unique strategy of flexible recognition of very long DNA target sites ranging from 12–40 bp (rare base-cutters). Homing endonucleases are similar to restriction endonucleases in that these require Mg^{2+} ions for cleavage to generate DNA fragments with a 5′-P and a 3′-OH; however, these are structurally unrelated. In contrast to restriction endonucleases, homing endonucleases do not fully exhaust the hydrogen bonding potential. Moreover, the mean GC content in homing enzyme binding sites is only 46% as compared to ~68% in case of restriction enzymes. Because these are encoded within the intervening sequence, there are interesting limitations on the position and length of their open reading frames and therefore on their structures (Table A).

Homing endonucleases are named in a fashion similar to the type II restriction enzymes except that intron-encoded endonucleases are given the prefix 'I-' and intein endonucleases have the prefix 'PI-'.

The heterogeneous family of homing endonucleases can be divided into four sub-families:

LAGLIDADG Family The LAGLIDADG sequence is the largest and the most common motif in the homing endonucleases. The LAGLIDADG family proteins have catalytic and DNA binding activities in the same domain. These either function as homodimers with one LAGLIDADG motif per polypeptide chain, e.g., I-*Cre* I and I-*Mso* I or as monomers with two motifs per polypeptide chain, e.g., PI-*Sce* I, PI-*Pfu* I, and I-*Dmo* I.

GIY-YIG Proteins These proteins have GIY-YIG domain that is 70–100 amino acids long and in most cases contains five distinct motifs. The proteins have catalytic and DNA binding activities in

Table A Examples of Homing Endonucleases

Enzyme	Source organism	Sequence around cleavage sites*	Positions of cleavage sites on two strands	Enzyme comments
I-*Ceu* I	*Chlamydomonas eugametos*	5′-CGTAACTATAACGGTC_CTAA↓GGTAGCGAA-3′	(−9/−13)	*Chlamydomonas eugametos* mitochondrial intron encoded endonuclease; Mol. wt. 24,889 Da
I-*Dmo* I	*Desulfurococcus mobilis*	5′-ATGCCTTGCCGG_GTAA↓GTTCCGGCGCGCAT-3′	(−14/−18)	Insertion site AA↓GT; Mol. wt. 22,002 Da
I-*Ppo* I	*Physarum polycephalum*	5′-TAACTATGACTCTC_TTAA↓GGTAGCCAAAT-3′	(−11/−15)	Insertion site CT↓TA; Mol. wt. 20,107 Da
I-*Sce* I	*Saccharomyces cerevisiae*	5′-AGTTACGCTAGGG_ATAA↓CAGGGTAATATAG-3′	(−13/−17)	In the yeast mitochondrial 21S rRNA gene; Mol. wt. 27,682 Da
PI-*Psp* I	*Pyrococcus* spp.	5′-TGGCAAACAGCTA_TTAT↓GGGTATTATGGGT-3′	(−13/−17)	Mol. wt. 62,301 Da
PI-*Sce* I	*Saccharomyces cerevisiae*	5′-ATCTATGTCGG_GTGC↓GGAGAAAGAGGTAAT-3′	(−15/−19)	Other Names: VMA1, VDE; Mol. Wt. 1,18,629
PI-*Tli* I	*Thermococcus litoralis*	5′-TAYGCNGAYACNG_ACGG↓YTTYT-3′	(−5/−9)	The original recognition sequence was 5′-GCTCTTTATGCGG_ACAC ↓TGACGGCTTTTA-3′; Mol. Wt. 45,285 Da

* The recognition sequence is loosely defined as the sequence that normally flanks the intron or intein encoding the endonuclease; (↓) indicates cleavage on the strand shown; (_) indicates cleavage on the complementary strand.

separate domains, suggesting modularity. Further, a flexible linker separates these domains. The second motif includes an invariant arginine and the third motif includes an invariant glutamate, both of which are essential for the catalytic activity, but not for the structure of the domain. Example: I-*Tev* I.

HNH Proteins The HNH proteins contain a conserved sequence domain of 30–33 amino acids. These proteins contain two pairs of conserved histidines flanking a conserved asparagine. The domain forms a metal binding catalytic site of the nuclease. These proteins act as monomers. In contrast to LAGLIDADG proteins, HNH proteins have catalytic and DNA binding activities in separate domains, suggesting modularity. Example: type II restriction enzyme *Nla* III from *Neisseria lactamica*.

His-Cys Box Containing Proteins These proteins act as homodimers. Similar to LAGLIDADG proteins, these proteins also contain the catalytic and DNA binding activities in the same domain. Example: I-*Ppo* I.

The structural information implies that the known sequence motifs in all homing endonuclease families define the catalytic (DNA cleaving) domain of the proteins.

Source: Jurica MS, Stoddard BL (1999) *Cell. Mol. Life Sci.* 55: 1304–1326; Roberts RJ, Belfort M, Bestor T, Bhagwat AS, Bickle TA, Bitinaite J, Blumenthal RM, Degtyarev SK, Dryden DT, Dybvig K (2003) *Nucleic Acids Res.* 31: 1805–1812.

modification (M), and restriction (R) subunits encoded by the *hsd* S, *hsd* M, and *hsd* R genes, respectively. For example, *Eco* KI with a molecular weight of ~400 kDa has a quaternary structure Hsd M_2·Hsd R_2·Hsd S comprising of two modification subunits (62 kDa each), two restriction subunits (135 kDa each), and one recognition subunit (55 kDa).

Type I restriction enzymes recognize specific DNA sequences that are quite long (8–16 bp) and asymmetric. The recognition sequences are bipartite structures, each consisting of a specific sequence of three base pairs separated by a few base pairs from a specific sequence of four base pairs. The sequence of the intervening region does not seem to be important. The separation of the two parts of the recognition site means that both lie on one face of the DNA. Neither side of the recognition site is symmetrical, but together these possess adenine residues that are methylated on opposite strands.

When a target site is recognized by S subunit, the binding of enzyme to DNA may be succeeded by either restriction or

Table 3.2 Differences Between Type I, II, and III Restriction Endonucleases

Property	Type I	Type II	Type III
Enzyme activities and structure	Single multisubunit, multifunctional enzyme having both endonuclease and methylase functions; Three different types of subunits are S (recognition), M (methylation) and R (restriction); Large in size (300–400 kDa)	Separate endonuclease and methylase enzymes; Small in size (50–100 kDa)	Separate subunits for restriction (R) and recognition/methylation (MS)
Recognition sequence	Asymmetric and bipartite	Palindromic with rotational (dyad) symmetry	Asymmetric and unipartite
Recognition site length	Long; 8–16 bp	Short; 4–9 bp	Short; 5–7 bp
Methylation site	Recognition site	Recognition site	Recognition site
Cleavage site	Possibly random (nonspecific); at least 1,000 bp away from recognition site	At or near recognition site	~25 bp downstream of the recognition site
Requirements for restriction	ATP, Mg^{2+}, SAM	Mg^{2+}	ATP, Mg^{2+}, SAM (stimulatory but optional)
Enzymic turnover	No	Yes	No
DNA translocation or loop formation	Yes	No	No
Types of fragments produced	Heterogeneous	Homogeneous	Homogeneous
Examples	*Eco* K; *Eco* B	*Eco* RI; *Eco* RV	*Eco* PI; *Eco* P15

modification. Thus, restriction and modification activities are mutually exclusive reactions. Because the methylation reaction is performed by the same enzyme which mediates cleavage, the target DNA may be modified before it is cut. The R and M subunits require ATP and *S*-adenosylmethionine (SAM or AdoMet), respectively, as cofactors. SAM acts as methyl group donor in the modification reaction and is itself converted to *S*-adenosyl homocysteine (SAH). Methylation takes place at the host specificity site (i.e., recognition site). The R subunit requires Mg^{2+} for activity. In the initial stage of the reaction, SAM acts as an allosteric effector that changes the conformation of S subunit to allow it to bind to DNA. After binding to DNA, the next step is a reaction with ATP. Whether the enzyme will catalyze methylation or restriction depends on the methylation status of its recognition sequence. The enzyme compares the methylation status of two adenines within the recognition sequence, and acts accordingly (Figure 3.4). Thus, (i) If the enzyme is bound at a completely methylated site, the arrival of ATP releases it from the DNA without exerting any action; (ii) If either one of the adenines is methylated, i.e., hemimethylated state (a signal that the DNA is host DNA) then the enzyme acts as a maintenance methylase and methylates the adenine on other strand, thereby perpetuating the state of methylation in DNA. This leads to generation of fully methylated DNA; and (iii) If both adenines are unmethylated (a signal that the DNA is not host DNA), the enzyme undergoes a conformational switch with the arrival of ATP thereby triggering the endonuclease activity. These enzymes cleave DNA following translocation of the DNA, which makes them important molecular motors. The cleavage of DNA appears to occur after blockage of the translocation activity, often following collision with another translocating enzyme. The DNA moves, and pulls through a second binding site on the enzyme, winding along until it reaches an undefined cleavage region. There is electron microscopic evidence that the enzyme generates loops in DNA (Figure 3.5). The enzyme seems to remain attached to its recognition site after cleavage. The SAM is released from the enzyme before the restriction step occurs. The cleavage of DNA strands occurs randomly, at least 1,000 bp away from the 5′-end of the recognition sequence, and produces heterogeneous populations of DNA fragments. However, the selection of a site for cutting does not seem to be entirely random, because same regions of DNA are preferrentially cleaved. The cleavage reaction itself involves two steps. First, one strand of DNA is cut, then the other strand is cut nearby. There may be exonucleolytic degradation in the regions on either side of the site of cleavage. These type I restriction

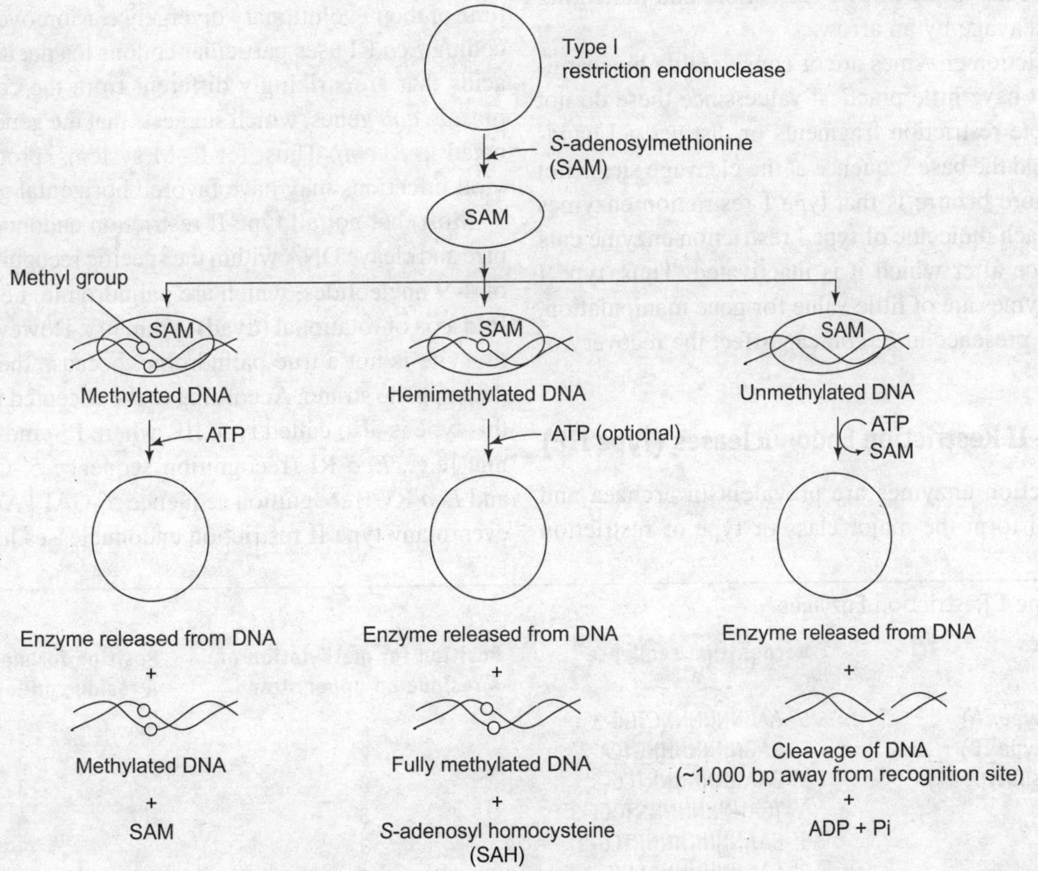

Figure 3.4 Methylation and cleavage by type I restriction endonuclease

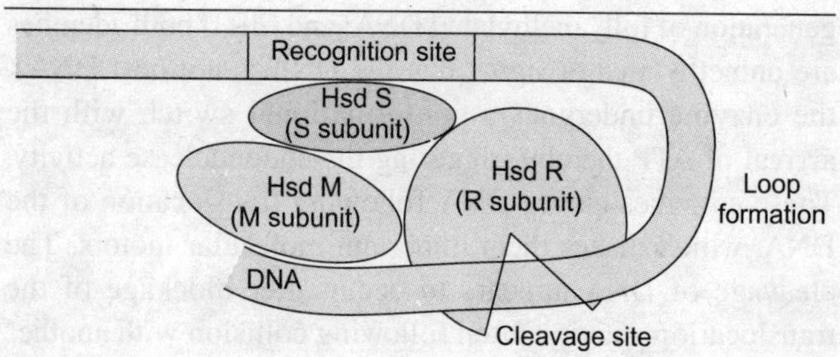

Figure 3.5 Loop formation during the action of type I restriction endonuclease allows cleavage ~1,000 bp away from the recognition site

enzymes exhibit a remarkably high ATPase activity and have an unusual property of promoting the hydrolysis of large amounts of ATP to ADP and Pi during cleavage. More than 10,000 molecules of ATP are hydrolyzed for the cleavage of each phosphodiester bond. The significance of these properties is however unknown.

Originally thought to be rare, it is now known from the analysis of sequenced genomes that these type I restriction enzymes are common. Few common examples of type I restriction endonucleases are listed in Table 3.3. For convenience it is usual practice to simplify the description of recognition sequences by showing only one strand of DNA, the one which runs in the 5′ → 3′ direction, and indicating the point of cleavage by an arrow.

Type I restriction enzymes are of considerable biochemical interest but have little practical value since these do not produce discrete restriction fragments or distinct gel banding patterns, and the base sequence at the cleavage site is not fixed. Even more bizarre is that type I restriction enzymes are suicidal. Each molecule of type I restriction enzyme cuts DNA only once after which it is inactivated. Thus, type I restriciton enzymes are of little value for gene manipulation, however, their presence in *E. coli* can affect the recovery of recombinants.

3.4.2 Type II Restriction Endonucleases (Type IIP)

Type II restriction enzymes are prevalent in archaea and eubacteria and form the major class or type of restriction enzymes. Type II restriction enzymes consist of separate proteins for restriction and modification. These are small 50–100 kDa proteins. Type II restriction enzyme is independent of its MTase, i.e., one enzyme recognizes and cuts DNA, the other enzyme recognizes and methylates the DNA. The subunit size ranges from 200 to 350 amino acids. These restriction enzymes usually exist as homodimers or tetramers, while MTases are monomeric in nature. At amino acid level, various type II restriction endonucleases do not reveal significant sequence similarity. However, analyses of 3-D structures (Exhibit 3.2) reveal a conserved core structure in different enzymes, which indicates evolutionary relatedness amongst type II restriction enzymes. These type II restriction enzymes have a catalytic core in common, and are possibly related by horizontal gene transfer. It is suggested that the bacteria may have obtained genes encoding these enzymes from other species by horizontal gene transfer that provides selective advantage in a particular environment. For example, the sequence of *Eco* RI (from *Escherichia*) and *Rsr* I (from *Rhodobacter*) share 50% identity, clearly indicating a close evolutionary relationship. However, these species of bacteria are not closely related, as is known from sequence comparisons of other genes and other evidences. Thus, it appears that these species obtained the genes for the restriction endonucleases from a common source more recently than the time of their evolutionary divergence. Moreover, the gene encoding *Eco* RI uses particular codons to specify given amino acids that are strikingly different from the codons used by most *E. coli* genes, which suggests that the gene has not originated in *E. coli.* Thus, for R–M systems, protection against viral infections may have favored horizontal gene transfer.

Most, but not all type II restriction endonucleases recognize and cleave DNA within the specific recognition sequences of 4–9 nucleotides, which are palindromic, i.e., having two-fold axis of rotational (dyad) symmetry. However, the restriction site is not a true palindrome, because the reverse is on the opposite strand. According to the accepted nomenclature, this type is also called type IIP, where P stands for palindromic [e.g., *Eco* RI (recognition sequence 5′-G↓AATTC-3′) and *Eco* RV (recognition sequence 5′-GAT↓ATC-3′)]. However, many type II restriction endonucleases have properties

Table 3.3 Type I Restriction Enzymes

Enzyme	Recognition sequence	Position for methylation of A residue on upper strand	Position for methylation of A residue on lower strand
Eco KI (subtype IA)	5′-AACNNNNNNGTGC-3′	2	3
Eco AI (subtype IB)	5′-GAGNNNNNNNGTCA-3′	2	3
Eco R124I (subtype IC)	5′-GAANNNNNNRTCG-3′	3	3
Eco BI	5′-TGANNNNNNNNTGCT-3′	3	4
Eco R124II	5′-GAANNNNNNNRTCG-3′	3	3
Sty LTIII	5′-GAGNNNNNNRTAYG-3′	2	4
Sty SPI	5′-AACNNNNNNGTRC-3′	2	3

Exhibit 3.2 Three-dimensional structure of type II restriction endonuclease

Type II restriction endonucleases are components of R-M systems that protect bacteria and archaea against invading foreign DNA. Most of these type II restriction enzymes are homodimeric or tetrameric, cleave DNA at defined sites of 4–9 bp in length, and require Mg^{2+} ions for catalysis; however, these enzymes exhibit a great deal of diversity in the process of recognition and the mode of cleavage. Despite this diversity, most of the type II restriction enzymes have a similar structural core, suggesting that these have evolved from a common ancestor, and for a long time it was believed that all of the type II restriction endonucleases belonged to the PD...(D/E)XK superfamily. Now, there is evidence from bioinformatics indicating that though most of the type II restriction enzymes belong to the PD...(D/E)XK superfamily, a few others do not. Some are rather related to the HNH and GIY-YIG families of homing endonucleases. Thus, there are at least four unrelated and structurally distinct superfamilies to which the type II restriction enzymes belong. These are PD...(D/E)XK (largest superfamily), Phospholipase D (PLD), HNH, and GIY-YIG. The first enzyme that unequivocally has been demonstrated not to be a member of this family was *Bfi* I, a type IIS enzyme that does not require Mg^{2+} for DNA cleavage. Based on sequence similarity with a nonspecific nuclease from *S. typhimurium*, *Bfi* I is considered to be a member of the phospholipase D superfamily. *Bfi* I is a homodimer with one catalytic center, which is used to cut two DNA strands within one binding event. Likewise, *Kpn* I has an HNH sequence motif, and unlike all other restriction enzymes known, it is active in the presence of Ca^{2+}, whereas it shows a degenerate specificity in the presence of Mg^{2+} or Mn^{2+}. Enzymes belonging to the same superfamily sometimes also have similar recognition sequences. For instance, *Eco* 29KI, *Ngo* MIII, and *Mra* I, which are related to the GIY-YIG superfamily, all bind to CCGCGG. *Hpy* I (CATG), *Nla* III (CATG), *Sph* I (GCATGC), *Nsp* HI (RCATGY), *Nsp* I (RCATGY), *Mbo* II (GAAGA), and *Kpn* I (GGTACC) belong to the HNH superfamily, and *Sso* II (CCNGG), *Eco* RII (CCWGG), *Ngo* MIV (GCCGGC), *Psp* GI (CCWGG), and *Cfr* 10I (RCCGGY) to the *Eco* RI branch. It has already been argued that these enzymes diverged early in evolution, presumably from a type IIP enzyme that recognized CCxGG or xCCGGx.

Based on structural differences, in particular the topology of secondary structure elements, and the arrangement of the subunits, the enzymes of the largest PD...(D/E)XK superfamily of type II restriction endonucleases can be divided into two major classes α (*Eco* RI-like) and β (*Eco* RV-like). Enzymes that belong to the *Eco* RI branch (*Bam* HI, *Bgl* II, *Bse* 634I, *Bso* BI, *Cfr* 10I, *Eco* RI, *Eco* RII, *Fok* I, *Mun* I, *Ngo* MIV) usually approach the DNA from the major groove, recognize the DNA mainly via an α-helix and a loop (α-class), and in general produce 5′ staggered ends. Enzymes of the *Eco* RV branch (*Bgl* I, *Eco* RV, *Hinc* II, *Nae* I, *Msp* I, *Pvu* II) usually approach the DNA from the minor groove, use a β-strand and a β-like turn for DNA recognition (β-class), and in general produce blunt or 3′ staggered ends.

Orthodox type II restriction endonucleases are dimers of identical subunits that are composed of a single domain (with

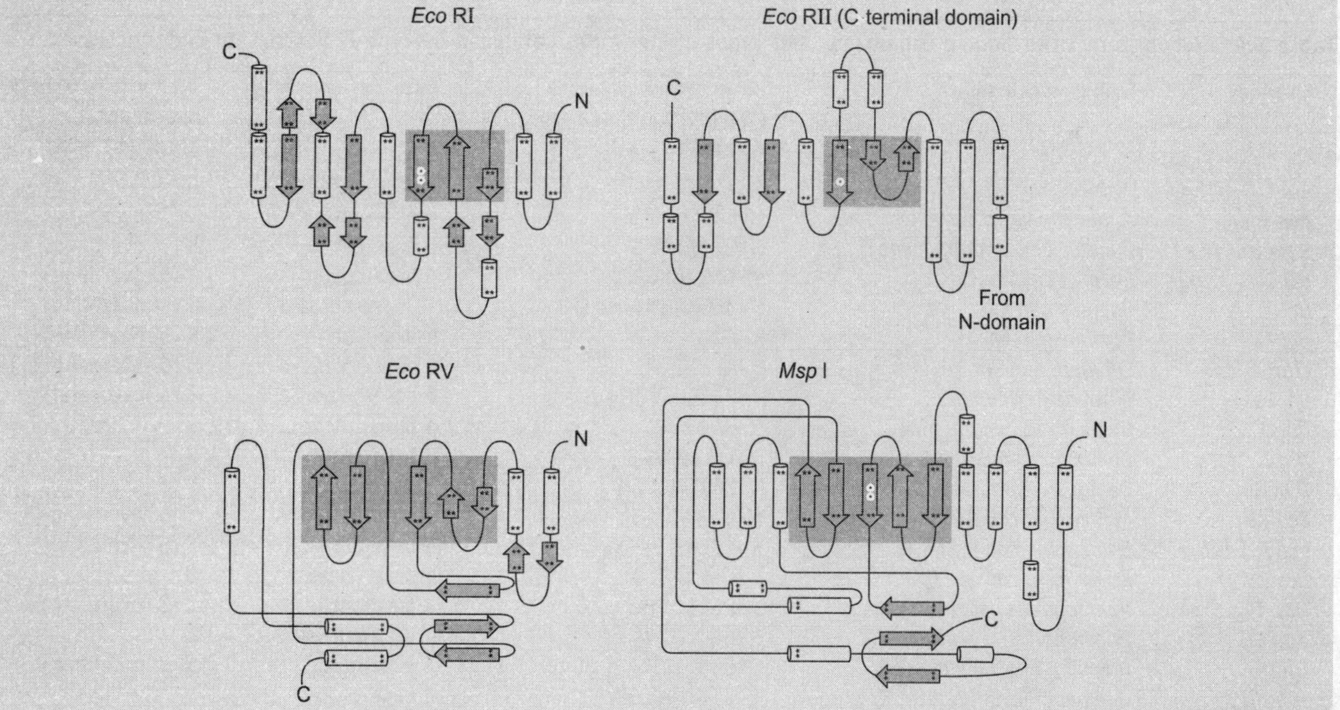

Figure A Topologies of the type II restriction endonucleases *Eco* RI, *Eco* RII (catalytic domain), *Eco* RV, and *Msp* I. *Eco* RI and *Eco* RII belong to the α-family (*Eco* RI family), *Eco* RV, and *Msp* I to the β-family (*Eco* RV family); both have a central five-stranded β-sheet. The amino acid residues of the PD...(D/E)XK motif are located on the second and third strand, of this β-sheet. Whereas in the α-family the fifth strand is parallel to the fourth strand, it is antiparallel in the β-family. The location of the PD...(D/E)XK motif is indicated in the topology diagram. α-helices are shown in light grey, β-strands in dark grey. The central five-stranded β-sheet is shaded grey.

subdomains being responsible for DNA recognition, DNA cleavage, and dimerization). Both *Fok* I and *Msp* I are monomers in the crystal. While *Fok* I is known to dimerize on the DNA, the quaternary structure of *Msp* I in its functional state is not yet known; however, analytical ultracentrifuge experiments show that in solution it exists in monomer-dimer equilibrium. The type IIE enzymes (*Eco* RII and *Nae* I), and the type IIS enzymes (*Fok* I) have a two-domain organization. These enzymes have a catalytic domain typical of the PD...(D/E)XK superfamily, and a DNA binding domain that in the case of the type IIE enzymes, serves as an effector domain, and in the case of the type IIS enzymes is responsible for DNA-recognition. The DNA-recognition domain of *Fok* I, and the effector domain of *Nae* I resemble the helix-turn-helix-containing DNA binding domain of the catabolite gene activator protein (CAP). Whereas it was originally believed that the effector domain of *Eco* RII has a novel DNA recognition fold, comparison with the structure of the type IIS enzyme *Bfi* I show that it rather resembles the DNA-recognition domain of *Bfi* I.

Some positive correlation between amino acid similarity and recognition sequence similarity of type II restriction enzymes has been reported. However, restriction enzymes are extremely divergent, and mostly structurally and evolutionarily unclassified. Even related enzymes binding to similar DNA sequences may differ much in the details of protein–DNA interaction. A comparison of crystal and co-crystal structures illustrates that the type II restriction enzymes have a similar core, which harbors the active site (one per subunit), and which serves as an important structure stabilization factor ('stabilization center'). This core comprises a five-stranded mixed β-sheet flanked by α-helices, the second and third strands of the β-sheet serve as scaffold for the catalytic residues of the PD...(D/E)XK motif. The fifth β-strand can be parallel (as in *Eco* RI), or antiparallel (as in *Eco* RV) to the fourth strand (Figure A).

Sources: Pingoud A, Jeltsch A (2001) *Nucleic Acids Res.* 29: 3705–3727; Pingoud V, Sudina A, Geyer H, Bujnicki JM, Lurz R, Luder G, Morgan R, Kubareva E, Pingoud A (2005) *J. Biol. Chem.* 280: 4289–4298; Pingoud A, Fuxreiterb M, Pingoud V, Wendea W (2005) CMLS, *Cell. Mol. Life Sci.* 62: 685–707; Roberts RJ, Vincze T, Posfai J, Macelis D (2005) *Nucleic Acids Res.* 33: D230–D232.

different from the type IIP enzymes (for details see Section 3.4.4). The type II restriction endonucleases require only Mg^{2+} for activity, and the corresponding MTases require only SAM. In contrast to other classes of enzymes, type II enzymes do not require ATP for action. Similar to type I and III restriction enzymes, methylation takes place at the host specificity site. The cleavage takes place within the recognition sequence and produces homogeneous populations of DNA fragments. The recognition sequences and cleavage sites of various type II restriction enzymes are summarized in Table 3.4.

Table 3.4 Recogntion Sites, Source Organisms, and Types of Cleavages Catalyzed by Type II Restriction Endonucleases

Enzyme	Source organism	Recognition or target site (cleavage site is marked as ↓)	Base-cutter	Sticky (S) (5′ or 3′ overhangs) or blunt (B) ends
Alu I	*Arthrobacter luteus*	5′-AG↓CT-3′	4 base-cutter	B
Ava I	*Anabaena variabilis*	5′-C↓YCGRG-3′	6 base-cutter	S (5′ overhang)
Bam HI	*Bacillus amyloliquefaciens* H	5′-G↓GATCC-3′	6 base-cutter	S (5′ overhang)
Bcl I	*Bacillus caldolyticus*	5′-T↓GATCA-3′	6 base-cutter	S (5′ overhang)
Bgl I	*Bacillus globigii*	5′-GCCNNNN↓NGGC-3′	11 base-cutter	S (3′ overhang)
Bgl II	*Bacillus globigii*	5′-A↓GATCT-3′	6 base-cutter	S (5′ overhang)
Cla I	*Caryophenon latum*	5′-AT↓CGAT-3′	6 base-cutter	S (5′ overhang)
Cfr 10I	*Citrobacter freundii*	5′-R↓CCGGY-3′	6 base-cutter	S (5′ overhang)
Dpn I	*Diplococcus pneumoniae*	5′-GA↓TC-3′	4 base-cutter	B
Dra I	*Deinococcus radiophilus*	5′-TTT↓AAA-3′	6 base-cutter	B
Dra II	*Deinococcus radiophilus*	5′-RG↓GNCCY-3′	7 base-cutter	S (5′ overhang)
Eco RI	*Escherichia coli* RY13	5′-G↓AATTC-3′	6 base-cutter	S (5′ overhang)
Eco RII	*Escherichia coli* R 245	5′-↓CCWGG-3′ (W = A, T)	5 base-cutter	S (5′ overhang)
Eco RV	*Escherichia coli* J62/pGL74	5′-GAT↓ATC-3′	6 base-cutter	B
Hae II	*Haemophilus aegyptius*	5′-RGCGC↓Y-3′	6 base-cutter	S (3′ overhang)
Hae III	*Haemophilus aegyptius*	5′-GG↓ CC-3′	4 base-cutter	B
Hha I	*Haemophilus haemolyticus*	5′-GCG↓C-3′	4 base-cutter	S (3′ overhang)
Hind II	*Haemophilus influenzae* R_d	5′-GTY↓RAC-3′	6 base-cutter	B
Hind III	*Haemophilus influenzae* R_d	5′-A↓AGCTT-3′	6 base-cutter	S (5′ overhang)
Hinf I	*Haemophilus influenzae* R_f	5′-G↓ANTC-3′	5 base-cutter	S (5′ overhang)
Hpa I	*Haemophilus parainfluenzae*	5′-GTT↓AAC-3′	6 base-cutter	B
Hpa II	*Haemophilus parainfluenzae*	5′-C↓CGG-3′	4 base-cutter	S (5′ overhang)
Kpn I	*Klebsiella pneumoniae*	5′-GGTAC↓C-3′	6 base-cutter	S (3′ overhang)

(*Contd.*)

Table 3.4 (*Contd.*)

Enzyme	Source organism	Recognition or target site (cleavage site is marked as ↓)	Base-cutter	Sticky (S) (5′ or 3′ overhangs) or blunt (B) ends
Mbo I	*Moraxella bovis*	5′-↓GATC-3′	4 base-cutter	S (5′ overhang)
Msp I	*Moraxella* spp.	5′-C↓CGG-3′	4 base-cutter	S (5′ overhang)
Nco I	*Nocardia corallina*	5′-C↓CATGG-3′	6 base-cutter	S (5′ overhang)
Nde I	*Neisseria denitrificans*	5′-CA↓TATG-3′	6 base-cutter	S (5′ overhang)
Not I	*Nocardia otitidis-caviarum*	5′-GC↓GGCCGC-3′	8 base-cutter	S (5′ overhang)
Pac I	*Pseudomonas alcaligenes*	5′-TTAAT↓TAA-3′	8 base-cutter	S (3′ overhang)
Pst I	*Providencia stuartii*	5′-CTGCA↓G-3′	6 base-cutter	S (3′ overhang)
Pvu I	*Proteus vulgaris*	5′-CGAT↓CG-3′	6 base-cutter	S (3′ overhang)
Pvu II	*Proteus vulgaris*	5′-CAG↓CTG-3′	6 base-cutter	B
Rsa I	*Rhodopseudomonas sphaeroides*	5′-GT↓AC-3′	6 base-cutter	B
Sac I	*Streptomyces achromogenes*	5′-GAGCT↓C-3′	6 base-cutter	S (3′ overhang)
Sac II	*Streptomyces achromogenes*	5′-CCGC↓GG-3′	6 base-cutter	S (3′ overhang)
Sal I	*Streptomyces albus* G	5′-G↓TCGAC-3′	6 base-cutter	S (5′ overhang)
Sau I	*Staphylococcus aureus*	5′-CC↓TNAGG-3′	7 base-cutter	S (5′ overhang)
Sau 3AI	*Staphylococcus aureus* 3AI	5′-↓GATC-3′	4 base-cutter	S (5′ overhang)
Sbf I	*Streptomyces* species Bf-61	5′-CCTGCA↓GG-3′	8 base-cutter	S (3′ overhang)
Sca I	*Streptomyces caespitosus*	5′-AGT↓ACT-3′	6 base-cutter	B
Sfi I	*Streptomyces fimbriatus*	5′-GGCCNNNN↓NGGCC-3′	13 base-cutter	S (3′ overhang)
Sma I	*Serratia marcescens*	5′-CCC↓GGG-3′	6 base-cutter	B
Sna BI	*Sphaerotilus natans*	5′-TAC↓GTA-3′	6 base-cutter	B
Sph I	*Streptomyces phaeochromogenes*	5′-G↓CATGC-3′	6 base-cutter	S (5′ overhang)
Ssp I	*Sphaerotilus* species	5′-AAT↓ATT-3′	6 base-cutter	B
Sst I	*Streptomyces stanford*	5′-GAGCT↓C-3′	6 base-cutter	S (3′ overhang)
Taq I	*Thermus aquaticus*	5′-T↓CGA-3′	4 base-cutter	S (5′ overhang)
Xba I	*Xanthomonas badrii*	5′-T↓CTAGA-3′	6 base-cutter	S (5′ overhang)
Xho I	*Xanthomonas holicicola*	5′-C↓TCGAG-3′	6 base-cutter	S (5′ overhang)
Xma I	*Xanthomonas malvacearum*	5′-C↓CCGGG-3′	6 base-cutter	S (5′ overhang)

Type II restriction enzymes offer several advantages over other types of restriction enzymes, owing to which these are widely used in DNA analysis and gene cloning experiments, and several type II enzymes are now commercially available. The advantages include: (i) Type II enzymes cut DNA at defined positions and produce discrete restriction fragments leading to distinct gel banding patterns. The results are predictable and reproducible; (ii) As the restriction and modification activities are mediated by separate enzymes, it is possible to cleave DNA in the absence of modification; (iii) The restriction activities do not require cofactors such as ATP or SAM, making them easier to use (note that SAM is required only by the modifying enzyme); and (iv) Most of the type II restriction enzymes make staggered cuts in the DNA. This increases the efficiency of ligation in gene cloning experiments.

3.4.3 Type III Restriction Endonucleases

Type III restriction enzymes consist of two subunits, M (or MS; responsible for DNA recognition and modification), and R (responsible for DNA cleavage). Active nucleases have a M_2R_2 stoichiometry. These enzymes recognize specific asymmetric, and unipartite DNA sequences of 5–7 bp length. The modification and restriction activities are expressed simultaneously. The enzyme first binds to its site on DNA, an action that requires ATP, subsequent dependence on ATP varies among the enzymes. Then the methylation and restriction activities compete for reaction with the DNA. Methylation occurs at host specificity site, consistent with the combination of methylation and recognition in the MS subunit (Figure 3.6). The enzyme methylates adenine residues, but the target sites have an intriguing feature, i.e., these can be

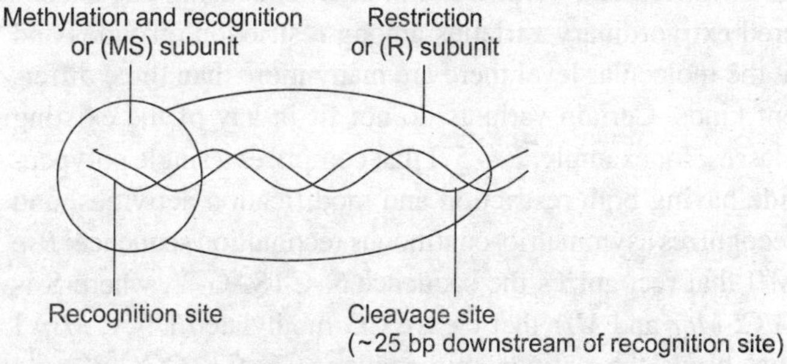

Figure 3.6 The positioning of two subunits of type III restriction enzyme (MS and R) allows separation of recognition site from the point of cleavage [The recognition and methylation occur by the MS subunit at the target site, while the restriction event occurs at a nearby site contacted by the R subunit.]

methylated only on one strand. Modification reaction is linked to the act of replication and the answer to the question whether the site will be methylated or restricted is unknown. Like type I enzymes the type III enzymes also require ATP and Mg^{2+}, however, in contrast to type I enzymes, these enzymes lack both the ATPase activity and absolute requirement of SAM. Thus, SAM is stimulatory but not essential. These enzymes cleave DNA at random sequences at variable distances from the recognition sites. The cleavage occurs downstream from the recognition sequence (e.g., ~25 bp in case of *Eco* P15) leading to generation of homogeneous population of DNA fragments. This distance of cleavage site from the recognition site is probably because the enzyme is large enough for the restriction subunit to contact DNA at this point. Two such recognition sequences in opposite orientations within the same DNA molecule are required to accomplish cleavage. First the cleavage takes place on a single strand followed by the cleavage of the second strand at some distance. Thus, restriction involves staggered cuts, 2–4 bases apart. Such staggered ends differ from each other and cannot be recombined at random. A whole set of such enzymes possessing characteristic non-palindromic target sequences is now available (Table 3.5). As the type III enzymes cut DNA at variable distances, these rarely give complete digests. Hence, no laboratory uses have been devised for them.

Table 3.5 Type III Restriction Enzymes

Enzyme	Recognition sequence	Position of methylating residue
Eco PI	5′-AGACC-3′	3
Eco P15	5′-CAGCAG-3′	5
Hinf III	5′-CGAAT-3′	–
Sty LTI	5′-CAGAG-3′	4

3.4.4 Variants of Traditional Classification System

The traditional classification of R-M systems into types I to III is convenient, but the amino acid sequencing has uncovered extraordinary varieties among restriction enzymes and at the molecular level there are many more than three different kinds. Certain variants do not fit in any of the existing classes, for example, *Eco* 57I that comprises a single polypeptide having both restriction and modification activities, and recognizes asymmetric, continuous recognition sequence; *Bse* MII that recognizes the sequence 5′-CTSAG-3′, where S is G/C; *Mcr* and *Mrr* that cleave CG-methylated DNA; *Msp* I that binds the palindromic sequence 5′-C↓*CGG-3′ as a monomeric protein (where * indicates methylation); homing endonucleases that recognize large asymmetric recognition sequence and tolerate some sequence degeneracy within their recognition sequence. Because of the discovery of such new restriction endonucleases that fall outside the traditional classes, reclassification has become mandatory. The current nomenclature tries to group the type II restriction enzymes according to the properties that are unique to the respective subtype. However, as will be seen, due to the great diversity among type II restriction endonucleases, overlap cannot be avoided. Variants of traditional classification are presently sub-classified as:

Type IIA Type IIA enzymes recognize asymmetric and unipartite recognition sequences and cleave outside of the recognition site, for example, *Bbv* CI: 5′-CCTCAGC (8/12)↓-3′. These enzymes bind to the DNA as heterodimers, and are ideal precursors for the generation of nicking enzymes.

Type IIB Type IIB enzymes cut twice on both the sides outside the recognition sequence generating stagerred ends, for example, *Bsp* 24I: 5′-↓(8/13)GACNNNNNN-TGG (12/7)↓-3′.

Type IIC Type IIC enzymes have both cleavage and modification domains within one polypeptide. One of the first discovered type IIC enzyme was *Bcg* I [5′-↓ (10/12) CGANNNNNNTGC(12/10)↓-3′], which has a very unusual functional organization. It has an A_2B quaternary structure with both endonuclease and MTase domains in the A subunit, and the target recognition domain located in the B subunit.

Type IIE These enzymes are dimeric, and bind to two recognition sites simultaneously. These enzymes cleave only one site during a single turnover but require a second DNA site for allosteric activation, for example, *Eco* RII (recognition sequence: 5′-↓CCWGG-3′) and *Nae* I (recognition sequence: 5′-GCC↓GGC-3′).

Type IIF These enzymes are homotetrameric in nature. Similar to type IIE enzymes, these enzymes bind to two recognition sites simultaneously and cleave both sites concertedly, for example, *Sfi* I (recognition sequence: 5′-GGCCNNNN↓NGGCC-3′) and *Cfr* 10I (recognition sequence: 5′-R↓CCGGY-3′).

Type IIG Type IIG enzymes, similar to and essentially a subgroup of Type IIC enzymes, have both cleavage and modification domains within one polypeptide. These are in general stimulated by SAM, but otherwise behave as typical type II enzymes, though most are also type IIS enzymes (see below). A well-studied example is *Eco* 57I [5′-CTGAAG (16/14)↓-3′]. Type IIG enzymes are very promising for the engineering of restriction endonucleases with new specificities.

Type IIH Type IIH enzymes behave like type II enzymes, but their genetic organization resembles type I R-M systems. *Ahd* I, for example, recognizes the sequence 5′-GACNNN↓NNGTC-3′, but its companion MTase consists of two modification and two specificity subunits.

Type IIM Many restriction enzymes are more or less tolerant to methylation, but for type IIM enzymes the methyl group is an essential recognition element. These enzymes recognize a specific methylated sequence, and cleave the DNA at a fixed site. The best-known representative is *Dpn* I (5′-GA↓TC-3′), which cleaves $G^{m6}ATC$, $G^{m6}AT^{m4}C$ and $G^{m6}AT^{m5}C$, yet not GATC, $GAT^{m4}C$, $GAT^{m5}C$ or certain hemimethylated sites. These enzymes differ from type IV enzymes, which do not cleave DNA at a fixed site.

Type IIS Type IIS enzymes cleave at least one strand of the target DNA outside of the recognition sequence. These enzymes are intermediate in size, 400–650 amino acids in length and recognize continuous and asymmetric sequences. Type IIS enzymes comprise two distinct domains, one for DNA binding or target recognition, and the other for catalysis or DNA cleavage (also serving as the dimerization domain). These enzymes are thought to bind to DNA as monomers for the most part, but cleave DNA cooperatively, through dimerization of the cleavage domains of adjacent enzyme molecules (i.e., these are active as homodimers). For this reason, some type IIS enzymes are much more active on DNA molecules that contain multiple recognition sites. For example, *Fok* I is monomeric in solution, but forms an active complex consisting of two proteins and two DNA sites. It recognizes non-palindromic sequence 5′-GGATG-3′ and cuts 9 and 13 base pairs to the right [i.e., 5′-GGATG(9/13)↓-3′]. Type IIS enzymes have been used for the creation of rare restriction sites ('Achilles' heel cleavage), and more recently for the generation of chimeric nucleases as well as strand-specific nicking endonucleases. These enzymes are also being used for cutting single stranded DNA (ss DNA) precisely at a desired point.

Type IIT Type IIT enzymes are heteromeric enzymes. A recently characterized representative is *Bsl* I (5′-CCNNNNN↓NNGG-3′), which is composed of two different subunits. The functional restriction endonuclease presumably is a $\alpha_2\beta_2$ tetramer. Several of these type IIT enzymes have been developed as nicking enzymes, e.g., *Bbv* CI, *Bsa* I, *Bsm* AI, *Bsm* BI, and *Bsr* DI.

Type IV Type IV enzymes are large in size ranging from 850–1,250 amino acids. The enzymes belonging to this type comprise of an N-terminal DNA cleavage domain joined to a DNA modification domain, and one or two DNA sequence-specificity domains forming the C-terminus or a separate subunit. The amino acid sequences of these enzymes are varied but their organization is consistent. When these enzymes bind to their substrates, these switch into either restriction mode to cleave the DNA or modification mode to methylate it. These enzymes recognize a specific sequence but cleave outside their recognition sequences at either one or both sides. Those that recognize continuous sequences [e.g., *Eco* 57I: 5′-CTGAAG(16/14)↓-3′] cleave on just one side; those that recognize discontinuous sequences [e.g., *Bcg* I: ↓(10/12)GCANNNNNNTCG(12/10)↓] cleave on both sides releasing a small fragment containing the recognition sequence. Type IV restriction enzymes also recognize and cleave methylated DNA. As such these are not part of an R-M system. The best-studied representative is *Mcr* BC, which consists of two different subunits, *Mcr* B and *Mcr* C, responsible for DNA recognition and cleavage, respectively. *Mcr* BC recognizes DNA with at least two recognition site sequences at a variable distance, containing methylated or hydroxy methylated cytosine in one or both strands. GTP and Mg^{2+} are required for DNA cleavage, which occurs close to one of the two methylated recognition sites.

As the type II R-M system is commonly used in gene cloning experiments, here onwards most of the discussion is with reference to type II enzymes unless otherwise mentioned.

3.5 RECOGNITION SITES

The restriction endonucleases recognize specific sequences on ds DNA called recognition sequences. These are also called cognate DNA/recognition site/binding site/DNA site/host specificity site/target site/restriction site (in case of type II enzymes the recognition and restriction sites are same).

3.5.1 Properties of Recognition Sites

The analyses of binding sites of all known type II restriction endonucleases reveal three important features, as is evident from the examples in Table 3.4. These include:

Significantly High GC Content As a generalization, one might hypothesize that an enhanced GC content may be an important property of protein binding DNA sequences whenever high specificity is needed. The enhanced GC content (68% GC and 32% AT) in all type II binding sites reflects biological functionality of the binding sites, as this value deviates significantly from the genome-wide average in host genomes as well as that of the bacteriophages (~50%). The facts associated with high GC content in recognition sites are: (i) Hosts protect themselves by methylating adenine(s) or cytosine(s) in the specific binding sites in their own genomes. As there are two different methylation sites in cytosine [yielding N^4-methylcytosine (m^4) and C5-methylcytosine (m^5)], but only one methylation site in adenine [yielding N^6-methyladenine (m^6)], evolution may therefore have favored cytosines over adenines in R–M binding sites; (ii) Like other DNA binding proteins, MTases and endonucleases recognize sequences on a ds DNA better than those on open DNA (without H-bonds between the two strands). Further, the GC rich sequences provide stability and higher binding strength to the DNA. This stability of GC rich sequences is due to better stacking interactions as compared to AT rich sequences. Moreover, as GC base pair involves three H-bonds, it has a higher binding strength than AT base pair, which has two H-bonds; and (iii) The restriction enzymes exert their action due to the H-bonding potential of their amino acid residues with the recognition sequence (protein–DNA interaction). As the number of canonical H-bonds formed between the bases and the recognizing amino acids are five in case of an AT base pair and six in case of a GC base pair, which may be beneficial for protein binding, the high GC content is favored for better interaction between the two. This fact, however, seems to be the most relevant reason for the high GC content in recognition sites.

Enhanced Occurrence of Adjacent R or Y Residues in DNA Binding Sites, i.e., RR/YY Dinucleotides The RR/YY dinucleotides are particularly accessible for specific protein–DNA interactions. This is because of following three reasons: (i) RR/YY dinucleotides provide on average stronger H-bond donor and acceptor clusters than other dinucleotides, and the average cluster strength of RR/YY steps is higher than that of all other steps. The only (very weak) exception is acceptor cluster in the minor groove, resulting from low strength of the GG/CC step; however, strong acceptor cluster in the major groove, and the donor cluster in the major and minor groove of the GG/CC dinucleotide counterbalance this. The close proximity of acceptor pairs (or donor pairs) on the DNA allows for the establishment of bifurcated H-bonds, which are stronger than canonical single donor-single acceptor interactions. This feature of RR/YY dinucleotides potentially facilitates the recognition by, and binding of interacting proteins; (ii) There is growing evidence that specific protein–DNA binding is accomplished not only by specific chemical contacts, but also by suitable geometrical arrangement of the DNA, and by its propensity to adopt a deformed conformation facilitating the protein binding. RR/YY dinucleotides allow for a special geometrical arrangement of the DNA, for example, minimal slide values, without exception, strong tilt in the negative direction, and a positive roll in all datasets, which implies positive bending towards the major groove; and (iii) RR/YY steps have a low stacking energy, and seem well-suited for the necessary conformational changes during specific protein binding. Moreover, the stacking energy of all RR/YY dinucleotides is anti-correlated with the statistical significance of the RR/YY subsequences. AA/TT has the highest stacking energy and the lowest significance, whereas GA/TC has the lowest stacking energy and the highest significance. Probably, all three possible reasons for an enhanced frequency of RR/YY dinucleotides in type II restriction enzyme binding sites together play a role in the corresponding specific DNA recognition. Asymmetric binding sequences of longer chains of purines or pyrimidines such as RRR, YYY, and YYYY are even more significant than RR/YY dinucleotides. This indicates that such substrings are preferred in binding sites. Some dinucleotide parameters, such as stacking energy, more or less add up in longer sequences. On the other hand, a negative correlation between motions at a given base pair step and neighboring steps has been found for most helical coordinates.

Under-representation of Recognition Sites The recognition sites for type II restriction enzymes are under-represented in host as well as phage genomes. If the composition of the four bases (A, T, G, C) in the genome is equal and as majority of the GC residues are confined to recognition sites of restriction endonucleases, the possibility of occurrence of such sites in the genome decreases, i.e., such sites are under-represented in the genome.

In contrast to type II restriction endonucleases, homing endonucleases do not fully exhaust the H-bonding potential. In support of this notion, the mean GC content in homing enzyme binding sites is only 46%.

3.5.2 Types of Recognition Sites

A variety of DNA sites recognized by restriction enzymes are as follows:

Palindromic Recognition Sequences

The sites on DNA that are recognized by type II restriction endonucleases are usually 4–9 bp in length, unambiguous, and non-degenerate. These sequences are palindromic having two-fold axis of dyad symmetry, i.e., the sequence on one strand reads the same as on complementary strand in the same polarity (say, 5′ → 3′ direction). For example, the sequence 5′-G↓AATTC-3′ (complementary strand 3′-CTTAA-G-5′), recognized by *Eco* RI is a palindrome. Such sequences are often referred to as palindromes because of their similarity to words or phrases that read the same backwards as forwards (e.g., Madam I'm Adam). Note that the recognition site is not a true palindrome, because the reverse is on the opposite strand. These palindromic sequences can be continuous or discontinuous.

Continuous where the two half-sites of the recognition sequence are adjacent (e.g., *Eco* RI: 5′-G↓AATTC-3′). Such continuous sequences are also called unipartite. Enzymes in homodimeric form mostly recognize these symmetric sites.

Discontinuous or Interrupted where the two half-sites are separated (e.g., *Bgl* I: 5′-GCCNNNN↓NGGC-3′; *Eco* NI: 5′-CCTNN↓NNNAGG-3′, where N is any nucleotide). Such restriction sites are bipartite and show hyphenated dyad symmetry. These sites are mostly recognized by enzyme in heterodimer form.

Degenerate or Ambiguous Recognition Sequences

Certain type II enzymes are not very specific for recognition sequence, and can recognize multiple recognition sequences (i.e., more than one sequence). Such sequences are called

Table 3.6 Degenerate or Ambiguous Recognition Sequences

Enzyme	Multiple recognition sequences
Bss KI	5′-↓CCNGG-3′
Btg I	5′-C↓CATGG-3′; 5′-C↓CGCGG-3′
Cfr I	5′-Y↓GGCCA-3′; 5′-Y↓GGCCG-3′
Eco RII	5′-↓CCAGG-3′; 5′-↓CCTGG-3′
Hae II	5′-RGCGC↓Y-3′
Hinf I	5′-G↓ANTC-3′
Xho II	5′-R↓GATCY-3′

Notes: R = purine; Y = pyrimidine; N = A/G/T/C.

degenerate recognition sequences. The ambiguity increases the frequency of occurrence of such sequences in DNA, and upon cleavage the fragments generated may or may not be compatible. However, like unambiguous recognition sequences, these are also 4–9 bp in length. Examples are listed in Table 3.6.

Non-palindromic Recognition Sequences

The recognition sequences can also be asymmetric or non-palindromic in nature. Further, the asymmetric recognition sequences can also be continuous/unipartite or discontinuous/bipartite. The enzymes in their heterodimeric form recognize such sequences. As discussed in Sections 3.4.1 and 3.4.3, the types I and III restriction enzymes recognize non-palindromic sites (Tables 3.3 and 3.5). However, recognition sites of some type II enzymes are also non-palindromic (Table 3.4).

The odd numbered recognition sites are also not true DNA palindromes, in that the central base pair reads differently in the two strands when read in same polarity (Table 3.4). For example, 5′-AANTT-3′ is a 5 base pair cutting site, where the N is any base. Thus, if N is A in coding strand, then the sequence is 5′-AAATT-3′ in coding strand and 3′-TTTAA-5′ (or 5′-AATTT-3′) in noncoding strand.

3.6 CLEAVAGE BY RESTRICTION ENDONUCLEASES

The central feature of a type II restriction enzyme is that it is specific in its action. It recognizes a specific sequence of nucleotides and cleaves DNA only at this particular sequence. As already mentioned, once a restriction enzyme finds its recognition sequence anywhere in a piece of DNA, it recognizes and cuts the two strands of DNA at a specific cleavage site that usually lies within the recognition sequence. For instance, the enzyme *Eco* RI recognizes the sequence 5′-GAATTC-3′ and cleaves between the G and A. Similarly, *Hae* III recognizes the sequence 5′-GGCC-3′ and does its surgical action between the adjacent G and C. The fragments produced are of precise length and nucleotide sequence. A list of restriction sequences for some of the most frequently used restriction endonucleases is presented in Table 3.4. Cleavage, however, does not occur unless the full-length of the recognition sequence is encountered. Thus, the enzyme 'scans' a DNA molecule, looking for recognition sequence and once it finds this sequence, it stops and cuts at both the strands. The positions of the two cuts made by a restriction endonuclease, both in relation to each other and to the recognition sequence itself, are determined by the identity of the restriction endonuclease used to cleave the molecule in the first place. Different endonucleases yield different sets of cuts, but one endonuclease will always cut a particular base sequence the same way, no matter what DNA molecule it is acting on. Once the cuts have been made, the DNA molecule will break into fragments. The number of fragments produced is the same as the number of recognition sites for the enzyme in a circular DNA, while the fragments produced in case of linear DNA is one more than the number of recognition sites for the enzyme. For example, the recognition sequence for *Hae* III occurs at 11 places in the circular DNA molecule of the virus ϕX174, hence the enzyme cleaves the viral DNA into 11 fragments. On the contrary, *Eco* RI cuts the λ DNA (linear) having five sites for *Eco* RI into six fragments.

3.6.1 Structure, Protein–DNA Interaction, and Cleavage by Type II Restriction Endonuclease

Protein–DNA selectivity is a central event in many biological processes, such as replication, transcription, and even restriction and modification. Type II restriction endonucleases are ideal systems for studying selectivity because of their high specificity and great variety. The enzyme in homodimeric (or tetrameric) form recognizes the palindromic recognition sequences such that the two-fold axis of DNA target site coincides with the two-fold axis of enzyme. The specificity of action is ensured by the specifically shaped and tightly packed active sites in enzyme–substrate complexes. In this section, specificity of action of type II enzymes is discussed taking the example of *Eco* RV (blunt end cutter), and *Eco* RI (sticky end cutter).

Eco RV

The determination of the structure of the complex between *Eco* RV and DNA fragments containing its recognition sequence (5′-GATATC-3′) (Exhibit 3.2) confirms the matching symmetry of the recognition sequence and the enzyme. The enzyme surrounds the DNA in a tight embrace in the form a homodimer. A unique set of interactions occurs between the enzyme and the cognate DNA sequence. Within the 5′-GATATC-3′ sequence, the G and A bases at the 5′-end of each strand, and their Watson–Crick partners, directly contact the enzyme by H-bonding with residues Gly 182, Ser 183, Gly 184, Asn 185, and Thr 186 that are located in two loops, one projecting from the surface of each enzyme subunit (Figure 3.7). Precise recognition is achieved even though the bases of target site remain fully paired with each other, as in classic Watson–Crick duplex. This enzyme can evaluate a

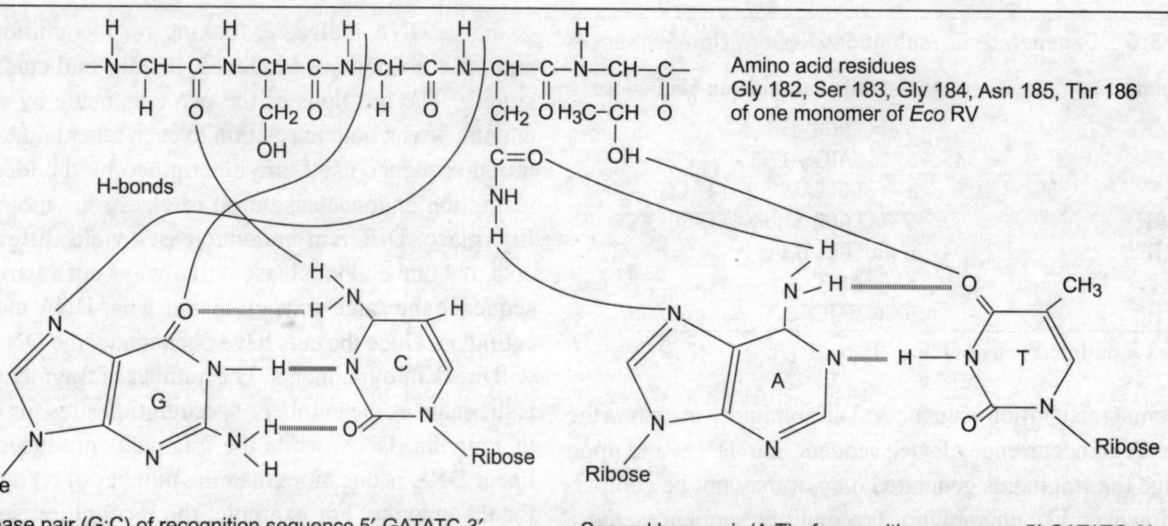

Figure 3.7 The interactions between *Eco* RV and DNA provide specificity of action [The H-bonding interactions shown above are for one monomer of the homodimeric enzyme. Fifth and sixth base pair of the recognition site forming A:T and G:C base pairs, respectively (A, G in bottom strand and T, C in upper strand), also exhibit such interactions, hence a total of 12 H-bonds are formed between the homodimeric enzyme and its cognate recognition site; Note that the H-bonds are not drawn up to scale and the figure just depicts the H-bond donor and acceptor groups involved in H-bonding.]

hexanucleotide sequence without denaturation of the ds DNA. A series of folds allows the protein to wedge into the large major groove of DNA, giving better access to the base pairs of the recognition sequence. The protein wraps almost completely around the DNA. The most striking feature of this complex is the distortion of the DNA that is substantially kinked in the center of the 5′-GATATC-3′ sequence. This abrupt disruption of the double helix is called type I neokink; the prefix 'neo' denotes the kink is imposed by the external agent (e.g., enzyme or drug). The structure of *Eco* RV bound to a substrate DNA shows a remarkable degree of distortion in the DNA, which is substantially bent towards the enzyme at the minor groove. Loops from each monomer wrap around the front of the DNA into the major groove, and make specific contacts with the residues of recognition sequence. The enzyme is only able to achieve this distortion when interacting with its target sequence. The central two TA base pairs in the recognition sequence play a key role in producing the kink. These do not make contact with the enzyme but appear to be required because of their ease of distortion. Note that 5′-TA-3′ sequence is known to be among the most easily deformed base pairs. Thus distortion and kinking are necessary for the formation of catalytically competent complex for cleavage, and for conferring specificity of action.

Once the enzyme recognizes and binds at the cognate recognition site, next action is the cleavage of phosphodiester backbone, and specifically the bond between the 3′ oxygen atom, and the phosphorus atom of the phosphodiester bond linking T and A (in the sequence 5′-GAT↓ATC-3′) is cleaved. The products of this reaction are DNA strands with a free 3′-hydroxyl group and a 5′-phosphoryl group. This hydrolysis step takes place by the direct attack of water on phosphorus atom by the mechanism involving 'in-line displacement of 3′ oxygen from phosphorus by Mg^{2+}-activated water'. Note that restriction enzymes, like many other enzymes that act on phosphate containing substrate require Mg^{2+} for activity. The Mg^{2+} binds to six ligands to form catalytically competent Mg^{2+}-binding pocket (Figure 3.8). These six ligands are: (i) Three water molecules; (ii) Two carboxylates of aspartate (Asp) residues (Asp 74 and Asp 90) at the active site of the enzyme; and (iii) One O atom of the phosphoryl group at the site of cleavage. Upon formation of Mg^{2+}-binding pocket, Mg^{2+} in conjunction with the two Asp residues (Asp 74 and Asp 90) and Lys 92 activates a water molecule by polarizing it towards deprotonation, and positions it so that it can exert a nucleophilic attack. Additional studies have revealed that a second Mg^{2+} must be present in an adjacent site to allow activated water to cleave its substrate.

The enzyme *Eco* RV binds with an almost equal affinity to both cognate and non-cognate sequences, however, in case of non-cognate DNA, conformation is not substantially distorted. Thus, no phosphate is positioned sufficiently close to the active site Asp residues to complete a Mg^{2+} binding site, and a complete catalytic apparatus is never assembled. The distortion of the substrate and the subsequent binding of Mg^{2+} account for the catalytic specificity of more than 10^6-fold despite very little preference at the level of substrate binding. In binding to the enzyme, the DNA is distorted in such a way that additional contacts are made between the enzyme and the substrate, which increases the binding energy. However,

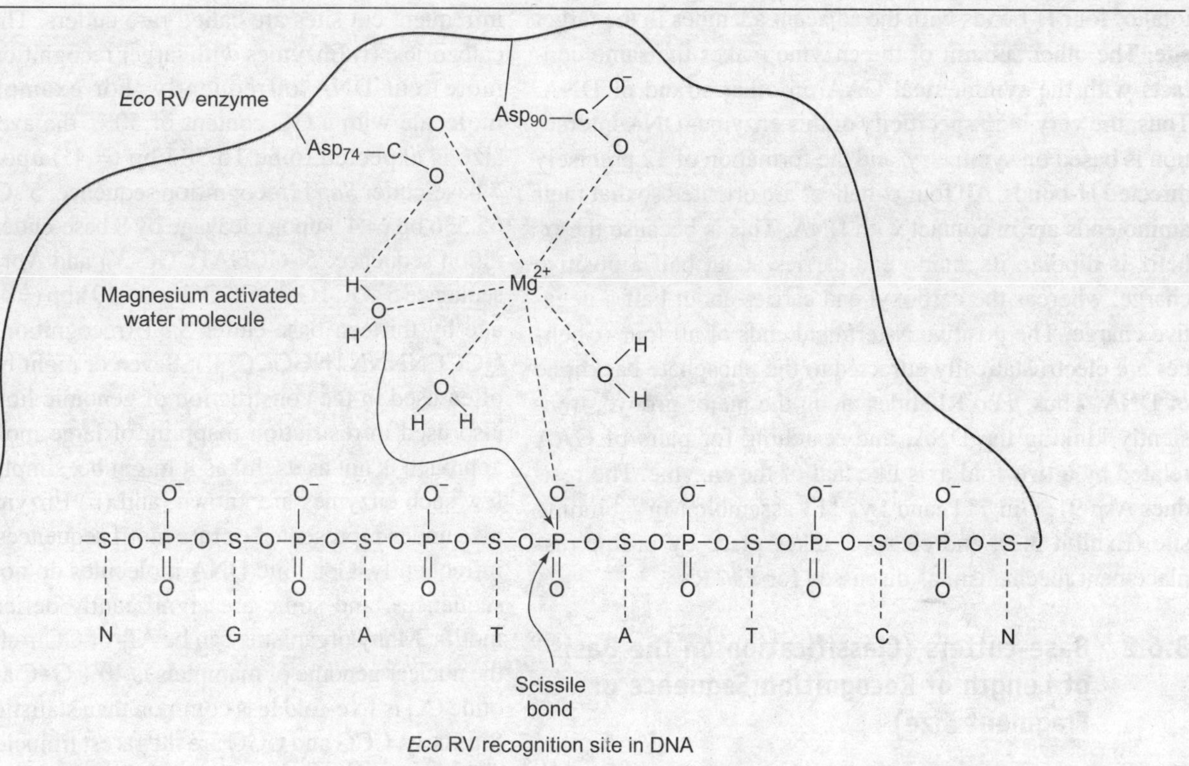

Figure 3.8 Formation of magnesium binding site that catalyzes the cleavage of phosphodiester bond between residues T and A of the recognition sequence 5′-GATATC-3′

this increase is cancelled by the energetic cost of distorting the DNA from its relaxed conformation. Thus, there is little difference in binding affinity for cognate and non specific DNA sequences. Interactions that take place within the distorted substrate complex stabilize the transition state leading to DNA hydrolysis.

Methylation of cognate recognition sequence blocks catalysis. This is because the addition of a methyl group to amino group of A nucleotide at 5′-end of recognition sequence, forming 5′-G*ATATC-3′, prevents formation of H-bond between amino group of A nucleotide, and side chain carbonyl group of Asn 185. Note that Asn residue is closely linked to another amino acid that forms specific contact with the DNA. The absence of H-bond disrupts other interactions between enzyme and substrate DNA, and the distortion necessary for cleavage does not take place. This is why methylation of recognition sequence protects host cell DNA from cleavage, whereas the unmethylated foreign DNA is cleaved.

Eco RI

Using computational resources at the Pittsburgh Supercomputing Center, Rosenberg and his colleagues built a detailed model of the structure of type II restriction enzyme *Eco* RI. The studies revealed a symmetrical structure of *Eco* RI. It is a dimer of identical subunits (31 kDa each), and binds DNA so that the two-fold axis of DNA target site coincides with the two-fold axis of the enzyme. It is found that this structure conforms closely to the structural pattern of a number of other DNA binding proteins. A high degree of discrimination requires intimate contact between enzyme and bases that leads to distortion and kinking of the DNA. The neokink widens the major groove by nearly 4 Å. It is believed that this folded structure, called the 'nucleotide binding fold', may be one of the keys to understanding enzyme–DNA recognition. Two hypothetical models have been proposed to explain the existence of a pronounced kink at the site of enzyme–DNA interaction. The first hypothesis suggests that the kink represents a 'structural form' of DNA not previously identified. The other model is that of 'molecular strain', in which the enzyme pushes the DNA into a distorted state. Molecular dynamic studies brought fundamental understanding showing that a kink in DNA is not inherent to DNA, but instead results from binding with the enzyme. The DNA in the complex is substantially unwound in the center of the recognition sequence by the insertion of the ends of two helices, one from each monomer, into the major groove. Loop structures from each monomer run along the major groove towards the end of the recognition sequence, and interact with the phosphate backbone at the point of cleavage. The entry of α-helices into major groove senses the identity of bases (four α-helices of enzyme enter widened major groove). One α-helix bears an Arg side chain that forms two H-bonds with a G of target site. A second α-helix of same subunit contains one Arg and one Glu residues that form a total of four H-bonds with the adjacent adenines in the target site. The other subunit of the enzyme makes the same contacts with the symmetrical GAA on other strand of DNA. Thus, the very high specificity of this enzyme–DNA interaction is based on symmetry, and the formation of 12 precisely directed H-bonds. All four α-helices are oriented so that their amino ends are in contact with DNA. This is because the α-helix is dipolar, its amino end carries about half a positive charge, whereas the carboxyl end carries about half a negative charge. The positive N-terminal ends of all four α-helices are electrostatically attracted to the phosphate backbone of DNA. Thus, *Eco* RI slides along the major groove, transiently kinking the DNA, and searching for pairs of GAA related by a two-fold axis like that of the enzyme. The residues Asp 91, Glu 111, and Lys 113 assemble Mg^{2+}-binding site (Exhibit 3.2), and cleavage takes place by in-line displacement mechanism as discussed for *Eco* RV.

3.6.2 Base-cutters (Classification on the Basis of Length of Recognition Sequence or Fragment Size)

The restriction enzyme recognizes the target site of specific length. For example, the enzymes *Eco* RI, *Sac* I, and *Sst* I each recognize a six base pair sequence of DNA, whereas *Not* I recognizes a sequence 8 bp in length and the recognition site for *Sau* 3AI is only 4 bp in length (Table 3.4). The frequency with which a class II restriction endonuclease cleaves DNA is dependent upon the size of its restriction site. The theoretical frequency of occurrence of any recognition sequence in the random sequence of DNA is calculated by the formula $1/4^n$, where n is the length (in bp) of the recognition sequence. This calculation is, however, based on the assumptions: (i) Four bases (A, T, G, and C) are randomly distributed throughout the length of substrate DNA; and (ii) There is equal proportion of all the four bases in the substrate DNA, i.e., G+C content is 50%. Based on the length of recognition site and cutting frequency, the restriction enzymes are thus classified as: four base-cutter that recognizes a tetranucleotide recognition sequence and tends to cleave DNA once in every $4^4 = 256$ bp; six base-cutter that recognizes a hexanucleotide recognition sequence and generates fragments with an average size of $4^6 = 4{,}096$ bp; eight base-cutter that recognizes an octanucleotide recognition sequence and generate fragments with an average size of $4^8 = 65{,}536$ bp, and so on. The above calculations thus indicate that the more the number of base pairs in the recognition sequence, the lesser are the chances of occurrence of that site to exist statistically in a given length of DNA. In simple terms, enzymes with short recognition sequences cut DNA frequently, whereas those with larger recognition sequences typically cut DNA very infrequently. Restriction endonucleases having infrequent cut sites are called rare cutters. These fall in two categories: (i) Enzymes with larger recognition sites (7 bp or more) cut DNA infrequently. For example, in a DNA molecule with a GC content of 50%, the average fragment size is expected to be 16,384 bp ($= 4^7$) upon cleavage by 7 base-cutter *Sap* I (recognition sequence: 5′-GCTCTTC-3′); 65,536 bp ($= 4^8$) upon cleavage by 8 base-cutters *Sgf* I (recognition sequence: 5′-GCGATCGC-3′) and *Not* I (recognition sequence 5′-GC↓GGCCGC-3′); 6,710 kbp ($= 4^{13}$) upon cleavage by thirteen base-cutter *Sfi* I (recognition sequence: 5′-GGCCNNNN↓NGGCC-3′). Seven or eight base-cutters are often used in the construction of genomic library. These are also used in restriction mapping of large molecules but the approach is not as useful as it might be, simply because very few such enzymes are known; and (ii) Enzymes that recognize under-represented or rare motif sequences also cut DNA infrequently. Genomic DNA molecules do not have random sequences, and some are significantly deficient in certain motifs. Many organisms can be AT- or GC-rich, for example, the nuclear genome of mammals is 40% G+C and the dinucleotide CG is five-fold less common than statistically expected. Similarly, CCG and CGG are the rarest trinucleotides in most AT-rich bacterial genomes and CTAG is the rarest tetranucleotide in GC-rich bacterial genomes. Take an example, the sequence 5′-CG-3′ is rare in human DNA because human cells possess an enzyme that adds a methyl group to carbon 5 of the C nucleotide in this sequence. The resulting 5-methylcytosine is unstable and tends to undergo deamination to give thymine. The consequence is that during human evolution many of the 5′-CG-3′ sequences that were originally in genome have become converted to 5′-TG-3′. Therefore, restriction enzymes that recognize a site containing 5′-CG-3′ cut human DNA relatively infrequently. For example, *Sma* I (recognition sequence: 5′-CCC↓GGG-3′) cuts human DNA on an average once every 78 kbp (instead of the expected once per ~4 kbp). Similarly, the recognition site for *Not* I (5′-GC↓GGCCGC-3′) containing two CpG motifs is slightly under-represented in mammalian genome (40% GC) leading to an estimated average fragment size of ~10 Mbp (instead of the expected ~65 kbp).

Certain restriction endonucleases show preferential cleavage of some sites in the same DNA molecule. For example, phage λ DNA has five sites for *Eco* RI, but the different sites are cleaved non-randomly. The site closest to the right terminus is cleaved ten-times faster than the sites in the middle of the molecule. Similarly, the enzymes *Nar* I, *Nae* I, and *Sac* II that require simultaneous interaction with two copies of their recognition sequence before catalyzing cleavage of DNA, rapidly cleave two of the four recognition sites on plasmid pBR322 DNA and seldom cleave the remaining two sites. Thus, the average practically produced fragment sizes may be significantly different from the expected values.

3.6.3 Sticky- and Blunt-End Cutters (Classification on the Basis of Types of Cuts or Types of Ends Generated)

Based on the types of cuts made, or the types of ends generated, there are two types of type II restriction enzymes.

Blunt-end Cutters

Some restriction endonucleases cut straight across the axis of symmetry of palindromic restriction site, producing blunt, or flush, or flat ended DNA. The two cuts lie at precisely opposite sites in the two strands of DNA. *Sma* I is an example of a restriction enzyme that cuts the DNA strands at the axis of symmetry, creating DNA fragments with blunt ends. As shown, it cuts at the center of the sequence 5′-CCC↓GGG-3′ (Figure 3.9). Only a few other enzymes cleave at the axis of symmetry and produce blunt ended fragments (Table 3.4).

Sticky End Cutters

Some restriction endonucleases cut the strand of DNA off the center of the palindrome sites, but between the same two bases on the opposite strands. The cleavage sites for such enzymes are displaced from the axis of symmetry and the cuts are asymmetric that do not lie exactly opposite to each other. This leaves one or more bases in single stranded form overhanging on each strand and such ends are called sticky (or cohesive or staggered) ends. Usually the size of overhang ranges from two or four nucleotides. Such ends are complementary to each other and can form intra- or intermolecular H-bonds. If a restriction enzyme has a non-degenerate palindromic cleavage site, all ends that it produces are compatible (i.e., these hydrogen bond to each other). Once the cuts have been made, the resulting fragments are held together only by the relatively weak H-bonds that hold the complementary bases to each other. The weakness of these bonds, however, allows the DNA fragments to separate from each other. The base pairing between two fragments with complementary sticky ends allows them to be efficiently annealed by DNA ligase. For example, in the base sequence 5′-G↓AATTC-3′, the restriction enzyme *Eco* RI cleaves the DNA between the G and A residues on each complementary strand, thus leaving the overhangs of 5′-AATT-3′. If an unpaired length of bases with the sequence 5′-AATT-3′ encounters another unpaired length of bases with the sequence 3′-TTAA-5′, both will bond with each other, i.e., these are 'sticky' for each other.

Generally, restriction fragments produced by the same enzyme are compatible (i.e., fragments produced by the same enzyme hydrogen bond to each other) and those produced by different enzymes are incompatible (i.e., these do not hydrogen bond to each other). However, there are examples where ends produced by different enzymes may also be compatible. For example, restriction endonucleases with different recognition sequences may produce the same sticky ends. Because recognition sequences and cleavage sites differ between different restriction enzymes, the length and the exact sequence of a sticky end 'overhang', as well as whether it is the 5′-end or the 3′-end that overhangs, depends on which enzyme produced it (Table 3.4). Depending upon the site of cleavage, two different types of overhangs may be produced by sticky end cutters. These are 5′ overhang and 3′ overhang.

5′ overhang When the restriction enzyme cuts asymmetrically within the recognition site at 5′-ends on both the strands, the fragments generated have 5′ overhanging termini (i.e., short single stranded segments extending or protruding from the 5′-ends on both the strands of DNA). For example, *Bam* HI cuts between G and G in the sequence 5′-G↓GATCC-3′ and generates 5′ protruding ends (Figure 3.10).

3′ overhang When the restriction enzyme cuts asymmetrically within the recognition site at 3′-ends on both the strands, the fragments generated have 3′ overhanging termini (i.e., short single stranded segments extending or protruding from the 3′-ends on both the strands of DNA). For example *Kpn* I cleaves between C and C in the sequence 5′-GGTAC↓C-3′ and generates 3′ protruding ends (Figure 3.11).

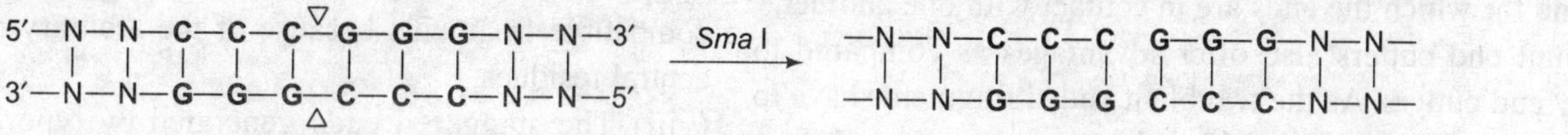

Figure 3.9 Cleavage by *Sma* I generates blunt ends [The recognition site is shown in bold and cleavage site as ∇.]

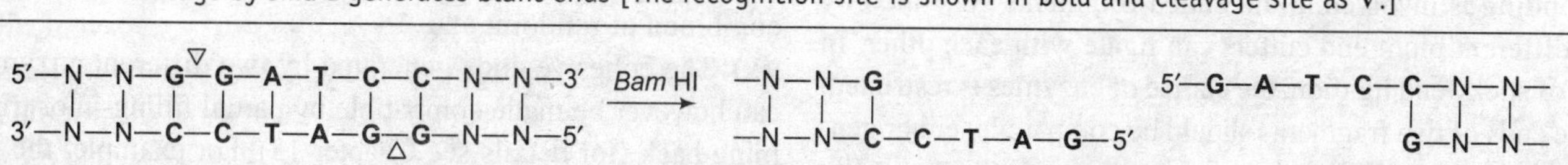

Figure 3.10 Cleavage by *Bam* HI creates sticky ends with 5′ overhangs [The recognition site is shown in bold and cleavage site as ∇.]

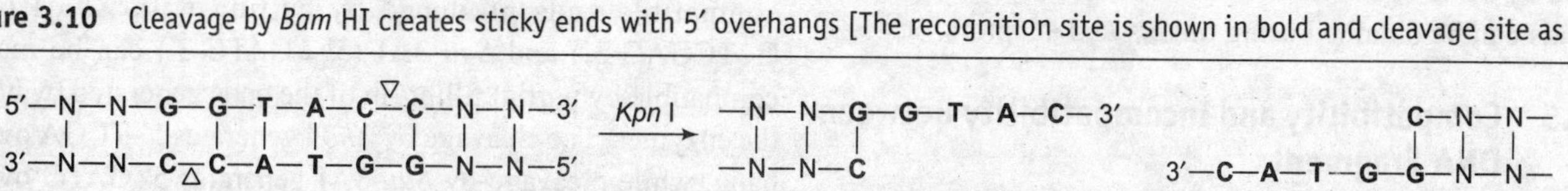

Figure 3.11 Cleavage by *Kpn* I creates sticky ends with 3′ overhangs [The recognition site is shown in bold and cleavage site as ∇.]

In general, the size of overhangs ranges up to five nucleotides. An example of enzyme producing a seven nucleotide 3′ overhang is *Tsp* RI (CASTGNN↓), where S = G/C, and N = A/T/G/C.

3.6.4 Sticky Ends are Easier to Join and Increase the Efficiency of Ligation

Under *in vitro* conditions, the fragments generated by restriction endonucleases are ligated in the desired order by DNA ligase. Together, these two enzymes are used to assemble customized genomes.

For obvious reasons, as compared to blunt ends, the sticky ends are much easier to join. Furthermore, as compared to sticky ended DNA ligation, blunt ended DNA ligation is inefficient, but is sufficient for most subcloning applications. As the blunt ends do not have any complementary sequence, the ligation of such fragments is based only on 'chance associations'. This not only makes the ligation process difficult and non-directional, but also decreases the efficiency of ligation. However, under *in vitro* conditions, in order to increase the chances of the ends of the molecules coming together in the correct way, the blunt end ligation should be performed at high DNA concentrations. The blunt end ligation can be made directional and efficient by changing blunt ends to sticky ends by employing high concentrations of linkers and adaptors (for details see Chapter 13). In contrast, ligation of complementary sticky ends is much easier and efficient. Being complementary, the single stranded tails (overhangs) can be made to form H-bonds with one another and the cohering fragments can then be ligated together with the help of DNA ligase. These H-bonds are weak, but as two to four H-bonds are formed, the structure is relatively stable for the ligase enzyme to form phosphodiester bonds. But, if the phosphodiester bonds are not synthesized fairly quickly the sticky ends will fall apart again. These transient, base paired structures, however, increase the efficiency of ligation by increasing the length of time for which the ends are in contact with one another.

Blunt end cutters also offer advantages as compared to sticky end cutters. As the two blunt ended fragments have to catch-hold of each other before ligation can occur and no H-bonding is involved, blunt ended fragments generated by two different blunt end cutters can ligate with each other. In case of sticky end ligation, the choice of enzymes is restricted, as the ends of two fragments should be compatible, either generated by same or different enzymes, however, the versatility of ends can be obtained using linkers and adaptors.

3.6.5 Compatibility and Incompatibility Between DNA Fragments

The fragments generated by restriction enzymes may be compatible or incompatible. The compatible fragments can ligate easily, while incompatible fragments do not ligate with each other. The knowledge of compatibility of ends allows molecular biologists to anticipate the ways in which the fragments can be joined and to choose enzymes appropriately. A few examples of compatible and incompatible fragments are as follows:

(i) The DNA molecules cleaved with same restriction enzyme exhibit compatible ends, for example, *Hind* III always cuts at the sequence 5′-A↓AGCTT-3′, and produces only single type of sticky ends (5′-AGCT-3′) that can ligate with one another easily;

(ii) The blunt ends generated by two different blunt end cutters are compatible. For example, fragments generated by *Sma* I (5′-GGG↓CCC-3′) can ligate with the fragments generated by *Hae* III (5′-GG↓CC-3′);

(iii) The digestion with restriction endonucleases with nested sites leads to the generation of compatible fragments (for details see Section 3.6.6). For example, *Bam* HI (5′-G↓GATCC-3′); *Bgl* II (5′-A↓GATCT-3′), and *Sau* 3A I (5′-↓GATC-3′), all produce 5′-GATC-3′ sticky ends;

(iv) Restriction site may contain one or more ambiguous nucleotides, which increases the frequency of the sequence in random DNA, but means that ends generated by the restriction endonuclease are not always compatible. For example, *Hinf* I (5′-G↓ANTC-3′, where N is any nucleotide) (Table 3.6);

(v) Cleavage with isoschizomers that recognize and cleave at the same target site also generate compatible fragments (for details see Section 3.7.1). For example, fragments generated by *Mbo* I (5′-↓GATC-3′) can ligate with fragments generated by *Sau* 3AI (5′-↓GATC-3′) (Table 3.7);

(vi) Isocaudomers recognize different target sequences but act on DNA to produce compatible cohesive ends (for details see Section 3.7.3). For example, *Bam* HI (5′-G↓GATCC-3′); *Bcl* I (5′-T↓GATCA-3′), and *Bgl* II (5′-A↓GATCT-3′), all produce 5′-GATC-3′ overhangs (Table 3.9);

(vii) Cleavage at a bipartite site does not generate universally compatible fragments because of the arbitrary nature of the central residues;

(viii) The staggered ends generated by type I and III restriction enzymes differ from each other, and cannot be recombined at random; and

(ix) The cohesive ends generated by two different enzymes can however be made compatible by partial filling-in or trimming back (for details see Chapter 13). For example, the incompatible ends produced by cutting with *Xho* I (5′-C↓TCGAG-3′) and *Sau* 3AI (5′-↓GATC-3′) can be made compatible by partial filling-in of the ends generated by both the enzymes. The cleavage by *Xho* I generates 5′-TCGA overhang, while cleavage by *Sau* 3AI generates 5′-GATC overhang. The partial filling-in one of the overhangs generated by *Xho* I leaves 5′-TC sticky end. This can be ligated with

5′-GA of the 5′-GATC-3′ sticky ends created by cleavage with *Sau* 3AI followed by partial filling-in.

3.6.6 Recleavable Ligation Products

The ligation of two fragments forms hybrid recognition site that can sometimes be recleaved with either the same or different enzyme. However, there is also a possibility that the hybrid sites formed are not recleavable. The formation of recleavable sites is beneficial in the recovery of the DNA of interest. Some of the important examples include: (i) The ligation products of cohesive ends produced by a higher base-cutter (say, six base-cutter) are often recleavable by lower base-cutters (say, four or five base-cutters). For example, the hybrid site ACCGGG can be cleaved by *Hpa* II (C↓CGG), *Nci* I (CC↓GGG), or *Scr* FI (CC↓NGG; where N = G); (ii) The recognition site for one enzyme may be embedded within the restriction site for another enzyme. Such sites are termed as nested sites. For example, the recognition and cleavage site for *Sau* 3AI is 5′-↓GATC-3′, while that for *Bam* HI is 5′-G↓GATCC-3′. Thus, the site for *Sau* 3AI lies within the site for *Bam* HI and both produce 5′-GATC-3′ overhangs that are compatible with each other. Joining of fragments generated by these enzymes forms a hybrid site (*Bam* HI/*Sau* 3AI), which may be cleaved only by *Sau* 3AI and the cleavage by *Bam* HI depends on the flanking residues. Since *Sau* 3AI recognizes a four base site, the adjacent bases are random. Thus, there is a one in four probability that the fusion of a *Sau* 3AI site to a *Bam* HI site will regenerate the *Bam* HI sequence; (iii) New restriction sites can be generated by filling-in the overhangs generated by restriction endonucleases and ligating the products together (for details see Chapter 13); and (iv) There are also many examples of combinations of blunt end restriction endonucleases that produce recleavable ligation products. For example, when molecules generated by cleavage with *Alu* I (5′-AG↓CT-3′) are joined to ones produced by *Eco* RV (5′-GAT↓ATC-3′), some of the hybrid ligation sites will have the sequence 5′-GATCT-3′, and others will have the sequence 5′-AGATC-3′. Both can be cleaved by *Mbo* I (↓5′-GATC-3′).

3.7 VARIANTS OF RESTRICTION ENDONUCLEASES

Occasionally, enzymes with novel DNA sequence specificities are found but most prove to have the same specificity as enzymes already known. Various forms of restriction endonucleases are described below.

3.7.1 Isoschizomers

Some restriction endonucleases from different sources may recognize the same restriction site and cleave identically at same position. Such restriction endonucleases are termed isoschizomers. The term is derived from Greek words; '*isos*' means 'equal' and '*schizein*' means 'to split'. At present, over a hundred isoschizomers are known and a few of them are listed in Table 3.7. The cleavage with isoschizomers gives distinct gel-banding patterns yielding same results. The fragments generated by cleavage with different isoschizomers are compatible and can be ligated with ease. However, isoschizomers often have different optimum reaction conditions, stabilities, and costs, which may influence the decision of which to purchase for the studies.

Table 3.7 Some Common Examples of Isoschizomers

Isoschizomers	Recognition and cleavage (↓) sites
Aha III; *Dra* I	5′-TTT↓AAA-3′
Cfr I; *Eae* I	5′-Y↓GGCCR-3′
Cla I; *Ban* III; *Bsc* I	5′-AT↓CGAT-3′
Dra III; *Ade* I	5′-CACNNN↓GTG-3′
Ege I; *Ehe* I	5′-GGC↓GCC-3′
Fnu DII; *Acc* II	5′-CG↓CG-3′
Hind II; *Hinc* II	5′-GTY↓RAC-3′
Hind III; *Hsu* I	5′-A↓AGCTT-3′
Hpa II; *Msp* I	5′-C↓CGG-3′
Mbo I; *Dpn* II; *Nde* II; *Sau* 3AI	5′-↓GATC-3′
Rsa I; *Afa* I	5′-GT↓AC-3′
Sac I; *Sst* I	5′-GAGCT↓C-3′
Sbf I; *Sda* I	5′-CCTGCA↓GG-3′
Tai I; *Tsc* I	5′-ACGT↓-3′
Xma III; *Eag* I	5′-C↓GGCCG-3′

Notes: R = A/G; Y = C/T; N = A/C/G/T.

3.7.2 Neoschizomers or Heteroschizomers

Restriction endonucleases from different sources that recognize the same target site but cleave at different positions are termed neoschizomers or heteroschizomers (Table 3.8). Similar to isoschizomers, the cleavage with neoschizomers also gives distinct gel-banding patterns yielding similar results, even though the ends of the fragments are different.

3.7.3 Isocaudomers

These enzymes recognize different target sequences but cleavage leads to the same termini. Thus, the ends produced by isocaudomers are compatible, such that the fragments generated can be ligated with ease (Table 3.9). The term is derived from the Greek word '*isos*' meaning equal, and the Latin word '*cauda*' meaning tail. For example, *Sal* I (5′-G↓TCGAC-3′), *Xho* I (5′-C↓TCGAG-3′), and *Ava* I (5′-C↓YCGRG-3′, where Y = T and R = A) recognize different sites but generate same 5′-TCGA-3′ termini upon cleavage.

Table 3.8 Some Common Examples of Neoschizomers or Heteroschizomers

Enzymes or isoschizomers	Recognition and cleavage site for enzyme	Neoschizomers	Recognition and cleavage site for neoschizomers
Cha I	5′-GATC↓-3′	*Mbo* I; *Dpn* II; *Nde* II; *Sau* 3AI	5′-↓GATC-3′
Fmu I	5′-GGNC↓C-3′	*Asu* I	5′-G↓GNCC-3′
Hae II	5′-RGCGC↓Y-3′	*Lpn* I	5′-RGC↓GCY-3′
Hha I	5′-GCG↓C-3′	*Hsp* AI	5′-G↓CGC-3′
Kpn I	5′-GGTAC↓C-3′	*Acc* 65I	5′-G↓GTACC-3′
Mae II	5′-A↓CGT-3′	*Tai* I	5′-ACGT↓-3′
Nae I	5′-GCC↓GGC-3′	*Mro* NI	5′-G↓CCGGC-3′
Nar I	5′-GG↓CGCC-3′	*Bbe* I	5′-GGCGC↓C-3′
		Ehe I; *Ege* I	5′-GGC↓GCC-3′
		Kas I	5′-G↓GCGCC-3′
Nhe I	5′-G↓CTAGC-3′	*Ace* II	5′-GCTAG↓C-3′
Sel I	5′-↓CGCG-3′	*Fnu* DII; *Acc* II	5′-CG↓CG-3′
Sma I	5′-CCC↓GGG-3′	*Xma* I	5′-C↓CCGGG-3′
Xho I	5′-C↓TCGAG-3′	*Sci* I	5′-CTC↓GAG-3′

Notes: R = A/G; Y = C/T; N = A/C/G/T.

Table 3.9 Some Common Examples of Isocaudomers or Different Enzymes Generating Fragments with Compatible Cohesive Ends

Enzyme	Recognition and cleavage (↓) sites	Compatible ends produced
Ava I	5′-C↓YCGRG-3′ (where Y = T and R = A)	
Sal I	5′-G↓TCGAC-3′	5′-TCGA-3′
Xho I	5′-C↓TCGAG-3′	
Bam HI	5′-G↓GATCC-3′	
Bcl I	5′-T↓GATCA-3′	
Bgl II	5′-A↓GATCT-3′	5′-GATC-3′
Mbo I	5′-↓GATC-3′	
Sau 3A	5′-↓GATC-3′	
Xho II	5′-R↓GATCY-3′	
Taq I	5′-T↓CGA-3′	
Acc I	5′-GT↓MKAC-3′ (where M = C and K = G)	
Acy I	5′-GR↓CGYC-3′ (where Y = A)	5′- CGA-3′
Asu II	5′-TT↓CGAA-3′	
Cla I	5′-AT↓CGAT-3′	
Age I	5′-A↓CCGGT-3′	
Ava I	5′-C↓YCGRG-3′ (where Y = C and R = G)	5′-CCGG-3′
Xma I	5′-C↓CCGGG-3′	

Notes: R = A/G; Y = C/T; N = A/C/G/T; K = G/T; M = A/C.

3.7.4 Heterohypekomers

As mentioned earlier, every restriction endonuclease has a cognate MTase that modifies recognition sites in the host genome by methylation and prevents restriction (Table 3.10). Thus, all restriction endonucleases are methylation sensitive to some degree. Some restriction endonucleases, however, due to the nature of their restriction sites, are also sensitive to genome-wide methylation by DNA adenine methyltransferase (Dam) and DNA cytosine methyltransferase (Dcm) in the *E. coli* genome, and the methylation of CG or CNG motifs in eukaryote genomes by CpG dinucleotide methyltransferase (Exhibit 3.3). The isoschizomers with different methylation sensitivity are called heterohypekomers. For instance, an MTase, M.*Sss* I, isolated from *Spiroplasma* methylates the dinucleotide CpG. This enzyme can be used to modify *in vitro* restriction endonuclease target sites which contain the CG sequence. Some of the target sequences modified in this way will be resistant to endonuclease cleavage, while others will remain sensitive. Thus, if the sequence 5′-CCGG-3′ is modified with M.*Sss* I, it will be resistant to restriction by *Hpa* II but sensitive to *Msp* I. Since 90% of the methyl groups in the genomic DNA of many animals, including vertebrates and echinoderms, occur as 5-methylcytosine in the sequence CG, M.*Sss* I can be used to imprint DNA from other sources with a vertebrate pattern. Both *Hpa* II and *Msp* I recognize the sequence 5′-CCGG-3′ but only the former is sensitive to methylation of the internal cytosine. Thus, 5′-C↓CGG-3′ sequence can be cleaved both by *Hpa* II and *Msp* I, while 5′-C↓*CGG-3′ sequence (where * indicates methylation) can be cleaved by methylation insensitive *Msp* I but not with methylation sensitive *Hpa* II. The availability of heterohypekomers is useful for mapping methylated DNA. By comparing the digestion patterns using such an isoschizomer pair, one can determine the extent to which the DNA sample is methylated. This can be used to measure percent methylation of a genomic DNA fragment or to reveal the ratio of transfected DNA to host DNA. These restriction endonucleases can thus be used to determine the positions of methylated CpG motifs in higher eukaryotic genomes.

Table 3.10 Some Common Examples of Heterohypekomers and their Sensitivity to Methylation

Enzyme	Recognition and cleavage sites (↓)	Sensitivity to methylation
Acc 65I	5′-G↓GTACC-3′	Overlapping Dcm or CpG methylation may influence DNA cleavage
Kpn I	5′-GGTAC↓C-3′	Not influenced by Dcm or CpG methylation
Apa I	5′-GGGCC↓C-3′	Overlapping Dcm or CpG methylation may influence DNA cleavage
Bsp 120I	5′-G↓GGCCC-3′	Blocked by overlapping Dcm or CpG methylation
Csp 6I	5′-G↓TAC-3′	Not influenced by CpG methylation
Rsa I	5′-GT↓AC-3′	Overlapping CpG methylation may influence DNA cleavage
Eco RII	5′-↓CCWGG-3′ (where, W = A/T)	Blocked by Dcm methylation, i.e., it will not cleave when the second cytosine in the recognition sequence is methylated to 5-methylcytosine
Mva I	5′-CC↓WGG-3′ (where, W = A/T)	Not influenced by Dcm methylation
Hpa II	5′-C↓CGG-3′	Blocked by CpG methylation
Msp I	5′-C↓CGG-3′	Not influenced by CpG methylation
Mbo I	5′-↓GATC-3′	Blocked by Dam methylated DNA, i.e., it will not cleave when the recognition sequence contains 6-methyladenosine
Sau 3AI	5′-↓GATC-3′	It will not cleave when its recognition sequence contains 5-methylcytosine
Bsp 143I	5′-↓GATC-3′	Not influenced by Dam, blocked by CpG methylation
Dpn I	5′-GA↓TC-3′	Cleaves only Dam methylated DNA
Sac I	5′-GAGCT↓C-3′	Not influenced by CpG methylation
Ecl 136II	5′-GAG↓CTC-3′	Overlapping CpG methylation may influence DNA cleavage
Sma I	5′-CCC↓GGG-3′	Blocked by CpG methylation
Cfr 9I	5′-C↓CCGGG-3′	CpG methylation may influence DNA cleavage

Exhibit 3.3 DNA methyltransferases or methylases (MTases)

DNA methyltransferases (MTases) are found in a wide variety of prokaryotes and eukaryotes. An understanding of DNA methylation is useful for troubleshooting cloning and transformation problems as well as for correct interpretation of experimental results, especially those obtained from digestions of eukaryotic genomic DNA. In addition, DNA methylation may be used as a way to limit the extent of cleavage of a target DNA. This may be useful for generating large fragments for megabase mapping, protecting specific sites, or directing digestion to a specific site. DNA MTases can be used to alter the apparent recognition specificity of restriction endonucleases. These altered specificities are unique, and increase the list of cleavage sequences that can be used by molecular biologists. Methylation of DNA at sequences that overlap the recognition sequence of a restriction endonuclease creates unique cleavage specificities *in vitro*. These modified sequences are resistant to cleavage by the restriction endonuclease. This methylation can be catalyzed by any one of the following MTases.

Cognate MTases The restrictive host strain must protect its own DNA from the potentially lethal effects of the restriction endonuclease, and so its DNA must be appropriately modified. In prokaryotes, DNA cleavage by a cognate restriction enzyme is prevented by the methylation of DNA by a sequence-specific methyltransferase, which is an integral component of every restriction–modification system. Modification involves methylation of certain bases at the recognition sequences (N^6 position of A and C5 position of C) by cognate methylases or methyltransferases. Depending on the particular type of bacteria, these methyl groups are added to adenine or cytosine bases. These methyl groups block the binding of restriction enzymes by protruding into the major groove of DNA at the binding site, thereby preventing the restriction enzyme from acting upon it. This, however, does not block the normal reading and replication of the genomic information stored in the DNA. This explains why phage that survive one cycle of growth upon the restrictive host can subsequently reinfect that host efficiently. Such restriction and modification processes can also occur whenever DNA is transferred from one bacterial strain to another. This modification differs fundamentally from mutation because it is imposed by the host cell on which the bacteriophage is growing but it is not inherited and the modification may be lost when the phage is grown in some other host.

Type I MTases The enzymes *Eco* K and *Eco* B are present in the *E. coli* strains K-12 and B, respectively. The genes *eco K* and *eco B*, encoding the *Eco* K and *Eco* B restriction–modification systems, respectively, occupy the same locus. The cognate methylases catalyze N^6 adenine methylation in the cognate recognition sequences described earlier.

Dam and Dcm MTases In prokaryotes, some genome-wide methylases may lead to prevention of restriction by restriction endonucleases, provided the specific A or C residues lying within the recognition sequence are modified. The genome-wide methylases are site-specific methylases. These can be *'de novo* methylases' that add methyl groups to unmethylated DNA at specific sites, and can therefore initiate a pattern of methylation or 'maintenance methylases' that add methyl groups to DNA that is already methylated on one strand (hemimethylated DNA), and thus perpetuate patterns of methylation through successive rounds of replication. The target sites for maintenance methylases often show dyad symmetry, so the same enzyme can methylate the nascent strand of both daughter duplexes.

In *E. coli*, adenine residues in the sequence 5′-GATC-3′ are methylated at the N^6 position (m^6N) by the enzyme DNA adenine methylase (Dam). The 5′-GATC-3′ site displays dyad symmetry, and adenine residues on both strands are methylated. The methylation thus yields a sequence symbolized as 5′-G*ATC-3′. The practical importance of this phenomenon is that a number of restriction sites may get methylated, and restriction endonucleases will not cleave them. The primary role of this modification is to allow the cell to discriminate between the parent and daughter strands following replication, when the newly synthesized strand is transiently unmethylated. This methylation also facilitates post-replicative mismatch repair.

Similarly, in *E. coli*, internal cytosine residues in the sequence 5′-CCAGG-3′ or 5′-CCTGG-3′ are methylated at the C5 position (m^5C) of the internal cytosine to form 5-methylcytosine by DNA cytosine methylase (Dcm). Their action generates the sequence 5′-C*CAGG-3′ or 5′-C*CTGG-3′. The function of this methylation system is unknown, although it might protect the genome from the restriction enzyme *Eco* RII. A DNA repair system encoded by the *vsr* gene corrects the G:T mismatches, which frequently occur within this target site caused by the deamination of 5-methylcytosine to thymine.

Eco KI sites (~1 site per 8 kbp) are much less common than Dam sites (~1 site per 256 bp) or Dcm sites (~1 site per 512 bp) in DNA of random sequence (GC=AT).

CpG Dinucleotide MTases Eukaryotic cells are not known to possess restriction/modification systems. In eukaryotes, methylation is catalyzed by CpG MTases, and the degree of methylation depends not only on the species, but also on the cell type and the developmental stage of the cell. In some mammals, methylation normally is limited to the m^5C position of cytosine in the dinucleotide CG, while in plants methylation also occurs at CNG sequences where N is any base. Up to 30% of the cytosine may be methylated in plants, while normally only 2–5% is methylated in mammals. In yeast and *Drosophila*, no methylation is detected. Other lower eukaryotes also have distinct DNA methylation patterns. The role of eukaryotic CG methylation is not fully understood. However, CpG MTases found in higher eukaryotes (e.g., *Dnmt* I) are not involved in restriction and modification. *In vivo* methylation by CpG MTase, M.*Sss* I has proven useful in overcoming methylation-based restriction during transformation, and in mapping chromatin structure *in vivo*.

Source: *Promega Notes Magazine* (1993) 42: 22.

3.7.5 *Mcr* and *Mrr* System

Mcr A, *Mcr* BC, and *Mrr* enzymes encoded by *mcr A*, *mcr BC*, and *mrr* genes, respectively, are methylation-requiring systems that attack DNA only when it is methylated at specific positions. The methylation requiring restriction systems are sequence specific. *Mcr* A cleaves *CG-methylated DNA, *Mcr* BC cleaves (A/G)*C-methylated DNA, and *Mrr* cleaves N^6-adenine methylated DNA. These three systems are of little concern when subcloning DNA from *E. coli*, since none of them cleaves Dcm or Dam-modified DNA. These systems however restrict DNA modified by CpG MTase (M.*Sss* I). In addition to restricting M.*Sss* I-modified DNA, *Mcr* A also restricts DNA modified by the *Hpa* II methylase (5′-C*CGG-3′). *Mcr* A and *Mcr* BC apparently do not distinguish between 5-methylcytosine and 5-hydroxymethylcytosine. *Mcr* BC also restricts DNA containing N^4-methylcytosine in appropriate sequences. *Mrr* also restricts DNA modified by a variety of adenine MTase and 5-methylcytosine MTases but no consensus recognition sequence has yet been deduced. *Mrr* does not restrict target sites modified by Dam, *Eco* KI, or *Eco* RI.

3.7.6 Unique Enzymes

Some enzymes do not fulfill the criteria of different types of restriction endonucleases and are considered to be unique in their action, for example:

Enzymes with Two Cleavage Sites

Certain enzymes cleave on both sides of the recognition sites (Table 3.11), such that these have four cleavage sites instead of two. Moreover, the cleavage on either side may or may not be at same distance from the recognition sequence.

Table 3.11 Some Common Examples of Enzymes that Cleave on Both sides of Recognition Site

Enzyme	Recognition and (cleavage) sites
Alf I	↓(10/12)GCA$(N)_6$TGC(12/10)↓
Bae I	↓(10/15)ACNNNNGTAYC(12/7)↓
Bcg I	↓(10/12)GCANNNNNNTCG(12/10)↓
Bsp 24I	↓(8/13)GACNNNNNNTGG(12/7)↓
Cje I	↓ (8/14)CCANNNNNNGT(15/9)↓
Cje PI	↓ (7/13)CCANNNNNNNTC(14/8)↓
Fal I	↓(8/13)AAG$(N)_5$CTT(13/8)↓

Enzymes that Recognize Two Altogether Different Sequences

Restriction enzymes that recognize two different recognition sequences have also been discovered. For example, *Taq* II recognizes two distinct sequences: 5′-GACCGA-3′ and 5′-CACCCA-3′.

Enzymes that Simultaneously Interact with Two Copies of their Recognition Sequence

Nar I, *Nae* I, and *Sac* II require simultaneous interaction with two copies of their recognition sequence before catalyzing cleavage of DNA.

Enzymes that Only Nick DNA

Some type II restriction enzymes only nick DNA. These DNA nicking endonucleases have been found in *Bacillus stearothermophilus*, for example, Nt.*Bst* NBI and Nt.*Bst* SEI (5′-GAGTCN4↓-3′), and in *Chlorella* viruses, for example, Nt.*Cvi* PII (5′-↓CCD-3′) and Nt.*Cvi* QXI (5′-R↓AG-3′). Nt stands for nicking enzyme with top strand cleavage activity. These nicking enzymes are very useful for the isothermal amplification of DNA.

3.8 APPLICATIONS OF RESTRICTION ENDONUCLEASES

The ability of restriction enzymes to reproducibly cut DNA at specific sequences has led to the widespread use of these tools in many molecular biology and biotechnology techniques, the major ones being isolation of DNA fragments for cloning experiments, restriction fragment length polymorphism and DNA fingerprinting, restriction mapping, and for studying protein–nucleic acid interactions and structure–function relationships. The application of restriction enzyme in the study of protein–DNA interactions has already been discussed in detail in Section 3.6.1, while the other applications are discussed below.

3.8.1 Isolation of DNA Fragments for Cloning Purpose (Restriction Digestion)

For molecular cloning, the first step is the isolation of DNA fragments and cutting of the cloning vector (for details see Chapter 1). The most commonly used method for a consistent and reproducible cutting is the usage of restriction enzyme. This *in vitro* cleavage catalyzed by restriction endonucleases is called restriction digestion. Further, it is possible to treat the cut ends with a variety of secondary enzymes to provide lots of flexibility with respect to subsequent cloning steps. A large number of restriction enzymes are now available commercially. Each restriction enzyme has its own optimal set of reaction conditions, which can be found on the information sheet provided by the supplier. To work with these enzymes, knowledge of its *in vitro* action is essential.

Setting Up Restriction Digestion Reaction

To set a restriction digestion reaction, the following constituents are added in a microfuge tube in the order described:

Sterile DNase-free Water A required volume of sterile DNase-free water is added to make up the final volume. Note that a single pipette tip may be used for adding water in all the reaction tubes set for restriction digestion of different DNA samples without any risk of contamination. Moreover, the addition of water and buffer in the reaction tube before restriction enzyme prevents the risk of irreversible denaturation of the enzyme.

An Appropriate Buffer Different enzymes cleave DNA optimally in different buffer systems, due to differing preferences for ionic strength and major cation. The buffer is usually commercially available as 10 X concentrate, and there is a 'unique buffer' for a particular enzyme or sometimes a 'standard buffer' can also be used. Buffer is diluted to get a final concentration of 1 X in the reaction mixture.

DNA Reliable cleavage by restriction enzymes requires DNA that is free from contaminants such as phenol or ethanol (EtOH). Excessive salt also interferes with digestion by many enzymes, although some are more tolerant of this problem. The amount of DNA to be used depends on the task at hand. Typically, a DNA band (0.5 cm wide) is easily visible under UV light in an ethidium bromide (EtBr) stained gel if it contains ~10 ng of DNA (for details see Chapter 6). For example, suppose we are digesting a plasmid that comprises 3 kbp of vector and 2 kbp of insert (i.e., total size of uncut plasmid 5 kbp). We are using *Eco* RI, and we expect to see three bands: the linearized vector (3 kbp), the 5′-end of the insert (0.5 kbp), and the 3′-end of the insert (1.5 kbp). In order to see the smallest band (0.5 kbp) we want it to contain at least 10 ng of DNA. The smallest band is $1/10^{th}$ the size of the uncut plasmid. Therefore we need to cut $10 \times 10 = 100$ ng of DNA. Then the three bands will contain 60 ng, 10 ng and 30 ng of DNA, respectively. All three bands will be clearly visible on the gel, and the band of the largest sized DNA fragment will be six times brighter than the band of the smallest sized DNA fragment. Now imagine cutting the same plasmid with *Bam* HI to linearize it, and that the enzyme cuts the plasmid only once. If we digest 100 ng of DNA in this case, the band will contain 100 ng of DNA, and will be very bright, and will probably be overloaded. Note that too much DNA loaded onto a gel is not advisable. In such a case the band appears to run fast (implying that it is smaller than it really is), and in extreme cases can mess up the electrical field for the other bands, making them also to appear in the wrong size. Too little DNA is only a problem if we are unable to see the smallest bands because these are too faint.

Restriction Enzyme It is added at the end of the reaction setup. It is always good to add water and buffer into the tube first, because adding the enzyme straight into the buffer may denature it irreversibly. Most of the restriction enzymes are commercially available at a concentration of 10–20 units (U) per μl. For a typical diagnostic digestion of 1 μg DNA, 1–2 units of restriction enzyme are sufficient. Moreover, the volume of the enzyme added should never be more than 10% of the final reaction volume. If several different DNA molecules are to be digested, it is advisable to make up a pre-mix containing buffer, water, and enzyme for the required number of reactions. This limits the number of pipetting steps. On addition of the enzyme, it is important to mix the reaction properly by sucking up and down with the pipette. Air-bubbles should be avoided during mixing, to prevent the enzyme from getting trapped at the air/liquid interface and becoming denatured. This is followed by brief centrifugation to settle down the contents of the reaction tube. The reaction tube is incubated at the recommended temperature for the desired period and analyzed by gel electrophoresis (for details see Chapter 6).

There are many variations from the typical above-mentioned reaction, for example, it is common practice to include a small quantity of a protein like bovine serum albumin (BSA) in the reaction, which often enhances the reaction, particularly, when the DNA is not exceptionally pure; a weak detergent may be included to reduce surface tension, etc. The restriction digestion reaction is performed for an hour (at least in case of plasmid DNA). If the DNA fragments produced by restriction are to be used in cloning experiments, the enzyme is destroyed so that it does not accidentally digest other DNA molecules that may be added at a later stage (for details see Section 3.8.1).

While working with restriction enzymes, some points that should always be kept in mind are (i) Restriction enzymes are expensive, so should be used with caution; (ii) Restriction enzymes are heat labile so should be kept at –20°C except for brief periods on ice or in a small freezer box at the time of use; (iii) Contamination not only spoils the experiment but also the enzyme stock, hence one enzyme should not be contaminated with another, i.e., tips should not be reused, even for pipetting the same enzyme twice. Gloves should be worn to avoid contamination as the enzyme stock may become contaminated with exonucleases from greasy hands; (iv) Micropipettes used for pipetting must be very accurate for pipetting minute quantities of reaction ingredients; and (v) The reaction should be set up quickly to avoid longer than required exposure of the enzyme to warm temperatures.

Unit of Enzyme Activity One unit (U) of restriction enzyme activity is the amount of enzyme required to fully digest 1 μg of bacteriophage λ DNA in 1 hour at 37°C, or under the conditions of the experiment.

Factors Influencing Efficiency and Extent of Restriction Digestion

It is not uncommon to have difficulties in digesting DNA with restriction enzymes. At times, the DNA does not appear to cut at all, and sometimes it cuts only partially. If the sequence is known, restriction sites can be predicted with accuracy, but in the laboratory, an enzyme may cut more often than it should, or at the wrong sites. There are a number of commonly encountered situations that influence how well restriction enzymes cut, and it is important to be aware of these for troubleshooting. Cleavage may depend on reaction conditions, and on the site being studied; cleavage of a substrate with multiple sites may yield a mixture of complete and partial digestion products. Major factors that affect restriction digestion under laboratory conditions are:

Incubation Temperature Reaction temperature is the most important reaction condition variable. The optimal temperature of incubation for most restriction enzymes is 37°C. However, there are several exceptions. Enzymes isolated from thermophilic bacteria give best results at temperatures ranging from 50 to 65°C, for example, restriction digests with *Taq* I must be incubated at 65°C to obtain maximum enzyme activity. On the other hand, some other enzymes have a very short half-life at 37°C, and it is recommended that these be incubated at 25°C.

Buffer Composition Most restriction endonucleases function adequately at pH 7.4, but different enzymes vary in their requirements for ionic strength (salt concentration), major cation (sodium or potassium), and Mg^{2+} concentration. It is also advisable to add a reducing agent, such as dithiothreitol (DTT), which stabilizes the enzyme, and prevents its inactivation. Providing the right conditions for the enzyme is very important as incorrect NaCl or Mg^{2+} concentrations may not only decrease the activity of the restriction endonuclease, but may also cause changes in the specificity of the enzyme. However, as compared to reaction temperature, the ionic strength is less stringent, and it is therefore permissible to broadly categorize restriction enzymes as requiring high, medium, or low salt. Certain enzymes can be digested in four standard buffers (A, B, C, D or 1, 2, 3, 4), but are also supplied with their own unique buffers, however, unique buffers should be preferred to ensure that the enzyme with the more specific buffer requirements work optimally and poor cleavage rates are avoided. When using restriction endonucleases in non-optimal buffers, more enzyme or longer digestion time may be needed to compensate for the slower rate of cleavage under those conditions.

A list of restriction enzymes and their optimal incubation temperatures and salt requirements is given in Table 3.12.

Enzyme Concentration and Quality Enzyme concentration can also affect the activity and specificity of action. A higher

Table 3.12 Optimal Conditions for Restriction Digestion of DNA by Some Commonly Used Restriction Enzymes

Enzyme	Incubation temperature	Salt requirement
Bam HI	37°C	Medium
Bgl II	37°C	Low
Eco RI	37°C	High
Hind III	37-55°C	Medium
Kpn I	37°C	Low
Mbo I	37°C	High
Pst I	21-37°C	Medium
Pvu II	37°C	Medium
Sau 3A	37°C	Medium
Taq I	65°C	Low
Xba I	37°C	High
Xma I	37°C	Low

concentration of enzyme is required for complete digestion, but a very high concentration (usually >100 U/μg of DNA) may lead to reduced specificity. Moreover, the volume of the enzyme added should never be more than 10% of the final reaction volume. It is also important to consider the quality of the enzyme supplied. High-quality enzymes are purified extensively to remove contaminating exonucleases and phosphatases, and tests for the absence of such contaminants form a part of routine quality control (QC) on the finished product. The absence of exonucleases is particularly important. If these contaminating exonucleases are present, these can nibble away the overhangs of cohesive ends, thereby eliminating or reducing the production of subsequent recombinants. Even where subsequent ligation is achieved, the resulting product may contain small deletions. Contaminating phosphatases can remove the terminal phosphate residues, thereby preventing ligation.

Purity of DNA Preparation The DNA should be as pure as possible. The presence of organic solvents such as phenol, EtOH, dimethyl sulphoxide, ethylene glycol, dimethyl acetamide, dimethyl formamide, and sulphalane in the DNA sample may lead to star activity or reduced specificity (for details see Section 3.8.1).

Methylation of Recognition Site Most restriction enzymes are sensitive to DNA methylation by various MTases (Exhibit 3.3). Thus, the pattern of DNA methylation can significantly affect the success of restriction digestions. The cleavage with restriction enzyme is blocked when the recognition sequence is methylated by the cognate methylase; while methylation at other bases can block cleavage, leave cleavage unaffected, or slower the rate or extent of cleavage.

Almost all strains used by molecular biologists for propagating cloned DNA are derivatives of K-12 (for details see Chapter 1), and possess two site-specific DNA MTases, Dam and Dcm, i.e., these are *dam*$^+$ *dcm*$^+$. If the recognition site of a restriction enzyme does not contain all or part of the Dam recognition sequence [5′-GATC-3′] or the Dcm sequence [5′-CC(A/T)GG-3′], it will be insensitive to Dam or Dcm methylation. DNA prepared from *dam*$^+$ cells will contain [5′-G*ATC-3′], and will not be restricted *in vitro* by enzymes blocked by this methylation pattern. DNA prepared from *dcm*$^+$ cells will contain [5′-C*C(A/T)GG-3′], and will not be restricted *in vitro* by enzymes blocked by this methylation pattern. Generally, Dam or Dcm methylation will be a problem only if the site contains one of these two sequences, or if it ends in GA, GAT, CC, or CC(A/T), and therefore potentially overlaps with a Dam or Dcm site. The Dam methylation site and target site for a restriction enzyme may be same, such that methylation of the MTase's recognition site will lead to prevention of cleavage at that site by the restriction enzyme. In DNA in which the GC content is 50%, the sites

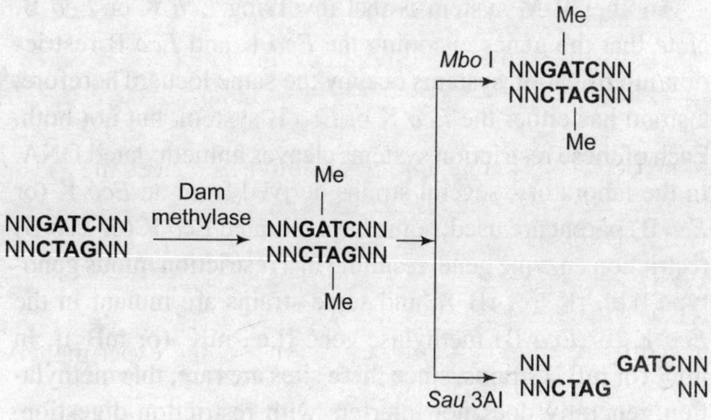

Figure 3.12 *Mbo* I is sensitive to Dam methylation while *Sau* 3AI is insensitive

for these two MTases occur every 256–512 bp on an average. Thus, some or all of the sites for a restriction endonuclease may be resistant to cleavage when isolated from strains expressing the Dcm or Dam MTases. For example, *Mbo* I and *Sau* 3AI are specific for the sequence 5′-GATC-3′, and as indicated this is precisely the sequence that is recognized by Dam MTase. Digestion of 5′-G*ATC-3′ by *Mbo* I is completely inhibited, while digestion by *Sau* 3AI is unaffected by methylation (Figure 3.12). In another example, *Cla* I recognizes 5′-AT↓CGAT-3′, which is not a substrate for Dam methylase, but when flanked by G on 5′-end or C on 3′-end, leads to generation of a Dam recognition site. The methylation at A residue then leads to inhibition of cleavage by *Cla* I (Figure 3.13). Similar blockage of restriction site is observed in case of Dcm catalyzed methylation of *Bss* KI recognition sequence 5′-CCWGG-3′ (where W is A/T), or *Acc* 65I recognition site 5′-GGTACC-3′ when preceded by CCW and followed by WGG. Thus, to cleave with a restriction enzyme that is sensitive to the Dam methylation, DNA should be purified from *dam*$^-$ *E. coli* strains, and to cleave with a restriction enzyme that is sensitive to Dcm methylation, DNA should be purified from *dcm*$^-$ *E. coli* strains. The *dam*$^-$ and *dcm*$^-$ strains should be used if it is required to cut the plasmid DNA grown in *E. coli* with a restriction endonuclease sensitive to *E. coli* K-12 methylation enzymes Dam or Dcm. Note that the derivatives of *E. coli* B, such as BL21(DE3), naturally lack Dcm.

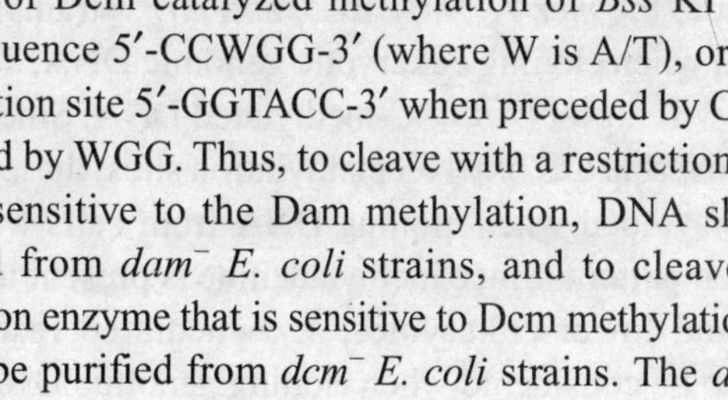

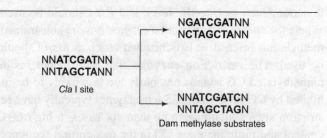

Figure 3.13 Flanking residues can convert a *Cla* I site, which is not a substrate for Dam methylase, to a Dam methylase site

Another R-M system is that involving *Eco* K or *Eco* B. Note that the genes encoding the *Eco* K and *Eco* B restriction/modification systems occupy the same locus. Therefore, a strain has either the *Eco* K or *Eco* B system, but not both. Each of these restriction systems cleaves unmethylated DNA. In the laboratory, several strains derived from an *Eco* K (or *Eco* B) parent are used; some have mutated *Eco* K (or *Eco* B) restriction enzyme gene, resulting in a restriction minus genotype [i.e., rK^- (or rB^-)], and some strains are mutant in the *Eco* K (or *Eco* B) methylase gene [i.e., mK^- (or mB^-)]. In mK^+ (or mB^+) strains, since these sites are rare, this methylation generally does not interfere with restriction digestion. DNA, if not methylated at the appropriate site before introduction into rK^+ (or rB^+) cells, will be cleaved. Restriction minus [rK^- (or rB^-)] strains may or may not produce the methylase [i.e., mK^+ or mK^- (or mB^+ or mB^-)]. If not, DNA propagated in these cells will not be protected when transformed into rK^+ (or rB^+) cells.

A further complication is that one or more of the three methylation-insensitive enzymes, *Mcr* A, *Mcr* BC, and *Mrr* are present in most commonly used strains of *E. coli*. The significance of these *Mcr* and *Mrr* restriction endonucleases is that DNA from many bacteria, and from all plants and higher animals, is extensively methylated and its recovery in cloning experiments will be greatly reduced if the restriction activity of these enzymes is not eliminated. There is no problem with DNA from *S. cerevisiae* or *D. melanogaster* since there is little methylation of their DNA. Because *Mcr* A and *Mcr* BC may cleave some sites containing m^5C in the CG dinucleotide, the *mcr* A^+, *mcr* BC^+, and mrr^+ strains should be avoided when cloning eukaryotic genomic DNA. The *mcr* BC^+ cells will digest (A/G)*C-methylated DNA. Since these sites may overlap CG or CNG methylation sites, these strains should be avoided when cloning DNA from cells with this methylation pattern. Since methyladenine is present in many bacteria and lower eukaryotes, *Mrr*-mediated restriction should also be considered when cloning genomic DNA from these organisms. The troubleshooting approach for genomic DNA is similar to that for cloned DNA, except that methylation affects the C in the CG dinucleotide and, in the case of plant DNA, the CNG trinucleotides. So the enzymes that have sites lacking CG dinucleotides or that are insensitive to methylation, such as *Hae* III, *Taq* I, and *Pst* I should be used. Where the choice of a restriction enzyme is more constrained, methylation-insensitive isoschizomers such as *Msp* I should be used. The restriction enzymes that tend to cut inside unmethylated CG islands obviously are less likely to be inhibited by CG methylation. Such enzymes typically have restriction sites containing more than six bases, a high G+C content and more than one CG in the recognition sequence, for example, *Not* I (5′-GCGGCCGC-3′).

The problem of methylation can be avoided by the use of strains in which the modifying ability is disabled by mutation. A strain completely disabled for restriction will be defective at the *hsd, mcr A, mcr BC*, and *mrr* loci. The foreign methylation pattern will be lost and the *E. coli* methylation pattern is acquired, upon replication of the clone in *E. coli*, unless the clone carries MTase activity. Once successfully introduced, clones can be moved freely among mcr^+, mrr^+ *E. coli* strains, since the methylation pattern will no longer be foreign.

Recognition Site Flanking Sequences The rate of cleavage may also be affected by the DNA sequences flanking the recognition site. The efficiency with which a restriction enzyme cuts its recognition sequence at different locations in a piece of DNA can vary 10- to 50-fold. This is apparently due to influence of sequences bordering the recognition site, which perhaps can either enhance or inhibit enzyme binding or activity.

A related situation is seen when restriction sites are located at or very close to the ends of linear fragments of DNA. Most enzymes require a few bases on either side of their recognition site in order to bind and cleave. Most of the companies dealing with restriction enzymes provide information about their 'end requirements' also.

Inactivation of Restriction Enzymes After Restriction Digestion

If the digested DNA is to be used in a cloning experiment, then the restriction enzyme should be destroyed so that it does not accidentally digest other DNA molecules that may be added at a later stage. There are different ways of 'killing' or 'inactivating' the restriction enzyme, which are as follows:

Heat Inactivation Incubation at 65°C for 20 min inactivates majority of restriction endonucleases that have optimal incubation temperature of 37°C. Enzymes that cannot be inactivated at 65°C can often be inactivated by incubation at 80°C for 20 min. Note that the temperature selected for inactivation of enzyme should not cause denaturation of DNA.

Deproteination Restriction enzymes can be removed by extraction with phenol. This procedure involves addition of an equal volume of phenol to the reaction mixture, mixing by inversion, followed by centrifugation to recover aqueous phase that contains DNA. Note that phenol denatures protein, which precipitates at the interface during centrifugation. DNA is then precipitated from the aqueous layer by EtOH precipitation (for details see Chapter 1).

Chelation of Mg^{2+} As all the restriction enzymes require Mg^{2+} ions for their activity, the addition of chelating agent such as ethylene diamine tetraacetic acid (EDTA; disodium salt) that binds Mg^{2+}, will prevent restriction endonuclease action.

Star Activity (Relaxed or Altered Specificity)

Under suboptimal reaction conditions, the restriction endonucleases are able to recognize and cleave nucleotide sequences that differ from the canonical site. In simple terms, under suboptimal conditions, the specificity of some restriction endonucleases is reduced so that only part of the normal recognition site is recognized. This altered or relaxed specificity or aberrant cutting is known as 'star' or 'relaxed' activity. The manner in which an enzyme's specificity is altered depends on the enzyme and the reaction conditions. The most common types of altered activity are single base substitutions, truncation of the outer bases in the recognition sequence, and single strand nicking. Under conditions of elevated pH and low ionic strength, *Eco* RI cleaves the sequence 5′-N↓AATTN-3′, i.e., any site which differs from the canonical recognition sequence by a single base substitution, provided the substitution does not result in an (A) to (T) or a (T) to (A) change in the central AATT tetranucleotide sequence. This reduced specificity is termed as *Eco* RI star activity (*Eco* RI*). Similarly, *Bam* HI (recognition sequence 5′-G↓GATCC-3′) under certain conditions is able to cleave the sequences 5′-N↓GATCC-3′, 5′-G↓PuATCC-3′ and 5′-G↓GNTCC-3′ also. *Sgr* AI, which recognizes and cleaves the sequence 5′-CR↓CCGGYG-3′, displays a new phenomenon of relaxation of sequence specificity. Under standard reaction conditions and in the presence of its cognate site, *Sgr* AI is capable of cleaving non-cognate sites, referred to as secondary sites, such as 5′-CR↓CCGGYN-3′ and 5′-CR↓CCGGGG-3′. Studies performed with *Sgr* AI reveal that the DNA termini generated by cleaving the cognate site are essential in the cleavage of secondary sites, as secondary sites are not cleaved on the DNA substrates that lack a cognate site. The star activity may be a general property of restriction endonucleases and any restriction endonuclease can be made to cleave non-canonical sites under certain extreme conditions. Thus, several other enzymes that have been shown to exhibit star activity include *Bam* HI, *Dpn* II, *Hae* III, *Hha* I, *Pst* I, *Pvu* II, *Sal* I, *Sca* I, *Ssp* I, and *Xba* I, etc.

Various suboptimal conditions that may lead to star activity are: (i) High pH (>8.0) or low ionic strength (<25 mM) (e.g., if one forgets to add the buffer); (ii) Glycerol concentrations (>5% v/v) (note that the restriction enzymes are usually available commercially as concentrates in 50% glycerol); (iii) Extremely high concentration of enzyme (varies with each enzyme, usually >100 U/μg of DNA); (iv) Presence of organic solvents in the reaction (e.g., EtOH, dimethyl sulphoxide, ethylene glycol, dimethyl acetamide, dimethyl formamide, sulphalane); (v) Substitution of Mg^{2+} with other divalent cations (e.g., Mn^{2+}, Cu^{2+}, Co^{2+}, Zn^{2+}); and (vi) Prolonged incubation times.

The relative significance of each of these altered conditions is dependent on the enzyme used. For example, *Eco* RI is much more sensitive to elevated glycerol concentrations than to elevated pH.

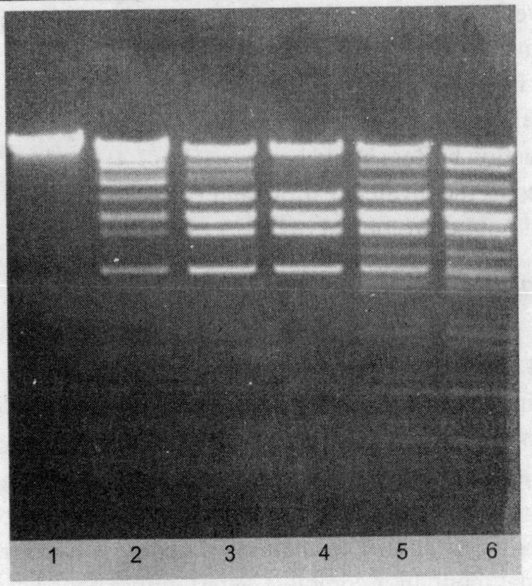

Figure 3.14 A typical electrophoresis pattern during incomplete digestion and enzyme star activity. [Lane 1: λ DNA; Lane 2: λ DNA incubated for 1 hour with 0.15 U of *Eco* RI (incomplete cleavage); Lane 3: λ DNA incubated for 1 hour with 0.4 U of *Eco* RI (incomplete cleavage); Lane 4: λ DNA incubated for 1 hour with 1 U of *Eco* RI (complete digestion); Lane 5: λ DNA incubated for 16 hours with 40 U of *Eco* RI (star activity); Lane 6: λ DNA incubated for 16 hours with 70 U of *Eco* RI (star activity)]

In most practical applications of restriction enzymes, star activity is not desirable. Any tendency of a restriction enzyme to exhibit star activity is indicated both in the product description, and in the certificate of analysis supplied with each enzyme. Star activity and incomplete DNA digestion result in atypical electrophoresis patterns (Figure 3.14). Incomplete DNA digestion results in additional low intensity bands above the expected DNA bands on the gel. No additional bands below the smallest expected fragment are observed. These additional bands disappear when the incubation time or amount of enzyme is increased. On the contrary, star activity results in additional DNA bands below the expected bands, and no additional bands above the largest expected fragment. With the increase of either the incubation time or the amount of enzyme, these additional bands become more intense, while the intensity of the expected bands decreases.

Star activity is completely controllable in a vast majority of cases, and is generally not of much concern during restriction digestion. It can, however, be inhibited by taking care of the following points: (i) For complete digestion, only required amount of the enzyme must be used. This avoids over-digestion, and reduces the final glycerol concentration in the reaction; (ii) The reaction must be free of any organic solvent such as alcohols that might be present in the DNA preparation; (iii) A standard 50 μl reaction volume should be used as it prevents evaporation during incubation (evaporation results in an increased glycerol concentration); (iv) The ionic strength of the reaction buffer should be raised to 100–150 mM (provided the enzyme is not inhibited by high salt); (v) The pH of the reaction buffer should be lowered to 7.4; (vi) Mg^{2+} should be used as the divalent cation; and (vii) The digestion period should not be prolonged.

Partial Digestion and its Significance

Either partial or complete digestion of the DNA may be required for different purposes. Partial digestion, i.e., cleavage at limited number of sites, with a four base-cutter, *Sau* 3AI, is a common practice in the construction of a genomic library, as it generates fragments compatible to that generated by *Bam* HI (its recognition site is common in vectors and nests *Sau* 3AI site). The chances of getting a large fragment (that may contain a complete gene) with a four base-cutter are very meagre if complete digestion is performed. Thus, partial digestion is performed to obtain large fragments, or fragments of various sizes (Figure 3.15). Alternatively, large fragments are also obtained with the application of rare base-cutters. Partial digestion can be achieved by any of the following ways: (i) By reducing the incubation period so that enzyme does not get time to cut all restriction sites within the DNA molecule; (ii) By reducing the amount of enzyme in the reaction mixture, so that there is not sufficient enzyme to restrict all sites; and (iii) By incubating at low temperature (performing the digestion at 4°C instead of 37°C) so that the activity of the enzyme is limited.

Double Digestion

To digest the DNA with two enzymes, it should be ensured that the buffers for both the enzymes are compatible, so that

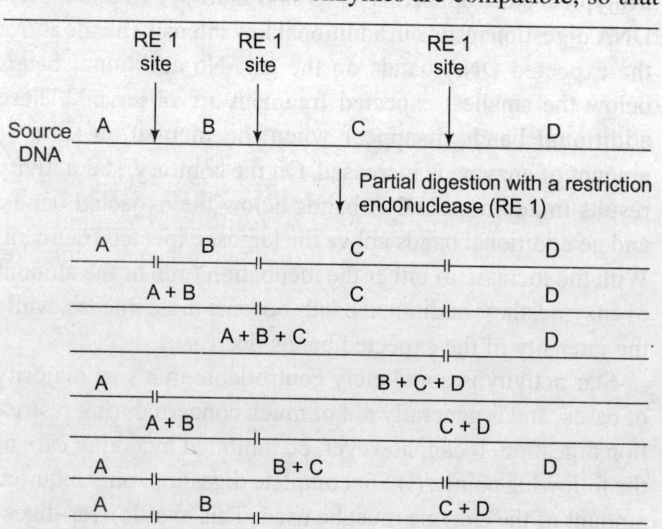

Figure 3.15 Partial digestion of DNA [In some of the DNA molecules, the restriction endonuclease cuts at all target sites. In other molecules, cleavages do not occur at all target sites. The desired outcome is a sample with DNA fragments of all possible lengths]

the digestion can be performed in a single tube (simultaneous digestion). At the same time, 10% volume rule should never be overlooked. However, for several enzymes that work under special conditions, a single buffer cannot be employed. This is due to inability of a single buffer to satisfy the buffer requirements of both the enzymes. In simple terms, the two enzymes work in incompatible buffers. While some enzymes require bovine serum albumin to be added into the mixture, some require weak detergents such as triton X-100 to reduce surface tension, and others require incubation temperatures other than 37°C (e.g., 50°C). In the case of reaction incompatibility (buffer composition or incubation temperature incompatibility), double digestion can also be performed in series (sequential digestion).

Thus, there are at least three ways to perform double digestion:

Digest with Both Enzymes in the Same Buffer (Simultaneous Digestion) Simultaneous digestion can be performed when the buffer and incubation temperature requirements for both enzymes are the same. Simultaneous digestion with two different enzymes may also be performed in a single standard buffer even though the buffer is not optimal for any one of these enzymes, but the cleavage rate of the enzyme for which the conditions are not optimal is sufficient. Enzyme manufacturer catalogues usually contain a reference table indicating the best single buffer for conducting specific double digests.

Cut with One Enzyme, Alter the Buffer Composition or Incubation Temperature and Cut with the Second Enzyme (Sequential Digestion) If the two enzymes have different salt requirements, first cleavage is carried out with the restriction endonuclease that requires the lower salt concentration. When the digestion with the first enzyme is complete, the salt concentration of the reaction is adjusted (using a small volume of a concentrated salt solution) to approximate the reaction conditions of the second restriction endonuclease, which is then added to perform the second digestion reaction.

Similarly, while working with two enzymes having different optimal incubation temperature requirements, first digestion is performed with the enzyme requiring lower optimal temperature (say, 25°C). This is followed by the addition of the second enzyme requiring higher optimal temperature (say, 37°C), and the temperature is raised to 37°C for digestion with the second enzyme. This allows the second enzyme to work, while the first enzyme is heat killed between the digests.

Change Buffers Between Digestions with Two Enzymes (Sequential Digestion) Digestion with two enzymes requiring totally incompatible buffers is done in sequential steps, i.e., after digestion with one enzyme, the DNA is recovered by EtOH precipitation, followed by resuspension in the buffer appropriate for the second enzyme.

For double digestion of 1–2 hour duration, the following should be kept in mind: (i) If BSA is required by either enzyme, it should be added to the double digest reaction as BSA does not inhibit any restriction enzyme; (ii) To minimize the possibility of star activity, the final concentration of glycerol in any reaction should be less than 5% v/v. However, to avoid this reduced specificity, an increase in total reaction volume may be necessary; (iii) Double digestion should not be performed overnight to avoid possible star activity; and (iv) Some enzymes such as *Bsg* I require SAM, which is an additive that does not have any negative effect on the activity of other enzymes.

Analyzing the Results of Restriction Digestion

A restriction digest results in a number of DNA fragments, the sizes of which depend on the exact positions of the recognition sequences for the restriction endonuclease in the original molecule. Clearly, if restriction endonucleases are to be used in gene cloning, a way of determining the number and sizes of the fragments is needed. It is fairly easy to determine if a DNA molecule has been cut, by simply testing the viscosity of the solution. This is because the cleavage is associated with the generation of smaller DNA fragments, and a decrease in viscosity (note that larger DNA molecules are more viscous than smaller DNA molecules). However, working out the number and sizes of the individual cleavage products is more difficult. In fact, for several years this was one of the most tedious aspects of experiments involving DNA. Eventually, the problem was solved in the early 1970s when the technique of gel electrophoresis to visualize DNA was developed (for details see Chapter 6).

Digestion of Agarose-embedded DNA

Mechanical shear forces can cause ds DNA breaks during manipulation of large DNA molecules (containing several million base pairs) in solution. To avoid this fragmentation, DNA can be embedded in an agarose matrix. Intact cells are first immobilized in agarose, and then treated to disrupt their cell walls and remove cellular protein (for details see Chapter 6). Subsequently, the DNA containing agarose plug is manipulated in much the same way as DNA in solution. Restriction digestion of agarose-embedded DNA has proved to be a valuable tool in the physical mapping of chromosomes. Most endonucleases can function in the presence of agarose. Since diffusion within agarose is more limited than in liquid reactions, complete DNA digestion generally requires longer incubation times and higher concentrations of restriction endonucleases. As a general rule, 20 units for 4-hours incubation or 5 units for 16 hours incubation are used. Most restriction endonucleases do not cut to completion in less than 2 hours, regardless of the number of units present. The extended times, together with the larger size of the target DNA, increase the necessity for highly purified restriction endonucleases lacking contaminating nuclease activity.

3.8.2 Restriction Fragment Length Polymorphism (RFLP) Analysis

RFLP refers to inherited differences in cleavage sites for restriction enzymes that result in differences in the lengths of the fragments produced by cleavage with the relevant restriction enzymes. This polymorphism is caused due to sequence variation that results in a restriction endonuclease site being changed or lost. In this technique, genomic DNA is isolated and digested with restriction enzymes, and the resulting fragments are separated according to size in an agarose gel followed by transfer on a membrane. The digested DNA on the membrane is allowed to bind to a radioactively or fluorescently labeled probe, which targets specific sequences that are bracketed by restriction enzyme sites. The size of these fragments varies in different individuals, generating a 'biological bar code' of restriction enzyme digested DNA fragments, a pattern that is unique to each individual. Thus, RFLP analysis is widely used for identification of individuals (humans and other species). RFLP is also used for genetic mapping to link the genome directly to a conventional genetic marker. RFLP is also used to infer phylogenetic relationship in both plants and animals, to prepare chromosome maps, and to study the inheritance or linkage relationship. The details of the process are discussed in Chapter 18.

3.8.3 Restriction Mapping

Once a novel clone has been isolated by the procedure outlined in Chapter 1, the first stage of analysis is the creation of a restriction map, which is a physical map showing the relative positions of the restriction sites for a number of different restriction endonucleases.

Applications of Restriction Mapping

Generating a restriction map is usually the first step in characterizing an unknown DNA, and a prerequisite to manipulate it for other purposes. Only when a restriction map is available, the sites for subcloning can be identified, and correct restriction endonucleases can be selected for cleavage. The pattern of restriction fragments generated may also be used to identify overlapping clones for contig mapping (for details see Chapter 19). Restriction maps are also used to determine the relationships between two different species at the molecular level. Another application of restriction maps is to compare mitochondrial DNA (mtDNA) of eukaryotic organisms. Besides being smaller in size as compared to nuclear genome, an added benefit with mtDNA is that it undergoes changes by mutation about 10 times faster than the nuclear genome thereby making it possible to sort out phylogenetic relationships between very closely related species, or even between different populations of the same species. Another use of restriction enzymes can be to find specific single nucleotide polymorphisms (SNPs). If a restriction enzyme can be found such that it cuts only one possible allele of a section of DNA (i.e., the alternate nucleotide of the SNP causes the restriction site to no longer exist within the section of DNA), this restriction enzyme can be used to genotype the sample without completely sequencing it. The sample is first run in a restriction digest to cut the DNA, and then gel electrophoresis is performed. If the sample is homozygous for the common allele, the result will be two bands of DNA, because the cut will have occurred at the restriction site. If the sample is homozygous for the rarer allele, the sample will show only one band, because it will not be cut. If the sample is heterozygous at that SNP, there will be three bands of DNA.

Generation of Restriction Maps

The resolution of the map depends upon the frequency of the restriction site in the DNA fragment, which reflects both its size and base composition. Typically, relatively inexpensive restriction enzymes that cleave DNA infrequently (e.g., those with 6 bp recognition sites) are used to generate restriction maps of small DNA molecules such as plasmids, PCR fragments, and λ inserts. For larger vectors, such as cosmids and artificial chromosomes, rare cutters are useful. This is termed long-range restriction mapping. Restriction maps of entire chromosomes can be prepared using such enzymes, although the DNA fragments produced must be separated by pulsed field gel electrophoresis (PFGE), or similar methods (for details see Chapter 6). However, because of the generation of so many fragments from the genome of a cell, restriction mapping is more practical for comparing smaller segments of DNA, usually a few thousand nucleotides long.

The DNA to be restriction mapped is usually contained within a well-characterized plasmid or bacteriophage vector, the sequence for which is known. In fact, there are usually multiple known restriction sites immediately flanking the uncharacterized pure DNA, which facilitates making the map. In the following discussion, it is assumed that the unknown DNA has been inserted into a plasmid vector, but the principles can readily be applied to other situations. Thus, two commonly used methods for generation of restriction maps are described as follows:

Restriction Map Generated by Digestion of Unknown DNA with Multiple Restriction Enzymes The most straightforward method for restriction mapping is to digest samples of the plasmid with a set of individual enzymes, and with pairs of those enzymes. The digests are then resolved on an agarose gel to determine sizes of the fragments generated.

Let us consider a plasmid vector (6,000 bp size) that contains a 3,000 bp fragment of unknown DNA (Figure 3.16). Within the vector, immediately flanking the unknown DNA are unique recognition sites for the enzymes *Kpn* I and *Bam* HI. Suppose separate digestion with *Kpn* I yields two fragments of the sizes 900 bp and 8,100 bp (vector plus the remaining 2,100 bp of the unknown DNA). Since there is a single *Kpn* I site in the vector, the presence of a 900 bp fragment indicates that there is also a single *Kpn* I site in the unknown DNA, and that it is 900 bp from the *Kpn* I site in the vector. Suppose, digestion with *Bam* HI yields three fragments of the sizes 600, 2,200, and 6,200 bp (vector plus 200 bp of unknown DNA). The presence of 600 and 2,200 bp fragments indicate that there are two *Bam* HI sites in the unknown DNA. Thus, it can be deduced that one *Bam* HI site is 2,800 bp (600 + 2,200 bp) from the *Bam* HI site in the vector. The second *Bam* HI site can be in either 600 or 2,200 bp from the *Bam* HI site in the vector. At this point, there is no way to know which of these alternative positions is correct. Thus, the digestion with individual enzymes indicates the presence or absence of recognition sites, and the number of sites for a particular enzyme in the unknown DNA.

To determine the exact position of the second *Bam* HI site, the plasmid is digested with *Kpn* I and *Bam* HI together. Now suppose, this double digest yields fragments of 600, 900, 1,300, and 6,200 bp. The 600 bp fragment is the same as that obtained by digestion with *Bam* HI alone. The 900 and 1,300 bp fragments give information that *Kpn* I cuts within the 2,200 bp *Bam* HI fragment. If the process outlined above is conducted with a larger set of enzymes, a much more complete map will result. In essence, the double digests are used to determine correct order and orientation of the fragments.

The success of this technique depends upon obtaining complete digestion of the DNA with each of the enzymes used, as partial digestion will yield fragments that are ultimately a great source of confusion. One way to avoid this problem is to add up the estimated sizes of all the fragments in each lane, which should roughly be similar to that of the intact DNA. Another source of confusion is the generation of two fragments of roughly the same size that appear as single fragment on an agarose gel. This situation is often suspected by observing an abnormally bright fragment on the gel, or by a fragment being broader than expected.

Restriction Map Generated by Partial Digestion of End-labeled DNA If a fragment of DNA is labeled with a radioisotope on only one end, it can be partially digested with restriction enzyme(s) to generate labeled fragments that directly reveal the locations of cleavage sites.

Let us consider an example (Figure 3.17), where the fragment of DNA to be mapped for *Kpn* I sites is contained within a plasmid (6,000 bp), and is flanked by restriction sites that are not present in the fragment itself, say for example, *Not* I and *Eco* RI. First, the plasmid is digested to completion with *Eco* RI, and then the ends of the linearized plasmid are radiolabeled. Secondly, the labeled DNA is digested with *Not* I

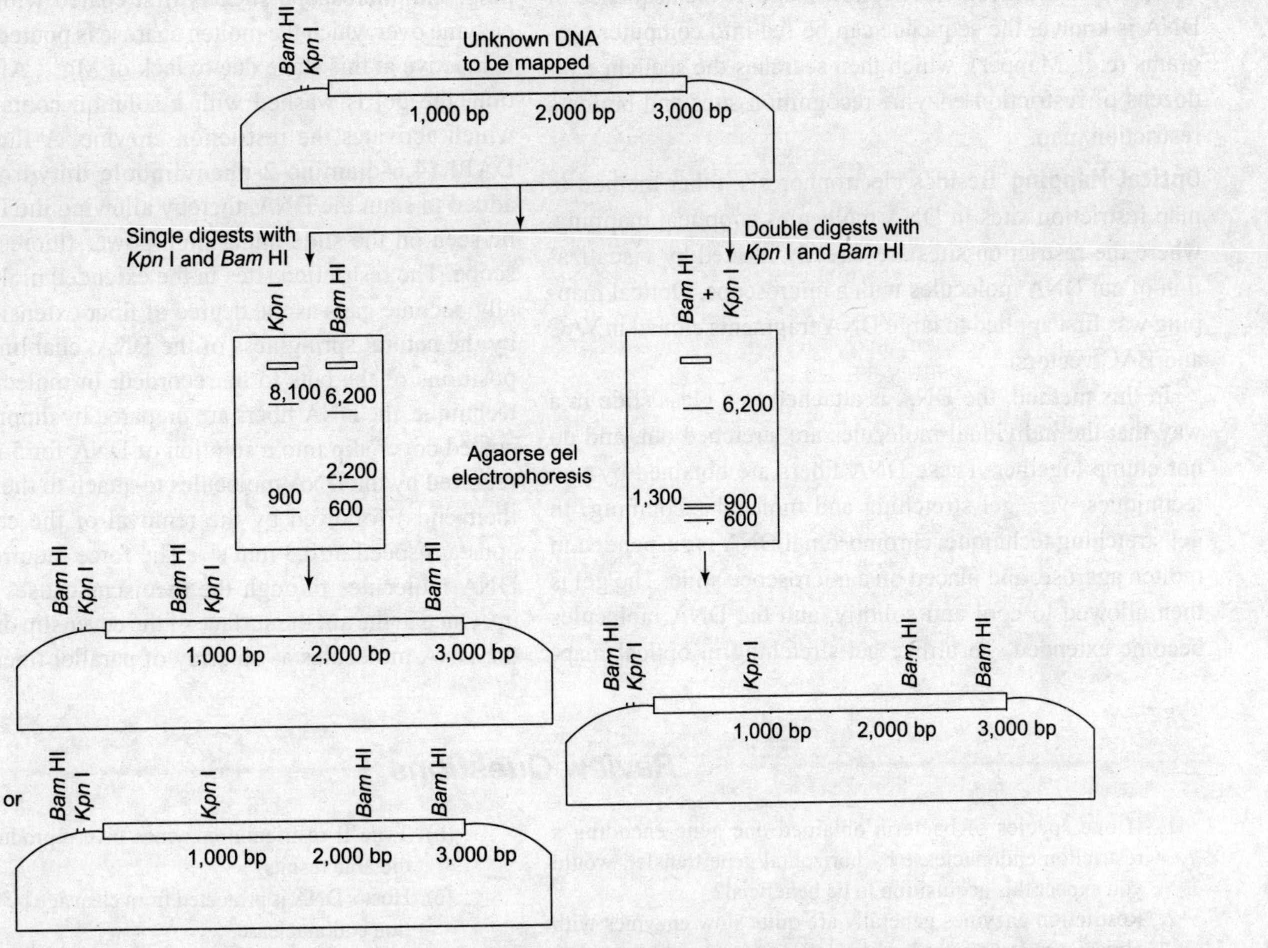

Figure 3.16 Restriction mapping by multiple restriction enzymes catalyzed digestion of unknown DNA

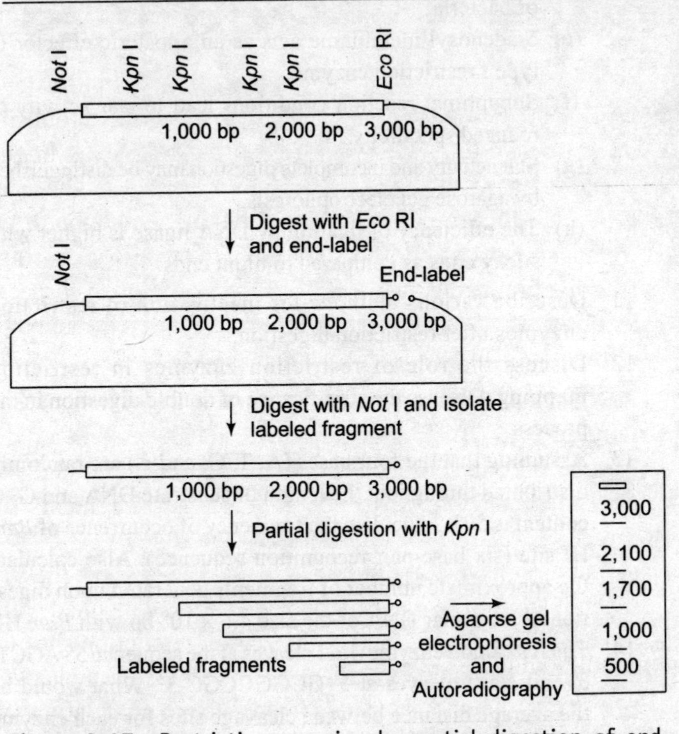

Figure 3.17 Restriction mapping by partial digestion of end-labeled DNA

and an agarose gel electrophoresis is performed. The fragment of interest is isolated, which now is labeled on only one end. This end-labeled fragment of DNA is the substrate to be used for partial digestion with *Kpn* I. Suppose in addition to the full-length fragment, four additional radiolabeled fragments are generated. The labeled partial digestion products are separated on an agarose gel followed by autoradiography to visualize the sizes of the labeled fragments, and hence the positions of all the *Kpn* I recognition sites.

A single preparation of end-labeled DNA can be used for mapping recognition sites for several different restriction enzymes, making this an efficient means of generating comprehensive maps. There are, however, at least two potential problems that can cause confusion in interpretation of results. For a given enzyme, some recognition sites can be cleaved much less efficiently than others. The cause of this problem is usually not known, but it can result in wrong mapping of certain sites. It is difficult to map sites near the ends of the fragment. For this reason, it is often best to perform the procedure twice, with preparations of fragment labeled at opposite ends, for example, in this case, one preparation labeled at the *Eco* RI end and the other labeled at the *Not* I end.

Computer Generated Restriction Maps If the sequence of DNA is known, the sequence can be fed into computer programs (e.g., Mapper), which then searches the sequence for dozens of restriction enzyme recognition sites and builds a restriction map.

Optical Mapping Besides electrophoresis, other method to map restriction sites in DNA molecules is optical mapping, where the restriction sites are directly located by visualization of cut DNA molecules with a microscope. Optical mapping was first applied to large DNA fragments cloned in YAC and BAC vectors.

In this method, the DNA is attached to a glass slide in a way that the individual molecules are stretched out, and do not clump together. These DNA fibers are obtained by two techniques, *viz.*, gel stretching and molecular combing. In gel stretching technique, chromosomal DNA is suspended in molten agarose, and placed on a microscope slide. The gel is then allowed to cool and solidify, and the DNA molecules become extended. To utilize gel stretching in optical mapping, the microscope slide is first coated with a restriction enzyme over which the molten agarose is poured. The enzyme is inactive at this stage due to lack of Mg^{2+}. After solidification, the gel is washed with a solution containing $MgCl_2$, which activates the restriction enzyme. A fluorescent dye, DAPI (4,6-diamino-2-phenylindole dihydrochloride), is added to stain the DNA, thereby allowing the DNA fibers to be seen on the slide under high-power fluorescence microscope. The restriction sites in the extended molecules gradually become gaps as the degree of fiber extension is reduced by the natural springiness of the DNA, enabling the relative positions of the cuts to be recorded. In molecular combing technique, the DNA fibers are prepared by dipping a silicone-coated cover slip into a solution of DNA for 5 min (the time required by the DNA molecules to attach to the cover slip by their ends) followed by the removal of the cover slip at a constant speed of 0.3 mm s^{-1}. The force required to pull the DNA molecules through the meniscus causes them to line up. Once in the air, the surface of the cover slip dries, retaining the DNA molecules as an array of parallel fibers.

Review Questions

1. If one species of bacteria obtained one gene encoding a restriciton endonuclease by horizontal gene transfer, would you expect this acquisition to be beneficial?
2. Restriction enzymes generally are quite slow enzymes with typical turnover numbers of 1 s^{-1}. Suppose that restriction enzymes were faster with higher turnover numbers, say 10^5 s^{-1}, and had similar levels of specificity, would this increased rate be beneficial to host cells?
3. Type I and III restriction endonucleases are not used in gene cloning experiments. Comment.
4. The restriction endonucleases are specific in recognition and cleavage of ds DNA. Comment.
5. What do you understand by the terms isoschizomers, neoschizomers, heterohypekomers, and isocaudomers? Give at least five examples of each. Also highlight the advantages of each in gene cloning.
6. Suppose we have isolated the gene of interest by cleavage with *Bam* HI, and we have cleaved the vector with *Sau* 3AI, can we ligate the two without the use of linkers or adaptors? If yes, can the hybrid site (*Bam* HI/*Sau* 3AI) generated be recleaved with *Sau* 3AI or *Bam* HI? Explain.
7. Describe the significance of *mcr*⁻ and *mrr*⁻ strains of *E. coli* in gene cloning experiments.
8. What is partial digestion and how is it performed? Explain its significance.
9. Enumerate various factors affecting restriction digestion. Also mention the optimal conditions for restriction digestion.
10. Explain the following:
    (a) Sterile water is added first to the reaction tube being set for a restriction digestion reaction.
    (b) Type II restriction enzymes give reproducible and predictable results.
    (c) Host's DNA is protected from cleavage by its own restriction endonuclease.
    (d) Restriction endonucleases act as self-protection system of bacteria.
    (e) *S*-adenosyl methionine acts as an allosteric effector of type I restriction enzyme.
    (f) Suboptimal reaction conditions lead to star activity or reduced specificity.
    (g) Star activity and incomplete digestion may be distinguished by agarose gel electrophoresis.
    (h) The efficiency of ligation by DNA ligase is higher with sticky ends as compared to blunt ends.
11. Describe various methods for inactivation of restriction enzymes after restriction digestion.
12. Discuss the role of restriction enzymes in restriction mapping. Discuss the significance of double digestion in the process.
13. Assuming that the four bases (A, T, G, and C) are randomly distributed throughout the length of substrate DNA and G+C content is 50%, calculate the frequency of occurrence of *Bam* HI site (six base pair recognition sequence). Also calculate the approximate number of fragments generated upon digestion of a circular DNA of the size 4.6×10^6 bp with *Bam* HI.
14. The restriction enzyme *Alu* I cleaves at the sequence 5′-AGCT-3′ and *Not* I cleaves at 5′-GCGGCCGC-3′. What would be the average distance between cleavage sites for each enzyme on ds DNA?

15. Given that a typical viral genome is 45,000 bp long, can a restriction endonuclease that cleaves at particular 10 bp target site be useful in protecting cells from viral infections? Explain.
16. In order to characterize a newly discovered plasmid, a restriction map is to be prepared. The sizes of the restriction fragments generated by using different enzymes individually and in combination, as determined by agarose gel electrophoresis are given in the table below. From the data, construct the restriction map of the plasmid.

Restriction enyzme(s)	Fragment sizes (kbp)
Eco RI	5.4
Hind III	2.1, 1.9, 1.4
Sal I	5.4
Eco RI + *Hind* III	2.1, 1.4, 1.3, 0.6
Eco RI + *Sal* I	3.2, 2.2
Hind III + *Sal* I	1.9, 1.4, 1.2, 0.9

17. Construct a restriction map from the given electrophoretic separation patterns obtained by single and double restriction digestions. The sizes of the restriction fragments generated (in bp) are indicated in the gel.

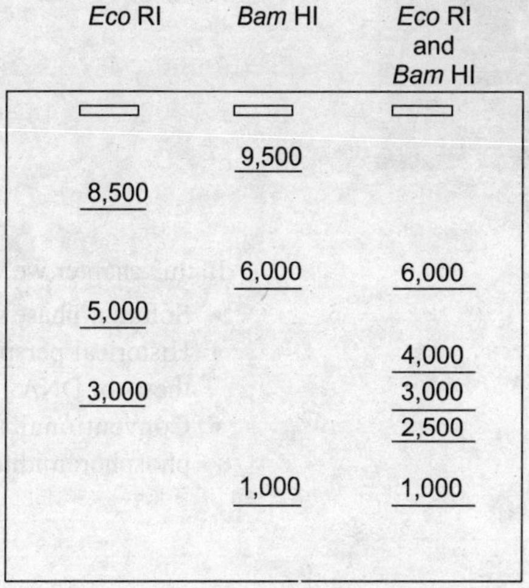

Reference

Smith, H.O. and D. Nathans (1973). A suggested nomenclature for bacterial host modification and restriction systems and their enzymes. *J. Mol. Biol.* **81**:419–423.

Chemical Synthesis of Oligonucleotides

Key Concepts

In this chapter we will learn the following:

- Solution-phase vs. solid-phase synthesis
- Historical perspective of chemical synthesis of DNA
- Conventional solid-phase synthesis: phosphoramidite method
- DNA synthesizers (automated solid-phase synthesis)
- Systematic evolution of ligands by exponential enrichment (SELEX)
- DNA chips

4.1 INTRODUCTION

The ability to chemically synthesize a strand of DNA with a specific sequence of nucleotides easily, inexpensively, and rapidly has contributed significantly in molecular cloning and characterization of DNA. Chemically synthesized, ss DNA oligonucleotides are used for assembling whole genes, and facilitating gene cloning, amplifying specific DNA sequences, introducing mutations into cloned genes, screening gene libraries, analyzing single nucleotide polymorphism, and sequencing DNA (for details see Chapters 1, 7, 13, and 16–19).

The basic strategy of oligonucleotide synthesis is analogous to that of polypeptide synthesis. It involves the repetitive formation of an ester linkage between an activated phosphoric acid function of one nucleotide, and the hydroxyl group of another nucleoside or nucleotide, thus forming the characteristic phosphodiester bridge. The major problem is that deoxyribonucleotides are very reactive molecules, having primary and secondary hydroxyl groups, a primary amino group, and a phosphate group. Thus, a suitably protected nucleotide is coupled to the growing end of the oligonucleotide chain. Consequently, the protecting groups are removed such that the chemistry involved does not result in scission or alteration of the phosphodiester backbone, the furanose rings, the sugar-purine/pyrimidine bonds or the bases themselves. The process is repeated until the desired oligonucleotide has been synthesized. Today, chemical synthesis is standardized to such an extent that the development of automated synthesis equipment has already led to commercially available 'gene synthesizing machines'.

Another miracle in the field of DNA synthesis is the development of DNA chips or microarrays with attached DNA that is either synthesized *in situ* (*on-chip*) or robotically deposited. These DNA chips are finding applications in gene discovery, detection of single nucleotide polymorphisms, monitoring patterns of gene expression, DNA resequencing, genotyping on a genomic scale, and identifying infectious agents.

This chapter highlights various processes used for the synthesis of oligonucleotides.

4.2 SOLUTION-PHASE vs. SOLID-PHASE SYNTHESIS

The synthesis of oligonucleotides can be done both in solution-phase and solid-phase. Solution-phase synthesis refers to the synthesis in solution form, while in solid-phase synthesis, one of the ends of the growing oligonucleotide is anchored or immobilized to solid support. The solid-phase synthesis simplifies manipulative (synthesis and purification) procedures. Thus, several advantages offered by solid-phase synthesis over solution-phase synthesis are: (i) Packing the support into a column, and passing the reagents through the column allows mechanization and automation of the synthesis process; (ii) All the reactions can be conducted in one reaction vessel. The appropriately blocked mononucleotides are added sequentially, and the reagents from one reaction step can be readily washed away before the reagents for the next step are added; (iii) The reagents can be used in excess in an attempt to drive the reactions to completion. After each reaction, the excess reagents can be removed simply by washing, thereby eliminating the need for purification steps (e.g., chromatography) between base additions; (iv) The different solubilities of the components can be neglected because the reactions take place in a heterogeneous phase; (v) Several oligonucleotides can be synthesized simultaneously in the same vessel; (vi) Individual reactions can be repeated as often as desired in order to increase yields; (vii) Another significant advantage is the ease of purification. (viii) Immobilization cuts out time-consuming purification steps following each cycle of condensation, and the filtration steps can be replaced by a simple washing procedure. Unreacted products are flushed out to waste. Thus, the growing biopolymer chains can be easily separated from other components of the reaction mixture as the oligonucleotides (products) are attached to the solid surface, while starting materials, excess reagents, and byproducts that are not attached to the support are removed simply by washing. These simple washing procedures allow the process to be automated; (ix) After synthesis, linkage to support is easily hydrolyzed by concentrated ammonium hydroxide to release the product. The base and phosphate protecting groups are then removed, and the full-length product is easily isolated; and (x) As compared to solution-phase synthesis, solid-phase synthesis allows synthesis of larger oligonucleotides.

4.3 HISTORICAL PERSPECTIVE OF CHEMICAL SYNTHESIS OF DNA

With the information on enzymatic synthesis of DNA, and the protocol for the chemical synthesis of polypeptides, several investigators began to try ways to accomplish chemical synthesis of DNA.

In nature, the formation of the phosphodiester linkages in DNA is catalyzed enzymatically by DNA polymerase in a reaction:

$$\text{dNTP} + (\text{dNMP})_n \rightarrow (\text{dNMP})_{n+1} + \text{PP}_\text{i}$$

i.e., the addition of a deoxyribonucleoside triphosphate to a growing deoxyribonucleoside monophosphate polymer by a DNA polymerase in the presence of Mg^{2+} results in the lengthened $(n + 1)$ polymer and a pyrophosphate. The above reaction is reversible, and favored in forward direction due to the pyrophosphorilytic activity of DNA polymerase.

Moreover, by that time, the chemistry involved in the chemical synthesis of polypeptides was also well established, which utilized a foundation of a single amino acid bound to a solid support (a resin bead), and a protected amino acid in the presence of a condensing agent, dicyclohexylcarbodiimide (DCC). The reaction would link the protected amino acid to the solid support, and the resulting dipeptide was then deprotected for the addition of the next amino acid. In this way, the synthetic polypeptide grew through a repeated cycle of steps.

As compared to enzymatic synthesis, the methods for chemical synthesis of oligonucleotides suffers from some drawbacks such as lesser efficiency than enzymatic synthesis; slower and lower yields; synthesis of single stranded sequences; size restricted to <~150 bases; costlier synthesis and purification. Though not as efficient as enzymatic synthesis, chemical synthesis of oligonucleotides is still a method of choice as it offers certain advantages as well. Thus, during chemical synthesis of oligonucleotides: (i) There are no constraints on sequence or structures; (ii) Cost-effective synthesis of DNA and RNA molecules is achieved in-house (cost is far over-weighed by benefits); and (iii) The modifications can confer new or enhanced functionality.

Besides, synthetic oligonucleotides are very valuable as reagents, diagnostics, DNA or RNA based therapeutics, and several other applications (for details see Section 4.4.14). Owing to various advantages, several researchers developed different strategies for oligonucleotide synthesis. These are:

Phosphodiester Method In this method, the 3′- and 5′-hydroxyl groups of the deoxyribose, and reactive amino groups of bases are suitably protected.

Triester Methods It includes two methods:

Phosphotriester method In this method, besides blocking the 3′- and 5′-hydroxyl groups of deoxyribose sugar, and reactive amino groups of bases, another protecting group is used for blocking the hydroxyl group at the internucleotide bond.

Phosphite-triester method It is basically a triester method, and the strategies for using protecting procedure are the same as for phosphotriester method.

However, in the pertinent scientific literature the term phosphotriester method is reserved exclusively for methods utilizing compounds of pentavalent phosphorus and phosphite-triester method for utilizing compounds of trivalent phosphorus.

As the above-mentioned approaches are associated with several disadvantages, only a brief outline of these approaches is given in the following sections.

4.3.1 Phosphodiester Method

Har Gobind Khorana and his group at the Institute for Enzyme Research at the University of Wisconsin developed the first practical technique for DNA synthesis, the phosphodiester approach (Figure 4.1). The scheme they developed involved three different protecting groups surrounding the nucleotide. The integrity of the ring structure of the base was protected by a benzoyl or an isobutyryl group, the oxygen on the 3′ carbon of the sugar was prevented from forming a reactive hydroxyl until needed by the addition of an acetyl group, and the 5′ carbon was blocked by a monomethoxytrityl (MMTr) group. The trityl group was synthesized on a polystyrene bead. The first step was attachment of a benzene ring on the polystyrene bead. Next, a second ring was attached via a Friedel–Crafts reaction (acetylation or alkylation of aromatic compounds by aluminum chloride). This structure was then exposed to Grignard reagent. Finally, the nearly completed trityl group was treated with acetyl chloride resulting in a polystyrene methoxytritylchloride species that could be easily attached to the

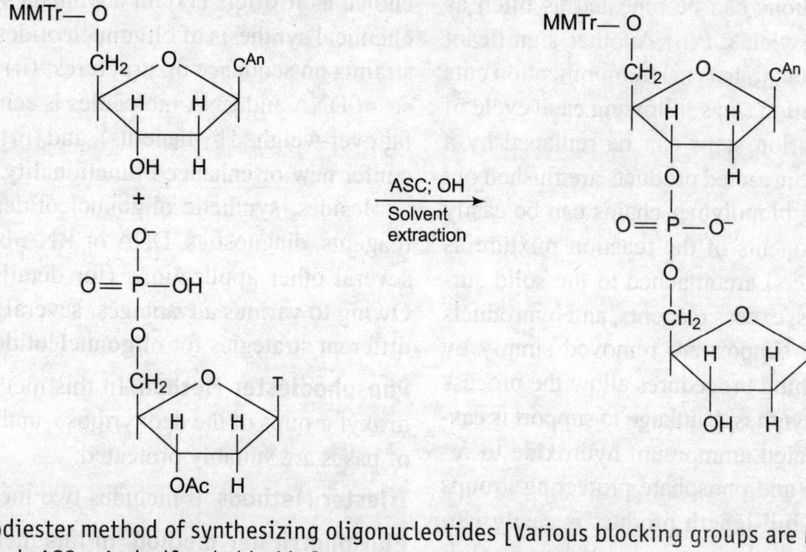

Figure 4.1 The phosphodiester method of synthesizing oligonucleotides [Various blocking groups are MMTr = Monomethoxytrityl; Ac = Acetyl; An = Anisoyl; ASC = Arylsulfonyl chloride.]

5′-carbon of the pentose sugar. The nucleotide to which the trityl was attached became the anchor of the synthesis. The oxygen on the 3′-carbon of the sugar was deprotected leaving the reactive hydroxyl group ready for coupling, and protected nucleotide without a trityl group was added via the condensing reaction of dicyclohexylcarbodiimide or arylsulfonyl chloride. The chain could thus be elongated by further step-wise condensations, and the synthesis of oligonucleotide proceeded in the 5′ → 3′ direction similar to that occurring in nature.

This method led to two major breakthroughs. First, the group was able to synthesize oligoribonucleotides that were used to confirm the genetic code. Second, the group realized that they could synthesize overlapping, complementary oligodeoxyribonucleotides, and assemble the gene using DNA ligase, and in 1970 they published the first completely synthetic and biologically active gene, the 77 bp yeast $tRNA^{Ala}$ gene. For his achievements, Khorana received the Nobel Prize in Physiology or Medicine in 1968.

The procedure was tedious and labor-intensive, and suffered from problems of solubility, long reaction times, only modest efficiency, rapidly decreasing yields with increasing chain length, and time-consuming purification procedures. Moreover, all the reactions were carried out in solution, and the products had to be isolated at each stage of the multistep synthesis. Owing to the drawbacks associated with this method, it was superseded almost completely by the triester approach.

4.3.2 Phosphotriester Method

The phosphotriester approach solves some of the problems faced in the phosphodiester method by blocking each internucleotide phosphodiester function during the course of building a defined sequence. The basic principle of the method is the use of a totally protected mononucleotide containing a fully masked 3′-phosphotriester group (compounds with phosphorus in pentavalent state). The procedure for synthesis of oligonucleotides is the same as in phosphodiester method (Figure 4.2).

Using this methodology, Edge and his colleagues (1981) synthesized 67 oligonucleotides of chain length 10–20, and spliced them together to generate a 517 bp interferon alpha gene. However, this method also suffered from the problems of long reaction times, and rapidly decreasing yields with the increasing chain length.

4.3.3 Phosphite-triester Method

An important alternative to the solid-phase phosphotriester method is the phosphite-triester method that utilizes the extreme reactivity of phosphite reagents (compounds with phosphorus in reactive trivalent state) (Figure 4.3). It is based on the use of phosphite, rather than phosphate intermediates, in which bifunctional derivatives of trivalent phosphorus of the *O*-alkylphosphochloridite type react quickly and quantitatively with the 3′-OH groups of suitably base-protected mononucleotides (Figure 4.4). The second phosphorylation step, involving, for example, a carrier-bound 5′ unprotected mononucleotide, is also more or less quantitative. The two building blocks are joined in a few minutes compared with the hours required in the phosphodiester and phosphotriester strategies. The phosphite-triester method works in solution for the condensation of 3 or 4 nucleotides, however, for the construction of large oligonucleotides, the 3′-end of the desired oligonucleotide has to be coupled to an insoluble support. Thus, the chemical synthesis does not follow the biological direction of DNA synthesis. Instead, each incoming nucleotide is coupled

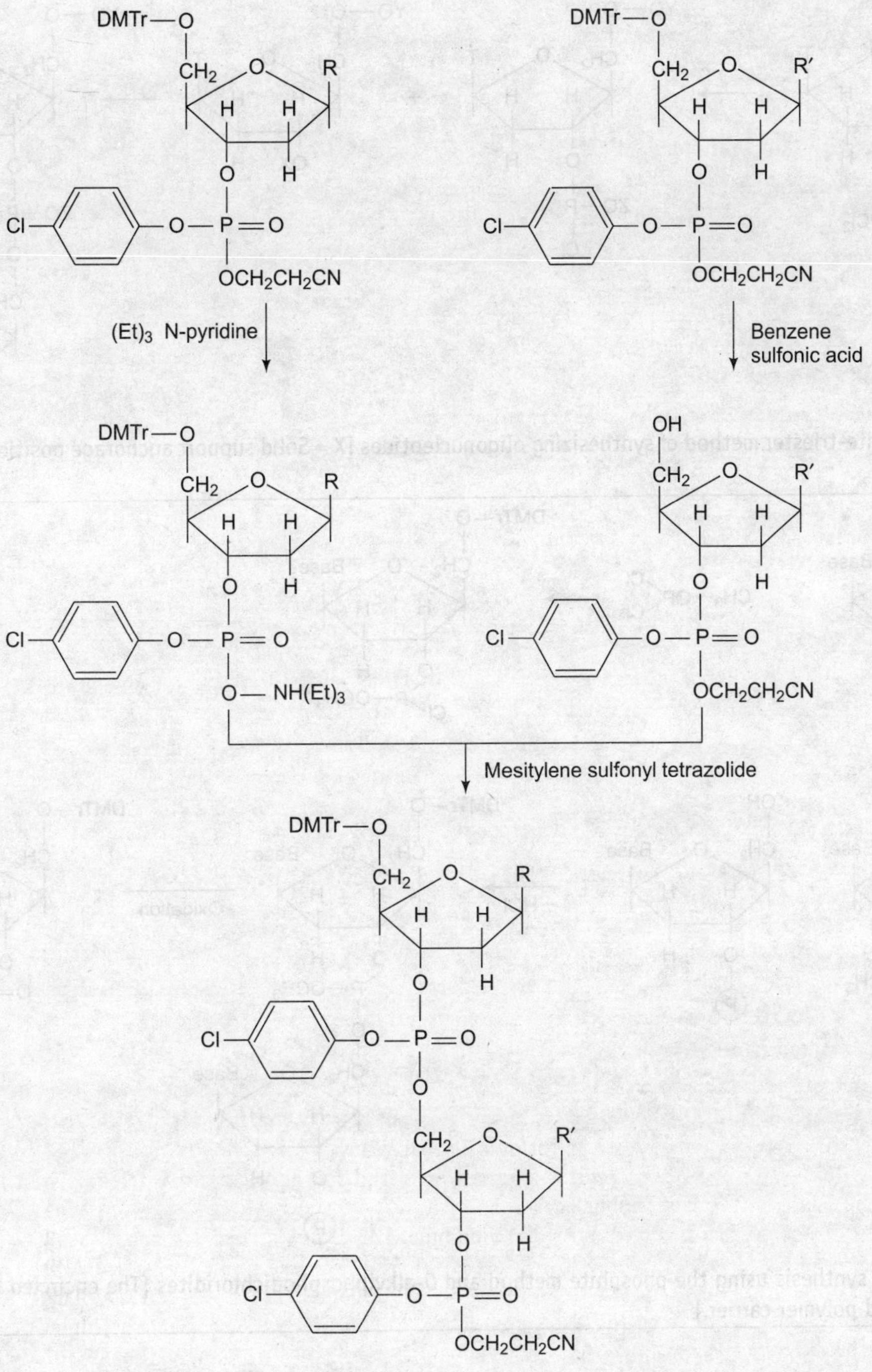

Figure 4.2 The phosphotriester method of synthesizing oligonucleotides [R or R = A/G/C/T; DMTr = Dimethoxytrityl.]

to the 5′-OH terminus of the growing chain. The phosphite approach has been well sited for mechanized solid-phase synthesis. However, a disadvantage of bifunctional phosphochloridites in their bifunctional as well as monofunctional forms, is the liability of the nucleoside phosphomonochloridites, as these compounds must usually be stored in hexane at –80°C.

The above methods are not generally used, and in the following sections, the commercially used methods are discussed in detail.

4.4 CONVENTIONAL SOLID-PHASE SYNTHESIS: PHOSPHORAMIDITE METHOD

In 1981, a breakthrough was achieved in synthetic chemistry that made it possible to make longer and longer oligonucleotides, and to make them much more efficiently with high yields. This faster solid-phase method also supplanted the old difficult and time-consuming processes, and permitted oligonucleotide synthesis to be automated. This most widely used chemistry, phosphoramidite method, was formulated by

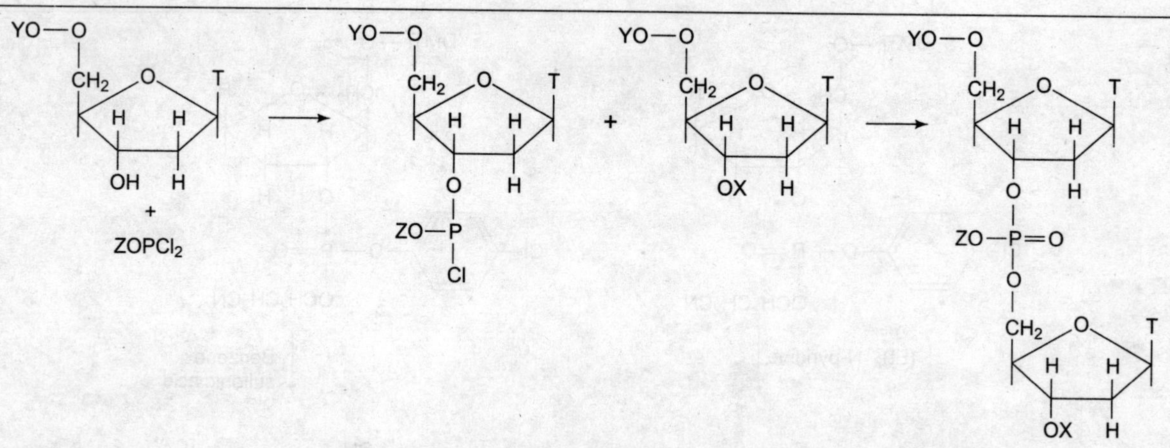

Figure 4.3 The phosphite-triester method of synthesizing oligonucleotides [X = Solid support anchorage position; Y = Dimethoxytrityl; Z = Methyl.]

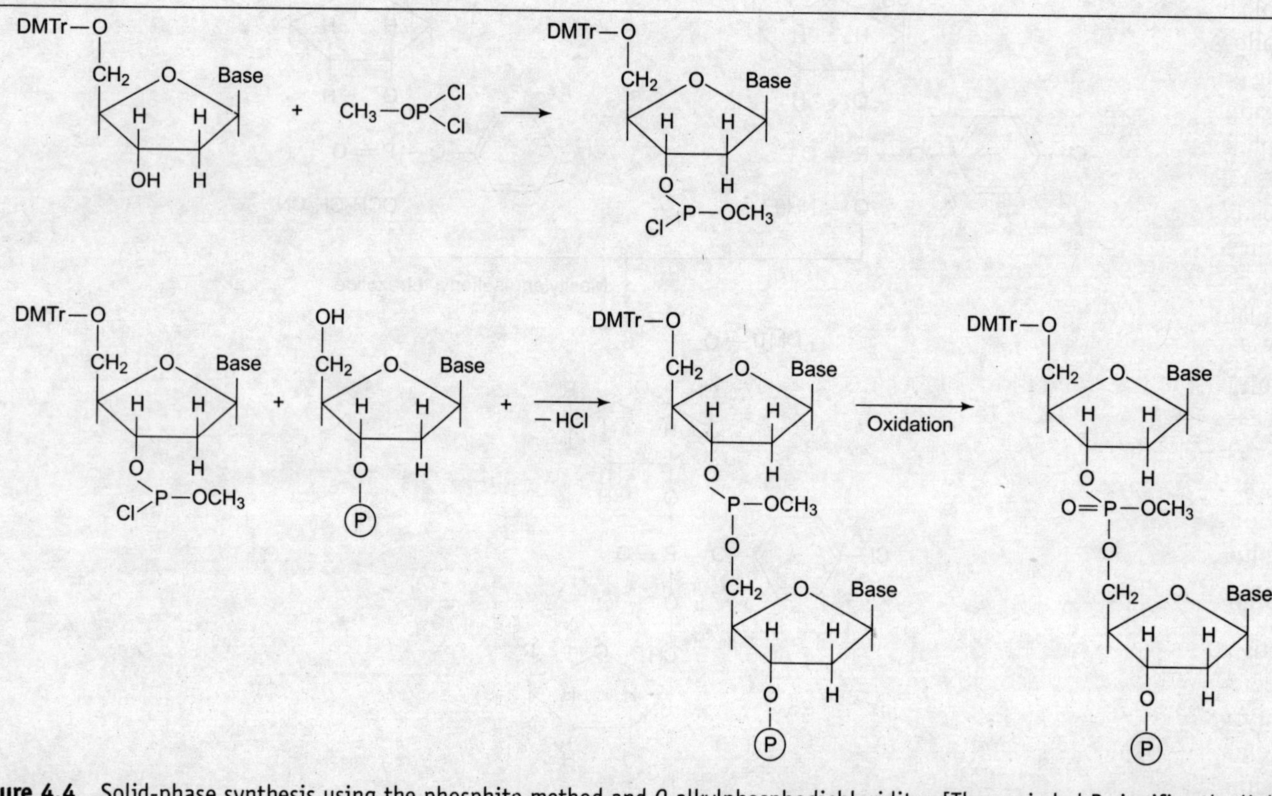

Figure 4.4 Solid-phase synthesis using the phosphite method and *O*-alkylphosphodichloridites [The encircled P signifies the linkage between reactants and polymer carrier.]

Letsinger, and further developed by Caruthers and Beaucage. This process is based upon the use of phosphoramidite monomers and tetrazole catalysis. This discovery revolutionized life sciences leading to billion-dollar economic activity since 1981. Currently, the phosphoramidite method is the procedure of choice for automated chemical synthesis of oligonucleotides in a DNA synthesizer.

4.4.1 Prerequisites for a Successful Synthesis

The following prerequisites have been established for the synthesis of desired biologically active oligonucleotides:

Solubility of Reactants in Non-aqueous Solvents The reactants should be soluble in non-aqueous solvents to be accessible to the repertoire of organic synthesis.

Blocking of Reactive Groups Suitable protecting groups should block the reactive hydroxyl- and amino-functions in nucleosides to avoid any side reaction.

Solid-phase Synthesis The solid-phase synthesis is useful to synthesize long oligonucleotides. Further, immobilization of growing oligonucleotide chain simplifies manipulative procedures, and cuts out time-consuming purification steps following each cycle of condensation.

Procedures Favoring High Yields Due to the costs of the starting material and the limitations presented by the various separation techniques, yield is an important consideration. The solid-phase synthesis in particular requires yields to be almost quantitative.

4.4.2 Blocking Agents or Protective Groups

For oligonucleotide synthesis, it is desired that the sequence must be in correct 5′–3′ frame without 5′–5′, 3′–3′, (5′–2′ in case of RNA) or branched linkages. Hence, every possible site of nonspecific side reaction (reactions at unwanted positions) is blocked before starting the extension or synthesis. The reactive 5′- and 3′-hydroxyl groups of sugar, and exocyclic amino functions of bases are thus blocked transiently by suitable protecting groups. While selecting protective groups, the following points have to be kept in mind: (i) These protecting groups should remain stable under conditions of chain elongation, i.e., during the formation of the internucleotide phosphodiester bonds; (ii) The 3′ and 5′ protective groups can be manipulated selectively. Hence acid- and alkali-labile groups have been developed to protect 3′- and 5′-OH functions, respectively. Similarly, the choice of protective group for amino function of the bases (A/C/G) is such that it is also alkali labile; (iii) The groups protecting the bases should be more stable than those at the deoxyribose sugar, because the former must be preserved during the entire synthesis; (iv) These groups should allow solubility in organic solvents; and (v) These groups should be labile enough to allow their removal at the end of the synthesis without damaging the reaction products. Thus, the protective groups used in phosphoramidite method of oligonucleotide synthesis are:

Benzoyl or Isobutyryl

Reactive exocyclic amino functions of the bases A and C are protected with benzoyl group, while that on G by *N*-2-isobutyryl group. Thymine does not have any reactive exocyclic amino group, and hence needs no protection. These protecting groups maintain the integrity of the ring structures of the bases, and are required throughout the synthesis. These groups are stable to acid, alkali-labile, and stripped off by heating at 65°C for 1 hour in ammonium hydroxide solution.

Dimethoxytrityl

The 5′-OH function of deoxyribose sugar is protected by 4,4′-dimethoxytrityl (DMTr) group. This is an acid labile, temporary blocking group that is removed at the beginning of synthesis to create a site for chain extension, and removed quantitatively by mild acid hydrolysis, without any risk of depurination.

Methyl or β-Cyanoethyl

Phosphate linkages at the 3′-end are protected as uncharged phosphotriesters using 2-cyanoethyl (β-cyanoethyl; β-CE) or methyl (Me) groups. These groups are kept throughout synthesis, are stable to acid, alkali-labile, and after synthesis these groups are removed by treatment with concentrated ammonium hydroxide to yield the negatively charged phosphodiester groups of naturally occurring nucleic acids.

Diisopropylamine

Phosphite group of the phosphoramidite derivative of nucleotide protected by 3′ Me/β-CE group has an attached diisopropylamine group. This group offers the following advantages: (i) The attachment of diisopropylamine to the 3′-phosphite of phosphoramidite preserves it in pure unreactive uncharged form till activated by protonation of its nitrogen atom; (ii) The protonated nitrogen atom of the diisopropylamine group serves as a site for nucleophilic attack during coupling reaction; and (iii) Diisopropylamine is easily released during activation.

4.4.3 Solid Support

The solid-phase synthesis offers advantages such as low-reaction and purification times, easy purification, and high yields, as already described in Section 4.2. However, several conditions have to be met, if these advantages are to be exploited. These are: (i) Yields must be quantitative, if, for example, the yield in each addition step is assumed to be of the order of only 80%, the total yield of a decanucleotide will be decreased to $0.8^{10} \times 100$ or 10%; (ii) The purity of the products, and the yields of the end products cannot be determined during the synthesis but only after its completion; (iii) Several oligonucleotide chains are usually attached to one polymer particle. A decreasing yield in each step will result in the formation of products, which will be heterogeneous with respect to composition and length. The application of the solid-phase method is, therefore, extremely limited by techniques available for subsequent separation; (iv) The chemical reactions take place in a heterogeneous phase, and are therefore usually slower than in solution. Since the reaction half-life, $t_{1/2}$, increases at least by a factor of 2 to 3, reaction conditions of the liquid phase can only be applied to solid-phase synthesis if reaction times are short, i.e., of the order of seconds or minutes; and (v) A number of additional and new reaction steps are required, for example, coupling of the monomers to the polymer carrier, and the removal of the end product from this solid support.

The recognized and accepted solid-phase of choice for oligonucleotide synthesis by phosphoramidite method is the controlled pore glass bead (CPG), a porous borosilicate material. It offers several advantages: (i) It provides uniform pore size allowing efficient synthesis of pure oligomers. This bead has a surface with holes and channels, and it is in these channels that the protected nucleotide is attached (Figure 4.5). CPG beads are available in different pore sizes that can be selected for the synthesis of different sized oligonucleotides.

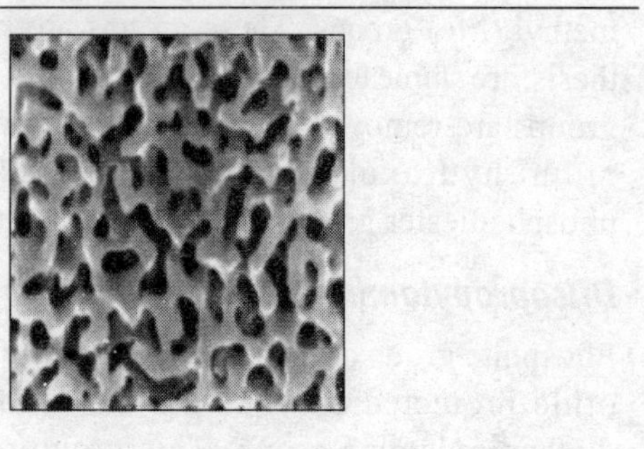

Figure 4.5 An electron photomicrograph of the surface of a CPG bead [The scale of this picture is 10 millionths of an inch square.]

Source: Behlke MA and Devor EJ (2005) *Molecular Genetics and Bioinformatics*, Integrated DNA Technologies.

Millipore produces DNA nucleoside CPG products [DNA nucleoside controlled pore glass (CPG) media®] for the solid-phase synthesis of oligonucleotides using the phosphoramidite chemistry on particles with 500 Å and 1,000 Å pore sizes. A CPG 500 Å matrix is recommended for synthesis of oligomers up to 50 bases in length, and a CPG 1,000 Å matrix for oligomers up to 150 bases; (ii) Surface loading of the support can be controlled to give good consistent coupling yields. Lower surface loading leads to higher coupling yields; (iii) Its high surface area and narrow pore size distribution provide high coupling efficiency and purity for oligonucleotide synthesis; (iv) High surface area also generates high yields of oligomers; (v) It is incompressible and rigid; (vi) The CPG matrices are inert and solvent compatible (compatible with aqueous and organic solutions), and hence unaffected by changes in a solvent system. This eliminates the risks of swelling and shrinking during synthesis cycles; and (vii) All CPG products are manufactured in a dedicated facility certified to internationally recognized standards, BS EN ISO9001.

4.4.4 Initial Deoxyribonucleoside Linked to CPG Bead Through Spacer Arm

In the initial deoxyribonucleoside, the 5′-end is blocked with DMTr, and exocyclic amino groups of bases are protected with benzoyl or isobutyryl groups. This initial deoxyribonucleoside (A/C/G/T) is covalently attached to the CPG bead through a long chain alkyl amine spacer (Figure 4.6). Deoxyribonucleoside can be obtained from either natural sources (for example, salmon sperm) or chemically synthesized. Covalently attached through the 3′-position to a CPG matrix, deoxyribonucleosides are available for solid-phase oligonucleotide synthesis (a derivatized CPG product).

4.4.5 Deoxyribonucleoside 3′-Phosphoramidite

The deoxyribonucleoside 3′-phosphoramidites, P (III) compounds with *P*-alkylamine bonds, are used as substrates for chain elongation. These have a 5′ DMTr group, a diisopropylamine group attached to a 3′-phosphite group protected

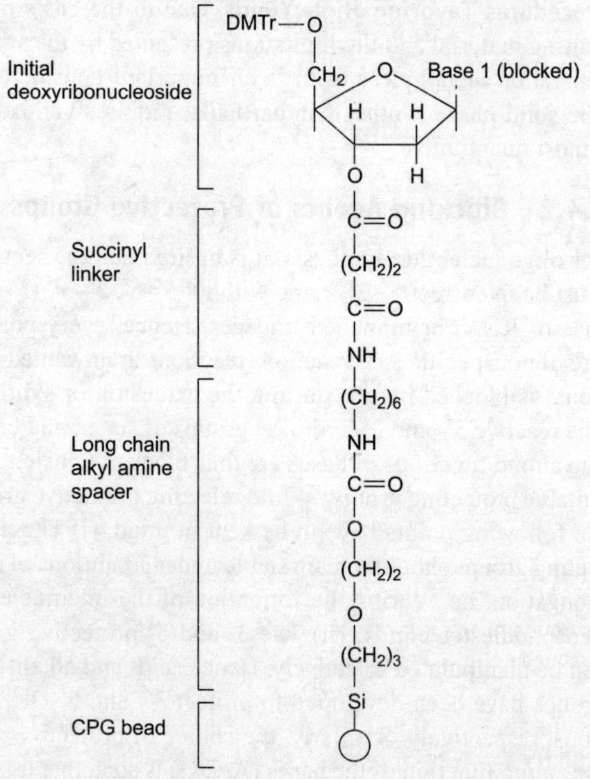

Figure 4.6 Starting complex for the chemical synthesis of a DNA strand

by a β-CE (or Me) residue, and a benzoyl/isobutyryl group blocking the exocyclic amino group of base (Figure 4.7). All four deoxyribonucleosides converted to 3′-phosphoramidites are available commercially. These phosphoramidite derivatives of deoxyribonucleosides possess the following properties: (i) These are much more stable in solution; (ii) There is no loss of activity when storing phosphoramidites in the freezer for even a period of 1 year. These are stable at elevated temperatures; (iii) These are less sensitive to oxidation; (iv) These are easy to handle until 'activated' or 'protonated'; and (v) The use of these phosphoramidites for the synthesis of DNA fragments simplifies deprotection and isolation of the final product. Cleavage of the oligonucleotide chain from the polymer support, *N*-deacylation, and deprotection of β-CE (or Me) group from the phosphate-triester moiety can be performed in one step with concentrated aqueous ammonia.

These new compounds allow a new approach to synthesis, in which the phosphorylating agent is generated in a two-step procedure. First, phosphotrichloride and methanol react to yield dichloromethoxyphosphine (I), which is then converted to chloro-*N,N*′-dimethyl amino methoxy phosphine (II) through a reaction with trimethyl silyl methylamine. The reaction with 5′-*O*-dimethoxy tritylnucleosides generates the corresponding nucleoside phosphoramidites (III). Apart from the dimethylamino compounds, a number of other secondary amines have also been used lately. Compounds derived from

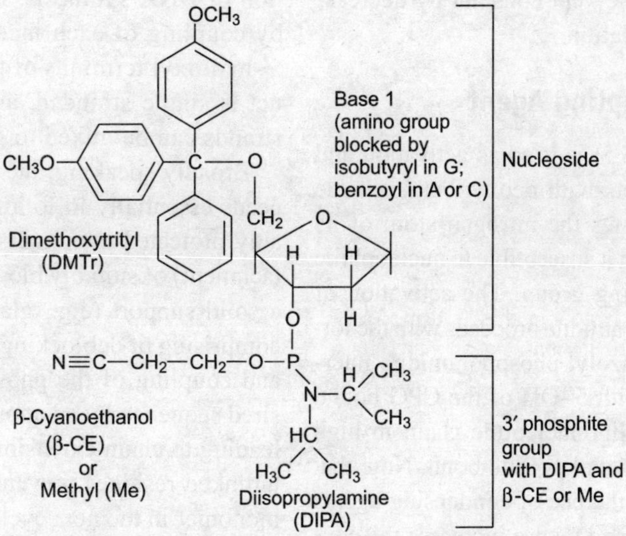

Figure 4.7 General structure of a deoxyribonucleoside 3′-phosphoramidite [Phosphoramidites are available for each of the four bases (A, C, G and T) that are used for the chemical synthesis of a DNA strand. A diisopropylamine group is attached to the 3′-phosphite group of the deoxyribonucleoside. A methyl (Me) or β-cyanoethyl (β-CE) group protects the 3′-phosphite group and a DMTr group is bound to the 5′-hydroxyl group of the deoxyribose sugar.]

diisopropylamine and morpholine have been shown to be particularly stable. The synthesis of the phosphorylating agents must be carried out with utmost care, purity controls should not only involve boiling point measurements, but especially a measurement of ^{31}P-NMR spectra, because only the latter method ensures that impurities, caused by phosphotrichloride or hydrolysis, and oxidation products, which would interfere with the reaction, are recognized with sufficient sensitivity.

4.4.6 Detritylating Agent

The 5′ DMTr groups from the CPG bound initial deoxyribonucleoside or other deoxyribonucleoside 3′-phosphoramidite (from growing chain) are removed by treatment with weak acid to yield a reactive 5′-hydroxyl group. Detritylation takes place in 50–60 seconds.

Depurination (cleavage of the glycosidic bond) under acidic conditions is an important side reaction that limits the synthesis of oligonucleotides. Originally, strong protic acids such as benzene sulphonic acid ($pK_a = 0.5$) or trichloroacetic acid (TCA; Cl_3CCOOH; $pK_a = 0.7$) were used. TCA [in dichloromethane (DCM)] has been successfully used for the syntheses of many millions of oligonucleotides, and was found to be adequate for most syntheses; however, TCA caused substantial depurination in silica bound oligonucleotides (3% TCA/DCM can cause 12–67% depurination in 1 hour, depending on the position of the deoxyadenosine). In 1983, two groups independently found that when CPG beads were used as the solid support, a weaker acid, dichloroacetic acid (DCA; $Cl_2CHCOOH$; $pK_a = 1.48$), could be satisfactorily employed. Indeed, when combined with phosphoramidite chemistry this combination allowed the synthesis of a 51 base long oligonucleotide, a length much greater than anything previously attempted at that time. DCA is the reagent of choice for detritylation in DNA synthesizers [e.g., Milligen/Biosearch, Pharmacia (1.3 mmol scale) and Applied Biosystems 390Z (large-scale) DNA synthesizers]. It offers several advantages over TCA. These include: (i) As compared to TCA, DCA is associated with reduced depurination; (ii) DCA solutions are also more conveniently prepared than TCA solutions because DCA is a liquid instead of a solid; (iii) DCA is also more stable towards decomposition to hydrochloric acid than TCA; (iv) Better results are obtained with DCA instead of TCA for making longer oligonucleotides (80–150 bases) as well as larger scale synthesis (10–15 mmol); and (v) It is not necessary to increase the total time for the detritylation step (which is usually about 50–60 seconds) because of the weaker acid.

Similar to TCA, the solution of DCA in required concentration may be prepared in DCM; however, 1, 2-dichloroethane (DCE) is preferred over DCM. For the synthesis of oligonucleotide up to ~40 bases, the use of 5% (v/v) DCA/DCE, and for still longer sequences, 2% (v/v) DCA/DCE is recommended. Note that DCE is advantageous over DCM as a solvent for DCA because it causes lesser depurination and has lower toxicity. Moreover, due to its more viscous nature, DCE is consumed in lower volume per cycle, and lasts almost twice as long on the synthesizer. The lower flow rate (due to higher viscosity) of DCE solvent is however compensated by a total time of 50–60 seconds for detritylation. This is easily achieved by extending the delivery time of the reagents to the column so that enough reagents reach the synthesis column. Usually the total detritylation time can be kept constant by decreasing any wait steps in the detritylation.

4.4.7 Activation and Coupling Agent

Tetrazole, a weak acid ($pK_a = 6.5$) is used as activation and coupling agent. It instantly reacts with deoxyribonucleoside 3′-phosphoramidite and protonates the nitrogen atom of its diisopropylamine moiety, making it susceptible to nucleophilic attack, and is also a good leaving group. The activation of deoxyribonucleoside 3′-phosphoramidite proceeds with the formation of a highly reactive tetrazolyl phosphoramidite intermediate, which reacts rapidly with 5′-OH of the CPG bound initial nucleoside or growing oligonucleotide chain in high yields to form new highly reactive phosphite bond. Note that the tetrazole activation replaces the use of condensing agents like DCC in earlier approaches, and its use increases the coupling efficiency to greater than 99%. This further opens the way for longer oligonucleotides to be synthesized.

4.4.8 Anhydrous Acetonitrile

The column is washed intermittently with anhydrous acetonitrile to maintain anhydrous conditions, and to dissolve unused phosphoramidites and tetrazole. The quality of anhydrous acetonitrile used is very important because even the trace amounts of moisture may lower the effective phosphoramidite concentration, and hence the coupling efficiency. Freshly opened commercially available anhydrous acetonitrile is good, but repeated sampling or prolonged storage of opened bottles can introduce moisture contamination. Hence in the laboratory, acetonitrile is kept in anhydrous state by removing large amounts of moisture through a two-stage distillation process (once from phosphorus pentaoxide, and once from calcium hydride) or a simple continuous reflux over calcium hydride, and under nitrogen.

4.4.9 Oxidizing Agent

Iodine in tetrahydrofuran (THF) acts as a mild oxidant and forms adduct with phosphate-triester linkage. Water acts as an oxygen donor, and decomposes the adduct leaving the phosphate-triester bond stabilized. Thus, I_2/H_2O mixture converts a highly reactive phosphite-triester with phosphorus in trivalent state into a stable phosphate-triester with phosphorus in a pentavalent state.

4.4.10 Outline of the Process

The chemical synthesis of DNA is a multistep process and takes place under non-aqueous (anhydrous) conditions. It involves addition of a single deoxyribonucleotide to a growing oligonucleotide chain immobilized onto a solid support. The DNA synthesis does not follow the biological direction of DNA synthesis. It takes place in 3′ → 5′ direction by coupling of each incoming deoxyribonucleotide to the 5′-hydroxyl terminus of the growing chain. The final product is single stranded, and the two complementary single strands can be mixed to get ds DNA.

Broadly speaking, the synthesis of oligonucleotides proceeds essentially in following steps: (i) Preparation of suitably protected monomers; (ii) Immobilization (covalent attachment) of suitably blocked initial deoxyribonucleoside on a solid support (e.g., glass); (iii) Four step coupling cycle comprising of deblocking of 5′-end (detritylation), activation and coupling of the phosphoramidite monomers in the desired sequence by an appropriate phosphorylation procedure leading to chain extension by one base at a time, capping of unlinked residues preventing them from linking to the next monomer in the next cycle, and oxidation of the highly reactive internucleotide phosphite-triester linkage to form stable phosphate-triester; and (iv) Deblocking, purification, and recovery of the synthesized oligonucleotides.

The flow sheet of oligonucleotide synthesis is depicted in Figure 4.8, and the details of each step are described as follows.

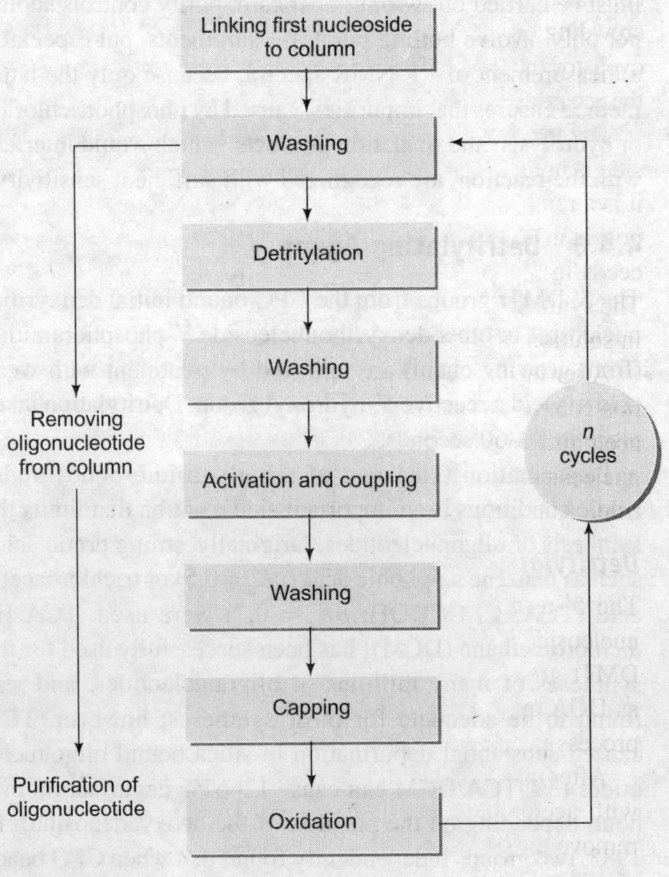

Figure 4.8 Flow chart for the chemical synthesis of oligonucleotides [After *n* coupling reactions (cycles), a single stranded oligonucleotide with *n* + 1 residues is produced.]

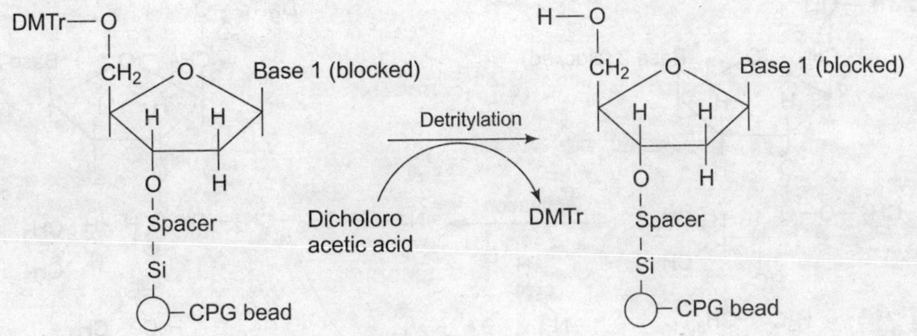

Figure 4.9 Detritylation step of phosphoramidite method [The 5′ DMTr group is removed by treatment with DCA. In this example, the detritylation of the first deoxyribonucleoside is depicted.]

Covalent Attachment of Initial Deoxyribonucleoside to Solid Support

The DNA synthesis cycle begins with the attachment of 3′-end of a suitably protected initial deoxyribo-nucleoside to a solid CPG support through a long alkyl amine spacer arm (Figure 4.6). The synthesis of long (50–150 bases) oligonucleotides can be greatly improved by using a CPG support with a much lower loading (~5 mmol/g) than the usual 30–40 mmol/g loading. The lower surface loading of the support gives much more consistent coupling yields, perhaps because of less surface crowding or just the greater excess of reagent present. Moreover, for the synthesis of 50–150 base long oligonucleotides, the recommended pore size of the support is 1,000 Å.

The trityl group of the protected initial deoxyribonucleoside is not used to link to the solid synthesis support, instead it is a removable protecting group. The link to the solid support is made through the 3′ carbon so that the synthesis proceeds in 3′ → 5′ direction rather than in 5′ → 3′ direction. The synthesis requires activation of the deoxyribonucleoside in solution rather than at the polymer carrier.

After the binding of initial deoxyribonucleoside, the reaction column is washed extensively with an anhydrous reagent, acetonitrile, to remove water and any unbound deoxyribonucleoside, and then flushed with argon to purge the acetonitrile.

Detritylation

The 5′-terminus of the support bound initial deoxyribonucleoside is set free for coupling reaction by removing the DMTr group (at 5′-end) by treatment with weak acid, such as DCA or TCA to yield a reactive 5′-hydroxyl group. The process is called detritylation (Figure 4.9).

After this detritylation step, the reaction column is washed with acetonitrile to remove TCA, and then with argon to remove the acetonitrile.

Addition of Deoxyribonucleoside 3′-Phosphoramidite, Activation, and Coupling

Once the initial deoxyribonucleoside is covalently attached to the CPG bead, the next step is the addition of next prescribed deoxyribonucleoside 3′-phos-phoramidite, and tetrazole in the reaction vessel for simultaneous activation and coupling steps. As described earlier in Section 4.4.7, tetrazole activates the nitrogen atom of the incoming phosphoramidite by protonation, which then serves as a site for nucleophilic attack by newly liberated 5′-end of the initial deoxyribonucleoside (due to detritylation). This leads to formation of a covalent bond (coupling) between 3′-phosphite of activated incoming phosphoramidite, and 5′-OH of the initial deoxyribonucleoside (Figure 4.10). Coupling is very fast (less than 30 seconds), and leads to the formation of a phosphite-triester with phosphorus in its highly reactive trivalent state (P III). The tetrazole is reconstituted, and the process continues. With the use of tetrazole, a coupling efficiency of the order of 99% can be achieved.

Of the first generation oligonucleotide analogs, the class that has resulted in the broadest range of activities, and about which the most is known is that of phosphorothioate oligonucleotides. This modification clearly achieves the objective of increased nuclease stability. In these oligonucleotides, one of the oxygen atoms in the phosphate group is replaced with a sulfur atom. Thus, for their synthesis, the internucleotide phosphite is sulfurized before the capping step. Modified deoxyribonucleosides (e.g., containing a fluorescent label) can also be incorporated into the growing oligonucleotide at this stage.

Unincorporated phosphoramidite and tetrazole are removed by washing the column with acetonitrile and flushing with argon.

Capping

Because not all of the support bound deoxyribo-nucleosides are linked to a phosphoramidite during the first coupling reaction, the unlinked residues must be prevented from linking to the next monomer during the following cycle. To do this, acetic anhydride and *N*-methylimidazole/dimethylaminopyridine are added to acetylate the unreacted 5′-OH groups and terminate any chains that have not undergone coupling (Figure 4.11). This reagent reacts only with free hydroxyl groups to irreversibly cap the oligonucleotides in which

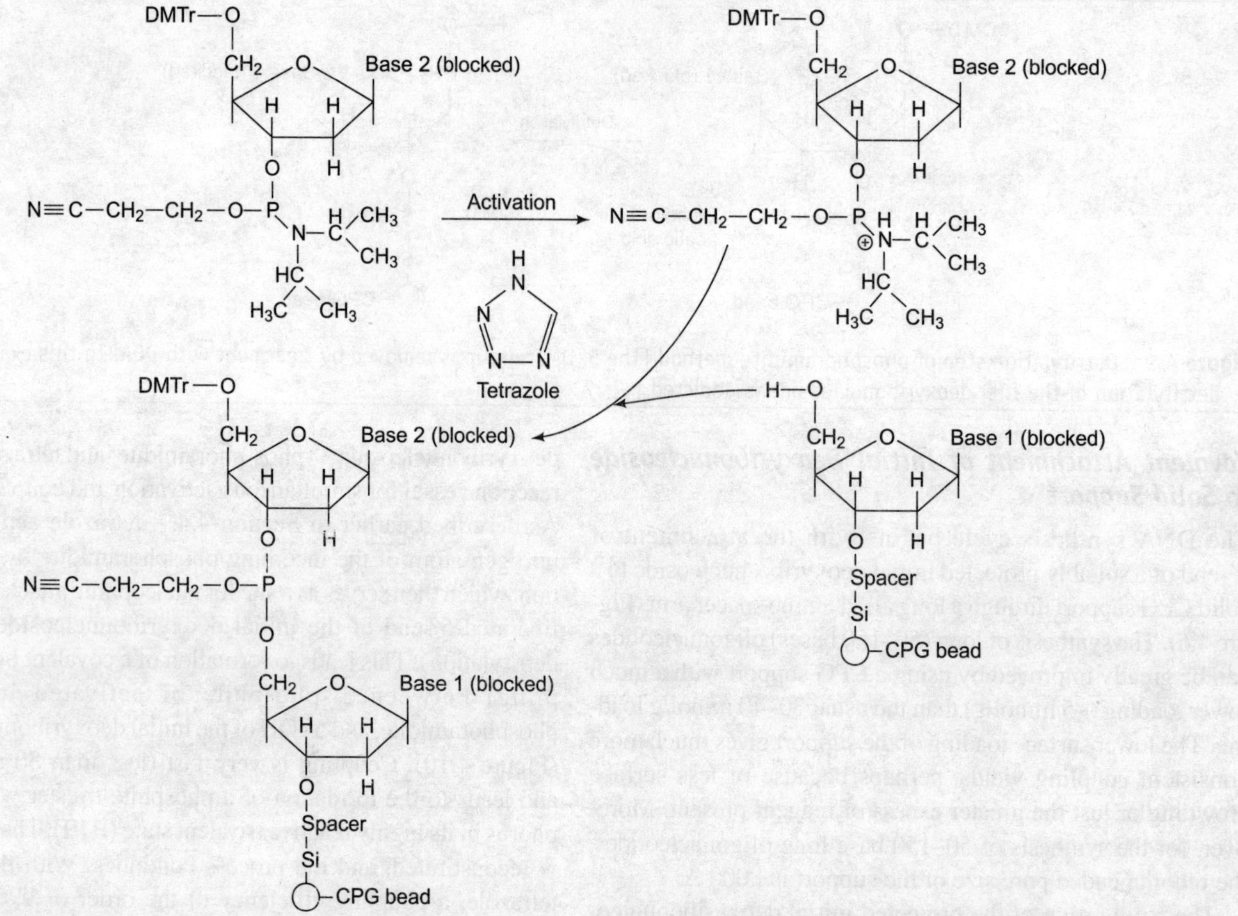

Figure 4.10 Activation and coupling steps of phosphoramidite method [The activation of a phosphoramidite enables its 3′-phosphite group to attach to the 5′-hydroxyl group of the bound detritylated deoxyribonucleoside.]

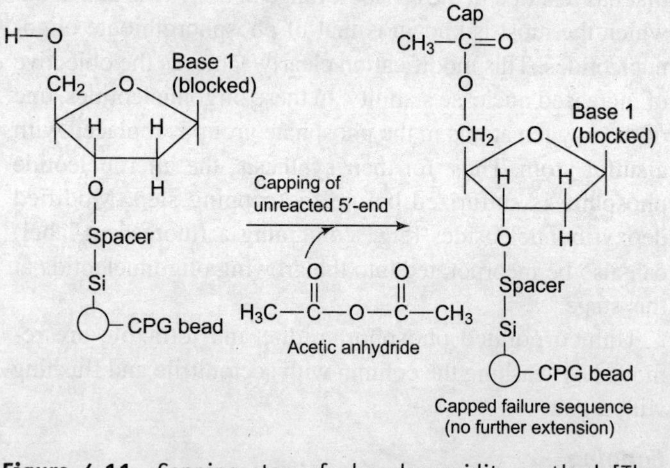

Figure 4.11 Capping step of phosphoramidite method [The available 5′-hydroxyl groups of unreacted detritylated deoxyribonucleosides are acetylated to prevent them from participating in the coupling reaction of the next cycle.]

the coupling failed. Thus, any unreacted 5′-end is capped by acetylation, and the extension of erroneous oligonucleotides (failure sequences) in subsequent coupling reactions is blocked, and these capped products are marked as failure products. In contrast, the DMTr group of the successful coupling step protects the 5′-OH end from being capped. The coupling failures thus result in unavoidable loss of mass relative to the initial synthesis scale. If this capping step is not carried out, then after a number of cycles, the growing chains will differ in both length and nucleotide sequence. Although this step is not absolutely necessary for DNA synthesis, it minimizes the length of impurities, and thus facilitates trityl-on HPLC purification (i.e., 5′ DMTr protecting group is retained on the deoxyribonucleotide and removed during purification).

Chemical capping by phosphitylation during oligonucleotide synthesis is also proposed. In this method, a phosphite-monoester is reacted with the 5′-OH or 3′-OH of the failure sequence between successive condensation steps in a synthesis procedure to form a 5′ or 3′-phosphite-diester with the failure sequence. The phosphite-diester substituent is inert with respect to subsequent reaction steps in the synthesis of the desired oligonucleotide product (U.S. Patent 4816571).

Oxidation and Stabilization

The internucleotide phosphite-triester linkage resulting from the coupling step is highly reactive, and prone to breakage in the presence of either acid or base. This highly reactive phosphite-triester (P III) is oxidized with I_2 (in THF)/H_2O mixture

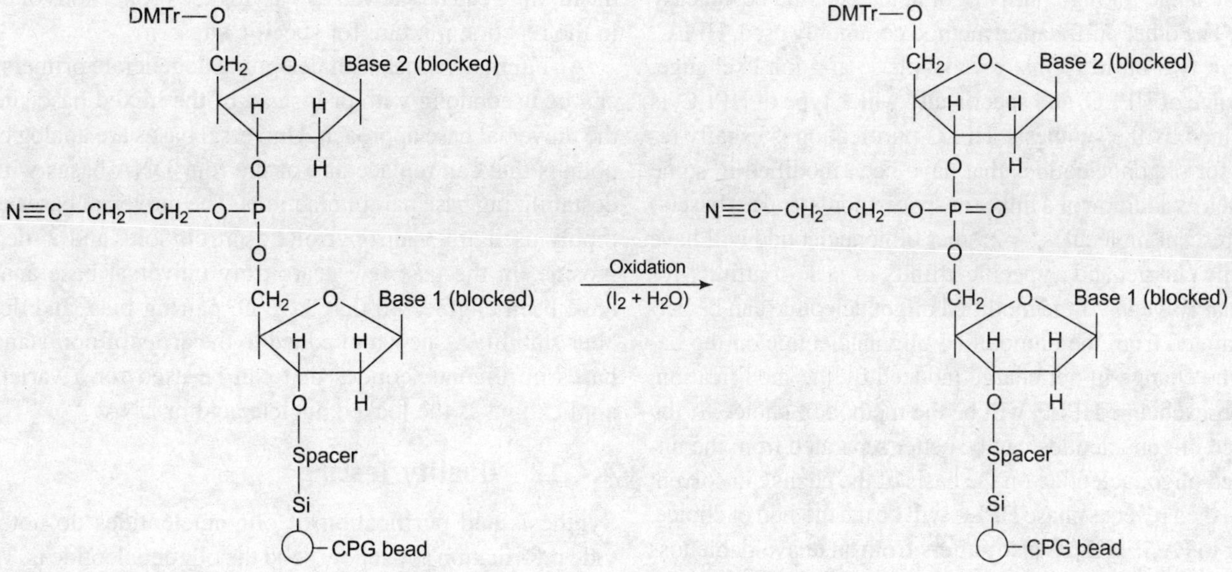

Figure 4.12 Oxidation step of phosphoramidite method

to form stable phosphate-triester (P V) (Figure 4.12), thereby yielding a chain that has been lengthened by one deoxyribonucleotide.

After this oxidation step and subsequent wash of the reaction column, the cycle of detritylation, phosphoramidite activation, coupling, capping, and oxidation is repeated to build up the desired sequence of oligonucleotide. This cycle continues with each successive phosphoramidite until the last programmed residue has been added to the growing chain.

Deprotection and Purification

Once an oligonucleotide of the desired sequence has been synthesized, its 3′-end is still bound to the CPG bead, all the bases (except T) of the synthesized oligonucleotide carry amino-protecting group, each internucleotide phosphate-triester linkage contains 3′ β-CE (or Me) groups, and the 5′-terminus of the oligonucleotide chain has a DMTr group. So to get a biologically active single stranded oligonucleotide, all these blocking groups should be removed after the synthesis is complete by a process called deprotection. The linkage of oligonucleotide to spacer is easily hydrolyzed by concentrated ammonium hydroxide (NH_4OH) to release the product with a free 3′-OH terminus. Ammonia treatment also removes the β-CE (or Me) protecting groups. The methoxy group at the internucleotide bond can also be removed by hydrolysis with the nucleophilic thiophenolate anion. Later the 5′-end of the oligonucleotide is detritylated by acid hydrolysis (weak acid or 80% acetic acid). The 5′-terminus of the oligonucleotide strand is phosphorylated either by a T4 polynucleotide kinase reaction or by a chemical procedure. Phosphorylation can also be carried out after detritylation, while the oligonucleotide is still bound to the support. The blocking groups protecting the exocyclic amines on the bases are also stripped off by heating at 65°C for 1 hour in ammonium hydroxide solution. However, deprotection removes the protecting groups but these remain with the oligonucleotide as organic salts that must be removed. The process of removing these contaminants is called desalting. The deprotected product is lyophilized, and can be used as such.

In many cases oligonucleotides must also be purified in an application-dependent manner after assembly. The desired product has to be separated from the incomplete synthesis products (failure products) and protecting groups. These methods are time saving, and the cost incurred is minimal, and far outweighed by the benefits. There are two methods for subsequent purification, and the choice of which purification method to use is largely dictated by the oligonucleotide itself. These are: (i) Polyacrylamide gel electrophoresis (PAGE) using 20% polyacrylamide/8 M urea gels; and (ii) High performance liquid chromatography (reversed-phase, ion exchange). If the oligonucleotide is unmodified, and the only issue is removal of the capped products, PAGE remains the most efficient means of purification. High percentage acrylamide gels permit isolation of full-length product from all shorter species with great efficiency. The full-length product is simply excised from the gel, and eluted from the gel slice. However, it suffers from one drawback. Similar to the coupling failures resulting in an unavoidable loss of mass relative to the initial synthesis scale, PAGE purification also results in an unavoidable loss of mass because it is physically impossible to recover every bit of full-length product from a gel slice. Thus, a PAGE-purified 50-mer will present with substantially lower mass relative to the starting synthesis scale than will a desalted 25-mer. However, given that the applications for which a 50-mer is utilized are most likely to be much more sensitive to non full-length species, the loss of mass exchange for high purity is, or at least should be, an easy choice. The other purification method commonly used, HPLC, comes in two basic forms: reverse-phase and ion exchange. The choice of HPLC, and specifically which type of HPLC, is determined by the synthesis. HPLC purification is usually reserved for oligonucleotides that have been modified in some way, such as addition of a linker or spacer, non-standard base(s) or fluorescent molecules. A 25-mer oligonucleotide will have a specific charge, and a specific affinity or lack of affinity for particular solvents. If the modified oligonucleotide can be better separated from the unmodified oligonucleotide on the basis of the change in net charge induced by the modification, then ion-exchange HPLC will be the method of choice. If the modified oligonucleotide can be better separated from the unmodified oligonucleotide on the basis of the change in solvent affinity, then reverse-phase HPLC will be the method of choice. Similar to PAGE, HPLC also suffers from an unavoidable loss of mass but again this should be more offset than by the gain in purity. Note that the synthesis scale refers to the amount of starting material that is composed solely of the 3′ most deoxyribonucleotide chemically linked to the CPG bead and synthesis yield refers to the cumulative loss of mass due to coupling failures and subsequent purification.

A new alternative strategy of oligonucleotide synthesis employs the use of phthaloyl protecting group for the exocyclic amino functions of the nucleobases. This approach combines the advantages of cheap, and easily accessible monomeric building blocks, standard machine-aided oligonucleotide synthesis, and a fast deprotection protocol, which is orthogonal to the cleavage procedure from the solid support. The crude oligonucleotides show high purity, and require no further chromatographic purification.

4.4.11 Synthesis of Degenerate Sequences

A sample of mixed probes or degenerate primers with a flexibility in specificity is required during hybridization with heterologous sequence or polymerase chain reaction (PCR) amplification using sequence from one species to amplify a homologous region in another species, respectively. Recent advances in combinatorial or 'irrational' drug design also require large pools of random sequences from which specific sequences can be retrieved. The synthesis of such mixed probes or degenerate primers is simple and straight-forward. Briefly, during chemical synthesis, instead of providing a specific phosphoramidite for a particular deoxyribonucleotide site, a mixture of different bases is added to the reaction. For example, with an equal concentration of four different bases for one deoxyribonucleotide position, four different oligonucleotides (probes/primers) are produced. If two sites are treated this way, $4^2 = 16$ different probes/primers will be synthesized, and for n sites, there are 4^n different probes/primers. Moreover, the frequencies of various probes/primers in the mixture can be skewed by varying the proportions of bases in the reaction mixture for specific sites.

An alternative means of designing degenerate primers that can be used along with or instead of the mixed base sites is the universal base approach. Universal bases are analog compounds that can replace any of the four DNA bases without destabilizing base pair interactions. The universal bases commonly used are 3-nitropyrrole, 5-nitroindole, and 2′-deoxyinosine. In the past few years truly universal base analogs have been engineered that have no pairing bias, and do not alter stability. A new tool added to the array of non-standard bases in oligonucleotides that can be used for a variety of applications is the locked nucleic acid or LNA.

4.4.12 Quality Testing

Synthesis and purification of oligonucleotides do not provide information about how good the oligonucleotide is. There are two methods by which the oligonucleotide can be tested for quality prior to its use in an experiment. These methods are capillary electrophoresis (CE) and mass spectrometry. Both of these methods provide excellent first order data about synthesis quality. Capillary electrophoresis requires a small amount of the final synthesis product, which is then subjected to a constant electrical field in a hair-thin capillary. As the product migrates in the capillary it is separated into component sizes in a manner exactly like gel electrophoresis. The fragments migrate past an optical window. An UV beam detector produces a series of peaks corresponding to material densities flowing past the detector, which then assesses the density of the fragments. This density profile is then made quantitative by establishing a base line, and integrating the area under the individual peaks. The purity of the product is calculated by the ratio of the main peak to the total area under all peaks. In practice, the main peak should be the last peak off since there should not be any species longer than the full-length product. If peaks do appear to the right of the main peak, it is indicative of residual impurities, and other potential contaminants.

The 'industry standard' method for assessing quality of oligonucleotides is a variant of mass spectrometry, MALDI-TOF (matrix assisted laser desorption ionization-time of flight). The principle of MALDI-TOF is that the speed at which an ion moves is inversely proportional to its mass. Thus, in the ion chamber of a MALDI instrument, materials are ionized, and given the same potential energy, eV, where V is the potential, and 'e' refers to the number of charges on the ion. As the ions emerge from the ion source, the potential is converted to kinetic energy in the moving ion. Ions in motion obey the rule $E = 1/2\ mv^2$ where m is the mass of the ion, v is velocity, and E is energy. Solving for m yields, $m = 2E/v^2$. Since the amount of energy is a constant in MALDI, the mass of the ion can be determined by velocity alone. Velocity is simply time over distance such that the time of arrival of the ion in the detector of the MALDI, i.e., time of flight, is directly converted to velocity because the distance is a constant as well. The details of the process are discussed in Chapter 19. MALDI-TOF has the added advantage that the mass of an oligonucleotide can be estimated with precision since the mass of the individual components (the deoxyribonucleotides) is fixed. For any given sequence, the expected arrival time will be given by the expected mass, and any deviation from that time of arrival will indicate a deviation from ideal size and/or purity.

4.4.13 Coupling Efficiency and Yield of Oligonucleotides

During synthesis of oligonucleotides, the release of orange colored trityl ions from each coupling step can be easily quantitated using either a colorimeter or spectrophotometer. These cations produced by the DMTr protecting groups in acid provide a convenient quantitative method for determining coupling efficiency. This is because of the one-to-one relationship between the number of trityl molecules and the number of new phosphate linkages formed. As this assay deals with relatively small changes in coupling efficiency, it is very important that the trityl ions are measured accurately, the other reagents are fresh, and the instrument is functioning properly. It is also possible that traces of moisture introduced in the packaging step or improper handling during shipment can affect the final coupling efficiency. Phosphoramidites are the most important reagents, and should be carefully purified or stored free of moisture to obtain good results.

To achieve a reasonable overall yield of an oligonucleotide, the coupling efficiency should be close to 100% because of the exponential relationship between overall yield, and average coupling yield. If, for example, the efficiency is 99%, at each cycle during the production of a 20-mer oligonucleotide, 82% (i.e., $0.99^{20} \times 100$) of the product will be 20 nucleotides long. If a 60-mer is synthesized with 99% efficiency at each cycle, then about 55% of the final product will contain all of the 60 nucleotides. With an average coupling efficiency consistently less than 98%, the yield of full-length oligonucleotides diminishes as a function of the required number of cycles. A coupling efficiency greater than 98% at each step is achieved using extremely pure reagent and chemicals. For very good preparations average yields of greater than 99% can also be acquired (Table 4.1). These requirements are not easily met in practice, and generally, the coupling efficiency may vary from 95 to 99% from cycle to cycle.

Table 4.1 Yield of Chemically Synthesized Oligonucleotides at Various Coupling Efficiencies for Each Cycle

Coupling efficiency (%)	Overall yield of oligonucleotide (%)				
	20-mer	40-mer	60-mer	80-mer	100-mer
90	12	1.5	0.18	0.02	0.003
95	36	13	4.6	1.7	0.6
98	67	45	30	20	13
99	82	67	55	45	37
99.5	90	82	74	67	61

As higher coupling efficiencies produce a higher overall yield, using good or better than average reagents can significantly improve oligonucleotide quality. This difference may not be noticeable with short oligonucleotides, but the effect becomes more pronounced as chain length increases, and the higher overall yield can determine whether a sequence will work without purification or even whether a long sequence can be isolated. Thus, the initial products of most chemical synthesis runs must be subjected to either reverse-phase HPLC or gel electrophoresis to purify the true products. This goal is easily achieved, because all the 'failure' sequences are shorter than the target oligonucleotide.

4.4.14 Applications of Oligonucleotides

Chemically synthesized oligonucleotides have a myriad of applications in gene technology that are as follows: (i) Partial or total gene synthesis; (ii) Primers synthesis for DNA and RNA sequencing; (iii) Synthesis of primers for PCR amplification; (iv) Synthesis for hybridization probes for the screening of cDNA or genomic libraries; (v) Synthesis of adaptors and linkers for restriction site modification during gene cloning; and (vi) Oligonucleotide synthesis for application in site-directed mutagenesis.

The application of these chemically synthesized oligonucleotides in gene synthesis is discussed in Exhibit 4.1. The applications of primers in PCR and sequencing, adaptors and link-

Exhibit 4.1 Assembly of complete gene

If chemically synthesized double stranded DNA is required for an application such as the synthesis of a gene or the formulation of an artificial coding sequence, then each complementary strand is made separately. The production of short genes (60–80 bp) is technically straight-forward and can be accomplished by synthesizing the complementary strands and then annealing them. For the production of longer genes (>300 bp), however, special strategies must be devised because the coupling efficiency of each cycle during chemical DNA synthesis is never 100%. For example, if a gene contains 1,000 bp and the average coupling efficiency is 99.5%, then the production of full-length single strands after the last cycle is a miniscule 0.007%. To overcome this problem, synthetic (double stranded) genes are assembled in modular form from (single stranded) fragments that are from 20–100 nucleotides in length.

One method for building a synthetic gene requires the initial production of a set of overlapping, complementary oligonucleotides, each of which is usually between 20 and 60 nucleotides long. The sequences of the strands are planned so that, after annealing, the two end segments of the gene are aligned to give blunt ends. Each internal section of the gene has complementary 3′- and 5′-terminal extensions that are designed to base pair precisely with an adjacent section (Figure A). Thus, after the gene is assembled, the only remaining requirement to complete the process is sealing the nicks along the backbones of the two strands with T4 DNA ligase. In addition to the protein coding sequence, synthetic genes can be designed with restriction endonuclease sites at their ends to facilitate insertion into a cloning vector and if necessary, additional sequences that contain signals for the proper initiation and termination of transcription and translation. To optimize translation, the codons of a gene from one organism can be changed to those that are preferred by the host cell without altering the amino acid sequence.

An alternative way to prepare a full-size gene is to synthesize a specified set of overlapping oligonucleotides. After the 3′ and 5′ extensions (6–10 nucleotides) are annealed, large gaps still remain, but the base paired regions are long and stable enough to hold the structure together. The duplex is completed and the gaps are filled by enzymatic DNA synthesis with *E. coli* DNA polymerase I.

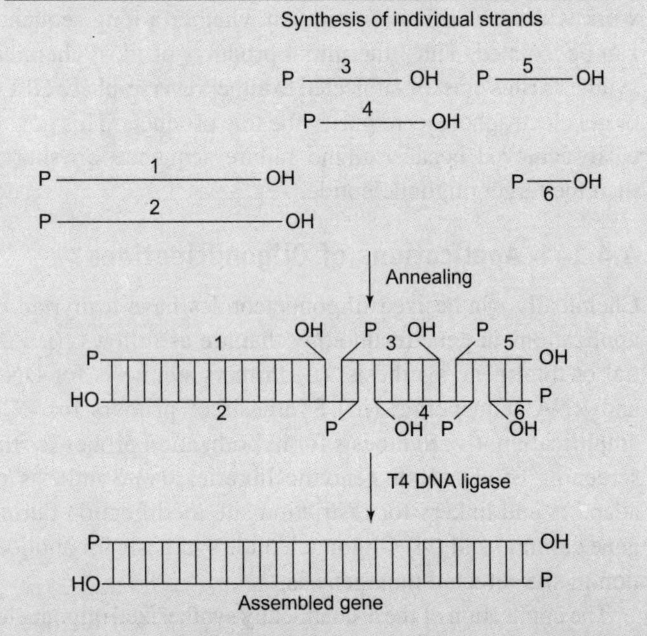

Figure A Assembly of a synthetic gene from short oligonucleotides

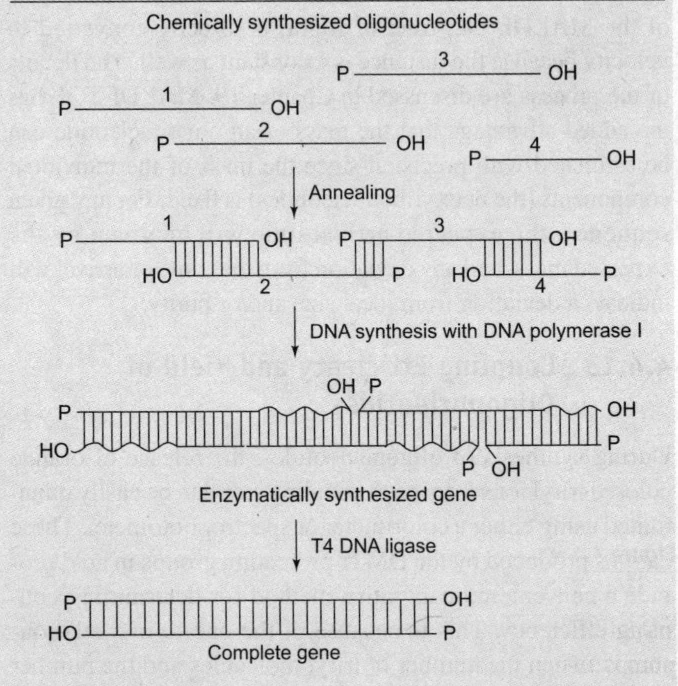

Figure B Assembly and *in vitro* enzymatic synthesis of a gene

This enzyme uses the 3′-hydroxyl groups as replication initiation points and the single stranded regions as templates. After the enzymatic synthesis is completed, the nicks are sealed with T4 DNA ligase (Figure B). For larger genes, (>1,000 bp), smaller sections of the gene are assembled and then combined to eventually form the complete coding sequence. Because it is absolutely essential that a chemically synthesized gene has the correct sequence of nucleotides, the complete sequence is characterized by DNA sequence analysis.

The somatostatin gene, the genes for the A and B chains of human insulin, the gene for human leukocyte interferon, and the human proinsulin gene are some of the examples of gene synthesis.

Sources: Itakura K, Hirose T, Crea R, Riggs AD, Heyneker HL, Bolivar F, Boyer HW (1977) Expression in *Escherichia coli* of a chemically synthesized gene for the hormone somatostatin, *Science* 198: 1056–63; Crea et al. (1978), Chemical synthesis of genes for human insulin, *Proc. Natl. Acad. Sci. USA* 75: 5765–5769; Brousseau R, Scarpulla R, Sung W, Hsiung HM, Narang SA, Wu R (1982) Synthesis of a human insulin gene V. Enzyme assembly, cloning and characterization of the human proinsulin DNA, *Nature* (1980) 292: 756–62; *Gene* (1982) 17: 279–289.

ers in increasing versatility in cloning experiments, oligonucleotides in site-directed mutagenesis, and probes in screening of libraries do not form the part of this chapter, and are dealt in other chapters (for details see Chapters 7, 13, 16, and 19).

4.5 DNA SYNTHESIZERS (AUTOMATED SOLID-PHASE SYNTHESIS)

Machines that automate the chemical reactions for DNA synthesis (DNA synthesizers or gene machines) have made the production of single stranded oligonucleotides (~100 deoxyribonucleotides) into a more or less routine procedure (Figure 4.13). Currently, the phosphoramidite method is the procedure used for automated chemical synthesis of DNA in a DNA synthesizer, and the synthesis takes place in 3′ → 5′ direction, opposite to the biological synthesis. Thus, each incoming deoxyribonucleotide is coupled to the 5′-hydroxyl terminus of the growing chain. The details of the method have already been described in previous section. The process in DNA synthesizers is fully automated as these machines con-

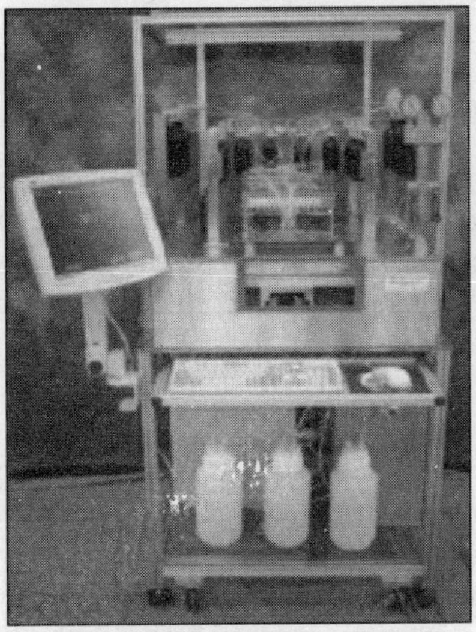

Figure 4.13 Automated DNA synthesizer

sist of reagent reservoirs equipped with a set of valves and pumps that are programmed to introduce specified deoxyribonucleotides, and the reagents required in the correct order for the coupling of each consecutive deoxyribonucleotide to the growing chain. These valves and pumps are controlled by a microcomputer, and all the reactions are carried out in succession in a single reaction column. The durations of all reactions and the washing steps are also computer controlled. Further the incorporation of modified deoxyribonucleosides (e.g., containing a fluorescent label) into the growing oligonucleotide eases the monitoring of the process. This reaction sequence can be repeated up to ~150 times with a cycle time of 40 min or less in commercially available automated synthesizers. The first commercially available DNA synthesizer was developed in Canada (McGill University) in 1981, since then the technique is advanced to a greater extent. Further, the performance of the synthesizer is monitored by trityl analysis (Exhibit 4.2).

For a core synthesis facility, the analysis of performance of DNA synthesizer on a daily basis is important because it provides an early indication of how well the coupling reactions are proceeding. This can be easily followed by trityl analysis, because of the one-to-one relationship between the number of trityl molecules released, and the number of new phosphate linkages formed. The spectrophotometric analysis of orange colored trityl cations is the most accurate method, but the method is much too labor intensive for routine use by any core facility. Hence a variety of alternative methods have been developed that are easier though less accurate (Exhibit 4.2). Although the trityl results cannot detect problems with poor oxidation or capping steps, these do warn of problems with moisture contamination, poor coupling reagents, blocked

Exhibit 4.2 Trityl analysis

During the detritylation step of oligonucleotide synthesis, orange colored trityl ions are released due to acid hydrolysis of DMTr protecting groups from each coupling step. Hence, the estimation of these cations provides a convenient quantitative method for determining coupling efficiency. These cations can be easily and accurately quantitated using a spectrophotometer. Besides, several other methods are also used, which are listed below in the order of increasing complexity and accuracy:

AutoAnalysis Detectors This option, available on Applied Biosystems 392/394 DNA synthesizers, consists of individual conductivity detectors (one per column) that measure the amount of charged trityl cation flowing out of each synthesis column. The ABI monitor is easy to use because it is integrated into the synthesizer. However, it is expensive and does not report the coupling yields for each individual step (although it measures and calculates them). Instead, an average stepwise yield, which masks individual step yields into one average value, is presented for each step. This makes it difficult to decipher the actual machine performance and so the detector is limited to the detection of rather severe failures.

The TritylTech on-line Trityl Detector These detectors, available from Ana-Gen Technologies, can be attached to almost any DNA synthesizer (including the ABI 380A/B series). This detector employs the more conventional colorimetric method for measuring trityl colors, but differs from the flow-through trityl detectors used by Pharmacia, Milligen etc. in using up to four quartz cuvettes to collect the trityl colours. The color measurements only takes place after the entire colours have been collected, partially diluted and mixed. Coupling data is presented in a clear, easily understood format (percent stepwise and overall yields) and the monitor is capable of detecting couplings that are only a few percentage points lower than normal. However, the detector is not integrated with the synthesizer and requires a separate PC controller that must be configured prior to each synthesis. Obtaining a printout of the results is difficult because the monitor uses a parallel port connection for data acquisition instead of a serial port and the time required to flush out the quartz cuvettes adds almost a minute to each coupling cycle.

PC-800 Fiber-optic 'Dipping Probe' Colorimeter This method is the most time consuming but offers accuracy similar to a spectrophotometer. This method utilizes a PC-800 fiber-optic 'dipping probe' colorimeter from Brinkmann Instruments, which eliminates the need to transfer cuvettes in and out of a spectrophotometer. This involves collection of trityl colors, diluting them to a constant volume with 5% DCA/DCE, vortexing and finally dipping in fiber optic probe to measure absorbance using either a 470 nm or 545 nm filter. The absorbance reading makes the measurements much faster although still not automated. This method is also the least expensive and the most effective for the facilities that may only want to perform the occasional trityl analysis.

Source: Richard TP, *Tips for Oligonucleotide Synthesis,* University of Calgary.

or partially plugged lines, and a variety of other instrument malfunctions. In addition, if the correct sequence has been entered into the synthesizer, and the correct number of good trityl colors has been verified, then one can be assured that the correct sequence has been made, and only a gel picture or chromatogram confirming sequence homogeneity (i.e., correct capping and oxidation) is required for complete quality control documentation.

4.6 SYSTEMATIC EVOLUTION OF LIGANDS BY EXPONENTIAL ENRICHMENT (SELEX)

Certain nucleic acids, for example tRNAs and rRNAs, have functions that require the maintenance of well-defined 3-D structures of these molecules. This has led to the development of a process called SELEX (systematic evolution of ligands by exponential enrichment), a technique devised by Gold, for generating single stranded oligonucleotides that specifically bind target molecules with high affinity and specificity. This technique involves the construction of a library of polynucleotides having fixed known sequences at their 5′- and 3′-ends and random sequences in their central region. Such sequences are synthesized using solid-phase methods by adding single fixed deoxyribonucleotides at the 3′- and 5′-ends, and a mixture of all four deoxyribonucleotides during the synthesis of the positions to be randomized. For a variable region that is 30 nucleotides long, this will yield a mixture of $4^{30} \approx 10^{18}$ different oligonucleotides. The next step is to select those sequences in the mixture that selectively bind a target molecule, say X. Those oligonucleotides that bind X with higher than average affinity can be separated from the other oligonucleotides in the mixture by affinity chromatography with X linked to a suitable matrix, and the precipitation of X-oligonucleotide complexes by anti-X antibodies. The oligonucleotides are then separated from X, and amplified by cloning or PCR. This selection and amplification process is repeated for usually 10–15 cycles. Such enrichment cycles are necessary because those few oligonucleotides in a library with the highest affinity for X compete for binding X with the far more abundant weak binders. The result is separation of oligonucleotides, known as aptamers, having high binding affinity for the target molecule X. If X is a protein, these oligonucleotide products are bound to X with dissociation constant typically of the order of ~10^{-9} M, and if X is a small molecule (for the reaction $A + B \rightleftharpoons A.B$, $K_d = [A][B]/[A.B]$), the dissociation constant lies in the range between 10^{-3} M and 10^{-6} M. Thus, aptamers selected for their affinity towards a particular member of a protein family do not bind other members of that family. Besides their application are used in the determination of structure of RNA or DNA, these aptamers are also potentially useful as diagnostic tools, and therapeutic agents, because of their ability to bind a wide variety of target molecules including simple ions, small molecules, proteins, organelles, and even whole cells.

4.7 DNA CHIPS

DNA chips are miniaturized 2-D arrays of different DNA molecules (oligonucleotides/cDNAs/PCR products) anchored to a flat surface of nylon wafer or silicon glass in a ~1 cm wide square grid (Figure 4.14). The different DNA molecules are synthesized or deposited at defined locations on the surface of support media where these remain attached, such that the identity of each spot is known. Simply speaking, microarray or chip represents an ordered collection of microscopic spots, each spot containing a single defined species of a DNA molecule attached to the glass surface. Microarray technology has evolved from Southern blotting, whereby fragmented DNA is attached to a substrate, and then probed with a known gene or fragment. Arrays of DNA can either be spatially arranged, or can be tagged, or labeled with specific DNA sequences such that these can be independently identified. The term 'DNA chip' came into being because of the use of photolithographic method for DNA synthesis, which was earlier known for integrated circuit ('computer chip') production. These are also commonly called GeneChips, or Gene Arrays, or BioChips. The terms slide, or array, or chip are used interchangeably to refer to the printed microarray.

Figure 4.14 DNA chip
Source: University of Calgary

4.7.1 Support Media

Chip represents a support media built of semiconducting material (material that is neither a good conductor of electricity nor a good insulator) either in the form of membrane, or a slide, or wafer. The membranes commonly used are commercially available nitrocellulose and charged nylon (similar to the ones used in standard blotting assays). These membranes being porous, allow relatively large amounts of DNA to be applied, resulting in strong signal intensities and good dynamic range, but because of the large surface area per spot, also give relatively high background. The nylon filter arrays can be reused, because the DNA sticks relatively well to the surface. However, spot sizes cannot be reduced to a level possible with non-porous media.

In contrast to nylon/nitrocellulose membrane, glass/silicon or polypropylene slides or wafer are advantageous, where the spot sizes can be reduced to a very high level. Moreover, the glass or polypropylene slides are chemically inert. This ability of miniaturization in combination with chemical inertness and low intrinsic fluorescence are the main advantages of polypropylene and especially glass permitting high DNA concentrations even from small samples. The planar surface structure, however, restricts the loading capacity. Moreover, the ability to reuse chips eliminates the variance between individual chips, thus increasing the experimental reliability of the analyses. Owing to various advantages described above, glass wafer in the form of microscopic slide is the support of choice in DNA chip technology.

4.7.2 Slide Chemistry

Owing to the inertness, DNA cannot couple directly to the surfaces of most plastics or to silicate glass through their silanol groups. Thus, DNA binding to these inert surfaces is a two-step process: (i) Derivatization of the surface: It is necessary to coat the surface with a group from which the growth of the oligonucleotide chain can be initiated. These groups are also called cross-linkers; and (ii) Binding of DNA to the cross-linkers: Immobilization of DNA is achieved through electrostatic (non-covalent) interaction or covalent bonding to the derivatized slides.

Thus, the following three strategies are adopted for cross-linking and further immobilization of DNA on the surface of support media:

Derivatization of Glass with Amino-reactive Silanes, Poly-*L*-lysine or Amino Silanes, and Derivatization of Polypropylene by Plasma-amination Such derivatization enhances both, the hydrophobicity of the slide and the adherence of the deposited DNA. The linker attachment to the support is achieved via amino groups (Figure 4.15). These provide positively charged surface groups that are exploited for immobilization of the negatively charged phosphate groups of DNA by electrostatic (non-covalent) interaction. The interaction takes

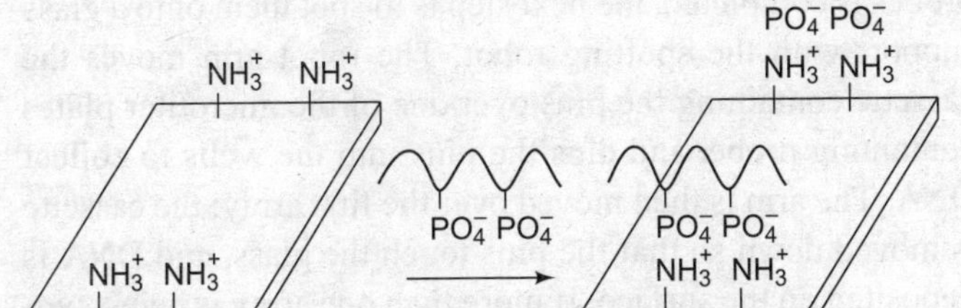

Figure 4.15 DNA binding by electrostatic interactions

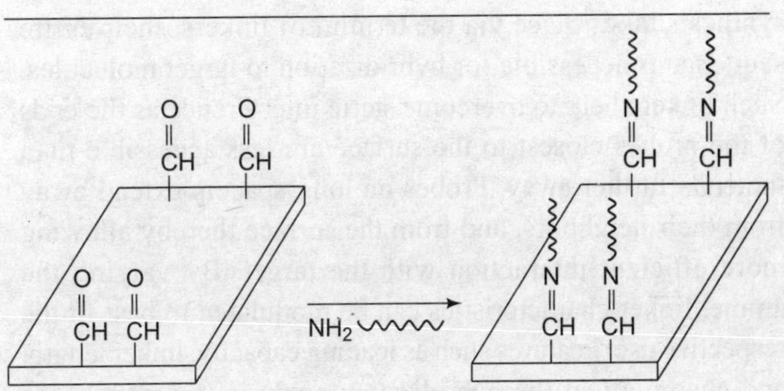

Figure 4.16 Aldehyde-derivatization of slides using Schiff's base reaction

place at several locations along the DNA backbone, so that the probe is tethered to the glass at many points. Such non-covalent interactions are used for robotic deposition of cDNA probes (for details see Section 4.7.3). Because most oligonucleotide probes are shorter than cDNAs, these interactions are not strong enough to anchor oligonucleotide probes to glass.

Derivatization of Glass with Aldehyde and Addition of Amino Cross-linkers Amino cross-linkers can be attached to the aldehyde-derivatized glass surface based on Schiff's base reaction (Figure 4.16). The amine group can be added to either end of the oligonucleotide during synthesis, although it is more usual to add it to the 5′-end of the oligonucleotide, which in turn is exploited for covalent bonding of DNA.

Derivatization of Glass with Epoxides and Addition of Amino Cross-linkers The glass surface is first derivatized with epoxides, and then amino groups are attached to them (Figure 4.17). In this case also the amine group can be added to either end of the oligonucleotide during synthesis, which is then exploited for covalent bonding of DNA.

In most cases, DNA is attached to the derivatized slide by baking or cross-linking with UV irradiation. As the attachment of prefabricated oligonucleotides (probes) or *in situ*

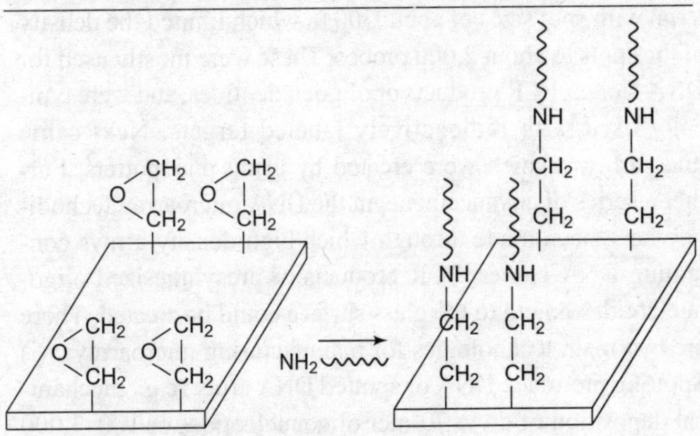

Figure 4.17 Epoxy-derivatization of slides [An amino cross-linker is used to covalently attach DNA to the epoxy-derivatized surface.]

synthesis takes place via the termini of linkers, their entire sequence is accessible for hybridization to target molecules. Such linkers help to overcome steric interference as the ends of the probes closest to the surface are less accessible than the ends further away. Probes on long spacers extend away from their neighbors, and from the surface thereby allowing more efficient interaction with the target. By varying the amine, linker characteristics can be modulated to best fit the respective use. Features such as loading capacity, linker length, and charge affect the hybridization analysis (for details see Chapter 19). Another factor that can be influenced during linker synthesis is the hydrophilic or hydrophobic character of the surface by incorporation of the respective form of amines. Most importantly, all linker derivatives permit covalent bonding of DNA to the support, and hence warrant reusability of the arrays. Further, the linker system is not restricted to the attachment of aminated biomolecules but amicable to alternative reactive groups such as thiol.

4.7.3 Fabrication Techniques

A DNA microarray consists of a solid surface, usually a microscope slide, onto which DNA molecules have been chemically bonded. The purpose of a microarray is to detect the presence and abundance of labeled nucleic acids in a biological sample, which will hybridize to the DNA on the array via Watson–Crick duplex formation, and which can be detected via the label. A good microarray slide should possess the following attributes: (i) It should possess high spot (feature) number and quality; (ii) It should have maximal DNA binding capacity; (iii) It should have consistent size; (iv) It should produce homogeneous signal; (v) It should give reproducible results; and (vi) It should possess minimal overall background.

Keeping the above points in mind, several DNA microarray fabrication techniques have been devised. The first arrays, created in the mid 1980s, were called macroarrays. These were fabricated by spotting DNA probes on a membrane-type material with spot sizes of about 300 μ, which limited the density of the spots to about 2,000 probes. These were mostly used for DNA clones, PCR products or oligonucleotides, and were typically used with radioactively labeled targets. Next came microarrays, which were created by using pin-spotters. Further a series of advancements in the DNA microarray technology have been made through which high-density arrays containing DNA clones, PCR products or presynthesized oligonucleotides bound to the glass surface could be created. There are two main technologies for manufacturing microarrays: (i) Spotting pre-made DNA or spotted DNA array (e.g., mechanical deposition of up to 70-mer oligonucleotides or 100–3,000 bp long PCR products); and (ii) Synthesis on slide (i.e., *in situ*, or '*on-chip*', or *de novo* synthesis on solid substrate) (e.g., synthesis using methods such as photolithography and printing).

Comparison of the two types of arrays, one with prefabricated oligonucleotides and the other with *in situ* synthesis reveals that the latter has some advantages over deposition of presynthesized oligonucleotides. It is not profitable to make large arrays using presynthesized oligonucleotides attached to the surface. Moreover, as the sequences for *de novo* synthesized arrays are stored electronically rather than physically in frozen DNA libraries, the costs, and the potential for errors in amplification, storage, and retrieval are eliminated. On the other hand, it is difficult to assess the quality of the oligonucleotides made on a surface. Therefore, this technique is used for quality control, but it is not available for most biological laboratories. In contrast, the presynthesized oligonucleotides can be assessed before these are attached to the surface. Further, the methods used also depend on the size of the DNA probe on the chip. In cDNA microarrays, relatively long DNA molecules are anchored on a solid surface by high-speed robots, and as already mentioned in Section 4.7.2, the cDNAs are preferably held by electrostatic interactions on derivatized slides. The oligonucleotide arrays can also be fabricated by robotic deposition of conventionally synthesized oligonucleotides on a glass substrate, but *in situ* chemical synthesis is preferred; moreover, covalent attachment methods (through aldehyde or epoxy-derivatization) are preferred.

The following section describes various techniques used for fabrication of DNA microarray:

Spotting Prefabricated DNA (Robotic Deposition of Presynthesized DNA)

This is the first technology by which first microarrays were manufactured, and such microarrays are also called spotted microarrays. DNA arrays are fabricated by high-speed robotics on glass or nylon substrates, where oligonucleotides are synthesized by conventional method (phosphoramidite synthesis), and later deposited (immobilized or spotted) on glass chip. The robot (arrayer) spots a sample of each oligonucleotide product onto a number of matrices in a serial operation. Several steps are involved in the synthesis of spotted arrays:

Production of the Probes First step is the production of probes to put on the array. The cDNA probes are made via highly parallel PCR, while oligonucleotide probes are presynthesized.

Spotting Step Once the synthesis of cDNA/oligonucleotide probes is completed, the next step is to spot them onto a glass support with the spotting robot. The robot arm moves the cassette containing the pins over one of the microtiter plates containing probe, and dips the pins into the wells to collect DNA. The arm is then moved over the first array; the cassette is moved down so that the pins touch the glass, and DNA is deposited on the surface. If more than one array is being synthesized, the cassette is moved to the subsequent arrays. The attachment to the derivatized glass surface is through covalent or non-covalent (electrostatic) interactions (for details see Section 4.7.2)

Washing Before collecting the next DNA to be spotted, the pins are washed to remove any residual solution, to ensure that there is no contamination of the next sample.

Post Spotting Processing on Glass Slide The final step of array production is fixing, in which the surface of the glass is modified. There are many fixing processes that depend on the precise chemistry on the surface of the glass. The purpose of fixing is to prevent the DNA target from the sample from sticking to the glass of the array during hybridization. It is also common to modify the surface so that the glass becomes more hydrophilic because this aids mixing of the target solution during the hybridization stage. Some microarray production facilities do not fix their arrays.

The schematic representation of the process is outlined in Figure 4.18 and a typical custom-made pin-spotter is depicted in Figure 4.19.

The first pin-based robotic system, pin-spotters (Stanford University), can dispense an accurate volume of DNA solution in a spot of about 150 μ onto a glass slide. DNA clones, PCR products or presynthesized oligonucleotides can be bound to the glass surface to create high-density arrays. The early release of the Stanford pin-spotting arrayer design also spurred commercial development of high quality microarrayers, which led to the availability of numerous models at competitive prices. Despite the problems and errors associated with collection of DNA libraries, the widespread availability of pin-spotted microarrays led to acceptance and standardization of protocols, terminology, data storage, and analysis techniques. In an example of Motorola gel pads, an array of prefabricated oligonucleotides is attached to patches of activated polyacrylamide. Several spotting robots are now available in the market. Genetix spotting robot located at the Mouse Genetics Unit in Harwell, Oxfordshire has the pins held in a cassette in a rectangular grid, which in turn is held on a robot arm that can be moved between the microtiter well plates and the glass arrays to deposit liquid. The number of pins in the cassette can vary. The more the pins, the greater the throughput of the robot, but greater the propensity for pin-to-pin variability. Each pin will spot a different grid on the array. Most pins in modern use have a reservoir that holds sample, and so can print multiple features, usually on different arrays, from a single visit to the well containing probe. Earlier robots use solid pins, which can only print one feature before the need to collect more DNA from the well.

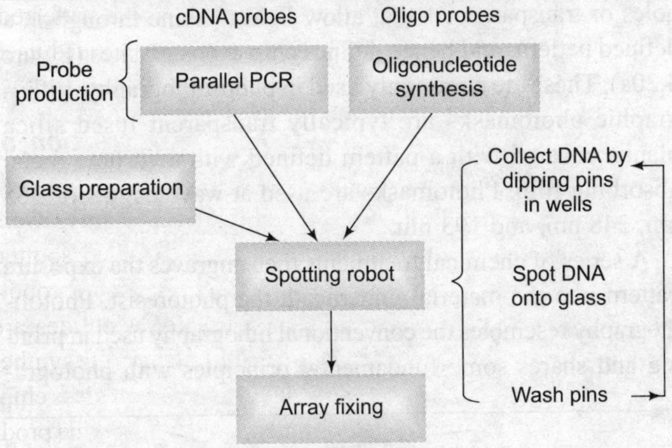

Figure 4.18 Scheme for manufacture of spotted cDNA or oligonucleotide microarrays

Figure 4.19 An example of a custom-made pin spotter
Source: Pat Brown Lab, Stanford University

There are a number of different designs of pins. The first spotting robots used solid pins. These can only hold enough liquid for one spot on the array, thus requiring the pin cassette to return to the plate containing probe before printing the next spot. Most array-making robots today have pins with a reservoir that holds liquid. This enables higher throughput production of arrays because each probe can be spotted on several arrays without the need to return the pins to the sample plates. However, not every spotting robot is a pin-based system, for example, Perkin Elmer robots use a piezoelectric system to fire tiny drops onto the arrays. Such spotting robots are a minority in the microarray field.

A spotted DNA array is thus made by transferring or spotting actual DNA clones (full-length or nearly full-length), more usually PCR products derived from them, or 20–70 residue oligonucleotides individually, onto a solid support (nylon membrane or glass slide), where these are immobilized. Nylon macroarrays produced by spotting are generally about 10–20 cm^2 in size, and the feature density is low, with typically 1–2 mm between targets (i.e., 10–100 targets/cm^2). These macroarrays are easy to manufacture and are therefore relatively inexpensive; these are also simple to use because standard hybridization procedures are applicable. For this reason, nylon macroarrays are still manufactured by a number of commercial suppliers, and the technology for in-house array production is readily available, involving simple robotic devices or even hand-held arrayers. Nylon macroarrays, however, suffer from certain disadvantages such as low probe density, and hence, limited number of sequences can be analyzed simultaneously; hybridization must be carried out in a large volume using a radioactive probe, the results are obtained by autoradiography or preferably using a phos-phorimager, and extensive miniaturization of nylon membranes has been difficult because resolution of the signal provided by radioactive probes is poor. A major advancement in spotted array technology was the development of microarrays on glass chips. Since glass is non-porous and has very little autofluorescence, fluorescent probes can be used, and these can be applied in very small hybridization volumes. The greater resolution afforded by fluorescent probes allows probe density to be increased significantly compared to nylon macroarrays, and the small hybridization volume improves the kinetics of the reaction. Together, these advantages signify that more probes can be assayed simultaneously with the same amount of probe without loss of sensitivity. Thus, glass arrays can routinely be manufactured with up to 5,000 features/cm^2. Because of the numerous advantages of glass microarrays, including their amenability for automated spotting, these have emerged as the most popular type of spotted array for expression profiling. However, spotted DNA array involves a high cost of production, and hence the technology is beyond the reach of all except for the best-funded laboratories. Earlier, researchers either purchased ready-made arrays designed for single use from a commercial source or invested in the resources required for in-house array manufacture, which was very expensive. The cost of a commercially available precision robot for array manufacture is $1,00,000 or more, even for those with the simplest specifications. Additionally, clone sets usually need to be purchased to provide the probes for a homemade array. Selected sources of clone sets for the manufacture of spotted arrays include Research Genetics, Incyte Genomics, and Genosys Biotech.

An alternative to a spotted DNA array is a high-density prefabricated oligonucleotide chip. These are similar to DNA arrays in that these consist of gridded DNA targets that are analyzed by hybridization. However, while DNA arrays consist of double stranded clones or PCR products that may be up to several hundred base pairs in length, oligo chips contain single stranded targets ranging from 25–70 nucleotides. Oligo chips can be made in the same way as spotted DNA arrays, by robotically transferring chemically synthesized oligonucleotides from microtiter dishes to a solid support, where these are immobilized. However, maximum array density is increased almost ten-fold if the oligos are printed directly onto the glass surface. Thus, up to 10,000 different oligonucleotides/cm^2 (lesser than *in situ* synthesis) are robotically deposited in an array on a coated glass surface.

In Situ (De Novo/On-Chip) Synthesis

These arrays are fundamentally different from spotted arrays, where instead of presynthesizing oligonucleotides, oligos are built up base-by-base on the surface of the array. The *in situ* synthesis is based on light-directed photolithography, printing, microfluidics technology, or microelectrodes. In several techniques, photolabile materials are used for protection or blocking of 5′-OH group. At the start of each coupling step, these groups have to be deprotected. Thus, the different methods for deprotection lead to the three main technologies for making *in situ* synthesized arrays. These are: (i) Photodeprotection using masks (used by Affymetrix® technology); (ii) Photodeprotection without masks (used by Nimblegen and Febit); and (iii) Chemical deprotection with synthesis via inkjet technology (used by Rosetta, Agilent, and Oxford Gene Technology).

Different methods for *in situ* synthesis of DNA on surface are discussed in detail in the following sections:

Photolithographically Controlled *In Situ* (*On-chip*) Synthesis Using Photomasks Photolithography, also called optical lithography, (Latin: Photo-litho-graphy: light-stone-writing) is a process used in microfabrication to selectively remove parts of a thin film or the bulk of a substrate. It uses light to transfer a geometric pattern from a photomask (or mirror in maskless photolithography) to a light-sensitive chemical ('photosensitive emulsion', or 'photoresist', or 'resist') on the substrate. A photomask is an opaque plate with holes or transparencies that allow light to shine through in a defined pattern, and hence defines chip exposure sites (Figure 4.20a). These are commonly used in photolithography. Lithographic photomasks are typically transparent fused silica blanks covered with a pattern defined with a chrome metal absorbing film. Photomasks are used at wavelengths of 365 nm, 248 nm, and 193 nm.

A series of chemical treatments then engraves the exposure pattern into the material underneath the photoresist. Photolithography resembles the conventional lithography used in printing and shares some fundamental principles with photogra-

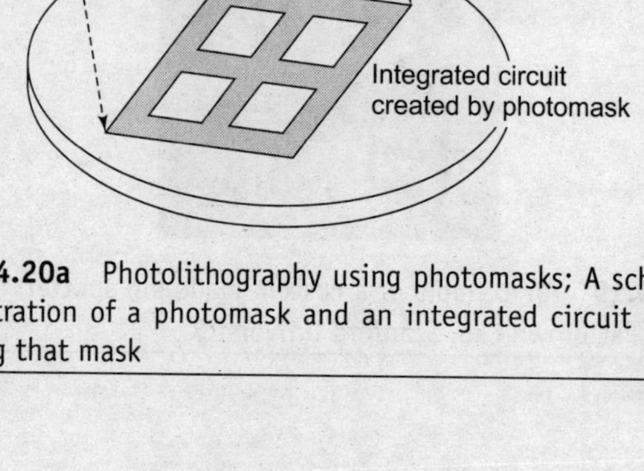

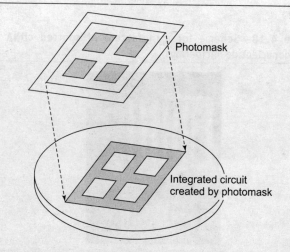

Figure 4.20a Photolithography using photomasks; A schematic illustration of a photomask and an integrated circuit created using that mask

phy. It is used because it affords exact control over the shape and size of the objects it creates, and because it can create patterns over an entire surface simultaneously. It requires a flat substrate to start with, and is not very effective in creating shapes on non-flat surfaces. Moreover, it also requires extremely clean operating conditions.

A combination of photolithography (the process used to fabricate electronic chips) and solid-phase DNA synthesis, commonly called *in situ* light-directed combinatorial synthesis, involves the synthesis of a large number of different oligonucleotides on a single chip. This technique, adopted from semiconductor industry, was developed by Fodor and colleagues at Affymetrix Inc. (Santa Clara, California, USA). By using photomasks and UV catalyzed base deprotection, large numbers of oligonucleotide arrays can be synthesized with a high feature density. In this method, direct '*on-chip*' light-directed oligonucleotide synthesis takes place on wafer with a very high density ranging from 3,00,000 to up to ~1 million oligonucleotides/cm^2. The oligonucleotides are synthesized at a high spatial resolution and at precise locations. Chips are now being made that contain all the genes predicted in the human genome. However, coupling chemistry using photolabile monomers is less efficient than standard phosphoramidite chemistries, resulting in arrays consisting of short oligonucleotides (25-mer) that require multiple features for the unambiguous identification of each gene. In addition, the design and construction of new arrays is slow and expensive because new masks need to be cut for each base changed in the array.

The *in situ* light-directed photolithographic process is complex and begins on a 5″ × 5″ glass or silicon wafer that is hydroxylated and silanized to attach DNA covalently attached to the surface in a simple chemical reaction. However, the covalent binding sites are blocked by photolabile protecting groups. A chromium photomask is then applied to the surface of the chip, which either blocks or transmits light onto specific locations of the wafer surface. Under illumination, the protecting groups at certain areas in the wafer are destroyed, allowing the addition of deoxyribonucleotides at specific locations. Thus, upon flooding the surface with a solution containing A/T/C/G, coupling occurs only in those regions on the wafer that have been deprotected through illumination. Any side reaction during oligonucleotide synthesis is prevented by blocking each deoxyribonucleotide at its 5′-end by photochemically removable (photosensitive/photolabile) protective group, for example, 2-(2-nitrophenyl) propoxycarbonyl (NPPOC) group and α-methyl-6-nitropiperonyloxycarbonyl protecting group (MeNPOC) (similar to DMTr in conventional solid-phase synthesis or phosphoramidite synthesis). Thus, for example, during synthesis, the oligonucleotides to which a particular deoxyribonucleotide (say G) is to be added are deprotected by shining light on them through a photomask that blocks the light from hitting those grid positions to which G is not to be added (Figure 4.20b). The chip is then incubated with a solution of activated deoxyguanosine. This leads to the coupling of G to the deprotected oligonucleotides only. The unreacted G is washed away and the process is repeated with photomasks of

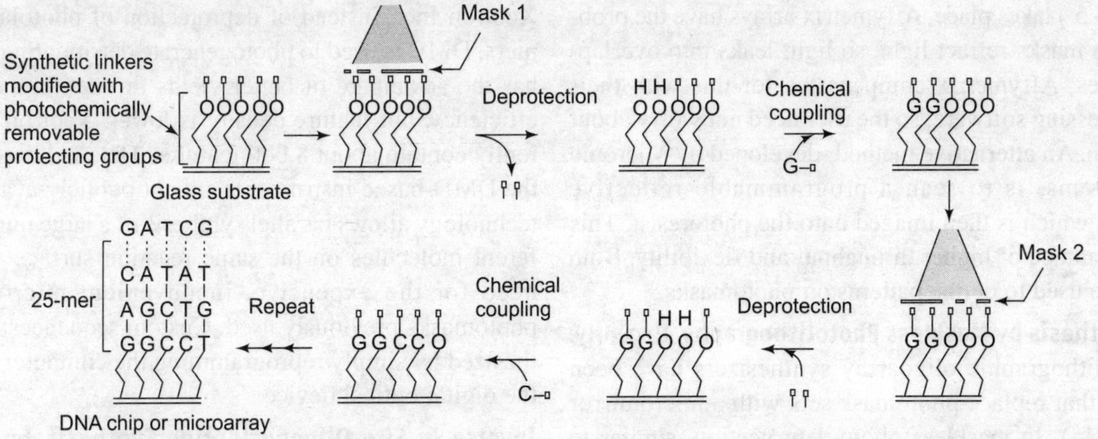

Figure 4.20b Photolithography using photomasks; The fabrication of GeneChip by *in situ* synthesis (photolithographic synthesis) by photoactivation and deprotection of nucleic acids [In step 1 of the process, oligonucleotides that are anchored to a glass substrate and each having a photosensitive protecting group at its 5′-end, are exposed to light through a mask that only permits the illumination of the oligonucleotides destined to be coupled, for example, to a G residue. Photomasks are used to pattern UV light at localized regions to selectively synthesize a patterned array. The light deprotects these oligonucleotides so that only these react with the activated G residue that is incubated with the chip in step 2. The entire process is repeated in steps 3 and 4 with a different mask for C residues and in subsequent reaction cycles for T and A residues, thereby extending all of the oligonucleotides by one residue. This quadruple cycle is then repeated for as many nucleotides as are to be added to form the final set of oligonucleotides. The method is developed by Affymetrix.]

different geometries that shield different positions of the grid from illumination, followed by addition of different deoxyribonucleotide solutions. As the coupled deoxyribonucleotide also bears a light-sensitive protecting group, the cycle can be repeated. In this way, the microarray is built as the oligonucleotides are synthesized through repeated cycles of deprotection and coupling. The process is repeated until the oligonucleotides reach their full-length, usually 25 nucleotides. The process is highly accurate, and depending on the demands of the experiment and the number of oligonucleotides required per array, each wafer is diced into tens or hundreds of individual arrays. This method allows the production of the currently available dense arrays [up to 64,000 features over an area slightly larger than 1 cm^2 in commercially available GeneChip® Arrays (from Affymetrix Inc.), but experimental chips with a density of $>10^6$ targets per cm^2 have been produced]. As the use of physical photomasks makes the GeneChip very expensive, several other modifications and other methods have been developed. The use of NPPOC yields near quantitative amounts of product in each individual condensation reaction. This is in contrast to the efficiencies of other established chemistries, including the commercially used MeNPOC chemistry of Affymetrix, which results in yields of 85%–90% per synthesis step. Hence, the accumulative total yield increases by more than three-fold in case of octamers, and this difference widens to a factor of considerably more than one order of magnitude for 20-mer oligonucleotides. For many applications, from approaches of highly parallel DNA sequencing to the creation of dsDNA microarrays, the availability of a free 3′-terminus is advantageous, and hence a reverse direction of oligonucleotide synthesis ($3' \rightarrow 5'$) takes place. Affymetrix arrays have the problem that the masks refract light, so light leaks into overlapping features; Affymetrix compensates for this with their image-processing software, so the user need not worry about this problem. An alternative method, developed by Micronic Laser Systems, is to scan a programmable reflective photomask, which is then imaged onto the photoresist. This has the advantage of higher throughput and flexibility. Both methods are used to define patterns on photomasks.

***In Situ* Synthesis by Maskless Photolithography** Recently, new photolithographic oligoarray synthesizers have been introduced that replace photomask sets with a micromirror (Figure 4.21a). In maskless photo-deprotection, similar to Affymetrix technology, light-mediated deprotection is done. Here, instead of using a physical mask, the array is synthesized using a computer-controlled micromirror. This technology consists of a large number of mirrors embedded on a silicon chip, each of which can move between two positions: one position to reflect light, and the other to block light. At each step, the mirrors direct light to the appropriate parts of the array. This maskless photolithography reduces the capital costs involved in chip manufacture by employing virtual masks. In maskless photolithography, light used to expose the photosensitive emulsion (or photoresist) is not projected from or transmitted through a photomask. Instead, most commonly, the radiation is focused to a narrow beam. The beam is then used to directly write the image into the photoresist, one or more pixels at a time. The main advantage of maskless photolithography is the ability to change lithography patterns from one run to the next, without incurring the cost of generating a new photomask. This may prove useful for double patterning. Using this technique new array designs can be produced in hours. NimbleGen Inc. (Houston, TX) employs UV light reflected from a miniature array of aluminium mirrors to focus selectively on the appropriate areas of the chip, and destroy the photolabile groups. This is known as the maskless array synthesizer (MAS) (Figure 4.21b).

One of the main advantages of micromirror technology over both Affymetrix technology, and spotted arrays is that the oligonucleotide being synthesized on each feature is entirely controlled by the computer input given to the array-maker at the time of array production. Therefore, this technology is highly flexible, with each array able to contain any oligonucleotide of operator's wish. However, this technology is also less efficient for making large number of identical arrays.

Another significant improvement in making photolithography a practical method for *in situ* combinatorial chemical synthesis is the introduction of the digital optical device or digital photolithographic device for light patterning projection onto a reaction surface. Xeotron Inc. and Febit GmbH have developed maskless technologies based on digital micromirrors device (DMD). In the technique developed by Xeotron Inc., instead of deprotection of photolabile monomers, DMD is used to photogenerate detritylating acids. This has the advantage of better yields through higher coupling efficiency, but feature density is lower. Xeotron arrays currently contain about 8,000 features. Detailed descriptions of the DMD-based instruments are not publicly available. This technology allows parallel synthesis of a large number of different molecules on the same reaction surface without the need for the expensive, inconvenient microfabricated photomasks previously used. Custom sequences can be synthesized by simply reprogramming the computer controlling the digital optical device.

Inverse *In Situ* Oligonucleotide Synthesis Inverse *in situ* synthesis is a variant of *in situ* oligonucleotide synthesis in which 5′-phosphoramidities, protected by 2-nitrophenylethyl (NPE), and 2-(4-nitrophenyl) ethoxycarbonyl (NPEOC) functions are employed for *in situ* synthesis of oligonucleotides in $5' \rightarrow 3'$ direction on flat glass surfaces. By this inverse synthesis format, the oligonucleotides are attached to the solid support via their 5′-ends, while the free 3′-OH groups are available as substrates for enzymatic reactions such as elon-

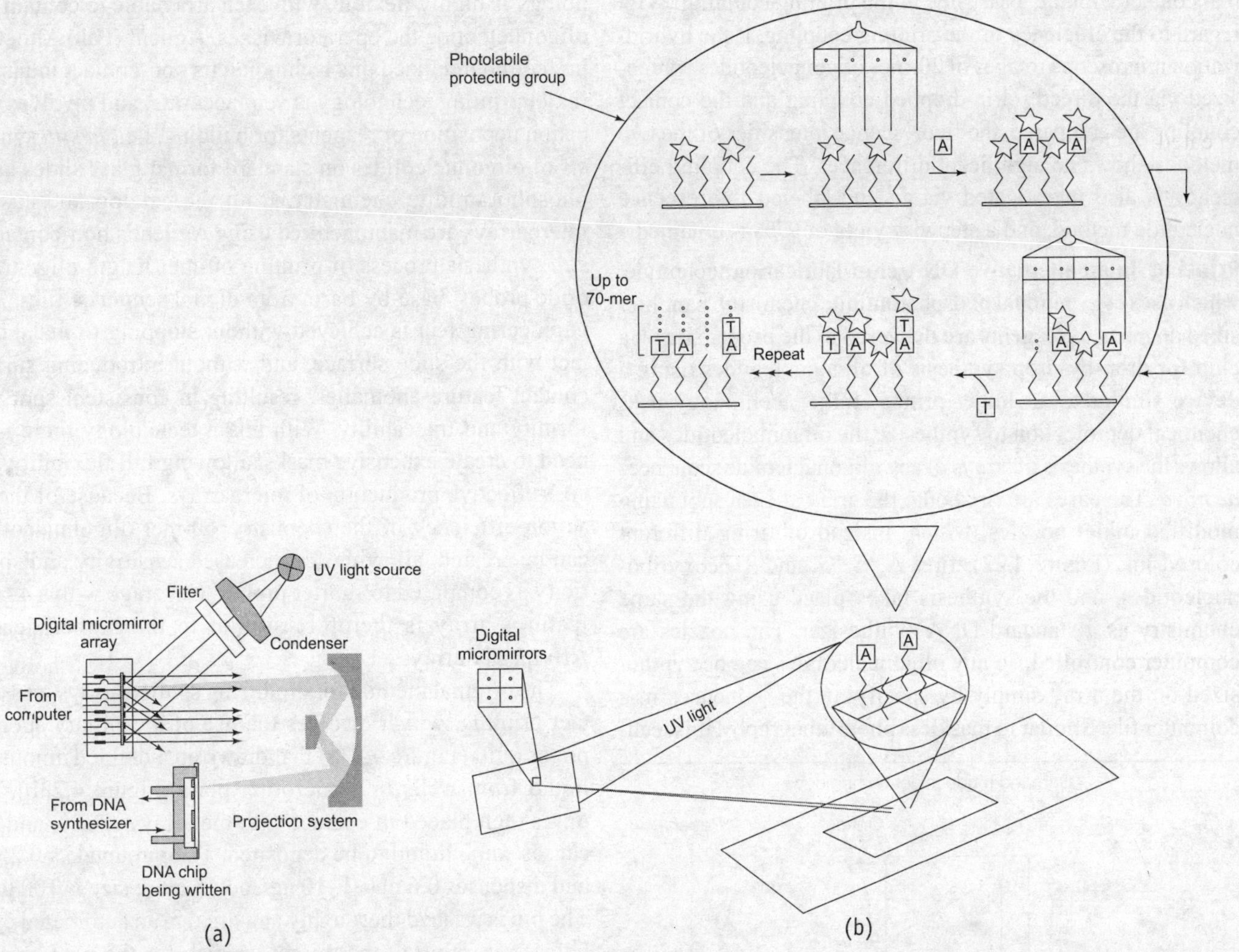

Figure 4.21 Maskless photolithography; (a) Maskless photodeprotection, (b) The synthesis of microarrays using MAS technology [The digital micromirrors reflect a pattern of UV light, which deprotects the nascent oligonucleotide and allows addition of the next base; This technology is developed by NimbleGen Systems.]

gation by polymerases, thereby adding another feature to the portfolio of chip-based applications. Having a fluorescence dye present at the first base during synthesis, the quality of the oligonucleotides can be analyzed quantitatively by capillary electrophoresis after release from the solid support. With about 95% yield per condensation, it is found to be equivalent to synthesis results achieved on CPG support. The chip-bound oligonucleotides can be extended enzymatically upon hybridization of a DNA template. Surprisingly, however, only 63% of the oligonucleotides are elongated in polymerase reactions, while oligonucleotides that are released from the support behave normally in standard PCR amplifications. This rate of 63% nevertheless compares favorably with an extension rate of only 50%, which is achieved under identical conditions, if prefabricated oligonucleotides of identical sequence are spotted to the glass support.

***In Situ* Synthesis of Oligonucleotide Arrays by Using Soft Lithography** In this method, based on the standard phosphoramidite chemistry protocol, the coupling is achieved by the glass slide being printed with a set of polydimethylsiloxane (PDMS) microstamps, on which is spread deoxyribonucleoside monomer, and tetrazole mixed solution. A high quality, high spatial resolution, and large-scale PDMS stamp is developed by integrating 168 different microstamps on one glass substrate for synthesizing oligonucleotide arrays. The stamp is modified to improve the surface wettability by plasma discharge treatments, so that microstamps can be used to fabricate oligonucleotide arrays. A motional printing head is developed to improve the contact effect of the glass slide with different microstamps. A higher boiling point solvent is used in the printing coupling to inhibit solvent volatilization, and to maintain the consistency of reagents on different features of the microstamp. A specific oligonucleotide array of four probes both matched and mismatched with the target sequence is fabricated to identify the perfect match and mismatch sequences. The elastic characteristic of PDMS allows it to make conformal contact with the glass slide in the printing coupling. With regard to the efficiency of the printing coupling, if the hybridization microscope images of 20-mer oligonucleotides synthesized via the directly drip-dropped coupling and the contact coupling are compared, the fluorescence intensities of the two methods shows no significant differences. The coupling efficiency is also investigated via an end-labeled fluorescence nucleotide method, and a stepwise yield of 97% is obtained.

Printing In an alternative DNA chip fabrication technique, which uses conventional phosphoramidite chemistry, nanoliter sized droplets of reagents are delivered to the proper site on a chip for drop-by-drop synthesis of oligonucleotides using a device similar to an inkjet printer. Inkjet technology uses chemical deprotection to synthesize the oligonucleotides and allows the synthesis of arrays of any oligonucleotide sequences *de novo*. The bases are fired onto the array at each spot using modified inkjet nozzles, which, instead of firing different colored ink (Figure 4.22), fires A, C, G, and T deoxyribonucleotides, and the synthesis takes place using the same chemistry as a standard DNA synthesizer. The nozzles are computer controlled, so any oligonucleotides can be synthesized on the array simply by specifying the sequences in a computer file. Similar to maskless photolithography, this technology is highly flexible, with each array able to contain any oligonucleotide the operator wishes. Agilent (Palo Alto, CA) has commercialized this technique. Its non-contact industrial inkjet printing technology is very accurate, and involves precision deposition of reagents for building, i.e., *in situ* synthesis of oligonucleotides on standard format glass slides using phosphoramidite chemistry. Both the catalog and custom microarrays are manufactured using Agilent's non-contact *in situ* synthesis process of printing 60-mer length oligonucleotide probes, base-by-base, from digital sequence files. This engineering feat is achieved without stopping to make contact with the slide surface, and without introducing surface contact feature anomalies, resulting in consistent spot uniformity and traceability. With inkjet technology there is no need to create expensive masks, allowing full flexibility, and cost-effective production of microarrays. Because of the reaction efficiency of the chemistry, 60-mer oligonucleotides can be created, allowing for increased sensitivity and specificity as compared to shorter probes. Coverage with a 44,000 features array is therefore similar to the higher density Affymetrix arrays.

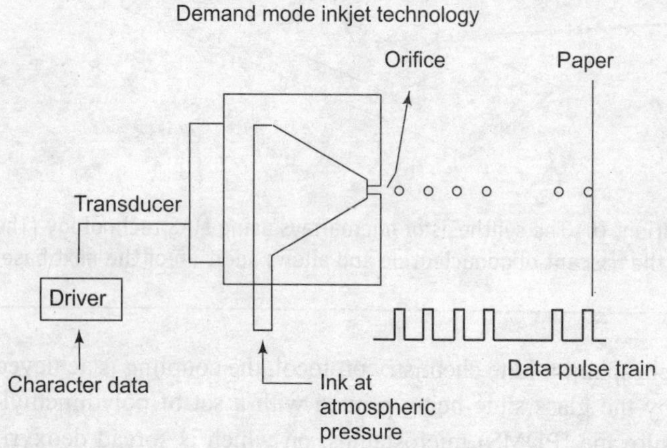

Figure 4.22 Principle of inkjet printing

The original method for producing spotted arrays was contact printing, which involves the use of a capillary spotting pin or quill (Figure 4.23a) that draws up a defined amount of liquid from wells in a microtiter plate (Figure 4.23b). The pin is then placed in contact with the array surface, and this causes some liquid to be deposited. The pin uptakes 0.25 µl, and dispenses 0.6 nl (~1–10 ng/spot; feature size is 100 µM). The pin is washed thoroughly and dried in an automated cycle before returning to the microtiter dish for the next sample. The application of multiplex print heads that deposit samples in a block may increase the speed of production of arrays.

Several laboratories have modified commercial ink-on-paper printers for use in spotting microarrays. These printers were usually based on 'bubble jet' print heads containing tiny heating elements that rapidly vaporize a water-based solution in a capillary to eject a droplet containing protein or DNA, onto a solid support. These printers, however, are difficult to

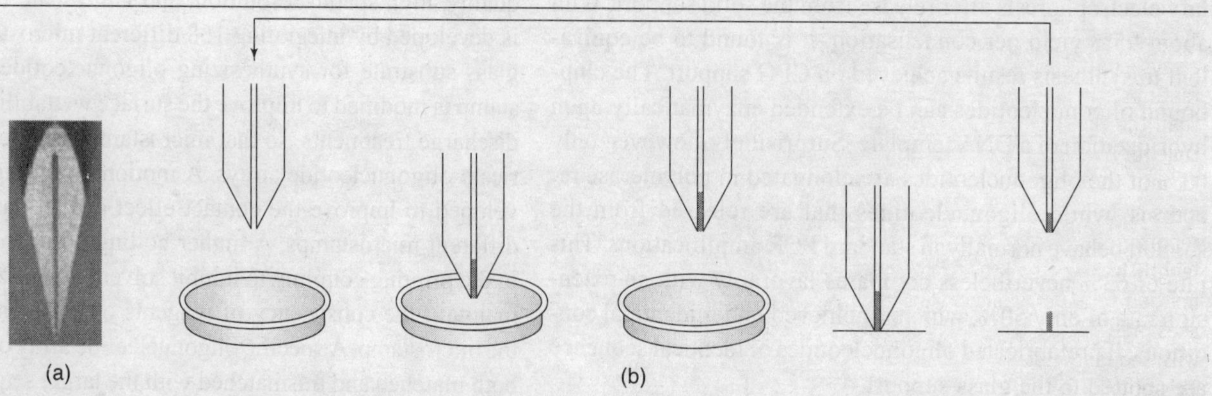

Figure 4.23 Contact printing; (a) Pin used in capillary contact printing, (b) Principle of array manufacture by capillary contact printing
Source: Stephen Wilcox, AGRF, Microarray Technologies

clean, and are not suitable for high-throughput production of oligoarrays or the parallel *de novo* synthesis of oligonucleotide arrays. Like pin-spotting microarrayers, these also require libraries of known nucleic acid reporters before arrays can be made.

Piezoelectric inkjet oligoarray synthesis is another method that is highly flexible, and allows rapid construction of oligonucleotide arrays containing any desired sequence. By dispensing DNA monomers from a multi-channel inkjet print head, large numbers of sequences can be chemically synthesized in parallel. *De novo* oligonucleotide synthesis using a piezoelectric inkjet oligoarray synthesizer overcomes several of the problems inherent in the pin-spotted arrays, and the conventional photolithographic mask arrays. First, as soon as the genomic DNA or expressed sequence tag (EST) library is even partially sequenced, oligonucleotide reporters, including intergenic regions, can be designed and synthesized on arrays without having to clone and store large libraries. Second, the use of standard phosphoramidite chemistry for oligoarray synthesis allows longer reporters to be synthesized, decreasing the number of reporters required for confident identification of the target molecules. For details of POSaM (piezoelectric oligonucleotide synthesizer and microarrayer), see Exhibit 4.3.

Exhibit 4.3 POSaM (Piezoelectric Oligonucleotide Synthesizer and Microarrayer)

The POSaM platform utilizes a low-cost piezoelectric print head with six fluid channels (Figure A). Four channels deliver phosphoramidite precursors and one delivers an activator (ethylthiotetrazole), leaving one channel available for an optional linker or modified base. The piezoelectric jets can deliver a wide range of nonvolatile solvents in volumes as low as 6 picolitre. Piezoelectric jetting, high quality motion controllers and standard phosphoramidite oligonucleotide synthesis chemistry allow users to synthesize arrays of any nucleic acid sequence at specific, closely spaced features on suitable solid substrates.

Acetonitrile, the preferred solvent for automated phosphoramidite synthesis, is not suitable for inkjet printing due to its high volatility; droplets may even evaporate before contact with the slide surface. In addition, phosphoramidite precipitates can accumulate on the print head and clog the nozzles. Propylene carbonate is one less-volatile alternative that has been shown to produce coupling efficiency of 94–98% for inkjet synthesis. A 1:1 mixture of methyl glutaronitrile (MGN) and 3-methoxypropionitrile (3MP) exhibits the most favorable combination of volatility, solubility, surface tension, and synthesis parameters. The mixture readily dissolves phosphoramidite monomers at a concentration of 250 mM and produces coupling efficiencies similar to acetonitrile or propylene carbonate. Moreover, the 1:1 MGN:3MP mixture jets easily, evaporates slowly and forms discrete 'virtual' reaction wells on epoxysilane-modified slides because of its favorable surface tension.

Key properties of the POSaM platform is that fabrication of a new array is flexible and rapid, and cost is practically independent of the number of sequences or the number of different arrays to be synthesized. Once the inkjet is running, it does not matter if the arrays contain 100 reporters or the current maximum of 9,800 features. There is no additional set up cost for a new design other than the time it takes for the reporter selection process. The cycle time for covalent attachment of each base in a reporter ranges from 11 min for one array to 20 min for eight arrays. Six arrays of 40-mers can be produced in about 13 hours. As each oligonucleotide reporter can be of a different length, it is possible, by varying the length of the reporter, to design arrays where all the theoretical melting temperatures are very similar. Inkjet oligoarray synthesis uses highly efficient, well-characterized phosphoramidite chemistry with sequences that are stored in a database rather than in a freezer. This eliminates the cost of amplification, replication, storage, and retrieval of the reporters, as well as minimizing the errors that are generated when cataloging many large DNA libraries. As the turnaround time for new arrays is so fast and the cost of printing 9,800 reporters is so small, many different oligonucleotide sequences can be tested to optimize the sensitivity and specificity of the reporters. The arrays may be stripped and reused, which is very helpful for optimizing hybridization conditions. Moreover, specificity and sensitivity can be measured under different buffer and temperature conditions on the same array. Because the same sample would be applied in every condition, residual signal of less than a few percent is not a problem.

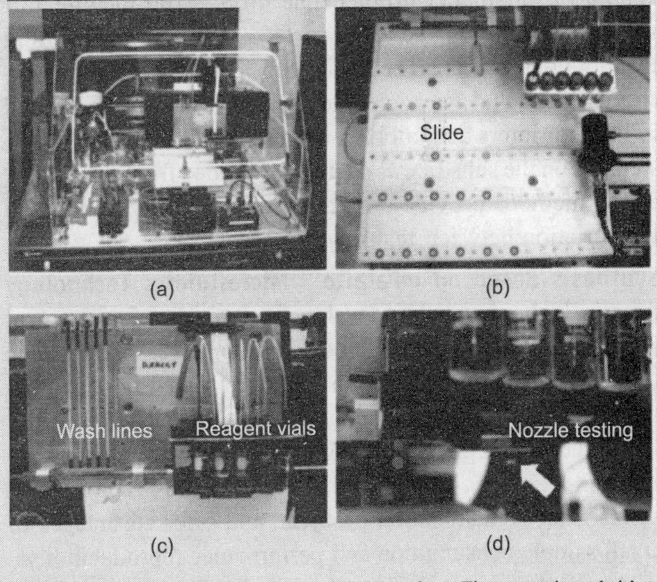

Figure A The POSaM platform; (a) Overview. The complete inkjet printing system is enclosed in an air-tight acrylic cover, 61 × 91 × 122 cm. (b) View from above showing the array holder. One slide is shown secured by the vacuum check with room for 26 additional slides. (c) Front view showing the print/wash head. Five PTFE wash lines deliver acetonitrile, oxidizer, and deprotecting acid in bulk. Six vials supply tetrazole and phosphoramidites to the inkjet print head. (d) Lower-front view of the inkjet print head showing droplets passing through the QC laser beam. The presence of a droplet produces forward scattered light as bright red flashes shown by arrow

Source: Lausted C, Dahl T, Warren C, King K, Smith K, Johnson M, Saleem R, Aitchison J, Hood L, Lasky SR (2004) POSam: a fast, flexible, open-source, inkjet oligonucleotide synthesizer and microarrayer, *Genome Biol.* (2004) 5 (8): article R58.

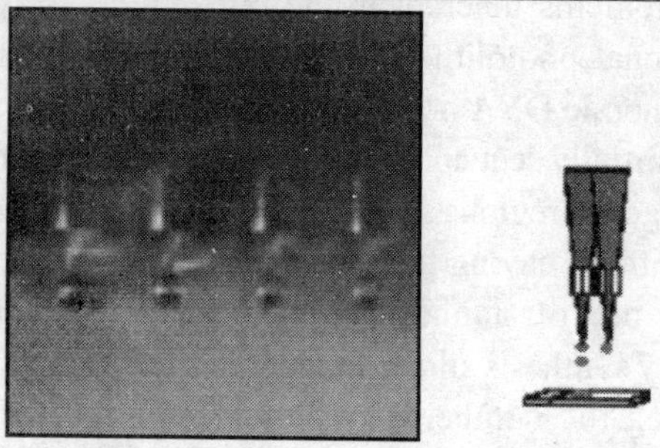

Figure 4.24 A non-contacting printing method (pin and ring system) for array manufacture [The method is developed by Genetic Microsystems and currently marketed by Affymetrix.]
Source: Stephen Wilcox, AGRF, Microarray Technologies

A number of alternative 'non-contact' printing methods are also available that are faster than established procedures with high reproducibility. The pin and ring system, devised at Genetic Microsystems and currently marketed by Affymetrix, is popular (Figure 4.24). The 'ring' is inserted into the well of a microtiter plate, and draws up a certain amount of liquid. The 'pin' then extends through the ring, and carries a smaller droplet (one drop = 100 picoliter) of solution down onto the array surface. Non-contact printing technologies have also been developed for microarray fabrication, and include piezoelectric devices similar to those found in inkjet printers and bubblejet printheads that deposit DNA samples on the substrate as a bubble extended from the nozzle. These methods provide a more uniform spot size, reducing the variation between features.

Synthesis Based on mParaflo™ Microfluidics Technology Atactic Technologies has developed approaches for chemical synthesis of oligonucleotides that encompass a microfluidic μParaflo™ reaction device, an advanced digital light synthesizer apparatus. This technology enables the massively parallel synthesis of high quality oligonucleotides in picoliter scale reaction chambers. The functionalized μParaflo™ chips do not suffer from contamination problem, and offer advantages of small sample consumption and performance reproducibility.

Another example is the microfluidic PicoArray reactor, which is made from silicon using standard microelectronic fabrication procedures (Figure 4.25). The reactor contains three topographical features: picoreaction chambers, fluid microchannels, and inlet/outlet through holes. The chip contains 128 × 31 (total 3,968) individual reaction chambers, each with an internal volume of 270 picoliter. The fluid microchannels are of a tapered shape derived from a fluid mechanical model to produce a uniform flow rate across all reaction chambers. In this method, a photo generated acid (PGA) precursor is fed into the microfluidic chamber prior to the light irradiation step to create the acid, which removes the acid labile DMTr protecting group. This process is simple in that it does not require an electrochemical surface or spe-

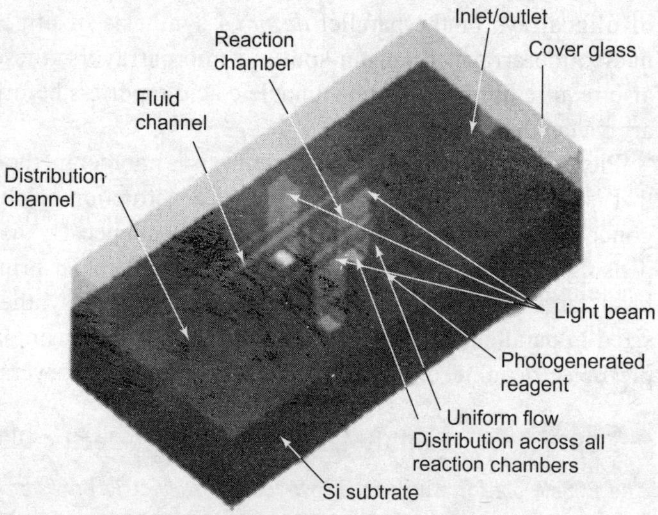

Figure 4.25 PicoArray Reactor

cialty monomers with photolabile protecting groups. Electrochemical deprotection methods require a complex circuit of electrodes that can withstand contact with strong organic reagents through multiple synthesis cycles. Another limitation of deprotection with electrodes is that side reactions can occur on the electrode surface. The need for specialty photolabile protecting group monomers means no flexibility for creating content variations in the sequences. Studies show the reaction efficiency is much lower than standard monomers, and these have been known to give rise to randomized misincorporation or insertion errors lowering sequence fidelity. PGA deprotection allows parallel synthesis with conventional chemicals and supplies, following well-established synthesis processes. The quality of the synthesis reaction is greater than 98.8%, and this approach is very flexible because virtually any modified monomer can be used creating a wide array of non-regular oligonucleotides. This microfluidics-based PicoArray method demonstrates several advantages: (i) It is simple to operate; (ii) It is programmable for synthesis of any desired sequence; (iii) It enables simultaneous synthesis and purification of oligonucleotides that are designed for multiplex gene synthesis; (iv) The method allows an ultrafast oligonucleotide parallel synthesis with a high density of uniform spots; (v) It is readily adopted by the existing automated synthesizers for parallel synthesis of oligonucleotides; (vi) It consumes only small amounts of reagents and solvents on a per sequence basis; (vii) It displays significant kinetic acceleration compared with that of conventional reactions; (viii) It allows easy recovery of sequences synthesized in a small volume for direct use in the subsequent enzymatic reactions; and (ix) It is a closed system that can protect the contained samples from contamination and air-oxidation.

Microelectrode Arrays The microelectrode technology using controlled electric fields for immobilization was also used for creating microarray (Nanogen). Nanogen's technology involves electronically addressing biotinylated DNA samples, hybridizing complementary DNA reporter probes, and applying stringency to remove unbound and nonspecifically bound DNA after hybridization. Density, however, is currently limited to 100 test sites.

Finally, CombiMatrix Inc. is leveraging this technology to synthesize nucleic acid compounds in a high throughput manner. They have developed a system using individually addressable microelectrode arrays to synthesize many different oligonucleotides *in situ* in parallel reactions. CombiMatrix's CustomArray™ product is a DNA microarray that allows scientists to understand the genetic causes of many complex diseases such as cancer, diabetes, infections, etc. This product also enables researchers to develop diagnostics for diseases, which are difficult to diagnose using other methods.

4.7.4 Applications of DNA Microarrays

DNA chip technology utilizes microscopic arrays (microarrays) of molecules immobilized on solid surfaces for biochemical analysis, and as discussed their applications depend strongly on the effectivity of the chemistry used to create the respective type of chip. DNA microarrays find applications in many fields, and the most common ones are listed below; and the details are discussed in Chapter 19. These include: (i) Large-scale gene discovery or identification of a sequence (gene); (ii) Determination of expression level (abundance) of genes, and expression profiling of gene or mRNA; (iii) Genotyping on genomic scale and characterization of whole genome; (iv) Single nucleotide polymorphism detection array (SNP array) for screening of polymorphism in the genome of populations, and to determine single base pair mismatch and gene mutation; (v) DNA resequencing; (vi) Comparative genomic hybridization array (CGH array) for assessing large genomic rearrangements; (vii) Pathogen analysis; (viii) Identification of complex genetic diseases; (ix) Detection of patterns of gene expression between tissues or disease states; and (x) Development of diagnostics, drug discovery, and toxicology.

Review Questions

1. Why is solid-phase synthesis preferred over solution-phase synthesis?
2. What advantages does enzymatic synthesis of DNA confer over chemical synthesis?
3. Why is controlled pore glass bead a preferable solid support for solid-phase synthesis by phosphoramidite method?
4. Discuss the advantages of phosphoramidite method over other historical methods of oligonucleotide synthesis.
5. Describe the significance and properties of various protecting groups (dimethoxytrityl; benzoyl/isobutyryl; β-cyanoethyl/methyl) in solid-phase synthesis.
6. Describe the significance of anhydrous conditions in conventional method of chemical synthesis of oligonucleotides.
7. Why is the use of dichloroacetic acid (DCA) preferred over trichloroacetic acid (TCA) for detritylation? What advantages does dichloroethane offer as compared to dichloromethane as a solvent for the preparation of DCA/TCA?
8. Draw the chemical structure of deoxynucleoside 3′-phosphoramidite. Describe the importance of deoxynucleoside 3′-phosphoramidite and tetrazole in phosphoramidite method of oligonucleotide synthesis.
9. How is a phosphite-triester (P III) converted to phosphate-triester (P V)? What is the advantage of this step in a DNA synthesizer?
10. Describe the significance of capping in the synthesis of oligonucleotides.
11. Describe the methods by which the contamination of protecting agents and failure products is removed from the required oligonucleotide.
12. If a DNA synthesizer is not working properly, what effect will it have on coupling efficiency? Describe several factors affecting the efficiency of coupling.
13. Give a brief account of various ways of trityl analysis. Also explain its significance.
14. If we are interested in the hybridization analysis using heterologous sequence, which type of probe DNA should we use? Describe the method of its synthesis.
15. How can chemical synthesis of oligonucleotides be used for the assembly of a complete gene?
16. Describe the synthesis of aptamers to be used as diagnostic and therapeutic agents.
17. Describe various methods by which oligonucleotide synthesis is performed on a chip.
18. Discuss the significance of photolabile protecting groups in the manufacture of DNA chips. Give two examples of photosensitive groups.
19. Differentiate between cDNA arrays and oligonucleotide arrays.
20. How does the photolithography using photomasks differ from the maskless photolithography?
21. Describe the advantages and disadvantages of spotted DNA array and prefabricated oligonucleotide chip.
22. Suggest two different strategies for synthesizing a 0.5 kbp gene. Discuss the advantages and disadvantages of both methods.
23. How is the yield of oligonucleotides calculated? Suppose a 30-mer oligonucleotide is synthesized with a coupling efficiency of 99%, how will you calculate its yield?
24. If the overall yield of a 40-mer oligonucleotide is 82%, calculate the coupling efficiency of the reaction.
25. If the new DNA synthesizer has an average coupling efficiency of 98.5%, what overall synthesis yield would you expect after the synthesis of a 50-mer DNA hybridization probe?

References

Beaucage, S.L. and M.H. Caruthers (1981). Deoxynucleoside phosphoramidites: a new class of key intermediates for deoxypolynucleotide synthesis. *Tetrahedron Lett.* **22**:1859–1862.

Caruthers, M.H. (1991). Chemical synthesis of DNA and DNA analogs. *Acc. Chem. Res.* **24**:278–284.

Caruthers, M.H. *et al.* (1992). Chemical synthesis of deoxynucleotides and deoxynucleotide analogs. *Methods Enzymol.* **211**:3–20.

Edge, M.D. *et al.* (1981). Total synthesis of a human leukocyte interferon gene. *Nature* **292**:756–762.

Fodor, S. (1997). DNA sequencing: Massively parallel genomics. *Science* **277**:393–395.

Gold, L. (1997). The SELEX process: A surprising source of therapeutic and diagnostic compounds. *Harvey Lectures* **91**:47–57.

Khorana, H.G. *et al.* (1972). Studies on polynucleotide. CIII. Total synthesis of the structural gene for an alanine transfer ribonucleic acid from yeast. *J. Mol. Biol.* **72**:209–217.

Khorana, H.G. *et al.* (1972). Studies on polynucleotide. CIII. Total synthesis of the structural gene for an alanine transfer ribonucleic acid from yeast. *J. Mol. Biol.* **72**:209–217.

Letsinger, R.L. and Lunsford, W.B. (1976). Synthesis of thymidine oligonucleotide by phosphate-triester intermediates. *J. Am. Chem. Soc.* **98**: 3655–3661.

Synthesis of Complementary DNA

Key Concepts

In this chapter we will learn the following:

- Enrichment and purification of mRNA
- cDNA synthesis

5.1 INTRODUCTION

A characteristic feature of most multicellular organisms is specialization of individual cells. For example, a human being is made up of a large number of different cell types such as brain cells, blood cells, liver cells, etc. Each cell contains the same complement of genes, but in different cell types different sets of genes are switched on, while others are silent. The fact that only a few genes are expressed in any one type of cell can be utilized in the preparation of a library if the material that is cloned is mRNA and not DNA. Only those genes that are being expressed are transcribed into mRNA, so if mRNA is used as the starting material then the resulting clones comprise only a selection of the total number of genes in the cell. A cloning method that uses mRNA is particularly useful if the desired gene is expressed at a high rate in an individual cell type. For example, the gene for gliadin is expressed at a very high level in the cells of developing wheat seeds. In these cells over 30% of the total mRNA specifies gliadin. Clearly, if mRNA is cloned from wheat seeds, a large number of clones specific for gliadin can be obtained. Similarly, if mRNA is cloned from chick oviduct, there will be abundance of ovalbumin. However, mRNAs are exceptionally labile molecules. These cannot be ligated into a cloning vector, and are difficult to amplify in their natural form. Hence, the information encoded by the RNA has to be converted into a stable DNA duplex, called complementary DNA (cDNA) or copy DNA, which is easily inserted into a self-replicating vector. The cDNA thus formed is representative of the expressed genes, and represents the information encoded in the mRNA of a particular tissue or organism, and at particular developmental stage. It also reflects mRNA levels, and the diversity of splice isoforms in particular tissues. Once the information is available in the form of a cDNA library, individual processed segments of the original genetic information can be isolated and examined with relative ease. In this chapter, details of the procedures for eukaryotic mRNA enrichment and preparation, commonly used methods for cDNA synthesis, and the strengths and limitations of each method are discussed. Once a double stranded cDNA (ds cDNA) is constructed, it is cloned into a suitable vector through strategies given in Chapter 15.

5.2 ENRICHMENT AND PURIFICATION OF mRNA

The underlying rationale for purifying mRNA is to increase the statistical representation of one, several, or all mRNAs in a sample as a fraction of the total purified RNA (polyadenylated and non-polyadenylated). Any approach of this nature is called enrichment strategy. The purposes for enriching samples in favor of mRNA include: (i) To reduce the amount of RNA necessary to assay specific transcripts by Northern analysis, S1 nuclease analysis, or RNase protection analysis; (ii) By increasing statistical representation, the likelihood to identify all elusive, rare mRNA molecules either on a blot, in a cDNA library, or by RT-PCR increases; (iii) To convert mRNA into cDNA; and (iv) As the mRNA fraction isolated from a cell or tissue contains many different mRNA sequences, the number of useless clones in a cDNA library increases consequently increasing the number of cDNA clones to be screened to isolate relevant clones. For example, if the percentage of the desired mRNA in the total poly (A) RNA of a eukaryotic cell is 1%, then only one out of 100 cDNA clones, on the average, may contain the desired genetic information. Thus, by enriching the mRNA of interest, the number of clones to be screened may be reduced.

The strategies routinely used for enrichment of mRNA of interest are based on: (i) Proper selection of tissues or cells for mRNA isolation; (ii) Manipulation of biological material before cellular disruption to superinduce or accumulate certain types of transcripts in the cell; and (iii) Manipulation of a previously purified RNA sample. In this section, various mRNA enrichment and purification techniques are described. Note that in all these procedures, careful precautions against degradation are necessitated. This is because mRNAs, like all other RNAs, are labile (much more than DNA), or highly reactive due to the presence of 2′-OH group of the ribose ring, and many ribonucleases (RNases) are very stable. Thus, for the isolation of intact, full-length mRNAs, an RNase-free environment should be maintained throughout. The general considerations to prevent degradation by RNases are the same as that followed in total RNA isolation procedures (for details see Chapter 1).

5.2.1 Enrichment of mRNA by Proper Selection of Tissues or Cells

The foremost mRNA enrichment strategy is careful selection of a tissue or cell so that a particular mRNA is abundantly produced, from where it can be easily isolated. The details of mRNA abundance are given in Exhibit 5.1. Some examples of the preferable sources of specific abundant mRNAs include chick oviduct cells for ovalbumin mRNA, reticulocytes for β-globin mRNA, pancreatic β-cells (Islets of Langerhans) for insulin mRNA, and wheat seeds for gliadin mRNA, etc.

5.2.2 Enrichment of Relevant mRNA by Manipulation of Biological Material

Besides using cells or tissues containing relevant mRNA in high abundance, another mRNA enrichment strategy is based on the manipulation of biological material. The strategies included are the following: (i) Induction of the expression of the gene of interest by treating tissues or cells with an agent; (ii) Using tumor tissue in which the particular mRNA is overexpressed; and (iii) The purification of certain mRNAs is also facilitated considerably if these possess a characteristic base composition. For example, silk fibroin and collagen have special composition of the corresponding mRNAs, and are isolated from total RNA preparations relatively easily by density centrifugation, because these mRNAs are guanine-rich and large (32S and 27S, respectively).

5.2.3 Enrichment of mRNA by Subtractive Hybridization

Subtractive hybridization procedure allows enrichment for RNA species that are absent in one cell type (say type A) but present in another (say type B). RNA is isolated from both the cell types. The RNA from type A cells is subjected to single stranded cDNA (ss cDNA) synthesis using oligo (dT) primers. The mRNA strand in the ss cDNA:mRNA hybrid obtained after first strand synthesis is destroyed, leaving a ss cDNA strand behind. These are then mixed, and allowed to anneal to the isolated RNA from type B cells. The RNA species that are unique to type B cells will not anneal to the cDNA from type A cells. This is because of the absence of that RNA in type A cells. Thus, unpaired RNAs remain as single strands. After annealing, the material is passed down a hydroxyapatite column, which binds more tightly to double stranded nucleic acids than to the ones that remain single stranded (Exhibit 5.2). The ss RNAs are then recovered, and used for cDNA synthesis and cloning (for details see Chapter 15).

5.2.4 Immunoprecipitation of Polysomes

This method uses antibodies for immunoprecipitation of polysomes, and is applicable for purification of both eukaryotic and prokaryotic mRNA as well as polyadenylated and non-polyadenylated mRNAs. Note that polysomes are active complexes of ribosomes and mRNA containing a certain fraction of nascent polypeptide chains. Thus, specific antibodies are added to each reaction tube, which bind to the corresponding proteins (antigens as polysome) wherever produced, provided

Exhibit 5.1 mRNA abundance

Eukaryotic mRNAs belong to an abundance class based on prevalence (i.e., the average number of copies of an RNA transcript) in the cell. Three classes are: high abundance, medium abundance, and low abundance mRNAs.

High Abundance mRNAs These mRNAs are present in hundreds of copies per cytoplasm. If a cell is highly specialized to perform a particular function or produces large amounts of a particular protein, then it is reasonable to expect that the cell transcribes a correspondingly elevated mass of the transcript. The example of high abundance mRNAs include β-globin mRNA in reticulocytes, ovalbumin mRNA in oviduct cell, and gliadin mRNA from wheat seeds, etc.

Medium Abundance mRNAs The mRNAs present in dozens of copies per cytoplasm are medium abundance mRNAs. Their expression is assayed frequently as control or reference genes in RNA-based analyses. For example, many standard house-keeping genes such as β-actin, histones, and GAPDH are medium abundance mRNAs.

Low Abundance or Rare mRNAs Low abundance mRNAs are those transcripts that are present in 14 or fewer copies per cytoplasm. These are also called rare mRNAs. Their detection is enhanced greatly through the application of supersensitive assay procedures such as PCR. It has been estimated that 30% of cellular mRNAs and 90% of the expressed sequences fall in the low abundance class.

Source: Farrell Jr. RE (2005) *RNA Methodologies—A Laboratory Guide for Isolation and Characterization*, Elsevier Acad. Press, NY, pp. 138–162.

Exhibit 5.2 Hydroxyapatite column chromatography

Hydroxyapatite $\{[Ca_{10}(PO_4)_6(OH)_2]$ or $[Ca_5(PO_4)_3(OH)]\}$, the most stable form of crystalline calcium phosphates precipitated from aqueous solution was originally developed as a matrix for protein chromatography. It is inert material with very large surface area in relation to particle size. It has widespread use in separating single stranded nucleic acids from double stranded nucleic acids. The method is also useful in the chromatographic purification and fractionation of complex nucleic acids by thermal elution according to their G+C content. It is used to investigate the reassociation kinetics of DNAs from many different sources, to construct transcription maps, and to measure the copy number of specific sequences in complex genomes. It also finds application in the preparation of cDNA for subtractive cloning, as well as for the removal of contaminants from DNA preparations. Despite its usefulness, with the availability of better ways to carry out almost all of these tasks, hydroxyapatite chromatography has disappeared from the standard repertoire of laboratory techniques.

The principle behind the technique is that the nucleic acids (ds DNA, ss DNA, ds cDNA, ss cDNA, RNA) bind to the hydroxyapatite by virtue of interactions between the phosphate residues of nucleic acid backbone and the calcium residues of matrix (Figure A). Binding of both single stranded and double stranded nucleic acids to hydroxyapatite is performed in 0.05 M sodium phosphate buffer (pH 6.8). Double stranded molecules bind with higher affinity to the matrix as compared to single stranded molecules. This is because the well-ordered and evenly spaced sets of phosphate residues in double stranded molecules make many regular contacts with the matrix, while single stranded molecules are more disordered, and a smaller proportion of their phosphate residues are available for contact with the matrix. This difference in binding pattern or affinity allows separation of double stranded and single stranded molecules. The bound nucleic acid molecules are then eluted in phosphate buffers. The nucleic acid molecules with higher affinity (i.e., double stranded molecules) are eluted in higher concentrations of phosphate (>0.36 M; pH 6.8), while those with lower affinity (i.e., single stranded molecules) are eluted in lower concentrations of phosphate (~0.12 M; pH 6.8). Partial duplexes and DNA:RNA hybrid molecules elute at intermediate concentrations. This elution step is usually carried out at 60°C, although there is no good reason to do so since the adsorption and elution profiles of nucleic acids are indistinguishable between 25°C and 60°C. As nucleic acids are often eluted in large volumes, these are later concentrated after removing phosphate ions from the solution. This is best achieved by concentrating the eluate by extraction with isobutanol and then removing the salt by chromatography, through Sephadex G-50 columns.

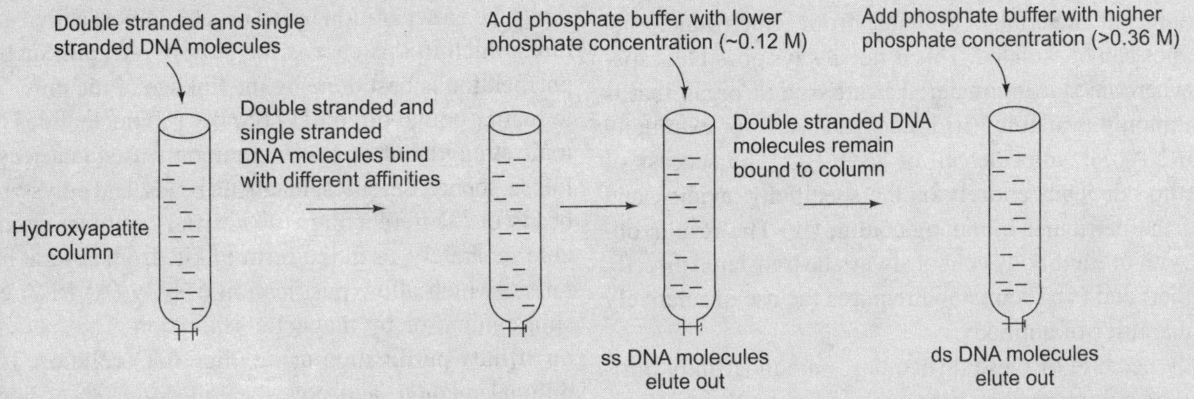

Figure A Hydroxyapatite column chromatography for separation of ds DNA from ss DNA molecules [Note that the same principle is applied for the separation of ds DNA and RNA; DNA:RNA hybrid and ss DNA; DNA:RNA hybrid and RNA; ss cDNA:mRNA hybrid and ss DNA; ss cDNA:mRNA hybrid and RNA.]

Source: Sambrook J and Russel DW (2001) *Molecular Cloning: A Laboratory Manual,* III Ed., Cold Spring Harbor Laboratory Press, Cold Spring Harbor, New York, pp. 8.32–8.34.

these polypeptides are sufficiently long. These antibody–antigen complexes are then precipitated (i.e., immunoprecipitation), and hence easily recovered. This technique works well for mRNAs encoding abundantly synthesized proteins such as albumin and immunoglobulin, but not for low abundance mRNAs. As a consequence the method is rarely employed. Moreover, the binding of a single antibody is not sufficient to precipitate the polysomes. Hence, certain modifications have been done to precipitate such antigen–antibody complexes. These modifications include indirect immunoprecipitation, i.e., reaction of antibody–antigen complex with a second antibody directed against the first antibody (i.e., anti-antibodies), and binding to a solid matrix, for example, immunoaffinity columns. Further the speed of immunoaffinity purification can be improved by adding protein A-sepharose (covalently coupled using cyanogen bromide), which binds antibodies, and can be precipitated easily. Note that the protein A is a component of the cell wall of *S. aureus* cells, which binds specifically to Fc fragments of immunoglobins IgG, and hence serves as a useful solid phase. The protein A-sepharose binds to the antibody–antigen complex, from where the antibody–antigen complex is pelleted and collected by centrifugation. The protein A-sepharose pellet thus recovered from each tube is denatured and electrophoresed in an SDS polyacrylamide gel. Note that as an alternative to protein A-sepharose, intact *S. aureus* cells can also be used, such that the polysomes get bound to the bacterial surface, from where these can be recovered.

The location and amount of recovered polypeptide is determined by autoradiography (or fluorography) provided radioactive (or fluorescent) amino acid is employed during synthesis. By this procedure, RNA fractions highly enriched for the mRNA of interest are identified as those from which large amounts of radioactive or fluorescent protein has been precipitated by the antibodies and protein A. The mRNA is then easily isolated from these precipitates by phenol extraction. The polysomes are then dissociated with EDTA, and poly (A) RNA is isolated by oligo (dT) chromatography (for details see Section 5.2.5). The immunoaffinity purified mRNA can be used to prepare cDNA probes, and to construct cDNA library enriched in cDNAs for particular mRNAs. Note that such a library is not representative of the total mRNA population of that tissue type.

Although a powerful technique, immunoaffinity purification of polysomes cannot be applied universally as enrichment technique for specific mRNAs because of the following reasons: (i) The technique does not work unless a reliable source of material is available from which functional polysomes can be isolated. This is not always possible, especially when the starting material is a tissue or organ that is not commonly available; (ii) It does not work for extremely rare mRNAs (1 molecule/cell or less); (iii) The success of the method depends entirely on the specificity, avidity, and type of the particular immunoglobulin; (iv) The results obtained with one antibody cannot always be translated directly to another; and (v) The method requires the use of relatively large quantities of antibody.

Partly because of these difficulties, immunoaffinity purification of polysomes has been superceded by the development of cDNA expression vectors (e.g., λgt 11 and λZAP II) that allow direct isolation of cDNA clones encoding specific antigens. Further, the ready access to PCR in virtually all laboratories has also decreased the application of this technique. Now, a standard approach is the precipitation of all polysomes to pull down the subpopulation of mRNA that is actively undergoing translation. Then, free from the ribosomes and ancillary translation factors, purified polysomal mRNA can be reverse transcribed to synthesize probe for microarray analysis, assayed by PCR with gene-specific primers, or enriched using universal primers to support the construction of a cDNA library.

5.2.5 Affinity Purification of Poly (A) RNA [Poly (A) Capture Techniques]

Most eukaryotic mRNAs, with few exceptions, contain a poly (A) tail comprising of 50–200 adenosine residues at their 3′-ends. *In vivo* functions of this poly (A) tail are the following: (i) To facilitate mRNA transport out of the nucleus (nucleocytoplasmic transport); (ii) To provide stability to mature mRNA by preventing nuclease attack; (iii) To increase the efficiency of translation by circularizing mRNA caused by the interaction between poly (A) tail binding proteins (PABP) and eIF-4G initiation factor bound at the 5′-end of the mRNA; and (iv) To regulate function of mature mRNA in the cytoplasm. Such poly (A) tail containing RNAs are called polyadenylated mRNAs or poly (A) RNAs. In contrast some RNAs, for example, histone mRNAs, some organelle mRNAs (e.g., chloroplasts and mitochondria), rRNAs, and tRNAs are non-polyadenylated, and hence are poly $(A)^-$ RNAs. Under *in vitro* conditions, this structural difference is exploited to distinguish most of the eukaryotic mRNAs from prokaryotic mRNAs, or to purify relatively small poly (A) RNA fraction from total cellular RNA, or to separate poly (A) RNA from poly $(A)^-$ RNA. The resolution of polyadenylated from non-polyadenylated RNA species is based on the natural H-bonding capability of A and T bases.

Philip Leder (1972) first reported eukaryotic mRNA purification by taking advantage of this 3′ poly (A) tail. Fractionation in this manner is a fine example of affinity chromatography, i.e., a separation based on a biological activity or structure, which in this case is the 3′ poly (A) tail. Such affinity purification is best done by the linkage of the poly (A) RNA to matrix bound oligo (dT) or poly (U) nucleotides in a high ionic strength buffer. Some commonly used matrices are cellulose, sepharose, silica, magnetic beads, and polystyrenelatex beads, etc. Different oligo (dT)/carrier combinations are available separately, or in the form of kits from several manufacturers, which allow purification of poly (A) RNA either on spin column or by magnetic separation. The method based on affinity purification using oligo (dT)-cellulose is the traditional method, and now the following newer approaches have replaced this traditional method. These are: (i) Mixing oligo (dT)-cellulose and total RNA together directly in a microfuge tube without any prior formation of some type of matrix geometry; (ii) Solution hybridization with oligo (dT) covalently linked to paramagnetic beads, and sequestration of hybrids using a magnet; (iii) Exploiting the natural affinity between biotin and avidin. Note that biotin can be conjugated to nucleic acids and various other molecules without changing their biological activities. These biotin-conjugated nucleic acids are then purified by affinity purification with streptavidin (biotin binds tightly with streptavidin; K_d 10^{-15} M). Thus, biotinylated oligo (dT) molecules are used to bind poly (A) RNAs, and the resulting conjugates are precipitated as biotin/streptavidin complexes. The conjugates may also be recovered using streptavidin linked to a paramagnetic bead and eluted in small, very convenient volumes without the need for further precipitation. Many permutations of this general approach are available as mRNA purification kits; and

(iv) Running an RNA sample over a column of poly (U)-sepharose to promote A:U H-bonding between the matrix and all poly (A) transcripts in a sample.

The general procedure for isolation of poly (A) RNA using any of the mentioned techniques is based on affinity purification. The purification is done either in a batch (i.e., suspending the matrix in the RNA solution) or by constructing a column by stuffing silanized glass wool into a silanized Pasteur pipette, and then pipetting the matrix onto this. Note that batch-binding technique is especially profitable for large preparations, and the problems frequently associated with conventional column technology such as slow flow rates and clogged columns are circumvented by this technique. In this section, the procedure for poly (A) RNA purification on oligo (dT)-cellulose, and its difference in separation mechanics using other matrices are discussed. Further details depending on the oligo (dT)/carrier combination should be obtained from the manufacturers' protocols.

Affinity Purification Using Oligo (dT)-Cellulose Spin Column

Poly (A) RNA can be enriched by affinity chromatography on oligo (dT)-cellulose columns [i.e., columns containing cellulose beads coated with oligodeoxythymidilic acid or short polymer comprising of 15–30 deoxythymidine residues $(dT_{15\text{-}30})$]. These oligo (dT) chains are covalently linked to the cellulose matrix via the terminal 5′-phosphates of oligonucleotide. Note that poly (U)-sepharose and oligo $(dT)_{25}$-silica are other polynucleotide/matrix combinations, and their principle and procedure is similar to oligo (dT)-cellulose. The preparation of poly (A) RNA using oligo (dT)-cellulose (Figure 5.1) consists of following steps:

Preparation of Oligo (dT)-Cellulose Column Oligo (dT)-cellulose (commercially available) is first washed and soaked in 0.1 M NaOH, washed with water, and ultimately equilibrated with binding buffer [0.5 M LiCl, 10 mM Tris-HCl (pH 7.5), 1 mM EDTA, and 0.1% SDS]. Note that 1 ml of bloated oligo (dT)-cellulose is sufficient for 5–10 mg of total RNA, but one should read the instructions from the manufacturer. The soaked oligo (dT)-cellulose is packed in a DEPC treated column by pipetting, followed by equilibration with a suitable buffer.

Loading of Total RNA Onto the Oligo (dT)-Cellulose Column During isolation of total RNA, phenolization has been observed to be associated with the formation of aggregates between mRNA and other RNA species. Such aggregates or any secondary structures formed between single stranded poly (A) RNAs due to intramolecular base pairing have to be destroyed by treating with formamide, or by heating. Thus, the total RNA is dissolved in a high salt buffer, and heated briefly to 65–70°C to disrupt secondary structures, and poured onto the oligo (dT)-cellulose column. In order to significantly increase the yield of poly (A) RNA bound specifically to the column, total cellular RNA preparations are usually recycled many times over the same column before elution.

Hybridization of Poly (A) RNA to Oligo (dT) Poly (A) RNA fraction of total cellular RNA anneals to oligo (dT) in a high ionic strength buffer at room temperature. This results in the formation of poly (A) RNA-oligo (dT) hybrids, while poly $(A)^{-}$ sequences (e.g., rRNA and tRNA) remain unbound. This is because at high ionic strength, monovalent cations act as counter ions neutralizing the net negative charge intrinsic to the phosphodiester backbone of polynucleotides, consequently

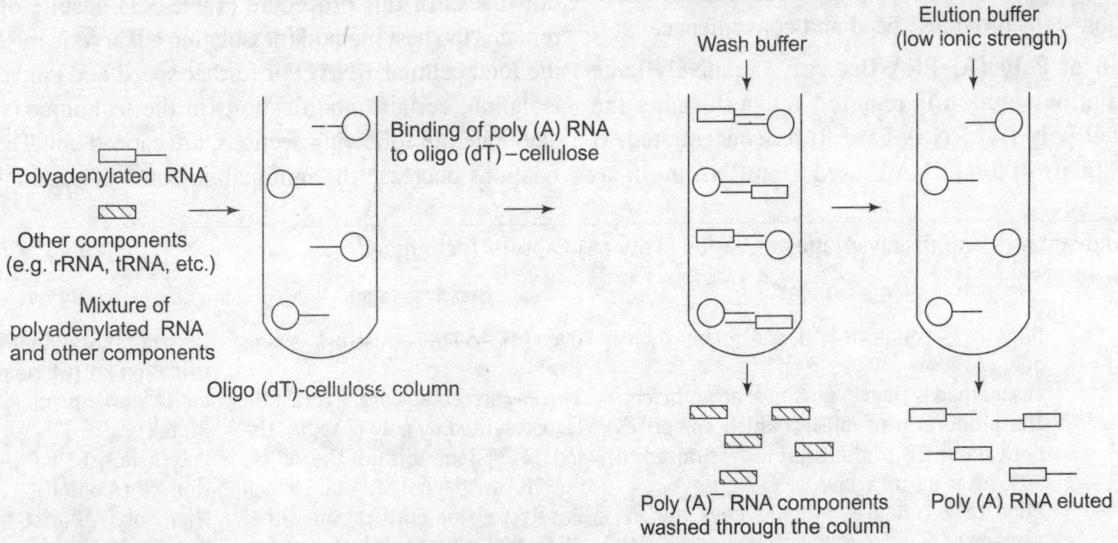

Figure 5.1 Poly (A) purification using oligo (dT)-cellulose column [Note that purification through oligo (dT)-silica, and poly (U)-sepharose follows the same principle; In these cases, oligo (dT)-cellulose will be replaced by oligo (dT)-silica, and poly (U)-sepharose, respectively.]

eliminating the electrostatic repulsion between the poly (A) RNA and the thymidylate residues linked to the matrix. Note that superior binding capacity results in high yields of mRNA from tissue/cells, and the binding capacity of oligo (dT)-cellulose increases with salt concentration up to about 500 mM NaCl, KCl, or LiCl, although the binding capacity is greater when KCl or LiCl is used in place of an equivalent amount of NaCl. Typical RNA binding capacity of the oligo (dT)-cellulose under standard assay conditions generally ranges from 50–80 OD (A_{260} units) per gram of matrix, depending upon the manufacturer, the grade of refinement, and the monovalent cation used to prepare the binding buffer. An alternative matrix configuration that supports the affinity separation of polyadenylated from non-polyadenylated RNA is known as microcrystalline oligo (dT)–cellulose (Monomer Sciences Inc., New Market, AL), which generally binds an excess of 100 OD per gram of matrix.

Washing Off Unbound Nucleic Acids and Elution of Poly (A) RNA As the oligo (dT) molecules are linked to a carrier, washing off of unbound nucleic acids while retaining the oligo (dT) bound poly (A) RNA is possible. Thus, several washing steps using high salt buffer or low temperature (low stringent conditions) are performed to wash off unbound poly $(A)^-$ RNA. Later, poly (A) RNA is eluted from the column under high stringent conditions using low salt Tris buffer or DEPC water, or by raising the temperature. Note that at low salt concentration, electrostatic repulsion between negative charges of oligo (dT) sequence, and the bound poly (A) RNAs is of sufficient magnitude to prevent base pairing between them, favoring rapid dissociation of the hybridized poly (A) RNAs from the matrix. Likewise, high temperature favors denaturation or separation of the two hybridized nucleic acids. Note that the application of spin column used in the washing and elution steps provides speed and convenience.

Precipitation of Poly (A) RNA Because a relatively large volume of elution solution is required for performing the elution, eluted poly (A) RNAs have to be concentrated. If more than ~800 µg of total RNA is used for mRNA purification, this is easily done by ethanol (EtOH) precipitation. Note that lesser starting material may result in poor recovery. EtOH precipitation is done with 2.5 volumes of EtOH at –20°C overnight in the presence of 1/10th volume of 3 M sodium acetate (pH 5.2). Poly $(A)^-$ RNA can also be concentrated from early fractions obtained after elution with high salt buffer.

DNase Treatment It should be noted that poly (A) RNA preparations may still contain some residual genomic DNA. For many applications, this is not a problem, but for some applications that involve a PCR step, it is necessary to perform RNase-free DNase treatment before using mRNA for downstream experiments. Several manufacturers offer DNases and buffer systems, which can be used prior to a reverse transcription and PCR step without the need for buffer removal (e.g., DNase I, amplification grade from Invitrogen). The purified poly (A) RNA is stored in EtOH at –70°C until required.

Note that the elution volumes vary from 20–250 µl for most kits accepting up to 1 mg of total cellular RNA. A yield of 2–30 µg of poly (A) RNA can be expected from 1 mg of total cellular RNA, but the value may vary considerably with cell growth conditions. Further, the mRNA comprises only a small percentage of all RNA species in a eukaryotic cell (<5% of the total cellular RNA), but after purification its proportion is raised to about 50%. Yields are almost never quantitative, probably because poly (A) tails may be rather short, and can be very heterogeneous in length. This method may also be useful to further fractionate poly (A) mRNAs to obtain a population enriched in poly (A) mRNA encoding a particular protein. Oligo (dT)–cellulose columns have the advantage of being reusable after washing with alkali. Moreover, the application of spin column used in the washing and elution steps provides speed and convenience. However, there are certain principal drawbacks of this procedure (Table 5.1), inspite of which it remains the best method for isolating eukaryotic mRNAs from the total cellular RNA. For further speed and convenience in isolation, certain modifications in the technique, which are elaborated in following sections, are carried out. These modifications decrease the minimal amounts of starting material,

Table 5.1 Advantages and Disadvantages of Various Poly (A) Capture Techniques

Matrix	Advantages	Disadvantages	Commercially available kits
Column purification by oligo (dT)–cellulose	Oligo (dT)–cellulose is useful for purifying poly (A) RNA; The method is simple, sensitive, and efficient; The proportion of mRNA, which comprises only about 2% of the total RNA is raised to 50% after purification; Oligo (dT)–cellulose columns have the advantage of being reusable after washing with alkali.	Oligo (dT)–cellulose matrix is expensive; Very low percentage of the total RNA is represented by poly (A) RNA. Using this system, it is not the mRNA, but RNA with a poly (A) tail, which is selected for purification. Other RNAs that also contain adenosine rich sequences are also isolated through this procedure, while the	Illustra™ QuickPrep™ mRNA purification kit (GE Health-care); mRNA enrichment kit (Omega Biotek); Poly(A)Purist™ mRNA purification kit (Ambion); MicroPoly(A)Purist™ purification kit (Ambion)

(Contd.)

Table 5.1 (*Contd.*)

Matrix	Advantages	Disadvantages	Commercially available kits
		mRNAs without a poly (A) tail are lost; The poly (A) RNA is eluted from the column in a relatively large volume, and at slow elution flow rates, making subsequent concentration of the RNA somewhat cumbersome; Due to large volumes, the procedure requires a step for precipitating the poly (A) RNA from the eluant; Because of the lack of sensitivity below 0.1 OD, this approach is unsuitable for smaller quantities of starting material (less than 100–150 μg total RNA); The preparation and running of the column is laborious; The mechanics of poly (A) selection often result in the loss, and further under-representation of very low abundance transcripts; Requires time consuming centrifugation step.	
Column purification by oligo (dT)-silica	Same as oligo (dT)–cellulose	Same as oligo (dT)–cellulose	Poly (U)–sepharose (Pharmacia)
Column purification by poly (U)-sepharose	Same as oligo (dT)–cellulose	Same as oligo (dT)–cellulose	GeNei™ mRNA purification kit (Bangalore Genei)
Column purification by oligo (dT) coated latex beads	Oligo (dT) coated latex bead system is useful for purifying poly (A) RNA; Perfectly spherical surface of latex beads allows uniform dispersion, and minimal centrifugation time, yielding increased recovery of purified mRNA.	Oligo (dT) coated latex beads are expensive; It leads to purification of poly (A) RNA only, and any non polyadenylated mRNA is lost.	NucleoTrap® mRNA kit (Clontech); Oligotex® mRNA purification kit (Qiagen)
Oligo (dT) magnetic beads	Process is rapid, highly sensitive, and efficient; No need of laborious step of preparation and running of the column; The technique is fast due to elimination of centrifugation step; Intact poly (A) RNA can be obtained in less than one hour; This method of poly (A) capture helps to concentrate the sample dramatically without the need to reprecipitate the sample; Rare mRNAs are not under-represented; Oligo (dT) magnetic beads are useful for purifying poly (A) RNA, and highly purified poly (A) RNA is recovered from supernatant; Magnetic particles can be repeatedly separated and resuspended without magnetically induced aggregation, as these particles become magnetic only when a magnetic field is applied to them;	Oligo (dT) magnetic beads are expensive; It leads to purification of poly (A) RNA only, and any non-polyadenylated mRNA are lost.	Poly(A)Purist Mag mRNA purification kit (Ambion); Absolutely mRNA™ purification kit (Stratagene); PickPen™ magnetic particle transfer device (Bio-Nobile); Magnetic mRNA isolation kit (New England Biolabs); Dynabeads® mRNA purification kit (Dynal Biotech)

(*Contd.*)

Table 5.1 (*Contd.*)

Matrix	Advantages	Disadvantages	Commercially available kits
	Magnetic separation is a rapid one-step procedure for purification of mRNA from tissues, and allows mRNA to be purified in a single tube by application of magnetic field thereby eliminating potential sample loss during liquid handling; The procedure permits manual processing of multiple samples, and can be adapted for automated high-throughput applications; Magnetic separation technology permits elution of intact mRNA, fully representative of the mRNA population of the original sample, in small volumes eliminating the need for precipitating the poly (A) RNA in the eluant; Oligo $(dT)_{25}$-magnetic beads can be reused up to three times, and the researcher has the option of eluting the isolated poly (A) RNA, or using the bound (dT) DNA as a primer in a first strand cDNA synthesis directly.		
Biotinylatedoligo (dT) in combination with streptavidin - coupled magnetic beads	Same as oligo (dT) magnetic beads; This procedure yields an essentially pure fraction of mature mRNA after only a single round of magnetic separation.	Same as oligo (dT) magnetic beads	PolyATtract® (Promega Corp.); MagNA Pure LC mRNA isolation kit II (Roche Molecular Biochemicals)
Magnetic porous glass (MPG)-streptavidin system	Same as oligo (dT) magnetic beads	Same as oligo (dT) magnetic beads	MPG® mRNA purification kit (CPG)

keep the volumes involved as small as possible, keep the sample as concentrated as possible, and increase the recovery of isolated poly (A) RNA.

Affinity Purification Using Oligo (dT) Coated Latex Beads

In this method, oligo (dT) coated on latex beads are employed for purification of poly (A) RNAs. When total cellular RNA is added, poly (A) RNAs hybridize with oligo (dT). These beads are pelleted by centrifugation, and washed in high salt washing buffer. This is followed by resuspension in low salt elution buffer, and the poly (A) RNAs are recovered from the supernatant. This procedure is advantageous because the perfectly spherical surface of latex beads allows uniform dispersion, and minimal centrifugation time, yielding increased recovery of purified mRNA. The procedure of mRNA purification using NucleoTrap® mRNA purification kit (Clontech) is outlined in Figure 5.2.

Biomagnetic Separation Using Oligo (dT)-Magnetic Beads

Magnetic separation technologies rely on supermagnetic particles that are made from polystyrene or iron oxide plus polysaccharides. These particles become magnetic only when a mag-

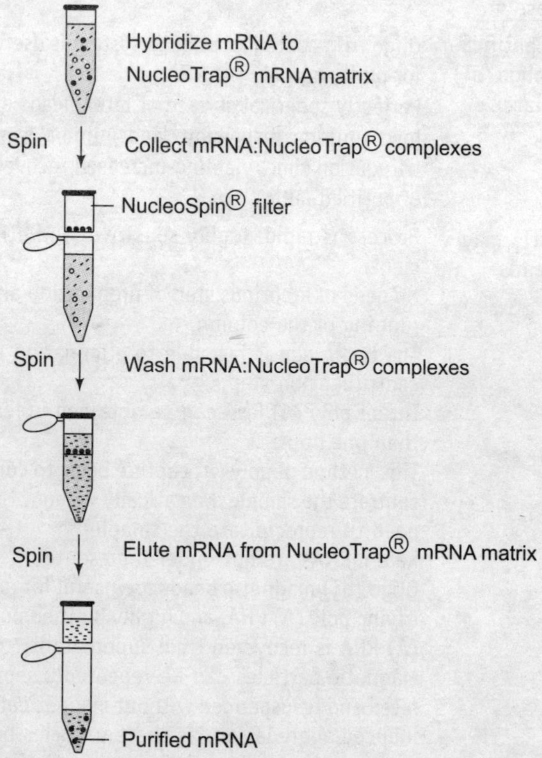

Figure 5.2 mRNA purification using NucleoTrap® mRNA purification kit

netic field is applied to them. These particles therefore can be repeatedly separated and resuspended without magnetically induced aggregation. The technology involves oligo $(dT)_{25}$ coupled to 1 μm paramagnetic beads [i.e., iron beads coupled to oligo (dT) molecules via functional groups that coat the particles] as the solid support for the direct binding of poly (A) RNAs. This is followed by the addition of total cellular RNA under conditions of high ionic strength. During this time, poly (A) RNAs anneal to oligo (dT) tract linked to the bead. With the application of magnetic field, the magnetic beads are captured to one side of the reaction tube by placing in a magnetic stand. This is followed by resuspension of isolated magnetic beads in buffer. The capturing/resuspending steps are repeated. Poly (A) RNAs are then purified from the beads by suspending them in RNase-free (DEPC treated) deionized water and capturing the magnetic beads. The detailed protocol is given in Figure 5.3.

The principal advantage of this procedure is that the magnetic separation is a rapid one-step procedure for purification of poly (A) RNAs from tissues, and allows poly (A) RNAs to be purified in a single tube by application of magnetic field, thereby eliminating potential sample loss during liquid handling. The procedure permits manual processing of multiple samples and can be adapted for automated high-throughput applications. Additionally, magnetic separation technology permits elution of intact poly (A) RNAs, fully representative of the poly (A) RNA population of the original sample in small volumes, eliminating the need for precipitating the poly (A) RNAs in the eluant. Moreover, the technique is fast due to elimination of centrifugation steps, and allows recovery of intact poly (A) RNAs in less than 1 hour. Oligo $(dT)_{25}$-magnetic beads can be reused two to three times, and the researcher has the option of eluting the isolated poly (A) RNAs, or using the bound (dT) DNA as a primer in a first strand cDNA synthesis directly.

Biomagnetic Capture using Biotinylated Oligo (dT) in Combination with Streptavidin-coupled Magnetic Beads

In this method, oligo (dT) coupled to biotin [i.e., biotinylated oligo (dT)] is used for purification of poly (A) RNAs. Poly (A) RNAs conjugate to biotinylated oligo (dT) resulting the formation of biotinylated oligo (dT)-poly (A) RNA complexes. Upon addition of streptavidin-coupled magnetic beads to the system, affinity association takes place between streptavidin and biotin, thereby forming a complex (comprising of five components) between streptavidin-magnetic beads and biotinylated oligo (dT)-poly (A) RNA complex. The system (Promega Corp.) uses streptavidin MagneSphere® paramagnetic particles (SA-PMPs) to capture poly (A) RNAs that have annealed to biotinylated oligo (dT) probes. This procedure yields an essentially pure fraction of mature poly (A) RNAs after only a single round of magnetic separation, and is represented in Figure 5.4. Furthermore, the procedure has all the advantages related to magnet-based separation

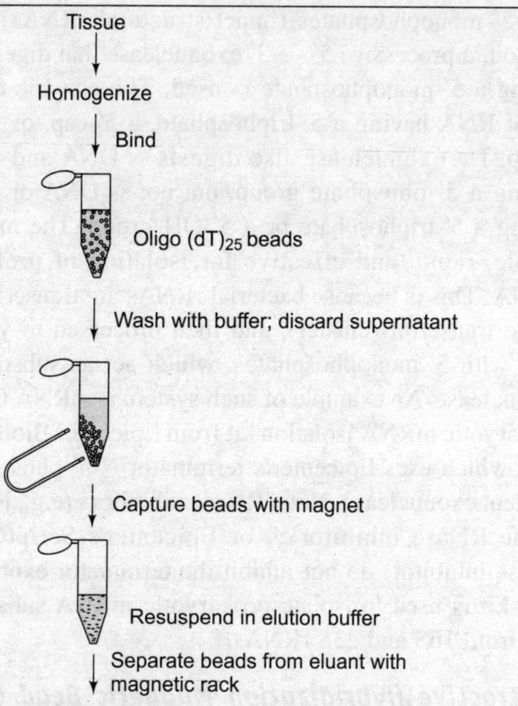

Figure 5.3 Biomagnetic separation of poly (A) RNA using oligo (dT)-magnetic beads

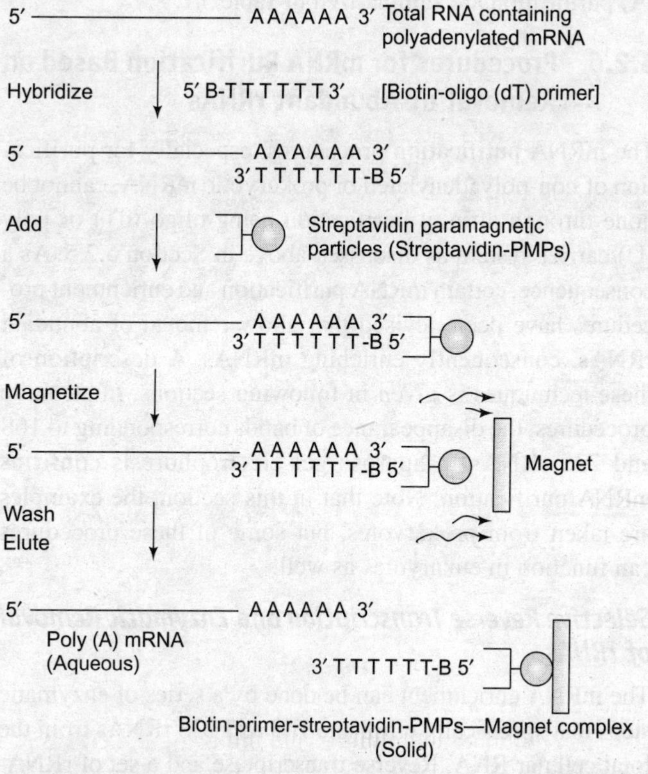

Figure 5.4 Biomagnetic capture using biotinylated oligo (dT) in combination with streptavidin-coupled magnetic beads

discussed above. Magnetic porous glass (MPG)-streptavidin system also works on the same principle.

Advantages and Disadvantages of Poly (A) Purification

Poly (A) purification offers several advantages, which are as follows: (i) Poly (A) purification increases poly (A) RNA (e.g. mRNA) mass as a percentage of the total RNA; (ii) Poly (A) purification results in minimal interference from rRNA and tRNA; (iii) Poly (A) purification increases assay sensitivity; (iv) Current methods of poly (A) capture help to concentrate the sample dramatically without the need to reprecipitate the sample; (v) Purified poly (A) RNAs are ready for immediate use; and (vi) The poly $(A)^-$ fraction, which can be collected and precipitated, is an excellent negative control, especially for blot analysis. The procedure, however, suffers from certain demerits, which include: (i) Poly (A) selection excludes mRNAs that do not possess poly (A) tail; (ii) Loss of poly $(A)^-$ mRNAs during the purification process may further under-represent rare mRNAs, especially when starting with a previously isolated sample of total cellular RNA; (iii) Depending on the lysis method, poly (A) selection may also yield polyadenylated heterogeneous nuclear RNA (hnRNA); (iv) Oligo (dT) matrices are expensive; and (v) Given the improvements in cloning and RT-PCR efficiency, poly (A) selection may not be necessary. Further advantages and disadvantages of individual nucleotide/carrier combination used for poly (A) purification are summarized in Table 5.1.

5.2.6 Procedures for mRNA Purification Based on Removal of Abundant rRNAs

The mRNA purification procedures, especially for purification of non-polyadenylated or prokaryotic mRNA, cannot be done through affinity purification using oligo (dT) or poly (U)/carrier system as discussed above in Section 5.2.5. As a consequence, certain mRNA purification and enrichment procedures have been devised that allow removal of abundant rRNAs, consequently enriching mRNAs. A description of these techniques is given in following sections. In all these procedures, the disappearance of bands corresponding to 16S and 23S rRNAs in agarose gel electrophoresis confirms mRNA purification. Note that in this section, the examples are taken from prokaryotes, but some of these procedures can function in eukaryotes as well.

Selective Reverse Transcription and Enzymatic Removal of rRNAs

The mRNA enrichment can be done by a series of enzymatic steps that specifically eliminate 16S and 23S rRNAs from the total cellular RNA. Reverse transcriptase and a set of rRNA-specific primers (i.e., primers specific for 16S, 23S, and 5S rRNAs) are used to synthesize ss cDNA:rRNA hybrids. Then

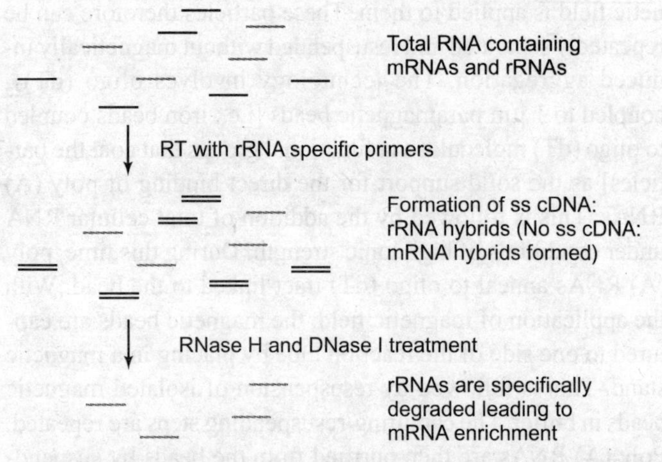

Figure 5.5 mRNA enrichment by selective reverse transcription and enzymatic removal of rRNAs

rRNAs are removed enzymatically by RNase H treatment (H stands for hybrid; the enzyme specifically digests RNA in ss cDNA:rRNA hybrid). Note, as the primers are specific for rRNAs, mRNAs do not participate in hybrid formation. The ss cDNA molecules are then removed by DNase I digestion, consequently enriching mRNAs. This method is somewhat complex and difficult, and is used by Affymetrix Inc. The details of the process are presented in Figure 5.5.

Digestion by 5′-Monophosphate Specific Exonuclease

This method called 'exonuclease digestion' is based on digestion catalyzed by exonucleases specific for rRNAs that bear 5′-monophosphates (intact structural rRNAs). In this method, a processive 5′ → 3′ exonuclease that digests RNA having a 5′-monophosphate is used. The enzyme does not digest RNA having a 5′-triphosphate, a 5′ cap, or a 5′-OH group. The exonuclease also digests ss DNA and ds DNA having a 5′-phosphate group, but not ss DNA or ds DNA having a 5′-triphosphate or a 5′-OH group. The method is simple, rapid, and effective for isolation of prokaryotic mRNA. This is because bacterial rRNAs are transcribed as a single transcript (cluster), and then processed to yield rRNAs with 5′-monophosphates, which act as substrates for exonuclease. An example of such system is mRNA ONLY™ prokaryotic mRNA isolation kit from Epicenter Biotechnologies, which uses Epicenter's terminator™ 5′-phosphate dependent exonuclease. Note, RNase inhibitors (e.g., RNasin®, Prime RNase inhibitor™, or Epicenter's ScriptGuard™ RNase inhibitor) do not inhibit the terminator exonuclease. This kit is used to isolate prokaryotic mRNA substantially free from 16S and 23S rRNAs.

Subtractive Hybridization Magnetic Bead Capture Using rRNA-specific Oligonucleotides as Capture Probes

This technique is used to remove abundant rRNAs from purified total cellular RNA, and enrich bacterial mRNA rapidly.

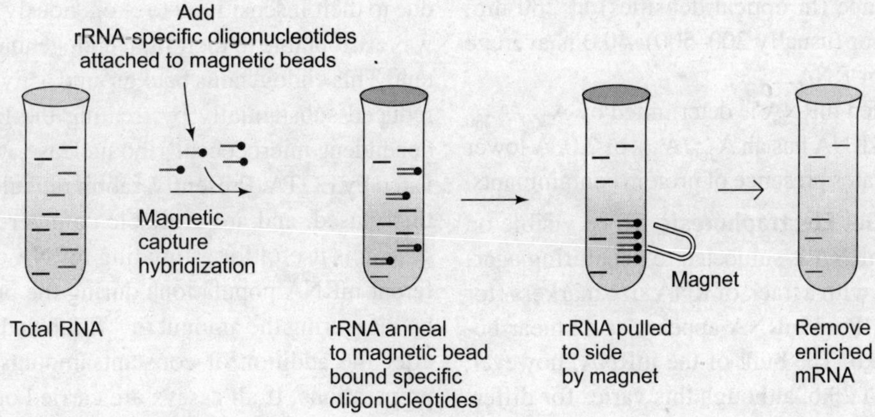

Figure 5.6 mRNA purification by magnetic capture-hybridization of rRNAs

An example employing this method is MICROBExpress™ bacterial mRNA purification kit (Ambion). In this method, the total RNA is incubated with the capture oligonucleotide mix in binding buffer. Magnetic beads derivatized with oligonucleotides that hybridize to the capture oligonucleotides are then added to the mixture, and allowed to hybridize. The magnetic beads, with the rRNAs attached, are pulled to the side of the tube with a magnet. The enriched mRNA in the supernatant is removed and precipitated with EtOH (Figure 5.6). Compared to conventional mRNA isolation procedures, this process is rapid, simple, and effective in the inhibition of rapid degradation of mRNA by cell indigenous or exogenous RNases.

Subtractive Hybridization Magnetic Bead Capture Using Biotin-labeled rRNA-specific Oligonucleotides as Capture Probes

This method utilizes biotin-labeled oligonucleotides complementary to conserved regions in rRNAs, and subtractive hybridization with streptavidin-coated magnetic beads. Thus in this method, magnetic particles bound biotin-labeled oligonucleotides specific for conserved regions of 5S, 16S, and 23S rRNAs of bacteria are used as capture probes. When total cellular RNA is added to these probes, rRNAs hybridize to them. The streptavidin-coated paramagnetic beads are then captured to the side of the tube by magnet, and the supernatant is carefully transferred to fresh tube. mRNA is then precipitated from the supernatant by EtOH, and recovered by centrifugation.

5.2.7 Purification and Enrichment of Prokaryotic or non-Polyadenylated mRNA

The concept of cDNA has much significance in eukaryotes, as the mature mRNA from which cDNA is synthesized lacks noncoding intronic sequences. In prokaryotes, genomic library is quite adequate, and bacterial cDNA libraries are rarely produced. It is not only technically difficult to isolate prokaryotic mRNA, but the prokaryotic cDNA libraries are less informative (for details see Chapter 15). However, cloning of bacterial cDNA plays an important role in identifying relatively abundant transcripts in the bacterial cells under selected conditions. Hence, for the purpose of analysis of gene expression in prokaryotes, mRNAs have to be isolated. Due to lack of poly (A) tails in prokaryotic mRNAs, oligo (dT) selection for poly (A) tails cannot be done. Hence, the methods based on removal of rRNAs as discussed in Section 5.2.6 are employed for non-polyadenylated mRNA enrichment. For simplifying the procedure of purification, polyadenylation of non-polyadenylated mRNA can also be done, followed by affinity purification strategies using oligo (dT)/matrices. Bacterial mRNA can also be separated from rRNAs and tRNAs by velocity sucrose density gradient centrifugation or gel electrophoresis.

5.2.8 Evaluating the Concentration, Integrity, and Purity of Isolated mRNA

The probability of finding a desired clone in a population of cloned molecules, and hence the amount of experimental work involved depends on the concentration and purity of the starting material. Hence, after preparing an mRNA fraction, it is necessary to check the concentration, integrity, and purity of the mRNA to be used as the template for cDNA synthesis, and to ensure that the mRNA of interest is present. A brief discussion of the techniques used for the purpose is given in this section. Note that some of the variants of the following methods used for the analysis of mRNA are same as for total cellular RNA (for details see Chapters 1 and 6).

Spectrophotometry The concentration and purity of mRNA samples are determined readily by taking advantage of their ability to absorb UV light, with an absorbance maximum at 260 nm (for details see Chapter 1). To estimate the mRNA concentration, it is assumed that a 40 μg/ml mRNA solution has an absorbance of 1 at 260 nm. Thus, absorbance measurements at this wavelength permit the direct calculation of mRNA concentration in a sample using the following formula:

$$[\text{RNA}]\ \mu\text{g/ml} = A_{260} \times \text{dilution} \times 40.0\ \mu\text{g/ml}$$

where A_{260} is absorbance (in optical densities) at 260 nm; dilution is dilution factor (usually 200–500); 40.0 is average extinction coefficient of RNA.

The purity of enriched mRNA is determined by A_{260}/A_{280} ratio. A pure sample of RNA has an A_{260}/A_{280} of 2.0. A lower A_{260}/A_{280} usually indicates presence of protein contaminants.

Denaturing Agarose Gel Electrophoresis To be visible on a gel, at least 1–2 μg mRNA is subjected to denaturing agarose gel electrophoresis with a track of RNA size markers (for details see Chapter 6). Pure mRNA appears as a smear between ~500 bp and 8 kb. The bulk of the mRNA, however, lies between 1.5 kb and 2 kb, although this varies for different tissues. As the amount less than 1 μg is not visible on an agarose gel, it has to be checked by Northern hybridization. The gel is also visualized to ensure the absence of residual amounts of rRNAs. Note that if mRNA appears to be even slightly degraded, a fresh preparation should be made.

Northern Blot Hybridization Quality of mRNA can also be determined by northern blot hybridization (for details see Chapter 16). For this, the gel is blotted and hybridized with a suitable probe representing a transcript known to be fairly abundant in the mRNA sample under test. Usually, β-actin is suitable for this purpose. Hybridization may also be carried out with the probe, which will be used to screen the library. This gives an indication of the usefulness of the probe in screening the cDNA library. As a rule of thumb, a transcript which gives a signal in a northern blot with 20 μg of total RNA should be easily visible as a clear band without any downward smear on 0.1 μg of mRNA. To determine size distribution of the eluted poly (A) RNAs, such a northern blot can be probed with oligo (dT), which should reveal a smear from ~0.2 to >3 kb.

***In Vitro* Translation and Analysis of Resulting Polypeptides** The assessment of integrity of isolated mRNA is typically done by *in vitro* translation of an aliquot of mRNA in each fraction in a cell-free system, and analyzing the resulting polypeptides by immunoprecipitation or SDS polyacrylamide gel electrophoresis. The most commonly used cell-free systems used for *in vitro* cell-free translation are those prepared from reticulocytes, wheat germ, or *Xenopus laevis* oocytes. These preparations contain the necessary ribosomes, tRNAs, and other components to translate the exogenous mRNA into proteins with a low background of translation products from endogenous mRNA. During *in vitro* translation, mRNA and twenty amino acids are added, one of which is radioactively labeled. This allows synthesis of radioactively labeled polypeptides, which are easily detected in very low quantities, and are distinguished from all the polypeptides present initially in the lysate or extract. Initially, reticulocyte lysates of anemic rabbits were considered disadvantageous due to their insensitivity to exogenously added mRNA, which was attributable to their high endogenous globin mRNA content. This endogenous background activity has however been reduced substantially by treating the lysates with calcium-dependent micrococcal ribonuclease, which is later inactivated by EGTA. Presently, rabbit reticulocyte lysates are routinely used, and are available commercially. This improved system is useful for estimating mRNA concentrations in different mRNA populations during the purification procedure by measuring the amount of ^{35}S-labeled proteins synthesized after the addition of constant amounts of different mRNA preparations. If all assays are carried out under non-saturating conditions, the amount of newly synthesized protein is proportional to the amount of exogenously added mRNA. After autoradiography, the radioactive band can be excised from the gel, and counted in a scintillation counter. In this case, the proportion of the total radioactivity contained in the band is an estimate of the relative concentration of the mRNA of interest. For example, bulk mRNA purified from mammalian cells should encode a large number of polypeptides with molecular weights ranging from <10,000 to >1,00,000. The polypeptides synthesized after *in vitro* translation are analyzed by immunoprecipitation reaction (for details see Section 5.2.3). Another strategy is to inject into *Xenopus* oocytes, and assay the resulting products either for biological activity, or by immunoprecipitation, or by gel electrophoresis.

Ability to Synthesize cDNA (Reverse Transcriptase Catalyzed Synthesis in Presence of Radiolabeled dNTPs or RT-PCR) The ability of the bulk mRNA preparation to direct the synthesis of long molecules of first cDNA strand (in presence of radiolabeled dNTPs) is another strategy to assess the quality and purity of isolated mRNA. If the preparation of mRNA is pure, cDNA synthesized from it should run on an agarose gel as a continuous smear from ~500 bases to 8 kb. When analyzed by autoradiography the bulk of the radioactivity should lie between 1.5 kb and 2 kb. No specific bands of cDNA should be visible unless the mRNA has been prepared from cells that express large quantities of specific mRNAs. In addition, the amount of radioactivity incorporated into the first strand of cDNA should be stimulated at least 20-fold by oligo $(dT)_{12-18}$ primers. Efficient cDNA synthesis in the absence of added primer is a sign that the mRNA preparation may be contaminated with fragments of DNA or RNA that can bind to mRNA at random sites, and serve as primers. If necessary, these fragments can be removed by denaturing the mRNA (100°C for 30 seconds in H_2O), cooling quickly in ice water, and immediately selecting poly (A) RNA by chromatography on oligo (dT)-cellulose. Note that the integrity of enriched mRNA can also be analyzed by RT-PCR (for details see Chapter 7).

5.3 cDNA SYNTHESIS

The synthesis of cDNA confers a number of advantages to the investigator interested in the characterization of gene structure or expression, developing nucleic acid probes, or expressing proteins that might otherwise be difficult to purify. The conversion of mRNA sequences to ds cDNA offers several advantages, which include: (i) mRNA is naturally labile single stranded molecule, and its conversion into more stable ds cDNA facilitates long-term storage of these sequences; (ii) cDNA is representative of the mRNA population (i.e., expressed genes) of the cell. Thus, the synthesis of cDNA is the creation of a permanent biochemical record of the cell at the time of lysis. Only those genes that are transcriptionally active are candidates for inclusion in a cDNA synthesis reaction, as cDNA cannot be synthesized from mRNA that is absent; (iii) By synthesizing and cloning the resulting cDNA from a single source, a method for propagating the cDNA is created. This approach is greatly facilitated by the huge variety of vectors compatible with an equally impressive variety of hosts; and (iv) cDNA molecules, both long and short, can be used to screen (hybridize to) members of much more complex genomic DNA libraries. This approach facilitates the isolation of exon and intron sequences from the structural portion of genes, and the flanking 5′ and 3′ sequences. This comparison also helps in the determination of 5′ and 3′ splice sites, and to get information about alternatively spliced genes. cDNA library also gives information about developmentally regulated genes as well as cell or tissue-specific genes.

Owing to the above advantages, the synthesis of ds cDNA from mRNA template was necessitated. Starting from mRNA, the synthesis of ds cDNA involves three major steps, which are listed below. The general scheme for ds cDNA synthesis is depicted in Figure 5.7.

First cDNA Strand Synthesis First strand cDNA synthesis using mRNA template is a reaction catalyzed by reverse transcriptase. Note that mRNA may be subjected to size fractionation before first cDNA strand synthesis. Reverse transcriptase requires primer to extend on, and the reaction leads to the formation of a hybrid between single stranded cDNA and mRNA (i.e., ss cDNA:mRNA hybrid).

Degradation of mRNA The mRNA template in ss cDNA:mRNA hybrid is degraded by alkali treatment or by RNase H-catalyzed nicking. This step provides primers for second cDNA strand synthesis (not in case of self-priming of second cDNA strand and random primers catalyzed second strand synthesis).

Second cDNA Strand Synthesis Second cDNA strand synthesis using the first strand as template is a reaction catalyzed by a DNA-dependent DNA polymerase such as *E. coli* DNA polymerase I or Klenow enzyme, etc. The reaction requires primers, and the end result of this step is the formation of ds cDNA.

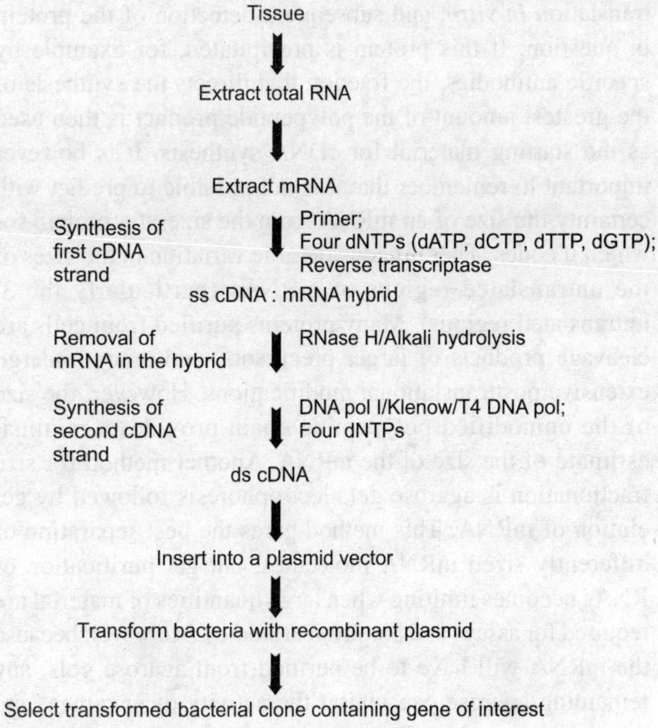

Figure 5.7 General scheme for cDNA synthesis and cloning

In this section, the primers used for first and second strand synthesis, and various strategies used for the synthesis of blunt ended ds cDNA are discussed. The cDNA once synthesized can be readily cloned into a suitable vector (for details see Chapter 15).

5.3.1 Size Fractionation of mRNA

The simplest method to enrich preparations of mRNA for sequences of interest is to fractionate them according to size. Size fractionation is used for enrichment of mRNAs that are much larger or smaller in size than the bulk mRNAs of the cell. Different ways of size fractionation include density gradient centrifugation and agarose gel electrophoresis (for details see Chapter 6). In density gradient centrifugation, different mRNAs present in a sample when applied to the top of a pre-poured density gradient, and subjected to centrifugation, are separated according to size. For instance, the model size of the mRNA population extracted from most types of mammalian cells is ~1.8 kb, but mRNAs smaller in size than 700 bases or larger than 4 kb can be enriched at least ten-fold by a single round of density gradient centrifugation carried out under denaturing conditions. Note that methylmercuric hydroxide is often used as a denaturing agent, but it is a poisonous and extremely toxic compound, and should be used with caution. The fractions collected are assayed for the presence of mRNA that encodes the relevant polypeptide through translation *in vitro*, and subsequent detection of the protein in question. If this protein is precipitated, for example by specific antibodies, the fraction that directs the synthesis of the greatest amount of the polypeptide product is then used as the starting material for cDNA synthesis. It is however important to remember that it is not possible to predict with certainty the size of an mRNA from the size of a protein for which it codes. There is considerable variation in the sizes of the untranslated regions of mRNAs (particularly the 3′ untranslated regions). Many proteins purified from cells are cleavage products of larger precursors, and many undergo extensive posttranslational modifications. However, the size of the unmodified polypeptide chain provides a minimal estimate of the size of the mRNA. Another method for size fractionation is agarose gel electrophoresis followed by gel elution of mRNA. This method gives the best separation of differently sized mRNA molecules, but gel purification of RNAs becomes limiting when large quantities of material are required for assays and biophysical studies. Moreover, because the mRNA will have to be purified from agarose gels, any remaining agarose can inhibit the activity of enzymes, and thereby reduce the efficiency of cDNA synthesis. Furthermore, the high susceptibility of mRNA to degradation by RNases reduces the chances of cloning a full-length corresponding cDNA; hence the method of choice is to size fractionate ds cDNA synthesized from mRNA template.

5.3.2 Effect of mRNA Length on cDNA Synthesis

The length of specific mRNA, and hence the cDNA that is to be cloned, influences the choice of primer and vector. The majority of mRNA species are between 1.5 and 2.0 kb in length. For these 'standard' mRNAs, the primer of choice is oligo (dT), which is 12–18 nucleotides in length. The oligo (dT) primer will anneal to the poly (A) tract, which is found at the 3′-terminus of most eukaryotic mRNAs (histone is a notable exception). There are, however, a number of mRNAs, which are extremely long, and it would be difficult to synthesize complete cDNA copies from these long mRNAs in one piece. This problem can be overcome by the use of mixed hexanucleotide primers to allow random priming of cDNA along the mRNA strand, which can then be cloned as a number of segments. A better option may be to use a specific primer for the mRNA of interest but this requires some knowledge of the sequence. This can be a useful approach when the mRNA of interest is a member of a superfamily of genes, which have short regions of high sequence conservation. It must be noted, however, that cDNA synthesis using such gene-specific primers is not always successful. If the size of the mRNA is larger than 2.5 kb, it may be worthwhile to enrich for either the starting mRNA, or preferably the ds cDNA synthesized from it. This will reduce both the size of library that has to be made and the number of clones that have to be screened.

5.3.3 Primers Used for First cDNA Strand Synthesis

The first step in the synthesis of ds cDNA is the formation of a ss cDNA:mRNA hybrid, which results from the synthesis of first cDNA strand using mRNA as a template. Note that for first cDNA strand synthesis, either total RNA or poly (A) RNA can be used depending on the purpose. If the cDNA is to serve as a template for PCR, total RNA is usually sufficient, while poly (A) RNA is employed if the cDNA is to be used as a hybridization probe. For first cDNA strand synthesis, mRNA template and primers are added together. The preparation is heated to 70°C to melt the secondary structure of mRNA, and then cooled slowly to room temperature to allow primer hybridization (called annealing of primers). This is followed by addition of buffer (neutral pH) containing Mg^{2+}, dNTPs, RNase inhibitors, reverse transcriptase, which is incubated for 1 hour at 37°C–42°C (depending upon the reverse transcriptase used). Note that at high temperature, secondary structure (intrastrand pairing) of the mRNAs is disrupted. This decreases the chances of pause of reverse transcriptase, and hence the likelihood of synthesis of incomplete molecules. The reverse transcriptase purified from an avian retrovirus (AMV RT) was used for the original cloning experiments, but this enzyme has largely been superceded by Moloney murine leukemia virus reverse transcriptase (Mo-MLV RT) because the latter has considerably lower levels of endogenous RNase H activity, which can cause degradation of the template before full-length cDNA has been synthesized (for details see Chapter 2). It is essential to add the RNase inhibitor (e.g., RNasin) to improve the length of first strand transcripts when using AMV RT, but this is optional if Mo-MLV RT is used. SuperScript II (Life Technologies), StrataScript (Stratagene), Transcript (AGS), and Mo-MLV RT RNase H Minus (Promega) are some modified reverse transcriptases that lack RNase H activity, and permit an increase in the length of first cDNA strand. These enzymes also have increased thermostability (up to 50°C), such that performing synthesis reaction at this temperature reduces secondary structure in the mRNA template. The reverse transcriptases are supplied with the cDNA synthesis kits. Some examples of commercially available kits for first strand synthesis are first cDNA strand synthesis kit (Novagen), ProtoScript® first cDNA strand synthesis kit (New England BioLabs), First strand synthesis kit (GE Healthcare), RevertAid™ first cDNA strand synthesis kit (Fermentas), etc.

The reverse transcriptase-catalyzed first strand synthesis requires primers to extend on. This is because reverse transcriptase, like other DNA polymerases, can only add residues at the 3′-OH group of an existing primer that is base paired with the template. There are many primers (oligonucleotides) for first strand synthesis to choose from, for example, oligo (dT) primers, random primers, oligo (dT) primer-adaptor, oligo (dT) primer-linker, and asymmetrically tailed plasmids. These primers are added to the reaction mixture in large molar excess (high primer:template concentration) so that each molecule of mRNA binds several molecules of primers. The important goals of first cDNA strand synthesis are to obtain a high yield of full-length cDNA, and low yield of small cDNA fragments (as reverse transcriptase has a tendency to terminate prematurely at mRNA secondary structures). The progress of first strand synthesis can be monitored by incorporation of ^{32}P-labeled dNTPs (for details see Section 5.3.9). In this section, different primers employed for first strand synthesis are described. The merits and demerits of each primer are summarized in Table 5.2.

Oligo (dT) Primers

Oligo (dT) primers that represent a chemically synthesized stretch of 16–20 adenosine residues are used when the mRNA species for which cDNA is required is polyadenylated at its 3′-end. If the mRNA is not polyadenylated, oligo (dT) primers cannot be used. The primer is added to the reaction mixture in large molar excess so that each molecule of mRNA binds several molecules of oligo (dT) primers at its long poly (A) tail. The annealed primer supplies free 3′-OH group, which is extended by reverse transcriptase. Thus, synthesis starts from the 3′-end of mRNA. It is shown from sequencing of cloned cDNAs that priming of first strand synthesis begins from the most proximal of these bound primers, and is very efficient. If however the synthesis is incomplete, the clones lack sequences from the 5′-end of the mRNA. Thus, the application of oligo (dT) primers may lead to 3′-end bias in the resulting library due to preferential accumulation of clones with sequences corresponding to the 3′-end of mRNA. Another disadvantage is that the oligo (dT) primers can be used only for cDNA synthesis from poly (A) RNA, leading to loss of sequences corresponding to poly (A)$^-$ RNAs.

Table 5.2 Merits and Demerits of Primers Used in Different Methods for First and Second cDNA Strand Synthesis

Primers used	Advantages	Disadvantages
First cDNA strand synthesis by reverse transcriptase		
Oligo (dT) primer	Method is efficient and simple; Used with poly (A) RNA; Allows use of the Gubler and Hoffman (1983) procedure for synthesis of the second strand of cDNA; Addition of different linkers or adaptors at each end of the cDNA allows directional cloning, and there is great flexibility in choice of vectors.	Method is not applicable for poly (A)$^-$ RNAs, thereby leading to loss of sequences corres ponding to poly (A)$^-$ RNAs; There may be 3′-end bias in the resulting library due to preferential accumulation of clones with sequences corresponding to the 3′-end of poly (A) RNA, and loss of sequences corresponding to the 5′-end of poly (A) RNA; Linkers, homopolymeric tails, or adaptors must be added before the cDNA is inserted into the vector.
Random primer	Higher efficiency of cloning 5′ sequences of a particular mRNA as compared to oligo (dT) primers; Readily used for both prokaryotic and eukaryotic mRNAs, as the priming is not based on the presence of 3′ poly (A) tail; Readily used for both poly (A) RNAs and poly (A)$^-$ RNAs; No loss of sequences complementary to mRNAs that might be non-polyadenylated; Increased chances of synthesis of sequences complementary to the 5′-end of mRNA; Random primers work very well with mRNAs that exhibit strong secondary structure at the 5′-terminus.	cDNAs generated by random priming tend to be small, and must be carefully fractionated according to size to eliminate the useless population of tiny molecules; Random hexamers are usually removed between synthesis of first and second cDNA strands, otherwise second strand synthesis will initiate by random priming; If one is interested in cDNA synthesis from poly (A) RNA, the product will be contaminated from cDNAs synthesized from poly (A)$^-$ RNAs as well; cDNA clones derived from random priming of first strand cDNA tend to be smaller than clones derived from oligo (dT) priming; The efficiency of priming varies from one mRNA to another, depending on the size of mRNA, and complexity and composition of the random primers pool.
Oligo (dT) primer-adaptor	Synthesis of the second strand can be carried out after the homopolymerically tailed cDNA has been ligated to a vector; Alternatively, second strand synthesis can be carried out using the Gubler Hoffman method. In this	Many protocols require the elimination of small oligonucleotides and unused primers from the first strand of cDNA by electrophoresis through alkaline agarose gels. The recovery of DNA at this stage is often poor.

(*Contd.*)

Table 5.2 (*Contd.*)

Primers used	Advantages	Disadvantages
First cDNA strand synthesis by reverse transcriptase		
	case, directional cloning of the cDNA can be achieved by the use of a primer-adaptor for the second restriction enzyme; such combination of first and second strand synthesis allows directional cloning; The background of empty clones is generally lower with primer–adaptors than with oligo (dT) primed synthesis of the first strand.	
Oligo (dT) primer-linker	Same as oligo (dT) primer-adaptor.	Same as oligo (dT) primer-adaptor; Such priming reaction requires methylation of internal restriction sites before primer ligation, and cleavage by restriction enzymes after primer attachment.
Asymmetrically tailed plasmid	As cDNA is immediately attached to the vector in the first step of its synthesis, and as homopolymeric tailing is done before second strand synthesis, direct and efficient cloning of full-length cDNA is possible; Selects for full-length or nearly full-length clones, perhaps because long cDNAs are preferred substrates for terminal transferase; The direct cloning of ds cDNA into the plasmid vector allows high overall efficiency; The number of steps involved in the synthesis and cloning of cDNA are reduced; Whereas the original protocol (Okayama and Berg, 1982) generated libraries of modest complexity, more recent versions are claimed to be extremely efficient and to generate almost 10^7 to 10^9 clones/μg of primer-linker DNA; The protocol is extremely valuable for cloning cDNA adjacent to promoter elements so that the cDNA can be expressed in its recipient cell.	The original protocol described by Okayama and Berg was difficult and lengthy; More modern versions are slightly easier but remain demanding; Being large in size, it is often not possible to achieve a high primer:template ratio, consequently resulting in less efficient utilization of mRNA templates.
Universal primer for RT-PCR	'Universal primers' may be used for amplification of all mRNAs, regardless of their sequence. These are used to generate cDNA libraries when only small amounts of poly (A) RNA (<200 ng) are available. Other PCR protocols rely on gene specific primers that are unique to or selective for particular target cDNAs. For example, the 5′ RACE method allows amplification of the 5′-end for which some sequence near the 5′-end is known.	Clones generated by standard PCR amplification are likely to contain a higher number of mutations than those produced by other means; Methods that require unique primer sequences can be used only if the partial sequence of a cDNA or its cognate protein is already available.
Second cDNA strand synthesis by Klenow or *E. coli* DNA polymerase I		
Hairpin loop catalyzed self-priming	A primer used for first cDNA strand synthesis initiates second strand synthesis; Method results in higher yields of cDNA.	There is no intrinsic reason why there should be complementarity between the 3′-end, and other parts of the first strand. So priming may not be able to take place at all, and may well be very inefficient; Hairpin loop has to be removed before cloning into a suitable vector. This is done by S1 nuclease treatment, which is an uncontrolled reaction, and frequently results in the loss of sequences corresponding to the 5′-terminus of the mRNA;

(*Contd.*)

Table 5.2 (*Contd.*)

Primers used	Advantages	Disadvantages
Second cDNA strand synthesis by Klenow or *E. coli* DNA polymerase I		
		It is technically extremely difficult to generate cDNA clones representative of the full-length of a given mRNA, and most cDNA clones contain sequences corresponding to the 3′-end of the mRNA.
RNA fragments generated by RNase H treatment (replacement synthesis)	This method produces longer cDNA molecules than self-priming method; The priming is very efficient; Such priming does not need any further treatment or purification, and can be carried out directly using the products of the first strand reaction; It eliminates the need to use S1 nuclease, and hence no loss of sequences corresponding to the 5′-end of mRNA The resulting cDNAs are often very nearly full-length, lacking only a few nucleotides corresponding to the 5′-terminus of the mRNA.	There is a loss of sequence complementary to the far 5′-end of the original mRNA; Hairpin priming to second strand synthesis may still occur during the replacement reaction if RNase H efficiently removes the mRNA sequences immediately adjacent to the cap structure.
Oligonucleotide primer	Such priming procedure prevents the loss of sequences complementary to the 5′-end of mRNA; It allows preferential synthesis of full-length cDNAs, when used in combination with oligo (dT) primers for first strand synthesis. This is because oligonucleotide tailing of first strand cDNA at 3′-end captures full-length, first strand cDNAs that are directly used as templates for second strand synthesis using oligonucleotide primers; Oligonucleotide tailing of first strand required for annealing of oligonucleotide primers used for second strand synthesis allows the cDNA to be ligated to a vector before synthesis of the second strand; The procedure is highly efficient, however, the efficiency of this method depends on the particular mRNA under study.	All of the clones generated by homopolymeric priming of second strand synthesis carry a tract of dG:dC residues immediately upstream of the sequences corresponding to the mRNA template, the presence of which may inhibit transcription of DNA both *in vivo* and *in vitro*; These extra nucleotides may form a barrier to the Klenow fragment of *E. coli* DNA polymerase I during DNA sequencing by dideoxy mediated chain termination method.
Oligonucleotide primer-adaptor	Facilitates construction of directional cDNA libraries; The ss cDNA:mRNA hybrid directly serves as the substrate for TdT reaction, consequently allowing generation of full-length cDNAs; The number of steps involved in the synthesis and cloning of cDNA is reduced, as the ss cDNA:mRNA hybrid can be attached to a vector by synthetic adaptors before synthesis of the second strand of cDNA; High efficiency; The method can be adapted to allow the amplification of first strand cDNA by PCR.	Same as oligonucleotide primer.
Oligonucleotide primer-linker	Same as oligonucleotide primer-adaptor.	Same as oligonucleotide primer; Any cDNA containing a restriction site for the same enzyme whose site is incorporated into the oligonucleotide (primer-linker) will be cleaved internally when the ends are cleaved, which will produce two truncated cDNA molecules, hence methylation of internal sites is mandatory.
Random primers	Random priming prevents 3′- or 5′-end bias in the resulting library.	Full-length cDNAs are hardly obtained.

Random Primers

Random primers (or random oligonucleotides) are a collection of chemically synthesized oligonucleotides of random sequence, usually hexamers. These are produced by oligomerization of equal quantities of mixed A, G, C, and T residues, so that all possible hexameric sequences are present. The hexamers bind throughout the length of the mRNA molecule (wherever complementary) resulting in internal priming of first cDNA strand instead of priming from the 3′-end of the template. The 3′-OH ends provided by these primers are simultaneously acted upon by reverse transcriptases, leading to the formation of first cDNA strand. This strategy thus solves the problems associated with oligo (dT) primers for first strand synthesis, for example, (i) These primers are readily employed in both prokaryotes and eukaryotes, as the priming is not based on the presence of 3′-poly (A) tail; (ii) The chances of synthesis of sequences complementary to the 5′-end of mRNA increases, or the chances of accumulation of sequences corresponding to the 3′-end of mRNA are reduced; (iii) The use of random primers overcomes the problem that some RNA molecules (e.g. bacterial mRNA and genomic RNA from viruses) are not polyadenylated, i.e., random primers also allow first cDNA strand syntheses using poly $(A)^-$ RNA templates. Consequently, there is no loss of sequences complementary to mRNAs that might be non-polyadenylated; (iv) Random primers also work very well with mRNAs that exhibit strong secondary structure at the 5′-terminus; and (v) Random priming also precludes the possibility that cDNA synthesis is primed from the 3′ most region of the poly (A) tail. In such an event, a significant portion of the 5′-end of the first cDNA strand would consist of a lot of Ts.

Random primers, however, create certain problems, for example, (i) If one is interested in cDNA synthesis from poly (A) RNA, the product will be contaminated from cDNAs synthesized from poly $(A)^-$ RNAs as well; (ii) cDNA clones derived from random priming of first strand cDNA tend to be smaller than clones derived from oligo (dT) priming; (iii) The efficiency of priming varies from one mRNA to another, depending on the size of mRNA, and complexity and composition of the random primers pool; and (iv) The primers have to be removed between first and second strand synthesis steps, otherwise second strand synthesis will also initiate by random priming.

From the above discussion it is clear that random primed cDNA is usually longer and more representative than cDNA primed by oligo (dT) alone. The efficiency of cDNA synthesis can be improved further by using a combination of oligo (dT) primers and random primers in the same reaction tube. This approach eliminates the need to worry about the possible exclusion of poly $(A)^-$ transcripts, and tends to produce a population of much larger cDNA molecules, many of which are close to full-length. The parameters governing the synthesis of the first strand need not be changed; the only difference is the simultaneous use of half of the recommended mass of each type of primer.

Oligo (dT) Primer-adaptor or Oligo (dT) Primer-linker

For first cDNA strand synthesis oligonucleotides such as oligo (dT) primer-adaptor may also be employed. These primers consist of oligo (dT) appended at their 5′-end with overhanging sequences that will support the addition of restriction enzyme sites or useful sequences on one end. The purpose of such primer-adaptor is directional cloning of ds cDNA, hence such first cDNA strand synthesis is done in combination with second strand synthesis with a similar primer-adaptor having different restriction enzyme site.

Oligo (dT) primer-linker works on the same principle for directional cloning. Note that while using such oligo (dT) primer-linker, internal restriction sites should be methylated. This is because linkers have inherent restriction enzyme site, and have to be digested with the cognate restriction enzyme before ligation to vector during cloning experiment.

Asymmetrically Tailed Plasmid (Oligomerically Tailed Plasmid)

Synthesis of the first strand is also primed from an oligo (dT) tail covalently attached to one end of the linearized plasmid. This strategy allows direct and efficient cloning of cDNA into the vector. Being large in size, it is often not possible to incorporate enough oligomerically tailed plasmid in the reaction mixture to achieve a high primer:template ratio. As a consequence, priming is usually the rate-limiting step during this kind of reaction, which may result in less efficient utilization of mRNA templates during first cDNA strand synthesis. However, the overall efficiency of the process of ds cDNA synthesis is higher due to direct cloning into the vector.

5.3.4 Primers Used for Second cDNA Strand Synthesis

In order to obtain a molecule that is stable, and can be manipulated and cloned into a vector, the mRNA in the ss cDNA:mRNA hybrid needs to be replaced with a cDNA strand of the same sequence. This second cDNA strand synthesis is conveniently carried out by *E. coli* DNA polymerase I (Kornberg enzyme), Klenow enzyme, reverse transcriptase, or T4 DNA polymerase. Thermostable DNA polymerases such as *Tth* and *Taq* DNA polymerases can also be used, albeit rarely. The reactivity of all these polymerases differs from one another. Klenow enzyme as well as T4 DNA polymerase are advantageous because these lack the $5' \rightarrow 3'$ exonuclease activity of DNA polymerase I, and hence do not degrade the newly synthesized ds cDNA. It has been observed that the use of AMV RT instead of DNA polymerase I generally produces full-length ds cDNA copies, and hardly ever any shorter reaction intermediates. The synthesis of the second strand directed by AMV RT, therefore, generally yields cDNA populations, which are more homogeneous than those produced by *E. coli* DNA polymerase I. On the other hand, certain cDNA molecules are good templates for *E. coli* DNA polymerase I but not for AMV RT. These molecules do not possess base paired 3′-ends, and in these cases the 3′ → 5′ exonuclease activities of the Klenow fragment and T4 DNA polymerase are required to eliminate protruding single stranded termini to generate active 3′-OH end. This is because T4 DNA polymerase has the property of both trimming-back and filling-in, which allows recovery of perfectly blunt ended ds cDNA. AMV RT is inactive in this case because it does not possess the required 3′ → 5′ exonuclease activity. In view of the experience reflected in various published reports, it is impossible to recommend solely the use of any one of these enzymes; rather all these enzymes possess their characteristic individual properties. Depending on the particular DNA polymerase used, the conditions employed for the synthesis of full-length second cDNA strand vary. Reactions that use *E. coli* DNA polymerase I are carried out at pH 6.9 to minimize the 5′ → 3′ exonuclease activity of the enzyme, and at 15°C to minimize the possibility of synthesizing 'snapback' DNA. These problems can be completely avoided by using either reverse transcriptase or Klenow fragment, both of which lack 5′ → 3′ exonuclease activity, and are available in pure form from commercial sources. *E. coli* DNA ligase is sometimes included in the reaction because of the evidence that it may improve the efficiency of cloning longer cDNAs. Frequently, T4 DNA polymerase or a thermostable polymerase such as *Pfu* DNA polymerase is added at the end of the second strand reaction to polish the termini of the completed ds cDNAs. However, generating blunt ends in this fashion sometimes results in the loss of 5′-terminal sequences so that the resulting cDNA clones are 5–30 nucleotides shorter than the original mRNA.

All these polymerases require free 3′-OH groups to extend on, which are provided by primers. Primers for second cDNA strand synthesis can be generated spontaneously by hairpin formation within the ss cDNA, i.e., by the 3′-end of the ss cDNA folding-back on itself. This is called self-priming, and initiates by the action of reverse transcriptase during first strand synthesis. Primers for second cDNA strand synthesis can also be provided in a separate reaction, for example, primers generated through RNase H-catalyzed nicking of mRNA strand in the hybrid, or annealing of oligonucleotides/oligonucleotide-adaptors/oligonucleotide-linkers complementary to the homopolymeric tail attached to the 3′-end of first cDNA strand or annealing of random primers after alkaline digestion of mRNA in the hybrid. Based on these primers, different ways for second cDNA strand synthesis are self-priming, replacement synthesis, and oligonucleotide priming (i.e., priming with oligonucleotide, oligonucleotide-adaptor, and oligonucleotide-linker). As with the synthesis of the first cDNA strand, high temperature diminishes intrastrand folding and increases the efficiency of synthesis of a full-length strand. The primers are used for second strand synthesis, and the yields of second cDNA strand achieved by these methods vary between 30% and 100%. The advantages and disadvantages of these primers are listed in Table 5.2. Note that careful consideration should be given to the method for the synthesis of second cDNA strand, because it can determine the choice of vector, and dictate the means used to link the cDNA to the vector.

Self-priming

This strategy takes advantage of the tendency of reverse transcriptase to form a hairpin loop during the first cDNA strand synthesis, which is used as a primer for second cDNA strand synthesis. This hairpin is formed due to spontaneous intramolecular base pairing during reverse transcriptase reaction. This technique is advantageous because a single primer that is used for first cDNA strand synthesis is used to synthesize the second cDNA strand as well. This method leads to higher yields of ds cDNA. For cloning ds cDNA into a suitable vector, hairpin loop is removed in a reaction catalyzed by S1 nuclease. As it is an uncontrolled reaction, there may be frequent loss of cDNA sequences corresponding to the 5′-end of mRNA.

RNA Fragments Generated by RNase H Treatment (Replacement Synthesis)

RNase H recognizes a ss cDNA:mRNA hybrid, and makes many nicks or gaps in the RNA strand. Many RNA fragments thus created along the length of the first cDNA strand provide free 3′-OH groups to be used as primers for extension by polymerase. The reaction has three main advantages: (i) The priming is very efficient; (ii) Such priming does not need any further treatment or purification, and can be carried out directly using the products of the first cDNA strand synthesis reaction; and (iii) It eliminates the need to use S1 nuclease, and hence is free from disadvantages such as loss of cDNA sequences that results from S1 nuclease treatment. However, three theoretical problems are associated with replacement synthesis, but these do not turn out to be serious in practice. These are: (i) Most of the eukaryotic mRNAs carry 'cap' structures at their 5′-termini, which have to be removed before ds cDNA is inserted into a prokaryotic vector, but how this cap is removed is unclear. However, in practice it has been observed that the products of the replacement reaction can be efficiently inserted into prokaryotic vectors without taking any special steps to remove the cap structure; (ii) The replacement reaction is incapable of synthesizing the second cDNA strand corresponding to the 5′-terminal region of the mRNA. However, the resulting cDNAs are often very nearly full-length, lacking only a few nucleotides corresponding to the 5′-terminus of the mRNA; and (iii) Hairpin priming to second cDNA strand synthesis may still occur during the replacement reaction if RNase H efficiently removes the mRNA sequences immediately adjacent to the cap structure. However, second cDNA strand synthesis is initiated by self-priming in only <10% of the cases studied.

Oligonucleotide Priming

In this method, homopolymeric tail is added to the 3′-end of the newly synthesized first cDNA strand. This tailing is done by exploiting the activity of terminal deoxynucleotidyl transferase (TdT), a template independent DNA polymerase from calf thymus, in the presence of only single type of dNTP (take for example, addition of oligo dC tail by TdT in presence of dCTPs). This oligo dC tail then acts as an annealing site for a chemically synthesized complementary oligonucleotide primer (oligo dG). This oligonucleotide tailing procedure allows the cDNA to be ligated to a vector before synthesis of the second cDNA strand, or it enables a second primer to be used to prime the synthesis of second cDNA strand. The annealed oligo dG then provides a free 3′-OH group for second cDNA strand synthesis. Such priming procedure prevents the loss of sequences complementary to the 5′-end of mRNA. It also allows preferential synthesis of full-length cDNAs, when used in combination with oligo (dT) primers for first cDNA strand synthesis.

Oligonucleotide-adaptor Priming or Oligonucleotide-linker Priming

In this method, once first strand is synthesized using oligo (dT) primer-adaptor, a homopolymeric tail [say oligo (dC)] is added to its 3′-end in a TdT-catalyzed reaction in the presence of dCTP. Note that during first cDNA strand synthesis, oligo (dT) anneals to poly (A) tail, and the adaptor provides a recognition site for a restriction endonuclease. This oligo (dC) tail provides complementary sequences for annealing of oligonucleotide primer. The mRNA strand in a ss cDNA:mRNA hybrid formed after first cDNA strand synthesis is subjected to alkali hydrolysis or denaturation. A chemically synthesized dual-function oligonucleotide, an oligonucleotide-adaptor primer comprising of an oligo (dG) at 3′-end, and a restriction enzyme site at its 5′-end is added to the system, which hybridizes to the oligo (dC) tail of the first cDNA strand. This oligo dG provides a free 3′-OH group for second cDNA strand synthesis. This is a good way of producing cDNA that facilitates construction of directional cDNA libraries. Moreover, as ss cDNA:mRNA hybrid directly serves as the substrate for TdT reaction, full-length cDNAs are preferentially obtained. This is because a cDNA that does not extend to the end of the mRNA will present a shielded 3′-OH group, which is a poor substrate for tailing.

Similar to above strategy, oligonucleotide-linker can also be employed as primers for second strand synthesis. The details are given in Section 5.3.5.

Random Priming

Similar to first cDNA strand synthesis by random primers, second cDNA strand synthesis by Klenow enzyme can also be primed by random primers. Random priming of the second cDNA strand is done after alkali digestion of RNA in ss cDNA:mRNA hybrid. Random priming prevents 3′ or 5′ bias in the resulting library, but with this method full-length cDNA is hardly obtained.

5.3.5 Procedures for Synthesis of ds cDNA from mRNA

In the earlier sections, a discussion of primers used for first and second cDNA strand syntheses has been made. A combination of any method for first and second cDNA strand syntheses can be used to get blunt ended ds cDNA, in a single- or two-step process. The single-step process involves self-priming initiated during first cDNA strand synthesis, and extension by polymerase. On the other hand, in the two-step synthesis procedure, the first step involves removal of RNA, either by alkaline hydrolysis or RNase H digestion, and the second step involves synthesis of second cDNA strand. Note that in all the procedures a high primer:template ratio is maintained. At the end of the procedure, a mixture of partial and complete (full-length) ds cDNA copies of the more prevalent mRNAs in the original sample are obtained, which can be efficiently ligated to vector DNA through linkers or adaptors, or by homopolymeric tailing (for details see Chapter 15). Some important ds cDNA synthesis strategies are discussed below, and their merits and demerits are tabulated in Table 5.2.

Oligo (dT) Priming of First cDNA Strand and Self-priming of Second cDNA Strand (Maniatis Strategy)

Maniatis strategy, the classical method of ds cDNA synthesis, takes advantage of the tendency of reverse transcriptase to form a hairpin loop after first cDNA strand synthesis, which is then used as a primer for second cDNA strand synthesis (Maniatis *et al.*, 1976). Thus in this method, the first cDNA strand synthesis catalyzed by reverse transcriptase is initiated by oligo (dT) primer, which anneals at the poly (A) tail of mRNA. The mRNA in ss cDNA:mRNA hybrid is denatured by boiling or hydrolyzed by alkali. The cDNA strand turns back on itself, and a few nucleotides are added to form a hairpin loop at the 3′-end due to folding-back or looping-back on itself as a consequence of hydrophobicity of bases. This hairpin loop allows self-priming of the second cDNA strand. Thus, the priming of second cDNA strand synthesis is initiated from the first cDNA strand synthesis reaction, as the growing first cDNA strand produced by the reverse transcriptase provides a 3′-OH group for the Klenow fragment or DNA polymerase I to complete the synthesis of the second cDNA strand, using the first cDNA strand as a template. The end product is a ds cDNA molecule, with a hairpin loop at one end. The hairpin loop is then cleaved with a single strand specific nuclease, S1 nuclease from *Aspergillus oryzae*, to generate a blunt ended ds cDNA (Figure 5.8a). Note that S1 nuclease cuts ss DNA, including exposed loops. The cleavage of a phosphodiester bond within the hairpin loop produces double stranded ends. In order to avoid the digestion of the double stranded parts of the cDNA, which correspond to the 5′-ends of the mRNA, this nuclease treatment has to be performed very carefully. The product thus obtained may be treated with T4 DNA polymerase to make the molecule blunt, and hence amenable to subsequent cloning.

Using this method, the reaction, and also the course of cDNA synthesis can be easily monitored if radioactively labeled precursors, for example, ^{3}H-dNTPs for the first reaction and ^{32}P-dNTPs for the second reaction are used. These radioactive labels are important in two ways. Firstly, these help in determination of the yield of ds cDNA by monitoring the extent to which the reaction products become S1 nuclease resistant during the reaction. Secondly, autoradiographic determination of gel fractionated cDNAs helps in identification of heterogeneous population of cDNAs (for details see Section 5.3.9). Clearly this technique has several disadvantages. For example, (i) There is no intrinsic reason why there should be complementarity between the 3′-end, and other parts of the first cDNA strand. So priming may not be able to take place at all, and may well be very inefficient; (ii) It is technically extremely difficult to generate cDNA clones representative of the full-length of a given mRNA, and most cDNA clones contain sequences corresponding to the 3′-end of the mRNA. This is due to several factors including the secondary structure of the RNA itself, which influences the extent of first cDNA strand synthesis; (iii) The second cDNA strand synthesis is a poorly controlled reaction probably because of the highly variable size and position of the terminal loop that acts as a primer for second cDNA strand synthesis; (iv) S1 nuclease, used to cleave hairpin loop, inevitably degrades some terminal nucleotides from the ds cDNA, resulting almost invariably to the loss of sequences corresponding to the 5′-terminus of the mRNA. This may lead to under-representation of the regions at the 5′-end of the mRNA in the cDNA library. This problem can however be countered by the application of TdT to add a

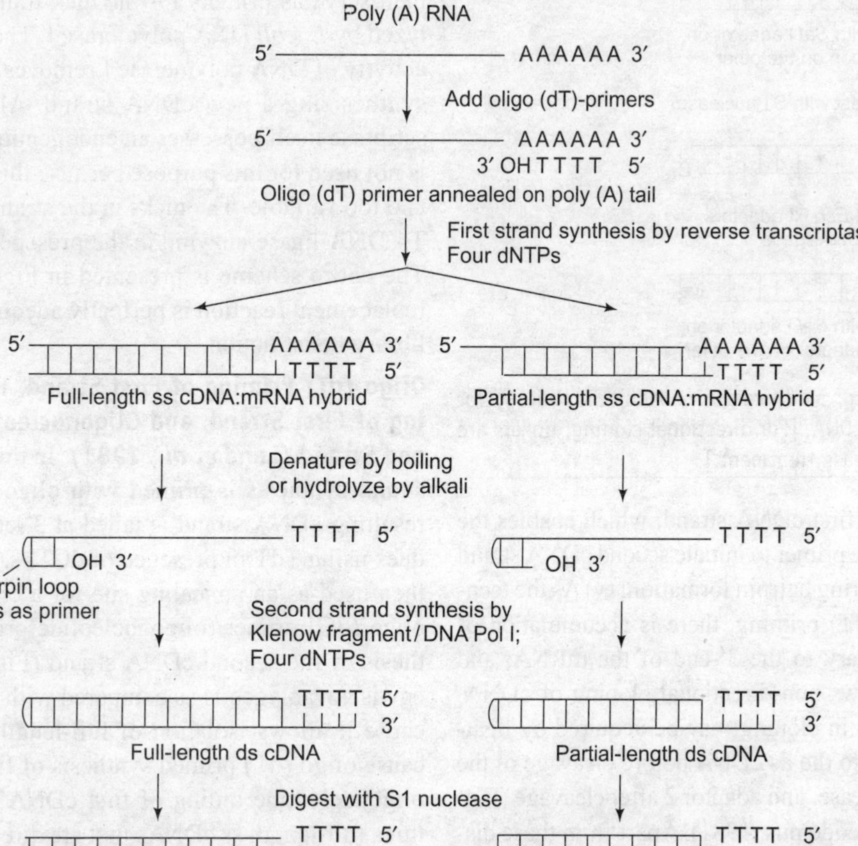

Figure 5.8a Self-priming classical method of ds cDNA synthesis (Maniatis strategy); In this 'traditional' cDNA synthesis procedure first strand synthesis is initiated by oligo (dT) primers, and second strand synthesis is initiated by self-priming

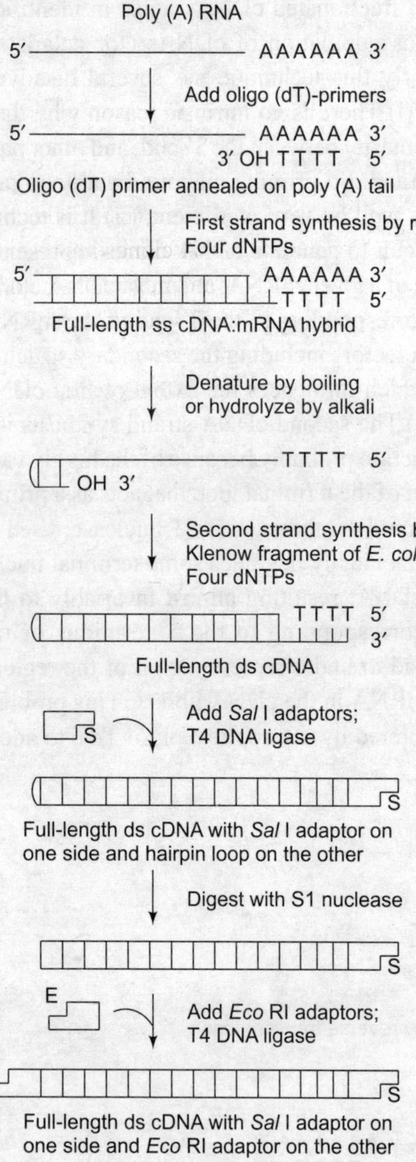

Figure 5.8b Self-priming; Modification of Maniatis strategy for directional cloning of cDNA [For directional cloning, linkers are ligated before S1 nuclease treatment.]

tail to the 3′-end of the first cDNA strand, which enables the use of an oligonucleotide primer to initiate second cDNA strand synthesis, without requiring hairpin formation; (v) As the technique employs oligo (dT) priming, there is accumulation of sequences complementary to the 3′-end of the mRNA; and (vi) The technique allows non-directional cloning of cDNA. However, directionality in cloning may be acquired by ligating synthetic adaptor 1 to the ds cDNA before cleavage of the hairpin loop by S1 nuclease, and adaptor 2 after cleavage (Figure 5.8b) (Kurtz and Nicodemus, 1981). Apart from these disadvantages, for many years there was no effective way other than self-priming to synthesize the second cDNA strand, and almost all clones of cDNA made before 1982 were obtained using this enzymatic reaction. This method was replaced later on by more efficient and less destructive methods where second cDNA strand synthesis was initiated in a separate reaction, and which yielded 10–100 times more cDNA clones as compared to self-priming.

Priming of Second cDNA Strand in a Separate Reaction

Besides self-priming, the priming of second cDNA strand synthesis can also be achieved in a reaction independent of first cDNA strand synthesis. Some common methods are discussed below.

Oligo (dT) Priming of First cDNA Strand and Replacement Synthesis of Second cDNA Strand (Gubler and Hoffman, 1983) In this method the product of the first cDNA strand synthesis, i.e., a ss cDNA:mRNA hybrid, is used as a template for a nick translation reaction catalyzed by *E. coli* DNA polymerase I. Thus in this method the first cDNA strand synthesis is initiated by oligo (dT) primers in a reverse transcriptase-catalyzed reaction. The mRNA in the resulting ss cDNA:mRNA hybrid is digested by RNase H, and not by alkali treatment. RNase H recognizes the RNA component of a ss cDNA:mRNA hybrid, and cleaves the RNA at a number of nonspecific sites leaving short oligoribonucleotides base paired to the cDNA. These short oligoribonucleotides then serve as primers for the nick translation reaction catalyzed by *E. coli* DNA polymerase I. The 5′ → 3′ exonuclease activity of DNA polymerase I removes RNA upstream while synthesizing a new cDNA strand. Although reverse transcriptase itself possesses an endogenous RNase H activity, it is not used for this purpose because this activity is too weak, and too variable. The nicks in the strands are then ligated by T4 DNA ligase enzyme in the presence of ATP as cofactor. The entire scheme is presented in Figure 5.9a. This simple replacement reaction is perfectly adequate for routine cDNA library construction.

Oligo (dT) Priming of First Strand, Homopolymeric Tailing of First Strand, and Oligonucleotide Priming of Second Strand (Land *et al.*, 1981) In this method, first cDNA strand synthesis is primed with oligo (dT) primer, and the resulting cDNA strand is tailed at 3′-end with cytidine residues using TdT in presence of dCTPs. The oligo (dC) tail is then used as an annealing site for a chemically synthesized oligo (dG) primer (oligonucleotide priming), allowing synthesis of the second cDNA strand (Figure 5.9b). This strategy is advantageous as compared with Maniatis method, because it allows isolation of full-length cDNAs. This is because oligo (dT) primed synthesis of first cDNA strand, and oligonucleotide tailing of first cDNA strand at 3′-end captures full-length ss cDNAs that are directly used as templates for second cDNA strand synthesis. Moreover, oligonucleotide tailing allows the cDNA to be ligated to a vector before synthesis of the second cDNA strand; however the cloning of cDNA is non-directional. The procedure is highly efficient,

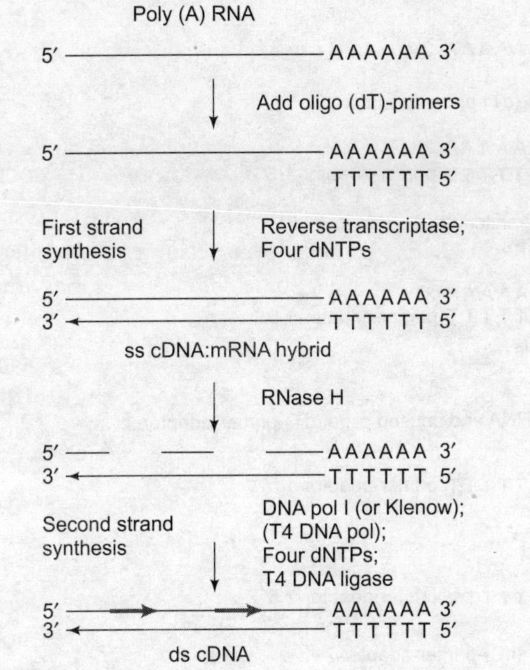

Figure 5.9a Poly (A) tail capture methods for synthesis of ds cDNA; ds cDNA synthesis by oligo (dT) priming of first strand and replacement synthesis of second strand

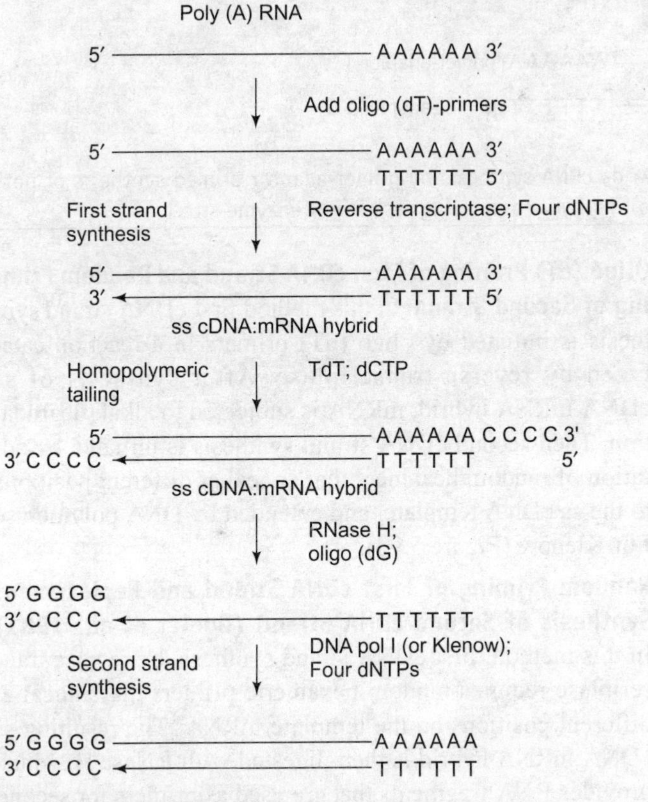

Figure 5.9b Poly (A) tail capture methods for synthesis of ds cDNA; Homopolymeric tailing to provide annealing site for oligonucleotide primers for second strand synthesis

however, the efficiency of this method depends on the particular mRNA under study. In the best cases (e.g., chicken lysozyme mRNA), full-length clones of cDNA can be obtained with high efficiency; other mRNAs, however, have provided more variable results, perhaps reflecting the relative efficiencies with which TdT uses the 3′-termini of different first cDNA strands as substrates for homopolymeric tailing. Yields of up to 10^6 clones per μg RNA have been described for instance in the cloning of cDNAs for the precursor of human growth hormone releasing factor and of rat thymosin β-4. The presence of oligomeric tails at each end of the first cDNA strand provides an opportunity to amplify the cDNA *in vitro* by PCR. This is useful when the amount of available mRNA is too small for producing cDNA by standard procedures. However, because PCR amplification of long cDNAs is inefficient, there is a strong selection against large cDNAs.

Synthetic Primer-adaptors (or Primer-linker) Primed Synthesis of Both Strands The strategy is based on the use of synthetic primer-adaptors that carry homopolymeric tails for priming the synthesis of both first and second strands of cDNA, and restriction sites for cloning into vectors. This strategy has the advantage of directional cloning. Thus, to allow directional cloning of first cDNA strand synthesis is initiated with oligo (dT) primer-adaptor 1. This is followed by TdT-catalyzed oligonucleotide tailing of first cDNA strand at its 3′-end (say oligo dC addition) (Figure 5.9c). Second cDNA strand synthesis is then initiated in a separate reaction by using oligonucleotide-adaptor 2 (say oligo dG-adaptor 2). This oligonucleotide tailing procedure besides providing 3′-OH group for extension by polymerase also allows the cDNA to be ligated to a vector before synthesis of the second cDNA strand. The ds cDNA thus synthesized contains one restriction site for one restriction enzyme (corresponding to adaptor 1) on one side, and a restriction site for second restriction enzyme (corresponding to adaptor 2) on the other side. The presence of two different restriction enzyme sites at the two ends favors directional cloning.

The use of primer-adaptors has following advantages: (i) The number of steps involved in the synthesis and cloning of cDNA is reduced. For example, in the protocol the ss cDNA: mRNA hybrid can be attached to a vector by synthetic adaptors before synthesis of the second strand of cDNA, thereby eliminating several steps that are required when ds cDNA is cloned by addition of synthetic linkers; (ii) The increased efficiency of priming of second cDNA strand yields libraries that contain a comparatively high proportion of full-length cDNA molecules; (iii) The presence of two different restriction enzyme sites at the two ends of the synthesized ds cDNA allows directional cloning; and (iv) The method can be adapted to allow the amplification of first cDNA strand by PCR. The method however has certain demerits as well. All of the clones generated by homopolymeric priming of second cDNA strand synthesis carry a tract of dG:dC residues immediately upstream of the sequences corresponding to the mRNA template. The

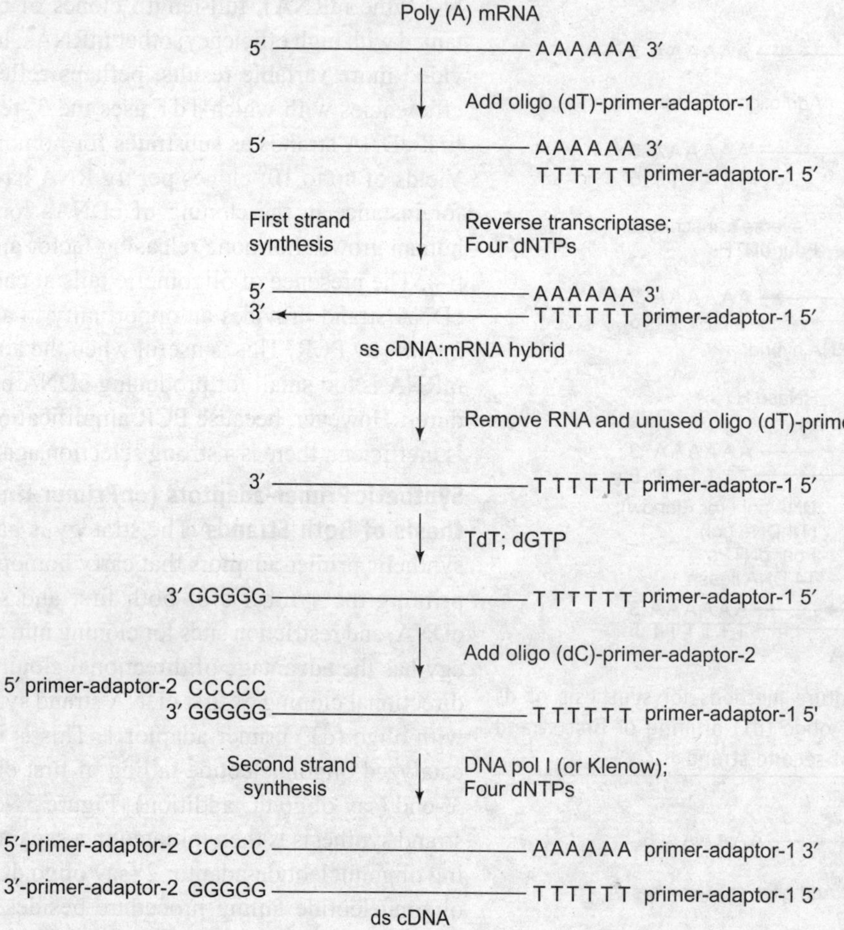

Figure 5.9c Poly (A) tail capture methods for synthesis of ds cDNA; ds cDNA synthesis by primer-adaptor primed synthesis of both strands favors directional cloning [Modification of homopolymer tailing by incorporating restriction enzyme site.]

presence of these additional sequences may inhibit transcription of DNA both *in vivo* and *in vitro*. Furthermore, these may form a barrier to the Klenow fragment during DNA sequencing by dideoxy chain termination method, requiring that another DNA polymerase (e.g., reverse transcriptase) be used instead, and the conditions adjusted accordingly.

Similar to the primer-adaptor approach discussed above, primer-linkers can also be employed. However, using primer-linkers any cDNA molecule that contains a restriction site for the same enzyme whose site is incorporated into the oligonucleotide (primer-linker) will be cleaved internally when the ends are cleaved. This will produce two or more truncated cDNA molecules depending on the number of restriction sites present for that enzyme within the cDNA. One way to avoid this is to protect the internal sites by cognate methylase (or methyltransferase). For example, if *Hin*d III oligonucleotides are being used, the cDNAs are methylated using *Hin*d III methylase that uses *S*-adenosyl methionine as methyl group donor, which protects any internal *Hin*d III site from subsequent restriction with *Hin*d III. After protection, the methylase is removed, and the oligonucleotides are attached as discussed for oligonucleotide-adaptors.

Oligo (dT) Priming of First cDNA Strand and Random Priming of Second Strand In this method first cDNA strand synthesis is initiated by oligo (dT) primers in a reaction catalyzed by reverse transcriptase. After synthesis of ss cDNA:mRNA hybrid, mRNA is subjected to alkali degradation. Then second cDNA strand synthesis is initiated by addition of random hexamers that anneal at different positions to the ss cDNA template, and extended by DNA polymerase I or Klenow (Figure 5.9d).

Random Priming of First cDNA Strand and Replacement Synthesis of Second cDNA Strand (Goelet et al., 1981) In this method, first cDNA strand synthesis by reverse transcriptase requires random hexameric primers that anneal at different positions on the template mRNA. The resulting ss cDNA:mRNA hybrid is then digested with RNase H, which provides RNA fragments that are used as primers for second cDNA strand synthesis by *E. coli* DNA polymerase I (Figure 5.10a). ds cDNAs synthesized in this fashion are unlikely to include sequence corresponding to the poly (A) tail unless an oligo (dT) random primer had been included in the primer mixture. Random priming generally produces a greater number of full-length cDNAs. However, the process requires an

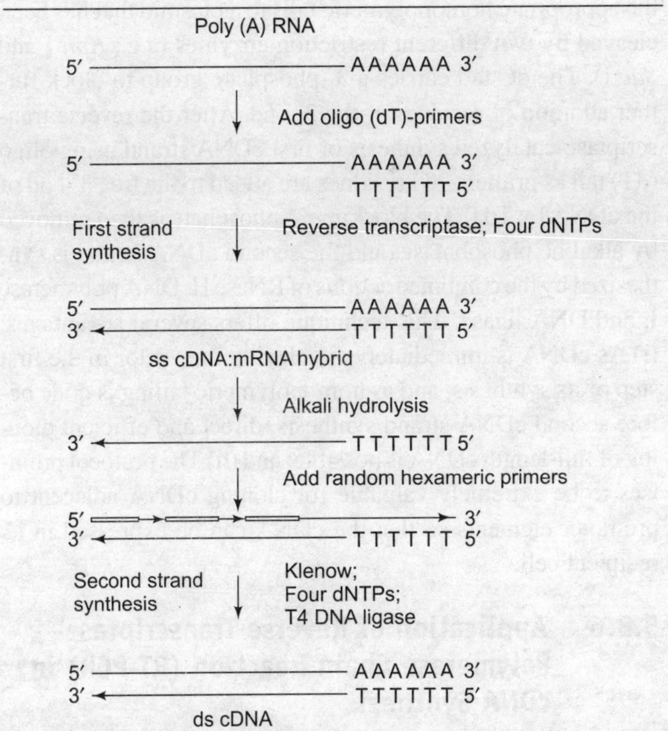

Figure 5.9d Poly (A) tail capture methods for synthesis of ds cDNA; ds cDNA synthesis by oligo (dT) priming of first strand and random priming second strand synthesis

intermittent purification step before second cDNA strand synthesis to remove random hexamers so as to prevent initiation of second cDNA strand synthesis by random primers.

Random Priming of First cDNA Strand, Homopolymeric tailing of First cDNA Strand, and Oligonucleotide Priming of Second cDNA Strand In this method, first cDNA strand synthesis is initiated by reverse transcriptase using random hexameric primers, which anneal at different positions on the template mRNA. The resulting ss cDNA:mRNA hybrid is then subjected to alkali denaturation, and a TdT reaction in the presence of single dNTP (say dCTP). The oligo (dC) homopolymeric tail thus added to the first cDNA strand is used as annealing site for the oligonucleotide primer [say oligo (dG) primer] to be extended in second cDNA strand synthesis step by Klenow fragment (Figure 5.10b). In this method random hexamers have to be removed by an intermittent purification step, otherwise random primers will also initiate second cDNA strand synthesis.

Random Priming of Both Strands In this method, first cDNA strand synthesis by reverse transcriptase requires random hexameric primers that anneal at different positions on the template mRNA. The resulting ss cDNA:mRNA hybrid is then subjected to alkali denaturation for degradation of mRNA

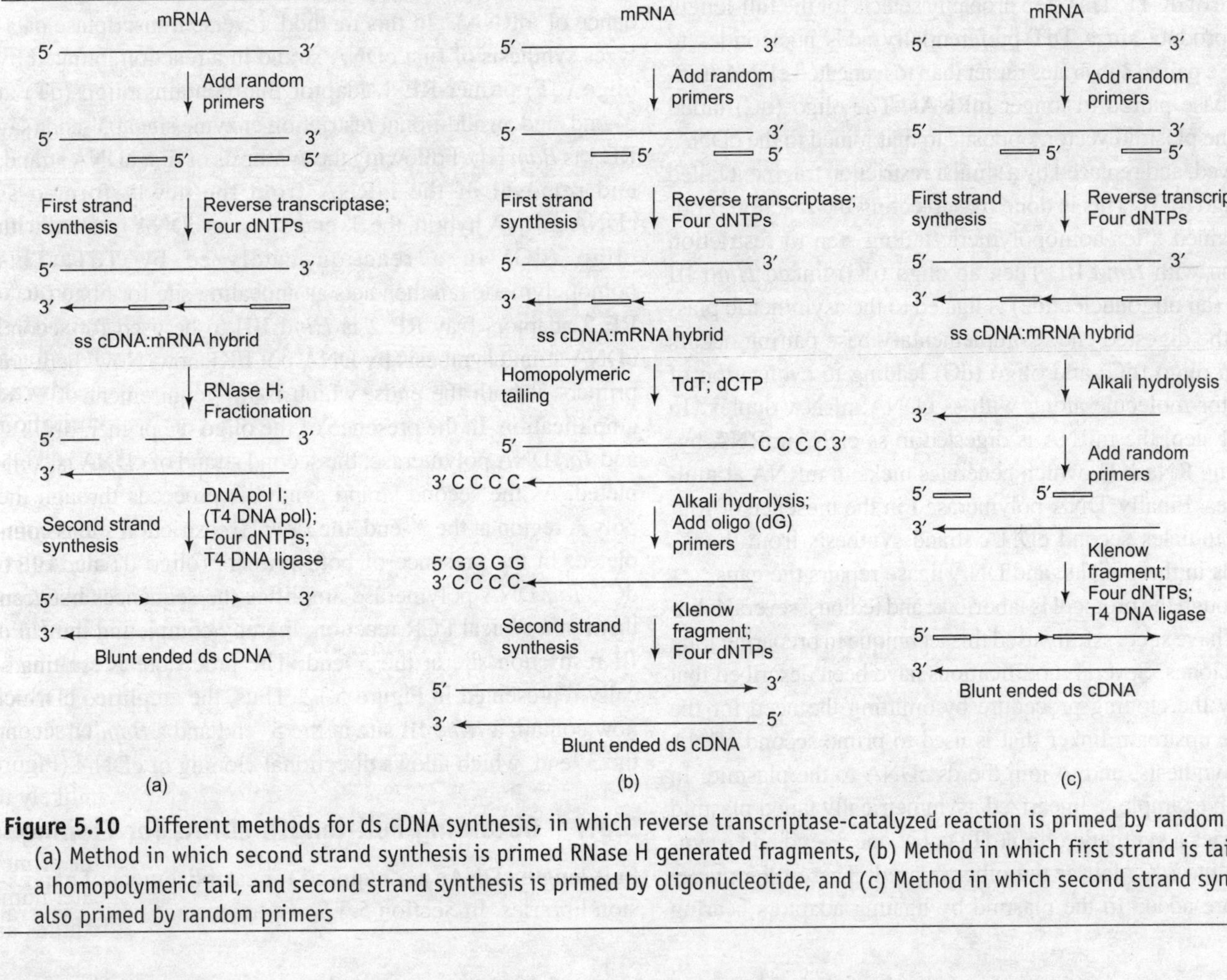

Figure 5.10 Different methods for ds cDNA synthesis, in which reverse transcriptase-catalyzed reaction is primed by random primers. (a) Method in which second strand synthesis is primed RNase H generated fragments, (b) Method in which first strand is tailed with a homopolymeric tail, and second strand synthesis is primed by oligonucleotide, and (c) Method in which second strand synthesis is also primed by random primers

in the hybrid. Then second cDNA strand synthesis is also initiated with random primers in a reaction catalyzed by Klenow fragment (Figure 5.10c). The advantage of this procedure is that it prevents any 5′ or 3′-end bias, i.e., preferential recovery of clones representing the 5′ or 3′-end of cDNA sequences in the resulting library is prevented. Moreover, the process does not require any intermittent purification step for the removal of random primers. However, the resulting clones are much smaller, and full-length cDNAs must be assembled from several shorter fragments.

Asymmetrically Tailed Plasmid Priming of First cDNA Strand, Homopolymeric Tailing of First cDNA Strand, and replacement synthesis of second cDNA strand (Okayama and Berg, 1982) An alternative to oligo (dT) or random primers for first cDNA strand synthesis is to use primers that are already linked to a plasmid. In this strategy, a linearized plasmid (pBR322) carrying a synthetic oligo (dT) tail at one of its 3′-ends is used (Figure 5.11). This oligo (dT) tail serves as the annealing site for poly (A) RNA. In a reverse transcriptase-catalyzed reaction, and using oligo (dT) tail attached to plasmid as primer, first cDNA strand is synthesized. The result of the reaction is a ss cDNA:mRNA hybrid. The synthesis of the second cDNA strand is not the next step, instead the hybrid is subjected to homopolymeric tailing in a reaction catalyzed by TdT in the presence of dCTP. This step probably selects for the full-length cDNA product since TdT preferentially adds nucleotides to fully base paired substrates rather than to truncated cDNA transcripts base paired to longer mRNAs. The oligo (dC) tailed end of the plasmid vector, opposite to that joined to the cDNA, is removed, and replaced by a similar restriction fragment tailed with oligo (dG). This is done by subjecting the reaction product obtained after homopolymeric tailing step to restriction digestion with *Hin*d III. Then an oligo (dG) linked *Hin*d III adaptor (an oligonucleotide) is ligated to the asymmetric plasmid at the digested end. Complementary base pairing occurs between oligo (dC) and oligo (dG) leading to cyclization of the vector molecule along with ss cDNA:mRNA duplex. In the next step, the mRNA is digested in ss cDNA:mRNA hybrid using RNase H, which generates nicks in mRNA at multiple sites. Finally, DNA polymerase I in the presence of four dNTPs initiates second cDNA strand synthesis from the 3′-OH ends in these nicks, and DNA ligase repairs the gaps.

Although the protocol is laborious, and tedious, several laboratories have successfully used this technique in preparing long cDNA clones. Several modifications have been described that simplify the cloning procedure by omitting the need for the separate upstream linker that is used to prime second cDNA strand synthesis, and to join the ds cDNA to the plasmid. In one such example, a linearized asymmetrically tailed plasmid that carries a synthetic oligo (dT) tail at one 3′-end and a synthetic oligo (dC) tail at the other is used. These oligonucleotides are added to the plasmid by ligating adaptors bearing the appropriate homopolymeric tails to a plasmid that has been cleaved by two different restriction enzymes (e.g., *Kpn* I and *Sac* I). The dC tail carries a 3′-phosphate group to block further addition of residues at the 3′-end. After the reverse transcriptase-catalyzed synthesis of first cDNA strand using oligo (dT) tail as primers, dG residues are added to the free 3′-end of the cDNA by TdT. The blocking 3′-phosphate is then removed by alkaline phosphatase, and the second cDNA strand is synthesized by the combined actions of RNase H, DNA polymerase I, and DNA ligase. This technique offers several advantages: (i) As cDNA is immediately attached to the vector in the first step of its synthesis, and as homopolymeric tailing is done before second cDNA strand synthesis, direct and efficient cloning of full-length cDNA is possible; and (ii) The protocol promises to be extremely valuable for cloning cDNA adjacent to promoter elements so that the cDNA can be expressed in its recipient cell.

5.3.6 Application of Reverse Transcriptase-Polymerase Chain Reaction (RT-PCR) in cDNA Synthesis

RT-PCR is also used in the production and cloning of cDNA particularly in situations where only very small quantities of mRNA are available (small amounts of tissue or low abundance of mRNA). In this method, reverse transcriptase catalyzes synthesis of first cDNA strand in a reaction initiated by oligo (dT) primer-RE 1 adaptor that contains oligo (dT) at 3′-end, and an additional restriction enzyme site at 5′-end (say RE 1 is *Bam* HI). Following the synthesis of first cDNA strand, and removal of the mRNA from the newly formed ss cDNA:mRNA hybrid, the 3′-end of the ss cDNA is tailed with oligo (dG) in a reaction catalyzed by TdT. This homopolymeric tail then acts as annealing site for oligo (dC)-RE 2 adaptors [say RE 2 is *Hin*d III] to be used for second cDNA strand synthesis by DNA pol I/Klenow. Now there are primers at both the ends, which is the requirement of PCR amplification. In the presence of the oligo dC primer, dNTPs, and *Taq* DNA polymerase, the second strand of cDNA is completed. As the second strand synthesis proceeds through the poly A region at the 3′-end, the *Bam* HI restriction site is completed. In the presence of both primers (oligo dT and oligo dC), *Taq* DNA polymerase amplifies the sequences between them in a typical PCR reaction, thereby completing the *Hin*d III restriction site at the 5′-end. The procedure is schematically represented in Figure 5.12. Thus, the amplified cDNAs now contain a *Hin*d III site at the 5′-end and a *Bam* HI site at the 3′-end, which allows directional cloning of cDNA.

5.3.7 Obtaining Full-length cDNAs for Cloning

Full-length cDNAs are required for establishment of expression libraries. In Section 5.3.5, we have come across several

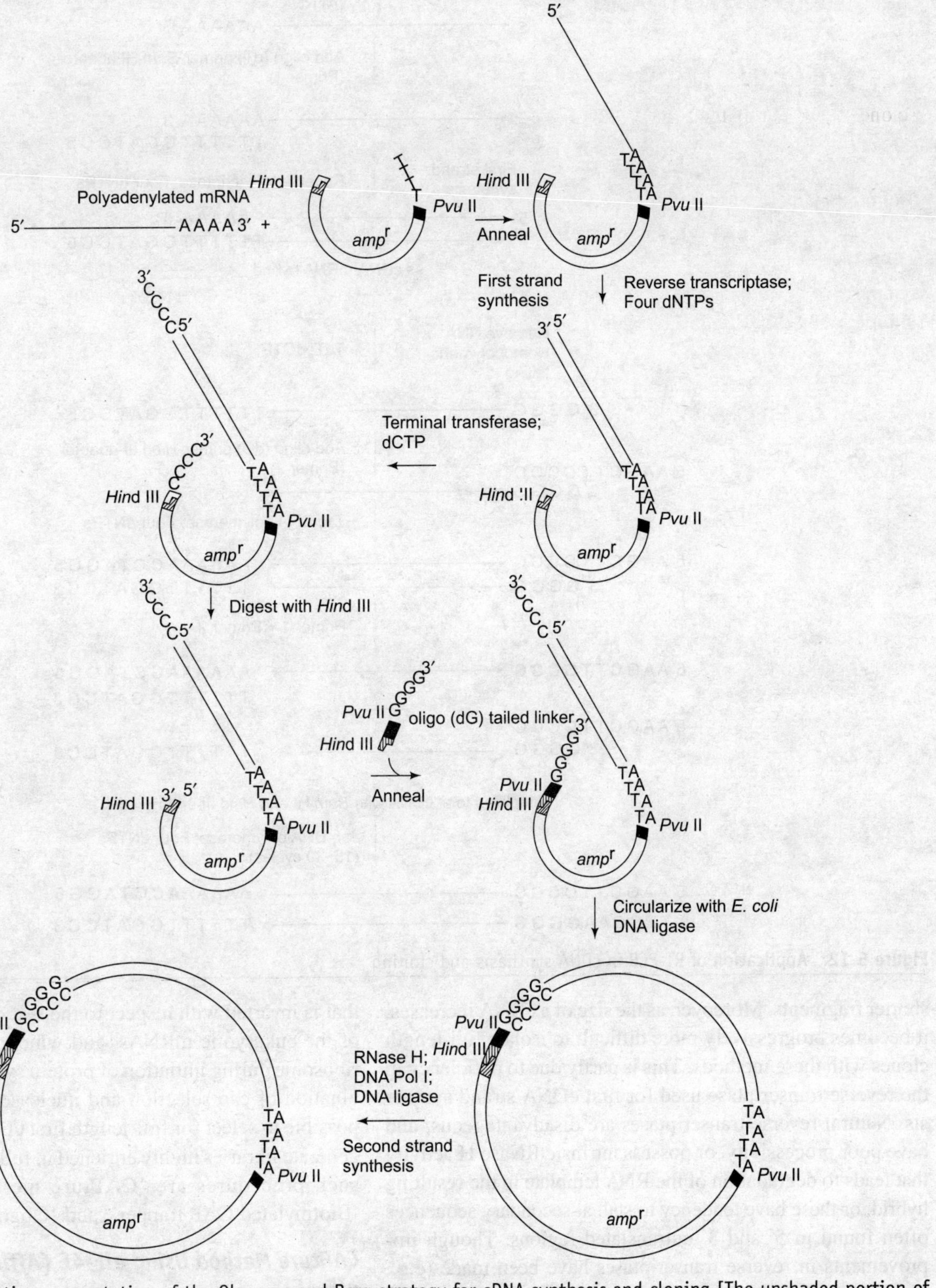

Figure 5.11 Schematic representation of the Okayama and Berg strategy for cDNA synthesis and cloning [The unshaded portion of each circle is pBR322 DNA, and the shaded segments represent the restriction endonuclease sites.]

approaches for cDNA synthesis. Some of these allow synthesis of full-length cDNAs, while rest of the methods when used for cDNA library construction show a bias of either 5′ or 3′-end, i.e., there is preferential accumulation of clones representing either 5′ or 3′-end of cDNA sequences in the resulting library. With certain approaches this problem is eliminated, but the resulting clones are much smaller, such that full-length cDNAs must be assembled from several

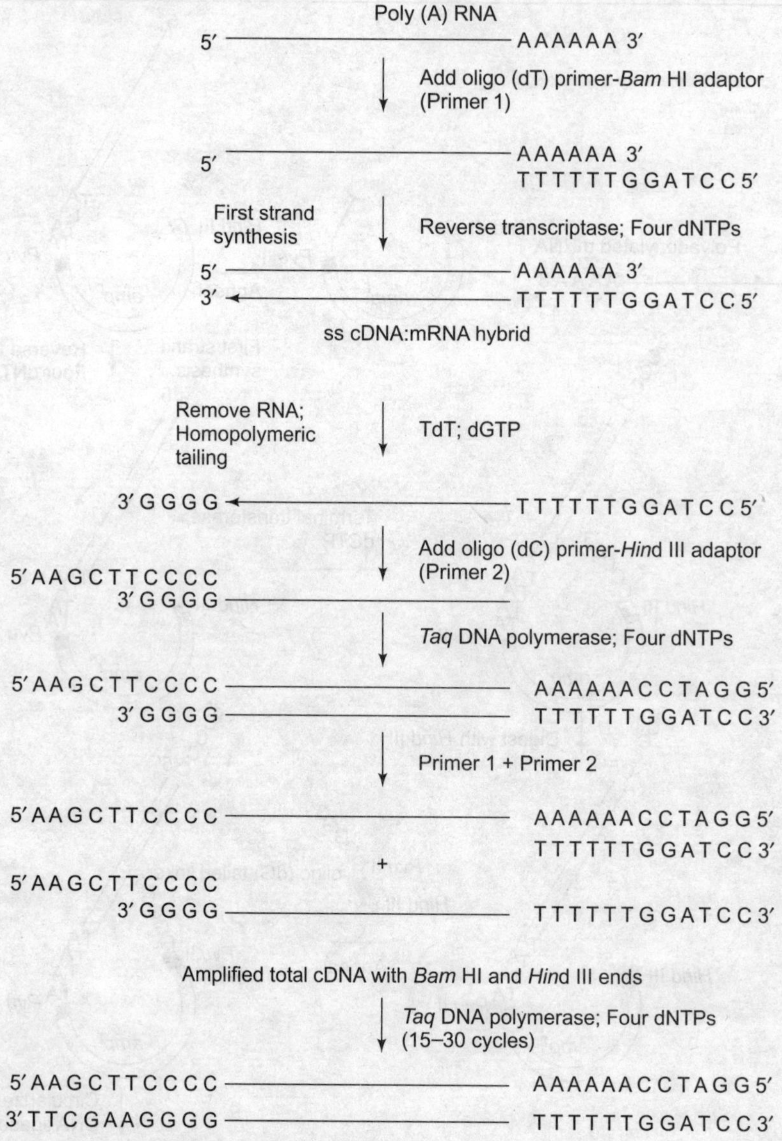

Figure 5.12 Application of RT-PCR in cDNA synthesis and cloning

shorter fragments. Moreover, as the size of a cDNA increases, it becomes progressively more difficult to isolate full-length clones with these methods. This is partly due to deficiency in the reverse transcriptase used for first cDNA strand synthesis. Natural reverse transcriptases are disadvantageous, and have poor processivity, or possess intrinsic RNase H activity that leads to degradation of the RNA template in the resulting hybrid, or these have tendency to stall at secondary sequences often found in 5′ and 3′ untranslated regions. Though improvements in reverse transcriptases have been made (e.g., SuperScript II, StrataScript, Transcript, Mo-MLV RT RNase H Minus, etc.), but the generation of full-length clones corresponding to large mRNAs has remained a problem for long. This problem has been addressed by the development of cDNA cloning strategies involving the selection of mRNAs with intact 5′-ends. Such processes exploit the presence of a 5′ cap [a specialized, methylated guanidine residue (m^7G) that is inverted with respect to the rest of the strand] in most of the eukaryotic mRNAs, and which is recognized by the ribosome during initiation of protein synthesis. Using a combination of cap selection and nuclease treatment, it is now possible to select for full-length first cDNA strands, and thus generate libraries highly enriched in full-length clones. Three such procedures are 'CAPture method using eIF-4E', 'Biotinylated CAP trapper', and 'Oligo-capping'.

CAPture Method Using eIF-4E (Affinity Capture Using eIF-4E)

In this strategy, described by Edery *et al.* (1995), oligo (dT) primer is used for first cDNA strand synthesis catalyzed by reverse transcriptase. Following the synthesis of first cDNA strand, ss cDNA:mRNA hybrids are formed. These hybrid molecules are treated with RNase A, a nuclease that digests only ss RNA, but not ss cDNA:mRNA hybrid. If the first cDNA

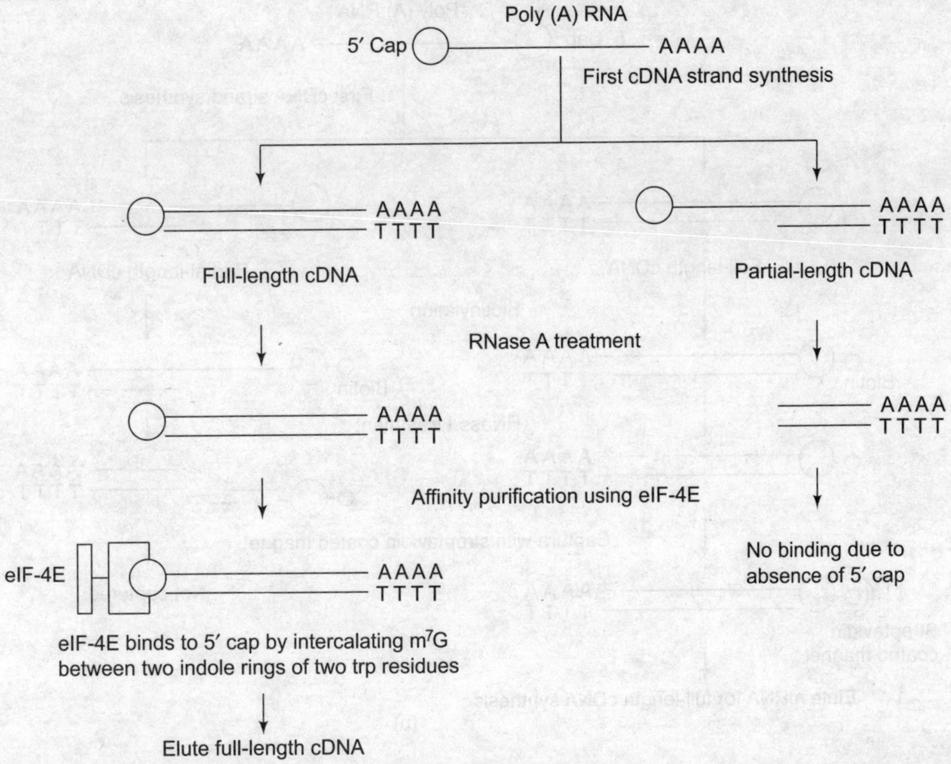

Figure 5.13a Techniques for obtaining full-length cDNAs; (a) Affinity capture of 5′ cap of eukaryotic mRNA using eIF-4E

strand is full-length, i.e., synthesis has occurred up to 5′ cap of the mRNA, the ss cDNA:mRNA hybrids will be protected from cleavage by RNase A. On the other hand, in case of partial-length first cDNA strand, a stretch of unprotected ss RNA is left between the end of the double stranded region (i.e., ss cDNA:mRNA hybrid) and the cap. This ss RNA region is digested by RNase A resulting in small sized ss cDNA:mRNA hybrids. The next step involves application of eukaryotic translational initiation factor eIF-4E for isolation of full-length molecules by affinity capture. Note that the m^7G of 5′ cap is recognized by eIF-4E because of its intercalation between the indole rings of two conserved Trp residues of eIF-4E. As incomplete cDNAs and cDNAs synthesized on partial-length template lack 5′ cap these do not bind eIF-4E and hence are not retained. On the other hand, the full-length cDNAs possess 5′ cap and hence are retained (affinity capture). The bound full-length cDNAs are then eluted from the column and used for cloning. The process is represented in Figure 5.13a. This method, however, co-purifies cDNAs resulting from the mispriming of first cDNA strand synthesis, which can account for up to 10% of the clones in a library.

Biotinylated CAP Trapper (CAPture Method Using Biotin)

Carninci *et al.* (1999) reported a method similar to CAPture method using eIF-4E for the isolation of full-length cDNAs (Figure 5.13b). This method is based on the biotinylation of mRNA. Similar to CAPture method, this method also co-purifies cDNAs resulting from the mispriming of first cDNA strand synthesis, which can account for up to 10% of the clones in a library. In this method, first cDNA strand synthesis is initiated in a reverse transcriptase-catalyzed reaction using oligo (dT) primers. After the first cDNA strand is synthesized, biotin is chemically attached to the ribose sugars of the cap nucleotide and the 3′-ends of the mRNA molecules. Note that deoxyribose is not biotinylated under the applied conditions. Next, the ss cDNA:mRNA hybrid is treated with RNase I, an enzyme that cleaves ss RNA, but not RNA base paired with DNA or ds DNA strands. In this reaction, mRNAs remain unaffected in full-length hybrids, but both the 5′ single stranded regions and the unpaired poly (A) tails of the mRNA molecules in partial-length hybrids are degraded. The sample is then mixed with streptavidin-coated magnetic beads. After RNase I treatment, the only biotinylated hybrids that remain are those with biotinylated cap. In other words, only full-length cDNAs are captured. A magnet is used to separate out the beads from solution. In the next step, the mRNAs in the streptavidin-bound ss cDNA:mRNA hybrids are hydrolyzed with RNase H, and full-length cDNA strands are released into the solution.

Oligo-capping Method

In contrast to the above method, where selection of full-length cDNA is done, oligo-capping allows selection of full-length mRNA (Figure 5.13c). This method, devised by Suzuki *et al.*

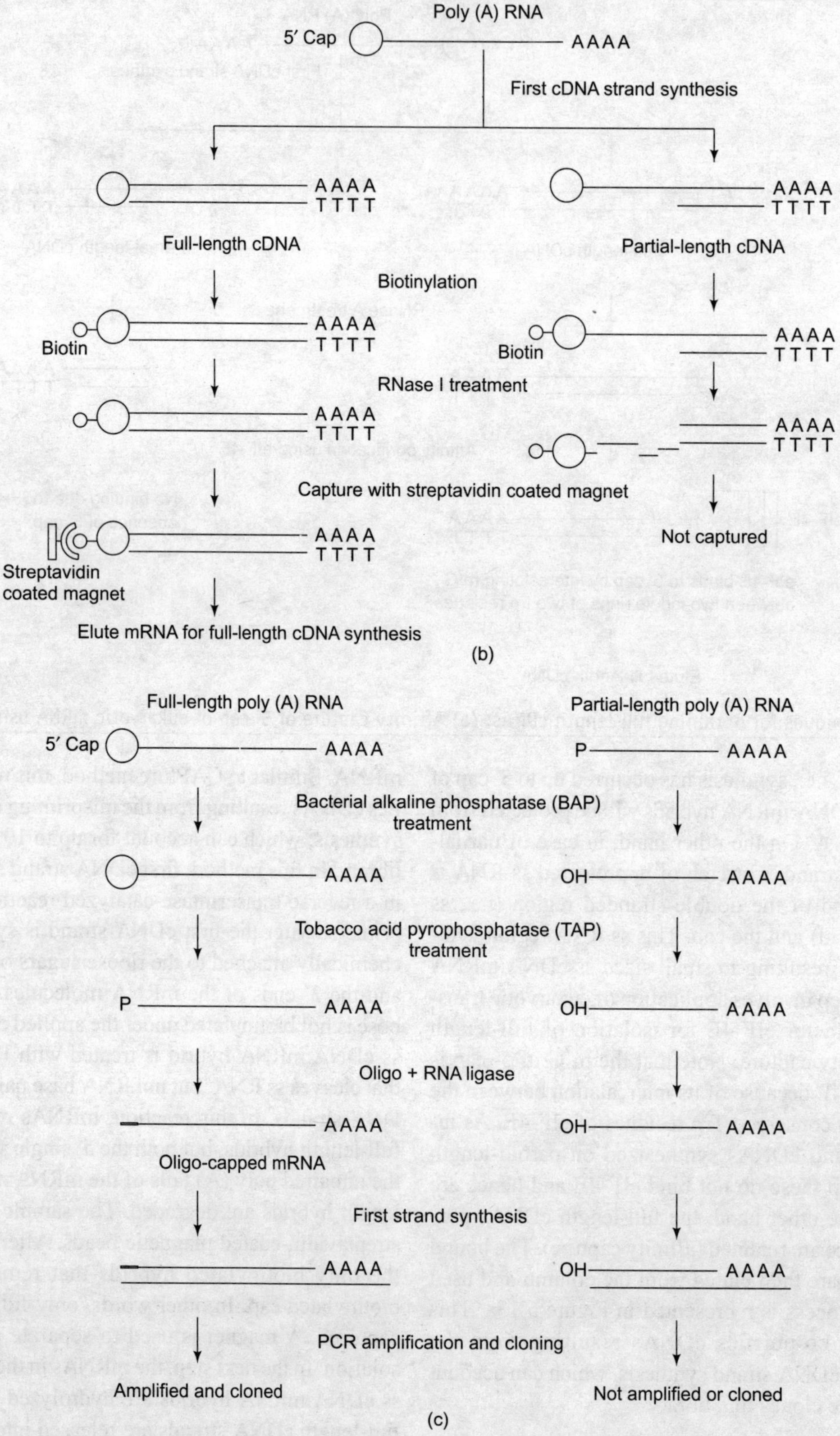

Figure 5.13b and c Techniques for obtaining full-length cDNAs; (b) Biotinylated CAP trapper, (c) Oligo-capping method

(1997), eliminates the problem of co-purification of cDNAs resulting from the mispriming of first cDNA strand synthesis, as observed in the above methods. In this procedure, mRNAs are sequentially treated with the enzymes bacterial alkaline phosphatase (BAP), and tobacco acid pyrophosphatase (TAP). BAP dephosphorylates 5′-ends of uncapped mRNA molecules (i.e., partial-length mRNA), while full-length mRNA molecules having 5′ cap remain unaffected. TAP treatment removes the 5′ cap from full-length mRNAs leaving a 5′-terminal residue with a phosphate group, while dephosphorylated partial-length mRNAs remain unaffected. In the next step, specific oligonucleotides are added in a reaction catalyzed by RNA ligase (oligo + RNA ligase). The specific oligonucleotides ligate to full-length mRNAs, but not to partial-length mRNAs (i.e., cap is replaced by oligonucleotides in case of full-length mRNAs only). This is because the full-length mRNAs contain 5′-phosphate group, while partial-length mRNAs are dephosphorylated. The result is an oligo-capped population of full-length mRNAs. This selected population is then subjected to first cDNA strand synthesis in a reaction catalyzed by reverse transcriptase using oligo (dT) primer. Second cDNA strand synthesis is then carried out by PCR using the oligo (dT) primer and a primer annealing to the oligonucleotide cap. This allows amplification of only full-length cDNAs that anneal to both primers. Thus, in this process, partial-length mRNAs are eliminated, as incomplete first cDNA strands lack a 5′ primer annealing site, and misprimed cDNAs lack a 3′ primer annealing site.

5.3.8 Polyadenylation of Non-polyadenylated RNA for Ease in cDNA Synthesis

To conveniently synthesize cDNA from non-polyadenylated RNA molecules, such as rRNA or the fragmented RNA genomes of certain RNA viruses, homopolymeric tails can be added to their 3′-ends. This can be accomplished with *E. coli* poly (A) polymerase, which like TdT acts in a template independent manner for the addition of homopolymeric deoxyribonucleotides. It is also possible to exploit the specific property of T4 RNA ligase, which efficiently joins oligoribonucleotides to the 3′-ends of ds RNA molecules. Both methods have been used for cloning the fragmented RNA genomes of reoviruses and rotaviruses.

5.3.9 Blunt Ending the cDNA

The 'ragged ends' on the cDNA obtained at the end of second cDNA strand synthesis may be subjected to T4 DNA polymerase-catalyzed filling-in and trimming-back reactions to get a final blunt ended ds cDNA. This blunt ending of ds cDNA is essential for ligation of linkers/adaptors during cDNA cloning (for details see Chapter 15), but this treatment may at times lead to loss of certain sequences.

5.3.10 Assessing cDNA Reactions

A typical cDNA synthesis reaction yields 40–60 ng of cDNA from 2 μg of mRNA. In order to determine the yield and integrity of synthesized cDNA, both first and second cDNA strand reactions should be assayed: (i) To determine the size of the cDNA by alkaline agarose gel electrophoresis and (ii) To calculate the yield of cDNA by measuring the incorporation of radiolabel. To follow the course of the first and second cDNA strand syntheses, radiolabeled dNTPs (α-^{32}P dCTPs) may be incorporated in the process, followed by precipitation of isotope-incorporated DNA with trichloroacetic acid (TCA). The radiolabel is then detected by agarose gel electrophoresis followed by autoradiography or by scintillation counting.

The quality of cDNA, which depends on the length of the cDNA, can be analyzed by alkaline agarose gel electrophoresis. The process involves radiolabeling [α^{32}P] of cDNA synthesis products (sample), as well as the DNA marker. An aliquot (1×10^5 dpm) of the sample is transferred into a separate tube, and diluted with water to 5 μl. Then 5 μl of 2 X loading buffer (60 mM NaOH, 2 mM EDTA, 6% Ficoll® 400, and 0.05% bromophenol blue) is added to the sample. The labeled DNA marker is also diluted with the same 2 X loading buffer. Note that 2 X loading buffer can be stored at –20°C, and the dye is added just before use. The prepared samples are loaded on 1.4% agarose gel [1.4% agarose gel is prepared in 30 mM NaCl, 2 mM EDTA, and equilibrated for at least 1 hour in alkaline electrophoresis buffer comprising of 30 mM NaOH and 2 mM EDTA]. Autoradiographic analysis of separated cDNA allows easy identification of heterogeneous populations of cDNA, if formed. Ideally, the exposure should show a lot of smearing toward the top of the gel. The closer the smearing is to the wells, the better. The result for the majority of cells and tissues is the appearance of few bands, and a smear from ~500 bp to ~8 kbp. The same size of the major bands for both first and second cDNA strands demonstrates that complete copying of the first cDNA strand has taken place. The band size also determines whether a hairpin is formed or not, as the hairpin formation leads to 'double sized' bands. Alternatively, it is also possible, albeit far more cumbersome, to electrophorese an aliquot of cDNA on a gel, stain it with SYBR Green, use image analysis software, quantify, and compare the resultant electrophoretic smears of cDNA observed among the samples on the gel (for details see Chapter 6). Whether to run an alkaline denaturing gel is really a matter of personal preference. One may also select to examine the products of cDNA synthesis by briefly denaturing a small aliquot with 0.1 volumes of 0.1 N NaOH followed by electrophoresis through a neutral 1.2% agarose gel. The alkaline agarose gel electrophoresis can also be employed for determination of mass of cDNA, as the measured amount of label incorporation is directly proportional to the mass of the newly synthesized cDNA. If the entire cDNA is not cloned directly, but digested with restriction endonucleases before cloning, very broad smears rather than discrete fragments are observed upon autoradiography. If, however, one or more mRNA species in the initial population have been particularly abundant, a pattern of specific bands will be observed, from which DNA can be isolated, and used for cloning. cDNA can be amplified and purified and particular fragment represents only a particular part of the desired mRNA. Clones obtained in this way can subsequently be used to isolate and identify intact mRNA molecules, which in turn, can be employed for cloning full-length cDNA.

The amount of synthesized cDNA can be quantified by adding radioactively labeled deoxyribonucleotide (5–10 μCi) to the reaction, and removing two 1 μl aliquots (aliquot 1 and 2) before adding reverse transcriptase. Then cDNA synthesis is carried out, and from the completed product another aliquot (aliquot 3) is removed. Aliquot 1 is diluted 1 to 100, and the total activity in the preparation is determined from 1 μl of this preparation. Aliquots 2 and 3 are precipitated with TCA, and the activity of the pellets is measured. The process involved is as follows: (i) Spot 1 μl aliquots of the sample on two Whatman® DE-81 filters (1.5 cm × 1.5 cm); (ii) Dry the filters under a heat lamp; (iii) Keep one filter aside, and use it directly for the determination of total radioactivity in the sample; (iv) Wash the other filter three times for 5 min in 10 ml of 7.5% (w/v) $Na_2HPO_4 \cdot 12\,H_2O$ to remove of unincorporated dNTPs; (v) Rinse with water, wash with acetone or 96% EtOH; (vi) Dry the washed filter under a heat lamp; and (vii) Transfer and count both filters (unwashed and washed) in a radioactivity counter. The difference between the values gives the quantity of additional activity. The amount of cDNA is then calculated by the following equation:

cDNA amount (ng) = incorporated activity (washed filter)/total activity (unwashed filter) × volume of preparation (μl) × 4 × dNTP concentration (nmol/μl) × 330 (ng/nmol)

where 330 represents the average molecular weight of nucleotide in ng/nmol (or 330 × 10^6 μg/mol); molecular weight is multiplied by a factor of 4, as all four dNTPs are involved in the reaction.

The proportion of ^{32}P incorporated in each synthesis reaction, multiplied by the mass of dNTPs present (66 μg) also gives a value for the mass of cDNA produced. Thus, mass of cDNA strand can be calculated according to the following formula:

Mass of cDNA strand (μg) = activity of TCA precipitation/ total activity (cpm) × 66 (μg)

cDNA yield (%) is then calculated by the following formula:

cDNA yield (%) = cDNA yield (μg)/RNA template (μg) × 100%

Review Questions

1. Enumerate the differences between genomic DNA and complementary DNA.
2. What is the significance of poly (A) tail in eukaryotic mRNAs?
3. Discuss the strategies used for isolation and purification of poly (A) RNA as well as non-polyadenylated mRNAs.
4. Discuss the advantages of biomagnetic separation of poly (A) RNA over conventional affinity purification followed by elution.
5. The presence of poly (A) tail in eukaryotic mRNAs facilitates its isolation as well as cDNA synthesis. Elaborate on the statement.
6. What are the precautions that should be taken during mRNA isolation? Discuss the importance of selection of cells and tissues in the process.
7. Poly (A) RNA bound to oligo (dT)-cellulose is eluted by low salt buffer. Comment.
8. Discuss the mRNA purification strategies based on the elimination of rRNAs.
9. Give a detailed account of procedures used for determination of integrity and purity of mRNA and cDNA synthesized from it. If A_{260}/A_{280} of an mRNA preparation is 2.0, comment upon the purity of preparation.
10. Why is it necessary to convert mRNA to cDNA for cloning into a vector?
11. Give a schematic representation of the general procedure of ds cDNA synthesis.
12. Enumerate various primers used for first and second cDNA strand syntheses. Discuss the advantages and disadvantages of each.
13. Discuss the reaction conditions for first and second cDNA strand synthesis. Also discuss the importance of T4 DNA polymerase in second cDNA strand synthesis.
14. Polyadenylation of non-polyadenylated mRNA can ease the process of cDNA synthesis. Discuss.
15. Elaborate upon the procedures of cDNA synthesis that allow directional cloning of cDNA into the vector.
16. Discuss the full-length cDNA synthesis procedures based on 5′ cap in eukaryotic mRNAs.
17. In the process of cDNA synthesis, homopolymeric tailing of the first cDNA strand may increase the chance of obtaining full-length cDNA. Comment.
18. If the A_{260} value of an mRNA preparation diluted 10 times is 1.8, calculate the concentration of mRNA in μg/ml.
19. 1 μg of mRNA is subjected to first cDNA strand synthesis reaction in the presence of 2×10^{-8} mol radiolabeled dNTPs, and the volume of preparation is 20 μl. If upon analysis, washed filter gives 3.2×10^4 dpm and the unwashed filter gives 2.2×10^6 dpm, calculate the amount of cDNA in μg and the yield (%) of cDNA.
20. The product of first cDNA strand synthesis reaction formed in the presence of ^{32}P labeled dNTPs was subjected to TCA precipitation. If the radioactivity in the pellet was found to be 20 cpm and the total activity was 60 cpm, calculate the proportion of ^{32}P incorporated in the synthesis reaction and the mass (in μg) of cDNA strand.

References

Aviv, H. and P. Leder (1972). Purification of biological active globin mRNA by chromatography on oligothymidylic acid cellulose. *Proc. Natl. Acad. Sci.* USA. **69**:1409–1412.

Carninci, P. and Y. Hayashizaki (1999). High-efficiency full-length cDNA cloning. *Methods Enzymol.* **303**:19–44.

Edery, I., L.L. Chu, N. Soneberg, and J. Pelletier (1995). An efficient strategy to isolate full-length cDNAs based on an mRNA cap retention procedure (CAPture). *Mol. Cell Biol.* **15**:3363–3371.

Goelet, P. *et al.* (1982). Nucleotide sequence of tobacco mosaic virus RNA. *Proc. Natl. Acad. Sci.* **79**:5818–5822.

Gubler, U. and B.J. Hoffman (1983). A simple and very efficient method for generating cDNA libraries. *Gene* **26**:263–269.

Kurtz, D.T. and C.F. Nicodemus (1981). Cloning of a globin cDNA using a high efficiency technique for the cloning of trace mRNAs. *Gene* **13**:145–152.

Maniatis, T. *et al.* (1976). Amplification and characterization of a β-globin gene synthesized *in vitro*. *Cell* **8**:163–182.

Okayama, H. and P. Berg (1982). High-efficiency cloning of full-length cDNA. *Mol. Cell Biol.* **2**:161–170.

Suzuki, Y. *et al.* (1997). Construction and characterization of full-length enriched and a 5'-end enriched cDNA library. *Gene* **200**:149–156.

6 Techniques for Nucleic Acid Analysis and Size Fractionation

Key Concepts

In this chapter we will learn the following:

- Gel electrophoresis
- Elution of DNA fragments from gel
- Gel filtration chromatography
- Density gradient centrifugation

6.1 INTRODUCTION

The nucleic acid isolated by methods described in previous chapters is visualized and analyzed for its purity and integrity. Moreover, differently sized fragments obtained by methods such as restriction digestion, mechanical shearing, sonication, etc., as discussed in Chapters 1 and 3, have to be size fractionated before ligating them into vector according to their insert capacity. In such a case, the techniques used for size fractionation should be such that the DNA fragments remain fully susceptible to further enzymatic restriction and modification reactions. In this chapter, the details of the techniques used for any or all of the above-mentioned purposes are presented; besides other applications of these techniques are also highlighted.

6.2 GEL ELECTROPHORESIS

Gel electrophoresis is a simple technique for rapid resolution of a mixture of charged macromolecules, including nucleic acids (DNA and RNA), proteins, and carbohydrates on a porous matrix under the influence of electric field. When a charged molecule is placed in an electric field, it enters the gel, and migrates towards the electrode with the opposite charge. In contrast to proteins, which can have either a net positive or net negative charge, nucleic acids have a consistent negative charge imparted by their phosphate backbone, and hence migrate towards the positive electrode (anode). The reactions that permit the passage of current from the cathode to the anode are:

Cathode reactions: $2e^- + 2H_2O \rightarrow 2OH^- + H_2$;
$HA + OH^- \rightleftharpoons A^- + H_2O$

Anode reactions: $H_2O \rightarrow 2\,H^+ + \tfrac{1}{2}\,O_2 + 2e^-$;
$H^+ + A^- \rightleftharpoons HA$

These reactions describe the electrolysis of water in the electrophoresis buffer, resulting in the production of hydrogen at the cathode and oxygen at the anode. For each mole of hydrogen produced, one-half mole of oxygen is produced.

The voltage applied between the two electrodes is the driving force, while the resistance offered by the solid medium is the opposing force. The resistance depends on the pore size of the solid medium, which can often be manipulated to some extent by changing the concentration of the matrix. The distance (d) traveled by the DNA molecules can be given by the expression:

$$d = a - b \log M$$

where a and b are the empirical constants that depend on the condition of electrophoresis, and M is the molecular weight of DNA.

Hence, larger the molecule, greater is the resistance, and slower is the mobility. Within a range, the higher the applied voltage, the faster the samples migrate.

In electrophoresis, one electrical parameter, voltage, current, or power, is always held constant. The consequences of an increase in resistance during the run because of electrolyte depletion, temperature fluctuation, etc., differ in the following ways:

(i) In constant current mode, velocity is directly proportional to the current, heat is generated, and the velocity of the molecule is maintained.
(ii) In constant voltage mode, there is a reduction in the velocity of the charged molecules, although no additional heat is generated during the course of the run.
(iii) In constant power mode, there is a reduction in the velocity of the molecules, but there is no associated change in heating.

As constant temperature throughout the electrophoretic process, particularly when running an agarose gel is very important, hence gels used for the separation of nucleic acids are run at constant voltage. This is necessary to prevent overheating, and consequently melting of the gel matrix.

In a gel, which consists of a complex meshwork of pores, the rate at which a nucleic acid molecule moves under the influence of electric field is determined by their ability to penetrate through this meshwork. A larger DNA molecule requires more force to move it; however, a longer DNA molecule also has more negative charge. In practice, these two factors cancel out because all fragments of DNA have the same number of charges per unit length. Consequently, all DNA molecules in free solution move towards the anode at the same speed, irrespective of their molecular weights. Hence, the penetration ability of these nucleic acid molecules, definitely, does not depend on the amount of charge that the molecule carries, rather, unlike some other applications of electrophoresis, it is either dependent on the size (or molecular weight or length) of the molecule, or its shape (or conformation). As nucleic acid molecules migrate towards the anode, these get separated according to their sizes as bands or zones, which are easily inspected visually by *in gel* staining. If a more precise confirmation of the nature of the sample is required, the gel is blotted onto a membrane support, and hybridized with a nucleic acid probe.

6.2.1 Choice of Matrix: Agarose and Polyacrylamide

Electrophoresis separates nucleic acids on the basis of their size and shape as does gel filtration (for details see Section 6.4). However, it possesses greater separation power, and is well-suited for the analysis of very small quantities of nucleic acids. Unlike the column matrix, the gel does not consist of separate beads; rather, it consists of submicroscopic pores that allow migration of charged particles under an electric field. Historically, the earliest applications of electrophoresis involved macromolecular chromatography in sucrose. Refinements in the methodology extended electrophoresis into starch gels, and finally led to the implementation of polyacrylamide, in which the structure of the gel influences the migration characteristics of the molecules under investigation. It was quickly realized that gels with large pores were required to efficiently electrophorese high molecular weight nucleic acids. At first, very low percentage polyacrylamide gels (2.5%) were used, although the instability and unreliability of such low percentage polyacrylamide gels precluded their widespread use. Agarose was later added to acrylamide gels to give them enhanced physical strength. Currently, the two standard matrices for the electrophoretic separation of nucleic acids are agarose and polyacrylamide. Both agarose and polyacrylamide gels can be poured in a variety of shapes, sizes, and porosities, and can be run in a number of different configurations. The choices within these parameters depend primarily on the sizes of the fragments being separated. During electrophoresis, as the samples of nucleic acids pass through the gel, the gel matrix acts as an anticonvective medium. This reduces convective transport as well as diffusion, and hence the separated sample components remain positioned in sharp zones or bands during the run. In addition, the gel consists of microscopic pores that act as a molecular sieve, allowing separation of nucleic acids according to molecular size. The chemical composition and properties of the two commonly used matrices in gel electrophoresis are described below.

Agarose Gels

Agarose is a highly purified linear polysaccharide isolated from seaweeds (red algae) such as *Gelidium* and *Gracilaria*. It is ~600–800 residues long polymer comprising of repeating units of *D*-galactose and 3,6-anhydro-*L*-galactose joined by β (1 → 4) linkage, and each repeating unit further linked by α (1 → 3) glycosidic linkage (Figure 6.1a). Some of the *D*- and *L*-galactose residues are esterified with sulfuric acid. Agarose forms a double helical structure when a suspension of agarose in water in heated and cooled. Two molecules in parallel orientation twist together with a helix repeat of three residues, and water molecules are trapped in the central cavity. These structures in turn associate with each other to form a gel, a 3-D matrix that traps large amount of water. Thus, chains of agarose form helical fibers that aggregate into supercoiled structures with a radius of 20–30 nm. Commercially prepared agarose polymers are believed to contain ~800

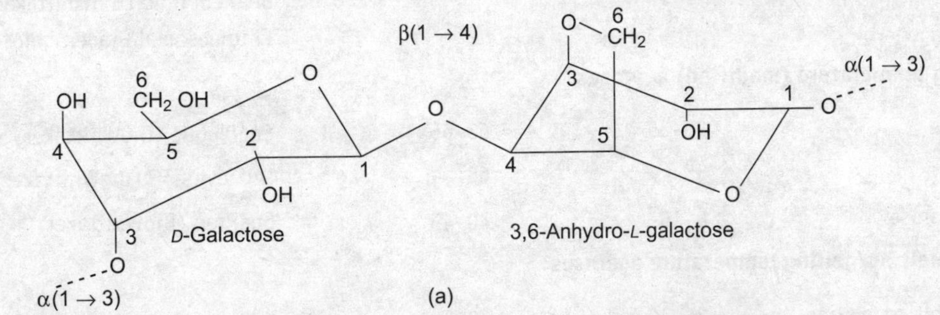

Figure 6.1a Gel matrices used in gel electrophoresis; Chemical structure of agarose [*D*-Gal β (1 → 4) 3,6-anhydro-*L*-gal]. [Disaccharide repeating unit comprising of *D*-galactose and 3,6-anhydro-*L*-galactose joined by β (1 → 4) linkage are linked to each other by α (1 → 3) glycosidic linkage to form a agarose polymer. Some of the *D*- and *L*-galactose residues are esterified with sulfuric acid.]

galactose residues per chain. However, agarose is not homogeneous, and the average length of the agarose chains varies from batch to batch, and from manufacturer to manufacturer. Hence, it is advisable to read the supplier's catalogue to obtain more precise information about specific brands of agarose.

Classes of Agarose The agarose used for gel electrophoresis is a more purified form of the agar used to make bacterial culture plates, and is available in various classes commercially. These include standard (high melting temperature) agarose, low melting/gelling temperature agarose, and low electroendosmosis (EEO) agarose. The properties of different types of agaroses are presented in Table 6.1, and their range of separation of DNA fragments is presented in Table 6.2.

Standard agarose Standard agaroses are manufactured from two species of red algae (seaweeds), namely *Gelidium* and *Gracilaria*. These agaroses differ in their gelling and melting temperatures, but agaroses from either source can be used to analyze, and isolate fragments of DNA ranging in size from 1–25 kbp. The lower grades of agarose may be contaminated with other polysaccharides, salts, and proteins, which affect the gelling/melting temperature of agarose solutions, the sieving of DNA, and the ability to recover DNA from the gel. These potential problems can be minimized by using special commercial grades of agarose, which exhibit features such as: (i) Minimal background fluorescence after staining with ethidium bromide (EtBr); (ii) No contamination of DNase and RNase; (iii) Minimal inhibition of restriction endonucleases and ligase; and (iv) Modest amounts of EEO. Newer types of standard agarose combine high gel strength with low EEO, allowing gels to be cast with agarose concentrations as low as 0.3%. These gels can be used in conventional electrophoresis to separate high molecular weight DNA (up to 60 kbp). At any concentration of these new agaroses, the speed of migration of the DNA is increased by 10–20% over that achieved using the former standard agaroses, depending on buffer type and concentration. This increase can lead to significant savings of time in pulsed field gel electrophoresis (PFGE) of megabase pair sized DNA.

Low melting/gelling point agarose (LMP agarose) LMP agaroses melt as well as gel at temperatures lower than that of standard agaroses. This property is attributable to hydroxyethylation, and the degree of solubilization determines the exact melting and gelling temperature. DNA of a given size runs faster through gels cast with LMP agarose than through conventional agarose gels. For this reason, the voltage applied to LMP agarose gels should be lower than

Table 6.1 Properties of Different Types of Agaroses

Type of agarose	Gelling temperature (°C)	Melting temperature (°C)	Commercial names
Standard agaroses			
Low EEO agarose isolated from *Gelidium* spp.	35–38	90–95	SeaKem LE (BioWhittaker); Agarose-LE (USB); Low EEO Agarose (Stratagene); Molecular Biology Certified Grade (Bio-Rad)
Low EEO agarose isolated from *Gracilaria* spp.	40–42	85–90	SeaKem HGT (BioWhittaker); Agarose-HGT (USB)
High-gel strength agaroses			
High-gel strength agarose	34–43	85–95	FastLane (BioWhittaker); SeaKem Gold (BioWhittaker); Chromosomal Grade Agarose (Bio-Rad)
Low melting/gelling temperature (modified) agaroses			
Low melting agaroses	25–35	63–65	SeaPlaque (BioWhittaker)
	35	65	NuSieve GTG (BioWhittaker)
Ultra low melting agarose	8–15	40–45	SeaPrep (BioWhittaker)
Low viscosity; low melting/gelling temperature agaroses			
Low viscosity; low melting/gelling temperature agaroses	25–30	70	InCert (BioWhittaker)
	38	85	NuSieve 3:1 (BioWhittaker)
	30	75	Agarose HS (BioWhittaker)

Table 6.2 Range of Separation of DNA Fragments Through Different Types of Agaroses

Agarose (%)	Size range of DNA fragments resolved by various types of agaroses			
	Standard	High gel strength	Low gelling/melting temperature	Low gelling/melting temperature, low viscosity
0.3	–	–	–	–
0.5	700 bp–25 kbp	–	–	–
0.8	500 bp–15 kbp	800 bp–10 kbp	800 bp–10 kbp	–
1.0	250 bp–12 kbp	400 bp–8 kbp	400 bp–8 kbp	–
1.2	150 bp–6 kbp	300 bp–7 kbp	300 bp–7 kbp	–
1.5	80 bp–4 kbp	200 bp–4 kbp	200 bp–4 kbp	–
2.0	–	100 bp–3 kbp	100 bp–3 kbp	–
3.0	–	–	500 bp–1 kbp	500 bp–1 kbp
4.0	–	–	–	100–500 bp
6.0	–	–	–	10–100 bp

that applied to standard agarose gels. LMP agaroses are used chiefly for rapid recovery of DNA, as most agaroses of tnis type melt at temperatures (65°C) that are significantly lower than the melting temperature of ds DNA. This feature allows for simple purification, enzymatic processing (restriction endonuclease digestion/ligation) of DNA, and bacterial transformation with nucleic acids directly in the remelted gel. LMP agarose can be held as liquids in the 30–35°C range and cells can be embedded without damage because of the lower gelling temperatures. This treatment is useful in preparing and embedding chromosomal DNA in agarose blocks before analysis by PFGE. Like standard agaroses, LMP agaroses are also available that display features such as minimal background fluorescence after staining with EtBr, no DNase and RNase contamination, minimal inhibition of restriction endonuclease and ligase, and modest amounts of EEO. Chemically modified agarose has significantly more sieving capacity than an equivalent concentration of standard agarose. This finding has been exploited to make agaroses that approach polyacrylamide in their resolving power, and are therefore useful for separation of PCR products, small DNA fragments, and small RNAs <1 kb in size. It is now possible to resolve DNA down to 4 bp, and to separate DNA molecules in the 200–800 bp range that differ in size by 2%.

Low electroendosmosis agarose (Low EEO agarose) EEO affects the rate of migration of nucleic acids toward anode. This process is due to ionized acidic groups (usually sulfates) attached to agarose. These acidic groups induce positively charged counter ions in the buffer that migrate through the gel toward the cathode, causing a bulk flow of liquid that migrates in a direction opposite to that of the DNA. The higher the density of negative charges on the agarose, the greater the EEO flow, and the poorer the separation of nucleic acid fragments. In such case, the retardation of small DNA fragments (<10 kbp) is minor, but larger DNA molecules can be significantly retarded, especially in PFGE. To avoid problems related to EEO, it is best to use low EEO agarose prepared by addition of few positively charged groups to neutralize a few sulfates on agarose. Note that commercially available 'zero' EEO agarose is undesirable because of two disadvantages, *viz.*, the chemical modification of agarose by adding positively charged groups to neutralize the sulfate groups on agarose may inhibit subsequent enzyme reactions, and the adulteration of this 'zero' EEO agarose by locust bean gum may retard the expulsion of water from the gel.

Properties of Agarose Gels Gelation of agarose results in a 3-D mesh of pores and channels whose diameters range from 50 nm to >200 nm. Agarose thus makes a large pore gel. A solid gel is formed upon melting of agarose in aqueous solution at concentrations ranging between 0.5–2% (w/v). Agarose gels are poured and electrophoresed as horizontal slabs. The horizontal configuration results in the full support of the gel by the casting tray beneath it while facilitating efficient separation of a wide range of nucleic acid samples, through relatively low percentage gels. This is desirable principally because of the low tensile strength that these gels exhibit. Some vertical apparatuses accommodate the weaker nature of agarose gels by the inclusion of a frosted glass plate to which the gels may cling.

Agarose gels have lower resolving power but greater range of separation as compared with polyacrylamide gels. DNAs from 50 bp to several Mbp in length can be separated on agarose gels of various concentrations, configurations, and the precise nature of the applied electric field (constant or pulse field) (for details see Sections 6.2.2 and 6.2.3). Thus, small DNA fragments (50–20,000 bp) are best resolved in conventional agarose gel electrophoresis, however, migration of DNA fragments >20 kbp in size through the conventional agarose gel becomes independent of the fragment length, and hence PFGE or its variants are used.

Agarose is nontoxic and biologically inert, and hence easy in handling, gel preparation and run on the bench top (or in the fume hood if toxic denaturants are added), and disposal. Moreover, the classical problems associated with prepara-

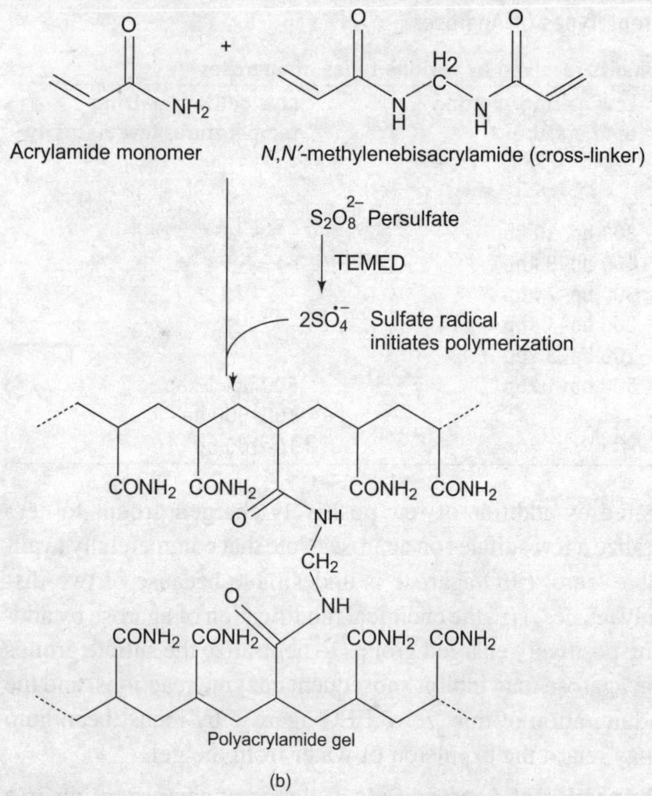

Figure 6.1b Gel matrices used in gel electrophoresis; Formation of polyacrylamide gel [A three-dimensional mesh is formed by co-polymerizing activated monomer and cross-linker.]

tive agarose gel electrophoresis, for example, co-elution of contaminants from the agarose, and difficulty in elution, have been overcome with the widespread use of extremely high-purity and low-melting temperature agarose formulations.

Polyacrylamide Gels

Polyacrylamide becomes cross-linked during the gelation process (Figure 6.1b). The process of gelation is initiated by generation of free radicals due to reduction of ammonium persulfate (APS) by TEMED (*N,N,N′,N′*-tetramethylene diamine). The generated free radicals then mediate vinyl polymerization of acrylamide monomers resulting in the formation of linear chains of polyacrylamide. When bifunctional cross-linking agent, *N,N′*-methylenebisacrylamide, or *bis* for short, is included, copolymerization reaction takes place, which generates 3-D ribbon-like networks of cross-linked polyacrylamide chains with a statistical distribution of pore sizes. Because the mean diameter of the pores formed in these networks is determined by the concentrations of the acrylamide monomer, and the bifunctional cross-linker, the pore size can be adjusted, and hence the separation range of the gel can be expanded (Table 6.3). A number of other factors that affect the efficiency of separation include gel thickness, Joulic heating, and electric field strength.

Properties of Polyacrylamide Gels In contrast to agarose gels, polyacrylamide gels owing to greater tensile strength are run in vertical configuration in a constant electric field. Polyacrylamide makes small pore gels, which are effective for separating small fragments of DNA (5–500 bp). The pore size can, however, be manipulated by adjusting the total percentage of acrylamide or by varying the amount of cross-linker added to the acrylamide to induce polymerization. When there is a wide range in the molecular weights of the material under study, a pore gradient gel in which the pore size is larger at the top of the gel than at the bottom may also be prepared. Thus, the gel becomes more restrictive as the electrophoretic run progresses. Conveniently, premade gels may now be purchased from a number of suppliers, which are available for most standard gel-box designs and formats.

Polyacrylamide gels offer several advantages over agarose gels: (i) The resolving power of polyacrylamide gels is extremely high, and fragments of DNA that differ in size by as little as 1 bp in length or by as little as 0.1% of their masses can be separated from one another, and hence these gels are used as sequencing gels; (ii) These gels can accommodate much larger quantities of DNA. Up to 10 μg of DNA can be applied to a single slot (1 cm × 1 mm) of a typical polyacrylamide gel without significant loss of resolution; and (iii) DNA recovered from polyacrylamide gels is extremely pure, which can be used for several applications including transgenesis. Despite these advantages, polyacrylamide gels have the disadvantage of being more difficult to prepare, and handle than agarose gels. Note that monomeric (unpolymerized form)

Table 6.3 Effective Range of Separation of DNA Molecules in Polyacrylamide Gels

S. No.	Concentration of acrylamide monomer (%)[a]	Effective range of separation (bp)	Size of ds DNA fragments that co-migrate with xylene cyanol FF (bp)	Size of ds DNA fragments that co-migrate with bromophenol blue (bp)
1.	3.5	1,000–2,000	460	100
2.	5.0	80–500	260	65
3.	8.0	60–400	160	45
4.	12.0	40–200	70	20
5.	15.0	25–150	60	15
6.	20.0	6–100	45	12

[a] *N,N′*-methylenebisacrylamide is included at 1/30th the concentration of acrylamide.

acrylamide is a very potent neurotoxin, and should be handled with caution. Furthermore, for blotting purpose the matrix of choice is agarose. This is because of highly efficient and complete transfer of nucleic acids from agarose gels.

6.2.2 Conventional Agarose Gel Electrophoresis (Constant Field Agarose Gel Electrophoresis)

Conventional or continuous agarose gel electrophoresis refers to an electrophoretic run in agarose gel in a horizontal configuration in an electric field of constant strength and direction. Under such conditions, the velocity of the DNA fragments decreases as their length increases, and is proportional to electric field strength. This electrophoresis is convenient for separating DNA fragments ranging in size from 50 bp to ~20 kbp. In this section, the principle and procedure of conventional agarose gel electrophoresis are described.

Mechanism of Separation

The agarose gel provides an anticonvective medium for the electrophoresis and resolves DNA fragments by acting as molecular sieve. In polynucleotides, the phosphate group of each nucleotide unit carries a strong negative charge that is much stronger than any of the charges on the bases above pH 7.0. The mass-to-charge ratio of the polynucleotides is independent of the base composition and, consequently, nearly the same for closely related species. For that reason, free-solution techniques (i.e., the electrophoretic medium containing no gel or polymer network solution) have not proven successful in the electrophoresis of oligonucleotides or ds DNA. Thus, DNA owing to a strong negative charge at neutral pH migrates through the gel towards the positive electrode in the presence of an electric field, and the pores in the gel affect its migration. Smaller molecules pass through the pores more easily, and migrate ahead of the larger molecules.

The actual mechanism involved in DNA separation in electrophoresis has been the topic of much discussion. The migration of the nucleic acids through the pores of agarose probably plays an important role in separation on the basis of molecular weights. Real time fluorescence video microscopy of stained molecules undergoing electrophoresis has revealed more subtle dynamics. However, no single model can fully account for the dependence of DNA mobility on its molecular size, and a number of experimental parameters such as field strength, gel concentration, etc. Various models have been proposed for unidirectional electrophoresis. The 'Ogston mechanism' treats DNA percolating through a network of polymer fibers as a random coil, which behaves as a globular (i.e., rigid and spherical) molecule interacting with the gel. Its electrophoretic mobility is proportional to the volume fraction of the pores of a gel that the DNA can enter. Since the average pore size decreases with increasing gel concentration, mobility decreases with increasing gel concentration and increasing molecular weight. The Ogston theory does not account for the fact that a relatively large DNA molecule may stretch or deform, and can squeeze its way through the pores. This model is unsatisfactory even for small DNA molecules as it produces a semi-logarithmic relationship between size and mobility rather than the reciprocal relationship that is observed. The 'reptation model' indicates that DNA is forced to squeeze through the tubes formed by the agarose network. The 'biased reptation model' represents the system as a 'snake moving through grass'. In this theory, the DNA molecule is thought to occupy a notational convoluted tube in the gel, through which it progresses led by one end. The forces acting to retard the progress of the molecule are the frictional interactions with the sides of the tube. Although this model successfully explains the reciprocal relationship between mobility and size, it does not explain some features of the system such as the fact that molecules above a certain size (>20 kbp) travel at the same rate, or may even travel more rapidly than smaller molecules under some conditions. Another model suggests that the DNA molecule is a long, highly flexible chain, which travels through the gel in fits and starts, becoming entangled, trapped, and hooked on the fibers, piling into tangled bundles in the vacuoles. This is the behavior observed by video microscopy. One of the more persuasive models suggests that DNA molecules display elastic behavior by stretching in the direction of the applied field, and then contracting into dense balls. The molecule moves in a stretched conformation, but occasionally, the front end becomes compressed and then trapped in a pore. When this happens, the rest of the molecule piles up behind it, and the molecule gets stuck in the pore, and initiates a new movement in the stretched state. And if the pore size is large, the ball of large size can pass through, thereby allowing separation of molecules. This model fits well with the main features of the conventional gel electrophoresis. Further, if the globular volume of the DNA molecules exceeds the pore size, for example DNA molecules >20 kbp in size, these molecules cannot get separated through conventional gel electrophoresis. This is because, once the size of DNA fragments exceeds a maximum value, the simple relationship between velocity of DNA fragments, their length, and electric field strength, as mentioned above, breaks down. This limit of resolution is reached when the radius of gyration of the linear DNA duplex exceeds the pore size of the gel. The DNA molecule can then no longer be sieved by the gel according to its size but instead migrates 'end-on' through the matrix as if through a sinuous tube, i.e., reptation. The separation of such large DNA fragments (>20 kbp) requires the recourse to pulsed electric fields (for details see Section 6.2.3).

Requirements for Conventional Gel Electrophoresis

In this section, a brief description of various equipments, solutions, and buffers required in conventional gel electrophoresis are discussed.

Electrophoresis Apparatus Horizontal slab gel electrophoresis assembly comprises of electrophoresis chamber (or gel-box or electrophoresis tank) available in diverse design and quality to carry electrophoresis or running buffer, a gel-casting tray (equipped with a holding chamber having rubber gaskets in some designs), comb(s) with teeth, a comb-holding stand, gel-sealing tape (e.g., Time tape or VWR lab tape), two electrodes (positive electrode or anode is red and negative electrode or cathode is black), electrical leads for connecting to power supply (red lead to connect to anode, and black lead to connect to cathode), and finally a power supply device capable of generating up to 500 V and 200 mA. The entire assembly and its components are depicted in Figures 6.2a and b, respectively. Note that better value electrophoresis tanks come with a variety of combs that can be used to generate wells of different sizes in different numbers. The teeth of typical combs generally are 1, 1.5, 2, or 3 mm thick, 0.5 cm long, and are capable of generating 7–20 wells per gel depending on the size of the electrophoresis apparatus.

Agarose For different purposes, different classes of agaroses are employed. These have already been discussed in significant details in Section 6.2.1.

Gel-loading Buffer (Tracking Dye) Samples are mixed with gel-loading buffer before loading them into the wells in the gel. There are different types of gel-loading buffers, and which type to use is a matter of personal preference. Different types of 6 X gel-loading dyes are: Type I [0.25% bromophenol blue (BPB), 0.25% xylene cyanol FF (XC FF), 40% (w/v) sucrose in water; storage at 4°C], Type II [0.25% BPB, 0.25% XC FF, 15% Ficoll (Type 400, Pharmacia) in water; storage at room temperature], Type III [0.25% BPB, 0.25% XC FF, 30% glycerol in water; storage at 4°C], and Type IV [0.25% BPB, 40% (w/v) sucrose in water; storage at 4°C].

These buffers serve three purposes: (i) Increase the density of the sample, ensuring that the DNA sinks evenly into the well, and to prevent diffusion or floating away of samples into the electrophoresis buffer; (ii) Add color to the sample, thereby simplifying the loading process, and also to keep a track of the extent of electrophoresis; and (iii) Contain dyes that move towards the anode at predictable rates under the influence of electric field. The increase in density is conferred to the sample by glycerol, sucrose, or Ficoll in the gel-loading buffer. BPB and/or XC FF in the gel-loading buffer impart the color to the sample. Both BPB and XC FF migrate towards the anode at

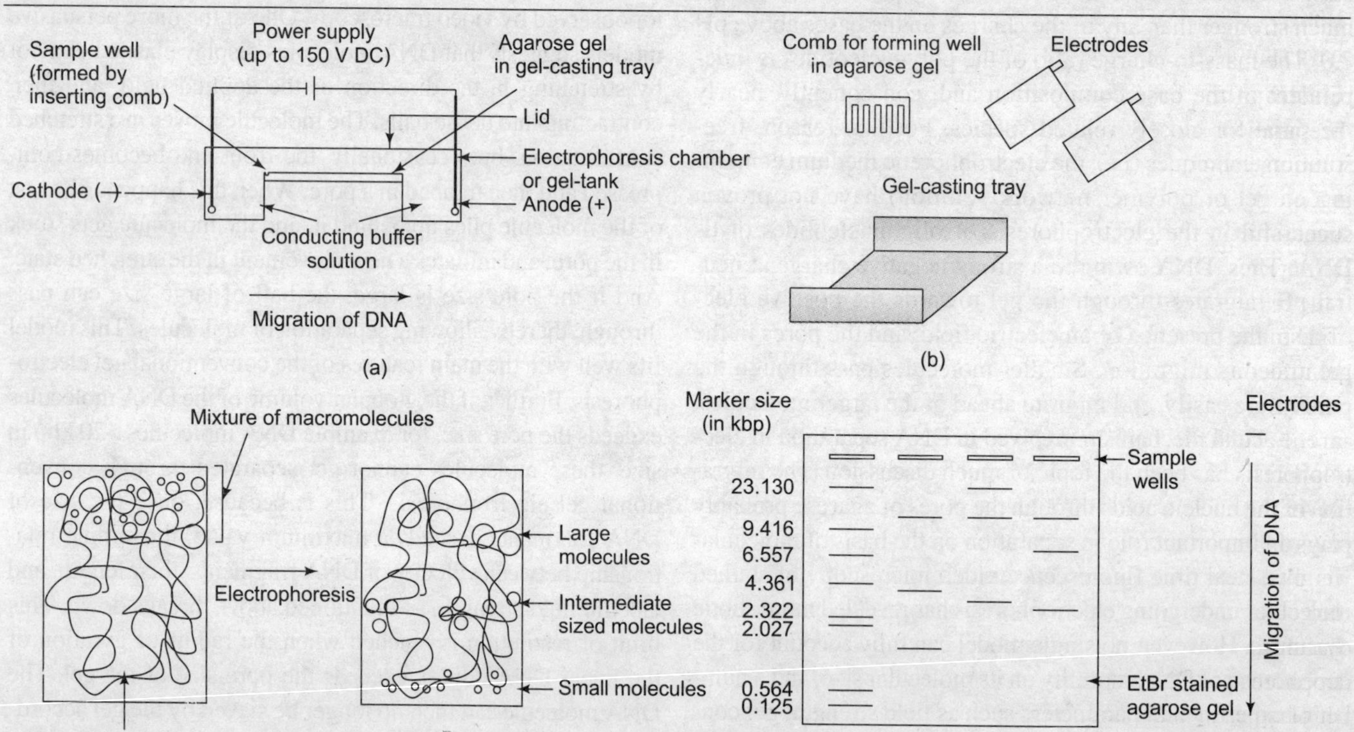

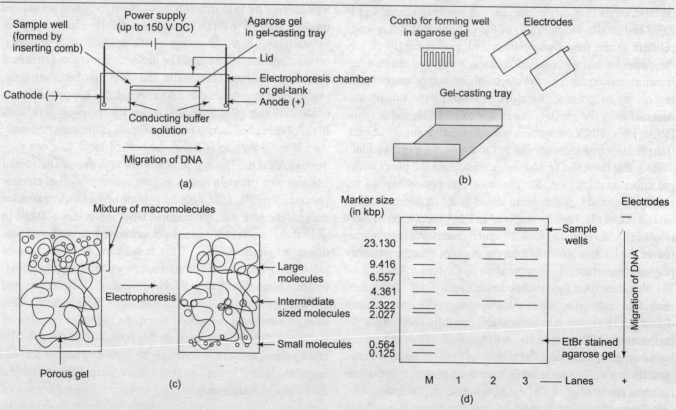

Figure 6.2 Agarose gel electrophoresis; (a) Schematic representation of agarose gel electrophoresis assembly, (b) Parts of agarose gel electrophoresis assembly: Comb, electrodes, and gel-casting tray. (c) The sieving action of a porous agarose gel separates nucleic acid molecules according to size, the large molecules move slowly, small molecules migrate rapidly, while intermediate sized molecules migrate a distance intermittent between the large and small molecules. (d) Gel showing DNA bands after agarose gel electrophoresis. [Lane M represents electrophoresis of marker DNA (λ DNA double digested with *Eco* RI and *Hind* III). Lanes 1, 2, and 3 indicate digested DNA samples. By comparing the size of DNA fragments in the sample DNA with marker DNA, the relative sizes of DNA fragments are estimated.]

predictable rates under the influence of electric field, independent of the agarose concentration. These tracking dyes generally run with the leading edge of the migrating sample, and are thus used to keep track of sample migration (or progress of electrophoresis). The choice of dye depends on its electrophoretic mobility relative to the DNA fragments. BPB migrates through agarose gels 2.2-fold faster than XC FF. The rate of migration of BPB through agarose gels run in 0.5 X TBE is approximately the same as linear ds DNA 300 bp in length, whereas XC FF migrates at approximately the same rate as linear ds DNA 4 kbp in length. These relationships are not significantly affected by the concentration of agarose in the gel over the range of 0.5–1.5%. An important point to emphasize here is that the inclusion of XC FF in the gel-loading buffer is usually not recommended when the interest is only in the DNA. This is because XC FF often interferes with the visualization of DNA after electrophoresis, particularly when a band of interest lies directly under the XC FF. Orange G is another example of a tracking dye that has a faster migration rate as compared to BPB. Thus, when comparing gels in which each dye is allowed to run to the end, the DNAs in the BPB gel will have run farther, and separated better than in the orange G gel. But if very small fragments are being investigated, these may have run off the gel with BPB, but not with orange G.

Electrophoresis Buffer Electrophoresis buffer is used in gel preparation as well as in running the gel (hence also called running buffer). Different electrophoresis buffers available for electrophoresis of native ds DNA include: (i) TAE buffer containing Tris-acetate and EDTA (pH 8.0) (also called E buffer); (ii) TBE buffer containing Tris-borate (pH ~8.3); and (iii) TPE buffer containing Tris-phosphate (pH 7.5–7.8). These electrophoresis buffers are usually made up as concentrated solutions (e.g., 10 or 50 X TAE, 5 X TBE, and 10 X TPE), and stored at room temperature. Working solutions are 1 X TAE comprising of 40 mM Tris-HCl, 1 mM Na_2EDTA, 5 mM glacial acetic acid; 0.5 X TBE comprising of 45 mM Tris-borate, 1 mM EDTA; and 1 X TPE comprising of 90 mM Tris-phosphate, 2 mM EDTA. Each buffer has its own merits or demerits, and the choice among these is largely a matter of personal preference. TAE has the lowest buffering capacity of the three, and becomes exhausted if electrophoresis is carried out for prolonged periods of time. When this happens, the anodic portion of the gel becomes acidic, and BPB migrating through the gel towards the anode changes in color from bluish-purple to yellow. This change begins at pH 4.6, and is complete at pH 3.0. The exhaustion of TAE can however be avoided by periodic replacement of the buffer during electrophoresis or by recirculation of the buffer between the two reservoirs. Both TBE and TPE have significantly higher buffering capacity, but both are slightly more expensive than TAE. Moreover, while performing cloning experiment, the presence of borate ions inhibit ligation reactions and can interfere with subsequent purification of the eluted DNA fragments on glass beads. Double stranded linear DNA fragments migrate ~10% faster through TAE than through TBE and TPE; the resolving power of TAE is slightly better than TBE or TPE for high molecular weight DNAs, but worse for low molecular weight DNAs. Thus, TAE yields better resolution of DNA fragments in highly complex mixtures such as mammalian DNA. For this reason, Southern blots used to analyze complex genomes are generally derived from gels prepared and electrophoresed in TAE. The resolution of supercoiled DNA is also better in TAE than in TBE.

Gel Staining Dye Gel staining dye is used for visualization of DNA in agarose gel by *in gel* staining. It is necessary that the gel staining dyes should be compatible with a variety of *in gel* enzymatic manipulations, and can be removed easily, if required. Some gel staining dyes may be added to the gel before casting as well as to the running buffer, or simply to the sample, while some dyes are used to stain the gel after the conclusion of electrophoresis. The most convenient and commonly used method to visualize DNA in agarose gels is staining with the fluorescent dye EtBr. Alternatives to EtBr are nucleic acid binding dyes such as SYBR dyes, GelStar®, acridine orange, and sometimes methylene blue. The properties, merits, and demerits of various DNA binding dyes are discussed here.

Ethidium bromide (EtBr) EtBr (3,8-diamino-6-ethyl-5-phenylphenathridium bromide) is a positively charged phenanthridinium intercalator that is structurally similar to propidium iodide, and contains a tricyclic planar group that intercalates between the two successive base pairs of DNA (Figures 6.3a and b). The dye is routinely used to detect ds DNA, however, the dye can also bind with highly variable stoichiometry to helical regions formed by intrastrand base pairing in RNA, or heat-denatured ds DNA, or ss DNA. However, the affinity of the dye for ss nucleic acid is relatively low, and the fluorescent yield is comparatively poor. The binding of dye to nucleic acids requires little or no sequence preference. At saturation in solutions of high ionic strength, approximately one EtBr molecule is intercalated per 2.5 bp in a ds DNA, independent of the base composition of the DNA. After insertion into the helix, the dye lies perpendicular to the helical axis, and makes van der Waals contacts with the base pairs above and below. The planar ring system of the dye is buried, its peripheral phenyl and ethyl groups project into the major groove of the DNA helix. The geometry of the base pairs and their positioning with respect to the helix are unchanged except for their displacement by 3.4 Å along the helix axis. This causes a 27% increase in the length of ds DNA saturated with EtBr. The intercalation of EtBr in closed circular DNA molecule causes local unwinding of the helix by ~26°, which leads to a reduction in twist and a corresponding increase in writhe. This property of intercalation of EtBr into the nucleic acids is used for their visualization in agarose (or polyacrylamide) gels under UV illumination. This is

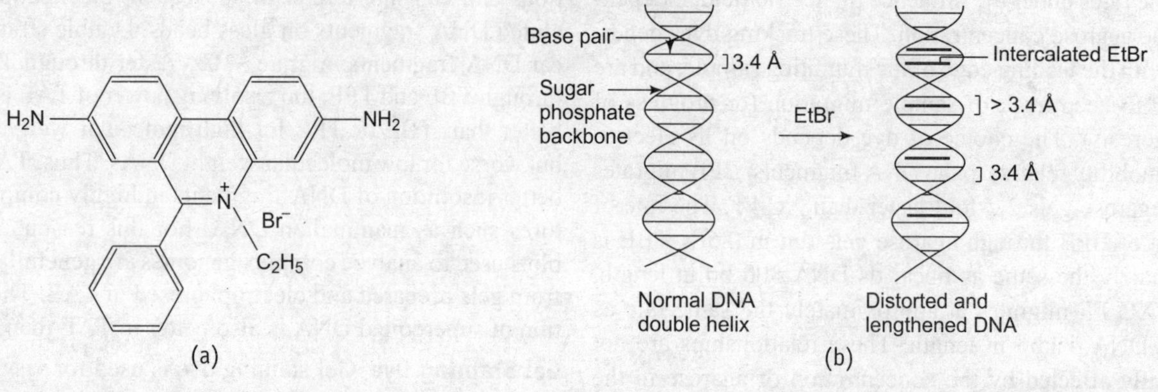

Figure 6.3 Role of EtBr in visualization of DNA; (a) Chemical structure of EtBr, (b) The intercalation of EtBr into a DNA molecule [EtBr increases the spacing of two successive base pairs, partially unwinds the double helical structure of DNA, distorts the regular sugar-phosphate backbone and decreases the pitch of the helix resulting in the 'stretched' structure.]

under UV illumination. This is because the fixed position of the planar group, and its close proximity to the bases cause dye bound to DNA to display a 20–25 fold increase in fluorescent yield compared to that of dye in free solution. As UV irradiation at 254 nm is absorbed by the DNA, and transmitted to the dye, and irradiations at 302 nm and 366 nm are absorbed by the bound dye itself, in both cases, the energy is re-emitted with a quantum yield of 0.3 at 590 nm in the red-orange region of the visible spectrum. Most of the commercially available UV light sources emit UV light at 302 nm. The fluorescent yield of EtBr–DNA complexes excited by irradiation is considerably greater at 302 nm than at 366 nm, but is slightly less than at 254 nm. However, the amount of photobleaching of the dye, and nicking of the DNA is much less at 302 nm than at 254 nm.

For visualizing nucleic acids, the dye is usually incorporated into the gel, and the electrophoresis buffer at a concentration of 0.5 µg/ml (EtBr is prepared as a stock solution of 10 mg/ml in water, which is stored at room temperature in dark bottles or bottles wrapped in aluminium foil). Note that the addition of dye into the buffer is essential along with the gel because the DNA runs from cathode to anode, and EtBr runs from anode to cathode, and gets depleted from the gel, however, the addition of EtBr to the electrophoresis buffer is not necessary, if the electrophoresis is short, for example, in checking for size, mass, or integrity. Rather than addition of the dye to the gel and the running buffer, EtBr may also be added to the sample. This reduces the total mass of the dye required, limits the fluorescence to the lanes in which a sample is loaded, and cuts back on the generation of EtBr waste. The addition of EtBr to the gel and running buffer, or to the sample helps in monitoring the progress of electrophoresis through UV irradiation simply after disconnecting the power. If greater separation is required, the gel can be placed back in the gel-box, and the power supply is reconnected for a suitable interval. Another alternative to visualize nucleic acids in the gel is to stain the gel with EtBr after electrophoresis. In this case, staining is accomplished by immersing the gel in electrophoretic buffer or water containing EtBr (0.5 µg/ml) for 30–45 minutes at room temperature. Because the fluorescent yield of EtBr–DNA complex is greater than that of unbound dye, small amounts of DNA (~10 ng/band) (0.5 cm wide band) can be detected in the presence of EtBr (0.5 µg/ml) in gels. Even smaller quantities of DNA can be detected if the DNA has previously been treated with chloroacetaldehyde, a chemical mutagen that reacts with adenine, cytosine, and guanine. Fluorescence can be enhanced by destaining the gel in a solution containing 10 mM Mg^{2+} before examining it under UV illumination. Unbound EtBr can be removed by soaking the stained gel in water or 1 mM $MgSO_4$ for 20 minutes at room temperature in a destaining step. Normally this destaining step is not essential, however, the detection of very small amounts (<10 ng) of DNA is made easier by reducing the background fluorescence caused by unbound EtBr. In case the DNA is required for further manipulations, EtBr can be easily removed from DNA by simple *n*-butanol extraction(s).

The use of EtBr is associated with certain disadvantages, which have led to the application of other gel staining dyes. These disadvantages include: (i) EtBr is a powerful mutagen. Hence, extreme care must be exercised when preparing and manipulating EtBr stock solutions and dilutions, gloves should be worn while handling EtBr containing solutions or equipment, contaminated buffers and equipment must be treated as toxic waste, and EtBr must be inactivated and disposed off through an activated charcoal filter; (ii) Though EtBr can be conveniently added into the agarose gel as well as buffer, or in the sample, the electrophoretic mobility of linear ds DNA is retarded by ~15% in its presence; sharper DNA bands are obtained when electrophoresis is carried out in the absence of EtBr. Thus, when an accurate size of a particular fragment of DNA is to be established (e.g. when a restriction endonuclease map is being constructed), the agarose gel should be run in the absence of EtBr, and stained after the conclusion of electrophoresis; (iii) Continuous exposure to ambient light or especially UV light for visualization nicks the RNA, making it difficult to work with; (iv) Upon exposure to ambient light or UV irradiation, EtBr exhibits photobleaching, making visualization and quantitative judgements difficult; and (v) The presence of EtBr in a gel reduces its transfer efficiency. If a traditional method of transfer is used, such as passive capillary diffusion, then the transfer period may need to be extended for unacceptably long periods. Destaining the gel may, however, facilitate transfer, although even minute amounts of EtBr remaining in the gel can result in poor transfer.

SYBR green I SYBR Green I is the trade name of an asymmetric cyanine dye marketed by Molecular Probes (http://www.probes.com) as a 10,000 X stock solution in anhydrous dimethyl sulphoxide (DMSO). The dye is essentially nonfluorescent in free solution but upon binding to ds DNA, displays greatly enhanced fluorescence, and a high quantum yield (0.8 upon binding to ds DNA), as a consequence of which the dye is used for visualization of ds DNA under UV illumination. The dye binds only to ds DNA by intercalation, and emits fluorescence at 520 nm only when bound.

For visualizing ds DNA in the gel, first the dye is diluted to 1 X in a Tris-containing buffer (pH 8.0), usually 1 X TBE, 1 X TAE, or TE buffer (10 mM Tris-HCl, 1 mM EDTA, pH 8.0) just before use. As optimal fluorescence occurs in Tris-based buffer around pH 8.0, the dilution of SYBR Green I in water is not recommended. SYBR Green I is used to stain the gel after electrophoresis, rather than adding the dye in the samples or gel or electrophoresis buffer. This is because its presence exerts a negative effect on the migration pattern of DNA, and leads to formation of bands that are fuzzy, wave-like, highly irregular shaped, and often indistinguishable from other bands in the vicinity. This phenomenon seems to occur at all concentrations of dye, and at all voltages, and makes mass and molecular weight determinations extremely difficult.

The dye offers significant advantages over EtBr, such as greater sensitivity, low background fluorescence, and reduced mutagenecity. Though not believed to be as mutagenic as EtBr, SYBR Green I should be handled with the same caution as EtBr. The level of background fluorescence is so low that no destaining is required; however, the dye can be removed from the nucleic acids simply by ethanol (EtOH) precipitation. Moreover, as the SYBR dyes generate strong signals with very little background, and have a high affinity for nucleic acids, these can be used in low concentrations as compared to conventional EtBr staining. Despite of these advantages, SYBR Green I is not commonly used to stain DNA in agarose gels, but is used chiefly to quantify DNA in solution, for example, in real-time PCR. This is because of some less desirable characteristics of SYBR Green I, for example, (i) The dye is not optimally stimulated by standard transilluminators that emit UV radiation at 300 nm, and the signal strength improve when illumination at 254 nm is used, but at this wavelength, damage to DNA is maximal; (ii) The dye penetrates agarose gels slowly, and post-electrophoretic staining can take 2 hours or more when the gels are thick or contain a high concentration of agarose; (iii) The dye is not particularly photostable; and (iv) The dye is only slightly more sensitive than EtBr in detecting ss DNA in agarose gels.

Other SYBR dyes (Molecular Probes) employed for *in gel* staining are SYBR Safe™ for DNA, SYBR Green II for RNA, SYPRO orange and SYPRO red for proteins. Note that SYBR Green II detects RNA in denaturing agarose gels with five-fold greater sensitivity than EtBr, and hence is useful for analyzing quantities of RNA that are too small to be detected by EtBr. Moreover, the dye does not interfere with Northern blotting.

SYBR gold SYBR Gold is a fluorescent asymmetrical cyanine dye marketed by Molecular Probes as a 10,000 X stock solution in anhydrous DMSO. The dye is used to stain single and double stranded nucleic acids (i.e., ss DNA, ds RNA, ss RNA) in gels. This dye has high affinity for DNA, and most probably binds to the backbone of charged phosphate residues, exhibit large fluorescence enhancement upon binding to nucleic acid. Like SYBR Green, SYBR Gold is essentially nonfluorescent in free solution but displays greatly enhanced fluorescence, and a high quantum yield upon binding to nucleic acids. The dye shows maximum excitation at 495 nm, and has a secondary excitation peak at 300 nm. Fluorescence emission occurs at 537 nm. When excited by standard transillumination at 300 nm, the SYBR Gold–DNA complexes generate bright gold fluorescent signals that can be captured on conventional black and white Polaroid film (type 667), or on charged couple device (CCD)-based image detection systems. The stained nucleic acids can be transferred directly to membranes for Northern or Southern hybridization.

DNA is stained by soaking the gel after separation of the DNA fragments (i.e., after electrophoresis) in a 1:10,000 fold dilution of the stock solution (in either 1 X TAE or 1 X TBE) before use. As the dye is sensitive to fluorescent light, working solutions containing SYBR Gold are freshly made daily in electrophoresis buffer. Note that the gels are stained post-electrophoresis, like SYBR Green I, because the binding of dye to the backbone of the charged phosphate residues of DNA markedly retards the electrophoretic mobility of stained DNA, and the bands of DNA are sometimes curved. Unlike SYBR Green I and II, SYBR Gold penetrates gels quickly, and can therefore be used to stain DNA both in conventional neutral polyacrylamide and agarose gels, and in gels containing denaturants such as urea, glyoxal, and formaldehyde. Moreover, unlike SYBR Green I, which has a quantum yield of 0.8 upon binding to ds DNA, SYBR Gold has a quantum yield of 0.7 and 1,000-fold enhancement of fluorescence.

As a result, <20 pg of ds DNA can be detected in an agarose gel. In addition, staining of agarose (or polyacrylamide) gels with this dye can reveal as little as 100 pg of ss DNA in a band or 300 pg of RNA. Similar to SYBR Green I, gel staining with SYBR Gold is associated with virtually no background fluorescence, requires the same type of photodocumentation system (photography is carried out with green or yellow filters), and handled with caution. The level of background fluorescence is so low that no destaining is required, and stained nucleic acid can be transferred directly to membranes for Northern or Southern hybridization. SYBR Gold is more sensitive (comparable in sensitivity to silver staining) than EtBr. It is ten-fold more sensitive than EtBr, and its dynamic range is greater. Although SYBR Gold does not inhibit many enzymatic reactions, PCR are sensitive to high concentrations of the dye. Inhibition can, however, be relieved by adjusting the concentration of Mg^{2+}, or be avoided by removing SYBR Gold from the template DNA by standard EtOH precipitation. The high cost of the dye precludes its use for routine staining of gels. However, the dye may be cost-effective as an alternative to using radiolabeled DNAs in techniques such as single strand conformation polymorphism (SSCP), and denaturing gradient gel electrophoresis (DGGE).

GelStar® Like the SYBR stains described previously, GelStar® stain (Cambrex BioScience) is a fluorescent nucleic acid stain that is suitable for staining ds or ss DNA as well as ds or ss RNA, and offers an increase in sensitivity compared with identical gels stained with EtBr. It can be added directly to the gel before electrophoresis, or the gel can be stained afterward. It is provided as a 10,000 X concentrate in DMSO, and functions best when diluted in a Tris-buffer solution between pH 7.0 and 8.5. GelStar® should be handled with same care as any of the other DNA binding dyes.

Acridine orange Staining with the cationic dye acridine orange is an older technique for visualizing nucleic acid molecules in a gel. Acridine orange either binds electrostatically to the phosphate groups of single stranded molecules and fluoresce red at about 650 nm, or intercalate into double helical molecules and fluoresce green at about 525 nm. Thus, the advantage of using this dye is that the single stranded nucleic acid molecules (e.g., ss DNA or RNA) appear red-orange, whereas double stranded molecules (e.g., DNA and non-denatured RNA) appear green in the same gel. Using this stain, it is easy to detect as little as 0.05 μg of double stranded nucleic acid and 0.1 mg single stranded nucleic acid on UV illumination. Given the versatility of this dye, the investigator may also detect incomplete denaturation or partial renaturation of an RNA sample before moving on to subsequent experiments. For staining, agarose gel is soaked in a solution containing 30 μg/ml acridine orange in 10 mM sodium phosphate buffer (pH 7.0) for 30 minutes. The main reason that acridine orange is not in widespread use is its associated high background staining. Furthermore, destaining the gel is an obligatory step, which is done by soaking the gel in 10 mM sodium phosphate buffer for 1–2 hours.

Methylene blue Methylene blue, also known as Swiss Blue, in recognition of the nationality of Caro who first synthesized the dye in 1876, is also used to stain bands of DNA in agarose gel. It has two absorption maxima (668 and 609 nm) in the visible spectrum, and is soluble in water. The staining is reversible and can be carried out before hybridization. This staining technique is generally used in school labs for demonstration purposes, where it is preferred over toxic and expensive EtBr and SYBR dyes, and to prevent UV exposure. Note that methylene blue does not pose any potential health hazards and disposal difficulties associated with EtBr and SYBR dyes. The application of methylene blue also prevents the risk of generation of UV induced pyrimidine dimers or lowering of the biological activity of DNA. This dye, however, has very poor sensitivity and produces a very high background. Moreover, it is difficult to completely remove all of the methylene blue from the gel after it has been stained. Thus, even after extensive destaining, methylene blue has little use as a DNA or RNA stain beyond being able to show that a band is present in the gel.

Silver staining Silver staining has been refined into a highly sensitive technique for post-electrophoretic detection of DNA bands in polyacrylamide gels. It has the advantages of visualizing the sample without any specialized equipment, and rendering gel backgrounds that are virtually colorless. Silver staining is a very sensitive method for detecting small amounts (<1.0 ng) of low molecular weight DNAs in polyacrylamide gels (for details see Section 6.2.6). In general, the method is several times more sensitive than EtBr for visualization of polynucleotides. The sensitivity is however drastically reduced in agarose gels, and hence usually not recommended.

DNA Size Standards DNA size standards allow accurate determination of the size (molecular weight) and relative concentrations of DNA species subjected to electrophoresis. These are also called molecular weight standards or marker DNA, and represent nucleic acid molecules of known sizes. These markers are typically generated by restriction digestion of a plasmid or bacteriophage DNA of known sequence. The most commonly used marker DNA is λ DNA double digested with *Eco* RI and *Hin*d III (as shown in Figure 6.2d, Lane M). The sizes (in kbp) of fragments generated are 23.130, 9.416, 6.557, 4.361, 2.322, 2.027, 0.564, and 0.125. Alternatively, DNA size standards are produced by ligating a monomer DNA fragment of known size into a ladder of polymeric forms [e.g. 100 bp DNA ladder containing fragments differing by 100 bp].

A proper use of these molecular weight standards requires certain key decisions to be made. These are the amount of the standard to use, the positioning of the standard in the gel, the size range of the standard, and the method of visualizing (staining/end-labeling/detection following hybridization). A stock solution of size standards can be prepared by diluting with gel-loading buffer, which is then used in individual electrophoresis experiments (usually 500 ng–1 μg of DNA size standards are loaded for proper visualization). Just as the mass of marker is an important parameter, so is the placement of the marker. The gel should always be loaded asymmetrically (in an unoccupied lane on the same gel as experimental samples), especially if one is running the gel for the first time. It is very frustrating to obtain a significant banding pattern, and not know which DNA sample is loaded in that lane. The confusion may also arise due to inversion of the gel during the transfer, and a failure in maintaining the orientation of the membrane after it is removed from the surface of the gel following the transfer. In addition, inversion of the exposed X-ray film during the developing step adds to yet another level of confusion. When the markers are loaded in the center of the gel, the distinction between left and right becomes further obscured. Consistently loading the marker DNA in an asymmetric fashion, either right hand corner or left hand corner, eliminates at least one source of ambiguity. Size standards for both agarose and polyacrylamide gel electrophoresis are commercially available. It is a good idea to have two size ranges of standards, including a high molecular weight range from 1 kbp to ~20 kbp, and low molecular weight range from 100 bp to 1,000 bp.

A size calibration curve can be constructed by plotting the $\log_{10}$ of the sizes of the DNA standards against distance migrated from the origin (the well into which the sample is loaded). A curve generated in this fashion can then be used to ascertain the sizes of DNA species observed directly in the gel or that may subsequently be detected by nucleic acid hybridization. Size determinations are now a standard feature of all image analysis software and workstations. Alternatively, a reasonable size calibration curve can be constructed manually with a couple of data points, a ruler, and a sheet of semilog paper.

UV Source A source of ultraviolet (UV) light is required to visualize a stained gel. UV light may be irradiated from below the gel through a transilluminator fitted with UV shield cover, or from above the gel with enhanced sensitivity (i.e., epi-illumination) through hand-held UV torch. Gel documentation system may also be used to visualize, analyze, and keep records in the form of photograph or gel scans. Note that the UV light sources emit intense, dangerous UV irradiation at 312 nm, or 302 nm, or 254 nm. Exposure to UV light from these sources can cause serious damage to the cornea, retina, and other structures of the eye, and to exposed skin. Moreover, as the damage is cumulative, minute amounts of exposure can be catastrophic in both the short- and long-term; hence, care should be taken to wear gloves, protective eye wear, or a face shield at all times, even when using transilluminators with a UV protective cover.

Photodocumentation Electrophoresis gels contain information critical for accurate interpretation of the outcome of an experiment; these data often suggest subsequent experiments that may give further definition to a scientific investigation. However, gels deteriorate in a relatively short time and the dyes used to stain nucleic acids in gels deteriorate even more rapidly. Hence, besides visualization, it is required to keep photographic records of gels (an electrophoretograms) or associated X-ray films generated on hybridization detecting (e.g., chemiluminescence or autoradiography). Photodocumentation is the only realistic option for providing a stable, long-term record of results for future analysis, for in-house presentations, and for publication. A variety of systems have gained widespread acceptance, and are currently in use for the purpose. These systems fall into two broad categories: (i) The traditional photograph generated on thermal paper or Polaroid film; and (ii) Digital imaging, in which the image of the gel or X-ray film is analyzed by, and stored in a computer. Digital image analysis can be done simply by mounting a digital camera above a gel, and by providing the proper filtration for the image and shielding for the user. Nowadays, digital image analysis is done by computer software and hardware, which allows for rapid, reproducible image analysis. It has replaced many of the traditional methods for the interpretation of electrophoretograms including visual inspection and densitometry. Useful measurements that are now automated include determination of molecular weight, concentration, relative abundance, and integrated optical density (IOD). The computer also provides a means of cataloging stored images. The image analysis software can be purchased separately, and used in conjunction with existing instrumentation. Many vendors now offer digital-imaging systems, including the gel documentation systems, Gel-Pro Analyzer (Media Cybernetics) or Archiver Eclipse workstation (FOTODYNE) etc. These systems are used to derive the desired information including identifying lanes, finding bands, determining intensity of bands for quantification, and performing user-defined macros. Although limited input is required for use, the capacity for complete override of all automated functions allows maximum flexibility. Although image analysis software and digital imaging systems often represent a significant investment, the purchase of such devices is a one-time cost; these offer various benefits such as: (i) To perform standard and customized measurements; (ii) To automate and unify image analysis in the laboratory; (iii) To save time and increase productivity; (iv) To provide reproducible data; (v) To perform intensity and background calibrations; (vi) To compensate for non-uniform background; (vii) To calculate integrated optical density of each band, relative abundance, or mass distribution among bands in the same lane, instead of the error-prone measurements associated with old-style densitometers; (viii) To export digital image data to

Excel or other commonly used spreadsheet software; (ix) To distribute digital images easily to e-mail as TIFF files; and (x) To import digital images easily into Word documents and Power Point presentations. Another imaging device is phosphorimager that is used to capture images from gels and blots for completely filmless, electronic analysis, and archiving (for details see Chapter 16).

Other Requirements Other requirements include microwave oven or boiling water bath for melting agarose, micropipettes with drawn-out tips for loading samples into the wells, and gloves to handle solutions, and pipette tips, etc.

Procedure of Conventional Gel Electrophoresis

The steps involved in conventional agarose gel electrophoresis:

Preparation of Gel and Assembly of Electrophoretic Apparatus The preparation of gel requires complete melting of agarose in 1 X electrophoresis buffer by boiling for 3–5 min in boiling water bath or microwave oven. Note that for routine use, usually 0.8% agarose gel is prepared (50 ml preparation is sufficient for 5 × 7 cm gel). Any decrease in the volume of agarose solution due to evaporation during boiling is replenished with water. Molten agarose solution is then cooled to <60°C and EtBr (0.5 μg/ml) is added to it. The solution is then poured into a gel-casting tray (with sealed open ends, and fitted with a comb held at appropriate height and position through comb-holding stand) to a thickness between 3 and 5 mm. Cooling step is essential because hotter solutions warp and craze the plastic trays, and EtBr is heat-labile. The open ends of the gel-casting tray are sealed to retain molten agarose until the matrix has solidified. In older style gel-casting trays, gel-sealing tape is used to seal the open ends. However, some newer designs incorporate built-in rubber gaskets that seal the casting trays when inserted into the holding chamber 90° to the intended direction of electrophoresis; this design precludes the taping requirement, and frequent leaks associated with older style trays. The positioning of comb is also important. It should not be placed too deep into the tray that it tears the gel when withdrawn, and causes leakage of samples between the gel and the tray. This problem is more common when low concentrations of agarose (~0.6%) or LMP agarose are used. It should not be too shallow that the wells are too small to accommodate sufficient sample. It should neither be too close to the edges that handling of gel leads to its breakage nor too distant from the edges that the samples do not get enough space for proper resolution. Air bubbles present in the molten agarose when poured in the gel-casting tray, if any, should be removed by poking them with the corner of a tissue paper. It should also be ensured that no air bubbles are entrapped under or between the teeth of the comb. The gel is then left undisturbed for 30–45 minutes at room temperature for complete solidification. Upon solidification, submicroscopic pores are formed in the gel, which allow the passage of DNA molecules under the influence of electric field. After solidification, small amount of 1 X electrophoresis buffer is added on the top of the gel just enough to cover the gel to a depth of ~1 mm, and the comb is carefully removed. The electrophoresis buffer is poured off and the gel-sealing tape is carefully removed. The insertion of comb in the gel at the time of setting of gel and its removal afterwards leads to the formation of 'wells' or 'slots' to be used for loading of multiple samples. The resulting horizontal slab gel is then placed carefully in an electrophoresis tank containing running buffer (with EtBr) in such a way that the entrapment of air bubbles beneath the gel-casting tray is prevented. Note that same running buffer should be added to both the chambers of the tank. Also note that the gel should be completely submerged in running buffer, however, a minimum volume necessary to completely cover the gel should be used so as to facilitate reasonable dissipation of heat when the gel is run at room temperature, and at sufficiently low and constant voltages.

Preparation of Samples and Loading into Wells The samples are prepared by addition of 0.2 volumes of the desired 6 X gel-loading buffer. Some investigators prefer to add 0.5 μg/ml EtBr to the sample instead of adding it to gel and running buffer. The samples are then loaded slowly into the wells of the submerged gel using a disposable micropipette. DNA size standard is also mixed with gel-loading dye and loaded into a slot on either right or left side of the gel. Different samples are loaded in different wells using different micropipette tips to prevent mixing of samples. Care should be taken not to insert the micropipette deep into the well or in a slanting position as this may puncture the well. For loading the sample, micropipette is kept straight just at the surface of the well, and upon pressing the pipette, the sample settles down the well due to high density conferred by glycerol in the sample. Moreover, not much time should be taken for loading, or else the samples will diffuse from the wells.

The maximum amount of DNA that can be applied to a slot depends on the dimensions of the slot, number of fragments in the sample, and their sizes. Overfilling of slots should be avoided because it results in trailing, smiling, and smearing problems that become more severe as the size of the DNA increases. Moreover, overfilling may lead to contamination of neighboring samples, hence it is best to make the gel a little thicker or to concentrate the DNA by EtOH precipitation rather than loading large amount of samples in the wells. When simple populations of DNA molecules (e.g., bacteriophage λ or plasmid DNAs) are to be analyzed, 100–500 ng of DNA should be loaded per 0.5 cm slot. When the sample consists of a very large number of DNA fragments of different sizes (e.g., restriction digests of mammalian DNA), 20–30 μg of DNA can be loaded per slot without any significant loss of resolution.

Electrophoresis, Sample Visualization, Gel Analysis, and Photodocumentation Once the samples are loaded, the gel-tank is fitted with electrodes (anode away from the wells), lid of the gel-tank is closed, and electrical leads from the electrodes are connected to power supply (Figure 6.2a). The power supply is switched 'on', and set at a constant voltage of 1–5 V/cm (measured as the distance between the positive and negative electrodes). If the leads have been connected correctly, bubbles are generated at the anode and cathode, and within a few minutes, BPB migrates from the wells into the body of the gel. Note that the amount of bubbles produced at anode is about half the amount produced at cathode. Though not the best way, but the observation of bubble formation due to electrolysis of water is a simple way to indicate that the electrodes are connected, and the current is flowing. The standard high concentration agarose gels are run at room temperature, while the gels cast with LMP agarose, and gels that contain less than 0.5% agarose should be poured and run at 4°C in a cold lab to reduce the chance of fracture. During electrophoresis, DNA being negatively charged migrates towards the anode. Electrophoresis is allowed to continue till the tracking dye front reaches $3/4^{th}$ of the way along the length of the gel. This ensures that small DNAs of possible interest are not lost over the leading edge (or distal edge) of the gel into the running buffer, produces very reasonable separation of DNA molecules, and simplifies accurate size determination. Thereafter, electric current is turned 'off', the leads are disconnected, and the lid of the gel-tank is removed. Then the gel is taken out carefully, visualized by UV irradiation, under which the DNA shows up as red-orange colored sharp bands (due to intercalation of EtBr), and photographed/analyzed. Note that instead of adding EtBr into the gel and buffer, or in the samples, DNA may be visualized by staining the gel with other gel stains after electrophoresis. The difference in mobility of different samples is based on their ability to fit through the pores. Larger molecules take longer time to fit through, and smaller ones travel more quickly. Thus the separation is according to their molecular sizes (Figures 6.2c and d). Another sensitive, albeit time-consuming, detection technique, autoradiography, is often used to identify specific DNA sequences in a procedure called Southern blotting and hybridization (for details see Chapter 16).

Effects of Various Factors on the Rate of Migration of DNA

The rate of migration of DNA through agarose gel run under constant voltage is affected by several factors, which are summarized in this section.

Molecular Size Range of the DNA Sample Molecules of ds DNA migrate through gel matrices at rates that are inversely proportional to the $\log_{10}$ of the number of base pairs. Larger molecules migrate more slowly because of greater frictional coefficient or drag, and because these worm their way through the pores of the gel (i.e., pulled through the pores of the gel) less efficiently as compared with smaller molecules.

DNA Conformation, Linking Number, and Secondary Structure The mobility of DNA is influenced greatly by the topology of the molecule. Take example of bacterial plasmids. Native plasmids are supercoiled circular molecules, but if a plasmid is nicked, it adopts a relaxed, open circular form. A double strand break produces a linearized plasmid. Although all three forms, *viz.* superhelical circular (form I), nicked circular or open circular (form II), and linear (form III), are the same size (in terms of base pair), these move differently in a gel. In all conditions, the migration rate of open circular form (form II) is the lowest. The relative mobilities of the forms I and III is more difficult to predict, and depend on their sizes, as well as agarose type and concentration used to prepare the gel. The strength of the applied current, the ionic strength of the buffer, and the density of superhelical twists in the form I DNA also influence the relative mobilities. Under some conditions form I DNA migrates faster than form III DNA, under other conditions the order is reversed. It is therefore quite normal for a purified plasmid preparation to show two or three bands in a gel, it does not necessarily mean that there is more than one plasmid. In most cases, the best way to distinguish between the different conformational forms of DNA is, simply, to include in the gel a sample of untreated circular DNA, and a sample of the same DNA linearized by digestion with a restriction enzyme that cleaves the DNA at a single site.

Gel electrophoresis also separates similar molecules on the basis of their compactness, so that the rate of migration of a circular duplex DNA increases with its degree of superhelicity. The agarose gel electrophoresis pattern of a population of chemically identical DNA molecules with different linking numbers therefore consists of a series of discrete bands. The molecules in a particular band have the same linking number and differ from those of adjacent bands.

It is important to note that the electrophoretic behavior of ss DNA or RNA is also influenced and modified by the presence of secondary structures known as hairpins, which are frequently assumed in non-denaturing conditions. Hence, to force identical species of ss DNA or RNA to co-migrate, the electrophoresis is routinely carried out under denaturing conditions (for details see Section 6.2.5).

Base Composition Unlike polyacrylamide gels, the migration characteristics of DNA in agarose are not influenced by their respective base compositions.

Agarose Concentration The concentration of agarose determines the 'stiffness' of the gel, and hence the rate of migration. The migration rate is linear over most of the length of the gel. This linear relationship is represented by the equation: $\log \mu = \log \mu_0 - K_r \iota$, where μ is electrophoretic mobility of the DNA, ι is the gel concentration, μ_0 is the free electrophoretic mobility of DNA, and K_r is the retardation coefficient, a constant related to the temperature of the gel, and the size and shape of the migrating molecules. Note that very large or very small molecules may end up outside its linear range. If this occurs, significant error may result from size determination based on comparisons to size standards that fall within the linear portion of the gel. In addition, there is a certain amount of smearing of low-molecular weight species frequently observed at the dye front or the 'leading-edge' of the gel. This is especially noticeable in gels made up of less than 1% (w/v) agarose. Agarose concentration also influences the pore size of the resulting gel. The usual working concentration of agarose is between 0.5 and 2.0% (w/v), although higher or lower concentrations of agarose may be adapted to specific circumstances, for example, 3% (w/v) agarose gel is used for the analysis of PCR products, and 0.3% (w/v) agarose gel is used for separation of very large molecules.

Purity of Agarose Various classes of agarose are used for specialized applications, which have already been discussed in Section 6.2.1.

Presence of EtBr in the Gel and Electrophoretic Buffer Intercalation of EtBr causes a decrease in the negative charge of the ds DNA, and an increase in both its stiffness and length. The rate of migration of the linear DNA–EtBr complex through gels is consequently retarded by a factor of ~15%.

Electrophoresis Buffer The electrophoretic mobility of DNA is affected by the composition and ionic strength of the electrophoresis buffer. Low ionic strength buffers (e.g., 1 X TAE, 0.5 X TBE, or 1 X TPE) are preferred because DNA migrates more quickly in such buffers. Moreover, low ionic strength buffers prevent heating during the run. In the absence of ions (e.g., if water is substituted for electrophoresis buffer in the gel or in the gel-tank), electrical conductivity is minimal and DNA migrates slowly, if at all. In buffer of high ionic strength (e.g., if 10 X electrophoresis buffer is mistakenly used), electrical conductance is very efficient and significant amount of heat is generated, even when moderate voltages are applied, in the worst case, the gel melts, and the DNA denatures.

Applied Voltage Conventional gel electrophoresis is run at constant voltage. The efficient dissipation of heat is directly related to the resolving power of a gel. When working with agarose gels, one must be certain that the applied voltage does not result in the overheating of the gel. In general, the slower the gel runs, the better is the resolution (2–10 kbp range), and the rate of migration of linear DNA fragments is proportional to the voltage applied. However, as the strength of the electric field is raised, the mobility of high molecular weight fragments increases differentially, and a noticeable loss of resolution is manifested. Moreover, the gel actually melts if the system becomes too hot. Thus, the effective range of separation in agarose gels decreases as the voltage is increased. To obtain maximum resolution of DNA fragments, agarose gels should be run at no more than 5–8 V/cm.

Temperature The electrophoretic behavior of DNA in agarose gels does not change appreciably between 4°C and room temperature, although as described earlier the overheating of gels is a matter of great concern. In conventional agarose gel electrophoresis, DNA samples are routinely electrophoresed at room temperature. Furthermore, gels can also be cooled as a direct function of gel-box configuration.

Field Direction Conventional electrophoresis is accomplished by separating DNA molecules in an electric field of constant direction, which however suffers from size limitations, allowing separation of molecules ranging in size from a few hundred base pairs to ~20 kbp. These limitations have now been overcome through the use of PFGE (for details see Section 6.2.3).

Analytical and Preparative Gel Electrophoresis

Based on the applications, there are two classes of gel electrophoresis, *viz.* analytical electrophoresis and preparative electrophoresis. Gel electrophoresis is routinely used for analyzing the composition and quality of a nucleic acid sample. It is invaluable for determining the size of DNA fragments from a restriction digest or the products of a PCR reaction. For this purpose it is necessary to calibrate the gel by running a standard marker containing fragments of known sizes, and it is observed that over much of the size range (few hundred bp to ~20 kbp) there is a linear relationship between the logarithm of the fragment size and the distance it has moved under the electric field applied in constant direction. This relationship, however, becomes non-linear for large DNA molecules (>20 kbp), and for a particular gel, all molecules above a certain size exhibit virtually the same mobility (for details see Section 6.2.3). The technique is also used to determine restriction digestion patterns, plays a role in the construction of restriction maps, and provides a reliable assessment of the integrity of the sample. Moreover, Southern and Northern blotting also require separation of DNA fragments and RNA, respectively, on agarose gel before transfer to membrane (for details see Chapter 16). Gel electrophoresis is also used for investigating protein–nucleic acid interactions in the gel retardation assay or band shift assay (for details see Chapter 19). It is based on the observation that binding of a protein to DNA fragments usually leads to a reduction in electrophoretic mobility. The assay typically involves the addition of protein to linear ds DNA fragments, separation of complex and naked DNA by gel electrophoresis and visualization.

Gel electrophoresis is not only used as an analytical method, it is also routinely used preparatively for the purification of specific DNA fragments from a complex mixture. In this case, LMP agarose is used. It is also a powerful technique for concentrating like species of RNA or DNA. The parameters governing electrophoresis favor co-migration of molecules of equivalent size, and in most instances specific bands can be recovered from the gel in a variety of ways for subsequent characterization and cloning (for details see Section 6.3).

6.2.3 Pulsed Field Gel Electrophoresis (PFGE) and its Variants

Conventional gel electrophoresis, run under constant electric field, and in constant direction is unable to effectively separate very large DNA molecules >20 kbp in size (i.e., larger than average agarose pore size), resulting in limiting of mobility. Thus, these large DNA molecules co-migrate through the conventional agarose gel in a size-independent manner, resulting in broad, unresolved, diffuse bands. Taking advantage of the elongated and oriented configuration of large DNA molecules in agarose gels at finite field strengths, Schwartz and Cantor (1984) developed a fundamentally different technique, namely pulsed field gel electrophoresis (PFGE), which provided a means for the routine separation of DNA fragments >20 kbp in size. In the first publications, PFGE stood for pulsed field gradient gel electrophoresis. However, later it was recognized that field gradients or inhomogeneous fields were not necessary to separate large DNA fragments, and nowadays, PFGE is generally used to mean pulsed field gel electrophoresis, i.e., all kinds of methods involving switching (pulsing) of fields to separate large DNA molecules. In this section, the general characteristics of PFGE, factors affecting PFGE, applications of PFGE in molecular biology, and variants of PFGE are discussed.

Principle of PFGE: Models for Migration of Large DNA Molecules

The limitation of conventional agarose gel electrophoresis is that the large DNA fragments above a critical size (>20 kbp) do not separate, but instead, co-migrate. This is because the globular forms of large DNA molecules are unable to fit into the matrix pores of even the lowest percentage agarose gels that can be easily handled (0.1–0.3%). This limitation could be overcome if the electric field is applied discontinuously (pulse), and even greater separation can be achieved by periodically altering the direction of migration by regular changes in the orientation of the electric field with respect to the gel. With each change in the electric field orientation, the DNA molecules realign their long axes to migrate in the new direction, and this process takes longer for larger molecules. This is because large DNA molecules become trapped in their reptation tubes every time the direction of the electric field is altered, and can make no further progress through the gel until these have reoriented themselves along the new axis of the electric fields. The larger the DNA molecule, the longer the time required for this realignment. DNA molecules whose reorientation times are less than the period of the electric pulse therefore get fractionated according to size. A number of models have been proposed to explain the migration of DNA molecules observed during PFGE. An early model of Southern *et al.* (1987) argues that DNA molecules move through the gel in a fully extended conformation, and that reorientation of the field by an angle >90° causes the molecules to reorient with their former trailing ends as the new leading ends. This simple model successfully predicts the separation of large DNA in a pulsed electric field. However, the model also predicts that the upper size limit of resolution is accompanied by zero DNA mobility, and is unacceptable. Another model pictures the motion of the DNA as a 'reptating' chain that crawls snakelike through a tube in the gel in response to the electric field. As the DNA moves forward, its leading end creates additional segments in the tube. Each step forward is in a random direction but with a bias in favor of the electric field direction. This relaxation from the rigid assumption of a fully extended conformation allows more complex DNA behavior. The reptation model successfully predicts that continuous fields will fail and pulsed fields will succeed in separating large DNA, and that in a pulsed field the size limit of separation increases proportionally to pulse time and is associated with a non-zero mobility. It also predicts two unexpected phenomena: First, under some conditions, transverse pulsed fields should cause larger DNA molecules to migrate faster than smaller molecules ('mobility inversion'). This has been experimentally observed in electrophoresis with periodic inversion of the electric field, but the reptation model predicts that it should also occur in transverse pulsed fields. Secondly, similar conditions should produce lateral band spreading, in which the DNA band remains sharply defined vertically but spreads laterally as it moves down the gel. Both mobility inversion and lateral band spreading have now been observed experimentally. The 'biased reptation model' (BRM) accounts for the experimentally observed fact that the mobility of DNA becomes independent of the molecular weight at high field strength. This is because at high field strengths, the DNA stretches, and hence the dependence of mobility on molecular size decreases. The 'reptating chain model' predicts that DNA molecules move through the gel like a flexible chain moving through a lattice of obstacles. The chain model relaxes the assumptions for DNA behavior further. The chain is not constrained to move inside a tube but is free to adopt any configuration as it interacts with the obstacles that represent the gel. Thus, DNA molecules move through a variety of extended and tangled configurations in response to the electric field. Like the reptation model, the chain model also predicts that the separation of large DNA molecules requires a pulsed field. However, chain model calculations require the power of supercomputers, and have not yet addressed the phenomena of mobility inversion and lateral band spreading. Direct observations in the light microscope indicate that DNA moves through agarose by oscillating among a complex repertoire of configurations. In a continuous field, DNA molecules fluctuate between extended and tangled configurations, but with a general orientation in line with the applied field. The leading end tends to accumulate DNA mass as the molecule encounters obstacles, slowing progress of the head in relation to the tail. At other times, DNA becomes looped over an obstacle forming an inverted U shape. When the field direction is suddenly changed, the DNA gradually realigns with the new orientation but initially retains a memory for the previous orientation. Subsegments first begin moving in line with the reoriented field before the entire molecule finally moves off in the new direction. This behavior clearly contradicts the reptation model but is in good agreement with the chain model. However, in the chain model, the chain responds to the reoriented field by developing a new set of configurations, retaining a memory that diminishes as it gradually deforms toward the new orientation. More recently, a 'bag model' has been proposed, which indicates that the DNA molecule moves through the gel as a deformable bag that moves with limiting mobility in a continuous electric field, adopts an orientation aligned with the field direction and reorients after a change in field direction in a size dependent manner. The model correctly predicts the resolution of large DNA in a pulsed field including the surprising phenomena of mobility inversion, lateral band spreading, and improved resolution for obtuse angles. A simple parameterization agrees with observations of two completely different aspects of DNA behavior, i.e., bulk mobility as measured during gel electrophoresis, and molecular reorientation as measured by linear dichroism.

PFGE Equipment

The basic components of a PFGE system include the following:

Gel-box (or PFGE-box) The basic design of PFGE boxes consists of an immobilized gel within an array of electrodes, and a means of circulating the electrophoresis buffer. The temperature of the buffer is controlled by a heat-exchange mechanism. Generally, the buffer is recirculated throughout the gel-box using inlet and outlet ports.

High Voltage Power Supply Precise control of the electric field gradient is necessary to obtain consistent PFGE separations. The output ratings of the power supply should therefore be high enough to meet both the voltage and current requirements of the gel-box. A typical PFGE gel-box has electrodes that are 25–50 cm apart. A power supply with a maximum voltage rating of 750 volts is used to achieve the commonly used range of voltage gradients of 1.5–15 V/cm. Note that the current drawn at this voltage in most PFGE gel-boxes is about 0.5 amp at 14°C using 0.5 X TBE as running buffer.

Switch Unit The ability to reproducibly control the switch interval is critical for the separation. This requires the application of relays, which are controlled by computer for obtaining high-speed switching necessary for the PFGE separation of small DNA molecules (2–50 kbp). High-voltage solid-state electronics has supplanted electromechanical relays in recently designed commercial PFGE systems. These switching units are commonly based on the use of metal oxide semiconductor field effect transistors (MOSFETs) in both switching and electrode voltage control circuits. These designs offer the advantages of improved reliability, the capability of high-speed switching (0.1 ms), and ample voltage (750 V) and current (0.5 amp) ratings. These apparatuses have the ability to control the reorientation angles between electric fields, however, these instruments cannot provide fast enough switching for the improvement of separation of DNA molecules <50 kbp in size.

Computer Program As described above, careful control of the switch interval is crucial in controlling the resolution in PFGE. A versatile switching unit should have software with the same characteristics. The algorithm should be fast enough so that switch times as short as 1 ms can be achieved, and switch interval increments should have at least 1 ms resolution. The algorithm should be slow enough to adjust run times as large as several weeks for separation of very large DNA molecules. Computer programmes are also used to control linear switch interval ramping.

Cooler Temperature variations within the gel during the run should be eliminated to alleviate buffer breakdown due to electrolysis. Moreover, DNA molecule migration is also sensitive to temperature, and thus a uniform temperature across the gel is needed to ensure uniform migration in each of the lanes. Thus, variation in temperature can be eliminated by recirculation of buffer through the gel-chamber by a reciprocating solenoid pump at a rate of about 450 ml/minute. The buffer is chilled in its reservoir tank by cold water (5°C) circulated through a glass tubing heat exchanger, and buffer temperature is maintained at 13–15°C throughout a typical run.

DNA Size Markers for PFGE

Another specific requirement for PFGE is DNA size marker. PFGE requires extremely large molecular weight markers. These are obtained by extracting DNAs from bacteriophages such as T7 (40 kbp), T2 (166 kbp), and G (78 kbp). Alternatively, DNA size standards are produced by ligating a monomer DNA fragment of known size into a ladder of polymeric forms [e.g. 250 kbp DNA ladder containing fragments differing by 250 kbp]. A better series of markers, which are evenly spaced over a wider range of molecular weights, are generated by ligation (through *cos* sites) of bacteriophage λ DNA into a nested series of concatemers (e.g., 20 tandemly arranged copies). Usually 400–600 ng of the concatenated DNA is loaded onto a single lane. Yeast chromosomes may also be used as even molecular weight markers.

Separation of DNA Fragments by PFGE

Large DNA is susceptible to easy shearing, and often difficult to pipette due to its high viscosity. Hence, preparation and manipulation of very large DNA molecules for PFGE requires more care than conventional gel electrophoresis normally used for smaller molecules. The steps involved in PFGE are:

Embedding of DNA in LMP Agarose Plugs The essential requirement in keeping the fragile large DNA molecules intact upon their release from intact cells (or isolated spheroplasts) is to embed them in agarose plugs or beads. The process of encapsidation of DNA involves harvesting of overnight grown cells, proper dilution of cells, formation of agarose plugs with immobilized cells, and treatment of embedded cells for releasing DNA. Thus, an overnight grown culture containing intact cells in exponential phase (with known cell count or cell density) is harvested by centrifugation, the cells are washed and diluted with suitable buffer (preferably TE), and the cell suspension is pipetted into equal volume of molten LMP agarose at required concentration (usually 1% w/v). After proper mixing, the agarose-cell suspension is poured in disposable plug molds, and allowed to set on ice for 10–20 minutes. These embedded cells are treated with different combinations of enzymes and detergents to remove cell walls, membranes, RNA, proteins, etc., to obtain naked DNA. This is possible because detergents and enzymes can diffuse by Brownian motion through the agarose pores of the plug, whereas the large pieces of DNA are unable to diffuse through agarose pores, and hence remain sequestered inside the plug.

Restriction (Digestion of Immobilized DNA within Agarose Plugs) The plugs are then washed with buffer, cut to size, treated with restriction enzymes in much the same way as the DNA in solution (20–30 units of enzyme are used for 4 hours to overnight), and sealed into place with agarose. It is advisable to use rare cutters, which have relatively few sites, and give larger fragments from the target DNA.

Running Conditions and Factors Affecting PFGE After digestion, next step is fractionation of DNA by PFGE, followed by either isolation from the gel, or analysis by Southern blotting. For electrophoresis, 1% agarose gel (in running buffer) is prepared, and mounted properly in the electrophoresis apparatus. All the plugs containing the DNA of interest are rinsed in TE (pH 8.0) buffer, embedded in individual wells in the gel, and sealed in the wells with molten 1% agarose (in running buffer) previously cooled to 14°C. Note that the DNA size standards are prepared and embedded in the same way. The gel apparatus is connected to the power supply set at appropriate pulse time, and electrophoresis is allowed to continue for the required time interval. Several parameters, which affect the migration and resolution of DNA fragments, act in concert during PFGE. These are as follows:

Electric field shape (number and configuration of electrodes) The number and configuration of the electrodes alter the shape of the applied electrical fields, and affect the resolution of DNA fragments through PFGE. The arrangement of electrodes and their alternate use generates an angle between the alternate electrical fields, which is a very critical parameter. Angles of 90° or smaller are not effective, probably because the DNA molecules easily become oriented midway between the two applied fields. Angles larger than 90° are more effective, and in cases of excellent resolution, field angles typically range from 120° to 150°. In PFGE, the hexagonal array of electrodes is used to generate two fields that intersect at 120° (Figure 6.4). Large molecules travel in the direction of the first field for a short time, and eventually get caught in the agarose matrix. When the field is changed, the DNA realigns by 120°, and then moves in this direction for a while. The two pulse times are kept equal, and overall movement is in the forward direction. Moreover, a progressive increase in angles along the direction of the net DNA motion produce band sharpening because the molecules at the front of each DNA zone always migrate more slowly than those at the rear.

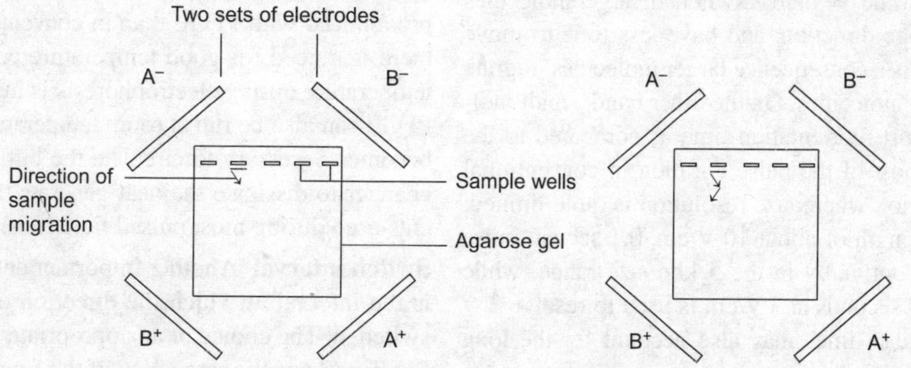

Figure 6.4 Electrode configuration in pulsed field gel electrophoresis (PFGE) [A (+ and –) and B (+ and –) represent two sets of electrodes. The DNA follows a zigzag path. When the A electrodes are on, the DNA is driven downward and to the right. When the A electrodes are turned off, the B electrodes are activated, which causes the DNA to move downward and to the left.]

Electrical field strength Electrophoretic mobility is defined as the velocity per unit field. Hence, field strength has a profound effect on pulsed field separations, and is a compromise between separation time and the resolution of molecules of a particular size. Increasing the field strength increases the migration of DNA molecules, but decreases the resolution obtained, and field strength higher than 10 V/cm results in poor recovery and smeared bands. Thus, resolution is much better at not too high field strength (less than 10 V/cm), and for separating DNA molecules in the range of 1–2,000 kbp, 5–7 V/cm electric fields are usually applied. Very large molecules are separated at low field strengths (1–3 V/cm), e.g., separation of *Schizosaccharomyces pombe* and *Neurospora crassa* chromosomes is done at field strengths of 1.3 and 2 V/cm, respectively. These low voltages, on the other hand, imply that long runs are required (as long as seven days for *N. crassa* chromosomes).

A progressive decrease in field strengths along the direction of the net DNA motion is also an effective strategy. It produces band sharpening because the molecules at the front of each DNA zone always migrate more slowly than those at the rear.

Pulse time (or pulse length) and reorientation time DNA is subjected alternately to two electrical fields at different angles for a time called the 'pulse time', and the time taken by the separating molecules to reorient or change direction is called 'reorientation time'. To obtain high resolution over separation ranges in a single run, different pulse times, or time ramping, should be applied. Moreover, the molecules must presumably change direction prior to net translational motion. Pulse times are selected so that DNA molecules of a targeted size spend most of the duration of the pulse reorienting rather than moving through the gel, and resolution in PFGE is likely to be optimal for molecules with reorientation times comparable to the pulse time.

The pulse time and reorientation time have significant effect on the mobility of the DNA molecules. Generally speaking, the larger the molecule, the longer the pulse time required for resolution. Each time the field is switched, larger molecules take longer to change direction, and have less time to move during each pulse. As a consequence larger molecules migrate slower than smaller molecules. On the other hand, small molecules that have short reorientation time as compared to the pulse time spend most of the pulse duration in conventional electrophoretic motion where size resolution is quite limited. At applied field strength of about 10 V/cm, 0.1 second pulse time resolve DNA optimally in the 5 kbp size range, while pulse time of 1,000 seconds at 3 V/cm is used to resolve 3–7 Mbp molecules. Pulse times may also account for the long periods of time, usually days or weeks, needed to fractionate large DNA molecules. The chromosomal DNA molecules of *Saccharomyces cerevisiae* in the 10 Mbp range require longer electrophoresis of approximately one week.

Pulse time also changes the size range of separation. Longer pulse times lead to separation of larger DNA. For example, at 5.4 V/cm, the 1.6 Mbp and 2.2 Mbp chromosomes from *S. cerevisiae* separate as a single band with 90 seconds pulse length, but upon increasing the pulse time to 120 seconds, two bands are resolved.

Reorientation angle The difference between the directions of migration induced by each of the electric fields is the reorientation angle, and corresponds to the angle that the DNA must turn as it changes its direction of migration each time the fields are switched. Studies have revealed that reorientation angles of less than 90° do not give effective separation of large DNA molecules. This is presumably because the DNA molecules easily become oriented midway between the two applied fields. The separation is superior at 120–150° reorientation angles.

The speed at which separation is attained differs greatly with the applied reorientation angle. When this angle is reduced, the velocity at which separation is obtained increases. With respect to the separation of *S. pombe* chromosomes, a 3.2-fold decrease of separation time has been reported with a reorientation angle of 96° as compared to 120°.

Voltage gradient or electrical potential applied to the gel The choice of the voltage used in PFGE must also be varied with the size of the DNA to be separated. Voltage gradients of 6–10 V/cm are used to separate molecules up to 1 Mbp, while still larger DNA samples require lower voltage gradients in order to migrate properly through the gel. However, when the voltage gradient is reduced to separate large DNAs, switching intervals must be lengthened. The practical effect is to increase the run times for larger DNA molecules.

Temperature As DNA mobility depends on the separation temperature, the temperature must be kept constant both during and between runs. The mobility of the DNA increases with increasing temperature of electrophoresis, thus permitting shorter run times. Moreover, at higher temperatures (25–35°C) the fractionation range of large molecules increases; however, the resolution of fragments is reduced resulting in blurred bands. Because the effect of temperature on fragment mobility is more pronounced with PFGE than in conventional electrophoresis, there is a need for good temperature control. In general, the temperature during electrophoresis is maintained at 10–15°C. PFGE can also be run at room temperature, but in this case, it becomes necessary to circulate the buffer through a heat exchanger to dissipate the heat generated by the voltage gradients used during most pulsed field runs.

Switch interval Another important determinant of mobility is the interval at which the direction of the electric field is switched. The choice of an appropriate switching interval for PFGE reflects the size range of the fragments to be resolved. The highest resolution for molecules of a given size is obtained by using the shortest switch intervals, which permit separation of the complete size range of the fragments. If the switch interval is increased beyond the time required for a fragment to reorient, the fragment will spend a large portion of the time in migrating, as in conventional electrophoresis, with a loss in resolution.

Degree of homogeneity of electric fields For separation of DNA into straight lanes so as to make lane-to-lane comparison easy, the electric fields should be homogeneous (i.e., electric field that has uniform potential differences across the whole field). The original pulsed field system, however, used inhomogeneous electric fields.

Agarose type and concentration When choosing agarose it is also important to keep the size of the sample in mind because different types of agarose influence the electrophoretic behavior of DNA molecules. Most standard electrophoresis grades of agarose are suitable for PFGE (e.g., SeaKem GTG), and agarose with minimal EEO provide a faster separation. Several low EEO 'pulsed field grades', which possess higher gel strength allowing low percentage gels to be made with high mechanical strength and reduced run times, are available, for example, FastLane and Gold (FMC BioProducts), and Megarose (Clontech).

The resolution and mobility of the DNA during PFGE is also affected by the agarose concentration. Higher concentrations of agarose yield sharper, but slower moving bands, and faster DNA migration occurs in gels of lower agarose concentration. Some band sharpening may also occur because of the difference in agarose concentration between the electrophoresis gel, and the agarose containing the digested sample. For most purposes 0.8–1% (w/v) gels are sufficient, which represent a compromise between speed and resolution. At agarose concentrations of more than 2% (w/v) separation is very poor. However, the effect of increased agarose concentration can be completely or partially reversed by using longer pulse times or higher temperatures.

Ionic strength of buffer DNA migrates more quickly in buffers of low ionic strength, for example, 1 X TAE and TBE. Moreover, low ionic strength buffers also prevent heating during the run. In the absence of ions (e.g., water) electrical conductivity is minimal, which leads to slow or no DNA migration, while in high ionic strength buffer (e.g., 10 X TAE) electrical conductance is very efficient, but it generates heat even at moderate voltage, and may result in melting of gel or denaturation of DNA.

DNA topology The electrophoretic behavior of circular and linear DNA molecules is completely different in PFGE; circular molecules are much less mobile than linear molecules. Thus, the circular bacterial chromosomes and the non-supercoiled form of plasmids are not amenable to analysis by PFGE unless the DNA is linearized by treatment with a restriction endonuclease. The supercoiled molecules are more mobile than relaxed circular molecules, which in turn are somewhat more mobile than nicked open circular molecules. Unlike linear molecules, their relative mobility is not or only slightly affected by pulse time (at least not for molecules up to 85 kbp). There is, however, an effect on absolute mobility. Supercoiled molecules smaller than 50 kbp show the lowest mobility at 10 seconds pulse time as compared with both shorter and longer pulse times. The pattern for nicked and relaxed circular molecules is not so clear. Because the mobilities of isometric linear and supercoiled circular DNA molecules are different and exhibit different responses to variations in pulse time, linear molecules cannot be used as standards in PFGE for determining the sizes of supercoiled circular plasmid molecules.

Although all electrophoresis parameters are discussed separately, one has to keep in mind that changing one parameter can have a strong influence on the effect of other parameters. For instance, separation obtained at 6 V/cm and pulse time of 90 seconds is almost similar to that at 3 V/cm with 300 seconds pulse time, while changing the temperature from 4–13°C has an effect comparable with lengthening the pulse time from 70 to 90 seconds. Therefore, applied conditions are always a compromise between the various parameters.

After electrophoresis is complete, power supply is disconnected, and the gel apparatus dismantled. The gel is visualized and photographed after staining with a suitable gel stain.

Applications of PFGE

Applications of PFGE are numerous and diverse. Some of these are: (i) Analysis of large DNA molecules from a variety of sources; (ii) Separation of large DNA fragments; (iii) Examining the elongated configuration of large DNA; (iv) Size fractionation of DNA fragments ranging between few kbp to over 10 Mbp; (v) Isolation and purification of large DNA molecules for sequencing (PFGE has formed an essential basis for projects aimed at total sequencing of eukaryote genomes notably and *S. cerevisiae* and human); (vi) Precise selection of large DNA fragments for cloning in high insert capacity vectors such as YACs and PACs, and construction of genomic libraries; (vii) Easy isolation of the individual restriction fragment for further restriction mapping, gene insertion, and functional gene mapping; (viii) Isolation of genes for production of transgenic plants and animals; (ix) Production of a discrete pattern of bands (restriction fragment length polymorphism; RFLPs) useful for genotyping or genetic fingerprinting, i.e., characterization of various strains at the DNA level and physical mapping of the chromosome; (x) More precise determination of the order of markers than possible with genetic linkage analysis; (xi) Identification of genetic defects causing hereditary diseases by detecting size differences in the normal and the defective genes due to chromosomal rearrangements; (xii) Study of diseases involving large deletions or duplications, e.g., duchenne muscular dystrophy; (xiii) Studying radiation-induced DNA damage and repair,

(xiv) Size organization and variation in mammalian centromeres; (xv) Mapping of new mutation by cloning the gene, followed by restriction analysis and hybridization to a set of ordered restriction fragments; (xvi) Establishment of the degree of relatedness among different strains of the same species; (xvii) Separation of whole chromosomes; (xviii) Rapid analysis of a large chromosomal region for detection of *in vivo* chromosome breakage and degradation; (xix) Providing insights into the genome organization, genome size, and characterization of organisms as diverse as bacteria and humans; and (xx) Separation and analysis of the number and size of chromosomes (electrophoretic karyotype) from yeasts, fungi, and parasites such as *Leishmania*, *Plasmodium*, *Trypanosoma*, etc.

Variants of PFGE

The original pulsed field systems used inhomogeneous electric fields, which did not separate DNA into straight lanes, making interpretation of gels difficult. To overcome the problem of inhomogeneity of electric fields, and to simplify lane-to-lane comparisons, the original apparatus has been modified. However, all systems are based on the same concept of using two electric fields applied alternatingly, and depend upon the fact that periodic changes in the direction of electric field retard the migration of larger DNA molecules through the agarose gel. These modifications include alteration in the number, design, and configuration of electrodes, field angles, and the consequent electric fields, pneumatic rotation of gel, conditions for higher field strengths and mobility inversion, homogeneity of electric fields, feedback clamping of voltages in a modified contour clamped homogeneous electric fields (CHEF) apparatus, ramping the switching parameters, simultaneous electrophoresis in a double-decker gel arrangement, faster separation by 'secondary' PFGE, etc. Note that all the variants of PFGE have their own specific requirements, but are affected by same parameters as described for PFGE. Some of the variants of PFGE are discussed below:

Orthogonal Field Alternation Gel Electrophoresis (OFAGE) OFAGE, the first PFGE method, was developed by Carle and Olson in 1984. The method uses double inhomogeneous electric fields (initially, a configuration with only one inhomogeneous field was used). This is a simple construction with two pairs of electrodes, placed to produce electric field diagonally across the gel, and at an angle greater than 90° to each other (Figure 6.5a). With this apparatus, a symmetrical separation pattern is obtained with straight lanes in the center of the gel, while the DNA in wells increasingly distal from the center of the gel migrates in increasingly curved and 'sidestepping' tracks (samples from one slot run off from the main track, and form a 'shadow') in the final migration pattern making lane-to-lane comparisons difficult. This problem results from the inhomogeneity of electric fields, and the variation of angle between the electric field across the gel. As compared with PFGE size estimates are more accurate, and DNA molecules from 1,000–2,000 kbp can be separated in OFAGE as sharp bands. Owing to the problem of DNA bending patterns, this apparatus is now largely redundant, and improved types have now largely replaced this system.

Transverse Alternating Field Electrophoresis (TAFE) and ST/RIDE™ TAFE is a simple, high-resolution pulsed field system that uses a vertical gel and simple electrode arrangement (straight lane geometry). An example of commercially available TAFE system is GeneLine (Beckman). The technique is used for the separation of fragments up to 9 Mbp in size. The apparatus comprises of a large gel-box (plexiglass box), in which the gel stands vertically, supported at each side by two thin plexiglass slots, and by the buoyancy of the buffer. The gels are shorter than many (10 cm), and the angle between fields is not readily altered. The electrodes, represented by dots in the Figure 6.5b, are wires stretched across the width of the box (in commercial designs, the electrodes are wired to removable plexiglass inserts). Electrodes are positioned on opposite sides of a vertically oriented gel (traverse positions), and hence the pulse directions of the DNA molecules are across the thickness of the gel. Thus, as the DNA moves first toward one electrode, and then towards the other, it forms a zigzag pattern, however, the vector of this oscillation is a straight line from the loading well to the base of the gel. Note that the angle between the two (A and B) fields is not constant down the length of the gel. At the wells, it is 115°, but is much greater at increasing distances from the wells. As the angle increases, the downward component of the field decreases. As in other pulsed field systems, if all other conditions are held constant, the pulse time can be used to optimize separation of the molecules of the size of interest. However, because of the unique geometry of the TAFE system, the pulse times and voltages required for resolution of a particular size of molecules cannot always be directly inferred from those used in other types of apparatus. In TAFE, as the electrodes are placed parallel to the gel face, the electric field is applied at an angle to the face of the gel plates, and the electric field is equivalent across all lanes of the gel. The advantages of TAFE are: (i) The placement of electrodes provides homogeneous fields produced across the width (depth) of the gel, and hence, the velocity of identical molecules does not vary from lane-to-lane. Consequently, DNA runs absolutely straight in all lanes, and the problem of 'bent' lanes seen with OFAGE is completely eliminated; (ii) Good resolution and very sharp bands are obtained for all size classes <1,600 kbp, in relatively short electrophoresis times (<18 hours); (iii) Band sharpening is observed, because the downward component of the field decreases down the gel; and (iv) Useful resolution of fragments is obtained in the Mbp range in three days. This system however has some drawbacks as well, (i) Because the reorientation angles are not constant

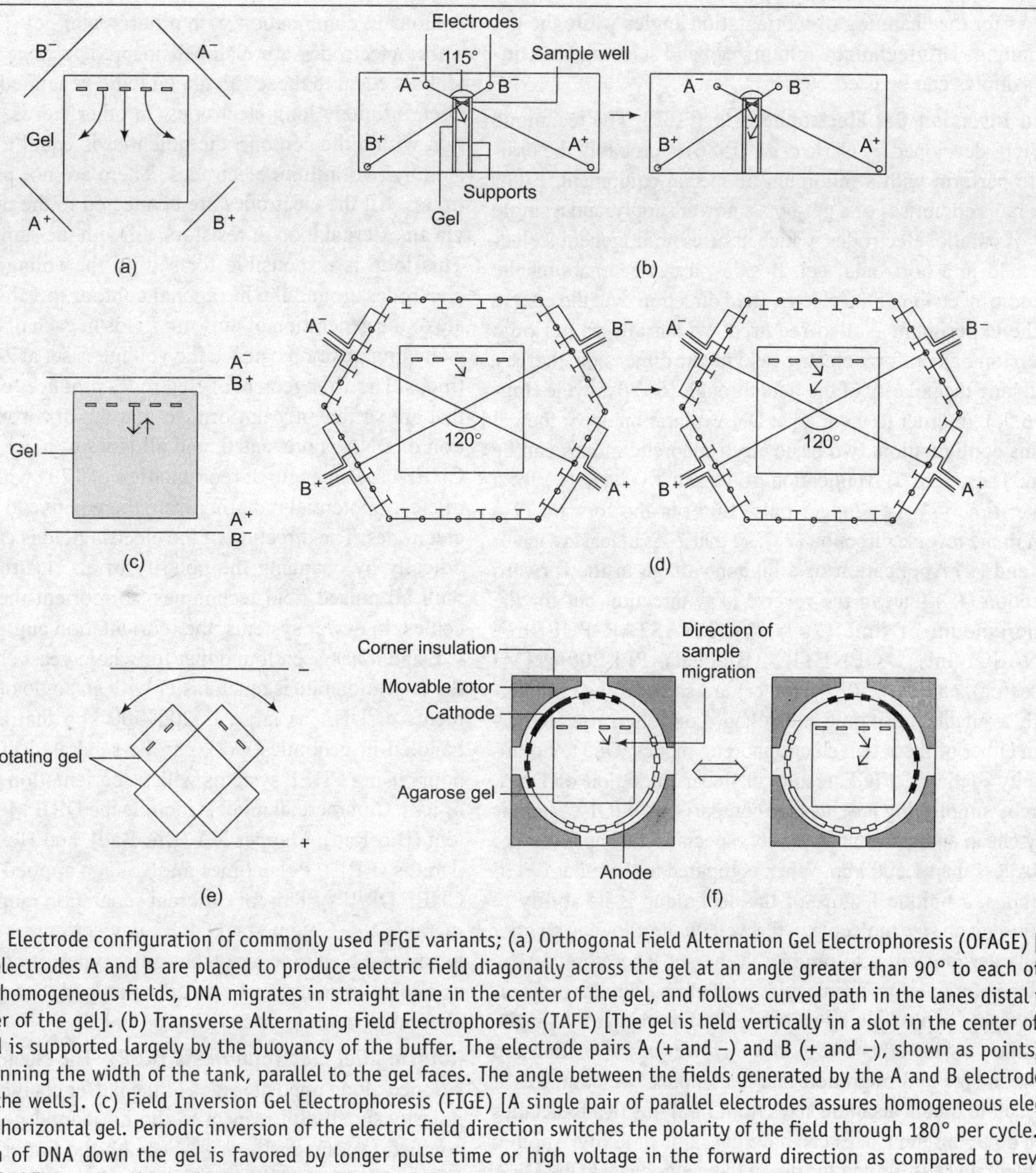

Figure 6.5 Electrode configuration of commonly used PFGE variants; (a) Orthogonal Field Alternation Gel Electrophoresis (OFAGE) [Two pairs of electrodes A and B are placed to produce electric field diagonally across the gel at an angle greater than 90° to each other. Due to inhomogeneous fields, DNA migrates in straight lane in the center of the gel, and follows curved path in the lanes distal from the center of the gel]. (b) Transverse Alternating Field Electrophoresis (TAFE) [The gel is held vertically in a slot in the center of the tank, and is supported largely by the buoyancy of the buffer. The electrode pairs A (+ and –) and B (+ and –), shown as points, are wires spanning the width of the tank, parallel to the gel faces. The angle between the fields generated by the A and B electrodes is 115° at the wells]. (c) Field Inversion Gel Electrophoresis (FIGE) [A single pair of parallel electrodes assures homogeneous electric field in a horizontal gel. Periodic inversion of the electric field direction switches the polarity of the field through 180° per cycle. Net migration of DNA down the gel is favored by longer pulse time or high voltage in the forward direction as compared to reverse direction]. (d) Contour Clamped Homogeneous Electric Fields (CHEF) [A horizontal gel is surrounded by 24 point electrodes equally spaced around the hexagonal closed contour, with no passive electrodes. Homogeneous electric fields at a reorientation angle of 120° are generated due to connection of all the electrodes to the power supply via an external loop of resistors, each one having the same resistance]. (e) Rotating Gel Electrophoresis (RGE) [A horizontal gel is rotated with the aid of magnetic drive. Electrodes are placed on opposite side of buffer chamber, and the polarity of electrodes with respect to gel is fixed]. (f) Rotating Field Electrophoresis (ROFE) [Cathode and anode are part of movable rotor. The direction of the field (long arrow) depends on the positions of the rotor relative to the gel in which the separation takes place. The movement of the DNA molecule out of a sample well is schematically depicted by short arrows. The field angle in this example is 110°.]

throughout the gel, molecules do not move at constant velocity throughout the gel, and liquid samples cannot be used; (ii) Due to the gel's vertical position, handling is more difficult, and gel size is restricted; and (iii) Though the molecules in all lanes moving down the gel are subjected equally to continual variations in field strength and reorientation angle, but due to variation of field strength and orientation angles along the length of the gel from top to bottom surfaces (the angle between the electric fields varies from 115° at the top of the gel to ~165° at the bottom) molecules still do not move at a constant velocity over the length of the gel. A modification of the TAFE system is the ST/RIDE™ system, which allows for the changing of reorientation angles while the gel is running. This technique minimizes band stacking, and liquid samples can be used.

Field Inversion Gel Electrophoresis (FIGE) The technique of FIGE developed by Carle *et al.* (1986) is probably the easiest to perform with a minimum of special equipment. FIGE apparatus comprises of a gel-box, a power supply, and a single pair of parallel electrodes, which assures homogeneous electric field in a horizontal gel. Besides, a device enabling the periodic inversion of the electric field direction over the course of the experiment is also required, which allows periodic inversion of a uniform electric field in one dimension, thereby switching the polarity of the field through 180° per cycle (Figure 6.5c). In order to favor a net DNA migration down the gel in this configuration, two basic electrophoretic modes can be used. These are: (i) Application of the same voltage (V) for a longer time (T) (i.e., longer pulse time) in the forward (T_F) than in the reverse direction (T_R), so that T_F is at least twice of T_R; and (ii) Application of a higher voltage in the forward direction (V_F) than in the reverse (V_R) direction but for the same amount of time ($T_F = T_R$). DNASTAR PULSE™ (DNAStar Inc.), 'GENETIC' (Biocent), PPI-200™ (MJ Research), and PC 750™ (Hoefer) are some of the commercially available pulse time and voltage controllers used to perform either of these two electrophoretic modes. Due to homogeneity of fields, FIGE results in linear migration of DNA, thereby simplifying lane-to-lane comparisons. FIGE system is very cheap and convenient to use, especially for separation of DNA less than 1,500 kbp. When compared to the other PFGE methods, a unique feature of this technique is its ability to retain a given size molecule in the well, while allowing smaller and larger molecules to migrate in the gel. Therefore, a very high resolution of two molecules with almost similar molecular weights is possible by stopping the first one at the origin, and allowing the other one to move. However, the relationship between size and migration rate is complex, which makes it difficult to obtain accurate sizes from FIGE, as two molecules with quite different sizes can have the same mobility. Another problem that has limited the use of the method is that the DNA bands are wider and more diffuse, and lack band sharpness for samples over 700 kbp, when compared with other PFGE methods. This problem necessitated the use of 'switch time ramping', i.e., progressively increasing the reorientation pulse time during separation. Now ramping is included in most commercial instrumentation to minimize inversions, nevertheless, ramping the switching parameters is not always enough, and DNA mobilities are not always monotonic with size for DNA molecules >2 Mbp, i.e., DNA molecules that differ markedly in size can at times co-migrate in FIGE.

Contour Clamped Homogeneous Electric Fields (CHEF) Chu *et al.* (1986) first developed CHEF. CHEF system utilizes 24 point electrodes equally spaced around the hexagonal closed contour in combination with a horizontal gel (Figure 6.5d). These electrodes are clamped to predetermine electric potentials equal to those calculated to be generated by two parallel, infinitely long electrodes. In other words, at any position within the contour, the potential is equal to that generated by two infinite electrodes. There are no 'passive' electrodes. All the electrodes are connected to the power supply via an external loop of resistors, all with the same resistance. This loop is responsible for setting the voltages of all the electrodes around the hexagonal contour to values appropriate to the generation of uniform fields in each of the alternate switching positions; hence the voltage is set at 24 point electrodes. The arrangement of electrodes produces electric fields that are sufficiently uniform so that the distortion of migration of DNA is prevented, and all lanes of a gel run straight. CHEF uses an angle of reorientation of 120° with gradations of electropotential radiating from the positive to the negative electrodes. The direction of the electric field is changed periodically by changing the polarity of an electrode array, as with all pulsed field techniques, to reorient the DNA molecules. In newer systems, the reorientation angle can be varied, and it has been found that for whole yeast chromosomes the migration rate is much faster with an angle of 106°. Fragments of DNA as large as 200–300 kbp that are routinely handled in genomics work can be separated in a matter of hours using CHEF systems with a reorientation angle of 90° or less. Commercial models include the DRII Megabase System (Bio-Rad), Mapper XA (Bio-Rad), and Hex-a-field Apparatus (BRL). Pulse times and voltage applied while using CHEF DR II system for different separation ranges are given in Table 6.4. Commercial CHEF devices currently employ a hexagonal electrode array, but other types of contours, such as circles or squares, if properly clamped, can also produce alternating homogeneous electric fields.

Rotating Gel Electrophoresis (RGE) The electrodes are positioned along opposite sides of the buffer chamber with their polarity fixed with respect to the horizontal gel. The gel is coupled to a magnetic drive beneath the buffer chamber to eliminate the possibility of leakage that a direct connection might cause. To force the migrating DNA to a new direction, the magnetic drive simply rotates the gel to the new angle (Figure 6.5e). This generates homogeneous electric fields, and hence allows straight migration of DNA molecules. Because the reorientation angle of the DNA is determined

Table 6.4 Pulse Times with Separation Ranges and Voltages When Using a CHEF DRII (Bio-Rad) Gel-box

Pulse time (s)	Separation range (kbp)	Voltage (V)
1–10	<100	200
50–100	100–200	150–200
120–180	2,000–4,000	125–175
600–3,600	>4,000	50–75

by a straightforward mechanical coupling, RGE offers a lot of flexibility at a reduced cost. Voltage, angle, and pulse times are varied with the help of a program stored into the memory of the unit. In contrast to most other PFGE systems which separate DNA over a relatively small area and limit the resolution of complex samples, RGE has a useful separation distance up to 20 cm, and a maximum gel size of 18 × 20 cm.

Rotating Field Electrophoresis (ROFE) Unlike RGE, in ROFE, electrodes are carried on a rotor that rotates around the stationary gel (Figure 6.5f). The electric field generated between two main electrodes, and stabilized with electronically regulated additional sets of electrodes, can be reoriented after predetermined intervals. In principle, this leads to separation of DNA molecules in the gel analogous to that obtainable with devices using purely electronic switching. ROFE has the same resolution characteristics as CHEF, but has additional engineering problems associated with the need to precisely reorient the electric field by mechanical means. This can be especially problematic when short pulse times (a few seconds or less) are required. Depending on the electrophoresis conditions, ROFE can separate short (0.5 kbp) or very long (10,000 kbp) nucleic acid molecules with excellent resolution. It is also possible to separate molecules in two dimensions, simply by turning the gel tablet by 90° after the first run and starting a new electrophoresis experiment, possibly under totally different separating conditions. This feature, for example, can be employed to separate closed circular from linear DNA molecules. An example of commercially available ROFE system is the ROTAPHOR apparatus (Whatman Biometra®).

Crossed Field Gel Electrophoresis (CFGE) In the crossed field apparatus, the gel is simply placed on a mobile platform that can be rotated in order to change the orientation of the electric field relative to the DNA. At each change in field direction, a DNA molecule takes off in the new direction of the field by a movement, which is led by what was formerly its back end. The effect of this ratcheting motion is to subtract from the DNA molecule's forward movement, at each step, an amount, which is proportional to its length. This separation method has the practical advantage over some others that the DNA molecules follow straight tracks. As with FIGE and PFGE, CFGE has little effect on the mobility of fragments below the limit mobility of conventional gel electrophoresis. The effect of the procedure is to slow down those molecules which move at the limit mobility, and to increase the limit mobility to a larger size. Within a size range dependent on the pulse regime and voltage, the separation between molecules is roughly proportional to the size difference. Between this size range and the new limit mobility is a range in which the separation is much greater. This latter effect has two practical consequences: (i) It makes size measurements inaccurate in this range unless there is a sufficient density of size markers; and (ii) It gives good resolution of fragments even if molecules are close in size. Thus, for optimal separation or accurate sizing it is important to know the characteristics of the system used for separation.

Zero-integrated Field Electrophoresis (ZIFE) Another more complex approach is ZIFE. It is slower than FIGE but has the ability to resolve larger molecules of DNA. In ZIFE, which is used for the 10–10,000 kbp range, the product of field strength and pulse times is approximately the same for backward and forward pulses.

Pulsed Homogeneous Orthogonal Field Gel Electrophoresis (PHOGE) PHOGE uses a 90° reorientation angle, but the DNA molecules undergo four reorientations per cycle instead of two, which helps in overcoming the problem of poor resolution associated with the use of electric fields at an angle of 90°. Thus, in PHOGE system, migration of DNA molecules proceeds in homogeneous electric fields placed perpendicularly to one another. This is a versatile system, allowing two degrees of resolution in different size ranges. Lower resolution is produced over a wide range of sizes (50 kbp – 1 Mbp), using ramped pulsed times (5–90 seconds), while maximal resolution is achieved for two-fold size differences. Owing to uniform fields, DNA tracks are straight.

Programmable Autonomously Controlled Electrodes (PACE) A computer-driven system known as PACE, designed by Lai *et al.* (1989) seems to be the ultimate PFGE device. The first developed PACE system allows independent regulation of 24 electrodes in a hexagonal array. With this system, all electric field parameters (number and angle of electric fields, voltage gradients, pulse time, and time ramping) are adjustable, and many types of PFGE can be chosen (OFAGE, FIGE, CHEF, PHOGE, etc.). Like CHEF, it is also used to generate voltage clamped homogeneous static fields. In contrast to FIGE, these systems require both a special gel-box with a specific electrode and gel geometry, and the associated electronic control for switching and programming the electrophoresis run. This system is very flexible, and allows the voltage gradient, magnitude, orientation, homogeneity, and duration of the electric field to be controlled precisely, and results in a better resolution of DNA fragments over an extensive size range, i.e., 100 bp to over 6 Mbp. Because of its precise and variable control of all parameters, it can be used to study the relationships of voltage, field angles, switch time, agarose concentration, and temperature with DNA mobility. The ability to alter the reorientation angle between the alternating fields permits an increased speed of separation for large DNA molecules.

6.2.4 Alkaline Agarose Gel Electrophoresis

Alkaline agarose gel electrophoresis was developed by McDonell *et al.* (1977) in Bill Studier's Laboratory at Brookhaven National Laboratory as a replacement for laborious alkaline gradient centrifugation of bacteriophage T7 DNA. In earlier times the technique was used to check for nicking activity in enzyme preparations used for molecular cloning, and to calibrate the reagents used in nick translation of DNA. Presently, the technique is chiefly used for speedy and accurate measurement of the size of first and second strands of cDNA, and the size of the DNA strand after S1 nuclease-catalyzed digestion of DNA:RNA hybrids.

Principle of Alkaline Agarose Gel Electrophoresis

Alkaline agarose gels are run at high pH (in presence of NaOH), which causes each thymine and guanine residue to lose a proton, and thus prevents the formation of H-bonds with their adenine and cytosine partners, respectively. Consequently, the denatured DNA is maintained in a single stranded state, which then migrates through an alkaline agarose gel as a function of its size. Denaturants such as formamide and urea do not work well because these cause the agarose to become rubbery. Denaturing gradient gel electrophoresis (DGGE) is another example, which combines gel electrophoresis with DNA denaturation. This technique allows separation of DNA molecules differing in sequence by only a single base (Exhibit 6.1).

Procedure of Alkaline Agarose Gel Electrophoresis

Agarose solution is prepared in the same way as described in the conventional agarose gel electrophoresis method. The solution is then cooled to 55°C, followed by the addition of 0.1 volume of 10 X alkaline agarose gel electrophoresis buffer (i.e., 10 X running buffer comprising of 500 mM NaOH, 10 mM EDTA, pH 8.0). Note that NaOH is not added to a hot agarose solution, because it causes hydrolysis of agarose. Moreover, EtBr is not added in the gel, as it does not bind to DNA at high pH. The gel is then poured immediately on a gel-casting tray. After the gel is completely set, it is mounted in the gel-tank. It is necessary to ensure that the gel is cooled to room temperature before installing it in the gel-tank. The sample is prepared by dissolving damp precipitates of DNA in 10–20 µl of running buffer, and adding 0.2 volumes of 6 X

Exhibit 6.1 Denaturing gradient gel electrophoresis (DGGE)

DGGE is combination of gel electrophoresis with DNA denaturation that allows separation of DNA molecules differing in sequence by only a single base. It is applicable to ds DNA fragments of a few hundred bases, which are often generated by PCR. The technique involves separation of two strands (melting) of ds DNA in stages by exposure to a gradient of increasing denaturation. As the ds DNA melts in stages, discrete zones known as 'melting domains' become unpaired along the piece of DNA. Partially melted DNA migrates more slowly as compared with ds DNA through an agarose gel during electrophoresis. This is because the small streamlined DNA fragments open into larger structures that get caught in the agarose meshwork. In practice, denaturation is done by a combination of moderately high but constant temperature (usually between 50 and 65°C) plus chemical denaturation with a mixture of urea and formamide. A concentration gradient of urea plus formamide is set up across the gel when it is formed. The chemical denaturation gradient may be arranged parallel or perpendicular to the direction of migration of the DNA, based on which there are two types of DGGE, *viz.* parallel DGGE and perpendicular DGGE. Parallel DGGE gives bands in unique positions, and the results resemble those of normal agarose gel electrophoresis. On the other hand, in perpendicular DGGE, a mixture of DNA molecules differing by one or a few base pairs are loaded across the whole gel slab, and a series of sigmoid bands are obtained.

DGGE is widely used in screening natural populations for genetic variability and/or relatedness. In particular, PCR of DNA extracted from soil or other natural habitat followed by DGGE has been used to analyze the phylogenetic relationships of microbial populations without the need to culture living microorganisms. In addition, DGGE is also used in the analysis of mutations, especially the ones that do not change the length of the DNA, for example, base substitutions. For example, DGGE has been used to screen mutations in the genes involved in causing breast cancer, *BRC* A1 and *BRC* A2. Since the melting temperature of each 'melting domain' is sequence specific, the melting profile of DNA with a single base mutation will differ significantly from the wild-type.

In Figures a and b, the results of standard electrophoresis and DGGE, respectively are presented. Under standard conditions, short pieces of ds DNA migrate fairly quickly, and are not hindered by the gel matrix. During DGGE, the agarose gel contains a gradient of a denaturing agent that melts the DNA into single strands. Close to the sample well, the concentration of denaturing agent is low, and the DNA does not melt. As the DNA migrates into higher concentrations of denaturing agent, it melts at one end, forming a Y-shaped molecule. This is much more likely to get caught in the agarose meshwork than the nondenatured version. The ease of DNA denaturation depends upon its base sequence. Therefore, wild-type and mutant versions of the same DNA sequence will denature at different places in the denaturation gradient. Small pieces of DNA that differ by as little as one base pair may be separated by this technique.

Figures c and d represent the gels of parallel and perpendicular DGGE, respectively.

In perpendicular DGGE, the gradient of denaturing agent is at right angles to the electric field. A mixture of DNA fragments (wild-type, mutant 1 and mutant 2 in this example) is loaded into a long well that spans the entire gel. On the left where the amount of denaturant is low, none of the three DNA fragments melt, and hence all migrate together. In the middle of the denaturation gradient, the three DNA fragments have melted to differing extents. Mutant 1 denatures most easily, and therefore migrates slower than the other fragments. Mutant 2 DNA melts the least and so migrates the farthest. On the far right of the gel, where the concentration of denaturant is greatest, all three DNA fragments are fully denatured, and run together as one band of ss DNA.

In parallel DGGE, the denaturing gradient runs from top to bottom, in the same direction as the electric field. Here, a mixture of mutant 1 and wild-type are loaded in one well, and a mixture of wild-type and mutant 2 in the second. Mutant 1 denatures most easily and migrates the least. Mutant 2 is most resistant to denaturation, and hence migrates the farthest.

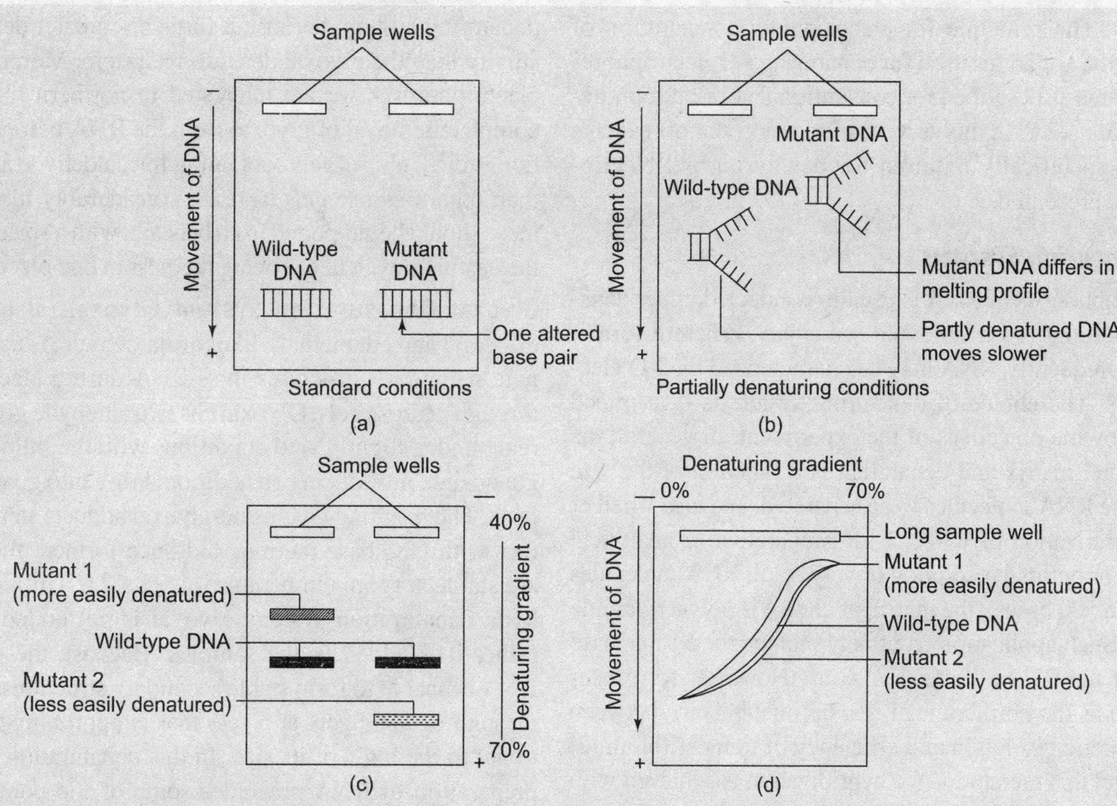

Source: Clark D (2005) Nucleic acids: isolation, purification, detection, and hybridization. *Molecular Biology*, Elsevier Inc., USA, pp. 567–598.

alkaline gel-loading buffer [300 mM NaOH, 6 mM EDTA, 18% (w/v) Ficoll (Type 400, Pharmacia), 0.15% (w/v) bromocresol green, and 0.25% (w/v) XC FF]. [Note that the exposure of the samples to the alkaline conditions in the gel is usually enough to render the ss DNA and it is not strictly necessary to denature the DNA with NaOH before electrophoresis. Also note that in alkaline agarose gels, bromocresol green is used as tracking dye instead of BPB. This is because incubation at high pH (alkaline conditions) quickly bleaches BPB]. Alternatively, if the volumes of the original DNA samples are small (<15 μl), 0.5 M EDTA (pH 8.0) may be added to a final concentration of 10 mM, followed by the addition of 0.2 volumes of the 6 X alkaline gel-loading buffer. Note that in solutions of high pH, Mg^{2+} forms insoluble $Mg(OH)_2$ precipitates that entrap DNA, hence it is important to chelate all Mg^{2+} with EDTA before adjusting the electrophoresis samples to alkaline conditions. The prepared samples are loaded into wells, and electrophoresis is performed at <3.5 V/cm. Note that alkaline gels draw more current than neutral gels at comparable voltages, and heat up during the run, hence alkaline agarose gel electrophoresis is carried out at a voltage lower than conventional gel electrophoresis. When bromocresol green migrates ~0.5–1 cm into the gel, the power supply is turned off, and a glass plate is placed on top of the gel. This is essential to slow the diffusion of bromocresol green dye out of the gel, and to prevent the gel from detaching and floating in the buffer. Electrophoresis is then continued until bromocresol green migrates ~$2/3^{rd}$ of the length of the gel. The gel is then stained with EtBr and visualized, or subjected to Southern hybridization, or wet gels are directly autoradiographed.

6.2.5 Denaturing Agarose Gel Electrophoresis

Virtually all RNA molecules contain short double helical regions consisting of standard Watson–Crick base pairs (A:U and G:C) as well as weaker G:U base pairs. These intramolecular base pairing interactions lead to the formation of secondary structures or 'hairpins'. As electrophoretic separation of nucleic acids in an electric field is also based on molecular conformation, these secondary structures impede electrophoretic separation. The identical species of RNA molecules exhibiting varying degrees of intramolecular base pairing migrate towards the anode at different rates, resulting in the smearing of distinct RNA transcripts, and hence making the analysis difficult. Thus, it is necessary that certain strategies be devised to prevent secondary structure formation, and allow co-migration of like species of RNA. This is routinely done by denaturing the RNA samples by heating at 65°C, or treating with a denaturant before loading into the wells, and performing the electrophoresis of RNA under denaturing conditions. The principle for electrophoretic separation of RNA, and the apparatus used for denaturing gel electrophoresis is the same as described for conventional gel electrophoresis in Section 6.2.2. In this section, a brief account of reagents or buffers specifically required for denaturing gel electrophoresis is presented.

RNA Denaturing Systems

The electrophoresis of RNA is usually conducted under denaturing conditions, so as to avoid secondary structure formation. Less frequently, RNA may be characterized by 2-D electrophoresis. The choice of denaturing system is determined primarily by the objectives of the experiment. In general, the choice of gel matrix and denaturant is a function of the size range of the RNA molecules to be separated, and also whether the characterization of RNA is for preparative or analytical purpose. For optimal resolution of very small RNA molecules (fewer than 500 bases), the matrix of choice is polyacrylamide (3–20%). Such applications routinely include the S1 nuclease assay, and the RNase protection assay. However, for procedures such as the northern analysis, optimal balance between electrophoretic resolution and efficiency of transfer (blotting) from the gel to a membrane for hybridization is achieved with a 1–1.2% denaturing agarose gel.

The most commonly used RNA denaturants for agarose gel electrophoresis include formaldehyde and glyoxal/dimethyl sulfoxide (DMSO). Historically, methylmercuric hydroxide system was the first to be used to denature RNA for agarose gel electrophoresis. Though methylmercuric hydroxide is the most efficient RNA denaturant, it is very toxic; hence its use is abandoned. For polyacrylamide gels, formamide and urea are used successfully as RNA denaturants. Note that formaldehyde and formamide are carcinogenic and teratogenic, and should be used with caution. Each of these denaturing systems has unique characteristics, safety requirement, and distinct advantages and limitations. In both denaturation systems, BPB migrates slightly faster than the 5S rRNA, whereas XC FF migrates slightly slower than the 18S rRNA. Methodologies pertaining to the use of formaldehyde and the glyoxal/DMSO system are presented in this section.

Formaldehyde Denaturation System To ensure that the RNA migrates only with respect to molecular weight, samples of RNA are denatured with formaldehyde or formamide before electrophoresis, and formaldehyde is added to the gel to maintain the denatured state during electrophoresis. Formaldehyde is routinely supplied as a 37% stock solution (12.3 M), containing 10–15% methanol as a preservative. Formaldehyde oxidizes when it comes into contact with air. As such, the pH of formaldehyde should be checked before each use, and the pH must be greater than 4.0 for RNA work. This is because at pH below 4.0, RNA experiences severe degradation. Most labs performing RNA electrophoresis use the formaldehyde denaturing system, because it furnishes greater detection sensitivity than the glyoxal denaturing system. Moreover, if after electrophoresis, we are interested in northern blotting, then complete removal of glyoxal from the RNA before hybridization is difficult. As gels containing formaldehyde are less rigid than other agarose gels, and are considerably more slippery, these should be supported from beneath, with a spatula or within the casting tray, when moving them from one place to another.

Glyoxal Denaturation System Glyoxal (also known as diformyl and ethanedial), like formaldehyde, is used to eliminate secondary structures in ss RNA during electrophoresis through agarose gels. Glyoxal has two aldehyde groups, which react under slightly acid conditions with the imino groups of guanosine, introducing an additional ring into guanosine residues. The resulting guanosine-glyoxal adducts sterically interfere with G:C base pairing, and once formed, these adducts are stable at room temperature at pH $\leq$ 7.0. Unlike formaldehyde denaturation system, glyoxal is not added to the gel, rather it is added to the samples. Because the glyoxalated RNA is unable to form stable secondary structures, it migrates through agarose gels at a rate that is approximately proportional to the $\log_{10}$ of its size. In this denaturation system, the preparation of RNA precludes some of the potential health hazards associated with the use of formaldehyde. Moreover, the banding of the glyoxalated RNA tends to be sharper than that observed in formaldehyde denaturing system. Although glyoxalation of RNA has advantages over formaldehyde denaturation, even then denaturation by formaldehyde is preferred. This is because glyoxalated RNA exhibits decreased electrophoretic mobility compared with an identical sample denatured with formaldehyde. Furthermore, the preparation of glyoxal, and subsequent handling of denatured samples, and northern membranes are major disincentives.

RNA Size Standards

The size of RNA of interest can be measured accurately only when markers of known molecular weights are loaded in the gel. There are two basic approaches for the size determination of RNAs of interest, *viz.* external size standards, and internal size standards. Similar to DNA size standards, key decisions regarding the amount to use, positioning in the gel, the size range, and the method of visualizing should be made with RNA size standards as well.

External Size Standards External molecular weight standards are nucleic acid molecules of known size that are electrophoresed in an unoccupied lane on the same gel as experimental samples. The marker is loaded asymmetrically, the reasons for which are discussed in Section 6.2.2. These standards are commercially available. There are two common examples of external size standards: (i) In RNA gels, a mixture of 5–10 ss RNAs that are produced by *in vitro* transcription of cloned DNA templates of known lengths is used. Because these markers are single stranded, these must be denatured by the same method used to denature the samples so as to prevent secondary structure formation. Moreover, care must be taken to protect RNA molecular weight markers from degradation, as these are just as susceptible to RNase degradation as the experimental RNA. One problem associated with these markers is the contamination of template DNA on which the RNA transcripts have been synthesized, and its associated plasmid sequences. Moreover, vector sequences present in the probe used in Northern hybridization may hybridize to these remnants, generating unwanted discrete bands, or more commonly, a smear on the autoradiogram. Furthermore, many of these templates exhibit limited sequence homology with plasmid vectors commonly used for molecular cloning; and (ii) Glyoxalated DNA and RNA molecules are electrophoretically equivalent, and hence small DNA molecules of known molecular weights can be used reliably as markers. Thus for glyoxal systems, a glyoxalated triple digest (*Pst* I, *Pvu* II, *Hin*d III) of the pBR322 plasmid containing three double stranded fragments of 0.779, 1.545, and 2.037 kbp length are used. If the hybridization probe is cloned into pBR322, or any vector that shares sequence homology with pBR322, then the markers will act as target sequences, hybridize with the probe, and produce a permanent marker image, autoradiographic or otherwise, during detection. Size markers generated in this fashion are also extremely well-suited for Southern analysis, and can be prepared from any plasmid or other DNA whose sequence is known.

In order to avoid problems due to strong signal of external standards, it is judicious to leave a blank lane in the gel between the size standards and the experimental samples, or to use nanogram quantities of the RNA marker. However, it is necessary to load about 2 μg marker per lane to have sufficient mass to photograph the marker after staining the gel. If no vector sequences are involved, or if it is known that the probe is cloned into a vector that does not hybridize to components of the molecular weight standard, then 4–5 μg per lane can be loaded. If vector sequences are involved, as little as 15 ng of marker will hybridize very efficiently to complementary vector sequences, and render sharp bands at the detection stage.

These external standards are also used in molecular weight determination. For this, a size calibration curve is constructed by plotting the $\log_{10}$ of the sizes of the RNA standards against distance migrated from the well. A curve generated in this fashion is then used to ascertain the sizes of RNA species observed directly in the gel, or that may subsequently be detected by nucleic acid hybridization.

Internal Size Standards Internal molecular weight standards are RNA species of known size that are actually part of the RNA experimental sample, and hence are not loaded separately. Most common examples are the 28S and 18S rRNAs. Depending on the method of lysis, a properly denatured RNA sample may also show a third, higher molecular weight band on post-electrophoresis staining. This less obvious transcript is the 45S rRNA, a nuclear precursor to the 28S and 18S rRNAs that accumulates in the cytoplasm. These highly abundant rRNAs (28S and 18S) are very useful as internal controls. However, their utility as markers is completely dependent on knowledge of their sizes either from the literature or from prior empirical determination. Thus, one should not use these rRNA species as internal size standards unless their molecular weights are known for the species under investigation. There is considerable variation in their sizes even among the mammals. There is extensive variability in the sizes of 28S rRNA in contrast to 18S rRNAs. The molecular weights of these rRNAs are known from various species, and in general, 18S rRNAs range from 1.7–1.9 kb, whereas the range of 28S rRNAs is between 4.6–5.2 kb. As these markers are a part of the sample, these are automatically denatured during preparation of the sample for electrophoresis. As a supplementary tool, the location of the tracking dye, most commonly bromophenol blue, may be used as an indicator of the progress of the migration of the samples. While electrophoresis is performed, bromophenol blue generally runs with the leading edge of the migrating sample. Another dye, xylene cyanol, runs just behind the 18S rRNA in 1.2% agarose gels. Xylene cyanol is not used routinely because of its inconvenient co-migration with the 18S rRNA, often obscuring the fluorescence coming from the bands of interest.

These rRNA species not only act as molecular weight standards, but also give an indication of the integrity of the sample by visual inspection, and provide evidence that an equivalent mass of total RNA has been loaded into each experimental lane. Note that equalization is otherwise done by densitometry or image analysis software. Having knowledge of the sizes of the 28S and 18S rRNAs, the distance migrated by each of these species can be measured, and a size calibration curve is constructed by drawing a straight line joining the two points.

Owing to the drawbacks of each system, the reliability of molecular weight data is enhanced when measurements are derived from both internal and external standards, especially if the RNA molecules of interest are very large or very small. This is true for two fundamental reasons. (i) While using external standards, the logarithmic separation of molecules in a gel may not be linear over the entire length of the gel. Very large or very small RNAs that end up outside the linear range of the gel may be sized inaccurately; and (ii) While using internal standards in conjunction with northern transfer for molecular weight determination, uneven transfer of sample material during northern transfer may occur due to entrapment of air bubbles between the gel and the filter membrane, or between the gel and the wicking material.

Formaldehyde Agarose Gel Electrophoresis

The solutions required for formaldehyde agarose gel electrophoresis are: (i) 10 X formaldehyde gel-loading buffer comprising of 50% glycerol (diluted in DEPC treated water),

10 mM EDTA (pH 8.0), 0.25% w/v BPB, and 0.25% w/v XC FF; and (ii) 10 X running buffer comprising of 200 mM 3-(*N*-morpholino) propane sulphonic acid (MOPS) (pH 7.0), 20 mM sodium acetate, and 10 mM EDTA (pH 8.0).

The procedure for formaldehyde agarose gel electrophoresis is almost the same as that described for conventional gel electrophoresis in Section 6.2.2. The steps involved in formaldehyde agarose gel electrophoresis are described in this section, and only those points that differ from conventional gel electrophoresis are elaborated. (i) The first step is the preparation of agarose gel containing 2.2 M formaldehyde. Thus, to prepare 100 ml agarose gel containing 2.2 M formaldehyde, 1.5 g of agarose is added to 72 ml of sterile water. Agarose is melted by boiling in a microwave oven or boiling water bath, resulting solution is cooled to 55°C and then 10 ml of 10 X running buffer and 18 ml of deionized formaldehyde are added to the solution. The solution is mixed thoroughly and poured into the gel-casting tray in a fume-hood. Note that 1.5% agarose gel is suitable for resolving RNA molecules in the 0.5–8.0 kb size range, and still larger RNA molecules are separated on gels cast with 1.0 or 1.2% agarose; (ii) Second step involves the preparation of samples. Thus, 2 μl of 10 X formaldehyde gel-loading buffer is added to each sample and the tubes are stored in ice; (iii) In the next step, the gel is mounted in gel-tank, and electrophoresis is performed under 1 X running buffer at 5 V/cm for 5 min. Then the RNA samples as well as RNA size standards are loaded into the wells. Note that for parallel analysis of samples, it is incumbent to ensure that the samples are normalized, meaning that equal quantities of RNA are loaded in each lane of the gel. The most widely accepted procedure for equalization (or normalization) is to measure the concentration of each RNA sample spectrophotometrically, and then to load equal (A_{260}) μg quantities of RNA in each lane on a gel. When purified poly (A) RNA is available, loading 2–5 μg per lane is sufficient to detect many low abundance mRNAs. In other cases, however, it may be desirable or even necessary to electrophorese 15–20 μg of total cellular RNA or total cytoplasmic RNA without prior selection of the poly (A) RNA; and (iv) In the final step, electrophoresis is performed under 1 X running buffer at 4–5 V/cm until BPB migrates to $3/4^{th}$ the way along the length of the gel. Note that application of higher voltage leads to smearing of bands. The gel is then stained with EtBr (or SYBR Green II or GelStar®), and visualized under UV illumination.

Glyoxal Agarose Gel Electrophoresis

The solutions required for glyoxal agarose gel electrophoresis are: (i) Glyoxal reaction mix comprising of 6 ml DMSO, 2 ml deionized glyoxal, 1.2 ml 10 X running buffer, 0.6 ml 80% glycerol in water, and 0.2 ml EtBr (10 mg/ml in water). As glyoxal oxidizes very rapidly, glyoxal stock solutions (40% glyoxal = 6 M) must be deionized to neutral pH before use. If not deionized, the oxidation product, glyoxylic acid, causes fragmentation of the RNA sample. For convenience, small aliquots of deionized glyoxal are stored at –20°C in tightly capped tubes, which are thawed at the time of use, and the unused portion is discarded; (ii) A stock solution of 10 X BPTE electrophoresis buffer (or 10 X running buffer) comprising of 100 mM PIPES [piperazine-1,4-*bis*(2-ethanesulfonic acid)], 300 mM *Bis* Tris-Cl, and 10 mM EDTA (pH 8.0) and a final pH of ~6.5 is used. The running buffer is diluted to 1 X before use. Note that *Bis* Tris-Cl, a component of running buffer, is *Bis* (2-hydroxyethyl) iminotris (hydroxymethyl) methane hydrochloride. It is a zwitterionic buffer, with a pK_a 6.5 at 25°C, and effective as a buffer between pH 5.8 and 7.2. Alternatively, glyoxal gels can also be electrophoresed in 1 X MOPS buffer; and (iii) RNA gel-loading buffer comprising of 95% deionized formamide, 0.025% w/v BPB, 0.025% w/v XC FF, 5 mM EDTA (pH 8.0), and 0.025% w/v SDS.

The procedure involves setting up of glyoxal denaturation reaction in a sterile microfuge tube by adding 1–2 μl RNA (up to 10 μg), 10 μl glyoxal reaction mix, and incubating at 55°C for 60 min. The sample is chilled for 10 minutes in ice water, and then centrifuged briefly to settle down the fluid. Note that RNA sample can also be heat denatured before electrophoresis at 65°C for 10 min. Gel is prepared, and mounted in gel-tank containing 1 X running buffer. Sample preparation involves the addition of 1–2 μl of RNA gel-loading buffer to the glyoxylated RNA samples, and loading them without delay into the wells of the gel. Size marker is also loaded into one of the wells. After loading is complete, electrophoresis is performed at 5 V/cm until the BPB reaches $3/4^{th}$ the way along the length of gel. Here also, higher voltage leads to smearing of bands. After staining of gel with EtBr or SYBR Green II or GelStar® dyes, RNAs are visualized under UV illumination.

6.2.6 Polyacrylamide Gel Electrophoresis (PAGE)

Raymond and Weintraub (1959) introduced cross-linked chains of polyacrylamide as matrix for electrophoresis (for details see Section 6.2.1). These serve as electrically neutral matrices for separation of ds DNA fragments according to size (under non-denaturing conditions), and ss DNAs according to size and conformation (under denaturing conditions).

Types of PAGE

There are two types of polyacrylamide gels: (i) Denaturing polyacrylamide gels; and (ii) Non-denaturing polyacrylamide gels. Denaturing polyacrylamide gels are used for the separation and purification of ss DNA fragments. These gels are polymerized in the presence of an agent (urea and/or less frequently formamide) that suppresses base pairing in nucleic acids, and denatured DNA migrates through these gels at a rate that is almost completely independent of its base composition and sequence. These gels are used for the isolation of radiolabeled DNA probes, analysis of the products of S1 nuclease digestion, and analysis of the products of DNA sequencing reactions. On the other hand, non-denaturing polyacrylamide gels are used for the separation and purification of fragments of ds DNA, and to detect protein–DNA complexes.

The principle for separation of nucleic acids through polyacrylamide gels has already been described in Section 6.2.1. As a general rule, ds DNAs migrate through non-denaturing polyacrylamide gels at rates that are inversely proportional to the $\log_{10}$ of their size. However, electrophoretic mobility is also affected by their base composition and sequence, so that duplex DNAs of exactly the same size may differ in mobility by up to 10%. An account of denaturing PAGE used in DNA sequencing is presented in Chapter 19, and in the following section, the apparatus required, and methods used for preparing and running polyacrylamide gels are presented.

Requirements for PAGE

In this section, a brief description of electrophoretic assembly, and other reagents/buffers required in PAGE are discussed.

Equipments Required Electrophoresis assembly comprises of electrophoresis tank or gel-box (with or without in-built thermal sensor), glass plates (one large and one notched), comb, spacers (usually Teflon, sometimes Lucite varying in thickness from 0.5 mm to 2.0 mm), binder or 'bulldog' paper clips, gel-sealing tape, electrical leads, power supply, and gel temperature monitoring strips (optional). A typical vertical electrophoretic unit used for PAGE is shown in Figure 6.6a. Other requirements include micropipettes with drawn-out plastic tips, petroleum jelly, syringe, and gloves.

Solutions/Buffers Required Various solutions/buffers required in polyacrylamide gel electrophoresis are:

Polyacrylamide gel Acrylamide:bisacrylamide (29:1 % w/v solution, stored in brown bottles), TEMED, ammonium persulfate (APS), and running buffer (TBE or TAE) for the preparation of gel. The cross-linker *N,N'*-methyle-nebisacrylamide is usually included at 1/30th the concentration of acrylamide monomer, and the percentage of acrylamide monomer to be used in preparing the gel is determined by the size of the DNA fragments to be resolved. Sometimes the acrylamide solution is deaerated, to reduce the chance of formation of air bubbles when thick gels (>1 mm) are poured, as well as to reduce the amount of time required for polymerization.

6 X gel-loading buffer This buffer is required for the preparation of sample prior to loading (for details see Section 6.2.2). Note that gel-loading buffer does not contain EtBr, or any other DNA binding dye, rather the gel is stained after electrophoresis.

Gel running buffer required for pouring and running gels It comprises of 0.5 X or 1 X TBE or 1 X TAE. Note that 0.5 X or 1 X TBE is used at low voltage (1–8 V/cm) to prevent denaturation of small fragments of DNA by Joulic heating, and while using 1 X TAE, the gel is run more slowly. Though 1 X TAE can also be used, it is not as good as TBE as it does not provide as much buffering capacity as TBE. Thus for electrophoresis runs greater than 8 hours, 1 X TBE buffer is recommended to ensure that adequate buffering capacity is available throughout the run.

Gel staining dyes The gel staining dyes include EtBr, SYBR dyes, etc. (for details see Section 6.2.2). Alternatively, a more sensitive silver staining may be performed. There are three general types of silver staining: (i) Photo development; (ii) Diammine staining; and (iii) Nondiammine staining. Like conventional photography and fluorography, photo development method uses photonic energy to reduce silver ions to the metallic silver element.

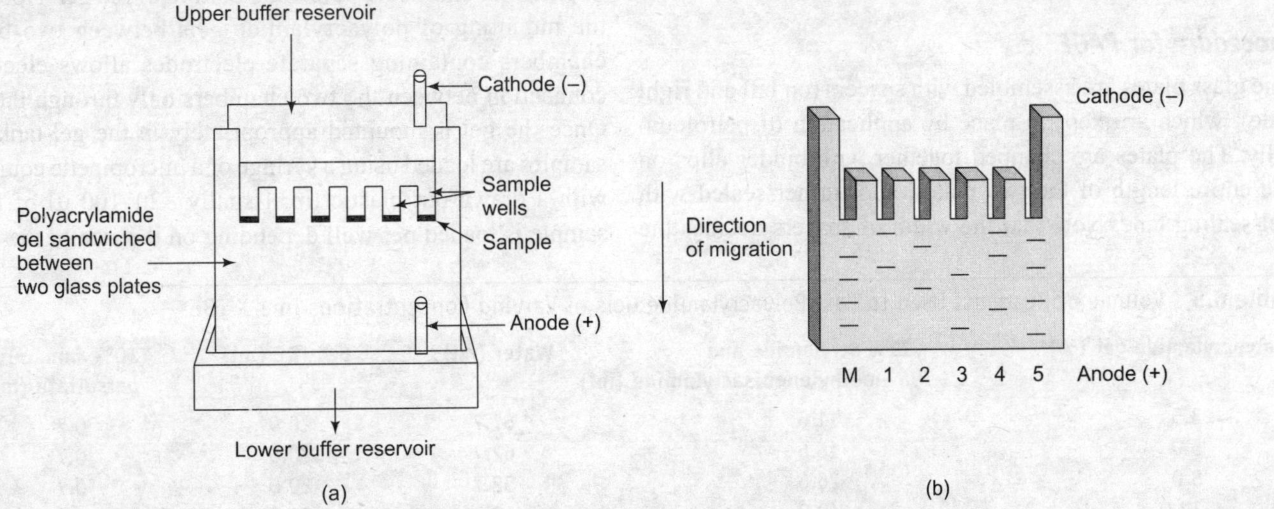

Figure 6.6 Polyacrylamide gel electrophoresis (PAGE); (a) Polyacrylamide gel electrophoresis assembly, (b) A gel showing the result of electrophoretic separation

The method is simple and rapid, and has sensitivity similar to that achieved by conventional staining with EtBr. Diammine staining uses ammonium hydroxide to generate silver diammine complexes, which bind to the nucleic acid. Silver ions are then liberated from the complexes by decreasing the concentration of ammonium ions with citric acid. The liberated silver ions are finally reduced to metallic silver by formaldehyde. The basic method established by Johansson and Skoog (1987) is both rapid, and reasonably sensitive (0.5–2 ng DNA/band), however, greater sensitivity (0.1–1 ng DNA/band) can be achieved using the modifications described by Vari and Bell (1996). Note that ammoniacal silver salts are potentially explosive, and must be handled with great care. Nondiammine staining involves acid fixation of DNA, sensitization of DNA with glutaraldehyde, impregnation of the gel or membrane with silver nitrate, i.e., soluble silver ions (Ag^+) at weakly acidic pH, and reduction of bound silver ions to metallic silver by alkaline formaldehyde. The initial deposit of the insoluble metallic silver initiates the autocatalytic deposit of more silver, manifested ultimately by visualization of the sample. Many variants of this technique have been developed, but most of them suffer to a greater or lesser extent from the problem of staining of DNA bands as gray or dog-yellow against a variable background of brownish surface staining. This problem can be minimized by carefully monitoring the gel during development so as to clearly discriminate specifically stained DNA (as band) and background staining of the gel. Differential reduction of silver ions is improved by adding sodium thiosulfate to the alkaline formaldehyde solution. Note that thiosulfate removes silver ions from the gel surface by forming soluble complexes with silver salts. When working well, the nondiammine staining method is simple and sensitive enough to detect a band containing 2–5 ng of DNA. A nondiammine silver stain is marketed by Promega as part of the Silver Sequence DNA sequencing kit.

Procedure for PAGE

The glass plates are assembled with spacers (on left and right side), which are kept in place by application of petroleum jelly. The plates are clamped together with binder clips on the entire length of the two plates, and further sealed with gel-sealing tape. Note that the width of spacers decides the thickness of the resulting gel. The thicker the gel, the hotter it will become during electrophoresis. Overheating results in 'smiling' bands of DNA. Thicker gels must be used when preparing large quantities of DNA (>1 μg per band), however, in general thinner gels are preferred, as these produce the sharpest and flattest bands of DNA. In the next step, the gel solution of required concentration is prepared by mixing the gel ingredients in the amounts indicated in Table 6.5, and poured through syringe/pipette into the space between the two plates. Note that unlike agarose gels, polyacrylamide gels cannot be cast in the presence of EtBr because the dye inhibits polymerization of acrylamide. Immediately after pouring the solution, a comb is inserted into the gel, clamped in place with bulldog paper clips in such a way that the upper part of the teeth of the comb remains slightly higher than the top of the glass. Precautions should be taken to avoid entrapment of air bubbles under the teeth while pouring the gel. The glass plates are then kept against a test tube rack at an angle of 10° to the bench top for 30–60 min at room temperature so as to allow the polymerization of acrylamide. Note that such positioning decreases the chance of leakage, and minimizes distortion of the gel. When polymerization is complete, a Schlieren pattern is formed just beneath the teeth of the comb. After solidification, the comb is carefully pulled out from the polymerized gel. The wells are washed thoroughly with 1 X running buffer, otherwise small amounts of acrylamide solution trapped by the comb will polymerize in the wells, producing irregularly shaped surfaces that will give rise to distorted bands of DNA. The gel-sealing tape is removed from the bottom of the gel, and the gel is mounted in the electrophoresis tank in such a way that the notched plate faces inward, i.e., toward the buffer reservoir. Now both the upper and lower reservoirs of electrophoresis tank are filled with 1 X running buffer. At this stage care is taken to prevent entrapment of air bubble below the bottom of the gel. Note that the mounting of polyacrylamide gels between two buffer chambers containing separate electrodes allows electrical connection between the two chambers only through the gel. Once the gel is mounted appropriately in the gel-tank, the samples are loaded using a syringe or a micropipette equipped with a drawn-out plastic tip. Usually ~20–100 μl of DNA sample is loaded per well depending on the size of the well.

Table 6.5 Volume of Reagents Used to Cast Polyacrylamide Gels of Varying Concentrations in 1 X TBE*

Polyacrylamide gel (%)	29% acrylamide and 1% *N,N*′-methylenebisacrylamide (ml)	Water (ml)	5 X TBE (ml)	10% ammonium persulfate (ml)
3.5	11.6	67.7	20.0	0.7
5.0	16.6	62.7	20.0	0.7
8.0	26.6	52.7	20.0	0.7
12.0	40.0	39.3	20.0	0.7
20.0	66.6	12.7	20.0	0.7

* Some investigators prefer to run acrylamide gels in 0.5 X TBE. In this case, the volumes of 5 X TBE and water are adjusted accordingly.

Note that not much time should be taken for loading, otherwise the samples will diffuse from the wells. Moreover, the syringe should be washed thoroughly between each loading. After loading of all the samples is complete, the electrodes are connected through electrical leads to a power pack (positive electrode connected to the bottom reservoir), power is turned-on, and electrophoresis is performed at voltages between 1–8 V/cm. If electrophoresis is carried out at a higher voltage, differential heating in the center of the gel may cause bowing of the DNA bands, or even melting of the strands of small DNA fragments. Therefore with higher voltages, gel-boxes that contain a metal plate or extended buffer chamber should be used to distribute the heat evenly. Many types of gel apparatuses are equipped with thermal sensors that monitor the temperature of the gel during the run. The electrophoresis is performed until the marker dye migrates the desired distance. After completion of electrophoresis, electric power is turned-off, and the leads are disconnected. The glass plates with the gel are taken out, and one of the glass plates is detached before subjecting the gel to staining. For staining step, the gel and its attached glass plate is gently submerged in the EtBr solution for 30–45 min. at room temperature. The staining solution should be just enough to cover the gel completely. The gel is then removed from the staining solution, washed with water, and excess liquid is carefully removed from the gel surface with tissue paper. Alternatively, staining with SYBR dyes, or silver staining may also be done. The gel is covered with a piece of Saran wrap, and any air bubbles or folds in the Saran wrap are smoothened out with the broad end of a comb, or a crumpled tissue paper. In order to visualize EtBr or SYBR dye stained gel, a piece of Saran wrap is placed on the UV transilluminator and the gel is inverted on it. The glass plate is carefully removed, leaving the gel on the Saran wrap. A typical polyacrylamide gel showing separation of large and small molecules is shown in Figure 6.6b. Note that in another method, bands of radioactive DNA separated by PAGE may be detected by autoradiography.

6.3 ELUTION OF DNA FRAGMENTS FROM GEL

It is sometimes necessary to isolate a particular DNA fragment for subsequent manipulation, for example, to use a specific fragment as a probe for a blot, or to clone a particular fragment from a genomic digest, or to directly inject into fertilized mouse embryos or transfect into murine embryonic stem cells, etc. The fragment of interest can be isolated by digesting the total DNA, cloning all of it, and identifying the clones that contain the fragment of interest; or, can enrich for the DNA fragment of interest prior to cloning. Enrichment for the DNA fragment of interest is generally accomplished by elution of the fragment from an agarose or polyacrylamide gel. While this sounds simple, it is often messy, and impurities in some agaroses can cause inhibition of the enzymes to be used later in the process of cloning. There are many methods for eluting DNA from a piece of gel for this purpose; some common ones are discussed in this section. In most of the methods used for isolation of specific DNA fragments from the gel, first a small piece of agarose/polyacrylamide containing the band(s) that need to be purified is excised out from the gel kept under or above a UV light source using a scalpel or razor blade. This gel piece containing the band(s) of interest is hereafter called gel slice. Note that the size of the gel slice should be as small as possible, so as to reduce the amount of contamination of DNA with inhibitors, to minimize the time required for elution of DNA from the gel, and to ensure an easy fit into the dialysis tubing (in dialysis method). The gel slice is then treated by one of a number of procedures to purify the DNA from the contaminating agarose/polyacrylamide, and EtBr stain, and the eluted DNA is generally free from contaminants that inhibit enzymes, or that are toxic to transfected or microinjected cells. The amount and quality of the purified fragment is finally checked by polyacrylamide or high-resolution agarose gel electrophoresis.

Organic Extraction

This method employs separation of DNA fragments according to size by electrophoresis through LMP agarose followed by cutting out of gel slice containing the band of interest. This gel slice is placed in a clean, disposable plastic tube, and ~5 volumes of low melting temperature (LMT) elution buffer is added to the tube. The gel slice is melted by incubation at 65°C for 5 min, and the resulting solution is cooled to room temperature. Equal volume of equilibrated phenol (pH 8.0) is added to the tube, the mixture is vortexed for 20 sec, and the aqueous phase is recovered by centrifugation at 4,000 g (5,800 rpm in a Sorvall SS-34 rotor) for 10 min at 20°C. Note that the white substance at the interface is agarose. The aqueous phase is extracted once with phenol:chloroform and once with chloroform. The resulting aqueous phase is transferred to a fresh tube, and 0.2 volume of 10 M ammonium acetate and 2 volumes of absolute EtOH are added. The mixture is incubated for 10 min at room temperature, and then DNA is recovered by centrifugation at 5,000 g (6,500 rpm in a Sorvall SS-34 rotor) for 20 min at 4°C. The DNA pellet is washed with 70% EtOH, and dissolved in an appropriate volume of TE (pH 8.0). The procedure works best for DNA fragments ranging in size from 0.5–5.0 kbp. Yields of DNA fragments outside this range are usually lower, but often are sufficient for many purposes. DNA purified from LMP agarose gels by this method is suitable for use in most enzymatic reactions of molecular cloning. Occasionally, more demanding experiments such as transfections, injections, or multifragment ligations require highly purified DNA, which require further purification of LMP agarose purified fragments by chromatography on DEAE-Sephacel columns or specialty resins available commercially as prepacked columns.

Electroelution into Dialysis Bags

This technique allows the recovery of ds DNAs of a wide range of sizes from slices of agarose and polyacrylamide gels in high yield. The method requires the insertion of individual gel slices into dialysis bags. Thus, a gel slice is placed on a square of Parafilm wetted with 0.25% TBE. A dialysis bag, sealed at one end, is filled with 0.25 X TBE, and the gel slice is transferred into it. The dialysis bag is sealed from the other side by placing the dialysis clip just above the gel slice without the entrapment of air bubbles, and immersed into a shallow layer of 0.25 X TBE in a horizontal electrophoresis tank. The dialysis bag is tied to a glass rod or pipette to prevent it from floating and to maintain the gel fragment in a parallel orientation to the electrodes. This is followed by the passage of electric current (7.5 V/cm) through the bag for 45–60 min, and the movement of DNA fragment out of the gel slice is visualized by hand-held long wavelength UV lamp. Under the mentioned conditions, ~85% of a 0.1–2.0 kbp DNA fragment is electroeluted from the gel slice. The polarity of current is reversed for 15–60 sec to release the DNA that is stuck to the dialysis bag. The electric current is switched off, the bag is recovered from the electrophoresis chamber, dialysis clip is removed, and the buffer surrounding the gel slice is transferred to a plastic tube. The DNA is purified from this buffer either by passage through DEAE-Sephacel, or by chromatography on commercial resins, or by extraction with phenol:chloroform and standard EtOH precipitation. Electroelution is more time consuming, tedious to perform, and is inefficient for recovering large numbers of fragments. Often this method requires vigilance, and the recovered DNA sample requires an additional concentration/desalting step. However, the technique works well, and is probably the best technique for recovery of large (>5 kbp) fragments of DNA.

Enzymatic Digestion with Agarase

In this technique, agarase enzyme is used to hydrolyze (or liquefy) agarose polymer to oligosaccharide and disaccharide subunits thereby releasing the DNA. The method is extremely gentle, and is particularly useful for the recovery of high molecular weight DNAs extracted from PFGE. However, the method also works well for recovery of smaller DNA fragments from agarose gels run in constant electrical fields.

The method involves incubation of the gel slice for 30 minutes at room temperature in 20 volume of gel equilibration buffer or agarase buffer [10 mM *Bis* Tris-Cl (pH 6.5), 5 mM EDTA (pH 8.0), and 0.1 M NaCl] followed by transfer to a fresh tube containing gel equilibration buffer (agarase buffer) approximately equal in volume to that of the gel slice. The gel slice is melted by incubation at 65°C for 10 min. The resulting solution is cooled to 40°C. DNase-free agarase (e.g., GELase from Epicenter Technologies, β-agarase I from New England BioLabs, and β-agarase I from Calbiochem) is added (1–2 units of agarase/200 μl gel slice), and the sample is incubated at 37–45°C for 1 hour according to the instructions of the manufacturer. During this time, agarose is digested to oligosaccharides and disaccharides. After digestion, agarase is inactivated by heating, or removed by phenol:chloroform extraction. Alternatively, agarase-digested agarose, which is in a liquid state, is passed through an ultrafiltration membrane as a small molecule. If desired, the DNA solution may be used directly at this stage for ligation, restriction enzyme digestion, and transformation. Alternatively, the DNA fragments may be purified and concentrated. Small DNA fragments (<20 kbp) are subjected to phenol extraction and EtOH precipitation, while large DNA fragments (>20 kbp) may be purified and concentrated by dialysis in a drop-dialysis apparatus (e.g., Microdialysis System, Life Technologies) (for details see Chapter 1). Note that to prevent shearing, larger DNA fragments are not subjected to phenol extraction, vortexing, or EtOH precipitation. Moreover, the inclusion of 30 μM spermine and 70 μM spermidine in the dialysis buffer can enhance recovery of large DNA fragments.

'Optimized Freeze-Squeeze' Method

Another method for purification of DNA fragments from LMP agarose gels is the 'optimized freeze-squeeze method'. In this method, gel slice is kept in tube, and elution buffer [0.5 M sodium acetate (pH 7.0) and 1 mM EDTA (pH 8.0)] is added to a volume until the level of the buffer is a few ml above the level of the gel slice. The tube is then heated in a 65°C water bath until the agarose melts (~5 min). The tube is subjected to fast-freezing by placing in a –70°C freezer for 10 minutes (or in an alcohol/dry ice bath, liquid nitrogen, or on a block of dry ice). The solution is immediately centrifuged for 10 minutes (i.e., before the solution can thaw), and the supernatant is transferred into a new tube. This is followed by the addition of fresh elution buffer to the pellet, and the process of elution is repeated. The two supernatants are pooled, and an equal volume of 1-butanol is added to the supernatant fraction, and mixed thoroughly. The tubes are then agitated for 15 min to remove EtBr from DNA. The top (1-butanol) phase is discarded, and extraction with 1-butanol is repeated one or two more times. The 50 μg of yeast tRNA (10 mg/ml stock) is added, and mixed thoroughly followed by addition of 2.5 volumes of cold 95% ethanol to precipitate DNA at –70°C for 30 minutes or overnight. Note, if PCR or cycle sequencing is to be performed in subsequent steps, tRNA should not be used. The tube is then centrifuged for 15 min, the resulting pellet is washed with 80% EtOH, and rehydrated in 20 μl of 0.1 X TE buffer.

This method has several advantages, for example, (i) It is not as messy as the other methods, which is desirable when working with small amounts of DNA, or where EtBr is used; (ii) It is very simple and fast to perform; (iii) The yields of DNA are excellent; 80–95% of small fragments (<5 kbp) are easily eluted from the gel, and even λ DNA (~50 kbp) can be recovered in yields up to 50%; and (iv) The DNA retains shape after precipitation.

Electrophoresis onto DEAE-Cellulose Membranes

This technique is based on the fact that at low concentrations of salt, DNA binds avidly to DEAE-cellulose membranes. This method involves cutting of the agarose gel kept in electrophoresis tank directly in front of the band of interest to create a slit. In this slit, a piece of paper/membrane (DEAE cellulose or DE-81 paper or membrane) is inserted that will bind DNA. A piece of dialysis membrane is also placed behind (side away from the DNA band) the paper to stop any DNA from passing completely through. Once DEAE-cellulose membrane and dialysis membrane are properly placed, electrophoresis is resumed until the band is stuck into the membrane. The DEAE-cellulose membrane is removed, washed free of agarose in low salt buffer (150 mM NaCl, 50 mM Tris, and 10 mM EDTA), and incubated for about 30 mins at 65°C in high salt buffer (1 M NaCl, 50 mM Tris, and 10 mM EDTA) to elute the DNA. Progress in binding DNA to the membrane and eluting it can be monitored with UV light to detect the EtBr bound to DNA. After elution, DNA is precipitated with EtOH. This procedure is simple, and provides very clean DNA. However, fragments >5 kbp do not elute well from the membrane, and the method yields inconsistent results.

Elution in Agarose Gel Wells

In this method, a well is cut in front of the band of interest in an agarose gel, and filled with buffer. A piece of dialysis membrane is placed on the far side (positive electrode side) of the well to act as a barrier to the DNA but not to the electric current. Then an electric current is applied so that the DNA moves into the well. Any DNA passing through the well is stopped by the dialysis membrane. After the band has moved out of the gel and into the well and membrane, the current is reversed for 15–60 seconds to remove any DNA from the membrane, and the solution in the well is removed. The DNA is then precipitated from this solution.

Affinity Purification Using Glass or Silica

In an environment of high salt and neutral or low pH, DNA binds avidly to glass, silica gel, or diatomaceous earth. This phenomenon is exploited to purify DNA from solutions containing impurities such as agarose. Typically, a gel slice containing the DNA of interest is melted by incubation in a solution containing chaotropic salt (e.g., sodium iodide, guanidine thiocyanate) at a pH of 7.5 or less. Glass powder or silica gel is then added, and the suspension is mixed to allow the adsorption of DNA. The particles are then recovered from the original liquid, and washed by centrifugation and resuspension in high salt EtOH buffer. Finally, the pellet is resuspended in a solution with low or no salt at basic pH, the free particles pelleted by another centrifugation, and the DNA-containing supernatant recovered.

This method is labor intensive, can cause shearing especially of fragments larger than 5 kbp, and leaves glass particles and EtOH in the DNA sample. However, this method is convenient, rapid, requires no organic denaturants or chloroform, and reliable for smaller DNA molecules. Moreover, the glass or silica particles used for this technique can be prepared in-house or, more conveniently, purchased from a number of suppliers. Several kits are now available commercially that contain premade spin columns designed to extract and purify DNA fragments or PCR products from agarose gel. These employ binding of DNA to glass beads, glass fiber matrix, or silica-based membranes in chaotropic salts. Some common examples are Gel/PCR DNA fragments extraction kit from Geneaid, Gel-M™ gel extraction system from Viogene, GFX PCR DNA and gel band purification kit from GE, QIAquick PCR purification and Gel DNA extraction kits from Qiagen, DNA gel extraction kit from Norgen Biotek Corp., PrepEase™ gel extraction kit from USB Corp., Zymoclean™ gel DNA recovery kit, and ZR-96 Zymoclean™ gel DNA recovery kit from Zymo Research, etc.

Crush and Soak Method

The 'crush' and 'soak' technique of DNA recovery from gel involves the transfer of gel slice into a tube, where it is crushed against the wall of the tube with disposable pipette tip or inoculation needle. It involves the transfer of gel slice into a tube, where it is crushed against the wall of the tube with disposable pipette tip or inoculation needle. Alternatively, the gel slice may be further sliced into smaller pieces with razor blade or scalpel prior to placement into a tube. One to two volumes of acrylamide gel elution buffer [0.5 M ammonium acetate, 10 mM magnesium acetate tetrahydrate, 1 mM EDTA (pH 8.0), and 0.1% (w/v) SDS (optional)] is added to the tube. Note that SDS improves the efficiency of recovery, most probably by blocking nonspecific adsorption of DNA to the walls of the tube. However, SDS is tenacious and difficult to remove from the eluted DNA, especially when purifying oligonucleotides on Sep-Pak columns. Hence, it is best to use SDS only when attempting to recover very small amounts (<20 ng) of DNA fragments which are >1 kbp in size, where recovery is already inefficient, and further losses may prejudice the experiment. Other buffers may be substituted for acrylamide gel elution buffer; for example, if the DNA fragment is radiolabeled and is to be used as a probe, hybridization buffer can be used. After addition of acrylamide gel elution buffer, the tube is incubated at 37°C on a rotating wheel or rotary platform. At this temperature, small fragments of DNA (<500 bp) are eluted in 3–4 hours, while larger fragments take 12–16 hours. The sample is then centrifuged at maximum speed for 1 min at 4°C in a microcentrifuge, and the supernatant is transferred to a fresh tube, being extremely careful to avoid transferring pieces of polyacrylamide. The elution steps are repeated with the pellet, and the two supernatants are pooled. DNA is precipitated from the supernatant by the addition of two volumes of EtOH at 4°C, and incubating the solution on ice for 30 min. DNA is recovered by centrifugation at maximum speed for 10 mins at 4°C in a microcentrifuge, and dissolved in 200 µl of TE (pH 8.0). DNA is reprecipitated by adding two volumes of EtOH in the presence of 25 µl of 3 M sodium acetate (pH 5.2). The pellet is carefully rinsed with 70% EtOH, and dissolved in TE (pH 8.0) to a final volume of 10 µl.

The method is very efficient, and the eluted DNA is generally free of contaminants that inhibit enzymes or that are toxic to transfected or microinjected cells. The method requires time but little labor, and results in recovery of <30–90% yield, depending on the size of the DNA fragment. The method is used to isolate both ds and ss DNA fragments from neutral and denaturing polyacrylamide gels, respectively. The method is widely used to isolate synthetic oligonucleotides from denaturing polyacrylamide gels, and occasionally to recover end-labeled DNA for chemical DNA sequencing. DNA recovered from polyacrylamide gels by this method is generally suitable for use as a hybridization probe, as a probe in gel-retention assays, or as a template in chemical sequencing and enzymatic reactions.

6.4 GEL FILTRATION CHROMATOGRAPHY

Gel filtration chromatography, also called size exclusion or molecular exclusion or molecular sieve chromatography, is used for separation of macromolecules (nucleic acid and proteins) on the basis of small differences in size and shape, or in more technical terms, their hydrodynamic volume. The technique (using starch gels) was invented by Lathe and Ruthven (1956), working at Queen Charlotte's Hospital, London, and these scientists later received the John Scott Award for this invention.

This method works independently of changes in the pH, concentration of metal ions, cofactors, etc. This is advantageous because under *in vivo* conditions, these biological macromolecules are controlled by small changes in the environment, and are profoundly affected by these factors. Thus, various solutions can be applied without interfering with the filtration process, while preserving the biological activity of the molecules to be separated. The technique can be readily combined with others that further separate molecules by other characteristics, such as acidity, basicity, charge, and affinity for certain compounds.

Principle of Gel Filtration Chromatography

The technique makes use of microscopic beads packed into columns, which form pores for separation of molecules. Both molecular weight and 3-D shape contribute to the degree of retention of the molecules in the column, and hence separation of different molecules on the basis of size and shape. Thus, when a DNA sample containing small (lower molecular weight) and relatively large molecules (higher molecular weight) is allowed to trickle down the packed column, small molecules diffuse or penetrate into the pores of gel beads from the surrounding medium, thereby spending more time in the maze of channels and pores in the bed, whereas relatively large molecules are prevented from entering into the pores of the beads to the same extent because of their size, and flow around and in between the beads taking a more direct path that involves less time in the beads. This results in a larger volume of liquid accessible to small molecules as compared with larger molecules, which remain confined to the solution outside. This larger volume of liquid to which small molecules are exposed comprises of the liquid surrounding the porous beads as well as that inside the beads. Consequently, the small molecules are delayed in their passage down the column, while large molecules move continuously down the column in the added eluant. The larger molecules thus leave the column first, followed by the smaller molecules in the order of their sizes (Figure 6.7a). In other words, the larger the particles, the less is the overall volume to traverse over the length of the column, and faster the elution.

Requirements for Gel Filtration Chromatography

The basic components of gel filtration chromatography are the gel filtration matrix or chromatographic medium, chromatography column, and the elution buffer.

The gel matrix is the material in the column that actually performs the separation of DNA fragments through formation of pores, which may be depressions on the surface or internal channels through the bead (Figure 6.7b). The gel matrix represents the stationary phase of chromatography. It is chosen for its chemical and physical stability, inertness (lack of adsorptive properties), and a carefully controlled range of sizes of the pores. The most widely used gel matrices are extremely small porous beads made up of insoluble, but highly hydrated polymers such as agarose, polyacrylamide, or dextran (a glucose polymer produced by bacterium *Leuconostoc mesenteroides*). Some common examples are: Sephadex (cross-linked dextran), Bio-Gel P (cross-linked polymers of acrylamide), Bio-Gel A, Sepharose CL, Sephacryl (cross-linked agarose), and Superdex (composite gel in which dextran chains are covalently bonded to a highly cross-linked agarose gel matrix). Note that the gel matrices may be polymers formed by cross-linking to form a 3-D network (e.g., Sephadex, Bio-Gel P, Sephacryl), or by spontaneous gel formation under appropriate conditions (e.g., agarose), or by grafting a second polymer onto a preformed matrix (e.g., Superdex). The porosity of dextran-based gels, sold under the trade name Sephadex, is controlled by the molecular mass of the dextran used, and the introduction of glyceryl ether

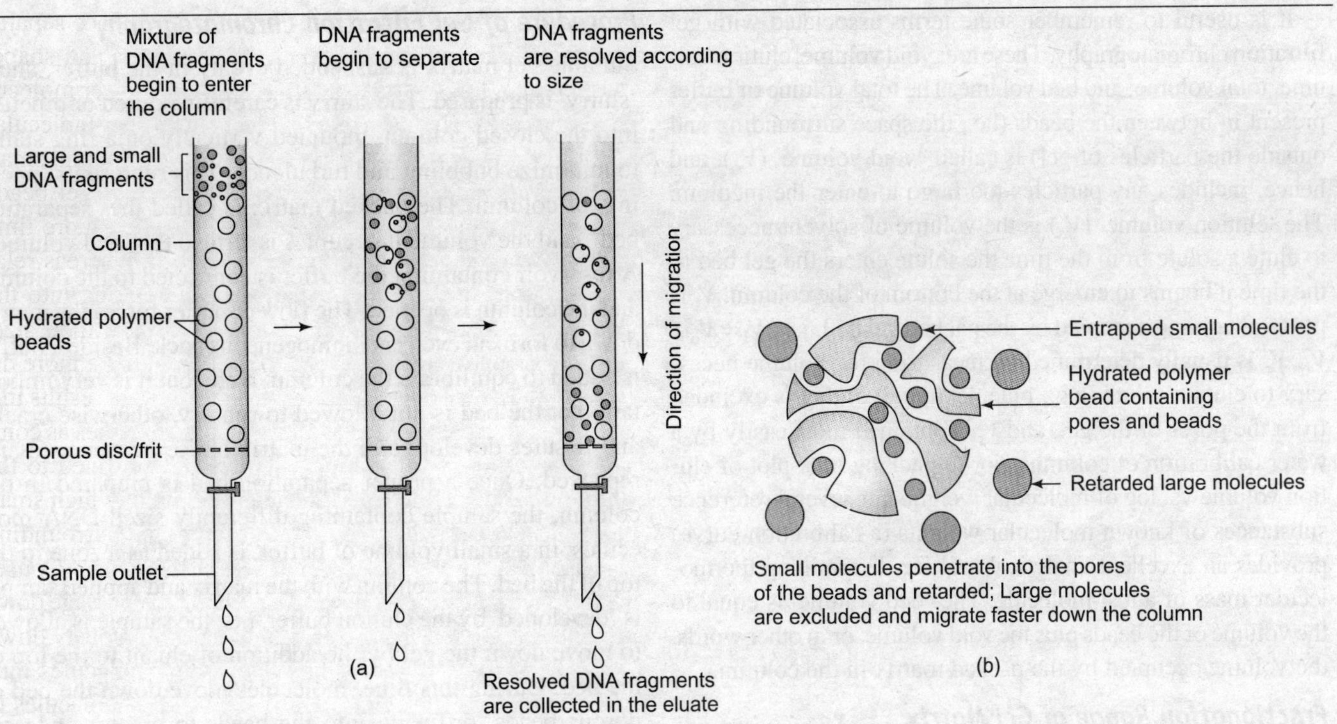

Figure 6.7 Principle of gel filtration chromatography; (a) Separation of DNA fragments of different sizes through gel filtration chromatography, (b) Magnification of gel beads showing the size exclusion process

units that cross-link the hydroxyl groups of the polyglucose chains. The control of porosities of agarose and polyacrylamide-based gels has already been described in Section 6.2.1. Composite gel matrices are of interest since these combine valuable properties from more than one gel-forming system.

Table 6.6 Some Commonly Used Gel Filtration Matrices

Name*	Type	Fractionation range (kDa)
Sepharose 6B	Agarose	10–4,000
Sepharose 4B	Agarose	60–20,000
Sepharose 2B	Agarose	70–40,000
Bio-Gel A-5	Agarose	10–5,000
Bio-Gel A-50	Agarose	100–50,000
Bio-Gel A-150	Agarose	1,000–1,50,000
Sephadex G-10	Dextran	0.05–0.7
Sephadex G-25	Dextran	1–5
Sephadex G-50	Dextran	1–30
Sephadex G-100	Dextran	4–150
Sephadex G-200	Dextran	5-600
Bio-Gel P-2	Polyacrylamide	0.1–1.8
Bio-Gel P-6	Polyacrylamide	1–6
Bio-Gel P-10	Polyacrylamide	1.5–20
Bio-Gel P-30	Polyacrylamide	2.4–40
Bio-Gel P-100	Polyacrylamide	5–100
Bio-Gel P-300	Polyacrylamide	60–400

* Sepharose and Sephadex gels are manufactured and marketed by Amersham Biosciences, while Bio-Gels are manufactured and marketed by BioRad Laboratories.

The properties of several gel matrices that are commonly employed in separating biological molecules are listed in Table 6.6. The choice of gel matrix is based on the required pore sizes, which should be comparable to the sizes of the molecules that are to be separated. Thus, a gel with higher degree of cross-linking will be less porous, and will be used for separation of small sized molecules, and vice versa, for example, Sephadex G-25 is used for separation of smaller sized molecules as compared with Sephadex G-75.

The column is a hollow vertical glass/polypropylene tube with a frit, and elution spout/sample outlet (Figure 6.7a), and holds homogeneously and tightly packed gels designed to have pores of different sizes. The frit is a membrane or porous disc that supports and retains the matrix in the column, but allows water and dissolved solutes to pass through. The sample outlet is fitted with a valve for the collection of samples.

The elution buffer is the mobile phase of the chromatography, and flows through the matrix and out of the column. In gel filtration chromatography, aqueous elution buffer is used as a mobile phase. Note that the technique is called gel permeation chromatography when the mobile phase is an organic solvent. The function of the elution buffer is to provide a means for developing the matrix with the applied sample contained in the column. This means that molecules in the sample are carried by the flow of buffer into the matrix where these get separated gradually. The filtered solution that is collected at the end is known as the eluate.

It is useful to remember some terms associated with gel filtration chromatography. These are: void volume, elution volume, total volume, and bed volume. The total volume of buffer present in between the beads (i.e., the space surrounding and outside the particles of gel) is called 'void volume' (V_o), and hence, includes any particles too large to enter the medium. The 'elution volume' (V_e) is the volume of solvent necessary to elute a solute from the time the solute enters the gel bed to the time it begins to emerge at the bottom of the column. V_t is the total volume occupied by the packed gel bed, i.e., $V_t = V_o + V_e$. V_o is usually determined by measuring the volume necessary to elute a solute (e.g., blue dextran dye) that is excluded from the pores of the gel, and V_t is obtained most easily by a water calibration of column prior to packing it. A plot of elution volume vs. log of molecular weights for several reference substances of known molecular weights (a calibration curve) provides an excellent analytical tool for determining the molecular mass of a test molecule. The 'bed volume' is equal to the volume of the beads plus the void volume, or in other words, the volume occupied by the packed matrix in the column.

Fractionation Range of Gel Matrix

Different types of gel matrices have different internal porosities, and hence different fractionation ranges. Thus, fractionation range of gel matrix is contributed by its pore size, and defines the 'exclusion limit' of the gel, i.e., the molecular mass of the smallest molecule unable to penetrate the pores of a given gel. This gives an indication of the molecular size of a molecule expected to elute at an elution volume equal to the bed volume, and the molecular size of a molecule expected to be totally excluded from the column, and to elute at the void volume.

The gel matrices have varying fractionation range, and hence allow separation of varying spectra of molecular weights (Table 6.6). In general, the exclusion limit of several classes of Sephadex is between 0.7 and 600 kDa, Bio-Gel P is between 0.1 and 400 kDa, and for Sepharose and Bio-Gel A, it ranges up to 1,50,000 kDa. These matrices are therefore selected according to the requirement. For example, a matrix having a fractionation range (in molecular weight) of 1,000–1,00,000 is used to separate molecules lying within this range. This is because molecules in the range of 1,000–1,00,000 enter the beads with varying efficiencies, and get partially or completely separated from one another. This matrix is not used for separation of molecules from each other with an average molecular weight of 1,000 or less, and 1,00,000 or above. The molecules of the size 1,000 or less penetrate the beads completely, and with equal efficiencies, and take the maximum volume of buffer for elution, which is equal to one bed volume. Molecules greater than 1,00,000 do not enter the beads, and are eluted in the void volume. Thus, all molecules having molecular weights of 1,00,000 or greater are not sieved by the matrix, and hence elute at the same time.

Procedure of Gel Filtration Chromatography

Suitable gel matrix is suspended evenly in the buffer, and a 'slurry' is prepared. The slurry is carefully poured or pipetted into the closed column, mounted vertically on a ring stand, to minimize bubbling and turbulence. This represents 'packing' of column. The packed matrix is called the 'separation bed', and the volume it occupies is termed the 'bed volume'. A reservoir containing the buffer is connected to the column, and the column is opened. The flow of buffer forces the matrix down to form an even and homogeneous pack. Besides, buffer is added to equilibrate the column. Note that it is very important that the bed is not allowed to run dry, otherwise cracks and fissures develop, and the matrix has to be removed and repacked. Once a proper separation bed is prepared in the column, the sample containing differently sized DNA molecules, in a small volume of buffer, is added as a zone to the top of the bed. The column with the matrix and applied sample is 'developed' by the elution buffer, i.e., the sample is allowed to move down the gel by the addition of eluant to the top of the bed. During this time, molecules move down the bed of porous beads, diffusing into the beads to greater or lesser degrees. As the pores in the gel matrix, which are filled by the liquid phase, are comparable in size to the molecules that are to be separated, the molecules in the sample carried by the flow of buffer into the gel matrix get gradually separated according to size and shape. In simple manual columns, as the solvent passes through the tube, and drips out from the end of the column, the eluant is collected in constant volumes known as fractions. The more the particles are similar in size, it is likely that these will elute in the same fraction, and not be detected separately. The collected fractions are often examined by spectroscopic techniques to determine the concentration of the particles eluted. Three common spectroscopic detection techniques are refractive index (RI), evaporative light scattering (ELS), and ultraviolet (UV) irradiation. When eluting spectroscopically similar species (such as during biological purification), other techniques may be necessary to identify the contents of each fraction. The elution volume decreases roughly linearly with the logarithm of the molecular hydrodynamic volume (often assumed to be proportional to molecular weight). Columns are often calibrated using 4–5 standard samples to determine the void volume, and the slope of the logarithmic dependence. This calibration may need to be repeated under different solution conditions. More advanced columns overcome this problem by constantly monitoring the eluant. It is thus possible to analyze the eluant flow continuously with techniques such as RI, ELS, UV irradiation, or by viscosity measurements.

Factors Affecting Gel Filtration Chromatography

For obtaining good results, the correct choice of gel, and operating conditions are critical. As with several experimental procedures, there are many variables and factors that can affect the results; for example, extreme levels of temperature, pressure, or pH can lead to adverse results. In real life situations particles in solution do not have a constant, fixed size, resulting in the probability that a particle, which would otherwise be hampered by a pore, may pass right by it. Also, the stationary phase particles are not ideally defined, both molecules and pores may vary in size. Elution curves therefore resemble Gaussian distributions. The stationary phase may also interact in undesirable ways with a molecule, and influence retention times, though great care is taken by column manufacturers to use inert stationary phases. As already discussed the shape and size of the DNA molecule is also a critical factor in the separation. Beads of different sizes allow nucleic acid molecules of different sizes to be effectively separated, hence a proper selection of matrix with respect to chemical nature as well as pore size, or fractionation range, is required. The technique is also used to separate molecules having the same molecular weights but radically different shapes. Molecules with a more compact shape, such as sphere, penetrate the beads more easily than those having elongated rod-like shape. Therefore, rod-like molecules elute before spherical ones of the same molecular weight; if the shape of molecule is markedly non-spherical, it elutes from the column in an unexpected position.

Like other forms of chromatography, the diameter and length of the column also influence resolution during the chromatographic process. Thus increasing the column length enhances the resolution, and increasing the column diameter increases the capacity of the column. Column packing is also an important factor to maximize resolution. An overpacked column can collapse the pores in the beads, resulting in a loss of resolution. An underpacked column can reduce the relative surface area of the stationary phase accessible to smaller species, resulting in those species spending less time trapped in pores.

Applications of Gel Filtration Chromatography

The special ability of gel filtration chromatography to separate macromolecules (nucleic acids and proteins) on the basis of small differences in size and shape makes it an essential fractionation tool. Besides, there are several other applications, which include: (i) During probe construction, this technique is used to separate unincorporated labeled dNTPs from DNA that have been labeled by nick translation or by filling of recessed 3′-termini or by random primer labeling. This is a rapid method used to free DNA from smaller molecules. The two most commonly used gel matrices are Sephadex and Bio-Gel, both of which are available in several porosities. Sephadex G-50 and Bio-Gel P-60 are ideal for purifying DNA larger than 80 nucleotides in length. Smaller molecules are retained in the pores of the gel, while the DNA is excluded, and passes directly through the column; (ii) The technique is also used to desalt macromolecules. This is because the macromolecules do not enter the beads, and are eluted early, whereas salt enters the porous beads, and is eluted later; and (iii) In addition, gel filtration chromatography can provide data such as molecular weights and sizes, molecular weight distributions of homopolymers, and data on binding equilibria. A linear relationship exists between the relative elution volume of a molecule (V_e/V_o), and the logarithm of its molecular mass. Thus, a 'standard' curve of V_e/V_o against $\log_{10}$ molecular mass can be estimated for the column using molecules of known mass. The elution volume of any sample molecule then allows its molecular mass to be estimated by reference to its position on the standard curve.

6.5 DENSITY GRADIENT CENTRIFUGATION

In density gradient centrifugation, the sample is centrifuged through a dense solution either as preformed gradient, or the one capable of forming gradient during centrifugation, so that individual components are separated. Meselson, Stahl, and Vinograd (1957) demonstrated the use of density gradient centrifugation in the separation of DNA molecules.

Principle of Density Centrifugation: Zonal and Isopycnic Centrifugation

A density gradient centrifugation has the advantages of both density gradient and the centrifugation procedures. A density gradient offers convenience in separating all the sizes of molecules in a small tube without sedimenting the large ones to the bottom, and also for getting sharp bands. The application of centrifugal force enhances the rate of sedimentation of a macromolecule by several folds. The acceleration imparted to a molecule by centrifugal force is given by $\omega^2 x$, where ω is the angular velocity of the centrifuge rotor (the angular velocity of the rotor is equivalent to the product of 2π, and the number of revolutions it is undergoing per sec), x is the distance of the molecule from the axis of rotation. The force exerted on the molecule is thus $m\omega^2 x$, where m is the mass of the molecule. The mass is corrected for its buoyancy, which is determined by the density of the liquid medium in which the molecule is present. There are two types of density gradient centrifugation, *viz.* velocity sedimentation and equilibrium sedimentation.

Velocity sedimentation [also called zonal or rate zonal centrifugation] is used to separate molecules on the basis of their differing 'effective' sizes, or on the basis of molecular weights. Thus, the technique is used to separate two DNA molecules of different or same base composition and structure (hence same density), but different chain lengths. For molecules having the same density, and differing only in size, the settling rate, S is related to molecular weight according to the relationship given by Svedberg: $S^2/s^2 = \text{MW/mw}$, where S and s = settling rates for large and small molecules, respectively, and MW and mw = molecular weights of large and small molecules, respectively. Since the force exerted on a molecule is proportional to its mass, the larger DNA molecule sediments faster, and hence occupies a position closer to the bottom of the tube, while the smaller molecule sediments slower, and occupies a position away from the bottom. In velocity sedimentation, the medium used for centrifugation should have a density lower than the density of the molecules to be separated, for example, a solution of sucrose, or Ficoll, or some other not very dense substance that has appreciable viscosity. This is a 'dynamic' situation because the molecules which started at the top, are settling through the gradient at rates roughly proportional to the square roots of their molecular weights. The centrifugation is interrupted after a suitable period of time. Note that if the tube is centrifuged for longer time period, all the molecules will eventually sediment at the bottom of the tube. A hole is punctured at the bottom of the centrifuge tube and samples are collected as drops in different tubes. Measurement of optical density of the fractions at 260 nm is used to detect the presence of DNA.

Equilibrium sedimentation [also called equilibrium density gradient centrifugation, or isopycnic ultracentrifugation (Greek: *isos* = equal; *pyknos* = dense)] is the technique of choice for separation of DNA molecules of different densities but having same or different masses. It is thus suitable for separation of a ds DNA from a triple stranded DNA irrespective of their chain lengths, or two ds DNA molecules with varying G:C contents (one rich in G:C, another rich in A:T). In this technique, the sample is dissolved in a relatively concentrated solution of a dense, fast-diffusing (and therefore low molecular mass) substance, and is spun at high speed until the solution achieves equilibrium. The medium used for centrifugation need to have a density higher than the density of the molecules to be separated, and the gradient used should be very steep. Moreover, at the highest point in the density gradient, density of the medium must exceed the density of the particles of interest. A concentrated solution of cesium chloride (CsCl)/cesium sulfate (Cs_2SO_4)/Cs formate/sodium bromide (NaBr) is generally used. Note that Cs_2SO_4 and Cs formate can be used to form a gradient twice as steep as that achievable with CsCl, the former is preferred for separating DNA molecules with widely different buoyant densities. Commonly employed gradients have densities (or specific gravities) of 1.10 g/cm^3 (top) to 1.80 g/cm^3 (bottom). Unlike velocity gradient centrifugation through sucrose, in this case there is no need to form a density gradient before-hand, rather the gradient is formed by itself during the centrifugation. The high centrifugal field causes the low molecular mass solute to form a steep density gradient in which the sample components band at positions where their densities are equal to that of the solution. The principle behind the operation is that during ultracentrifugation under equilibrium condition, the rate at which the molecules sediment down becomes equal to the rate at which the molecules diffuse back because of thermal motion and Brownian motion. Hence, each molecule floats or sinks to position where its density equals density of solution, after which there is no net sedimenting force on molecules. Consequently, the molecules separated stay at their positions in the centrifuge tube once the equilibrium is attained, irrespective of the time of centrifugation. This is a 'static' system as the molecules come to rest at points in the tube at which these are in density equilibrium with the surrounding solvent at that point. The tube is then punctured to isolate specific bands. Rotors are also available with quartz windows, and the position of bands can be photographed while the centrifugation is on.

Sucrose Zonal Density Gradient Centrifugation

DNA fragments migrate through a linear sucrose gradient at a rate that is dependent on their size. This procedure provides good resolution for DNA fragments of 5–60 kbp in size. Sucrose gradients are also useful for purification of bacteriophage λ vector arms. Using this technique, partially digested genomic DNA can be fractionated for the production of cosmid or bacteriophage libraries, and completely digested DNA can be fractionated for subgenomic DNA libraries.

In this technique, preformed continuous density gradient of sucrose is required, which is prepared in plastic centrifuge tube by placing layer after layer of sucrose solutions of different concentrations, and hence densities (Figure 6.8a). This is done with the aid of a device that mixes concentrated sucrose solution and water in decreasing ratio as the tube is filled, so that the density of the medium is greatest at the bottom, and lowest at the top of the tube. Note that instead of manual layering, programmable density gradient systems (Figure 6.9) (e.g., Isco) can reproducibly form density gradients in centrifuge-ready condition, onto which the sample is loaded directly. The mixture of macromolecules to be resolved is layered on top of the gradient, and centrifuged for up to 20 hours at about 20,000 rpm. Note that centrifugation of the tube in a horizontal position in a rotor at a high speed causes each type of macromolecule to sediment down the density gradient at its own rate, determined primarily by its particle weight but also by its density and shape, in the form of separate bands or zones. Thus, the molecule sinks under gravitational force if the density of the particle is higher than that of the immediate surrounding solution, and continues to sink until a position is reached where the density of the surrounding solution is exactly the same as the density of the molecule. As large and small molecules move down at different rates, separate bands are formed. Usually centrifugation is stopped before equilibrium is reached. The positions of the separated bands are located optically at a wavelength of 260 nm, at which DNA absorbs strongly, or by draining off the contents of the tube carefully through a pinhole in the bottom, and analyzing successive small samples. Note that programmable density gradient systems

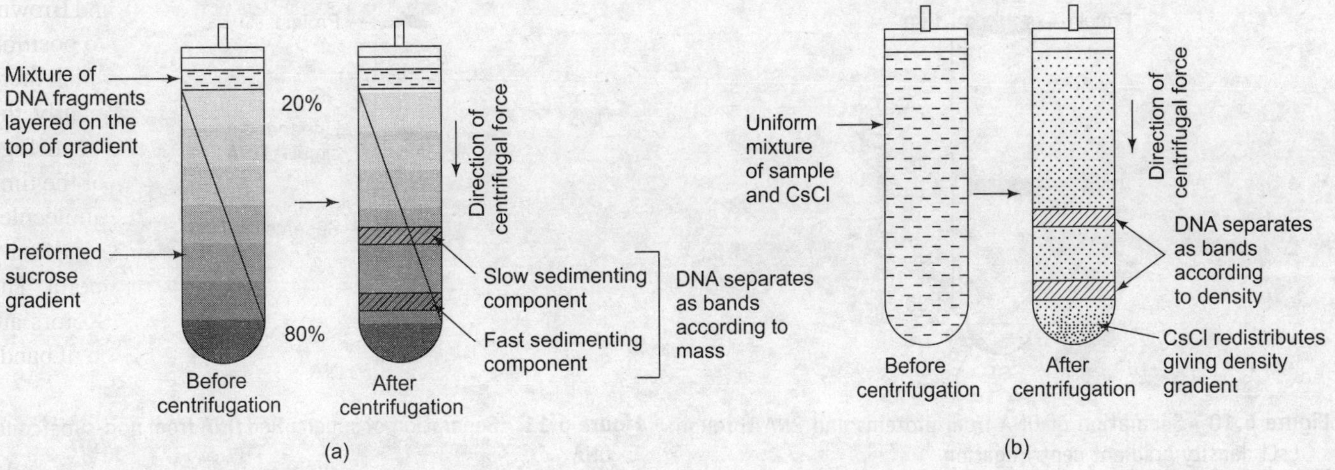

Figure 6.8 Separation of DNA fragment by density gradient centrifugation; (a) Separation of DNA fragments in a sucrose density gradient, (b) Separation of DNA fragments in a CsCl density gradient

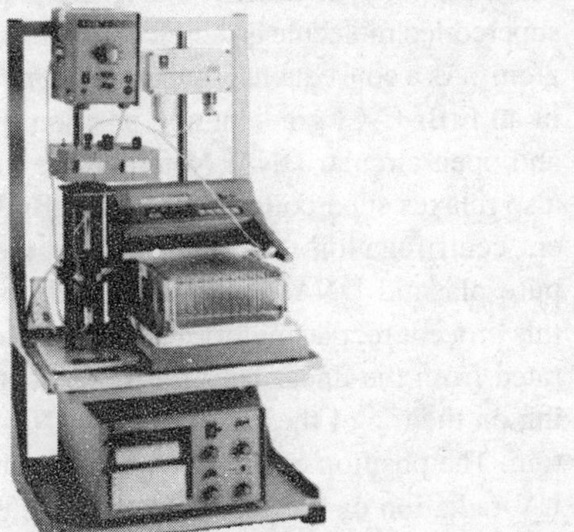

Figure 6.9 Programmable density gradient system

produce a continuous absorbance profile as the gradient is collected in precisely measured fractions. Alternatively, the plastic tube can be frozen solid, and then cut into thin slices for analysis. This technique can also be used for the determination of molecular weights of DNA molecules by comparing their rates of sedimentation with that of a DNA sample of known size, and sedimentation coefficient.

CsCl Equilibrium Density Gradient Centrifugation

CsCl is an example of a medium having the capability of producing 'self-forming' gradient, which works on the principle of isopycnic centrifugation. CsCl, historically the salt used for the classical studies of semiconservative replication, is routinely used to establish gradients with densities ranging up to about 1.8 g/cm^3.

In case of CsCl density gradient centrifugation, a preformed gradient is not required; rather, solid CsCl is added to a nucleic acid mixture, which is then subjected to ultracentrifugation. Under the centrifugal force experienced during ultracentrifugation, the gradient material redistributes, resulting in part from the intrinsic density of cesium salts (Figure 6.8b). The gradient develops because the high centrifugal force pulls the cesium and chloride ions towards the bottom of the tube. Their downward migration is counterbalanced by diffusion, so a concentration gradient is set up, with the CsCl density greater towards the bottom of the tube. These self-forming gradients usually require several hours at ultracentrifugation speeds (above 30,000 rpm) before macromolecules become isopycnically banded; however, the required centrifugation time can be significantly reduced through the use of microultracentrifuge. In the linear gradient generated during the run, components of the sample either sediment or float to their isopycnic locations, based only on their densities. Thus, in contrast to sucrose zonal centrifugation, molecules are separated on equilibrium position, and not by rates of sedimentation. The tube is pierced for fractionating the contents, each fraction containing a different size class of DNA.

Applications of Density Gradient Centrifugation

Both CsCl equilibrium density gradient centrifugation and sucrose zonal density gradient centrifugation have several applications in the field of molecular biology, which include: (i) Density gradient centrifugation is used to separate macromolecules according to their densities, molecular weights, or shapes. CsCl density gradient centrifugation separates macromolecules according to their densities irrespective to their chain length, and hence is used to separate ds DNA from a triple stranded DNA, or even two ds DNAs with varying G:C contents. On the other hand, sucrose density gradient centrifugation is used to separate two DNA molecules having different sizes but same base composition and structures, and hence same densities. Thus, density gradient centrifugation serves as an effective fractionation tool (Figures 6.8a and b); (ii) EtBr–CsCl density gradient centrifugation is also used to separate DNA from RNA and protein, and is an alternative to

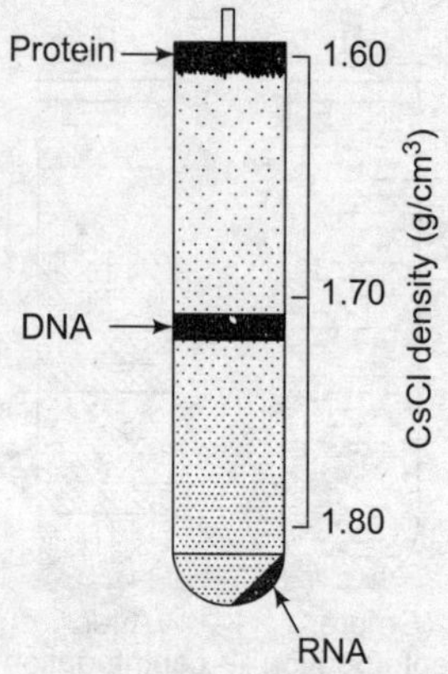

Figure 6.10 Separation of DNA from proteins and RNA through CsCl density gradient centrifugation

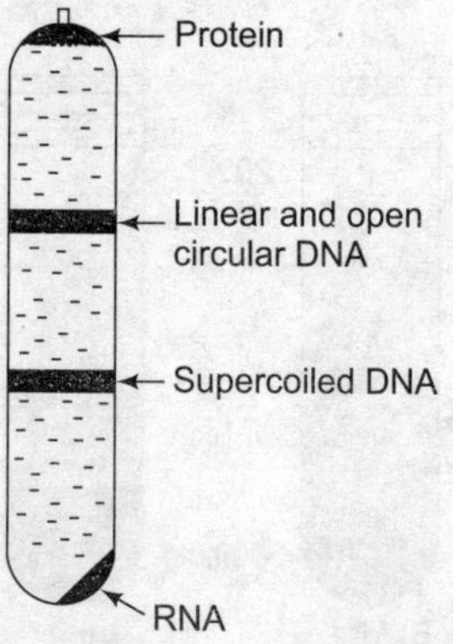

Figure 6.11 Separation of supercoiled DNA from non-supercoiled DNA

phenol extraction and ribonuclease (RNase) treatment for DNA purification. This is because DNA has a buoyant density of about 1.7 g/cm^3, and therefore migrates to the point in the gradient where the CsCl density is also 1.7 g/cm^3. In contrast, protein molecules have much lower buoyant densities, and so float at the top of the tube, whereas RNA being denser than DNA forms a pellet at the bottom (Figure 6.10); (iii) CsCl density gradient centrifugation is also used in the purification of λ phage DNA for removal of PEG precipitate, which contains a certain amount of bacterial debris, possibly including unwanted cellular DNA (for details see Chapter 9). λ particles band in the CsCl gradient at 1.45–1.50 g/cm^3, and can be withdrawn from the gradient as described previously for DNA bands. Removal of CsCl by dialysis leaves a pure phage preparation from which the DNA can be extracted by either phenol or protease treatment to digest the phage protein coat; (iv) EtBr–CsCl density gradient centrifugation is used to separate supercoiled DNA from non-supercoiled molecules (Figure 6.11). EtBr binds to DNA molecules by intercalating between adjacent base pairs, causing partial unwinding of the double helix. This unwinding results in a decrease in the buoyant density, by as much as 0.125 g/cm^3 for linear DNA. However, supercoiled DNA, with no free ends, has very little freedom to unwind, and can only bind a limited amount of EtBr. The decrease in buoyant density of a supercoiled molecule is therefore much less, only about 0.085 g/cm^3. As a consequence, supercoiled molecules form a band in an EtBr-CsCl gradient at a position different from linear and open-circular DNA. Note that the intercalation of EtBr also relaxes supercoiled DNA; (v) EtBr–CsCl density gradient centrifugation is a very efficient method for obtaining pure plasmid DNA. When a cleared lysate is subjected to this procedure, plasmid forms bands at a distinct point, separated from the linear bacterial DNA, with the protein floating on the top of the gradient, and RNA pelleted at the bottom. The position of the DNA bands can be seen by shining UV radiation on the tube, and the pure plasmid DNA is removed by puncturing the side of the tube, and withdrawing a sample with a syringe. EtBr bound to the plasmid DNA is then extracted with *n*-butanol, and the CsCl is removed by dialysis (Figure 6.12). The resulting plasmid preparation is virtually 100% pure, and ready for use as a cloning vehicle; (vi) Besides separating two molecules having different densities, equilibrium sedimentation in CsCl gradient is widely used to determine the buoyant density of DNA molecules, and to get information about base composition. The precise

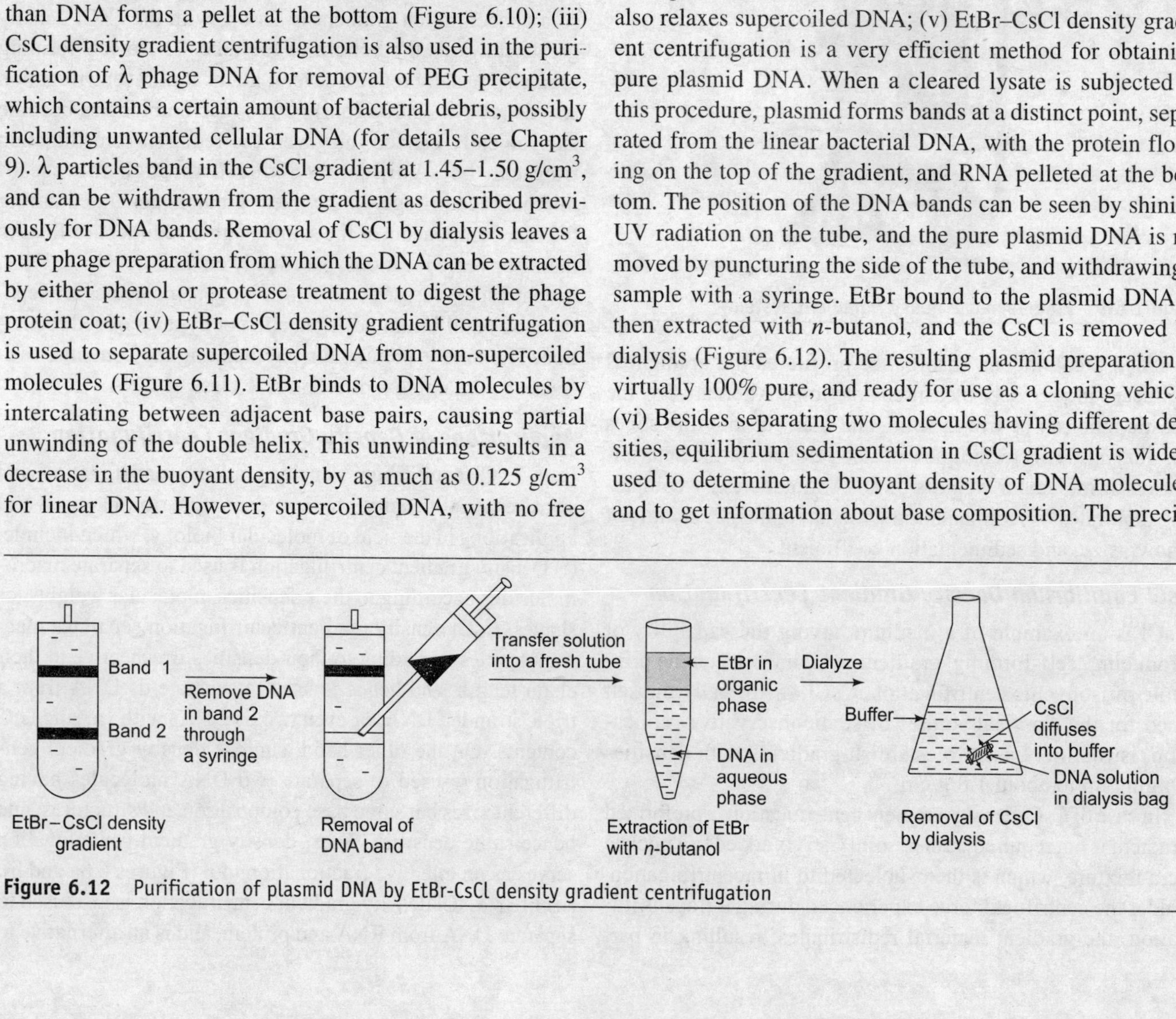

Figure 6.12 Purification of plasmid DNA by EtBr-CsCl density gradient centrifugation

buoyant density of the DNA (ρ) is a linear function of its G+C content represented by the equation: $\rho = 1.66 + 0.098\%$ (G+C). When a concentrated CsCl solution (8.0 M) is centrifuged to equilibrium in a high gravitational field, CsCl becomes distributed in a linear gradient down the tube, at the top of a 1 cm column the density of the solution is about 1.55 g/cm^3 and at the bottom about 1.80 g/cm^3. When DNA is added to the CsCl solution during formation of such a gradient, it concentrates into a stable band at that position in the tube at which its buoyant density is exactly equal to the density of the CsCl solution. The density of the DNA can be calculated directly, or compared with the density of a known standard DNA specimen centrifuged in the same gradient. ssDNA is denser in such a CsCl gradient than ds DNA, which in turn is denser than proteins in general. RNA can be distinguished from DNA since it is denser than either ss or ds DNA. Most important, however, is the fact that buoyant density measurements provide information on the base composition of the DNA specimen, because GC base pairs, which are joined by three H-bonds, are more compact and denser than AT base pairs, which are joined by only two H-bonds; (vii) Denaturing density gradient centrifugation is used to separate mRNA molecules according to size. Thus, in this technique, the mRNA sample is applied to the top of a pre-poured gradient, and subjected to centrifugation. The mRNA molecules separate according to size, and the larger molecules occupy the lowest position in the centrifuge tube due to the highest density. The bottom of the tube is pierced for fractionating the contents. Each fraction thus contains a different size class of mRNA; and (viii) Density gradient centrifugation is very useful in separating and purifying organelles, different cell types, viruses, membranes, etc.

Review Questions

1. Differentiate between the following:
   (a) Agarose and polyacrylamide gels used in gel electrophoresis.
   (b) EtBr and SYBR dyes.
   (c) Formaldehyde and glyoxal denaturation systems.
   (d) Zonal and isopycnic density gradient centrifugation.
   (e) PFGE and CHEF.
   (f) External and internal size standards used in RNA gels as size markers.
   (g) Conventional agarose gel electrophoresis and denaturing agarose gel electrophoresis.
   (h) LMP agarose and standard agarose.
   (i) Continuous and pulsed field gel electrophoresis.
2. Explain the mechanism of separation of DNA molecules through a porous gel under the influence of electric field. What factors affect this separation process?
3. Assign reasons for the following:
   (a) Electrophoresis of RNA is done under denaturing conditions.
   (b) SYBR dyes are not added to the gel before electrophoresis.
   (c) Larger molecules migrate faster than smaller molecules in gel filtration chromatography.
   (d) Conventional agarose gel electrophoresis is not suitable for separation of larger molecules >20 kbp in size.
   (e) EtBr is mutagenic in nature.
   (f) Bromophenol blue is the most suitable tracking dye in agarose gel electrophoresis.
   (g) Horizontal slab gel is submerged completely, but only with a minimum volume just enough to cover the gel.
   (h) Unlike agarose gels, polyacrylamide gels are not cast in the presence of EtBr.
   (i) During elution of DNA from gel, the size of cut gel slice should be as small as possible.
   (j) BPB is not used as tracking dye in alkaline gels.
   (k) Low ionic strength buffer is used in conventional agarose gel electrophoresis.
   (l) High ionic strength buffer and water are not suitable as running buffers in electrophoresis.
4. Elaborate upon the role of CsCl density gradient centrifugation in the purification of plasmid DNA and phage arms.
5. Write short notes on the following:
   (a) Fractionation range of a gel matrix used in gel filtration chromatography
   (b) Field inversion gel electrophoresis
   (c) RNA size standards
   (d) Alkaline agarose gel electrophoresis
   (e) Gel-stains
   (f) Low EEO agarose
   (g) Preparative agarose gel electrophoresis
   (h) Gel-loading buffers
6. Describe the principles of the following techniques used for separation of nucleic acid molecules:
   (a) Density gradient centrifugation
   (b) Gel filtration chromatography
   (c) Polyacrylamide gel electrophoresis
   (d) Denaturing density gradient electrophoresis
   (e) Alkaline agarose gel electrophoresis
7. Give a descriptive account of PFGE and its variants. Also describe the applications of PFGE.
8. Discuss the application of organic extraction and enzymatic digestion with agarase in the elution of DNA fragments from the gel.
9. Give an elaborative account of the electrophoretic technique for speedy and accurate measurement of the size of first and second strands of cDNA synthesized by reverse transcriptase.

References

Carle G.E., M., Frank and M.V. Olson (1986). Electrophoretic separations of large DNA molecules by periodic inversion of the electric field. *Science* **232**:65–68.

Chu, G., D. Vollrath, and R.W. Davis (1986). Separation of large DNA molecules by contour clamped homogeneous electric fields. *Science* **234**:1582–1585.

Johansson, S. and B. Skoog (1987). Rapid silver staining of polyacrylamide gels. *J. Biochem. Biophys. Meth.* **14**: 33.

Lai, E. *et al.* (1989). Pulsed field gel electrophoresis. *Biotechniques* **7**:34–42.

Lathe, G.H. and C.R.J. Ruthven (1956). The separation of substances and estimation of their relative molecular sizes by the use of columns of starch in water. *Biochem J.* **62**:665–674.

McDonell, M.W., M.N. Simon, and F.W. Studier (1977). Analysis of restriction fragments of T7 DNA and determination of molecular weights by electrophoresis in neutral and alkaline gels. *J. Mol. Biol.* **110**:119–146.

Raymond, S. and L. Weintraub (1959). Acrylamide gel as a supporting medium for zone electrophoresis. *Science* **130**:711–712.

Schwartz, D.C. and C.R. Cantor (1984). Separation of yeast chromosome-sized DNAs by pulsed field gradient gel electrophoresis. *Cell* **37**:67–75.

Southern, E.M. *et al.* (1987). A model for the separation of large DNA molecules by crossed field gel electrophoresis. *Nucleic Acids Res.* **15**:5925–5943.

Vari, F. and K. Bell (1996). A simplified silver diamine method for the staining of nucleic acids in polyacrylamide gels. *Electrophoresis* **17**:20–25.

Polymerase Chain Reaction

Key Concepts

In this chapter we will learn the following:

- Setting up a PCR reaction
- Thermostable DNA polymerases
- Primer designing
- Cycle number
- PCR product yield
- Verification of PCR product
- Factors affecting PCR amplification
- Variants of PCR
- Isothermal amplification of DNA
- Applications of PCR

7.1 INTRODUCTION

Polymerase chain reaction (PCR) is a technique that results in exponential amplification of a selected region of a DNA molecule. PCR is widely held as one of the most important inventions of the 20th century in molecular biology. With this technique, small amounts of the genetic material can be amplified (i.e., to make a huge number of copies of a DNA) to be able to identify and manipulate DNA, detect infectious organisms including the viruses that cause AIDS, hepatitis, tuberculosis and detect genetic variations including mutations in human genes and numerous other tasks. The idea of PCR is credited to Kary Mullis, a Research Scientist at a California Biotech Company, Cetus, in 1983. Mullis and five other researchers in the Human Genetics Department at Cetus demonstrated that oligonucleotide primers could be used to specifically amplify defined segment of genomic DNA or cDNA. For this work, Mullis received the 1993 Nobel Prize in Chemistry jointly with Michael Smith.

7.2 SETTING UP PCR REACTION

7.2.1 Constituents of PCR Reaction

The basic components of a PCR reaction are the following: (i) One or more molecules of target DNA; (ii) Two oligonucleotide primers (forward and reverse primers); (iii) All four deoxyribonucleoside triphosphates (dNTPs; where N is A/T/G/C); and (iv) Thermostable DNA polymerase. The above components are added in a PCR tube and mixed well.

7.2.2 Steps in PCR Reaction

After mixing all the components, the PCR reaction comprising of the following three temperature sensitive steps, in the given order, is performed in a thermocycler (Figure 7.1). The temperature profile of PCR is shown in Figure 7.2.

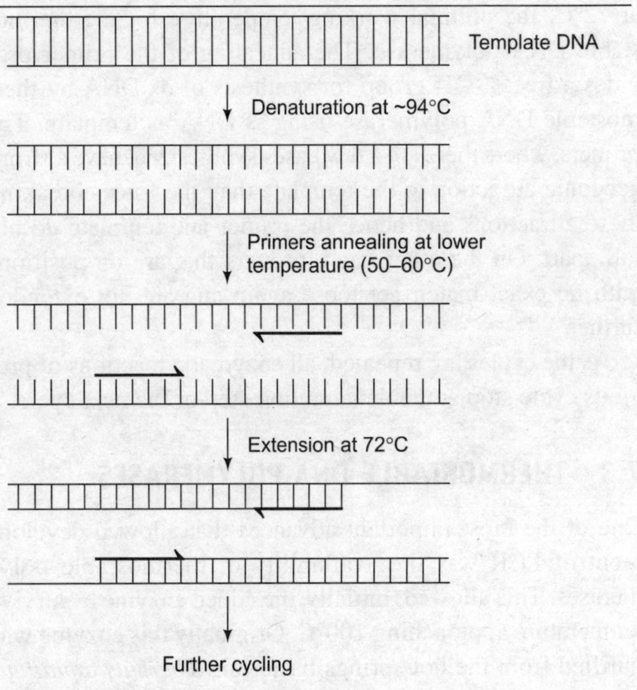

Figure 7.1 Different steps of PCR

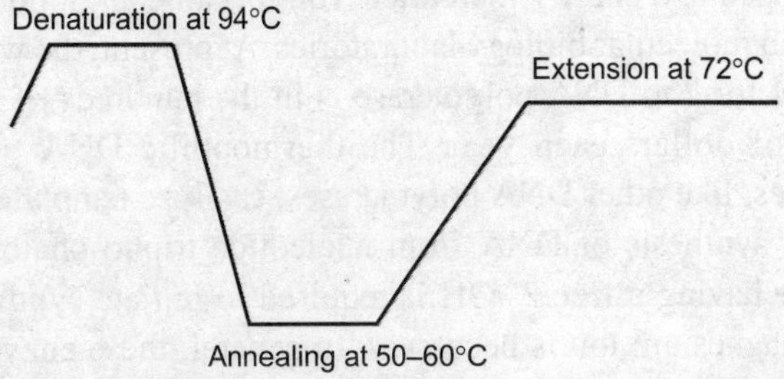

Figure 7.2 Temperature profile of PCR

Denaturation of Double Stranded DNA (ds DNA)

The genetic material is denatured, converting the ds DNA molecules to single strands. This reaction is usually performed at 94°C.

Annealing of Primers to Single Stranded DNA (ss DNA) Template

The complementary base pairing of the custom-made single stranded primers to the complementary regions of the ss DNA molecules is known as annealing. The common choice of temperature range for this reaction is 55–60°C. The primers are jiggling around, caused by the Brownian motion, and H-bonds are constantly formed and broken between the single stranded primers and the ss DNA template. The more stable bonds are formed in case of primers with higher complementarity to the ss DNA template and this annealing lasts for a little longer time interval.

Extension of Primers or Synthesis of ds DNA

Extension signifies the synthesis of DNA by a thermostable DNA polymerase using 3′-OH ends of the primers. It is done at 72°C, the optimal working temperature for the thermostable DNA polymerase. The annealing of the primer provides a free 3′-OH group for synthesis of ds DNA by thermostable DNA polymerase using ss DNA as template. The primers, where there are a few bases synthesized, have a stronger ionic attraction to the template than the forces breaking these attractions and hence the primer and template do not fall apart. On the other hand, primers that are on positions with no exact match get loose again and are not extended further.

As the cycles are repeated, all enzymatic reactions of previous cycle stop at the denaturation step of the next cycle.

7.3 THERMOSTABLE DNA POLYMERASES

One of the most important advances that allowed development of PCR was the availability of thermostable polymerases. This allowed, initially, the added enzyme to survive temperature approaching 100°C. Originally this enzyme was purified from the hot springs bacterium *Thermus aquaticus* (published in 1976). Roughly 10 years later, the PCR was developed and shortly thereafter '*Taq*' became an important word in molecular biology laboratories. At present, the world market for *Taq* DNA polymerase is in the hundreds of millions of dollars each year. The thermophilic DNA polymerases, like other DNA polymerases, catalyze template-directed synthesis of DNA from nucleotide triphosphates. A primer having a free 3′-OH is required to initiate synthesis and magnesium ion is necessary. In general, these enzymes have maximal catalytic activity at 75°C–80°C, and substantially reduced activities at lower temperatures. At 37°C, *Taq* DNA polymerase has only about 10% of its maximal activity.

The crystallization and structure determination of full-length *Taq* DNA polymerase as well as a Klenow like 5′ → 3′ deficient protein have been reported. The structure of polymerase domains of *Taq* DNA polymerase/Klenow are identical and structure of corresponding 3′ → 5′ exonuclease domain is altered in *Taq* DNA polymerase resulting in the absence of this activity. 5′ → 3′ exonuclease active site of *Taq* DNA polymerase is positioned ~70 Å from polymerase active site in a separate domain. Mutational analyses of the residues comprising the 5′ → 3′ exonuclease of *Taq* DNA polymerase have identified several amino acids that alter 5′ → 3′ exonuclease activity. These include: (i) Mutations at Arg 25/Arg 74 to Ala result in reduction in nuclease activity; (ii) Arg 74/Lys 82/Arg 84 mutated to Ala result in 80–90% reduction of 5′ → 3′ exonuclease activity; and (iii) Gly 46 mutated to Asp shows 1,000-fold reduction in 5′ → 3′ exonuclease activity and high processivity. This mutated enzyme is known as AmpliTaq and is used in sequencing reaction.

In addition to *Taq* DNA polymerase, several other thermostable DNA polymerases have been isolated and expressed from cloned genes. These DNA polymerases are thermoactive and thermostable, and are capable of catalyzing polymerization at the high temperature required for stringent and specific DNA. One of the most discussed characteristics of thermostable polymerases is their error rate. Error rates are measured using several different assays, and as a result, estimates of error rate vary, particularly, when different labs perform the assays. As would be expected from first principles, polymerases lacking 3′ → 5′ exonuclease activity generally have higher error rates than the polymerases with exonuclease activity. The total error rate of *Taq* DNA polymerase has been variously reported between 1×10^{-4} and 2×10^{-5} errors per base pair. *Pfu* DNA polymerase appears to have the lowest error rate at roughly 1.5×10^{-6} error per base pair, and Vent is probably intermediate between *Taq* and *Pfu* DNA polymerases.

The most commonly used polymerases are described in the Table 7.1.

7.4 PRIMER DESIGNING

The important considerations during designing of primers are discussed below, which are a key to specific amplification with high yield.

7.4.1 Primer Length

Generally, the optimal length of PCR primers is 18–30 nucleotides (18–30 mer). This length is sufficient for adequate specificity and short enough for primers to bind simply to the template at the annealing temperature. Shorter primers could lead to amplification of nonspecific PCR products.

Table 7.1 Properties of Different Thermostable DNA Polymerases

Enzyme	Source	M.W. (kDa)	Optimum temperature (°C)	Exonuclease activity	Fidelity	Stability (Half-life)	Remarks
Taq DNA polymerase (Natural)	*Thermus aquaticus*	94	74	5′ → 3′	Low	40 min at 90°C	Used in routine PCR experiment
Platinum *Taq* DNA polymerase (Recombinant + proprietary antibody)	*Thermus aquaticus*			3′ → 5′	High (6-fold of *Taq*)		Polymerase activity is blocked at ambient temperature and restored after denaturation
AmpliTaq (Stoffel fragment) (Recombinant)	*Thermus aquaticus*		75–80	None	Low	21 min at 97.5°C (half life)	Processivity is lower than full-length *Taq* DNA polymerase
Hot Tub (AMR) (Natural)	*Thermus ubiquitus*		–	None	Low		
Pyro stase (Natural)	*Thermus flavis*						
Vent (Recombinant)	*Thermococcus litoris*		70–80	3′ → 5′	High (5–15 fold of *Taq*)	Half-life of 23 hours at 95°C	Works with difficult templates: ideal for GC-rich or looped sequences
Deep Vent (Recombinant)	*Pyrococcus* strain *GB-D*		70–80	3′ → 5′	High (5–15 fold of *Taq*)	~500 min at 100°C	
Tth DNA polymerase (Recombinant)	*Thermus thermophilus*	94	75–80	5′ → 3′	Low	20 min at 95°C	PCR, RT-PCR, and primer extension in the presence of Mn^{2+}, reverse transcriptase activity enhanced
Pfu DNA polymerase (Natural)	*Pyrococcus furiosus*	90	75	3′ → 5′	High	~240 min at 95° (half life)	Used in PCR of high fidelity and primer extension
UL *Tma* (Recombinant)	*Thermotoga maritima*		75–80	3′ → 5′	High	50 min at 95°C (half life)	–
Tfl DNA polymerase (Recombinant)	*Thermus flavis*	~94	74	5′ → 3′	Low	40 min at 95°C	Used in PCR and primer extension reactions at elevated temperatures, In the presence of Mn^{2+} reverse transcriptase activity enhanced
Tli DNA polymerase (Recombinant)	*Thermococcus litoris*	90	74	3′ → 5′	High	400 min at 95°C	Primer extension and high fidelity PCR
Pwo DNA polymerase	*Pyrococcus woesei*	90	60–65	3′ → 5′	High	>2 hours at 100°C	Generated blunt-end product best-suited for cloning
Tbr (AM<FINNz) DNA polymerase	*Thermus brockianus*		75–80	5′ → 3′	Low	150 min at 95°C	–

7.4.2 Melting Temperature (T_m)

Melting temperature of a DNA molecule is defined as the temperature at which one-half of the duplex DNA will dissociate to become single stranded and indicates the duplex stability.

The specificity of PCR depends strongly on the melting temperature of the primers. Both primers in a PCR reaction should have similar range of T_m to ensure that these will have the same hybridization kinetics during the annealing step. For the maximum amplification both the primers should have closely matched melting temperatures, a difference of 5°C or more can reduce the efficiency.

The T_m of primer hybridization can be calculated using various formulae:

(i) $T_m = 4\,(G + C) + 2\,(A + T)$ °C

This is the most commonly used formula. This formula was determined originally from oligonucleotide assays, which were performed in 1 M NaCl and appears to be accurate in lower salt conditions only for primers less than ~20 nucleotides in length. T_m is 3–5°C lower than the value calculated from this formula:

(ii) $T_m = 22 + 1.46\,[2 \times (G + C) + (A + T)]$

This formula is useful for primers of 20–35 bases in length.

(iii) $T_m = 81.5 + 16.6\,[\log_{10}(J^+)] + 0.41\,(\%GC) - (600/l)$

where (J^+) = the molar concentration of monovalent cation; l = length of oligonucleotide. This formula is useful for primers of 14–70 bases in length.

7.4.3 Primer Dimer

Usually a large amount of primers are used in PCR compared to the amount of target gene. This increases the chances of formation of primer dimers by intermolecular interactions.

Primer dimers are generally of two types:

Self Self dimer is formed by intermolecular interactions between the two same primers, where the primer is homologous to itself;

Cross If primer dimer is formed by intermolecular interaction between complementary regions of two different primers, i.e., sense and antisense, it is known as cross primer.

The formation of primer dimers prevents the hybridization of primers to the template DNA, thereby reducing the product yield. Moreover, the homology at the 3′-end of primers should be avoided to prevent dimer formation.

7.4.4 GC Content

The GC content of primer should be 40–60%. The presence of G or C bases within the last five bases from the 3′-end of primers helps promote specific binding at the 3′-end due to the stronger bonding of G and C bases. More than three G or C should be avoided in the last 5 bases at the 3′-end of the primer.

7.4.5 Runs and Repeats

Primers should lack stretches of polynucleotide sequences, i.e., runs (e.g., poly dA) or repeating motifs, because these can hybridize at wrong places on the template.

7.4.6 Distance Between Primers

Theoretically, the least distance between primers on template DNA should be 150 bp and utmost 10 kbp. Typically, yield is reduced when the primers extend from each other beyond ~3 kbp.

7.4.7 Secondary Structures

Presence of the secondary structures produced by intermolecular or intramolecular interactions can lead to poor or no yield of the product. These adversely affect primer template annealing and thus the amplification. These greatly reduce the availability of primers to the reaction.

7.5 CYCLE NUMBER

The number of amplification cycles necessary to produce a band visible on a gel depends largely on the starting concentration of the template DNA. It is recommended that to amplify 50 target molecules 40–45 cycles are needed, while 25–30 cycles are required to amplify 3×10^5 molecules to the same concentration. This non-proportionality is called plateau effect (Figure 7.3), which is the decrease in the exponential rate of product accumulation in late stages of a PCR. This may be caused because of: (i) Degradation of reactants (dNTPs, DNA polymerase); (ii) Reactant depletion (primers, dNTPs—former a problem with short products, latter for long products); (iii) End-product inhibition (pyrophosphate formation); (iv) Competition for reactants by nonspecific products; and (v) Competition for primer binding by reannealing of concentrated (10 nM) product.

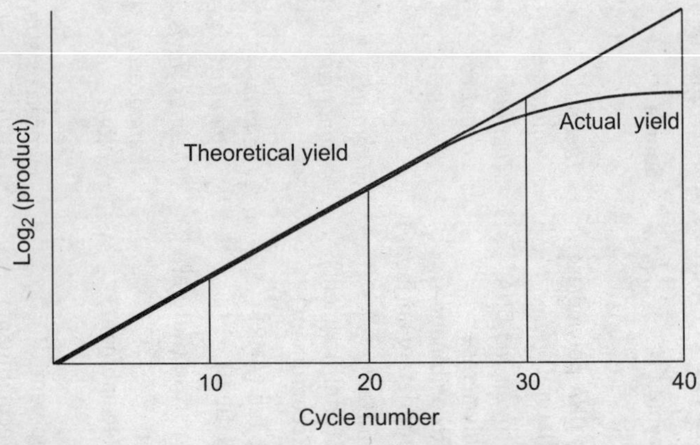

Figure 7.3 'Plateau effect' in PCR amplification

If desired product is not obtained in 30 cycles, small sample (1 μl) of the amplified mix is taken and reamplified 20–30 times in a new reaction mix rather than extending the run to more cycles. In some cases where template concentration is limiting, this can give good amount of product where extension of cycling to 40 or more does not.

7.6 PCR PRODUCT YIELD

Because both strands are copied during PCR, there is an exponential increase in the number of copies of DNA. Suppose at the start of the process, there is only one copy of DNA, the number of copies of DNA after one cycle will be two, after two cycles, it will be four and so on (Figure 7.4). Thus, after n cycles there will be 2^n copies of DNA.

The predicted yield of PCR product can be calculated by a simple 'invested equation' that incorporates terms for the number of target molecules at the onset of reaction, efficiency of each cycle, and the number of cycles. The PCR investment formula can be expressed as:

PCR product yield = (input target amount) × (1 + % efficiency) × number of cycles

Thus, ~26 cycles are required to produce 1 mg of PCR product from 1 pg of a target sequence (million fold amplification) with an efficiency value of 70%, i.e.,

1 mg PCR product = (1 pg target) × (1 + 0.7) × 26

7.7 VERIFICATION OF PCR PRODUCT

After a PCR reaction is complete, the agarose gel electrophoresis is performed to determine the following:

Whether or Not a Product is Formed The possibility that a product is not formed is observed when the quality of the DNA is poor, or when one of the primers does not base pair stably with the template DNA, or when there is too much or too less starting template;

Whether or Not the Product Formed is of Right Size The size of the product is determined by comparison with known molecular weight marker;

Whether or Not a Single Band of Right Size is Formed Multiple bands may be formed when one of the primers base pairs at different sites.

7.8 FACTORS AFFECTING PCR AMPLIFICATION

Several factors affect the PCR reaction at various steps. These include:

7.8.1 Buffer Composition

Recommended buffers for PCR reaction generally contain: (i) 10–50 mM Tris-HCl, pH 8.3; (ii) Up to 50 mM KCl; (iii) 1.5 mM or higher $MgCl_2$; (iv) Primers (forward and reverse) 0.2–1 μM each; (v) 50–200 μM each dNTP; (vi) Gelatin or BSA up to 100 μg/ml; and (vii) Non-ionic detergents (in some cases) such as Tween-20/Nonidet P-40/Triton X-100 (0.05–0.10% v/v). Different buffer constituents play definite roles in the process and affect the reaction at some or the other stage.

[Mg^{2+}] *Taq* DNA polymerase requires Mg^{2+} in buffer to form soluble complex with dNTP that is essential for dNTP incorporation. Mg^{2+} ion concentration affects primer annealing. In the presence of Mg^{2+}, dNTPs, primers and template should be allocated for the reaction, all of which chelate and sequester the cation; among these, dNTPs are the most concentrated, so [Mg^{2+}] should be 0.5–2.5 mM greater than [dNTP]. A titration should be performed with varying [Mg^{2+}] with all new template–primer combinations, as these can differ distinctly in their requirements, even under the same conditions of

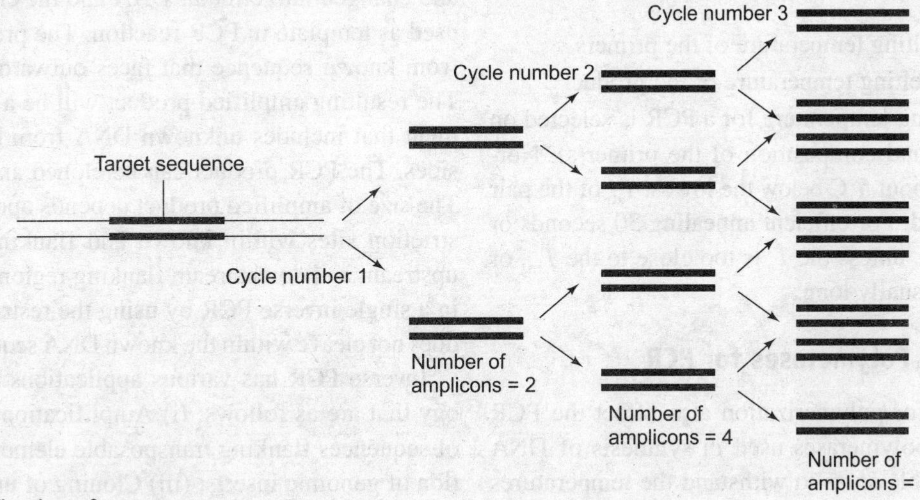

Figure 7.4 Amplification of target gene

concentrations and cycling times/temperatures, it increases the T_m of primer/template interaction. Generally low Mg^{2+} concentration leads to low yields and high concentration leads to accumulation of non-specific products.

[KCl] or [NaCl] Higher than 50 mM KCl or NaCl inhibits *Taq* DNA polymerase, but a little amount is necessary to facilitate primer annealing.

Protein and Detergent Some polymerases do not need added protein and detergent, others are dependent on it. Some enzymes work noticeably better in the presence of detergent, probably because it prevents the natural tendency of the enzyme to aggregate.

7.8.2 Primer and its Concentration

Good primer design is indispensable for successful reactions. The primers anneal to the complementary sequences on the template DNA and thereby determine the boundaries of the amplified product. The designing of a suitable primer is described in Section 7.4. Primer concentration is also important. It should not go above 1 µM except for degenerate primers. 0.2 µM is sufficient for homologous primers.

7.8.3 Nucleotide Concentration

Nucleotide concentration should not exceed 50 µM each. Long products may need higher concentration.

7.8.4 Primer Annealing Temperature

The annealing temperature (T_a) of the PCR reaction is determined on the basis of T_m of primers. Too high T_a will produce insufficient primer–template hybridization resulting in low PCR product yield. If the T_a is too low, it may lead to the formation of nonspecific products because of high number of mismatches. T_a can be calculated by the following formula:

$$T_a = 0.3 \times T_m \text{ (primer)} + 0.7\, T_m \text{ (product)} - 14.9$$

where

T_m (primer) = Melting temperature of the primers

T_m (product) = Melting temperature of the product

Thus, the annealing temperature for a PCR is selected on the basis of length and composition of the primer(s). Normally T_a is chosen about 5°C below the lowest T_m of the pair of primers to be used. For efficient annealing 30 seconds or less time is required, unless the T_a is too close to the T_m, or unless these are unusually long.

7.8.5 Choice of Polymerases for PCR

The enzymes used for polymerization also affect the PCR amplification. The polymerases used in synthesis of DNA should be thermostable so as to withstand the temperatures up to 94°C. The polymerases lacking 3′ → 5′ exonuclease activity generally have higher error rates than the polymerases with exonuclease activity. The properties of various enzymes used for synthesis of DNA have already been discussed in Table 7.1.

7.8.6 Cycle Number

As already discussed in Section 7.5, the number of amplification cycles necessary to produce a band visible on a gel depends largely on the starting concentration of the target DNA.

7.9 VARIANTS OF PCR

PCR is a highly versatile technique. It has been customized in a diversity of ways to suit conditions and applications. Several of the important variations of PCR are as follows:

7.9.1 Inverse PCR

Inverse PCR (or Inverted PCR or Inside-out PCR) is used to amplify unknown DNA that flanks one end of known DNA sequence for which no primers are available. For standard PCR it is required to have information about 5′ and 3′ flanking regions of the DNA fragment of interest while in inverse PCR, information of only one internal sequence of the target DNA is required. It is therefore very useful in identifying flanking DNA sequences of genomic inserts. Similar to other PCR methods, inverse PCR amplifies target DNA using DNA polymerase.

Inverse PCR uses standard PCR, however, it has the primers oriented in the reverse direction of the usual orientation. The template for the reverse primers is a restriction fragment that has been ligated upon it to form a circle.

This technique involves digestion by restriction enzyme of a DNA preparation having known DNA sequence and its flanking sequence. The restriction fragments are self-ligated and changed into circular DNA and the circularized DNA is used as template in PCR reaction. The primers are designed from known sequence that faces outward from that region. The resulting amplified product will be a single linear fragment that includes unknown DNA from both left and right sides. The PCR product can be cloned and then sequenced. The size of amplified product depends upon allocation of restriction sites within known and flanking sequences. The upstream and downstream flanking regions can be amplified in a single inverse PCR by using the restriction enzyme that does not cleave within the known DNA sequence (Figure 7.5).

Inverse PCR has various applications in molecular biology that are as follows: (i) Amplification and identification of sequences flanking transposable elements; (ii) Identification of genomic inserts; (iii) Cloning of unknown cDNA sequences from total RNA; (iv) Construction of end-specific

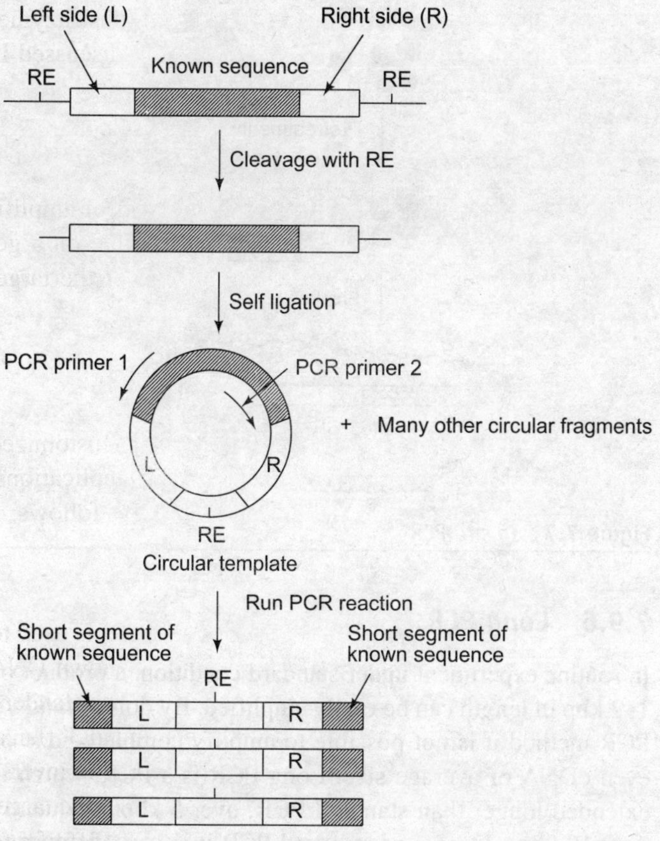

Figure 7.5 Inverse PCR

probes for chromosome walking; and (v) Amplification of integration sites used by viruses and transgenes.

7.9.2 Colony PCR

Colony PCR is used for the screening recombinants from bacterial, bacteriophage or yeast transformation products. Selected colonies of bacteria or yeast are picked with a sterile toothpick or pipette tip from a growth plate. This is then inserted into autoclaved water and denatured for 5 min at 94°C followed by chilling on ice. This heating helps in liberation of template DNA and inactivation of nucleases and proteases. After that, the PCR master mix containing the primers targeted to vector sequences flanking the cloned DNA is added. PCR is then conducted to determine if the colony contains the DNA fragment or plasmid of interest (Figure 7.6).

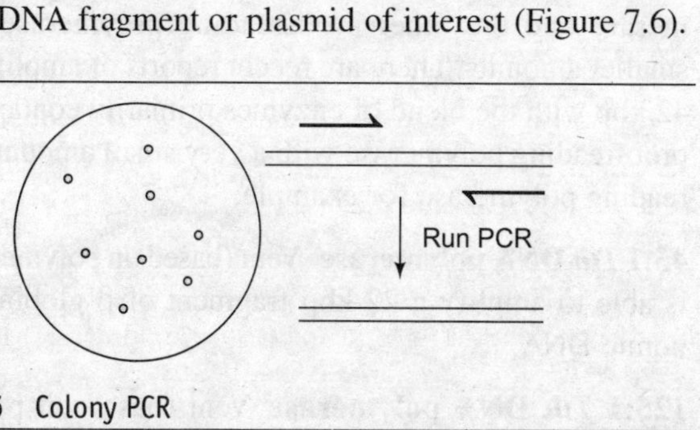

Figure 7.6 Colony PCR

Colony PCR is a fast and reliable method for the screening of recombinants. A number of colonies or plaques can be assayed simultaneously and there is no need to store large number of transformed clones for long periods. This method can easily be used for cDNA library screening and gene disruption in yeasts.

7.9.3 Hot Start PCR

Hot Start PCR permits the inhibition of polymerase activity during setting of PCR reaction, by limiting polymerase activity prior to PCR cycling. Initially, Hot Start PCR was performed manually, i.e., by adding an essential component of reaction mixture only after heating to an elevated temperature. Semi-automated means of performing Hot Start PCR have now been developed, where wax beads are used, which are first melted and then cooled forming a solid wax barrier to separate an essential reaction component from the bulk of the reagents. The primers, Mg^{2+}, buffer and dNTPs can be mixed at room temperature in the bottom of the PCR tube and then covered with melted wax (e.g., Ampliwax PCR Gems from Perkin-ELMER). The wax solidifies on cooling and limits the reagents to bottom of the tube. The remaining components are then added on top of the barrier. Wax layer melts upon heating during the denaturation step and mixes the two aqueous layers, resulting in a fully active reaction. The melted wax floats to the top of the reaction mixture where it acts as a barrier to evaporation.

More recently, an alternate method of automated Hot Start PCR has been developed where *Taq* DNA polymerase-directed monoclonal antibodies, as thermolabile inhibitors of enzymes, are used. At low temperature, i.e., at ambient temperature antigen–antibody interaction results in potent inhibition of DNA polymerase; when temperature rises at the start of thermocycling process, the antibodies are denatured and fully active *Taq* DNA polymerase is released. The preincubation of *Taq* DNA polymerase and antibodies results in highly specific PCR reaction with increased sensitivity. Sometimes ligands are used instead of antibodies, and in presence of ligand *Taq* DNA polymerase is no longer active at room temperature but retains its full activity at temperature >40°C. Nowadays, *Taq* DNA polymerase with antibodies or ligands are commercially available, e.g., AmpliTaq™ Gold DNA polymerase, TaqStart, TthStart, Platinum Taq polymerase.

Hot Start PCR reduces nonspecific amplification and increases the sensitivity, specificity, precision of amplification of low copies of target DNA and yield of PCR product. It is used in multiplex PCR, where multiple primer pairs are used to simultaneously amplify multiple sequences in a single reaction. It is the best-suited method for multiplex PCR as it prevents the dimer formation.

7.9.4 Multiplex PCR

It is often advantageous to amplify all sequences of interest simultaneously in a multiplex reaction. Multiplex PCR has significant template, time and cost saving advantage, especially when a large number of individual sequences need to be analyzed. Single aliquot of DNA or RNA is required rather than an aliquot for each marker to be analyzed. Generally up to eight primer pairs can be used in a standard multiplex reaction, otherwise the yield of some amplicons is reduced and not visible on agarose gel. This type of PCRs is important for forensic application, prenatal diagnosis, and for clinical applications in which the tissue/DNA samples are limited. It is also used for genotyping applications where simultaneous analysis of multiple markers is required.

Multiplex assays can be tedious and time consuming and require lengthy procedure. During optimization, the principal challenge is to overcome primer dimer formation; the T_m of primers should be in similar range, without stable secondary structure (hairpin). The amplified products must be in similar range but can be distinguished from one another on gel electrophoresis, otherwise multiplex reactions are partial in favor of smaller products.

The following points should be considered for setting multiplex PCR: (i) Make sure that all target templates can be amplified in separate reactions by the same PCR program; (ii) Standardize the amount of primer pair (generally 100–400 mM) to get maximum yield in separate reaction; (iii) Optimize the amount of primer pairs in the reaction to attain amplification of target region; (iv) Mg^{2+} concentration is kept higher than optimal range, if the reaction is partial for smaller products; (v) K^+ concentration is kept higher than optimal range, if the reaction is partial for larger products; and (vi) If all of the products are not amplified, attempt with increasing concentration of template and primers.

7.9.5 *In situ* PCR (IS-PCR)

It is a collective term used to describe primer driven amplification of a DNA or RNA template by PCR and its subsequent detection within the histological tissue section or cell preparation. *In situ* PCR amplification can be performed on fixed tissue or cells or on a slide (Figure 7.7).

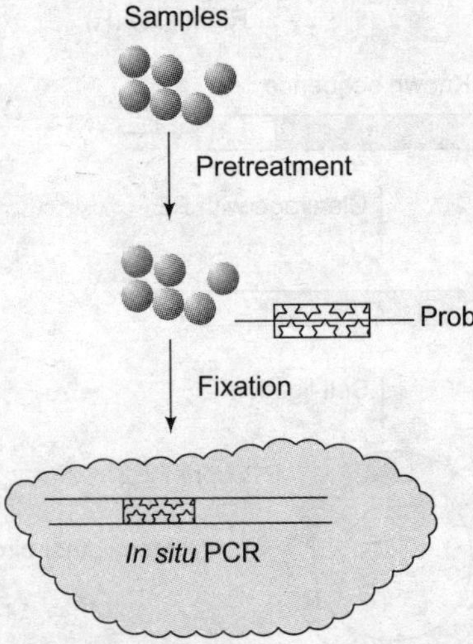

Figure 7.7 *In situ* PCR

The protocol for IS-PCR is divided into three steps: (i) Preparation of sample for PCR, i.e., fixation and permeabilization; (ii) Actual PCR; and (iii) Detection of amplified signal.

It is somewhat difficult to detect the genes of low copy number by *in situ* PCR as it is below the detection limit. Hence, due to copy number limitations, hybridization of RNA is more sensitive than DNA detection.

The factors that affect *in situ* PCR sensitivity are (i) The strandedness of the target molecule and (ii) The lack of a complementary sequence proximal to the target sequences.

7.9.6 Long PCR

In routine experiment under standard condition, a product of 1–2 kbp in length can be easily amplified. By doing standard PCR method it is not possible to amplify complete gene or even cDNA of average size. Long PCR is a PCR, which is extended longer than standard PCR, over 5 kbp (frequently over 10 kbp). In case of standard PCR it is very difficult to get long PCR product because of damage to the template and product DNAs due to exposure to high temperature, presence of Mn^{2+} and difficulties in denaturing of long DNA molecules. These problems can be fixed by simple alteration to the reaction conditions. However, PCR is limited beyond 5 kbp due to incorporation of incorrect bases at the 3′-end of newly synthesized strand. Long PCR is usually only useful if it is accurate. Polymerases with proofreading activity demonstrate lower misincorporation rates, but it is limited to perform due to long extension times, and exposing primers and template to extensive degradation. Secondary structure of template may also limit extension with proof reading DNA polymerase due to ability of the enzyme to idle between polymerase and exonuclease mode rather than performing strand displacement. Therefore, efficient long PCR results from the use of two polymerases, an efficient non-proofreading polymerase and the other proofreading polymerase in much smaller amounts. There are recent reports of amplification of 42 kbp with the blend of enzymes primarily containing non-proofreading polymerase with a very small amount of proofreading polymerase for example,

45:1 *Tth* DNA polymerase Vent (based on polymerase units) is able to amplify a 22 kbp fragment of β globin from genomic DNA.

125:1 *Tth* DNA polymerase Vent (based on polymerase units) is able to amplify 39 kbp of λ DNA.

160:1 to 640:1 Klentaq (deletion mutant of *Taq*) *Pfu* DNA polymerase is able to amplify 35 kbp fragment of λ DNA.

These blends having more fidelity of amplification and their ability to amplify long sequences make these polymerase mixtures useful for variety of PCR applications. Now many polymerases blends are commercially available, e.g., *rTth* DNA polymerase, XL, Expand DNA polymerase, *Taq* plus DNA polymerase, Elongase DNA polymerase, and LA *Taq* DNA polymerase.

7.9.7 Nested PCR

In case of nested PCR, two or more pairs instead of one pair of PCR primers are used to amplify a fragment. The first pair of PCR primers amplifies a fragment similar to a standard PCR. Internal primers, called nested primers as these hybridize to the sites nested within the first fragment, are used to amplify PCR products formed by external primers. The binding of nested primers inside the first PCR product fragment allows amplification of a second PCR product that is shorter than the first one.

The advantage of nested PCR is that if the wrong PCR fragment is amplified, the probability is quite low that the region is amplified a second time by the second set of primers. Thus, nested PCR is used to increase magnitude and specificity of amplification.

7.9.8 Touchdown PCR

Touchdown PCR is a simple method that is used to optimize yield of amplified product at different annealing temperature. In some cases it is very tricky to amplify DNA because of mismatches between oligonucleotide primers and the template DNA and it is not possible to calculate annealing temperature. This situation arises in the following circumstances, e.g., when the desired gene belongs to multigene family or the primers are designed from amino acid sequence, or when heterogeneous primers are used.

In touchdown PCR, the T_a during the first two cycles is set at ~3°C above the calculated T_a. The annealing temperature is then reduced by one degree centigrade for every one or two cycles. When the T_a reaches at an optimum point, then specific amplification of the target DNA starts. The onset of nonspecific amplification is delayed for few additional cycles until the T_a is lowered to the point where nonspecific amplification priming can occur. However, by this time the specific amplification product has supremacy over the reaction and will successfully restrain nonspecific amplification.

Touchdown PCR is the quickest method to optimize PCR when it is required to use new template and primer combinations, particularly in the amplification of template with heterologous primers. Nowadays, modern PCR machines which have the facility of gradient setting are easily programmed to run touchdown PCR.

7.9.9 Reverse Transcriptase-PCR (RT-PCR)

RNA is easily converted into cDNA (complementary DNA) by performing reverse transcription reaction. In eukaryotes during posttranscriptional modification, introns are removed and mRNA has only coding sequences, hence it can be used as template that is much more suitable for expression and cloning studies. RT-PCR is a multipurpose and sensitive technique which is used to find and clone 5′- and 3′-ends of mRNAs and to generate large cDNA libraries from small amounts of mRNA. It can be easily adapted to identify mutations and polymorphisms in transcribed sequence and to measure the strength of gene expression when the amounts of available mRNA are limited and when the RNA is poorly expressed. RT-PCR is widely used in the diagnosis of genetic diseases.

Procedure of RT-PCR

RT-PCR is a technique used to amplify cDNA copies of RNA (Figure 7.8). It is a two step process:

First Strand Reaction In the first step which is known as 'first strand reaction', RNA strand is first reverse transcribed into a ss cDNA template using dNTPs and an RNA-dependent DNA polymerase (reverse transcriptase) through the process of reverse transcription. An oligodeoxynucleotidyl primer is hybridized to 3′-end of mRNA and is then extended by a reverse transcriptase to make a cDNA copy that can be amplified by PCR. This reaction is usually carried out at 37°C (depends on type of reverse transcriptase).

Second Strand Reaction After the reverse transcription reaction is complete, and cDNA has been generated from the original ss mRNA, standard polymerase chain reaction, termed the 'second strand reaction' is initiated. In this step, standard PCR is performed in the presence of gene-specific reverse and forward primers. After ~35 cycles, millions of copies of the sequence of interest are generated. The original RNA template is degraded by RNase H, leaving pure cDNA.

This process can be simplified into a single-step process by the use of wax beads containing the required enzymes for the second stage of the process, which are melted, releasing their contents on heating for primer annealing in the second strand reaction.

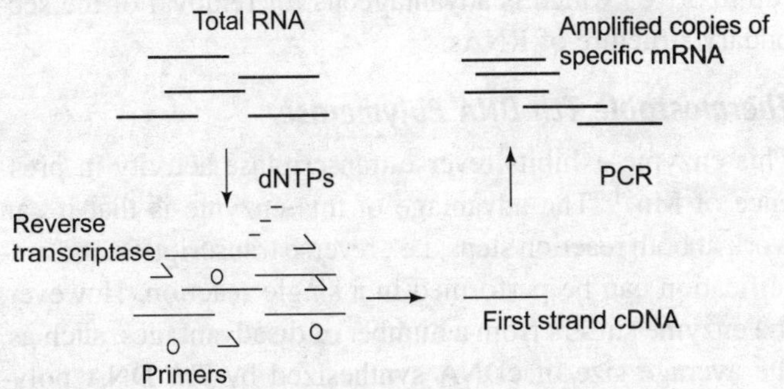

Figure 7.8 Reverse transcriptase-PCR

Depending on the requirement, the primer for first cDNA strand synthesis can be of two types: (i) Oligo dT is used as universal primer which binds to poly (A) tail of eukaryotic mRNA; and (ii) Reverse primers are specifically designed which can hybridize to a particular target gene or defined mRNA family.

Reverse Transcriptase

Reverse transcriptase is RNA-dependent DNA polymerase which is used to catalyze synthesis of DNA complementary to an RNA template. The most commonly used and commercially available reverse transcriptases are discussed below:

AMV (Avian Myeloblastosis Virus) RT AMV RT was first isolated from purified avian myeloblastosis virus. These days, the enzyme is extracted from a strain of *E. coli* which expresses a cloned copy. This enzyme uses RNA or DNA as template with an RNA or DNA primer having 3′-OH end. The avian enzyme consists of two polypeptide chains that carry both activities *viz.* polymerase (5′ → 3′ DNA polymerase) and RNase H. It, however, lacks 3′ → 5′ exonuclease, i.e., proofreading activity. High levels of associated RNase H activity suppress the yield of cDNA and restrict its length. The optimum temperature and pH are 42°C and 8.5, respectively.

Mo-MLV (Moloney Murine Leukemia Virus) RT The enzyme has a single polypeptide chain of M_r 84 kDa. The enzyme exhibits a 5′ → 3′ DNA polymerase activity and a weak RNase H activity that offers considerable advantage when cDNAs of long mRNAs are to be synthesized. The enzyme lacks 3′ → 5′ exonuclease activity and is thus prone to error. Its optimal temperature is 37°C, which is a slight disadvantage if the RNA template used has a high degree of secondary structure. Its optimal working pH is 7.6.

Variant of Mo-MLV RT Some improved reverse transcriptases have been developed, for example, Superscript (Life Technologies) and StrataScript (Stratagene).

These improved reverse transcriptases lack inherent RNase H activity; hence there is no possibility of degradation of RNA template at the 3′-end of the synthesizing DNA. Moreover, these reverse transcriptases can synthesize longer cDNA than the wild-type and work at higher temperature (up to 50°C) which is advantageous for removal of the secondary structure of RNAs.

Thermostable Tth DNA Polymerase

This enzyme exhibits reverse transcriptase activity in presence of Mn^{2+}. The advantage of this enzyme is that it can work at both reaction steps, i.e., reverse transcription and amplification can be performed in a single reaction. However, the enzyme suffers from a number of disadvantages, such as, the average size of cDNA synthesized by *Tth* DNA polymerase is ~1–2 kbp as compared with ~10 kbp by Mo-MLV RT. Second the fidelity of reaction is very low in the presence of Mn^{2+}. Third it cannot work well with oligo (dT) primer and random hexamer as the hybrids are unstable.

7.9.10 Band-stab PCR

If during amplification the yield is very low, then the desired fragment can be recovered by gel electrophoresis and reamplified, with which is known as band-stab PCR. In band-stab PCR, ethidium bromide (EtBr) stained agarose gel is analyzed by UV illumination and excess fluid is removed by placing a piece of Whatman 3 MM paper on the surface of the gel. Each band of interest is sampled carefully with the help of hypodermic needle. The needle is withdrawn, the tip is washed in PCR mixture, and the DNA of that particular band is reamplified by using nested primers.

The bands from polyacrylamide gels can also be recovered easily. However, in this case, the surface of the gel is dried by gentle wiping with gloved hands and not by Whatman paper. Then with the help of a needle, a small fragment of polyacrylamide containing the band of interest is transferred into fresh PCR mixture. The template is kept in the PCR reaction mixture for an hour or two before commencement of reaction so that the DNA diffuses into the mixture.

7.9.11 Degenerate PCR

Degenerate PCR is similar to standard PCR, except, that instead of using specific primers of a given sequence, mixed primers are used in degenerate PCR.

This PCR variation has proven to be a very powerful technique to find new genes or gene families. Most of the genes that belong to a gene family share structural and functional homologies. If the sequence for a gene from one organism is known and the gene from another closely related organism is desired, the DNA sequence for the particular gene in both organisms will be close, though rarely similar. 'Clustal' alignment of the protein sequences from a number of related proteins can be carried out to find the conserved and variable domains. The primers are designed on the basis of conserved protein motifs.

An amino acid is coded by triplet codon and the genetic code is degenerate, i.e., several codons can code for the same amino acid. Usually many codon families share the first two bases and vary only at the third position of the codon. In view of the fact that the sequence of the protein, rather than the DNA, is most significant for function, most of the variations in closely related genes occur in the third codon position; the degenerate or redundant DNA primers can be designed such that these have a mixture of all possible bases in every third position. Several possibilities exist for the sequence of DNA that corresponds to any particular polypeptide sequence due to the Wobble hypothesis of genetic code (third base of

Exhibit 7.1 Mixed oligonucleotide-primed amplification of cDNA (MOPAC)

By using MOPAC, it is easy to clone the gene encoding a protein of which only a small portion is known. The N-terminal and C-terminal amino acid sequences of the peptide are used for designing of primers, which are used to amplify a segment of cDNA. The amplification products are loaded on agarose gel and the DNA of accurate size is isolated, cloned and sequenced. After confirming the clone, it can be used as probe to screen libraries. It is a very tedious job to design primers for MOPAC and usually degenerate primers are used for amplification and sometimes a universal nucleotide inosine residue is added in the primer sequence.

Source: cshprotocols.cshlp.org; Sambrook J and Russel DW (2001) *Molecular Cloning: A Laboratory Manual*, 3rd edition, Cold Spring Harbor Laboratory Press, Cold Spring Harbor, New York.

the codon is 'wobble' position). Such degenerate primers, which have a number of options at several positions in the sequence, allow annealing and amplification of a variety of related sequences. In simple words, a degenerate primer is a mixture of closely related primers, and the gene of interest is recognized by one of the primers from this mixture. Moreover, a perfect match is not really required for amplification. For example, if 18–19 bases of 22 bases hybridize, a primer works quite well. After verification of PCR reaction, the band of the expected size is excised from agarose gel and the product can be cloned and sequenced.

Degenerate PCR Primer Designing

If protein motif is: Met-Trp-Asp-Arg-Lys-Glu-Ala-Cys

The probable codons will be: ATG-TGG-GAC(T)-CGT(CAG)-AAA(G)-GAA(G)-GCT(CAG)-TGT(C)

This gives a mix of 256 different oligonucleotides.

Thus, corresponding primer is: ATG-TGG-GAC/T-CGT/C/A/G-AAA/G-GAA/G-GCT/C/A/G-TGT/C

The degeneracy of the primer is created during DNA synthesis; there is no need to order 256 different primers to get a 256 mix.

Deoxyinosine (dI) can be used instead of mixed oligos at degenerate positions. This makes complementary base pairing with any other base, effectively giving a four-fold degeneracy at any position. This minimizes the problem, but may result in too high a degeneracy where there are four or more dIs in an oligonucleotide.

After successful isolation of a protein of interest, the terminals of this protein are sequenced. Sequences amplified by degenerate primers can then be sequenced to confirm that the sequence is correct. If the primers produce a truncated or partial sequence, these can further be utilized as the probes to fish out the gene of interest from a genomic library or a cDNA library.

For the degenerate PCR reaction, the following should be kept in mind: (i) cDNA template should be used for eukaryotes; (ii) Primer concentration must be increased to compensate for the degeneracy; if the primers used are reasonably degenerate, 50 pmoles can be used as a starting point and optimized from there; (iii) The setting up of a thermocycle requires a lot of experimentation and optimization. First, the annealing temperature is standardized. The first 4–5 PCR cycles can be set to run at annealing temperatures 5–10°C lower than the set annealing temperature to extend primers with multiple mismatches; (iv) It is desirable to start with about 35 cycles and increase to 50 cycles, if required; and (v) DNA polymerase with 3′ → 5′ exonuclease activity should not be used as these degrade the primers (*Taq* DNA polymerase however works well).

The degenerate PCR can be used to 'solve' a number of problems, some of examples are given below: (i) To find the corresponding gene when the amino acid sequence of a protein is known; (ii) To clone the homologous gene from another source (say rice) when the gene of interest from one source (say *Arabidopsis*) has been found; (iii) To find the human homologue when an interesting gene in *S. cerevisiae* or *C. elegans* has been found; (iv) To perform phylogenetic and evolutionary studies of genes, i.e., to find specific orthologous genes from a number of related species and compare them; and (v) To study gene families.

A variant of degenerate PCR is mixed oligonucleotide-primed amplification of cDNA (Exhibit 7.1).

7.9.12 Anchored PCR

In some type of PCR where only enough information to make a single primer is known, a known sequence is added to the end of the DNA by enzymatic addition of a polynucleotide stretch or by ligation of known sequence, and the second primer is designed by sequences of anchored DNA. This technique of amplification with single sided specificity has been known as one-sided PCR or anchored PCR. The anchored or single-sided PCR allows specific amplification of DNA where the 5′ sequence of the molecule of interest is unknown. This approach is based on homopolymer tailing of cDNA catalyzed by the terminal deoxynucleotidyl transferase. The amplification is performed by using one primer specific for the molecule of interest (gene-specific primer), and a second primer containing a defined 'anchor' sequence attached to a homopolymer sequence complementary to the tail. Finally, reamplification is done by using one gene-specific primer (nested) and one anchor-specific primer.

This technique, though very useful, has a number of disadvantages that have limited its application, for example: (i) It is a relatively difficult protocol; (ii) Large amounts of starting template are required; (iii) The process needs multiple purification steps; and (iv) The process generates nonspecific products due to use of homopolymer-containing primers in the PCR.

7.9.13 Ligation-anchored PCR (LA-PCR)

An innovative technique of anchored PCR has been developed in which an anchor of defined sequence is directly ligated to first cDNA strand, and the resulting product is amplified by using primers specific for both the cDNA of interest and the anchor. Because of its simplicity and sensitivity, ligation-anchored PCR (LA-PCR) has several advantages over previously described anchored PCR. These include: (i) This advanced technique avoids TdT tailing, which has often proved technically demanding, requiring assessment of tailing efficiency and optimization of reaction conditions; (ii) Since LA-PCR employs direct ligation of the anchor to cDNA, it eliminates the use of homopolymer-containing primers with limited specificity and the necessity of performing two rounds of PCR. Another means of increasing both specificity and sensitivity of LA-PCR is the use of nested primers in two rounds of amplification; (iii) The LA-PCR protocol involves only three steps (Figure 7.9) and is simpler than the anchored PCR; and (iv) Finally, the LA-PCR approach can yield specific product from as little as 1 ng of total RNA.

Thus, LA-PCR is very simple and sensitive technique, which has many important applications. It is used for: (i) The production of general cDNA libraries from very small amounts of starting material; (ii) The analysis of T-cell receptor and immunoglobulin V regions; (iii) To clone developmentally regulated mRNAs detected with gene trap vectors; (iv) To characterize alternative promoter usage and splice products; and (v) To clone specific genes that share a single region of sequence similarity by amplification with one primer specific for the anchor and a set of degenerate primers designed to hybridize with the conserved domain.

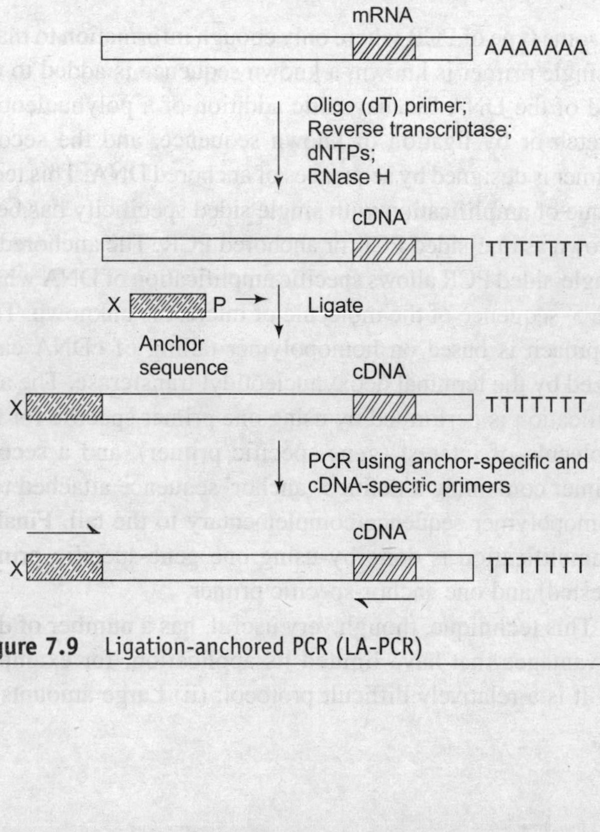

Figure 7.9 Ligation-anchored PCR (LA-PCR)

7.9.14 Asymmetric PCR

In standard PCR, as amplification reaches plateau, the total amplicon yield becomes independent of the starting template amount and varies significantly among replicate reactions. Amplification of ds DNA occurs exponentially during the early stages of symmetric PCR, but in the end slows down and plateau is formed because of negative feedback between the double stranded products and the *Taq* DNA polymerase. The plateau value of symmetric PCR is unsuitable for end-point analysis of starting target numbers because slight differences in reaction components, thermal cycling conditions, and early mispriming events cause individual replicate samples to exit exponential amplification at slightly different times. As a result, at the end of a symmetric PCR amplification, the amplicon yield varies significantly among replicates, and the amount of accumulated amplicon at plateau does not reflect the amount of DNA present in the initial sample.

A PCR technique in which the predominant product is a ss DNA as a result of unequal primer concentration is known as asymmetric PCR. Asymmetric PCR is carried out as usual, but with a great excess of the primers for the chosen strand. As asymmetric PCR proceeds, the lower concentration primer (P_L) is quantitatively incorporated into ds DNA. The higher concentration primer (P_X) continues to synthesize DNA, but only of its template strand. This generates one of the strands by linear amplification, and a fraction of its total product as ds DNA, limited by the concentration ratio of the primers used. Thus, single stranded target strand along with the ds DNA can be generated by an asymmetric PCR. There is linear amplification of one of the strands and the amount of amplified product is expected to be less than that obtainable by a conventional symmetric PCR. Thus, asymmetric PCR provides lower intensity yield, hence less sensitivity than symmetric PCR on agarose gel analysis, as expected. Due to the slow (arithmetic) amplification later in the reaction after the limiting primer has been used up, extra cycles of PCR are required. Asymmetric PCR is best-suited for end-point analysis as it gives linear amplification of one of the strands. Nowadays it is coupled with real-time PCR technique (for details see Section 7.9.16). The advanced form of asymmetric PCR, Linear-After-The-Exponential PCR (LATE-PCR) uses a limiting primer and an excess primer that differ 10–50 fold in their relative concentrations.

7.9.15 Differential Display Reverse Transcriptase PCR (DDRT-PCR)

It is a powerful technique for comparing the patterns of gene expression in RNA sample of different types, or under different biological conditions. It produces partial cDNA fragments by a combination of reverse transcription and PCR of randomly primed RNA. Changes in the expression level of genes are identified after separation of cDNAs on sequencing type gels (Figure 7.10). The DDRT-PCR method consists of two major steps: (i) Reverse transcription of RNAs isolated from different sample with a set of degenerate, anchored oligo (dT) primers to generate cDNA pools; and (ii) PCR amplification of random partial sequences from the cDNA pools.

This method is based on two different types of primers, i.e., anchored antisense and arbitrary sense primers. The typical anchored primer is 11–12 mer oligo (dT) XY which is complementary to nucleotides of poly (A) tail of mRNA, and X may be dG, dA or dC, and Y may be any one of the dNTPs. As these primers anneal to the junction of poly (A) tail and the 3′ UTR of the template, the synthesis of first cDNA strand starts. Arbitrary primer of ~10 nucleotides is added to the reaction mixture and second strand of cDNA is synthesized by using standard PCR method at low stringency. The annealing temperature of the PCR is kept low to maximize the number of amplified mRNA species. The products are separated on denatured polyacrylamide gel electrophoresis (PAGE) and visualized by autoradiography. By comparing the banding patterns of cDNA products derived from different mRNAs, differentially expressed genes can be detected easily. The band of interest can then be recovered from the gel, amplified, and cloned, and can be used as probe for screening of libraries or for northern blots.

DDRT-PCR provides several advantages: (i) The process requires very small amount of mRNA (0.1–0.5 pg); (ii) Analysis of several samples in parallel is possible and amplification products from different sources are displayed in the same gel. It is promising to identify both quantitative and qualitative changes in gene expression; (iii) Simplicity of the technique allows detection of up and downregulated genes in the same set of experiment; (iv) Identification of both rare and abundant messages as PCR is clubbed with RT; and (v) The

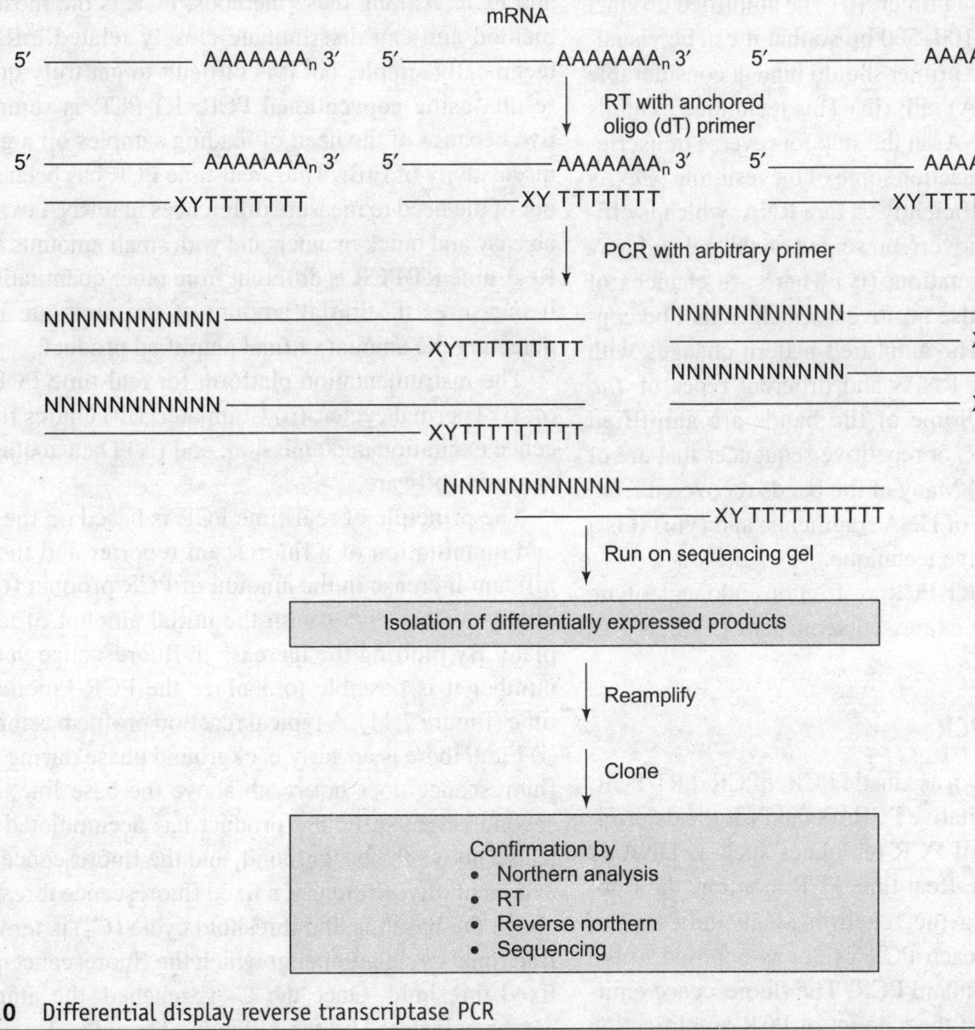

Figure 7.10 Differential display reverse transcriptase PCR

Exhibit 7.2 Restriction endonucleolytic analysis of differentially expressed sequences (READS)

This is a modification of DDRT-PCR and was previously called ordered differential display. Reverse transcription is performed with dinucleotide oligo (dT) with a 20-base heel such that all cDNAs will have a common 3′-end. These are digested with a restriction enzyme and ligated to a Y-shaped adapter. The Y-shaped adapter consists of three sections:

- A 3′ base overlap compatible with the restriction enzyme.
- A complementary middle region.
- A noncomplementary 5′ base end.

In the amplification step, the 3′-base primers anneal to the heel and the 5′-base primer anneals to the upper branch of the Y-shaped adapter. Hence, only the 3′-base ends of restriction digested cDNA are amplified under stringent conditions. The results of READS are dependent on the enzyme used to digest the cDNA, as a result of which, the lengths of the cDNAs differ. Therefore, cDNA prepared from total RNA can be systematically resolved into a pattern of 3′-end restriction fragments by using different restriction enzymes. The result of this technique is highly reproducible. This technique is further modified: the adapter-specific primers are extended by two randomly chosen bases at the 3′-end. The ds cDNA is digested with *Rsa* I, and restriction fragments are ligated to adapter-specific primers.

Source: Sturtevant J (2000) *Clinical Microbiology Reviews* 13(3): 408–427.

major advantage of DDRT-PCR is that it can detect changes in expression of closely related genes that is not possible from subtraction libraries.

In spite of a number of advantages this technique suffers from the following drawbacks: (i) It is quite frustrating that only a part of mRNA in a cell can be amplified, because for amplification the template should be hybridized with a particular sense and antisense primer; (ii) The amplified product must be of the length of 100–500 bp so that it can be visualized on gel, i.e., the sense primer should bind at considerable distance from the poly (A) tail; (iii) This technique exhibits preference for some mRNAs at the step for reverse transcription and during the PCR reaction some of the resulting cDNAs will be amplified more efficiently. A rare RNA, which is efficiently amplified may be overrepresented on the gel and may lead to incorrect interpretation; (iv) There are chances of obtaining a number of false positives, which cannot be confirmed by blotting; (v) The amplified pattern changes with different preparations of RNAs and different types of *Taq* DNA polymerases; (vi) Some of the bands are amplified from introns, or 3′-UTR, or repetitive sequences that are of no use for the study; (vii) Many of the bands recovered from the gel contain a mixture of DNA fragments; and (viii) It is a very lengthy and expensive technique.

A modification of DDRT-PCR, restriction endonucleolytic analysis of differentially expressed sequences (Exhibit 7.2) is highly reproducible.

7.9.16 Real-time PCR

Real-time PCR, also known as kinetic PCR, qPCR, qRT-PCR, and RT-qPCR, is a quantitative PCR method for the determination of copy number of PCR templates such as DNA or cDNA in a PCR reaction. Real-time PCR monitors the fluorescence emitted during the reaction as an indicator of amplicon production at each PCR cycle, as opposed to the endpoint detection in standard PCR. The fluorescence emitted acts as an indicator of the amount of PCR amplification that occurs during each PCR cycle. Thus, in real-time PCR machines, one can visually see the progress of the reaction in 'real-time'. It is the method of choice for measuring changes in gene expression.

There are various methods for the quantitative measurement of mRNA expression. These include, northern blotting, *in situ* hybridization, ribonuclease protection assays (RPA) and PCR. Among these methods, PCR is the most sensitive method and can discriminate closely related mRNAs. It is technically simple, but it is difficult to get truly quantitative results using conventional PCR. RT-PCR is semi-quantitative because of the need of loading samples on a gel and the insensitivity of EtBr. Thus, real-time PCR has been developed out of the need to measure differences in mRNA expression in an easy and quick manner, and with small amounts of mRNA. Real-time RT-PCR is different from other quantitative PCR as it measures the initial amount of the template instead of detecting the amount of final amplified product.

The instrumentation platform for real-time PCR consists of: (i) Thermal cycler; (ii) Computer; (iii) Optics for fluorescence excitation and emission; and (iv) Data acquisition and analysis software.

The principle of real-time PCR is based on the detection and quantitation of a fluorescent reporter and the first significant increase in the amount of PCR product (C_T-threshold cycle) correlates with the initial amount of target template. By plotting the increase in fluorescence versus cycle number it is possible to analyze the PCR kinetics in real-time (Figure 7.11). A typical reaction profile has three stages: (i) First, there is an early background phase during which the fluorescence does not reach above the base line; (ii) In the second stage, sufficient product has accumulated to be detected above the background, and the fluorescence increases exponentially. Normally, a fixed fluorescence threshold is set above the baseline and threshold cycle (C_T) is termed as the fractional cycle number at which the fluorescence passes the fixed threshold. Once the C_T is reached, the amplification can be calculated by the following formula:

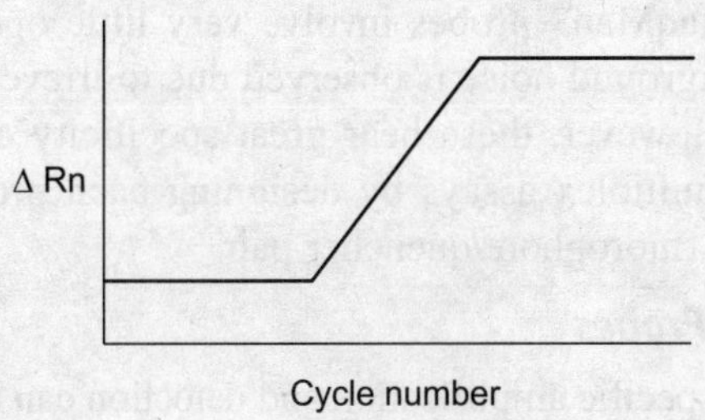

Figure 7.11 Kinetics of real-time PCR

$$T_n = T_0 (E)^n$$

where

T_n = amount of target sequence at cycle number n

T_0 = initial amount of template

E = the efficiency of amplification

In the final stage the reaction efficiency decreases until no product accumulates.

Different methods used for the quantitative assays are described in detail below:

DNA-binding Agents

Various DNA-binding agents used in quantitative assays include the following:

Intercalation of EtBr Initially EtBr was used to quantify PCR products as these accumulated. Amplification produces increasing amounts of ds DNA, which binds EtBr, resulting in an increase in fluorescence. However, this method is no longer used.

SYBR® green SYBR® Green method uses a florescent dye that can bind to any ds DNA and its fluorescence is enormously increased upon binding to ds DNA. As the dye binds to ds DNA, there is no need to design a specific probe for any particular target being analyzed. During the extension phase, more and more SYBR® Green will bind to the PCR product, resulting in an increased fluorescence (Figure 7.12). Therefore, during each subsequent PCR cycle more fluorescence signal will be detected. This method examines the total amount of ds DNA, but cannot differentiate between different sequences.

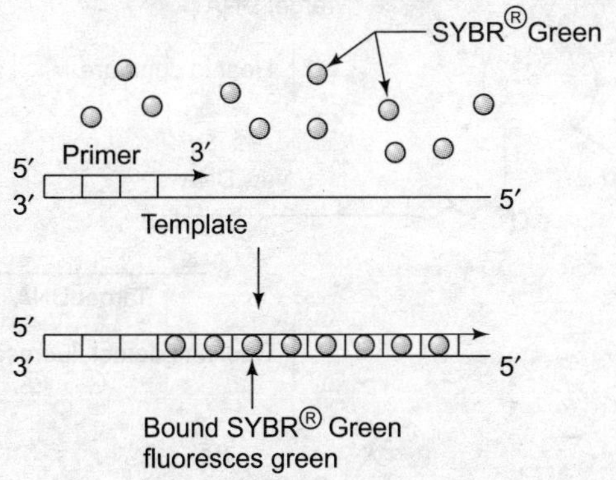

Figure 7.12 Extension step of SYBR® Green PCR (single strand shown)

Various steps of the process are: (i) At the beginning of amplification, the reaction mixture contains the denatured DNA, the primers, and the dye. The unbound dye molecules weakly fluoresce, producing a minimal background fluorescence signal which is subtracted during computer analysis; (ii) After annealing of the primers, a few dye molecules can bind to the double strand. DNA binding results in a dramatic increase of the SYBR Green I molecules to emit light upon excitation; (iii) During elongation, more and more dye molecules bind to the newly synthesized DNA. If the reaction is monitored continuously, an increase in fluorescence is viewed in real-time; and (iv) Upon denaturation of the DNA for the next heating cycle, the dye molecules are released and the fluorescence signal falls.

SYBR® Green method is the simplest, easy to use and most reasonable method for detecting and quantitating PCR products in real-time reactions. The technique is a relatively cheap and does not require probe designing. It can be used in following applications: (i) Assays that do not require specificity of probe; (ii) Detection of thousands of molecules; and (iii) General screening of transcripts prior to moving to probe-based assays.

This method, however, suffers from certain disadvantages. The SYBR Green will bind to any ds DNA in the reaction, including primer dimers and other nonspecific reaction products, which results in an overestimation of the target concentration. Moreover, detection by SYBR Green requires extensive optimization. Since the dye cannot distinguish between specific and nonspecific products accumulated during PCR, follow up assays are needed to validate results. Nonspecificity can lead to false positives and is not attuned for complex protocols. When the PCR system is fully optimized, no primer dimers or nonspecific amplicons, e.g., from genomic DNA are formed. This method is not suitable in following assays such as SYBR® Green allelic discrimination assays, multiplex reactions, amplification of rare transcripts, and low-level pathogen detection.

Hydrolysis Probes

This type of probe is linked with a nonfluorescent chromophore (quencher), which absorbs the fluorescence of the fluorescent chromophore (reporter fluorochromes) as long as the probe is intact. However, upon amplification of the target sequence, the hydrolysis probe is displaced and subsequently hydrolyzed by the 5′ → 3′ exonuclease activity of *Taq* DNA polymerase. This results in the separation of the reporter and quencher flurochromes, and as a result the fluorescence of the reporter fluorochrome becomes quantifiable. This fluorescence will increase with the PCR cycle because

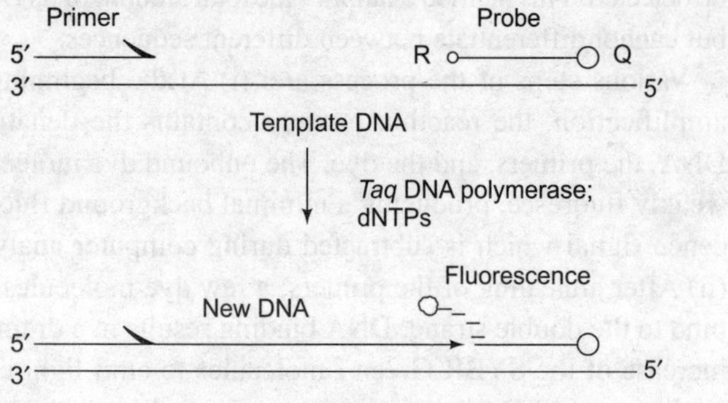

Figure 7.13 Extension step of real-time PCR with TaqMan® probe (single strand shown)

of the exponential accumulation of free reporter fluorochromes. For example,

TaqMan® probe TaqMan® uses a fluorogenic probe which is a single stranded oligonucleotide of 20–26 bases and is designed to bind only the target DNA sequence between the two PCR primers (Figure 7.13). Therefore, only specific PCR product can generate fluorescent signal in TaqMan PCR.

To perform TaqMan PCR, in addition to reagents required for standard PCR, the following additional things are required: (i) Real-time PCR machine; (ii) Two PCR primers with a preferred product size of 50–150 bp; (iii) A probe having: Fluorescent reporter (R) or reporter dye or fluorophore [It is covalently attached to the 5′-end of the probe. It is usually a short-wavelength colored dye (green color), for example, 6-carboxyfluorescein (FAM), Tetrachlorofluorescein (TET) and Green fluorescent protein (GFP)], and Quencher (Q) [It is covalently attached to 3′-end of the probe. It is usually a long-wavelength colored dye (red color), for example, Tetramethyl rhodamine (TAMRA) and Texas Red].

The process is outlined as follows: (i) At the start of the PCR reaction, the quencher reduces fluorescence of the reporter by the use of Fluorescence (or Förster) Resonance Energy Transfer (FRET), which is the inhibition of one dye caused by another without emission of a proton; (ii) When PCR is performed in the presence of TaqMan® probe, the probe hybridizes with the target sequence as the two strands separate after the denaturation step; (iii) During annealing and extension of PCR reaction the primers anneal to the DNA and then *Taq* DNA polymerase adds nucleotides; (iv) The 5′ → 3′ exonuclease activity of *Taq* DNA polymerase cleaves the TaqMan® probe from the template DNA. This separates the quencher from the reporter and fluorescence increases; (v) The increase in fluorescence which is directly related to the amount of the specific target sequences that have been amplified is then captured by the sequence detection instrument and displayed and quantified by the software using a computer attached with real-time PCR machine.

It is an expensive technique as a separate probe is needed for each mRNA template that is being analyzed, but correctly designed TaqMan® probes involve very little optimization. Some background noise is observed due to irreversibility of reaction. However, these bear great specificity and can be used for multiplex assays by designing each probe with a distinctive fluorophore/quencher pair.

Scorpion Probes

Sequence-specific amplification and detection can be done by means of Scorpion probes. Scorpion primers are bifunctional molecules in which a primer is covalently linked to the probe. The Scorpion primer carries a Scorpion probe element at the 5′-end. The Scorpion probe has a hairpin loop structure in the unhybridized state. The loop portion of the stem has sequences complementary to the target DNA and comprises the probe segment. The fluorophore is attached to the 5′-end and is quenched by a moiety attached to the 3′-end. The gene-specific Scorpion primer is attached at the 3′-end and the sequence is modified at the 5′-end. It contains a PCR blocker (e.g., HEG monomers) at the start of the hairpin loop. In the absence of the target, the quencher virtually absorbs the fluorescence emitted by the fluorophore. In the initial PCR cycles, the primer hybridizes to the target and extension occurs due to the action of polymerase. During the Scorpion PCR reaction, in the presence of the target, the fluorophore and the quencher separate leading to an increase in the fluorescence emitted (Figure 7.14). The fluorescence can be detected and measured in the reaction tube.

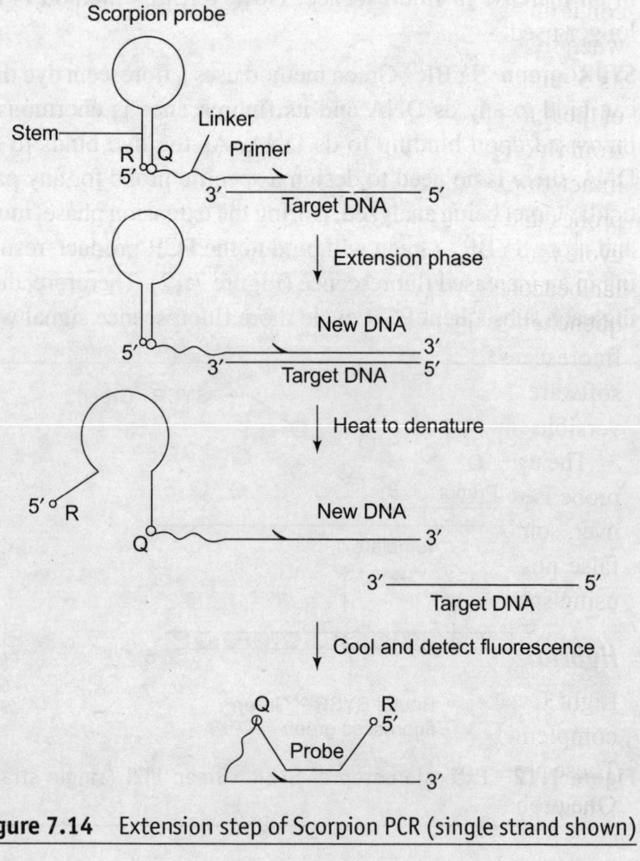

Figure 7.14 Extension step of Scorpion PCR (single strand shown)

Scorpion primers can be used for the examination and identification of point mutations using multiple probes. Each probe can be tagged with a different fluorophore to produce different colors.

A good scorpion primer is the key requirement for a successful real-time PCR reaction. The important characteristics considered during designing of Scorpion primers are as follows: (i) The length of probe sequences should be ~18–27 bases and the probe target should be at a distance of 11 bases or less from the 3′-end of the Scorpion® probe; (ii) The primer pair should be designed to give an amplification product of ~100–200 bp; (iii) The primers should not have secondary structure. While designing, primers should be tested for hairpins and secondary structures; (iv) The melting temperature of the two primers should be in similar range. The T_m of stem should be 5–10°C higher than the probe T_m; (v) The Scorpion® primer should be written as the reverse complement of the target; (vi) The stem can be of 6–7 bases in length, and the 5′ stem sequence should begin with a C, as G may quench the FAM; and (vii) There is always a possibility of the primer hybridizing to the probe element. This could lead to linearization of the probe in an amplification-independent manner, causing significant target-independent fluorescence.

Molecular Beacons

The basic principle of molecular beacons is quite similar to TaqMan® probe, as this technique also uses FRET to detect and measure the synthesized PCR product. Molecular beacon is an oligonucleotide probe having a stem-loop structure when free in solution, in which the fluorophore is attached to its 5′-end and a quencher to its 3′-end. The close proximity of the fluorophore and quencher molecules prevents the probe from fluorescing. Molecular beacons are designed to remain intact during the amplification reaction unlike TaqMan® probes and these hybridize to target DNA sequence in every cycle for signal measurement (Figure 7.15). When a molecular beacon hybridizes to a target, the fluorescent dye and quencher are separated, FRET does not take place, and the fluorescent dye emits light, which can be quantified by the software. Molecular beacons provide greater specificity. Reversible fluorescence in this case means lower background.

The use of molecular beacons is a costly affair as a specific probe is required for each target like TaqMan® probe. Moreover, some nonspecific interactions of the hairpin can lead to false positives. It can also be used for multiplex assays by using spectrally separated fluor/quench dyes on each probe.

Hybridization Probes Technique

Light Cycler is a technique in which two probes that have complementary sequences to the template DNA are used. These probes hybridize in close vicinity on the target sequence. One probe is labeled with a donor fluorochrome at the 3′-end

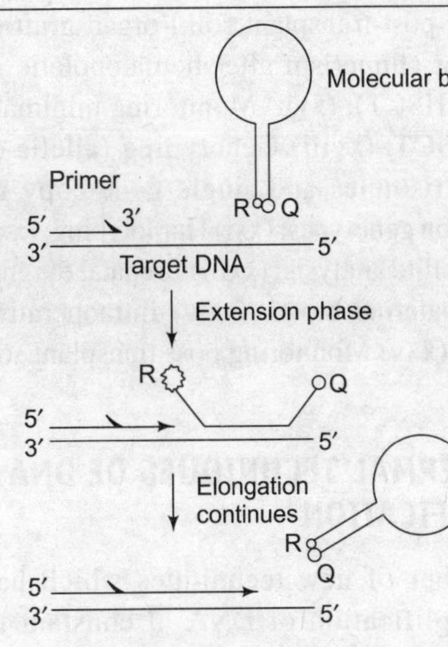

Figure 7.15 Extension step of molecular beacon PCR (single strand shown)

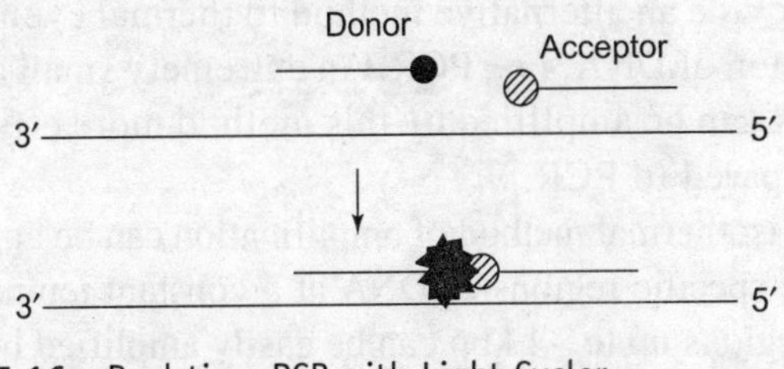

Figure 7.16 Real-time PCR with Light Cycler

and a second probe is labeled with an acceptor fluorochrome at the 5′-end. When the two fluorochromes approach each other in close vicinity during PCR, for example, these are one to five nucleotides apart, the emitted light of the donor fluorochrome will excite the acceptor fluorochrome (FRET). This results in the emission of fluorescence, which can subsequently be detected during the annealing phase and the first part of the extension phase of the PCR reaction (Figure 7.16). During PCR more hybridization probes can anneal, resulting in higher fluorescence signals which can be detected by computer attached to a real-time PCR machine.

Advantages of Real-time PCR

Real-time PCR can be used for same applications as conventional PCR. It is also used for new applications that would have been less effective with conventional PCR. With the ability to collect data in time, the real-time PCR technique can be used for following purposes: (i) Viral quantitation; (ii) Quantitation of gene expression; (iii) Array verification; (iv) Drug therapy efficacy; (v) DNA damage measurement; (vi) Quality control and assay validation; (vii) Pathogen detection; (viii) Genotyping; (ix) Radiation exposure assessment; (x) *In vivo* imaging of cellular processes; (xi) Mitochondrial DNA studies; (xii) Methylation detection; (xiii) Detection of inactivation at X-chromosome; (xiv) Determination of identity at highly polymorphic human leukocyte antigen system;

(xv) Monitoring post-transplant solid-organ graft outcome; (xvi) Monitoring chimerism after hematopoietic stem cell transplantation (HSCT); (xvii) Monitoring minimal residual disease after HSCT; (xviii) Genotyping (allelic discrimination); (xix) Trisomies and single-gene copy numbers; (xx) Microdeletion genotypes; (xxi) Haplotyping; (xxii) Quantitative microsatellite analysis; (xxiii) Prenatal diagnosis from foetal cells in maternal blood; (xxiv) Intraoperative cancer diagnostics; and (xxv) Monitoring post-transplant solid organ graft outcome.

7.10 ISOTHERMAL TECHNIQUES OF DNA AMPLIFICATION

There are number of new techniques which have been devised for amplification of DNA at constant temperature in the presence of DNA polymerase. These new methods bypass the thermal cycling steps of PCR. This method may provide an alternative method to thermal cycling amplification of DNA, i.e., PCR. An extremely small amount of DNA can be amplified by this method more efficiently as compared to PCR.

This isothermal method of amplification can be applied to amplify specific regions of DNA at a constant temperature. DNA regions up to ~4 kbp can be easily amplified by using this method.

7.10.1 Components of Isothermal Techniques of DNA Amplification

The reaction of this method requires the following components: (i) Chimera-primers composed of RNA (3′-end) and DNA (5′-end); (ii) DNA polymerase with strand displacement activity (e.g., *Bca* BEST™ DNA polymerase); (iii) RNase which hydrolyzes the part of DNA–RNA hybrid (RNase H); and (iv) dNTPs as substrates for DNA synthesis.

7.10.2 Helicase-dependent Amplification (HDA) Technology

Helicase-dependent amplification (HDA) is a method for isothermal amplification of nucleic acids. HDA reaction selectively amplifies a target sequence defined by two primers similar to PCR. In this method, the two strands of DNA are separated by an enzyme known as helicase and not by heat denaturation. The ss DNA is amplified isothermally with the help of DNA polymerase, avoiding thermocycling step (Figure 7.17). This reaction can also be coupled with reverse transcription for RNA analysis.

This new technique can be used to amplify and detect short DNA sequences (70–120 bp) at a constant temperature. It can be used for amplification of different types of templates, e.g., microbial genomic DNA, viral DNA, plasmid DNA, human genomic DNA, and cDNA. A single copy of target DNA can be amplified by HDA and visualized by agarose gel electrophoresis in the presence of optimized primers and buffers.

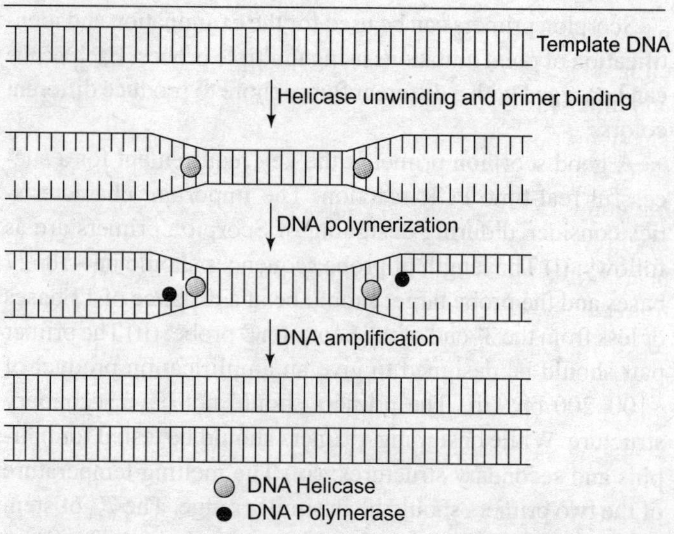

Figure 7.17 Helicase-dependent amplification (HDA) technology

7.10.3 Whole Genome Amplification (WGA)

The WGA technique is used to amplify genomic DNA for down-stream analysis including genotyping. In this technique total DNA is amplified nonspecifically. This technique is based on naturally existing cellular DNA replication system that is redesigned *in vitro* for isothermal amplification. The two strands of DNA are separated with the help of the enzyme helicase. The primers are synthesized on site in the presence of primase, and unlike PCR, no primers are added in this reaction. The primase creates multiple initiation sites for random, whole genome amplification. The fast amplification is performed at constant temperature without the need for prior heat denaturation and thermocycling.

WGA method is best-suited for cancer and genetic research and also used in forensics, comparative genomic hybridization, and single cell analysis.

7.10.4 Loop-mediated Isothermal Amplification of DNA (LAMP)

Loop-mediated isothermal amplification (LAMP) amplifies DNA with high specificity, efficiency and rapidity under isothermal conditions. This method uses a DNA polymerase and a set of four specifically designed primers that can hybridize at six distinct sequences on the target DNA. The sense and antisense primers which are internally located initiate LAMP. The following strand displacement DNA synthesis primed by an outer primer releases a ss DNA. This serves as template for DNA synthesis primed by the second inner and outer primers that hybridize to the other end of the target producing a stem-loop DNA structure. In subsequent LAMP cycling, one inner primer hybridizes to the loop on the product and initiates displacement DNA synthesis, yielding the original stem-loop DNA and a new stem-loop DNA with a stem twice as long. The cycling reaction continues with the accumulation of 10^9 copies of target in less than an hour. The final products are stem-loop DNAs with several inverted repeats of the target, and cauliflower-like structures with multiple loops, formed by annealing between alternately inverted repeats of the target in the same strand. Because LAMP recognizes the target by six distinct sequences initially and by four distinct sequences afterwards, it is expected to amplify the target sequence with high selectivity.

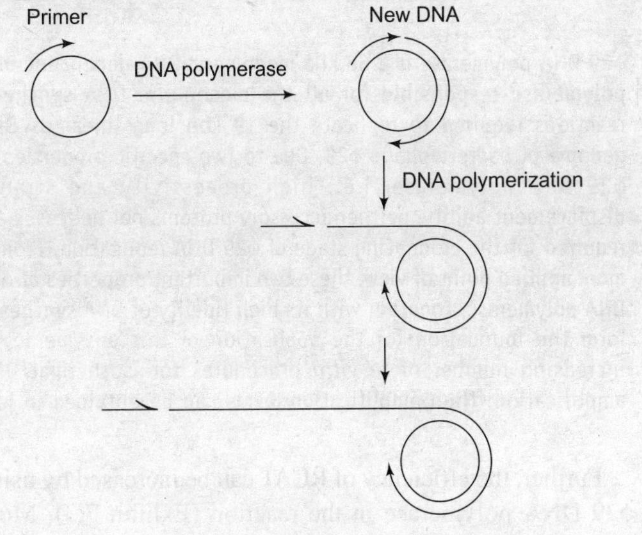

Figure 7.18 Linear rolling circle amplification technology

7.10.5 Rolling Circle Amplification Technology (RCAT)

Rolling circle is the most promising method of isothermal amplification of DNA. The principle of this technique is based on rolling circle mechanism of DNA replication. Two types of RCAT have been devised, *viz*., Linear RCAT and Exponential RCAT.

Linear RCAT

In linear RCAT, a primer binds to circular DNA, which is then copied many times to give a long ss DNA consisting of ~10,000 tandem repeats of target sequence (Figure 7.18). This method may be used in combination with microarray. A linear mode of RCAT can generate 10^5-fold signal amplification during a brief enzymatic reaction that has been demonstrated in microarray assays.

Exponential RCAT (E-RCAT)

In E-RCAT two primers are used and the second primer binds to the newly synthesized strands, which results in strand displacement (Figure 7.19). The single strand branches are generated which can bind to the first primer forming ds DNA. E-RCAT is capable of generating amplification in excess of 10^{12}-fold in an hour and can detect a single target DNA.

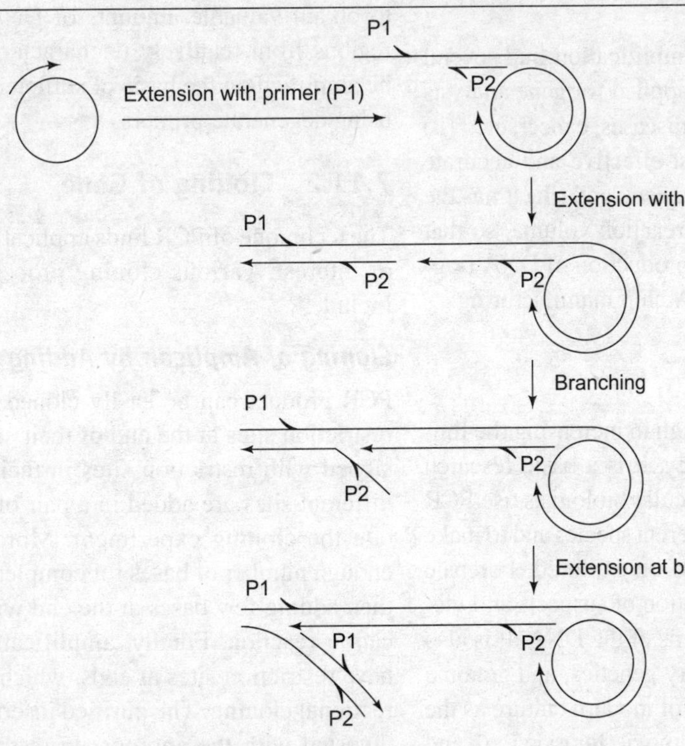

Figure 7.19 Exponential rolling circle amplification technology

Exhibit 7.3 ϕ29 DNA polymerase

ϕ29 DNA polymerase is a 66 kDa monomeric DNA-dependent DNA polymerase responsible for all the mesophilic DNA synthesis reactions required to replicate the 19 kbp long linear ds DNA genome of bacteriophage ϕ29. Due to two specific properties of ϕ29 DNA polymerase, i.e., high processivity and strand-displacement ability, neither accessory proteins nor helicases are required for the elongating stage of ϕ29 DNA replication. From a more applied point of view, these two important properties of ϕ29 DNA polymerase, together with its high fidelity of DNA synthesis, form the foundation for the application of this enzyme in an increasing number of *in vitro* procedures for isothermal DNA amplification. High amplification levels can be obtained in less than 4 hours using random hexamers to initiate rolling circle amplification. ϕ29 DNA polymerase can initiate synthesis from many primers on each circle, simultaneously advancing multiple replication forks around the circle in a highly processive manner producing as much as 10^7-fold amplification of the input circular DNA. Typically, reactions require modest levels of input DNA (1 pg–1 ng) and the amplification products of such reactions will become viscous after only a few hours of incubation at 30°C. Afterwards, the products can be added directly into sequencing reactions.

Source: www.horizonbioscience.com

Further, the efficiency of RCAT can be increased by using ϕ29 DNA polymerase in the reaction (Exhibit 7.3). More recently, it has been established that RCAT can be used to increase the sensitivity of immunoassays. In a manner analogous to immunopolymerase chain reaction it has been possible to couple the specificity of antibodies to the signal amplification of RCAT. However, in contrast to immunopolymerase chain reaction, signal amplification via immuno-RCAT results in an easily detectable high molecular weight nucleic acid molecule at the site of antibody binding. Furthermore, because the product of immuno-RCAT reaction remains attached to the immune complex, it is compatible with localization of the product on a microarray or within a biological structure.

7.10.6 Advantages of Isothermal Techniques of DNA Amplification

The isothermal techniques of DNA amplification find several applications: (i) This method can be applied for gene analysis dependent diagnoses, e.g., infectious diseases, cancer, etc.; (ii) This technology can be used for cost effective and accurate detection of genetically modified organism; and (iii) It has the advantage of easiness in scale-up of reaction volume, so that this method can be applied for mass-production of DNA fragments, which would be used for DNA chip manufacturing.

7.11 APPLICATIONS OF PCR

The discovery of PCR is a breakthrough in increasing the limits of knowledge in molecular biology. It is a basic research tool in a molecular biology lab. Molecular biologists use PCR for finding specific genes in many different species and to make many copies of the piece of DNA to be explored. Forensic technicians use it to help in identification of suspects and victims based on the amplification patterns of the DNA. It is also used in disease diagnosis, evolutionary genetics, and genome sequencing. It has also become a part of modern culture as the theoretical basis of the science fiction movie Jurassic Park and by producing key evidence in many real life court cases. The advent of PCR has dramatically impacted many areas of our lives and promises to yield many more exciting discoveries. PCR can also be exploited in both site-directed and localized mutagenesis and the generation of specific gene (and hence protein) fusions through the choice of 'hybrid primers'. There is little doubt that this procedure and its variants will play an ever-increasing role in our future science. Several important applications of the PCR are as follows:

7.11.1 Isolation of Gene

Using standard PCR method, a particular gene can be isolated and amplified within the known boundaries. Now, several methods have been developed for the amplification of DNA sequences that flank regions of known sequences, e.g., inverse PCR and gene walking. By this method, one is able to obtain valuable amounts of DNA without cloning. Even regions from relatively uncharacterized organisms can also be amplified on the basis of sufficient sequence homology by using degenerate primers.

7.11.2 Cloning of Gene

The technique of PCR finds application in the cloning of gene of interest. Various cloning procedures that are followed include:

Cloning of Amplicon by Adding Restriction Sites

PCR product can be easily cloned by adding sequences of restriction sites at the end of their sequence. Primers are designed with restriction sites in their 5′-end. Generally, two different sites are added in a pair of primers that will facilitate the cloning experiment. Moreover, the primers have enough number of bases for complementary base pairing and then adding few bases at the end will not effect the amplification reaction. Finally, amplification generates fragments have restriction sites at ends, which can be exploited for directional cloning. The purified insert and the vector DNA are digested with the appropriate restriction enzymes, ligated, and transformed in *E. coli*.

This reaction, however, suffers from a number of limitations: (i) After the PCR reaction, the presence of remaining DNA polymerase and dNTPs after digestion of amplicon by restriction enzyme can generate blunt ended DNA; (ii) The efficiency of restriction digestion of amplicon may also decrease because of the presence of residual primers and primer dimers as these compete for restriction reaction; and (iii) Various restriction enzymes fail to digest recognition sequences located close to the ends of amplicon, e.g., *Hin*d III, *Not* I, *Sal* I, *Xba* I, and *Xho* I.

To solve the above-mentioned problems, the following points are considered: (i) Residual primers, polymerase, dNTPs should be eliminated before setting the restriction reaction; (ii) The pair of primers is designed having sites of different restriction enzymes that help in directional cloning; (iii) The primers should lack restriction sites that are internally located somewhere in the insert; (iv) The restriction enzymes, which fail to digest if located at the end of amplicon, should be avoided for the incorporation in the primers; (v) Restriction enzymes having star activity (for details see Chapter 3) should also be avoided; and (vi) Clamp sequence is added at the 5′-end of the primers to help in holding the ends of amplicon and restriction.

TA Cloning

TA cloning is one of the most popular methods of cloning the amplified PCR product by exploiting the properties of *Taq* and other polymerases. As these polymerases have a non-template-dependent terminal transferase activity that adds a single dATP residue to the 3′-ends of the double stranded amplicon, this type of amplified product can be directly cloned in a linearized vector having complementary 3′ T-overhangs. Such vectors are called T-vectors. T-vectors may be generated in the following ways:

- The vector DNA is digested with *Xcm* I, *Hph* I, and *Mbo* I, which generates 3′-terminal deoxythymidine.
- T residue is also added in the presence of dideoxy TTP with the help of terminal transferase at the 3′-end.
- T-vectors are also created by utilizing the template independent terminal transferase activity of *Taq* DNA polymerase that catalyzes the addition of T residues at the 3′-end of linearized vector.
- T vectors can be supplied by many suppliers as a component of T-vector cloning kit for example, pCR Script (SK$^+$) (Stratagene), pCR II (Invitrogen), and pGEM-T (Promega).

The PCR product with 'A' overhang is mixed with this vector in high proportions. The complementary overhangs of a T-vector and the PCR product hybridize. The result is a recombinant DNA, the recombination being brought about by DNA ligase. This cloning procedure does not rely on sequence of template or primers and is successfully applied when appropriate restriction sites are not accessible for cloning.

TOPO TA Cloning

TOPO TA cloning is designed for cloning PCR products directly from a PCR reaction in just 5 min. This method of cloning uses a pCR®-TOPO® vector (Invitrogen) with covalently bound topoisomerase I for fast cloning and obtaining >95% recombinants.

This vector has: (i) 3′ T overhangs for direct ligation of *Taq* DNA polymerase amplified PCR products; (ii) Ampicillin and kanamycin resistance gene for selection and X-gal, IPTG for screening in *E. coli*; (iii) *Eco* RI sites flanking the PCR product insertion site for easy excision of insert; (iv) Topoisomerase I covalently bound to the vector (referred to as 'activated' vector); and (v) M13 reverse and forward primers for sequencing of cloned product.

As discussed earlier, *Taq* DNA polymerase has a nontemplate-dependent terminal transferase activity that adds a single deoxyadenosine (A) to the 3′-ends of PCR products. The linearized vector has single, overhanging 3′ deoxythymidine (T) residue. This allows PCR insert to ligate efficiently with the vector.

Topoisomerase I from *Vaccinia* virus binds to duplex DNA at specific sites and cleaves the phosphodiester backbone after 5′-CCCTT-3′ in one strand. The energy from the broken phosphodiester backbone is conserved by formation of a covalent bond between the 3′-phosphate of the cleaved strand and a tyrosyl residue (Tyr 274) of topoisomerase I. The phosphotyrosyl bond between the DNA and enzyme can subsequently be attacked by the 5′-hydroxyl of the original cleaved strand, reversing the reaction and releasing topoisomerase.

7.11.3 cDNA Synthesis and Rapid Amplification of cDNA Ends (RACE)

As discussed in Chapter 5, RT-PCR can be used for the synthesis of cDNA. However, there are chances of getting partial clones during cloning by RT-PCR method or by screening cDNA libraries. This happens when reverse transcriptase reaction does not succeed to synthesize full-length first cDNA strand from mRNA. By applying rapid amplification of cDNA ends (RACE) technique, the sequences of 5′- and 3′-ends of any partial clone can be completed. RACE is a procedure for amplification of DNA sequences from an mRNA template between a distinct internal site and unknown sequences at either the 5′- or the 3′-end of the mRNA. Using these methods searches for and sequencing of the 5′- and 3′-ends of any mRNAs of interest can be sped up , provided some sequence is known from the internal portion of the mRNA.

5′ RACE

5′ RACE, i.e., rapid amplification of 5′-ends of cDNA, is a method of amplification and is actually a type of 'anchored' PCR. In general, standard PCR amplification of few target molecules in a complex mixture requires two sequence-specific primers that flank the region of the sequence to be amplified. However, to amplify and characterize regions of unknown sequences is greatly restricted via standard PCR technique. RACE technique offers possible solution to this problem. 5′ RACE is a technique that makes possible the isolation and characterization of 5′-ends from low copy messages. This method was reviewed independently by Frohman and Loh in 1989. Even though the particular protocol varies among different users, the general strategy remains constant. The process (Figure 7.20) is outlined as follows:

- Total or poly (A) RNA is reverse transcribed and first strand of cDNA is synthesized by using a reverse gene-specific primer termed GSP 1 and SuperScript™ II, a derivative of Mo-MLV RT with reduced RNase H activity.
- After first cDNA strand synthesis, the original mRNA template is removed by treatment with the RNase Mix (mixture of RNase H, which is specific for RNA:DNA heteroduplex molecules and RNase T1).
- cDNA is purified to get rid of unincorporated dNTPs, GSP 1, and proteins.
- A homopolymeric tail is then added to the 3′-end of the cDNA using TdT (terminal deoxynucleotidyl transferase) and a dNTP.
- In the original protocol, the tailed cDNA is then amplified by PCR using a mixture of three primers: a nested gene-specific primer (GSP 2), which anneals 3′ to GSP 1; a combination of a complementary homopolymer containing anchor primers and corresponding adapter primer which permit amplification from the homopolymeric tail. This allows amplification of unknown sequences between the GSP 2 and the 5′-end of the mRNA.
- In the modification of the original protocol, PCR amplification is performed by using a nested gene-specific primer (GSP 2) that anneals to a site located within the cDNA molecule and a novel deoxyinosine-containing abridged anchor primer.
- Finally, the primary PCR product can be reamplified by GSP 3 (nested primer) and universal amplification primer. Generally, two universal amplification primers are used in RACE system *viz.*

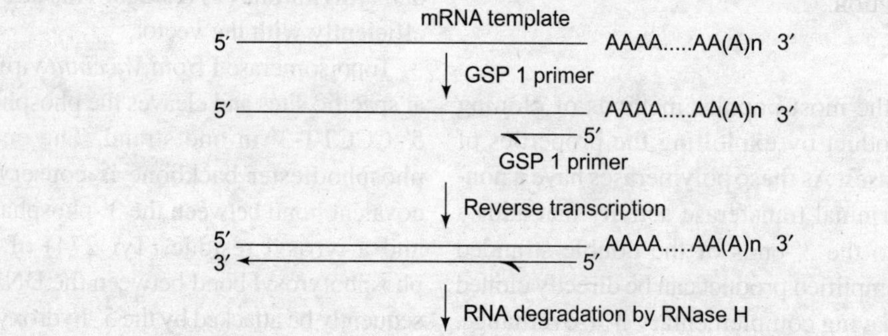

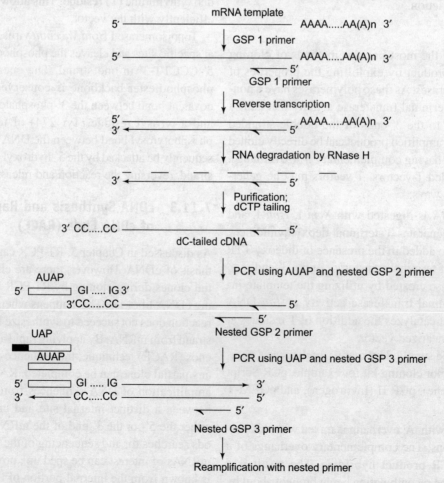

Figure 7.20 Rapid amplification of 5′-ends of cDNA

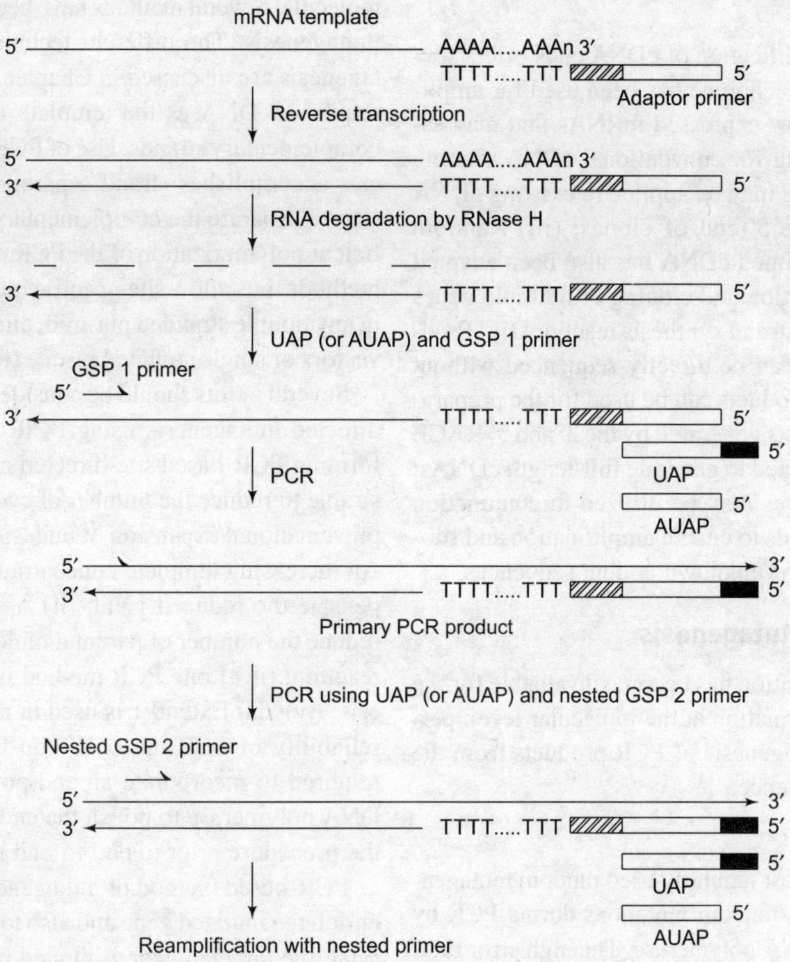

Figure 7.21 Rapid amplification of 3′-ends of cDNA

– Universal amplification primer (UAP): This primer is designed for the rapid and efficient cloning of RACE products using the uracil DNA glycosylase (UDG) cloning method.

– The abridged universal amplification primer (AUAP): This primer is homologous to the adapter sequence used to prime first cDNA strand synthesis.

• Following amplification, 5′ RACE products can be cloned into an appropriate vector for subsequent characterization procedures, which may include sequencing, restriction mapping, and preparation of probes to detect the genomic elements associated with the cDNA of interest or *in vitro* RNA synthesis.

3′ RACE

3′ RACE, i.e., rapid amplification of 3′-ends of cDNA, takes benefit of the poly (A) tail of mRNA as a general priming site for PCR amplification. In this procedure, mRNAs are converted into cDNA using reverse transcriptase, and an oligo (dT) adapter primer. Specific cDNA is then directly amplified by PCR using a gene-specific primer (GSP 1) that anneals to a region of known sequences and an adapter primer that targets the poly (A) tail region. This permits the amplification of unknown 3′ mRNA sequences that lie between the known coding region and the poly (A) tail. The standard protocol of 3′ RACE is simpler than 5′ RACE. Different steps of 3′ RACE (Figure 7.21) are as follows.

• First cDNA strand is synthesized by reverse transcription of mRNA, which is initiated at the poly (A) tail of mRNA using the adapter primer (AP).

• After first cDNA strand synthesis, the original mRNA template is degraded with RNase H, which is specific for RNA:DNA heteroduplex molecules.

• Then amplification is performed, without intermediate phenol:chloroform extractions or ethanol precipitations, with the help of two primers:

– One is a user-designed GSP 1 that anneals to a site located within the cDNA molecule.

– The other is a universal amplification primer that targets the mRNA of the cDNA complementary to the 3′-end of the mRNA as discussed in 5′ RACE.

Applications of RACE

The process of rapid amplification of cDNA ends offers several advantages: (i) This technique has been used for amplification and cloning of low expressed mRNAs that may escape, or prove challenging for conventional cDNA cloning methodologies; (ii) RACE may be applied to existing cDNA libraries to complete the 5′-end of clones; (iii) Random hexamer or oligo (dT) primed cDNA has also been adapted to 5′ RACE for amplification and cloning of multiple genes from a single first cDNA strand synthesis reaction; (iv) Products of RACE reactions can be directly sequenced without any cloning; (v) RACE products can be used for the preparation of probes; (vi) Products generated by the 3′ and 5′ RACE procedures may be combined to generate full-length cDNAs; and (vii) RACE procedures may be utilized in conjunction with exon trapping methods to enable amplification and subsequent characterization of unknown coding sequences.

7.11.4 PCR-based Mutagenesis

PCR-based gene manipulation has become invaluable for the alteration of genetic information at the molecular level, permitting site-directed mutagenesis of PCR products from the ends of the template sequence.

Error-prone PCR

Error-prone PCR is the most regularly used random mutagenesis method. It introduces random mutations during PCR by reducing the fidelity of DNA polymerase. The high error rates may be introduced with the use of DNA polymerase lacking proofreading activity, e.g., *Taq* DNA polymerase, which causes misincorporation of incorrect nucleotides during the PCR reaction, yielding randomly mutated products. The fidelity of DNA polymerase can also be reduced by adding Mn^{2+} ions or increasing the Mg^{2+} concentration or by biasing the dNTP concentration. Moreover, with DNA up to 10 kbp size, it is possible to introduce alterations per gene ranging from ~1 to ~20. This form of mutagenic PCR also finds a crucial role in *in vitro* selection methods (a protocol in use before PCR itself was used to generate the initial, randomized libraries). With *in vitro* selection, a random oligonucleotide library is expressed and screened, high-performing variants from this library are then retained and accurately amplified and the expression/screening cycle is started again, in analogy with Darwinian selection. This method has proven useful both for generation of randomized libraries of nucleotide sequences and also for the introduction of mutations during the expression and screening processes in a mutagenesis step.

PCR-based Oligonucleotide Directed Mutagenesis

Site-directed mutagenesis is a technique of molecular biology in which a mutation is created at a defined site in a DNA molecule. Several methods have been devised for site-directed mutagenesis. The different techniques of site-directed mutagenesis are discussed in Chapter 17. A number of methods require ss DNA as the template to prevent reannealing of complementary strands. Use of PCR in site-directed mutagenesis accomplishes strand separation by using a denaturing step to separate the complementary strands and allowing efficient polymerization of the PCR primers. PCR site-directed methods thus allow site-specific mutations to be incorporated in any double stranded plasmid, eliminating the need for M13 vectors or single stranded rescue (for details see Chapter 10).

Several points should be considered while performing site-directed mutagenesis using PCR. These are: (i) When performing PCR-based site-directed mutagenesis, it is often desirable to reduce the number of cycles (5–10) during PCR to prevent clonal expansion of undesired second-site mutations. An increasing template concentration (1,000-fold) can compensate the reduced yield; (ii) A selection must be used to reduce the number of parental molecules coming through the reaction; (iii) Long PCR method is best-suited for mutagenesis; (iv) *Taq* Extender is used in the PCR mix for increased reliability of PCR up to 10 kbp DNA; and (v) It is often required to incorporate an end-polishing step by using *Pfu* DNA polymerase to polish the ends of the PCR product into the procedure prior to end-to-end ligation.

PCR-based method of mutagenesis has been developed to enrich the mutated gene and also to avoid M13 phage system. First, the targeted gene is cloned in plasmid vector and then divided into two parts. Two specific primers are added to each reaction, one primer is completely complementary to a sequence within or adjacent to the insert and the other is complementary to a different part of insert, except for one nucleotide, which has been targeted for change. In both of the reactions the primers (A and C) anneal to opposite strands, so both the nucleotides have been targeted. The primers (B and D) are complementary to a DNA fragment. The positioning of the hybridization regions of the primers in the two reactions is such that these have different ends (Figure 7.22). After PCR, both of the reaction mixtures are combined, denatured and renatured. As these have different ends, a fragment from one reaction hybridizes with complementary fragment from the other reaction and a circular DNA is formed with two nicks. The nicks are sealed and repaired *in vivo* after transformation in *E. coli*. There are chances of formation of linear DNA molecules because of hybridization of two complementary sequences from the same reaction mix; moreover, the circular form and not the linear ones are stably maintained in *E. coli*. Finally, this procedure introduces a specified point mutation into a cloned gene without the need of introduction into M13 vector.

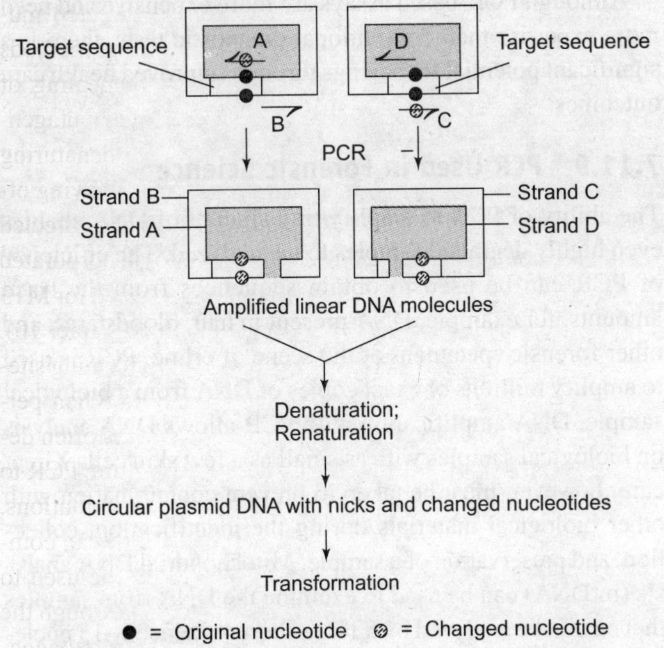

Figure 7.22 PCR-based oligonucleotide-directed mutagenesis

7.11.5 PCR-based Molecular Markers

Heritable traits that can be assayed are known as markers. A molecular marker may be defined as a DNA sequence used for chromosome mapping as it can be located on a specific chromosome. These are segments of an organism's DNA that show genetic variability between individuals in the same population or species. Molecular markers are very useful in genome mapping and analyzing genetic variation within and between specific populations. The science of mapping genetic traits, including those of agronomic interest, is well established and many genetic marker systems are available which are useful for molecular breeding program of plants. Several molecular marker methods are based on DNA amplification, for example, Randomly amplified polymorphic DNA (RAPD) and its variants such as Allele specific PCR (AP-PCR) and DNA amplification fingerprinting (DAF), Sequence characterized amplified regions (SCAR), Single strand conformational polymorphism (SSCP), Amplified fragment length polymorphism (AFLP), Minisatellites or Variable number tandem repeats (VNTR), Microsatellites or Simple tandem repeats (STR), Expressed Sequences Tags (EST), Sequence tagged sites (STS), and Cleaved amplified polymorphic sequence (CAPS). These molecular markers are discussed in detail in Chapter 18.

7.11.6 Study of Fossil DNA Using PCR

The knowledge of our past life forms is mainly dependent on the study of fossils, which is the only available evidence of extinct species.

In fact, traces of DNA molecules preserved in fossil bones are usually analyzed by PCR amplification. This method, however, is adversely affected by inhibitors commonly present in fossil bone extracts and by the chemical modifications of the preserved DNA molecules. Moreover, increased efficiency of amplification is observed with shorter fragments. Recently, quantitative real-time PCR is used to determine the optimal extract amount that results in maximum PCR product yield with minimum PCR inhibition. Further, the PCR amplification can also be improved by performing PCR reaction in the presence of dUTP and the cloning of PCR products is performed in *E. coli* strains that incorporate dUTP instead of dTTP in DNA. Moreover, putative contamination with products of previous PCR amplification or DNA cloning experiments is degraded prior to each PCR with uracil-*N*-glycosylase (UNG). Therefore, this procedure decreases the contaminants by ~10,000-fold. This procedure permits amplification of reliable sequences from fossils and ancient material, and is relevant for applications in paleogenetics, forensics, conservation biology, and DNA tracing.

7.11.7 Analysis of Environment by PCR

It is very easy to isolate the DNA directly from soil and water for study. Conventional culture is generally used to detect and count microbes present in environmental sample, but it can take up to 10 days to obtain a firm result. In addition, the sensitivity of culture is poor especially when samples also contain microorganisms that inhibit growth of particular microbe. Moreover, the cells that are viable but unculturable are not detected by conventional culture yet are potentially pathogenic. PCR is an alternative tool for rapid detection of the presence of microorganisms in environmental samples. The detection rate of contaminant DNA can be increased by using qualitative real-time PCR.

As the classification of microbes is based on prokaryotic rRNAs of small subunit, i.e., 16S rRNA, the primers designed from sequences of rRNA can be used for the identification of particular microbes in the environmental sample. It is also easy to quantify the amount of microbes by amplifying the gene of 16S rRNA. The microbial diversity can be also studied by amplifying 16S rRNA and with the help of PCR-based molecular markers. Genetic diversity has been shown to identify populations that may benefit from the addition of relocated individuals or that should remain as such because these represent different subspecies.

Eco-trawling approach is used to amplify and clone the genes encoding useful proteins directly from the environmental sample without knowing their source. For example, *Taq* DNA polymerase can be amplified from water of hot spring.

7.11.8 Medical Diagnosis by PCR

As PCR is a highly sensitive technique that is used to amplify and detect minute quantities of specific DNA sequences in biological material, the PCR-based assay can be used in diagnosis to detect unculturable or particular microorganisms that cannot be identified by conventional cultural techniques. Usually three types of tests are performed, *viz.*, qualitative test, quantitative test, and genotype test. These diagnostics tests are used for detection of the quantity and subtype of particular infectious microorganisms. The infection of common strains that cause epidemics can easily be identified within hours, as compared to the weeks it can take to culture. This technique is able to detect the presence of viral DNA well before the virus has reached the levels required to initiate disease response. It is commonly used to measure the amount of HIV in the blood as a marker of disease progression and thus aids in determining the most effective antiviral therapy. PCR-based assays are really fast and are often completed within 2–5 hours. It is because of their speed, the treatment can be implemented earlier. PCR is often the only test that can provide a diagnosis of HIV in neonates in the first week of life. The resulting early antiviral treatment may help in prevention of devastating neurological complications. In the study of HIV and AIDS, where such evidence exists, PCR-based assays have been embraced, and novel assays developed at Laval University may soon lead to widespread acceptance of this approach in bacteriology. Now, RT-PCR has been applied as a test to measure viral load with HIV. It may also be used with other RNA viruses such as measles, mumps etc. HCV PCR is a test used for the detection of hepatitis C virus.

Although PCR-based assays are more expensive and need more expertise than conventional diagnostic tests, there is a significant potential for savings through improved health care outcomes.

7.11.9 PCR Used in Forensic Science

The ability of PCR to amplify tiny amount of DNA enables even highly degraded samples to be analyzed. The efficiency of PCR can be used to obtain sequences from the trace amounts, for example, DNA present in hair, bloodstains, and other forensic specimens at the scene of crime. PCR is used to amplify millions of exact copies of DNA from a biological sample. DNA amplification with PCR allows DNA analysis on biological samples with as small as a few skin cells. Great care, however, must be taken to prevent contamination with other biological materials during the identification, collection, and preservation of a sample. Mitochondrial DNA analysis (mtDNA) can be used to examine the DNA from samples that cannot be analyzed by STR (a short sequence 1–13 nucleotides in length that can be repeated several times in random array; for example, in the human genome the most common STR is the [CA]*n*, where *n* is the number of repeats, and usually ranges from 5 to 20). Nuclear DNA must be extracted from samples for use in RFLP, PCR, and STR; however, such analysis is performed using mtDNA in older biological samples lacking nucleated cellular material, such as hair, bones, and teeth. Thus, in the investigation of cases that have gone unsolved for many years, mtDNA is extremely valuable. PCR is a much more sensitive technique and most importantly, PCR assays can be performed properly and are more agreeable to standardized quality control (QC) protocols as compared to others.

Review Questions

1. The stringency of PCR amplification can be controlled by altering the temperature at which the hybridization of the primers to the target DNA occurs. How does altering the temperature of hybridization effect the amplification? Suppose that you have a particular yeast gene *A* and you wish to see if it has a counterpart in humans, how does controlling the stringency of the hybridization help you?
2. PCR is typically used to amplify DNA that lies between two known sequences. Suppose that you want to explore DNA on both sides of a single known sequence. Devise a variation of the usual PCR protocol that would enable you to amplify an entirely new genomic terrain.
3. A successful PCR experiment often depends on designing the correct primers. In particular, the T_m for each primer should be approximately the same. What is the basis of this requirement?
4. A gel pattern displaying PCR products shows four strong bands. The four pieces of DNA have lengths that are approximately in the ratio of 1:2:3:4. The largest band is cut out of the gel and PCR is repeated with the same primers. Again, a ladder of four bands is observed in the gel. What does this result reveal about the structure of the encoded protein?
5. What is the role of each of the following components of PCR?
   - Template
   - Primers
   - Polymerase
   - Nucleotides
6. Discuss the polymerase chain reaction (PCR), including the method itself and the components of the PCR reaction mixture. Finally, describe two applications of PCR in Modern Molecular Biology.
7. What is a primer? What are the parameters one must take into account when designing the primer for PCR amplification?
8. What is a molecular beacon probe? How does it work?

9. Explain the real-time fluorescent PCR with TaqMan® probe. How does this technique differ from real-time PCR with SYBR® Green?
10. Discuss the advantages and disadvantages of oligonucleotide-directed mutagenesis using PCR.
11. What is error-prone PCR and how is it useful?
12. What special property must be possessed by the DNA polymerase used in PCR, and why is this so important for doing PCR? How can you clone PCR products by taking advantage of the properties of the *Taq* DNA polymerase itself?
13. What is DDRT-PCR, and what is its main application?
14. In the process of cDNA library screening if you have ended up getting partial clones, how would you complete these partial clones with the help of PCR?
15. PCR invention has been credited to Karry Mullis, who was awarded the Noble Prize in Chemistry in 1993. Why was the discovery of a thermostable DNA polymerase so important for the development of PCR?

References

Frohman, M.A., M.K. Dush and G.A. Martin (1988). Rapid production of a full-length cDNAs from rare transcripts; Amplification using a single gene-specific oligonucleotide primer. *Proc. Natl. Acad. Sci.* **85**:8998–9002.

Loh, E. *et al.* (1989). Polymerase chain reaction with single-sided specificity: Analysis of T cell receptor delta chain. *Science* **243**:217–220

Mullis, K. (1990). The unusual origin of the polymerase chain reaction. *Sci. Am.* **262**:56–61.

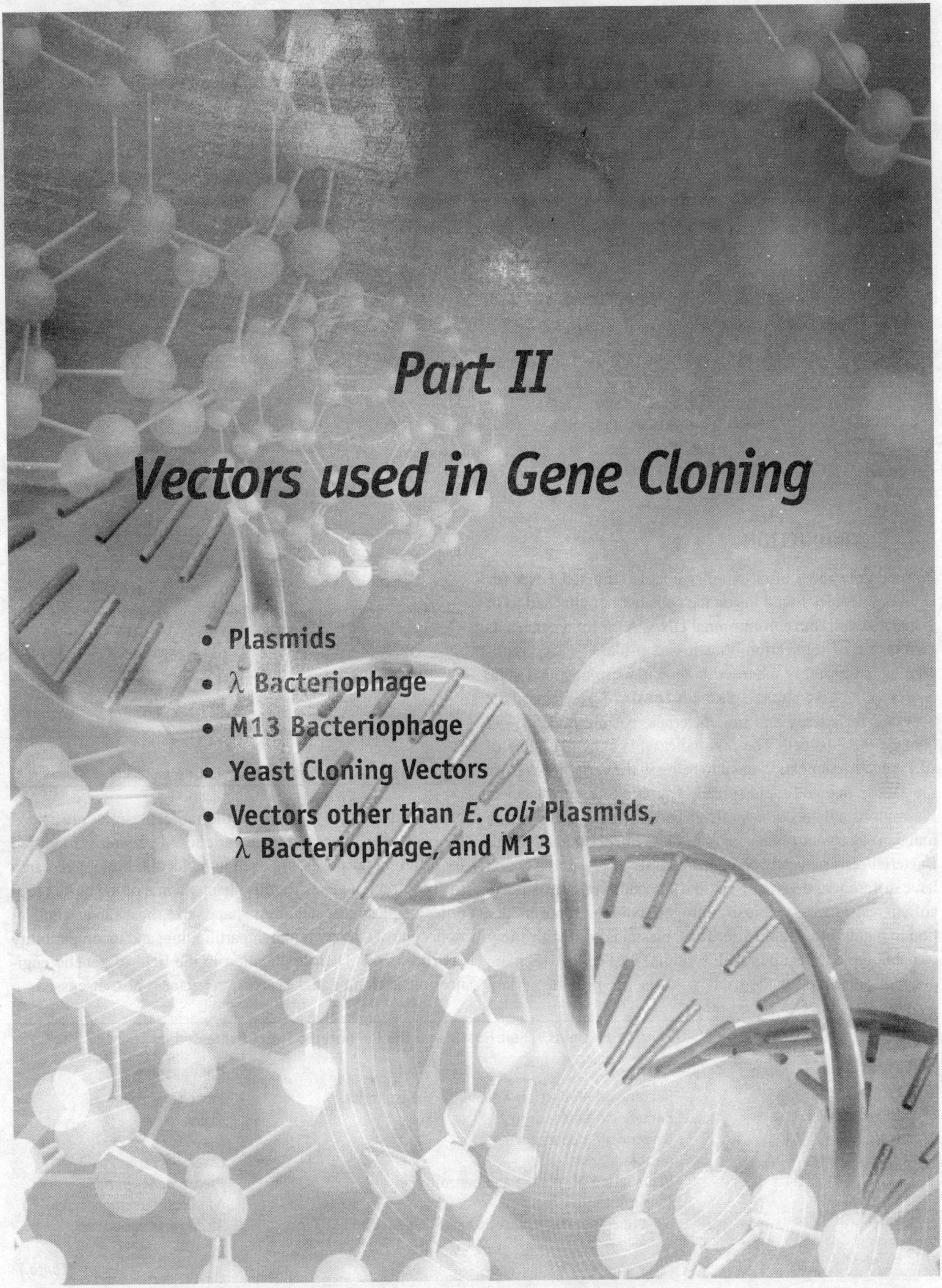

Part II

Vectors used in Gene Cloning

- Plasmids
- λ Bacteriophage
- M13 Bacteriophage
- Yeast Cloning Vectors
- Vectors other than *E. coli* Plasmids, λ Bacteriophage, and M13

8 Plasmids

Key Concepts

In this chapter we will learn the following:

- Definitions of plasmid
- Why is plasmid not considered genome?
- Plasmid vs. virus
- Plasmid size range
- Shapes of plasmid
- Replication of plasmids
- Plasmid host range
- Stable maintenance of plasmids
- Control of plasmid copy number
- Plasmid incompatibility
- Plasmid classification
- Plasmid purification
- Plasmids in genetic engineering

8.1 INTRODUCTION

Plasmids are most often circular double stranded DNA (ds DNA) molecules found inside the cells but not attached to or associated with the chromosomal DNA. A plasmid carries its own origin of replication, thus it is considered a true replicon, which is stably inherited in an extrachromosomal state (Figure 8.1). Like viruses, plasmids are also dependent on the host cell for energy and raw materials, but plasmids do not damage the host cell. The first indication of the existence of such genetic elements came into being in the early fifties from the appearance and rapid spread of multiple drug resistance. Plasmids usually occur naturally in bacteria, but are sometimes found in eukaryotic organisms (e.g., the 2-μm plasmid of yeast). Bacterial cells naturally contain a number of plasmids, which have different and varied functions like antibiotic resistance, colicin resistance, symbiosis, nitrogen fixation, tumor induction in plants, etc. (Table 8.1). Thus, plasmids confer a variety of phenotypic traits upon their bacterial host cells, which are not indispensable for the bacterial cells, but provide certain growth advantages to them. Furthermore, plasmids are a boon for genetic engineering; wide varieties of plasmids, modified for different purposes, are used in molecular biology research to carry foreign genes. In this chapter some of the basic properties of plasmids including shape, size, host range, replication, control of copy number, partitioning, and incompatibility are discussed. Their application in the field of genetic engineering as cloning and expression vector is also discussed.

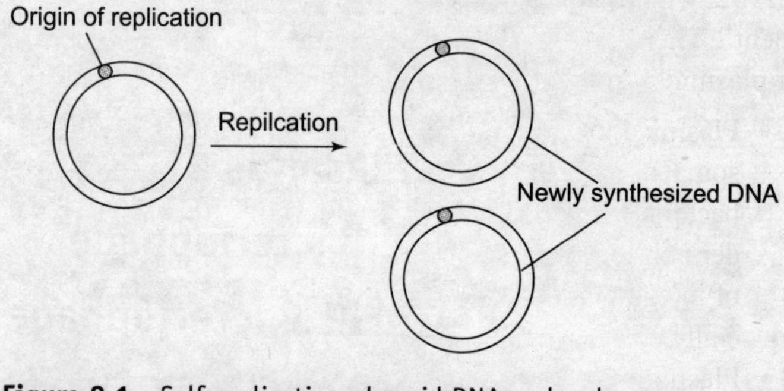

Figure 8.1 Self-replicating plasmid DNA molecule

Table 8.1 Examples of Naturally Occurring Plasmids, Their Hosts, and the Phenotypic Traits Expressed by Them

Plasmid	Trait	Host species
Col E1	Bacteriocin production which kills the bacteria	*Escherichia coli*
Col E1-K30	Bacteriocin production	*E. coli*
Col V	Bacteriocin production	*E. coli*
R6K	Antibiotic resistance	Enterobacteria
Tol	Degradation of toluene and benzoic acid	*Pseudomonas putida*
Ti	Tumor initiation in plants	*Agrobacterium tumefaciens*
pJP4	2,4-D(dichlorophenoxyacetic acid) degradation	*Alcaligenes eutrophus*

(Contd.)

Table 8.1 (*Contd.*)

Plasmid	Trait	Host species
pSym	Nodulation on roots of legume plants	*Rhizobium meliloti*
SCP1	Antibiotic methylenomycin biosynthesis	*Streptomyces coelicolor*
RK2	Resistance to ampicillin, tetracycline, and kanamycin	*Klebsiella aerogenes*
FP2	Heavy metal tolerance	*Pseudomonas* sp.
2-μm plasmid	Cryptic plasmid	*Saccharaomyces cerevisiae*
Mauriceville plasmid	Cryptic	*Neurospora* mitochondria
3-μm plasmid	Carries host genes	*S. cerevisiae*
pAl 2-1	Longevity	*Podospora anserina*
OCT plasmid	Metabolite utilization	*Pseudomonas* sp.
N3	Restriction modification system	Promiscuous plasmid
Kalilo	Senescence	*Neurospora crassa*
Satellite DNA plasmids	Gas vacuole synthesis	*Halobacterium* sp.
Col V	Siderophore synthesis (iron transport)	Enterobacteria
S1 and S2	Sterility	Mitochondria of *Zea mays*
Ent P307	Toxin synthesis	Enterobacteria
R46	UV protection	Promiscuous plasmid

8.2 DEFINITIONS OF PLASMID

The term plasmid was first introduced by Joshua Lederberg in 1952, who defined it as an 'extrachromosomal genetic element'. Thereafter, various definitions have been proposed for plasmids, some of which are as follows:

- Plasmids are autonomously replicating extrachromosomal circular DNA molecules, distinct from the normal bacterial genome and nonessential for cell survival under nonselective conditions. Some plasmids are capable of integrating into the host genome. A number of artificially constructed plasmids are used as cloning vectors.
- Plasmids are independent, free-floating circular pieces of DNA in a bacterium, capable of making copies of itself in the host cell. Plasmids can be used in recombinant DNA experiments to clone genes from other organisms and make large quantities of DNA.
- Plasmid is a structure in cells consisting of DNA that can exist and replicate independently of the chromosomes. In various organisms that have been studied, it appears that plasmids interfere with gene activity.
- Plasmids are small circular pieces of DNA in bacteria that resembles the bacterial circular chromosome, but is dispensable. Some bacterial strains contain many plasmids and some contain none. Plasmids are often used in genetic engineering as cloning vectors.
- Plasmids are self-replicating (autonomous) circle of DNA distinct from the chromosomal genome of bacteria. A plasmid contains genes normally not essential for cell growth or survival. Some plasmids can integrate into the host genome, be artificially constructed in the laboratory, and serve as vectors in cloning.
- Plasmids are small self-replicating ring of DNA found in many bacteria and some yeasts. These are widely used in genetic modification because these are able to pass easily from one cell to another.
- Plasmids are extrachromosomal genetic elements found in bacteria, which are not essential for growth. These usually contain genetic information for resistance to an antimicrobial agent or for degradation of additional substrates.
- Plasmids are genetic particles physically separate from the chromosome of the host cell (chiefly bacterial) that can stably function and replicate and are not essential to the cell's basic functioning.
- Plasmids are DNA molecules that can replicate independently of the chromosome, and are often used in cell regulation.
- Plasmids are loops of DNA in bacteria, which float in the cytoplasm.
- Plasmids are extrachromosomal genetic material that is not essential for growth and has no extracellular form.
- Plasmids are autonomously replicating DNA found in a wide variety of bacterial species; most plasmids have a narrow host range and can be maintained only in a limited set of closely related species.
- Plasmids are extrachromosomal elements, which behave as accessory genetic units that replicate and are inherited independently of the bacterial chromosome.
- Plasmids are genetic elements with a variety of mechanisms to maintain a stable copy number of the plasmid in their bacterial hosts and to partition plasmid molecules accurately to daughter cells.
- Plasmids are extrachromosomal DNA molecules that are dependent, to a greater or lesser extent, on the enzyme and proteins encoded by the host for their replication and transcription.

- Plasmids are DNA molecules that frequently contain genes coding for enzymes advantageous to the bacterial host. These genes specify a remarkable diverse set of traits, many of which are of great medical and commercial significance. Among the phenotypes conferred by plasmids are resistance to and production of antibiotics; degradation of complex organic compounds; and production of colicins, enterotoxins, and restriction and modification enzymes.

8.3 WHY IS PLASMID NOT CONSIDERED GENOME?

Plasmids are extrachromosomal, autonomous self-replicating ds DNA molecules (seldom RNA). Although plasmids reside in living cells and carry genetic information, these are not regarded as part of the cell's genome or chromosome because of the following reasons:

- A particular plasmid can be found in cells of different species, and can easily mobilize from one host species to another.
- A plasmid may sometimes be present or absent from the cells of a particular host species.
- The genetic information present on plasmids is not an essential part of the cell's genetic make-up, and the plasmids are not needed for cell growth and division under normal conditions.

8.4 PLASMID vs. VIRUS

Plasmids can be considered to be independent life-forms like viruses because both are capable of autonomous replication in the suitable host, which provides the replication enzymes, energy, and raw materials. However, unlike viruses, plasmids do not possess protein coats, cannot leave the host cell, and lack the ability to move from cell to cell. Moreover, unlike virus–host relationship, the plasmid–host relationship tends to be more symbiotic than parasitic. This is because plasmids can also endow their hosts with useful packages of DNA to assist mutual survival in stress conditions. For example, plasmids can confer antibiotic resistance to host bacteria, which can then survive along with their life-saving guests, i.e., plasmids, and carry them to future generations. Furthermore, viruses replicate in the host cell and usually destroy them to be released as infectious particles and infect fresh host cells. In contrast, plasmids replicate along with their host cell from generations to generations. When the host cell divides, the plasmid also divides, and each daughter cell gets a copy of the plasmid (Figure 8.2).

8.5 PLASMID SIZE RANGE

Plasmids are widely distributed throughout the prokaryotes, vary enormously in size from 1 to over 200 kbp, and are generally nonessential for cell growth and division. The F plasmid of *E. coli* is average in size and is ~1% of the size of the *E. coli* chromosome. However, multicopy plasmids are much smaller (Col E plasmids are about 10% of the size of the F plasmid); a few large F plasmids can be up to 10% of the size of a chromosome. Note, as it is difficult to work with large plasmids, only a few have been properly characterized.

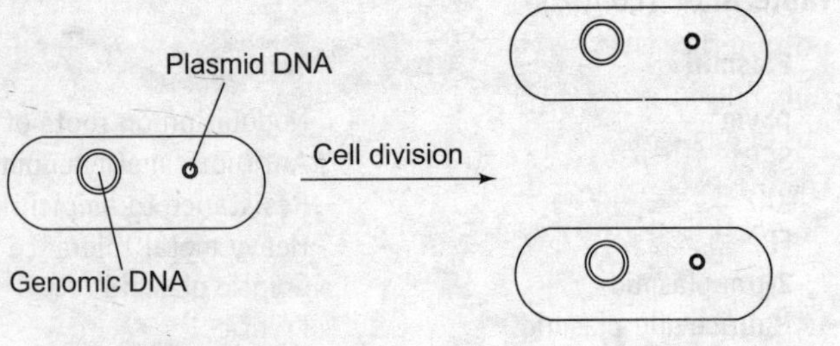

Figure 8.2 Plasmid replicates with the genomic DNA

Some higher organisms also contain plasmids, although these are less common than in bacteria. For example, most strains of yeast comprise of a circular ds DNA plasmid, the 2-μ circle or 2-μm plasmid (for details see Chapter 11).

8.6 SHAPES OF PLASMIDS: COVALENTLY CLOSED CIRCULAR AND LINEAR PLASMIDS

A majority of the plasmids are circular ds DNA molecules, while some plasmids are linear ds DNA molecules. The linear plasmids have been found in various bacteria, fungi, and higher plants. The linear plasmids of *Borrelia burgdorferi* and *Streptomyces* sp. are best-characterized; these bacteria also contain linear chromosomes. Unlike linear chromosomes of eukaryotes, these linear DNA plasmids are not protected by telomeres; rather two different mechanisms are adopted for their protection. For example, in *Borrelia*, repeated sequences are present that end in terminal single stranded DNA (ss DNA) hairpin loop, and in *Streptomyces*, inverted repeats are present at the ends which are covalently bound by proteins. The occurrence of linear plasmids have also been reported among eukaryotes. For example, the fungus *Flammulina velutipes,* (enoki mushroom) has two very small linear plasmids within its mitochondria. One single linear plasmid is also found in the cytoplasm of *Kluyveromyces lactis* (dairy yeast). The physiological roles of these plasmids are not yet established.

Some RNA plasmids have also been discovered in plants, fungi, and even animals; their presence is rare and is mostly poorly characterized. For example, some strains of yeast, *S. cerevisiae,* contain linear RNA plasmids. Similarly, RNA plasmids are found in the mitochondria of some varieties of maize plants. Both ss RNA and ds RNA plasmids are found, and these replicate in a manner similar to certain RNA viruses. The RNA plasmid encodes RNA-dependent RNA polymerase that directs its own synthesis. RNA plasmids do not contain genes for coat proteins unlike RNA viruses. It has been proposed that these RNA plasmids may have evolved from RNA viruses that have lost their ability to move from one cell to the other as virus particles.

Plasmid DNA of a particular size may occur in one of the five different conformations, which move at different speeds during electrophoresis in a gel. The different conformations, in the order of electrophoretic mobility from slowest to fastest, are as follows: (i) Open-circular DNA (OC DNA) with one nicked strand; (ii) Relaxed covalently closed circular DNA (CCC DNA) with both the strands fully intact, but enzymatically relaxed (supercoils removed); (iii) Linear DNA with free ends, either because both strands have been cut or because the DNA is linear *in vivo*; (iv) A fully intact supercoiled (or covalently closed circular) DNA with both strands uncut, and with a twist built in, resulting in a compact form; (v) Supercoiled denatured DNA with unpaired regions that make it slightly less compact; this can result from excessive alkalinity during plasmid preparation. The OC DNA, relaxed CCC DNA, and supercoiled CCC DNA are interconvertible (Figure 8.3). At low voltage, the rate of migration for small linear fragments is directly proportional to the voltage applied, while large linear fragments (>20 kbp or so) migrate at a certain fixed rate regardless of the length. At a specified, low voltage, the migration rate of small linear DNA fragments is a function of their length. However, the larger fragments (>20 kbp) move at continuously yet different rates, at higher voltages, and the resolution of a gel decreases with

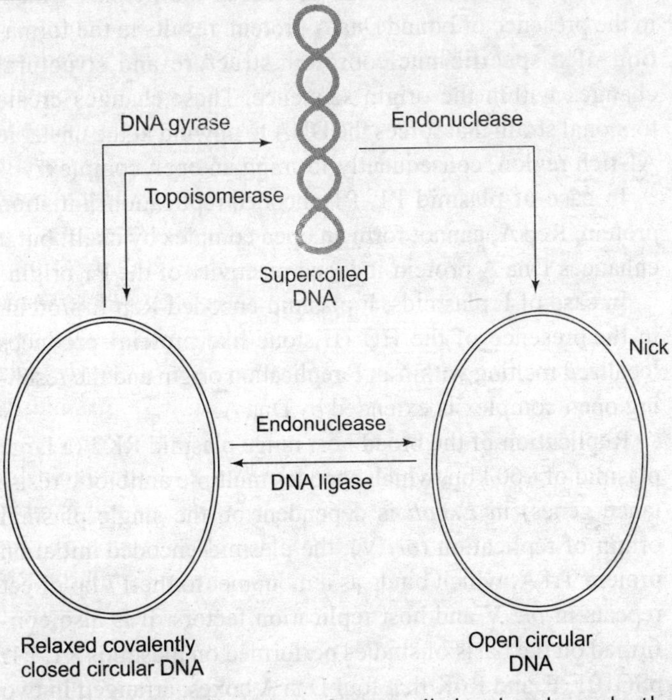

Figure 8.3 Interconversion of supercoiled, relaxed covalently closed and open circular DNA

increased voltage. Because of its tight conformation, supercoiled DNA migrates faster through a gel than linear or OC DNA. Note that the purified plasmids are usually analyzed after linearization by digesting with restriction enzymes.

8.7 REPLICATION OF PLASMIDS

During the division of a bacterial cell, chromosomal DNA duplicates and the replication machinery of the cell also duplicates the plasmid DNA. Plasmid must be able to replicate independently from the host chromosome and the copies of the plasmid are divided equally between the daughter cells. The replication of the plasmid does not harm the cell. As the plasmid carries its own origin of replication (*ori*), it is considered to be a true replicon. The origin of replication of a plasmid is a distinct segment of DNA that is associated with *cis*-acting controlling elements. Plasmids rely on the host cell for their energy requirement, raw materials, and many enzyme activities. Hence, a plasmid replicon is simply defined as the smallest piece of plasmid DNA that is able to replicate autonomously and maintain normal copy number. There are a number of replicons that have been identified in different plasmids. Generally pMB1 or its close relative, the colicin E1 (Col E1) replicon, are present in plasmids, which are commonly used in genetic engineering. These plasmids maintain 15–20 copies in each bacterial cell; however, these replicons are modified extensively over the period to increase their copy number. Replicons present in different plasmids and their corresponding copy numbers are given in Table 8.2.

Table 8.2 Replicons Present in Plasmids

Plasmid	Replicon	Copy number
pBR322	pMB1	15–20
pUC	Modified pMB1	500–700
pMOB45	pKN402	15–118
pACYC	p15A	18–22
pSC101	pSC101	~5
Col E1	Col E1	15–20

8.7.1 Modes of Replication of Circular ds DNA Plasmid

The two alternative mechanisms of replication of plasmids containing circular ds DNA are as follows:

Theta or Bidirectional Mode of Replication

This type of replication occurs in most of the common Gram negative bacterial plasmids, for example, Col E1, RK2, and F. The enzymes involved in replication recognize an origin of replication, unwind the DNA, and replication begins. Then two replication forks move around the circular plasmid DNA in opposite directions until collision with each other (Figure 8.4).

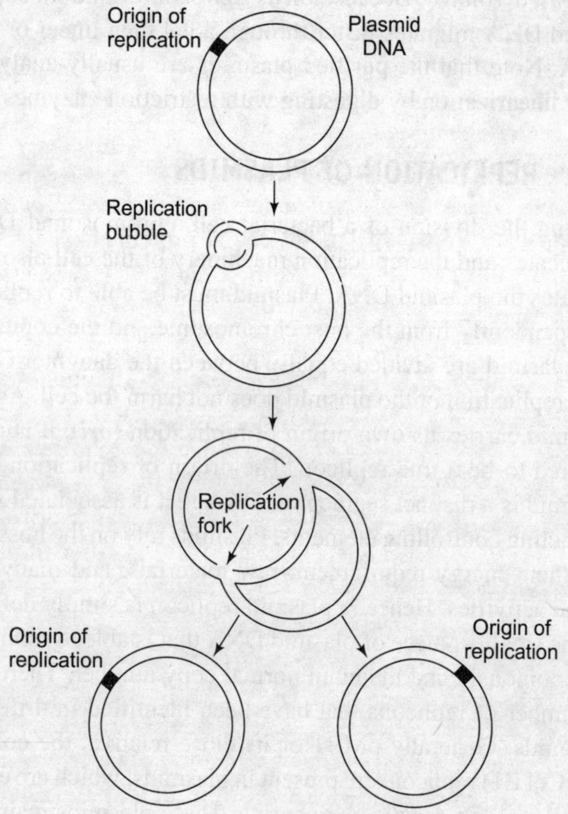

Figure 8.4 θ or bidirectional mode of replication

Finally, when both ds DNA circles have been completed, two distinct plasmids are generated. Some smaller plasmids have only one replication fork that travels around the circle until it returns to the origin.

Rolling Circle Mode of Replication

This occurs in some Gram positive bacterial plasmids. The process of replication is presented in Figure 8.5. Here, one strand of the ds DNA molecule is nicked at the origin of replication, and the broken strand separates from the circular strand. The gap generated by the separation is filled with new DNA, starting at the origin of replication. Newly synthesized DNA displaces the linear strand until the circular strand is completely synthesized. The processes of rolling and filling continue. Finally, the original broken strand is completely unrolled and the circular strand is fully paired with a newly synthesized strand of DNA.

8.7.2 Process of Replication

The plasmid DNA is most commonly circular ds DNA, and replicates like miniature bacterial chromosomes. Some undergo bidirectional mode of replication, while some follow rolling circle mechanism. Linear plasmids, however, follow altogether different modes of replication.

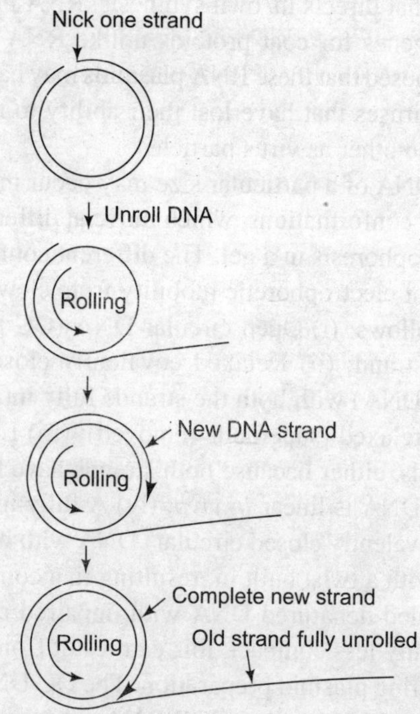

Figure 8.5 σ or rolling circle mode of replication

Process of Replication of Circular ds DNA Plasmids

The replication origin (*ori*) in Gram negative bacterial plasmid contains one or more clusters of repeats known as iterons. These include one or more essential binding sites for the Dna A protein (Dna A box) and an AT-rich region. For these plasmids, replication initiation involves binding of a plasmid-encoded replication initiation protein to the iterons, which, in the presence of bound Dna A protein, results in the formation of a specific nucleoprotein structure and structural changes within the origin sequence. These changes create torsional strain that forces the DNA to unwind at the unstable AT-rich region, consequently forming an open complex.

In case of plasmid P1, P1-encoded replication initiation protein, Rep A, cannot form an open complex by itself, but it enhances Dna A protein-induced reactivity of the P1 origin.

In case of F plasmids, F plasmid-encoded Rep E protein, in the presence of the HU (Histone-like protein) produces localized melting within an F replication origin and the resulting open complex is extended by Dna A.

Replication of the broad host range plasmid RK2 (a large plasmid of ~60 kbp, which encodes multiple antibiotic resistance genes) in *E. coli* is dependent on the single plasmid origin of replication (*ori* V), the plasmid-encoded initiation protein Trf A, which binds as a monomer to the 17 bp direct repeats in *ori* V and host replication factors. It is also confirmed on the basis of studies performed on plasmids R1, P1, pSC101, F, and R6K that four Dna A boxes, arranged in two pairs in an inverted orientation with respect to each other are present within the RK2 minimal origin. It has been shown that the Dna A protein binds to Dna A consensus sequences within the RK2 origin. Thus, specific binding of *E. coli* Dna A protein to four Dna A boxes localized upstream of the Trf A binding sites takes place. It is of interest that the requirement of Dna A protein for *in vitro* replication of *ori* V using a reconstituted system with purified components is substantially lower than the concentration of Dna A protein required for *ori* C (*E. coli* origin) replication. This observation suggests different structural features of the Dna A–DNA nucleoprotein complex for the two origins of replication. The formation of a cruciform-like structure at the minimal RK2 origin is worth considering since the internal Dna A boxes (box 2 and box 3) can form a hairpin structure (duplex length of 5 bp with a 5-nucleotide loop; ΔG of –3.8 kcal). It has been investigated that the binding of Dna A protein to a DNA hairpin structure in the pBR322 origin is essential. Furthermore, deletion of one of the Dna A boxes significantly reduces Dna A-dependent replication of pBR322, and deletion of both Dna A binding sites abolishes template activity. This confirms the possibility that for the initiation of RK2 replication, binding of the Dna A protein to a hairpin structure is important for helicase delivery to the plasmid origin. The RK2 initiation protein, Trf A, in the presence of HU protein can produce localized opening at the AT-rich region of *ori* V. In simple words, Trf A protein opens the RK2 origin. The Trf A-induced opening does not absolutely require the presence of nucleotides; however, the formation of the open complex is somewhat enhanced by ATP or ATPγS. On the contrary, ATP is required for *E. coli ori* C open complex formation mediated by the Dna A protein. The Dna A protein stimulates and/or stabilizes this RK2 open complex formation, but cannot, on its own, form an open complex. *E. coli* Dna A initiator protein, Dna B helicase, Dna C helicase loader, Dna G primase, DNA gyrase, DNA polymerase III holoenzyme, and SSB proteins are required for RK2 replication in crude extracts as well as in an *in vitro* replication system reconstituted with purified components. The binding of the Trf A protein to the iterons at the minimal RK2 origin of replication has been characterized in detail; however, little is known about the nature of binding of the Dna A protein to *ori* V. The HU and IHF (integration host factor) proteins have been shown to stimulate *ori* C replication (chromosomal) *in vivo* and *in vitro*. It has been reported that Trf A-mediated *ori* V opening is strictly dependent on HU in the absence of Dna A protein. In contrast, HU is not necessary when both Trf A and Dna A proteins are present. Also, at any HU concentration tested, the Dna A protein alone cannot open the RK2 origin. Thus, both the Dna A and the HU proteins can stimulate and/or stabilize Trf A-mediated origin melting; the two proteins show differences in their ability to enhance the formation of the open complex. The ribonucleotides other than ATP are not required for the opening at the RK2 origin. Dna A or HU protein is required for full opening of *ori* V by Trf A. Similarly, the formation of an open complex at the replication origin of F plasmid by Rep E protein requires HU protein and is enhanced by Dna A. HU protein has also been shown to enhance Dna A-mediated opening of the origin of *E. coli*. The mechanism by which HU or Dna A protein enhances open complex formation is not known. Recently, it has been shown that binding of Dna A protein induces a bend in *ori* C DNA. Similarly, HU protein binds nonspecifically to ds or ss DNA, and this binding either induces bending and folding or alters the superhelicity of supercoiled DNA. Thus, it is possible that Dna A or HU-dependent DNA bending is responsible for the observed increase and/or stabilization of Trf A-mediated *ori* V opening. Interestingly, although Dna A can perform a role similar to that of HU in the formation of the open complex at the RK2 origin, it is unable to open *ori* V by itself or in the presence of HU. This is in contrast to *ori C* and origin region of plasmid P1. The addition of the Rep A protein increases Dna A-induced strand opening at the P1 origin region, but Rep A alone does not form an open complex. The ability of the plasmid-specified initiation protein, in the case of RK2, to bring about the formation of an open complex in the absence of a specific host protein may be a contributing factor, accounting for the different replication activities observed for plasmids RK2 and P1. Thus, RK2 is a broad host range replicon, whereas the P1 replicon appears to have a more limited host range. Although Dna A protein by itself does not induce open complex formation, it is required for RK2 replication *in vivo* and *in vitro*, and it enhances or stabilizes Trf A-induced open complex formation. More recently, it has been shown that Dna A protein plays a critical role in the delivery of *E. coli* Dna B to RK2 origin. In the absence of Trf A protein, an *ori* V · Dna A · Dna B · Dna C complex can be isolated using a gel filtration method. These observations indicate at least two roles for the Dna A protein during RK2 replication initiation: (i) Enhancement of the formation of an open complex; and (ii) Helicase delivery to the origin region. Trf A protein in turn forms an open complex at the 13-mer region and in concert with the Dna A protein induces specific nucleoprotein structural changes at the origin, which allows proper positioning of the Dna B helicase within the open complex. The helicase is activated by these events, and replication is allowed to proceed. Finally, the formation of a specific nucleoprotein structure containing four Dna A boxes (which potentially can form a cruciform structure) that bind Dna A protein and 17 bp iterons that bind the Trf A protein, appear to be critical factors in the initiation of RK2 replication and may also contribute to the unique broad host replication properties of this plasmid.

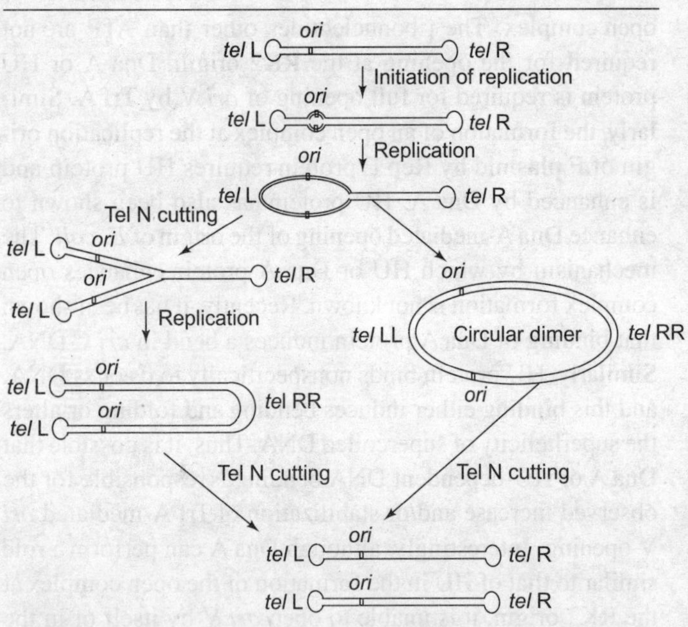

Figure 8.6 Replication mechanism of linear plasmid of *Borrelia burgdorferi*

Process of Replication of Linear ds DNA Plasmids

Linear plasmids have an altogether different mode of replication. Because of the 5′ → 3′ polarity of DNA replication, the propagation of all linear replicons requires a solution to the 'end replication problem', i.e., the need to provide both a template and a primer for DNA synthesis at telomeres. The replication of linear plasmid of *Borrelia burgdorferi* is well-characterized, which is described in Figure 8.6.

As will be evident in following sections, replication is central to the control of a number of important plasmid properties: host range, copy number, incompatibility, and mobility.

It was proposed that the replication mechanism of linear plasmids found in these bacteria is quite similar to replication mechanism of coliphage N15, a temperate phage, which, as a prophage, replicates as a linear plasmid with covalently closed ends. Basically, replication mechanism of N15 is considered as a model for the replication of other replicons of similar structure, such as *Borrelia* chromosome and linear plasmids. The prophage of coliphage N15 is not integrated into the bacterial chromosome but exists as a linear plasmid molecule with covalently closed ends. Upon infection of an *E. coli* cell, the phage DNA circularizes via cohesive ends. A phage-encoded enzyme, protelomerase, then cuts at another site, *tel* RL, and forms hairpin ends (telomeres). Thus, N15 protelomerase acts as a telomere-resolving enzyme responsible for processing of replicative intermediates during prophage DNA replication. Replication is initiated at an internal *ori* site located close to the left end of plasmid DNA (located within *rep* A) and proceeds bidirectionally. *In vitro*, purified Tel N processes circular and linear plasmid DNA containing the proposed target site *tel* RL to produce linear ds DNA with covalently closed ends and exerts its activity as cleaving-joining enzyme in a concerted action. *In vivo tel* N has been proved to be necessary and sufficient to maintain linear, N15 mini-plasmids. After replication of the left telomere, protelomerase cuts this sequence and forms two hairpin loops, *tel* L. Electron microscopy data suggest that after duplication of the left telomere, protelomerase cuts this site generating Y-shaped molecules. After duplication of the right telomere (*tel* R), the same enzyme resolves this sequence producing two linear plasmids. Thus, full replication of the molecule and subsequent resolution of the right telomere then results in two linear plasmid molecules. Alternatively, full replication of the linear prophage to form a circular head-to-head dimer may precede protelomerase-mediated formation of hairpin ends. N15 prophage replication thus appears to follow a mechanism that is distinct from that employed by eukaryotic replicons with this type of telomere and suggests the possibility of evolutionarily independent appearances of prokaryotic and eukaryotic replicons with covalently closed telomeres.

8.8 PLASMID HOST RANGE

The host range of plasmids, i.e., range of bacteria in which plasmid replicates, varies widely. The *ori* region and replication proteins encoded by plasmids determine their host range. Note that the plasmids encode only one or a few of the proteins required for their own replication. All the other proteins required for replication, e.g., DNA helicases, DNA polymerases, and DNA ligase, etc., are encoded by host chromosome. Note that as the plasmid-encoded replication proteins are located very close to the *ori* site, it is possible to delete other parts, and instead, insert foreign sequences into the plasmid without affecting plasmid replication. This feature of plasmid has greatly simplified the construction of versatile cloning vectors.

On the basis of host range, there are two types of plasmids: (i) Narrow host range plasmids; and (ii) Broad or wide host range plasmids.

8.8.1 Narrow Host Range Plasmids

Some plasmids because of specificity of their origin of replication can replicate in only one species of host cell. These are called narrow host range plasmids. For example, plasmids in which *ori* region is derived from Col E1 have a restricted host range: these only replicate in enteric bacteria, such as *E. coli, Salmonella*, etc. Most of the plasmids used in gene cloning are more restricted in their host range.

8.8.2 Broad Host Range Plasmids

Some plasmids have less specific origins of replication and can replicate in a number of bacterial species. These are called broad host range plasmids. Plasmids with a broad host range encode most, if not all of the proteins required for replication. For example, plasmids of the P-family can easily survive in hundreds of different species of bacteria. Because of their unusually wide host range, P-family plasmids are called promiscuous plasmids; however, these plasmids were originally named after *Pseudomonas*, the bacterium in which these were first discovered. Plasmids like RSF1010 and RK2 are noncon-jugative, but can be transformed into a wide range of Gram negative and Gram positive bacteria, where these are stably maintained. As discussed above, replication of RK2 plasmid requires *ori* V and a plasmid-encoded protein Trf A. RK2 plasmid is replicated with the help of host-encoded proteins/enzymes, which are different in different species. While RSF1010 is more independent, it encodes many of the proteins for replication of the plasmids and requires very few of the host-encoded proteins. These genes can be easily expressed, so their promoter and ribosome binding sites must have evolved such that these can be recognized in a diversity of bacterial families. These are often responsible for resistance to multiple antibiotics, including penicillin. Such promiscuous plasmids are very important for genetic engineering; these offer the potential of readily transferring cloned DNA in a wide range of genetic backgrounds.

8.9 STABLE MAINTENANCE OF PLASMIDS

The stable maintenance of plasmids in cells requires a specific partitioning mechanism; the loss of plasmids due to defective partitioning is called segregative instability. All naturally occurring plasmids are stably maintained as these have a partitioning function, *par*, which is responsible for their stable maintenance at each cell division. These *par* regions are indispensable for stability and maintenance of single copy number plasmids (i.e., plasmids having one copy per cell). Partition systems require both *trans*-acting factors and *cis*-acting elements located on the plasmid and may involve attachment of the plasmid to the cell membrane. The bacterial chromosome uses a similar partition mechanism. High copy number plasmids (i.e., plasmids having >1 copy per cell) lack specific partition mechanisms and rely on the high probability that each daughter will receive at least one copy. However, the *par* region is also found in high copy number plasmids like Col E1, although this has been deleted in many Col E1 derived cloning vectors, e.g., pBR322. To maintain a high copy number, *par* region from a plasmid such as pSC101 is cloned into pBR322; otherwise plasmid-free cells arise under nutrient limitation or other stress conditions. Reports suggest that DNA superhelicity is involved in the partitioning mechanism; pSC101 derivatives lacking the *par* locus show decreased overall superhelical density when compared with wild-type pSC101. Furthermore, partition-defective mutants of pSC101 and similar mutants of unrelated plasmids are stabilized in *E. coli* by a mutation (*top* A) which increases negative DNA supercoiling *top* A. On the other hand, DNA mutations in DNA gyrase and gyrase inhibitors increase the rate of loss of *par*-defective pSC101 derivatives.

Plasmid instability may also occur due to the formation of multimeric forms of the plasmid. The mechanism that controls the copy number of a plasmid ensures a fixed number of plasmid origins per bacterium. Cells containing multimeric plasmids have the same number of origins of replication but fewer plasmid molecules, which leads of segregative instability in the absence of a partitioning function. However, some plasmids like Col E1 have a natural method of resolving multimers back to monomers. The plasmid encodes a site-specific recombination system, which counteracts homologous recombination events leading to multimerization, and ensures that monomers are available for partition. It contains a highly recombinogenic site (*cer*). If the *cer* sequence occurs more than once in a plasmid, as in a multimer, the host's Xer protein promotes recombination, thereby regenerating monomers. A similar function encoded by *xer* D and *xer* C genes of *E. coli* ensures chromosome monomerization. The yeast 2-μm plasmid also encodes a site-specific recombination system, which increases its copy number (for details see Chapter 11).

8.10 CONTROL OF PLASMID COPY NUMBER

Copy number of a plasmid is defined as the average number of plasmids per bacterial cell or per chromosome under normal growth conditions. Plasmids vary widely in their copy number, and on this basis there are two types of plasmids: low copy number plasmids (or single copy plasmids) and high copy number plasmids (or multicopy plasmids). In ideal conditions, the copy number of a plasmid remains constant as the population of the plasmid doubles at exactly the same rate as the population of the host cell. The copy number depends on the origin of replication, which determines whether these plasmids are under relaxed or stringent control, and the size of the plasmid and its associated insert, and is thus determined by the regulatory mechanisms controlling replication. Origins of replication and copy numbers of various plasmids are given in Table 8.3. Both high copy and low copy plasmids regulate their copy numbers carefully. This indicates that the mechanisms that regulate the copy numbers of plasmids are of major importance for the maintenance of plasmids in bacteria. On the basis of copy number and replication regulatory mechanisms, plasmids are categorized into two groups, which are explained in the following sections:

Table 8.3 Origins of Replication and Copy Numbers of Various Plasmids

DNA construct	Origin of replication	Copy number	Classification
pUC vectors	pMB1	500–700	High copy
pBlueScript vectors	Col E1	300–500	High copy
pGEM vectors	pMB1	300–400	High copy
pTZ vectors	pMB1	>1,000	High copy
pBR322 and derivatives	pMB1	15–20	Low copy
pACYC and derivatives	p15A	10–12	Low copy
pSC101 and derivatives	pSC101	~5	Very low copy

8.10.1 High Copy Number Plasmids (Relaxed Plasmids)

Plasmids of this category are present in multiple copies (>1 copy/cell) in the bacterial cell. These plasmids are also known as relaxed copy number plasmids, and the controlling mechanism is relaxed control. Multicopy control systems allow multiple initiation events per cell cycle, with the result that there are several copies of the plasmid per bacterium. Once the number of plasmids in the cell reaches a certain level, the initiation of plasmid replication is controlled. High copy number plasmids are normally nonconjugative and have low molecular weights. These plasmids can replicate without plasmid-encoded *de novo* protein biosynthesis. These are entirely dependent on functional, long-lived enzymes and proteins supplied by the host, including DNA polymerase I, DNA-dependent RNA polymerase, ribonuclease H (RNase H), DNA gyrase, topoisomerase I, and chaperone. Several multicopy plasmids illustrate quite unusual behaviour, as these continue to replicate while protein synthesis is inhibited by amino acid starvation or by addition of an antibiotic such as chloramphenicol. As *de novo* protein synthesis is required for the initiation of each round of host DNA synthesis but not for plasmid replication, the content of plasmid DNA in cells exposed to chloramphenicol amplifies relative to the amount of the chromosomal DNA. After several hours of incubation, thousands of copies of a relaxed plasmid may accumulate in the cell; at the end of the process, plasmid DNA may account for 50% or more of the total cellular DNA. Examples included in this type are plasmids carrying the pMB1/Col E1 replicon.

Usually antisense RNA acts as positive regulatory molecule in regulating the copy number of relaxed plasmids. The details of this mechanism are best-studied for the multicopy plasmid pMB1/Col E1 (Figure 8.7). Initiation of replication occurs within a 600-nucleotide region that contains all of the *cis*-acting elements required for replication with the transcription of a preprimer, which can later on, after processing, work as a primer for DNA synthesis. This primer is called RNA II. The synthesis of precursor of RNA II is initiated by RNA polymerase-catalyzed transcription initiated at a promoter 550 bp upstream of the origin, and proceeding through the origin terminating at one of a number of closely-spaced sites located ~150 nucleotides downstream of the origin. The 5′-end of the ~750 nucleotide primary transcript folds into a complex secondary structure that aligns a G-rich loop in RNA II with a C-rich stretch of plasmid DNA located 20 nucleotides upstream of the origin on the template strand. Thereafter, preprimer RNA, called RNA II, is processed by RNase H, which cleaves the RNA within a sequence of five A residues at the origin. The resulting 555 nucleotide mature RNA II with a free 3′-OH group is used by DNA polymerase I as a primer to initiate leading strand synthesis. The stable DNA:RNA hybrid [pairing between RNA and DNA occurs just upstream of origin (around position –20) and also farther upstream (around position –265)] is elongated on the complementary strand of DNA at which discontinuous synthesis of the lagging strand is initiated. Since lagging strand synthesis is blocked ~20 nucleotides upstream of the origin by the unhybridized segments of RNA II, replication occurs unidirectionally in Cairns or theta structures. RNA II can only act as primer if it is cleaved by RNase H to leave a free 3′-OH group; if RNase H fails to cleave the preprimer RNA II, no free 3′-OH is available and replication cannot proceed. Replication process is controlled by another small (108 nucleotides) antisense RNA molecule called RNA I, which is encoded by the same region of DNA as RNA II but by the complementary strand. The complementary regions of RNA II and RNA I begin to base pair, forming regions of bubbles. Eventually, the entire sequence aligns and a ds RNA molecule is formed, which is resistant to RNase H action. Therefore, RNA II is not processed, and does not function as a primer thereby restricting replication to proceed. When the copy number of the plasmid is high, RNA I is synthesized. Furthermore, a plasmid-encoded protein called Rom (RNA I modulator)/Rop (Repressor of primer) facilitates copy number maintenance. Rom/Rop is a homodimer of a 63 amino acid polypeptide encoded by a gene located 400 nucleotides downstream from the Col E1 origin of replication. Each subunit of the dimer comprises of two α helices joined by sharp bend, and the dimer is a tight bundle of four α helices having two-fold symmetry. The sequences and structural elements in both RNA I and RNA II are recognized by Rom proteins. This protein enhances the pairing between RNA I and RNA II so that processing of primer is inhibited even at reasonably low concentrations of RNA I.

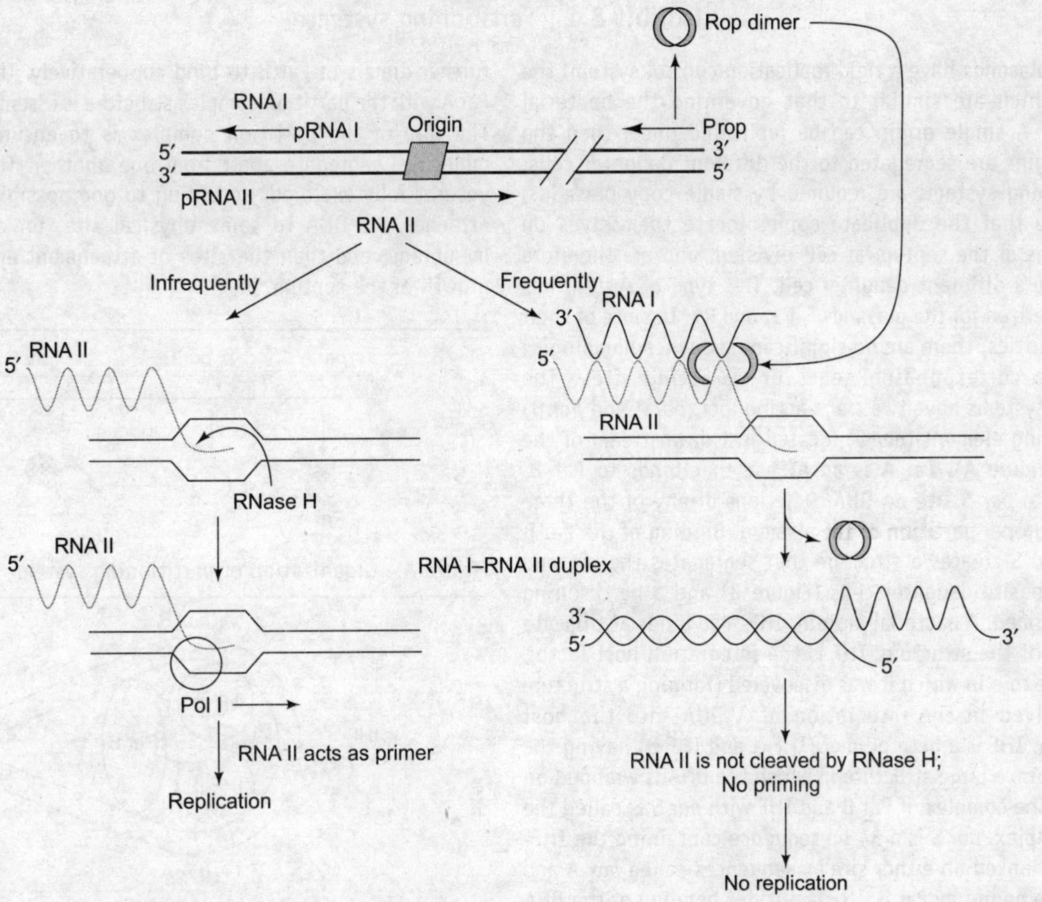

Figure 8.7 Regulation of replication of Col E1-derived plasmids

As Rom binds to RNA I and RNA II with similar affinities, the unstable intermediates of two complementary RNAs are driven into a more stable structure. In conclusion, Rom proteins bind to the stem of the RNA hybrid and stabilize the kissing complex formed between RNA I and RNA II. There are reports showing that deletion of *rom/rop* gene increases the copy number of Col E1 plasmids by two orders of magnitude. For instance, deletion of the *rom/rop* genes increases the copy number of pBR322 from 15–20 copies to >500 copies per bacterial cell. The relative copy numbers of pBR322 and pUC18 illustrate the role of Rop/Rom perfectly. There are 15–20 copies of pBR322 per cell; however, there are 50–100 copies of pUC18, which is derived from pBR322. The relevant difference between the two plasmids is the fact that pBR322 contains the *rop* gene and pUC18 does not. Furthermore, pUC plasmids have a single mutation (G $\rightarrow$ A), one nucleotide upstream of the initiation of RNA I, which decreases negative regulation of RNA II resulting in high copy number (500–700 copies per cell).

The mechanism of control of replication described above allows increase in copy number of relaxed plasmids to 1,000–3,000 copies/cell after inhibition of protein synthesis by chloramphenicol. On this basis, a relaxed plasmid is also defined as a plasmid, which does not require continued protein synthesis for its replication, and its copy number increases if protein synthesis is blocked, due to the removal of negative regulator protein.

8.10.2 Low Copy Number Plasmids (Stringent Plasmids)

Copy numbers of some plasmids are strictly controlled and correlated with the number of chromosomal DNA molecules, and these are maintained at a limited number of copies per cell. With some exceptions plasmids with low copy numbers usually have a relatively high molecular weight and are conjugative. Low copy number plasmids have stringent copy control as their division is more tightly regulated and such plasmids replicate only once during the cell cycle. Single copy control systems resemble that of the bacterial chromosome and result in one replication per cell division. Single copy plasmids have a partitioning system (Exhibit 8.1). These require active protein synthesis but no DNA polymerase I for their replication. Many of the broad host range and low copy number plasmids contains three to seven copies of an iteron sequence, which is 17–22 bp long at their *ori* region. The *rep*

Exhibit 8.1 Partitioning system

Single copy plasmids have a rigid replication control system, the effects of which are similar to that governing the bacterial chromosome. A single origin can be replicated once; then the daughter origins are segregated to the different daughter cells. The partitioning systems are required by single copy plasmids, which ensure that the duplicate copies locate themselves on opposite sides of the septum at cell division, and are therefore segregated to a different daughter cell. This type of system has been characterized for the plasmids F, P1, and R1. In spite of their overall similarities, there are no significant sequence homologies between the corresponding genes or *cis*-acting sites. The partitioning systems have two *trans*-acting loci (*par* A and *par* B) and a *cis*-acting element (*par* S) located just downstream of the two genes (Figure A). Par A is an ATPase that binds to Par B, which binds to *par* S site on DNA. Deletions of any of the three loci prevent proper partition of the plasmid. Binding of the Par B protein to *par* S creates a structure that segregates the plasmid copies to opposite daughter cells (Figure B) and a partitioning complex is formed. A bacterial protein, IHF, also binds at this site to form part of the structure. IHF is the integration host factor, named for the role in which it was discovered (forming a structure that is involved in the integration of λ DNA into the host chromosome). IHF is a heterodimer (IHF α and IHF β) having the capacity to form a large structure in which the DNA is wrapped on the surface. The complex of Par B and IHF with *par* S is called the partition complex. *par* S is a 34 bp sequence containing the IHF-binding site flanked on either site by sequences called *box* A and *box* B that are bound by Par B. IHF facilitates bending of the DNA so that Par B can bind simultaneously to the separated *box* A and *box* B sites. Complex formation is initiated when *par* S is bound by heterodimer of IHF together with a dimer of Par B. This enables further dimers of Par B to bind cooperatively. The interaction of Par A with the partition complex structure is essential but transient. The role of the partition complex is to ensure that two DNA molecules segregate apart from one another. The mechanism is yet not fully explored, according to one possibility the complex attaches the DNA to some physical site, for example, on the membrane, and then the sites of attachment are segregated by growth of the septum.

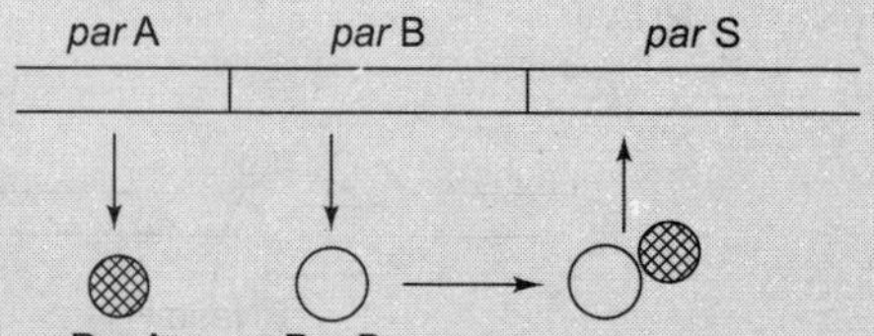

Figure A Organization of partitioning system

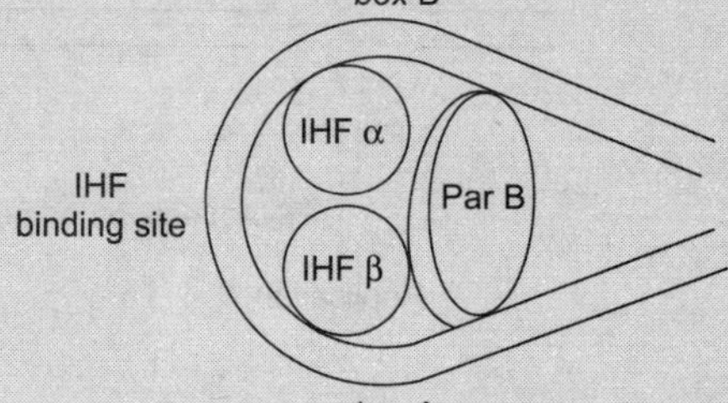

Figure B Partitioning complex

Source: Benjamin L (2005) *Genes VIII*, Pearson Prentice Hall, Pearson Education, Inc. NJ.

A gene, which is located closer to the *ori* region, plays an important role in regulation of copy number by encoding the Rep A protein. This protein acts positively on the origin of replication and negatively regulates the transcription of its own gene. Rep A is very unusual because it acts only *in cis*. Rep A protein binds to the iterons and initiates DNA synthesis. These plasmids require ongoing synthesis of the Rep A protein for replication; their copy number cannot be amplified, nor their yield increased, by inhibiting cellular protein synthesis. Such plasmids are said to replicate under stringent control. This indicates that the plasmids replicating under stringent control maintain harmony between their rate of replication and that of the host by rationing the supply of a molecule affecting the frequency of initiation of plasmid synthesis. Examples included in this category are pSC101 and R1.

Stringent control of copy number is regulated by two mechanisms related to Rep A. Firstly, the Rep A protein represses its own synthesis by binding to its own promoter region and blocking transcription of its own gene. If the copy number is high, synthesis of Rep A is repressed. The copy number and concentration of Rep A is decreased after cell division and replication is initiated. Mutations in the Rep A protein can lead to increased copy number. Secondly, two plasmids can be linked together with the help of Rep A protein by binding to their iteron sequences, thus preventing them from initiating replication. This mechanism which is known as 'handcuffing' the replication of iteron plasmids is regulated both by the concentration of Rep A protein and plasmids. pSC101 is the best-studied example of a plasmid whose replication is controlled by iterons. Among them two sites are inverted repeat sequences which are located near the *rep* A promoter. Rep A binding to these two sites inhibits its own expression. Three other 18 bp tandem repeats are also present, which may serve to sequester Rep A so that it is not available for replication.

Besides Rep A-mediated regulation of replication as described above, antisense RNA is also involved in the regulation of copy number. This is best-illustrated for R1 plasmid, a large (100 kbp) broad host-range plasmid that replicates unidirectionally from a single origin of replication (*ori* R). Rep A protein specifically initiates replication *in vitro* of plasmid DNA bearing *ori* R. R1 plasmid replication is triggered by

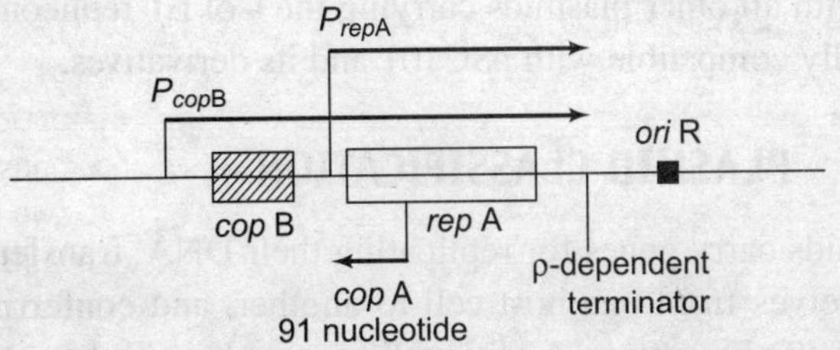

Figure 8.8 Organization of regulatory elements of R1 plasmids

interaction of Rep A and Dna A proteins with the *ori* R sequence. Rep A binding is required to permit Dna A to bind to the origin, which in turn allows Dna B to bind and start the assembly of a replisome. The replication process is strictly dependent on the added Rep A protein, and is independent of host RNA polymerase, but requires other host replication functions such as Dna B, Dna C, SSBs, and DNA gyrase. The replication is also completely dependent on the host Dna A function. Dna A protein specifically binds to a 9 bp Dna A recognition sequence (Dna A boxes) within *ori* R only, when Rep A is bound to the sequence immediately downstream to the Dna A box. The plasmid copy number is regulated by *rep* A expression, as Rep A is essential for replication. Expression of *rep* A occurs primarily as a result of transcription from the *rep* A promoter. However, when the R1 plasmid enters a cell for the first time, *rep* A can also be expressed due to transcription from the *cop* B promoter. The organization of these elements is shown in Figure 8.8. The Cop B (small; 10 kDa protein) regulates negatively transcription from the *rep* A promoter. In R1 plasmid, antisense RNA also controls the copy number, however, the mechanism is different from that of Col E1. Expression from the *rep* A promoter is also regulated by the *cop* A RNA. This small (91 nucleotide) antisense RNA (*cop* A) is complementary to the leader sequence of *rep* A transcript (called *cop* T) and hybridizes with it. Both *cop* A and *cop* T RNAs form secondary structures and mutational analysis has shown that the interaction between them may involve these secondary structures rather than full duplex formation (a so-called kissing complex). Thereafter, Shine–Dalgarno ribosome binding site is sequestered, as a consequence of which 16S rRNA in the ribosome is unable to be positioned correctly and translation of the mRNA does not take place.

A stringent plasmid (low copy number plasmid) does not undergo plasmid amplification upon inhibition of protein synthesis and may be defined as a plasmid, the replication of which requires continued protein synthesis, and its copy number does not increase if protein synthesis is inhibited.

8.11 PLASMID INCOMPATIBILITY

Plasmid incompatibility is the inability of two different plasmids to coexist in the same cell in the absence of selection pressure. The term incompatibility can only be used when it is certain that entry of the second plasmid has taken place and that DNA restriction is not involved. Groups of plasmids, which are mutually incompatible, are considered to belong to the same incompatibility (Inc) group. Based on incompatibility, plasmid families are designated by alphabets (F, P, I, X, etc.). Table 8.4 lists some plasmids belonging to different incompatibility groups. It is quite possible to have two or more types of plasmids in the same cell as long as these belong to different families. More than 30 incompatibility groups have been defined in *E. coli* and one in plasmids of *S. aureus*. Plasmids of one incompatibility group are related to each other, but cannot survive together in the same bacterial cell. This ensures that the cell contains a wider selection of plasmids, since only plasmids not closely related can live in the same cell.

Table 8.4 Examples of Plasmids Belonging to Different Incompatibility Groups

Incompatibility group	Plasmids
FI	F, R386
FII	R1, R100
FIII	Col B-K99, Col B-K166
FIV	R124
I	R62, R64, R483 (at least 5 subgroups)
J	R391
N	R46
O	R724
P	RP4, RK2
Q	RSF1010
T	R401
W	R388, S-a

Plasmids are incompatible if these share elements of the same replication machinery, as these compete with each other during replication and the subsequent step of partitioning into daughter cells. Plasmids containing the same replicon belong to the same incompatibility group and are unable to coexist within the same bacterial cell. However, plasmids having replicons whose components are not interchangeable belong to different incompatibility groups and can be easily maintained in the same cell. It has been discovered that incompatible plasmids share more sequence homology than compatible plasmids, and in the case of conjugative plasmids, usually specify similar transfer systems, hence incompatibility between independently isolated heterologous plasmids reflects their similarity or evolutionary relatedness. As a purpose of characterizing plasmids, incompatibility provides a feature of distinction that applies to both conjugative and nonconjugative plasmids.

The genetic control of incompatibility is complex and several interactions are involved, resulting in not so clear-cut method of categorizing plasmids. Variations of incompatibility properties and interactions have been discovered, such as unidirectional incompatibility, where a plasmid can displace another plasmid, although these are compatible if the later plasmid is to act as the superinfecting agent. Examples of dual compatibility have also been reported, where a plasmid is incompatible against more than one Inc groups, and a few more cases of low levels of incompatibility have occurred with members of one or several groups. Plasmids, which fall into these categories pose a problem for their classification into Inc groups, and are thus left unassigned in terms of incompatibility grouping. Incompatibility may have different causes including randomization during replication and partition, and systems that stabilize plasmid maintenance by killing plasmid-free segments from within may give rise to incompatibility effects. Randomization takes place during replication as well as during partition and leads to distortion of the ratio between the numbers of molecules of each of the two derivatives in individual cells. Partition incompatibility, however, is weaker than replication incompatibility and the copy number of each of the two plasmids is unaffected by randomization and contributes nearly equally to the degree of incompatibility observed. As discussed in Section 8.10, *trans*-acting molecules control copy numbers of plasmids. For example, in the case of Col E1 plasmids, RNA I acts *in trans*, and in the case of R1 plasmids, *cop* A RNA and Cop B both act *in trans*. If a cell contains two different plasmids, each of which has the same mechanism of control, then because control is exercised through *trans*-acting molecules, each plasmid is able to control the replication of the other. The expected consequence of this is that one of the two plasmids is eventually lost from the cell, simply as a result of random partitioning of plasmids into daughter cells during cell division. Thus, the two plasmids appear to be incompatible. Plasmids, which have different control mechanisms, replicate independently of one another and each is partitioned between daughter cells. Thus, both plasmids are maintained together. By introducing segments of the DNA into an unrelated multicopy replicon, and determining the ability of a test plasmid to coexist with the hybrid, the region of plasmid DNA conferring incompatibility can be identified. Several well-known plasmids and their negative control elements, which are responsible in regulating their replication, are given in Table 8.5. Most of the plasmid vectors that are used in genetic engineering carry a replicon derived from the plasmid pMB1. These vectors are incompatible with all other plasmids carrying the Col E1 replicon, but are fully compatible with pSC101 and its derivatives.

Table 8.5 Controlling Elements of Different Incompatibility Groups

Incompatibility groups	Negative control element	Remark
Col E1, pMB1	RNA I	Processing of pre-RNA II into primer is controlled
Inc FII, pT181	RNA II	Synthesis of Rep A protein is controlled
P1, F, R6K, pSC101, p15A	Iterons	Rep A protein is sequestered

8.12 PLASMID CLASSIFICATION

Plasmids carry genes for replicating their DNA, transferring themselves from one host cell to another, and conferring a variety of phenotypes. Additionally some plasmids have killer (*kil*) systems or 'post-segregational killing system (PSK)', or addiction systems, which are discussed in Exhibit 8.2. There are number of ways by which naturally occurring plasmids are classified into different groups depending upon their properties. The major characteristics on the basis of which plasmids are classified into different groups are described in this section.

8.12.1 Conjugative Ability

The naturally occurring plasmids are classified into two major groups termed as conjugative or nonconjugative, depending upon whether or not these carry a set of transfer genes. Besides, some plasmids exhibit the property of being mobilized into other cell in the presence of conjugative plasmids. These plasmids are called mobilizable plasmids.

Conjugative Plasmids

Conjugation is a process in which DNA is directly transferred from a donor cell to a recipient cell. This transfer process is facilitated by plasmid-encoded *tra* (transfer) genes. The *tra* genes consist of a cluster of ~40 different genes, which encode for the synthesis of pili and other surface components. The formation of pili between donor and recipient cells, which permit physical contact between the cells, is essential for conjugation. For example, F factor and other plasmids belonging to the incompatibility classes FI, FII, FIII, and FIV are conjugative. In general, conjugative plasmids (Tra^+) are of relatively high molecular weight and are present as one to three copies per cell, whereas nonconjugative plasmids are of low molecular weight and are present as multiple copies per cell. An exception is the conjugative plasmid R6K, which has high molecular weight and multiple copies.

Transfer of F Plasmid During Conjugation

The conjugation process is mediated by the F (fertility) plasmid, which is the classic example of an episome, an element that may exist as a free circular plasmid, or that may become integrated into the bacterial chromosome as a linear sequence. The F factor or F plasmid is a large circular DNA molecule of ~100 kbp in length. The F plasmid can integrate at several sites in the *E. coli* chromosome, often by a recombination event involving certain sequences (IS sequences) that are present on both host chromosome and F plasmid. In its free (plasmid) form, the F plasmid utilizes its own replication origin (*ori* V)

Exhibit 8.2 Plasmid addiction system

Some plasmids have killer (*kil*) systems or postsegregational killing system (PSK), or addiction systems working on the basis that 'we hang together or we hang separately', which ensure that a bacterium carrying a plasmid can survive only as long as it retains the plasmid. So, bacteria that lose the plasmid certainly die, and the bacterial population is anticipated to retain the plasmid indefinitely. There are several ways to ensure that a cell dies if it is cured of a plasmid (cells lacking plasmid), all sharing the principle demonstrated in Figure A. The plasmid produces both a poison and an antidote. The poison is a killer substance (a protein which is lethal to host cell) that is relatively stable, whereas the antidote consists of a substance that blocks the killer substance, but is relatively short-lived, acting as an antagonist to killer protein itself or an inhibitor of its synthesis. These systems take various forms. One specified by the F plasmid consists of killer and blocking proteins. F plasmid consists of two genes, *ccd* A coding for an antidote and *ccd* B encoding a toxin. Ccd A is readily degraded by the host cell proteases. As long as the plasmid is present, it constantly produces Ccd A, which binds Ccd B and blocks its action. But as the plasmid is lost, Ccd A is readily degraded; consequently, Ccd B kills the cell by inhibiting DNA gyrase and generating double stranded breaks in the bacterial chromosome. Another interesting example of this system is *hok*/*sok* (host killing/suppressor of killing) system of plasmid R1 in *E. coli*. The plasmid R1 has a killer in the form of mRNA, which is translated into a toxic protein, while the antidote is a small antisense RNA that prevents expression of the mRNA. During bacterial cell division, daughter cells that retain a copy of the plasmid survive, while a daughter cell that fails to inherit the plasmid dies or suffers a reduced growth-rate because of the lingering poison from the parent cell.

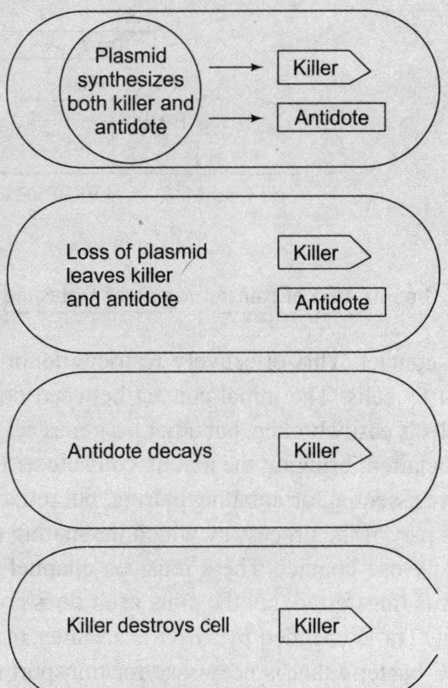

Figure A Plasmid addiction system

Source: Benjamin L (2005) *Genes VIII*, Pearson Prentice Hall, Pearson Education, Inc. NJ.

and control system, and is maintained at a level of one copy per bacterial chromosome. When it is integrated into the bacterial chromosome, this system is suppressed, and F DNA is replicated as a part of the chromosome. The presence of the F plasmid, whether free or integrated, has important consequences for the host bacterium. Bacteria that are F-positive (F^+) are able to conjugate with bacteria that are F-negative (F^-). Conjugation involves a contact between the donor F^+ and recipient F^- bacteria, which is followed by transfer of the F factor. If the F factor exists as a free plasmid in the donor F^+ bacterium (male), it is transferred as plasmid, and the infective process converts the F^- recipient (female) into F^+ state (transconjugant). If the F factor is present in an integrated form in the donor, the transfer process may also cause some or the entire bacterial chromosome to be transferred. Many plasmids have conjugation systems that operate in a similar manner, but the F factor is the first to be discovered and remains the paradigm for this type of genetic transfer.

Mechanism of Conjugation

During conjugation, transfer of the F factor is initiated at a site called *ori* T, the origin of transfer, which is located at one end of the transfer region, and a ss DNA is transferred. In this section, the mechanism of conjugation is described in detail.

The F plasmid (~100 kbp) comprises of a large transfer region (~33 kbp) that includes ~40 genes. The *tra* locus is shown in Figure 8.9. The loci includes *tra* and *trb* genes. The *tra* Y-I unit is expressed in coordination as part of a single 32 kbp transcription unit while *tra* M and *tra* J are expressed separately. Tra J encoded by *tra* J acts as a regulator that turns on both *tra* M and *tra* Y-I. On the opposite strand, *fin* P (fertility inhibition) is a regulator that codes for a small RNA antisense to *tra* J, thereby turning off *tra* J activity. Its activity requires expression of another gene, *fin* O. Four of the *tra* genes in the major transcription unit *tra* Y-I are concerned directly with the transfer of DNA during conjugation; most are involved with the properties of the bacterial cell surface and in maintaining contacts between mating bacteria. The *tra* A codes for the single subunit, pilin, that is polymerized into the pilus. At least 12 *tra* genes are required for the modification and assembly of pilin into the pilus. The F pili are hair-like structures, 2–3 μm long, that protrude from the bacterial surface and a typical F^+ cell has 2–3 pili. The pilin subunits are polymerized into a hollow cylinder ~8 nm in diameter, with a 2 nm axial hole. Mating is initiated when the tip of F pilus contacts the surface of the recipient F^- cell. A donor cell does not contact other cells carrying the F factor, because the genes *tra* S and *tra* T code for surface exclusion proteins, which make the cell a poor recipi-

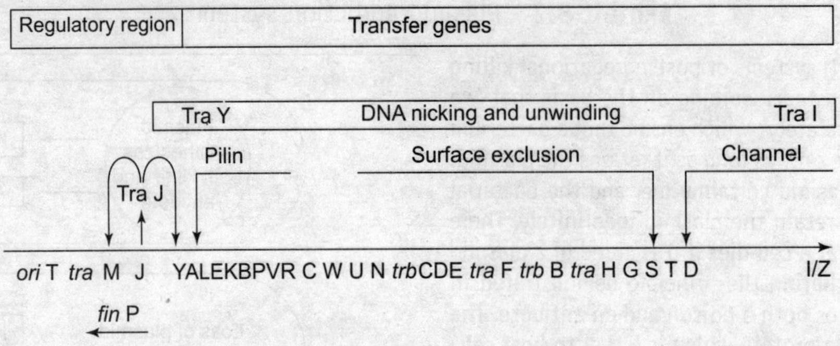

Figure 8.9 Organization of transfer region of F plasmid

ent in each contact. This effectively restricts donor cells to mating with F^- cells. The initial contact between donor and recipient cells is easily broken, but other *tra* genes act to stabilize the association, bringing the mating cells closer together. The F pili are essential for initiating pairing, but retract or disassemble as part of the process by which the mating cells are brought into close contact. There must be channel through which DNA is transferred, but the pilus itself does not appear to provide it. Tra D encoded by *tra* D is an inner membrane protein in F^+ bacteria that is necessary for transport of DNA and it may provide or be part of the channel. The transfer process is initiated when Tra M recognizes that a mating pair has formed. Then Tra Y binds near *ori* T and causes Tra I to bind. Tra I is a relaxase, which generates nick at a unique site (called *nic*) in *ori* T, and then forms a covalent link with the 5′-end. Tra I also catalyzes unwinding of ~200 bp of DNA. The free 5′-end initiates the transfer of ss DNA from the donor F^+ bacterium (male) into the recipient F^- bacterium (female), which as a result is converted to F^+ (transconjugant). Only a single unit length of the F factor is transferred to the recipient bacterium, as some important event terminates the process after one revolution, after which the covalent integrity of F plasmid is reestablished. Subsequently, a complementary strand is synthesized in the donor F^+ bacterium to replace the strand that has been transferred; the process is called donor conjugal DNA synthesis. Thus, conjugating DNA usually appears like a rolling circle, but replication as such is not necessary to provide the driving energy, and single strand transfer is independent of DNA synthesis. In the recipient cell, the newly transferred ss DNA is used a template to generate a ds DNA molecule, which is then recircularized; the process is called repliconation. An outline of the process of conjugation is presented in Figure 8.10.

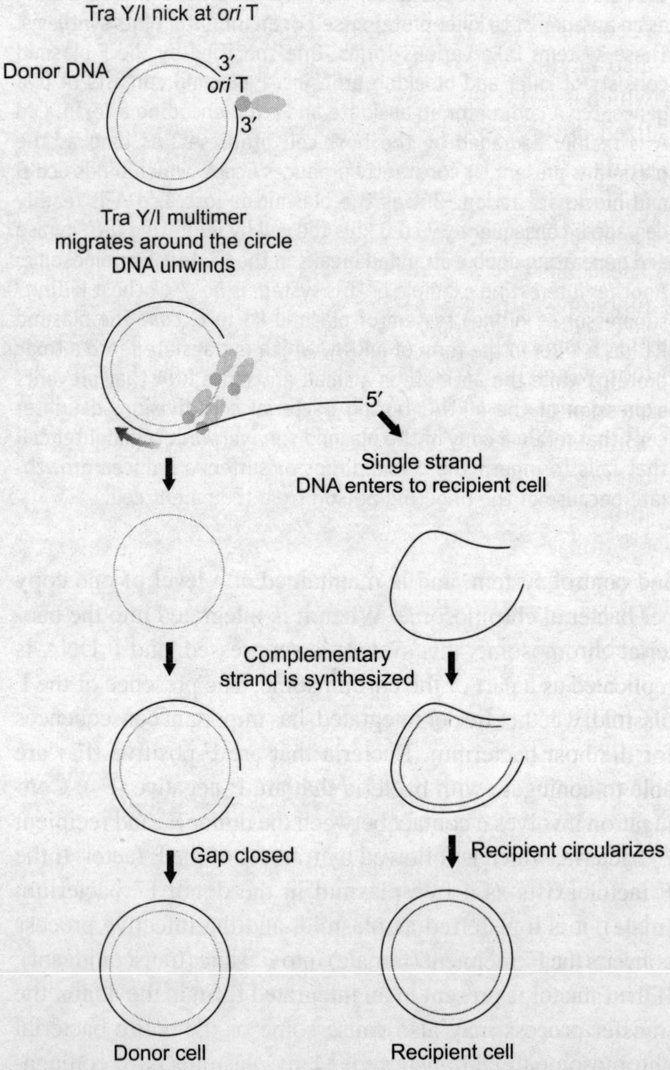

Figure 8.10 Transfer of ss DNA during conjugation
Source: Lewin B (2004) *Genes VIII*, Pearson Prentice Hall, Pearson Education Inc., pp 367.

In the case of integrated F plasmid, transfer of DNA begins with a short sequence of F DNA and continues until prevented by loss of contact between the bacteria. It takes ~100 min to transfer the entire bacterial chromosome, and under standard conditions, contact is often broken before the completion of transfer. The donor DNA that enters a recipient bacterium is converted to double stranded form, and may recombine with the recipient chromosome. Thus, conjugation process facilitates the exchange of genetic material between bacteria. A strain of *E. coli* with an integrated F factor supports such recombination at relatively high frequencies compared with strains that lack integrated F factors. Such strains are described as Hfr (high frequency recombination). Each position of integration for the F factor gives rise to a differentiated Hfr strain, with a characteristic pattern of transferring bacterial markers to a recipient chromosome. As contact between conjugating bacteria is usually broken before transfer of DNA is complete, the probability that a region of the bacterial chromosome will be transferred depends upon its distance from *ori* T. Bacterial genes located close to the site of F integration enter into recipient bacteria first, and are therefore found at greater frequencies than those located farther away and enter later.

Mobilizable Plasmid

A conjugative plasmid is not only self-transmissible, but may also facilitate the conjugal transfer of other plasmid DNA, which is not self-transmissible. Such plasmids that are mobilized by conjugative plasmids are called mobilizable plasmids. This is because mobilizable plasmids are nonconjugative in nature due to deficiency of transfer (*tra*) functions, but contain a specific region of DNA, called *mob*. These plasmids are thus phenotypically Tra^-, Mob^+. The *mob* functions are expressed by at least two regions on the plasmid DNA. The first region, *mob*, encodes a mobilizing protein, which binds the *nic/bom* region (*nic* = nick; *bom* = basis of mobility). The formation of this complex is coupled with a strand break at *nic/bom* region. Mobilization may occur *in cis* if the mobilizable plasmid DNA becomes covalently joined to the conjugative plasmid (*cis* mobilization or conduction) or *in trans* if another plasmid is mobilizable, i.e., it can respond to *trans*-acting mobilization factors such as the Tra Y and Tra I but does not encode such factors itself (*trans* mobilization or donation).

Examples of mobilizable plasmids are Col E1 and CloDF13, which can be mobilized effectively by F factor and F-like plasmids. Note that F plasmid is capable of mediating the transfer of chromosomal as well as plasmid DNA. The F plasmid is also capable of mediating the transfer of other plasmids, which are not capable of self-transfer. If a cell containing Col E1 is mixed with one that lacks it, no transfer occurs; but if Col E1 containing cell also contains the F plasmid, Col E1 is transferred to the recipient at almost the same frequency. The mechanism of the F plasmid-mediated transfer of Col E1 is analogous to, but not mechanistically same as, the transfer of chromosomal DNA by an Hfr (high frequency recombination) strain because Col E1 and F occur in recipient cell as separate circular molecules of DNA; some more subtle interactions must occur between the F plasmid and the other plasmid. The cotransformed plasmid is said to be mobilized by F. Thus, the loss of *mob* region is associated with the loss of F-mediated transfer of Col E1 plasmid.

The plasmids of Tra^-, Mob^- phenotypes are neither conjugative nor mobilizable. However, if these plasmids have lost only the *mob* region, which encodes for the mobilizing protein, their mobilization can be facilitated by another plasmid having functional Mob protein, which can act *in trans* on the Mob^- plasmid. Furthermore, the *nic/bom* region is a *cis*-acting element, which should be present on that particular plasmid. Nowadays various designer nonconjugative and transfer-deficient plasmids have gained significance as cloning vectors because of biological safety concerns. On the contrary, the plasmids having both functional regions, i.e., Tra^+, Mob^+, are not preferred, because these are both conjugative and mobilizable.

8.12.2 Functional Properties

The plasmids can also be categorized on the basis of their functional properties. There are six main classes and plasmids can belong to more than one of these functional groups.

Cryptic Plasmid

A plasmid that confers no identified characteristics or phenotypic properties on the host cell is referred to as cryptic plasmid. These types of plasmids presumably carry genes whose characteristics are still unknown. For example, yeast 2-μm plasmid (for details see Chapter 11) does not carry gene that may confer any phenotypic traits. Another example of cryptic plasmid is the mauriceville plasmid, a novel retroelement found in the mitochondria of certain *Neurospora* strains. This plasmid, termed a mitochondrial retroplasmid (mRP), is a small (3.6 kbp), closed circular ds DNA that encodes a reverse transcriptase.

Fertility (F) Plasmid

F plasmids are conjugative plasmids, which contain *tra* (for transfer) genes. The fertility factor also known as F factor or sex factor is a bacterial DNA sequence that allows a bacterium to produce a pilus necessary for conjugation. The details of conjugative plasmids have already been described in Section 8.12.1. It contains clusters of *tra* genes and a number of other genetic sequences responsible for incompatibility, replication, and other functions. The F factor is an episome, i.e., it may integrate into the bacterial genome, and be stably maintained in this form through numerous cell divisions, but at some stage exist as independent plasmid.

Resistance (R) Plasmids

Plasmids that carry and transmit antibiotic resistance genes from one bacterium to another are known as resistance (R) plasmids or resistance factors. These plasmids provide protection to bacteria both from human medicine and from antibiotics produced naturally in the soil. In the late 1950s, just after World War II, an epidemic of dysentery caused by *Shigella*, which was resistant to sulfonamides, broke out in Japanese hospitals. The genes that provide resistance to sulfonamide were found on plasmids, which were able to transfer their copies from one bacterial cell to another, and hence sulfonamide resistance also. Over that period it was observed that 80% of the dysentery-causing *Shigella* in Japan had become resistant to sulfonamides. Later, 10% of the *Shigella* in Japan was found to be resistant to four antibiotics, namely, sulfonamides, chloramphenicol, tetracycline and streptomycin. Although this resistance plasmid does not effect the survival of *Shigella*, but the transferable antibiotic resistance is very dangerous from the human medical viewpoint. Nowadays bacterial strains having resistance against multiple antibiotics have been naturally developed. Moreover, these strains carry resistance genes for different antibiotics on one single plasmid. Now, the transfer of multiple antibiotic resistance plasmids between bacteria has become a major concern from the medical perspective. This clinical problem can be critical in case of patients suffering from severe burns or infections after surgery, or having immunodeficient systems.

R plasmids are of moderate to large size, are present in one to two copies per host cell, and are self-transmissible at a low frequency; some derepressed mutants illustrate high transfer frequency and belong to a wide variety of incompatibility groups. R plasmids comprise of genes that confer resistance to one or more antibiotics on the host cell. The antibiotic resistance genes do not alter the vital cell component; rather these inactivate the antibiotic by covalent modification or expel it. The genes that provide antibiotic resistance are frequently located in or are derived from transposons. These genes are flanked by 0.8–1.4 kbp repetitive IS elements (insertion sequence), which can integrate into various sites of the plasmid or host chromosome. Therefore, in addition to mobilization of the plasmid by itself, these insertion sequences also play a very important role in the spread of antibiotic resistance.

The antibiotic resistance provided by plasmids usually does not modify vital cell components, however, at times, plasmids do provide an altered but still functional target component. In most of the cases the antibiotic may be inactivated by covalent modification or pumped out of the cell. Numerous antibiotic resistance genes that are used as selectable markers in genetic engineering have been derived from plasmids (for details see Chapter 16). The mechanisms of the most studied antibiotic resistances are discussed below.

Resistance to β-Lactam Antibiotics The β-lactam or penicillin family of antibiotics that includes penicillin, cephalosporin, ampicillin, and carbenicillin is the best-known and the most widely used group of antibiotics. These antibiotics have the β-lactam structure, which is a four-membered ring containing an amide group. The antimicrobial activity of these compounds is targeted to the cell wall of susceptible organisms, which leads to disintegration of the cell wall and prevents the cells from growing or dividing. Penicillin inhibits the final stage of transpeptidase-catalyzed cross-linking of peptidoglycans, which occurs outside the cell. In view of the fact that peptidoglycans are distinctive to bacteria, apart from some allergies, this group of antibiotics has almost no side effects in humans. Additionally, penicillins inhibit enzymes necessary for the rod-like structures of *E. coli* and for septum formation during division.

Resistance plasmids contain the gene (*bla*) encoding the enzyme, β-lactamase, which catalyzes the hydrolysis of the cyclic amide bond of the β-lactam ring, with concomitant detoxification of penicillins (Figure 8.11). The most prevalent β-lactamase in Gram negative bacteria, the TEM β-lactamase, is named after the initials of the Athenian from whom a strain of *E. coli* expressing the enzyme was first isolated. The TEM β-lactamase, which is widely used as selectable marker, is a 286 amino acid residue protein encoded by *bla* or *amp*^{r} gene. The 23 amino acids from the N-terminal function as a signal sequence and are cleaved during translocation of the protein in periplasmic space. Certain strains of *Pseudomonas* carrying the R plasmid RP1 that encodes a high activity, broad-spectrum β-lactamase can actually grow on ampicillin as the sole carbon and energy source.

An enormous number of improved penicillin derivatives have been manufactured by the pharmaceutical industry to make them less susceptible to breakdown by β-lactamase. However, their improvement has led to the appearance of

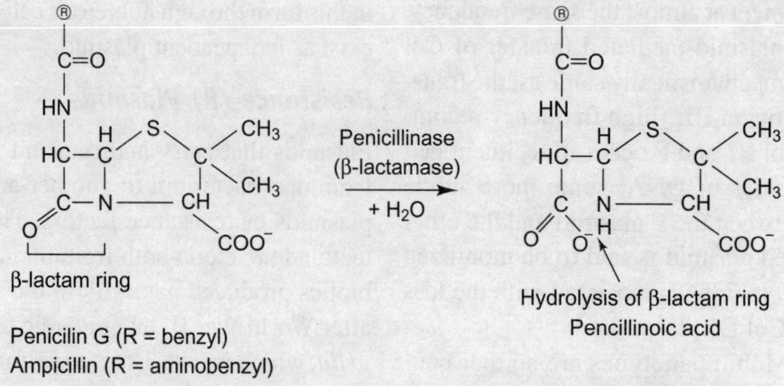

Figure 8.11 Inactivation of penicillin by β-lactamase

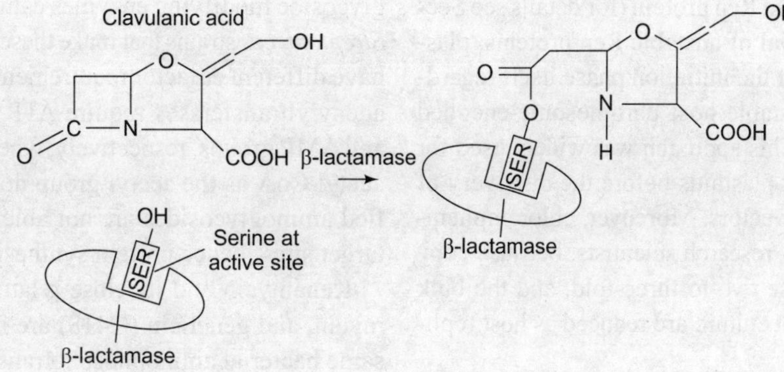

Figure 8.12 Inactivation of β-lactamase by clavulanic acid

altered and improved β-lactamases among bacteria carrying the R plasmids. Furthermore, another strategy has been devised to inactivate β-lactamase where a mixture of a β-lactam antibiotic and a β-lactam analog, clavulanic acid, that inhibits β-lactamase, are added (Figure 8.12). Clavulanic acid and its derivatives bind covalently to the active site of β-lactamase. As a result, the administered penicillin can now kill the bacteria, even though the bacteria contain the resistance genes.

Resistance to Chloramphenicol Chloramphenicol was first isolated from soil actinomycetes and was later synthesized and used as a broad-spectrum antibiotic. The clinical use of this antibiotic has been reduced as it induces bone-marrow toxicity and the bacteria readily develop resistance to chloramphenicol. Chloramphenicol has two hydroxyl groups, which are important for binding to the bacterial ribosomes. Chloramphenicol decreases the catalytic rate constant of peptidyl transferase positioned on 23S rRNA of 70S ribosome, and hence blocks protein synthesis.

R plasmids provide protection to bacteria by synthesizing the enzyme chloramphenicol acetyl transferase (CAT; encoded by *cat* gene). This enzyme catalyzes the addition of one acetyl group from acetyl coenzyme A to the C-3 hydroxyl group of the antibiotic (Figure 8.13). The resulting product, 3-acetoxychloramphenicol, cannot bind to the 70S ribosome. The *cat* gene is constitutively expressed and is usually carried

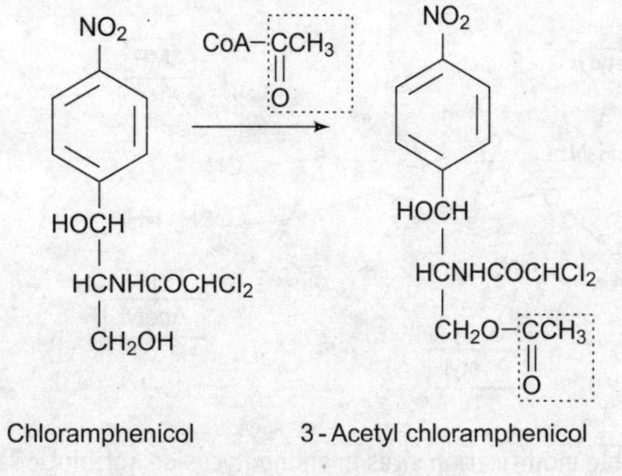

Figure 8.13 Inactivation of chloramphenicol

by plasmids that confer multiple drug resistance in some strains of Enterobacteriaceae and Gram negative bacteria. All products of the *cat* gene form trimers of identical subunits of M_r ~25,000. The type I variant is widely used as a reporter gene, which is encoded by 1,102 bp fragment of transposon *Tn* 9. The most characterized variant is type III as it forms a crystal suitable for X-ray analysis. Both the substrates reach the active site located on opposite sides of the molecule. The active site of the enzyme is positioned at the subunit interface, which has a conserved histidine residue that probably acts as a general base catalyst in the acetylation reaction.

Chloramphenicol also plays a role in plasmid amplification, an important requirement in genetic engineering experiments. The aim of amplification is to increase the copy number of a plasmid, i.e., to obtain good yields of plasmid. Some relaxed plasmids (multicopy plasmid with copy number of ≥20) have the useful property to replicate in the absence of protein synthesis, while bacterial chromosome cannot. This property can be made use of during the growth of a bacterial culture for plasmid DNA purification. After obtaining satisfactory cell density, chloramphenicol is added, and the culture is incubated for further 12 hours. During this time, chloramphenicol prevents protein synthesis thereby ceasing the replication of bacterial chromosome and blocking the cell division; however, the replication of relaxed plasmid, including all vectors having wild-type pMB1 or Col E1 replicon, continues until 1,000–3,000 copies have been synthesized. However, stringent plasmids (low copy number plasmids) do not undergo plasmid amplification upon inhibition of protein synthesis. This reflects that host chromosome-encoded DNA replication proteins (Dna A, Dna B, Dna C, DNA polymerase, DNA ligase, etc.) are stable, while plasmid-encoded regulators (e.g. Rop/Rom or Rep A, etc.) are unstable. As in the case of relaxed plasmids, replication is negatively-regulated by plasmid-encoded Rop/Rom protein (for details see Section 8.10.1), which being unstable is removed immediately after inhibition of protein synthesis, but the replication continues using stable host chromosome-encoded DNA replication proteins. On the other hand, in case of stringent plasmids, the initiation of replication is positively-regulated by plasmid-encoded Rep protein (for details see Section 8.10.2); with the removal of unstable Rep proteins, plasmid replication is blocked at the initiation phase itself regardless of the availability of stable host chromosome-encoded DNA replication proteins. This approach was widely used for the amplification of relaxed plasmids before the discovery of high copy number plasmid vectors. Moreover, chloramphenicol treatment is still used by research scientists, because copy numbers of plasmid increase two to three-fold, and the bulk and viscosity of the bacterial culture are reduced as host replication is blocked.

Resistance to Aminoglycosides The aminoglycoside family of antibiotics includes streptomycin, kanamycin, neomycin, tobramycin, and gentamycin. These antibiotics consist of three or more sugar rings attached to one or more amino groups. These antibiotics are polycations that diffuse readily through porin channels in the outer membranes of Gram negative bacteria. Thereafter, the antibiotic is transported from the periplasmic space into the cytosol by an energy-dependent process driven by the negative membrane potential of the inner periplasmic membrane. Upon entering the cytosol, these antibiotics interact with three ribosomal proteins of the 30S ribosomal subunit resulting in inhibition of protein synthesis and an increased frequency of induced translational errors. *In vitro*, kanamycin and other aminoglycoside antibiotics that lack guanido groups (e.g., neomycin and gentamycin) also inhibit the splicing of group I introns.

The antibiotics are inactivated by covalent modifications performed by enzymes encoded by genes present on the plasmids. There are many different aminoglycosides, and a correspondingly wide range of modifying enzymes, which catalyze different types of modifications, for example, phosphorylation or adenylation (i.e., addition of AMP) of hydroxyl group, or acetylation of amino groups (Figure 8.14). As aminoglycosides are naturally synthesized by the *Streptomyces* group of soil bacteria, which need to protect themselves against the antibiotics produced by them, hence the aminoglycoside modifying enzymes came originally from the same *Streptomyces* strains that make these antibiotics. These enzymes have different cofactor requirement; phosphotransferases and adenylyltransferases require ATP as a source of phosphate and AMP groups, respectively, whereas acetyltransferases use acetyl CoA as the acetyl group donor. The covalently modified aminoglycosides are not able to inhibit their ribosomal target sites; hence protein synthesis is not inhibited.

Kanamycin and its close relatives like gentamycin, neomycin, and geneticin (G418) are inactivated by many of the same bacterial aminophosphotransferases (APHs). There are seven groups of well-characterized APHs, which have been distinguished on the basis of substrate specificities. The APHs encoded by these genes inactivate kanamycin by transferring the γ-phosphate of ATP to the hydroxyl group in the 3′ position of the pseudosaccharide. Among them, two have been extensively used as selectable markers for kanamycin resistance in prokaryotic vectors. For example, APH (3′)-I is isolated from *Tn* 903, and APH (3′)-II is isolated from *Tn* 5. APH (3′)-II efficiently inactivates geneticin (G418) and is used as dominant selectable marker in eukaryotic cells.

Resistance to Tetracycline Tetracyclines are made of an identical four-ring carbocyclic skeleton that supports a variety of groups. It can readily enter inside the bacterial cells by passive diffusion across the outer membrane through porin channels, which are composed of the Omp F protein. Transport of the antibiotic across the cytoplasmic membrane and into the cytoplasm is facilitated by pH or electropotential gradients. There are five well-characterized tetracycline efflux genes, namely *tet* A (A) to *tet* A (E), which encode hydrophobic proteins of homologous sequence and similar structure known as Tet protein pump. These proteins import tetracycline from the environment and actively pump the antibiotic and a proton into the cell. When tetracycline enters inside the cell, it makes complexes with Mg^{2+} ion (Tet-Mg^{2+} complex), which binds to a single site on the 16S RNA of the 30S ribosomal subunit and prevents the attachment of

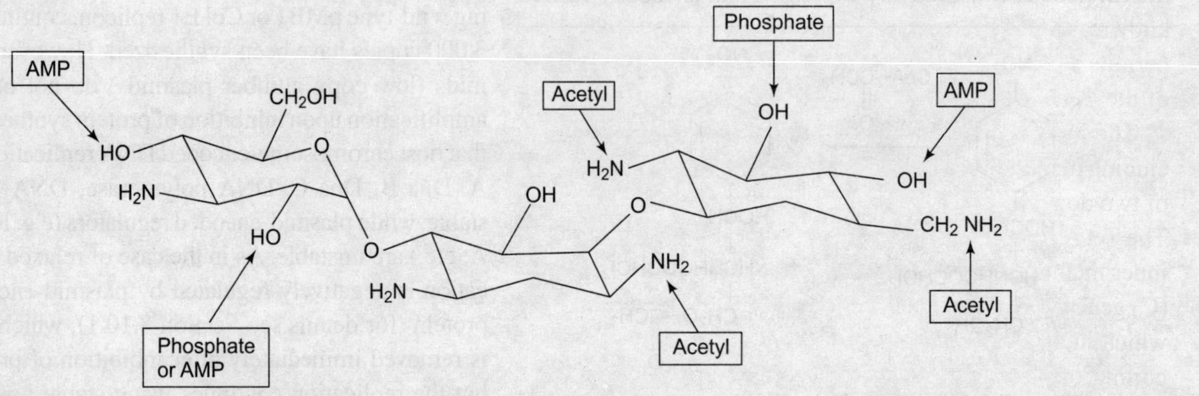

Figure 8.14 Inactivation of aminoglycoside antibiotics [In the figure possible modification sites in aminoglycoside antibiotic kanamycin B are shown.]

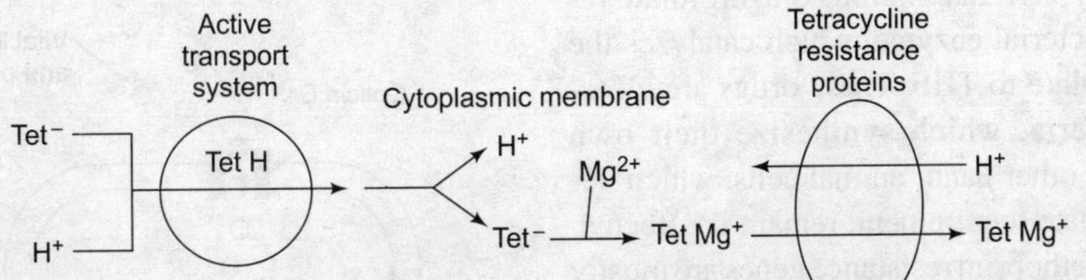

Figure 8.15 Expulsion of tetracycline from bacteria having R-plasmid for tetracycline

aminoacyl tRNA to the acceptor (A) site on the ribosome. It also blocks protein synthesis by disrupting codon:anticodon interactions. Though tetracycline has the ability to bind to both prokaryotic and eukaryotic ribosomes, bacteria are more sensitive than animal cells because tetracycline is actively taken up by bacterial cells, but not by eukaryotic cells. Moreover, eukaryotic cells actively export tetracycline.

The mechanism of plasmid encoded tetracycline resistance is quite unique: instead of inactivation by modification, tetracycline resistance is due to energy-driven export of the antibiotic which is governed by antiporter proteins known as Tet proteins, which are located in the inner bacterial membrane and in the exchange of proton, expel intracellular tetracycline metal complexes against a concentration gradient. Thus, resistant cells grow in the presence of tetracycline because these maintain a low concentration of antibiotic. There are reports suggesting that bacterial cells containing R plasmid harbouring tetracycline resistance gene take up tetracycline by a different transport mechanism from that of sensitive cells. Generally, plasmid-encoded tetracycline resistance is expressed at two levels, i.e., basal constitutive and induced. The basal constitutive level of resistance provides protection by five to ten-folds as compared to sensitive bacteria. Additionally, when tetracycline is administered inside the cells, resistance of a higher level is induced. Both resistance levels are due to production of proteins that are found in the plasma membrane, and actively expel tetracycline from the cell. In bacterial cells consisting of an R plasmid harbouring tetracycline resistance gene, an additional transport protein known as tetracycline resistant protein, is synthesized, which permits a proton to enter the cell to produce energy for export of the Tet^{-}–Mg^{2+} complex (Figure 8.15).

The *tet* A (C) gene which is present in various vectors including pBR322, encodes a 392 residue polypeptide composed of two domains each containing six transmembrane segments. The Tet A (C) protein assembles into a multimeric form in the inner membrane of the bacteria. When the expression of *tet* A (C) gene has been induced, cells are affected in multiple ways, which include reduced growth and viability, increased supercoiling of plasmid DNA, complementation of defects in potassium uptake, and increased susceptibility to other antibacterial agents, including aminoglycoside antibiotics and lipophilic acids. Inactivation of *tet* A (C) gene provides growth advantage to bacteria that are exposed to such agents. Earlier this effect was used to select for bacteria carrying recombinant plasmids in which the *tet* A (C) gene had been inactivated by insertion of foreign DNA sequence.

Resistance to Sulfonamides and Trimethoprim Sulfonamides and trimethoprim act as antagonists of the vitamin, folic acid. The reduced form of folate, tetrahydrofolate (THF), is used as a cofactor by several enzymes whose function is to add a one-carbon fragment, and is involved in the synthesis of metabolites like methionine, adenine, thymine, etc. Sulfonamides are synthetic antibiotics and are analogs of *p*-aminobenzoic acid (Figure 8.16), a precursor of the vitamin folic acid, which inhibits dihydropteroate synthetase, an enzyme in the biosynthetic pathway for folate. Trimethoprim acts as an analog of

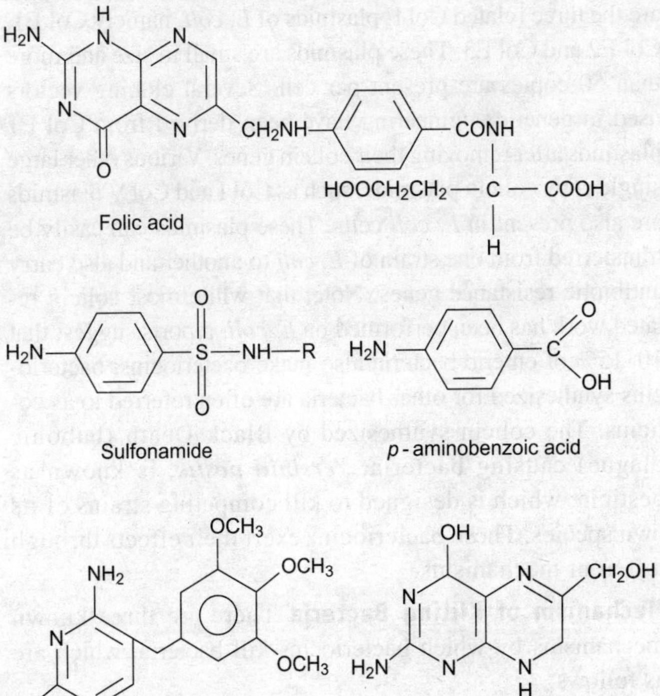

Figure 8.16 Structure of folic acid, sulfonamide, trimethoprim, folate cofactor [Sulfonamide is an analog of *p*-amino benzoic acid Trimethoprim is an analog of dihydropteridine.]

the pterin ring portion of THF that inhibits dihydrofolate reductase (DHFR), the bacterial enzyme, which catalyzes the conversion of dihydrofolate to THF. These drugs are effective only against bacteria, which synthesize their own tetrahydrofolate. On the other hand, animal cells, which depend on diet for their folate requirement, remain unaffected.

Sulfonamide and trimethoprim resistance genes are mostly localized on the same R plasmid. These R plasmids provide resistance to both sulfonamides and trimethoprim by encoding folic acid biosynthetic enzymes that are not able to bind the antibiotics. Dihydropteroate synthetase, which is encoded by R plasmid, has the same affinity for *p*-aminobenzoic acid as the chromosomal enzyme, but is resistant to sulfonamides. Likewise, R plasmid-encoded DHFR is resistant to trimethoprim.

Bacteriocinogenic Plasmid

Several strains of bacteria are known for their aggressive properties, rather than defensive, and comprise of plasmids that encode toxic proteins to kill closely related bacteria. As more closely related strains of bacteria compete for the same resources, these synthesize proteins, which are known as bacteriocins to kill their relatives. A particular bacteriocin is named after the species that makes them. For example, many strains of *E. coli* manufacture a diverse variety of colicins, which kill other strains of the same species. The best-known examples are the three related Col E plasmids of *E. coli*, namely Col E1, Col E2 and Col E3. These plasmids are small in size and more than 50 copies are present per cell. Several cloning vectors used in genetic engineering have been derived from Col E1 plasmids after removing their colicin genes. Various other large single copy colicin plasmids, such as Col I and Col V plasmids are also present in *E. coli* cells. These plasmids can easily be transferred from one strain of *E. coli* to another and also carry antibiotic resistance genes. Note, that while most colicin related work has been performed on *E. coli,* reports suggest that 10–15% of enteric bacteria also make bacteriocins; bacteriocins synthesized for other bacteria are often referred to as colicins. The colicin synthesized by Black Death (bubonic plague) causing bacteria, *Yersinia pestis*, is known as pesticin, which is designed to kill competing strains of its own species. These bacteriocins exert their effects through different mechanisms.

Mechanism of Killing Bacteria There are three known mechanisms by which bacteriocins kill bacteria, which are as follows:

Effect on cell membrane Some colicins, for example, colicin E1 encoded by a gene of Col E1 plasmid, attack their victims, punctures a hole through the cell wall and plasma membrane creating a channel (Figure 8.17). After the channel is formed, an influx of protons and an efflux of vital

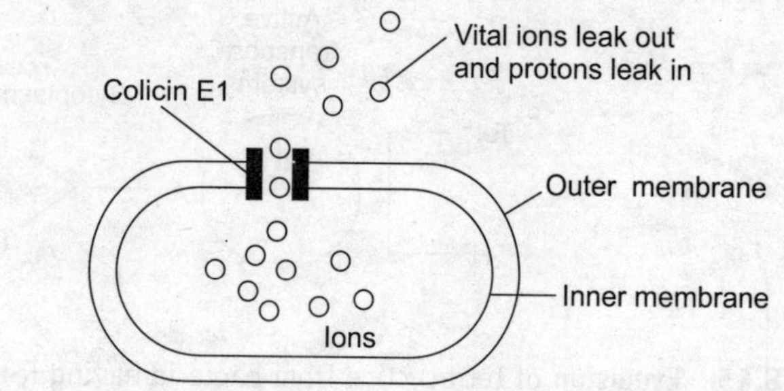

Figure 8.17 Col E1 plasmid encodes the colicin E1 protein that damages the cell membrane

contents starts, because of which, all the vital cell contents, including essential ions, leak out from the bacterial cell. The protons are flooded into the cell through this channel because of the collapse of proton motive force. The energy generated by the proton motive force is required for the production of ATP and the uptake of many nutrients, which are essential for the survival of bacteria. Even a single molecule of colicin E1 that penetrates the membrane is enough to kill the target cell. Colicin I and colicin V also follow this mechanism for killing of related bacteria.

Effect on cell wall In the second mechanism, some bacteriocins penetrate the outer surface of the cytoplasmic membrane site, i.e., the site of peptidoglycan assembly, and degrade the peptidoglycans of the cell wall. In the absence of peptidoglycans, the bacterial cell loses shape and eventually bursts. Colicin M and pesticin A1122 follow this mechanism for killing of related bacteria. For example, pesticin A1122, which is synthesized by *Yersinia pestis, kills Y. pseudotuberculosis, Y. enterocolitica,* plasmid-free *Y. pestis*, and many strains of *E. coli*.

Effect on nucleic acids In this mechanism, nucleic acids of the victim are degraded. The plasmids Col E2 and Col E3 encode nucleases, enzymes that degrade nucleic acids. Colicin E2 and E3 proteins show ~75% homology at their N-terminal, which suggests that both of them bind to the same receptor on the surface of sensitive bacteria, while both differ in the C-terminus and have different nucleic acid targets. Colicin E2 is a deoxyribonuclease that degrades the chromosome of the target cell while colicin E3 is a ribonuclease that recognizes specific sequence of the 16S rRNA of the small ribosomal subunit (30S) and hydrolyzes the backbone releasing a fragment of 49 nucleotides from the 3′-end. Thus, protein synthesis is blocked; even a single colicin molecule that enters is enough to kill the target cell.

Immunity Against Colicins Bacteriocin synthesizing bacteria develop immunity for self-protection from their own bacteriocin. This is because a particular colicin producing bacterial cell synthesizes some specific immunity proteins, called antidotes, which are immune to their own product, but not to

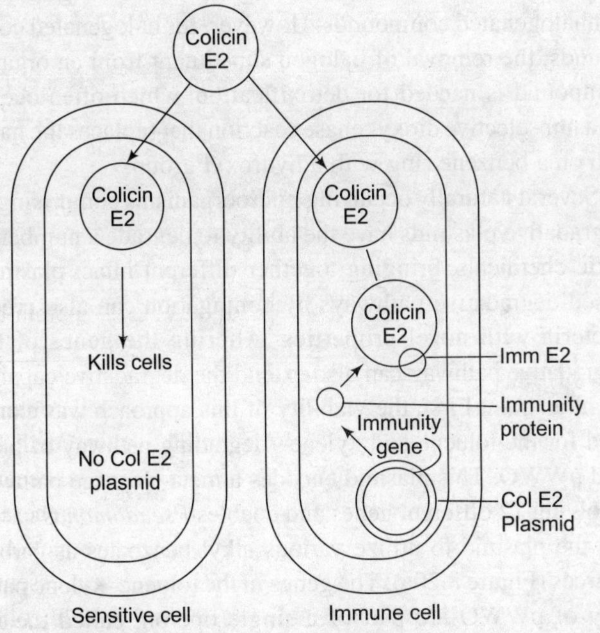

Figure 8.18 Colicin immunity system

other brands of colicins. These immunity proteins bind to active sites of the corresponding colicin proteins (Figure 8.18). For example, the Col E2 plasmid comprises of genes for both colicin E2, and a soluble immunity protein that binds colicin E2. This immunity protein has the ability to only protect against homologous, but not heterologous colicin, not even the closely related colicin E3. In case of membrane active colicin, immunity is provided due to a plasmid-encoded inner membrane protein that blocks the colicin from forming a pore in the host cell. For instance, the Ia immunity protein provides protection against colicin Ia but not against the closely related colicin Ib, even though colicins Ia and Ib have extensive sequence homology, share the same receptor, and have the same mode of action.

Synthesis and Release In a population of Col E plasmid-carrying bacteria, about 1 in 10,000 cells actually produce colicin in each generation. Colicins are normally produced by the suicidal mechanism, in which a rare cell sometimes produces a large amount of colicin, which then bursts and releases the colicin into the medium. The burst and release mechanism kills the producer cell, and the released toxin destroys all the sensitive bacteria in the area. Col E plasmid containing cells that have the immunity protein remain protected. Expression of two plasmid genes, *cea* (colicin protein) and *kil* (lysis protein) are needed for colicin E production. These proteins, however, remain repressed by Lex A, the protein of SOS DNA repair system. The production of colicin is induced by DNA damage and those cells that sacrifice themselves are probably injured anyway. However, the suicidal mechanism is not responsible for colicin production in all the cases. For example, various colicins (e.g., colicin V, colicin I) are synthesized continuously in smaller amounts. Unlike the colicins E, these colicins have the tendency to attach to the surface of the producer cell rather than being released as freely soluble proteins. Finally, when the producer cell bangs into a sensitive bacterium, the colicin may be transferred to them, resulting in their death.

Virulence Plasmids

Virulence plasmids carry genes for various characters, which help bacteria to infect higher organisms, both plants and animals, including humans. Bacteria can infect the target organism by a variety of mechanisms. Bacteria having virulence plasmids are able to attack animal cells by attaching to their cellular membranes and releasing toxins (Figure 8.19). The plasmids consist of genes that encode protein filaments, i.e., adhesins. Note that adhesins are similar to pili but vary in length and thickness, and recognize cell-surface receptors available on target cells and attach to them. After attachment, the bacteria secrete toxins, which can infiltrate in the target cell and then destroy that cell. For example, pathogenic *E. coli* strains usually rely on plasmid-borne virulence factors. A broad range of toxins is available in different pathogenic strains of *E. coli* and includes heat-labile enterotoxin (similar to choleratoxin), heat-stable enterotoxin, hemolysin (lyses red blood cells), and Shiga-like toxin (resembles to the toxin of dysentery-causing *Shigella*). However, similar types of adhesins or 'colonization factors' are synthesized by virulence plasmids, which facilitate the attachment of bacteria to the animal cells. Another prevalent example of virulent plasmids is tumor-inducing plasmid (pTi) of *A. tumefaciens* that causes crown gall tumour on wounded plant cells. The *vir* genes on pTi plasmid are responsible for this virulence (for details see Chapter 12).

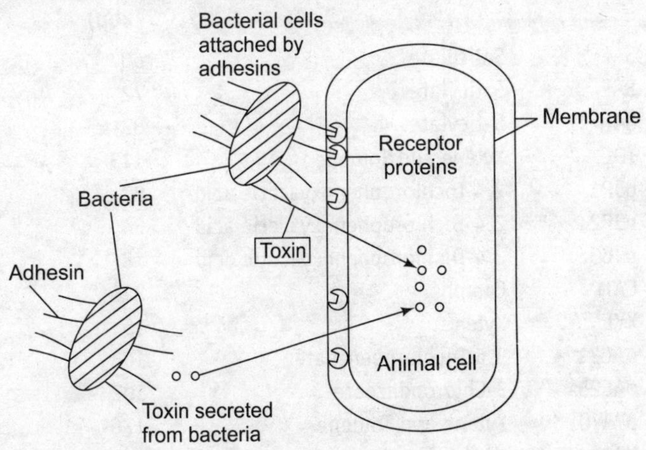

Figure 8.19 Toxins and adhesins of virulence plasmids

Degradative Plasmids

Degradative plasmids are the group of plasmids which enable the digestion of unusual substances, e.g., toluene, xylene, naphthalene, camphor, chlorobenzoate, and salicylic acid, etc. These plasmids provide genes that allow bacteria to grow by breaking down various unusual metabolites or industrial chemicals, including the components of petroleum or herbicides. Degradative plasmids are thus very important for bioremediation. The microbial degradation provides an economic and effective means of disposing off toxic chemical wastes. During the 1960s, a number of soil microorganisms were revealed that were able to degrade xenobiotic ('xenos' meaning foreign) chemicals such as herbicides, pesticides, refrigerant solvents, and other chemical compounds. Several bacteria belonging to the *Pseudomonas* group are the major soil microorganisms that degrade and detoxify more than 100 different organic compounds.

Biodegradation of complex organic molecules requires the collaborative effort of several different enzymes involved in the biodegradation pathway. The genes encoding these enzymes are located in the genomic DNA, although these are more often found on large ~50–200 kbp plasmids. *Pseudomonas* plasmids and their degradative pathways are enlisted in Table 8.6. Degradative bacteria, in most cases enzymatically convert xenobiotic, nonhalogenated aromatic compounds to either catechol or procatechuate; thereafter, through a series of oxidative cleavage steps, catechol and procatechuate are processed to yield acetyl CoA and succinate, or pyruvate and acetaldehyde (Figure 8.20a). Almost all microorganisms readily metabolize these end products. Halogenated aromatic compounds, which are the main components of most pesticides and herbicides, are also converted to catechol, procatechuate, hydroquinone, or the corresponding halogenated derivatives by the same enzymes, which degrade the nonhalogenated compounds. However, for halogenated compounds, the removal of halogen substituent from an organic compound is needed for detoxification, which often occurs by a nonselective dioxygenase reaction that replaces the halogen on a benzene ring with a hydroxyl group.

Table 8.6 Properties of *Pseudomonas* Plasmids

Plasmid	Compound(s) degraded	Plasmid size (kbp)
SAL	Salicylate	60
SAL	Salicylate	72
SAL	Salicylate	83
TOL	Xylene and Toluene	113
pJP1	2,4-Dichlorophenoxyacetic acid	87
pJP2	2,4-Dichlorophenoxyacetic acid	54
pJB3	2,4-Dichlorophenoxyacetic acid	78
CAM	Camphor	225
XYL	Xylene	15
pAC31	3,5-Dichlorobenzoate	108
pAC25	3-Chlorobenzoate	102
pWWO	Xylene and Toluene	176
NAH	Naphthalene	69
XYL-K	Xylene and Toluene	135

Several naturally occurring microorganisms comprising of degradative plasmids have the ability to degrade a number of toxic chemicals, bringing together different intact plasmid-based degradative pathways by conjugation can also create bacteria with novel properties. Altering the genes of the degradative pathway can also extend the degradative capability of a strain. First, the viability of this approach was examined for the toluene and xylene– degrading pathway of plasmid pWWO. This plasmid encodes a meta-cleavage pathway involving 12 different genes and enables *Pseudomonas* carrying the plasmid to utilize various alkyl benzoates as carbon sources (Figure 8.20a). The genes in the toluene–xylene pathway of pWWO are part of a single operon, called the *xyl* operon, under the control of the P^{m} promoter. The transcription from P^{m} promoter is positively regulated by the *xyl* S gene product, which is activated by most of the initial substrates such as benzoates and 3-methylbenzoate of the pathway.

Similary, TOL plasmids contain *tod* genes (*tod* A, *tod* B, *tod* C1 and *tod* C2), which encode proteins/enzymes that catalyze degradation of toluene. The *tod* A encodes a flavoprotein that accepts electrons from NADH and transfer them to a ferredoxin encoded by *tod* B, which reduces the terminal dioxygenase encoded by *tod* C1 and *tod* C2. Toluene is thus converted to *cis*-toluene dihydrodiol by the concerted enzymic activities of Tod proteins (Figure 8.20b).

Various genetic engineering efforts are made to meet the challenge of environmental contamination by designing bacterial strains with expanded degradative capabilities (for details see Chapter 20).

8.13 PLASMID PURIFICATION

Various methods have been devised for purification of plasmids from bacterial cells. Each of these methods include three major steps: (i) Growth of bacterial cultures; (ii) Harvesting and lysis of bacteria; and (iii) Purification of plasmid DNA or removal of contaminants.

8.13.1 Growth of Bacterial Cultures

Plasmids are generally prepared from bacterial cultures grown in the presence of a selective agent such as an antibiotic. The yield and quality of plasmid DNA may depend on factors such as plasmid copy number, host strain, inoculum, antibiotic, and type of culture medium. Note that plasmids vary widely in their copy number per cell (Table 8.3), depending

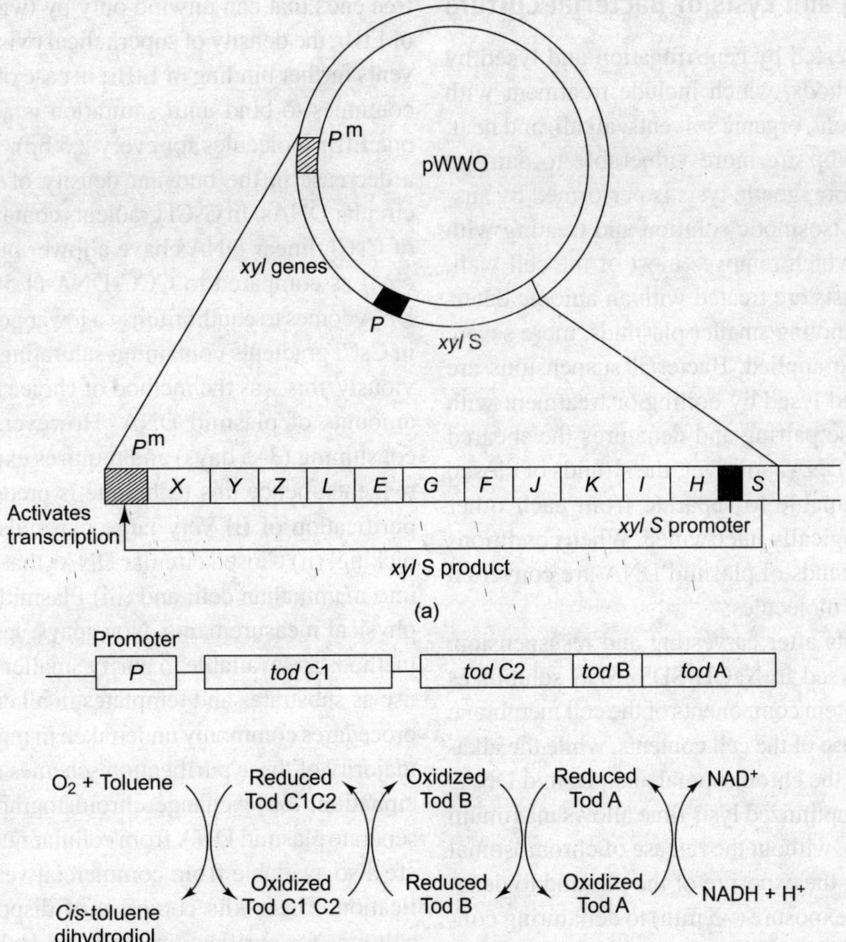

Figure 8.20 Degradative plasmids; (a) pWWO plasmid and *xyl* operon, (b) TOL plasmid and *tod* genes

on their origin of replication. Although the strain used to propagate a plasmid has an effect on the quality of the purified DNA, yet most *E. coli* strains can be used successfully to isolate plasmid DNA. Host strains such as DH1, DH5α, and C600 give high-quality DNA. The slow-growing strain XL1-Blue also yields DNA of very high quality, which works extremely well for some specific purposes like sequencing. XL1-Blue and DH5α are highly recommended for reproducible and reliable results. Luria Bertani broth (LB) is the recommended culture medium for plasmid purification (for details see Chapter 1). Bacterial cultures for plasmid preparation should always be grown from a single colony picked from a freshly streaked selective plate. A single colony should be inoculated into 1–5 ml of medium containing appropriate selective agent, and grown with vigorous shaking for 12–16 hours. Note that growth for more than 16 hours is not recommended since cells begin to lyse and plasmid yields may be reduced. Aliquots of appropriately grown culture can be used to prepare small amounts of plasmid DNA (minipreparation; miniprep) for analysis, or as inoculum for medium-scale culture (midiprep) or large-scale culture (maxiprep). It is advisable to grow the bacterial culture in selective conditions, i.e., in the presence of antibiotics. Concentrations of commonly used antibiotics are given in Table 8.7.

Table 8.7 Concentrations of Commonly Used Antibiotics

Antibiotic	Stock solutions		Working concentration (dilution)
	Concentration	Storage	
Ampicillin (sodium salt)	50 mg/ml in water	–20°C	100 μg/ml (1/500)
Chloramphenicol in ethanol	34 mg/ml	–20°C	170 μg/ml (1/200)
Kanamycin in water	10 mg/ml	–20°C	50 μg/ml (1/200)
Streptomycin in water	10 mg/ml	–20°C	50 μg/ml (1/200)
Tetracycline HCl in ethanol	5 mg/ml	–20°C	50 μg/ml (1/200)

8.13.2 Harvesting and Lysis of Bacterial Culture

Bacterial cells are harvested by centrifugation and lysed by any of the various methods, which include treatment with ionic or nonionic detergent, organic solvents, alkali, and heat. Plasmids of size >15 kbp are more vulnerable to damage during cell lysis, therefore, gentle lysis is performed by suspending the bacteria in isosmotic solution and treating with EDTA and lysozyme, which removes most of the cell wall. The resulting spheroblasts are treated with an anionic detergent like SDS. When handling smaller plasmids, more severe methods of lysis can be applied. Bacterial suspensions are exposed to detergent and lysed by boiling or treatment with alkali. This disrupts base pairing and denatures the sheared bacterial chromosomal DNA; however, the strands of closed circular plasmids are unable to separate from each other because these are topologically intertwined. When conditions return to normal, the strands of plasmid DNA are converted into native superhelical molecules.

In routine experiments after harvesting and resuspension, the bacterial cells are lysed in NaOH/SDS. SDS solubilizes the phospholipid and protein components of the cell membrane leading to lysis and release of the cell contents, while the alkaline conditions denature the chromosomal and plasmid DNAs as well as proteins. The optimized lysis time allows maximum release of plasmid DNA, without the release of chromosomal DNA, while minimizing the exposure of the plasmid to denaturing conditions. Long exposure (>5 min) to denaturing conditions causes closed circular DNA to enter an irreversibly denatured state. This denatured form of plasmid runs faster on agarose gels, is resistant to restriction enzyme digestion, and stains poorly with intercalating dyes such as ethidium bromide (EtBr).

8.13.3 Purification of Plasmid DNA

The resulting lysate (after SDS/NaOH treatment) is contaminated with considerable amount of RNA and bacterial genomic DNA. Therefore, it is required to remove the contaminants or at least reduce them to some extent. All schemes for purification of plasmids take advantage of the relatively small size and the covalently closed circular structure (CCC) of plasmid DNA. For a long time, buoyant density centrifugation in gradient of cesium chloride (CsCl) and EtBr has been the most valued method for separating CCC plasmid DNA from contaminating bacterial genomic DNA. In this technique, separation depends upon differences in the amount of EtBr that can be bound to linear and CCC DNA molecules. EtBr, which is an intercalating dye (for details see Chapter 6), has the capacity to bind very tightly to DNA in high concentrations of salt. As the dye intercalates between the bases, it unwinds the ds DNA molecule and leads to an increase in length of the linear or relaxed circular ds DNA molecules. Since CCC DNA has no free ends that can unwind only by twisting and after addition of EtBr, the density of superhelical twists increases, which prevents further binding of EtBr; in case of linear molecules, EtBr continues to bind until saturation is achieved (an average of one EtBr molecules for every 2.5 bp). Binding of EtBr causes a decrease in the buoyant density of both linear and closed circular DNAs. In CsCl gradients containing saturating amount of CsCl, linear DNAs have a lower buoyant density (1.54 g/cm^3) as compared to CCC DNA (1.59 g/cm^3). Hence, CCC DNA comes to equilibrium at a lower position than linear DNAs in CsCl gradients containing saturating amounts of EtBr . Previously, this was the method of choice for preparation of large amounts of plasmid DNA. However, this process is time-consuming (3–5 days) and requires expensive equipment and reagents; hence this technique is predominantly used for the purification of (i) Very large plasmids that are vulnerable to nicking; (ii) Closed circular DNAs that are to be microinjected into mammalian cell; and (iii) Plasmids that are used for biophysical measurements. Nowadays, less expensive and faster methods are available to purify smaller plasmids (<15 kbp) for use as substrates and templates in all enzymatic reactions and procedures commonly undertaken in molecular cloning. A great majority of these purification schemes rely on differential precipitation, ion exchange chromatography, or gel filtration to separate plasmid DNA from cellular nucleic acids. Various kits are also available from commercial vendors for plasmid purification. These kits comprise of disposable chromatography columns for purification of plasmids. However, these kits are unnecessary for minipreps of plasmid DNA to be used in routine experiments.

Among the several available methods, the alkaline lysis method developed by Birnboim and Doly (1979) is the most popular because of its simplicity, relatively low cost, and reproducibility. This method is based on the principle that there is a narrow range of pH (12.0–12.5) within which only the linear DNA is denatured. After lysis, the lysate is neutralized with acidic potassium acetate, which results in renaturation and aggregation of chromosomal DNA to form an insoluble network. The high salt concentration causes denatured proteins, chromosomal DNA, cellular debris, and SDS to precipitate, while the smaller plasmid DNA renatures correctly and remains in the clear supernatant. The precipitate can be removed by centrifugation and the plasmid DNA, if required, can be purified further by gel filtration (for details see Chapter 6).

8.14 PLASMIDS AS VECTORS IN GENETIC ENGINEERING

Plasmids are used as ideal cloning vectors (for details see Chapter 1) in recombinant DNA technology due to following characteristics:

Autonomous Replication Plasmids contain their own origin of replication and have the ability to replicate autonomously in host cells.

Small Size The size of plasmids is small. There are several advantages of having a low molecular weight plasmid. First, the plasmid DNA is much easier to handle and is readily isolated from host cells, i.e., it is more resistant to damage by shearing. Secondly, low molecular weight plasmids are usually present as multiple copies, thereby facilitating their isolation. This, however, leads to gene dosage effects for all cloned genes. Thirdly, with a low molecular weight there is less chance that the vector will have multiple recognition sites for any restriction endonuclease. Finally, the transformation efficiency decreases with plasmids that are more than 15 kbp long.

Presence of Selectable Markers Plasmids possess genes which encode for selectable markers (at least one or several), such as resistance to antibiotics, which confers a new selectable phenotype upon the cells.

Presence of Unique Restriction Enzyme Site Plasmids possess a single restriction site for one or more restriction endonucleases. Insertion of a linear molecule at one of these sites does not alter its replication properties. Some of these sites are present in a marker gene so that this gene is insertionally inactivated during cloning (for details see Chapter 16). A second marker may also be present on the same plasmid molecule, which is allowed to remain intact in order to provide a selectable phenotype. Plasmid vectors may also be manipulated to contain a multiple cloning site (MCS) or polylinker, which is a short DNA sequence that carries sites for several restriction endonucleases, preferably in the genes with a readily scorable phenotype. An MCS increases the number of potential cloning strategies available by extending the range of enzymes that can be used to generate restriction fragments suitable for cloning. By combining them within an MCS, the sites are made contiguous, so that any two sites within it can be cleaved simultaneously without excising vector sequences.

Nonconjugative and Nonmobilizable Due to biological safety considerations, plasmids should neither be transmissible nor mobilizable. Many plasmids are naturally nonconjugative and nonmobilizable; however, such plasmid vectors may also be constructed artificially by deleting the relevant genes or loci (for details see Section 8.12.1).

Exhibit 8.3 Runway Plasmid Vectors

One important reason for cloning a gene on a multicopy plasmid is to increase expression and hence facilitate purification of the protein it encodes. However, some genes cannot be cloned on high copy number vectors because excess gene production is lethal to the cell. The use of low copy number vectors avoids cell killing but this may be self-defeating since expression of the cloned gene will be reduced and solution to this problem is to use runway plasmid vectors. Copy numbers of these plasmids can be altered by mutagenesis in such a way that plasmid replication becomes temperature-sensitive and either ceases after the temperature has been raised or becomes relaxed. Certain temperature-sensitive mutants of plasmid R1, for example, have a low copy number (1-2) at 30°C, however, replication control is lost and the plasmids begin to replicate much faster than the chromosome of the host cell at 42°C. This phenomenon, known as runaway replication, is responsible for the fact that up to 50% of the total DNA in a bacterial cell may consist of plasmid molecules after only 4-5 cell divisions. In contrast to the chloramphenicol-induced amplification of plasmid DNA, which requires inhibition of protein biosynthesis, runaway replication can proceed in the presence of normal protein biosynthesis. Since this leads to a considerable overproduction of plasmid-encoded gene products, runaway plasmids play a certain role as expression vectors. Later, a system has been devised which consists of two plasmids with temperature-sensitive replicons, one of which replicates at 30°C, but not at 42°C, and carries a gene coding for lac repressor. The second plasmid carries a runaway replicon, which allows high copy numbers at the elevated temperature (42°C), and a gene coding for the protein to be expressed in this system. The *lac* operator/promoter region controls the expression of this gene. At low temperatures, this gene will not be expressed, because the first plasmid provides large amounts of lac repressor which blocks transcription by binding to the *lac* operator. When the temperature is raised, replication of the first plasmid ceases and, at the same time, the second plasmid which carries the *lac* I control region is amplified. Since the cell soon runs out of repressor, the expression of the desired gene is derepressed. Such selective expression systems always play a particular role if the heterologous protein is toxic for the host cell, because production of the protein can be repressed in actively growing bacteria and switched on at will by a simple temperature shift, *e.g.*, when the cell density has reached the desired level.

The two plasmids, which were developed as runway plasmid vectors, are now called as pBEU1 and pBEU2; each carries *Bam* HI site and a single antibiotic resistance marker. Later, improved vectors have been constructed, like a runway plasmid vector carrying a kanamycin resistance gene and having unique restriction sites for five restriction enzymes. One more runway plasmid vector has been devised comprising of one gene encoding resistance to ampicillin and tetaracycline, while the later marker has unique sites to permit insertional inactivation. Later cloning vectors which are present in one copy per chromosome at temperature below 37°C and which display uncontrolled replication at 42°C have also been designed. In addition, these also carry partitioning function, which stabilizes the plasmid at low temperature when grown in the absence of selection pressure.

Source: Uhlin B E Molin, *et al.* (1979) Plasmids with temperature-dependent copy number for amplification of cloned genes and their products. *Genes* 6: 91–106; Uhlin B E and Clark A J (1981) Overproduction of the *E. coli rec* A protein without stimulation of its proteolytic activity. *J. Bacteriol.* 148: 386–390.

Replicon under Relaxed Control Several plasmids carry a replicon, which is under relaxed control as it is often desirable to maintain plasmids in multiple copies per cell. Sometimes runway plasmid vectors are used which are discussed in Exhibit 8.3.

The basic strategy of gene cloning has already been described in Chapter 1. Most of the routine gene cloning experiments use one or other plasmid vectors.

8.14.1 Natural Plasmid Vectors for *E. coli*

Previously natural plasmids of *E. coli* were used as cloning vectors. A few natural plasmids such as pSC101, pSF2124 (or RSF2124) and Col E1 have been used as cloning vectors.

pSC101

This plasmid was first used for *in vitro* cloning of eukaryotic DNA, when *Eco* RI fragments carrying the genes for rRNA from *Xenopus laevis* were inserted into the pSC101 replicon. It is 9 kbp in size and has low copy number (1–2 copies). This plasmid has the advantage of a single *Eco* RI site at which DNA can be inserted. pSC101 plasmid also carries a strong selectable marker for tet^r. pSC101 is derived from the conjugative plasmid R6-5. This R factor shares homology with F plasmid in the region containing the *tra* operon, i.e., the genes necessary for conjugal transfer of DNA, and carries the genes for resistance to several antibiotics. pSC101 was devised by chance from an experiment in which R6-5 DNA was hydrodynamically sheared and then used to transform *E. coli* to tet^r. It was probably developed by cyclization of a segment of R6-5 DNA and the concomitant activation of a tet^r gene, which is not active in parental plasmid. In fact, pSC101 might have co-existed with R factor as a small shear resistant, supercoiled DNA molecule. pSC101 was only used for insertion of *Staphylococcus* and *Xenopus* DNA. This plasmid has disadvantages of large size, stringent replicative control, low copy number, and low insert capacity. Consequently, the yields of plasmid DNA from cells carrying pSC101 are low as compared to the plasmid vectors that are in current use. Furthermore, it is very difficult to distinguish between chimera and nonrecombinants, unless the insert DNA confers a new phenotype on transformants.

pSF2124 (RSF2124)

pSF2124 was produced by the transfer of the amp^r genes by *Tn* 1. Thus, pSF2124 has ability for colicin biosynthesis as well as ampicillin resistance. This plasmid has high copy number, and hence, the yields from bacterial cells having pSF2124 are high. It has single sites for *Bam* HI and *Eco* RI, which facilitate the cloning process. This plasmid is not currently used as vector, as it does not provide easy selection by insertional inactivation. Moreover, pSF2124 is a mobilizable plasmid.

Col E1 Plasmid

Col E1 is a small, circular, and colicinogenic plasmid, that codes for a 57 kDa protein toxin (colicin E1) that can kill other *E. coli* cells by depolarizing the bacterial membrane. The size of this plasmid is 6,466 bp. This plasmid has relaxed control mechanism of replication and exists in multiple copies (~15–20 copies) per cell in *E. coli.* Plasmid can be amplified up to 1,000-3,000 copies by the addition of chloramphenicol, an antibiotic that inhibits protein synthesis. The vector map of Col E1 is presented in Figure 8.21, and the genes and their encoded functions are given in Table 8.8. Col E1 has *cea* gene for colicin production; *imm* provides immunity against its own colicin; *kil* gene encodes for killer or lysis protein; *inc* transcribes into RNA I that is determinant of incompatibility; *rop*/*rom* gene is involved in copy number control (for details see Section 8.10.1); *mob* gene facilitates mobilization during conjugation with the help of F plasmid; *cer* gene assists in maintenance of protein as monomer, and *exc* encodes for exclusion protein. It also carries an origin of replication *ori* V and a region *bom* responsible for mobilization of plasmid. Replication of Col E1 proceeds unidirectionally from *ori* V and copy number depends on RNA II, which is transcribed from a promoter 555 bp away from *ori* V (for details see Section 8.10.1). Previously, this plasmid was explored as vector in genetic engineering as it has high copy number and hence gives high yields. As this

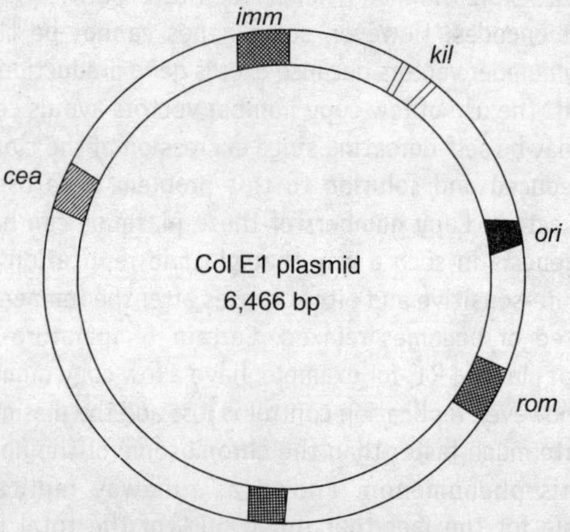

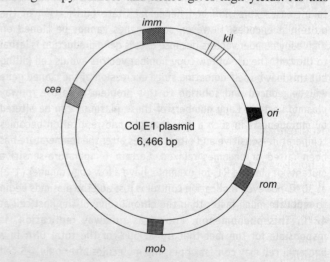

Figure 8.21 Gene map of Col E1 plasmid

Table 8.8 Genes of Col E1 Plasmid and Their Functions

Gene	Function
cea	Colicin toxin
imm	Immunity protein
kil	Lysis protein
inc	RNA I incompatibility determinant
rop/rom	Protein that regulates priming and copy number
mob	Proteins for mobilization during conjugation
cer	Maintains plasmid as monomers
exc	Exclusion protein

plasmid has a single *Eco* RI site in *cea* gene, transformants can be selected by inactivation of colicin production. This selection strategy is however technically difficult because cell resistance to colicins arises spontaneously at quite high frequency in a bacterial population, and the selection system has to be applied with great care.

8.14.2 Artificially Constructed Plasmid Vectors for *E. coli*

Naturally occurring plasmids often lack several important features that are required for a high-quality cloning vector. Therefore, a range of superior plasmid cloning vectors containing several elements from naturally occurring plasmids have been genetically engineered. During vector designing the size of natural plasmids are reduced, nonessential and toxin producing genes are deleted, antibiotic resistance genes are inserted either by *in vitro* ligation of restriction fragment or by *in vivo* transposition, multiple cloning site for the purpose of cloning is inserted usually within a selectable marker gene for ease in selection of recombinants, and the *tra* or *mob* genes, which are responsible for conjugation or mobilization, are removed (for details see Chapter 1). Three important artificially constructed plasmid cloning vectors are described below.

pBR322

The earliest, the best-studied, and the most frequently used general-purpose plasmid cloning vector is pBR322. This vector was named according to traditional rules of naming a plasmid. For example, in pBR322, 'p' (lowercase) stands for plasmid; 'BR' identifies the laboratory in which the vector was originally constructed (BR stands for F. Bolivar and R. Rodriguez, the two researchers who created pBR322); '322' distinguishes this plasmid from others developed in the same laboratory (there are also plasmids called pBR325, pBR327, pBR328, etc.).

Molecular Origin of pBR322/Geneology pBR322 is 4,361 bp in size and comprises of valuable DNA fragments derived from three different naturally occurring plasmids. The two antibiotic resistance genes, *amp*^{r} and *tet*^{r} were derived from plasmids R1 and R6-5, respectively. Both these genes were ligated with a part of pMB1 (Col E1-like plasmid), which carries the origin of DNA replication (Figure 8.22). A summary of the scheme used for construction of pBR322 is shown in Figure 8.23. The plasmid R1 was first isolated in London in 1963, which was earlier named as R7268. Thereafter, a variant, R1drd19, which was derepressed for mating transfer, was isolated. In a separate reaction, the *Tn* 3 of R1drd19 was transposed to Col E1 to form pSF2124. The *amp*^{r} transposon *Tn* 3 from this plasmid was transposed onto pMB1 to form pMB3. This plasmid was reduced in size by *Eco* RI rearrangement to form a tiny plasmid, pMB8, which

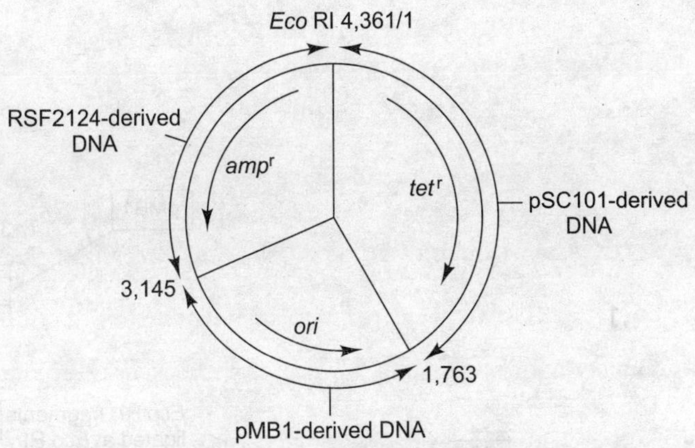

Figure 8.22 Origins of plasmids pBR322—Boundaries between the pSC 101, pMB1, and RSF 2124-derived DNA

carried only colicin immunity. Simultaneously, R6-5, a donor of *tet*^{r} gene, was recircularized to generate pSC101. In the next step, a chimeric molecule pMB9 coding for *tet*^{r} as well as colicin immunity (*col*^{imm}) was constructed. For this, *Eco* RI* fragments of pSC101 were mixed with plasmid pMB8 (linearized at its unique *Eco* RI site). The resulting chimeric molecule was rearranged by *Eco* RI* activity to generate pMB9. The *Tn* 3 element was then transposed to pMB9 to form pBR312, which was later converted into pBR313 by *Eco* RI rearrangement. Two separate fragments were isolated from pBR313 and ligated to form pBR322. These construction schemes were based on two principles. Firstly, due to transposability of transposon *Tn* 3, the *amp*^{r} gene was permitted to jump unspecifically between DNA molecules. This transposability was represented by the *amp*^{r} gene jumping from plasmid Col E1 to a new DNA molecule to produce pSF2124. Secondly, these constructions demonstrated that DNA rearrangements could be accomplished efficiently by random or direct reassociation of mixtures of certain restriction fragments, for example, pBR322 was generated from pBR313.

The phenotype of plasmid pBR313 was *amp*^{r} *tet*^{r} *col*^{imm}, which had three *Pst* I sites, one of which is located in the *amp*^{r} gene. During the generation of pBR322 from pBR313, two new plasmids were constructed. In the first attempt it was tried to remove two *Pst* I sites, and to maintain a functional *amp*^{r} gene. Plasmid pBR318 was obtained by selecting for tetracycline resistance after transformation of *E. coli* strain RRI with a mixture of three pBR313 *Pst* I fragments. A new molecule which was tetracycline resistant but ampicillin sensitive was generated by removing two small *Pst* I fragments. Further, it was attempted to reconstruct *amp*^{r} gene in pBR318 in a separate reaction and a second plasmid pBR320 was constructed. In view of the fact that the *amp*^{r} gene of pBR313 did not contain an *Eco* RII site, bacteria were transformed with unligated *Eco* RII fragments of pBR313. Ampicillin resistant cells were selected and screened

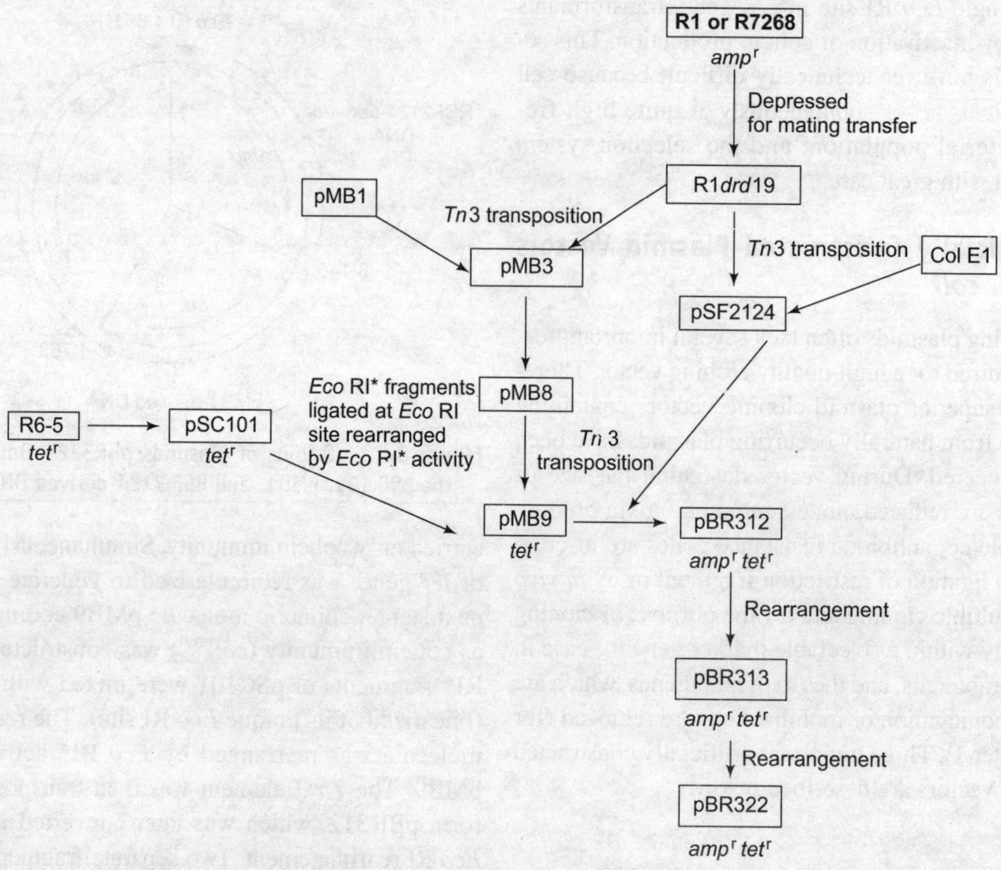

Figure 8.23 Geneology of plasmid pBR322

for tetracycline and colicin sensitivity. Plasmid pBR320 was one of the sixteen clones with the desired phenotype, containing a unique *Pst* I site in the *amp*^{r} gene. Finally, this plasmid, in combination with pBR318, was used to obtain plasmid pBR322. For this purpose, plasmid pBR318 was simultaneously digested with *Pst* I and *Hpa* I, yielding two fragments with molecular weights of 1.95 and 2.2 × 10^6 Da, respectively; the smaller fragment contained the entire *tet*^{r} gene and a part of the *amp*^{r} gene. Plasmid pBR320 was digested with *Pst* I and *Hinc* II, which yielded three fragments, the largest of which contained the origin of DNA replication and that particular part of the *amp*^{r} gene lacking in pBR318. After transformation of *E. coli* strain RRI with a mixture of the DNA fragments obtained from pBR320 and pBR318, an ampicillin and tetracycline resistant, colicin sensitive phenotype was selected. This selection strategy regenerated a new functional *amp*^{r} gene in the new plasmid, and at the same time, blocked the parent plasmids and their fragments, which could not be replicated, because these were either *tet*^{r} *amp*^{s} (pBR318) or *amp*^{r} *tet*^{s} (pBR320). Through this experiment, pBR322 with a molecular weight of 3.1 × 10^6 Da was obtained, which was 0.5 × 10^6 Da smaller than expected plasmid.

Properties of pBR322 pBR322 is an ideal cloning vector. It is small in size (4,361 bp) and hence can be easily purified without degradation. The recombinant pBR322 with additional 6 kbp of DNA can be easily handled. pBR322 has been completely sequenced. The original sequence was 4,362 bp long. The '0' position was randomly set between the A and T residues of the *Eco* RI recognition sequence (5′-GAATTC-3′). The sequence was revised by including an additional CG base pair at position 526, thus increasing the size of plasmid to 4,363 bp. The fragment from position 1 to 1,763 bp is taken by pSC101 having *tet*^{r}. pMB1 provides the fragment from 1,764 to 3,147 bp, which carries the origin of replication and *rop* gene. However, it does not provide the *mob* gene, but a *nic* site is present which acts as a site of action of Mob proteins. Note that pBR322 can be mobilized by a third plasmid, e.g., Col K. Col K provides mobility proteins, which interact *in trans* with a specific DNA sequence '*nic*' on pBR322. The fragment from 3,148 to 4,363 bp has been taken from RSF2124, which provides *amp*^{r}. The size of this vector was later revised by eliminating base pairs at coordinates 1,893 and 1,915. Finally, the size is 4,361 bp, and there are unique recognition sites for 40 restriction endonucleases. Eleven of these enzymes (*Eco* RV, *Nhe* I, *Bam* HI, *Sgr* I, *Sph* I, *Eco* NI, *Sal* I, *Psh* A, *Eag* I, *Nru* I, and *Bsp* MI) lie within the *tet*^{r} gene, and two sites (*Cla* I and *Hind* III) lie within the promoter of *tet*^{r} gene. There are unique sites for six enzymes

(*Ahd* I, *Bsa* I, *Ase* I, *Pst* I, *Pvu* I, and *Sca* I) within the amp^r gene. Thus, cloning in pBR322 with the aid of any one of the 19 restriction enzymes for which a unique site is present results in insertional inactivation of either the amp^r or the tet^r markers. However, cloning in the other unique sites does not permit the easy selection of recombinants, because neither of the antibiotic resistance determinants is inactivated. This plasmid has high copy number (~24 copies/cell) and can be increased by amplification to up to 1,000–3,000 copies in the presence of chloramphenicol.

β-Lactamase gene (bla or amp^r) of pBR322 The β-lactamase gene, which is responsible for ampicillin resistance, plays an important role in pBR322. The 5′-ends of two mRNA species coding for this enzyme have been mapped to positions 4,189 and 37. The first base of the methionine initiation codon AUG (ATG) is localized at 4,362. A Shine–Dalgarno box, which, in this case, shows a pronounced homology with the 3′-end of 16S rRNA, is found at the usual distance from the AUG codon. The counter-clockwise transcription of a peptide, 286 amino acids in length, initiates at the AUG codon at position 4,155. The 23 N-terminal amino acids are regarded as hydrophobic secretion signal and are not found on the mature periplasmic enzyme. pBR322 DNA contains a unique *Pst* I site at position 3,609. Therefore, cleavage of the plasmid by *Pst* I interrupts the *bla* gene in front of the codon for amino acid 182 (Ala). This *Pst* I site is essentially useful, because the 3′-tetranucleotide extensions formed on digestion are ideal substrates for terminal transferase. Thus, this site is ideal for cloning by the homopolymer tailing method. If oligo (dG.dC) tailing is used, the *Pst* I site is regenerated and the insert may be cut out. Furthermore, if the reading frame is conserved in spite of the homopolymeric tails, suitably transformed cells express a fusion protein, which consists of β-lactamase and the protein coded by the inserted DNA. The antigenic determinants of a rat insulin fusion protein in the periplasmic space of transformed cells were first detected by employing this technique. Practically it is important to note that cells transformed with plasmids expressing such type of fusion proteins may show a certain degree of ampicillin resistance even after the insertion of DNA. Sometimes researchers also perform cloning into the β-lactamase gene on unique *Pvu* I site (5′-CGAT↓CG-3′) at position 3,734.

Tetracycline resistance gene The tetracycline resistance cassette of pBR322 contains three open reading frames which are partially overlapping and span ~1,200 bp. The major reading frame (position 86–1,273 encodes the most important peptide (386 amino acid) which is linked with the tetracycline resistance phenotype. If the DNA is interrupted either by small deletion mutations or by the insertion of DNA fragments in the unique *Sal* I site (5′-G↓TCGAC-3′; position 650) of pBR322, then the synthesis of this protein ceases, resulting in the inactivation of tet^r gene. Two other useful restriction sites are also present that can be used for marker inactivation, which are *Bam* HI site (5′-G↓GATCC-3′) at position 375 and the *Sph* I site (GCATG↓C) at position 561. The neighboring *Hind* III (5′-A↓AGCTT-3′; position 29) and *Cla* 1 (5′-AT↓CGAT-3′; position 23) sites are only of limited use for marker inactivation since these lie in the immediate vicinity of the promoter region for the tet^r gene. This promoter region has Pribnow box between positions 33 and 39 and an initiation site for mRNA synthesis at position 45, and an insertion of foreign DNA at positions 23 or 29 does not necessarily influence this region.

Derivatives of pBR322 Several derivatives, which are smaller than pBR322, have been devised and are discussed below:

pBR324 This plasmid vector carries Col E1 structural and immunity genes derived from pMB9. It has unique *Eco* RI and *Sma* I sites in Col E1 structural gene which provides easy selection for *Eco* RI and *Sma* I generated DNA fragments.

pBR325 It carries chloramphenicol resistance (cm^r) gene derived from P1 phage. To construct this vector, pBR322 was digested with *Eco* RI, treated with S1 nuclease, and then *Hae* III fragment containing amp^r gene from P1 phage was added to it. It has unique *Eco* RI site in cm^r gene and the recombinant phenotype is amp^r tet^r cm^s.

pBR327 This vector was constructed by deleting a 1,089 bp segment from pBR322 (Figure 8.24). pBR322 possesses six *Eco* RII sites and contains a long nonessential region between the origin of DNA replication and the tet^r gene, which is flanked by two *Eco* RII cleavage sites at positions 1,442 and 2,502. The pBR322 DNA was partially digested with *Eco* RII for deletion of this particular DNA fragment. Since the two *Eco* RII sites present at positions 1,442 (5′-CCTGG-3′) and 2,502 (5′-CCAGG-3′) differ from each other, the resulting noncomplementary ends were trimmed with the single strand-specific S1 nuclease. It is also important to note that S1 nuclease not only digests single stranded regions, but also removes double stranded stretches of DNA, up to ~20 bp. This can be prevented only by using specially purified enzyme preparations. Subsequently, the DNA mixture was subjected to gel electrophoresis. The desired 3.3 kbp DNA fragment was isolated, circularized with T4 DNA ligase, and used for transformation. Thereafter, pBR327 plasmid DNA was isolated from amp^r tet^r transformants. However, this deletion left the amp^r and tet^r genes intact, but changed the replicative and conjugative abilities of the resulting plasmid. Therefore, pBR327 differs from pBR322 in two important ways. Firstly, pBR327 has a higher copy number than pBR322, being present at ~30–45 molecules per *E. coli* cell. The higher copy number makes this vector more suitable to study the function of the cloned gene. However, this is not of

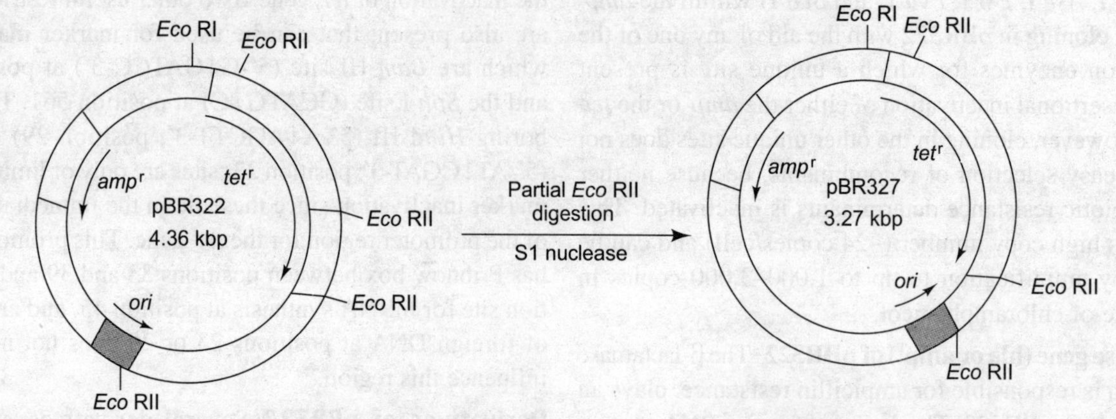

Figure 8.24 Construction of pBR327 from pBR322

great significance as far as plasmid yield is concerned, as both plasmids can be amplified to copy number >1,000. In the case of functional analysis, pBR327 with its high copy number is a better choice, as more copies of the cloned gene are obtained, which give detectable effects of the cloned gene on the host cell. Secondly, the deletion makes pBR327 nonmobilizable, i.e., it cannot be mobilized by conjugative plasmid R64drd11. This aspect is important for biological safety concerns, as it prevents the possibility of a recombinant pBR327 molecule escaping from the test tube and contaminating the gut of a careless molecular biologist. Earlier, pBR322 had been certified as an EK2 vector, because it could not be mobilized by the conjugative plasmid R64drd11; however, later on it has been shown that pBR322 can be mobilized by another plasmid, Col K. Therefore, as the special relaxation (*bom*) site between positions 2,207 and 2,263 of pBR322 is missing from pBR327, it offers additional safety and is considered an EK2 vector. As far as biological safety is concerned, pBR327 even exceeds the c1776/pBR322 system.

pBR328 The *Pst* I/*Bam* HI from pBR327 has been ligated to *Pst* I/*Bam* HI fragments of pBR325, which carries the *cm*^{r} gene to produce a recombinant pBR328. Plasmid pBR328, like pBR325 not only carries *cm*^{r} *amp*^{r} *tet*^{r} genes, but also has additional single sites for *Pvu* II and *Bal* I.

pBR329 Like pBR328, this plasmid also carries *cm*^{r} gene.

pHP34 The vector pHP34 is derived from pBR322 and contains a linker with an additional *Sma* I (5′-CCC↓GGG-3′) ligated at *Eco* RI site (Figure 8.25). This site may be used for cloning of blunt ended DNA molecules. The *Sma* I site is destroyed by the insertion, but the insert can be recovered from the plasmid by *Eco* RI digestion. Thus, pHP34 permits cloning of DNA fragments with blunt ends.

pAT153 pAT153 lacks two *Hae* II fragments of pBR322 (positions 1,644–2,349). This deletion, which is only 705 bp in length, also removes sequences responsible for conjugational transfer, and hence pAT153 is no longer mobilizable.

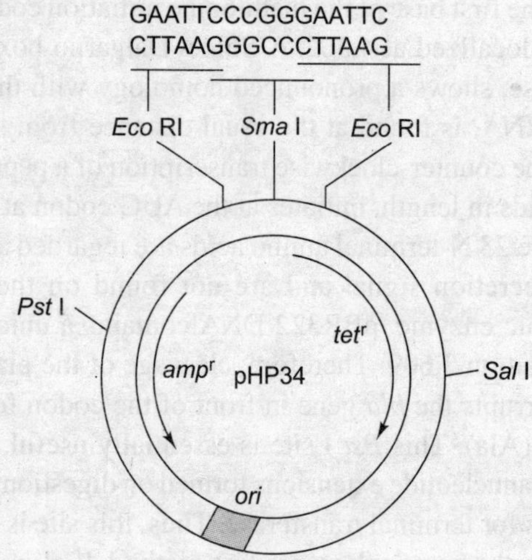

Figure 8.25 Gene map of plasmid vector pHP34

In comparison to pBR322 and pBR327, pAT153 shows a two- to four-fold increase in copy numbers; presumably, the deleted DNA codes for a repressor of replication, which controls copy number in relaxed plasmids.

Cloning Strategies with pBR322 For cloning into pBR322, DNA is isolated and restriction digestion is performed (say, with *Bam* HI, which cuts pBR322 at just one position, within the clusters of genes that code *tet*^{r}). The gene of interest is then ligated at this position and the ligated mix is used for transformation of *E. coli* cells. Note that any of the unique restriction enzyme sites present within the selectable marker genes (*amp*^{r} or *tet*^{r}) can be used for cloning.

Recombinant selection with pBR322 The transformants are easily detected because the *E. coli* cells are transformed from ampicillin and tetracycline sensitive *amp*^{s} *tet*^{s} to ampicillin and tetracycline resistant *amp*^{r} *tet*^{r}. Thus, plating onto a selective medium containing ampicillin and tetracycline facilitates the screening of transformants from nontransformants (Figure 8.26). However, it is quite tricky to distinguish the recombinant from nonrecombinant colonies. In this case

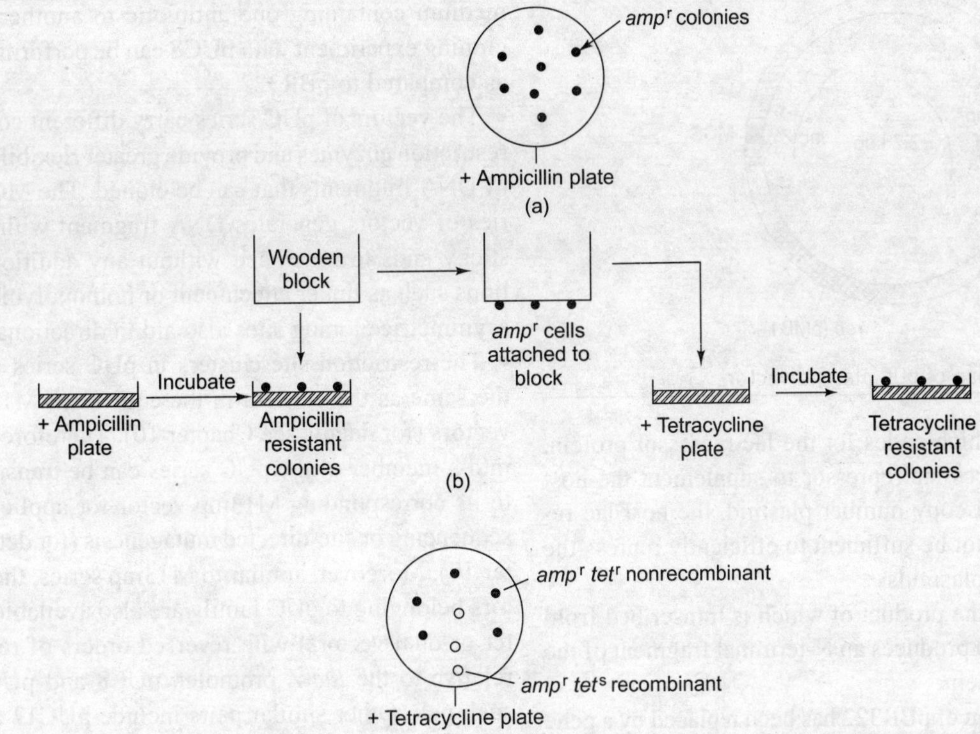

Figure 8.26 Strategy for selection of recombinant selection with pBR322. (a) Colonies grown on ampicillin agar plate. (b) Replica plating. (c) *amp*^{r} *tet*^{r} colonies grow on tetracycline agar plate

recombinants can be identified by insertional inactivation because the host cells no longer displays the character encoded by the inactivated gene (Figure 8.27). A recombinant pBR322 molecule, one that carries an extra piece of DNA in the *Bam* HI site, is no longer able to confer tetracycline resistance on its host, as the inserted DNA disrupts the necessary *tet*^{r} genes. Cells containing these recombinant pBR322 molecules are still resistant to ampicillin, but sensitive to tetracycline (*amp*^{r} *tet*^{s}). After transformation, the cells are plated onto ampicillin containing medium and incubated until the appearance of colonies. Only the transformed colonies grow, as the untransformed cells are *amp*^{s} and as a result do not produce colonies on selective medium. However, while only a few cells contain recombinant pBR322 molecules, most contain

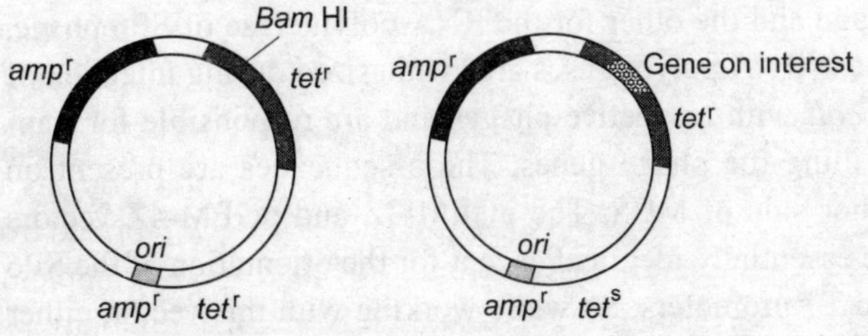

Figure 8.27 Insertional inactivation of tetracycine resistance gene during cloning in pBR322; (a) non-recombinant pBR322 DNA molecule, (b) Recombinant pBR322 molecule

only the normal, self-ligated plasmid. Finally, the recombinants are identified by plating (in replica) onto agar media containing tetracycline. After incubation, some of the original colonies regrow, but others do not. Only nonrecombinant colonies are able to grow on tet$^+$ plates as these carry the nonrecombinant or reconstituted pBR322 with no inserted DNA (*amp*^{r} *tet*^{r}). The colonies that do not grow on tetracycline agar medium (*amp*^{r} *tet*^{s}) are those of recombinant. Once their positions are localized, colonies for further study can be recovered from the original ampicillin agar plate.

pUC Series

Vieira and Messing (1982) constructed the pUC series of plasmids, all of which are derived from pBR322. This vector was named after the place of their initial preparation, i.e., the University of California. These plasmids are ~2,700 bp long, have high copy number and possess the *amp*^{r} gene. The origin of replication is derived from pBR322 and the *lac* Z′ gene is derived from *E. coli* chromosome (Figure 8.28). The nucleotide sequence of the *amp*^{r} gene has been altered so that it no longer contains the unique restriction sites; all these sites are now clustered into a short segment. This family of vectors also contains an inducible promoter, the *lac* promoter (P_{lac}) along with the associated *lac* operator region (O_{lac}). Besides these, pUC series of vectors contain a couple of other elements that provide improved qualities to this vector. These elements are as follows:

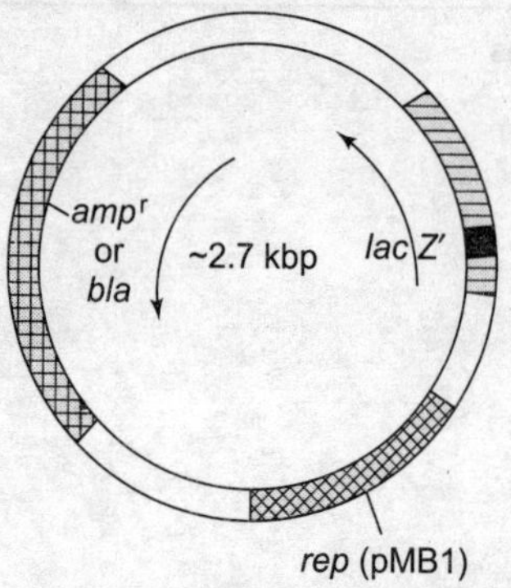

Figure 8.28 Gene map of pUC plasmid vector

• The *lac* I gene which codes for the lac repressor protein. The plasmid produces lac repressor to supplement the host levels. Being a high copy number plasmid, the host lac repressor levels may not be sufficient to efficiently repress the *lac* operator on the plasmids.

• The *lac* Z′ gene, the product of which is transcribed from the *lac* promoter and produces an N-terminal fragment of the β-galactosidase protein.

• The *tet*^{r} gene region of pBR322 has been replaced by a gene segment coding for part of the β-galactosidase enzyme into which a series of unique restriction sites have been designed in the form of MCS or polylinker. In pUC series of vectors, MCS is located just downstream from the *lac* promoter and after the first few codons of the *lac* Z′ gene (between positions 396 and 454). The usual site for insertion of the MCS is between the initiator ATG codon and a region that encodes a functionally nonessential part of the α-complementation peptide. The insertion of the MCS into the *lac* Z′ fragment does not affect the ability of the α-peptide to mediate complementation, but cloning DNA fragments into the MCS does. Therefore, recombinants can be detected by blue-white screening (for details see Chapter 16) on growth medium containing X-gal and IPTG.

pUC series of vectors are one of the most popular *E. coli* cloning vectors as these have some significant advantages as compared to other vectors. A few of them are described below:

• A chance mutation within the origin of replication during construction of pUC8 resulted in the plasmid having a copy number 500–700 even before amplification. As a result high yield of cloned DNA can be easily obtained from *E. coli* cells transformed with recombinant pUC8 plasmids. The copy number can be increased further (up to 1,000–3,000) upon addition of chloramphenicol.

• The identification of recombinant cells can be accomplished by a single step process by plating onto agar medium containing X-gal and IPTG by blue-white screening, while in the case of pBR322 series of vectors, recombinants are selected in a two-step process, requiring replica plating from medium containing one antibiotic to another. Therefore, a cloning experiment with pUC8 can be performed in less time as compared to pBR322.

• The vectors of pUC series carry different combinations of restriction enzymes and provide greater flexibility in the types of DNA fragments that can be cloned. The MCS of pUC series of vectors generates DNA fragment with two different sticky ends to be cloned without any additional manipulations such as linker attachment or homopolymer tailing. The asymmetric cloning sites also aid in directional cloning.

• The restriction site clusters in pUC series of vectors are the same as the clusters in the equivalent M13mp series of vectors (for details see Chapter 10). Therefore, DNA cloned into a member of the pUC series can be transferred directly to its corresponding M13mp vector for application in DNA sequencing or site-directed mutagenesis (for details see Chapter 19). Moreover, similar to M13mp series, the cloning vectors belonging to pUC family are also available in pairs (sister or dual vectors) with reversed orders of restriction sites relative to the *lac* Z promoter. pUC8 and pUC9 make one such pair. Other similar pairs include pUC12 and pUC13 or pUC18 and pUC19 (Figure 8.29). Note that pUC18 is a translational fusion vectors, however, it is not used in this way.

pGEM^R Series

pGEM^R series of vectors, comprising of pGEM-3Z and pGEM-4Z, are projected for use as standard cloning vectors, as well as for the highly efficient synthesis of RNA *in vitro*. pGEM^R series of vectors provides an example of expression vectors (transcriptional fusion vectors) (for details see Chapter 12). This vector series is very similar to pUC series of vectors. The size of pGEM is ~2,750 bp and it carries the *amp*^{r} and *lac* Z′ genes having polylinker at the 5′-end to be used for introduction of new DNA into the pGEM molecule. Beside these, pGEM has two additional promoter sequences, each of which acts as a recognition site for the attachment of an RNA polymerase enzyme (Figure 8.30). The promoters carried by these vectors are not the standard sequences recognized by the normal *E. coli* RNA polymerase. One of the promoters is specific for the RNA polymerase coded by T7 phage and the other for the RNA polymerase of SP6 phage. These RNA polymerases are synthesized during infection of *E. coli* with respective phages and are responsible for transcribing the phage genes. These sequences are present on either side of MCS. The pGEM-3Z and pGEM-4Z vectors are essentially identical except for the orientation of the SP6 and T7 promoters. So while working with this vector, either T7 or SP6 RNA polymerase is supplied *in vitro*, depending upon the strand-specific RNA of interest. The genes for these RNA polymerase may also be engineered into the host chromosome. These RNA polymerases are selected because these

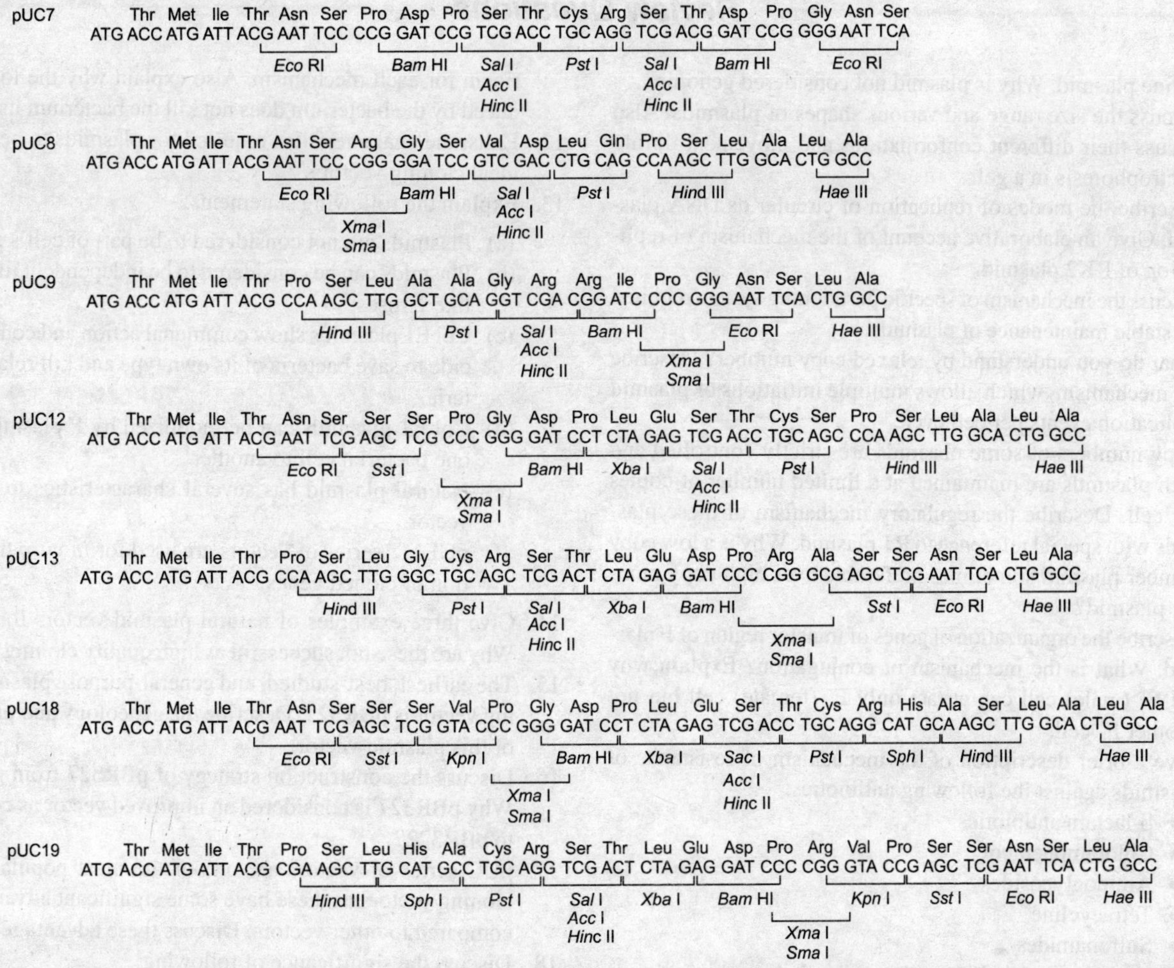

Figure 8.29 The polylinker regions (MCS) of pUC series of vectors

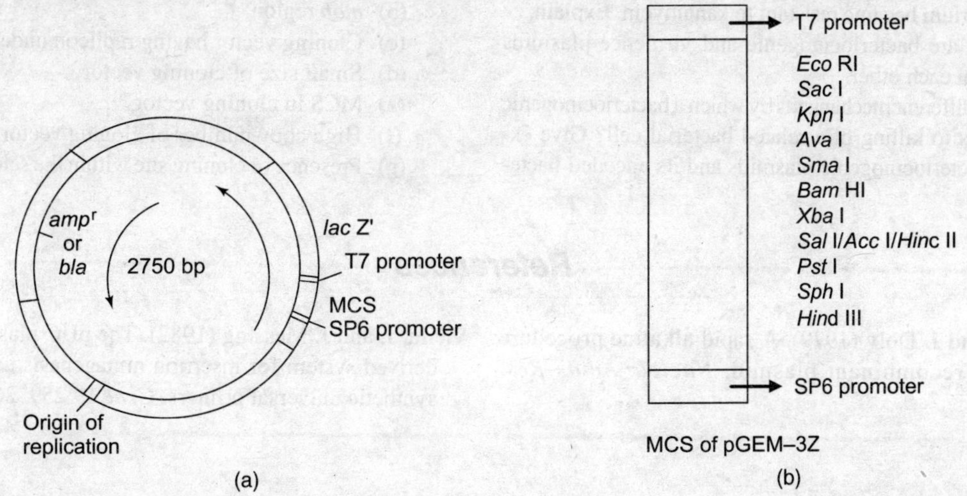

Figure 8.30 pGEM-3Z vector; (a) Vector map, (b) MCS

are very active (entire lytic infection cycle takes only 20 min so phage genes must be transcribed very quickly) and are able to synthesize 1–2 μg RNA/minute, which is significantly more than that produced by standard *E. coli* RNA polymerase. Besides getting higher yield of RNA, the other advantage is that the RNA complementary to both the strands can be synthesized by using two different promoters at different times. The RNA thus synthesized can be used as probe or used in experiments designed for studying RNA processing (e.g., splicing) or protein synthesis.

Review Questions

1. Define plasmid. Why is plasmid not considered genome?
2. Discuss the size range and various shapes of plasmids. Also discuss their different conformations and movement during electrophoresis in a gel.
3. Describe the modes of replication of circular ds DNA plasmid. Give an elaborative account of the mechanism of replication of RK2 plasmid.
4. Discuss the mechanism of specific partitioning system required for stable maintenance of plasmids.
5. What do you understand by relaxed copy number? Describe the mechanism, which allows multiple initiations of plasmid replication events per cell cycle.
6. Copy numbers of some plasmids are strictly controlled and such plasmids are maintained at a limited number of copies per cell. Describe the regulatory mechanism of these plasmids with special reference to R1 plasmid. Why is a low copy number plasmid sometimes preferable over high copy number plasmid?
7. Describe the organization of genes of transfer region of F plasmid. What is the mechanism of conjugation? Explain why the F^+ (male) cell can attract only F^- (female) cell but not another F^+ cell.
8. Give a brief description of the mechanism of resistance of plasmids against the following antibiotics:
   - β-lactam antibiotics
   - Chloramphenicol
   - Aminoglycosides
   - Tetracycline
   - Sulfonamides
9. A bacterium was sensitive to the antibiotic kanamycin. It was then transformed with a plasmid containing *amp*^{r} and *kan*^{r} gene and the bacterium became resistant to kanamycin. Explain.
10. Explain how are bacteriocinogenic and virulence plasmids different from each other.
11. What are the different mechanisms by which a bacteriocinogenic plasmid leads to killing of a related bacterial cell? Give examples of bacteriocinogenic plasmids and its encoded bacteriocin for each mechanism. Also explain why the toxin produced by the bacterium does not kill the bacterium itself.
12. Enlist the characteristics required for plasmids to be used as ideal cloning vectors.
13. Explain the following statements:
    (a) Plasmids are not considered to be part of cell's genome.
    (b) Plasmids can be considered to be independent life-forms like viruses.
    (c) Col E1 plasmids show communal action and commit suicide to save bacteria of its own type and kill related bacteria.
    (d) Col E1 plasmids can be mobilized by F plasmids from one bacterial cell to another.
    (e) Natural plasmid has several characteristics to act as a vector.
    (f) pGEM® series of vectors are used for *in vitro* transcription of cloned genes.
14. Give three examples of natural plasmid vectors for *E. coli*. Why are these not successful as high quality cloning vectors?
15. The earliest, best-studied, and general-purpose plasmid cloning vector is pBR322. Describe the geneology and properties of this plasmid vector.
16. Discuss the construction strategy of pBR327 from pBR322. Why pBR327 is considered an improved vector as compared to pBR322?
17. pUC series of vectors are one of the most popular *E. coli* cloning vectors as these have some significant advantages as compared to other vectors. Discuss these advantages.
18. Discuss the significance of following:
    (a) *tra* locus
    (b) *mob* region
    (c) Cloning vector having replicon under relaxed control
    (d) Small size of cloning vector
    (e) MCS in cloning vector
    (f) High copy number of cloning vector
    (g) Presence of cloning site within the selectable marker gene

References

Birnboim, H.C. and J. Doly (1979). A rapid alkaline procedure for screening recombinant plasmid. *Nucleic Acids Res.* **7**:1513–1523.

Vieira, J. and J. Messing (1982). The pUC plasmids, an M13-mp7 derived system for insertion mutagenesis and sequencing with synthetic universal primers. *Gene* **19**:259–268.

9 λ Bacteriophage

Key Concepts

In this chapter we will learn the following:

- Biology of λ bacteriophage
- λ bacteriophage growth and preparation of λ DNA
- Methods of introduction of λ DNA into bacterial cells
- λ bacteriophage as cloning vehicle

9.1 INTRODUCTION

Ester Lederberg found *E. coli* K 12 to be lysogenic for λ phage, which marked the discovery of λ phage as well as phenomenon of lysogeny. λ phage is a temperate phage. Its genome may integrate into the host genome in a more or less stable relationship, or the phage may enter a productive lytic cycle resulting in lysis of the host cell with the liberation of a number of phage particles.

λ phage is a type of lambdoid phage, and of all the phages that infect *E. coli*, it is the most extensively studied and genetically complex phage. Because it has been the object of molecular genetic research, right from the beginning of gene manipulation, it was investigated and developed as a vector for transgenes. The high rate of infectivity makes it ideal for creating large number of clones that are often needed for construction of gene libraries that represent large populations of molecules. In this chapter, the basic biology of λ phage, and the nature and use of λ cloning vectors are summarized.

9.2 BIOLOGY OF λ BACTERIOPHAGE

The λ phage is among the most extensively characterized complex phages, and its genetic regulatory mechanism forms one of the best paradigms for the control of development in higher organisms. In order to trace the development of a versatile series of phage λ based vectors and to exploit them to full capacity for gene cloning, information of morphological features, gene organization, and morphogenesis are necessary. In this section, the structure of the λ genome and its interactions with host cell are discussed in detail.

9.2.1 Composition and Structure of λ Phage

λ phage is a coliphage (phage that infects *E. coli*), and is a typical example of a head-and-tail phage belonging to the family Siphoviridae. The phage has ~50 nm diameter icosahedral head ($T = 7$ icosadeltahedron), a flexible tubular protein tail ~150 nm long bearing a single thin protein tail fiber at its end (Figure 9.1). In wild-type as well as laboratory strains of λ, tail fibers, if at all visible, are stubs that project from the junction between the tail shaft and the tail spike. The cylindrical head–tail connector is present at one of the five-fold vertices of the head, and breaks the icosahedral symmetry of phage head. This connector serves as a site for attachment of preformed tail to the head. The tail aids in attachment of the phage to the bacterial surface, and for injecting its DNA into the cell.

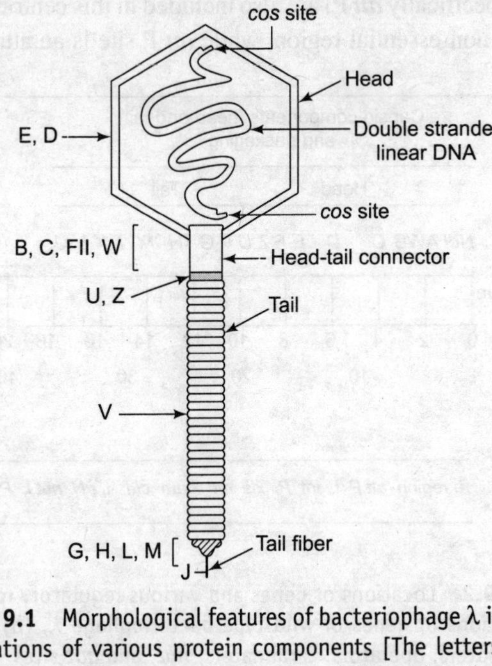

Figure 9.1 Morphological features of bacteriophage λ indicating locations of various protein components [The letters refer to specific proteins or gene products as indicated in the text.]

The protein head of the wild-type virion contains a 48,502 bp double stranded (ds) linear B-DNA molecule (MW 31 × 10^6 Da) with short complementary ('sticky' or 'cohesive') single stranded (ss) projections of 12 nucleotides at its 5′-ends, called cohesive termini or *cos* sites. The presence of these *cos* sequences on either end of the wild-type λ DNA ensures that each head receives the correct amount (48.5 kbp) of DNA, and because of complementariness of *cos* sites, DNA adopts a circular structure when injected into host cell (for details see Section 9.2.5). The entire genome sequence of λ phage has been determined and is available online.

9.2.2 Gene Organization

λ DNA has been extensively studied by the techniques of gene mapping and DNA sequencing. As a result the positions and identities of most of the genes on λ DNA molecule are known. The studies reveal that the λ genome carries a complement of ~50 genes, which are organized into functionally related clusters placed together on the λ genome map (Figure 9.2a), which are as follows:

Left-hand Region

The genes on the left of the conventional linear map includes genes *Nu* 1 through *J* that code for head and tail proteins, or those required for assembly of infectious phage particles by packaging of the λ DNA into λ phage heads (Figure 9.2a).

Central Region

The central region includes genes *J* through *gam* that code for integration and recombination functions (*int, xis, red, gam*). Besides, two genetic loci, namely, *b*2 region and *att* site (or more specifically *att* P) are also included in this central region. *b*2 is a non-essential region, while *att* P site is an attachment site that plays a role in integration of λ genome into the host chromosome (for details see Section 9.2.6). An important feature of central region is that many genes of this region are not essential for lytic growth, and can be deleted or replaced to make space for foreign DNA during construction of λ vectors without seriously impairing the infectious growth cycle.

Right-hand Region

Genes to the right of the central region comprise of regulatory genes (*N* and *Q*), genes controlling lysogeny and lysis (*cI*, *cII*, *cIII*, and *cro*), two genes essential for DNA replication during lytic growth (*O* and *P*), and genes required for host cell lysis (*S*, *R*, and R_Z).

The properties and functions of various λ genes and genetic sites are tabulated in Table 9.1. Further, host genes and genetic loci that play significant roles in phage infection are summarized in Table 9.2. Note that the organization of functionally related genes in clusters enables these genes to be transcribed together, i.e., as an operon. Clustering of related genes is also profoundly important for controlling expression of the λ genome, as it allows the genes to be switched 'on' and 'off' as a group rather than individually. Moreover, as we shall see later in Section 9.5, it is also advantageous during construction of λ-based vectors.

9.2.3 Regulatory Regions

A thorough knowledge of λ physiology and control circuits affecting λ gene expression is essential for construction of efficiently replicating λ cloning vectors, and those allowing efficient expression of DNA cloned under the control of λ promoters. In this section, the locations and functions of various regulatory elements (Figure 9.2b) present in the λ genome are discussed.

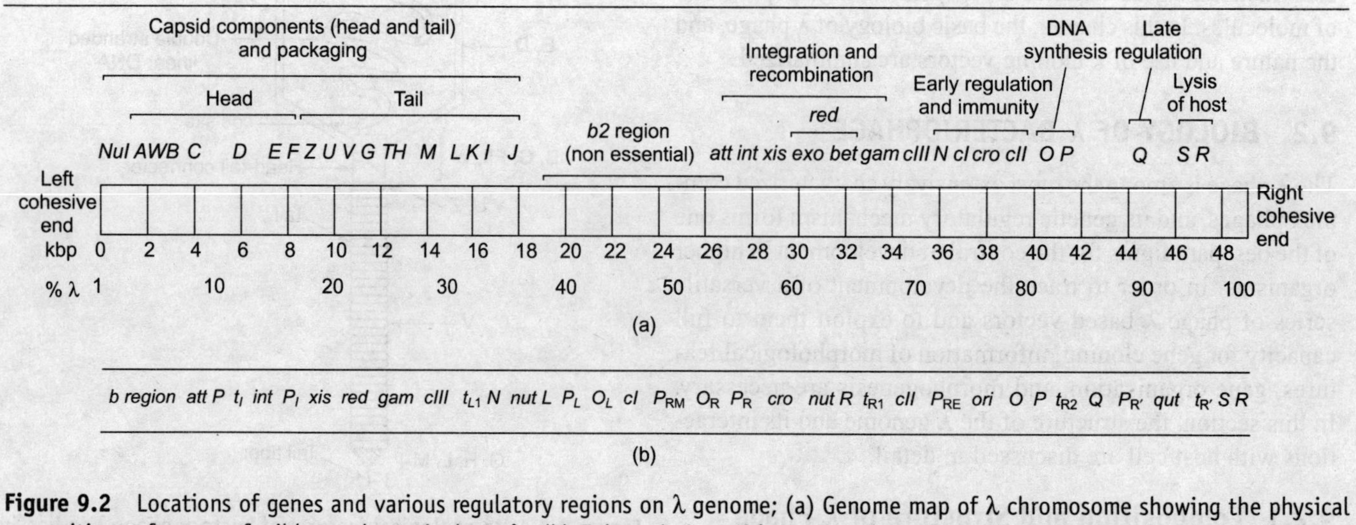

Figure 9.2 Locations of genes and various regulatory regions on λ genome; (a) Genome map of λ chromosome showing the physical positions of genes of wild-type bacteriophage λ, (b) Enlarged view of λ genome from *b*2 region to *R* gene showing positions of promoters, operators, terminators, *nut*, and *qut* sites

Table 9.1 Properties and Functions of Various Genes and Genetic Sites of λ Bacteriophage

Phage genes or genetic sites	Properties and functions of encoded proteins
N	Position on map: 35,040–35,360 bp; pN (107 amino acid), an RNA binding protein, is an immediate early gene product expressed from the transcript initiated at P_L promoter; It is a positive regulator of delayed early genes; The transcription of delayed early genes commences as soon as a significant quantity of the pN accumulates. It is thus responsible for immediate early-to-delayed early switch; It acts in *trans*, but it also requires the presence of *nut* L and *nut* R sites. But pN does not bind to these sequences within DNA, rather it binds to corresponding *nut* sequences embedded in the leftward and rightward immediate early RNAs; It acts as transcriptional antiterminator. pN modulated RNA polymerase passes over many different terminators that it encounters. Thus, once RNA polymerase has passed a *nut* site, the *box* B RNA (of *nut* site) is recognized by dimers of pN via its Arg rich N-terminal helix (~18 residue), which binds against the major groove of *box* B hairpin via electrostatic and hydrophobic interactions, forming pN-*box* B complex. From the complex, pN is loaded on to the polymerase itself. This antitermination at t_{L1}, t_{R1}, and t_{R2} allows transcription initiated at P_L and P_R to proceed in leftward and rightward directions, and eventually causes the transcription of delayed early genes, *viz. cII*, *cIII*, *O*, *P*, and *Q*; The activity of pN is necessary for the lytic growth of bacteriophage carrying t_{R2}. However, λ *nin* carries a deletion of t_{R2}, and can grow independently of pN.
cro	Position on map: 38,041–38,238 bp; *cro* stands for control of repressor and other things; pCro is an immediate early gene product expressed from the transcript initiated at P_R; It is a single domain protein that competes with pCI repressor for binding to three subsites in each of the two operator regions, O_L (O_{L1}, O_{L2}, O_{L3}), and O_R (O_{R1}, O_{R2}, and O_{R3}) with different preferences of binding and affinity. It binds as a dimer to operator O_L and O_R in the order: $O_{R3} > O_{R2} \approx O_{R1}$ This order is opposite to that of binding of pCI repressor to these subsites. Unlike pCI, the binding is non-cooperative; As pCro accumulates, it binds to O_{R3} so as to abolish transcription from overlapping promoter P_{RM}. It is repressor of pCI and early genes. It helps in establishment of lysis.
cII	Position on map: 38,360–38,650 bp; pCII is activator of transcription of λ genes that repress lytic functions, and catalyzes integration of the viral DNA into the host chromosome; It is delayed early gene product expressed from the transcript initiated at P_L promoter; Its concentration is high during poor nutritional conditions, and higher multiplicity of infection; It coordinately regulates transcription from three separate leftward promoters: P_{RE}, P_I, and P_{aQ}. It thus aids in establishment of lysogeny by regulating synthesis of λ repressor from P_{RE} promoter, integrase (pInt) from P_I promoter, and antisense RNA for *Q* from P_{aQ} promoter.
O	Position on map: 38,686–39,582 bp; pO (333 amino acid) aids in DNA replication initiated in lytic infection cycle at *ori* λ; It, presumably as dimer, specifically binds by its N-terminus to each of the four repeated 18 bp palindromic segments within *ori* λ region, leading to formation of *O*-some. The interaction of pO with *ori* λ results in more extensive bending of the repeated region, and induces helix destabilization of the adjacent AT-rich region.
P	Position on map: 39,582–40,280 bp; pP (233 amino acid) plays a role in DNA replication; It interacts with the C-terminus of pO in *O*-some, and also associates with the host Dna B helicase
Q	Position on map: 43,886–44,506 bp; pQ (207 amino acid) is product of delayed early transcription from P_R promoter; It is regulator of late gene expression. Like pN, pQ acts as an antiterminator and is therefore a genuine positive regulator of transcription. It is only in its presence that transcripts, which have been prematurely terminated, are extended into the late genes; Synthesis of large amounts of pQ marks the beginning of late phase of transcription, and the synthesis of head and tail proteins.
cIII	Position on map: 33,302–33,463 bp; pCIII inhibits host Hfl protein (cellular protease that degrades pCII), and hence stabilizes pCII; It is delayed early gene product expressed from the transcript initiated at P_L promoter; Its concentration is high during poor nutritional conditions, and higher multiplicity of infection.

(*Contd.*)

Table 9.1 (*Contd.*)

Phage genes or genetic sites	Properties and functions of encoded proteins
gam or γ	Position on map: 32,819–33,232 bp; Gam protein inhibits host Rec BCD (exonuclease V). This, in turn promotes rolling circle replication and formation of concatemeric DNA, which is a good substrate for packaging; It is delayed early gene product expressed from the transcript initiated at P_L promoter.
red	Position on map: *red* A (31,351–32,028 bp); *red* B (32,028–32,810 bp); Red protein is responsible for recombination leading to formation of dimeric (or multimeric) DNA, which is also a good substrate for packaging; It is delayed early gene product expressed from the transcript initiated at P_L promoter.
xis	Position on map: 28,863–29,078 bp; pXis (72 amino acid) is an excisionase that catalyzes excision of prophage from lysogen upon induction of lysis; It is delayed early gene product expressed from the transcript initiated at P_L promoter.
int	Position on map: 27,815–28,882 bp; pInt (356 amino acid) is integrase or Int recombinase, which stimulates prophage integration and excision; It is a type I topoisomerase and its preferred substrate is closed circular DNA with negative supercoils. It catalyzes a two-strand exchange at a precise nucleotide by way of a transient DNA–protein bond on a tyrosine residue thereby generating a Holliday structure. Branch migration then occurs across a 7 bp segment of the core, and the Holiday structure is resolved by exchange between the two other strands. Proper positioning of integrase is facilitated by integration host factor. The cross-over point of pInt is included within λ *att* P; It is expression product of the transcript initiated from P_I promoter, stimulated by binding of pCII; pInt is not expressed from the leftward transcript initiated at P_L promoter. This signifies that pInt is not formed in the absence of pCII.
cI	Position on map: 37,230–37,940 bp; pCI (236 amino acid monomer) is λ repressor with two roughly equal sized domains connected by ~30 residue hinge, which is susceptible to cleavage by proteolytic enzymes. The N-terminal domain is DNA binding domain (a helix-turn-helix domain) that binds operator. The C-terminal domain provides the contacts for dimer formation; It competes with pCro binding to three subsites in each of the two operator regions, O_L (O_{L1}, O_{L2}, O_{L3}), and O_R (O_{R1}, O_{R2}, and O_{R3}). It binds to DNA as a dimer so that its two-fold symmetry matches those of the operator subsites to which it binds. Thus, a single dimer recognizes a 17 bp DNA sequence at operator, each monomer recognizing one half-site. It binds to operator region in cooperative manner; It is repressor of the lytic genes and blocks the synthesis of proteins pN, pO, pP, and pQ by binding to operators O_L and O_R, thus preventing transcription initiating at P_L and P_R. This in turn prevents the entry into the lytic pathway. In other words, it helps in establishment of lysogeny; Despite its name, it can both activate and repress transcription. When functioning as a repressor, it works in the same way, as does Lac repressor, it binds to sites that overlap the promoter and excludes RNA polymerase. As an activator, it works like catabolite activator protein (CAP), by recruitment. Its activating region is in the N-terminal domain of the protein. Its target on polymerase is a region of the σ subunit adjacent to the part of σ that recognizes –35 region of the promoter (i.e., region 4 of σ)
S	Position on map: 45,186–45,506 bp; pS (107 amino acid; 8.5 kDa) is an inner membrane protein that plays role in host lysis by forming pores/holes in the cell membrane. These pores allow pR to enter the periplasmic space and access their peptidoglycan substrate; It is late gene product, which is expressed from the transcript initiated at $P_{R'}$ promoter.
R and R_Z	Position on map: R (45,186–45,506) bp; pR and pR_Z play role in host lysis; pR (17.5 kDa) is a transglycosylase that cleaves the glycosidic bonds between *N*-acetyl glucosamine (NAG) and *N*-acetyl muramic acid (NAM) in the host cell wall peptidoglycan, generating 1,6-disaccharide; pR_Z (~29 kDa) is a membrane protein that plays role in host lysis, possibly an endopeptidase that cleaves oligopeptide cross-links in the peptidoglycan cell wall; Both are late gene products expressed from the transcript initiated at $P_{R'}$ promoter.
A,Nu 1	Position on map: A (711–2,633); These two leftmost genes on the bacteriophage λ genome code for terminase or endonuclease, a hetero-oligomer of polypeptides [pNu 1 (181 amino acid); pA (641 amino acid)]; Its function is to recognize and cut DNA at *cos* sites as it is being inserted into the head. The DNA sequence that is recognized and cleaved by terminase consists of two subsites, the *cos* N (*cos* nicking) site and the adjacent *cos*

(*Contd.*)

Table 9.1 (*Contd.*)

Phage genes or genetic sites	Properties and functions of encoded proteins
	B (*cos* binding) site. Cleavage at *cos* N is carried out by pA, which is accompanied by ATP hydrolysis. pA acts both as an allosteric effector of terminase and to melt the *cos* ends after cleavage. Following nicking at *cos* N, terminase remains tightly bound to the left end of λ genome forming complex I. This complex I then binds to the portal protein pB, which serves as the site of DNA entry into the prohead. Terminase remains bound to pB as DNA is reeled from the cleaved concatemer into the prohead. Packaging ceases when terminase introduces staggered nicks at the next *cos* site, by which time the prohead contains a complete λ genome in linear form; These are late genes products, and synthesized during lytic infection cycle from $P_{R'}$ promoter.
W,B,Nu 3,C, D,E,FI,FII	These code for proteins required for the assembly of head and connector; These are late genes products, and synthesized during lytic infection cycle from $P_{R'}$ promoter; The proteins pE (6,135–7,157 bp) and pD (5,747–6,076 bp) are the major capsid proteins, which account for 72% and 20% of the total head protein, respectively. pE forms polyhedral shell, and are arranged on the surface of a *T* = 7 icosadeltahedron. pD decorates capsid surface. pB, pC form a cylindrical structure, called precursor of connector or initiator, which attaches the tail to the head. pFII and pW form tail binding site. pFI possibly helps terminase pA in recognition of *cos* sites and pumping of DNA into the prehead by an ATP-dependent process; The other minor head components are pB*, derived by proteolytic cleavage of the pB. pX1 and pX2, two similar polypeptides are derived from the fusion of part of pE with pC. The formation of pB*, pX1, and pX2 requires the putative core protein pNu 3.
Z,U,V G,T,H, M,L,K,I,J	These genes encode proteins required for tail assembly; These are late genes products, and synthesized during lytic infection cycle from $P_{R'}$ promoter; pV is the major tail protein. pG, pH, pL, pM, pJ form adsorption organelle. pI and pK are not the components of mature tail. Immature tail is activated by the action of pZ before joining the head.
ori λ and *ice* site	*ori* λ is origin of DNA replication, responsible for autonomous DNA replication during lytic infection cycle. This region is recognized by pO forming *O*-some; *ori* λ comprises of two structurally distinct domains. First domain consists of four 18 bp direct repeats, each with dyad hyphenated symmetry. Periodically phased tracts of adenine residues separate each repeat, which gives this domain a bent character. The second domain, lying immediately adjacent to the first, comprises of a highly AT rich segment in which almost all of the purines are on the same strand. At least two of the direct repeats, and 35–40 base pairs of the adjacent AT-rich region are required for any initiation of DNA replication; Inceptor (*ice*) site lies in gene *cII*, which having a particular secondary structure promotes the switch from RNA synthesis from P_O promoter to DNA synthesis in replication initiation.
b region	Position on map: 19,000–27,700 bp; It is accessory gene region (non-essential region); This region specifies the so-called accessory proteins that, although not essential for lytic growth, increase its efficiency.
att P site	Position on map: 27,723–27,742 bp; *att* P sequence on phage λ is the site of action of pInt for integration of phage DNA into the homologous sequence *att* B on the host chromosome; *att* P spans ~240 bp of phage DNA centered on a 15 bp core that is identical between phage and host chromosomes (*att* B), and includes the cross-over point of pInt. During integration, *att* P recombines with *att* B to form a prophage flanked by recombinant *att* L and *att* R sites.
nut sites N utilization sites)	These are left and right binding sites for pN that leads to anti-termination of transcription, and hence switch over from immediate early gene transcription to delayed early gene transcription; *nut* L is located between P_L and *N*, while *nut* R is located between *cro* and t_{R1}. Precisely, *nut* L and *nut* R are located 60 and 200 nucleotides downstream of P_L and P_R, respectively; *nut* sites contain *box* B whose transcripts can form H-bonded hairpin loops and *box* A.
qut site (Q utilization site)	This is binding site for pQ that leads to transcription antitermination, and hence switch over from delayed early gene transcription to late gene transcription.
cos sites	These are single stranded extensions of 12-nucleotide at 5′-end of DNA that are complementary (cohesive) to each other. These are located 48.5 kbp apart (representing one genome) ensuring that head receives the correct amount of DNA; These sites are recognized and cleaved by terminase. The DNA between two *cos* sites, provided it lies within the range 37–52 kbp, is packaged in the phage head after cleavage by pNu 1/pA; Upon injection into host cell, the λ DNA circularizes by base pairing between complementary *cos* sites at two ends.

Table 9.2 Host Genes and Genetic Sites Required for λ Phage Infection Cycle

Host genes and genetic sites	Function
lam B	Host recognition protein (outer membrane protein)
dna A	DNA replication initiation
dna B	DNA helicase for unwinding of DNA during replication
primosome	Primer synthesis
pol C	DNA polymerase III for DNA replication
pol A	DNA polymerase I for excision of primers and filling of gaps
lig	DNA ligase for sealing of nicks during replication
gyr A, *gyr* B	DNA gyrase for supertwisting
rpo A, *rpo* B, *rpo* C	RNA polymerase core enzyme
rho	Transcription termination factor
nus A, *nus* B, *nus* E	Necessary for pN function
gro EL, *gro* ES	Head assembly
him A, *him* D (*hip*)	Integration host factor α and β
hfl A, *hfl* B	Host protease that degrades pCII
cap, *cya*	Catabolite repressor system
att B	Prophage integration site
rec A	Induction of lytic growth by stimulation of self-cleavage of pCI
rec BCD (*exonuclease* V)	Exonuclease that degrades the concatemeric meric (consisting of tandemly linked identical units) linear duplex DNA

Promoters P_R and P_L

The promoters P_R (promoter rightward) and P_L (promoter leftward) are particularly important in the early phase of infection in both lysogenic and lytic infection cycles. P_R and P_L promoters are located to the left and right side of the *cI* gene, respectively. These are strong, constitutive promoters, i.e., these bind host RNA polymerase efficiently, and direct transcription without the help of an activator. Thus, immediately after infection, bacterial RNA polymerases bind to promoters P_R and P_L, and generate early rightward and leftward transcripts, respectively (for details see Section 9.2.7). The sequence of P_L promoter overlaps with operator O_L, and that of P_R overlaps with O_R (Figure 9.3).

Promoter $P_{R'}$

$P_{R'}$ (promoter rightward for late genes) lies close to the *Q* gene, and has overlapping sequences with *qut* (*Q* utilization) site (for details see Section 9.2.9). Immediately after infection, bacterial RNA polymerase stimulates transcription from $P_{R'}$ promoter leading to synthesis of a 194 nucleotide long RNA. In the late phase of infection, this transcript is extended by the intervention of antiterminator protein pQ, such that the transcript initiating at $P_{R'}$ spans the entire late region. The extended transcript serves as mRNA for lysis proteins and phage structural proteins in the late phase of infection during lytic infection cycle.

Promoter P_{RM}

P_{RM} (promoter for repressor maintenance), in contrast to P_R and P_L promoters, is a weak promoter, and directs efficient transcription only when an activator is bound just upstream. It resembles *lac* promoter in this regard. The sequences of P_{RM} promoter overlap with P_R promoter (Figure 9.3). Transcription from this promoter is stimulated for the maintenance of lysogeny and transcribes only *cI* gene.

Promoter P_{RE}

P_{RE} (promoter for repressor establishment) lies several hundred base pairs upstream from P_{RM}. It is a weak promoter because it has a very poor –35 region. pCII binds to a site that overlaps the –35 region but is located on the opposite face of the DNA helix. By directly interacting with polymerase, pCII helps polymerase to bind to this promoter. Thus, it is under the positive regulation by pCII, the concentrations of which are influenced by the physiological and nutritional state of the cells, and the transcription from this promoter is stimulated for establishment of lysogeny.

Promoter P_I

P_I (promoter for integrase) has a sequence similar to that of P_{RE} and is located in front of the phage gene *int* (within the gene *xis*). The transcription from this promoter is stimulated during lysogeny. Upon activation of P_I promoter by pCII, transcription of *int* gene is initiated resulting in the production of integrase (pInt), but not excisionase. Note that *int* gene, though transcribed from P_R promoter, does not lead to formation of pInt during lytic infection cycle (for details see Section 9.2.9).

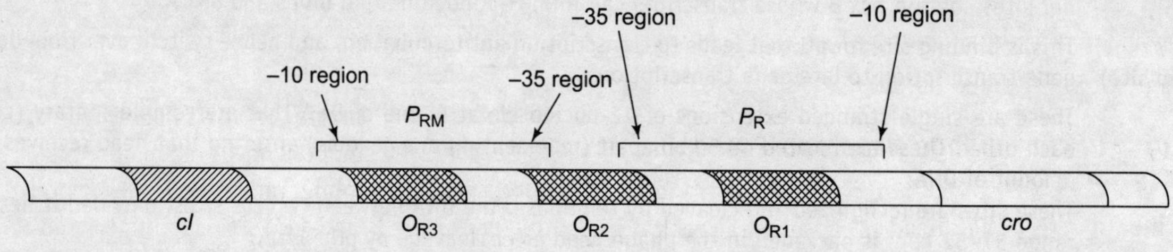

Figure 9.3 Relative positions of two promoters, P_R and P_{RM}, with respect to operator O_R [Note that O_{R2} overlaps the –35 region of P_R by three base pairs, and that of P_{RM} by two. This difference is enough for P_R to be repressed, and P_{RM} activated by repressor bound at O_{R2}.]

Promoter P_{aQ}

P_{aQ} (promoter for anti-Q) is located in the middle of gene *Q*. It is also stimulated by pCII, leading to the synthesis of antisense RNA for Q. This, in turn, forms RNA:RNA hybrid with Q mRNA, which is readily degraded by nucleases. This results in cessation of synthesis of late gene products. This promoter thus functions during lysogeny.

Operator O_R

O_R (right operator) lies upstream from its promoter P_R and shares some overlapping sequences with it. O_R comprises of three subsites, namely O_{R1}, O_{R2}, and O_{R3}. Each of these subsites consists of a similar 17 bp segment with an approximate palindromic symmetry (i.e., not identical), and is separated from the others by 6 or 7 nucleotides long spacers. O_{R2} overlaps the –35 region of P_R by three base pairs, and that of P_{RM} by two base pairs (Figure 9.3). These three subsites serve as binding sites for either a dimer of λ repressor (pCI) or a dimer of pCro (control of repressor and other things). Note that pCro and pCI repressor bind to the three subsites with different preferences (order of binding, cooperativity, and affinity).

Operator O_L

O_L (left operator) is the left counterpart of O_R. It also comprises of three palindromic sites, namely O_{L1}, O_{L2}, and O_{L3}. It lies upstream from its promoter P_L, with some overlapping sequences. O_L serves as binding site for either a dimeric λ repressor or a dimeric pCro. Similar to O_R, here also pCro and λ repressor bind to three subsites with different preferences. However, O_L plays only a minor role in the λ switch relative to that of O_R.

9.2.4 Assembly of Infectious λ Phage Particles (Morphogenesis)

The assembly of the heads and tails and packaging of DNA (during lytic cycle) are highly coordinated, and take place in following order:

- Assembly of empty head shells and tails in separate steps.
- Packaging of λ DNA in the empty head shells.
- Attachment of preformed tails to the filled heads.

A model for the assembly of infectious phage including separate assemblies of head and tail is shown in Figure 9.4. The details of the mechanism through which these phage particles are assembled are highlighted below:

Assembly of λ Bacteriophage Head and Packaging of λ DNA

As already mentioned, phage head is icosahedral with ~50 nm diameter encompassing 48.5 kbp linear ds DNA. The assembly of phage head and packaging of λ DNA occurs in five steps in an obligatory order given below:

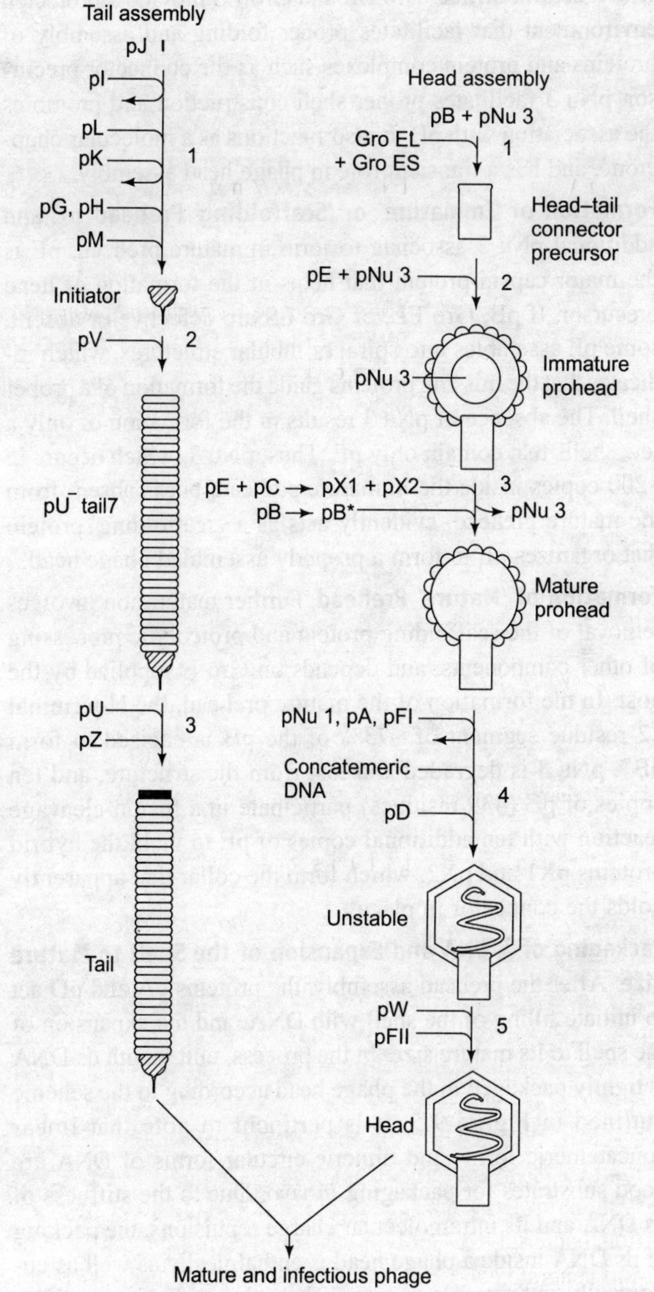

Figure 9.4 Assembly of infectious bacteriophage λ [The numbered steps are described in the text. The head and tails are assembled in separate pathways before joining to form the mature phage particle. Within each pathway the order of the various reactions is obligatory for proper assembly to occur.]

Formation of 'Initiator' or 'Precursor of Head–tail Connector' The components of the mature phage head are not entirely self-assembling; rather the *E. coli* encoded chaperonin proteins, Gro EL and Gro ES, facilitate head-tail connector assembly. Thus, two phage proteins pB and pNu 3 together with two host proteins Gro EL and Gro ES interact to form an 'initiator' that consists of 12 copies of pB arranged in a ring with a central orifice. Gro EL and Gro ES provide a protected environment that facilitates proper folding and assembly of proteins and protein complexes such as the connector precursor. pNu 3 facilitates proper shell construction and promotes the association with pB. It also functions as a molecular chaperone, and has a transient role in phage head assembly.

Formation of 'Immature' or 'Scaffolding' Prehead pE and additional pNu 3 associate to form immature prehead. pE is the major capsid protein that helps in the formation of head precursor. If pB, Gro EL, or Gro ES are defective or absent, some pE assembles into spiral or tubular structures, which indicates that the missing proteins guide the formation of a proper shell. The absence of pNu 3 results in the formation of only a few shells that contain only pE. Thus, pNu 3, which occurs in ~200 copies inside the immature prehead, but is absent from the mature prehead, evidently acts as a 'scaffolding' protein that organizes pE to form a properly assembled phage head.

Formation of 'Mature' Prehead Further maturation involves removal of the scaffolding protein and proteolytic processing of other components, and depends on Gro E supplied by the host. In the formation of the mature prehead, the N-terminal 22 residue segment of ~75% of the pB is excised to form pB*, pNu 3 is degraded and lost from the structure, and ten copies of pC (439 residues) participate in a fusion-cleavage reaction with ten additional copies of pE to yield the hybrid proteins pX1 and pX2, which form the collar that apparently holds the connector in place.

Packaging of λ DNA and Expansion of the Shell to Mature Size After the prehead assembly, the proteins pA and pD act to initiate filling of the shell with DNA, and the expansion of the shell to its mature size. In the process, unit length ds DNA is tightly packaged in the phage head according to the scheme outlined in Figure 9.5. It is pertinent to note that linear concatemeric forms and dimeric circular forms of DNA are good substrates for packaging *in vivo*. Due to the stiffness of ds DNA and its intramolecular charge repulsions, the packing of ds DNA inside a phage head is enthalpically as well as entropically unfavorable process. Thus, the packaging requires ATP, which actively pumps the ds DNA into phage head. Further, the packaging of λ genome into preheads requires two phage-encoded proteins, pNu 1 and pA, which bind to the concatenated linear ds DNA near left *cos* site. Two adjacent *cos* sites of the concatenated linear DNA are brought close together at the entrance to the head, where a staggered double stranded scission (endonucleolytic cleavage) is made by the terminase function of the phage protein pA to generate 12 nucleotide long cohesive termini (Exhibit 9.1). λ phage therefore contains a unique segment of DNA (in contrast to some phages in which the amount of DNA packaged is limited) by a 'headful' mechanism that results in their containing somewhat more DNA than an entire chromosome.The cleaved DNA–protein

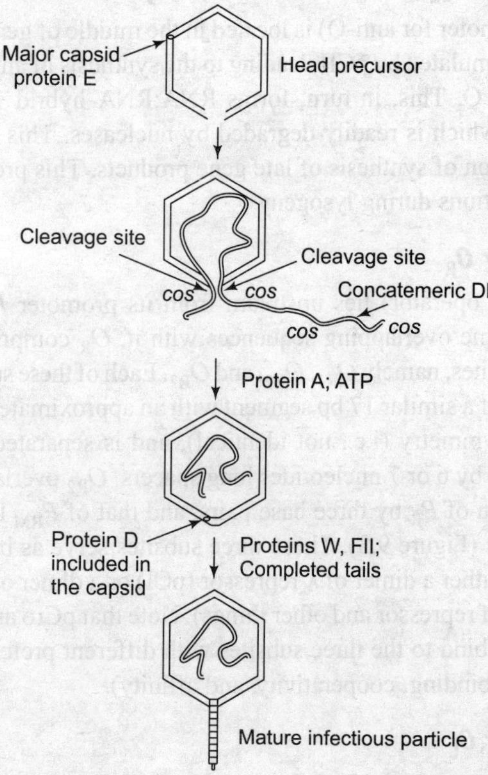

Figure 9.5 Packaging of λ DNA into phage heads and formation of infectious phage particles under natural conditions

complex then becomes attached to a defined area on the prehead. In the presence of protein pFI, the DNA is pumped into the prehead by an ATP-dependent process. The 'left' end of the DNA chromosome enters the prehead first as is indicated by the observation that only this end of the chromosome is packaged by an *in vitro* system when λ DNA restriction fragments are used. Whether the cutting of the initial *cos* site precedes or follows the initiation of packaging is unknown. However, at least *in vitro*, this process requires the binding to *cos* of the *E. coli* histone like protein known as integration host factor (IHF). IHF binds specific sequences of ds DNA, which it wraps around its surface, thereby inducing a sharp bend in the DNA. During the packaging process, the capsid proteins undergo a conformational change that results in an expansion of the phage head by ~11–45% (a process that occurs in 4 M urea in the absence of DNA). Final packaging in phage head is completed by pD or 'decoration' protein. pD attaches to the outside of the filled capsid at a newly exposed binding sites on pE, thereby partially stabilizing the expanded structure of the capsid. This locks the head in place around the DNA. The head is also stabilized by the addition of one final protein, pFII, which forms at least a portion of the site to which the tail binds. Note that due to the exact locations of the two *cos* sequences in the multiple length linear λ DNA (at the ends of one complete genome), it is ensured that each head receives

Exhibit 9.1 λ Terminase

λ terminase is a packaging enzyme that fashions concatemers of phage DNA into unit length genomes with protruding 5′-termini, 12 bp in length. The site of action of terminase is called the cohesive site or *cos* site. Unit length λ genomes are therefore normally generated by cleavage of linear concatemers at two *cos* sites spaced one genome length apart.

Terminase is a hetero-oligomer of polypeptides encoded by the two leftmost genes on the bacteriophage λ genome: the *Nu* 1 gene encoding a 181 amino acid polypeptide and the *A* gene encoding a 641 amino acid polypeptide. Both subunits can hydrolyze ATP, and mutants of either gene display the same phenotype, i.e., the accumulation of concatemers and empty proheads.

The DNA sequence that is recognized and cleaved by terminase consists of two subsites, the *cos* N (*cos* nicking) site, previously called *cos*, and the adjacent *cos* B (*cos* binding) site. Under normal conditions, cleavage at *cos* N is carried out by the larger subunit of terminase (pA) and is accompanied by hydrolysis of ATP, which acts both as an allosteric effector of terminase and to melt the *cos* ends after cleavage. Following nicking at *cos* N, terminase remains tightly bound to the left end of the λ chromosome in a complex called complex I. Complex I then binds to the portal protein (encoded by bacteriophage λ *B* gene), which serves as the site of DNA entry into the prohead. Terminase remains bound to the portal protein as DNA is reeled from the cleaved concatemer into the prohead. Packaging ceases when terminase introduces staggered nicks at the next *cos* site, by which time the prohead contains a complete linear bacteriophage λ genome 48.5 kbp in length. The filled heads then associate with tail units assembled by a separate pathway.

Source: Sambrook J and Russel DW (2001) *Molecular Cloning: A Laboratory Manual*, III ed., Cold Spring Harbor Laboratory Press, Cold Spring Harbor, New York, pp. 2.15.

the correct amount of DNA. The packaging range is 75–105% of the wild-type genome (i.e., 75% or 105% of 48.5 kbp). Thus, if less than 37 kbp of DNA is packed into a head, a non-infective phage particle is produced, and more than 52 kbp of DNA cannot fit into the phage head.

Formation of Tail Binding Site In the final stage of phage head assembly, pW and pFII are added in the given order to stabilize the head and form the tail-binding site.

Assembly of Tail

λ phage contains a flexible tubular protein tail ~150 nm long bearing a single thin protein tail fiber at its end. Tail assembly, which occurs independently of head assembly, proceeds from 200 Å long tail fiber toward the head-binding end. This strictly ordered series of reactions involves the following three stages.

Formation of 'Initiator' The formation of the 'initiator', which ultimately becomes the adsorption organelle, requires the sequential actions of pJ (the tail fiber protein; 1,133 residue), pI, pL, pK, pG, pH, and pM proteins. Of these, pI and pK are not components of the mature tail. The distal tip of the tail is involved in attachment to the *lam* B receptor of host during adsorption.

Formation of Immature Tail The initiator forms the nucleus for the polymerization of the major tail protein pV to form a stack of 32 hexameric rings, each containing six polypeptide subunits that represent the non-contractile tail shaft. These subunits form a hollow tube with 9 nm diameter through which DNA is injected upon infection. The length of this stack is probably regulated by pH, which becomes extended along the length of the growing tail, and somehow limits its growth.

Formation of Mature Tail In the termination and maturation stage of tail assembly, pU attaches to the growing tail, thereby preventing its further elongation. The resultant immature tail has the same shape as the mature tail, and can attach to the head so as to form an infectious phage particle. However, the immature tail must be activated by the action of pZ before joining the head.

Assembly of Infectious Phage

The completed tail then spontaneously attaches to a mature phage head to form an infectious λ phage particle, the exact mechanism of which is however not properly understood.

9.2.5 Significance of *cos* Sites

As mentioned earlier, the λ genome is linear ds DNA, base paired according to the Watson–Crick rules, with a length of 48.5 kbp. On either 5′-end of the ds DNA is a short 12 nucleotide single stranded stretches of DNA that are complementary to one another. These complementary single stranded ends are often referred to as 'sticky ends' or 'cohesive ends', because base pairing between them can 'stick' together the two ends of a DNA molecule (or the ends of two different DNA molecules during cloning experiments). The λ cohesive ends of a DNA molecule are called the *cos* sites, and these play distinct roles during the λ infection cycle. The various *in vivo* and *in vitro* roles of *cos* sites are as follows.

Circularization of λ Genome

First, the presence of single stranded complementary *cos* sites at two ends of genome allow linear λ DNA to circularize by intramolecular base pairing (Figure 9.6) when injected into host cell. This is also a necessary prerequisite for integration of λ DNA into the bacterial genome during lysogeny, or for θ replication, followed by rolling circle replication during lytic infection cycle (for details see Section 9.2.6). The circularized ds DNA is covalently closed and supertwisted by the host DNA ligase and DNA gyrase, respectively.

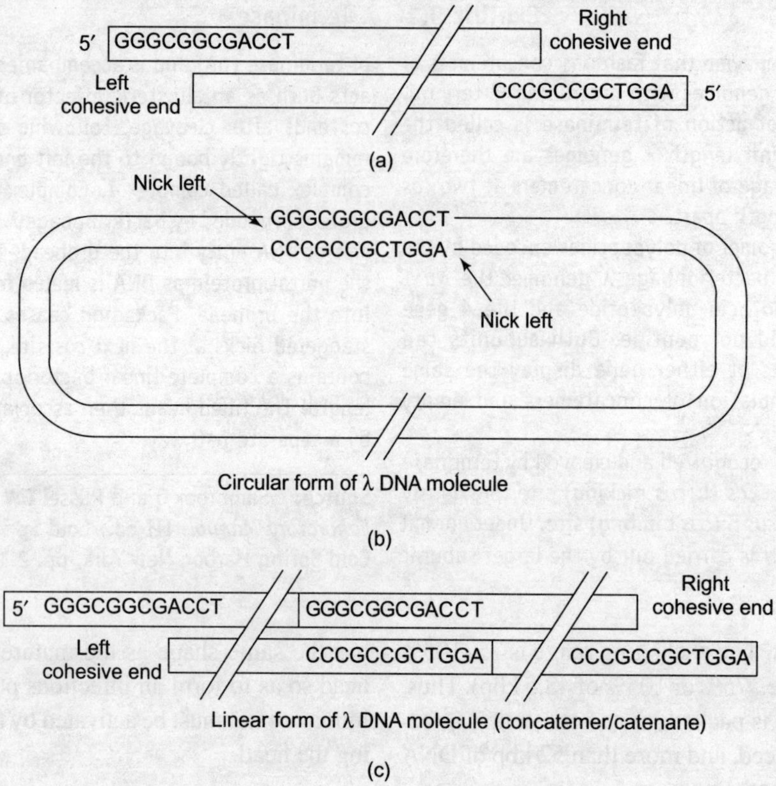

Figure 9.6 Structures of circular and linear forms of λ DNA; (a) λ DNA showing the complementary sequences of left and right cohesive ends, (b) Left and right cohesive ends undergo annealing by intramolecular circularization (base pairing between overlapping protruding complementary termini) to form a circular DNA inside host, (c) Linear concatemeric forms are generated by intermolecular reactions between different DNA molecules

Formation of Catenane or Dimer, and Formation of Site for Endonucleolytic Cleavage of λ DNA for Packaging into λ Phage Head

The second role of the *cos* sites comes into play after the prophage has excised from the host genome, or λ phage is following lytic infection cycle. At this stage a large number of new λ DNA molecules are produced by rolling circle replication resulting in a catenane consisting of a series of linear λ genomes joined together at the *cos* sites. The role of the *cos* sites is now to act as recognition sequences for endonuclease or terminase pA that cleaves the catenane at the *cos* sites, producing individual λ genomes (Exhibit 9.1). pA creates the single stranded sticky ends, and also acts in conjunction with other proteins to package each λ genome into a λ phage head. This property of *cos* cites is also exploited for *in vitro* packaging of genome into λ phage head (for details see Section 9.4.2). Thus, DNA with at least two *cos* sites can be packaged. The distance between two *cos* sites is also important for packaging. Only the two *cos* sites separated by 37–52 kbp (i.e., 75–105% of λ DNA) aid in correct cleavage during *in vivo* morphogenesis. The nature of the DNA between these *cos* sites is of little or no significance. Thus, linear concatemeric forms, and dimeric circular forms of DNA are good substrates for packaging *in vivo*. On the contrary, linear monomeric form of DNA cannot be packaged. Similarly, monomeric circular DNA molecules with a single *cos* site are also not suitable substrates for packaging, and in such cases, packaging takes place if a second *cos* site is introduced into the DNA either *in vitro* or by a recombinational event *in vivo*.

Construction of Cosmids

Further, the *cos* sites can be inserted into plasmid vectors for construction of chimeric plasmids, i.e., cosmids. These cosmids are high capacity cloning vectors (for details see Chapter 12).

9.2.6 Infection Cycle

Generally, phage λ is grown in K-type strains of *E. coli*, although other strains of *E. coli* can also serve as hosts. The infection of *E. coli* with λ phage is initiated when the phage specifically adsorbs to the host cell and injects its DNA into it. The fate of linear DNA entering the host cell is diagrammatically represented in Figure 9.7. After penetrating the host cell, the phage follows either lysogeny or the lytic cycle. The nutritional conditions and multiplicity of infection govern the mode of life cycle to be followed (for details see Section 9.2.9). During lysogeny, the phage DNA is stably integrated at a specific site in the host chromosome as prophage so that

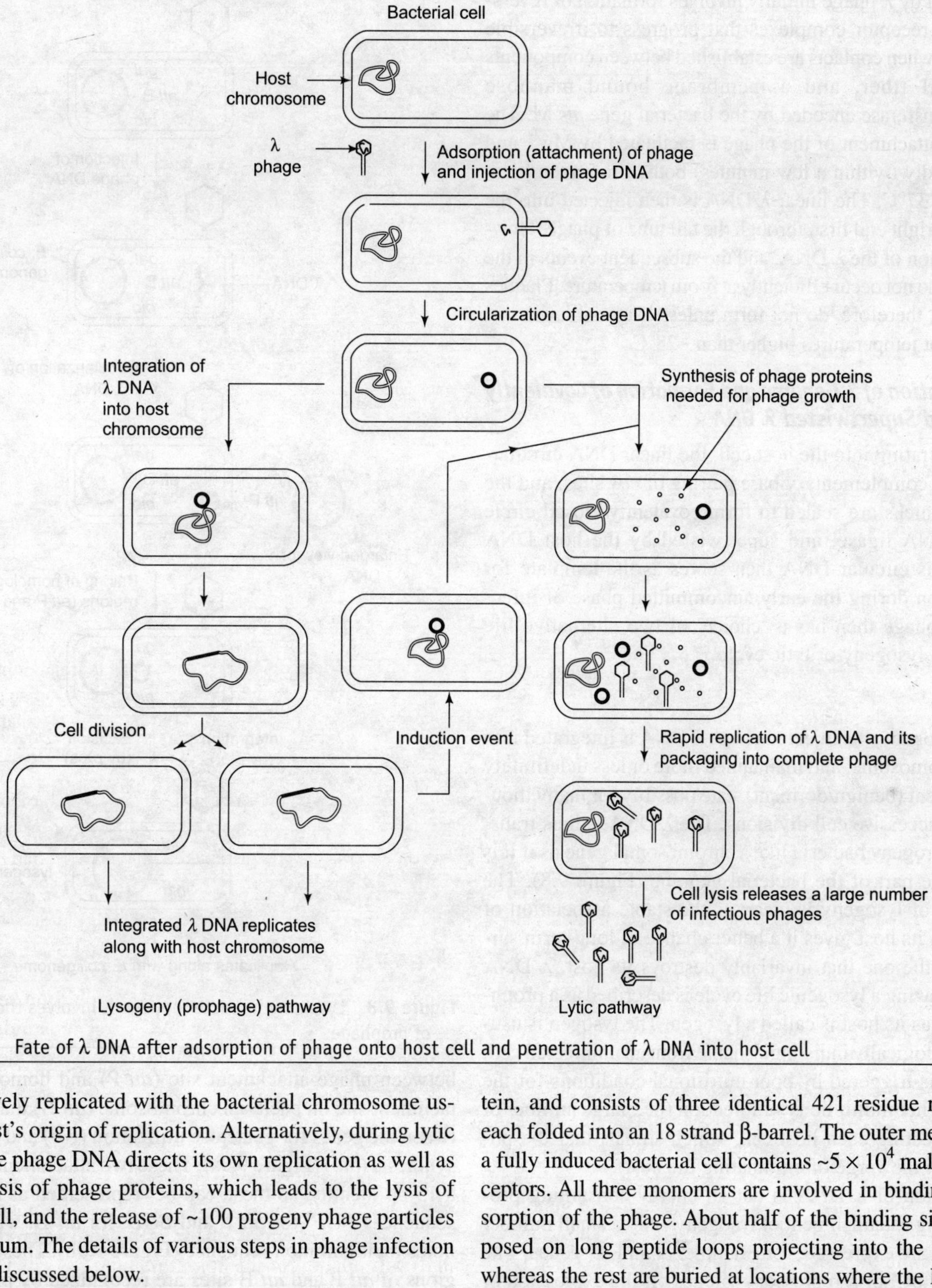

Figure 9.7 Fate of λ DNA after adsorption of phage onto host cell and penetration of λ DNA into host cell

it is passively replicated with the bacterial chromosome using the host's origin of replication. Alternatively, during lytic growth, the phage DNA directs its own replication as well as the synthesis of phage proteins, which leads to the lysis of the host cell, and the release of ~100 progeny phage particles per bacterium. The details of various steps in phage infection cycle are discussed below.

Adsorption of λ Bacteriophage on Host Cell and Injection of λ DNA

λ phage adsorbs to the trimeric maltoporin receptor of the host through a specific interaction with the C-terminal residues of pJ located at the tip of the tail fiber. The maltoporin receptor, encoded by the *E. coli lam B* gene, is an outer membrane protein, and consists of three identical 421 residue monomers, each folded into an 18 strand β-barrel. The outer membrane of a fully induced bacterial cell contains $\sim 5 \times 10^4$ maltoporin receptors. All three monomers are involved in binding and adsorption of the phage. About half of the binding sites are exposed on long peptide loops projecting into the periplasm, whereas the rest are buried at locations where the loops pack together into the β-barrel. As its name suggests, the maltoporin receptor is normally used to facilitate diffusion of maltose and maltodextrins into the cell (maltose binding protein). The synthesis of these receptors is repressed by glucose and induced by maltose. When high expression of this *lam* B gene is accomplished by growing the bacteria in maltose, λ adsorbs to the bacteria with high efficiency via the central tail spike.

Infection by λ phage initially involves formation of reversible phage receptor complexes that progress to irreversible complexes when contacts are established between components of the tail fiber, and a membrane bound mannose phosphotransferase encoded by the bacterial gene *pts* M. The reversible attachment of the phage is facilitated by Mg^{2+} and occurs rapidly (within a few minutes) both at room temperature and at 37°C. The linear λ DNA is then injected into the bacterium, right end first, through the tail tube of phage. However, injection of the λ DNA, and the subsequent events in the lytic cycle do not occur efficiently at room temperature. Plaques of λ phage, therefore, do not form unless bacterial lawns are incubated at temperatures higher than ~28°C.

Circularization of λ Genome and Formation of Covalently Closed and Supertwisted λ DNA

After penetrating into the host cell, the linear DNA circularizes due to complementary base pairing of *cos* sites, and the remaining nicks are sealed to form covalently closed circle by host DNA ligase, and supertwisted by the host DNA gyrase. This circular DNA then serves as the template for transcription during the early, uncommitted phase of infection. The phage then has a 'choice' of two alternative life styles, *viz.*, lysogeny or lytic cycle.

Lysogeny

During lysogeny, the injected phage λ DNA is integrated into *E. coli* chromosome, and maintained more or less indefinitely in a quiescent (benign/dormant) state possibly for many thousands of successive cell divisions. The λ DNA is thus transmitted to progeny bacteria like a chromosomal gene as if it is a legitimate part of the bacterial genome (Figure 9.8). The advantage of lysogeny is clear, as the stable association of phage with its host gives it a better chance of long-term survival than the one that invariably destroys its host. λ DNA that is following a lysogenic life cycle is described as a prophage, whereas its host is called a lysogen. The lysogen is usually physiologically indistinguishable from an uninfected cell. Lysogeny is triggered by poor nutritional conditions for the host, or higher multiplicity of infection (i.e., large number of phages infecting each host cell, which signals that the phages are on the verge of eliminating the host).

Integration of λ DNA is not an energy-consuming process. During establishment of lysogeny, only a small number of λ genes are expressed, for example, *cI* and *int* genes. The *cI* encoded pCI or λ repressor switches off the expression of lytic functions, and positively regulates transcription of its own gene, and *int* encoded pInt or λ integrase plays a role in the integration of λ DNA into the bacterial chromosome. The control circuit leading to synthesis of only pCI is discussed in detail in Section 9.2.9. As explained by the Campbell model, integration takes place through a site-specific recombination process, similar to Cre recombinase of phage P1,

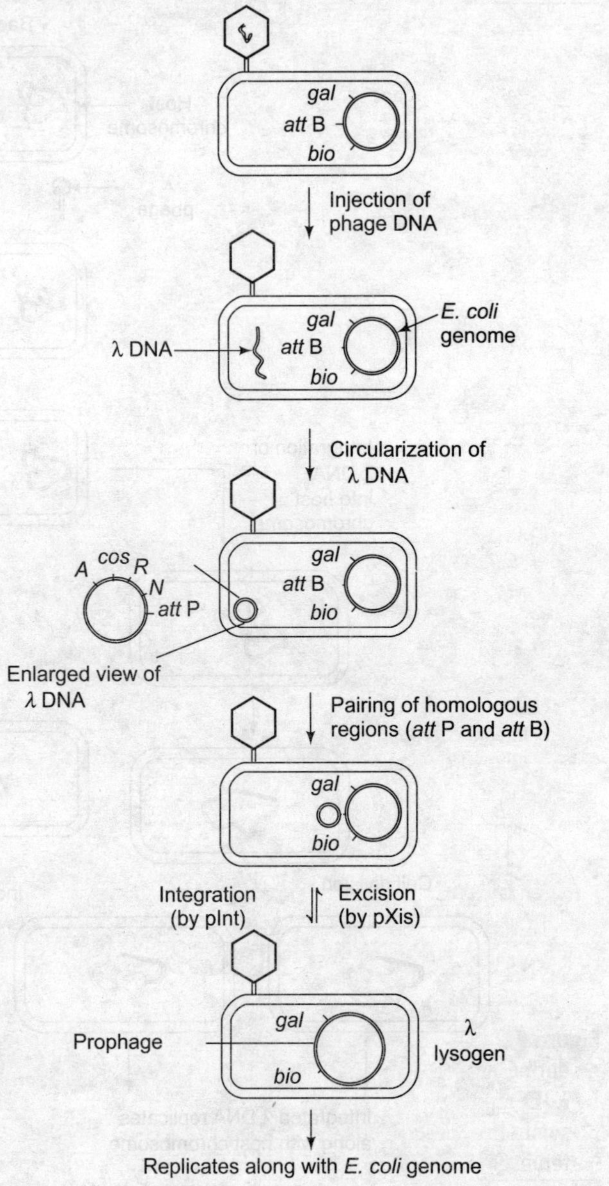

Figure 9.8 Lysogeny in bacteriophage λ involves the formation of prophage

between phage attachment site (*att* P) and homologous attachment site on bacterial chromosome (*att* B) that maps between the genes for galactose utilization (*gal*) and biotin biosynthesis (*bio*) (Figure 9.9a). These two attachment sites have a 15 bp identity, so these can be represented as having the sequences *POP′* for *att* P and *BOB′* for *att* B, where *O* denotes their common sequence (Figure 9.9b). These 15 bp regions of *att* P and *att* B sites are recognized by pInt in concert with a heterodimer host protein IHF. Note that IHF has no demonstrable endonuclease or topoisomerase activity, but specifically binds to a ds DNA bearing an *att* sequence. IHF presumably facilitates the action of λ integrase by bending the DNA in a U-turn so as to bring the two DNA segments of *att* sites to which λ integrase binds into close proximity. The action of pInt and IHF at *att* sites for integration of λ genome

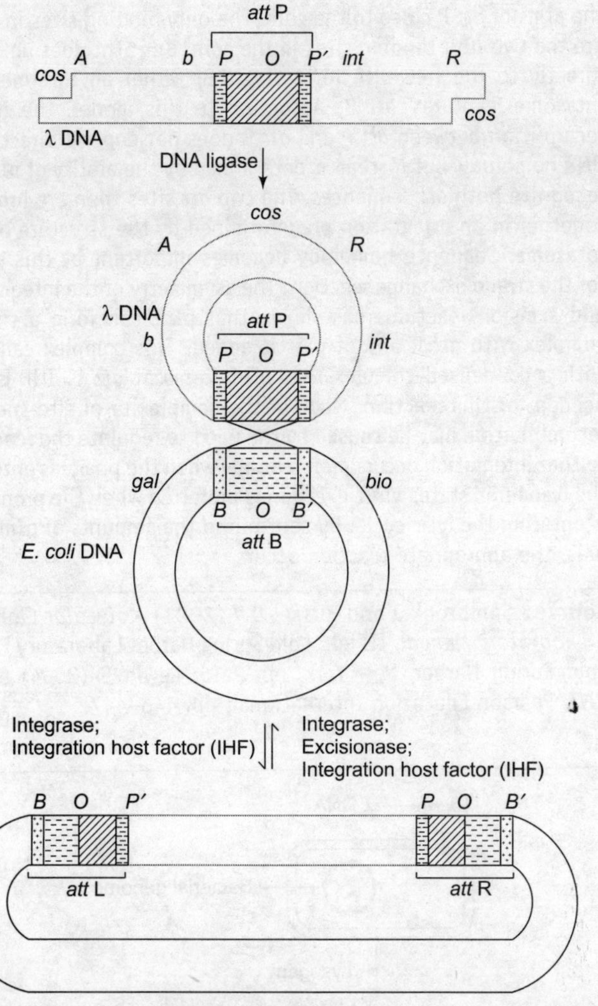

Figure 9.9a The details of the attachment and integration steps during lysogeny; Site-specific recombination in bacteriophage λ [O represents 15 bp cross-over sequences, whereas B and B′ symbolize unique sequences of bacterial origin, while P, P′ represent unique sequences of phage origin.]

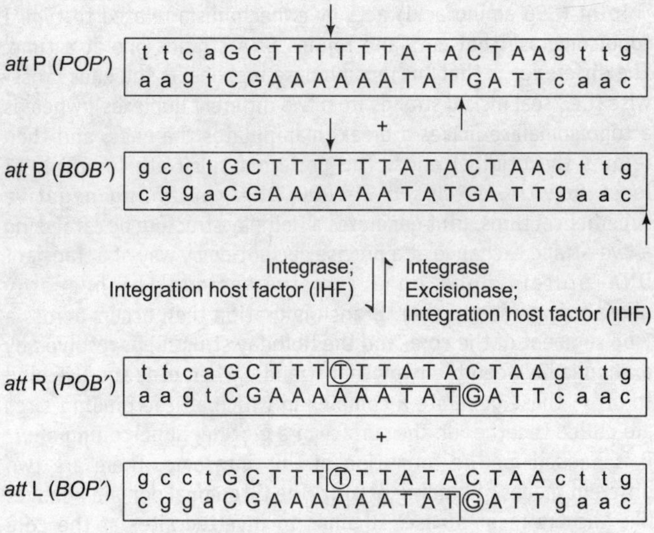

Figure 9.9b The details of the attachment and integration steps during lysogeny: Homologous sequences of *att* P and *att* B sites. [During integration, exchange occurs between the phage *att* P site and bacterial *att* B site, while during excision exchange takes place at the prophage *att* L and *att* R sites. The strand breaks at approximate positions are indicated by the short arrows. The sources of the circled bases in *att* R and *att* L are uncertain. The uppercase letters represent bases in the O region common to the phage and bacterial DNAs, whereas lowercase letters symbolize bases in the flanking B, B′, P, and P′ sites]

into host chromosome is indicated in Exhibit 9.2. Bacteriophage integration occurs through a process that yields the inserted phage chromosome flanked by the sequence *BOP′* on the left (*att* L site), and *POB′* on the right (*att* R site). The cross-over site occurs at a unique position on each strand that is displaced with respect to its complementary strand so as to form a staggered recombination joint. Since the locations of the *cos* sites at the extreme ends of the linear λ DNA, and of the *att* site are not identical, the genetic maps of the

Exhibit 9.2 Attachment site (λ *att* P), λ integrase (pInt), and integration host factor (IHF)

Unlike the Cre/*lox* recombination system of P1 phage, which requires only the enzyme and the two recombining sites, phage λ recombination occurs in a large structure, and has different components for integration and excision. Integration of phage λ genome into host chromosome occurs by site-specific recombination between two sites, *att* P (phage) and *att* B (bacteria), with the help of a host protein integration host factor (IHF) and phage protein pInt (Integrase; Int recombinase).

The λ *att* P spans ~240 bp of phage DNA centered on a 15 bp core that is identical between phage genome and host chromosomes, and includes the cross-over point of pInt. The bacterial *att* B is 21 bp long, and is composed chiefly of the core region. During integration, *att* P recombines with *att* B to form a prophage flanked by recombinant *att* L (left) and *att* R (right) sites. The inserted prophage is therefore bracketed by a 15 bp repeat in direct orientation. During excision, these flanking repeats serve as substrates for recombination, regenerating the *att* B and *att* P sites. The *in vitro* reaction requires supercoiling in *att* P, but not in *att* B.

IHF is a 20 kDa heterodimeric protein comprising of two different subunits, IHFα and IHFβ, encoded by *him* A and *him* D (*hip*) genes, respectively. Mutation analysis has shown that mutations in *him* genes prevent phage λ site-specific recombination, which can however be suppressed by mutations in λ *int*, thereby suggesting interaction between IHF and pInt. IHF binds to sequence of ~20 bp in *att* P, and bends the DNA duplex into structures compatible with the required DNA–protein interactions, and hence facilitate proper positioning of pInt. The IHF binding sites are approximately adjacent to sites where pInt binds. pXis binds to two sites located close to one another in *att* P, so that the protected region extends over 30–40 bp. pInt, pXis, and IHF cover virtually all of the *att* P. The binding of pXis changes the organization of the DNA so that it becomes inert as a substrate for the integration reaction.

pInt (356 amino acid) acts by a mechanism related to type I topoisomerase that cuts and rejoins DNA strands one at a time. The difference is that Int recombinase reconnects the ends cross-wise (i.e., seal nicked strands from two different duplexes), whereas a topoisomerase makes a break, manipulates the ends, and then rejoins the original ends. The preferred substrate for pInt, at least *in vitro*, is closed circular DNA containing negative superhelical turns. pInt generates a Holliday structure by catalyzing a two-strand exchange at a precise nucleotide by way of a transient DNA–protein bond on a tyrosine residue (high energy phosphotyrosine linkage). Branch migration then occurs across a 7 bp segment of the core, and the Holliday structure is resolved by exchange between the two other strands. When pInt and IHF bind to *att* P, these generate a complex in which all the binding sites are pulled together on the surface of a protein. Supercoiling of *att* P is needed for the formation of this intasome. There are two different modes of binding of pInt. The C-terminal domain behaves like the Cre recombinase. It binds to inverted sites at the core sequence, positioning itself to make the cleavage and ligation reactions on each strand at the positions illustrated in Figure 9.9b. The N-terminal domain binds to sites in the arms of *att* P that have a different consensus sequence. This binding is responsible for the aggregation of subunits into the intasome. The two domains probably bind DNA simultaneously, thus bringing the arms of *att* P close to the core. The only binding sites in *att* B are the two pInt binding sites in the core. But pInt does not bind directly to the free DNA form of *att* B; rather an intermediate intasome 'captures' *att* B. According to this model, the initial recognition between *att* P and *att* B does not depend directly on DNA homology, but instead is determined by the ability of pInt to recognize both *att* sequences. The two *att* sites then are brought together in an orientation predetermined by the structure of the intasome. Sequence homology becomes important at this stage for the strand exchange reaction. The asymmetry of the integration and excision reactions has shown that pInt can form a similar complex with *att* R only if pXis is added. This complex can pair with a condensed complex that pInt forms at *att* L. IHF is not needed for this reaction. Much of the complexity of site-specific recombination may be caused by the need to regulate the reaction so that integration occurs preferentially when the phage is entering the lysogenic state, while excision is preferred when the prophage is entering the lytic cycle. By controlling the amounts of pInt and pXis, the appropriate reaction occurs.

Source: Sambrook J and Russel DW (2001) *Molecular Cloning: A Laboratory Manual*, III ed., Cold Spring Harbor Laboratory Press, Cold Spring Harbor, New York, pp. 2.16; Lewin B (2004) *Genes VIII*, Pearson Education International, pp. 446–447.

integrated prophage and free phage DNA become cyclically permutated.

There are two intriguing properties of lysogeny:

Immunity to Superinfection Lysogens are resistant to further attack by other λ phage. This means that the lysogens cannot be reinfected by phages of the type with which they are already lysogenized, i.e., these are immune to superinfection. This immunity to superinfection is thus a characteristic of lysogenic cells that results due to the presence of λ repressor in their cytoplasm. However, λ repressor does not repress other lambdoid phages (or λ-like phages), such as ϕ434, ϕ21, and ϕ80. This is because λ repressor cannot bind to the operators of other lambdoid phages.

Lysogenic Induction (Induction of Lysis) The integration of λ DNA into bacterial chromosome is not readily reversible, but the stability of integrated prophage is not absolute too. Thus, prophage is not readily excised out from the bacterial chromosome, and under normal conditions lysogens spontaneously induce only about once per 10^5 cell divisions. This property is apparently due to the inherent asymmetry of λ recombination system that ensures the kinetic stability of the lysogenic integration product, pInt. However, transient exposure to inducing conditions triggers lytic growth in almost every cell of a lysogenic bacterial culture. For example, DNA damage (e.g., creation of single stranded regions) caused by UV radiation or exposure to chemical agents induces the excision of the prophage DNA from the lysogenic bacterial chromosome, and causes the phage to take up lytic mode (Figure 9.10).

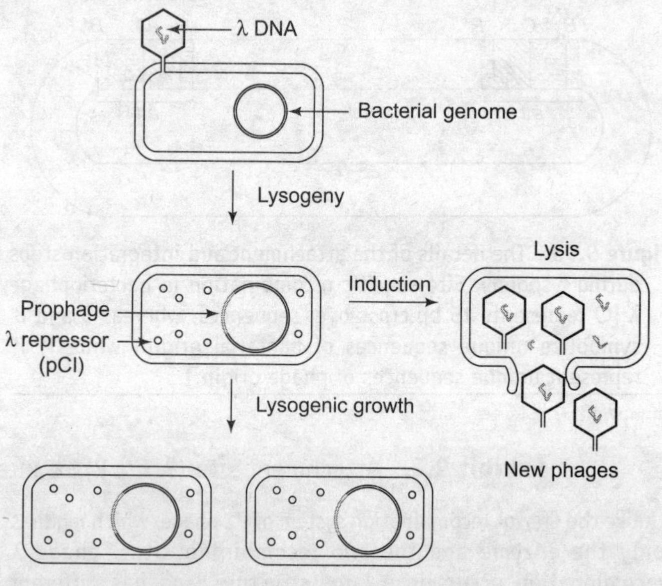

Figure 9.10 Lysogenic phage can be induced to revert to lytic cycle (induction of lysis)

Lytic Growth

As an alternative to lysogeny, after the phage infects *E. coli* by injection of its DNA, it can enter a lytic cycle, where the circular DNA directs the synthesis of ~30 proteins required for its replication, the assembly of phage particles, and host cell lysis. Bacteriophage replicates lytically only in a host actively growing in rich nutritional medium, or when the multiplicity of infection is low (for details see Section 9.2.9)

Early in the infection cycle, the circular DNA molecules initially replicate in a plasmid-like manner (θ or bidirectional replication), to produce more circular DNA molecules. This bidirectional replication originates at *ori* λ region lying between *cII* and *O* (Figure 9.2), and requires the activity of the phage genes *O* and *P*. Eventually, however, replication switches to an alternative mode (rolling circle replication), which generates long concatemers, composed of several linearly arranged genomes joined end-to-end in a continuous structure. The multimeric λ genomes, which are products of recombination, may also serve as substrate for cleavage. While all this is going on, the genes carried by the phage are being expressed to produce the components of the phage particle. These proteins are assembled first of all into head, and tail in two separate assembly steps as discussed in Section 9.2.4. The concatemers or multimers are cleaved to unit length genomes by the action of pA at the *cos* sites, and packaged into preformed head, which later associates with preformed tail, and an infectious phage particle is assembled. Phage particles thus contain monomeric linear DNA molecules with protruding ends. In lytic cycle, the phage DNA molecule is never maintained in a stable condition in the host cell. After 40–45 minutes, the lysis of the host cell, and the subsequent release of phage particles is mediated by pS, pR, and pR_Z, and ~100 infectious phage particles per infected bacterium are released. The details of the lytic cycle are discussed in Figure 9.11. Bacteriophage following lytic cycle is said to reproduce by vegetative growth. It is interesting to note that during lytic cycle, proper timing is essential, i.e., the DNA must be replicated in sufficient quantity before it is made unavailable by packaging into phage particles, and packaging must be completed before the host cell is enzymatically lysed.

9.2.7 Immediate Early, Delayed Early, and Late Phases of Transcription

The lytic infection cycle comprises of three phases, *viz.*, immediate early, delayed early, and late phases.

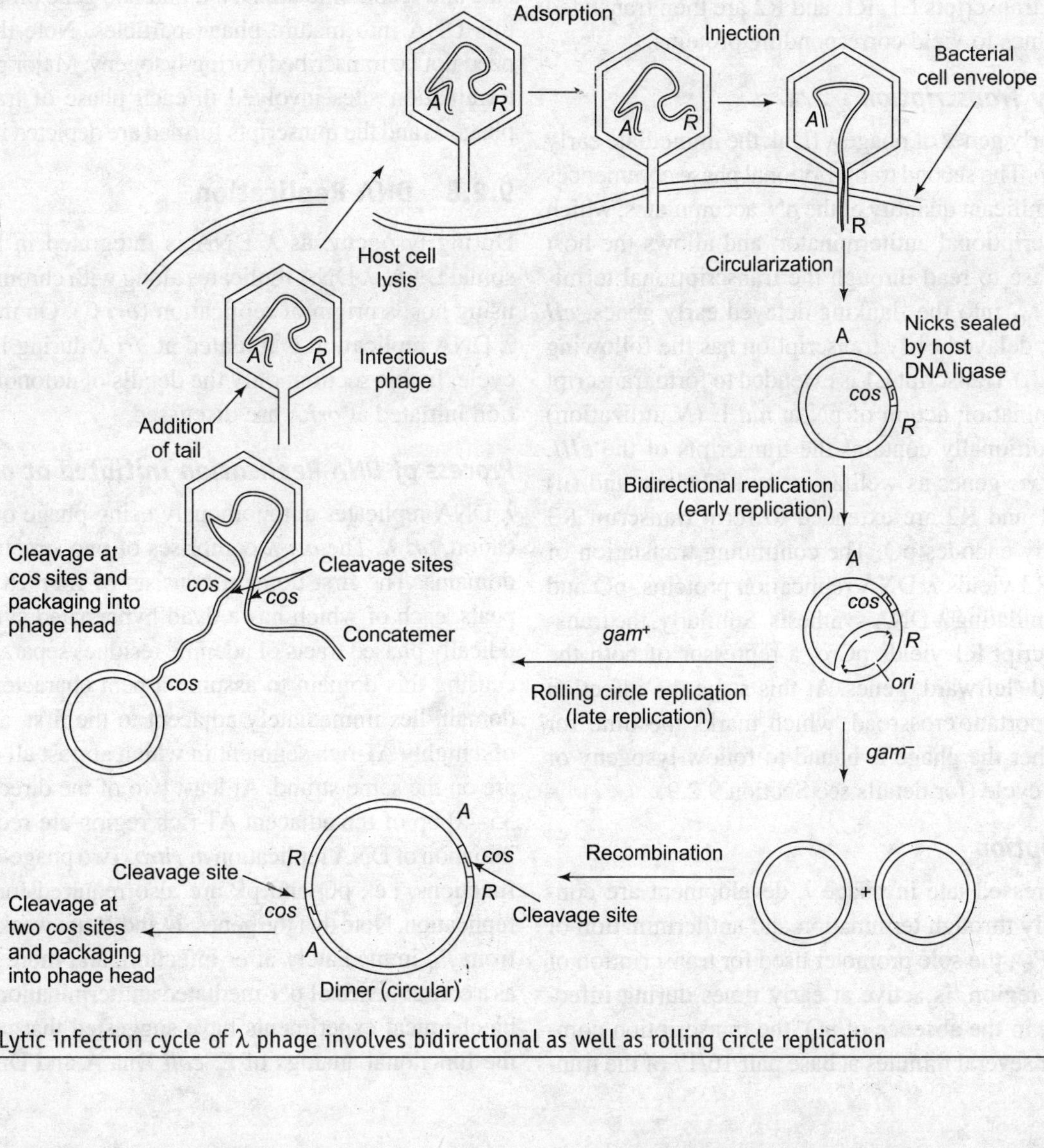

Figure 9.11 Lytic infection cycle of λ phage involves bidirectional as well as rolling circle replication

Immediate Early Transcription

Soon after phage infection, when the linear λ DNA has been converted to a superhelical, covalently closed circle, early transcription is initiated at two divergent promoters, P_L and P_R. Thus, *E. coli* RNA polymerase commences 'leftward' transcription of the λ DNA starting at the promoter P_L, and 'rightward' transcription from the promoter P_R (from the opposite DNA strand) resulting in 'immediate early RNAs' that are as follows: (i) The formation of 12S leftward transcript L1 (850 nucleotides) initiating at P_L, and terminating at t_{L1}. This transcript contains only *N* gene transcript; and (ii) Rightward transcription from P_R terminates at t_{R1}, to yield transcript R1 (310 nucleotides), although ~40% of the rightward transcripts continue up to t_{R2} to yield transcript R2. Transcript R1 contains only *cro* gene transcript, whereas transcript R2 comprises of transcripts of *cII*, *O*, and *P* genes also. However, the transcription up to *P* gene (i.e., formation of transcript R2) is weak. Note that the host RNA polymerase initiates transcription from P_L and P_R promoters, and the termination of RNA synthesis at t_{L1}, t_{R1}, and t_{R2} is mediated by the *E. coli* encoded protein ρ. The transcripts L1, R1, and R2 are then translated by host ribosomes to yield corresponding proteins.

Delayed Early Transcription

The delayed early genes of phage λ flank the immediate early genes *N* and *cro*. The second transcriptional phase commences as soon as a significant quantity of the pN accumulates, which acts as a transcriptional antiterminator, and allows the host RNA polymerase to read through the transcriptional terminators t_{L1} and t_{R1} into the flanking delayed early genes, *cII* and *cIII*. Thus, delayed early transcription has the following consequences: (i) Transcript L1 is extended to form transcript L2 by antitermination action of pN at *nut* L (*N* utilization) site, which additionally contains the transcripts of the *cIII*, *gam*, *red*, and *xis* genes as well as *att* and *b*2 sites; and (ii) Transcripts R1 and R2 are extended to form transcript R3 that additionally encodes pQ. The continuing translation of R2, and later R3 yields λ DNA replication proteins, pO and pP, thereby stimulating λ DNA synthesis. Similarly, the translation of transcript R1 yields pCro, a repressor of both the 'rightward' and 'leftward' genes. At this stage the infection reaches an important crossroad, which marks the time for deciding whether the phage is bound to follow lysogeny or lytic infection cycle (for details see Section 9.2.9).

Late Transcription

Functions expressed late in phage λ development are controlled primarily through termination and antitermination of transcription. $P_{R'}$, the sole promoter used for transcription of the entire late region, is active at early times during infection. However, in the absence of pQ, the transcription complex pauses for several minutes at base pair 16/17 of the transcript, and then terminates at a strong terminator, $t_{R'}$, located at base pair 194. Under the influence of pQ, and in the presence of the host Nus (*N* utilization substance) proteins, transcripts initiated at $P_{R'}$ are rapidly extended through both the pause site at base pair 16, and the terminator at $t_{R'}$. This is because pQ acts as antiterminator, and prevents termination of transcription at $t_{R'}$ (for details see Section 9.2.9). Transcription then proceeds around the circular genome through the late genes, and terminates within the *b*2 region to form transcript R4. The 'gene dosage' effect of the ~30 copies of λ DNA that have accumulated by the beginning of this stage results in the rapid synthesis of the following proteins: (i) Proteins conferring host cell lysis (pS, pR, and pR_Z); and (ii) λ DNA packaging and morphogenesis proteins (pNu 1/pA for terminase function, pW through pF for head assembly, and pZ through pJ for tail assembly). The first phage particle is completed ~22 min postinfection, and within 40–45 min, ~100 progeny particles per bacterium are liberated.

Basically, early gene transcription establishes the lytic cycle (in competition with lysogeny), middle gene products replicate and recombine the DNA, and late gene products package this DNA into mature phage particles. Note that late genes need not be transcribed during lysogeny. Major promoters and termination sites involved in each phase of transcription in phage λ, and the transcripts formed are depicted in Figure 9.12.

9.2.8 DNA Replication

During lysogeny, as λ DNA is integrated in host chromosomal DNA, λ DNA replicates along with chromosomal DNA using host's origin of replication (*ori* C). On the other hand, λ DNA replication is initiated at *ori* λ during lytic infection cycle. In this section, only the details of autonomous replication initiated at *ori* λ are discussed.

Process of DNA Replication Initiated at ori λ

λ DNA replicates autonomously using phage origin of replication, *ori* λ. The *ori* λ comprises of two structurally distinct domains. The first domain consists of four 18 bp direct repeats, each of which has a dyad hyphenated symmetry. Periodically phased tracts of adenine residues separate each repeat causing this domain to assume a bent character. The second domain lies immediately adjacent to the first, and comprises of a highly AT-rich segment in which almost all of the purines are on the same strand. At least two of the direct repeats, and 35–40 bp of the adjacent AT-rich region are required for any initiation of DNA replication *in vitro*. Two phage-encoded *trans* functions, i.e., pO and pP are also required for autonomous replication. Note that the genes, *O* and *P*, are weakly transcribed from P_R immediately after infection, and more strongly later as a consequence of pN-mediated antitermination. Genetic and biochemical experiments have suggested that pO and pP are the functional analogs of *E. coli* Dna A and Dna C proteins,

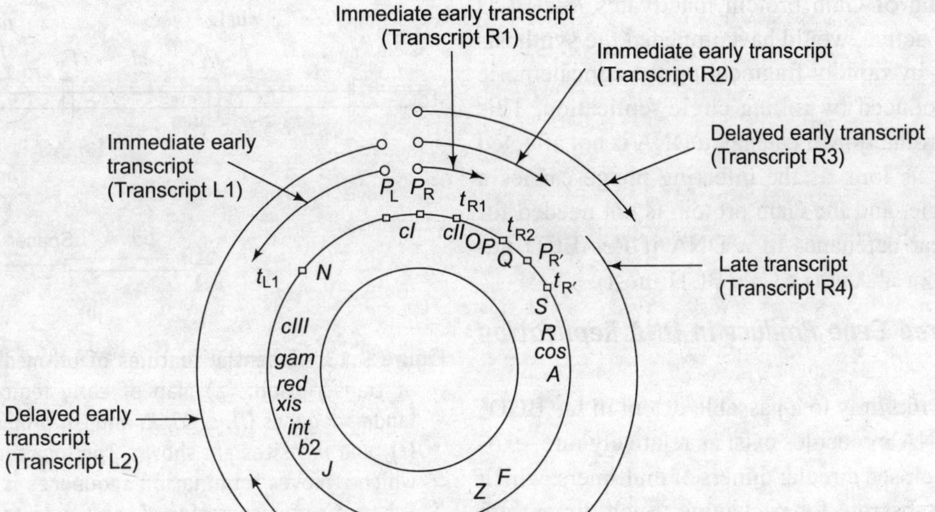

Figure 9.12 Major promoters, and termination sites involved in each phase of transcription in phage λ and the transcripts formed

that serve to divert and direct the host replicative machinery to *ori* λ via a series of protein–protein interactions.

The initiation of λ DNA replication requires an ordered assembly and partial disassembly of specialized nucleoprotein structures. The stages of protein association and dissociation reactions involved in the prepriming pathway are as follows: (i) pO, presumably as dimer, specifically binds by its N-terminus to each of the four repeated 18 bp palindromic segments within *ori* λ region. This leads to formation of an organized nucleoprotein structure termed the *O*-some. The *O*-some serves to localize and initiate a sequential reaction that provides Dna B helicase for localized unwinding of the origin region, which is the critical pre-priming step for precise initiation of DNA replication at *ori* λ. The interaction of pO with *ori* λ results in more extensive bending of the repeated region, and induces helix destabilization of the adjacent AT rich region; (ii) pP interacts with the C-terminus of pO in *O*-some. pP also associates with the host Dna B protein to form a pP.Dna B complex, which then adds to the *O*-some to generate tetrameric complex comprising of pO.pP.Dna B.*ori* λ; and (iii) Other host proteins, such as Dna G, Dna J, and Dna K add to the above complex to initiate DNA replication at *ori* λ. With the addition of ATP, and single stranded binding proteins (SSBs), pP is largely removed. Dna B then acts as a helicase to generate locally unwound, SSB-coated ss DNA. Thus, an origin specific unwinding reaction, presumably catalyzed by the helicase activity of Dna B is initiated. The unwinding is unidirectional, proceeding 'rightward' from the origin.

The three-stage pre-priming pathway to unwinding of *ori* λ appears to resemble that of the *E. coli* host, in which Dna A, Dna C, and Dna B proteins interact at the *E. coli* origin (*ori* C) to generate an unwound pre-priming structure. However, the two systems differ in that unwinding from the host *ori* C region is bidirectional, and requires neither Dna J nor Dna K.

Once the prepriming complex is assembled, replication is initiated at *ori* λ using primer formed by leftward transcription starting at promoter P_O that yields a transcript of 77 nucleotides, called oop RNA. This oop RNA is 350 bases away from *ori* λ region. The *ice* site (inceptor site in gene *cII*) having a particular secondary structure promotes the switch from RNA synthesis to DNA synthesis in replication initiation. As evident from electron microscopy, during early stages of lytic infection, λ DNA replication occurs via the bidirectional θ mode (circle to circle form) from a single replication origin (*ori* λ). However, by the late stage of the lytic cycle, when ~50 λ DNA circles have been synthesized, θ mode of DNA replication ceases, probably due to exhaustion of one or more of the required host proteins. At this point, DNA replication takes place by rolling circle σ mode (circle to concatemer form) accompanied with the synthesis of complementary strand, although the mechanism of switch-over between the two modes of DNA replication is unclear. Rolling circle replication yields DNA concatemers that are cleaved exactly at the *cos* L and *cos* R sites by phage-encoded terminase, pA, and packaged into phage heads as described in Section 9.4.2. The course of DNA replication in lytic phage is diagrammatically presented in Figure 9.11.

Significance of gam Gene Product in DNA Replication

It is postulated that phage Gam (γ) protein plays a crucial role during the switch-over between two modes of replication. The Gam protein binds to the host heterotrimeric and multifunctional exonuclease V (also called *Rec* BCD nuclease), encoded by the bacterial *rec* B, *rec* C, and *rec* D genes. This binding of Gam protein inactivates *Rec* BCD nuclease, which if active, would have impeded the synthesis of catenated DNA by rapidly fragmenting the concatemeric linear ds DNA produced by rolling circle replication. This suggests that the production of catenated DNA is not affected in *rec* BCD^+ host as long as the infecting phage carries a functional *gam* gene, and the Gam protein is not needed for production of linear catenanes of λ DNA if *Rec* BCD nuclease is defective or absent (i.e., *rec* BCD^- host).

Significance of red Gene Product in DNA Replication in gam⁻ Phage

A gam^- phage can multiply to a passable extent in *rec* BCD^+ host cells, if the DNA molecules exist as relatively rare, exonuclease resistant, closed circular dimers or multimers, which is also a suitable substrate for packaging. Such dimers are generated by bidirectional replication, and subsequent recombination of two monomeric forms, catalyzed by the products of the λ *red* gene and the host *rec* A gene. During packaging, these circular forms, like the head-to-tail tandem polymers produced by rolling circle replication, are cleaved at two *cos* sites by terminase (Exhibit 9.1).

9.2.9 Regulation of Lysogeny and Lytic Cycle

To appreciate the functioning of phage λ cloning systems, an understanding of the molecular aspects of lysogeny and lytic cycle is necessary. This section describes the regulatory mechanism through which phage λ selects and maintains its life cycle, and the genetic system that controls the orderly formation of phage particles in the lytic mode.

Immediate Early Transcription Involves Formation of pN and pCro

As mentioned in Section 9.2.7, the leftward immediate early transcription initiates at P_L and terminates at t_{L1} leading to synthesis of pN. Similarly, the rightward immediate early transcription initiates at P_R and terminates at t_{R1} leading to synthesis of pCro.

pN acts as Antiterminator and Allows Transcription of Delayed Early Genes

As mentioned in Section 9.2.7, the transcription of delayed early genes commences as soon as a significant quantity of the pN accumulates. pN then acts as transcriptional antiterminator by an improperly understood mechanism, and stimulates immediate early-to-delayed early switch. In principle, pN acts in *trans*, but it also requires the presence of a specific short DNA sequence, designated *nut* L and *nut* R that are 60 and 200 nucleotides downstream of P_L and P_R, respectively (Figure 9.13a). Thus, *nut* L is located between P_L and *N*, while *nut* R is located between *cro* and t_{R1}. But pN does not bind to these sequences within DNA; rather it binds to corresponding *nut* sequences embedded in the leftward

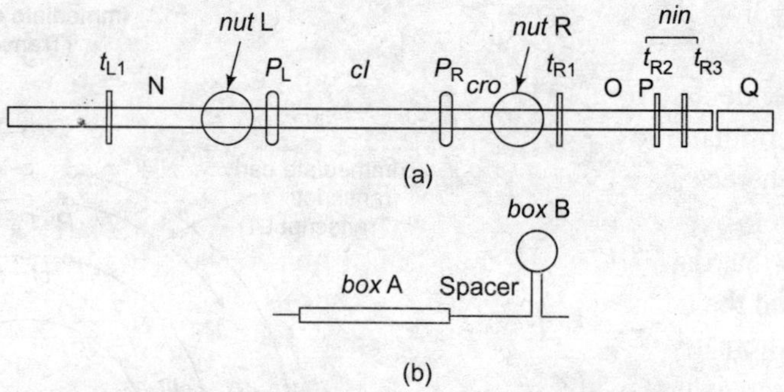

Figure 9.13 Essential features of pN-mediated antitermination of transcription; (a) Map of early region of λ genome [The landmark genes (*N*, *cI*, *O*, *P*, and *Q*), promoters (*P*), terminators (*t*), and *nut* sites are shown. The location of the *nin* deletion, which removes termination sequences is also indicated.], (b) Schematic representation of the λ *nut* sites showing the relative positions *box* A and *box* B, and the intervening spacer region. The hairpin structure formed by *box* B is also shown

Table 9.3 The *nut* L and *nut* R Sites of λ Bacteriophage

	nut R	*nut* L
box A	CGCTCTTAC	CGCTCTTAA
Spacer	ACATTCCA	AAATTAA
box B	GCCCTGAAAAAGGGC	GCCCTGAAGAAGGGC

and rightward immediate early RNAs. The *nut* sites have two conserved sequences, *box* A and *box* B. The sequences of *box* A, spacer, and *box* B of both *nut* L and *nut* R sites are tabulated in Table 9.3. The transcripts of *box* B form H-bonded hairpin loops (Figure 9.13b). The observation that *E. coli* defective in antitermination has mutations that map in the *rpo* B gene (which encodes the RNA polymerase β subunit) suggests that pN acts at *nut* sites to render core RNA polymerase (lacking a σ subunit) resistant to termination. Indeed, pN modulated RNA polymerase passes over many different terminators that it encounters either naturally or by experimental design. Thus, once RNA polymerase has passed a *nut* site, the *box* B RNA is recognized by dimers of RNA binding protein, pN, via its ~18 residue, Arg rich, N-terminal segment forming pN–*box* B complex, and from there pN is loaded on to the polymerase itself. This allows the host RNA polymerase to read through the transcriptional terminators t_{L1} and t_{R1} into the flanking delayed early genes, *cII* and *cIII* (transcripts L2 and R2).

Full activity of pN, however, requires a set of host encoded accessory antitermination factors namely, Nus proteins. These include Nus A (cellular transcription factor, which binds to both pN and RNA polymerase), Nus E (small ribosomal subunit protein S10), Nus G (protein which binds to RNA polymerase), and Nus B (protein which binds to S10). The pN-*box* B complex interacts with these factors and the RNA polymerase. Note that pN forms a helix that binds against the major groove of the *box* B hairpin via electrostatic and hydrophobic interactions. This presumably orients the opposite face of RNA for interaction with host factors. At ρ-independent terminators, the release of the transcript at the terminator's weakly bound poly (U) segment is deterred, whereas at ρ-dependent terminators, rho factor is prevented from overtaking RNA polymerase, the consequence of which is that the RNA polymerase is not stopped from unwinding, and the transcript is not released at the transcription bubble. Alternatively, since Nus G binds directly to rho, this interaction, as modulated by pN, may inhibit rho protein from releasing the nascent transcript. The interaction allows RNA polymerase to elongate nascent RNA chains at an increased rate and to skip through sites of transcriptional pausing, thereby allowing RNA synthesis to proceed through several ρ-dependent (t_{L1} and t_{R1}) and ρ-independent (t_{R2}) terminators. This antitermination at t_{L1}, t_{R1}, and t_{R2} allows transcription initiated at P_L and P_R to proceed in leftward and rightward directions, and eventually causes the transcription of *cII*, *cIII*, *O*, *P*, and *Q*, i.e., formation of transcripts L2 and R3. The synthesis of large amounts of pQ marks the beginning of late phase of transcription, and the synthesis of head and tail proteins. Note that pN if present at high concentration can work in the absence of Nus proteins, suggesting that it is pN itself that promotes antitermination.

We now know that pN is a positive regulatory element whose activity is necessary for the lytic growth of phage carrying t_{R2}. However, mutants of phage λ that carry a deletion of t_{R2} can grow, albeit poorly. Such phages are known as *nin* mutants (*N* independent), the details of which are given in Section 9.5.6.

Significance of pCro in Host Cell Lysis

pCro, a single domain protein, is expressed from the rightward transcript initiated at P_R. It binds as a dimer to the three subsites (17 bp each) of each of the two operators O_L and O_R. The binding preference of pCro to the three subsites of O_R is $O_{R3} > O_{R2} \approx O_{R1}$. Thus, pCro binds O_{R3} with highest affinity, and only binds O_{R2} and O_{R1} when present at tenfold higher concentrations. The binding is non-cooperative. Similar order of binding is followed for O_L subsites. The binding of pCro to O_L and O_R prevents binding of λ repressor to these sites. As P_L and P_R have overlapping sequences with O_L and O_R, these promoters are free for binding of RNA polymerase to initiate transcription (Figure 9.14). Moreover, occupancy of O_{R3} prevents RNA polymerase binding at P_{RM}, thereby blocking the synthesis of pCI. Thus, pCro acts as a repressor of lysogeny, and favors lytic infection cycle.

Significance of λ Repressor in Lysogeny

λ repressor binds to DNA as a dimer to the three subsites of O_L and O_R operators. The two-fold symmetry of λ repressor is such that it matches the symmetry of the operators to which

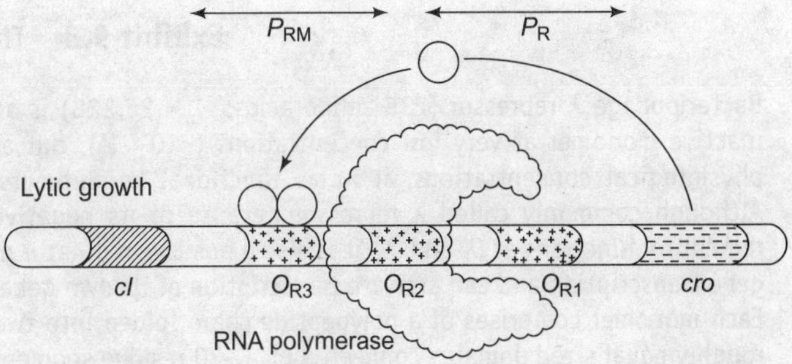

Figure 9.14 The binding of pCro to O_R stimulates transcription from P_R promoter that leads to lytic infection cycle

it binds. Each monomer comprises of a polypeptide chain folded into two roughly equal sized domains connected by a ~30 residue segment that is susceptible to rapid cleavage by proteolytic enzymes. The two domains are C-terminal dimerization and tetramerization domain and N-terminal DNA binding domain. Important features of λ repressor are also highlighted in Exhibit 9.3. The isolated N-terminal domains retain their ability to bind specifically to operators (although with only half of the binding energy of the intact repressor) but do not dimerize in solution. The isolated C-terminal domains can still dimerize but lack the capacity to bind DNA. This has significance in cleavage of λ repressor upon DNA damage.

The establishment and maintenance of lysogeny by λ repressor is explained as follows:

pCII and pCIII Aids in Establishment of Lysogeny At the time of higher multiplicity of infection (≥10), delayed early transcription occurs at high rate from P_R and P_L promoters, which when followed by translation leads to high rate of synthesis of pCII and pCIII, respectively. These proteins play significant role in the establishment of lysogeny. This is because oligomeric forms of pCII coordinately stimulate transcription from three different promoters, *viz*., P_{RE}, P_I, and P_{aQ} (Figure 9.15). Thus, when sufficient concentration of pCII is present in an infected cell, transcription of *cI* gene from P_{RE} promoter, and *int* gene from P_I promoter is activated. The pCII stimulated transcription leads to synthesis of λ repressor, which binds to three 17 bp subsites in O_L and O_R operators (that overlap with P_L and P_R promoters, respectively), thereby denying access of RNA polymerase to these promoters. This blocks the transcription of phage early genes, whose products are essential for onward progression of the lytic cycle. Moreover, as transcription proceeds from P_{RE}, antisense *cro* RNA is formed, which base pairs with *cro* mRNA, and hence decreases the levels of pCro and consequently chances of lysis (Figure 9.16). Simultaneously, transcription from P_{aQ} leads to formation of antisense RNA for *Q*. Base pairing between antisense RNA for *Q* formed from P_{aQ} and *Q* mRNA formed from P_R (RNA:RNA hybrid) prevents transcription of late genes. pInt, synthesized from P_I, catalyzes breaking and joining events that

Exhibit 9.3 The λ repressor or pCI

Bacteriophage λ repressor (236 amino acids; M_r = 26,228) is an inactive monomer at very low concentrations (<10^{-9} M), but at physiological concentrations, it forms functional homodimers. Although commonly called λ repressor because of its negative regulatory functions at O_L and O_R, it is also a positive regulator of gene transcription and can activate transcription of its own gene. Each monomer comprises of a polypeptide chain folded into two roughly equal sized domains connected by a ~30 residue segment (hinge) that is susceptible to rapid cleavage by proteolytic enzymes. The C-terminal domain of the λ repressor contains the major sites for dimerization. In addition to providing the dimerization contacts, the C-terminal domain mediates interactions between dimers thereby leading to tetramerization. The DNA binding domain lies within the N-terminal region of the molecule and contains five stretches of α-helix, of which two helices (helices 2 and 3) form a helix-turn-helix (HTH) motif. These HTH motifs are involved in sequence-specific binding to the major groove of DNA at operator region. The λ repressor binds symmetrically to DNA, so that each N-terminal domain contacts a similar set of bases. A single dimer thus recognizes a 17 bp DNA sequence, each monomer recognizing one-half site, as in *lac* system. This region also contains activating region for interaction with RNA polymerase.

λ repressor aids in lysogeny. The operators, O_L and O_R both contain three binding sites for λ repressor. The two-fold symmetry of λ repressor is such that it matches the symmetry of the operators to which it binds. In each case, site 1 (i.e., O_{L1} and O_{R1}) has a ~10-fold greater affinity than the other sites for λ repressor. The λ repressor, therefore, always binds first to O_{L1} and O_{R1}, and then binds to the other sites in the operator in a cooperative manner. The C-terminal domains of repressor dimers mediate this cooperative behavior. The binding of λ repressor to O_L and O_R thus prevents transcription from promoters P_L and P_R that have overlapping sequences with O_L and O_R, respectively.

When incubated at high pH *in vitro*, λ repressor undergoes an autocatalytic cleavage at Glu-Ala peptide bond located between the two domains. *In vivo* as well as after transient exposure of a lysogenic strain to UV light or a chemical agent, bacterial Rec A protein is activated, which acts as a coprotease that stimulates autocleavage of λ repressor between Ala 111–Gly 112 bond located in the hinge region. This cleaves N-terminal domain from C-terminal domain, and hence there is immediate loss of cooperative behavior of λ repressor. This in turn leads to dissociation of λ repressor from O_L and O_R operators. λ repressor is also negatively autoregulated through P_{RM} promoter.

Source: Sambrook J. and Russel D.W. (2001) *Molecular Cloning: A Laboratory Manual*, III ed. Cold Spring Harbor Laboratory Press, Cold Spring Harbor, New York, pp. 2.8.

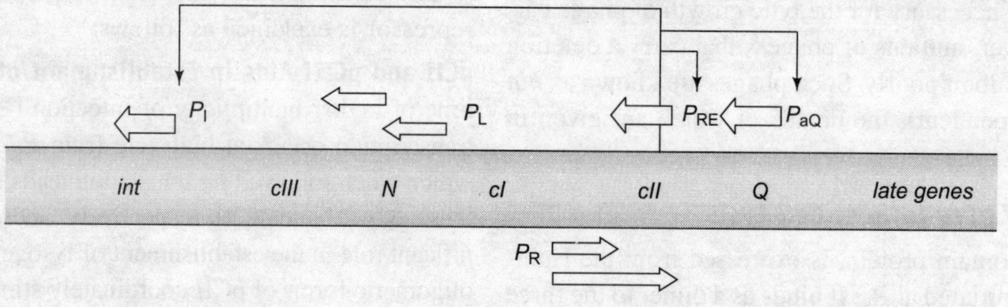

Figure 9.15 The bacteriophage λ pCII is a transcriptional activator of three promoters, P_I, P_{RE}, and P_{aQ} [The products from these promoters are represented by thick arrows, and the products of early transcription initiated from the promoters P_L and P_R are represented by thin arrows.]

lead to insertion of the phage DNA into the host chromosome. All these products thus favor in establishment of lysogeny. Note that these three pCII-activated promoters are weak, the sequences of which are given in Table 9.4.

pCII is in turn metabolically unstable with a half-life of ~1 min. This is preferentially due to proteolysis of pCII by host proteases, Hfl A and Hfl B. As pCII is required for stimulation of transcription from P_{RE}, its instability will result in immediate cessation of synthesis of pCI, pInt, and antisense *Q*. However, during high multiplicity of infection, pCIII transcribed from P_L promoter, protects pCII from the action of Hfl A protease by binding and inactivating it. Note that high multiplicity of infection can activate synthesis of pCII even in a phage following lytic infection cycle. This in turn increases the activity of pCII, which would otherwise have been cleaved proteolytically by Hfl. Note that inactivation of Hfl can also be achieved by a mutation in the *hfl* loci (high frequency of lysogenization), the details of which are given in Section 9.5.6. The pCII stimulates synthesis of pCI that binds to operators O_L and O_R, and hence represses the synthesis of immediate and delayed early proteins pN, pO, pP, and pQ, thereby preventing lytic cycle. On the other hand, at low concentrations of pCIII, i.e., when the multiplicity of infection is low, host protein Hfl is not inhibited. This reduces the activity or the amounts of pCII, and hence of pCI. Host RNA polymerase can therefore initiate transcription from P_L and P_R promoters leading to synthesis of immediate and delayed early proteins, *viz*., pN, pO, pP, and pQ. pQ then acts as antiterminator, and allows synthesis of late proteins. This sequence of events precipitates the lytic infection cycle.

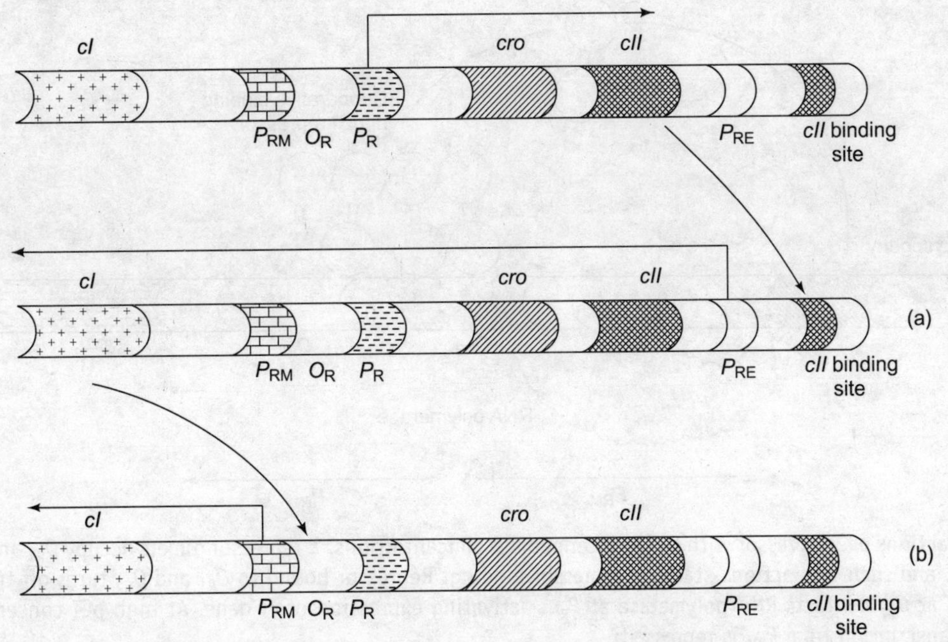

Figure 9.16 λ repressor is autoregulatory and favors lysogeny; (a) pCII stimulates transcription from P_{RE} and hence aids in establishment of lysogeny, (b) At intermediate concentration, pCI prevents transcription initiation at P_R, while at high concentration pCI prevents its own synthesis from P_{RM}. [P_{RM} transcribes only the *cI* gene. Lytic growth proceeds when P_L and P_R remain switched on, while P_{RM} is kept off; Lysogenic growth, in contrast, is a consequence of P_L and P_R being switched off, and P_{RM} switched on. Occupancy of O_{R2} by λ repressor stimulates transcription from P_{RM}, whereas occupancy of O_{R3} prevents transcription from P_{RM}.]

Table 9.4 pCII Activated Promoters

Promoter	*Sequence*
P_{RE}	**TTGC** GTTTGT **TTGC**........13 bp .... *AAGTAT*
P_I	**TTGC** GTGTAA **TTGC**.........13 bp .... *TGTACT*
P_{aQ}	**TTGC** GAGCAC **TTGC**.........13 bp .... *TAGTAT*

* *Conserved pCII binding sites are shown in bold font; −35 sequences are underlined −10 sequences are in italicized.*

λ Repressor Prevents Transcription from P_R and P_L Promoters and Blocks Transcription of Genes Required for Lytic Infection Like Lac repressor, λ repressor binds to two operators O_R and O_L that overlap the promoters P_R and P_L, respectively. Here only the repressor binding to O_R is considered. Note that the binding preference of λ repressor to the three subsites of O_R is opposite to that of pCro, and in contrast to pCro, the binding of λ repressor to operator sites is cooperative. At intermediate concentrations of pCI, the binding of λ repressor at O_{R1} stimulates cooperative binding of another molecule of λ repressor at lower affinity site O_{R2}. In this way, two dimers of repressor can bind cooperatively to adjacent sites on DNA. O_{R3}, however, remains unbound (Figure 9.17). As O_{R1}, O_{R2} sites overlap with P_R promoter, the binding of λ repressor to these sites excludes host RNA polymerase from binding to P_R promoter, thereby repressing rightward transcription. Similar is the case for leftward transcription from P_L promoter. Thus, during establishment of lysogeny, almost all the phage genes including *cro*, *cII*, and *cIII* are repressed (i.e., these genes are inactive in lysogens). A few λ genes that are expressed include *cI* (the product of which inhibits expression of lytic functions and positively regulates transcription of its own gene), and *int* (the product of which is required for integration of the λ DNA into the bacterial chromosome).

Transcription from P_{RM} Promoter Leads to Maintenance of Lysogeny and Regulation of λ Repressor Synthesis (λ Repressor Synthesis is Autoregulated) The λ repressor not only blocks transcription of early genes from P_R and P_L promoters, but also positively regulates its own synthesis (Figure 9.16). As discussed earlier, binding of λ repressor to O_R prevents transcription from P_R promoter, and hence prevents further synthesis of pCII. As levels of pCII decrease, pCI synthesis no longer remains under the control of P_{RE}. However, up to this point, λ repressor has accumulated to sufficient amount. At intermediate concentrations of λ repressor, the binding to DNA is stabilized by cooperative binding of λ repressor dimers at O_{R1} and O_{R2}, while O_{R3} remains free, and the sequences of O_{R3} overlap with P_{RM} promoter (Figure 9.17). Thus, the λ repressor bound at O_{R1} and O_{R2} is able to make direct contact with RNA polymerase at P_{RM}, and hence synthesis of more of λ repressor. As an activator, λ repressor works like catabolite activating protein (CAP), and its target on RNA polymerase is region 4 of the σ subunit, which lies adjacent to the part of σ subunit that recognizes the −35 region of the promoter. As pCI concentration increases, O_{R3} is also occupied,

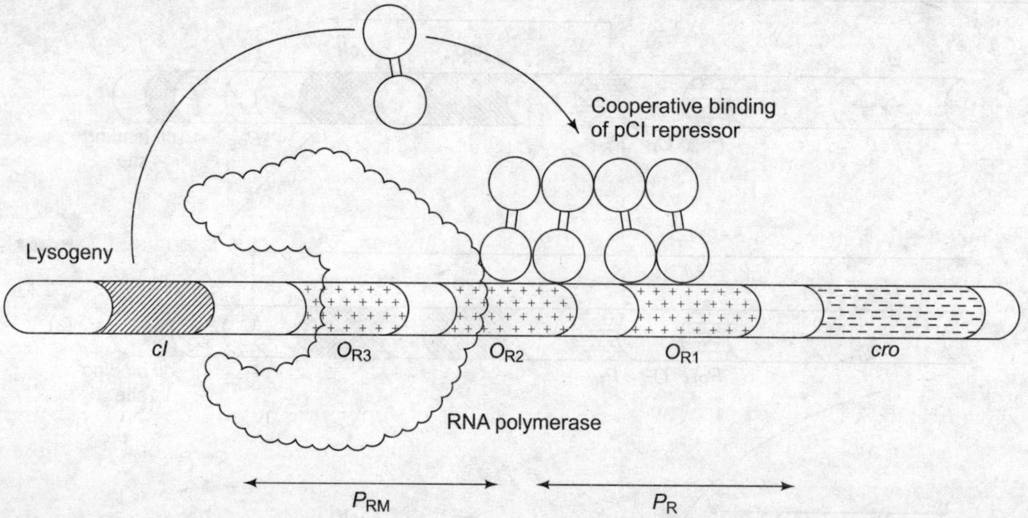

Figure 9.17 Interactions of λ repressor with O_R [At intermediate concentrations, λ repressor dimers occupy O_{R1} and O_{R2} subsites, while O_{R3} remains free, and such interactions stabilize repressor binding. Repressor bound to O_{R1} and O_{R2} turns off transcription from P_R. Repressor bound at O_{R2} contacts RNA polymerase at P_{RM}, activating expression of *cI* gene. At high pCI concentrations, O_{R3} is also occupied, and transcription from P_{RM} is repressed]

and the transcription from P_{RM} becomes repressed. The stimulation of transcription from P_{RM} is abolished by mutations in O_{R2} that prevent repressor binding, whereas its repression at high repressor concentrations is relieved by mutations in O_{R3}. Thus, occupancy of O_{R2} by λ repressor stimulates transcription from P_{RM}, whereas occupancy of O_{R3} prevents transcription from P_R. In this way, λ repressor prevents the synthesis of all phage gene products, except itself. Yet, at high repressor concentration, its own synthesis is also repressed, thereby maintaining the repressor concentration within reasonable limits. Thus, the synthesis of λ repressor is autoregulated.

Thus, during the establishment of lysogeny, P_{RE} promoter controls the synthesis of λ repressor, and as the phage gets integrated in the host chromosome, lysogeny is stably maintained from generation to generation by λ repressor synthesized from the P_{RM} promoter. In fact, λ repressor is synthesized in sufficient excess to repress transcription from superinfecting λ phage, thereby accounting for the phenomenon of immunity to superinfection, as discussed in Section 9.2.6. The prophage remains integrated as long as it synthesizes active λ repressor, but all events that inactivate the repressor induce the lytic growth cycle.

Synthesis and Stability of pInt

pInt (λ integrase) plays significant role in both integration of λ genome into host chromosome and excision of prophage upon induction of lysis.

On one hand, the promoter P_I that directs expression of *int* gene is activated by pCII. As discussed above, upon infection, conditions favoring pCII protein activity give rise to a burst of pInt along with λ repressor. RNA initiated at P_I stops at t_I terminator ~300 nucleotides after the end of the *int* gene. On the other hand, the *int* gene is also transcribed from P_L promoter as part of transcript L2. This suggests that pInt should be made even in the absence of pCII protein, but this does not happen. The interplay of controls on RNA synthesis and stability determines *int* gene expression. The *int* mRNA initiated at P_L is degraded by cellular nucleases, whereas mRNA initiated at P_I is stable, and can be translated into pInt. This occurs because the two messages have different structures at their 3′-ends. The t_I terminator has a typical stem-and-loop structure followed by six uridine nucleotides, while the longer mRNA (transcript L2) goes beyond t_I terminator (due to antitermination by pN at *nut* L site), and can form a stem that is a substrate for nucleases. Because the site responsible for this negative regulation is downstream of the gene it affects, and because degradation proceeds backward through the gene, this process is called retroregulation. The biological function of retroregulation is clear. It is considered that when pCII activity is low and lytic development is favored, there is no need for pInt, and hence its mRNA is destroyed. But when pCII is high and lysogeny is favored, *int* gene is expressed to promote recombination of the repressed λ DNA into the bacterial chromosome.

There is yet another intricacy in this regulatory device. When a prophage is induced, it needs to make pInt together with pXis to catalyze reformation of free λ DNA by recombination out of the bacterial DNA, and it must do this whether or not pCII activity is high. Thus, under these circumstances, the phage must make stable *int* mRNA from P_L despite the antitermination activity of pN. When the prophage genome is integrated into the bacterial chromosome during the establishment of lysogeny, the phage *att* site lies between the end of the *int* gene, and the sequences encoding the extended stem from which mRNA degradation is begun. Thus, in the integrated form, the site causing degradation is removed from the end of the *int* gene, and hence *int* mRNA formed from P_L is stable.

Crossroad Where Decision is Taken for Lysogeny of Lytic Cycle

The outcome of infection, lysogeny, or vegetative growth, remains unresolved until the end of the delayed early phase. In wild-type *E. coli*, when the infected cell contains the proteins pN, pCro, pCIII, pCII, pO, pP, and small amounts of pQ, the cell must decide between the lysogenic or lytic pathways depending on the circumstances. This decision between lysogeny and lytic cycle is influenced by the following factors:

The Multiplicity of Infection (Ratio of Infecting Phages to Bacteria) Higher multiplicity of infection favors lysogenization, while lytic mode is favored at lower multiplicity of infection. As discussed below, this is attributable to relative levels of pCII, pCIII, and host Hfl protease.

Nutritional State of the Cell The frequency of lysogenization is higher in case of poor nutritional state of cell, while rich medium favors lytic infection cycle. As discussed below, this is related to intracellular cAMP concentrations.

Thus, the effects of multiplicity of infection and nutritional state of the cell are regulated in following ways:

Relative amounts of pCII, pCIII, and host protein Hfl A key element in the decision between lysis and lysogeny is the phage-encoded pCII protein. As discussed earlier, a high intracellular concentration of pCII favors lysogeny, whereas a low concentration favors lysis.

Relative levels of proteins pCro and λ repressor pCro and λ repressors help in establishment of lysis and lysogeny, respectively. The basis of a genetic switch that stably maintains phage in either the lytic or the lysogenic state is the conceptually simple difference between the actions of both repressors. pCro and λ repressor compete with each other for binding to three subsites in each of the two operator regions, O_L and O_R, with different preferences of affinity and binding. The outcome of this competition determines whether the cell will become lysogenic or will advance to the late stages of lytic infection.

On one hand, at low levels, pCro binds to O_{R3} so as to abolish transcription from overlapping promoter P_{RM}. This blocks the synthesis of even basal levels of λ repressor thereby committing irreversibly to the lytic pathway. When sufficient pCro has accumulated, it binds to O_{R2} and/or O_{R1}, and turns off transcription from P_R. Similarly, transcription from P_L promoter is also turned off. However, by this stage sufficient amounts of another positive control protein, pQ, have been synthesized to ensure efficient transcription of late phage genes (encoding lysis proteins and phage head and tail proteins). There is thus no mechanism for selectively inactivating pCro, and the phage irreversibly enters the lytic mode. On the other hand, as discussed, at only marginal concentrations, λ repressor blocks transcription initiating at P_L and P_R, while stimulates transcription from P_{RM}. In this way, λ repressor prevents the synthesis of all phage gene products, but itself. Transcription from P_{RM} only becomes repressed at high levels of λ repressor, when O_{R3} is also occupied. This is how pCI controls its own level. Simultaneously, pInt, synthesized from P_I, catalyzes a breaking and joining event that leads to insertion of the λ DNA into the host chromosome.

Each repressor protein, pCro and pCI, thus turns off the gene encoding its competitor as well as appropriate downstream genes required for lysogeny or lytic growth, respectively. The two repressors therefore not only compete for the same sites, but also have mutually antagonistic physiological effects. The λ switch, once thrown, cannot be reset.

Intracellular levels of 3′–5′ cAMP The intracellular concentration of 3′–5′ cyclic AMP alters in response to changes in nutritional conditions. This in turn serves as biochemical mediator of lysogeny or lysis. When bacteria are grown in rich medium, the intracellular concentrations of cAMP are low, and the lytic pathway is favored. On the other hand, when bacteria are grown under poor nutritional conditions, the intracellular concentration of cAMP is high, and hence lysogeny is favored.

Because none of the known phage promoters are responsive to cAMP, it seems likely that the decision between lysogeny and lytic infection is influenced in part by a bacterial gene that is regulated by cAMP. It is proposed that the activity of host Hfl A is dependent on the host cAMP activated catabolite repression system as is indicated by the observation that *E. coli* mutants defective in this system lysogenize with less than normal frequency. Yet, if these mutant strains are also *hfl* A^-, these lysogenize with greater than normal frequency (for details see Section 9.5.6). Apparently the *E. coli* catabolite repression system, which is known to regulate the transcription of many bacterial genes, controls *hfl* A, perhaps by directly repressing Hfl A protein synthesis at high cAMP concentrations. This explains why poor host nutrition, which results in elevated cAMP concentrations, stimulates lysogenation.

pQ Acts as Antiterminator and Regulates Transcription of Late Genes Required for Host Cell Lysis

The synthesis of large amounts of pQ expressed from the distal portion of the extended P_R transcript is responsible for early-to-late switch. This marks the beginning of late phase of transcription by extension of short P_R transcript into the late genes, across the cohered *cos* region (i.e., genes responsible for host cell lysis, cleavage and packaging of DNA, and head and tail assembly are transcribed), so that lysis is established and many mature phage particles are ultimately produced. Like pN, pQ is a positive regulator of transcription, and acts as an antiterminator that modifies RNA polymerase so that the enzyme no longer recognizes downstream terminators. Unlike pN, pQ mediated antitermination occurs via a quite different mechanism. In contrast to pN, which binds to a unique hairpin in nascent RNA, pQ does not bind to RNA, instead it binds to a specific DNA sequence called *qut* located ~20 bp downstream of the late gene promoter $P_{R'}$ (i.e., between –10 and –35 regions of $P_{R'}$), or simply speaking, the *qut* site overlaps $P_{R'}$. At *qut* DNA site, pQ together with Nus A binds to RNA polymerase holoenzyme that is paused at $P_{R'}$ during the initiation phase, thereby accelerating it out of this promoter site, and somehow inducing it to antitermination state so as not to terminate transcription at $t_{R'}$. This modification, the nature of which is unknown, may lower the K_m of the enzyme for nucleoside triphosphates (NTPs). On the other hand, in the absence of pQ, RNA polymerase binds $P_{R'}$ and initiates transcription but terminates when it reaches the $t_{R'}$ terminator located ~200 bp downstream of $P_{R'}$.

DNA Damage Induces Lysis in a Prophage

Lysogeny can be interrupted by transiently exposing a lysogenic strain to UV light or to a chemical that damages DNA, thereby creating single stranded regions. The consequences of DNA damage are the following.

Inactivation of λ repressor Upon DNA damage, SOS response is initiated by ss DNA stimulated activation of host Rec A protein. The activated Rec A protein is a protease, which then stimulates self-cleavage of λ repressor (λ repressor has evolved to resemble Lex A, ensuring that repressor too undergoes autocleavage in response to activated Rec A). Cleavage of λ repressor occurs between an alanine and a glycine residue (Ala 111–Gly 112 bond) in the hinge region linking the DNA binding N-terminal domain, and the C-terminal dimerization domain. It has been shown that cleavage of the λ repressor occurs autocatalytically at pH 10 in an intramolecular reaction that displays first order kinetics, and is independent of protein concentration. Due to autocleavage, the N-terminal domain may bind specifically to operators (although with only half of the binding energy of the intact repressor) but does not dimerize in solution. This leads to immediate loss of cooperative behavior of λ repressor (at concentrations of repressor found in a lysogen). As these functions are critical for λ repressor binding to O_{R1} and O_{R2}, loss of cooperativity ensures that repressor dissociates from those sites as well as from O_{L1} and O_{L2}, with a consequent reduction in the concentration of intact free monomers. This in turn shifts the monomer–dimer equilibrium such that the operator-bound dimers dissociate to form monomers, which are then cleaved through the influence of activated Rec A before these can rebind to their target DNA. Note that for induction to work efficiently, the level of λ repressor in a lysogen must be tightly regulated. If its level is too low, the lysogen is spontaneously induced, and if its level is too high, appropriate induction is inefficient. The reason for the latter is that more λ repressor has to be inactivated by Rec A for the concentration to drop enough to vacate O_{R1} and O_{R2}. Loss of repression due to inactivation of λ repressor then triggers transcription from P_R and P_L. Besides other proteins, this leads to synthesis of the phage protein pXis, which excises the prophage at the *att* site.

Excision of Prophage from Host Chromosome The excision of prophage requires the participation of excisionase pXis in concert with pInt, IHF, and Fis (a DNA binding host protein that also stimulates Hin-mediated gene inversion). The mechanism by which excisionase reverses the integration process is unknown, although it has been shown that pXis specifically binds to *POB′*, where it induces a sharp bend in this DNA, and excises the prophage.

Expression of Phage Genes, Initiation of Phage Replication, and Host Cell Lysis The excision of prophage allows expression of phage genes, and the phage then undergoes normal vegetative growth. The cell lyses and progeny virions are liberated. Thus, a culture of an induced bacterial lysogen normally contains phage particles in the supernatant, due to a low level of spontaneous failure of the repression mechanism.

The induction of lysis is thus described as the 'lifeboat' response. The prophage remains integrated as long as it synthesizes active λ repressor, but escapes a doomed host through the formation of infectious phage particles that have at least some chance of further replication. The process of induction is depicted in Figure 9.10.

Besides UV or chemical agents, the temperature-induced inactivation of λ repressor can also lead to induction of lysis. The repressor having temperature sensitive mutation in *cI* (*cIts* 857) can be inactivated by a temperature shift to the non-permissive temperature leading to excision of the prophage from the host chromosome, and initiation of lytic growth phase (for details see Section 9.5.6).

As *att* B site is located between *gal* and *bio* operons, the abnormal excisions of the prophage can result in the incorporation of genes from one or other of these operons into the λ genome, with the concomitant deletion of some λ DNA. Depending upon the extent of this deletion, these transducing phages may or may not be defective for vegetative growth or lysogenization. If the normal attachment site is deleted from the *E. coli* chromosome then the phage can integrate at secondary attachment sites. Abnormal excisions of prophage from these other regions of the genome can result in the transducing phage carrying a number of other *E. coli* genes. A description of transducing phages is presented in Exhibit 9.4.

Exhibit 9.4 Transducing phages

Transduction is a phenomenon in which bacterial DNA is transferred from one bacterial cell to another by a phage particle, and the phage particles that contain host DNA are called transducing particles. Such transducing particles are not always formed and usually constitute only a small fraction of a phage population. Such particles can arise by two mechanisms, namely, aberrant excision of a prophage and packaging of bacterial DNA fragments. There are two types of transduction: generalized transduction and specialized transduction.

Generalized transduction phages are produced by an error during packaging of DNA into a phage head, resulting in the insertion of random, phage-size fragments of bacterial DNA into the phage head instead of phage DNA. Thus, in a lysate of generalized transducing phage, some particles contain DNA obtained from the host cell rather than phage DNA. The bacterial DNA fragment packaged within a phage head can be derived from any part of the host genome, although certain regions of the genome may be packaged at varying frequencies. When a transducing particle adsorbs to sensitive recipient, the ds DNA is injected into the recipient cell. Stable inheritance of the donor DNA requires either that it integrates into the recipient chromosome via homologous recombination or that it is replicated (e.g., following transfer of plasmids). Such generalized transducing particles can be produced during lytic growth of either virulent or lysogenic phage.

Generalized transduction is an efficient method to move mutations between strains to construct derivatives with different genotypes. In order to determine the genetic or biochemical effects of a particular mutation, it is necessary to compare the mutant with a strain that only differs by a single mutation (an 'isogenic strain'). If other uncharacterized mutations are present, it is not possible to determine whether the mutation of interest is responsible for the observed phenotypic change. The most common way to ensure that two strains are isogenic is to transfer a small region of DNA carrying the mutation into the parental strain by recombination. If a mutation confers a selectable phenotype, it can be transferred easily from a donor to an appropriate recipient strain. If the mutation cannot be selected directly, linkage to an adjacent, selectable marker can be used to move the mutation into a recipient strain. For example, an auxotrophic mutation may be brought into a recipient strain by selecting for inheritance of a nearby gene, and then screening for co-inheritance of the auxotrophic mutation.

Specialized transducing bacteriophages result from aberrant excision of λ prophages from the bacterial chromosome. In contrast to generalized transducing phage, only regions of the host DNA that flank the prophage are packaged in specialized transducing phage, allowing transduction of a contiguous DNA fragment with both host and phage DNA. Thus, specialized transducing particles derived from lysogens carrying the λ prophage at its normal integration site in the bacterial chromosome can carry the host's neighboring *gal* or *bio* genes. Because lysogenic phage can be made to integrate at a variety of sites in the genome by genetic tricks, specialized transducing phage can be isolated from many different regions of a bacterial genome. Stable inheritance of specialized transducing fragments may occur either by site-specific recombination mediated by the phage integrase or by homologous recombination mediated by bacterial recombinases.

Source: Blot M (2004) *Prokaryotic Genomics*, Springer (India) Private Limited, New Delhi.

9.3 λ BACTERIOPHAGE GROWTH AND PREPARATION OF λ DNA

Under conditions of lysis or induction of lysis, phage particles accumulate in extracellular medium in broth culture, while on solid medium they form clear plaques on bacterial lawn. Unlike bacteria, the growth of λ phage and isolation of λ DNA requires high skill, experience, and instrumentation.

9.3.1 Obtaining High λ Titre in Broth Culture

The naturally occurring λ phage is lysogenic, and an infected culture consists mainly of cells carrying the prophage integrated into the bacterial DNA. The extracellular λ titre is extremely low under these circumstances. In this case, to get a high yield of λ in extracellular medium, the culture must be induced, so that all the bacteria enter the lytic phase of the infection cycle resulting in cell death, and release of the λ particles into the medium. Many cloning vectors derived from λ are modified by deletions of the *cI* and other genes, so that lysogeny never occurs. These phages follow lytic infection cycle, and do not integrate into the bacterial genome. With these phages the key to obtaining a high titre (the number of phage particles per ml of culture) lies in the way in which the culture is grown, in particular the stage at which the bacterial cells are infected by adding phage particles (Figure 9.18). For example: (i) If phages are added before the bacterial cells are dividing at their maximal rate, then all the bacterial cells are lysed very quickly, resulting in a low titre; (ii) If the bacterial cell density is too high when the phages are added, then the bacterial culture will never be completely lysed, and again the phage titre will be low; and (iii) The ideal situation is when the age of the culture and the size of the phage inoculum are balanced such that the culture continues to grow, but eventually all the cells are infected and lysed.

Thus, skill and experience are needed to judge the matter to perfection.

After complete lytic infection cycle, the phages are released into the extracellular medium. However, the remains of lysed bacterial cells along with any intact cells that are inadvertently left contaminate the phage preparation. This contamination can be removed by centrifugation, leaving the phage particles in suspension (Figure 9.19). In the next step, the size of the suspension is reduced, so that the size is manageable for further processes, for example, DNA extraction.

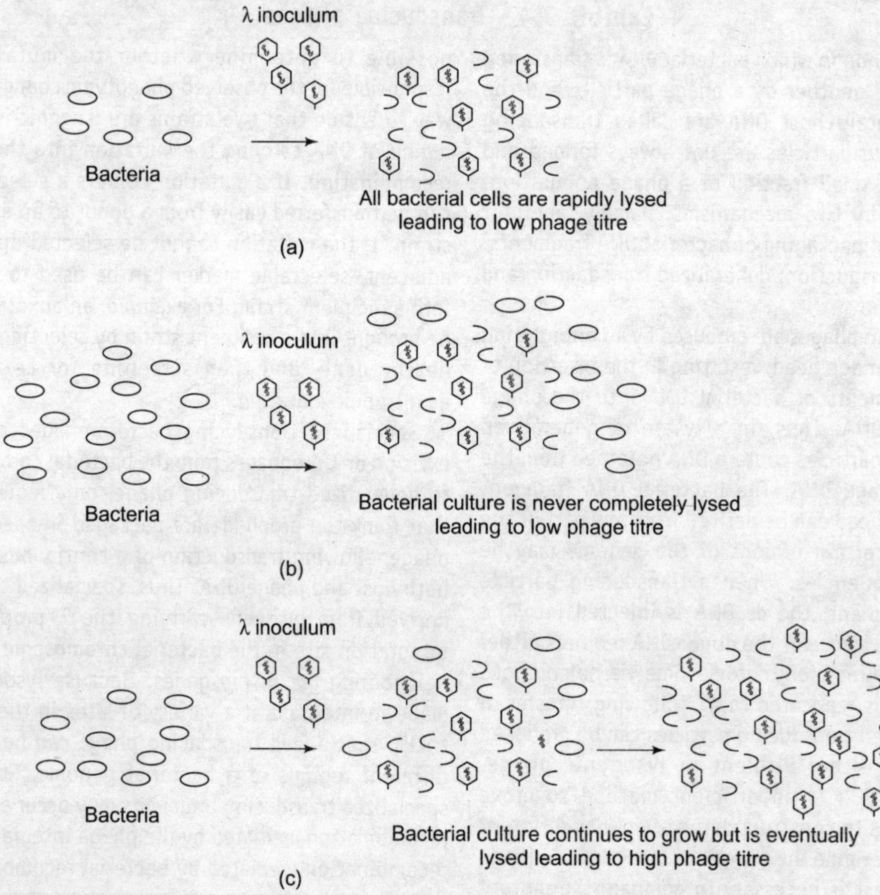

Figure 9.18 Achieving the right balance between bacterial culture age and phage inoculum size when preparing a sample of a nonlysogenic phage; (a) Bacterial cell density is too low, i.e., the phage inoculum is added before the cells are dividing at their maximal rate, (b) Bacterial cell density is too high when the phage inoculum is added, (c) Bacterial culture density is just right at the time of addition of phage inoculum

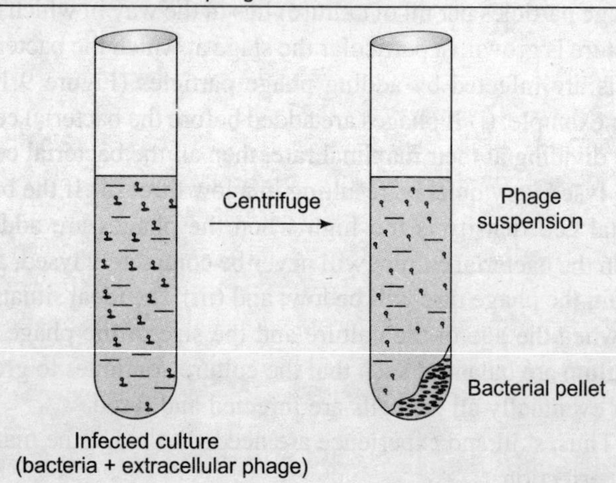

Figure 9.19 Preparation of a phage suspension from an infected culture of bacteria

Bacteriophage particles are so small that they are pelleted by very high-speed CsCl density gradient centrifugation with polyethylene glycol (PEG), a long chain polymeric macromolecular assembly (Figure 9.20a and b). The precipitate is then collected and redissolved in a suitably small volume. Removal of CsCl by dialysis leaves a pure phage preparation.

9.3.2 λ Bacteriophage Forms Plaques on Solid Medium

The final stage of the lytic infection cycle is host cell lysis. If infected bacterial cells are spread onto a solid agar medium immediately after addition of phage or immediately after transfection with λ DNA, cell lysis is visualized as plaques on a lawn of bacteria. Each plaque is a zone of clearing produced as the phage lyses the bacterial cells, and the resulting progeny move on to infect and eventually lyse neighboring bacteria. The plaque is thus represented as a clear circular area in an otherwise turbid background of bacterial growth that is visible to the naked eye after several hours of incubation (Figure 9.21). Note as the diffusion of the progeny of phage particle is limited in solid or semi-solid medium, the size of plaque is also restricted. Because each plaque is derived from a single transfected or infected cell, the phages derived from a single plaque are essentially genetically identical to one another.

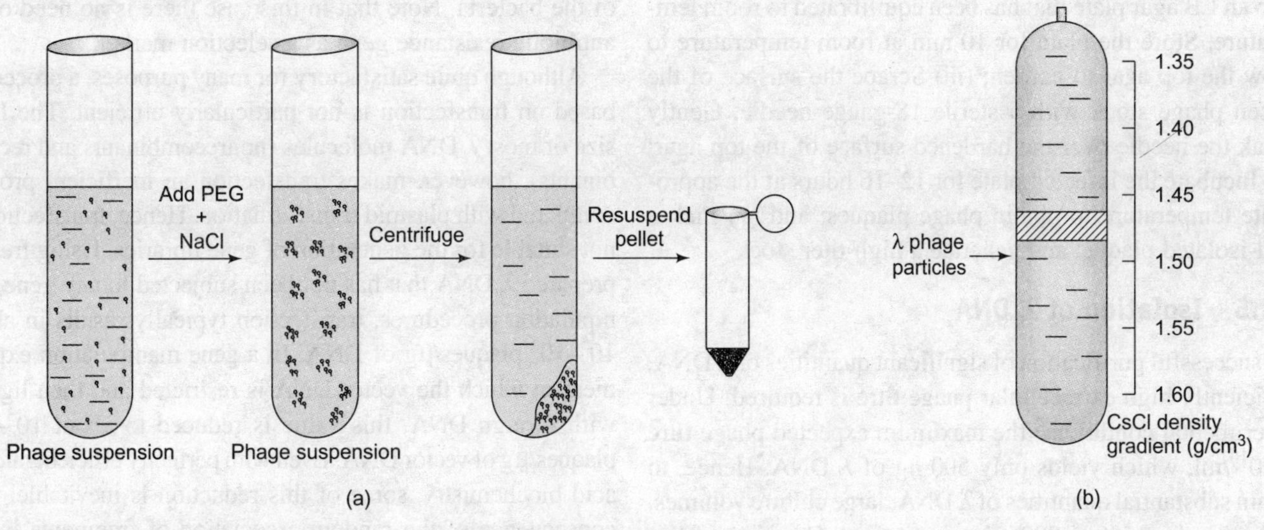

Figure 9.20 Isolation and purification of λ phage; (a) Collection of phage particles by polyethylene glycol (PEG) precipitation, (b) Purification of λ phage particles by CsCl density gradient centrifugation

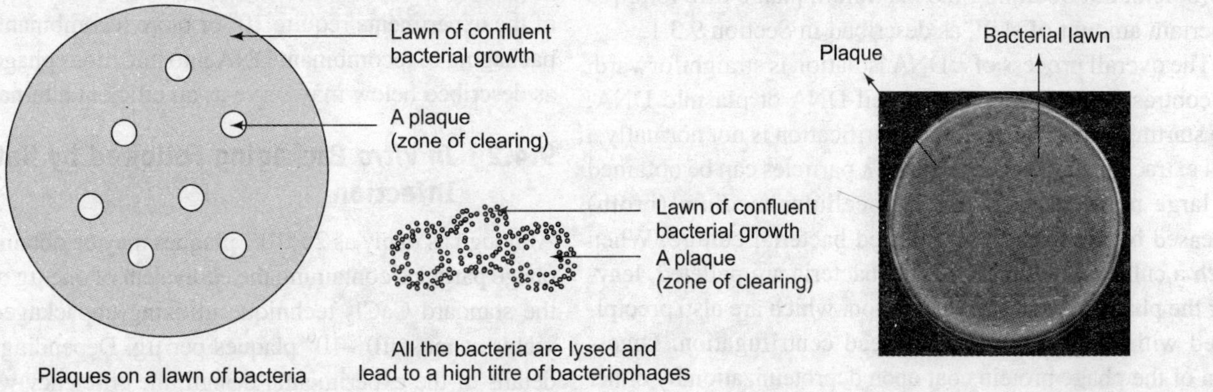

Figure 9.21 The appearance of clear plaques on a lawn of bacteria formed by the phages [A photograph of the petriplate showing plaques is also shown.]

9.3.3 Importance of Maltose in Medium

Maltose is often added to the medium used to grow and plate λ phage. The presence of 0.2% maltose leads to a substantial induction of maltose operon including the *lam* B gene, which encodes the cell surface receptor to which phage λ binds (for details see Section 9.2.6). This induction should theoretically increase the efficiency of infection, and hence the yield of phage. The use of maltose is a double-edged sword, since florid growth of some strains of *E. coli* in such rich medium may lead to cell lysis, and the accumulation of cellular debris laden with Lam B protein. Binding of phage particles to this debris leads to futile release of the λ DNA, and an unproductive infection.

9.3.4 Long-term Storage of λ Bacteriophage Stocks

Stocks of most strains of λ phage are stable for several years when stored in SM medium [sodium chloride and magnesium sulfate medium; 5.8 g/l NaCl, 2 g/l $MgSO_4.7H_2O$, 50 ml/l Tris–HCl (1 M; pH 7.5), and rest H_2O, autoclaved for 20 minutes at 15 psi on liquid cycle] containing a small amount of chloroform (0.3% v/v). However, many investigators have found that some λ phage recombinants lose viability when stored in this way for a few months. This problem can usually be avoided by using freshly distilled chloroform or chloroform that has been extracted with anhydrous sodium bicarbonate to remove products of photolysis. However, it is also recommended that the titer of master stocks be checked every 2–3 months, and that master stocks of all important λ phage strains be stored at –70°C. Before storage, dimethysulfoxide (DMSO) is added to the λ phage stock to a final concentration of 7% (v/v) followed by gentle inversion of the tube into liquid nitrogen. When the liquid has frozen, the tube is transferred to a –70°C freezer for long-term storage. The λ phage can be recovered from stock culture by following steps: (i) Add 0.1 ml of plating bacteria to 3 ml of molten top agar at 47°C. Mix the contents of the tube by tapping or gentle vortexing; (ii) Pour the contents of the tube onto an LB agar plate that has been equilibrated to room temperature. Store the plate for 10 min at room temperature to allow the top agar to harden; (iii) Scrape the surface of the frozen phage stock with a sterile 18-gauge needle. Gently streak the needle over the hardened surface of the top agar; (iv) Incubate the infected plate for 12–16 hours at the appropriate temperature to obtain phage plaques; and (v) Pick a well-isolated plaque, and generate a high-titer stock.

9.3.5 Isolation of λ DNA

For successful purification of significant quantities of λ DNA, sufficiently high extracellular phage titre is required. Under experimental conditions, the maximum expected phage titre is 10^{10}/ml, which yields only 500 ng of λ DNA. Hence, to obtain substantial quantities of λ DNA, large culture volumes, in the range of 500–1,000 ml, are required. Growing a large volume culture (bacterial cultures of 50 liters and above) is not a problem, but obtaining the maximum phage titre requires a certain amount of skill, as described in Section 9.3.1.

The overall process of λ DNA isolation is straightforward. In contrast to isolation of total cell DNA or plasmid DNA, the starting material for λ DNA purification is not normally a cell extract. This is because phage λ particles can be obtained in large numbers from the extracellular medium (broth) released by the lysis of an infected bacterial culture. When such a culture is centrifuged, the bacteria are pelleted, leaving the phage particles in suspension, which are also precipitated with PEG through high-speed centrifugation. Digestion of the phage protein coat upon deproteinization by either phenol or protease treatment of the redissolved PEG precipitate is sufficient to extract pure λ DNA.

9.4 METHODS OF INTRODUCTION OF λ DNA INTO BACTERIAL CELLS

There are two different methods by which a purified or recombinant λ DNA is introduced into a bacterial host cell. These are transfection and *in vitro* packaging followed by natural infection.

9.4.1 Transfection

Naked λ DNA can be introduced into a host bacterial cell by transfection (counterpart of bacterial transformation) (for details see Chapter 14). Just as with a plasmid, the purified λ DNA (or recombinant DNA) is mixed with competent *E. coli* cells, and the DNA uptake is induced by heat shock. However, in this case instead of plating on a selective agar medium, and counting bacterial colonies, the transfection mix is allowed to adsorb on a phage-sensitive indicator bacterium in molten soft agar. When this mix is overlaid onto an agar plate, zones of clearing, i.e., plaques are formed due to lysis of the bacteria. Note that in this case there is no need of an antibiotic resistance gene as a selection marker.

Although quite satisfactory for many purposes, a procedure based on transfection is not particularly efficient. The large size of most λ DNA molecules (nonrecombinants and recombinants), however, makes transfection an inefficient process compared with plasmid transformation. Hence, transfection is not suitable for the generation of gene libraries. Using freshly prepared λ DNA that has not been subjected to any gene manipulation procedures, transfection typically results in about 10^5–10^6 plaques/μg of DNA. In a gene manipulation experiment in which the vector DNA is restricted and then ligated with foreign DNA, this value is reduced to about 10^3–10^4 plaques/μg of vector DNA. Even with perfectly efficient nucleic acid biochemistry, some of this reduction is inevitable. It is consequence of the random association of fragments in the ligation reaction, which produces molecules with a variety of fragment combinations, many of which are non-viable. As most of the experiments require 10^6 or more recombinants, *in vitro* packaging of recombinant DNA into infectious phage particles as described below may serve as an efficient alternative.

9.4.2 *In vitro* Packaging Followed by Natural Infection

Although as many as 2×10^{10} plaques may be obtained from λ phage particles containing the equivalent of one μg of λ DNA, the standard $CaCl_2$ technique utilizing unpackaged λ DNA yields at most 10^5–10^6 plaques per μg. Depending upon the details of the experimental design, the efficiency with which hybrid clones are generated is of the order of 1–5×10^5 plaques per μg of inserted DNA, while nonrecombinant λ DNA may yield up to 10^7–10^8 plaques per μg of DNA. Since *in vitro* packaging of λ DNA is much more effective than transfection, it is the method that is almost always used.

In vitro packaging means encapsidation of λ DNA using packaging extract under *in vitro* conditions. Packaging of λ DNA into phage head leads to assembly of infectious phage particle, which is capable of naturally infecting the bacterial host cell. One of the most important features of this process is that the length of DNA that is packaged into the phage head is determined by the distance between two *cos* sites. Another important feature is that the head has a fixed size, and hence it puts a limit on the size of DNA to be accommodated into it. Thus, DNA of the size below 37 kbp, though packaged, does not lead to formation of infectious phage particle, and that above 52 kbp is not accommodated into phage head (for details see Section 9.5.1).

The principle of packaging *in vitro* is to supply λ DNA or ligated recombinant DNA (whichever may be the case) with high concentrations of phage head precursor proteins, packaging or assembly proteins, and phage tail proteins. Two different systems are in use for *in vitro* packaging. With both systems, packaged molecules are introduced into *E. coli* cells simply by adding the assembled phage to the bacterial culture, and allowing the normal λ infective process to take place.

Single Mutant Strain System

In the single strain system, the defective λ phage carries a mutation in the *cos* sites, so that these are not recognized by the endonuclease that normally cleaves the λ catenanes during phage replication. Consequently, the defective λ phage cannot replicate, though it directs the synthesis of all the proteins needed for packaging and morphogenesis. The proteins accumulate in the bacterium, and can be purified from cultures of *E. coli* infected with mutated λ strain. These proteins are then used for *in vitro* packaging of recombinant λ DNA molecules (Figure 9.22).

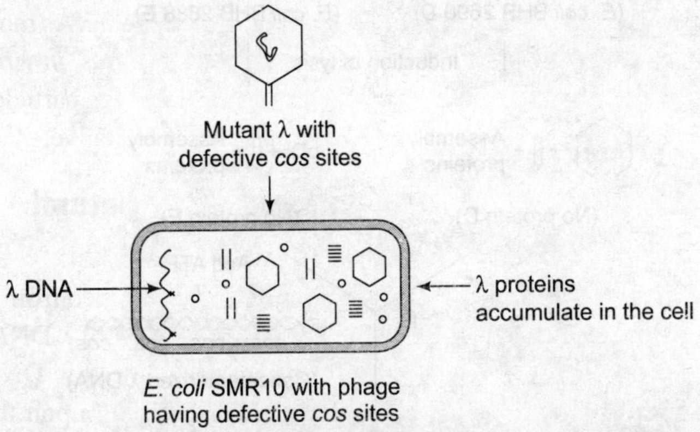

Figure 9.22 *In vitro* packaging using single strain packaging system [*E. coli* strain SMR 10 that carries a λ phage having defective *cos* sites is used for the synthesis of λ proteins]

Two Defective Strains System

Practically, *in vitro* packaging is most efficiently performed in a very concentrated mixed lysate, commonly called packaging extract or mix, of two induced lysogens, each one harboring mutant phage strain defective in one of the steps of morphogenesis. For example: (i) The first of the two lysogens (*Dam* lysogen) contains a λ prophage (λ D^- strain) with an amber mutation in gene *D* (*Dam*), encoding decoration protein that is located on the outside of mature phage particles, and participates in the maturation of head structures and packaging of DNA into phage heads. *Dam* lysogen thus supplies all proteins required during lytic cycle, except pD; and (ii) A second lysogen (*Eam* lysogen) harbors a prophage (λ E^- strain) that is amber mutant in gene *E* (*Eam*). These are deficient in pE, a major structural protein of λ phage head required in an early phase of head assembly. This lysogen, however, supplies pD and all other proteins required during lytic cycle, except pE.

Both the above strains are used as temperature sensitive lysogens carrying the temperature sensitive repressor *cIts*, due to which lysogenic state is maintained stably at 30°C, while lytic growth, and hence synthesis of the desired structural proteins ensues at 42°C (for details see Section 9.5.6). The two prophages also carry an amber mutation in *S* gene, *Sam* 7.

If extracts from λ D^- and λ E^- lysogens are used for packaging the recombinant DNA (or any exogenous DNA), it may impose two potential problems. These problems along with their solutions are discussed below:

Packaging of Endogenous DNAs Endogenous DNAs from the phage lysates themselves compete with exogenously added (recombinant) DNA, and gives a background of 10^2–10^3 plaques per packaging experiment. This background problem can be avoided by one of the following ways: (i) Select appropriate genotype for these prophages, for example, use λ D^- and λ E^- strains having *b2* deletions so that excision upon induction is inhibited, or use λ D^- and λ E^- strains having *imm* 434 repressor region so that plaque formation is prevented if an *imm* 434 lysogenic bacterium is used for plating the complex reaction mixture; and (ii) If the vector does not contain any amber mutation, a non-suppressing host bacterium can be used at the time of plating the complex reaction mixture so that endogenous DNAs of λ D^- and λ E^- strains do not give rise to plaques.

Recombination Between Exogenous DNA and Induced Prophage Markers The second potential problem arises from recombination in the lysate between exogenous DNA and induced prophage markers. In such cases, phage particles obtained by packaging of endogenous DNA may require containment procedures to ensure their biological safety, because of the generation of wild-type (Am^+) recombinants from endogenous and exogenous DNAs. Such recombinants may arise because mutations existing in exogenous and endogenous DNAs may not map at exactly the same site within the same gene. In this case recombination between the two loci may lead to Am^+ revertants, albeit with a very low frequency. If troublesome, such recombinational events can be overcome by: (i) Using recombination deficient (i.e., red^- rec^-) λ D^- and λ E^- lysogens; and (ii) By UV irradiating the cells used to prepare the lysate after induction of lysis, thereby eliminating the biological activity of endogenous DNA. Note that the enzymatic packaging activities are fully retained if the lysates are irradiated at a dose that corresponds to 40 lethal hits per λ DNA molecule.

Thus, for *in vitro* packaging, *E. coli* strains used are BHB 2688 *E* amber mutant [N205 *rec* A^-, *Eam 4*, *Sam* 7, λ *imm* 434, *cIts* 857, *b2*, *red* 3], and BHB 2690 *D* amber mutant [isogenic with BHB 2688 *E* amber mutant apart from the fact that it carries *Dam* 15 instead of *Eam* 4 mutation]. Cultures of the two lysogens are induced by temperature shift, however, host cell lysis remains blocked due to a mutation in the *S* gene. Thus, combination of *imm* 434 *cIts* and *Sam* 7 mutation allows accumulation of various proteins required during lytic cycle (except the one for which the phage is mutant), and also prevents their premature release by the lysis of bacterial cell. Thus, upon heat induction, *Dam* lysogen leads to the accumulation of empty prehead particles, while *Eam* lysogen allows accumulation of all components of head structures, including pD, in a free form without assembly of phage preheads. Besides, all other proteins required for lytic growth, for example, head proteins, packaging proteins, and tail proteins are produced by both the lysogens. Both type of cells are then concentrated by centrifugation. Extracts or lysates are prepared by either sonication of cell pellet, or by subjecting the cells to freeze thaw cycle. Neither strain is able to complete an infection cycle in *E. coli* on its own, because a complete capsid structure cannot be made in the absence of the product of the mutated gene. Thus, when the concentrated lysates of both the induced lysogens are mixed, they genetically complement each other *in vitro* forming an *in vitro* packaging mix or packaging extract that contains all the necessary components required for packaging. Such *in vitro* packaging extracts are also commercially available under different trade names, for example, Gigapack® III Gold Packaging Extract from Stratagene and Packagene® from Promega. The exogenous DNA of appropriate length (37–52 kbp), when added to the packaging extract in the presence of exogenous ATP and biogenic amines, is encapsidated or packaged by cleavage at *cos* sites, and mature viable phages are assembled. These *in vitro* packaged phages have infectivity indistinguishable from that of normal phage particles. Although concatemeric DNA is the substrate for packaging (covalently joined concatemers are produced in the ligation reaction by association of the natural cohesive ends), the *in vitro* system also packages added monomeric DNA, which presumably first concatemerizes non-covalently. Monomeric circular molecules with a single *cos* site are packaged very poorly. This feature is markedly different from plasmid vectors, where the ideal ligation product is a monomeric circular plasmid consisting of one copy of the vector plus insert (for details see Chapter 8). Thus, for λ vectors it is advantageous to adjust the ligation conditions in such a way that multiple end-to-end ligation of λ DNA molecules together with the insert fragments through sticky *cos* ends are achieved. The importance of *cos* sites in packaging has already been discussed in Section 9.2.5. Figure 9.23 outlines the events involved in the *in vitro* packaging process. The resulting λ phage particles can then be assayed by addition of a sensitive bacterial culture and plating as an overlay.

9.5 λ BACTERIOPHAGE AS CLONING VEHICLE

The primary use of λ phage is to clone large pieces of DNA, from 5–23 kbp, much too big to be handled by plasmid or

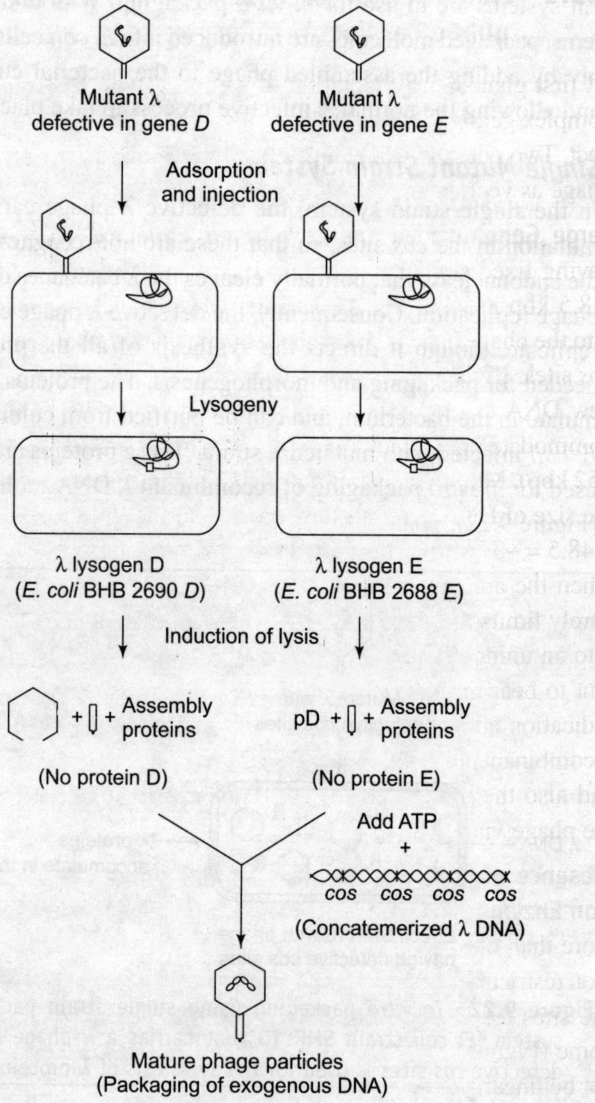

Figure 9.23 *In vitro* packaging of concatemerized phage λ DNA in a mixed lysate using two strains packaging system [Induced λ lysogen *E. coli* BHB 2688 *E* amber mutation produces λ tails, pD, assembly proteins, but no pE and hence no pre-heads. Induced λ lysogen *E. coli* BHB 2690 *D* amber mutation produces λ tails, pre-heads, assembly proteins, but DNA packaging blocked due to lack of pD.]

M13 vectors. It is pertinent to note that the wild-type λ DNA is not directly used for cloning; rather it is manipulated in several ways to construct derivatives for use as cloning vectors. λ vectors should therefore satisfy following requirements: (i) Firstly, these should be small in size, and allow cloning of DNA molecules of a broad size range; (ii) Secondly, these should contain useful target sites for restriction enzymes; (iii) Thirdly, it should be possible to distinguish recombinant and parent phages by plaque morphology or marker inactivation; (iv) Fourthly, recombinant phages should be obtainable in high yields; and (v) Finally, such vectors should guarantee a sufficient level of biological safety.

9.5.1 Manipulation of Wild-type λ Bacteriophage for Vector Construction

At first glance, the adaptation of λ phage, with its large and complex genome, for use as a vector seems to be a bleak prospect. Two potential problems related to the use of wild-type λ phage as vectors are:

Large Genome Size and Low Insert Capacity The λ head having fixed size can accommodate certain amount of DNA (48.5 kbp in wild-type λ), and the length of DNA packaged into the phage head is determined by the distance between two *cos* sites. If wild-type λ DNA is used as vector, only 3 kbp of new DNA can be inserted into the head, as the head can accommodate only about 5% extra DNA ($48.5 + 5/100 \times 48.5 = \sim 52$ kbp). Moreover, infectious phage particle is not formed, if the size of DNA is reduced by more than 25% ($48.5 - 25/100 \times 48.5 = \sim 37$ kbp). Thus, effective packaging takes place only when the amount of DNA ranges from 37–52 kbp. This severely limits the size of a DNA fragment that can be inserted into an unmodified λ vector. These factors are clearly important to bear in mind when designing λ vectors, as these give indication about the limitations imposed upon the lengths of recombinant phage DNAs that can be packaged into the λ head, and also the limitations in size diminution without affecting the phage viability.

Presence of Multiple Sites for a Commonly Used Restriction Enzyme The wild-type λ DNA is so large that it contains more than one recognition sequence for virtually every common restriction endonuclease. For example, there are five cleavage sites for *Eco* RI, and six for *Hin*d III in wild-type λ genome (Figure 9.24). As a consequence, the vector DNA cannot be linearized by a common restriction enzyme for ligation of insert DNA. It is rather cut into several fragments depending on the number of restriction sites for that enzyme, and it is almost impossible to join all these fragments and the insert together in right order and orientation. Hence, wild-type λ DNA itself is not suitable as a vector. Moreover, these sites are often located in regions of the genome essential for lytic growth of the phage, and hence cannot be deleted.

Thus, for the construction of vectors, wild-type λ genome is manipulated as follows:

Size Diminution and Increasing Cloning Capacity

The way forward for the development of λ cloning vectors was provided by the discovery that only about 50% of genes of wild-type λ phage are essential for its replication and for the lysis of the host cell, and most of the non-essential genes are located together in a cluster around the middle of the genome. This signifies that a large segment in the central region of the λ DNA can be removed without affecting the ability of the phage to infect *E. coli* cells. Thus, deletion of all or part of this non-essential region, between positions 20 and 35 on the genome map shown in Figure 9.2, decreases the size of the resulting λ DNA by up to 15 kbp, or in other words, only 15 kbp DNA (upper limit minus lower limit, i.e., $52 - 37 = 15$ kbp) can be deleted without affecting the phage viability. As already stated, for the stability and viability of the phage, a minimum of 37 kbp of DNA (about 75% of wild-type λ DNA) between the two *cos* sites should be present, hence too much from the wild-type λ genome cannot be deleted. Moreover, as 5% extra DNA (~3 kbp) can be accommodated in the phage head, a total of 18 kbp (due to space provided by deletion of 15 kbp non-essential segment of λ genome + 3 kbp extra space) can be added before the cut-off point for packaging is reached. The existence of these packaging limits is a very important feature of the design and application of λ vectors and also of cosmids (for details see Chapter 12). As this non-essential region contains *b*2 region, integration and excision genes, and genes required for establishment of lysogeny, a deleted λ genome is non-lysogenic, and can follow only lytic infection cycle. This in itself is desirable for a cloning vector as it means induction is not needed before clearly distinguishable plaques are formed.

The deletion of certain region of λ DNA carrying the restriction sites also increases cloning capacity. However, even a deleted λ genome, with the non-essential region removed, may have multiple recognition sites for many restriction enzymes.

Alteration (Reduction/Introduction) in the Number of Cleavage Sites for a Restriction Enzyme(s)

Before using λ phage as a cloning vehicle, it is necessary to decrease the number of the restriction sites of common restriction enzymes used for cloning. Thus, the following methods

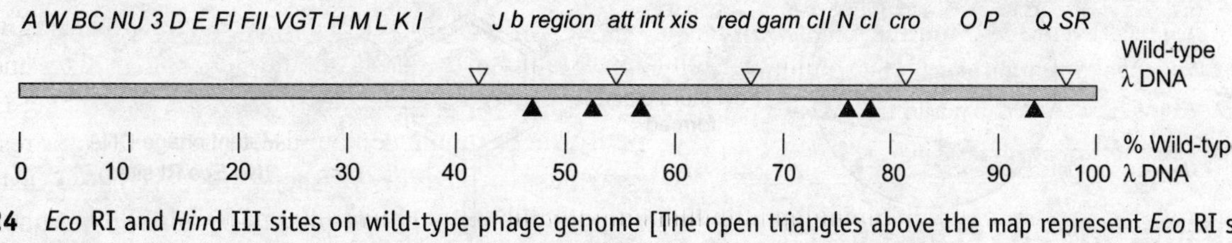

Figure 9.24 *Eco* RI and *Hin*d III sites on wild-type phage genome [The open triangles above the map represent *Eco* RI sites (five in number), and the solid triangles below the map indicate the *Hin*d III sites (six in number).]

are adopted for the construction of λ derivatives having either a single or two target sites for a restriction enzyme:

Natural Selection of Modified λ Bacteriophage Lacking Certain Restriction Sites Natural selection signifies *in vivo* selection of the derivatives of λ phage from which few or all sites for a restriction enzyme (here *Eco* RI is considered as an example) have been eliminated from the essential portion of the λ genome. This was the first method to be performed for the elimination of restriction enzyme sites. Natural selection can be brought into play by using an *E. coli* host strain that produces *Eco* RI. *Eco* RI cleaves most of the λ DNA molecules invading the cell, but a few phages survive and produce plaques. Thus, in an infection of a host carrying the *Eco* RI restriction-modification (R–M) system by unmodified phage, the progeny phages are vastly reduced in number, and have arisen from DNA molecules that have undergone spontaneous alteration in their recognition sequence at one or more sites early in infection. The *Eco* RI sites may in addition have mutations such that these are not cleaved by *Eco* RI. Mutants lacking some sites are selected by cycling the phage between hosts containing and hosts lacking the *Eco* RI R–M system. Several cycles of infection eventually result in mutant phages that lack all or most of the *Eco* RI sites (Figure 9.25). The cycling is continued until the efficiency of plating indicates that the DNA can no longer be restricted. The *in vivo* selection method, however, requires the growth of phage in hosts that synthesize the restriction enzyme of interest. This is not acceptable as this may lead to development and spread of resistant strains.

Natural Selection Followed by Genetic Crossing with λ Bacteriophage In this method, the first step is to look for λ mutants, which do not contain a target site for a particular restriction enzyme. This is because the complete elimination of restriction sites by growing the phages on restricting hosts is comparatively easy. Only phage DNA devoid of susceptible *Eco* RI sites will be able to survive in host cells containing, for example, the *Eco* RI R–M system. In the second step, such restriction resistant phages are genetically crossed with suitable susceptible phages in order to introduce the desired cleavage sites. For example, the mutant phages that completely lack *Eco* RI sites are crossed *in vivo* with phage containing all the *Eco* RI sites and recombinants are selected, which contain sites marked 1, 2, and so on as indicated in Figure 9.24.

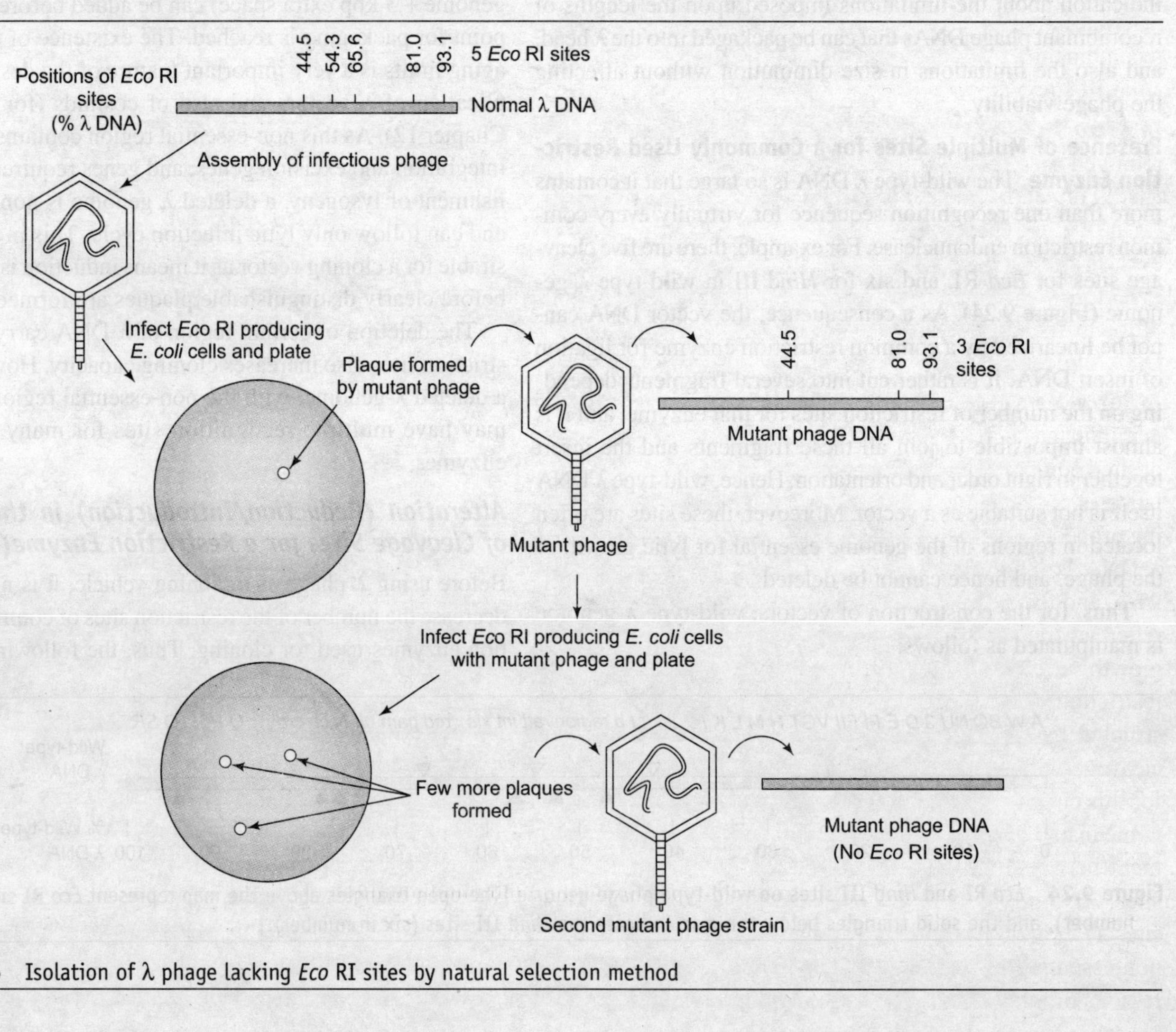

Figure 9.25 Isolation of λ phage lacking *Eco* RI sites by natural selection method

Recombinational Insertion of Regions from other Lambdoid Phages (i.e., Construction of Hybrid Immunity Phage) All the lambdoid phages (e.g., ϕ434, ϕ80, and ϕ21) are active in *E. coli*, and their genomes have considerable regions of sequence homology with λ DNA. However, their immunity regions differ, because of which these phages display different immunities. It is possible to construct recombinant λ phage, which carries the immunity region (the operators, *cI*, and *cro* genes) of ϕ434, for example. This is done by *in vivo* genetic recombination that leads to exchange of immunity region of these two phages. Such a phage is called λ *imm* 434. Similarly, λ *imm* 80 or λ *imm* 21 can be constructed. Such hybrid phage provides a useful means of varying the number of restriction sites for a given enzyme in λ vectors. Such recombinants either contain or lack recognition sites for any particular restriction enzyme. First, all the target sites can be removed, and then the desired ones replaced by *in vivo* genetic recombination to get unique target sites. In this way, λ vectors have been constructed for the restriction fragments produced by *Hin*d III. The deletion of the restriction fragment between *Eco* RI sites 1 and 2 removes two *Hin*d III sites, and *b* 538 deletion removes *Hin*d III sites 1, 2, and 3. *In vivo* genetic recombination leading to formation of λ *imm* 434, removes *Hin*d III sites 4 and 5, while site 6 is lost in λ *imm* 80.

***In Vitro* Mutagenesis** If just one or two sites need to be removed then the technique of *in vitro* mutagenesis can be used. For example, an *Eco* RI site, G**A**ATTC, is changed to G**G**ATTC, which is not recognized by the enzyme. However, *in vitro* mutagenesis was in its infancy when the first λ vectors were developed. Even today this technique is not efficient for alteration of more than a few sites in a single DNA molecule.

***In Vitro* Manipulation** λ vectors nowadays are genetically manipulated to remove unwanted restriction sites or to get desired combination of restriction sites. This *in vitro* manipulation is possible due to the development of *in vitro* packaging procedures for the assembly of infectious particles. Thus, λ DNA is digested with a restriction enzyme, packaged *in vitro*, and the surviving molecules are propagated in *E. coli*. After several successive cycles of digestion, packaging, and growth, infective phage genomes that have lost some of the restriction sites can be obtained. λ vectors have been constructed by joining the restriction fragments produced by cleavage with *Hin*d III. The wild-type λ DNA has six sites for this enzyme, the positions of which are shown on the genome map in Figure 9.24. The deletion of the restriction fragment between *Eco* RI sites 1 and 2 removes two *Hin*d III sites. Also phage having *b* 538 deletion lacks the *Hin*d III sites 1, 2, and 3. Further, any target site located within the non-essential region can be deleted simply by deleting that region, provided the size does not reduce below the limit required for the assembly of infectious phage. The digested fragment can also be replaced by foreign DNA segment.

Other Manipulations

After size diminution and the reduction of restriction sites, several other manipulations need to be carried out to construct suitable vectors. These include: (i) Synthetic oligonucleotides comprising several restriction sites are inserted into the λ genome to be used for cloning. In this way commonly used restriction sites are placed at desired locations with precision and high efficiency; (ii) Manipulations are also done in a way that leads to positive selection of recombinants, for example by size, Spi phenotype (sensitivity to P2 interference or sensitivity to P2 inhibition), etc. (for details see Section 9.5.6); Besides, in some vectors, selectable markers may also been introduced; (iii) Early work with recombinant DNA molecules was beset with difficulties because of the worry that a potentially hazardous gene could be inadvertently cloned, and expressed within *E. coli*. In order to provide λ vectors with a level of biological containment, most λ vectors contain genetic markers, which are important for their biological safety. The guidelines for EK2 vector require that only one phage particle in 10^8 should survive under natural conditions if phages are employed as cloning vehicles. For this reason amber mutations are introduced into a number of late genes (capsid or lysis genes). For example, *S* amber mutation renders the burst of mature phage particles more difficult. Such phage can be propagated only on a host containing an amber suppressor mutation, which does not occur in the natural environment, for example, bacterial host strain DP50 *sup* F, a derivative of a strain χ 1953 into which the *sup* F58 (*su* III) suppressor mutation has been introduced (for details see Section 9.5.6). *nin* deletions also force phages into the lytic growth cycle, and hence provide an additional safety factor; (iv) As an amber mutation in late genes ensures lysogeny, the vector may be constructed that has amber mutation in one of its late genes, for example *S*, *A*, *B*, etc. Such phage DNA replicates along with host chromosome, which leads to increase in copy number. When such lysogenic phages are induced for lysis, burst of mature phage particles is achieved; and (v) For expression of foreign DNA, transcriptional or translational fusion vectors (for details see Chapter 12) that contain regulatory regions may be constructed.

Together, these manipulations have led to the development of a large number of vectors that can accept and propagate fragments of foreign DNA generated by a variety of restriction enzymes. Since early 1970s, when phage λ was first used as a cloning vehicle, more than 400 different vectors have been described. Some of these are direct descendants of wild-type strains of lambdoid phages, while others are far more esoteric. These λ vectors with different features are also available commercially, and can be selected according to the requirement.

9.5.2 Cloning in λ Vectors

λ phage has many uses as a cloning vehicle, ranging from subcloning of genomic DNA sequences initially cloned in vectors with larger capacity (e.g., bacterial artificial chromosomes or P1 phages) to construction of complex cDNA or genomic DNA libraries. A cloning experiment with a λ vector can proceed along the same lines as with a plasmid vector. Thus, cloning in λ vectors involves several steps that are summarized below: (i) Restriction digestion of λ vector as well as genomic DNA; (ii) Ligation of the DNA fragments into λ vector to form recombinants; (iii) Introduction into host by transfection of competent *E. coli* host or *in vitro* packaging of recombinant DNA by addition of packaging extract followed by natural infection of the host; and (iv) Selection and screening of recombinants.

Further details of cloning strategies and the selection procedures are summarized along with the details of each vector.

9.5.3 Choice of λ Vectors

There is no single λ vector that is suitable for cloning DNA fragments for all purposes. Therefore, a careful choice must be made among the commercially available vectors for the one best-suited to the particular task at hand. This has been possible because extensive information is available on the genetics and molecular biology of the phage that is required for wise selection of vectors, and to use them effectively. Based on this information, following considerations influence the selection of a vector: (i) Vectors are selected after considering the restriction enzyme(s) to be employed, the size of the foreign DNA to be inserted, and the type of screening system to be used; (ii) Expression vectors are used for expression of cloned DNA sequences in *E. coli*; (iii) Phasmids are used if the foreign DNA is to be rescued from the λ vector in the form of a plasmid or phagemid (for details see Chapter 12); and (iv) Vectors with large insert capacity are employed if the foreign DNA carried in the vector is to be assembled into a large overlapping contig of cloned DNA in sequencing experiments (for details see Chapter 19).

The development of vectors containing multiple cloning sites, and the availability of a variety of restriction enzymes that cleave DNA to produce compatible termini have greatly simplified the mechanics of cloning into λ vectors. However, the size of the insert remains an important factor to be considered when choosing a vector. Only ~60% of the vector DNA (the left arm, ~20 kbp in length, including the head and tail genes *A–J*, and the right arm, ~9 kbp in length, from P_R through the *cos* R site) is necessary for lytic growth of the phage, while the central one-third of the wild-type λ genome can be replaced by foreign DNA. This signifies that the central one-third region can be easily deleted without affecting the viability of phage. However, the viability of phage λ decreases dramatically when the length of genome is greater than 105% or less than 75% of that of wild-type λ DNA. Hence, the combination of vector and foreign DNA must result in a recombinant λ DNA that falls within the acceptable size limits. This also indicates that there is restriction to the deletion of λ DNA and insert capacity. This information has lead to the construction of two classes of vectors.

9.5.4 Classes of λ Vectors

Once the problem posed by packaging constraints and by the multiple restriction sites had been solved, the way was opened for the development of different types of λ based vectors. The first two classes of vectors to be produced were λ insertional and λ replacement (or substitution) vectors (Figure 9.26). Many vector derivatives of both types are now available. Most of these vectors are constructed for use with common restriction enzymes such as *Eco* RI, *Bam* HI, or *Hind* III. However, their application can easily be extended to other endonucleases by the use of linkers or adaptors (for details see Chapter 13). Salient features of these two classes of λ vectors are summarized in the following sections.

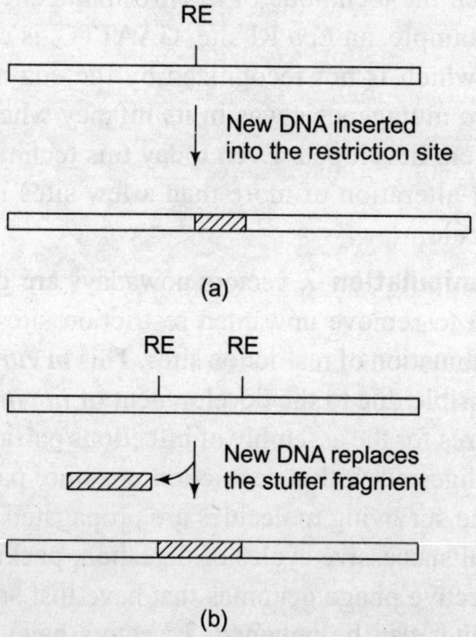

Figure 9.26 Principles of two types of λ vectors; (a) Insertion vector, (b) Replacement vector [RE represents restriction enzyme sites.]

Insertional Vectors

Vectors containing at least one unique restriction enzyme site for the insertion of foreign DNA are designated as insertional vectors (Figure 9.26a). These are the simplest form of λ vectors, and have a similar concept as plasmid vectors. In insertional vectors, the vector size is decreased by deleting ~20% of the wild-type genome comprising of several genes not required for lytic growth. Unwanted restriction sites are also removed by any of the strategies described above in Section 9.5.1. Note that the deletion of the regions of DNA carrying unwanted restriction sites also increases the cloning capacity. Insertion of an appropriate segment of foreign DNA into the cloning site (or unique restriction enzyme site) restores the genome size to its full length, and facilitates *in vitro* packaging and assembly of infectious phage particles. The size of the DNA fragment that an individual vector can carry thus depends on the extent to which the non-essential region has been deleted. As the packaging limits for λ DNA are between 37 kbp and 52 kbp, the size of insertional vector should not be below 37 kbp, and the size of the foreign DNA should be in the range that its insertion does not increase the size of the vector above 52 kbp. Thus, the maximum cloning capacity of an insertion vector is 15 kbp (52 – 37 = 15 kbp). The cloning capacity of insertional vectors is thus higher than plasmid cloning vectors, but is still unsuitable for genomic library construction.

Examples of Insertional Vectors Two popular insertional vectors are λgt 10 and λgt 11. Besides, there are several other derivatives of λgt 11. These vectors are discussed in detail below:

λgt 10 λgt 10 has a unique *Eco* RI site located within the *cI* gene. Note that gt stands for generalized transducing (Exhibit 9.4). The deletions and other manipulations that this phage has undergone have removed some of the unwanted sites, and have also reduced the overall size of the vector DNA to 43.3 kbp (which is still large enough to produce viable phage particles, and still contains the genes that are needed for viability), and hence allows the insertion of foreign DNA up to a maximum of ~8 kbp. The vector map of λgt 10 is shown in Figure 9.27. The vector has *b* 527 deletion, and contains immunity region from ϕ434 (instead of λ) having unique *Eco* RI site that is employed for cloning. The

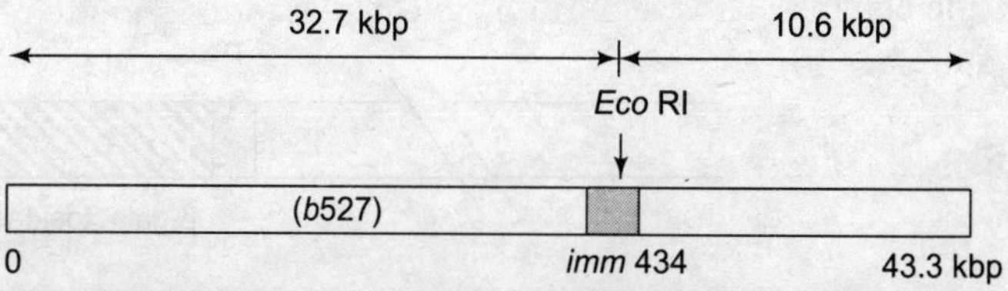

Figure 9.27 Vector map of insertion vector λgt 10

insertion of foreign DNA into this *Eco* RI site leads to insertional inactivation of *cI* gene, which destroys the ability of recombinants to form functional repressor (*Eco* RI cleaves the vector into two fragments of 32.7 and 10.6 kbp). On the other hand, a functional repressor is formed in religated λgt 10 that establishes lysogeny leading to formation of turbid plaques. This allows easy distinction between recombinants and nonrecombinants (parental vector formed by religation of the arms without an insert), as recombinants form clear plaques, while nonrecombinants form turbid plaques. This distinction is made more pronounced by plating on hfl^- host (for details see Section 9.5.6).

λgt 11 λgt 11 (gt stands for generalized transducing) is a useful cDNA cloning and translational fusion vector described by Young and Davis. The vector is 43.7 kbp in length and contains a *lac* promoter driven *E. coli lac Z′* gene, encoding β-galactosidase, with a single *Eco* RI restriction site in its C-terminal coding region, precisely 53 bp upstream from the termination codon. The vector map of λgt 11 is shown in Figure 9.28. The cleavage with *Eco* RI cuts the vector into two fragments of 19.6 and 24.10 kbp. The cloning capacity of the vector is ~8 kbp. The presence of a unique *Eco* RI site within *lac Z′* gene confers two advantages: (i) Firstly, insertion of DNA at *Eco* RI site inactivates the β-galactosidase gene, so that recombinants form white plaques on a medium containing X-gal (5-bromo 4-chloro 3-indolyl β-*D*-galactoside), while non-recombinants form blue plaques (for details see Chapter 16); and (ii) Secondly, if the insertion maintains the correct orientation and translational reading frame with

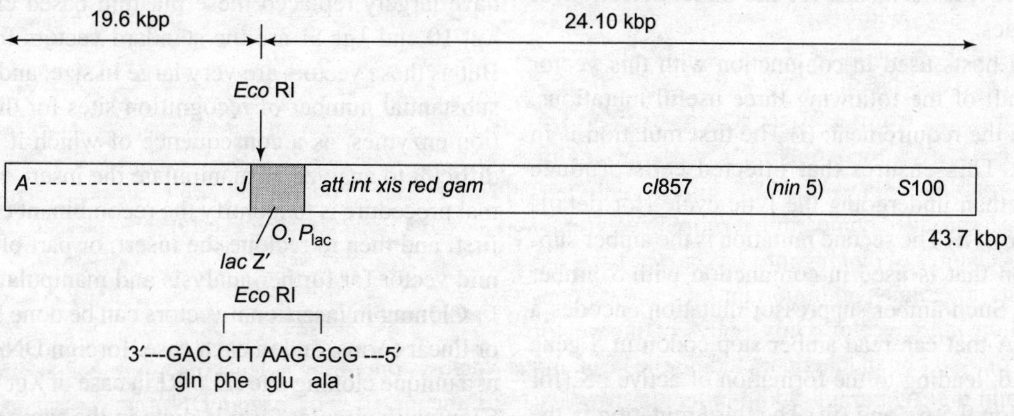

Figure 9.28 Vector map of insertion vector λgt 11 [DNA sequence shows codons 1003 (Ala) and 1006 (Gln) of β-galactosidase with the relevant *Eco* RI site]

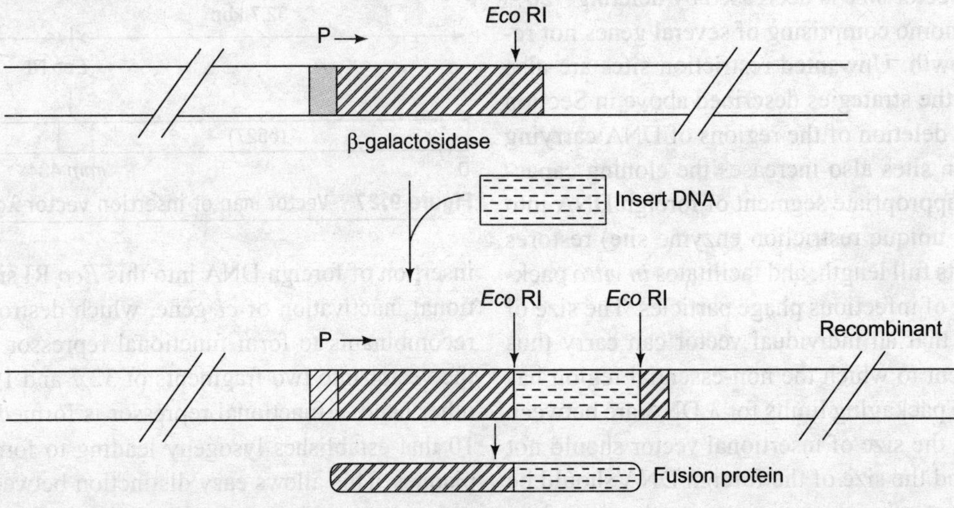

Figure 9.29 Application of λgt 11 for generation of fusion proteins

the β-galactosidase gene, then the recombinant phage directs the synthesis of fusion proteins containing the product encoded by the insert fused to the β-galactosidase protein (Figure 9.29). The fusion proteins thus formed may not show any biological functions associated with the cloned gene, as these contain a large amount of extraneous material, however their interaction with specific antibodies forms easy detection system for the clone of interest (for details see Chapters 12 and 16).

In order to permit efficient screening, the phage genome carries the *cI* 857 gene so that the shifting of temperature from 30°C to 42°C can induce the lysogens (for details see Section 9.5.6). The thermo-induced colonies can be lysed by exposure to chloroform vapor. The vector also has *nin* 5 deletion making the vector *N* independent, and *S* 100 mutation for biological containment. The amber mutation in the *S* gene renders the cells to be filled with large amounts of phage products before lysis. Kemp and his colleagues further modified λgt 11 to incorporate the β-*lactamase* gene from the plasmid pBR322. This specifies resistance to ampicillin, and therefore provides a means for the direct selection of lysogenic colonies.

The bacterial hosts used in conjunction with this vector contain any or all of the following three useful mutations, depending upon the requirement: (i) The first mutation is in the *hfl* A gene. This ensures that infected cells produce lysogens rather than undergoing the lytic cycle (for details see Section 9.5.6); (ii) The second mutation is the amber suppressor mutation that is used in conjunction with *S* amber mutant vectors. Such amber suppressor mutation encodes a suppressor tRNA that can read amber stop codon in *S* gene as an amino acid, leading to the formation of active pS (for details see Section 9.5.6); and (iii) The third mutation is the *lon*$^-$ mutation, which causes a defect in protein degradation pathways. This prevents degradation of fusion proteins (for details see Section 9.5.6).

λgt 18 to λgt 23 These are derivatives of λgt 11 that accept larger inserts, and allow the use of up to four additional restriction enzymes. Like λgt 11, λgt 18–λgt 23 also carry the *cIts* 857 mutation in addition to an amber mutation in the *S* gene, *Sam* 7. At 42°C, where the *cIts* 857 repressor is only partially functional, these vectors form plaques on strains of *E. coli* (such as Y1090 *hsd* R) that carry the amber suppressor *sup* F. At temperatures where the repressor is active (30°C), these phage strains form lysogens. Like λgt 11, its derivatives λgt 18–λgt 23 are expression vectors that allow expression of the cloned fragment. The chimeric gene can be induced in appropriate lytic or lysogenic hosts resulting in the synthesis of a fusion protein consisting of the N-terminal portion of the β-galactosidase gene fused to sequences encoded by the downstream open reading frame.

Strategy Used for Cloning in Insertional Vectors Most early cDNA libraries were constructed using plasmid vectors. However, λ phage libraries prepared by using insertional vectors have largely replaced these plasmid based cDNA libraries; λgt 10 and λgt 11 are the standard vectors for the purpose. But as these vectors are very large in size, and may contain a substantial number of recognition sites for different restriction enzymes, as a consequence of which it becomes more difficult to analyze or manipulate the insert. Hence, the normal procedure is to identify the recombinant clone of interest first, and then to reclone the insert, or part of it, into a plasmid vector for further analysis and manipulation.

Cloning in insertional vectors can be done both in circular or linear forms, and in each case, foreign DNA is introduced at a unique cloning site (*Eco* RI in case of λgt 10 and λgt 11). Cloning in circular form is done in the same way as in plasmids, by just cutting open the vector and inserting the gene

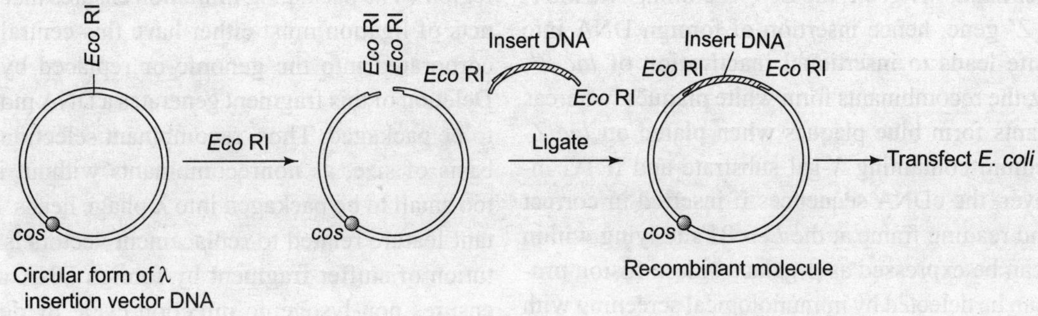

Figure 9.30 Cloning strategy with circular forms of λ insertion vectors is similar to plasmids

of interest. The recombinants obtained are directly transfected into host cells (Figure 9.30). On the other hand, cutting the linear form of insertional vector with restriction enzyme having unique recognition site produces two DNA fragments, which are referred to as the left and right arms. As one end of each of the two fragments is derived from the cohesive ends of the λ DNA, it allows quite stable annealing of the two arms at 37°C, which though not covalently joined can be considered as a single DNA fragment. The insert DNA is then ligated between the two vector arms, and the resulting recombinants are *in vitro* packaged to assemble infectious phage particles (Figure 9.31). In either method, recombinants are selected through appropriate strategies.

In λgt 10, as the *Eco* RI cloning site interrupts the phage *cI* gene, the recombinants can be selected on the basis of plaque morphology. This is because the recombinants undergo lytic cycle, whereas nonrecombinants follow lysogenic cycle, and hence recombinants form clear plaques, while nonrecombinants form turbid plaques. Moreover, this distinction is made more pronounced by plating on *hfl* A^- host (for details see Section 9.5.6). In such a strain, pCII is stable that leads to efficient establishment of lysogeny in religated vector producing functional λ repressor. This does not affect the recombinants, which are unable to make functional repressor, and therefore do not establish lysogeny. Thus, in parental phage, lytic cycle is prevented, and the plaques if formed are turbid, while in case of recombinants, clear plaques are formed. This also allows substantial enrichment of recombinant phages over the religated vector molecules without having to resort to dephosphorylation with alkaline phosphatase. λgt 10 libraries are screened by nucleic acid hybridization. λNM 1149 vector is also based on the insertional inactivation of *cI* gene.

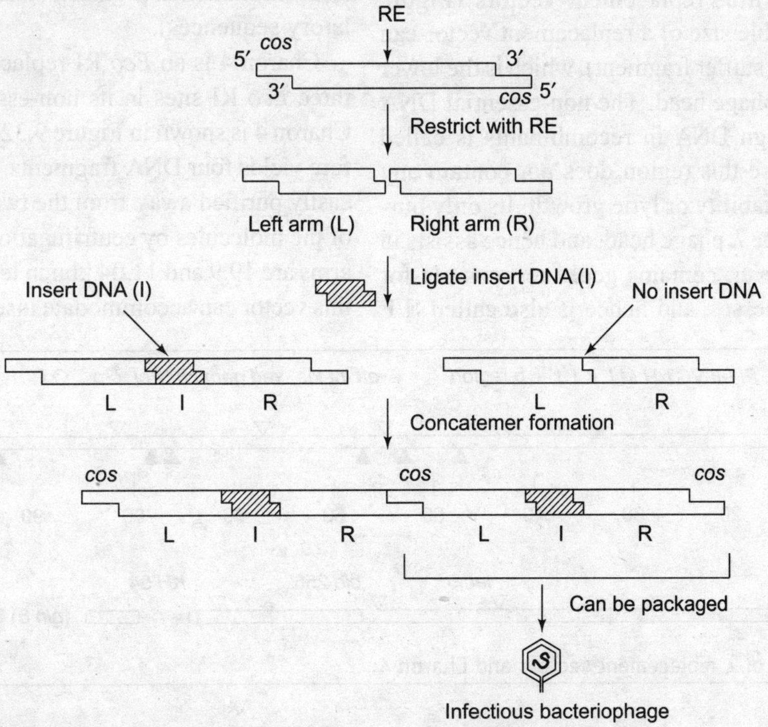

Figure 9.31 Cloning in λ insertion vectors is also done in linear form of the vector

On the other hand, in λgt 11, the *Eco* RI cloning site interrupts the *lac* Z′ gene, hence insertion of foreign DNA into this unique site leads to insertional inactivation of *lac* Z′. Consequently, the recombinants form white plaques, whereas nonrecombinants form blue plaques when plated on *lac* Z^- host on a medium containing *X*-gal substrate and IPTG inducer. Moreover, the cDNA sequences if inserted in correct orientation and reading frame at the *Eco* RI site lying within *lac* Z′ gene, can be expressed as β-galactosidase fusion proteins, which can be detected by immunological screening with other ligands. This approach has been remarkably successful, and has led to the isolation of many genes encoding proteins using specific antisera probes. Note that λgt 11 libraries can also be screened by nucleic acid hybridization, although λgt 10 is more appropriate for this screening strategy because high titres are possible.

Replacement (or Substitution) Vectors

The packaging limits that restrict the cloning capacity of insertion vectors are imposed by the physical requirements of the λ phage head rather than by the nature of the genes needed. There are more genes that are not essential for lytic growth and could be deleted, except that it would make the λ DNA too small to produce viable progeny. This suggests the possibility of substituting this non-essential region with any DNA of interest. Thus, instead of merely inserting extra DNA vectors with a pair of cloning sites flanking a segment of non-essential DNA, which allow foreign DNA to be substituted for the DNA sequences between these sites, have been designed. These are known as replacement vectors (Figure 9.26b). The lowest possible size of a replacement vector can thus be 37 kbp (including stuffer fragment), which is the lower limit of packaging in λ phage head. The non-essential DNA that is replaced by foreign DNA in recombinants is called 'stuffer fragment' because this region does not contain any genes essential for cell viability or lytic growth. Its only purpose is to help to fill up the λ phage head, and hence assists in its propagation. It, however, contains genes responsible for integration/excision processes, and hence is also called 'I/E region'. The packaging limitation ensures that the viable products of ligation must either have this central fragment reincorporated into the genome or replaced by foreign DNA. Deletion of this fragment generates a DNA molecule too small to be packaged. Thus, recombinant selection is often on the basis of size, as nonrecombinants without insert DNA are too small to be packaged into λ phage heads. Another important feature related to replacement vectors is that the substitution of stuffer fragment by foreign DNA in recombinants ensures non-lysogenic infection cycle by deletion of genes involved in integration and recombination. Thus, the recombinant phage λ undergoes lytic cycle and is maintained by growth on *E. coli*.

As evident above, replacement vectors are generally designed to carry larger pieces of DNA (up to 23 kbp) than insertional vectors. As such, replacement vectors are routinely used for genomic library construction.

Examples of Replacement Vectors Some common examples of replacement vectors are Charon series, λEMBL series, λgtWES.λB′, λgt.λC, λGEM series, λDASH, and λFIX. These are described in detail below

Charon series These phages are named for the ferryman of Greek mythology who conveyed the spirits of the dead across the river Styx, which separated the realm of the living from Hades, the underworld. The series includes several vectors, out of which only a few exhibiting different properties are described. Besides, some Charon insertional vectors have also been developed. Many vectors in Charon series carry the *lac* 5 substitution (i.e., β-galactosidase gene together with its regulatory sequences).

Charon 4 is an *Eco* RI replacement vector which contains three *Eco* RI sites in its non-essential region. Vector map of Charon 4 is shown in Figure 9.32. Cleavage with *Eco* RI therefore yields four DNA fragments. The central fragments can be easily purified away from the two other fragments at the ends of the molecules by centrifugation. Since the left and the right arms are 19.9 and 11.04 kbp in length (30.94 kbp total length), this vector can accommodate insertions between 6 and 21 kbp.

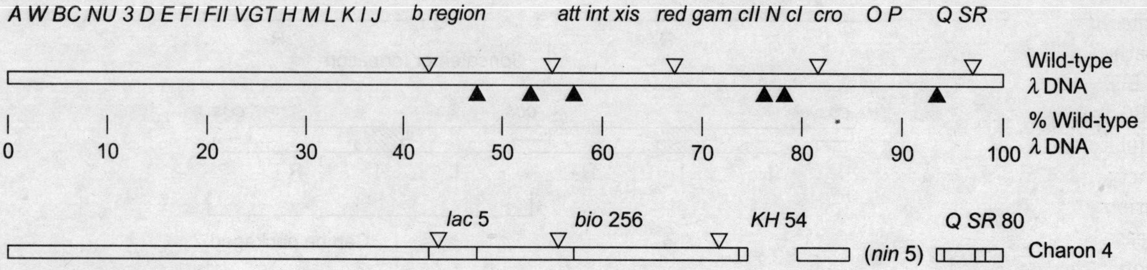

Figure 9.32 Vector maps of λ replacement vectors and Charon 4

Because of the high insert capacity, the vector can be used in genomic library construction (for details see Chapter 15). It has been found that this vector is unable to grow well on many hosts. This is because of the absence of *chi* sequences (χ site; cross-over hotspot instigator) on the vector (for details see Section 9.5.6 and Exhibit 9.5). However, if the foreign DNA fortuitously provides these *chi* sequences, recombinants may grow well. As the recombinant molecules lack *int* and *att*, these are unable to form lysogens. These vectors are also confined to the lytic mode by deletions in the immunity regions and by the *nin* 5 deletion which results in the *N* independent activation of gene *Q*, and hence the genes of the late transcriptional pathway. In this phage, the separation of the two internal fragments eliminates the markers *lac* 5 and *bio* 256 (*lac* 5 and *bio* 256 lie on separate fragments). In the presence of X-gal and suitable inducers, the intact vector forms blue plaques, while recombinants containing both vector arms, and inserted foreign DNA yield colorless plaques. However, this selection is only of minor importance, because the size of the insert DNA itself exerts a much more powerful positive selection pressure for recombinant DNA molecules.

Charon phages are however not restricted to *Eco* RI target sites. A second generation of vectors has been developed to allow cloning at a number of other restriction sites. Charon 30, for example, is a *Bam* HI replacement vector. The vector is 46.8 kbp in size, i.e., 96.5% of λ. The length of DNA between arms is 14.86 kbp. Like Charon 4, Charon 30 also lacks χ site. The *att* site is also missing due to *b* 1007 deletion. The insert capacity while cloning at *Bam* HI site ranges between 6.1 and 19.1 kbp, and the recombinants are selected on the basis of size. Cloning in this vector gives rise to gam^- recombinants, and hence this vector should only be used for construction of libraries using *rec* BC^- or *rec* D^- hosts.

The presence of an active bacterial recombination system occasionally leads to rearrangements, and hence deletion of segments of foreign DNA that carry repeated sequences. To

Exhibit 9.5 χ (*chi*) Site

χ (chi) site is an octameric (8 bp) sequence of double stranded genomic DNA with a sequence 5′-GCTGGTGG-3′. χ sequences were first discovered in bacteriophage λ as mutations creating recombinational hot spots for recombination promoted by *Rec* BCD, a heterotrimeric enzyme encoded by the *rec* B, *rec* C, and *rec* D genes. There are, however, no χ sites in wild-type λ DNA. χ sequences have been created by single base change at several widely-spaced sites in the bacteriophage λ genome.

The *E. coli* genome contains a total of ~103 χ sites, ~1 per 4-5 kbp. Of these sites in the *E. coli* chromosome, 90% are oriented such that these will protect against degradation of DNA proceeding toward the origin of replication. χ sequence, in *E. coli*, causes increased recombination over a region of several kilobase pairs located asymmetrically around it. This stimulatory effect of χ sequences on recombination is due to the formation of a nick about five nucleotides 3′ to the χ site by the *E. coli Rec* BCD protein. The nick only occurs when the enzyme passes through the sequence in one direction (from right to left), unwinding the DNA as it goes. The combination of unwinding and nicking the DNA generates a single stranded 'tail' with a χ sequence near its 3′-end. This tail is believed to be a potent substrate for *rec* A gene encoded Rec A, which catalyzes the formation of recombinant molecules between the single stranded tail and homologous ds DNA. Of the several major pathways of recombination in *E. coli*, χ sites stimulate only the Rec BCD pathway. χ sites are therefore inactive in *rec* B, *rec* C, and *rec* BC double mutants. Normally, the Rec BCD protein expresses a powerful exonuclease V activity that destroys foreign DNA cleaved by restriction enzymes. However, χ sites protect linear DNA from degradation both *in vivo* and *in vitro*, perhaps by inactivating the exonuclease V activity of the Rec BCD nuclease. With the help of the Rec A protein, the frayed end of DNA invades a homologous sequence to form a branched structure that can be converted into a replication fork and resolved by recombination.

The final genetic factor to consider in choosing a vector is the presence of a χ site. Because recombination is essential for efficient replication and packaging of bacteriophage λ genomes, the presence or absence of χ sites, and the state of the host recombination machinery are important factors to consider when selecting a host cell for a particular bacteriophage λ vector. The growth of the red^- gam^- bacteriophages on wild-type *E. coli* is limited by the availability of concatenated DNA molecules that can be efficiently packaged into phage particles. However, the presence of a χ site increases the efficiency of the recombination process that gives rise to packaging substrates. As a result, the plaques formed by red^- gam^- χ^+ bacteriophages are almost as large as those formed by red^+ gam^+ bacteriophages. The requirement for a χ site in red^- gam^- bacteriophages is, however, eliminated by using *rec* D mutant bacterial strains as hosts (*rec* D^-). These strains, which are superproficient in recombination but defective in exonuclease V production, generate concatemers of bacteriophage λ DNA by both recombination and rolling circle replication. The recombination system in *rec* D^- mutants is so active that a χ site is no longer required in red^- gam^- bacteriophages. Because some bacteriophage λ vectors are red^- gam^- and do not carry a χ site, recombinants would normally be expected to display a small plaque phenotype on wild-type *E. coli*. However, eukaryotic DNAs contain sequences that can mimic χ sites. Thus, some recombinant bacteriophages will carry insertions of foreign DNA that contain χ sequences and others will not. The recombinants that contain χ sequences will grow to higher titre, form larger plaques, and become over-represented during amplification of libraries. This problem can be overcome by avoiding vectors (such as Charon 28 and 30), which give rise to gam^- recombinants that do not contain a χ site. These vectors should only be used for construction of libraries if the recombinants are propagated on *rec* BC^- or *rec* D^- hosts.

Source: Sambrook J and Russel DW (2001) *Molecular Cloning: A Laboratory Manual*, III ed., Cold Spring Harbor Laboratory Press, Cold Spring Harbor, New York, pp. 2.13.

overcome this problem, Charon 32–Charon 35 and 40 have been designed that carry the *gam* gene on one of the arms of the λ DNA. Thus, cloning in these vectors generates *red*$^-$ *gam*$^+$ recombinants. Since each of these vectors is *red*$^-$, infected *rec* A cells are phenotypically *Rec* A$^-$, *Rec* BC$^-$, and *Red*$^-$. The use of such vectors has in some instances overcome the problems of cloning segments of foreign DNA containing recombinogenic sequences.

The guidelines for EK2 vector require that only one phage particle in 10^8 should survive under natural conditions if phages are employed as cloning vehicles. For this reason phage vectors frequently carry suppressor mutations in genes coding for capsid proteins pA and pB, which provide biological safety or containment. Charon 4A, a derivative of Charon 4, is designed to contain such genetic markers. Such phages with amber mutations in *A* and *B* genes grow on suppressor positive (*sup*$^+$) bacterial host strain, e.g., DP50 sup F, a derivative of a strain χ 1953 into which the *sup* F58 (*su* III) suppressor mutation has been introduced. Note such suppressor positive host strains do not occur in the natural environment. The vector is 45.3 kbp in size, i.e., 93.4% of λ. The length of DNA between arms is 14.38 kbp. The vector contains *Eco* RI and *Xba* I cloning sites. Insert capacity while cloning at *Eco* RI site is 7–20 kbp, and the recombinants are *lac*$^-$ *bio*$^-$. On the other hand, insert capacity while cloning at *Xba* I site is up to 5.63 kbp, and the recombinants are *red*$^-$ *gam*$^-$.

λEMBL series Loenen and Brammar constructed a vector, L47, which is used as a replacement vector for fragments generated by *Bam* HI, *Eco* RI, or *Hin*d III. The recombinants are selected by their Spi$^-$ phenotype (for details see Section 9.5.6), and the left phage arm contains a *chi* site to facilitate propagation of the recombinant (for details see Section 9.5.6 and Exhibit 9.5). A second such vector, λ1059, carries a Col E1 plasmid with cloned λ *att* sites. Such recombinants are capable of growing lytically as phages or as plasmids in the presence of λ repressor. The plasmids can be released from the phage arms by infecting *E. coli* that is constitutively producing λ integrase. One problem with such a vector is that a minor proportion of parental phage does survive the Spi$^-$ selection (for details see Section 9.5.6), and these are subsequently picked up during screening for foreign genes by nucleic acid hybridization. This is because such screening is almost invariably carried out with probes made from recombinants in which the plasmid vectors have homology to Col E1. To circumvent this problem, λEMBL 1 was constructed in which the *Hin*d III fragment of λ1059, which contains the plasmid sequence, was replaced by a fragment carrying the *E. coli trp* E gene. Variants of λEMBL 1 have been selected, which have lost the *Eco* RI sites, and the *Bam* HI sites are replaced by linkers having the recognition sequences for *Eco* RI:*Bam* HI:*Sal* I. Most common of such variants are λEMBL 3 and λEMBL 4, λEMBL 3A, and 4A, the vector maps of which are given in Figure 9.33. It is important to note that *Bam* HI is the most common cloning site in λEMBL series of vectors, as for many other vectors. This is advantageous because, during genomic library construction, DNA fragments are obtained by partial digestion of genomic DNA with a frequent base cutter, *Sau* 3AI, which produces sticky ends that are compatible to that produced by *Bam* HI (for details see Chapter 3). Such partially digested DNA is size fractionated and cloned into the *Bam* HI site of vectors.

λEMBL 3 and λEMBL 4 vectors contain stuffer fragment flanked on both the ends by polylinkers comprising of *Eco* RI, *Bam* HI, and *Sal* I sites. The orientation of linker containing these three restriction enzyme sites is opposite in these two vectors (Figure 9.33). Any of these three restriction endonucleases can be used to remove the stuffer fragment, and replace it with DNA fragment generated by partial digestion of genomic DNA followed by size fractionation. These vectors are high capacity vectors, and permit cloning of large fragments of DNA ranging from 9 to 23 kbp. These vectors offer an advantage of isolation of insert DNA for subcloning purposes and prevention of vector reconstitution by digestion with the site lying in the stuffer fragment. Thus, if the middle *Bam* HI site is used for cloning DNA fragment generated by digestion of genomic DNA with *Sau* 3AI, the *Bam* HI site is destroyed in this reaction, but the foreign DNA may still be recovered by digesting with outermost restriction enzyme (*Sal* I in λEMBL 3, and *Eco* RI in λEMBL 4), and the inner most site (*Eco* RI in λEMBL 3, and *Sal* I in λEMBL 4) is used to cleave the stuffer fragment to prevent vector reconstitution. Recombinants are conveniently selected on the basis of size or Spi phenotype (for details see Section 9.5.6). This is because, similar to other replacement vectors, λEMBL vectors also carry the *red* and *gam* genes on the stuffer fragment and a *chi* site on one of the vector arms (for details see Section 9.5.6 and Exhibit 9.5). The recombinants follow lytic infection cycle and form clear plaques, while nonrecombinants are lysogenic. Further plating on *hfl*$^-$ host provides clear distinction between recombinants and nonrecombinants.

λEMBL 3A and λEMBL 4A vectors are similar to λEMBL 3 and λEMBL 4, respectively, but are biologically contained due to the presence of amber mutations in *A* and *B* genes. The cloning strategies in λEMBL 3A and λEMBL 4A are also similar to λEMBL 3 and λEMBL 4, respectively, however, screening of recombinants after cloning in these amber mutant vectors is done by plating on *sup*$^+$ P2 lysogenic hosts (i.e., amber suppressor host that is lysogenic for P2 phage) (for details see Section 9.5.6).

λgtWES.λB′ This vector, developed by Thomas *et al.*, is 40.4 kbp in size. This is an EK2 vector, and has the advantage of biological containment. This is because the amber mutations

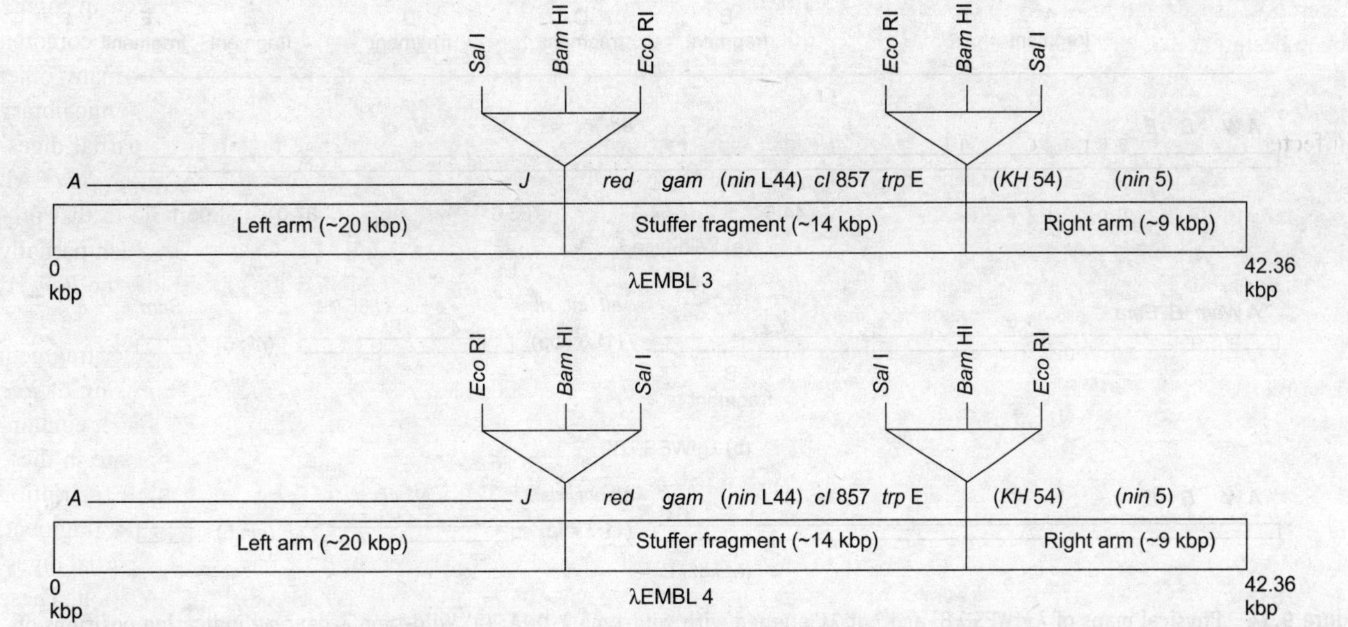

Figure 9.33 Vector maps of λEMBL series of vectors [The polylinker sequence is present in opposite orientation in the λEMBL 3 and λEMBL 4 vectors.]

in genes *S* (*Sam* 1000 mutation), *W*, and *E* reduce the likelihood of recombinants escaping from laboratory environment. Note that appropriate amber suppressor strains are very uncommon in nature. Moreover, as pS is required for host cell lysis, *S. amber* mutation renders the burst of mature phage particles more difficult. The vector is constructed by deleting the *Eco* RI fragment C lying between sites 2 and 3 in wild-type λ DNA, by restriction and religation *in vitro*. Consequently, the vector is *att*⁻ and *int*⁻, and hence loses the capability to lysogenize. The B′ fragment (i.e., an inadvertent inversion of B fragment during construction of the vector) is the replaceable *Eco* RI fragment, and it contains the only cleavage sites for *Sst* I within the phage genome. The background of parental phage to recombinant phage is therefore greatly reduced if the vector DNA is cleaved with *Eco* RI and *Sst* I before carrying out the ligation reaction. In addition, the two most right-hand *Eco* RI sites have been eliminated by *nin* 5 deletion. This deletion increases the capacity of vectors for the insertion of foreign DNA, and also removes the strong early terminator site t_{R2} thereby making the phage *N* independent. The *nin* deletion increases the synthesis of pQ immediately after infection. Since pQ is an important regulator of the late phase of infection, *nin* deletions force phages into the lytic growth cycle, and hence provide an additional safety factor (for details see Section 9.5.6). The vector contains unique *Eco* RI site for cloning (Figure 9.34). For cloning, the vector DNA is digested with *Eco* RI, and the B′ fragment is removed by preparative gel electrophoresis, or other physical methods. Alternatively, this fragment can be destroyed by treatment of *Eco* RI digest with *Sst* I. The length of DNA between arms is 4.85 kbp, and the insert capacity while cloning at *Eco* RI site is 2.2–15.1 kbp. The *Eco* RI treated foreign DNA is then added to a mixture of vector arms, the mixture is ligated, and used to transfect an appropriate amber suppressor strain of *E. coli* (e.g., *sup* F suppressor *E. coli* host DP50 sup F) so that viable recombinant phages are recovered. Note that the nonrecombinants formed by joining of the two DNA arms without insertion of foreign DNA are too short [9.8% (B fragment) + 11.3% (C fragment) + 6.1% (*nin* deletion) = 27.2% less than λ] to produce viable phages even though these contains all of the genes necessary for lytic growth.

λgt.λC This vector was also developed by Thomas and coworkers, but unlike λgtWES.λB′, it does not contain amber mutations in *W*, *E*, and *S* genes, and the replacement fragment lies between *Eco* RI sites 2 and 3 (fragment C) (Figure 9.34). This difference, however, changes the properties a lot. As the fragment C contains *att*, *int*, and *xis*, this gives the phage the capability to form stable lysogens, but when foreign DNA replaces fragment C, the phage becomes integration defective. Thus, nonrecombinants are lysogenic in nature, while recombinants follow lytic infection cycle. The recombinants (*red*⁻ *gam*⁻) are positively selected on the basis of size or Spi phenotype (for details see Section 9.5.6).

λGEM 11 and λGEM 12 The λ replacement vectors belonging to λGEM series, λGEM 11 and λGEM 12, are high insert capacity vectors. These have a cloning capacity of 23 kbp (near the theoretical maximum), and the insert is cloned into these vectors by replacing the stuffer fragment, which is flanked by polylinkers containing seven different restriction sites. The

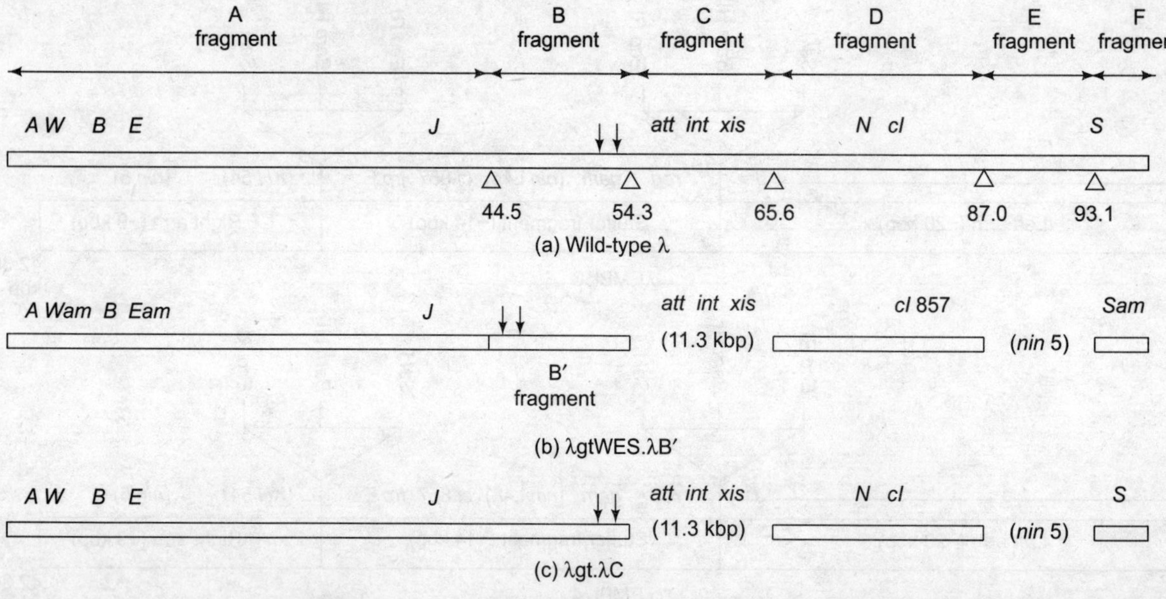

Figure 9.34 Physical maps of λgtWES.λB′ and λgt.λC aligned with wild-type λ DNA; (a) Wild-type λ genome indicating positions of *Eco* RI and *Sst* I sites, (b) λgtWES.λB′ (regions aligned with λ genome), (c) λgt.λC (regions aligned with λ genome) [Parenthesis indicates deletions. Downward arrows are *Sst* I sites. Upward triangles are *Eco* RI sites. Numbers under *Eco* RI sites indicate the positions of the sites as percentages of wild-type genome length.]

polylinkers are slightly different in the two vectors, but in both cases the outermost sites on either side are for *Sfi* I, which recognizes the 13 bp sequence 5′-GGCCNNN-NNGGCC-3′. This sequence is very rare and unlikely to be present in the cloned DNA fragment. Restriction of the recombinant vector with *Sfi* I can therefore be used to cut out the cloned fragment, with a high likelihood of recovering the fragment in intact form. Moreover, as discussed later in this section, the risk of religation of stuffer fragment can be prevented by partial filling-in of the *Xho* I site. These vectors carry the *red* and *gam* genes on the stuffer fragment, and a χ site on one of the vector arms (not shown in vector map) (Figure 9.35). This allows selection of the recombinants on the basis of size and Spi phenotype (for details see Section 9.5.6).

λDASH and λFIX λDASH, λFIX, and their derivatives are particularly versatile because the multiple cloning sites flanking the stuffer fragment contain opposed promoters for the T3 and T7 RNA polymerases. Vector maps of λDASH and λFIX are shown in Figure 9.36. These vectors also have χ site in one of their arms (not shown in vector map). If the recombinant vector is digested with a restriction endonuclease that cuts frequently, only short fragments of insert DNA are left attached to these promoters. This allows RNA probes to be generated that correspond to each end of the insert. Such probes are used to identify overlapping clones, and have the great advantage that these can be made conveniently, directly from the vector, without recourse to subcloning. λFIX differs from λDASH in containing an additional *Xho* I sites flanking the stuffer fragment, which as discussed below in this section, may be filled-in partially to prevent vector religation. The partially filled-in *Sau* 3AI fragments are then ligated to the partially filled-in *Xho* I ends in vector DNA. This strategy prevents the ligation of vector arms without genomic DNA, and also prevents the insertion of multiple fragments (Figure 9.37). Because of their high insert capacity ranging up to 23 kbp, both these vectors are used for

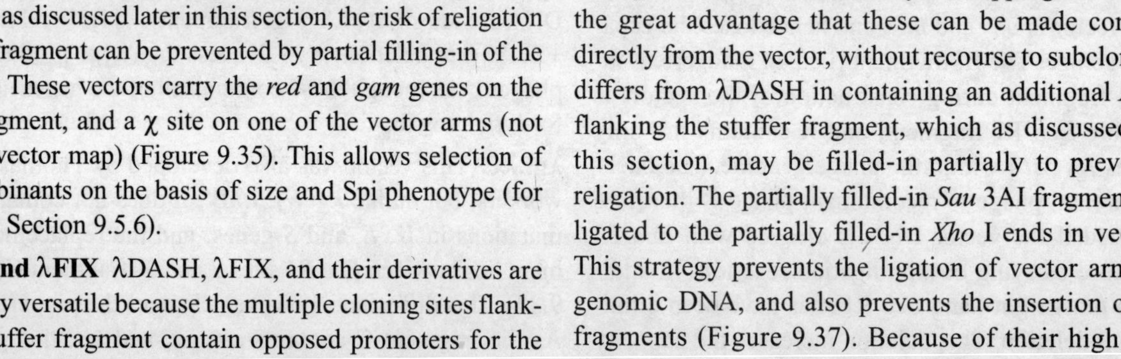

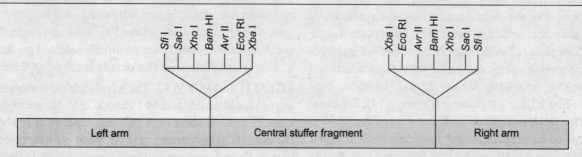

Figure 9.35 Vector map of λ replacement vector λGEM 11

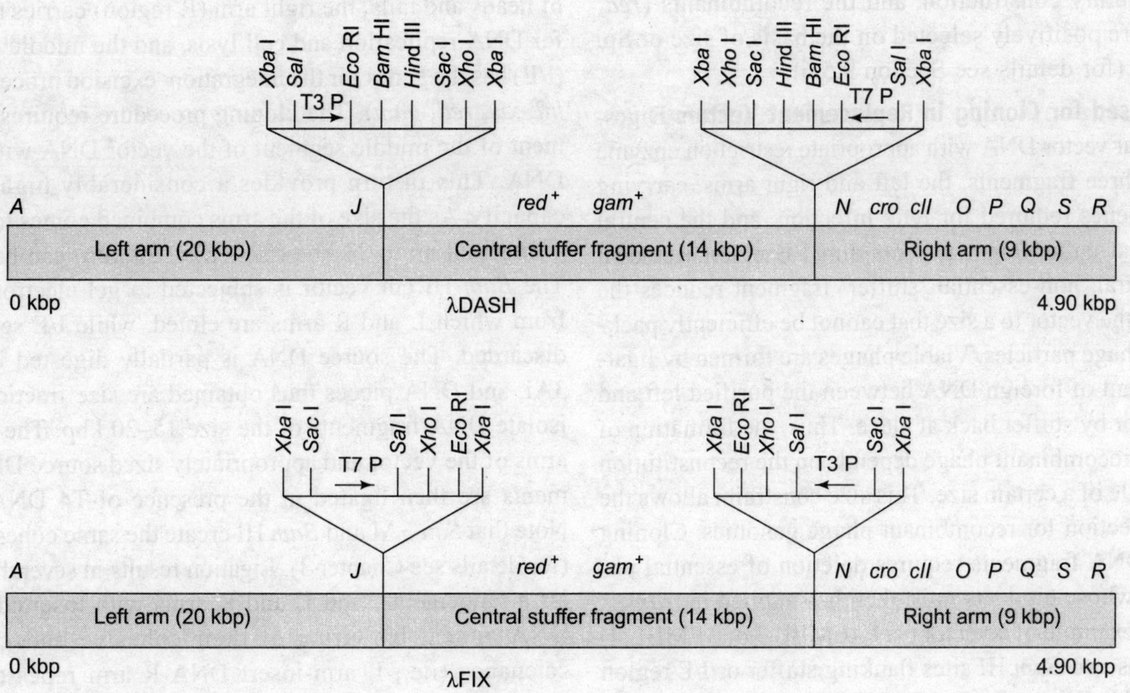

Figure 9.36 Vector maps of λ replacement vectors λDASH and λFIX

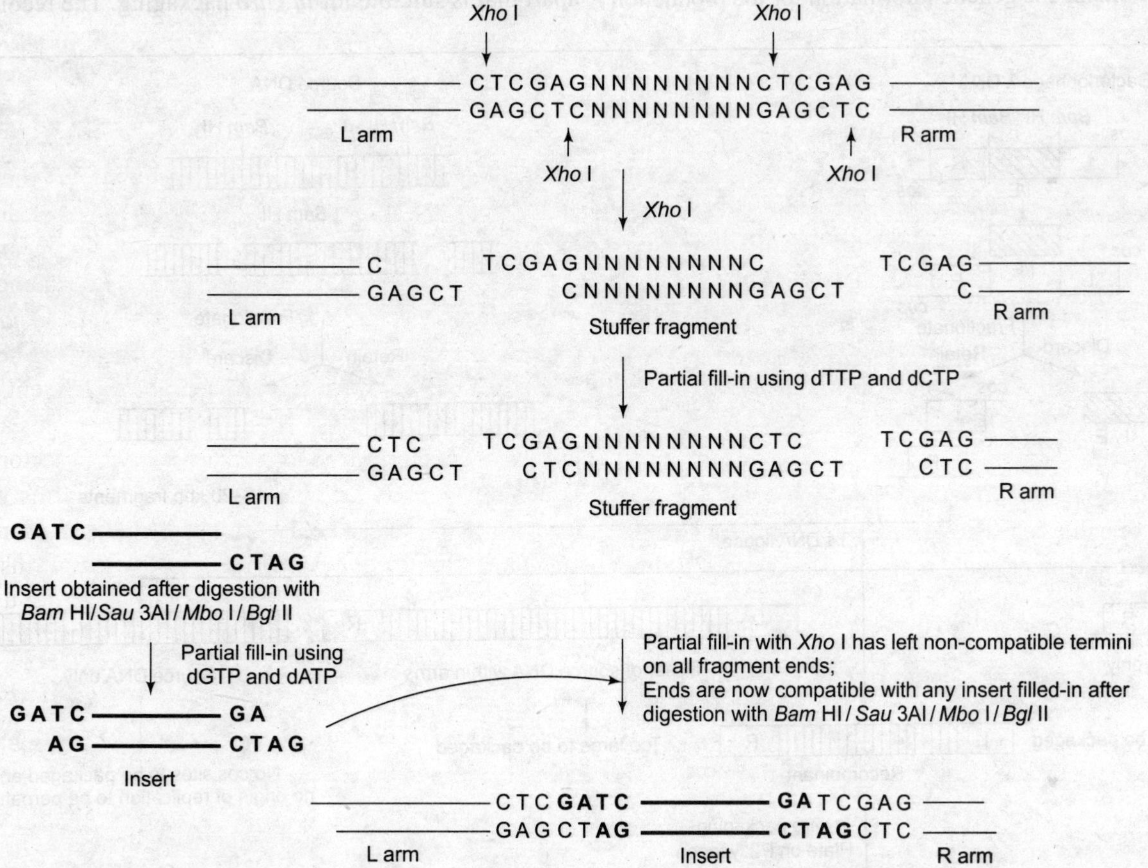

Figure 9.37 Partial filling-in of *Xho* I site in case of λFIX replacement vector prevents religation of stuffer fragment [Only the partially filled-in *Bam* HI/*Sau* 3A/*Mbo* I/*Bgl* II insert DNA can now be ligated within the vector arms.]

genomic library construction, and the recombinants (*red*$^-$ *gam*$^-$ χ^+) are positively selected on the basis of size or Spi phenotype (for details see Section 9.5.6).

Strategy Used for Cloning in Replacement Vectors Digestion of linear vector DNA with appropriate restriction enzyme generates three fragments, the left and right arms, carrying all of the genes required for lytic infection, and the central non-essential stuffer fragment containing I/E region. Removal of this central, non-essential 'stuffer' fragment reduces the genome of the vector to a size that cannot be efficiently packaged into phage particles. Viable phages are formed by ligating a segment of foreign DNA between the purified left and right arms or by stuffer back at place. Thus, the formation of plaques by recombinant phage depends on the reconstitution of a molecule of a certain size. This size constraint allows the positive selection for recombinant phage genomes. Cloning of longer DNA fragments requires deletion of essential phage genes, whose products must then be supplied in *trans*.

Take an example of a vector (say λEMBL 3 or λEMBL 4) that contains two *Bam* HI sites flanking stuffer or I/E region (Figure 9.38). When purified DNA from this phage is cut with *Bam* HI, three DNA segments are created. The left arm (L region) contains the genetic information for the production of heads and tails, the right arm (R region) carries the genes for DNA replication and cell lysis, and the middle fragment (I/E) has the genes for the integration–excision processes (i.e., *int*, *xis*, *red*, *gam*). The cloning procedure requires replacement of the middle segment of the vector DNA with cloned DNA. This in turn provides a considerably high cloning capacity. As the size of the arms combined comes to 29 kbp, a fragment up to 23 kbp (52 – 29 = 23 kbp) can be cloned. The *Bam* HI cut vector is subjected to gel electrophoresis, from which L and R arms are eluted, while I/E segment is discarded. The source DNA is partially digested with *Sau* 3AI, and DNA pieces thus obtained are size fractionated to isolate DNA fragments of the size 15–20 kbp. The L and R arms of the vector and appropriately sized source DNA fragments are then ligated in the presence of T4 DNA ligase. Note that *Sau* 3AI and *Bam* HI create the same cohesive ends (for details see Chapter 3). Ligation results in several molecular arrangements, and L and R arms with inserted foreign DNA anneal by virtue of their cohesive ends forming catenanes (i.e., L arm-insert DNA-R arm repeated many times). If the insert DNA is of the correct size then the *cos* sites separating these structures are placed at the right distance apart that is suitable for *in vitro* packaging. The recombinant

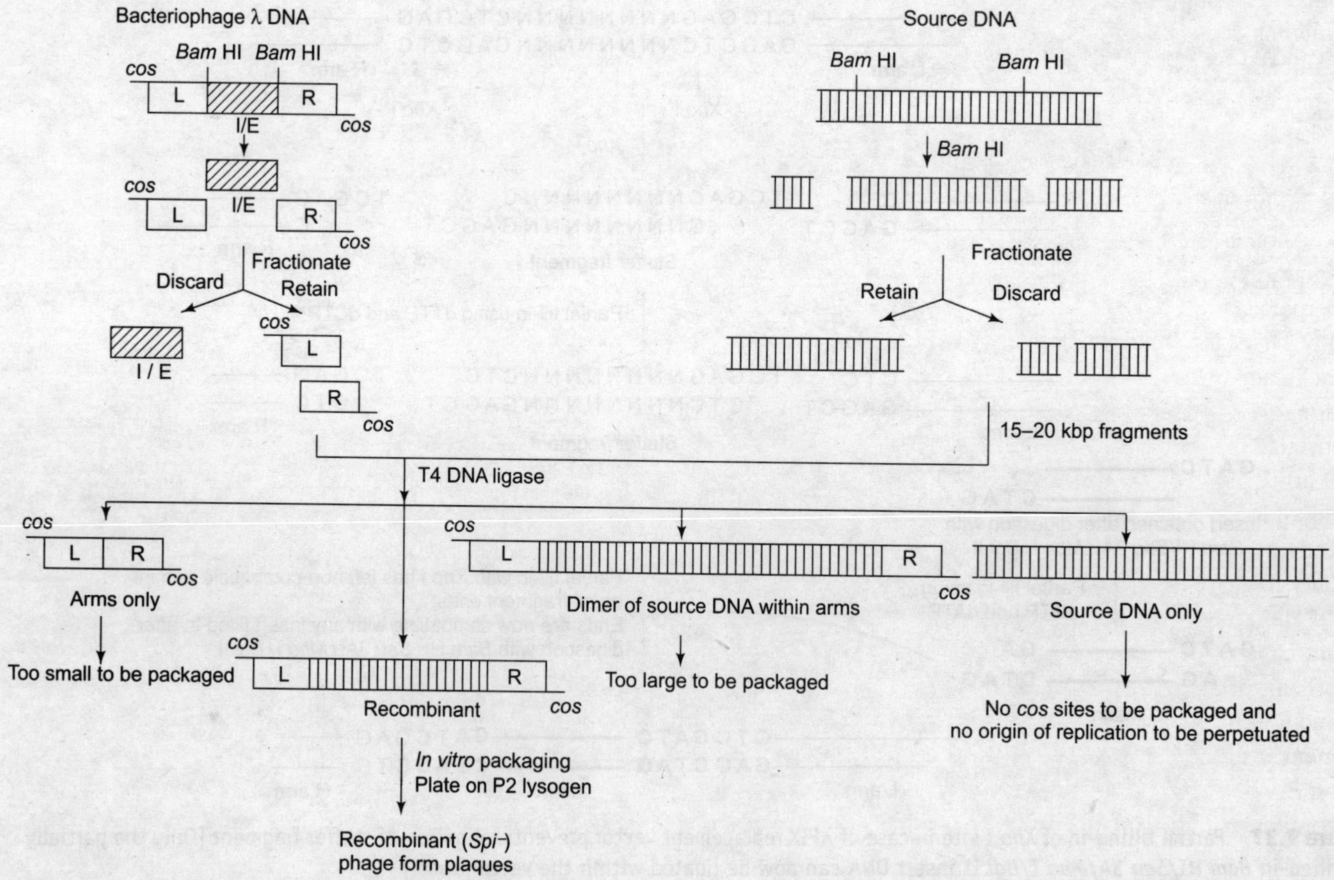

Figure 9.38 Cloning system in replacement vector

DNA (37–52 kbp) is then packaged *in vitro* by adding packaging extract comprising of empty λ phage head, tail, and assembly proteins, and result in the formation of infective phage particle. The DNA molecules that can or cannot be packaged are described below.

Ligation product of L and R arms without stuffer or foreign DNA cannot be packaged As the combined size of L and R arms is only 29 kbp, which is less than the minimum required for packaging, any pair of arms that are ligated without an insert or stuffer itself is therefore too small (<37 kbp) to produce viable phage particle.

Ligation product where foreign DNA is inserted in dimeric form between L and R arms cannot be packaged Other products from the ligation reaction, such as recombinants with insert dimers are very large [20 kbp L arm + 9 kbp R arm + 2 × (~20 kbp as obtained by size fractionation) insert dimer = >52 kbp], and cannot be packaged.

Recombinant DNA comprising of single foreign DNA fragment inserted between L and R arms can be packaged Viable particles are produced if ligation results in an insert of at least 8 kbp (37 – 29 kbp = 8 kbp). The vector thus provides a positive selection on basis of size for recombinants that contain an insert of at least 8 kbp.

Reconstituted vector without foreign DNA can be packaged Ideally there should not be any contamination of stuffer fragment, but practically there may be some stuffer DNA contaminating the preparation of the vector arms. Thus the reconstituted vector, where stuffer fragment is inserted between L and R arms, may contaminate the recombinants during cloning experiment. This is because their size lies within the packaging size range.

Another important finding is that the packaging of some ligation products results in assembly of phages that cannot be perpetuated, and hence do not form plaques. These include any DNA molecule ranging in size from 37 to 52 kbp that lacks any of the following: (i) Functional origin of replication, *ori* λ (i.e., those recombinants containing foreign DNA inserted between two L arms, or in other words those lacking R arm); and (ii) The *cos* ends (i.e., ligation product comprising of L and R arms joined by their *cos* ends instead of sticky ends generated by restriction digestion or those containing only ligated inserts).

The above findings indicate that the phage particles that are capable of forming plaques include recombinants comprising of single copy of foreign DNA inserted between L and R arms, and nonrecombinants comprising of stuffer fragment inserted between L and R arms. However, by plating on *E. coli* P2 lysogen strain, the contamination from phages containing reconstituted vector DNA can be prevented (for details see Section 9.5.6). Replacement vectors thus provide a positive selection for recombinants as opposed to parental phage, and the gene library remains free of nonrecombinant phages. Moreover, during cloning experiments, the following strategies play crucial role in either avoiding reconstitution of vector or formation of insert dimers, and a combination of these is used to increase efficiency of cloning.

Size fractionation of source DNA fragments The source DNA, after partial digestion with a suitable restriction enzyme is size fractionated by gel filtration chromatography, and the fragments of desired size are collected for ligation into vector. As there is a packaging constraint while cloning in λ phage, (i.e., the DNA of the size >52 kbp and <37 kbp is not packaged, or if packaged does not form infectious phage particle), this size fractionation step prevents chances of cloning of insert dimers, and allows selection of recombinants on the basis of size.

Fractionation of digested vector by agarose gel electrophoresis The vector DNA after digestion with suitable restriction enzyme is subjected to agarose gel electrophoresis. From the gel, L and R arms are eluted out and used in cloning experiment, while the stuffer fragment is discarded. (Note that in most of the replacement vectors, the size of L arm is 20 kbp, R arm is 9 kbp, and stuffer fragment is 14 kbp, and hence these fragments form separate bands during electrophoresis). This prevents contamination with stuffer fragment during cloning.

Alkaline phosphatase treatment Alkaline phosphatase treatment of the vector is done to prevent recircularization (Chapter 2), and hence to ensure formation of recombinant progeny rather than reconstituted parental vector molecules. Moreover, both vector and foreign DNAs cannot be dephosphorylated in a same experiment. This is because DNA ligase requires that one of the two DNA molecules should contain 5′-P, and the other should have 3′-OH end for ligation. However, in case of replacement vectors, reconstituted vector is prevented from forming plaques by plating on P2 lysogen, and hence phosphatase treatment of the vector is not necessary. Thus, here by dephosphorylating the insert DNA by alkaline phosphatase treatment, insert–insert ligation can be prevented. This is necessary because in the construction of a gene library, the insertion of more than one fragment into the same vector molecule can give rise to anomalies in characterizing the insert in relation to the source genome.

Cleavage of stuffer fragment with internal restriction sites Often the replaceable stuffer fragment carries additional restriction sites that can be used to cleave it into small pieces, so that its own reinsertion during a cloning experiment is very unlikely. For example, the vector λEMBL 3 is digested with *Bam* HI and *Eco* RI prior to ligation with foreign DNA fragments produced with *Bam* HI. This allows cleavage of stuffer fragment forming small *Bam* HI–*Eco* RI fragments, which if removed, prevent reincorporation of stuffer fragment.

Partial filling-in of sticky overhangs of vector and foreign DNA molecules followed by ligation The vectors having *Xho* I sites flanking the stuffer fragment are first digested with *Xho* I followed by partial filling-in of the sticky ends. This partial filling-in prevents vector religation, as the two ends no longer remain complementary. These are then ligated to partially filled-in *Sau* 3AI sticky ends of source DNA. Though, *Xho* I and *Sau* 3AI produce fragments with incompatible ends, their partial filling-in makes these fragments compatible, which can then be ligated with one another. As we have seen, such strategy is adopted in replacement vectors λGEM 11 and λFIX. Figure 9.37 clearly depicts this strategy for λFIX.

It is interesting to note that if the above strategies are not performed, even then the packaged nonrecombinants can easily be distinguished from other viable phages with recombinant DNA. This is achieved by Spi phenotype selection by plating on P2 lysogen. The reconstituted vectors are unable to grow on P2 lysogens, whereas the recombinant phages outgrow the parental vector (for details see Section 9.5.6). It therefore seems unnecessary to follow the mentioned steps to minimize the number of nonrecombinants formed when constructing genomic DNA libraries. In practice, however, removal of the stuffer fragment has been found to improve the efficiency with which fragments of foreign DNA can be cloned and propagated in vectors of this type. Besides, the placing of relevant genes onto the stuffer fragment may also be beneficial, as their loss gives rise to a detectable phenotypic signal. This provides an immediate check for any contamination of stuffer fragment. For example, when *lac* Z′ gene is inserted within the stuffer fragment, the reconstituted vectors containing stuffer fragment form blue plaques, while recombinants lacking stuffer fragment form white plaques.

9.5.5 λ Bacteriophage as Expression Vector

Beginning in the mid 1980s and continuing to the present day, several phage expression vectors have been developed that permit not only the propagation of foreign sequences, but also their expression in bacterial cells, their transcription *in vitro* into RNA, and in some cases, their automatic recovery as autonomously replicating phagemids. The prototype λ expression vector is λgt 11, which carries a portion of the *E. coli* β-galactosidase gene, including the upstream elements that are essential for its expression. Other examples include λgt 18–23, λDASH, and λFIX, etc. The details of these vectors have already been discussed in Section 9.5.4.

9.5.6 Significance of Mutant Host Strains and Vectors (Genetic Markers)

While cloning in λ vectors, a combination of vectors with mutations in certain genes, and mutant host strains is used for either growth or screening of recombinants from nonrecombinants. The significance of such mutants (genetic markers) during cloning is discussed in this section.

P2 lysogen and Spi Phenotype

P2 lysogen (host infected with P2 lysogenic phage) and Spi phenotype are two terms related to selection of recombinants generated by cloning in λ replacement vectors.

The Spi selection takes advantage of the fact that growth of wild-type λ phage is restricted in P2 lysogens, while mutant phages deficient in *gam* and *red* (both *red* B and *red* A) genes can grow well in P2 lysogens. Simultaneous inactivation of both *gam* and *red* genes is induced by deletions in the region, and such double deletion mutants are called λ *red*⁻ *gam*⁻ or *Fec*⁻ or *Feckless*. Wild-type λ phages form no plaques, i.e., these are either *Spi*⁺, or form small plaques and are partially *Spi*⁻. On the contrary, mutant phages form normal sized plaques, i.e., these are *Spi*⁻ (insensitive to P2 interference). Thus, Spi selection is used to detect deletion mutations of λ phage generated in *E. coli* cells. This principle, when extended to replacement vectors, allows distinction between recombinants and nonrecombinants. As replacement vectors have been designed to replace the central stuffer fragment carrying the *red* and *gam* genes of λ phage by the gene of interest, cloning in such replacement vectors allows replacement of phage *red* and *gam* genes with foreign DNA. In such cases, the recombinant phage genomes are selected directly from the products of a ligation mixture by plating on a P2 lysogen. The loss of the *red* and *gam* genes confers the *Spi*⁻ phenotype upon the phage, so enabling it to form plaques on P2 lysogens. On the other hand, nonrecombinants are unable to form plaques on P2 lysogen, and like wild-type λ phage, these are *Spi*⁺. This suggests that the screening of all replacement vectors is easily done on the basis of Spi phenotype. As the reconstituted vectors are unable to grow on P2 lysogens, Spi phenotype selection offered by replacement vectors also removes the necessity for purifying the phage arms from the central stuffer fragment before ligating the foreign DNA.

As will be evident below, the screening of recombinants by Spi phenotype also requires the vectors to be χ^+, and the host to be P2 lysogen as well as *rec* A⁺ or *rec* BCD⁻.

Rec A⁺ Hosts and χ^+ Vector

The *gam*⁻ phages (i.e., recombinants) cannot switch over to rolling circle replication, and hence concatemers are not formed. As discussed in Section 9.2.7, in such phages, the only substrate DNA that can be packaged is a dimeric or multimeric circular form. But as the phage is also *red*⁻, the only way of recombination is by using *rec* A⁺ host encoding recombination gene product, Rec A. However, the presence of an active bacterial recombination system can sometimes lead to instability in sequences cloned in λ phage vectors, particularly in genomic sequences that contain repetitive elements. This problem can be avoided in the following three ways: (i) A *gam*⁻ phage can multiply if DNA molecules exist as circular dimers formed by recombination catalyzed by the products of the λ *red* gene or the host *rec* A gene. However, the presence of an active bacterial recombination system occasionally leads to rearrangements (in particular to deletion) of segments of foreign DNA that carry repeated sequences. To avoid this problem, several vectors have been designed that carry the *gam* gene on one of the vector arms. Examples of such vectors are Charon 32–35 and 40. The use of such vectors has in some instances overcome the problem of cloning segments of foreign DNA containing recombinogenic sequences; (ii) Gam protein can be supplied in *trans* from a plasmid. In this system, the expression of *gam* is controlled by pQ of the incoming phage. Inactivation of exonuclease V can therefore occur only after infection; (iii) A number of mutant strains of *E. coli* are available that are recombination-proficient but deficient in exonuclease V. These include strains that are defective in *rec* B or *rec* C, as well as strains carrying the mutations *sbc* A or *sbc* B, both of which suppress mutations in *rec* BC; and (iv) λ DNA is a poor substrate for *rec*-mediated exchange. Thus, red^- gam^- phages (i.e., recombinants) make vanishingly small plaques when plated on *rec* A^+ host, because synthesis of catenated DNA is impeded by exonuclease V. Hence, for propagation of red^- gam^- recombinants in *rec* A^+ host, vectors are designed to contain a χ site (*chi* site), an octameric sequence that stimulates Rec A mediated exchange or recombination, on one of its non-replaceable arms (Exhibit 9.6). Such red^- phages carrying χ site produce plaques that are close to normal in size. The presence of the χ sequence in the vector DNA, and an active *rec* BCD system provided by the host leads to an increase in the efficiency of recombination events that generate closed circular dimers and multimers from θ forms. The red^- gam^- phages package these multimers efficiently, and red^- gam^- χ^+ phages grow to a reasonable yield in wild-type *E. coli*. Plaque phenotypes of wild-type and red^- gam^- λ phages are presented in Table 9.5.

Table 9.5 Plaque Phenotypes of Wild-type and red^- gam^- λ Phages

Phage	Host	Plaque phenotype
λ red^- gam^-	*rec* A^-	–
	rec A^-	
	rec BC^-	+
	P2 *lysogen*	*Minute*
λ red^- gam^- χ	*rec* A^-	–
	rec A^-	
	rec BC^-	+
	P2 *lysogen*	*Small*
λ^+	*rec* A^-	+
	rec A^-	
	rec BC^-	+
	P2 *lysogen*	–

(–) indicates no plaque formation; (+) indicates normal size plaque formation.

hfl A^- Host

Some bacterial host strains carrying a mutation known as *hfl* (high frequency of lysogenization) produce a much higher proportion of lysogens when infected with wild-type λ, which can be useful if a more stably altered host strain is required, for example, if the aim is to study the expression of genes carried by the phage. This is because mutations in the *E. coli* *hfl* locus result in a high concentration of pCII in the absence of Hfl protein, and the phages are quantitatively forced into the lysogenic pathway. Cell lysis and plaque formation in *hfl* A^- strains is only observed when the λ repressor is inactivated, for example, by temperature sensitive mutation or the insertion of recombinant DNA. Thus, if cloning site lies within the *cI* gene, by plating on *hfl* A^- host, recombinants can be easily distinguished from nonrecombinants, for example, λgt 10 (for details see Section 9.5.4). This is because nonrecombinants or parental insertion vectors, when plated on *hfl* A^- strain, do not undergo the lytic infection cycle, and hence no plaque formation. Recombinants in which the *cI* gene is inactivated, on the other hand, form plaques.

sup^+ Host

The phages belonging to EK2 series are amber mutant phages (e.g., λgt II, Charon 4, λEMBL 3A, λEMBL 4A, and λgtWES.λB′) basically designed for biological containment (for details see Section 9.5.4). Such phages require amber mutation suppressive hosts for growth, i.e., *sup* F suppressor host. The amber suppressor mutation codes for suppressor tRNA that reads amber stop codon in phage gene as amino acid, leading to synthesis of functional protein corresponding to the mutated gene. The most suitable *sup* F laboratory strain is *E. coli* DP50 sup F. Note such suppressor positive (sup^+) host strains do not occur in the natural environment. Moreover, as the growth of these sup^+ strains is absolutely dependent upon the presence of diaminopimelic acid and thymidine in the culture medium, and as diaminopimelic acid is not found in the mammalian gastrointestinal tract, it is highly unlikely that such mutant phage would be able to grow in the wild-type *E. coli* that populates the gastrointestinal tract. Moreover, the strains are also sensitive to bile salts and detergents, and are quite difficult to grow even in the laboratory. The potential risk from the accidental ingestion of recombinant DNA molecules is therefore very much reduced.

lon^- host

ATP dependent Lon protease degrades specific short-lived regulatory proteins as well as defective and abnormal proteins in

Exhibit 9.6 Lon Protease

Rapid proteolysis plays a major role in post-translational cellular control by the targeted degradation of short-lived regulatory proteins, and also serves an important function in protein quality control by eliminating defective and potentially damaging proteins from the cell. In all cells, protein degradation is predominantly carried out by ATP dependent proteases, which are complex enzymes containing both ATPase and proteolytic activities expressed as separate domains within a single polypeptide chain or as individual subunits in complex assemblies. Five ATP dependent proteases, Lon, FtsH (Hfl), ClpAP, ClpXP, and HslUV, have been discovered in *E. coli,* and homologous proteases have been found in all eubacteria and in many eukaryotes. Alone among them, Lon protease has been found in virtually all the living organisms, from Archaea to eubacteria to humans.

Lon protease (*E. coli*) is the first ATP dependent protease to be identified. It is an oligomeric multidomain enzyme whose single polypeptide chain is composed of 784 amino acids. Comparison of the amino acid sequences of various members of the Lon family suggests that Lon consists of three functional domains: a variable N-terminal domain, an ATPase domain, and a C-terminal proteolytic domain. The domain organization has been confirmed by expression of the functional domains of the *E. coli*, the yeast mitochondrial Lon, and by the limited proteolysis of the *E. coli* and *Mycobacterium* Lons. Despite extensive studies of this enzyme, many of its structural characteristics remain undetermined, although its function in selective energy-dependent proteolysis has been characterized in considerable detail. However, the structure confirms the presence of a Ser-Lys catalytic dyad in the active site, and reveals a unique structural fold distinct from both the classical serine proteases containing active site catalytic triads, and from other hydrolytic enzymes that are utilizing Ser-Lys catalytic dyads. The catalytic domain of Lon in the crystal assembles into hexameric rings, which provides strong support for a hexameric structure of the holoenzyme.

Like other ATP dependent proteases, the Lon ATPase domain belongs to the superfamily of AAA^+ proteins (ATPases associated with a variety of cellular activities). The characteristic AAA^+ domain consists of 220–250 amino acids that include hallmark Walker A and B motifs, and several other regions of high sequence conservation. An AAA^+ module has two structural domains: a Rec A-like α/β domain, and an α domain that may interact with protein substrates. In response to ATP hydrolysis, these domains undergo changes in conformation and orientation. Transduction of the mechanical motions within the AAA^+ module to bound substrates provides the driving force for various functions, including binding and unfolding of target proteins, translocation of proteins to an associated functional domain (in this case, the protease), and coordinated activation of the functional domain.

Source: Botos I *et al.* (2004) The catalytic domain of *E. coli L* on protease Ras a unique fold and a Ser-Lys dyad in the active site. *Journal of Biological Chemistry* **279**(9):8140–8148.

the cell. The protein synthesized using expression vectors (e.g., λgt 11, where fusion protein is synthesized) is thus susceptible to cleavage by host Lon protease (for details see Section 9.5.4). Hence in such cases, recombinants are plated on the hosts defective in protein degradation pathway (i.e., *lon*⁻), so as to prevent degradation of the protein of interest. The structure and functions of Lon protease are highlighted in Exhibit 9.6.

cI Gene With Temperature Sensitive Mutation (cIts or cIts 857) in λ Vectors

Generally, it is preferred that the recombinant phages carrying the cloned genes follow lytic cycle. This is because lytic phages form easily visible clear plaques, give high yields of cloned DNA or protein encoded by cloned gene. On the other hand, lytic infection cycle is not preferable due to the risk of spread of phages, i.e., lytic phages are not biologically contained. Thus, the vector should be designed in such a way that the phage follows lysogeny, and when required, lysis is induced. Induction is normally very difficult to control, but most laboratory strains of λ carry a temperature sensitive (*ts*) mutation in the *cI* gene. Recall that *cI* is one of the genes that are responsible for maintaining the phage in the integrated state. If inactivated by a mutation, the *cI* gene no longer functions correctly, and the switch to lysis occurs (for details see Section 9.2.9).

Several widely used λ vectors carry *cIts* mutation. A good example is insertion vector, λgt 11 (for details see Section 9.5.4). Bacteriophages carrying *cIts* 857 gene are able to establish and maintain lysogeny state as long as the cells are propagated at low temperature (30°C). At this temperature, the repressor retains its ability to inhibit transcription from P_L and P_R. However, heating the bacterial culture for 10–15 min at 45°C can induce lytic growth. This is because at such a high temperature partial inactivation of λ repressor occurs (Figure 9.39). Although, λ repressor is capable of renaturation upon cooling to 37°C, its concentration is too low to allow reestablishment of lysogeny. Instead transcription resumes from P_L and P_R when ~80% of λ repressor has been inactivated. At this stage only ~40% monomers of repressors remain in the cell, too small a number to sustain cooperative binding to O_{R1} and O_{R2}. In addition, because pCI stimulates transcription of its own gene, synthesis of new repressor molecules decreases as the concentration of pCI falls. The first protein synthesized when transcription resumes from P_R is pCro, which binds to O_{R3}, blocking any further synthesis of repressor, and therefore committing the phage genome to a cycle of vegetative growth. The integration and excision proteins act together to cause the *att* sites to recombine, resulting in excision of the prophage DNA from the host chromosome. The lytic cycle then follows its usual course. A culture of *E. coli* infected with λ *cIts* can therefore be induced

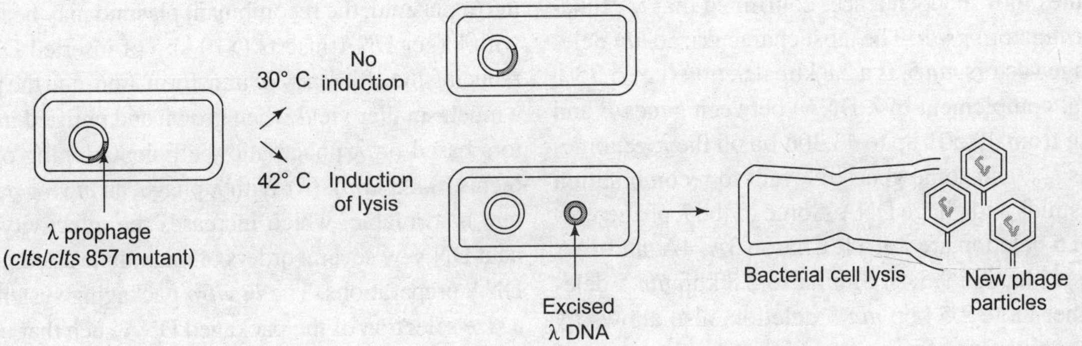

Figure 9.39 Temperature-regulated induction of lysis mediated by temperature sensitive mutant repressor, *cIts* or *cIts* 857

to produce extracellular phage by transferring from 30°C to 42°C. This is called thermoinduction of lysogen.

S Amber Mutant Vectors

For biological containment of recombinant phages carrying the cloned genes, it is preferred that the vector follows lysogeny till required, and then can be easily induced to lysis. One way is the application of *cIts* mutation. Another strategy is to block the stage of morphogenesis by mutating the genes encoding head, tail, and assembly proteins, for example, in λEMBL 3A and λEMBL 4A (for details see Section 9.5.4). A yet another strategy, which is also used commonly, is to block or delay lysis by amber mutations in the genes responsible for cell lysis. Thus, as lysis of the bacterial cell is accomplished largely through the action of λ-encoded pS, the product of gene *S amber* mutation in this gene can cause a delay or failure of lysis. This is also advantageous in increasing the yield of phages, as late transcription is drastically reduced, and the replication of the phage continues for a longer time. This allows the assembly of progeny particles to continue for an extended period of time, instead of being interrupted by lysis of the host cell. This also allows a high yield of phage DNA, and retention of all determined polypeptides within the cell. The accumulated intracellular phage particles can later be liberated by plating on amber suppressor host (sup^+) or by artificially lysing the infected cell with chloroform. The gut bacteria are not naturally sup^+; hence the potential risk from the accidental ingestion of recombinant DNA molecules is also reduced. *S amber* mutants can also be applied to selection procedures, such as recombinational selection, or tagging DNA with a sup^+ gene. Similarly, lysis is abolished in *R* mutants, but the infected cells die at the usual time of lysis. Likewise, in R_Z mutants, lysis in liquid culture is unaffected unless 5–10 mM divalent cations are present, in which case, spheroplasts form at the normal time for lysis.

nin (N independent) Mutant Vectors

As discussed in Section 9.2.9, temporal regulation of gene expression in λ phage is done by transcription termination and antitermination systems. For maximal expression of delayed early genes, RNA polymerase modification is required, which is effected through *nut* L and *nut* R sites located downstream of the early P_L and P_R promoters, respectively. About 50% of the P_R-initiated rightward transcription continues through t_{R1} but terminates in the *nin* region without reaching gene *Q*. Note that *nin* region is 3 kbp in length located between genes *P* and *Q*, and contains rho dependent transcription termination signals as well as 10 open reading frames, some having identified function, but none being essential for phage growth under laboratory conditions. Termination in the *nin* region blocks expression of the *Q* gene, the product of which is a second transcription antitermination function required for effective levels of late gene expression. In the presence of the *N* antitermination complex that includes the phage encoded pN, phage RNA site *nut,* and several host Nus factors, the RNA polymerase is modified into a transcription termination-resistant form. The modified polymerase is able to pass through t_{R1} and the terminators in the *nin* region, and extends into the *Q* gene.

From the above facts, it is concluded that *nin* deletions, which remove the *nin* termination signals, obviate the need for pN mediated antitermination of transcription thereby allowing sufficient P_R read-through into *Q*, and hence leading to synthesis of pQ and late genes. Thus, deletions in the *nin* region frees phage growth from dependence on the *N* transcription antitermination system, conferring a *nin* phenotype. Such phages are known as *nin* (*N* independent) mutants. Simply, *nin* deletions increase the synthesis of the pQ immediately after infection (i.e., '*N* independent' activation of gene *Q*). Since pQ is an important regulator of the late phase of infection, *nin* deletions force phages into the lytic growth cycle, and hence provide an additional safety factor. A N^- phage can produce a small plaque if the termination site t_{R2} is removed by a small deletion termed *nin* as in λ N^- *nin*. Mutants of λ phage that carry a deletion of t_{R2} can grow, albeit poorly. The *N*-independence conferred by *nin* deletions suggests that the 50% termination occurring at t_{R1} in the absence of pN is not sufficient to prevent lytic growth. Subsequently, the identification of a mutation in the t_{R2} sequence that contributes to N independence, confirmed that t_{R2} functions as a terminator *in situ*. The most characterized *nin* deletion of λ phage vectors *nin* 5, is a 2.8 kbp deletion (i.e., 5.75% of the normal complement of λ DNA) between genes *P* and *Q* (extending from 40,501 bp to 43,306 bp on the λ genome) that removes t_{R2}, and some genes relevant to recombination between plasmids and phage DNA. Some of the λ phage vectors with *nin* 5 deletion are λgt 11, Charon 3A, 4A, and 16A (for details see Section 9.5.4). Similar to 2.8 kbp *nin* 5 deletions, the other large 2.5 kbp *nin* 3 deletion, also allows '*N* independent' activation of gene *Q*, and thereby the genes of the late transcriptional pathway. Such deletions allow the growth of λ derivatives defective in *N* antitermination.

A subregion of *nin*, namely *roc*, is defined by a 1.9 kbp deletion (λ*roc*), which partially frees λ growth from the requirement for *N* antitermination. The *roc* region has strong rho dependent transcription termination activity as assayed by a plasmid-based terminator testing system. Unlike *nin* 5 deletion, the *roc* deletion leaves the t_{R2} terminator intact and does not render λ completely *N* independent.

9.5.7 Advantages of λ Vectors

λ phage based vectors extend the cloning capacity over that readily obtainable with plasmid vectors, and can easily generate very large number of recombinants that are required for a gene library. Even today, these are considered to be the most suitable cloning vehicles for cloning genomic eukaryotic DNA, because these have a number of advantages over plasmids. These include: (i) Plasmid vectors are best for cloning relatively small fragments of DNA. Although there is probably no fixed limit to the size of a DNA fragment that can be inserted into a plasmid, the recombinant plasmid may become unstable with larger DNA inserts (>10 kbp of inserted DNA), thereby reducing the efficiency of transformation, and the plasmid gives a much smaller yield when grown and purified in *E. coli*. Vectors based on λ phage allow efficient cloning of larger fragments (8–23 kbp); (ii) With λ phage, an *in vivo* packaging system is available, which increases the infectivity of recombinant DNA by several orders of magnitude as compared to pure DNA preparations. The *in vitro* packaging system also allows a size selection of the packaged DNA such that under suitable conditions only recombinant DNA molecules will be packaged; (iii) Its very high rate of infectivity makes it ideal for creating large numbers of clones that are often needed for construction of gene libraries that represent large populations of molecules; (iv) Millions of independently cloned phage particles constituting a gene library, which represents the entire genetic information of a complex organism, can be stored in a few milliliters of broth; (v) Bacteriophage libraries consist of plaques, and the main advantage is that the plaque libraries are easy to sort through, as λ phage clones are particularly easy to array spatially at high density, and the clone density can reach 10^8–10^9 plaque forming unit (*pfu*) per μg of ligated DNA; (vi) As compared with the colony library generated by cloning in plasmids, plaque library is easy to store and has longer shelf-life; and (vii) λ vectors also have advantages in screening gene libraries. Several thousand phage plaques can be screened and characterized, e.g., by DNA:DNA hybridization, on a single petridish. It is much easier to screen large libraries when using phage vectors, because larger the inserts, fewer the clones to be screened. Moreover, the results with phage plaques are much cleaner than those obtained with bacterial colonies.

Review Questions

1. Give an elaborative account of morphology of λ phage. Also describe the process of morphogenesis of λ phage giving details of the various proteins involved.
2. Categorize the λ genes according to phase of transcription. Enumerate the functions of various λ genes.
3. Describe the factors affecting the decision for lysogeny or lytic infection cycle in λ phage. How is this decision regulated?
4. Explain how pCII aids in establishment of lysogeny. Also explain the role of pCIII in the process.
5. Give a detailed account of the process of integration of λ genome into host chromosomal DNA during lysogeny.
6. Explain the process of DNA replication during lytic infection cycle. Also explain the roles of *gam* and *red* genes in the process.
7. Assign reasons for the following:
   (a) Transcription from P_{RE} promoter leads to establishment of lysogeny.
   (b) λ repressor maintenance requires transcription from P_{RM}, and not P_{RE}.
   (c) λ repressor synthesis is negatively autoregulated.
   (d) Poor nutritional conditions and higher multiplicity of infection favor lysogeny.
   (e) Turbid or no plaques are formed during lysogeny, while clear plaques are formed during lysis.
   (f) Late genes are not transcribed during lysogeny.
   (g) Replacement vectors have higher cloning capacity as compared with insertional vectors.
   (h) *nin* deletion frees phage growth from dependence on the *N* transcription antitermination system.
   (i) Mutation in host *hfl* locus favors lysogeny.
   (j) *gam* gene product plays a crucial role in switch-over from bidirectional mode to rolling circle mode of replication.
   (k) Exposure of a lysogen to UV rays causes induction of lysis.
   (l) Recombinants formed while cloning in replacement vectors are Spi^-.

(m) A *gam*⁻ *red*⁻ χ⁺ phage can be propagated in *rec* A⁺ host.
(n) Maltose is added in the medium used to plate and grow λ phage.

8. Explain the regulatory roles of pN and pQ.
9. pInt is synthesized from P_I promoter and not from P_L promoter indicating that pInt is not formed in the absence of pCII. Explain.
10. Explain the process of 'induction of lysis' in detail.
11. Discuss how the stage of infection of bacterial cells with phage particles affects phage titre.
12. Give a detailed account of 'two defective strains system of *in vitro* packaging'. How are the potential problems of 'packaging of endogenous DNA' and 'recombination between exogenous DNA and induced prophage markers' circumvented?
13. Differentiate between transfection and *in vitro* packaging. Which one of these two procedures is more efficient?
14. What do you understand by 'immunity to superinfection'? Explain the advantages of introduction of immunity region of other lambdoid phages in a λ vector.
15. Explain how is natural selection process used to reduce the number of target sites for a restriction enzyme.
16. What are various advantages of λ cloning vectors over plasmids? How do λ based vectors differ from plasmid vectors in terms of substrates (DNA ligation products) for introduction into host?
17. What feature of λ genome has led to the construction of λ replacement vectors?
18. Describe how is the packaging constraint (or packaging limit) imposed by λ phage beneficial during cloning in λ vectors.
19. Differentiate between insertional vectors and replacement vectors. Also describe the strategy of cloning in linear as well as circular forms of insertional vectors.
20. λgt II is a translational fusion vector (or expression vector). Explain.
21. What are the advantages of RNA probes? Give one example of a vector that can be used for preparation of RNA probes.
22. If we are interested in cloning a eukaryotic gene in a λ vector, which type of vector should we use? Describe the cloning strategy to be adopted.
23. What are the various strategies that increase the efficiency of cloning in a replacement vector? Taking suitable example, give a detailed account of the process of 'partial filling-in of cohesive ends in vector as well as source DNA' before ligation.
24. Give the significance of following:
   (a) Clustered arrangement of functionally related genes on λ genome.
   (b) *cos* sites.
   (c) λ terminase.
   (d) λ *att* P and bacterial *att* B sites.
   (e) *nut* and *qut* sites in λ genome.
   (f) *S* amber mutation in vector DNA.
   (g) χ site in replacement vector.
   (h) Size fractionation of source DNA before ligation to vector.
   (i) *lon*⁻ host for production of fusion proteins.
   (j) Presence of cloning site within *lac* Z′ gene.
   (k) Use of *hfl*⁻ host while cloning in λgt 11.
   (l) Temperature sensitive mutation in *cI* gene in vector DNA.
   (m) Stuffer fragment in replacement vectors.
   (n) DP50 sup F host.
25. Describe the salient features and vector maps of λgt 10, λgt 11, λEMBL 3, λEMBL 3A, λgtWES.λB′, λDASH, λFIX, and λGEM 11. Also discuss the cloning strategies and selection procedures for each of them.

References

Loenen, W.A.M. and W.J. Brammar (1980). A bacteriophage l vector for cloning large DNA fragments made with several restriction enzymes. *Gene* **20**:249–259.

Murray, K. and N.E. Murray (1975). Phage l receptor chromosomes for DNA fragments made with restriction endonuclease III of *Haemophilus influenzae* and restriction endonuclease I of *Escherichia coli*. *J. Mol. Biol.* **98**:551–564.

Murray, N.E., S.A. Bruce, and K. Murray (1979). Molecular cloning of the DNA ligase gene from bacteriophage T4. II. Amplification and preparation of the gene product. *J. Mol. Biol.* **132**:471–491.

Murray, N.E. (1983). Phage l and molecular cloning. In: R.W. Hendrix *et al.* (eds.), *The Bacteriophage* λ, Vol. 2, Cold Springer Harbor Laboratory, CSH, New York.

Thomas, M., J.R. Cameron, and R.W. Davis (1974). Viable molecular hybrids of bacteriophage l and eukaryotic DNA. *Proc. Natl. Acad. Sci.*, USA **71**:4579–4583.

M13 Bacteriophage

Key Concepts

In this chapter we will learn the following:

- Biology of filamentous bacteriophages
- M13 bacteriophage as cloning vector
- Cloning in M13 vectors

10.1 INTRODUCTION

M13 is a filamentous rod-shaped *E. coli* bacteriophage, which belongs to family Inoviridae. The 'M' in M13 stands for the city of Munich, where M13 was discovered. It is one of the smallest known phage, and has circular single stranded DNA (ss DNA) of ~6,400 nucleotides in length. The DNA of M13 phage behaves quite unusually as it produces both double stranded (ds DNA) and ss DNA in different phases of its replication cycle. During replication inside the host cell, M13 is present in the double stranded replicative form [ds DNA (RF)] like a plasmid; however, during morphogenesis, the DNA is converted into single stranded form. Unlike λ phage, M13 is a non-lytic phage. The infected host cells continue to grow and divide, and extrude ss DNA containing M13 phage particles at the rate of several hundred phages/cell/generation. These phages have been developed as cloning vectors, as these have a number of advantages over plasmid and λ phage vectors. M13 phage is basically used as a vector for subcloning of single stranded copies of DNA fragments cloned in other vectors. The recombinant M13 phages are used instead of older physical and enzymatic techniques of denaturation of ds DNA. The ss DNA are primarily used as templates for site-directed mutagenesis, sequencing of DNA fragments by the dideoxy chain termination method, construction of subtractive cDNA libraries, and synthesis of strand-specific probes. The basic biology, M13 vectors, and cloning strategies used with M13 vectors are summarized in this chapter.

10.2 BIOLOGY OF FILAMENTOUS BACTERIOPHAGES

The genome of M13 has been completely sequenced. More than 98% sequences are identical to other filamentous coliphages belonging to its family, *viz.*, f1 and fd. The sites of differences are scattered around the genome, mostly in the third position of redundant codons. These actively complement and recombine with one another and have identical properties, which are as follows:

- The phage particles have dimensions of ~900 × 6–7 nm. These phages are sex-specific, and infect only enteric male bacteria harboring F pili. The adsorption site appears to be the end of the F pilus, but exactly the mechanism of entry is not known. These phages do not lyse their host cells like λ phage. Instead, these are easily released from infected cells even as the cells continue to grow and divide though at slower rate than uninfected cells. Several hundred particles may be released into the medium per cell per generation.
- The ss DNA of phage is transformed into a circular ds DNA (RF) when injected in host cells. The ds DNA (RF) is replicated as a θ structure, i.e., bidirectional mode of replication. The ds DNA (RF) multiplies rapidly until ~100 ds DNA (RF) molecules are formed inside the cell. Replication of ds DNA (RF) then becomes asymmetric due to the accumulation of phage-encoded single strand-specific DNA binding proteins pV. This protein binds to the phage plus (+) strand and prevents synthesis of the complementary strand. From this point on, only phage single strands are synthesized. These single stranded plus (+) strand progeny DNA molecules are produced by a rolling circle mechanism.
- ds DNA (RF) molecules also provide templates for transcription of 11 genes of M13. Among these, three of the phage gene products (pII, pV, and pX) are responsible for the replication of the phage DNA, five phage-encoded transmembrane structural proteins (pIII, pVI, pVII, pVIII, and pIX) are involved in the formation of capsid, and three phage proteins (pI, pIV, and pXI) play role in secretion or morphogenesis of phage.
- First, the coat proteins are congregated into bacterial membrane, where these are assembled around the (+) strand progeny DNA molecules and the progeny phage particles are generated. As the DNA passes through the membrane, the DNA-binding proteins are stripped off and replaced with capsid

proteins. The resulting protein–DNA complex is extruded from the cell in the form of a filamentous phage particle. Secretion and assembly of the filamentous phage particle occur simultaneously in a synchronized manner. These progeny single strands are released from the cell as filamentous particles following morphogenesis at the cell membrane.

- As genomes of these phages are unstable, the recombinant filamentous phages are not used for cloning and long-term propagation of segments of foreign DNA.

10.2.1 Genome Map of M13

The M13 genome is comprised of 11 genes that are separated by only a few nucleotides. The genetic map of M13 phage is shown in Figure 10.1. All structural proteins of the phages are inserted into the bacterial membrane prior to phage assembly. The comprehensive organization of genes and structures and functions of proteins encoded by M13 genes are presented in Table 10.1.

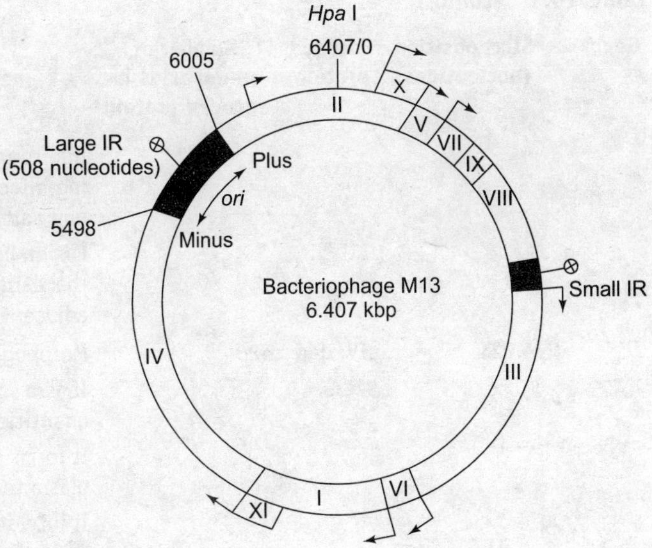

Figure 10.1 Genome map of M13 phage. [Arrows indicate major promoters; cross in a circle represents transcription terminator.]

Table 10.1 Functions of Proteins Encoded by Genes of M13 Bacteriophage

Gene	Start position (nucleotides)	Encoded protein	Number of amino acids in encoded protein	Functions of encoded protein	Copy number of encoded protein per virion or per cell
I	3,197	pI; g1p	348	Morphogenetic protein; Span the inner membrane of M13 infected bacteria; Initiates phage assembly by interacting with the phage pIV and host-encoded thioredoxin.	—
II	6,007	pII; g2p	410	Replication protein; Introduces a nick at a specific site in the intergenic region of the plus (+) strand of ds DNA (RF); Initiates rolling circle replication, which generates plus (+) strand progeny DNA molecules; Cleaves the single stranded product of rolling circle replication into monomeric molecules that are packaged into progeny phage particles.	—
III	1,579	pIII; g3p	427	Structural protein (Minor coat protein); Their location on phage particle marks the tail or proximal end of M13 phage; pIII is linked with pVI at the proximal tip, and is required for adsorption of the phage to the sex pili of new hosts and for penetration of the M13 DNA, hence also called pilot protein; pIII interacts with the host Tol Q, Tol R, and Tol A proteins, which mediate the removal of the major capsid protein pVIII, and facilitate the M13 DNA to penetrate into the cytoplasm of the bacterium; Before morphogenesis, pIII is anchored to the bacterial membrane by a single membrane-spanning domain near its C-terminus; As the M13 filament emerges from the infected cell, three to five copies of pIII are attached to the proximal tip of phage particle;	3–5

(*Contd.*)

Table 10.1 (*Contd.*)

Gene	Start position (nucleotides)	Encoded protein	Number of amino acids in encoded protein	Functions of encoded protein	Copy number of encoded protein per virion or per cell
				Due to pIII, the tip enters first into new host cell during infection and leaves the membrane last when progeny particles are extruded; Fusion peptides in phage display libraries are located in the early mature region of pIII, near or immediately adjacent to the signal peptide cleavage site.	
IV	4,221	pIV; g4p	426	Morphogenetic protein; It is a multimeric protein that is synthesized in large quantities in infected cells; It forms a gated channel connecting the bacterial cytoplasm to the exterior; It interacts with pI, and is required for the induction of the *E. coli psp* (phage shock protein) operon during M13 infection.	—
V	843	pV; g5p	87	Replication protein; It is a ss DNA-specific protein; It is synthesized in large amounts in infected cells; It binds strongly and in cooperative manner to newly synthesized plus (+) strands to form pV-DNA complex; The resulting pV–DNA complex moves to specialized packaging sites on the bacterial membrane, where pV is stripped off from the phage DNA and is recycled into the cell; In addition to DNA binding, this protein also acts as a translational repressor by binding specifically to the leader sequences of phage mRNAs coding for gene *II* and other phage proteins; These two properties of pV work in combination to regulate both the expression and replication of the M13 genome; When the intracellular concentration of pV reaches critical levels (10^5 to 10^6 molecules/cell/generation), the conversion of progeny plus (+) strand to ds DNA (RF) form is suppressed and the rolling circle replication of M13 DNA therefore proceeds at an essentially constant pace.	10^5–10^6
VI	2,856	pVI; g6p	112	Structural protein (Minor coat protein); Few molecules of pVI are associated with pIII at proximal or tail end; It is a membrane protein that is located in the bacterial cytoplasmic membrane before incorporation into phage particles.	3–5
VII	1,108	pVII; g7p	33	Structural protein (Minor coat protein); It interacts with the packaging signal located in the intergenic region of M13 DNA; Five molecules of pVII are located at the cap or distal end of the phage particle that emerges first from the infected cell during secretion and morphogenesis.	5

(*Contd.*)

Table 10.1 (*Contd.*)

Gene	Start position (nucleotides)	Encoded protein	Number of amino acids in encoded protein	Functions of encoded protein	Copy number of encoded protein per virion or per cell
VIII	1,301	pVIII; g8p	50 (precursor = 73)	Structural protein; These are major capsid protein; It is synthesized as precursor known as precoat, which binds to the inner surface of the plasma membrane and subsequently translocates as a loop structure across the membrane in the presence of a transmembrane potential; These proteins are arranged in a cylindrical sheath around the M13 DNA.	~2,300
IX	1,206	pIX; g9p	32	Structural protein (Minor coat protein); Five copies of pIX are located at the distal or cap end of the phage particle where assembly begins; These proteins interact with the packaging signal located in the intergenic region.	5
X	496	pX; g10p	111	Replication protein; Translation of pX begins at the *in frame* AUG triplet (codon 300) of pII, hence pX is identical in sequence to C-terminal of pII; It is required for efficient accumulation of ss DNA; Like pV, it is a powerful repressor of phage-specific DNA synthesis *in vivo* and is thought to limit the number of ds DNA (RF) molecules in infected cells.	—
XI (also known as *I**)	—	pXI; g11p		Morphogenetic protein; It is produced by internal initiation of translation within gene *I* transcript at methionine codon 241; During secretion and assembly, it spans the bacterial cytoplasmic membrane but lacks a cytoplasmic domain; pXI, along with pI, may be involved in forming the site for assembly of M13 phage particles.	—
Small *IR*	—	—	—	It is noncoding intergenic region located between genes *VIII* and *III*; It contains transcription terminator; This site can be employed for cloning foreign DNA.	—
Large *IR*	5,501–6,007		—	It is noncoding intergenic region located between genes *II* and *IV*; It contains origin for replication for plus (+) and minus (–) strands, terminators of replication, transcription terminator, and packaging signals; This site can be employed for cloning foreign DNA.	—

There are two intergenic noncoding regions. The smaller intergenic region (*IR*) lies between genes *VIII* and *III*, whereas a considerably large intergenic region (508 nucleotides) is located between genes *II* and *IV*. All essential *cis*-acting elements of the M13 genome are sequestered into this larger intergenic region. These sequences are involved in the regulation of packaging and orientation of the DNA within phage particles, sites for the initiation and termination of synthesis of plus (+) strand and minus (–) strand of the DNA molecules, and also provides signal for ρ-independent termination of transcription. These *cis*-acting elements, which are tightly packed into a small region of DNA, can easily be cloned *en bloc* to other types of cloning vehicles such as plasmids. These phage-plasmid vectors are known as phagemids (for details see Chapter 12). The promoters and terminators of the phage genes display unique organization. The initiation of transcription process occurs at any one of series of promoters in ds DNA (RF) and proceeds unidirectionally to one of the two terminators, which are located immediately downstream from gene *VIII* and *IV*. Due to this unique organization of promoters and terminators, gradients of transcription take place. The genes located closest to a termination site (e.g., gene *VIII*) are transcribed much more frequently than the genes located further upstream (e.g., gene *II*).

10.2.2 M13 Phage Life Cycle

The basic steps of M13 phage life cycle include infection, i.e., adsorption of phage and entry of DNA, replication of M13 DNA, assembly of new phage particles, and release of progeny particles from the host.

Adsorption of Phage on Host Cell and Entry of DNA

The phage particles are adsorbed by the bacterial cell, which requires interaction between a sex pilus and the phage minor coat protein III (pIII), three to five copies of which are positioned at proximal end (tail) of the filamentous phage. Thereafter, the rod-shaped phage penetrates the pilus and pIII interacts with the host Tol Q, Tol R, and Tol A proteins. These Tol proteins mediate the removal of the major coat protein pVIII, and facilitate the M13 DNA to penetrate into the cytoplasm of the bacterium.

Replication of M13 DNA

The M13 replication cycle is divided into three stages (Table 10.2). In the first stage, immediately after infection, the plus (+) strand directs the synthesis of its complementary strand or minus (–) strand to form the circular ds DNA (RF). The ds DNA (RF) may either be nicked (RFII) or supercoiled (RFI) with the help of host replicating machinery during 0–1 min of postinfection. The ds DNA (RF) produces daughter strands in the second stage. In the third, i.e., final stage, ds DNA (RF) produces ss DNA, i.e., phage plus (+) strands. The origins of replication of both strands are located in the intergenic space between genes *II* and *IV*. The minus (–) strand serves two important functions as it acts as a template for all the genes of M13 and also for plus (+) strand synthesis.

Although there are three stages of replication, however, there are only two mechanisms, one for production of (–) strand and another for (+) strand synthesis. The mechanism of M13 replication is depicted in Figure 10.2. The first step of (–) strand replication is the synthesis of a 20–30 nucleotides long RNA primer by *E. coli* RNA polymerase at the (–) origin on a single stranded (+ strand) phage DNA molecule. M13 proteins are not required for synthesis of (–) strand. *E. coli* DNA polymerase III holoenzyme adds nucleotides to the free 3′-OH terminus. Finally the primer is removed by DNA polymerase I-catalyzed nick translation, thereby forming RFII, which is converted to RFI by the sequential action of DNA ligase and DNA gyrase.

The ds DNA (RF) molecules serve as templates for further rounds of transcription and synthesis of additional plus (+) strands. M13 pII protein is essential for phage daughter strand synthesis. The daughter strands are synthesized from RFI templates by a variation of the rolling circle model. Protein pII introduces a nick at the origin of the plus (+) strand of parental ds DNA (RF). Thereafter the plus (+) strand is extended from the 3′-OH end of the nick with the help of *E. coli* replication proteins. As the elongating new plus (+) strand displaces the old plus (+) strand, it is coated with single strand binding protein (SSB). As soon as the new plus (+) strand reaches the origin, pII again generates a nick, freeing the old plus (+) strand, the 3′-OH and 5′-P ends of which are joined by pII. Finally, after one round of replication, a single stranded plus (+) circle and a ds DNA (RF) molecule are produced. At early times during infection, the daughter plus (+) strand serves as template for the synthesis of minus (–) strand, accounting for RF→RF replication. During this period, ds DNAs (RFs) are also transcribed, producing a coat protein pVII that associates with the cytoplasmic membrane.

Table 10.2 Replication Cycle of M13 Bacteriophage

Stage	Postinfection time (min) at 33°C	Event
ss→RF	0–1	Adsorption and penetration; ss DNA molecules of phage are converted to parental ds DNA (RF); Transcription of ds DNA (RF).
RF→RF	1–20	Replication of parental ds DNA (RF); ~100–200 progeny ds DNA (RF) are formed; Replication of ds DNA (RF) continues.
RF→RF	25	Replication of ds DNA (RF) molecules ends.
RF→ss	20–30	Rolling circles form progeny of M13 DNA; ~500 ss DNA molecules are produced; Morphogenesis, i.e., DNA is packaged into phage particles, which are then excreted from the bacterial cell without causing bacterial cell lysis.

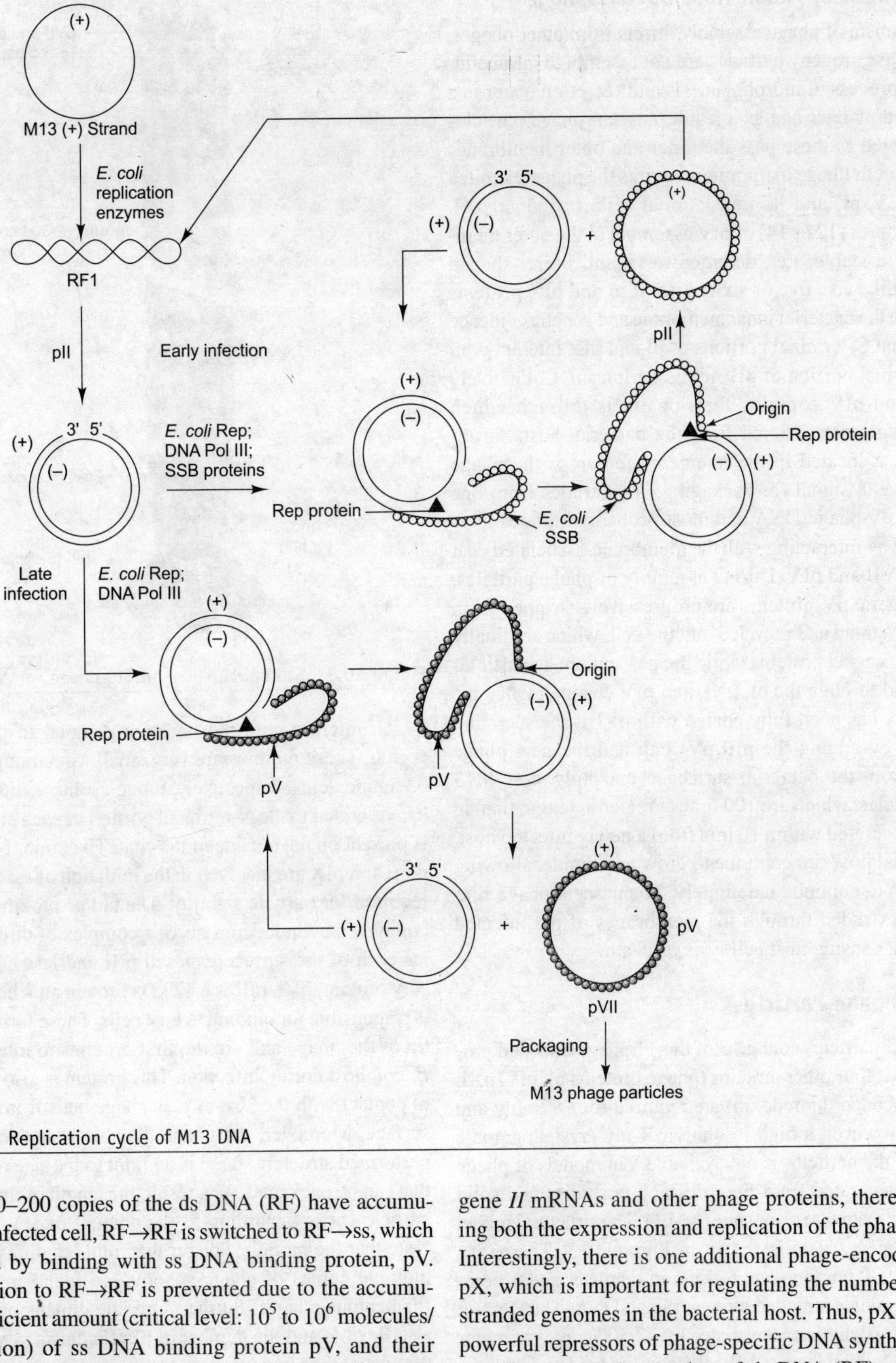

Figure 10.2 Replication cycle of M13 DNA

When 100–200 copies of the ds DNA (RF) have accumulated in the infected cell, RF→RF is switched to RF→ss, which is facilitated by binding with ss DNA binding protein, pV. The conversion to RF→RF is prevented due to the accumulation of sufficient amount (critical level: 10^5 to 10^6 molecules/cell/generation) of ss DNA binding protein pV, and their strong and cooperative binding to the newly synthesized plus (+) strands. Additionally this protein also acts as repressor of translation by binding specifically to the leader sequences of gene *II* mRNAs and other phage proteins, thereby regulating both the expression and replication of the phage genome. Interestingly, there is one additional phage-encoded protein, pX, which is important for regulating the number of double stranded genomes in the bacterial host. Thus, pX and pV are powerful repressors of phage-specific DNA synthesis and are thought to limit the number of ds DNA (RF) molecules in infected cells and the rate of production of progeny plus (+) strands are kept within reasonable limit.

Morphogenesis of Filamentous Bacteriophage

The mechanism of phage assembly differs from other phages, as in this case progeny particles are not assembled intracellularly. The process of morphogenesis and secretion occur in a coordinated manner and as a result, nascent phage particles are assembled as these pass the inner and outer membranes of the host cell. Phage maturation requires the phage-encoded proteins pIV, pI, and its translational restart product pXI. Multiple copies (12 or 14) of pIV assemble in the outer membrane into a stable, i.e., detergent-resistant, barrel-shaped structure. Likewise five or six copies of pI and pIX proteins assemble in the bacterial inner membrane and genetic evidence suggests that C-terminal portions of pI and pIX interact with the N-terminal portion of pIV in the periplasm. Collectively pI, pXI, and pIV complex form channels through which mature phages are secreted from the bacterial host. An inverted repeat located in the intergenic region of the phage DNA acts as a signal for packaging; it protrudes from one end of the pV-phage DNA and most probably initiates morphogenesis by interacting with the membrane-associated coat proteins, pVII and pIX. During assembly of phage particles, ~1,500 dimeric pV proteins are progressively stripped from the plus (+) strand and recycled into the cell, which are finally replaced by capsid proteins while the nascent phage particles are threaded through the pI, pXI, and pIV channel. Once the phage DNA has been fully coated with pVIII, the secretion terminates by adding the pIII/pVI cap and the new phage detaches from the bacterial surface. Amazingly, new M13 phage particles, which are 100 times greater in length than in width, are secreted within 10 min from a newly infected host. The bacterial host can continue to grow and divide, allowing this process to continue indefinitely. The nascent phage filaments are extruded through the membranes of the infected cell without causing host cell lysis or death.

10.2.3 Phage Particles

Mature M13 particles contain 5 of the 11 phage-encoded proteins. At least four other proteins (phage proteins pI, pIV, pXI, and host-encoded thioredoxin) are required for assembly and secretion. However, a high-resolution X-ray crystallographic structure of the particles is not available, but models of phage have been generated from fiber diffraction studies and solid state nuclear magnetic resonance (NMR). The structural model of M13 phage particle is shown in Figure 10.3. The native particle is ~900 nm long and 6–7 nm in diameter and is composed of the single stranded circular phage DNA. The DNA at core of the particles is encapsulated by ~2,300 copies of major coat protein, which is an α-helical 50 amino acid residue protein, pVIII. The cap or distal end of the phage cylinder extruded from the membrane consists of five copies each of surface exposed pIX and more buried companion protein pVII,

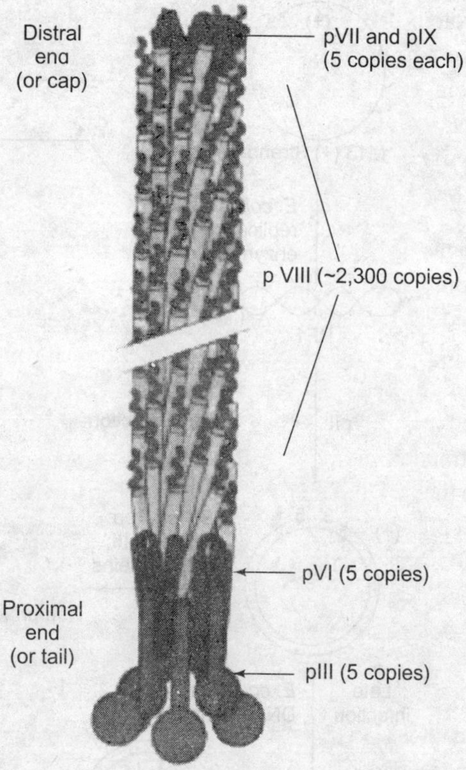

Figure 10.3 Structural model of M13 phage

which interact with the packaging signal in the intergenic region. These proteins are very small, containing only 32 and 33 amino acids, respectively, though some additional residues can be added to the N-terminal portion of each amino acid that is present on the outside of the coat. The minor phage proteins pVII and pIX are involved in the initiation of assembly and are required for particle stability. The tail or proximal end of the completed cylinder consists of a complex of three to five copies each of the surface exposed pIII and less exposed accessory protein, pVI. pIII is a 42 kDa protein attached by pVI and is responsible for binding to host cells. These form the rounded tip of the phage and are the first proteins to interact with the *E. coli* host during infection. This protein is also the last point of contact with the host as new phage buds from the bacterial surface. Moreover, the phage genome is not inserted into a preformed structure; there is no limit to the size of the ss DNA that can be packaged. As a result, the length of the filamentous particle varies according to the amount of DNA it contains. Note that the particles longer than unit length and containing multiple copies of phage genome are found in stocks of all filamentous phages. Furthermore, the dimensions of coat are also flexible and the number of pVIII adjusts to accommodate the size of the single stranded genome. For example, when the phage genome is mutated to reduce its number of DNA bases (from 6,400 nucleotides to 221 nucleotides), the number of pVIII copies decreased to less than 100, causing the pVIII coat to shrink in order to fit the reduced genome. Deletion of a phage protein pIII prevents full escape from the host *E. coli* and phage that is 10–20 times larger than the normal length with several copies of the phage genome are seen shedding from the *E. coli* host.

10.3 M13 BACTERIOPHAGE AS CLONING VECTOR

M13 phages are small in size and have their own origin of replication, which are prime requisites for a DNA molecule to be used as a cloning vector (for details see Chapter 1). Besides, M13 phages also have several unique properties that make them suitable as cloning vectors. First, the phage is replicated via circular ds DNA (RF) intermediate. Thus, double stranded segments of foreign DNA can be inserted *in vitro* into the ds DNA (RF), which can be easily purified from infected cells, manipulated in a manner similar to plasmid DNA, and easily reintroduced into cells by standard transformation techniques. Secondly, both ds DNA (RF) and ss DNA transfect competent *E. coli* cells to yield either plaques or infected colonies depending on the assay method. Thirdly, the size of phage particle is governed by the size of phage DNA and therefore there are no packaging constraints. In fact, foreign DNA inserts having size seven times longer than the wild-type M13 genome can be easily cloned and propagated in M13. Finally, determination of orientation of an insert is very easy with the help of these phages. Usually the relative orientation can be easily obtained from the restriction analysis of ds DNA (RF). However, comparatively an easier method has been opted for mass screening of cultures. If two clones carry the insert in opposite directions, the ss DNA from these will hybridize, which can be detected by agarose gel electrophoresis. In a nutshell, M13 phages have all the advantages of plasmids while producing particles containing ss DNA in an easily available form.

Despite the advantages described above, M13 vectors are used only for subcloning of single stranded copies of DNA fragments cloned in other vectors, and not for primary cloning and long-term propagation. This is because of higher frequency and extent of deletions and rearrangements in ss DNA. The DNA fragment has to be subcloned from plasmids to M13 vectors, which is a time-consuming process. The insert capacity of M13 vectors is very small (up to <1,000 nucleotides). This is because M13 DNA carries several essential genes encoding components of the phage-specific DNA replicative enzymes and phage protein coat, the alteration or deletion of which may impair or destroy the replicative ability of the resulting molecule. Thus, there is much less freedom to modify phage DNA molecules and generally M13 vectors are only modestly different from the parent molecule. These problems can however be overcome by using phagemids (for details see Chapter 12).

10.3.1 Types of M13 Vectors Based on Location of Cloning Site

Based on the location of the cloning site, three types of M13 vectors have been devised, which are discussed in the following sections.

Cloning into the Large Intergenic Region

As discussed above, all genes of filamentous phage are essential for phage DNA replication and its propagation in the host cell. However, scientists have attempted the insertion of segments of foreign DNA into the 508 nucleotides intergenic region located between genes *II* and *IV*. Though insertion of foreign DNA segments within this intergenic region severely affects the replication of the phage genome, but by chance, most vectors carry mutations in gene *IV* or *V* that partially compensate for the disruption of *cis*-acting elements.

Cloning into the Small Intergenic Region

The second relatively smaller intergenic region, located between genes *VIII* and *III*, has also been used as the cloning site in another series of vectors. Furthermore, a number of vectors carrying selectable markers like β-lactamase (*bla*) gene and chloramphenicol acetyl transferase (*cat*) gene, which may be used for cloning by insertional inactivation, have been devised. However, cloning in this vector system is much slower and more tedious process as compared to visual examination of plaques, which requires the establishment of colonies of infected cells that must be screened by replica plating.

Cloning into Gene X

As gene *X* encoded protein acts as a repressor, its overproduction completely blocks phage-specific DNA synthesis. For example, M13-100 carries a large number of the gene *X* in the large intergenic region of M13. This gene is under the control of T7 promoter and hence occurs only in host cells that express T7 RNA polymerase, for instance, *E. coli* strain JM109 (DE). Once foreign DNA fragments are cloned into the supernumerary copy of the gene *X*, the gene is inactivated and the resulting recombinants can therefore replicate on JM109 (DE) cells, which obviously are nonpermissive hosts for empty vectors. In fact, this vector system was devised to reduce the cost and manual labor of screening of large numbers of plaques by histochemical staining and so far has not been widely acknowledged.

10.3.2 Modified Vectors

The wild-type filamentous phages are not very promising as vectors because these include very few unique sites within the large intergenic region, for example, *Asu* I in the case of fd and *Asu* I and *Ava* I in the case of M13. Hence modified M13 vectors with large number of unique cloning sites were developed. Vector M13mp1 (mp signifies Max Planck Institute where these vectors were first developed) was developed first in this series of M13 vectors; thereafter more advanced asymmetrical M13mp series of vectors were devised. These M13mp vectors are usually available as pairs (e.g., M13mp8 and M13mp9, M13mp10 and M13mp11, M13mp18 and M13mp19) that vary only in the orientation of the asymmetrical polylinker region within *lac* Z′ region, and are called dual vectors or sister vectors.

Advantages of Modified M13 Vectors

The modified M13 vectors offer following advantages:

- As it is always the plus (+) strand that is packaged, ss DNA can be directly isolated from the M13 phages. This ss DNA is useful as templates for sequencing of DNA fragments by Sanger's method, site-directed mutagenesis, construction of subtractive cDNA libraries, and synthesis of strand-specific probes.
- Cloning in M13 vectors containing asymmetrical restriction sites (e.g., M13mp8, M13mp9, M13mp10, M13mp11, M13mp18, and M13mp19) facilitates directional cloning by using two different enzymes for cleavage.
- Cloning in dual vectors is advantageous. Here, dual vectors M13mp18 and M13mp19 are taken as example. After cleavage of ds DNA (RF) with two different restriction enzymes, M13mp18 and M13mp19 cannot circularize easily unless a double stranded foreign DNA fragment with compatible termini is present in the ligation mixture. Cleavage with two different restriction enzymes in asymmetrical polylinker favors directional cloning. The foreign DNA is inserted in opposite orientation and different strands of DNA fragment are attached to the plus (+) strand in the two vectors of the pair. Thus, the progenies of M13mp18 and M13mp19 recombinants will contain different strands of the foreign DNA. This indicates that by using M13mp18 and M13mp19 as dual vectors, it is possible to use single primer (universal primer) to determine the sequence of nucleotides on opposite strands from each end of the inserted DNA and to generate probes that are complementary to one strand or the other of the foreign DNA.
- The *lac* Z′ fragment present in M13 series of vectors is similar to that present in pUC series of vectors. Moreover, the *lac* Z′ fragments in M13 vectors carry the same constellation of restriction sites in their multiple cloning sites (MCS) as the analogous vectors of pUC series. Thus, the cloned DNA fragments can be moved between M13mp vectors and pUC vectors with great ease.
- In addition, the intergenic region of M13 is now a standard fixture of most plasmid vectors, which can therefore be used as conventional plasmids or as phagemids (for details see Chapter 12).

Geneology of M13 Vectors

The large intergenic region, which is 508 nucleotides long and is located between genes *II* and *IV*, was targeted for the development of a series of M13 vectors. As mentioned in Section 10.2.1, this region contains the origin of replication, but it was revealed that the integrity of the whole region is not necessary for phage development by Messing and his colleagues who introduced a segment of the *E. coli lac* operon into this site. This was the first stage in the development of a series of M13 vectors. The phage DNA has 10 cleavage sites for *Bsu* I, one of which is located in the intergenic region. For cloning, ds DNA (RF) was partially digested with *Bsu* I, and linear full-length molecules were isolated by agarose gel electrophoresis. A 789 nucleotide *Hind* II fragment containing the regulatory region of *E. coli lac* operon and the first 146 codons of the β-galactosidase gene (*lac* Z′) was inserted into this site (Figure 10.4). The complete ligation mixture was used to transform a strain of *E. coli* with deletion of α-fragment of β-galactosidase. Only those recombinants in which the *lac* DNA was inserted into a nonessential region were viable. The recombinants were detected by intragenic complementation on media containing IPTG and X-gal. One of the blue plaques was selected and the phage in it was designated M13mp1. However, M13mp1 lacked unique restriction endonuclease recognition sites, hence it was not a suitable cloning vector for the

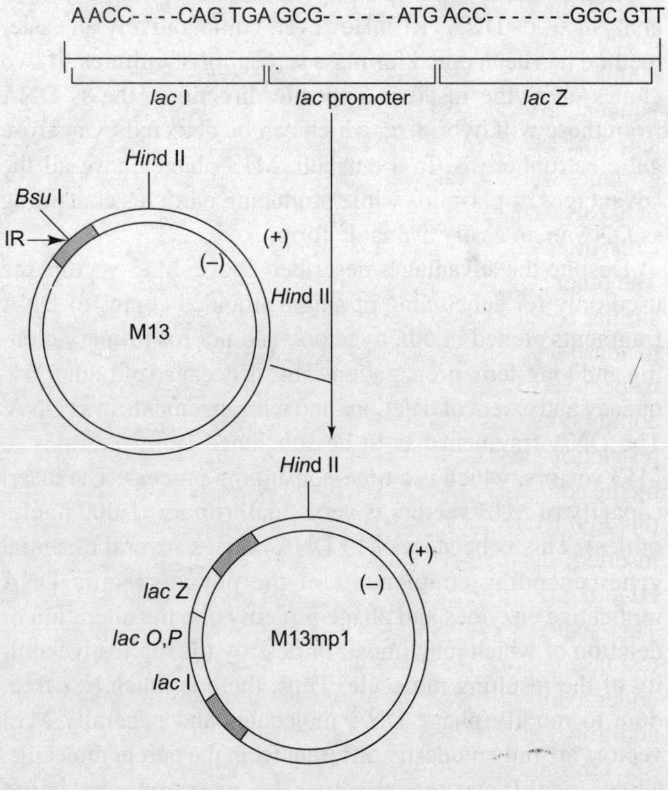

Figure 10.4 Construction of M13mp1 from M13

insertion of foreign DNA into the *lac* Z′ region. The *lac* region only contained unique sites for *Ava* II, *Bgl* I, and *Pvu* I, and three sites for *Pvu* II. Sequence analysis of *lac* Z′ fragment revealed the sequence GGATTC at a site corresponding to the sixth amino acid of β-galactosidase. Thus, it was speculated that a single mutation changing G to A would generate an *Eco* RI site in *lac* Z′ gene without affecting its biological function (Figure 10.5). This was achieved by methylating guanine residue in DNA with *N*-methyl-*N*-nitrosourea. As the resulting O^6-methylated guanine residue base pairs with T, the progeny contained the sequence GA**A**TTC instead of GGATTC. Sufficient alkylating agent was used to methylate two to four guanine residues per genome and the DNA was then transferred into bacterial cells. After several cycles of infection, ds DNA (RF) was recovered and cleaved with *Eco* RI. Linearization of DNA was possible only in effectively mutagenized ds DNA (RF) molecules.

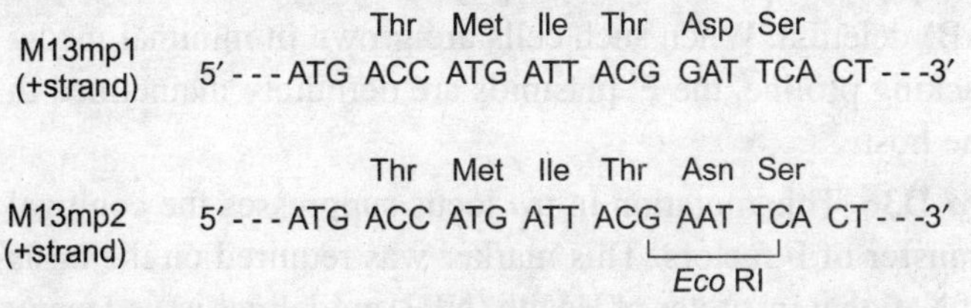

Figure 10.5 Introduction of an *Eco* RI site into M13mp1 vector

The linear molecules were recircularized, religated, and retransferred into *E. coli.* Three mutant phages were isolated in this way: M13mp2, M13mp3, and M13mp4. M13mp2 contained a new *Eco* RI site within the codons for amino acids 5 and 6 of *lac* Z′. M13mp2 thus exhibited slightly altered *lac* Z′ gene, where the sixth codon specified asparagine instead of aspartic acid, but the β-galactosidase enzyme produced by cells infected with M13mp2 was still perfectly functional. The other two vectors, M13mp3 and M13mp4, comprised of similar changes at the position of amino acid 119 of β-galactosidase and in the gene *X* of the M13 region of the phages, respectively. One problem with M13mp2 vector was that it could be used for cloning only at *Eco* RI site. Therefore the restriction site range of M13mp2 was subsequently extended by the addition of a set of chemically synthesized linkers into this *Eco* RI site (Figure 10.6). Thus, an oligonucleotide was inserted, which had the cleavage site for the enzymes *Bam* HI–*Sal* I–*Pst* I–*Sal* I–*Bam* HI and was flanked by *Eco* RI. This sequence contributes 14 extra codons to the β-galactosidase gene but does not affect the ability of the phage to undergo intracistronic complementation. There are, however, additional sites for *Bam* HI in gene *III* and for *Hinc* II and *Acc* I in gene *II* (note that *Acc* I and *Hinc* II are isoschizomers of *Sal* I). These sites were removed in a manner analogous to that described above for the introduction of *Eco* RI site. The

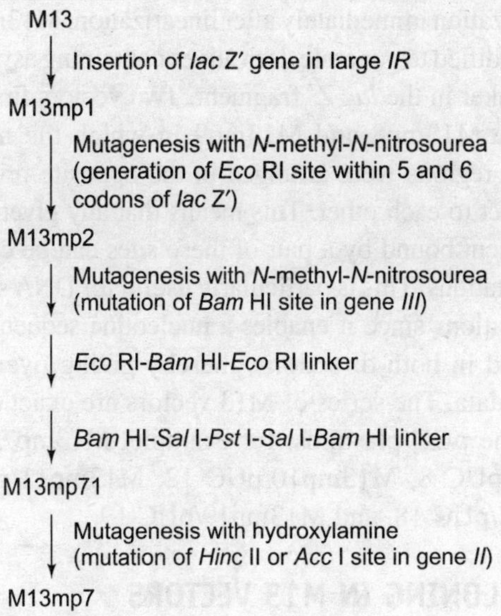

Figure 10.6 Geneology of M13 vectors

phage DNA was mutagenized, propagated in *E. coli*, and the mutant phage ds DNA (RF) was selected from molecules resistant to the appropriate restriction enzyme. The mutagenesis was planned in such a way that the nucleotide change was in the third base positions of the codons concerned and so did not change the amino acid. First, the *Bam* HI site in gene *III* was mutated by *N*-methyl-*N*-nitrosourea treatment, subsequently synthetic linker containing *Sal* I, *Pst* I was inserted in the *Bam* HI recognition site and M13mp71 DNA molecules were produced. Finally, *Hinc* II or *Acc* I site in gene *II* were altered by mutagenesis with hydroxylamine (NH_2OH) and M13mp7 DNA was obtained. Note that hydroxylamine is a hydroxylating agent and it specifically reacts with C residues leading to the formation of hydroxylamine cytosine, which then base pairs with A, consequently changing a G:C base pair to A:T base pair in next round of replication. These mutations were then transferred onto phage genome carrying the synthetic 17-mer cloning site. This was achieved by isolating a restriction fragment containing the mutated phage genes *II* and *III* from ds DNA (RF), followed by denaturation and annealing to the plus (+) strand of the other phage DNA. The partial heteroduplexes formed in this way replicated following their introduction into *E. coli.* The resulting vector M13mp7 was still infectious and directed the synthesis of final α-peptide, unless the reading frame was disrupted by the insertion of foreign DNA into the oligonucleotide. A big advantage of M13mp7, with its symmetrical cloning sites, is that the foreign DNA inserted into either the *Eco* RI, *Bam* HI, or *Pst* I sites can be excised from the recombinant molecule using *Eco* RI. Very few vectors allow cloned DNA to be recovered so easily. However, the symmetrical cloning vector M13mp7 does not allow directional cloning and requires alkaline phosphatase treatment to prevent vector recircularization immediately after linearization, M13mp7 was further modified to generate derivatives containing asymmetrical polylinker in the *lac* Z′ fragment. Two vectors first developed were M13mp8 and M13mp9 in which the modified polylinker regions were arranged in the opposite orientation with respect to each other. This means that any given restriction fragment bound by a pair of these sites can be cloned in both orientations. This is particularly useful for DNA sequencing applications since it enables a nucleotide sequence to be determined in both directions, thereby giving overlapping sequence data. The series of M13 vectors are exact counterparts of the pUC plasmids, for example, M13mp7/pUC 7, M13mp8/pUC 8, M13mp10/pUC 12, M13mp11/pUC 13, M13mp18/pUC 18, and M13mp19/pUC 19.

10.4 CLONING IN M13 VECTORS

M13 vectors are ideal for generating ss DNA, which is required for several specialist applications of cloned DNA, for example, sequencing by the original dideoxy chain termination method, oligonucleotide-directed mutagenesis, construction of subtractive libraries, and some methods of probe preparation. In this section, suitable strains of *E. coli* used as host cell, plating, growing in liquid culture, preparation of DNA, and strategy for cloning into M13 vectors are presented.

10.4.1 Bacterial Hosts for M13 Vectors

Only male (F^+) bacteria are used for the proliferation of the M13 phage because it enters the host cell through sex pili encoded by an F factor. However, female (F^-) cells can also be infected by introducing M13 DNA through transfection, but in this case, progeny particles produced by the transfected cells are unable to infect other cells in the culture, resulting in low yields of phage. Therefore, bacterial strains having F′ plasmids and several other genetic markers or traits have been constructed, which should be appropriately selected for working with M13 vectors. The most important markers are discussed below.

***lac* ZΔM15** This is a deletion mutant of *lac* Z gene lacking the coding sequences from the N-terminal portion of β-galactosidase. The peptide that is expressed by ΔM15 has the capacity to demonstrate α-complementation. Various host cells carrying this deleted version of *lac* Z gene on F′ plasmids are used for the proliferation of M13 vectors.

Δ(*lac pro* AB) The chromosomal segment that spans the *lac* operon and neighboring genes coding for enzymes involved in proline biosynthesis are deleted. The bacterial host cells carrying this deleted marker are unable to use lactose as a carbon source and in the absence of F′ plasmid comprising of *pro* AB, proline is required for growth.

***lac* I^q** This is a mutant of *lac* I gene, resulting in the overproduction of lac repressor, which is approximately ten-fold higher as compared to wild-type. Excessive concentration of lac repressor suppresses the already low level of transcription of the *lac* operon in the absence of inducer. Therefore, synthesis of potentially toxic proteins is minimized when foreign coding sequences are placed under the control of *lac* Z promoter in host cells carrying a *lac* I^q mutation. If the host cells having a *lac* I^q mutation are grown in the presence of glucose, catabolite repression decreases the level of transcription across the intergenic region from the *lac* Z promoter, and therefore the production of ss DNA is increased. In most of the cases *E. coli* strains, which are used with M13 vectors carry *lac* I^q with *lac* ZΔM15 on an F′ plasmid.

***pro* AB** The *pro* AB region of the bacterial chromosome encodes enzymes required for proline biosynthesis. Usually the *pro* AB genes are often present on an F′ plasmid, which complement proline prototrophy in a host with a (*lac–pro* AB) deletion. When such cells are grown in minimal media lacking proline, the F′ plasmids are definitely maintained in the host.

***tra* D36** This mutation in *tra* locus suppresses the conjugal transfer of F factors. This marker was required on the basis of National Institutes of Health (NIH) guidelines issued many years ago. However, these guidelines are no longer in force, but the bacterial strains having this trait are still in use.

***hsd* R17 and *hsd* R4** These mutations lead to loss of restriction, but not modification, by the type I restriction/modification system of *E. coli* strain K. The foreign DNA directly cloned into M13 vectors and propagated in bacterial strains carrying this mutation will be modified and therefore protected against restriction if the vector is subsequently introduced into an *hsd*$^+$ strain of *E. coli* K.

mcr* A and *mrr Basically *mcr* A and *mrr* mutations demonstrate the loss of restriction of methylated DNA and hence eukaryotic genomic DNA fragments can be easily cloned in M13 vectors.

***rec* A1** The *rec* A gene of *E. coli* codes for a DNA-dependent ATPase that is essential for genetic recombination in *E. coli*. The strains that carry the *rec* A1 mutation are defective in recombination and therefore have two advantages. First, plasmids propagated in these strains remain as monomers and do not form multimeric circles. Second, segments of foreign DNA propagated in M13 vectors may experience fewer deletions in *rec* A^- strains.

***sup* E** An amber suppressor mutation codes for suppressor tRNA that inserts glutamine at UAG codons. Formerly, there were NIH guidelines requiring M13 vectors to carry an amber mutation. However, this is not necessary now and most vectors that are commonly used do not carry such mutations although several of the strains of *E. coli*, which were developed at that time as host cells for propagation of M13 phage, are still available.

10.4.2 Choosing and Maintaining a Suitable Strain of *E. coli* for Propagation of M13

As infection of *E. coli* by M13 phage requires an intact F pilus, the bacterial cell used for propagating M13 phage should retain the F′ plasmid encoding the pilus structure. The master stocks of F′ strains of *E. coli* used for propagation and plating of M13 phage should be stored at –70°C in Luria Bertani (LB) medium containing 15% glycerol. There are two types of selection strategies that are normally used for screening of suitable host for proliferation of M13 phage.

- The F′ plasmids present in all strains of *E. coli*, which are commonly used for the propagation of male-specific phages, have genes that encode enzymes for proline biosynthesis *pro* AB$^+$. When bacteria are grown on minimal medium lacking proline (e.g., M9, see Chapter 1), only bacteria carrying F′ plasmid will be a prototroph of proline and hence form colonies because genes involved in proline biosynthesis have been deleted from the host chromosome. Since bacterial cells are grown much more slowly on minimal media than on rich media, these are not viable for prolonged periods of storage at 4°C. Therefore, during experiments, fresh cultures of host bacteria should be prepared on M9 minimal medium by using a master stock stored at –70°C as inoculum.
- A number of F′ plasmids most frequently carry kanamycin resistance (*kan*^{r}) or tetracycline resistance (*tc*^{r}) gene. The host cells, which are able to maintain the F′ plasmids, will grow on antibiotic containing media. This selection strategy is a better alternative, because it eliminates the use of minimal medium and thus greatly speeds up the work.

Besides the presence of the F′ plasmid, the chromosomal markers carried by the bacterial strain can greatly influence the stability and yield of M13 recombinants and the purity of the ss DNA templates prepared from bacterial cells. Usually, two types of problems related to strain development are encountered while cloning in M13 vectors. First, when cloning is performed in the intergenic region of filamentous phages, deletions of the DNA fragments are observed. The larger the cloned segment the greater the extent of deletion. This kind of obstruction can be minimized by using *rec* A mutant host strains and using fresh stocks of recombinant phages stored at –70°C for plating on a suitable host. The propagation of phage by serial growth of infected cells in liquid culture should be avoided.

Secondly, the foreign DNA fragments are inserted only in one orientation while using the M13 vector. This situation arises as sequences on opposite strands of the foreign DNA interfere to different extents with the functioning of the vector's intergenic region. This problem can be solved by inserting the foreign DNA at a different site within the multiple cloning sites by changing host strains or by performing directional cloning by using combination of restriction enzymes. It was observed that if undesirable sequences of foreign DNA are forcibly cloned in M13 vectors, the resulting recombinants are often unstable and give rise to progeny that carry deleted or rearranged versions of the foreign sequences. In this case, a better alternative is to propagate the foreign DNA in phagemid vectors (for details see Chapter 12) that carry an origin of replication derived from a filamentous phage.

The yield and stability of recombinant DNA cloned in M13 phage are greatly influenced by several factors like the structure and size of the foreign DNA, the type of vector, and the properties of the host strain, including its efficiency of transformation and ability to produce high yields of ss DNA. Therefore several combinations of vector and host cell should be attempted. First, experiment should be started by using standard host strains such as TG1 or XL1-blue followed by switching over to a host cell such as DH11S that contains several additional markers. DH11S carries several mutations, which include: (i) The *rec* A gene to enhance stability of inserts; (ii) The restriction systems that attack methylated mammalian genomic DNA; and (iii) The *deo*^{r} gene that increases the efficiency of transformation by larger plasmids.

10.4.3 Plating of M13 Bacteriophage

M13 phages are propagated by infecting susceptible *E. coli* strains and plating the infected cells on soft agar. Infection is carried out with either a phage stock or freshly grown colonies of the appropriate *E. coli* strain. For plating of phage in semisolid medium containing agar or agarose, master culture of a bacterial strain carrying F′ plasmid is streaked and incubated at 37°C. Thereafter single, well-isolated colony is picked from the agar plate; 5 ml of LB medium is inoculated, and incubated at 37°C for 6–8 hours. Then different amounts of phage are taken into a sterile tube containing plating bacteria and mixed by vortexing. This culture is mixed with the top agar/agarose and poured on the LB-agar plate supplemented with 5 mM $MgCl_2$ and incubated at 37°C. After 4–8 hours, plaques or zones of reduced cellular growth are visible in the top agar. These plaques are turbid in contrast to the clear plaques formed by other phages (e.g. λ phage) that lyse the bacterial cells. The replication occurs in harmony with that of host cell and the infected cells are therefore not lysed but continue to grow, albeit at 1/2 to 3/4 of normal rates, while producing several hundred phages per cell per generation. The reason for the formation of zones of reduced growth is simple. If a single phage particle infects a single bacterial cell, a single plaque is obtained. The progeny phage particles then infect neighboring bacteria, which in turn release another generation of daughter phage particles. However, as the bacteria are growing on agar/agarose, the diffusion of the progeny phage is limited. Moreover, the bacterial cells infected with M13 phage have a longer generation time as compared to uninfected *E. coli*; therefore plaques are identified as areas of slower growing cells on a faster growing lawn of bacterial cells.

Minimal M9 agar medium is also used for the growth of M13 phage. Growing M13 for plaque formation on this minimal medium is advantageous as these can be clearly identified and the probability of finding of female cells diminishes. When plaques are formed on richer medium (YT or LB), auxotrophic female cells show overgrowth as these are resistant to infection by M13 phage, but this danger is minor as the spontaneous loss of F′ plasmid from bacteria grown for a limited period in rich medium is rarely high enough to cause problems. Moreover, plaques form more rapidly on richer medium than on minimal medium. Hence, YT or LB medium is recommended for routine plating of M13 phage stock.

10.4.4 Growing M13 Bacteriophage in Liquid Culture

Usually stocks of M13 phage are grown in liquid culture. The infected bacteria grow at a slower than normal rate to form a dilute suspension. The inoculum of phage should either be a freshly picked plaque or a suspension of phage particles obtained from a single plaque.

10.4.5 Preparation of M13 DNA

Infected bacterial cells contain up to 200 copies of ds DNA (RF) and extrude several hundred phage particles per generation. Thus, enough amounts (1–2 μg) of ds DNA (RF) are obtained from 1 ml culture, which is sufficient for restriction mapping and recovery of cloned DNA inserts. Approximately 5–10 μg of ss DNA can also be isolated, which is sufficient for specific experiments. The experiments, which are performed for manipulations involving M13 phages are started with small-scale liquid cultures. Generally, a plaque is picked from agar plates and added to a 1–2 ml aliquot of medium containing uninfected *E. coli* cells. After a 4–6 hours period of growth, the titer of phage particles in the medium reaches ~10^{12} pfu/ml, which is sufficient for the subsequent isolation of phage DNA or for storage as a stock solution. If required, larger volumes of supernatant can be used to scale-up the production of phage, ds DNA (RF), and ss DNA.

Preparation of Double Stranded DNA (RF)

The ds DNA (RF) can be isolated from small cultures of infected cells using procedures similar to those used to purify plasmid DNA (for details see Chapter 8). Several micrograms of ds DNA (RF) can be isolated from a 1–2 ml culture of infected cells, which is enough for subcloning and restriction enzyme mapping. However, the preparation of ds DNA (RF) is often contaminated with ss DNA of phage that can confuse the pattern of DNA fragments obtained by digestion with restriction enzymes. The ss DNA is visible as fuzzy, faster migrating bands in agarose gel electrophoresis. This confusion can however be prevented by analyzing in parallel an aliquot of undigested ds DNA (RF). Moreover, linear and circular ss DNAs are not cleaved by most of the restriction enzymes commonly used for cloning in M13 phage. The fragments of ds DNA (RF) with recessed 3′-termini can also be distinguished from ss DNAs by end-labeling using the Klenow fragments of *E. coli* DNA polymerase I.

Preparation of Single Stranded M13 DNA

M13 DNA in single stranded form (ss DNA) is prepared from phage particles secreted from the infected bacterial cells into the surrounding medium. The phage is first concentrated by precipitation with polyethylene glycol 8000 (PEG) in the presence of high salt (2.5 M NaCl). Thereafter, ss DNA is extracted with phenol and then collected by precipitation with ethanol. The resulting ss DNA preparation is pure enough to be used for various applications. The recombinant phages having DNA of 300–1,000 nucleotides provide ~5–10 μg of ss DNA/ml of infected cells, which is enough for 10–20 cycles of sequencing reactions.

10.4.6 Cloning Strategy with M13 Vectors

Foreign DNA fragments can be easily cloned into M13 vectors by following standard protocol of cloning. The ds DNA fragments are ligated with compatible cohesive or blunt termini into ds DNA (RF) molecules. The ligation mixture is then used to transform competent male *E. coli* cells, the products of which are plated on top agar containing IPTG and X-gal and incubated at 37°C for 6–8 hours. The blue (nonrecombinant) and white (recombinant) plaques are picked and propagated. Note that the blue-white screening (for details see Chapter 16) is the best strategy, but not a perfect one, opted for the selection of recombinants. In the case of very small (<100 nucleotides) segment of foreign DNA, which is inserted *in frame* into the *lac* Z′ of the M13 vector, the resulting fusion peptide retains sufficient α-complementing activity so that the recombinant generates a blue plaque in medium having IPTG and X-gal. Theoretically, there is no prescribed limit for the size of the fragment of foreign DNA that can be cloned in M13 phage. However, larger segments of foreign DNA (>5,000 nucleotides) are more prone to deletions and rearrangements than smaller fragments. Hence, it is best to clone the DNA up to 1,000 nucleotides in length into M13 phage. Additionally, this length of DNA is ideal for DNA sequencing from both directions, which is primed by the universal forward and reverse sequencing primers. For minimizing the deletion and other rearrangements of cloned DNAs the strategies that can be adopted include: (i) The length of foreign DNA is restricted; (ii) Recombinant phages are not propagated for more than one or two serial passages in culture; and (iii) While performing site-directed mutagenesis, ds DNA fragments are recovered as early as possible.

Basically there are two different approaches that are commonly used to construct M13 recombinants which are discussed below.

Cloning by Ligation of Insert to Linearized Vector

The vector DNA is linearized by digesting with a single unique restriction enzyme. After linearization, the vector DNA is dephosphorylated by alkaline phosphatase treatment so as to reduce its ability to recircularize during ligation and to increase the proportion of recombinant clones. The dephosphorylation step must be included when the amount of foreign DNA is less or when ligating blunt ended DNA fragments into M13 vectors, for example, when creating libraries of M13 subclones for sequencing. When ligation has to be performed with complementary cohesive termini, ~50 ng of linearized vector DNA and one- to five-fold molar excess of the target DNA are mixed in a microfuge tube. The resulting ligation mix is used for the transformation and the recombinant plaques are analyzed by blue-white screening (for details see Chapter 16). This technique is best for cloning a single segment of the DNA, which is purified from a gel, in preparation for site-directed mutagenesis, DNA sequencing, or synthesis of radiolabeled probes.

Hypothetically, recombinant phages generated by this approach can carry the foreign DNA sequences in two possible orientations. However, using this strategy, populations of recombinants having the insert in the same orientation are mostly obtained. This situation arises as sequences on opposite strands of the foreign DNA interfere in the functioning of the intergenic region. This problem can be avoided by cloning the foreign DNA at a different site within the multiple cloning sites or by performing directional cloning with the help of combination of restriction enzymes. Instead, ds DNA (RF) from the original recombinant is isolated, digested with the same restriction enzyme used for cloning, and the resulting DNA fragments are religated and used to transform *E. coli*. Therefore screening of the resulting white plaques may possibly yield clones carrying the insert in the opposite orientation. The best answer to this type of technical difficulty is to propagate the foreign DNA in phagemid vectors (for details see Chapter 12).

Forced or Directional Cloning

For directional cloning, vector DNA with asymmetric polylinker is digested with two different restriction enzymes, producing molecules with incompatible ends that are not able to be recircularized and hence do not require dephosphorylation step. Simultaneously foreign DNA is also prepared by digesting with same set of restriction enzymes. Thereafter, the vector is ligated to one- to three-fold molar excess of foreign DNA in a desired orientation within M13. This strategy of cloning is excellent when cloning of a single segment of foreign DNA is required in a desired orientation within M13. In contrast, it is not appropriate for the subcloning in M13 of large DNAs cloned in high-capacity vectors like cosmids, yeast artificial chromosomes (YACs), or P1-derived artificial chromosome (PAC) vectors. Directional cloning is also a type of additional burden when single stranded templates are required for sequencing or for production of radiolabeled strand-specific probes. If a universal primer is used for the sequencing reaction, the sequences obtained will be restricted to 200–700 nucleotides of foreign DNA immediately downstream from the primer-binding site. In the case of radioactive labeling of probes, only one of the strands of the foreign DNA is labeled. This problem can be easily solved by inserting the foreign DNA individually into each member of a pair of M13 vectors (e.g., M13mp18 and M13mp19) that carry the MCS in opposite orientations. The resulting single stranded template DNAs prepared from these recombinant phages definitely provide the terminal sequences of each of the two strands of the foreign DNA and radiolabeled probes can be generated that will hybridize specifically to one strand or the other of the foreign DNA.

10.4.7 Analysis of Recombinant M13 Bacteriophage Clones

The size and orientation of foreign DNA carried in M13 recombinants are analyzed by numerous methods. These include the following:

- Screening *lac*$^-$ (recombinants; white) and *lac*$^+$ (nonrecombinants; blue) M13 phage plaques by blue-white screening procedure by plating on medium containing X-gal and IPTG.
- Analysis of small-scale ds DNA (RF) by hybridization with sequence-specific probes (for details see Chapter 16).
- Analysis of *lac*$^-$ plaques by polymerase chain reaction (for details see Chapter 7) with the help of specific primers.
- Analysis of small-scale ds DNA (RF) preparations by restriction enzyme digestion (for details see Chapter 3).
- The size of ss DNA is determined by electrophoretic analysis (for details see Chapter 6) of putative recombinant clones. Thus, recombinant M13 phage clones carrying sequences of foreign DNA >200–300 nucleotides are detected by gel electrophoresis of ss DNA released from infected bacteria into the culture medium.

Review Questions

1. Give a brief account of the biology of filamentous phage.
2. Describe the genome map of M13 phage. Enumerate the functions of various M13 genes.
3. Give details about the process of DNA replication during the life cycle of M13 phage. Also explain its regulatory mechanism.
4. Give an elaborative account of the process of morphogenesis of M13 phage giving details of various proteins involved.
5. Describe the structural model of M13 phage.
6. Write a short note on unique properties of M13 phages that make them suitable as vectors.
7. Write in brief about following:
   - Development of M13mp1 vector from M13
   - Development of M13mp2 vector from M13mp1
   - Development of M13mp7 vector from M13mp2
8. Give an explanatory account of several genetic markers that are valuable for working with M13 vectors.
9. Enumerate different strategies normally used for screening of suitable host for proliferation of M13 phages.
10. Write the basic principle involved in the preparation of ss DNA and ds DNA from M13 phage.

Reference

Messing, J. et al. (1977). Filamentous coliphages M13 as a cloning vehicle: Insertion of a Hind II fragment of the lac regulatory region in M13 replicative form *in vitro*. *Proc. Natl. Acad. Sci.* **74**:3342–3246.

Yeast Cloning Vectors

Key Concepts

In this chapter we will learn the following:

- Why is yeast system required?
- Nutritional auxotrophic complementation for recombinant selection
- Transformation of yeast cells
- Yeast 2-μm plasmid
- Yeast cloning vectors

11.1 INTRODUCTION

Eukaryotic hosts, like yeasts, are preferred for studying the behavior and effects of eukaryotic gene products in an environment more closely related to their original source to understand the cellular functions unique to eukaryotes such as mitosis, meiosis, signal transduction, obligate cellular differentiation, etc., and for functional expression of eukaryotic genes. The latter has received much emphasis as far as genetic engineering is concerned. Though prokaryotic expression systems have been used for cloning eukaryotic cDNAs for the production of corresponding heterologous proteins, but in many instances, authentic versions of eukaryotic proteins are not obtained. Thus, some eukaryotic proteins synthesized in bacteria are reported to be unstable or lacking biological activity. This is because of lack of eukaryotic protein-specific folding and other posttranslational modifications that include proteolytic cleavage and chemical modifications of amino acids (e.g., phosphorylation, acetylation, glycosylation, hydroxylation, sulfation, γ-carboxylation, myristoylation, palmitoylation, etc.). Furthermore, if the starting material to be cloned is eukaryotic genomic DNA rather than cDNA, then eukaryote-specific posttranscriptional modifications including splicing are also problematic. In addition, despite careful purification procedures, bacterial compounds that are toxic or pyrogenic may contaminate the final product. However, when the heterologous protein is intended for medical use (e.g., therapeutic agents), it is especially important that the expressed protein is stable, biologically active, pyrogen-free, and essentially identical to the natural protein in all its properties. These requirements have led to the development of eukaryotic vectors with a major emphasis on gene expression; however, the current interest in the application of genetic engineering in yeast is the ability to clone very large pieces of DNA using yeast artificial chromosomes. This chapter provides a detailed account of all the yeast vectors, their properties, main concepts of cloning, and applications. Special emphasis is laid on yeast artificial chromosome, which is the highest capacity vector till date to be used in several whole-genome projects.

11.2 WHY IS YEAST SYSTEM REQUIRED?

Yeast constitutes an effective model system for understanding fundamental cellular processes and metabolic pathways in humans and has facilitated the molecular analysis of many human genes. Thus, several eukaryotic proteins, e.g., cell cycle proteins, signaling proteins, protein processing enzymes, and disease genes have first been discovered as yeast homologs. This has been possible due to our knowledge from long-standing expertise in yeast biotechnology, its ease of manipulation, genetic tractability, and its comparatively similar structure to eukaryotic cells in contrast to the prokaryotes (bacteria and archaea). It has been reported that at least 31% of the proteins encoded by yeast genes have human homologs and conversely, nearly 50% of human genes implicated in heritable diseases have yeast homologs. The following features of yeast make it an organism of choice for detailed analysis of eukaryotes as well as for cloning eukaryotic genes:

(i) Yeast is a single-cell eukaryote. It represents the simplest form of eukaryotic cell that shares the complex internal cell structure of plants and animals;

(ii) It has a short life cycle of 90 min and hence it is easy to grow and study. Its culture is also inexpensive to maintain;

(iii) Yeast strains are genetically well-characterized. Yeast genome was the first eukaryotic genome to be completely sequenced and detailed genetic maps for *S. cerevisiae* and *Schizosaccharomyces pombe* are readily available. It is estimated that yeast shares ~23% of its genome with that of humans;

(iv) Yeast is physiologically well-characterized. It is stable in both the haploid and diploid states. This particular feature is an essential prerequisite for detailed genetic analyses;

(v) Its haploid genome is of relatively low complexity with a size of ~12 Mbp and is packaged into 16 well-characterized metacentric chromosomes. Its 6,466 open reading frames (ORFs) exist in readily usable form, and annotated information is available through several databases;

(vi) Owing to the small genome size, it is easy to isolate and manipulate;

(vii) Yeast cells are much easier to grow in tissue culture than cells from higher eukaryotes;

(viii) It is grown readily in both small culture vessels as well as large-scale bioreactors;

(ix) Many auxotrophic and other markers of yeasts are known. These can serve as markers to select transformed clones simply by complementation (for details see Section 11.3);

(x) Several strains of yeast harbor a 2-μm plasmid. These naturally occurring plasmids can be used as part of an endogenous yeast expression vector system to attain high copy number or to mediate Flp/*frt*-based recombination (for details see Section 11.5);

(xi) It is a source of several strong promoters, which are used for efficient transcription of heterologous genes in yeast vectors (for details see Section 11.6.9). Several constitutive, repressible, and hybrid promoters that combine the features of different promoters are available from yeast;

(xii) It is possible to introduce naked DNA into yeast strains with ease by techniques such as generation of yeast protoplasts and their transformation using $CaCl_2$ or electroporation;

(xiii) Many yeast genes are functionally expressed in *E. coli*. This allows the isolation of a number of yeast genes for metabolically important enzymes by their ability to complement cognate genetic defects in *E. coli* mutants. The bacterial genes can also be expressed in yeasts, a step required for gene manipulation studies;

(xiv) It is capable of carrying out many posttranscriptional and posttranslational modifications (splicing, protein folding, covalent modification, glycosylation in case of glycoproteins, etc.) specific for a eukaryotic protein;

(xv) It normally secretes very few proteins. Hence, when yeast is engineered for extracellular release of a heterologous (recombinant) protein, it can be readily purified. The most commonly used signal sequence for yeast is derived from the mating α-factor gene. Also, synthetic leader sequences have been created to increase the amount of secreted protein;

(xvi) As yeast is used in baking and brewing industries for several years, U.S. Food and Drug Administration (FDA) has classified it as GRAS (generally recognized as safe) microorganism. Therefore, the use of this organism for the production of human therapeutic agents (drugs or pharmaceuticals) does not require the same extensive experimentation demanded for unapproved host cells. A number of vaccines, pharmaceuticals, and diagnostic agents produced in yeast are commercially available; and

(xvii) Yeast has a relatively high rate of recombination between homologous DNA sequences as compared with other eukaryotic model systems. This allows precise insertion of DNA sequences at specific locations within the yeast genome. This phenomenon enables a simple gene knock-out technique to be performed in yeast using a PCR-mediated approach and the integration of a selectable marker, and/or a mutant allele. Similar approaches can be used to introduce regulated promoters upstream of a given ORF or to introduce different epitope tags.

For further information on *S. cerevisiae*, see Exhibit 11.1. The utility of yeast as a model organism for eukaryotes is further advanced with the development of several yeast-based functional genomics and proteomics technologies (for details see Chapter 19). These approaches include utilization of the yeast deletion strain collection, large-scale determination of protein localization *in vivo*, synthetic genetic arrays, variations of the yeast two-hybrid system, protein microarrays, and tandem affinity purification (TAP) tagging.

Exhibit 11.1 *S. cerevisiae*: The simplest eukaryotic model organism

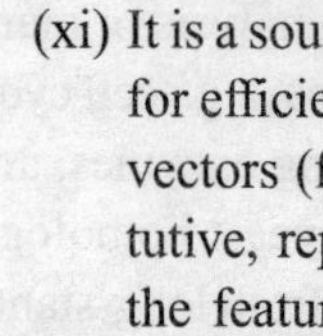

S. cerevisiae is one of the most intensively studied eukaryotic model organisms in molecular and cell biology, much like *E. coli* as the model prokaryote. The word '*Saccharomyces*' derives from Greek and means 'sugar mold'; '*cerevisiae*' comes from Latin and means 'of beer'. *S. cerevisiae* is perhaps the most important yeast owing to its use since ancient times in baking and brewing. It is one of the major types of yeast used in wine fermentation and ethanol production. This species is also the main source of nutritional yeast and yeast extract. Depending on its applications, other names for the organism are Brewer's yeast, Ale yeast, Top-fermenting yeast, and Baker's yeast. It is also commonly known as budding yeast because of its asexual reproduction by budding. Yeasts have also revolutionized the field of genetic engineering, where these play a significant role in gene expression and synthesis of biologically active eukaryotic proteins, genome-proteins, genome-wide analysis by cloning large insert DNA in yeast artificial chromosome and in detection of protein–protein interactions using yeast two-hybrid system.

S. cerevisiae is a unicellular fungus (eukaryote) belonging to family Saccharomycetaceae. It is widespread in distribution. These are frequently found growing saprophytically on substrates that contain sugar, such as decaying vegetables, ripe fruits, grains, sugary exudates of trees, and nectar of flowers. These are also found in soil containing abundance of humus, in decaying organic matter, and milk products. It can also be grown easily under laboratory conditions. Single yeast cells are colorless, but when grown on solid medium, colonies are formed by yeast cells which may be white or cream colored.

Yeast cells are very polymorphic and are capable of assuming different forms and shapes depending upon the growth medium. The cells are round to ovoid with 5–10 μm diameter. The cell wall is thin and delicate and is made up of two layers, an outer dense layer ~0.05 μm thick and an inner less dense layer ~0.2 μm thick, containing microfibrils. Enclosed within the cell wall is the granular cytoplasm, which in the old cells can be differentiated into an outer ectoplasm and inner endoplasm. Embedded within the cytoplasm are mitochondria, endoplasmic reticulum, nucleus, and ribosomes. The reserve food materials in the cytoplasm are in the form of glycogen, oil globules, and protein particles.

The development of yeast cells involves stable haploid as well as diploid stages. Both forms undergo a simple life cycle of mitosis and growth. However, under conditions of high stress, haploid cells generally die, while diploid cells can undergo sporulation, entering meiosis and producing a variety of haploid spores, which can mate (conjugate) reforming the diploid. There are two mating types, a and α, which show primitive aspects of sex differentiation and hence are of great interest.

The genome of wild-type haploid *S. cerevisiae* consists of 12,057,495 bp of linear ds DNA, i.e., approximately three times that of *E. coli* and constitutes ~90% of total DNA in the cell. There are two other independent genetic elements, viz. mitochondrial DNA and 2-μm plasmid. Some strains contain a third independent replicon, known as the killer plasmid that are ds RNA molecules coding for a toxin, which kills other susceptible yeast strains. The genomic DNA is packaged into 16 metacentric chromosomes ranging in size from 230 kbp to over 1,700 kbp. The genome comprises of more than 6,000 genes, but all of them are not believed to be true functional genes. Genetic crosses have established that the genome of *S. cerevisiae* has a total length of ~4,900 cM. This value has been confirmed by sequence analysis of cloned yeast DNA comprising the region between *leu* 2 and *cdc* 10 loci on the left arm of chromosome III.

The genome of *S. cerevisiae* is the first eukaryotic genome to be completely sequenced. It is striking that, with some exceptions, functionally related genes are generally not grouped together in yeasts, whereas such genes are often linked in prokaryotes. Thus, the three gene complexes atypical of genomic organization of yeast are *GAL* 1, 7, and 10 (coding for galactokinase, galactose transferase, and galactose epimerase, respectively) located on chromosome II and the five genes of the *ARO* 1 complex (coding for enzymes involved in aromatic amino acid biosynthesis) located on chromosome IV. Annotated information on the functions of 6,466 open reading frames and their corresponding protein products is available through several databases. These include the *Saccharomyces* Genome Database (SGD; www.yeastgenome.org), the Yeast Protein Database (YPD; www.proteome.com), the Munich Information Center for Protein Sequences (MIPS), Comprehensive Yeast Genome Database (CYGD; mips.gsf.de/genre/proj/yeast/index.jsp), and the Yeast Resource Center (depts.washington.edu/~yeastrc).

Source: Suter B, Auerbach D, Stagljar I (2006) *BioTechniques* 40(5): 625–644.

11.3 NUTRITIONAL AUXOTROPHIC COMPLEMENTATION FOR RECOMBINANT SELECTION

For bacterial vectors, a large number of antibiotic resistance genes are exploited for the selection of transformants, however, only a few antibiotics are available to which yeasts are sensitive. Thus, only a few yeast selection systems based on *E. coli* antibiotic resistance genes have been developed in the past several years. Since aerobically grown cells of yeast are sensitive to chloramphenicol, chloramphenicol acetyl transferase (*cat*) gene is used for direct selection. Resistance to aminoglycoside antibiotics is only rarely exploited. Some yeast cloning vectors also exploit genes conferring resistance to inhibitors such as methotrexate for selection of transformants. Nevertheless, in yeast with its extensive collection of mutants, selection based on complementation of mutant strains is preferred over dominant markers of the kind described earlier. Thus, most of the popular yeast vectors make use of the markers that can be selected readily in yeast, which often complement the corresponding mutations in host cell. Note that the mutations in the cognate chromosomal markers are recessive and nonreverting mutants are available. This complementation cloning approach is based on the selection of yeast DNA sequences in bacteria and exploits the remarkable fact that ~20% of all yeast genes can complement mutations of bacterial host. Although such experiments using bacterial hosts were very helpful in the initial stages of gene manipulation in yeasts, today it is preferable to characterize yeast genes by complementation in yeast itself. Thus, many yeast selection schemes rely on the functional expression of yeast genes in the host defective in biosynthetic pathway encoding products required for the growth of yeast. Such strains are called auxotrophic mutants because minimal growth medium must be supplemented with a specific nutrient for their growth, which the mutants cannot synthesize. In practice, the vector containing a functional (wild-type) version of the gene complements the function of mutated gene in the host strain growing in a specific nutrient drop-out minimal medium (i.e., agar medium lacking specific nutrient).

Table 11.1 Yeast Selectable Marker Genes Complement Nutritional Auxotrophy in Host Cells

Yeast selectable marker genes	Enzyme encoded by selectable marker gene	Complemented host mutations
HIS 3	Imidazole glycerol phosphate (IGP) dehydratase	*his* B
TRP 1	Tryptophan synthase	*trp* AB
LEU 2	β-Isopropyl malate dehydrogenase	*leu* B
URA 3	Orotidine-5′-phosphate decarboxylase	*pyr* F
ARG 4	Argininosuccinate lyase	*arg* H

Thus, in a medium lacking the specific nutrient, only the transformants harboring the functional gene are able to grow, while nonrecombinants do not survive. This selection system, called nutritional auxotrophic complementation (or rescue of nutritional auxotrophy), allows positive selection of transformants. In principle, every yeast gene for which there is a mutation available can be used as a selection marker in a transformation experiment. Table 11.1 lists some yeast genes and the corresponding mutants, which have been used for this type of complementation analysis. The details of some commonly used selection systems are summarized below.

leu 2$^-$ Host and *Leu* 2$^+$ Plasmid

Isopropyl malate dehydrogenase encoded by *leu* 2 gene is one of the enzymes involved in the conversion of pyruvic acid to leucine. In order to use *leu* 2 as a selectable marker, the host used is an auxotrophic mutant that has a nonfunctional *leu* 2 gene. Such *leu* 2$^-$ yeast is unable to synthesize leucine on its own (Figure 11.1). However, the host can survive if this amino acid (leucine) is supplied as a nutrient in the growth medium or the vector is *Leu* 2$^+$ that complements the function of the mutated gene by providing the enzyme isopropyl malate dehydrogenase. This is a good selection system because only the transformants contain a plasmid-borne copy of *Leu* 2 gene and hence are able to grow in the absence of leucine in the medium.

trp 1$^-$ Host and *TRP* 1$^+$ Plasmid

The gene locus *TRP* 1 for tryptophan synthase is involved in the biosynthesis of tryptophan and complements the *E. coli* mutation *trp* 1. Thus, transformants containing a plasmid-borne copy of *TRP* 1 gene are able to grow in the absence of tryptophan in the medium.

his 3$^-$ Host and *HIS* 3$^+$ Plasmid

The gene locus *HIS* 3 for imidazole glycerol phosphate (IGP) dehydratase is involved in the biosynthesis of histidine and complements the *E. coli* mutation *his* 3. Thus, the mutant host is incapable of growing in the absence of histidine in the medium, while transformants containing a plasmid that harbors a *HIS* 3 gene can grow on such nutritionally depleted medium.

lys 2$^-$ Host and *LYS* 2$^+$ Plasmid

LYS 2 encodes α-aminoadipate reductase, an enzyme that is required for the biosynthesis of lysine. The *LYS* 2 yeast gene has a marked advantage because both positive and negative selections are possible. Positive selection is carried out by auxotrophic complementation of the *lys* 2 mutation. On the other hand, negative selection is based on a specific inhibitor, α-aminoadipic acid (αAA), that prevents the growth of the prototrophic strains but allows growth of the *lys* 2 mutants (i.e., *lys* 2$^-$). Thus, *lys* 2$^-$ and *lys* 5$^-$ mutants, but not normal strains

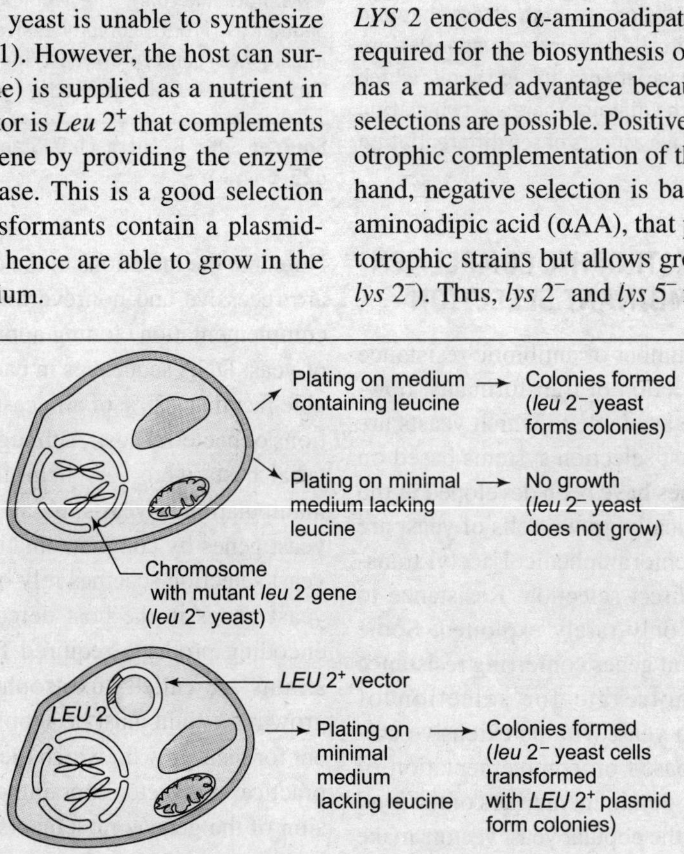

Figure 11.1 *LEU* 2 gene as a selectable marker in a yeast cloning experiment

grow on a medium containing lysine and αAA and lacking the normal nitrogen source. Apparently, *lys* 2 and *lys* 5 mutations cause the accumulation of a toxic intermediate of lysine biosynthesis that is formed by high levels of αAA, but these mutants still can use αAA as a nitrogen source. Numerous *lys* 2^- mutants and low frequencies of *lys* 5^- mutants can thus be conveniently obtained by simply plating high densities of normal cells on αAA medium. *LYS* 2-containing plasmids can be conveniently expelled from *lys* 2^- hosts. Because of the large size of *LYS* 2 gene and the presence of numerous restriction sites, this system is less commonly used in yeast vectors, as will be evident in Section 11.6.

ura 3^- Host and *URA* 3^+ Plasmid

URA 3 encodes orotidine-5′-phosphate decarboxylase, an enzyme required for the biosynthesis of uracil. Similar to *LYS* 2 system, both positive and negative selections are possible with *URA* 3 yeast gene. Positive selection is carried out by auxotrophic complementation of the *ura* 3 mutation in host strains by *URA* 3^+ plasmid. On the other hand, negative selection is based on specific inhibitor, 5-fluoro-orotic acid (5-FOA), which prevents growth of the prototrophic strains but allows growth of the *ura* 3 mutants (i.e., *ura* 3^-). Thus, *ura* 3^- (or *ura* 5^-) cells can be selected on media containing 5-FOA. The *URA* 3^+ cells are killed because 5-FOA is converted to the toxic compound 5-fluorouracil by the action of orotidine-5′-phosphate decarboxylase, whereas *ura* 3^- cells are resistant. The negative selection on 5-FOA medium is highly discriminating and usually less than 10^{-2} 5-FOA-resistant colonies are *URA* 3^+. This 5-FOA selection procedure can be used to produce *ura* 3 markers in haploid strains by mutation and more importantly for expelling *URA* 3-containing plasmids.

ade Mutant Host and *SUP* 4 Plasmid Vector (Red–White Screening)

The *ade* 1^- and *ade* 2^- yeast strains are defective in adenine biosynthesis pathway due to nonsense mutations in *ade* 1 and *ade* 2 alleles. These mutants are therefore unable to synthesize phosphoribosyl amino imidazole succinocarbozamide synthetase and phosphoribosyl amino imidazole carboxylase, respectively. These mutant yeast strains are grown on complete acid-hydrolyzable casein medium (AHC) in order to inhibit reversion of mutation (note for the growth of *ade* mutants, AHC medium containing high concentrations of adenine, up to 100 mg/l, is used). These *ade*$^-$ host strains form the basis of red–white screening, which further requires the presence of *SUP* 4 gene carried in vector, e.g., pYAC 4 (for details see Section 11.6.6). *SUP* 4 gene codes for suppressor tRNA and is designed to contain restriction sites for cloning. If no insert is introduced in the vector, the *SUP* 4 gene remains intact (i.e., forms active suppressor tRNA) and suppresses a mutation at the *ade* locus of appropriate strains of yeast, resulting in a change in the color of colonies from red to white in the presence of limiting concentrations of adenine in AHC medium (note for the initial construction of libraries to select for insert-containing vectors, AHC medium containing low concentrations of adenine, up to 20 mg/l, is used). Conversely, if an insert is present in the vector, the *SUP* 4 gene undergoes insertional inactivation and the nonsense mutation in the *ade* 2 allele prevents synthesis of the phosphoribosyl aminoimidazole carboxylase enzyme in the host. This leads to accumulation of phosphoribosyl glycinamidine in yeast cells, giving red or pink color to yeast colonies growing on AHC medium. Thus, yeast cells containing parent vector grow as white colonies under limiting adenine concentrations on AHC medium, while transformants carrying recombinant plasmids, in which the *SUP 4* gene is interrupted by insertion, grow as red colonies. This forms the basis of red–white screening, which was first reported by Herschel Roman (1957). Since that time, the scheme has been used with great elegance to monitor gene recombination events and to examine chromosome instability. Only the *ade* 1 and *ade* 2 mutants, but no other *ade*$^-$ mutants (e.g. *ade* 3^- or double mutant *ade* 2^- *ade* 3^-), produce a red pigment. These facts have been incorporated as a detection scheme in a number of diverse genetic screens. Examples of *ade*$^-$ red genetic screens include the detection of conditional mutations (plasmid shuffle), isolation of synthetic lethal mutations (synthetic enhancement and epistatic relationships), detection of yeast artificial chromosome (YAC) transformants, and the isolation of mutations that affect plasmid stability.

lac Z Reporter Gene Containing Translated Region of *HIS* 3 is Used to Identify *trans*-Acting Factors

Activities of promoters, protein–protein, and protein–DNA interactions involving promoter regions can be readily converted into selectable and quantifiable traits by fusing the promoter regions into reporter genes. Besides determining the levels of transcription or translation under various physiological conditions and identifying elements required for transcription by systematically examining series of mutations in the promoter regions, reporter genes are also used to identify *trans*-acting factors that modulate expression by transcription or translation. For such purposes, combining the *HIS* 3 selection with a *lac* Z screen is a commonly used strategy. For positive selection, the *E. coli lac* Z reporter gene includes the translated region of *HIS* 3 gene lacking the upstream-activating sequence (UAS). *HIS* 3^+ colonies arise when active promoters are formed, such as in the cloning of heterologous components required for the activation of a defined DNA segment. This approach of using two different reporters in parallel with the same promoter region is an efficient means for identifying *trans*-acting factors.

11.4 TRANSFORMATION OF YEAST CELLS

Like *E. coli*, fungi are not naturally transformable and artificial means are required to introduce exogenous DNA. Three techniques are commonly used for yeast transformation.

Spheroplasts Technique

This method was first developed for *S. cerevisiae*. In this method, the cell wall is removed enzymatically by lyticase or zymolyase and the resulting spheroplasts are fused with polyethylene glycol (PEG) in the presence of DNA and $CaCl_2$ (Ca^{2+} ions). The spheroplasts are then allowed to generate new cell walls in a stabilizing medium containing 3% agar. This is a highly efficient but laborious method, as the latter step makes subsequent retrieval of cells difficult.

Lithium Acetate Treatment

This is an efficient method for the introduction of DNA into intact yeast cells. The cells are first suspended in 0.1 M lithium acetate and then DNA and PEG are added. This is followed by a brief heat shock.

Electroporation

Electroporation provides a simpler and more convenient alternative to the use of spheroplasts. It involves subjecting cells to short pulses of electric current, thus creating transient pores through which DNA enters the cell (for details see Chapter 14). Cells transformed by electroporation can be selected on the surface of solid media, thus facilitating subsequent manipulation.

The intact cell methods are suitable for most applications, but spheroplasts are required for the transformation of YAC vectors (for details see Section 11.6.6). DNA can also be introduced into yeasts and filamentous fungi by conjugation. It has been found that enterobacterial plasmids, such as R 751 (Inc Pβ) and F (Inc F), facilitate plasmid transfer from *E. coli* to *S. cerevisiae* and *S. pombe*. The T-DNA of *A. tumefaciens* pTi plasmid is also conjugally transferable to protoplasts of *S. cerevisiae* and a range of filamentous fungi. T-DNA can also be transferred to hyphae and conidia.

11.5 YEAST 2-μM PLASMID

Plasmids are found in higher organisms although these are less common than in bacteria. Most strains of *S. cerevisiae* harbor a plasmid known as 2-μm circle or 2-μm plasmid. Owing to its presence in yeast, this plasmid is also called scp plasmid (*S. cerevisiae* plasmid). Similar plasmids are found in other species of yeast.

Genetic Organization

The 2-μm plasmid of yeast is a circular covalently closed molecule consisting of 6,318 bp of double stranded DNA (ds DNA). It is found in almost all strains of *S. cerevisiae* in 50–100 copies per haploid cell, which corresponds to ~2–3% of the entire chromosomal DNA. It is located in the nucleus of yeast cell, where like chromosomal DNA, it is bound by histones and forms nucleosomes. The plasmid has two perfect inverted repeats (*IVR* 1 and *IVR* 2) of 599 bp that separate the plasmid into two regions of 2,774 and 2,346 bp. These two inverted repeats can align. The plasmid contains one origin of replication (*ori*) of the size ~350 bp located predominantly on one of the inverted repetitions, although the origin also extends into the neighboring unique DNA sequences. The presence of this *cis* function is sufficient to allow replication. The primary structure of the 2-μm plasmid reveals different ORFs for proteins (Figure 11.2). The plasmid *rep* 1

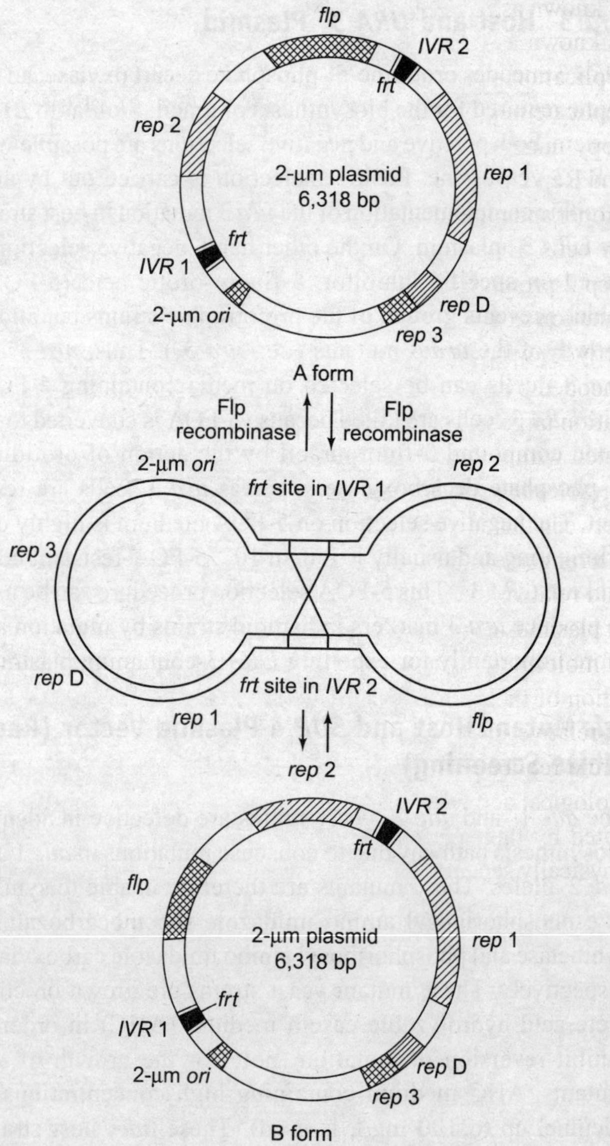

Figure 11.2 2-μm circle or 2-μm plasmid of yeast [Two forms of 2-μm plasmid formed by Flp/*frt*-mediated recombination are shown. The plasmid 'A form' has *ori* close to the *rep* 2, whereas the plasmid 'B form' has *ori* close to *flp* gene. Flp catalyzed site-specific recombination at *frt* sites lying within intervening regions (IVR 1 and 2) is shown in central figure, which converts A form of plasmid to B form, and vice versa.]

and *rep* 2 genes encode replication initiation proteins Rep 1 and Rep 2, respectively, which bind to the *ori* region. Besides, several host proteins are also employed in replication. The *rep* 3 gene is required in *cis* to the *ori* region for mediating the action of the *trans*-acting *rep* 1 and *rep* 2 genes. Rep 1 and Rep 2 proteins thus bind to the *rep* 3 DNA sequence and promote partitioning of the plasmid between cells at division. The replication occurs in the S-phase in the nucleus and is coupled to genome replication by unknown mechanism. However, the analysis of tetrads obtained from a cross of plasmid-containing strains (*cir*$^+$) with plasmid-free strains (*cir*^{0}) reveals that all haploid cells in an ascus contain 2-μm plasmids (note that a yeast strain containing a 2-μm plasmid is known as *cir*$^+$, while a yeast strain lacking a 2-μm plasmid is known as *cir*^{0}). The presence of *ori* is sufficient to allow replication and chimeric plasmids containing this region can replicate in *cir*^{0} strains, albeit with low copy numbers. High copy numbers are observed only in the presence of Rep 1 and Rep 2 proteins. These proteins are probably required for proper and random segregation of the plasmid into the daughter cells. Besides regulating plasmid replication, Rep 1 and Rep 2 proteins also regulate the expression of *flp* gene. The function of Rep D protein encoded by *rep* D gene is not exactly known. The Flp protein (Flp recombinase or flippase) encoded by plasmid *flp* gene recognizes 48 bp Flp recombination target sites (*frt* sites or flip recombination targets) located within the inverted repeats and catalyzes intra- and intermolecular site-specific genetic recombination or cross-over between them, which is very efficient in yeast. This results in rearrangement of genes and inversion of one-half of the plasmid relative to the other. These processes yield a population of plasmids consisting of roughly equal amounts of two types of molecules in the yeast cell, which differ only in the orientation of their unique sequences (Figure 11.2). This process of relative inversion controls DNA amplification by rolling circle replication. Although this plasmid shows a variety of biological activities, *cir*$^+$ and *cir*^{0} strains cannot be differentiated by their phenotypes; however, cloning in *E. coli* can physically separate these two forms. It has been demonstrated that these plasmids can be easily transferred into strains of *S. carlsbergensis* and other strains of *S. cerevisiae*, which normally do not contain endogenous 2-μm sequences. The Flp$^-$ strains, which have been transformed with suitably deleted plasmids, contain only one of these molecular species. The biological function of this autonomously replicating 2-μm plasmid is still unclear.

Applications of 2-μm Plasmid

The discovery of 2-μm plasmid has stimulated the development of cloning vector for yeast owing to its small size and autonomous replication. Due to high copy number, the plasmid also forms the basis of multicopy eukaryotic cloning vectors. The plasmid, however, lacks a selection marker gene, which can be inserted into it by genetic manipulation. The vector is then transformed into an amino acid/nucleic acid-defective auxotrophic mutant host cell. The transformed yeast cells are selected by plating onto minimal medium deprived of amino acid/nucleotide for which the host is defective. In such a medium, only transformed cells are able to survive and form colonies. Further, the incorporation of either a full copy of the 2-μm plasmid or a region encompassing the 2-μm *ori* and the *rep* 3 gene into a plasmid vector leads to the construction of yeast episomal plasmid (YEp) vectors (for details see Section 11.6.2).

The plasmid-encoded Flp recombinase is also used in genetic engineering to control the expression of a variety of genes by inverting segments of DNA. The two forms can be interconverted by recombination. In addition to the inversion reaction, Flp recombinase also promotes site-specific insertion and deletion reactions of segments flanked by *frt* sites. This site-specific Flp recombinase catalyzes a reciprocal ds DNA exchange between two DNA segments provided both segments carry very specific sequences. The target DNA sequences can be on the same or different DNA molecules. If these are on different molecules and one of the molecules is circular, then an insertion event will occur. However, if the two DNA molecules are linear, then segments of DNA will be exchanged (Figure 11.3). Yeast Flp recombinase also enhances

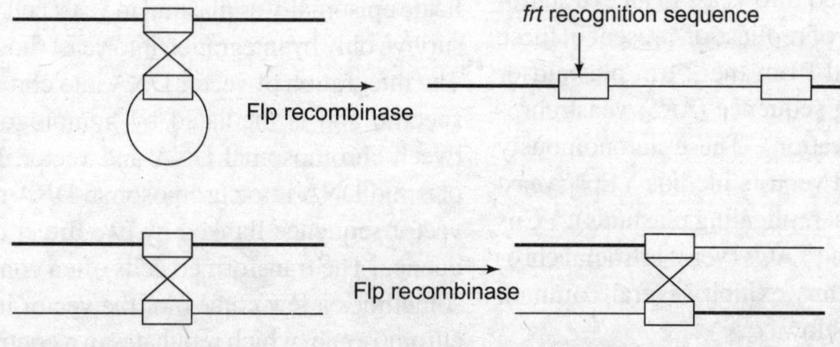

Figure 11.3 The action of Flp recombinase when the *frt* target sites are on different DNA molecules [Cre/*lox* recombinase system of P1 phage works on the same principle]

mitotic recombination in somatic tissues and allows efficient production of genetic mosaics. Note that the Flp/*frt* system is similar to the widely used Cre/*lox* P recombinase system of P1 phage. Flp is functional in bacteria, insects, plants, and animals provided the correct recognition sites are present. If *frt* sequences are inserted near the centromere of each of the four chromosomes of *Drosophila* using the method of P-element transposon-mediated transformation, and the Flp recombinase is introduced through transgenic *Drosophila* strains that contain the yeast Flp recombinase coding sequence under the control of the heat-inducible *hsp* 70 promoter, then Flp is synthesized in all the cells upon heat shock. Flp thus formed then binds to *frt* motifs in the two homologous chromosomes containing the gene of interest. This efficiently catalyzes mitotic recombination. By providing short pulses of heat, heterozygous flies are produced, which contain a null allele in a specific gene on one chromosome, while the homologous chromosome still contains the wild-type copy of the same gene. As there is no endogenous Flp recombinase in *Drosophila*, the heterozygous flies are stable and viable.

11.6 YEAST CLONING VECTORS

The standard procedure for cloning in yeast is to perform the initial cloning experiment with *E. coli* and to select recombinants in *E. coli* first. Most primary cloning experiments (i.e., the initial isolation of a gene or other DNA fragment from a target organism, its purification, and characterization) are carried out using bacterial hosts (usually *E. coli*) and the correct molecule is introduced into yeast. This is because it is easy to manipulate bacteria and a range of powerful techniques has been developed for the bacterial system. On the other hand, it is technically difficult to genetically manipulate yeast DNA as well as to recover recombinant DNA molecule from a transformed yeast colony. As such, all the yeast vectors, except yeast integrative plasmids (YIps), have been constructed as shuttle vectors (for details see Chapter 12) that are capable of replicating autonomously both in *E. coli* and yeast cells. Except for YIps, which lack yeast origin of replication and integrate into yeast genome, other cloning vectors are designed for autonomous replication so as to maintain the heterologous DNA introduced into yeast in an extrachromosomal state. The yeast origin of replication present in these shuttle vectors is either derived from the 2-μm plasmid or yeast autonomously replicating sequence (*ARS*; yeast chromosomal DNA origin of replication). These autonomously replicating yeast-based plasmid vectors include YEps (yeast episomal plasmids), YRps (yeast replicating plasmids), YCps (yeast centromeric plasmids), and YACs (yeast artificial chromosomes). The yeast vectors thus exhibit several common features, which are described below.

Presence of both *E. coli* and Yeast Origin of Replication (Shuttle Vectors) All the yeast vectors, except YIps, are shuttle vectors (for details see Chapter 12). Thus, all the vectors are capable of autonomous replication in *E. coli* either at high or low copy number, depending on the source of origin of replication. Besides, all the yeast vectors, except YIps, also contain yeast origin of replication (either from 2-μm plasmid or *ARS*) and can replicate in yeast as well. As YIps lack yeast origin of replication, these cannot multiply autonomously in yeast cells. Consequently, the only way for their multiplication is by integrating into the yeast genome (for details see Section 11.6.1).

Presence of Unique Restriction Enzyme Sites for Cloning All these vectors contain unique target sites, usually for a number of restriction endonucleases, which are used for cloning.

Presence of Selectable Markers for Selection Both in *E. coli* and Yeast All these vectors contain two separate selectable markers genes for easy and rapid analysis of respective *E. coli* and yeast transformants. The yeast selectable marker genes are often the ones that complement the corresponding mutations in host cells (*E. coli* and yeast strains) and check the reversion of auxotrophy, while antibiotic resistance genes are the most commonly used *E. coli* selection markers.

All these yeast vectors are readily distinguishable by their interaction with recipient cells. The principal features with respect to gene organization, cloning strategies, and selection criteria of different yeast vectors are discussed in the following sections. The properties, advantages, and disadvantages of yeast vectors are also presented in Table 11.2.

11.6.1 Yeast Integrating Plasmids

Yeast integrating plasmids are the first yeast vectors to be discovered. These are basically bacterial plasmids carrying a yeast gene with a selectable phenotype. These plasmids thus contain an *E. coli* origin of replication, selectable marker gene for selection in *E. coli*, yeast selection marker gene, and unique restriction enzyme site(s) for cloning, but lack yeast origin of replication (2-μm plasmid origin or *ARS*). The vector map of a yeast integrative plasmid, YIp 5 (size 5.5. kbp), which is basically pBR322 with an inserted *URA* 3 selectable marker gene, is shown in Figure 11.4a.

Due to lack of yeast origin of replication, YIps cannot replicate episomally as plasmid in yeast cell. These plasmids rather survive only by integration into yeast (host) chromosomal DNA. The integration of vector DNA into chromosomal DNA is site-specific and is mediated by homologous recombination between chromosomal DNA and vector. Integration of circular plasmid DNA into chromosomal DNA results in a copy of the vector sequence flanked by two direct copies of the yeast sequence. The transformed cells often contain only one copy (or sometimes a few copies) of the vector integrated into the host chromosome, which replicates in a controlled manner together with chromosomal DNA. The integration at multiple sites, however, occurs at low frequencies. YIps with two yeast

Table 11.2 Properties of Yeast Vectors

Vector	Bacterial origin of replication	Yeast origin of replication		Yeast *CEN* sequences	*TEL* sequences	Copy number per cell	Transformation frequency (Number of transformants per μg DNA)	Loss in nonselective medium (percentage/generation)	Disadvantages	Advantages
		2-μm plasmid origin	Yeast chromosomal *ARS* element							
YIp	+	–	–	–	–	1	10^2	Much less than 1%	Cannot be maintained episomally; Transformation frequency low; Plasmid copy number is low, and hence low yield; Recovery of plasmid from yeast is difficult. It can be recovered only by cutting chromosomal DNA with restriction endonuclease that does not cleave original vector containing cloned gene	It allows most stable maintenance of cloned genes (as compared to all other vectors); Integrated YIp behaves as an ordinary genetic marker, e.g., a diploid heterozygous for an integrated plasmid segregates the plasmid in a Mendelian manner; Most useful for surrogate genetics of yeast, e.g., can be used to introduce deletions, inversions, and transpositions; Suitable for cloning genes whose products inhibit cell growth, and are toxic for cell.
YEp	+	+	–	–	–	25–200	10^3– 10^5	1%	New recombinants can be generated *in vivo* by recombination with endogenous 2-μm plasmid.	The plasmid can be readily recovered from yeast; Plasmid copy number is high, hence high yield; Transformation frequency is high; Very useful for complementation studies.
YRp	+	–	+	–	–	1–20	10^4	Much greater than 1%; May integrate into chromosomal DNA	Transformants are usually unstable except for the ones that integrate into yeast genome.	The plasmid can be readily recovered from yeast; Copy number is high, but usually less YEp. This may be useful if cloning gene whose product is deleterious to the cell if produced in excess; Transformation frequency is high; Very useful for complementation studies;

(*Contd.*)

Table 11.2 *(Contd.)*

Vector	Bacterial origin of replication	Yeast origin of replication		Yeast *CEN* sequences	*TEL* sequences	Copy number per cell	Transformation frequency (Number of transformants per µg DNA)	Loss in nonselective medium (percentage/generation)	Disadvantages	Advantages
		2-µm plasmid origin	Yeast chromosomal *ARS* element							
										Can integrate into the chromosome, and may offer advantages similar to YIps.
YCp	+	–	+	+	–	1–3	10^4	Less than 1%	Low copy number; Recovery from yeast more difficult than that with YEps or YRps.	Low copy number is useful if product of cloned gene is deleterious to cell; Transformation frequency is high; Stable episomal maintenance; Very useful for complementation studies; Transformants are mitotically stable; At meiosis generally shows Mendelian segregation.
YAC	+	–	+	+	+	1–2	–	Depends on length; Stability increases with the length of YAC; Stable maintenance as chromosome if length is above 50 kbp.	Difficult to map by standard techniques; Transformation frequency is low; Copy number is low; Classical YACs had low insert capacity and suffered from the problems of instability and chimerism.	Highest capacity vector that permits DNA molecules greater than 40 kbp to be cloned (routinely employed for cloning ~600 kbp fragments, or even up to ~2 Mbp); Useful for generating libraries of large eukaryote genome. Widely used in genome projects; Circular versions available that are advantageous as compared to classical YACs; Can amplify large DNA molecules in a simple genetic background.

Note: + indicates present; – indicates absent.

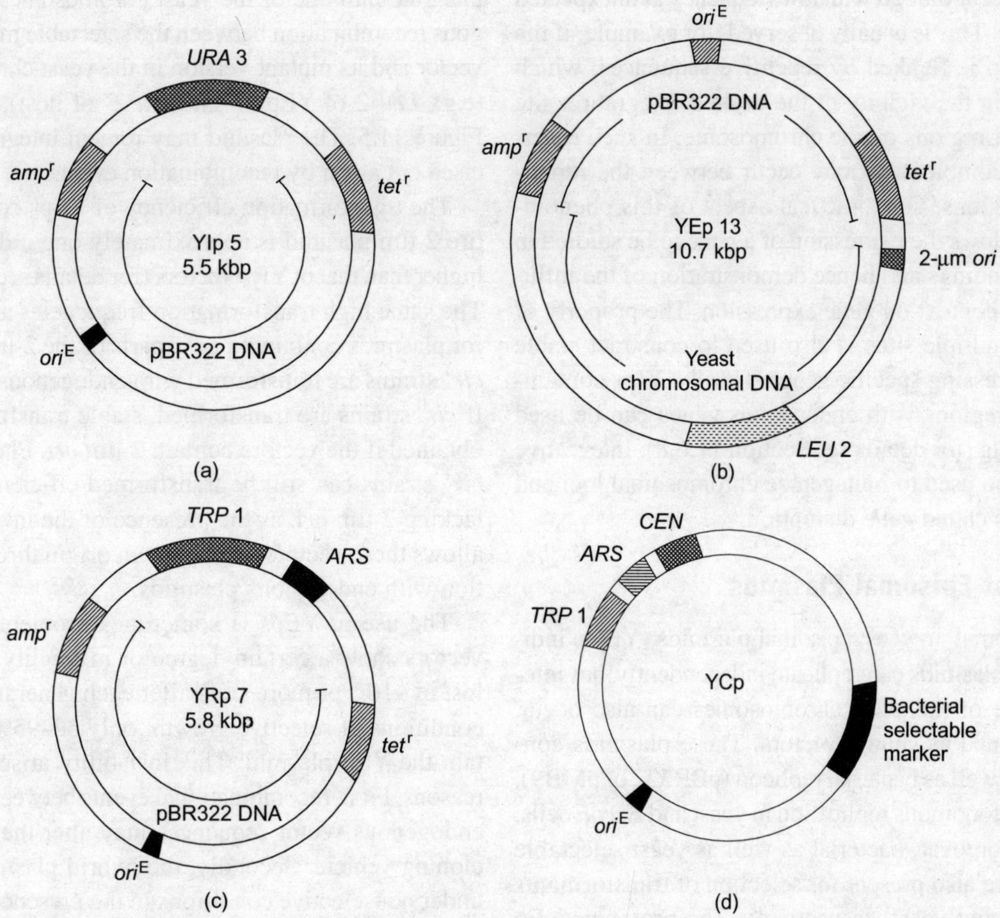

Figure 11.4 Vector maps of common yeast plasmids; (a) YIp 5, (b) YEp 13, (c) YRp 7, (d) generalized YCp [The arrangement of genes in the circular vector as well as important sequence elements are shown on the vector map. *amp*^{r} signifies ampicillin resistance gene; *tet*^{r} represents tetracycline resistance gene; *ori*^{E} is origin of replication for *E. coli*; 2-μm *ori* indicates origin of replication of 2-μm plasmid for replication in yeast at high copy number; *ARS* indicates autonomously replicating sequence for replication in yeast; *CEN* indicates centromeric sequence; and *URA* 3, *TRP* 1, *LEU* 2 are yeast selection markers. Note that the yeast origin of replication is absent in YIp 5.]

segments, for example, the gene of interest and selectable marker gene, have the potential to integrate at either of the two genomic loci, whereas vectors containing repetitive DNA sequences, such as Ty elements, can integrate at any of the multiple sites within the host genome. The site of integration is easily confirmed by examining the strains constructed with YIps by PCR analysis or other methods. Cutting the yeast segment in the YIps with a restriction endonuclease and transforming the yeast strain with the linearized plasmid can target the site of integration. The linear ends are recombinogenic and direct the integration to the site in the genome that is homologous to these ends. In addition, linearization increases the efficiency of integrative transformation to fifty-fold from ten-fold.

Transformation yields obtained with these vectors are comparatively low, which is ~1 transformant per μg of DNA per 10^7–10^8 cells. As the YIps integrate into host genome, strains transformed with YIp plasmids are extremely stable even in the absence of selective pressure. However, plasmid loss can occur at ~10^{-3}–10^{-4} frequencies by homologous recombination between tandemly repeated DNA, leading to looping out of the vector sequence and one copy of the duplicated sequence.

To recover the integrated YIps from the chromosome, special techniques are required. Chromosomal yeast DNA must be cleaved with a suitable restriction endonuclease, which does not cleave within the cloned sequence or the prokaryotic vector sequences. The mixture of yeast DNA fragments obtained after digestion is ligated at low DNA concentrations since this favors intramolecular reactions and desired plasmids are selected in *E. coli* cells.

Owing to various properties described earlier, YIps offer certain advantages. First, these vectors allow the introduction of genes into yeast cells. Moreover, their integration can be done at exactly their normal chromosomal locations. This particular feature permits the introduction of specific mutations into the yeast genome through substitution of normal alleles for mutated genes. Secondly, the integration of vector

DNA can also occur though with low frequency at unexpected or multiple sites. This is usually observed, for example, if the gene in question is flanked by repetitive sequences, which occur not only in the vicinity of the desired integration site but also in other regions on the chromosome. In such cases, homologous recombination may occur between the repetitive flanking regions. One practical aspect of this phenomenon is that it allows the expression of a gene to be studied in different surroundings and hence demonstration of the influence of genetic context on gene expression. The property of integration at multiple sites is also used to construct stable strains overexpressing specific genes. Thirdly, YIps containing homology regions with endogenous genes can be used for gene targeting (for details see Section 11.6.8). Integrative plasmids are also used to mutagenize chromosomal loci and the procedure is called gene disruption.

11.6.2 Yeast Episomal Plasmids

The word 'episomal' in yeast episomal plasmids (YEps) indicates that these plasmids can replicate independently, but integration into one of the yeast chromosomes can also occur. YEps are designed as shuttle vectors. These plasmids contain 2-μm *ori* as well as bacterial replicon (pBR322 or pMB9), which ensure autonomous replication in yeast and *E. coli* cells, respectively. Moreover, bacterial as well as yeast selectable marker genes are also present for selection of transformants in both *E. coli* and yeast, respectively. The prototype YEp vectors contain either a full copy of the 2-μm plasmid, or a region encompassing the 2-μm *ori* and the *rep* 3 gene. In most of the YEps, the latter case is prevalent and such vectors can propagate in cir^{+} hosts (i.e., hosts harboring the native 2-μm plasmid). One of the most common examples of YEps is YEp 13 (size 10.7 kbp), the vector map of which is presented in Figure 11.4b.

The 2-μm *ori* in YEps is responsible for the high copy number and high frequency of transformation. The copy number of most YEps ranges from 10 to 40 copies per cell of cir^{+} hosts. However, YEps are not equally distributed among the cells and there is a high variance in the copy number per cell in populations. Several systems have been developed for producing very high copy numbers of YEps per cell. These include the use of the partially defective mutation *leu* 2-d, whose expression is several orders of magnitude less than the wild-type *Leu* 2^{+} allele. The copy number per cell of such YEp *leu* 2-d vectors range from 200 to 300 and the high copy number persist for many generations after growth in leucine-containing medium without selective pressure. These YEp *leu* 2-d vectors are useful in large-scale cultures with complete media where plasmid selection is not possible. Due to high copy number, YEps are employed for the overproduction of gene products in yeast. Besides autonomous replication, entire YEp may also integrate into one of the yeast chromosomes due to homologous recombination between the selectable marker gene in the vector and its mutant version in the yeast chromosomal DNA (e.g., *Leu* 2 of YEp 13 and *leu* 2^{-} of host), as illustrated in Figure 11.5. The plasmid may remain integrated or later excised out again by recombination event.

The transformation efficiency of YEps containing the entire 2-μm plasmid is approximately one order of magnitude higher than that of YRp vectors (for details see Section 11.6.3). The same high transformation frequencies are also observed for plasmids containing only parts of the 2-μm plasmid if the cir^{+} strains are transformed with endogenous 2-μm plasmids. If cir^{0} strains are transformed, stable transformants are only obtained if the vectors contain 2-μm *ori*. Plasmid containing cir^{+} strains can still be transformed efficiently with vectors lacking 2-μm *ori*, as the presence of the inverted repetitions allows these vectors to acquire an origin through recombination with endogenous plasmids.

The use of YEps is sometimes problematic since these vectors show a certain degree of instability. Most YEps are lost in ~10^{-2} or more cells after each generation. Even under conditions of selective growth, only 60–95% of the cells retain the YEp plasmid. This instability arises due to certain reasons. First, recombinational events between exogenous and endogenous vector sequences may alter the structure of the cloning vehicle. Secondly, the hybrid plasmids may be lost under nonselective conditions in the presence of endogenous 2-μm plasmids with a frequency of 5–40% per cell division. This phenomenon is presumably due to the incompatibility of identical replicons competing for the same enzymes. The reason for the predominant loss of hybrid rather than endogenous plasmid vectors is still unknown. The problem can however be reduced by using strains, which do not contain

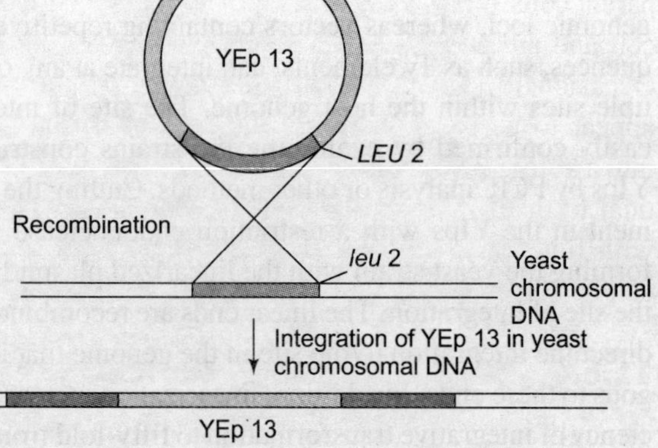

Figure 11.5 Recombination between plasmid *LEU* 2 and chromosomal *leu* 2 genes leads to integration of YEp 13 into yeast chromosomal DNA [After integration, two copies of the *leu* 2 gene are formed, usually one is functional, and the other is mutated.]

endogenous plasmids. Taking advantage of a better understanding of the biology of the 2-μm plasmid, newer versions of YEps are being constructed that are more stable.

11.6.3 Yeast Replicating Plasmids

In contrast to YEps that are 2-μm *ori*-based vectors, YRps are *ARS*-based vectors. The *ARS* sequences are believed to correspond to the natural replication origins of yeast chromosomes and contain signal for bidirectional mode of replication. Their presence allows autonomous replication as extrachromosomal elements presumably in the nucleoplasm of yeast cells. However, the *ARS* sequences can be replaced by origins of DNA replication of other lower eukaryotes and higher plant cells. *ARS* sequences are located very close to several yeast genes, including the ones that can be used as yeast selectable markers. The vector also contains ori^{E} and bacterial selectable marker genes for autonomous replication and selection in bacteria, respectively. The prototype of YRp vector, YRp 7 (size 5.8 kbp) contains the entire pR322 sequence and a 1.4 kbp *Eco* RI fragment of yeast DNA coding for the *TRP* 1 gene with a neighboring *ARS* region (Figure 11.4c). Since the yeast components of this plasmid have been cloned into the *Eco* RI sites, the cloning in *Eco* RI site is rendered more difficult.

In comparison to YEps, YRps exist at a lower copy number. YRps exhibit very high transformation efficiency (10^4 transformants per μg DNA), but the resulting transformants have been found to be exceedingly unstable. Studies have indicated that during selection of YRp 7 transformants in the medium lacking tryptophan, the plasmid is found in only 10–20% of cells after a few cell generations. In the presence of tryptophan, i.e., under conditions that also allow the growth of plasmid-free transformants, less than 1% of the cells retain the TRP^{+} phenotype. The rapid loss of these plasmids in a cell population is due to preferential segregation into the parent cells. YRps tend to remain associated with the mother cell and are not efficiently distributed to the daughter cell. Occasionally stable transformants may be obtained, and these appear to be cases in which the entire YRp has integrated into a homologous region on a chromosome in a manner identical to that of YIps. However, by providing YRp plasmids with functional centromeric DNA sequences, this problem can be solved (for details see Section 11.6.4).

11.6.4 Yeast Centromeric Plasmids

Though autonomously replicating YRps containing *ARS* transform yeast at high frequency, but the transformants are unstable and are lost at very high frequency, i.e., more than 10% per generation. This makes YRps undesirable as general cloning vectors. However, some studies have indicated that a few YRps that integrate into the yeast genome can be maintained stably through mitosis and meiosis. This stability was attributable to centromeric sequence (e.g., 1.6 kbp region lying between the *leu 2* and *cdc* 10 loci on chromosome III), which resulted in the plasmids behaving like minichromosomes. Genetic markers on the minichromosomes acted as linked markers, segregating in the first meiotic division as centromere-linked genes and were unlinked to genes on other chromosomes. Such centromeric sequence (*CEN*) contains a central AT-rich sequence of ~90 bp flanked by highly conserved regions of 11 and 14 bp. Taking advantage of this sequence, yeast centromeric plasmids (YCps) were constructed by incorporation of *CEN* sequences, having chromatin structure similar to the centromere region of yeast chromosomes, into plasmids containing *ARS*. *ARS* 1, which is in close proximity to a selectable marker *TRP* 1, is the most commonly used *ARS* element for YCp vectors although others have also been used. *CEN* 3, *CEN* 4, and *CEN* 11 are commonly used centromeres that can be conveniently manipulated. The prototype of YCps is YCp 50, which contains *CEN* 4, *ARS* 1, *TRP* 1, ori^{E}, and bacterial selection marker gene.

The YCps are thus autonomously replicating vectors that possess *ARS* and *CEN* sequences and hence behave as true chromosomes (Figure 11.4d). Thus, YCps show four characteristics of chromosomes. First, YCps replicate only once per cell cycle. Secondly, as compared to YRps, YCps are mitotically more stable in the absence of selective pressure. The loss amounts to ~10^{-2} cells per generation without selective pressure. Note that the mitotic stabilization function of *CEN* in YCp vectors is dependent on its three conserved domains designated I, II, and III; however, YCps do not possess the same stability as intact functional chromosomes. Moreover, remaining residual instability decreases with increasing length of such minichromosomes. Thirdly, YCps are evenly distributed to daughter cells, i.e., these plasmids segregate during meiosis in a Mendelian manner. In many instances, YCps segregate to two of the four ascospores from an ascus, indicating that these mimic the behavior of chromosomes during meiosis, as well as during mitosis. Finally, YCps are typically present at low copy number, from 1 to 3 per cell, and possibly more, in the host cell. These findings indicate that a 6–10 kbp DNA fragment called *CEN* is required for DNA molecules to segregate correctly in both meiosis and mitosis. This is because *CEN* sequences serve as attachment sites for mitotic spindle and 'pulls' one copy of each duplicated chromosome into each new daughter cell, resulting in efficient segregation of chromosomes and maintaining the plasmid at a low copy number (1–3) in the cell. The above-mentioned characteristic features make YCps ideal for construction of yeast genomic DNA libraries and for investigating the functions of genes altered *in vivo*.

11.6.5 Yeast Linear Plasmids

The vectors that replicate as plasmids in *S. cerevisiae* are often unstable under large-scale (≥10 l) growth conditions even in the presence of selection pressure. These plasmid-based yeast vectors tend to be lost from the culture as plasmid-free daughter cells accumulate. This is due to erratic partitioning during mitosis. Moreover, all the autonomously replicating plasmid vectors described in previous sections are maintained in yeast as circular DNA molecules and none of these vectors resembles the normal yeast chromosomes, which have a linear structure. Hence, an attempt was made to construct yeast linear plasmid (YLp), which was similar to YCps. YLps consist of telomeric sequences at both ends of a cleaved *ARS*-based plasmid. Telomeres preserve the integrity of the ends of DNA molecules, as these prevent end-shortening problem during replication and protect chromosomal ends from exonuclease action. This plasmid is however unable to replicate in *E. coli* and consequently it has drawn less attention for cloning. Nevertheless, YLps have raised the possibility to put together the complete set of genes in artificial chromosome of eukaryotes (for details see Section 11.6.6).

11.6.6 Yeast Artificial Chromosomes

Yeast artificial chromosomes (YAC) are yeast-based high-capacity vectors or supervectors constructed from the components of yeast chromosomes. These were first designed to study various aspects of chromosome structure and behavior, for example, to examine the segregation of chromosomes during meiosis. Later these artificial chromosomes were recognized for cloning DNA fragments >40 kbp in size. As with yeast plasmid vectors referred to earlier (YEps, YRps, and YCps), YAC vector is also propagated as a circular plasmid in *E. coli*. In this section, the gene organization, cloning strategy, operational problems, examples, and applications of the YAC vectors are discussed.

Gene Organization of YAC Vector

YAC vector mimics a chromosome and contains three key components of yeast chromosome, the significance of which has already been discussed in previous sections. These are: (i) The *ARS*, which is the position along the chromosome at which bidirectional DNA replication initiates in yeast. There may be one or more *ARS* in YAC vector; (ii) The centromeric sequence (*CEN*) required for proper segregation of chromosome in a Mendelian manner to daughter cells during cell division. *CEN* thus maintains the artificial chromosome at a low copy number (1–2) in the cell; and (iii) Two telomeric repeat sequences (*TEL*) to preserve the integrity of DNA ends. These *TEL* sequences are incomplete, but once inside the yeast nucleus, these incomplete sequences act as seeding sequences onto which telomeres will be built. In most YAC vectors, the *TEL* sequences are derived from *Tetrahymena*.

Besides *ARS*, *CEN*, and two *TEL* sequences, YAC also contains yeast selectable marker(s), ori^E, and bacterial selectable marker (amp^r). Because of the presence of ori^E, YACs can propagate as conventional plasmids in *E. coli*. In circular plasmid, a stuffer fragment separates two *TEL* sequences, which is removed when YAC is linearized. This generates linear molecules, the structure and topology of which resemble that of an authentic yeast chromosome. The two telomeric sequences thus appear at both ends of linearized plasmid, which help in maintenance of YAC in yeast as linear structures resembling a chromosome by providing stability to chromosome ends. Further, recognition site for the restriction enzymes, one for linearization of artificial chromosome and at least one for cloning purpose are also present.

There is a minimum size requirement for a linear centromeric plasmid to function as an artificial chromosome. This is because the size affects the segregation behavior of a chromosome. It has been found that linear plasmids containing the *LEU* 2 gene as a selectable marker, and the elements *CEN* 3 and *ARS* 1, flanked by the terminal 0.7 kbp *Tetrahymena* ribosomal DNA termini as telomeres, are less stable than any circular plasmid shorter than 10–16 kbp. However, with increasing length, mitotic stability of these constructs increases considerably. The estimated lengths of genuine yeast chromosomes range from 300 to 3,000 kbp.

Examples of YAC Vectors

Several YAC vectors have been developed but each one is constructed along the same lines. A typical example is given below:

pYAC 3 is essentially a pBR322 plasmid (i.e., contains Col E1 origin of replication and amp^r marker gene for selection in bacteria) into which a number of yeast genes have been inserted (Figure 11.6a). Two of these genes are yeast selectable markers *URA* 3 and *TRP* 1 that confer on *ura* and *trp* auxotrophs of yeast the ability to grow on media lacking uracil and tryptophan, respectively. As in YRp 7, the DNA fragment that carries *TRP* 1 also contains an origin of replication, *ARS*, but in pYAC 3, this fragment is extended even further to include the sequence called *CEN* 4, which is the DNA from the centromere region of chromosome 4. The *TRP* 1, *ARS*, *CEN* 4 fragment therefore contains two of the three components of the artificial chromosome. Two *TEL* sequences from *Tetrahymena* are also present at the ends. A stuffer fragment, flanked on either side by *Bam* HI restriction sites, is located between two *TEL* sequences. Further, a *SUP* 4 gene is also present that contains cloning site (*Sna* BI) into which new DNA can be inserted.

Another example of artificial chromosome, pYAC 4 (Figure 11.6b), is essentially similar to pYAC 3. It contains Col E1 origin of replication, amp^r, *ARS*, *CEN*, and two *TEL* sequences separated by a stuffer fragment similar to pYAC 3. The stuffer fragment is flanked on either end by *Bam* HI site.

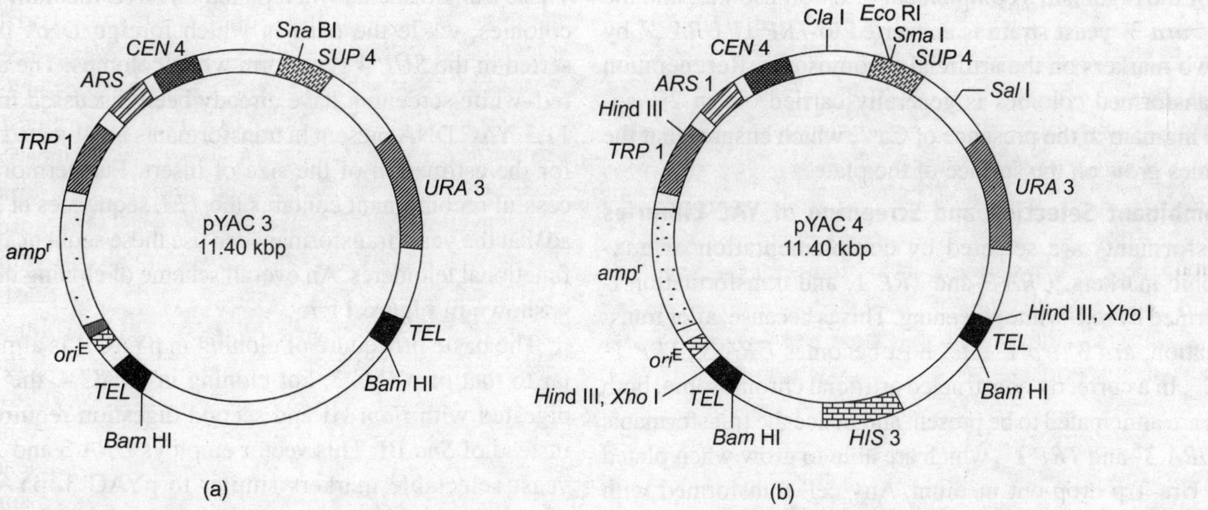

Figure 11.6 Vector maps of yeast artificial chromosomes, (a) pYAC 3 and (b) pYAC 4 [The general features important for construction of YAC libraries are depicted. The arrangement of genes in the circular vector, important sequence elements, and key restriction sites are shown on vector map. The genes or loci have same significance as indicated in Figure 11.4. *TEL* indicates telomeric sequences, *CEN* 4 is centromeric sequence from chromosome 4.]

The difference from pYAC 3 is that this stuffer fragment carries a *HIS* 3 gene that is employed to check the contamination of stuffer fragment during the cloning experiment. A unique *Sma* I site located within the *SUP* 4 gene is used for cloning.

Construction of YAC Library or Strategy for Cloning in YAC Vector

YAC is the highest insert capacity vector, and is widely used in whole genome projects. Cloning in YAC vectors is simple and involves following steps. pYAC 3 is taken as an example to describe the cloning procedure in YAC vectors.

Linearization of Vector DNA and Removal of Stuffer Fragment After amplification in *E. coli*, vector DNA is digested with *Bam* HI to remove the stuffer fragment located between two *TEL* sequences. This linearizes the vector DNA.

Separation of Left and Right Vector Arms The linearized vector is then digested with *Sna* BI at its unique site located within the *SUP* 4 gene, which serves as cloning site. This cuts the vector into two linear arms, namely, left and right arms. The left arm thus formed contains one of the *TEL* sequences, *CEN 4*, *ARS*, *amp*^{r}, Col E1 origin of replication, and *TRP* 1. The right arm contains the second *TEL* sequence and *URA* 3. Note that the presence of two yeast selectable markers on two different arms ensures that the recombinant contains both arms.

Dephosphorylation of Vector Arms The vector arms are dephosphorylated by alkaline phosphatase treatment to prevent their religation.

Restriction Digestion and Size Fractionation for Obtaining Insert DNA The high molecular weight genomic DNA extracted from the target organism is partially digested with a restriction enzyme, which generates blunt ends that can easily ligate with *Sna* BI-digested vector (note that *Sna* BI used for restriction of vector is a blunt end cutter that recognizes and cleaves the sequence TAC↓GTA). DNA fragments thus generated are fractionated by pulsed field gel electrophoresis (PFGE). Note that as the natural yeast chromosomes range from 230 kbp to over 1,700 kbp, YACs have the potential to clone Mbp-sized DNA fragments.

Ligation of Vector and Insert DNA Size fractionated DNA fragments of genomic DNA are ligated to a large molar excess of dephosphorylated vector arms. This leads to the production of artificial chromosome. Note that the presence of an extra chromosome in the form of YAC increases the size of genome by ~1 Mbp or more (depending on insert size).

Transformation of Host Cells Artificial chromosome is then introduced into yeast cell by spheroplast transformation. For this, the recipient *ade*$^-$ host strain, which is also double auxotrophic mutant, *trp* 1$^-$ *ura* 3$^-$ (e.g., *ade* 2-1 *ura* 3-52 *trp* 1-D host strain) is first grown in rich medium containing uracil and tryptophan and converted to spheroplasts by digestion of cell walls with lyticase or Zymolyase. The spheroplasts are then transformed with the products of ligation reaction in the presence of polyamines (to prevent degradation of DNA). The transformants are plated on Ura-Trp drop-out medium (agar medium lacking uracil and tryptophan). After regeneration of the cell walls, the transformed cells form colonies that carry a recombinant YAC and are prototrophic for uracil and tryptophan. Because pYAC 3 vector contains all the three elements required for autonomous replication, segregation of chromosomes between daughter cells, and chromosome stability, recombinant pYAC 3 introduced into yeast becomes part of the organism's complement of chromosomes, and the *trp* 1^- *ura* 3^- yeast strain is converted to *TRP* 1^+ *URA* 3^+ by the two markers on the artificial chromosome. Regeneration of transformed colonies is generally carried out in 2% sodium alginate in the presence of Ca^{2+}, which ensures that the colonies grow on the surface of the plate.

Recombinant Selection and Screening of YAC Libraries Transformants are selected by complementation of auxotrophic markers, *URA* 3 and *TRP* 1, and transformation is confirmed by red–white screening. This is because, after transformation, *ura* 3^- *trp* 1^- *ade*$^-$ host becomes *URA* 3^+ *TRP* 1^+ *ADE*$^-$. In a correctly constructed artificial chromosome, both arms are anticipated to be present and hence the transformants are *URA* 3^+ and *TRP* 1^+, which are able to grow when plated onto Ura-Trp drop-out medium. Any cell transformed with an incorrect artificial chromosome, containing two left or two right arms rather than one of each, is not able to grow on minimal medium as one of the markers is absent. The presence of the insert DNA in the vector can also be confirmed by testing for insertional inactivation of *SUP* 4, which is carried out by a simple red–white screening. This is possible because the transformants harboring recombinant pYAC 3, where foreign DNA is inserted in *SUP* 4 gene, are *ADE*$^-$. These transformants when plated on AHC medium form red colonies, while the ones in which foreign DNA is not inserted in the *SUP* 4 gene form white colonies. The details of red–white screening have already been discussed in Section 11.3. YAC DNA present in transformants is subjected to PFGE for the estimation of the size of insert. Furthermore, a successful recombinant contains the *TEL* sequences at each end so that the yeast transformant can use these sequences to build functional telomeres. An overall scheme of cloning in pYAC 3 is shown in Figure 11.7.

The basic procedure of cloning in pYAC 4 is almost similar to that of pYAC 3. For cloning in pYAC 4, the vector is digested with *Bam* HI and second digestion requires *Sma* I instead of *Sna* BI. This vector employs *URA* 3 and *TRP* 1 as yeast selectable markers similar to pYAC 3. pYAC 4 has another yeast marker gene, *HIS* 3, located within the stuffer fragment. This has an added advantage, where analyzing auxotrophic complementation in *his* 3^- host checks the contamination of stuffer fragment in cloning experiments (i.e., formation of nonrecombinants due to vector religation). Thus, the host strain used in combination with pYAC 4 is *ura* 3^- *trp* 1^- *his* 3^- *ade*$^-$, which after transformation becomes *URA* 3^+ *TRP* 1^+ *HIS* 3^- *ADE*$^-$.

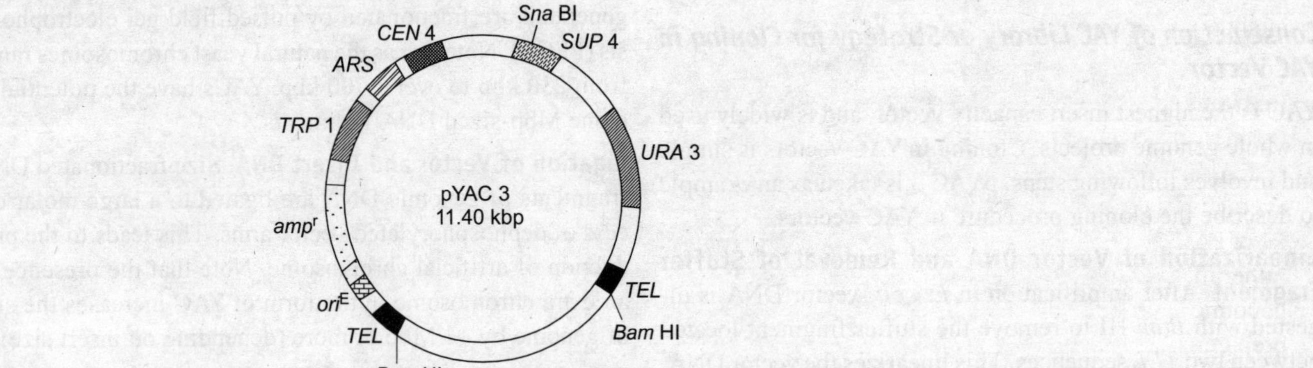

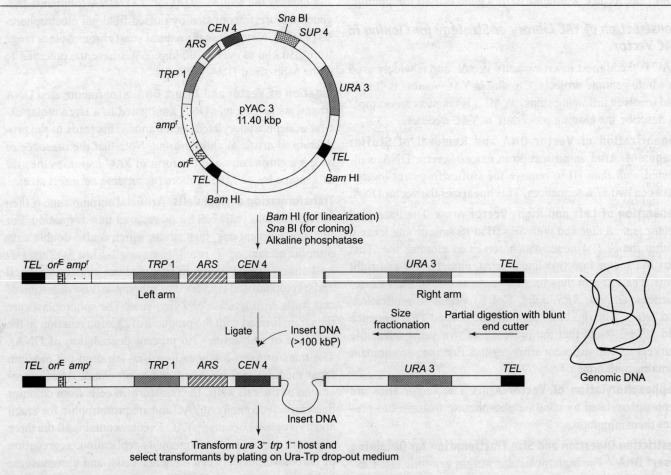

Figure 11.7 Cloning strategy with pYAC 3

In earlier times, unordered YAC libraries were screened by hybridization of radiolabeled probes to the DNA of lysed transformed colonies immobilized on filters. Today, transformants are generally picked, either manually or by robot, and assembled into ordered arrays in the wells of microtiter dishes. Copies of the arrayed library may be stored at –70°C, replica-plated at high density to membranes as needed, and screened by hybridization. Alternatively, the arrayed transformants may be organized into a hierarchy of pools that can be screened by multiple rounds of PCR. The number of individual clones decreases with every round of screening.

Characterization of YAC Libraries

Isolating a few dozen random clones and measuring the size of the inserts by PFGE generally assess the quality and complexity of a YAC library. Multiplying the average insert size by the number of full clones in the library and dividing the product by the size of the genome is used to calculate the extent of genome coverage (for details see Chapter 15). A library containing four genome equivalents is required to cover 95% of the genomic DNA sequence.

Operational Complexity Associated with YAC Vectors

A number of operational problems are associated with the use of YACs, many of which have now been overcome. These include:

Construction of YAC Libraries Demands Skill and Instrumentation Construction of genomic libraries in YACs is time-consuming, expensive, and demanding.

Earlier many YAC libraries contained inserts with an average size of only 100–200 kbp, not much bigger than a P1 or PAC clone. This problem was overcome by using PFGE rather than sucrose gradient centrifugation to fractionate the genomic DNA before cloning, by devising gentle methods to prepare and manipulate large fragments of genomic DNA, and by including polyamines and/or high molecular weight carrier DNA in the ligation and transformation reactions. At present, using these techniques, YACs with insert capacity up to 2 Mbp have been developed.

Due to fragility of large target DNA, it is susceptible to breakage while handling. The inclusion of carrier DNA, the ends of which are incompatible with those of the cloning vector, increases the viscosity of the solution and thereby offers passive protection to the target DNA molecules against damage by shearing. Furthermore, circular YACs are less susceptible to shear forces.

Efficient storage of genomic YAC libraries involves arraying clones into the wells of microtiter plates, whereas library screening often necessitates generating a hierarchy of pools of YAC clones from within the library.

As YAC clones are carried at low copy number, the recombinant molecules are also not easy to recover and purify. It is sometimes difficult to separate YACs from endogenous yeast chromosomes. Purification of intact YAC clones can be achieved only by specialized electrophoresis systems such as PFGE. Alternatively, the cloned segment of genomic DNA can be recovered in fragmented form by subcloning the total DNA extracted from an individual yeast transformant into cosmid or bacteriophage vectors. Subclones carrying genomic sequences are then identified, for example, by hybridization to probes containing highly repetitive segments of mammalian DNA.

This combination of skills and resources, expertise in handling large DNA molecules, manipulation of fragile molecules of DNA, fluency in yeast genetics, and access to robotic devices is rarely found in a single laboratory. The construction of new YAC libraries is therefore best left to specialists, whereas screening of existing YAC libraries is generally carried out in collaboration with Genome Centers or with increasing frequency on a commercial basis. YAC libraries of the genomes of many species are now available from these sources.

Instability and Rearrangement of Cloned Genomic Sequences YACs have problems with the stability of the insert, especially with very large fragments. This instability can be attributable to rearrangement by recombination between repetitive sequences in the cloned DNA. This recombination may be associated with high-frequency spontaneous deletions ranging in size from 20 to 260 kbp. These unstable clones thus tend to delete internal regions from their inserts, which usually results in the appearance of submolar bands of DNA of lower molecular weight when YACs from individual clones are analyzed by PFGE. Such deletions are generated both during the transformation process and during subsequent mitotic growth of transformants. Approximately 40% of the YACs from most libraries are deleted. It is reported that the frequency of deletion can be reduced by use of a strain rendered recombination-deficient as a result of a *rad* 52 mutation. However, such strains grow more slowly and transform less efficiently than RAD strains, and therefore are not ideal hosts for YAC library construction. Later, it has been shown that the *rad* 54-3 allele significantly stabilizes YAC clones containing human satellite DNA sequences. Strains carrying this allele can undergo meiosis and have growth and transformation rates comparable with wild-type strains.

Chimerism YACs also suffer from clone chimerism. Chimeric YACs are artifacts in which DNA fragments from distant regions of the target genome become joined together and propagated in a single YAC (i.e., inserts consist of more than one fragment of genomic DNA). Chimeras can be generated either during *in vitro* ligation of genomic DNA to YAC arms (e.g., coligation of DNA inserts *in vitro* prior to yeast transformation) or more rarely by mitotic recombination between different YAC clones introduced into the same yeast spheroplasts by cotransformation. Depending on the methods used to construct YAC libraries, between 10% and 60% of clones may be chimeric. However, the intact yeast cells are much less recombinogenic as compared with spheroplast. Chimerism is best detected by fluorescent *in situ* hybridization (FISH) of individual YACs to preparations of metaphase chromosomes. Hybridization to more than one chromosomal location, i.e., to two or more chromosomes, or to distinct regions of the same chromosome is indicative of a chimera.

Low Copy Number YACs are maintained at low copy number (usually one per cell) and hence the yield of DNA is not very high. Moreover, owing to low copy number, recombinant clones are difficult to purify from yeast chromosomes.

Low Transformation Efficiency YAC offers low transformation efficiency and hence it becomes difficult to recover the recombinant vector after transformation.

Despite these problems, YAC is the highest capacity vector till date and is used for a variety of applications indicated below.

Circular YACs have a Number of Advantages Over Classical YACs

To overcome the disadvantages of classical YACs, a method was developed to convert linear YACs into circular chromosomes that can propagate in *E. coli* as bacterial artificial chromosomes (BACs). The method involves a specialized BAC that contains a yeast centromeric sequence, a marker ($G418^r$) for selection in yeast, and two sequences homologous to those flanking the *Eco* RI site in a standard YAC (e.g., pYAC 4). These two sequences are known as hooks, which are separated by *Sal* I sequence, thereby enabling the BAC to be linearized (Figure 11.8). The linearized BAC is electroporated

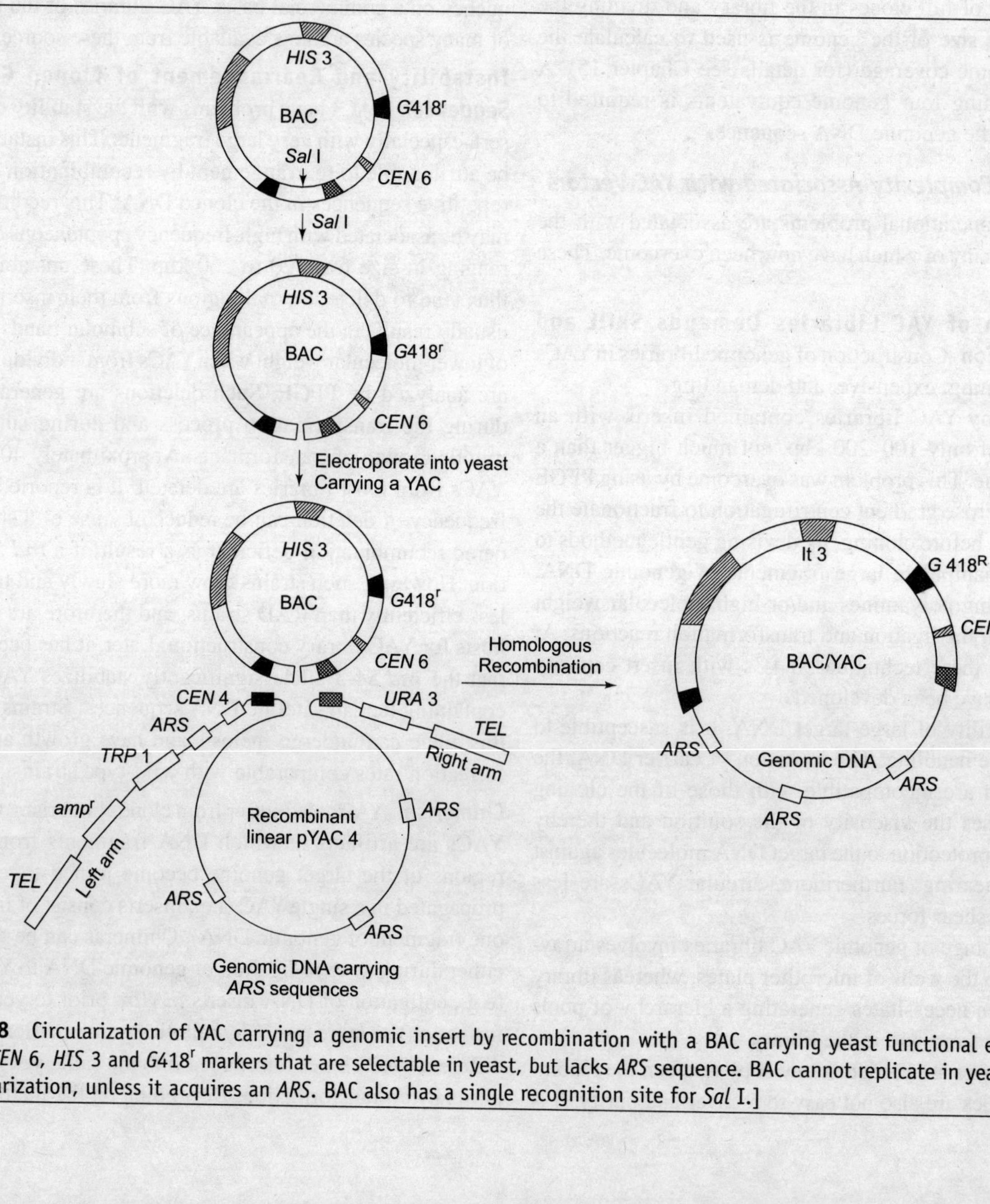

Figure 11.8 Circularization of YAC carrying a genomic insert by recombination with a BAC carrying yeast functional elements [BAC carries *CEN* 6, *HIS* 3 and $G418^r$ markers that are selectable in yeast, but lacks *ARS* sequence. BAC cannot replicate in yeast, even after recircularization, unless it acquires an *ARS*. BAC also has a single recognition site for *Sal* I.]

into a yeast cell carrying a standard linear YAC, and recombination results in the generation of a circular BAC/YAC and loss of telomeres of the original YAC. Note that the use of recombination in this application is identical to the principle of recombinogenic engineering. Such circular YACs are advantageous as compared with classical YACs. These include: (i) Circular YACs can be separated easily from linear yeast chromosomes using standard alkaline lysis methods and are much less susceptible to shear forces. Thus, molecules >250 kbp in size can be isolated intact; (ii) Circular YACs exhibit far greater structural stability than linear YACs and in the size range 100–200 kbp have comparable stability to BACs. In the larger size range, up to 40% of circular YACs are chimeric but it is possible to isolate stable clones; and (iii) Circular YACs up to 250 kbp can be electroporated into *E. coli*, where these behave as BACs.

Half-YACs

The difficulty of cloning DNA-containing centromeres and telomeres has led to some of the gaps in complete sequences. The absence of restriction sites in the telomeric repeats [e.g., $(TTAGGG)_n$ in case of human telomere] means that these are unlikely to be inserted into cloning vector. This problem can be solved by using half-YACs. These are circular yeast vectors containing a single telomere. On cleavage with the appropriate restriction enzyme, a linear molecule containing a single telomere at one end (half-YAC) is generated. This molecule is incapable of replicating in a linear form in yeast unless another telomere is added. Using such half-YACs, 32 telomere regions were linked to the draft sequence of the human genome.

Applications of YAC Vectors

Several important mammalian genes are >100 kbp in length (e.g., the human cystic fibrosis gene is ~250 kbp), a size beyond the capacity of all the most sophisticated *E. coli* cloning systems. Construction of YACs raised the possibility of using these vectors as vehicles for genes that are too long to be cloned as a single fragment in an *E. coli* vector. Standard YACs are now routinely used to clone ~600 kbp fragments and specialized versions called 'mega-YACs' are also available, which can accommodate inserts close to 2 Mbp. Note that by using the vectors with higher insert capacity for genomic library construction, the number of clones to be analyzed decreases. Unfortunately, the mega-YACs suffer from the problem of insert instability when the cloned DNA sometimes becomes rearranged by intramolecular recombination. Nevertheless, currently YAC is the highest capacity vector. The cloning of large fragment of DNA has also opened the way to studies of the functions and modes of expression of genes that was previously intractable to analysis by recombinant DNA techniques. YACs are employed for various purposes and some common ones are listed follows:

Large-scale Sequencing and Genome Analysis Because of large insert capacity, YAC vectors can easily accommodate any eukaryotic gene in its entirety along with the distant regulatory sequences and hence YAC clones are employed extensively in several genome projects for large-scale sequencing and genome analysis.

For DNA sequencing, YAC DNA subcloned into plasmid or bacteriophage M13 vectors can also be used. In this case, the aim is to produce YAC DNA that is free of contaminating yeast sequences. The YAC DNA is therefore purified by PFGE and then sheared into fragments of suitable size either by sonication or nebulization (for details see Chapter 1). The proportion of clones containing yeast sequences is greatly reduced if the YAC DNA is subjected to two cycles of purification by PFGE. The termini of the sheared DNA molecules are repaired and the fragments are fractionated according to size and ligated into the desired vector. Overlapping sequences obtained from single sequencing reactions of subclones chosen at random from these libraries can be used to construct small contigs, which then can be confirmed and extended by further rounds of directed sequencing (for details see Chapter 19).

Mapping Genomic Inserts Cloned in Individual YACs In most cases, screening a YAC genomic library yields several positive clones carrying overlapping inserts that span the region corresponding to the probe. These clones are checked for chimerism and their sizes are measured by PFGE. Groups of YACs are sometimes aligned by their content of polymorphic markers, or by their content of recognition sites for rare restriction enzymes, or more frequently by DNA fingerprinting. The latter involves digestion of total DNA (yeast plus YAC) isolated from each of the positive clones with several restriction enzymes. The products of digestion are then analyzed by Southern hybridization using repetitive DNA as a probe [e.g., small interspersed nuclear elements (SINEs) such as *Alu* DNA in case of YACs containing human sequence and *B*1 DNA in case of mouse]. Clones to be analyzed in detail are mapped with additional restriction enzymes. This involves partial digestion and Southern hybridization using probes specific to the pBR322 sequences in the left and right vector arms. In order to ensure accuracy of location of sites, most of the restriction sites are mapped from both ends of YAC vector. Because yeast DNA is unmethylated, the restriction map of YAC does not necessarily correspond to that of the homologous region of genomic DNA.

Identification of Ends of Genomic DNA Sequences Several well-validated techniques exist to identify and subclone the ends of genomic sequences cloned in YACs. All these techniques take advantage of the YAC vector sequences that lie immediately proximal to the cloned DNA. For example, the left arm of pYAC 4 and its derivatives contain the pBR322 origin of replication and a selectable marker. The genomic sequences immediately adjacent to the left arm can therefore be recovered by end-rescue subcloning. Alternatively, libraries of subclones can be screened by hybridization using probes specific for the plasmid sequences carried in the right and left ends of the vector. Among the PCR-based methods that can be used to amplify and clone genomic sequences adjacent to the left and right arms are inverse PCR, which uses vector sequences in outward orientation as primers and vectorette or bubble PCR, which uses an oligonucleotide cassette as a primer. Vectorette PCR is the better method since it not only amplifies insert sequences immediately proximal to each arm, but also provides template for DNA sequencing.

Creating Partial Diploids or Polyploids YACs and all other yeast vectors can be used to create partial diploids or partial polyploids and the extra gene sequences can be integrated or extrachromosomal.

Introduction of Mutation in Yeast Cell Homologous recombination can be used to introduce specific mutations at defined sites in segments of genomic DNA cloned in YACs. Thus, deletions, point mutations, and frameshift mutations can be introduced *in vitro* into cloned genes. The altered genes are then returned to yeast and used to replace the wild-type allele.

Generation of Transgenic Animals and Functional Analysis of Mammalian Genes An advantage of YACs over other genomic vectors is the ability to accommodate complete sequences of large mammalian genes. This has led to the growing use of YACs as vehicles to transfect mammalian cells and to generate transgenic animals. The capability of YACs to propagate in mammalian cells allows the functional analysis to be carried out in the organism in which the gene normally resides. In some cases, YACs are employed to complement mutations in mouse chromosomal genes.

11.6.7 Choice of Vectors

The availability of different kinds of yeast vectors offers great flexibility. Following four factors come into play when deciding which type of yeast vector is most suitable for a particular cloning experiment.

Transformation Frequency Vector choice is based on transformation frequency, a measure of the number of trans-formants that can be obtained per μg of plasmid DNA. A high transformation frequency is necessary if a large number of recombinants are needed or if the starting DNA is in short supply. Thus, YIps are the least preferred as the frequency of transformation is very low and it is difficult to recover the recombinants.

Copy Number The copy number, i.e., the number of vector DNA molecules per cell also influences the vector choice. This is important if the objective is to obtain protein from the cloned gene and to study its regulation. The greater the number of gene copies, the greater the expected yield of the protein product. Thus, to get higher yield of protein, YEps and YRps are preferred over YIps and YCps, while YIps and YCps are preferred for the isolation of genes whose products are toxic when overexpressed, e.g., the genes for actin and tubulin. Furthermore, low copy number vectors are employed if the overexpression of genes other than the gene of interest can suppress the mutation used for selection.

Stability of Transformants YIp produces very stable recombinant and loss of a YIp that has become integrated into a chromosome occurs at only a very low frequency. YRp recombinants are extremely unstable and the plasmids tend to congregate in the mother cell when a daughter cell buds off, so the daughter cell is nonrecombinant. YEp recombinants suffer from similar problems though an improved understanding of the biology of the 2-μm plasmid has enabled more stable YEps to be developed in recent years. YCps that contain *CEN* sequences are more stable than YRps. Nevertheless, YIps are the vector of choice if the need of the experiment dictates that the recombinant yeast cells must retain the cloned gene for long periods in culture.

Insert Capacity Among the cloning vectors, YACs have the highest insert capacity. Hence, YACs are preferred for the isolation of large and full-length genes, and have been successfully used in several genome projects.

11.6.8 Yeast Gene Targeting Vectors and Yeast Transplacement Plasmid

YIps containing homology regions with endogenous genes can be used for gene targeting. Two vector types are used for this purpose. Standard YIps possess a unique restriction site within the homology region, which favors a single cross-over and hence insertional interruption at the target locus. Multicopy integration vectors have been developed using this principle by targeting the reiterative rRNA genes. The second type of vector, yeast transplacement plasmid, stimulates a double cross-over and replaces a segment of chromosomal DNA, producing stable, single copy integration at a defined site. Similar principles are used in mammalian targeting vectors although homologous recombination is rare compared with random integration events in mammalian genomes.

11.6.9 Yeast Expression Vectors

One of the purposes of using yeast as a host is to express the cloned gene; hence yeast vectors are commonly designed as expression vectors. The importance of yeast for production of proteins by recombinant DNA methods is illustrated by the fact that the first approved human vaccine, hepatitis B core antigen, and the first food product, rennin, were produced in yeast. This is because yeast is considered as GRAS microorganism and cloning in yeast expression vectors offers some attractive features in contrast to bacterial expression systems. First, the proteins produced in yeast, unlike those produced in *E. coli*, lack endotoxins. Secondly, in certain special cases, the products produced in yeast have higher activity than those produced in *E. coli*, e.g., hepatitis B core antigen. Thirdly, in contrast to using *E. coli*, several post-translational processing mechanisms are available in yeast. As such, it has been possible to express several human or human pathogen-associated proteins with appropriate authentic modifications. In addition, heterologous proteins secreted from specially engineered strains are correctly cleaved and folded and are easily harvested from yeast culture media. Finally, the use of either homologous or heterologous signal peptides has also allowed authentic maturation of secreted products by the endogenous yeast machinery.

In principle, eukaryotic expression vectors do not differ from their prokaryotic counterparts except that the expression signals involved are applicable to *S. cerevisiae*. A basic eukaryotic expression vector has a eukaryotic promoter that drives the transcription of the DNA insert, eukaryotic transcriptional stop signals, a sequence that enables polyadenylation of mRNA, and a eukaryotic selectable marker gene (Figure 11.9). Numerous normal and altered yeast promoters are being used in expression vectors and these are chosen because of their high activity and sometimes because of their regulatory properties. Some of the promoters are derived from genes encoding alcohol dehydrogenase isozyme I ($P_{ADH\,1}$), phosphoglycerate kinase (P_{PGK}), glyceraldehyde-3-phosphate dehydrogenase (P_{GAPD}), enolase ($P_{Enolase}$), triose phosphate isomerase (P_{TPI}), galactokinase (P_{GAL1}), acid phosphatase ($P_{PHO\,5}$), α-mating factor, etc. (Table 11.3). Note that the latter promoter is usually employed along with the signal sequence of α-mating factor for the secretion of heterologous protein into extracellular medium. In expression vectors, these promoters include a transcription initiation site and variable amounts of DNA encoding the 5′ untranslated region. Because most of the yeast

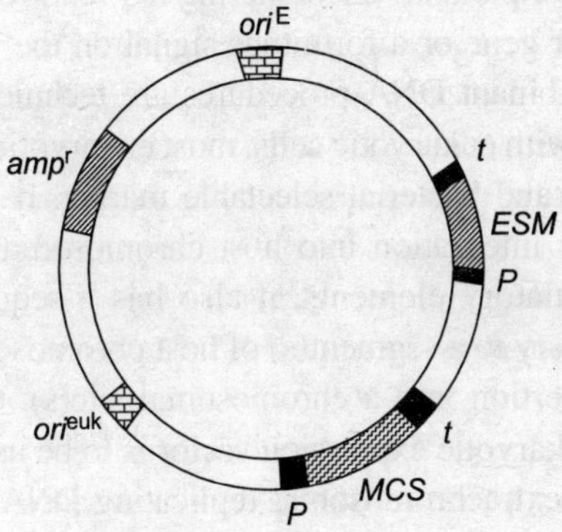

Figure 11.9 Generalized eukaryotic expression vector [Eukaryotic expression vector includes a eukaryotic transcription unit with a promoter (*P*), a multiple cloning site (*MCS*) for cloning of a gene of interest, a DNA segment with termination and polyadenylation signals (*t*), a eukaryotic selectable marker (*ESM*) gene system, an origin of replication that functions in eukaryotic cell (ori^{euk}), and origin of replication that functions in *E. coli* (ori^{E}), and an *E. coli* selectable marker gene (amp^{r}).]

expression vectors do not contain an ATG in the transcribed region of the promoter, the heterologous gene must provide an ATG in order to establish the correct reading frame corresponding to the N-terminus of the protein. It is essential that this ATG corresponds to the first AUG of mRNA because translation almost always initiates at the first AUG on mRNAs from yeast as well as from other eukaryotes. The 5′ untranslated region of the vector should also be similar to the naturally occurring leader region of abundant mRNAs by lacking secondary structures, and being A-rich and G-deficient, and by having an A at position -3 relative to the ATG translational initiator codon (i.e., Kozak sequence). Many of the expression vectors include a known signal for 3′-end formation of yeast mRNA although vectors lacking such defined signals syn-

Table 11.3 Common Yeast Promoters Used for Manipulation of Eukaryotic Gene Expression

Promoter	Gene	Properties
P_{PHOS}	Acid phosphatase	Expression in phosphate-deficient medium Phosphate repressed
$P_{ADH\,1}$	Alcohol dehydrogenase I	Expression in 2–5% glucose
$P_{ADH\,2}$	Alcohol dehydrogenase II	Glucose repressed
$P_{CYC\,1}$	Cytochrome c_1	Glucose repressed
$P_{GPGP\text{-}UT}$	Gal-1-P Glc-1-P uridyltransferase	Galactose induced
$P_{GAL\,1}$	Galactokinase	Galactose induced
P_{GAPD}, P_{GAPDH}	Glyceraldehyde-3-phosphate dehydrogenase	Expression in 2–5% glucose
$P_{CUP\,1}$	Copper metallothionein	Expression in 0.03–0.1 mM copper Copper induced
P_{PGK}	Phosphoglycerate kinase	Glucose induced
P_{TPI}	Triose phosphate isomerase	Expression in 2–5% glucose
$P_{GAL\,10}$	UDP galactose epimerase	Galactose induced
$P_{MF}\,\alpha_1$	Mating factor α-1	Constitutive Temperature-induced variant also available

thesize transcripts until encountering a 3′-end forming signal from another gene or a fortuitous signal on the plasmid. Because recombinant DNA procedures are technically difficult to carry out with eukaryotic cells, most eukaryotic vectors also contain ori^E and bacterial selectable marker. If the vector is designed for integration into host chromosomal DNA, then besides regulatory elements, it also has a sequence that is complementary to a segment(s) of host chromosomal DNA to facilitate insertion into a chromosomal site(s). On the other hand, if a eukaryotic expression vector is to be used as a plasmid, i.e., as extrachromosomal replicating DNA, then it also has a eukaryotic origin of replication. Numerous varieties of expression vectors are currently available for producing heterologous proteins in yeast and these are derivatives of YIp, YEp, YRp, and YCp. The choice of an expression system depends primarily on the quality of the recombinant protein that is produced, but the yield of product, ease of use, and cost of production and purification are also important considerations.

In most cases for expression of cloned DNA sequence, it is better to use a cDNA or synthetic DNA or genomic DNA lacking introns, with coding sequences under the control of yeast promoters. The cloning of specific cDNAs from other organisms and the study of their function using yeast as a surrogate does not necessarily require high-level expression of the foreign protein. In these instances, the aim is just to produce physiological quantities of the protein in a form that is correctly modified and localized in the cell such that the activity accurately reflects the activity in the original organism. However, commercial and laboratory preparations of proteins generally require high expression vectors.

Yeast expression vectors have been used successfully for the production of intracellular proteins as well as extracellular proteins. The details of production of secretory proteins are given in Section 11.6.10. For the production of intracellular proteins in *S. cerevisiae*, take an example of human Cu/Zn–SOD cloned in YEp expression vector (Figure 11.10). The vector comprised of *LEU* 2 as yeast selectable marker gene, 2-μm *ori*, amp^r for selection in bacteria, *E. coli* origin of replication, human Cu/Zn–SOD cDNA inserted between the promoter region of the yeast glyceraldehyde phosphate dehydrogenase gene (P_{GAPD}), and a sequence containing the signals for transcription termination and polyadenylation of mRNA from the same gene (GAPD*t*). After cloning, *leu* 2^- yeast strain was transformed with the vector and the cells were plated onto the medium deprived of leucine. Under these conditions, only cells with the functional *Leu* 2 gene supplied by the vector could grow. Continuous expression from P_{GAPD} during cell growth led to production of high levels of intracellular Cu/Zn–SOD having an acetylated N-terminal alanine residue, much like the authentic protein from human cells.

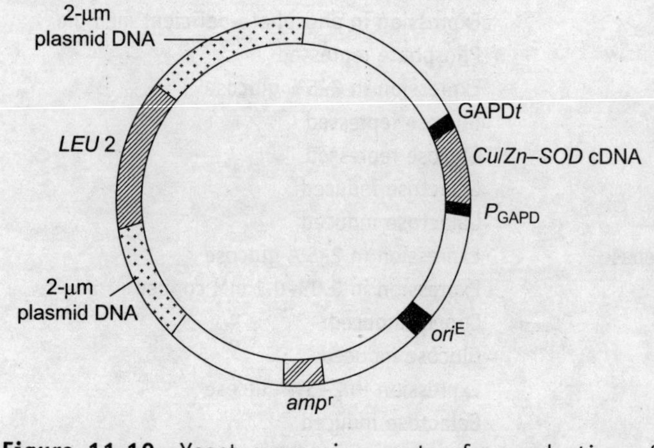

Figure 11.10 Yeast expression vector for production of intracellular protein

Yeast expression vectors have successfully been used for the commercial production of a number of vaccines, pharmaceuticals, and diagnostic agents (Table 11.4).

Table 11.4 Recombinant Proteins Produced by Yeast Expression Vectors

Vaccines
Hepatitis B virus surface antigen
Malaria circumsporozoite protein
HIV-1 (human immunodeficiency virus type 1) envelope protein
Diagnostics
Hepatitis C virus protein
HIV-1 antigens
Human therapeutic agents
Epidermal growth factor
Insulin
Insulin-like growth factor
Platelet-derived growth factor
Proinsulin
Fibroblast growth factor
Granulocyte-macrophage colony-stimulating factor
α_1-antitrypsin
Blood coagulation factor XIIIa
Hirudin
Human growth factor
Human serum albumin

11.6.10 Specialist Yeast Vectors

Different special-purpose yeast vectors have also been developed that incorporate the useful features found in the corresponding *E. coli* vectors, e.g., an F1 origin to permit sequencing of inserts, production of the cloned gene product as a fusion protein, secretion of heterologous protein in extracellular medium, surface display of the expression product corresponding to the cloned gene, etc. In this section, detailed description of the heterologous proteins secreted in the extracellular medium and displayed on surface is presented.

Secretion of Heterologous Proteins in Extracellular Medium for Easy Purification

The secretion of heterologous proteins encoded by cloned genes into the extracellular medium is usually preferred (for details see Chapter 12). This is because it is easy to isolate and purify proteins from extracellular medium rather than from cell lysate, and purification takes lesser time and is cost-effective. Furthermore, of the yeast-genome encoded proteins, only a few secretory proteins are released in the extracellular medium. As such, the protein encoded by cloned gene faces very little contamination and hence is easy to purify. The yeast-genome encoded secretory proteins released into the extracellular medium include pheromone α-mating factor and killer toxin. In the case of the α-mating factor, an 89 amino acid leader sequence is present, out of which the initial ~20 amino acids are similar to the conventional hydrophobic signal sequence. The presence of hydrophobic extension at their N-terminal end targets it to endoplasmic reticulum (ER). From there the protein reaches Golgi apparatus, where these sequences are cleaved after a Lys–Arg sequence at positions 84 and 85. Thus, to make a cloned gene-encoded heterologous protein secretory in nature, the leader DNA sequence of yeast α-mating factor (*MAT*α signal peptide) is inserted immediately upstream of the cloned cDNA in an expression vector. Furthermore, the Lys–Arg codons are located adjacent to the cDNA sequence. Thus, apart from the necessary leader sequence, the fusion proteins also contain the Lys–Arg residues that are required for endoproteolytic cleavage of the precursor by an endoprotease and the Glu–Ala sequence required for exoproteolytic processing by a dipeptidylaminopeptidase. Under these conditions, an active recombinant protein is secreted, which contains correct disulfide bonds and appropriate posttranslational modifications. Following removal of the leader peptide, the recombinant protein has the correct amino acid residue at its N-terminus. Using this methodology, a properly processed and active form of the protein hirudin has been synthesized and secreted by *S. cerevisiae* strain containing a YEp expression vector.

Additional strategies have been devised to enhance the secretion of recombinant protein by *S. cerevisiae* strains that require additional posttranslational modifications. For example, the overproduction of a naturally occurring secretory system enzyme such as protein disulfide isomerase (PDI), which promotes the proper folding of proteins during secretion, might increase the release of a recombinant protein, especially those with disulfide bonds. To test this hypothesis, the constitutive glyceraldehyde phosphate dehydrogenase promoter and a transcription terminator sequence were placed at the 5′- and 3′-ends, respectively, of cloned yeast *PDI* gene in a YIp expression vector and the entire gene construct (transcription unit) was integrated into a chromosome site. The modified strain showed a 16-fold increase in PDI production compared with the wild-type strain. When PDI overproducing cells were transformed with a YEp expression vector carrying the gene for human platelet-derived growth factor B, there was a 10-fold increase in the secretion of recombinant protein with disulfide bonds.

Surface Display

Phage-display system is used to detect protein–ligand interactions and to select mutant proteins with altered binding capacity. However, phage-display system often cannot display secreted eukaryotic proteins in their native functional conformation. Hence, yeast surface display has been developed in a manner similar to phage-display system.

Yeast surface display makes use of the cell surface receptor α-agglutinin (Aga), which is a two-subunit glycoprotein. The Aga 1p subunit (725 amino acid residues) anchors the

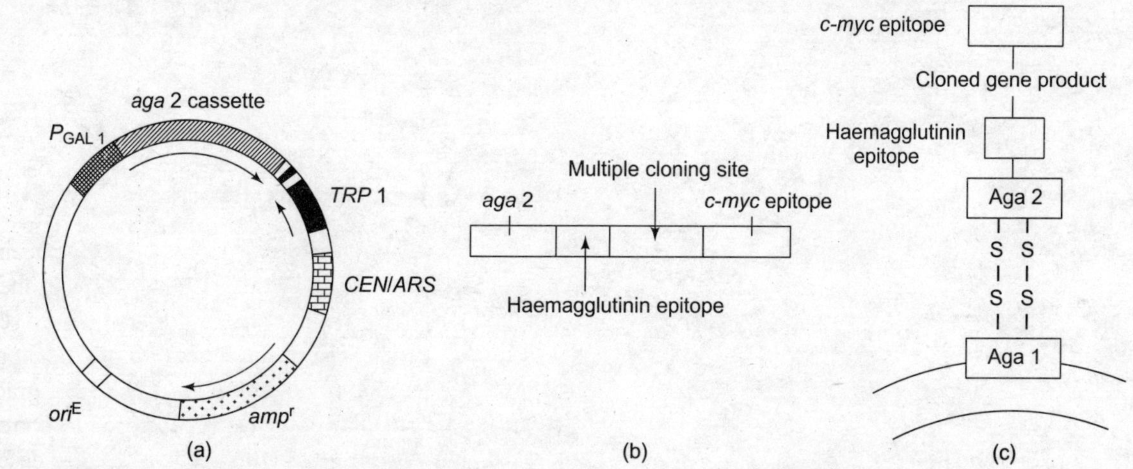

Figure 11.11 Surface display of heterologous protein in yeast; (a) vector used for surface display, (b) details of *aga* 2 cassette, and (c) schematic representation of surface display

assembly to the cell wall via a covalent linkage, while the Aga 2p binding subunit (69 amino acid residue) is linked to Aga 1p by two disulfide bonds. To achieve surface display, the appropriate gene is inserted at the C-terminus of a vector borne *aga* 2p gene under the control of the promoter, $P_{GAL\ 1}$. The construct is then transformed into a yeast strain carrying a chromosomal copy of the *aga* 1p gene, also under the control of $P_{GAL\ 1}$. If the cloned gene has been inserted in the correct translational reading frame, its gene product is synthesized as a fusion with Aga 2p subunit. The fusion product associates with Aga 1p subunit within the secretory pathway and is exported to the cell surface (Figure 11.11).

Review Questions

1. Describe the purpose of cloning in yeast. What properties of yeast make it model system for higher eukaryotes?
2. Write in brief about:
   (a) Red-white screening
   (b) Yeast selectable markers and complementation of nutritional auxotrophy
   (c) Significance of *lac* Z reporter gene containing translated region of *HIS* 3
   (d) Yeast promoters
   (e) Recombinogenic engineering and retrofitting
   (f) Significance of *ARS* and *CEN* sequences in yeast vectors
   (g) Methods of yeast transformation
   (h) Significance of Flp/*frt* recombinase system *in vivo* and *in vitro*
3. Comment on following statements:
   (a) The most commonly used selection system of bacteria is based on antibiotic resistance, but not in yeast.
   (b) YACs behave as chromosomes.
   (c) YEps are high copy number plasmids.
   (d) YCps are mitotically and meiotically more stable than YRps.
4. Discuss the properties, advantages, and disadvantages of autonomously replicating yeast vectors.
5. Discuss in detail the strategy used for cloning in YACs. How are the recombinants selected? Also describe the strategy for construction of circular YAC, and its advantage over classical YACs.
6. Enumerate various applications of cloning in YACs. Also discuss operational problems while working with YAC vectors.
7. Describe the features of eukaryotic expression vector. How can a heterologous protein be made to secrete by yeast in extracellular medium?
8. What are the disadvantages of traditional methods of isolation of gene sequences? Give an outline of a faster strategy based on recombination that is used for isolation of large genes.

Reference

Roman, H.L. and F. Jacob (1957). Effet de la lumière ultraviolette sur la récombinaison génétique entre alleles chez la levure. *C. R. Acad. Sci.* **245**:1032–1034.

12 Vectors other than *E. coli* Plasmids, λ Bacteriophage, and M13

Key Concepts

In this chapter we will learn the following:

- Chimeric vectors
- Gram negative bacteria other than *E. coli* as cloning vectors
- Gram positive bacteria as cloning vectors
- Plant and animal viral vectors
- P1 phage as vector
- Fungal systems other than yeast
- Artificial chromosomes
- Shuttle vectors
- Expression vectors
- Advanced gene tagging/trapping vectors

12.1 INTRODUCTION

Previously, only a limited number of vectors based on plasmids, bacteriophage λ, and M13 were available. In course of time, a series of other vectors were designed and constructed. An enormous range of vectors are now available, developed by a combination of elements known as chimeric vectors, which includes cosmids, phasmids, and phagemids. Furthermore, viral vectors and vectors derived from Gram negative bacteria (other than *E. coli*, e.g., *Agrobacterium tumefaciens*) and Gram positive bacteria have been devised for special purposes. The vectors derived from artificial chromosomes are high-capacity vectors and ideal for cloning large pieces of DNA and therefore best-suited for genome sequencing projects. Further, some expression vectors having different features that facilitate cloning and expression in a single vector have been developed. Nowadays, various vendors supply vector DNAs and associated reagents. This chapter offers an elaborative account of the properties, main concepts of cloning, and applications of all these vectors.

12.2 CHIMERIC VECTORS

The vectors containing DNA sequences from more than one source are called chimeric or hybrid vectors. These hybrid vectors have been constructed to combine useful properties of different vectors. Some common examples of chimeric plasmid vectors are cosmids, phagemids, phasmids, and fosmids. In this section, the properties of these hybrid vectors and the advantages offered by each are described in detail.

12.2.1 Cosmids

Plasmids are not suitable for cloning large DNA fragments. This is due to decrease in transformation frequency beyond the acceptable limit and cloned fragments or their parts very often undergo rearrangement or deletion. To overcome the problems associated with plasmids and taking advantage of *in vivo* and *in vitro* packaging characteristics of bacteriophage λ (i.e., DNA between *cos* sites in a concatemer is packaged into phage head), chimeric plasmids called as cosmids were constructed. Thus, cosmids are hybrid vectors comprising of a part of plasmid sequences and one or two copies of *cos* sites of bacteriophage λ. Plasmid sequences include origin of replication (Col E1 replicon), one or two selectable marker genes, and unique restriction enzyme sites for cloning. The *cos* site(s) contain all of the *cis*-acting elements required for packaging of phage DNA into bacteriophage particles. Recall that the presence of *cos* sites in bacteriophage λ aids in circularization of λ genome when internalized into host cell and forms concatemers in the presence of insert DNA at high concentrations (for details see Chapter 9). The concatemers of cosmid molecules linked at their *cos* sites act as substrate for *in vitro* packaging because the *cos* site is the only sequence that a DNA molecule needs in order to be recognized as a 'λ genome' by the proteins that package DNA into infectious λ phage. Note that the particles containing cosmid DNA are as infectious as real λ phages, but once inside the cell, the cosmid cannot direct the synthesis of new phage particles and instead replicates as a plasmid without the expression of the phage functions. The hosts containing cosmids do not form plaques; rather such hosts form colonies. Packaging the cosmid recombinants into phage coat imposes a desirable selection upon their size. Thus similar to λ-based vectors, the upper limit for the size of the cloned DNA is set by the space available within the λ phage head or simply between 37 and 52 kbp. As a cosmid is small (~5–8 kbp), it has high insert capacity (i.e., 30–45 kbp).

Types of Cosmids

Depending on the number of *cos* sites, cosmids can be single *cos* cosmids or dual *cos* cosmids.

Single *cos* Cosmids Cosmids containing one *cos* site are single *cos* cosmids. These contain at least one unique restriction enzyme site. The examples of single *cos* cosmids are MUA3, Homer I, Homer II, pJC79, and pJB8. The structural features of these plasmids are presented in Table 12.1 and vector map of pJB8 in Figure 12.1.

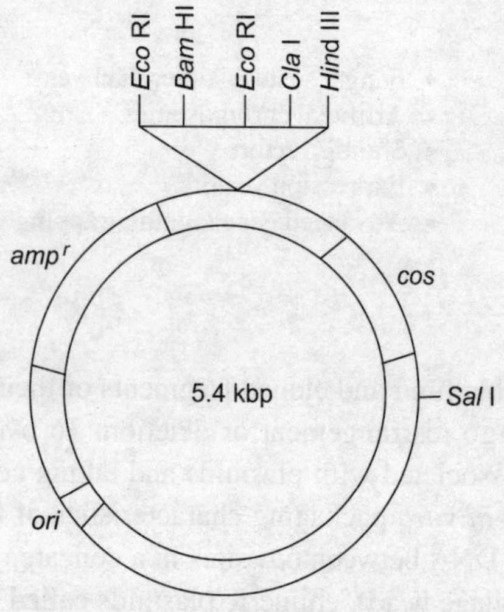

Figure 12.1 Vector map of single *cos* cosmid, pJB8 [The vector contains single *cos* site, unique restriction site for *Bam* HI that is used both for linearization and cloning, and *amp*^r gene for selection of recombinants.]

Table 12.1 Structural Features of Few Single *cos* Cosmids

Cosmid	Size (kbp)	Cleavage sites	Size of insertion (kbp)
MUA3	4.76	*Eco* RI, *Pst* I, *Pvu* II, *Pvu* I	40–48
Homer I	5.4	*Eco* RI, *Cla* I	30–47
Homer II	6.38	*Sst* I	32–44
pJC9	6.4	*Eco* RI, *Cla* I, *Bam* HI	32–44
pJB8	5.4	*Bam* HI	32–45

Dual *cos* Cosmids Dual *cos* cosmids contain two *cos* sites, which are separated by a recognition site for a restriction enzyme that cleaves the vector only once. These vectors have been constructed to reduce the number of steps involved in a cloning experiment. Moreover, cloning in dual *cos* cosmids does not require fractionation of partial digests of genomic DNA before ligation and packaging. The examples of dual *cos* cosmids are c2XB, Supercos-1, and pLFR-5 (Figure 12.2).

Many different single or dual *cos* cosmid vectors are now available, which may have specialized functions. Among these is the insertion of selectable marker genes for drug resistance, which may be used to establish mammalian cell lines that have incorporated cosmid sequences after transfection. Other specialized functions include insertion of bacteriophage promoters for the production of RNA probes complementary to the termini of the cloned genomic DNA sequences or recognition sites for restriction enzymes that cleave mammalian DNA very rarely (e.g., *Not* I, *Sal* I, *Sac* I, *Pac* I), thereby

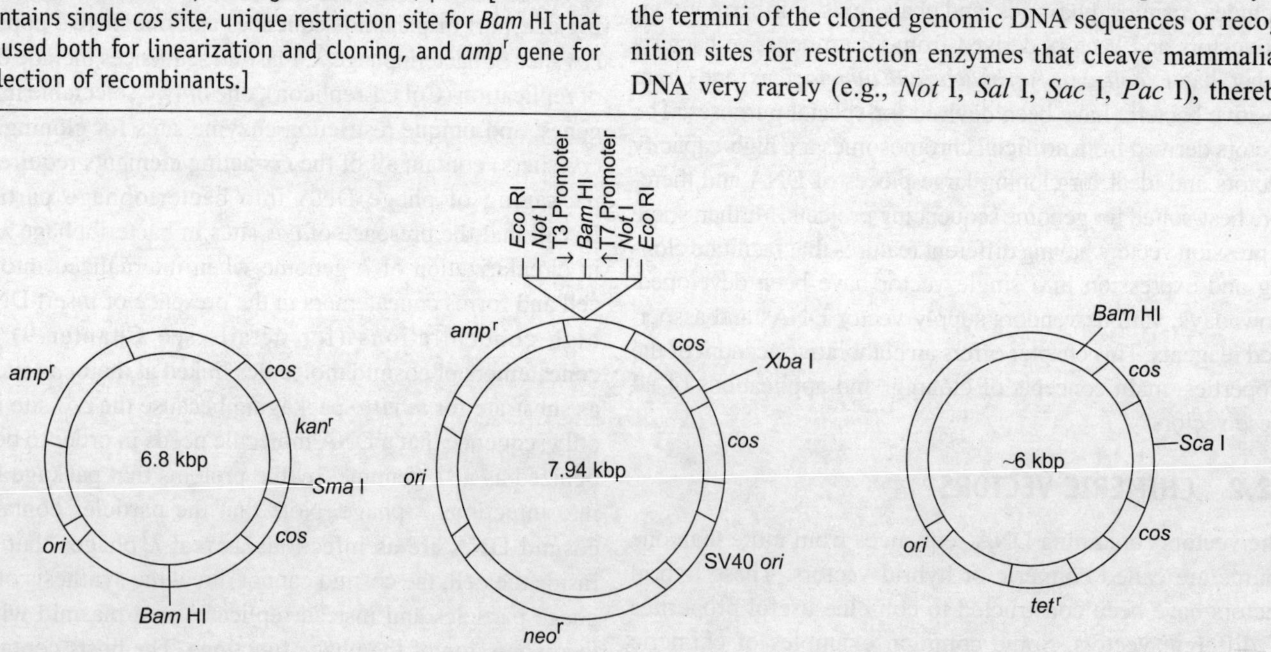

Figure 12.2 Vector maps of dual *cos* cosmids (a) c2XB [The vector carries a *Bam* HI insertion site; two *cos* sites are separated by a site for blunt end cutter *Sma* I; *kan*^r and *amp*^r genes are used for selection of recombinants]; (b) Supercos-1 [The vector contains two *cos* sites separated by a unique *Xba* I site; SV40 *ori* indicates an SV40 origin of replication; *neo*^r and *amp*^r represents two antibiotic resistance genes; *ori* is Col E1 origin of replication; the MCS is flanked by bacteriophage promoters]; (c) pLFR-5 [The vector contains two *cos* sites closely flanking a *Sca* I site and a *Bam* HI site near, but outside one of the *cos* sites; *tet*^r gene represents tetracycline resistance gene.]

allowing isolation of the cloned segment from the cosmid in one piece or sequences that facilitate homologous recombination between cosmids, multiple cloning sites (MCSs), or replicons of several different types.

Cloning Strategies used with Cosmid Vectors

The processes used to construct genomic DNA libraries in bacteriophage λ vectors and cosmids are essentially the same. In each case, segments of genomic DNA are ligated *in vitro* to vector DNA, forming concatemers that can be packaged in phage head. These infectious phage particles merely serve as Trojan horses to deliver recombinant DNA molecules efficiently into bacteria, where the DNA circularizes, and propagated as large plasmids without the expression of the phage functions. For cloning in cosmid vectors, genomic DNA of size appropriate for cloning (30–45 kbp) are generally obtained by partial digestion of high molecular weight chromosomal DNA with a four base-cutter, for example, *Sau* 3AI, *Mbo* I, which generates DNA fragments that can be cloned into a *Bam* HI site in a vector.

Cloning in Single *cos* Cosmids High molecular weight segments of the foreign DNA, generated by restriction enzyme cleavage, are incubated with the linearized cosmid molecule (e.g., by *Bam* HI in pJB8) at high DNA concentration to favor the production of concatemers (Figure 12.3). A complex mixture of ligation products is produced, but amongst these are the recombinant DNA molecules that have ~40 kbp insert DNA flanked by two cosmid molecules and hence *cos* sites in the same orientation. In order to increase the transformation efficiency, the number of nonrecombinants or recombinants with wrong inserts is decreased by techniques such as alkaline phosphatase treatment, partial filling-in, or size fractionation (for details see Chapter 9). These structures resemble the molecules produced late in phage infections and are suitable substrates for the terminase function of the phage protein pA. The foreign DNA that intervenes between the *cos* sites is packaged into bacteriophage particle in an *in vitro* packaging reaction, whereupon the structure is cleaved to give a linear molecule with cohesive ends. Upon infection of *E. coli*, the injected recombinant cyclizes through the *cos*

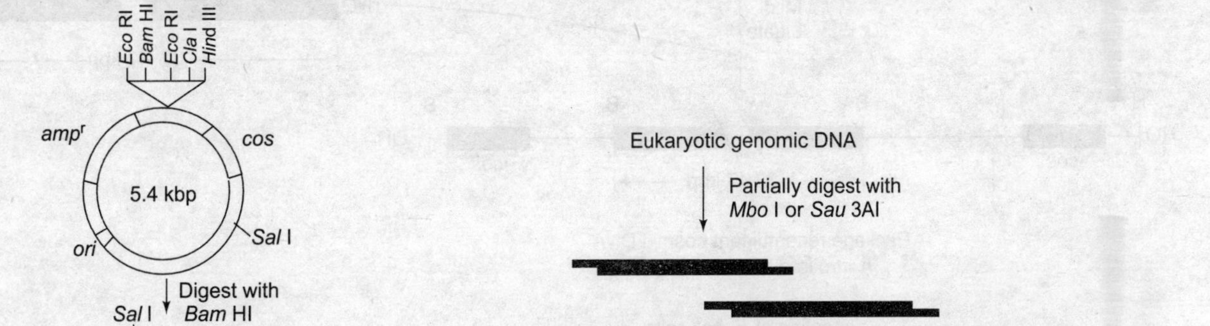

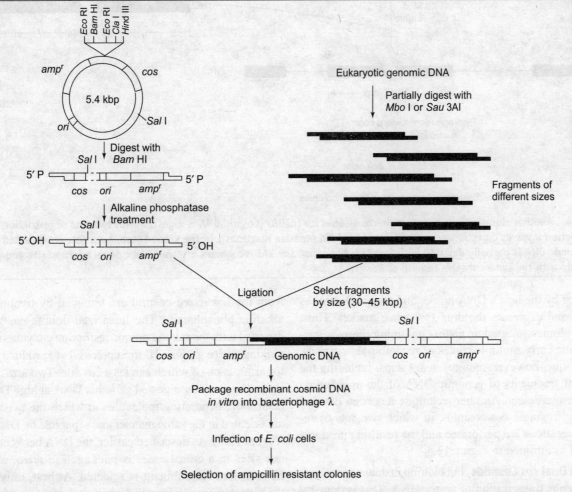

Figure 12.3 Cloning strategy for single *cos* site vectors (pJB8) [Cosmid DNA is linearized by digesting with *Bam* HI, dephosphorylated with alkaline phosphatase, and finally a vector with protruding 5′-termini is generated, which can be ligated with 30–45 kbp fragments of eukaryotic DNA produced by partial digestion with *Mbo* I or *Sau* 3AI. The resulting concatemers are *in vitro* packaged into bacteriophage particles followed by infection of *E. coli* cells. Recombinants are selected on ampicillin.]

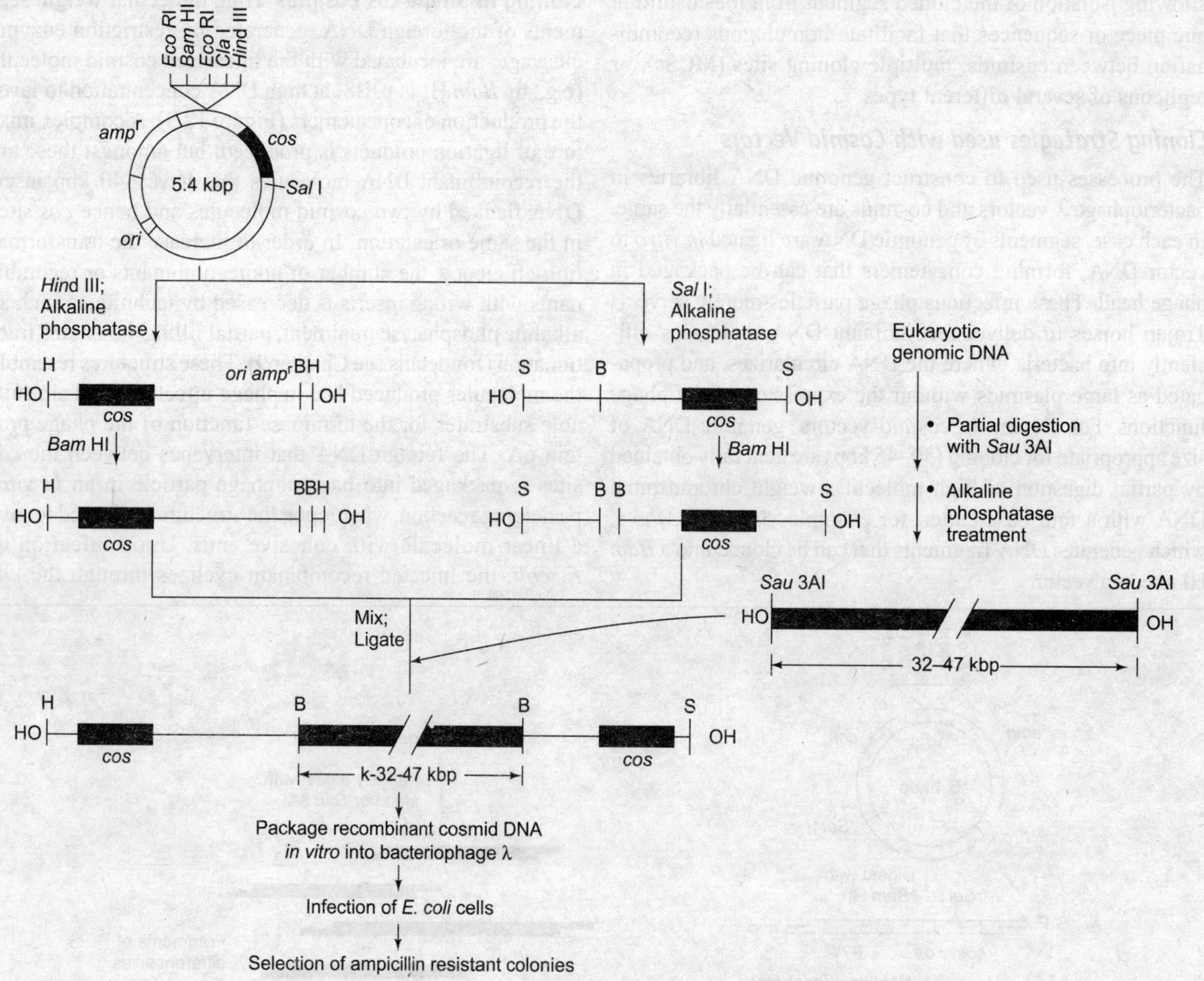

Figure 12.4 Another cloning strategy for single *cos* site vector (pJB8) [Cosmid DNA is digested with two sets of restriction enzymes. First reaction involves digestion with *Hind* III and *Sal* I (in separate reactions). In the second step, both sets are digested with *Bam* HI. Genomic DNA is partially digested with *Sau* 3AI. Note that *Sau* 3AI recognizes a tetranucleotide recognition site and generates fragments with the same cohesive termini as *Bam* HI.]

sites, sealed by the host's DNA ligase, and then replicates as a plasmid and expresses the drug resistance marker. Thus, bacterial colonies selected on plates containing the appropriate antibiotic carry multiple copies of recombinant cosmids. This technique, however, requires many steps, including the isolation of fragments of genomic DNA of the appropriate size and is inefficient. Another technique has been devised for cloning in single *cos* cosmids, in which two sets of restriction digestions are performed and the resulting products are ligated as outlined in Figure 12.4.

Cloning in Dual *cos* Cosmids For cloning in dual *cos* cosmid, the first step is linearization of vector DNA. This is done by digesting the vector DNA using restriction enzyme, the recognition site of which is present between two *cos* sites (e.g., *Xba* I in Supercos-1). Then the 5′-terminal phosphate residues from the linearized cosmid are removed by treatment with alkaline phosphatase. The linearized double *cos* vector is digested with the second unique restriction enzyme site called cloning site (e.g., *Bam* HI in Supercos-1) to produce two vector arms, each of which carries a *cos* site. Two arms are then ligated to partially digested genomic DNA at high DNA concentration, generating molecules in which the two *cos* sites are oriented in the same manner and separated by DNA inserts (Figure 12.5). As described earlier, the DNA between the two *cos* sites in a concatemer is packaged *in vitro*, plated on *E. coli*, and recombinant is selected. At best, only 50% of the concatemers can have the correct arrangement of *cos* sites. Such molecules are packaged into λ heads with very high efficiency, provided these are between 37 and 52 kbp in size. If all works well, 10^5–2×10^7 transformed colonies/

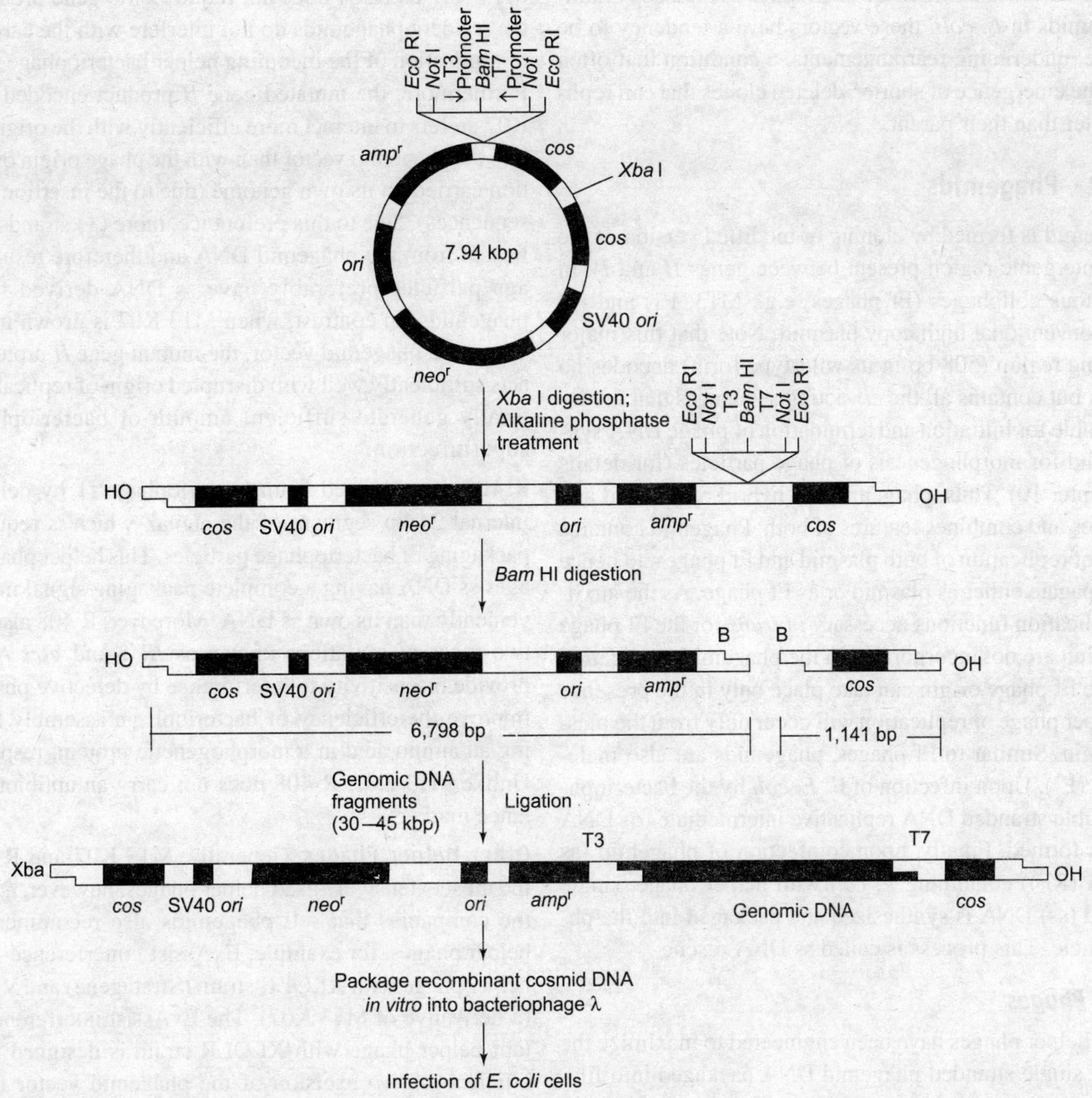

Figure 12.5 Cloning strategy for double *cos* site vector (SuperCos-1) [SuperCos-1 DNA is digested with *Xba* I, treated with alkaline phosphatase, and further subjected to *Bam* HI digestion, the resulting vector DNA is ligated to partially digested (with *Sau* 3AI) and dephosphorylated genomic DNA. The recombinant DNA is then packaged into λ phage particles and finally *E. coli* cells are infected]

μg of genomic DNA are generated. Such a high efficiency is advantageous because only 5 μg of partially digested genomic DNA is required to generate a library that reasonably represents a mammalian genome. Similar to λ bacteriophage and single *cos* cosmids, there are chances of formation of nonrecombinants or concatemers with wrong insert DNA, but their numbers can be reduced by alkaline phosphatase treatment, partial filling-in, or size fractionation (for details see Chapter 9).

Advantages of Cosmids

Cosmids offer several advantages. For example, a few of them are presented: (i) Cosmids can be used as gene cloning vectors in conjunction with *in vitro* packaging; (ii) Cosmids are high insert capacity vectors with insert capacity ranging between ~30 and 45 kbp. Gene clusters and large genes are easier to clone. Larger insert means that fewer clones of a genomic library have to be screened for a specific gene; (iii) Cosmids are attractive for constructing libraries of eukaryotic genomic fragments, and hence used for the analysis of complex genomes; (iv) Large genes can be studied intact and genetic linkage studies can be carried out at molecular level; (v) Background molecules, which do not have the insert DNA or have smaller inserts, are eliminated during packaging; and (vi) The transformation frequency of cosmids is much higher than that of plasmids because of *in vitro* packaging.

However, since cosmids are maintained as high copy number plasmids in *E. coli*, these vectors have a tendency to be unstable, undergoing rearrangements, a condition that often favors the emergence of shorter, deleted clones that can replicate faster than their parent.

12.2.2 Phagemids

A phagemid is formed by cloning of modified version of the major intergenic region present between genes *II* and *IV* in filamentous coliphages (Ff phages; e.g., M13, F1, and fd) into a conventional high copy plasmid. Note that this major intergenic region (508 bp in its wild-type form) encodes no proteins but contains all the *cis*-acting sequences that are indispensable for initiation and termination of phage DNA synthesis and for morphogenesis of phage particles (for details see Chapter 10). Thus, phagemid is a hybrid of plasmid and Ff phages and combines features of both. Phagemid contains origins of replication of both plasmid and Ff phage and hence can propagate either as plasmid or as Ff phage. As the auxiliary replication functions necessary *in trans* for the Ff phage replication are not incorporated in the phagemid, replication from the Ff phage origin can take place only in the presence of a helper phage or replication will occur only from the plasmid origin. Similar to Ff phages, phagemids are also male-specific (F^+). Upon infection of F^+ *E. coli* by the bacteriophage, double stranded DNA replicative intermediate (ds DNA (RF)) is formed. Finally, upon coinfection of phagemid (as ds DNA (RF)) containing *E. coli* with helper phage, single stranded (ss) DNA is synthesized and packaged into the phage particle. This process is called ss DNA rescue.

Helper Phages

Several helper phages have been engineered to maximize the yield of single stranded phagemid DNA packaged into filamentous particles after superinfection of a phagemid-transformed culture. When working well, the ratio of phagemid to helper genomes in the bacteriophage particles released into the medium should be 20:1 and a small-scale (1.5 ml) culture should provide enough single stranded phagemid DNA for four to eight sequencing reactions.

M13 K07 It is a derivative of M13 bacteriophage that carries a origin of replication of p15A plasmid, the kanamycin resistance gene from *Tn* 903, and a mutated version of gene II, which is derived form M13mp1 wherein G residue at 6,125 position has been replaced by a T. When the cells having phagemids are infected by M13 K07, the incoming ss DNA of helper bacteriophage is transformed to ds DNA with the help of cellular enzymes, which then replicates by using origin of plasmid p15A. Gradually, the pool of double stranded M13 K07 genomes expresses all of the proteins needed to generate progeny ss DNA, as the accumulation of M13 K07 ds DNA does not require viral gene products and the resident phagemids do not interfere with the early stages of replication of the incoming helper bacteriophage genome. Furthermore, the mutated gene *II* product encoded by M13 K07 prefers to interact more efficiently with the origin cloned into the phagemid vector than with the phage origin of replication carried on its own genome (due to the insertion of *lac* Z sequences). Due to this preference, more (+) strands are replicated from the phagemid DNA and therefore resulting phage particles preferably have ss DNA derived from the phagemids. In contrast, when M13 K07 is grown in the absence of a phagemid vector, the mutant gene *II* protein interacts sufficiently well with disrupted origin of replication, and finally generate sufficient amount of bacteriophage for superinfection.

R 408 It is derived from bacteriophage f1 by deleting an internal 24 bp segment of the signal, which is required for packaging of bacteriophage particles. This helper phage packages ss DNA having a complete packaging signal more conveniently than its own ss DNA. Moreover, R 408 also carries two types of mutations known as *IR* 1 and *gtrx* A, which provide insensitivity to interference by defective phages and improve the efficiency of bacteriophage assembly by altering an amino acid in a morphogenetic protein, respectively. Unlike M13 K07, R 408 does not carry an antibiotic resistance marker.

Other Helper Phages Generally, M13 K07 and R 408 are the most extensively used helper phages, however, nowadays the companies that sell phagemids also recommend other helper phages, for example, ExAssist® interference-resistant helper phage with XLOLR strain (Stratagene) and VCS M13 (a derivative of M13 K07). The ExAssist interference-resistant helper phage with XLOLR strain is designed to allow efficient *in vivo* excision of the phagemid vector from the phasmid while preventing problems associated with helper phage coinfection. The ExAssist helper phage contains an amber mutation that prevents replication of the phage genome in a nonsuppressing *E. coli* strain (e.g., XLOLR cells). Only the excised phagemid can replicate in the host, removing the possibility of coinfection from the ExAssist helper phage. Because ExAssist helper phage cannot replicate in the XLOLR strain, ss DNA rescue cannot be performed in this strain using ExAssist helper phage. XLOLR cells are also resistant to λ infection, preventing λ DNA contamination after excision.

Bacterial Hosts

Principally, bacteriophage particles that contain ss DNA can be propagated in any male strain of *E. coli* (for details see Chapter 10). However, the product yield of ss DNA can be greatly affected by the bacterial strain used to propagate the plasmid. Moreover, several strains including MV 1184, DH 11S, TG 2, and XL1 Blue MRF′ have been used by many laboratories that generally yield enough amount of ss phagemids DNA. Note that with all strains, maximum yields of phagemids DNA are obtained when infected cultures are well aerated.

Examples of Phagemids

The particular strand of the foreign DNA that is packaged into the bacteriophage particles depends on its orientation in the polylinker site of the phagemid vector and on the orientation of the bacteriophage origin of DNA replication carried in the vector. Hence, most commercially available phagemid vectors come in four chiralities in which the orientation of the polylinker sequence is opposite in one pair of vectors.

Some examples of phagemids are listed below. For first time users of phagemids, the safest choices for helper phage, phagemid vector, and host is a well tested and reliable helper phage such as M13 K07 and a dependable phagemid (e.g., one or more of the Stratagene SK or Promega pGEMZf series) in an *E. coli* strain such as DH 11S.

pEMBL 8 The first generation of vectors in this category are the pEMBL vectors. For example, one member, namely, pEMBL 8 is 3,997 bp in size and is made by transferring a 1,300 bp fragment of M13 genome into pUC8. This piece of M13 contains the signal sequence recognized by enzymes that convert M13 ds DNA into ss DNA before secretion of new phage particles. This signal sequence is still functional even though detached from rest of the M13 genome, so the resultant pEMBL 8 molecules are also converted into ss DNA and secreted as defective phage particles. pEMBL 8 contains *amp*^{r} gene for the selection of transformants and a polylinker site similar to pUC8 is also present in *lac* Z′ gene. *E. coli* cells used as hosts for a pEMBL 8 cloning experiment are infected with normal M13 to act as a helper phage providing the necessary replicative enzymes and phage coat proteins. pEMBL 8 can generate single stranded versions of cloned DNA fragments up to 10 kbp in length. However, pEMBL 8 gave phagemids a bad name because, after superinfection with a helper phage, the yield of ss DNA was generally poor and was influenced by many factors, including the density of the culture at the time of infection, the multiplicity of infection, the length of time after infection, and the size and nature of foreign DNA (larger the fragment, poorer the yield). Even when superinfection was performed under optimal conditions, the yield of progeny phage often consisted predominantly of helper phages rather than packaged ss phagemid DNA. This problem is, however, solved up to some extent by using helper phages encoding a mutated version of gene *II* that preferentially activates the phagemid origin of replication or engineering novel strains of *E. coli* (e.g., DH 11S, TG 2, and MV 1184) that are efficiently transformed by plasmids, are easily infected by commonly used helper phages, and yield preparations of ss phagemid DNA that are free from contamination with bacterial DNA and helper phages.

pBluescript II Series The pBluescript II series (Stratagene) includes four chiralities, *viz.*, pBluescript II KS (+), pBluescript II KS (–), pBluescript II SK (+), and pBluescript II SK (–) (Figure 12.6). All these four pBluescript II vectors are 3.0 kbp in size and contain the intergenic (IR) region of a filamentous f1 phage that encodes all of the *cis*-acting functions of the phage required for packaging and replication. These four chiralities differ with respect to the orientation of the polylinker sequence and the orientation of the intergenic region. In each vector, the polylinker is derived from pUC19 and is present in *lac* Z′ gene. The *lac* Z′ region is under the control of *lac* Z promoter and *lac* I. The presence of polylinker in *lac* Z′ gene allows blue-white screening (for details see Chapter 16). For antibiotic selection of recombinants, *amp*^{r} is used as selection marker. The polylinker has 21 unique cloning sites flanked by T3 and T7 promoters in opposite directions on the two strands. The prefixes 'KS' or 'SK'

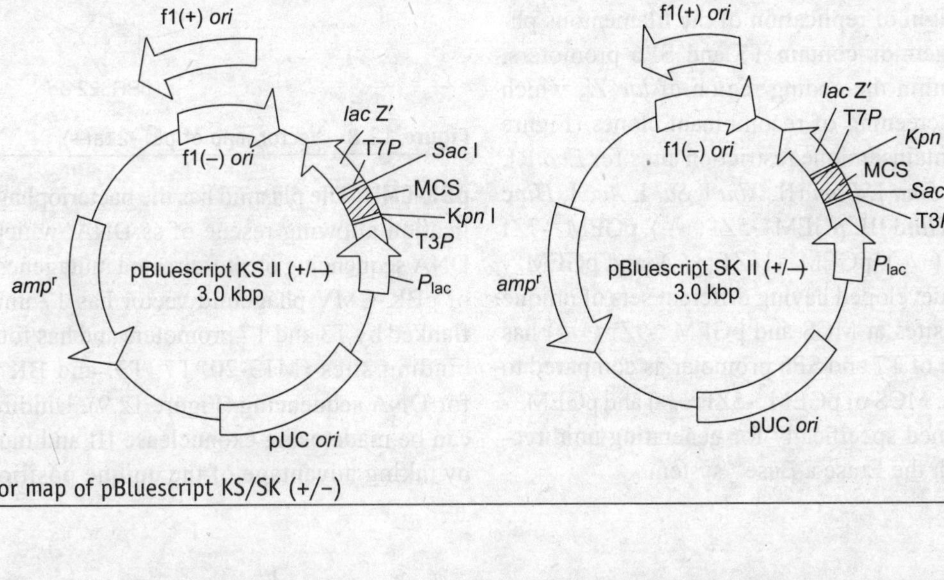

Figure 12.6 Vector map of pBluescript KS/SK (+/–)

denote the orientation of polylinker (K signifies *Kpn* I and S signifies *Sac* I). Thus, in 'KS' vectors transcription of *lac* Z′ gene proceeds from the *Kpn* I restriction enzyme site toward *Sac* I site, while in 'SK' vectors, transcription proceeds from the *Sac* I restriction enzyme site toward *Kpn* I site. Due to the (+) and (–) orientations of the intergenic region, sense and antisense ss DNAs can be rescued by bacteriophages after coinfecting with the helper phage. Note that in the absence of helper phage, origin of replication of plasmid (Col E1 *ori*) is utilized. One pair of vectors, *viz.*, pBluescript II SK and pBluescript II KS have opposite orientation of polylinker sequence, while another pair, *viz.*, pBluescript II SK (+) and pBluescript II SK (–), have opposite orientation of intergenic region. Similarly, pBluescript II KS (+) and pBluescript II KS (–) have opposite orientation of intergenic region.

The polylinker provides a choice of six different primer sites for DNA sequencing. These include binding sites for M13–20 primer, T7 primer, KS primer, SK primer, T3 primer, and M13 reverse primer. The presence of bacteriophage f1 origin of replication allows rescue of ss DNA, which can be used for DNA sequencing or site-directed mutagenesis. Unidirectional deletions can be made with exonuclease III and mung bean nuclease by taking advantage of the unique positioning of 5′ and 3′ restriction sites. Transcripts made from the T3 and T7 promoters generate riboprobes useful in Southern and northern blotting and the *lac* Z promoter may be used to drive expression of fusion proteins suitable for western blot analysis or protein purification.

pGEMZf Series Six types of pGEMZf are supplied from Promega and each of them are available in two chiralities, which differ with respect to orientation of intergenic sequences. These phagemids serve as standard cloning vectors, templates for *in vitro* transcription, and for the production of circular ss DNA. The (+) and (–) orientations of the intergenic region allow the rescue of sense and antisense ss DNAs by coinfecting helper phages. The pGEM®-3Zf (+/–) vector is derived from pGEM®3Z vector (for details see Chapter 8) and contains the origin of replication of the filamentous phage f1. These phagemids contain T7 and SP6 promoters, flanking a MCS within the coding region of *lac* Z′, which allows blue-white screening of recombinant clones (Figure 12.7). The MCS contains unique restriction sites for *Eco* RI, *Sac* I, *Kpn* I, *Ava* I, *Sma* I, *Bam* HI, *Xba* I, *Sal* I, *Acc* I, *Hinc* II, *Pst* I, *Sph* I, and *Hind* III. pGEM®-5Zf (+/–), pGEM®-7Zf (+/–), pGEM®-9Zf (+/–), pGEM®-11Zf (+/–), and pGEM®-13Zf (+/–) are also developed having different sets of unique restriction enzyme sites at MCS and pGEM®-9Zf (+/–) has opposite orientation of T7 and SP6 promoter as compared to others. Furthermore, MCS of pGEM®-5Zf (+/–) and pGEM®-7Zf (+/–) are designed specifically for generating unidirectional deletions with the Erase a Base® system.

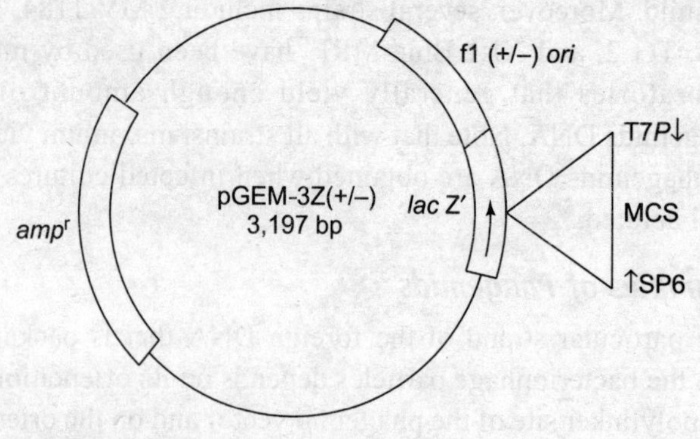

Figure 12.7 Vector map of pGEM®-3Z (+/–)

pET Vectors pET vectors are supplied by Novagen and in pET-21 series, four vectors are available, *viz.*, pET-21a (+), pET-21b (+), pET-21c (+), and pET-21d (+). The pET-21a-d (+) vectors carry an N-terminal T7•Tag® sequence plus an optional C-terminal His•Tag® sequence. The size of pET-21a (+) is 5,443 bp and has both pBR322 and f1 origins (Figure 12.8). The f1 origin is oriented so that infection with helper phage produces virions containing ss DNA that corresponds to the coding strand. These vectors differ from pET-24a-d (+) only by their selectable marker (ampicillin vs. kanamycin resistance). The MCS has eight unique cloning sites flanked by T7 promoter and T7 terminator. Therefore, single stranded sequencing should be performed using primer complementary to T7 terminator.

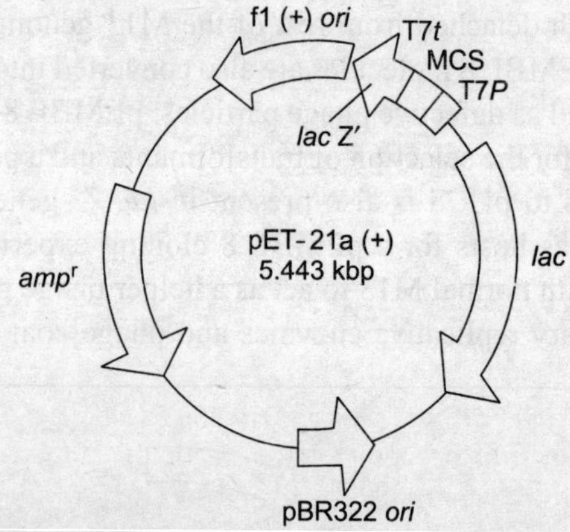

Figure 12.8 Vector map of pET-21a(+)

pBK-CMV The plasmid has the bacteriophage f1 origin of replication allowing rescue of ss DNA, which can be used for DNA sequencing or site-directed mutagenesis. The polylinker of pBK-CMV phagemid vector has 17 unique cloning sites flanked by T3 and T7 promoters and has four standard primer binding sites (M13-20, T7, T3, and BK reverse primers) for DNA sequencing (Figure 12.9). Unidirectional deletions can be made using exonuclease III and mung bean nuclease by taking advantage of the unique positioning of 5′ and 3′

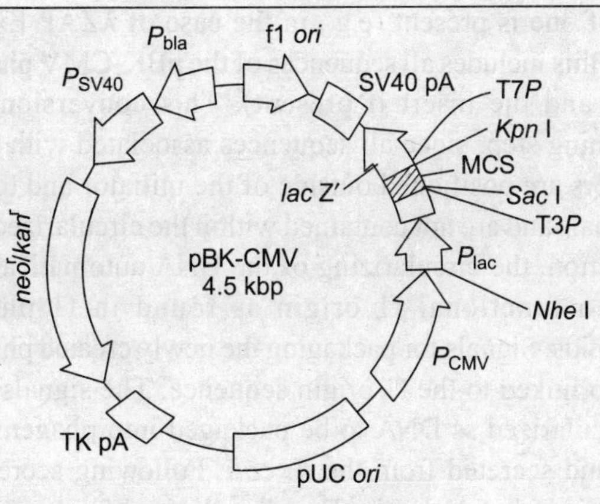

Figure 12.9 Vector map of pBK-CMV

restriction sites. Transcripts made from the T3 and T7 promoters generate riboprobes useful in Southern and northern blotting, and the *lac* Z promoter may be used to drive expression of fusion proteins suitable for western blot analysis or protein purification. The cytomegalovirus (CMV) immediate early (IE) promoter, SV40 transcription terminator and polyadenylation signal control the eukaryotic expression of inserts. Stable selection of clones in eukaryotic cells is made possible by the presence of the neo^r and kan^r resistance gene, which is driven by the SV40 early promoter with *tk* transcription polyadenylation signals to render transfectants resistant to G418 (geneticin).

Applications of Phagemids

Segments of foreign DNA can be cloned in phagemid and propagated as plasmids in the usual way. However, when a male strain of *E. coli* carrying a phagemid is infected with a suitable filamentous phage, the mode of replication of the phagemid changes in response to gene products expressed by the incoming phage. The gene *II* protein encoded by the helper phage introduces a nick at a specific site in the intergenic region of phagemid and hence initiates rolling circle replication, which generates copies of one strand of the plasmid DNA. These single stranded copies of the plasmid DNA are packaged into progeny phage particles, which are then extruded into the medium. The secreted phage particles are easily recovered by precipitation with polyethylene glycol and the ss DNA is purified by extraction with phenol.

The ss DNA produced by phagemid is used for the same purposes as ss DNAs of recombinant M13 phage, i.e., for DNA sequencing, synthesis of strand-specific radiolabeled probes, subtractive hybridization, and oligonucleotide-directed mutagenesis. As seen above, vectors permitting packaging of either strand of the cloned gene have been constructed and widely used for sequencing both the strands of any given gene independently. In addition, phagemid vectors can be used in appropriate strains of *E. coli* (e.g., BW 313) to produce ss DNAs that contain uracil in place of a proportion of the thymidine residues. These uracil-substituted DNAs are excellent substrates for certain types of oligonucleotide-directed mutagenesis. In addition, it is possible to construct a complete cassette in a phagemid containing, for example, a strong promoter, the gene or cDNA of interest, and a transcription terminator. Expression phagemids of this type can be isolated in ss DNA form, subjected to site-directed mutagenesis, and then used to transform *E. coli* or yeast for phenotypic expression.

In phagemids, as in M13 vectors, the yield of ss DNA can vary over a 5- to 10-fold range depending on the size and nature of the foreign DNA. In general, the larger the fragment size, the poorer the yield. Furthermore, for reasons that are not understood, foreign DNAs of equivalent size can suppress the yield of ss DNA to varying extents. For example, most segments of yeast DNA seem to be amenable to propagation in phagemids, whereas human genomic DNAs of equivalent size may produce disappointing yields of ss DNA. The phagemid vector can also dramatically affect yields. Thus, recloning a fragment in the opposite orientation or in a vector with the bacteriophage origin of replication in the opposite orientation sometimes solves a problem of low yields.

Advantages of Phagemids Over M13 Bacteriophage

Phagemids have several attractive features that overcome problems commonly encountered during cloning in M13 bacteriophage. These include: (i) Rapid growth as plasmids and hence higher yield of ds DNA; (ii) Ease of isolation, manipulation, and reinsertion into host as plasmids; (iii) Phagemid can be induced to produce phage particles containing ss DNA by coinfection with a fully functional helper phage, as and when required. Hence, phagemid eliminates the need for time-consuming recloning and subcloning procedures for transfer of foreign DNA fragments from plasmids to filamentous phage vectors for sequencing; (iv) Higher insert capacity (similar to plasmids) than M13; (v) A positive selective marker present that can be used to select bacteria transformed by the phagemid; (vi) Cloning in phagemids allows segments of DNA several kilobase pairs in length to be isolated in ss form; and (vii) Significant reduction in the frequency and extent of deletions and rearrangements in ss DNA (no or low instability problems).

12.2.3 Phasmids

Phasmid is another example of chimeric vector. It consists of λ genome into which a phagemid (or plasmid) vector carrying λ attachment site (*att* P) has been inserted by site-specific recombination mechanism of phage. Note that *att* P is normally responsible for recombinational insertion of the phage into bacterial chromosome during lysogen formation. But in the case of phasmid, a reversible recombinational insertion of phagemid (or plasmid) into the phage occurs, which is referred to as 'lifting' of phagemid (or plasmid) into the phage genome. These novel genetic combinations contain functional origins of replication of λ phage and phagemid (or plasmid) and may be propagated as a phagemid (or plasmid) or as a phage in appropriate *E. coli* strains. Reversal of the lifting process may, however, release the plasmid (or phagemid) vector.

In vivo Excision of Phagemid and ss DNA Rescue

The phasmids are designed to allow simple and efficient *in vivo* excision and recircularization of any cloned insert contained within the λ vector to release the uplifted phagemid containing the cloned insert. This *in vivo* excision depends on the placement of the DNA sequences within the λ phage genome and on the presence of a variety of proteins, including filamentous (e.g., M13) bacteriophage-derived proteins. The M13 phage proteins recognize a region of DNA normally serving as the f1 bacteriophage origin of replication. This origin of replication can be divided into two overlying parts: (i) The site of initiation; and (ii) The site of termination for DNA synthesis. These two regions are subcloned separately into the phasmid vector. The λ phage (target) is made accessible to the M13-derived proteins by simultaneously infecting a strain of *E. coli* with both the λ vector and the M13 helper phage. Inside *E. coli*, the helper proteins (i.e., proteins from M13 helper phage) recognize the initiator DNA that is within the λ vector. The gene *II* encoded protein then nicks one of the two DNA strands. At the site of this nick, new DNA synthesis begins and duplicates the existing DNA in the λ vector downstream (3′) of the nicking site. DNA synthesis of a new single strand of DNA continues through the cloned insert until a termination signal (positioned 3′ of the initiator signal) is encountered within the constructed λ vector. The gene *II* product from the M13 phage circularizes the ss DNA molecule and this circular DNA molecule contains the DNA between the initiator and the terminator. This comprises the sequences of phagemid vector and the insert if one is present (e.g., in the case of λZAP Express® vector, this includes all sequences of the pBK-CMV phagemid vector and the insert if present). This conversion is the subcloning step, since all sequences associated with normal λ vectors are positioned outside of the initiator and terminator signals and are not contained within the circularized DNA. In addition, the circularizing of the DNA automatically recreates a functional f1 origin as found in f1 phage or phagemids. Signals for packaging the newly created phagemid are also linked to the f1 origin sequence. The signals permit the circularized ss DNA to be packaged into phagemid particles and secreted from the *E. coli*. Following secretion of the phagemid particle, the *E. coli* cells used for *in vivo* excision of the cloned DNA are killed and the λ phage is lysed by heat treatment at 70°C. The phagemid is not affected by the heat treatment. *E. coli* is infected with the phagemid and can be plated on selective media to form colonies. DNA from excised colonies can be used for analysis of insert DNA, including subcloning, and mapping. Colonies from the excised phagemid vector can also be used for subsequent production of ss DNA suitable for dideoxy sequencing and site-specific mutagenesis.

Examples of Phasmid Vectors

Some examples of phasmids are as follows:

λZAP II® The original λZAP® vector has only limited choice of suitable host strains (containing a *sup* F genotype) as it contains the *Sam* 100 mutation. Later λZAP II, a new variation of the λZAP vector, has been derived, which does not contain the *Sam* 100 mutation, therefore, highly efficient growth can be obtained on various non-*sup* F strains, including XL1-Blue MRF′ cells. The λZAP® II system, covered by Stratagene's U.S. Patent 5,128,256, combines the high efficiency of library construction in λ vector and the convenience of a plasmid system with improved blue-white selection. The use of the XL1-Blue MRF′ host strain with λZAP II enhances

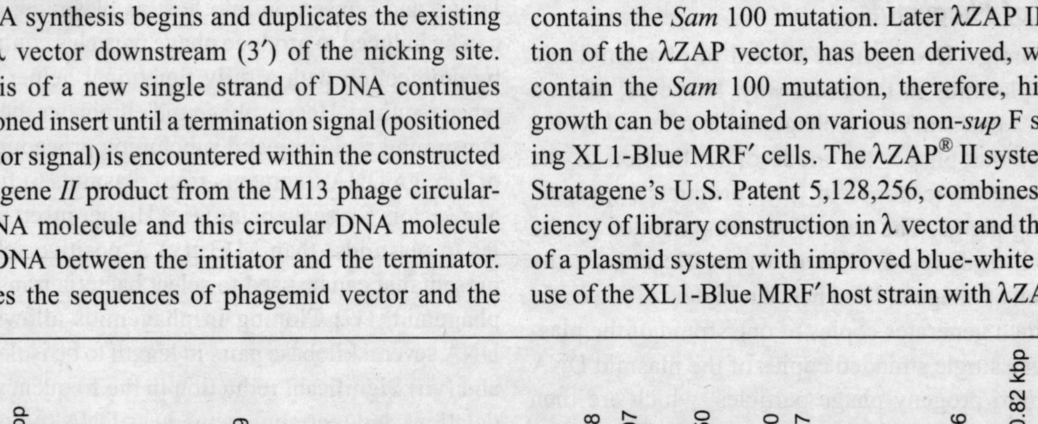

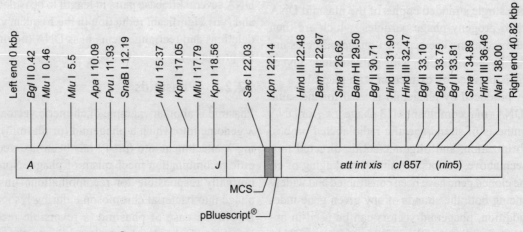

Figure 12.10 Vector map of λZAP II®

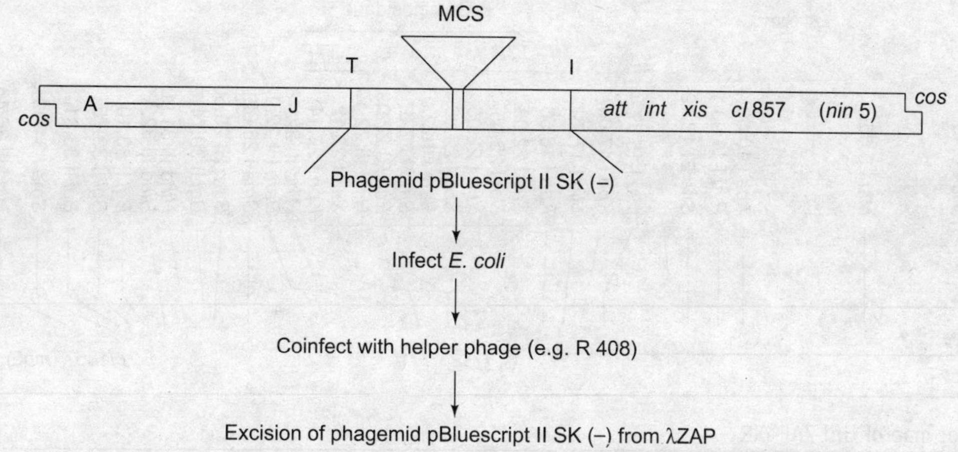

Figure 12.11 ss DNA rescue of phagemid pBluescript II SK (–) from λZAP II® [The phagemid pBluescript II SK (-) may be excised from λZAP II® by addition of helper phage. The helper phage provides the necessary proteins and factors for transcription between the initiation (I) and termination (T) sites in the parent phage to produce the phagemid with the DNA cloned into the parent vector.]

the blue color produced by nonrecombinant phage, thereby improving blue-white selection. λZAP II vector has six unique cloning sites (*Sac* I, *Not* I, *Xba* I, *Spe* I, *Eco* RI, and *Xho* I) similar to the original λ ZAP, which can accommodate DNA inserts up to 10 kbp in length. Clones in the λZAP II vector can be screened with either DNA probes or antibody. When infected with helper phage, rapid *in vivo* excision of the phagemid pBluescript II SK (–) occurs (Figure 12.11), thus allowing the insert to be characterized in a phagemid system. The vector map of λZAP II® is drawn in Figure 12.10.

The advantageous features of this vector are: (i) High capacity (up to 10 kbp of foreign DNA can be cloned, which is large enough to encompass most cDNAs); (ii) Presence of a polylinker with six unique restriction sites, which increases cloning versatility and also allows directional cloning, (iii) Availability of T3 and T7 promoters flanking the polylinker, which allows sense and antisense RNA to be prepared from the insert; and (iv) Most importantly, all these features are included within a plasmid vector called pBluescript, which is inserted into the phage genome. Thus, the cDNA clone can be recovered from the phage and propagated as a high copy number plasmid without any subcloning, simply by coinfecting the bacteria with a helper f1 phage that nicks the λ ZAP II vector at the flanks of the plasmid and facilitates excision.

λZAP Express® The λZAP Express® vector (Figure 12.12) is designed and covered by Stratagene's U.S. Patents 5,128,256 and 5,286,636. It contains pBK-CMV phagemid and exhibits all its properties. This vector allows both eukaryotic and prokaryotic expression, while also increasing both cloning capacity and the number of unique λ cloning sites. The λ ZAP Express® vector has 12 unique cloning sites, which can accommodate DNA inserts up to 12 kbp in length. The 12 unique cloning sites are *Apa* I, *Bam* HI, *Eco* RI, *Hind* III, *Kpn* I, *Not* I, *Sac* I, *Sal* I, *Sma* I, *Spe* I, *Xba* I, and *Xho* I. As λZAP Express® includes the human CMV promoter and SV40 terminator, fusion proteins can be expressed in mammalian cells as well as bacteria. Thus, cDNA libraries can be cloned in the phage vector in *E. coli*, rescued as plasmids,

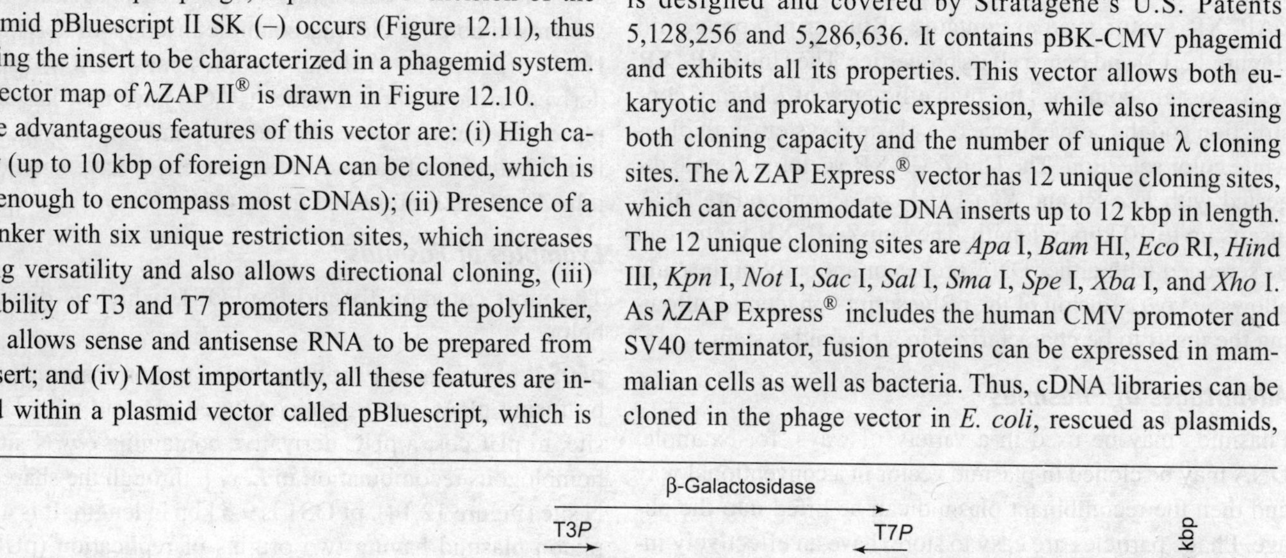

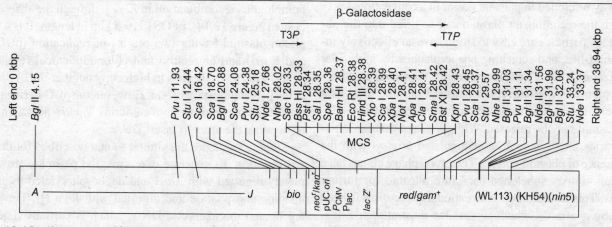

Figure 12.12 Vector map of λZAP Express® vector

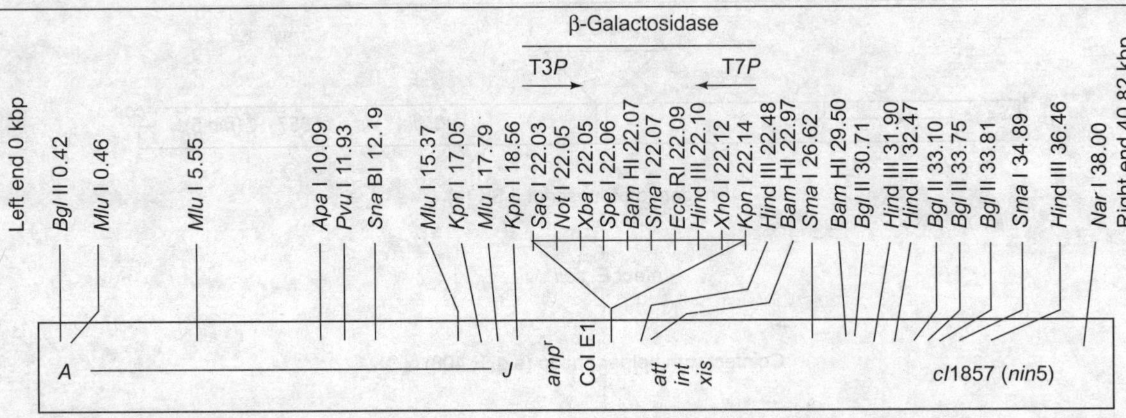

Figure 12.13 Vector map of Uni-ZAP®XR

and then transfected into mammalian cells for expression cloning. Inserts cloned into the λZAP Express® vector can be excised out of the phage in the form of the *kan*^{r}-resistant pBK-CMV phagemid vector by the same excision mechanism found in the λZAP® vectors. Clones in the λZAP Express® vector can be screened with either DNA probes or antibody probes, and *in vivo* rapid excision of the pBK-CMV phagemid vector allows insert characterization in a plasmid system.

Uni-ZAP®XR Vector System The Uni-ZAP®XR vector is covered by Stratagene's U.S. Patent 5,128,256. The Uni-ZAP®XR vector system contains pBluescript® phagemid (Figure 12.13) and hence all its properties. The Uni-ZAP®XR vector system combines the high efficiency of λ library construction and the convenience of a plasmid system with blue-white color selection. The Uni-ZAP®XR vector is double digested with *Eco* RI and *Xho* I and can accommodate DNA inserts up to 10 kbp in length. The Uni-ZAP®XR vector can be screened with either DNA probes or antibody probes and allows *in vivo* excision of the pBluescript® phagemid, allowing the insert to be characterized in a plasmid system.

Advantages of Phasmids

Phasmids may be used in a variety of ways, for example, DNA may be cloned in plasmid vector in a conventional way and then the recombinant plasmid can be lifted into the phage. Phage particles are easy to store, have an effectively infinite shelf-life, and screening phage plaques by molecular hybridization often gives cleaner results than screening bacterial colonies. Alternatively, a phasmid may be used as a phage cloning vector, from which subsequently a recombinant plasmid may be released. While λ phage generates better libraries, these cannot be manipulated *in vitro* with the convenience of plasmid vectors. Therefore, phage clones have to be laboriously subcloned back into plasmid for further analysis. This limitation of conventional λ vectors has been solved by phasmids, which possess the most attractive features of both λ phage and plasmids.

12.2.4 Fosmids

Fosmids contains the F plasmid origin of replication and a λ *cos* site. These are similar to cosmids, but have a low-copy replicon (F factor). In simple terms, fosmids have low copy number in *E. coli*, which means that these are less prone to instability problems as compared to cosmids and plasmids. Fosmid system may be useful for constructing stable libraries from complex genomes. Fosmid clones are very helpful in the assessment of the accuracy of the genome sequence. Thus, fosmid is a bacterially propagated phagemid vector system suitable for cloning genomic inserts ~40 kbp in size. However, bacterial artificial chromosomes (BACs) and P1-derived artificial chromosomes (PACs) have now largely replaced fosmids for genome mapping and sequencing, but a limited number of fosmid insert sequences have been determined and are represented in the genome.

Examples of Fosmids

The most common fosmid is pFOS1, which is discussed below.

pFOS1 pFOS1 is an F replicon-based vector. It is constructed by fusing pBAC, an F replicon-based plasmid with λ*cos* N site, to pUCcos, a pUC derivative containing *cos* N site, by homologous recombination in *E. coli* through the shared *cos* N site (Figure 12.14). pFOS1 is 9.5 kbp in length. It is a bireplicon plasmid having two origins of replication (pUC *ori* and F *ori*) and ampicillin and chloramphenicol resistance genes. This fosmid exists in high copy number in *E. coli* due to the pUC replication origin. However, the pUC-derived portion of the plasmid is removed during *in vitro* packaging after ligating the arms to insert DNA.

Cloning strategies are similar to that described for double *cos* cosmids. To generate two arms, the plasmids are completely digested with *Aat* I and dephosphorylated by using alkaline phosphatase and digested with *Bam* HI. The arms are ligated to eukaryotic DNA, which is partially digested with *Mbo* I followed by dephosphorylation. The resulting

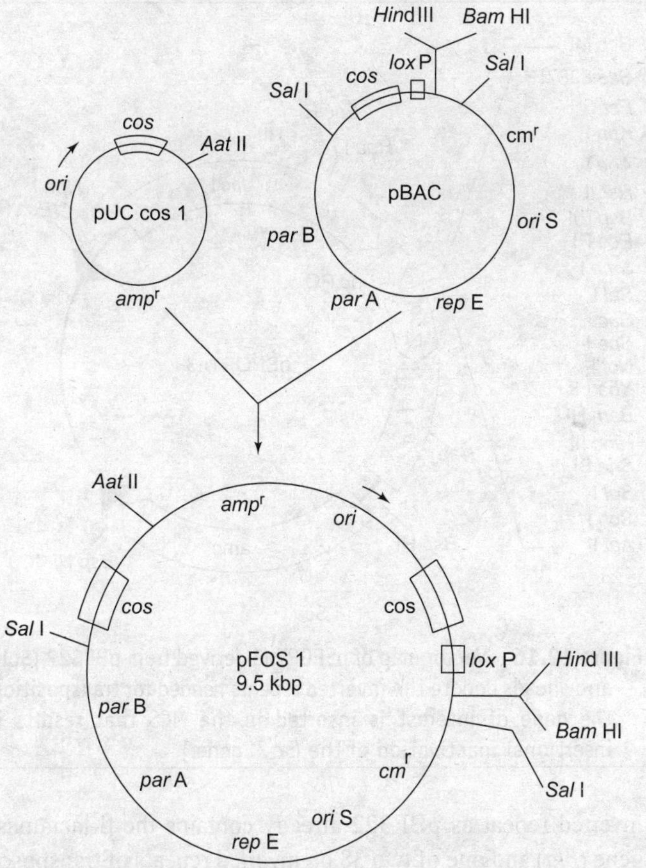

Figure 12.14 Construction of pFOS1

ligated DNA is *in vitro* packaged by using packaging system and finally the fosmid particles are transfected to *E. coli* strain and the cells are plated onto appropriate selective media.

12.3 GRAM NEGATIVE BACTERIA OTHER THAN *E. coli* AS CLONING VECTORS

The need for cloning in Gram negative bacteria other than *E. coli* had arisen due to following reasons:

- *E. coli* plasmids can be propagated only in *E. coli* and closely related enteric bacteria.
- There are many nonenteric Gram negative bacteria of considerable commercial interest, which do not serve as hosts for *E. coli* vectors. For example, strains of *Pseudomonas*, which can degrade a variety of organic chemicals, strains of *Rhizobium* and *Azotobacter*, which fix nitrogen, possess photosynthetic activities, and *Agrobacterium*, which are capable of symbiosis with plant cells. Such types of symbiotic relationships are of practical importance for the transfer of genes into plant cells. Furthermore, few methylotrophic bacteria are used for the production of single cell proteins (SCP).
- *E. coli* lacks some auxiliary biochemical pathways that are essential for phenotypic expression of certain functions, for example, degradation of aromatic compounds or plant pathogenicity.

For cloning of DNA in these nonenteric bacteria, plasmid vectors are required which can replicate in them. However, in most of the cases, recombinant plasmids are screened in *E. coli* cells. Thus, the vector must be able to replicate in both cells. Two types of vectors are used for this purpose, which includes shuttle vector and broad host range plasmid as a vector. Shuttle vectors are described in Section 12.9. For broad host range, various natural plasmids have been exploited and extensively modified, but only a few approaches the level of standard *E. coli* vectors. In addition, for cloning in *Agrobacterium*, specialist vectors have been derived from the natural tumor-inducing (pTi) plasmid.

12.3.1 Examples of Broad Host Range Plasmid Vectors

Following are the examples of broad host range plasmid vectors used as cloning vectors for nonenteric Gram negative bacteria.

Plasmid RSF1010 or pSF1010 (IncQ Group)

Plasmid RSF1010 belonging to the IncQ group is a multicopy replicon ~8,684 kbp long, which has been completely sequenced. A detailed physical and functional map has been constructed, which includes restriction endonuclease recognition sites, RNA polymerase binding sites, antibiotic resistance genes, genes for plasmid mobilization (*mob*), three genes encoding replication proteins (*rep* A, B, and C), and the origins of vegetative replication (*ori*) and transfer (*nic*) (Figure 12.15). This plasmid has resistance for two antimicrobial agents, sulfonamide (su^r) and streptomycin (sm^r). Plasmid RSF1010 has unique cleavage sites for *Eco* RI, *Bst* EII, *Hpa* I, *Dra* II, *Nsi* I, *Sac* I, *Afl* III, *Ban* II, *Not* I, *Sac* II, *Sfi* I, and *Spl* I. Among them, none of the unique cleavage sites is located within the antibiotic resistance genes and none is particularly useful for cloning. Additionally, there are two *Pst* I sites flanking the sulfonamide resistance genes, located ~750 bp apart. As su^r and sm^r genes are transcribed from the

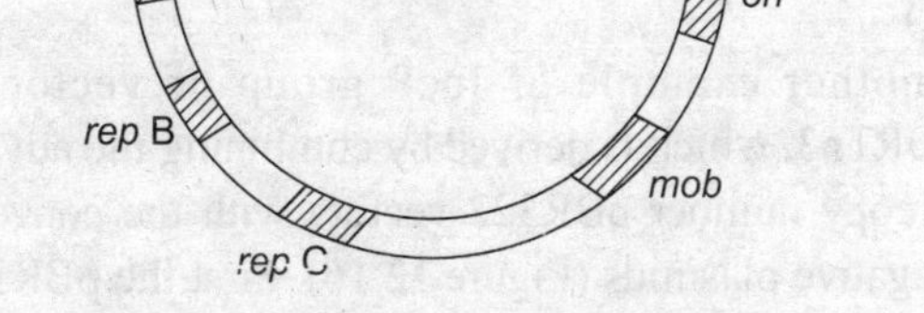

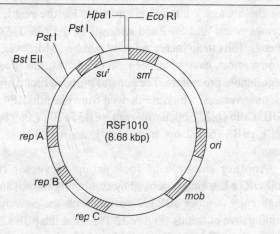

Figure 12.15 Vector map of plasmid RSF1010

same promoter, insertion of a DNA fragment between the *Pst* I sites inactivates both resistance determinants. Later, a series of improved vectors has been derived from RSF1010 by introduction of another selective marker. For example, plasmids KT230 and KT231 encode kanamycin and streptomycin resistance and have unique sites for *Hind* III, *Xma* I, *Xho* RI, and *Sst* I, which can be used for cloning by marker inactivation. These vectors have been utilized for cloning of *Pseudomonas putida* genes involved in the catabolism of aromatic compounds. Further, some expression vectors having *tac* or phage T7 promoter, which are used for the regulated expression of cloned genes have also been devised from this plasmid. By using these vectors, various cloned genes can be expressed in a wide range of Gram negative bacteria.

Plasmid RP4 (IncP Group)

Plasmid RP4 belongs to IncPα group plasmids of 60 kbp in size. Other examples of this group include R18, R68, RK2, RP1, and RP4. A relatively smaller (52 kbp) plasmid such as IncPβ (e.g., R751) also belongs to this category. These plasmids have been completely sequenced and as a result, location of restriction sites and genes are known. Due to their large size, the IncP group plasmids are not widely used as vectors. However, various mini-IncP plasmids have been developed as vectors ranging between 4.8 and 7.1 kbp in size, but can be easily maintained in a wide range of Gram negative bacteria. All the vectors comprise of a *lac* Z′ region having common polylinker (anticlockwise direction, *Hind* III, *Sph* I, *Pst* I, *Sal* I/*Hinc* II/*Acc* I, *Xba* I, *Bam* HI, *Xma* I/*Sma* I, *Kpn* I, *Sac* I, and *Eco* RI) and thus facilitate the cloning and identification of inserts by blue-white screening. Most of the vectors carry two antibiotic resistance genes. The vectors of these series have the *ori* T (origin of transfer) gene, which permits them to be conjugally transferred in those recipients, which are unable to get vector by any of the gene transfer techniques including transformation or electroporation. Some of the vectors have *par* DE region that greatly enhances their segregative stability in certain hosts. In addition, *trf* A locus on the vectors contains unique sites for the restriction enzymes *Nde* I and *Sfi* I and removal of the *Nde* I–*Sfi* I fragment results in an increased copy number. Moreover, expression vectors have also been developed by the insertion of regulatable promoters. The general purpose broad host range cloning vectors, which are derived from plasmid RP4, include pJB3Cm6 (6,227 bp, *amp*^{r}, *cm*^{r}), pJB3Tc20 (7,069 bp, *amp*^{r}, *tet*^{r}), pJB3 (6,052 bp, *amp*^{r}, *kan*^{r}), and pJB321 (5,594 bp, *amp*^{r}).

Another example of IncP group of vector includes pSPORTn3, which is derived by combining the advantages of high copy number pBR322 vectors with the convenience of conjugative plasmids (Figure 12.16). First the pBR322 vector is transformed into a transposable element by adding a second

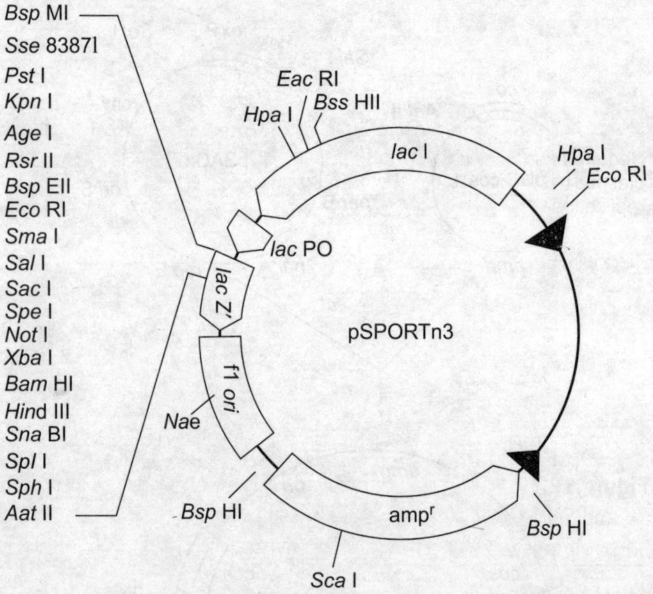

Figure 12.16 Vector map of pSPORTn3 derived from pBR322 [Solid arrowheads denote the inverted repeats needed for transposition; the gene of interest is inserted in the MCS that results in insertional inactivation of the *lac* Z′ gene.]

inverted repeat as pBR322 already contains the β-lactamase gene (*bla*) and one of two 38 bp inverted repeats of transposon *Tn* 2, while the missing transposase activity is provided by pSC101 carrying the *tnp* A gene. This system is used for cloning gene of interest into the transposition vector. Thereafter, recombinant molecules are transformed into an *E. coli* strain carrying the IncP group plasmid (e.g., R751) and the pSC101 *tnp* A derivative and recombinants are selected. After selection of a suitable transformant, it is mobilized by conjugation to desired Gram negative bacterium and finally selection is carried out on the basis of *amp*^{r} selectable marker gene carried on the transposon.

pSa (IncW Group)

This plasmid derived from group IncW infects a wide range of Gram negative bacteria. It has been developed mainly as a vector for use with the oncogenic bacterium *A. tumefaciens*. The size of this plasmid is 29.6 kbp, carrying *kan*^{r}, *sm*^{r}, *su*^{r}, and *cm*^{r} resistance genes (Figure 12.17). pSa *ori* is localized within a 4 kbp DNA fragment, which permits the replication of the plasmid in *E. coli* and *A. tumefaciens*. In addition, two regions of the plasmid have been identified, one involved in conjugal transfer of the plasmid and the other between *Sst* I and *Sal* I responsible for suppression of oncogenicity by *A. tumefaciens*. Later, a series of broad host range vectors of smaller size have been derived from pSa, carrying single-target sites for a number of the common restriction endonucleases and at least one selectable marker in each is subject

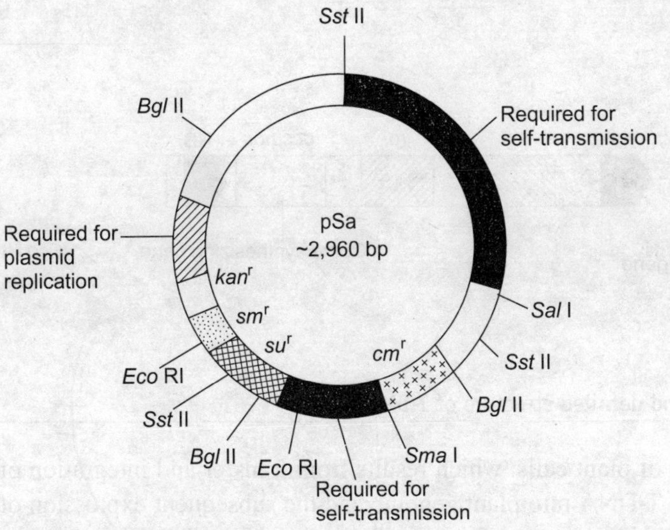

Figure 12.17 Vector map of pSa [The region between the *Sst* II and *Sal* I sites is responsible for suppression of tumor induction by *Agrobacterium tumefaciens*.]

to insertional inactivation. These Sa derivatives are nonconjugative and can be mobilized by other conjugative plasmids.

pBBR1

pBBR1 is a broad host range plasmid, which was first isolated from *Bordetella bronchiseptica*. It is compatible with IncP, IncQ, and IncW plasmids and replicates in a wide range of bacteria. A series of relatively small (~5.3 kbp) vectors have been developed from pBBR1 having different MCSs, allow direct selection of recombinants in *E. coli* by blue-white screening, and are mobilizable by IncP plasmids. Further, some expression vectors were also developed from one of these pBBR1 derivatives by incorporating the *ara* BAD/*ara* C cassette from *E. coli* and it was shown that the promoter is controllable in *Agrobacterium* and *Xanthomonas*.

12.3.2 *A. tumefaciens*-based Vectors

A. tumefaciens, a rod-shaped spore-forming Gram negative pathogenic soil bacteria, has played a major role in the development of plant genetic engineering and basic research in molecular biology. It is capable of genetically transforming a plant cell naturally and has long been exploited to introduce DNA into plants. Initially, it was believed that only dicots, gymnosperms, and a few monocot species could be transformed by this bacterium, but later, this view was totally changed as many 'recalcitrant' species not included in its natural host range such as monocots and fungi could also be transformed. The simple *in vivo* transformation of tissue in intact plants by cocultivation and the 'agrolistic' methods to transform recalcitrant plants are the two novel technical achievements related to *Agrobacterium*. Combined with powerful vectors such as bacterial artificial chromosome, very large DNA fragment can be transformed into the plant genome by *Agrobacterium*.

The property of natural tendency of gene transfer to plants comes from the genes present on pTi plasmid of *Agrobacterium*, which catalyzes the natural transfer of T-DNA (i.e., transferred DNA) to plants. This property is exploited in plant transformation, where an exogenous DNA inserted into T-DNA is naturally transferred to plants. Thus, in order to understand the mechanism of *Agrobacterium*-mediated gene transfer, the structure of pTi, functions of various pTi-encoded genes, and their role in T-DNA transfer to plants should be properly understood.

Tumor-inducing Plasmid (*pTi*)

Tumor-inducing plasmid (pTi) is a megaplasmid of the size greater than 200 kbp. During infection by *A. tumefaciens*, a portion of DNA, called T-DNA (transferred DNA), from the pTi is transferred to the wounded plant cell. The functions of various pTi-encoded gene products are tabulated in Table 12.2. The T-DNA is ~23 kbp in size and is flanked by 25 bp direct repeats, namely right border (RB) and left border (LB) (Figure 12.18). These border sequences are *cis*-elements necessary to direct T-DNA processing and any DNA between these borders can be transferred to a plant cell. Wild-type T-DNAs contains shoot inducing loci (*shi*) and root-inducing loci (*roi*), which comprise of genes encoding for the synthesis of the plant growth regulators. The *roi* locus contains auxin-synthesizing genes *iaa* M, encoding tryptophan 2-monooxygenase and *iaa* H encoding indole acetamide hydrolase [tryptophan is converted to indole 3-acetamide in a tryptophan 2-monooxygenase-catalyzed reaction; indole 3-acetamide is hydrolyzed to indole 3-acetic acid (auxin) in an indole 3-acetamide hydrolase-catalyzed reaction]. The *shi* locus contains *ipt* (*tmr*) gene encoding isopentenyl transferase, which plays a role in cytokinin synthesis [isopentenyl transferase adds 5′-adenosine monophos-

Table 12.2 Products of Various *Agrobacterium* pTi-encoded Genes

Gene	Function
Produced by bacteria and required by plants	
vir	DNA transfer to plants
shi	Shoot induction
roi	Root induction
Produced by plants and required by bacteria	
nos	Nopaline synthesis
ocs	Octopine synthesis
occ	Octopine catabolism
noc	Nopaline catabolism
Produced and required by bacteria	
tra	Bacterial transfer
inc	Incompatibility
ori V	Origin of replication

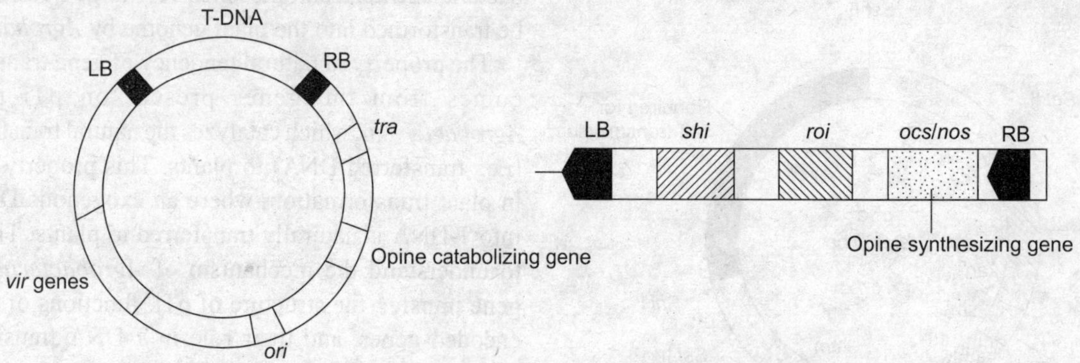

Figure 12.18 Location of useful genes and sites on pTi plasmid and detailed structure of T-DNA

an isoprenoid side chain to form isopentenyladenine and isopentenyl adenosine; hydroxylation of these two molecules by plant enzymes generates the cytokinins called transzeatin and transribosylzeatin, respectively]. The excessive production of these plant growth regulators in infected plant cells is responsible for tumorous phenotype (crown gall). In addition, wild-type T-DNA also encodes enzymes for the synthesis of novel amino acid derivatives called opines [unique condensation products of either an amino acid and a keto acid or an amino acid and a sugar, for example, the condensation product of arginine and pyruvic acid is called octopine; arginine with α-ketoglutaraldehyde is nopaline; agropine is bicyclic sugar derivative of glutamic acid]. These compounds serve as sources of carbon and/or nitrogen and energy and also stimulate plasmid conjugative processes as well as virulence gene (*vir* genes) expression. The genes encoding enzymes for opine catabolism (*occ*/*noc*) are located outside the T-DNA. Based on the presence of the type of opine synthesizing and catabolizing genes on pTi, the plasmid is named as octopine, nopaline, agropine, argininopine, succinamopine, mannopine plasmids, etc. Two best-characterized examples of wild-type pTi used in vector construction are pTiAch5 (octopine type) and pTiC58 (nopaline type).

In contrast to the other mobile DNA elements, T-DNA does not itself encode the products that mediate its transfer. Thus, processing and transfer of T-DNA is stimulated by the products of *vir* genes (i.e., Vir proteins), which occupy ~40 kbp of pTi outside T-DNA. Their location outside the T-DNA indicates that the *vir* genes are not themselves transferred during infection. The detailed description of process of T-DNA transfer to plants is provided in Chapter 14. Besides, the pTi plasmid is capable of autonomous replication using its own origin of replication (*ori*).

pTi-based Vectors

T-DNA transfer to plant cells demonstrates certain important facts for its practical use in plant transformation. Firstly, the tumor formation is a consequence of transformation process of plant cells, which results from transfer and integration of T-DNA into plant genome and the subsequent expression of T-DNA genes. Thus, the disarming of the T-DNA to remove tumor-inducing genes prevents pathogenicity or oncogenicity associated with *Agrobacterium* and also increases the space for insertion of exogenous DNA. Secondly, any foreign DNA placed between the T-DNA borders can be transferred to plant cells, irrespective of its source. Thirdly, while growing *A. tumefaciens* under *in vitro* conditions, opine synthesizing and catabolizing genes may also be deleted from pTi. Rather other nitrogen and carbon sources required for their growth can be supplied in the medium. This also increases the insert capacity. Finally, the T-DNA genes are transcribed only in plant cells and do not play any role during the transfer process. Thus, the genes essential for the transfer of T-DNA (i.e., *vir* genes) need not be present on the same vector and can be supplied *in trans*.

The above facts have allowed the construction of pTi-based vectors and bacterial strain systems that can be employed for plant transformation in order to acquire a particular trait. Thus, pTi-based vector construction requires manipulation of wild-type pTi in the following respects: (i) Size reduction by disarming: For recombinant DNA experiments, a much smaller version is preferred, so large segments of DNA that are non-essential for a cloning vector can be removed. Thus, genes governing auxin and cytokinin production (the oncogenes) as well as opine synthesizing genes may be deleted from the T-DNA in a process called disarming of T-DNA. Further, the opine catabolizing genes may also be deleted from the wild-type pTi plasmid. The *tra* loci may also be deleted to prevent conjugative ability; (ii) Ability to replicate in *E. coli* (shuttle vector system): Because the pTi does not replicate in *E. coli*, the convenience of perpetuating and manipulating pTi carrying inserted DNA sequences in these bacteria does not exist. Therefore in developing pTi-based vectors, an origin of replication that can be used in *E. coli* must be added. As *E. coli* and *Agrobacterium* both are Gram negative bacteria, a broad host range origin of replication may also be used, which allows replication in both *E. coli* and *Agrobacterium*; (iii) Insertion of a selection marker gene: A selection marker gene such as an antibiotic resistance gene conferring antibiotic resistance to plants is inserted into plant transformation vector. As the selection marker gene is prokaryotic in origin, it is put under the control of plant (eukaryotic) transcriptional signals (promoter and termination polyadenylation sequence) to allow its efficient expression in plants; (iv) Insertion of polylinker: MCS is added to facilitate insertion of the cloned gene into the region between T-DNA border sequences (RB and LB). In an ideal vector, the MCS for cloning purpose should lie close to RB and selectable/screenable marker gene should lie closer to LB within the T-DNA; and (v) Deletion of *tra* loci: In order to make the vector suitable for biological containment, the conjugative ability of the bacterium should be prevented by deleting the *tra* loci present outside the T-DNA. As described above, this deletion also decreases the size of the pTi-based vector.

Depending on the *trans* or *cis* supply of *vir* gene products, two vector systems, *viz.*, cointegrate vector system and binary vector system have been developed, the details of which are as follows:

Cointegrate Plant Transformation Vector This system comprises of a single large plasmid formed by homologous recombination of two plasmids in an agrobacterial cell. These two plasmids are *E. coli* plasmid (cloning vector or cointegrate cloning vector) and disarmed helper pTi. The cloning vector has a plant selectable marker gene, the polylinker site for insertion of exogenous DNA, RB, an *E. coli* origin of replication, bacterial selectable marker gene, and homologous sequences. The disarmed helper pTi is deletion derivative of wild-type pTi and comprises of *vir* genes, origin of replication (*Agrobacterium*), LB, and homologous sequences. After all the manipulations, the cloning vector is introduced into agrobacterial cell harboring disarmed helper pTi, where the cloning vector recombines with disarmed helper pTi. As a result, entire cloning vector is integrated into disarmed helper pTi forming a recombinant pTi called cointegrate vector. This is because both cointegrate cloning vector and the disarmed helper pTi carry homologous sequences, which provide a shared site for *in vivo* homologous recombination. Following recombination, the cloning vector becomes the part of disarmed helper pTi, which provides the *vir* genes necessary for the transfer of T-DNA to the host plant cells. The cointegrate vector system suffers from certain drawbacks. These include large size and difficulty in handling, manipulation, and insertion (relative to binary vectors). Moreover, the requirement of homologous sequences limits the integration of *E. coli* plasmid into limited number of pTi. However, once formed, this plasmid is stable in *Agrobacterium*. The best-known example is pGV2260. The strategy used for cointegrate construction and a general vector map of a cointegrate vector are given in Figure 12.19.

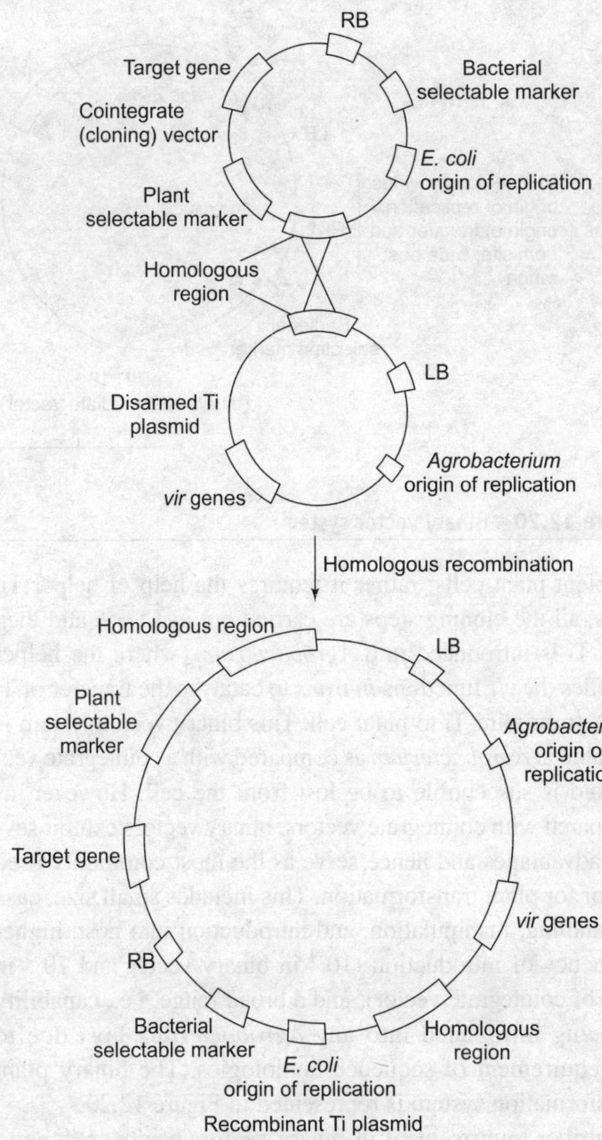

Figure 12.19 Cointegrate vector system

Binary Plant Transformation Vector In order to overcome the problems associated with using cointegrate vector system, binary vector system has been developed. Binary plant transformation vector system is a two-vector system, which comprises of a pair of plasmids, *viz.*, mini-Ti (or intermediate vector; iv; micro-Ti) and helper Ti. In mini-Ti, disarmed T-DNA flanked by RB and LB contains plant selection marker gene and a site for insertion of exogenous DNA. It is a shuttle vector capable of replication both in *Agrobacterium* and *E. coli* and also contains bacterial selection marker gene. The helper strain acts as recipient of mini-Ti and harbors a modified (defective or disarmed) pTi that contains a complete set of *vir* genes but lacks T-DNA. It is however capable of autonomous replication in *Agrobacterium*. The example of a commonly used helper is pAL4404. As mini-Ti lacks *vir* genes, it is incapable of catalyzing the transfer and integration of T-DNA region into

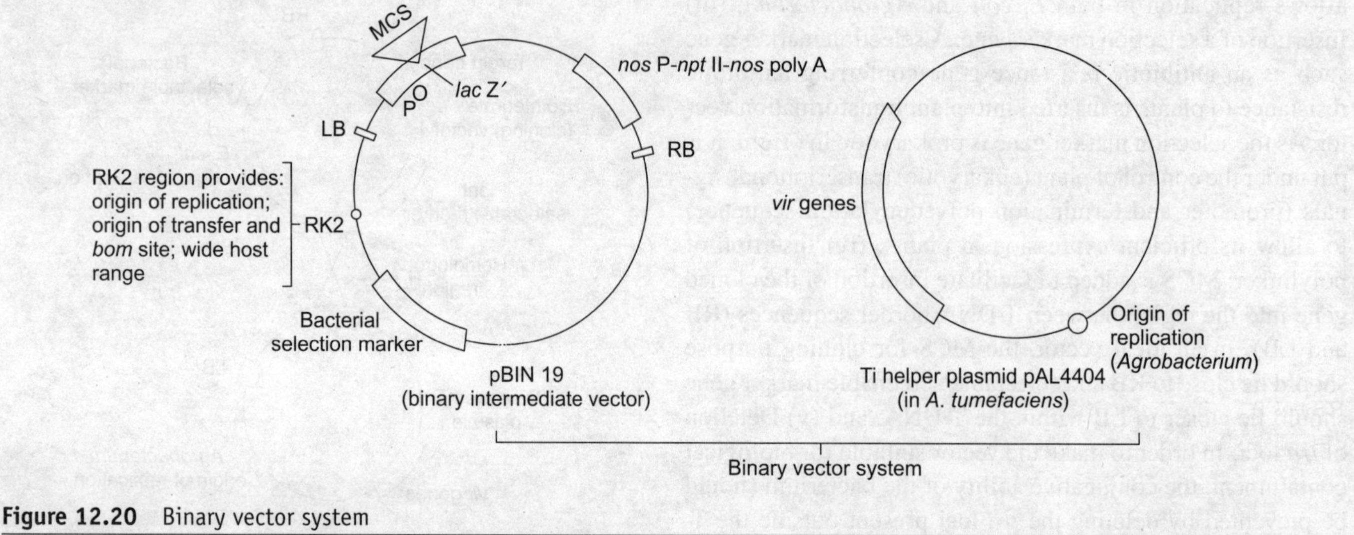

Figure 12.20 Binary vector system

recipient plant cells; rather it requires the help of helper Ti. Thus, all the cloning steps are carried out in *E. coli* and then mini-Ti is introduced into *Agrobacterium*, where the helper supplies the *vir* functions *in trans* to catalyze the transfer of T-DNA from mini-Ti to plant cell. This binary vector system is unstable in *Agrobacterium* as compared with a cointegrate vector and is susceptible to be lost from the cell. However, as compared with cointegrate vectors, binary vectors exhibit several advantages and hence serve as the most commonly used vector for plant transformation. This includes small size, ease of handling, manipulation, and introduction into host, higher frequency of introduction (10^{-1} in binary vector and 10^{-5} in case of cointegrate vector), and a broad range, i.e., capability of being introduced into any *Agrobacterium* host due to nonrequirement of sequence homologies. The binary plant transformation system is represented in Figure 12.20.

Further improvement in binary vectors has been done to increase insert capacity (>50 kbp). These high-capacity binary vectors, for example BIBAC 2, have been constructed on the basis of artificial chromosome type vectors of *E. coli*. These contain an F plasmid origin of replication and are modeled on the BAC.

Examples of some commonly used mini-Ti vectors are discussed below:

pBIN19 pBIN19 is an example of one of the first binary vectors for plant transformation (developed in early 1980s) (Figure 12.21). It has been one of the widely used plant transformation vectors, and is a progenitor for several binary vectors developed later. It is large, 11,777 bp in size (hence manipulation is difficult). It has kanamycin resistance gene for selection in plants. It also has kanamycin resistance gene for selection in bacteria. It has eight unique restriction sites (MCS) in *lac* Zα (*lac* Z′) gene. Hence, insertion of gene of interest at any of these sites leads to insertional inactivation of *lac* Zα, thereby providing an easy way of selection of recombinants by blue–white screening. Note that MCS is located between *Eco* RI and *Hind* III. Origin of replication is derived from pRK252, a broad host range plasmid. This leads to low copy number of this plasmid in bacterial cell and hence low yield of product. The *kan*ʳ gene (plants) is located close to right border (RB), which may lead to possibility of false transformants due to truncation of the T-DNA.

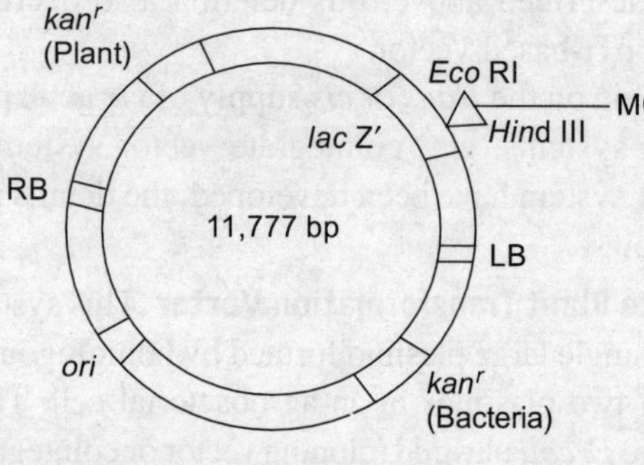

Figure 12.21 pBIN19 vector map

pBI101 It is a promoterless plant transformation binary vector (Figure 12.22). It is derived from pBIN19 and is 12.2 kbp in size. The origin of replication is derived from plasmid RK2, a broad host range low copy number plasmid. This vector allows cloning and testing of promoters using β-glucuronidase (*gus*) expression. There are two expression cassettes, one for *npt* II and the other for *gus* gene. The one for *npt* II is complete (i.e., both promoter and terminator are present), while that for *gus* gene is incomplete (i.e., lacks promoter, while terminator is present). 'Promoterless' *gus* gene is 1.87 kbp in size. A 260 bp *Sst* I-*Eco* RI fragment containing nopaline synthase polyadenylation signal (*nos* poly A) (from *Agrobacterium* pTi) is inserted downstream of *gus*. Thus, promoters are easily cloned upstream of the *gus* gene and then transferred to the host. The *gus* gene cassette is taken from pRAJ260 ligated into *Kpn* I site of pBIN19 polylinker. There is no spurious ATG in the *gus* fragment. Any promoter cloned into the *Hind* III site

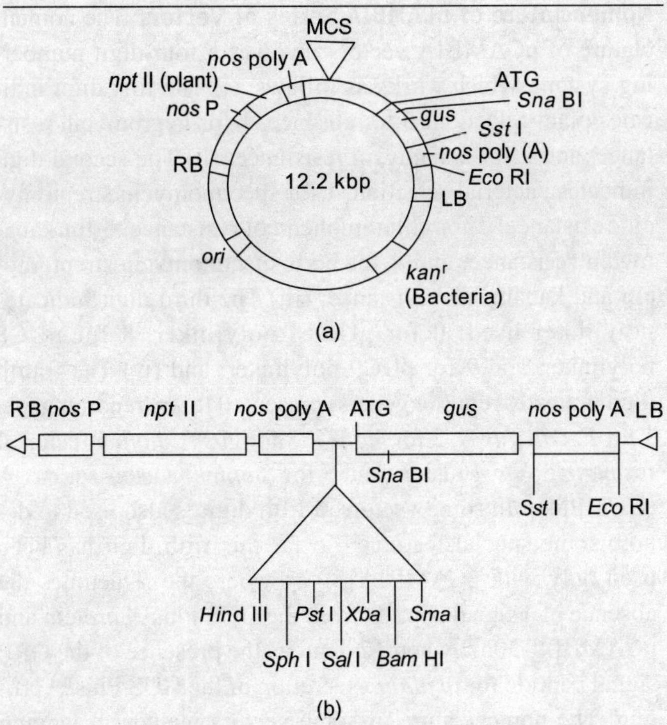

Figure 12.22 pBI101 binary plant transformation vector; (a) Vector map, (b) T-DNA

may encounter problems with the ATG from the polylinker's *Sph* I site (CCATGC). It has kanamycin resistance gene for selection in plants that lie within the T-DNA region. It also has kanamycin resistance gene for selection in bacteria. The pBI101.2 and pBI101.3 are identical to pBI101, except that their reading frames are shifted one and two nucleotides, respectively, relative to the polylinker region.

pBI121 It is an example of widely used binary expression vector for plant transformation. It is 13.0 kbp in size and is derived from pBIN19. It is similar to pBI101 with an 800 bp *Hin*d III–*Bam* HI fragment containing the CaMV 35S promoter (CaMV 35S P) cloned upstream of the *gus* gene (Figure 12.23). The CaMV 35S P expresses high levels of GUS activity in plants after pTi-mediated transformation. It has kanamycin resistance gene for selection in plants that lie within the T-DNA region. It has kanamycin resistance gene for selection in bacteria. The origin of replication is derived from plasmid RK2, a broad host range low copy number plasmid. There are two complete expression cassettes, one for *npt* II and the other for *gus* gene. The gene of interest can be inserted in place of *gus* gene and can be used to check its level of expression under CaMV 35S P. The *gus* gene is 1.87 kbp in size. A 260 bp *Sst* I–*Eco* RI fragment containing nopaline synthase polyadenylation signal (*nos*-poly A from *Agrobacterium* pTi) is inserted downstream of *gus*. The *gus* gene cassette is taken from pRAJ 260 ligated into *Kpn* I site of pBIN19 polylinker. There is no spurious ATG in the *gus* fragment. Any promoter cloned into the *Hin*d III site may encounter problems with the ATG from the polylinker's *Sph* I site (CCATGC).

pGreen This is one of a large family of plant transformation vectors. It is much smaller than pBIN19, 4,632 kbp in length (Figure 12.24). The polylinker has 15 restriction enzymes, which are situated in *lac* Z′ (between *Kpn* I and *Sac* I) and so

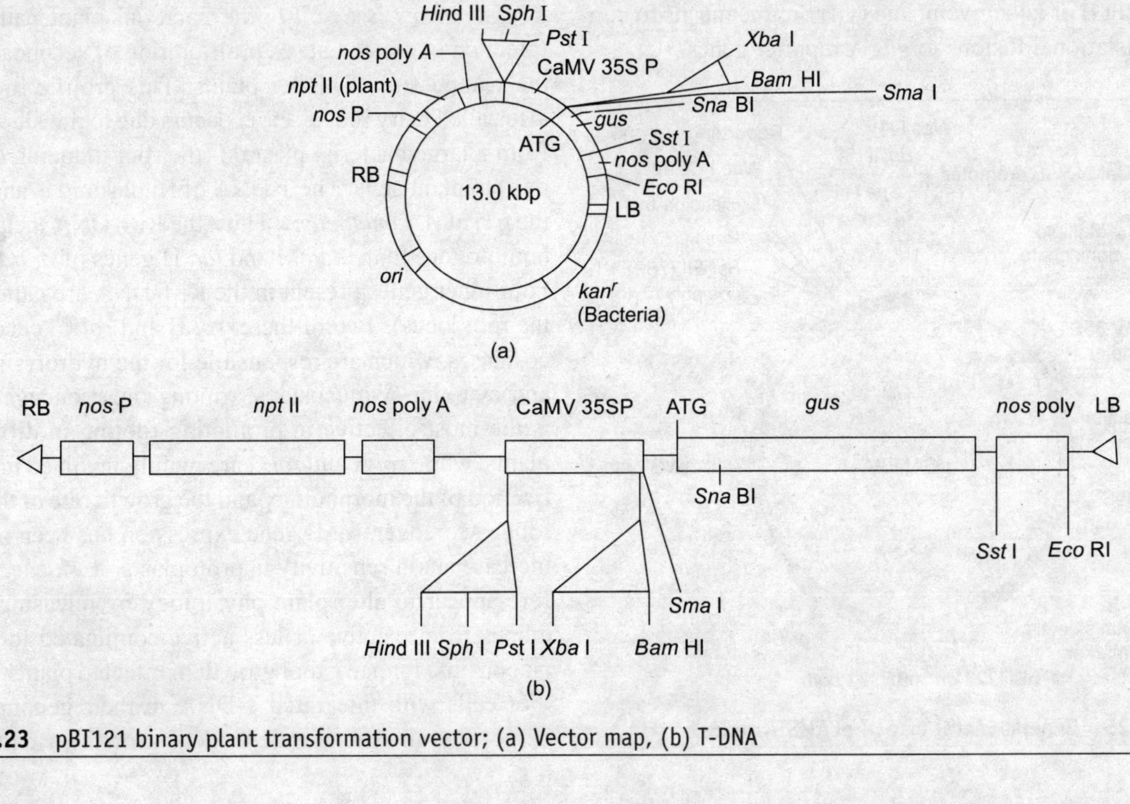

Figure 12.23 pBI121 binary plant transformation vector; (a) Vector map, (b) T-DNA

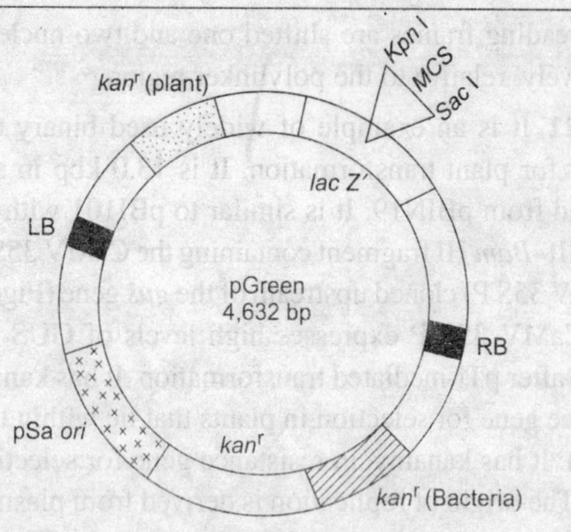

Figure 12.24 General vector map of pGreen series

blue-white screening is possible. This vector has pSa origin and kanamycin resistance, which contributes to the overall small size and antibiotic resistance, respectively.

pCAMBIA series The pCAMBIA contains a series of binary plant transformation vectors, which differ in several characteristics. A general vector map of pCAMBIA series of vectors is drawn in Figure 12.25. These vectors are derived from the pPZP vectors. While these are not perfect and still have technical limitations, these offer (i) High copy number in *E. coli* for high DNA yields; (ii) pVS1 replicon for high stability in *Agrobacterium*; (iii) Small size, 7–12 kbp depending on the plasmid; (iv) Restriction sites designed for modular plasmid modifications and small but adequate polylinkers for introducing DNA of interest; (v) Bacterial selection with chloramphenicol or kanamycin; (vi) Plant selection with hygromycin B or kanamycin; and (vii) Simple means to construct translational fusions to *gus* A reporter genes.

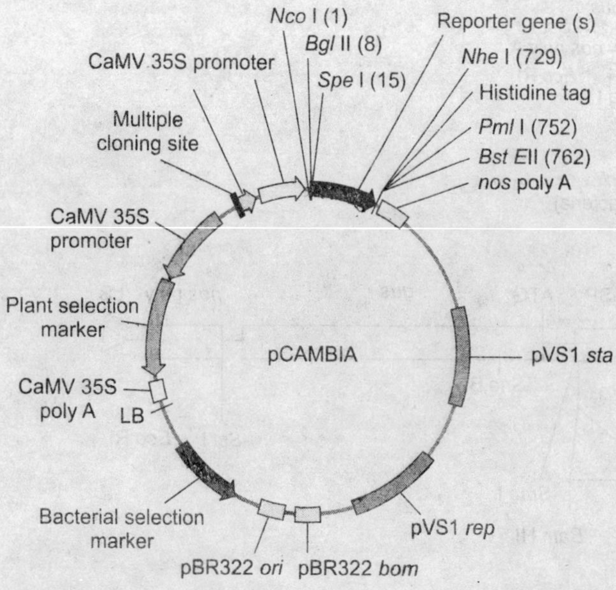

Figure 12.25 General vector map of pCAMBIA series

Nomenclature of pCAMBIA Series of Vectors The nomenclature of pCAMBIA vectors involves a four-digit numbering system, which works as follows: (i) The first digit indicates plant selection: 0 for absence, 1 for hygromycin resistance, and 2 for kanamycin resistance; (ii) The second digit indicates bacterial selection: 1 for spectinomycin/streptomycin resistance, 2 for chloramphenicol resistance, 3 for kanamycin resistance, and 4 for both spectinomycin/streptomycin and kanamycin resistance; (iii) The third digit indicates polylinker used: 0 for pUC18 polylinker, 8 for pUC8 polylinker, and 9 for pUC9 polylinker; and (iv) The fourth digit indicates reporter gene(s) present: 0 for no reporter gene, 1 for *E. coli gus* A, 2 for *mgfp* 5, 3 for *gus* A:*mgfp* 5 fusion, 4 for *mgfp* 5:*gus A* fusion, and 5 for *Staphylococcus* sp. *gus* A (GUS Plus). In some vectors, a fifth digit is also used to denote some special features. So far this fifth digit has been used only with pCAMBIA1305.1 where the .1 denotes the absence of a signal peptide from the GUS Plus™ protein and pCAMBIA1305.2 where .2 denotes the presence of the GRP signal peptide for *in planta* secretion of the GUS Plus™ protein. The nomenclature of some vectors includes a lagging letter. The lagging letter X indicates that the reporter gene lacks its own start codon and the vector is for creating fusions to the reporter; lagging letter Z indicates presence of a functional *lac* Zα for blue–white screening, and lagging letter a/b/c indicates the reading frame for fusions with the Fuse and Use vectors.

12.3.3 *A. rhizogenes*-based Vectors

A. rhizogenes is a soil-borne bacterial plant pathogen, the infection of which causes proliferation of secondary roots at the wound site of higher plants. This prolific root growth, also called hairy root disease, occurs due to transfer of T-DNA from a large bacterial plasmid (the root inducing or Ri plasmid) to plant cells. The T-DNA of Ri plasmid is analogous to the pTi of *A. tumefaciens*. Thus, the Ri T-DNA includes genes homologous to the *iaa* M and *iaa* H genes of *A. tumefaciens*. Four other genes present in the Ri T-DNA are named *rol* (for the root locus). Two of these, *rol* B and *rol* C, encode *P*-glucosidases, which are responsible for the hydrolysis of indole and cytokine-*N*-glucosides. Among these oncogenes, *rol* B is the most effective in promoting rooting in different host plants, while *rol* A and *rol* C are mainly involved in the modification of the morphology and the growth rate of the induced roots. Moreover, *rol* B gene expression has been reported to increase auxin sensitivity in protoplasts. *A. rhizogenes* therefore appear to alter plant physiology by releasing free hormones from inactive or less active conjugated forms and is responsible for hairy root growth in infected plants. The hairy root cells with integrated T-DNA in their genome can undergo phytohormone-independent growth in culture. This property of hairy root cells is routinely exploited in several laboratories for the production of secondary metabolites and opines from plants by maintaining them as hairy root cultures. The hairy root cultures are advantageous as compared to suspension cultures, as these are much more stable genetically and the yield of secondary metabolites is higher. However, a major limitation for the commercial use of hairy root cultures is the difficulty involved in scale-up, since each culture comprises a heterogeneous mass of interconnected tissue, with highly uneven distribution. *A. rhizogenes* also finds place in plant transformation. The Ri plasmid appears to be equivalent to disarmed pTi and many of the principles explained in the context of disarmed pTi are applicable to Ri plasmid as well. Now several *A. rhizogenes*-based vectors are available. When foreign DNA is introduced into T-DNA region, it can be carried via the Ri plasmid into host plant cells, making it a natural vector for rapid introduction of genes into higher plants. Further it is reported that *Agrobacterium* containing both Ri plasmid and a disarmed pTi can frequently cotransfer both of them. If such a dual system is used, no drug resistance marker on the T-DNA is necessary for selection of transformants. In contrast, the transformants can be easily selected on the basis of hairy root formation in the target tissue, thereby indicating cotransfer of recombinant T-DNA. Intact plants are then regenerated by manipulation of phytohormones in culture medium.

12.4 GRAM POSITIVE BACTERIA AS CLONING VECTORS

The base composition of the different genomes of Gram positive bacteria shows inconsistency in GC content, which ranges from <30% to >70%. Therefore, no universal cloning vehicles can be used for cloning in Gram positive bacteria. In this case, two sets of vector systems have been developed, one for low-GC organisms (e.g., *Bacillus, Clostridium, Staphylococcus, Streptococcus, Lactococcus*, and *Lactobacillus*) and another for high-GC organisms (e.g., *Streptomycetes*). The vectors used for these two systems are described in the following sections.

12.4.1 *B. subtilis* as Vector

Bacillus sp. have attracted interest in genetic engineering because of following reasons: (i) *Bacillus* sp. can be used as model organisms for studying Gram positive bacteria having low GC content; (ii) *Bacillus* sp. is generally obligate aerobe compared with *E. coli*, which is Gram negative facultative anaerobe and hence can be used to study behavior of obligate aerobes; (iii) *Bacillus* sp. have the ability to sporulate and consequently are used as models for prokaryotic differentiation; (iv) *E. coli* lacks some secondary biochemical pathways that are essential for phenotypic expression of certain functions. *Bacillus* sp. are widely used in fermentation industries particularly for production of exoenzymes and can be tailored to secrete the products of cloned genes; and (v) It is safe to use *Bacillus* sp., as these are not known for any pathogenic interactions with human and animals.

Table 12.3 Properties of Plasmids of *S. aureus* used as Vectors in *B. subtilis*

Plasmid	Size (bp)	Copy no.	Antibiotic resistance	Remarks
pC194	2,906	15	Chloramphenicol	Large amount of high molecular weight DNA are produced
pE194	3,728	10	Erythromycin	Temperature-sensitive replication
pUB110	4,548	50	Kanamycin	Site-specific plasmid recombination

Plasmids from *S. aureus* were first used as vectors for the transformation of *B. subtilis* (Table 12.3). None of the natural *S. aureus* plasmids were ideal as cloning vector as these lacked unique restriction sites and contained not more than one selectable marker. As a result, improved vectors were devised by gene manipulation and the resulting plasmids were stable in *B. subtilis*, though segregative stability was greatly reduced after insertion of exogenous DNA. Later, cryptic *Bacillus* plasmid (pTA1060) was used as a vector. Finally, chimeric plasmids were constructed by performing fusion between pBR322 and pC194 or pUB110. These chimeric plasmids can replicate in both *E. coli* and *B. subtilis*; hence *E. coli* is used as an efficient intermediate host for cloning with such hybrid plasmid vectors. Plasmid preparations extracted from *E. coli* clones are subsequently used to transform competent *B. subtilis* cells. The resultant DNA preparations contain sufficient amounts of multimeric plasmid molecules sufficient enough for the transformation of competent cells of *B. subtilis* (for details see Chapter 14). These hybrid vectors have *E. coli lac* Z α-complementation fragment and MCS. Some also have the phage f1 origin for subsequent production of ss DNA in a suitable *E. coli*. Following are commonly used vectors for cloning in *B. subtilis*:

pAMβ1 and pTB19 The major problem associated with cloning in plasmid is that only short DNA fragments could be efficiently cloned and longer DNA segments often undergo rearrangements. This structural instability is due to the mode of replication of the plasmid vector as all the *B. subtilis* vectors replicate by a rolling circle mechanism thereby generating ss DNA. Moreover, ss DNA is known to be a reactive intermediate in every recombination process. Therefore, if structural instability is an outcome of rolling circle replication, then vectors that replicate by the theta mechanism may possibly be more stable. Two potentially useful plasmids, pAMβ1 and pTB19 are derived from *Streptococcus (Enterococcus) faecalis* and *B. subtilis*, respectively. These are large (26.5 kbp) natural plasmids. Replication of these plasmids does not lead to accumulation of detectable amounts of ss DNA, whereas the rolling circle mode of replication does. Also, the replication regions of these two large plasmids share no sequence homology with the corresponding highly conserved regions of the rolling circle-type plasmids. It is worth noting that the classical *E. coli* vectors, which are derived from plasmid Col E1, replicate via theta-like structures. Later a series of cloning vectors have been developed from pAMβ1 that carry a gene essential for replication, *rep* E, and its regulator, *cop* F. The *cop* F gene can be inactivated by inserting a linker into its unique *Kpn* I site. Since *cop* F downregulates the expression of *rep* E, its inactivation leads to an increase in the plasmid copy number per cell. This new replicon has been used to build vectors for making transcriptional and translational fusions and for expression of native proteins. Few shuttle vectors based on pAMβ1 have been constructed for the production of transcriptional fusions, which can be transferred via conjugation between *E. coli* and a wide range of Gram positive bacteria.

A range of expression systems has also been devised by employing controllable elements from *E. coli*. For instance *lac* I gene and the promoter of phage SPO-1 coupled to the *lac* operator are inserted into the vector known as Pspac. In addition, T7 RNA polymerase gene (*rpo* T7) is inserted into the chromosome under the control of a xylose-inducible promoter and cloning the gene of interest coupled to a T7 promoter on a *B. subtilis* vector.

pMUTIN Vectors pMUTIN vector is used for systematic gene inactivation in *B. subtilis* that facilitates functional analysis of uncharacterized ORFs, which is performed by directed insertional mutagenesis in the chromosome. pMUTIN vector has erythromycin resistance (em^r), *lac* I, and *lac* Z genes expressed in *B. subtilis* and amp^r gene expressed in *E. coli* (Figure 12.26).

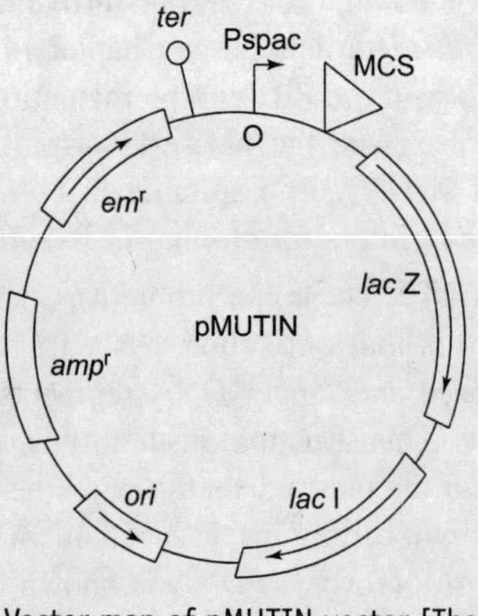

Figure 12.26 Vector map of pMUTIN vector [The em^r, *lac* I, and *lac* Z genes are expressed in *B. subtilis*; the amp^r gene is expressed in *E. coli*; *ter* is the terminator that prevents run-through transcription from the em^r gene; O indicates the *lac* I operator; Mcs contains *Hind* III, *Eco* RI, *Not* I, *Sac* I, and *Bam* HI sites.]

It also carries the inducible Pspac promoter to allow controlled expression of genes downstream of it and found in the same operon as the target gene. A terminator sequence (*ter*) is located after em^r gene, which prevents run-through transcription of that gene. An internal fragment of the target gene is amplified by polymerase chain reaction (PCR) and cloned in a pMUTIN vector; the resulting plasmid is used to transform *B. subtilis*. After integration, the target gene is interrupted and a fusion product is produced between its promoter and the reporter *lac* Z gene. If the targeted gene is part of an operon, then any gene downstream of the operon is placed under the control of the Pspac promoter. Note that simultaneously two types of mutants are generated, *viz.*, an absolute (null) in *orf* 2 and a conditional mutation in *orf* 3, which can be alleviated by induction with isopropyl-β-*D*-thiogalactoside (IPTG).

12.4.2 *Streptomyces* as Vector

This species has attracted interest as vector because of following reasons: (i) Large number of antibiotics are prepared by members of this genus, thus genetics and regulation of antibiotic synthesis can be analyzed; (ii) *Streptomyces* vectors serve as model organisms for studying Gram positive bacteria having high GC content; (ii) *Streptomyces* can express genes of low GC organisms, however, *Streptomyces* genes are usually difficult to express in *E. coli* because most promoters do not function and translation may be inefficient unless the initial amino acid codons are changed to lower GC alternatives by applying codon usage approach.

All the cloning vectors used in *Streptomyces* are derived from plasmids and phages that occur naturally in them as no plasmid (except RSF1010) from any source has the ability to replicate in *Streptomyces*. Various replicons that have been used as vectors are given in Table 12.4. All *Streptomyces* plasmids are equipped with transfer abilities that permit conjugative transfer and provide different levels of chromosome-mobilizing activity. Plasmid SCP2*, a derivative of the sex plasmid SCP2, is 31.4 kbp in size and progenitor of various stable and very low copy number vectors. High copy number vectors are derived from pIJ101 and pSG5. The two *Streptomyces* plasmids SLP1 and pSAM2 are examples that normally reside integrated into a specific highly conserved chromosomal transfer RNA (tRNA) sequence.

Table 12.4 Plasmids from *Streptomyces* Used in the Development of Vectors

Plasmid	Size (kbp)	Mode of replication	Copy number
pIJ101	8.8	Rolling circle	300
pJV1	11.1	Rolling circle	—
pSG5	12.2	Rolling circle	20-50
SCP2	31	Theta	1–4
SLP1	17.2	Rolling circle	Integrating
pSAM2	10.9	Rolling circle	Integrating

Furthermore, several specialist vectors have been derived from these plasmids, including cosmids, expression vectors, vectors with promoterless reporter genes, positive selection vectors, and temperature-sensitive vectors, etc. For instance, the temperate phage ϕC31 is comparable to λ phage and has been derived as a vector. These ϕC31-derived vectors can clone up to an average insert size of 8 kbp. On the contrary, no such size constraints are observed with plasmid vectors although recombinant plasmids of size >35 kbp are rare with the usual vectors. However, phage vectors have one important advantage as plaques can be obtained overnight, whereas plasmid transformants can take up to 1 week to sporulate. By incorporating the integration functions of ϕC31 into plasmids, hybrid vectors have also been generated. For cloning of large gene clusters, BACs are generated that can accommodate up to 100 kbp of *Streptomyces* DNA. These vectors can replicate autonomously in *E. coli* and integrate site-specifically into the chromosome.

12.5 PLANT AND ANIMAL VIRAL VECTORS

Since viral infection of a cell results in the addition of new genetic material, which is expressed in the host, viruses provide natural examples of genetic engineering. In both plant and animal systems, viruses have played important roles in vector development. Additional genetic material incorporated in the genome of a plant or animal virus might be replicated and expressed in the host (plant/animal) cell along with the other viral genes. Various viral vectors have been engineered for their specific applications but generally share a few key properties. These include: (i) Although viral vectors are occasionally created from pathogenic viruses, these are modified in such a way so as to minimize the risk of handling them. This usually involves the deletion of a part of the viral genome critical for viral replication. Such a virus can efficiently infect cells but, once the infection has taken place, it requires a helper virus to provide the missing proteins for production of new virions; (ii) The viral vector should have a minimal effect on the physiology of the cell it infects; (iii) Some viruses are genetically unstable and can rapidly rearrange their genomes. This is detrimental to predictability and reproducibility of the work conducted using a viral vector and is avoided in their design; and (iv) Most viral vectors are engineered to infect a wide range of cell types. However, sometimes the opposite is preferred. The viral receptor can be modified to target the virus to a specific kind of cell.

In the following sections, vectors derived from plant and animal viruses are described.

12.5.1 Plant Viral Vectors

Like *Agrobacterium*, plant viruses exhibit natural tendency of gene transfer to plants and thus have the potential for use as gene transfer and expression vectors. The construction of plant virus-based vectors requires modification of viral genomes to accommodate the insertion of foreign sequences, which are transferred, multiplied, and expressed in a plant as part of the recombinant virus genome. Thus, for vector construction, following points should be taken into consideration: (i) The genome size should be as small as possible so as to allow easy handling during isolation and manipulation. Further in a small genome, there are chances of occurrence of unique restriction site(s) for cloning. Thus to use plant virus as vector, nonessential genes from its genome are deleted. Besides reducing the genome size, this also increases the space for insertion of exogenous DNA; (ii) For use as vector, recombinant plant viruses are designed to mimic their wild-type counterparts in all respects except for the genes conferring pathogenicity. The removal of pathogenicity-conferring genes further leads to size reduction and increase in insert capacity; (iii) Vector should have a broad host range, ease of mechanical transmission, and rate of seed transmission; (iv) Vector should have the potential to carry additional genetic information since there are strict packaging limitations. Thus, viruses whose capsid are filamentous or rod-shaped or viruses that possess a multipartite genome or a helper or satellite component offer the potential for carrying exogenous DNA; (v) Virus stability as a vector depends on the fact that genetic material must be able to be manipulated in such as way that these are expressed in the plant; and (vi) Finally, there should not be any loss of infectivity of viruses.

Certain plant viruses (DNA and RNA plant viruses) having the biological characteristics as described above have been assessed for their plant transformation capability. These are summarized below.

DNA Viruses as Vectors

Two groups of DNA viruses that infect plants are the cauliflower mosaic virus (CaMV) and gemini virus (GV).

Cauliflower Mosaic Virus (CaMV) CaMV is the extensively available member of caulimovirus. It is the first plant virus to be discovered that contains DNA and is the first plant viral genome to be sequenced. It is also the first plant virus to be manipulated by the use of recombinant DNA technology and is cited as the most likely potential vector for introducing foreign genes into plants. It is spherical, isometric, and ~50 nm in diameter. It is a ds DNA virus with the genome size of 8,024 bp (as relaxed circular molecule), which comprises of an unusual structure characterized by the presence of three discontinuities in the duplex (Figure 12.27). The DNA exists in linear, open circular, and twisted, or knotted forms; however, none of the circular forms is covalently closed due to the presence of site-specific single stranded breaks. The genome is organized into eight tightly packed genes that are expressed as two major transcripts, *viz.*, the

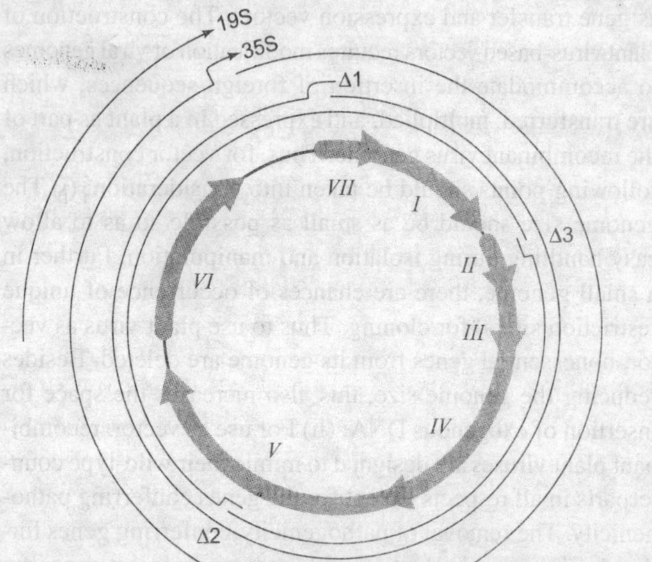

Figure 12.27 Schematic diagram of the CaMV genome [Circular ds DNA (8 kbp) is represented by thin lines with sequence discontinuities (Δ1–3); ORFs are shown by arrows; Seven genes are indicated as Roman numerals (I codes for cell-to-cell movement protein, II and III are responsible for aphid transmission, IV codes for the precursor of the capsid protein, V encodes the precursor of aspartic proteinase, reverse transcriptase and RNase H, VI codes for an inclusion body protein/translational activator, VII function unknown); the solid black lines of the inner circle are the long and small intergenic regions, which contain the 35S and 19S promoters, respectively; the two external arrowed lines correspond to the 35S and 19S RNAs.]

35S RNA corresponding to almost the whole genome and the 19S RNA that corresponds to the coding region of gene *VI*. Genome also contains large and small intergenic regions, which contain 35S and 19S promoters, respectively.

The properties that make CaMV as one of the most likely potential vector for introducing foreign genes into the plants are as follows: (i) CaMV genome is ds DNA, which lends itself more readily to the manipulations; (ii) These viruses are naturally transmitted by aphids and are also capable of easy mechanical transmission; (iii) Virion DNA alone or cloned CaMV DNA is infectious when simply rubbed on the surface of susceptible leaves; (iv) These viruses can be found at a high copy number (up to 10^6 virions per cell) and spread rapidly in 3–4 weeks in a systemic manner throughout the entire plant; (v) These viruses have only few genes, which can be expressed to high levels by a variety of means; (vii) Because these viruses accumulate in the cytoplasm as inclusion bodies, which consist of a protein matrix with embedded virus particles, these may be isolated from the inclusion body by simple method using urea and nonionic detergents; and (viii) CaMV 35S promoter and terminator (or polyadenylation sequences) are routinely utilized in plant expression vectors for efficient and constitutive expression of the cloned gene.

CaMV has been used successfully as expression vector for the expression of very small transgenes, for example, the 240 bp bacterial *dhfr* gene encoding dihydrofolate reductase, the 200 bp murine metallothionein cDNA, and a 500 bp human interferon cDNA. However, it is not suitable for cloning large-sized inserts. DNA sequence analysis has revealed that six major and two minor ORFs are present on one coding strand in a very tightly packed arrangement. As a consequence, there is very little space for the insertion of transgene and the insert capacity cannot be increased greatly without affecting the efficiency of packaging. It has been found that even after the removal of nonessential regions, a maximum insert capacity of <1 kbp is obtained, indicating that the maximum DNA-holding capacity of the icosahedral capsid of CaMV can be exceeded only up to 8.3 kbp. There is also no scope of increasing insert capacity by developing two separate vectors, i.e., replicon vectors for cloning and helper viruses providing the essential functions *in trans*. This is because of occurrence of high-level of recombination that leads to rapid excision of the foreign DNA.

Gemini Virus (GV) Gemini viruses are also valuable as expression vectors. These are characterized by their twin (geminate) virions, comprising two partially fused icosahedral capsids. The small ss DNA genome is circular and in some species is divided into two segments called DNA A and DNA B, where DNA B supplies viral movement functions. Thus, systemic infection occurs only if DNA B is also present in the plant. Two genera of GV, begamovirus and mastrevirus, have been developed as vectors. The begamoviruses have predominantly bipartite genomes and are transmitted by the whitefly *Bemisia tabaci* and infect dicots. On the other hand, the mastreviruses have monopartite genomes, transmitted by leafhoppers, and predominantly infect monocots. The properties of these viruses that make them ideal for cloning purpose include the following: (i) These viruses use a ds DNA (RF), which suggests that these can be more stable than CaMV, whose RNA-dependent replication cycle is rather error-prone. Further, it makes *in vivo* manipulation more convenient; (ii) There appears to be no intrinsic limitation to the size of the insert, although larger transgenes tend to reduce the replicon copy number. Generally, it appears that mastrevirus replicons can achieve a much higher copy number in protoplasts than replicons based on begamovirus. A wheat dwarf virus (WDV) shuttle vector capable of replicating in both *E. coli* and plants was shown to achieve a copy number greater than 3×10^4 in protoplasts derived from cultured maize endosperm cells, whereas the typical copy number achieved by tomato golden mosaic virus (TGMV) replicons in tobacco protoplasts is less than 1,000. This may, however, reflect differences in the respective host cells, rather than the intrinsic efficiencies of the vectors themselves; (iii) GV replicon vectors can facilitate the high-level transient expression of foreign genes in protoplasts; and (iv) A number of GVs have been developed as expression vectors because of the possibility of achieving high-level recombinant protein expression as a function of rapid viral replication.

In the case of begamoviruses, which have bipartite genomes, the coat protein gene is located on DNA A along with all the functions required for DNA replication. Replicons based on DNA A are therefore capable of autonomous replication in protoplasts. Species that have been developed as vectors include African cassava mosaic virus (ACMV) and TGMV. The coat protein genes of ACMV and TGMV are nonessential for systemic infection, but these are required for insect transmission. Thus, an attractive feature of bipartite GV is its ability to contain a deletion or a replacement of virus coat protein sequence by foreign gene without interfering with the replication of the viral genome. On the other hand, all viral genes appear to be essential for systemic infection in the case of the mastreviruses and hence coat protein replacement vectors cannot be used. Therefore, replicon vectors based on these viruses provide an *in planta* contained transient expression system. Species that have been developed as vectors include maize streak virus (MSV) and WDV.

RNA Viruses as Vectors

The assets of RNA viruses that make them the ideal choice for certain applications as vectors include ease of use, perhaps the highest levels of gene expression and the vector's ability to infect and spread in differentiated tissue. There are two basic types of ss RNA viruses, *viz.*, monopartite and multipartite viruses. The monopartite viruses have undivided genomes containing all the genetic information and are usually fairly large, for example, tobacco mosaic virus (TMV). The multipartite viruses, as the name suggests, have their genome divided among small RNAs either in the same particle or separate particles, for example, the brome mosaic virus (BMV) contains four RNAs divided between three separate particles. The RNA components of multipartite genomes are small and appear to be able to self-replicate in plants. With some members of the group, the genes encoding the coat protein may be dispensable, as their loss does not affect viral DNA multiplication. The second group, subgenomic RNA (e.g., RNA IV of BMV), is unlikely to find application as cloning vector as it is unable to self-replicate in infected plant. The third group, satellite RNA, has perhaps the greatest potential, as it is totally dispensable to the virus. Satellite RNAs vary in size from 270 bases (e.g., tobacco ringspot virus satellite) to 1.5 kbp (e.g., tomato black ring virus satellite). These satellites appear to share little homology with the viral genomic RNAs, templates for their own replication, and utilize the machinery for replication, but are capable of altering the pathogenicity of the viral infection. These satellite RNAs have a number of other unusual properties, including the ability to code for proteins and stability in the plant in the absence of other viral components.

As most plant RNA viruses have a filamentous morphology, the packaging constraints as observed in case of DNA viruses should not present a limitation in vector development. Some of these RNA viruses have been extensively developed as vectors for foreign gene expression. Two such examples are TMV and potato virus X (PVX; potato X potexvirus).

Tobacco Mosaic Virus (TMV) TMV is one of the most extensively studied plant viruses. The virus has a monopartite RNA genome of 6.5 kb. The subgenomic RNAs are translated into at least four polypeptides, which include a movement protein and a coat protein. TMV is a natural choice for vector development since there is no packaging constraint. With certain modifications, very stable TMV expression vectors have been developed that are used to synthesize a variety of valuable proteins in plants, for example, ribosome-inactivating protein and scFV antibodies.

Potato Virus X PVX is the type member of the Potexvirus family. Like TMV, it has a monopartite RNA genome of ~6.5 kb packaged in a filamentous particle. The genome of PVX contains genes for replication, viral movement, and the coat protein, the latter expressed from a subgenomic promoter.

In terms of vector development, it is notable that PVX-based vectors are probably most widely used to study virus-induced gene silencing and related phenomenon and to deliberately induce silencing of homologous plant genes. PVX has also been used for the synthesis of valuable proteins, such as antibodies and for the expression of genes that affect plant physiology (e.g., the fungal avirulence gene *avr* 9). Reporter genes, such as *gus* A and *gfp*, when added to PVX genome under the control of duplicated coat protein subgenomic promoter, are expressed at high levels in infected plants. Further, the stable transformation of plants with cDNA copies of PVX genome potentially provides a strategy for extremely high-level transgene expression because transcripts are amplified to a high copy number during viral replication cycle. PVX, however, has certain associated drawbacks: (i) There is a tendency for the transgene to be lost by homologous recombination; (ii) In the case of PVX, no alternative virus has been identified whose coat protein gene promoter can functionally substitute for the endogenous viral promoter. For this reason, PVX is generally not used for long-term expression, but has been widely employed as a transient expression vector; (iii) Instead of high-level expression, the transformation strategy employing PVX leads to potent and consistent transgene silencing; (iv) Plants may also develop resistance to viral infection. However, in terms of vector development, it is notable that PVX-based vectors are probably most widely used to study virus-induced gene silencing and related phenomena and to deliberately induce silencing of homologous plant genes.

Besides their use as plant expression vectors, certain plant viruses have also been developed to present short peptides on their surfaces. Epitope-display systems based on alfalfa mosaic virus, cowpea mosaic virus, PVX, and tomato bushy stunt virus have been developed as potential source of vaccines, especially against animal viruses.

12.5.2 Animal Viral Vectors

Animal viruses are used as vectors for transduction, i.e., the introduction of genes into animal cells by exploiting the natural ability of the virus particle within which the transgene is packaged. Viruses can deliver their nucleic acids with great efficiency into cells and replicate and express at high levels. Thus, viruses have been used as vectors not only for gene expression in cultured cells but also for gene transfer to living animals. Various classes of viral vectors have been devised for use in human gene therapy including retrovirus, adenovirus, herpesvirus, and adeno-associated virus (AAV) vectors. Transgenes are inserted into whole genome of viral vectors or by replacement of one or more viral genes, which is carried out either by ligation (at unique restriction sites) or homologous recombination. In the case of replacement, cointroduction of helper phage is required. The availability of a wide variety of animal virus vectors enables a foreign gene to be expressed in different cell types at different levels and with different consequences for the host cell. Factors influencing the choice of virus vectors include the types of cells or animals to be infected, whether infection leading to cell transformation or lysis is required, the number of genes to be expressed, and whether virus infectivity is to be retained. Small DNA viruses may have restricted host range and limited capacity for foreign DNA due to severe packaging constraints imposed by the icosahedral virus capsid. On the other hand, rod-shaped viruses, for example, the baculoviruses have no such type of size constraints as capsid is formed around the genome. In conclusion, none of the animal virus is considered as ideal vector for gene transfer as each one is associated with certain disadvantages. Recently, several hybrid viral vectors have been developed by including the favorable features of two or more viruses.

The animal viruses, which are generally used as vector for animal transformation, are explained in this section.

Retrovirus as Vector

Retroviruses are a class of enveloped viruses containing two copies of single strand sense (+) RNA genome that replicates via a ds DNA intermediate. Following infection, the viral genome is reverse transcribed into ds DNA intermediate, which precisely integrates into the host genome, where it is transcribed to yield daughter genomes that are packaged into virions.

The viral genome is 7–10 kb in size, containing at least three genes: *gag* (coding for core proteins), *pol* (coding for reverse transcriptase, integrase, and protease), and *env* (coding for the viral envelope proteins) (Figure 12.28a). At each end of the genome are long terminal repeats (LTRs), which include promoter/enhancer regions and sequences involved in integration. In addition, there are sequences required for packaging the viral DNA (ψ) and RNA splice sites in the *env* gene. The viral envelope interacts with the host cell's plasma membrane, delivers the particle into the cell, and infection cycle is initiated. After infection, RNA (genome) is reverse transcribed to produce a cDNA copy, which involves two template switches and the terminal regions of the RNA genome are duplicated in the DNA as LTRs. The resulting ds DNA intermediate then integrates into the host genome at an essentially random site by a transposition-like event. Viral genomic RNA is synthesized by transcription from a single promoter located in the left LTR and ends at a polyadenylation site in the right LTR. Thus, the full-length genomic RNA is shorter than the integrated DNA copy and lacks the duplicated LTR structure (Figure 12.28b). The genomic RNA comprises of cap and poly (A) tail, permitting the *gag* gene to be

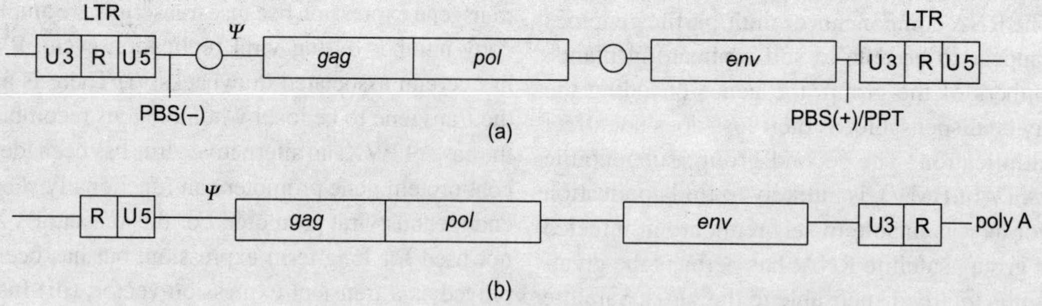

Figure 12.28 Generic map of a retrovirus genome (a) Structure of an integrated provirus, with long terminal repeats (LTRs) comprising three regions U3, R and U5, enclosing the three open reading frames: *gag, pol* and *env*; (b) Structure of a packaged RNA genome, which lacks the LTR structure and possesses a poly (A) tail. [PBS indicates primer binding sites; PBS(–) indicates primer binding site for minus strand synthesis; PBS(+) indicates primer binding site for plus strand synthesis; PPT represents polypurine sequence, which serves as PBS(+); ψ represents packaging signal; Circles represent splice sites.]

translated (the *pol* gene is also translated by read-through, producing a Gag–Pol fusion protein, which is later processed into several distinct polypeptides). Some of the full-length RNAs also undergo splicing, eliminating the *gag* and *pol* genes and allowing the downstream *env* gene to be translated. Finally, two copies of the full-length RNA genome and reverse transcriptase/integrase are packaged into each capsid, which requires a specific *cis*-acting packaging site ψ.

Due to following reasons, retroviruses have been developed as vectors; (i) Retroviruses have the natural ability to act as replication defective gene transfer vectors; (ii) The majority of retroviruses does not kill the host, but produce progeny virions over an indefinite period, and hence retroviral vectors can be used to make stably transformed cell lines; (iii) Strong promoters drive retroviral gene expression. These promoters can also be utilized to control the expression of transgenes; (iv) Some retroviruses have a broad host range, allowing the transduction of many cell types; and (v) As retroviral genome is small, it is easily manipulated *in vitro* in plasmid cloning vectors, resulting in the development of efficient and convenient vectors for gene transfer, which can be propagated to high titers (~10^8 pfu/ml) and with 100% *in vitro* infection.

Besides the above-described advantages, retroviral vectors also have some disadvantages as most of them only infect dividing cells, which restrict their use for gene therapy applications. However, some complex retroviruses like lentiviruses (e.g., human immunodeficiency virus; HIV) have the ability to infect nondividing cells. Lentiviral vectors permit the stable transduction of multiple cell types.

Replication-competent retroviral vectors are made by adding sequences to existing viruses, but most commonly used vectors are replication-defective as these provide the maximum capacity for foreign DNA (~8 kbp). A single promoter in the 5′-LTR expresses retroviral proteins in most of the naturally occurring oncogenic retroviruses and the expression of multiple viral coding regions is achieved by alternative splicing. However, multiple promoters, insertion of genes in the reverse orientation, and the use of internal ribosome entry sites (IRESs) are the different strategies used during development of vectors. Furthermore, efficient gene transduction and integration depend on the insertion of a number of *cis*-acting viral elements in the retroviral vector. These *cis*-acting elements include: (i) A promoter and polyadenylation signal in the viral genome; (ii) A viral packaging signal (ψ) to direct incorporation of vector RNA into virions; (iii) Signals required for reverse transcription, including a tRNA primer-binding site (PBS) and polypurine tract (PPT) for initiation of first and second DNA strand synthesis, respectively, and a repeated (R) region at both ends of the viral RNA required for transfer of DNA synthesis between templates; and (iv) Short, partially inverted repeats located at the termini of the viral LTRs required for integration. The replication process of retrovirus involves an RNA intermediate; therefore every approach used for inserting genes into a retroviral vector must permit synthesis of full-length copies of the vector genome. For example, insertion of a polyadenylation signal between the LTRs should be avoided; otherwise premature polyadenylation within the vector will reduce full-length vector transcription, which interferes with the production of vector-containing virions. Besides, introns enclosed within the insert may be removed during vector replication. This property allows the use of retroviral vectors to generate complementary DNAs from genes with introns, but it can also be a disadvantage if an intron contains information that is required for the appropriate regulation of gene expression. Vectors that rely on alternative splicing for gene expression retain their introns because critical elements of the packaging signal are contained within the intron in these vectors and RNAs without this signal are not efficiently encapsulated into virions. The replication-competent retroviral vectors are usually derived from avian viruses [e.g., HIV, Rous sarcoma virus (RSV), murine viruses (e.g., MLVs), and human foamy virus (HFV)].

In most of the applications, replication-defective vectors are used that can be accomplished by replacement of most or all of the coding regions of retroviruses with the gene(s) or sequence elements to be transferred, so that the vector by itself is incapable of making proteins required for additional rounds of replication. Viral proteins needed for the initial infection can be provided *in trans* by a retroviral packaging cell; there is no need for retroviral protein synthesis in recipient cells for proviral integration. The process of gene transfer and expression by retroviral vectors is referred to as transduction. Most of the replication-defective retroviral vectors have been derived from the murine and avian retroviruses, but vectors have also been derived from bovine leukemia virus, simian retroviruses, HIV, virus-like 30S, and HFV. The construction of replication-defective vector involves the generation of a 'provirus' that contains all of the signals needed *in cis* for vector packaging, reverse transcription, and integration, but that lacks the coding regions for most or all of the viral proteins. Furthermore, a corresponding packaging cell line is also required that produces viral proteins needed for virus assembly and transduction. Although it is in principle possible to derive vectors from any retrovirus, the complexity, toxicity, and regulatory elements of some retroviruses, especially the lentiviruses such as HIV, make construction of vectors more difficult and the titers of such vectors are typically low. New vector systems are constantly being developed to take advantage of particular properties of the parent retroviruses such as host range, usage of alternative cell surface receptors, and levels of tissue-specific expression. Retroviral vectors are typically used to induce the production of a specific protein in particular cells. Changes can be made in the enhancer/promoter of the LTR to provide tissue-specific expression. The presence of retroviral sequences between the viral promoter and the translational start codon of the inserted gene appears not to noticeably affect gene expression even though several ATG start codons are present in the 5′-untranslated regions of commonly used vectors. Alternatively, a single coding region can be expressed by using an internal promoter, which allows more flexibility in promoter selection. These strategies for expression can be most easily implemented when the gene of interest is also a selectable marker, as in the case of hypoxanthine-guanine phosphoribosyl transferase (*hprt*), which allows simplistic selection of vector-transduced cells. If the vector contains a gene that is not a selectable marker, the vector can be introduced into packaging cells by cotransfection with a selectable marker present on a separate plasmid. This strategy has the appealing advantage for gene therapy that a single protein is expressed in the ultimate target cells and possible toxicity or antigenicity of selectable markers is avoided.

Three general strategies have been used to design retroviral vectors that express two or more proteins: (i) Expression of different proteins from alternatively spliced mRNAs transcribed from one promoter; (ii) Use of the promoter in the LTR and internal promoters to drive transcription of different cDNAs; and (iii) Use of IRES elements to allow translation of multiple coding regions from a single mRNA. An alternative approach is to design fusion proteins that can be expressed from a single ORF. Furthermore, deleted vectors can be propagated only in the presence of a replication-competent helper virus or a packaging cell line. The former strategy leads to the contamination of the recombinant vector stock with nondefective helper virus. Conversely, packaging lines can be developed where an integrated provirus provides the helper functions but lacks the *cis*-acting sequences required for packaging. Many different retroviruses have been used to develop packaging lines and since these determine the type of envelope protein inserted into the virion envelope, these govern the host range of the vector (pseudotype of vector). For example, packaging lines based on amphotropic MLVs allow retroviral gene transfer to a wide range of species and cell types, including human cells. It is still possible for recombination to occur between the vector and the integrated helper provirus, resulting in the production of wild-type contaminants. The most advanced 'third-generation' packaging lines limit the extent of homologous sequence between the helper virus and the vector and split up the coding regions so that up to three independent cross-over events are required to form a replication-competent virus.

The simplest strategy for the high-level constitutive expression of single gene in retroviral vectors is to delete all coding sequences and place the foreign gene between the LTR promoter and the viral polyadenylation site. On the other hand, an internal heterologous promoter can be used to drive transgene expression. However, interference between the heterologous promoter and the LTR promoter are observed, which can be avoided by devising self-inactivating vectors containing deletions in the 3′-LTR that are copied to the 5′-LTR during vector replication, thus inactivating the LTR promoter. This approach also solves additional problems associated with the LTR promoter such as: (i) Adjacent endogenous genes may be activated following integration; and (ii) The entire expression cassette may be inactivated by DNA methylation after a variable period of expression in the target cell. As retroviral vectors are used for the production of stably transformed cell lines, selectable marker genes are cointroduced along with the gene of interest. Moreover, the expression of two genes can be achieved by arranging the transgene and marker gene in tandem, each under the control of a separate promoter, one of which may be the LTR promoter. This leads to the production of full-length and subgenomic RNAs from the integrated provirus. If splice sites flank the first gene, only a single promoter is necessary because the RNA is spliced like in the typical retroviral life cycle, allowing translation of the downstream gene. Vectors in which the downstream gene is controlled by an IRES have also been used. The viral replication cycle involves transcription and splicing, thus during construction of vector the foreign DNA must not contain sequences that interfere with these processes.

Adenovirus as Vector

Adenovirus belongs to the family Adenoviridae. Adenoviruses are nonenveloped viruses containing a linear ds DNA genome of ~36 kbp. While there are over 40 serotype strains of adenovirus, most of them cause benign respiratory tract infections in humans, however, subgroup C serotypes 2 or 5 are predominantly used as vectors. Most of the adenovirus vectors are derived from serotype 5, the genome of which is shown in Figure 12.29. There are six early-transcription units, most of which are essential for viral replication and a major late transcript that encodes components of the capsid.

Due to some helpful features, which are available with adenovirus like stability, a high-capacity for cloning, a wide host range that includes nondividing cells, and the ability to produce high-titre stocks (up to 10^{11} pfu/ml), adenoviruses have been widely used as gene transfer and expression vectors. Moreover, the life cycle does not normally involve integration into the host genome rather these viruses replicate as episomal elements in the nucleus of the host cell, consequently making them suitable for transient expression.

The first generation adenovirus-based vectors were replication-deficient E1 replacement vectors, which lacked the essential E1a, E1b, and nonessential E3 (sometimes) regions. These vectors had a maximum capacity of ~7 kbp and were propagated in the human embryonic kidney line 293. This cell line was transformed with the leftmost 11% of the adenoviral

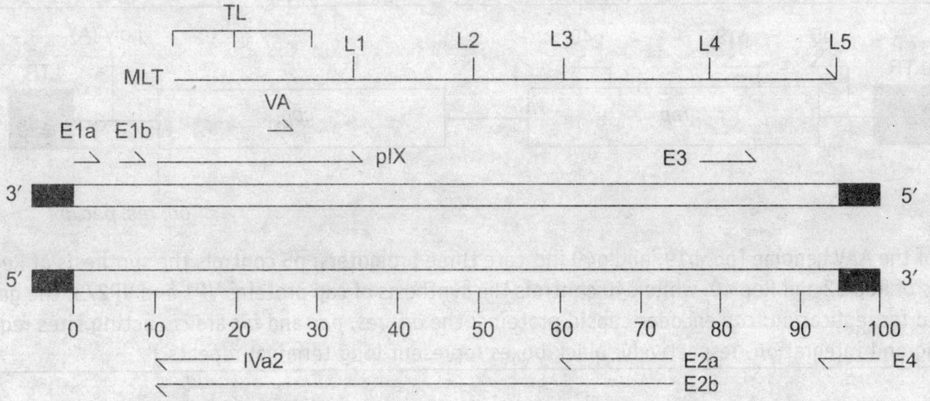

Figure 12.29 Map of the adenovirus genome [E indicates early transcription units; MLT indicates the major late transcript; TL indicates tripartite leader; VA, pIX, and IVa2 are other genes; dark-shaded boxes indicate terminal repeats.]

genome, comprising the E1 transcription unit and hence supplied these functions *in trans*. These vectors had been used for transient expression analyses, however, two types of problems were observed while working with them: (i) Cytotoxic effects resulted from low-level expression of the viral gene products; and (ii) The tendency of recombination between the vector and the integrated portion of the genome, which led to the revival of replication-competent viruses.

High-capacity vectors have been devised by deleting the E2 or E4 regions as well as E1 and E3, giving a maximum cloning capacity of ~10 kbp. The complementary cell lines with multiple functions, which are used for propagation of these vectors, have been developed in several laboratories. The vectors comprising of E1/E4 deletions is useful, as the E4 gene is responsible for many of the immunological effects of the virus. The problem of unwanted recombination has been solved through the use of a refined complementary cell line transformed with a specific DNA fragment corresponding exactly to the E1 genes. According to another strategy a large fragment of 'stuffer DNA' is inserted into the nonessential E3 gene, so that recombination yields a genome too large to be packaged. Gutless adenoviral vectors are favored for *in vivo* gene transfer, as these have a large insert capacity (up to 37 kbp) and minimal cytotoxic effects. However, presently complementary cell lines supplying all adenoviral functions are not available, so gutless vectors must be packaged in the presence of a helper virus, which presents a risk of contamination.

As Ad5 vectors are able to transduce a wide range of cell types, these vectors are the most commonly used adenovirus vectors. Ad5 vectors utilize the Coxsackie-adenovirus receptor (CAR) to enter cells and the transduction efficiency is related to the level of CAR expression on the cell membrane. Further, a new adenovirus vector has been developed where the fiber gene from adenovirus type 35 has been substituted for the Ad5 fiber. Ad35 fiber does not bind CAR and enters the cell by a different mechanism and hence does not need expression of CAR. Adenoviruses are particularly used as gene therapy vectors because cells take up the virions efficiently *in vivo* and adenovirus-derived vaccines have been used in humans. Though, generally, these do not cause any side-effect, there has been report of death of a patient after an extreme inflammatory response to adenoviral gene therapy. This emphasizes the requirement for a thorough safety testing.

Adeno-associated Virus as Vector

Adeno-associated viruses (AAV) belong to the family Parvoviridae. AAV is not related to adenovirus, but is so named as it was first discovered as a contaminant in an adenoviral isolate. Eleven serotypes of AAV have been isolated till date. AAV2 is the best-characterized serotype and the most gene transfer studies have been based upon this serotype.

AAV are nonenveloped, nonpathogenic ss DNA viruses that can only replicate in the presence of a helper virus, which includes adenovirus (Ad), herpes virus, *Vaccinia* virus, or cytomegalovirus (CMV). The gene map of AAV genome is given in Figure 12.30, which is small (~5 kb) and comprises a central region containing *rep* (replicase) and *cap* (capsid) genes flanked by 145 base inverted terminal repeats. Wild-type AAV may integrate into the host cell genome (preferentially into human chromosome 19) and remains latent until a helper virus supplies the necessary genes for replication. However, in the absence of these helpers, the AAV DNA integrates into the host cell's genome and successive infection by adenovirus or herpesvirus can 'rescue' the provirus and induce lytic infection.

AAV vector characteristics include ability to be produced in high titers, ability to infect a broad range of cells and stable expression from randomly integrated sequences or episomal sequences. Moreover, the dependence of AAV on a heterologous helper virus provides a remarkable degree of control over vector replication, making AAV theoretically one of the safest vectors for use in gene therapy. The only disadvantage associated with adeno-associated viral vectors is their limited cloning capacity (~4.5 kbp).

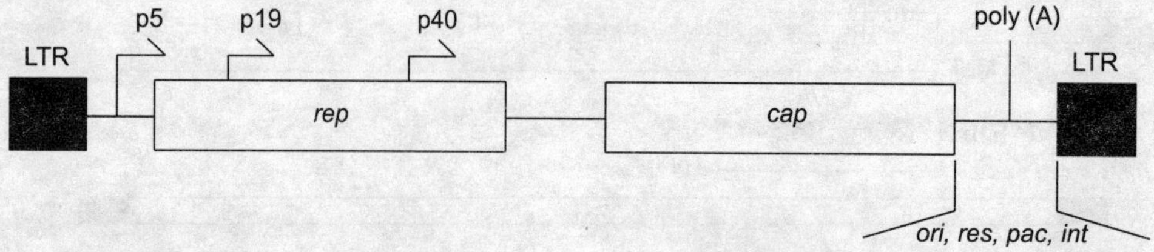

Figure 12.30 Map of the AAV genome [p5, p19, and p40 indicate three promoters; p5 controls the synthesis of Rep 78 and Rep 68, p19 regulates synthesis of Rep 52 and Rep 40, while p40 controls the synthesis of cap proteins VP1 and VP2/3; the gene *rep* and *cis*-acting site *ori* are required for replication; *cap* encodes capsid proteins; the *ori*, *res*, *pac* and *int* are *cis*-acting sites required for replication, excision, packaging and integration, respectively; black boxes represent long terminal repeats.]

In the early developed AAV vectors, foreign DNA replaced the cap region and was expressed from an endogenous AAV promoter. However, heterologous promoters were also used, although in many cases, expression of transgenes was inefficient because the Rep protein inhibited their activity. The interference of Rep with endogenous promoters is also responsible for many of the cytotoxic effects of the virus. Thereafter, more advanced vectors were developed in which both genes were deleted and the transgene was expressed from either an endogenous or a heterologous promoter and it was concluded that the repeats were the only elements required for replication, transcription, proviral integration, and rescue. The entire ranges of recently used AAV vectors are based on this principle. For *in vitro* manipulation of AAV, the inverted terminal repeats are cloned in a plasmid vector and the transgene is inserted between them. Usually, recombinant viral stocks are produced by cotransfecting this construct into cells with a helper plasmid to furnish AAV products and then infecting the cells with adenovirus to stimulate lytic replication and packaging. The resulting recombinant AAV is of low titre and contaminated with helper AAV and adenovirus thus is not suitable for use in human gene therapy. Recently developed transfection-based adenoviral helper plasmids, packaging lines, and the use of affinity chromatography to isolate AAV virions have helped to solve such problems. AAV vectors have been used to successfully introduce genes into many wide ranges of cell types. Further, deletion of the *rep* region eliminates the site specificity of proviral integration, so the vector integrates at essentially random positions, which may increase the risk of insertional gene inactivation. In addition, AAV uses concatemeric replication intermediates that have been used to avoid the disadvantage of AAV vectors, i.e., the limited capacity for foreign DNA. This problem can be solved by cloning a large cDNA as two segments in two separate vectors, which are cointroduced into the same cell. The 5′ portion of the cDNA is cloned in one vector, downstream of a promoter and upstream of a splice donor site. The 3′ portion is cloned in another vector, downstream of a splice acceptor. By following this strategy, cDNAs of up to 10 kbp can be easily expressed as concatemerization results in the formation of heterodimers and transcription across the junction yields an mRNA that can be processed to splice out the terminal repeats of the vector.

Epstein-Barr Virus and Herpes Simplex Virus (Herpes Viruses) as Vectors

The herpes viruses (HSVs) are large ds DNA viruses that include Epstein–Barr virus (EBV) and the HSVs (e.g., HSV-I, *Varicella zoster*). EBV is a herpes virus with a large ds DNA genome (~170 kbp) used as replicon vector. This virus mostly infects primate and canine cells and naturally lymphotrophic, infecting B cells in humans. This virus can establish as an episomal replicon with ~1,000 copies per cell in cultured lymphocytes. Only lymphocytes are infected naturally with this virus; however, the viral genome is maintained in a wide range of primate cells if introduced by transfection. Two relatively small regions of the genome, the latent origin (*ori* P) and a gene encoding a *trans*-acting regulator called Epstein–Barr nuclear antigen 1 (EBNA1) are only required for episomal maintenance.

A series of latent EBV-based plasmid expression vectors have been derived on the basis of two sequences, *viz.*, *ori* P and EBNA1. These vectors are maintained at a copy number of ~2 to 50 copies per cell. The earliest EBV vectors including the *ori* P element cloned in a bacterial plasmid can replicate only if EBNA1 was supplied *in trans*. Later shuttle vector, *viz.*, pBamC, consisting of *ori* P, a bacterial origin and amp^r gene derived from pBR322 and the neomycin phosphotransferase marker for selection in animal cells were developed. In pHEBO, which is derivative vector, neomycin phosphotransferase is replaced with hygromycin resistance marker. Afterward, EBNA1 gene was also added to pHEBO, generating a construct known as p201, which could replicate independently. In addition, EBV replicons have also been used to express a wide range of proteins in mammalian cell lines that have resulted in high-level and long-term gene expression. Moreover, EBV-derived vectors can carry large DNA fragments, including mammalian cDNAs and genes; hence can be used for the preparation of episomal cDNA libraries for expression cloning and genomic libraries.

HSV has been developed as a transduction vector. Generally these viruses are transmitted without symptoms and causes prolonged infections. The whole genome of HSV has been sequenced and extensively studied. The viral genome is a linear ds DNA molecule of 152 kbp. There are two unique regions, long and short (termed UL and US), which are linked in either orientation by internal repeat sequences (IRL and IRS). At the nonlinker end of the unique regions are terminal repeats (TRL and TRS). There are ~81 genes in the genome, of which about half are not essential for growth in cell culture. The HSV-1 genes are categorized into three main classes, namely, the immediate-early (IE or α) genes, early (E or β) genes, and late (L or γ) genes. When susceptible cells are infected, lytic replication is regulated by a temporally coordinated sequence of gene transcription. The early genes encode for nucleotide metabolism and DNA replication, and activate late genes that code for structural proteins. The entire cycle takes less than 10 hours and invariably results in cell death.

With the in depth knowledge of the HSV genome, it is now clearly understood which genes and DNA sequences may be deleted and which sites are suitable for insertion of foreign DNA. It has been found that by deleting nonessential genes, ~40–50 kbp of foreign DNA can be accommodated within the virus. HSV vectors are predominantly suitable for gene therapy in the nervous system because the virus is extremely neurotropic; in contrast the wild-type herpes viruses can replicate in many cell types in a wide range of species if the genome is introduced by transfection. HSV-based vector strategies rely on the ability of HSV to infect neuronal cells and to establish a latent infection. Latency is defined as a state in which viral DNA is maintained within the cell nucleus in the absence of any viral replication. During latency, viral gene expression is largely absent with the exception of the latency-associated transcripts (LATs), which may remain transcriptionally active. The two main strategies for HSV-based vectors in recent use are genetically engineered viruses and plasmid-derived 'amplicon' vectors, which carry only those *cis*-acting elements required for replication and packaging. The first strategy involves the construction of recombinant viruses containing deletions in one or more viral genes whose expression is essential for viral replication. Although therapeutic use of herpes virus vectors has been restricted, but several genes have been successfully transferred to neurons *in vivo*.

Sindbis Virus and Semliki Forest Virus (Alphaviruses) as Vectors

The alphaviruses are enveloped viruses with a single strand sense (+) RNA genome belonging to family Togaviridae. These viruses are never integrated into the host genome. Additionally, alphaviruses replicate in the cytoplasm and produce a large number of daughter genomes, allowing very high-level expression of any transgene. In this category, Sindbis virus and Semliki forest virus (SFV) have been extensively used for vector development and transgenic research. These viruses exhibit a broad host range and cell tropism. Moreover, mutants have been isolated with reduced cell toxicity effect. The wild-type alphaviruses comprise of two genes: a 5′ gene encoding viral replicase and a 3′ gene encoding a polyprotein from which the capsid structural proteins are autocatalytically derived. Since the genome is made of RNA, it can act as a substrate for protein synthesis. Soon after the synthesis of replicase protein, antisense strand is synthesized, which acts as a template for the production of full-length daughter genomes. However, the negative strand also contains an internal promoter, which allows the synthesis of a subgenomic sense (+) RNA containing the capsid polyprotein gene. This subgenomic RNA is subsequently capped and translated.

Alphaviruses are mostly used as expression vectors and various strategies have been utilized to express recombinant proteins. For example, replication-competent vectors have been constructed in which an additional subgenomic promoter is placed either upstream or downstream of the capsid polyprotein gene. If foreign DNA is introduced downstream of this promoter, the replicase protein produces two distinct subgenomic RNAs, one of them corresponding to the transgene. Such type of insertion vectors are unstable and have been largely superseded by replacement vectors in which the capsid polyprotein gene is replaced by the transgene. Furthermore, the first 120 bases of both the Sindbis and the SFV structural polyprotein genes include a strong enhancer of protein synthesis, which is also included in many vectors. This allows significant increase in the yield of recombinant protein and expresses the foreign gene as an N-terminal fusion protein as it is located downstream of its translational initiation site.

Both plasmid replicon and viral transduction vectors have been developed from the alphavirus genome. A common example of alphavirus-based replicon vector is pSinRep5, which is derived from Sindbis virus and marketed by Invitrogen (Figure 12.31). The vector is a bacterial plasmid containing the Sindbis nonstructural replicase genes (*ns* 1–4) and packaging site, which are required for *in vitro* replication of recombinant RNA and its packaging into virus, respectively. An expression cassette having a Sindbis subgenomic promoter, a MCS, and a polyadenylation site is also available in this vector. SP6 promoter is located upstream of the replicase genes and expression cassette for generating full-length *in vitro* transcripts. A second set of restriction sites is localized downstream from the polylinker, allowing the vector to be linearized prior to *in vitro* transcription. The gene of interest is cloned in the expression cassette, the vector is linearized and transcribed, and the resulting infectious recombinant Sindbis RNA is transfected into cells from which recombinant protein can be recovered. Alternatively, the entire alphavirus

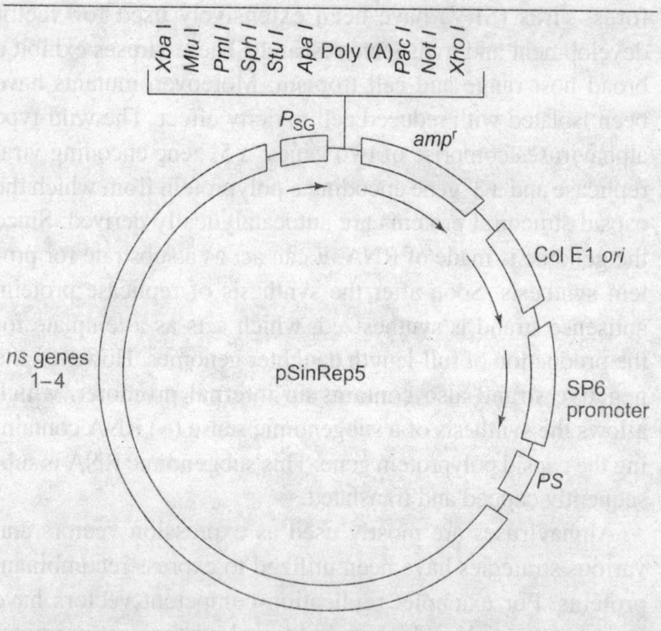

Figure 12.31 Map of pSinRep5, a Sindbis virus-based vector [*ns* genes 1–4 are nonstructural genes required for *in vitro* replication of recombinant RNA; P_{SG} indicates subgenomic promoter to drive expression of inserted genes; SP6 promoter indicates promoter for binding of SP6 RNA polymerase; *PS* indicates packaging signal.]

genome has been placed under the control of a standard eukaryotic promoter, such as SV40, and cells are transfected with the DNA. The DNA is expressed in the nucleus and the recombinant vector RNA is exported into the cytoplasm. As transduction is a more suitable delivery procedure for gene therapy applications, in this case pSinRep5 is cotransfected with a defective helper plasmid supplying the missing structural proteins. The recombinant viral particles are produced, which can be isolated from the extracellular fluid. Sometimes the replicon and helper vectors undergo recombination to produce moderate amounts of contaminating wild-type virus, which can however be alleviated by supplying the structural protein genes on multiple plasmids and by the development of complementary cell lines producing the structural proteins.

Vaccinia Virus and Other Pox Viruses as Vectors

Vaccinia and other pox viruses belonging to family Poxviridae have a complex structure and a large linear ds DNA genome ~300 kbp in size. These viruses encode their own DNA replication, transcription, and packaging machinery, which is unique feature of DNA viruses as most of the DNA viruses use them from the host cell nucleus. Moreover, the pox viruses replicate in the cytoplasm of the infected cell rather than their nucleus unlike other DNA viruses. Due to extraordinary replication strategy and large size of the *Vaccinia* genome, the designing and construction of expression vectors are more complex as compared to other viruses. Recombinant viruses are produced by homologous recombination using a targeting plasmid transfected into virus-infected cells since the virus normally packages its own replication and transcription enzymes. Thus, recombinant genomes introduced into cells by transfection are noninfectious. Later direct ligation vectors have been devised and these are transfected into cells having a helper virus to supply replication and transcription enzymes *in trans*. Finally, recombinant clones are screened by hybridization, which is performed on plaques produced by permissive cells that can be selected. However, the efficiency of this process can be further improved by applying various screening systems (for details see Chapter 16). In view of the fact that *Vaccinia* vectors have a high capacity for foreign DNA, selectable markers, such as *neo*[r], or screenable markers, such as *lac* Z or *gus* A, can be cointroduced with the experimental transgene to identify recombinants. For expression of transgene, it should be driven by an endogenous *Vaccinia* promoter, since transcription relies on proteins supplied by the virus. The highest expression levels are observed by late promoters such as P11, allowing the production of up to 1 μg of protein per 10^6 cells, but other promoters such as P7.5 and 4b are used, especially where early expression is desired. Up to 2 μg of protein per 10^6 cells is produced by utilizing a synthetic late promoter. *Vaccinia* vectors cannot be used to express genes with introns as these lack not only host transcription factors but also the nuclear splicing apparatus. Additionally, TTTTTNT must be removed from all foreign DNA sequences expressed in *Vaccinia* vectors as the virus uses this motif as a transcriptional terminator sequence. Furthermore, a useful binary expression system has been developed in which the T7 promoter drives the transgene and the T7 RNA polymerase itself is expressed in a *Vaccinia* vector under the control of a *Vaccinia* promoter. Initially, the transgene was placed on a plasmid vector and transfected into cells infected with the recombinant *Vaccinia* virus, but higher expression levels (up to 10% total cellular protein) can be achieved using two *Vaccinia* vectors, one carrying the T7-driven transgene and the other expressing the T7 RNA polymerase.

Simian Virus as Vector

Simian viruses belonging to family Retroviridae replicate to a very high copy number, resulting in cell lysis and release of thousands of progeny virions. These follow lytic infections, i.e., during the infection cycle, high concentrations of viral gene products are accumulated; therefore, this strategy has been exploited for the production of recombinant proteins. The first animal virus, which is molecularly characterized in detail, is SV40 and consequently it was also the first to be exploited as a vector. SV40 has a small icosahedral capsid and a circular ds DNA genome of ~5 kbp. The host range of this virus is very limited; it can propagate only in certain simian cells. The genome has two transcripts (Figure 12.32), known as the early and late regions, facing in opposite directions. These transcription units generate multiple products by alternative splicing.

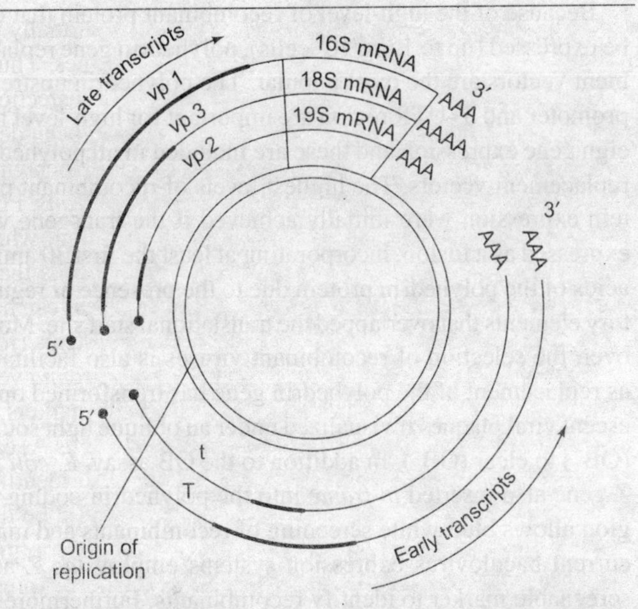

Figure 12.32 Transcripts and transcript processing of SV40 [Thick lines are intron sequences that are spliced out of the transcripts.]

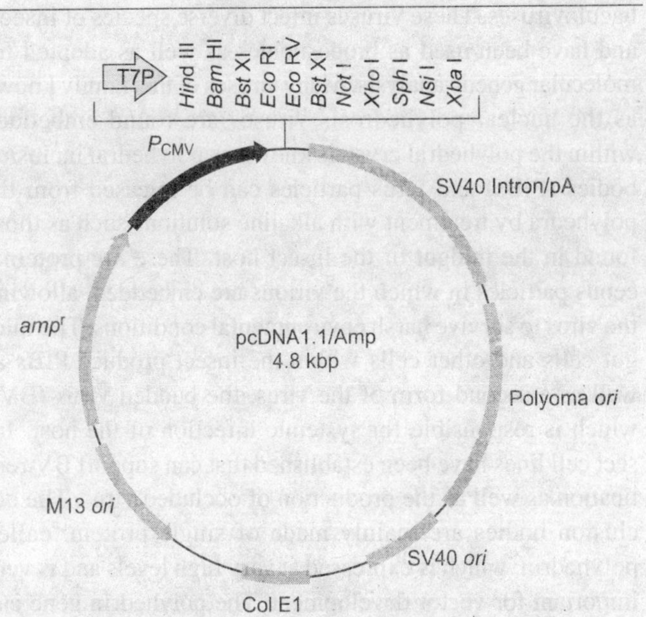

Figure 12.33 Vector map of pcDNA1.1/Amp

The regulatory proteins are produced by early region, whereas the late region produces components of the viral capsid. Moreover, the transcription is regulated by a complex regulatory element located between the early and late regions, including early and late promoters, an enhancer and the origin of replication. The early transcript produces two proteins, known as the large T and small t tumor antigens during the first stage of the SV40 infection cycle. T-antigen is absolutely required for replication because this protein binds to the viral origin of replication. Thus, vectors derived from SV40 should comprise of functional T-antigen for replication. This protein also works as an oncoprotein, interacting with the host's cell cycle machinery and resulting in uncontrolled cell proliferation.

Firstly, SV40 viral vectors were used to introduce foreign genes into animal cells by transduction. Due to the small size of the viral genome, *in vitro* manipulation is straightforward, as either the late region or the early region can be replaced with foreign DNA. Since both these regions are essential for the infection cycle, it is essential to provide their functions *in trans* by a cointroduced helper virus. For this, COS cell lines, derivatives of the African green monkey cell line CV-1 containing an integrated partial copy of the SV40 genome have been developed. The entire T-antigen fragment has been integrated in this cell line, which contains coding sequence and offers this protein *in trans* to any SV40 recombinant. These initial SV40 vectors have very low capacity as a maximum of only ~2.5 kbp of foreign DNA can be incorporated. The major breakthrough occurred during the development of vectors based on SV40 when it was observed that plasmids having the SV40 origin of replication produced high copy number DNA in permissive monkey cells since the problem of size constraints were solved. Various vectors were devised based on this principle, consisting of a small SV40 DNA fragment (containing the origin of replication) cloned in an *E. coli* plasmid vector. Furthermore, some vectors also enclosed a T-antigen coding region and could be easily used in any permissive cell line, although others contained the origin alone and could only replicate in COS cells. Moreover, permanent cell lines were not established in case of transfection of SV40 replicons to COS cells, as the immense vector replication ultimately led to cell death. A high copy number (10^5 genomes per cell) was obtained even though only a low proportion of cells were transfected, permitting the transient expression of cloned genes and harvesting of large amounts of recombinant proteins. More recently, similar vectors have been designed that incorporate the murine-polyomavirus origin, which is functional in mouse cells. High-level recombinant protein is transiently expressed in permissive mouse cells, such as MOP-8 (having an integrated polyomavirus genome and supplies T-antigen *in trans*) and WOP (which is latently infected with the virus).

Nowadays, a series of vectors currently marketed by Invitrogen are more versatile, having both the SV40 and the murine-polyomavirus origins, allowing high-level protein expression in cells that are permissive for either virus such as the vector pcDNA1.1/Amp containing the SV40 and polyoma origins, a transcription unit comprising the human CMV promoter, and SV40 intron/polyadenylation site, a polylinker to insert the transgene, and the *amp*^{r} selectable marker for selection in *E. coli* (Figure 12.33).

Baculoviruses as Vectors

Baculoviruses are ds DNA viruses having rod-shaped capsids belonging to family Baculoviridae comprising of >500 baculoviruses. These viruses infect diverse species of insects and have been used as biopesticides as well as adopted for molecular genetic analysis. Many viruses of this family known as the nuclear polyhedrosis viruses are found embedded within the polyhedral crystals known as polyhedral inclusion bodies (PIBs) and virus particles can be released from the polyhedra by treatment with alkaline solutions such as those found in the midgut of the insect host. These are proteinaceous particles in which the virions are embedded, allowing the virus to survive harsh environmental conditions. The midgut cells and other cells within the insect produce PIBs as well as a second form of the virus, the budded virus (BV), which is responsible for systemic infection of the host. Insect cell lines have been established that can support BV replication as well as the production of occluded virus. The occlusion bodies are mainly made of single protein, called polyhedrin, which is expressed at very high levels and is very important for vector development. The polyhedrin gene can be replaced with foreign DNA, which can be expressed at high-levels under the control of the endogenous polyhedrin promoter as the nuclear-occlusion stage of the infection cycle is nonessential for the productive infection of cell lines. The baculovirus most commonly is used for high-level transient protein expression in insects and insect cells. Baculovirus has large (~80–200 kbp) circular DNA genomes, several of which [including *Autographa californica* multiple polyhedrosis virus (AcMNPV) 1,33,894 bp] have been extensively sequenced.

Two baculoviruses have been extensively developed as vectors, namely the AcMNPV and the *Bombyx mori* nuclear polyhedrosis virus (BmNPV). The AcMNPV genome contains ~154 open reading frames, most of which have been assigned at least a basic function. Approximately 20 genes appear to encode viral structural proteins, whereas a similar number have been assigned roles in gene regulation. AcMNPV genes are transcribed in there basic phases: early, late, and very late phases. AcMNPV is used for protein expression in insect cell lines, particularly those derived from *Spodoptera fruiperda* such as Sf9 and Sf21. Among them, Sf21 infects the silkworm and has been used for the production of recombinant protein in live silkworm larvae. This expression system has limitation that the glycosylation pathway in insects differs from that in mammals, so recombinant mammalian proteins may be incorrectly glycosylated and hence immunogenic. This limitation has been solved by using insect cell lines chosen specifically for their ability to carry out mammalian-type posttranslational modifications, e.g., those derived from *Estigmene acrea*. A novel strategy to solve this problem is to exploit the indefinite capacity of baculovirus vectors to coexpress multiple transgenes. Thus, the glycosylation process can be modified in the host cell line by expressing appropriate glycosylation enzymes along with the transgene of interest.

Because of the high-level of recombinant protein that can be expressed (up to 1 mg/10^6 cells), polyhedrin gene replacement vectors are the most popular. The polyhedrin upstream promoter and 5′-UTR region are important for high-level foreign gene expression and these are included in all polyhedrin replacement vectors. The highest levels of recombinant protein expression were initially achieved if the transgene was expressed as a fusion, incorporating at least the first 30 amino acids of the polyhedrin protein due to the presence of regulatory elements that overlapped the translational start site. Moreover, the selection of recombinant viruses is also facilitated as replacement of the polyhedrin gene has transformed opalescent viral plaques if visualized under an oblique light source (OB^+) to clear (OB^-). In addition to the OB assay, *E. coli lac* Z gene also inserted *in frame* into the polyhedrin-coding region allows blue-white screening of recombinants and many current baculovirus expression systems employ *lac* Z as a screenable marker to identify recombinants. Furthermore, in some advanced vectors, *lac* Z substituted with the gene for green fluorescent protein allows rapid identification of recombinants by exposing the plaques to UV light (for details see Chapter 16).

The construction of baculovirus expression vectors involves inserting the transgene downstream of the polyhedrin promoter. Thus, the transfer vector contains a MCS adjacent to and downstream from the polyhedrin promoter so that the cloned gene is transcribed from that promoter. Since the genome is large, this is usually achieved by homologous recombination using a plasmid vector carrying a baculovirus homology region flanking the MCS, on either side. A problem with homology-based strategies for introducing the gene of interest into large viral genomes is that recombinants are generated at a low efficiency; recombinant vectors are recovered at a frequency of 0.5–5% of total virus produced. This problem of low recombination efficiency can however be overcome by using linear derivatives of the wild-type baculovirus genome containing large deletions (instead of intact viral DNA), which can be repaired only by homologous recombination with the targeting vector. As the linearized viral DNA lacks a portion of an essential gene, the production of nonrecombinant viruses is eliminated. Compatible targeting vectors span the deletion and provide enough flanking homologous DNA to sponsor recombination between the two elements (Figure 12.34). Recombination with the compatible targeting vector, after cotransfection of insect cells, restores this essential gene; this results in the production of up to 90% recombinant plaques. Additional features can be built into this system, including the incorporation of a marker such as β-galactosidase into the transfer vector so that blue recombinant plaques can be readily identified.

An alternative system has been devised in which the baculovirus genome is maintained as a low copy number replicon in bacteria or yeast, allowing the powerful genetics of

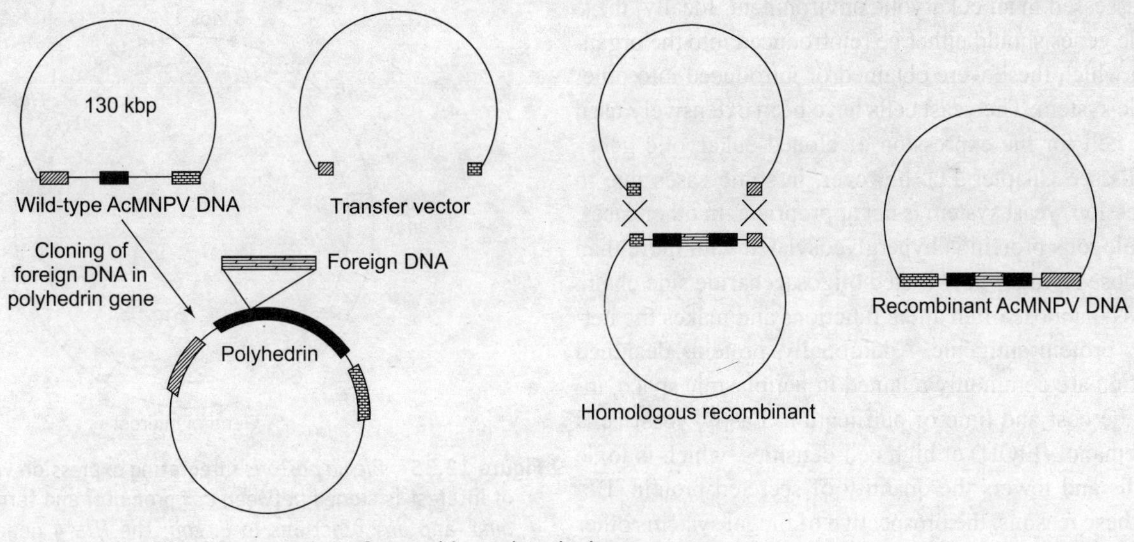

Figure 12.34 Procedure for the generation of recombinant baculovirus vectors

these microbial systems to be exploited. 'Bac-to-Bac' is bacterial system in which the baculovirus genome is engineered to contain an origin of replication from the *E. coli* F plasmid. The resulting hybrid replicon, known as bacmid, also contains the target site for the transposon *Tn* 7, inserted *in frame* within the *lac* Z gene, downstream of the polyhedrin promoter. The gene of interest is cloned in another plasmid between two *Tn* 7 repeats and introduced into the bacmid-containing bacteria, which also contains a third plasmid expressing *Tn* 7 transposase. Transposase synthesis is induced, resulting in a site-specific transposition of the transgene into the bacmid, which generates a recombinant baculovirus genome that can be isolated for transfection into insect cells. Transposition of the transgene into the bacmid interrupts the *lac* Z gene, allowing recombinant bacterial colonies to be identified by blue–white screening. Though insect cells are efficiently infected with baculoviruses, mammalian cells can also take up these baculoviruses without producing progeny virions. Now the number of mammalian cell lines reported to be transduced by baculoviruses is growing and this suggests that recombinant baculoviruses can be developed as vectors for gene therapy. Various baculovirus-borne transgenes have been expressed using constitutive promoters such as the CMV IE promoter and the RSV promoter. Some mammalian cell lines have also been generated that are stably transformed with baculovirus vectors.

12.6 P1 PHAGE AS VECTOR

P1 phage lives as plasmid and divides in steps with the host cell. The genome of P1 phage is circular ds DNA of ~100 kbp size, which undergoes bidirectional replication like a typical plasmid. Each descendant of the infected bacterial cell gets a single copy of the P1 DNA. The cell is unharmed and no phage particles are made. This state is known as lysogeny and a host cell containing such a phage is called a lysogen. Changing conditions may stimulate a lysogenic phage to return to destructive phage mode. This tends to happen if the host cell is injured, in particular if there is severe damage to the host cell DNA. The phage decides to make as many phage particles as possible before the cell dies. If on the other hand, the host cell is growing and dividing in a healthy manner, the phage most likely decides to lie dormant and divide in steps with its host.

P1 vectors are very similar to λ vectors and are based on a deleted version of a natural phage genome. Similar to λ phage, packaging of P1 DNA is also limited to the 'headful' and packaging is followed by plasmid propagation. Thus, the capacity of the cloning vectors is determined by the size of the deletion and the space within the phage particle. In comparison with λ-based vectors, P1 vectors have high insert capacity up to 125 kbp using current technology. This is because the P1 genome is larger than the λ genome and the P1 phage particles are bigger in size. P1 phage possesses Cre/*lox* recombination system, which promotes circularization of phage DNA during infection by recognizing *lox* sites on phage DNA (for details see Chapter 17).

12.7 FUNGAL SYSTEMS OTHER THAN YEAST

Eukaryotic DNA sequences are analyzed routinely after cloning in vectors derived from prokaryotic system. Moreover, cloned DNA can be obtained easily in large amounts and manipulated *in vivo* by bacterial genetic techniques and *in vitro* by specific enzyme modifications. The effects of these experimentally induced changes on the function and expression of eukaryotic genes are studied by isolating DNA from the host bacteria and reintroducing into a eukaryotic organism. As there are many functions common to eukaryotic cells, which are absent from prokaryotes, the genetic control of such functions must be assessed in an eukaryotic environment. Ideally, these eukaryotic genes should either be reintroduced into the organism from which these were obtained or introduced into other eukaryotic system. The yeast cells have been extensively used as a host cell for the expression of cloned eukaryotic genes (for details see Chapter 11), however, in some cases due to low expression, yeast system is not appropriate. In other cases, the heterologous protein is hyperglycosylated with more than 100 mannose residues in *N*-linked oligosaccharide side chain. The excess mannose often alters functions and makes the heterologous protein antigenic. Additionally, proteins designed for secretion are commonly retained in periplasmic space, increasing the cost and time of purification. Lastly, yeast cells produce ethanol (EtOH) at high cell densities, which is toxic to the cells and lowers the quantity of secreted protein. Because of these reasons, the prospective of fungal systems other than yeast cells for cloning have been investigated, which is explained in this section.

12.7.1 *Pichia pastoris* as Vector

P. pastoris are methylotrophic yeasts that are particularly very important hosts for the overexpression of heterologous proteins because of following reasons: (i) *P. pastoris* has a highly efficient and tightly regulated promoter of methanol-inducible gene that encodes alcohol oxidase, the first enzyme of the methanol utilization pathway. In the presence of methanol, as much as 30% of the cellular protein is alcohol oxidase whereas in its absence, the *AOX* 1 gene encoding alcohol oxidase is completely turned off. In addition, the *AOX* 1 gene promoter responds rapidly to the addition of methanol to the medium. In conclusion, the *AOX* 1 promoter is an excellent candidate for both driving the transcription of cloned genes and producing large amounts of recombinant protein. Moreover, the induction of cloned gene can be timed to maximize recombinant protein production during large-scale fermentations; (ii) *P. pastoris* does not synthesize EtOH and therefore the strains can be cultivated to very high density; (iii) It is possible to stably integrate expression plasmids at specific sites in the genome in either single or multiple copies; and (iv) *P. pastoris* normally secretes very few proteins, thus simplifying the purification of secreted recombinant proteins.

Various *P. pastoris* expression vectors have been devised. The basic features of these vectors include a gene of interest under the control of promoter and transcription termination sequences from *P. pastoris AOX* 1 gene, an *E. coli* origin of replication, a bacterial selectable marker gene, and a yeast selectable marker (Figure 12.35). The signal sequence either from *P. pastoris* phosphatase *PHO* 1 gene or from another yeast gene is also inserted into the vector to facilitate secretion of a recombinant protein. *P. pastoris* vectors are designed to be integrating plasmid vectors to avoid the problems of plasmid instability during long-term growth. Both the engineered gene of interest and a yeast selectable marker gene are inserted together into a specific chromosome site by either a single or a double recombination event. *P. pastoris* expression system has been used to produce more than 300 different biologically active proteins from bacterium, fungi, invertebrates, plants, and mammals, including humans. Many of these recombinant proteins, for example, the hepatitis B surface antigen, human serum albumin, and bovine lysozyme are indistinguishable from the native protein.

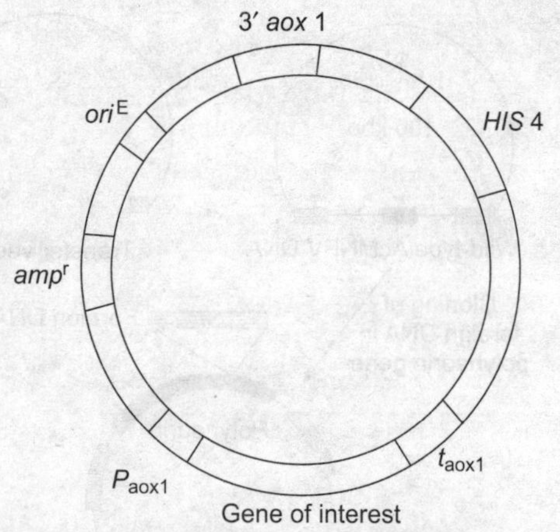

Figure 12.35 *Pichia pastoris* integrating expression vector. [Gene of interest is cloned between *aox* promoter and terminator; the *amp*^{r} and *ori*^{E} functions in *E. coli*; the *HIS* 4 gene encodes a histidinol dehydrogenase of histidine biosynthesis pathway; 3′ *aox* 1 is a DNA fragment from the 3′-end of *aox* 1 gene of *P. pastoris*; a double recombination event between the P_{aox1} and 3′ *aox* 1 regions of the vector and the homologous segments of chromosome DNA results in the insertion of DNA carrying the gene of interest and the *HIS* 4 gene.]

12.7.2 Other Fungi as Vectors

Heterologous genuine proteins for industrial and pharmaceutical importance have also been produced in other fungi. For instance, the cDNAs for the α- and β-globin chains of human hemoglobin A were cloned between the methanol oxidase promoter (P_{MOX}) and transcription terminator (t_{MOX}) sequences of the methylotrophic yeast, *Hansenula polymorpha*, and placed in tandem in an expression vector. Large amounts of animal feed enzyme supplement phytase have also been synthesized by transformed *H. polymorpha*. Moreover, an expression vector for *Schizosaccharomyces pombe* equipped with mammalian promoters controlling the transcription of both yeast selectable marker gene and cloned human gene produced large quantities of recombinant proteins from several human genes. Furthermore, the food yeast *Candida utilis*, which grows exclusively on inexpensive sugar sources such as molasses, produced a single-chain version of the protein sweetener monellin for a long period of time. In addition, some of species of the filamentous fungus *Aspergillus* have been extensively used for the commercial production of enzymes such as α-amylase, lipase, amyloglucosidase, catalase, and cellulase for the food, beverage, and pulp and paper industries. These fungi secrete large amounts of native proteins, grow rapidly on inexpensive media, process eukaryotic mRNA, and carry out many posttranslational modifications. Numerous vectors have been devised fitted with fungal transcription and translation control elements for the expression of recombinant proteins.

12.8 ARTIFICIAL CHROMOSOMES

Vectors that accept larger fragments of DNA than phage or cosmids have the distinct advantage that fewer clones need to be screened when searching for foreign gene of interest. Additionally, these have an enormous impact in the mapping of genomes of organisms. The important properties of some commonly used artificial chromosomes are presented in Table 12.5. The properties and cloning capacities of various artificial chromosomes are discussed in this section.

12.8.1 Yeast Artificial Chromosome

Yeast artificial chromosomes (YACs) are high insert capacity vectors, which contain yeast *ARS*, centromere, and telomeres (for details see Chapter 11). However, YACs are associated with certain problems, for example, (i) ~40–60% of the YACs from most libraries are chimeric; (ii) ~40% of the YACs from most libraries are deleted; (iii) Transformation efficiency is low; and (iv) YACs are very difficult to manipulate.

To overcome these problems, other high insert capacity vectors have been designed, which are discussed in following sections.

12.8.2 Bacterial Artificial Chromosome

BACs are based on the naturally occurring F plasmid of *E. coli*. Unlike the plasmids used to construct the early cloning vectors, the F plasmid is relatively large and vectors based on it have a higher capacity for accepting inserted DNA. Thus, BACs can be used to clone fragments of 300 kbp or longer. BACs have low copy number origin of replication and allow replication of clones at one copy per cell. BACs replicate clones faithfully across 60–100 generations. The recombinant DNA molecules are introduced into host cell by electroporation.

In comparison with YACs, BACs offer certain advantages such as the following: (i) These are bacterial based systems that are easy to manipulate; (ii) Libraries are generated using bacterial hosts with well-defined properties; (iii) Transformation efficiency is higher than that obtained for YACs; (iv) BACs are nonchimeric; and (v) BACs have very stable inserts and do not delete sequences.

12.8.3 P1-derived Artificial Chromosome

PACs combine features of P1 vectors and BACs and have insert capacity in the range of 70–300, and average size ~150 kbp. PACs have a low copy number origin of replication based on P1 bacteriophage which is used for propagation. Similar to BACs, PACs allow replication of clones at one copy per cell and replicate clones faithfully across 60–100 generations. The recombinant DNA molecules are introduced into host cells by electroporation. In contrast to BACs, PACs also have a negative selection against nonrecombinants. PACs also have an IPTG-inducible high copy number origin of replication that can be utilized for DNA production.

In comparison with YACs, PACs offer certain advantages: (i) These are bacterial based systems that are easy to manipulate; (ii) Libraries are generated using bacterial hosts with well-defined properties; (iii) Transformation efficiency is higher than that obtained for YACs; (iv) PACs are nonchimeric; and (v) PACs have very stable inserts and do not delete sequences.

12.8.4 Mammalian Artificial Chromosome

Mammalian artificial chromosomes (MACs) represent powerful tools for human gene therapy and animal transgenesis. The functional elements of such an artificial chromosome are telomeres, a centromere, and a replication origin.

Table 12.5 Properties of Some Large Fragment Cloning Vectors (Artificial Chromosomes)

Property	P1AC	pBAC	pUCBAC	pCYPAC	YAC
Vector size (kbp)	31	6.5	7.2	19.3	11.5
Vector copy	Single	Single	Multiple	Multiple	Multiple
Insert size (kbp)	75–95	0–300	0–300	0–300	0–2,000
Cloning strategy	Two arms *Bam* HI/*Sca* I	Single digest *Bam* HI or *Hind* III	Single digest *Bam* HI/*Sca* I	Single digest *Bam* HI or *Hind* III link	Double digest *Bam* HI/*Eco* RI
Cloning method	Packaging	Electroporate	Electroporate	Electroporate	Spheroplast
Chimeric clones (%)	0	2	2	0 (24/24)	20-60
Positive selection	Yes	No	No	Yes	Yes
Copy induction	Yes	No	No	Yes	No

12.8.5 Human Artificial Chromosome

A human artificial chromosome (HAC) is a microchromosome that can act as a new chromosome in a population of human cells. Thus, instead of 46 chromosomes, the cells have 47 chromosomes, with the 47th chromosome being very small, roughly 6–10 Mbp in size and able to carry newly introduced genes. HACs are usually constructed by recombineering. Previously, two models were established for construction of HACs. The first 'top-down' approach or TACF (telomere-associated chromosome fragmentation) involved modifying natural chromosomes into smaller defined minichromosomes in cultured cells. Following recombination and subsequent breakage between homologous sequences on the endogenous host chromosome and an incoming telomere containing the targeting vector, engineered minichromosomes as small as 450 kbp in size in avian cells have been generated. This approach has been important for studying the structure, sequence organization and size requirements of the human X and Y chromosomes. Second approach, the 'bottom-up' or assembly approach involved generating HACs in human cells by introducing defined chromosomal sequences as naked DNA including human telomeres, alpha satellite (alphoid) DNA and genomic fragments containing replication origins. The *de novo* HACs are generated following recombination and some amplification of the input DNA within the host cell. Together, the generation of minichromosomes and *de novo* HACs has identified alphoid DNA as the major sequence element of the centromere and determined the minimum size (~70–100 kbp) required for centromere function and stability. Other approaches include the generation of SATACs (satellite DNA-based artificial chromosomes) following integration of repetitive DNA into pre-existing centromeric regions of host chromosomes and modifying small human marker chromosomes (minichromosomes derived from naturally occurring chromosomes).

The *de novo* HACs have been generated in human cells following the introduction of BACs or PACs containing large arrays of cloned or synthetic alphoid DNA repeats from chromosomes 5, 13/21, 14/22, 17, 18 and X. The introduced DNA undergoes a process of recombination and amplification, forming large (1–10 Mbp) circular molecules (usually at one or two copies per cell) which are mitotically stable in the absence of selection for at least 9 months in some cells. The efficiency of *de novo* HAC formation and stability depends on the presence of a centromere protein B-binding sequence (CENP-B box) and, to some extent, on the chromosomal origin of the alphoid template and the longer length of the alphoid array (>100 kbp).

HAC vectors have been used successfully for gene transfer and expression in human cells which have focused on using standard cell-culture models, including studies of the human *HPRT* 1 (hypoxanthine guanine phosphoribosyl transferase 1) and the *GCH* 1 (GTP cyclohydrolase 1) genes. Two transfection approaches are used in these studies, yielding similar results. In the first strategy, a single vector containing chromosome 17 alphoid DNA and the *HPRT* 1 gene is introduced into human fibroblast (HT1080) cells containing the *HPRT* 1 deficiency. The vector is constructed in *E. coli* following recombination between a 17 alphoid BAC and an *HPRT* 1 PAC containing the entire *HPRT* 1 genomic locus, generating a large stable HAC vector of 404 kbp. HACs containing the *HPRT* 1 genes are generated which complemented the genetic deficiency. In another strategy, an alphoid PAC and an HPRT 1 BAC are introduced simultaneously into HPRT 1-deficient HT1080 cells by cotransfection, generating similar HPRT 1-expressing HACs. Later this approach was also used to introduce an alphoid BAC and GCH 1 BAC to generate GCH 1-expressing HACs. The *de novo* HACs are viable transfer vectors for gene expression studies and can complement genetic deficiencies in human cells. Similar to normal chromosomes, these HACs contain regions of both euchromatin and heterochromatin and therefore provide a transcriptionally permissive environment for gene expression. For efficient and stable *de novo* HAC formation, centromeric chromatin formed over alphoid DNA should probably be flanked at least unilaterally by heterochromatin.

12.9 SHUTTLE VECTORS

It is possible to insert two origins of replication into plasmid, one for *E. coli* and the other for the chosen host. Such vectors are thus able to replicate in *E. coli* using one origin and in chosen host using alternative replication origin. Such a vector is called a shuttle vector because it can be transferred back and forth between two species. The first such shuttle vector was derived from fusions between the *E. coli* vector pBR322 and the *S. aureus*/*B. subtilis* plasmids pC194 and pUB110. The advantage of shuttle plasmids is that *E. coli* can be used as an efficient intermediate host for cloning. Moreover, *E. coli* can be used as an intermediate host for the transformation of the ligation mix and screening for recombinant plasmids, which is very useful if transformation or electroporation of the alternative host is very inefficient. Shuttle vectors have been developed for nonenteric bacteria, other prokaryotic and eukaryotic systems.

12.10 EXPRESSION VECTORS

The insertion of a gene into cloning vector does not essentially ensure that it will be successfully expressed in a selected host. However, for the biotechnological applications, expression of the cloned gene in a selected host organism and that too at high rate is required. Hence to allow expression of the cloned gene, an expression vector is used, which allows transcription and translation of the cloned gene. Several specialized prokaryotic and eukaryotic expression vectors have now been constructed that provide genetic elements for controlling transcription, translation, protein stability, and secretion of the product of cloned gene from the host cell. A detailed description of both prokaryotic and eukaryotic expression vectors is presented here.

12.10.1 Prokaryotic Expression Vectors

A wide range of prokaryotic organisms can express foreign genes and several commercially important proteins are produced by genetic engineering usually synthesized in *E. coli.* The strategies that have been elaborated for *E. coli*, in principle, are applicable to all systems.

Basic Features of Expression Vectors

The basic features of expression vectors are as follows.

Transcriptional Promoter Prokaryotic expression vector contains a promoter, which serves as binding site for RNA polymerase to initiate transcription of the cloned gene. There are two types of promoters, *viz.*, constitutive promoter and inducible promoter. The constitutive promoter is the one that causes expression of the genes under its control irrespective of the time and developmental stage and such promoters cannot be regulated (i.e., cannot be induced or repressed). The promoters that do not express a transgene unless induced by an inducer (chemical agent/environmental stimuli/stage of development) are called inducible promoters. The inducible promoter gives finer control over transgene expression than a constitutive promoter. The inducible promoters allow the timing of transgene expression to be carefully controlled. Generally, tightly regulatable, inducible promoters are preferred for producing large amounts of recombinant protein at a specific time during large-scale growth. Such promoters allow separation of growth phase from induction phase and hence there is minimization of selection of nonexpressing cells and can permit the expression of proteins normally toxic to cell. The protein is kept uninduced till sufficient growth is achieved and later the protein synthesis is achieved by inducing the promoter through some substance or environment.

In *E. coli*, a wide range of suitable promoters is available. The standard (σ^{70}) promoter of *E. coli* comprises of –10 region (Pribnow box) and –35 region separated by an intervening distance of ~17 bp. The sequences of –10 and –35 regions are conserved, 5′-TATAAT-3′ and 5′-TTGACA-3′, respectively. The sequences of these conserved regions determine the strength of promoter. Amongst other naturally occurring promoters, the *lac* and *trp* promoters that normally drive expression of the *lac* (lactose utilization) and *trp* (tryptophan biosynthesis) operons, respectively, and the P_L promoter from λ bacteriophage are examples of commonly used promoters; the P_L promoter being one of the strongest natural promoters of *E. coli* or its associated genetic elements. However, none of these three natural promoters actually represents a perfect match with the consensus sequence obtained by comparing the sequence of a large number of standard *E. coli* promoters.

Table 12.6 Comparison of Selected *E. coli* Promoters

Promoter	–35 region	–10 region
Consensus	TTGACA	TATAAT
lac	TTTACA	TATATT
trp	TTGACA	TTAACT
P_L	TTGACA	GATACT

The –10 and –35 regions are not the sole determinants of the strength of a promoter, a region of ~70 bp that makes contact with the RNA polymerase also influences the strength of that contact. Nevertheless, these two regions and the distance separating them are the most highly conserved and play the major role. Since none of these sequences is a perfect match with the consensus, it can be inferred that further manipulation of the sequence can produce even stronger promoters and this is in fact the case. One artificial promoter that is commonly used is known as the *tac* promoter, since it represents a hybrid between the *trp* and *lac* promoters and is capable of higher levels of transcription than any of the natural promoters.

Not all promoters conform to the consensus shown (Table 12.6). In particular, some bacteriophage RNA polymerases (notable from T7, T3, and SP6) have a virtually absolute requirement for a sequence of bases specific to that enzyme. These enzymes are very useful in ensuring expression of the cloned gene only at the time of requirement. These promoters allow firmer control of expression as compared with *lac* promoter (which is leaky).

Translational Signals (Ribosome Binding Site, Start Codon, and Stop Codon) The vectors containing promoter *in frame* with the MCS allows proper transcription of the cloned gene. But for product formation corresponding to the cloned gene, an *in frame* ribosome-binding site (RBS) should also be present. A ribosome-binding site is a sequence of six to eight nucleotides (e.g., UAAGGAGG) in mRNA that can base pair with a complementary sequence (AUUCCUCC for *E. coli*) of the 16S RNA of small ribosomal subunit. The stronger binding of mRNA to the ribosomal RNA exhibits greater efficiency of translation initiation. Various *E. coli* expression vectors have been devised having strong RBS; thus heterologous prokaryotic and eukaryotic genes can be translated readily in *E. coli*. Furthermore, some other conditions must be fulfilled for this approach to function properly: (i) RBS must be located a precise distance from the translational start codon (ATG) of the cloned gene; (ii) DNA sequence that includes the RBS and first few codons of gene of interest must not contain sequences that after transcription can form intrastrand loops; and (iii) Transcription terminator sequences must also be included.

Plasmid Copy Number In addition to the choice of promoter in the expression vector, the plasmid copy number is also an important consideration. It is considered that more the copies of the plasmid, more the amount of cloned gene and hence more the product formation. Hence, most routine plasmid cloning vectors in *E. coli* are 'multicopy' plasmids. The early cloning vector such as pBR322 normally occurs in 15–20 copies per cell, although under conditions of inhibition of protein synthesis, this plasmid can be amplified to 1,000–3,000 copies per cell in a process called plasmid amplification (for details see Chapter 8). Subsequent development of pUC series of plasmid vectors removed some control elements so that hundreds of copies per cell could be obtained. However, the relationship between copy number of the plasmid and the amount of product formed is not linear. Though more products may be obtained from a high copy number plasmid (say, 200 copies/cell) as compared with that obtained from the one with only 20 copies/cell, it may not necessarily be ten times. Another problem with high copy number plasmids may be that excessive amounts of product may kill the cells producing it.

Thus to maximize gene expression, both strategies should be incorporated, i.e., optimization of promoter and high copy number plasmid. Under these conditions (assuming translation works perfectly), bacterial clone, which represents up to 50% of desired product of the total protein of the cell, is obtained. For commercial production, in addition, the proportion of contaminating protein (and other material) that has to be removed is lower, thus reducing the costs of downstream processing.

Conditional Expression

There is a downside to such high levels of product formation as the production of vast quantities of a protein not useful to cell certainly results in a reduction in cell growth rate. This is because of the diversion of significant amount of resources in a manner that is nonproductive from the cell's perspective. This applies even if the protein itself has no damaging effects. Moreover, the protein may be directly damaging, which makes the problem much more acute. The slower growth of the producing cells means that there is a very strong selective pressure in favor of any of a wide range of potential mutants that are nonproductive. This includes cells, which have either lost the plasmid altogether or have any mutation in the plasmid that reduces or prevents product formation. Slower growth rates in turn reduce the efficiency of the process. This can be a problem both for a few milliliters in the laboratory or many thousands of liters in an industrial fermentor. However, on a laboratory scale, it is possible to include antibiotic selection in the culture to ensure that any mutants that have lost the plasmids are unable to grow. On a large-scale, this is not only an expensive solution, but also the disposal of large volumes of antibiotic containing waste is a problem. Thus, there should be certain means to ameliorate this problem. These include the following:

Application of Regulatable or Inducible Promoters One such way is to use controllable promoters, i.e., promoters whose activity can be altered by changes in the culture conditions. The *lac* and *trp* promoters can be regulated by changing the culture conditions in the following ways:

***lac* promoter** The *lac* promoter is naturally expressed only if *E. coli* is growing on lactose as a carbon and energy source. Note that normally bacterial geneticists use IPTG instead of lactose, as it is a gratuitous inducer (i.e., not broken down by the cell's β-galactosidase) and more convenient to use. In the absence of an inducer, a repressor protein binds to a DNA sequence known as operator, which in this case overlaps with the *lac* promoter and prevents transcription initiation. The inducing agent binds to the repressor protein, altering its conformation so that it no longer binds to the operator. Thus to obtain high levels of gene expression, the culture is first grown to an appropriate density in the absence of an inducer thereby removing the selective pressure imposed by excessive product formation, and after getting enough cells, inducing agent is added to switch on the gene expression. In case of multicopy plasmid vectors, which are mostly in use, enough of the repressor protein is produced to switch off the single copy of the promoter that it has in the chromosome, but this is not enough to switch off several hundred copies of this promoter. This is because the repressor is titrated out by the presence of so many copies of the promoter. Hence there should be some strategy to increase the production of the repressor protein. As the *lac* I gene encoding the repressor has its own promoter, one way of increasing production of the Lac I repressor protein is by using a mutated version of the *lac* I gene, which has a more active promoter (an up-promoter mutant). This altered *lac* I gene is known as *lac* I^q. Alternatively, *lac* I gene is put onto the plasmid itself, subjecting it to the same gene dosage effect and therefore increasing the production of Lac I repressor.

***trp* promoter** Adding IPTG to a laboratory culture is appropriate, but on a commercial scale, it is not an ideal solution as its addition in an industrial scale fermentor would be very expensive as well as create other problems including the disposal of the waste material. An alternative strategy is to use *trp* promoter, which controls transcription of the *trp* operon and is subjected to repression by tryptophan (trp). *E. coli* switches off expression of the *trp* operon when the enzymes encoded by it are not needed, i.e., if there is a plentiful supply of trp. It is possible to monitor and control the availability of trp so that there is an adequate supply during the growth phase and then limit the supply of trp when expression is required. However, it is advisable to supply a low level of trp, which is not enough to switch off the *trp* promoter but enables production of the required protein.

The regulatable promoters described may solve the problem in part by removing the selective pressure imposed by excessive product formation; but there is still some element of selection imposed by the presence of so many copies of the plasmid, which may also slow growth rates. This can be countered by using a plasmid with a different replication origin, so that replication is tightly controlled at only one or two copies per cell. However, the level of expression achieved with a low copy vector will be less than that achievable with a multicopy plasmid.

Application of Runaway Plasmid Vectors Another strategy for conditional expression is to program gene expression by using a runaway plasmid (for details see Chapter 8). If the control of plasmid copy number is temperature sensitive, then growing the culture initially at say 30°C leads to production of cells with only a few copies of the plasmid. Then, once sufficient growth has been achieved, the culture can be shifted to a higher temperature, say 37°C, control of plasmid replication is lost and the copy number increases dramatically until it represents perhaps 50% of the DNA of the cell. If the gene expression is switched 'on' at the same time, a very substantial amount of the product is obtained. Eventually the cells will die, but by that time enough of the product will be generated.

An alternative way of achieving a similar effect is by providing the vector with two origins of replication: one that results in many copies of the plasmid and the other that produces only one or two copies per cell. Initially origin is controlled and conditions are switched on at an appropriate stage so that the culture starts off with only a few copies of the plasmid per cell and then at the desired time, gene expression is switched on by inducing the promoter and switch to the other replication origin, consequently increasing the copy number of the plasmid.

Expression of Lethal Genes

Some genes encode very damaging products or products that are lethal to the growing bacterial cell. As the *lac* promoter is a leaky promoter, it remains active at a lower level even in the absence of induction though full activity is observed in the presence of an inducing agent (e.g., IPTG). If the gene product is very damaging, the cell will not be able to tolerate even this low level of expression. To avoid such problem, other promoters with tighter control, for example, T7 promoter should be used. Note that T7 RNA polymerase does not initiate transcription from the usual promoters, or in other words, T7 promoters are not recognized by *E. coli* RNA polymerases. Thus, if desired DNA fragment is cloned downstream of a T7 promoter using an 'ordinary' *E. coli* host (lacking a T7 RNA polymerase gene), no expression is observed. Once right construct is prepared, then the plasmid is isolated and put into another *E. coli* strain engineered to contain a T7 RNA polymerase gene so as to allow transcription of the cloned gene. The expression of the T7 RNA polymerase gene itself can also be regulated, for example, by putting it under the control of a *lac* promoter. In such a case, the expression of T7 RNA polymerase can be regulated by IPTG (i.e., switched on in its presence and switched off in its absence), which in turn controls the level of expression of the cloned gene.

12.10.2 Expression in Eukaryotic Cells

Although bacteria are convenient hosts for many purposes and *E. coli* is usually the host of choice for initial gene cloning (including the production of primary gene libraries), bacterial systems have many limitations for the expression of cloned genes, especially for large-scale production of proteins of eukaryotic origin. In particular, the posttranslational modifications needed for obtaining biologically active product encoded by eukaryotic gene more likely occur in a eukaryotic host. There are a wide variety of eukaryotic systems; a few of them are discussed here.

Yeast Expression System

Yeasts have many advantages for the expression of cloned genes. These grow rapidly in simple defined medium, and as unicellular organisms, these are relatively easy to manipulate and enumerate. *S. cerevisiae* is used as an experimental organism in microbial genetics and there is now a wealth of biochemical and genetic information available to support its use for gene cloning including the genome sequence. Powerful and versatile systems are also available for several other yeast species, notably *P. pastoris*. Some common fungal promoters, which are used for manipulation of gene expression, are listed in Table 12.7.

***S. cerevisiae* System** Vectors for cloning and expression of genes in the yeast *S. cerevisiae* were introduced in Chapter 11. In many respects, if using episomally replicating vectors, the concepts are similar to those involved in the design of plasmid-based expression vectors in bacteria. There is a choice between vectors carrying the 2-μm origin of replication, which are maintained episomally at high copy number (up to 200 copies per cell), and centromere vectors, which are maintained at low copy number (1–2 copies per cell). Both types of vectors are designed as shuttle vectors as these also carry

Table 12.7 Common Fungal Promoters Used for Manipulation of Gene Expression

Species	Promoter	Gene	Regulation
General			
S. cerevisiae	PGK	Phosphoglycerate kinase	Glucose-inducible
S. cerevisiae	*GAL* 1	Galactokinase	Galactose-inducible
S. cerevisiae	*PHO* 5	Acid phosphatase	Phosphate-inducible
S. cerevisiae	*CYC* 1	Cytochrome C1	Glucose-repressible
S. cerevisiae	*ADH* II	Alcohol dehydrogenase II	Glucose-repressible
S. cerevisiae	*GAL*	Galactose-1-phosphate-glucose-1-phosphate uridylyltransferase	Galactose-inducible
S. cerevisiae	*GAL* 10	UDP galactose epimerase	Galactose-inducible
S. cerevisiae	*CUP* 1	Copper metallothionein	Copper-inducible
S. cerevisiae	*MF*α1	Mating factor α1	Constitutive but temperature-inducible variant available
S. cerevisiae	*TPI*	Triose phosphate isomerase	Constitutive
Candida albicans	*MET* 3	ATP sulfur lyase	Methionine- and cysteine-repressible
Methanol utilizers			
Candida boindnii	*AOD* 1	Alcohol oxidase	Methanol-inducible
Hansenula polymorphia	*MOX*	Alcohol oxidase	Methanol-inducible
Pichia methanolica	*AUG* 1	Alcohol oxidase	Methanol-inducible
Pichia pastoris	*AOX* 1	Alcohol oxidase	Methanol-inducible
P. pastoris	*GAP*	Glyceraldehyde-3-phosphate dehydrogenase	Strong constitutive
P. pastoris	*FLD* 1	Formaldehyde dehydrogenase	Methanol- or methyl amine–inducible
P. pastoris	*PEX* 8	Peroxin	Methanol-inducible
P. pastoris	*YPT* 1	Secretory GTPase	Medium constitutive
Lactose utilizers			
Kluveromyces lactis	*LAC* 4	β-Galactosidase	Lactose-inducible
Starch utilizers			
Schwanniomyces occidentalis	*AMY* 1	α-Amylase	Maltose- or starch-inducible
Xylose utilizers			
P. stipitis	*XYL* 1	–	Xylose-inducible
Alkane utilizers			
Yarrowia lipolytica	*XPR* 2	Extracellular protease	Peptone-inducible
Y. lipolytica	TEF	Translation elongation factor	Strong constitutive
Y. lipolytica	*RPS* 7	Ribosomal protein S7	Strong constitutive

an *E. coli* replication origin, which enables the initial construction and verification of the recombinant plasmid to be carried out in *E. coli* before transferring the finished construct into yeast cells. As with bacterial expression vectors, these *S. cerevisiae*-based vectors are designed with a controllable promoter adjacent to the cloning site to enable expression of the cloned gene to be switched on or off. Most commonly, this involves the promoter and enhancer sequences from the *GAL* 1 (galactokinase) gene, which is strongly induced by the addition of galactose.

***P. pastoris* System** As described in Section 12.7.1, *P. pastoris* is able to use methanol as a carbon source, the first step in the pathway being catalyzed by alcohol oxidase encoded by *AOX* 1 gene. This gene is tightly controlled so that in the absence of methanol, no alcohol oxidase is detectable. On addition of methanol to the culture, the *AOX* 1 gene is expressed at a very high level. The use of the *AOX* 1 promoter in the expression vectors, adjacent to the cloning site, therefore provides vectors that are capable of generating substantial levels (up to several grams per liter) of the required product.

Expression in Insect Cells (Cloning using a Baculovirus System)

As already described abovebaculoviruses, such as the AcMNPV that infect insect cells are exploited as the basis of systems for gene expression in such cells. The polyhedrin promoter (*polyh*), which is a very strong promoter, can be used to drive the expression of the cloned gene. However, the viral DNA is itself too large (>100 kbp) for direct manipulation to be carried out easily. The gene to be expressed is therefore first inserted into the polyhedrin gene in a smaller transfer vector. The resultant plaques can be distinguished from those produced by wild-type virus. Following transfection, viral plaques can be picked and the virus characterized to verify the presence of the cloned gene. The characterized recombinant virus can then be used to infect a large-scale culture of insect cells; high-levels of expressed protein are usually obtained before the cells lyse. The levels of production are generally lower than those obtainable with *Pichia* expression systems, but insect cells are claimed to provide post-translational modification that is more closely similar to that in mammalian cells. On the other hand, yeast cells are much easier and cheaper to grow.

Expression in Mammalian Cells

There are many varieties of vectors and systems for expression in mammalian cells. Although the details are more complex than the systems described so far, the general principles remain similar. A summary of major expression systems used in animal cells is presented in Table 12.8.

Most vectors use the enhancer/promoters from the human CMV, the SV40 virus, or the HSV thymidine kinase (HSV-*tk*) to drive transcription. These give high-level, constitutive expression. As in prokaryotic systems, it is sometimes desirable to control the onset of expression. One way of achieving such regulation is by interposing the operator sequence (*tet* O) from the bacterial tetracycline resistance operon between the promoter and the cloned gene. If the mammalian cells are cotransfected with a second plasmid containing the tetracycline repressor gene (*tet* R), driving expression by using the CMV promoter, the Tet R protein will bind to the *tet* O site, thus preventing transcription. When tetracycline is added to the culture medium, it will bind to the Tet R protein, altering its conformation and releasing it from the DNA, thus derepressing transcription of the cloned gene. The system is advantageous; being prokaryotic in origin, its activation does not affect the induction of native mammalian genes. [Note that this system is based on the bacterial tetracycline repressor-operator system wherein the Tet R binds to the *tet* O and negatively regulates its expression. The derepression or inactivation of gene expression is regulated by tetracycline; binding of tetracycline modifies and releases Tet R from the operator]. Similarly, an insect ecdysone-responsive element

Table 12.8 Summary of Major Expression Systems Used in Animal Cells

System	Host cells	Major applications
Nonreplicating plasmid vectors		
No selection	Many cell lines	Transient assays
Dominant selectable markers	Many cell lines	Stable transformation; Long-term expression
DHFR/methotrexate	CHO	Stable transformation; High-level expression
Plasmids with viral replicons		
SV40 replicons	COS	High-level transient expression
BPV replicon	Various murine	Stable transformation (episomal)
EBV replicons	Various human	Stable transformation (episomal); Library construction
Viral transduction vectors		
Adenovirus E1 replacement	293 cells	Transient expression
Adenovirus amplicons	Various mammalian	*In vivo* transfer
Adeno-associated virus	Various mammalian	*In vivo* transfer
Baculovirus	Insects	High-level transient expression
	Various mammalian	*In vivo* transfer
Herpes virus	Various mammalian	Stable transformation
Oncoretrovirus	Various mammalian and avian	Stable transformation
	ES cells	Transgenic mice
Lentivirus	Non-dividing; Mammalian	*In vivo* transfer
Sindbis, Semliki forest virus	Various mammalian	High-level transient expression
Vaccinia virus	Various mammalian	High-level transient expression

is sometimes used to induce gene expression in mammalian cells. Like the tetracycline system, ecdysone does not have any effect on native mammalian genes and the induction is therefore specific to the genes placed under its control.

Additionally, signal sequences can be added to enable targeting of the product to specific cellular locations such as the nucleus, mitochondria, endoplasmic reticulum, or cytoplasm, or secretion into the culture medium.

In contrast to bacterial cells, the introduction of DNA into mammalian cells does not depend on the independent replication of the vector. The introduced DNA can be stably integrated into the nuclear DNA. However, some expression vectors can be stably maintained at high copy extrachromosomally, such as those containing the origin of replication from the EBV; with a suitable promoter system, these are capable of allowing high-levels of protein expression.

The advantage of using mammalian cells for expression of eukaryotic genes, especially those from mammalian sources, rather than the other systems described, lies in the greater likelihood of a functional product being obtained. This is especially relevant for studies of structure–function relationships and the physiological effect of the protein on cell function. However, the relative difficulty and cost of large-scale production, compared with either *Pichia* or baculovirus systems, makes mammalian cells less attractive if the objective is the large-scale production of recombinant proteins.

Expression in Plant Cells

The expression of foreign genes in transgenic plants is under the control of regulatory elements, most commonly promoters. Unless a promoter is well-characterized and widely used in the plant species to be transformed, characterization of the expression pattern conferred by the promoter (may be as promoter:reporter gene fusions) should be carried out. This is particularly so as it becomes apparent that many *cis*-acting regulatory sequences are, in fact, found in regions other than the 'promoter'. There are two types of promoters, *viz.*, constitutive promoters and inducible promoters. For producing large amounts of recombinant protein at a specific time during large-scale growth, generally, tightly regulatable, inducible promoters are preferred. Inducible plant expression systems are divided into two categories, namely, nonplant-derived systems and plant-derived systems.

Constitutive Promoters The examples of constitutive plant promoters include CaMV 35S, rice actin promoter/first intron sequence, and maize ubiquitin I promoters.

Nonplant-derived Inducible Promoters The nonplant-derived system can be induced by the application of external/exogenous chemical agent (inducer). These systems are advantageous, as these are independent of the normal plant processes. Such systems require the application of a specific substance to induce expression. This can also be their weakness, as in some cases the application of inducers on an agricultural scale may be expensive or not feasible or economically nonviable.

Nonplant-derived systems should have the following features: (i) The promoter should not be leaky, i.e., the transgene should not be expressed in the absence of the inducer; (ii) The system should be specific to the inducer, i.e., it should respond to only one inducer or one class of inducer; (iii) Gene expression should be induced very rapidly following the application of the inducer; (iv) Gene expression should cease rapidly following withdrawal of the inducer; (v) The inducer should not cause nonspecific changes in gene expression; and (vi) The inducer should not be toxic. Thus, in this system, a chimeric promoter is used to drive transgene expression. In addition to CaMV 35S promoter region, the promoter also contains operator/binding sites for a transcription factor, which in turn is regulated by an external inducing agent. Moreover, the repressor/transcription factor is allowed to be constitutively overexpressed. Some examples of nonplant-derived systems are described as follows:

Tetracycline-inducible promoter The bacterial tetracycline repressor-operator system is also adapted to function in plants, where Tet R is constitutively overexpressed from a CaMV 35S promoter. The transgene is under the control of a chimeric promoter consisting of a core CaMV 35S promoter and several copies of the *tet* operator. The Tet R normally represses expression of the transgene, but with the application of tetracycline, it is released from the operator resulting in transgene expression. This system was one of the first examples in all plant species.

Tetracycline can also be used to inactivate transgene expression. A similar promoter is used to control the expression of the transgene, but the Tet R is modified to convert it to an activator (tetracycline transactivator or tTA). The tTA binds to the operator and induces gene expression in the absence of tetracycline. When tetracycline is added to the system, the tTA is released, consequently ceasing the transgene expression.

Alcohol (ethanol)-inducible promoter This system is based on binding sites for the Alc R transcription factor (from *Aspergillus nidulans*) in the promoter. A hybrid promoter is constructed in which core 35S promoter region is combined with Alc R-binding sites. Upon application of EtOH to the system, Alc R binds to the Alc R-binding site and activates transcription.

Copper-inducible promoter In this system, the yeast metallothionein regulatory system is used to regulate expression. The chimeric promoter comprises of CaMV 35S constitutive promoter plus elements (binding site) for binding transcription factor. The transcription factor is constitutively expressed. Upon binding copper, the transcription factor activates and binds to elements in a chimeric promoter and activates expression.

Steroid-inducible promoter Various systems conferring steroid inducibility on transgene expression are developed. Binding sites for the modified transcription factor are included in a chimeric promoter, which controls transgene expression Thus, most of them comprises of a modified transcriptional activator capable of binding a steroid hormone or its analogue. In the absence of an inducer, the transcription factor is inactive, but on application of the inducer, it binds to the promoter and activates expression. Systems that respond to different steroidal inducers have been developed, for example, systems responding to glucocorticoids (the synthetic analogue dexamethasone is used), estrogen, and ecdysone (an insect hormone). The dexamethasone system appears to have some serious drawbacks that may limit its usefulness. The development of the ecdysone-based system is, however, particularly interesting as it also responds to the nonsteroidal ecdysone agonist RH5992 (tebufenozide agrochemical used in field).

Plant-derived Inducible Promoters Systems based on plant-derived components depend on normal plant processes. This is disadvantageous as compared with nonplant-derived systems; however, as this system does not require the application of an inducer, its use in agriculture is potentially simpler. Other disadvantages associated with plant-derived systems are that there is no single system, which suits all situations, and some systems have proved to be difficult to use or have proved to be 'leaky'.

The plant-derived systems are further subdivided into two subcategories, *viz.*, those responding to environmental signals and those based on developmental control of gene expression.

Plant-derived systems responding to environmental signals Some plant-derived systems have been developed that respond to a variety of environmental signals. These systems are either dependent on the construction of chimeric promoters or on the use of complete promoters to control the expression of transgenes. The chimeric promoters usually consist of a core promoter element (usually from the CaMV 35S promoter) and sequence elements that bind transactivating factors, which respond to particular environmental signals. Some of the examples of plant-derived systems that respond to the environmental signals include promoters inducible by wound and heat-shock.

Wound-inducible Promoters Wound-inducible promoters can be used to drive the expression of pest resistance genes after insect damage. It has been established that the promoters of genes encoding some proteinase inhibitors confer wound inducibility on transgene expression. Furthermore, the wound inducibility may be mediated by other factors such as methyl-jasmonate, which can be used to mimic the wound response in some cases. The *vir* genes of *Agrobacterium* are also wound-inducible.

Heat-shock-inducible Promoters If heat-shock elements (HSEs) are included in chimeric promoters, these confer heat-shock inducibility on transgene expression. Several heat-shock proteins (e.g., chaperones) or protective proteins are inducible by heat-shock.

Plant-derived systems based on developmental control of gene expression Various temporally expressed (development-specific) genes have been identified that are expressed at particular stages in plant development. The promoters from some of these genes have been studied in detail. Either chimeric promoters or complete promoters are used to drive transgene expression. Some of the examples of plant-derived systems based on developmental control are those induced by senescence, abscissic acid, and auxin.

Senescence-specific Promoters The promoters from senescence-associated gene *viz.*, SAG12 and SAG13 obtained from *Arabidopsis* genes, have been shown to confer senescence-inducible expression on transgenes.

Abscissic Acid (ABA)-inducible Promoters The chimeric promoter resulting from combining ABA-inducible sequence elements with the core CaMV 35S promoter confers ABA inducibility on transgene expression.

Auxin-inducible Promoters By combining these auxin-response elements (AuxREs) with a core CaMV 35S promoter, auxin inducibility on transgene expression is attained.

12.10.3 Tissue-specific Expression

For expression of the protein encoded by the cloned gene in only a particular tissue, the gene should be cloned under the control of tissue-specific promoters. These promoters have been successfully used to drive transgene expression in the predicted pattern in a tissue-specific manner. These promoters are used when the expression of any potentially harmful substance in limited tissue is required (in this case, the expression of that substance is limited to those tissues not consumed by animals or humans) and when the expression of genes involved in specific processes needs to be limited to tissues in which that process occurs originally (i.e., expression in restricted tissues). For example, ripening genes are required to be expressed in fruits only and not in stems or roots, so a fruit-specific promoter is used.

12.10.4 Fusion Vectors

Fusion vectors are of two main types: transcriptional fusion vector and translational fusion vector.

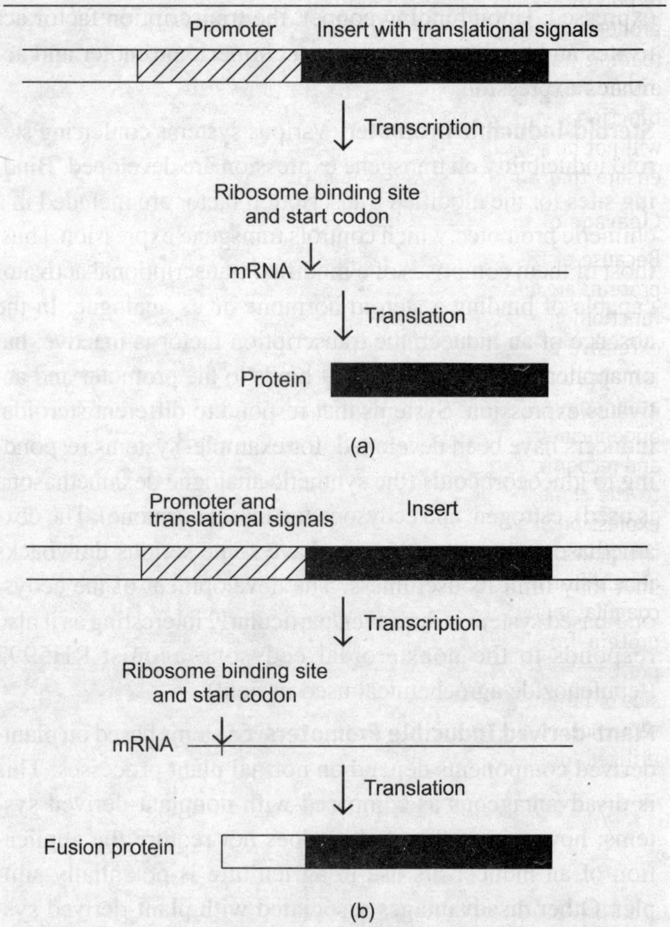

Figure 12.36 Transcriptional and translational fusions; (a) transcriptional fusions, (b) translational fusions

Transcriptional Fusion Vector

If the vector just carries a promoter and relies on the translational signals present in the cloned DNA, it is referred to as a transcriptional fusion vector. When attempting to express a foreign gene in a bacterial host, the first parameter to be considered is the requirement for a fully functional promoter attached to the cloned gene. In principle, this can be addressed by simply adding a known, characterized promoter in a separate cloning step. A more convenient procedure is to use a ready-prepared vector, usually a plasmid that already carries a suitable promoter adjacent to the cloning site. This forms a simple type of expression vector. Insertion of the cloned fragment at the cloning site in the correct orientation puts that fragment under the control of the promoter carried by the vector. In other words, transcription initiated at the promoter site continues through the cloned gene. Since this means a fusion of the gene and the promoter into a single transcriptional unit, it is referred to as a transcriptional fusion (Figure 12.36a). This is very similar to the concept of a reporter gene, but for a different purpose; a reporter gene is used to study the activity of the promoter. Example of transcriptional expression vector is pGEM® series of vectors (for details see Chapter 8).

In case of transcriptional vector, the only point that has to be ensured by the researcher is that the insert is the right way round. Moreover, it does not matter too much where it is; some untranslated leader mRNA can be tolerated, so it does not have to be precisely located with respect to the promoter.

Translational Fusion Vector

If the gene has to be provided with translational signals (ribosome binding site and start codon) as well as promoter, an expression vector different from transcriptional fusion vector is required. Such vector must supply the translational signals along with the promoter and should give rise to a translational fusion. In these vectors, the target DNA is cloned into the coding region of a vector gene, i.e., the insert must be *in frame* with the start codon. In this case, part of the translational product (the protein or polypeptide) is derived from the insert and part from the vector (Figure 12.36b). This is referred to as fusion protein (Exhibit 12.1). An example of translational fusion vector is λgt 11 (for details see Chapter 9). Note that the construct should be designed very carefully while using translational fusion vector.

In contrast to transcriptional fusion, location is important with a translational fusion. Since translation starts at the initiation codon in the vector sequence and the ribosomes then read the sequence in triplets, it has to be ensured that the correct reading frame is maintained at the junction. If one or two bases are out in either direction, the insert will be read in the wrong frame, giving rise to a completely different amino acid sequence and probably resulting in premature termination, as the ribosomes may soon come across a stop codon in this frame.

Exhibit 12.1 Fusion proteins

Small foreign proteins are obtained in minute amounts when produced in heterologous host cells due to the degradation of the foreign protein. Engineering a DNA construct encoding a target protein that is *in frame* with a stable host protein can solve this problem. The resulting single protein (combined) is known as a fusion protein. It has been reported in various studies that cloned gene proteins have been found to be stable when these are part of a fusion protein. In contrast, when these are expressed as separate intact proteins, these are susceptible to proteolysis. Generally, fusion proteins are stable because the target proteins are fused with proteins that are not especially susceptible to proteolysis.

The constructs for fusion proteins are prepared at the DNA level by ligating together portions of coding regions of two or more genes. Simply, a fusion vector system is required, which permit the insertion of a target gene or gene segment into the coding region of a cloned host gene. After ligation, reading frame of the products should be verified if the combined DNA has an altered reading frame, i.e., a sequence of successive codons that yields either an incomplete or an incorrect translation product, then a functional version of protein encoded by the cloned target gene will not be produced. Various approaches have been developed to ensure that a proper reading frame is achieved.

Cleavage of fusion proteins

Because of the presence of the host protein segment, most fusion proteins are unsuitable for clinical use and also affect the biological functioning of target protein. Moreover, fusion proteins require more extensive testing for approval by regulatory agencies. Therefore it is mandatory to devise strategies for removal of the unwanted amino acid sequences from the target protein. In general, the oligonucleotide linkers that code for the protease recognition site and recognized only by a specific non-bacterial protease are ligated to the cloned gene before the construct is inserted into a fusion expression vector. For example, an oligonucleotide linker encoding the amino acid sequence Ile-Glu-Gly-Arg can be joined to the cloned gene. After synthesis and purification of the fusion protein, a blood coagulation factor protein called X_a can be used to release the target protein from the fusion partner because factor X_a is a specific protease that cleaves peptide bonds uniquely on the C-terminal side of Ile-Glu-Gly-Arg sequence (Figure A). Furthermore, this peptide sequence occurs rarely in native proteins. This approach can be utilized for many different cloned gene products.

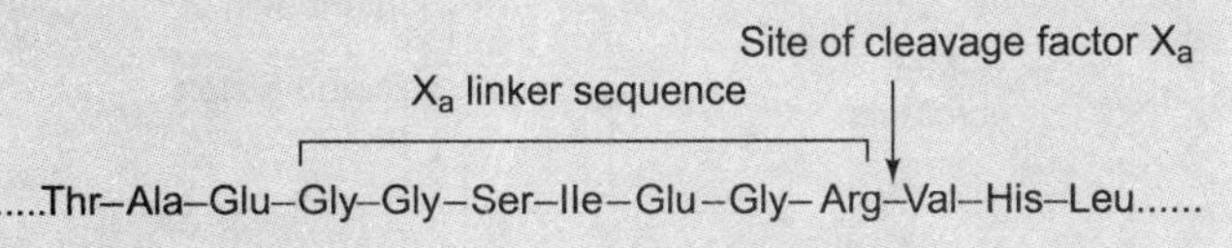

Figure A Cleavage of fusion protein by blood coagulation factor X_a

Uses of fusion proteins

A fusion protein can be a satisfactory end product for some applications such as a specific antigenic site that is required in large amounts and part of a fusion protein may be used for research and diagnostic purposes as long as the stabilizing protein does not interfere with the correct folding of the antigenic site. In this example, the fusion protein can be used as an antigen and any antibodies that are directed against the stabilizing protein can be removed by absorption with this protein alone, thus leaving in the antiserum only those antibodies that bind to the targeted proteins sequence.

One example of a fusion cloning vector is presented in Figure B. It includes the 5′-terminal segment of the *E. coli omp* F gene, which directs the synthesis of an outer membrane protein, and a portion of *E. coli lac* Z (β-galactosidase) gene, and used to generate antibodies against selected target proteins. The *omp F* gene segment contributes the signals for the initiation of transcription and translation and for secretion of the fusion protein. Although the truncated *lac* Z gene lacks the codons for the first eight amino acids, the shortened protein encoded by this gene fragment is still enzymatically active. This truncated form of β-galactosidase is able to function with any N-terminus localized peptide fused to it. The *lac* Z gene is cloned on the vector at a location that puts it in an altered reading frame with respect to the *omp* F leader sequence;

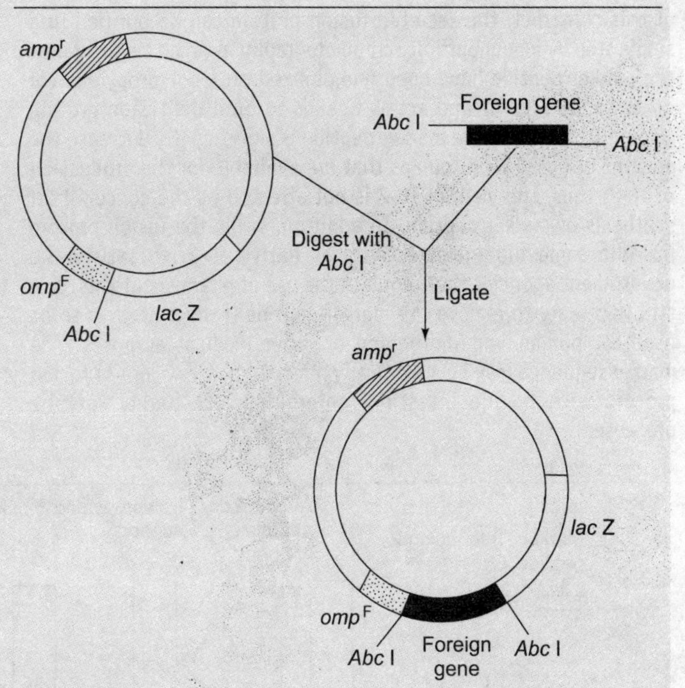

Figure B Fusion protein cloning vector [Plasmid contains an amp^r gene as selectable marker, *omp* F fragment for outer membrane protein, and a truncated *lac* Z. Foreign gene is inserted at *Abc* I site; after transcription and translation, a tribrid protein is produced]

consequently no functional β-galactosidase will be produced. However, after cloning of target DNA, if *in frame omp* F and *lac* Z are generated, that means 'tribrid' (a three-part hybrid) protein that comprises the Omp F amino acid sequence, the protein encoded by the cloned target gene, and the functional C-terminal portion of β-galactosidase, this hybrid protein can be used either as an antigen to produce antibody that will cross-react with the cloned gene protein or as a means of producing large amounts of small important portions of specific proteins.

A number of fusion proteins have been developed to simplify the purification of recombinant proteins. This strategy is utilized with both prokaryotic and eukaryotic host organisms. For example, a vector that contains the human interleukin-2 gene joined to DNA encoding the marker peptide sequence Asp-Tyr-Lys-Asp-Asp-Asp-Asp-Lys has the dual function of reducing the degradation of expressed interleukin-2 gene product, and then enabling the product to be purified (Figure C). IL-2 is a biological factor that stimulates both T-cell growth and B-cell antibody synthesis. Following expression

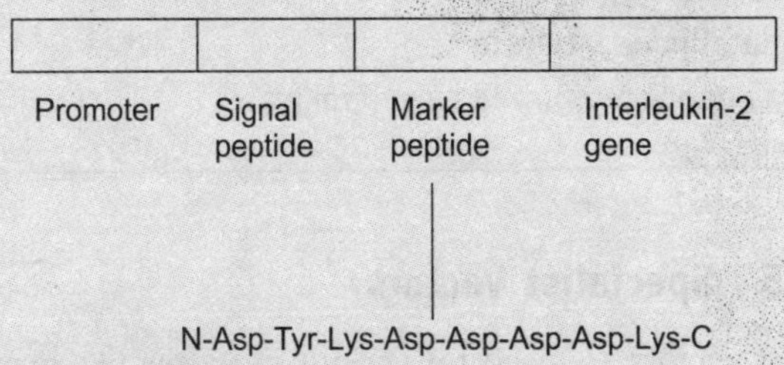

Figure C Example of genetic construct used to produce a secreted fusion protein having a marker peptide and interleukin-2

of this construct, the secreted fusion protein can be purified in a single step by immunoaffinity chromatography in which McAb against the marker peptide have been immobilized on a polypropylene (or other solid) support and act as ligands to bind the fusion protein (Figure D). Because the marker peptide is significantly decrease the amount of host cell resources that are available for the production of IL-2; thus, the yield of IL-2 is not affected by the concomitant synthesis of marker peptide. In addition, while the fusion protein has the same biological activity as native IL-2, to satisfy the government agencies that regulate the use of pharmaceuticals, it is still necessary to remove the marker peptide if the protein is to be used for human immunotherapy or other medical purposes. The marker sequence may be specifically removed by treatment of fusion protein with bovine intestinal enterokinase, a highly specific protease.

Most often antigen–antibody complexes are difficult to separate without the use of denaturing chemicals. In these cases, it has become very popular to generate a fusion protein containing six or eight histidine residues attached to either the N- or C-terminal of target protein. This His-tagged protein, along with other cellular proteins, is then passed over an affinity column of nickel nitrilotriacetic acid. His-tagged protein, but not the other cellular proteins bind tightly to the column. The bound protein is eventually eluted from the column by the addition of imidazole. With this protocol, some cloned and overexpressed proteins have been purified up to 100-fold with greater than 90% recovery in a single step. Some fusion systems used to facilitate the purification of foreign proteins produced in *E. coli* are listed in Table A.

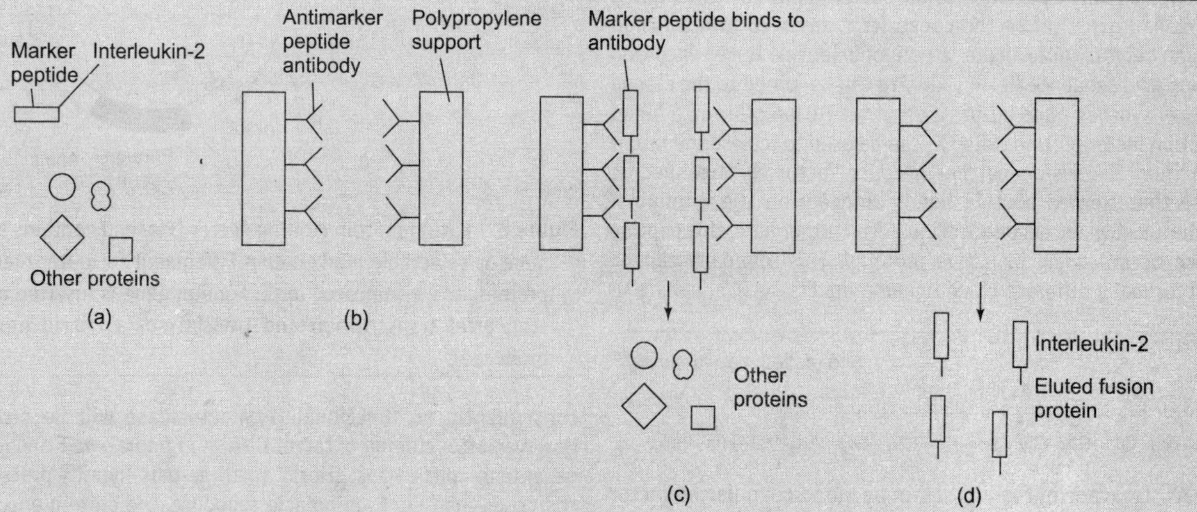

Figure D Immunoaffinity chromatographic purification of a fusion protein; (a) Secreted protein mixture is concentrated, (b) Immunoaffinty column is prepared, (c) Secreted protein mixture is added to the column, (d) Fusion protein is eluted

Table A Some Fusion Systems Used to Facilitate the Purification of Foreign Proteins Produced in *E. coli*

Fusion partner	Size	Ligand	Elution condition
ZZ (a fragment of *Staphylococcus aureus* protein A)	14 kDa	IgG	Low pH
His tail (histidine tail)	6–10 amino acids	Ni^{2+}	Imidazole
Strep-tag (a peptide with affinity for streptavidin)	10 amino acids	Streptavidin	Iminobiotin
Pinpoint (a protein fragment, which is biotinylated *in vivo* in *E. coli*)	13 kDa	Streptavidin	Biotin
Maltose binding protein (MBP)	40 kDa	Amylose	Maltose
β-Lactamase	27 kDa	Phenylboronate	Borate
GST (glutathione *S*-transferase)	25 kDa	Glutathione	Reducing agent
Flag (a peptide recognized by enterokinase)	8 amino acids	Specific monoclonal antibody (McAb)	Low calcium

12.10.5 Specialist Vectors

Specialist vectors are used for various purposes like protein tagging, secretion signals, surface display, and production of RNA probes.

Protein Tagging

Purification of the recombinant product by conventional means can often be tedious and inefficient. The fact that a translational fusion vector can add a short stretch of amino acids to the N- or C-terminal end of the polypeptide product can be turned to advantage. If the vector contains not just a start codon but also a short sequence coding for a few amino acids, known as a tag, the resulting fusion protein carries this tag at the N-terminus.

To remove the tag after protein purification, the vectors are designed to contain not only the tag but also a site at which the product can be cleaved by highly specific peptidases. If the peptidase is sufficiently specific, it will cleave the product only at this site and not elsewhere. Therefore, the product can be affinity purified, cleaved with the peptidase, and separated from the cleaved tag.

Epitope Tagging The tag may constitute a recognition site (an epitope) that can be recognized by a monoclonal antibody (McAb). The protein is recovered from the cell extract in a single step by affinity purification using the ability of the McAb to bind to the epitope tag (Figure 12.37). A further application of this type of vector is that specific McAb can be used to detect the expression of product. A variety of such tags and corresponding McAbs are now available commercially, and these can be added to either end of the insert (N- or C-terminus), using modified PCR primers rather than by incorporation into the vector.

Histidine Tagging Designing the expression vector or manipulating the insert so that a sequence coding for a number of histidine residues is added to one end of the sequence to be expressed provides an alternative to epitope tagging. This produces a His-tagged protein, which can be purified using a Ni^{2+} resin (e.g., nickel nitrilotriacetic acid). The His-tagged protein, but not the other cellular proteins bind tightly to the column. The bound protein is eventually eluted from the column by the addition of imidazole (the side chain of histidine). Thus the affinity of the histidine residues for nickel

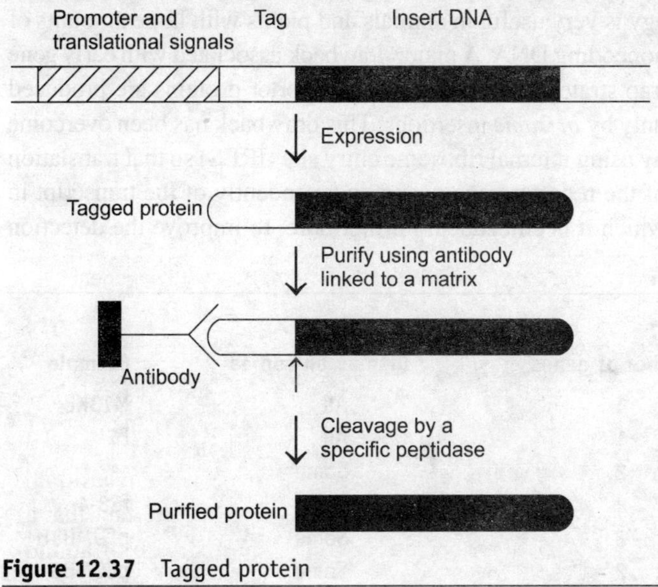

Figure 12.37 Tagged protein

results in the tagged protein being retained by the resin, while other proteins are washed through. The His-tag can also be used as an epitope tag, since antibodies that specifically recognize this sequence of histidine residues are available facilitating the detection of the tagged product.

Secretion Signals

In a similar way to the incorporation of tags into the product, secretion signals can be added to target the proteins to definite locations.

E. coli does not secrete many proteins into the culture supernatant and most of the proteins secreted across the cytoplasmic membrane remain in the periplasm, trapped by the outer membrane. In some other bacteria, particularly Gram positive bacteria such as *B. subtilis*, many enzymes are secreted into the culture supernatant. This can be advantageous for several reasons. First, if a protein is synthesized at high-levels and accumulates within the cytoplasm, it results in a very high concentration of that protein. This can cause aggregation of the protein into insoluble inclusion bodies. These may be damaging to the cell and are often very difficult to resolubilize. However, secretion of the protein into the culture medium prevents this problem since the volume of the supernatant is greater than that of the total volume of the cytoplasm of all the cells. Even with a very dense culture, the space occupied by the cells is only a small proportion of the culture volume. Furthermore, although *Bacillus* secretes a number of enzymes, the culture supernatant contains a much simpler mixture of proteins than the cell cytoplasm. The task of purifying the required product is therefore much simpler.

Secretion of proteins by bacteria depends normally on the presence of a signal peptide at the N-terminus. This labels it as a secreted protein; hence it is recognized by the secretion machinery and transported across the cytoplasmic membrane. Therefore, if a sequence coding for a signal peptide is incorporated into the vector, or into the insert, the final product may be secreted. It is not inevitable, as it also depends on the overall structure of the protein. If the protein is naturally secreted in its original host, then there are chances of success. If it is normally a cytoplasmic protein, the chances of getting it secreted successfully are very much lower.

In mammalian systems, specific signals can be added to direct the product to specific cellular locations or to obtain a secreted product.

Surface Display

Specialized fusion protein systems have been devised for screening cDNA libraries that contain very large numbers of different clones (sometimes up to 5×10^{10}) for proteins that are encoded by rarely occurring cDNAs. Generally, of these libraries, cDNA molecules are cloned into a surface protein (filamentous protein or pilus protein) gene of either a filamentous bacteriophage such as M13 or a bacterium. After transcription and translation, the fusion protein is incorporated into a surface structure of bacteriophage or bacteria, respectively, where it can be identified by an immunological assay. More specifically, fusions are often made with the M13 gene encoding protein pIII. The pIII is normally found at the tip of this tubular shaped bacteriophage and is responsible for initiating phage infection of *E. coli* by binding to F pili. A plasmid that contains a small piece of M13 DNA that allows the plasmid (phagemid) to be packaged *in vitro* into M13 phage particles, a *pIII* gene under the control of a regulatable bacterial promoter such as *E. coli lac* promoter and a cloning site near the 5′-end of the *pIII* gene has been constructed. The target protein is fused to M13 phage protein pIII near its N-terminus. After M13 replication in *E. coli* cells, the plaques are assayed immunologically for the presence of target protein. Recombinant phagemids isolated from positive plaques can then be used as source of target cDNA. This is an extremely powerful selection system that has the capability of finding cDNAs for very rarely expressed but important proteins.

Finally, it is also possible, although less straightforward and therefore less common to create recombinant phage in which a target peptide is fused to protein pVIII, the bacteriophage coat protein. The classification of major phage display vectors is presented in Table 12.9.

Alternatively, libraries with bacterial surface structures composed of fusion proteins can be screened for clones that carry specific coding sequences. To export proteins to the surface of a Gram negative bacteria, such as *E. coli*, fusions between surface protein are created. Bacterial fusion partners that have been used in these types of constructs include outer membrane protein A (Omp A) and peptidoglycan-associated protein (PAL) from *E. coli*, as well as *Pseudomonas aeruginosa* outer membrane protein (Opr F). With most bacterial surface fusion proteins, the target protein is located at either the N- or C-terminus of the fusion protein.

In addition to facilitating the screening of large cDNA libraries, surface displayed proteins can provide an effective means of overexpressing peptides and proteins. For example, in one study the amino acid repeating epitope (antigenic determinant) Asn-Ala-Asn-Pro of a protein from the parasite *Plasmodium falciparum*, the causative agent of malaria, was inserted into the regions that encode surface exposed loops of the major outer membrane protein from *P. aeruginosa* (Opr F). Whole bacterial cells expressing this fusion protein reacted positively when challenged with McAbs against *P. falciparum*. It may, therefore, be possible to use some surface displayed fusion proteins as vaccines.

Production of RNA Probes

RNA probes are produced by cloning of relevant gene under the control of a phage promoter in a plasmid vector. After purification, the plasmid is linearized with a suitable restriction enzyme and then incubated with the phage RNA polymerase and the four rNTPs (for details see Chapter 2). Transcriptional fusion vectors (e.g., pGEM® series, LITMUS), in which two phage promoters flank the polylinker, are commonly used for the synthesis of RNA probes.

12.11 ADVANCED GENE TRAPPING VECTORS

Gene trapping is the technique applied to mark interrupted genes with a unique DNA sequence by using an insertional mutagen. This particular DNA sequence is then further used as a target for hybridization. This can be simply achieved by inverse PCR (for details see Chapter 7) with primers designed on the basis of flanking sequences. Moreover, some gene trapping vectors are also designed, which facilitate cloning and provide more information about the interrupted genes. Few of them are discussed briefly in this section.

12.11.1 Gene Trap Vector

The insertion element of this type of vector contains a visible marker gene, e.g., *lac* Z or *gus* A, which is localized downstream to the splice acceptor site. Thus, the marker gene is activated only if element is inserted within the transcription unit of a gene and generates a transcriptional fusion. This strategy is very useful in animals and plants with large amounts of noncoding DNA. A major drawback associated with early gene trap strategy is that functional reporter proteins are produced only by *in frame* insertions. This drawback has been overcome by using internal ribosome entry site (IRES) so that translation of the reporter gene occurs independently of the transcript in which it is embedded. Furthermore, to improve the detection

Table 12.9 Classification of Phage Display Vectors

Vector type	Coat protein for display	Number of genes	Display on copies	Example
Type 3 (phage)	pIII	1	All	M13KE
Type 8 (phage)	pVIII	1	All	F1
Type 33 (phage)	pIII	2	Some	–
Type 88 (phage)	pVIII	2	Some	f88-4
Type 3+3 (phage)	pVII	2	Some	pCOMB3H
Type 8+8 (phage)	pVIII	2	Some	pCOMB8

of unexpressed genes while using gene trap vectors, second marker gene (driven by its own promoter) is incorporated that carries a downstream splice donor making it dependent on the surrounding gene for polyadenylation.

12.11.2 Plasmid Rescue Vector

Plasmid rescue vector carries the insertion element having the origin of replication and antibiotic resistance from a bacterial plasmid. This plasmid rescue technique is described in Figure 12.38. Genomic DNA from a tagged organism is digested with restriction enzyme that does not cut in the insert and the resulting linear fragments are self-ligated to generate circles. Thereafter, bacteria are transformed with complex mixture of resultant circles and positive transformants are selected under antibiotic selection pressure. The circle comprising of the origin of replication and resistance gene is propagated as a plasmid while others are lost. The genomic sequences flanking the insert can be selectively amplified in a single step. Although this technique is more time-consuming than the direct amplification of flanking sequences by PCR, rescued plasmids can however be maintained as a permanent resource library.

12.11.3 Enhancer Trap Vector

In this type of construct, a visible marker gene is used, which is driven by a minimal promoter. In normal conditions, the promoter is too weak to activate the marker gene, which remains unexpressed. However, if the construct integrates in the vicinity of an endogenous enhancer, the marker is expressed, which is driven by the enhancer (Figure 12.39). As enhancers are most frequently located quite distant from the corresponding gene, the enhancer trap cannot be used to directly clone genes by tagging, however, it can be exploited in other ways, such as to drive the expression of a toxin (such as ricin or diphtheria toxin). Enhancer trap lines have been widely used to identify and clone novel *Drosophila* genes.

Furthermore, a modified strategy (Figure 12.40) has been developed in order to facilitate the use of enhancer trapping as a general method for driving cell-specific expression. For this purpose, the ability of the yeast transcription factor GAL 4,

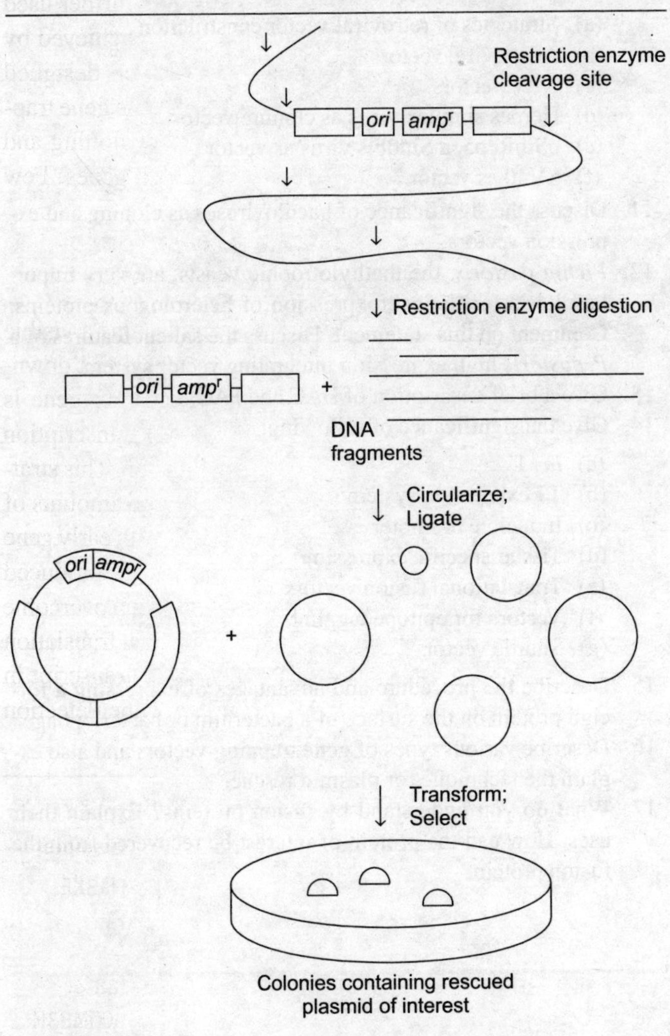

Figure 12.38 Principle of plasmid rescue technique

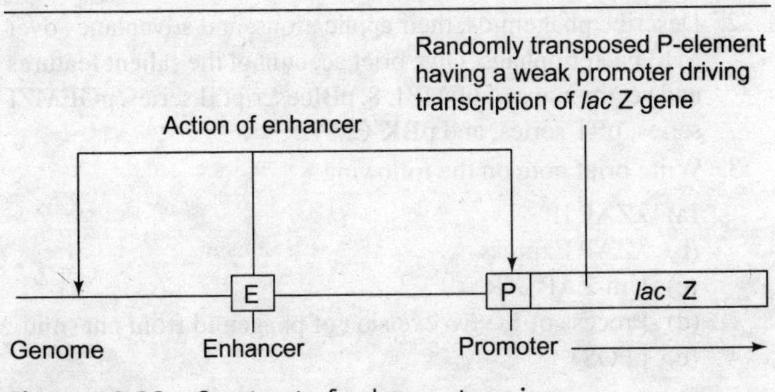

Figure 12.39 Construct of enhancer trapping

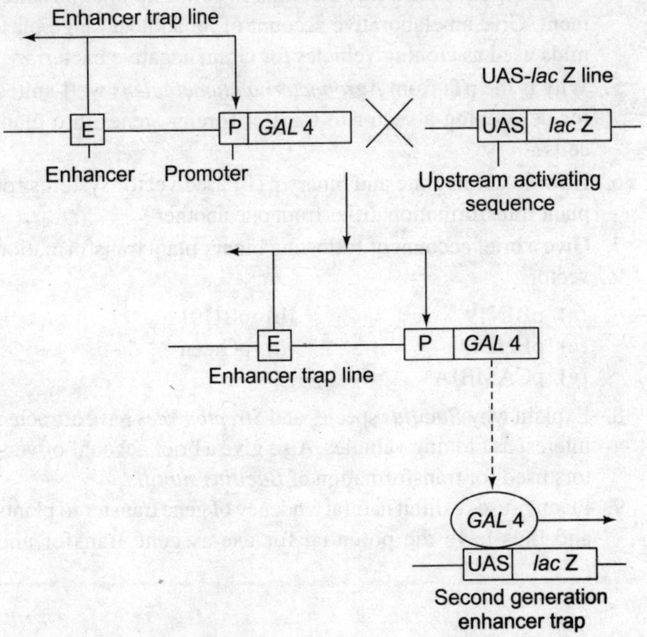

Figure 12.40 Second-generation enhancer trap

which activates transgenes containing its recognition site in the heterologous environment (e.g., fly) has been exploited. In this case, the *lac* Z gene is replaced by the coding region for GAL 4. In resulting transgenic flies, GAL 4 is expressed in the pattern dictated by a local enhancer. Moreover, the pattern of GAL 4 expression can be analyzed by crossing the enhancer trap line to flies carrying a reporter transgene in which the *lac* Z gene is coupled to a promoter containing GAL 4-binding sites. This system is used for generation of a bank of fly stocks with different trapped enhancers, each with a definite pattern of GAL 4 expression; such bank can be utilized to analyze expression of desired gene. Similar kinds of *lac* Z enhancer trap systems have also been used in mice.

Enhancer trap vectors have also been developed for plants using T-DNA insertions carrying a *gus* A gene driven by a weak promoter adjacent to the RB repeat. The activity of endogenous enhancers can be determined by screening the plants for GUS activity.

12.11.4 Activation Tagging

The insertion element having a strong outward-facing promoter is used in this technique. During gene tagging, if the element integrates adjacent to an endogenous gene, that particular gene is activated by the promoter unlike other insertion vectors. Therefore, an activation tag results in a gain of function through overexpression or ectopic expression.

Review Questions

1. Give a detailed account of cosmids, types of cosmids, and cloning strategies of both types of cosmids with suitable examples of each. What are the advantages of using cosmid DNA as a vector?
2. Describe phagemids, their applications, and advantages over M13 bacteriophage. Give brief account of the salient features and vector maps of pEMBL 8, pBluescript II series, pGEMZf series, pET series, and pBK-CMV.
3. Write brief note on the following:
   - (a) λZAP II®
   - (b) λZAP Express
   - (c) Uni-ZAP®XR
   - (d) Process of *in vivo* excision of phagemid from phasmid
   - (e) pFOS1
4. For some important experiments, it is required to clone in Gram negative bacteria other than *E. coli*. Explain this statement. Give an elaborative account of broad host range plasmids used as cloning vehicles for Gram negative bacteria.
5. Why is the pTi from *Agrobacterium tumefaciens* well-suited for developing a vector to transfer foreign genes into plant cells?
6. How do cointegrate and binary pTi-based vector systems for plant transformation differ from one another?
7. Give a brief account of following binary plant transformation vectors:
   - (a) pBIN19
   - (b) pBI101
   - (c) pBI121
   - (d) pGreen
   - (e) pCAMBIA
8. Explain why *Bacillus* species and *Streptomyces* have attracted interest as cloning vehicles. Also give a brief account of vectors used for transformation of *Bacillus subtilis*.
9. Plant viruses exhibit natural tendency of gene transfer to plants and thus have the potential for use as gene transfer and expression vectors. Explain this statement. Give a brief account of RNA and DNA viruses, which are used as expression vectors for plant transformation.
10. Write in brief about the following:
    - (a) Strategies of retroviral vector construction
    - (b) Adenoviral vectors
    - (c) AAV vectors
    - (d) Herpes simplex virus as cloning vector
    - (e) pSinRep5, a Sindbis virus as vector
    - (f) SV40 as vector
11. Discuss the significance of baculoviruses as cloning and expression vectors.
12. *Pichia pastoris,* the methylotrophic yeasts, are very important hosts for the overexpression of heterologous proteins. Comment on this statement. Discuss the salient features of a *P. pastoris* high expression integrating vector system.
13. Give a brief description of BAC and PAC.
14. Give the significance of following:
    - (a) *lac* I^q
    - (b) T7 expression system
    - (c) Inducible promoter
    - (d) Tissue-specific expression
    - (e) Translational fusion vectors
    - (f) Vectors for epitope tagging
    - (g) Shuttle vector
15. Describe the procedure and advantages of expressing a foreign protein on the surface of a bacterium or bacteriophage.
16. Describe various types of gene trapping vectors and also explain the technique for plasmid rescue.
17. What do you understand by fusion proteins? Explain their uses. How can the protein of interest be recovered from the fusion protein?

Part III

Generation and Screening of Recombinants

- Joining of DNA Fragments
- Introduction of DNA into Host Cells
- Construction of Genomic and cDNA Libraries
- Techniques for Selection, Screening, and Characterization of Transformants

13 Joining of DNA Fragments

Key Concepts

In this chapter we will learn the following:

- Ligation of DNA fragments using DNA ligase
- Ligation using homopolymer tailing
- Increasing versatility and efficiency of ligation

13.1 INTRODUCTION

One of the fundamental steps in recombinant DNA technology is the joining of foreign DNA molecules to the vector DNA. In the previous chapters, the first step of cloning experiment, i.e., isolation of DNA fragments was described. In this chapter, the ways in which these fragments can be joined to the vector in order to create artificial recombinant molecules will be explained. Currently, following three methods are mostly used for joining DNA fragments *in vitro*:

Ligation by *E. coli* DNA Ligase This enzyme has the ability to join covalently the annealed cohesive ends produced by certain restriction enzymes.

Ligation by T4 DNA Ligase from T4 Phage-infected *E. coli* This enzyme has the ability to catalyze the formation of phosphodiester bonds between blunt ended fragments.

Homopolymer Tailing Catalyzed by Terminal Deoxynucleotidyl Transferase (TdT) TdT has the ability to synthesize homopolymeric single stranded tails at the 3′-ends of the DNA fragments. Thus, by adding complementary tails to two different DNA fragments, the two fragments can be annealed together.

In this chapter, these methods are explained in detail. Besides, the procedure for independent ligation with DNA ligase is also discussed. The versatility of ends and ligation efficiency can be increased by the modification of DNA ends. The techniques employed include trimming back, filling-in, and ligation of short oligonucleotides, *viz.* linkers and adaptors. These techniques aid in converting one restriction enzyme site into another, and may also help in creating new restriction enzyme sites for the introduction of insert DNA into the vector molecule, or for easy recovery of the insert DNA for subcloning.

13.2 LIGATION OF DNA FRAGMENTS USING DNA LIGASE

DNA ligase is ubiquitous and essential to all cells. The term 'Ligase' is derived from the Latin word *Ligare*, which means 'to tie'. The natural role of DNA ligase is to join the short Okazaki fragments during DNA replication, or to repair single stranded breaks (nicks) resulting due to DNA damage in the sugar phosphate backbone of a ds DNA. This ability of DNA ligase can be exploited under *in vitro* conditions for joining DNA fragment to the vector in cloning experiments or in the assembly of a gene (for details see Chapter 4). DNA ligase catalyzes the formation of a phosphodiester bond between a 3′-OH and a 5′-phosphate group of DNA; a total of four phosphodiester bonds are formed during ligation of DNA fragment in the vector molecule. As discussed in Chapter 3, for DNA ligase-catalyzed ligation both the DNA fragments and the vector should have compatible ends, or else compatibility between them may be acquired by partial filling-in or trimming back or with the addition of linkers and adaptors.

The two compatible DNA molecules may anneal to each other through H-bonds (in case of molecules having cohesive ends) or may remain as blunt ended DNA molecules in the solution. DNA ligase then catalyzes the formation of phosphodiester bonds between cohesive ends that are 'annealed and held together by H-bonding' (Figure 13.1) or between blunt ends 'by random association based just on chance' (Figure 13.2). As compatible sticky ends are annealed by the formation of H-bonds between the complementary cohesive ends, these are easier to join by DNA ligase and efficiency of ligation is higher as compared to blunt end ligation, which is based just on chance. However, the efficiency of blunt end ligation is high enough for most subcloning applications. The sticky end ligation by DNA ligase is restricted due to compatibility of

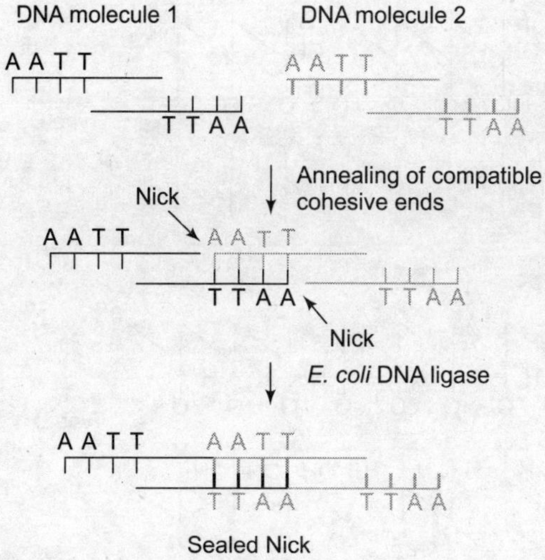

Figure 13.1 Sticky end ligation

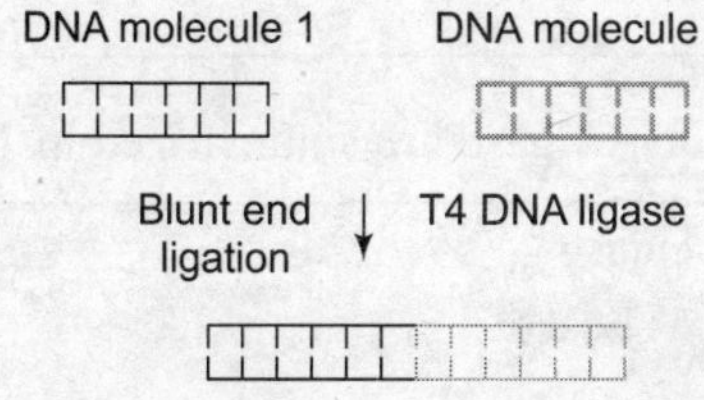

Figure. 13.2 Blunt end ligation

ends, whereas there is no such restriction in case of blunt end ligation; the only requirement is two blunt ends, irrespective of how these are generated. Hence, at times, blunt ended ligation is also preferred.

13.2.1 Types of DNA Ligases

The two types of DNA ligases that find application in cloning experiments are: (i) *E. coli* DNA ligase; and (ii) T4 DNA ligase. Although both these ligases catalyze the formation of phosphodiester bonds, these differ in few respects:

Source and Structure The *E. coli* DNA ligase from *E. coli* is a single polypeptide (monomer) of ~75 kDa comprising of 671 amino acid residues and a sedimentation coefficient of 3.9S. The enzyme is susceptible to adventitious proteolysis as well as to controlled cleavage by trypsin. The proteolytic product, significantly smaller in size than the intact enzyme, can still form the first covalent intermediate, DNA ligase-adenosine monophosphate (DNA ligase-AMP), at an unaltered rate and to the full extent, but it cannot subsequently transfer the AMP group to DNA for phosphodiester bond formation. The number of DNA ligase molecules per *E. coli* cell (estimated from the extracted activity divided by the specific activity per molecule of pure enzyme) is about 300.

The T4 DNA ligase from T4-infected *E. coli* cells resembles the *E. coli* enzyme in size and shape. It is a ~68 kDa polypeptide with a sedimentation coefficient of 3.5S. The enzyme shows marked inhibition by salt, being virtually inert at 0.2 M KCl. Spermine inhibits T4 DNA ligase by raising the K_m for the DNA substrate.

Cofactor Requirement Free energy required for phosphodiester bond formation (ligation reaction) is obtained in a species-dependent manner. The cofactor or energy source (AMP donor) for the *E. coli* enzyme is NAD^+, while that for the T4 enzyme is ATP. In each case, the cofactor is split and forms an enzyme–AMP complex, which is discussed later in this chapter in Section 13.2.3.

Substrate Specificity The DNA ligases isolated from *E. coli* and T4-infected *E. coli* differ not only in their coenzyme requirements but also in their substrate specificities. The physiological substrate for both enzymes is thought to be the breakage point at a phosphodiester bond between neighboring 3′-OH and 5′-phosphate ends still held together by an intact complementary strand. Another substrate for both these enzymes contains the open and staggered phosphodiester bonds formed through annealing of the cohesive protruding termini of different DNA molecules generated by digestion with certain type II restriction enzymes. In this case also, the strands may be held together by base pairing between protruding nucleotides, but leaving four nicks a few base pairs apart in opposite strands. In addition, the T4 phage-encoded T4 DNA ligase catalyzes a number of other ligation reactions. It is capable, for example, of joining nicks in the RNA chains of RNA:DNA hybrids and also anneals RNA termini with DNA strands. Such reactions may play an important biological role in the RNA-primed enzymatic synthesis of DNA. Another remarkable property of T4 DNA ligase, which distinguishes it from the *E. coli* DNA ligase, is its ability to accomplish blunt end ligation of ds DNA molecules. The joining of two blunt ended linear DNA complexes is noteworthy because it was once assumed that an intact complementary strand is needed to orient the ends of a broken strand for joining. Little is known about the mechanism of this reaction, apart from the fact that the reaction requires high concentrations of ligase and that it is stimulated by T4 RNA ligase. The *E. coli* DNA ligase does not catalyze blunt end ligation except under special reaction conditions of macromolecular crowding. The T4 DNA ligase is thus capable of catalyzing the reactions involving substrates of types A, B, and C by itself (Figure 13.3).

Neither of the two enzymes is capable of ligating single stranded polynucleotides, however, this reaction can be carried out by another enzyme, RNA ligase, which does not require a complementary strand for its activity.

Basis of Ligation *E. coli* DNA ligase ligates mostly the fragments that are annealed and unstably held by H-bonds (not more than four). On the other hand, T4 DNA ligase can also catalyze blunt end ligation, which is based just on 'chance association'.

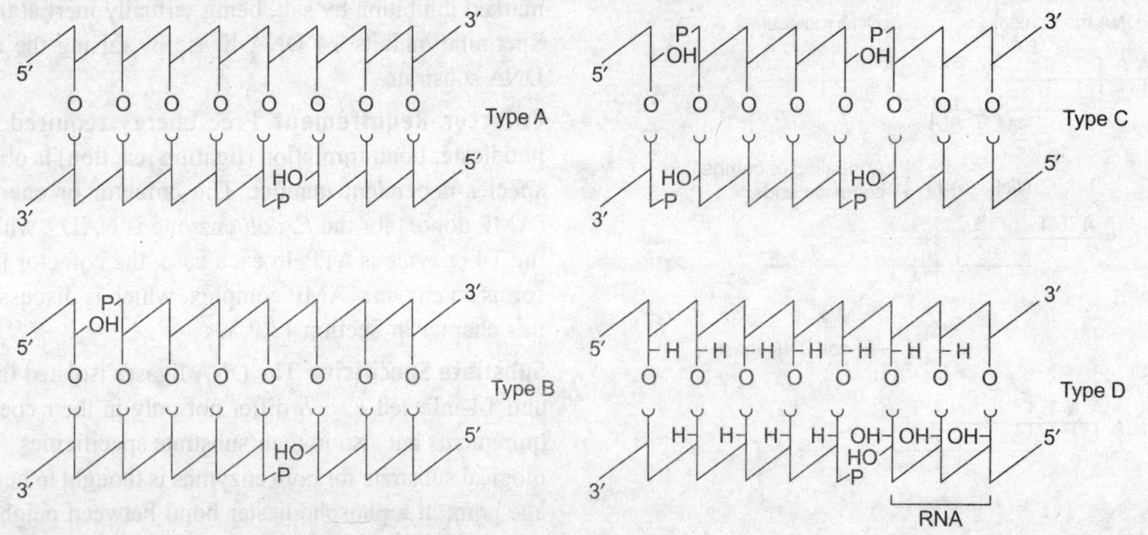

Figure 13.3 Four substrates for DNA ligases [*E. coli* DNA ligase only catalyzes reactions with substrates of types A and B, while T4 DNA ligase is capable of using all four substrates.]

The differences and similarities between *E. coli* DNA ligase and T4 DNA ligase are summarized in Table 13.1.

Table 13.1 Differences or Similarities Between *E. coli* DNA Ligase and T4 DNA Ligase

Property	*E. coli* DNA ligase	T4 DNA ligase
Source	*E. coli* cells	T4-infected *E. coli* cells
Structure	Single polypeptide (Monomeric enzyme)	Single polypeptide (Monomeric enzyme)
M_r	~ 75 kDa	~ 68 kDa
Sedimentation coefficient	3.9S	3.5S
Cofactor	NAD^+	ATP
Substrates	Nick in one strand of a ds DNA; Open and staggered phosphodiester bonds formed through annealing of the cohesive protruding termini of different DNA molecules generated by digestion with certain type II restriction enzymes	Nick in one strand of a ds DNA; Open and staggered phosphodiester bonds formed through annealing of the cohesive protruding termini of different DNA molecules generated by digestion with certain type II restriction enzymes; Nicks in the RNA chains of RNA:DNA hybrids; It can also anneal RNA termini with DNA strands.
Type of ligation	Sticky end ligation	Sticky end ligation; Blunt end ligation.
Capability of ligating single stranded polynucleotides	Absent	Absent
Basis of ligation	Two DNA molecules must be annealed due to complementary base pairing of cohesive ends, i.e., unstably held by H-bonds	Random and based just on 'chance association' for blunt end ligation; Two DNA molecules must be annealed due to i.e., complementary base pairing of cohesive ends, unstably held by H-bonds during sticky end ligation.
Active site residue	Lysine	Lysine
Reaction mechanism	Reaction involves three steps: • Adenylylation of DNA ligase; • Adenylyl activation of 5′-phosphoryl terminus of the nick; • Sealing of the nick by formation of phosphodiester bond	Reaction involves three steps: • Adenylylation of DNA ligase; • Adenylyl activation of 5′-phosphoryl terminus of the nick or one of the two blunt ended DNA molecules; • Joining of two DNA molecules by formation of phosphodiester bond or sealing of the nick.

13.2.2 Isolation of DNA Ligase

DNA ligases from *E. coli* and T4 phage-infected *E. coli* can be obtained in highly purified form, because suitable overproducers have been developed. Overproduction of ligase has been achieved using either T4 mutants defective in DNA replication or recombinant λ phages containing a ligase gene. The T4 mutants, due to their genetic defect, overproduce large quantities of 'early' T4 ligase enzymes. The recombinant λ phages carry the genes for *E. coli* ligase or T4 DNA ligase, under the control of phage promoters. Thus, T4 DNA ligase has been efficiently produced by a recombinant containing T4 DNA ligase gene (*gene* 30) under the control of the λ late promoter P'_R. In this case, the T4 insertion is ~3.2 kbp in length. It comprises the ligase gene with its own promoter and containing four *Eco* RI sites. In the phage genome itself, the ligase gene is transcribed from a 0.4 kbp *Eco* RI fragment toward another 2.2 kbp *Eco* RI fragment. Note that the T4 *gene* 30 promoter is an efficient promoter, but T4 DNA ligase production is far excellent by λ promoter P'_R, which predominantly drives synthesis in the λ NM989 recombinant. In this construction, up to 10% of the total protein in infected bacteria may be T4 DNA ligase. This corresponds to an overproduction by a factor of 500–2,000. Only a few steps are therefore required for the purification of ligase produced in this manner. Similar protocols have been worked out for *E. coli* DNA ligase. Thus, amplified levels of the *E. coli* and T4 DNA ligases, induced by cloning the genes, have provided ample sources for studies of the enzymes and their widespread use (particularly T4) in recombinant DNA technology.

13.2.3 Reaction Catalyzed by DNA Ligase

DNA ligase is a monomeric enzyme with a deep cleft containing predominantly positively charged amino acid residues. This cleft forms the DNA binding site. A highly negative patch formed by the highly conserved side chains of Asp and Glu residues lies in the vicinity of positively charged invariant Lys residue, thereby making the ε-NH_2 group of this active site Lys a nucleophile to initiate the reaction. The enzyme further catalyzes the formation of a phosphodiester bond between intact 3′-OH and 5′-phosphate groups at the ends of DNA molecules to be joined. The ligation process involves the formation of four phosphodiester bonds, i.e., two at each end of the molecule. In principle, such bonds can be formed either *in vitro* or *in vivo*. While the *in vitro* reaction is catalyzed by enzymes and is carried out under carefully controlled reaction conditions, *in vivo* ligation takes place in the organism that has been chosen to harbor and replicate the recombinant DNA. It is now evident that in cloning experiments, *E. coli* DNA ligase is employed for sticky end ligation, in which the sticky ends are already annealed due to complementary base pairing, and T4 DNA ligase is used for blunt end ligation of two separate DNA molecules. In either case, the mechanism of action is the same, where the phosphodiester bond is formed utilizing the group transfer potential of the phosphoanhydride bonds of NAD^+ (in case of *E. coli* DNA ligase) or ATP (in case of T4 DNA ligase) through the three discrete steps defined in the sections that follow.

Adenylylation of DNA Ligase (E. coli or T4)

This step is the enzyme activation step. In this step, the enzyme makes use of the ε-amino group of a Lys residue at its active site as a nucleophile that catalyzes a nucleotidyl transfer reaction. Thus ε-NH_2 group of active site lysine exerts a nucleophilic attack on phosphorus atom bonded to 5′ oxygen of the adenosine group of NAD^+ or α-P of ATP, leading to the transfer of adenylyl group (AMP) onto it. This reaction leads to the formation of the energy-rich phosphoamide adduct called adenosine monophosphate-DNA ligase (AMP–DNA ligase) adduct, which is the first covalent intermediate in the reaction. The complex formation is accompanied by the release of NMN^+ (in case of *E. coli* DNA ligase using NAD^+ as AMP donor) or pyrophosphate (in case of T4 DNA ligase using ATP as AMP donor).

Adenylyl Activation of 5′-Phosphoryl Terminus of the DNA Molecule/Nick

In the next step, the phosphoryl group of energy-rich phosphoamide linkage in AMP–DNA ligase adduct is attacked by the free 5′-phosphate group of the nick or the other DNA molecule (whichever may be the case). This leads to the displacement of the enzyme. In this case, AMP is linked to the 5′ nucleotide via a pyrophosphate rather than the usual phosphodiester bond resulting in the formation of an energy-rich second covalent intermediate, adenosine diphosphate-DNA (ADP–DNA) adduct.

Joining by the Formation of Phosphodiester Bond

The activated 5′-phosphate of ADP–DNA adduct then undergoes nucleophilic attack by the free 3′-OH group of the nick or on the other DNA molecule, resulting in the formation of a new phosphodiester bond. Thus, in case the two DNA molecules are blunt ended, these are simply joined and if the two DNA molecules are annealed due to compatible sticky ends with four nicks left, these are sealed (one in each reaction step) and AMP is released as a reaction product.

The above reaction mechanism is based on the isolated covalent intermediates, reversal of each step, and steady-state kinetic analysis, and is diagramatically depicted in Figure 13.4.

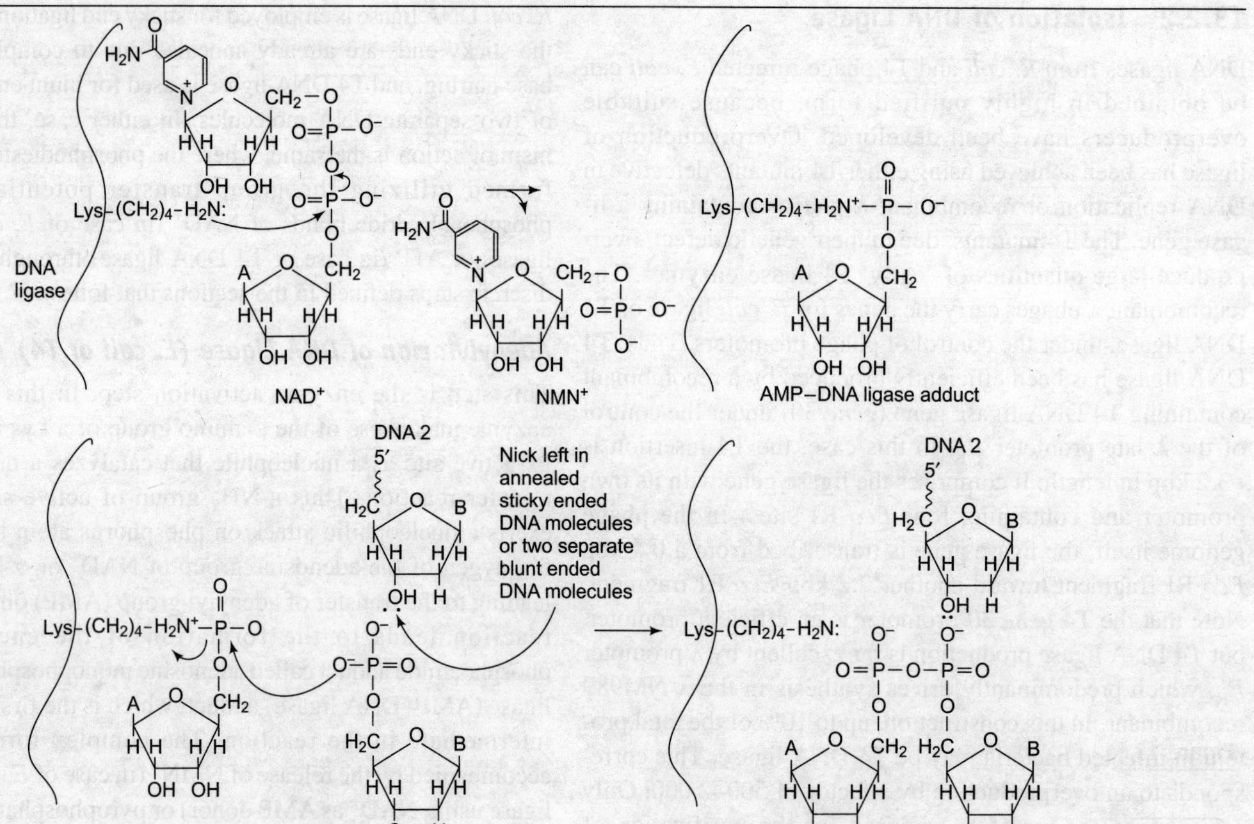

Figure 13.4 Reaction mechanism of DNA ligase [A is adenine; B is any base; cofactor is NAD^+ for *E. coli* DNA ligase, and ATP (not shown here) for T4 DNA ligase; leaving group is NMN^+ in *E. coli* DNA ligase-catalyzed ligation and PP_i (not shown here) in T4 DNA ligase-catalyzed reaction.]

The overall reaction catalyzed by DNA ligase (in terms of cloning experiment) is thus explained by the following equation:

$$2\,(\text{DNA molecules}) + \text{NAD}^+\,(\text{or ATP}) \xrightarrow[\text{DNA ligase}]{E.\ coli\ (\text{or T4})} (\text{DNA})_{\text{ligated}} + \text{NMN}^+(\text{or PP}_i) + \text{AMP}$$

The absolute requirement for the 5′-phosphate is extremely important; by removing the 5′-phosphate, the occurrence of unwanted ligation could be prevented.

13.2.4 *In vitro* Reaction Conditions

Under *in vitro* conditions, DNA ligase is employed in cloning experiments for recombinant DNA formation by either

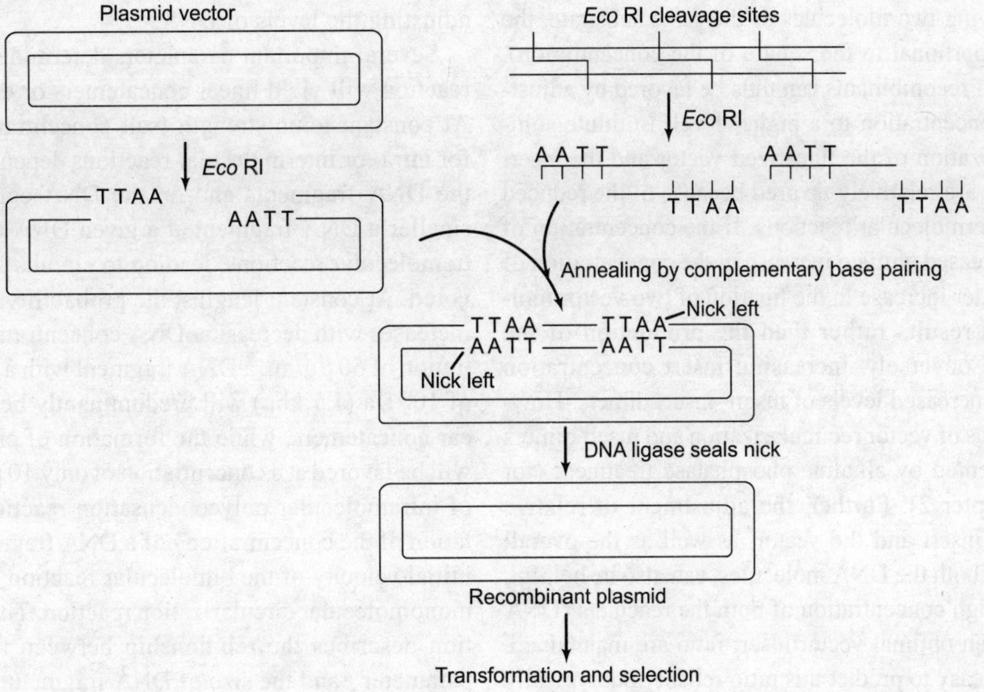

Figure 13.5 Recombinant formation using sticky end ligation

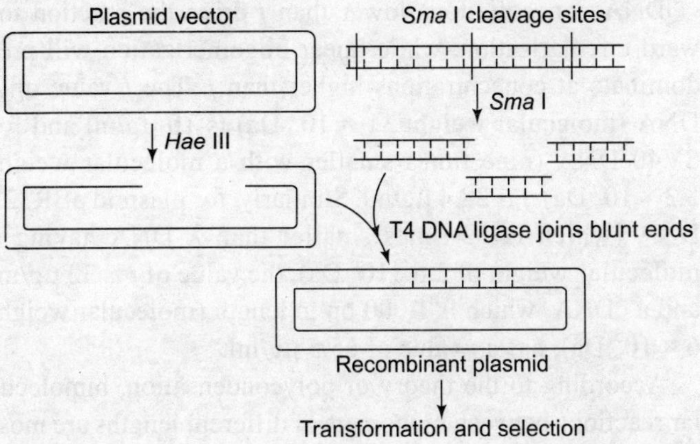

Figure 13.6 Recombinant formation using blunt end ligation

sticky end ligation or blunt end ligation using *E. coli* or T4 DNA ligase, respectively (Figures 13.5 and 13.6). Several factors listed in the following sections, however, affect these ligation reactions.

Temperature

Reaction temperature is an important parameter, which influences ligase reactions. The optimal ligation temperature is 37°C. At this temperature, however, base pairing between complementary protruding ends of DNA fragments generated by restriction enzymes is very unstable, because such paired structures involve, at most, only four base pairs. For example, *Eco* RI generated termini associate through only four AT base pairs, and these are not sufficient to resist thermal disruption at such a high temperature. As a compromise, a much more favorable temperature will therefore be the one between the rate of enzyme action and sufficient stability of base pairing between the protruding ends. This optimum temperature also increases with increasing G+C content at the site of ligation. Hence for sticky end ligation, a lower temperature range from 4°C to 16°C is preferable. Many protocols now recommend 16°C, but this is inconvenient unless cooled water bath is available. Thus, 4°C, which is achieved simply by performing the ligation reaction in the refrigerator, may sometimes be preferred; however, enzyme activity is low at this temperature, and longer incubation times are required. Blunt ends have less affinity for one another and their ligation is aided by very low temperature (4°C) and the presence of additional ATP in the reaction.

DNA Concentration and Length of Insert DNA

As the cloning experiments involve bimolecular reaction, in which one end of a DNA molecule (insert) reacts more frequently with one end of another DNA molecule (vector) than with its own other end (i.e., intermolecular ligation), these procedures are expected to be extremely sensitive to concentration of both the reactants. Such reactions thus follow second-order kinetics, indicating that the fraction of recombinants should increase with increasing concentrations of both the reactants (DNA molecules). At low concentrations of DNA, it is more likely that the two ends of the same molecule join (an intramolecular ligation), since the rate of a reaction involving one component will be linearly related to its concentration, whereas a reaction involving two different components will be proportional to the product of the same concentrations (or for a reaction involving two molecules of the same substrate, the rate will be proportional to the square of the concentration). The formation of recombinants can thus be favored by adjusting the DNA concentration to a higher level. In dilute solutions, recircularization of the linearized vector and the insert dimer formation are relatively favored because of the reduced frequency of intermolecular reactions. If the concentration of the vector is increased with no increase in the concentration of the insert, a greater increase in the ligation of two vector molecules together results rather than the production of the recombinants. Conversely, increasing insert concentration would result in increased levels of insert–insert dimers. However, the problems of vector recircularization and insert dimers can be circumvented by alkaline phosphatase treatment (for details see Chapter 2). Further, the adjustment of relative amounts of the insert and the vector, as well as the overall concentration of both the DNA molecules, can also be helpful, i.e., an overall high concentration of both the reactants (DNA molecules) and an optimal vector:insert ratio are maintained. Though it is not easy to predict this ratio reliably but typically the molar ratios (that take account of the relative size of the vector and insert) should range from 3:1 or 1:3. For example, if the vector is 4 kbp (4,000 bp) and the insert is 400 bp, then a 1:1 molar ratio would involve 10 times as much vector, by weight, as the insert (e.g., 400 ng of vector and 40 ng of insert). Conversely, for the same 4 kbp vector but a 40 kbp insert, if 400 ng of vector is used, 4,000 ng of insert is needed to achieve a 1:1 molar ratio. In general, to convert the amount of DNA by weight into a value that can be used to calculate the molar ratio, the amount used (by weight) is divided by the size of DNA (in bp or kbp). Thus, if W_V and W_I are weights of the vector and the insert DNA, respectively, and S_V and S_I are the sizes of vector and insert DNA, respectively, then the vector:insert ratio is given by

$$W_V / S_V = W_I / S_I \qquad (13.1)$$

A further complication arises if a heterogeneous collection of potential insert fragments is used, for example, while making a gene library. Intramolecular ligation of the vector molecules, or formation of vector dimers, results in the formation of transformants lacking the insert DNA, while addition of too much insert DNA results in the formation of insert dimers (which cannot be cloned) or recombinant vectors carrying multiple inserts. In other words, recombinant vectors that carry two or more completely different pieces of DNA are produced, leading to seriously misleading results when characterizing the clones in the library. So, adjusting the relative amount of the vector and insert not only influences successful ligation of the insert and the vector, but also has an effect on the nature of the products formed. Fortunately, as discussed earlier, alkaline phosphatase can be used to prevent such problems instead of relying entirely on adjusting the levels of DNA.

Several important parameters determine whether a ligase reaction will yield linear concatemers or circular structures. At constant ionic strength (salt concentration), preferences for intra- or intermolecular reactions depend on the length of the DNA fragments and on the DNA concentrations. The smaller a DNA fragment at a given DNA concentration, intramolecular reactions, leading to circularization, will be favored. At constant lengths, the probability of circularization increases with decreasing DNA concentrations. At a concentration of 50 μg/ml, a DNA fragment with a molecular weight of 10^6 Da (1.7 kbp) will predominantly be converted to linear concatemers, while the formation of circular monomers will be favored at a concentration of only 10 μg/ml. The theory of intramolecular polycondensation reactions allows calculation of the concentration *j* of a DNA fragment, at which the initial velocity of the bimolecular reaction equals that of the monomolecular circularization reaction. The following equation describes the relationship between the concentration parameter *j* and the size of DNA fragment:

$$j \text{ (in g/l)} = 51.1 \times M_r^{-1/2} \qquad (13.2)$$

DNA concentrations lower than *j* drive the reaction toward circularization, while linear oligomerization will predominate at concentrations higher than *j*. The *j* value of λ DNA (molecular weight 31×10^6 Da) is 10 μg/ml and for SV40 DNA (nine times smaller with a molecular weight 3.3×10^6 Da), *j* is 28.4 μg/ml. Similarly, for plasmid pBR322 DNA (which is 12 times smaller than λ DNA having a molecular weight of 2.6×10^6 Da), the value of *j* is 32 μg/ml and a cDNA, which is 1,000 bp in length (molecular weight 6×10^5 Da), has a *j* value of 65.5 μg/ml.

According to the theory of polycondensation, bimolecular reactions between monomers of different lengths are most favored when the two reactants are present in equimolar amounts. Hence, it follows that the length of the shortest DNA fragment will always determine the concentration of vector ends required to compete with the monomolecular circularization of the shortest fragment. The relationship between the concentration of vector DNA and a foreign DNA to be cloned is expressed as

$$\text{Foreign DNA (in μg/ml)} = \frac{M_r \text{ (foreign DNA)} \times \text{vector DNA (in μg/ml)}}{M_r \text{ (vector DNA)}} \qquad (13.3)$$

The following example from a cDNA cloning experiment in plasmid pBR322 clarifies this proportion. The minimal vector concentration *j* for pBR322 is 32 μg/ml. An equivalent molar concentration of a cDNA, which is 1,000 bp in length, will be $(0.6 \times 10^6)/(2.6 \times 10^6) \times 32 = 7.38$ μg/ml. This amount, however, is considerably less than the *j* value obtained by applying

Table 13.2 Minimal Vector DNA Concentrations for the Cloning of Foreign DNA

Insert size (kbp)	Vector DNA concentration (μg/ml)	
	λ DNA	pBR322 DNA
0.5	9,750	813
1	3,447	286
2	1,218	102
3	663	55
4	430	36
5	308	26
10	109	9
15	59	5
20	38	3.2
30	21	1.8

equation 13.2 (65.6 μg/ml). Note that at concentrations lower than 65.6 μg/ml, the undesired circularization of this cDNA will be favored, so its concentration must be raised to at least 65.6 μg/ml. In order to obtain equimolar conditions again, the concentration of vector DNA will accordingly have to be raised to $65.5/7.38 \times 32 = 284$ μg/ml. Table 13.2 lists the concentrations of λ and pBR322 DNAs that are required for an efficient competition with the free ends of foreign DNA fragments if reaction conditions are such that the rates of the monomolecular circularization reactions and the rates of the bimolecular reactions between vector and insert DNA molecules are equal. Thus, for obtaining necessary concentration of such ends to drive the reaction, DNA concentration must be higher for larger vectors. The vector concentrations obtained through these calculations are usually much higher than the j values (e.g., 284 μg/ml as compared to 32 μg/ml for pBR322 DNA) and hence the bimolecular reaction is favored. By definition, the j values for foreign DNA obtained by applying Eq. (13.1) specify DNA concentrations at which bimolecular and monomolecular reactions occur at the same rate. Experimental data demonstrate that the calculated j values should be at least two times higher to ensure that an efficient bimolecular reaction between vector DNA and foreign DNA is accomplished. In practice, however, such high concentrations often are unnecessary because a number of procedures have been developed that allow competing monomolecular circularization reactions to be eliminated. DNA fragments smaller than 200 bp, for example, circularize very inefficiently. It is therefore quite unnecessary to use extremely high vector concentrations in order to compensate for the self-circularization of the foreign DNA when the insert DNA is small. An alternative approach is to prevent self-circularization of the vector itself. Since ligase reactions require 5′-phosphate and 3′-OH ends, the method of choice is to remove the 5′-phosphate termini of linearized vector DNAs by treatment with alkaline phosphatase (for details see Chapter 2). Such DNA molecules can neither perform intermolecular nor intramolecular reactions with themselves. The foreign insert DNA can only provide the lacking 5′-phosphate residues and hence circularization can only be achieved by means of a bimolecular recombination event. Although this leaves two open phosphodiester bonds in each molecule, these molecules still transform bacteria quite efficiently because cellular enzymes subsequently complete the double strand *in vivo*.

ATP Concentration (In Case of T4 DNA Ligase)

The T4 DNA ligase is capable of catalyzing the reactions involving substrates of types A, B, and C by itself (Figure 13.3). This has been demonstrated by cloning of the T4 DNA ligase gene in λ phages and by the subsequent identification of the gene product in thermally induced *E. coli* lysates, which did not contain any phage-encoded proteins other than T4 DNA ligase. The specificity of T4 DNA ligase for type A, B, and C molecules can be switched to type A and B substrates simply by increasing ATP concentrations from 0.5 to 5 mM. At the elevated ATP concentration, the specificity of the enzyme for blunt ends is reversibly inhibited. An increase of the ATP concentration to 7.5 mM reversibly inhibits both activities. Therefore, DNA fragments with protruding and blunt ends can be ligated sequentially in the same assay mixture by altering the ATP concentrations appropriately.

Inhibitory Materials

Factors that may compromise the success of ligation reaction include the presence of inhibitory materials contaminating the DNA preparations and degrading the enzyme or the DNA (including loss of 5′-phosphate group). For a successful ligation, the 3′-OH and 5′-phosphate groups should be intact in the DNA molecules/nick.

13.3 LIGATION USING HOMOPOLYMER TAILING

Tailing involves using an enzyme terminal deoxynucleotidyl transferase (Tdt), commonly called terminal transferase (purified from calf, rat, or mouse thymus), to add a series of nucleotides onto 3′-OH termini of ds DNA molecule (for details see Chapter 2). If this reaction is carried out in the presence of just one dNTP, a homopolymer tail is produced. For example, Tdt adds a tail made entirely of deoxyguanosine to the 3′-end of ds DNA in a reaction catalyzed in the presence of only dGTP. This is an example of a homopolymer and is referred to as polydeoxyguanosine or poly (dG). Typically, 10–40 homopolymeric residues are added to each 3′-end of a ds DNA.

This homopolymer tailing serves as a general method of annealing (base pairing) two DNA molecules having complementary homopolymer tails. For example, by adding oligo (dA) sequences to the 3′-ends of one population of DNA molecules and oligo (dT) tracts to the 3′-ends of another population, the two types of molecules can anneal to form mixed dimeric circles (Figure 13.7). This further opened the avenue for ligation of insert to vector DNA in

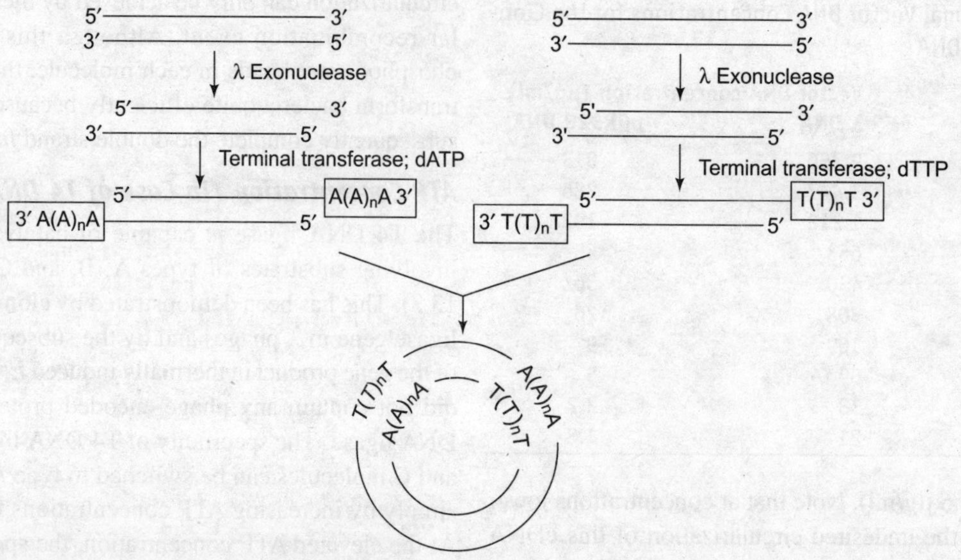

Figure 13.7 Cloning using homopolymer tailing method

cloning experiments. One of the earliest known examples that employed homopolymer tailing for the construction of recombinant molecules was the insertion of a fragment of λ DNA into SV40 viral DNA. Subsequently, the homopolymer method, using either dA.dT or dG.dC homopolymers was used extensively to construct recombinant plasmids for cloning in *E. coli*. Studies have indicated that DNA with exposed 3′-OH groups, such as arising from pretreatment with phage λ exonuclease or restriction with an enzyme such as *Pst* I (a sticky end cutter that generates 3′ overhangs), is a very good substrate for Tdt enzyme. However, conditions have been found in which the enzyme will extend even the shielded 3′-OH of 5′ cohesive termini generated by *Eco* RI. If the cleavage sites possess 5′ protruding ends, which is the case, for example, with ends generated by *Hind* III, *Eco* RI, and *Sal* I, etc., it is advisable to fill-in these ends with T4 DNA polymerase in the presence of dNTPs (for details see Chapter 2). The technique can also be employed if the enzymes used in the experiment generate blunt ends, such as *Hae* III, *Eco* RV, *Alu* I, and *Sma* I, etc.

In a gene cloning experiment, it is not possible to ensure that the two homopolymer tails (added on insert and vector DNA) are exactly of the same length. This leads to nicks or discontinuities/gaps in the base paired recombinant DNA molecule. Under *in vitro* conditions, the recombinant DNA molecule is repaired by a two-step process, in which Klenow polymerase fills the gaps and DNA ligase catalyzes the formation of final phosphodiester bonds. However, this repair reaction need not always be performed under *in vitro* conditions. This is because the recombinant DNA molecule with nicks or gaps (i.e., not completely ligated) contains the two components that are held together by base pairing between complementary homopolymer tails longer than ~20 nucleotides in size, which provide it stability at room temperature. When this recombinant DNA molecule is introduced into host cell (i.e., transformation), the gaps and nicks are repaired *in vivo* by bacterial DNA polymerase and DNA ligase, respectively (Figure 13.8).

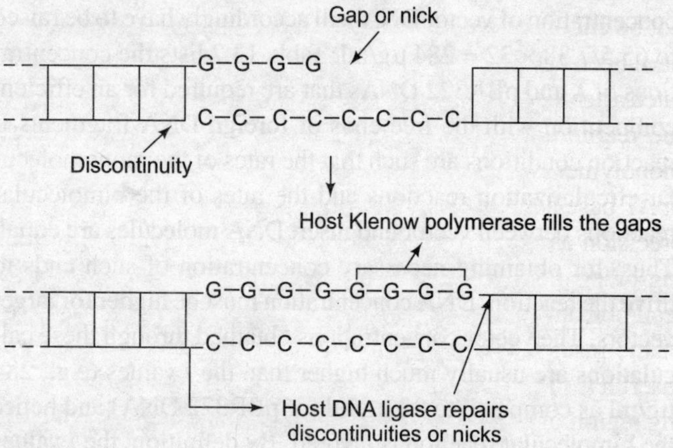

Figure 13.8 *In vivo* repair steps during cloning with homopolymer tailing method

Cloning by homopolymer tailing approach offers several advantages. The use of different but complementary homopolymeric tails in vector and foreign DNA clearly eliminates the competition between monomolecular and bimolecular reactions. The relative and absolute DNA concentrations in the ligase assay are thus of lesser importance. The advantage of using this strategy is that the vector does not reform without insertion of foreign DNA, as the ends of the vector are not complementary to one another. This technique also offers a radically different approach to the production of sticky ends on a blunt ended DNA molecule, thereby increasing the versatility of cloning strategies. It should be noted that homopolymeric tailing also suffers from certain disadvantages.

Exhibit 13.1 Application of Vaccinia DNA Topoisomerase in Joining DNA Fragments

A novel approach developed for the synthesis of recombinant molecules involves the application of a single enzyme, Vaccinia DNA topoisomerase. This ligase-free technique requires placement of the CCCTT cleavage motif for Vaccinia topoisomerase near the end of a ds DNA, which permits efficient generation of a stable, highly recombinogenic protein DNA adduct that can only religate to acceptor DNAs that contain complementary single strand extensions. A single enzyme thus catalyzes both the cleavage and rejoining steps. This technique has been successfully employed for insertion of linear DNAs containing CCCTT cleavage sites at both ends, activated by topoisomerase, into a plasmid vector. This topoisomerase-mediated joining offers certain advantages over DNA ligase-mediated ligation. It requires low ligation times (5 min) as compared with overnight incubation (at 4°C) for DNA ligase. The method is particularly suited for cloning PCR fragments. A linearized vector with single 3′-T extensions is activated with the topoisomerase. On addition of the PCR product with 3′ overhangs, ligation is very rapid. In addition, the high substrate specificity of the enzyme means that there is a low rate of formation of vectors without inserts.

Sources: Shuman S (1994) *J. Biol. Chem.* 269: 32678–32684; Heyman JA *et al.* (1999) *Genome Res.* 9: 383–92.

Firstly, the vector DNA must be completely intact, because Tdt enzyme also attacks free internal 3′-OH groups. This leads to the formation of branched structures and inactivates the vector. Secondly, tailing frequently destroys the restriction enzyme cleavage site on the vector DNA so that the inserted DNA cannot be cut out in an inert form, however, now several methods have been devised that allow the restriction site to be reconstituted after the addition of tails and cloning of the DNA. Thirdly, the recombinant DNA molecule constructed by using homopolymer tailing approach contains a variable number of GC or AT base pairs at either end of the insert, thereby lacking the precision associated with the formation of a recombinant using other methods. This is a disadvantage if the insert from recombinant vector is to be recovered. For example, if recloning of the insert DNA in another vector is to be performed, the insert has to be released by using restriction sites in the flanking region of the vector. However, the amplification of the insert DNA is not a problem, where the ends of the insert need to be sequenced and this information is used to design polymerase chain reaction (PCR) primers for amplification. In recent years, homopolymer tailing has been largely replaced as a result of the availability of a much wider range of restriction endonucleases and other DNA-modifying enzymes; however, the method is still important for cDNA cloning.

Besides, the approaches that utilize DNA ligase for the synthesis of recombinant molecules, a ligase-free technique using Vaccinia DNA topoisomerase has also been developed by Shuman (1994) and Heyman *et al.* (1999), the details of which are presented in Exhibit 13.1.

13.4 INCREASING VERSATILITY AND EFFICIENCY OF LIGATION

In a DNA ligase-mediated ligation reaction, the restriction fragments with sticky ends are useful as these are readily ligated with higher efficiency. However, there is a limitation on their usefulness, as these can only be ligated to another fragment with compatible ends. So an *Eco* RI fragment can be ligated to another *Eco* RI fragment, but not to the fragment generated by *Bam* HI. Thus, the advantage of cohesive ends is not only lost, but the unpaired DNA also intervenes, and so, at times, blunt end ligation may also be preferred. This is a potential disadvantage, as the vector used will only have a limited number of possible sites into which DNA can be inserted, and it may not be possible to generate a sensitive insert with the same enzyme and very often, it is necessary to ligate DNA fragments having different and incompatible ends, or blunt ends with staggered ends. In order to overcome such problems in cloning, changing a specific restriction site at will may be advantageous, i.e., one sticky end is altered into another sticky end, or one blunt end into another blunt end, or one sticky end into blunt end or vice versa. Such manipulations involve filling-in missing bases complementary to the unpaired ends, or digesting away the single stranded regions, or by using linkers and adaptors that add new restriction enzyme sites to the ends of DNA fragments. These modifications of restriction fragment ends make the cloning process easier and more versatile. The strategies that enable even the incompatible DNA fragments generated in different ways to be joined together include: (i) Trimming back and filling-in; (ii) Use of linkers; (iii) Use of adaptors; and (iv) Homopolymer tailing. The modification of DNA ends by homopolymer tailing and its applicability in gene cloning has already been discussed in Section 13.3, while a discussion of other techniques is given below.

13.4.1 Trimming Back and Filling-in

Trimming back and filling-in strategies are employed for converting a sticky end into a blunt end.

Enzymes Used in Trimming Back and Filling-in

Incompatible DNA fragments with recessed ends can be ligated if the termini are first converted into blunt ends. This can be achieved either by digesting protruding strands or by filling-in protruding tails with complementary nucleotides. Single stranded DNA (ss DNA) regions are most conveniently digested with the single strand-specific endonuclease S1 from *Aspergillus oryzae*, which is capable of digesting 3′ as well as 5′ protruding ends and generates 3′-OH and 5′-phosphate termini, respectively. The resulting flush ends are suitable substrates for ligases. Instead of removing nucleotides from the ends of a DNA molecule, it is also possible to fill-in single stranded 5′-ends with T4 DNA polymerase. Although this enzyme possesses a 3′ → 5′ exonuclease activity, which is active under these circumstances, there will be a turnover of only the terminal nucleotide in the presence of suitable (complementary) deoxyribonucleoside triphosphates. T4 DNA polymerase is preferred to S1 nuclease, because, due to contamination with double strand-specific activities, the S1 reaction often leads to ill-defined reaction products. Alternatively trimming back can be done by Klenow fragment of *E. coli* polymerase I or even by avian myeloblastosis virus (AMV) reverse transcriptase. The processes of trimming back and filling-in have already been discussed in detail in Chapter 2.

Application in Cloning

This filling-in strategy for joining of two incompatible fragments has been successfully used in cloning experiments. For example, filling-in of an *Eco* RI cohesive end leads to formation of a blunt end that can be easily ligated to a vector DNA molecule cut with a blunt end cutter (Figure 13.9).

The filling-in strategy can also be used for creating new restriction enzyme sites. Figure 13.10 shows that after filling-in the cohesive ends produced by *Eco* RI, and ligating the products together produces restriction sites recognized by four other enzymes. Many other examples of creating new target sites by filling-in and ligation are known.

In another example, for the cloning of a blunt DNA fragment generated by cleavage with *Fnu* DII (CG↓CG) and *Rsa* I (GT↓AC) into a pBR322 derivative pDD52 (containing unique sites for the enzymes *Xba* I, *Eco* RI, *Xho* I, and *Hind* III), the vector is cleaved with *Eco* RI and *Xba* I (Figure 13.11), the ends are then filled-in and ligated with the *Fnu* DII–*Rsa* I fragment. The desired recognition sequences (*Eco*

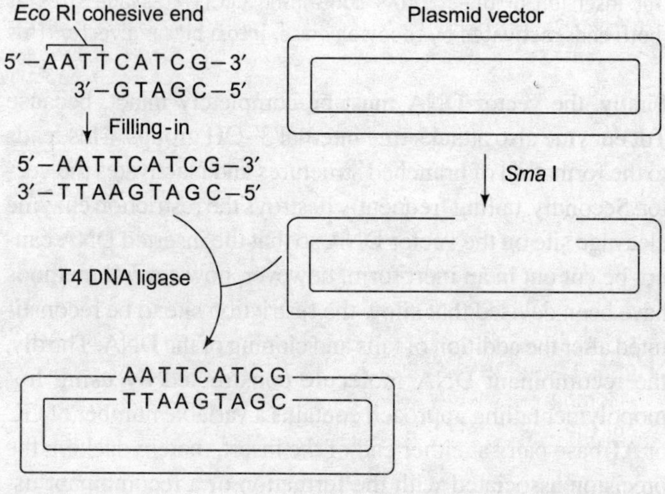

Figure 13.9 Conversion of sticky end into blunt end by filling-in

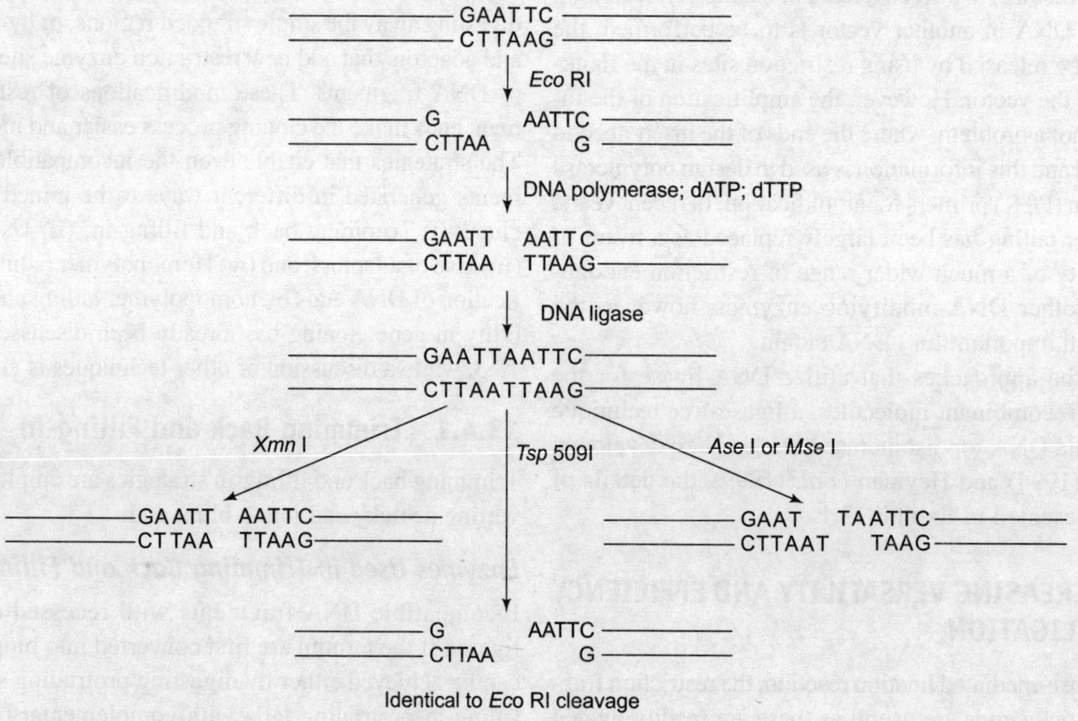

Figure 13.10 Creation of new restriction enzyme sites by filling-in [Three new restriction enzyme sites are created by filling-in of the cohesive ends generated by *Eco* RI and ligating the products together. There are two target sites for *Tsp* 5091 that lie 4 bp apart in the reconstituted molecule.]

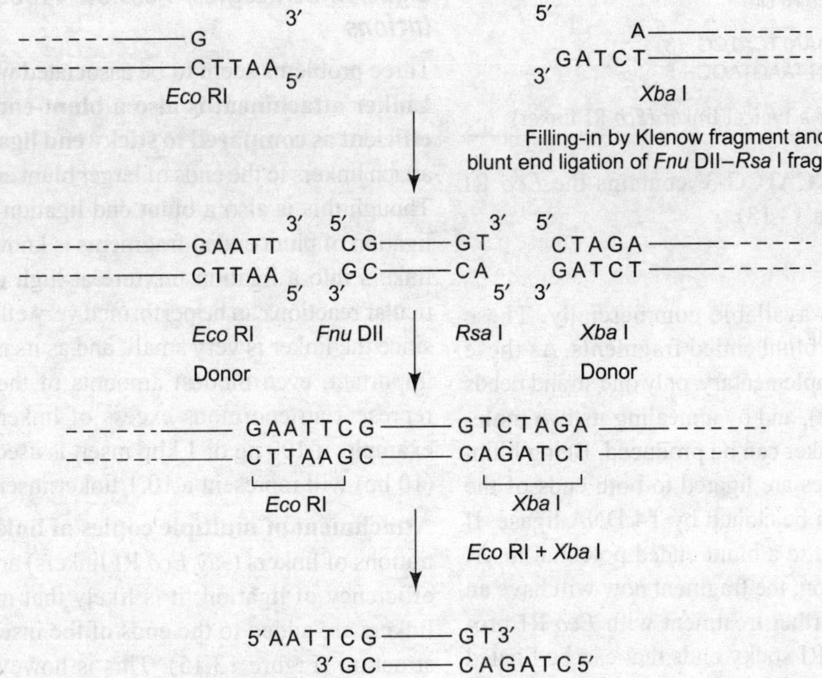

Figure 13.11 Cloning of a DNA fragment with blunt *Fnu* DII–*Rsa* I ends in a donor plasmid with filled-in *Eco* RI or *Xba* I termini [After cloning, the inserted DNA fragment can be cut out from the plasmid with *Eco* RI or *Xba* I. This approach does not alter the structure of the 3′-ends of the inserted DNA fragment.]

RI and *Xba* I) are restored if the DNA fragment is inserted in the correct orientation. The cloned fragment can then be cut out from the vector DNA using these two enzymes. It should be noted, however, that the two 5′-ends of the cloned DNA have now been structurally altered and carry the extensions derived from the *Eco* RI/*Xba* I cleavage. The 3′-termini remain unchanged, single strands of fragments cloned this way can thus be used as primers for DNA sequencing without any restrictions.

For certain applications, it becomes mandatory to isolate the cloned DNA fragment from a recombinant DNA molecule, thereby requiring the restriction site to be reconstructed during the end manipulation and ligation reactions. For example, if the four protruding bases of a hexameric cleavage site are filled-in, five of the six base pairs are restored. The full restoration depends upon a careful selection of the terminal base pair of the DNA used for ligation in such a way that it supplies the missing sixth base pair (Figure 13.12). The example shown in the figure involves filling-in of *Eco* RI site (5′-G↓AATTC-3′) followed by the annealing of the flush ends thus generated with a double stranded molecule carrying a 5′-terminal C provided by DNA fragments digested with *Hae* III (5′-GG↓CC-3′), leading to the restoration of *Eco* RI sites. Flush ends required for ligation can also be obtained by filling-in of donor DNAs. There are many other combinations of such donors and acceptors that allow cleavage sites to be reconstituted.

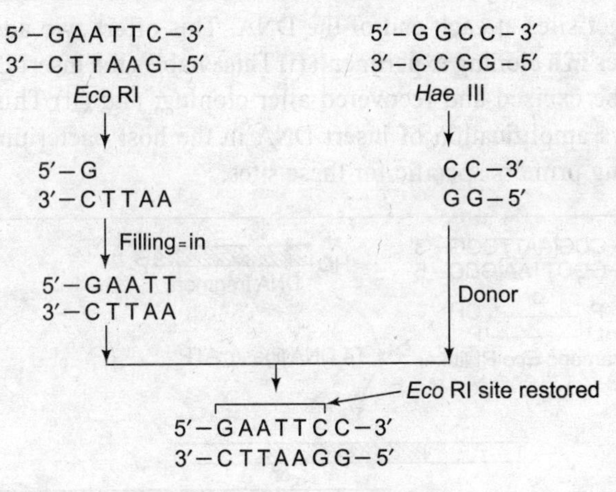

Figure 13.12 Filling-in of the protruding ends converts a sticky end into blunt end and its ligation with another blunt ended fragment may reconstitute the original restriction site [Filling-in of *Eco* RI site converts a sticky end into a blunt end, which can be easily ligated with the fragments generated by a blunt end cutter, *Hae* III. The ligation of filled-in *Eco* RI acceptor site and *Hae* III restores the *Eco* RI recognition site.]

13.4.2 Linkers

In order to make ligation much more efficient and versatile, the use of linkers is recommended. These are chemically synthesized short pieces of DNA molecules of known nucleotide sequence. These are blunt ended, self-complementary, and contain an inherent restriction enzyme site. For example, the

Eco RI site

5′– CGATGAATTCATCG – 3′
3′– GCTACTTAAGTAGC – 5′

Figure 13.13 An example of a typical linker (*Eco* RI linker)

sequence 5′-CGAT**GAATTC**ATCG-3′ contains the *Eco* RI site 5′-**GAATTC**-3′ (Figure 13.13).

Application in Cloning

A wide range of linkers is available commercially. These linkers are used for joining blunt ended fragments. As these synthetic linkers are self-complementary, only one strand needs to be synthesized (or bought), and by annealing its two molecules together, a ds DNA linker can be produced. Being blunt ended, these linker molecules are ligated to both ends of the blunt ended foreign DNA to be cloned by T4 DNA ligase. If the linker molecule is joined to a blunt ended potential insert fragment by blunt end ligation, the fragment now will have an *Eco* RI site at each end. Further treatment with *Eco* RI produces a fragment with *Eco* RI sticky ends that can be ligated with an *Eco* RI cut vector (Figure 13.14). Linkers can also be attached to the vector DNA in order to increase the versatility of use of restriction enzymes in cloning.

Insertion by means of the linker creates restriction enzyme target sites at each end of the DNA. This offers two advantages in a cloning experiment: (i) This enables the insert DNA to be excised and recovered after cloning; and (ii) This allows amplification of insert DNA in the host bacterium by using primers specific for these sites.

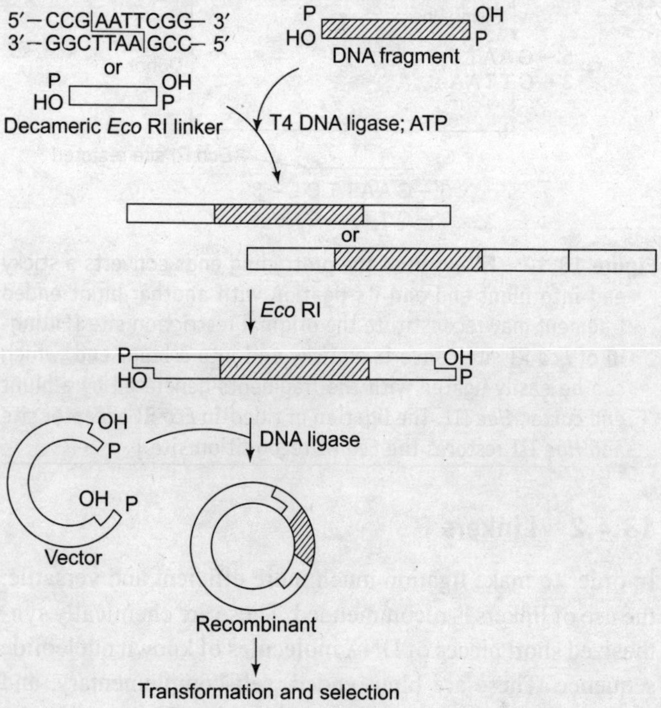

Figure 13.14 Cloning using linkers

Ligation Strategies, Possible Problems, and their Solutions

Three problems seem to be associated with the use of linkers: **Linker attachment is also a blunt-end ligation that is less efficient as compared to sticky end ligation** DNA ligase can attach linkers to the ends of larger blunt ended DNA molecules. Though this is also a blunt end ligation and the efficiency of ligation of blunt ended fragments is known to be low, by using linkers into a ligation mixture at high concentrations, a particular reaction can be performed very efficiently. Furthermore, since the linker is very small, and as its molar concentration is important, even modest amounts of the linker by mass will represent an enormous excess of linker in molar terms. For example, if 100 ng of 1 kbp insert is used, then 10 ng of linker (10 bp) will represent a 10:1 linker:insert ratio.

Attachment of multiple copies of linkers As high concentrations of linkers (say *Eco* RI linkers) are used to increase the efficiency of ligation, it is likely that multiple copies of the linkers are added to the ends of the insert, producing a chain structure (Figure 13.15). This is however not really a problem as the subsequent restriction digestion will remove them. The digestion produces a large number of cleaved linkers and the original DNA fragment, now carrying *Eco* RI sticky ends. The cleaved linkers are size-fractionated and the modified fragment is ready for ligation into a cloning vector restricted with *Eco* RI.

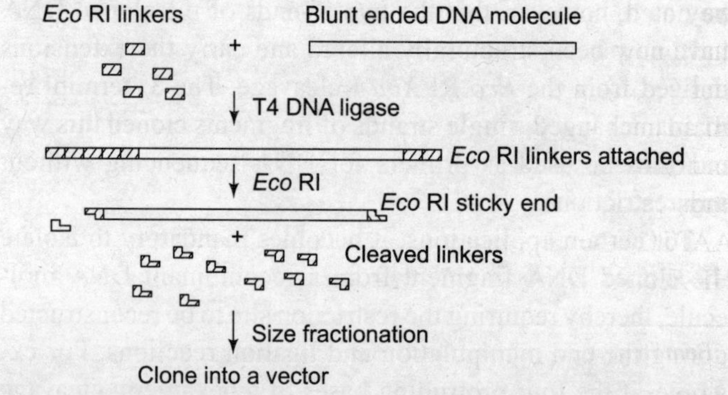

Figure 13.15 Dimer or multimer formation is not a problem with the use of linkers

Cleavage at internal sites within the DNA upon digestion after linker attachment Another problem arises if the insert DNA already has internal *Eco* RI site(s), then cloning using linkers is not possible. In the situation of possibility of cleavage of insert DNA at internal sites during digestion of attached linkers, the insert DNA will be cloned as two or more subfragments (Figure 13.16). This problem can however be overcome by any of the following three procedures: (i) Linkers with restriction enzyme sites other than the ones present in the insert DNA should be selected. This is a solution of choice if the insert DNA is small and its sequence is known.

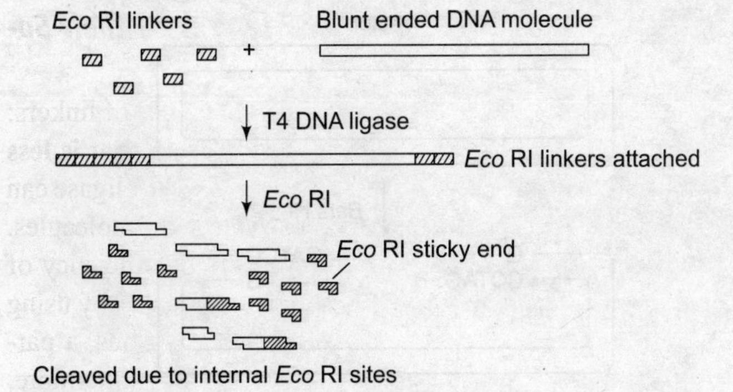

Figure 13.16 Internal sites for the same restriction enzyme in insert DNA lead to its cleavage while using linkers in cloning

This solution is however not suitable if the foreign DNA is large and has sites for several restriction enzymes; (ii) Internal restriction sites in insert DNA should be methylated. This can be done with appropriate modification methylase, for example, *Eco* RI MTase is used for methylation of internal sites while using *Eco* RI linkers (see Maniatis strategy for genomic library construction, Chapter 15); and (iii) Chemically synthesized adaptors with preformed cohesive ends should be used instead of linkers (for details see Section 13.4.3).

13.4.3 Adaptors

Adaptors, like linkers, are short synthetic oligonucleotides that also increase the versatility of ligation. Two pairs of such oligonucleotides are designed to anneal together in such a way as to create a short ds DNA fragment with one sticky and one blunt end [e.g., the sequences 5′-GATCCCGG-3′ and 5′-CCGG-3′ anneal to produce a double stranded *Bam* HI adaptor with a *Bam* HI sticky end on one side and a blunt end on the other (Figure 13.17a)], or with two different sticky ends [e.g., the sequences 5′-GATCCCGGG-3′ and 5′-AATTCCCGGG-3′ anneal to produce a double stranded *Bam* HI–*Eco* RI adaptor with a *Bam* HI sticky end at one end and an *Eco* RI sticky end on the other (Figure 13.17b)]. Both these adaptors have dephosphorylated 5′ sticky ends. Note that the only difference between an adaptor and a linker is that the former has cohesive end(s) and the latter has blunt ends. Hence, adaptors help in overcoming the problem of cleavage at internal sites associated with linkers, as no restriction digestion has to be performed after the ligation of adaptors to the DNA of interest. A wide range of adaptors is available commercially.

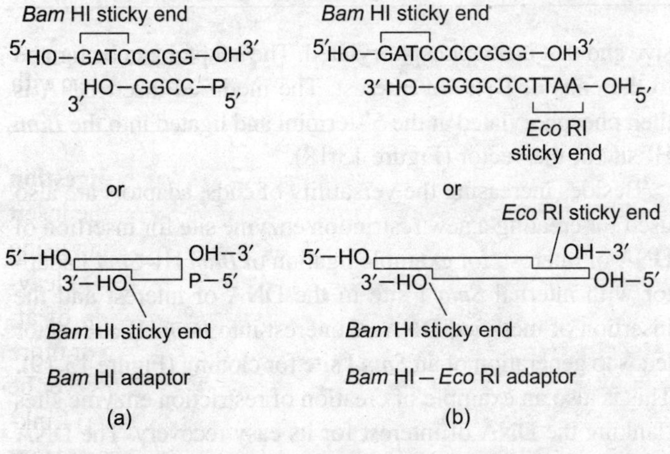

Figure 13.17 Examples of a typical adaptors; (a) *Bam* HI adaptor, (b) *Bam* HI–*Eco* RI adaptor

Application in Cloning

Adaptors are used for increasing the versatility of ends and cloning efficiency. These are also used for the creation of a novel restriction enzyme site for the introduction of insert DNA and for the creation of restriction enzyme sites flanking the insert DNA for its easy recovery from the recombinant DNA. Like with linkers, high molar concentrations of adaptors are used to drive ligation very efficiently. As the adaptors have nonphosphorylated sticky ends, it is ensured that multiple additions of the adaptors to the ends of the DNA of interest are prevented. This is because DNA ligase is unable to form a phosphodiester bond between 5′-OH and 3′-OH ends. The result is that, although base pairing always occurs between the sticky ends of two adaptor molecules, the association is never stabilized by ligation. Adaptors can therefore be ligated only to a DNA molecule but not to themselves. The adaptor with one blunt end and one sticky end, e.g., *Bam* HI adaptor, ligates the blunt ends of DNA fragments, such as those generated by

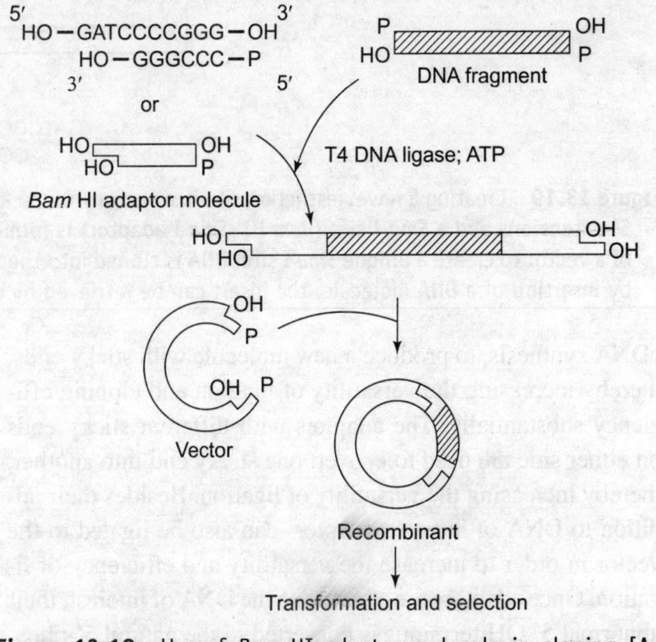

Figure 13.18 Use of *Bam* HI adaptor molecule in cloning [A synthetic adaptor molecule is ligated to the foreign DNA. The adaptor is used in the 5′-hydroxyl form to prevent self-polymerization. The foreign DNA plus ligated adaptors is phosphorylated at the 5′-termini and ligated into the vector previously cut with *Bam* HI.]

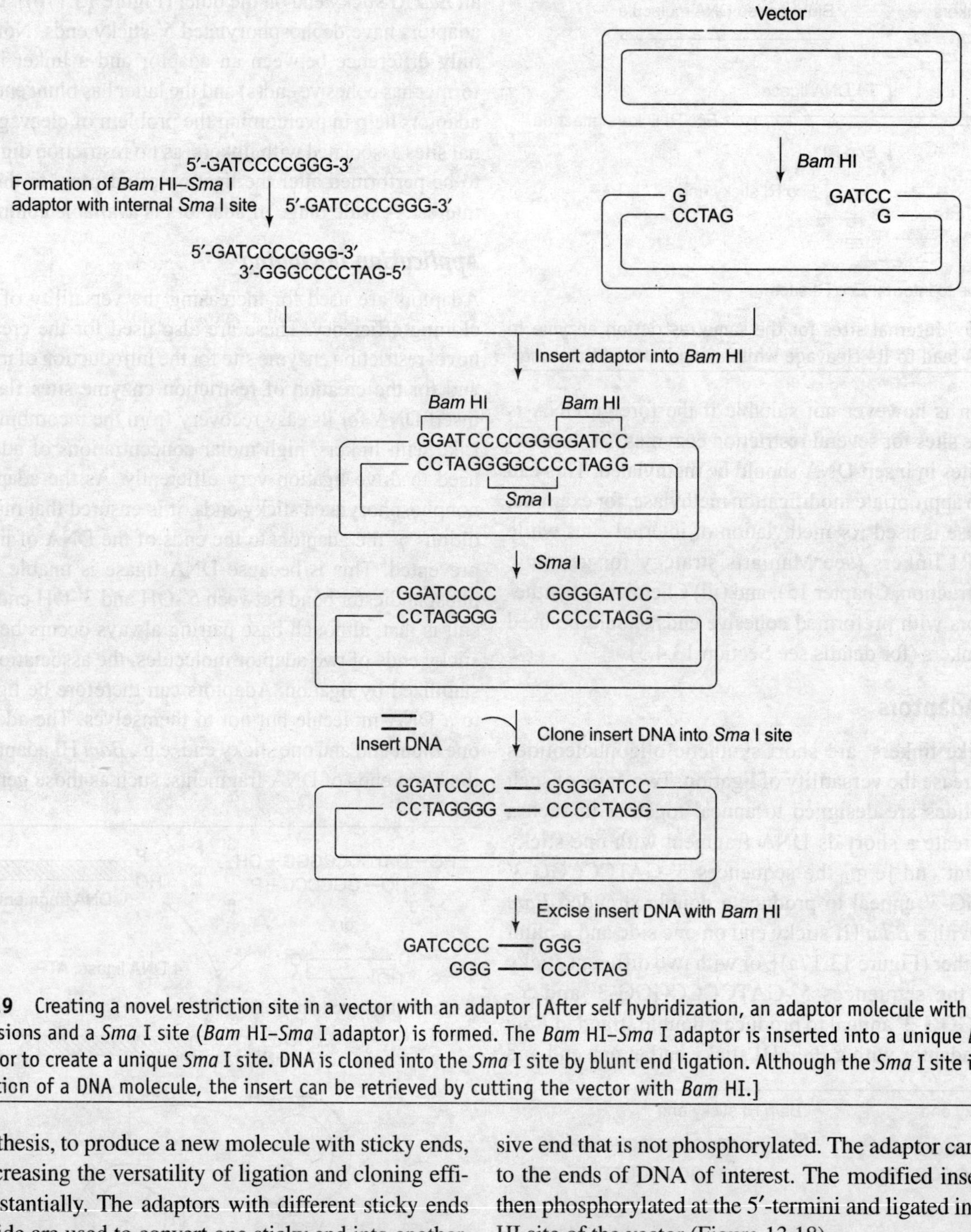

Figure 13.19 Creating a novel restriction site in a vector with an adaptor [After self hybridization, an adaptor molecule with two *Bam* HI 5′-extensions and a *Sma* I site (*Bam* HI–*Sma* I adaptor) is formed. The *Bam* HI–*Sma* I adaptor is inserted into a unique *Bam* HI site of a vector to create a unique *Sma* I site. DNA is cloned into the *Sma* I site by blunt end ligation. Although the *Sma* I site is destroyed by insertion of a DNA molecule, the insert can be retrieved by cutting the vector with *Bam* HI.]

cDNA synthesis, to produce a new molecule with sticky ends, thereby increasing the versatility of ligation and cloning efficiency substantially. The adaptors with different sticky ends on either side are used to convert one sticky end into another, thereby increasing the versatility of ligation. Besides their addition to DNA of interest, adaptors can also be ligated to the vector in order to increase the versatility and efficiency of ligation. Once adaptors are attached to the DNA of interest, their abnormal 5′-OH terminus is converted to the natural 5′-phosphate by treatment with polynucleotide kinase, producing a sticky ended fragment that can be inserted into appropriately cut vector. Consider a *Bam* HI site, which is to be cloned in a *Bam* HI cut vector. The *Bam* HI adaptor molecule has one blunt end bearing a 5′-phosphate group and a *Bam* HI cohesive end that is not phosphorylated. The adaptor can be ligated to the ends of DNA of interest. The modified insert DNA is then phosphorylated at the 5′-termini and ligated into the *Bam* HI site of the vector (Figure 13.18).

Besides increasing the versatility of ends, adaptors are also used for creating a new restriction enzyme site for insertion of DNA of interest, for example ligation of *Bam* HI–*Sma* I adaptor with internal *Sma* I site to the DNA of interest and the insertion of modified DNA of interest into *Bam* HI cut vector leads to generation of an *Sma* I site for cloning (Figure 13.19). This is also an example of creation of restriction enzyme sites flanking the DNA of interest for its easy recovery. The DNA of interest introduced in the modified vector can be recovered for subcloning by digestion with *Bam* HI.

Review Questions

1. Differentiate between *E. coli* DNA ligase and T4 DNA ligase.
2. Describe in detail the mechanism of action of T4 DNA ligase.
3. Why is adenylylation of 5′-end phosphoryl group required in a DNA ligase-catalyzed reaction?
4. What is the prerequisite for a sticky end ligation? Why is efficiency of blunt end ligation lower than sticky end ligation?
5. Enumerate various factors that affect the activity of DNA ligase. How can intramolecular ligation be prevented?
6. It is known that the size of homopolymer tails formed by terminal transferases is not always the same. Thus, in a cloning experiment, the size of dG tail added to the insert DNA may differ from the dC tail added to the vector molecule. Now, supposing the size of tail in each case exceeds 20 nucleotides and the annealed recombinant is stable, though it lacks few nucleotides and two nicks also remain unsealed, can this recombinant be transformed into a host cell? How does filling of gaps and sealing of nicks take place in this case?
7. What are the various advantages and disadvantages of cloning by homopolymer tailing?
8. What is the advantage of modifying restriction fragment ends? Enumerate various strategies of modification.
9. With suitable examples, explain the application of filling-in:
   - For modification of DNA ends, such as converting a sticky end into blunt end
   - For creation of new restriction enzyme sites
   - For reconstitution of the original restriction site
10. Why is T4 DNA polymerase preferred for trimming back?
11. What is a linker? How is it used in cloning experiments?
12. Differentiate between linkers and adaptors. Also explain the advantages of adaptors over linkers in cloning experiments.
13. Why are linkers and adaptors used in high molar concentrations in a cloning experiment?
14. What are the various problems associated with the use of linkers? How can these be circumvented?
15. Describe in detail the cloning strategies mediated by homopolymer tailing, linkers, and adaptors.
16. How can adaptors be employed for the creation of a novel restriction enzyme site for the introduction of the insert DNA? Explain with the help of suitable example.
17. What are the various strategies adopted for the recovery of the insert DNA from the recombinant DNA molecule?
18. Suppose the size of vector DNA is 5 kbp and that of insert DNA is 500 bp. In a ligation reaction, if 500 ng of vector DNA is used, calculate the amount of insert DNA for 1:1 molar ratio of vector and insert.

References

Heyman, *et al.* (1999). Genome-scale cloning and expression of individual open reading frames using topoisomerase I-mediated ligation. *Genome Res.* **9**:383–392.

Shuman, S. (1994). Novel approach to molecular cloning and polynucleotide synthesis using vaccinia DNA topoisomerase. *J. Biol. Chem.* **269**:32678–32684.

Introduction of DNA into Host Cells

Key Concepts

In this chapter we will learn the following:

- Introduction of DNA into bacterial cells
- Introduction of DNA into yeast cells
- Genetic transformation of plants
- Introduction of DNA into insect cells
- Genetic transformation of animals

14.1 INTRODUCTION

The introduction of biologically active recombinant DNA into host cells is a very important step in genetic engineering and holds a key position among the cloning protocols. Several techniques have been devised for the mobilization of recombinant DNA molecules into various types of host cells. These techniques fundamentally differ in their mechanisms, and have their own merits and demerits (Table 14.1). The term 'transformation' is used to describe nonviral DNA transfer in bacteria and nonanimal eukaryotic cells such as fungi, algae, and plants, while the term 'transfection' defines the process of introducing nucleic acids into cells by nonviral methods. The original meaning of transfection is 'infection by transformation', i.e., introduction of DNA (or RNA) from a eukaryote virus or bacteriophage into cells, resulting in an infection. Because the term transformation had another sense in animal cell biology (a genetic change allowing long-term propagation in culture, or acquisition of properties typical of cancer cells), the term transfection is acquired for animal cells that means: a change in cell properties caused by introduction of DNA. In this chapter, for convenience, the host cell receiving the foreign DNA by either transformation or transfection is referred to as 'transformant'.

Table 14.1 Advantages and Disadvantages of Transformation/Transfection Procedures

Transformation/ transfection procedure	Organism transformed	Advantages	Disadvantages
		Biological methods	
In vitro packaging and phage infection	Bacteria	The efficiency of the process of *in vitro* packaging is high (more than transfection).	There is host range limitation; Conditions should be maintained such that concatemers and multimers are formed.
Agrobacterium tumefaciens-mediated transformation	Plants	Exhibits natural tendency of gene transfer; Excellent, simple, convenient, and highly effective system; Technically simple (considered as poor man's vector); Ability to transfer >50 kbp long stretches of T-DNA; High frequency of transformation (0.1–5%); Wounded explants are readily transformed; Precision of transfer and integration of DNA sequences with defined ends; Excellent stability of expression of introduced transgene; Rapid production of transgenic plants;	Requirement of biological vector system; *Monocot plants were originally outside the host range of *Agrobacterium tumefaciens*, but now it is considered 'universal'; Cannot transform chloroplast and mitochondria.

(*Contd.*)

Table 14.1 (*Contd.*)

Transformation/ transfection procedure	Organism transformed	Advantages	Disadvantages
		Biological methods	
		Reliability and reproducibility of results; Linked transfer of gene(s) of interest along with the transformation marker; Foreign genes delivered by this method are usually transmitted to progeny plants in Mendelian manner; Transformed cells usually carry single or low copy number of T-DNA; Very little or no rearrangement; Reasonably low incidence of transgene silencing; Low chances of chimera formation; Cost-effective process with no requirement of skilled workers, and sophisticated instrumentation; *Earlier used to transform dicots that are natural hosts of *Agrobacterium*. Now it is suitable for a wide range of dicot plants as well as monocots.	
Plant virus-mediated transformation	Plants	Exhibits natural tendency of gene transfer; High efficiency of gene transfer; Amplification of transferred genes in lesser time (due to fast growth rate); Introduction of gene(s) into all the plant cells (due to systemic nature); High transgene copy number (due to high growth rate); High yield (due to high growth rate); Biological containment (due to nonintegrative nature and non-germ line transfer).	Requirement of biological vector; Low insert capacity, hence effective for cloning small cDNAs or genes.
Transposon-mediated transformation	Insects	No specific host factors are required for stable transformation; High frequency of transformation.	Requirement of biological vector.
Baculovirus-mediated transformation	Insects	High efficiency of gene transfer; It is an insect-specific agent and hence does not infect any other species.	Requirement of viral vector.
Viral transduction	Animals	High efficiency of gene transfer; Only a single copy of the viral provirus is integrated; The genomic DNA surrounding the transgenic locus generally remains intact.	Size constraints; Possible interference of viral regulatory elements with the expression of surrounding genes; Susceptibility of the virus to *de novo* methylation; Only useful for generating transgenic sectors of embryos and this is not the preferable method for producing transgenic animals.
YAC transgenesis	Animals	Large DNA segments can be transformed; No intrinsic cytotoxic effects; Expression and regulation of animal gene under the control of its own *cis*-acting regulatory element can be studied.	Requirement of biological vector.

(*Contd.*)

Table 14.1 (*Contd.*)

Transformation/ transfection procedure	Organism transformed	Advantages	Disadvantages
		Direct gene transfer by chemical and biochemical means**	
$CaCl_2$/heat shock	Bacteria, yeast	No requirement of biological vector; Circumvent the host range limitation.	The standard $CaCl_2$ technique utilizing unpackaged λ DNA yields at most 10^5–10^6 plaques per μg. This yield is lower than *in vitro* packaging procedure.
Ca–DNA coprecipitation	Plants, animals	No requirement of biological vector; Circumvent the host range limitation.	Usually allows transient expression; Cell toxicity; Prone to mutation.
Liposome packaging and fusion or endocytosis	Plants, animals	Free from host range limitation; No requirement of biological vector; Efficient in terms of DNA encapsulation and transfer; Enhanced delivery of encapsulated DNA by membrane fusion; Protection of nucleic acids from nuclease activity; Delivery into a variety of cell types besides protoplasts entry through plasmodesmata; Delivery of intact small organelle; Protect the genetic material from immunogenic reaction (liposomes are nonimmunogenic); Composition of lipids can be manipulated to confer specific properties to liposomes; These are easy to manipulate, their lipid contents can be varied at will, and many substances may be trapped inside the interlamellar spaces.	Although it is efficient in terms of DNA encapsulation and transfer, the method requires protoplasts, and its application is limited to species for which protocols are available for protoplast regeneration; Requires bath-type sonicator and rotary evaporator; Cell toxicity; Susceptible to interference by fats and lipoproteins in serum and by charged components on the extracellular matrix such as chondroitin sulfate (in case of animal lipofection).
DEAE–dextran transfection	Plants, animals	No requirement of biological vector; Circumvent the host range limitation.	Transient expression; Cell toxicity; Prone to mutation.
Polycation-mediated DNA uptake	Plants, animals	No requirement of biological vector; Circumvent the host range limitation; Polycations such as polyethylene glycol (PEG), Poly-*L*-lysine, Poly-*L*-ornithine, polybrene, etc. increase the adsorption of DNA to the surface, and hence efficiency of DNA uptake; Polybrene is less toxic than other polycations; The efficiency of transformation is high, which allows very small quantities of plasmid DNA to be used; The polybrene method can be used for both stable and transient expression.	PEG is toxic; Usually allows transient expression.
		Direct gene transfer by physical means**	
Microprojectile bombardment	Plants, animals	Simple, easy, and safe method; Reliable and reproducible method; DNA delivery appears to be genotype independent; Method is used successfully to transform monocots that were once impossible to transform by *Agrobacterium tumefaciens* based system. Method is used for wide range of monocots and dicot plants;	Gene transfer leads to nonhomologous integration into the chromosome, and is characterized by multiple copies, and some degree of rearrangement; Unpredictable pattern of integration of foreign DNA (due to integration by nonhomologous recombination); Possibility of emergence of chimera plants;

(*Contd.*)

Table 14.1 (*Contd.*)

Transformation/ transfection procedure	Organism transformed	Advantages	Disadvantages
		Direct gene transfer by physical means**	
		DNA delivery is tissue-independent. There is no need for protoplast isolation and DNA is able to penetrate intact cell walls; No requirement of any biological vector; Microprojectile is shot through the physical barriers of the cell; One shot leads to multiple hits; All the cells are not in the same physiological state after transformation; Particle reaches deeper cell layers allowing stable transformation.	Low transformation rate (0.01–1%); Lack of control over the velocity of bombardment often leads to substantial damage to the target cells; Transformed cells may be at a dis-advantage due to damage from the transformation procedure; Costly process, and requires specific sophisticated instrumentation, gold or tungsten microparticles, and skilled workers.
Electroporation	Animals, plants, bacteria, insects	Convenient, simple, efficient, safe, and fast method; Host range (genotype) independent; Applicable to wide range of monocots and dicots; Great degree of control and reproducibility; Free from biological obstacles as there is no requirement of biological vector; Even under optimal conditions the amount of DNA delivered to each cell is low, which has the benefit of producing transformants with a low transgene copy number; The delivery rate (the proportion of electroporated cells that actually receive DNA) can be high, in some experiments it has been shown that between 40% and 60% of the incubated cells receive DNA; Causes lower damage to cells as compared to gene gun. 50% of the cells survive the treatment; All the cells are in the same physiological state after transformation, unlike the situation with particle bombardment;	Requires sophisticated electroporator; Technically difficult to perform; Damage to 50% of the cells; Earlier the technique was restricted to transfection of protoplasts, which required systems for isolation of protoplasts and their regeneration to calli or whole plants (which is difficult). As a consequence the method was not applicable to some plant systems. However, now the method can be used for immature embryos also.
Microinjection	Animals, plants, insects	Convenient, simple, and safe method; Overcomes biological obstacles. The form of the DNA applied to the cells is controlled entirely by the experimenter and not by an intermediate biological vector; Independent of host range. Used to transform monocots and dicots with equal efficacy; No cell type restriction. DNA is able to penetrate intact cell walls; The amount of DNA delivered per cell is not limited by the technique and can be optimized thereby improving the chance for integrative transfor-mation; Delivery is precise into the cell and the results are predictable; Delivery can be done either in cytoplasm or in nucleus of target cell. Number and localization of recipient cell is known;	Technically difficult to perform. Handling requires specialized skill and instrumentation; Has limited usefulness because only one cell can be injected at a time; However, if hundreds of microinjections can be performed per working day, low transformation rate is yet not a serious disadvantage; Low numbers of manipulated cells; Slower process, and 150–200 cells can be injected in 10 min.

(*Contd.*)

Table 14.1 (*Contd.*)

Transformation/ transfection procedure	Organism transformed	Advantages	Disadvantages
		Direct gene transfer by physical means**	
		Direct delivery in nuclei of the target cell increases the chance for integrative transformation; No limitation for material to be transferred. Small structures can be injected containing only a few cells and with high regeneration potential, for example, microspore-derived and zygotic preembryos; DNA does not normally move from cell to cell, hence recombination or rearrangement occurs at relatively low frequency by this method; The process does not necessarily require a protoplast regeneration system; Very little sample volume (2 μl) required; Possibility of visual control on injection of target cell.	
Silicon carbide whiskers (silicon carbide fibers)-mediated transformation	Plants	Technically the method is very simple and easy to perform; Free from host range limitation; No requirement of biological vector; Very low equipment costs as only microfuge tubes and a vortex is sufficient equipment. No requirement of skilled personnel; Ability to transform walled cells (thus avoiding protoplast isolation).	Overall efficiency is around one-tenth that of conventional microprojectile bombardment; Requires careful handling as there are inherent dangers of the fibers; Possibility of damage to target cells.
Ultrasonication	Plants	Free from host range limitation; No requirement of biological vector; Equipment for ultrasound transformation is simple, cheap and multifunctional; Versatile cell types such as plant protoplasts, cells in suspension, and intact pieces of plant tissue can be readily transformed.	Costly and requires ultrasound generating machines.
UV laser microbeam irradiation	Plants	Free from host range limitation; No requirement of biological vector; Laser beam can be used inside the tissues, cells and organelles, thus increasing the rate of transformation, and the speed of operation; Wide controllability of the laser beam intensity, flexibility in optical manipulation, wide choice for laser wavelength, and pulse width are other advantages; Technique can be used for genetic manipulation of plant tissue, individual cells, protoplasts and organelles. Technique is also applicable to cells with hard walls; It permits selective fusion in a part of a subcellular organelle present inside a single living cell; It has the ability to transfer cytoplasmic material.	Costly and requires a UV laser microbeam source.

* For long, the inability of *Agrobacterium* to transfer DNA to monocotyledonous plants was considered its major limitation. Though, monocots produce inducers of *vir* gene expression and are competent to import T-complex into the nucleus, these are recalcitrant to infection. It may be due to that the proteins on the surface of *Agrobacterium* are unable to recognize receptors on the surface of monocot cell, thereby

preventing attachment and transfer of T-DNA. However, with effective modifications in pTi vectors and finer modifications of transformation conditions, a number of monocot plants including rice, wheat, maize, sorghum, and barley have now been transformed. In fact, *Agrobacterium* T-DNA transfer is now viewed as 'universal' based on successful transformation of yeast, *Aspergillus*, and human cells.

** Direct gene transfer methods show unpredictable pattern of foreign DNA integration. During their passage into the nucleus, the DNA is subjected to nucleolytic cleavage resulting in multicopy insertion, chimera formation, truncation, recombination, rearrangement, and silencing. Direct gene transfer methods do not require any biological vectors, and are free from host range limitation.

The DNA introduced into the host cell is either transiently expressed (where DNA introduced into the cell is eventually lost by dilution and degradation) or stably integrated into the host genome (where DNA introduced into the cell is passed on to the offsprings). The introduced DNA can be maintained conveniently in bacteria and yeast, but not in higher eukaryotes. In some eukaryotic cases, this is not important because it is unnecessary for DNA to be stably maintained; many experiments, such as reporter gene assays, can be carried out relatively quickly and transient expression is sufficient. Stable transfection, required for protein overexpression, is generally achieved by the stable integration of DNA into the genome in the absence of episomal vectors. DNA transfected directly into higher eukaryotic cells frequently integrates randomly into the genome, providing a relatively easy mechanism for the genetic transformation of animal and plant cells in culture. In the following sections, detailed descriptions of successful methods for the introduction of DNA into various host cells are provided.

14.2 INTRODUCTION OF DNA INTO BACTERIAL CELLS

Theoretically, cloning in plasmid vector is a very straightforward process. The plasmid DNA is digested with one or two restriction enzymes and ligated *in vitro* to the gene of interest having compatible termini. Because the new recombinant DNA molecule cannot reproduce on its own, it is necessary to introduce it into a host, where it can replicate and produce many copies of itself. However, DNA molecules do not easily enter bacteria on their own, but require assistance for passing through the outer and inner bacterial membranes and in reaching an intracellular site where these can be replicated and expressed. Plasmid transformation into bacterial cells is a key technique in molecular cloning and bacteria that are able to take up DNA are called 'competent'. Though some bacteria are naturally conjugative, such strains are not preferred hosts in recombinant DNA technology experiments. The process of conjugation has already been described in Chapter 8. After transformation, the cells are transferred to a nutrient medium and incubated for growth to permit expression of phenotypic properties conferred by the plasmid, for example, antibiotic resistance. Thereafter, the cells are plated on selective medium. Finally, bacteria transformed by recombinant plasmids are distinguished from bacteria carrying empty wild-type vector DNA by using any of the procedures devised for this purpose, which includes blue-white screening, nucleic acid hybridization (for details see Chapter 16), and colony polymerase chain reaction (PCR) (for details see Chapter 7). The methods devised for transformation of *Escherichia coli* cells fall into two classes: chemical and physical.

14.2.1 Transformation by Chemical Method (Preparation of Competent Cells and Transformation by Heat Shock)

Although certain types of bacteria have DNA transport systems for DNA uptake, the efficiency of DNA uptake can be improved by making them competent. *E. coli*, the most commonly used host for the propagation of recombinant molecules are naturally incompetent; however, these can be made competent for DNA uptake by treating a rapidly growing culture with a solution containing high concentrations of calcium chloride ($CaCl_2$). The solution of $CaCl_2$ is prepared in a hypotonic buffer, a liquid, which contains fewer dissolved solids than the cell it surrounds. As water always tends to move away from the hypotonic area, the cells swell in response to the $CaCl_2$ solution as water leaves the hypotonic buffer and enters the cell, and become spheroblasts. This treatment combined with the heat shock, a brief period (1–2 min) during which the temperature of the cells and their surrounding fluid is raised from 37°C (98°F) to 42°C (102°F), makes the cells competent for DNA uptake. Heat shock is a crucial step, and it is very important that the cells be raised to exactly the right temperature at the correct rate. During this heat shock, the spheroblasts contract, and the bacteria take up any bound DNA. This forms the basis of competent cells. After the heat shock, bacterial cells are incubated with nonselective growth medium for the recovery of cells. The cells containing exogenous DNA can be plated on a growth medium that selects for cells containing the plasmid vector, and inhibits the growth of other cells (e.g., medium containing an antibiotic or blue-white screening, etc.) (for details see Chapter 16). The principle and procedure of the preparation of competent cells and their transformation by heat shock method is outlined in Figure 14.1.

The original method of treatment with $CaCl_2$/heat shock, which was devised by Cohen, is an easy, but an inefficient process, and generally results in up to 1×10^6 transformation frequency, which is best for routine experiments. Apart from this technique, two more protocols were developed, which provide better efficiency. In one protocol, Hanahan prepared

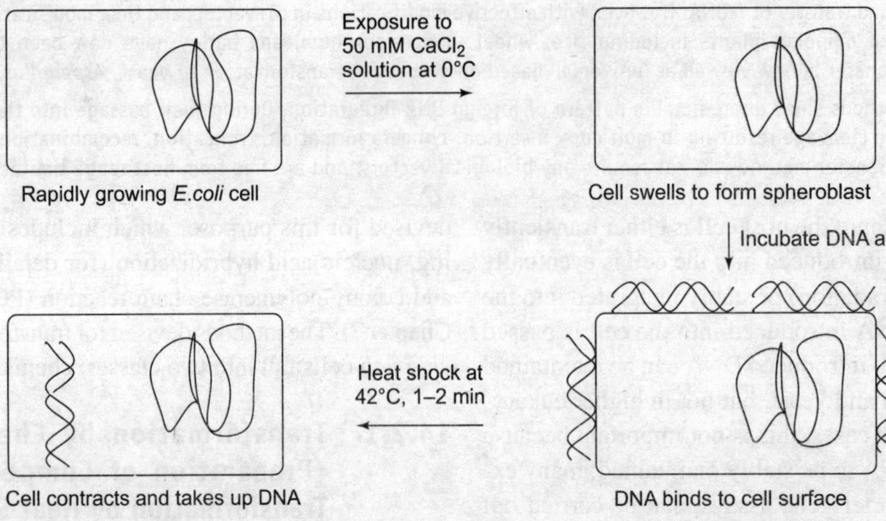

Figure 14.1 Principle of chemical transformation of *E. coli* by calcium chloride treatment and heat shock

competent cells of high competency (5×10^8 transformants/µg DNA) for immediate use in standard transformation buffer comprising of $MnCl_2$, $CaCl_2$, KCl, and hexamine cobalt chloride in MES [2-(*N*-morpholino) ethanesulfonic acid] buffer. However, Hanahan prepared competent cells for storage at –70°C in frozen storage buffer (FSB) having similar salts in potassium acetate buffer. In the second protocol, Inoue grew bacterial culture at 18°C as opposed to the conventional 37°C. The competent cells prepared by Inoue's protocol are less fussy, more reproducible, and therefore more reliable than the Hanahan method. Most probably the composition or the physical characteristics of bacterial membranes synthesized at 18°C are more favorable for uptake of DNA, or perhaps the phases of the growth that favor efficient transformation are extended.

Transformation Efficiency

Transformation efficiency is defined as the number of colony forming unit (cfu) produced by 1 µg of plasmid DNA. It is measured by performing a control transformation reaction using a known quantity of DNA. The transformation efficiency decreases rapidly as the size of the DNA molecule increases and is almost insignificant when the size exceeds 50 kbp.

Equation for Transformation Efficiency (cfu/mg)

Transformant

cfu = No. of bacterial colonies × dilution ratio × original transformation volume/plated volume

Thus, if 25 colonies were observed on the plate, and before plating the transformed competent cells were diluted 10,000 times, the original transformation volume was 100 µl, and 50 µl was used for plating, then the transformant cfu is

$25 \times 10{,}000 \times 100/50 = 5.0 \times 10^5$.

Transformation efficiency = Transformant cfu/plasmid DNA (µg)

Thus, if 0.5 µl of the plasmid DNA (1 µg/µl) was added, the transformation efficiency is

$$5.0 \times 10^5/0.5 = 10 \times 10^5 \text{ cfu/µg}$$

Factors Affecting Transformation Efficiency High transformation efficiency is very important in genetic engineering experiments and is affected by many factors, which are listed in the following discussion.

Growth period of bacteria For higher transformation efficiency, the bacterial cells must be in their early logarithmic growth period. Moreover, as different growth characteristics are observed with different strains of *E. coli*. Therefore, optimal OD_{600} for preparation of competent cells of different strains also varies, for example, the optimal OD_{600} value for XL1 blue is 0.15–0.45; for TG1 0.2–0.5; and for DH5α 0.145–0.45. It was observed that competent cells prepared from the overgrown or undergrown bacterial cultures exhibited reduced or no transformation capacity.

Temperature of competent cells during preparation, storage, and transformation Temperature is the second important factor that has an effect on the transformation efficiency. The competent cells must be maintained in a cold environment during preparation, storage, and transformation. Competent cells can be stored at –70°C for 15 days without a significant reduction in their transformation capacity. However, if the competent cells are stored at –20°C, the highest transformation efficiencies appear at 2–7 days and thereafter, the transformation efficiency is reduced considerably, while competent cells stored at 4°C lose their competency in just 3 days. Competent cells cannot be stored for a long time in liquid N_2 and should not be thawed more than once.

It is well-known fact that the results of the $CaCl_2$ treatment can be enhanced if a heating step follows the treatment, although there is some debate about whether the heat shock step is critical for the uptake of DNA. When *E. coli* is subjected to a temperature of 42°C, a set of genes called the heat shock genes are expressed, which enable the bacteria to survive at such temperatures. However, at temperatures above 42°C, the ability of bacteria for DNA uptake becomes reduced and at more extreme temperatures, the bacteria die.

Concentration of $CaCl_2$ The concentration of $CaCl_2$ is also a significant factor having an impact on the transformation efficiency of competent cells. By using 50–100 mM $CaCl_2$ for preparation of competent cells 1×10^5 to 1×10^6 transformation efficiency is achieved. This efficiency can be increased up to 1×10^7 to 9×10^7 by using TB solution (transformation buffer containing $CaCl_2$; 10 mM Pipes, 55 mM $MnCl_2$, 15 mM $CaCl_2$, 250 mM KCl, pH 6.7) under the same conditions. The addition of dimethylsulfoxide (DMSO) or polyethylene glycol 8000 (PEG 8000) during bacterial transformation can also affect transformation efficiency. It is observed that addition of DMSO or PEG 8000 during transformation process can result in 100 to 300-fold higher transformation efficiency as compared to the original Cohen's method.

Bacterial culture medium The bacterial culture medium can also affect the transformation efficiency. The growth of bacteria is higher in richer media. Therefore, transformants can be observed earlier in the SOC medium (Tryptone (20 g/l), Yeast extract (5 g/l), NaCl (0.5 g/l), 5 ml of 2 M $MgCl_2$, 20 mM glucose, pH 7.0), appearing after 12 hours as opposed to 24 hours in the Luria Bertani (LB) medium (note that SOC is a richer medium than LB medium). Moreover, SOC provides 10–30 times higher transformation efficiency than LB.

14.2.2 Transformation by Electroporation

Bacterial cells can also be transformed directly by electroporation, which helps in avoiding possible contamination with other strains. Electroporation is a direct physical method of gene transfer. It is a fast and efficient method of transformation. This procedure provides ~100-fold more efficiency as compared to chemical induction method. However, electroporation is an expensive process, requiring costly electrical equipment and highly priced specially designed cuvettes.

Principle of Electroporation

Cell membrane is an electric capacitator that is unable to pass current (except through ion channels). Subjecting cell membranes to electrical pulses of high field strength (high voltage) reversibly permeabilizes them, which facilitates uptake of macromolecules such as DNA. This is because an electric field, delivered as exponentially decaying pulse, results in the temporary breakdown of the cell membrane and induces the formation of transient pores in the cell membrane (perforation) through which the DNA molecules are able to enter the cell (Figure 14.2). It has been hypothesized that the mechanism of DNA entry into the cell during electroporation may

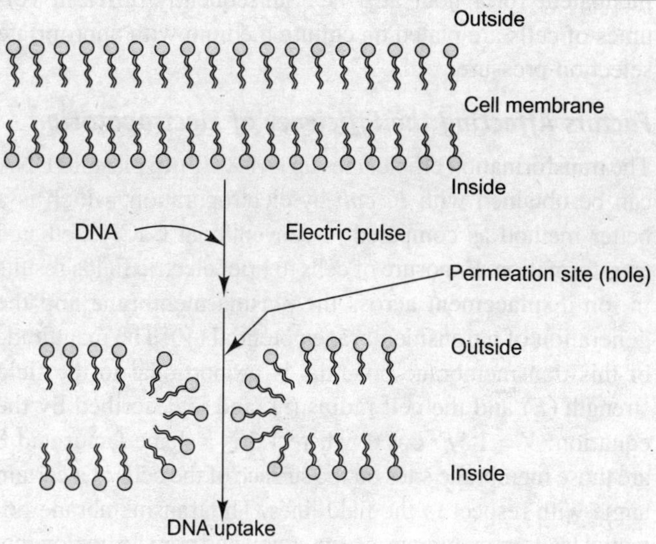

Figure 14.2 Principle of electroporation

be either due to nonspecific membrane process, or it may be specifically mediated by permease. The activation of permease may be due to high electric field, and thus may lead to thinning of membrane. As a result pores may be created, which are stabilized by a favorable dipole interaction of lipid dipoles in an electric field. The pores allow DNA uptake by diffusion, however, it is also suggested that DNA binding through electrostatic bridging to plasma membrane facilitates DNA uptake. The direct introduction of genes by electroporation allows both rapid transient expression of introduced genes as well as their stable integration.

Procedure of Electroporation

For electroporation of *E. coli* cells, it is required to make electrocompetent cells, which are simply prepared by growing bacteria to mid-log phase, chilling, centrifuging, washing thoroughly with chilled buffer or water to reduce the ionic strength of cell suspension, and then finally suspended in an ice-cold buffer containing 10% glycerol. Electroporation is done in the chamber of an electroporator, which is cylindrical in shape, and comprises of a disposable 1 ml cuvette and two parallel stainless steel electrodes placed at a distance of ~1 cm. The electrocompetent cells and plasmid DNA are transferred to cuvette containing electroporation buffer, which is kept in the electroporator, and electrodes are inserted into it. This is followed by the delivery of a single exponential decaying pulse (shock of desired voltage) by discharge of a capacitator across the cell. Usually an electric field with very high field strengths (12.5–15 kV) is used for *E. coli* cells. After pulse delivery, cuvettes are removed and 1 ml SOC or LB medium is added at room temperature. Note that most probably addition of medium at room temperature fires a heat shock that increases the efficiency of transformation. Thereafter, the cells are transferred to polypropylene tube and incubated for 1 hour at 37°C. Subsequently different volumes of cells are plated on culture medium with appropriate selection pressure.

Factors Affecting the Efficiency of Electroporation

The transformation efficiencies up to 1×10^9/μg plasmid DNA can be obtained with *E. coli* by electroporation, which is a better method as compared to conventional $CaCl_2$-mediated transformation. Exposure of cells to brief electric fields results in ion-displacement across the plasma membrane and the generation of a transmembrane potential (V). The magnitude of this transmembrane potential is proportional to the field strength (E) and the cell radius (r), and is described by the equation: $V = 1.5Er \cos \theta$, where 1.5 is a shape factor and θ are those membrane sites on the surface of the cells at a certain angle with respect to the field lines. The transmembrane potential leads to membrane compression and pore formation that facilitates DNA uptake. Efficiency of electroporation is affected by electroporation conditions such as field strength (voltage applied across the electrodes), capacitance and resistance of circuit, pulse duration, voltage, and electroporation buffer. The efficiency of electroporation is also affected by other factors including temperature of the cells before, during, and after electroporation, concentration and topological form of the DNA, plating efficiency, and state of the cell. Electroporation is a temperature-dependent technique and is best performed at 0–4°C. The efficiency drops as much as 100-fold when electroporation is carried out at room temperature. The size of DNA also affects the efficiency of electroporation; it is possible to get transformation efficiencies of 10^6 cfu/μg DNA with molecule as big as 240 kbp. Even molecules three to four times this size also can be electroporated successfully. More recently, it has been reported that transformation efficiencies can be increased 10-fold with plasmid DNAs by coprecipitating the DNA with transfer RNA before electroporation. Additionally, linear plasmid DNAs transformed very inefficiently than the corresponding closed circular DNAs, because the exposed termini of linear DNAs are susceptible to attack by intracellular nucleases. Moreover, the cell type or strain, growth conditions, and postpulse treatment are also very significant.

14.2.3 Phage-mediated Procedures

The recombinant DNA molecules can also be introduced into bacterial cells with the help of phages. The procedures included are *in vitro* packaging followed by natural infection or transfection.

In vitro Packaging Followed by Natural Infection

The recombinant DNA molecule (produced as concatemers or multimers) can be packaged under *in vitro* conditions when supplied with phage packaging extracts comprising of head, tail, and assembly proteins. The phage thus assembled is naturally infectious that transfers the exogenous DNA into bacterial cell (for details see Chapter 9).

Transfection

The process of transfer of naked bacteriophage DNA into a host bacterial cell in much the same way as described for bacterial transformation is called transfection. It is thus the counterpart of bacterial transformation. The process requires the mixing of recombinant DNA molecule with competent *E. coli* cells, and the DNA uptake is induced by heat shock (for details see Chapter 9).

14.2.4 Introduction of DNA into Bacteria Other than *E. coli*

E. coli is used as a universal host for genes cloned from eukaryotes or other prokaryotes for many experiments. However, use of *E. coli* is not always feasible because it lacks some auxiliary biochemical pathways that are essential for the phenotypic expression of certain functions, for example, degradation of aromatic compounds, antibiotic synthesis, pathogenicity, sporulation, etc. In such circumstances, the genes have to be cloned back into species similar to those from where these were derived. The transformation event in these organisms is quite different from *E. coli* in the ways as listed: (i) Transformation occurs in these organisms naturally, while transformation in *E. coli* is artificially induced. Competence for transformation is a transient phenomenon with few exceptions, e.g. *Neisseria gonorrhoeae*; (ii) Transformation in some species is sequence-independent, as in *Bacillus subtilis* and *Acinetobacter calcoaceticus*, but in other species (*Haemophilus influenzae, N. gonorrhoeae*) it is dependent on the presence of specific uptake sequences; and (iii) The mechanism of natural transformation involves breakage of DNA duplex and degradation of one of the two strands so that a linear single strand can enter the cell. This mechanism is not compatible with efficient plasmid transformation.

Electroporation provides a much simpler method for introduction of DNA for these types of bacteria; however, with a very low efficiency, for example introduction of DNA by electroporation into *Methanococcus voltae* produced only 10^2 transformants/μg plasmid DNA. On the basis of biochemical and genetic analysis of various model organisms, it is observed that the mechanisms of DNA uptake is conserved in Gram positive and Gram negative bacteria and have some common structural features, whereas the regulation of competence is not conserved. Now within the bioinformatics domain, it is possible to search for homologous genes in the organisms whose genome has been completely sequenced, and on the basis of presence of these genes in particular bacterium, the ability of transformation can be analyzed. As plasmid transformation is not possible in many nonenteric bacteria, conjugation is also an alternative method of choice (for details see Chapter 8). However, after DNA manipulation, it has to be mobilized into a host cell by transformation at some stage usually in *E. coli*. Once in *E. coli*, or any other organism for that matter, it may be moved to other bacteria directly by conjugation, as an alternative to purifying the DNA and moving it by transformation or electroporation.

Some specialist methods have been devised for transforming non-*E. coli* bacteria such as *A. tumefaciens, B. subtilis,* and *Streptomyces* sp. with plasmid DNA. These are discussed in detail below.

Transformation of Agrobacterium tumefaciens

A. tumefaciens, a Gram negative soil bacterium, exhibits a natural tendency of gene transfer to plants. The cells of *A. tumefaciens* are transformed by following procedures.

Transformation by Triparental Mating Triparental mating (also called three-way mating) is a type of bacterial conjugation where a conjugative plasmid present in one bacterial strain assists the transfer of a mobilizable plasmid present in a second bacterial strain into a third bacterial strain. This method is usually used for transforming non-*E. coli* bacterium that cannot be transformed, for example, *Agrobacterium, Pseudomonas*, etc. This procedure helps in overcoming some of the barriers to efficient plasmid mobilization. For instance, if the conjugative plasmid and the mobilizable plasmid are members of the same incompatibility group, these do not need to stably coexist in the second bacterial strain for the mobilizable plasmid to be transferred. The requirements for the procedure are a helper strain, a donor strain, and a recipient strain. Helper strain (strain B) carries a conjugative plasmid (e.g., F-plasmid) that codes for genes required for conjugation and DNA transfer, for example, *E. coli* C199 containing the helper plasmid pRK2013. It contains the conjugation and transfer genes (*tra*), an origin of replication allowing maintenance in a narrow range of hosts. A donor strain (strain A) carries a mobilizable plasmid (a small cloning vector), which can utilize the transfer functions of the conjugative plasmid. This small cloning vector contains an origin of replication that allows the plasmid to replicate in a wide variety of bacterial hosts (broad host range), a selectable marker gene (antibiotic resistance), one or more restriction sites for the insertion of DNA fragments, and an origin of transfer (*ori* T) that allows the plasmid to be mobilized and transferred during conjugation. A recipient strain (strain C) is a bacterium that cannot be readily transformed, and in which the mobilizable plasmid is intended to be introduced. To transfer the cloning vector to a bacterium that cannot be readily transformed, the strain of *E. coli* containing the vector (A), mobilizing helper plasmid (B), and the recipient bacterium (C) are mixed together. A three-way or tripartite mating occurs, and the mobilizing plasmid is transferred to the bacterium containing the vector. It then promotes transfer of vector as well as itself, and both plasmids can transfer to the non-*E. coli* recipient bacterium (C). The vector with the broad host range origin can continue to replicate in the new host, but the mobilizing plasmid with the narrow host range origin cannot replicate in the new host and is rapidly lost. The diagrammatic representation of triparental mating is outlined in Figure 14.3. Antibiotic selection or nutrient conditions are generally used to eliminate the *E. coli* cells (A and B) and select the non-*E. coli* bacterium (C) containing the vector. Five to seven days are required to determine whether or not the plasmid has been successfully introduced into the new bacterial strain and to confirm that there is no carryover of the helper or donor strain.

Transformation by Electroporation Highly competent *A. tumefaciens* or *A. rhizogenes* cells are easily transformed by electroporation. The principle and procedure is similar to *E. coli* as already described in Section 14.2.2.

Transformation of Bacillus subtilis

Bacilli are Gram positive and the prototype of this is *Bacillus subtilis. B. subtilis* have been used in fermentation industry for a long time for large-scale production and synthesis of important proteins like amylase, β-lactamases, and a series of proteases, which are excreted into the growth medium. These organisms have also gained great importance in genetic engineering. Although *Bacillus* is naturally competent, it is very easy to transform *B. subtilis* with fragments of chromosomal DNA, while transformation by plasmid DNA molecules is not very easy. There are three types of procedures for transformation of *B. subtilis*, which are described in the following discussion.

Transformation by Competent Cells It was first reported that competent cells of *B. subtilis* could be transformed with covalently closed circular (CCC) plasmid DNA from *Staphylococcus aureus*, which could replicate and express in its new host. However, the transformation of *B. subtilis* with plasmid DNA is very inefficient as compared to chromosomal transformation; only one transformant is obtained per 10^3–10^4 plasmid DNA molecules. Moreover, the transformability of plasmid DNA molecules is dependent on its degree of oligomerization during transformation of *B. subtilis*, as purified monomeric CCC forms of plasmids transform *B. subtilis* less efficiently in contrast to impure plasmid preparations having multimers. The efficiency of transformation increases with an increasing degree of oligomerization. In a transformation experiment performed with a recombinant plasmid (e.g., pHV14) capable of replicating in both *E. coli* and *B. subtilis*, it has been demonstrated that plasmid transformation of *E. coli* occurs irrespective of the degree of oligomerization, on the contrary to the condition with *B. subtilis*. Furthermore, DNA ligase-catalyzed oligomerization of linearized plasmid DNA results in a substantial increase of spe-

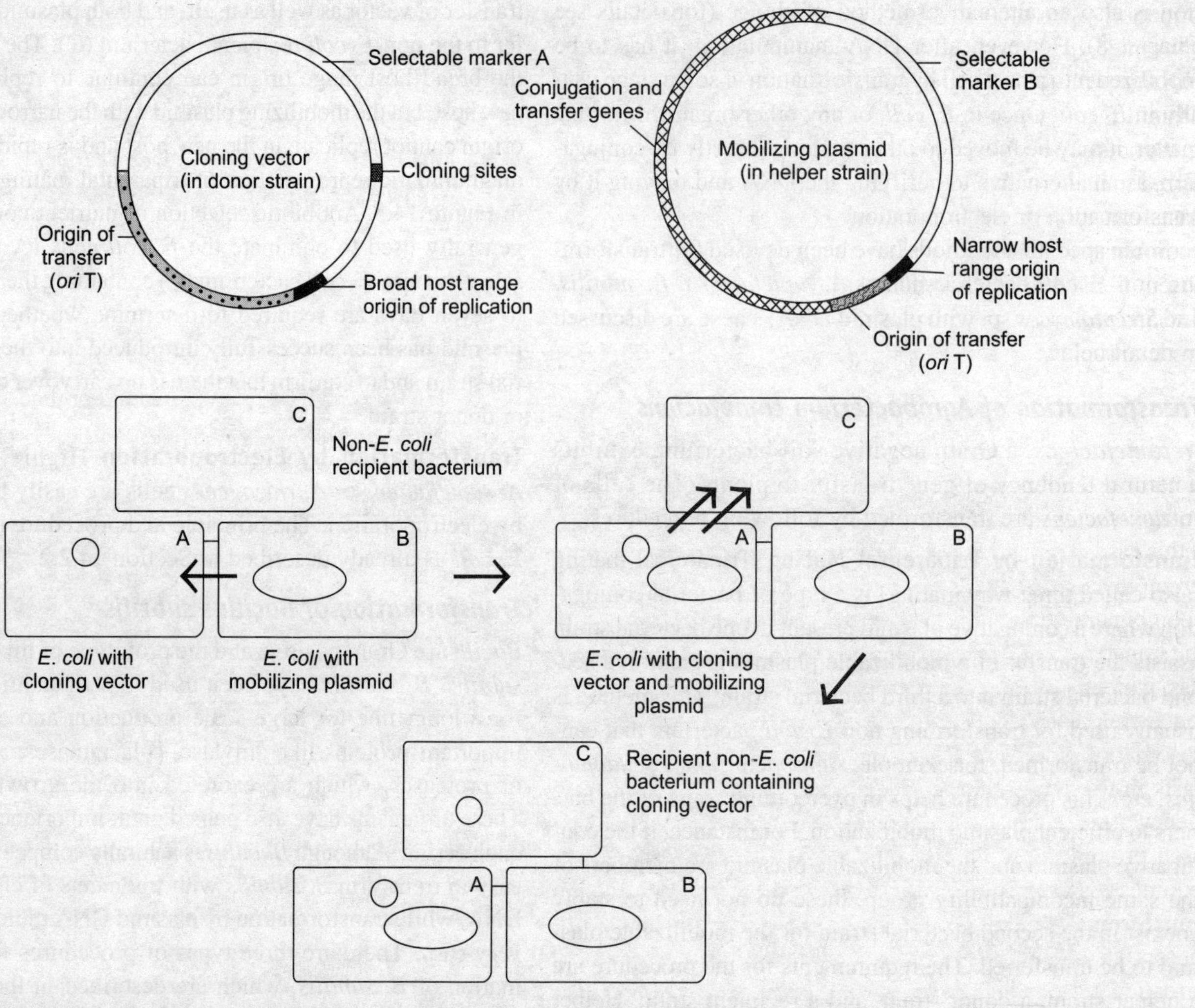

Figure 14.3 Bacterial transformation by triparental mating

cific transforming activity when assayed with *B. subtilis* and causes a decrease when used to transform *E. coli*. During transformation in *Bacillus,* like chromosomal DNA, the plasmids are cleaved into linear molecules upon contact with competent cells. When the linear single stranded form of the plasmid DNA enters the cell, it needs to circularize for replication, facilitating the recombination of the requisite multimers, which provide regions of homology. The recombinant plasmids having direct internal repeats (260–2,000 bp long) or having an insert of *B. subtilis* genomic DNA can actively mobilize into *B. subtilis* cells, though with low efficiency as compared to oligomers. The plasmids introduced into competent cells of *B. subtilis* by transformation are unstable and frequently show deletions, as DNA appears to be extremely labile during its conversion into single stranded form and the subsequent recombinational events.

Transformation by Plasmid Rescue Unfractionated plasmid DNA can be transformed by this alternative strategy. This procedure does not require oligomers for transformation. Practically, foreign DNA is ligated to monomeric vector DNA and *in vitro* recombinants are used to transform *B. subtilis* cells carrying a homologous plasmid. In this case, the recipient must be Rec^+ as there is recombination between the linear donor DNA and the resident plasmid. Using such a 'plasmid rescue' system, various genes from *B. licheniformis* have been cloned in *B. subtilis*. The resulting transformants contain resident plasmid and incoming plasmid. These have to be segregated by segregation or retransformation; hence several subculturing steps are required. The observed transformation efficiency by this procedure is 2×10^6 transformants/µg DNA and transformation is possible even with linear DNA.

Transformation of Protoplasts in the Presence of PEG A third alternative strategy has been devised for transforming cells of *B. subtilis*, which is a milestone in the development of cloning techniques for this bacterium. In this technique, the protoplasts that are obtained by lysozyme treatment can be transformed with plasmid DNA in the presence of PEG. The procedure is highly efficient (>10^7 transformants/µg DNA) and yields up to 80% transformants, making the method suitable for the introduction of even cryptic plasmids. The efficiency of transformation is size-dependent and decreases sharply as size of DNA molecule increases. The additional advantage associated with this procedure is that nonsupercoiled monomeric circular plasmid DNA molecules constructed by ligation *in vitro* can be introduced at high efficiency. Linear DNA can also be mobilized in the cell by using this procedure. Furthermore, protoplasts can also be transformed in the absence of functional recombination systems. A major weakness linked with this system is that the regeneration medium is nutritionally complex and the system requires a two-step selection procedure for auxotrophic markers.

Transformation of Streptomyces

Streptomyces are Gram positive, mycelial and spore-forming bacteria, and these produce more than 60% of the known antibiotics, several extracellular enzymes, and enzyme inhibitors. In this case, genes are naturally exchanged between hyphae of *Streptomyces* through conjugation, which is mediated by recombination systems based on the activity of sex plasmids. The transfer of sex plasmids between cells is an efficient process, which is often accompanied by chromosomal mobilization; therefore these cells demonstrate a high frequency of chromosomal recombination. Furthermore, fusing bacterial protoplasts in the presence of PEG can also induce the high frequency of recombination between chromosomal markers. The fused protoplasts, which contain the two complete parental genomes, can regenerate efficiently into substrate mycelia. Transformants are generally identified by the selection of appropriate phenotypes (e.g., antibiotic resistance) and transformation frequencies up to 10^7 transformants/μg DNA are easily obtained. Clones carrying conjugative plasmids can also be detected by the visualization of so-called 'pocks'. The property of pock formation is exhibited if a strain containing plasmid is replica plated onto a lawn of the corresponding plasmid-free strain. The clones containing plasmids are surrounded by a narrow zone in which growth of the plasmid-free strain is hindered. This particular feature has been used to identify the presence of cryptic plasmids.

14.3 INTRODUCTION OF DNA INTO YEAST CELLS

Simple eukaryotes such as yeasts are not naturally transformable, like *E. coli*. Hence transformation of yeasts can often be accomplished with chemical treatments very similar to those for transforming bacteria. These include techniques such as spheroplasts technique, lithium acetate treatment, and electroporation. The transformation procedures for yeast cells have already been discussed in Chapter 11.

14.4 GENETIC TRANSFORMATION OF PLANTS

Engineering plants with useful quality attributes is gaining importance to overcome the problems associated with existing strategies involving environmentally damaging chemicals, erosion-causing tillage, or costly equipments and purification steps, etc. Thus, for genetic manipulation of plants, a suitable plant transformation vector, plant explants, and a source of transgene (exogenous DNA) are required. General procedure for vector construction and recombinant DNA preparation used for plant transformation are the same as discussed in Chapter 12. Thus, vector construction involves cloning of gene of interest under the control of a strong and organ-specific plant promoter to allow high-level expression of cloned gene in plants. Further targeting of the protein to appropriate cellular compartment may be helpful in stabilizing the protein. For this, a plant-specific ER-targeting sequence is preferred. Likewise for expression of protein in chloroplast, chloroplast plant transformation vectors harboring chloroplast sequences are used. In expression vectors, optimized translation start site context is created and the codons are altered to suit the expression of prokaryotic genes in a plant. After vector construction (i.e., formation of recombinant vector DNA), it is transformed into suitable host (biological system) such as *Agrobacterium* or plant viruses, which then transfers the recombinant DNA into target explant. The explants selection is done on the basis of method used for plant transformation. Thus, protoplasts/cells/calli/auxillary nodes, etc. can be used. Once selected, the explant is surface sterilized and transformed with DNA by any of the plant transformation methods. Some of the plant transformation techniques require biological system, while others employ direct transfer. The targeted explants are finally transferred to selection medium for selection of transformants exploiting the property of plant selection marker gene present on plant transformation vectors. Assay of reporter gene products can also be done to determine the extent of expression of the cloned gene. In the method, a suitably transformed explant is regenerated into a plant (putative transformant), which is acclimatized to environmental and soil conditions, and then transferred to the field. Following transformation and plant regeneration, F_0 and progeny plants are analyzed to ensure stable integration and proper functioning of the transgene using techniques such as PCR, reverse transcriptase-PCR (RT-PCR), Southern, northern hybridization, and protein assay.

Currently a variety of techniques are available for easy and successful transformation of almost all the plant species, which provide the basis for the advances in plant biotechnology. These techniques fall into two main categories.

Biological Method or Vector-mediated Gene Transfer Gene transfer techniques that involve biological vectors for the transfer of genetic information to plants from other organisms, *viz.*, bacteria, fungi, and animals as well as between plants are called biological methods or vector-based methods. Examples included in this category are *A. tumefaciens*, *A. rhizogenes*, and plant viruses.

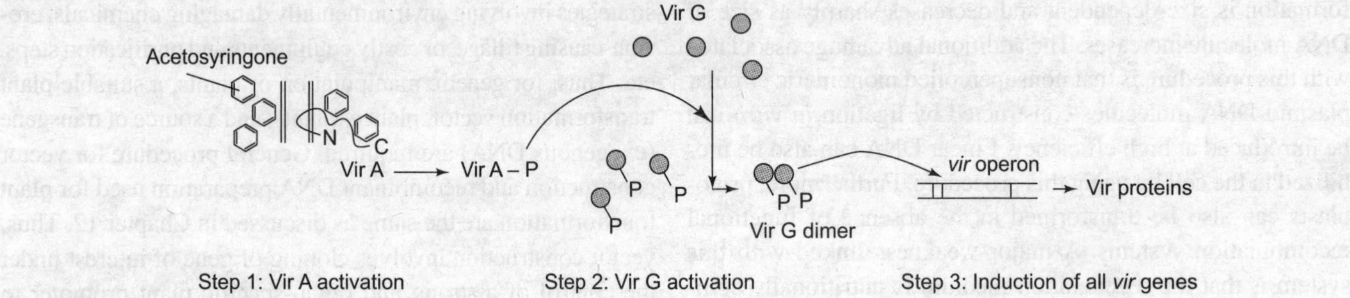

Figure 14.4 Induction of *vir* genes

Direct DNA Transfer or Vectorless Gene Transfer The initial difficulties while using *Agrobacterium* to transform monocot plants (grasses and cereals) has spurred on the development of 'direct gene transfer' methods, however now *Agrobacterium*-mediated gene transfer in monocot species is also possible. These direct gene transfer techniques do not employ any biological vector, and are species- and genotype-independent. These methods serve as simple and effective means for the introduction of foreign DNA into plant genomes. These techniques are further divided into two types:

By physical means These methods are based on direct delivery of naked DNA to the plant cells. This is also called DNA-mediated gene transfer (DGMT). Various methods included in this subcategory are microprojectile bombardment, electroporation, microinjection, silicon carbide whiskers, ultrasonication, UV laser microbeam irradiation.

By chemical and biochemical means These methods employ direct gene transfer with the aid of chemical agents. For example, calcium phosphate coprecipitation, polycation-mediated gene transfer, and liposome-mediated transformation.

14.4.1 *A. tumefaciens*-mediated Gene Transfer

A. tumefaciens has a natural tendency of gene transfer to the plant cells. This property is attributable to its pTi plasmid, which contains T-DNA flanked by left and right border sequences (LB and RB) and *vir* genes (for details see Chapter 12). In this section, the events involved in T-DNA transfer to plants and method used for plant transformation using *A. tumefaciens*-based vectors are discussed.

Process of T-DNA Transfer to Plants

The process of T-DNA transfer to plants and its integration into plant genome has been studied by various researchers. Three genetic elements, *Agrobacterium* chromosomal virulence genes (*chv*), T-DNA flanked by RB and LB, and pTi-encoded *vir* genes constitute the T-DNA transfer machinery. The process of gene transfer from *A. tumefaciens* to plant cells involves several essential steps, which are as follows:

Induction of *vir* Genes The *vir* region encodes *vir* operons (*vir* A to *vir* J; depending on the plasmid type), which are tightly

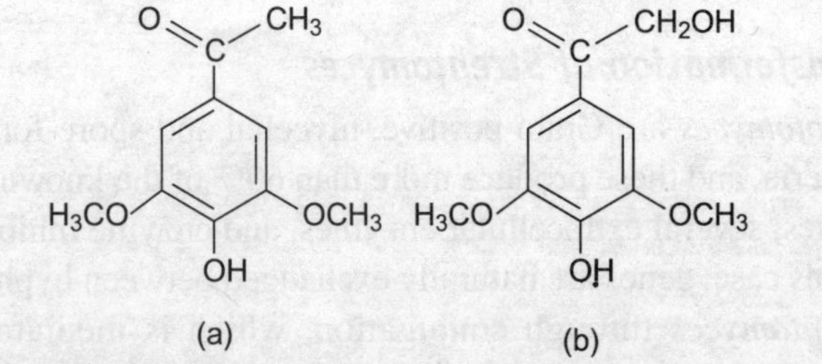

Figure 14.5 Phenolic compounds released from wounded plant cells; (a) Acetosyringone, (b) α-hydoxyacetosyringone

regulated. The number of genes per operon differs: *vir* A, *vir* G, and *vir* F have only one gene, *vir* E, *vir* C, *vir* H have two genes each, *vir* D has four genes, while *vir* B has 11 genes. The expression of *vir* genes occurs only in wounded plant cells, i.e., the targets of infection. These *vir* genes are activated first and their products then directly involve in T-DNA processing and transfer. The steps involved in the induction of *vir* genes (Figure 14.4) are as follows: (i) Vir A, a transmembrane dimeric sensor protein encoded by *vir* A gene, is activated by autophosphorylation after detecting the signals such as small phenolic compounds, for example, acetosyringone or α-hydroxyacetosyringone (Figure 14.5), released by wounded plant. Besides phenolic compounds, other signals for *vir* A activation include acidic pH and certain class of monosaccharides, which act synergistically with phenolic compounds. The activation of *vir* system also depends on external factors like temperature and pH. At temperatures greater than 32°C, *vir* genes are not expressed because of a conformational change in the folding of Vir A induces inactivation of its properties; (ii) Activated Vir A has the capacity to transfer its phosphate to a conserved aspartate residue of the cytoplasmic DNA-binding protein Vir G, which is then activated by dimerization; and (iii) Activated Vir G binds specifically to the upstream of other *vir* genes called *vir* box (*vir* B, C, D, E, F, H, and J) to induce their expression.

Bacterial Colonization and their Attachment to Plant Cell Bacterial colonization is an essential and one of the earliest steps in tumor induction, which takes place when *A. tumefaciens* is attached to the plant cell surface (Figure 14.6). Different chromosomal genes or loci are involved in the

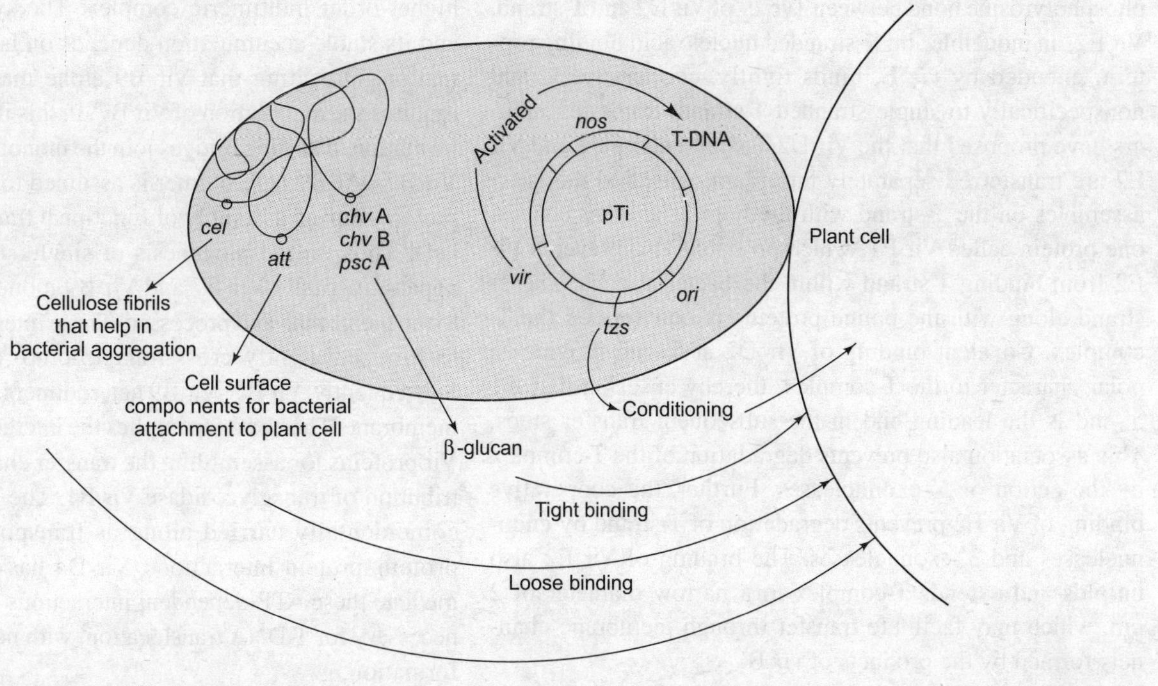

Figure 14.6 Bacterial aggregation, chemotaxis and attachment to plant cell [The transmembrane sensing protein Vir A senses the release of phenolic compounds from wounded plant cells. The activated Vir A in turn phosphorylates Vir G, which results in the chemotaxis of the bacteria to the plant cell. Expression of the bacterial chromosomal genes *att*, *chv A*, *chv B*, and *psc A* result in the binding of the bacterial cell to the plant cell, whereas the product of *cel* is involved in the formation of cellulose fibrils, which assist in bacterial aggregation. The activated Vir G serves as a transcriptional activator of other *vir* genes. Trans-zeatin, encoded by the *tzs* gene of the pTi probably conditions the plant cell for transformation by inducing cell division.]

attachment of *A. tumefaciens* to the plant cell and bacterial colonization, for example, *chv* A and *chv* B involved in the synthesis and excretion of the β-1,2 glucan; *chv* E required for the sugar enhancement of induction of *vir* genes and bacterial chemotaxis; *cel* responsible for the synthesis of cellulose fibrils; *psc* A (*exo* C) playing role in the synthesis of both cyclic glucan and acid succinoglycan; and *att* involved in the cell surface proteins.

Generation of T-complex Once various Vir proteins are formed, the production of a transfer intermediate begins with the generation of a single stranded copy of T-DNA, the T-strand. The synthesis of T-strand begins with the endonucleolytic cleavage at RB. LB may also act as a starting site for single stranded T-strand synthesis, but the efficiency is much lower. The difference may be a consequence of the presence of an enhancer or 'overdrive' sequence next to RB. This overdrive sequence is specifically recognized by Vir C1 protein. Together, Vir D1/D2 recognize the 25 bp RB and produce a single stranded endonucleolytic cleavage in the bottom strand of each border. The cleavage occurs at the T-DNA RB between the third and fourth bases. Both nick sites contain a 12 nucleotide consensus sequence that has also been found among various rolling circle-type replicons. Thus the two nicks at RB and LB are used as the initiation and termination sites for T-strand production (Figure 14.7). Vir D2 remains tightly associated even after nicking with the 5′-end of the T-strand by a high-energy

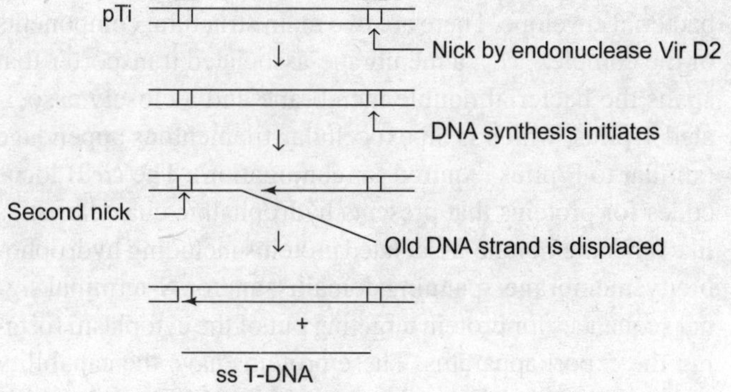

Figure 14.7 T-strand production by displacement of the bottom strand of the T-DNA [A nick is made at RB by endonuclease activity of Vir D2 which provides a priming end for synthesis of a single stranded DNA (ss DNA). Synthesis of new strand displaces old strand, which is used in transfer process. Transfer process is terminated when DNA synthesis reaches a nick at LB. This model accounts for the polarity of the process. If the LB fails to be nicked, transfer could continue farther along the pTi. The transfer process involve ss DNA. This model for transfer of T-DNA closely resembles the events involved in bacterial conjugation, when *E. coli* chromosome is transferred from one cell to another in single stranded form. A difference is that in bacterial conjugation there is no predetermined end to the process and up to the whole chromosome may be transferred, depending on the conditions.]

phosphotyrosine bond between Tyr 29 of Vir D2 and T-strand. Vir E2, an inducible single stranded nucleic acid binding protein, encoded by *vir* E, binds tightly, cooperatively, and nonspecifically to single stranded T-strand. Some researchers have proposed that the Vir D2/T-strand complex and Vir E2 are transferred separately into plant cells, and the latter assembles on the T-strand with the help of another chaperone protein called Vir E1, which probably also prevents Vir E2 from binding T-strand within the bacterial cells. The T-strand along with the bound proteins is now termed the T-complex. Covalent binding of Vir D2 at 5′-end provides a polar character to the T-complex, thereby ensuring that the 5′-end is the leading end in the subsequent transfer steps. This association also prevents degradation of the T-complex by the action of 5′-exonucleases. Further, the cooperative binding of Vir E2 prevents degradation of T-strand by endonucleases and 3′-exonucleases. The binding of Vir E2 also unfolds and extends T-complex to a narrow diameter of 2 nm, which may facilitate transfer through membrane channels formed by the products of *vir* B.

Transport of T-complex to Plant Cell The T-complex of an appropriate diameter exits the bacterial cell passing through the inner and outer membranes as well as the bacterial cell wall. It then crosses the plant cell wall and membrane. The bacterium uses a Type IV secretion system (T4SS), a complex of 11 Vir B and Vir D4 proteins, to transfer the T-complex and other Vir proteins across the double membrane of the bacterial envelope. There are two main structural components of the complex, *viz.*, a membrane-associated transporter that spans the bacterial double membrane and a closely associated T-pilus, which is an exocellular filamentous appendage (similar to F-pilus required for conjugation). The *vir* B locus codes for proteins that presents hydrophathic characteristics like other membrane-associated proteins including hydrophobicity, membrane-spanning domains, and/or N-terminal signal sequences for protein targeting out of the cytoplasm forming the export apparatus. These proteins have the capability of interkingdom delivery. Vir B4 and Vir B11 are hydrophilic ATPases localized in the inner membrane and provide the energy for translocation of the T-complex. Vir B11 is reported to associate with the cytoplasmic face of the inner membrane exhibiting both ATPase and protein kinase activity. It lacks continuous sequence of hydrophobic residues, forming periplasmic domains. Vir B4 has a nucleotide-binding site that is essential for virulence and tightly associates with the cytoplasmic membrane. Its two putative extracellular domains confer transmembrane topology to this protein, which presumably allows the ATP-dependent conformational change in the conjugation channel. Vir B4 synthesis is well-correlated with the accumulation and distribution of Vir B3. Vir B7 seems to be crucial for the conformation of the transfer apparatus and interacts with Vir B9, forming a heterodimer and probably a higher-order multimeric complex. The synthesis of Vir B9 and its stable accumulation depends on heterodimer conformation, indicating that Vir B9 alone may be unstable and requires the association of Vir B7. In this intermolecular conformation, disulfide bridges join the monomeric subunits. The Vir B7–Vir B9 heterodimer is assumed to stabilize other Vir proteins during assembly of functional transmembrane channels. Thus, in the biogenesis of single stranded T-complex apparatus, firstly Vir B7 and Vir B9 monomers are exported to the membrane and processed. These interact with each other to form covalently cross-linked homo- and heterodimers. Subsequently, Vir B7–Vir B9 heterodimer is sorted to the outer membrane. The next step implies the interaction with the other Vir proteins for assembling the transfer channel with the contribution of transglycosidase Vir B1. The T-complex may be coincidentally carried along as transport is mediated by protein–protein interactions. Vir D4 has been suggested to mediate these ATP-dependent interactions of protein complex necessary for T-DNA translocation, with no role in T-complex formation.

Nuclear Import of T-complex Once inside the plant cell, the T-complex targets the plant cell nucleus and crosses the nuclear membrane. This nuclear uptake is a tightly regulated process and can occur through the nuclear pore. Besides, the nuclear uptake is mediated by the presence of nuclear localization signal (NLS). The T-strand itself does not possess a signal for nuclear import and any DNA between RB and LB can be transferred. These NLS are localized on Vir proteins, one on Vir D2 and two on each of the cooperatively bound Vir E2. The multiple NLSs of Vir E2 are functionally important for uninterrupted nuclear import of T-complex by keeping the cytoplasmic and nucleoplasmic sides of the nuclear pore simultaneously open. Vir D2 localized at the 5′-end of the T-complex ensures that the 5′-end of the T-strand enters the nucleus first. The role of Vir F also seems to be related with the nuclear targeting of the ss T-DNA complex but its contribution is less important. Vir H1 and Vir H2 proteins are not essential in T-DNA transport but probably enhance the transfer efficiency by detoxifying certain plant compounds that can affect bacterial growth. Thus, it is proposed that Vir H protein plays a role in the host-range specificity of bacterial strain for different plant species.

The nuclear uptake of T-complex by the plant cell is also regulated by the interaction of Vir D2 with plant host proteins, namely importin-α, VIP1, KAP-α, cyclophilins, karyophilins, pp2C, and Histone H2A. Importin-α enables transport through the plant nuclear pore complex (NPC). Cyclophilins may cause conformational changes in Vir D2 or may be involved in targeting to chromatin and integration of the T-DNA. Serine/threonine protein phosphatase type 2C (pp2C) negatively affects nuclear import of the T-DNA.

Integration of T-DNA into Plant Genome Integration of T-strand into a host chromosome, after entry into plant cell, is the culmination step of the T-DNA transfer process. After nuclear import, bound Vir D2, Vir E2, and host importin proteins (e.g., VIP 1) are removed before (or during) T-DNA integration into the host genome. Vir F localizes to the plant cell nucleus along with the T-complex, destabilizes both VIP1 and Vir E2, presumably targeting them for proteosomal degradation. Vir F does not interact with or destabilize Vir D2. Hence, Vir D2 remains bound to the T-DNA until a later stage.

The mechanics of integration step is, however, not very clearly understood. The T-DNA does not encode enzymatic activities required for integration; rather the protein components of the T-complex, i.e., Vir D2 and Vir E2, and/or host nuclear factors such as AtKu80 and plant DNA ligases provide these functions. Two models have been proposed for the integration of T-DNA into plant genome. These are defined in the following.

'Illegitimate recombination' of ss T-DNA followed by second strand synthesis A model for the integration step has been proposed that compares T-DNA integration to illegitimate recombination, i.e., a mechanism that joins two DNA molecules not sharing extensive homology. According to this model, pairing of a few bases, known as microhomologies, are required for a preannealing step between T-DNA strand coupled with Vir D2 and plant DNA. These homologies are very low and provide just a minimum specificity for the recombination process by positioning Vir D2 for the ligation. Two models proposed for illegitimate recombination are described.

- ***Model 1*** It is suggested that the Vir D2 has an active role in the precise integration of T-strand in the plant chromosome. The release of Vir D2 protein may provide the energy contained in its phosphodiester bond at the Tyr 29 residue, with the first nucleotide of T-strand providing the 5′-end of the T-strand for ligation to the plant DNA. This phosphodiester bond can serve as electrophilic substrate for nucleophilic 3′-OH from nicked plant DNA. Vir D2 bound to the 5′-end of the T-strand thus joins a nick in plant DNA. This leads to further unwinding of the plant DNA, thereby forming a gap. The 3′-end of the T-strand may pair with another region of plant DNA in proximity, or may bind covalently to the plant DNA by plant repair and recombination enzymes. Displaced plant DNA is subsequently cut at the 3′-end position of the gap by endonucleases. These reactions result in the introduction of the T-strand into one strand of plant DNA. This in turn leads to torsional strain resulting in the introduction of a nick into the opposite strand of plant DNA. This situation activates the repair mechanism of the plant cell and the complementary strand is synthesized using the early inserted T-DNA strand as a template. Thus, following strand invasion, DNA polymerase is thought to synthesize the complementary strand of the T-DNA and remove the Vir E2 molecules. *In vitro* studies of T-DNA integration shows that it is dependent on DNA ligase, suggesting that the T-DNA exploits the host DNA repair machinery. The sites of integration appear to be randomly selected and sites of DNA damage at one or both strands probably serve as T-DNA entry points. Small deletions, base substitutions, and duplicated border and genomic DNA sequences are found around the T-DNA/plant DNA junctions.
- ***Model 2*** The T-DNA integration in the plant genome is favored at sites rich in AT content. The sequence immediately downstream of the plant AT-rich region is the master element for setting up the DNA duplex and deletions into the left end of the integrated T-DNA depend on the location of a complementary sequence on the T-DNA. The integration process is thought to commence with the 3′-end of the T-DNA invading the plant DNA through limited base pairing. The formation of a short DNA duplex between the host DNA and the left end of the T-DNA sets the frame for the recombination. Recombination at the right end of the T-DNA with the host DNA involves another DNA duplex, 2–3 bp long, which preferentially includes a G close to the right end of the T-DNA. Recognition of a bended DNA region might, therefore, be a common feature in the integration of foreign DNA in plant genome.

Nonhomologous end joining of ds T-DNA to double strand breaks Alternatively, a double stranded form of T-DNA might be generated first and then integrated into the chromosome. This process is called nonhomologous end joining (NHEJ) to double strand break created in the plant genome by an endonuclease (e.g., I-*Ceu* I or I-*Sce* I). AtKu80 is required for the initiation of NHEJ.

A frequent outcome of the process of integration by any model is the random integration of a copy of the T-DNA in the plant genome. The integration preferentially occurs in the transcription active and/or repetitive regions of the genome. Random integration of T-DNA into the plant genome occurs as single or multiple copies (but low) and produces various forms of rearrangements such as small duplications, deletions, and fillings. The causative factors of single or low copy number integration are not clearly known. There probably exists some kind of negative feedback system during infection, which prevents too many copies of T-DNA transfer and integration. After integration into the plant genome, bacterial genes are expressed, the products of which disrupt the hormonal balance within the plant cells and induce their proliferation to form crown gall tumors. In addition, it also produces enzymes to synthesize opines, which can be used by bacteria as their nutrition. Once a plant incorporates the T-DNA with its inserted gene, the encoded character passes on to future generations of the plant with a normal pattern of Mendelian inheritance.

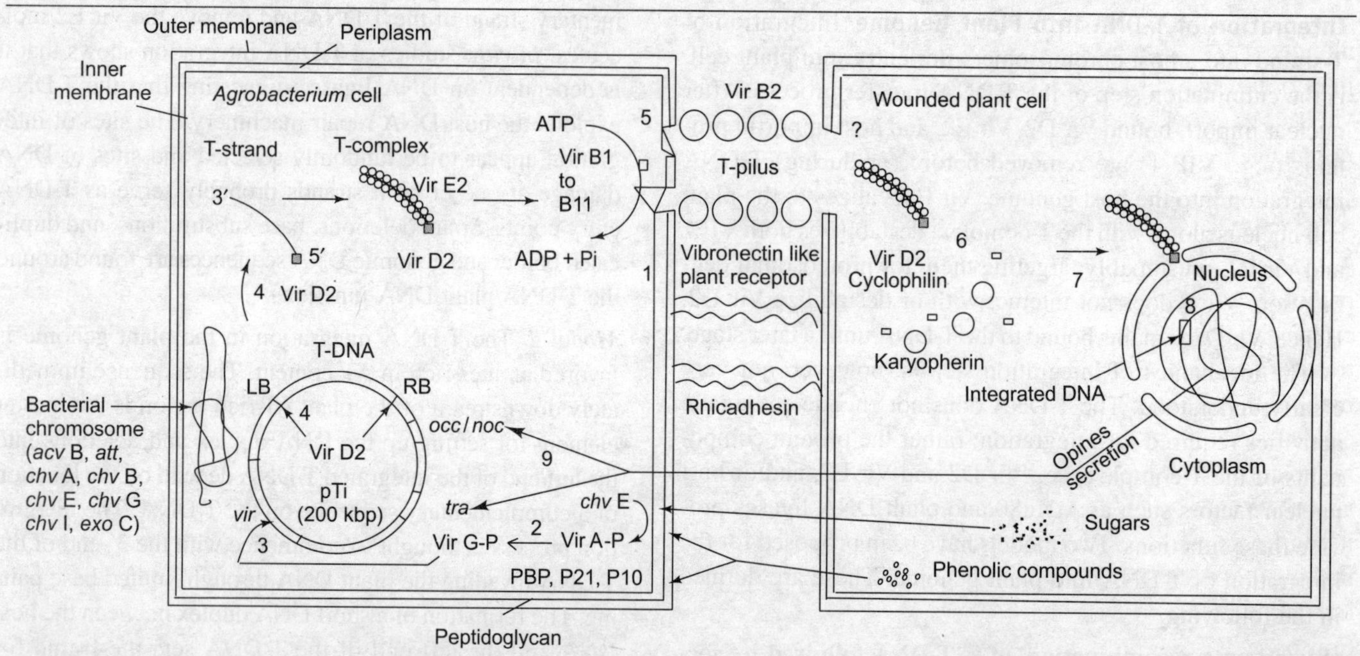

Figure 14.8 Basic steps in the transformation of plant cells by *A. tumefaciens* [Step 1: Binding of *Agrobacterium* to the plant cell surface receptors; Step 2: Direct or indirect mode of recognition of plant signal molecules by the bacterial Vir A/Vir G two-component sensortransducer system; Step 3: Induction of bacterial *vir* genes on pTi; Step 4: Generation of the T-strand; Step 5: Formation of the T-complex and its transport into the host plant cell through the T-pilus; Step 6: Recognition and interaction of T-complex with host cell cytoplasmic proteins (karyopherins and cyclophilins); Step 7: Nuclear transport of T-complex; Step 8: T-DNA integration into plant chromosome and opine secretion; Step 9: Activation of bacterial opine catabolism genes (*occ/noc*) and pTi conjugal transfer genes (*tra*); *acv* B, *att*, *chv* A, *chv* B, *chv* E, *chv* G, *chv* I, and *exo* C are the chromosome-encoded virulence genes; '?' indicates that precise steps are yet to be established; P10, P21, and PBP are chromosome-encoded putative phenol-binding proteins.]

The above-described procedure of *A. tumefaciens*-mediated plant transformation is outlined in Figure 14.8.

Plant Transformation using pTi-based Vectors

For plant transformation, either of the two vectors, i.e., binary or cointegrate vectors, are employed (for details see Chapter 12). The gene of interest for required trait is inserted into the polylinker site lying within T-DNA portion of the vector and a recombinant vector is constructed. This recombinant vector is then used for transformation of plant explant. This can be achieved simply by coculture (or cocultivation) of recombinant vector harboring *Agrobacterium* with various explants, such as protoplasts, leaf disks, and the tissues in whole plants. Cocultivation is routinely used in most of the laboratories for transformation of dicot as well as monocot species. The protocol has to be standardized for an individual plant species. A general protocol of cocultivation is summarized in Figure 14.9. Besides cocultivation, agroinfection/agroinoculation (integrating non-*Agrobacterium* factors such as the systematic virus infection) and agrolistics (the biolistic method for gene transfer) can also be employed. However, some of the plant viruses (e.g., mastreviruses) are not mechanically transmissible and hence cannot be effectively introduced into plant cells. *Agrobacterium* can also introduce a complete copy of a viral genome into isolated plant cells or whole plants. In this

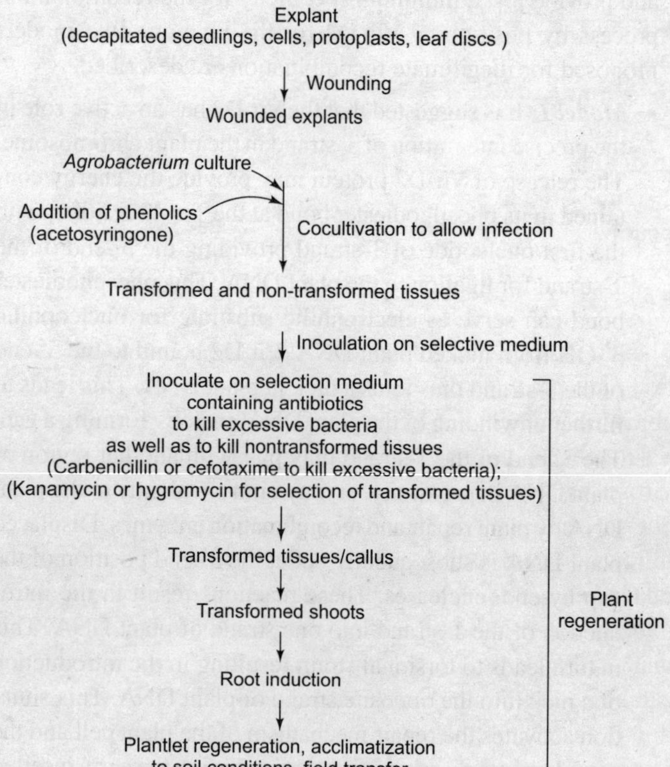

Figure 14.9 Protocol for *A. tumefaciens*-mediated transformation of explants

manner, it is possible to generate transiently transformed cell lines or transgenic plants carrying an integrated recombinant viral genome and the viral symptoms are visible within 2 weeks of inoculation.

14.4.2 *A. rhizogenes*-mediated Gene Transfer

A. rhizogenes also infects plants, however, the resulting pathology is different as it produces hairy roots instead of crown gall tumor. This system also finds some place in plant transformation. The Ri plasmid, which is responsible for formation of hairy roots at the site of infection, appears to be equivalent to disarmed pTi and many of the principles explained in the context of disarmed pTi are applicable to Ri plasmid as well. When foreign DNA is introduced into T-DNA region, it can be carried via the Ri plasmid into host plant cells, making it a natural vector for rapid introduction of genes into higher plants. Previously, *A. rhizogenes* transformation procedure was considered as an alternative strategy to *A. tumefaciens* for gene transfer as it led to the production of defined tissues (hairy roots) that could be regenerated into whole plants. However, later this strategy was discarded, as more efficient *A. tumefaciens* systems were developed. Further it is reported that *Agrobacterium* containing both Ri plasmid and a disarmed pTi can frequently cotransfer both of them. If such a dual system is used, no drug resistance marker on the T-DNA is necessary for the selection of transformants. In contrast, the transformants can be easily selected on the basis of hairy root formation in the target tissue, thereby indicating cotransfer of recombinant T-DNA. Intact plants are then regenerated by manipulation of phytohormones in culture medium. Moreover, hairy roots have been used as a source of secondary metabolites and for the production of pharmaceutical proteins.

14.4.3 Plant Viruses-mediated Gene Transfer

The plant (DNA and RNA) viruses used in plant transformation have already been discussed in Chapter 12. These have ability to transfer gene(s) to plants and thus have the potential for genetic transformation of plants. This potential of plant viruses to perform stable plant transformation is speculated due to certain properties possessed by them, which are as follows: (i) Plant viruses (DNA or RNA viruses) exhibit natural tendency of gene transfer to plant, hence transgene(s) when integrated into their genome may replicate and express in the plant cell along with the other viral genes; (ii) Plant viruses can be easily transmitted by direct adsorption and introduction of their genome into intact plant cells. Besides this normal infection route, naked DNA or RNA is also infectious in many viruses. Hence recombinant vectors can be introduced directly into plants by simple methods such as leaf rubbing. Further, grafting infected scions onto free hosts also serves as an easy mode of transmission; (iii) Vectors based on viruses are desirable because of the high efficiency of gene transfer that can be obtained by infection; (iv) The plant viruses have their own origin of replication, and unlike *A. tumefaciens*, these are nonintegrative in nature. Thus, plant viruses do not integrate into the host genome; rather these accumulate to high copy numbers in their respective target cells. Thus, recombinant viral vectors have the potential for high-level transgene expression. Moreover, as the viruses persistently infect transgenic plants, their use as vectors may produce large amounts of transgene-encoded recombinant protein; (v) Plant viral infections are often systemic. The virus spreads throughout the plant, allowing transgene expression in all cells; (vi) Viral infections are rapid and accumulate to high copy number in their respective target cells. Thus, when used as expression vectors, these may lead to production of large amount of recombinant proteins in a few weeks' time; (vii) All known plant viruses replicate episomally, hence the transgene carried by them is not subjected to the position effects that often influence the expression of integrated transgenes; (viii) Plant viruses neither integrate into the plant genome nor are transmitted through seeds (germ line). This is advantageous in terms of biological containment of the virus; and (ix) In the case of DNA viruses, *Agrobacterium*-mediated transient or stable transformation with T-DNA containing a partially duplicated viral genome can lead to the 'escape' of intact genomes, which then replicates episomally. The latter process known as 'agroinfection' or 'agroinoculation' provides a very sensitive assay for gene transfer. More recently, the *Agrobacterium*-mediated delivery of viral genomes has been enhanced through a process called 'magnifection', in which amplification of the vector occurs in all infected leaves.

14.4.4 Plant Transformation by Microprojectile Bombardment

Monocot plants and conifers were at one time considered intractable to *A. tumefaciens*-mediated transformation due to inappropriate wound responses and hormonal differences between dicots and monocots. At that time, microprojectile bombardment was developed as a method to introduce transgene into plant cell suspensions, callus cultures, meristematic tissues, immature embryos, protocorms, coleoptiles, and pollen in a wide range of plants, including monocots, and conifers. Furthermore, this method is used to deliver genes into organelles such as chloroplasts and mitochondria, thereby opening the possibility of introducing transgenes into these organelles. Note that this is not possible with *A. tumefaciens*-mediated transformation procedure, as it targets the DNA only into the nucleus of plant cell. Thus, microprojectile bombardment serves as one of the most important alternative to

A. tumefaciens-based pTi DNA delivery systems for plants. Today microprojectile bombardment is the method of choice in most of the laboratories and is routinely used to create transgenic monocot as well as dicot plants. Microprojectile bombardment is also called 'biolistics' (a combination of biological and ballistics), 'particle gun', 'particle acceleration', 'particle bombardment', 'gene gun', and 'bioblaster'.

Principle of Microprojectile Bombardment

The principle involved in gene gun is that when small high-density microparticles (microprojectiles) are accelerated to high velocity, these particles acquire sufficient kinetic energy and can penetrate cells and membranes. When DNA-coated microprojectiles are used, these particles penetrate the outer cell layers and target the DNA into the cells.

Basic Equipment Design

After its first development in 1987, biolistic devices have become much more sophisticated and advanced. The basics of all of these designs remain the same and involve coating of DNA onto small dense microparticles followed by their high-velocity acceleration toward a target tissue. Thus the microprojectile bombardment equipment comprises of high-pressure source, microprojectiles, macrocarriers, stopping screen, vacuum pump, and baffles/vents/meshes.

High-pressure Source for Microparticle Propulsion There are three different sources for generation of high pressure, which are used to provide the propelling force for targeting microparticles to target tissues. These include: (i) Chemical explosion (use of gun powder); (ii) Discharge of compressed air; and (iii) Helium shock. Different designs of particle gun employ any one of these three high-pressure sources for particle acceleration. The gun powder-driven system employs nail-gun cartridges. This blast system from the gun powder cartridge is not only dangerous, but also leads to uncontrolled generation of dirty gases and debris within apparatus. This necessitates frequent cleaning of the chamber and the polyethylene macrocarrier as well as stopping plate has to be discarded after each transformation process. Moreover, the pressure cannot be regulated and the unregulated pressure may lead to target variation, and nonreproducibility of results after transformation. As a consequence of these drawbacks, this system has been abandoned. Today, the most common system is helium-driven system. Helium is a light gas and expands faster, and is superior to other gases such as nitrogen or compressed air. This leads to better dispersal of microparticles, resulting in more uniform field of transformation. It is more effective, safer, and cleaner than gun powder-driven system. Another advantage of using a helium-driven system is that it reduces drag on microparticles. The pressure thus generated is monitored by pressure gauge and is used to accelerate the DNA-coated microparticles with high velocity and kinetic energy sufficient to penetrate the target cell. Note that helium-driven system is functional in pressure ranging from 600 to 2,400 psi, but ~1,000 psi is optimal. Pressure below 600 psi is suboptimal for transformation and at pressure above 2,400 psi, though there is an increase in velocity, gas shock that impacts biological target also increases.

Microprojectiles/Microparticles Microprojectiles are high-density microparticles used for carrying coated DNA. Ideally these microparticles should have good initial affinity for the DNA, yet should be able to freely release DNA once in the cytoplasm or nucleus of the target cell. Moreover, these microparticles should be small enough to enter the plant cell without too much damage, yet large enough to have the mass to penetrate the cell wall and carry an appropriate amount of DNA on its surface into the interior of the plant cell. As such, gold or tungsten microparticles are used for coating of DNA. These microparticles are high-density microparticles because the low-density particles may have ability to penetrate cell walls, but their reduced momentum decreases the efficiency of penetration. These microparticles are spherical in shape ranging in diameter from ~0.5–2 μm, or about the size of some bacterial cells. As a rule of thumb, microparticles should be roughly one-tenth the diameter of the cell. Gold microparticles offer certain advantages as compared to tungsten microparticles. Gold microparticles are biologically inert and do not produce cytotoxic effects. Moreover, gold is more malleable and therefore can be shaped into uniform particles. This uniformity in size of gold microparticles allows size optimization for a cell type. Furthermore, gold microparticles are chemically inert, and do not attack the DNA bound to them. Because of their properties, Food and Drug Administration (FDA) of the United States has approved gold microparticle as a human therapeutic agent. One problem while using gold microparticles is that these tend to agglomerate irreversibly in aqueous suspensions over a period of time, i.e., these are unstable. This problem can be overcome by preparing them at the time of use. On the contrary, tungsten microparticles are somewhat cytotoxic and vulnerable to oxidation, which promotes degradation of bound DNA. As compared to gold microparticles, the tungsten microparticles tend to be heterogeneous in size, as a consequence of which size optimization for a cell type is a problem. Moreover, tungsten microparticles are subject to surface oxidation, which can alter their DNA-binding capacity. However, the cost of gold particles may limit their accessibility and the tungsten particles being more affordable are sufficient for most studies. Besides gold and tungsten, other microparticles reported with some success include glass needles, silica, and dried cells of *E. coli* and *A. tumefaciens* (biological capsules).

Macrocarrier/Macroprojectiles/Flying Disk Many of the particle guns utilize a macrocarrier. These support or carry the microparticles, and thus are accelerated under high pressure along with the microparticles toward the target explant.

A stopping plate usually retains the macrocarrier before it could collide with the target explant, whereas the microparticles continue along their course. The macrocarriers are of different types. For gun powder-driven system, cylindrical shaped, high-density polyethylene macroprojectiles are used, while for helium-driven system, circular Kaplon membranes (2.5 cm diameter, 0.06 mm thickness, DuPont) are best-suited. Membranes require very little time to gain speed; hence short flight distance is required. The less massive membrane can be stopped with a screen, rather than Lexan disk as in the case of polyethylene macrocarriers. As a result, more microparticles can be delivered without any associated high velocity 'debris', which can be generated from the macrocarrier or stopping screen.

Rupture Disk Some designs of gene gun include plastic rupture disk, which is fitted in a rupture disk holder, and placed at an appropriate position in the equipment chamber. This rupture disk bursts due to passage of high-pressure gas and allows it to hit the macrocarrier loaded with DNA-coated microprojectiles. These have a pressure cut-off value and are available at the operating pressures of 450, 650, 1,100, and 1,350 psi.

Stopping Screen or Plate These are metallic plates with a small opening at the center. This obstructs the movement of macrocarrier ahead of it, while DNA-coated microparticles travel through the opening, and penetrate the cell wall and membrane of the target explant.

Vacuum Pump Microprojectiles are rapidly decelerated as these pass through gas (note that helium exerts reduced drag on microparticles as compared to any other gas). Hence, in most cases, the microparticles are accelerated under partial vacuum in a vacuum chamber to reduce drag due to gas. As gas can transmit potentially damaging shock wave, evacuation of the chamber by vacuum pump reduces the severity of gas shock. Standard vacuum gauge reading inside the chamber is 25–29 mmHg.

Baffles, Vents, or Meshes Acoustic shock and gas blast are reduced by proper positioning of baffles, vents, or meshes between the microprojectile launch site and the target explant.

Particle Acceleration Devices

The original gene gun employs DNA-coated tungsten microparticles in aqueous slurry loaded on the end of a small plastic bullet, a macrocarrier. The plastic macrocarrier is placed into a 0.22 caliber barrel in front of a gun powder cartridge. Once the cartridge is fired, the plastic bullet is propelled toward a stopping plate containing a small opening in the center. The microprojectiles continue through the hole toward the target tissue while the macrocarrier is retained. Despite various drawbacks associated with gun powder-driven system of particle acceleration, this first unit was used successfully in different laboratories. However, now several instruments have been developed where particle acceleration is controlled by pressurized gas. This results in greater control over particle velocity, leading to greater reproducibility of transformation conditions. Different modified acceleration devices, which differ in one or more features, are now available and are listed:

Helium-modified Bombardment Device The first important modification of the original device includes the substitution of a helium blast for the gun powder discharge. Helium pressure continuously builds up in a reservoir and is released using rupture disks. The release of helium produces a shock wave, which travels to a macrocarrier. Macrocarrier that holds the DNA-coated microparticles flies a short distance into a stopping screen, which retains the macrocarrier, while the microprojectiles continue traveling and ultimately penetrate the target explant held in a partial vacuum. The most common example in most of the laboratories is PDS-1000/He device (Figure 14.10).

Accel Particle Gun This gun uses a higher voltage electrical discharge to vaporize a water droplet, which produces a controlled shock wave. The device is constructed so that the initial shock wave is reflected to produce secondary shock waves. This, in turn, accelerates a Mylar sheet coated with microparticles. The Mylar sheet is accelerated toward a retaining screen, which stops the macrocarrier and allows the microparticles to continue onto the target tissue. Adjusting the intensity of the voltage passing through the water droplet accurately modifies particle speed and the resulting depth of particle penetration. It is the most efficient particle bombardment device based on electric discharge in current use that is particularly useful for the development of variety of independent gene transfer methods for the more recalcitrant cereals and legumes. However, it is a costly device due to manufacturing expenses.

Particle Inflow Gun (PIG) In this device, a solenoid controlled by a timer relay is used to generate a burst of low-pressure helium. The particles, supported by a screen in a reusable syringe filter unit, are accelerated directly in a stream of helium without the need for a macrocarrier. The target tissue is held in a chamber under a partial vacuum. To minimize the localized impact of the particles, a nylon mesh baffle may be placed over the tissue to aid in dispersal. For this device, macrocarriers or rupture membranes need not be purchased. It requires less cleanup and minimal down time between shots. Moreover, it is affordable, simple, safe, and effective device.

Microtargeting Device This device was originally developed to allow precise delivery of particles to the shoot meristem. In this method, DNA is not precipitated on the gold particles but is used as a DNA–particle mixture. This device accelerates small amounts of this DNA–microparticle mixture in a focused stream of high-pressure nitrogen.

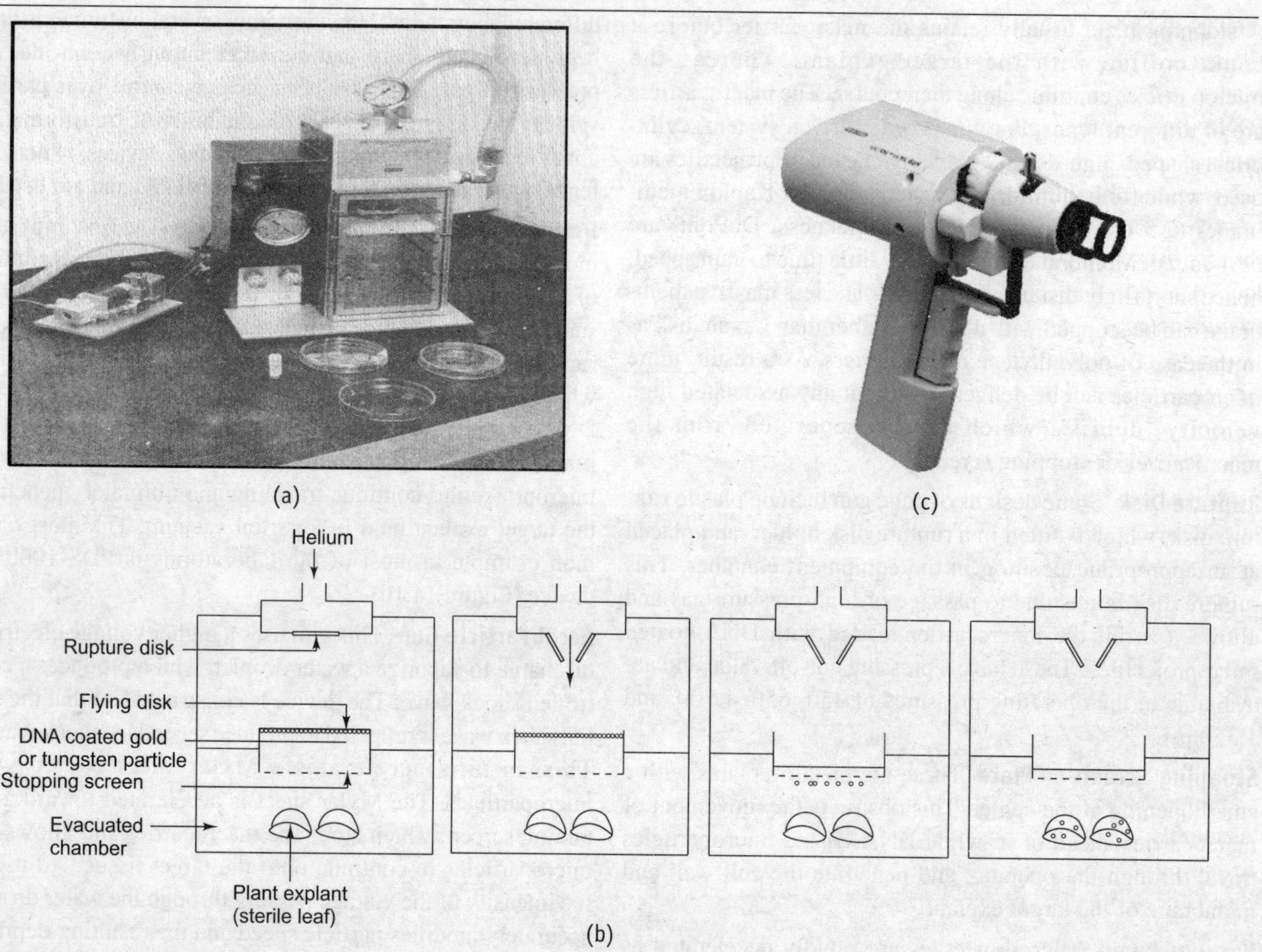

Figure 14.10 Technique of microprojectile bombardment; (a) DNA particle acceleration apparatus, (b) Schematic representation indicating functioning of PDS-1000/He device (Bio-Rad), and (c) Helios gene gun (Bio-Rad)

Helicos Gene Gun or Handheld Gene Gun In the gene guns discussed so far, targeted tissue is held in a vacuum and apparatuses are not highly portable. Moreover, these devices cannot be easily used to transform target tissues sensitive to conditions created by evacuated chamber. The Helios gene gun is a handheld device, which uses low-pressure helium to accelerate DNA-coated microparticles from the interior of a small plastic cartridge toward the target explant (Figure 14.10c). A spacer at the tip of the gun maintains optimal target distance and minimizes cell damage by venting the helium gas away from the explant. The Helios gene gun can accommodate up to 12 loaded cartridges at one time, thus allowing for multiple firing of the device before reloading is required. The greatest utility of this pressurized handheld gene gun seems to be *in situ* transformation of cell and tissues and the DNA-coated gold or tungsten particles can be directly shot into the meristematic tissue of plants growing in experimental fields.

Procedure

In this section, the procedure for gene gun-mediated direct gene transfer is described. PDS-1000/He device (BioRad) is used for discussion, as it is the most commonly used device in most of the laboratories. This biolistic transformation system uses an evacuated sample chamber that positions the target cells directly in the path of DNA-coated tungsten or gold particles traveling at a velocity of 1,100 mph. Various steps involved in PDS-1000/He transformation system are as follows:

Preparation of Microparticle Suspension First, the gold or tungsten microparticles are suspended in sterile water (30 mg/500 ml) and vortexed briefly. The microparticles are then washed once in 70% ethanol (EtOH) and twice in absolute EtOH. Finally, the microparticles are suspended in sterile water. While vortexing slowly, the particles are aliquoted into very small volume (50 µl each) in clean tubes for storage at 4°C (gold) or –20°C (tungsten). Note that this preparation is good for 2–3 months.

Precipitation of DNA on Microparticles (Preparation of DNA-coated Microparticles) At the time of use, one tube of microparticle (prepared in first step) is taken and water is discarded by brief spin (15 seconds). With slow and continuous vortexing, 5–10 µl of DNA (1–2 mg/ml), 50 µl of 2.5 M $CaCl_2$, and 20 μl of 0.1 M spermidine-free base are added in the mentioned order. Spermidine probably protects DNA from degradation and/or prevents conformational changes in DNA. The DNA-coated microparticles are spun at 10,000 rpm for 10 seconds in cold and supernatant is removed to the maximum possible extent. The microparticles are washed with EtOH and resuspended in 60 volumes of ultrapure EtOH, vortexed very briefly at low speed, and the tube is then kept on ice till required. Note that this preparation of DNA-coated microparticles should be used within few hours of preparation.

Application of DNA-coated Microparticles on Macrocarrier Membrane On one hand, macrocarrier membrane, stopp-ing screen, and macrocarrier holder are soaked in EtOH for 15 min. EtOH is then removed and the apparatuses are air-dried under the laminar flow hood. Caution is taken to avoid touching of these apparatuses while drying. On the other hand, the DNA-coated microparticles are briefly sonicated, vortexed, and suspended with the help of a tip and 10 μl of the solution is pipetted, and the suspension is immediately but slowly released onto the center of macrocarrier without touching its surface. The drop is allowed to air-dry under laminar flow hood. It is pertinent to note here that the particle density used should be such that it does not significantly damage the cells.

Setting of Macrocarrier and Rupture Disk Holders in PDS Assembly The stopping screen is placed in the assembly followed by macrocarrier in the macrocarrier holder. Rupture disk is rinsed once in isopropanol (optional) and placed into rupture disk holder.

Steps for Using PDS-1000/He First stage involves unlocking of helium cylinder and setting the first stage meter reading to more than 2,200 psi. The PDS assembly is then switched on and 'vent' switch is kept in the on position. Then second stage pressure regulator is opened and the pressure is kept at least 200 psi more than the grade of rupture disk used. For example, if 1,350 psi rupture disk is used, the meter should read at least 1,550 psi. The rupture disk holder containing rupture disk is placed at appropriate position and tightened into the PDS assembly. Then macrocarrier holder with fitted metallic stopping screen and macrocarrier is mounted on the PDS launch chamber in the required slot. The petridish that contains the target to be transformed is then placed under the assembly in the desired slot and the door of the chamber is closed. The chamber is then evacuated to 25–29 mmHg and held at the same pressure during transformation. When the helium pressure builds up to a required value, the 'fire switch' is turned on and kept in position till a 'phut' sound is heard in the launch area, indicating the bursting of plastic rupture disk. During this shot, released gas accelerates the macrocarrier with the DNA-coated microparticles on its side. Propelled under high pressure, the DNA-coated microprojectiles pass the stopping screen and penetrate plant cell walls and membranes, while stopping screen holds back the macrocarrier. Once inside a cell, the DNA is removed from the microparticles and integrates into plant genome in some cells. After the shot, the 'fire switch' is switched off and the vacuum is quickly released. This step is important and should be performed within 2 seconds of firing. The door is opened after release of the entire vacuum. The petriplate containing target explant on suitable medium is removed. It is estimated that ~10,000 transformed cells are formed per bombardment if proper considerations are taken into account. The macrocarrier assembly, and rupture disk holder are also removed and dismantled. The PDS assembly is wiped with EtOH and switched off. Helium cylinder is closed with the help of key, and by noticing that needles on both the meters of pressure gauges attached to helium cylinder have fallen to zero, its closure is ensured.

Factors Affecting Particle Bombardment-mediated Transformation

DNA delivery by particle bombardment is governed by several parameters, which must be optimized for each species and the type of explant. These include following:

Particle Accelerator Parameters (Gene Gun Settings) These parameters include the following.

High-pressure source or power source Helium-driven system is superior to gun powder-driven system in effectiveness.

Intensity of explosive burst or gas pressure or power load This factor is important to regulate the extent of microparticle penetration into target plant cells. An increase in power load can compensate for increased gap distance (i.e., the distance between rupture disk and macroprojectile launch site), decreased macroflight distance (i.e., the distance between macroprojectile launch site and stop screen), or target distance (i.e., distance that the microparticles have to travel to reach the target cells).

Gap distance The gap distance plays an important role in controlling the velocity and hence penetration of target plant cells. Short gap distance increases velocity and also variability and off-centered macroflight. Thus, 6–12 mm is optimal for transfer with minimum variability.

Macroprojectile flight distance (macroflight distance) Macroflight distance is also crucial in transformation procedure. As macroflight distance increases, velocity increases, but variability and off-centered hits also result. Macroflight distance of 10 mm is considered optimal without serious variations in flight orientation or transformation frequency.

Target distance Besides gap distance and macroflight distance, the target distance is also important. By altering the flight distance, the extent of microparticle penetration into target plant cells can be controlled. This distance is not very critical when large microparticles are used. However, for small microprojectiles, target distance must be minimized to maintain adequate impact velocity for penetration into the target cells.

Microparticle Parameters Microparticle parameters that affect the microprojectile bombardment-mediated transformation include the following:

Choice of the microparticle Either gold or tungsten microparticles are used. Gold particles have several advantages over tungsten microparticles.

Size of the microparticles By using different sized particles, the extent of microparticle penetration into target plant cells can be controlled.

Preparation of the microparticle The use of aminosiloxanes, instead of $CaCl_2$ and polyamines, to coat the microparticles with DNA may lead to higher transformation efficiencies at lower DNA concentrations.

Microparticle density Particle density used in the transformation procedure should be such that it does not significantly damage the cells.

Microparticle acceleration Microparticle acceleration, which depends on the pressure or burst, affects the depth of penetration and the amount of tissue damage.

Biological Parameters These parameters include the following.

Amount, conformation, and purity of DNA It is possible to increase the transformation frequency by increasing the amount of plasmid DNA. However, too much plasmid DNA can be inhibitory and may lead to a high copy number and rearrangements of the transgene constructs, whereas too little DNA may lead to low transformation frequencies. The conformation and size of the vector that is used for biolistic delivery of foreign genes to plants influences both the integration and expression of those genes. For example, transformation is more efficient when linear rather than circular DNA is used. Moreover, large plasmids (>10 kbp) in contrast to small ones may become fragmented during microprojectile bombardment and therefore produce lower levels of foreign gene expression. However, large segments of DNA may be introduced into plants using yeast artificial chromosomes.

Nature of target explant (state of the target, target type, cell age, and physiology) The nature of the transformation target is one of the most important variables in the success of gene transfer. It is very important to target the appropriate cells that are capable of cell division and are competent for transformation. It should not be damaged during penetration and there must be high rate of particle penetration, cell survival, and growth after bombardment. Target type is important from the point of view of intact tissues penetrated at high particle velocity, while cell cultures require gentle treatment. Cell age and physiology are also important and hence healthy cells that can withstand stress of bombardment process should be employed.

Target cell size Target cell size is crucial because particle size and target distance are based on cell size. The particle size is kept one-tenth the size of the target cell.

Osmoticum The pretreatment of explants with an osmoticum has often been shown to improve transformation efficiency, probably by preventing the deflection of particles by films or droplets of water. Moreover, elevated osmoticum protects cells from leakage and bursting and improves penetration. Thus to protect the plant tissue from damage sustained during the bombardment procedure, treatments to induce limited plasmolysis or culture on high-osmoticum medium should be used.

This indicates that for successful transformation via particle bombardment, the vector DNA should be carefully designed and selected, and special attention also needs to be placed on the nature, state, and receptivity of the target tissues. Moreover, after the DNA-coated microparticles are introduced, there are numerous factors that influence gene expression and the subsequent recovery of transgenic plants.

14.4.5 Plant Transformation by Electroporation

Electroporation is suitable for both monocot and dicot plant cells, or protoplasts, and successful gene transfer using electroporation has been reported in tobacco, petunia, maize, rice, wheat, sorghum, and carrot. Initially protoplasts were used for transformation, which employs a difficult plant regeneration system to generating calli or whole plants. But now one of the advantages of the system is that intact plant cells and tissues, such as callus culture, immature embryos, and inflorescence, can also be used. However, the plant material may require specific treatments, such as pre- and post-electroporation incubation in high-osmotic buffers. In many cases, the target cells are wounded or penetrated with enzymes in order to facilitate gene transfer. Currently electroporation using organized and easily regenerable tissues such as immature embryos is possible, and immature rice, wheat, and maize embryos can be easily transformed without any form of pretreatment. The procedure for transformation of plant by electroporation is essentially the same as described for bacteria (for details see Section 14.2.2). In this case after pulse delivery, cuvettes are kept on ice for 10–15 min. Explants are taken out, rinsed in medium, and plated on MS medium with appropriate selection pressure.

Factors Affecting Efficiency of Electroporation

Factors affecting electroporation have been already described in Section 14.2.2.3. During electroporation formation in plasma membrane or protoplast fusion only occurs if the membrane voltage exceeds a certain threshold or critical voltage (V_c). Pores are first formed at those membrane sites, which are positioned on the 'equator' of the protoplasts. Larger protoplasts are exposed to a greater transmembrane voltage than smaller protoplasts in response to a given field strength. If field strength (E) is increased so that transmembrane potential (V) is greater than critical voltage (V_c), a corresponding increase in the number and size of pores occurs, leading eventually to membrane rupture and protoplast lysis. The latter aspect of membrane rupture and protoplast lysis (i.e., excessive damage to protoplasts) reduces protoplast density and releases metabolites that inhibit subsequent division and growth. It is because of this reason that somewhat milder electroporation conditions and modified posttransfection culture conditions are recommended.

Moreover, the cell type or strain, condition of plant material used, and the tissue treatment conditions are significant. It has been reported that by using linear DNA rather than circular DNA, field strength of 1.25 kV/cm, and employing PEG can increase efficiency of transformation. Addition of spermidine to the incubation buffer also improves the efficiency of electroporation. Furthermore, keeping the required DNA fragment in direct contact with the explant facilitates the entry of DNA molecules into the cells.

14.4.6 Plant Transformation by Microinjection

It is a direct physical approach for mechanical introduction of DNA into a specific target under microscopical control. This technique was originally developed for animal cells, but is now applied successfully to plant cells as well. The possible targets in plants include defined cells within a multicellular structure such as embryo, ovule, and meristem. Both intact cells and protoplasts can be employed for introduction of foreign DNA by this technique. The introduction of DNA is so precise that besides intracellular injection, direct intranuclear injection is also possible.

Principle of Microinjection

The principle involved in microinjection technique is that the appropriate volume of biological sample can be injected directly into the cytoplasm or nucleus of cell with the aid of gentle air pressure applied to the capillary through a manual or electronic air pulse system (syringe or micropipette) (Figure 14.11).

Procedure of Microinjection

Microinjection technique is performed under phase contrast microscope and uses a glass capillary needle (or micropipettes) with a tip diameter ranging from 0.5 to 10 μm for precise transfer of the DNA into the cytoplasm or the nucleus of a recipient target cell or protoplast tightly held or immobilized on solid support (i.e., glass cover slips or slides imprinted with numbered squares). The immobilization of the recipient cells is done by using methods such as agarose embedding, agar embedding, polylysine treatment of glass surfaces, and suction holding with the aid of a micromanipulator that positions the holding pipette to hold the cell in position. The capillary is filled from the tip with the biological sample and directed into immobilized cells. An appropriate sample volume is transferred by gentle air pressure exerted by a syringe connected to the capillary. Recipient cells are grown on squared glass slides for convenient localization of the cells injected. The procedure is such that the cells survive the introduction of DNA and can develop further into clones. In order to acquire rapid injection rates, protoplasts with partially reformed cell walls are attached to solid support with poly-*L*-lysine or artificially bound substrate without damaging the cells. After one day in culture, the cell walls of protoplasts become firm enough to provide support and protection and at the same time, these are easily penetrable with micropipettes with a tip diameter ranging from 0.5 to 10 μm. Each protoplast is then microinjected, segregated, and transferred to optimal conditions to obtain transformed cell lines. An alternative method employs a blunt holding pipette to hold each protoplast under suction and segregating it directly into microdrop culture after microinjection.

14.4.7 Plant Transformation Using Silicon Carbide Whiskers or Silicon Carbide Fibers

This method is technically very simple and does not require any sophisticated equipment. This technique only requires silicon carbide fibers (0.3–0.6 μm diameter and 10–100 μm length), vortex, target tissue, and sample DNA. Presently this method is used with cells in suspension culture, embryos, and embryo-derived calli of only a few plant species; however, it offers a great potential for plant transformation.

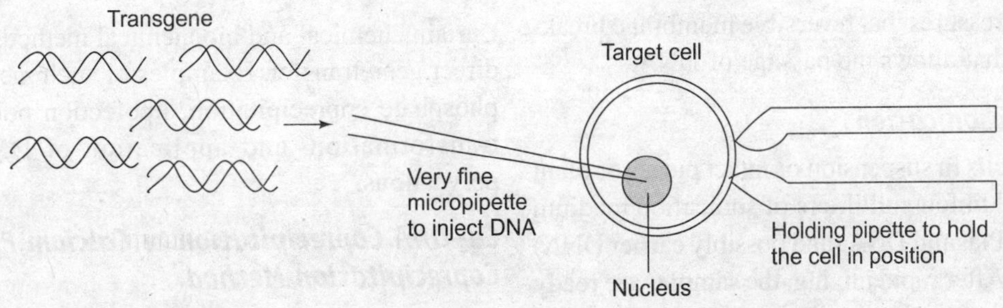

Figure 14.11 Technique of microinjection

Principle of Silicon Carbide-mediated Transformation

Silicon carbide forms microscopically small needles (silicon whiskers) that can be used to perforate plant cell walls and plasmalemma by forceful intercellular collisions during vortexing. This allows the DNA adsorbed to them to gain access inside of the cell by diffusion. Thus the silicon carbide fibers apparently act as microinjection needles facilitating DNA delivery.

Procedure of Silicon Carbide-mediated Transformation

In this method, DNA, silicon carbide whiskers and tissue of the species to be transformed are mixed in a suitable buffer in a microfuge tube. This mixture is vortexed to facilitate cellular penetration by enhancing frequency and forceful intercellular collisions. During these collisions, DNA adsorbed on silicon carbide fibers diffuses into the cell.

14.4.8 Plant Transformation by Ultrasonication

The latest method developed for plant transformation uses ultrasound to facilitate the uptake of nucleic acids into plant cells. By this technique, versatile cell types such as plant protoplasts, cells in suspension, and intact pieces of plant tissues can be readily transformed. Furthermore, the equipment for ultrasound transformation is simple and multifunctional. This method has been successfully used to transform tobacco, wheat, and sugar beet, and holds prospects for substantial future applications in plant transformation.

Principle of Ultrasonication

The procedure is based on the permeabilization of membrane by ultrasonication, leading to uptake of DNA. Two hypotheses have been given for its explanation. First, the violent collapse of activation bubbles, generating high pressure and temperature shock waves, possibly causes localized rupture of plasmalemma and leads to uptake of exogenous DNA. The second hypothesis predicts the existence of critical hydrostatic pressure at which the intrinsic membrane potential is sufficiently high to induce mechanical breakdown of membrane (60–100 MPa) in a hyperbaric chamber. Consequently, it is possible that either the high oscillating pressure generated by the ultrasonic field, and/or the high pressure shock waves originating from collapse cavitation could produce such higher hydrostatic pressures that reversible membrane breakdown would occur that allows the passage of DNA.

Procedure of Ultrasonication

The protoplasts or cells in suspension or intact pieces of plant tissue are suspended in few milliliters of sonication medium in a microfuge tube. Plasmid DNA (and possibly carrier DNA) is added to the tube. After rapid mixing, the samples are ready for sonication. A microtip, generally tapered, is immersed 2–3 mm into sonication medium, and pulses of ultrasound of selected intensity and duration are delivered to the tube by ordinary machines such as ordinary homogenizing machines. In the process, reversible membrane damage takes place, which serves as a route for DNA diffusion.

14.4.9 Laser Microbeam Irradiation-mediated DNA Uptake

This method uses finely focused laser beam for introduction of transgenes into whole plant cells. This technique is an ideal tool for the manipulation of cell components and organelles because it offers several advantages. First, this technique can be used for the genetic manipulation of plant tissue, individual cells, protoplasts, and organelles. The technique is also applicable to cells with hard walls. Secondly, laser beam can be used inside the tissues, cells, and organelles, thus increasing the rate of transformation and the speed of operation. Thirdly, laser beam intensity can be widely controlled and optical manipulation is very flexible. A wide choice of laser wavelength and pulse width is another advantage. Fourthly, it permits selective fusion in a part of a subcellular organelle present inside a single living cell. Finally, it has the ability to transfer cytoplasmic material.

Principle and Procedure of Laser Microbeam Irradiation-mediated Transformation

Plant cell transformation by laser irradiation is based on the delivery of high-energy light pulses focused through a microscope phase contrast objective onto regions of the target tissue of less than 1 μm in diameter. A continuously adjustable optical attenuator placed in the light path controls the laser energy. The beam can be focused very accurately so that the targeted area can be finely controlled. It is suggested that the focused high energy removes components from the plant cell walls or perforate self-healing holes of defined dimensions in the plant cell wall and membrane, which allow DNA uptake through passive diffusion. By creating an osmotic gradient between the target cells and an external hypotonic buffer, the efficiency of the technique for plant transformation can be increased.

14.4.10 Direct Gene Transfer by Chemical and Biochemical Methods

Certain chemical and biochemical methods are also used for direct gene transfer. Examples of such methods are calcium phosphate coprecipitation, lipofection polycation-mediated transformation, and application of DMSO along with polycations.

Ca–DNA Coprecipitation or Calcium Phosphate Coprecipitation Method

Calcium phosphate coprecipitation method can be used for transfection purpose.

Principle and Procedure In this method, DNA is mixed with $CaCl_2$ solution and isotonic phosphate buffer to form a DNA–calcium phosphate coprecipitate (also called Ca–DNA coprecipitate). The Ca-DNA coprecipitate enters the cell by endocytosis and a part of the coprecipitate escapes from endosomes or lysosomes to enter the cytoplasm, and finally the coprecipitate is transferred to the nucleus.

When protoplasts are contacted with Ca–DNA coprecipitate, protoplast/Ca–DNA complex is formed (instead of protoplasts, other actively dividing cells can also be used). In order to encourage the endocytic uptake of the coprecipitate, the protoplast/Ca–DNA complex is treated with polyvinyl alcohol and high pH calcium. A physiological shock with DMSO can also increase the efficiency of transformation to a certain extent. Upon several hours of incubation, the Ca–DNA coprecipitate is transferred across the plasma membrane in a calcium-requiring process. The protoplasts are washed and then implanted on fresh culture medium. Relative success depends on high DNA concentration and its apparent protection in the coprecipitate. This method was first used for transformation of tobacco leaf protoplast.

Factors affecting Ca–DNA Coprecipitation Several variables have been reported to affect the efficiency of transfection mediated by Ca–DNA coprecipitation. These are described below.

Speed of mixing of DNA and calcium phosphate Coprecipitates can be formed gently by bubbling air from an electric pipetting device or by slowly adding DNA followed by gentle vortexing. However, the formation of coarse precipitates should be avoided.

Size and concentration of the DNA High molecular weight genomic DNA is added to the coprecipitate to increase the efficiency of transformation by small DNAs. After transfection, the small DNAs integrate in the carrier DNA, frequently forming an array of head-to-tail tandems. The resulting complex subsequently integrates into the chromosome of the transfected cell.

pH of the buffer Exact pH of the buffer and the concentration of calcium and phosphate ions are also important.

Use of facilitators Transfection efficiency and transient expression level can be increased by exposing the cells to glycerol, transfection maximizers, or certain inhibitors of cysteine proteases. In most cases, these agents are toxic to cells and their effects on viability and transfection efficiency vary from one type of cell to another. Therefore, the optimal time, length, and intensity of treatment with facilitators should be standardized for each cell.

Enhancement of extent of transcription The level of transient expression is primarily dependent on the promoter and its associated *cis*-acting regulatory elements. In some cases, the level of expression increases by exposing the transfected cells to hormones, heavy metals, or other substances that activate the appropriate cellular transcription factors. Nowadays, transfection kits are commercially available from a number of vendors having some kind of modifications of the original protocol.

Lipofection

Liposomes serve as ideal carrier systems and are now used in genetic engineering to deliver foreign DNA molecules into plant protoplasts. This use of liposomes as a transformation or transfection system is called lipofection.

Principle of Lipofection Liposomes represent special type of lamellar phase in which water is self-contained and the lipid molecules are disposed in bimolecular layers attached by their nonpolar interfaces. Thus, liposomes are closed, self-sealing, solvent-filled vesicles bound by a single bilayer and possess many properties similar to those of biological membranes. These are easy to manipulate, their lipid contents can be varied as needed, and many substances may be trapped inside the interlamellar spaces. Owing to their properties, a wide variety of molecules including DNA can be encapsulated within their aqueous interior and can be delivered to cells in a biologically active form via endocytosis or membrane-membrane fusion (Figure 14.12).

Procedure of Lipofection A stable water-in-oil emulsion is produced in a screw-capped glass tube by brief sonication of phospholipid (phosphatidyl serine or phosphatidyl choline or phosphatidyl glycerol) plus an aqueous buffer [Tris-HCl or HEPES (*N*-(2-hydroxyethyl) piperzazine-*N*′-(2-ethanesulfonic acid)] or phosphate buffer, pH 7.8] containing DNA (i.e., recombinant plasmid DNA) to be packaged in

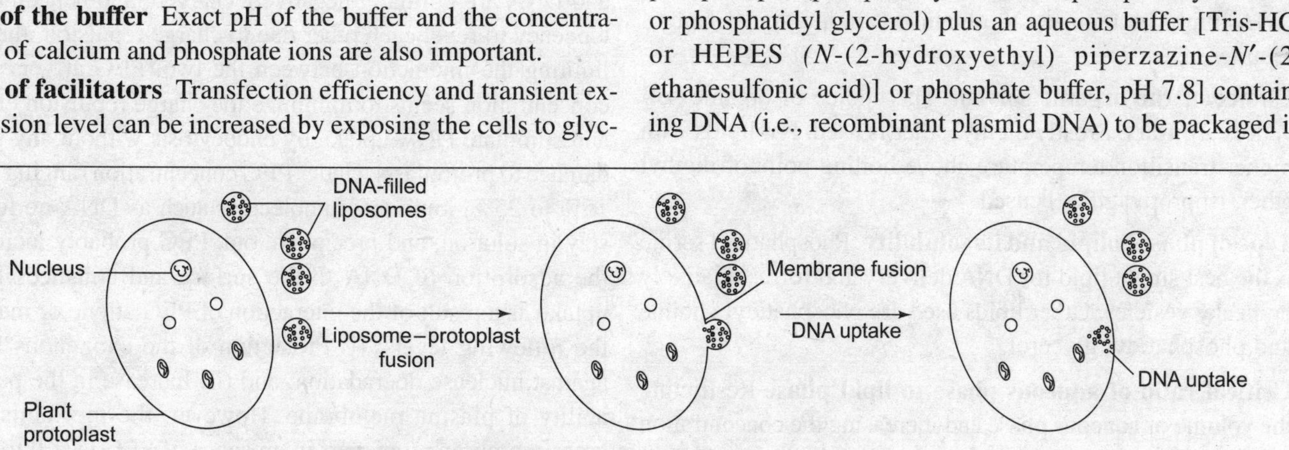

Figure 14.12 Fusion of DNA-filled liposomes with protoplast

a large volume of organic solvent (diethyl ether or isopropyl ether). The sonicator is purged with inert gas to keep the preparation sterile. The phases are reversed subsequently by evaporation of the organic solvent (reversed phase evaporation; REV) using rotary evaporator till a viscous gel is obtained. The system is vortexed briefly to promote breakdown of the gel and then vacuum is increased to 500–600 mmHg. At this point, encapsulation of DNA is complete and the preparation is stable for months at inert atmosphere and 4°C. These liposomes with encapsulated or internalized DNA then serve as carrier of DNA, which when incubated for ~30 min with plant protoplasts, transfer DNA into them by either direct fusion with the naked plasma membrane or through endocytosis of the liposome. At the end of incubation, the cells are washed with buffer and plated on growth medium. Liposomes can also be used in combination with PEG or DMSO for the delivery of plasmid DNA into protoplasts, which enhances effective delivery of liposome-entrapped DNA.

Liposome strategy is also useful for transferring DNA into vacuoles of walled plant cells. The targeting of DNA to vacuole is done for delivering the encapsulated DNA into the cytoplasm. This strategy is advantageous because the liposome protects the DNA from being damaged by the acidic pH and proteases present in the vacuole. To achieve this, DNA is encapsulated into liposome and pressure-injected into the vacuole. Other way is to construct liposomes from materials compatible with the lipids of the tonoplast, encapsulate DNA into it, and allow liposome–tonoplast fusion.

Factors Affecting Liposome-mediated Gene Transfer Several factors affect liposome-mediated gene transfer, which are listed below.

Purity and amount of DNA preparation DNA should be dissolved in water rather than buffers containing EDTA. It should be purified by chromatography on anion exchange resins or CsCl–ethidium bromide equilibrium density centrifugation. The amount of DNA concentration depends on the sequences of interest, as little as 50 ng and as much as 40 µg of DNA might be required to obtain maximum signal from a reporter gene.

Choice of the organic solvent The choice of organic solvent is arbitrary. Mostly diethyl ether is used. For lipids with higher transition temperature above boiling point of diethyl ether, isopropyl ether is used.

Type of phospholipid and its solubility Phosphatidyl serine is the best single lipid for DNA delivery and forms relatively nonleaky vesicles. Other lipids used are phosphatidyl choline and phosphatidyl glycerol.

Critical ratio of aqueous phase to lipid phase Restricting the volume of aqueous phase and increasing the concentration of DNA reduces the amount of material required and enhances the probability that each vesicle produced will contain DNA.

Size and surface charge of liposome Negatively charged vesicles are generally superior to positively charged or neutral vesicles in functional delivery of DNA. More negatively charged liposomes adhere to cells than do vesicles of other charge and infectivity of encapsulated DNA can be explained on the basis of the capacity to bind cells.

Liposome concentration For most cells studied, the capacity for liposome uptake saturates at about 100–200 nmol (negatively charged) phospholipid per 10^6 cells and there is likely to be little advantage to adding more material. In addition, high concentrations of liposomes (>500 µmol/ml), particularly those made with charged lipids can be toxic to plant cells.

Period of incubation between liposome and target cells For negatively charged liposomes, the interaction relevant to nucleic acid delivery takes place within 30 min of incubation. Longer incubations may be undesirable because of effects on cellular metabolism.

Use of PEG or DMSO The presence of PEG and DMSO can enhance the delivery of liposome-encapsulated DNA. These are added just before postincubation washes. The usual range of PEG concentrations that cell can tolerate is 25–50% w/v for 60–90 seconds, although toxicity is dependent on dose and time. It should be noted that PEG of molecular weight 6,000 Da or more is difficult to wash away and may be toxic.

Polycation-mediated DNA uptake (Use of PEG or Poly-L-lysine or Poly-L-ornithine or Polybrene)

Another method of chemical transfection employs the application of PEG.

Principle and Procedure Direct DNA uptake by protoplasts can be stimulated by using extremely hydrophilic, high molecular weight, long chain polycations such as PEG with or without metal ions such as Zn^{2+}, Li^+, Cs^+, Rb^+, K^+, Na^+, Ca^{2+}, or Mg^{2+}. When placed in a solution, PEG removes 'free water', i.e., molecules that interact with charged (usually ionic) molecules soluble in water. Both protoplast membrane and DNA are normally negatively charged, and hence have a tendency to repel each other due to charge repulsion, thereby limiting the interaction between the two. PEG at very high concentration seems to minimize the charge repulsion effects and stimulate DNA uptake by endocytosis without any gross damage to protoplasts. Under PEG concentration ranging from 15% to 25%, ionic macromolecules such as DNA no longer stay in solution, and precipitate out. PEG probably increases the adsorption of DNA to the surface and enhances DNA uptake as a result of the interaction of PEG at one or more of the following levels: (i) Protection of the exogenous DNA against nuclease degradation; and (ii) Increase in the permeability of plasma membrane. However, the mechanism of cross-membrane transport in presence of PEG is still to confirm. The technique is efficient and can be applied virtually to every protoplast system and for transfer of even small quantities of DNA. As evident in above sections, PEG enhances the efficiency of transformation by other procedures such as electroporation, liposome-mediated gene transfer, and Ca–DNA coprecipitation procedures.

Other polycations such as poly-*L*-lysine, poly-*L*-ornithine, or polybrene (1,5-dimethyl-1,5-diazaundecamethylene polymethobromide) also work on the same principle as that of PEG. Polybrene possesses lesser toxicity as compared to PEG and is widely used in plant transformation.

Factors Affecting Polycation-mediated DNA Uptake Several variables have been reported to affect the efficiency of transfection mediated by polycation-mediated DNA uptake. These are described below.

Amount of DNA 5 ng to 40 μg DNA is used for transformation, DNA toxicity is generally not a problem in the polybrene method of transfection. However, the linear relationship between DNA concentration and transformation efficiency breaks down when using very high concentration of DNA.

Concentration of polycations In experiments using polybrene, 30 μg is used for transformation.

Length of incubation time of cells and polycation–DNA mixture The cells are incubated for 6–16 hours and gentle shaking during the early stages of this period is recommended to ensure even exposure of the cells to the DNA–polybrene mixture.

Addition of DMSO and its concentration Further improvement of the method involves the use of a polycation (which increases the adsorption of DNA to the surface) followed by a brief treatment with 25–30% DMSO, which increases membrane permeability and therefore enhances DNA uptake. The factors affecting DNA uptake mediated by polycation–DMSO include the concentration of DMSO and polycations, the amount of DNA, the temperature of incubation with DMSO, and the length of incubation time of cells and polycation–DNA mixture.

DNA uptake by Diethyl Aminoethyl-Dextran (DEAE-Dextran) Treatment

DEAE–dextran treatment also facilitates DNA uptake. DEAE–dextran can also be used as an adjuvant to enhance the efficiency of electroporation. Although the outcomes appear to vary from one cell to another, the combination of electroporation and DEAE–dextran in some cases can improve the efficiency of transfection by a factor of 10–100.

Principle and Procedure The principle of mechanism of action of DEAE–dextran is, however, not clearly understood. Most probably this high molecular weight, positively charged polymer serves as a bridge between the negatively charged DNA and negatively charged surface of the cell. The DEAE–dextran/DNA complexes thus formed are internalized by endocytosis. The DNA somehow manages to escape from the increasingly acidic endosomes and is transported by unknown mechanisms across the cytoplasm and into the nucleus.

Factors Affecting Efficiency of Transfection by DEAE–Dextran Several variables that affect the efficiency of transfection have been reported. These include the concentration and molecular weight of DEAE-dextran, length of time of exposure of DNA, use of facilitators such as DMSO or glycerol, concentration and type of facilitating agents, concentration of transfecting DNA, density of cells and their state of growth, the way of addition of DEAE–dextran and DNA, length and temperature of the posttransfection facilitation, and cell type. Numerous variants of DEAE–dextran mediated transfection have also been published, and in most of the cases, the cells are exposed to a preformed mixture of DNA and high molecular weight DEAE–dextran (>50,0000). In some of the modified procedures, cells are exposed first to high molecular weight DEAE–dextran and then to DNA. The aim of all these methods is to maximize the uptake of DNA and to minimize the cytotoxic effects of DEAE–dextran.

Differences from Ca–DNA Coprecipitation Method This method of transfection is different from calcium phosphate coprecipitation in following aspects: (i) It is used for transient expression of cloned genes and not for stable transformation of cells; (ii) Smaller amounts of DNA are used for transfection with DEAE–dextran than with calcium phosphate coprecipitation; and (iii) Carrier DNA is rarely used with the DEAE–dextran transfection method.

DNA transfected into cells by DEAE–dextran method and calcium phosphate coprecipitate is prone to mutation. However, this effect is confined to the transfected sequences and does not affect the chromosomal DNA of the host cell.

14.4.11 *In planta* Transformation

The gene transfer methods described earlier involve the use of cells or explants as transformation targets and also require an obligatory tissue culture step for the regeneration of whole fertile transgenic plants. These extensive periods of tissue culture have several drawbacks such as loss of time (typically many months), requirement of space and labor, and tendency of induction of somaclonal variations (i.e., the variation in plants derived from tissue culture). There is therefore considerable interest in developing transformation methods that minimize or obviate the need for tissue culture steps. Thus recently *in planta* transformation system has been developed in which DNA is introduced into intact splants, with no need for plant regeneration under tissue culture conditions. Such *in planta* transformation methods have been demonstrated to work in a limited number of plant species, opening up the possibility of widespread application to crop improvement.

The *in planta* transformation protocols are of two types, *viz.*, germ line transformation and meristem transformation. Germ line transformation is done at an appropriate time in the life cycle of plant so that the DNA becomes incorporated into cells that will contribute to the germ line. Thus DNA is transformed directly into the germ cells themselves, often around the time of fertilization or into the very early plant embryo. However, the germ line transformation techniques are not reproducible; hence an alternative is to introduce DNA directly into meristems *in planta*, followed by the growth of transgenic shoots. Some commonly used methods for *in planta* transformation are as follows.

DNA Imbibition

The first *in planta* transformation system involved imbibing *Arabidopsis thaliana* (a model dicot plant) seeds overnight in an *Agrobacterium* culture, followed by germination. This resulted in a large number of transgenic plants containing T-DNA insertions. In case of *A. thaliana*, probably its small size and its ability to produce over 10,000 seeds per plant are advantageous. Generally, *in planta* transformation methods have a very low efficiency and low reproducibility. These limitations have so far prevented *in planta* techniques from being widely adopted for other plant species.

Vacuum Infiltration of Bacteria

A more reliable method for *in planta* transformation is vacuum infiltration, in which the bacteria are vacuum-infiltrated into flowers or ovules.

Incubation of Dry Seeds or Embryos in DNA

This approach has potential for the recovery of transgenic cereals and grain legumes. Dry embryos are split off from the endosperm, creating a giant wound across scutellum. As the tissues are dry, cell contents do not leak out. These are incubated in DNA solution. A microenvironment is created in the open cell that enables *in vitro* transcription, translation, and replication.

Macroinjection

This method involves the injection of naked DNA solution (5–10 μl) with the aid of micropipettes into the developing floral tillers of plants such as rye tillers and postfertilization cotton flowers. Within the floral tillers are archesporial cells that give rise to pollen in the developing sac by two meiotic cell divisions. It is suggested that such large cells are capable of taking up large molecules of DNA. Using this procedure, a plasmid harboring *npt* II gene was introduced into the tillers of rye plants. Two plants exhibiting resistance to kanamycin were obtained, but these experiments were not reproducible.

Incubation of DNA with Cut Pistil or Pollen Transformation

This is based on the fact that if DNA is applied to the cut pistil, it will reach zygote and hence seed. The plant produced from this seed will be transgenic.

Floral Dip

This is an extremely simple and popular technique, in which at the time of fertilization, plants with young flowers are dipped (with or without a vacuum being applied) into a culture of *Agrobacterium*, which also contains a surfactant. The plant is subsequently allowed to set seed, whereupon a small proportion of the seeds produced are transgenic. This technique was used with *Arabidopsis* flowers, but the efficiency was very low and the transformed plants thus obtained were chimeric. However, the vast number of seeds produced resulted in acceptable overall transformation efficiency.

Inoculation of Cut Apical Shoots

Meristem transformation is achieved simply by severing apical shoots at their bases, followed by inoculation of the cut tissue with suspension. Using this process, transgenic *Arabidopsis* plants have been recovered from the transformed shoots at a frequency of about 5%. In rice, explanted meristem tissue has been transformed using *Agrobacterium* and particle bombardment, resulting in the proliferation of shoots that were regenerated into transgenic plants.

14.4.12 Chloroplast Transformation

Chloroplast transformation is an important technique for effectively producing transgenic plants. This technique offers opportunities for creating environment-friendly and more efficient crops. Various advantages of the technique include the following: (i) There is no risk of gene transfer to weeds or wild-type relatives by pollination. In many crop species, chloroplasts display only maternal inheritance so there is no danger of any gene transfer to related weedy species through pollen; (ii) Chloroplast transformation allows high-level gene expression and results in a high yield. Through this technique, very large transgene copy number may be obtained. Transgene insertion may occur into a single genome contained within an individual chloroplast, however, by using several rounds of selection during the plant regeneration process, it is possible to develop a homoplasmic population of chloroplasts (where all the chloroplasts are transformed). As there may be as many as 100 chloroplasts per cell, each containing up to 100 copies of the genome, there may be as many as 10,000 copies of the transgene per cell; (iii) There is no risk of gene silencing. As the integration of recombinant vector into the chloroplast genome occurs by homologous and site-specific

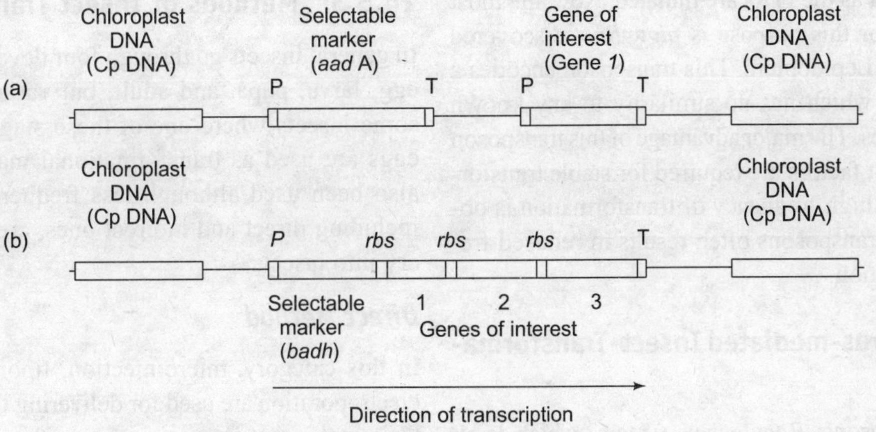

Figure 14.13 Vectors for chloroplast transformation; (a) A construct designed for the expression of one gene of interest (Gene *1*). Vector contains two expression cassettes, one for the gene of interest and the other for selectable marker gene (shown here is *aad* A conferring spectinomycin resistance). Each gene is driven by promoter (P) and terminator (T) sequences. Flanking the two expression cassettes are regions of chloroplast DNA (Cp DNA) that plays role in homologous recombination for site-specific integration into chloroplast genome. (b) A construct designed for the expression of multiple transgenes (Genes *1*, *2*, and *3*). The complete polycistronic unit or operon [comprising of selectable marker gene (shown here is *badh*, which codes for betaine-aldehyde dehydrogenase), and three genes of interest] is driven by single promoter and terminator sequences. Each transgene is preceded by ribosome binding site (*rbs*) to ensure efficient translation. Flanking the two expression cassettes are regions of chloroplast DNA (Cp DNA) that plays role in homologous recombination for site-specific integration into chloroplast genome

integration, any problems associated with random insertion into the nuclear genome are avoided. This is in contrast to that observed with the methods mentioned earlier where nonhomologous and random integration into nuclear genome takes place. Thus, gene silencing is not a problem during chloroplast transformation; and (iv) There is a tremendous potential in chloroplast transformation for very high-level gene expression and synthesis of biologically active proteins because chloroplast correctly folds and cross-links protein expressed from exogenous DNA.

Procedures used for Chloroplast Transformation

The major strategy for introducing recombinant DNA molecule into plastids is microprojectile bombardment (for details see Section 14.4.4). For chloroplast transformation, specifically designed vectors are used. Thus, a basic chloroplast expression vector for single transgene contains two expression cassettes, one for selection of transformants (e.g., *aad* A for spectinomycin resistance; betaine-aldehyde dehydrogenase gene) and another for expression of transgene. Transgene is kept under the control of chloroplast-specific regulatory sequences, *viz.*, promoter and ribosome binding site of *psb* A, which ensures proper transcription and translation of the transgene, respectively. Regions of chloroplast DNA (Cp DNA) flank these two expression cassettes (Figure 14.13). After introduction into the chloroplast by microprojectile bombardment, homologous recombination occurs between the plastid sequences on the vector and those on the genome.

14.5 INTRODUCTION OF DNA INTO INSECTS

Insects are one of the most abundant groups of living organisms and play very important role in human life. Several insects are beneficial for the mankind and some are quite harmful. Therefore, researchers have attempted to solve the problems of harmful insects like crop pests and also to exploit the potential of beneficial ones. With the advancements made in field of genetic engineering, insects are modified genetically for various purposes. It was first reported for stored-grain pest (*Ephestia khuniella*) wherein DNA from external source was introduced. Thereafter, several attempts have been made, resulting into transient expression.

14.5.1 Insect Transformation by Transposon-based Vectors

The real genetic transformation of *Drosophila melanogaster* was observed after the discovery of the P-element, a transposon. Afterward the discovery of other transposons like *Hermes, Hobo, Minos, MosI*, and *piggyBac* (TTAA-specific element), and more advanced DNA delivery devices like microinjection, electroporation, and particle bombardment facilitated the genetic transformation of various insects. For a stable integration of the transgene, two separate plasmids are used, *viz.*, a donor plasmid, carrying the transgene of interest and a visible transformation marker within the functional inverted terminal repeats (ITRs) and helper plasmid that codes for a functional transposase enzyme with defective ITRs. The helper plasmid cannot transpose itself as the ITRs are mutated. Now, the most utilized transposon for this purpose is *piggyBac*, discovered from insects of order Lepidoptera. This transposon encodes a transposase enzyme, which has no similarity to any known eukaryotic transposases. The major advantage of this transposon is that no specific host factors are required for stable transformation and generally high frequency of transformation is obtained. Use of other transposons often results in reduced frequency of transformation.

14.5.2 Baculovirus-mediated Insect Transformation

In addition to transposons, *Baculovirus* system, which is an insect-specific agent and unable to infect any other species, is also used for genetic transformation of insects, Viral vectors are derived either from *Autographa california* nuclear polyhedrosis virus (*Ac*NPV or *Ac*MNPV) or *Bombyx mori* nuclear polyhedrosis virus (*Bm*NPV) (for details see Chapter 12) for this purpose.

14.5.3 Methods of Insect Transformation

In general insects go through four developmental stages, *viz.*, egg, larva, pupa, and adult, but variations are observed in some insects where one of these stages is lacking. Usually eggs are used as transformational material but adults have also been used although less frequently. Several methods, including direct and indirect ones, are used for DNA delivery into insects.

Direct Method

In this category, microinjection, lipofection, biolistics, and electroporation are used for delivering the gene of interest into the target insect species.

Microinjection is the most commonly used direct method (for details see Section 14.4.6). The procedure involves aligning the needle with the micromanipulator and moving the mechanical stage to orient the micropyle end of the egg to the needle (Figure 14.14). For genetic modification of insects, the transgene of interest has to be introduced into the

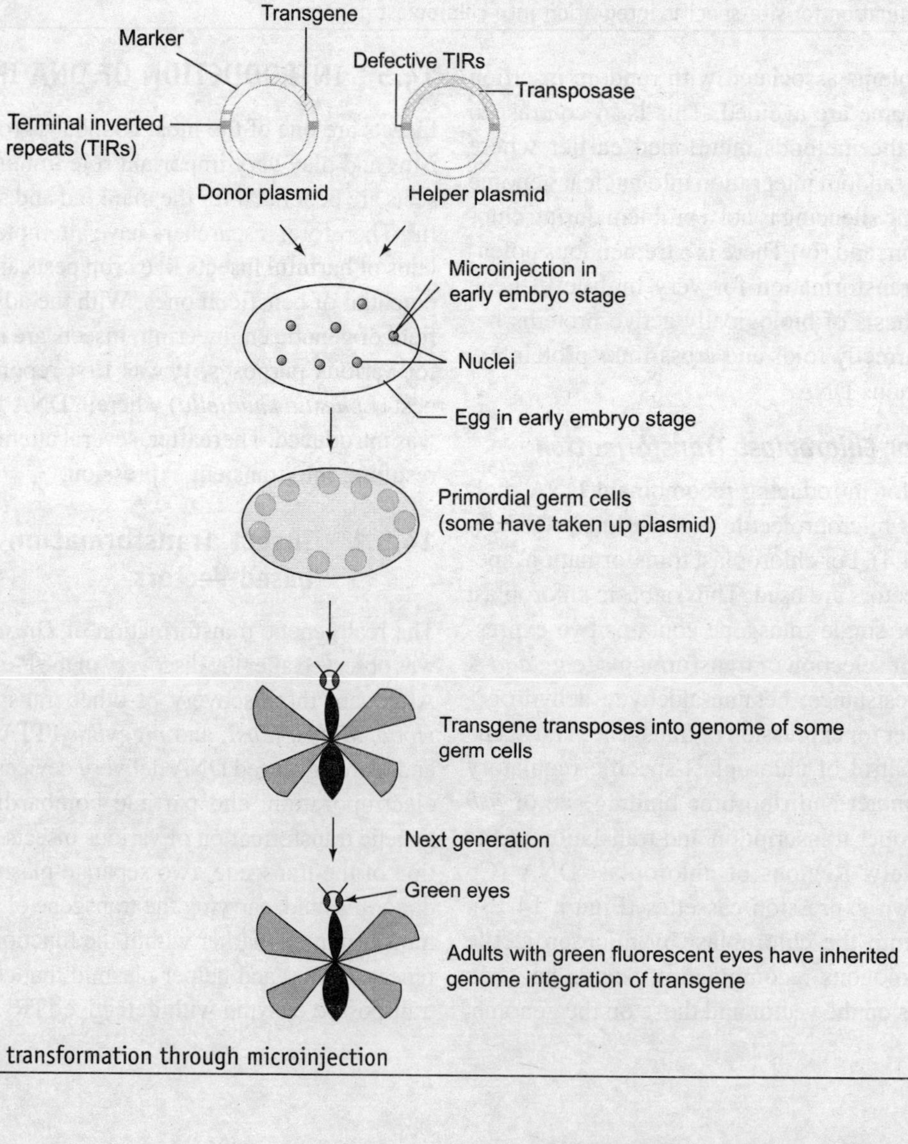

Figure 14.14 Insect transformation through microinjection

germ line of an egg and transposons are generally employed for delivering the gene of interest in insects. Mixture of two plasmids (donor and helper) is then microinjected into the egg in the early embryo stage. During embryonic development, the transposase mediates the transposition of the transgene onto a chromosome that results in the production of genetically modified insects. Stable genetic transformation occurs when the transgene is inherited in the next generation. Microinjection is also performed into ovarian egg follicles prior to oviposition and into female haemocoel for the uptake of DNA into egg follicles along with vitelline.

The next most widely used method is lipofection, which is routinely used for DNA delivery into *in vitro* cultured cells. Other less common methods are biolistics that involves coating DNA with gold or tungsten microparticles and bombarding into cells or tissues (for details see Section 14.4.4), and electroporation (for details see Sections 14.2.2 and 14.4.5) that is employed to genetically engineer insects belonging to the orders Lepidoptera and Diptera.

Indirect Methods

In addition to direct methods, indirect method like paratransgenesis is also employed for the transformation of insects, which are less amenable for laboratory rearing or have a long generation time. In this method, two bacterial endosymbionts *Wolbachia* sp. and *Rhodococcus* sp. are commonly employed, though *Wolbachia* sp. is the most widely used for delivering genes to a whole population.

Note that the genetically engineered insects (transformed either by direct or indirect methods) are selected using either *npt* II (neomycin phosphotransferase II) or organophosphorus dehydrogenase (*opd*) (which confers resistance to paraoxan) or the gene for dieldrin resistance (*rdl*). Visual marker like green fluorescent protein (GFP) and its spectral variant, enhanced green fluorescent protein (EGFP), are also used. For multiple transgene delivery, a reporter gene that is distinct from the marker is needed, where yellow and blue fluorescent proteins are very useful. The advantage with these markers is their easy visualization by naked eye in the live insect.

14.6 Genetic Transformation of Animals

Animal systems are genetically transformed by introducing the transgene in an early embryonic stage, sometimes even at the stage of freshly fertilized egg, either by viruses or by microinjection of the transgene directly in the nucleus. Certain stages of early development of animals are characterized by the presence of cells known as 'embryonic stem cells', which have the capability to differentiate into different types of tissues. Such transfected stem cells or fertilized eggs are then placed in the uterus of foster-mother in case of mammals like mice or cattle for full development. Thereafter, the newborn young ones are screened for the presence of the transgene by Southern hybridization or PCR. Then their transgenic siblings are mated to obtain homozygous transgenic animals. Similar procedure is also followed to obtain transgenic birds like chicken and transgenic fish like salmon. The applications of transgenic animals are extensively discussed in Chapter 20.

Basically two types of strategies are opted for transformation of eukaryotic cells, *viz*., transient and stable transfection. During transient transfection, DNA is introduced into a recipient cell line for temporary but high level of expression of the target gene as the transfected DNA does not essentially become integrated into the host chromosome. This method is preferable when a large number of samples are to be analyzed within a short period of time. In contrast, stable or permanent transfection approach is utilized to develop cells lines in which the transfected target gene is integrated into chromosomal DNA. Generally, efficiency of stable transfection is lower than the efficiency of transient transfection. Like other transformation procedures, the isolation of the rare stable transformants is facilitated by the use of a selectable genetic marker. The marker either may be located on the recombinant plasmid DNA having gene of interest or may be carried on a separate vector. The selectable marker, which is present on separate vector, is introduced with the recombinant plasmid into the desired cell line by cotransfection.

The transfection methods, which are commonly used for introducing DNA into animal cells, are categorized into three categories: transfection by chemical and biochemical method, transfection by physical methods, and vector-mediated transformation. The details of these procedures are described in the following section and summarized in Table 14.1.

14.6.1 Transfection by Chemical and Biochemical Induction

Chemical and biochemical methods of transfection include transfection mediated by calcium phosphate, DEAE–dextran, lipofection, and polybrene.

Transfection with Calcium Phosphate–DNA Coprecipitation

Mammalian cells grown in a culture dish can be transformed with DNA that has been allowed to form a coprecipitate with calcium phosphate (Ca–DNA coprecipitate). The principle and procedure has already been discussed above. The transfection performed by this method provides ~50% of the population expressing transfected genes in a transient manner, while transformed cell lines carrying integrated copies of the transfected DNA can be obtained at a much lower frequency. The transformation rates vary widely from one cell type to another. Calcium-phosphate-mediated DNA transfection was firstly devised to introduce adenovirus and simian virus 40 (SV40) DNA into adherent cultured cells. Later on, this method was used for various purposes like transformation of genetically marked mouse cells with cloned DNAs, for transient expression of cloned genes in a variety of mammalian cells, and for the isolation and identification of cellular oncogenes, tumor-suppressing genes, and other single-copy mammalian genes.

The variables that affect the efficiency of transfection mediated by Ca–DNA coprecipitation include the speed of mixing of DNA and calcium phosphate, size and concentration of the DNA, pH of the buffer (usually HEPES-buffered saline in the pH range 6.90–7.15 is used), use of facilitators such as glycerol, transfection maximizers, or certain inhibitors of cysteine proteases, or enhancement of the extent of transcription by exposing the transfected cells to hormones, heavy metals, or other substances. Additionally, treating transfected simian and human cells with sodium butyrate can enhance expression of genes carried on plasmids under the control of SV40 enhancer.

DNA Transfection by Lipofection

Lipofection comprises of set of techniques used to introduce exogenous DNAs into cultured eukaryotic cells. First the DNA to be transfected is coated by a lipid, which either interacts directly with the plasma membrane of the cell or is taken into the cell by nonreceptor-mediated endocytosis. When lipofection is performed in ideal conditions, lipofection can deliver DNA into cells more efficiently than precipitation with polycations such as calcium phosphate and at lower cost than electroporation. However, like other transfection techniques, lipofection is not universally successful. The efficiency of both transient expression and stable transfection by exogenously added genes varies widely from one cell line to another. Moreover, different protocols used for the same cell line may provide a wide range of results. However, lipofection is the method of choice as it generates positive results in various difficult situations like transfection of primary culture(s) of differentiated cells or introduction of very high molecular weight DNA into standard cell lines. There are two general classes of liposomal transfection reagents, *viz.*, anionic and cationic.

Anionic Liposomes Lipofection with anionic liposomes was first used to deliver DNA and RNA in a biologically active form and requires that the DNA be trapped in the internal aqueous space of large artificial lamellar liposomes. Because of its time-consuming nature and lack of reproducibility, this technique is not widely used.

Cationic Liposomes The commonly used lipofection techniques are based on the fact that cationic lipids react spontaneously with DNA to form a unilamellar shell, which can ultimately fuse with cell membranes. The DNA–lipid complexes are formed because of ionic interactions between the head group of the lipid, which carries a strongly positive charge that neutralizes the negatively charged phosphate groups on the DNA. First generations of cationic lipids were monocationic double chain amphiphiles with a positively charged quaternary amino head group linked to the lipid backbone by ether or ester linkages. These monocationic lipids however suffer from two major problems: these are toxic to many cell types and their ability to promote transfection is restricted to a small range of cell lines. Hence polycationic lipids replaced the monocationic lipids. These polycationic lipids have wider host range and are noticeably less toxic as compared with monocationic lipids. Generally, preparations of cationic lipids used for transfection comprises of a mixture of synthetic cationic lipid and a fusogenic lipid [e.g., phosphatidyl ethanolamine or dioleoylphosphatidylethano-lamine (DOPE)].

Several monocationic and polycationic lipids, which are actively used in transfection, are listed in Table 14.2. The DNA to be transfected is incorporated either into multilamellar structures composed of alternating layers of lipid bilayer and hydrated DNA or into hexagonal columns depending on the composition of the lipid mixture. In the case of honeycomb structure, each column comprises of a central core of hydrated DNA molecules and a surrounding hexagonal shell of lipid monolayers. Moreover, honeycomb arrangements deliver

Table 14.2 Examples of Some Lipids Used in Lipofection

Abbreviation	IUPAC Name	Type
DOTMA	*N*-[1-(2,3-dioleoyloxy) propyl]-*N*,*N*,*N*-trimethylammonium chloride	Monocationic
DOTAP	*N*-[1-(2,3-dioleoyloxy) propyl]- *N*,*N*,*N*-trimethylammonium methyl sulfate	Monocationic
DMRIE	1,2-Dimyristyloxypropyl-3-dimethyl hydroxyethylammonium bromide	Monocationic
DDAB	Dimethyl dioctadecylammonium bromide	Monocationic
Amidine	*N*-*t*-butyl, *N*′–tetradecyl-3-aminopropionamide	Monocationic
DC-Cholesterol	3β-[*N*-(*N*′*N*′-Dimethylaminoethane) carbamoyl]-cholesterol	Monocationic
DOSPER	1,3-Dioleoyloxy-2-(6-carboxyspermyl) propylamide	Dicationic
DOGS	Sperimine-5-carboxy-glycine dioctadecyl–amide	Polycationic
DOSPA	2,3-dioleoyloxy-*N*-[2(sperminecarboxamido) ethyl]–*N*,*N*-dimethyl-1-propane-aminium trifluroacetate	Polycationic
TM-TPS	*N*,*N*′*N*,*N*′-tetramethyl-*N*,*N*′*N*″,*N*‴-tetrapalmitylspermine	Polycationic

DNA across lipid bilayers more efficiently than multilamellar structures.

Furthermore, several other variables affect the efficiency of lipofection. These are already described above and the specific ones include: (i) Initial density of the cell culture: Cell monolayers should be in mid-log phase and should be between 40% and 75% confluent; (ii) Amount of DNA: As little as 50 ng and as much as 40 μg DNA is required to give maximum signal from a reporter gene, which depends on the concentration of the DNA of interest; and (iii) Medium and serum: The culture medium and serum, which are used to grow the cells, are also important for lipofection.

Note that lipids that are used as facilitators for the formation of transfection-competent structures also have undesirable side effects on DNA, for example, generalized toxicity. Lipofection is also susceptible to interference by fats and lipoproteins in serum and by charged components of the extracellular matrix such as chondroitin sulfate. Various modifications of the cationic and neutral lipids have been made in an effort to overcome these drawbacks.

DNA Transfection Using Polycations

Animal cells, which are insensitive to transfection by other methods, are easily transfected with the help of several polycations, including polybrene and poly-*L*-ornithine, in the presence of DMSO. Although the mechanism by which DMSO increases the efficiency of transfection is not clearly understood, but it may perhaps involve a combination of permeabilization of the cell membrane and the osmotic shock. In some cases, DMSO has been replaced by solution of 5–7% NaCl. The most important benefit of this method is that it does not pose toxicity to the cells. This procedure is very efficient for stable transformation, yielding more transformants as compared to calcium phosphate–DNA coprecipitation method. However, there is no difference between the two methods in the efficiency of transformation of cells by high molecular weight DNA, as the linear relationship between DNA concentration and transformation efficiency breaks down when using very high concentrations of DNA.

Transfection-mediated by DEAE–Dextran

This procedure is widely used for transfection of cultured cells with viral genomes and recombinant plasmids. The principle and procedure of DEAE–dextran-mediated transfection has already been discussed above. This technique was originally used for introducing poliovirus RNA, SV40 DNA, and polyomavirus DNA into cells.

For consistent high efficiencies of transfection with a typical cell line, the following factors should be standardized: (i) Concentration of transfecting DNA; (ii) Concentration and molecular weight of DEAE–dextran; (iii) Density of cells and their state of growth; (iv) Lengths of time of exposure of DNA; (v) Use of facilitating agent, its type and concentration; (vi) The way of addition of DEAE–dextran and DNA; (vii) Length and temperature of the posttransfection facilitation; (viii) Cell type; and (ix) Whether the cells are transfected while growing on a solid support or first removed from the solid support and transfected in suspension.

The method works very efficiently with lines of cells such as BSC-1, CV-1, and COS but is unsatisfactory with other types of cells. Usually the cells are exposed to high concentration of DEAE–dextran (1 μg/ml) for short periods (0.5–1.5 hours) or a lower concentration (250 μg/ml) for longer periods of time (up to 8 hours). Among these, the second technique is less demanding and more reliable, but slightly less efficient. Furthermore, it can be combined with shock treatments, which can raise the efficiency of transfection to very high levels. DMSO, chloroquine, or glycerol is used as facilitators to increase the efficiency of transient expression of transgenes. If cells are exposed to DMSO, glycerol, polyethyleneimine, or other substances such as Starburst dendrimers, these act as facilitators and hinder osmosis and increase the efficiency of endocytosis. A similar increase in efficiency of transfection of some lines of cultured cells may be obtained by exposing the transfected cells to chloroquine, which prevents acidification of endosomes and promotes early release of DNA into the cytoplasm. However, the efficiency of DNA transfection using DEAE–dextran with a facilitator varies greatly from one cell line to another.

Moreover, DEAE–dextran can also be used as an adjuvant to enhance the efficiency of electroporation. Although the outcomes appear to vary from one cell line to another, for example, the combination of electroporation and DEAE–dextran in some cases can improve the efficiency of transfection by a factor of 10–100. DNA transfected into cells by DEAE–dextran and calcium phosphate coprecipitation method is prone to mutation. However, this effect is confined to the transfected sequences and does not affect the chromosomal DNA of the host cell.

Transformation by Physical Methods

Physical methods of transfection of animal cells include microinjection, particle bombardment, and electroporation.

Microinjection The principle and procedure of this technique have already been discussed in Section 14.4.6. This technique of microinjection has been successfully used for somatic cell nuclear transfer (SCNT) (Exhibit 14.1) and embryonic stem (ES) cell technology (Exhibit 14.2), and their applications are summarized in Chapter 20. The techniques of pronuclear microinjection and sperm-mediated gene transfer for animal transfection are discussed below.

Pronuclear Microinjection This was the first approach attempted to generate transgenic mice. SV40 DNA was directly injected into the blastocoele cavities of preimplanta-

Exhibit 14.1 Somatic cell nuclear transfer (SCNT)

SCNT is a laboratory technique used for creating an ovum with a donor nucleus. The principle of nuclear transfer was first established in amphibians by transplanting nuclei from the blastula of the frog *Rana pipens* to an enucleated egg, obtaining a number of normal embryos in the process. Later similar results were obtained in *Xenopus laevis*. In this case, nuclei from various types of cell in the swimming tadpole were transplanted to an egg, which was UV-irradiated to destroy the peripheral chromosomes. This procedure is based on the principle that during development of animal cells, the cytoplasm becomes irreversibly dedicated to their ultimate fate, while the nuclei of most cells retain all the genetic information required for the entire developmental programme. Therefore nuclei can be reprogrammed by the cytoplasm of the egg to recapitulate development under appropriate circumstances. Moreover in all species, it appears that the earlier the developmental stage at which nuclei are isolated, the greater their potential to be reprogrammed. This technique can be exploited for the production of clones of animals with the same genotype by transplanting many somatic nuclei from the same individual into a series of enucleated eggs, which permit animals with specific and desirable traits to be propagated. SCNT can create clones for both reproductive and therapeutic purposes (for details see Chapter 20). In SCNT, the nucleus of a somatic cell, which contains the organism's DNA, is removed and the rest of the cell is discarded (Figure A). At the same time, the nucleus of an egg cell is removed. The nucleus of the somatic cell is then inserted into the enucleated egg cell. After being inserted into the egg, the somatic cell nucleus is reprogrammed by the host cell. The egg, now containing the nucleus of a somatic cell, is stimulated with a shock and begins to divide. After many mitotic divisions in culture, this single cell forms a blastocyst (an early stage embryo with about 100 cells) with almost identical DNA to the original organism.

In a highly publicized case, a sheep named Dolly was cloned by transfer of a nucleus from a mammary (udder) cell of an adult organism. Since the cloning of Dolly, somatic cell nuclei have been used to clone cattle, goats, and pigs. In these cases, the nuclear transfer procedures are similar. Briefly, embryonic, fetal, or adult donor cells are isolated, cultured, and genetically modified. Although not always feasible with adult cells, prolonged culture is preferred because experimenters have additional time to carry out successive genetic alterations, such as inactivating both alleles of a locus or creating multiple gene changes. After establishing a cell line with a specific genetic modification(s), individual donor cells are fused to enucleated oocytes with short-duration electric pulses. For example, 2.5 kV/cm pulses for 10 ms each are used to fuse adult cattle fibroblast cells with enucleated oocytes. The pulses simultaneously

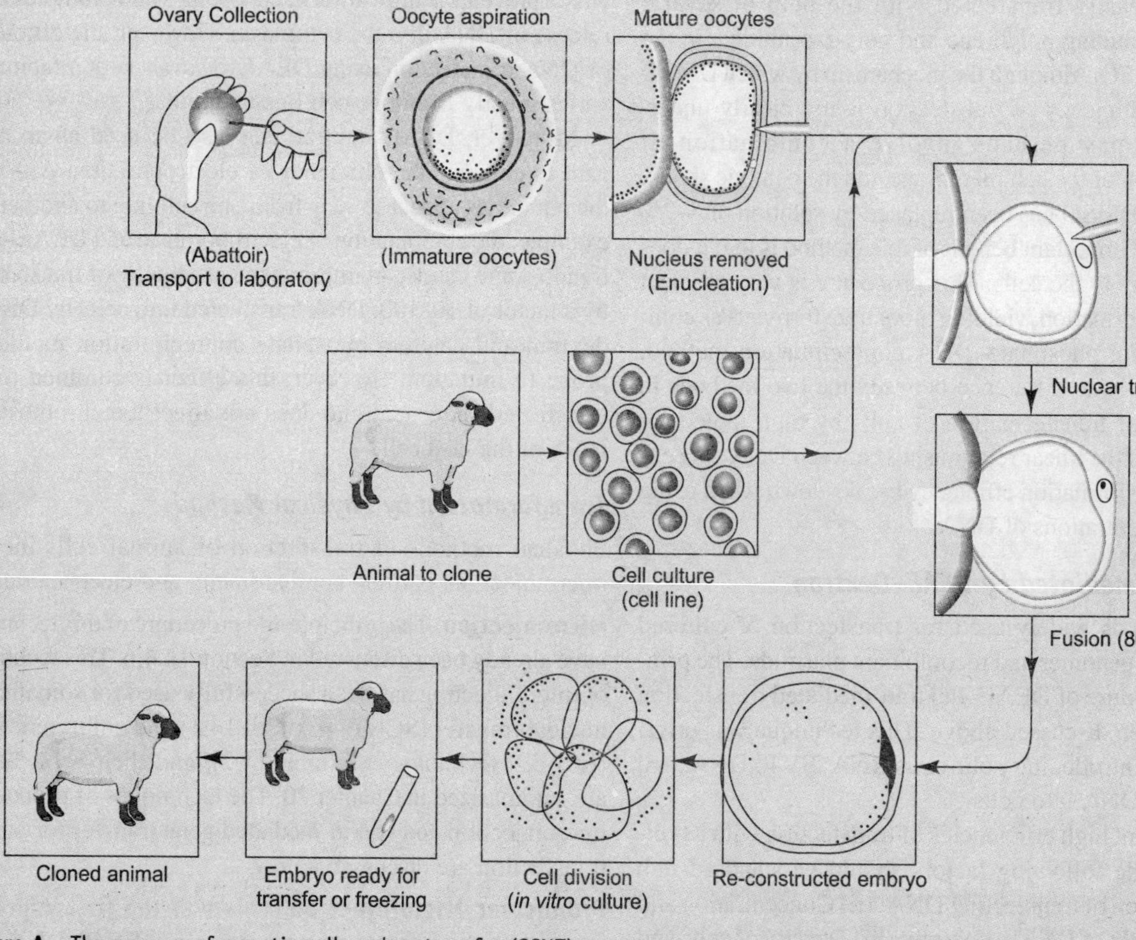

Figure A The process of somatic cell nuclear transfer (SCNT)

induce cell fusion and activation of oocytes. Each fused cell is cultured to the blastocyst stage before being transferred into the uterus of a pseudopregnant female. At birth, genotype analysis is used to confirm the presence of the transgene.

Some researchers use SCNT in stem cell research. The aim of carrying out this procedure is to obtain stem cells that are genetically matched to the donor organism, although at present no human stem cell lines have been derived from SCNT research. A potential use of genetically customized stem cells would be to create cell lines that have genes linked to the particular disease. For example, if a person with Parkinson's disease donates somatic cells, then the stem cells in the resulting SCNT would have genes that contribute to Parkinson's disease. In this scenario, the disease-specific stem cell lines would be studied in order to better understand the disease. In another case, genetically customized stem cell lines would be generated for cell-based therapies to transplant to the patient. The applications of this technique are described in Chapter 20.

Source: Campbell KH, McWhir J, Ritchie WA, Wilmut I (1996) Sheep cloned by nuclear transfer from a cultured cell line. *Nature* 380: 64–66.

tion embryos. Thereafter, the embryos were implanted into the uteri of foster-mothers and allowed to develop. If the DNA was taken up by some of the embryonic cells that ultimately develop into the germ line resulted in transgenic mice containing integrated SV40 DNA in the next generation. Transgenic mice were also developed through injection of viral DNA into the cytoplasm of the fertilized egg. However, the technique that has become accepted is the injection of DNA into one of the pronuclei of the egg. Usually, this technique is carried out immediately after fertilization. At this stage, the small egg nucleus (female pronucleus) and the large sperm nucleus (male pronucleus) are discrete. The male pronucleus is usually selected as the target for injection because it is larger. The egg is held in position with a suction pipette and DNA solution (~2 pl) is transferred into the nucleus with the help of a fine needle. The resulting embryos are cultured *in vitro* to the morula stage and then transferred to the uteri of pseudopregnant foster-mothers. This technique requires

Exhibit 14.2 Embryonic stem (ES) cell technology

ES cell technology is based not on the manipulation of individual fertilized eggs or embryos, but on cells that can be grown in culture, for example, certain types of cells taken from an adult animal. In some cases, these can give rise to established cell lines, which can be maintained indefinitely through serial subculture. Moreover, manipulation of animal cells in culture is considerably easier than introducing genes into a fertilized egg. The differentiation of such cells is not usually reversible; these cannot be used to produce a whole animal. However, the cells at an early stage in the development of an embryo still have the potential to develop into many types of cell in the adult mouse. These are known as embryonic stem cells (ES cells).

ES cells are derived from the inner cell mass of the mouse blastocyst and thus have the potential to contribute to all tissues of the developing embryo. The ability of ES cells to contribute to the germ line has been observed and requires culture conditions that maintain the cells in an undifferentiated state. Since these cells can be serially cultured, like any other established cell line, DNA can be introduced by transfection or viral transduction and the transformed cells can be selected using standard markers. This is an important advantage, since there is no convenient way to select for eggs or embryos that have taken up foreign DNA, and each potential transgenic mouse has to be tested by Southern blot hybridization or polymerase chain reaction (PCR) to confirm transgene integration. ES cells are also particularly efficient at carrying out homologous recombination, and therefore, depending on the design of the vector, DNA introduced into ES cells may integrate randomly or may target and replace a specific locus. Whichever strategy is chosen, the recombinant ES cells are introduced into the blastocoele of a host embryo at the blastocyst stage, where these mix with the inner cell mass. This creates a true chimeric embryo. ES cells are harvested from donor mouse and constructs are prepared by inserting transgene having selectable marker (Figure A). Thereafter, DNA is introduced by transfection or viral transduction and the transformed cells can be selected using standard markers. The positive cells are inserted into the uterus of foster-mother and finally offspring is screened for the presence of the transgene. The progeny will be chimeric, since the blastocyst still contains some of the unmanipulated cells, and heterozygous, since normally only one of the chromosomes will be altered. Subsequent breeding and selection of the progeny will give rise to stable homozygous mutant mice. ES cell technology is used for gene knock-out, gene knock-in, and gene targeting (for details see Chapter 17).

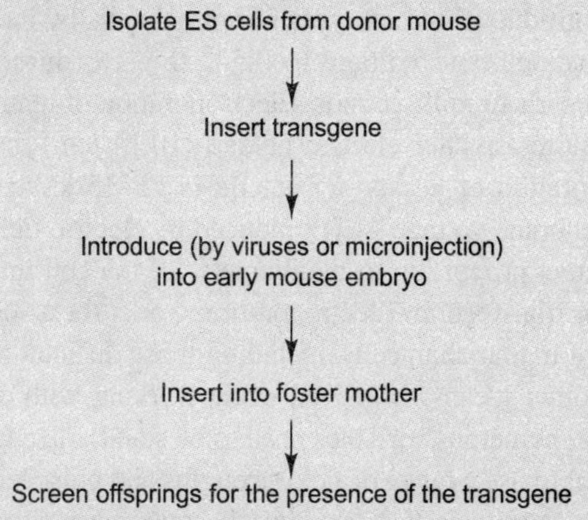

Figure A Embryonic stem (ES) cell technology

Source: Evans M, Kaufman M (1981) Establishment in culture of pluripotential cells from mouse embryos. *Nature* 292: 154–156.

specialized microinjection equipment. The resulting animal may be transgenic or may be chimeric for inserted transgene as it either integrates immediately or may remain extrachromosomal for one or more cell divisions. This technique is reliable, and the transfection efficiency varies in the range of 5–40%. Foreign DNA tends to form head-to-tail arrays prior to integration, and the copy number varies from a few copies to hundreds. The site of integration appears random and may depend on the occurrence of natural chromosome breaks. There are reports of extensive deletions and rearrangements of the genomic DNA with transgene integration. Furthermore, once the transgene is transmitted through the germ line, stable inheritance is reported.

Sperm-mediated Gene Transfer The sperm heads are directly injected into the cytoplasm of the egg, which is known as intracytoplasmic sperm injection (ICSI). This technique is a boon, because with the help of this technique, infertility in humans can be defeated. Moreover, it has been observed that sperm heads bind spontaneously to naked plasmid DNA *in vitro*, suggesting that sperm injections can be used for transformation. This was first demonstrated by mixing mouse sperm with plasmid DNA carrying the gene for GFP. Thereafter, the resulting sperm was injected into unfertilized oocytes and the resulting embryos (94%) showed GFP activity. Finally, these embryos were transferred to pseudopregnant females and 20% transgenic mice were generated. Later this method was adopted for transformation of other animal species. For example, Rhesus monkey oocytes fertilized in the similar way gave rise to a number of embryos with GFP activity, but this only lasted until the blastula stage, suggesting that there was no stable integration. However, in some cases, it was observed that this procedure was compatible with normal development process.

Electroporation The principle and method of electroporation has already been described in Section 14.2.2. Thus, the DNA can be introduced into all types of animal cells by increasing the transmembrane voltage to 0.5–1.0 V for durations of microseconds to milliseconds. Electroporation of mammalian cells requires smaller electric fields (<10 kV/cm) than does electroporation of yeasts or bacteria (12.5–15 kV/cm). The transmembrane voltage $\Delta U(t)$, induced by electric fields, varies in direct proportion to the diameter of the cell that is the target for transfection. Electroporation works for a very wide variety of mammalian cells, including those difficult to transfect by other means. However, while working with different cell lines, numerous variables need to be standardized: (i) It is important to use a range of field strengths and pulse lengths to optimize conditions that generate the maximum numbers of transformants; (ii) The proportion of cells that survive after exposure to the electric field is usually measured by determining plating efficiency; (iii) The temperature of the cells before, during, and after electroporation is a very important parameter. Usually, electroporation is carried out on cells that have been prechilled to 0°C, the cells are held at 0°C after electroporation, and are diluted into warm medium for plating; (iv) The concentration and conformation of the DNA should be considered for efficient transfection of animal cells. Linear DNA is preferred for stable transformation; circular DNA is preferred for transient transfection. Preparations containing DNA at a concentration of 1–80 µg/ml are optimal; and (v) The state of the cells is also an important factor for high efficiency of electroporation. The best results are obtained with cell cultures in the mid-log phase of growth that are actively dividing.

Microprojectile Bombardment Microprojectile bombardment or biolistics is used to transform some specific cell types, tissues, and intracellular organelles, which remain impermeable to foreign DNA. The principle and mechanism of biolistics has been already discussed in Section 14.4.4. Like other methods of transfection, biolistic transformation also requires optimization of the array of variables that affect the transfection efficiency of each cell type. These variables are categorized as particle accelerator parameters (gene gun settings), microparticle parameters, and biological parameters. Various cell types including hepatocytes within the livers of living rodents have been successfully shot with DNA-coated microprojectiles. In each case, the conditions that provide a positive result are different. The transfection frequency is greatly affected by the density at which cultured cells are bombarded. In some cases, at low cell density, cells are transfected with greater efficiency (e.g. for transfection of intracellular organelle), while in others, optimum efficiency is obtained at higher cell density. Some cells are best transfected in the early log phase of their growth cycle, whereas other cells are best transfected after being grown to saturation. High osmolarity of the culture medium increases the efficiency of transformation and agents such as sorbitol and/or manitol are used in differing concentrations (0.05–1.5 M) for different species and cell type. For each application, the optimum particle size also differs between cell types and ranges from 0.6 µm for subcellular organelles to 1.6 µm for cultured mammalian cells.

Biological Methods of Gene Transfer to Animal Cells

Biological methods of gene transfer to animal cells includes virus transduction and yeast artificial chromosome (YAC) transgenesis.

Virus Transduction Virus transduction is a process by which genes are introduced into animal cells by utilizing the natural gene transfer ability of the virus particle within which the transgene is packaged. As already discussed in Section 14.4.3, viruses can deliver their nucleic acid into cells with high efficiency, which results in high levels of replication and gene expression. Therefore, viruses have been used as vectors not only for gene expression in cultured cells but also for gene transfer to living animals. Various classes of viral vectors (for details see Chapter 12) have been developed to facilitate transduction in animal cells. Gene of interest may be incorporated into these viral vectors either by addition to the whole genome or by replacing one or more viral genes, which is performed either by ligation or homologous recombination. If the transgene is cloned in replication-competent or helper-independent vector, then it can propagate independently. In contrast, if the transgene replaces an essential viral gene, then the vector is replication-defective or helper-dependent, and requires cointroduction of a helper virus or transfecting the cells with a helper plasmid carrying missing genes. In this case, care is taken to prevent the helper virus completing its own infection cycle so that only the recombinant vector is packaged. Furthermore, it is also advisable to prevent chances of recombination between the helper and the vector; otherwise wild-type replication-competent viruses are again produced. Complementary cell line or 'packaging line', which is transformed with the appropriate genes, is also used as a better substitute for cointroduction of helper viruses. The viral vectors can infect early embryos and ES cells, i.e., recombinant viral vectors can be used for germ line transformation. The embryos are infected with recombinant virus before implantation. Finally, the infected embryos are implanted in the uterus of foster-mother and the transgene is transmitted into germ line. This technique is better than microinjection as only a single copy of the viral provirus is integrated. Moreover, the genomic DNA surrounding the transgenic locus generally remains intact. The practical limitation of this method includes size constraints, the possible interference of viral regulatory elements with the expression of surrounding genes, and the susceptibility of the virus to *de novo* methylation. Viral transduction is only useful for generating transgenic sectors of embryos and this is not the preferable method for producing transgenic animals.

Artificial Chromosomes for Gene Transfer: YAC Transgenesis

The expression of small transgenes often fails to follow the normal temporal and spatial patterns of expression or matches the expression level of the endogenous homologue. Moreover, mammalian genes are very large, and even in the case of small genes, important regulatory elements that are required for proper expression may be located many nucleotides (several kbp) upstream of the coding sequence. Thus, in order to study the expression and regulation of a human gene under the control of its own *cis*-acting regulatory elements, it is necessary to establish transfection conditions, which allow the transfer of large DNA clones. First, the transfer of large DNA segments to the mouse genome was achieved by transformation with YAC vectors (for details see Chapter 11). This procedure was first used for the transformation of ES cells with a YAC vector via fusion with yeast sphaeroplasts. The original method had some drawbacks, for example, the endogenous yeast chromosomes were cointroduced into the host cell along into the vector. Thereafter, alternative strategies involving isolation of the vector DNA by pulsed field gel electrophoresis (for details see Chapter 6), followed by introduction of the purified YAC DNA into mouse eggs by pronuclear microinjection or transfection into ES cells have been devised. The technique involving ES transfection is more appropriate as microinjection involves shear forces that break the DNA into fragments. Additionally, YAC transfer to ES cells can also be achieved by lipofection. YAC transgenics have been used to study gene regulation of long-range regulatory elements and to introduce the entire human immunoglobulin locus into mice for the production of fully humanized antibodies. Microcell-mediated fusion technique is also used to introduce chromosomes and chromosome fragments into ES cells. In this technique, cultured human cells are arrested at the mitotic stage by treating with an inhibitor such as colchicine. The nucleus is then broken up into vesicles containing individual chromosomes, which can be rescued as microcells comprising a nuclear vesicle surrounded by a small amount of cytoplasm and a plasma membrane. Finally, transgenic mice have been generated using ES cells already fused to human microcells. Such vectors have high capacity for foreign DNA and have no intrinsic cytotoxic effects.

Review Questions

1. Describe the principle and mechanism of chemical transformation of *E. coli* cells. How is transformation efficiency calculated? Also give an account of factors affecting the efficiency of competent cells.
2. Give an elaborative account of electroporation with special reference to principle, procedure, and factors affecting electroporation.
3. Explain the procedure and mechanism of triparental mating used for transformation of *Agrobacterium*.
4. *B. subtilis* has great importance in genetic engineering and three types of procedures have been exploited for transformation of this important organism. Give a brief account of these methods.
5. Give a detailed note on the mechanism of plant transformation by *A. tumefaciens* with the help of suitable diagram.
6. Write the role of each *vir* operon involved in the process of T-DNA transfer from *A. tumefaciens* to plant.

7. Due to certain properties possessed by plant viruses, these have the potential to perform stable plant transformation. Discuss these properties in detail.
8. Write a short note on various components of microprojectile bombardment equipment and different particle acceleration devices required for bombardment.
9. Illustrate the procedure for gene gun-mediated direct gene transfer by taking the example of PDS-1000/He device (BioRad).
10. Describe various parameters affecting particle bombardment-mediated transformation.
11. Give a brief account of the principle and procedure of plant transformation by microinjection.
12. Write short notes on the following with special reference to principle and procedure:
    (a) Plant transformation using silicon carbide whiskers
    (b) Plant transformation by ultrasonication
    (c) Laser microbeam irradiation-mediated DNA uptake
    (d) Direct gene transfer by chemical and biochemical methods
13. Liposomes are used in genetic engineering to deliver foreign DNA molecules into plant protoplasts. Describe the principle and procedure of this technique.
14. Discuss various advantages of chloroplast transformation in the production of transgenic plants.
15. Discuss the significance of *in planta* transformation. Give an elaborative account of various methods for *in planta* transformation.
16. Give an elaborative account of principle and procedure of insect transformation.
17. Discuss the principle and procedure of animal transfection mediated by calcium phosphate, DEAE–dextran, polybrene, and lipofection.
18. Write short notes on the following:
    (a) Virus transduction
    (b) YAC transgenesis
    (c) Somatic cell nuclear transfer
    (d) Sperm-mediated gene transfer
    (e) Pronuclear microinjection
    (f) Embryonic stem cell technology
19. Plasmid DNA 0.5 μl (1 μg/μl) was added to 200 μl of competent cells. The transformed competent cells were diluted 1,000 times from which 50 μl was plated and finally 35 colonies were found. Calculate the transformation efficiency.

References

Alwine, J.C., D.J. Kemp, and G.R. Stark (1977). Method for detection of specific RNAs in agarose gels by transfer to diazobenzylmethoxy methyl-paper and hybridization with DNA probes. *Proc. Natl. Acad. Sci.* **74**:5350–5354.

Bender, W., P. Spierer, and D.S. Hogness (1983). Chromosomal walking and jumping to isolate DNA from the *Ace* and *rosy* loci and the bithorax complex in *Drosophila melanogaster*. *J Mol. Biol.* **168**:17–33.

Benton, W.D. and R.W. Davis (1977). Screening lambda gt recombinant clones by hybridization to single plaques *in situ*. *Science* **196**:180–182.

Cohen, S.N., A.C.Y. Chang, and L. Hsu (1972). Nonchromosomal antibiotic resistance in bacteria: Genetic transformation of *E. coli* by R-factor DNA. *Proc. Natl. Acad. Sci.* **69**:2110–2114.

Grunstein, M. and D.S. Hogness (1975). Colony hybridization: A method for the isolation of cloned DNAs that contain a specific gene. *Proc. Natl. Acad. Sci.* **72**:3961–3965.

Hanahan, D. (1983). Studies on transformation of *E. coli* with plasmids. *J. Mol. Biol.* **166**:557–580.

Inoue, H., H. Nojima, and H. Okayama (1990). High efficiency transformation of *E. coli* with plasmids. *Gene* **96**:23–28.

Construction of Genomic and cDNA Libraries

Key Concepts

In this chapter we will learn the following:

- Genomic library
- cDNA library
- PCR as an alternative to library construction

15.1 INTRODUCTION

Recombinant DNA library or gene library is a collection of random clones that is representative of the entire starting population of DNA (genomic DNA/cDNA) in an organism in a stable form. Thus, during library construction, the nucleotide sequences of interest are preserved as inserts to a vector, which are then used to infect host cells. Gene libraries are of two types: genomic library and cDNA (complementary DNA) library. Genomic library is prepared from total genomic DNA of an organism, which is a set of clones, packaged in the same vector that together represents all regions of the particular genome. The number of clones that constitute a genomic library depends on the size of the genome, and the insert size further depends on the capacity of the particular cloning vector system employed. For small genomes like that of bacteria, yeast, and fungi, the number of clones needed for a complete genomic library is not so large and hence manageable. While for plant and animal genomes, complete libraries contain so many different clones that the identification of the desired gene may prove to be a massive task. With these organisms, a second type of library, called cDNA library, which is made from DNA copies of its RNA sequences, is more useful. The cDNA library is specific to a particular cell type and not to the whole organism.

A cDNA library represents all the mRNA present in a particular tissue, which has been reverse transcribed to a DNA template by the use of the enzyme reverse transcriptase. It represents all the genes that are transcribed in particular tissues under particular physiological, developmental, or environmental conditions. cDNA library should not be confused with a genomic library, as it does not represent the entire genome. A genome sequence provides only partial information about a given gene. However, in contrast, cDNA reveals expression profiles in different cell types, developmental stages, and in response to natural or experimentally stimulated external stimuli. The cDNA sequence also gives useful information about splice isoforms and their abundance in different tissues and developmental stages in higher organisms. Although up till now, various genomes including human, all model plants, animals, and prokaryotic organisms have been completely sequenced and annotated, but there are several important organisms whose genomes are not sequenced. Besides this, there are numerous other reasons for single gene cloning. Further gene cloning strategies are also very important for the elucidation of gene function.

The cloning strategies in both types of libraries consist of similar steps (Figure 15.1). The genomic DNA or cDNA is first prepared and then ligated into appropriate vector. Thereafter, the ligated vector is introduced into host cells either by *in vitro* packaging into λ phage heads, or by direct transformation, or by transfection. The library once constructed is titrated, characterized, and screened for desired clone with the help of suitable probe. It can be further amplified for long-term storage. In this chapter, the procedures for the construction of genomic and cDNA libraries, representations and randomness of the size of both libraries, a comparison of genomic and cDNA libraries, and subtractive libraries are described.

15.2 GENOMIC LIBRARY

The process of subdividing genomic DNA into clonable fragments and inserting them into host cells leads to generation of a large number of random clones of recombinant donor DNA molecules, the collection of which is called a genomic library or clone bank or gene bank. A complete genomic library, by definition, contains all the genomic DNA of the source organism. Physically it is a collection of plasmid clones, or phage lysates, which contain recombinant DNA molecules. The genome of any organism particularly eukaryotic is remarkably complex and a particular fragment of interest comprises only a small fraction of the whole genome. Therefore, construction of a useful and representative

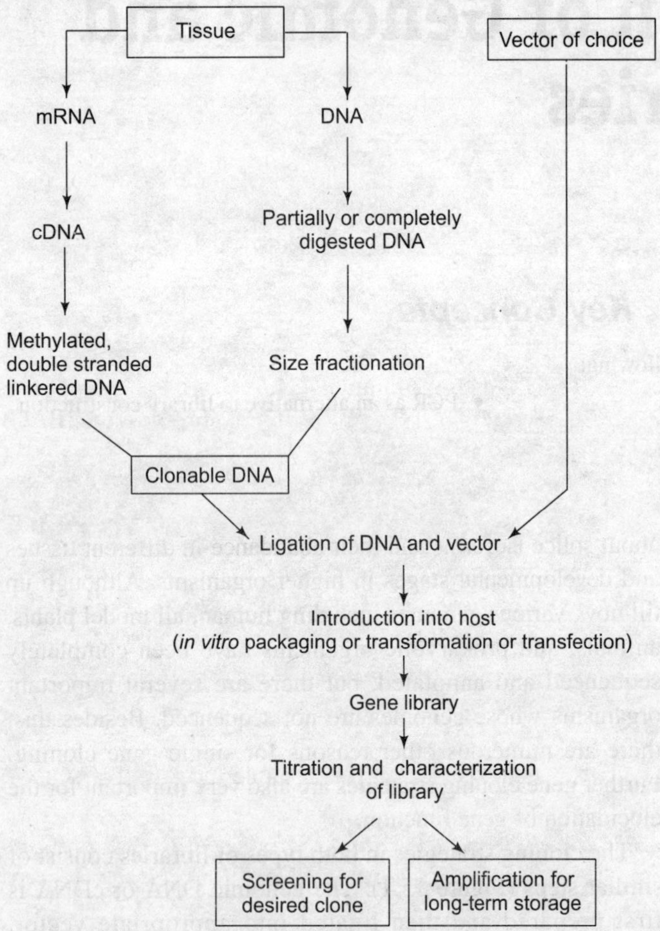

Figure 15.1 Overview of cloning strategies [In this figure, cDNA is methylated and linkered. For preparation of clonable DNA, cDNA may also be ligated with adapters or subjected to homopolymer tailing.]

recombinant genomic library is dependent on the generation of a large population of clones. This is indispensable to make certain that the library contains at least one copy of every sequence of DNA in the genome.

15.2.1 Construction of Genomic Library

Genomic library construction involves following steps:

- Isolation of purified total genomic DNA and generation of suitably sized DNA fragments.
- Joining or ligation of DNA fragments into suitable vectors to generate recombinant DNA molecules (i.e., cloning into vector).
- Introduction of recombinant DNA molecules into host cells and plating on suitable selection medium.

Isolation and Purification of Total Genomic DNA and Generation of Suitable Size DNA Fragments

The primary step in genomic library construction is the isolation and purification of total genomic DNA of an organism (for details see Chapter 1). Since genomic DNA has high molecular weight, gentle handling is required to minimize the extent of shearing and nonspecific damage. Besides necessary steps should be included in the purification procedure so that the preparation of total genomic DNA is free from organelle DNA, RNA, and other macromolecules. Genomic DNA ranging in size between 100 and 150 kbp is adequate for the construction of genomic library.

Once total genomic DNA is isolated and purified, the next step is to generate appropriately sized random DNA fragments. Note that the random fragmentation of DNA allows systematic inclusion of all the DNA sequences of the genome. Several methods have been devised for this purpose including (i) Partial digestion of genomic DNA with a suitable restriction endonuclease (sticky or blunt end cutters); (ii) Mechanical shearing of large genomic DNA; (iii) Subjecting genomic DNA to sonication; (iv) Treating genomic DNA with nonspecific endonucleases; and (v) Chemical degradation of genomic DNA. Note that for cloning a small target gene, the sequence of which is known, methods such as chemical synthesis or amplification through polymerase chain reaction (PCR) can also be followed.

Mechanical shearing, sonication, and digestion with nonspecific endonuclease have the advantage of allowing random breakage with generation of controlled average fragment size. If a high molecular weight DNA is homogenized in a blender at 1,500 rpm for 30 min, then fragments with an average size of ~8 kbp are obtained. Likewise, with the application of intense ultrasound, ~300 bp fragments are obtained. Thus, with these procedures the size of fragments can be controlled, but the resulting fragments contain short single stranded regions, which may hybridize with complementary sequences in the other fragment. Consequently, ligation of the resulting fragments into vectors needs further modifications. For example, the resulting fragments are made blunt by single strand specific DNase before ligation step.

Partial digestion with a four base-cutter [e.g., *Sau* 3A (5′-↓GATC-3′), *Alu* I (5′-AG↓CT-3′), *Hae* III (5′-GG↓CC-3′)], which theoretically cleaves the DNA nearly once every 256 bp, is the most preferable method. In this way, all possible fragment sizes (random fragments) are generated and the chances of obtaining a complete target gene (with its flanking control regions) as a single fragment are improved. Complete digestion is avoided since it generates fragments that are too heterogeneous in size (for details see Chapter 3). If complete digestion is performed, there is a possibility of cutting the target gene internally one or more times so that the target gene is not obtained as a single fragment. This is more likely if the target gene is large in size. Moreover, the chances of obtaining even a small target gene along with its flanking control regions as a single fragment are also lost. One problem associated with the application of partial restriction digestion for the generation of random DNA fragments is the construction of incomplete genomic library. This is because the restriction enzyme sites are not randomly located, and hence some fragments may be too large to be cloned. Consequently it may become difficult or even impossible to find a specific target DNA sequence in the library. This problem can, however, be overcome by forming libraries with different restriction enzymes.

The resulting large-sized, high molecular weight DNA fragments are checked by pulsed field gel electrophoresis parenthesis (PFGE) (for details see Chapter 6).

Joining of DNA Fragments into Vector

The fragmented genomic DNA is then mixed and ligated to a suitable vector, which has previously been linearized by suitable restriction enzyme(s). Note that a number of pilot ligation reactions are performed using a set amount of vector and varying amounts of insert prior to setting of large-scale ligation. The number of clones in different ligations is compared and the optimum vector to insert ratio is then used to set a large-scale ligation reaction. The ligation of DNA fragments to vector molecules can simply be done by DNA ligase or may require a prior end modification step, for example, homopolymer tailing or addition of linkers/adaptors (for details see Chapter 13). This ligation step may or may not be associated with a prior DNA size fractionation step depending upon the vector employed (for details see Chapter 6).

Introduction of Recombinant DNA Molecules into Host Cells and Plating on Suitable Selection Medium

The resulting recombinant DNA is then introduced into the host cells by transformation, or transfection, or *in vitro* packaging (for details see Chapters 9 and 14) and the transformants are selected by plating on a selection medium (for details see Chapter 16).

In the resulting genomic library, the clones with the target sequence are then identified by methods such as DNA hybridization with a DNA probe followed by screening for the probe label, immunological screening for the protein product, assaying for protein activity, and functional (genetic) complementation (for details see Chapter 16).

15.2.2 Representative Genomic Library and Randomness of DNA Fragments

In prokaryotic organisms genes form a continuous coding domain in the genomic DNA, whereas in eukaryotes the coding regions of genes are interrupted by noncoding regions called introns. The human genome contains ~50,000 unique genes within 3 to 4 billion base pairs of DNA, scattered in 23 pairs of chromosomes. The size of library of completely random fragments of genomic DNA is needed to guarantee representation of a particular sequence of interest, which is dictated by the size of the cloned fragments and size of the genome. To ensure that the entire genome, or most of it, is contained within the clones of a library, the sum of the inserted DNA in the clones of the library should be three to four times more than the amount of DNA in the genome. Thus, for a genome size 4×10^6 bp and the average insert size 1,000 bp, ~12,000 clones are required for three-fold coverage, i.e., 3 [$4 \times 10^6/10^3$]. Likewise for human genome, which has a size of 3.3×10^9 bp, ~80,000 bacterial artificial chromosome (BAC) clones having an average insert size of 150 kbp compose a library with four-fold coverage, i.e., 4 [(3.3×10^9)/1,50,000]. The exact probability of having any given DNA sequence in the library can be calculated from the equation given by Clarke and Carbon:

$$N = \frac{\ln[1-P]}{\ln[1-(1/n)]} = \frac{\ln[1-P]}{\ln[1-f]}$$

where N is the necessary number of recombinants to be screened in order to have a reasonable chance of including the desired sequence; P the desired probability or the extent of coverage; f the ratio of the insert length (average) to the entire genome, and n (= $1/f$) is the size of genome relative to a single cloned fragment.

The equation indicates that the number of clones in a genomic library depends on the extent of coverage, the size of the genome of the organism, and the average size of the insert in the vector. If insert length remains constant, the number of clones required to find a particular gene increases with increasing complexity of the genome. Hence, for larger and more complex genome, more clones of a specific size are needed for complete representation of the entire genetic information. Approximately 65,000 clones are required for a 99% probability ($P = 0.99$) of discovering a particular sequence in a human genomic library with an average insert size of 20 kbp. This simple analysis assumes that the cloned DNA segments randomly represent the sequences present in the genome, which is true only if the target DNA is cleaved randomly prior to insertion into the vector. This randomness of DNA fragments does not allow systematic exclusion of any DNA sequence of the genome. Moreover, random clones overlap one another, allowing the sequence of very large genes to be assembled and an opportunity of 'walking' from one clone to an adjacent one. This randomness can be easily approached by mechanical shearing, sonication, or by careful partial restriction digestion (for details see Section 15.2.1). One limitation of this approach is that fragments that are larger than the capacity of vector size will be excluded from the library.

15.2.3 Strategies Used for Genomic Library Construction Using Different Vectors

Principally all vectors that are used in genetic engineering can be used for construction of genomic libraries. However, vectors with high cloning efficiency and relatively large insert capacity are employed for genomic library construction. Because of the larger size of each cloned DNA fragment, fewer clones are required for a complete or nearly complete library (for details see Section 15.2.1). Moreover, by using large insert capacity vectors, the possibility of cloning the entire target gene within a single clone also increases. The examples of vectors used for genomic library construction include λ replacement vectors, cosmid vectors, bacteriophage P1 vectors, and artificial chromosomes [yeast artificial chromosome (YAC), bacterial artificial chromosome (BAC), P1-derived artificial chromosome (PAC), and mammalian artificial chromosome (MAC)] are generally used to construct genomic DNA libraries. Plasmid vectors are not suitable for genomic library construction because these cannot be used for inserts of >15 kbp size. Further the efficiency of transformation decreases with increasing length of the inserts. On the other hand, most λ replacement vectors can easily accommodate up to 23 kbp inserts and cosmids can be used to insert fragments up to 45 kbp in size. Besides, artificial chromosomes allow cloning of inserts >100 kbp in size. The insert capacities of various cloning vectors are given in Table 15.1. Note that unless the target sequence is known to be very large and needs to be isolated as a single clone, λ replacement vectors and cosmids remain the most appropriate choice for many experiments. The main applications of libraries in artificial chromosomes are genome mapping, sequencing, and assembly of clone contigs.

Genomic Library Construction Using Bacteriophage λ Vectors

Various phage vectors are used for library construction, all of which have three basic features in common that aid in cloning into them: (i) Ability to accept fragments generated by several restriction enzymes; (ii) Ease of selection owing to presence of biochemical and/or genetic selection markers; and (iii) Capability of *in vitro* packaging of the DNA segment (37–52 kbp in size) between two *cos* sites.

For the first time, Maniatis *et al.* (1978) used Charon 4A, a λ replacement vector, for genomic library construction. The strategy, called Maniatis strategy, is presented in Figure 15.2. It involves partial digestion of target DNA with a mixture of two unrelated restriction enzymes, *viz.*, *Hae* III (5′-GG↓CC-3′) and *Alu* I (5′-AG↓CT-3′). Note that these two enzymes are four base-cutters, which cut frequently, and create blunt ends. Partial digestion leads to generation of large random fragments, ranging in size between 10 and 30 kbp. In the next step, DNA molecules of ~20 kbp are isolated by sucrose gradient centrifugation, which are suitable for insertion into Charon 4A. The preexisting *Eco* RI sites in DNA fragments are methylated by *Eco* RI methylase. This is followed by ligation of *Eco* RI linkers to the ends of DNA fragments and then the resulting fragments are digested with *Eco* RI. Note that the previous methylation step is essential to prevent further fragmentation of the DNA upon digestion with *Eco* RI. The insert is now ready for ligation with *Eco* RI-digested cloning vector. Charon 4A is prepared for cloning by circularization at the *cos* sites and *Eco* RI digestion, which produces a large 31 kbp fragment containing the *cos* site and two smaller fragments of 7 and 8 kbp, respectively. The large fragments are purified by gradient centrifugation. The target DNA molecules are then ligated with the vector under conditions, allowing concatemer formation. Once recombinant DNA molecule is formed, it is subjected to *in vitro* packaging and finally a large number of independent recombinants are recovered, which generate a representative library. The recombinants are selected by plating on sup^+ P2 lysogen for Spi^- phenotype (for details see Chapter 9).

Currently, the most widely used λ vectors for library construction are λEMBL series, λ2001, λDASH, λFIX, and their derivatives (for detail see Chapter 9). λEMBL series of vectors are advantageous for genomic library construction because these have high insert capacity, permits both genetic and biochemical strategy to avoid contamination of stuffer fragment, and have several useful cloning sites in the polylinker. For cloning in λEMBL series of vectors (for

Table 15.1 Insert Capacities of Cloning Vectors

Cloning vector	Host	Insert capacity (kbp)
Plasmid	*E. coli*	Up to 10
M13	*E. coli*	Up to 5
λ bacteriophage vector	*E. coli*	5–23
Cosmid	*E. coli*	35–45
Phagemid	*E. coli*	Up to 10
Fosmid	*E. coli*	40
P1 bacteriophage vector	*E. coli*	70–125
P1-derived artificial chromosome (PAC)	*E. coli*	100–300
Bacterial artificial chromosome (BAC)	*E. coli*	≤300
Yeast artificial chromosome (YAC)	Saccharomyces cerevisiae	200–2,000

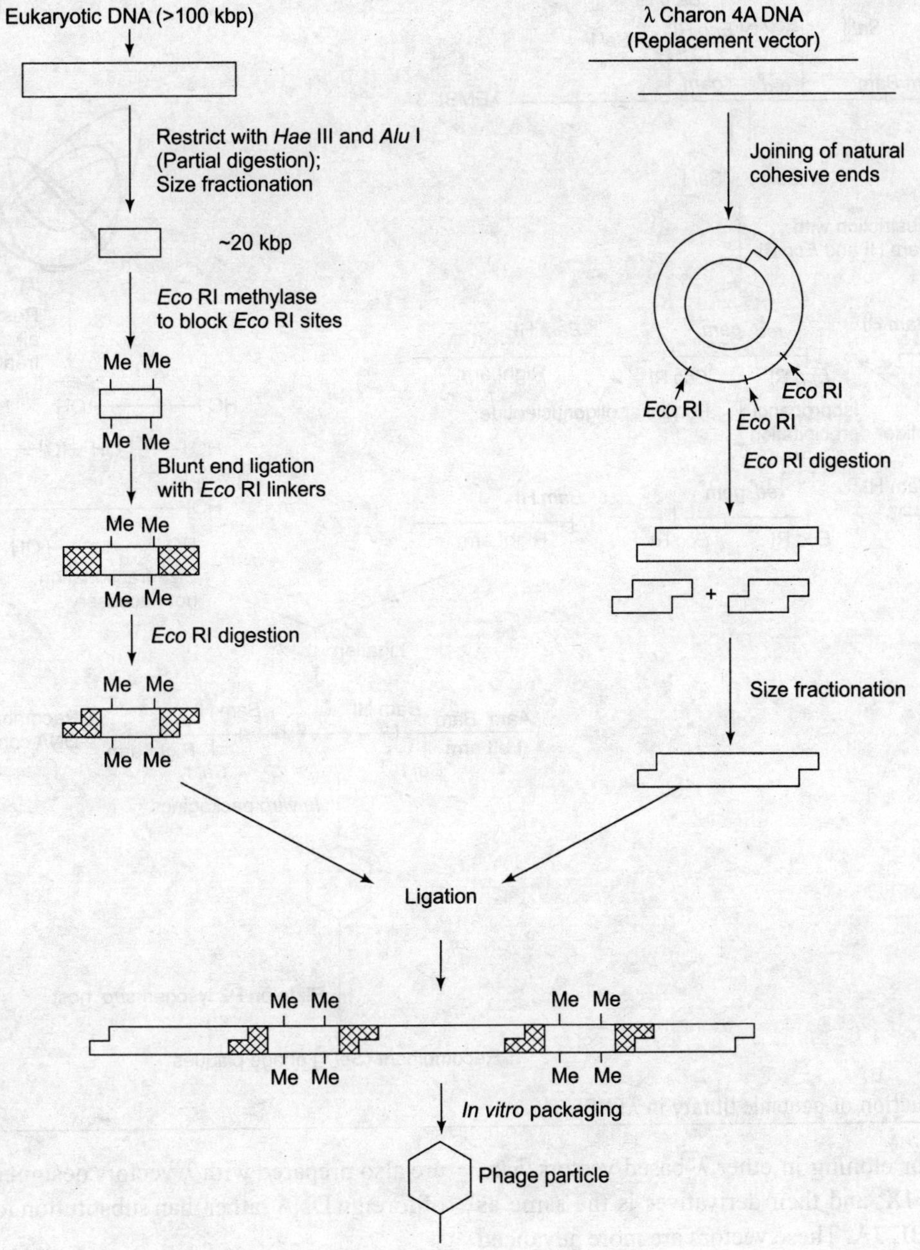

Figure 15.2 Maniatis strategy for generating genomic library

example, λEMBL 3A), the genomic DNA is partially digested with single restriction endonuclease that cuts more frequently, such as *Sau* 3AI, and less random fragments are obtained. These *Sau* 3AI fragments have the benefit of being capable to be readily inserted into *Bam* HI-digested λEMBL vectors. Note that *Sau* 3AI and *Bam* HI have nested sites and produce compatible ends (for details see Chapter 3). 5′-Phosphate groups of the fragments are removed by treating with alkaline phosphatase to prevent dimer formation. Note that in this case an intermittent size fractionation step is not necessary. The vector DNA is double digested with *Bam* HI and *Eco* RI, and the vector arms are purified either by isopropanol precipitation or by preparative agarose gel electrophoresis. Note that this double digestion step (i.e., cleavage at internal sites lying in stuffer fragment) prevents vector religation by preventing precipitation of polylinker oligonucleotide having *Bam* HI site on one end and *Eco* RI site on the other end (for details see Chapter 9). The vector arms are then mixed and ligated with the dephosphorylated genomic DNA fragments (~20 kbp). Thereafter, the recombinant is subjected to *in vitro* packaging and plaques are obtained after plating on P2 lysogen of a sup^+ *E. coli* (for details see Chapter 9). Finally, for the purpose of subcloning, the foreign DNA is excised from the vector by digesting with flanking *Sal* I sites in the polylinker. The entire strategy for cloning in λEMBL 3A vector is presented in Figure 15.3.

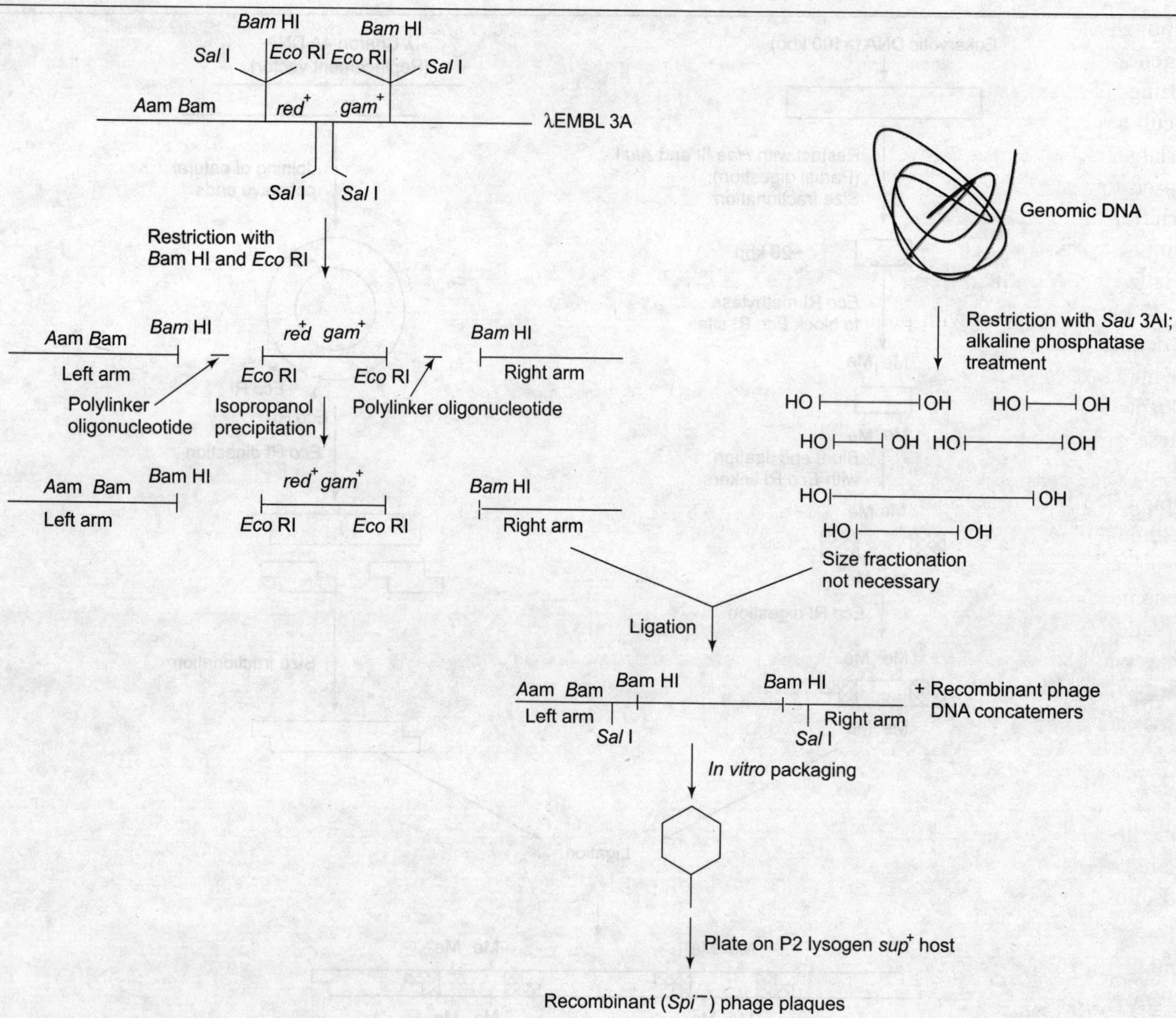

Figure 15.3 Construction of genomic library in λEMBL 3A

The procedure for cloning in other λ-based vectors, *viz.*, λ2001, λDASH, λFIX, and their derivatives is the same as described for λEMBL 3A. These vectors are more advanced and are more advantageous than λEMBL series of vectors. The polylinkers of these vectors, flanking the stuffer fragment, contain promoters for the T3 and T7 RNA polymerases. The availability of promoter sites facilitates the production of short RNA probes corresponding to the ends of any genomic insert. Note that if the recombinant vector is digested with any frequent cutter restriction enzyme, only short fragments of insert DNA are left attached to these promoters. These RNA probes are best for probing the genomic library for screening of overlapping clones and these can be prepared conveniently, directly from the vector, without subcloning. In these vectors, the *Xho* I site in the polylinker flanking the stuffer fragment is used to prevent vector religation by partial filling-in procedure (instead of cleavage at internal sites as used in λEMBL 3A) (for details see Chapter 9). Subgenomic libraries are also prepared with λ vectors designed for direct insertion of foreign DNA rather than substitution for a stuffer fragment.

Application of High Insert Capacity Vectors in Genomic Library Construction

Genomic libraries for higher eukaryotes (large genomes) are usually constructed using a number of higher insert capacity cloning vectors instead of bacteriophage λ vectors. These include cosmids, YACs, BACs, PACs, MACs, and bacteriophage P1 vectors (for details see Chapter 11 and 12). These vectors offer certain advantages as compared to bacteriophage λ vectors: (i) Average insert size of such vectors is much larger than λ replacement vectors; (ii) The number of recombinants that need to be screened to identify a particular gene of interest is reasonably lower; (iii) Large genes are most probably present within a single clone; and (iv) Fewer clones are required to assemble a contig by chromosome walking.

Though the high insert capacity vectors seem to be ideal for genomic library construction because of capacity of inserting very large inserts, such libraries are generally more difficult to prepare. This is because of difficulty in handling such larger inserts. Generally strategies similar to the λ vectors are used to construct such libraries, except that the partial restriction digestion conditions are optimized for larger fragment sizes, and size fractionation is performed by specialized electrophoresis methods that separate fragments over 30 kbp in length. These include PFGE and its variants (for details see Chapter 6). The procedures for cloning in these high insert capacity vectors have already been discussed in Chapters 11 and 12. Cosmids are favored over λ vectors because of their larger insert capacity (up to 45 kbp). The BAC, PAC, YAC, and MAC libraries are used for genome mapping, sequencing, and the assembly of clone contigs.

Although the development of modern vectors and cloning strategies have simplified library construction, the production of genomic DNA libraries is still a painstaking job. Nowadays pre-made libraries are available from many commercial sources and the same companies often offer custom library services. These libraries are often of high quality and such services are becoming increasingly popular.

15.2.4 Plaque Library vs. Colony Library

Plaque libraries (e.g., those formed by bacteriophage λ) are preferred over colony libraries (e.g., those formed by plasmids or cosmids). This is because of the following reasons:

Higher Insert Stability As plasmids or cosmids are maintained at high copy number in *E. coli*, these have a tendency to become unstable due to deletions and rearrangements. Moreover, instability increases with the increase in the insert size. On the other hand, phage libraries are more stable than plasmid libraries.

Ease of Amplification Effective amplification of plasmid and cosmid libraries is more difficult as compared to amplification of phage library. The procedures for amplification of colony and plaque libraries are discussed in Section 15.2.5.

Lesser Background Hybridization and Cleaner Results Plaques give less background hybridization than colonies. Hence screening libraries of λ phage recombinants by plaque hybridization gives cleaner results as compared to screening plasmid or cosmid libraries by colony hybridization.

Longer Shelf-Life and Ease of Storage It is desired to retain and store the amplified genomic library. With phage, the initial recombinant DNA population is packaged and plated out, and can be screened at this stage. Alternatively, the plates containing the recombinant plaques can be washed to give amplified library of recombinants indefinitely because of the long shelf-life of phage. With bacterial colonies containing cosmids, it is also possible to store an amplified library, but bacterial populations cannot be stored as readily as phage populations, and there is often an unacceptable loss of viability. Thus as compared to colony library, plaque library is easy to store and has longer shelf-life.

Reuse of Library The amplification of phage library is so high that samples of the amplified library can be plated out, and screened with different probes on hundreds of occasions.

15.2.5 Amplification of a Genomic Library

All the recombinants in a genomic library do not propagate equally well. This is because variations in target DNA size or sequence may affect replication (or growth rate) of a recombinant phage, plasmid, or cosmid. Hence it is advisable to amplify a genomic library. This amplification step may be carried out as soon as possible after the library is packaged, and after this step, particular recombinants may be increased in frequency, decreased in frequency, or lost altogether. This change in composition is due to differences in growth rate. Such amplified library can be stored for an indefinite period and screened as many times as necessary. Suitable *E. coli* host strains are used for amplifying λ libraries (Table 15.2).

The amplification step thus involves minimization of differences in growth rate by preadsorbing the library to the vector using a high plating density and short incubation period. For amplification, 250 ml LB medium containing 0.2% maltose and 10 mM $MgSO_4$ is inoculated with 2.5 ml of a fresh overnight culture of host bacteria. The culture is incubated in incubator shaker for 2–4 hours at 37°C with moderate agitation until OD_{600} reaches 0.5. The cells are then prepared for plating. The number of nonviable cells must be minimized since a phage incorporated into a dead cell is lost from the library. Approximately 10^5 phages from packaged library are added into host bacterial culture and incubated for 15 min at 37ºC. The entire mixture is plated by combining 0.5 ml adsorbed host/phage per 8 ml of ~47°C top agarose onto fresh prewarmed petriplates. The culture is spread evenly and allowed to solidify. The plates are inverted and incubated at ~37°C (incubation temperature depends upon type of phage vector) until clear plaques are visible. The plates are removed and overlaid with SM buffer (NaCl, $MgSO_4$, and Tris–HCl, pH 7.5), and plates are stored for 2 hours at room temperature or overnight at 4°C. The suspensions from all plates are pooled carefully and cellular

Table 15.2 Suitable *E. coli* Host Strains for Amplifying λ Libraries

Vector	*E. coli* host	Relevant host genotype
λgt 10	C600 *hfl* A	*hfl* A
λgt 11	Y1088	*sup* F, *lac* I^q
λEMBL 3 or 4	P2392, Q359, NM539	P2 lysogen
Charon 4A	LE392	*sup* F

debris is removed by centrifugation in a single tube at 2,800 *g* for 5 min. Chloroform (0.5 volume) is added and mixed by inverting, and amplified library is titrated. The expected titer of amplified library is of 10^{10} to 10^{11} pfu/ml. The library in this form is stable for many years. Dimethyl sulfoxide (7%) is added to the amplified library and 1 ml aliquots are stored in screw-capped microcentrifuge tubes at –80°C.

Effective amplification of plasmid and cosmid libraries is more difficult because a disproportionate growth of recombinant bacteria results in under-representation of particular clones. This concern must be considered against the time required to create a library *de novo* for each screening. The antibiotic-resistant bacteria are plated on nitrocellulose filters on LB plates containing the appropriate antibiotic. The colonies are rubbed off by using a sterile rubber policeman. The bacterial suspensions from all plates are combined into one 50 ml plastic tube, followed by addition of sterile glycerol. The contents are mixed thoroughly and 500 µl aliquots are prepared. The amplified libraries are then kept at –80°C.

15.2.6 Subgenomic Libraries

Genomic libraries have been prepared from single human chromosomes separated by flow cytometry. Even greater enrichment for particular regions of the genome is possible using the technically demanding technique of chromosome microdissection. In *Drosophila*, it has been possible to physically excise a region of the salivary gland chromosome by micromanipulation. This is followed by digestion of DNA and cloning it in λ vectors, all within a microdrop under oil. Similarly, this has been achieved with specific bands of human chromosomes using either extremely fine needles or a finely focused laser beam. Regardless of the species, this is a laborious and difficult technique, and is prone to contamination with inappropriate DNA fragments. This technique has, however, been rendered obsolete with the advent of high insert capacity vectors.

15.3 cDNA LIBRARY

A cDNA library is a collection of clones containing cDNA and represents the genes that are expressed within a given cell or tissue type at a particular period. As the entire sequence of eukaryotic genes is often very large due to presence of interrupting introns (noncoding sequences) within the exons (coding sequences), and as the introns can increase the genomic space taken up by the gene to a point where prokaryotic genetic manipulation schemes cannot handle the length of the DNA, it becomes essential to synthesize DNA copies of functional mRNA molecules that lack introns, i.e., cDNAs, and create tissue-specific cDNA libraries from eukaryotic species. Thus, cDNA libraries represent not just the collection of expressed sequences from that organism, but also those sequences after any posttranscriptional modification (e.g., splicing of introns). Before embarking on the synthesis and cloning of cDNA, it is essential to consider carefully which source material, methods, vectors, and screening procedures offer the best chance of success. The choice of vector, the steps involved in library construction, and the method by which the cDNA molecules are inserted into the vector are all factors that are heavily influenced by the method to be used for library screening.

On the contrary, the arguments in favor of cDNA rather than genomic libraries carry much less force with bacterial targets and hence when generating gene libraries from prokaryotic species, genomic DNA libraries are often preferred. This is because of the following reasons: (i) Prokaryotes generally lack introns and hence DNA is much shorter, and the RNA does not require much post-transcriptional processing events, e.g., splicing, to be expressed correctly. Prokaryotes also lack the apparatus necessary for splicing; (ii) Prokaryotic genome is small in size and hence easy to isolate. As compared with eukaryotic gene, the chance of cloning entire prokaryotic gene in a single vector is higher. Hence there is hardly any need for cDNA library construction in prokaryotes; (iii) Introns are rare in prokaryotic genes; hence the information contained in cDNA libraries is almost the same as in genomic DNA libraries. On the other hand, introns are present in eukaryotic genome, while these are spliced out in mature mRNAs. This forms the basis of difference between eukaryotic genomic and cDNA libraries; (iv) Prokaryotic mRNA is remarkably unstable and many mRNA species have a short *in vivo* half-life of only 1–2 min. Hence it is difficult to isolate prokaryotic mRNA. On the other hand, eukaryotic mRNA is relatively stable; (v) Prokaryotic genes are organized into polycistronic operons (groups of genes that are transcribed into a single long mRNA). As a consequence, a prokaryotic mRNA is 10–20 kbp in length. It is not only difficult to isolate this long mRNA intact, but it is also very difficult to produce a full-length cDNA from it; and (vi) It is technically difficult to construct cDNA libraries with bacteria, as prokaryotic mRNA is difficult to isolate. This is because prokaryotic mRNA is rarely polyadenylated. Few exceptions of mRNAs that are polyadenylated are those from *B. subtilis*, *B. brevis*, *B. polymyxa*, *E. coli*, *Rhodospirillum rubrum*, *Rhodopseudomonas capsulata*, *Caulobacter cresentus*, *Hypomicrobium* B-522, *Anabaena variabilis*, *Nostoc* MAC, etc. Thus, only 2–60% of the molecules of a given mRNA species is polyadenylated, and these poly (A) tails are much shorter than that of eukaryotes, ranging from 15 to 60 A residues. Moreover, the sites of polyadenylation of bacterial mRNA are diverse. These include the 3′-ends of primary transcripts, the sites of endonucleolytic processing in the 3′ untranslated and intercistronic regions, and the sites within the coding regions of mRNA degradation products. This diversity of polyadenylation sites suggests that mRNA polyadenylation in prokaryotes is a relatively indiscriminate process that can occur at 3′-ends of all mRNAs and does not require specific consensus sequences as in eukaryotes. Thus, affinity chromatography using oligo (dT)–cellulose, poly (U)–sepharose, or oligo (dT)–silica does not provide a reliable way of isolating bacterial mRNA, and hence it becomes technically difficult to isolate mRNA from total RNA in prokaryotes. On the contrary, most of the eukaryotic mRNAs contain poly (A) tail, which allows easy isolation through affinity chromatography.

15.3.1 cDNA Library vs. Genomic DNA Library

Genomic library is constructed for application in genome mapping, sequencing, or contig assembly. cDNA library also offers several applications. As cDNA library differs from genomic library in several respects, a comparison of both may also provide valuable information. The applications of cDNA library are mentioned below: (i) As cDNA reflects only those genes expressed in the particular tissue under the chosen conditions, the complete cDNA library of an organism gives the total proteins it can possibly express; (ii) cDNA library is representative of the RNA population from which it is derived. Hence the contents of cDNA libraries vary widely according to developmental stage and tissue. A given cDNA library becomes enriched for abundant mRNA in a particular developmental stage, or tissue, and hence is used for easy isolation of abundant mRNAs or corresponding polypeptide products. On the contrary, regardless of the cell type or developmental stage from which the DNA has been isolated, genomic libraries are essentially the same; (iii) Eukaryotic genes are large in size (due to presence of introns), hence it is difficult to clone entire gene in a single clone in case of genomic library. Moreover, there are increased chances of cloning of the target gene as fragments into several clones. On the other hand, the probability of cloning the entire cDNA is high. This is particularly essential when the expressed protein is of interest; (iv) The problem with genomic DNA is that cloning large segments of DNA is technically difficult and may suffer from lower transformation efficiency. On the contrary, cDNA, being smaller in size as compared to genomic DNA, is easier to handle and manipulate; (v) The advantage of cDNA cloning is that in many cases the size of cDNA clone is significantly lower than that of the corresponding genomic library. This reduces the number of clones to be analyzed to isolate the target gene; (vi) Since removal of eukaryotic intron from pre to posttranscript by splicing does not occur in bacteria, eukaryotic genes containing introns cannot be expressed in prokaryotes. As such, cDNA clones find application where bacterial expression of foreign eukaryotic DNA is necessary, either as a prerequisite for detecting the clone or if the expression of the polypeptide product is the primary objective. A cDNA of an eukaryotic organism can be cloned into a prokaryotic organism, and expressed there. This is because no posttranscriptional processing event that is lacking in prokaryotes is required; (vii) Comparison of cDNA sequences between libraries constructed from cells derived from different organisms can provide insight into the genetic and evolutionary relationship between organisms through the similarity of their cDNAs; (viii) As cDNAs lack introns, a comparison of genomic library with cDNA library can provide information about positions of introns and locations of exon–intron boundaries; (ix) As cDNAs are obtained by reverse transcribing mRNAs, a comparison of genomic library with cDNA library can provide information about polyadenylation sequences. Note that eukaryotic mRNA contains 3′ poly (A) tail; and (x) Where the gene is differentially or alternatively spliced, a cDNA library will contain different clones representing alternative spliced variants. On the contrary, genomic library does not give any information about differentially spliced genes.

15.3.2 Enrichment Methods

Any cDNA library is likely to represent only a fraction of the RNA species of any one organism, determined by the particular type, developmental stage, physiological state, etc. of the tissue from which the RNA is isolated. The library is therefore expected to contain cDNAs for abundant mRNAs, such as the general 'house-keeping' genes and for those genes whose expression is specific to that particular tissue (for details see Chapter 5). If cDNA encoding a specific protein is needed, it is wise to use a cDNA library from a tissue with a high quantity of RNA for that protein relative to other RNAs. In other words, a cDNA library from tissues producing large amounts of required protein should be used.

As in the chicken oviduct, mRNA-encoding ovalbumin is superabundant, hence the starting population is naturally so enriched in ovalbumin mRNA. Thus isolation of ovalbumin mRNA can be achieved without the use of a library. An appropriate strategy for obtaining such abundant cDNAs is to clone them directly into a suitable vector. For the isolation of cDNA clones in moderately and low abundance mRNA classes, it is usually necessary to construct a cDNA library. Like genomic library, in cDNA library also, *in vitro* packaging in λ bacteriophage is an attractive strategy.

It is pertinent to note that the abundance of mRNA determines the size of a cDNA library that has to be constructed to ensure the presence of an appropriate clone. Thus, the number of clones required to give a 99% probability that a particular rare mRNA is present in a cDNA library can be calculated according to the Clarke and Carbon equation:

$$N = \frac{\ln[1-P]}{\ln[1-(1/n)]}$$

where N is the number of clones required; P the desired probability (99% or 99/100); n the fractional proportion of the total mRNA that is represented by a single type of rare (or low abundance) mRNA, or $1/n$ is the fractional proportion of the rare mRNA in the total pool of mRNA.

Take an example of a rare mRNA in a human fibroblast cell line. In this case, rare mRNAs (<14 copies per cell) constitute ~30% of the mRNA and ~11,000 different mRNAs belong to this class. The minimum number of cDNA clones required to obtain a complete representation of mRNAs of this class is therefore 11,000/0.30 or 37,000. Thus $1/n$ = 1/37,000. The number of clones (N) required to give a 99% probability that a particular rare mRNA is present in a cDNA library from a human fibroblast cell that contains ~12,000 different transcribed sequences will be: $\ln(1 - 0.99)/\ln(1 - 1/37,000) = 1,70,000$.

In practice, a larger number of recombinant clones have to be screened to ensure isolation of a rare cDNA. A number of factors influence the representation of a particular cDNA clone in a library. These include mRNA instability, the tendency of some sequences to be preferentially cloned, and the presence of multiple cell types in a particular tissue, some of which do not contain the required RNA. These numbers have been calculated for a number of cell types. Typically 10^5–10^6 clones are sufficient for the isolation of low abundance mRNAs from most cell types.

Some mRNAs are even less abundant than the sufficient number (1 copy per cell) and may be further diluted if these are expressed in only a few specific cells in a particular tissue. Under these circumstances, mRNA preparation should be enriched prior to library construction, e.g., by size fractionation and testing the fractions for the presence of desired molecules. One way in which this can be achieved is to inject mRNA fractions into *Xenopus* oocytes and test for the production of the corresponding protein (for details see Chapter 5). Additionally, it is often necessary to clone cDNAs from populations of mRNAs isolated from tissues that consist of several cell types. In such cases, the frequency at which the sequences of interest are represented in the initial preparation of mRNA may be reduced still further and it then becomes necessary to construct and screen libraries that contain several million independent cDNA clones. Currently, the efficiency with which cDNA can be synthesized and cloned has gradually increased, and cDNA libraries of comprehensive size can be generated routinely from 1 to 5 μg or less of poly (A) mRNA. On the other hand, screening large number of cDNA clones is both tedious and expensive activity. Therefore, enrichment methods have been devised to enrich either the starting population of mRNA molecules or ds cDNA synthesized from it for sequences of interest. Enrichment allows the size of the cDNA library to be reduced and decreases the cost and labor involved in screening for the desired cDNA clones. Different mRNA enrichment strategies have already been described in Chapter 5.

15.3.3 Vectors Used for cDNA Cloning

Previously plasmid vectors were used for construction of cDNA libraries which were usually maintained as a collection of $>10^5$ independently transformed bacterial colonies. These colonies were sometimes pooled, amplified in liquid culture, and stored at –80°C. These were also stored in immobilized form on nitrocellulose filters. In both cases, results were unsatisfactory. The libraries were not easily preserved and screening by hybridization with multiple radioactive probes required replication of colonies from one nitrocellulose filter to another. This was a painstaking process that could be repeated only a few times before the colonies on the master filter were smeared beyond recognition. With the advent of refined linkers, adaptors, methylase, and packaging mixtures, it became possible to use λ phage as a vector, and to take advantage of the high efficiency and reproducibility of *in vitro* packaging of λ DNA into virus particles. The resulting libraries are often large enough to be screened directly without amplification. Alternatively these are amplified to be stored for indefinite period without loss of viability, and depending on the particular vector, screened with nucleic acid probes, antibodies, or ligands. During 1980s, the standard vectors for construction of cDNA libraries were λgt 10 and λgt 11. However, λgt 10 was used for the preparation of libraries that were to be screened only by nucleic acid hybridization, while for cDNA to be screened with antibodies or other ligands for isolation of DNA sequences encoding specific proteins, libraries were constructed in λgt 11. Presently, the most commonly used vectors for cDNA library construction are λZAP series of vectors (marketed by Stratagene) including λZAP II, λZAP Express, and λZipLox (marketed by Life Technologies). Using these vectors, the cDNA inserts are recovered as plasmids from the respective libraries and hence the tedious jobs like preparation of λ DNA and subcloning into plasmids can be avoided. Although the plasmids that were used in 1970s for construction of cDNA libraries are no longer in use, their derivatives (phagemids) are employed to construct and propagate cDNA libraries that are used for functional screening, expressed sequence tagged (EST) sequencing, and generation of subtracted libraries and probes. The examples of such plasmid vectors are pSPORT 1, pCMV-Script, and pcDNA3.1. These have variety of sequence elements, including versatile polylinkers, promoters for phage T3, T7 and/or SP6 RNA polymerase, and intergenic region of filamentous f1 phage that permits DNA to be rescued in single stranded form (for details see Chapter 12).

15.3.4 Construction of cDNA Library

A wide range of technical and theoretical advances has been made in cDNA cloning and now cDNA cloning is well within the range of any competent laboratory. Comprehensive cDNA libraries can be routinely established from small quantities of mRNA and a variety of reliable methods are available to identify cDNA clones corresponding to extremely rare species of mRNA. As the enzymatic reactions used to synthesize cDNA have improved, the sizes of cloned cDNAs have increased and it is now possible to isolate full-length cDNAs corresponding to mRNAs (for details see Chapter 5).

The first stage of cDNA library construction is the synthesis of ds cDNA using mRNA as the template. The details of the procedures used for mRNA isolation and blunt ended ds cDNA synthesis have already been discussed in Chapter 5. After the construction of ds cDNA, it is ligated to a suitable vector. The ligation step is facilitated by the addition of linkers, adaptors, or homopolymer tails to the end of blunt ended ds cDNA. The details of linkers, adaptors, and homopolymer tails have already been described in Chapter 13.

Application of Linkers and Adaptors in cDNA Cloning

The traditional protocol for construction of cDNA library is divided into six steps.

Synthesis of First and Second cDNA Strand Different protocols for first two steps have already been described in detail in Chapter 5. In brief, the first cDNA strand is synthesized by reverse transcriptase using mRNA template, and the second cDNA strand is synthesized by DNA polymerase I/Klenow fragment using the first cDNA strand as template. These two steps result in the formation of ds cDNA. The ds cDNA is then treated with T4 DNA polymerase or DNA polymerase I, enzymes that remove protruding single stranded 3′-termini with their 3′ → 5′ exonuclease activities, and fill-in their recessed 3′-OH termini by their polymerase activities. Finally, because of combinations of these steps, phosphorylated, blunt ended ds cDNA molecules are generated.

Methylation of cDNA at Internal Sites For ligating the blunt ended ds cDNA molecules into vector DNA, attachment of linkers or adaptors is required. If it is essential to clone into a restriction site that occurs as frequently as *Eco* RI, it is advisable to modify the internal *Eco* RI sites in the ds cDNA by methylation using cognate methylases (*Eco* RI methylase). Methylated bases protect internal restriction sites in the ds cDNA against cleavage by cognate restriction enzyme (*Eco* RI), a step required to generate cohesive termini after addition of double stranded *Eco* RI linkers to the ds cDNA. There are two ways to produce methylated cDNA: (i) 5-methyl dCTP is used as a precursor instead of dCTP during first cDNA strand synthesis. Second cDNA strand is synthesized in the presence of four normal dNTPs. The resulting hemimethylated ds cDNA is partially resistant to cleavage by *Eco* RI. If the cDNA is methylated, then a strain of *E. coli* must be used that is deficient in the *mcr* restriction system, which cleaves DNA at methylated cytosine residues; and (ii) Before addition of synthetic linkers, ds DNA can also be methylated by using a cognate methylase enzyme (*Eco* RI methylase). *Eco* RI methylase catalyzes the transfer of methyl groups from donor such as *S*-adenosyl methionine to the second adenine residues in the recognition sequence of *Eco* RI (5′-G↓A*ATTC-3′) leading to the formation of 6-methyl amino adenine (N^6-methyladenine). The modified base protects the DNA from digestion by *Eco* RI at the modified sites. As Mg^{2+} inhibits this enzyme, it is essential to precipitate the ds cDNA with ethanol to remove Mg^{2+} before proceeding with the methylation step. Note that the methylation step is essential with linkers, but this modification is not required while using adaptors. When cDNA libraries are constructed from small amounts of DNA, many researchers simply skip the methylation step to generate the maximum number of cDNA clones.

Attachment of Linkers and Adaptors, and Restriction Digestion Linkers having one or more restriction sites offer a valuable technique for joining ds cDNA to vector and have replaced other methods for cloning cDNA populations. The methylated blunt ended ds cDNAs generated after step III are then incubated with large molar excess of linker molecules in the presence of T4 DNA ligase, an enzyme capable of catalyzing blunt end ligation. The simplest of all the linkers are 8- to 12-mer oligonucleotides, consisting of recognition site for a restriction enzyme such as *Eco* RI (5′-pCG*GAATTC*CG-3′). These blunt ended oligonucleotides are easily ligated to the polished end of ds cDNA (Figure 15.4). The products of the ligation reaction are cDNA molecules carrying polymeric linker sequence at their termini. These ligation products are then cleaved by restriction enzyme (in this example *Eco* RI) by which excess of linkers is removed and cohesive termini are exposed at the ends of cDNA. Thus, the end products obtained after ligation and restriction digestion steps are cDNA molecules carrying a single linker at each end and free linker molecules (for details see Chapter 13).

Usually the ds cDNA contains one or more internal recognition sites for the restriction enzyme, which hinders the isolation and analysis of full-length cDNAs. Note that any internal unmethylated site is susceptible to digestion by restriction enzyme (*Eco* RI). Various efforts have been made to circumvent this problem, which includes the following: (i) The average distance between *Eco* RI sites is ~4 kbp and so there is good probability that a cDNA of average size will be cleaved into two or more parts. Hence to avoid such a problem, restriction enzyme sites (in linkers) that cleave the target cDNA very rarely (e.g., *Not* I and *Sal* I) are used. For example, the likelihood of finding the GC-rich octanucleotide sequence recognized by *Not* I in a 5 kbp clone of mammalian cDNA are less than 1 in 200. For *Sal* I, the sites of which occur on an average once every 100 kbp in mammalian DNA, the possibility increases to 1 in 20. Moreover, the internal sites are methylated by cognate methylases to avoid cleavage at internal *Not* I or *Sal* I sites, if present. The usage of *Not* I and *Sal* I enzymes allows recovery of intact cDNA from the vector.

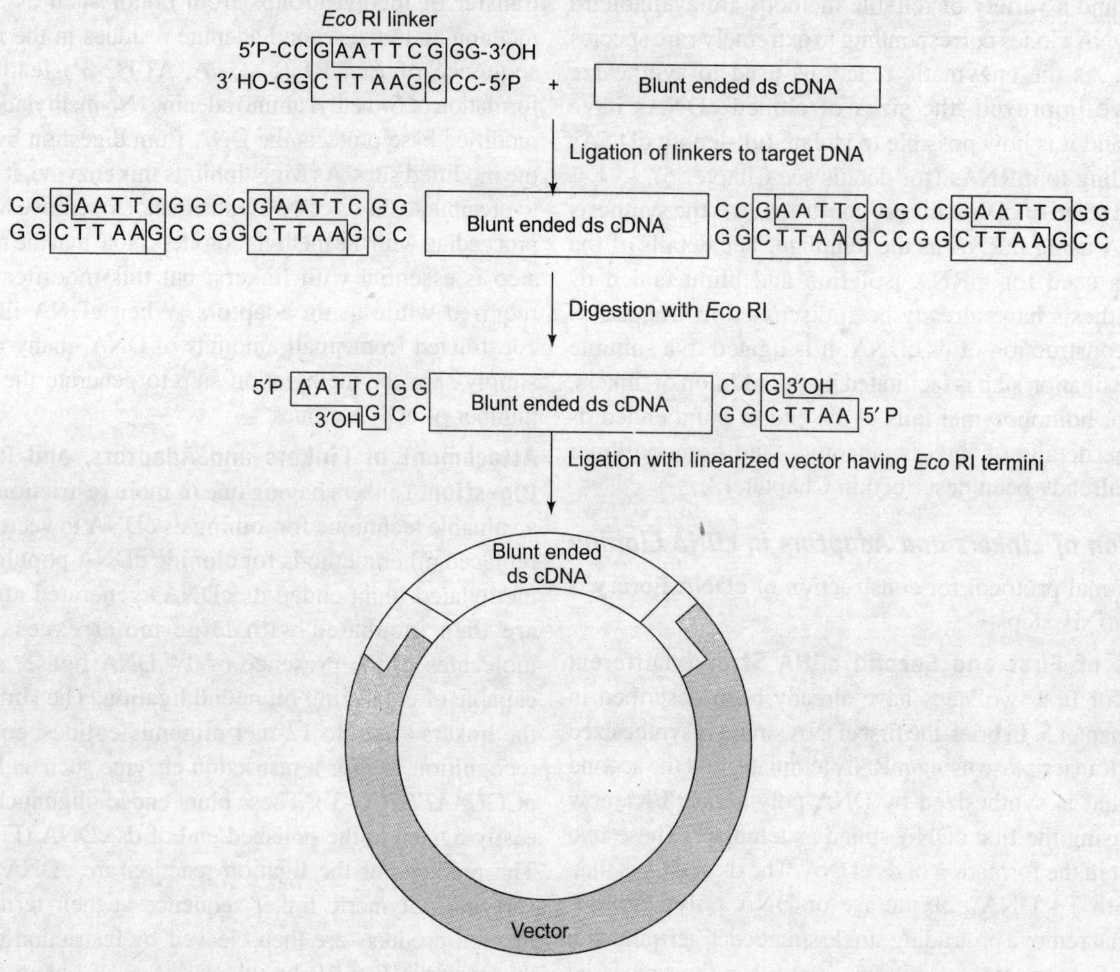

Figure 15.4 Ligation of linker to ds cDNA products in the presence of high concentrations of T4 DNA ligase

This is because *Not* I and *Sal* I recognition sites are rarely present within cDNA sequence; and (ii) Synthetic adaptors can be used instead of linkers.

Adaptors are short, double stranded oligonucleotides having an inherent restriction endonuclease recognition site, one blunt end, and one cohesive end, which can be ligated at compatible termini in the vector. Various types of adaptors are used in cDNA cloning. The example of most commonly used adaptor is as follows:

HO 5′-AATTCGCGGCCGCGTCGAC-3′ HO

HO 3′-GCGCCGGCGCAGCTG-5′P

This adaptor consists of cohesive *Eco* RI terminus, internal *Not* I and *Sal* I sites, and one phosphorylated 5′-terminus. When adaptors are ligated with blunt ended ds cDNA, then only single adaptor is added, and thus the resulting molecules have cohesive termini at both ends without the requirement of *Eco* RI digestion. Note that similar to linkers, the attachment of adaptors is performed in very small volume, as it is very important to maintain high concentration of adaptors. The molar concentration of adaptors is required to be at least 100 times greater than the concentration of ends of cDNA to minimize blunt end ligation of cDNA molecules. Another important point is that after ligation with ds cDNA, the adaptors need not be digested with restriction enzymes. However, as cDNA molecules carrying phosphorylated adaptors can form covalently closed circular molecules, which cannot be cloned, or chimeric linear molecules, which are highly undesirable during the subsequent ligation reaction in the presence of dephosphorylated vector DNA, cDNAs are phosphorylated by T4 polynucleotide kinase (T4 PNK) at both ends after the addition of adaptors. The end products obtained after ligation step are cDNA molecules carrying a single adaptor at each end and dimer adaptor molecules. As already mentioned, in contrast to linkers, the application of adaptors does not require any methylation step, i.e., step III is not required while using adaptors. Moreover, any restriction digestion step after ligation of adaptors to ds cDNA is also not required.

It is pertinent to note here that in addition to facilitating the modification of termini of cDNA for cloning, linkers and adaptors can also be used as binding sites for primers in PCRs. By incorporating a PCR step, large cDNA libraries can be established from a very small amount of cDNA.

Fractionation of cDNA Subsequently, before the cDNA is inserted into the vector, unutilized adaptors and the low molecular weight products produced by restriction enzyme digestion of polymerized linkers should be efficiently removed by size fractionation techniques (for details see Chapter 6). Fractionation is usually carried out through agarose or polyacrylamide gel electrophoresis, however, purification by gel filtration via a long narrow column (27 × 0.3 cm) of Sepharose CL-4B is the best choice. Currently, some biotech manufacturers supply prepacked columns of Sepharose CL-4B (e.g., SizeSep 400 Spun columns). The removal of debris increases the number of recombinants that contain cDNAs and reduces the number of false recombinants. The cDNAs <500 bp in length, which are undesired products of cDNA synthesis, are also eliminated by size fractionation. Therefore, the massive exercise of screening procedures is also reduced and the probabilities of getting nearly full-length cDNA clones are increased.

Ligation of Blunt Ended ds cDNA to Greek Letter Lamda Arms or Linearized Plasmids Finally the target cDNA molecules with linkers or adaptors attached to either ends are ligated with dephosphorylated vector DNA carrying cohesive termini compatible with those of the linker or adaptor. Before ligation, λ arms or linearized plasmid vectors are prepared. If large library is to be constructed, then it is much cheaper to isolate phage arms by agarose gel electrophoresis or by sucrose density gradients rather to purchase them. Some commercial suppliers manufacture dephosphorylated arms. During this ligation step, the ds cDNA molecules containing the synthetic cohesive termini are ligated with each other, as well as to vector DNA. Hence, it is essential to adjust the ligation conditions to minimize the formation of chimeric molecules. Therefore, the ligation mixture should always contain a high molar excess of vector DNA to cDNA. However, these conditions strongly favor reformation of vector by self-ligation, leading to unacceptably high backgrounds of nonrecombinant clones. This problem can be reduced by dephosphorylation of cleaved vector by treating with alkaline phosphatase before ligation to cDNA.

Before setting of actual ligation reaction, it is suggested to perform pilot ligation reactions to determine the amount of cDNA that is required to produce a cDNA library of reasonable size. The pilot ligation reactions are set to minimize the chance of formation of compound cDNAs. This is done by (i) Using a molar ratio of phosphorylated λ arms to cDNAs such that only 5% of the resulting bacteriophages or plasmids are recombinants; and (ii) If dephosphorylated λ arms or plasmid vectors are used, then the amount of cDNA should be used in the concentration that generates ten-fold more plaques or colonies, respectively, in contrast to the vector ligated in the absence of cDNA. Different λ vectors require different strains of *E. coli* for plaque formation (Table 15.3). Since λ vectors do not have any selection system against nonrecombinant bacteriophages, and if cDNA libraries are created in these vectors, then it is advisable to use dephosphorylated arms. This strategy reduces ~100-fold the number of nonrecombinant bacteriophages in the cDNA library. In some of these vectors such as λgt 11, it is feasible to approximate the ratio of nonrecombinant to recombinant bacteriophages in the library by plating on strains of *E. coli*, for example Y1090 *hsd* R, through blue-white selection. In the presence of the inducer IPTG, and the chromogenic substrate X-gal, nonrecombinant plaques are blue and recombinants are white. In the case of λgt 10, C600 (BNN 93) is used for growth and BNN 102 (C600 *hfl* A) for screening of cDNA libraries. The single round of growth on host *E. coli hfl*$^-$ mutant strain (BNN 102) eliminates the nonrecombinants. Recombinants form clear plaques, as insertion of the cDNA inactivates the *cI* gene encoding λ repressor (for details see Chapter 9). Finally, *in vitro* packaging is carried out with the help of packaging extracts. Sometimes plasmid vectors may be preferable, especially when preparing a cDNA expression library, which are screened after transformation into cultured cells.

Table 15.3 Suitable *E. coli* host strains to plate bacteriophage λ vectors for plaque formation

Vector	*E. coli* host	Medium for growth
λgt 10	C600 and BNN 102	NZCYM + 0.2% maltose
λgt 11 and λgt 18–23	Y1090 *hsd* R	NZCYM + 0.2% maltose
λZAP	BB4	NZCYM + 0.2% maltose
λZAP II	XL1-Blue	NZCYM + 0.2% maltose
λZipLox	Y1090 (ZL)	NZCYM + 0.2% maltose

Application of Homopolymer Tailing in cDNA Cloning

Homopolymeric tailing is also used to ligate ds cDNA to vector DNA in cloning experiments. The procedure for obtaining ds cDNA remains essentially the same as described above. This blunt ended ds cDNA is then tailed by adding just one dNTP (e.g., dCTP) using the enzyme terminal deoxynucleotidyl transferase at the 3′-ends and complementary homopolymer tails are generated in vector DNA (for details see Chapter 13). Thereafter, target DNA and vector DNA having complementary tails are ligated and *in vitro* packaged as described in Step VI in the discussion before.

15.3.5 Amplification of cDNA Library

The primary libraries are usually used for cDNA screening, but amplification of primary library can provide an unlimited source of cDNA clones to screen. In spite of the fact that the single round of amplification misrepresents the mRNAs in a cDNA library, even then amplification is the best choice when the amount of given library reduces. Amplification involves plating entire library on agar plates followed by preparation of lysates from each plate. The lysates are then used as the source of recombinant bacteriophages for screening with nucleic acid, antibodies, or other ligands. The basic procedure of amplification of cDNA library is quite similar as discussed above in Section 15.2.5.

15.3.6 Construction of Eukaryotic cDNA Libraries in Expression Vectors

The eukaryotic cDNA that encodes functional proteins is cloned in particular types of expression vectors. An eukaryotic expression vector essentially contains a promoter that drives the transcription of gene of interest, eukaryotic transcriptional and translational stop signals, a sequence that enables polyadenylation of mRNA and a eukaryotic marker gene. Most of the eukaryotic expression vectors are shuttle vectors having two origins of replication and selectable marker genes, one for *E. coli* and other functions in eukaryotic host cells as yeasts and mammalian cells.

Strategies for cDNA Library Construction Based on Location and Function of Encoded Proteins

Basically three types of strategies, which depend on location and function of particular proteins, are available for construction of cDNA libraries in eukaryotic expression vectors. These are:

Receptor, Channel, and Transporter Proteins For receptor, channel, and transporter proteins, a cDNA library is prepared in a plasmid or λ vector equipped with promoters for λ-encoded DNA-dependent RNA polymerase. Pools of cDNA clones chosen at random from the library are transcribed *in vitro* and the resulting mRNA is injected into *Xenopus* oocytes. The individual oocytes are then scored for the expression of the functional protein. Pools of cDNAs that generate a positive signal are divided into subpools and retested until a single clone encoding the target protein is identified and isolated. This kind of selection system is termed as sib-selection.

Cell Surface and Secreted Proteins A cDNA library is prepared in a λ vector equipped with signals for expression in eukaryotic cells. These signals include a promoter/enhancer as well as splice donor and acceptor sites for expression in particular eukaryotic cells. Pools of cDNA clones are then transfected into a line of cultured cells. In the case of secreted protein, supernatant medium from the cells is assayed for functional activity. Binding to a specific ligand or antibody then identifies transfected cells expressing a target surface protein. The individual cDNAs encoding the target proteins may be identified by sib-selection as discussed above. The cells expressing the target protein at the cell surface may be isolated by panning or by use of radiolabeled ligands or by fluorescence activated cell sorting (FACS). The plasmid DNA is then recovered from the selected population of transfected cells and amplified in *E. coli*.

Intracellular Proteins In this case, pools of cDNA clones are transfected into cultured mammalian cells. The proteins from cultured cells are extracted and then screened for an activity distinctive of the target protein.

Host/Vector Systems for Amplification of Transfected cDNA Clones

For reducing the extensive labor, which is involved in traditional sib-selection systems, various host/vector systems are developed in which transfected cDNA clones replicate as episomes in mammalian cells. cDNA libraries are prepared in plasmids, which have a wild-type origin of DNA replication derived from SV40, polyomavirus, and sometimes Epstein-Barr virus or bovine papillomavirus. Pools of cDNA clones are then transfected into lines of mammalian cells that express all of the *trans*-acting factors required to drive replication of the transfected cDNA clones. Cells that express the target protein may be identified by immunological screening, and in some cases, isolated by panning on petriplates coated with appropriate ligands or antibodies. The cells expressing the target protein are lysed and the episomal DNAs are isolated and reestablished in *E. coli*. The gene of interest may be isolated after three to four times of transient expression and selection which depends on: (i) Efficiency of panning; (ii) Amount of amplification of the transfected cDNA clones in the mammalian hosts; and (iii) Frequency with which the target clone is represented in the cDNA library.

Various types of host-vector systems have been designed to broaden the host range of cells that can be utilized for expression cloning (Table 15.4). However, the most commonly used host cells are permanent lines of African green monkey kidney cells that express SV40 T-antigen from chromosomally integrated copies of an origin-minus mutant of SV40, which are known as COS lines. The integrated copies of the viral DNA lack viable origin of replication and are not activated by the endogenously expressed T-antigen. Thus, the yield of episomal DNA totally reflects the amplified transfected plasmids. Note that in this case, an immediate early cytomegalovirus (CMV) promoter, which is a strong promoter, drives the expression of cloned DNA.

The success of any expression library depends upon various factors like the assay system that is used to detect the target protein must be robust, sensitive, and specific. As most cDNAs isolated by expression cloning encode rare proteins, it is usually required to generate a library consisting of 10^6 or more independent cDNA clones. It is also important to optimize the efficiency of transfection of the host cells using plasmids with reporter genes.

15.3.7 cDNA Libraries Generated by RT-PCR

Reverse transcriptase polymerase chain reaction (RT-PCR) can be used to provide the cDNA for library construction

Table 15.4 Vector System for Expression Cloning

Vector	Promoter(s)	Expression in	
		Xenopus oocytes	Cultured mammalian cells
pcDNA3.1	CMV, T7	No	Yes
pcDNA4	CMV with enhancer QBI SP163	No	Yes
pSPORT plasmid	*lac*, SP6, T7	Yes	No
pCMV-Script plasmid	CMV	Yes	Yes
λZipLox	*lac*, SP6, T7	Yes	No
λZAP-CMV	T3, T7, CMV	No	Yes
λZAP Express	*lac*, T3, T7, CMV	Yes	Yes
λEcCell	SP6, T7	Yes	No

when the source of mRNA is unsuitable for conventional approaches, for example, a very small amount of starting material or fixed tissue. In this technique, instead of gene-specific primers, universal primers can be used that lead to the amplification of all mRNAs, which can then be subcloned into suitable vectors. The advantages and disadvantages of RT-PCR are discussed in Section 15.3.9.

15.3.8 Subtractive Cloning

Subtractive cloning is a type of representational difference analysis technique. It is carried out for removal of common sequences from the two different cDNA preparations before library construction. Initially, this method was developed for the comparative analysis of genomes, but later it has been modified for cloning differentially expressed genes. As discussed in Chapter 5, this procedure is used for enrichment of differentially expressed RNA species. A library is generated that is rich in differentially expressed clones by removing sequences that are common to two sources. This is termed as subtracted cDNA library, which is ideal for isolation of both up- and downregulated genes. Removal of common sequences is achieved by hybridizing ss cDNA prepared from mRNA (tester) extracted from the tissue of interest (Type A) to a 10- to 20-fold excess of DNA or RNA (driver) prepared from other source (Type B) that does not express the genes of interest. The DNA:RNA hybrids are then separated from unhybridized cDNA by hydoxyapatite chromatography. After hybrids containing the cDNA have been removed, the residual cDNA molecules are cloned after addition of linkers or adaptors. For the generation of highly enriched subtracted libraries, several rounds of extraction with driver mRNA are performed. Few interesting examples are available where subtractive cloning has been used with genomic libraries to identify differentially represented sequences. The most significant example of this approach is the cloning of the human gene for muscular dystrophy using normal DNA and DNA from an individual with a deletion spanning the DMD gene (chromosome deletion in the region *Xp*21 known as DMD locus).

Nowadays, more advancements have been made like bacteriophage promoters can be used to generate DNA drivers from cDNA cloned in single stranded vectors, and hybrids can be captured by biotin:avidin affinity chromatography on latex beads. More effective PCR-based techniques have also been devised for subtractive cloning, i.e., common sequences between two sources are eliminated prior to amplification. Basically, the technique involves the same principle, in which a large excess of DNA from one source, the driver, is used to hybridize the common sequences in the other source, the tester DNA. The strategy generally used for this purpose is shown in Figure 15.5. The ds cDNA is prepared from two sources, digested with restriction enzymes, to obtain shorter blunt ended fragments. The tester cDNA is divided into two portions. Each is ligated with different adaptor that provides annealing sites for a unique pair of PCR primers. These adaptors are not added to the driver cDNA. A large excess of driver cDNA is then added to the tester cDNA and the populations are mixed. Hybridization kinetics lead to equalization and enrichment of differentially expressed sequences among tester molecules. Driver/driver fragments possess no adaptors and cannot be amplified, while driver/tester fragments possess only one primer annealing site and are amplified in a linear fashion. cDNAs that are present only in the tester possess adaptors on both strands and are amplified exponentially and can therefore be isolated and cloned. Finally, templates for PCR are generated for amplification from differentially expressed sequences, which are thus amplified exponentially.

15.4 PCR AS AN ALTERNATIVE TO LIBRARY CONSTRUCTION

PCR can be used as an alternative to genomic library construction, while RT-PCR can be used as an alternative to cDNA library construction. The procedures for both these techniques have already been described in detail in Chapter 7.

15.4.1 PCR as an Alternative to Genomic Library Construction

It is not feasible to construct the genomic libraries from small amounts of starting material, like single cells, or from problematic sources such as fixed tissues. PCR with specific primers is the only technique that facilitates gene isolation from small amounts of genomic DNA. However, the

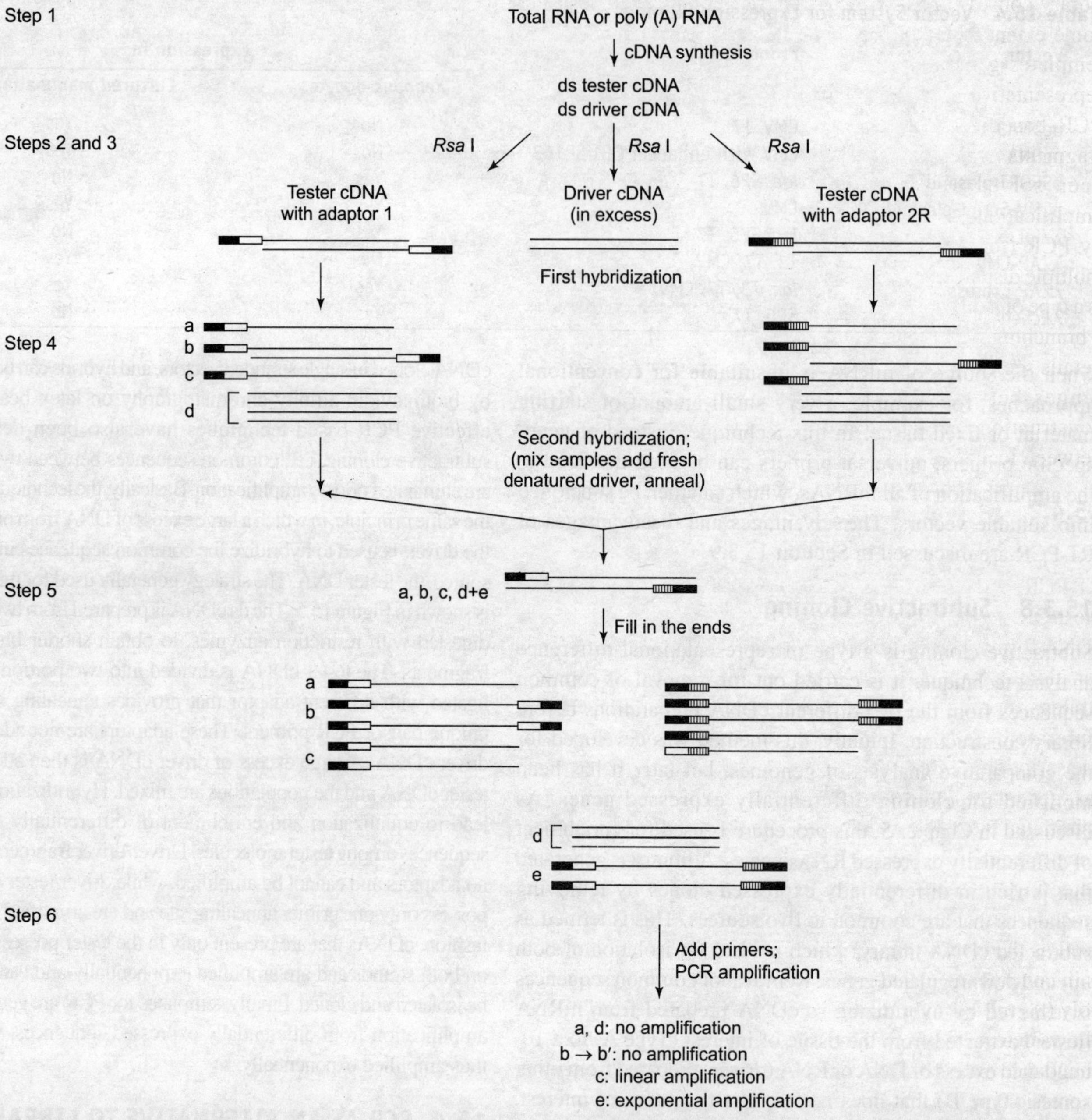

Figure 15.5 A generalized strategy for cDNA subtraction technique [Step 1: cDNA is prepared from either total RNA or poly (A) RNA; Step 2: Tester and driver ds cDNAs are digested separately by *Rsa* I to get shorter blunt ended fragments; Step 3: Tester cDNA is divided into two parts and each is ligated to a different adaptor (1 and 2R), while driver cDNA is not ligated to any adaptor; Step 4: First hybridization leads to equalization and enrichment of differentially expressed sequences among tester molecules; Step 5: Templates are generated for PCR amplification from differentially expressed sequences; and Step 6: Differentially expressed sequences are amplified exponentially.]

maximum product size that can be obtained by PCR is ~5 kbp, although the typical size is more likely to be 1–2 kbp in a PCR reaction. In this case, long PCR is performed, which utilizes the blend of polymerases to amplify long DNA templates (for details see Chapter 7). Furthermore, PCR can also be used to generate libraries, i.e., by amplifying a representative collection of random genomic fragments. Usually, this can be performed by two strategies. In the first strategy, random primers are used for PCR amplification followed by size selection for accurate PCR products. In the second strategy, genomic DNA is digested with restriction enzymes and then linkers are ligated to the ends of the DNA fragments, which provide annealing sites for one specific type of primer. These PCR-based techniques permit genomic fragment libraries to be prepared from material that could not yield DNA of suitable quality or quantity for conventional library construction. As PCR-based techniques demonstrate some extent of favoritism towards particular sequences, the templates generally do not allow the production of truly representative libraries. In spite of several advancements in PCR-based techniques, usually the products are short fragments, which have limited usefulness. This problem has been solved in a technique known as whole genome amplification (WGA) in which whole genome is amplified by PCR (for details see Chapter 7). Another alternative is multiple displacement amplification (MDA or E-RCAT). This is a type of isothermal amplification technique, which involves a branching reaction, and utilizes λ29 DNA polymerase (for details see Chapter 7). The product length obtained in MDA ranges between 20 and 200 kbp, which is ideal for construction of genomic library, and the discrimination between loci is also reduced.

15.4.2 RT-PCR as an Alternative to cDNA Library Construction

RT-PCR, which involves reverse transcription followed by the PCR, leads to the amplification of RNA sequences in cDNA form and helps in obtaining a specific cDNA sequence for cloning. The procedure is almost similar to basic PCR strategy, except that the template for PCR amplification is generated in the same reaction tube in a prior reverse transcription reaction. It involves usage of specific 3′ primer that generates first cDNA strand and then PCR amplification is initiated by addition of 5′ primer to the reaction mix. Using gene-specific primers, RT-PCR is a sensitive means for detecting, quantifying, and cloning specific cDNAs. The procedure is highly sensitive because (i) Total RNA can be used as the starting material rather than poly (A) RNA [this is in contrast to conventional cDNA cloning, which requires poly (A) mRNA as starting material]; and (ii) This technique allows detection of a specific mRNA in a single cell, and low abundance mRNA (the technique is also used to analyze gene expression in cells that are difficult to obtain in large numbers).

RT-PCR has several advantages as compared with cDNA cloning, or in other words, several disadvantages or problems are associated with cDNA library construction. These include: (i) RT-PCR requires very small amount of starting material or fixed tissue (<200 ng RNA), and as mentioned above even total RNA can be used as starting material; (ii) RT-PCR is faster than cDNA cloning procedure; (iii) cDNA library construction is a labor-intensive process; (iv) cDNA library may not yield a cDNA clone with a full-length coding region due to technical difficulties with long mRNAs; (v) RT-PCR may allow amplification of even very low abundant mRNA; and (vi) The sought-after cDNA may be very rare even in specialized cDNA libraries.

Despite the advantages associated with RT-PCR, there are still reasons for constructing cDNA libraries. Some examples are given: (i) The cDNA library may be used as a source from which a specific cDNA can be obtained by PCR amplification; (ii) The sought-after mRNA may occur in a source that is not readily available, perhaps a small number of cells in a particular human tissue. A good quality cDNA library from such tissue has to be constructed only once to give a virtually infinite resource for future use; (iii) The specialized cDNA library can be stored for further use and is permanently available for screening; (iv) DNA polymerases used for PCR are error-prone and hence may result in large number of mutations; (v) In RT-PCR, there is likely to be a certain amount of distortion due to competition among templates and a bias toward shorter cDNAs; (vi) In RT-PCR, false results may be obtained from the amplification of contaminating genomic sequences in the RNA preparation and even trace amounts of genomic DNA may be amplified. However, this problem can be circumvented by certain manipulations. In the study of eukaryotic mRNAs, it is desirable to choose primers that anneal in different exons such that the products expected from the amplification of cDNA and genomic DNA are of different sizes. Moreover, if the intron is suitably large, genomic DNA will not be amplified at all. In case of bacterial system, RNA can be treated with DNase prior to amplification to destroy any contaminating DNA; and (vii) PCR-based approaches for screening are dependent upon specific probes and hence a knowledge of complete or at least partial sequence is required. On the other hand, with cDNA libraries, screening strategies are based on immunological screening rather than nucleic acid hybridization.

Review Questions

1. A DNA library is a collection of clones, each containing a different fragment of DNA, inserted into a cloning vector. Write the differences between a genomic library and a cDNA library. Clearly explain the starting material, the types of vectors, and the strategies used for the construction of these libraries.
2. A genomic library of a prokaryotic organism is often constructed by cloning the products of a *Sau* 3AI partial digest of the genomic DNA into a *Bam* HI site of the vector. Why are two different enzymes used in this experiment? Why is partial digestion often used for constructing genomic libraries?
3. What is a genomic library and what is its significance? How can the size of the genomic library be estimated?
4. Describe the Maniatis strategy of genomic library construction with the help of suitable diagram.

5. Plaque libraries are preferred over colony libraries. Justify this statement.
6. Why is it important to amplify the genomic library? Write the protocol for effective amplification of the library.
7. Genomic DNA libraries are often preferred over cDNA libraries for prokaryotic species. Write some important reasons to explain this statement.
8. Describe the different steps of construction of cDNA library. What are the applications of cDNA library?
9. Give a detailed account of different strategies used for construction of eukaryotic cDNA libraries in expression vectors. Write host/vector systems for amplification of transfected cDNA clones.
10. What do you understand by subtractive cloning? Describe the generalized strategy for cDNA subtraction technique with the help of suitable diagram.
11. Explain how PCR and RT-PCR are used as alternatives to library construction.
12. The yeast genome is 12,000 kbp in size. How many yeast DNA fragments of average length 5 kbp must be cloned so that a genomic DNA contains a particular segment with 90%, 99%, and 99.9% probability?
13. Calculate and compare the number of required recombinants in genomic library for a probability of 99% and an average insert size of 20 kbp of *E. coli* (4.6×10^6 bp) and human (3×10^9 bp).
14. How large a library (or how many clones) would you need in order to have a 99% probability of finding a desired sequence represented in a human genome library created by digestion with a six base-cutter? Note that on an average, a six base-cutter will cut at approximately every 4.1 kbp.

Reference

Maniatis, T. *et al.* (1978). The isolation of structural genes from libraries of eukaryotic DNA. *Cell* **15**:687–701.

Techniques for Selection, Screening, and Characterization of Transformants

Key Concepts

In this chapter we will learn the following:

- Selectable marker genes
- Reporter genes
- Screening of clone(s) of interest
- Nucleic acid blotting and hybridization
- Protein structure/function-based techniques

16.1 INTRODUCTION

In recombinant DNA technology, after introduction of DNA into host cell, it is very important to select the host cell that takes up the DNA construct (i.e., transformed cell) from those that do not. Modern cloning vectors comprising of selectable markers permit only cells in which the vector, but not necessarily the insert, has been transformed to grow. Therefore, plating onto selective media enables transformants to be distinguished from nontransformants. Additionally with most cloning vectors insertion of a DNA fragment into them destroys the integrity of one of the genes (i.e., insertional inactivation) present in molecules and recombinants can be identified because the phenotype or characteristic encoded by the inactivated gene is no longer displayed by the host cells. Nevertheless, these selection steps do not absolutely guarantee that the DNA insert is present in the cells. Further screening and investigation of the resulting colonies is required for confirmation and characterization. A genetic screen (often shortened to screen) is a procedure or test to identify and select individuals that possess a phenotype of interest. Using this method of analysis, researchers can search through entire genomes to find genes of interest without having prior knowledge of their location. Positional cloning is a method of gene identification in which a gene for a specific phenotype is identified, with only its approximate chromosomal location or candidate region (but not the function) known; whereas functional cloning uses a known gene sequence as a probe to look for new gene sequences that may have similar functions, based on the similarity of sequence (Exhibit 16.1). In the present chapter, all the procedures employed for selection, screening, and analysis of gene of interest are presented.

16.2 SELECTABLE MARKER GENES

In cloning experiments, only one in a several million or billion cells take up DNA. Selectable marker genes are present on vector itself into which the DNA of interest has been cloned and thus provide an easy way to select and screen positive recombinant clone(s), i.e., to determine that a piece of DNA

Exhibit 16.1 Positional and functional cloning

Positional cloning employs mapped position of a gene to obtain a clone containing it. Simply, positional cloning is done when there is no biological information about a gene (often the case for human disease genes), but its position can be mapped relative to other genes or markers. Using this information, the nearest physical markers can be located, which is followed by chromosome walking (for details see Chapter 19). In this technique, overlapping clones are obtained spanning the region of interest by using each successive clone as a probe to detect the next. Thus, one can 'walk' from one flanking marker to the next, knowing that at least one of the clones will contain the gene of interest. For large distances, libraries based on high-capacity vectors, such as BACs and YACs, should be used so as to reduce the number of steps involved. In one of the first applications of this technology, Bender and his coworkers cloned DNA from the *Ace* and *rosy* loci and the homeotic *Bithorax* gene complex in *Drosophila*.

In contrast to positional cloning, functional cloning does not require any prior knowledge about the gene in the genome, for example, there is no requirement of knowledge about the nucleotide sequence of the clone or the amino acid sequence of its product. Rather functional cloning depends on the full biological activity of the protein. As long as the expressed protein is functional and the function can be exploited to screen an expression library, the corresponding clone can be identified.

Source: en.wikipedia.org/wiki/Positional_cloning, en.wikipedia.org/wiki/Functional_cloning.

has been successfully inserted into the host organism. It protects the organism from a selective agent that would normally kill it or prevent its growth. Therefore, all cells that do not contain the foreign DNA are killed in the presence of selective agents provided in the growth medium, and only the desired ones are left behind. However, there are examples of 'positive selection' systems as well. Some examples of selectable marker genes are discussed in this section.

In bacteria, resistance to antibiotics are used almost exclusively, for example, genes encoding resistance to ampicillin, chloramphenicol, tetracycline, streptomycin, and kanamycin (for details see Chapter 8). Reversal of auxotrophy is another type of positive selection, which forms the basis of selection of recombinants in yeast (for details see Chapter 11). Some other markers are useful because their inactivation can be positively selected. For example, *sac* B encodes levansucrase and its activity is lethal to cells growing on medium containing 7% sucrose; *ccd* B gene is lethal to host cells unless these carry the DNA gyrase mutation *gyr* A.

Plant transformation is in many cases a very low-frequency event. It is therefore vital that some means for selecting the transformed plant tissue is provided by the plant transformation vector. In most cases, this selection is based on the inclusion into the culture medium of a substance that is toxic to plants. The selectable marker on the vector confers resistance to the toxic substance when expressed in transformed plant tissue. In many cases, antibiotic resistance genes are used as selectable markers in plants. Although plants are eukaryotic, antibiotics efficiently inhibit protein synthesis in the organelles, particularly the chloroplasts. Perhaps the most widely selectable marker gene is the neomycin phosphotransferase (*npt* II) gene that confers resistance to the antibiotic kanamycin. Dominant marker selection provides a direct means of obtaining only transformed cells in culture, for example, in the presence of the antibiotic kanamycin, only plant cells with a functional *npt* II gene can grow. The selectable marker allows the transformed cells to proliferate in the presence of a selective agent, while untransformed cells either do not grow or multiply at a very reduced rate. Other antibiotic resistance genes have also been used with some considerable success in plant transformation vectors. In part, this use of alternative selectable marker genes was driven by the observations that some plant species exhibited a very high degree of natural resistance to kanamycin (such as cereals) and that some species (some soft fruits) were too sensitive to kanamycin for it to be successfully used. This made the selection of transformed tissues very difficult, leading to a high number of false-positives or the inability to recover transformed plants. Thus, the use of any selectable marker has to be closely controlled, with appropriate concentrations being determined by kill-curves on a case-by-case basis. The development of alternative selectable marker genes also allows for the retransformation of plant tissue that already expresses one or more different selectable markers. Thus, genes conferring resistance to antibiotics such as bleomycin, spectinomycin, and hygromycin are used quite widely. Hygromycin B phosphotransferase (plasmid-borne bacterial gene) encodes a 341 amino acid that inactivates the antibiotic. This gene has been used as a selectable marker in *E. coli,* and chimeric genes constructed with the appropriate promoters act as dominant selectable markers in yeast, mammalian cells, and plants. These selective agents can be used at lower concentrations than kanamycin and therefore usually result in a cleaner selection of transformed tissue. Other resistance genes can also be used as selectable markers in plants. Amongst those widely used are genes that confer resistance to herbicides such as chlorsulfuron and bialaphos. The resistance genes can act either by modifying the selective agent itself to inactivate it or by producing an insensitive form of the normal target of the selective agent. Some of the most interesting selectable markers are based on the principle that allows the use of mannose or xylose as carbon sources. These genes have also generally proved to be superior to standard antibiotic resistance genes in some transformation protocols. These are examples of 'positive-selection' methods and thus termed because untransformed tissues do not die (due to the presence of low concentrations of sucrose in the medium used for regeneration), but do not proliferate much. Some examples of plant selectable markers are summarized in Table 16.1.

In the case of mammalian cells, the first gene, which is used extensively for selection, was a herpes simplex virus (HSV) gene encoding thymidine kinase (*tk*). However, thymidine kinase is expressed in various mammalian cell lines, therefore several TK^- cell lines were produced by selection for growth in the presence of 5-bromodeoxyuridine (Brd U). The HSV *tk* gene is an example of a class of genes known as endogenous markers because these confer a property that is already present in wild-type cells. Some of the endogenous genes that act in redundant metabolic pathways have been exploited and used as markers are given in Table 16.2. The major disadvantage associated with endogenous markers is that these can only be used with mutant cell lines in which the corresponding host gene is nonfunctional, which restricts the range of cell lines used for transfection.

Nucleotides are produced via two alternative routes, the *de novo* and salvage pathways in mammalian cells. In the *de novo* pathway, nucleotides are synthesized from basic precursors, such as sugars and amino acids, while the salvage pathway recycles nucleotides from DNA and RNA. When the *de novo* pathway is blocked, nucleotide synthesis becomes dependent on the salvage pathway, and this procedure can be exploited for the selection of cells carrying functional *hprt* (hypoxanthine-guanine phosphoribosyl transferase) and *tk* genes. For example, cells grown in the presence of aminopterin (an analog of dihydrofolate) are unable to catalyze *de novo* synthesis

Table 16.1 Selectable Marker Genes used in Plant Transformation, Their Source, and Mode of Action

Selectable marker	Abbreviation	Source of gene	Selection medium	Selective agent
Antibiotic resistance				
Aminoglycoside adenyltransferase	*aad A*	*Shigella flexneri*	Antibiotic resistance	Streptomycin Spectinomycin
Bleomycin resistance	*ble*	*E. coli*	Antibiotic resistance	Bleomycin
Dihydropteroate synthase	*sul/dhps*	*E. coli*	Antibiotic resistance	Sulfonamides
Dihydrofolate reductase	*dhfr*	Mouse	Antibiotic resistance	Methotrexate
Hygromycin phosphotransferase	*hpt/aph* IV/*hyg*	*E. coli*	Antibiotic resistance	Hygromycin
Neomycin phosphotransferase II	*npt* II/*neo*	*E. coli*	Antibiotic resistance	Kanamycin Geneticin (G418)
Neomycin phosphotransferase III	*npt* III	*Streptococcus faecalis*	Antibiotic resistance	Kanamycin Geneticin (G418)
Herbicide resistance				
Acetolactate synthase	*als*	*Arabidopsis* spp./ Maize/Tobacco	Herbicide resistance	Sulfonylurea
Enol pyruvyl shikimate phosphate synthase	*epsps/aro* A	*Petunia hybrida/ Agrobacterium* spp.	Herbicide resistance	Glyphosate
Glyphosate oxidoreductase	*gox*	*Streptomyces hygroscopicus/S. viridochromogenes*	Herbicide resistance	Bialaphos Glufosinate *L*-phosphinothricin
Cyanamide hydratase	*cah*	*Myrothecium verrucaria*	Herbicide resistance	Cyanamide
Bromoxynil nitrilase	*bxn*	*Klebsiella pneumoniae*	Herbicide resistance	Bromoxynil
Others				
Lysine-threonine aspartokinase	*lys* C	*E. coli*	Resistance to high levels of lysine and threonine	Lysine and threonine
Mannose-6-phosphate isomerase	*pmi/man* A	*E. coli*	Alternative carbon source	Mannose
Xylose isomerase	*xyl* A	*Thermoanaerobacterium thermosulfurogenes/ Streptomyces rubiginosus*	Alternative carbon source	Xylose
β-Glucuronidase	*gus/uid* A	*E. coli*	Complementation of missing medium component	Cytokinin glucuronide
Betaine aldehyde dehydrogenase (used as selectable marker in chloroplast transformation)	*badh*	*Spinach*	Detoxification	Betaine aldehyde
Green fluorescent protein	*gfp*	*Aequoria victoria* (jelly fish)	Visual screening	None
Isopentyl transferase (If used with a constitutive promoter, plants have poor root formation and are often sterile, thus limiting the use of the *ipt* gene as a selectable marker. However, use with an inducible promoter has resulted in the development of an antibiotic-free selectable marker system)	*ipt*	*Agrobacterium* spp.	Growth in absence of exogenous cytokinin	None
Leafy cotyledon (Leafy cotyledon gives the transformed cells a competitive growth advantage and when coupled with a visual marker to identify transformed cells can be used as an anti-biotic-free selectable marker system)	*lec*	Maize	Growth	None

Table 16.2 Endogenous Selectable Markers Commonly Used in Animal Transformation

Selectable marker	Abbreviation	Selection medium
Adenosine deaminase	*ada*	Xyl-A (9-β-*D*-xylofuranosyl adenosine) and 2′-deoxycoformycin
Adenine phosphoribosyltransferase	*aprt*	Adenine plus azaserine to block *de novo* dATP synthesis
Multifunctional enzyme	*cad*	PALA (*N*-phosphonacetyl-1-aspartate) inhibits the aspartate transcarbamylase activity of carbamyl phosphate synthetase/aspartate transcarbamylase/dihydroorotase (CAD)
Hypoxanthine-guanine phosphoribosyltransferase	*hprt*	Hypoxanthine and aminopterin to block *de novo* IMP synthesis; Selected on HAT medium
Thymidine kinase	*tk*	Thymidine and aminopterin to block *de novo* dTTP synthesis; Selected on HAT medium

of both inosine monophosphate (IMP) and thymidine monophosphate (TMP) by inhibiting key enzymes. However, cells treated with aminopterin can survive only if these have functional *hprt* and *tk* genes and a source of hypoxanthine and thymidine. The *hprt* and *tk* positive transformants can be selected using HAT (hypoxanthine, aminopterin, thymidine) medium, which contains 100 μM hypoxanthine, 0.4 μM aminopterin, 16 μM thymidine, and 3 μM glycine. This approach was utilized for the introduction of foreign DNA in mammalian cells by cotransfection with a plasmid encoding the *tk* gene as most genes did not produce a conveniently selectable phenotype and the isolation of transformants in such experiments problematic. Nowadays, endogenous markers have been largely replaced by dominant selectable markers, which confer a phenotype that is entirely novel to the cell and hence can be used in any cell type (Table 16.3). Examples of such markers include drug resistance genes of bacterial origin and transformed cells are selected on a medium that contains the drug at an appropriate concentration. The most commonly used dominant selectable markers in animal cells are genes encoding aminoglycoside phospho-transferase (resistance to G418 or neomycin), hygromycin B phosphotransferase (resistance to hygromycin B), xanthine-guanine phosphoribosyl transferase (resistance to mycophenolic acid and aminopterin), and puromycin acetyltransferase (resistance to puromycin). These selectable markers can be linked with promoters such as early promoter of simian virus 40 (SV40) or the HSV Tk promoter and expressed in many cell types.

16.3 REPORTER GENES

An alternative to a selectable marker is a screenable marker (or reporter gene), which allows the researcher to distinguish between wanted and unwanted cells. It is essential to be able to detect the foreign DNA that has been integrated into host genomic DNA so that those cells that have been transformed could be identified. Furthermore, in studies of transcriptional regulatory signals and the functioning of these signals in

Table 16.3 Commonly Used Dominant Selectable Marker Genes for Animal Transformation

Selectable marker	Abbreviation	Source	Principle of selection
Asparagine synthase	*as*	*E. coli*	Toxic glutamine analog albizziin
Glycopeptide-binding protein	*ble*	*Streptoalloteichus hindustantus*	Resistance to glycopeptide antibiotics bleomycin, pheomycin, Zeocin™
Blasticidin deaminase	*bsd*	*Aspergillus terreus*	Resistance to blasticidin S
Guanine-xanthine phosphoribosyltransferase	*gpt*	*E. coli*	Analogous to *hprt* in mammals, but possesses additional xanthine phosphoribosyl transferase activity, allowing survival in medium containing aminopterin and mycophenolic acid
Histidinol dehydrogenase	*his* D	*Salmonella typhimurium*	Resistance to histidinol
Hygromycin phosphotransferase (*E. coli*)	*hpt*	*E. coli*	Resistance to hygromycin B
Neomycin phosphotransferase (*E. coli*)	*neo* (*npt* I)	*E. coli*	Resistance to aminoglycoside antibiotics (e.g., neomycin, kanamycin, G418)
Puromycin *N*-acetyltransferase	*pac*	*Streptomyces alboniger*	Resistance to puromycin
Tryptophan synthase	*trp* B	*E. coli*	Resistance to indole

specific tissues and cell lines, it is often important to be able to quantify the level of expression of a gene with a readily identifiable product. Quantification and other applications require the use of reporter genes or screenable marker that can be easily assayed. To this end, a number of different genes have been tested as reporters for transformation, including gene that can be used as dominant selectable marker and gene encoding protein that produces a detectable response to a specific assay. Many of these reporter genes are from bacteria and have been equipped with plant-specific regulatory elements for expression in plant cells. Reporter genes are widely used in transformation vectors both as means of assessing gene expression by promoter analysis and as easily scored indicators of transformation. Reporter genes are widely used for *in vitro* assays of promoter activity. However, reporters that can be used as cytological or histological markers are more versatile because these allow gene expression profiles to be determined in intact cells and whole organisms.

Ideally, reporter genes should have following properties: (i) Its product should be easy to assay, preferably with a non-destructive assay system; (ii) There should be little or no endogenous activity (of the reporter gene-encoded protein/enzyme) in the cell to be transformed; (iii) It should be non-toxic i.e., it should not interfere with normal cellular activities and exert no effect on survival and growth of transformants; and (iv) It should tolerate N-terminal fusions.

In contrast to selectable marker genes, the transformants are not selected by growth in the medium containing selection pressure (e.g., antibiotic/herbicide); rather the products encoded by reporter genes are analyzed separately either *in vitro* or *in situ*. Various reporter genes that are in widespread use in transformation vectors are described in this section, and their advantages and disadvantages are presented in Table 16.4.

Chloramphenicol Acetyltransferase Chloramphenicol acetyltransferase (CAT), encoded by gene *cat*, inactivates antibiotic chloramphenicol by acetylation. CAT is not normally

Table 16.4 Commonly Used Reporter Genes

Reporter gene	Source	Advantages	Disadvantages
Chloramphenicol acetyltransferase (*cat*)	Bacterial	No endogenous activity; Tolerates N-terminal fusions and remains enzymatically active; Automated ELISA available.	Narrow linear range; Use of radioisotopes; Assay procedure difficult, tedious, and expensive; Endogenous esterases, phosphatases, transferases, and other enzymes catalyze competing reactions and make quantitation difficult; Stable and persists in cell, hence not useful for assaying transcriptional repression or rapid changes in gene activity.
β-Galactosidase	Bacterial	Well-characterized and stable; Simple colorimetric assay; Sensitive bio- or chemiluminescent assay available; Tolerates N-terminal fusions.	Endogenous activity; Stable and persists in cell, hence not useful for assaying transcriptional repression or rapid changes in gene activity.
Nopaline synthase	*Agrobacterium tumefaciens*	No endogenous activity; N-terminal fusion possible.	Assay procedure difficult and cumbersome.
Octopine synthase	*Agrobacterium tumefaciens*	No endogenous activity.	Cannot tolerate N-terminal fusions; Assay procedure difficult and cumbersome.
Neomycin phosphotransferase (*npt* II)	Bacterial	N-terminal fusion possible.	Assay procedure difficult, tedious, and expensive; Variable endogenous activities in plant cells; Endogenous esterases, phosphatases, transferases, and other enzymes catalyze competing reactions and make quantitation difficult;
Green fluorescent protein (*gfp*)	*Aequoria victoria* (jelly fish)	Autofluorescent endogenous activity;	Requires posttranslational modification;

(*Contd.*)

Table 16.4 (*Contd.*)

Reporter gene	Source	Advantages	Disadvantages
		Easy nonradioactive and nondestructive assay; Assay cellular processes in real-time; Nontoxic; N-terminal fusions possible; Mutants with altered spectral qualities available.	Stable and persists in cell, hence not useful for assaying transcriptional repression or rapid changes in gene activity; Requires expensive equipment.
Luciferase (*lux* A and *luc* B)	*Vibrio harveyi*	Nontoxic; Good for analyzing prokaryotic gene transcription.	Less sensitive than firefly; Not suitable for mammalian cells.
Luciferase (*luc*)	*Photinus pyralis* (firefly)	Ideal *in situ* marker; High specific activity; No endogenous activity; Nontoxic; Easy nonradioactive and nondestructive assay; Useful for assaying transcriptional repression or rapid changes in gene activity; Variants of other colors available.	Requires substrate (luciferin) and presence of O_2 and ATP; Requires expensive equipment; Labile with rapid decay of light emission.
Alkaline phosphatase	Human placenta	Secreted protein; Inexpensive colorimetric and highly sensitive chemiluminescent assays available.	Endogenous activity; Interference with compounds being screened.
β-Glucuronidase (*gus*; *uid* A)	*E. coli*	Simple histochemical, fluorometric and colorimetric assays; No or low endogenous activity; Tolerates N-terminal fusions.	Very stable and persists in cell, hence not useful for assaying transcriptional repression or rapid changes in gene activity.

found in plant or animal cells, i.e., there is no endogenous activity in plants and animals. CAT is analyzed by radioactive assay, which is a very sensitive and semiquanti-tative method. The assay procedure of enzyme is difficult, tedious, and expensive. Competing reactions catalyzed by endogenous esterases, phosphatases, transferases, and other enzymes make quantitation of CAT difficult. The enzyme is stable and persists in cell that expresses them. N-terminal fusion is possible, i.e., the gene can tolerate N-terminal fusions and remain enzymatically active. This reporter system is widely used in mammals. *cat* was the first bacterial gene to be expressed in plants; however the availability of *gus* and *gfp* as reporter genes has limited the use of *cat* as reporter gene in plant transformation.

β-Galactosidase The source of gene *lac* Z that encodes β-galactosidase (Lac Z) is *E. coli*. β-galactosidase is a hydrolytic enzyme that hydrolyzes a range of β-*D*-galactopyrano-sides such as lactose and various analogs. There is high level of endogenous activity of β-galactosidase in plants. The assay procedure for the enzyme is nonradioactive and sensitive. The cell lysates can be assayed spectrophotometrically using chromogenic substrate ONPG (*o*-nitrophenyl-β-*D*-galactopyra-noside), which yields *o*-nitrophenol that is yellow in alkaline solution (λ_{max} 420 at pH 10.2). Alternatively, a more sensitive fluorescent assay may be preferred using the substrate MUG (4-methyl umbelliferyl-β-*D*-galactopyranoside), which yields blue fluorescence. In yet another method, *viz*., histological staining, the substrate X-gal (5-bromo-4-chloro-3-indolyl-β-*D*-galactopyranoside) yields an insoluble blue precipitate. β-Galactosidase is also an ideal histochemical marker. The *lac* Z gene was first expressed in mammalian cells. Moreover, β-galactosidase is often used as an internal reference in transfection studies. It is a stable enzyme and its expression in cultured mammalian cells, yeasts, *Drosophila*, or transgenic mammals does not appear to be deleterious or advantageous to the host. The enzyme is stable and persists in cell that expresses it. The gene can tolerate N-terminal fusions, i.e., fusion protein construction is possible.

Nopaline Synthase The gene *nos* encodes an enzyme, nopaline synthase (NOS), and used as reporter gene in plant system. Its source is *Agrobacterium tumefaciens*. NOS is not found in normal plant cells, i.e., it has no endogenous activity. The assay procedure for the enzyme though nonradioactive is difficult and cumbersome. The gene can tolerate N-terminal fusions.

Octopine Synthase The gene encoding octopine synthase (OCS) is abbreviated as *ocs*. Like NOS, its source is *A. tumefaciens*, and there is no endogenous activity of OCS in normal plant cells. The assay procedure for the enzyme though nonradioactive is difficult and cumbersome. The gene cannot tolerate N-terminal fusions.

Neomycin Phosphotransferase The gene encoding neomycin phosphotransferase enzyme (NPT II) is abbreviated as *npt II*. The gene *npt II* encodes NPT II with specificity not normally found in plant tissues. NPT II phosphorylates neomycin or kanamycin and hence inactivates it. The addition of signal sequence at N-terminal end (N-terminal fusions) is possible, thereby making it suitable for studying organelle transport in plants. The assay procedure for the enzyme though nonradioactive is difficult, tedious, and expensive. The enzyme suffers from variable endogenous activities in plant cells (generally caused by enzymes with broader substrate specificity), which limits both their sensitivity and the validity of quantitation. Competing reactions catalyzed by endogenous esterases, phosphatases, transferases, and other enzymes make quantitation of NPT II difficult.

Green Fluorescent Protein The gene encoding green fluorescent protein (GFP) is abbreviated as *gfp*. Source of *gfp* is *Aequorea victoria* jellyfish, brightly luminescent organisms. GFP is a bioluminescent marker that causes cells to emit green fluorescence when excited with UV or blue light. GFP is easy to assay and the assay procedure is nonradioactive and nondestructive. GFP has the advantage that it is even easier to assay than GUS (given the correct equipment); the fluorescence assay requires expensive equipment. It is an ideal *in vivo* or *in situ* marker and works efficiently in bacteria, yeasts, *Drosophila*, zebra fish, plants, and cultured mammalian cells. GFP was first used as a heterologous marker in *Caenorhabditis elegans*. GFP is noncatalytic and hence there is no requirement of any substrate or cofactor and can be used as a vital marker to assay cellular processes in real-time. GFP is nontoxic i.e., it does not interfere with normal cellular activities and hence no effect on survival and growth of transformants under field conditions. The enzyme is stable even under harsh conditions. The gene can undergo N-terminal fusions to form fusion proteins; thereby facilitating the investigation of intracellular protein trafficking. GFP can be used in situations where GUS cannot be used, for example, in screening primary transformants in time-course experiments or in analyzing segregation in small seedlings.

Now in transformation experiments, a modified *gfp* (*mgfp*) is used, which can resist aberrant splicing. Modified form of *gfp* is obtained by removing a cryptic splice site recognized in plants; the original *gfp* gene has also been extensively modified to alter various properties of the protein such as the excitation and emission wavelengths to increase the signal strength and to reduce photobleaching. Derivatives of the gene that encode the GFP having different spectral properties and increased levels of fluorescence are also available (e.g., red fluorescent protein from Anthozoa).

Luciferase The gene encoding luciferase from *Photinus pyralis* (firefly) is abbreviated as *luc* gene, and those from *Vibrio harveyi* (bacteria) are *lux* A and *lux* B genes. The *luc*-encoded luciferase is a single polypeptide of 550 amino acids, which converts luciferin to light in the presence of oxygen, ATP, and Mg^{2+}. This oxidation results in emission of light. The *lux* A and *lux* B encodes luciferase, which catalyzes oxidation of long chain fatty aldehydes and causes emission of light. Luciferase is an ideal, nontoxic, *in situ* marker, i.e., can be detected in intact plant tissues. The assay procedure requires oxidation of luciferin/long chain fatty aldehydes to emit luminescence, which is analyzed using luminometer; the assay is nonradioactive and nondestructive but requires expensive instrumentation. Highly sensitive assays (based on photomultipliers, luminometers, or film exposure) have been developed to detect the extremely rapid emission of light; other assay systems are also available commercially. Luciferase is labile and there is rapid decay of light emission; hence it is difficult to assay with accuracy. However, this property is advantageous as well as it does not remain for long in the cell that expresses them. Note that CAT, GUS, and Lac Z are stable proteins and persist in the cells, and these stable reporter proteins though serve as useful markers for gene activation, are less useful for assaying transcriptional repression or rapid changes in gene activity. The reaction is complex. The reporter has little, if any, potential for meaningful histochemical analysis or fusion genetics. Firefly luciferase is not a widely used marker gene as assaying the gene product is difficult, but it is useful for the detection of low-level or highly localized expression; detection technique limits the flexibility of the process. The reporter has high sensitivity (100 X than Lac Z).

The amenability of the luciferase system has been expanded by the isolation of alternative luciferases from other organisms, which bioluminesce in different colors. For example, luciferase genes have been isolated from diverse marine organism such as sea pansy (*Renilla reniformis*), which utilizes substrate form other than that used by firefly luciferase. Together the firefly and sea pansy luciferase enzymes form a dual reporter assay system that is supplied from Promega, which permits measurement of both enzymes in the same test tube.

Alkaline Phosphatase *Se*AP (secreted alkaline phosphatase) is used as reporter enzyme and can be assayed by nonradioactive systems. This reporter system has been useful in many cases because various sensitive colorimetric, fluorometric, and chemiluminescent assay procedures are available. It can be assayed in the growth medium because the reporter protein is secreted so there is no need to kill the cells. Alkaline phosphatase is conjugated to a ligand such as streptavidin that specifically interacts with a biotinylated target molecule.

β-Glucuronidase The gene encoding β-glucuronidase (GUS; EC 3.2.1.31) is abbreviated as *gus* or *uidA*. The source of *gus* is *E. coli*. GUS is a hydrolytic enzyme that catalyzes the cleavage of a range of glucuronides. This system is the easiest to analyze, most widely used, most popular, and preferred in plants and has widespread acceptance. It has minimal background activity of endogenous enzyme as compared to other reporters (hence it is the most preferred reporter) with the possible exception of reproductive tissues. The enzyme is very stable and persists in cell that expresses them; hence it is less useful for assaying transcription repression and rapid changes in gene activity. GUS is remarkably resistant to protease action with a very long half-life in living cells and in most extracts. There is no cofactor or ionic requirement by the enzyme. The enzyme shows thermal resistance. The optimal pH for the enzyme is 5.2–8.0. The enzyme is easily assayed by nonradioactive and destructive method. Though nondestructive methods are also reported, but these are not widely adopted. The enzyme can be assayed extremely sensitively and is most reactive in the presence of thiol-reducing agents such as β-mercaptoethanol (β-ME) and dithiothreitol (DTT). The enzyme is inhibited by heavy divalent metal ions as Cu^{2+} and Zn^{2+} and hence EDTA is added in the assay system.

The enzymatic activity can be analyzed by histochemical/fluorometric/spectrophotometric methods (similar to *lac* Z). The enzymatic activity is analyzed by histochemical method in stem or leaf cross-sections (*in situ* method) using 5-bromo-4-chloro-3-indolyl-β-*D*-glucuronide (X-gluc) as substrate. The reaction as mentioned below results in blue colored product.

X-gluc (Colorless) $\xrightarrow{\text{GUS}}$ Indoxyl derivative (Colorless)

↓ O_2 Oxidative dimerization stimulated by oxidation catalyst potassium ferrocyanide/ferricyanide mixture

Indigo derivative (Blue colored)

The enzymatic activity can also be quantified in plant extract by fluorometric method using 4-methyl-umbelliferyl-β-*D*-glucuronide (MUG) as substrate giving blue fluorescence under UV illumination. For quantitative analysis, fluorescence is measured by fluorometer. It is the most sensitive assay. The reaction catalyzed is as follows:

MUG (Nonfluorescent) $\xrightarrow{\text{GUS}}$ 4-Methyl umbelliferone or 7-Hydroxy 4-methyl coumarin

(Soluble; fluorescent only when the hydroxy group is ionized)

The GUS activity can be analyzed by spectrophotometric method using *p*-nitrophenyl-β-*D*-glucuronide (PNPG) or Naphthol AS-BI glucuronide as substrates. It is moderately sensitive assay procedure.

PNPG $\xrightarrow{\text{GUS}}$ *p*-Nitrophenol (measure absorption at 415 nm)

One unit of enzyme activity is defined as the amount of enzyme that produces one nanomole of product per minute at 37°C.

16.4 SCREENING OF CLONE(S) OF INTEREST

Once the recombinant clones are selected, the next step in genetic engineering experiment is to identify a specific clone from a genomic DNA or cDNA library. There are two types of screening strategies: (i) Sequence-dependent screening; and (ii) Protein structure/function-dependent screening (i.e., screening of expression library).

16.4.1 Sequence-dependent Screening

Sequence-dependent screening can be achieved by exploiting the sequence of the clone that can be applied to any type of library, genomic or cDNA, and involves either nucleic acid hybridization or the polymerase chain reaction (PCR). In both cases, the design of the probes or primers can be used to home in on one specific clone or a group of structurally related clones. These techniques therefore require sufficient information about the sequence of interest to make suitable probes and primers.

Nucleic Acid Hybridization Techniques

Colony and plaque blotting and hybridization strategies are employed for the detection of recombinant clones directly in the library. Other applications of blotting and nucleic acid hybridization techniques include determination of the presence or absence of the gene of interest, location of the cloned gene, information about the coding sequences, extent of expression, and size of transcript. These are discussed in detail in Section 16.5.

PCR

PCR is a useful technique for library screening. It is possible to identify any clone by PCR provided sufficient information about its sequence to make suitable primers is available. PCR is carried out with gene-specific primers that flank a unique sequence in the target. In this screening technique, pools of clones are maintained in multiwell plates and each well is screened by PCR and positive wells are identified. The clones from each positive well are diluted and then subjected to secondary screening. This screening process is repeated until wells having homogeneous clones corresponding to the gene of interest have been identified.

In several applications, degenerate primers are favorable such as when the primer sequences have to be deduced from amino acid sequences. Moreover, degenerate primers may also be employed to search for novel members of a known family of genes or to search for homologous genes between species. The amino acids with low codon degeneracy are selected for primer designing. It has been reported that 128-fold degeneracy in each primer can be successful in amplifying a single-copy target from the human genome.

Note that PCR screening can also be used to isolate DNA sequences from uncloned genomic DNA and cDNA.

16.4.2 Protein Structure/Function-dependent Screening (Expression Screening)

Besides nucleic acid sequences, the structure/function of its expressed product can also be analyzed. Screening the product of a clone applies only to expression libraries, i.e., libraries where the DNA fragment is expressed to yield a protein. In this case, the clone can be identified because its product is recognized by an antibody or a ligand of some nature or because the biological activity of the protein is preserved and can be assayed in an appropriate test system. The screening techniques based on analysis of protein structure/function are discussed in detail in Section 16.6.

The details of both the nucleic acid hybridization and protein structure/function-based screening methods are discussed in separately in Sections 16.5 and 16.6, respectively. The details of PCR have already been discussed in Chapter 7.

16.5 NUCLEIC ACID BLOTTING AND HYBRIDIZATION

A range of experimental techniques in genetic engineering is based on nucleic acid blotting (immobilization) on membranes, probe labeling, and hybridization. These techniques are employed for isolation and quantification of specific nucleic acid sequences, the study of their organization, intracellular localization, expression and regulation, and some more specific applications like diagnosis of infectious and inherited diseases. In this technique, the nucleic acid is isolated and either separated on gel or subjected to blotting in denatured form directly onto the membrane; the blotted nucleic acids are then hybridized with labeled probes. An outline of the basic steps involved in nucleic acid blotting and membrane hybridization procedures is given in Figure 16.1. A quite different hybridization technique that uses fluorochromes of different colors to enable two or more genes to be located within chromosome preparation in a single *in situ* experiment is known as fluorescence *in situ* hybridization (Exhibit 16.2). The important variables of these techniques include type of membrane, method of immobilization, method of probe labeling, type of

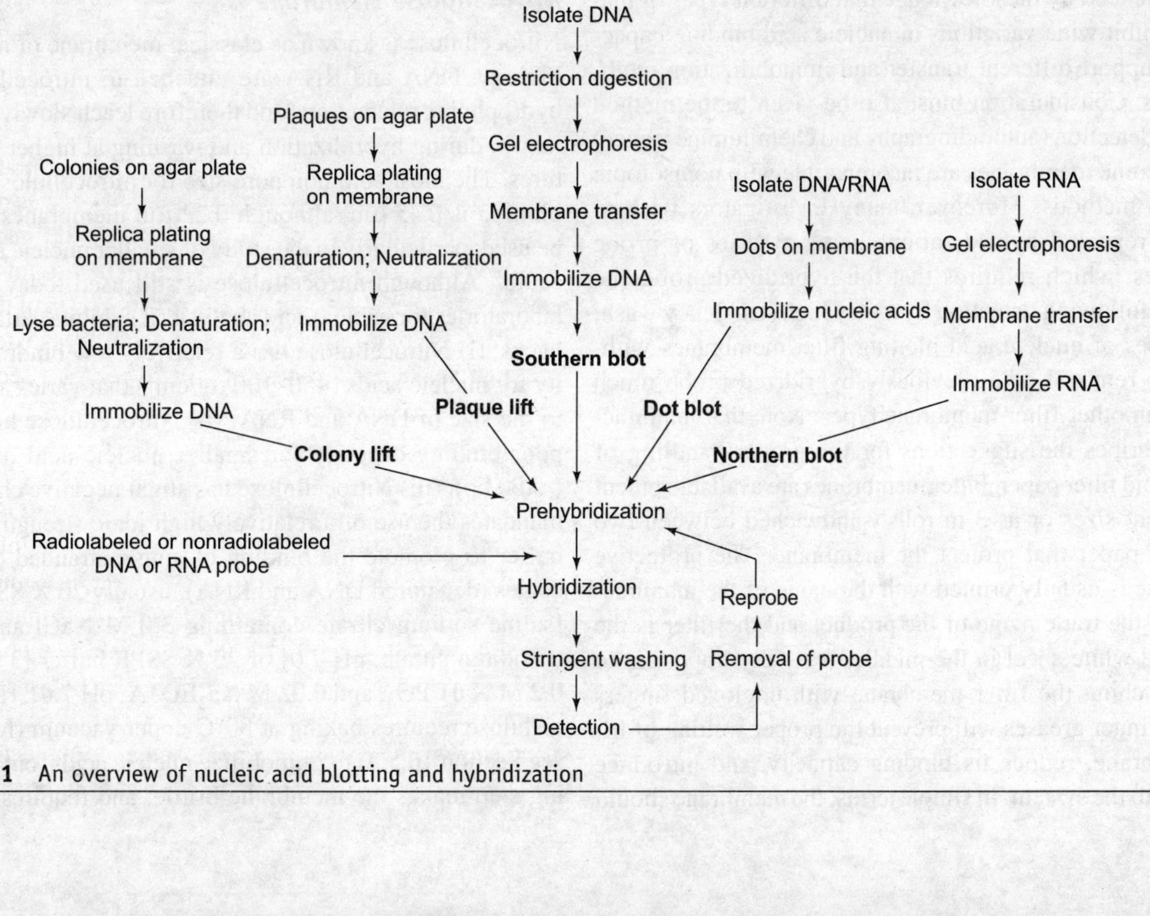

Figure 16.1 An overview of nucleic acid blotting and hybridization

Exhibit 16.2 Fluorescence *in situ* hybridization (FISH)

Fluorescence *in situ* hybridization (FISH) is a cytogenetic technique that can be used to detect and localize the presence or absence of specific DNA sequences on chromosomes. This technique uses fluorescent probes that bind to only those parts of the chromosome with which these show a high degree of sequence similarity, and the location of binding of fluorescent probe to the chromosomes are detected by fluorescence microscopy. FISH is often used for finding specific features in DNA for use in genetic counseling, medicine, and species identification. FISH is also useful for detection and localization of specific mRNAs within tissue samples; hence it facilitates the analysis of the spatial-temporal patterns of gene expression within cells and tissues.

To perform FISH, a probe of optimum length is constructed first. Note that the probe must be large enough to hybridize specifically with its target but not too large to impede the hybridization process. The probe having targets for antibodies or biotin is tagged directly with fluorophores. Then, an interphase or metaphase chromosome preparation is produced. The chromosomes are firmly attached to a glass substrate. Repetitive DNA sequences are blocked by adding short fragments of DNA to the sample. The probe is then applied to the chromosomes on the glass substrate and incubated for approximately 12 hours to allow hybridization. All unhybridized or partially hybridized probes are removed by several wash steps. The results are then visualized and quantified using a fluorescent microscope that is capable of exciting the dye and recording images. If the fluorescent signal is weak, amplification of the signal may be necessary in order to exceed the detection threshold of the microscope. Fluorescent signal strength depends on many factors such as probe labeling efficiency, the type of probe, and the type of dye. These secondary components, i.e., the fluorescently tagged antibodies or streptavidin bound to the dye molecule, are selected so that a strong signal is achieved. Furthermore, FISH experiments designed to detect or localize gene expression within cells and tissues rely on the use of a reporter gene, such as one expressing green fluorescent protein, to provide the fluorescence signal.

Source: http://en.wikipedia.org; http://science.jrank.org

probe, hybridization recipe, and method of detection, which are discussed in this section.

16.5.1 Choice of Membrane

The selection of solid support matrix to which nucleic acid (electrophoretically resolved or denatured) is transferred is very important before hybridization. This selection is therefore influenced by the knowledge that different types of matrices exhibit wide variations in nucleic acid binding capacity and support different transfer and immobilization methodologies. Consideration must also be given to the method of probe detection (autoradiography and chemiluminescence) because some membranes are incompatible with nonisotopic detection methods. Moreover, many investigators wish to perform repeated hybridizations using a series of probe sequences, which requires that the hybridized probes be removed following detection by a very high stringency wash; some types of nucleic acid blotting filter membranes withstand the removal of a previously hybridized probe much better than other filter membrane types. Note that manufacturer describes the suggestions for the proper handling of nucleic acid filter paper. Filter membranes are available precut in different sizes or as 3 m rolls, sandwiched between two sheets of paper that protect the membrane. The protective membrane is usually printed with the name of the manufacturer and the trade name of the product and the filter is the unmarked white sheet in the middle. It is very important to avoid touching the filter membrane with ungloved fingers because finger greases will prevent the proper wetting of filter membrane, reduce its binding capacity, and introduce RNase into the system. In simple terms, the membrane should be picked up from the corner with nuclease-free forceps or gloved hands. Filter membranes should be stored in a cool dry location away from direct sunlight and should not be refrigerated, frozen, or heated above room temperature before their first use. If stored correctly, filters are stable for many months. The different types of membranes used for nucleic acid blotting are presented below.

Nitrocellulose Membrane

Nitrocellulose is known as classical membrane of molecular biology. DNA and RNA are attached to nitrocellulose by hydrophobic interactions and therefore leach slowly from the matrix during hybridization and washing at higher temperatures. The most common pore size for nitrocellulose used in blotting is 0.45 μm, although 0.22 μm membranes can also be used, particularly in the study of smaller nucleic acid molecules. Although nitrocellulose is still used today in many laboratories for nucleic acid analysis, it exhibits certain drawbacks: (i) Nitrocellulose has a relatively low binding capacity for nucleic acids (~50–100 $\mu g/cm^2$) that varies according to the size of DNA and RNA; (ii) Nitrocellulose has a very poor binding capacity for smaller nucleic acid molecules (<400 bp); (iii) Nitrocellulose has slight negative charge that mandates the use of a relatively high ionic strength transfer buffer to promote the binding of single stranded (ss) molecules (denatured DNA and RNA), usually 20 X SSC buffer [saline sodium citrate containing 3.0 M NaCl and 0.3 M trisodium citrate; pH 7.0] or 20 X SSPE buffer [3 M NaCl, 0.2 M NaH_2PO_4, and 0.02 M Na_2EDTA; pH 7.4]; (iv) Nitrocellulose requires baking at 80°C under vacuum (for details see Section 16.5.3) to immobilize nucleic acids, but this baking step makes the membrane brittle, and requires extreme care when handling the membranes; (v) The brittleness of this membrane after the baking step makes the subsequent manipulations very difficult and repeated probing impractical; and (vi) Nonisotopic detection methods generate excessive background with the use of nitrocellulose. Due to associated drawbacks other membranes have also been invented for blotting purposes.

Nylon Membrane

Nylon membrane was devised to avoid many of the difficulties inherent in nitrocellulose and have largely replaced nitrocellulose for most routine nucleic acid blotting applications. Two types of nylon membranes are commercially available: (i) Unmodified (or neutral); and (ii) Charge-modified nylon that carries amine group and therefore known as positively charged or (+) nylon. Like nitrocellulose, many of the nylon membranes used for blotting have a pore size of 0.45 μm, although 0.22 μm filters are also available. Both types of nylon membranes can bind single stranded and double stranded (ds) nucleic acids and retention is quantitative in solvents as diverse as water, 0.25 N HCl, and 0.45 N NaOH. Nylon membranes are supplied under a variety of trade names (e.g. Amersham Hybond, Fermentas SensiBlot™ Plus Nylon, GE Nylon, Magna Nylon, MagnaProbe Nylon, Zeta-Probe GT (Bio-Rad) membrane, etc.). The advantages associated with nylon membrane are as follows: (i) These membranes exhibit great tensile strength, which enables repeated probing of the same filter. However, this also requires very sensitive handling of the filter membrane; (ii) Nylon filters, particularly nylon (+) variety, have also proven to be very compatible with chemiluminescence-related technologies; (iii) Nylon membranes exhibit an enhanced nucleic acid binding capacity compared to nitrocellulose, in some cases as much as 400–500 μg/cm^2, depending on the manufacturer and the exact buffers used by investigator; (iv) Nylon membranes also show a particular affinity for smaller molecules (~500 bases), so routine blotting of small molecules, including PCR products, can be readily performed; (v) Compared with nitrocellulose, the transfer buffers used for nucleic acid transfer are of lesser ionic strength, often as low as 5 X SSC or 5 X SSPE; (vi) Although the positive charge associated with nylon (+) membranes certainly increases the electrostatic attraction between target nucleic acid and the membrane, adequate blocking of the membrane during prehybridization and, if applicable, during the detection process normally produces blots with an exceptionally clean background. However, failure to block and prehybridize properly produces excessive background because of nonspecific interaction between the positively charged membrane and the negatively charged sugar-phosphate backbone of the probe; and (vii) Immobilization of nucleic acid can be accomplished either by baking or by irradiation with UV light (for details see Section 16.5.3). The baking of nylon filters does not require a vacuum oven and can be performed in any instrument capable of generating 80°C for 2 hours.

Polyvinylidene Difluoride

Polyvinylidene difluoride (PVDF) is a fluorocarbon polymer that historically has been a favorite membrane for western blotting. The use of this membrane for Southern and northern analyses has also been reported. PVDF is a highly durable and chemically stable membrane sold under a number of different trade names (e.g. Hoefer, Biodyne Membranes). These membranes possess an average pore size of 0.45 μm. Nucleic acids can be permanently immobilized by cross-linking with UV light or by baking (vacuum oven is not required). PVDF membranes have gained acceptance in many nonisotopic applications as well.

16.5.2 Transfer of Nucleic acids from Gel to Membrane

The electrophoretically separated nucleic acid is transferred from the gel to a solid support by using any of the transfer techniques that differ primarily in the method by which nucleic acids are drawn out of the gel matrix. There are six different methods that are used to transfer DNA fragments or RNA from agarose gels to solid supports. These are described below.

Upward Capillary Transfer

The traditional passive capillary transfer method is the least expensive and requires the fewest accessories; however, the method is the most time-consuming. In this method, a stack of adsorbent material, such as paper towels or blotting papers, serve as wick to draw the transfer buffer from reservoir, through the gel, and finally into the dry stack of paper towels. The buffer is drawn through the gel by capillary action that is established and maintained by a stack of dry absorbent paper towels. As a result, DNA fragments or RNA are carried from the gel in an upward flow of liquid and deposited on the surface of the solid support (Figure 16.2a). Note that the size of the membrane does not allow DNA fragments or RNA to pass beyond it. The rate of transfer depends on the size of the DNA fragments or RNA and the concentration of agarose in the gel. As elution proceeds, fluid is drawn not only from reservoir, but also from the interstices of the gel itself and the gel is transformed to a rubbery substance through which DNA or RNA molecules cannot pass. Two common complaints associated with this transfer method are poor transfer efficiency, especially of larger DNA/RNA molecules, and transfer takes too long (18–20 hours).

Downward Capillary Transfer

In this method, DNA fragments/RNA are carried in a downward direction in a flow of alkaline buffer and are deposited

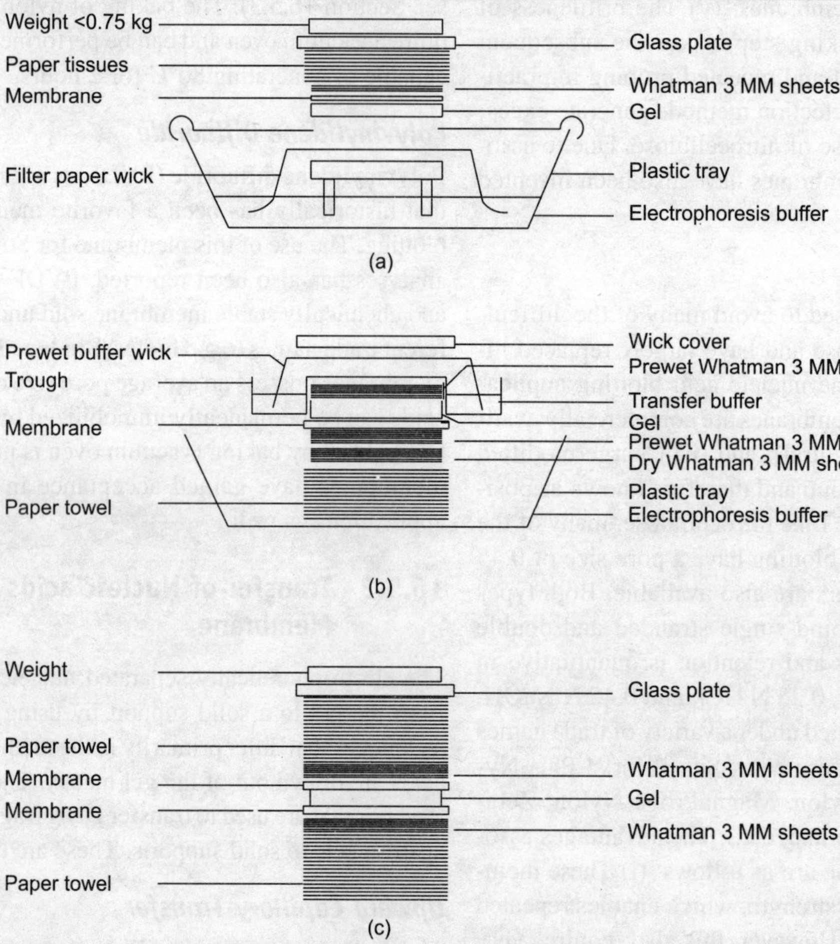

Figure 16.2 Capillary transfer of nucleic acids from gel to the membrane; (a) upward capillary transfer, (b) downward capillary transfer, (c) simultaneous transfer to two membranes

onto the surface of a charged nylon membrane. Different arrangements of wicks, reservoirs, and various formulations are used for this transfer; best results are obtained with 0.4 M NaOH and a setup in which the transfer buffer is drawn from reservoirs at the top of the gel through wicks and pulled through the gel by an underlying stack of paper towels (Figure 16.2b). Downward transfer systems take advantage of gravity and offer enhanced transfer efficiency in a much shorter time (<4 hours) as compared to older upward capillary transfer method. Transfer of DNA fragments/RNA is rapid and the intensity of signal is ~30% greater than upward capillary transfer. Nucleic acid fragments are efficiently migrated through the interstices of the gel, as these are not under pressure from weights placed on top. This method likewise eliminates the need for heavy weights on top of the capillary stack, thereby minimizing compression and concomitant collapsing of the pores of the gel. The technique does not require the use of any expensive equipment, such as the vacuum blot devices or electroblotting apparatus described later, and is extremely simple to assemble. However, TurboBlotter apparatus may be used to facilitate the upward capillary transfer.

Simultaneous Transfer to Two Membranes

If the DNA fragments/RNA are present in high concentrations in the gel, the capillary method can be used to transfer DNA fragments/RNA simultaneously and rapidly from a single gel to two nitrocellulose or nylon membranes (Figure 16.2c). Usually this method is used for analysis of plasmids, bacteriophages, cosmids, or the genomes of simple organisms by Southern hybridization.

Electrophoretic Transfer

Electrophoretic transfer, a completely different approach, is used with the charged nylon membrane and has become the method of choice for analysis of small fragments of DNA or RNA separated by electrophoresis through polyacrylamide gel electrophoresis (PAGE). Although ss DNA and RNA can be transferred directly, while fragments of ds DNA must first be denatured *in situ*, thereafter the gel is neutralized and soaked in electrophoresis buffer (1 X TBE) before being mounted between porous pads aligned between parallel electrodes in a large tank of buffer (Figure 16.3). The time for complete transfer depends on the size of DNA fragments/RNA, the porosity of gel, and the strength of applied field.

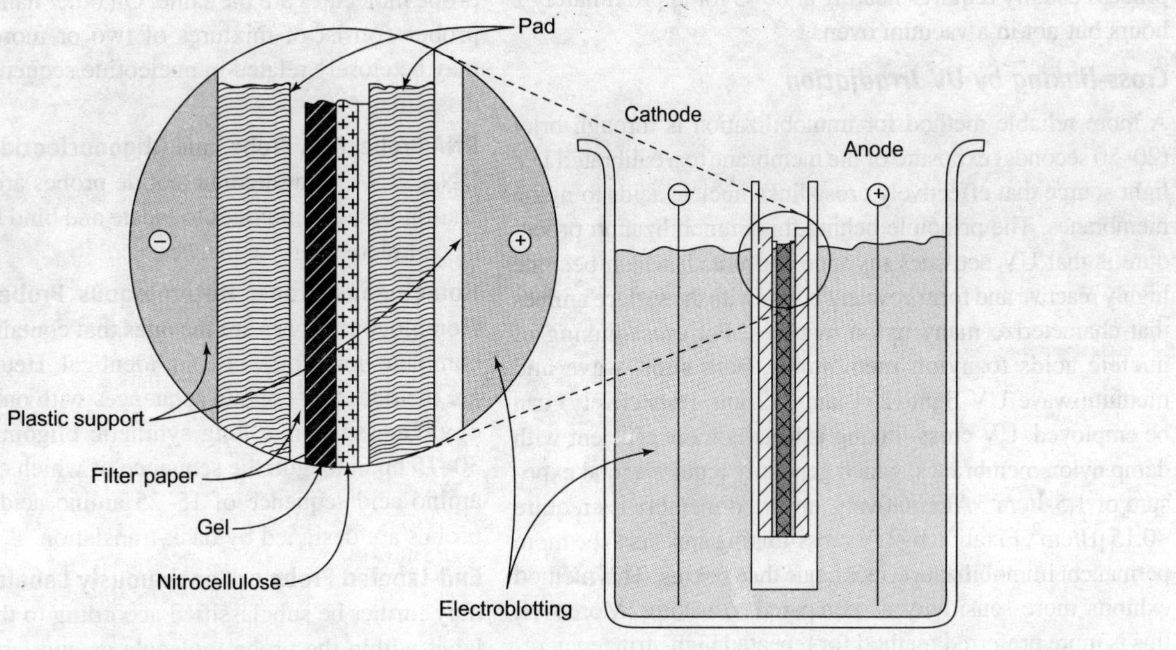

Figure 16.3 Electroblotting with a tank blotting unit

Note that very small fragments of nucleic acids (~50 bp) bind to charged nylon membranes in buffers of very low ionic strength. As high molecular weight nucleic acid migrates relatively rapidly from the gel, depuriniation step is not needed and transfer is completed within 2–3 hours. During electroblotting, a significant amount of heat is produced; therefore many protocols strongly recommend cooling the electrolyte buffer before use and controlling the temperature during the run. Usually electroblotting is the standard methodology for western blotting that is used in conjunction with PAGE for the transfer of proteins and less frequently used for the transfer of nucleic acids from agarose gels.

Vacuum Transfer

Nucleic acid can also be transferred rapidly and quantitatively from the agarose gels under vacuum. In a vacuum blotting system, negative pressure is applied to accelerate the transfer process. This technique is also known as vacuum-assisted capillary transfer, in which buffer drawn from upper reservoir, elutes nucleic acids from the gel, and deposits them on the membrane. There are a number of different instruments for vacuum blotting that are available from different vendors (e.g. Bio-Rad Vacuum Blotter Appligene Vacuum Blotter). The major advantage of vacuum blotting is that the transfer occurs far more rapidly than by the traditional capillary transfer, and takes just ~30–60 min. This accelerated process reduces molecular diffusion of nucleic acid molecules on the way out of the gel therefore minimizing the lateral spreading resulting into very sharp bands. However, great care must be taken so that hydrostatic pressure of the gel is in limited range (vacuum between 3 and 5 inches are optimal).

Positive Transfer

In contrast to vacuum blotting, positive pressure can also be used to push nucleic acids from the gel. The PosiBlot uses positive pressure to drive the transfer buffer through the gel and to complete the transfer process in much less time, reportedly as little as 15 min for the DNA/RNA. Like vacuum blotting, the accelerated transfer process reduces lateral molecular diffusion of nucleic acid molecules, thereby maintaining the sharpness of bands. Moreover, the positive pressure greatly reduces the problem of gel collapse that sometimes occurs when an excessive vacuum is applied.

16.5.3 Method of Immobilization

After the transfer of nucleic acid molecules from the gel to the membrane or directly spotted onto the membrane, two methods of immobilization are used for fixation of DNA/RNA on membranes, which are presented below.

Baking

The classical technique for nucleic acid immobilization involves baking the sample onto the surface of nitrocellulose in a vacuum oven. Although the exact nature of interaction between the matrix and nucleic acids is not clear, it is believed to be hydrophobic. The procedure involves complete air-drying of the nitrocellulose filters followed by baking the membrane for 2 hours at 80°C *in vacuo*. This method of immobilization is not a completely permanent process; after a few reprobings, some of the sample is lost from the surface of the filter, thereby compromising the quantitativeness of all subsequent assays. With nylon membranes, the baking process usually requires heating at 80°C for approximately 2 hours but not in a vacuum oven.

Cross-linking by UV Irradiation

A more reliable method for immobilization is through brief (20–30 seconds) exposure of the membrane to a calibrated UV light source that effectively cross-links nucleic acids to nylon membranes. The principle behind this immobilization procedure is that UV activates thymine and uracil, which become highly reactive and form covalent bonds with the surface amines that characterize many nylon matrices. For cross-linking of nucleic acids to nylon membranes, both short-wave and medium-wave UV light (254 and 312 nm, respectively) can be employed. UV cross-linking is usually more efficient with damp nylon membranes, which generally require a total exposure of 1.5 J/cm^2. Alternatively, air-dried membranes require ~0.15 μJ/cm^2. Fixation by UV cross-linking appears to be more permanent immobilization technique than baking. This method exhibits more sensitivity as compared to baking. Moreover, this is more preferred method for repeated high-stringency removal of a hybridized probe (for details see Section 16.5.6). Note that in case the filters are to be hybridized with several different probes and stripped between probing reactions, then it is better to cross-link the DNA to the filters with UV light. Moreover, cross-linking is faster than baking, requiring 20–30 seconds per filter compared to baking for 2 hours. The practice of UV irradiation of nitrocellulose is however avoided because it does not result in the comparable retention of nucleic acids that is achievable by cross-linking to nylon. More importantly, over irradiation of nitrocellulose poses a serious risk of fire.

16.5.4 Nucleic Acid Probe Technology

Nucleic acid probes, i.e., labeled DNA, RNA, and oligonucleotide probes are used in hybridization-based techniques to locate and bind DNAs and RNAs of complementary sequences. The probe labeling systems are selected on the basis of required sensitivity and resolution, method of label incorporation, probe stability after labeling, type of hybridization, and desired method of detection. Different types of probes used in hybridization-based methods of screening are described below.

Types of Probes

Different types of probes are used in hybridization-based techniques. These are as follows:

Radioisotope Labeled Probe and Nonisotope Labeled Probe The probes can be labeled with a radioisotopic or a nonisotopic method.

Homogeneous Probe and Heterogeneous Probe Probes can be either homogenous or heterogeneous. In a hybridization solution, a probe preparation is said to be homogenous if all probe molecules are the same. On other hand, heterogeneous probes consist of mixtures of two or more sequences that may be closely related in nucleotide sequence or completely dissimilar.

DNA Probe, RNA Probe, and Oligonucleotide Probe Labeled DNA, RNA, and oligonucleotide probes are used in hybridization-based techniques to locate and bind DNAs and RNAs of complementary sequences.

Homologous Probe, Heterologous Probe, or Guessmers Homologous probes are the ones that contain sequences with complete homology, i.e., are identical. Heterologous probes are the ones that contain sequences with only partial homology. Guessmer is a long synthetic oligonucleotide usually 30–75 bp in length, the sequence of which corresponds to an amino acid sequence of 15–25 amino acid residues. These probes are designed by back-translation.

End-labeled Probe or Continuously Labeled Probe Probes may further be subclassified according to the distribution of label within the probe molecule as end-labeled or continuously labeled. End-labeled probes are those to which the label is added at either the 5′-end or 3′-end. Continuously labeled probes are those in which the labels are added along the length of the backbone of the molecule at regular intervals such as probes generated by random priming, PCR, *in vitro* transcription, or cross-linkage to biotin and digoxigenin.

Labeling (Radioisotopic and Nonisotopic) and Detection Systems

There are two basic methods for labeling nucleic acids: (i) Radioisotopic labeling for detection by autoradiography; and (ii) Nonisotopic labeling for detection by chemiluminescence, chemifluorescence, or chromogenic techniques. Both these radiolabeling and nonisotopic labeling techniques are compatible with the labeling of DNA, RNA, and oligonucleotide for most applications.

Radioisotope Labeling Radioisotopes provide excellent sensitivity and compatibility with many labeling techniques. The most frequently used isotopes for nucleic acid probe synthesis are ^{32}P, ^{3}H, and ^{35}S; ^{33}P is used less frequently. Nucleoside triphosphate precursors are labeled in the appropriate position (α or γ) and facilitate enzyme-mediated transfer of the portion of the nucleotide having the isotope to the molecules being labeled. Radioisotope labeling system has some inherent disadvantages that include short half-life of the isotope, potential health hazards, requirement for containment of radioactivity, high purchase and disposal cost, no possibility of repeated probing, and the need for relatively long detection periods.

Radiolabeled nucleotides are shipped in dry ice and stored frozen below –20°C or at –80°C. The specific activity of ^{32}P nucleotides on any day before the calibration date can be calculated according to the following equation:

$$SA\ (Ci/mol) = \frac{SA_{cal}}{Df + [SA_{cal}\ (1 - Df)/9120]}$$

where SA is specific activity; SA_{cal} is the specific activity on the calibration date; Df is the decay factor obtained from a decay chart and is the fraction of current radioactivity that will remain on the calibration date; and 9120 represents theoretical specific activity of carrier-free ^{32}P.

The SA of ^{32}P nucleotides on any day after the calibration date can be calculated according to the following formula:

$$SA\ (Ci/mol) = \frac{Df}{Df + [SA_{cal}\ (1 - Df)/9120]}$$

The probe labeled with radioisotope is detected by autoradiography (for details see Exhibit 16.3).

Exhibit 16.3 Autoradiography

Autoradiography is a simple and sensitive photochemical technique used to record the spatial distribution of radiolabeled compounds within a specimen or an object. It produces permanent images of the distribution of radioactive atoms on a 2-D surface of photographic film. Autoradiography is subdivided into two broad groups, commonly referred to as macroautoradiography and microautoradiography, based on type of specimen containing the radioactivity, the type of emulsion necessary for image formation, and the method of examining the results. The most common applications of autoradiography in molecular biology include the quantitative analysis of nucleic acid hybridization events. If used in conjunction with electrophoresis, the resultant image provides a qualitative aspect as well. For example, in northern analysis of gene expression (macroautoradiography), the extent of hybridization, and proved specificity to RNA target molecules is assessed via the magnitude and location of radioisotope emission from the filter membranes. In some applications, such as the nuclease protection assays, isotopes can be detected directly on polyacrylamide or agarose gels. Microautoradiography, on the other hand, is performed by coating a thin section of larger specimen with a light sensitive emulsion.

The film base is a flexible piece of polyester that has been coated on one or both sides with an emulsion. Among the more commonly used types of X-ray films for autoradiography and chemiluminescence are the Kodak X-OMAT AR (XAR), X-OMAT LS (XLS), Fuji RX, and Bio MAX film series. The radio- and light-sensitive emulsion consists partly of sensitive crystals of silver halide (bromide, chloride, or mixed halides). During film exposure, energy is absorbed by silver halide grains within the film emulsion with the release of electrons. The resulting negatively charged 'sensitivity specks' i.e., faults in the crystal lattice attract positively charged silver ions, thus forming an atom of metallic silver (Figure A). Exposure of X-ray film generates latent image, consisting only of developable silver grains, which is not visible in the dark room and the emulsion is still extremely light sensitive. Formation of the latent image can be improved by preflashing the film and by low temperature exposure. Once formed, the latent image is not stable for long exposure period. The formation of a visible autoradiographic image can be obtained manually or by machine. In either case, there are distinct stages in the processing of film: development, washing, fixation, washing, and drying. When immersed in a photographic developer, the exposed silver halide grains in the emulsion of the film become reduced to metallic silver, thus amplifying the latent image. The amount of time required to complete the developing process is variable depending on the degree of exposure of the film, the temperature of the developer, the age of the developer, and the tolerance of the investigator for background. The action of the developer is arrested by immersion of the film in a chemical 'stop bath'. Although a variety of stop bath solutions may be purchased, a homemade version can be concocted by preparing 1% acetic acid solution. Arresting the activity of the developer at the precise time is very important because overdevelopment can obscure any image that may have been captured on the film. It is advisable to immerse the film after development into water to wash off developing solution. When developing manually, it is necessary to develop, wash, and fix each film for the same length of time. The final developed image may be assessed by visual inspection or with greater precision by digital image analysis. This forms the basis of direct autoradiography in which the sample is placed in intimate contact with the film and is best-suited for detection of weak- to medium-strength β-emitting radionuclides (^{3}H, ^{14}C, ^{35}S). On the other hand, indirect autoradiography describes the technique for improved autoradiographic detection in which emitted energy is converted to light by means of intensifying screens or fluorography.

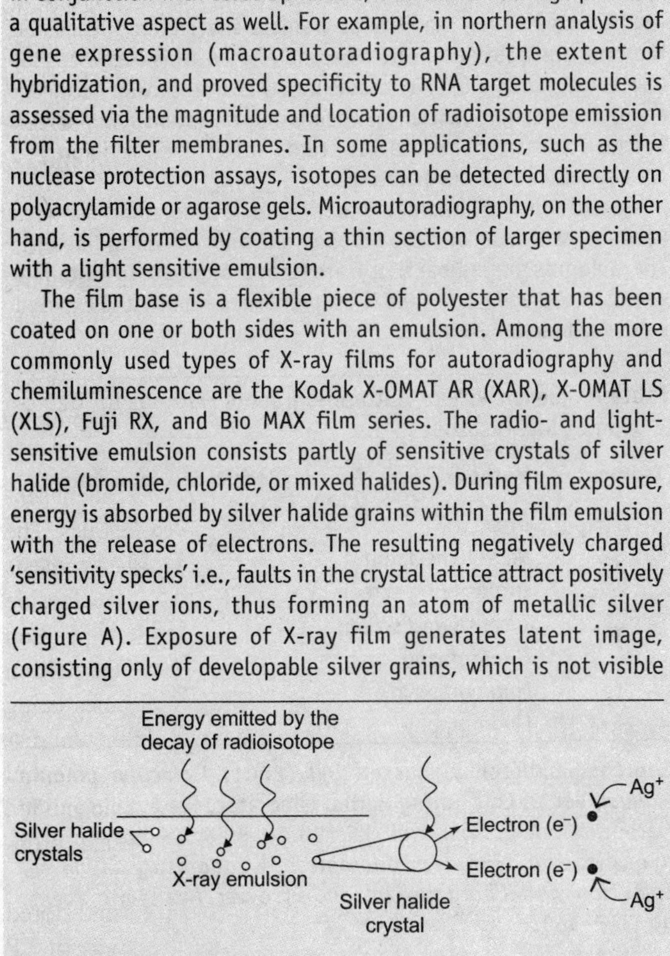

Figure A Principle of autoradiography

Intensifying Screen Intensifying screen is a solid inorganic scintillator that consists of a flexible polyester base coated with inorganic phosphors, and is placed behind the X-ray film. The most commonly used intensifying screen for macroautoradiography is the calcium tungsten ($CaWO_4$) phosphor-coated screen, such as DuPont Cronex lightening plus intensifying screen and Fuji Mach 2, which are excellent for exposures longer than one day. The $CaWO_4$ phosphor coating provides one side of the screen a white appearance and the uncoated, nonfunctional side has an off-white appearance. The organic phosphors convert ionizing radiation to ultraviolet or blue light; these screens emit blue light, exhibit very low-level noise (background signal), and are extremely compatible with standard autoradiography use. [Note that X-ray films are most sensitive to blue light]. In simple terms, any emissions passing through the photographic emulsion are absorbed by the screen and converted to light, effectively superimposing a photographic image upon the direct autoradiographic image. Intensifying screens thus improve the detection of the high-energy emitters such as ^{32}P and γ-ray emitters such as ^{125}I, but are not effective with ^{3}H, ^{35}S, or ^{14}C. Note that ^{32}P emits a β-particle with sufficient energy (1.71 MeV) to penetrate water or plastic to a depth of 6 mm, and to pass completely through an X-ray film. Gels and filters therefore need not be completely dried, although the sharpness of autoradiographic image is much improved if these are covered with Saran wrap before these are exposed to the film. This is in contrast to ^{35}S. Radioactive (^{32}P) particles that pass through the film hit the intensifying screen, and cause it to emit photons that are captured by silver halide crystals in the emulsion. This leads to a ~5-fold enhancement in the intensity of an autoradiographic image when the film is exposed at low temperature (–70°C). A further 2-fold enhancement can be obtained by prexposing the film to a short (≤ 1 ms) flash of light emitted by a stroboscope or a photographic flash unit. This activates the silver halide crystals and increases the probability that any single crystal exposed to light emitted from the intensifying screen is reduced to Ag metal during the developing process. Preflashed film has another advantage; the intensity of the image on the film becomes proportional to the amount of radioactivity in the sample. An intensified autoradiographic image is shown in Figure B.

Fluorography Fluorography is not a commonly used methodology for molecular biology applications. It is an intensification process by which some of the energy associated with the decay process of the isotope is converted to light by the interaction of radiodecay particles with a compound known as flour, which then exposes the X-ray film. It is used mainly when low-energy emitters, especially ^{3}H, are used as radiolabels. Fluorography is carried out through the impregnation of a specimen, such as a polyacrylamide gel, with a fluorographic solution (a scintillant), after which the gel is dried down for efficient autoradiographic detection. Commer-cially available fluorographic formulations include mixtures of 2,5-diphenyloxazole, also known as PPO, or sodium salicylate. Although fluorographic impregnation does not improve detection performance with ^{32}P or ^{125}I, it may enhance detection with ^{14}C and ^{35}S and it is almost always used with ^{3}H. In this way, fluorographic detection differs from signal amplification through the use of intensifying screens, which may be viewed as a type of external flour. Flour-impregnated samples should be exposed to preflashed film (Kodak X-OMAT AR) at –80°C to ensure that the response of film is linear.

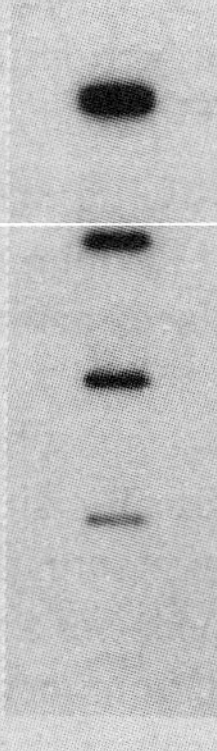

Figure B An intensified autoradiographic image

Preflashing and Low Temperature Autoradiography The gain in sensitivity, which is achieved by use of indirect autoradiography, is offset by nonlinearity of film response. A single hit by a β-particle or γ-ray can produce hundreds of silver atoms, but a single hit by a photon of light produces only a single silver atom. Although two or more silver atoms in a silver halide crystal are stable, a single silver atom is unstable and reverts to a silver ion very rapidly. This means that the probability of a second photon being captured before the first silver atom has reverted is greater for large amounts of radioactivity than for small amounts. Hence small amounts of radioactivity are underrepresented with the use of fluorography and intensifying screens. This problem can be overcome by a combination of preexposing a film to an instantaneous flash of light (preflashing) and exposing the autoradiograph at –80°C. Preflashing provides many of the silver halide crystals of the film with a stable pair of silver atoms. Lowering the temperature to –80°C increases the stability of a single silver atom, increasing the time available to capture a second photon. This suggests that low temperature (–80°C) autoradiography is necessary for maximum energy conversion efficiency when using an intensifying screen because film response to low light intensities is improved dramatically at low temperatures.

Sensitivities of autoradiographic and fluorographic methods used for detection of radioisotopes are summarized in Table A. The amounts of radioactivity are those required to obtain a detectable image ($A_{545} = 0.02$) on preflashed film that is exposed to the sample for 24 hours.

Table A Sensitivity of Autoradiographic Methods for Detection of Radioisotopes

Isotope	Method	Sensitivity (dpm/min^2)
^{32}P	Direct	2–5
^{35}S	No enhancement	15–25
^{32}P	Intensifying screen	0.5
^{125}I	Intensifying screen	1–2
^{3}H	Fluorography	10–20
^{14}C	Fluorography	2
^{35}S	Fluorography	2

Source: Sambrook J, Russell DW (2001) *Molecular Cloning Manual*, Vol. 3. Cold Spring Harbor Laboratory Press, Cold Spring Harbor, NY, A 9.14; Farrell RE (2005) Practical nucleic acid hybridization. *RNA Methodologies: A Laboratory Guide for Isolation and Characterization*, Elsevier Academic Press, pp. 317–360.

Nonisotopic Labeling Various nonisotopic labeling and detection methodologies have been devised in order to minimize the use of radioisotopes and the inherent dangers associated with their presence in the laboratories. Various available nonisotopic labeling and detection systems provide similar levels of sensitivity and resolution, although these differ noticeably in their complexity and the number of different steps required during analysis. The most commonly used nonisotopic labeling systems are described below.

Biotin Biotin is a small, water-soluble vitamin that can be readily conjugated to a number of biological molecules. The structure of biotin is given in Figure 16.4. The attachment of biotin to the probe is called biotinylation. Probe biotinylation can be accomplished enzymatically; photochemically through the use of biotinylated nucleotides; chemically using photoactivatable biotin and during the synthesis of oligonucleotides. Although 5′-biotinylation of oligonucleotides is one of the easiest and most efficient methods to accomplish labeling but from detection point of view, it really does not matter how the biotin becomes part of the probe. Biotinylation offers several advantages: (i) Biotinylation does not interfere with the biological activity of the probe; (ii) In the case of biotinylated nucleotides such as Bio-11 dUTP[6], linker arms between the biotin and the backbone of the probe effectively minimize steric interference; (iii) Biotinylated probes are stable at –20°C for at least one year after labeling; and (iv) After hybridization, the biotinylated probe is easily detected via interaction with streptavidin that has been conjugated to a reporter enzyme. Note that streptavidin, a tetrameric protein (M_r 60,000), is isolated from the bacteria *Streptomyces avidinii* and has a neutral isoelectric point at physiological pH with few charged groups and contains no carbohydrate. These properties reduce nonspecific binding and background problems that enhance the sensitivity of many forms of this assay. In most systems, alkaline phosphatase is used as reporter enzymes, which exist as a conjugate, while in some cases streptavidin–horseradish peroxidase (HRP) conjugate is also useful. Depending on the specific substrate to which alkaline phosphatase or HRP is subsequently exposed, hybridization events are then localized and quantified by any of the available detection systems, which include chromogenic detection, fluorescent assays, and chemiluminescence. Examples of these three detection methods in an alkaline phosphatase-catalyzed reaction are given below.

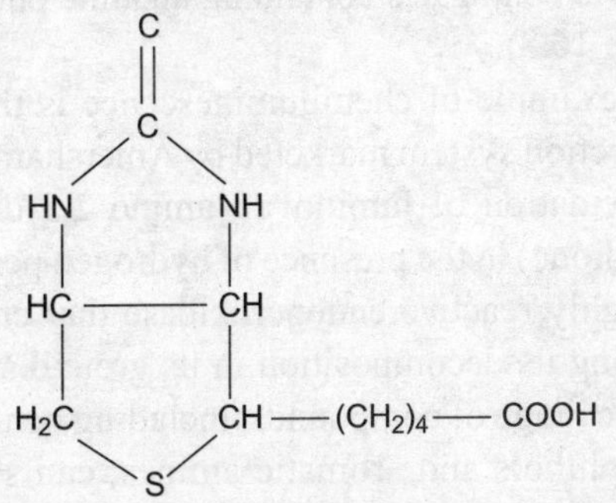

Figure 16.4 Chemical structure of biotin

The most sensitive indicator system for chromogenic *in situ* detection of alkaline phosphatase conjugate is the 5-bromo-4-chloro-3-indolyl phosphate (BCIP) and nitroblue tetrazolium (NBT) system. In this reaction, alkaline phosphatase catalyzes the removal of the phosphate group from BCIP, generating 5-bromo-4-chloro-3-indolyl hydroxide, which dimerizes to form insoluble blue compound, 5,5′-dibromo-4,4′-dichloro indigo. The two reducing equivalents produced during the dimerization reaction reduce one molecule of NBT to the insoluble, intensely blue purple dye diformazan (Figure 16.5), which becomes visible at sites where the labeled probe has hybridized

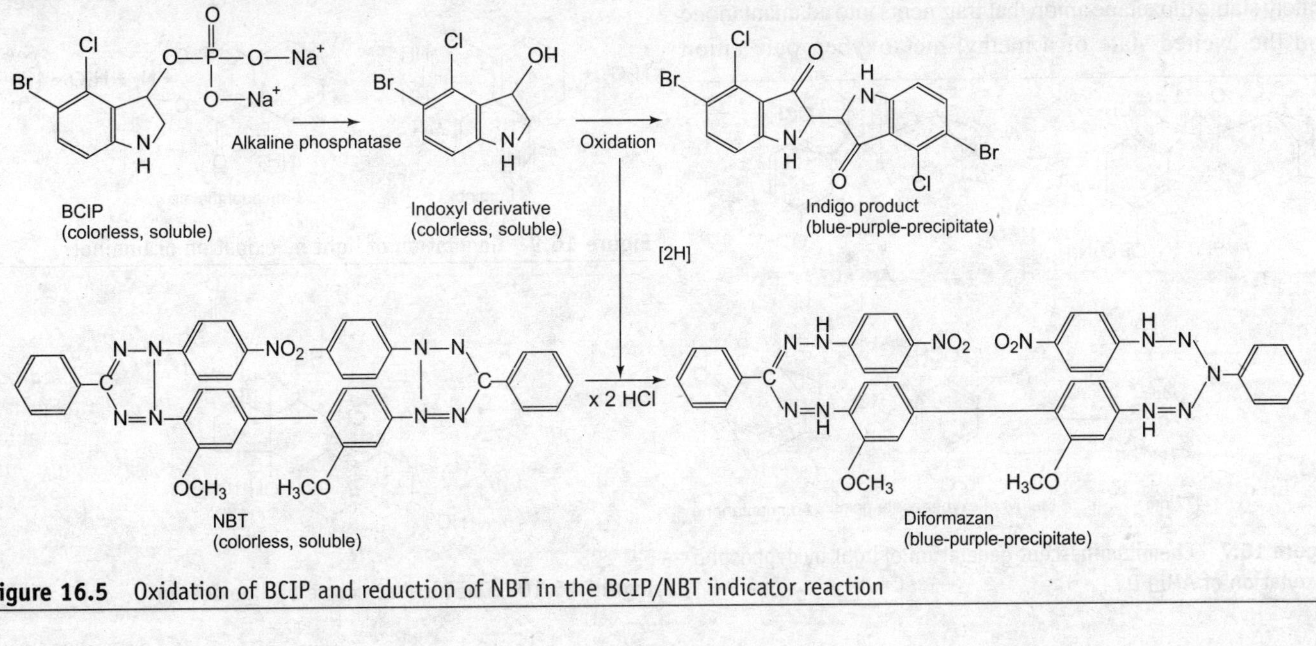

Figure 16.5 Oxidation of BCIP and reduction of NBT in the BCIP/NBT indicator reaction

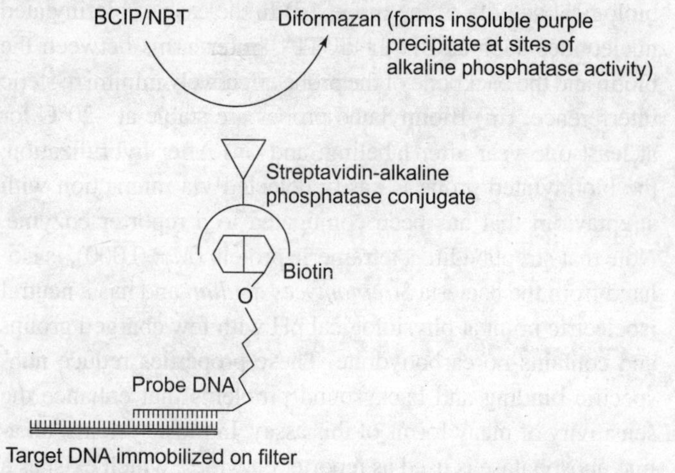

Figure 16.6 Detection of biotin-labeled nucleic acid probes with BCIP/NBT

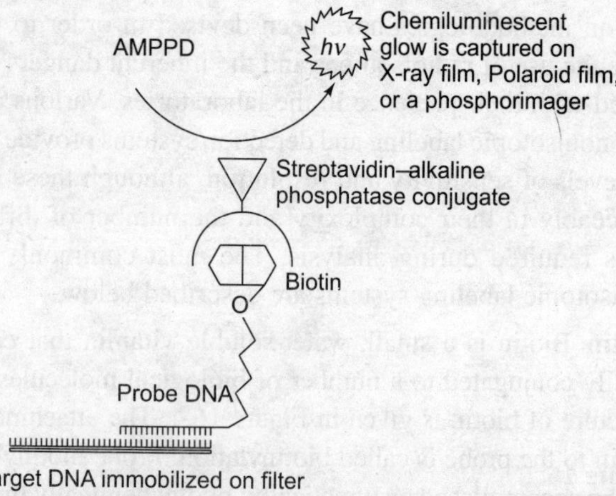

Figure 16.8 Detection of biotin-labeled nucleic acid probes with AMPPD

to its target. The results can be analyzed visually or recorded on photographic film. BCIP/NBT is also widely used as an indicator in the nonradioactive systems to detect nucleic acids or proteins (Figure 16.6).

In fluorescent assays, the substrate for alkaline phosphatase is 2-hydroxy-3-naphthoic acid 2′-phenylanillide phosphate (HNPP). After dephosphorylation by alkaline phosphatase, HNPP generates a fluorescent precipitate on membranes that can be excited by irradiation at 290 nm from a transilluminator. Light emitted at 509 nm can be captured by a charged-coupled device (CCD) camera or on Polaroid film using the same photographic setup.

Adamantyl-1,2-dioxetane phosphate (AMPPD) is used as substrate in alkaline phosphatase–triggered chemiluminescence for the detection of biopolymers immobilized on membranes (Figure 16.7). Alkaline phosphatase catalyzes the removal of the single phosphate residue of AMPPD, generating a moderately stable dioxetane anion that fragments into adamantanone and the excited state of a methyl-metaoxybenzoate anion (Figure 16.7). On return to ground state, the anion emits visible, yellow-green light. The dephosphorylation of dioxetane by alkaline phosphatase is quite efficient with a turnover rate of ~4.0×10^3 molecules/second. Note that chemiluminescence is the fastest and most sensitive assay to detect DNA labeled with biotin via conjugates containing alkaline phosphatase or HRP (Figure 16.8).

Another example of chemiluminescence is that of HRP-luminol detection system marketed by Amersham. HRP catalyzes the oxidation of luminol (5-amino-2,3-dihydro–1,4-phthalazinedione) in the presence of hydrogen peroxide, generating a highly reactive endoperoxidase that emits light at 425 nm during its decomposition in its ground state (Figure 16.9). A wide range of compounds, including benzothiazoles, phenols, naphthols and aromatic amines, can significantly enhance the amount of light generated during the reaction.

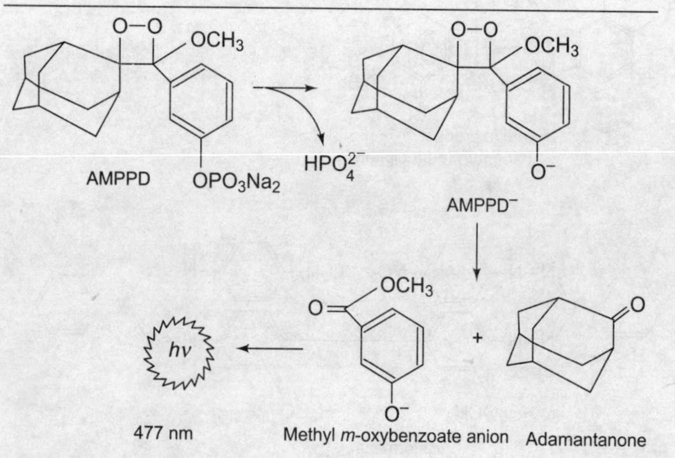

Figure 16.7 Chemiluminescent generation of light by dephosphorylation of AMPPD

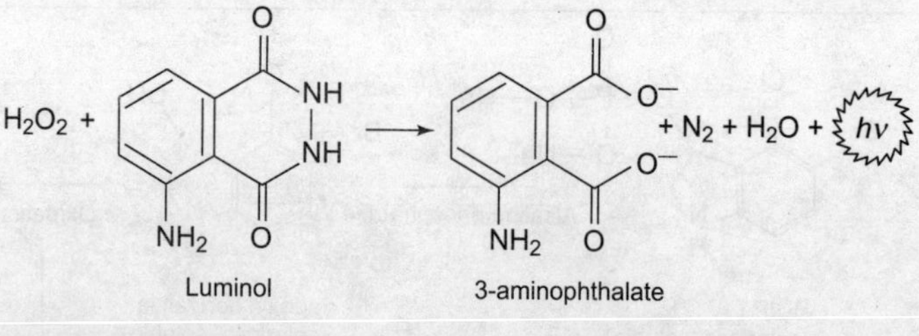

Figure 16.9 Generation of light by oxidation of luminol

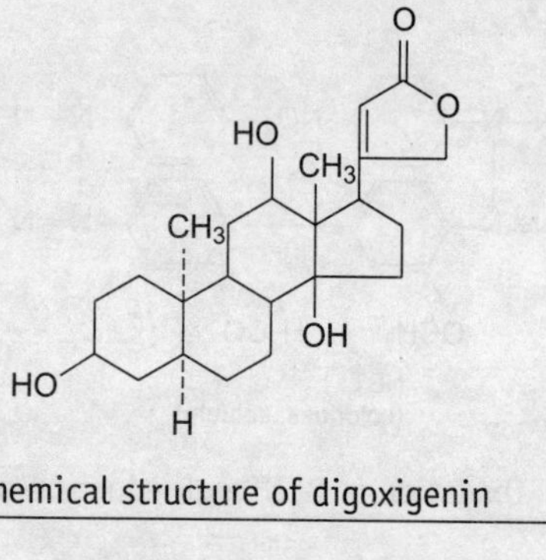

Figure 16.10 Chemical structure of digoxigenin

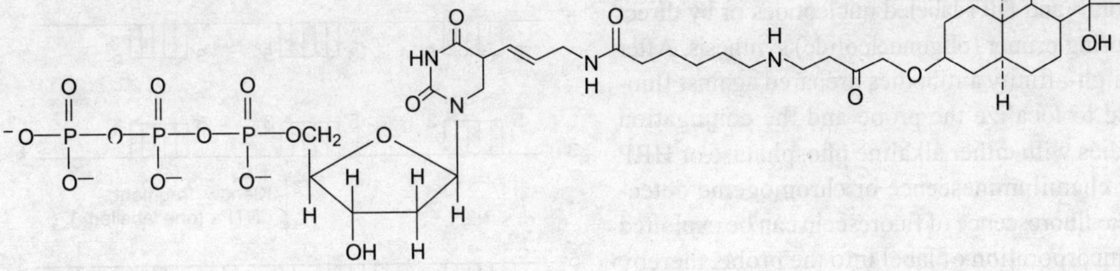

Figure 16.11 Structure of DIG-11-dUTP

Under ideal conditions, upto ~1 pg of target DNA can be detected on Southern blot. Furthermore, various commercial vendors sell detection systems to detect alkaline phosphatase with dioxetane-based chemiluminescent substrates such as AMPDD and Lumigen-PPD, and their more recent derivatives CSPD and CDP-Star, which are the most sensitive chemiluminescent substrates.

Digoxigenin Digoxigenin (DIG) (Roche Applied Science) is a steroid hapten derived from plants of the genus *Digitalis* (purple foxglove) (Figure 16.10). This is widely used for nucleic acid labeling and is supported by both chromogenic and chemiluminescence detection. In this system, DNA probes are enzymatically labeled; usually by random priming with DIG-dUTP (or DIG-11-dUTP) (Figure 16.11) and RNA probes are synthesized by *in vitro* transcription with DIG-11-UTP. These nucleotides are linked via a spacer arm to DIG. Oligonucleotides can also be DIG-labeled during their synthesis. Like biotinylation, DIG labeling does not interfere with biological activity and probes are stable at –20°C for one year after labeling. Moreover, DIG labeling supports posthybridization detection by chemiluminescence or by formation of an insoluble colored precipitate. After the posthybridization stringency washes, DIG-labeled probes are detected by enzyme-linked immunoassay using an antibody conjugated alkaline phosphatase (i.e., anti-DIG-alkaline phosphatase). As with systems involving biotin, the emission of light or the formation of precipitate is mediated by alkaline phosphatase-catalyzed dephosphorylation of a substrate compatible with the method of detection, either colorimetric or chemiluminescence (Figures 16.12 and 16.13). HRP-labeled antibody conjugates are also used in this system.

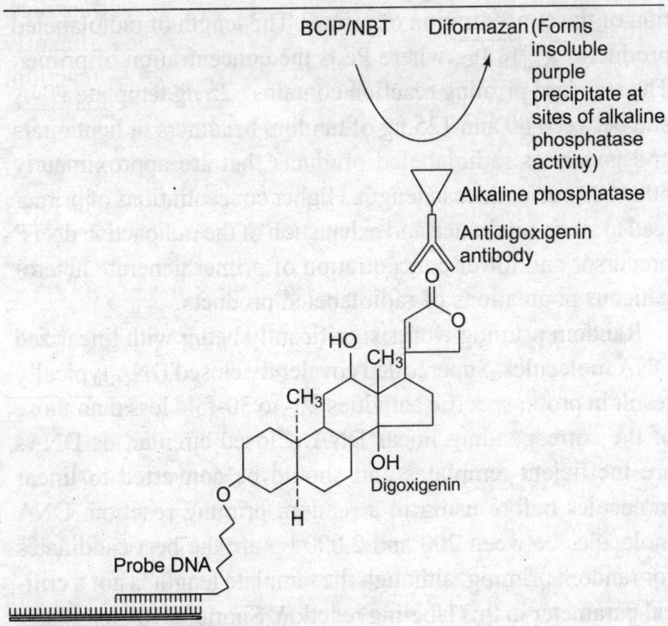

Figure 16.12 Detection of DIG-labeled nucleic acid probes with BCIP/NBT

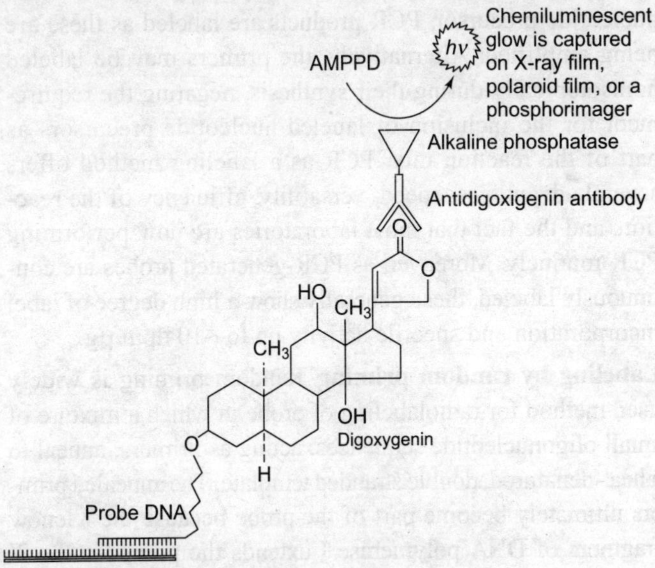

Figure 16.13 Detection of DIG-labeled nucleic acid probes with AMPPD

Fluorescein More recently, the fluorescent nucleotides readily available for either DNA or RNA probe synthesis are fluorescein labeled (FL): FL-12-dUTP for the synthesis of DNA probes and FL-12-UTP for the synthesis of RNA probes. These nucleotides are used to generate probes in much the same way as biotin- and DIG-labeled nucleotides or by direct incorporation during primer (oligonucleotide) synthesis. After hybridization, high-affinity antibodies prepared against fluorescein are used to localize the probe and the conjugation of these antibodies with either alkaline phosphatase or HRP provides either chemiluminescence or chromogenic detection. The intrinsic fluorescence of fluorescein can be exploited to monitor the incorporation of label into the probe, thereby permitting the evaluation of the efficiency of the labeling reaction.

Furthermore, fluorescent dyes like Cy3 and Cy5 are used to label DNA probes by direct label incorporation via labeled nucleotides or by indirect labeling via aminoallyl dUTP or aminoallyl dCTP, followed by a secondary reaction in which Cy3 or Cy5 is linked to the probe. The most common application of these fluorescent nucleotides is the synthesis of cDNA for microarray analysis.

DNA Probes

DNA probes are of diverse types, such as ds cDNA, genomic DNA, and single stranded oligonucleotides.

Techniques for Preparation of DNA Probes Various approaches are used for labeling of DNA, a brief description of more common techniques are as follows.

Labeling by PCR PCR is an excellent method for probe synthesis, requiring small quantities of template DNA. In the presence of the appropriate radiolabeled or hapten-labeled nucleotide precursor, PCR products are labeled as these are being amplified. Alternatively, the primers may be labeled nonisotopically during their synthesis, negating the requirement for the inclusion of labeled nucleotide precursors as part of the reaction mix. PCR as a labeling method offers several advantages: speed, versatility, efficiency of the reaction, and the fact that most laboratories are now performing PCR routinely. Moreover, as PCR-generated probes are continuously labeled, these generally show a high degree of label incorporation and specific activity up to ~10 dpm/μg.

Labeling by random priming Random priming is widely used method for radiolabeling of probe in which a mixture of small oligonucleotide sequences, acting as primers, anneal to a heat-denatured, double stranded template. The annealed primers ultimately become part of the probe because the Klenow fragment of DNA polymerase I extends the primer in the 3′ direction and incorporate the labeled nucleotide (Figure 16.14). In this method, one (α-^{32}P) dNTP and three unlabeled dNTPs are used as precursors, generating probes with specific activities of 5×10^8 to 5×10^9 dpm/μg. Oligonucleotide primers can be produced either by DNase I digestion of calf thymus DNA or synthesized chemically. The average size of the probe generated during priming reactions is an inverse function of the concentration of primer. The length of radiolabeled product $= K\sqrt{\ln \text{Pc}}$, where Pc is the concentration of primer. The standard priming reaction contains ~25 ng template DNA and between 60 and 125 ng of random hexamers or heptamers and generates radiolabeled products that are approximately 400–600 nucleotides in length. Higher concentrations of primer lead to steric hindrance and exhaustion of the radioactive dNTP precursor and lower concentration of primer generate heterogeneous populations of radiolabeled products.

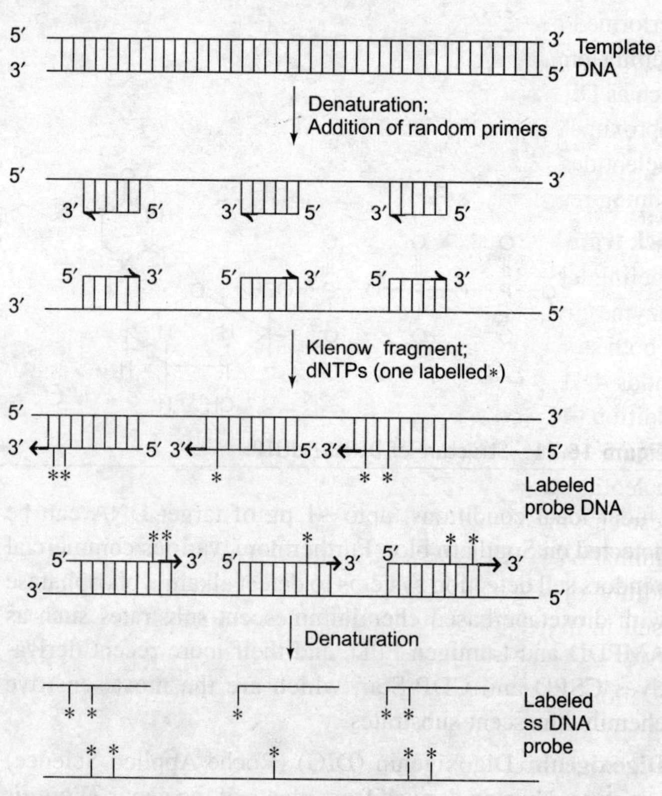

Figure 16.14 Preparation of radiolabeled probe by the random primer method

Random priming works significantly better with linearized DNA molecules. Supercoiled covalently closed DNA typically result in probe-specific activities 20- to 30-fold less than those of the corresponding linear DNA. Closed circular, ds DNAs are inefficient templates and should be converted to linear molecules before using in a random priming reaction. DNA molecules between 200 and 2,000 bp are the best candidates for random priming, although the template length is not a critical parameter in this labeling reaction. Shorter DNA templates produce probes of low specific activity that do not hybridize under stringent conditions. Labeling usually requires incubation of 10–30 minutes. Furthermore, random priming is also performed on the DNA in slices cut from gels cast with low melting temperature agarose. dNTPs that carry reporter groups such as DIG can be substituted for radioactive nucleotide and approximately one modified base in incorporated per 25–35 nucleotides of DNA synthesized during a standard random priming reaction.

Nick translation Nick translation is one of the oldest probe labeling techniques. It involves the activity of two different enzymes: (i) DNase I is used to create nicks at random sites in both strands of a ds target DNA; and (ii) DNA polymerase I adds dNTPs to the 3′-OH termini created by DNase I. In addition to its polymerizing activity, DNA polymerase I carries a $5' \rightarrow 3'$ exonuclease activity that removes deoxyribonucleotides from the 5′ side of the nick. The details of both these enzymes are discussed in Chapter 2. The simultaneous elimination of deoxyribonucleotides from the 5′ side and the addition of radiolabeled deoxyribonucleotides to the 3′ side result in movement of the nick (nick translation) along the DNA, which becomes labeled to high specific activity. Optimization of concentrations of DNase I, DNA polymerase I, primer, and dNTPs is very important for nick translation. Like random priming, it is adequate to use one labeled α-^{32}P (8,000 Ci/mole) and three unlabeled dNTPs for most purposes. Nick translation is efficient for both linear and covalently closed circular (CCC) DNA molecules and labeling requires about one hour. When labeling with ^{32}P, nick translation produces probes with specific activities of 3 to 5×10^8.

5′ end-labeling The 5′-end of the DNA molecule can be labeled with the addition of a radiolabeled phosphate. This method of 5′-end labeling is known as forward kinasing reaction; it specifically involves the transfer of γ-phosphate of ATP (not dATP) to a 5′-OH substrate group on dephosphorylated DNA molecules. Thus, 5′-phosphates are first removed by alkaline phosphatase treatment (for details see Chapter 2) and then radiolabeled phosphates are added by bacteriophage T4 polynucleotide kinase (PNK). When this reaction is performed efficiently, 40–50% of the protruding 5′-termini in the reaction becomes radiolabeled. The forward kinasing reaction is far more efficient than the exchange reaction, which involves the substitution of 5′-phosphates. This T4 PNK-catalyzed labeling reaction is an excellent method for labeling short nucleotides. The 5′-ends of DNA molecules with blunt ends, recessed 5′-termini, or internal nicks can also be labeled but less efficiently. The specific activity of the resulting probes is not as high as that obtained by other radiolabeling methods since only one radioactive atom is introduced per DNA molecule. However, the availability of γ-^{32}P labeled ATP with specific activities in the 3,000–7,000 Ci/mmole range allows the synthesis of probes suitable for many purposes. Moreover, by incorporating polyethylene glycol 8000 (PEG 8000) and high concentrations of ATP in the reaction mix, the efficiency of labeling can be enhanced. Finally, various nonisotopic labeling kits are available from several commercial vendors for 5′-end modification.

3′ end-labeling The ss DNA and ds DNA molecules can also be labeled by the addition of dNTPs to 3′-OH termini of double stranded, blunt ended fragments or fragments with 3′-overhang structure. Probe synthesis by 3′-end labeling involves the addition of nucleotides at 3′-end of DNA. The simplest way to label ds DNA is to use the Klenow fragment to catalyze the addition of one or more [α-^{32}P] dNTPs into a recessed 3′-terminus. The choice of which α-^{32}P dNTP to use for the reaction depends on the sequence of the protruding 5′-termini at the ends of the DNA and objective of the experiment. For example, ends created by cleavage of DNA with *Eco* RI can be labeled with [α-^{32}P] dATP and termini created by cleavage of DNA with *Bam* HI can be labeled with [α-^{32}P] dGTP. The Klenow enzyme is then used to catalyze the attachment of dNTPs to the recessed 3′-OH groups. Furthermore, restriction enzymes such as *Pst* I, *Pvu* I, and *Sac* I cleave asymmetrically to produce fragments with protruding 3′-ends, and these types of termini are efficiently labeled with T4 DNA polymerase by adding one radiolabeled dNTP during filling reaction. The protruding 3′-termini of ds DNA, such as those produced by *Pst* I or *Sac* I can be labeled by using terminal deoxynucleotidyl transferase (Tdt) to catalyze the transfer of (α-^{32}P) cordycepin 5′-triphsphate or (α-^{32}P) dideoxy ATP.

Direct Enzyme Labeling This method involves cross-linking an enzyme capable of supporting chemiluminescence directly to the backbone of heat-denatured DNA using glutaraldehyde. The method is extremely efficient and requires only 20 min to label the probe. Furthermore, there are no subsequent steps required to remove unincorporated label. Briefly, ds DNA is boiled for 5 min and then rapidly cooled on ice to minimize renaturation. Then glutaraldehyde links a modified alkaline peroxidase or HRP to the ss DNA. Because the probe cannot be boiled again after the labeling has been completed and because the denatured DNA reanneals in a relatively short time, only a required mass of probe should be labeled, which is then added to the hybridization mix immediately after labeling. As the enzyme is present throughout the hybridization and subsequent washes, stringency must be regulated by modifying ionic strength and including destabilizing agents such as sodium dodecyl sulfate (SDS) or urea, rather then by raising the temperature.

RNA Probes

RNA probes as hybridization tools have several key advantages and these probes are synthesized by *in vitro* transcription or end-labeling and can be substituted for DNA probes in nearly all hybridization reactions.

Characteristics of RNA Probes Some important properties of RNA probes are as follows: (i) RNA probes are single stranded; (ii) All RNA molecules are available for hybridization; (iii) Being single stranded, there is no problem of intermolecular base pairing; however intramolecular base pairing may reduce the effective concentration of the probe. Although RNA probes do not require boiling before use, heating them briefly may disturb intramolecular base pairing; (iv) RNA probes are continuously labeled as these are being transcribed, thereby generating probes with a high degree of label incorporation; (v) RNA probes show greater thermodynamic stability with both DNA and RNA target molecules as compared to DNA probes; (vi) RNA probes are synthesized by *in vitro* transcription in a template-dependent fashion as discussed below; therefore, all probe molecules are of uniform length; (vii) Large quantities of probes can be synthesized in a single *in vitro* transcription reaction; (viii) In the construction of the transcription template, the cDNA to be transcribed is generally flanked by highly efficient bacteriophage RNA polymerase promoters such as SP6 and T7 or T3, thus both sense and antisense RNA probes can be synthesized as needed; (ix) The SP6, T3, and T7 bacteriophage RNA polymerase promoters reveal virtually no cross-reactivity; therefore, transcription reactions initiated from one promoter or the other are virtually free of transcripts initiated at other promoters; (x) As with all RNA procedures, RNA transcribed *in vitro* must be treated with RNase-free reagents. Failure to do so may result in rapid degradation of the probe; (xi) RNA probes often produce unacceptable, high levels of background; thus at the conclusion of the hybridization period, it is common practice to digest all probes with RNase A and RNase T1. This treatment results in degradation of all probe molecules that did not participate in duplex formation; and (xii) Radiolabeled RNA probes often have such a high specific activity that these may experience radiolysis if stored for extended period.

RNA Probe Synthesis RNA is labile and susceptible to RNase degradation, and hence RNA probes must be treated with the same care as any other RNA preparations. In contrast to the diverse methods for DNA probe synthesis; there are only two types of methods for RNA probe synthesis, namely, *in vitro* transcription and end-labeling. A brief description of these RNA labeling techniques is as follows.

In vitro transcription *In vitro* transcription is the most reliable method for generating RNA probes. The method is based on the fact that large amounts of efficiently labeled probes of uniform length can be generated by transcription of a DNA sequence ligated next to a promoter. In the construction of the transcription template, the cDNA to be transcribed is usually ligated between two different RNA polymerase promoters from bacteriophage, which flank a multiple cloning site in opposite orientations. An integral part of the preparation of RNA probes is the transcription of sense and antisense strands from the same template. The resulting mRNA and antisense RNA are complementary and hence base pair with each other. Note that in addition to the ability to function as a probe in transcription studies, antisense RNA can also inhibit gene expression by forming untranslatable ds RNA (for details see Chapter 17). Many plasmid and phagemid vectors are commercially available that carry various combinations of bacteriophage promoters and polylinker sites (e.g., pGEM series) (for details see Chapters 8 and 12) that facilitate *in vitro* transcription to obtain RNA probes in the presence of corresponding bacteriophage RNA polymerase (for details see Chapter 2). Transcripts are elongated by the addition of nucleoside monophosphate into the nascent backbone of the probe and isotopic labeling with ^{32}P-UTP requires radiolabel in the α position, as with all continuously labeled probes. Because the probe is labeled as it is synthesized, the degree of label incorporation is high. Furthermore, bacteriophage RNA polymerases can also accept biotinylated nucleotides as precursors; however, compared to radiolabeled NTPs, incorporation is less efficient and the products of the reaction contain a higher proportion of truncated RNAs.

5′ end-labeling Like DNA, RNA can also be labeled at 5′-end by kinasing reaction.

3′ end-labeling 3′ end-labeling of RNA can also be performed by a reaction catalyzed by poly (A) polymerase (for details see Chapter 2). Note that in the presence of ATPs, poly (A) polymerase adds A residues at the 3′-end of RNA in a template independent manner.

Oligonucleotide Probes

Oligonucleotide probes are used for screening cDNA or genomic libraries, identifying or detecting specific genes, examining genomic DNA and PCR-amplified DNA, screening expression libraries, and identification of clones that carry specific base change.

Characteristics of Oligonucleotide Probes An ideally designed oligonucleotide probe should form perfect duplex only with its target sequence, and the resulting duplex should be sufficiently stable to withstand the posthybridization washing steps applied to remove probes nonspecifically bound to non-target sequences (for details see Section 16.5.6). For oligonucleotide of 200 residues or less, the reciprocal of the melting temperature T_m^{-1} (measured in Kelvin) is approximately proportional to n^{-1}, where n is the number of bases in the oligonucleotide. Note that the T_m is the temperature at which the probe and target are 50% dissociated. Several equations are available to calculate the T_m of hybrids formed between an oligonucleotide probe and its complementary target sequence. The equations for calculating T_m of oligonucleotides have been already given in Chapter 7. Because of the degeneracy of the genetic codons, many thousands of possible nucleotide sequences could code for such a sizeable tract of amino acids. However, because most amino acids are specified by codons that differ only in the third position, at least two of the three nucleotides of each codon are likely to be perfectly matched. In addition, the number of mismatches can be minimized by selecting codons that are used preferentially by particular organism, organelle, or cell type. In this way, it is possible to make guessmers of sufficient length so that the detrimental effects of mismatches are outweighed by the stability of extensive tracts of perfectly matched bases. There are no defined rules that guarantee selection of the correct codon at position of ambiguity. Moreover, on the basis of mathematical calculations, a probe would be expected to have only 76% identity with the authentic gene even if all codon choices were made on random basis. It is sometimes possible to synthesize a small pool containing 2–8 guessmers that includes all possible codon choices at certain amino acid positions. In this way, it may be possible to generate a continuous sequence within the guessmer that is a perfect match for the target gene. However, when using a mixture of guessmers as probes, the strength of the hybridization signal generated by the correct probes is usually reduced. Therefore, the most critical step in the use of guessmers is the choice of conditions for hybridization. The temperature should be high enough to suppress hybridization of probe to the incorrect sequence, but must not be too high that it prevents hybridization to the correct sequence, even though it may be mismatched. Therefore, before performing the experiment, pilot reactions should be carried out under different degrees of stringency. Hybridization with oligonucleotide probes is performed in aqueous solutions and blots are washed with buffers containing quaternary ammonium salts.

Preparation of Oligonucleotide Probes After chemical synthesis, oligonucleotides are usually labeled by phosphorylating the 5′-ends in the presence of T4 PNK.

Probe Purification

After labeling of probes, it is very important to purify the labeled probe from unincorporated precursors; otherwise especially in the case of radioisotopes, unacceptably high levels of background signals may be obtained. There are three basic methods for probe purification, which are as follows.

Ethanol Precipitation In the first approach, probes are precipitated with a combination of salt (0.1 volume 3 M sodium acetate) and 2.5 volumes of 95% EtOH. After EtOH precipitation, the unincorporated label remains in the supernatant and discarded. This is an efficient purification strategy for larger probes; however, precipitation of oligonucleotides is not possible, especially if small masses of nucleic acids are involved. If the unincorporated probes are small in size, then EtOH precipitation is done in the presence of ammonium ions. This allows small probes to remain in the supernatant and hence discarded.

Purification by Gel Filtration Chromatography The second strategy involves the purification of probes by gel filtration chromatography. The most common approach is to make or purchase a spun column, which consists of a 1 ml syringe, packed with Sephadex; Sephadex G-50 is used for larger probes and Sephadex G-25 is reserved for the purification of labeled oligonucleotides. During spun-column chromatography, which involves centrifugation of the column at 1,000 × g to 1,200 × g for ~4 min, the probe is rapidly eluted while the unincorporated nucleotides remain behind in the column. The gel filtration probe purification is compatible with both isotopic and nonisotopic labeling methods.

Purification by Concentration or Absorption The third general method for probe purification utilizes one of the available concentration or absorbent devices, for example, Elutip-d minicolumns (Schleicher and Schuell, Keene, NH), ultrafree filtration units, and Centricon devices (Millipore, Bellerica, MA). All these devices provide rapid and convenient alternatives to traditional column chromatography or alcohol precipitation for separating radiolabeled nucleic acids from unincorporated nucleotides. Moreover, these devices can be easily disposed off, thereby minimizing the spread of radiolabeled waste products.

Probe Storage

Radioisotope-labeled probes are useful only as long as there is sufficient activity remaining and the probe molecules remain intact. In general, radiolabeled probes should be used immediately after labeling. Radiolysis of probes, especially those labeled to extremely high specific activity, becomes more problematic as the probe ages. Probes are stored at –20°C and –80°C until use. Repeated freezing and thawing should be avoided. Nonisotopic labeling systems vary with respect to the recommended postlabeling storage and handling of probes. Biotinylated, DIG-, and FL-labeled probes can be stored at –20°C for up to 1 year.

16.5.5 Hybridization Procedure

Unlike solution hybridization, membrane hybridization tend not to proceed to completion. One reason for this is that some of the bound nucleic acids are embedded in the membrane and are inaccessible to the probe. Prolonged incubations may not generate any significant increase in detection sensitivity. Hence, the composition of the prehybridization/hybridization buffer can greatly affect the speed of the reaction and the sensitivity of detection. The key components of these prehybridization/hybridization buffers are: (i) Rate enhancers (e.g., dextran sulfate and other polymers such as PEG) that act as volume excluders to increase both the rate and the extent of hybridization); (ii) Detergents (e.g., SDS) and blocking agents (e.g., Denhardt's reagent, BLOTTO (Bovine Lacto Transfer Technique Optimizer), dried milk, and heparin) that suppress nonspecific binding of the probe to the membrane; (iii) Heterologous DNA (e.g., salmon sperm DNA) that reduces nonspecific binding of probes to nonhomologous DNA on the blot; and (iv) Denaturants (e.g., urea or formamide) that are used to depress the T_m of the hybrid so that reduced temperatures of hybridization can be used.

The blotted membrane is placed in hybridization container (e.g., heat sealable bag, plastic container, and roller bottle) and first the prehybridization is done in a prehybridization buffer.

Prehybridization/Hybridization in Aqueous Buffer

Solution for prehybridization/hybridization in aqueous buffer consists of 6 X SSC or 6 X SSPE, 5 X Denhardt's reagent [Blocking agent; 50 X contains 1% w/v Ficoll 400, 1% w/v bovine serum albumin (BSA), and 1% w/v polyvinylpyrrolidone (PVP)], 0.5% SDS, 1 µg/ml poly (A) RNA and 100 µg/ml salmon sperm DNA. Prehybridization/hybridization solution may be prepared with or without poly (A) RNA. Note that poly (A) RNA prevents the probe from binding to T-rich sequences commonly found in eukaryotic DNA. Approximately 0.2 ml of prehybridization buffer is required for each square centimeter of membrane while using heat–sealable bags or plastic containers. Smaller volumes (~0.1 ml/cm^2) can be used when hybridizing in roller bottles. For prehybridization, the hybridization chamber is incubated for 1–2 hours at a controlled temperature (68°C). Nucleic acid probes have a tendency to bind nonspecifically to other materials on the filter or even to filter itself. To minimize this nonspecific probe binding, some blocking agents like detergents, BSA, and nonhomologous DNA (salmon sperm DNA) are added in prehybridization solution. Thereafter, denatured probe is added and hybridization is performed for 18–20 hours. Note that if the probe is a ds DNA, it is heat-denatured by boiling prior to hybridization and made accessible to the target in hybridization solution. After hybridization, nonspecifically bound or unbound probes are removed by washing. The hybridization and washing conditions depend on percent sequence identity, which can be analyzed by using Web-based bioinformatics tools. Very short nucleotides can be compared manually, but computer algorithms are required to find the best alignments when the sequences are longer than about 10–15 residues. The two principal algorithms are BLAST (Basic local alignment search tool) and FASTA. There are several variants of each algorithm that are adapted for different types of searches depending on the query sequence and the database (Table 16.5). If, for example, the intention is to hybridize a target and probe (homologous) from same source, then high-stringency (i.e., high temperature and low salt concentration) conditions should be chosen. This ensures that the probe will remain bound to the membrane only where it has annealed to the correct complementary sequences. If on the other hand, the investigator wishes to detect with a probe made from different source (heterologous probe), then low-stringency (i.e., low temperature and high salt concentration) should be chosen in order to protect the expected partially mismatched hybrids. Members of a gene family from single species or orthologous genes from different species can be isolated by low-stringency hybridization and washing. The identification of genes that share <65% sequence is trickier but can be accomplished by: (i) Using an RNA probe prepared by *in vitro* transcription; (ii) Using a ss DNA probe prepared from bacteriophage M13 template; (iii) Rinsing and washing the hybridized membranes as described above; and (iv) Including crowding agent in the hybridization reaction such as 10% dextran sulfate or PEG 8000.

Table 16.5 Variants of the BLAST and FASTA Algorithms

Program	Compares
BLASTN	A nucleotide sequence against a nucleotide sequence database
BLASTX	A nucleotide sequence translated in all six reading frames against a protein sequence database
BLASTP	An amino acid sequence against a protein sequence database
TBLASTN	An amino acid sequence against a nucleotide sequence database translated in all six reading frames
EST BLAST	A cDNA/EST sequence against cDNA/EST sequence databases
FASTA	A nucleotide sequence against a nucleotide sequence database or an amino acid sequence against a protein sequence database
TFASTA	An amino acid sequence against a nucleotide sequence database translated in all six reading frames

Prehybridization/Hybridization in Formamide Buffers

The solution for prehybridization/hybridization in formamide buffer has the same composition as that described for aqueous buffer, but also contains 50% v/v formamide. The prehybridization/hybridization procedure is essentially the same as described above however, the hybridization temperature is 42°C. Furthermore the identification of genes that share <65% sequence can also be accomplished by decreasing the formamide concentration to 20% and hybridizing at 34°C.

Formamide (50% v/v) eliminates the requirement of elevated temperature in the hybridization process, as aqueous solutions of formamide denature DNA. Furthermore, increased retention of immobilized DNA by membrane filters and decreased nonspecific background absorption is observed by adding formamide. Due to these two factors, an increased reproducibility of replicates is obtained in the hybridization procedure. Hybridization in formamide solution at low temperature is also helpful in minimizing scission of nucleic acid molecules during prolonged period of incubation. Additionally, increased flexibility is also introduced into the design of reaction conditions for a given experiment. It is more convenient to control the stringency of hybridization with formamide rather than through adjustment of the incubation temperature. Depression of the T_m of duplex DNA is a linear function of the formamide concentration. For DNAs with 30–75% G+C content, the T_m is depressed by 0.63°C for each percentage of formamide in the hybridization mixture.

The T_m of hybrid formed between a probe and its target may be estimated from the following equation:

$$T_m = 81.5°C + 16.6 (\log_{10} [Na^+]) + 41 (\text{mole fraction } [G + C]) – 0.63 (\%\text{formamide}) – 500/n$$

where n is the length of the DNA in nucleotides.

In the case of RNA:RNA hybridization,

$$T_m = 79.8°C + 18.5 (\log_{10} [Na^+]) + 58.4 (\text{mole fraction } [G + C] + 11.8 (\text{mole fraction}) [G+C]^2 – 0.35 (\%\text{formamide}) – 820/n$$

For DNA:RNA hybrids:

$$T_m = 79.8°C + 18.5 (\log_{10} [Na^+]) + 58.4 (\text{mole fraction } [G + C] + 11.8 (\text{mole fraction}) [G+C]^2 – 0.50 (\%\text{formamide}) – 820/n$$

By comparing these reaction equations, the relative stability of nucleic acid hybrids in high concentrations of formamide decreases in the following order: RNA:RNA (most stable), RNA:DNA (less stable), DNA:DNA (least stable). The rate of DNA:DNA hybridization in 80% formamide is slower than in aqueous solution. This effect is the consequence of increased viscosity of the hybridization solution at the temperatures used for renaturation. The breakdown of formamide that occurs during prolonged incubation at temperatures in excess of 37°C can cause the pH of the hybridization buffer to drift upward. When formamide is included in the hybridization buffer, 6 X SSPE is preferred to 6 X SSC because of its greater buffering power.

Rapid Hybridization Buffers

Several rapid hybridization buffers have also been devised that dramatically enhance the rate of hybridization of two complementary strands of nucleic acid, thereby decreasing the hybridization time from 16 hours to 1–2 hours. Although the chemical composition of these premade solutions is a trade secret, it seems likely that some of them contain cationic detergents in the millimolar range [e.g., quaternary amine compounds such as dodecyltrimethylammonium bromide (DTAB) and cetyltrimethylammonium bromide (CTAB)], whereas others contain volume excluders [e.g., 10% dextran sulfate or 5% PEG 35000]. These increase the rate of renaturation of two complementary strands of DNA by >10,000-fold. Moreover, the increase in hybridization rate is specific and occurs in the presence of as much as 10^6-fold excess of noncomplementary DNAs. In addition, these hybridization accelerators improve the efficiency of hybridization when low concentrations of probe are used (~1 ng/ml). Radiolabeled probes are added to preheated rapid hybridization buffer before adding it to the membrane. These buffers work efficiently well for Southern hybridization, however, when used in northern analysis, the background of hybridization to rRNAs increases greatly.

16.5.6 Washing

After hybridization, the membrane is removed from hybridization container and hybridization solution is poured into a container suitable for disposal of radioactivity (in case of radiolabeled probe). Membrane is submerged in a tray containing 2 X SSC and 0.5% SDS (1 ml/cm^2 membrane) at room temperature for 5 minutes to rinse the membrane, and the first rinse solution is also poured off into a radioactivity disposal container. A solution (several hundred milliliters) containing 2 X SSC and 0.1% SDS is then added to the tray containing the membrane and incubated for 15 min at room temperature with occasional gentle agitation; after incubation, the rinse is discarded as described above. Finally, several hundred milliliters of fresh 0.1 X SSC with 0.1% SDS are added and incubated for 30 min to 4 hours at 65°C with gentle agitation; after incubation, the rinse is discarded as described above. During the washing step, it is required to periodically monitor the amount of radioactivity on the membrane using a handheld minimonitor. After that, most of the liquid from the membrane is removed by placing it on pad of paper towels and then damp membrane is placed on a sheet of Saran wrap and subjected to detection system depending upon the labeling method used [e.g., exposed to X-ray film for 16–24 hours at –70°C with an intensifying screen to obtain an autoradiographic image by autoradiography or sometimes by fluorography (Exhibit 16.3), or chromogenic assay, or chemiluminescence, or fluorescence analysis. Hybridization data can also be analyzed by phosphorimager, a more sophisticated and very sensitive digital imaging device (Exhibit 16.4).

16.5.7 Stringency Control

Stringency is an evaluation of the probability that a double stranded nucleic acid molecule will or will not dissociate into its constituent single strands. This phenomenon measures the ability of single stranded nucleic acid molecules to discriminate between those molecules that share a high degree of homology and those that exhibit a lesser degree of homology. Practically, higher stringency conditions (lower salt or higher temperature) favor stable hybridization only between nucleic acid molecules with a high degree of homology (i.e., hybrid-

Exhibit 16.4 Phosphorimager: A sophisticated imaging device

Phosphorimager is a form of solid state liquid scintillation where radioactive material can be both localized and quantified. It produces digital images of the gels and blots by capturing images from them and allows complete filmless, electronic analysis, and archiving.

Two types of phosphorimaging devices have become available that create images of radiation sources on computer screens rather than on conventional photographic film. One type of device (area detector) scans the gel or filter in small windows with a Gieger counter, compiling a contour map of the number of radioactive disintegrations per unit area. The other device uses plates coated with a light-responsive phosphor. The gel or film or filter is directly exposed to the phosphorimager screen or plate and the energy emitted is stored in a europium-based coating. Scanning of the screen is done by laser and the released photons are collected to form an image. The intensity of photon emissions released from the storage phosphor screen during scanning is measured as an arbitrary unit called 'phosphorimager count'. The linear dynamic range of phosphorimager systems begins at one phosphorimager count. Given the sensitivity of the phosphorimager system, if the average pixel value for the band is below one phosphorimager count, there are insufficient particle emissions from the band for statistically valid quantitation. Both devices present the image on a computer screen. The images are captured electronically and can be stored and prepared for publication using programs such as Adobe Photoshop (Adobe Systems Incorporated).

The first model of phosphorimager became available in the late 1980s from Molecular Dynamics Company, which was originally designed for the quantification of radioactivity in gels. Present models include the Phosphorimager 400, Phosphorimager 425, Phosphorimager 445SI, Phosphorimager SI, Typhoon, and Storm series of imaging instrumentations. Depending on the model, phosphorimagers are capable of quantification of fluorescence (e.g., EtBr, SYBR Green, SYBR Gold), chemiluminescence, and radioactive decay. Present systems are also capable of imaging Coomassie blue and silver-stained gels, TLC plates, and Western blots. Analysis has also been extended to the realm of microarrays.

Though phosphorimagers are more expensive than those required for conventional autoradiography and the images have a lower resolution, there are numerous advantages associated with them. These are as follows:

- In blot analysis by phosphorimager, X-ray films are not required but images are generated on special screens that can be used again and again and last indefinitely with proper care. Because of this advantage, it has become an increasingly popular choice for several applications in molecular biology.
- Though similar to traditional autoradiographic techniques in that it relies on high-energy particle decay, phosphorimaging is more sensitive, and has a greater dynamic range than X-ray film. The sensitivity associated with phosphorimaging is generally 5- to 10-fold greater than is possible using X-ray film.
- Phosphorimager systems are linear over a wide dynamic range so exposure time does not matter as much as it does with X-ray film. The exposure time is reduced to just one-tenth of the time required to expose the samples to X-ray films and the images can be detected in approximately 10% to 20% of the time required by conventional autoradiography.
- Unlike autoradiography, there is no requirement of a dark room.
- The phosphorimager detection technology is completely compatible with the qualitative and quantitative assays of nucleic acids, proteins, and any other macromolecule that carries a fluorescent or radioactive tag.
- The elimination of X-ray film also means the elimination of the very narrow linear response that is an intrinsic shortcoming of all photographic films. Typically, phosphorimagers offer a linear response of five logs, resulting in quantitative data that highlights the true variations among samples.

Source: Sambrook J, Russell DW (2001) *Molecular Cloning Manual*, Vol. 3 Cold Spring Harbor Laboratory Press, Cold Spring Harbor, NY, A 9.14; Farrell RE (2005) Practical nucleic acid hybridization. *RNA Methodologies: A Laboratory Guide for Isolation and Characterization*, Elsevier Academic Press, pp. 317–360.

ization between homologous probe and target sequence). When the stringency in a system is lowered, a proportional increase in nonspecific hybridization is favored, meaning that so-called mismatches will be tolerated (i.e., hybridization between heterologous probe and target sequence or guessmer and target sequence).

The variables that are commonly manipulated to either promote or prevent nucleic acid hybridization *in vitro* include pH, ionic strength, temperature, and the presence of formamide. Stringency is most commonly controlled by the temperature and salt concentration in the posthybridization washes, although these parameters can also be utilized in the hybridization step. The addition of formamide permits maintenance of high-stringency conditions at temperatures lower than would otherwise be required. The melting temperature (T_m) of a probe–target hybrid can be calculated to provide a starting point for the determination of correct stringency. For probes longer than 100 base pairs:

$$T_m = 81.5°C + 16.6 \log M + 0.41 (\% G + C)$$

where M is the ionic strength of buffer in moles/liter.

Oligonucleotides can give a more rapid hybridization rate than long probes as these can be used at a higher molarity. Also, in situations where target is in excess to the probe, for example, dot blots, the hybridization rate is diffusion-limited and longer probes diffuse more slowly than oligonucleotides. The availability of the exact sequence of oligonucleotides allows conditions for hybridization and stringency washing to be tightly controlled so that the probe only remains hybridized when it is 100% homologous to the target. Adjusting the temperature of the wash buffer commonly controls stringency. The 'Wallace rule' is used to determine the appropriate stringency wash temperature:

$$T_m = 4 \times (\text{number of GC base pairs}) + 2 \times (\text{number of AT base pairs})$$

In filter hybridizations with oligonucleotide probes, the hybridization step is usually performed at 5°C below T_m for perfectly matched sequences. When the probe is used to detect partially matched sequences, the hybridization temperature is reduced by 1°C for every 1% sequence divergence between probe and target. For every mismatched base pair, a further 5°C reduction is necessary to maintain hybrid stability.

16.5.8 Probe Stripping

After detection, probes are removed from nylon blots to allow subsequent hybridization with a different probe by heating at 65°C in the presence of 0.1% SDS.

16.5.9 Techniques Based on Nucleic Acid Blotting and Hybridization

In this section, various techniques are described that are based on nucleic acid blotting and hybridization.

Colony Blotting and Hybridization

Colony blot hybridization, devised by Grunstein and Hogness (1975), is applied to nucleic acid released from blotted microbial colonies. Colony blotting is used to screen transformed bacterial colonies (100–200) from agar plates. It is the most commonly used and rapid method of library screening as it can be applied to very large numbers of clones. In the first step, the microbial colonies to be screened are transferred (blotted) to a membrane by applying the filter membrane to the upper surface of agar plates containing the colony library, making direct contact between the colonies and filter. Usually three types of solid supports, *viz.*, nitrocellulose filters, nylon filters, and Whatman 541 filter papers are used for *in situ* hybridization of lysed bacterial colonies. Previously, Whatman 541 filter paper was used to screen bacterial colonies as it possessed a high wet strength. However, it is now used mainly for screening of arrayed libraries that are stored as cultures of individual transformed colonies in separate wells of microtiter plates. These ordered libraries are duplicated on the surface of agar medium and resulting colonies are then transferred to Whatman 541 paper and lysed either by alkali or by a combination of alkali and heat. Nylon filters are the most durable among the three and easily withstand several rounds of hybridization and washing at elevated temperatures; nylon membranes are preferred when colonies are to be screened sequentially with a number of different probes. The filter bearing the colonies is removed and treated with SDS, denaturing solution, and neutralizing solution sequentially so that the bacterial colonies are lysed and their DNA is denatured. In the meantime, the master plate is stored at 4°C until the results of the screening procedure become available. The DNA is fixed firmly either by baking the filter at 80°C or by UV cross-linking to the membrane. The molecules become attached to the membranes through their sugar–phosphate backbones, so the bases are free to pair with complementary nucleic acid molecules. The labeled probe is denatured by heating, and applied to the membrane in a prehybridization/hybridization solution. Traditionally, the probe is labeled with radioactive nucleotide a technique that results in a probe with higher activity and therefore able to detect smaller amounts of membrane-bound DNA. However, the hybridization probe may be labeled in a nonradioactive manner to avoid the hazardous effects of radioactivity. After a period to allow hybridization to take place, the filter is washed to remove unbound probe, dried, and the positions of the bound probe detected. A colony whose DNA print gives a positive result is then picked from the reference master plate. The protocol is summarized in Figure 16.15.

Plaque Lift and Hybridization

Plaque lift and hybridization, devised by Benton and Davis (1977), is a procedure used to identify and isolate specific recombinants from libraries of bacteriophage λ. This method is commonly used for mass screening of plaques by *in situ* hybridization. This procedure permits the rapid screening of several hundred thousand plaques and so facilitates the isolation of single copy genes from complex eukaryotic genomes.

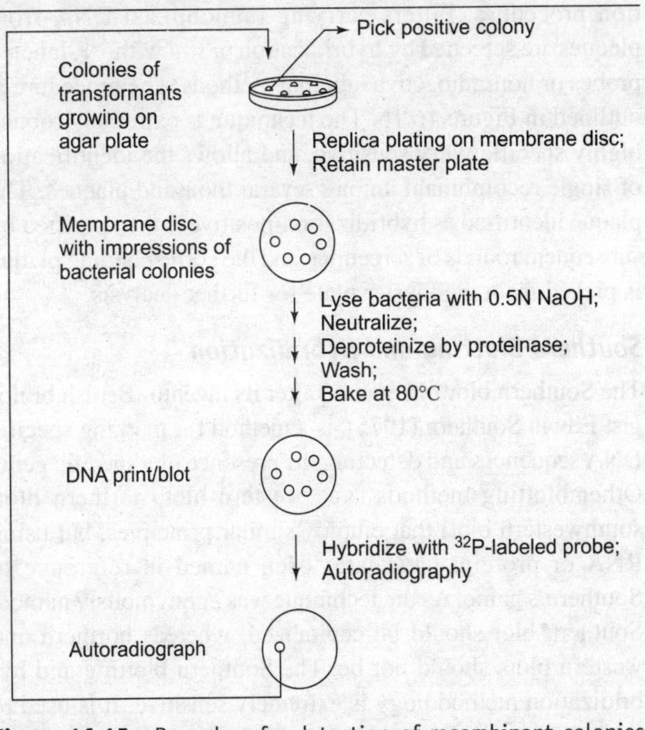

Figure 16.15 Procedure for detection of recombinant colonies by colony blot hybridization

This method has the advantage that several identical DNA prints can easily be made from a single-phage plate; this allows the screening to be performed in duplicate and hence with increased reliability, and also allows a single set of recombinants to be screened with two or more probes.

First, packaging mixture, bacteriophage stock, or library is diluted. Aliquots of the diluted bacteriophage stock are mixed with an appropriate amount of freshly prepared host cells and incubated for 20 min at 37°C. Molten (~47°C) top agarose is added to each aliquot of infected cells and the contents are poured onto separate numbered agar plates. The plates are closed and top agarose is allowed to be hardened and incubated at 37°C in an inverted position until plaques appear and are just beginning to make contact with one another (~10–12 hours). The plates are chilled for at least 1 hour at 4°C to allow the top agarose to harden. The plaques are then lifted on filter membrane (nitrocellulose or nylon) by applying them to the upper surface of agar plates, making direct contact between plaques and filter. This procedure is called plaque lift. The plaques contain phage particles as well as a considerable amount of unpackaged recombinant DNA. Both phage and unpackaged DNA bind to the filter. The filters are transferred, with the plaque side up to a sheet of Whatman 3 MM paper impregnated with denaturing solution and finally to neutralizing solution for 1–5 min in each case. The released DNA is fixed on membranes either by baking or UV cross-linking. Large numbers of recombinant phages can be easily screened for complementarity toward a given radiolabeled sequence by the nucleic acid hybridization procedure. Filters carrying immobilized DNA from plaques are screened by hybridization *in situ* with ^{32}P-labeled probes or nonradioactive labeling methods. The procedure is outlined in Figure 16.16. The technique is extremely robust, highly specific, very sensitive, and allows the identification of single recombinant among several thousand plaques. The plaque identified as hybridization-positive is then purified by subsequent rounds of screening, and the corresponding plaque is picked from the master plate for further analysis.

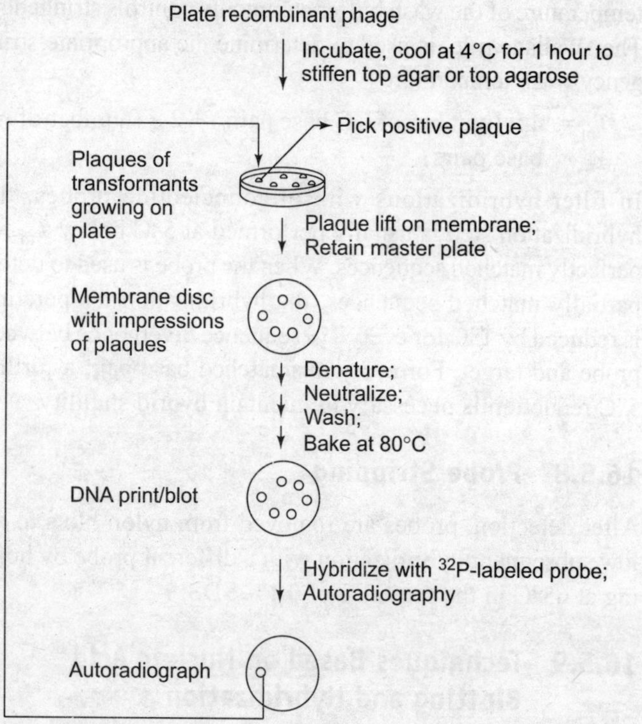

Figure 16.16 Procedure for detection of recombinant plaques by plaque lift and hybridization

Southern Blotting and Hybridization

The Southern blotting, named after its inventor British biologist Edwin Southern (1975), is a method for marking specific DNA sequences and detecting the presence of a specific gene. Other blotting methods (i.e., western blot, northern blot, southwestern blot) that employ similar principles, but using RNA or protein, have later been named in reference to Southern's name. As the technique was eponymously named, Southern blot should be capitalized, whereas northern and western blots should not be. The Southern blotting and hybridization methodology is extremely sensitive. It is used to study how genes are organized within genomes by mapping restriction sites in and around segments of genomic DNA. It can also be applied to mapping restriction sites around a single-copy gene sequence in a complex genome such as that of humans, and when a 'minisatellite' probe is used it can be applied forensically to minute amounts of DNA.

Procedure of Southern Blotting and Hybridization The procedure involves isolation and purification of DNA, restriction digestion, agarose gel electrophoresis, gel treatment (optional), transfer of DNA fragments to membrane, immobilization of DNA on membrane, hybridization with labeled probes, and washing to remove unincorporated probes.

Agarose Gel Electrophoresis DNA is first digested with one or more restriction enzymes and the resulting fragments are separated according to size by agarose gel electrophoresis. DNA size standards are run in one lane (for details see Chapter 6).

Gel Treatment and Transfer of DNA Fragments to Membrane The DNA is then subjected to *in situ* or *in gel* depurination, denaturation, and neutralization followed by transfer from the gel to solid support, i.e., hybridization membrane. The procedure of Southern blotting and hybridization is diagrammatically presented in Figure 16.17. In the beginning, the nitrocellulose membrane was used as hybridization membrane; thereafter nylon membranes have been developed, which have greater binding capacity for nucleic acids in addition to high tensile strength. Gel pretreatment is important for efficient transfer of DNA to uncharged membranes; large DNA fragments (>15 kbp) require incorporation of depurination step

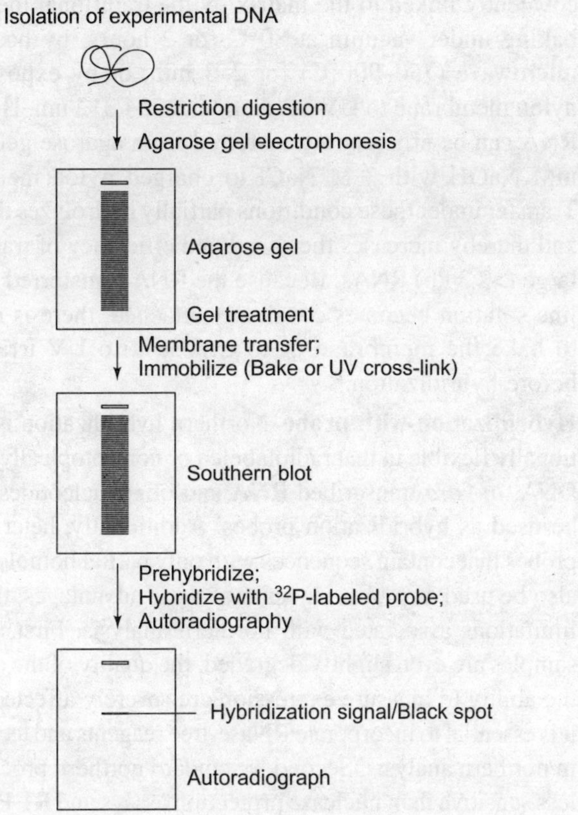

Figure 16.17 Procedure of Southern blotting and hybridization

in which gel is treated with weak acid (0.2 N HCl) solution that permits uniform transfer of a wide size range of DNA fragment. In the case of transfer to uncharged membranes, denaturation is performed by incubating the gel into 10 gel volumes of denaturation solution (1.5 M NaCl, 0.5 M NaOH) for ~45 min or even more at room temperature with constant agitation that denatures the DNA fragments so that these are available in single stranded state and accessible for probe hybridization. Alkali treatment also hydrolyzes the phosphodiester backbone at depurinated sites. Finally, the gel is equilibrated in 10 gel volumes of neutralizing solution [1.5 M Tris (pH 7.4), 1.5 M NaCl] prior to blotting. Thereafter, DNA is transferred to uncharged membrane in the presence of neutral transfer buffer (10 X SSC or 10 X SSPE by upward capillary transfer or following any of the procedures described in Section 16.5.2). Moreover, DNA can be simultaneously transferred from opposite sides of a single agarose to two membranes, which is useful when the need arises to analyze the same set of restriction fragments with two different probes. The method works best when the target DNA sequences are present in high concentration, for example, analyzing cloned DNAs [plasmid, bacteriophages, cosmids, yeast artificial chromosomes (YACs), bacterial artificial chromosomes (BACs), P1-derived artificial chromosomes (PACs), or less complex genome such as *Drosophila* and yeast]. For transfer to positively charged nylon membranes, extended gel pretreatment is not required; in this case alkaline transfer buffer (0.4 N NaOH, 1 M NaCl) is used to transfer the DNA to the membrane. After transfer, DNA is fixed to the membrane by baking or UV cross-linking. These steps lead to formation of Southern blot, and the procedure is called Southern blotting.

Hybridization with Probe Once the Southern blot is ready, next the immobilized ss DNA molecules are subjected to Southern hybridization. For this, the blot is treated with a labeled (say, for example, radiolabeled) hybridization probe, which is simply a ss DNA, RNA, or single stranded oligodeoxyribonucleotide molecule with a known sequence that pairs with the blotted DNA's sequences. It is important to choose the conditions, which maximize the rate of hybridization, compatible with a low background of nonspecific binding on the membrane. A common approach is to carry out the hybridization under conditions of relatively low stringency (i.e., lower temperature or more commonly high ionic strength). The membrane is then subjected to posthybridization washes of increasing stringency (i.e., higher temperature or more commonly lower ionic strength to remove unbound probes (i.e., radioactivity). The regions of hybridization are then detected autoradiographically by placing the membrane in contact with X-ray film (Exhibit 16.3). By examining the pattern of hybridization with autoradiography, the investigator can determine which fragment contains specific DNA sequences or genes.

Northern Blotting and Hybridization

The northern blotting is a fundamental technique used in molecular biology research to study the gene expression, i.e., the amount and size of RNAs transcribed from eukaryotic genes and estimation of their abundance. It takes its name from the similarity of the procedure to the Southern blotting with the key difference that RNA, rather than DNA, is the substance being analyzed by electrophoresis. This technique was developed in 1977 by Alwine and coworkers at Stanford University. Northern analysis provides a direct relative comparison of message abundance between samples on a single membrane. It is the preferred method for determining transcript size and for detecting alternatively spliced transcripts. Northern and Southern hybridizations have much in common, including, for example, the mechanism of hybridization, the types of probes, and the posthybridization processing of the membranes. All of these topics are discussed in depth in this section.

Procedure of Northern Blotting and Hybridization Although many variations and improvements have been reported after the development of this technique, but the basic steps in northern analysis are the same. These include isolation of intact total RNA or mRNA, probe generation, denaturing agarose gel electrophoresis, transfer of RNA to a solid support in a way that preserves its topological distribution within the gel, fixation of the RNA to solid support, hybridization of the immobilized RNA to probes complementary to the sequences of interest, removal of nonspecifically bound probes to solid matrix by washing, detection capture, and analysis of an image of the specifically bound probe molecules. Stripping and reprobing may be done if required.

Denaturing agarose gel electrophoresis and equalization Total RNA or mRNA is isolated according to the procedures described in Chapters 1 and 5, respectively. The purified preparation is subjected to electrophoresis under denaturing conditions, i.e., by including formaldehyde or glyoxal (for details see Chapter 6). Markers of known molecular weights are loaded in the gel for accurate measurement of the size of the RNA of interest. These include RNA standard, DNA standard, highly abundant rRNAs (28S and 18S) (for details see Chapter 6). Note that 18S rRNAs range in size from 1.8 to 2.0 kb whereas 28S rRNAs range between 4.6 and 5.3 kb. Tracking dye bromophenol blue migrates slightly faster than 5S rRNA, and xylene cyanol migrates slightly slower than the 18S rRNA.

When a number of different samples are to be compared in northern analysis, then it is important to equalize the amounts of RNA loaded into lanes of northern gels. Equalizing the amounts of RNA in northern gel is a tricky and important issue (For details see Chapter 6).

Loading of equal amounts of poly (A)$^+$ RNA The amount of poly (A)$^+$ RNA can be compared by slot or dot–blot hybridization to a radiolabeled poly (dT) probe. Finally, equivalent amounts of poly (A)$^+$ RNA can be loaded into each lane of a northern gel.

Using a synthetic psneudomessage as a standard In some cases, RNAs synthesized *in vitro* can be utilized as externally added standards to calibrate the expression of the gene of interest in different preparations of cellular RNA. After that, the expression of the experimental gene is obtained by analyzing relative intensity of the hybridization signals obtained from the authentic and pseudomessages.

Transfer of RNA to membrane After electrophoresis, RNA is transferred to solid support. Initially, activated cellulose papers were used for immobilization of RNA and subsequently nitrocellulose paper was the support of choice for northern hybridization for several years. Since the introduction of various types of nylon membranes as solid support, nylon membranes are the most preferred solid supports for northern analysis. Transfer of denatured RNA from the gel to the surface of a membrane is a crucial step, which must be performed in a way that preserves the distribution of the molecules along the length of the gel. Although RNA transfer can be carried out by any of the procedures described in Section 16.5.2, the simple and economical technique used for northern blotting is upward capillary transfer.

Transfer to uncharged nylon membrane is performed at neutral pH with 10 X SSC or 20 X SSC. The RNA is then covalently linked to the matrix by the traditional method of baking under vacuum at 80°C for 2 hours, by heating in microwave (750–900°C) for 2–3 min, or by exposing the nylon membrane to UV irradiation at 254/312 nm. However, RNA can be efficiently transferred from agarose gels in 8.0 mM NaOH with 3 M NaCl to charged nylon membrane. Transfer under these conditions partially hydrolyzes the RNA and thereby increases the speed and efficiency of transfer of large (>2.3 kb) RNAs. Because the RNA transferred in alkaline solution becomes covalently attached, there is no need to bake the membrane or to expose it to UV irradiation before hybridization.

Hybridization with probe Northern hybridization is exceptionally flexible in that radiolabeled or nonisotopically labeled DNA, *in vitro* transcribed RNA and oligonucleotides can all be used as hybridization probes. Additionally, heterologous probes that contain sequences with only partial homology may also be used as probes. In spite of these advantages, there are limitations associated with northern analysis. First, if RNA samples are even slightly degraded, the quality of the data and the ability to measure expression are severely affected. Thus, it is essential to incorporate RNase-free reagents and techniques in northern analysis. Second, a standard northern procedure is less sensitive than nuclease protection assays and RT-PCR (for details see Chapter 7) although improvements in sensitivity can be achieved by using high specific activity antisense RNA probes, optimized hybridization buffers, and positively charged nylon membranes.

Reverse Northern Blot A variant of the procedure known as the reverse northern blot is an occasionally used technique. In this procedure, the substrate nucleic acid (that is affixed to the membrane) is a collection of isolated DNA fragments and the probe is RNA extracted from a tissue and radioactively labeled.

DNA and RNA Dot Blot

Dot blot (or slot blot) is a technique in molecular biology used to detect nucleic acids. It can replace Southern or northern blot. Note that dot blot can also replace western blotting to detect the presence or absence of particular protein by using antibody probes (similar to western blot). The procedure of dot blotting and hybridization is presented in Figure 16.18.

For several years, dot blotting and slot blotting were viewed with disfavor by many investigators chiefly because of variability in the hybridization signal obtained from identical samples applied to the same membrane, especially when analyzing complex populations of RNA or DNA. Although this problem has not been entirely solved, the advent of positively charged nylon membranes has led to a marked improvement in minimizing sample to sample variation. In the case of DNA, purified preparations or alkaline lysates of cells and tissue samples can be loaded onto the membrane under alkaline conditions. The technique offers significant savings in time as

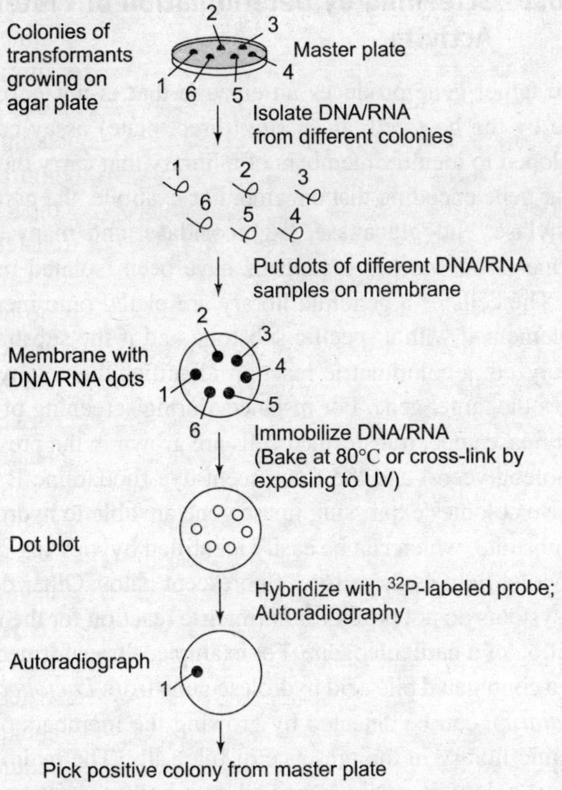

Figure 16.18 Technique of dot blot and hybridization

gel electrophoresis and complex transfer procedure are not required. However, it offers no information on the size of the biomolecule. Furthermore, two molecules of different sizes appear as a single dot, and hence cannot be detected. Dot blot only confirms the presence or absence of a biomolecule, which can be detected by the nucleic acid probes or the antibody. The amount of target sequence is estimated by comparing the intensity of signals emitted by dots containing the test samples with standards containing known concentrations of the target sequence.

Dot blot analysis of RNA is slightly trickier than dot blotting of DNA. At one stage, investigators experimented with dot and slot blots of crude cytoplasm prepared from freshly or frozen cultured cells or animal tissues. However, the results obtained from dot blotting of crude preparations of RNA do not always match those obtained from northern blots of purified RNA. For this reason, dot blotting and slot blotting are generally carried out with purified preparations of RNA denatured with glyoxal or formaldehyde.

Procedure of Dot Blotting and Hybridization In dot blot, the nucleic acid molecules to be detected are not separated by electrophoresis. Instead a mixture possibly containing the molecule to be detected is applied directly on a membrane as a dot. Thus, several preparations of nucleic acids are immobilized on the same membrane. This is then immediately followed by detection with nucleotide probes similar to northern blot and Southern blot.

Applying the samples and standards to the membrane Although samples of DNA/RNA can be applied to the membrane manually with an automatic pipetting device, the spacing and size of resulting dots are often variable. The resulting images may be blurred, misshaped, and quantitation will be hampered. Therefore, the preferred method of applying samples to a membrane is by vacuum manifold. Many of the commercially available manifolds are supplied with a choice of molds that deposit the samples on the membrane as dots or slots in various geometries. This ensures that all the immobilized samples have the same shape, area, and spacing, which facilitate comparison of the intensity of hybridization.

To obtain quantitative results, it is essential to include positive and negative controls that have physical properties similar to those of nucleic acid under test. For example, when analyzing RNAs, the negative control should consist of RNA from a cell or tissue that does not express the target sequences. Positive controls should consist of preparations of RNA from a cell or tissue that is mixed with known quantities of RNA standards that are complementary to the probe. These standards and radiolabeled probes are best synthesized *in vitro* from DNA templates that have been cloned into plasmid vectors in which two different bacteriophage promoters in opposite orientations flank the cloning site. Sense strand RNA, which is used as a hybridization standard, can be synthesized using one promoter; radiolabeled (antisense) probe can be synthesized using another promoter. When creating standards, the synthetic sense strand RNA is mixed with an irrelevant RNA so that the resulting mass is equal to that of the test samples. The irrelevant RNA should be prepared in the same manner as the test RNA.

Normalization To avoid overloading the membrane, not more than 5 μg of total RNA should be used in a slot of standard size. As a routine, the same amount of RNA is loaded into each slot. However, there is always some uncertainty about the actual amounts of RNA that are retained on the membrane. This problem can be solved by staining the membrane with methylene blue after the RNA has been cross-linked to the positively charged membrane by UV irradiation. Alternatively, the amount of poly $(A)^+$ RNA retained on the membrane can be measured by hybridization with radiolabeled oligo (dT). This method is especially useful when loading small amounts of purified poly $(A)^+$ RNA in each slot.

Hybridization and measuring the intensity of the signal Once the DNA/RNA samples are immobilized on the membrane, next step is hybridization with probes. For many purposes, visual assessment of the intensity of hybridization is sufficient. However, densitometric scanning, direct phosphor imaging, or luminometry can provide more accurate estimates of the amount of target sequence in each sample. Liquid scintillation counting also provides direct quantitation of the concentration of target DNA/RNA. However, this method requires that the sample be cut into pieces and placed in scintillation flour, thereby eliminating any possibility of reprobing the dot blots.

Reverse Dot Blot A variant of dot blot known as reverse dot blot is also devised for some experiments in which the probes are immobilized on the membrane and hybridized with the sample. In this way, only one hybridization step is required because each probe occupies a unique position on the membrane.

Zoo Blot Hybridization

Zoo blot hybridization is a type of an assay for DNA sequences that are highly conserved between species. Zoo blotting is not a distinct method but a variant of Southern blot that can be used for analysis of DNA of different species. This is based on the fact that eukaryotic genome contains both coding and noncoding base sequences and during evolution, the base sequence of noncoding region undergoes mutation and changes rapidly, whereas coding sequences change much more slowly and can easily be recognized after millions of years of divergence between two species. For example, zoo blot can be prepared from DNA of related mammalian species such as a human, monkey, mouse, rabbit, and cow, etc. DNA is then digested with suitable restriction enzyme and the fragments are separated on a gel and transferred to a nylon membrane. Thereafter, human DNA is labeled and used as a probe. DNA sample containing a coding sequence will definitely hybridize with some fragment of DNA from most other closely related animals. Basically, this procedure is used to detect similar or exact relationships between the DNA in question and from the other organisms and so the technique takes advantage of nonexact hybridization. Moreover, the locations of introns and exons can also be detected as the latter are far more conserved than the former.

16.6 PROTEIN STRUCTURE/FUNCTION-BASED TECHNIQUES

When a DNA library is constructed using expression vectors, it permits a range of alternative techniques to be employed, each of which exploits some structural or functional property of the gene product as each individual clone can be expressed to yield a polypeptide. These approaches are very important in cases where the DNA sequence of the target clone is completely unknown and enough information is not available to design a suitable probe or set of primers. Some of these strategies used for screening of expression libraries are presented in this section. Note that some of protein-based blotting techniques, for example, western, southwestern, and northwestern blotting, are also used for detection of specific proteins in any system.

16.6.1 Screening by Determination of Protein Activity

If the target gene produces an enzyme that is not normally made by the host cell, an *in situ* (direct plate) assay can be developed to identify members of a library that carry the particular gene encoding that enzyme. For example, the gene for α-amylase, endoglucanase, β-glucosidase, and many other enzymes from various organisms have been isolated in this way. The cells of a genomic library are plated onto medium supplemented with a specific substrate and if the substrate is hydrolyzed, a colorimetric reaction identifies the colony that carries the target gene. For instance, during screening of bacterial lipase gene, transformed cells are grown in the presence of trioleoglycerol and the fluorescent dye rhodamine B. The positive colonies expressing lipase gene are able to hydrolyze the substrate, which can be easily identified by viewing under ultraviolet light as these form fluorescent halos. Other detection systems do not provide colorimetric reaction for the identification of a particular gene. For example, a transformed cell with a conjugated bile acid hydrolase gene from *Lactobacillus planatarum* can be detected by growing the members of the genomic library in the presence of bile salts. The hydrolase-positive colony is easily identified as it becomes surrounded with a ring of precipitated free bile acids.

16.6.2 Screening by Functional Complementation

Functional complementation is powerful method of isolating genes that encode enzymes. It is defined as the process by which a particular DNA sequence compensates for a missing function in a mutant cell and thus restores the wild-type phenotype. This procedure involves transforming host cells that have particular genetic defect with plasmids of a DNA library constructed from a wild-type organism and selecting the transformed cells that function normally. The functional complementation can be homologous or heterologous, which means that the cloned gene may come from either the same or a different species, respectively. The functional complementation may be genetic or nutritional.

Functional (Genetic) Complementation

Earlier, insertional inactivation of antibiotic resistance or *cI* gene provided a convenient means of recombinant selection. This is based on the fact that the insertion of a foreign gene into the cloning site within the antibiotic resistance or *cI* gene disrupts its reading frame, and hence loss of function. Thus, the recombinants lose the function that is exhibited by the nonrecombinants. In simple terms, in case the cloning is done within the antibiotic resistance gene, the recombinants become antibiotic sensitive, while nonrecombinants are antibiotic resistant. This is a common strategy for pBR322 (for details see Chapter 8). However, several other systems are now available for screening of positive clones.

α-Complementation of β-Galactosidase (Blue-white Screening) The most convenient and widely used phenomenon for screening is α-complementation of β-galactosidase. In this complementation system, two inactive fragments of *E. coli* β-galactosidase combines with each other and produces a functional enzyme. The 5′ region of the *lac* Z gene encoding the initiating methionine residues is deleted that results in a translation to initiate at a downstream methionine residue, producing a C-terminal fragment of the enzyme (the ω or α-acceptor fragment). An N-terminal fragment (the α-donor fragment or α-peptide) is generated by deletion or chain-terminating mutations in the structural gene. The resulting α-donor fragment and ω-acceptor fragment are enzymatically inactive, however, the two parts of the enzyme can associate to form an active β-galactosidase enzyme *in vivo* as well *in vitro*. Although various α-donor fragments of different lengths are functional for complementation, the minimum requirement is reported to be α-peptide containing amino acid residues 11–41.

The α-complementation system is explained by analyzing the 3-D structure of *E. coli* β-galactosidase. This enzyme is a tetramer comprised of four identical monomers of 1,023 amino acids, which are folded into five sequential domains. The complementation region (the N-terminal segment of the monomer) directly contributes to the interfacial association among monomers that associate with domains 1, 2, and 3 of its own monomer. Moreover, the complementation peptide also stabilizes an interfacial four-helix bundle composed of two helices from each of the two monomers.

The vector DNA usually carries a short segment of *E. coli lac* Z gene along with its regulatory sequence (*lac* I) and the coding information for the first 146 amino acids of the β-galactosidase gene. Vectors of this type are propagated in host cells that express the C-terminal portion of the β-galactosidase. Although neither the host encoded nor the plasmid-encoded fragments of β-galactosidase are active, these can associate to form as enzymatically active enzyme. This type of complementation, in which deletion mutants of the operator-proximal segment of *lac* Z gene (*lac* ZΔM15) are complemented with β-galactosidase negative mutants that have the operator-proximal region intact (*lac* Z′), is called α-complementation. The *lac* Z^+ bacteria that result from α-complementation are easily recognized because these form blue colonies in the presence of the chromogenic substrate X-gal (5-bromo-4-chloro-3-indolyl-β-*D*-galactoside). This compound is colorless but when cleaved by β-galactosidase a blue indigo derivative is released (Figure 16.19).

In order to utilize this unique protein for cloning, a polylinker is inserted into the *lac* Z′ gene without disrupting the reading frame in the vector; note that the small addition does not affect the enzyme. The gene of interest is then cloned into this polylinker site. positioned in *lac* Z gene of the plasmid; note that the insertion of the foreign gene almost invariably results in the production of an N-terminal fragment that is no longer capable of α-complementation. After cloning the gene of interest, the plasmid vector is introduced into the host cell. The bacterial host strains, which are used for α-complementation, carry the ω-fragment encoded by the deletion mutant *lac* ZΔM15, lacking 11–41 amino acid residues of β-galactosidase, which is usually carried on an F′ plasmid. For screening, the transformants are plated on media containing the substrate X-gal. As strains of bacteria commonly used for α-complementation do not synthesize significant quantities of lac repressor, there is no need to induce synthesis of ω and α fragments. However, if required, the synthesis of both fragments can be fully induced by IPTG (isopropyl-β-*D*-thiogalactoside), a nonfermentable lactose analog that inactivates the lac repressor. If the *lac* Z gene remains intact after digestion and ligation, blue colony is obtained whereas the recombinant colonies are white because of insertional inactivation. This screening procedure is referred as blue-white screening. Using this procedure, many thousands of putative recombinant clones are easily screened, which can be further verified by restriction analysis of minipreps or by colony PCR (for details see Chapters 3 and 7, respectively).

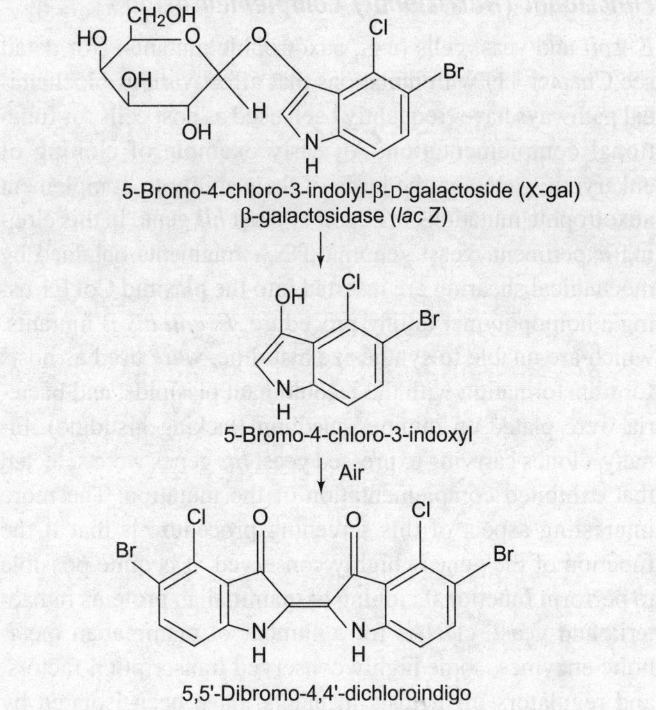

Figure 16.19 Principle of blue-white screening

Red-white Screening Red–white screening method is another example of functional (genetic) complementation based on *sup*$^+$ vector and *ade*$^-$ host strains of yeast cells (for details see Chapter 11).

Functional (Nutritional) Complementation

E. coli and yeast cells (e.g., auxotrophic mutants) (for detail see Chapter 11) with mutations that affect various biochemical pathways have frequently been used as host cells for functional complementation. An early example of cloning of eukaryotic genes on the basis of their ability to complement auxotrophic mutations in *E. coli* is yeast *his* gene. In this cloning experiment, yeast genomic DNA fragments obtained by mechanical shearing are inserted into the plasmid Col E1 using a homopolymer tailing procedure. *E. coli his* B mutants, which are unable to synthesize histidine, were used as hosts for transformation with the recombinant plasmids, and bacteria were plated on minimal medium (lacking histidine). Finally, clones carrying expressed yeast *his* genes were selected that exhibited complementation of the mutation. The more interesting aspect of this screening procedure is that if the function of the gene is highly conserved, it is quite possible to perform functional cloning of mammalian proteins in bacteria and yeast. cDNAs for a number of mammalian metabolic enzymes, some highly conserved transcription factors, and regulators of meiosis in plants, have been isolated by functional complementation in yeast. Most commonly, a pool system is employed where cells are transfected with a complex mixture of up to 1,00,000 clones. Thereafter, the pools that successfully complement the mutant phenotype are then subdivided for a further round of transfection and the procedure is repeated until the individual cDNA is isolated. Functional complementation is also possible in transgenic animals and plants. The advantage of using this screening procedure is that no sequence information is required; without the functional assay, a mouse or a human gene is only identified through a laborious chromosome walk from a linked marker.

16.6.3 Screening by Gain of Function

In many cases, the function of the target gene is too specific, which does not work in a bacterial or yeast expression host and even in a higher eukaryotic system. Therefore, an alternative method has been devised in which positive clones can be identified on the basis of gain of function conferred on the host cell. This gain of function may be a selectable phenotype that permits cells having the corresponding clone to be positively selected. It has been reported in early experiments where the mammalian *dhfr* gene was expressed in *E. coli* host cells, a population of recombinant plasmids containing cDNA derived from unfractionated mouse mRNA were constructed. This population of mRNA molecules was used to identify clones that express *dhfr* gene encoding enzyme dihydrofolate reductase (DHFR). As mouse DHFR is much less sensitive to inhibition by the drug trimethoprim as compared to *E. coli* DHFR, growing transformants in medium containing the drug permitted selection for those cells containing the mouse *dhfr* cDNA. The clone of interest can also be detected on the basis of visible appearance in phenotype. For example, clones of cellular oncogenes have been identified on the basis of their ability to stimulate the propagation of quiescent mouse fibroblast cells either in culture or after transplantation into mice.

Furthermore, various specific assays have also been devised for the functional cloning of cDNAs encoding particular types of gene product. For instance, *Xenopus* melanophores (dark cells containing melanosomes) have been used for the functional cloning of G-protein-coupled receptors or tyrosine kinase receptors. These organelles exhibit a useful characteristic as these disperse when adenyl cyclase or phospholipase C are active and aggregate when these enzymes are inhibited. Thus, the expression of cDNAs encoding G-protein-coupled receptors and tyrosine kinase receptors directs the redistribution of pigmentation within the cell, which can be used as an assay for the identification of receptor cDNAs.

Gain of function strategy of screening can also be employed for analysis of transgenes.

16.6.4 Phage display

In phage display system, a library of proteins or peptides is expressed as fusion proteins with a surface-displayed coat protein of a filamentous phage such as M13. This technique is called phage display because it involves the display of proteins on the surface of M13 bacteriophage. Proteins expressed in this manner can be probed using strategies similar to those utilized for screening conventional bacterial expression libraries. Two types of vector systems are used for phage display: (i) First is the bacteriophage display system, in which a segment of foreign DNA is inserted into gene *III* or gene *VIII* of M13 (for details see Chapter 10), a few nucleotides downstream from the cleavage site. Thereafter, *E. coli* cells are transfected with the recombinant viral DNA that synthesize and secrete a pure population of infectious fusion phage particles, and the amino acid encoded by the foreign DNA are displayed on their surface. Every copy of pIII or pVIII on the surface of an infectious bacteriophage particle carries the sequences encoded by the foreign DNA, which are therefore displayed in a densely packed multivalent fashion; and (ii) Second the phagemid display system consists of a plasmid that carries a single copy of gene *III* or gene *VIII* and the filamentous phage origin of DNA replication. In this system, a segment of foreign DNA is inserted into gene *III* or gene *VIII* just downstream from the cleavage site that separates the hydrophobic signal sequence from the mature protein. The recombinant plasmid is then used to transform an appropriate strain of *E. coli.* Thereafter, bacteriophage particles displaying the amino acid sequences encoded by the segment of foreign DNA are obtained by superinfecting the transformed cells with helper phages. Specialized phagemid display vectors having an amber (UAG) chain-terminating mutation immediately downstream from the inserted segment of foreign DNA and upstream of the body of pIII or pVIII are commercially available. When the recombinant phagemid is used to transform nonsuppressing strains of *E. coli* (e.g., HB2151), the protein encoded by the foreign DNA terminates at the amber codon and is secreted into the culture medium. In the case of phagemids that carry fragment of immunoglobulin genes, the supernatant medium from individual suppressor-minus (*sup*$^-$) transformants can be screened for soluble antibody fragments with the capacity to bind antigen. However, when the phagemid is used to transform cells expressing an amber suppressor, the entire fusion protein is synthesized and fragments of antibody are displayed on the surface of secreted bacteriophage particles in the normal way.

It is possible to generate large libraries of recombinant bacteriophages that display millions of peptides by inserting into gene *III* or *VIII* populations of synthetic oligonucleotides that differ in sequence but are of equal length. Bacteriophage carrying a specific target peptide for a receptor, an antibody or another protein can be purified from the alternating rounds of affinity enrichment and bacteriophage growth. Since the gene encoding the target peptide is part of the bacteriophage genome, bacteriophage peptide display can be used to isolate a specific DNA sequence by affinity selection of its protein product. Phage display is used: (i) To identify and analyze features on the surface of proteins; (ii) To isolate new ligands that bind to particular amino sequences; and (iii) To improve the affinity and specificity of the interaction between ligands and their target structures. The strength of the phage display system results from the powerful combination of affinity selection and biological amplification. Furthermore, bacteriophage peptide display facilitates epitope mapping, analysis of protein–protein interaction and the isolation of inhibitors, agonists, and antagonists.

16.6.5 Membrane Immobilization or Blotting Techniques

Similar to nucleic acid blotting, protein blotting techniques are also used for detection of proteins. These include immunological screening, plus–minus screening, western blotting, southwestern blotting, and northwestern blotting. The details of these procedures are discussed below.

Immunological Screening

Immunological screening involves the use of antibodies that specifically recognize antigenic determinants on the polypeptide synthesized by a target clone. This is one of the most versatile expression cloning strategies because it can be used for any protein provided an antibody is available. Moreover, if a cloned DNA sequence is transcribed and translated, the presence of the protein, or even part of it, can be determined by an immunological assay. For this type of assay, there is also no need for that protein to be functional. Basically, the molecular target for recognition is an epitope, a short sequence of amino acids that folds into a particular 3-D conformation on the surface of the protein. Epitopes can fold independently from the rest of the protein and even when the polypeptide chain is incomplete or when expressed as a fusion with another protein. More importantly, it is possible to generate many epitopes under denaturing conditions when the overall conformation of the protein is abnormal. The first immunological screening techniques were developed in the late 1970s, when expression libraries were generally constructed using plasmid vectors. The first devised method was based on two facts that (i) Certain types of plastics, such as polyvinyl can very strongly adsorb antibodies; and (ii) IgG antibodies can be readily labeled with ^{125}I by iodination *in vitro*. During screening, transformed cells were plated out on petridishes and resulting colonies were lysed, e.g., by using chloroform vapor or by spraying with an aerosol of virulent phage. Note that a replica plate should be maintained as this procedure kills the bacteria. A sheet of polyvinyl already coated with the appropriate antibody was then applied to the surface of the plate, allowing antigen–antibody complexes to form. The sheet was then removed and exposed to ^{125}I-labeled IgG specific to a different determinant. Note that this determinant was not involved in the initial binding of the antigen to the antibody-coated polyvinyl sheet (Figure 16.20). The sheet was then washed and exposed to the X-ray film. Finally, the positive clones identified by this procedure were isolated from the replica plate. This type of sandwich immunological technique can be used only where two antibodies recognizing different determinants of the same protein are available. In the case of fusion protein, antibodies that bind to each component of the fusion can be used efficiently selecting for recombinant molecules.

Later, λ insertion vectors were subjected to expression screening by immunological assay. A simplified procedure was adopted by which recombinant phages were directly screened. In this method, the library was plated out at moderately high density with *E. coli* Y1090 as the host. Note that

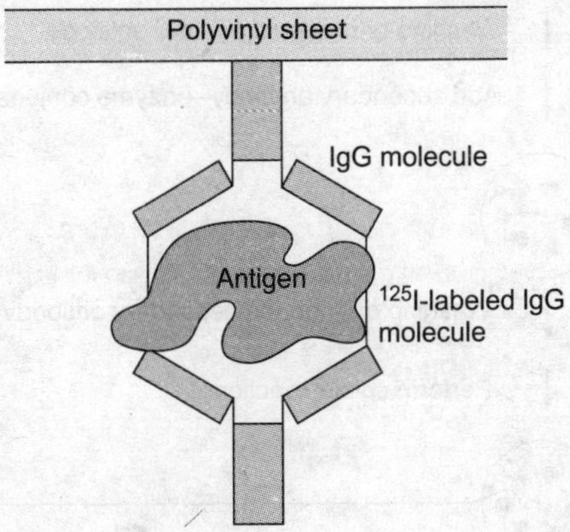

Figure 16.20 Antigen–antibody complex formation in the immunochemical detection

Y1090 overproduces the *lac* repressor and ensures that no expression of cloned gene takes place until the inducer IPTG is added to the infected cells. This strain is also deficient in the lon protease, consequently increasing the stability of recombinant fusion proteins. The expressed fusion proteins in plaques were absorbed onto a nitrocellulose membrane previously soaked with IPTG and after plaque lift, this membrane was processed for antibody screening. After identifying a positive signal on the membrane, the positive plaque was picked from the original agar plate and the recombinant phage DNA was isolated and further characterized.

In recent times, the original detection method using iodinated antibodies has been superseded by more convenient methods using nonisotopic labels, which are also more sensitive and have a lower background of nonspecific signal. Theoretically, this procedure is very similar to DNA hybridization assay (Figure 16.21). All the clones of the library are grown separately on master plates. A sample of each colony is transferred to a known position on a solid support by replica plating where the cells are lysed and the released proteins are attached to the matrix. The membrane with the bound proteins is treated with an antibody (primary antibody) that specifically binds to the protein encoded by target gene. Note that the recombinants contain either an intact gene or portion of the gene that is large enough to produce a protein product to be recognized by the primary antibody. After the interaction of the primary antibody with the target protein (antigen), any unbound antibody is washed away and the membrane is treated with a second antibody that is specific for the primary antibody (secondary antibody). The secondary antibody recognizes the species-specific constant region of the primary antibody and is conjugated to either HRP or alkaline phosphatase, each of which can in turn be detected using a simple colorimetric assay carried out directly on the membrane. This method eliminates the need for isotopes and also incorporates an amplification step since two or more secondary antibodies bind to the primary antibody. The positive colonies are then picked up from the master plate for further characterization.

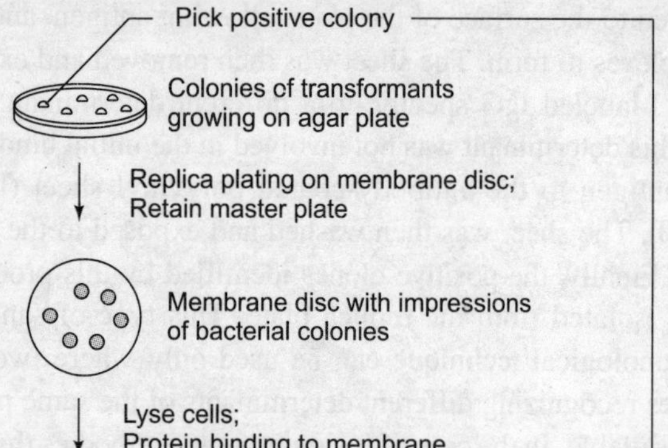

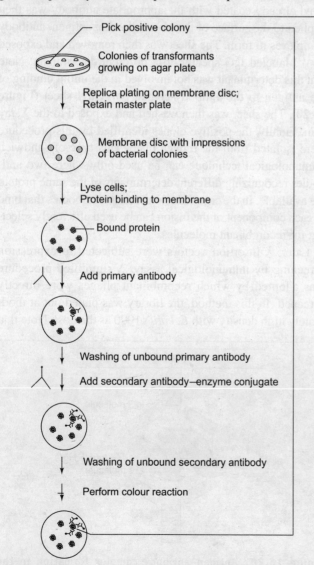

Figure 16.21 Immunological screening of a gene library

Plus–Minus Screening (Differential Screening)

Plus–minus screening, also known as differential screening, involves construction of cDNA library from cells expressing the gene of interest. Duplicate copies of the library are then screened separately with labeled cDNA probes synthesized from two preparations of mRNA. First, mRNAs are isolated from a cell type or tissue that expresses the gene(s) of interest in high abundance and the second from a cell type or tissue with low abundance of the target mRNA(s). Recombinant clones that hybridize equally to both probes that correspond to genes equivalently expressed in both cell types are not studied further. Recombinant clones that hybridize differentially to the two probes are isolated and analyzed by northern hybridization to determine whether these contain cDNAs derived from differentially expressed mRNAs.

Take an example of the isolation of cDNAs derived from mRNAs, which are abundant in the gastrula embryo of the frog *Xenopus* but which are absent, or present at low abundance, in the egg (Dworkin and Dawid, 1980). A cDNA clone library is prepared from gastrula mRNA. Replica membranes carrying identical sets of recombinant clones are then prepared (Figure 16.22). One of these membranes is probed with ^{32}P-labeled mRNA (or cDNA) from gastrula embryos, and one with ^{32}P-labeled mRNA (or cDNA) from the egg. Some colonies that give a positive signal with both probes represent cDNAs derived from mRNA types that are abundant at both stages of development. Some colonies that do not give a positive signal with either probe correspond to mRNA types present at undetectably low abundance in both tissues. Note this is a feature of using probes derived

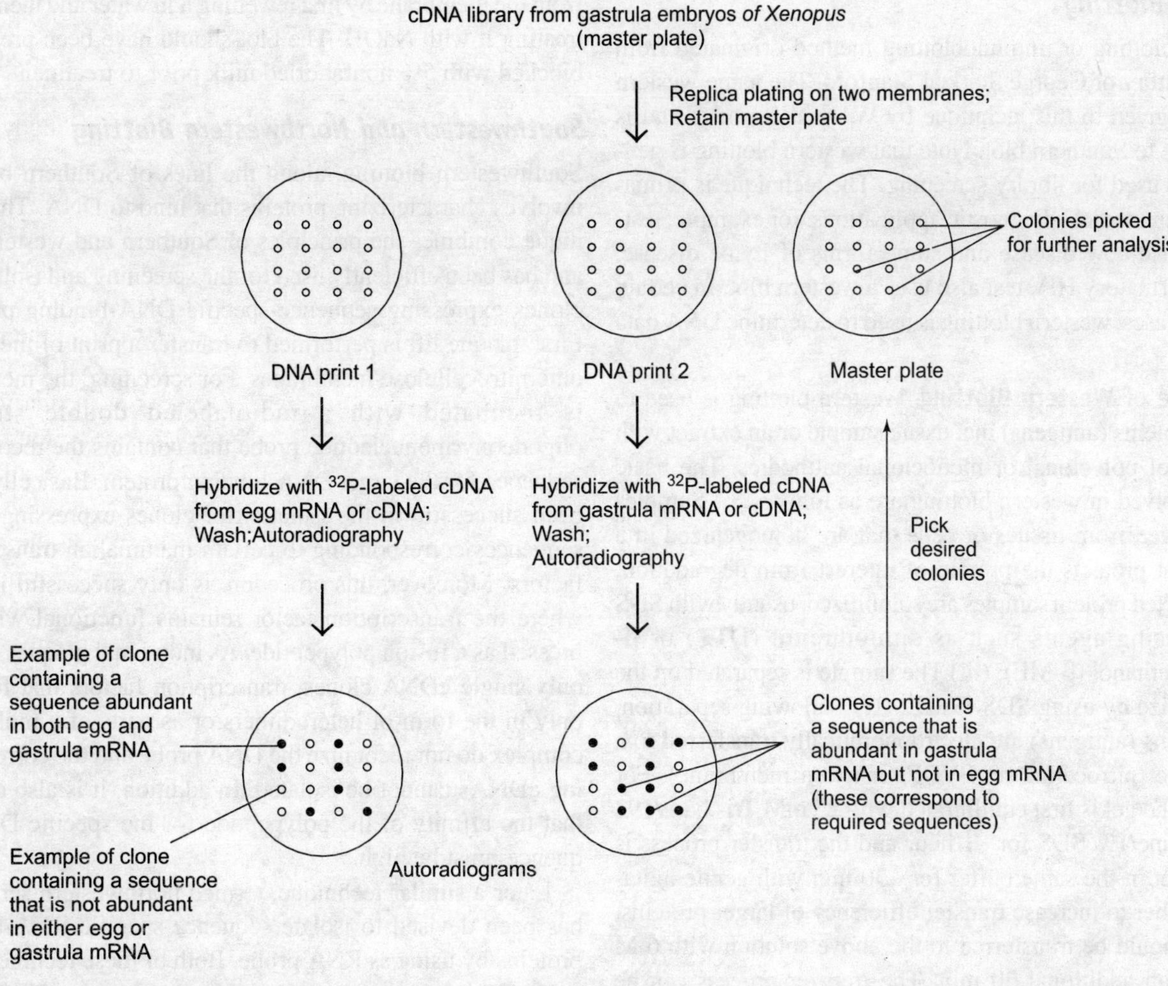

Figure 16.22 Plus-minus screening

from mRNA populations; only abundant or moderately abundant sequences in the probe carry a significant proportion of the label and are effective in hybridization. Importantly, some colonies give a positive signal with the gastrula probe, but not with the egg probe; these correspond to the required sequences. This procedure allows detection of 5-fold differences in mRNA abundance.

Plus-minus screening and subtractive library (for details see Chapter 15) screening are the methods of choice for the differential analysis of temporal and topographical expression patterns. Principally, plus-minus screening method of comparative screening can be used to identify differentially regulated genes. As cDNA probes are copied from the entire population of mRNAs, these probes used in differential screening are of very high complexity, with each mRNA represented in proportion to its abundance. Therefore, plus-minus screening is successful only when the concentration of the target mRNAs in the two preparations differs by factor of 5 or more and when the abundance of the mRNAs in one of the preparations exceeds 0.05%. The success of differential screening depends on three important points: (i) Depth and quality of the cDNA library; (ii) Use of probes of high sequence complexity, which are capable to detect cDNAs corresponding to low-abundance mRNAs; and (iii) The ability to capture and quantify hybridization data efficiently. Moreover, the efficiency of the technique is greatly enhanced if the cDNA libraries are organized into a format such as wells of microtiter dishes, which is amenable to automation. There are reports of identification of more than 100 differentially expressed genes by using some form of plus-minus screening with laborious approach. With the introduction of automated method, more complex and more efficient forms of differential screening have been developed using densely arrayed libraries of cDNAs or synthetic oligonucleotides. The arrays are probed with cDNAs that are prepared from two different sources and labeled with different chromophores. Finally, an image of the pattern of hybridization of the cDNAs is captured electronically and normalized by computer to compare the efficiency of hybridization of each of the probes to each of the immobilized targets. When differentially expressed genes have been identified, full-length cDNAs can be obtained from a repository.

Western Blotting

Western blotting or immunoblotting method originated from the laboratory of George Stark at Stanford. The name western blot was given to this technique by W. Neal Burnette and is analogous to Southern blot. Note that western blotting is generally not used for library screening. The technique is primarily used in medical diagnostic applications, for example, testing for mad-cow disease and some forms of Lyme disease. The confirmatory HIV test also uses a western blot. In certain forensic cases, western blotting is used to determine DNA data as well.

Procedure of Western Blotting Western blotting is used to detect proteins (antigens) in a tissue sample or an extract with the help of polyclonal or monoclonal antibodies. The basic steps involved in western blotting are as follows: (i) Samples are prepared from tissues or cells that are homogenized in a buffer that protects the protein of interest from degradation; (ii) Extracted protein samples are solubilized, usually with SDS and reducing agents such as dithiothreitol (DTT) or β-mercaptoethanol (β-ME); (iii) The sample is separated on the basis of size by using SDS-PAGE; (iv) Following separation, the proteins (antigens) are electrophoretically transferred to a membrane (nitrocellulose, PVDF, or nylon membrane). For transfer, the gel is first equilibrated with 25 mM Tris base/192 mM glycine/1% SDS for ~1 hour and the transfer process is carried out in the same buffer for ~30 min with gentle agitation. Further to increase transfer efficiency of larger proteins, the gel should be transferred to the above solution with 6 M urea for an additional 30 min. The transfer process can be monitored by reversible staining or by Ponceau S staining; (v) The blotted membrane is placed in heat-sealable bag or plastic incubation tray with 5 ml blocking buffer [TBS (Tris-buffered saline) for nylon membrane and TTBS (0.1% v/v Tween 20 in TBS) for PVDF and nitrocellulose membrane] and incubated for 30 min to 1 hour at room temperature with agitation on an orbital shaker or rocking platform. All remaining nonspecific binding sites are blocked by immersing the membrane in a solution containing either a protein (casein or nonfat dried milk) or detergent (Tween 20); (vi) Thereafter, the membrane is incubated with diluted primary antibody (1/100 to 1/1,000 for a polyclonal antibody; 1/10 to 1/100 for hybridoma supernatants) prepared in TBS or TTBS for 30 min to 1 hour at room temperature with constant agitation; (vii) After probing with the primary antibody, the membrane is washed and conjugated secondary anti-IgG antibody (e.g., goat anti-rabbit IgG) is added to the antibody–antigen complexes. Secondary antibody is conjugated with HRP or alkaline phosphatase enzymes that are attached directly or via an avidin-biotin bridge; (viii) Chromogenic or luminescent substrates are then used to visualize the enzyme activity, and hence to detect antigen-antibody complex; and (ix) Finally, membranes may be stripped and reprobed. For stripping, all residual antibodies are removed from the membrane by first rewetting it in water and then briefly treating it with NaOH. The blot should have been previously blocked with 5% nonfat dried milk prior to treatment.

Southwestern and Northwestern Blotting

Southwestern blotting, along the lines of Southern blotting, involves characterizing proteins that bind to DNA. This technique combines the principles of Southern and western blots and has been efficiently used for the screening and isolation of clones expressing sequence-specific DNA-binding proteins. First, plaque lift is performed to transfer a print of the library onto nitrocellulose membranes. For screening, the membrane is incubated with a radiolabeled double stranded oligodeoxyribonucleotide probe that contains the recognition sequence for the target DNA-binding protein. Basically, it has been successful in the isolation of clones expressing cDNA sequences corresponding to certain mammalian transcription factors. Moreover, this procedure is only successful in cases where the transcription factor remains functional when expressed as a fusion polypeptide. As individual plaques contain only single cDNA clones, transcription factors that function only in the form of heterodimers or as parts of a multimeric complex do not recognize the DNA probe and the corresponding cDNAs cannot be isolated. In addition, it is also obvious that the affinity of the polypeptide for the specific DNA sequence must be high.

Later a similar technique, termed northwestern screening, has been devised to isolate sequence-specific RNA-binding proteins by using ss RNA probe. Both of these techniques are most efficient when the oligonucleotide contains the binding sequence in multimeric form. Thus, several fusion polypeptides on the membrane should bind to each probe, thereby greatly increasing the average dissociation time. In both cases, along with the specific probe, a large excess of unlabeled ds DNA (or ss RNA) is added to minimize nonspecific binding. It is usually indispensable to confirm the specificity of binding in secondary screening with specific oligonucleotide probe and one or more alternative negative probes.

16.6.6 Methods Based on Cell-free Translation

Hybrid release translation (HRT) and hybrid arrest translation (HART) are two related techniques that are used for characterization of the translation product encoded by a cloned gene. The principles of these techniques are based on the ability of purified mRNA to direct synthesis of proteins in cell-free translation system containing ribosomes, tRNAs, and all the other molecules needed for active protein synthesis. Usually two types of cell-free translation systems are used: cell extracts prepared from germinating wheat seeds and rabbit reticulocyte cells. When the purified mRNA sample is added to the cell-free translation system along with a mixture of 20 amino acids found in proteins, one of which is labeled (preferably

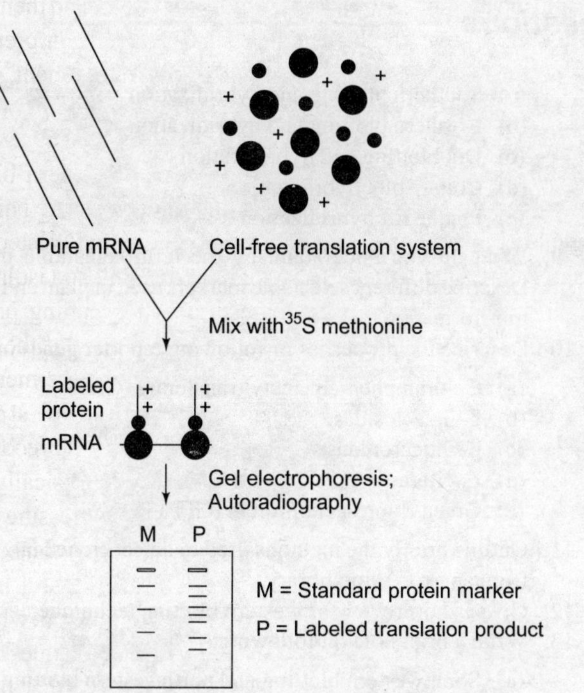

Figure 16.23 Principle of cell-free translation system

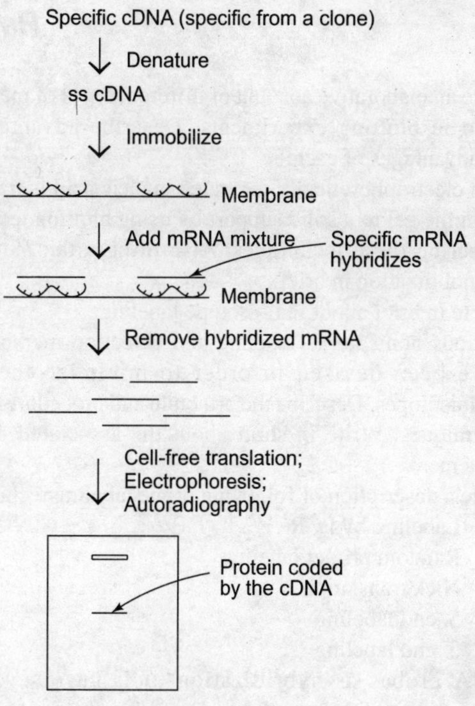

Figure 16.24 Hybrid release translation

^{35}S-methionine), the different mRNA molecules are translated into a mixture of radioactive proteins (Figure 16.23), which can be separated by gel electrophoresis and visualized by autoradiography. The individual band thus obtained represents a single protein coded by one of the mRNA molecules present in the sample.

In both techniques, best results are obtained when a cDNA clone prepared directly from the mRNA sample is available.

Hybrid Release Translation

In the case of HRT, the cDNA is denatured, immobilized on a nitrocellulose or nylon membrane, and incubated with the mRNA sample (Figure 16.24). The specific mRNA complementary of the cDNA hybridizes and remains attached to the membrane. Afterward, the unbound molecules are removed by washing and the hybridized mRNA is recovered and translated in a cell-free system. This provides a pure sample of the protein coded by the cDNA.

Hybrid Arrest Translation

HART is slightly different from HRT because the denatured cDNA is added directly to the mRNA sample (Figure 16.25). Hybridization occurs between the cDNA and its complementary mRNA, but here the unbound mRNA is not discarded. Instead after hybridization, the entire sample is translated in the cell-free system. The resulting hybridized mRNA is unable to direct translation, so all the proteins except the one coded by the cloned gene are synthesized. Finally, the cloned gene's translation product is identified as the protein that is absent from the autoradiograph.

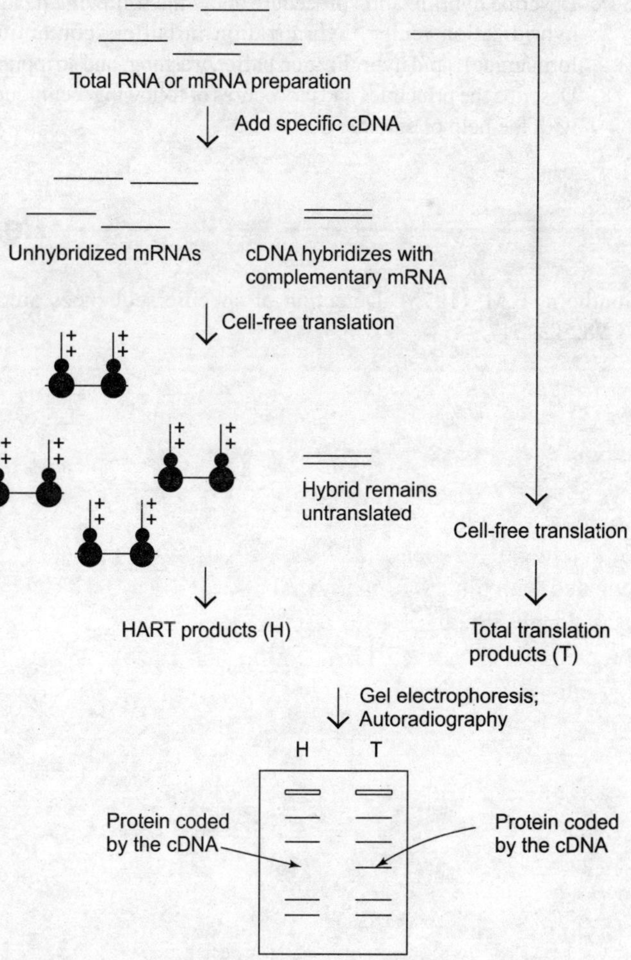

Figure 16.25 Hybrid arrest translation

Review Questions

1. Give an elaborative account of different types of membranes used in blotting experiments. Describe advantages and disadvantages of each.
2. The electrophoretically separated nucleic acid is transferred from the gel to a solid support by using blotting techniques. Describe these techniques. Also discuss the methods of immobilization in brief.
3. Write in brief about radioisotope labeling.
4. Various nonisotopic labeling and detection methodologies have been devised in order to minimize the use of radioisotopes. Describe the principle and procedures of these techniques. Write in short about the associated detection systems.
5. Give a description of following probe labeling techniques:
   (a) Labeling by PCR
   (b) Random primer labeling
   (c) Nick translation
   (d) 5′ end labeling
   (e) 3′ end labeling
6. RNA probes as hybridization tools have several key advantages. Discuss them. Describe the principle and procedure of RNA labeling by *in vitro* transcription.
7. Describe hybridization procedure under the following heads: hybridization recipe, hybridization in buffers containing formamide, rapid hybridization buffer, washing, and stripping.
8. Describe the principles and procedures of following techniques with the help of suitable diagrams:
   (a) Southern blotting and hybridization
   (b) Northern blotting and hybridization
   (c) Dot blotting and hybridization
   (d) Colony blot hybridization
   (e) Plaque lift hybridization
9. What do you understand by the term selectable marker? Describe different selectable markers used in plant and animal transformation.
10. Describe the properties of following reporter genes in brief:
    (a) Chloramphenicol acetyltransferase (CAT)
    (b) β-Galactosidase
    (c) β-Glucuronidase
    (d) Luciferase
    (e) Green fluorescent protein (GFP)
11. Outline briefly the methods used to detect cloned target gene from an expression library.
12. Give a short review of western blotting technique.
13. Write a brief note on following:
    (a) Southwestern blotting and northwestern blotting
    (b) Hybrid arrest translation (HART) and hybrid release translation (HRT)
    (c) Plus-minus screening
    (d) Autoradiography
    (e) Positional and functional cloning
    (f) Phosphorimager

Reference

Southern, E.M. (1975). Detection of specific sequences among DNA fragments separated by gel electrophoresis. *J. Mol. Biol.* **98**:503–517.

Part IV

Applied Genetic Engineering

- Other Techniques for Genetic Manipulation
- Molecular Markers
- Techniques Used in Genomics and Proteomics
- Applications of Cloning and Gene Transfer Technology
- Safety Regulations Related to Genetic Engineering

Other Techniques for Genetic Manipulation

Key Concepts

In this chapter we will learn the following:

- *In vitro* mutagenesis
- Gene manipulation by recombination
- Techniques based on gene silencing

17.1 INTRODUCTION

Various novel techniques used in the field of genetic engineering have made possible the construction of new designer genes with specific properties and their expression in specific host organisms, determination of the biological function of a protein encoded by a particular gene, specific targeting of gene in the host organisms, specific manipulation of endogenous gene as well as transgene in a regulated manner, and many other functionalities. The recombinant DNA technology for genetic manipulation have already been discussed in Chapter 13. Other techniques such as *in vitro* mutagenesis, recombination, antisense RNA technology, RNA interference (RNAi), cosuppression, and ribozyme technology also play significant roles in genetic manipulation. Thus, *in vitro* mutagenesis techniques can be employed to determine the biological function of a gene-encoded protein. This involves intentional mutation of a gene in a directed manner and evaluation of its effect. A diverse type of *in vitro* mutagenesis methods can also be used to improve the physical and chemical properties of proteins. Combinations of gene targeting, site-specific recombination, and inducible transgene expression are also being exploited for specific manipulation of endogenous gene as well as transgene in a regulated manner. Apart from these, several other methods for regulated gene silencing have been explored for use in genetic manipulation. In this direction, RNA engineering has been shown to be a potent strategy for the development of molecular tools that have potential for a variety of applications in genetic engineering. For example, antisense RNA, RNAi, cosuppression as well as a number of ribozymes have been used to silence undesired gene expression, however, RNAi dominates the field.

17.2 *IN VITRO* MUTAGENESIS

In vitro mutagenesis is used at present as a basic tool for gene manipulation techniques. This technique allows different types of alterations of base sequences of a segment of DNA. The methods for *in vitro* mutagenesis fall into two categories: random (or nonspecific) and targeted (or site-specific or localized). The random mutagenesis methods are used to analyze regulatory regions of genes. On the other hand, site-directed mutagenesis methods, which allow defined alteration in a gene sequence under *in vitro* conditions, are used to understand the contribution of amino acid(s) in a protein. Methods involving site-directed mutagenesis also find application in protein engineering to change the structure and possibly the activity of a protein. Both types of methods generate mutants *in vitro* without the need of phenotypic selection. Initially, site-directed mutagenesis was considered to be a difficult and time-consuming process, but with the development of different procedures of recombinant DNA technology such as polymerase chain reaction (PCR), DNA synthesizer, and DNA hybridization techniques, specific mutations can be generated in a more rapid, precise, and efficient way. Some of the important methods of *in vitro* mutagenesis are described in the following sections.

17.2.1 Oligonucleotide-directed Mutagenesis

In 1978, an important breakthrough in the field of genetic engineering was reported, which described *in vitro* synthesis of mutant DNA by using oligonucleotides. This oligonucleotide-directed mutagenesis developed by British-born Canadian biochemist Michael Smith (winner of the 1993 Nobel Prize for Chemistry with Kary Mullis) is central to all the available techniques of site-directed mutagenesis.

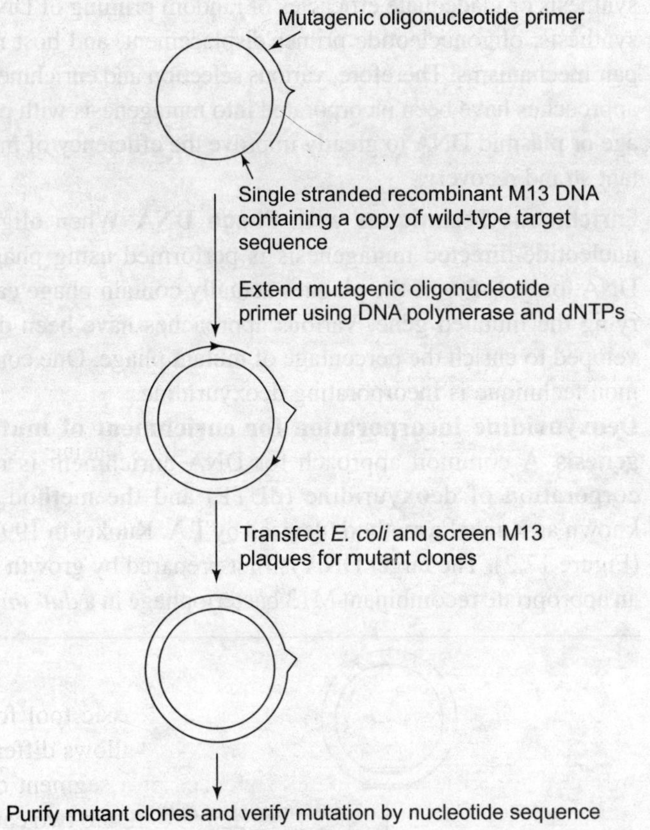

Figure 17.1 Generalized scheme for oligonucleotide-directed mutagenesis

All the methods devised for oligonucleotide-directed mutagenesis are based on the same principle and involves the cloning of the gene followed by the introduction of new coding information into the cloned genes. The principle is outlined below and presented in Figure 17.1.

- For oligonucleotide-directed mutagenesis, the information of precise nucleotide sequence in the region of DNA encoding the mRNA codon that is to be altered should be known.
- Based on the information of DNA sequence to be mutated, synthetic oligonucleotide complementary to the relevant region of the gene but containing the desired mutation is prepared. This oligonucleotide anneals to the target region of the wild-type template DNA and acts as a primer.
- A DNA polymerase is then used to extend the primer and finally double stranded DNA (ds DNA) is generated that carries the desired mutation.
- The mutated DNA is then inserted at the appropriate location of the target gene and the mutant protein is expressed. Large numbers of novel proteins thus generated are then assayed to determine whether a particular property has been created.

The procedure for oligonucleotide-mediated mutagenesis typically involves following sequence of steps: (i) Clone the target DNA into M13 phage or phagemid vector; (ii) Design and synthesize mutagenic oligonucleotide; (iii) Hybridize mutagenic oligonucleotides to the already cloned target DNA; (iv) Extend hybridized oligonucleotide in presence of all four dNTPs by DNA polymerase; (v) Ligate circular DNA with T4 DNA ligase; (vi) Insert circular DNA into bacteria by transfection; (vii) Screen resulting clones by hybridization technique using mutagenic oligonucleotide as probe; (viii) Prepare single stranded DNA (ss DNA) from the mutagenized clones; (ix) Sequence target DNA for confirmation of desired mutation; and (x) Recover mutated fragment of DNA.

The technique has enabled scientists to analyze the structure of gene and redesign them by inducing specifically tailored mutations at defined sites within a gene in a very straightforward manner. The technique is used to test the role of particular amino acid residues in the structure, catalytic activity, and ligand-binding capacity of the protein. The determination of specific amino acids in a protein that should be changed to attain a specific property is easier if the 3-D structure of the protein is well-characterized by analytical procedures. However for many proteins, such detailed information is lacking, so directed mutagenesis is a trial and error approach in which most probable nucleotides are altered to yield a particular property. The protein encoded by each mutated gene is tested to establish whether or not the mutagenesis process has certainly generated the desired change. The success of this technique depends on the availability of reliable assay techniques. And finally, if reliable assay system is available to confirm the result of mutagenized protein, new protein can be engineered, functions of proteins can be mapped to specific domains, undesirable activities of enzymes can be eliminated, and their desirable catalytic and physical properties can be enhanced.

Presently, several strategies have been devised for oligonucleotide-directed mutagenesis. Some of the important approaches are discussed below.

Oligonucleotide-directed Mutagenesis with M13 DNA

This method involves insertion of the target gene into the double stranded replicative form, ds DNA (RF), of an M13 phage or a phagemid vector. The small fragment (<500 bp) carrying the target sequence is cloned into an appropriate M13 vector such as M13mp18 or M13mp19. The ss DNA template and ds DNA (RF) is isolated from a freshly grown plaque generated from recombinant bacteriophage. The mutagenic oligonucleotide and universal sequencing primer are phosphorylated by T4 polynucleotide kinase. Both phosphorylated primers are mixed with single stranded M13 DNA containing the target sequence and incubated for annealing at 20°C above the theoretical T_m of mutagenic primer for 5 min in high salt concentration followed by slow cooling to room temperature. The molar ratio of primer to template should be in range of 10:1 and 50:1. Klenow fragment, T4 DNA ligase, dNTPs, and ATP are added to this reaction mixture and incubated for 6–15 hours at 16°C. DNA synthesis is catalyzed by Klenow fragment, which utilizes four dNTPs. T4 DNA ligase is added to join the last nucleotide of the synthesized strand to the 5′-end of the primer. As *in vitro* DNA synthesis is often incomplete, partially ds DNA (RF) must be removed from the mixture by sucrose gradient centrifugation. Competent *E. coli* cells are transfected with reaction mixture. Infected cells produce M13 phage particles, which eventually lyse the cells and form plaques. Screening is performed by hybridization under highly stringent conditions with a radiolabeled oligonucleotide probe to detect mutants. The ds DNA (RF) of M13 is isolated, followed by excision of the mutated gene by digestion with restriction enzymes and splicing onto an *E. coli* plasmid expression vector. For further study, the altered protein is expressed and purified from the *E. coli* cells. Finally, the putative mutant plaques are confirmed by sequencing of ss M13 DNA.

Oligonucleotide-directed Mutagenesis with Plasmid DNA

This method is an alternative to the M13 system, which reduces the time-consuming steps of the above system that are performed to eventually isolate mutated form of the target gene. The template of interest is cloned into plasmid vector, which is transformed into host *E. coli* cells. After growing the transformed cells, the ds DNA of plasmid is extracted and denatured by alkali treatment to produce a single stranded circular DNA molecule. The principle of this mutagenesis process is the same as described for M13 DNA. Thus, a synthetic mutagenic oligonucleotide primer having a desired mutation is annealed to the target strand, T4 DNA polymerase is used to synthesize a new complementary strand, and finally the resulting nick between the end of the new strand and the oligonucleotide is sealed by T4 DNA ligase. The resulting chimeric DNA molecule is propagated by transformation in *E. coli*. Note that T4 DNA polymerase is considered to be the most suitable enzyme for site-directed mutagenesis. This is because it does not displace the mutagenic primer after the completion of mutant strand synthesis and its usefulness in reducing undesirable secondary mutations. These two advantages of T4 DNA polymerases are respectively due to lack of any strand displacement activity, i.e., 5′ → 3′ exonuclease activity and 1,000-fold higher fidelity as compared to other polymerases lacking proofreading activity (for details see Chapter 2).

Mutant Enrichment Techniques

Theoretically due to the semiconservative nature of DNA replication, 50% of the transformants (phage or plasmid) should contain wild-type sequence and the remaining 50% should have mutated sequence with the specified nucleotide change. However, the observed efficiencies of mutagenesis are much lower because of some basic reasons such as incomplete strand synthesis or inadequate efficacies of random priming of DNA synthesis, oligonucleotide primer displacement, and host repair mechanisms. Therefore, various selection and enrichment approaches have been incorporated into mutagenesis with phage or plasmid DNA to greatly improve the efficiency of mutant strand recovery.

Enrichment Techniques with Phage DNA When oligonucleotide-directed mutagenesis is performed using phage DNA, only 1–5% of the plaques actually contain phage carrying the mutated gene. Various approaches have been developed to enrich the percentage of mutant phage. One common technique is incorporating deoxyuridine.

Deoxyuridine incorporation for enrichment of mutagenesis A common approach for DNA enrichment is incorporation of deoxyuridine (dUTP) and the method is known as Kunkel's method (devised by T.A. Kunkel in 1991) (Figure 17.2). The target DNA is first prepared by growth of an appropriate recombinant M13 bacteriophage in a dut^{-} ung^{-}

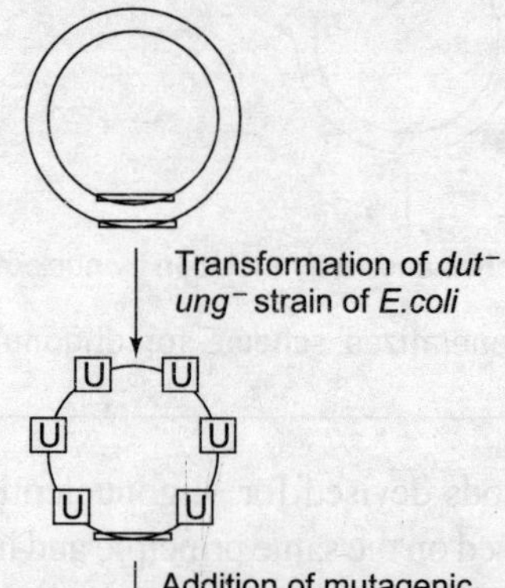

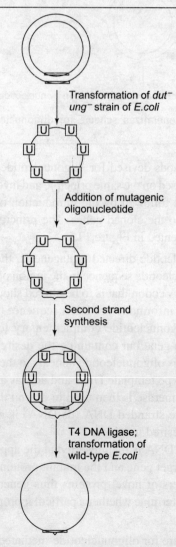

Figure 17.2 Enrichment of mutated M13 by deoxyuridine incorporation

F' E. coli strain (e.g., *E. coli* CJ 236). This strain is defective in two genes encoding two different enzymes involved in DNA metabolism. These are dUTPase (DUT) and uracil *N*-glycosylase (UNG). In the absence of functional dUTPase, intracellular level of dUTP elevates in the cell, and as a result, a few dUTP residues get incorporated into replicating DNA instead of dTTP. In the absence of functional uracil *N*-glycosylase, the cells are unable to remove the spuriously incorporated dUTP from DNA. The ss DNA, which now contains uracil, is used as the template for oligonucleotide-directed mutagenesis, and a chimeric molecule is generated containing uracil in the template strand and thymine in the newly synthesizing strand. This resulting heteroduplex DNA is then introduced into a normal wild-type *E. coli* strain having functional ung^+ strand, resulting in the destruction of the template strand. This reduces the generation of wild-type bacteriophage up to 80%. Since the synthesis of this strand is primed by the mutagenic oligonucleotide, a high proportion of the progeny bacteriophages carry the desired mutation. This significantly enriches the yield of M13 bacteriophage carrying a gene with a site-specific mutation.

Enrichment Techniques with Plasmid DNA Similar to phage DNA, various enrichment techniques are employed for increasing the frequency of mutant formation using plasmid system for oligonucleotide-directed mutagenesis.

Antibiotic selection Site-directed mutagenesis is performed easily by antibiotic selection. In this method, selection oligonucleotide(s) is simultaneously annealed with the mutagenic oligonucleotide to repair an antibiotic resistance gene. Antibiotic resistance of the mutated DNA and the sensitivity of the nonmutated strand facilitate selection of the mutant strand. In one of the plasmid-based mutagenesis protocols, the target DNA is inserted into an MCS on a plasmid vector containing functional tetracycline resistance gene (tet^r) and nonfunctional ampicillin resistance gene (amp^r) having a single nucleotide substitution in the middle of MCS. The recombinant vector is transformed into *E. coli* host cells. Double stranded plasmid DNA is extracted from transformed cells and then denatured by alkali treatment to form a single stranded circular DNA molecule. Three different oligonucleotide primers, four dNTPs, T4 DNA polymerase, and DNA ligase are mixed with ss DNA and incubated for DNA synthesis and ligation. The three oligonucleotides, which are added in this reaction, are designed for specific purposes: the first oligonucleotide is used to alter the target DNA, the second oligonucleotide is designed to correct the substituted nucleotide in the nonfunctional ampicillin resistance gene, and the third oligonucleotide is designed to change a single nucleotide in the tet^r gene to make it nonfunctional. The resulting reaction mixtures are used to transform *E. coli* cells and the transformants are selected on the basis of ampicillin resistance and tetracycline sensitivity. Finally, the cells with the specified mutation in the target gene are identified by DNA hybridization and confirmed by sequencing. This approach increases the frequency of mutagenesis by approximately 90% and hence provides a very efficient way to generate a large number of the desired mutations in limited time.

Deoxyuridine incorporation As discussed above, deoxyuridine incorporation can enrich the mutant frequency by up to ~90%.

Unique restriction site elimination (USE) In this approach, two oligonucleotide primers are hybridized to the same strand of the denatured double stranded recombinant plasmid. One primer is mutagenic primer whereas the other is a selection oligonucleotide containing a mutated sequence for a unique restriction site in the plasmid. The 5′-end of each primer is phosphorylated. Both primers anneal to the same strand, which are extended by T4 DNA polymerase in the presence of four dNTPs, followed by nick sealing by T4 DNA ligase. This leads to generation of two types of plasmids. These are wild-type molecules and heteroduplex molecules (molecules that have lost the unique restriction enzyme site after gaining the desired mutation). The resulting heteroduplex plasmid DNA consists of a wild-type parental strand and a new full-length strand with unique restriction enzyme site disrupted. The concentration of NaCl in the reaction mixture is adjusted to an optimal level for the selected unique restriction endonuclease. Thereafter, the enzyme is added to the reaction mixture and incubated at the appropriate temperature. During this reaction, the wild-type molecules are linearized, while the mutated plasmids are resistant to digestion. The mixture of DNA is then used for transformation of a strain of *E. coli* deficient in repair of mismatched bases (e.g., *mut* S^- strain *E. coli* BMH 71-18). The transformed DNA is then isolated and digested with the same enzyme to linearize the wild-type plasmid and then used to transform a standard strain of *E. coli*. However, the theoretical frequency of mutant produced by this method is ~50%, but usually 5–30% mutant frequencies are obtained. The rate of mutant recovery can be further improved by combining this method with Kunkel method or by increasing the concentration of mutagenic primer so that the molar ratio of two primers is 10:1 in favor of the mutagenic primer.

Restriction with Dpn I This method of enrichment utilizes a supercoiled ds DNA vector with an insert of interest and two synthetic oligonucleotide primers containing the desired mutation. The basic procedure is, however, the same as discussed for USE. The oligonucleotide primers, each complementary to opposite strands of the vector, are extended during temperature cycling by *Pfu Turbo* DNA polymerase. [Note that *Pfu Turbo* DNA polymerase has six-fold higher fidelity in DNA synthesis than *Taq* DNA polymerase and it lacks strand displacement reaction, which protects oligonucleotide primers from displacement]. Incorporation of the oligonucleotide primers generates a mutated plasmid containing staggered nicks.

Following temperature cycling, the product is treated with *Dpn* I (target sequence: 5′-G^{m6}ATC-3′), an endonuclease specific for methylated and hemimethylated DNA. As DNA isolated from almost all *E. coli* strains is deo-xyadenosine methylase (DAM) methylated, it is susceptible to *Dpn* I digestion. Thus *Dpn* I is used to select for mutation containing synthesized DNA. This is possible because *Dpn* I digests the parental DNA template, while mutation containing synthesized DNA remains undigested. The nicked vector DNA containing the desired mutations is then transformed into supercompetent cells. The small amount of starting DNA template required for this method, the high fidelity of the *Pfu Turbo* DNA polymerase, and the requirement of low number of thermal cycles contribute to high mutation efficiency and decreased potential for generating random mutations during the reaction.

Phosphorothioate incorporation This method of site-directed mutagenesis relies on the ability of a dNTP analog containing a thiol group to generate heteroduplex DNA resistant to restriction enzyme digestion. This approach requires the use of ss DNA and a dNTPαS nucleotide analog in addition to the mutagenic oligonucleotide. The mutant strand is extended from the mutagenic oligonucleotide and synthesized in the presence of a dNTPαS. Unused template DNA is removed by digestion with T5 exonuclease. Theoretically, only circular heteroduplex DNA remains undegraded. The heteroduplex DNA is then nicked, but not cut, at the restriction site(s). Exonuclease III is then added to digest the nicked strand (the nonmutant strand) of the heteroduplex and fragment of DNA is generated. This fragment then acts as a primer for repolymerization with DNA polymerase I and ligation with T4 DNA ligase. This creates a mutant homoduplex.

PCR-based Oligonucleotide-directed Mutagenesis

The principle involved in the PCR-based techniques is that mutations can be introduced with the help of mutagenic primer consisting of centrally located mismatches during PCR. PCR-based mutagenesis methods are easy and fast, resulting in high mutant frequency. ds DNAs are used as templates and mutations can be introduced at almost any site. PCR-based methods are also associated with some disadvantages like high error rates in PCR products, which can be reduced by using thermostable enzyme having high fidelity and limiting the number of amplification cycles. Although a number of variants of PCR-based mutagenesis have been published, two methods are more effective.

Megaprimer PCR Method This is simplest and most cost-effective method of PCR-based oligonucleotide-directed mutagenesis. In this approach, three oligonucleotides are used and two rounds of amplifications are performed wherein a product strand from the first amplification serves as a primer in the second amplification (Figure 17.3). One mutagenic primer is used with two flanking primers, which lie upstream and downstream from the binding site for the mutagenic oligonucleotide. These flanking primers can be complementary to sequences in the cloned gene or to adjacent vector sequences. However, theoretically, the mutagenic primer can be oriented toward either one of the flanking primers, while practically the mutagenic primer is oriented toward the nearer of the two flanking primers, hence the length of the megaprimer is in limit. The first PCR reaction is performed with mutagenic primer (M) and the closer of the external primers (R1) to generate mutated fragment of DNA. This amplified fragment is termed as the megaprimer, which is then used in the second PCR reaction in combination with the second external primer (F2) to amplify a longer region of the template DNA.

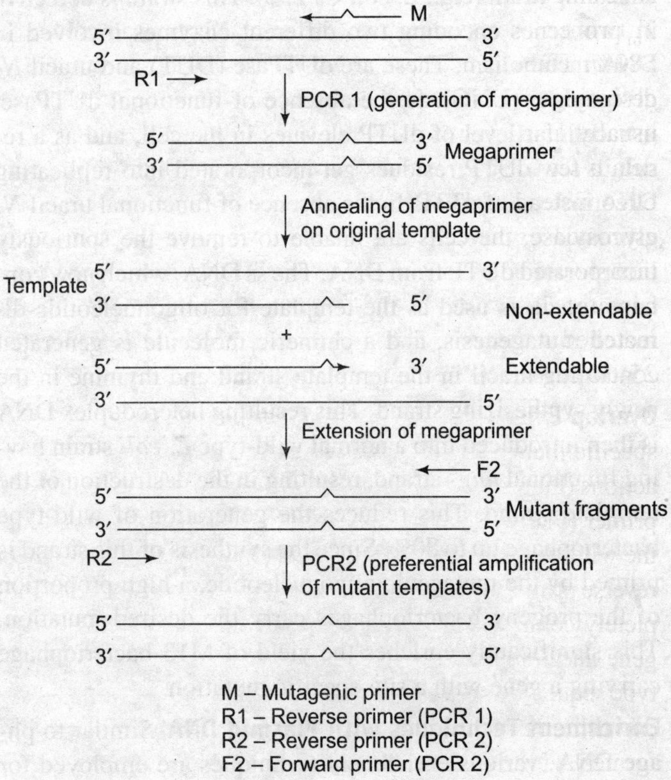

Figure 17.3 Megaprimer PCR method of oligonucleotide-directed mutagenesis

In early methods, megaprimer was purified before second amplification reaction by polyacrylamide gel electrophoresis (PAGE). As this purification was tedious and time-consuming, new approaches have been devised that do not require DNA purification between two rounds of PCR reactions. One such method requires the application of forward and reverse external primers with significantly different T_m values. Thus a megaprimer is generated in the first PCR reaction at a low annealing temperature in the presence of a mutagenic primer and an external primer with a low T_m. The second PCR is performed in the same tube at a higher annealing temperature (~72°C) in the presence of the megaprimer and an external primer with a high T_m. During second PCR reaction, high annealing temperature prevents mispriming by the residual low T_m primer. The average mutagenic efficiency of this method is >80%. The success of this method is thus based on careful designing of oligonucleotides. The short external primers are usually 15–16 bases long and consist of calculated T_m (for details see Chapter 7) between 42°C and 46°C. The mutagenic oligonucleotide primer should also be designed to have a low estimated T_m and a length of 16–25 bases. The desired mutation should be located in the middle of the primer with 8–10 bases of correctly matched sequences on each side of mismatched region. The longer external primer should be of 25–30 bases with estimated T_m between 72°C and 85°C.

Overlap Extension The overlap extension technique for site-specific mutagenesis requires four primers and three PCR reactions (Figure 17.4). The four primers are forward mutagenic primer (FM) containing the mutation(s) to be introduced into the wild-type template DNA, reverse mutagenic primer (RM), reverse primer (R) containing a wild-type sequence that is completely complementary to a sequence upstream to the cloned gene and forward downstream primer (F) containing a wild-type sequence that is completely complementary to a sequence downstream to the cloned gene. Two PCR reactions are performed with different sets of primers. First PCR reaction is done with R and FM primers, while the second PCR reaction is performed with RM and F primers. RM and FM primers contain the mutation to be introduced into the template DNA and at least 15 bases including mutagenic sequences of both primers are exactly complementary. Since the mutation of interest is located in the region of overlap, it will be included in both sets of amplified fragments resulting from the two PCR reactions. The overlapping fragments resulting from first and second rounds of amplification are mixed, denatured, and annealed to generate chimeric DNA. This chimeric DNA is extended into a full-length heteroduplex DNA in a third PCR reaction using two primers that bind to downstream and upstream of two initial fragments. Restriction site can be included in these primers to facilitate subcloning of the mutated segment of DNA. This technique is unexpectedly effective. The resulting heteroduplex DNA is introduced into *E. coli* by transformation and the mutated sequences are confirmed by sequencing. In some cases, a simpler version of the method can be used (one mutagenic primer and two sequential PCRs) if strategically placed restriction sites are available to clone the segment of amplified DNA containing the mutation. Deoxyuridine residues can also be incorporated for mutant enrichment.

17.2.2 Error-prone PCR-mediated Mutagenesis

Error-prone PCR is the technique used to generate randomized genomic libraries. By this method, a large amount of mutated DNA can be produced with small amounts of template DNA. The principle of this technique is that under imperfect PCR conditions, low-fidelity thermostable DNA polymerase (e.g., *Taq* DNA polymerase) is capable of annealing noncomplementary base. Thus, PCR is performed with *Taq* DNA polymerase (lacks proofreading activity) and a library of mutants of the target gene is constructed. The randomly mutagenized DNA produced by error-prone PCR is cloned into expression vectors and screened for altered or improved

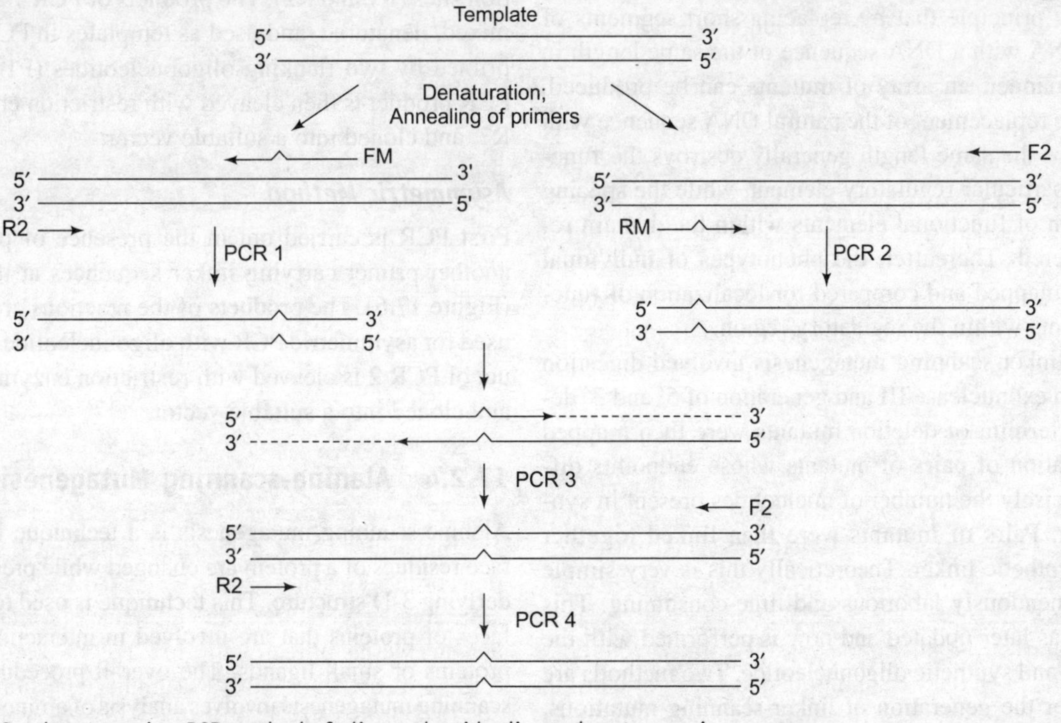

Figure 17.4 Overlap extension PCR method of oligonucleotide-directed mutagenesis

protein activity. The DNA from positive clones that encode the desired activity is isolated and confirmed by sequencing. Error-prone PCR has been used to create enzymes with improved solvent and temperature stabilities and with enhanced specific activity. The introduction of error in the target gene can be improved by employing several conditions, some of which are as follows: (i) By modifying concentration of the DNA template up to 10 kbp in size, it is possible to vary the number of alterations per gene from about 1 to 20; (ii) Extension step of PCR is carried out with unequal amounts of four dNTPs. Usually dATP concentration is lowered as compared to the other dNTPs; (iii) By performing *in vitro* synthesis reaction under limiting amounts (usually 1–10% of the normal concentration) of one of the four dNTPs. This leads to misincorporation of bases. The DNA polymerase starved of the correct nucleotide stalls briefly and then inserts one of the three available nucleotides; (iv) Error rate can be further increased by adding Mn^{2+} at a concentration of 0.5 mM in the PCR buffer; and (v) Performing PCR in the presence of urea, isopropanol, and butanol in the reaction can also incorporate random mutations. The addition of 7.0–8.0% (v/v) propanol in PCR reaction can generate mutation frequencies up to 9.8×10^{-3} mutation/bp/PCR by using Vent (exo-) DNA polymerase. The decrease in fidelity promoted by propanol is due to partial destabilization of the polymerase.

17.2.3 Linker-scanning Mutagenesis

Linker-scanning mutagenesis technique is used to locate and identify *cis*-acting regulatory elements such as functional elements in promoters and replication origins. The method is based on the principle that by replacing short segments of the target DNA with a DNA sequence of the same length in systematic manner, an array of mutants can be produced. However, the replacement of the natural DNA sequence with a sequence of the same length generally destroys the function of that particular regulatory element, while the spacing or orientation of functional elements within the domain remains unaltered. Thereafter, the phenotypes of individual mutants are mapped and compared for localization of functional elements within the regulatory region.

Initially, linker-scanning mutagenesis involved digestion of DNA with exonuclease III and generation of 5′ and 3′ deletions. The termini of deletion mutants were then mapped for identification of pairs of mutants whose endpoints differed by precisely the number of nucleotides present in synthetic linker. Pairs of mutants were then linked together through a synthetic linker. Theoretically this is very simple but it is tremendously laborious and time-consuming. This procedure was later updated and now is performed with the help of PCR and synthetic oligonucleotide. Two methods are developed for the generation of linker-scanning mutations, which are discussed below

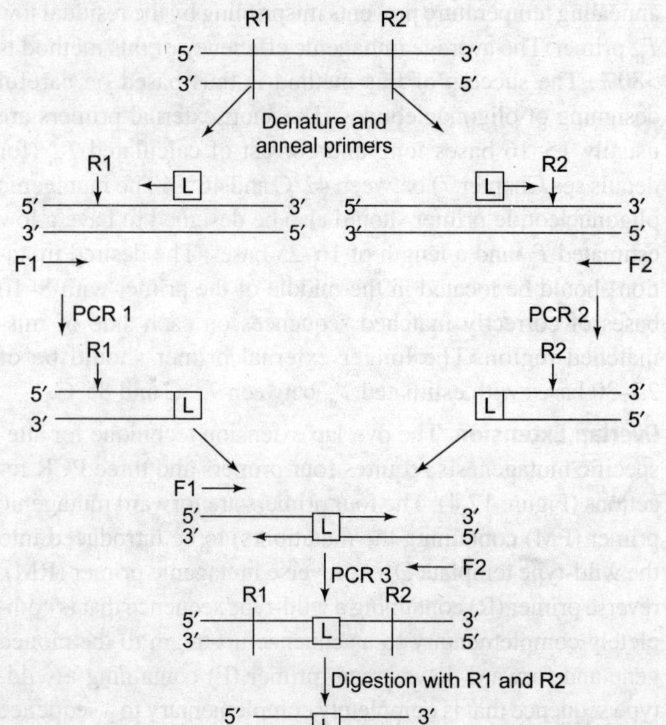

Figure 17.5 Linker-scanning mutagenesis by overlap method of PCR

Overlap Method

Two complementary oligonucleotides carrying linker sequences at the 3′-ends are used to prime PCR 1 and PCR 2 (Figure17.5). In each case, the second primer (F1 or F2) is complementary to sequences lying outside the unique restriction sites (R1 and R2). The products of PCR 1 and PCR 2 are mixed, denatured, and used as templates in PCR 3, which is primed by two flanking oligonucleotides (F1 and F2). The PCR product is then cleaved with restriction enzyme R1 and R2, and cloned into a suitable vector.

Asymmetric Method

First PCR is carried out in the presence of primer Fl and another primer carrying linker sequences at its 3′-terminus (Figure 17.6). The products of the reactions are purified and used for asymmetric PCR with oligonucleotide F2. The product of PCR 2 is cleaved with restriction enzymes R1 and R2 and cloned into a suitable vector.

17.2.4 Alanine-scanning Mutagenesis

Alanine-scanning mutagenesis is a technique by which surface residues of a protein are changed while preserving its underlying 3-D structure. This technique is used to map the surfaces of proteins that are involved in interactions with other proteins or small ligands. The overall procedure of alanine-scanning mutagenesis involves analysis of amino acid sequence of the protein for the presence of clusters of highly charged

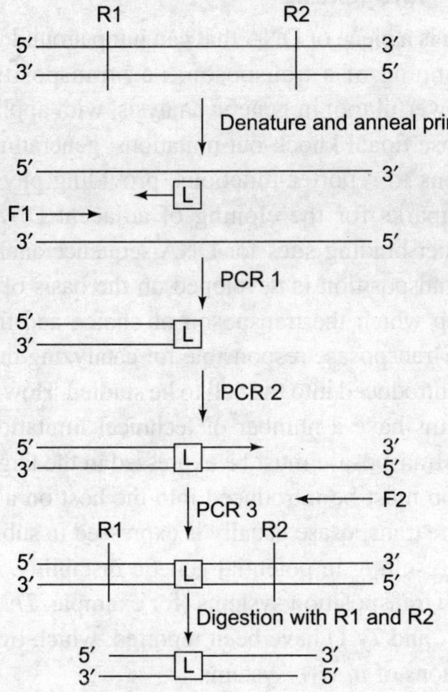

Figure 17.6 Linker-scanning mutagenesis by asymmetric PCR

amino acid residues like arginine, aspartic acid, and glutamic acid (i.e., the amino acids whose side chains when exposed to the surface of the protein contribute to functional properties of the proteins such as interactions with substrates, inhibitors, and other ligands), their deletion by site-directed mutagenesis, and replacement with amino acids having chemically inert side chain that cannot extend beyond β-carbon. The preferred amino acid for this type of substitution is alanine. This is because the replacement of charged amino acids localized at the surface with alanine does not disrupt the folding of core protein, while the absence of polar groups from critical locations may hamper the functions of the protein that involve surface residues. Alanine-scanning mutagenesis of charged amino acids thus generates a complete set of mutant proteins that can be assayed for loss of function and avoids the necessity of generating, sequencing, and characterizing a large library of random mutants.

17.2.5 Cassette Mutagenesis

Cassette mutagenesis involves removal of a small fragment of DNA from the wild-type gene and its replacement with a synthetic DNA segment carrying single or multiple mutations. Thus one or more mutations can be introduced within the target region by this technique, leading to the generation of hybrid mutants and mutant families. Theoretically, this is the procedure in which individual DNA codon is systematically altered, producing all possible mutants at that particular site to determine the effect of such alterations on protein folding or function. The technique is quite useful, particularly in cases where a targeted region is present on a small DNA fragment flanked by unique restriction endonuclease cleavage sites that do not cut elsewhere in the target plasmid. Cleavage with this enzyme thus excises a small segment of DNA, which is then replaced with a ss DNA or ds DNA segment containing a defined mutation or a mixture of mutations. *Bsp* MI cassettes can be more frequently used for cassette mutagenesis. The targeted region is replaced with a DNA segment containing two *Bsp* MI recognition sequences (5′-ACCTGT-3′) arranged in opposite orientations. As *Bsp* MI cleaves outside its recognition sequence, the cleavage removes a segment of DNA, leaving an acceptor molecule into which an oligonucleotide can be inserted to restore the normal sequence except for the mutagenized position. The designing of pools of oligonucleotides carrying mutations at the desired site(s) is very crucial. This is accomplished by a number of methods. An example of particularly efficient method is to synthesize one strand of the cassette as a pool of oligonucleotides with equal amounts of all four nucleotides in the first two positions of the target codon and an equal mixture of G and C in the third position. The resulting DNA contains 32 different codons at particular site, which then encodes the entire set of amino acids. The second strand of the cassette is synthesized with inosine at each target position. The two strands are then annealed and the population of mutagenic cassettes is ligated into the vector. The method, however, suffers from certain drawbacks. First, the cost of oligonucleotide synthesis and screening is very expensive for cassette mutagenesis because a different set of oligonucleotide cassettes harbouring the codon changes is produced for each mutagenized region. Second, theoretically the resulting population of degenerate cassettes will comprise of codons specifying each of the 20 amino acids at each target site, however, because of redundancy in the codon, some amino acids will be represented more frequently than others. Further it is required to sequence large number of candidates to obtain a complete collection of cassettes. Third, the frequency of particular mutations and of double and triple mutations usually deviates from theoretical predictions. This deviation is probably a result of inequalities in the efficiency with which the four-nucleotide precursors are incorporated into the mutagenic cassettes.

To overcome these above drawbacks, new methods of codon cassette mutagenesis have been developed that utilize a set of 11 universal mutagenic cassettes to insert or replace individual codons at blunt end target sites in DNA molecules. The desired codon change is the only residue of the cassette left in the target molecule, so the same set of cassettes can be used to introduce mutations at all blunt end targets. Blunt end DNA is generated at any position in DNA, which thus permits codon cassette mutagenesis to be employed at any position in a gene regardless of the DNA sequence context. Thus, a single series of cassettes can be used to introduce codons encoding all possible amino acids at targets that can be constructed at all positions in any gene. An interesting example of cassette mutagenesis is the work on subtilisin by Estell *et al*. (1985) to create all 19 possible amino acid substitutions at given position. By applying cassette mutagenesis approach, Sung *et al*. (1986) have inserted oligoribonucleotide in front of a synthetic gene to enhance accumulation of the human proinsulin in *E. coli*. By systematically varying the codes, the level of expression of synthetic human proinsulin gene reached around 25% of the total protein of bacteria.

17.2.6 Insertional Mutagenesis

Mutagenesis of DNA by the insertion of new DNA fragment (one or more bases) is known as insertional mutagenesis. By this technique, defined fragment of DNA is randomly inserted throughout the genome. This mutagenic approach is quite attractive as it has ability to target any gene within a complex genome and the potential to apply the same mutagenic construct to a diverse set of experimental systems. This is the most powerful mutagenesis strategy to functional genomics. The different ways by which insertional mutagenesis is performed are described in the following sections.

Virus Insertional Mutagenesis

The phenomenon of natural insertion of infecting retroviruses into the host genome has been exploited for the purpose of *in vitro* mutagenesis. However, the insertion of an inert DNA fragment in diploid cells is masked by the remaining intact allele. To overcome this problem, a retroviral long terminal repeat (LTR) promoter, which drives transcription of the adjacent host DNA, is utilized. The hybrid transcript derived from the retrovirus and the target gene may encode a full-length or truncated protein or an antisense RNA, depending on the position of the insertion. The resulting products, if functional, can confer a dominant phenotype. The random promoter insertion event is similar to the combined introduction of several types of expression libraries, but without the difficulty of constructing, maintaining, and delivering them. Some researchers have experimented with the random insertion of a regulatable promoter so that the link between the hybrid transcript and the phenotype can be verified by modulating the promoter. However, the regulation of exogenous promoters is sensitive to the genomic microenvironment, and multiple copies of the same promoter within a cell would be regulated similarly, preventing the identification of the one responsible for the phenotype. Furthermore, an alternative improved approach is devised that is based on physical removal of the inserted promoter by site-specific recombination and the relevant gene is easily identified from multiple insertions of same cells.

Transposon Mutagenesis

A transposon is a piece of DNA that can jump around within a genome. Jumping of a transposon, i.e., transposition has become a powerful tool in genetic analysis, with applications in creating insertional knock-out mutations, generating gene-operon fusions to reporter functions, providing physical or genetic landmarks for the cloning of adjacent DNAs, and locating primer binding sites for DNA sequence analysis. *In vitro* DNA transposition is developed on the basis of *in vivo* approaches in which the transposon of choice and the gene encoding the transposase responsible for catalyzing the transposition are introduced into the cell to be studied. However, all *in vivo* systems have a number of technical limitations. For example, the transposase must be expressed in the target host, the transposon must be introduced into the host on a suicide vector, and the transposase usually is expressed in subsequent generations, resulting in potential genetic instability. A number of *in vitro* transposition systems (for example, *Tn* 5, *Tn* 7, *Mu*, *Himar* 1, and *Ty* 1) have been reported, which overcome many limitations of *in vivo* systems.

Transposon mutagenesis has been developed as a simple, robust technology by which high-efficiency transposition is generated in bacterial species (*E. coli*, *Salmonella typhimurium*, and *Proteus vulgaris*), and in *Saccharomyces cerevisiae*. *Tn* 5 is a bacterial genetic element, which transposes via a cut-and-paste mechanism. The macromolecular components required for *Tn* 5 transposition are transposase, the transposon, which can apparently be any sequence defined by two specific inverted 19 bp sequences and the target DNA into which the insertions are made. The steps involved in *Tn* 5 transposition are as follows (i) Binding of transposase monomers to the 19 bp end sequences; (ii) Oligomerization of the end-bound transposase monomers, forming a transposition synaptic complex; (iii) Blunt end cleavage of the transposition synaptic complex from adjoining DNA, resulting in formation of a released transposition complex or transposome (this step requires Mg^{2+}); (iv) Binding to target DNA; (v) Strand transfer of the transposon 3′-ends into a staggered 9-bp target sequence (this step also needs Mg^{2+}). After strand transfer, transposase is removed from the product DNA and the 9-base gaps at the ends are repaired by host machinery. The central DNA sequence between the 19 bp end sequences plays no mechanistic role in this process as long as it is of a size that permits synaptic complex formation. Thus any sequence (gene of interest) can make up this region. The 'transposed strain' can be treated as a mutant pool like any other. The mutated gene is then identified by screening a gene library by either hybridization or PCR. After a round of transposition, progeny are screened by PCR using transposon and gene-specific primers for the proximity of the transposon sequence to the gene of interest. As PCR can only produce products up to 1–2 kbp, a large fraction of progeny identified as positive by PCR will have a transposon close enough to the gene to inactivate or otherwise alter its pattern of expression.

Further, if transposon jumps into the middle of a coding gene, the phenotype of that particular gene is disrupted. If transposons carry gene(s) responsible for certain phenotypic character, its transposition will confer the phenotype to the host.

T-DNA Mutagenesis

A. tumefaciens is a natural genetic engineer for plants and is routinely used for gene transfer to plants. During transformation, T-DNA is inserted into the plant genome (for details see Chapter 12). This T-DNA can also act as an insertional mutagen. The method of insertional mutagenesis involving the use of T-DNA from *A. tumefaciens* is used for generating a large population of tagged lines and for identifying the insert carrying the gene of interest. T-DNA insertional mutagenesis has been applied for genome-wide mutagenesis programs in *Arabidopsis thaliana* and rice. T-DNA tagged *Arabidopsis* lines are currently available by the University of Wisconsin Arabidopsis knock-out facility. These lines comprise two populations; one generated by the insertion of a simple T-DNA construct and another is generated by the insertion of an activation tag. These populations can be used for screening and a mutation in particular target gene is identified in several rounds of PCR by using gene-specific primer. As T-DNA is not a transposon, stable insertions are produced during primary transformation. One disadvantage associated with this technique is its tendency to produce complex, multicopy integration, and sometimes deletion or rearrangement of surrounding genomic DNA, thereby complicating subsequent analysis.

Use of Plant Transposons in Mutagenesis

Insertional mutagenesis in plants can also be achieved by using endogenous and heterologous plant transposons. The application of plant transposons has led to the discovery of many new genes. Various transposons that have been used include *Activator (Ac), Suppressor mutator (Spm), and Mutator (Mu)* from maize, *Tam* 3 from *Antirrhinum,* and *Tph* 1 from petunia. Similar to T-DNA from *A. tumefaciens*, plant transposons can also be used to generate a large population of tagged lines and to identify the insert carrying the gene of interest. In contrary to T-DNA insertion, transposons tend to produce simple, single copy insertions, however most of the plant transposons show pronounced target-site preference, which can make it difficult to achieve whole genome saturation. Since transposons have intrinsic ability to mobilize unless their source of transposase is removed, additional crossing steps are also required to stabilize the insertions, which are produced by transposons. This kind of control is generally achieved by the use of two-component transposon systems, consisting of an autonomous (self-mobilizing) element, and a nonautonomous derivative. The most common example to this effect is maize *Ac*/*Ds* system. The maize *Ac* transposon is autonomous because it encodes for its own transposases, but *Ds* lacks the transposase gene. However, *Ds* elements can also transposase in the presence of transposase from Ac. Thus, when *Ac* and *Ds* are present in the same genome, both elements can be mobilized. However, if Ac is removed by crossing, progeny plants can be recovered with stable Ds insertions. The use of maize *Ac*/*Ds* system has been reported to be potentially efficient and has been widely used for genome-wide mutagenesis.

17.2.7 Random Mutagenesis

Random mutagenesis is carried out when detailed information regarding the role of particular amino acid residues in the functioning of the protein is not known. By this technique, some novel mutants encoding proteins with a range of interesting and useful properties may be produced because the introduced changes are not limited to one amino acid. Random mutations can be created easily by several molecular biological techniques. The drawback in producing a large number of random mutations within a cloned gene, regardless of the method employed, is that each clone must be assayed to determine if it produces a protein with desired properties. This sort of testing is not a simple task, but often it is the only way to find proteins with novel features. Once a mutant of potential interest has been identified, the sequence of the cloned gene can be determined to ascertain the mutated site. Some common methods employed for random mutagenesis are explained.

Application of Degenerate Oligonucleotide Primers

Random mutations can be generated by the application of degenerate primers in PCR. The degenerate primers contain heterogeneous set of DNA sequences that will generate a series of mutations clustered in a defined position of the target gene (for details see Chapter 7). Degenerate primers are incorporated by various ways to generate random mutagenesis (Figure 17.7). According to one strategy, the target gene is inserted into a plasmid between two restriction sites (A and B) and then PCR is performed to incorporate random mutations in separate reactions. The left fragment is amplified with one degenerate oligonucleotide primer, which binds to lower strand of target DNA along with regular, completely complementary primer that hybridizes to a region of the upper strand flanking the left unique restriction endonuclease (A). The right fragment is generated by using degenerate oligonucleotide primer for upper strand of the target DNA along with the primer that is complementary with a region of the lower strand lying outside the second unique restriction endonuclease site (B). The products of both reactions are purified and pooled

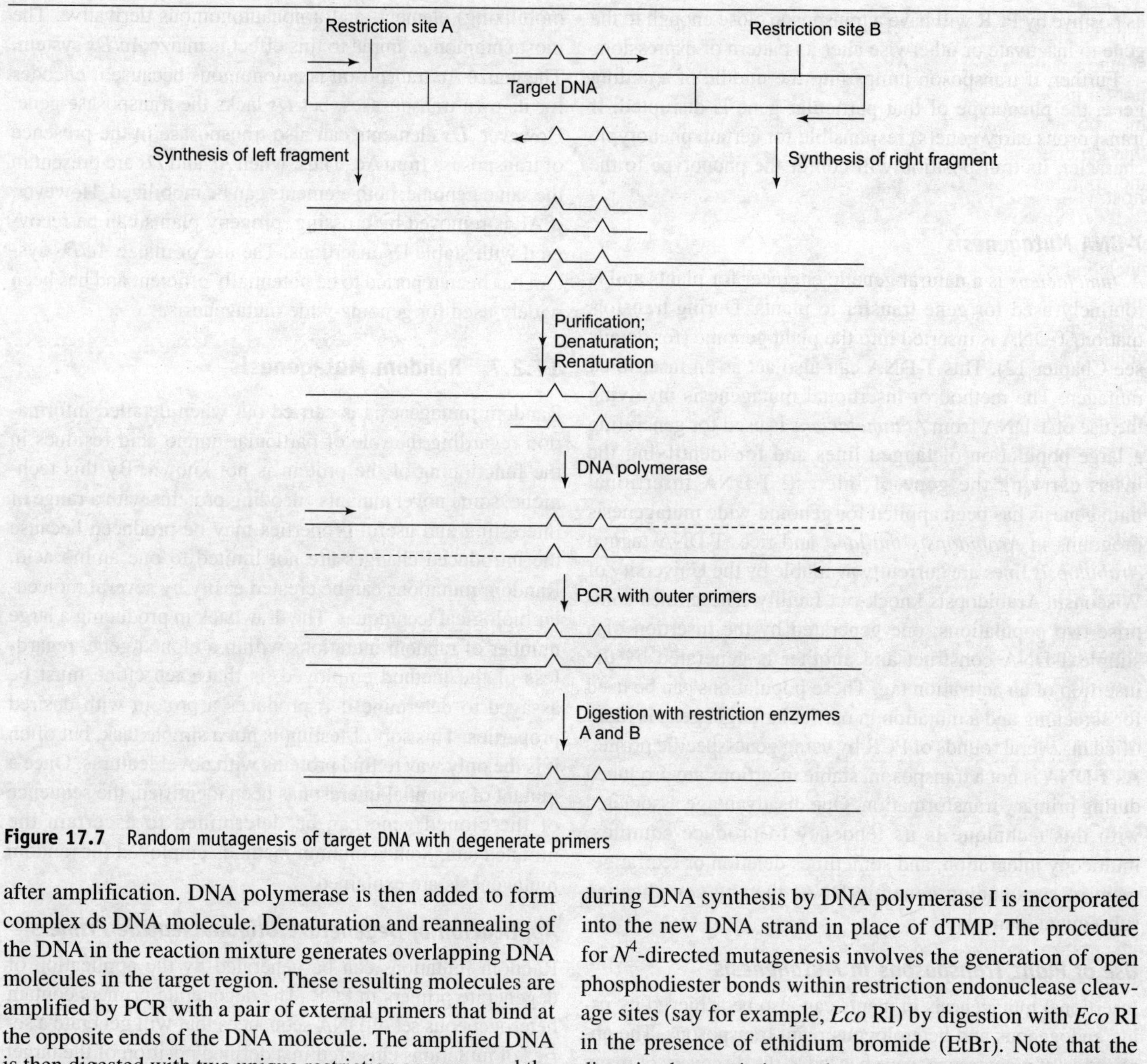

Figure 17.7 Random mutagenesis of target DNA with degenerate primers

after amplification. DNA polymerase is then added to form complex ds DNA molecule. Denaturation and reannealing of the DNA in the reaction mixture generates overlapping DNA molecules in the target region. These resulting molecules are amplified by PCR with a pair of external primers that bind at the opposite ends of the DNA molecule. The amplified DNA is then digested with two unique restriction enzymes and the DNA is then cloned into a suitable plasmid vector. The procedure results in the production of an altered gene that has mutated sites in the region of the overlap of the original oligonucleotides. However, if in the first attempt desired mutation is not achieved, it may be necessary to repeat the entire procedure with a set of degenerate primers complementary to a different region of the gene.

Application of Nucleotide Analogues

Random mutagenesis is easily performed by the incorporation of nucleotide analogues into DNA. Presently several techniques are utilized for this purpose. The three most importantly used techniques are discussed in the following:

Application of Nucleotide N^4-analogue dCTP N^4-hydroxy dCTP is a nucleotide analogue of dTMP, which when added during DNA synthesis by DNA polymerase I is incorporated into the new DNA strand in place of dTMP. The procedure for N^4-directed mutagenesis involves the generation of open phosphodiester bonds within restriction endonuclease cleavage sites (say for example, *Eco* RI) by digestion with *Eco* RI in the presence of ethidium bromide (EtBr). Note that the digestion by restriction endonuclease in the presence of EtBr results in the cleavage of only one of the two phosphodiester bonds within the restriction site. This nick is than used for *in vitro* initiation of DNA synthesis by DNA polymerase I, which will proceed only along one of the two complementary strands in 5′ → 3′ direction. This yields two different sets of molecules in which dTMP residues are replaced by N^4-hydroxy dCTPs in either of the two strands. The resulting DNA is introduced into *E. coli* by transformation. The incorporated N^4-hydroxy dCTPs residue directs the incorporation of G and A residues, thereby generating the expected transitions from TA to CG. In view of the fact that all mutations occur in region of an *Eco* RI cleavage site, the transitions can be easily recognized or selected because such mutations generates *Eco* RI resistant DNA molecules. The mutation frequency of this method is ~5–15%.

Application of α-Thiophosphate Analogues α-Thiophosphate analogues act as substrates for DNA polymerases, but not for its 3′ → 5′ exonuclease activity. These analogues when incorporated in growing DNA chains act as false nucleotides that generate random errors. Once this nucleotide analogue is incorporated into DNA chain, the 3′ → 5′ exonuclease activity of polymerase cannot excise it. However, its terminus can be used as primer for further synthesis. Thus, the misincorporated nucleotides are presumably held frozen in a state in which these are resistant to the 3′ → 5′ exonucleolytic correcting activities of the enzyme. Upon transformation, the wild-type and mutant DNAs segregate so that half of the clones are of mutant type. This method yields up to 40% mutant clones, which is very close to the one predicted theoretically.

Generation of 3′ and 5′ Recessed Ends In this procedure, a cloned gene is inserted into a plasmid cloning site next to two closely placed restriction sites. These sites are specifically selected for the purpose to generate 3′ and 5′ recessed ends. The cloned DNA is digested with both restriction enzymes. The resulting DNA has a 3′ recessed end and a 3′ protruding end. This is followed by cleavage with *E. coli* exonuclease III, an enzyme that degrades DNA exclusively from the 3′ recessed ends, one nucleotide at a time. After a specified time, the exonuclease III reaction is terminated and the gap is filled with Klenow fragment. A dNTP mixture that contains the four normal dNTPs and one dNTP analogue is supplied during the Klenow fragment-catalyzed DNA synthesis step. This leads to incorporation of nucleotide analogue into the newly synthesized DNA strand. During subsequent plasmid replication, the nucleotide analogue directs the incorporation of a nucleotide that is different from the one present in wild-type DNA into the complementary strand of the cloned gene.

Variant of Oligonucleotide-directed M13 Mutagenesis

Random mutagenesis can also be achieved by a variant of the oligonucleotide-directed M13 mutagenesis procedure. In this case, an oligonucleotide mixture in which the nucleotide sequences of some positions are randomly varied is used as a primer during DNA synthesis. As a result, libraries containing clones with many different mutated sites are generated, which requires extensive screening to identify the mutant of interest.

17.2.8 DNA Shuffling

In DNA shuffling, several of the genes or cDNAs for particular proteins are isolated and mixed to allow recombination. This may generate proteins with greatly improved performances, for example, some of the recombinant proteins may combine important features of two or more of the original proteins such as high activity and thermostability. Thus DNA shuffling is used to rapidly propagate beneficial mutations in a directed evolution experiment.

DNA Shuffling Amongst Gene Families

Most of the important proteins/enzymes are encoded by genes belonging to the multigene families. These proteins have slightly different nucleotide sequences and biological activities. It can be performed by the following ways: (i) The simplest technique for DNA shuffling of similar genes involves digestion by common restriction enzyme. Two or more DNAs that encode native form of similar proteins are digested with one (or more) restriction enzymes that cut DNA at the same place followed by ligation of a mixture of DNA fragments. The technique potentially generates a large number of hybrids. For example, two DNAs, each having three unique restriction enzyme sites can be recombined to produce 14 different hybrids in addition to original DNA; and (ii) The DNA of several members of a gene family are isolated, combined, and digested with the enzyme deoxyribonuclease I (DNase I) into random 10–50 bp fragments. These small fragments are separated, subjected to denaturation by heating, and then allowed to reanneal and amplify by PCR. The DNA fragments bind to one another in regions of high homology as gene fragments from different members of a gene family cross-prime each other. In the final step, full-length products are obtained by including terminal primers in the PCR reaction. After 20–30 cycles, full-length hybrids DNAs are generated. The hybrids are cloned into a suitable vector, and introduced into bacteria to create an expression library, which is then screened for desired activity.

DNA Shuffling Amongst Genes of Different Gene Families With Little or no Homology

Although DNA shuffling works well with gene families having a high degree of homology, this technique is also useful when proteins have little or no homology. For the production of recombinant proteins from the genes of different families, several variants of the DNA shuffling protocol have been developed. Thus, many variants of the gene can be recombined rapidly by using DNA shuffling techniques and even variants from different species can be employed. Even nonhomologous DNA, totally unrelated species can be recombined. Larger pieces of DNA such as entire plasmids, viral, and bacterial genomes can also be recombined. To increase the variety of recombinants, short pieces of synthetic DNA can be added to the mixture. The first report of this strategy was on recombinant of a β-lactamase gene. A recombinant of a β-lactamase that increased resistance to a β-lactam antibiotic by 16,000-fold has been generated. When this recombinant was 'backcrossed' to the parental gene using the same DNA shuffling technique, the enzyme obtained was 32,000 times as effective as the wild-type enzyme.

17.2.9 Bisulfite Mutagenesis

Bisulfite mutagenesis technique developed by Shortle and Nathans (1978) is based on the principle that sodium bisulfite treatment deaminates exposed cytosine residues of ss DNA to form uracil, which results in cytosine-to-thymidine transition mutations following DNA replication. The single stranded regions are generated in ds DNA molecules by various techniques, which then become susceptible to bisulfite treatment. Some techniques are described below: (i) If any ds DNA molecule is treated with restriction endonuclease in the presence of EtBr, it hydrolyzes only one phosphodiester bond in the cleavage site (Figure 17.8). The single strand break thus generated is further enlarged by exonucleolytic treatment. Usually DNA polymerase I having 5′ → 3′ polymerase and 3′ → 5′ exonuclease activity is used for this purpose in the presence of only one dNTP. After treatment of DNA with bisulfite, the same DNA polymerase fills the gap by utilizing its polymerase activity. For screening, digestion with the initially used restriction enzyme is required. This technique is restricted to mutagenesis of regions located around restriction enzyme cleavage sites; (ii) Another strategy involves the application of D-loops in the desired DNA region generated by the action of *E. coli* Rec A protein, a protein that catalyzes the formation of D-loops in mixtures of superhelical plasmid DNA and a complementary single strand. The single strand displaces a DNA region of the same polarity in the double strand. Thereafter, controlled S1 nuclease treatment induces a single strand break in the displaced single stranded region, and the unstable relaxed intermediate displaces the single stranded short DNA fragment within fractions of a second, yielding a relaxed circular molecule with a single strand break in the desired region. The single stranded region is then mutagenized by bisulfite treatment as described above. Since *E. coli* Rec A protein is commercially available, this technique can be used practically. However, one should be careful while performing S1 nuclease digestion and the reaction conditions must be chosen to prevent further digestion of the desired single stranded region; (iii) An alternate and simple approach for generating ss DNA is cloning in M13 vector system followed by mutagenesis of this DNA by bisulfite. The first step in this protocol is the isolation of a recombinant ss DNA and the corresponding replicative form (RF). A DNA fragment containing the desired region, i.e., the site to be mutagenized is then cut out from the RF DNA. This linearized DNA fragment is denatured and reassociated in the presence of the original ss DNA, which leads to the formation of a heteroduplex having a single stranded region at the desired site. The desired site is then mutagenized with bisulfite as described earlier in this section; and (iv) Another technique utilizes the previously existing deletions. The wild-type DNA and DNA containing a deletion are cloned separately in suitable plasmid vectors and linearized with two different enzymes, namely A and B. The mixtures of linearized plasmids are then denatured and renatured, which then yields heteroduplex and homoduplex molecules. The resulting heteroduplex is circular because of the staggered position of cleavage sites, while homoduplexes are linear. The circular heteroduplexes contain a single stranded region, the size of which is exactly the same as the size of the original deletion. This region can be mutagenized with bisulfite. The bacterial ung^{-} mutant cells are then transfected by mutagenized DNA. Approximately half of the clones thus obtained are of wild-type size and the other half are of the size of deletion mutants. Moreover, more than 80% of the full-size clones show the expected transitions from GC to AT.

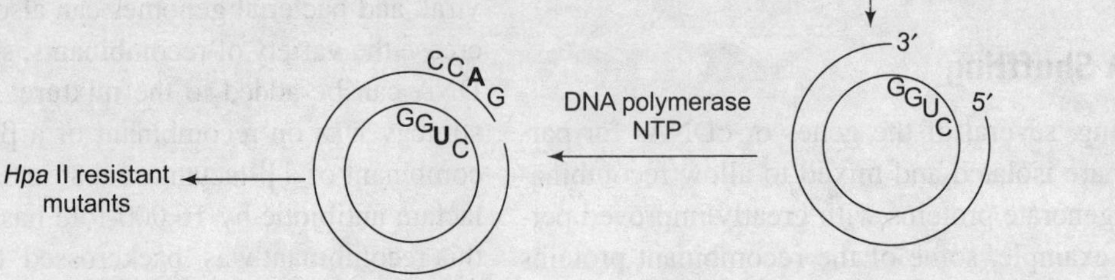

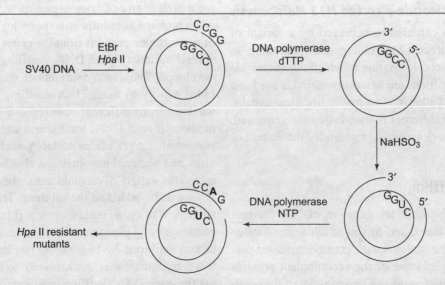

Figure 17.8 Bisulphite mutagenesis by generating single stranded break with restriction enzyme in presence of EtBr

17.2.10 Mutant Proteins with Unusual Amino Acids

By following one of the site-directed mutagenesis approaches, any protein can be easily modified. The diversity of proteins can be further increased after mutagenesis by introducing synthetic amino acids with unique side chains at specific sites. This approach utilizes engineered *E. coli*, which produces both a pair of novel tRNA and a new aminoacyl tRNA synthetase that aminoacylates only its cognate tRNA. The gene of novel tRNA and its corresponding aminoacyl tRNA synthetase are isolated from archaebacteria *Methanococcus jannaschii* and introduced into *E. coli*. The tyrosinyl tRNA synthetase from *M. jannaschii* has ability to add an amino acid to an amber suppressor tRNA, which is mutant of its tyrosine-tRNA. The amino acid specificity of the tyrosine-tRNA synthetase from *M. jannaschii* can be modified by random mutagenesis of its gene, so that instead of tyrosine, it places *O*-methyl-*L*-tyrosine onto tRNA. An amber suppressor tRNA is modified that can insert an amino acid into a protein against an amber codon (UAG; stop codon) in the mRNA. The cloned target gene is modified by oligonucleotide-directed mutagenesis so that it contains a 5′-TAG-3′ in that portion of DNA that encodes the amino acid targeted for change to *O*-methyl-*L*-tyrosine. Modified DNA is selected and used to transform an *E. coli* strain that is previously engineered to produce *O*-methyl-*L*-tyrosine into proteins that contain UAG stop codon, resulting in full-length protein. This approach can be manipulated for the insertion of a variety of different amino acid analogues into specified sites within protein for the production of functional proteins with altered activities as compared with the native form. An example of this technique is the modification of valine-tRNA synthetase gene so that the altered enzyme inserts the nonstandard amino acid aminobutyrate into proteins. Although the full prospective of this strategy has yet to be recognized, it is however obvious that proteins containing unusual chemical structures, and possibly having unique properties are produced.

17.2.11 Applications of Site-directed Mutagenesis

Site-directed mutagenesis is one of the basic tools for genetic manipulation and protein engineering. It is easy to perform and requires less hard work. By using any of the techniques described earlier, specific amino acid encoded by a cloned gene can be altered. Site-directed mutagenesis is performed with the objective of generating proteins that are better suited than wild-type counterparts for therapeutic and industrial applications. Thus site-directed mutagenesis is used for following purposes: (i) Most widely used application of site-directed mutagenesis is to get mutated forms of gene encoding a protein; (ii) To introduce specific single base changes in order to study the effects of changed amino acid on structure, function, and stability of the protein; (iii) To increase protein stability by increasing its resistance to cellular proteases. This may simplify protein purification and increase the recoverable yield; (iv) To enhance thermal tolerance and/or pH stability of the protein, enabling the altered protein to be used under conditions that would inactivate the native version; (v) To increase the catalytic efficiency of enzymes by altering K_m and V_{max} values; (vi) To introduce new active sites and hence generate new catalytic activity in enzymes; (vii) To eliminate the cofactor requirement of an enzyme so that the cofactor is no longer required for certain continuous industrial production processes; (viii) To modify the substrate-binding site of an enzyme in order to increase its specificity, thereby decreasing the extent of undesirable side reactions; (ix) To alter the allosteric regulation of an enzyme in order to reduce the impact of metabolite feed back inhibition and increase the product yield; (x) To modify the reactivity of an enzyme on nonaqueous solvent so that chemical reactions can be catalyzed under nonphysiological conditions; (xi) To produce designer enzymes in large quantities by increasing the expression of the gene through modification of regulatory elements; (xii) To improve the effect or function of antibodies and their affinity. It can thus help in antibody engineering in the development of hybrid antibodies; (xiii) To change the amino acid pattern of storage proteins in grain so as to remove the amino acid deficiency; and (xiv) To introduce deletions or insertion of sequences.

An important industrial application of site-directed mutagenesis is the production of genetically modified protease enzyme (used in detergents for oxidation-resistant proteases) having greater stability. In 1982, Genetics developed and commercialized these oxidation-resistant proteases. When methionine (position 222) located at active site of protease subtilisin isolated from *Bacillus amyloliquefaciens* was replaced by alanine, the altered protein exhibited 80% of the specific activity of the wild-type. Further replacement of this methionine with serine or glutamine gave 100% of the specific activity of the wild-type. Storage stability of this enzyme was also greater as compared with the original wild-type enzyme.

17.3 GENE MANIPULATION BY RECOMBINATION

A number of recombination-based techniques have been formulated for gene transfer in plants and animals, which are now considered as routine experiments in many species, particularly in the case of model organisms. These recombination-based genetic manipulation techniques are discussed in the following sections.

17.3.1 Gene Targeting Using Embryonic Stem Cell

Gene targeting is a genetic technique that utilizes homologous recombination to alter an endogenous gene. Mario R. Capecchi, Martin J. Evans, and Oliver Smithies won the 2007 Nobel Prize in Physiology and Medicine for their work on

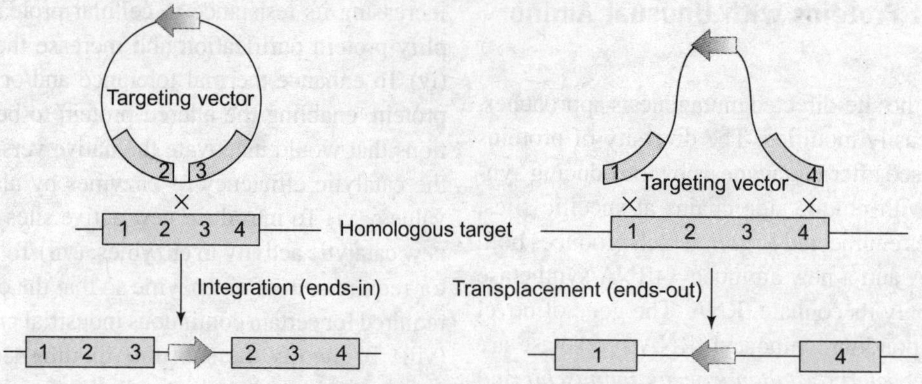

Figure 17.9 Structure and integration mechanisms of two types of targeting vector

'principles for introducing specific gene modifications in mice by the use of embryonic stem cells', or 'gene targeting'. It is applied to modify the endogenous genome in a specific manner, for example, for deletion of a gene, removal of exons, and introduction of a defined mutation in a selected gene. In yeast, gene targeting by homologous recombination occurs with high efficiency (for details see Chapter 11). Previously gene targeting was performed for the repair of mutations in selectable markers like *neo*^{r} with lower efficiency in animal cells. Later targeting of an endogenous gene was carried out into a fibroblast cell × mouse erythroleukemia cell line. The observed frequency of this experiment was quite low as compared to random integration. However, gene targeting was performed with significantly higher efficiency in some cell lines like mouse embryonic stem (ES) cells. Gene targeting can be permanent or conditional (e.g. developmental stage and specific tissue).

Procedure

The procedures of gene targeting vary depending on the organisms. In this section, the basic procedure for gene targeting in mice is described. The steps involved are (i) A construct for targeting is prepared in bacterial system containing target gene, a reporter gene, and a (dominant) selectable marker; (ii) This construct is transfected into mouse embryonic stem cells in culture. It is preferable to linearize the vector prior to transfection; (iii) Positive transformants with the genuine insertion are selected; and (iv) Finally, chimeric mice where the modified cells make up the reproductive organs are selected via breeding. After this step, the entire body of the mouse is based on the previously selected embryonic stem cell. Note that this procedure is modified for successful gene targeting to cattle, sheep, swine, and many fungi and mosses.

Types of Gene Targeting Vectors

Gene targeting requires the creation of a specific vector for each gene of interest. However, it can be used for any gene, regardless of transcriptional activity or gene size. Targeting vectors are specialized plasmid vectors that promote homologous recombination when introduced into ES cells. A region that is homologous to the target gene is inserted into the vector to facilitate synapse of the targeting vector with the endogenous DNA. The size of the homology region and the level of sequence identity play very important role in the efficiency of gene targeting.

Generally, two types of gene targeting vectors have been developed that are used for gene knock-out experiments. These are insertion vectors and replacement vectors (Figure 17.9).

Insertion Vectors During targeting, the entire vector is inserted into the target locus as it is linearized within the homology regions. Although the target gene is disrupted by insertion, some of the sequences are duplicated during this process, which can restore the wild-type genotype by consequent homologous recombination event.

Replacement Vectors In these types of vectors, homology region is colinear with the target and the vector is linearized outside the homology region, resulting in cross-over events in which the endogenous DNA is replaced by the incoming DNA. When the gene targeting event is carried out by replacement vector, sequences localized only within the homology regions are inserted.

Both vectors are equally efficient, but majority of knockout experiments are performed by using replacement vectors.

Screening Strategies for Targeting

For efficient screening of positive transformants, it is required to include a selectable marker in the targeting vector. In case of insertion vectors, selectable markers can be placed anywhere on the vector backbone, but in replacement vectors, the marker must interrupt the homology region. Usually, *neo*^{r} (neomycin phosphotransferse, which confers resistance to aminoglycoside antibiotics) has been used and selection is carried out on G418-containing medium. In some cases, other dominant marker that is equally used is *hprt* (hypoxanthine guanine phosphoribosyl transferase, which blocks *de novo* IMP synthesis). Selection is done on HAT medium (hypoxanthine, aminopterin, and thymidine) in combination with *hprt*$^-$ mutant cells. However, sometimes efficient screening is not possible with the help of single marker as a random integration event may also give positive signal. This situation can be circumvented by combined positive-negative selection using neo^r and the herpes simplex virus thymidine kinase (HSV *tk*) gene, wherein the HSV *tk* gene is located outside the homology region. Thus only recombinant positive transformants will survive in the presence of toxic thymidine analogs ganciclovir or 1-(2-deoxy-2-fluoro-β-*D*-arabinofuranosyl)-15-iodouracil (FICU), while in the cells having randomly integrated copies of the *tk* gene, the analogs will be incorporated into the DNA leading to cell death. However, a different approach is also developed in which expression of the neo^r gene is linked only with homologous recombination. Thus neo^r genes are first located on the vector having no promoter, and after genuine recombination event, the genes are positioned under the control of promoter of endogenous gene and expressed. PCR can also be used for screening of large numbers of G418-resistant transfected cells to identify authentic recombinants.

17.3.2 Site-specific Recombination

Site-specific recombination (SSR) or site-specific recombinase technology is a genetic technique that utilizes homologous recombination for the precise *in vivo* manipulation of DNA sequences in animal and plant genomes. In this technique, the transgene is integrated into the genome of an organism at sites where its expression can be properly regulated.

Procedure

In site-specific recombination, DNA strand exchange takes place between segments having short, specific recognition sites, also called target sites. When the target sites for recombination are introduced into transgenes, recombination can occur in a heterologous cell provided a source of recombinase is also supplied. This is possible because recombinase recognizes the target sites, allowing excision of DNA backbone between two sites, followed by exchange of two DNA helices, and finally their religation.

Types of Recombinases

Various types of SSR systems have been recognized in different systems. Although in some SSRs, recombinase enzyme and recombination sites are sufficient to perform this event, while in other systems a number of accessory proteins and accessory sites are also needed. Most site-specific recombinases are grouped into two families, *viz.*, the tyrosine recombinase family and the serine recombinase family, the names of which come from the conserved nucleophilic amino acid residue that these enzymes use to attack DNA and which becomes covalently linked to DNA during strand exchange.

Tyrosine Recombinase Family The recombinases belonging to tyrosine family contains a conserved tyrosine residue that is responsible for attacking DNA and this tyrosine residue becomes covalently linked to DNA during strand exchange. Tyrosine recombinase breaks and rejoins single strands in pairs to form a Holliday junction intermediate. This family of recombinases is also called integrase family. The examples of tyrosine recombinases are the well-known enzymes such as Cre (Cyclic recombinase) from the P1 phage, Flp (flippase) from 2-μm plasmid of yeast *S. cerevisiae* (for details see Chapter 11), and λ integrase from λ phage (for details see Chapter 9).

Cre recombinase recognizes 34 bp DNA sequence called *lox* P (locus of X-over P1) (Table 17.1) consisting of a pair of 13 bp inverted repeats flanking an 8 bp central element. The 34 bp sequence *frt* (Flp recombination target) is recognized by Flp. In addition to this, it also possesses an additional copy of 13 bp repeat sequence, which is nonessential for recombination (i.e., total size 48 bp). Cre works optimally at 37°C and has been used extensively in mammalian cells. The optimal temperature for Flp is 30°C.

Serine Recombinase Family The recombinases belonging to serine family contains a conserved serine residue that is responsible for attacking DNA, and this serine residue becomes covalently linked to DNA during strand exchange. Serine recombinase cuts all strands prior to strand exchange and religation. This family of recombinases is also called invertase/resolvase family. The serine recombinases include enzymes such as gamma-delta resolvase from the *Tn* 1000 transposon and *Tn* 3 resolvase from *Tn* 3 transposon.

Among all these recombinases, Cre and Flp recombinases are the most widely used ones and have been shown to be successful in many heterologous eukaryotic systems such as mammalian cells, transgenic animals, and plants. In the following section, the Cre/*lox* P system is described in detail and the Flp/*frt* system has already been discussed in Chapter 11.

Cre/lox P System for SSR

The Cre/*lox* P system is used as a genetic tool to control SSR events in genomic DNA. This recombination system was discovered by Brian Sauer. This system can be used to control gene expression in a number of genetically modified organisms, delete undesired DNA sequences, and modify chromosome architecture. Cre/*lox* P recombination system has also found application in gene knock-out technology, and in the development of P1-derived artificial chromosomes (PACs) (for details see Chapter 12).

Table 17.1 Sequences of *lox* P (locus of X-over P1) Site

Sequence of 13 bp region	Sequence of 8 bp region	Sequence of 13 bp region
ATAACTTCGTATA–	GCATACAT	–TATACGAAGTTAT

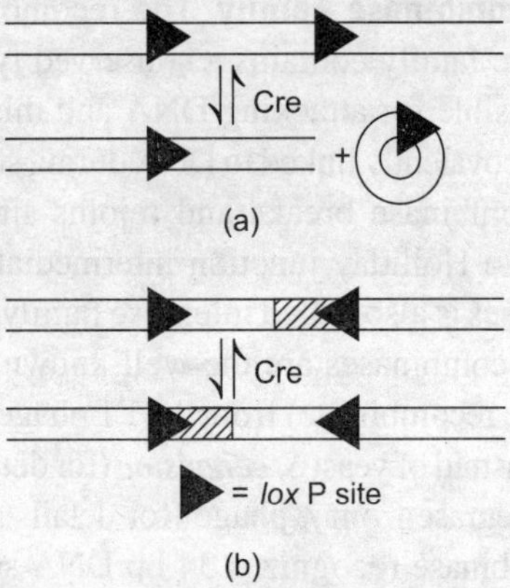

Figure 17.10 Structure of the *lox* P site and reactions catalyzed by Cre recombinase; (a) Direct repeats lead to deletion. (b) Inverted repeats lead to inversion

When cells having *lox* P sites in their genome express Cre recombinase, a reciprocal recombination event occurs between the *lox* P sites. The ds DNA is cut at both *lox* P sites by the Cre recombinase and then rejoined together. It is a quick, efficient, and reversible process. However, the efficiency of recombination depends on the orientation of the *lox* P sites (Figure 17.10). Cre/*lox* P system also plays a significant role in inversion as well as deletion of DNA segment between two *lox* P sites located on the same chromosome arm. When inverted *lox* P sites are present, it leads to inversion, while the presence of a direct repeat of *lox* P sites causes a deletion event. If *lox* P sites are on different chromosomes, Cre-induced recombination may catalyze translocation events.

Applications

SSR has numerous applications, some of which are discussed below:

Introduction and Excision of Selectable Markers The ability of flanking *lox* P for site-specific deletion can be used to delete unwanted transgenes. For example, it is applied for the replacement of wild-type allele of a given endogenous gene with an allele having a point mutation with concurrent introduction of marker genes (*neo*^{r} and *tk*) followed by marker excision. First positive and negative markers (*neo*^{r} and *tk*) flanked by *lox* P sites are introduced by homologous recombination. After screening of positive transformants on G418, the cells are transfected with a plasmid expressing Cre recombinase. Cre recombinase is used to excise both markers, leaving a single *lox* P site remaining in the genome. The excision event can be identified by selection for the absence of *tk* using ganciclovir or FIAU. This strategy can also be used for the development of marker-free transgenic plants. Vectors in which the selectable marker gene is placed between recognition sites for site-specific recombinase can be used to transform plants. Transgenic plants are developed and leaf explants from these plants are then transformed with a second construct, in which Cre recombinase is driven by CaMV 35S promoter. Consequently, the recombinase is expressed, which results in the excision of selectable marker gene from the plant genome. *R*/*RS* recombinase system from *Zygosaccharomyces rouxii* is also used for excision of selectable marker in clean gene technology of plants.

Recombinase-activated Gene Expression (RAGE) Site-specific recombinase can be used to activate transgene expression or switch between two alternative transgenes. This is called recombinase-activated gene expression (RAGE), in which a blocking sequence, such as polyadenylation site is placed between the transgene and its promoter, such that the transgene cannot be expressed. If this blocking site is flanked by *lox* P sites, Cre recombinase can be used to excise the sequence and activate the transgene. This approach has been applied successfully both in animal and plant systems. In mice, Cre recombinase was expressed under the control of developmentally regulated promoter, and in tobacco regulated by seed-specific promoter. However, Cre-mediated precise transgene integration has been successfully accomplished in plants while in mammalian cells integration is performed by using Flp recombinase (for details see Chapter 11).

Recombinant-mediated Cassette Exchange (RMCE) RMCE is performed routinely in animal cells. Successful targeted replacement occurs when the recognition sites between the donor and the target recombine and the intervening DNA is exchanged. The use of RMCE has improved our basic understanding of how chromatin and higher-order nuclear structure influence gene expression within specifically targeted regions of the genome. RMCE is also possible even in the presence of constitutively expressed recombinase.

Deletion, Inversion, and Translocation SSR between widely separated target sites or target sites on different chromosomes can be used to produce large deletions, translocations, and other types of mutations. This has been achieved in Drosophila, mice, and plants.

Cre-estrogen Receptor-mediated Targeting Cre expression is placed under the control of a specific promoter sequence, which permits localized and temporal expression of Cre. For regulating temporal activity of the Cre excision reaction, forms of Cre which take advantage of various ligand binding domains have been devised. One successful strategy for inducing temporally specific Cre activity involves fusing the enzyme with a mutated ligand-binding domain of the human estrogen receptor (ERt). Upon introduction of the drug tamoxifen (an estrogen receptor antagonist), the Cre-ERt construct enters the nucleus and induces targeted mutation. ERt binds tamoxifen with greater affinity than endogenous estrogens, which allows Cre-ERt to remain cytoplasmic in animals untreated with tamoxifen. The temporal control of SSR activity by tamoxifen permits genetic changes to be induced later in embryogenesis and/or in adult tissues. This allows researchers to bypass embryonic lethality while still investigating the function of targeted genes.

Role in Gene Knock-out (Targeted Disruption) This can be achieved by inserting a cassette anywhere in the integration vector, or within the homology domain of a transplacement vector. This cassette is usually a dominant selectable marker, such as the bacterial *neo*^{r} gene, which allows selection of targeted cells (Exhibit 17.1).

Role in Gene Knock-in This is a novel application where one gene is replaced by another (nonallelic) gene. This is achieved by inserting the incoming gene as a cassette within the homology domain, and is most readily achieved when swapping alternative members of multigene families.

Gene Therapy In this case, a mutant nonfunctional allele is replaced by a normal allele (for details see Chapter 20).

Development of PACs Cre/*lox* P recombination system is used in the development of bacteriophage P1-based vectors, PACs (for details see Chapter 12).

17.4 TECHNIQUES BASED ON GENE SILENCING

Gene silencing phenomenon was originally perceived as a problem or obstacle for further progress in genetic engineering, but unexpectedly, it has provided scientists with novel tools for regulating gene expression in many eukaryotes.

Two general classes of transgene silencing phenomenon have been demonstrated. The first type includes position effects, in which either flanking DNA or chromosomal location negatively influences the expression of single copy transgene loci. The second type of silencing results from interactions between homologous or complementary nucleic acid sequences, and is known as homology-dependent gene silencing (HDGS) that occurs when multiple copies of a particular sequence are present in a genome. There are two types of HDGS phenomena, which are distinguished according to the level of gene expression. These are transcriptional gene silencing (TGS) and posttranscriptional gene silencing (PGTS).

Exhibit 17.1 Gene Knock-out and Gene Knock-in

A gene knock-out is a genetic technique in which an organism is engineered to carry genes that have been knocked out (non-functional). The knock-out organisms are used in learning about a gene that has been sequenced, but which has an unknown or incompletely known function. Mice are currently the most closely related laboratory animal species to humans, for which the knock-out technique can easily be applied. A knock-out mouse is a genetically engineered mouse in which one or more genes have been turned off through a gene knock-out. By causing a specific gene to be inactive in the mouse, and observing any differences from normal behavior or condition, researchers can infer its probable function. Knock-out mice are widely used in investigation of genetic questions related to human physiology. The first knock-out mouse was created by Mario R. Capecchi, Martin Evans and Oliver Smithies in 1989, for which they were awarded the Nobel Prize for Physiology and Medicine in 2007. Examples of research in which knockout mice have been useful include studying and modeling different kinds of cancer, obesity, heart disease, diabetes, arthritis, substance abuse, anxiety, aging, and Parkinson's disease. Knock-out mice also offer a biological and scientific context in which drugs and other therapies can be developed and tested.

Knock-out is accomplished through a combination of techniques, beginning in the test tube with DNA construct, and proceeding to cell culture. Individual cells are genetically transformed with the DNA construct. Often the goal is to create a transgenic animal that has the altered gene. The effect of the absence of a gene can be very informative about the normal function of the gene. Eliminating a gene (gene Knock-out) completely from a diploid organism requires knocking-out both copies of the gene in the cells. There are a variety of methods for producing gene knock-outs in different model organisms.

In the mouse:

- DNA that has been engineered to contain a mutant copy of the gene is introduced into special ES cells that were growing in tissue culture,
- Cells that take up the DNA are tested to find those in which the mutant copy has replaced one good copy of the gene,
- Cells with one mutant copy are introduced into an early embryo (blastocyst) that will take up these cells,
- Mice that are born from this manipulation (and contain the one mutant copy in their germ cells) are mated to each other,
- One in four mice from this mating will contain two mutant copies of the gene.

While knock-out mice technology represents a valuable research tool, some important limitations exist. About 15% of gene knock-outs are developmentally lethal, which means that the genetically altered embryos cannot grow into adult mice. This problem is often overcome through the use of conditional mutations. The lack of adult mice limits studies to embryonic development and often makes it more difficult to determine a gene's function in relation to human health. In some instances, the gene may serve a different function in adults than in developing embryos. Knocking-out a gene also may fail to produce an observable change in a mouse or may even produce different characteristics from those observed in humans in which the same gene is inactivated. For example, mutations in the p53 gene are associated with more than half of human cancers and often lead to tumors in a particular set of tissues. However, when the p53 gene is knocked-out in mice, the animals develop tumours in a different array of tissues.

Knock-in is similar to knock-out, but instead it replaces a gene with another instead of deleting it. Knock-in (or Gene knock-in) refers to a genetic engineering method that involves the insertion of a protein coding cDNA sequence at a particular locus in an organism's chromosome. It is a technique by which scientific investigators may study the function of the regulatory machinery that governs the expression of the natural gene being replaced. This is accomplished by observing the new phenotype of the organism in question. This technique is essentially the opposite of a gene knock-out.

Source: www.en.wikipedia.org/wiki/Knock-out_mouse

Transcriptional Gene Silencing It is generally observed in plants but has also been seen in animals. TGS usually occurs when genes share homology in their promoter regions and hence no mRNA is produced from the targeted gene. Thus, gene expression is reduced by a blockade at the transcriptional level. Evidence indicates that transcriptional repression might be caused by chromatin modification or DNA methylation.

Posttranscriptional Gene Silencing It is commonly known as RNA silencing. In this type, protein complexes are involved that target specific mRNAs. Thus, in PTGS, the transcript of the silenced gene is synthesized but does not accumulate because of its rapid degradation. Different forms of RNA silencing include antisense RNA technology, cosuppression, quelling, and RNA interference (RNAi). All these forms depend on the presence of double stranded RNA (ds RNA). These techniques are being widely used in genetic manipulation and are discussed here in detail.

17.4.1 Antisense RNA Technology

In this technology, RNA having an opposite sense has been created and stable duplex of sense and antisense RNA molecules is formed in the cell, which may interfere with gene expression. Naturally occurring antisense transcripts are reported in prokaryotic and eukaryotic systems for the regulation of gene expression.

General Mechanism

Antisense RNA technology involves the designing of construct in which the target gene sequence is introduced in an antisense orientation with respect to the promoter. The resulting gene sequence is transcribed in reverse orientation using the opposite strand as the template, thereby generating antisense transcript. As antisense transcript from the particular gene has a sequence complementary to the sense mRNA, stable duplex is formed between antisense RNA and sense mRNA, the formation of which inhibits the expression of that particular gene (Figure 17.11).

Antisense inhibition mechanism may vary from case to case (Figure 17.12). Various mechanisms have been proposed which are categorized into two groups, nuclear or primary and cytoplasmic or secondary inhibition. Nuclear inhibition includes blocking of transcription, splicing, and transport of duplex mRNA from nucleus to the cytoplasm. In cytoplasm, antisense RNA blocks translation, as ribosome cannot gain access to the nucleotides in the mRNA or duplex RNA. Antisense RNA also blocks ribosome assembly and ribosome migration. The duplex RNA hybrids are recognized by cell-defense mechanisms and finally degraded.

The efficiency of the technique varies widely and the effect can in some cases be nonspecific. Using this technique, it is possible to shut down endogenous gene activity almost completely. However, conditional gene silencing can also be achieved by placing antisense constructs under the control of an inducible promoter. A number of other points to be considered include construct length (full-length vs. partial-length), locations of the partial-length fragments, antisense gene copy number, and use of proper gene switches. By increasing the copy number, the inhibitory effect of an antisense gene can be enhanced.

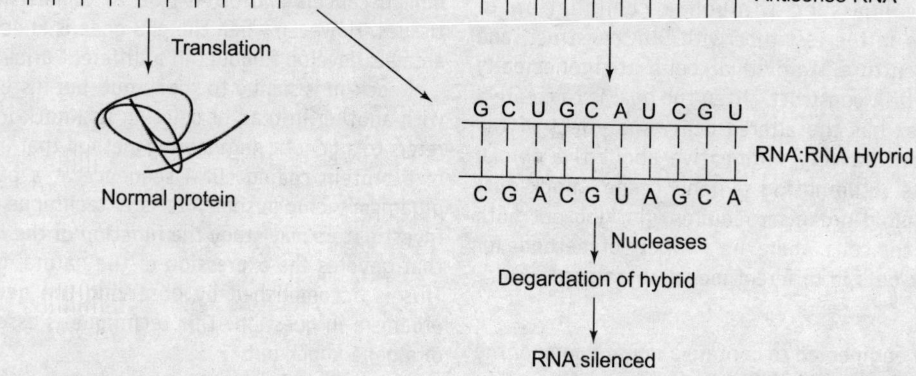

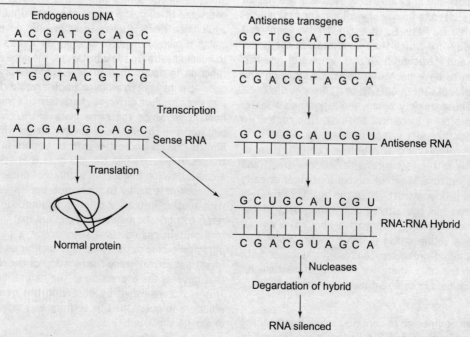

Figure 17.11 Principle of antisense RNA inhibition

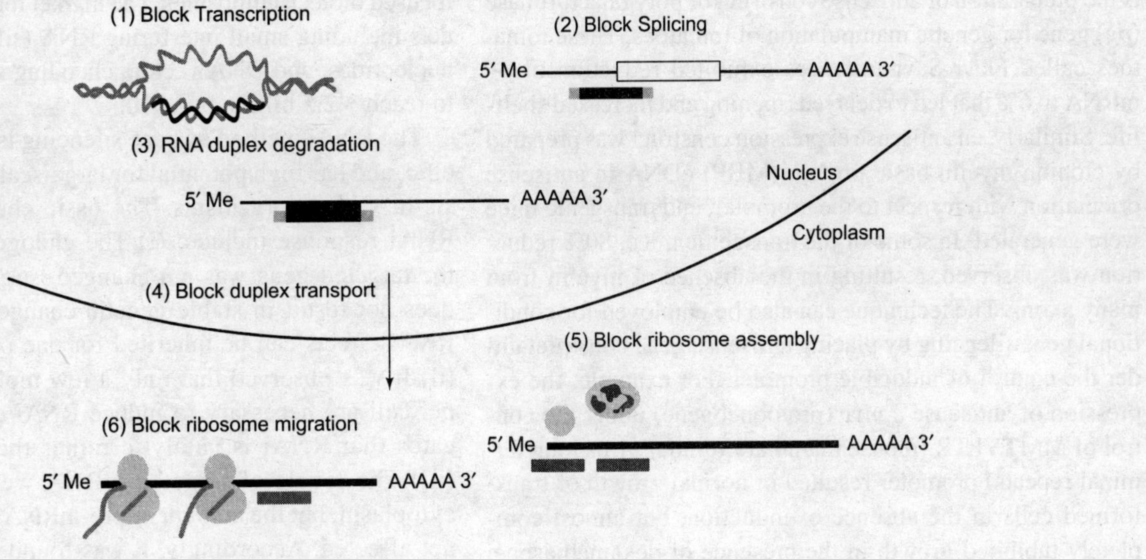

Figure 17.12 Different mechanism of antisense RNA inhibition

Applications

Antisense RNA technology is used to study transient inhibition of particular genes by directly introducing antisense oligonucleotide into cells. However, for the generation of stable inhibition of gene expression, antisense transgenes are used and the cells are transformed. The antisense RNA technology is successful for the genetic manipulation of both plants and animals. First application of antisense RNA technology

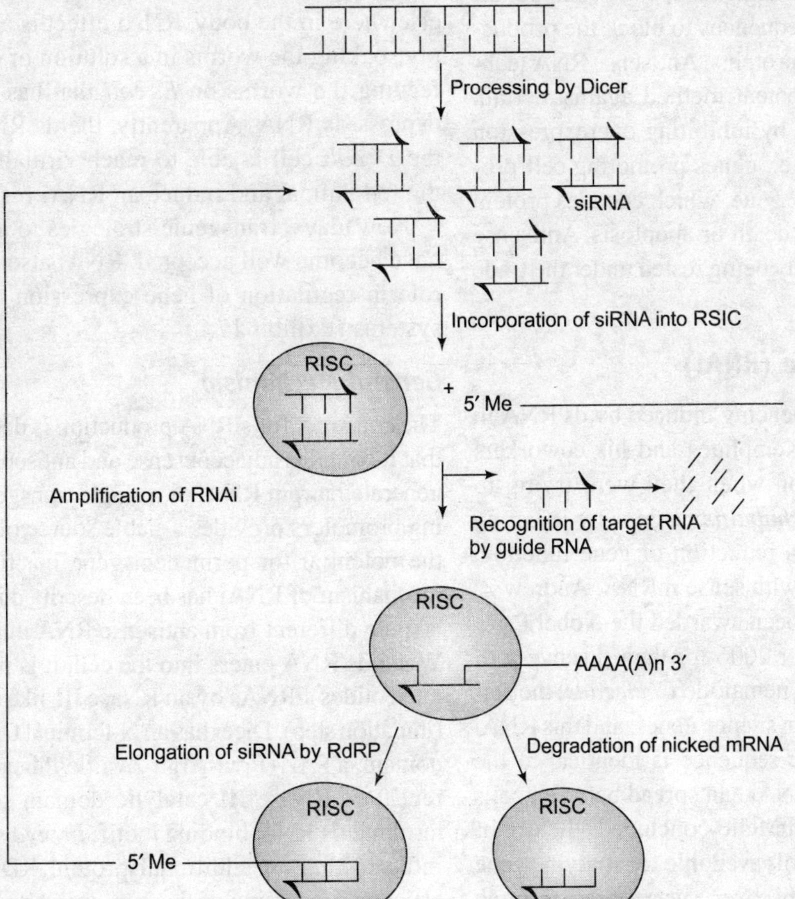

Figure 17.13 Mechanism of RNA interference

is the preparation of antisense construct of polygalacturonase (*pg*) gene for genetic manipulation of tomatoes. These tomatoes called Flavr Savr tomatoes exhibited reduction of *pg* mRNA to 6% that led to delayed ripening and increased shelf-life. Similarly, an antisense expression construct was prepared by cloning myelin basic protein (MBP) cDNA in antisense orientation with respect to the promoter, and transgenic mice were generated. In some of the transgenic mice, 80% reduction was observed, resulting in the absence of myelin from many axons. The technique can also be employed for conditional gene silencing by placing antisense gene construct under the control of inducible promoter. For example, the expression of antisense *c-myc* (protooncogene) under the control of MMTVLTR (mouse mammary tumour virus long terminal repeats) promoter resulted in normal growth of transformed cells in the absence of induction, but almost completely inhibited growth in the presence of dexamethasone. Nowadays, antisense constructs are widely used in gene inhibition for the generation of transgenic plants and animals (for details see Chapter 20). This technology has been used for improving the shelf-life of fruits and flowers, controlling viral diseases, develop insect, herbicide, disease resistance, metabolic engineering, and cancer treatment, etc. Currently, antisense technology is used to design therapeutic compounds, which target specific mRNA sequences to block the production of certain disease-causing proteins. Antisense RNA technology is also explored as a potent method against certain forms of cancer, for example, by inhibiting overexpression of tumour-suppressor genes (i.e., genes promoting cell proliferation) or by targeting *bcl*-2 gene, which encodes protein that prevents programmed cell death or apoptosis. Antisense therapy against *bcl*-2 is currently being tested under the trade name Genazyme.

17.4.2 RNA Interference (RNAi)

RNAi is the process of gene silencing induced by ds RNA in a sequence-specific manner. Kemphues and his coworkers first revealed this phenomenon when they were trying to downregulate genes in *Caenorhabditis elegans* by injecting antisense RNA, which led to a reduction of gene function. Similar results were observed with sense mRNA. Andrew Z. Fire and Craig C. Mello have been awarded the Nobel Prize in Physiology and Medicine for 2006 for their discovery of RNAi. While working with the nematode *C. elegans*, the scientists observed that ds RNA can silence genes, and this RNAi is specific for the gene whose sequence is identical to the injected RNA molecule, and RNAi can spread between cells and even be inherited. Fire and Mello concluded 'the use of ds RNA injection adds to the tools available for studying gene function in *C. elegans*'. This discovery was a breakthrough in molecular biology research and provided innovative possibilities, leading to the establishment of over a dozen RNAi-focused biotech companies. The market for RNAi-based products including small interfering RNA (siRNA), RNA oligonucleotides, and DNA vectors encoding siRNA is estimated to reach $ 2.5 billion by 2010.

The RNAi method of gene silencing is quite simple, specific, and has high potential for large-scale functional analysis in various organisms. The basic characteristics of the RNAi response include: (i) The endogenous sequence of the targeted gene was not changed, suggesting that RNAi does not result in stable genetic changes. However, some RNAi effects can be inherited for one or two generations; (ii) It was observed that only a few molecules of ds RNA per cell are necessary to induce RNAi effect, which indicates that RNAi is catalytic rather than stoichiometric; (iii) The levels of targeted mRNA were reduced in the cytoplasm, but the amount of pre-mRNA in the nucleus was not affected. Accordingly, it was found that genes located in operons usually do not cross react in RNAi experiment. The effect of interference can be only attained if the introduced ds RNA is homologous to the exons of a target gene, demonstrating that it is a posttranscriptional process; and (iv) RNAi effect is capable of spreading within the nematode. In *C. elegans*, *in vitro* synthesized ds RNA is injected into the germ line of adult worms, which exerts an effect elsewhere in the body. RNAi effect is also induced by simply soaking the worms in a solution of ds RNA, or even by feeding the worms on *E. coli* that has been engineered to express ds RNA. Apparently, the ds RNA produced inside the *E. coli* cell is able to reach virtually all tissues within the fed animal and induce an RNAi response.

Nowadays, transgenic strategies to achieve RNAi effect have become well accepted. RNAi also play very important role in regulation of gene expression naturally in various systems (Exhibit 17.2).

General Mechanism

The construct for siRNA production is designed in such a way that it contains adjacent sense and antisense transgenes, which generate hairpin RNA, or a single transgene with dual opposing promoters provides a stable source of ds RNA, and hence the potential for permanent gene inactivation. The general mechanism of RNAi has been described in Figure 17.3 which is quite different from antisense RNA inhibition (Table 17.2). When ds RNA enters into the cell, it is processed into 21–28 nucleotides siRNAs by an RNase III-like enzyme called Dicer (initiation step). Dicer has an N-terminal DExH/DEAH helicase domain, a PAZ (Piwi-Argo-Zwiille/Pihead) domain, a tandem repeat of RNase III catalytic domain sequences, and a C-terminal ds RNA-binding motif. *In vivo* studies in *C. elegans* indicate that an additional protein RDE-4 is required for efficient processing of the introduced ds RNA. The RDE-4 is found in complex with DCR-1, together with RDE-1, and an RNA helicase, DRH-1, indicating that these proteins together

Exhibit 17.2 Natural Occurrence of RNAi

Various studies have been performed in several systems that proved that RNAi plays a very important role in the regulation of gene expression. This response provides protection to genomes against invading sequences such as viruses and transposable elements required for proper development, cellular regulation, and belong to a large class of evolutionarily conserved RNAs. There are three different types of small RNAs that occur naturally. These are siRNAs, microRNAs, and repeat-associated short-interfering RNAs (rasiRNAs), which can be distinguished by their origin. siRNAs are processed from ds RNA precursors made up of two distinct strands of perfectly base paired RNA. Many of the endogenous siRNAs are derived from repetitive sequences within the genome, hence the term repeat-associated siRNAs, or rasiRNAs. miRNAs originate from a single, long transcript that forms imperfectly base paired hairpin structures. miRNA gene products most probably regulate gene expression by regulating mRNA, either destroying the mRNA when the sequences match exactly or repressing its translation when the sequences are only a partial match. To repress translation, several miRNAs bind simultaneously in the 3′-untranslated region (3′-UTR). These 'riboregulators' are best-suited for this purpose as these are too small, these can be rapidly transcribed from their genes and these do not need to be translated into a protein product to act.

The processing of the small RNAs is performed either by Dicer or Drosha. The genes of miRNAs are first transcribed as long primary transcripts and then processed by Drosha in the nucleus. Once Drosha has excised the miRNA precursor, a 2-nucleotide 3′ overhang and a 5′-phosphate are left over at the stem base. This miRNA precursor is then moved to the cytoplasm by a nuclear export receptor called exportin-5. After entering into cytoplasm, it is processed further to RNA duplexes of ~21 nucleotides in length having 2-nucleotide 3′ overhangs and 5′-phosphates. A novel study to regulate gene expression was discovered when *lin*-4 and *let*-7 hairpin RNAs were identified in *C. elegans*. The *C. elegans lin*-4 regulatory gene, a 22-nucleotide RNA processed from a precursor hairpin RNA, was identified during the screening of mutations that affected the timing and sequence of postembryonic development. Multiple *lin*-4 RNAs bind to multiple regions of the 3′-UTR of the *lin*-14 gene (heterochronic genes), which represses the translation. Later on *let*-7 gene was identified, which acts as a posttranscriptional negative regulator targeting the 3′-UTR of *lin*-41 and *lin*-14 genes (heterochronic genes). However, the precise function of these RNAs is unknown, but it is clear that this gene shows a developmentally regulated timing of expression. On this basis, it is termed as small temporal RNA (stRNA).

Source: Matzke *et al.* (2003) In: Gregory J. Hanon (Ed.), *RNA I-A Guide to Gene Silencing*, CSHL Press.

are required for siRNA production during RNAi in *C. elegans*. In *Drosophila*, the rate of siRNA formation is ATP-dependent and siRNA produced in the absence of ATP is one nucleotide longer than in the presence of ATP. However, the involvement of ATP during processing of ds RNA yet remains to be explored. Formation of siRNA has now been recognized for plants, filamentous fungi, *C. elegans, Trypanosoma brucei*, and mouse ES cells. The length of siRNA produced varies between 21 and 28 nucleotides apparently reflecting the structural differences of various Dicer orthologs. The distinct size and structure of siRNAs reflect the geometric spacing between the active sites of ds RNA-bound dimers of Dicer during processing. Then the siRNA duplexes assemble into a multisubunit protein complex, known as RNA-induced silencing complex (RISC). RISC forms complex with siRNA duplexes prior to target RNA recognition that is referred to as siRNP. The formation of RISC on siRNA duplexes requires ATP, but once formed, RISC can mediate robust, sequence-specific cleavage of its target in the absence of ATP. One strand of siRNA is integrated into the RISC complex, which is known as the guide strand and is selected on the basis of the stability of the 5′-end. The remaining strand is known as the antiguide or passenger strand, which is degraded as a RISC complex substrate. The siRNA strands subsequently guide the RISCs to complementary RNA molecules, which induces a break in the target mRNA in the region covered by the siRNA molecule. The target RNAs are cleaved precisely 10 nucleotides upstream of the target position complementary to the 5′ most nucleotide of sequence-complementary guide siRNA. Importantly, the 5′-end, and not the 3′-end, of the guide siRNA sets the ruler for target RNA cleavage. Furthermore, the presence of a 5′-phosphate at the target-complementary strand of a siRNA duplex is required for siRNA function and ATP is used to maintain the 5′-phosphates of the siRNAs. The mechanism of siRNA-mediated target RNA cleavage is conserved between *Drosophila melanogaster* and mammals.

Table 17.2 Comparison of Antisense RNA Technology and RNAi

Antisense RNA inhibition	RNA interference
Single strands of RNA are targeted to bind with the mRNA	RNAi mimics a natural process by using ds siRNAs
No RISC complex	RISC complex forms
Antisense technology is hard to target and unstable in the body	Relatively easy to target
Antisense therapies seek to block a key cellular biochemical process	RNAi 'directs' the cell's own biology (RISC complex) to detect and shut off rouge genetic activity and corresponding protein synthesis

Some of the components of RISC have been explored. One of the components of RISC has been identified as Ago2, that is, fly homolog of the *C. elegans* RDE-1 gene product. These proteins are members of a large family, with other members such as Argonaute (*Arabidopsis*), and the QDE-2 (*Neurospora crassa*) proteins, both of which are required for RNA-induced gene silencing. It has been proposed that RDE-1 protein acts as an adaptor between Dicer and RISC. Some other protein factors have been identified in the germ line tissue of *C. elegans* through genetics, such as MUT-7 (an exori-bonuclease), MUT-14 (an RNA helicase), EGO-1 (an RNA-directed RNA polymerase; an RdRP), and MUT-8 and MUT-15 (two proteins with no recognizable protein motifs). These factors are required for RNAi, but tested only in the germ line tissue of *C. elegans*.

Amplification of RNAi

In *C. elegans*, plants, and *Neurospora*, the introduction of a few molecules of ds RNA has a potent and long-lasting effect. In plants, the gene silencing spreads to adjacent cells through plasmodesmata and even to other parts of the plant via phloem. RNAi within a cell can continue after mitosis in the progeny of that cell. Triggering of RNAi in *C. elegans* can even pass through the germ line into its descendants. Such amplification of an initial trigger signal suggests a catalytic effect. The siRNA population is amplified, but only if there is a suitable target RNA that can be used as a template. When the sequence of an siRNA molecule is complementary to the 5′-end of the original ds RNA, it is used as a primer and ds RNA will be produced with the help of enzyme RNA-dependent RNA polymerase. This ds RNA will also be processed by Dicer, leading to secondary siRNA, which again mediates gene silencing. These secondary siRNAs target additional areas of the original mRNA and mRNAs of other genes that may carry the same sequence of nucleotides. This phenomenon is called 'transitive RNAi'. Usually, transitivity works only in the $5' \rightarrow 3'$ direction. It is noteworthy that transitivity is also observed in plants and the effect is not unidirectional. RNAi process is capable of building up a powerful response when both ds RNA and homologous target RNA are present, all of which bear a striking resemblance to the immune response.

Mammalian RNAi

RNAi phenomenon seems to be quite general and has been applied for gene silencing experiments in many other organisms, including *Drosophila*, plants, and even mammalian cells. The application of RNAi in vertebrate animals, including mammals, has proven to be more difficult because of the presence of additional ds RNA-triggered pathways that mediate nonspecific suppression of gene called the interferon response, which shuts down protein synthesis in the presence of ds RNA molecules greater than 30 bp in length and masks any specific effects of gene silencing. Fortunately, these responses to ds RNA in vertebrates are not triggered by siRNAs. siRNA can target genes as effectively as long ds RNAs and are widely used today for assessing gene functions in cultured mammalian cells or early developing vertebrate embryos. Chemically synthesized siRNAs are easily transfected into mammalian cells that induces specific RNAi effects without showing specific interferon response. Mammalian cells also lack intrinsic amplification of the triggering RNA, which prolongs the effect, and makes it more potent. Because of the absence of amplification in mammalian cells, there has been much interest in the development of transgenic systems for the expression of siRNA. Expression cassettes have been constructed and transcribed by RNA polymerase III, since this enzyme transcribes the naturally occurring short RNA genes in mammalian genomes. Three types of strategies have been developed for the production of siRNA in mammalian cells, which are as follows: (i) In one approach, a plasmid is constructed, which contains two RNA polymerase III transcription units in tandem, each producing one of the siRNA strands. The two separate strands are assembled spontaneously into the siRNA duplex *in vivo*; (ii) A second very similar strategy has been developed, in which RNA polymerase III transcription units are present on separate vectors. The individual RNA strands assemble in the same manner; and (iii) In a third strategy, the plasmid DNA produces a hairpin RNA, which assembles into siRNA by self-pairing. Either RNA polymerase II or III can be used to produce hairpin siRNAs because the transgene is longer.

Basic Requirements of RNAi Experiments

RNAi experiments have four basic requirements: (i) A specific ds RNA that targets a particular gene transcript to induce the RNAi pathway. In mammalian cultured cells, RNAi is typically induced by siRNA or short hairpin RNA (shRNA). The comparison between siRNA and shRNA for RNAi are presented in Table 17.3; (ii) An efficient delivery method for the trigger. The two most commonly used methods are transfection and electroporation; (iii) A robust phenotypic assay for the RNAi is critical to the success of screening experiments using RNAi tools; and (iv) Positive and negative controls for both siRNA delivery and for the phenotypic assay.

Applications

The discovery of RNAi adds a promising tool to the toolbox of molecular biologists, as introducing ds RNA corresponding to a particular gene knocks-out the cell's own expression of that gene. Some of the advantages and disadvantages associated with this technique are given in Table 17.4. The exciting result was obtained when ds RNAs corresponding to 20,326 of *C. elegans* genes (98% of the total) were injected. It was observed that at least 661 different genes altered some process during this period, about half of them involved in cell division, and half in general cell metabolism.

Table 17.3 Comparison of siRNA and shRNA for RNAi

siRNA	shRNA
Expensive	Economical
Easier to design	More difficult to design
Produced by *in vitro* or *in vivo* transcription	Primarily expression-based but can be produced by chemical synthesis or *in vitro* transcription
Toxicity is an important issue, especially in primary cells and neurons	Toxicity usually not a problem
	Can be used to generate stable cell lines
	High level of purity required
Suppression effect lasts for 5–7 days	Sustained suppression of gene expression
Less time from start to finish	More time from start to finish
Focuses on short-term effects	Focuses on long-term effects

Because RNAi can be performed in particular tissues at a chosen time, RNAi has been useful for different purposes in plant and animal genomes. The medical applications of RNAi are potentially very exciting. RNAi phenomenon also has wide scope in the field of plant biotechnology. Some common examples of RNAi technology are: (i) Screening genes for their effect on drug sensitivity; (ii) Human cells in culture can be protected from poliovirus and HIV-1 by siRNAs with sequences that match RNA molecules encoded by the viruses; (iii) RNAi has high potential in human therapy, as its target is very specific. The possibility of using RNAi to shut down the expression of a single gene has created great excitement that a new class of therapeutic agents is on the horizon. Antisense RNA that is complementary to the proto-oncogene *bcl*-2 is being examined as a possible therapy for certain B-cell lymphomas and leukemias. In mice and monkeys, an intravenous injection of a chemically modified siRNA specific for the mRNA of apolipoprotein B reduces the levels of the apolipoprotein B mRNA with an accompanying reduction in the level of the protein, which results in lowering their cholesterol and LDL levels. Introducing siRNAs targeting genes of herpes simplex virus-2 (HSV-2) protects mice from lethal infection by the virus; (iv) Coffee plants have been engineered to express a transgene that makes siRNA, which interferes with the expression of a gene needed to make caffeine; (v) Two transgenic plants cotton and corn are developed that express a ds RNA, allowing resistance to cotton bollworm and western corn rootworm, respectively. When these herbivorous pests feed on the transgenic plants, RNAi inhibits synthesis of an enzyme that would otherwise give the animal resistance to the chemical defenses of the host plant; and (vi) RNAi is also used to overcome genetic redundancy in polyploids, modifying plant height in rice grains by interfering gibberellin metabolism, changing the glutein content of rice grains and the oil content of cotton seeds, and controlling the development of leaves.

17.4.3 Cosuppression

Cosuppression is defined as the mutual suppression of transgene and homologous endogenous gene expression. This is one example of PTGS, in which expression of a transgene, even though intact and stably integrated, is suppressed. Researchers transformed *Petunia* with chalcone synthase sense gene under the control of strong promoter, with the aim to raise gene expression levels but were surprised at the result. Surprisingly

Table 17.4 RNAi Advantages and Disadvantages

Advantages	Disadvantages
Analyze function(s) of one gene at a time	RNA oligos are more expensive than DNA oligos
Highly specific to the gene(s) of interest	Skilled-hand worker can perform
Systematic approach to investigate gene function and which genes are not involved in which pathways	ds RNA >30 bp upregulates interferon response genes in mammalian cells; leads to apoptosis
Highly evolutionarily conserved process more efficient at mRNA destruction than ribozymes	Some cells, especially neurons, tend to be resistant to RNAi
Less toxic than phosphorothioate-based oligonucleotides	Reproducibility is sometimes a problem
Easy to deliver to mammalian cells	Some siRNAs, no matter how rationally designed, do not work, mandating the trial and error approach
Well-suited for drug-target identification	Interlaboratory comparisons often difficult because of variations in protocols
ds RNA is effective at lower concentrations than antisense oligos	Lack of positive and negative controls
	Synthetic RNAi oligos must be extremely pure

~50% of the transgenic petunias were either variegated or completely white instead of a darker flower. This phenomenon was termed cosuppression, since both the expression of the existing gene (the initial purple color) and the introduced gene (to deepen the purple color) were suppressed. Similar results were reported by using a transgene encoding another pigment biosynthesis pathway enzyme, dihydroflavonol-4-reductase. Like antisense technology, the tomato ripening system also acts as one of the pioneers to disclose the event of cosuppression, which was observed in fruits having sense *pg* construct as transgene. In several transgenic fruits, the PG expression was less as compared to nontransgenic. This suggested that plant cells have some mechanism by which the presence of extra gene copies of the endogenous gene is detected, and expression of both endogenous gene and transgene has been suppressed. RNA viruses can also induce cosuppression in plants as long as there is a region of homology between the virus genome and an integrated gene. This was clearly shown by nuclear run-on assays, which indicated that transcript was being made but that it failed to accumulate in the cytoplasm. Furthermore, silencing was a systemic phenomenon, which could spread from the source plant by grafting tissue onto another host plant that did not contain the transgene. This phenomenon was observed when transgenic plants were generated by transforming potato virus X (PVX) genome having the *gus* A reporter gene. It was expected that extremely high levels of transgene expression could be obtained through amplification of the viral RNA by its own RNA-dependent RNA polymerase. Surprisingly plants infected with virus vector (containing a reporter) had moderate levels of reporter gene expression, combined with viral resistance. This response was also effective against genes that had sequence homology to the viral RNA. Later this was performed with replication- defective viruses, which was unable to produce this response, indicating that ds RNA intermediate involved in viral replication was the trigger for silencing. PVX vectors have been used very successfully to generate functional knock-outs in plants. PVX vectors having cellulose synthase cDNA was used to raise transgenic plants with reduced levels of cellulose.

The transgene cosuppression phenomenon has also been observed in *Drosophila*, *C. elegans*, and rodent fibroblasts. Transgene silencing in the ascomycete fungus, *N. crassa*, is termed 'quelling' and occurs at the posttranscriptional level. Furthermore, in *Drosophila*, single copy of a *white-adh* (for alcohol dehydrogenase) fusion transgene can induce silencing of the endogenous *adh* gene. This silencing response was more prominent in flies that were homozygous for the transgene than those that were homozygous, because of interaction between the two alleles. This result suggested that silencing was most probably occurring at the transcriptional level. In another study, it was observed that transgenes that were present as inverted repeats were more effectively silenced than those arranged in direct repeats. It was probable that, at least in the latter case, silencing was triggered by a ds RNA that induces PTGS through the RNAi pathway. However, links between transgene cosuppression and RNAi have come from genetic analysis, search for mutant organisms that are resistant to either RNAi or transgene cosuppression has revealed a common set of molecules. Additionally, some mutations that prevent RNAi in *C. elegans* also relieve transgene cosuppression. It is unclear whether cosuppression that occurs at the transcriptional level or that is provoked by unlinked and single copy transgenes is related to RNAi.

17.4.4 Ribozymes

The statement that all enzymes are proteins was proven to be wrong by Thomas Robert Cech and Sidney Altman who shared the 1989 Nobel Prize in Chemistry for their demonstration that RNA could act as an enzyme. These antisense RNA molecules with enzymatic properties are termed ribozymes. The ribozyme action is generated by the formation of particular secondary and tertiary structures that create active sites. These function by binding to the target RNA moiety through base pairing and inactivate it by cleaving the phosphodiester backbone at a specific cutting site. Various types of reactions performed by ribozymes are based on transesterification. These reactions include splicing, oligonucleotide chain extension, RNA ligation, endonuclease action, and phosphatase action. A ribozyme has two sites: a substrate binding site and a guanosine-binding site. Ribozymes have the potential to become useful therapeutic agents and currently the vast majority of effort has been expended in the development of *trans* cleaving hammerhead and hairpin ribozymes as inhibitors of viral gene expression, in particular to cleave and destroy HIV-I RNAs to inhibit viral replication in infected cells. RNA enzymes can potentially be quite useful for a variety of gene therapy applications. A number of research laboratories around the world are now using these ribozymes to regulate the gene expression in a precise manner. However, the use of ribozymes is associated with hurdles, since cell is producing a large number of RNAs from a huge number of different genes. Obviously when the ribozyme is introduced into the cell, the researcher does not want the ribozyme to cut all the RNA messengers since a large number of genes will be turned off. This problem of specificity was solved by incorporating the ribozyme catalytic centers into antisense RNA, which permitted the ribozyme to be targeted to particular mRNA molecules, which could then be cleaved and degraded. The ribozyme approach has advantage over antisense RNA because of its catalytic activity. Ribozymes are recycled after the cleavage reaction and can therefore inactivate many mRNA molecules, while antisense inhibition relies on stoichiometric binding between antisense and sense RNA molecules.

The use of ribozyme constructs for specific gene inhibition in eukaryotes has first been reported in *Drosophila*, where eggs were injected with P-element vector containing a ribozyme construct targeted against the white gene. The resulting transgenic flies have reduced eye pigmentation, demonstrating that expression of endogenous gene has been repressed. Ribozymes have been also been used in mammalian cell lines, mostly for the study of oncogenes and in attempts to confer resistance to viruses. Intensive research has been performed in the field of ribozyme-mediated inhibition of HIV and incredible success has been achieved using retroviral vectors, particularly carrying multiple ribozymes. Mice expressing three different ribozymes targeted against β2-macroglobulin mRNA have been developed, which have succeeded in reducing endogenous RNA levels by 90%. Tissue-specific expression has also been reported. Thus, a ribozyme targeted against glucokinase mRNA has been expressed in transgenic mice under the control of insulin promoter, resulting in specific inhibition of the endogenous gene in pancreas. The ribozyme approach has also been used in plants as an antiviral tool, but has not been very successful.

Review Questions

1. What physical and chemical properties of enzymes are targets for enhancement by directed mutagenesis?
2. You have cloned a bacterial gene that is expressed in *E. coli* and now you want to alter its activity. However, because of technical problems with the original M13 protocol, only a very small fraction of mutagenized clones of this gene actually carry the modified gene. How would you perform site-specific mutagenesis so that a much larger proportion of the clones have the desired mutation?
3. You have isolated a gene for an enzyme that is expressed in *E. coli.* Describe how you would alter the catalytic activity of the enzyme. Assume that you know the DNA sequence of the gene but do not know anything about which regions of the enzyme are important for catalytic activity.
4. Describe a strategy for oligonucleotide-directed mutagenesis with plasmids. How can degenerate oligonucleotides be used to generate randomly mutagenized DNA?
5. What do you understand by error-prone PCR and why is it useful?
6. Give a brief explanation of DNA shuffling and outline two ways in which this technique may be used to generate hybrid genes.
7. Bisulfite mutagenesis is used for introducing localized point mutagenesis. To perform this, you need to generate the single stranded regions that will be susceptible to bisulfite treatment. Give an explanatory description of various known techniques used for this purpose.
8. Mario R. Capecchi, Martin J. Evans, and Oliver Smithies won the 2007 Nobel Prize in physiology and medicine for their work on gene targeting. Describe the procedure of gene targeting by discussing the vectors and screening strategies.
9. What do you understand by site-specific recombination? Describe various recombinase systems and also mention some applications of this gene manipulation strategy.
10. What is the principle of mechanism of antisense RNA technology? Give a brief account of engineering of fruit ripening by utilizing this approach. Give a brief description of strategy by which you can enhance the antisense inhibition by using ribozyme approach.
11. What do you understand by the term RNA interference and how can you relate it to RNA silencing in *C. elegans*?
12. Define cosuppression and explain how this approach is used for gene silencing in plants.

References

Altman, S. (2007). A view of RNase P. *Mol Biosyst* **3**(9):604–607.

Capecchi, M.R. (2000). Choose your target. *Nature Genet.* **26**: 159–161.

Cech, T.R. (1987). The chemistry of self-splicing RNA and RNA enzyme. *Science* **236**:1532–1539.

Estell, D.A., T.P. Graycar, and J.A. Wells (1985). Cassette mutagenesis of lysine 130 of human glutamate dehydrogenase. *Eur. J Biochem.* **268**:3205–3213.

Evans, M.J. *et al.* (1997). Gene trapping and functional genomics. *Trends Genet.* **13**:370–374.

Fire, A.Z. *et al.* (1998). Potent and specific genetic interference by ds RNA in C. elegans. *Nature* **391:**806–811.

Guo, S. and K.J. Kemphues (1995). *par-1*, a gene required for establishing polarity in *C. elegans* embryos, encodes a putative ser/thr kinase that is asymmetrically distributed. *Cell* **81**:611–620.

Kunkel, T.A., K. Bebenek, and J. McClary (1991). Efficient site-directed mutagenesis using uracil-containing DNA. *Methods Enzymol.* **204**:125–139.

Sauer, B. (1994) Site-specific recombination: developmemts and applications. *Curr Opin. Biotechnol.* **5**:521–527.

Shortle, D. and D. Nathans (1978). Local mutagenesis: A method for generating viral mutants with base substitutions in preselected regions of the viral genome. *Proc. Natl. Acad. Sci.* **75**:2170–2174.

Smith, M. (1985). *In vitro* mutagenesis. *Annu. Rev. Genet.* **19**: 423–462.

Smithies, O. (1993). Animal models of human genetic diseases. *Trends Genet.* **9**:112–116.

Sung-Woo C. *et al.* (1986). Engineering an enzyme by site-directed mutagenesis to be resistant to chemical oxidation. *J Biol. Chem.* **260**:6518–6521.

Molecular Markers

Key Concepts

In this chapter we will learn the following:

- Biochemical markers
- Molecular markers
- Restriction fragment length polymorphism (RFLP)
- Random amplified polymorphic DNA (RAPD)
- Amplified fragment length polymorphism (AFLP)
- Microsatellite or short tandem repeat (STR)
- Single strand conformation polymorphism (SSCP)
- Single nucleotide polymorphism (SNP)
- Map-based cloning
- Marker-assisted selection

18.1 INTRODUCTION

In the 1860s, Gregoire Mendel carried out plant breeding experiments on the garden pea and developed the law of inheritance. Mendel's work has set the foundation for the molecular marker technologies that are in use at present. The markers are heritable characteristics associated with and useful for the identification and characterization of specific genotypes.

In the past, morphological markers were used in crop improvement projects, but these have certain limitations. These include limited availability of easily scorable markers, difficulty in scoring homozygous from heterozygous individuals, influence of environment, and difficulty in comparing phenotypes with genotypes.

On the contrary, molecular markers offer several advantages. These are more adaptable, exhibit abundant polymorphism, display no pleiotropic effect, less affected by environment, and rapidly detected. Molecular markers can be used to select individual plants or animals carrying genes that affect economically important traits such as fruit yield, wood quality, disease resistance, milk and meat production, or body fat. Measuring such characteristics by conventional methods is much more difficult, time-consuming, and expensive, since it requires the organism to grow to maturity. Molecular markers are useful to measure the extent of variation at the genetic level, within and among populations. Molecular markers are normally divided into two classes: biochemical molecular markers and molecular genetic markers derived from direct analysis of polymorphism in DNA sequences. Antibodies are also used as molecular probes to recognize specific protein sequences as the information inherent in the DNA sequence is ultimately translated into amino acid sequence of protein.

18.2 BIOCHEMICAL MARKERS

Biochemical markers is derived from study of products of gene expression, i.e., proteins. These can be isolated and characterized by electrophoresis and staining. Two major types of biochemical markers are isozymes and allozymes.

18.2.1 Isoenzymes

Isoenzymes (isozymes) are different variants of the same enzyme having identical functions, i.e., catalyze the same chemical reaction. The isozymes are encoded by different genes and thus represent different loci. These enzymes usually show different kinetic parameters or different regulatory properties. These have different molecular weight and hence show different banding pattern during electrophoresis. The presence of isozymes permits the fine-tuning of metabolism to meet the specific requirement of a given tissue or developmental stage. These are coded by homologous genes that have diverged over time.

18.2.2 Allozymes

Allozymes are different molecular forms of an enzyme that correspond to different alleles of a gene (locus). These are quite different from isozymes, which are derived from separate gene per loci. Allozymes are detected using electrophoresis. This is because different allozymes migrate differently owing to slight difference in amino acid composition and as a result of different electric charges. Allozymes have traditionally been used to assess genetic variation within a population or species, but these are also used in phylogenetic analysis of closely related species.

Isozymes and allozymes have been among the most widely used molecular markers to study the causes and effects of genetic variation within and between populations. Although these have now been largely superseded by more informative DNA-based approaches, these are still amongst the quickest and cheapest marker systems and serve as excellent choice for projects that only need to identify low levels of genetic variation.

18.3 MOLECULAR MARKERS

Molecular markers, commonly known as DNA-based markers, are specific segments of DNA that can be identified within the whole genome at specific locations. Because of their specific locations on a chromosome, these genetic molecular markers are used for chromosome mapping. These are used to 'flag' the position of a particular gene or the inheritance of a particular characteristic. In a genetic cross, the characteristics of interest are usually linked with the molecular markers. Thus, individuals can be selected in which the molecular markers are present since the markers indicate the presence of the desired characteristics. Nowadays, in plant and animal breeding projects, the analysis of genomes with the help of molecular markers plays an important role in the rapid development of improved crops and livestock with enhanced productivity. Molecular markers provide enough information about the genome organization and several practical applications like variety identification through DNA fingerprinting, development of genetic map, which facilitate individual selection of economic traits such as disease resistance without cumbersome screening, cloning of important genes, and evolutionary and phylogenetic studies. The DNA fingerprints of a particular species verify the uniqueness of the species that makes it different from others. The sequence of DNA is different in every individual; therefore a specific sequence of DNA can be utilized as a tool to screen variations within a given population. Moreover, DNA-based molecular marker techniques provide advanced and reliable approaches to study variability, most likely applicable to any genome with a widespread array of promising studies. The desirable properties of ideal molecular marker are as follows: (i) The marker should show high polymorphic behavior, as it is the polymorphism itself, which is measured for genetic diversity studies. However, the level of polymorphism detected can vary depending on the method of measurement; (ii) Generally codominant marker is preferred, as it can easily discriminate between homozygotic and heterozygotic states in diploid organisms. Codominance can be defined as the absence of intralocus interactions; (iii) The marker must show nonepistatic behavior, i.e., its genotype can only be inferred from its phenotype, whatever the genotype at other loci may be. Epistatis is intergenic interaction in which one gene masks the effect of another gene; (iv) The allelic substitutions at the marker locus should not have phenotypic effects. It must be neutral; (v) It should be insensitive to the environment; (vi) The marker should be evenly and frequently distributed throughout the genome; (vii) It should be easy, fast, and inexpensive to detect; and (viii) It should show high reproducibility.

The most commonly used DNA marker systems are compared in Table 18.1.

Most biochemical markers also have all these qualities. However, morphological markers do not fulfill these requirements, as these are insufficiently polymorphic and are generally dominant. Furthermore, these often interfere with other traits and can be influenced by the environment.

18.3.1 Types of Molecular Markers

Various types of molecular markers are utilized to evaluate DNA polymorphism and are generally categorized as follows:

PCR-based Markers

PCR-based markers involve *in vitro* amplification of particular DNA sequences with the help of specifically or arbitrarily chosen oligonucleotide primers. The amplified fragments are separated by electrophoresis and banding patterns are detected by different methods such as ethidium bromide (EtBr) staining and autoradiography.

Table 18.1 Comparison of Commonly Used Marker Systems

Feature	RFLPs	RAPDs	AFLPs	Microsatellite	SNPs
Template requirement (μg)	10	0.02	0.5–1.0	0.05	0.05
DNA quality	High	High	Moderate	Moderate	High
PCR based	No	Yes	Yes	Yes	Yes
Number of loci tested	1.0–3.0	1.5–50	20–100	1.0–3.0	1.0
Ease of use	Not easy	Easy	Easy	Easy	Easy
Amenable to automation	Low	Moderate	Moderate	High	High
Reproducibility	High	Unreliable	High	High	High
Development cost	Low	Low	Moderate	High	High
Cost per analysis	High	Low	Moderate	Low	Low

Hybridization-based Markers

During hybridization-based marker analysis, DNA profiles are visualized by hybridizing the restriction enzyme– digested DNA to a specific labeled probe, which is a DNA fragment of known origin or sequence.

18.4 RESTRICTION FRAGMENT LENGTH POLYMORPHISM

Restriction fragment length polymorphism (RFLP) is a genetic marker technique in which organisms are differentiated by analysis of patterns derived from cleavage of their DNAs. This technique takes the advantage of differences in DNA sequences generated by cutting with restriction enzyme. If two organisms differ in the distance between sites of cleavage of a particular restriction endonuclease, the length of the DNA fragments produced upon digestion will differ. The similarity of the patterns generated can be used to differentiate species and even strains from one another. Restriction enzymes digest double stranded DNA (ds DNA) at specific recognition sites, generating a collection of DNA fragments of precisely defined length. If the bases A, T, G, and C of DNA are evenly distributed with same frequency, the enzyme having a recognition site of six bases cuts the DNA on an average every 4,096th [or $(4)^6$] base (for details see Chapter 3). A genome of 10^9 bases will thus produce ~2,50,000 fragments of variable lengths. The specificity is such that the replacement of a single base in a site is enough to prevent the enzyme from cutting the DNA at that site. It is this specificity that is exploited for the detection of a polymorphism. Thus, the presence or absence of the restriction site leads to a restriction fragment length polymorphism Since polymorphism is a basic property of living organisms, digestion of the DNA of any two individuals in a given species produces a great number of differences in fragment length. This difference is then detected by using molecular probes.

18.4.1 Procedure for RFLP Analysis

RFLP analysis of genomic DNA is performed in the following steps (Figure 18.1):

DNA Isolation, Purification, and Restriction Digestion DNA from different genotypes that might contain polymorphism due to variation in the DNA base sequence is isolated and purified. DNA samples are then digested with restriction enzymes.

Agarose Gel Electrophoresis and Visualization of DNA Bands Digested DNA samples are subjected to agarose gel electrophoresis separated by loading different samples in wells on the same gel. The DNA fragments of different lengths resulting from digestion are separated by gel electrophoresis. However, the difference in the distribution pattern of the fragments of different DNA samples cannot be detected directly. This is because the enormous fragments are produced and the range in size is rather continuous so that these form continuous smear on the gel.

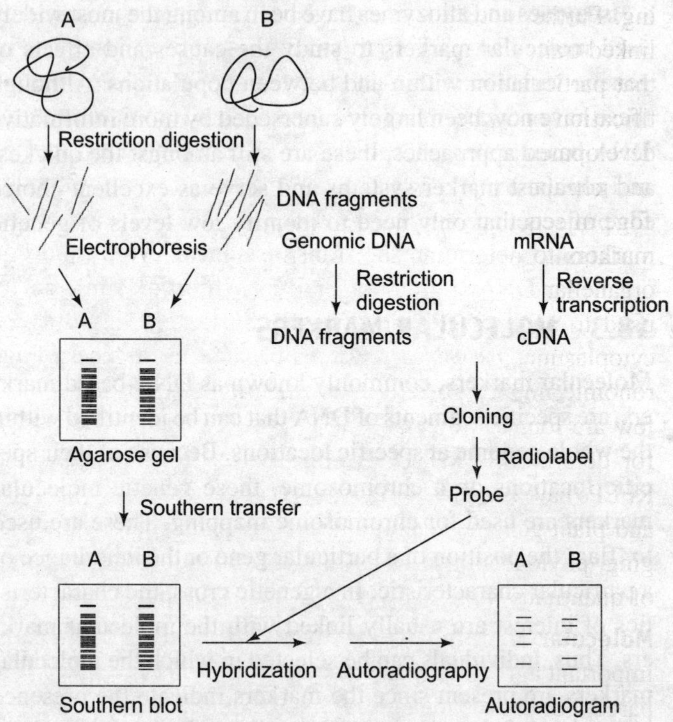

Figure 18.1 Procedure of RFLP

Southern Blotting and Hybridization DNA fragments are then subjected to Southern blotting by transferring the DNA fragments on a nylon or nitrocellulose membrane from the gel. Blotted DNA is hybridized with specific radiolabeled probes and hybridization signals are detected by autoradiography (for details see Chapter 16). Short, single- or low-copy genomic DNA or cDNA clones are typically used as RFLP probes.

18.4.2 Limitations of RFLP

RFLP has reduced usefulness due to following reasons: (i) Large amount of pure DNA is required for restriction digestion and Southern blotting; (ii) The requirement of radioactive isotope makes the analysis relatively time-consuming, expensive, and health hazardous; (iii) The assay is time-consuming and labor-intensive, and only out of several markers the detection of fewer alleles may be possible, which is highly inconvenient especially for crosses between closely related species; and (iv) These are unable to detect single base changes, which restrict their use in detecting point mutations occurring within the region at which these are detecting polymorphism.

18.4.3 Applications of RFLP

RFLP is a type of codominant marker, which enables heterozygotes to be distinguished from homozygotes, as DNA fragments from all homologous chromosomes are detected. These are very trustworthy markers in linkage analysis and breeding. Furthermore, these can easily determine the status of linked trait, which is either homozygous or heterozygous in that particular individual. RFLP marker allows direct identification of a genotype or cultivar from any tissue at any developmental stage. It can also discriminate between species and population. RFLP marker is simple and no prior knowledge of sequence is required. Initially, RFLPs were used as markers to determine the transmission of cytoplasmic or organellar DNA (mitochondrial and chloroplastic) that is used to analyze genetic diversity. However, the utility of cytoplasmic DNA is very limited because most genes of agronomic importance are located on nuclear chromosomes, few are found in cytoplasmic DNA. RFLPs are very useful for determining the geographic structure of populations. RFLPs have been used in various types of analyses of animal and plant genomes including molecular mapping, gene tagging, phylogenetic relationship, identification and mapping of quantitative trait loci (QTLs), and DNA fingerprinting.

Molecular Mapping RFLP maps of various economically important agricultural and horticultural crops have been prepared. For probe preparation, a single copy DNA from a species of interest is cloned and labeled. The labeled probe is used to follow the segregation of the homologous region of the genome in individuals from segregating populations (F2 or backcross). For example, two parents are crossed to produce an F1 (homozygous at all loci) and the F1 is selfed to produce an F2 population. When the F1 plants undergo meiosis to produce gametes, their chromosomes undergo recombination by crossing-over, and the chromosomes are mosaics with segments from each of the two parental chromosomes and no two chromosomes have similar arrangement of segments. This type of genetic map is constructed with the help of RFLP markers as these directly tag on chromosome segments during recombination. Therefore, one can directly analyze the RFLP map, i.e., the genotype instead of phenotype.

Gene Tagging Gene tagging of major genes of economic value can be performed with the help of RFLP markers as these markers are tightly linked with genes of interest. For breeding, various traits that are coded by single gene, for example disease resistance gene, are transferred from one variety to another. Incorporation of disease resistance genes into sensitive varieties requires crosses with stocks that carry the resistance genes followed by screening among the progeny for having the desired gene. The selection of disease resistance variety can easily be performed by detecting the disease resistance gene with the help of RFLP probes linked to that particular disease resistance gene. In human beings, sickle cell anemia can be detected by screening human DNA for the presence of potentially deleterious sickle cell anemia gene with the help of linked RFLP marker.

Phylogenetic Relationship RFLP of genomic DNA are useful for phylogenetic analysis in several groups of plants. For establishment of phylogenetic relationship among different species of a genus, data on RFLP markers are collected with the help of a large number of probes and restriction enzyme combinations. The restriction fragments, which are obtained on Southern blotting are numbered 1 to n and a distance matrix is computed using the presence or absence of these fragments in different species. This information in then used for computer analysis of the data and phylogenetic relationship is prepared.

Identification and Mapping of QTLs The heritable characters, which are a consequence of the joint action of the several genes, are known as polygenic or quantitative traits, for example, yield, maturity date, and drought tolerance in plants, and height, weight, etc. in animals. The genetic loci of such characters are known as QTL (Exhibit 18.1). For detection of QTLs, a cross-pollination is performed between two plants that are genetically different for one or more characters of interest and a number of progeny are assessed for the character of interest and their genotypes at RFLP marker loci throughout the genome. Further, an analysis is done to find out the association between RFLP markers and the characters of interest; if these characters are associated with each other, it should be due to linkage between RFLP markers to that gene.

DNA Fingerprinting DNA fingerprinting provides evidence for the establishment of innocence or a probability of the guilt of a crime suspect. RFLP is one of the original applications of DNA analysis to forensic investigation. With the development of newer, more efficient DNA analysis techniques, RFLP is not used as much as it once was because it requires relatively large amounts of DNA. In addition, samples degraded by environmental factors, such as dirt or mold, do not work well with RFLP.

18.4.4 Variant of RFLP: Cleaved Amplified Polymorphic Sequence or PCR–RFLP

Isolation of sufficient DNA for RFLP analysis is time-consuming and labor-intensive. RFLP can be coupled with PCR, which can be used to amplify very small amounts of DNA, usually in 2–3 hours, to the levels required for RFLP analysis. Therefore, more samples can be analyzed in a shorter time. This technique is known as cleaved amplified polymorphic sequence (CAPS). This method is used for analyzing polymorphism within a product of amplification. In CAPS, amplified fragments are digested with one or several restriction enzymes having four-base recognition sites. The four base-cutter restriction enzymes digest the DNA after ~256 bp [i.e., (4^4)] and the length of products normally range between 0.5 and 2 kbp, which results in a high probability of cutting. Finally, polymorphisms at the restriction sites are revealed by agarose gel electrophoresis. CAPS is thus a vari-

Exhibit 18.1 Quantitative trait loci (QTL)

Some phenotypic characters depend upon expression of two or more genes and their interaction with the environment. These do not follow patterns of Mendelian inheritance (qualitative traits) and their phenotypes typically vary along a continuous gradient depicted by a bell curve. A quantitative trait locus (QTL) is a region of DNA that is associated with a particular phenotypic trait. QTLs are either genes or stretches of DNA that are closely linked to the genes that underlie the trait in question. These QTLs are often found on different chromosomes. QTLs can be molecularly identified (for example, with PCR) to help map regions of the genome that contain genes involved in specifying a quantitative trait. Most phenotypic characteristics are the result of the interaction of multiple genes (polygenic traits) like skin color, height, and body mass. All of these phenotypes are complicated by a great deal of interplay between genes and environment. Many disorders are also example of polygenic traits including autism, cancer, diabetes, etc. Typically, QTLs are continuous traits, i.e., the trait can have any value within a range. Moreover, a single phenotypic trait is usually determined by many genes. Consequently, many QTLs are associated with a single trait. Knowledge of a number of QTLs that explains variation in the phenotypic trait provides information on the genetic architecture of a trait. It also denotes that plant height is controlled by many genes of small effect or by a few genes of large effect. Another use of QTLs is to identify candidate genes underlying a trait. Once a region of DNA is identified as contributing to a phenotype, it can be sequenced. The DNA sequence of any genes in this region can then be compared to a database of DNA for genes whose function is already known. In a recent development, classical QTL analyses are combined with gene expression profiling, i.e., by DNA microarrays. Such expression QTLs (e-QTLs) describes *cis*- and *trans*-controlling elements for the expression of often disease-associated genes. Observed epistatic effects have been found to be beneficial to identify the gene responsible by a cross validation of genes within the interacting loci with metabolic pathway and scientific literature databases.

QTL mapping is performed by statistical analyses of the alleles that are present in a locus and the phenotypes that these produce. As most traits of interest are governed by more than one gene, analyzing the entire locus of genes linked to a trait helps in understanding the relationship between the genotypes and phenotypic expression. Statistical analysis is essential to explain that different genes interact with one another and to determine whether these produce a significant effect on the phenotype. QTLs identify a particular region of the genome as containing a gene that is associated with the trait being assayed or measured. These are shown as intervals across a chromosome, where the probability of association is plotted for each marker used in the mapping experiment. To begin with, a set of genetic markers must be developed for the species in question. QTL segregate, recombine, and exhibit the linkage theory of association of marker loci with them. Usually RFLP markers are used for QTL mapping, as segregation of these markers of all chromosomes can be easily followed during a cross-hybridization. The basic theme behind this mapping technique is to establish the correlation between the QTL of interest and specific chromosome segments marked by RFLPs. The QTL techniques can be performed in self-pollinated species and on inbred strains of any cross-pollinated species. When a QTL is found, it is often not the actual gene underlying the phenotypic trait, but rather a region of DNA that is closely linked with the gene. For organisms whose genomes are known, one might now try to exclude genes in the identified region whose function is known with some certainty not to be connected with the trait in question. If the genome is not available, it may be an option to sequence the identified region and determine the putative functions of genes by their similarity to genes with known functions usually in other genomes. Another interest of statistical geneticists using QTL mapping is to determine the complexity of the genetic architecture underlying a phenotypic trait. For example, they may be interested in knowing whether a phenotype is shaped by many independent loci, or by a few loci, and do these loci interact? This can provide information on how the phenotype is evolving.

Source: www. nature.com/reviews/genetics, November 2003, vol 4, pp 916; *Euphytica* (2005) 142: 169–196.

ant of RFLP. Like RFLP, this technique provides codominant markers that are detected individually. CAPS have been used for classification of rice genotypes, gene mapping in barley, screening of sequence tagged sites (STS) linked to resistance genes in lettuce and population genetics.

18.5 RANDOM AMPLIFIED POLYMORPHIC DNA

Random amplified polymorphic DNA (RAPD) is a PCR-based molecular marker, which exhibits differential amplification of a DNA sample with short oligonucleotide primers. This PCR-based RAPD method was developed independently in two laboratories (Welsh and McClelland, 1990; Williams *et al.*, 1990). Eukaryotic nuclear DNA is quite complex and when PCR is performed with some randomly chosen single short primer (8–10 mer) (Figure 18.2), sometimes by chance, pairs of sites complementary to that primer may be present in correct orientation and close enough to one another for PCR amplification. Since primers used in RAPD analysis are short, these have the possibility of annealing at a number of locations in the genome, and for amplification the binding must be to inverted repeat sequences generally 150–4,000 base pairs apart. The number of amplification products is directly related to the number and orientation of the sequences that

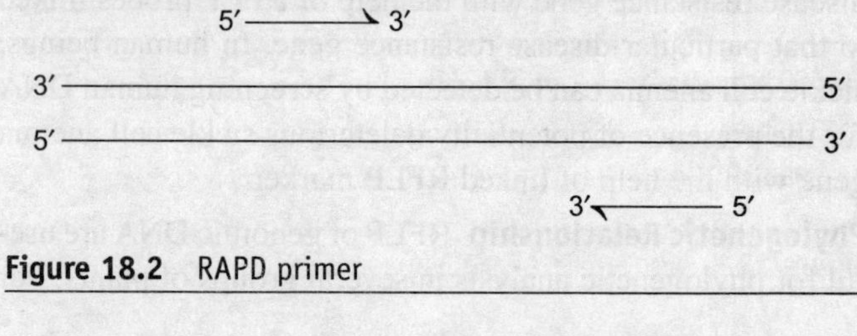

Figure 18.2 RAPD primer

are complementary to the primer in the genome. While performing analysis, some primers do not provide amplification; others generate the amplicon of the same size with the DNA of different individuals. With still others, different patterns of bands are obtained for every individual in a population. The variable bands are commonly referred to as randomly amplified polymorphic DNA (RAPD) bands. For example, as shown in Figure 18.3, three of the bands are RAPD bands. The RAPD amplification reaction is performed on a genomic DNA template and primed by an arbitrary oligonucleotide primer, resulting in the amplification of several discrete DNA products. Each amplicon is derived from a region of the genome that contains two short DNA segments with some homology to the primer. These segments must be present on opposite DNA strands and be sufficiently close to each other to permit DNA amplification. The PCR reaction is performed usually with short primers and at low annealing temperature, which ensures that amplification occurs. Genomic DNA of two individuals produces different RAPDs. A specific DNA segment (generated for one individual but not for other) shows DNA polymorphism. Hence, it can be used as a genetic marker. Many different gene loci can be analyzed because each random primer anneals to different regions of the DNA. These are usually separated on agarose gels and visualized by EtBr staining.

If an assumption is made that complete complementarities between primer and target is required for efficient amplification, it becomes possible to derive a general equation to predict the approximate number of expected amplicons, *A*.

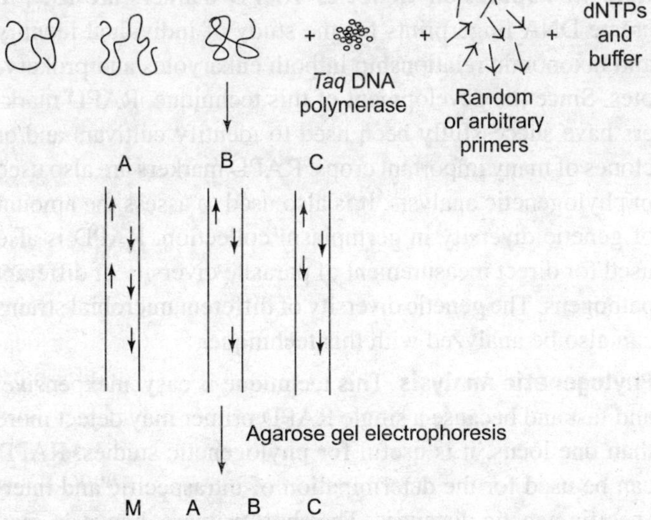

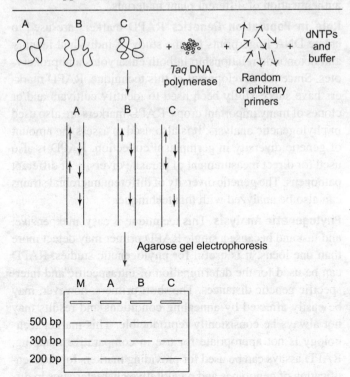

Figure 18.3 Procedure of random amplified polymorphic DNA

For amplified fragments of 2 kbp or smaller in size the equation is:

$$A = (4{,}000\ C/16^{N})$$

where *N* is length of primer; and *C* is complexity of genome.

18.5.1 Procedure for RAPD Analysis

RAPD amplification using genomic DNA is performed in the following steps (Figure 18.3):

Genomic DNA Isolation Genomic DNA is isolated from the individuals of interest (for details see Chapter 1).

PCR Amplification using Random Primers PCR conditions are standardized according to species. Several parameters have to be considered, such as concentrations of $MgCl_2$, primer, target DNA, dNTPs, and quality of the target DNA. Similar amounts of target DNA, primer, and dNTP concentration are taken in each reaction. This requires consistency between wells and also between electrophoretic runs. This is extremely significant because without this control, it cannot be ensured that the amplification polymorphisms are the result of population variability or reaction variability.

Usually the PCR amplification steps are as follows: denaturation at 94°C, 1 min; annealing at 35–40°C, 30 seconds to 2 min; extension at 72°C, 1.5 to 2 min. This cycle is repeated 35–45 times and then the products are generated by a single extension for 7 min at 72°C.

Agarose Gel Electrophoresis and Visualization of DNA Bands The amplification products are then resolved on a 2% agarose gel, stained, and photographed (for details see Chapter 6).

Analysis of Data Variability is then scored as the presence or absence of a specific amplification product.

18.5.2 Advantages of RAPD

RAPD has been proved to be a better marker than RFLP in many ways. The advantages of RAPD over RFLP are listed below: (i) RAPD is quicker than RFLP; (ii) Smaller amounts of DNA (15–25 μg) is required for analysis as compared to RFLP; (ii) Crude DNA preparation may be used for RAPD analysis of whole genome, while RFLP requires pure DNA; (iv) RAPD analysis does not need any species-specific probes. Different types of species can be screened with the help of similar sets of primers, while in RFLP sequence-specific probes are required for screening; (v) RAPD protocols are relatively simple. (RAPD and RFLP procedures are compared in Table 18.2); and (vi) It does not require blotting or hybridization.

18.5.3 Limitations of RAPD

RAPD technique suffers from certain disadvantages, for example, the following: (i) Nearly all RAPD markers are dominant, i.e., it is not possible to distinguish whether a DNA

Table 18.2 Comparison of Procedures of RFLP and RAPD

RFLP	RAPD
Isolation of genomic DNA	Isolation of genomic DNA
DNA digestion with restriction enzyme	Amplification by PCR using arbitrary primers
Separation of DNA fragments on agarose gel	Separation of DNA fragments on agarose gel
Southern transfer	Visualization of markers on the gel and photography of the gel
Hybridization of Southern blots with probe	–
Detection by auto-radiography or using a non-radioactive system	–

segment is amplified from a locus that is heterozygous (one copy) or homozygous (two copies). Codominant RAPD markers, observed as different-sized DNA segments and amplified from the same locus, are detected only rarely; (ii) PCR is an enzymatic reaction, therefore the quality and concentration of template DNA, concentrations of PCR components, and the PCR cycling conditions may greatly influence the outcome. Thus, the RAPD technique is highly laboratory-dependent and needs carefully developed laboratory protocols to be reproducible; and (iii) Mismatches between the primer and the template may result either in the total absence of PCR product or in a merely decreased amount of the product. Thus, the RAPD results can be difficult to interpret.

18.5.4 Applications of RAPD

RAPD exhibits polymorphism and thus can be used as genetic markers. By identifying RAPD bands closely linked to the marker of interest to be transferred and by scoring individuals or groups of individuals for the linked RAPD marker, the process of breeding can be sped up. RAPD markers, which are linked to the genes of interest, can serve as starting points for chromosome walking for the isolation of these genes. RAPD markers have been useful for different purposes in plant and animal genome, which are summarized below:

Preparation of Genetic Maps As RAPD is a fast and an efficient technique, it is used for the construction of genetic maps. Till date, genetic maps of many organisms have been prepared by using RAPD. Since RAPD polymorphism is the result of either a nucleotide base change that alters the primer binding site or an insertion or deletion within the amplified region, polymorphism is usually noted by the presence or absence of an amplicon from a single locus. RAPD technique provides only dominant markers, which signifies that the homozygote of the parental type from which RAPD band is amplified cannot be distinguished from the heterozygote in a segregating population. This is because the heterozygote also produces a RAPD band. The only definitely assigned genotype is the homozygote of the recessive parental type (no RAPD band). Therefore, the segregating F2 population may be scored as follows: band present AA or Ab; band absent bb. However, population like backcross populations, F2 populations or recombinant inbred populations, haploid populations, or somatic embryos are best-suited for the construction of genetic maps with RAPD markers.

Targeting Markers to Specific Regions of Genome Several groups have used the RAPD assay as an efficient tool to identify molecular marker that lies within regions of a genome introgressed during the development of near isogenic lines. By definition, any region of the genome that is polymorphic between two near isogenic plants is potentially linked to the introgressed traits. Another advantage of this technology is that genetic map of entire genome is not required to identify markers linked to trait of interest and instead specific regions of the genome can be focused upon.

Mapping of Traits As RAPD markers are targeted to the smaller locus within the genome and the likelihood of identifying false-positive markers is small, this method is used to target RAPD markers to regions of some economically important traits. It is used for indirect selection of segregating population through the tagging of a gene of high economic value such as resistance against pathogen in many crop plants, e.g., potato, tomato, pea, maize, rice, wheat, etc. Molecular polymorphisms can be identified at random and used for genetic mapping. RAPD can be effectively used for the authentication of different plant materials.

Role in Population Genetics RAPD markers are used to create DNA fingerprints for the study of individual identity and taxonomic relationship in both eukaryotes and prokaryotes. Since the development of this technique, RAPD markers have successfully been used to identify cultivars and/or clones of many important crops. RAPD markers are also used in phylogenetic analysis. It is also used to assess the amount of genetic diversity in germplasm collection. RAPD is also used for direct measurement of parasite diversity of different pathogens. The genetic diversity of different microbial strains can also be analyzed with this technique.

Phylogenetic Analysis This technique is easy, inexpensive, and fast and because a single RAPD primer may detect more than one locus, it is useful for phylogenetic studies. RAPD can be used for the determination of intraspecific and interspecific genetic distances. The short primers, however, may be easily affected by annealing conditions and results may not always be consistently reproducible. This marker technology is not appropriate for use in comparative mapping. RAPD assays can be used for providing markers for the identification of genotypes and quantitative characteristics in different plants.

DNA Fingerprinting RAPD-PCR needs very small amount of sample of the subject's DNA, which is then selectively amplified a million-fold, thereby producing a large amount of DNAs. PCR products can be separated into bands of different molecular weight DNAs by agarose gel electrophoresis. Banding patterns from an individual or a population can then be used as a 'fingerprint' that may distinguish it from other genetic types. Unique fingerprint profiles generated by the RAPD techniques can be exploited for strain identification and these can be useful in epidemiological studies to determine the origin of a bacterial population.

18.5.5 Variants of RAPD

Various types of methods based on PCR with arbitrary primers have been developed, which are used to detect polymorphism. The principles of these variant methods are similar to RAPD and the only variation is in length of the primers and the medium used for migration of bands.

DNA Amplification Fingerprinting

In DNA amplification fingerprinting (DAF), a single 5–8 nucleotide arbitrary primer is used for PCR reaction. In DAF, higher primer concentration is required and PCR is performed only at two temperature cycles. As very complex banding pattern is produced by DAF, the product of DAF is analyzed by polyacrylamide gel electrophoresis (PAGE) and detected by silver staining. DAF may reveal many bands, sometimes even up to 100. DAF can be automated by fluorescent tagging of primers, which helps in easy and fast determination of amplified products.

Arbitrarily Primed PCR

In arbitrarily primed PCR (AP-PCR), the PCR amplification of genomic DNA is carried out with a single primer of 20–50 bases in length. Initially annealing is performed at lower temperature and with higher primer concentration. Annealing temperature of subsequent cycles is increased to increase the stringency of reaction. Usually primers of variable lengths are randomly selected for this purpose. Amplified products are resolved on PAGE and analyzed by autoradiography.

Sequence Characterized Amplified Regions

New, longer, and RAPD-based specific primers are designed for the DNA sequence, which is called sequence characterized amplified region (SCAR) which can increase industrial application of the molecular techniques. In SCAR, pairs of 20–25 bp oligonucleotide primers specific to the sequence of polymorphic bands can be used to amplify the characterized regions from genomic DNA under stringent conditions, which makes these markers more specific and dependable as compared to RAPD markers. The genetic polymorphism observed among the cultivars is interesting and can be used to develop markers for cultivar identification. These markers have been used for the authentication of different commercially important crop plants.

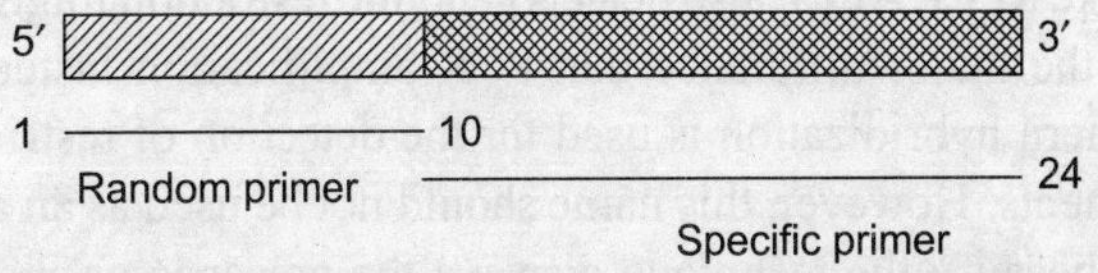

Figure 18.4 Primer of sequence-characterized amplified region

SCAR is a genomic DNA fragment at a single genetically defined locus, i.e., identified by oligonucleotides. At first, RFLP markers were converted into SCARs by sequencing two ends of genomic DNA clones and designing oligonucleotide primers based on the end sequences. The primers were used directly on genomic DNA in a PCR reaction to amplify the polymorphism region. If no polymorphism is noticed after PCR, then the PCR fragments can be subjected to restriction to detect RFLPs within that amplified product.

When a RAPD marker is located near a gene to be cloned, or a gene that needs to be followed through generations, specific SCAR marker can be developed from that particular locus. Usually RAPD markers are converted into SCARs.

For the development of SCAR, following steps should be performed: (i) The particular band is excised, cloned, and DNA is sequenced; (ii) Two 24-base oligonucleotide primers are designed corresponding to the ends of fragment (Figure 18.4); and (iii) PCR is performed with the help of these primers.

SCAR primer has increased the specificity of a reaction. SCAR offers several advantages over RAPD and some of them are listed below: (i) The dominant RAPD marker is converted into codominant SCARs that increase the information at F2 generation; (ii) SCAR detects only single locus, as SCAR is performed in stringent conditions; (iii) As larger and specific primers are used in SCAR, the results are more reproducible; and (iv) SCAR is applied to commercial breeding projects as no radioactivity is used in SCAR.

18.6 AMPLIFIED FRAGMENT LENGTH POLYMORPHISM

The amplified fragment length polymorphism (AFLP) technique, developed in the early 1990s by Vos *et al.*, is based on the selective PCR amplification of restriction fragments from a total digest of genomic DNA. This technique is a combination of both RFLP and RAPD techniques, which are very sensitive in detecting polymorphism throughout the genome. The AFLP technique is robust and consistent because stringent reaction conditions are used for primer annealing. The reliability of the RFLP technique is thus combined with the power of the PCR technique. It is applied universally because of its reproducibility at high level.

Like RFLP, AFLP also detects genomic restriction fragments with the major difference that PCR amplification instead of Southern hybridization is used for the detection of restriction fragments. However, this name should not be used as an acronym because the technique displays the presence or absence of restriction fragments rather than length differences.

18.6.1 Procedure for AFLP Analysis

AFLP involves three basic steps (Figure 18.5):

Isolation and Purification of DNA, its Restriction Digestion, and Ligation of Oligonucleotide Adaptors Genomic DNA is isolated and purified and then digested with two restriction enzymes for 1 hour at 37°C. Restriction enzyme that cuts frequently, e.g., *Mse* I (4 base-cutter) and the one that cuts less frequently, e.g., *Eco* RI (6 base-cutter) are used for digestion. The resulting fragments are ligated to complementary, double stranded, end-specific adaptors with the help of T4 DNA ligase; in this example, two adaptors used are *Eco* RI adaptors and *Mse* I adaptors.

Selective Amplification of Sets of Restriction Fragments A subset of the restriction fragments is amplified using two primers complementary to the adaptor and restriction site fragments.

A preselective PCR amplification is done by using primers complementary to each of the two adaptor sequences, except for the presence of one additional base at the 3′-end, which is chosen by the user. This step is normally performed for 20 cycles with the following cycle profile: a 30 seconds DNA denaturation step at 94°C, a 1 min annealing step at 56°C, and a 1 min extension step at 72°C. After this preamplification step, the reaction mixtures are diluted 10-fold with 10 mM Tris–HCl, 0.1 mM EDTA (pH 8.0), and used as templates for the second amplification reaction.

The second amplification reaction is performed for AFLP reactions with primers having longer selective extensions. One of the two primers is radioactively labeled, preferably the *Eco* RI primer. The primers are generally end-labeled by using (γ^{32}P) ATP and polynucleotide kinase. AFLP reactions with primers having two or more selective nucleotides are performed for 36 cycles with the following cycle profile: a 30 seconds DNA denaturation step at 94°C, a 30 seconds annealing step, and a 1 min extension step at 72°C. Usually gradient PCR is performed for AFLP. For example, the annealing temperature in the first cycle is 65°C, which is subsequently reduced in each cycle by 0.7°C for the next 12 cycles, and is continued at 56°C for the remaining 23 cycles.

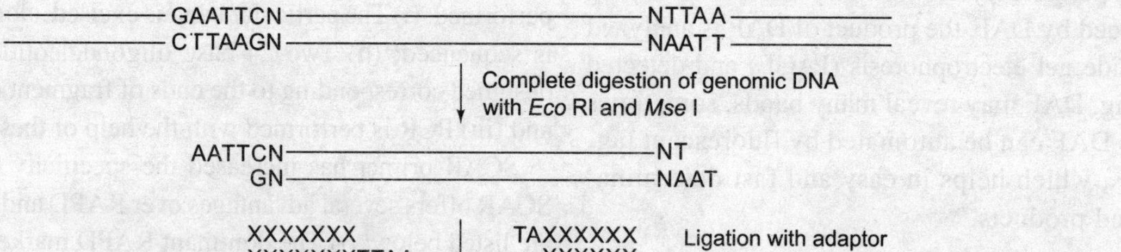

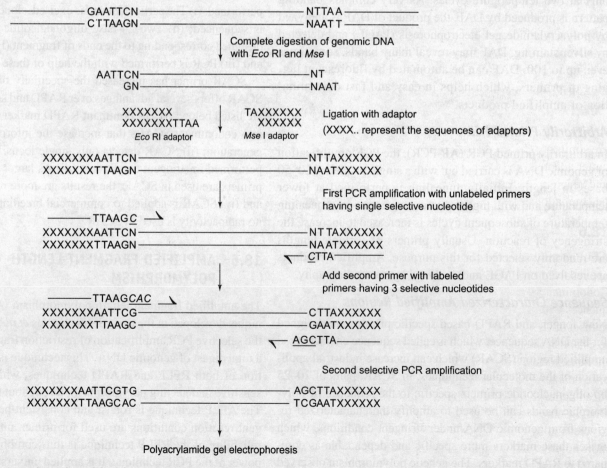

Figure 18.5 Procedure of amplified fragment length polymorphism

Gel Analysis of the Amplified Fragments The DNA products, which have been amplified, are separated on denaturing polyacrylamide gel. Fingerprint patterns are visualized through autoradiography. Fluorescent or silver staining is used to visualize the products in cases where radiolabeled nucleotides are not used. After PAGE, several tens of fragments (up to 100) are detected, the polymorphism of which originates from restriction sites and/or hybridization sites of arbitrary bases.

18.6.2 AFLP Primers and Adaptors

All oligonucleotides are prepared on DNA synthesizer and the quality of the crude oligonucleotides is determined by end-labeling with polynucleotide kinase and (γ^{32}P) ATP and subsequent electrophoresis on denaturing PAGE. AFLP adaptors consist of a core sequence and an enzyme-specific sequence.

The structure of the *Eco* RI adaptor is

5′-CTCGTAGACTGCGTACC
CATCTGACGCATGGTTAA-5′

The structure of the *Mse* I adaptor is

5′-GACGATGAGTCCTGAG
TACTCAGGACTCAT-5′

AFLP primers consist of three parts: a core sequence, an enzyme-specific sequence (ENZ), and a selective extension (EXT). This is illustrated below for *Eco* RI and *Mse* I primers with three selective nucleotides (selective nucleotides shown as NNN):

	CORE	ENZ	EXT
Eco RI	5′-GACTGCGTACC	AATTC	NNN-3′
Mse I	5′-GATGAGTCCTGAG	TAA	NNN-3′

AFLP adaptors and primers for other 'rare cutter' enzymes are similar to the *Eco* RI adaptors and primers, and frequent cutter adaptors and primers are similar to the *Mse* I primers and adaptors, except the enzyme-specific parts, which should be compatible to the respective enzymes.

18.6.3 Advantages of AFLP

AFLP is extremely sensitive, highly reproducible, and widely applied technique. AFLPs are codominant markers like RFLPs. Codominance results when the polymorphism is due to sequences within the amplified region. However, because of the number of bands seen at one time, additional evidence is needed to establish that a set of bands result from different alleles at the same locus. If, however, the polymorphism is due to presence or absence of a priming site, the relationship is dominant. It can permit detection of restriction fragments in any pool of DNA. Compared to RAPD, fewer primers should be needed to screen all possible sites. AFLP procedure typically detects more polymorphisms per reaction than RFLP or RAPD analysis. The primary reason for the superiority of AFLP is that it detects very large number of DNA bands, enabling the identification of many polymorphic markers. AFLP also enables rapid creation of high-density genetic maps, as 50–100 bands are observed in each lane of the gel. It detects more point mutations than RFLPs, enables detection of very large number of polymorphic markers than RAPD or RFLP, and is simpler than microsatellites, as no prior sequence information is needed for amplification.

18.6.4 Limitations of AFLP

Although this approach is highly informative, a criticism of this technique includes the following points: (i) AFLP requires purified genomic DNA; (ii) AFLP uses multiple procedures in the protocol. Hence it is a cumbersome and laborious technique; (iii) As compared to RAPD and RFLP, AFLP is an expensive technique; and (iv) The technique requires radioactivity to detect DNA, which is one of the major drawbacks that restricts its use. However, nonradioactive silver staining protocols to detect AFLP markers with no major loss in sensitivity have now been developed.

18.6.5 Applications of AFLP

The AFLP technology has the potential to identify various polymorphisms in different genomic regions simultaneously. It is also highly sensitive and reproducible. This technique is widely accepted for the following purposes:

Construction of Genome Map Most AFLP fragments correspond to unique positions on the genome and therefore can be exploited as landmarks in the construction of genetic and physical maps, each fragment being characterized by its size and primers required for amplification. AFLP markers can be used to construct high-density genetic maps of genomes or genome segments. In most organisms, AFLP proves to be the most effective way to construct genetic maps as compared to other marker techniques. The AFLP technique is also used for mapping of cloned DNA segments like cosmids, P1 clones, bacterial artificial chromosomes (BACs), or yeast artificial chromosomes (YACs). By simply using no or few selective nucleotides, restriction fragment fingerprints are produced, which are subsequently used to line up individual clones and make contigs.

Population Genetics AFLP is a very sensitive technique and hence it is widely used for the identification of genetic variation among closely related species of plants, fungi, animals, and bacteria. As a result, AFLP has become extremely beneficial in the study of taxa including bacteria, fungi, and plants. The markers have immediate applications in supportive research for advanced breeding projects mainly in relation to quality control, e.g., clonal identification and orchard contamination within orchard mating patterns, etc.

Detection of Restriction Fragment AFLP technique is a very efficient tool to reveal restriction fragment polymorphisms.

It can allow detection of restriction fragments in any background or complexity, including pooled DNA samples and cloned DNA segments.

DNA Fingerprinting AFLP is used to produce fingerprints of any DNA irrespective of its origin or complexity. AFLP fingerprinting of simpler genome, e.g., yeast, can be prepared with the help of *Eco* RI and *Mse* I primers having one or two selective nucleotides. For fingerprinting of complex genomes, AFLP primers with at least three selective nucleotides at both the *Eco* RI and *Mse* I primers are required to generate useful band patterns. For example, by following the two-step amplification strategy, AFLP fingerprints have been obtained for DNAs of *Arabidopsis thaliana*, tomato, and maize as well as human DNA. For *Arabidopsis*, DNA primer combinations with a total of five selective nucleotides were used because of the small genome of this plant species. For tomato, maize, and human DNA, a total of six selective nucleotides were used, three each for *Eco* RI and *Mse* I primers. However, fingerprints of these complex DNAs predominantly consist of unique AFLP fragments, but it also demonstrates occurrence of more intense repeated fragments. The fingerprints generated by AFLP are used in plant and animal breeding, medical diagnostics, forensic analysis, and microbial typing. AFLP fingerprinting is especially useful in screening backcross individuals. AFLP fingerprinting is also used in criminal and paternity tests.

Genomic Library Screening AFLP markers are also used to detect corresponding genomic clones, e.g., YACs. This is most effectively achieved by working with libraries, which are pooled to allow for rapid PCR screening and subsequent clone identification. For example, an AFLP marker is able to detect a single corresponding YAC clone in pools of as much as 100 YAC clones.

Mapping of Traits AFLP markers have recently been employed to identify many DNA markers tightly linked to the tomato genes for resistance to the leaf mold pathogen. Similarly, 29 AFLP markers linked to R1 gene for resistance to late blight diseases in potato and two markers for the resistance to root cyst nematode have been identified.

18.6.6 Variant of AFLP: Transposon Display or Sequence-specific Amplification Polymorphism

Transposon display or sequence-specific amplification polymorphism (SSAP) is a modification of the AFLP technique, which is based on anchoring AFLP with transposon. This technique, like AFLP, involves digestion of DNA with restriction enzymes, followed by ligation of adaptors. However, in this technique, PCR amplifications are performed with one primer complementary to the adaptors and another one complementary to a conserved sequence of transposon. This approach generates genetic marker similar to AFLP, but on an average. These markers are more polymorphic.

18.7 MICROSATELLITE OR SHORT TANDEM REPEATS

Eukaryotic genomes contain repetitive heterogeneous classes of noncoding DNA that are located either within or between genes. These short identical segments of DNA aligned head to tail in a repeating fashion are interspersed in the eukaryotic DNA genome (Figure 18.6). These repeat sequences exhibit length variation, which is termed as simple sequence length polymorphism (SSLP). SSLP is multiallelic, as each SSLP can have a number of different length variants. These arise due to polymerase (replication) slippage or unequal crossing-over during meiosis. There are two types of SSLPs:

Minisatellites Minisatellites, also known as variable number of tandem repeats (VNTRs), are tandem repeats with a repeat length of ~10–100 bp.

Microsatellites Microsatellites, also known as simple sequence repeats (SSRs) or simple tandem repeats (STRs), are tandem repeats with a repeat length of 2–10 base pairs, and the cluster size up to 150 bp. The most frequent microsatellite DNAs are $(A)_n$, $(TC)_n$, $(CA)_n$, $(TAT)_n$, $(AAAC)_n$, $(GATA)_n$, $(CATG)_n$, etc., the value of '*n*' ranging from a few to 100 units.

Microsatellites are more predominantly used as DNA markers than minisatellites for two important reasons. First, minisatellites are not evenly spread around the genome but

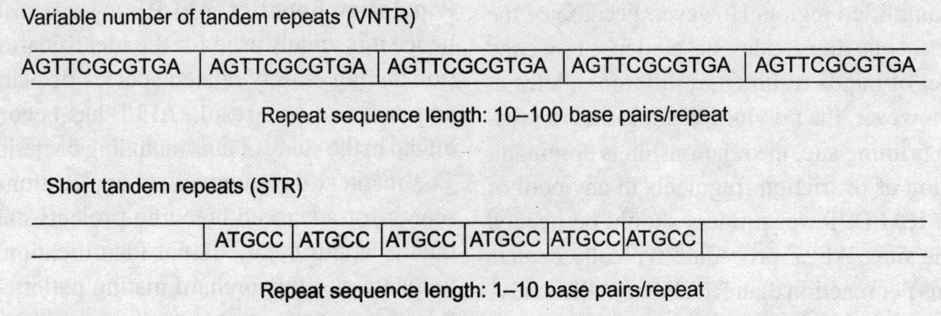

Figure 18.6 Types of simple sequence length polymorphism (SSLP)

tend to be found more frequently in the telomeric regions while microsatellites are evenly distributed throughout the genome. Second, microsatellites are much more amenable to analysis by PCR as compared to minisatellites; however, minisatellite technology relies on probe-based hybridization.

Microsatellites constitute excellent genetic markers with the advantages of PCR for routine detection. These are random and frequently distributed throughout the eukaryotic genomes. These show large, stable polymorphism due to variation in the number of repeat units and are ideal as molecular markers for genome mapping. While the repeated sequences themselves are usually the same from person to person, the number of times these are repeated tends to vary. In a population, many alleles of a single microsatellite locus may exist. Microsatellite alleles differ in the number of repeats. For example, one allele may have seven repeats of a CT motif and another allele may have eight repeats. In a population, many alleles may perhaps exist (up to 70 or 80) at a single locus, with each allele having a different length. An individual homozygous for a locus has the same number of repeats on both chromosomes, whereas a heterozygous individual has different numbers of repeats on the two chromosomes. The regions surrounding the microsatellite locus, called the flanking regions, may still have the same sequence. This is important because the flanking regions can therefore be used as PCR primers when amplifying microsatellite loci and can be conserved across genera or sometimes even in families.

In an example shown below, the two lines represent the sequences on two homologous chromosomes in a diploid organism. For clarity, only one strand of each chromosome is shown.

Homozygous alleles: (Both chromosomes have 7 CT repeats)

5′-CGTAGCCTTGCATCCTT **CTCTCTCTCTCTCT** ATCGGTACTACGTGG-3′
5′-CGTAGCCTTGCATCCTT **CTCTCTCTCTCTCT** ATCGGTACTACGTGG-3′
5′ flanking region Microsatellite locus 3′ flanking region

Heterozygous alleles: (One chromosome has 7 CT repeats, and the other has 8 CT repeats)

5′-CGTAGCCTTGCATCCTT **CTCTCTCTCTCTCT** ATCGGTACTACGTGG-3′
5′-CGTAGCCTTGCATCCTT **CTCTCTCTCTCTCTCT** ATCGGTACTACGTGG-3′
5′ flanking region Microsatellite locus 3′ flanking region

Interestingly, it is estimated that microsatellites mutate 100 to 10,000 times as fast as base pair substitutions. This makes microsatellites useful for studying evolution over short time spans (hundreds or thousands of years). As mentioned microsatellites are mutated because of polymerase slippage or unequal crossing over during meiosis.

Microsatellite is a polyallelic marker, which can be very useful for mapping the simple Mendalian traits as well as polygenic traits in segregating populations. PCR can be useful in detecting them individually, providing codominant and highly polymorphic locus-specific markers for a given locus containing microsatellite PCR primers for sequences flanking the repeats. Thus, microsatellite alleles are usually based on length polymorphism. The most common way to detect microsatellites is to design PCR primers that are unique to one locus in the genome and that base pair on either side of the repeated portion (Figure 18.7). Therefore, a single pair of PCR primers will work for every individual within the species and produce amplicons of different sizes for each of the different length microsatellites. Various approaches have been devised based on microsatellite sequences.

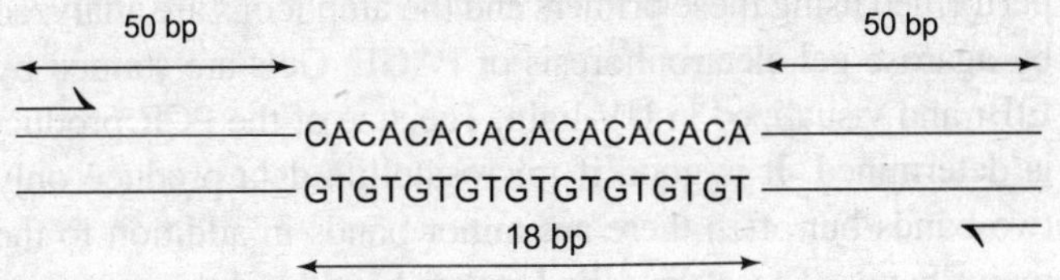

Figure 18.7 Microsatellites primers

Microsatellite Primed PCR (MP-PCR) This approach is also known as single primer amplification reaction (SPAR). This is very simple reaction similar to RAPD in that exponential amplification occurs from single primer reactions only when the particular repeat used as a primer is represented in multiple copies, which are closely spaced (2–3 kbp) and inversely oriented in the template DNA. Multiple loci can be identified from a genome using a single PCR reaction. Normally, 15–20 bp oligonucleotides are used as primers. No other DNA sequence information is required. After PCR amplification, the DNA pattern is easily resolved by agarose gel electrophoresis. The same set of primers can be used in different species.

Anchored Microsatellite-directed PCR (AMP-PCR) This technique is also known as inter SSR amplification (ISA or ISSR). In this approach, radiolabeled di- or trinucleotide repeats are modified by the addition of either 3′ or 5′ anchor sequence of two to four nucleotides composed of nonrepeat sequences. The anchored sequences can be RY, RG, and RTCY (where Y = pyrimidine; R = purine). These sequences help in anchoring the primers during PCR reaction at fixed points.

Random Amplified Microsatellite Polymorphisms (RAMP) In this approach, amplification is performed with the help of 5′-anchored primer and a RAPD primer. The RAPD primer binding site serves as an arbitrary endpoint for microsatellite-based amplification.

Selective Amplification of Microsatellite Polymorphic Loci (SAMPL) This is microsatellite-based modification of AFLP technique. In this method, same adaptor-modified restriction fragment is used as template but the amplification is carried out with ^{32}P-labeled STR primer combined with an unlabeled adaptor primer. During second amplification, one AFLP primer with three selective nucleotides is used with STR primer that is at least partially complementary to microsatellite sequences.

18.7.1 Procedure for Microsatellite Analysis

The most common way to detect microsatellites is to design PCR primers that are unique to one locus in the genome and that base pair on either side of the repeated portion (Figure 18.8). The procedure for microsatellite analysis is as follows:

Designing of PCR Primers Two PCR primers (forward and reverse) are designed from flanking regions of microsatellite.

PCR Amplification and Analysis of Results PCR reaction is performed using these primers and the amplicons are analyzed by agarose gel electrophoresis or PAGE. Gels are stained by EtBr and visualized in UV light. The size of the PCR product is determined. It is good if microsatellite data produce only two bands but often there are minor bands in addition to the major bands. These are called stutter bands and these usually differ from the major bands by two nucleotides.

18.7.2 Advantages of Microsatellite Analysis

Microsatellites offer certain advantages, which are as follows: (i) Microsatellites are abundant, multiallelic, and segregate as codominant marker; (ii) Microsatellites display high levels of genetic variation based on difference in the number of randomly repeating units at a locus; (iii) Very minute amount of DNA is required for microsatellite analysis; (iv) The analyses of microsatellites are easily performed by PCR; (v) During microsatellite analysis, use of radioactivity can be avoided as it can be easily visualized on agarose gel; (vi) It is highly reproducible marker; and (vii) These types of analyses can be shared among the different laboratories by exchanging primer DNA sequences; even same primers can be used for analysis in different genomes.

18.7.3 Limitations of Microsatellite Analysis

Despite several advantages, microsatellite analysis suffers from few disadvantages as well: (i) It is quite costly and time-consuming procedure; (ii) Specific oligonucleotides are required; (iii) Polymorphic primer sites have to be established; and (iv) There is a requirement for cloning and sequencing of microsatellite loci.

18.7.4 Applications of Microsatellite Marker Analysis

Because microsatellites are widely dispersed in eukaryotic genomes, are highly variable, and are PCR-based, these have been used in many different areas of research such as the following:

Conservation Biology Microsatellites can be used for detection of sudden changes in population, effects of population disintegration and interaction between different populations. Microsatellites are useful for the identification of new and early populations.

Forensics Microsatellite loci are predominant genetic markers that are widely used for forensic studies. In forensic identification cases, usually the aim is to link a suspect with a sample taken from a crime scene. Another application involves linking DNA samples with the relatives of a missing

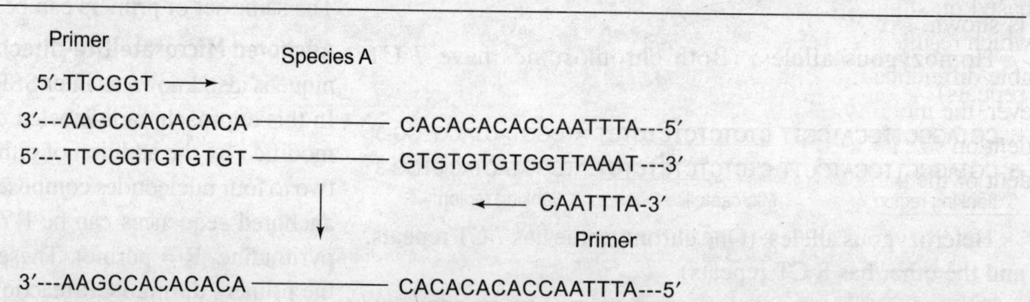

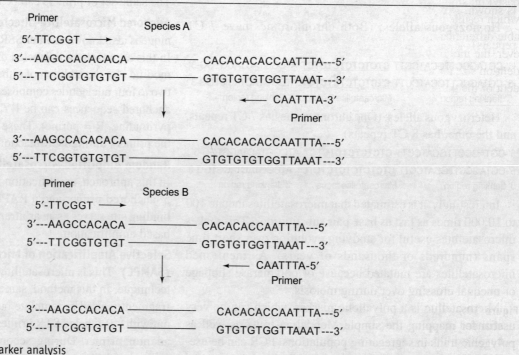

Figure 18.8 Microsatellites marker analysis

person. As the lengths of microsatellites may vary from one person to the other, it can be used to identify criminals and for paternity testing by performing DNA fingerprinting. The ability to employ PCR to amplify small samples is particularly valuable in this setting, since in a criminal case, only minute samples of DNA might be available.

Diagnosis and Identification of Human Diseases Because microsatellites change in length early in the development of some cancers, these are useful markers for early cancer detection. Being polymorphic in nature, these are useful in linkage studies in an attempt to locate genes responsible for various genetic disorders.

Population Studies By looking at the variation of microsatellites in populations, assumption can be made about population structures and differences, genetic drift, genetic bottlenecks, and even the date of the last common ancestor.

Genome Mapping Microsatellites have been adopted for mapping studies in maize, rice, barley, soyabean, and *Arabidopsis*.

Diversity Determination Microsatellites have been implicated in analysis of diversity in grape, rice, and soybean.

18.8 SINGLE STRAND CONFORMATION POLYMORPHISM

A technique for detecting polymorphisms using electrophoresis of single stranded DNA (ss DNA) whose conformation differs due to point substitutions, deletions, or insertions is known as single strand conformation polymorphism (SSCP). SSCP is the electrophoretic separation of single stranded nucleic acids based on slight differences in sequence (even a single base), which results in a different secondary structure and a measurable difference in mobility through a gel (Figure 18.9). However, the mobility of ds DNA in gel electrophoresis is dependent on size and length of the strand but is relatively independent of the particular nucleotide sequence but the mobility of single strands is markedly affected by very small changes in sequence, possibly one changed nucleotide out of several hundred. Because of the absence of a complementary strand the single strand display intrastrand base pairing, resulting in loops and folds that give the single strand a unique 3-D structure regardless of its length. A single nucleotide change could dramatically affect the strand's mobility through a gel by altering the intrastrand base pairing and its resulting 3-D conformation. Single strand conformation polymorphism analysis takes advantage of this quality of ss DNA. SSCP analysis offers an inexpensive, convenient, and sensitive method for determining genetic variation.

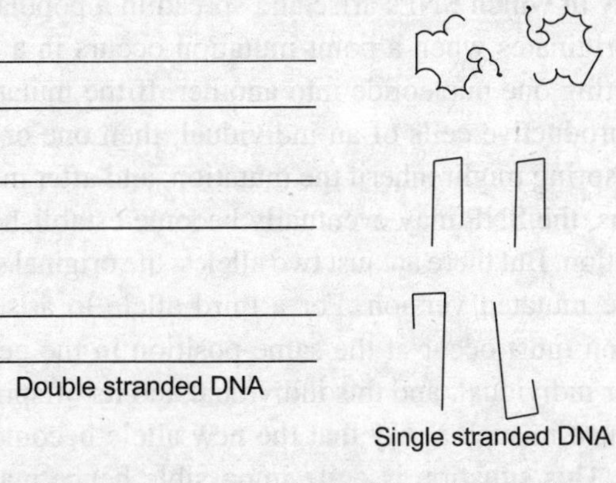

Figure 18.9 Single strand conformation polymorphism [Single stranded DNA with slight differences in sequence (even a single base) exhibits different secondary structures and hence a measurable difference in mobility]

18.8.1 Procedure of SSCP Analysis

The procedure SSCP analysis is as follows: (i) Digestion of genomic DNA with restriction endonucleases; (ii) Denaturation in an alkaline solution; (iii) Electrophoresis on a neutral polyacrylamide gel; (iv) Transfer to a nylon membrane; and (v) Hybridization with either DNA fragments or more clearly with RNA copies synthesized on each strand as probes.

Most SSCP protocols are designed to analyze the polymorphism at single loci. To this end a specific pair of PCR primers bracketing the target region is used to amplify DNA from individuals. ss DNA is produced by asymmetric PCR, in which one primer is present in excess over the other. After the low concentration primer is used up, the reaction continues, producing only the target strand (Figure 18.10). The mobilities of single strands are compared by gel electrophoresis.

PCR products are simply denatured at 94°C, cooled to ice temperature to prevent hybrid formation, and then electrophoresed on standard nondenaturing polyacrylamide gels.

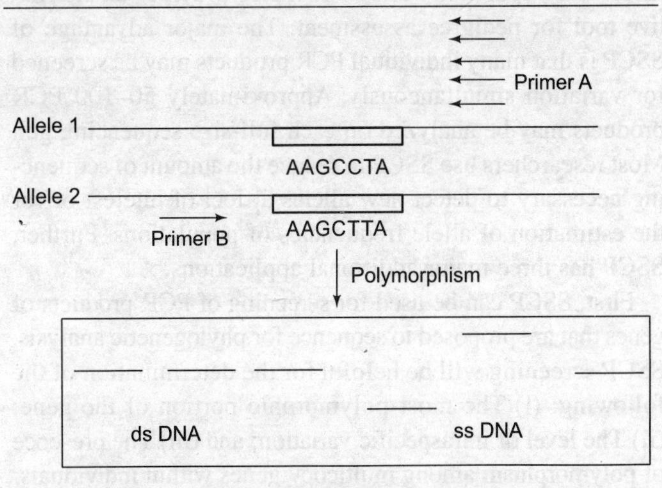

Figure 18.10 Asymmetric PCR for the production of ss DNA

18.8.2 Advantages of SSCP

SSCP offers following advantages: (i) ss DNA mobility is sequence-dependent because of varying degrees of intrastrand base pairing and the resulting looping and compaction;

(ii) SSCPs are codominant markers like RFLPs. In this regard, the technique differs from allele-specific PCR in which primers are designed to amplify from one allele but not from the other; (iii) Precise knowledge of the sequence polymorphism is not necessary. Indeed, polymorphisms at any of several sequence positions may be detectable with SSCP; (iv) PCR primers can be designed directly from cDNA sequences; and (v) If one amplified region does not show polymorphism, one can always move upstream or downstream to another.

18.8.3 Limitations of SSCP

There are certain demerits of SSCP, which include: (i) SSCP is sensitive technique and the mobility of ss DNA depends on temperature. While performing SSCP, the gel electrophoresis must be performed at constant temperature; (ii) The resolution of bands is also affected by pH. ds DNA fragments are usually denatured by exposure to basic conditions and addition of glycerol to the polyacrylamide gel lowers the pH of the electrophoresis buffer. Tris-borate buffer is preferred for electrophoresis; and (iii) The length of fragment also affects SSCP analysis. For optimal results, DNA fragment size should fall within the range of 150–300 bp, although SSCP analysis of RNA allows for a larger fragment size. The presence of glycerol in the gel may also allow a larger DNA fragment size at acceptable sensitivity.

18.8.4 Applications of SSCP

SSCPs have been widely applied in medical diagnosis; however, few studies have been reported in population genetics. The usefulness and convenience of SSCP is quite distant from molecular population biologists. It also provides an alternative tool for pedigree assessment. The major advantage of SSCP is that many individual PCR products may be screened for variation simultaneously. Approximately 50–100 PCR products may be analyzed on each full-size sequencing gel. Most researchers use SSCP to reduce the amount of sequencing necessary to detect new alleles at loci of interest or for the estimation of allele frequencies of populations. Further, SSCP has three major additional applications.

First, SSCP can be used for screening of PCR products of genes that are proposed to sequence for phylogenetic analysis. SSCP screening will be helpful for the determination of the following: (i) The most polymorphic portion of the gene; (ii) The level of intraspecific variation; and (iii) The presence of polymorphism among multicopy genes within individuals.

Second, SSCP is a much more efficient method for obtaining information about levels of polymorphisms within unidentified nuclear loci as compared to RFLP analyses. The advantages of using SSCP in this perspective are as follows: (i) The PCR amplicons are smaller and thus easier to amplify; (ii) As the length of amplicons are only 200–300 bp hence the entire sequence needed for PCR primer design can be obtained from a single sequencing reaction; (iii) Similar amounts of information are obtained by simply running the amplicons on two SSCP gels (e.g., one with glycerol and one without) rather than doing a number of restriction digestions and assuming that the restriction sites are independent; and (iv) When polymorphisms are detected, the amplicons are of optimal size to sequence using an ABI automated sequencer (so getting sequence level information is easy).

Third, SSCP is quite useful for detection of single base substitutions. SSCP analysis has proven to be a simple and effective technique for the detection of single base substitutions. Researchers have used SSCP to analyze different mutations like globin, p53, and rhodopsin mutations in mouse. As the sensitivity of SSCP varies dramatically with the size of the DNA fragment being analyzed, therefore the optimal size of fragment should be ~150 bp for this type of analysis.

18.9 SINGLE NUCLEOTIDE POLYMORPHISM

A single nucleotide polymorphism (SNP) is a DNA sequence variation occurring when a single nucleotide A, T, C, or G in the genome (or other shared sequences) differs among members of a species or between paired chromosomes in an individual. For example, two sequenced DNA fragments from different individuals, AAGC**C**TA to AAGC**T**TA, contain a difference in a single nucleotide. In this case, there are two alleles: C and T. SNPs are frequently present in the genome with a density of at least one common (>20% allele frequency) SNP per kbp. Few of SNPs give rise to RFLP; however, many of these do not lie between recognition sequences for restriction enzymes. For example, in human genome, there are 1.42 million SNPs, only 1,00,000 of which result in RFLP. As there are four types of nucleotides in the DNA sequence, which could be present at any position in the genome, there are possibilities of four alleles. However, most SNPs have only two alleles and are thus easy to assay. This is because of the way in which SNPs arise and spread in a population. An SNP originates when a point mutation occurs in a genome, converting one nucleotide into another. If the mutation is in the reproductive cells of an individual, then one or more of the offspring might inherit the mutation, and after many generations, the SNP may eventually become established in the population. But there are just two alleles: the original sequence and the mutated version. For a third allele to arise, a new mutation must occur at the same position in the genome in another individual, and this individual and its offspring must reproduce in such a way that the new allele becomes established. This situation is quite impossible hence majority of SNPs are biallelic. SNP may lie within coding sequences of genes, noncoding regions of genes, or in the intergenic regions between genes. SNPs within a coding sequence do not necessarily change the amino acid sequence of the protein that is produced. This is due to degeneracy of the genetic code. A SNP in which both forms lead to the same polypeptide sequence is known as synonymous (silent mutation) and if a different polypeptide sequence is produced, these are nonsynonymous. SNPs that are not in protein-coding regions may still have consequences for gene splicing, transcription factor binding, or the sequence of noncoding RNA. SNP can be used as molecular markers on the chromosome.

18.9.1 Procedure for SNP Analysis

The detection of SNP is based on oligonucleotide hybridization analysis. Note that the oligonucleotides are usually synthesized by techniques such as photolithography (Chapter 5). Hybridization is performed at highly stringent conditions. The temperature for incubation must be just below the T_m of the oligonucleotide. At this temperature, a stable hybrid is formed only if the oligonucleotide is able to form a completely base paired structure with the target DNA. If there is single mismatch, no hybrid is formed (Figure 18.11). At temperatures above the T_m, the fully base paired hybrid is unstable and at more than 5°C below the T_m, mismatched hybrids might be stable. Oligonucleotide hybridization easily discriminates between the two alleles of an SNP. Generally two types of strategies have been used for screening. These are explained below.

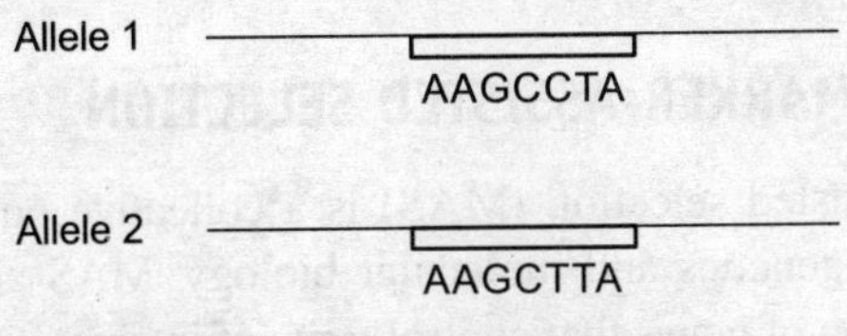

Figure 18.11 Single nucleotide polymorphism

Solution Hybridization Technique This is performed in the wells of a microtiter tray, each well containing a different oligonucleotide and use a detection system that can discriminate between unhybridized ss DNA and the double stranded product that results when an oligonucleotide hybridizes to the test DNA. The two ends of oligonucleotide probe are labeled with fluorescent dye and quencher. Normally there is no fluorescence because the oligonucleotide is designed in such a way that the two ends base pair, placing quencher next to the fluorophore, thereby forming a hairpin structure. Hybridization between oligonucleotide and test DNA disrupts this base pairing, moving the quencher away from the fluorophore and fluorescent signal is generated (Figure 18.12).

DNA Chips A DNA chip (for details see Chapter 4), carrying many different oligonucleotides in a high-density array, is also used for screening SNPs. The target DNA is labeled with a fluorescent marker and pipetted onto the surface of the chip.

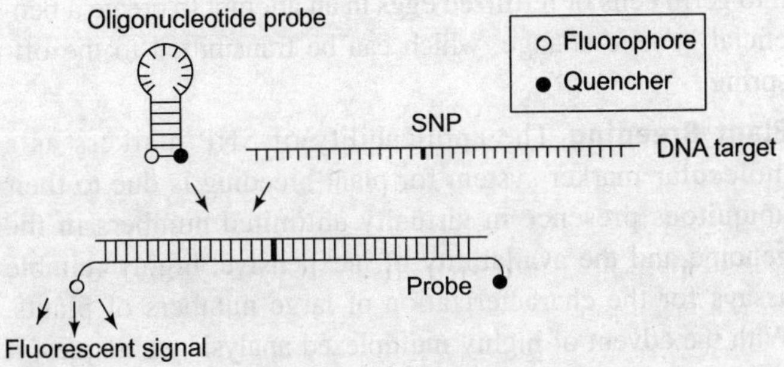

Figure 18.12 Detection of SNP by solution hybridization

Hybridization is detected by examining the chip with a fluorescence microscope. The positions at which the fluorescent signal is emitted indicates which immobilized DNA has hybridized with the test DNA. Many SNPs can therefore be scored in a single experiment. For example, a DNA chip with 2,50,000 oligonucleotides per cm^2 can be used to type 1,25,000 polymorphisms in a single experiment, presuming there are oligonucleotides for both alleles of each SNP.

18.9.2 Advantages of SNPs

For genetic mapping, SNPs provide several advantages over other markers. This is because SNPs are highly abundant, exist mostly in only two codominant variants, and are phenotypically neutral.

18.9.3 Applications of SNPs

The following fields utilize SNP information and haplotype maps. The major application of SNP information is toward improved and futuristic health care. Genomics and specifically SNP research can be used to improve health care through gene therapy, to yield new targets for drug discovery, to refine the process of drug development, and to discover new diagnostics.

Biomedical Research SNPs are of great interest to biomedical industry. Variations in the DNA sequences of humans can affect how humans develop diseases or respond to pathogens, chemicals, drugs, etc. SNP genotyping identifies SNPs that are common DNA variants present across the human genome. Over the past few years, SNPs have been proposed as the next generation markers to identify the loci associated with complex diseases and for pharmacogenomics. SNPs have been shown to be responsible for differences in genetic traits, susceptibility to disease, and response to drug therapies, pathogens, and chemicals. Genotyping of SNPs has become extremely important to researchers working on understanding and treating disease. An individual's genotype can be determined and then analyzed according to a haplotype map to determine the patient's disease risk or reception to different treatments. Germ line gene therapy involves the insertion of normal genes into germ cells or fertilized eggs in an attempt to create a beneficial genetic change, which can be transmitted to the offspring.

Plant Breeding The applicability of SNP markers as a molecular marker system for plant breeding is due to their ubiquitous presence in virtually unlimited numbers in the genome and the availability of inexpensive, highly reliable assays for the characterization of large numbers of plants. With the advent of highly multiplexed analysis techniques, it is possible to characterize individual plants for a large number of SNPs in a very short timeframe. However, as a biallelic system, SNP markers usually have low polymorphism information content so that large numbers of markers have to be developed in order to saturate the entire genome of a crop plant with informative markers, which can be used for plant breeding purposes.

18.10 MAP-BASED CLONING

Map-based cloning or positional cloning is a direct approach utilized for cloning of gene of interest with the help of closely linked molecular marker. The approach of map-based cloning is to find molecular markers tightly linked to the gene of interest, which serves as the starting point for chromosome walking or jumping to the gene (Figure 18.13). The steps of map-based cloning can be summarized as follows: (i) Identification of a molecular marker that lies in the vicinity of the gene of interest (few cm away) in a smaller population; (ii) The next step is to saturate the region around that original molecular marker with other markers in a relatively larger population; (iii) After the screening of a large insert genomic library (BAC or YAC) with marker, clones that hybridize to that particular molecular marker are isolated (iv) New markers are then created from the large insert clone, which are analyzed for cosegregation with the gene of interest; (v) If necessary, large insert genomic library is rescreened for other clones and searched for cosegregating markers; (vi) Candidate genes, whose markers cosegregate with the gene, are identified from large insert clone; (vii) DNA fragments between the flanking markers are cloned and introduced by plant transformation approach into a genotype that is mutant for the gene of interest; (viii) The gene fragment is sequenced to find a potential open reading frame (ORF); and (ix) Once this ORF is shown to rescue the mutant phenotype, then in-depth molecular and biochemical analysis of newly cloned gene is performed.

The steps outlined above for map-based cloning demonstrate the general approach needed to clone a gene. The specific steps may vary between this and other map-based cloning experiments, but the same general steps must be met. To speed the cloning process, it is best to start with a marker that is tightly linked to the gene of interest, which curtails a good amount of additional screening. If effective transformation system is not developed for species of interest, the confirmation is much more difficult. The most powerful approach in this case is recombinational or mutant analysis. For example, if two mutants are present in the gene of interest, then crossing is done in these two mutants and an individual with a restored normal phenotype is searched in a large segregating population. In the next step, gene is cloned from each of these three individuals, i.e., the two mutants and the restored line. If it can be shown that the restored line contains the same sequence as the predicted gene and that the two mutants have unique changes in the gene sequence not found in the normal gene, compelling evidence can be obtained, which indicate that the putative sequence is indeed the gene of interest. The first example of map-based cloning is that for gene from tomato plant, which provides resistance against bacterial speck disease of tomato caused by *Pseudomonas syringae* pv. tomato, i.e., Pto.

18.11 MARKER-ASSISTED SELECTION

Marker-assisted selection (MAS) is a collective outcome of traditional genetics and molecular biology. MAS allows for the selection of genes that control traits of interest. Combined with traditional selection techniques, MAS has become a valuable tool in selecting organisms for traits of interest, such as color, disease resistance, etc. Nowadays, researchers can use molecular markers to find genes of interest that phenotypically and functionally affect plants and animals. Some of the molecular markers are positioned very close to major genes of interest, which can be easily located by using well-designed experiments. Linked markers are present only near the gene of

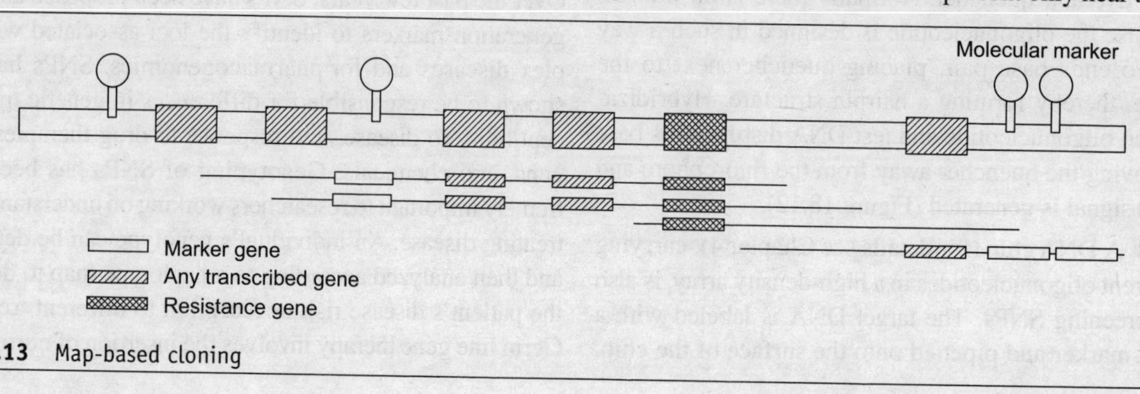

Figure 18.13 Map-based cloning

interest on the chromosome and are not parts of the DNA of the gene of interest. A second kind of molecular marker is one that is part of the gene of interest. However, once detected, direct markers are easier to work with, but as compared to linked markers, these are often much difficult to find.

The fundamental prerequisites for MAS technique are as follows: (i) Linked or direct molecular marker should be available for particular gene of interest; (ii) The screening for the molecular marker should be easy and fast. Generally, PCR-based molecular markers are used for this purpose; (iii) The screening methods should be highly reproducible among different laboratories; and (iv) The technique should be user-friendly and economical.

A number of strategies have been developed for screening a large number of random, unmapped molecular markers in a reasonably short time and to select few markers that are located in the vicinity of the target gene. The high volume marker technologies *viz.*, RAPD, RFLP, AFLP, SSR, and SNP, etc. are highly efficient techniques. These procedures are dependent on following two principles:

- Production of hundreds or even thousands of potentially polymorphic DNA segments and rapid detection from single preparation of DNA.
- Identification of those few fragments, which are derived from a region adjacent to the targeted gene by using genetic stocks.

In the past few years, by using one or more approach, several loci, which are scattered throughout the genome, have been assayed in very short tome. The next step is to determine which of the amplified loci lies near the targeted gene.

18.11.1 Strategies for Marker-assisted Selection

For MAS, two strategies have been proved most successful. These are nearly isogenic line (NIL) strategy and bulk segregant analysis (BSA).

Nearly Isogenic Line Strategy

Nearly isogenic lines (NILs) are created when a donor line (P1) is crossed to a recipient line (P2). The resulting F1 hybrid is then backcrossed to the P2 to generate backcross 1 (BC1). From BC1, a single individual containing the dominant allele of the target gene from P1 is selected on the basis of phenotype. This selected BC1 individual is again backcrossed to P2, and the cycle of backcross selection is repeated for a number of generations. In the BC7 generation, most, if not all, of the genome will be derived from P2 except for a small chromosomal segment containing the selected dominant allele, which is derived from P1. The homozygous lines for the target gene can be selected from that particular generation, which is said to be nearly isogenic with the recipient parent P2.

Bulk Segregant Analysis

When P1 and P2 are hybridized, the F2 generation derived from the cross segregates for alleles from both parents at all loci throughout the genome. If the F2 population is divided into two pools of contrasting individuals on the basis of screening at a single target locus, these two pools differ in their allelic content only at loci contained in the chromosomal region close to the target gene. Bulk 1 individuals selected for recessive phenotype will contain only P2 alleles near the target gene, whereas Bulk 2 plants selected for dominant phenotype will contain alleles from both P1 and P2.

High-volume marker techniques, such as those based on easy selection, can be used to compare the genetic profiles of pairs of NIL or pairs of bulks to search for DNA markers near the target gene.

18.11.2 Applications of MAS

Nowadays, MAS has great potential for combining the molecular biology approach with conventional and traditional breeding. However, there are currently few examples of MAS for economically important traits used by animal and plant breeders. Here, molecular markers can be used as chromosome landmarks to facilitate the selection of chromosome segments including useful agronomic traits during the breeding process. An additional advantage of the incorporation of MAS into breeding programs is that very different types of traits, e.g., a disease resistance gene or a gene to increase grain protein content can be manipulated using the same technology. Some of the examples of application of MAS in breeding are illustrated below:

MAS in Animal Breeding A number of different approaches have been used to identify the genes that control the genetic variation for a wide range of complex traits, including most of the economically relevant traits for animal production that are used for the improvement of animal breeds, for example, Mastitis resistance, double muscling, control of milk fat in bovine milk, carcass composition in pigs, and bovine leukocyte adhesion deficiency.

MAS in Plant Breeding Plant breeders are also utilizing MAS approach in some basic issues. MAS are particularly useful for genes that are highly affected by the environment, genes for resistance to diseases that cannot be easily screened, and accumulation of multiple genes for resistance to specific pathogens and pests within the same cultivar (gene pyramiding). MAS are also useful for the improvement of qualitative traits, hybrid rye breeding, male sterility restoration, resistance to abiotic stress, etc.

Review Questions

1. What are biochemical markers? Differentiate between isozymes and allozymes.
2. Define a molecular marker. Give a brief account of desirable properties of an ideal molecular marker.
3. Name the molecular marker technique based on restriction digestion and hybridization of DNA. Describe its principle in brief.
4. Describe the steps involved in restriction fragment length polymorphism (RFLP) study. Comment on the statement: Each RFLP data is unique to a specific enzyme–probe combination. Give a brief account of variation of RFLPs.
5. What are RAPDs and how is this data obtained? Give an elaborative account of principle, procedure, and applications of RAPD. Also write its advantages and limitations.
6. Describe the variants of RAPD technique. How can you develop SCAR from RAPD data?
7. Illustrate the technique for generating AFLPs with the help of suitable diagram. Also mention the primers and adaptors used in the technique. How does this technique differ from RFLP?
8. Give a brief account of different types of simple sequence length polymorphism in eukaryotic genome. Describe the different approaches of microsatellite-primed PCR.
9. Describe the molecular marker technique for detecting polymorphisms using electrophoresis of ss DNA. Also write its advantages and limitations in comparison to other techniques.
10. A majority of SNPs are biallelic. Explain this statement. Describe the two types of strategies for SNP analysis.
11. Map-based cloning is a direct approach utilized for cloning of gene of interest with the help of closely linked molecular marker. Give a brief description of different steps of map-based cloning.
12. Write a short note on marker-assisted selection. What are the fundamental prerequisites for marker-assisted selection technique?

References

Vos, *et al.* (1995). AFLP: A new technique for DNA fingerprinting. *Nucleic Acids Res.* **23**:4407–4414.

Welsh, J. and M. McClelland (1990). Fingerprinting genomes using PCR with arbitrary primers. *Nucleic Acids Res.* **18**:7213–7218.

Williams, J.G.K. *et al.* (1990). DNA polymorphism amplified by arbitrary primers are useful as genetic markers. *Nucleic Acids Res.* **18**:6531–6535.

Techniques Used in Genomics and Proteomics

Key Concepts

In this chapter we will learn the following:

- Techniques used in genomics
- Techniques used in proteomics

19.1 INTRODUCTION

The systematic study of the complete DNA sequences, i.e., genome, of organisms is known as genomics. In genomics, various techniques of molecular biology and bioinformatics are used to sequence and analyze the DNA sequences. Several genomes, including human, have been sequenced and now the dream of having the complete genome sequence is a reality. Proteomics is much more complicated than genomics. This term is analogous with genomics and considered as the next step. The genome is a somewhat constant entity while the proteome is constantly changing through its biochemical interactions with the genome. The organism exhibits differential expression patterns in different parts of its body and in different stages of its life cycle. Proteomics refers to any procedure that characterizes large sets of proteins. Proteomics focuses on identifying when and where proteins are expressed in a cell and to ascertain their physiological roles in an organism. 'If the genome is a list of the instruments in an orchestra, the proteome is the orchestra playing a symphony,' according to R. Simpson. The molecular biological techniques that are commonly used for studying genomes and proteomes are presented in this chapter.

19.2 TECHNIQUES USED IN GENOMICS

The acquisition of DNA sequence data is known as genomics. Determining the DNA sequence is useful in fundamental research and in applied fields such as diagnostic or forensic research. Usually, the sequence data are obtained in the form of many individual sequences of 500–800 bp. The advent of DNA sequencing has significantly accelerated biological research and discovery. The fundamental reasons for knowing the sequence of DNA molecule are the following: (i) To characterize the newly-cloned DNA; (ii) For predictions about its functions; (iii) To facilitate manipulation of the molecule; (iv) To confirm the identity of a clone or a mutation; (v) To check the fidelity of newly created mutation and ligation junction; (vi) Screening tool to identity polymorphisms and mutation in genes of particular interest; and (vii) To confirm the product of a polymerase chain reaction (PCR)

19.2.1 Conventional DNA Sequencing

The information content of DNA is encoded in the form of four bases (A, C, G, and T) and the process of determining the sequence of these bases in a given DNA molecule is referred to as DNA sequencing. The technique of DNA sequencing is a great achievement for molecular biologists and lies at the heart of molecular biology revolution. The first DNA to be sequenced was that of yeast ala-tRNA having 77 bases, which was sequenced by Holley's method. In Holley's method, RNA was digested with sequence-specific RNases followed by fractionating the resulting oligoribonucleotides by ion exchange chromatography or 2-D chromatography. Finally the order of base within each fragment was established by exonuclease digestion. As a result of using two different methods of fragmentation, Holley was able to establish overlaps between fragments and thus succeeded in assembling the entire 77 nucleotide sequence of a yeast ala-tRNA. Further, numerous attempts were made to use this approach of DNA sequencing by cleaving DNA into smaller fragments with different types of endonucleases or chemicals. Even there were few reports in which DNA was first transcribed into RNA and subsequently sequenced. However, none of these approaches were fully successful. Finally, the techniques of DNA sequencing were developed in the Gilbert's laboratory at Harvard University and in the Sanger laboratory at Cambridge, England. With these techniques at hand, the rate of obtaining new sequence information has increased rapidly over the last 30 years. Thus, the two seq-uencing techniques currently in use are as follows:

- Chain termination method or Sanger's dideoxy method (1977)
- Chemical degradation method or Maxam and Gilbert's method (1977)

Both these methods, though based on different principles, generate oligonucleotide sequence that begins with fixed point and terminates at particular kind of nucleotide residue. The result is the generation of four populations of oligonucleotides terminating at A/C/G/T residues. These populations of fragments are then resolved by polyacrylamide gel electrophoresis (PAGE) under conditions that distinguish between individual DNA fragments differing in length by as little as one nucleotide (for details see Chapter 6). The four populations of oligonucleotides are loaded into adjacent lanes of a sequencing gel and the order of nucleotides can be read directly from the image of the gel. In the following sections, the details of these two DNA sequencing methods are discussed.

Chemical Degradation Method (Maxam and Gilbert's Method)

The chemical degradation method was developed by Allan Maxam and Walter Gilbert and involves the base-specific chemical cleavage of an end-labeled DNA segment to generate a nested set of labeled molecules.

Principle The partially cleaved DNA fragment is subjected to five separate chemical reactions, each of which is specific for a particular base or type of base. The resulting fragments terminate at that specific base followed by high-resolution denaturing gel electrophoresis and detection of the labeled fragments by autoradiography (for details see Chapter 16). This method has not changed much since its initial development while some more chemical cleavage reactions have been formulated and other reactions have been simplified.

Steps Involved in Chemical Degradation Method The cleavage reactions are performed in following steps.

Restriction digestion and radiolabeling The starting point of the DNA to be sequenced is defined by a site of restriction enzyme. The double stranded DNA (ds DNA) fragment is first radiolabeled at 5′-end with α-^{32}P. Dimethyl sulfoxide (DMSO) is then added and the sample is heated to 90°C, which results in separation of strands. The two strands are separated by PAGE on the basis of difference in their mobility. One strand is purified from the gel and distributed into four parts.

Base-specific cleavage reactions The base-specific cleavage is performed by using reagents that modify a specific base or bases (Table 19.1) in such a way that subsequent treatment with hot piperidine (volatile secondary amine; 1 M in water, 90°C) cleaves the sugar phosphate backbone at the site of modification (Figure 19.1). These base-specific cleavage reactions are carefully designed to modify on an average only one of the target bases in each DNA molecule. This yields a set of end-labeled molecules ranging from one to several hundred nucleotides. Subsequently piperidine is removed either by ethanol (EtOH) precipitation or vacuum drying or a combination of both.

Table 19.1 Different Types of Base-specific Reactions Used in the Chemical Degradation Method

Reagents	Base	Specific Modification
Dimethyl sulfate (pH 8.0)	G	Methylation of N7 renders the C8–C9 bond susceptible to cleavage.
Piperidine formate (pH 2.0)	A + G	Weakens the glycosidic bonds of adenine and guanine residues by protonating nitrogen atoms in the purine rings, resulting in depurination.
Hydrazine	C + T	Opens pyrimidine rings, which recyclize in a five-membered form, which is vulnerable.
Hydrazine + 1.5 M NaCl	C	Only cytosine reacts with hydrazine.
Hot (90°) 1.2 M NaOH	A>C	Strong cleavage at A residues and weaker cleavage at C residues.

Polyacrylamide gel electrophoresis and autoradiography The resulting end-labeled DNA molecules are resuspended in a buffer containing 90% formamide and denatured by heating to 90°C. These are separated by PAGE and subjected to autoradiography. The autoradiogram thus developed is read in order to determine the DNA sequence.

Advantages and disadvantages of chemical degradation method Chemical degradation method has a number of advantages and disadvantages as compared to chain termination method.

The advantages are as follows: (i) Sequencing by this method can be initiated anywhere in the target DNA by labeling a restriction site or by end-labeling via PCR; (ii) Original DNA molecules can be sequenced rather than the complementary copy; and (iii) The sequence of DNA very close to the labeled site (2–3 bases) can also be obtained.

The disadvantages are as follows: (i) A smaller length of DNA sequence is usually determined by this method; (ii) The reactions of this method are normally slower and less consistent; (iii) The chemicals used for cleavage reactions are hazardous; and (iv) It is labor-intensive, as this method cannot be automated.

However, for most specific purposes, these disadvantages are insignificant than the advantages. For example, this method is used for special application where it is crucial to sequence the original DNA molecule rather than complementary copy that includes the study of DNA secondary structure and the interaction of proteins with DNA. The chemical degradation is sometimes performed on a DNA fragment of known sequence to generate a set of size markers. The sequence obtained by this method can be used as an access point to generate oligonucleotides that can be used as primers for enzymatic sequencing reactions. Additionally, the chemical degradation method has a crucial role in ascertaining the sequence of oligonucleotides in the functional dissection of transcriptional control signals and in the DNase footprinting.

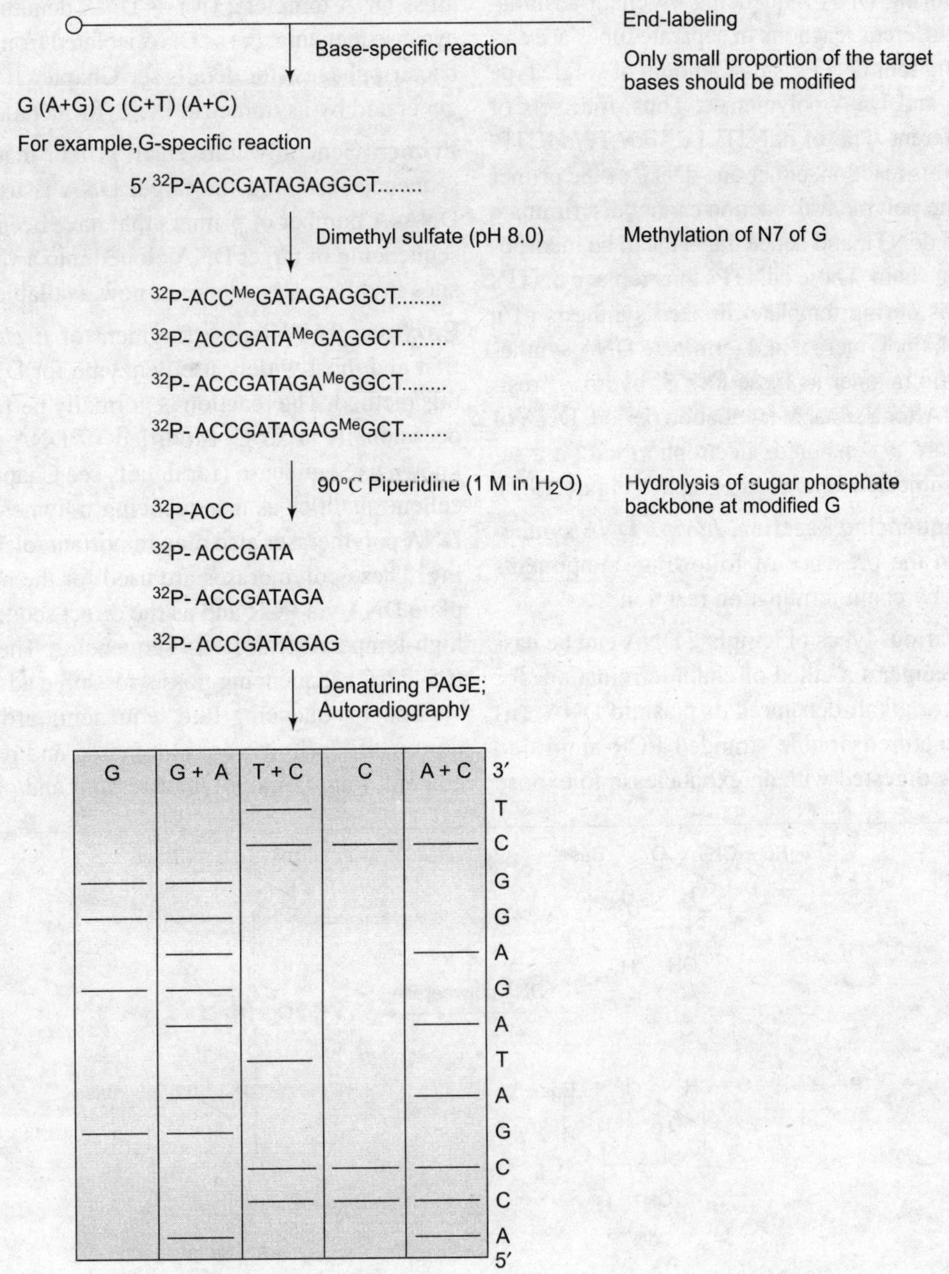

Figure 19.1 Maxam–Gilbert method of chemical degradation for DNA sequencing

Chain Termination Method (Sanger's Dideoxy Method)

Developed by Frederick Sanger, it is the most widely accepted and used method for sequencing DNA. This method involves controlled synthesis of DNA to generate fragments terminating at specific points. This idea was originally developed for partial repair in the presence of one, two, or three deoxyribonucleoside triphosphates (dNTPs) to sequence the 12 nucleotide cohesive ends of bacteriophage λ DNA. Further, by this method, DNA of bacteriophage ϕX174 (5,386 bases) was successfully sequenced. This system was termed as plus and minus system and developed by Sanger and Coulson, which was the immediate predecessor of the present dideoxy-mediated chain termination method. Chain termination is based on the use of a specific primer for extension by DNA polymerase, base-specific chain termination, and polyacrylamide gels to discriminate between single stranded DNA (ss DNA) chains differing in length by a single nucleotide.

Principle The underlying principle of chain termination method of DNA sequencing is that the replacement of dNTPs with 2′,3′-ddNTPs (dideoxyribonucleoside triphosphates) in the DNA chain terminates DNA synthesis. This is because these ddNTPs are nucleotide analogs that lack the 3′-OH group that is necessary for phosphodiester bond formation and chain elongation.

Steps Involved During DNA sequencing by chain termination method, four different reactions in separate tubes are carried out, each having four dNTPs, small amount of single type of ddNTP, primer, and DNA polymerase. Thus, four sets of reactions have different type of ddNTP, i.e., ddATP/ddCTP/ddGTP/ddTTP. In this reaction, either one dNTP or the primer is α-^{32}P labeled. The polymerase reaction cannot discriminate between dNTP and ddNTP and hence ddNTP can be incorporated in the growing chain. These ddNTPs thus replace dNTPs at random positions during template-directed synthesis of a DNA strand by DNA polymerase and terminate DNA synthesis in a base-specific manner as these lack 3′-hydroxyl residues (Figure 19.2). After a suitable incubation period, DNA of each reaction mixture is denatured, electrophoresed on a sequencing gel, and subjected to autoradiography (Figure 19.3).

Components of Sequencing Reaction *In vitro* DNA synthesis is performed in the presence of following components during sequencing by chain termination reaction.

Template DNA Various types of template DNA can be easily sequenced by Sanger's method of chain termination, for example, (i) Heat- or alkali-denatured ds plasmid DNA; (ii) Heat- or alkali-denatured double stranded PCR-amplified DNA; (iii) ds DNA digested with an exonuclease to expose an ss DNA template; (iv) ds DNA denatured by heat during cycle sequencing; (v) ss DNA isolated from recombinant M13 bacteriophages (for details see Chapter 10); and (vi) ss DNA generated by asymmetric PCR (for details see Chapter 7).

Primer Generally 'universal' primer that anneals to vector sequences flanking the target DNA is used for sequencing DNA. A number of primers that have been designed to allow sequencing of target DNA cloned into a variety of restriction sites in different vectors are now available.

Enzymes The Klenow fragment of *Escherichia coli* is the first and most widely used enzyme for DNA sequencing by this method. The reaction is normally performed at 37°C and occasionally at 50°C. Modified T7 DNA polymerase, better known as Sequenase (for details see Chapter 2), also has excellent qualities as a sequencing polymerase. Thermostable DNA polymerases also play important roles in DNA sequencing. These polymerases are used for the preparation of template DNA via PCR and as the direct sequencing enzymes in high-temperature dideoxy sequencing. The high-temperature (65–72°C) sequencing possesses some advantages over conventional sequencing like sequencing artifacts arising from nonspecific priming are eliminated and regions of high GC content, long homopolymer region, and/or secondary struc-

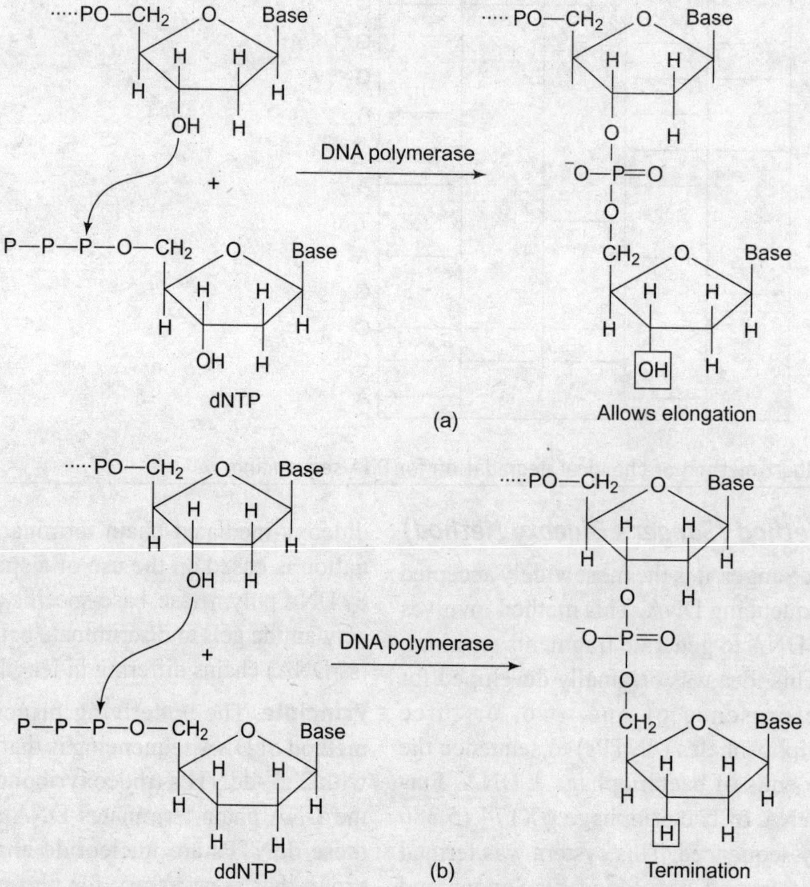

Figure 19.2 Action of DNA polymerase; (a) chain elongation by dNTP, (b) termination of DNA synthesis by ddNTPs

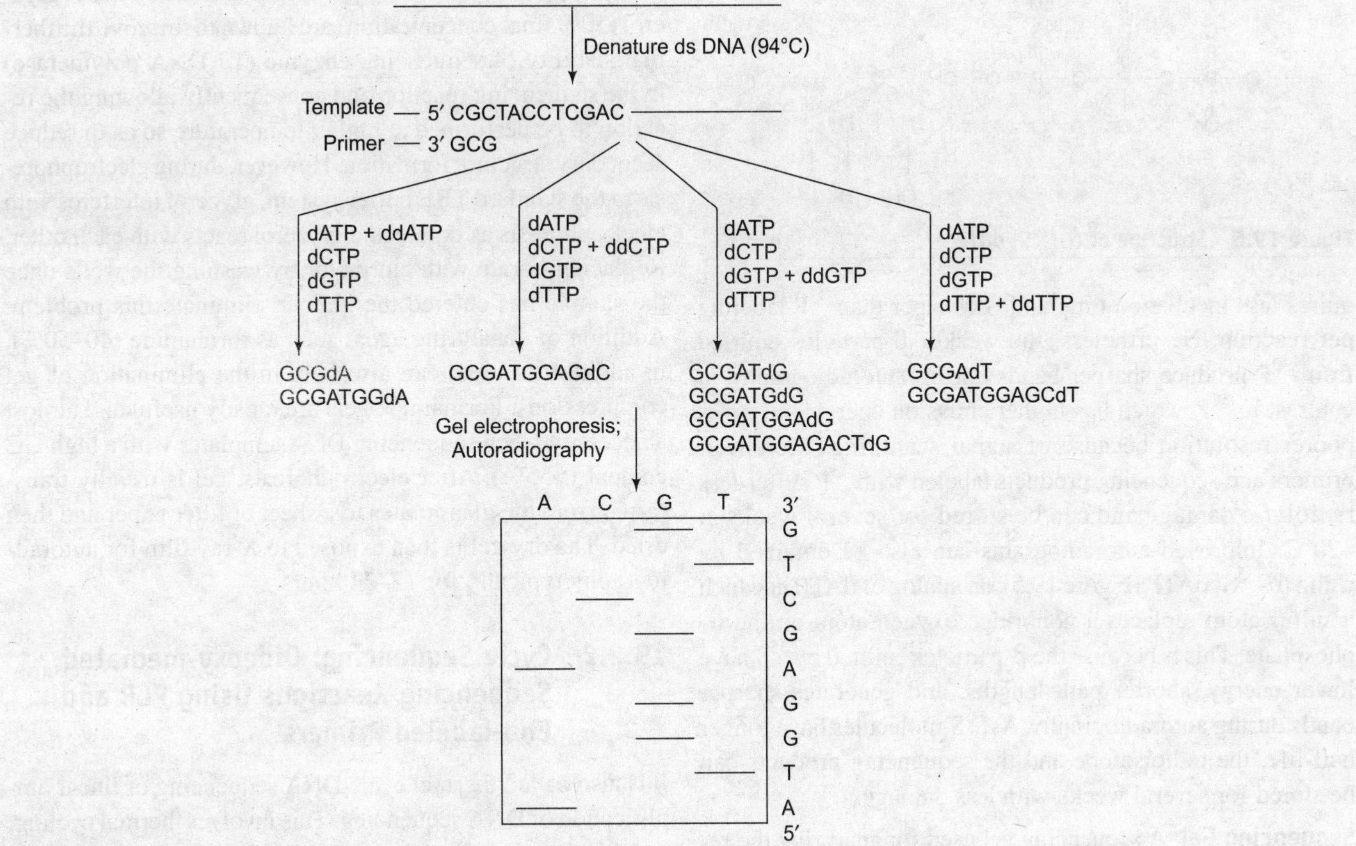

Figure 19.3 Chain termination method of DNA sequencing

tures can be easily sequenced. AMV and Mo-MLV reverse transcriptase (for details see Chapter 2) are frequently used for dideoxy sequencing of DNA and are indispensable tools for sequencing of RNA.

Substrates As already discussed above, dNTPs (dATP, dCTP, dGTP, dTTP) and ddNTPs (ddATP, ddCTP, ddGTP, ddTTP) are required as substrates for DNA sequencing. The dNTP:ddNTP ratio is carefully selected so that the resulting labeled strands form a nested set of DNA molecules up to few thousands bases long, each having a common 5′-end and terminating at a specific base.

Nucleotide analogs Compressions are major sources of error in analyzing sequencing gels. These artifacts arise as a result of intramolecular base pairing, leading to a folded or hairpin structure. This type of structure may migrate faster through the gel than an equivalent linear molecule, which disrupts the even spacing of the bands. GC hybrids are significantly more stable and hence the frequency of occurrence of compressions is more in GC-rich regions. These artifacts can be eliminated or reduced by using nucleotide analogs, such as dITP and 7-deaza-dGTP instead of dGTP, which destabilize GC base pairs. Inosine and 7-deaza-dGTP form only two H-bonds with cytosine and hence these base pairs are less stable than GC base pairs. Secondary structure is therefore

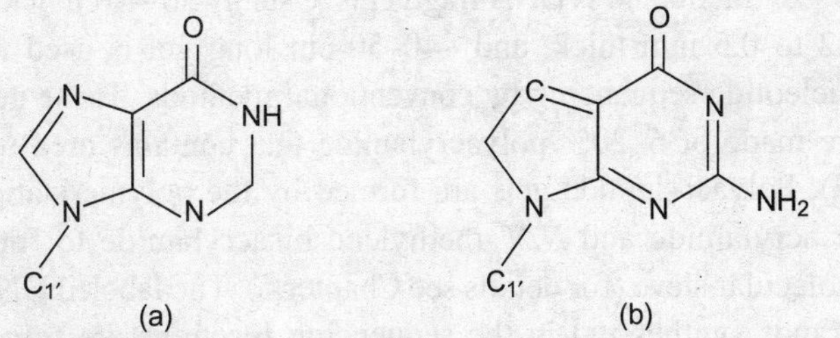

Figure 19.4 The structure of base analogs used to resolve compressions; (a) Inosine, (b) 7-deaza-guanine

less stable when dITP or 7-deaza-dGTP is used instead of dGTP (Figure 19.4).

Labeling of primers Usually, end-labeled primers are used for sequencing, which reduce the frequency and intensity of sequencing artifacts. This is because the end-labeled primers equally label all the DNA chains and hence the intensity of the bands is also equal throughout the entire length of the gel. Sequencing ladders obtained from reactions primed by labeled oligonucleotides are therefore far easier to read than ladders produced in conventional sequencing reactions containing unlabeled primers and radiolabeled dNTPs.

Primers can be labeled at their 5′-ends by transfer of the γ-phosphate from [γ-^{32}P] ATP or [γ-^{33}P] ATP in a reaction catalyzed by T4 polynucleotide kinase. The labeling with ^{32}P re-

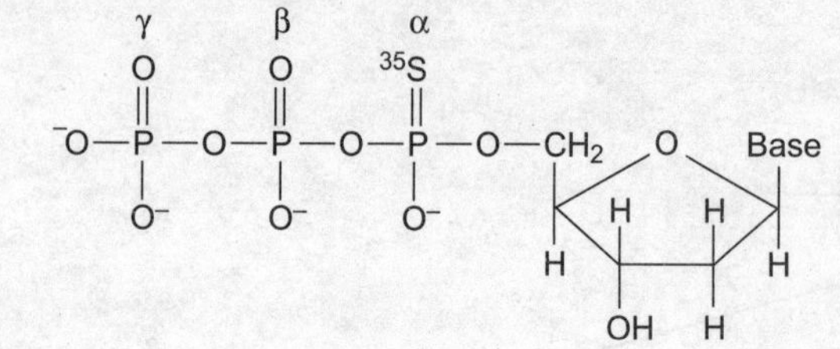

Figure 19.5 Structure of α-[^{35}S] dNTP

quires less incubation time and is cheaper than ^{33}P labeling per reaction. Nevertheless, the weaker β-particles emitted from ^{33}P produce sharper bands during autoradiography in contrast to ^{32}P, which has higher emission energy and gives poorer resolution because of signal scattering. Moreover, primers and sequencing products labeled with ^{33}P suffer less radiolytic damage and can be stored for several weeks at –20°C. Improved autoradiograms can also be obtained by using [α-^{35}S] dATP (Figure 19.5), an analog of dATP in which a sulfur atom replaces a nonbridge oxygen atom on the α-phosphate. This is because the β-particles emitted by ^{35}S have lower energy, shorter path lengths, and generates sharper bands during autoradiography. As ^{35}S molecules have longer half-life, the radioisotope and the sequencing products can be stored for several weeks with less damage.

Sequencing Gel A sequencing gel used for analyzing the reaction is long and thin having high-resolution quality designed to size fractionate ss DNA fragments. Usually 30–40 cm wide, 0.2 to 0.5 mm thick, and ~40–50 cm long gel is used for nucleotide sequencing by conventional methods. These gels are made of 6–20% polyacrylamide that contains urea (~7 M). Polyacrylamide gels are formed by the polymerization of acrylamide and *N,N′*-methylene bisacrylamide to form molecular sieve (for details see Chapter 6). The labeled DNA strands synthesized in the sequencing reactions are forced through this sieve by electrophoresis. Urea denatures the DNA and reduces artifacts due to presence of secondary structure of DNA. This further ensures that the separated DNA strands remain apart and migrate through the gel as linear molecules. This type of gel is capable of resolving fragments differing in length by a single base. Electrophoretic separation takes place in 3–10 hours. Electrophoresis tends to set up a temperature gradient across the gel. This leads to uneven band mobility across the gel, giving an effect known as 'smile'. This gradient system can be dissipated by attaching a metal plate to the gel plates or by using the buffer reservoir as a heat sink. An active cooling system such as a fan permits higher power to be used, giving shorter run times. A thermostat system, allowing accurate control of the gel temperature, aids reproducibility of the results. Furthermore, the temperature of gels needs to be maintained at ~40–50°C during the run. For maximum resolution, 0.6 X TBE is used in running buffer and 1.2 X TBE in the gel. High concentrations of glycerol (50% final concentration) are known to improve the thermal stability of sequencing enzyme (T7 DNA polymerase) in the sequencing reaction and consequently allowing the reaction to be performed at higher temperature so as to reduce secondary structure formation. However, during electrophoresis in the standard TBE buffer system, glycerol interferes with electrophoresis as borate and glycerol reacts with each other. Replacing borate with taurine or by washing the wells once the sample has entered the gel can eliminate this problem. Addition of denaturing agent such as formamide (40–50%), in addition to urea, can also help in the elimination of gel compressions. Formamide gels are mostly useful and almost a necessity when sequencing DNA templates with a high GC content (>55%). After electrophoresis, gel is usually transferred from the glass plates to a sheet of filter paper and then dried. The dry gel is then exposed to X-ray film for autoradiography typically for 12–24 hours.

19.2.2 Cycle Sequencing: Dideoxy-mediated Sequencing Reactions Using PCR and End-labeled Primers

It is also called thermal cycle DNA sequencing or linear amplification of DNA sequencing. This involves thermal cycling, i.e., heating the reaction mix to 94°C to denature the template, cooling below the melting temperature of the primer to allow annealing, and repeating the sequencing reaction (Figure 19.6). In principle, this procedure can be repeated until one of reaction components is exhausted.

Steps Involved and Cycling Programs

In this technique, four separate amplification reactions are set up, each having the same primer and a different ddNTP. Usually two cycling programs are used in cycle sequencing.

In one program, reaction mixtures are subjected to 15–40 cycles of conventional PCR cycling, i.e., amplification cycle of normal three steps: denaturation of the ds DNA template, annealing of a ^{32}P-labeled sequencing primer to its target sequence, and then extension of the annealed primer by a thermostable DNA polymerase. Finally, termination of the extended strand is done by the incorporation of a ddNTP. The resulting partially double stranded hybrid comprises of the full-length template strand and its complementary chain-terminated product. This hybrid is denatured during the first step of the next cycle, thereby liberating the template strand for another round of priming, extension, and termination. Therefore, the radiolabeled chain terminated products accumulate in a linear fashion during the entire first phase of the cycle sequencing reaction.

In second cycling program, the annealing step is eliminated so that no further extension of primers is possible. However,

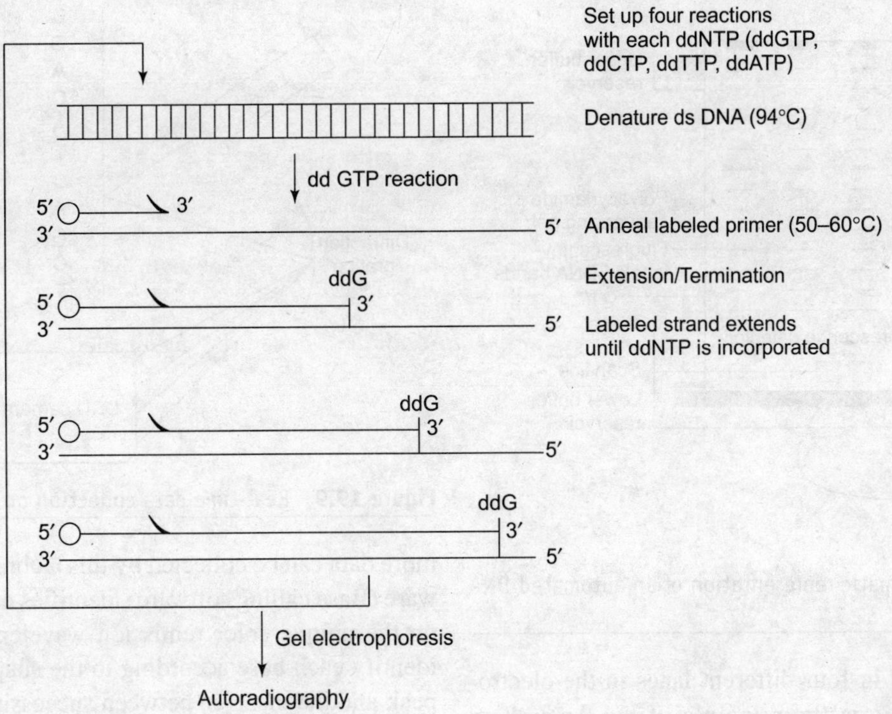

Figure 19.6 Outline of method for cycle sequencing

the reaction products that were not terminated by incorporation of a ddNTP during the initial rounds of conventional thermal cycling are further extended.

The radiolabeled products of cycle sequencing reactions are finally resolved on a denaturing polyacrylamide gel and visualized by autoradiography.

Advantages

Cycle sequencing offers a number of advantages, some of which are mentioned below: (i) Cycle sequencing requires only small amounts (50 femtomole; the mass of 50 fmol of ds DNA is ~33 ng/kbp) of template; (ii) Cycle sequencing works well with ds DNA as well as ss DNA templates; (iii) It can be easily set up in either microtiter plates or microfuge tubes; (iv) Cycle sequencing can be modified for use with internal labeling by α-^{32}P, α-^{33}P, or α-^{35}S labeled nucleotides or with 5′ labeled primers; (v) Cycle sequencing can also accommodate nonradioactive biotin and fluorescent labels. The fluorescent labels can be attached to either the 5′-end of the primer or to the ddNTPs; (vi) It can be used with commercially available robotic workstations and reactions are easily automated; (vii) It can work with nucleotide analogs, e.g., 7-deaza-dGTP or dITP; and (viii) It can be modified to obtain the sequence of each strand of a ds DNA template.

The most important precaution that should be taken care during cycle sequencing is that the template must be pure. This can be easily achieved by chromatography through spun columns of Sepharose CL-6B or Sephacryl S-400.

19.2.3 Automated DNA Sequencing

A key advantage of automated sequencing is the automated data collection in an easy way and in lesser time. As the dideoxy chain termination method of DNA sequencing with enzymatic reactions is performed in aqueous solvents at moderate temperatures, it was easily automated. The procedure used in manual sequencing by chain termination, i.e., radiolabeling of DNA fragments, resolution on the sequencing gel in four lanes, and finally detection by autoradiography, is not well-suited for automation. Thus for automation of this process, it is desirable to acquire sequence data in real-time by detecting the DNA bands during the gel electrophoresis. However, this is not possible as there are only ~10^{-15} to 10^{-16} moles of DNA per band, which is not easy to detect. Fluorescence methods can solve this problem of detection where the DNA bands are detected by fluorescence as these electrophorese past a detector (Figure 19.7). The sequence can be printed as chromatograms (Figure 19.8) or entered directly into a storage device for analysis.

Types of Fluorescence Labeling Systems

Two types of fluorescence labeling systems are used in automated DNA sequencing.

Four Reaction/One Gel System The fluorescent primers can be used with nonlabeled ddNTPs. The primers used in each of four chain extension reactions are 5′-linked to different fluorescence dyes. The resulting fluorescent-labeled DNA

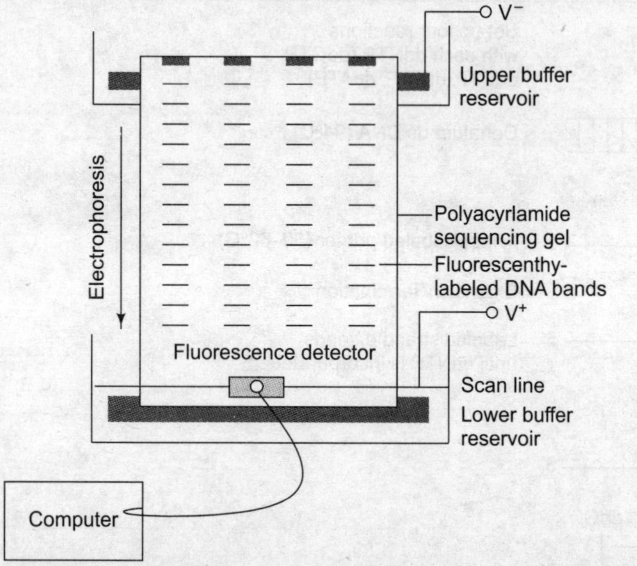

Figure 19.7 Diagrammatic representation of an automated DNA sequencer

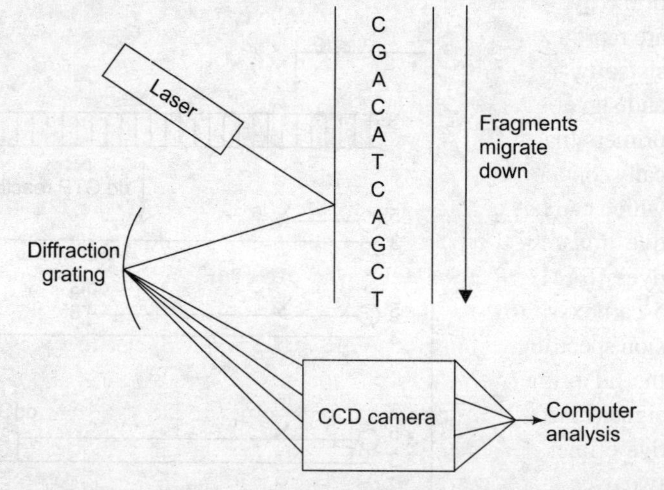

Figure 19.9 Real-time data collection on ABI 377

strands are separated in four different lanes in the electrophoresis system. This is the principle of the Amersham Pharmacia Biotech ALF sequencer. It has fixed argon laser that emits light, which passes through the width of the gel and is sensed by detectors in each of the lanes.

One Reaction/One Gel System Each of the four ddNTPs is labeled with spectrally different fluorophore. The chain extension reaction is carried out in a single tube. The resulting fragment is subjected to gel electrophoresis in a single lane. The tag is incorporated into the DNA molecule by DNA polymerase and completes two processes in one step, i.e., termination of DNA chain synthesis and attachment of fluorophore at the end of DNA molecules. Because of their higher throughput, better consistency, and lower sensitivity to electrophoretic artifacts, the single lane instruments have gradually become the approach of choice. This is the principle on which the original Applied Biosystems instruments (ABI) operate. The gel is illuminated with an argon beam, and the fluorescent signals are amplified and detected by photomultiplier tubes [or a charge-coupled device (CCD) camera] (Figure 19.9). Since fluorescent signals can be detected and processed in real-time, sequencing gels can be run for longer times and more data can be collected by this technique. Computer software (Base calling software) identifies each nucleotide based on the unique color (emission wavelength) of each dye to identify each base according to the shape of the fluorescent peak and the distance between successive peaks.

Chemistry of Automated Sequencing

Two types of chemistries are available for the ABI model automated sequencer, which is the currently used equipment by most sequencing facilities, i.e., dye primers and dye terminators. The difference between these two systems is simply the point at which the fluorescent label is incorporated into the newly synthesized strand.

Dye Primer System In dye primer system, the fluorescent dye is linked to the primer. Four separate reactions are carried out for each DNA sample as in conventional sequencing, with each reaction containing a different dye-labeled primer. This set of four reactions is then mixed and loaded into a single lane for gel electrophoresis. The sequencing reagent 'kit' has a set of four primers, each containing a different dye incorporated into an oligonucleotide, ~20 bases in length. The single labeled primer is more expensive than the single unlabeled primer, therefore the synthesis of four primers are more than four times costlier than the primer of dye-terminator system. Moreover, the dye interacts with first five bases of 5′-end of primer sequence that affects the mobility

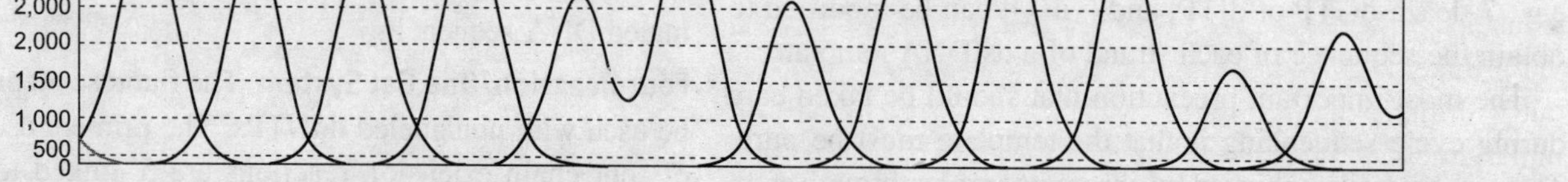

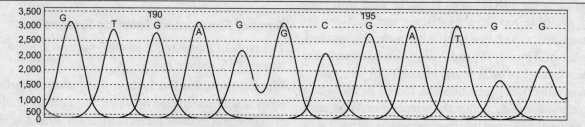

Figure 19.8 Print out from an automated sequencer

of the dye-labeled sequencing fragments. This mobility shifts are reproducible and can be compensated for by the gel analysis software. The sensitivity of this system can be improved and signal intensity enhanced by using 'energy transfer' (ET) primers that carry two separate dyes. These ET primers typically contain a single donor dye [6-carboxyfluorescein (FAM)], which can be easily excited by the argon ion laser. The donor dye transfers the energy to one of the secondary acceptor dyes [FAM, 5-carboxytetramethylrhodamine (TAMRA), or 5-carboxy-X-rhodamine (ROX)], each with a distinctive emission spectrum. Compared to the earlier single dye primer sets, the ET primers typically provide three- to four-fold stronger signals. Amersham and PE Bio-systems are two suppliers of dye primer reagents. Amersham (ET primer kit) provides the two dyes attached at two different bases, about five nucleotides apart. PE Bio-systems (Big dye primer kit) provide the two dyes connected by an amino benzoic acid linker (big dye), both attached to the 5′-end of the primers.

Usually *Taq* DNA polymerase or Sequenase (Chapter 2) can be used for sequencing under cycle sequencing conditions. However, *Taq* DNA polymerase discriminates between dNTPs and ddNTPs, so that it incorporates ddNTPs and typically labeled ddNTPs inefficiently. This means that high concentration of labeled ddNTPs have to be used for sequencing reaction. This problem, i.e., inefficient incorporation of ddNTPs, can be solved by using *Taq* DNA polymerase mutant (e.g., AmpliTaq FS, which has one mutation that eliminates the 5′ → 3′ nuclease activity, and a second that allows much more efficient incorporation of ddNTPs) for cycle sequencing. With the use of thermostable enzymes, the signal is linearly amplified and hence <0.1 μg of template is needed. The standard sequencing kit replaces dGTP with deaza–dGTP to reduce gel compression. However, for unknown reasons, the use of the deaza compound sometimes results in peak broadening, which reduces the readable length of sequencing tracts. The average read length with dye primer sequencing has achieved up to ~650–700 bp at >99% accuracy in 11–12 hours runs (2,400 V with 48 cm slab gel) on the ABI 377 Sequencer.

Dye Terminator System Although the dye primer method is suitable for sequencing projects that use universal primers, projects that require custom-designed primers become cumbersome and expensive because each primer must be modified in four separate dye-labeling reactions. This problem can be solved by attaching the fluorescent dyes to each of the four ddNTPs, which become incorporated at the 3′-end of the products of sequencing reactions. Note that each of the four ddNTPs is labeled with a different dye linked to the nitrogenous base via a linker. Four chain-extension reactions are then carried out with the same primer in a single tube. This requires low molar amounts of template DNA and there is a considerable reduction in labor and cost. In the reaction, the chains terminated for other reasons are not labeled and so are not detected, which eliminates noise. This means that dye terminator chemistry is less prone to false stops. As incorporation of dITP increases the false stops, dye terminator type of chemistry is well-suited for dITP incorporation to reduce compressions, because these false stops do not interfere in sequencing.

Dye terminators are frequently used with cycle sequencing. By using a commercial kit (ABI or PE), the sequencing reaction simply requires the mixing of DNA, primer, and water with an aliquot of premixed reagents from the kit, followed by 20 cycles in PCR machine. The unincorporated labeled nucleotides are removed by EtOH precipitation and the sample is air-dried. The reaction after this step is stable up to 1 week at –20°C. The sample is resuspended in a gel loading buffer and heated to denature the DNA.

As in dye primer system, AmpliTaq FS is also best-suited enzyme for this system. Convenient cycle sequencing conditions are normally used and the sequencing reactions can be performed with any primer and with a wide variety of templates (ss DNA, ds DNA, or PCR-generated DNA). Presently, two sets of second generation dye terminators are available from PE Biosystems. One terminator incorporates dichlororhodamine dyes and the other incorporates the big dyes terminator. The big dyes contain a fluorescein isomer as the donor dye and four dichlororhodamine dyes as the acceptors.

Types of Sequencers

Sequencers can be of different types. Some examples are given in the following.

LI-COR Sequencers LI-COR instruments of automated sequencing also detect fluorescent-labeled DNA molecules as a laser near the bottom of gel excites them. This range of instruments is equipped with a scanning fluorescence microscope for detection of the fluorescent label. Thus focusing can be controlled automatically based on differential fluorescence of the gel relative to the glass plates. There is single detection wavelength, so data can be collected more rapidly during each scan, which means four lanes are required per sample. Labeled primers are commercially available, but the standard labeling system is internal labeling using dATP labeled with an infrared dye. Additionally, LI-COR two-dye near-infrared DNA analysis system can detect the products of two different sequencing reactions in parallel, enabling pooling reactions and simultaneous bidirectional sequencing. The LI-COR 4000 LS system can read more than 800 bases per sample at 99% accuracy, but can run 7 to 11 samples per gel compared with 36 or 64 on the ABI 377. Longer read lengths can save time and money in primer walking strategies as fewer steps and fewer custom primers are required. This instrument is also cheaper than ABI sequencers.

Capillary-based Sequencers Practically DNA sequencing by capillary electrophoresis is very simple and is usually undertaken in high-purity silica capillaries with an internal diameter of 50–100 μm. A number of different materials can be used for the gel matrix but it is usually linear polyacrylamide. The sample is applied on top of capillary gel and the labeled DNA fragments travel via the capillary. They come out at the end in a vertical stream and are finally detected. A capillary electrophoresis system has other advantages over traditional slab gel. These are as follows: (i) Capillary-based sequencers provide six to eight runs a day and eliminate the tedious gel preparation and sample loading steps; (ii) As each capillary is very small in diameter, heat generated during electrophoresis can be rapidly dissipated. Very high voltages can therefore be applied to achieve separation of the products of sequencing reactions in a shorter period of time; (iii) Silica capillaries are very flexible and are easily incorporated into automated instruments that reduce the time of operation; and (iv) Because each sample is loaded into a discrete capillary, there is no need for time-consuming tracking of lanes on gel images.

19.2.4 Pyrosequencing

Pyrosequencing is more rapid 'minisequencing' method because it does not require electrophoresis or any other fragment separation procedure. This method determines which of the four bases is incorporated at each step in the copying of DNA template. In this method, dNTPs are used and ddNTPs are not required. As the new strand is being made, the order in which the dNTPs are incorporated is detected, so the sequence can be read as the reaction proceeds. In the reaction, all four dNTPs are not added at one time; rather each dNTP is added individually in a sequential manner. If the particular dNTP is not incorporated into the growing chain, then it is rapidly degraded by nucleotidase or washed before the addition of next dNTP. The addition of a nucleotide to the end of the growing chain is detectable because it is accompanied by the release of pyrophosphate, which is detected in an enzyme cascade that emits light (Figure 19.10). The released pyrophosphate converts adenosine phosphosulfate (APS) into ATP in the presence of enzyme ATP sulfurylase. Finally, on addition of luciferin and enzyme luciferase, ATP is degraded into AMP that is accompanied with a flash of chemiluminescence. Since both ATP and dATP are used by luciferase, dATP cannot be used for nucleotide incorporation; otherwise false results may be obtained. Instead α-thio-dATP, an analog is used in place of dATP, which is used as a substrate by DNA polymerase but not by luciferase. As indicated above, all the four dNTPs are not added at one time because in such a case, flashes of light would be seen all the times, and no useful sequence information would be obtained. The template for pyrosequencing can be easily generated or amplified by PCR. After PCR, the product is simply purified by treating with alkaline phosphatase or apyrase and exonuclease I to destroy the dNTPs and primers, respectively. The sequencing primer is then added and the mixture is rapidly heated and cooled, which inactivates the enzymes, denatures the DNA, and facilitates the primers to anneal to the templates. In addition to this, an alternative method is also used for template preparation in which biotinylated PCR product is captured with the help of magnetic beads, washed, and denatured with alkali.

Types of Pyrosequencing

There are two methods for pyrosequencing, which are as follows:

Solid-phase Pyrosequencing In this type of pyrosequencing, template and primer DNA are immobilized on a solid support. All the four dNTPs are added stepwise and the incorporation of a particular dNTP is detected by addition of ATP sulfurylase and luciferase. A washing step is carried out after addition of each dNTP for removal of the excess substrate (Figure 19.11).

Liquid-phase Pyrosequencing The template DNA, oligonucleotide primer, polymerase, ATP sulfurylase, luciferase, and most importantly nucleotidase (apyrase) are placed in the well of a microtiter plate. The four dNTPs are then added sequentially and incorporation is detected because of chemiluminescence. The unincorporated dNTPs are continuously degraded by nucleotidase. Thus, the addition of nucleotidase in the reaction mixture eliminates the requirement of immobilization and washing steps of solid-phase pyrosequencing, and reaction is easily performed in a single tube (Figure 19.12).

Applications

Pyrosequencing makes it possible to follow the order in which the dNTP are incorporated into the growing strand. This technique simply requires that a repetitive series of additions be made to the reaction mixture, which is easily automated. As

$$\underset{\text{(Oligonucleotide)}}{(\text{dNMP})_n} + \underset{\text{(Nucleotide)}}{\text{dNTP}} \xrightarrow{\text{DNA polymerase}} \underset{\text{(Lengthened oligonucleotide)}}{(\text{dNMP})_{n+1}} + \underset{\text{(Pyrophosphate)}}{\text{PPi}}$$

$$\text{PP}_i + \text{APS} \xrightarrow{\text{Sulfurylase}} \text{ATP} + \text{Sulfate}$$

$$\text{ATP} + \text{Luciferin} + \text{O}_2 \xrightarrow{\text{Luciferase}} \text{AMP} + \text{PPi} + \text{Oxyluciferin} + \text{Light}$$

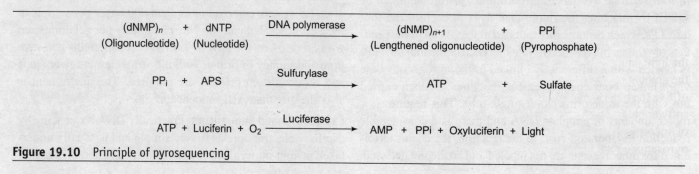

Figure 19.10 Principle of pyrosequencing

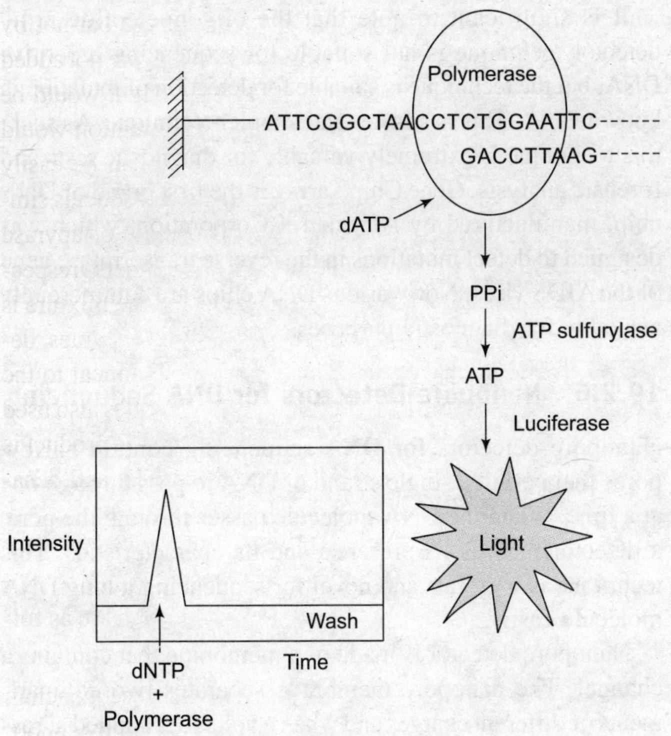

Figure 19.11 Diagrammatic representation of solid-phase pyrosequencing

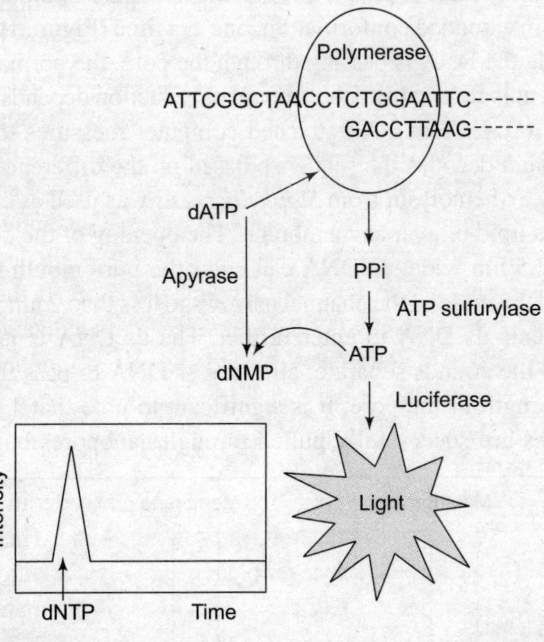

Figure 19.12 Diagrammatic representation of liquid-phase pyrosequencing

the light emitted by enzyme cascade is directly proportional to the amount of pyrophosphate released, it is easy to detect runs of five to six identical bases. For longer runs of a single base, software algorithms must be applied for exact quantification of incorporated nucleotides. The acceptable read length of pyrosequencing is ~200 bases, which are much less than Sanger's methods of sequencing. Many modifications are being introduced in this system and the availability of automated systems for pyrosequencing enables the use of the technique for high-throughput analyses. Roche has developed new sequencing technique based on the principle of pyrosequencing named as 454 (also called 'Roche 454'). In this sequencing technique, ss DNA is annealed to an excess of DNA capture beads and then beads and PCR reagents are emulsified in water-in-oil microreactors. Beads that contain amplified template DNA are deposited into the wells of a picotiter plate (fused fiber-optic reaction vessels arranged in tandem). 454 Sequencing™ is performed to obtain reads with average read lengths of 250 bases. The major advantage of using this sequencing technique is that labeling of test DNA and gel separation are not required. Pyrosequencing is quite often used for short regions in multiple individuals rather than for sequencing long regions of unknown DNA. It is specifically used for comparative analysis of known DNA sequences that differ in few bases. Roche has launched its next generation system, GSFLX (Genome Sequencer FLX system) in December 2006 where FLX stands for flexibility and reflects the outstanding versatility of the system. It has unique combination of read length and reads per run, which significantly improves the sequence accuracy. This system provides applications for research fields such as cancer research, genetic diseases, infectious diseases, plant genomics, and metagenomics. The second generation Roche 454 sequencer is capable of generating 4,00,000 reads in an 8 hours run. The first individual human genome (the genome of DNA pioneer and Nobel Laureate James D. Watson) has been sequenced by using this technique.

19.2.5 Gene Chip Array-based DNA Sequencing

As mentioned in Chapter 4, DNA chips are used for simultaneously detecting and identifying many short DNA fragments by DNA:DNA hybridizations. This is also known as DNA array or oligonucleotide detector that can be used for large-scale sequencing or mutations or detecting single nucleotide polymorphisms.

For DNA sequencing, the DNA sequence is first denatured into single strands. Then one strand is tested for hybridization to a known probe sequence (octanucleotide), usually to all possible stretches of eight bases, one at a time. Every possible 8-base sequences, which have to be used as probes, are arranged in a square array and anchored to or synthesized on the surface of a glass chip (for details see Chapter 4). An array about 1 cm^2 usually contains up to a million nucleotide probe sequences. The glass chip is subsequently dipped in a solution of the test DNA, which will hybridize simultaneously to all those probe sequences complementary to that particular sequence, for example, the sequence of test sequence is GCTA TGCT GGCT TACG and its complementary sequence will

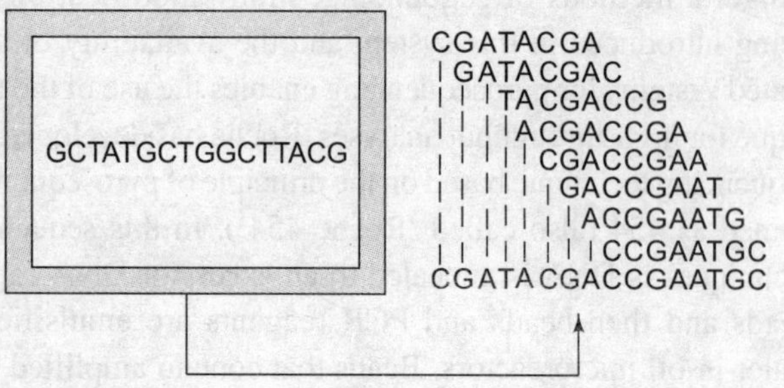

Figure 19.13 Oligonucleotide array detector-sequencing from oligonucleotide overlaps

therefore be CGAT ACGA CCGA ATGC. If this test sequence hybridizes to DNA chip having all possible 8-base sequences, the base sequences that will hybridize are CGATACGA, GATACGAC, ATACGACC, TACGACCG, ACGACCGA, CGACCGAA, GACCGAAT, ACCGAATG, and CCGAATGC (Figure 19.13). By analyzing all the possible 8-base probe sequences, sequencing of test DNA can be performed by hybridization and analysis by computer software. For sequencing of larger DNA, it must be broken into smaller pieces and tagged with a fluorescent dye. The DNA binds only with complementary octanucleotide probes on the DNA chip. Then the chip is scanned by a laser to locate the fluorescently-tagged DNA. Analysis is done with the help of an attached computer. The sequences are arranged into correct overlapping order by computer software (Figure 19.14).

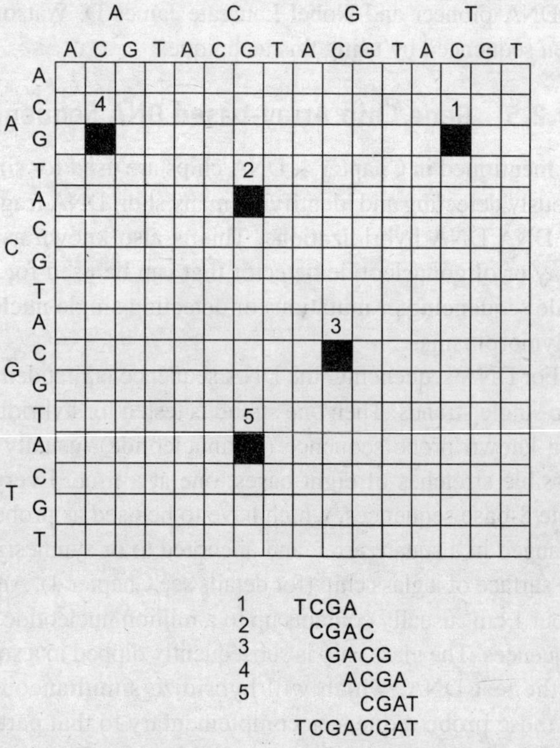

Figure 19.14 Oligonucleotide array hybridizes with five fragments and sequence is assembled by computer

It is significant to note that the oligonucleotide array detector technique is not suitable for sequencing repetitive DNA, but the technique is suitable for detection of mutation in known DNA. This is a faster and simpler technique. As such, this technique is extremely valuable for diagnostic tests and forensic analysis. Gene Chip® array is the first brand of DNA chip, manufactured by Affymetrix Corporation, which was designed to detect mutations in the reverse transcriptase gene of the AIDS virus. Now various DNA chips are commercially available for diagnostic purposes.

19.2.6 Nanopore Detectors for DNA Sequencing

Nanopore detectors for DNA sequencing contain narrow pores that permit a single strand of DNA to pass through one at a time. When the DNA molecule passes through the pore, a detector records its presence and its characteristics. This technique is very fast and novel for sequencing a long DNA molecule easily.

Nanopore detector is made of a membrane that contains a channel. The nanopore membrane separates two compartments of different charges and when a voltage is applied across the membrane, ions flow through the open channel (Figure 19.15a). Because of this charge separation, negatively charged DNA molecule is pulled through the nanopore to the positive side in extended conformation, one at a time (Figure 19.15b). While the DNA is passing through the pore, the normal ionic current is reduced and the amount of reduction depends on the base (G>C>T>A). An attached computer measures the current and decodes the sequence based on the differences. Initially, α-hemolysin from *Staphylococcus* was used as channel and a lipid bilayer as membrane. The opening of the channel is ~2.5 nm wide. ds DNA can enter the pore mouth but toward the middle, the channel narrows to less than 2 nm, which prevents ds DNA to enter further. The ds DNA is jammed until the strands separate, allowing ss DNA to pass through the length of the pore. It is significant to note that 1 kbp ss DNAs are successfully pulled through nanopores but these

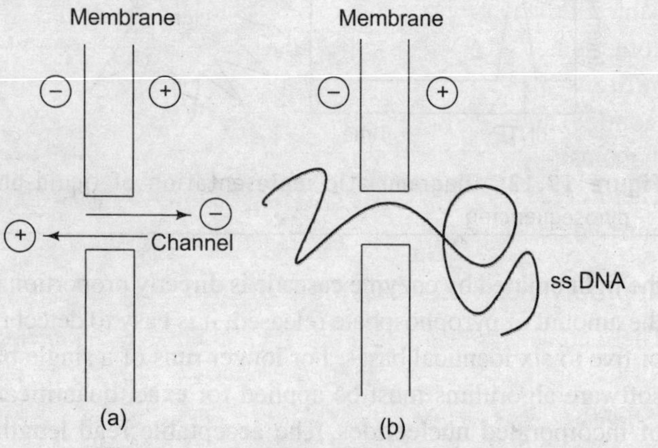

Figure 19.15 Principle of nanopore detector

move so fast through pores that it is difficult to detect individual bases. Presently, nanopore detectors can analyze two 20 nucleotides DNA strands that differ in a single base. Theoretically, it can be possible to sequence whole bacterial genome in ~1 min and the entire human genome in less than 2 hours provided technical improvements will be incorporated in this technique such as the chips are used consisting of 500 pores, each having a capacity to read 1 kbp/second.

19.2.7 RNA Sequencing

RNA can be easily sequenced by modifying DNA sequencing techniques. The RNA to be sequenced is first transcribed into a cDNA with the help of reverse transcriptase enzyme and then the sequencing of resulting cDNA is performed.

19.2.8 Genome Sequencing

The first completely sequenced genome accomplished in 1975 was that of ϕX174. Thereafter, the genome of bacterium *Haemophilus influenzae* was sequenced in 1995. Subsequently numerous bacterial genomes were sequenced and data published. Eukaryotic genome sequencing is still a developing field. Now, genome sequencing is turning toward species that are of interest because of their key position in the evolutionary tree or because of their economic value. The eukaryotic genomes sequenced are that of yeast, *Caenorhabditis elegans*, *Drosophila melanogaster*, *Arabidopsis thaliana*, and mouse.

For genome sequencing, usually the sequence data have been obtained in the form of many individual sequences of 500–800 bp using the most advanced automated sequencers equipped with robotic systems for sample preparation and sample loading, 96 capillary tubes for separation, and fully automated electrophoresis and data analyses systems. With the help of these types of sequencers, 96 samples are simultaneously sequenced averaging ~700 bases every ~2 hours and thus can identify ~8,00,000 bases per day. Therefore, the major sequencing centers have over 100 advanced sequencing systems, all working round the clock and generating abundant data. The major technical hurdle in sequencing a genome is not the DNA sequencing itself, but assembling the millions of sequenced fragments into contiguous blocks (contigs) and finally assigning them to their proper positions that must be assembled into the contiguous genome sequence. Therefore, an approach for correctly assembling the sequences must be developed.

Approaches for the Assembly of DNA Sequences

There are two different approaches that have been developed for assembling the sequences obtained by automated DNA sequencers.

Shotgun Sequencing Shotgun sequencing is the strategy of choice for large-scale sequencing projects and does not require any genetic map or prior information about the organism whose genome is to be sequenced.

Steps involved In this technique, the genome to be sequenced is randomly broken into short fragments (1–2 kbp), which are suitable for sequencing. These fragments are ligated into a suitable vector and then partially sequenced. In a single sequencing run, 400–500 bp of sequence can be determined from each fragment. In some cases, both ends of a fragment are sequenced. Computerized searching for overlaps between individual sequences then assembles the complete sequence. Overlapping sequences are assembled to produce contigs. The term contig refers to known DNA sequences that are contiguous and lack gaps. Since fragments are cloned at random, duplicates will quite often be sequenced. To get full coverage, the total amount of sequence obtained must therefore be several times that of the genome to allow for duplications. For example, 99.8% coverage requires a total amount of sequence six to eight times the genome size. It should be possible to deduce the complete genome sequence by arranging together the individual sequence reads. Practically, while performing shotgun sequencing, two types of problems arise. First, the presence of dispersed repetitive sequences creates confusion during sequence assembly. Second, assembly of sequences into contigs is possible but there are gaps between contigs that need closing by individual approach. Thus, after obtaining contiguous sequence data, it is required to close each of the gaps individually by most direct approach in time effective manner. If end sequences of contigs are present within the same clone, further sequencing of this insert with internal primers helps in closing of gaps (Figure 19.16a). However, a number of strategies are applied for gap closures though the most successful is the hybridization analysis of a clone library, in which the library is screened with a series of probes whose sequences correspond with the ends of each of the segments. If two probes are hybridized to the same clone, it specifies that the two segment ends represented by these two clones are adjacent to one another in the genome (Figure 19.16b). Finally, sequencing the appropriate part of the clone closes the gaps between these two ends and the task is known as finishing. Many of the gaps between contigs are due to regions of DNA that are degraded during cloning, which are known as physical gap, particularly in multicopy vector. To overcome these gaps, a second library is prepared in a different vector, often a single copy vector. For example, λ vector is often used during final stages of shotgun sequencing. Thereafter, pairs of oligonucleotide probes that are designed from the ends of contigs are used for screening of second library. This bridges the gap between the two contigs. In another approach, PCR reaction is performed on whole genomic sequence by using primers corresponding to contig ends. Amplification occurs only if the two ends of contig are within a few kilobases of each other (Figure 19.16c).

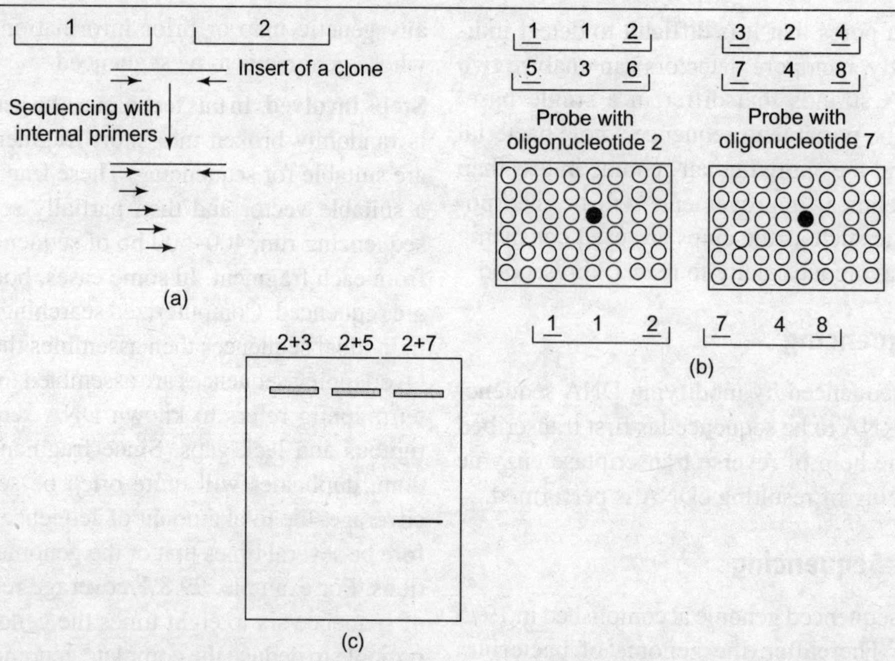

Figure 19.16 Assembly of genome sequence; (a) Closing sequence gap by direct approach, (b) Closing sequence gap by oligonucleotide hybridization. Oligonucleotides 2 and 7 hybridize the same clone demonstrating that contigs 1 and 4 are adjacent, (c) PCR with pairs of oligonucleotides confirming that contig ends are close together in the genome. The amplified product from the clone can be sequenced for closure of physical gap between two contigs

Case The whole genome of bacterium *H. influenzae* was sequenced by this approach (Figure 19.17). The genomic DNA was prepared and fragmented into small pieces for preparation of template having manageable size for sequencing. This was possible by mechanically shearing or by digesting the DNA with restriction enzymes. Sonication was preferred for fragmentation due to its more randomness, thereby minimizing the possibility of gaps appearing in the genome sequence. The resulting fragments were size fractionated by agarose gel electrophoresis. The fragments ranging between 1.6 and 2.0 kbp in size were purified from the gel (for details see Chapter 6). Two DNA sequences were produced with this size of fragments, one from each end, reducing the amount of cloning and DNA preparation. Additionally, with maximum size of 2.0 kbp, the chances of occurence of complete genes on a DNA fragment was reduced so that the possibility of their loss through expression of deleterious gene products was minimized. The selected fragments were ligated to a sequencing vector and again size fractionated to reduce contamination from double inserts, chimeras or free vectors. All cloning was carried in host cells deficient in all recombination and restriction functions to prevent deletion and rearrangement of insert. After cloning, 28,643 sequence experiments were performed with 19,687 of the clones, of which 4,339 sequences were rejected because these were less than 400 bp in length. The remaining 24,304 sequences were entered into computer that took 30 hours analyzing the data. The result was contiguous sequences, each being a different segment of *H. influenzae* genome. Finally, 11,631,485 bp of sequence was generated, which meant that huge amount of extra work would have been performed before the genome were correctly sequenced.

Bacterial artificial chromosome (BAC) walking As bacterial genomes contain no repeated DNA sequences, many bacterial genomes have been sequenced using the shotgun sequencing approach. However, for sequencing of eukaryotic genome, a number of modifications are included in this sequencing approach. Using this approach without having the knowledge of physical maps whole genome of eukaryotes can be sequenced. This is performed in stages. First a bacterial artificial chromosome library of ~150 kbp inserts is generated. The individual clones of the library are then arrayed in micotiter wells. The insert in each of this clone is sequenced starting at vector-insert points ~500 from each end and known as sequence-tagged connectors (STC) or BAC ends. STCs are normally scattered every 5 kbp across the genome and make up 10% of the sequence. STCs can help in the connection of one BAC to other BACs, and each BAC clone can be linked with approximately 30 other BACs. One BAC insert is then fragmented, shotgun cloned into M13 vectors (~3,000 clones), and the fragments are sequenced and assembled into contigs. The restriction enzyme fingerprinting of each clone is made by which size of insert is detected. The fingerprints of overlapping clones are compared, which help in selection of contigs. Thereafter, the sequence of the particular seed BAC is then compared with the database of

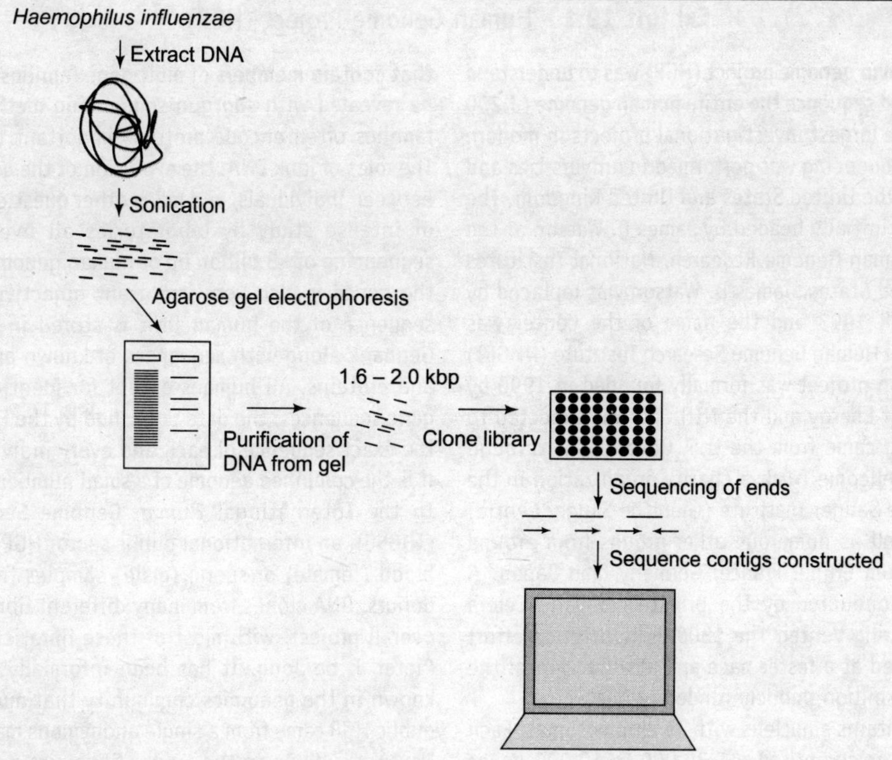

Figure 19.17 Shotgun sequencing of *H. influenzae* genome

STCs to identify ~30 overlapping BAC clones. The two clones with minimal overlap at both ends and also showing internal consistencies among the fingerprints are then selected, sequenced, and the process is repeated until the entire chromosome is sequenced. This is called BAC walking (Figure 19.18). This shotgun strategy is readily automated through robotics and hence is faster and less expensive than the conventional strategy. The majority of the identified whole genome sequencing has been performed by this shotgun strategy in a very short time. By applying this approach with some modification, chromosome 2 (19.6 Mbp) of *Arabidopsis thaliana* has been sequenced. The entire human genome has been also been sequenced with the help of modified approach of shotgun sequencing (Exhibit 19.1).

Clone Contig Approach The clone contig approach is the conventional method for sequencing of a eukaryotic genome and has also been used with the microbial genomes that have already been mapped. This is also known as clone-by-clone approach or map-based approach and it has been effectively used to produce the complete genome sequence of *S. cerevisiae* and the nematode *C. elegans*. This approach does not suffer from the limitation of shotgun sequencing and provides an accurate sequence of a large genome that even contains repetitive DNA. The major drawback of this method is that it involves much more work and hence longer time and more money. Additional time and effort is needed to construct the overlapping series of cloned DNA fragments.

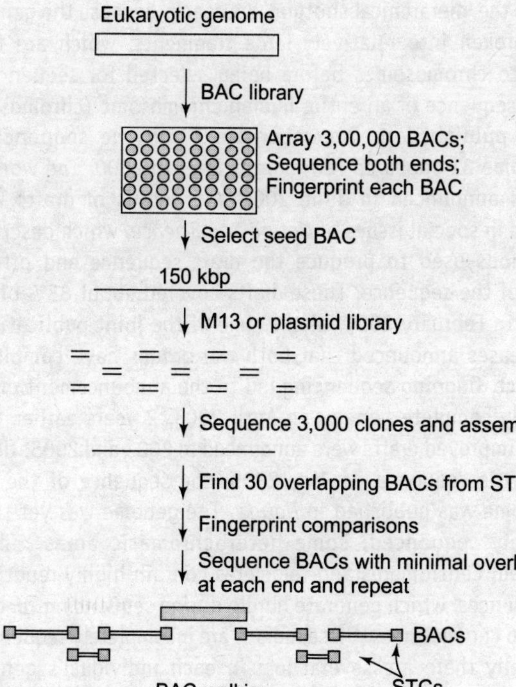

Figure 19.18 Shotgun sequencing strategy (BAC walking) for eukaryotic genome

Exhibit 19.1 Human Genome Project (HGP)

The original goal of human genome project (HGP) was to understand the genetic make up and sequence the entire human genome (3,200 Mbp). It was one of the largest investigational projects in modern science. Most of the sequencing was performed in universities and research centers from the United States and United Kingdom. The project began in 1990 initially headed by James D. Watson at the National Center for Human Genome Research, National Institutes of Health (NIH), United States. James D. Watson was replaced by Francis Collins in April 1993 and the name of the Center was changed to the National Human Genome Research Institute (NHGRI) in 1997. The $ 3 billion project was formally founded in 1990 by the U.S. Department of Energy and the NIH and was expected to take 15 years. Funding came from the U.S. Government through the NIH and the the Wellcome Trust, a charity organization in the U.K., which funded the Sanger Institute (then the Sanger Centre) in Great Britain, as well as numerous other groups from around the world primarily from China, France, Germany, and Japan. A parallel project was conducted by the private company Celera Genomics headed by Craig Venter. The $300 million Celera effort was intended to proceed at a faster pace and at a fraction of the cost of the roughly $3 billion publicly funded project.

Each human cell contains a nucleus with 46 chromosomes. Each of these chromosomes is comprised of ~30,000 to 50,000 genes and intervening sequences. The genome was broken into smaller pieces, ~1,50,000 bp in length. A library of 3,00,000 BAC clones was generated. The insert from each BAC clone was completely sequenced by shotgun method and then assembled. The larger fragment, 1,50,000 bp helped in creating chromosomes. This is known as the 'hierarchical shotgun' approach, because the genome is first broken into relatively large fragments, which are then mapped to chromosomes before being selected for sequencing. The first sequence of an entire human chromosome (chromosome 22) was published in December 1999 and the sequence of chromosome 21 appeared few months later in 2000. The working draft was announced in June 2000 and details of drafts were published in special issues of *Nature* and *Science*, which described the methods used to produce the draft sequence and offered analysis of the sequence. These drafts covered about 83% of the genome. In February 2001, at the time of the joint publications, press releases announced that both the groups have completed the project. Ongoing sequencing led to the announcement of the essentially complete genome in April 2003, 2 years earlier than planned. Improved drafts were announced in 2003 and 2005, filling ~92% of the sequence. In May 2006, the sequence of the last chromosome was published in *Nature*. The genome was yet to be completely sequenced; some heterochromatic areas remain unexplored. Centromeres and telomeres contain highly repetitive DNA sequences, which generate hurdle during sequencing, and for most of 46 chromosomes, these regions are incompletely sequenced. Additionally there are several loci in each individual's genome that contain members of multigene families, which are difficult to be revealed with shotgun sequencing methods. These multigene families often encode proteins important for immune functions. The roles of junk DNA, the evolution of the genome, the differences between individuals, and many other questions are still the subject of intense study by laboratories all over the world. Besides sequencing of ~3 billion bp of human genome, identification of all the genes is also very important objective of this project. The sequence of the human DNA is stored in a database known as Genbank along with sequences of known and hypothetical genes and proteins. All humans except for identical twins have unique gene sequences; the data published by the HGP does not represent the exact sequence of each and every individual's genome, rather it is the combined genome of a small number of anonymous donors. In the International Human Genome Sequencing Consortium (IHGSC), an international public sector HGP, researchers collected blood (female) or sperm (male) samples from a large number of donors. DNA clones from many different libraries were used in the overall project, with most of those libraries being created by Dr. Pieter J. de Jong. It has been informally reported and is well-known in the genomics community that much of the DNA for the public HGP came from a single anonymous male donor from Buffalo, New York. While in the Celera Genomics project, DNAs from five different individuals were used for sequencing. Craig Venter of Celera Genomics later acknowledged that his DNA was also one of those in the pool. On September 4, 2007, a team led by Craig Venter published his complete DNA sequence, unveiling the genome of an individual for the first time. It is anticipated that detailed knowledge of the human genome will provide new avenues for advances in the field of medicine and biotechnology. The outcomes of HGP are summarized as follows:

- Half of the human genome comprises of repeating sequences of various types.
- Approximately 28% of the genome is transcribed to RNA.
- Approximately 1.1% to 1.4 % of the genome encodes protein.
- The human genome is made of 20,000 to 25,000 genes as compared to earlier predicted number 30,000 to 50,000 (announced by IHGSC of the HGP, 2004). This number continues to fluctuate and it is now expected that it will take many years to confirm the precise value for the number of genes in the human genome.
- Small fraction of human protein families is only unique to vertebrates as most of them are also available in other life forms.
- Any two people are likely to be >99.9% genetically identical as the sequences of two randomly selected human genomes are different from each other by only 1 nucleotide per 1,250 on average.

Source: www.genome.gov/HGP

Steps involved In this approach, low-resolution physical maps of each chromosome are prepared by identifying shared landmark on overlapping ~250 kbp inserts that are cloned in high-capacity vectors like BAC, yeast artificial chromosome (YAC), or P1-derived artificial chromosome (PAC). These landmarks often take the form of 200–300 bp segments known as sequence tagged sites (STSs), whose exact sequences are unique and occur at specific sites in the genome. Thus, STS is a DNA sequence whose position has been mapped in a genome and is present at just one position in the genome. Hence, two clones that contain the same STS must overlap. The STS-containing inserts are then randomly fragmented into ~40 kbp segments that are subcloned into cosmid vectors so that a high-resolution map can be prepared by identifying their landmark overlaps. The cosmid inserts are then randomly fragmented into overlapping 5–10 kbp or 1 kbp segments for insertion into plasmid or M13 vectors. These inserts (800 M13 clones/cosmid) are then sequenced (~400 bp per clone) and assembled into contigs to yield the sequence of the parent cosmid insert (with a redundancy of 400 bp per clone × 800 clones per cosmid/40,000 bp per cosmid = 8) Finally, the cosmid inserts are assembled through cosmid walking, which is an analog of chromosome walking, using their landmark overlaps to yield the sequences of the YAC inserts, which are then assembled using their STSs to yield the chromosome's sequence.

Assembly of clone contigs Assembling all of these sequence information into chromosome-specific contigs is a massive task and is performed by methods such as chromosome walking, combinatorial screening of clones, and other rapid techniques such as clone fingerprinting by restriction fingerprinting, repetitive DNA fingerprinting, or repetitive DNA PCR.

Chromosome walking This is the simplest and first method devised for assembly of clone contigs. In this method, overlapping series of cloned DNA fragments are prepared, which begin with one clone from a library, a second clone is identified whose insert overlaps with the insert in the first clone, then the third clone is identified whose insert overlaps with the second clone, and so on. First, a clone is selected at random from the cloned library, labeled, and used as hybridization probe for all the other clones in the library. Clones whose insert overlaps with the probe gives positive hybridization signals and now their inserts can be used as new probes to continue the chromosome walk. Gradually the clone contig is prepared in a step-by-step manner. The main problem associated with this approach is that if the probe contains a genome-wide repetitive sequence, then it can also hybridize with nonoverlapping clones having that repeat. This nonspecific hybridization can be reduced up to some extent by blocking the repeat sequences by prehybridization with unlabeled genomic DNA. However, this problem is not fully solved in case of walking through long inserts from high insert capacity vectors. Because of this reason, instead of full-length insert, chromosome walk is usually performed with the fragment from the end of an insert, as there is less possibility of occurrence of genome-wide repeats in a short-end fragment as compared to the full-length insert. End fragment is sequenced prior to use as a probe to verify the absence of repetitive fragments.

After sequencing the end fragments, the speed of the walk to identify clones with overlapping inserts can be increased by using PCR rather than hybridization. Primers are designed from the sequence of the end fragment and PCR is performed with all the other clones in library. A clone that gives a PCR product of the correct size contains an overlapping insert. The chromosome walking by PCR is further speeded up by performing a PCR of mixed templates, prepared from groups of clones rather than individual clone in such a manner that overlapping ones can be easily identified. This technique is known as combinatorial screening of clones in microtiter plates by PCR. For example, the library of 960 clones has been prepared in ten microtiter plates, each plate consisting of 96 wells in an 8 × 12 array, with one clone per well, and PCR is performed as follows: (i) Samples of each clone in row 1 of first microtiter plate are mixed together and a single PCR is performed, which is repeated for every row of every plate. That means all together 80 amplification reactions are to be performed (Figure 19.19a); and (ii) Each clone of column A of the first microtiter plate are mixed together and a single PCR is carried out, which is repeated for every column of each plate. One hundred and twenty PCRs are carried out (Figure 19.19b); clones from well 1 A of each of the 10 microtiter plates are mixed together and a single amplification reaction is performed. This is repeated for every well, requiring 96 PCR reactions in all (Figure 19.19c); finally 296 PCR reactions are performed, which provide enough information for identification of positive clones among 960 clones. Ambiguity arises only if a large number of clones turn out to be positive.

Clone fingerprinting As chromosome walking is slow and laborious technique for clone contig assembly and it is rarely possible to assemble contigs of more than 15–20 clones by this method, this procedure is only valuable for positional cloning, where the purpose is to walk from a mapped site to a particular gene of interest that is located at not more than few Mbp distance. It starts at a fixed starting point and builds up the clone contig step-by-step. Consequently, chromosome walking is slow and less valuable for assembling clone contigs of the entire genome. Rapid techniques for clone contig assembly do not use a fixed starting point and instead aim to identify pairs of overlapping clones. Various techniques can be used and are collectively known as clone fingerprinting. This technique gives information on the physical structure of cloned DNA fragment, fingerprint being compared with equivalent data from other clones. The clones having similarities, i.e., consisting of overlaps can be identified. The techniques that are used in clone fingerprinting are discussed below.

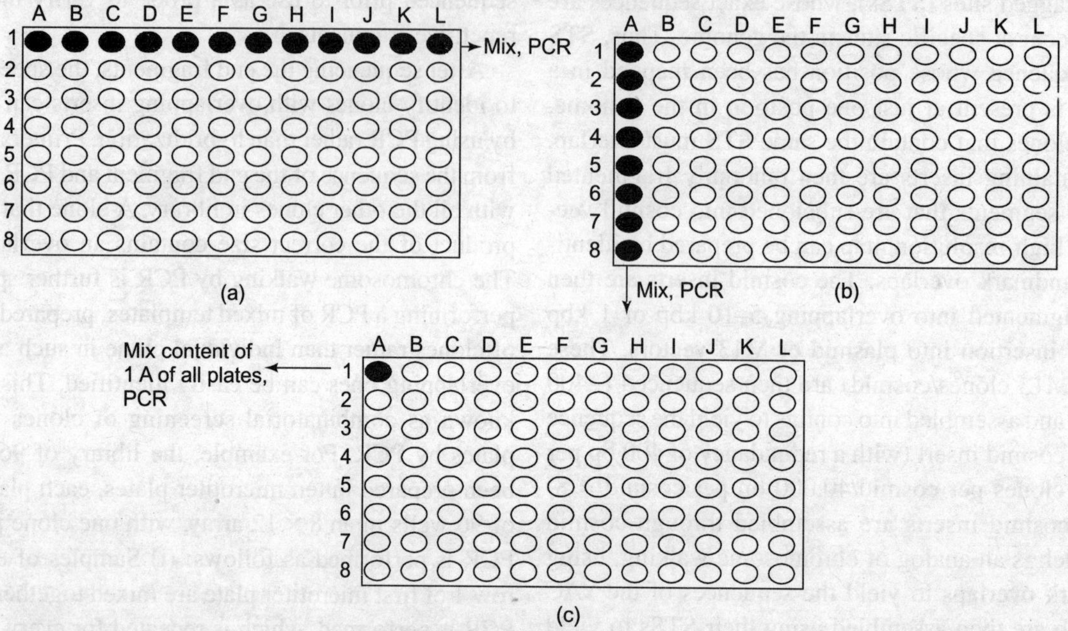

Figure 19.19 Combinatorial screening of clones in microtiter plates by PCR

Restriction fingerprinting For the production of restriction fingerprints, clones are digested with different restriction enzymes and the samples resolved on agarose gel. Two clones having overlapping inserts observe result in common restriction fragments. However, this technique appears to be easy to perform, but practically it takes lot of time to scan the agarose gel for shared fragments. Sometimes the clones that do not actually overlap produce restriction fragment whose size is impossible to differentiate by agarose gel electrophoresis (Figure 19.20a).

Repetitive DNA fingerprinting The repetitive fingerprints can be prepared by blotting a set of restriction fragments and performing Southern hybridization with probes specific for one or more types of genome-wide repeats and overlaps are identified by observing common hybridizing bands for two clones (Figure 19.20b).

Repetitive DNA PCR This is also known as interspersed repeat element PCR (IRE-PCR). PCR is carried out with primers that are designed to anneal within repetitive DNA sequence and amplify the single-copy DNA between adjacent repeats. As repeats of a particular type are distributed fairly randomly in eukaryotic genomes, with varying distances between them, a variety of amplicons are produced when these primers are used with clones of eukaryotic DNA, thereby generating fingerprints. If a pair of clones generate PCR products of the same size, these must contain repeats that are identically spaced, possibly because the cloned DNA fragments overlap. With human DNA, the genome wide repeats, i.e., *Alu* elements are used, as these occur on an average once every 4 kbp. An *Alu*-PCR of a human BAC insert of ~150 kbp generates ~38 PCR products of varying sizes (Figure 19.20c).

STS content mapping According to this sequencing strategy, the first stage in sequencing a genome is to identify the appropriate number of evenly spaced STSs. PCRs directed at individual STSs are performed with each member of a clone library. As STS is a DNA sequence present at just one position in the genome, all clones that give PCR products must contain overlapping inserts. However, the genomes of most complex eukaryotes contain various stretches of repetitive sequences that create difficulty in finding STSs. To solve this particular problem, STS-like sequences, known as expressed sequence tags (ESTs) are used instead of STSs. Finally, the rapid assembly of clone contigs requires efficient application of the above fingerprinting techniques for screening of clones in combinations perfectly with the help of bioinformatics for analyzing the resulting data (Figure 19.20d).

19.2.9 High-throughput Sequencing

Nowadays majority of sequencing experiments are performed by automated sequencer based on dye terminator method and DNA separation by capillary electrophoresis. However, this technique is quite costly for sequencing of DNA libraries, therefore high-throughput sequencing techniques are developed, which require less expenses. These new high-throughput methods include techniques that parallelize the sequencing process, producing thousands or millions of sequences at a time. *In vitro* cloning is carried out to produce many copies of each individual molecule. Emulsion PCR is one method of choice used for this purpose where individual DNA molecules are isolated along with primer-coated beads in aqueous bubbles within an oil phase. The amplification reaction

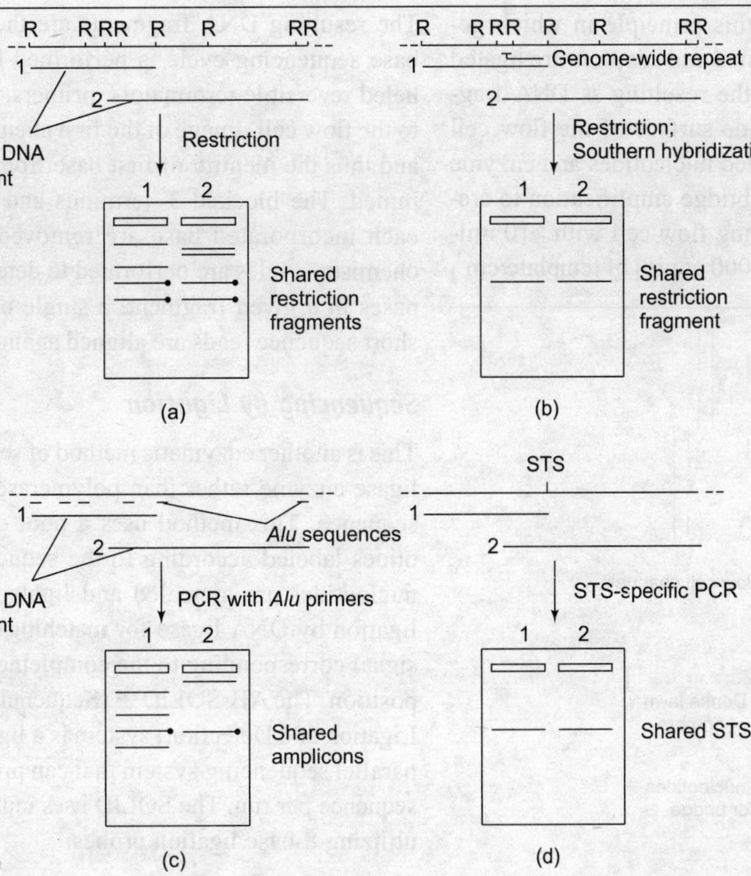

Figure 19.20 Clone fingerprinting techniques. (a) restriction fingerprint; (b) repetitive DNA fingerprint; (c) repetitive DNA PCR; (d) STS content mapping

then coats each bead with clonal copies of the isolated library molecule and these beads are subsequently immobilized for later sequencing. Another method for *in vitro* clonal amplification is 'bridge PCR', where fragments are amplified using primers attached to a solid surface. Both of the above-described methods produce many physically isolated locations, each containing many copies of a single fragment. New single molecule methods are also developed, which skips this amplification step, directly fixing DNA molecules to a surface. Finally, once clonal DNA sequences are physically localized to separate positions on a surface, various sequencing approaches may be used to determine the DNA sequences of all locations in parallel.

Sequencing by Synthesis

This method like the popular dye termination electrophoretic sequencing uses the process of DNA synthesis by DNA polymerase to identify the bases present in the complementary DNA molecule. In this approach, modified dNTPs are used as reversible terminators, in which four different fluorophore having a distinct fluorescent emission is linked to each of the four bases through a photocleavable linker and the 3′-OH group is capped by a small chemical moiety. Recently it is discovered that 3′-OH group is modified by adding allyl moiety and a fluorophore is tethered to the base using allyl as a linker, forming chemically cleavable fluorescent nucleotide reversible terminators. The resulting nucleotides are 3′-*O*-allyl-dATP-allyl–ROX, 3′-*O*-allyl-dCTP-allyl–Bodipy-FL-510, 3′-*O*-allyl-dGTP-allyl–Bodipy-650, 3′-*O*-allyl-dUTP-allyl–Bodipy-FL-R6G. The 3′-*O*-Allyl-dNTPs-allyl-fluorophore is incorporated in the growing DNA chain by polymerase reaction and then the fluorophore and the 3′-*O*-allyl groups are removed simultaneously in 30 seconds by Pd-catalyzed deallylation in aqueous buffer solution. This one-step dual deallylation reaction thus allows the reinitiation of the polymerase reaction and increases the efficiency of the reaction. Millions of different DNA templates are immobilized on the surface of a chip by using emulsion PCR on microbeads. These high-density DNA templates, coupled with four color sequencing by synthesis approach, generate a high-throughput (>20 million bases per chip) and highly accurate platform for a variety of sequencing and digital gene expression analysis projects. Using this class of fluorescent nucleotide analogs on a DNA chip and a four-color fluorescent scanner, DNA templates consisting of homopolymer regions are accurately sequenced.

Illumina/Solexa is based on this principle in which genomic DNA is randomly fragmented and adaptors are ligated to both ends of fragments, and the resulting ss DNA fragments bind randomly to the inside surface of the flow cell channels (Figure 19.21). Unlabeled nucleotides and enzyme are added to initiate solid-phase bridge amplification to create an ultrahigh-density sequencing flow cell with >10 million clusters, each containing ~1,000 copies of template/cm^2.

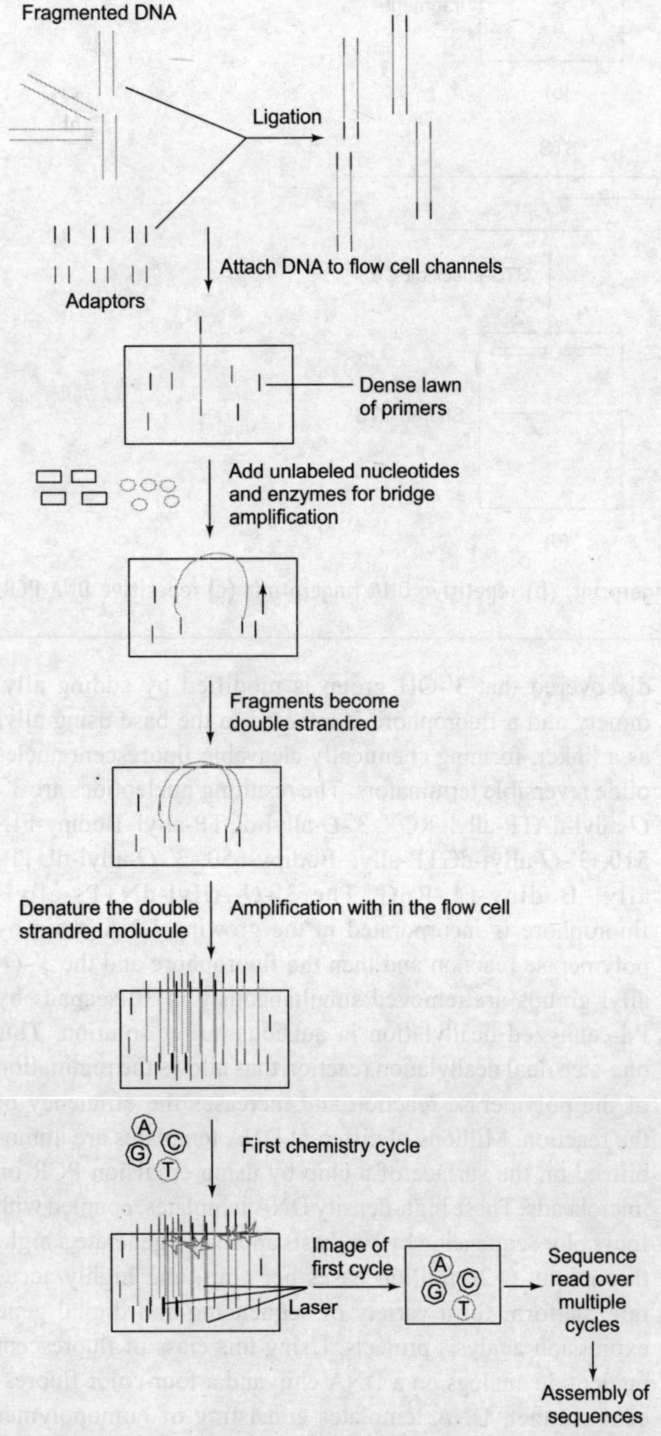

Figure 19.21 DNA sequencing by Solexa/Illumina technique

The resulting DNA fragments are then denatured and first base sequencing cycle is performed by adding all four labeled reversible terminators, primers, and DNA polymerase to the flow cell. Image of the first chemistry cycle is captured and thus the identity of first base from each cluster is determined. The blocked 3′-terminus and the fluorophore from each incorporated base are removed; thereafter, multiple chemistry cycles are performed to determine the sequence of bases in a given fragment, a single base at a time. Finally, short sequence reads are aligned against a reference genome.

Sequencing by Ligation

This is another enzymatic method of sequencing using a DNA ligase enzyme rather than polymerase to identify the target sequence. This method uses a pool of random oligonucleotides labeled according to the sequenced position. Oligonucleotides are annealed and ligated and the preferential ligation by DNA ligase for matching sequences results in a signal corresponding to the complementary sequence at that position. The AB-SOLiD™ (Sequencing by Oligonucleotide Ligation and Detection) system is a ligation-based massively parallel sequencing system that can produce >1 Gbp of DNA sequence per run. The SOLiD uses unique ligation chemistry, utilizing 8-base ligation probes.

Helicos

A new sequencing method has been proposed termed as 'Helicos', which is single-molecule dideoxy sequencing technique used for genome sequencing.

19.3 TECHNIQUES USED IN PROTEOMICS

The entire complement of proteins synthesized by a given cell or organism is known as proteome and the branch of science that deals with proteins is called proteomics. The word 'proteomics' was developed in 2004 by Twyman. It is equivalent to transcriptomics, i.e., study of protein expression and abundance. Proteins are functional molecules found in every cell and are at the center of every biological process. These are richer source of information as compared to nucleic acid. Moreover, these are obvious candidates for drug targeting, as a consequence of which proteomics is enjoying a rapidly increasing level of attention. The structure of protein is very complex and determination of a protein's structure first requires protein sequencing, i.e., determining the amino acid sequences of its constituent peptides, and then determining their abundance, conformation, modification, localization, biochemical and physiological function, and finally interaction with proteins and other nonpeptide molecules. Thus, the study of proteome involves a variety of techniques and some of these are discussed below.

19.3.1 Protein Sequencing

Protein sequencing, i.e., primary structure (sequence of amino acids) determination of any protein is very important because of following reasons: (i) It is prerequisite for the determination of X-ray and NMR structure of proteins. This has helped in establishment of the fact that proteins have unique covalent structure; (ii) It is important to understand the molecular mechanism of action of that particular protein; (iii) It is essential for establishment of evolutionary relationship among the proteins and the organisms that produce them. After getting the information of primary structure of proteins, sequences of analogous proteins are compared from the same individuals, from members of the same species, and from members of related species; and (iv) Sequence analyses have important clinical applications because many genetically inherited diseases are caused due to mutations, leading to change in amino acid(s) of a protein. This has led to the development of valuable diagnostic tests for such diseases and also used to develop symptom-relieving therapy.

The major landmark in the history of protein sequencing is the first complete sequence determination of bovine polypeptide hormone, namely insulin by Frederick Sanger in 1953. For his work, Sanger was awarded the Nobel Prize in Chemistry in 1958. Subsequently, the amino acid sequences of numerous proteins from many species have been determined using same principle developed by Sanger. The determination of the primary structure of insulin, which comprises of 51 residues, was the hard work of many scientists for a decade that altogether consumed ~100 g of protein. However, procedures for primary structure determination have since been so advanced and automated that an experienced technician can sequence proteins of similar size in a few days using only a few micrograms of protein. In spite of automation of sequencing procedure, the basic principle for primary structure determination is based on Sanger's method.

Steps Involved

The protein sequencing involves several steps, *viz.*, (i) Separation, purification, and characterization of the individual polypeptide chains of the protein; (ii) Determination of the terminal amino acids of each chain; (iii) Breaking of disulfide bonds; (iv) Determination of amino acid composition of each polypeptide chain; (v) Specific peptide cleavage reactions to produce manageable sized fragments; (vi) Determination of the sequence of each fragment; (vii) Repetition with a different pattern of cleavage; and (viii) Construction of the sequence of the overall protein. All these steps are discussed in detail:

Separation, Purification, and Characterization of the Individual Polypeptide Chains of the Protein The unidentical polypeptides of proteins should be separated and purified prior to sequencing, which is usually performed under acidic or basic condition, at low salt concentration and at high temperature. Treatment of protein with denaturing agents such as urea, guanidinium ion, detergents such as sodium dodecyl sulfate (SDS) dissociates the protein subunits. The dissociated subunits can then be separated by ion exchange or gel filtration chromatography. However, molecular mass of protein can be routinely measured by gel filtration chromatography and SDS-polyacrylamide gel electrophoresis (SDS-PAGE) with an accuracy of 5–10%, but mass spectrometry can determine the molecular masses of picomoles quantity of the protein sample (>100 kDa) up to the accuracies of ~0.01%. After knowing the molecular mass, the number of amino acid residues on the polypeptide to be sequenced can be estimated (~110 Da/residue).

Determination of the Terminal Amino Acids of Each Chain Each polypeptide chain has an N-terminal residue and a C-terminal residue. Identification of these end groups can establish the number of chemically distinct polypeptides in a protein.

N-terminus identification A generalized method for N-terminal amino acid analysis is as follows. The protein is reacted with a reagent, which selectively labels the terminal amino acid. The protein molecule is then hydrolyzed to its constituent amino acids and the labeled amino acid is analyzed.

Various reagents are available to label and identify the N-terminal amino acid residue. These include 1-fluoro-2,4-dinitrobenzene (FDNB), 1-dimethylaminonaphthalene-5-sulfonyl chloride (dansyl chloride), dabsyl chloride, and phenylisothiocyanate (PITC; Edman reagent). FDNB reagent, developed by Sanger, reacts with primary amines (including ε-amino group of lysine) and form 2,4-dinitrophenyl derivatives (Figure 19.22a). Dansyl chloride yields dansylated derivatives after acid hydrolysis that are more easily detectable and sensitive than the dinitrophenyl derivative (Figure 19.22b). This is because dansylated derivatives exhibit intense yellow fluorescence that can be chromatographically identified from as little as 100 picomoles of sample. The chemical structure of dabsyl chloride, another reagent used for recognizing N-terminal amino acid residue is given in Figure 9.22c. After the N-terminal residue is labeled with one of these reagents, the polypeptide is hydrolyzed to its constituent amino acids and the labeled amino acid is identified. As the reagents produce colored derivatives, staining is not needed and only qualitative analysis is required, so the amino acid does not have to be eluted from the chromatography column and analysis is performed by comparing with standard amino acid. Since any amine group can react with the labeling reagent, ion exchange chromatography should be avoided. Either thin layer chromatography or high-pressure liquid chromatography should be used for identification. Because the hydrolysis stage destroys

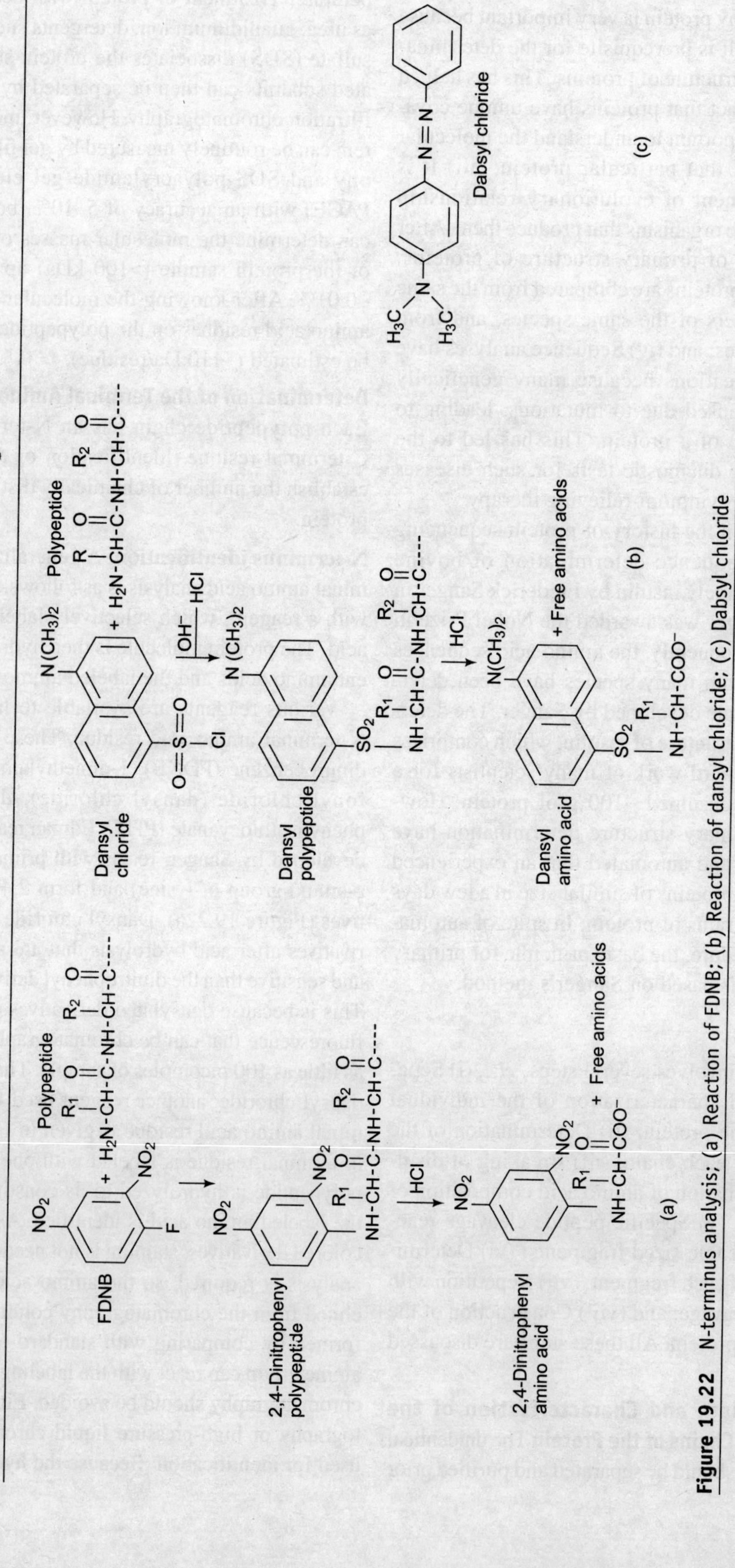

Figure 19.22 N-terminus analysis; (a) Reaction of FDNB; (b) Reaction of dansyl chloride; (c) Dabsyl chloride

the polypeptide, this procedure cannot be used to sequence a polypeptide beyond its N-terminal residue. However, it can help to determine the number of chemically distinct polypeptides in a protein, provided each has a different N-terminal residue. The most commonly used reagent for N-terminal analysis is PITC, also called as *Edman reagent* after the name of its inventor Pehr Edman. The method comprises of three steps, *viz.*, coupling, cleavage, and conversion, and is called *Edman degradation method*, which also forms the basis of automated protein sequencing (for details see Section 19.3.3).

Besides chemical methods, enzymatic analysis of N-terminal residue can also be done by using aminopeptidase, an exopeptidase. The enzyme sequentially cleaves the amino acids from the N-terminal of polypeptide, which can be easily analyzed.

C-terminus identification The number of procedures available for C-terminal amino acid analysis is much less than that for N-terminal analysis. There is no reliable chemical procedure as compared to the Edman degradation for the sequential end group analysis from the C-terminus of polypeptides. This can be performed enzymatically by using exopeptidases. For example, carboxypeptidases catalyze the hydrolysis of the C-terminal residues of polypeptide. These carboxy-peptidases are highly specific for the chemical characteristics of the substrate. The specificities of commonly used exopeptidases are listed in Table 19.2. As carboxypeptidases show selectivity toward side chains, so their use, either singly or in mixtures, do not exhibit the correct order of more than the first few C-terminal residues of a polypeptides. Furthermore, C-terminal residues with preceding Pro residues are not subjected to cleavage by some carboxypeptidases such as A and B. Chemical methods are therefore usually employed to identify their C-terminal residue. The most commonly used chemical methods are hydrazinolysis, in which a polypeptide is treated with anhydrous hydrazine at 90°C for 20 to 100 hours in the presence of a mild acidic ion exchange resin (catalyst) (Figure 19.23). All the peptide bonds are thereby cleaved, yielding the aminoacyl hydrazides of all amino acid residues except that of the C-terminal amino acid, which is released in the free form and is identified chromatographically. However, hydrazinolysis is associated with many side reactions and therefore is only applied to carboxypeptidase-resistant polypeptides.

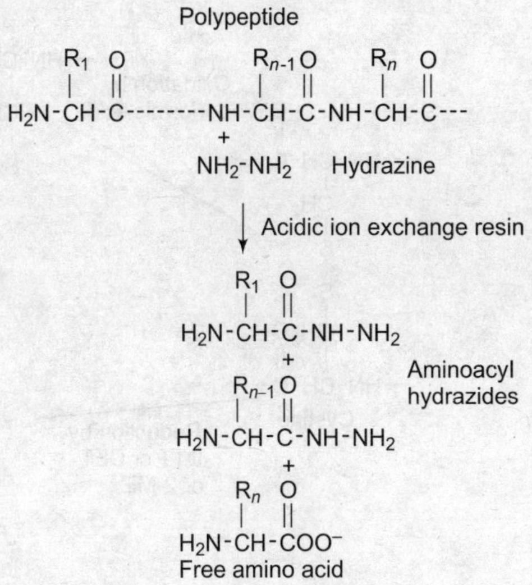

Figure 19.23 C-terminus analysis by hydrazinolysis

Breaking Disulfide Bonds The native conformation of the protein is stabilized by disulfide bonds that hinder with the enzymatic or chemical cleavage of the polypeptide used in primary structure determination or sequencing procedures. A cysteine residue that has one of its peptide bonds cleaved by the Edman procedure may remain attached to another polypeptide strand via its disulfide bond. Therefore, cleavage of disulfide bonds is required for separation of polypeptide chains. Two approaches are usually applied for irreversible breakage of disulfide bonds (Figure 19.24). Oxidation of cysteine residues with performic acid produces two cysteic acid residues. Disulfide bonds are most often cleaved by treatment with reducing agents such as β-mercaptoethanol (β-ME), dithiothreitol (DTT), or dithioerythritol (DET; Cleland's reagent). Reduction by DTT to form Cys residues must be followed by further modification of the reactive SH groups to prevent reformation of disulfide bond through oxidation by O_2. The resulting free SH groups are alkylated, usually by treatment with iodoacetic acid. The *S*-alkyl derivatives are stable in air and under the conditions used for the subsequent cleavage of peptide bonds.

Table 19.2 Specificities of Various Exopeptidases Used for Protein Sequencing

Name of enzyme	Source	Specificity
Carboxypeptidase A	Bovine pancreas	$R_n \neq$ Arg, Lys, Pro; $R_{n-1} \neq$ Pro
Carboxypeptidase B	Bovine pancreas	$R_n \neq$ Arg, Lys; $R_{n-1} \neq$ Pro
Carboxypeptidase C	Citrus leaves	All free C-terminal residues; pH optimum = 3.5
Carboxypeptidase Y	Yeast	All free C terminal residues, but slowly with R_n = Gly
Leucine aminopeptidase	Porcine kidney	$R_1 \neq$ Pro
Aminopeptidase M	Porcine kidney	All free N-terminal residues

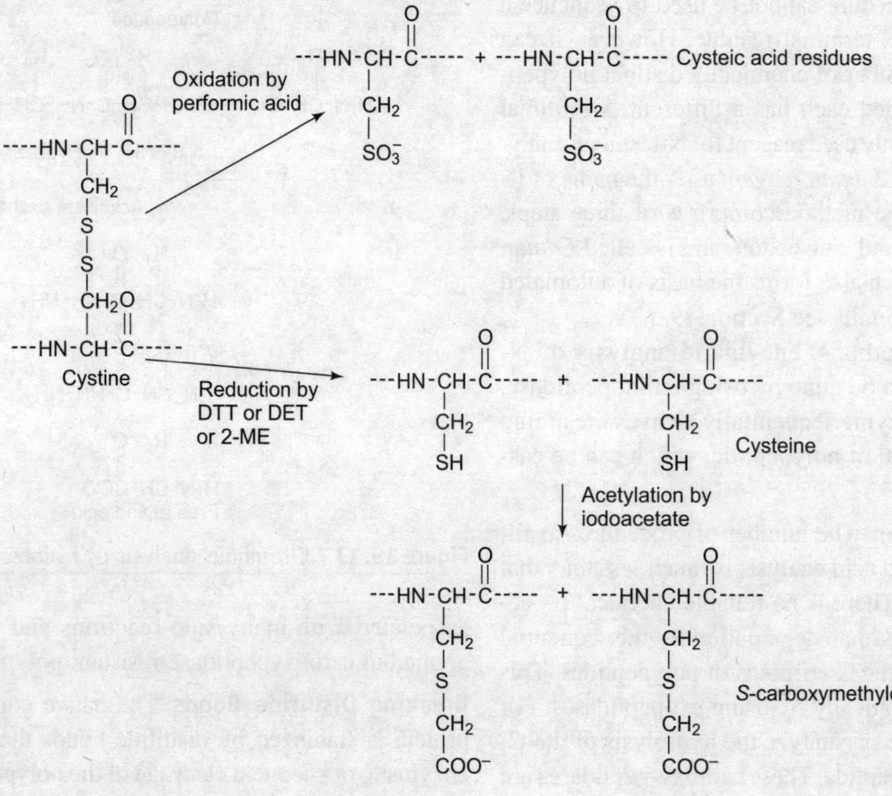

Figure 19.24 Breaking of disulfide bonds of polypeptide chains

Determination of Amino Acid Composition of Each Polypeptide Chain It is important to have an idea about the amino acid composition (number of each type of amino acid residue) of polypeptide chain prior to sequencing reaction. The amino acid composition of polypeptide chain is determined by its complete hydrolysis followed by the quantitative analysis of the liberated amino acid. The hydrolysis of polypeptide can be accomplished either by chemical (acid or base) or enzymatic methods.

Acid-catalyzed hydrolysis The polypeptide is dissolved in 6 M HCl, sealed in an evacuated tube to prevent air oxidation of S-containing amino acids, and heated at 100–120°C for 10 to 100 hours. Such harsh conditions of hydrolysis are required for complete liberation of aliphatic amino acid like Val, Leu, and Ile. This condition of hydrolysis often leads to partial degradation of Ser, Thr, and Tyr, and mostly destroys Trp residues. Moreover, Gln and Asn are converted to Glu and Asp. Thus that amount of Asx (Asp + Asn) and Glx (Glu + Gln) can be independently measured after acid hydrolysis.

Base-catalyzed hydrolysis In base-catalyzed hydrolysis, polypeptides are dissolved in 2–4 M NaOH and heated at 100°C for 4 to 8 hours. This hydrolysis procedure is more problematic because it leads to decomposition of Cys, Ser, Thr, and Arg, and partially deaminates and racemizes other amino acids. Therefore, alkaline hydrolysis is predominantly used to determine Trp content.

Enzymatic hydrolysis As individual peptidases are not able to cleave all peptide bonds, the complete hydrolysis of polypeptide requires mixture of peptidases. Pronase, which is a mixture of relatively nonspecific proteases from *Streptomyces griseus*, is often used for this purpose. Since proteolytic enzymes are proteins, therefore these are self-degrading and it is very important to use limited amount of enzyme, i.e., ~1% by weight of the polypeptide to be hydrolyzed, otherwise the enzymes hydrolyze themselves, and significantly contaminate the final digest. Enzyme digestion is frequently applied for determining the quantities of Trp, Gln, and Asn in a polypeptide, which are destroyed by harsher chemical methods.

Specific Peptide Cleavage Reactions to Produce Manageable Sized Fragments Polypeptides that are longer than ~50 residues cannot be directly sequenced. Larger polypeptides must therefore be cleaved, either enzymatically or chemically to smaller fragments, which are of manageable size for sequencing.

Cleavage by trypsin Trypsin specifically cleaves peptide bonds at the C-terminal side of positively charged residues such as arginine and lysine, if the next residue is not proline. Since it exhibits the greatest specificity, it is the most valuable member of the armory of peptidases used to fragment polypeptides. As trypsin cleaves peptide bonds that follow positively charged residues, trypsin cleavage sides may be added or deleted from a polypeptide by chemically adding or deleting positive charges

Table 19.3 Specificities of Various Endopeptidases Used for Protein Sequencing

Name of enzyme	Source	Specificity	Remarks
Trypsin	Bovine pancreas	R_{n-1} = Arg, Lys $R_n \neq$ Pro	Highly specific
Chymotrypsin	Bovine pancreas	R_{n-1} = Phe, Trp, Tyr, $R_n \neq$ Pro	Cleaves slowly for R_{n-1} = Asn, His, Met, Leu
Thermolysin	*Bacillus thermoproteolyticus*	R_n = Ile, Met, Phe, Trp, Tyr, Val $R_{n-1} \neq$ Pro	Rarely cleaves R_n = Ala, Asp, His, Thr
Elastase	Bovine pancreas	R_{n-1} = Ala, Gly, Ser, Val $R_n \neq$ Pro	–
Pepsin	Bovine gastric mucosa	R_n = Leu, Phe, Trp, $R_{n-1} \neq$ Pro	Quite nonspecific; pH optimum 2
Endopeptidase Arg-C	Mouse submaxillary gland	R_{n-1} = Arg	May hydrolyze at R_{n-1} = Lys
Endopeptidase Asp–N	*Pseudomonas fragi*	R_n = Asp	May hydrolyze at R_{n-1} = Glu
Endopeptidase Glu-C	*Staphylococcus aureus*	R_{n-1} = Glu	May hydrolyze at R_{n-1} = Gly
Endopeptidase Lys-C	*Lysobacter enzymogenes*	R_{n-1} = Lys	May hydrolyze at R_{n-1} = Asn

to or from its side chains. For example, the positive charge on Lys is eliminated by treatment with dicarboxylic anhydride such as citraconic anhydride. The reagent forms a negatively charged derivative of the Lys ε-amino group that trypsin does not recognize. After trypsin hydrolysis of the polypeptide, Lys residue can be deblocked for the identification by mild acid (pH 2–3) hydrolysis. On the other hand, Cys may be aminoalkylated by β-halomine (e.g., 2-bromoethylamine) to form positively charged residue, i.e., subject to tryptic cleavage. Other endopeptidases (Table 19.3) show broader side chain specificity than trypsin and frequently form a series of peptide fragments with overlapping sequences. Through limited proteolysis, i.e., by adjusting reaction conditions and limiting reaction times, less specific endopeptidases can yield useful peptide fragments. Limited proteolysis is often employed to generate peptide fragments of useful size from subunits that have too many or too few Arg and Lys residues.

Cleavage by cyanogen bromide Several chemical reagents promote peptide bond cleavage at specific residues. The most useful among these reagents, cyanogen bromide (CNBr), causes specific and quantitative cleavage on the C-terminal side of Met residues to form peptidyl homoserine lactone. The reaction is performed in an acidic solvent (0.1 M HCl or 70% formic acid) that denatures most proteins so that cleavage normally occurs at all Met residues.

A peptide fragment generated by a specific cleavage process may still be too large to sequence. In that case, after its purification, it can be subjected to a second round of fragmentation using a different cleavage process.

Determination of the Sequence of Each Fragment Once the manageable sized peptide fragments formed through specific cleavage reactions have been isolated, their amino acid sequences can be determined. This is done through repeated cycles of Edman degradation. An automated device for doing so was first developed by Edman and Geoffrey Begg. In modern sequencers, the peptide sample is absorbed into a polyvinyl difluoride (PVDF) membrane or dried onto glass fiber paper that is impregnated with polybrene. In both cases, the peptide is immobilized but is readily accessible to Edman reagents. Accurately measured quantities of reagents, either in solution or as vapours in a stream of argon, are then delivered to the reaction cell at programmed intervals. The thiazolinone amino acids are automatically removed, converted to the corresponding PTH-amino acid, and identified by UV detector equipped with reverse-phase HPLC system. Such instruments are very sensitive and are able to identify as little as 0.1 pmol of PTH-amino acid.

Repeat with a Different Pattern of Cleavage and Construct the Sequence of the Overall Protein Once the sequencing of individual peptide fragment is completed, the last and major step is to assemble the sequence of overall protein in the order in which these are present in original polypeptide. The amino acid sequence of a polypeptide chain is determined by comparing the sequences of two sets of mutually overlapping peptide fragments. For this, one set is cleaved by trypsin and the other set by CNBr. The amino acid sequences of each fragment obtained by the two cleavage procedures are examined for the overlapping sequences. Overlapping fragments obtained from the second cleavage reaction facilitate in the arrangement of the correct order of the peptide fragments generated in the first reaction. N-terminal amino acid has already been identified before the first cleavage of the protein, which helps in the confirmation regarding the fragment that is derived from N-terminus. The two sets of fragments can also be compared for possible inaccuracy in determining the amino acid sequence of each fragment. If the second cleavage procedure fails to establish continuity between all peptides from the first cleavage, a third or even a fourth cleavage must be performed to obtain a set of peptides that can provide the necessary overlaps.

Finally, to complete the protein's structure, positions of original disulfide bonds are determined by cleaving a sample of the native protein under conditions that leave its disulfide bonds intact. The peptide fragments are separated by reverse-phase HPLC and cys residues are identified by determining amino acid composition. The Cys residues having free –SH groups can be identified by treating them with radioactive iodoacetate. The Cys-containing fragments are then subjected to Edman degradation. Although disulfide-linked peptides yield two PTH–amino acids in each step, the locations of Cys within the predetermined amino acid sequence are recognized and positions of the disulfide bonds are established. Alternatively, this information can also be obtained by using a diagonal electrophoresis technique to isolate the peptide sequences containing such bonds. First, the protein is specifically cleaved into peptides under conditions in which the disulfide bonds remain intact. The mixture of peptides is applied to a corner of a sheet of paper and subjected to electrophoresis in a single lane along one side. The resulting sheet is exposed to vapours of performic acid, which cleaves disulfides and converts them into cysteic acid residues. Peptides originally linked by disulfide bonds are now independent and more acidic because of the formation of an SO_3^- group. This mixture is subjected to electrophoresis in the perpendicular direction under the same conditions as those of the first electrophoresis. Peptides that are devoid of disulfide bonds have the same mobility as before and consequently all are located on a single diagonal line. In contrast, the newly formed peptides containing cysteic acid usually migrate differently from their parent disulfide-linked peptides and hence lie off the diagonal. These peptides are then isolated and sequenced and the location of the disulfide bonds can be established.

19.3.2 Predicting Protein Sequence from DNA/RNA Sequences

The amino acid sequence of a protein can also be deduced by determining the sequence of nucleotides in the gene that code for it. The mRNA, which codes for that particular protein is isolated, reverse transcribed and then amplified by PCR. The amplified DNA is then sequenced. The amino acid sequence of the protein is then deduced from this amplified sequence.

19.3.3 Automated Sequencing by Automated Amino Acid Analyzer

The sequencing of peptide chain is performed from the N-terminus to C-terminus by subjecting the polypeptide to repeated cycles of the Edman degradation and, after every cycle, identifying the newly liberated product (i.e., PTH-amino acid). This technique has been automated, resulting in great savings of time and materials. Automated Edman sequencers known as sequenators are now in widespread use and are able to sequence peptides up to ~50 amino acids long. The amino acid content of a polypeptide hydrolyzate can be quantitatively determined through the use of an automated amino acid analyzer, which is based on the principle of Edman's degradation (Figure 19.25).

Steps Involved in Edman Degradation

The Edman degradation procedure is divided into three steps: coupling, cleavage, and conversion (Figure 19.26).

Coupling In the coupling reaction, PITC (Edman reagent) chemically modifies the free α-amino group of N-terminal of a polypeptide to form phenylthiocarbamoyl polypeptide (PTC adduct). Under mild alkaline conditions (pH 9.0), coupling is favored at α-amino group and takes place within 15–30 min at a temperature of 40–55°C. Modifying the N-terminal residue by formylation, acetylation, and fatty acid acylation can inhibit coupling reaction.

Cleavage For the cleavage reaction, the PTC adduct is treated with anhydrous strong acid, e.g., trifluoroacetic acid (TFA), which cleaves N-terminal amino acid as thiazolinone derivative, anilinothiazolinone (ATZ). During this treatment, the PTC adduct with N-terminal amino acid is cleaved from the polypeptide chain. This process is facilitated because of the proximity of nucleophilic sulfur atom of the derivatized N-terminus amino acid to the carbonyl carbon of the first peptide bond to yield a five-membered heterocyclic derivative,

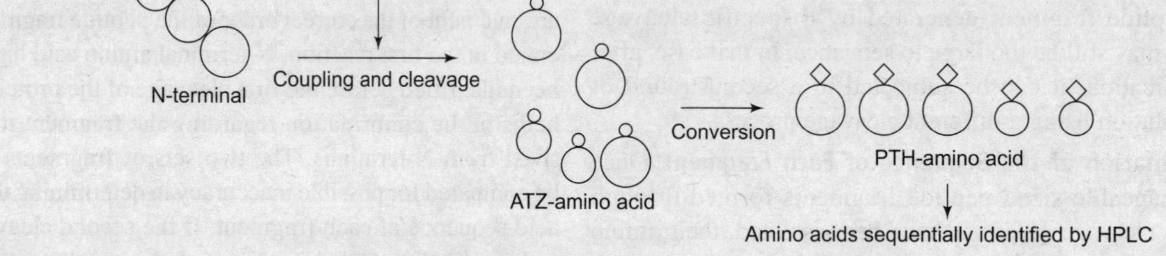

Figure 19.25 A conceptual view of the principle of Edman degradation

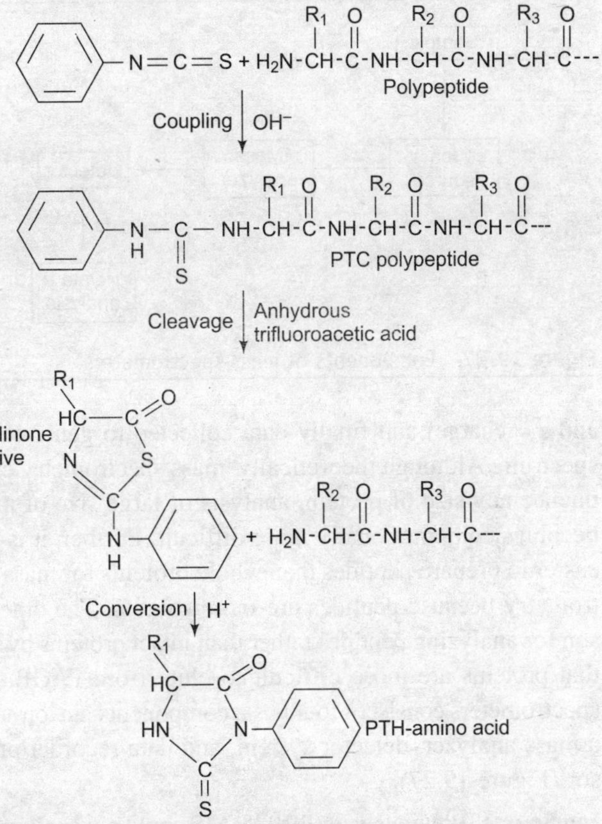

Figure 19.26 Edman degradation reaction for protein sequencing

ATZ-amino acid, and the n–1 polypeptide (where n represents total number of amino acid residues in the polypeptide chain). Now, the shortened n–1 polypeptide has a reactive N-terminal α-amino group, which can undergo another cycle of coupling and cleavage. This procedure is then repeated. The solubility of the small, hydrophobic ATZ-amino acid is significantly different from that of the hydrophilic polypeptide and can be extracted selectively by a nonpolar (organic) solvent such as chlorobutane or ethyl acetate.

Conversion In this step, the unstable ATZ is converted into more stable phenylthiohydantoin derivative (PTH-amino acid), which occurs in two-step reaction by treatment with aqueous acid. Although cleavage and conversion can be completed in one step by treating with just one aqueous acid, the anhydrous acid optimizes the specific cleavage of the peptide bond of the N-terminal amino acid. First the unstable ring structure of the ATZ-amino acid is opened by aqueous acid at increased temperature (~60°C) to form a PTC-amino acid and finally converted into more stable PTH-amino acid by rearrangement. It has been suggested that the thiohydantoin is actually formed preferentially when cleavage is performed in the presence of a thiol such as DTT. Upon comparison of the retention time on HPLC with those of known PTH-amino acid, PTH-amino acid generated in the reaction can be easily identified. Since Edman degradation occurs in separate three stages and each requires quite different conditions, therefore amino acid residues can be sequentially removed from the N-terminus of a polypeptide in controlled stepwise manner.

Types of Sequencers

Different types of sequencers have been developed.

Automation of Edman Degradation by Spinning-cup Sequencer Edman's technical assistant, Geoffrey Begg, made first effort to automate the repetitive reaction, and produced prototype protein sequenators in 1961. Coupling and cleavage steps are automated, which enabled researchers to determine the sequence of first 30–40 amino acid residues in a protein (0.3 μmole). Edman and Begg's sequencer consisted of a solvent delivery system that delivered solvents and reagents by nitrogen pressure to the spinning cup via an electronically operated valve pack. In the spinning cup reaction vessel, the protein was held against the inner wall by centrifugal force. Reagents and solvents delivered to the spinning cup were precisely measured to wet only the protein film and waste, i.e., excess of reagents and solvents, were removed by evaporation in vacuum. After coupling and cleavage, the released ATZ was delivered to a cooled fraction collector. Thereafter, collected ATZ-amino acids were manually converted to the PTH derivatives. Initially, PTH-amino acids were separated and identified by paper chromatography, thin layer chromatography, or partition chromatography. Afterward gas chromatography and HPLC techniques were applied for the separation of PTH-amino acids. Later devices for automatic conversion of the separated thiazolinones to the PTH-amino acids were developed, which were followed by online detection of the PTHs using HPLC. The sensitivity limit of the original spinning cup sequencer is 0.3 μmole and repetitive yield 92–95%.

Microsequencing with a Gas-phase Sequencer The gas-phase sequencer used miniaturized components and gaseous coupling and cleavage reactions of the protein immobilized on a glass fiber disc. This technique has ~1,000-fold more sensitivity as compared to original spinning cup instrument. Sample wash out was minimized by the addition of polybrene to the glass support. The two key reagents, a coupling base (trimethylamine) and a cleavage acid (TFA), were delivered to the reaction vessel (glass cartridge) in the gaseous phase via argon or nitrogen stream. Organic solvent that was automatically converted to PTH derivatives following their transfer from the reaction vessel to a conversion flask selectively extracted ATZ-amino acids. The resulting PTH derivatives were separated by online HPLC and detected. The sensitivity of these types of sequencers is ~10 pmoles, which has been extended to the femtomole level by introducing capillary columns.

Pulsed Liquid-phase Sequencer Although the basic principle of this type of sequencer is the same as gas-phase sequencer, however the cleavage acid, i.e., TFA, is delivered as a liquid pulse. The amount of acid delivered is precisely controlled, sufficient to wet the sample but not enough to wash the sample from the reaction vessel. Liquid-phase cleavage results in faster cleavage times hence shorter cycle times (<30 min), and accelerates sequencing. After cleavage, the volatile TFA is removed. Further, the microcolumn has been introduced with these type sequencers, which increases the sensitivity up to 0.5 pmole with a repetitive yield of ~95%.

Biphasic Column Sequencer Hewlett Packard has introduced a new generation of sequencer that uses a different approach for immobilizing proteins for Edman chemistry. The reaction vessel in this instrument comprises an adsorptive biphasic column. One half of the column contains a solid hydrophobic support and other half contains a hydrophilic support. A very large volume of sample is applied to hydrophobic portion of biphasic sequencers under dilute acidic conditions. The sample is retained at the top of the column and any inorganic salts and buffers are washed away. The hydrophobic and hydrophilic halves of the column are then reassembled and resulting biphasic column is positioned in the sequencers. The solvents are permitted to flow in reverse direction that minimizes wash out of the sample. Aqueous solvent flows are directed toward the hydrophobic half of the column, which subsequently immobilizes the sample via hydrophobic interactions. Otherwise, organic solvents are flown toward the hydrophilic half of the column, which retains the sample when organic solvents are employed to elute hydrophobic contaminates/reaction products (ATZ-amino acids).

19.3.4 Mass Spectrometry for Peptide Characterization and Sequencing

During sequencing, Edman degradation proceeds from the N-terminus of the protein, which is most often prevented if N-terminal amino acid is chemically modified or concealed within the protein. Ordering of fragments and determining the position of disulfide bonds also require the use of either guesswork or a separate procedure to determine the positions of disulfide bridges. The other direct method for sequencing and characterizing polypeptides is mass spectrometry (MS). MS is an instrumental procedure that generates information about the mass of molecule and the masses of product ions derived from the parent molecule after fragmentation. The liberated ions are easily detected and provide valuable information about the amino acid sequence of polypeptides. MS is much more sensitive, copes up with different protein mixtures, and offers much higher sample throughput.

Principle and Components of Mass Spectrometers

A mass spectrometer is capable of forming molecular ions from the sample, and liberated ions are separated and detected on the basis of mass to charge ratio (m/z) (where m = ion's mass and z = charge) and finally data collected to generate mass spectrum. Although theoretically, mass spectrometry can sequence any size of protein, analysis of large size of protein becomes computationally more difficult. Further, it is much easier to prepare peptides than whole proteins for mass spectrometry because peptides are more soluble. The other reason for analyzing peptides rather than intact proteins by MS is that proteins are more difficult to elute from PAGE. Mass spectrometers consist of four basic components: an ion source, a mass analyzer, detector system, and data recorder/processor (Figure 19.27).

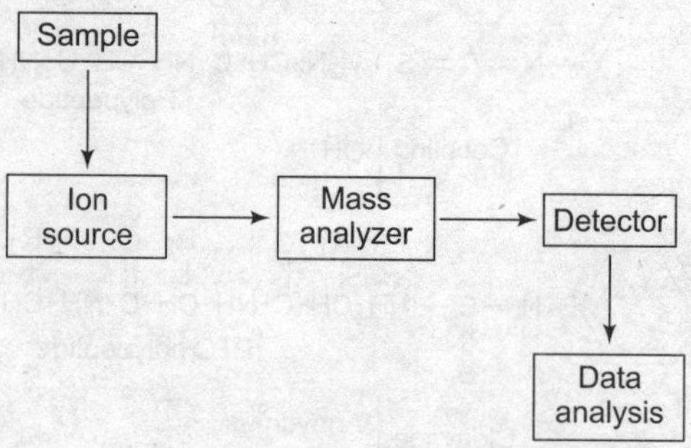

Figure 19.27 Components of mass spectrometers

Ion Source Previously until 1985, MS could not analyze macromolecules like proteins and nucleic acids because the method by which mass spectrometers produced gas phase ions destroyed macromolecules. Later, more sophisticated MS techniques have been designed, which eliminated this barrier. These techniques are both extremely valuable and easily automated, and are presented in the following:

Fast atom bombardment (FAB) FAB along with liquid secondary ion-MS (LSIMS) is the soft-ionization technique (i.e., ionization without causing significant amounts of fragmentation) first used to ionize proteins and peptides in mass spectrometer. The sample to be analyzed is dissolved in a nonvolatile matrix and placed under vacuum in the instrument. The sample is bombarded with a stream of fast neutral ions, which are generated by ionization/neutralization of argon or xenon. This stream induces a shock wave within the solution that ejects the ions of sample into gas phase (Figure 19.28). The examples of common matrices are glycerol, thioglycerol, 3-nitrobenzyl alcohol (3-NBA), 18-Crown-6 ether, 2-nitrophenyloctyl ether, sulfolane, diethanolamine, and triethanolamine. FAB produces primarily intact protonated molecules denoted as $[M+H]^+$ and deprotonated molecules such as $[M-H]^-$. Following ionization, the selected positive or negative ions are extracted, accelerated, and then mass analyzed. The practical mass limit of FAB is ~7 kDa and therefore it is used for polypeptides having <70 residues.

Electrospray ionization (ESI) This method of producing ions in MS was established by John Bennet Fenn for which he was

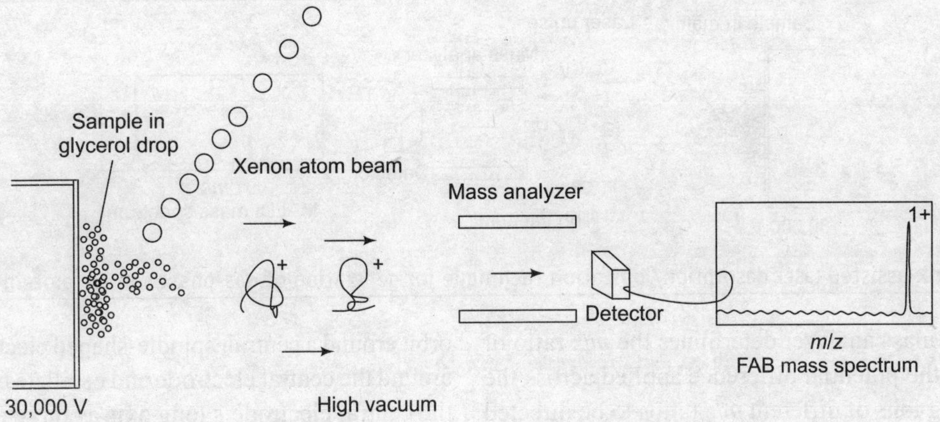

Figure 19.28 Fast atom bombardment technique for generation of gas-phase ion for protein analysis

awarded Noble Prize for Chemistry in 2002. In this method, the protein is digested by an endoprotease and the resulting solution is passed through a HPLC column, and at the end of this column, the solution is sprayed from a narrow capillary tube, which is maintained at high voltage (~4,000 V). The resulting droplets have positive charge and the most remarkable characteristic of ES is that a family of multiple charge states of ion can arise from a single precursor (Figure 19.29). The extent of multiple charging of an ion is influenced by factors such as the composition and pH of electrospray solvent. This unique charging phenomenon has the effect of lowering the *m*/*z* values of the protein sample to a range that can be readily measured by many different types of mass analyzers. The mass spectrum is analyzed by computer and often compared against a database of previously sequenced proteins in order to determine the sequences of the fragments. This process is then repeated with a different cleavage system and the overlaps in the sequences used to construct a sequence for the protein. The practical mass limit of ESI is ~70 kDa and sensitivity ranges between femtomole (10^{-15}) to picomoles (10^{-12}). This is the softest ionization technique, which is capable of observing biologically native noncovalent interactions. The advantage of this technique is that it can be directly coupled to liquid separation techniques like capillary electrophoresis or HPLC.

Matrix-assisted laser desorption/ionization Matrix-assisted laser desorption/ionization (MALDI) is a laser-based soft ionization technique and has proven to be one of the most successful ionization methods for MS analysis and investigation of large molecules. MALDI is carried out in an excess of specific wavelength-absorbing crystalline matrix (UV 337 nm). The sample is directly added to appropriate matrix for analysis and ~1:1000 molar ratio of sample to matrix is used (Figure 19.30). Finally, drying a drop of solution containing the sample to be analyzed generates cocrystallized mixture. The matrix plays a key role in this technique by absorbing the laser light energy and causing a small part of the target substrate to vaporize. The examples of commonly used matrices for protein analysis are α–cyano-4-hydroxy-cinnamic acid and 3,2-dimethoxy-4-hydroxy-cinnamic acid. Ions are produced by bombarding the sample with short pulse (1–10 ns) of UV light from a nitrogen laser. The interaction of the laser pulse with the sample results in ionization of both matrix and sample via an energy transfer mechanism from the matrix to the embedded sample. A high potential electric field (30 kV) is applied between the samples and sampling orifice and the desorbed ions are then accelerated to the mass analyzer by a series of lenses. The practical limit is ~300 kDa and sensitivity ranges from low femtomoles to low picomoles. According to some reports, attomole (10^{-18}) sensitivity is also possible. Posttranslational modification of proteins may also be detected by shifts in molecular weights. The proteins separated by 2D-PAGE can be routinely detected by MALDI/TOF (time of flight) (for details see Section 19.3.4).

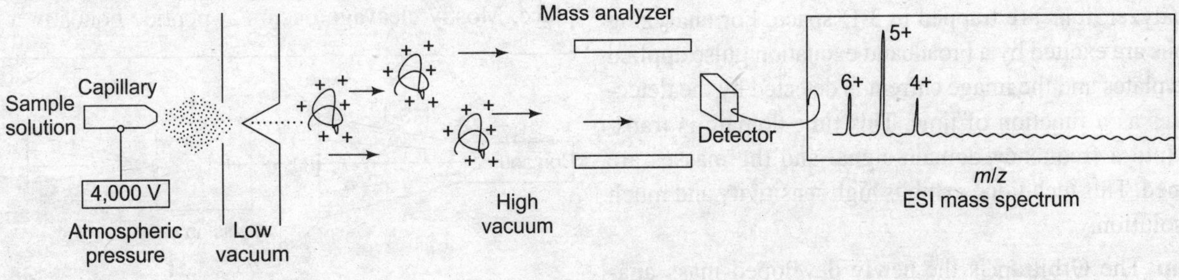

Figure 19.29 Electrospray ionization technique for generation of gas-phase ion for protein analysis

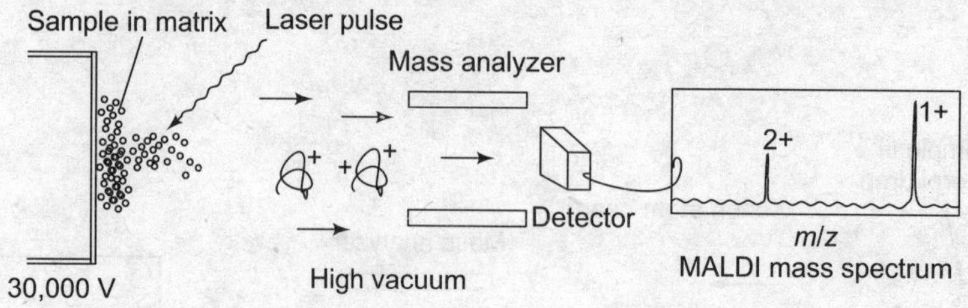

Figure 19.30 Matrix-assisted laser desorption/ionization technique for generation of gas-phase ion for protein analysis

Mass Analyzer A mass analyzer determines the *m/z* ratio of an ion by varying the potential difference applied across the ion steam, allowing ions of different *m/z* ratios to be directed toward the detector. The mass analyzers used with MS are time of flight (TOF), quadrupole, triple–quadrupole, or quadrupole TOF (qTOF), Fourier transform ion cyclotron resonance (FT-ICR), and Orbitrap.

Time of flight (TOF) analyzer The simplest to understand is the time of flight (TOF) analyzer. It uses an electric field to accelerate the ions through the same potential and then measures the time for an ion to fly from the ion source to the detector, which is proportional to the square root of *m/z*.

Quadrupole It comprises of four metal rods, pairs of which are electrically connected and carry opposing voltages that can be controlled by the operator. This type of mass analyzers use oscillating electrical fields to selectively stabilize or destabilize ions passing through a radio frequency (RF) quadrupole field. More than one quadrupole may be connected in series, as in triple-quadrupole MS. The variation in the voltage gradually over time detects a MS. Till date only ESI sources are available for commercial triple–quadrupole instruments. A number of hybrid quadrupole/TOF mass analyzer have also been developed for tandem MS/MS. qTOF has been developed by introducing a device for pulsed orthogonal ion injection into TOF mass analyzers. The qTOF is advantageous as both ESI and MALDI sources can be successfully coupled to it. This type of mass analyzer overcomes many of the limitations of triple–quadrupole and TOF-based MS/MS methods.

Fourier transform ion cyclotron resonance (FT-ICR) The reactions in the FT-ICR are carried out in a cell bound by electrodes located in a magnetic field and the *m/z* value of an ion is directly related to its cyclotron frequency. In FT-ICR mass analyzer, ions are trapped in 3-D space. For analyzing mass, ions are excited by a broadband excitation pulse applied to excite plates and the image current is detected by the detection plates as a function of time. This time domain is transformed into a frequency domain signal and the masses are determined. This technique exhibits high sensitivity and much high resolution.

Orbitrap The Orbitrap is the newly developed mass analyzer. In the Orbitrap, ions are electrostatically trapped in an orbit around a central, spindle-shaped electrode. The ions orbit around the central electrode and oscillate back and forth along the central electrode's long axis as these confine by the electrode. An image current is generated due to oscillation in the detector plates, which is recorded by the instrument. The frequencies of these image currents depend on the *m/z* of the ions in the Orbitrap. Finally, mass spectra are obtained by Fourier transformation of the recorded image currents. Orbitraps have a high mass accuracy, high sensitivity, and a good dynamic range like FT-ICR.

Tandem Mass Spectrometer for Peptide Sequencing

Mass spectrometry can be used to directly sequence the short stretches of polypeptide through the use of a tandem mass spectrometer. Tandem MS/MS is performed by using an instrument with two mass analyzers arranged in tandem, which may be of same or different types and a collision cell that contains inert gas (i.e., a small amount of noble gas such as He or Ar, also called 'collision gas') separates these two mass analyzers. MS/MS is generally used for fragment ion analysis. For performing tandem MS/MS, the first step is the enzymatic or chemical cleavage of the sample protein followed by fragmentation. The mixture is then injected into tandem MS/MS. In the first mass analyzer, the peptide ion of interest is selected from other peptide ions or any other contaminants on the basis of their *m/z* values. The ionized fragments are manipulated so that only one of the several types of peptides produced by cleavage emerges at the other end. The selected peptide ion is then passed into a collision cell, where the peptide is further fragmented by high-energy impact with the 'collision gas' (Figure 19.31). A number of the peptide molecules in the sample are fragmented while this procedure is designed in such a manner that each individual peptide is broken on an average at one place. Mostly cleavage occurs at peptide bonds in a vacuum

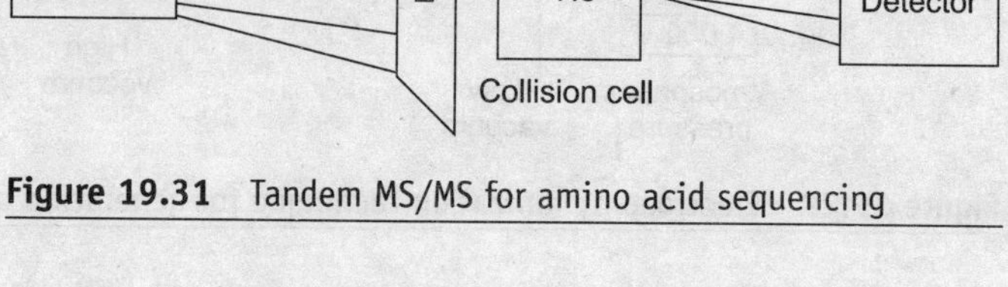

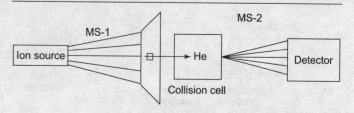

Figure 19.31 Tandem MS/MS for amino acid sequencing

and the products may include molecular ion radicals such as carbonyl radicals. The charge on the original peptide is retained on one of the fragments generated from it. There are various methods for fragmenting molecules for tandem MS, including collision-induced dissociation (CID), collisionally activated dissociation (CAD), electron capture dissociation (ECD), electron transfer dissociation (ETD), infrared multiphoton dissociation (IRMPD), and blackbody infrared radiative dissociation (BIRD). The *m/z* ratios of all the charged fragments are then measured by second mass spectrometer that produces one or more sets of peaks. However, a set of peaks comprises all the charged fragments that are produced by cleaving the same type of bond but at different points in the peptide. These fragments are generated by bond breakage from the same side of peptide bond either at C- or N-terminal side. Each successive peak in a given set has one less amino acid than the peak before and the difference in mass from peak to peak identifies the amino acid that is lost in each case. The sequence of entire polypeptide can thus be determined except the isomeric residues Ile and Leu (same mass) and Gln and Lys (difference 0.36 kDa) residues. This whole process can be easily computerized, which reduces the time needed to sequence a polypeptide to only a few minutes as compared to 30 to 50 min required per cycle of Edman degradation. The other advantages of this method of sequencing are that sequences of several polypeptides in a mixture can be easily determined even in the presence of contaminants by sequentially selecting the corresponding polypeptide ions in the first MS of the tandem instrument. Less effort is needed for sample preparation on MS-based sequencing as compared to Edman degradation. MS can be used to sequence peptides even with chemically blocked N-termini and to characterize other posttranslational modifications such as phosphorylation and glycosylation. However, protein sequencing by Edman's degradation is still the foundation of amino acid sequencing, while MS has become an important tool in the characterization of polypeptides.

19.3.5 Two-dimensional Gel Electrophoresis

For proteins, there is no amplification technique like PCR. The most widely used technology for protein analysis is separation by two-dimensional gel electrophoresis (2DGE) and/or multidimensional liquid chromatography followed by annotation using high–throughput MS. 2DGE was first described by P. H. O' Farrell and J. Klose in 1975, who used it for resolution of 1,000 proteins from *E. coli* on a single gel. They separated proteins of *E. coli* first by isoelectric focussing (IEF) in tube gel. The tube was cracked open and the proteins exposed to SDS by immersion of the gel in SDS solution. Tube gel was then attached to an SDS-PAGE slab gel and the proteins were separated according to mass at low pH (Figure 19.32). During the past several years, a number of developments have improved the resolution and reproducibility of the technique and expanded the ability to interpret results. Thus, the original method has been improved; however, the basic principle of 2DGE is same. The tube gel, which is not easily managed, has been replaced by strip gel, which is easy to handle and gives more reproducible results.

Steps Involved

Presently, 2DGE is powerful technique for separating complex mixture of proteins such as those extracted from cells and tissues into many more components. As all proteins in an electric field migrate on basis of their conformation, size, and electric charge, 2DGE uses two of these characteristics to allow high-resolution separation of proteins. The method involves at first dimension step IEF, which separates proteins according to charge, whereas second dimension step SDS-PAGE separates proteins according to molecular mass. Details of the process are described here.

Protein Solubilization Protein solubilization is very important for separation by 2DGE. The protein must be completely solubilized, disaggregated, denatured, and reduced prior to loading on IEF. The standard lysis buffer includes a chaotropic

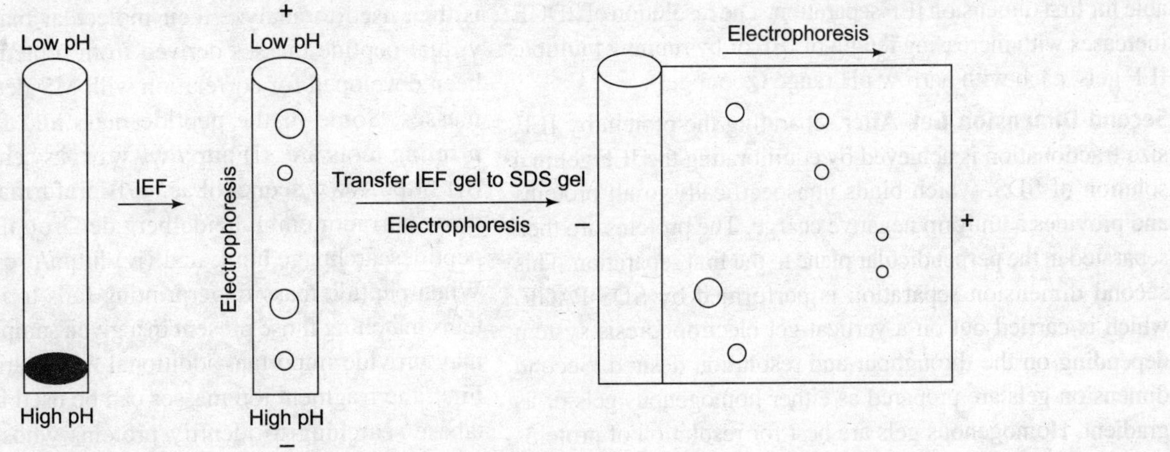

Figure 19.32 Two-dimensional gel electrophoresis for protein analysis

agent to disrupt H-bonds (8 M or 7 M urea and 2 M thiourea), nonionic detergent such as NP-40, and reducing agents like β-ME or DTT. These conditions are not suitable for the solubilization of membrane proteins and this is why a membrane protein is underrepresented on standard 2-D gels. Choosing detergents such as CHAPS can increase the recovery of membrane proteins.

First Dimension Gel The first dimension step is IEF, which is performed on a pH gradient allowing each protein to migrate to its isoelectric point, i.e., the point at which its isoelectric point (pI) is equivalent to surrounding pH and its net charge is zero. The original IEF was performed by carrier ampholyte-generated pH gradients in polyacrylamide tube gels. Carrier ampholytes are small amphoteric molecules, which have high buffering capacities around their pIs. When a voltage is applied across the mixture of these molecules, the carrier ampholytes align themselves according to their pIs and create a continuous pH gradient across the gel. However, the use of carrier ampholyte in 2DGE has several limitations like less reproducibility, instability of carrier ampholytes generated pH gradients, and tendency to drift toward the cathode over time. Later, an alternative method has been devised, in which immobilized pH gradients (IPGs) are generated for IEF by incorporating acrylamide buffers into a PAGE gel at the time of its casting. During polymerization, the acrylamide portion of the buffers copolymerizes with the acrylamide and bisacrylamide monomers to form a polyacrylamide gel. The two acrylamido buffer solutions contain a relatively acidic mixture and basic mixture, and thus pH gradient is formed. For improved handling, these gradients are cast onto a plastic support, and after polymerization the gel is washed to remove catalysts and unpolymerized acrylamide, which can interfere with the separation of proteins and reduce the sample loading capacity. Though acrylamide buffers are available from some manufacturers, commercial preparations of immobilized pH gradient, i.e., precast gel strips are simpler but an expensive alternative. The gel strips can be rehydrated in any desired solution suitable for first dimension IEF separation. The resolution of 2DGE increases with increasing length of IEF or by running multiple IEF gels, each with narrow pH range (zoom gel).

Second Dimension Gel After separating the protein by IEF, size fractionation is achieved by equilibrating the IEF gel in a solution of SDS, which binds nonspecifically to all proteins and provides a uniform negative charge. The proteins are then separated in the perpendicular plane to the first separation. This second dimension separation is performed by SDS-PAGE, which is carried out on a vertical gel electrophoresis system depending on the throughput and resolution desired. Second dimension gels are prepared as either homogenous gels or as gradient. Homogenous gels are best for resolution of proteins belonging to similar molecular weight range and easiest to pour reproducibly. However, a gradient gel is best when proteins over a wide range of molecular weights are to be analyzed, which provide a wider linear separation interval and the resulting spots are sharper, because the decreasing pore size functions to minimize diffusion. Vertical 2-D is generally run on 1.0 or 1.5 mm thick gels. Precast vertical SDS-PAGE gels are also commercially available. Large-sized SDS-PAGE gels can increase 2DGE resolution. Thus, the second dimension consists of four steps: (i) Preparation of the SDS-PAGE; (ii) Equilibration of the IPG strips in SDS buffer; (iii) Placement of the equilibrated IPG gels on the SDS-PAGE gel; and (iv) Electrophoresis of the second-dimensional gel.

Staining 2-D Gels Various methods are developed for staining proteins separated by 2-D electrophoresis. Depending on the objective of the experiment, different types of staining procedures can be used, for example, colloidal coomassie method, ammoniacal silver staining, mass spectrometry–compatible silver staining, fluorescent staining of proteins with SYPRO Ruby, or phosphoprotein staining. Usually, SYPRO reagent is used to stain and compare protein spots on different gels, but the gels can also be stained with additional reagents that identify specific classes of proteins. Numerous types of stains have been developed by manufacturers such as Molecular Probes Inc. that recognize various structurally or functionally related proteins, e.g., glycoproteins and phosphoproteins, oligohistidine-tagged proteins, calcium binding proteins, and in particular drug-binding proteins.

Analysis of Results The result is an array of protein spots that are assigned *x* and *y* coordinates rather than the protein bands seen in one-dimension separations. The large number of spots on typical gels must be analyzed by image analysis software, e.g., MELANIE II and CAROL.

19.3.6 Peptide Mass Fingerprinting

A MS with an ESI or MALDI TOF can be used for peptide mass fingerprinting. Protein spots are excised from a 2-D gel and digested with a specific endopeptidase like trypsin. MS is then used to analyze their molecular mass. A number of virtual peptide masses derived from protein databases have been developed for correlation with MS-determined peptide masses. Some of the peptide mass and fragment fingerprinting tools are: (i) http://www.expasy.ch/tools/proteome; (ii) http://www.seqnet.dl.ac.uk/Bioinformatics; (iii) http://www.narrador.embl–heidelberg.de/GroupPages/PageLink/peptidesearchpage.html; and (iv) http://prospector.ucsf.edu. When peptide mass fingerprinting fails to identify any proteins matching those present in a given sample, the fragments may provide important additional information in two ways. First, the fragment ion masses can be used in correlative database searching to identify proteins whose peptide yields similar CID spectra under same fragmentation condition. In probability-based matching, virtual CID spectra are derived from the peptides of all protein sequences in the database and these are compared with the observed data to derive a list of potential matches. Second, the peaks of MS can be interpreted, either manually or automatically, to derive the partial *de novo* peptide sequence that can be used as standard database search queries.

19.3.7 Multidimensional Liquid Chromatography

Multidimensional liquid chromatography (MDLC) is a sensitive and flexible technique used in proteomics and is compatible with MS. This is a serial analysis technique, where multiple gels having related samples can be run simultaneously. Liquid chromatography (LC) can be used either upstream to prefractionate samples before loading on 2DGE or downstream to separate peptides isolated from individual gel spots. Different liquid chromatography separation techniques can be combined in a different order to increase the utility of this method. MDLC has solved some of the practical limitations associated with 2DGE. For example, large sample volume can be loaded on HPLC columns, which is concentrated on the column and detection of low abundance proteins becomes easier. Many of the proteins like membrane proteins that are difficult to analyze by 2DGE can be separated easily using an appropriate column matrix. Proteins separated by LC methods are not stained for detection. LC methods can separate peptides as well as proteins, and LC columns can easily couple to MS. The entire analytical process from sample preparation to peptide mass profiling can be automated. The sequential application of different chromatographic techniques exploiting different physical or chemical separation principles can provide sufficient resolution for the analysis of very complex protein or peptide mix. The sequential use of ion exchange chromatography (charge-based) and reversed-phase HPLC (RP-HPLC; mass-based) can achieve the same resolution as 2DGE with the added advantage of automation, increased sensitivity, and better representation of basic proteins. Nevertheless, MDLC technique is associated with some practical limitations, which should be considered before performing this analysis. The first important concern is compatibility of the buffer and solvent used in different steps of each procedure. The elution buffer used in first dimension step has to be a suitable solvent for second dimension step and the elution buffer for second dimension step needs to be compatible with solvent used in MS. Usually ion exchange or size exclusion chromatography is used as a first dimension separation technique in combination with HPLC. Sometimes affinity chromatography step is associated prior to HPLC, forming 3-D separation techniques. Previously MDLC was performed by discontinuous process in which fractions were collected from first dimension column and then manually injected to HPLC. Thereafter, continuous MDLC has been developed with column switching. The third and most advanced strategy for MDLC is the use of biphasic columns, in which distal part is filled with reversed phase resin and the proximal part with another type of matrix (Figure 19.33).

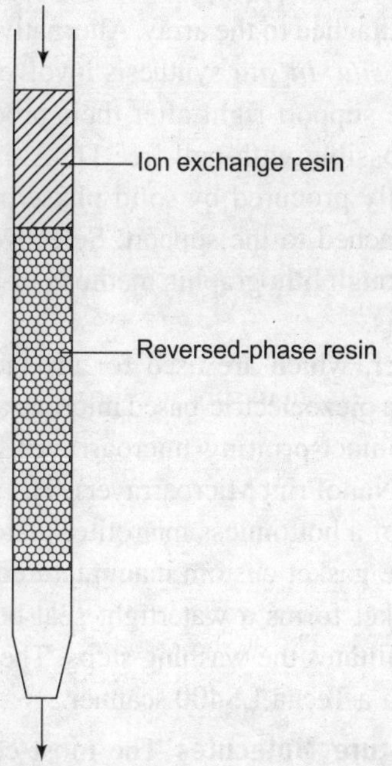

Figure 19.33 Biphasic column used in multidimensional liquid chromatography for protein analysis

19.3.8 Protein Microarray

Protein microarray (biochip, protein chip) technique has the potential to be an important tool for proteomics research and is used to identify protein–protein interactions to identify the substrates of protein kinases or to identify the targets of biologically active small molecules. Protein chips serve as measurement devices used in biomedical applications to determine the presence and/or amount (referred to as quantitation) of proteins in biological samples, e.g., blood. The technology is well-suited to large-scale, system-wide investigation because it enables many different samples to be analyzed simultaneously in a rapid and economical fashion and it enables such experiments to be performed hundreds or even thousands of times with different cells under different conditions. These two features of the technology apply equally well to the system-wide study of protein function. The protein microarray technology provides a robust way to study protein function in a rapid, economical, and system-wide fashion. This technology can be applied to the comprehensive and quantitative study of whole families of protein interaction domains.

Fabrication of Protein Chip Protein chip is a piece of glass or silicon on which different molecules of protein have been spotted at separate locations in an ordered manner, thus forming a microscopic array. Besides these, nanowell arrays are also available. The protein chips usually have a multitude of different capture agents deposited on a chip surface in a miniature array. The proteins are externally synthesized, purified, and attached to the array. Alternatively these can be synthesized *in situ*. *In situ* synthesis involves attachment of proteins to the support right after their production and the synthesis is possible with cell-free DNA expression. Peptides chemically procured by solid-phase peptide synthesis are already attached to the support. Selective deportation is carried out through lithographic methods or by the so-called SPOT synthesis.

Microarrayer, which are used for production of protein array, is either a piezoelectric-based microarrayer from Perkin Elmer or a contact-printing microarrayer from TeleChem International (NanoPrint Microarrayer). The glass is attached to the bottom of a bottomless microtiter plate using an intervening silicone gasket custom manufactured by Grace BioLabs. The gasket forms a watertight seal between adjacent arrays and facilitates the washing steps. The arrays are then visualized with a Tecan LS400 scanner.

Types of Capture Molecules The most common protein microarray is the antibody microarray, where antibodies are spotted onto the protein chip and are used as capture molecules to detect proteins from cell lysates. Though antibodies have several problems including the fact that there are no antibodies for most proteins and also problems with specificity in some commercial antibody preparations, these still represent the well-characterized and effective protein capture agents for microarray. However, nowadays there has been a great demand for other types of capture molecules, which are quite similar such as peptides or aptamers. Nucleic acids, receptors, enzymes, and proteins have also been spotted onto chips and used as capture molecules. This allows a vast variety of experiments to be conducted on protein–protein interactions and all other protein-binding substrates.

Detection Methods Although protein microarrays may use similar detection methods as DNA microarrays, a problem is that protein concentrations in a biological sample may be different from that for mRNAs. Therefore, protein chip detection methods must have a much larger range of detection. The preferred method of detection, which is used currently, is fluorescence detection and is safe, sensitive, and has high resolution. Moreover, the fluorescent detection method is compatible with standard microarray scanners; however, some minor alterations to software may need to be incorporated. In order to reduce variation introduced at the probing and washing steps, as well as to facilitate processing thousands of arrays, microtiter plates with 96 and 384 wells have been generated. To maximize the number of spots in each well, proteins are arrayed onto microtiter plate-sized pieces of aldehyde-derivatized glass.

Protein microarrays have been used for the biochemical and enzymatic analysis of proteins as well as to review protein–protein interaction. Most proteome arrays have used yeast *S. cerevisiae* as model organism. A complete proteome analysis needs an array of ~6,000 proteins in this case. Such arrays are assembled using proteins tagged with groups, allowing binding to solid supports such as 96-well microtiter plates. Library that includes nearly 90% of the yeast proteins are fused to glutathione *S*-transferase (GST) tag, which binds to a solid support via glutathione or to the His tag, which allows binding via nickel. These constructs are expressed under control of the promoters, $P_{GAL\,1}$ (galactose-inducible) or $P_{CUP\,1}$ (copper-inducible). Such protein libraries are pooled or distributed individually into the wells of microtiter dishes. The functional assays must be designed so that the arrays can be screened conveniently usually for fluorescence, less often for radioactivity. For example, the yeast proteome has been screened for those proteins that bind with calmodulin or phospholipids. His-tagged proteins are attached to Ni-coated slides. Both calmodulin and phospholipid are tagged with biotin. After binding of calmodulin or phospholipids to the proteome array, the biotin is detected by streptavidin carrying a Cy3 fluorescent label.

19.3.9 Yeast Two-hybrid System to Detect Protein–Protein Interaction

The association of protein molecules and the study of these associations from the perspective of biochemistry, signal transduction, and networks are referred to as protein-protein interactions. This is sometimes termed as the protein interactome or interaction trap. The study is of significance because the interactions between proteins are important for many biological functions. Further, many protein–protein interactions can lead to diseases when disrupted. Useful information about protein–protein interactions are available at several databases. These include: database of interacting proteins (DIP; dip.doe-mbi.ucla.edu), database of ligand receptor partners (DLRP; http://dip.doe-mbi.ucla.edu/dip/DLRP.cgi), biomolecular interaction network database (BIND; http://www.blueprint.org/bind/bind.php), protein-protein interaction and complex viewer (MIPS-CYGD; http://mips.gsf.de/proj/yeast/CYGD/interaction), hybrigenics (PIM; www.hybrigenics.fr), general repository for interaction datasets (GRID; http://biodata.mshri.on.ca/grid), molecular interactions database (MINT; http://cbm.bio.uriroma2.it/mint), curagen *Drosophila* interactions database (http://portal.curagen.com/cgi-bin/interaction/yeastHome.pl), and *Saccharomyces* genome database (SGD; http://www.yeastgenome.org).

Experimentally, interactions between pairs of proteins are inferred from several methods. The most useful methods are phage display (for details see Chapter 16) and the yeast two-hybrid system.

Principle of Classical Yeast Two-hybrid System

Fields and Song pioneered yeast two-hybrid system in 1989. The principle behind this system is based on the modular organization of many eukaryotic proteins that consist of several discrete structurally and functionally independent domains, which can function when brought together through noncovalent interaction. For example, transcription factor comprises of a DNA-binding domain (DBD), which recognizes and binds to a defined promoter sequence upstream of a gene, and an activation domain (AD), which interacts with the RNA polymerase II complex (Figure 19.34). Both of these domains are required for transcription. If either of these domains is absent, the gene is not transcribed. DBD of transcription factor helps in locating itself to promoter of its target gene and AD then recruits the RNA polymerase II complex at the start of the gene, thereby activating its transcription. However, when DBD and AD are expressed as separate polypeptides, the function of the transcription factor is lost. This is because DBD alone can bind to its cognate promoter sequence, but cannot activate transcription, and AD alone can interact with the RNA polymerase II complex, but cannot activate transcription from promoter. On this basis, it is possible to use the yeast two-hybrid system to confirm interactions between known proteins and to screen for unknown proteins that interact with a given protein of interest.

Steps Involved in Classical System

In the first step, a cDNA encoding the protein of interest (X) is cloned into a bait vector, creating a fusion of the DBD and protein of interest termed as bait. This bait is then translocated to the nucleus of the yeast cell, where it binds to the promoter located upstream of a reporter gene. As this bait lacks an activation domain, no activation of the downstream reporter genes takes place. A second cDNA encoding an interacting protein (Y) is cloned into a prey vector, producing a fusion of the AD and the interacting protein termed as prey. Theoretically, the prey is able to activate yeast genes because of the presence of AD, but is unable to do so as it is not in the proximity of a reporter gene. Finally, if proteins X and Y have capability of interaction with each other, i.e., if the bait captures the prey, the gene will be activated, creating a hybrid transcription factor. A convenient downstream reporter gene is used to monitor for a successful interaction, for example, an auxotrophic marker (e.g.; *HIS* 3, *ADE* 2, or *URA* 3) or a selectable marker *lac* Z. If the reporter is transcribed, the result is histidine (or adenine or uracil) prototrophy (for details see Chapter 11) or blue coloration of the yeast cells. Thus, the interaction of two proteins is easily studied by constructing hybrid transcription factor and the resultant activation of a set of specific reporter genes. The success of yeast hybrid system is thus dependent on joint strategy; first, the positive clones are selected on the basis of auxotrophic markers, and second, the selected clones are analyzed further using the convenient color assay.

Classical Yeast Two-hybrid System to Determine Interactions Between Yeast Proteins

The yeast two-hybrid system developed in yeast was successfully used to produce a complete list of interactions between more than 6,000 yeast proteins. The steps involved in classical yeast two-hybrid system are presented in Figure 19.35. For this, each ORF of the yeast genome was amplified by PCR and cloned into two separate vectors, i.e., one having the DBD and another AD domain, and the probability of each yeast protein was tested both as bait and prey for all possibilities of interactions. The vectors were designed to give *in frame* gene fusions of each ORF with the DBD domain and AD domain of a transcription factor like GAL 4. One vector

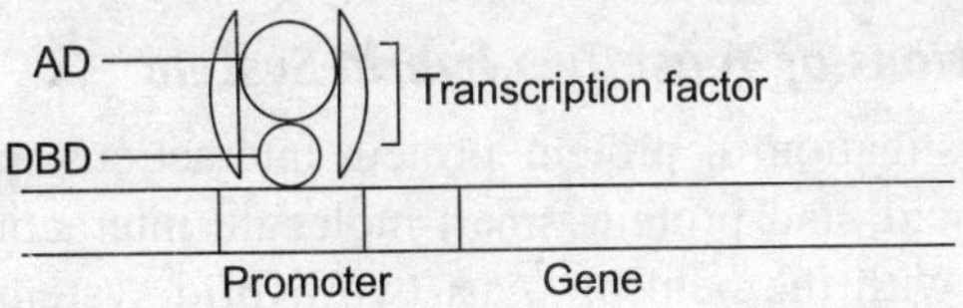

Figure 19.34 Principle of classical yeast two-hybrid system

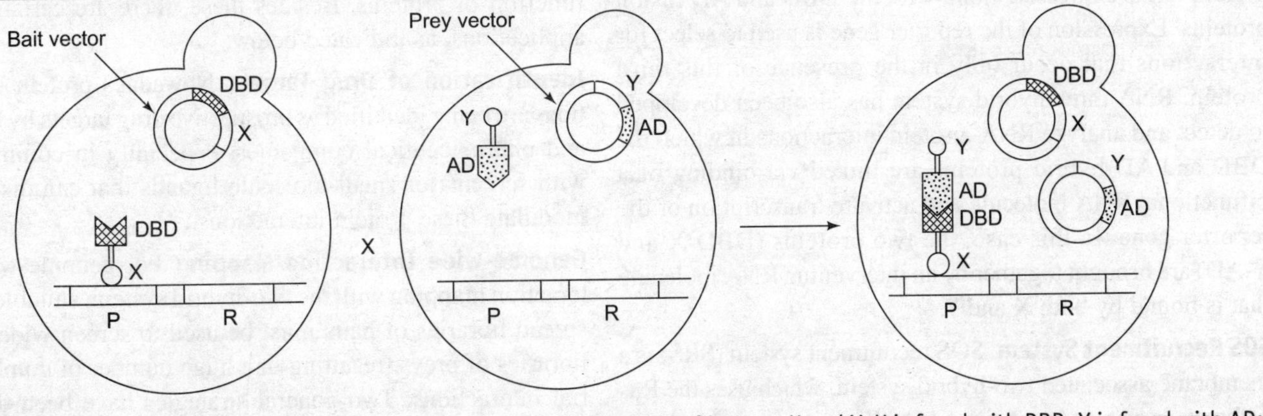

Figure 19.35 Steps involved in classical yeast two-hybrid system [Protein of interest X and Y; X is fused with DBD; Y is fused with AD; P is promoter; R represents reporter]

had a MCS downstream of the GAL 4-DBD and so generated a 3′ fusion of GAL 4-DBD to protein (GAL 4-DBD-X). The other vector had its MCS upstream of the GAL 4-AD and gave a 5′ fusion of GAL 4-AD and protein (Y-GAL 4-AD). Both fusion constructs were transformed into yeast cells of different mating types and two sets of ~6,000 transformants were formed. The transformants were mated, and the resulting diploid cell had a bait vector and a prey vector. Consequently, the diploid cells from 6,000 × 6,000 matings were selected on minimal medium. This is because if the two proteins X and Y had interacted with each other, the reporter gene would have been switched on.

Variations of Yeast Two-hybrid System

The original yeast two-hybrid system has several limitations, e.g., it relies on the proteins interacting within nucleus and proteins having hydrophobic transmembrane domains are not transported to the nucleus. Moreover, if transported, these are not properly folded in the nucleus. On the contrary, other proteins may need modification by cytoplasmic or membrane associated enzymes for interaction with binding partner. Some large proteins demonstrate toxic and steric problems, which may overlook some interactions. Moreover, many proteins bind RNA and/or rely on small molecules to alter their conformation thus promoting protein–protein interactions. Thus, in view of dealing with the limitations of the classical yeast two-hybrid system, various modified systems have been developed. Some important variations are described below.

Reverse Two-hybrid and Split-hybrid Systems Sometimes classical two-hybrid interactions are disrupted due to mutations, drugs, or competing proteins. In this system, the interaction of X and Y proteins induces the transcription of a reporter gene that confers toxicity to the yeast. For example, selection can be induced by the addition of 5-fluoro-orotic acid (FOA), which is converted to the toxic compound 5-fluorouracil by the *URA* 3 gene product.

Three-hybrid System When performing this variation, a third protein (Z) is expressed along with the DBD and AD fusion proteins. Expression of the reporter gene is used to select for interactions that occur only in the presence of this third protein. RNA three-hybrid system has also been developed to detect and analyze RNA–protein interactions in which the DBD and AD hybrid proteins are linked via binding of a bifunctional RNA molecule and activate transcription of the reporter gene. In this case, the two proteins (DBD-X and Y-AD) are brought together by an intervening RNA molecule that is bound by both X and Y.

SOS Recruitment System SOS recruitment system (SRS) is a membrane-associated two-hybrid system, which uses the Ras pathway in yeast. Aronheim and colleagues formulated this system in 1997. The yeast Ras guanine nucleotide exchange factor (RGEF), *cdc* 25, when positioned at the plasma membrane, stimulates GDP/GTP exchange on Ras. Downstream signaling events are then promoted that eventually lead to cell growth. A mutant yeast strain having the temperature-sensitive *cdc* 25-2 allele is still able to grow at 25°C but fails to grow at 36°C. The human RGEF (hSOS), which efficiently complements the yeast mutation, when targeted to the plasma membrane leads to cell growth at 36°C. In the SRS, the translocation of hSOS is dependent on a protein–protein interaction. The bait X is fused to C-terminally truncated hSOS, which is active but unable to target to the plasma membrane. The bait X is coexpressed with a prey Y, which can either be an integral membrane protein or a soluble protein that is anchored to the membrane by myristoylation. Interaction between X and Y recruits SOS to the membrane, where it stimulates guanine nucleotide exchange on Ras and ultimately GTP-bound Ras stimulates cell growth.

Split-ubiquitin System A cytoplasmic two-hybrid assay based on ubiquitin was developed by Johnsson and Varshavsky (1998). As ubiquitin is a small protein of 76 amino acids, which acts as a 'tag' for protein degradation, so that proteins fused to ubiquitin are rapidly cleaved *in vivo* by ubiquitin-specific proteases (UBPs). If the C-terminal of ubiquitin (Cub) is fused to a reporter protein and coexpressed with the N-terminal fragment of ubiquitin (Nub), the two halves reconstitute the native ubiquitin, resulting in the cleavage of the reporter protein. For its adaptation to detect protein–protein interactions, a mutant Nub that is unable to interact with Cub on its own is fused to one protein and a Cub reporter hybrid is fused to its prospective interaction partner. Interaction between the two proteins allows ubiquitin to be reconstituted, leading to cleavage and release of the reporter gene.

Applications of Yeast Two-hybrid System

The investigation of protein–protein interactions, protein–nucleic acid, and protein–small molecule interactions are possible with the help of yeast two-hybrid systems. This technique is basically used in research for studying the function of proteins. Besides these, there are certain other applications, as indicated below:

Identification of Drug Targets Nowadays protein interactions are being identified as prospective drug targets by biotech and pharmaceutical companies, especially in combination with screens for small-molecule ligands that can disrupt or modulate these protein interactions.

Genome-wide Interaction Mapping For genome-wide interaction mapping with the two-hybrid system, suitable widespread libraries of baits must be used to screen widespread libraries of preys, resulting in a huge number of combinatorial interactions. Two general strategies have been devised for this purpose. These are the matrix approach and the random library method.

Matrix interaction screening This screening procedure involves panels of defined bait and prey, i.e., constructs derived from known ORFs, which are mated systematically in an array format. Since this approach depends on the availability of sequence data corresponding to each protein, it can only be used for predefined proteins. The advantage of this approach is that it is fully extensive and can provide exhaustive proteome coverage. However, it is also laborious because each bait and prey construct must be prepared individually by PCR followed by subcloning in the appropriate expression vector. Haploid yeast cells of opposite mating types are then transformed with the bait and prey constructs, respectively, and arrayed in microtiter plates. Mating generates specific pair-wise combinations and candidate interactions are assayed in the resulting diploid cells.

Random library method In the random library method, random clones from a highly complex expression library represent baits and preys. The prey can be screened by using defined ORFs as baits. If comprehensive proteome analysis has to be performed, random libraries are prepared for bait and prey. Unlike the matrix method, where all constructs are predefined and candidate interactors can be traced on the basis of their grid positions on the array, interacting clones in the random library must be characterized by sequencing and then these are compared to sequence databases for annotation.

19.3.10 Gel Retardation Assay to Detect Protein–Nucleic Acid Interactions

The binding of many regulatory proteins immediately upstream of the genes is required for proper genome expression. A number of experimental methods have been devised for establishing protein-binding sites within DNA fragments. Deletion analysis of these upstream sites determines how these affect gene expression, but do not confirm the actual binding of a protein at the site. Therefore, after finding a binding site, the binding of the regulatory protein must be confirmed experimentally by different techniques. The first and foremost technique is gel retardation assay. It is also called mobility shift assay or band shift assay or electrophoretic mobility shift assay (EMSA). This is a common procedure that is used to study protein–DNA or protein–RNA interactions. By this technique, the capacity of binding the protein or mixture of proteins with a given DNA or RNA sequence can be determined.

Principle of Gel Retardation Assay

Gel retardation method is based on considerable difference between the electrophoertic mobility of DNA fragments that carry a bound protein and that do not contain bound proteins. Usually, DNA fragments are separated by agarose gel electrophoresis. The speed of migration of different molecules through the gel is determined by their size and charge and to some extent by their shape. Moreover, if a DNA fragment binds to protein, then its mobility during the gel electrophoresis is hindered, thus the DNA–protein complex forms a band at a position closer to the well (Fig. 19.36).

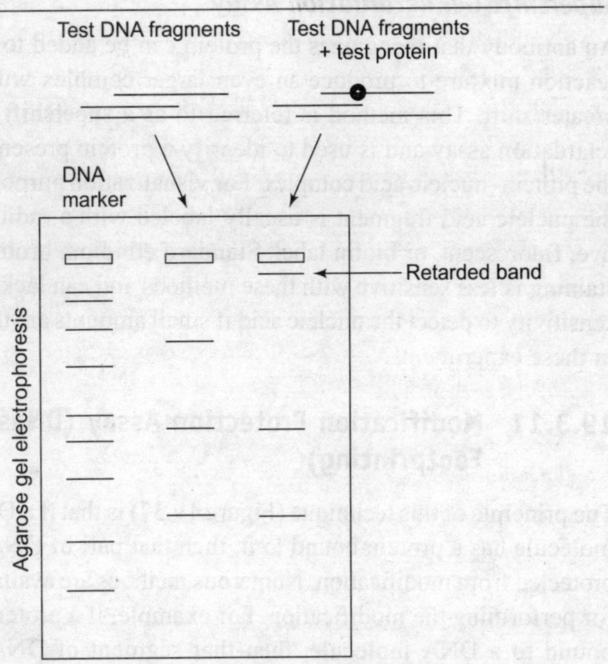

Figure 19.36 Principle of gel retardation assay

Steps Involved

An average protein has molecular weight of ~40 kDa and segment of DNA of 1 kbp has a molecular weight of ~700 kDa. Further, if a typical protein is bound to a length of DNA much bigger that this, the relative change in size and hence mobility would be 5% or less, which cannot be observed. Therefore, for gel retardation analysis, restriction is chosen to give segments of DNA in the range of 250–1,000 base pairs. A series of fragments thus obtained are radioactively (or fluorescent or biotin)-labeled. Sometimes PCR with primers from the upstream region of a gene are used instead of restriction enzymes to generate fragments from that particular region. The PCR fragments or radioactively-labeled DNA fragments are examined one at a time for binding to the test protein. A series of incubations containing DNA fragment and no protein (control) or increasing amounts of the protein are performed to allow binding to occur. The samples are run on native polyacrylamide gel or agarose gels (large fragments). The gel buffer should contain a low concentration of salt to stabilize protein–DNA interactions. If the protein binds to one of the DNA fragments, the complex formed will be larger and run slower than the original DNA, i.e., the fragments are retarded (Figure 19.36). The gel is then visualized by autoradiography or fluorescence or chemiluminescence, depending upon the label on DNA fragments.

Supershift Gel Retardation Assay

An antibody that recognizes the protein can be added to the reaction mixture to produce an even larger complex with a greater shift. This method is referred to as a supershift gel retardation assay and is used to identify a protein present in the protein–nucleic acid complex. For visualization purposes, the nucleic acid fragment is usually labeled with a radioactive, fluorescent, or biotin label. Standard ethidium bromide staining is less sensitive with these methods and can lack the sensitivity to detect the nucleic acid if small amounts are used in these experiments.

19.3.11 Modification Protection Assay (DNase I Footprinting)

The principle of this technique (Figure 19.37) is that if a DNA molecule has a protein bound to it, then that part of DNA is protected from modification. Numerous methods are available for performing the modification. For example, if a protein is bound to a DNA molecule, then that segment of DNA is prevented from cleavage by a nuclease, while rest of the phosphodiester bonds in the DNA molecule are cleaved. Similarly as dimethylsulfate (DMS), which adds methyl group to G nucleotides in DNA, is unable to methylate Gs protected by the bound protein. These procedures are done experimentally by footprinting. The most commonly used technique based on modification protection assay is DNase I footprinting. This technique is used to locate a protein-binding site on a particular DNA molecule. As the name suggests, this method uses deoxyribonuclease (DNase I), a relatively nonspecific

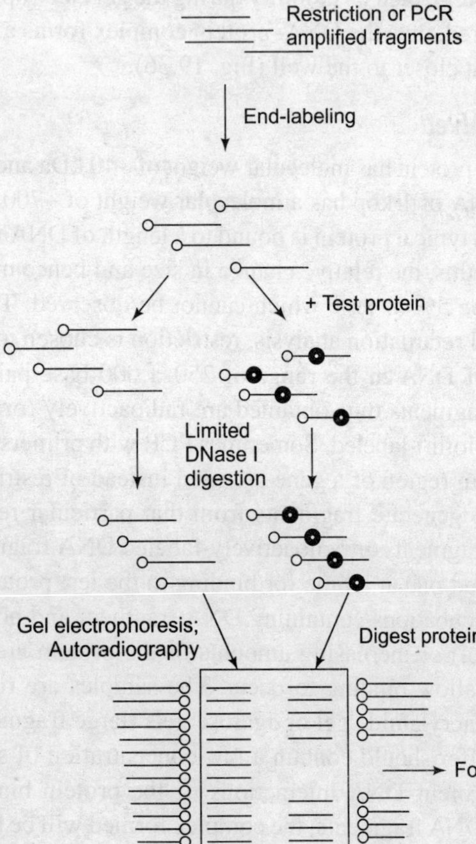

Figure 19.37 Principle of DNase I footprinting

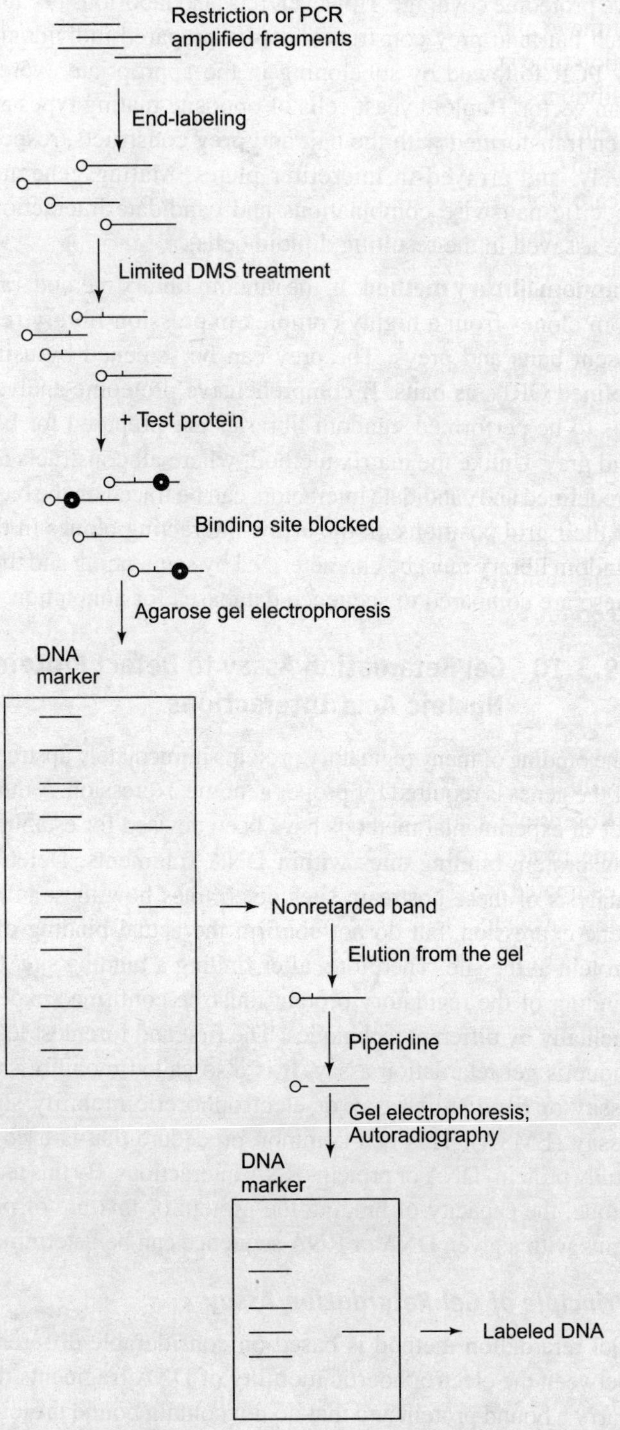

Figure 19.38 Principle of modification interference assay by treatment with dimethyl sulfate

endonuclease, and hydrolyzes phosphodiester bonds between any nucleotide (for details see Chapter 2). The underlying principle is that DNase I degrades all the regions of DNA except those covered by bound proteins (Figure 19.37). The first step of the procedure is the generation of a target DNA fragments ~100–300 bp in length either by PCR with specific primers or by restriction digestion from a vector. Next, these DNA fragments are labeled at 5′-end and incubated with protein for some time to allow binding. As binding is not instant, more incubation time is required to reach equilibrium, particularly if equilibrium constants have to be measured. In another set, no protein binding is done. Thereafter, both the samples are treated with DNase I. The most important point, which should be taken care while performing the digestion with DNase I, is that the DNA molecules are treated lightly with DNase I so that there is an average of <1 nick/strand. The reaction is stopped, the protein is digested, and DNA is denatured. Finally, both the reaction mixtures are run on a denaturing polyacrylamide gel (similar to sequencing gel). These gels are capable of resolving molecules differing in length by a single nucleotide. The cleavage pattern of the DNA in the absence of a DNA-binding protein is compared to the cleavage pattern of DNA in the presence of a DNA-binding protein by autoradiography or by the use of a phosphorimager (for details see Chapter 16). The result indicates that the sample of protected DNA lacks certain bands and hence fragments. In contrast, cutting a sample of unprotected DNA gives rise to a series of fragments of all possible lengths, varying by a single base pair. This is because, if the protein binds DNA, the binding site is protected from enzymatic cleavage and this protection results in a clear area on the gel, which is referred to as the 'footprint'. By varying the concentration of the DNA-binding protein, the binding affinity of the protein can also be estimated according to the minimum concentration of protein at which a footprint is observed. Practically, the footprint is run side by side with a sequencing reaction, which allows matching the footprint with the DNA sequence.

19.3.12 Modification Interference Assay

This is quite different technique from modification protection assay. This technique is based on the principle (Figure 19.38) that if a particular nucleotide of DNA fragment, which is critical for protein binding is altered, e.g., by addition of a methyl group, then binding may be prevented. There are four nucleotides and it is possible to specifically modify each type of nucleotide by chemicals and thereafter the protein binding is inhibited. The DNA fragments either amplified by PCR or cut by restriction enzyme are end-labeled and treated with the modification reagent under limiting conditions so that only one nucleotide per fragment is modified, for example, methylated by DMS. The DNA-binding protein is then added and the resulting fragments electrophoresed. The results indicate two bands, one corresponding to the DNA–protein complex and one containing the DNA without protein. The latter represents the DNA molecule that is prevented from attaching to the protein because methylation has modified one or more Gs that are crucial for binding. To identify which particular Gs are modified, the fragments are purified from the gel and treated with piperidine, a compound that cleaves at methyl- G. Each fragment is cut into two fragments, one of which carries the label. The resulting fragment is electrophoresed and autoradiographed by which the length of the labeled fragments is determined. The particular nucleotides of the original fragment, which are methylated, are known and therefore, the positions in the DNA sequence (Gs) that participate in the binding reaction can be identified. Similarly A, C, and T nucleotides can also be modified by base-specific chemicals, followed by piperidine treatment and then modification interference assay is performed.

Review Questions

1. Outline how is DNA sequenced by the chain termination method. How is this method different from chemical degradation method?
2. Draw a sequencing gel showing the sequence that would be obtained if the antisense strand of the DNA (*) were used as template and a primer that hybridized to the DNA just beyond the end of the coding sequence.

   * 5′-ATGAGCGCAACGACG-3′
3. Explain the following statements:
   (a) The sequence read from the gel in Sanger dideoxy sequencing method does not begin with the sequence of the primer used.
   (b) ddNTP terminates the chain growth of DNA synthesis and all growing chains are not terminated at the same length.
   (c) dNTPs labeled by ^{35}S are better for DNA sequencing as compared to those labeled by ^{32}P.
   (d) The fluorescent labels have more advantages as compared to radioactive labels for sequencing.
4. Give an elaborative account of the principle and advantages of following techniques:
   (a) Pyrosequencing
   (b) DNA microarray

5. Describe with the help of suitable diagram the shotgun approach of genome sequencing taking the example of *Heamophilia influenzae.*
6. Explain the principle and steps involved in high-throughput sequencing by using Solexa/Illumina technique.
7. A polypeptide is specifically fragmented by following treatments, resulting in polypeptide fragments with the following amino acid sequences. Derive the amino acid sequence of the entire polypeptide.

   (a) **Cyanogen bromide treatment**

   Asp-Ile-Lys-Gln-Met
   Lys-Met
   Lys-Phe-Ala-Met
   Arg- Gly-Met

   (b) **Trypsin hydrolysis**

   Gln-Met-Lys
   Gly-Met-Asp-Ile-Lys
   Phe-Ala-Met-Lys
   Tyr-Arg

8. Following are the results obtained after subjecting a polypeptide to varying treatments. Determine the primary structure of the polypeptide.

   (a) **Acid hydrolysis**
   Ala, Arg, Cys, Glx, Gly, Lys, Leu, Met, Phe, Thr

   (b) **Carboxypeptidase A + Carboxypeptidase B**
   No fragments

   (c) **Trypsin hydrolysis followed by Edman degradation of the separated products**
   Cys-Gly-Leu-Phe-Arg
   Thr-Ala-Met-Glu-Lys

9. What type of assay should be performed to study protein–protein interaction? Give an elaborative account of the variation of that assay.
10. Describe how gel retardation assay is used to study DNA–protein interaction. What are the limitations of this technique?
11. Draw diagrams to illustrate the modification protection and modification interference techniques. Indicate the key differences and describe how these differences underlie the specific applications of the two techniques.

References

Aronheim, A. *et al.* (1997). Isolation of an AP-1 repressor by a novel method for detecting protein- protein interactions. *Mol. Cell. Biol.* **17**:3094–3102.

Edman, P. (1950). Method for determination of the amino acid sequence in peptides. *ACTA Chemica Scandinavica* **4**:283–293.

Fields and Song (1989). A novel genetic system to detect protein-protein interactions. *Nature* **340**:245–246.

Holley, R.W. *et al.* (1965). The base sequence of yeast alanine transfer RNA. *Science* **147**:1462–1465.

Johnsson, N. and A. Varshavsky (1996). Split-ubiquitin protein sensor No. 5,503,977.

Klose, J. (1975). Protein mapping by combined isoelectric focusing and electrophoresis of mouse tissues: a novel approach to testing for induced point mutations in mammals. *Humangenetik* **26**:231–243.

Maxam, A.M. and W. Gilbert (1977). A new method of sequencing DNA. *Proc. Natl. Acad. Sci.* **74**:560–564.

O'Farrell, P.H. (1975). High-resolution two-dimensional electrophoresis of proteins. *J. Biol. Chem.* **250**:4007–4021.

Sanger, F. and E.O.P. Thomson (1953). The amino acid sequence in the glycyl chain of insulin 1. The investigation of lower peptides from partial hydrolysates. *Biochem. J.* **53**:353–366.

Sanger, F. and A.R. Coulson (1975). A rapid method for determining sequences in DNA by primed synthesis with DNA polymerase. *J Mol. Biol.* **94**:441–448.

Sanger, F., S. Nicklen, and A.R. Coulson (1977). DNA sequencing with chain terminating inhibitors. *Proc. Natl. Acad. Sci.* **74**:5463–5467.

Applications of Cloning and Gene Transfer Technology

Key Concepts

In this chapter we will learn the following:

- Applications of recombinant microorganisms
- Applications of transgenic plant technology
- Applications of animal cloning and transgenic technology

20.1 INTRODUCTION

Recombinant DNA technology (rDNA technology) or genetic engineering has brought about a complete revolution in the way living organisms are exploited. By transferring new DNA sequences into microbes, plants, and animals, or by removing or altering DNA sequences in the endogenous genome, completely new strains or varieties can be created to perform specific tasks. An organism whose genetic composition has been altered by the addition of exogenous DNA is said to be 'transgenic'. Such an organism is also called transformant, genetically modified organism (GMO), or genetically engineered organism (GEO). The DNA introduced is called a transgene and the overall process of incorporation of a gene or part of a gene from one individual into the genome of another individual is called 'transgenesis' (also called 'transgenic technology' or 'transformation'). The introduction of a transgene into an organism means insertion of foreign gene into the genome, which may come from the same or different species. Owing to the virtual universality of the coding properties of DNA, the huge evolutionary distances can be bridged in transgenesis. The transgenic organisms are usually normal in appearance and character and differ from the parent only with respect to the function and influence of the inserted gene. Thus, transgenic organisms exhibit at least one new and useful trait and transmit their newly acquired genetic determinants through the germ line as simple Mendelian traits. In order to be of any use in a multicellular organism, the transgene must be inserted into germ line (cells involved in reproduction of the organism) DNA so that it can be propagated in subsequent generations. On the other hand, genes are inserted into somatic cell (cells not involved in reproduction of the organism) DNA for the purpose of gene therapy.

In contrast to transgenesis, 'cloning' in essence means 'making an exact copy' and the collection of cells or organisms (animal/plant/microorganism) that are genetically identical to the original ancestor are referred to as a 'clone'. Cloning is also defined as a process of asexually producing clones or to produce multiple copies of a single gene or segment of DNA or to produce a cell or an organism from a somatic cell of an organism with the same nuclear genomic (genetic) characters without fertilization.

The potential applications of genetic engineering are innumerable and some of these are discussed in this chapter. Emphasis is given to applications of transgenic organisms for purposes other than research. Note that the term genetic engineering, as used in this context, is restricted to cloning, genetic experimentation, and genetic enhancements.

20.2 APPLICATIONS OF RECOMBINANT MICROORGANISMS

A recombinant microorganism (also called transgenic microorganism) harbors genetically engineered DNA in all its cells and through the usual processes of protein synthesis, the gene of interest gives rise to proteins, which endow the bacteria with a new trait. To develop transgenic bacteria, the gene of interest can be easily incorporated into a plasmid, which is then incubated with bacteria under specific conditions that favor plasmid uptake by the bacteria. A bacterium that absorbs and retains the plasmid is transgenic. As a bacterial cell gives rise to a colony of identical cells, a bacterium that contains recombinant DNA gives rise to a colony of identical transgenic bacteria within a few hours. Most recombinant microorganisms are unicellular bacteria or yeasts (unicellular fungi). Making a unicellular recombinant organism is relatively easier than producing a multicellular transgenic plant or animal. Moreover, in contrast to transgenic multicellular organisms, plants, and animals, there is no risk of chimera formation (i.e., recombinants lacking new DNA in all the cells).

20.2.1 Applications in Genetic Research

Transgenic microorganisms find extensive applications in genetic research. Developing bacteria that possess and/or express a certain gene is often a routine procedure carried out as part of the study of that gene. These transgenic microorganisms can be used as sources of DNA or proteins, which can be used to understand biochemical processes or gene regulation and function.

20.2.2 Production of Recombinant Therapeutic Proteins or Recombinant Biopharmaceuticals

When rDNA technology was first developed, it was heralded as a means of producing a whole range of possible human therapeutic agents in sufficient quantities for both efficacy testing and eventual human use. The early commercial eukaryotic recombinant proteins of medical use produced in *E. coli* are relatively simple aglycosylated proteins, which accumulate as inclusion bodies within the bacterial cell and are subjected to *in vitro* denaturation and refolding. Because of the importance of glycosylation, the addition of sugar chains to proteins during their synthesis, the *E. coli* system has been superceded in many cases by mammalian cells because the latter can produce more complex proteins and achieve correct folding and glycosylation *in vivo*. However, *E. coli* is much cheaper to cultivate than mammalian cells and low gene dosage systems, which facilitate the expression of soluble proteins and targeting systems, which direct recombinant proteins to the periplasm, have been very successful. Note that the mammalian proteins targeted to the bacterial periplasm are more likely to fold correctly because this compartment has the ability to form and isomerize disulfide bonds. An alternative strategy, which has also been successful, is the expression of protein disulfide isomerase in the bacterial cytosol along with the recombinant protein of interest. Another alternative is to use yeast cells, which grow in a similar manner to bacterial cells and like bacteria require simple and relatively inexpensive media for growth. However, being eukaryotes, these have the ability to fold and assemble and secrete recombinant proteins much more efficiently than bacteria. The secretion of recombinant proteins from cultured yeast cells allows the formation of disulfide bonds, proteolytic maturation, *N*- and *O*-linked glycosylation, and other posttranslational modifications that are either absent or occur very inefficiently in bacteria. *S. cerevisiae* was the first yeast used for recombinant protein production and it is now being used for the commercial production of several approved drugs and vaccines. Unfortunately, as a general production system for recombinant pharmaceuticals, *S. cerevisiae* suffers from a number of limitations including low product, plasmid instability, difficulties in scaling up production, the hyperglycosylation of recombinant human glycoproteins, and inefficient secretion. Although improved *S. cerevisiae* strains have been described, some of which generate glycan chains compatible with humans, other yeast species have been developed as alternative production hosts. These include *Schizosaccharomyces pombe*, the methylo-trophic yeasts *Pichia pastoris* and *Hansenula polymorpha*, the dairy yeast *Kluyveromyces lactis*, and others such as *Schwanniomyces occidentalis* and *Yarrowia lipolytica*. These organisms often outperform *S. cerevisiae* in terms of yield, reduced hyperglycosylation, and secretion efficiency. The methylotrophic yeasts in particular are now emerging as competitive production systems. Filamentous fungi, for example, *Aspergillus nidulans*, *Aspergillus niger*, *Aspergillus oryzae*, *Trichoderma reesei*, *Acremonium chrysogenum*, and *Penicillium chrysogenum*, also exhibit high capacity for protein secretion.

Currently the cloning in bacteria and yeasts has resulted in the production of a number of therapeutic proteins (Table 20.1). The process of production of recombinant insulin is highlighted in Exhibit 20.1. Genentech founded by Herbert

Table 20.1 Some Examples of Biopharmaceuticals Produced Commercially in Microorganisms (Approved in the United States and Europe)

Products		Trade names
Category	**Examples**	
Recombinant hormones	Insulin	Actrapid; Velosulin; Monotard; Insulatard; Protaphane; Mixtard; Actraphane; Ultratard; Novolog; Novolog mix 70/30; Novolog 30; NovoRapid; Novolin; Lantus; Optisulin; Liprolog; Humalog; Humulin; Insuman
	Human growth hormone	Somavert; Nutropin AQ; Serostim; Saizen; Genotrophin; Norditropin; Bio Tropin; Nutropin; Humatrope; Protropin
	Follicle-stimulating hormone	Follistim; Puregon; Gonal F
	Parathyroid hormone	Forsteo; Forteo
	Choriogonadotrophin	Ovitrelle
	Thyroid-stimulating hormone	Thyrogen

(*Contd.*)

Table 20.1 (Contd.)

Products		Trade names
Category	**Examples**	
	Luteinizing hormone	Luveris
	Calcitonin	Forcaltonin
	Glucagon	Glucagen
Recombinant anticoagulants	Tissue plasminogen activator	Tenecteplase (also marketed as Metalyse); TNKase; Ecokinase; Rapilysin; Retavase; Activase
	Hirudin	Refludan; Revasc
Recombinant blood factors	Factor VIII	Helixate NexGen; ReFacto; Kogenate; Bioclate; Recombinate
	Factor VIIa	NovoSeven
	Factor IX	Benefix
Recombinant hematopoietic growth factors	Erythropoietin	Aranesp; Nespo; Neorecormon; Procrit; Epogen
	Granulocyte-macrophage colony stimulating factor	Neulasta (also marketed as Neupopeg); Leukine; Neupogen
Recombinant interferons and interleukins	Interferon-α	Pegasys; PegIntron A; Viraferon; ViraferonPeg; Alfatronol; Intron A; Rebetron; Infergen; Roferon A
	Interferon-β	Rebif; Avonex; Betaferon; Betaseron
	IL-1	Kineret
	IL-2	Proleukin
	IL-11	Neumega
	IFN-γ-1b	Actimmune
Recombinant vaccines	Hepatitis B	Ambrix; Pediarix; HBVAXPRO; Twinrix; Infanrix-Hexa; Infanrix-Penta; Hepacare; Hexavac; Procomax; Primavax; Infanrix Hep B; Comvax; Tritanrix; Recombivax
	Pertussis	Tricelluvax
Monoclonal antibody-based products	Against CD 20 antigen on the surface of B lymphocytes	Mabthera; Ritunax
	Against CD 52	Mabcampath; Campath
	Against CD 33	Mylotarg
	Against human epidermal growth factor receptor 2	Herceptin
	Against TNF-α	Remicade
	Against epitope on the surface of respiratory syncytial virus	Synagis
	Against α-chain of IL-2 receptor	Zenapax; Simulect
	Against cytokeratin tumor-associated antigen	Humaspect
	Against carcinoma-associated antigen	Verluma
	Against tumor-associated antigen CA 125	Indimacis
	Against platelet surface receptor GPIIb/IIIa	ReoPro
	T-lymphocyte surface antigen CD 3	Orthoclone OKT3
Recombinant enzymes	Galactosidase	Fabrazyme; Replagal
	Laronidase	Aldurazyme
	Urate oxidase	Fasturtec
	β-Glucocerebrosidase	Cerezyme
	DNase	Pulmozyme
Other recombinant products	Bone morphogenetic proteins	Inductos; Infuse; Osteogenic protein 1
	Protein C	Xigris
	Tumor necrosis factor	Beromun
	Platelet-derived growth factor	Regranex

Exhibit 20.1 Microbial production of recombinant insulin

Insulin, a hormone responsible for the control of glucose levels in the blood, is synthesized by the β-cells of the Islets of Langerhans in the pancreas. Its deficiency leads to diabetes mellitus, a complex of symptoms, which may lead to death if untreated. Many forms of diabetes can be alleviated by a continuing program of insulin injections, thereby supplementing the limited amount of hormone synthesized by the patient's pancreas. Traditionally, the insulin has been obtained from the pancreas of pigs and cows slaughtered for meat production. Though animal insulin is satisfactory, problems may arise in its use to treat human diabetes. These include side effects in some patients due to slight differences between the animal and the human proteins, difficulty in purification, and contamination with potentially dangerous contaminants. Presently, recombinant microorganisms are used to produce insulin on a commercial scale. Moreover, the structure of insulin facilitates its production by recombinant DNA techniques. As the human insulin does not undergo glycosylation after translation, recombinant insulin synthesized by a bacterium should be active. Note that bacteria lack posttranslational modification abıuty. The second advantage is its small size, which comprises of two polypeptides (A chain of 21 amino acids and B chain of 30 amino acids). In humans, these chains are synthesized as a precursor called preproinsulin, which contains the A and B segments linked by a third chain (C), and preceded by a leader sequence. After translation, the leader sequence is removed, and the C chain is excised leaving the A and B polypeptides linked to each other by two intrachain disulfide bonds (Figure A).

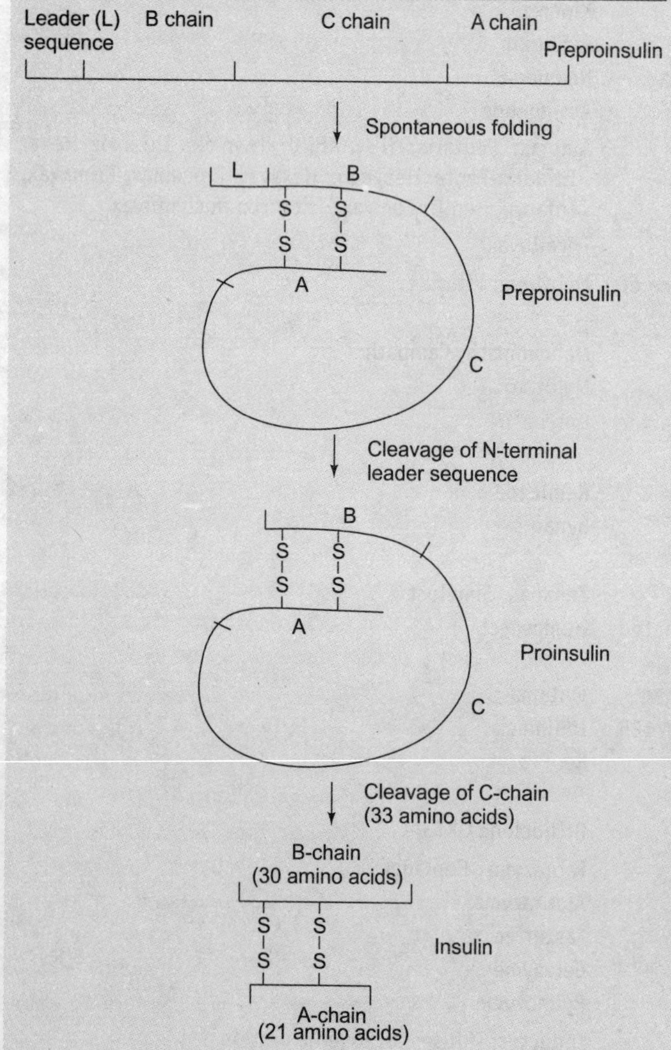

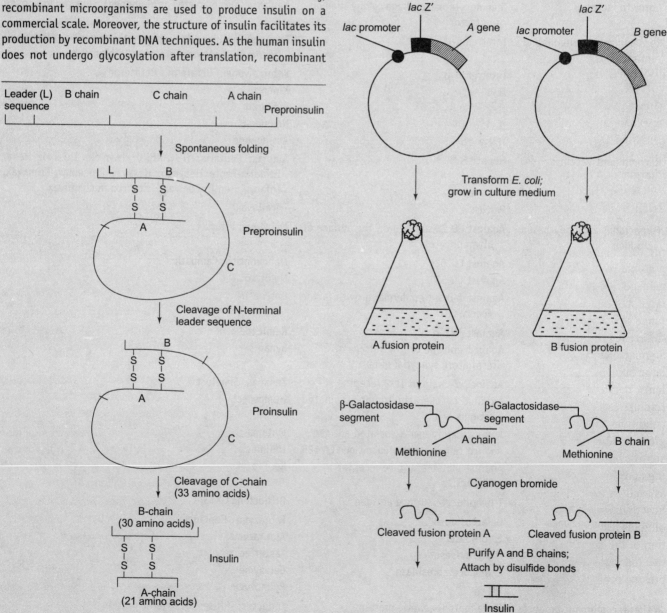

Figure A Synthesis by processing from preproinsulin to form mature insulin

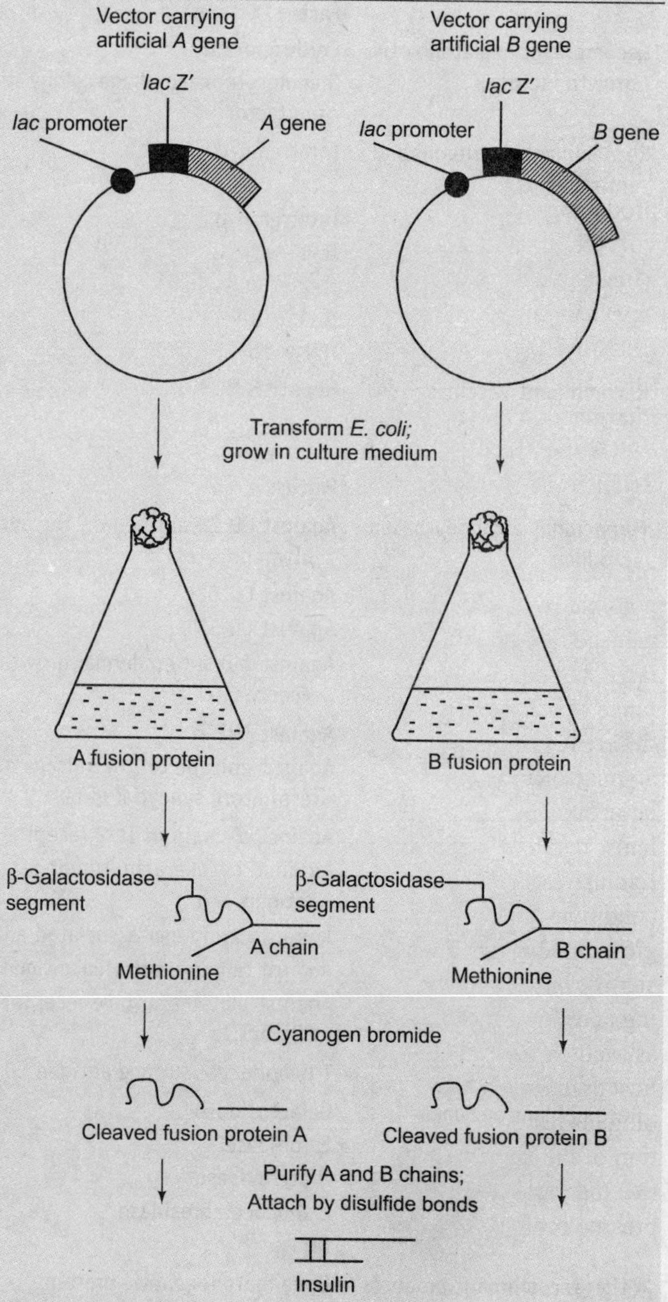

Figure B The synthesis of recombinant insulin from artificial *A* and *B* chain genes

One of the first strategies to obtain recombinant insulin involved artificial synthesis of genes encoding A and B chains followed by production of fusion proteins in *E. coli*. The procedure involved synthesis of trinucleotides representing all the possible codons, and joining them together in the order dictated by the amino acid sequences of the A and B chains. Two recombinant plasmids (pBR322-type vector) were constructed, one carrying the artificial gene for the A chain, and the other contained B chain. In each case, the artificial gene was ligated to a *lac Z′* (that encodes β-galactosidase) reading frame. The insulin genes were therefore under the control of the strong promoter and were expressed as fusion proteins, consisting of the first few amino acids of β-galactosidase followed by the A and B polypeptides. Each gene was designed in such a manner that a methionine residue separated its β-galactosidase and insulin segments, so as to allow the cleavage of insulin polypeptides from the β-galactosidase segments by cyanogen bromide treatment. The purified A and B chains were then attached to each other by disulfide bond formation in the test tube; this final step is actually rather inefficient. The process is shown in Figure B.

An improvement in the above procedure involved synthesis of the entire proinsulin reading frame, specifying B chain–C chain–A chain, rather than the individual A and B chains. The C chain segment could then be excised relatively easily by proteolytic cleavage. Though large in size, the prohormone has the advantage of spontaneous folding into the correct disulfide-bonded structure.

Source: Goeddel DV, Kleid DG, Bolivar F (1979) *Proc. Natl. Acad. Sci., USA*, 76: 106–110; Brown TA (2001) *Gene Cloning and DNA Analysis*, Blackwell Science, pp. 293–312.

Boyer was the first company to use rDNA technology and in 1978, the company announced the creation of an *E. coli* strain producing human insulin. Other companies or institutes involved in marketing of these biopharmaceuticals are Monsanto, Aventis, Baxter Healthcare, Bayer, Biotechnology General, Boehinger Mannheim/Centocor, Centeon, Ciba, Eli Lilly, Europharma, Galenus Mannheim, Genetics Institute, Hoechst Marion Roussel, Novartis, Novo Nordisk, Pharmacia, Serono Laboratories, and Upjohn, etc.

20.2.3 Production of Restriction Enzymes

rDNA technology would not be possible without a ready supply of a range of different restriction endonucleases. This is possible by fermentative production using different microorganisms. These processes, however, require maintenance of a large number of different microorganisms, stock a very wide range of microbial growth medium components, designing of different types of fermentors, and spending an inordinate amount of time developing optimal growth conditions for a large number of different organisms. To overcome these problems, restriction enzyme genes are often cloned into *E. coli*. Exclusive use of *E. coli* allows standardization of production conditions for all restriction enzymes. In addition, *E. coli* cells grow rapidly to high cell densities and can be engineered to significantly overexpress each target restriction enzyme. Cloning and expressing the genes encoding the restriction enyzme as well as its specific (cognate) modification enzyme in the host organism prevents the degradation of host DNA by heterologous restriction enzymes. In addition, to prevent the digestion of the host DNA by the restriction enzyme, it is imperative that, after transformation, the methylation activity be expressed prior to the production of the restriction enzyme.

20.2.4 Production of Antibiotics

The discovery of antibiotics is done through labor-intensive research programs, which involve screening of thousands of different microorganisms. Moreover, with the high costs of development and clinical testing, only the compounds that show significant therapeutic and economic promise are marketed and only about 1–2% of newly discovered antibiotics are added annually to the disease-fighting arsenal. In addition, to date, nearly all of the genetic improvement of industrially important antibiotic-producing strains has been achieved by the use of classical mutagenesis and selection. Though the yield of antibiotics from many strains has been significantly improved. rDNA technology can have a positive impact on this endeavor in two ways. First, this technology can be used to develop new, structurally unique antibiotics with increased activities against selected targets and decreased side effects. Second, genetic manipulation can be used to rapidly and inexpensively enhance yields and hence lower the cost of production of existing antibiotics. Some examples of role of genetic manipulation in antibiotic production are discussed below:

Cloning antibiotic biosynthesis genes There are a number of examples of the cloning and transfer of large fragments of DNA encoding antibiotic biosynthesis genes. For example, bacterial artificial chromosomes (BACs) have been used to transfer of large DNA fragments encoding entire antibiotic biosynthetic pathways into *Streptomyces* host cells.

Production of novel antibiotics With the knowledge about the biochemistry of various antibiotic pathways, it has become possible to design new antibiotics with unique properties and specificities by genetic manipulation of the genes encoding relevant enzymes involved in the biosynthesis of existing antibiotics, for example, introducing the *act* VA gene from *S. coelicolor* into a strain that makes meder-mycin leads to the synthesis of mederrhodin A.

In another example, a novel 7ACA biosynthetic pathway has been constructed in the fungus *A. chrysogenum*, which normally synthesizes only cephalosporin C. This included cloning of cDNA encoding D-amino acid oxidase from *Fusarium solani* and genomic DNA encoding cephalosporin acylase from *Pseudomonas diminuta*.

Improving antibiotic production Genetic engineering has also been used to enhance the yield and rate of production of known antibiotics. Two important examples are discussed in this section.

In case of antibiotic production by aerobic fermentation (e.g. by *Streptomyces*), the manipulation may include development of microbial strains that are better able to utilize the available oxygen. Thus to cope with oxygen-poor environments, the gene for the *Vitreoscilla* hemoglobin was isolated and subcloned onto a *Streptomyces* plasmid vector. When both transformed and untransformed *S. coelicolor* cultures were grown in the presence of a low level of dissolved oxygen (i.e., ~5% saturation), the transformed cells with a functional *Vitreoscilla* hemoglobin produced 10 times more actinorhodine per gram (dry weight) of cells and had greater cell densities as compared with untransformed cells.

Many antibiotic-producing organisms are slow growing, require special growth conditions, or yield only low levels of cells. To overcome these problems, *E. coli* was engineered to produce polyketide antibiotics at rates that were potentially useful for drug production. Thus three genes (each 10–12 kbp in length) encoding the components of the polyketide synthase from the bacterium *S. erythraea* were expressed in *E. coli*. Then a *Bacillus subtilis* gene encoding an enzyme that attaches the cofactor phosphopantetheine to the polyketide synthase was cloned into the engineered *E. coli*. In addition, to supply the polyketide synthase with sufficient building blocks for polyketide synthesis, *viz.*, propionyl-coenzyme A (CoA) and methylmalonyl-CoA, the *E. coli* gene encoding an enzyme that breaks down propionyl-CoA was inactivated and an *S. coelicolor* gene for propionyl-CoA carboxylase was introduced.

20.2.5 New Routes for Production of Small Molecules by Metabolic Engineering

A branch of rDNA technology, called metabolic engineering, allows modification of metabolic pathways of organisms either by introducing new genes or by altering existing ones. The goal is to create an organism with a novel enzymatic activity, which can convert an existing substrate into a commercial compound that with current technology can be produced only by a combination of chemical treatments and fermentation steps. Note that *E. coli*, owing to much better characterized metabolic pathways, procedures for genetic manipulation, and relative ease of manipulation, is an attractive host organism for metabolic engineering. There are many examples of *E. coli* metabolic engineering to develop new routes for synthesis of small molecules, some of which are discussed here.

Synthesis of Indigo

E. coli has been engineered to produce indoxyl through two routes. In one route, a single gene from naphthalene-degrading (NAH) plasmid of *Pseudomonas putida*, encoding naphthalene dioxygenase, was cloned in *E. coli* strain. To develop another route, a single gene from toluene-degrading (TOL) plasmid of *P. putida*, encoding xylene oxidase, was cloned in *E. coli* strain. Another strategy involved modification of *trp* B gene, encoding the subunit of tryptophan synthase, which resulted in the release of indole for conversion by the dioxygenase. Moreover, indoxyl is susceptible to spontaneous oxidation to isatin and indirubin, hence to make textile-quality indigo, indirubin must be absent. This was achieved by cloning the gene for isatin hydrolase, an enzyme that degraded isatin to isatic acid in the indigo-overproducing strains.

Synthesis of L-Ascorbic Acid

The steps involved in conventional synthesis of Vitamin C are: catalytic reduction of *D*-glucose to *D*-sorbitol, sorbitol dehydrogenase (NAD-requiring) catalyzed reduction of *D*-sorbitol to *L*-sorbose, chemical oxidation of *L*-sorbose to 2-keto-*L*-gulonic acid (2-KLG), formation of sodium salt of 2-KLG (enol form), and lastly acid-catalyzed conversion of 2-KLG (sodium salt) to *L*-ascorbic acid. Biochemical studies of the metabolic pathways of a number of different microorganisms have shown that it may be possible to synthesize 2-KLG by a different pathway, for example, some bacteria such as *Acetobacter*, *Gluconobacter*, and *Erwinia* can convert glucose to 2,5-diketo-*D*-gluconic acid (2,5-DKG), and other bacteria such as *Corynebacterium*, *Brevibacterium*, and *Arthrobacter* have the enzyme 2,5-DKG reductase, which converts 2,5-DKG to 2-KLG. In an attempt to engineer pathway for vitamin C synthesis in microbes, *Corynebacterium* gene encoding 2,5-DKG reductase was cloned in *Erwinia*.

Synthesis of Amino Acids

Some progress has been made in increasing the amino acid output of *C. glutamicum*. For example, the synthesis of the essential amino acid tryptophan was enhanced by introducing into wild-type *C. glutamicum* cells a second copy of the gene encoding anthranilate synthetase. An even higher level of tryptophan production was achieved when modified genes for the three key enzymes, 3-deoxy-*D*-arabino-heptulosonate 7-phosphate synthase, anthranilate synthase, and anthranilate phosphoribosyl transferase, were introduced into *C. glutamicum* cells. The genes encoding these enzymes were mutagenized to render them insensitive to feedback inhibition. An alternative to producing amino acids in *Corynebacterium* and *Brevibacterium* spp. is to produce them in *E. coli* by cloning genes coding for enzymes involved in their biosynthesis from other sources.

20.2.6 Production of Biopolymers

The ability to genetically engineer organisms has stimulated researchers to design new biopolymers, replace synthetic polymers with biological equivalents, modify existing biopolymers

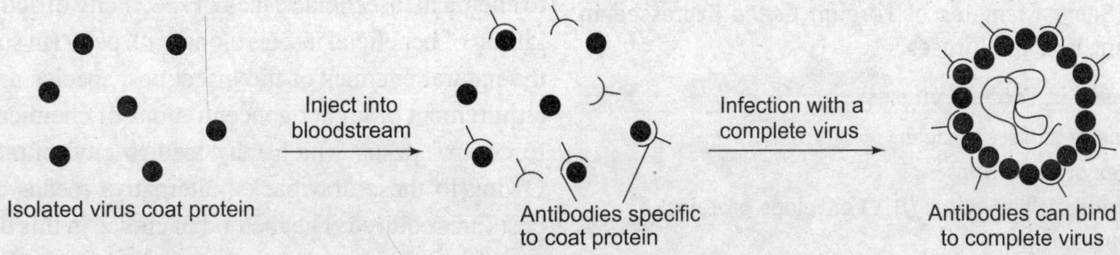

Figure 20.1 The principle of using a preparation of isolated virus coat proteins as vaccine

to enhance their physical and structural characteristics, and find ways to increase the yields and decrease the costs of biopolymers produced by industrial processes. Some common examples are: xanthan gum, melanin, adhesive biopolymer from blue mussel *Mytilus edulis*, and rubber. Other interesting example is that of biodegradable polymers such as poly-(3-hydroxybutyric acid), its copolymer poly-(3-hydroxybutyrate-co-3-hydro-xyvalerate), and poly-(3-hydroxyvaleric acid).

20.2.7 Combating Human Diseases

Genetically modified microorganisms may play a significant role in combating human diseases by production of recombinant vaccines, gene therapy, etc. Some of these are listed in this section.

Production of Recombinant Vaccines

The antigenic material present in a vaccine that elicits an immune response is normally an inactivated form of the infectious agent. For example, antiviral vaccines often consist of virus particles that have been attenuated by heating or a similar treatment. In the past, two problems have hindered the preparation of attenuated viral vaccines: (i) The inactivation process must be 100% efficient, as the presence in a vaccine of just one live particle can result in infection; and (ii) The large amounts of virus particles needed for vaccine production are usually obtained from tissue culture and unfortunately some viruses, notably hepatitis B virus, do not grow in tissue culture. Thus to circumvent these problems, the concept of recombinant vaccines, i.e., producing vaccines as recombinant proteins, came into existence. It is believed that these recombinant vaccines would be free of intact virus particles and could be obtained in large quantities.

The use of gene cloning in the field of vaccine preparation centers on the discovery that virus-specific antibodies are sometimes synthesized in response not only to the whole virus particle, but also to isolated components of the virus. This is particularly true of purified preparations of the proteins present in the virus coat. Thus if the genes encoding antigenic proteins of a particular virus could be identified and inserted into an expression vector, recombinant vaccines could be produced using gene cloning procedures (Figure 20.1). Unfortunately, this approach has not been entirely successful, mainly because recombinant coat proteins often lack the full antigenic properties of the intact virus. The one notable success has been with hepatitis B virus, whose coat protein (the major surface antigen) has been synthesized in *S. cerevisiae*, using a vector based on the 2-μm plasmid. The protein was obtained in reasonably high quantities and when injected into monkeys, it provided protection against hepatitis B. This recombinant vaccine has been approved for use in humans.

Another idea is that of using recombinant *Vaccinia* viruses as live vaccines against small pox and other diseases. Thus, if a gene coding for a virus coat protein, for example, the hepatitis B major surface antigen, is ligated into the *Vaccinia* genome under the control of a *Vaccinia* promoter, then the gene will be expressed. After injection into the bloodstream, replication of the recombinant virus results not only in new *Vaccinia* particles, but also in significant quantities of the major surface antigen (Figure 20.2). Immunity against both smallpox and hepatitis B would result. This remarkable technique has considerable potential and recombinant *Vaccinia*

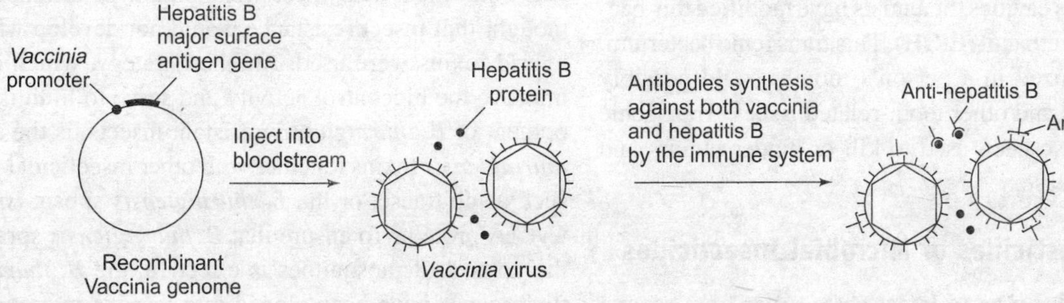

Figure 20.2 The principle of using a recombinant *Vaccinia* virus

Table 20.2 Some Examples of Foreign Genes Expressed in Recombinant *Vaccinia* Viruses

Genes expressed in *Vaccinia* viruses
Hepatitis B major surface antigen
Herpes simplex glycoproteins
Human immunodeficiency virus (HIV) envelope proteins
Influenza virus coat protein
Influenza virus hemagglutinin
Plasmodium falciparum (malaria parasite) surface antigen
Rabies virus G protein
Sindbis virus proteins
Vesicular stomatitis virus coat proteins

viruses expressing a number of foreign genes have been constructed (Table 20.2) and shown to confer immunity against the relevant diseases in experimental animals. The possibility of broad-spectrum vaccines is raised by the demonstration that a single recombinant *Vaccinia*, expressing the genes for influenza virus haemagglutinin, hepatitis B major surface antigen, and herpes simplex virus glycoprotein, confers immunity against respective diseases in monkeys.

Some of the recombinant vaccines under development include recombinant anthrax vaccine, recombinant HIV vaccine, and recombinant cholera vaccine, etc.

Gene Therapy

Genetically modified viruses make possible gene therapy a relatively new area of medicine. A virus reproduces by injecting its own genetic material into an existing cell. The targeted cell then follows the instructions of introduced genetic material and produces more viruses. In medicine, this process is adapted to deliver a gene that could cure disease into human cells. Although gene therapy is still relatively new, it has some successes. It has been used to treat genetic disorders such as severe combined immunodeficiency and a range of other incurable diseases, such as cystic fibrosis, sickle cell anemia, and muscular dystrophy.

Other Strategies

Streptococcus mutans, a tooth decay causing bacterium, consumes sugars in mouth and produces acid that eats away tooth enamel and causes cavities. Scientists have modified this bacterium to produce ethanol (EtOH). This transgenic bacterium, if properly colonized in a person's mouth, could possibly eliminate cavities and other tooth-related issues. Transgenic microbes have also been used to kill or hinder tumors and fight Crohn's disease.

20.2.8 Biopesticides or Microbial Insecticides

Chemical pesticides used to target pests suffer from several drawbacks, for example, development of increasing resistance to chemical insecticides, lack of specificity of action, and hence killing of beneficial insects along with pests (in some instances, the natural enemies of the insect pest species are killed), and requirement of higher concentrations of chemical insecticides to control pests, which may lead to environment pollution. Owing to these drawbacks, alternative means of controlling pests are required. The method of choice in this direction is the use of insecticides produced naturally by microorganisms, for example, the application of insecticidal activities of the bacterium *Bacillus thuringiensis* and insect baculovirus.

Bacillus thuringiensis-based Biopesticides

The insecticidal activity of *B. thuringiensis* is due to *cry* gene-encoded insecticidal protein (Bt toxin or Cry protein). Thus based on the action of Bt toxin, microbial insecticides, called biopesticides, have been developed as environmentally friendly biological substitutes for chemical pesticides. These biopesticides are highly specific for a limited number of insect species, nontoxic to nontarget species, and biodegradable. Consequently, these are unlikely to cause significant biological selection for resistant forms under normal conditions. These attributes make these biological insecticides good candidates for controlling insect damage to certain crops and prevent the proliferation of insects. Details of insecticidal action of *B. thuringiensis* are given in Exhibit 20.2. The *cry* genes for various *B. thuringiensis* toxins have been cloned and characterized. By expressing a *B. thuringiensis cry* gene in a nonsporulating *Bacillus* strain, the production of the insecticidal protein has been achieved during vegetative growth, bypassing the need for parasporal crystal formation. To expand the specificity of a *B. thuringiensis* toxin to other pests, toxin genes from different subspecies were cloned into plasmids and introduced into a host strain, either on a broad host range plasmid or by integration into the chromosomal DNA of the host cell. In addition to dual toxicity, the bacterium with two different toxin genes sometimes showed an effect against a nontarget insect pest. In one instance, a fusion protein consisting of two domains from different *B. thuringiensis* toxin genes was constructed by genetic manipulation and the fusion protein retained both the toxic activities. In another study, the receptor-binding domain of one insecticidal toxin was combined with the toxin domain of another. It was thought that insect resistance would not develop when such hybrid toxins were used. Another strategy, which can both improve the biocontrol activity and serve to limit the development of *B. thuringiensis*-resistant insects, is the use of *B. thuringiensis* toxins together with other insecticidal proteins, such as chitinase, or the *B. thuringiensis* subsp. *israelensis* Cyt 1A protein. To ensure that *B. thuringiensis* spraying for the control of mosquitoes is effective, the *B. thuringiensis* toxin genes have been cloned into various microorganisms that live near the surface of ponds and are eaten by mosquito

Exhibit 20.2 Insecticidal action of *B. thuringiensis*

Bacillus thuringiensis, a Gram positive bacterium (Kingdom: Eubacteria, Phylum: Firmicutes, Class: Bacilli, Order: Bacillales, Family: Bacillaceae, Genus: *Bacillus*, Species: *thuringiensis*), is pathogenic for a number of insect pests. This insecticidal activity is attributed to the production of a crystalline protoxin on sporulation, which is converted to a toxin in an alkaline environment. The common names of this crystalline protoxin are δ-endotoxin, Bt toxin, Cry protein, or crystal protein (Figure A).

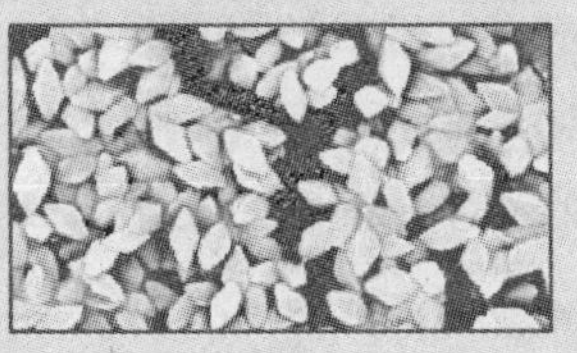

Figure A Spores and bipyramidal crystals of *B. thuringiensis* morrisoni strain T08025 (magnified view)

The toxin is encoded by *cry* gene and is active against lepidopteran (including most significant corn- and cotton-damaging insects), dipteran, and coleopteran insects. The extent of mortality (ppm) of target insects is reported to the level of 100%.

A range of insecticidal crystal proteins has been isolated and characterized from *B. thuringiensis* strains (Table A). Previously these were grouped into four major classes: Cry I, Cry II, Cry III, and Cry IV based on the insecticidal activity of the toxin. The Cry I proteins are toxic to Lepidoptera, Cry II proteins are toxic to both Lepidoptera and diptera, Cry III proteins are toxic to coleopteran, and Cry IV proteins are toxic to diptera. These proteins are further organized into subclasses (A, B, C, etc.) and subgroups (a, b, c, etc.) according to the DNA sequence of the toxin gene. In the past few years, as increasing numbers of *B. thuringiensis* strains were isolated and their genes were characterized, it became clear that the original classification was unable to accommodate many of the newly discovered *B. thuringiensis*. In the new classification scheme, Cry proteins are assigned designations based on their degree of evolutionary divergence, as estimated by certain mathematical algorithms. This scheme is readily visualized by constructing a phylogenetic tree based on the amino acid sequences of Cry proteins. Basically, amino acid sequences of the proteins are compared and if the proteins are identical, then these are 100% homologous. If only 50% of the amino acids are the same, then the proteins have 50% identity. The relationship among a set of protein sequences can be deduced and represented as a branched tree. The nodes (branch points) of the tree represent points of divergence. For the classification of the Cry proteins, a four-part naming system was devised. Demarcations, set at 95, 78, and 45% homology, show the boundaries that define the different nomenclature ranks. The name that is given to a particular toxin depends on the location of the node where the toxin enters the tree relative to the set boundaries. A toxin that joins the tree to the leftmost boundary is assigned a new primary rank (an Arabic number), one that joins the tree between the central and left boundaries is assigned a new secondary rank (an uppercase letter), one that joins the tree between the central and the right boundaries is assigned a new tertiary rank (a lowercase letter), and one that joins to the right of the rightmost boundary is assigned a new quaternary rank (an Arabic number). For example, Cry proteins that are less than 45% homologous are given a number (e.g., Cry 1 and Cry 2), and assigned to the primary rank. Cry proteins, which are 45–78% identical to proteins of the primary rank, are further designated with an uppercase letter (e.g., Cry 1A and Cry 1F). The complete Cry protein tree consists of the positions of all Cry proteins. This new classification system will be utilized throughout this chapter, even when referring to the work that was published prior to the development of this system.

The ingestion of protoxin produced by a number of subspecies of *B. thuringiensis* as a part of a parasporal crystal leads to killing of specific insects. The parasporal crystal does not usually contain the active form of the insecticide. Rather, once the crystal has been solubilized, the protein that is released is generally a protoxin, a precursor of the active toxin. The protoxins of many of the Cry toxins that are directed against lepidoptera have a molecular mass of ~130 kDa. The proposed mode of actions of all the Bt proteins

Table A Insecticidal Crystal Proteins from Various Strains of *Bacillus thuringiensis*

***Bacillus* subspecies and strains**	**Crystal protein**
aizawai	Cry 1Aa, Cry 1Ab, Cry 1Ad, Cry 1Ca, Cry 1Da, Cry 1Eb, Cry 1Fa, Cry 9Ea, Cry 40Aa
entomocidus	Cry 1Aa, Cry 1Ba, Cry 1Ca, Cry 1Ib
galleriae	Cry 1Ab, Cry 1Ac, Cry 1Da, Cry 1Cb, Cry 7Aa, Cry 8Da, Cry 9Aa, Cry 9Ba
israelensis	Cry 10Aa, Cry 11Aa
japonensis	Cry 8Ca, Cry 9Da
jegathesan	Cry 11Ba, Cry 19Aa, Cry 24Aa, Cry 25 Aa
kenyae	Cry 2Aa, Cry 1Ea, Cry 1Ac
kumamotoensis	Cry 7Ab, Cry 8Aa, Cry 8Ba
kurstaki HD-1	Cry 1Aa, Cry 1Ab, Cry 1Ac, Cry 11a, Cry 2Aa, Cry 2Ab
kurstaki HD-73	Cry 1Ac
kurstaki NRD-12	Cry 1Aa, Cry 1Ab, Cry 1Ac
morrisoni	Cry 1Bc, Cry 1Fb, Cry 1Hb, Cry 1Ka, Cry 3Aa
tenebrionis	Cry 3Aa
tolworthi	Cry 3Ba, Cry 9Ca
wuhanensis	Cry 1Bd, Cry 1Ga, Cry 1Gb

are same and involves following steps: (i) Ingestion of spore containing protoxin by insects, dissolution, and solubilization of protoxin; (ii) Proteolytic processing of crystal protein to smaller 'active' form (molecular mass ~68 kDa) by gut hydrolytic enzymes (digestive proteases) at alkaline pH (7.5–8.0); (iii) Interaction of active toxin with cell receptor in the midgut epithelium; (iv) Conformational change exposing α-4–5 helical hairpin; (v) Oligomerization and insertion of toxin in membrane, making it permeable to ions (i.e., formation of ion channels in the gut cells) and small molecules so that the cell bursts forming a pore; and (vi) Escape of ATP molecules and hence decrease in cellular metabolism after about 15 min of ion channel formation, cessation of feeding on plants by the insect, dehydration, and eventual death of insect. The proposed mode of action of Bt toxin is schematically represented in Figure B.

Because the conversion of the protoxin to the active toxin requires both alkaline pH and presence of specific proteases, it is unlikely that nontarget species such as humans and farm animals will be affected. Moreover, the lack of persistence in the natural environment means that selection of resistant insects is highly unlikely. Furthermore, Bt toxin is a safe means of protecting plants as it does not persist in the environment and is nonhazardous to mammals. Due to its insecticidal nature, biodegradability, and nontoxicity to mammals, Bt toxin has been developed as a biopesticide (biocontrol agent). *B. thuringiensis* subsp. kurstaki is applied by spraying ~1.3×10^8 to 2.6×10^8 spores/ft^2 of the target area. Administration of the spores is timed to coincide with the peak of the larval population of the target organism because the parasporal crystals, being sensitive to sunlight, are short-lived in the environment. In simulated conditions, sunlight degrades over 60% of the tryptophan residues of the parasporal crystal within 24 hours. Depending on the amount of sunlight present, parasporal crystals may persist in the environment for as little as a day or as long as 1 month. The negative side of Bt toxins that has limited their use for a variety of applications includes their low potency and high cost of production.

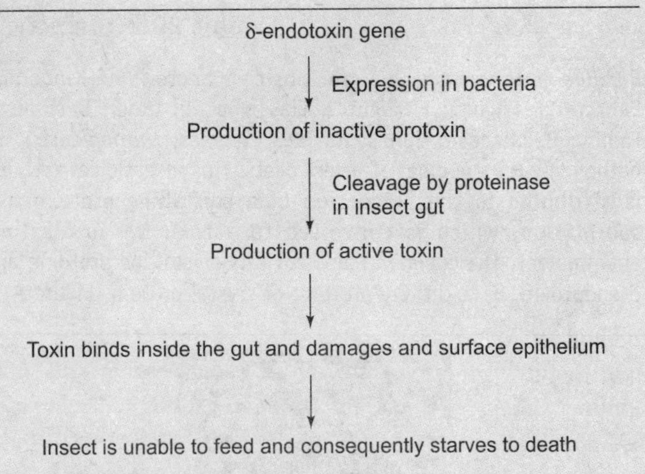

Figure B Mode of action of Bt toxin

Source: De Maagd RA, Bravo A, Crickmore N (2001) *Trends Genet.* 17: 193–199; Schnepf HE, Wtiteley HR (1998) *Proc. Natl. Acad. Sci., USA* 78: 2893–2897.

larvae. This strategy appears to be an effective means for delivering the *B. thuringiensis* toxin to the targeted insect. Also rhizosphere bacteria that have been engineered with *B. thuringiensis* toxin genes lessen the damage caused by insects attacking the roots of plants.

Baculovirus-based Biopesticides

Another example of biopesticide involves the use of baculovirus, a rod-shaped, double stranded (ds) DNA virus that infects and kills a wide range of invertebrates. Subgroups of this family of viruses are pathogenic to several orders of insects including the Lepidoptera, Hymenoptera, Diptera, Neuroptera, Trichoptera, Coleoptera, and Homoptera. In nature, some of these baculoviruses are important for the control of certain pest insects and several have been registered for use as biological insecticides. A positive feature of using baculoviruses as biocontrol agents is their limited host range and no effect on nontarget organisms. But these baculoviruses are relatively slow in killing target insects (few days to several weeks depending upon the conditions). To remedy this effectiveness, several attempts have been made to enhance the virulence of baculoviruses by introducing foreign genes that either severely impair or kill the targeted insect species (Table 20.3).

Table 20.3 Some Examples of Genes Expressed in Baculoviruses to Increase Insecticidal Activity

Gene	Effect on host insect
Diuretic hormone	Reduced hemolymph volume
Juvenile hormone esterase	Feeding cessation
Bacillus thuringiensis toxin	Feeding cessation
Scorpion toxin	Paralysis
Mite toxin	Paralysis
Wasp toxin	Premature melanization; Low weight gain

Cloning of Juvenile Hormone Esterase During insect development, a reduction in the level of juvenile hormone in larvae initiates metamorphosis into pupae and leads to cessation of larval feeding. The reduction in the juvenile hormone level is due to an increase in the amount of juvenile hormone esterase, an enzyme that converts the biologically active methyl ester form of juvenile hormone to an inactive acid form. Inhibition of juvenile hormone esterase activity leads to an *in vivo* accumulation of active juvenile hormone, so the larvae remain in the feeding stage longer, continue to grow, and eventually become giant larvae. Therefore, researchers reasoned that an experimentally induced increase in the supply of juvenile hormone esterase should lower the endogenous level of active juvenile hormone and cause a premature cessation of feeding. Thus, cDNA for juvenile hormone esterase from tobacco budworm (*Heliothis virescens*) has been inserted into the genome of a baculovirus under the control of baculovirus transcription signals, and this genetically modified baculovirus was used to treat cabbage looper (*Trichoplusia ni*) at the first instar stage. The result of cloned juvenile hormone esterase was a reduction in the amount of juvenile hormone and a consequent reduction in the larval feeding and growth rates. This approach was however suitable for the reduction of first instar larvae and had to be applied when majority of the target insect population was in its first larval instar stage, a strategy that is difficult to achieve under natural conditions.

Cloning of Genes Encoding Toxins Another approach was to incorporate an insect-specific toxin gene into the viral genome, which when expressed during the larval lifecycle, would yield a potent insect neurotoxin. Thus, a gene encoding insect-specific neurotoxin produced by the North African fat-tailed scorpion (*Androctonus australis* Hector) was cloned into a baculovirus strain and the genetically engineered virus was found to be an effective biopesticide. Note that this neurotoxin disrupted the flow of sodium ions in neurons of targeted insects and eventually led to paralysis and death. In another example, a cDNA for the toxin from Israeli yellow scorpion (*Leiurus quinquestriatus hebraeus*) was cloned and expressed in the baculovirus *Autographa californica* nuclear polyhedrosis virus, which hastened the demise of the infected insect larvae and also significantly decreased the ability of the insects to damage plants. Mite toxins are also potent neurotoxins that have been cloned in baculoviruses to increase insecticidal activity. Wasp and Bt toxins have also been cloned in baculoviruses to increase insecticidal activity. Bt toxin leads to cessation of feeding (for details see Section 20.2.8), while wasp toxin causes premature melanization and low weight gain.

20.2.9 Destaining of Fatty Stains

The removal of lipid stains is a persistent problem for the laundry industry. Though a combination of high temperature and high alkalinity can effectively emulsify and remove many fatty stains, these conditions often damage fabrics and also require large amounts of energy. An effective solution to this problem can be to add lipases produced by *Pseudomonas alcaligenes* that are compatible with the wash conditions. Unfortunately, this enzyme is produced at such low levels that it is prohibitively expensive to use in laundry. Moreover, it is extremely difficult to overproduce this enzyme in a variety of heterologous hosts, including *Bacillus licheniformis*, *E. coli*, *Streptomyces lividans*, *A. niger*, and *K. lactis*. The difficulty in overexpressing the *P. alcaligenes* lipase may reflect the requirement for the simultaneous expression of another gene product that is involved in either the secretion or stabilization of the bacterial lipase. Thus the DNA fragment encoding the lipase gene (*lip* A) and the second (helper) gene *lip* B were cloned into a broad host range expression vector and used to transform *P. alcaligenes*. When the vector was derived from a low copy number plasmid, the lipase activity of the transformants was four- to five-fold greater than the nontransformants, regardless of the presence or absence of the *lip* B. However, with a high copy number plasmid, the lipase activity of the transformants was about 20-fold greater than the wild-type in the absence of the *lip* B gene and ~35-fold greater than the wild-type in its presence. Since the lipase was secreted to the growth medium, the purification step simply involved removal of cells and concentration of the growth medium.

20.2.10 Efficient Utilization of Carbohydrates

rDNA technology is also employed in efficient utilization of starch, glucose, and cellulose, etc. Some common examples are as follows.

Improved Production of Alcohol or Fructose

The use of milled grain for the production of alcohol or fructose requires a number of enzymatic steps. The enzymes that are used in these processes are often used only once and then discarded. To enhance enzymatic conversions and decrease costs, bacterial genes encoding enzymes that are thermostable, highly efficient catalytically, or tolerant to alcohol have been cloned, characterized, and tested. Thus, innovative approaches to the inexpensive large-scale production of these enzymes could lower the cost of alcohol or fructose production. These include: (i) Overproduction of each of the enzymes in a fast-growing recombinant microorganism that utilizes an inexpensive substrate, thereby lowering the cost of production from native organisms; (ii) Utilization of variants of α-amylase, either naturally occurring or genetically manipulated that function efficiently at 80–90°C and allow the liquefaction step to be performed at this temperature (note that heat-resistant α-amylase speeds up the hydrolysis of gelatinized starch and decreases the amount of energy required to cool the gelatinized starch to a temperature suitable for starch hydrolysis); (iii) Alteration of α-amylase and glucoamylase genes so that each enzyme has the same temperature and pH optimum, thereby enabling the liquefaction and saccharification steps to be performed under the same conditions; (iv) Isolation and engineering an enzyme that could efficiently degrade raw starch, obviating the need for the gelatinization step and thereby saving a large amount of energy; and (v) Developing a fermentation organism that can synthesize and secrete glucoamylase, eliminating the need to add this enzyme during fermentation. Some of the achievements in the direction of efficient production of alcohol or fructose by genetic manipulation include expression of α-amylase or glucoamylase.

Glycerol Overproduction

To produce wine with a lower concentration of alcohol and a slightly increased level of sweetness, *S. cerevisiae* yeast strains were engineered to overexpress *GPD* 1 gene encoding glycerol-3-phosphate dehydrogenase. This resulted in overproduction of glycerol-3-phosphate dehydrogenase in different commercial wine yeast strains and led to an increase in the glycerol concentration of the wine from 7–9 to 12–18 g/l. Moreover, the amount of acetate that the engineered strains produced remained quite low.

Improving Conversion of Glucose to Fructose

The isomerization of glucose to fructose is a reversible reaction catalyzed by glucose isomerase (or xylose/glucose isomerase). The final fructose content is dependent on the reaction temperature; at higher temperatures, the fructose content in the final product is high. Most commercial processes use conversion temperatures of around 60°C. Consequently, increasing the temperature optimum for the enzymatic activity and thermostability of xylose/glucose isomerase is one strategy to make the process more efficient. The xylose/glucose isomerase from the thermophilic bacterium *Thermus thermophilus* is a good candidate for use in industrial processes. Note that this enzyme is not only active at 95°C, but is also very stable at high temperature. Unfortunately, wild-type *T. thermophilus* does not produce large amounts of this enzyme. To circumvent this problem, the *T. thermophilus* xylose/glucose isomerase gene was isolated and expressed in *E. coli* and *Bacillus brevis* under the control of various promoters and ribosome-binding sites. One of the constructs overproduced xylose/glucose isomerase more than 1,000-fold relative to the amount found in the original organism.

Efficient Utilization of Cellulose

For efficient utilization of cellulose, several genetic manipulation strategies have been employed. These include expression of genes encoding cellulase, endoglucanase, exoglucanase, or β-glucosidase.

20.2.11 Plant Growth Promotion

Under natural environmental conditions, successful plant development and high crop yields depend on the genetic constitution of the crop species, the availability of nutrients, the presence of certain beneficial microorganisms, and the absence of pathogenic ones in the surrounding soil. Some beneficial indigenous soil bacteria and fungi act directly by providing a plant growth-enhancing product and others act indirectly. The latter organisms inhibit the growth of pathogenic soil microorganisms, thereby preventing them from hindering plant growth.

In order to create microbial strains with augmented plant growth-promoting activities, the strategies adopted include: (i) Increasing the level of microbial nitrogen fixation and consequently lessen the dependency on chemical fertilizers for crop plants; (ii) Genetically engineering root nodule formation capability in bacteria to outcompete naturally occurring symbiotic bacteria; (iii) Microbial synthesis of iron-sequestering compounds (siderophores) to produce beneficial strains that prevent the growth of phytopathogenic microorganisms; (iv) Cloning enzymes encoding phytohormones to produce and release specified levels of selected phytohormones that in turn stimulate plant proliferation; (iv) Engineering better biocontrol strains of bacteria to decrease the damage to plants from a variety of pathogens, and hence to replace some of the environment-polluting chemical pesticides; and (vi) Using bacteria to prevent high levels of ethylene from accumulating in plants, and thereby decreasing the damage to the plant from a variety of environmental stresses including drought, flooding, salt stress, and the presence of pathogens. In this section, various plant growth promotion strategies using recombinant microbes are discussed.

Nitrogen Fixation

Nitrogen fixation requires the concerted action of a large number of different proteins. As a consequence, it is difficult to clone an intact single DNA fragment containing all the genetic information for nitrogen fixation from a diazotrophic microorganism into a nondiazotrophic organism. It is also difficult for a recipient organism to maintain the physiological conditions needed for nitrogenase activity. The strategies adopted for efficient nitrogen fixation are as follows: (i) Expression of *nif* genes; (ii) Modulation of oxygen levels within the bacterial cells; (iii) Expression of hydrogenase; and (iv) Increasing the efficiency of nodulation by expressing *nod* genes.

Expression of *nif* Genes The first *nif* genes were isolated from diazotroph *Klebsiella pneumoniae*. These genes are arranged in a single cluster that occupies ~24 kbp of the bacterial genome. The cluster contains seven separate operons that together encode 20 distinct proteins and all of the *nif* genes must be transcribed and translated in a concerted fashion under the regulatory control of the *nif* A and *nif* L genes to produce a functional nitrogenase. Most diazotrophic organisms have a similar array of genes encoding their nitrogen-fixing apparatus. Thus it seems possible to increase the amount of nitrogen fixed by diazotrophic organisms by manipulating *nif* A and *nif* L genes. As nitrogen fixation by microorganisms is a very complicated process, it is evident that simply by adding one or two *nif* genes, the ability to fix nitrogen cannot be conferred on a nondiazotrophic recipient cell. Moreover, genetic modification of plants with the entire *nif* gene cluster would not be effective because the normal level of oxygen in the host cell would inactivate nitrogenase.

And, if this level were reduced, the host plant cell would probably die. In addition, it is difficult to regulate nitrogen fixation since there are no plant promoters that respond to Nif A protein. As a consequence, *nif* genes would remain switched off in such a transgenic plant. Furthermore, all of the *nif* genes have to be under the control of separate promoters to be able to respond to the level of fixed nitrogen in the cell because plant cells are unable to process multigene transcripts. Thus, the introduction of a functional nitrogen fixation capability into plants is at present extremely unlikely.

Modulation of Oxygen Levels within the Bacterial Cells Based on the observation of a dramatic increase in nitrogenase activity upon addition of exogenous leghemoglobin, which binds free oxygen tightly to isolated bacteroids, more efficient strains of *Rhizobium* have been engineered for overproducing leghemoglobin. Alternatively, since the plant produces the globin portion of leghemoglobin, it may be more efficient to transform rhizobial strains with genes encoding a bacterial equivalent of leghemoglobin. Following the transformation of a strain of *R. etli* with a broad host range plasmid carrying the *Vitreoscilla* hemoglobin gene, at low levels of dissolved oxygen (0.25–1.0%) in the growth medium, the rhizobial cells had a two- to three-fold higher respiratory rate than the untransformed strain.

Expression of Hydrogenase The side reaction of nitrogen fixation, i.e., reduction of H^+ to H_2 (hydrogen gas) by nitrogenase is highly undesirable because the energy in the form of ATP is wasted on the production of hydrogen, which is eventually lost to the atmosphere. As a result of this side reaction, only 40–60% of the electron flux through the nitrogenase system is transferred to N_2 and hence there is significant lowering of the overall efficiency of the nitrogen fixing process. By introducing a hydrogenase gene into *R. legumino-sarum* (a bacterium that forms symbiotic relationship with crop plants), i.e., conversion of a Hup^- strain of *Rhizobium* to a Hup^+ strain and using this genetically modified bacterium (Hup^+) to transform plants, it has been shown that the transformed plants grew larger and contained more nitrogen than the plants inoculated by the Hup^- parental strain.

Increasing the Efficiency of Nodulation It has been reasoned that enhancing nodulation by genetic engineering will enable inoculated rhizobial strains to be more effective competitors for sites on the roots of target plants than indigenous strains. Detailed biochemical and genetic studies have revealed that nodulation and its regulation require the functioning of a large number of genes (*nod* genes), which are grouped into three separate classes, *viz.*, common genes, host-specific genes, and the regulatory *nod* D gene. The *nod* ABC genes are common to all *Rhizobium* species and are structurally interchangeable, and in most species, the *nod* ABC genes are located on a single operon. The first event in nodulation is the recognition of a flavonoid molecule, which is excreted by roots of host plants, by the product of constitutively expressed *nod* D gene. This activates Nod D and enables the flavonoid–Nod D complex to attach to a nodulation promoter element, *nod* box, which turns on the transcription of all the other *nod* genes. It is speculated that the transfer of a *nod* D gene from a broad specificity rhizobial strain to one with a narrow specificity can alter the host specificity. However, as the process of nodulation is quite complicated, at present, there is no simple way to manipulate this process genetically.

Biocontrol of Pathogens

Currently the problem of phytopathogen is generally dealt with the use of chemical agents. Many of the chemicals used to control phytopathogens are hazardous to animals and humans and persist and accumulate in natural ecosystems. It is therefore desirable to replace these chemical agents with biological control agents that are more ecofriendly. One approach for the control of phytopathogens is to use plant growth-promoting bacteria as biocontrol agents. Plant growth-promoting bacteria can produce a variety of substances that limit the damage to plants by phytopathogens, which include siderophores, antibiotics, other small molecules, and a variety of enzymes. A brief description of each is given below.

Production of Antibiotics One of the most effective mechanisms by which a plant growth-promoting bacterium can prevent phytopathogen proliferation is the synthesis of antibiotics. For example, some of the antibiotics synthesized by biocontrol pseudomonads include agrocin 84, agrocin 434, 2,4-diacetyphloroglucinol, herbicolin, oomycin, phenazines, pyoluteorin, and pyrrolnitrin. Presently there is only one commercially available genetically engineered biocontrol bacterial strain, which is a modified version of *Agrobacterium radiobacter* K84 that has been marketed in Australia since 1989 as a means of controlling crown gall disease caused by the bacterium *Agrobacterium tumefaciens*.

Production of Fungus-degrading Enzymes The genes encoding chitinase, β-1,3-glucanase, protease, and lipase, if transferred to plant growth-promoting bacteria, produce both antibiotics and fungus-degrading enzymes that may cause lysis of fungal cells. In one series of experiments, a chitinase gene was isolated from the bacterium *Serratia marcescens* and transferred to *Trichoderma harzianum* and *R. meliloti* cells. In both cases, the transformed microorganisms produced chitinase and displayed increased antifungal activity. When the *S. marcescens* chitinase gene was introduced into a strain of *Pseudomonas fluorescens*, the transformant stably expressed and secreted active chitinase, and effectively controlled the phytopathogen *R. solani*.

Synthesis of Siderophores In metal contaminated soils, plants are generally unable to obtain enough iron because iron uptake is inhibited by the metal contaminant(s). Plant siderophores bind to iron with a much lower affinity than bacterial siderophores, so plants are often unable to accumulate a sufficient amount of iron unless bacterial siderophores are present. One bacterial siderophore, called pseudobactin, has been estimated to bind to Fe (III). All fluorescent pseudomonads produce structurally related siderophores. For accumulation of sufficient amount of iron, one strategy can be to genetically engineer bacteria to produce modified siderophores. Further, the cloning of genes encoding iron-siderophore receptors from one plant growth-promoting bacterium to other may be done to extend the range of iron-siderophor complexes, so that a genetically altered plant growth-promoting strain may take up and use siderophores synthesized by other soil microorganisms.

Cloning gene Encoding Antifreeze Protein Antifreeze proteins (AFPs) have been identified in many organisms that have evolved resistance to freezing temperatures, including bacteria, fungi, fishes, insects, and plants. AFPs technically called thermal hysteresis proteins lower the freezing point of a liquid below the melting point, inhibit ice recrystallization, and prevent freezing damage to cellular structures by slowing or stopping the formation of ice crystals inside a cell. Once the ice crystals melt, ice recrystallization (growth of larger ice crystals at the expense of smaller ones) is inhibited by the adsorption of AFPs to the crystal surface. Genes encoding AFPs have been isolated from certain marine teleosts (fishes), spruce budworm (*Choristoneura fumiferana*), common mealworm beetle (*Tenebrio molitor*), and overwintering or frost-hardy plants. A genetically engineered bacterium, called Frostban has been developed to protect strawberries and other fruits from frost damage. This bacterium lacked the genes coding for the ability to form ice crystals on the leaves of crop plants, and was field-tested in 1987–88. Certain pathogenic leaf bacteria, for example, *Pseudomonas syringae* damage plants by synthesizing ice nucleation proteins. Note that ice nucleation proteins are produced at low temperatures, are present on the bacterial surface, and act as sites that facilitate the formation of ice crystals at freezing temperatures. These ice crystals grow, and may pierce the plant cells leading to irreparable damage. The bacteria benefit from this damage by gaining direct access to the nutrients from the lysed plant cells. In the absence of ice nucleation proteins on the leaf surface, a brief overnight frost would not damage the plant because the water in a plant cell must usually be several degrees below the freezing point before ice crystals could be formed, i.e., it must be supercooled. In this direction, a bacterial AFP that regulates the formation of ice crystals outside the bacterium may be helpful in providing protection for the bacterium in the soil. This is because in the presence of bacterial AFP, ice crystals still form, but their size is limited. While in the absence of AFPs, ice crystals can grow to a large size, and eventually puncture the bacterial cell wall and membrane, causing cell lysis. Ice crystals do not form inside the bacterium to any great extent. This is because at low temperatures bacteria decrease their volume by pumping some of their water from inside to outside of the cell. Thus the isolation and cloning of genes encoding bacterial AFPs in certain strains of plant growth-promoting bacteria may allow them to persist and proliferate at cold temperatures.

Root Colonization A biocontrol bacterium should be able to bind tightly to the plant root to reduce the damage caused by pathogenic microorganisms to plants. This process, called root colonization, can be improved by overexpressing the bacterial *sss* gene, a gene that probably plays a role in DNA rearrangements that regulate transcription of a gene(s) involved in the biosynthesis of cell surface components. The introduction of *sss* gene into plasmids of poor and good root colonizing strains of *P. fluorescens* has been reported to result in an increase in colonization of tomato roots by ~28- and 12-fold, respectively.

Reduce Inhibitory Levels of Ethylene Fungal pathogens not only directly inhibit plant growth, but these also cause the plant to synthesize stress ethylene, which causes much of the damage sustained by infected plants. Thus to overcome the problem of fungal phytopathogens, several biocontrol bacterial strains have been transformed to regulate ethylene synthesis. In this direction, the *Enterobacter cloacae* UW 4 ACC deaminase (1-amino cyclopropane-1-carboxylic acid deaminase) gene has been cloned in several biocontrol bacterial strains and the effect was assessed on the damage to cucumbers caused by *Pythium ultimum*. ACC deaminase containing biocontrol bacterial strains were found to be significantly more effective in lessening the damage as compared to the wild-type strain.

20.2.12 Bioremediation or Environment Clean-up

New and improved methods have currently been developed to deal with both inorganic and organic environmental contaminants. In this direction, certain microorganisms with the ability to thrive on hazardous or toxic materials such as hydrocarbons (e.g., oil), methylene chloride, detergents, creosote, pentachlorophenol, sulfur, and polychlorinated biphenyl (PCBs) to harmless products have been isolated and characterized. The use of living organisms to degrade and filter wastes before it is introduced into the environment is termed bioremediation. Bioremediation programs involving the use of microorganisms are currently in progress to clean-up contaminated air, tracks of land, lakes, and waterways. There are two strategies for bioremediation: (i) Manipulation by transfer of plasmids; and (ii) Manipulation by gene alteration.

Despite the ability of many naturally occurring microorganisms to degrade a number of different xenobiotic chemicals, there are limitations to the biological treatment of these waste materials. For example, (i) No single microorganism can degrade all organic wastes; (ii) High concentrations of some organic compounds can inhibit the activity or growth of degradative microorganisms; (iii) Most contaminated sites contain mixtures of chemicals and an organism that can degrade one or more of the components of the mixture may be inhibited by other components; (iv) Many nonpolar compounds adsorb onto particulate matter in soil or sediment and become less available to degradative microorganisms; and (v) Microbial biodegradation of organic compounds is often quite slow. One way to address some of these problems is to transfer plasmids that carry genes for different degradative pathways by conjugation into recipient strain. If two resident plasmids contain homologous regions of DNA, recombination can occur and a single, larger 'fusion' plasmid with combined functions can be created. Alternatively, if two plasmids do not contain homologous regions and belong to different incompatibility groups, these can coexist within a single bacterium. However, bringing together different intact plasmid-based degradative pathways by conjugation is not the only way to create bacteria with novel properties. It is also possible to extend the degradative capability of a strain by altering the genes of the degradative pathway. In this direction, several researchers have determined enzymes naturally occurring in bacteria that are effective in degrading toxic substances and have developed transgenic microorganisms by cloning the genes encoding these enzymes for waste treatment and prevention of pollution. For example, transgenic *Pseudomonas* bacterium has been developed that is able to degrade polyhalogenated compounds, a large and important class of pollutants, to harmless products. Also, a company called Envirogen at Princeton, New Jersey, has developed transgenic PCB-degrading microorganisms. Some applications of genetically manipulated microbes (simply by transfer of plasmids or by genetic engineering) in environment cleanup are listed below.

Superbug' for Cleaning Oil Spills

Oil consists of a variety of hydrocarbons, the main being xylenes, naphthalenes, octanes, and camphors. Certain strains of *P. putida* can consume each of these hydrocarbons, but no single strain found in nature can consume all four types. The capability to utilize one or a few of the many different types of hydrocarbons present in oil by *Pseudomonas* strain depends on the presence of genes encoding enzymes catalyzing the degradation of these hydrocarbons. Thus the genes, which enable these strains to feed on hydrocarbons, are found on four types of plasmids, referred to as xylene-degrading (XYL), naphthalene-degrading (NAH), octane-degrading (OCT), and camphor-degrading (CAM) plasmids. In 1979, Anand Mohan Chakrabarty, an India-born US scientist, and his colleagues created a single strain of *Pseudomonas* with expanded degradative capabilities permitting degradation of xylene, naphthalene, octane, and camphor. They introduced plasmids from different strains of *Pseudomonas* into a single cell and the resulting bacterium was capable of consuming these four types of hydrocarbons present in oil. This unique bacterial strain with increased metabolic capabilities was named 'superbug'. The idea behind the creation of a superbug was to mix them with straw, dry, and store. When required, this superbug-laden straw could be scattered over the oil spills where the straw would soak up the oil and the superbug would act on oil to convert it into harmless nonpolluting products. However, several scientists doubted the usefulness of Chakrbarty's superbug in coping up the problem of large oil spills produced by tankers.

To prepare superbug, first the CAM plasmid was transferred by conjugation into a strain carrying the OCT plasmid. But CAM and OCT plasmids being incompatible could not coexist inside the same cell as two separate plasmids. Hence the relevant genes from these two plasmids were first joined into a single plasmid by recombination between the two plasmids. The resulting single plasmid was able to perpetuate and carried both CAM and OCT degradative activities. On the other hand, the NAH plasmid was transferred by conjugation into a strain carrying the XYL plasmid. Note that the NAH and XYL plasmids were compatible and could therefore coexist within the same host cell. Finally, the CAM/OCT fusion plasmid was transferred by conjugation into the strain carrying the NAH and XYL plasmids. The final result of these manipulations was the generation of a strain that grew better on crude oil than any of the single plasmid strains either alone or in combination. The diagrammatic representation of the entire process of creation of superbug is illustrated in Figure 20.3. Although this particular multiple degradative strain has not been used to clean up oil spills, it has played a critical role in the development of the biotechnology industry. The inventor of this 'superbug' was granted a US patent describing its construction and use in March 1981. This was the first patent ever granted for a genetically engineered microorganism and represented a landmark decision by the US Supreme Court. Since then, it was ruled that biotechnology companies could protect their inventions in the same way as the chemical and pharmaceutical industries had in the past.

Manipulating Degradative Bacteria for Temperature Tolerance

Most of the degradative bacteria that have been genetically engineered by plasmid transfer are mesophiles, organisms that grow well only at temperatures between 20°C and 40°C. However, temperatures of polluted rivers, lakes, and oceans generally range from 0°C to 20°C. To test whether bacteria with enhanced degradative abilities could be created for cold

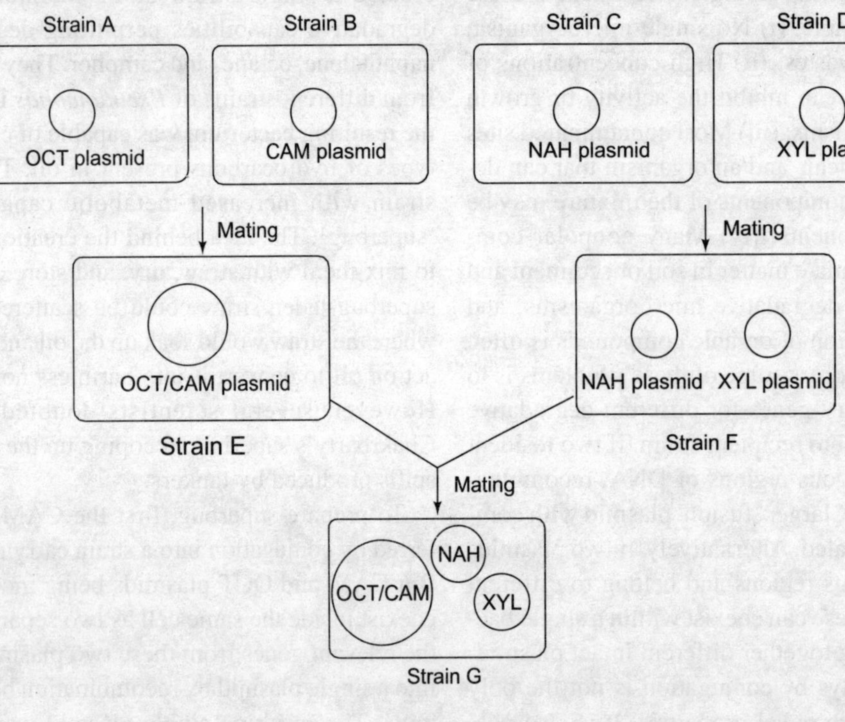

Figure 20.3 The process of creation of superbug

environments, a TOL plasmid from a mesophilic *P. putida* strain was transferred by conjugation into a physchrophile, an organism with a low temperature optimum that was able to degrade salicylate, but not toluene, and used it as a sole carbon source at temperatures as low as 0°C. The transformed strain carried the introduced TOL plasmid and its own SAL (salicylate-degrading) plasmid and was able to use either toluene or salicylate as its sole carbon source at 0°C.

Detoxification of Organophosphate Pesticides

The approaches for detoxification of organophosphate pesticides (e.g., methyl and ethyl parathion, paraoxon, Dursban, coumaphos, cyanophos, and Diazinon) in the environment, such as chemical treatment, incineration, or burial in landfill sites, suffer from serious environmental drawbacks. This problem can, however, be circumvented by the application of bacteria capable of degrading these pesticides, for example, soil bacteria such as *P. diminuta* MG and *Flavobacterium* spp. This property is attributable to an organophosphorus hydrolase-catalyzed hydrolysis of these pesticides to environmentally innocuous compounds. Moreover, *E. coli* has also been engineered to express organophosphorus hydrolase. Another approach can be to engineer *E. coli* cells to express organophosphorus hydrolase as part of a fusion protein that contains the *E. coli* lipoprotein signal peptide, the N-terminal portion of the lipoprotein, and outer membrane protein A on the outer surface of the bacterium, thereby eliminating the problem of a slow rate of pesticide uptake into the engineered bacterium.

Degradation of Trichloroethylene

When the *bph* A1 gene encoding the large subunit of biphenyl dioxygenase from *P. putida* KF 715 was replaced by homologous recombination with the *tod* C1 gene encoding the large subunit of the toluene dioxygenase from *P. putida* F1, the resultant strain was able to degrade trichloroethylene, demonstrating that it was possible to engineer bacterial strains that could degrade a number of different compounds. Using chimeric versions of the *bph* A1 gene, a degradative bacterium with novel biological activities has also been created. In this case, one *bph* A1 gene was from a strain of *P. pseudoalcaligenes* with the ability to degrade only a narrow range of PCBs, while the other was from *Burkholderia cepacia*, which could degrade a very wide range of PCBs.

Cleanup of Radioactive Environments

Most microorganisms are sensitive to damaging effects of radiation, however, a nonpathogenic soil bacterium *Deinococcus radiodurans* is naturally resistant to quite high levels of ionizing radiations. This resistance has been attributed to exceptionally effective DNA repair processes. It has been found that any DNA introduced into *D. radiodurans*, either as part of a plasmid or inserted into the chromosome, is protected against high levels of potentially damaging radiation. Since this bacterium can express foreign genes while growing in the presence of continuous radiation, it can serve as an ideal candidate for the expression of bioremediating proteins in toxic environments containing radioactive contaminants. Thus to develop a system to remediate organic pollutants present in radioactive environments, four genes encoding toluene dioxygenase were placed on a plasmid under the control of a constitutive *D. radiodurans* promoter. The entire plasmid was then inserted into the chromosome of *D. radiodurans* by a homologous recombination (a single cross-over between the chromosomal DNA and a chromosomal DNA fragment adjacent to the toluene dioxygenase genes). The integrated toluene dioxygenase was active and conferred upon *D. radiodurans* the ability to degrade toluene, chlorobenzene, and 3,4-dichloro-1-butene irrespective of the presence or absence of high levels of ionizing radiations.

Detoxification of Toxic Metals

Bacteria have been developed that are capable of converting toxic heavy metal ions to less toxic elemental forms, which are easier to remove from the environment. One of the most studied biological systems for detoxifying organometallic or inorganic compounds involves the *mer* operon. The *mer* determinants, *RTPCDAB*, in the bacteria are often located on plasmids, transposons, or chromosomes. There are two classes of mercury resistance, *viz.*, narrow-spectrum specifying resistance to inorganic mercury and broad-spectrum including resistance to organomercurials, encoded by *mer* B. The regulatory gene *mer R* is transcribed from a promoter that is divergently oriented from the promoter for the other *mer* genes. Mer R regulates the expression of the structural genes of the operon in both a positive and a negative fashion. Resistance is due to Hg^{2+} being taken up into the cell and delivered to the NADPH-dependent flavoenzyme mercuric reductase, which catalyzes the two-electron reduction of Hg^{2+} to volatile, low-toxicity Hg^0. Thus, the microbial *mer* operon has potential for bioremediation applications; researchers have reported the development of transgenic *E. coli* that is capable of cleaning up environments contaminated with mercury (Hg^{2+}).

Bioluminescors

rDNA technology is also being used in development of bioindicators where microorganisms have been genetically modified as 'bioluminescors' that give off light in response to several chemical pollutants. For example, transgenic microorganisms have been developed harboring genes for bioluminescence coupled to the naturally occurring genes that allow them to degrade contaminants. The result of this genetic linking is that the bacteria light up while working at decontamination. For example, genes for light production (*lux* genes) have been coupled to a microbial gene encoding for naphthalene degradation. The microbial activity is measured on-site by changes in light level recorded through fiber-optic sensors. Based on the intensity of light emitted, this tracking system determines that the bacterium is exerting its action. The efficiency of microbe can be analyzed rapidly and the conditions (like nutrient levels) can be optimized for maximum light emission and hence biodegradation activity.

20.3 APPLICATIONS OF TRANSGENIC PLANT TECHNOLOGY

Transgenic plant technology is defined as an *in vitro* gene transfer technique for transferring refined desirable gene(s) across taxonomic boundaries into plants from other plants, animals, and microbes, or even to introduce artificial, synthetic, or chimeric genes into plants. The technique, also called plant genetic transformation, thus involves recombinant DNA techniques, allowing artificial insertion of gene(s) to plants (excluding insertion of genes through pollination), its subsequent stable integration into the plant genome, and expression of foreign gene. The inserted gene sequence is known as transgene and the plant developed after a successful gene transfer is known as transgenic plant (or genetically modified or GM plant). As the transgene may come from another unrelated plant or even from an entirely different kingdom, there is possibility of not only bringing in desirable characteristics from other varieties of the plant, but also of adding characteristics from other unrelated species. These transgenes are introduced into plant cells, tissues or organs by a variety of methods, which include biological vector-mediated transformation or direct gene transfer mechanisms. The transgene once introduced into the plant genome by any of these methods is expressed in plants. This allows production of new plant variety, which is usually normal in appearance and differs from the parent only with respect to the function of the inserted transgene. This directed genetic engineering of plants requires that gene(s) of interest is available, that the gene be introduced into plant cells capable of regenerating into intact plants, and that the gene carries with it a selectable marker so that the transformed plant cells can be isolated from a large population of untransformed normal cells. Finally, the transformed plant cell must retain its capacity to regenerate. Producing transgenic plants is much simpler, as plant cells, unlike animal cells, are totipotent.

The aim of plant transgenic technology is to create diversity of plants serving human needs and hence to benefit producers, processors, and consumers. In this section, the applications of plant transformation in various fields are presented.

20.3.1 Development of Insect-resistant Plants

Every year crop plants in the field and harvested products in transit or storage are infested by insects consequently leading to heavy economical losses. In addition, the insecticides used to prevent these insects (often six to eight times during a growing season) lack specificity of action, are not cost-effective, and are health- and environment-damaging. As a consequence of health and environmental concerns, the registration of many promising chemical pesticides has been restricted. Hence alternative opportunities to reduce the use of chemical pesticides in many important crops are required. One such strategy is to deploy genes for pest resistance carefully in crops. In this field of transgenics, extensive research has been done and several transgenic plants have been developed that are capable of producing functional insecticides and are intrinsically resistant to insect predators. These transgenic plants do not need to be sprayed with costly and hazardous chemical pesticides, and the cost of maintaining such insect-resistant crops is lower than that for nonresistant crops. These transgenic plants are usually highly specific in action and act on a limited number of insect species. Moreover, these biological insecticides are also nonhazardous to humans and other higher animals. In many cases, these have also benefited the environment because of reduced pesticide usage or by providing the means to grow crops with less tillage.

The strategies employed for generation of insect-resistant transgenic plants include the introduction of genes encoding the following through rDNA technology: (i) Insecticidal protoxin produced by one of several subspecies of *B. thuringiensis*; and (ii) Proteins that protect plants against a variety of insects, which exert their action by interfering with insect development, for example, protease inhibitors, α-amylase inhibitors, lectins, cholesterol oxidase, tryptophan decarboxylase, and avidin. It is significant to note that presently *Bt* technology remains the most effective and the safest method for controlling plant pests, and hence the best choice for pest control.

Increasing Expression of *Bacillus thuringiensis cry* (*Bt*) Gene As certain economically significant pests of crop plants feed on internal plant tissues, the application of Bt toxin as a biopesticide as discussed in Section 20.2.8 may not be a useful insect control strategy. To overcome this problem, the best way is to express *cry* genes encoding Bt toxin in plant itself. Such transgenic plants are insect-resistant and do not require spray of insecticidal toxin, consequently limiting the environmental distribution of the toxin. For the production of commercially viable insect-resistant transgenic plants, high yield of *cry* gene-encoded protoxin is required. Several researchers worked in this direction to device strategies to increase the level of expression of cloned *cry* gene. It has been determined that all of the insecticidal toxin activity resides within the first 646 amino acids from the N-terminus of the 1,156 amino acid protoxin. This segment of *cry* gene that encodes the highly conserved amino acid sequence (i.e., N-terminus) has been cloned under the control of strong plant promoter. The plants transformed with this construct led to the expression of shortened Bt protein that protected the transgenic plants against damage from lepidopteran insect predation. The practical results obtained with *Bt* crops in several countries have demonstrated that these are safe and beneficial and few of the *Bt*-transgenic plants have been commercialized. In 1995, potato crop containing modified *cry* 3A gene for resistance against Colorado beetle became the first transgenic insect-resistant crop to reach commercial production as NewLeaf™ potato marketed by Monsanto. The same company also released insect-resistant cotton and maize under the trade names Bollgard® (Figure 20.4a) and YieldGard™, respectively.

(a) *Bt* cotton (b) Round-up Ready soybean (c) PRSV-resistant papaya (d) Freedom II squash

(e) Golden rice (f) BetaSweet® carrot (g) Blue rose (h) Flavr Savr tomato (i) Long life carnation

Figure 20.4 Some examples of transgenic plants

Three transgenic *Bt* cotton hybrids containing lepidopteran-specific Bollgard® *Bt* gene, *cry* 1Ac offered protection against all the major species of Indian bollworms, *viz.*, *Helicoverpa armigera, Pectinophora gossypiella, Earias vittella*, and *Earias insulana*. Maize YieldGard™ expressed *cry* 1Ab and was resistant to the European corn borer. The status of commercialization of *Bt* cotton in India is summarized in Exhibit 20.3.

Exhibit 20.3 *Bt* cotton in India

In India alone, bollworm causes losses worth US $555 million despite yearly spraying of insecticides worth almost as much. Cotton, grown by more than four million farmers, occupies only 5% of the country's total crop area, yet half the pesticides used in India are applied to combat the bollworm. Due to hazardous and environment-damaging nature of pesticides, *Bt* cotton was produced. It has been commercialized in six countries, namely, the United States (1996), Australia (1997), South Africa (1997), Argentina (1998), Mexico (1996), China (1998), and Indonesia (2000).

In India, after a thorough assessment of biosafety of *Bt* cotton through various scientific studies conducted by National Level Institutes, the Government of India through GEAC (on the recommendations of RCGM), Ministry of Environment and Forests (MoEF) considered the proposal for the commercial release of *Bt* cotton in its 32nd meeting held on March 26, 2002, and after the careful and in-depth consideration, accorded approval for release into the environment of three transgenic *Bt* hybrid cotton varieties, developed by Maharashtra Hybrid Seed Company (Mahyco), namely, MECH-12 *Bt*, MECH-162 *Bt*, and MECH-184 *Bt*. The insect resistance in these hybrids was introgressed from *Bt* containing Cocker-312 (event MON 531) developed by Monsanto, USA, into parental lines of Mahyco propriety hybrids. The Government of India, MoEF, issued the letter to this effect, vide letter no. D.O. No.10/1.IV/2002-CS, dated April 5, 2003. The approval given to three Mahyco *Bt* hybrids for commercial release was accompanied by 15 conditions stipulated by GEAC (for details see Chapter 21). MoEF also reserved the right to stipulate additional conditions and the right to revoke the approval if the implementation of the conditions was not satisfactory.

Subsequent to approval to Mahyco in 2002, several other companies have taken sublicenses from Mahyco and integrated *cry* 1Ac gene into their hybrids. Taking into consideration the need for introducing diversity in the gene as well as germplasm as a tool to contain the development of insect resistance, hybrids containing three new *Bt* cotton genes/events have been approved in 2006. These are: (i) *cry* 1Ac gene (event 1) by M/s J.K. Agri Seeds Ltd.; (ii) fusion genes (*cry* 1Ab + *cry* 1Ac) GFM by M/s Nath Seeds, and (iii) stacked genes *cry* 1Ac and *cry* 1Ab by M/s Mahyco. As of now 62 hybrids of *Bt* cotton are approved for commercial cultivation in the country as listed in Table A and the total area under *Bt* cotton has increased from 38,000 ha in 2002 to 38 lakh ha.

Table A *Bt* Cotton Hybrids Approved for Commercial Cultivation in India

Zone	Company	Hybrid
North	Ankur Seeds Ltd.	Ankur 2534 *Bt*
	J.K. Agri Genetics Seeds Ltd.	JKCH-1947 *Bt*
	Mahyco	MRC-630 *Bt*, MRC-6029 *Bt*, MRC-6025 *Bt*
	Nath Seeds Ltd.	NCEH-6R *Bt*
	Nuziveedu Seeds Ltd.	NCS 138 *Bt*
	Rasi Seeds Ltd.	RCH-134 *Bt*, RCH-308 *Bt*, RCH-317 *Bt*, RCH-314 *Bt*
Central/North	Ankur Seeds Ltd.	Ankur-651 *Bt*
	Mahyco	MRC-6301 *Bt*
Central	Ajeet Seeds Ltd.	ACH-11-2 BG II
	Ankur Seeds Ltd.	Ankur-09
	Ganga Kaveri Seeds Pvt. Ltd.	GK 205 *Bt*, GK 204 *Bt*
	J.K. Agri Genetics Seeds Ltd.	JK Varun *Bt*
	Krishidhan Seeds Pvt. Ltd.	KDCHH 9821 *Bt*, KDCHH-441 BG II
	Mahyco	MECH 12 *Bt**, MRC 7301 BG II, MRC-7326 BG II, MRC-7347 BG II
	Nath Seeds Ltd.	NCEH-2R
	Pravardhan Seeds Ltd.	PRCH-102 *Bt*
	Rasi Seeds Ltd.	RCH 377 *Bt*, RCH-144 *Bt*, RCH-118 *Bt*, RCH-138 *Bt*
	Vikki Agrotech Pvt. Ltd.	VCH-111 *Bt*
North/Central South	Nuziveedu Seeds Ltd.	NCS-913 *Bt*
South	Ganga Kaveri Seeds Pvt. Ltd.	GK-209 *Bt*, GK-207 *Bt*
	J.K. Agri Genetics Seeds Ltd.	JK Durga *Bt*, JKCH-99 *Bt*
	Mahyco	MRC-6322 *Bt*, MRC-6918 *Bt*, MRC-7351 VG II, MRC 7201 BG II
	Nath Seeds Ltd.	NCEH-3 R
	Prabhat Seeds Ltd.	PCH-2270 *Bt*
	Rasi Seeds Ltd.	RCH-20 *Bt*, RCH-368 *Bt*, RCH 111 BG I, RCH-371 BG I, RCHB-708 BG I

* Approval not renewed for Andhra Pradesh.

[North zone: Haryana, Punjab, and Haryana; Central zone: Gujarat, Madhya Pradesh and Maharashtra; South zone: Andhra Pradesh, Karnataka, and Tamil Nadu].

The chronology of events related to *Bt* cotton in India is as follows:

1995: • Department of Biotechnology (DBT), Government of India permitted to Mahyco import of 100 g of transgenic Cocker-312 variety of cottonseed cultivated in the United States by Monsanto.

1996: • Mahyco imported 100 g of Cocker-312 seed containing the *cry* 1Ac gene from Monsanto, USA. Crossing with Indian cotton breeding lines to introgress *cry* 1Ac gene was carried out and 40 elite Indian parental lines were converted for *Bt* trait.

1996–1998: • Greenhouse, risk assessment studies, and limited field trials (1 location) were conducted using *Bt* cotton from converted Indian lines for pollen escape studies, aggressiveness, persistence studies, biochemical analyses, toxicological studies, and allergenicity studies.

1998–1999: • Monsanto-Mahyco tied up. • Monsanto was given permission for small trials of *Bt* cotton 100 g per trial by DBT. On the recommendations of RCGM, two sets of replicated field trials were conducted to test the performance (agronomic benefits and biosafety) of the three *Bt* hybrids, MECH-12 *Bt*, MECH-162 *Bt*, and MECH-184 *Bt*. • Thousands of farmers occupied and burned down *Bt* cotton trial fields in Karnataka as part of Operation Cremation Monsanto. • Vandana Shiva's Research Foundation for Science, Technology, and Ecology went to the Supreme Court challenging the 'illegality' of the field trials authorized by the DBT.

2000: • Field trials repeated at 10 locations in six states. • GEAC gave approval for conducting large-scale field trials on 85 ha and also to undertake seed production on 150 ha. • Mahyco discovered illegal commercial *Bt* cotton farming over several thousand hectares in Gujarat. Source of the cotton was traced back to Navbharat Seeds Pvt. Ltd. • Mahyco allowed conducting large-scale field trials including seed production at 40 sites in six states. The permission was granted based on the 'totally confidential' data from the small trials that allowed regulators to infer that *Bt* cotton was 'safe'. • The DBT set up a committee to 'independently' monitor and evaluate large-scale field trials.

2001: • An open dialogue was held between Monsanto and Greenpeace to discuss *Bt* cotton with scientists, Ministry of Environment representatives, and farmers. No data on field trials was presented though farmers vociferously demanded *Bt* cotton to be commercialized. Technical questions and concerns raised by Greenpeace remained unanswered. • GEAC ordered *Bt* cotton fields to be burnt in Gujarat. • Gene Campaign filed a case in the Delhi High Court charging the Government with negligence in allowing large-scale field trials to be conducted without appropriate monitoring, regulation, and safety precautions. • Mahyco conducted large-scale field trials on 100 ha in seven states. • All India Coordinated Crop Improvement Project of the ICAR also conducted field trials. ICAR conducted multilocation field trials on the three hybrids especially to make a cost–benefit analysis of *Bt* cotton. The results proved the effectiveness of *Bt* technology in reducing bollworm infestation, decreasing number of insecticide sprays, increasing cotton yields, and net incomes.

• GEAC extended field trials of *Bt* cotton by another year. • The large field trials at 21 sites (four trials were damaged) yielded similar results with *Bt* hybrids.

2002: • The ICAR submitted a positive report to the Ministry of Environment on the field trials of *Bt* cotton. • Dr Manju Sharma, secretary of DBT, declared that the latest round of *Bt* cotton trials were satisfactory and that it was up to the GEAC and the Ministry of Environment to decide on a date of commercial release. • GEAC approved three *Bt* cotton hybrids for commercial cultivation after taking into account the data on their performance. Among the nine states in which *Bt* cotton was cultivated, Maharashtra, Andhra Pradesh, Gujarat, and Madhya Pradesh were leading with 49.7%, 18.1%, 11.8%, and 11.4% of the national *Bt* cotton acreage, respectively. Central zone showed the highest adoption of *Bt* cotton followed by south and north zones.

2003: • Cotton farmers in Andhra Pradesh, where the state government had aggressively promoted GM technology, suffered severe agricultural and financial losses and many including entire families committed suicide.

2004–2006: • The hybrids released up to 2004 were approved for cultivation in central and south zones, while six *Bt* hybrids were for the first time approved for cultivation in north zone in 2005. Zone-wise, 14, 24, and 9 hybrids were approved for growing in north, central, and south zones, respectively. • There was no evidence to firmly link the cotton losses to *Bt* technology, but in May 2005 Andhra Pradesh revoked permission to grow three varieties of *Bt* cotton. • In 2005, the area covered by *Bt* cotton plantations in India was raised to 3.44 million ha (30 times larger as compared to that in 2002). Mahyco reported that in 2005, planting *Bt* cotton cut the use of pesticides by 42% and yield losses went down, and net returns increased by US $373/ha compared to non-*Bt* crops.

2006: • Forty *Bt* cotton hybrids developed by 13 seed companies were approved for commercial cultivation after going through a similar process of GEAC approval as prescribed for the three Mahyco hybrids. All the hybrids, except four, had the Monsanto-Mahyco *Bt* technology (event MON 531) that has been sublicensed to the respective seed companies. JKCH-1947 *Bt* and JK Varun *Bt* contained *cry* 1Ac event 1 developed by the Indian Institute of Technology, Kharagpur while NCEH-2R *Bt* and NCEH-6R *Bt* contained fusion genes *cry* 1Ab/*cry* 1Ac from China. • The repeated failures of cotton crops and another spate of farmers' suicides in Maharashtra once again put *Bt* cotton in the spotlight. • Hybrids containing three new *Bt* cotton genes/events were approved. These were *cry* 1Ac (event 1) by M/s J.K. Seeds Ltd., fusion genes (*cry* 1Ab + *cry* 1Ac) GFM by M/s Nath Seeds, and stacked genes *cry* 1Ac and *cry* 1Ab by M/s Mahyco. • The government's efforts to usher in GM crops suffered more setbacks in September, when the Supreme Court temporarily banned further trials on new GM crops. This followed a public interest petition urging the regulatory system to be foolproof before permitting trials. • Angry farmers' unions burnt trial plots of GM rice in two northern Indian states. The protestors alleged that an Indian company leased the farmers' plots for trials without explaining the full implications of GM crops to them.

Source: *Bt cotton in India—A status report* (2006) by Asia-Pacific Consortium on Agricultural Biotechnology, New Delhi; *Primer on genetically modified organisms* by Project coordinating and monitoring unit, GEF-World bank capacity building project on biosafety, MoEF, in association with Biotech Consortium India Ltd. (BCIL) (2007) Marshall Advertising Co. 25-26 pp.; http://www.envfor.nic.in/divisions/csurv/geac/bgnote.doc

Though a range of *B. thuringiensis* protoxins are available, no single protoxin is effective against a broad range of insect species, which may limit their overall usefulness. Further modifications in the technique have been reported that resulted in dramatically increased levels of expression. These are as follows: (i) In the first strategy, bacterial *cry* gene was isolated and modified to change any DNA sequences that could inhibit efficient transcription or translation of a plant host by site-directed mutagenesis (96.5% unchanged, 3.5% changed). This partially modified gene gave 10-fold higher level of Bt toxin; (ii) Second strategy involved chemical synthesis of a fully modified gene, which contained codons more commonly used by plants as opposed to those favored by *B. thuringiensis*. This chemically synthesized gene was also modified to eliminate any potential mRNA secondary structure or plant polyadenylation sequences, which might decrease gene expression. The G+C content amounted to 49% (note that the wild-type gene contains 37% G+C), with only 78.9% identity to the wild-type gene. The plants transformed with such highly modified gene showed ~100-fold higher level of protein and hence increased insecticidal activity; (iii) Third strategy designed to increase the effectiveness of relatively low levels of *B. thuringiensis* insecticidal toxin activity entailed combining truncated *cry* gene with a serine protease inhibitor gene; and (iv) Finally, chloroplast transformation procedure, as discussed in Chapter 14, was employed for the expression of Bt toxin in chloroplast. This technique not only led to an increase in the yield of Bt toxin, but also prevented the risk of generation of 'superweeds' (for details see Chapter 21). A brief overview of the transgenic strategies used to develop insect-resistant plants is presented in Table 20.4. A list of ten food crops (besides *Bt* cotton) that were under contained limited field trials in India in 2006 have been listed in Table 20.5.

Table 20.4 Some Examples of Genetically Engineered Pest Resistant Plants

Genes	Encoded protein	Origin (plant or bacteria)	Target insects	Transformed plants
Cry proteins (Bt toxin)				
cry genes	Bt toxin	*Bacillus thuringiensis*	Lepidoptera, Diptera, Coleoptera	Cotton, corn, soybean, brinjal, cabbage, castor, cauliflower, okra, rice, and tomato
Protease inhibitors from plants				
c-II	Serine protease	Soybean	Coleoptera, Lepidoptera	Oilseed rape, poplar, potato, tobacco
CMe	Trypsin	Barley	Lepidoptera	Tobacco
CMTI	Trypsin	Squash	Lepidoptera	Tobacco
CpTI	Trypsin	Cowpea	Coleoptera, Lepidoptera	Apple, lettuce, oilseed rape, potato, rice, strawberry, sunflower, sweet potato, tobacco, tomato, wheat
14K-CI	Bifunctional serine protease and α-amylase	Cereals	–	Tobacco
MTI-2	Serine protease	Mustard	Lepidoptera	Arabidopsis, tobacco
OC-1	Cysteine protease	Rice	Coleoptera, Homoptera	Oilseed rape, poplar, tobacco
PI-IV	Serine protease	Soybean	Lepidoptera	Potato, tobacco
Pot PI-I	Proteinase	Potato	Lepidoptera, Orthoptera	Petunia, tobacco
Pot PI-II	Proteinase	Potato	Lepidoptera, Orthoptera	Birch, lettuce, rice, tobacco
KTi3, SKTI	Kunitz trypsin	Soybean	Lepidoptera	Potato, tobacco, tomato
PI-I	Proteinase	Tomato	Lepidoptera	Alfalfa, tobacco, tomato
PI-II	Proteinase	Tomato	Lepidoptera	Tobacco, tomato
α-Amylase inhibitors from plants				
a-AI-Pv	α-Amylase inhibitor	Common bean	Coleoptera	Azuki bean, pea, tobacco
WMAI-I	α-Amylase inhibitor	Cereals	Lepidoptera	Tobacco
14K-CI	Bifunctional serine protease and α-amylase inhibitor	Cereals	–	Tobacco

(*Contd.*)

Table 20.4 (*Contd.*)

Genes	Encoded protein	Origin (plant or bacteria)	Target insects	Transformed plants
Lectins from plants				
GNA	Lectin	Snowdrop	Homoptera, Lepidoptera	Grapevine, oilseed rape, potato, rice, sweet potato, sugarcane, sunflower, tobacco,
p-lec	Lectin	Pea	Homoptera, Lepidoptera	Potato, tobacco
WGA	Agglutinin	Wheat germ	Lepidoptera, Coleoptera	Maize
Jacalin	Lectin	Jack fruit	Lepidoptera, Coleoptera	Maize
Rice lectin	Lectin	Rice	Lepidoptera, Coleoptera	Maize
Plant enzymes				
bch	Chitinase	Bean	Homoptera, Lepidoptera	Potato
pod	Anionic peroxidase	Tobacco	Lepidoptera, Coleoptera, Homoptera	Sweet gum, tobacco, tomato
tdc	Tryptophan decarboxylase	*Catharanthus roseus*	Homoptera	Tobacco
Others				
Avidin	Avidin	Chicken eggs	Coleoptera	Corn
Cholesterol oxidase	Cholesterol oxidase	Streptomyces	Coleoptera (*Anthonomus grandis grandis*)	Tobacco, Cotton

Table 20.5 Transgenic Crops Approved for Conducting Contained Limited Field Trials (Including Multilocation Field Trials) (Source: Department of Biotechnology, Government of India)

Crop	Organization	Transgene
Brinjal	IARI (New Delhi)	*cry* 1Ac and *cry* 1Aabc
	Mahyco (Mumbai)	*cry* 1Ac
	Sungro Seeds Research Ltd. (Delhi)	*cry* 1Ac
Cabbage	Nunhems India Pvt. Ltd. (Gurgaon)	*cry* 1Ba and *cry* 1Ca
Castor	Directorate of Oilseeds Research (DOR; Hyderabad)	*cry* 1Aa and *cry* 1Ec
Cauliflower	Sungro Seeds Research Ltd. (Delhi)	*cry* 1Ac
	Nunhems India Pvt. Ltd. (Gurgaon)	*cry* 1Ba and *cry* 1Ca
Corn	Monsanto India Ltd. (Mumbai)	*cry* 1Ab (Mon 810 event)
Groundnut	ICRISAT (Hyderabad)	*chitinase gene* from rice (*Rchit*)
Okra	Mahyco (Mumbai)	*cry* 1Ac, *cry* 2Ab
Potato	Central Potato Research Institute (CPRI; Shimla)	*RB* gene derived from *Solanum bulbocastanum*
Rice	Mahyco (Mumbai)	*cry* 1Ac and *cry* 2Ab
	Tamil Nadu Agricultural University (TNAU)	Rice *chitinase* (*chi 11*) or tobacco *osmotin* gene
	IARI (New Delhi)	*cry* 1B-*cry* 1Aa fusion gene
Tomato	IARI (New Delhi)	Antisense *replicase* gene of tomato leaf curl virus
	Mahyco (Mumbai)	*cry* 2Ab

Expression of Plant Proteinase Inhibitor Insects use diverse proteolytic or hydrolytic enzymes in their digestive guts for digestion of food proteins and other food components. Certain plants have natural insect defense mechanisms that are sufficient for plant survival, for example, some plants produce proteinase inhibitors (protease inhibitors) that upon ingestion by the feeding insect prevent the hydrolysis of plant proteins, thereby effectively starving the predator insect. However, the level of these proteinase inhibitors in plants is not always effective enough to keep the damage by insects to a level that would be acceptable for crop plants. Thus in order to produce sufficiently high levels of proteinase inhibitor, it is reasonable to isolate the plant proteinase inhibitor gene, clone it into an expression vector under the control of a strong plant promoter, and create transgenic insect-resistant plants. In this direction, cowpea trypsin inhibitor (CpTI) and potato proteinase inhibitor II have been successfully used.

Expression of Plant α-amylase Inhibitor The seed feeding beetles such as *Callosobruchus maculates* (cowpea weevil) and *C. chinensis* (azuki bean weevil) infest crops and cause tremendous economic losses. In this direction, the gene for α-amylase inhibitor from the common bean was placed under the control of strong seed-specific promoter for the bean phytohemagglutinin gene, and used to transform pea plants susceptible to damage by both these insects.

Expression of Plant Tryptophan Decarboxylase In an observation, the expression of enzyme tryptophan decarboxylase from *C. roseus* (periwinkle) in tobacco led to generation of tobacco plants protected against damage by whitefly (*Bemisia tabaci*). It was suggested that in the reaction catalyzed by tryptophan decarboxylase, tryptophan is decarboxylated to form tryptamine, which is used for the production of plant alkaloids that inhibit insects.

Expression of Plant Lectins The activity of lectins (carbohydrate-binding proteins) found in seeds and storage tissues of a variety of plant species has also been used to protect plants from insect predation. These lectins are toxic to certain species of insects. With this in mind, the snowdrop lectin gene has been introduced into approximately a dozen different plants. The expression of lectin in transformed plants resulted in the reduction in the extent of damage caused by aphids.

Expression of Bacterial Cholesterol Oxidase Cholesterol oxidase, an enzyme present in a variety of bacteria, is responsible for the oxidation of 3-hydroxysteroids to ketosteroids and hydrogen peroxide. This enzyme, even at low-levels, has a high-level of insecticidal activity against larvae of *Anthonomus grandis grandis* (boll weevil, a coleopteran pest of cotton). The enzyme also has lower activity levels against some lepidopteran pests. This enzyme, similar to Bt toxin, probably kills the insect by disrupting the midgut epithelial membrane. The enzyme cholesterol oxidase has been successfully used for the generation of transgenic insect-resistant plants. In an example, a cholesterol oxidase gene encoding the enzyme plus leader peptide was isolated from a *Streptomyces* strain and cloned into a vector under the control of a plant virus (figwort mosaic virus) promoter, the introduction of which into tobacco cell protoplasts led to the generation of transformed cells that actively expressed cholesterol oxidase.

Expression of Avidin Transgene Avidin, a glycoprotein isolated from chicken eggs or artificially synthesized, has been successfully used to transform plants. As this glycoprotein has an extremely high affinity for coenzyme biotin, its expression in transgenic plants causes biotin deficiency, which leads to stunted growth and death in a number of different insect species.

20.3.2 Development of Herbicide-tolerant Plants

In agriculture, two strategies in common use for controlling weeds are tillage (ploughing) and application of herbicides; however, these are not effective strategies. Extensive tillage may lead to soil erosion, thereby causing a serious loss of water content. Moreover, many herbicides are unable to discriminate between weeds and crop plants, some have to be applied early before the establishment of weeds, and some persist in the environment due to their nonbiodegradable nature. Besides these drawbacks, chemical herbicides incur heavy expenditure and despite their application on crop plants, ~10% of the global production is lost per year through weed infestation. Thus to overcome these drawbacks, herbicide-resistant plants have been developed. Several strategies have been envisioned to develop herbicide-tolerant plants, which involve any one of the following: (i) Overproduce the target protein; (ii) Reduce the ability of herbicide sensitive target protein to bind to a herbicide; (iii) Inhibit the uptake of the herbicide; (iv) Detoxify herbicide; and (v) Endow plants with the capability to metabolically inactivate the herbicide. Details of various strategies for the development of herbicide-tolerant transgenic plants are given in this section and a list of herbicide-tolerant transgenic crops in the field is given in Table 20.6.

Overexpression of Herbicide-resistant Target Protein Some of the herbicides exert their action by inhibiting target protein (a metabolic pathway enzyme) in weeds. To overcome the effect of this inhibition in crop plants, the strategy of effectively titrating out the herbicide from the crop plants can be employed. This can be achieved by the overproduction of the target protein. As a consequence, the target protein is supplied in excess amounts for carrying out its cellular function and hence to partially overcome the effect of inhibition by herbicide. Overexpression can be achieved by integration of multiple copies of the target protein-encoding gene and/or the use of a strong promoter plus translational enhancer to drive expression of the target protein-encoding gene. Using this strategy, resistant plants have been developed for herbicide glyphosate, which acts as an inhibitor of an enzyme 5-enol pyruvyl shikimate-3-phosphate synthase (EPSPS) involved in the shikimate pathway for aromatic amino acid biosynthesis. The EPSPS-encoding gene (*epsps*) was isolated from a glyphosate-resistant strain of *E. coli*, cloned under the control of plant promoter and transcription polyadenylation sequences, and transformed into plant cells. The transgenic plants (for tobacco, petunia, tomato, potato, and cotton) have been successfully raised that produced an amount of *E. coli* EPSPS, which was sufficient to replace the inhibited plant enzyme, and were resistant to the effects of glyphosate. Thus in these cases, the weeds are affected by glyphosate treatment, while the crop plants remain unaffected. Using this strategy, 'Round-Up Ready' soybean plants (Figure 20.4b) resistant to Round-Up™ herbicide (glyphosate) were grown in the United States by Monsanto Co. These transgenic soybean plants contained gene encoding Roundup®-tolerant enzyme (CP4-EPSPS).

Table 20.6 Some Examples of Herbicide-resistant Transgenic Plants

Class of herbicide	Herbicide	Mode of action of herbicide	Transgene introduced	Crops transformed	Mode of herbicide resistance	Companies marketing transgenic variety
Glycine	Glyphosate (Roundup)	Inhibits enzyme EPSPS involved in aromatic amino acid biosynthesis	*Agrobacterium* CP4-*epsps* resistant gene	Soybean, rape, tomato	Overproduction of EPSPS; Effect of inhibition of EPSPS by herbicide is overcome its overproduction	Monsanto
			Glyphosate oxidoreductase gene	Maize, rape, soybean	Detoxification (or degradation) of this herbicide by glyphosate oxidoreductase	Monsanto
Phosphinic acid	Phosphinothricin (Basta), (Liberty), (Glufosinate)	Inhibits glutamine synthase, an important enzyme involved in ammonia assimilation, and regulation of nitrogen metabolism	*bar* gene from *Streptomyces hygroscopicus* or *pat* gene from *S. viridochromogenes*	Maize, rice, wheat, cotton, rape, potato, tomato, sugar beet	Detoxification (or degradation) of this herbicide by phosphinothricin acetyltransferase	Hoechst/AgrEvo/Aventis, Novartis/Syngenta
Sulfonylurea	Chlorsulfuron (Glean)	Noncompetitively inhibit acetolactate synthase involved in the biosynthesis of branched amino acids	Mutant plant acetolactate synthetase selected from tissue culture	Rape, rice, flax, tomato, Sugar beet, maize	Altered enzyme is herbicide insensitive	DuPont/Pioneer/Hi-Bred
Imidazolinone	Arsenal	Noncompetitively inhibit acetolactate synthase involved in the biosynthesis of branched amino acids	Mutant plant acetolactate synthetase selected from tissue culture	—	Altered enzyme is herbicide insensitive	American Cyanamid
S-triazines	Atrazine (Lasso)	Targets Q_B of photosystem II	Mutant plant chloroplast *psb* A gene or Q_B gene or introduction of gene for glutathione-*S*-transferase	Soybean	*psb* gene encodes chloroplast protein D-1; Mutant psb protein is insensitive to the action of herbicide; Glutathione-*S*-transferase detoxifies atrazine	DuPont, Ciba Geigy/Novartis
Nitriles	Bromoxynil (Buctril)	Inhibits photosynthesis	Bacterial nitrilase gene	Cotton, rape, potato, tomato, tobacco	Detoxification (or degradation) of this herbicide by nitrilase	Calgene
Phenoxy-carboxylic acids	2,4-D	Action like indole acetic acid (synthetic auxin); Decrease acetate metabolism	*tfd* A gene from *Alcaligenes*	Maize, cotton, tobacco	Detoxification (or degradation) of this herbicide by oxygenase	Schering/AgrEvo

Mutation of the Target Protein This strategy involves the functional substitution of the native protein by a modified target, which is resistant to inhibition by the herbicide, and to incorporate the resistant target protein-encoding gene into the plant genome. Several sources of resistant proteins can be exploited. Similar to the overexpression strategy, this strategy also requires knowledge of the mode of action of the herbicide.

Inactivation of the Herbicide using a Single Gene from a Foreign Source This strategy is based on the conversion of herbicide to a less toxic form and/or to remove it from the system. In this case, a detailed knowledge of the mode of action of herbicide is not required. An example of resistance due to herbicide inactivation has been developed for bromoxynil (3,5-dibromo-4-hydroxy benzonitrile), a herbicide that acts by inhibiting photosynthesis. In this case, nitrilase-encoding gene was isolated from soil bacteria *Klebsiella ozaenae*, which encodes an enzyme nitrilase that inactivates bromoxynil. This gene was cloned in a plant transformation vector under the control of light-regulated promoter from rubisco small subunit and used for the transformation of tobacco plants. As expected, the transgenic plants expressed nitrilase in their shoots and leaves, but not in their roots, and were resistant to the toxic effects of the herbicide.

Enhance Plant Detoxification The aim of this strategy is to improve the natural plant defenses against toxic compounds. This requires detailed information about endogenous plant detoxification pathways and the mechanisms by which compounds are recognized and targeted for detoxification by the plant.

Increased Superoxide Dismutase (SOD) Level By increasing the SOD level in higher plants, resistance to the herbicide methyl viologen (paraquat) can be increased.

20.3.3 Development of Fungal or Bacterial Disease-resistant Plants

Many phytopathogenic fungal or bacterial infections resulting simultaneously, or as a consequence of infestation of plants by insects often result in extensive damage and loss of crop productivity, accounting for enormous economic losses. Presently chemical agents such as fungicide or bactericide are used for the control of fungi or bacteria, respectively. However, the use of chemicals is disadvantageous because these may persist and accumulate in the environment and may pose threat to animals or humans. To overcome these problems associated with the application of chemicals, fungal- or bacterial pathogen-resistant transgenic plants have been developed as a simple, inexpensive, effective, and environment-friendly nonchemical means. In this direction, several approaches have been tested, which are grouped into five categories: (i) The expression of gene product that is directly toxic to pathogens or that reduces their growth. These include pathogenesis-related (PR) proteins such as hydrolytic enzymes [e.g., chitinases, glucanases], antifungal proteins [e.g., osmotin and thaumatin-like proteins (thaumatin is a small and very sweet protein)], antimicrobial peptides [e.g., thionins, defensins, and lectin], ribosome-inactivating proteins (RIP), and phytoalexins; (ii) The expression of gene product that destroys or neutralizes a component of the pathogen arsenal such as polygalacturonase, oxalic acid, and lipase; (iii) The expression of gene product that can potentially enhance the structural defenses in the plant, for example, elevated levels of peroxidase and lignin; (iv) The expression of gene product releasing signals that can regulate plant defenses, for example, production of specific elicitors, hydrogen peroxide, salicylic acid (SA), and ethylene (C_2H_4); and (v) The expression of resistance gene product (R) involved in the hypersensitive response (HR) and in interactions with avirulence (Avr) factors. A brief description of these strategies is given below and some common examples of fungal disease-resistant transgenic plants using any of these strategies are listed in Table 20.7.

Expression of Chitinase and Glucanases To combat fungal pathogens, many plants have been transformed with genes that code for chitinase and glucanase enzymes, which degrade polymers in the cell walls of many, but not all fungi, exerting no effect on the host plants or animals. The genes for these enzymes have been isolated from a number of sources, including plants (rice, barley), bacteria (*S. marcescens*), and even fungi (*T. harzianum*). Transgenic plants that constitutively expressed high levels of chitinase have been engineered. Thus transgenic tobacco seedlings constitutively expressing a bean chitinase gene under the control of the CaMV 35S promoter showed an increased ability to survive in soil infested with the fungal pathogen *R. solani* and delayed the development of disease symptoms. Glucanases have also been used as a tool to enhance resistance to fungal infection. Doubly transgenic plants obtained by crossing-over two transgenic plants, one transformed with chitinase gene and the second with β-glucanase gene showed synergistic effects and thus were found to be more resistant to fungal damage as compared to ones transformed with single gene constructs.

Overproduction of Salicylic Acid Plants often respond to fungal or bacterial pathogen invasion or other environmental stresses by converting a conjugated storage form of salicylic acid (salicylic acid 2-*O*-β-*D*-glycoside) to salicylic acid, which induces the synthesis of PR proteins, such as β-1,3-glucanases, chitinases, thaumatin-like proteins, and protease inhibitors. Thus one strategy to engineer plants with broad-spectrum disease resistance may involve overproduction of salicylic acid. Theoretically, this can be achieved by transforming plants with

Table 20.7 Some Examples of Transgenic Plants Exhibiting Resistance to Fungal Diseases

Expressed gene product	Resistance against	Plant species engineered
Expression of hydrolytic enzymes		
Tomato chitinase	*Cylindrosporium concentricum* and *Sclerotinia sclerotiorum*	Canola
Alfalfa glucanase	*Phytophthora megasperma*	Alfalfa
Trichoderma harziamum endochitinase	*Venturia inaequalis*	Apple
Baculovirus chitinase	*Alternaria alternata*	Tobacco
Barley chitinase	*Blumeria graminis and Puccinia recondita*	Wheat
Tobacco chitinase	*Cercospora arachidicola*	Peanut
Expression of PR proteins		
Pea PR-10 gene	*Verticillium dahliae*	Potato
Tobacco PR-1a	*Peronospora tabacina* and *Phytophthora parasitica*	Tobacco
Potato osmotin-like protein	*Phytophthora infestans*	Potato
Pea defense response gene, defensin	*Leptosphaeria maculans*	Canola
Rice thaumatin-like protein (TLP)	*Fusarium graminearum*	Wheat
Rice *Rir* 1b defense gene	*Magnaporthe grisea*	Rice
Aspergillus giganteus antifungal protein	*Blumeria graminis and Puccinia recondita*	Wheat
Expression of antimicrobial proteins, peptides, or compounds		
Barley ribosome inactivating protein (RIP)	*Rhizoctonia solani*	Tobacco
Bacillus amyloliquefaciens barnase (RNase)	*Phytophthora infestans*	Potato
Arabidopsis thionin	*Fusarium oxysporum*	*Arabidopsis*
Human lysozyme	*Erysiphe cichoracearum*	Tobacco
Mistletoe thionin viscotoxin	*Plasmodiophora brassicae*	*Arabidopsis*
Onion antimicrobial protein	*Botrytis cinerea*	*Geranium*
Human lysozyme	*Erysiphe heraclei* and *Alternaria dauci*	Carrot
Alfalfa defensin	*Verticillium dahliae*	Potato
Synthetic cationic peptide chimera	*Fusarium solani* and *Phytophthora cactorum*	Potato
Sarcotoxin peptide from *Sarcophaga peregrina*	*Rhizoctonia solani, Pythium aphanidermatum*, and *Phytophthora nicotianae*	Tobacco
Chloroperoxidase from *Pseudomonas pyrocinia*	*Colletotrichum destructivum*	Tobacco
Synthetic antimicrobial peptide	*Colletotrichum destructivum*	Tobacco
Radish defensin	*Alternaria solani*	Tomato
Barley RIP	*Blumeria graminis*	Wheat
Antifungal (killing) protein from virus infecting *Ustilago maydis* (ds RNA)	*Ustilago maydis* and *Tilletia tritici*	Wheat
Synthetic magainin-type peptide	*Peronospora tabacina*	Tobacco
Expression of phytoalexins		
Grape stilbenes (Resveratrol) synthase	*Botrytis cinerea*	Barley
Alfalfa isoflavone *O*-methyltransferase	*Phoma medicaginis*	Alfalfa
Peanut resveratrol synthase	*Phoma medicaginis*	Alfalfa
Inhibition of pathogen virulence products		
Pear polygalacturonase inhibiting protein	*Botrytis cinerea*	Tomato
Collybia velutipes oxalate decarboxylase	*Sclerotinia sclerotiorum*	Tomato
Wheat oxalate oxidase	*Septoria mustiva*	Poplar
Alteration of structural compounds		
Wheat germin (no oxalate oxidase activity)	*Blumeria graminis*	Wheat

(*Contd.*)

Table 20.7 (*Contd.*)

Expressed gene product	Resistance against	Plant species engineered
Regulation of plant defense responses		
Tobacco catalase	*Phytophthora infestans*	Potato
Arabidopsis *NPR* 1 protein	*Peronospora parasitica*	*Arabidopsis*
Phytophthora infestans elicitor (β-cryptogein)	*Phytophthora parasitica*	Tobacco
Talaromyces flavus glucose oxidase	*Rhizoctonia solani* and *Verticillium dahliae*	Cotton
Bacterial enzymes generating salicylic acid	*Oidium lycopersicon*	Tobacco
Expression of combined gene products		
Rice chitinase + alfalfa glucanase	*Cercospora nicotianae*	Tobacco
Barley chitinase + β-1,3-glucanase, or chitinase + RIP	*Rhizoctonia solani*	Tobacco
Tobacco chitinase + β-1,3-glucanase	*Fusarium oxysporum*	Tomato
Tobacco chitinase + β-1,3-glucanase, osmotin	*Alternaria dauci*, *Alternaria radicina*, and *Erysiphe heraclei*	Carrot

bacterial genes encoding isochorismate synthase and isochorismate pyruvate lyase. In practice, since chorismic acid, which is the end product of the shikimate pathway and an intermediate in the biosynthesis of the amino acid tryptophan, is produced in large amounts in the chloroplast, both bacterial genes were fused to chloroplast targeting sequences from the gene for the small subunit of rubisco. The result of this gene manipulation was that when both these enzymes were localized in the tobacco chloroplast, salicylic acid was produced constitutively. Consequently, these plants constitutively expressed a number of PR proteins. The plants appeared normal but showed enhanced resistance to both viral and fungal pathogens. In yet another strategy, the *NPR* 1 gene that could be activated or induced by the addition of salicylic acid was employed. In *Arabidopsis*, overexpression of *NPR* 1 gene has led to the generation of broad-spectrum disease resistance against both fungal and bacterial pathogens.

Expression of Ribosome-inactivating Proteins RIPs are also employed as a defense system against fungal infections. These proteins usually remove an adenine residue from a specific site in the large rRNA of eukaryote and prokaryote ribosomes, consequently inhibiting protein synthesis with varying specificities. Thus type-1 barley RIP, expressed constitutively, has been used to provide resistance to fungal attack. This RIP when used in tandem with the chitinase gene led to a synergistic effect against *R. solani* in transgenic plants.

Expression of Antimicrobial Proteins Many other genes that code for proteins with both antifungal and antibacterial activities have been introduced into plants with varying success. The proteins used are lysozyme (enzyme that degrades peptidoglycan), thionin proteins, cationic antimicrobial peptides, and defensins. Examples include the following: (i) Bacteriophage T4 lysozyme gene was successfully employed for the protection of potato against pathogenic soil borne *E. carotovora*. The strategy involved high-level expression and targeting of lysozyme for secretion into the apoplast (part of the plant where *E. carotovora* enters and spreads) in potato plants; (ii) Various thionin proteins have been introduced into plants for generation of fungal- or bacterial-resistant transgenic plants. Barley α-thionin expressed in transgenic tobacco was shown to increase resistance to *P. syringae*; (iii) Another approach includes engineering plants to express cationic antimicrobial peptides. In this direction, the genes encoding peptide cecropin A (35 amino acid peptide from the giant silk moth, *Hyalophora cecropia*) and melittin (26 amino acid peptide, which is the major component of bee venom) have been employed. Individually these peptides were unable to provide the desired activity. With an aim to combine the properties of both these peptides, a hybrid antibacterial peptide, namely CEMA, was developed by fusion of N-terminus from melittin and C-terminus from cecropin A. CEMA had reduced hemolytic activity compared with melittin while retaining most of the antibacterial activity and unfortunately it was toxic to plants. To overcome this problem, six additional amino acids were added to the N-terminal end of CEMA. The resultant peptide, called Msr A1, had a lower, although still potent, level of antibacterial activity and was nontoxic to plants. Expression of the gene for Msr A1 under CaMV 35S promoter and introduction of this construct into potato and tobacco protected the plants against bacterium *E. carotovora* as well as fungi *Phytophthora cactorum* and *F. solani*; and (iv) Another exciting transgenic approach to microbial resistance is the use of defensins, the small antimicrobial peptides found in all living cells. An artificial defensin gene has been made that incorporates the cecropin and mellitin genes, obtained from a giant silk moth and bee, respectively. This gene has been introduced into potatoes and transgenic plants challenged in tissue culture conditions with the bacterium *E. carotovora* were found to be resistant.

Expression of Plant Proteinase, Bromelain Plant extract such as bromelain comprises a variety of proteolytic enzymes. Thus the cloning of *BAA* 1 gene encoding the fruit bromelain into a plant expression vector under the control of CaMV 35S promoter has been successfully used to transform *Brassica rapa*. The transformants exhibited enhanced resistance to the soft rot pathogen *Pectobacterium carotovorum* ssp. *carotovorum*.

20.3.4 Development of Virus-resistant Plants

Plant viruses often cause considerable crop damage and significantly reduce yields. Moreover, there is no effective chemical treatment to control plant viruses. To overcome the problems associated with plant viruses, plant breeders have attempted to transfer naturally occurring virus resistance genes from one plant strain (cultivar) to another. However, resistant cultivars often revert to virus sensitivity and resistance to one virus does not necessarily confer resistance against other similar viruses. Plants also exhibit certain natural viral resistance mechanisms, for example, (i) Blockage of viral transmission; (ii) Prevention of establishment of virus; and (iii) Bypassing of viral symptoms. Genetic engineering has also been used to develop nonconventional types of virus-resistant transgenic plants. Such virus-resistant transgenic plants will benefit farmers as well as the environment by a reduction in the use of pesticides. The type of interaction between transgenic sequences and virus life cycles is categorized into two groups, *viz.*, those involving the synthesis of viral proteins and those involving viral RNA. Based on these two types of interactions, various strategies have been developed for raising virus-resistant plants, which are as follows:

Coat Protein Gene Approach (or CP-mediated Resistance) The first antiviral transgenic approach used pathogen-derived resistance (PDR), originally known as parasite-derived resistance. This involves engineering of pathogen genomic sequences into the plant genome on a nonscientific basis so that the sequence may be expressed at an inappropriate time, in inappropriate amounts, or in an inappropriate form during the infection cycle, thereby inducing some form of resistance in the plant. The mechanisms of PDR are varied and complex. Probable explanations for PDR-based resistance are that the presence of the pathogen sequence may directly interfere with the replication of the pathogen or may induce some host defense mechanism. The most successful PDR transgenic approach to viral resistance is the transgenic expression of viral coat protein (CP-mediated resistance), which usually is the most abundant protein of a virus particle. The precise mechanism of CP-mediated resistance is not clearly understood, but it is not antibody-mediated. When transgenic plants express the gene for a CP protein of a virus that normally infects those plants, the ability of the virus to subsequently infect the plants and spread systemically is often greatly diminished. The antiviral effect occurs early in the viral replication cycle and as a result, prevents any significant amount of viral synthesis. This offers an advantage because it decreases the probability of selecting for spontaneous viral mutants that can overcome this resistance and replicate in the presence of viral coat protein. This approach was first reported with a tobacco mosaic virus (TMV)-tobacco model system in 1986. Later this strategy was used with several viruses to develop transgenic plants resistant to particular viruses (Table 20.8). Although complete protection is not usually achieved, high levels of virus resistance have been reported. Moreover, a coat protein gene from one virus sometimes provides tolerance to a broad-spectrum of unrelated viruses. There are several important features of the CP strategy. First, there is some level of cross-protection against infection by related viruses. Another important point is that for TMV, alfalfa mosaic alfamovirus (AIMV), potato X potexvirus (PVX), and several viruses from other groups, the level of resistance is related to the level of transgenic coat protein produced. Using this strategy, transgenic papaya plants resistant to papaya ringspot potyvirus (PRSV) (Figure 20.4c) have been commercialized, and grown in Hawaii. Zucchini yellow mosaic virus and watermelon mosaic virus resistant 'Freedom II' squash plants have also been produced using this strategy (Figure 20.4d).

Table 20.8 Some Examples of GM Plants Created to have Viral CP-mediated Protection Against Viral Infection

Viral source of coat protein	Transgenic plant
Alfalfa mosaic virus	Alfalfa, tobacco, tomato
Arabis mosaic virus	Tobacco
Beet necrotic yellow vein virus	Sugar beet
Cucumber mosaic virus	Cucumber, tobacco
Cymbidium ringspot virus	Tobacco
Grapevine chrome mosaic virus	Tobacco
Maize dwarf mosaic virus	Sweet corn
Papaya ringspot virus	Papaya, tobacco
Plum pox virus	Tobacco
Potato aucuba mosaic virus	Tobacco
Potato leaf-roll virus	Potato
Potato virus S	Potato
Potato virus X	Potato, tobacco
Potato virus Y	Potato, tobacco
Rice stripe virus	Rice
Soybean mosaic virus	Tobacco
Tobacco etch virus	Tobacco
Tobacco mosaic virus	Tobacco, tomato
Tomato mosaic virus	Tomato
Tomato rattle virus	Tobacco
Tomato streak virus	Tobacco
Tomato spotted wilt virus	Tobacco
Watermelon mosaic virus II	Tobacco
Zucchini yellow mosaic virus	Muskmelon, tobacco

Antisense RNA for Viral Coat Protein The RNA molecule complementary to mRNA (normal gene transcript) is called antisense RNA (for details see Chapter 17). This antisense RNA hybridizes to complementary mRNA, which has two different consequences: (i) Prevention of translation of mRNA and decrease in the synthesis of the gene product; and (ii) Rapid degradation of antisense RNA:mRNA duplex by nucleases and hence rapid diminution of the amount of that particular mRNA in the cell. Theoretically, by creating transgenic plants that synthesize antisense RNA complementary to virus coat protein mRNA, the plant viruses can be prevented from replicating and subsequently damaging plant tissues. One such example is that of transgenic potato plant expressing antisense RNA to potato leaf roll luteovirus (PLRV) coat protein, which was resistant to PLRV infection.

Expression of *E. coli* RNase III Gene Application of viral coat protein gene in transgenic plants is usually an effective strategy only against closely related viruses. However, as a large number of different viruses infect a crop, it is advantageous to engineer plants in such a way that it is resistant to a broad spectrum of viruses. In this direction, ribonuclease *RNase* III gene (*rnc*) from *E. coli* has been employed for the protection of plants against damage from infective viruses. The enzyme is responsible for the cleavage of ds RNA, which is the genetic material of most of the plant viruses. When tested, transgenic plants that expressed the *rnc* gene were resistant to several different plant RNA viruses. Unfortunately, plants that expressed this gene were often stunted and did not exhibit normal development. This was probably attributable to the interaction between the plant RNA and the enzyme. To overcome this problem, a mutant gene (*rnc* 70) was used, which encoded RNase III that had lost substrate cleavage ability, but not ds RNA binding capability. When introduced into wheat under the control of a corn ubiquitin gene, normal transgenic plants were developed that expressed mutant RNase III and exhibited a high level of resistance to infection by barley stripe mosaic virus. In this instance, binding of the mutant RNase III to replicating barley stripe mosaic virus prevented viral replication. This approach is also effective for the elimination of viriod infection of plants.

Expression of Antiviral Plant Protein Expression of antiviral plant proteins can confer protection against viruses. For example, *Phytolacca americana* (pokeweed) has three antiviral proteins in its cell wall: pokeweed antiviral protein (PAP) in spring leaves, PAP II in summer leaves, and PAP-S in seeds. PAP and PAP II are good candidates for developing transgenic plants resistant to a broad-spectrum of plant viruses. This is because these are RIPs and lead to the removal of a specific adenine residue of rRNA component from 60S subunit of eukaryotic ribosome. American pokeweed produces an antiviral protein called dianthrin that functions as RIP. The cDNA for this protein has been cloned and expressed in tobacco, providing resistance against African cassava mosaic virus (ACMV).

Expression of Virion-specific Antibodies Antibodies specific for virion proteins have also been used to protect plants from viruses. The first experiment in this direction is that of expression of a single chain Fv fragment (scFV) specific for ACMV in *Nicotiana benthamiana* and the resulting transgenic plants were found to be resistant to viral infection.

Expression of Other Nonviral Gene Products Besides RNase, which destroy viral genome, several other classes of nonviral gene products can be used to provide specific protection against viruses, including RIPs, which block protein synthesis, 2′,5′-oligoadenylate synthetases, which interfere with replication, and ribozymes, which cleave viral genomes, causing them to be degraded by cellular enzymes.

20.3.5 Development of Abiotic Stress-tolerant Plants

Plants have evolved several physiological strategies to circumvent stresses such as oxidative stress and those induced by high levels of light, UV irradiation, heat, cold, high salt concentrations, or drought. Taking advantage of these strategies, several transgenic plants resistant to abiotic stresses have been generated. Some of them are summarized below and a list of various abiotic stress-tolerant transgenic plants is provided in Table 20.9.

Table 20.9 Engineering Abiotic Stress-tolerant Transgenic Plants

Transgene	Product encoded	Crop plant (host)	Stress tolerance
Arthrobacter globiformis cod A	Choline oxidase	*Arabidopsis* chloroplast	Salt, chilling, freezing, heat, strong light
Arthrobacter pascens cox	Choline oxidase	*Arabidopsis*	Freezing, salt
Annexine-like protein	Annexine like protein	*Brassica*	Water deficit, salinity
Antifreeze gene from fish, and overwintering plants	Antifreeze protein	Tobacco, potato	Cold

(*Contd.*)

Table 20.9 (*Contd.*)

Transgene	Product encoded	Crop plant (host)	Stress tolerance
Anti-ProDH	Proline dehydrogenase [Proline osmoprotectant]	*Arabidopsis*	Salt, freezing
Apple *s6pdh*	Sorbitol-6-phosphate dehydrogenase [Sorbitol osmoprotectant]	Tobacco	Oxidative stress
Apple *s6pdh*	Sorbitol-6-phosphate dehydrogenase [Sorbitol osmoprotectant]	Persimmon	Salt
apx 1	Ascorbate peroxidase	Tobacco	Oxidative stress
apx 1	Ascorbate peroxidase	*Arabidopsis*	Heat
AtHSF 1	Heat-shock transcription factor HSF1::GUSfusion	*Arabidopsis*	Heat
Bacillus subtilis sac B	Fructans osmoprotectant	Tobacco	Drought
B. subtilis sac B	Fructans osmoprotectant	Sugar beet	Drought
bet A	Choline dehydrogenase [Glycine betaine osmoprotectant]	Rice	Drought, salt
bet A/*bet* B	Choline dehydrogenase with betaine aldehyde dehydrogenase [Glycine betaine osmoprotectant]	Tobacco	Chilling, salt
CBF 1	CRT/DRE element binding protein	*Arabidopsis*	Cold
Chloroplast Cu/Zn-SOD pea	Cu/Zn-SOD	Tobacco chloroplast	High light, chilling
cod A	Choline oxidase	Rice	Salt, chilling
cox	Choline oxidase	*Brassica napus*	Drought, salt
cox	Choline oxidase	Tobacco	Salt
Cytosolic Cu/Zn-SOD	Cu/Zn-SOD	Tobacco cytosol	Acute ozone exposure
E. coli bet A	Choline dehydrogenase	Tobacco cytosol	Salt
E. coli mt1D	Mannitol-1-phosphate dehydrogenase [Mannitol osmoprotectant]	*Arabidopsis*	Salt
E. coli mt1D	Mannitol-1-phosphate dehydrogenase [Mannitol osmoprotectant]	Tobacco	Salt
E. coli ots A + *ots* B	Trehalose-6-phosphate synthase with Trehalose-6-phosphate phosphatase [Trehalose osmoprotectant]	Tobacco	Drought
Fe-SOD (Arabidopsis)	Fe-SOD	Tobacco	Acute ozone exposure
Glutathione reductase (*E. coli*)	Glutathione reductase	Tobacco cytosol	Paraquat
Glycerol-3-phosphate acyl-transferase	Glycerol-3-phosphate acyl-transferase	Tomato	Cold
GS 2	Chloroplastic glutamine synthetase [Glutamine osmoprotectant]	Rice	Salt, chilling
GST/GPX	Glutathione *S*-transferase with glutathione peroxidase	Tobacco	Oxidative stress
Hsp 101	HSP100 class heat-shock protein	*Arabidopsis*	Heat
Hsp 17.7	SmHSP small heat-shock protein family	Carrot	Heat
Hsp 70	HSP70 class heat-shock protein	*Arabidopsis*	Heat
Ice plant *imt* 1	*Myo*-inositol *o*-methyltransferase [*D*-Ononitol osmoprotectant]	Tobacco	Drought, Salt
lea	Late embryogenesis abundant protein	*Arabidopsis*	Cold

(*Contd.*)

Table 20.9 *(Contd.)*

Transgene	Product encoded	Crop plant (host)	Stress tolerance
Mitochondrial Mn-SOD tobacco	Mn-SOD	Alfalfa chloroplast	Drought
Mitochondrial Mn-SOD tobacco	Mn-SOD	Tobacco chloroplast	Ozone
Mn SOD	Mn-SOD	Canola	Aluminium
Mothbean *P5 CS*	Pyrroline carboxylate synthetase [Proline osmoprotectant]	Tobacco	Salt
Mothbean *P5 CS*	Pyrroline carboxylate synthetase [Proline osmoprotectant]	Rice	Drought, salt, oxidative stress
Mothbean *P5 CS*	Pyrroline carboxylate synthetase [Proline osmoprotectant]	Soybean	Osmotic stress, heat
MsFer (alfalfa)	Ferritin	Tobacco	Oxidative damage caused by excess iron
Nt 107	Glutathione-*S*-transferase	Tobacco	Cold, salinity
Nt pox	Glutathione peroxidase	*Arabidopsis*	Aluminium toxicity, oxidative stress
osm 1 to *osm* 4 (protein accumulation)	Osmotin osmoprotectant	Tobacco	Drought, salt
P5 CS^{F127A} (feedback inhibition insensitive)	Pyrroline carboxylate synthetase [Proline osmoprotectant]	Tobacco	Salt
Par B	Glutathione-*S*-transferase	*Arabidopsis*	Aluminium toxicity, oxidative stress
TLHS 1	Class I sm HSP	Tobacco	Heat
Yeast *tps* 1	Trehalose-6-phosphate synthase [Trehalose osmoprotectant]	Tobacco	Drought
Yeast *tps* 1	Trehalose-6-phosphate synthase [Trehalose osmoprotectant]	Potato	Drought

Oxidative Stress Tolerance

Several abiotic stresses, such as salt, freezing, heat, drought, exposure to pollutants as ozone, SO_2, lead to generation of reactive oxygen species (ROS) in plant cells. These molecules result in oxidative stress by damaging membranes, membrane-bound structures, and macromolecules including proteins and nucleic aids, especially in the mitochondria and chloroplast. Thus, plants engineered to tolerate ROS (i.e., oxidative stress) can withstand such environmental stresses. This can be easily accomplished by targeting oxygen radical scavengers in the following ways:

Application of SOD A common type of potentially damaging oxygen radical is the superoxide anion. Within a cell, under oxidative stress, the enzyme SOD detoxifies superoxide anion by converting it to hydrogen peroxide, which in turn is broken down to water by various cellular peroxidases or catalases. In a study, tobacco plants transformed with a *SOD* gene cloned under CaMV 35S promoter had reduced oxygen radical damage under stress conditions compared with control plants. There are different isoforms of SOD, *viz*., Cu/Mn SOD primarily present in chloroplast and to some extent in cytosol, Mn SOD located in mitochondria, and Fe isoform also present in some plants. Transgenic tobacco plants transformed with a vector carrying a cDNA for Cu/Mn *SOD* under the control of CaMV 35S promoter were found to be more resistant to damage by high light as compared to untransformed plants. In another study, targeting of Mn SOD into the chloroplast led to generation of transgenic plants, which were three to four times less sensitive to ozone damage as compared to untransformed plants.

Application of Oxidized Glutathione Oxidative stress may also be reduced if the level of oxidized glutathione (GSSG) is increased within a plant by using glutathione-*S*-transferase and glutathione peroxidase. In addition, an optimal strategy for lowering oxidative stress in plants is the introduction of both glutathione-*S*-transferase/glutathione peroxidase and SOD.

Salt and Submergence Stress Tolerance

Growth of many plants is severely impaired by either drought or high salinity. Moreover, a large fraction (nearly one-third) of the world's irrigated land is laden with excessive salt and has become unsuitable for growing crops. Thus the development of crops with an inbuilt salinity resistance can stabilize annual production and use of marginal lands for cultivation. The resistance to salt stress can be achieved by any one of the following strategies:

Increase in Cellular Accumulation of Osmoprotectants To survive under high salt conditions, many plants naturally synthesize low molecular weight nontoxic compounds collectively called osmoprotectants (or osmolytes), for example, sugars, alcohols, proline, and quaternary ammonium compounds, etc. These compounds function by facilitating water uptake as well as water retention. These compounds also protect and stabilize cellular macromolecules from damage by high levels of salt. Thus, a useful strategy for the acquisition of salt tolerance is the development of transgenic plants with increased cellular accumulation of osmoprotectants. A list of osmoprotectants used to engineer salt-tolerant plants is provided in Table 20.9.

Compartmentalization of Salts into Vacuole by Overproduction of Na^+/H^+ Antiport Protein Plants that live in saline environment suffer from Na^+ toxicity and also have to contend with water loss caused by osmotic stress. Thus sequestering sodium ions in the large intracellular vacuole can be beneficial. This not only protects plants from toxic effects of high Na^+ ions, but also results in plants that use water more efficiently. This is because when Na^+ ions are concentrated in vacuole (i.e., compartmentalized), water becomes salt-free, and is readily driven into plant cells. This strategy has been successfully used in *Arabidopsis thaliana* by overexpression of the gene encoding Na^+/H^+ antiport protein, which transports Na^+ into the vacuole using the electrochemical gradient of protons generated by vacuolar H^+ translocating enzymes. Thus the transgenic plants overproducing Na^+/H^+ antiport protein thrived in soil watered with 200 mM salt.

Cold/freezing Tolerance

Injury inflicted by frost and freezing is found in all plants exposed to such damaging temperatures. Different plants vary enormously in their ability to withstand cold and freezing temperatures. Most tropical plants have virtually no capacity to survive freezing conditions. On the other hand, many temperate plants can survive a range of freezing temperatures from –5°C to –30°C depending upon the species. Plants from colder regions routinely withstand temperatures even lower than this. Considering this difference, efforts were made to engineer cold stress-tolerant plants. In this direction, 50 different cold-induced proteins have been identified in different plant species. These include 'late embryogenesis abundant' (LEA) proteins and 'cold responsive' (COR) proteins (Table 20.9). Both these proteins contribute directly to freezing tolerance by mitigating the potentially damaging effects of dehydration associated with freezing.

Overexpression of *COR* and *CBF* 1 Genes Overexpression or ectopic expression of these cold-induced proteins could be a possible route to the specific engineering of cold or freezing stress tolerance. There are some examples of the expression of cold-induced proteins in transgenic plants. For example, constitutive expression of small, hydrophilic, chloroplast-targeted COR protein COR15a in *Arabidopsis* improved the freezing tolerance of chloroplasts frozen *in situ* and of protoplasts frozen *in vitro*. Other strategy is based on engineering plants with the overexpression of a common regulatory protein. Note that several cold tolerance-related genes form a *COR* regulon, where all the cold tolerance-related genes contain a similar regulatory element in their promoters (*CRT/DRE* element), which is bound and activated by CBF 1 transcription factor. Thus by overexpressing *CBF* 1 gene, the entire group of *COR* regulon can be induced. In this direction, transgenic *Arabidopsis* plants carrying a CaMV 35S promoter::*CBF* 1 gene construct have been produced. These plants expressed a number of *COR* genes and have been shown to be freezing-tolerant.

Overexpression of Glycerol 3-Phosphate Acyltransferase The overexpression of tomato glycerol-3-phosphate acyltransferase gene (*LeGPAT*) under the control of constitutive promoter CaMV 35S has also been shown to improve chilling tolerance in tomato. Chilling treatment induced less ion leakage from the transgenic plants than from the wild-type. The photosynthetic rate and the maximal photochemical efficiency of PS II (Fv/Fm) in transgenic plants decreased more slowly during chilling stress and recovered faster than in wild-type under optimal conditions.

Cloning of Genes Encoding Antifreeze Proteins The expression of AFPs in a plant is a possible means of increasing the frost resistance and freeze tolerance of plants. In 1991, researchers at DNA Plant Technology developed a genetically engineered variety of tomato that expressed a gene identified in an Arctic flounder. The flounder gene encoded a protein conferring cold resistance to the fish. The goal was to develop tomato plants that could withstand frost in the field and fruits that resisted cold damage in storage. This particular experiment, however, was a failure and did not produce frost-resistant plants, but its story is continued. Later a tale about strawberries with fish genes, the 'fishberry' was heard though there was no such scientific report to this effect. Several plants have also been engineered for cold tolerance using antifreeze genes from overwintering plants.

Heat Tolerance

Similar to cold/freezing tolerance, heat-shock proteins (Hsp) are inducible by heat stress. In a way analogous to cold/freezing tolerance, individual heat-shock proteins have been transformed into plants in order to enhance heat tolerance. Similar to CBF 1, heat-shock transcription factor (HSF) also binds and activates the promoter element (heat-shock element; *HSE*) contained in all the heat-shock genes. Thus, fusion of *AtHSF* 1 gene to N- or C-terminus of the *gus* A reporter gene was used for *Arabidopsis* transformation. The resulting transgenic plants constitutively expressed heat-shock proteins and demonstrated enhanced thermotolerance (Table 20.9).

20.3.6 Improvement of Plant Nutritional Status

Improvement of plant nutritional content can be done in a variety of ways, for example, increase the amino acid content (specifically methionine and lysine) of some seed storage proteins, modify lipid (or fatty acid) composition of both edible and nonedible oil-producing crops so that the oil becomes better suited for intended end use, synthesis of Vitamin E and β-carotene, and increasing the levels of available iron, etc. Such modifications increase the nutritional status of the foods and may help to improve human health by addressing malnutrition, undernutrition, and micronutrient deficiencies. Traditionally, plant nutritional improvement was achieved by plant breeding, but these approaches were difficult, slow, and intrinsically limited by the existing genetic content and crossbreeding strains. Presently, genetic engineering strategies are being routinely adopted for obtaining nutritionally enriched plants.

Improvement in Amino Acid or Protein Content

Seed storage proteins are used as sources of both carbon and nitrogen during seed germination and are the major protein source for most of the world's population. But most often, these seed storage proteins are deficient in one or the other essential amino acids. Cereals are usually deficient in lysine. Rice, though has higher lysine content compared to other cereals, has low protein content. Recently genetic engineering techniques are being employed to alter the amino acid composition of seed storage proteins and hence to improve their nutritional quality. This also decreases the need for amending grains with costly feed supplements.

Expression of a Gene Encoding a Protein Rich in Particular Amino Acid Engineering plants with a gene that encodes a seed protein rich in a particular amino acid can be used as a strategy to enrich its own seed proteins with the lacking amino acids. For example, lupine, a grain legume used as feed for cattle, pigs, and chicken in Australia lacks methionine and cysteine. Thus lupine has been successfully engineered to express sunflower seed albumin, a rich source of sulfur-containing amino acids methionine and cysteine. The protein was stable in rumen and was available for digestion and absorption in the lower gut. These transgenic lupine plants when used as rat feed as sole nitrogen source, rats made significantly higher weight gains. Similarly, seed-specific 2S albumin gene from Brazil nut (*Bertholletia excelsa*) has been used to transform several plant species, e.g., *Vicia*, tobacco, alfalfa, *Trifolium*, etc. to increase methionine or sulfur content.

Deregulation of Amino Acid Biosynthetic Pathway One novel way to increase the lysine content of seeds is to increase the production of lysine in transgenic plants by deregulating lysine biosynthesis pathway. The amino acids lysine, threonine, methionine, and isoleucine are all derived from aspartate. The first step in the conversion of aspartate to lysine is phosphorylation of aspartate by aspartokinase (AK) to produce β-aspartyl phosphate. The condensation of aspartic β-semialdehyde with pyruvic acid to form 2,3-dihydrodipicolinic acid, which is catalyzed by dihydrodipicolinic acid synthase (DHDPS), is the first reaction in the pathway committed to lysine biosynthesis. Thus to overproduce lysine it is necessary to abolish the feedback inhibition of these two enzymes. This was accomplished by cloning lysine feedback-insensitive genes for DHDPS and AK from *Corynebacterium* and *E. coli*, respectively.

Expression of Albumin In a study in potato, nutritional enhancement by deploying a nonallergenic seed albumin gene *ama* I derived from the plant *Amaranthus hypochondriacus* has also been achieved.

Alteration of Lipid or Fatty Acid Composition

Soybean, palm, rapeseed (canola), and sunflower accounts for ~75% of worldwide plant oil production, and for most part, these oils consist of fatty acids such as palmitic, stearic, oleic, linoleic, and linolenic acids. In addition, some vegetable oils contain fatty acids with conjugated double bonds. This is in contrast to the more usual case in which the typical polyunsaturated fatty acids (PUFA) of plant seed oils contains double bonds that are separated by methylene ($-CH_2$) groups. Different types of fatty acids, *viz.*, saturated fatty acids or fatty acids with conjugated double bonds, or PUFA, are required for different applications. Besides chain length is also significant in different applications. For example, for use as drying agents in paints and inks, oils used should be such that less oxygen is required for the polymerization reactions occurring during drying, hence oils containing conjugated double bonds (i.e., methylene interrupted double bonds) are most suitable. This is because the presence of conjugated double bonds increases the rate of oxidation (in contrast to PUFA). Production of shortening, margarine, and confectionery goods requires large amounts of stearate. Thus for application of fatty acids for different purposes, its degree of unsaturation (i.e., the number of C–C double bonds), and chain length should be modified. This modification is easily done in plants by genetic engineering. In this direction, a number of canola and other transgenic plants with modified fatty acid content and composition have been created that produced oils with better cooking and nutritional properties or for better suitability in detergents, soaps, cosmetics, lubricants, and paints. A list of enzymes targeted for engineering plants with altered lipid content and composition is provided in Table 20.10, and examples of transgenic canola varieties with modified seed lipid contents are presented in Table 20.11.

Table 20.10 Some Examples of GM Plants with Altered Lipid Content and Composition

Trait	Gene	Source	Crop
Increased lauric acid	Lauryl acyl carrier protein (ACP) thioesterase	*Umbellularia californica*	Canola
Increased ricinolic acid	Oleate hydroxylase	*Ricinus communis*	Canola
Increased seed oil	Acetyl-CoA carboxylase	Rapeseed, *Triticum aestivum*, Alfalfa	Canola
Increased stearic acid	Antisense stearoyl-ACP desaturase	*Brassica rapa*	Canola
Increased petroselinic acid	Palmitoyl-ACP desaturase	*Coriandrum sativum*	Canola

Table 20.11 Examples of Transgenic Canola with Modified Seed Lipid Contents

Seed product	Commercial use(s)
40% Stearic	Margarine, cocoa butter
40% Lauric	Detergents
60% Lauric	Detergents
80% Oleic	Food, lubricants, inks
Petroselinic	Polymers, detergents
'Jajoba' wax	Cosmetics, lubricants
40% Myristate	Detergents, soap, personal care items
90% Erucic	Polymers, cosmetics, inks, pharmaceuticals
Ricinoleic	Lubricants, plasticizers, cosmetics, pharmaceuticals

Increase in Vitamin E Content

It has been evidenced through several studies that dietary supplement with a lipid-soluble antioxidant vitamin E [400 International units or ~250 mg of (*R*,*R*,*R*)-α-tocopherol daily] results in a decreased risk for cardiovascular diseases and cancer, assists in immune function, and prevents or slows a number of degenerative diseases in humans. The oils extracted from seeds have a relatively high level of total tocopherols, but in most of the cases, the fraction of α-tocopherol is not high enough. In an example of sunflower oil, which contains high fraction of α-tocopherol, intake should be ~400 g per day so as to obtain sufficient amount of Vitamin E. This is a large amount to be consumed daily, hence an alternative approach is to engineer plants to produce a greater percentage of α-tocopherol than γ-tocopherol or to increase the synthesis of α-tocopherl. Some approaches in this direction are as follows.

Expression of γ-Tocopherol Methyltransferase (γ-TMT) Genetic manipulation using γ-*TMT* gene encoding an enzyme catalyzing methylation of γ-tocopherol to α-tocopherol is a useful strategy in increasing the percentage of α-tocopherol. Thus cloning of *Arabidopsis* γ-*TMT* under the transcriptional control of a seed-specific promoter from carrots (DC3 promoter) has been used to transform *Arabidopsis* plants. As compared to untransformed plants, the α-tocopherol level of these transgenic plants was ~80-fold greater.

Expression of Homogentisic Acid Prenyltransferase (HPT) Tocopherol synthesis in plants requires the input from two metabolic pathways. The shikimate pathway generates homogentisic acid, which forms the aromatic ring of the compound, whereas the side chain is derived from phytyldiphosphate, a product of the methylerythritol phosphate (MEP) pathway. These precursors are joined together by HPT enzyme to form the intermediate 2-methyl-6-phytylbenzoquinol (MPBQ). Thus, overexpression of HPT in *A. thaliana* has been reported to produce twice the level of vitamin E found in normal seeds.

Enhancement of Vitamin A Precursor

In mammals, vitamin A is synthesized from the precursor provitamin A (β-carotene), a common carotenoid pigment normally found in plant photosynthetic membranes. But few plants, including rice, are a poor source of several nutrients and vitamins, including vitamin A. Traditional breeding methods have been unsuccessful in producing crops containing a high vitamin A concentration and most national authorities rely on expensive and complicated supplementation programs to address the problem. To overcome this problem, transgenic approach was developed to engineer plants with inbuilt β-carotene pathway. Examples include 'Golden rice' and 'Beta-sweet® carrot', which are discussed below.

Expression of β-Carotene Biosynthetic Pathway Enzymes in Rice As rice is the staple food of approximately half of the world's population, one way to circumvent the symptoms associated with vitamin A deficiency (VAD) would be to engineer rice to produce β-carotene. In this direction, *Agrobacterium*-mediated transformation was used to introduce the entire β-carotene biosynthetic pathway into rice to catalyze the biosynthesis of β-carotene from geranyl geranyl diphosphate (GGDP; the progenitor of vitamin A in rice endosperm). Rockefeller Foundation, Swiss Federal Institute of Technology, and European Community Biotech Program granted the funding for the research. Three genes of vitamin A biosynthesis pathway, *viz.*, *psy* (phytoene synthase), *crt* (phytoene desaturase), and *lcy* (lycopene β-cyclase) were fused to transit peptides for targeting encoded proteins into the plastids. Phytoene synthase converts GGDP to phytoene, phytoene desaturase causes desaturation of phytoene to lycopene, and lycopene β-cyclase catalyzes cyclization of lycopene to form β-carotene. For plant transformation, *psy* and *lcy* genes were isolated from daffodil (*Narcissus pseudo-marcissus*) and cloned separately under glutelin promoter (i.e., promoter of rice seed storage protein). The *crt* gene isolated from bacteria *Erwinia uredovora* was put under the control of CaMV 35S promoter. The *psy* and *crt* genes were introduced on a construct lacking selectable marker. The *lcy* gene was part of a separate construct that contained a selectable marker. All three genes were expressed in endosperm (major part of the rice grain) and the frequency of insertion of all the three genes into the genomic DNA and their subsequent expression was quite high. Thus the engineered rice produced β-carotene, which, after ingestion by mammals, could be converted to vitamin A. This strategy facilitated the eventual development of transgenic strains of rice that not only produced high levels of β-carotene, but could also be easily manipulated so that these no longer contained antibiotic resistance marker genes. This β-carotene producing transgenic rice was transformed with gene encoding an iron storage protein ferritin (from bean), which allowed overexpression of this protein in endosperm of rice. As this transgenic rice was yellow or golden color, it was named 'golden rice' (Figure 20.4e).

Preliminary studies indicated that this golden rice synthesized enough β-carotene to provide individuals with more than the minimal daily requirement of vitamin A. Moreover, in contrast to vitamin A, excess dietary β-carotene had no harmful effects. Unfortunately production was too low and normal serving of rice (300 g) provided a few percentage of daily diet. Further transformation of rice with phytase gene from *Aspergillus fumigatus* fungus or overexpression of endogenous cysteine-rich metallothionein protein (e.g., cysteine peptidase having opposite effect to phytic acid) has been reported to favor efficient iron absorption.

Expression of β-Carotene Biosynthetic Pathway Enzymes in Carrot Beta Sweet® carrot (Figure 20.4f) was produced at Texas A&M University, which contained inbuilt β-carotene biosynthesis pathway to combat cancer (β-carotene is also a potent cancer-fighting antioxidant). The transgenic carrot contained ~50% more β-carotene than normal carrot. It was dark maroon-purple colored, which was due to addition of another antioxidant, anthocyanin (most common flower pigment belonging to the class of flavonoids). Its taste was similar to wild-type carrots, but had a very crispy texture, which was easier to chew.

Modulation of Carotenogenesis in Tomato Tomato fruit is the most important accumulator of the carotenoid pigment lycopene, which can be readily cyclized to β-carotene by the plant enzyme lycopene β-cyclase (β-Lcy). Thus, overexpression of β-Lcy by cloning a transgene conveying the tomato β-*lcy* cDNA under the control of fruit-specific promoter for the phytoene desaturase gene has led to the production of transgenic tomato plants whose fruits contained high amounts of β-carotene as a result of the almost complete cyclization of lycopene.

Increase in Iron Content

According to WHO estimates, iron deficiency affects ~30% of world's population and the problem is most severe where vegetable-based diets are the primary food source. Although a number of crops are rich in iron, its absorption is often prevented by the presence of phytic acid. Hence, iron fortification of cereal grains or food crops is required. As a first step toward developing food crops with a high enough level of iron to prevent iron deficiency anemia, cloning of gene encoding ferritin (an iron storage protein that is found in animals, plants, and bacteria that carries up to 4,500 iron atoms in its central cavity formed from the interaction of 24 monomeric ferritin subunits) was done. Rice was transformed with ferritin-encoding genes from soybean and a gene encoding an enzyme that facilitated iron availability in the human diet. These transgenic rice plants contained two to four times the levels of iron normally found in nontransgenic rice. Another example is that of double-transgenic golden rice, which was designed to prevent and treat vitamin A and iron deficiencies.

Prevention of Oxalic Acid-induced Nutritional Stress

Oxalic acid is an important stress factor. Some green leafy vegetables (e.g., *Amaranthus*, spinach, rhuburb) are rich sources of vitamins and minerals but these contain oxalic acid, a nutritional stress factor. This is because oxalate chelates calcium and precipitation of calcium oxalate in kidney leads to hyperoxaluria and destruction of renal tissues. Similarly, consumption of *Lathyrus sativus* (chicken vetch) causes neurolathyrism, which is characterized by spasticity of leg muscles, lower limb paralysis, convulsions, and death. This is the result of formation of β-*N*-oxalyl-*L*-α,β-diamino propionic acid (ODAP), a neurotoxin, from oxalic acid. Hence, development of transgenic plants with low oxalic acid content may be beneficial. Several studies have been conducted in this direction by targeting oxalate decarboxylase gene.

Modification of Sweetness

Certain food products though having high nutritional values are not relished by human beings and need to be made more appetizing. In addition, with the increasing health consciousness, a desire to get products having lesser calories has increased. Thus transgenic plants with a modification of sweetness have been generated. The purpose of this modification can be varied. For example, the purpose can be to prepare a sugar substitute (e.g., low calorie sweetener or nondigestible sweetener), or to make sweet-tasting vegetables more palatable and with longer shelf-life. Thus genetic manipulation to alter sweetness of fruits and vegetables can be done by any of the following ways:

Expression of Sweet-Tasting Proteins as Low-calorie Sugar Substitutes Given the demand for low-calorie sweeteners as well as interest in healthy and natural food products, it is likely that many plants can be genetically engineered to produce sweet-tasting proteins to increase their sweetness. Examples of such sweet-tasting proteins are monellin from African plant *Dioscoreophyllum cumminsii* Diels, thaumatin from *Thaumatococcus danielli* Benth, mabinlin from *Capparis masakai* Levl, Pentadin from *Pentadiplandra brazzeana* Baillon, brazzein from *P. brazzeana* Baillon, curculin from *Curculingo latifolia*, and miraculin from *Richadella dulcifica*. Monellin is approximately 3,000 times sweeter than sucrose on a weight basis. Moreover, owing to its proteinaceous nature, it does not exert the same metabolic impact as sugar. Though it is a likely candidate as a sugar substitute, its dimeric nature (A chain of 45 amino acid residues and a B chain of 50 amino acid residues held together by weak noncovalent bonds) restricts its usefulness as a sweetener. This is because it is readily dissociated (denatured) upon heating during cooking or upon exposure to acid (e.g., lemon juice or vinegar) and hence loses its sweetness. Furthermore, the need to clone two separate genes and express both of them in a coordinated manner complicates the production of protein in transgenic plants. This problem was, however, circumvented by chemically synthesizing monellin gene, which was used to produce fusion protein in transgenic tomato and lettuce plants. In addition to monellin, genes for other sweet-tasting proteins have also been isolated and characterized.

Expression of 1-sucrose:sucrose Fructosyl Transferase to Produce Fructans Fructans, the naturally occurring polymers of fructose, are not usually degraded in the human digestive tract. However, some beneficial bacteria in human intestine can utilize fructans. Small fructans, i.e., up to five monosaccharide units, have a sweet taste, and can be used as a natural low calorie sweetener instead of sucrose. On commercial level, fructans are generally produced from sucrose by the action of fungal invertases or extracted from the roots of chicory plants or Jerusalem artichoke tubers. But these processes are expensive and hence there is a need to develop an inexpensive method of producing fructans. Transgenic plants capable of converting sucrose to fructans can be a useful strategy. Thus sugar-beet plants have been engineered to convert sucrose normally stored in the vacuole of taproot parenchyma cells into fructans. For this, sugar-beets were transformed with a genetic construct containing 1-sucrose:sucrose fructosyl transferase cDNA from Jerusalem artichoke under transcriptional control of CaMV 35S promoter. These transgenic plants produced a mixture of fructans, which was essentially the same as enzymatically (natural) produced fructans.

Antisense Inhibition of Sucrose Phosphate Synthase (SPS) to Reduce Sucrose Content and Sweetness Most studies on the plant leaves have indicated that during lowered light intensity or darkness, synthesized starch is mobilized to sucrose. A key enzyme for sucrose metabolism in higher plant is SPS, which is responsible for catalyzing the formation of sucrose-6-phosphate from fructose-6-phosphate and UDP-glucose. The reaction is essentially irreversible *in vivo* due to its close association with the sucrose phosphate phosphatase. Thus, downregulation of SPS may be a useful strategy for reducing sucrose content and hence sweetness. In this direction, antisense expression of gene encoding SPS under the control of tuber specific patatin promoter has been reported to produce sweet-resistant potato. These potatoes exhibited longer shelf-life as compared to wild-type potatoes.

Modification of Starch Content and Composition

The starch found in potatoes and most other crop plants consists of 20–30% straight chain amylose having α-1,4 linkage and 70–80% branched chain amylopectin having α-1,6 linkage. The stepwise synthesis of starch requires enzymes ADP–glucose pyrophosphorylase (that catalyzes the transfer of ADP from ATP to glucose-1-phosphate to form ADP-glucose) and starch synthase (that catalyzes the transfer of glucose from ADP-glucose to the nonreducing end of a preexisting glucan chain via an α-1,4 linkage, i.e., amylose chain). Branching is catalyzed by starch-branching enzyme (SBE) (that joins two glucan chains by α-1,6 linkage). This suggests that the physical and chemical properties of starch are greatly influenced by amylose/amylopectin ratio. For many industrial applications, it is useful to have starch that is highly enriched in either amylose or amylopectin. The modification of starch content and composition can be achieved by any of the following ways:

Antisense Inhibition of Starch-branching Enzyme (SBE) to Increase Amylose Content Potato plants having a high percentage of amylose and a corresponding low percentage of amylopectin have been produced by transformation using antisense versions of SBE under transcriptional control of CaMV 35S promoter. In transgenic lines exhibiting only ~1% of normal amount of SBE activity, the fraction of amylose increased from around 28% to 60–89% of starch content.

Genetic Manipulation of Plastidic ADP-Glucose Pyrophosphorylase (AGPase) to Alter Starch Content The pathway of sucrose to starch conversion has been intensively investigated in growing potato. After entering the cytosol of tuber parenchyma cells via plasmodesmata, sucrose is converted by sucrose synthase (SuSy) to fructose and UDP-glucose. Fructose is subsequently phosphorylated to fructose-6-phosphate and UDP-glucose is converted to glucose-1-phosphate (Glc 1-P) via a pyrophosphate-dependent reaction catalyzed by UDP-glucose pyrophosphorylase. After interconversion of hexose-phosphates, Glc 6-P or Glc 1-P enter the amyloplast by the action of a hexose phosphate/phosphate antiporter. Within the amyloplast, Glc 1-P serves as the substrate for the first committed step in starch synthesis, catalyzed by AGPase. Thus, overexpression of AGPase will lead to accumulation of more starch, while its antisense inhibition leads to reduction in starch content. In one study, high starch potatoes were produced by expression of an *E. coli AGPase* gene in potato, which contained 60% more starch. Owing to the presence of less water, the finger chips from these potatoes exhibited uniform frying and consumed lesser frying medium. In another study, the small subunit of *Vicia faba AGPase* cDNA was expressed in antisense orientation in *Vicia narbonensis* under the control of the seed-specific legumin B4 promoter. This manipulation led to a moderate reduction in starch content and an increase in sucrose content.

Genetic Manipulation Using Plastidic ATP/ADP-Transporter Protein (AADP) to Alter Starch Content Conversion of sucrose to starch is ATP-dependent. The ATP required for the AGPase reaction is imported into the plastid via AATP located on the inner-envelope membrane. In a recent study, it has been indicated that the rate of ATP import exerts considerable control on the rate of starch synthesis and affects the molecular composition of starch in potato tubers. Thus in contrast to wild-type plants, antisense plants with decreased activity of the plastidic AATP exhibited drastically reduced levels of starch content and a reduced level of amylose/amylopectin ratio, whereas sense plants with increased transporter possessed higher starch and a higher amylose/amylopectin ratio.

Engineering Polyamine Accumulation

Polyamines, ubiquitous organic aliphatic cations, have been implicated in a myriad of physiological and developmental processes in many organisms. The enhanced expression of a yeast *S*-adenosylmethionine decarboxylase gene (sped fused with a ripening-inducible E8 promoter) in tomato fruit during ripening has been shown to result in increased conversion of putrescine into higher polyamines and thus to ripening-specific accumulation of spermidine and spermine. This led to an increase in lycopene, prolonged vine life, and enhanced fruit juice quality. Furthermore, the rates of ethylene production in the transgenic tomato fruit were consistently higher than those in the nontransgenic control fruit.

20.3.7 Genetic Manipulation of Flower Pigmentation

Flower industry is continuously attempting to improve flower appearance. In this direction, traditional breeding techniques have been used to create thousands of new varieties differing from one another in color, shape, and architecture. However, these techniques are slow, tedious, and are limited by gene pool of a particular species. Now uniquely colored flowers have been developed by manipulation of genes encoding enzymes involved in anthocyanin and carotenoid biosynthetic pathway. Some common examples are as follows:

Genetic manipulation using chalcone synthase (CHS) Anthocyanin, the most common flower pigment, is synthesized from phenylalanine. The first step in the biosynthesis is catalyzed by CHS. Thus, genetic manipulation of plants targeting this enzyme is expected to alter flower pigmentation. In this direction, constitutive expression of antisense *CHS* gene in transgenic petunia with high frequency resulted in an altered flower pigmentation due to reduction in level of both mRNA for enzyme and enzyme itself.

Expression of dihydroflavonol-4-reductase The color of flowers is determined by the chemical side chain substitutions of different chemical structures. For example, cyanidin derivatives lead to production of red color, while delphinidin derivatives give blue color. The wild-type petunia enzyme dihydroflavonol-4-reductase is responsible for the conversion of colorless dihydroquercetin to red cyanidin-3-glucoside, and colorless dihydromyricetin to blue delphinidin-3-glucoside, however it cannot use colorless dihydro-kaempferol as a substrate. The latter property was conferred to petunias by transforming them with a dihydroflavonol-4-reductase gene from maize. The maize enzyme led to utilization of dihydrokaempferol to form pelargonidin-3-glucoside and the transformants thus produced exhibited unique brick red-orange flowers.

Expression of β-carotene ketolase The characteristic pink color of salmon, trout, shrimp is due to accumulation of a carotenoid called astaxanthine, which is synthesized by marine bacteria and microalgae and then passed on to fish through food chain. It is a powerful antioxidant and protects salmon and trout eggs from damage by UV irradiation and improves the survival and growth rate of juveniles. Taking clue from this, cDNA of β-carotene ketolase (β-*C*-4-oxygenase) from unicellular algae *Haematococcus pluvialis* was cloned and expressed in tobacco flowers. Once inside the chromoplast, the algal β-carotene ketolase worked in concert with endogenous β-carotene hydroxylase to convert β-carotene to astaxanthin, which accumulated in flower nectaries.

Expression of delphinidin Roses lack *delphinidin* gene, which encodes for the primary plant pigment responsible for the production of true blue flowers. Australian company Florigene and Japanese company Suntory used CSIRO's gene silencing technology and created a blue rose (Figure 20.4g). The strategy involved three steps: (i) Silencing of endogenous dihydroflavonol reductase (DFR) responsible for red color; (ii) Insertion of *delphinidin* gene from pansy (responsible for production of blue pigment); and (iii) Replacement of rose *DFR* gene with iris *DFR* gene. The resulting rose was pale violet rather than true blue. The color of other 'blue' roses currently in the market is a chemical modification of red pigment.

20.3.8 Delaying of Postharvest Softening and Discoloration of Fruits

One of the problems faced by food industry is postharvest softening and discoloration of fruits and vegetables during marketing and transport. As these rotten and discolored foods are not acceptable by consumers, food industry has made heavy investment in the addition of food additives to prevent their discoloration and ripening. However, the safety of some of these food additives, especially sulfites, is questionable. It was therefore planned to engineer plants in such a way that the fruits and vegetables have increased shelf-life (i.e., long-term storage without discoloration as well as delayed ripening or softening). In this direction, following strategies have been adopted:

Antisense inhibition of polyphenol oxidase (PPO) The enzyme thought to be responsible for the initial discoloration of fruits and vegetables is nucleus-encoded PPO that is localized in chloroplast and mitochondrial membranes. PPO catalyzes the oxidation of monophenols and *O*-diphenols to *O*-quinones. This results in discoloration of fruits and vegetables. Thus the transgenic potato plants harboring antisense *PPO* gene construct have been reported to exhibit reduced discoloration. Phytoene synthase gene has also been used for development of red color in fruits as it is an important enzyme for lycopene synthesis.

Expression of cell wall-degrading enzymes Postharvest discoloration of fruits and vegetables is due to premature ripening and softening and is a part of natural aging (or senescence) process. Some of the genes that are induced during ripening encode the enzymes cellulase and polygalacturonase (PG). Thus by interfering with the expression of one or more of these genes using sense or antisense RNA, ripening process can be delayed. As PG is responsible for cell wall (pectin) degradation and fruit softening, a simple strategy to delay fruit ripening is antisense inhibition of PG production. In this direction, transgenic tomato plants constitutively expressing antisense *PG* gene were produced and marketed under the trade name 'Flavr Savr' (pronounced flavor saver) (Figure 20.4h). The Flavr Savr tomato was the first transgenic fruit developed by antisense RNA technology to be commercialized in United States (released into the market in 1994 by Calgene in the Unites States). This strategy lowered PG production and inhibited tomato ripening. The idea behind this product was to produce a healthy tomato that could vine ripen and yet still withstand the bumps and bruises of shipping. Thus these tomatoes exhibited longer shelf-life as compared to normal tomatoes, while the morphology, red pigment (lycopene) accumulation, and appearance of these tomatoes were retained. As such, these transgenic tomatoes were considered to be safe for human consumption. But as the cultivar lacked consistent production qualities, these transgenic tomatoes could not be sold well and were off the market in 1997.

Pectin methylesterase (PME) affects the integrity of middle lamella, which controls cell-to-cell adhesion and thus influences fruit texture. Diminished PME activity considerably increases the viscosity of tomato juice or paste, which is correlated with reduced polyuronide depolymerization during processing.

Several other reports available for delaying fruit ripening or softening and increasing shelf-life include manipulation of enzymes such as β-galactosidase, endo (1 → 4) β-*D*-glucanase (EGase), xyloglucan endotransglycosylase (XET), pectate lyase, and expansin.

Antisense inhibition of ethylene biosynthesis pathway In climacteric fruits (e.g., tomato), synthesis of plant growth regulator ethylene induces expression of a number of genes encoding enzymes involved in fruit ripening. If such already ripened fruits are marketed, these usually get spoiled before reaching the customer. Hence, such climacteric fruits have to be harvested before ripening. In the direction of preventing premature ripening, an effective strategy can be blocking ethylene biosynthetic pathway. As ethylene biosynthesis proceeds from methionine in the reactions catalyzed by *S*-adenosyl synthetase (SAM synthetase), 1-amino cyclopropane-1-carboxylic acid synthase (ACC synthase), and ACC oxidase, targeting these enzymes may delay fruit ripening. In this direction, transgenic plants containing antisense RNA of any of these three enzymes have been engineered that contained much lower than normal levels of ethylene. However, the antisense effect may be reversed by ethylene treatment.

The amount of 1-amino cyclopropane-1-carboxylic acid (ACC), and hence the amount of ethylene may also be decreased by increasing the activity of either malonyl ACC transferase or glutamyl ACC transferase, the enzymes which convert ACC into dead-end storage compounds.

Another strategy to inhibit synthesis of ethylene is to transform plants with the bacterial enzyme ACC deaminase that converts ACC to α-ketobutyrate.

20.3.9 Delaying Flower Wilting, Discoloration, and Senescence

Similar to food industry, flower industry also suffers from extensive economic losses due to premature wilting of flowers. The strategies for delaying wilting and senescence are as follows:

Antisense inhibition of ethylene production Genetic manipulations resulting in decreased levels of ethylene by any of the strategies described in the section of delayed fruit ripening may be used for delaying wilting in flowers, thereby extending their storage life. Thus SAM synthetase, ACC synthase, ACC oxidase, malonyl ACC transferase, glutamyl ACC transferase, ACC deaminase may be employed. In an example, Melbourne-based Company, Florigene Pvt. Ltd. had produced transgenic long-life carnations by antisense inhibition of ACC synthase. These genetically eternal transgenic flowers neither required expensive refrigeration nor silver thiosulfate (STS) or EthylBlock compound (Figure 20.4i).

Overexpression of SOD Petal deterioration (wilting) is due in part to oxygen radicals produced during flower senescence. Hence a *SOD* gene can also be used to maintain the quality of cut flowers during shipping. By cloning *SOD* gene under the control of a flowering-specific promoter, the endogenous oxygen scavenging capacity in cut flowers is increased, thereby increasing the shelf-life.

20.3.10 Detoxification of Toxic Metals by Transgenic Plants

Toxic metals such as mercury, lead, cadmium, and selenium pose a great threat to environment and humankind. To overcome the associated problems, systems for detoxification and/or chelation of such toxic compounds are required. Presently, phytoremediation (i.e., use of green plants to remove, contain, or render harmless environmental pollutants) offers an effective, environmentally nondestructive, and cheap remediation method. However, the use of genetic engineering to modify plants for metal uptake, transport, and sequestration has opened up new avenues for enhancing efficiency of phytoremediation. Such strategies include the following:

Expression of phytochelatin synthase (PC synthase) or glutathione synthetase PC synthase and glutathione synthetase enzymes are responsible for the synthesis of phytochelatin (PC), a metal chelator. The expression of these enzymes in plants thus aid in improved metal uptake and sequestration. In this direction, several transgenic plants have been generated that harbor genes encoding these enzymes.

Metal detoxification Plants have also been engineered to detoxify toxic metals. For example, transgenic plants containing genes for bacterial detoxifying enzymes have been used for bioremediation of contaminated soils. Thus in a study, *Arabidopsis* plants engineered with *mer* A gene encoding mercuric reductase were found to be capable of growing on mercurium-contaminated soils. These plants were found to be 50 times more tolerant to mercury than wild-type plants. Further the introduction of *mer* B gene encoding organomercurial lyase into *Arabidopsis* plants led to generation of plants, which were 10 times more tolerant to mercury than *mer* A plants.

20.3.11 Increase Plant Biomass and Yield of Components

It is always desirable to increase the yield of plant, or its components for maximal exploitation of plant biomass for any application. Theoretically, this may be achieved by a number of ways as exemplified below.

Increasing Iron Availability

Despite of abundance of iron on the surface of the earth, plant growth is often limited by its availability. At physiological pH, under aerobic conditions, iron is sparingly soluble in the soil solution and hence it is not readily available to plants. Therefore some plants have developed mechanisms to acquire iron from the soil, for example, by secreting iron-binding phytosiderophores, which are linear hydroxy- and amino-substituted iminocarboxylic acids, such as mugineic acid and avenic acid. Mugineic acid family phytosiderophores (MAs) are natural iron chelators that graminaceous plants secrete from their roots to solubilize iron in the soil. Rice plants, which secrete very low amount of mugineic acid, are highly susceptible to growth inhibition from iron deficiency. Thus by engineering rice plants to increase the secretion of mugineic acid, their capability to take up iron can be enhanced, which further increases the plant yield. This has been achieved by transforming rice plants with a 11 kbp fragment of barley genomic DNA containing two *naat* genes, *naat* A and *naat* B, encoding subunits of nicotianamine aminotransferase (NAAT), an enzyme responsible for catalyzing amino group transfer of nicotianamine (an intermediate in the biosynthesis of MA phytosiderophores).

Increasing Oxygen Content

O_2 is an essential substrate for plant respiratory metabolism. Hence the amount of available O_2 may limit a number of plant biochemical reactions. Thus to increase O_2 concentration inside a plant, an O_2 sequestering protein may be provided where needed. To achieve this, the gene encoding a dimeric Hb (that binds oxygen tightly) has been inserted from a Gram negative bacterium *Vitreoscilla* into tobacco.

Increasing Photosynthesis

In terms of yield enhancement, photosynthesis is perhaps the most obvious target for genetic intervention because it determines the rate of carbon fixation and hence the overall size of the organic carbon pool. Strategies used for increasing photosynthetic activity include the modification of light-harvesting phytochromes and key photosynthetic enzymes. Attempts have been made to introduce components of the energy-efficient C_4 photosynthetic pathway into C_3 plants, which lose a proportion of their fixed carbon through photorespiration. In this direction, maize gene encoding phosphoenol pyruvate carboxylase (PEPC) of C_4 pathway has been transferred into several C_3 plants. This manipulation increased the overall level of carbon fixation. Transgenic rice plants expressing pyruvate orthophosphate dikinase (PPDK) and NADP-malic enzyme have also been produced.

Preliminary field trials in China and Korea have demonstrated 10–30% and 30–35% increase in yield, respectively.

Decreasing Lignin Content

The presence of lignin in forest trees is a major impediment in obtaining cellulose for pulp and paper industry. The presence of residual lignin results in inferior quality paper that yellows with age. Presently lignin is removed from cellulose by harsh physical and chemical treatments, which are energy-intensive, environmental-damaging, and expensive. These processes are also associated with a reduction in pulp yield. Cellulose degree of polymerization is also reduced due to action of bleaching agents. Another negative implication of lignin in plants is the decrease in nutritional value of forage crops to be used as cattle feed. In forage crops, lignin being intimately associated with cellulose and hemicellulose also limits their digestibility and hence energy yields. This is because enormous lignification limits the accessibility of digestive hydrolytic enzymes to these cell wall polysaccharides. Thus for optimal utilization of plant biomass in pulp and paper industry and as nutritious cattle feed, lignin content should be reduced. In this direction, lignin downregulation has been successfully obtained by inhibition of any one (or more) of the enzymes involved in monolignol biosynthesis, their transport to cell wall, and their polymerization using antisense RNA or cosuppression or RNAi technology. The candidate genes are *O*-methyltransferase (*omt*), ferulate 5-hydroxylase (*f5h*), cinnamoyl CoA *O*-methyltransferase (*ccoaomt*), cinnamoyl CoA reductase (*ccr*), or cinnamyl aldehyde dehydrogenase (*cad*), coniferin β-glucosidase, and laccase, etc., either as single or multiple constructs. Besides a reduction in lignin content, the results of some of the studies on downregulation of lignin also indicated a concomitant increase in cellulose content. As structural genes in lignin biosynthetic pathway are regulated through transcriptional control, another strategy for genetic manipulation of lignin involves manipulation of transcription factor genes, *myb*.

Bioenergy

Realizing the potential of bioenergy crops necessitates optimizing the dry matter and energy yield of these crops per area of land and obtaining efficient conversion. Thus, by genetically engineering these plants for better yield and cell wall digestion, the economics and efficiency of the conversion into a liquid fuel, bioethanol, could be significantly enhanced. Currently, employed biological process for converting the lignocellulose to EtOH involves a series of steps, *viz.*, delignification to liberate cellulose and hemicellulose from their complex with lignin, depolymerization of the carbohydrate polymers to produce free sugars, and fermentation of mixed hexose and pentose sugar to produce EtOH; this is however a costly procedure. Thus, plant transgenic technology may provide *in planta* alternative to pretreatment by rendering the plants more amenable to bioprocessing. This can be achieved by altering plant biochemical composition (*i.e.*, starch content and composition and lignocellulosic matter characteristics to improve the processing characteristics), by modifying of plant stature and architecture, by ameliorating the negative biotic and abiotic stresses on plants, and by increasing the capacity of plants to produce harvestable yield. Presently, in order for cellulases to access the cellulose for degradation, costly and harsh heat or acid pretreatment of biomass is required to remove lignin and hemicellulose from the lignocellulosic matter. Transgenic manipulation of different lignin biosynthetic pathway genes may be targeted to decrease lignin content or alter lignin composition and thus reduce the pretreatment costs. It has been established that reducing lignin biosynthesis can lead to lower recalcitrance and higher saccharification efficiency. Transgenic plants can also be designed to overexpress cellulases, endoglucanases, exoglucanases, and β-glucosidases, or produce recombinant (microbial) cellulases within the biomass crop. Recently, transgenic plants constitutively producing either a hyperthermophilic α-glucosidase or β-glycosidase using genes derived from the archeon *Sulfolobus solfataricus* have been produced. *In planta* expression of expensive cellulases and cellulosomes could potentially reduce the cost of enzymatic saccharification of lignocellulosic biomass at the biorefinery by providing the enzymes needed for cell wall degradation.

20.3.12 Plants as Bioreactors for Large-scale Production

Another major goal for production of transgenic plants is their use as bioreactors (or biofactories) for the production of specialty chemicals and pharmaceuticals.

Plant system is relatively more advantageous as compared to recombinant microbial system because of following reasons: (i) Transgenic plants are relatively easy to grow on large-scale, resulting in extensive biomass. Hence, a large amount of biomass can be easily produced by cultivation in fields with relatively few inputs. In addition, transgenic plants capable of producing several different products can be created at any given time by crossing plants producing different products; (ii) Plants are easy to transform and allow high and controlled level of expression so that proteins are obtained in high quantity; (iii) Transgenic plants can be easily subjected to scale-up by simply planting seeds. Moreover, their production is not limited by fermentation capability; (iv) Transformation of plants generally results in the stable integration of foreign DNA into plant genome, while most microorganisms are transformed with plasmids that can be lost during prolonged or large-scale fermentation; (v) The processing and assembly of foreign proteins in plants are similar to those in animal cells, whereas bacteria do not readily process, assemble, or posttranslationally modify eukaryotic proteins. Thus glycoproteins and properly folded proteins can be made in plant system, while bacteria cannot do so; (vi) Transgenic material, in the form of seed or fruit, can be easily and stably stored for long periods under ambient conditions and transported from one place to another without fear of its degradation or damage; (vii) The final cost incurred in the production of any value-added product depends on the level of foreign protein expression, the cost of generating sufficient quantities of recombinant organism, and the cost and yield of purification process. Taking these into account, cultivation of transgenic plants on a large-scale is relatively inexpensive and does not require highly trained personnel or expensive equipments. Plants therefore require low cost of investment, dispersed capital requirement, and low cost of production; and (viii) While using plant system, contaminating pathogens are not likely to be present. Hence plants face fewer concerns about product safety and public acceptance, consequently providing opportunities as a safe method of delivery of recombinant proteins for therapeutic use.

Only problem while using transgenic plants for large-scale production of high-value products is their purification from an enormous amount of harvested transgenic plant. However, this is easily achieved either by expressing foreign protein at a high level or by ensuring its expression in a manner that facilitates its purification. These strategies include: (i) Chloroplast targeting of foreign protein; (ii) Directing the foreign protein to cell apoplast, thereby facilitating its exudation, and (iii) Fusing foreign protein to plant oleosins.

In view of various aforesaid advantages of plant system, considerable attention has been drawn in using transgenic plants as bioreactors for the large-scale production of biopolymers, vaccines, therapeutic proteins, nutraceuticals, hormones, and industrial enzymes.

Production of Biopolymers

Considering the above advantages of using plants as bioreactors and to overcome the problems associated with the fermentative production of biopolymer poly-(3-hydroxy butyric acid) (PHB), which is used in the synthesis of biodegradable plastics (bioplastics), research has been conducted to produce this biopolymer in plants. As plants lack enzymes involved in PHB biosynthesis, engineering plants with these genes from some other source (e.g., bacteria) may be a useful strategy. It is speculated that plant seeds may be a potential source for plastics that could be produced and easily extracted. In *A. eutrophus*, PHB is synthesized from acetyl-CoA in three enzyme-catalyzed reactions (encoded by genes in a single operon). These three enzymes are 3-ketothiolase, acetyl-CoA reductase, and PHB synthase. Thus to produce transgenic *A. thaliana* plants, genes for each of these three enzymes were isolated, cloned separately into a plasmid, and targeted to chloroplast.

Production of Antibodies, Antibody Fragments, and Therapeutic Agents

As mentioned earlier, crops have unmatched opportunities of biomass production.

Besides the common advantages of exploiting plants as bioreactors (as discussed earlier), certain other advantages are as follows: (i) As biopharmaceuticals are intended for human use, these should be biologically active (posttranscriptionally and posttranslationally processed). Hence eukaryotic proteins can be properly expressed in plants; (ii) The production in transgenic plants significantly reduces the risk of mammalian viral contamination in comparison to the ones produced in animal cells grown in culture. There is no danger from mammalian cells and tissue culture medium that might be contaminated with infectious agents; and (iii) Large-scale cultivation is inexpensive. The cost–benefit analysis has revealed that antibodies production costs \$5,000/g from hybridoma cell culture, \$1,000/g from transgenic bacteria, and \$10–100/g from transgenic plants. Thus the exploitation of transgenic plants is envisaged for producing large quantities of biomolecules of therapeutic use to combat various diseases in developing countries in a convenient, effective, and inexpensive way. In this direction, various antigens, antibodies, antibody fragments, hormones, enzymes, therapeutic proteins, etc. have already been expressed successfully in plants and have been shown to retain their native functional forms.

Table 20.12 Some Examples of Antigens Produced in Transgenic Plants

Host plant	Antigen
Tobacco	Phosphonate ester
Tobacco	(4-Hydroxy-3-nitrophenyl) acetyl
Tobacco	Phytochrome
Tobacco	Artichoke mottled crinkle virus
Tobacco	Human creatine kinase
Tobacco	*Streptococcus mutans* cell surface antigen SA I/II
Tobacco	Fungal chitinase
Tobacco	Oxazolone
Tobacco	Abscissic acid
Tobacco	Cell surface protein from mouse
Tobacco	B-cell lymphoma
Tobacco	Human carcinoembryonic antigen
Tobacco	TMV
Tobacco	Gibberellin
Tobacco	Beet necrotic yellow vein virus coat protein
Tobacco	Stolbar phytoplasma membrane protein
Tobacco	Root rot nematode surface glycoprotein
Petunia	DHFR
Soybean	Herpes simplex virus
Pea	Abscissic acid
Pea	Human cancer cell surface antigen

Table 20.13 Some Examples of Antibodies and Antibody Fragments Produced in Transgenic Plants

Antibody	Antigen	Plant
IgG (κ)	Transition stage analog	Tobacco
IgM (l)	NP (4-hydroxy-3-nitrophenyl) acetyl hapten	Tobacco
Single domain (dAb)	Substance P	Tobacco
Single chain Fv	Phytochrome	Tobacco
Single chain Fv	Artichoke mottled crinkle virus coat protein	Tobacco
Fab; IgG (κ)	Human creatine kinase	*Arabidopsis*
IgG (κ)	Fungal cutinase	Tobacco
IgG (κ) and IgG/ hybrid	*S. mutans* adhesin	Tobacco
Single chain Fv	Abscissic acid	Tobacco
Single chain Fv	Nematode antigen	Tobacco
Single chain Fv	β-Glucuronidase, β-1,4 endoglucanase	Tobacco
Single chain antibody fragment	Atrazine, paraquat	Tobacco
IgG	Glycoprotein B of herpes simplex virus	Soybean

These antibodies produced in plants are called plantibodies, which are used for diagnostic and therapeutic purposes. Plants producing recombinant sIgA against the oral pathogen *Streptococcus mutans* have been generated, and these plantibodies are being commercially developed as the drug CaroR$_x$^{TM}, marketed by Planet Biotechnology Inc. Various other examples of transgenic plants used for the production of antigens, antibodies, antibody fragments, and therapeutic agents are given in Tables 20.12 through 20.14. For some recent examples and information, the readers are advised to visit http://www.molecularfarming.com/plantigens.html.

Production of Edible Vaccines

Vaccine development is essential to effectively combat infectious diseases and to overcome increased occurrence of antibiotic-resistant pathogens. Many commercial vaccines developed through conventional ways are administered through injections, which require trained personnel. Moreover, use

Table 20.14 Some of the Therapeutic Recombinant Proteins (Biopharmaceuticals) Produced in Transgenic Plants

Protein	Application	Plant(s)
Human protein C	Anticoagulant	Tobacco
Human hirudin variant C	Anticoagulant	Tobacco, canola, Ethiopian mustard
Human granulocyte-macrophage colony stimulating factor	Neutropenia	Tobacco
Human erythropoietin	Anemia	Tobacco
Human enkephalins	Antihyperanalgesic by opiate activity	Thale cress, canola
Human epidermal growth factor	Wound repair/control of cell proliferation	Tobacco
Human α-interferon	Hepatitis C and B	Rice, turnip
Human serum albumin	Liver cirrhosis	Potato, tobacco
Human hemoglobin	Blood substitute	Tobacco
Human homotrimeric collagen I	Collagen synthesis	Tobacco
Human α-antitrypsin	Cystic fibrosis, liver disease, hemorrhage	Rice
Human growth hormone	Dwarfism, wound healing	Tobacco
Human aprotinin	Trypsin inhibitor for transplantation surgery	Corn
Angiotensin-1-converting enzyme	Hypertension	Tobacco, tomato
α-Tricosanthin	HIV therapy	Tobacco
Glucocerebrosidase	Gaucher disease	Tobacco
Hirudin Leech	Anticoagulant	Canola
Interleukin-2,4,10,12,18	Antiviral, anticancer	Potato, tobacco
Lactoferrin	Antimicrobial	Potato tubers, rice, tobacco
Lysozyme	Antimicrobial	Rice
Pancreatic lipase	Exocrine pancreatic deficiency	Tobacco, maize
Synthetic elastin	Tissue repair	Tobacco
Collagen Human	Tissue repair	Tobacco
Factor XIII (A-domain)	Bleeding	Tobacco
Angiotensin-1 converting enzyme	Hypertension	Tobacco, tomato

of antigens for inducing oral tolerance requires production of human antigens in large amounts, which is generally difficult to achieve by conventional means. Furthermore, traditionally produced vaccines are expensive to produce and package. These vaccines are costly and are not available for widespread distribution in the developing countries of the world, where infectious diseases are still the primary cause of death. These vaccines are destroyed (or denatured) rapidly due to lack of physical infrastructure such as roads and refrigeration. Thus to eliminate and simplify the process of vaccine production on large-scale, vaccine administration, and vaccine dissemination in developing countries, a relatively newer concept of edible vaccines was developed.

These edible vaccines represent production of protective antigens in transgenic plants, such that eating raw fruits or vegetables can induce neutralizing antibodies, and trigger immune response to pathogen. Thus, food crops can play significant new roles in promoting human health by serving as vehicles for both production and delivery of vaccines. The advantages and strategy used for the production of edible vaccines in presented in Exhibit 20.4. This concept was first used for the production of surface protein antigen A (*spa* A) from *Streptococcus* in tobacco. In 1992, Charles Arntzen and coworkers expressed hepatitis B surface antigen (HBsAG) in tobacco and an accumulation of 0.01% of soluble protein level has been reported. Since then, several other plants are being considered as effective vaccine delivery systems. These include tomato, potato, banana, papaya, lettuce, carrot, peanut, cucumber, rice, wheat, soybean, and corn (Table 20.15).

Exhibit 20.4 Edible Vaccines

Edible vaccines represent production of protective antigens in transgenic plants such that eating raw fruits or vegetables can induce neutralizing antibodies and trigger immune response to pathogen. Thus food crops can play significant new roles in promoting human health by serving as vehicles for both production and delivery of vaccines.

Edible vaccine technology involves several steps (Figure A). These include (i) Identifying an antigen that provides the desired immune

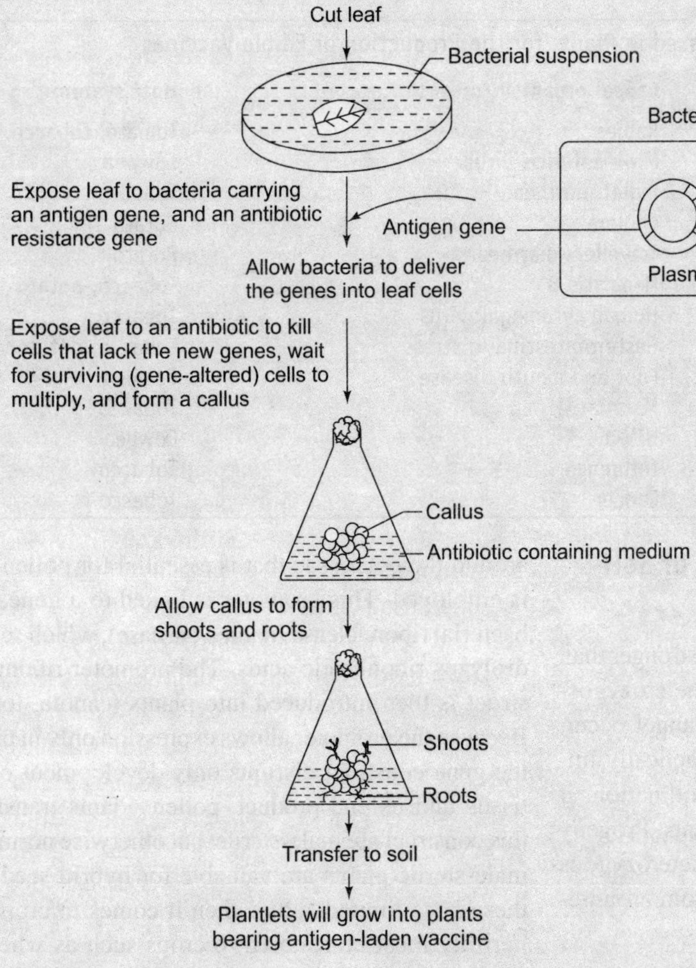

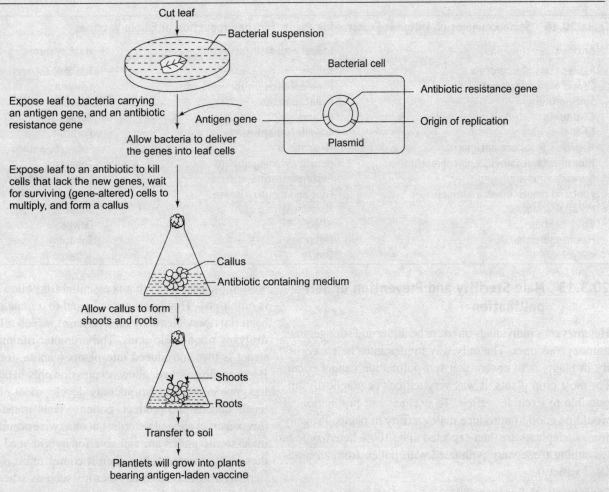

Figure A Production of edible vaccines

response; (ii) Cloning gene from a human pathogen into a suitable vector; (iii) Infection of explants and high-level expression of the antigen in an edible part of the plant such as grain or fruit (using tissue-specific promoters); (iv) Plant regeneration from transformed explant; and (v) Analyzing immune response in an animal model system and extension to human clinical trials. This indicates that the success of edible vaccine production, dissemination, and administration depends on factors such as antigen selection, efficacy in model system, choice of plant species for vaccine delivery, delivery and dosing issues, safety issues, and public perception and attitudes to genetic modification and quality control and licensing.

In contrast to traditionally produced vaccines, edible vaccines offer several advantages: (i) Edible vaccines provide a fail-safe immunization system against diseases for which a protective antigen has been defined. It generates systemic and mucosal immunity, which are essential to avoid respiratory and digestive tract infections; (ii) Plant-based vaccines may be subjected to mass production, i.e., on large scale; (iii) Antigens or antibodies expressed in plants can be administered orally as any edible part of the plant or by parenteral route such as intramuscular or intravenous injection after isolation and purification from the plant tissue. The edible part of the plant to be used as a vaccine is fed raw to experimental animals or humans to prevent possible denaturation during cooking and avoid cumbersome purification protocols. Edible vaccines can be improved for their oral immunogenicity by the use of appropriate adjuvant, which could be used either as a fusion to the candidate gene, or as an independent gene; (iv) Edible means of administration gives excellent safety compared to injection; (v) The procedure for edible vaccine production is well-established and is much cheaper than production of conventional vaccines. Thus edible vaccines provide an economically efficient immunization system in developing countries and do not require elaborate production facilities, purification, sterilization, packaging, or specialized delivery systems; (vi) Edible vaccines are heat-stable at room temperature and do not require refrigeration. Hence these are easy-to-store; (vii) It is easy and cheaper to transport agricultural products throughout the world. This adds to the cost-effectiveness of the procedure; and (viii) Edible vaccine serves as socioculturally acceptable vaccine delivery system. There is growing acceptance of transgenic crops in both industrial and developing countries. Initially thought to be useful only for preventing infectious diseases, it has also found application in prevention of autoimmune diseases, birth control, and cancer therapy, etc. Edible vaccines are currently being developed for a number of human and animal diseases. This new technology may contribute to global vaccine programs and may have a dramatic impact on health care in developing countries.

Source: Langridge WHR (2000) *Scientific American*, pp. 66–71.

Table 20.15 Some Examples of Antigens Expressed in Plants for the Production of Edible Vaccines

Antigen	Causal organism/disease	Host system
Rabies virus glycoprotein	Rabies	Tomato, tobacco, spinach
Capsid protein epitope	Mink enteritis virus	Cowpea
Spike protein	Piglet diarrhea	Tobacco
CT-B toxin	Cholera	Potato
ET-B	Travellers diarrhea	Potato
Hepatitis B surface antigen	Hepatitis B	Tobacco, potato
Human cytomegalovirus glycoprotein B	Human cytomegalovirus	Tobacco
Norwalk virus antigen	Gasterointestinal distress	Tobacco, potato
Foot and mouth disease antigen	Foot and mouth disease	Cowpea
Malarial antigens	Malaria	Tobacco
Gp41 peptide	HIV-1	Cowpea
Haemagglutinin	Influenza	Tobacco
c-*Myc*	Cancer	Tobacco

20.3.13 Male Sterility and Prevention of Self-pollination

Heterozygous individuals often are healthier and stronger than homozygous ones. The only way to guarantee heterozygosity in plants is to ensure that self-pollination cannot occur. For most crop plants, it was very tedious or practically impossible to exclude selfing. To exclude self-pollination, it would be good to introduce male sterility in plants. Progeny from such plants are then expected to be 100% heterozygous (assuming these were pollinated with pollen from an unrelated variety).

Expression of Barnase To introduce male sterility, a promoter that is turned on exclusively in tapetum cells (a tissue around the pollen sac that is essential for pollen production) is employed. This promoter is linked to a gene coding for a bacterial ribonuclease (named barnase), which selectively hydrolyzes ribonucleic acids. The promoter/ribonuclease construct is then introduced into plants (canola, tobacco, etc.). Because the promoter allows expression only in tapetum cells, the gene construct disrupts only development of the tapetal tissue and its end product, pollen. Plants transformed with this construct are male-sterile, but otherwise normal. Although male-sterile plants are valuable for hybrid seed production, these have limited value when it comes to crop production. Fertility must be restored to crops such as wheat, rice, and tomato, in which the seed or fruit is the harvested product. Fortunately, the ribonuclease is inhibited very much by a

Table 20.16 List of Companies/Institutions Engaged in Plant Transgenic Technology and the Products Commercialized by them

Name of company	Trade name of transgenic plant	Crops	Property
AgrEvo Canada, Inc.	Innovator Liberty link	Canola, corn	Glufosinate tolerance
Asgrow Seed Co.	Freedom II squash	Squash	Virus resistance
Aventis	StarLink	Maize (Cry 9C)	Resistance to European corn borer
Bejo-Baden	–	Chicory	Male sterility/Glufosinate tolerance
Calgene	Flavr Savr™	Tomato	Delayed fruit ripening
	BXN	Cotton	Bromoxynil tolerance
DeKalb	Bt-Xtra	Maize (Cry 1Ac)	Resistance to European corn borer
DuPont	–	Cotton	Sulfonylurea tolerance
	–	Soybean	High oleic acid
Florigene	–	Carnations	Modified flower color/Increased vase life
Monsanto	New-Leaf	Potato	Resistance to Colorado beetle
	Bollgard	Cotton	Resistance to tobacco budworm, cotton bollworm, pink bollworm
	YieldGard	Maize (Cry 1Ab)	Resistance to European corn borer
	YieldGard Rootworm	Corn (Cry 3Bb)	Insect resistance
	Round-Up Ready	Cotton, soybean, canola	Glyphosate tolerance
Mycogen	Herculex 1	Maize (Cry 1F)	Resistance to European corn borer
	NaturGard	Corn	Insect resistance
Novartis	YieldGard	Corn (Cry 1Ab)	Insect resistance
	Knockout	Corn (Cry 1Ab)	Insect resistance
Pioneer	STS Liberty Link	Soybean	Insect resistance
Plant Genetic Systems	–	Oilseed rape	Male sterility/Glufosinate tolerance
Rockefeller Foundation, Swiss Federal Institute of Technology, and European Community Biotech Program	Golden rice	Rice	Overcome Vitamin A and iron deficiencies
Texas A & M University	Beta Sweet® carrot	Carrot	β-Carotene synthesis
University of Saskatchewan	Triffid	Flax	Sulphonylurea tolerance
Zeneca/Petoseed	–	Tomato	Improved ripening

Note that some of these GM crops are now out of the market.

simple protein, named barstar. One can thus cross the male sterile plant with a male fertile variety in which the gene for barstar has been introduced and the result is progeny with viable pollen and restored fertility. Several genetically modified crops with barstar and barnase are available.

Gene Protection Technology A closely related approach is called 'gene protection technology' or 'terminator technology' (for details see Chapter 21). D & PL Company claimed that the technology has been developed to ensure biosafety by preventing the escape of transgenic to the wild-type and untargeted plants. This technology was, however, criticized as a way for companies to protect and enforce their patents.

In total, the plant transgenic technology offers tremendous potential in various areas discussed above. A list of companies or institutions engaged in plant transgenic technology and the products launched by them in the market is provided in Table 20.16. A very useful database of genetically modified crops is also available at http://www.agbios.com/dbase.php.

20.4 APPLICATIONS OF ANIMAL CLONING AND TRANSGENIC TECHNOLOGY

Genetic modification of animals by rDNA technology (transgenesis) entails the introduction of a cloned gene(s) into the genome of a cell that might, after proliferation and embryonic development, be present in the germ lines of some of the progeny and from which it may be possible to establish true-breeding transgenic lineages. These genetic manipulation strategies for animals involve the introduction of the transgene in an early embryonic stage, sometimes even at the stage of freshly fertilized egg, either by viruses or by microinjection of the transgene directly in the nucleus (for details see Chapter 14). The earliest method for cloning is 'embryo splitting', in which embryos split into two before implantation. Another method namely 'parthenogenesis' is possible only in females to give female progeny. The main technique in current cloning experiments is 'nuclear transplantation', which includes embryonic stem (ES) cell technology and somatic cell nuclear transfer (SCNT) (for details see Chapter 14). Using these processes, a number of transgenic animals such as mice, rats, rabbits, pigs, sheep, and cows have been produced during the last decade with the aims directed toward obtaining new or improved products to develop disease model, to understand particular diseases, and to allow the testing of drugs that is not always possible on humans. Also, there is an interest to increase basic knowledge about mammalian genetics and physiology, including complex traits controlled by many genes such as many human and animal diseases. In this section cloning of adult animals and its significance is discussed. The potential applications of animal transgenesis are also discussed.

20.4.1 Cloning in Adult Animals

Cloning creates an identical copy of an organism by copying its genetic material. The process occurs naturally in single-celled organisms, such as yeast and bacteria, as these reproduce by cell division and in identical twins, which are clones derived from a single fertilized egg. In animals, two very different procedures are referred to as 'cloning'. These are reproductive cloning and therapeutic cloning. However, it should be noted that in true sense, the term 'cloning' refers to reproductive cloning only.

Cloning is highly desirable in many situations since this allows indefinite multiplication of an elite desirable genotype without the risk of segregation and recombination during meiosis, which must precede sexual reproduction. The technique holds a great promise in genetic research and studying the impact of epigenetic changes such as imprinting and telomere shortening. This technique makes it feasible to target transgenes in livestock by nuclear transfer from transgenic cell populations developed *in vitro* into enucleated oocytes to recover nonchimaeric transgenic animals. Based on the applications, there are two types of cloning, *viz.*, reproductive cloning and therapeutic cloning.

Reproductive Cloning

Use of cloning technology to produce one or more complete and genetically identical individuals (apart from the genes in mitochondria and chloroplasts) is termed as reproductive cloning. It is also called 'adult DNA cloning'. It relies on the technique of cell nuclear replacement (CNR), which leads to the production of an individual, by transferring the nucleus of differentiated adult cells into an oocyte from which the nucleus has been removed.

Procedure of reproductive cloning Reproductive cloning involves removing the DNA from an embryo and replacing it with the DNA of a cell removed from an individual, implanting the resulting embryo in the womb of a foster mother and developing a new individual whose DNA is identical to that of the original individual. This method is called 'nuclear transfer' (for details see Chapter 14). It is significant to note here that adult DNA cloning is ethically not allowed to produce a human clone because experiments on animals have sometimes produced defective specimen.

Applications of reproductive cloning Cloning is routine in plants, but in case of animals, only a limited success had been achieved so far. The important potential applications of cloning in animals include animal breeding to produce animals to perfect new therapies for human diseases, conservation of biodiversity, cloning of elite animals, and preservation of endangered species. It may also play a role in animal transgenic technology using gene transfer procedures. Use of cloning in animal genetic improvement may increase the rate of selection progress in certain cases, particularly in situations where artificial insemination is not possible, such as in pastoral systems with ruminants. Production based on clones of the best animals of the population may allow for a one-time large 'jump' in breeding value, so the commercial animals might be very close to those in the nucleus. However, further genetic improvement must be based in the continued use of the genetic variation by selection programs. In early experiments, nuclei from tadpole were transplanted into the cytoplasm of an enucleated fertilized frog egg and normal frogs were obtained. Since that time, the technique has been extended to mammalian cloning as well for application in several industries.

In 1995, Megan and Morag (Figure 20.5a), two Welsh mountain sheep, were cloned from embryonic cells by nuclear transfer into an enucleated egg cell by Ian Wilmut and colleagues at the Roslin Institute at Scotland. These were the first cloned mammals and their genes were not modified in any way, but these were the result of a successful attempt to show that it

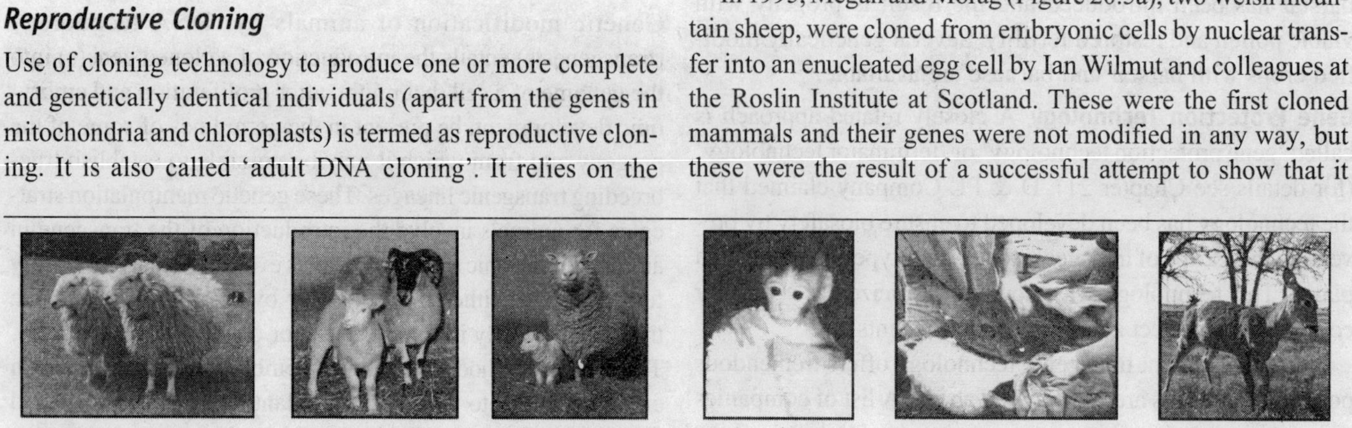

(a) Megan and Morag (b) Dolly with her surrogate mother (c) Bonnie with her mother Dolly (d) Tetra (e) Noah (e) Dewey

Figure 20.5 Some examples of cloned animals

was possible to derive live lambs from embryo cells that had been cultured for several months in the laboratory. In 1996, Ian Wilmut and colleagues at the Roslin Institute in collaboration with Pharmaceuticals Proteins Ltd. in Edinburgh, Scotland successfully cloned a sheep, called 'Dolly' (Exhibit 20.5). Dolly was bred with a Welsh Mountain ram and produced a

Exhibit 20.5 Dolly: The Cloned Sheep

The first animal clone from an adult somatic cell was a ewe, named Dolly, which was created by Ian Wilmut, Keith Campbell, and colleagues at the Roslin Institute in collaboration with PPL Therapeutics in Edinburgh, Scotland.

Dolly, a genetic replica, was born on July 5, 1996 by the technique of somatic cell nuclear transfer (SCNT), which involved cloning of a single cell taken from her mother's udder. This was the first time that genetic information from a fully differentiated and a vegetative mammalian cell was used to give rise to a new, fully differentiated organism. The same team had produced cloned sheep from embryonic cells in 1995, but this was not considered a breakthrough since adult cloned animals had already been produced from embryonic tissue of the frog *Xenopus laevis* in 1958. Though cell division of vegetative cells to eventually yield a new organism with the same genetic makeup as the parent is pretty usual among lower organisms and certain plants, but until 1997, it had never been shown for mammals. As a consequence, the announcement of cloning of sheep in early 1997 caused a fairly large stir in the popular media.

For the production of Dolly, cells from the mammary gland of an adult sheep (6-year-old ewe) were cultured. The growth of cultured cells were inhibited and induced to enter the G_0 phase (quiescent phase) by reducing the concentration of serum in the medium from 10% to 0.5% for 5 days. The cell nucleus from an adult udder cell (G_0) was transferred into an unfertilized enucleated oocyte (developing egg cell) under *in vitro* conditions. Note that oocytes were recovered between 28 and 33 hours after injection of gonadotropin-releasing hormone and enucleated as soon as possible with the aid of pipette. The fused cell (hybrid cell) was grown in culture or in ligated oviducts and its division was stimulated by short-duration electric pulses and allowed to develop into a normal early stage embryo. The blastocyst formed was then implanted in a surrogate mother and it developed into a healthy lamb. The technique is shown in Figure A. This cloning process was, however, highly inefficient, because out of 277 attempts, only 29 reached morula/blastula stage, and were transferred into surrogate mothers, leading to 13 pregnancies, but only one lamb survived to adulthood. She was named Dolly after the name of country singer Dolly Parton.

Dolly was a normal sheep in most respects. She was fertile, and in April 1998, it was announced that she was mated with a Welsh Mountain ram at the Roslin Institute. There she produced six lambs in total and her first lamb was named Bonnie. The next year Dolly produced twin lambs and in the subsequent year, triplets were born.

In 2001, Dolly developed arthritis, which was successfully treated with anti-inflammatory drugs. But on February 14, 2003, Dolly died at the age of six years. Her relatively early death fueled the debate about the ethics of cloning research and the long-term health of clones. This was because a Finn Dorset such as Dolly has a life expectancy of around 12–15 years. An autopsy showed she had a form of lung cancer called Jaagsiekte, a fairly common disease of sheep caused by the retrovirus JSRV. A group of Roslin scientists stated that her death had no connection with Dolly being a clone and that other sheep in the same flock had also died of the same disease. Dolly was most susceptible to this disease, as the sheep kept indoors are particularly at risk and she had to sleep inside for security reasons. Another group of workers believed that a contributing factor to Dolly's death was that she could have been born with a genetic age of six years, the same age as the sheep from which she was cloned. This was based on the finding that her telomeres were shorter than usual for an animal of her age, which typically was a result of the aging process. However, whether this

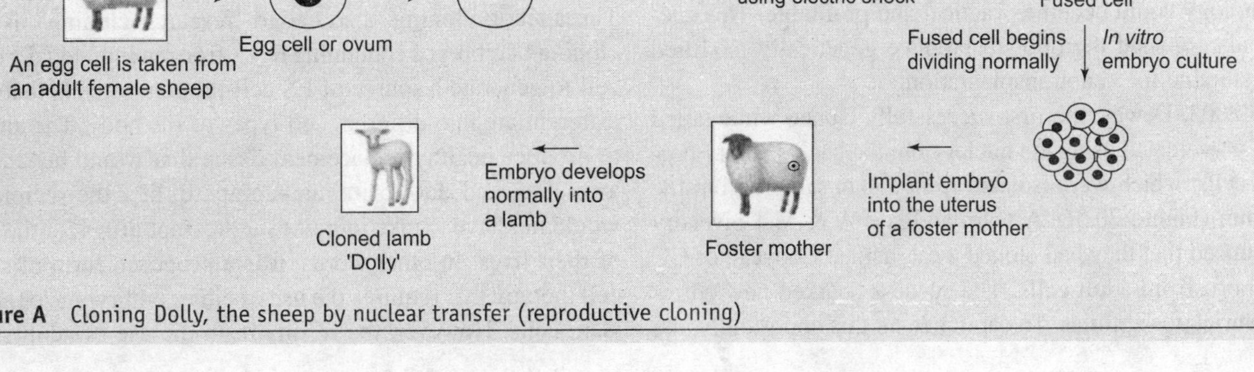

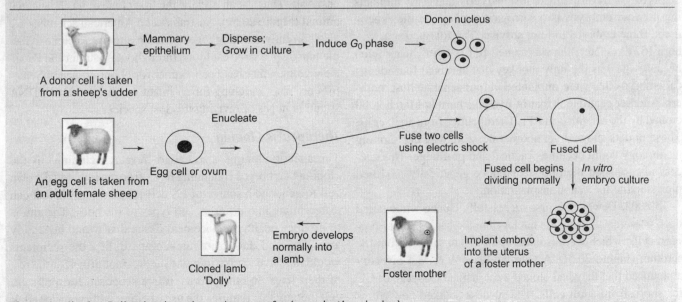

Figure A Cloning Dolly, the sheep by nuclear transfer (reproductive cloning)

had an impact on her health and longevity was unclear. The scientists at the Roslin Institute have stated that intensive health screening did not reveal any abnormalities in Dolly that could have come from advanced aging.

Prior to Dolly, scientists believed that once a cell became specialized as a certain type of cell, the change was permanent. Some scientists believed that errors in reprogramming during reproductive cloning led to the high numbers of dead or deformed animal clones. The production of Dolly revealed certain interesting findings. Firstly, as the donor cell was taken from a mammary gland, it was suggested that a specialized adult cell, such as an udder cell, which was programmed to express only udder cell genes, could be reprogrammed to create an entirely new creature. Thus, it was the first demonstration of pluripotency of a nucleus of a differentiated adult cell. Secondly, mature differentiated somatic cells in the body of an adult animal could revert back to an undifferentiated pluripotent form under some circumstances and then develop into any part of an animal. Thirdly, the differentiation of the adult cell did not involve the irreversible modification of genetic material required for development to term. Finally, the birth of lambs from differentiated fetal and adult cells also reinforced previous speculation that by inducing donor cells to become quiescent it would be possible to obtain normal development from a wide variety of differentiated cells.

Source: Campbell KH, McWhir J, Ritchie WA, Wilmut I (1996) *Nature* 380: 64–66; Wilmut I, Schnieke AE, McWhir J, Kind AJ, Campbell KH (1997) *Nature* 385: 810–813; Wilmut I, Campbell K, Tudge C (2000) In: *The Second Creation: Dolly and the Age of Biological Control*, Harvard University Press, Cambridge, MA.

named 'Bonnie'. Pictures of Dolly, her surrogate mother, and her lamb Bonnie are shown in Figures 20.5b and c. Since the publication of the original paper on cloning of Dolly, there are several other reports on adult-cloned animals involving mice, cattle, cats, goats, pigs, sheep, and rabbits involving the same and other cloning techniques. However, there are no documented cases of successful human cloning.

In January 2000, scientists at the Oregon Regional Primate Research Center cloned the bright-eyed rhesus macaque female, named Tetra, (Figure 20.5d) using embryo splitting (i.e., splitting an early-stage embryo and implanting the pieces into mother animals) with an aim to use monkeys to perfect new therapies for human diseases. Note that Tetra is not the first monkey to be cloned, but she is the first using the embryo-splitting technique. The cloning procedure involved creation of a fertilized egg by fusion of egg from a mother and sperm from a father, growing the resulting embryo into eight cells, splitting an eight-celled embryo into four two-celled embryos, each consisting of just two cells, and lastly implanting the four embryos into surrogate mothers. In the experiment, three embryos did not survive. The fourth, Tetra, was born 157 days later. She was named Tetra ('tetra' means four) because she was the only monkey that survived four identical embryos that were implanted in four separate host mothers. Another example is that of five pigs born in March 2000 cloned by the scientists at PPL Therapeutics from adult cells. These piglets showed yet another possible way that cloning technology could become practical and profitable. This success also opened the door to produce genetically modified pigs suitable for xenotransplantation.

In 2003, Dewey, the first successfully cloned white-tailed deer, was created using the nuclei from the buck's adult fibroblast cells, which were isolated from a skin sample from the scrotum (Figure 20.5f). A team at Texas A & M University announced that they had cloned a cat, named Copycat, or CC for short, from adult cells. The success sparked new entrepreneurial possibilities. To capitalize on this opportunity, the company Genetic Savings & Clone, Inc. (GSC), based in Sausalito, California, that funded the first successful domestic cat and dog cloning had gone commercial.

In yet another example, a team at the Laboratory of Reproductive Technologies in Cremona, Italy, cloned the first horse from adult cells. The foal, named Prometea, is also the first clone of any species to be carried to term by its twin mother.

Cloning may also before provide one method of saving species from extinction because intact DNA can be taken from these endangered species now. Cloning may enable population growth of currently endangered species and therefore permit their reintroduction into the wild. It may also enhance genetic diversity within these species. These benefits are particularly keen for animals, such as giant pandas and tigers, which are notoriously difficult to breed in captivity. Scientists have already adopted aggressive reproductive cloning practices. In 2001, at Iowa, a host cow gave birth to a baby bull guar, named Noah, an endangered species of ox native to India and Southeast Asia. Thus, Noah became the first clone of an endangered animal (Figure 20.5e). Australian and American scientists are also currently working together in an attempt to clone a Tasmanian tiger, a large cat found throughout Australia and Papua New Guinea that has been extinct for about 70 years. Scientists are also searching for an intact woolly mammoth DNA strand in hopes of reviving this lost species.

Therapeutic Cloning

Therapeutic cloning, also called 'research cloning', is the cloning of embryos containing DNA from an individual's own cell to generate a source of ES cell-progenitor cells that can differentiate into different cell types of the body. The aim is to produce healthy replacement tissue that would be readily available, and due to immunocompatibility, the recipients would not have to take immunosuppressant drugs for the rest of their lives. In other words, it is a proposed form of stem cell therapy that requires the use of either embryonic or adult stem cells. However, the results from the use of adult stem cells have to date proved generally less favorable than those with ES cells. There is a general consensus that ES cells offer significantly more scientific chances of success than limiting research to adult cells. It finds application in human development research and therapeutic remedies for several diseases. It is pertinent to note that this cloning technique is capable of creating cloned human cells, but not whole embryos beyond the 14-day stage. Thus, in therapeutic cloning, no sperm fertilization is involved nor is there implantation into the uterus to create a child. Therapeutic cloning thus differs from reproductive cloning in the sense that the end result is not a complete individual (human being or animal), rather the end product is a replacement organ, a piece of nerve tissue, or a quantity of skin. One problem with therapeutic cloning is that many attempts are often required to create a viable egg. The stability of the egg with the infused somatic nucleus is poor and it can require hundreds of attempts before success is attained.

Procedure of Therapeutic Cloning Therapeutic cloning aims to combine CNR, stem cell culture, and stem cell therapy. Its goal is to remove healthy adult cells from a patient, reprogramme the cell's nuclei by CNR, collect and grow pluripotent ES cell clones from the resulting blastocyst, and then induce these to differentiate into the stem cell or mature cell types required for transplantation to treat diseases. Note that at present CNR is the only practical means of reversing the differentiation of adult cells to restore their embryonic potential and of generating perfect match transplant tissue. Therapeutic cloning procedure starts with the same procedure as the one used in adult DNA cloning, but differs at later stage (Figure 20.6). The procedure involves the following steps: (i) Woman's ovum is taken and its DNA is removed. This step converts it into a form of human (or animal) life that serves as a factory for creating a pre-embryo; (ii) The DNA is removed from a cell taken from a human (or animal) and inserted into the ovum; (iii) The resulting ovum is given an electrical shock to start up its embryo making operation. Note that in a small percentage of cases, a pre-embryo will be formed; (iv) Unlike reproductive cloning, the pre-embryo is not implanted in a woman's womb in order to try to produce a pregnancy, rather

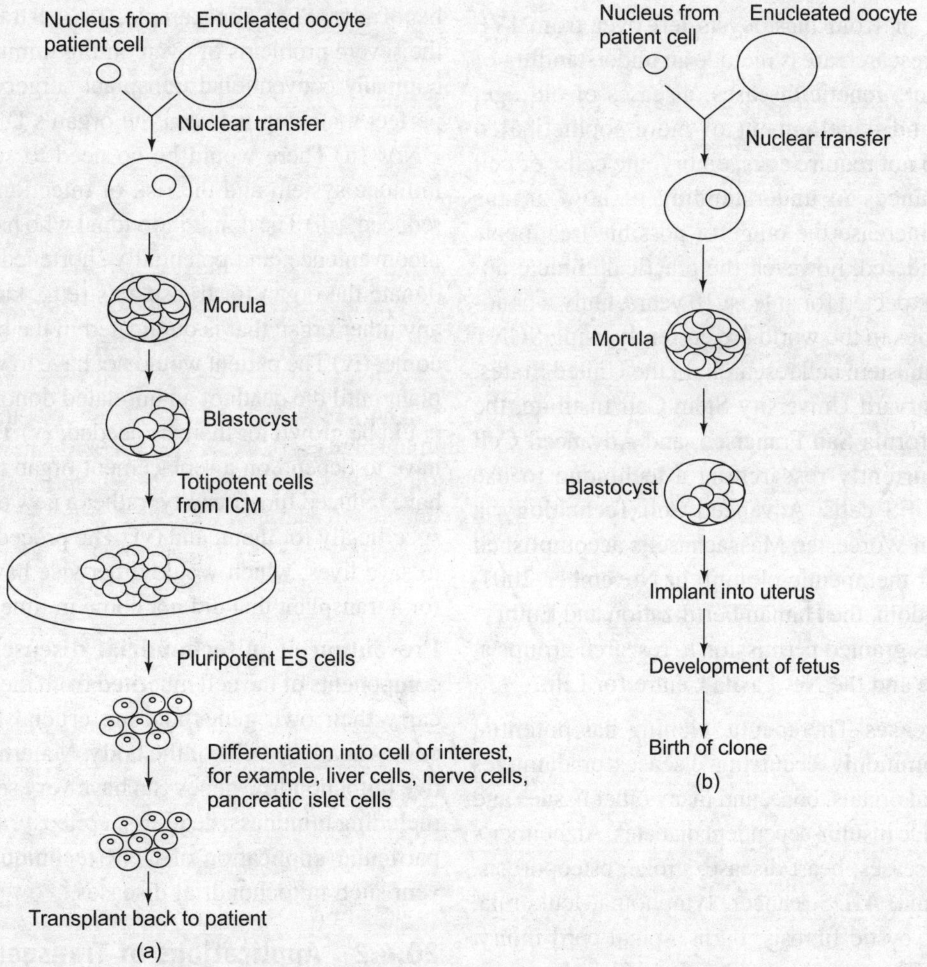

Figure 20.6 The process of therapeutic cloning and its comparison with reproductive cloning; (a) Therapeutic cloning, (b) reproductive cloning

it is allowed to grow and develop for ~14 days and produce many stem cells; (v) The stem cells are extracted from the pre-embryo resulting in its death; and (vi) The stem cells are grown into a piece of human (or animal) tissue or a complete organ for transplant in the patient. Note that the stem cells are a unique form of cells that can theoretically develop into many organs or body parts. Theoretically, these stem cells can be used to develop into replacement organs (heart, liver, pancreas, and skin, etc.).

Benefits of Therapeutic Cloning In theory, therapeutic cloning has enormous potential medical benefits, but extensive research would be required to realize this promise. The clinical applications of therapeutic cloning depend on the advances in stem cell therapy, the availability of eggs, learning to correct gene defects in inherited diseases, and in the shorter term, therapeutic cloning could provide a valuable transitory research tool to study differentiation process. Some possible applications are discussed in brief.

Stem cell therapy research Therapeutic cloning is a potential source of human pluripotent ES cells for stem cell therapy research. At present, these cells are isolated from aborted and miscarried fetuses or from blastocysts left over from IVF programmes. The research areas include an understanding of human development, genetic diseases, diseases of old age, serious injuries, and development of more sophisticated alternatives that do not require eggs, embryonic cells, or cell cloning. The advances in understanding of how organs regenerate would increase the range of possible treatments that could be considered, however, the practical clinical applications are not expected for at least 10 years. Only a handful of the laboratories in the world are currently using SCNT techniques in human stem cell research. In the United States, scientists at the Harvard University Stem Cell Institute, the University of California San Francisco, and Advanced Cell Technology are currently researching a technique to use SCNT to produce ES cells. Advanced Cell Technology, a biotech company in Worcester, Massachusetts accomplished the first successful therapeutic cloning in November 2001. In the United Kingdom, the Human Fertilization and Embryology Authority has granted permission to research groups at the Roslin Institute and the Newcastle Centre for Life.

Treatment of diseases Therapeutic cloning has potential applications in commonly occurring diseases or damages of the brain, internal organs, bone, and many other tissues and organs. These include insulin-dependent diabetes, Alzheimer's and Parkinson's diseases, heart disease, stroke, osteoporosis, blindness, glaucoma, AIDS, cancer, lymphoma, leukemia, Down's syndrome, cystic fibrosis, burns, spinal cord injury, and infertility, etc. Therapeutic cloning also offers the possibility of development of new and successful treatments for diseases. A stem cell scientists at the Memorial Sloan-Kettering Cancer Centre in New York, and her team used the same technique as that used for creation of Dolly, and extracted skin cells from the tails of mice with Parkinson's disease. They removed the nucleus from each cell and implanted these into egg cells, which had their own nuclei removed. The resulting cells were genetically identical to the donor mouse. By allowing these to develop, the scientists could extract stem cells that developed into dopamine neurons, which are missing in Parkinson's disease. The team injected the stem cells into the affected region of the brains of the donor mice and monitored the animals' behavior for 11 weeks. The treatment brought about a marked improvement in behavioral symptoms of the disease.

Organ transplantation The most attractive feature of therapeutic cloning is its potential to create perfectly matched transplant material, genetically identical to that of the patient. The procedure involves isolation of healthy cells from a patient and creation of copies that can be used as a source of perfectly matched transplants to replace or repair damaged or diseased tissues and organs. As compared to regular organ transplant donated by a second person, this procedure would offer a number of advantages, for example, (i) Such transplants would avoid the severe problems of overcoming immune rejection that accompany conventional transplant surgery. This is because of perfect matching between the organ's DNA and the patient's DNA; (ii) There would be no need to suppress the patient's immune system and the risk of infection would therefore be reduced; (iii) The donor individual who has to experience pain, inconvenience, and potentially shortened life span in order to donate the organ for transplants (e.g., kidney or theoretically any other organ that is duplicated in the body) would be overcome; (iv) The patient would not have to wait to obtain a transplant until the death of an unrelated donor; rather a new organ could be grown for them as needed; (v) The patient would not have to depend on a replacement organ that is old and might have reduced functionality; rather a new organ could be grown specifically for them; and (vi) The procedure has the potential to save lives, which would otherwise have been lost, waiting for a transplant that did not come in time.

Prevention of mitochondrial diseases Mitochondria are components of the cell inherited from the mother's egg. These carry their own genes and are crucial to fundamental processes in all the cells of the body. Maternally inherited defective mitochondrial genes can have very serious consequences, including blindness, deafness, epilepsy, and infant death. One particular application of CNR techniques would be to prevent such mitochondrial diseases.

20.4.2 Applications of Transgenic Animals

The underlying principle in the production of transgenic animals is the introduction of transgene(s) into an animal purposely by the human intervention, which must be transmitted through the germ line, so that every cell, including germ cells (i.e., cells that function to transmit genes to offspring), of the animal contains the same modified genetic material. Animal transgenic technology holds great potential in many fields, including agriculture, medicine, and industry. Thus, transgenic animals may have genetic modifications with potential use in studying mechanisms of gene function, changing attributes of the animal in order to synthesize products of high value, create models for human disease, or to improve productivity or disease resistance in animals. Some common applications are discussed in this section.

Transgenic Animals as Basic Research Models

Transgenic animals have been extremely important for analyzing human genes and have helped greatly in our understanding of a variety of fundamental biological processes, notably in immunology, neurobiology, cancer, and developmental studies. Some major applications are given below.

Investigation of gain-of-function and determination of biological function of expressed protein In principle, any mammalian gene that produces a dominant negative effect or gain-of-function can be investigated by introduction of an appropriate transgene. In some cases, this can provide proof of a suspected biological function. Of the different transgenic animals that have been made so far, transgenic mice, *Drosophila*, frogs, and fish have been very important for understanding aspects of gene function and development in these species. There are several examples of such studies and a few are described here: (i) A classical example is that of determination of the role of growth hormone (GH). The overexpression of GH in transgenic mice created huge mice that were almost twice the size of their litter mates (Figure 20.7a). The details are presented in Exhibit 20.6. Similarly, overexpression of GH in *Xenopus laevis* has been reported to stimulate growth of tadpoles and frogs; and (ii) Another example concerns the *sry* gene (sex-determining region of the Y chromosome) involved in sex determination in mice. A variety of different genetic analyses had implicated this gene as a major male-determining gene but convincing proof was obtained using a transgenic mouse approach. The experiment consisted of transferring a cloned *sry* gene into a fertilized 46, XX mouse oocyte. When *sry* gene was microinjected into mouse embryos, the resulting transgenic mice were all male. Indeed, even in the mice that had two X chromosomes and were thus genetically female, the presence of the *sry* gene was sufficient to cause them to develop testes and led to complete sex reversal. As a result of this artificial intervention, the resulting mouse, which nature had intended to be female, turned out to be male.

Investigation of loss-of-function Gene targeting has been used to identify the function of hundreds of mouse genes. One dramatic example was the deletion of the *Lim* 1 gene. The mice carrying this deletion died during embryonic development because of a complete lack of brain and head structure development. This demonstrated that the *Lim* 1 gene was critical for head development.

Gene disruption or gene knock-out provides clues to gene function In order to probe the function of a gene, it is inactivated and the resulting abnormalities are investigated. Transgenic organisms can be produced to inactivate a particular gene using 'gene knock-out technology', which relies on the process of homologous recombination (for details see Chapter 17). Through this process, regions of strong sequence similarity exchange segments of DNA. Foreign DNA inserted into a cell thus can disrupt any gene that is at least in part homologous by exchanging segments. Specific genes can be targeted if their nucleotide sequences are known. The resulting knock-out organisms display mutant phenotype, indicating the function of the original gene and hence are very important tools in understanding the functions of a gene and its role in the overall life process of the organism. For example, (i) The gene knock-out approach has been applied to the genes encoding transcription factors (i.e., gene regulatory proteins) that control the differentiation of muscle cells. When both copies of the gene for the regulatory protein myogenin are disrupted, an animal dies at birth because it lacks functional skeletal muscle. Microscopic inspection reveals that the tissues from which muscle normally forms contain precursor cells that have failed to differentiate fully. Heterozygous mice containing one normal myogenin gene and one disrupted gene appear normal, indicating that the level of gene expression is not essential for its function. Analogous studies have probed the function of many other genes to generate animal models

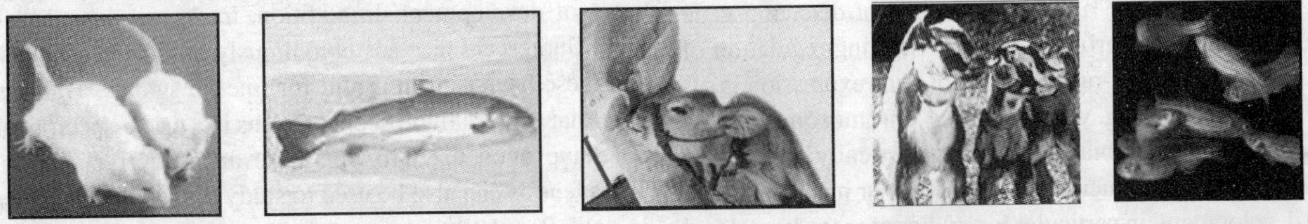

(a) Transgenic GH mice (b) Antifreeze fish (c) Patagonia (d) Webster and Peter (e) GloFish™

Figure 20.7 Some examples of transgenic animals

Exhibit 20.6 Supermice or giant mice: Transgenic mice harboring rat growth hormone

The growth hormone (somatotropin), a 21 kDa protein, is normally synthesized by the pituitary gland. A deficiency of this hormone produces dwarfism and its excess leads to gigantism. Growth hormone (GH) has been valuable tool to study structure–function relationships in the GH molecule. Much of the research in this area has centered on developing transgenic mice.

In 1982, Ralph Brinster and colleagues at the University of Pennsylvania's School of Veterinary Medicine successfully injected the gene encoding rat GH into mouse embryos. These genetically engineered giant mice illustrated the expression of foreign genes in mammalian cells for the first time. The gene for rat growth hormone was placed on a plasmid next to the mouse metallothionein promoter. This promoter site is normally located on a chromosome, where it controls the transcription of metallothionein, a cysteine-rich protein that has high affinity for heavy metals. Metallothionein binds to and sequesters heavy metals, many of which are toxic for metabolic processes. The synthesis of this protective protein by the liver is induced by heavy metal ions such as cadmium. Hence, if mice contain the new gene, its expression can be initiated by the addition of Cd to the drinking water. The construct used for transformation is drawn in Figure A. Several hundred copies of the above plasmid containing the promoter and *GH* gene were microinjected into the male pronucleus of a fertilized mouse egg, which was then inserted into the uterus of a foster mother mouse. A number of mice that developed from such microinjected eggs contained the gene for rat GH, as shown by Southern blots of their DNA. These transgenic mice, containing multiple copies (~30 copies/cell) of the rat *GH* gene, grew much more rapidly than the control mice. In the presence of Cd, the level of GH in these mice was 500 times as high as in normal mice. The transgenic mice were huge, twice the size of their litter mates (in figure shown above, left is transgenic, and right is nontransgenic). Owing to their large size, these transgenics mice were dubbed as 'supermice' by the press. The foreign DNA was transcribed and its five introns correctly spliced out to form functional mRNA. These experiments thus strikingly demonstrated that a foreign gene under the control of a new promoter could be integrated and efficiently expressed in mammalian cells. The supermice also passed the *GH* gene on to their offspring. Later this strategy was used to create transgenic mice harboring human *GH* gene.

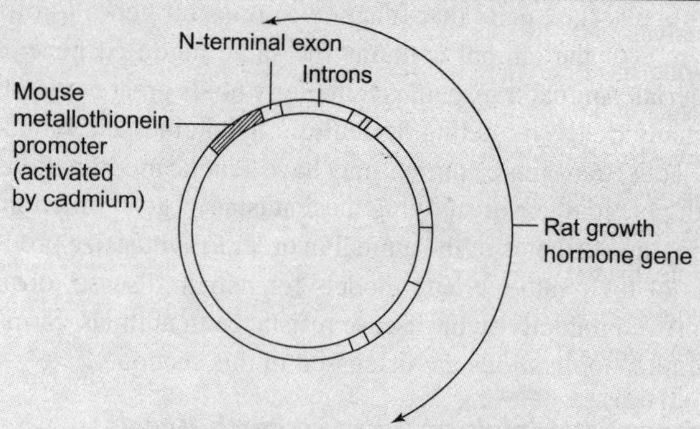

Figure A Rat growth hormone metallothionein gene construct [The gene for rat GH is inserted into a plasmid next to the metallothionein promoter, which is activated by the addition of heavy metals, such as cadmium ion]

Source: Brinster RL, Chen HY, Warren R, Sarthy A, Palmiter RD (1982) *Nature* 296: 39–42; Brinster RL (2002) *Science* 296: 2174–2176.

for known human genetic diseases; and (ii) Knock-out mice with inactivated tumor suppressor genes, which suppress the spontaneous development of cancer, were seen to be highly prone to the development of cancers of various kinds. This gave the unambiguous evidence of the role of the tumor suppressor genes.

Investigation of gene expression and its regulation Although evidence for *cis*-acting regulatory elements is often inferred initially from studies using cultured cells, these need to be validated in whole animal studies. Transgenes consisting of the presumptive regulatory sequence(s) coupled to a reporter gene, such as *lac* Z, provide a sensitive method of detecting gene expression and a powerful way of investigating regulation of gene expression. Long-range control of gene expression is often investigated using yeast artificial chromosome (YAC) transgenes. The amphibian embryo has classically been one of the best systems for elucidating the molecular mechanisms of early development, in particular for studies of mesodermal and neural induction. Amphibian embryos develop externally and are large and robust. Therefore, tissues can be dissected, isolated, or transplanted with high precision and ease in these embryos. In addition, it is relatively easy to manipulate the expression of gene products by injecting *in vitro* transcribed RNAs into developing embryos. However, since RNAs are translated soon after injection, this method has been used mainly for studying early stages of development. Manipulating genes specifically during later stages of development requires fine control over the time and place of expression, which can be achieved only through transgenic technology. Transgenesis has been used to express the wild-type form as well as the mutant form of genes in specific regions of the embryo or at distinct times of development. In addition, transgenic lines that express fluorescent proteins ubiquitously have been established and these have proven useful for lineage studies. Transgenic lines that express fluorescent proteins in a tissue-specific manner have been used to visualize an inductive response. Transgenesis can also be used to study the regulation of genes *in vivo*. Regulatory elements from various species can drive appropriate expression in transgenic *Xenopus* embryos. By comparison with mouse transgenesis, transgenic *Xenopus* embryos can be obtained rapidly, at low cost, and in large numbers. Therefore, transgenesis is useful for large-scale functional analysis of *cis*-regulatory regions of genes. The transgenic technique has enormous potential for addressing the function of genes in late development and organogenesis.

Investigation of dosage effects and ectopic expression In some cases, valuable information can be gained by overexpressing a transgene or by expressing it ectopically. Note that the transgene is coupled to a tissue-specific promoter, which causes expression in specific cells, and the phenotypic consequence may provide valuable clues to function.

Cell lineage ablation The elimination of certain cell lineages in the animals is called cell ablation. Transgenes can be designed consisting of a tissue-specific promoter coupled to a sequence encoding a toxin, for example, diphtheria toxin subunit A or ricin. When the promoter becomes active at the appropriate stage of tissue differentiation, the toxin is produced and kills the cells. Thus, certain cell lineages in the animal can be eliminated and the phenotypic consequences monitored.

Modeling human disease Human mutant genes may be inserted into mice, causing them to suffer from human diseases, which act as disease models (i.e., animals genetically manipulated to exhibit disease symptoms so that effective treatment can be studied). These disease models decrease the number of animals used in such experimentation and permit to undertake studies not always possible on humans. These disease models also allow the testing of drugs to devise treatments without experimenting on humans. These models of human diseases are created with the application of gene targeting. Insertional inactivation is often used to model loss-of-function mutations while gain-of-function mutations can often be modeled by inserting a mutant transgene. Some common examples of disease models are as follows: (i) Cystic fibrosis is caused by defective genes and can be mimicked in a transgenic mouse. For this, mutations have been made in the mouse version of the cystic fibrosis transmembrane conductance regulator gene. Although mice with the mutated gene do not develop the devastating symptoms of cystic fibrosis in their lungs, they do develop the intestinal and pancreatic duct defects associated with the disease and thereby provide a model to study at least part of the disease; (ii) Transgenic mice overexpressing the amyloid precursor protein form deposits in the brain that resemble the amyloid plaques found in Alzheimer's patients. Mouse models can potentially be used to test drug therapies and to gain information about the progression of the disease; (iii) Harvard scientists created a genetically engineered mouse, called OncoMouse® or the Harvard mouse, to test potential cancer drugs. This mouse contained exogenously derived oncogenes and hence was capable of developing a cancer. For the genetically engineered mouse, the scientists received a US patent (the company DuPont holds exclusive rights to its use). The transgenic mouse model for the investigation of breast cancer led to two discoveries, *viz.*, the identification of a regulating element in the mouse mammary tumor virus (MMTV) that is active specifically in cells of the mammary gland and the involvement of *myc* and *ras* oncogenes in breast cancer in mice transformed with these genes; (iv) Another transgenic mice model for gain-of-function disease Gerstmann-Straussker-Scheinker (GSS) syndrome, a neurodegenerative disease caused by a dominantly acting mutated prion protein genes was developed, which carried the mutant form of the prion protein gene in codon 102 in addition to the wild-type locus. These transgenic mice developed similar neurodegenerative pathology to their human counterparts; (v) AnyGene, the pharma company in Manchester, detected, analyzed, and cloned a gene for a certain form of brain cancer (called *brac* 1). This gene was incorporated into mice, and the resulting transgenic mice was used as a disease model for investigations into the development of brain tumor and to try out medicines to prevent its growth; (vi) Simple transgene addition has also been used to model diseases caused by dominant negative alleles, as shown for the premature aging disease, Werner's syndrome; and (vii) Recessively inherited diseases are generally caused by loss-of-function and these can be modeled by 'gene knockout'. The earliest report of this strategy was a mouse model for hypoxanthine-guanine phosphoribosyl transferase (HPRT) deficiency, generated by disrupting the gene for HPRT. A large number of genes have been modeled in this way, including cystic fibrosis, fragile-X syndrome, β-thalassemia, and mitochondrial cardiomyopathy. Gene targeting has been widely used to model human cancers caused by the inactivation of tumor suppressor genes such as TP53 and RB1.

While the studies above provide models of single gene defects in humans, attention is now shifting toward the modeling of more complex diseases that involve multiple genes. In some cases, the crossing of different modified mouse lines has led to interesting discoveries. For example, undulated mutant mice lacks the gene encoding the transcription factor Pax-1 and Patch mutant mice are heterozygous for a null allele of the platelet-derived growth factor gene. Hybrid offspring from a mating between these two strains were shown to model the human birth defect spina bifida occulta. In other cases, such crosses have pointed the way to possible novel therapies. For example, transgenic mice overexpressing human α-globin and a mutant form of the human β-globin gene that promotes polymerization provide good models of sickle cell anemia. However, when these mice are crossed to those ectopically expressing human fetal hemoglobin in adulthood, the resulting transgenic hybrids show a remarkable reduction in disease symptoms. Similarly, crossing transgenic mice overexpressing the antiapoptotic protein Bcl-2 to *rds* mutants that show inherited slow retinal degeneration resulted in hybrid offspring in which retinal degeneration was strikingly reduced. This indicates that Bcl-2 could possibly be used in gene therapy to treat the equivalent human retinal degeneration syndrome. The most complex diseases involve many genes and transgenic models would be difficult to create. However, it is often the case that such diseases can be reduced to a small number of 'major genes' with severe effects and a larger number of minor genes. Thus, it has been possible to create mouse models of Down's syndrome, which in humans is generally caused by the presence of three copies of chromosome 21. Trisomy for the equivalent mouse chromosome 16 is a poor model because the two chromosomes do not contain all the same genes. However, studying Down's syndrome patients with partial duplications of chromosome 21 has identified a critical region for Down's syndrome. The generation of YAC transgenic mice carrying this essential region provides a useful model of the disorder and has identified increased dosage of the *Dyrk* 1a (minibrain) gene as an important component of the learning defects accompanying the disease.

Transgenic Animals as Bioreactors for Large-scale Production of Substances for Human Welfare and Improvement of Production Traits

Conventional properties of farm animals, such as their rate of growth or milk or wool production, have been manipulated by selected breeding for thousands of years, resulting in the specialized breeds available today that bear little resemblance to their original ancestors. Today many of the transgenic animal experiments are designed to increase speed of growth, raise weight, reduce fat, improve milk, and wool yield, etc. Besides, animals can be used as sources of a variety of useful products by introduction of additional genes from other sources. Animal genetic engineering has the capability to benefit humans in several ways, including advancing health care research and treatment and satisfying the food needs of the growing population. Thus, animal transgenic technology holds great potential in agriculture, medicine, and industry. Following is a brief discussion of animal transgenesis for betterment of livestock and for human welfare.

Generation of herds with specific traits in short time and improvement in the size of livestock Farmers have always used selective breeding to produce animals that exhibit desired traits (e.g., increased milk production, high growth rate). Traditional breeding is a time-consuming and difficult task. With the advent of animal transgenic technology, it has become possible to develop traits in animals in a shorter time and with more precision. In addition, it offers the farmer an easy way to increase yields. Transgenic cows have been created that produce more milk or milk with less lactose or cholesterol, pigs and cattle that have more meat on them, and sheep that produce more wool.

Improvement in growth rates and food production Genetic engineering offers the potential to increase this food supply by improving growth rates of fishes, pigs, sheep, and cattle that could greatly enhance the food supply. Some common examples include the following.

Application of GH In most of the earlier work in domestic species (pig, sheep, cattle, and goat) for their application as bioreactors, a high-level expression of GH was achieved under the control of metallothionein promoter. The technique of transgenic mice as mentioned in Exhibit 20.6 was also extended to livestock animals. These GH transgenic animals were robust, with more flesh, and the expression of GH offered an alternative to the labor-intensive means of administering GH by daily injection in livestock. Generally, the expression of transgenes in the mammary glands of sheep, cows, pigs, rabbits, and goats has no ill effects on either lactating females or nursing progeny. However, this is not always so. For example, when the transgene for bovine *GH* under the control of the metallothionein promoter was introduced into pigs, adverse results were observed. Although the levels of GH varied among these transgenic pigs, the group as a whole showed an increased ability to convert into body weight.

In 1985, a transgenic fish harboring *GH* gene was created. Using a mouse metallothionein promoter ligated to a human *GH* structural gene, transgenic loach, goldfish, and silver carp were successfully produced. On the average, the transgenic fish was one to three times larger than control. Since then, several reports using similar gene constructs are available that have been used successfully for the transformation of species including salmon, trout, carp, and tilapia, etc. Canadian geneticists from the Vancouver Fisheries and Ocean Department (British Columbia) jointly with two researchers from America and Singapore have created a line of giant transgenic salmon, the famous Sumosalmon. When one year old, they were on average 11 times as big as normal salmon of the same age. One even reached a growth rate 37 times the normal. In another study, transgenic channel catfish with the *rtGH* cDNA had more protein, less fat, and less moisture in their edible muscle than nontransgenic full-siblings (about a 10% change). All fish gene constructs have also been developed and used for transformation of fishes. In one study, transgenic Atlantic salmon (*Salmo salar* L.) was produced using a gene construct (*OPAFPesGH*) containing a *GH* gene from chinook salmon (*Oncorhynchus ischawytcha*) driven by regulatory sequence from an antifreeze gene of the ocean pout (*Macrozoarrces americanus*) and the resulting transgenic fish grew two to six times more than the transgenic siblings. In another study, up to 11-fold more growth as compared to nontransgenic siblings was observed in coho salmon (*Oncorhynchus kisutch*) when a gene construct (*OnMTGH* 1) consisting of a sockeye salmon (*O. nerka*) *GH* gene spliced to metallothionein promoter of the same origin was used.

Application of oncogene Another effort to genetically alter growth rates and patterns included the production of transgenic swine and cattle expressing a foreign *c-ski* oncogene, which targets skeletal muscle.

Application of insulin-like growth factor I In a report, there is evidence for the use of transgenesis allowing significant reduction in body fat in transgenic pigs and increased diameter of muscle fiber by increasing insulin-like growth factor I (IGF-I) levels without serious pathological side effects.

Application of fatty acid desaturase gene Transgenic pigs have been generated that carry the fatty acid desaturation 2 gene for a 12-fatty acid desaturase from spinach. Levels of linoleic acid in adipocytes that differentiated *in vitro* from cells derived from the transgenic pigs were 10 times higher than those from wild-type pigs. In addition, the white adipose tissue of transgenic pigs contained 20% more linoleic acid than that of wild-type pigs. These results demonstrated the functional expression of a plant gene for a fatty acid desaturase in mammals, opening up the possibility of modifying the fatty acid composition of products from domestic animals by transgenic technology.

Improvement in milk yield and its quality Genes encoding milk proteins can be altered to modify milk composition and properties. Among the different applications of milk modification in transgenic animals, the following can be highlighted: (i) As compared to bovine milk, human milk lacks β-lactoglobulin, has higher contents of lactoferrin and lysozyme and has a higher serum proteins to casein ratio. Note that β-lactoglobulin acts as a major allergen in cow's milk and should be eliminated. Moreover, human lactoferrin protein is involved in iron transport and inhibits the bacterial growth, and hence its inclusion in cow's milk is preferable. Thus to make the bovine milk more appropriate for consumption by infants, transgenic cows have been created with the introduction of human lactoferrin gene and elimination of β-lactoglobulin from their milk; (ii) It is considered that 70% of the world population lacks intestinal lactase (lactose hydrolyzing enzyme) and suffers from lactose intolerance. Hence for such people, it is desirable to reduce lactose content from milk. This can be easily achieved by expressing β-galactosidase in the milk or by diminishing the content of α-lactalbumin. This probability has been tested in transgenic mouse. However, a serious practical drawback of this method is that this milk is very viscous and it is not secreted to the exterior of the mammary gland due to the importance of the lactose in the osmoregulation of the milk; (iii) Another field of interest in milk industry is alteration of milk casein content to increase its nutritive value, cheese yield, and processing properties. Researchers have intended to increase the number of copies of the gene of the κ-casein to reduce the size of the micelles and modifying the κ-casein to make it more susceptible to digestion with chymosin. Thus female bovine fetal fibroblasts have been engineered to express additional copies of transgenes encoding two types of casein, i.e., bovine β-casein and κ-casein. The modified cell lines of fibroblasts were used to create 11 cloned calves. Milk from the cloned animals was enriched for β- and κ-casein, resulting in a 30% increase in the total milk casein or a 13% increase in total milk protein, demonstrating the potential of this technology to make modified milk. The overexpression of casein also increased the thermal stability of milk, reducing protein breakdown during manufacturing; (iv) A transgenic mice has been produced that expressed human lysozyme or a modified bovine casein (a protein used in cheese-making and for some industrial purposes) in the milk. The production of human lysozyme in milk of transgenic mice also increased the antimicrobial properties of milk. It is suggested that if this technique is extended to cows, it could reduce infections in the mammary gland and perhaps eliminate undesirable pathogens in the gut of humans who consume the milk; and (v) Another example of animal transgenesis is the expression of antibacterial substances in the milk such as proteases to increase mastitis (mammary gland abscesses) resistance. The objective here is to alter the concentrations of antibacterial proteins such as lysozyme or transferrin in the milk.

Production of cysteine rich wool Animal transgenic technology has been used to improve production of sheep wool and to modify the properties of the fiber. Because cysteine seems to be the limiting amino acid for wool synthesis, the first approach was to increase its production through transfer of genes encoding enzymes involved in cysteine biosynthesis from bacterial to sheep genome. This approach, however, did not achieve the efficient expression of these enzymes in the rumen of transgenic sheep.

Acquisition of new and useful characteristics by animals The animal transgenic technology can be employed for acquisition of new and useful characteristics such as disease resistance and cold tolerance, etc.

Generation of disease-resistant livestock Currently, vaccination, drugs, physical isolation, and careful monitoring are used to control infectious diseases of domestic animals. These are costly processes and the cost of disease prevention can be as much as 20% of the total production value. Scientists are attempting to produce animals with inherited resistance to bacterial, viral, and parasitic diseases, for example, genetic resistance to bacterial diseases such as mastitis in dairy cattle, neonatal scours (dysentery) in swine, and fowl cholera would be likely targets. If the basis of resistance in each case is due to a single gene, it may be possible to create transgenic animals with specific protection against a bacterial disease after these genes have been isolated and characterized. Some common examples of developing disease-resistant animals are as follows: (i) *S. aureus* bacteria that destroy milk-secreting cells in the animal's mammary gland cause about 30% of all mastitis cases in dairy cows. Antibiotics are only effective in about 15% of cows infected with *S. aureus*, so dairy producers are forced to sacrifice these cows from their herds. US Department of Agriculture and University of Vermont researchers produced a clone of a pure-bred Jersey cow, named 'Annie' in March 2000, named after Annie Powell, a contributing team member. In the procedure, genes for lysostaphin, a green fluorescent protein (GFP) tag, an antibiotic marker, and gene for β-lactoglobulin were inserted into a fibroblast cell using a mild electric current. The altered fibroblast was fused to an unfertilized cow egg, whose nuclear DNA contents had been removed. The surviving fused cells were allowed to divide like a normal fertilized embryo. The healthiest embryos, which included Annie, were cultured and implanted into surrogate mothers. The gene for lysostaphin was taken from a benign species *S. simulans* that competes with its virulent cousins. The purpose of raising Annie was to develop cows resistant to mastitis disease by secreting lysostaphin protein. Moreover, β-lactoglobulin helped in targeting the protein expression in milk. Though not the first cow clone, Annie is the first to be genetically altered with a gene for an agricultural application; (ii) Another approach that may be used to develop lines of animals that are resistant to infectious agents entails creating inherited immunological protection by transgenesis. A number of candidate genes that contribute to the immune system (e.g., genes encoding major histocompatibility complex, T cell-receptor, and lymphokine) are being studied from this perspective. However, the most favorable preliminary results to date have come from research in which the genes encoding the H and L chains of a McAb have been transferred to mice, rabbits, goats, and pigs. The rationale behind this strategy is to provide a built-in, inherited biological protection mechanism for the transgenic animal that eliminates the need for immunization by vaccination. The concept of introducing into a recipient organism the transgenes for an antibody that binds to a specific antigen has been called *in vivo* immunization. To test this idea, the genes for the immunoglobulin chains of a mouse McAb against an antibody that binds to 4-hydroxy-3-nitrophenylacetate were cloned in tandem and microinjected into fertilized eggs of mice, rabbits, and pigs. In each case, McAb activity was found in the serum of the transgenic animals. However, the level of McAb containing both of the cloned H and L chains was low; and (iii) Scientists are working to develop disease resistance in fishes through genetic enhancement. Insertion of lytic peptide cecropin B construct enhanced the resistance to bacterial diseases two- and four-fold in channel catfish. Producers have begun to push for approval of transgenic fish. The FDA received an application for the production of genetically engineered Atlantic salmon from A/F Protein, Inc., making them the first regulatory agency in the world to receive an application to approve the commercial development of transgenic fish.

Cold tolerance in fishes As natural fisheries become depleted, the production of worldwide resources will depend more on aquaculture. Early research involved the transfer of a full-length *AFP* gene encoding the major liver secretory AFP from the winter flounder into the fertilized eggs of Atlantic salmon (Figure 20.7b). The *AFP* transgene successfully integrated into the salmon chromosomes, expressed, and exhibited Mendelian inheritance. Expressed levels of AFP in the blood of these fishes were quite low (100–400 μg/ml) and were insufficient to confer any significant increase in freeze resistance to the salmon. In another study, antifreeze promoter of a fish called ocean pout has been cloned with the *GH* cDNA from salmon. The resulting transgenic salmon grew faster and survived cold waters as compared to nontransgenic controls.

Xenotransplantation or Xenografting Human-to-human organ transplants (allotransplantations or allografts) of hearts, livers, kidneys, and lungs are 70–80% effective for the first year, and 50–70% effective for 5 years. However, throughout the world, the demand for donated organs far exceeds the available supply. As a lack of supply of a replacement heart, liver, or kidney, several patients die every year. Given shortages in most countries for organs, the development of other medical options has become critical. Some experts believe that xenotransplantation, i.e., transplanting organs from animals to humans, may be able to solve the organ shortage problem. In this context, swine has been considered the most likely source of organs for xenotransplantation because they are raised for food and as a result (but not necessarily), they might be socially acceptable as organ donors. Moreover, their organs are similar in size and physiological functions to humans and none of the known porcine infectious agents are thought to be human pathogens. Other advantages of using pigs are that they have a small gestation period (only four months), acquire fertility at 12 months of age, and produce large litter sizes (usually 10 to 12 piglets). The 'knock-out' pigs with the deletion of the gene responsible for the human rapid immune rejection response and their cloning to generate large herds may thus serve as source of organs to transplant in humans. The major impediment to organ transplantation between species is hyperacute rejection of the animal organ. It entails the binding of preexisting antibodies of the host organism to a carbohydrate epitope (α-Gal) on the surface of the cells of the grafted organ. These bound antibodies elicit an inflammatory response (complement cascade) that destroys the antibody-coated cells and lead to the loss of the transplanted organ within hours. Under natural conditions, proteins on the surface of the cells lining the blood vessels protect the cells from the inflammatory response. These complement-inhibiting proteins are species-specific. Therefore, it was reasoned that if the donor animal carried one or more of the genes for a human complement-inhibiting protein, a transplanted organ would be protected from the initial inflammatory response. To this end, transgenic pigs with different human complement inhibitor genes have been produced. It is believed that in future, patients with kidney, heart, and lung failure are likely to be benefited from animal organ transplants. Biotech companies such as Nextran and Alexion are trying to insert human genes into the germ lines of animal embryos to make their organs more compatible with the human genome and less likely to be rejected.

Recombinant proteins from live animals (pharming) Another application of animal transgenesis is the production of nutritional supplements and human therapeutic proteins. Transgenic animals are expensive to produce but the technology is cost-effective because once a transgenic animal has been made, it can reproduce and pass its cloned gene to its offspring according to standard Mendelian principles. As milk is the staple food for ages among children, adolescents, and adults and as it is a renewable secreted body fluid produced in substantial quantities, and collected frequently without harming the animal, mammary glands of dairy cows, sheep, pigs, and goat serve as excellent protein production factories. Further, the expression in milk leads to large-scale production of proteins encoded by cloned genes. To convey the idea that milk from transgenic farm ('pharm') animals can be a source of authentic human pharmaceuticals, the term 'pharming' was coined. As the expression of proteins in all cells may be fatal for animal, the expression of foreign proteins is restricted only to mammary glands. This is possible by attaching the foreign genes encoding pharmaceutically important proteins to a promoter active in mammary tissue, e.g., casein or β-lactoglobulin promoters. Moreover, to prevent the movement of proteins into the bloodstream, the foreign gene is usually modified such that the protein produced in the mammary gland cannot work unless it is subjected to proteolysis, which is carried out during processing and purification of the milk extracts. This strategy is particularly advantageous because a novel drug protein confined to the mammary gland does not have any effect on normal physiological processes of the transgenic animal and it can be easily purified. This strategy opened the field not only for enriching the milk itself with various proteins, but also for the use of cattle or sheep as bioreactors for producing large amounts of proteins with various uses in an efficient way (Table 20.17).

It is pertinent to note that before animal transgenesis, the pharmaceutically important proteins had to be painstakingly isolated from tissues or blood, but since these are produced in such minute quantities, the isolation of significant quantities required the processing of large amounts of material. The advent of animal transgenesis has offered substantial advantages over the use of recombinant bacterial cultures: (i) It avoids the need to build and run expensive industrial-scale fermentors; (ii) Using animal hosts means that the product is likely to have the appropriate posttranslational modifications, which is often not the case for products obtained from bacteria; (iii) The purification of the protein from the milk is relatively straightforward; and (iv) Producing pharmaceutical proteins can be prohibitively expensive when it is done through the large-scale culture of human cells. Some bacteria and plants can be used as production means, but in some instances, the use of animals is the only way to ensure appropriate levels of biological activity. For example, the protein α_1-antitrypsin, normally produced in liver, is required for the treatment of lung diseases such as emphysema or cystic fibrosis. While this protein can be produced in transgenic plants, the product obtained does not have certain carbohydrate elements and is processed in the bloodstream at much faster rates than that produced by the liver. Animals modified to secrete the human protein in their milk produce the more effective protein version.

On September 8, 1981, a team of scientists from the University of Ohio, jointly with the Jackson Laboratory at Bar Harbor, Maine, achieved the first successful animal transgenesis by transplanting a rabbit's β-globin gene into a mouse embryo. Although proteins can be produced at high concentrations in mouse milk (e.g., 50 ng/ml for tPA), the system is not ideal due to the small volume of milk produced. Therefore, other animals, such as cattle, sheep, and goats, have been investigated as possible bioreactors. Such animals not only produce large volumes of milk, but the regulatory practices regarding the use of their milk are more acceptable. Human proteins that have been expressed in milk of transgenic cows, sheep, or goats include insulin, GH, lactoferrin, α-lactalbumin, serum albumin, blood anticlotting factors, and tissue plasminogen activator, etc. (Table 20.17). GTC Biother-apeutics (Framingham, Massachusetts) used both goats and cows to produce more than 60 therapeutic proteins, including plasma proteins, McAbs, and vaccines. GTC is also using animal transgenic technology for the production of pharmaceutically important proteins in collaboration with various companies such as Abbott, Centocor (a subsidiary of Johnson & Johnson), Alexion Pharmaceuticals, Immunogen, BASF, Bristol Myers Squibb, and Eli Lilly.

Some examples of animal transgenesis in the field of pharming are the following: (i) Pharming, NV (Leiden, The Netherlands) developed the first transgenic bull, named Herman in the late 1980s, and developed a line of transgenic cows to produce several proteins including human lactoferrin and α-glucosidase; (ii) Normally α_1 proteinase inhibitor (αPI) is extracted from human blood plasma, but the amounts available are small and cost excessive. To overcome this problem, the Roslin Institute in collaboration with PPL Therapeutics at Scotland developed transgenic sheep, named Tracey,

Table 20.17 Some Examples of Human Recombinant Proteins Expressed in the Milk of Transgenic Animals

Gene	Transgenic species	Promoter region	
		Gene	Source
α_1-Antitrypsin	Sheep	β-Lactoglobulin	Sheep
	Mice	β-Lactoglobulin	Sheep
α-Glucosidase	Mice	α_{S1}-Casein	Bovine
α-Lactalbumin	Rats	α-Lactalbumin	Human
Antithrombin III	Goat	J-Casein	Goat
β-Interferon	Mice	WAP	Mouse
γ-Interferon	Mice	β-Lactoglobulin	Sheep
CFTR	Mice	β-Casein	Goat
EPO	Mice	β-Lactoglobulin	Bovine
	Rabbits	β-Lactoglobulin	Bovine
Factor VIII	Sheep	β-Lactoglobulin	Sheep
Factor IX	Mice	β-Lactoglobulin	Sheep
	Sheep	β-Lactoglobulin	Sheep
Fibrinogen	Mice	WAP	Mouse
GM-CSF	Mice	α_{S1}-Casein	Bovine
Growth hormone	Mice	J-Casein	Rat
Hepatitis B surface antigen	Goat	α_{S1}-Casein	Bovine
IGF-1	Rabbits	α_{S1}-Casein	Bovine
Interleukin-2	Rabbits	β-Casein	Rabbit
Lactoferrin	Mice	α_{S1}-Casein	Bovine
	Cattle	α_{S1}-Casein	Bovine
Lysozyme	Mice	α_{S1}-Casein	Bovine
Protein C	Mice	WAP	Mouse
	Pigs	WAP	Mouse
Serum albumin	Mice	β-Lactoglobulin	Sheep
Superoxide dismutase (SOD)	Mice	β-Lactoglobulin	Sheep
	Mice	WAP	Mouse
Surfactant protein B	Mice	WAP	Rat
TAP	Mice	WAP	Rat
tPA	Mice	WAP	Rat, Mouse
	Rabbits	α_{S1}-Casein	Bovine
	Goats	WAP	Mouse
Urokinase	Mice	α_{S1}-Casein	Bovine

using the technique of pronuclear microinjection (for details see Chapter 14). This procedure involved the microinjection of artificially inseminated eggs with a DNA construct containing an *aat* gene encoding αPI fused to a β-lactoglobulin promoter (200–300 copies of the transgene were inserted). The eggs were implanted into surrogate mothers, of which 112 gave birth. Four females, including Tracey, and one male were found to have incorporated intact copies of the gene and all five developed normally. Tracey was thus a 'furry bioreactor' that produced human α_1 proteinase inhibitor in her milk in a cost-effective way (30 g of αPI/l milk). The production potential of Tracey was significant and during her first lactation, she produced 1.5 kg of αPI; (iii) Using similar protocols as that used for creation of Tracey, Ebert *et al.* (1991) demonstrated the production of a variant of human tissue plasminogen activator (tPA) in goat milk. Of 29 offsprings, one male and one female contained the transgene. The transgenic female underwent two pregnancies and one out of five offsprings was transgenic. Milk collected over her first lactation contained only a few milligrams of tPA per liter, but improved expression constructs have since resulted in an animal generating several grams per liter of the protein; (iv) Recombinant human antithrombin III, which is used to treat intravascular coagulation in patients who have undergone heart bypass operations, was the first protein expressed in transgenic animal milk to reach commercial production and is currently marketed by Genzyme Transgenics Corporation; (v) In 1996, Genzyme Transgenics announced the birth of Grace, a transgenic goat carrying a gene that produces BR-96, an antibody being developed and tested by Bristol-Meyers Squibb to deliver conjugated anticancer drugs;

(vi) In July 1997, the scientists at the Roslin Institute, Edinburgh, Scotland headed by Ian Wilmut developed Polly and Molly, the first transgenic sheep. The creation of Polly and Molly was based on the same procedure as that for Dolly, i.e., SCNT using adult cell. However, unlike Dolly, these were not simply clones, but were also genetically transformed. Moreover, the source cell type of the nucleus that was transferred was mammary gland cell from a 6-year-old ewe for the creation of Dolly, while it was fibroblast cell in the case of Polly and Molly. The procedure involved a single, diploid cell originating from the adult sheep, which was genetically altered by introducing a human gene coding for blood clotting factor IX before fusing with an enucleated egg cell. Note that the deficiency of factor IX leads to haemophilia B and its treatment requires intravenous infusion of factor IX. The advantage of the production of factor IX in milk is the reduction in cost and freedom from potential risk of infection associated with the current source of this protein, i.e., human blood. The fused cell was made to divide and to develop into an embryo, which was implanted into a surrogate mother. In the procedure, a cell type PDFF was used. PDFF5 produced male animals and were not transduced, while cell type PDFF2 produced female animals and were transduced. Of the gestations that occurred, three PDFF2 animals were born, two of which survived and were named 7LL8 and 7LL12. These animals were transfected but contained a marker gene, which was not the cloned gene of interest. These were named 'Holly' and 'Olly'. Another subset of female-producing PDFF2 cells, PDFF2-12 and PDFF2-13, also produced animals that had the gene of interest together with the marker. Of these lambs, 7LL12, 7LL15, and 7LL13 were born alive and healthy. Two of these were named Polly and Molly. These lambs contained the human factor IX in every cell of their body. The generation of Polly harboring a human gene provided a proof that this technique could eventually make human proteins, new drugs, and even replacement organs; (vii) Using SCNT procedure, the first transgenic cow, Rosie, created in 1997, contained the human gene α-lactalbumin in her milk (2.4 g/l). This transgenic milk is a more nutritionally balanced product than natural bovine milk and can be given to babies or the elderly with special nutritional or digestive needs; (viii) Argentine scientists have successfully produced a transgenic cow, named Patagonia (Figure 20.7c) that carries in its chromosomes the human gene of insulin. Bio-Sidus, the biotechnology company based in Argentina is a pioneer in this kind of technology in South America. The gene was allowed to express only in mammary glands, and not in all cells; (ix) Currently, whole human blood from donors is the only source for replacing blood loss from trauma. Active programs exist for both recruiting donors to prevent shortages and testing donated blood for human immunodeficiency virus and other viruses. Much of this effort could be alleviated if there was a safe and reliable supply of whole human blood or a cell-free blood substitute from a nonhuman source. Accordingly, with a construct that consisted of the regulatory region from the human β-globin gene joined to two human α_1-globin genes and one human β^A-globin gene, healthy transgenic pigs that expressed human hemoglobin in their blood cells were produced. After the human β-globin promoter was replaced with the pig β-globin promoter, higher levels of human hemoglobin having same properties as natural human hemoglobin were produced. The transgenic human hemoglobin was then separated from the pig hemoglobin by a simple chromatographic procedure. Although preliminary, these results indicated that human hemoglobin could be purified from the blood of transgenic pigs. However, free hemoglobin may not be the answer for replacing whole blood because it does not conduct oxygen exchange as effectively as the hemoglobin within erythrocytes. Also, free hemoglobin decomposes after administration to animals and its breakdown products cause kidney damage. Thus, the formulation of a human whole blood substitute by transgenesis may be a long way off; and (x) Research is also underway to manufacture milk through transgenesis for treatment of debilitating diseases such as phenylketonuria (PKU), hereditary emphysema, and cystic fibrosis.

Departing from the trend line of using large farm animals, Bio-Protein Technologies (Paris) specialized in producing therapeutic proteins in rabbit milk. The main advantage of using rabbits is the shorter development time and time-to-market. The gestation time for rabbits is only 1 month (as opposed to 9 months for goats and cows) and the female rabbits mature sexually in just 4 months. Rabbits are also prolific breeders. These two features compensate for the low milk production from each rabbit. However, a cow can produce milk 20 l per day compared to 0.25 l per day for the rabbit. Bio-Protein's potential products include antibodies, plasma proteins, and hormones. In yet another example, a transgenic mouse was created that was capable of synthesizing an extensive array of different human antibodies. For the creation of human antibody-producing transgenic mice, a 1,000 kbp YAC that carried 66 V_H segments, about 30 D_H segments, six J_H segments, $C\mu$, $C\delta$, and $C\gamma$ was formed by combining segments from four human H chain YACs. Similarly, an 800 kbp YAC with 32 $V\kappa$ segments, 5 $J\kappa$ segments, and $C\kappa$ was created from three YACs that carried different $V\kappa$ segments. These constructs were introduced into ES cells and transgenic mice that synthesized only human antibodies were eventually established. These transgenic mice produced a full range of human antibodies against every antigen. For validation, the human antibody-producing transgenic mice were immunized with three different antigens, and in each case, the hybridomas secreted a human McAb that had a strong affinity for the immunizing antigen. In animal studies, a human McAb derived from a transgenic mouse that binds to the epidermal growth factor receptor effectively blocked the proliferation of human tumors that overexpressed this protein. Moreover, in a human clinical trial, a mouse-derived human McAb was not immunogenic. It is very likely that this transgenic system will be used routinely to develop human therapeutic McAbs. A commercialized version of the human antibody-producing mouse has been designated the XenoMouse.

Besides using milk as a source of pharmaceutical proteins, it is also suggested that the egg with its high protein content could also be used. By analogy to the mammary gland of livestock, the expression of a transgene in the cells of the reproductive tract of a hen that normally secretes large amounts of ovalbumin could lead to the accumulation of a transgene-derived protein that becomes encased in the eggshell. The recombinant protein could either be fractionated from these sterile packages or consumed as nutraceuticals. The expected annual yield of recombinant protein from one hen is 0.25 kg. Transgenic chickens are now being used to synthesize McAbs, GH, insulin, human serum albumin, and α-interferon.

Production of therapeutic proteins by mammalian cells While bacteria are easy to manipulate, these are not always ideal as producers of human proteins, and mammalian cells have dominated the biopharmaceutical industry since the mid-1990s. This is because only mammalian cells can glycosylate human proteins in the correct manner. There are several principal cell lines of choice: Chinese hamster ovary (CHO) cells, the murine myeloma cell lines NSO and SP 2/0, BHK and HEK-293 cells, and the human retinal line PER-C6. These mammalian cell lines have been used to produce most of the recombinant therapeutic products licensed by the FDA.

Application in human gene medicine While disease modeling uses gene manipulation to create diseases in model organisms, gene medicine refers to the use of the same technology to ameliorate or even permanently cure diseases in humans. Gene medicine has a wide scope, and to name a few, included under gene medicine are targeted killing of disease cells (e.g., cancer cells), use of DNA vaccines (Exhibit 20.7), targeted ablation of specific cells, use of oligonucleotides as drugs, gene-directed enzyme prodrug therapy (GDEPT), and use of gene transfer to correct genetic defects (gene therapy) (Exhibit 20.8).

Exhibit 20.7 DNA vaccines

DNA vaccines (also called genetic vaccines), the third generation vaccines, represent the expression constructs the products of which stimulate the immune system to protect the organism against diseases. The technique of protecting an organism against disease by injecting it with genetically engineered DNA to produce an immunological response is called DNA vaccination or DNA immunization. DNA vaccines are still experimental and have been applied to a number of viral, bacterial, and parasitic models of disease, as well as to several tumor models.

DNA vaccines comprise a bacterial plasmid carrying a gene encoding the appropriate antigen under the control of a strong promoter that is recognized by the host cell. The DNA vaccines may be administered by injection using liposomes or by particle bombardment where the inner machinery of the host cells reads the DNA and converts it into pathogenic proteins, i.e., antigen(s). These proteins are recognized as foreign, processed by the host cells, and displayed on their surface to alert the immune system, which then triggers a range of immune responses. In principle, DNA vaccination has much in common with gene therapy since both involve DNA transfer to humans using a similar selection of methods. However, while the aim of gene therapy is to alleviate disease by either replacing a lost gene or blocking the expression of a dominantly acting gene, the aim of DNA vaccination is to prevent disease by causing the expression of an antigen that stimulates the immune system. In the original demonstration of DNA vaccination, DNA corresponding to the influenza virus nucleoprotein was introduced and protection against influenza infection was achieved. Since then, many DNA vaccines have been used to target viruses (e.g., measles, HIV, Ebola virus), other pathogens (e.g., tuberculosis bacterium), and even the human cellular prion protein in mice. In June 2006, positive results were announced for a bird flu DNA vaccine and a veterinary DNA vaccine to protect horses from West Nile virus has been approved. In August 2007, a preliminary study in DNA vaccination against multiple sclerosis was reported as being effective.

DNA vaccination approach has several advantages over conventional vaccines, including the following: (i) The method is simple and widely applicable; (ii) Certain bacterial DNA sequences have the innate ability to stimulate the immune system; (iii) It is able to induce the expression of antigens that resemble native viral epitopes more closely than standard vaccines do since live attenuated and killed vaccines are often altered in their protein structure and antigenicity; (iv) DNA vaccines have the ability to induce a wider range of immune response types; (v) DNA vaccines encoding several antigens or proteins can be delivered to the host in a single dose, only requiring a microgram of plasmids to induce immune responses. Other genes encoding proteins influencing the function of the immune response can be cointroduced along with the vaccine; (vi) DNA vaccination can be used to treat diseases that are already established as chronic infections; (vii) DNA vaccines are very temperature stable, making storage and transport much easier; (viii) It is easy to produce large quantities of the DNA vaccines rapidly; and (ix) Rapid and large-scale production are available at costs considerably lower than traditional vaccines.

Another approach to DNA vaccination is expression library immunization (ELI). Using this technique, potentially all the genes from a pathogen can be delivered at one time, which may be useful for pathogens that are difficult to attenuate or culture. ELI can be used to identify the pathogen gene that induces a protective response. This has been tested with *Mycoplasma pulmonis* and it was found that even partial expression libraries can induce protection from subsequent challenge.

Source: Weiner DB, Kennedy RC (1999) *Scientific American* 281(1): 34–41; Lewis PJ, Babiuk LA (1999) *Adv. Virus Res.* 54: 129–188; Robinson HL, Pertmer TM (2000) *Adv. Virus Res.* 55: 1–74.

Exhibit 20.8 Gene Therapy

Genetic engineering has provided many new approaches to fight against human diseases. Besides providing new diagnostic tools and animal disease models, a number of novel therapeutic strategies have resulted directly from the ability to clone and manipulate human genes. These include the expression cloning of human gene products, the development of novel vaccines, and the engineering of therapeutic antibodies. Different gene therapy methods used to replace or repair nonfunctional or mutated genes include the following: (i) Insertion of a normal gene into a nonspecific location within the genome to replace a nonfunctional gene; (ii) An abnormal gene could be swapped for a normal gene through homologous recombination; (iii) The abnormal gene could be repaired through selective reverse mutation, which returns the gene to its normal function; and (iv) The regulation (i.e., the degree to which a gene is turned on or off) of a particular gene could be altered. Gene therapy is thus used to treat or prevent diseases caused by mutations in the patient's own DNA (inherited disorders, cancers), as well as infectious diseases, and is particularly valuable in cases where no conventional treatment exists, or that treatment is inherently risky. This technique may allow treatment of diseases by inserting a gene into the patient's cells instead of using drugs or surgery.

Vectors used for DNA Delivery in Gene Therapy Approaches

In most gene therapy studies, a carrier molecule called a vector is used to deliver the therapeutic gene to the patient's target cells. Currently, the most common vector is a virus that has been genetically altered to carry normal human DNA. Some of the different types of viruses used as gene therapy vectors are retroviruses, adenoviruses, adeno-associated viruses, and herpes simplex viruses. Besides virus-mediated gene delivery systems, there are several nonviral options for gene delivery. The simplest method is the direct introduction of therapeutic DNA into target cells by particle bombardment. Another nonviral approach involves liposomal packaging of therapeutic DNA and its delivery through target cell's membrane. Therapeutic DNA can also get inside target cells by chemically linking the DNA to a molecule that binds to special cell receptors. Once bound to these receptors, the therapeutic DNA constructs are engulfed by the cell membrane and passed into the interior of the target cell. This delivery system however tends to be less effective than other options. Researchers are also experimenting with introducing an artificial human chromosome into target cells. This chromosome would exist autonomously alongside the standard 46 chromosomes with no effect on their working or without causing any mutations. It is considered that such a vector would not be attacked by the body's immune systems. However, a problem with this potential method is the difficulty in delivering such a large molecule to the nucleus of a target cell.

Gene Therapy Approaches on the Basis of Cells Modified

Germ line gene therapy In this case, germ cells, i.e., sperm or eggs, are modified by the introduction of functional genes, which are ordinarily integrated into their genomes. In germ line therapy, a fertilized egg is provided with a copy of the correct version of the relevant gene, which is then reimplanted into the mother. If successful, the gene is expressed in all cells of the resulting individual. Germ line therapy is usually carried out by microinjection of DNA into the isolated egg cell and theoretically could be used to treat any inherited disease. Gene therapy in germ line cells has the potential to affect the individual being treated and these changes are heritable and are passed on to next generations. Theoretically, this approach should be highly effective in counteracting genetic disorders; however, this option is prohibited for application in human beings, at least for the present, due to ethical and technical reasons.

Somatic cell gene therapy In this case, the gene is introduced only in somatic cells, especially of those tissues in which expression of the concerned gene is critical for health and those that can be removed from the organism. Expression of the introduced gene relieves or eliminates symptoms of the disorder, but this effect is not heritable as it does not involve the germ line. The technique is most promising for inherited blood diseases such as hemophilia and thalassaemia, which involves introduction of the genes into bone marrow stem cells, giving rise to all the specialized cell types in the blood. The strategy involves preparation of a bone extract containing several billion cells, transfecting these with retrovirus vector and then reimplanting the cells. Subsequent replication and differentiation of transfectants leads to the added gene being present in all the mature blood cells. The advantage of using retrovirus vector is its extremely high transfection frequency, enabling a large proportion of the stem cells in a bone marrow extract to receive the new gene. Somatic cell therapy also has potential to treat lung diseases such as cystic fibrosis, as after introduction into the respiratory tract via an inhaler, the epithelial cells in the lungs take up the DNA cloned in adenovirus vector or contained in liposome. However, gene expression occurs for only a few weeks and as yet this has not been developed into an effective means of treating cystic fibrosis.

Gene Therapy Approaches on the Basis of Manipulation of Cells of Living Patient or *in vitro* Cultured Cells

***In vivo* gene therapy** Gene therapy involving genetic modification of cells in a living patient is called *in vivo* gene therapy. In the *in vivo* method, the delivery agent is introduced within the body of the patient. For most of the *in vivo* methods, viral vectors are used. The transgene is introduced where it replaces the defective gene by homologous recombination. Note that the transgene in this case is the corrected gene, which needs to replace the defective disease-causing gene. The choice of the viral vector often depends upon the target tissues or organs. *In vivo* gene therapy encompasses both genetic modification of target cells (somatic transgenesis) and the therapeutic use of DNA as an epigenetic treatment (i.e., without changing the nucleotide sequence).

***Ex vivo* gene therapy** Gene transfer can be carried out in cultured cells, which are then reintroduced into the patient. This forms the basis of *ex vivo* gene therapy. This approach can be applied only to certain tissues, such as bone marrow, in which the cells are amenable to culture. In *ex vivo* gene therapy method, some of the body cells, usually blood, are withdrawn and treated with the delivery agent. After the gene transfer is completed, the treated cells are returned to the patient's body, which then proliferate and spread to the entire body, taking with them the introduced gene. For *ex vivo* methods, besides animal virus-mediated delivery, electroporation, lipofection, etc. can also be used.

Gene Therapy Approaches to Treat Loss-of-function or Gain-of-function Diseases

The genetic diseases in which the defect has arisen due to mutation of gene, the therapy involves introduction of the correct version of the gene into the cell without the need to remove defective genes. In contrast, in case of dominant genetic diseases, where defective gene product itself is responsible for the disease state, the therapy involves addition of the correct gene as well as the removal of the defective version. Thus a gene delivery system is required that promotes recombination between the chromosomal and vector-borne versions of the gene so that the defective chromosomal copy is replaced by the gene from the vector. Gene targeting, a form of *in vivo* site-directed mutagenesis involving homologous recombination between a targeting vector containing one allele and an endogenous gene represented by a different allele can be used for either the integration of introduced gene into the host chromosome or replacing the mutated gene (for details see Chapter 17). Two types of targeting vectors are used. Integration vectors (ends-in vectors) where cleavage within the homology domain stimulates a single cross-over, resulting in integration of the entire vector and transplacement vectors (ends-out vectors), where linearization occurs outside the homology domain and a double cross-over or gene conversion event within the homology domain replaces part of the genome with the homologous region of the vector.

For treatment of loss-of-function and gain-of-function diseases, gene therapy approaches used are gene augmentation therapy and gene inhibition therapy, respectively (Figure A):

Gene augmentation therapy (GAT) The traditional approach to gene therapy involves addition of a normal copy of a gene (transgene) to the genome of a person carrying defective copies of the gene. In simple words, DNA is added to the genome to replace a lost function or to correct a loss-of-function effect (e.g., caused by a deletion)

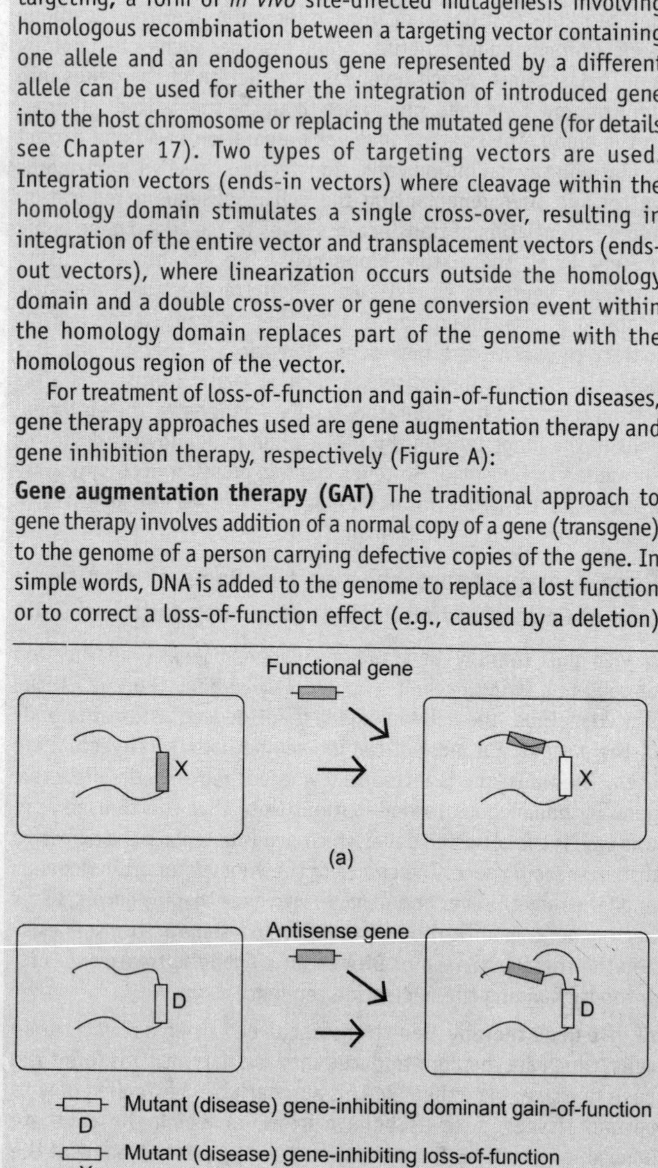

Figure A (a) Gene augmentation therapy, (b) Gene inhibition therapy

by adding the functional allele. Transferred genes may be stably integrated into the genome (in which case there is the potential to permanently correct the defect, especially if stem cells are transformed) or may be maintained episomally (in which case there is an inevitable delay in the maintenance of gene expression, and treatment may need to be repeated). Several GAT clinical trials are currently underway including cystic fibrosis, adenosine deaminase deficiency, and familial hypercholesterolemia.

Gene inhibition therapy For the treatment of dominant negative loss-of-function mutations and gain-of-function mutations, GAT is less powerful and additional functional copies of a dominantly malfunctioning gene are unlikely to affect the phenotype. Thus a valid approach to the treatment of such diseases is targeted correction (i.e., allele replacement) or gene 'knock-out' to remove the mutant allele. This is a very inefficient process in practice. A novel approach, which may play a role in gene therapy in the future, is targeted correction at the RNA level by using ribozymes or RNA editing enzymes to correct pathogenic mRNAs. An alternative strategy, which is presently undergoing clinical trials for the treatment of several types of cancer, is the use of nucleic acids to inhibit gene expression. The advantage of this approach is that the inhibitor can be manipulated to inhibit a specific allele, so expression of any normal functioning allele is not affected. The introduction of antisense genes allows the stable and permanent expression of antisense RNA, which binds to (mutant) mRNA and prevents translation (it may also target the mutant mRNA for degradation). Furthermore, increasing use is being made of antisense constructs containing ribozymes, which degrade the mRNAs to which these bind. Oligonucleotides can be used for epigenetic gene therapy (i.e., therapy which does not involve changes to the genome) and can act in two ways. Antisense oligonucleotides act in the same way as antisense RNA by binding to mRNA and causing inhibition and degradation probably by recruiting RNase H. Secondly, pyrimidine-rich oligonucleotides can potentially form triple helix structures with purine-rich strands of DNA (by Hoogsteen base pairing), which blocks transcription (triple helix therapy). Peptide nucleic acid (PNA), an analog of DNA where the phosphate backbone is replaced by a neutral peptide backbone, forms very stable triple-helix structures and can be used as agents for gene therapy.

Besides gene inhibition therapy at nucleic acid level (as described above), targeted inhibition of gene expression can also occur at the protein level by the expression within a cell of genetically engineered antibodies (intrabodies), which bind to and inactivate mutant proteins. Antibodies are not the only molecules being developed for protein-level gene therapy. Instead any protein acting as multimer is a potential target for dominant negative inhibitory proteins and this strategy is being explored to prevent viral coat protein assembly, including that of the HIV. A novel approach is to use degenerate oligonucleotides to identify specific oligonucleotide sequences, which interact with proteins. These oligonucleotides, or aptamers, can then be used to inactivate specific mutant proteins.

Problems Associated with Gene Therapy

Although gene therapy is a promising option for the treatment of a number of diseases (including inherited disorders, some types of cancer, and certain viral infections), the technique remains risky and is still under study to make sure that it is safe and effective. Gene therapy is currently only being tested for the treatment of diseases that have no other cures. The problems associated with gene therapy include its short-lived nature (in case of somatic gene therapy), risk of stimulating an immune and inflammatory response, toxicity, gene control and targeting issues associated with virus vectors as gene delivery systems, the risk of virus recovering its disease-causing ability once inside the patient cell, difficulty in treating multigene or multifactorial disorders such as heart disease, high blood pressure, Alzheimer's disease, arthritis, and diabetes. Note that conditions or disorders that arise from mutations in a single gene are the best candidates for gene therapy. Moreover, some ethical concerns have also been associated with germ line therapy, because it would change the genetic pool of the entire human species and future generations would have to live with that change.

Some Case Studies of Human Gene Therapy

Some important examples of human gene therapy are discussed here:

Gene augmentation therapy for severe combined immune deficiency The first clinical trial using a therapeutic gene transfer procedure involved a 4-year-old female patient, Ashanthi DeSilva, suffering from severe combined immune deficiency, resulting from the absence of the enzyme adenosine deaminase (ADA). This disease fitted many of the ideal criteria for gene therapy experimentation. The disease was life-threatening, therefore making the possibility of unknown treatment-related side effects ethically acceptable. The study aimed at cloning of the corresponding gene and understanding the biochemical basis of the disease. Importantly, since ADA functions in the salvage pathway of nucleotide biosynthesis, cells in which the genetic lesion had been corrected had a selective growth advantage over mutant cells, allowing them to be identified and isolated *in vitro*. Conventional treatment for ADA deficiency involves bone marrow transplantation from a matching donor. Essentially the same established procedure could be used for gene therapy, but the bone marrow cells derived from the patient and would be genetically modified *ex vivo*. Cells from the patient were subjected to leukopheresis and mononuclear cells were isolated. These were grown in culture under conditions that stimulated T-lymphocyte activation and growth and then transduced with a retroviral vector carrying a normal *ADA* gene and *npt* II gene. Following infusion of these modified cells, the patient showed an improvement in clinical condition as well as to study *in vitro* and *in vivo* immune functions. However, the production of recombinant *ADA* in the patient is transient, so regular infusions of recombinant T-lymphocytes were required.

Gene augmentation therapy for cystic fibrosis (CF) Cystic fibrosis, a recessive disorder that predominantly affects the lungs, liver, and pancreas, is caused by the loss of a cAMP-regulated membrane spanning chlorine channel. This results in an electrolyte imbalance and the accumulation of mucus, often leading to respiratory failure. This loss-of-function disorder can be corrected by introducing a functional copy of the gene, i.e., the epithelial cells isolated from CF patients can be restored to normal by transfecting them with the cloned cystic fibrosis transmembrane regulator (*CFTR*) cDNA. Unlike ADA deficiency, the cells principally affected by CF cannot be cultured and returned to the patient, so *in vivo* delivery strategies must be applied, for example, adenoviral vectors, which have a natural tropism of the epithelial lining of the respiratory system or by liposomes. Targeting into patients has also been done using an inhaler.

Gene therapy strategies for cancer Most cancers result from activation of oncogene or inactivation of tumor suppressor gene. In both cases, gene therapy can be envisaged to treat the cancer. For example, in the first case, the introduction of antisense RNA copy of the oncogene could reduce or prevent oncogene expression and reverse its tumorigenic activity. In the latter case, gene therapy would involve introduction of an active version of tumor suppressor gene. Gene therapy can also be used to improve the natural killing of cancer cells by the patient's immune system, for example, with a gene that causes the tumor cells to synthesize strong antigens that are efficiently recognized by the immune system. Another most effective and general approach to the treatment of many types of cancer is the introduction of a gene encoding toxic protein that selectively kills cancer cells and the normal cells remain unaffected. Yet another strategy is the activation of prodrug through an enzyme encoded by the targeted gene.

Cancer gene therapy was initially an extension of the early gene-marking experiments. The tumor-infiltrating leukocytes were transformed with a gene for *TNF* and *npt* II, with the aim of improving the efficiency with which these cells kill tumors by increasing the amount of TNF secreted. Although TNF is highly toxic to humans at levels as low as 10 μg/kg body weight, there have been no side effects from the gene therapy and no apparent organ toxicity from secreted TNF. One alternative strategy is to transform the tumor cells themselves, making them more susceptible to the immune system through the expression of cytokines or a foreign antigen. Another is to transform fibroblasts, which are easier to grow in culture, and then coinject them together with tumor cells to provoke an immune response against the tumor. A number of such 'assisted killing' strategies (Figure B) have been approved for clinical trials. Note that assisted killing approach does not require a detailed understanding of the genetic basis of the disease, however, a very accurate delivery system or some means of ensuring the expression of gene only in the cancer cells (e.g., placing it under the control of a promoter that is active only in cancer cells) is required.

Direct intervention to correct cancer-causing genes is also possible. Dominantly acting oncogenes have also been targeted using antisense technology, either with antisense transgenes, oligonucleotides, or ribozymes. An early report of cancer gene therapy with antisense oligonucleotides was that for the treatment of chronic myeloid leukemia. In this study, two 18-mers specific for the *BCR–ABL* gene junction generated by the chromosomal

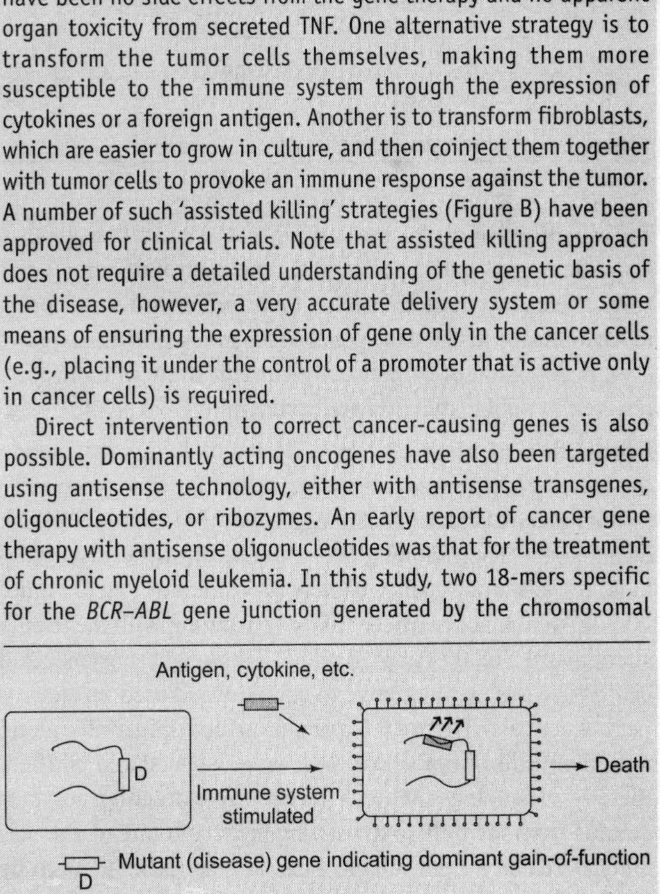

Figure B Assisted killing strategy

translocation that causes this particular cancer were used for suppression of colony formation in cells removed from cancer patients. Cancers caused by loss of tumor-suppressor gene function have been addressed by replacement strategies in which a functional copy of the appropriate gene is delivered to affected cells.

Gene-directed enzyme prodrug therapy (GDEPT) or prodrug activation therapy involves the activation of a particular enzyme specifically in cancer cells, which converts a nontoxic 'prodrug' into a toxic product, thereby killing the cancer cells. This can be achieved by driving the expression of a so-called 'suicide gene' selectively in cancer cells. Note that a large number of prodrugs including novel self-immolative prodrugs, have been designed that are converted to a range of different classes of drugs. Thus the prodrug/drug system selected can be tailored for the tumor type. An example includes the use of HSV *thymidine kinase* gene in combination with the prodrug ganciclovir. Thymidine kinase converts ganciclovir into a nucleotide analog, which is incorporated into DNA, and blocks replication by inhibiting the DNA polymerase (Figure C). Activation of the enzyme specifically in cancer cells is through the use of oncoretroviruses. Another way is to use transcriptional regulatory elements that are active only in cancer cells.

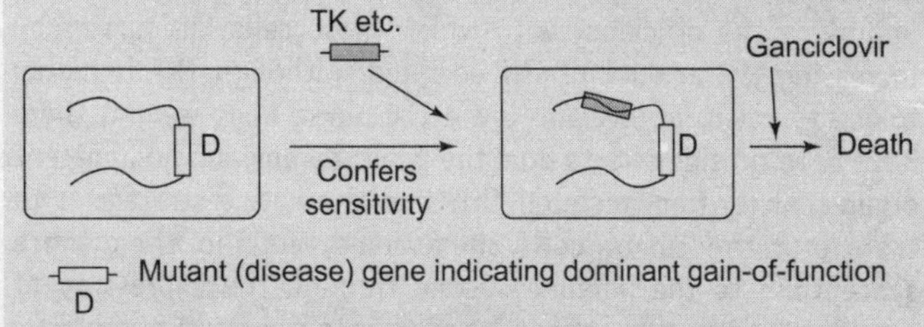

Figure C Prodrug activation therapy

GDEPT also involves targeting of the gene for the exogenous enzyme carboxypeptidase G2 (CPG2) to the tumor by the use of a selective vector, followed by administration of a prodrug that is activated by the enzyme. CPG2 has been expressed in a variety of forms, both intracellular and tethered to the outer surface of mammalian cells. The former requires intracellular activation of prodrugs, whereas the latter allows this to be extracellular. Excellent cytotoxicity differentials are obtained between those cell lines that express the enzyme compared with the parent cell lines without the CPG2. Good bystander cytotoxicity is obtained *in vitro*, when only 2% of the cells need to express the enzyme in order to achieve total cell ablation. Tumor xenografts of breast and colorectal carcinoma cell lines stably expressing the enzyme CPG2 have been developed for *in vivo* analyses of the GDEPT systems. A range of prodrugs causes a dramatic decrease in xenograft tumor volume and many cures are observed in both xenograft models. Tumors that regrow after initial treatment respond to further administration of the prodrug, indicating that the gene is stable and active in the long term. In collaboration with Cell Genesys Inc. (South San Francisco, CA), adenoviral vectors expressing the gene for CPG2 have been engineered to be under the control of a cancer-selective promoter. This restricts expression of CPG2 and thus prodrug sensitivity to cancer cells. *In vivo* models of human hepatocellular carcinoma treated with a single intravenous systemic dose of adenoviral vectors showed tumor-selective replication of adenoviruses and expression of CPG2, not found in normal tissues. Levels of CPG2 greatly exceeded those previously shown to be sufficient to cause regression of the tumors following administration of the prodrug. Thus the adenoviral vectors show great promise for efficacy in GDEPT protocols.

Source: http://www.the-scientist.com; http://en.wikipedia.org; www.scientistlive.com; genomics.energy.gov; http://biobasics.gc.ca.

Transgenic animals for disease control Animal transgenesis also offers tremendous potential for the control of animal diseases. Scientists have controlled mice, which are major pests in Australia, by altering the genes of the mousepox virus. However, the creation of an unusually virulent strain of mousepox resulted in public attention and concern.

Production of biosteel Nexia Biotechnologies in Canada implanted a single spider gene (regulated for expression only in mammary glands) into the egg cells of lactating Nigerian dwarf goats. Their cloning led to the birth of the first 'web-kids' or 'silk milk goats', namely Webster and Peter (Figure 20.7d). Note that the spider silk is very strong with the tensile strength of 1,36,000 kg/in^2, flexible, lightweight substance and can be used as an alternative to compounds based on steel or petrochemicals. These GM goats produced spider silk along with their milk, from where these were isolated and purified, thereby producing a whitish liquid. By extracting polymer strands from the milk and weaving them into thread, the scientists created a light, tough, flexible material that could be used as biosteel and for the production of ultralight bullet-resistant vests, military uniforms, medical sutures (surgical thread), fishing line that is stronger and better for the environment (because it eventually decomposes), tennis racket strings, artificial ligaments, repairing broken limbs and torn tissue, coatings of space stations, aircraft, racing vehicles, and possibly bridges, etc. The silk milk technique works because the way mammals produce milk proteins and spiders make silk proteins are broadly similar.

Transgenic Animals for Mutagenicity Testing The assessment of the potential genotoxicity of chemicals *in vivo* is important for both the verification and confirmation of intrinsic mutagenicity and for establishing the mode of action of chemical carcinogens. Although the present trend is to reduce animal testing, *in vitro* data must be confirmed by testing under *in vivo* conditions, which take into account whole animal processes like absorption, tissue distribution, metabolism, excretion of the chemical and its metabolites, and overall toxicity. Advances in genetic engineering have created opportunities for improved understanding of the molecular basis of carcinogenesis. The mouse spot test, an *in vivo* mutation assay, has been used to assess a number of chemicals. More recently transgenic animal mutagenicity assays using transgenic animals have been developed for testing *in vivo* gene mutagenicity, and for mechanistic research. Thus toxicity-sensitive transgenic animals have been produced for routine testing of chemicals to determine their carcinogenic potential. The two transgenic mouse models with the best database available are the *lac* I model (commercially available as the Big Blue® mouse) and the *lac* Z model (commercially available as the Muta™ Mouse). The study involves comparison of the results of *in vivo* testing of a number of chemicals using the mouse spot test with the results from these two transgenic mouse models.

Ornamental fish production Ornamental fish distributors and retail locations report unprecedented consumer demand of ornamental fishes. To capitalize on this bourgeoning market, a US Company developed GloFish™ (Figure 20.7e), the first commercialized genetically modified pet in North America. The GloFish™ was common Zebra Danio aquarium fish, which was fluorescent red in color due to the insertion of red fluorescent coral gene. Noting that the GloFish™ researchers used genetic engineering on fish for frivolous purposes and the risks were not all identified, the Department of Fish and Game banned the possession, sale, and transport of GloFish™ in California, but it was legal in the rest of the country. However, the FDA approved GloFish™ sales, finding no evidence that the fish pose any more threat to the environment than their unmodified counterparts, which have long been widely sold in the United States.

Environmental applications Transgenesis is also being used to address certain environmental concerns. For example, (i) Livestock production, particularly intensive systems like dairy, swine, poultry, and aquaculture, needs to reduce the amount of minerals excreted by animals. Reducing phosphate pollution of water is a major challenge for swine, poultry, and finfish producers. This is because pigs and poultry, unlike ruminants, are unable to digest and utilize phytate (*myo*-inositol 1,2,3,4,5,6-hexakisdihydrogen orthophosphate or phytic acid), the predominant storage form of phosphorus in plant-based animal feeds and hence their feed has to be supplemented with phosphorus. The excretion of large amounts of phosphorus may run off into water systems and accumulate at elevated levels in surface water and groundwater. These minerals may contaminate the drinking water. Their release into water bodies may lead to excessive growth of cyanobacterial populations (algal blooms), which in turn deplete the oxygen supply, and subsequently kill fish and other aquatic organisms. In addition, large amounts of phosphorus in the environment are implicated in the production of gases that enhance the greenhouse effect and contribute to global warming. The 'enviro-pig' is an interesting example of transgenesis as a way of overcoming the detrimental environmental impact of high-phosphate pig manure. In this case, a phytase transgene that is expressed in the salivary gland efficiently removes the phosphate residues from phytate. As a consequence, the phosphorus in phytate becomes metabolically accessible, which enhances growth and concomitantly significantly reduces the amount of phosphorus that is excreted; and (ii) Another example is that of transgenic fish, which is being considered as an economical and efficient bioassay system. In this context, the bacteriophage γ *cII* gene protocol that detects mutagens has been incorporated into fish for determining the presence of pollutants in water system or testing the mutagenicity of newly developed compounds.

Review Questions

1. Give a detailed account of biopesticides.
2. Discuss the roles of genetically engineered microorganisms in the following fields:
   (a) Therapeutic protein production
   (b) Designing new routes to the production of small molecules
   (c) Plant growth promotion
   (d) Environmental cleanup
3. Discuss the applications of plant transgenic technology in the field of insect and disease resistance. Give a descriptive account of commercialization of *Bt* cotton in India.
4. What are edible vaccines? How are these produced? What are the advantages of edible vaccines over conventional vaccines?
5. Discuss the role of plant transgenic technology in the field of improvement of nutritional content and composition with respect to the following:
   (a) Enhancement of vitamin A precursor
   (b) Modification of sweetness
   (c) Alteration of starch content and composition
   (d) Modification of lipid content and composition
   (e) Increase in vitamin E content
6. Write short notes on the following:
   (i) Superbug
   (ii) Recombinant vaccines
   (iii) Bollgard® cotton
   (iv) Golden rice
   (v) Flavr Savr tomato
   (vi) Blue rose
   (vii) Cold-tolerant plants
   (viii) Round-up Ready soybean
   (ix) CP-mediated virus resistance in plants
   (x) BetaSweet® carrot
   (xi) Supermice
   (xii) GloFish™
   (xiii) Dolly, the cloned sheep
   (xiv) Polly and Molly sheep
   (xv) Annie, the cow

(xvi) Tetra monkey
(xvii) Enviro-pig
(xviii) Xenomouse
(xix) Silk milk goats
(xx) Xenotransplantation
(xxi) Human disease models

7. Define the term 'cloning'. How does it differ from transgenesis? Also describe in detail the procedure for cloning of Dolly, the sheep. How does Dolly differ from Polly?
8. Differentiate between therapeutic cloning and reproductive cloning. Also enumerate their applications.
9. How are transgenic mice harboring growth hormone gene created?
10. Write a note on sheep produced at the Roslin Institute in collaboration with PPL Therapeutics, Scotland.
11. Write an essay on 'transgenic plant and animals as bioreactors or biofactories'.
12. Give a descriptive account of application of animal transgenic technology in human gene therapy.
13. Discuss how transgenesis could be used to improve organ transplantation.
14. What do you understand by the term 'pharming'? Explain the role of animal transgenesis in 'pharming'. What strategy is adopted to target the exogenous protein production in milk?
15. Discuss various applications of animal transgenesis in agriculture, medical, and industrial sectors.
16. Explain the following:
    (a) Role of transgenic technology in overcoming the detrimental impact of high phosphate pig manure on the environment.
    (b) Mechanism of action of *Bt* toxin from *B. thuringiensis* in killing lepidopteran insects.
    (c) Milk is preferred for the production of pharmaceutically important proteins.
    (d) Role of proteinase inhibitors in the development of insect-resistant transgenic plants.
    (e) Animal transgenesis is advantageous as compared to bacterial recombinant cultures for the production of pharmaceutically important proteins.
    (f) Role of osmoprotectants in salt tolerance.
    (g) Overexpression of 5-enol pyruvyl shikimate-3-phosphate synthase provides herbicide tolerance in crop plants.

References

Baetge, E.E. (2006). Production of pancreatic hormone expressing endocrine cells from human embryonic stem cells. *Nat. Biotechnol.* **24**:1392.

Campbell, K.H. *et al.* (1996). Sheep cloned by nuclear transfer from a cultured cell line. *Nature* **380**:64–66.

Ebert, K.M. *et al.* (1991). Transgenic production of a variant of human tissue-type plasminogen activator in goat milk: Generation of transgenic goats and analysis of expression. *Biotechnol.* **9**:835–838.

Mason, H.S., Lam, D.M.K. and C.J. Arntzen (1992). Expression of hepatitis B surface antigen in transgenic plants. *Proc. Natl. Acad. Sci. USA* **89**:11745–11749.

Wagner, T.E. *et al.* (1981). Microinjection of a rabbit β-globin gene into zygotes and its subsequent expression in adult mice and their offspring. *Proc. Natl. Acad. Sci. USA* **78**:6376–6380.

Wilmut, I. *et al.* (1997). Viable offspring derived from fetal and adult mammalian cells. *Nature* **385**:810–813.

Wilmut, I., K. Campbell, and C. Tudge (2000). *The Second Creation: Dolly and the Age of Biological Control*, Harvard University Press, Cambridge, MA; http://library.thinkquest.org/C0122429/history/1995.htm

Safety Regulations Related to Genetic Engineering

Key Concepts

In this chapter we will learn the following:

- National regulatory mechanism for implementation of biosafety guidelines for handling GMOs
- Salient features of recombinant DNA biosafety guidelines, revised guidelines for research in transgenic plants, and risk assessment
- Regulations for GM plants: *Bt* cotton in India—a case study
- Regulations related to stem cell research and human cloning

21.1 INTRODUCTION

Genetic engineering is sometimes described by phrases such as 'playing God', 'manipulation of life', 'the most threatening scientific research ever undertaken', and 'human-made evolution', owing to the concerns raised by scientists, public, and government officials about the safety, ethics, and unforeseen consequences of genetic engineering. These issues had far-reaching implications that resulted in both the establishment of official guidelines for the conduct of research and commercialization of genetically engineered products. The major apprehension related to genetically engineered microorganisms (GMM) is that either inadvertently or possibly deliberately for the purposes of warfare, unique microorganisms that had never previously existed would be developed and would cause epidemics or environmental catastrophes. Likewise, multiple concerns against genetically modified (GM; transgenic) plants and animals are prevalent even today. These include human health, biosafety, environmental protection, corporate control of industries, world trade monopolies, trustworthiness of public institutions, integrity of regulatory agencies, and loss of individual choice, etc. It is therefore recommended that scientific research aimed at risk analysis, prediction, and prevention, combined with adequate monitoring and stewardship, must be done so that negative impact from GM products, if any, may be kept to a minimum. This suggests that the problems raised by science can be solved by additional science itself. In this chapter, various aspects of the regulation of genetically modified organisms (GMOs; includes GMM, GM plants, GM animals) are highlighted. Further, regulations related to human stem cell research and reproductive cloning are also discussed.

21.2 NATIONAL REGULATORY MECHANISM FOR IMPLEMENTATION OF BIOSAFETY GUIDELINES FOR HANDLING GMOs

The new capabilities to manipulate the genetic material (i.e., recombinant DNA technology; rDNA technology) present tremendous potential in the betterment of humankind. However, it is also recognized that the GM technology may entail rare unintended risks and hazards. These biosafety concerns have led to the formulation of safety guidelines and regulations in various countries for research, testing, safe use and handling of GMOs and products thereof. The Government of India has also adopted a policy of careful assessment of the benefits and risks of GMOs at various stages of their development and field release to ensure biosafety. Two nodal agencies, Ministry of Environment and Forests (MoEF) and Department of Biotechnology (DBT) in the Ministry of Science and Technology are responsible for the preparation and implementation of the regulations on GMOs. In this section, a brief overview of rules and regulations in India relevant for foods derived for GM crops (GM foods) is presented.

21.2.1 Rules 1989

The MoEF notified the rules and procedures for the manufacture, import, use, research, and release of GMOs as well as products made by the use of such organisms on December 5, 1989 under the Environmental Protection Act (EPA) 1986. These rules and regulations, commonly referred as 'Rules 1989' cover the areas of research as well as large-scale applications of GMOs and products made thereof throughout India. A copy of the rules can be accessed at http://envfor.nic.in. The 'Rules 1989' order compliance of the safeguards through regulatory approach and any violation and noncompliance including nonreporting of the activity in this area attracts punitive action provided under the EPA. The two main agencies responsible for implementation of the rules are the MoEF and the Department of Biotechnology (DBT), Government of India.

The 'Rules 1989' have defined six competent authorities to handle various issues related to rDNA technologies, which are listed below. In general, these authorities are vested with nonoverlapping responsibilities. The overall mechanism and functional linkages among various committees and departments concerned with the approval of GMOs for commercial release are summarized below.

Recombinant DNA Advisory Committee

The Recombinant DNA Advisory Committee (RDAC) is constituted by DBT to monitor the developments in biotechnology at national and international levels. RDAC submits recommendations from time to time that are suitable for implementation of upholding the safety regulations in research and applications of GMOs and products thereof. This Committee prepared the first 'Recombinant DNA Biosafety Guidelines' in 1990, which were adopted by the Government of India for handling of GMOs and conducting research on them. The guidelines were revised in 1998.

Institutional Biosafety Committee

It is necessary that the institutions or organizations intending to carry out research activities involving genetic manipulation of microorganisms, plants, and animals should constitute the Institutional Biosafety Committee (IBSC). All the ISBCs need to have one nominee from the DBT. It serves as the nodal point for interaction within the institution for implementation of the guidelines. Any research project, which is likely to have biohazard potential (as envisaged in the guidelines) during the execution stage or which involves the production of either microorganisms, or biologically active molecules that may cause biohazard should be notified to IBSC. The institution or occupier for each of these activities should prepare the on-site emergency plan with the help of IBSC. Authorization for interstate exchange of etiologic agents, diagnostic specimens and biological products is also done by IBSC. Manipulation of plants under containment is performed under the regulatory clearance of IBSC. IBSC has the mandate to approve low-risk (category I and II) experiments and to ensure adherence to rDNA safety guidelines. IBSC recommends category III or above experiments to Review Committee on Genetic Manipulation (RCGM) for approval. It thus acts as a nodal agency for interaction with various statutory bodies.

Review Committee on Genetic Manipulation

The Review Committee on Genetic Manipulation (RCGM) is constituted by DBT to bring out manuals of guidelines specifying producers for regulatory process on GMOs in research, use and applications including industry with a view to ensure environmental safety. The Committee reviews all ongoing rDNA projects involving high-risk (category III and above) and controlled field experiments. RCGM approves applications for generating research information on transgenic plants in contained greenhouse or small pots. The small experimental field trials are limited to a total area of 20 acres in multilocations in one crop season. In one location where the experiment is conducted with transgenic plants, the land used should not be more than one acre. RCGM approval is granted for one season and applicant must provide entire details of the experimentation to the Committee. RCGM visits site of experimental facilities periodically where projects with biohazard potential are being pursued and also at a time prior to the commencement of the activity to ensure that adequate safety measures are taken as per the guidelines (for details see Section 21.3). Monitoring-cum-Evaluation Committee (MEC) of RCGM carries out monitoring of field trials. The latter also directs the generation of toxicity, allergenicity and any other relevant data on transgenic materials in appropriate systems. RCGM lays down procedures restricting or prohibiting production, sale, importation and use of GMOs both for research and applications. It also issues clearances for import/export of etiologic agents, and vectors, transgenic germplasm including transformed calli, seed and plant parts for research use only.

Genetic Engineering Approval Committee

The Genetic Engineering Approval Committee (GEAC) functions as a body under the MoEF and is responsible for approval of activities involving large-scale use of hazardous microorganisms and recombinant products in research and industrial production from the environment point of view. GEAC also permits the use of GMOs and products thereof for commercial applications. Large-scale experiments beyond the limits specified within the authority of RCGM are authorized by GEAC. Thus, testing of GMOs against pathogens and products in the environment should follow regulatory guidelines for seeking field use permit from GEAC. GEAC can authorize approval and prohibitions of any GMOs for import, export, transport, manufacturing, processing, use or sale both for research and applications under the EPA 1986. The large-scale planned release of organisms into the environment both for environmental and agricultural applications should be done under license issued by GEAC. GEAC also authorizes agencies or persons to have powers to take punitive actions under the EPA 1986.

State Biotechnology Coordination Committee

The State Biotechnology Coordination Committee (SBCC) is constituted in each state where research and application of

GMOs are contemplated. The Chief Secretary of the State heads SBCC. SBCC has the authority to inspect, investigate, and take punitive actions in case of violations of the statutory provisions through the State Pollution Control Board or the Directorate of Health, etc. The Committee periodically reviews the safety and control measures in various institutions handling GMOs. It also acts as nodal agency at State level to assess the damage, if any, due to release of GMOs and to take on site control measures. The Committee coordinates the activities related to GMOs in the State with the Central Ministries. The Committee also nominates State Government representatives in the activities requiring field inspection concerning GMOs.

District Level Committee

The District Level Committee (DLC), constituted at the district level, is considered to be the smallest authoritative unit to monitor the safety regulations in installations engaged in the use of GMOs in research and applications. The District Collector heads the Committee, and can induct representatives from State agencies to enable smooth functioning and inspection of the installations with a view to ensure the implementation of safety guidelines while handling GMOs, under the Indian EPA 1986. The Committee has the power to inspect, investigate and report to the SBCC or the GEAC about compliance or noncompliance of rDNA guidelines or violations under the EPA 1986. It also acts as a nodal agency at District level to assess the damage, if any, due to release of GMOs and to take on site control measures.

21.2.2 Recombinant DNA Safety Guidelines, 1990; 1994

With the advancement of research work initiated in biotechnology in the country by various Indian institutions and industry, DBT formulated 'Recombinant DNA Biosafety Guidelines' in 1990. These guidelines were further revised in 1994. The revised guidelines include guidelines for the research and development activities on GMOs, transgenic crops, large-scale production and deliberate release of GMOs, plants, animals, and products into the environment, shipment and importation of GMOs for laboratory research. Besides prescribing the criteria for assessment of the ecological aspects on a case-by-case basis for planned introduction of GMOs into the environment, these guidelines also suggest regulatory measures to ensure safety for import of GMOs and various quality control methods needed to establish the safety, purity, and efficacy of GM products. The issues relating to genetic engineering of human embryo, use of embryos and fetuses in research and human germ line, and gene therapy areas have not been considered while framing the guidelines.

21.2.3 Guidelines for Research in Transgenic Plants, 1998

Most of the genetic manipulation studies pertain to the plants; hence an elaborative account of the regulatory framework in India governing GMOs and their applications particularly in agriculture is given here. In 1998, DBT framed separate guidelines for regulation of transgenic plants in India. These 'Revised Guidelines for Research in Transgenic Plants', also includes guidelines for toxicity and allergenicity of transgenic seeds, plants, and plant parts. These guidelines cover areas of rDNA research on plants including the development of transgenic plants and their growth in soil for molecular and field evaluation. The guidelines also deal with import and shipment of genetically modified plants of research use. To monitor over a period of time, the impact of transgenic plants on the environment, a special MEC has been set up by the RCGM. The Committee undertakes field visits at the experimental sites and suggests remedial measures to adjust the trial design, if required, based on the on-the-spot situation. This Committee also collects and reviews the information on the comparative agronomic advantages of the transgenic plants and advises the RCGM on the risks and benefits from the use of transgenic plants put into evaluation. The guidelines include complete design of a contained green house suitable for conducting research with transgenic plants. Besides, it provides the basis for generating food safety information on transgenic plants and plant parts. Details of the guidelines may be seen at www.dbtbiosafety.nic.in.

21.2.4 Seed Policy, 2002

The Seed Policy, 2002 on transgenic plant varieties issued by Ministry of Agriculture, Government of India contains a separate section (No. 6) on transgenic plant varieties. The policy states that all genetically engineered crops/varieties have to be tested for environment and biosafety before their commercial release as per the regulations and guidelines under the EPA 1986. According to this policy, seeds of transgenic plant varieties for research purposes will be imported only through the National Bureau of Plant Genetic Resources (NBPGR) as per the EPA 1986. Transgenic crops/varieties will be tested to determine their agronomic value for at least two seasons under the All India Coordinated Project Trials of ICAR, in coordination with the tests for environment and biosafety clearance as per the EPA 1986 before any variety is commercially released. Once the transgenic plant variety is commercially released, its seed will be registered and marketed in the country as per the provisions of the Seeds Act. The Ministry of Agriculture and State Departments of Agriculture will monitor the performance of the commercially released variety in the field for at least 3–5 years. It has also been mentioned that transgenic varieties can be protected under the Plant Varieties and Farmers Rights Protection (PVP) legislation in the same manner as nontransgenic varieties after their release for commercial cultivation. The details of the 'Seed Policy, 2002' are available at http://agricoop.nic.in/seedpolicy.htm).

21.2.5 Plant Quarantine Order, 2003

The provisions of 'Plant Quarantine (regulation of import into India) Order, 2003', which came into force from April 1, 2004, are also applicable to import of transgenic seeds. According to this order, NBPGR has been designated as the Competent Authority to issue import permits for import of seeds by public and private sector agencies for research purposes after getting permission from DBT and MoEF as the case may be under the Rules 1989. Moreover, no transgenic material should be permitted for experimentation in open environment without prior authorization from the Government of India. All precautions should be taken to prevent the escape of the genetic material into the open environment and shall follow the 'Recombinant DNA Safety Guideline, 1990' of the Government of India. The supplier of the transgenic material shall certify that the transgenic has the genes as has been prescribed in the permission. The supplier shall also certify that these transgenic materials do not contain any embryogenesis deactivator gene sequence. The details of this order are available at http://www.plantquarantineindia.org.

21.2.6 Task Force on Application of Agricultural Biotechnology, 2005

In 2005, the Ministry of Agriculture set up a 'Task Force on Application of Agricultural Biotechnology' to formulate a draft long-term policy on applications of biotechnology in agriculture and suggest modifications in the existing administrative and procedural arrangements for the approval of GM crops (for details see http://agricoop.nic.in/TaskForce/tf.htm). The report suggests that the issues in regard to the release of GM crops are not understood correctly owing to the lack of information on this subject amongst the otherwise well-informed members of the public; hence an information campaign needs to be conducted to generate public awareness on the benefits and risks associated with biotechnology and the social, ethical, economic, scientific, environmental, and health issues, which are addressed by regulatory bodies before allowing the cultivation of GM crops. Post-release monitoring and management of GM crops and their products, such as insect resistance management, transgene stability at the farm level, use of transgenic diagnostic kits, and maintenance of transgenic seed quality, should be organized with effective involvement of State Level and District Level Coordination Committees of the existing transgenic biosafety evaluation and management mechanism.

21.2.7 The Food Safety and Standards Bill, 2005

In 2005, the Ministry of food processing industries introduced 'The Food Safety and Standards Act, 2005', which seeks to consolidate the laws relating to food and establish the 'Food Safety and Standards Authority of India'. There is a provision for a separate scientific panel on GMOs. According to it, no person should manufacture, process, export, import, or sell GM articles of food, organic foods, functional foods, nutraceuticals, health supplements etc. except in accordance with the regulations made under this Act. The full text of the Food Safety and Standards Bill, 2005 can be accessed at http://www.mfpi.nic.in.

21.2.8 National Biotechnology Development Strategy

In 2005, DBT brought out a 'National Biotechnology Development Strategy', which also covers regulatory mechanism for transgenic plants (for details see http://dbtindia.nic.in/biotechstrategy/biotech_strategy.htm). The strategy suggests involvement of Panchayati Raj institutions in the process of analysis and understanding the risks and benefits associated with GMOs, as these will play an important role in the local level management of biodiversity and access to benefit sharing, etc.

21.2.9 Regulation for Import of GM Products Under Foreign Trade Policy, 2006–07

On April 7, 2006, the Ministry of Commerce and Industry through Director General of Foreign Trade (DGFT) has notified new regulation for import of GM products by amending Schedule-I (Imports) of the ITC(HS) Classifications of Export and Import Items, 2004–09 under the Foreign Trade Policy (2004–09) effective from April 1, 2006. As a result of this notification, import of GM food, feed, GMOs, and living modified organisms (LMOs) is subjected to the following conditions: (i) The import of GMOs and LMOs for the purpose of research and development, food, feed, processing in bulk, and for environment release will be governed by the provisions of the EPA 1986 and Rules 1989; (ii) The import of any food, feed, raw or processed or any ingredient of food, food additives or any food product that contain GM material (e.g., GM soya oil) and is being used either for industrial production, environmental release, or field application will be allowed only with the approval of the GEAC; (iii) Institutes/companies that wish to import GM material for research and development purposes will have to submit their proposal to the RCGM. In case the Companies/institutes use these GM materials for commercial purposes, approval of GEAC is also required; and (iv) At the time of import, all consignments containing products, which have been subjected to genetic modification, will have to carry a declaration stating that the product is genetically modified. In case a consignment does not carry such a declaration and is later found to contain GM material, the importer is liable to penal action under the Foreign Trade (Development and Regulation) Act, 1992.

21.2.10 Guidelines and Standard Operating Procedures for Confined Field Trials of Regulated, Genetically Engineered Plants, 2008

In view of the enormous progress made during the last decade in the research and development of GM crops a need was felt to revisit the 'Revised Guidelines for Research in Transgenic Plants, 1998' to streamline the procedures for the safe conduct of confined field trials and methodical evaluation of the same. MoEF and DBT jointly prepared the 'Guidelines and Standard Operating Procedures (SOPs) for Confined Field Trials of Regulated, Genetically Engineered (GE) Plants, 2008'. These guidelines were finalized by incorporating the comments received after a series of regional consultations and public review (by placing the guidelines on the websites of MoEF and DBT), and adopted in July 2008. These guidelines supplement the biosafety measures for field trials given in the 'Revised Guidelines for Research in Transgenic Plants, 1998'.

The objective of these guidelines is to ensure that confined field trials are conducted under appropriately controlled conditions and in workable and efficient manner. A new application format has been designed to seek detailed information at the beginning of the process itself. The SOPs have been formulated to ensure quality and safety at each stage of a field trial, including transportation and storage of experimental GM plant material. To keep pace with the advancements in developments in GM crops, it is proposed that guidelines and SOPs be reviewed at least once in two years and suitably updated. SOPs have been prepared to provide guidance for the following aspects of conducting field trails of regulated GM crops in India: (i) Transport of regulated GM plant material; (ii) Storage of regulated GM plant material; (iii) Management of confined field trails; (iv) Management of harvest or termination of controlled field trails; and (v) Post-harvest management of controlled field trials.

The above-mentioned regulatory systems in India have effectively managed the approval process at various stages of development of *Bt* cotton, the only GM crop commercialized so far. However, with several crops under various stages of development, continuous improvement in the management of field trials is required in line with increasing knowledge in this area as well as to build public confidence in this technology.

21.3 SALIENT FEATURES OF RECOMBINANT DNA BIOSAFETY GUIDELINES, REVISED GUIDELINES FOR RESEARCH IN TRANSGENIC PLANTS, AND RISK ASSESSMENT

Based on the above-mentioned regulatory mechanisms, the research activities and environmental release of GMOs involve the following: (a) GMOs should be evaluated for potential risk on a case-by-case basis prior to application in agriculture and environment. As there are three components of genetic engineering, *viz.*, (i) The selected sequence of DNA of the donor (any living species or even synthetic sequences); (ii) The vector (usually a virus or a plasmid that carries the ligated donor sequences into the recipient host); and (iii) The host (invariably a microbial cell or a cultured cell) to achieve the required biotechnological potential, all three of them are manipulated. As a consequence, prior to introduction of microorganisms, hazards posed by them, their properties, possible interaction with other disease causing agents, and the infected wild plant species should be evaluated as per recombinant DNA safety guidelines; (b) Depending on the type of organism handled and the assessment of potential risks involved, appropriate containment facilities must be provided to ensure safety and to prevent unwanted release in the environment; (c) Prerelease tests of GMOs in agricultural applications should include elucidation of genetic markers, host range, requirement for vegetative growth, persistence, stability in small plots, and experimental field trials for two years. Soil samples in experiments under controlled containment conditions should be tested for the absence of viable cells before disposal into the environment; (d) Biowastes resulting from laboratory experiments or industrial operations should be properly treated. The microorganisms should be either destroyed or rendered harmless before disposal into the environment; and (e) Special facilities should be created for disposal of experimental animals. All refuse and carcasses must be incinerated. Exemption/relaxation of safety measures on specific cases may be considered based on the risk assessment criteria.

In the following sections, a brief description of risk assessment procedures and safety measures taken during GM research and applications is provided.

21.3.1 General Scientific Considerations for Risk Assessment of Microorganisms

Basically, microorganisms belong to four categories, *viz.*, intergeneric organisms, well-defined organisms with noncoding regulatory regions, biological agents whose source of DNA is a pathogen, and organisms that are generally recognized as nonpathogenic and may imbibe the characteristics of a pathogen on genetic manipulation. In genetic engineering experiments, the hazards are mostly posed by donor microorganisms, hence it is appropriate to consider the classification of donor microorganisms according to the hazards posed by them to the handlers, their ease of transmission in the society, and the respective containment measures that are required to be followed. Depending upon the conditions prevalent in the country, the microorganisms are classified as high-risk and low-risk categories. The widely prevalent microorganisms require lower degree of safety measures and are assigned low-risk category, while the ones not widely prevalent require higher degree of safety measures and are categorized under high-risk group. Further, the ones not present in the country are assigned a special category requiring highest degree of safety. The infective microorganisms are classified under four risk groups, namely, I, II, III, and IV, in increasing order of risk based on the parameters such as (i) Pathogenicity of the agent; (ii) Modes of transmission and host range of the agent; (iii) Availability of effective preventive treatments or curative medicines; (iv) Capability to cause diseases to humans/animals/plants; and (v) Epidemic causing strains in the country. These parameters are influenced by certain factors including levels of immunity, density, and movement of host population, presence of vectors for transmission, and standards of environmental hygiene.

Handling of Pathogenic Microorganisms

For the assessment of potential risks in handling of pathogenic organisms, the following scientific considerations are taken into consideration:

Characteristics of Donor and Recipient Organisms Under this category, the characteristics to be considered are: (i) Taxonomy, identification, source, culture; (ii) Genetic characteristics; and (iii) Pathogenic and physiological traits.

Characteristics of the Modified Organism This category includes (i) Description of the modification; (ii) The nature, function, and source of the inserted donor nucleic acid, including regulatory or other elements affecting the function of the DNA and the vector; (iii) The method(s) by which the vector with insert(s) has been constructed; (iv) Method(s) for introducing the vector-insert into the recipient organism and the procedure for selection of the modified organism; (v) The structure and amount of any vector and/or donor nucleic acid remaining in the final construction of the modified organism; (vi) Characterization of the site of modification of the recipient genome; (vii) Stability of the inserted DNA; and (viii) Frequency of mobilization of inserted vector and/or genetic transfer capability.

Expression and Properties of the Gene Product This includes (i) Rate and level of expression of the introduced genetic material; (ii) Method and sensitivity of method; (iii) Activity of the expressed protein; (iv) Allergenic hazard of the product; and (v) Toxic hazard of the product.

Certain quantitative values are also assigned to the potential risks determined by evaluating the above scientific findings. These include (i) Access factor of the organism (i.e., the probability of entry and survival of the manipulated organism in the target tissue or cell if it escapes by chance); (ii) Expression factor of DNA (i.e., the probability of translation of the gene in the manipulated organism and secretion of the cloned gene product from the altered organism); and (iii) Damage factor of the biologically active substance (i.e., the probability that the expressed product cause physiological damage to the individual. Note that only approximations are done).

Host–Vector Systems

The host-vector system is devised on the basis of access factor of the microorganisms employed. Thus, there are three categories of host–vector systems, *viz.*, normal, disabled, and especially disabled systems that have access factors of 10^{-3}, 10^{-6}, and 10^{-9}, respectively. Usually, disabled or especially disabled categories of host–vector system are selected. This is because such a system facilitates biological containment of the microorganism and naturally brings down the physical containment level of the experiment (for details see Section 21.3.2). A general consideration for the vectors is that it must be safe not only to human beings but also to domestic animals and there should not be any neoplastic effect. Besides these aspects, other scientific considerations taken into account while selecting a host–vector system are summarized below.

Plasmids The disabled or especially disabled plasmids have certain characteristics (for details see Chapter 8). These are: (i) The plasmids must not be self-transmissible; (ii) The plasmids must be nonmobilizable or only very inefficiently mobilizable; and (iii) The plasmids should not code for the mobilization proteins and also must be deficient in *nic* site on which the mobilization proteins act. Such plasmids have an access factor of 10^{-6} even on normal *E. coli* host.

Bacteriophage λ The disabled or especially disabled bacteriophage λ-based vectors have certain characteristics (for details see Chapter 9). These are: (i) Bacteriophage λ must have reduced host range, which can be achieved by the incorporation of amber mutations (reversion frequency 10^{-5} or less) in two different genes not involved in lysis; (ii) The phage must be nonlysogenic, which can be achieved by deletion of phage *att* site and defective *cI* gene; (iii) The phage must not be able to propagate in the plasmid mode; (iv) If the repressor is temperature-sensitive, the host strains must be *rec* A mutants; and (v) If a lysogenic phage vector is used, then the host must be disabled, like *E. coli* strains DP50 sup F or MRCI.

M13 Vector Systems The disabled or especially disabled bacteriophage M13-based vectors have certain characteristics (for details see Chapter 10), for example, (i) F-factor in the host must be defective for mobilization; and (ii) The vector must have amber mutations in at least two genes.

21.3.2 Containment

The term 'containment' is used to describe the safe methods for managing infectious agents belonging to four different biosafety levels in the laboratory environment where these are being handled or maintained. The purpose of containment is to reduce exposure of laboratory workers, other persons, and outside environment to potentially hazardous agents. At the laboratory level, there are basically two different types of containment, *viz.*, biological containment and physical containment. When transfer of GM plants to the field conditions is to be done, glasshouse containment is also required.

Biological Containment

In consideration of biological containment (BC), the vector (plasmid, organelle, or virus) for the recombinant DNA and the host (bacterial, plant, or animal cell) in which the vector is propagated in the laboratory are considered together. Any combination of vector and host, which is to provide biological containment must be chosen or constructed to limit the infectivity of vector to specific hosts and control the host-vector survival in the environment. These have been categorized into two levels, one permitting standard biological containment and the other even at a higher level that relates to normal and disabled, or especially disabled host–vector systems, respectively.

Physical Containment

The objective of physical containment (PC) is to confine pathogenic and recombinant organisms, thereby preventing the exposure of the researcher and the environment to the harmful agents. This is achieved by way of good laboratory practices, safety equipments, and laboratory design and facilities. Based on the protection procedure employed, there are two types of physical containment.

Primary Containment The protection of personnel and the immediate laboratory environment from exposure to infectious agents is provided by good microbiological techniques and the use of appropriate safety equipment.

Secondary Containment The protection of the environment external to the laboratory from exposure to infectious materials is provided by a combination of facility design and operational practices.

The above levels of containment indicate that there are three elements of containment, *viz.*, (i) Laboratory practice and techniques (i.e., strict adherence to standard microbiological practices and techniques, awareness of potential hazards, providing appropriate training of personnel, selection of safety practices in addition to standard laboratory practices, adopting a biosafety or operations manual, which identifies the hazards); (ii) Primary barriers such as containment or safety equipments [safety equipment includes Class I, II, and III biological safety cabinets (BSC provides containment of infectious aerosols), a variety of enclosed containers (e.g., safety centrifuge cup), items for personal protection such as gloves, coats, gowns, shoe covers, boots, respirators, face shields, and safety glasses, etc.]; (iii) Secondary barriers such as special laboratory or facility design (the purpose of facility design is to provide a barrier to protect persons working in the facility, outside of the laboratory, and those in the community, from infectious agents, which may be accidentally released from the laboratory. Note that there are three types of facility designs, *viz.*, the basic laboratory (for risk groups I and II), the containment laboratory (for risk group III) and the maximum containment laboratory (for risk group IV). However, if an increase in the laboratory-acquired infections is seen despite of advances in containment techniques, it may be due to the volume of microbiological research or the broadened spectrum of infectious agents under investigation.

Glasshouse Containment

According to the 'Revised Guidelines for Research in Transgenic Plants, 1998' the genetic engineering experiments on plants have been grouped under three categories. Category I includes routine cloning of defined genes, defined noncoding stretches of DNA and open reading frames in defined genes in *E. coli* or other bacterial/fungal hosts, which are generally considered as safe to humans, animals and plants. The category II experiments include experiments carried out in laboratory and green house/net house using defined DNA fragments nonpathogenic to humans and animals for genetic transformation of plants, both model species and crop species. Category III includes experiments having high risk where the escape of transgenic traits into the open environment could cause significant alterations in the biosphere, the ecosystem, plants and animals by dispersing new genetic traits, the effects of which cannot be judged precisely. Further, this also includes experiments conducted in green house and open field conditions having risks mentioned above.

Thus, the growth of whole plants requires special environmental conditions, which are categorized as two glasshouse containment facilities.

Glasshouse Containment A This is appropriate for plant experiments involving no plant pathogen or nonpathogen DNA vector systems and regeneration from single cells. Such experiments require notification to competent authority.

Glasshouse Containment B This is recommended for experiments involving genetically manipulated plant pathogens including plant viruses, for example, propagation of genetically manipulated organisms in plants and growth of plants regenerated from cells transformed by genetically manipulated pathogen vector systems, which still contain the pathogen. Approval of competent authority is required before commencement of activity.

Biosafety Levels

Biosafety is ensured by the combination of laboratory practices and techniques, safety equipments, and laboratory facilities appropriate for the operations performed and the hazards posed by the infectious agents. The guidelines for microbiological and biomedical laboratories suggest four biosafety levels in incremental order depending on the nature of work. Additional flexibility in containment levels can be obtained by combination of physical barriers with biological barriers. The proposed safety levels for work with recombinant DNA technique take into consideration the source of the donor DNA and its disease-producing potential. These four levels correspond to facilities approximate to four risk groups assigned for etiologic agents and are categorized as PC1<PC2<PC3<PC4. These levels and the appropriate conditions are enumerated as follows:

Biosafety Level 1 These practices, safety equipments, and facilities are appropriate for undergraduate and secondary educational training, teaching laboratories, and for other facilities in which work is done with defined and characterized strains of viable microorganisms not known to cause disease in a healthy adult human. No special accommodation or equipment is required but the laboratory personnel are required to have specific training and to be supervised by a scientist with general training in microbiology or a related science.

Biosafety Level 2 These practices, safety equipments, and facilities are applicable in clinical, diagnostic, teaching, and other facilities in which work is done with the broad spectrum of indigenous moderate risk agents present in the community and associated with human diseases of varying severity. Laboratory workers are required to have specific training in handling pathogenic agents and to be supervised by competent scientists. Accommodation and facilities including safety cabinets are prescribed, especially for handling large volume and high concentrations of agents when aerosols are likely to be created. Moreover, access to the laboratory is controlled.

Biosafety Level 3 These practices, safety equipments, and facilities are applicable to clinical, diagnostic, teaching research, or production facilities in which work is done with indigenous or exotic agents where the potential for infection by aerosols is real and the disease may have serious or lethal consequences. Personnel are required to have specific training in work with these agents and to be supervised by scientists experienced in this kind of microbiology. Specially designed laboratories and precautions, including the use of safety cabinets, are used and the access is strictly controlled.

Biosafety Level 4 These practices, safety equipments, and facilities are applicable to work with dangerous and exotic agents, which pose a high individual risk of life-threatening diseases. Strict training and supervision are required and the work is done in specially designed laboratories under stringent safety conditions, including the use of safety cabinets and positive pressure personnel suits. Access is strictly limited. A specially designed suit area may be provided in the facility. Personnel who enter this area wear a one-piece positive pressure suit that is ventilated by a life support system. The life support system is provided with alarms and emergency break-up breathing air tanks. Entry to this area is through an airlock fitted airtight doors. A chemical shower is provided to decontaminate the surface of the suit before the worker leaves the area. The exhaust air from the suit area is filtered by two sets of high-efficiency particulate air (HEPA) filters installed in the series. A duplicate filtration unit, exhaust fan, and an automatically starting emergency power source are provided. The air pressure within the suit area is lower than that of any adjacent area. Emergency lighting and communication systems are provided. All penetrations into the inner shell of the suit area are sealed. A double door autoclave is provided for decontamination of disposable waste materials from the suit area.

21.3.3 Categorization of rDNA Research Activities

Biosafety practices correspond to the related incremental risk involved with the use of microorganisms, plants, and animals in experiments. Hence, the rDNA activities have been classified into three categories based on the level of the associated risk (e.g., pathogenicity, local prevalence of disease, and epidemic causing strains in the country), and the requirement for the approval of competent authority.

Category I

This category includes those activities that are exempted for the purpose of intimation and approval of competent authority. It is also called 'exempt category'. The activities included are as follows: (i) The experiments involving self-cloning using strains and also interspecies cloning belonging to organism in the same exchanger group; (ii) Organelle DNA including those from chloroplasts and mitochondria; and (iii) Host–vector systems consisting of cells in culture and vectors (either nonviral or viral) containing defective genomes, except from cells known to harbor class III, IV, and special category etiologic agents. Note that the strains used for self-cloning exempted from notification are: *E. coli* K12, other well-characterized nonpathogenic laboratory strains of *E. coli, B. subtilis, B. stearothermophilus, B. thuringiensis*, nonpathogenic strains of *Streptomyces*, nonpathogenic strains of *Micromonospora, strains of Nocardia mediterranei, Klebsiella pneumoniae,*

Acremonium chrysogenum, Penicillium chrysogenum, nonpathogenic strains of *Haemophilus, Saccharomyces cerevisiae, Neurospora crassa* with selected vectors, and mouse cells with polyoma virus.

Category II

This category includes those research activities that require prior intimation to competent authority. For example, (i) Experiments falling under containment levels II, III, and IV; (ii) Experiment wherein DNA or RNA molecules derived from any source except for eukaryotic viral genome may be transferred to any nonhuman vertebrate or any invertebrate organism and propagated under conditions of physical containment PC1 and appropriate to organism under study; (iii) Experiments involving nonpathogen DNA vector systems and regeneration from single cells; and (iv) Large-scale use of recombinants made by self-cloning in systems belonging to exempt category.

Category III

Category III includes research activities that require review and approval of competent authority before commencement. (i) Cloning of toxin genes (e.g., *B. subtilis* and *B. sphericus* toxin genes in *B. subtilis*, cloning of cholera toxin genes and *B. thuringiensis* crystal protein genes in *E. coli* K12). Note that the cloning of these toxin genes is done under PC 1 and BC 1 containment conditions, and all toxin gene cloning experiments producing LD_{50} less than 50 µg/kg of body weight of vertebrates or large-scale growing may be referred to IBSC for clearance; (ii) Cloning of genes for vaccine production, e.g., Rinderpest and leprosy antigens. The containment is decided by RCGM on a case-by-case basis on experiment utilizing DNA from nondefective genomes of organisms recognized as pathogen. In view of no demonstrated risk during handling free *Mycobacterium leprae* antigens, inactivated whole cells as well as antigens can be assigned to risk group I; (iii) Cloning of mosquito and tick DNA experiments should be prescribed on a case-by-case basis since these are natural vectors for certain endemic viral and parasitic diseases; (iv) Genes coding for antibiotic resistance into pathogenic organisms, which do not naturally possess such resistance; (v) Introduction of recombinant DNA molecules containing complete genes of potentially oncogenic viruses or transformed cellular genes into cultured human cells; (vi) Introduction of unidentified DNA molecules derived from cancer cells or *in vitro* transformed cells into animal cells; (vii) Experiments involving the use of infectious animal and plant viruses in tissue culture systems; (viii) Experiments involving gene transfer to whole plants and animals; (ix) Cell fusion experiments of animal cells containing sequences from viral vectors if the sequence leads to transmissible infection either directly or indirectly as a result of complementation or recombination in the animals. For experiments involving recombinant DNA of higher-class organisms using whole animals, approval depends on case-by-case basis following IBSC review; (x) Transgenosis (a method used to transform animal cells with foreign DNA by using viruses as vectors or by microinjection of DNA into eggs and preembryos) in animal experiments. The expression of an inserted gene can be influenced both by the regulatory sequences associated with the gene and the sequences present at the site of integration of host genome; (xi) All experiments involving genetic manipulation of plant pathogens and the application of such genetically manipulated plant pathogens requires approval of IBSC; (xii) Transfer of genes with known toxicity to plants using *A. tumefaciens* or other vectors. Case-by-case clearance is needed though exemption may be made for the use of well-characterized vectors and nontoxic genes; (xiii) In case of plant viruses, permission may be obtained only when it is known that there is a chance of nonspecies specific spread of infection to plants that could produce changes in pathogenicity, host range, or vector transmissibility; (xiv) Experiments requiring field-testing and release of rDNA-engineered microorganisms and plants; (xv) Experiments involving engineered microbes with deletions and certain rearrangements; (xvi) No major risk can be foreseen on diagnostics involving *in vitro* tests, but for diagnostics involving *in vivo* tests, specific containment levels have to be prescribed on case-by-case basis. For example, tuberculin moiety can be cloned and used for *in vivo* hypersensitivity test as a diagnostic method; and (xvii) Gene therapy for hereditary diseases of genetic disorders.

21.3.4 Large-scale Experiments

Large-scale production (i.e., above 20 liters) of biomolecules from GMMs is not done presently in the country, however, regulations for large-scale production and field-testing including environmental release have been laid down under statutory provisions of the EPA 1986. For such activities, it is recommended that one should seek approval of the GEAC by furnishing relevant details. For good large-scale practice (GLSP) as well as levels of containment, principles of occupational safety and hygiene have to be applied. Besides, certain safety criteria are to be adopted for GLSP. These include the following: (i) The host organism should not be a pathogen and should have an extended history of safe use and built-in environmental limitations that permit optimum growth in the bioreactor, but limited survival with no adverse consequences in the environment; (ii) The vector/insert should be well-characterized and free from known harmful sequences; (iii) The DNA should be limited in size as much as possible (with insert <10 bp or molecular weight <30,000) to perform the intended function and should be poorly mobilizable. In cases where the insert sequence exceeds the above limit, toxicity screening should be made; (iv) The DNA should not increase the stability of the recombinant in the environment unless that is a requirement of the intended function; (v) There should not be transfer of resistance markers to microorganisms not known to acquire them naturally; (vi) Like host organism, the GMOs should not be a pathogen and should be assessed as being as safe in the bioreactor as the host organism and without adverse consequences in the environment; and (vii) For large-scale operations, measures such as proper engineering for containment, quality control, personnel protection, and medical surveillance are recommended. Offsite contingency plans in event of unanticipated effects of novel organisms/products on accidental release are to be worked out and appropriate control measures are to be developed in consultation with the competent authority (SBCC and DLC) to meet any exigency.

21.3.5 Product Safety Assessment

To provide assurance that this technology will generate food 'as-safe-as' that produced by traditional breeding programs, safety assessment strategies have been developed for products of plant genetic engineering, which are more thorough than those used to evaluate new foods using conventional breeding techniques. The overall goal of these assessments is to determine that the GM plant is 'substantially equivalent' to food derived from a conventional source, which has a history of safe use. Besides molecular characterization of the genetic modification and agronomic characterization, other key steps that are recommended for assessing the safety of GM plants by many international organizations are as follows:

Nutritional Assessment (i.e., Assessment of Key Nutrients) Key nutrients are those components in a particular food product that may have a substantial health impact in the overall diet. These may be major constituents (e.g., fats, proteins, carbohydrates) or minor components (e.g., essential minerals, vitamins). Critical nutrients to be assessed are determined, in part, by knowledge of the function and expression product of the inserted gene. Further the critical nutrients to be addressed are determined using consumption data for the target region.

Toxicological Assessment (i.e., Assessment of Key Antinutrients and Toxicants) Clinical toxicants and antinutrients are those compounds known to be inherently present in the crop variety whose potency could have an impact on health if the levels are increased significantly, e.g., solanine glycoalkaloids in potatoes, trypsin inhibitors in soybeans. Knowledge of the biologic function of the expression product of the inserted gene provides clues as to which toxicants or antinutrients are examined.

Safety Assessment of the Gene Expression Product When a GM food crop has been shown to be substantially equivalent to conventional crop with the exception of the introduced trait(s), which may impart one or more characteristics such as pest resistance, selectivity to preferred herbicides or modification of the ripening process, the safety assessment focusses on the introduced trait, and the protein expression product of the cloned gene. The biological function/specificity and mode of action of the protein determines the key assessment undertaken. If the protein is an enzyme, the potential effects of the enzyme on metabolic pathways and levels of endogenous metabolites based on the mode of action and specificity are assessed. If the newly introduced protein is functionally or structurally related to proteins that are known toxins, antinutrients, derived from a food with a history of allergy, structurally similar to known protein allergens, not degraded by digestive enzymes, or the biological function of the protein has not been characterized and there are no structurally related proteins identified in protein databases, then additional testing may be undertaken after a case-by-case assessment. Toxicological and nutritional endpoints are evaluated in rat feeding studies to determine if an adequate safety margin exists. Human nutritional studies have also been considered.

21.3.6 Release of GMOs into the Environment

The release of GMOs into the environment is a very sensitive issue. Before their release into the environment, potential risk factors, safety measures, quality control, etc. need to be evaluated.

Risk Assessment for Environmental Release of GMOs

The following factors should be taken into account when initial local risk assessment is being made. It is not expected that for any particular release proposal, all the points will be relevant, however, the extent of the information to be provided will depend on the type of organism and release proposed. Thus, in general, following factors should be taken into account for risk assessment: (i) The nature of the organism or the agent to be released, i.e., the species (or culture), its host range, and pathogenicity (if any) to humans, animals, plants, or microorganisms; (ii) The procedure used to introduce the genetic modification; (iii) The nature of any altered nucleic acid, its source, its intended function, and the extent to which it has been characterized; (iv) Verification of the genetic structure of the novel organism; (v) Genetic stability of the novel organism; (vi) Predicted effects of manipulation on the behavior of the organism in its natural habitat; (vii) The ability of the organism to form long-term survival forms, e.g., spores, seeds etc., and the possible effects that the altered nucleic acid may have on this ability; and (viii) Details of any target biota (e.g., pest in the case of a pest control agent), known effects of unmanipulated organism, and predicted effects of manipulated organism.

Information on the nature, method, and magnitude of the release is also important in assessing potential risk. This involves following considerations: (i) Geographical location, size, and nature of the site of release and physical and biological proximity to humans and other significant biota. In the case of plants, proximity to plants that might be cross-pollinated; (ii) Details of the target ecosystem and the predicted effects of release on that ecosystem; (iii) Method and amount of release, rate frequency, and duration of application; (iv) Monitoring capabilities and intentions; how many novel organisms be traced, e.g., to measure effectiveness of application; (v) On-site worker safety procedures and facilities; and (vi) Contingency plans in the event of unanticipated effects of novel organism.

Confined Field Trials, Survival, and Dissemination

Confined field trials, i.e., field experiments of growing a regulated, GM plant in the environment under specified terms and conditions that are intended to mitigate the establishment and spread of the plant, are important components of the process for approval of a GM crop for commercial cultivation. Confined field trials have been regularly conducted for almost 20 years across the world and more than 10 years in India. These trials represent the first controlled introduction of a GM crop into the environment falling in between experiments in contained facilities and commercial release to farmers. A single confined field trial may be comprised of one or more events of a single plant species that are subject to the same terms and conditions of confinement which include, but are not limited to, reproductive isolation, site monitoring, and post-harvest land use restrictions. It is understood that the experimental plant species in confined trials are those that have yet to receive regulatory approval for environmental release from GEAC. This confinement is also to be understood in terms of confinement of a particular GM plant in a particular region, state, village or a research farm of the applicant and is not accessible by other parts of the country in environmental terms. Embodied in this definition of confinement are three important considerations. First, confined field trials are typically carried out on a small-scale, usually to a maximum of one hectare. There may be exceptions to this, e.g., the cultivation of larger areas so that sufficient plant material may be harvested for livestock feeding trials. Secondly, a confined trial is an experimental activity conducted to collect data on potential biosafety impacts. The collection of such field trial data is a prerequisite for safety assessment of the GM crop under evaluation. Additionally, field trials are carried out to produce sufficient plant material so that the developer can undertake research to address the information and data requirements for livestock feed and human food safety assessment. Finally, the trial is conducted under conditions known to mitigate: (i) Pollen- or seed-mediated dissemination of the experimental plant; (ii) Persistence of the GM plant or its progeny in the environment; and (iii) Introduction of the GM plant or plant products into the human food or livestock feed pathways.

The survival, persistence, and dissemination of a released novel organism clearly have a major bearing on environmental consequences. This is especially so if the organism persists beyond the time required for its intended purpose. To evaluate this aspect, the following points should be considered: (i) Growth and survival characteristics of the host organism and the effects of manipulation; (ii) Susceptibility to temperature, humidity, desiccation, ultraviolet radiation, and ecological stresses; (iii) Details of any modification to the organism designed to affect its ability to survive and to transfer genetic material; (iv) Potential for transfer of inserted DNA to other organisms including methods for monitoring survival and transfer; and (v) Methods to control or eliminate any superfluous organism, or nucleic acid surviving in the environment or possibly in a product.

21.3.7 Import and Shipment

The import or receipt of etiologic agents and vectors of human and animal disease or their carriers is subject to the quarantine regulations. Permits authorizing the import or receipt of regulated materials for research (e.g., toxin genes, hybridomas, cell cultures, organelle) and specifying conditions under which the agent or vector is shipped, handled, and used are issued by RCGM, while large-scale imports for industrial use are regulated by GEAC. Safety testing may be required to ensure that the agent or the vector is risk-free.

21.4 REGULATIONS FOR GM PLANTS: *BT* COTTON IN INDIA—A CASE STUDY

Mahyco (Maharashtra Hybrid Seed Company), a leading seed company in India, was granted permission for commercial cultivation of three *Bt* cotton hybrids in India. After several years of field trials and based on the recommendations of RCGM, GEAC in its 32nd meeting on March 26, 2002 approved the commercial cultivation of three cotton hybrids, *viz.*, MECH-12 *Bt*, MECH-162 *Bt*, and MECH-184 *Bt* that were developed by introgression of insect resistance from *Bt*-containing Cocker-312 (event MON 531) developed by Monsanto Corporation, USA, into parental lines of Mahyco propriety hybrids. This approval was accompanied by 15 conditions as mentioned in Section 21.4.2. MoEF also reserved the right to stipulate additional conditions and the right to revoke the approval if the conditions were not implemented. Till date, 40 *Bt* cotton hybrids developed by 13 seed companies have been approved for commercial cultivation after going through a similar process of GEAC approval as prescribed for the three Mahyco hybrids. All the hybrids, except four, have the Monsanto-Mahyco *Bt* technology (event MON 531) that has been sublicensed to the respective seed companies. JKCH-1947 *Bt* and JK Varun *Bt* contain *cry* 1Ac event 1 developed by the Indian Institute of Technology, Kharagpur, while NCEH-2R *Bt* and NCEH-6R *Bt* contain fusion genes *cry* 1Ab/*cry* 1Ac from China. The hybrids released up to 2004 were approved for cultivation in Central and South zones, while in 2005 six *Bt* hybrids were approved for the first time for cultivation in North Zone. Zonewise, 14, 24, and 9 hybrids are presently approved for growing in North, Central, and South zones, respectively.

21.4.1 Concerns About *Bt* Cotton

Bt cotton has evoked extraordinary interest and emotion among a large section of Indian public comprising biotechnologists, plant breeders, social scientists, environmentalists, and civil society organizations (CSOs). The non-governmental organizations (NGOs) like Gene Campaign, Center for Sustainable Agriculture, and Research Foundation for Science, Technology and Ecology have expressing opinions about the performance and desirability of GM plants in general and *Bt* cotton in particular for Indian agriculture and social environment. These concerns are summarized here.

Low Bt Toxin Level

The critical minimum level of toxin (Cry 1Ac) in plant tissue essential for bollworm mortality is reported to be 1.9 μg/g. The leaves of *Bt* cotton expressed the highest level of toxin, while the expression in boll-rind, square, and ovary of plants was inadequate to confer full protection to the fruiting parts. Further, the *cry* 1Ac gene expression decreased consistently as the plant grew old, the decline being more rapid in some hybrids than others. This report was interpreted as the failure of *Bt* cotton in India.

Poor Performance Under Certain Environmental Conditions

Some *Bt* cotton hybrids have been reported to perform poorly under unirrigated conditions, while others yield inferior quality cotton staple.

Building up of Pest Resistance and Non-Compliance of Proposed 'Refuge Strategy'

The hybrids carrying *cry* 1Ac gene may become susceptible to bollworm, thereby posing a serious threat of widespread breakdown of resistance to the insect. Certain strategies, such as 'refuge strategy' and 'pyramiding strategy', have been proposed for resistance management by delaying adaptation to *Bt* crops by pest populations. However, due to small land holdings and lack of information, generally these strategies are not followed in practice, which may lead to rapid build-up of resistance to Bt toxin in bollworm. Likewise, cross-resistance among the toxins is a potential risk to the use of pyramids and it is suggested that 100% mortality of susceptible insects on the *Bt* crop is more critical to delay the onset of resistance.

Proliferation of Illegal or Spurious Bt Cotton

Owing to the high demand of *Bt* cotton, unapproved *Bt* cotton seeds of dubious origin and quality have been produced. Even though the production, sale, and use of such illegal and unapproved seeds is a violation of rules and liable to punitive action under the EPA 1986 , illegal *Bt* cotton seeds were in the market even before GEAC first approved the commercial cultivation of *Bt* cotton. A news report stated that against 90,000 seed packets of legal *Bt* cotton sold in Yavatmal district of Maharashtra the number of illegal packets sold was 2,50,000. According to field reports of Research Foundation for Science Technology and Ecology, illegal *Bt* cotton sold under 32 different names was sown in 2004 season. The testing of 10 packets of such seeds from Gujarat at Central Institute for Cotton Research, Nagpur, by polymerase chain reaction (PCR) and enzyme-linked immunosorbent assay (ELISA) revealed the presence of *cry* 1Ac gene.

Intellectual Property Rights (IPR) issues

The Plant Variety Protection and Farmers' Rights Act 2001 of India has a crucial provision according to which farmers are allowed to save, use, sell, and exchange seeds of a protected variety, the restriction being that the seed cannot be sold under the breeders' registered name. The *cry* 1Ac MON 531 technology is patented in the United States. While the non-*Bt* hybrid seed is sold at ~Rs. 450 per 450 g packet, the *Bt* hybrid seed is sold at Rs. 1,500 to Rs. 1,800 per packet, of which Rs. 1,250 is charged towards 'trait value' or 'technology fees'. Andhra Pradesh government has filed a case under Monopolies and Restrictive Trade Practices Act, claiming that the Indian farmers have to pay unusually high rates for the 'trait value' of *Bt* seed. On the other hand, the seed companies claim that the *Bt* seed saves four to five insecticide sprays and gives a higher net profit to the farmer. Recent reports indicate an agreement on a 30% reduction in 'technology fees' of *Bt* cotton seeds. Another intellectual property-related issue of concern is the restriction imposed on commercialization of 'gifted' or 'borrowed' *cry* genes. Substantial research has been done in public sector laboratories using 'gifted' *cry* genes, but the efforts have not culminated in release of commercial varieties since the genes were available for academic and experimental purposes only and the required authorization for their commercialization could not be negotiated.

Inadequate Testing, Genetic Illiteracy, and Recommendations of the National Commission on Farmers

The National Commission on farmers headed by Dr. Swaminathan held consultations with farmers on September

22, 2005 after receiving information regarding certain concerns about *Bt* cotton from the Indian Society for Sustainable Agriculture. Although none of the farmers reported cases of any health, food, or environmentally negative effects of *Bt* cotton, some expressed concerns about the possible risks. Several farmers emphasized the need for a cautious approach while exploiting GM technology and asked for a science-based pre and postrelease testing and monitoring. The Commission recorded that inadequate testing under the major cotton-growing agroclimatic conditions is a serious problem. The Commission also observed "genetic literacy (amongst farmers) has been generally low as most of the *Bt* cotton farmers grew no refugia, and did not provide recommended isolation distances needed for preventing cross-pollination between *Bt* and non-*Bt* strains". A general misgiving prevails, may be partly due to aggressive advertisement by seed companies, that the *Bt* cotton needs no pesticide applications, forgetting that the *Bt* provides protection (often not 100%) only against bollworms. For controlling other pests such as aphids and whitefly, which at times assume serious proportions, pesticides should be applied as per recommendations. The Commission noted that some participants reported failure of *Bt* cotton due to drought and multiple pest epidemics, while reporting additional net profit of at least about Rs. 12,000/ha and about 40–50% savings in the pesticide use and in the number of sprays. The Commission also expressed grave concern over proliferation of spurious *Bt* cotton seeds and suggested that in order to curb this trend, the company must compensate the losses incurred by the farmer. It also suggested insurance cover to be provided along with the sale of GM seeds.

21.4.2 Conditions Stipulated by the MoEF for Release of MECH-12 *Bt*, MECH-162 *Bt*, and MECH-184 *Bt*

Mahyco was instructed to follow certain rules and regulations for the release of three *cry* 1Ac MON 531 *Bt* cotton hybrids, *viz*., MECH-12 *Bt*, MECH-162 *Bt*, and MECH-184 *Bt* in India. The instructions included the following: (i) The approval was valid for three years, *viz*., April 2002 to March 2005; (ii) Every field with planted *Bt* cotton hybrids should be fully surrounded by 'refuge'; (iii) To facilitate 'refuge strategy', each packet of seeds of the approved varieties should contain a separate packet of the seeds of the same non-*Bt* cotton variety sufficient for planting in the prescribed refuge area; (iv) Each packet should be appropriately labeled indicating the contents and the description of the *Bt* hybrid including the name of the transgene, the GEAC approval reference, physical and genetic purity of the seeds. The packet should contain detailed directions for use including sowing pattern, pest management, suitability of agroclimatic conditions, etc., in vernacular language; (v) Mahyco should enter into agreements with their dealers/agents that would specify the requirements from dealers/agents to provide details about the sale of seeds, acreage cultivated, and state/regions where *Bt* cotton had to be sown; (vi) Mahyco should prepare annual reports by March 31 each year on the use of *Bt* cotton hybrid varieties by dealers, acreage, locality (state and region), and submit the same in electronic form to GEAC, if asked by the GEAC; (vii) Mahyco should develop plans for *Bt*-based integrated pest management (IPM) and to include this information in the seed packet; (viii) Mahyco should monitor annually the susceptibility of bollworm to *Bt* gene vis-à-vis baseline susceptibility data and to submit data related to resistance development, if any, to GEAC; (ix) The Ministry entrusted the Central Institute for Cotton Research, Nagpur, to carry out the monitoring of susceptibility of bollworms to the *Bt* gene, and that the MoEF should undertake monitoring at applicant's cost; (x) Mahyco should undertake an awareness and education programme, *inter alia* through development and distribution of educational material on *Bt* cotton, for farmers, dealers, and others; (xi) Mahyco should undertake studies on possible impacts on nontarget insects and plants and report back to GEAC annually; (xii) The label on each packet of seeds and the instruction manual inside the packet should contain all relevant information; (xiii) Mahyco should deposit 100 g seed each of the approved hybrids as well as their parental lines with the National Bureau of Plant Genetic Resources (NBPGR); (xiv) Mahyco should develop and deposit the DNA fingerprints of the approved varieties to NBPGR; and (xv) Mahyco should provide the testing procedures for identifying transgenic traits in the approved varieties by DNA and protein methods to NBPGR.

21.4.3 Trial Results of MECH-12 *Bt*, MECH-162 *Bt*, and MECH-184 *Bt*

Before release, Mahyco conducted the following biosafety, risk management, and field performance trials on the *Bt* hybrids submitted for approval of GEAC. These studies were carried out in the laboratories and experimental fields designated by RCGM/GEAC. Besides these studies, the socioeconomic impact of *Bt* cotton cultivation was also assessed.

Biosafety Assessment

Bt cotton hybrids were evaluated for environment and food safety, the results of which are indicated here.

Studies on Environmental Safety The studies on environmental safety included the following:

(i) Pollen escape/out-crossing studies Multilocation experiments conducted in 1996, 1997, and 2000 revealed that outcrossing occurred only up to 2 m and only 2% of the pollen reached a distance of 15 m. It is significant to note that as the pollen in heavy and sticky, the range of pollen transfer is limited. The studies demonstrated that there was no chance for the transfer of *Bt* gene from cultivated tetraploid species (e.g., *Bt* hybrids) to traditionally cultivated diploid species.

(ii) Evaluation of aggressiveness and weediness In order to assess weediness, the rate of germination and vigor were compared with untransformed parental lines by laboratory test and in soil. The results indicated no substantial differences between *Bt* and non-*Bt* cotton, suggesting that there is no difference between them with regard to their weediness potential.

(iii) Evaluation of effects of Bt on nontarget organisms Further studies to determine any toxic effects on nontarget species, such as parasites or pest predators reared on herbivore prey that had ingested corn leaves expressing Bt toxins (e.g., aphids, jassids, whitefly, and mites) revealed no such effects. The population of secondary lepidopteran pests, e.g., tobacco caterpillar, remained negligible during the study period. Moreover, the beneficial insects (lady bird beetle and spiders) remained active in both *Bt* and non-*Bt* varieties.

(iv) Assessment of presence of Cry 1Ac toxin in soil For accessing the possible risk of accumulation of Cry 1Ac protein in the soil, insect bioassays and ELISA were performed. Cry 1Ac protein was not detected in soil samples, indicating its rapid degradation in soil in both the purified form and as part of the cotton plant tissue. Moreover, the half-life for the plant tissue was 41 days, which was comparable to the degradation rates reported for microbial formulations of Cry 1Ac protein.

(v) Determination of effects of Cry 1Ac protein on soil microflora *B. thuringiensis* (a ubiquitous soil-borne bacterium) naturally releases *Bt* toxin into soil, which can bind to elements within the soil (such as clay particles or humic acids), become stabilized, and remain active for possibly hundreds of days. Further, *Bt* plants express toxins from their roots during their entire life cycle, which are released from dead plant material after harvest and incorporated into the soil. Thus, the soil microorganisms in the rhizosphere of *Bt* crops are additionally exposed to these toxins. As the toxin needs to be ingested to be active against insects, it is likely that toxins bound to soil particles would potentially threaten only those organisms that feed on soil, for example, earthworms. The analysis of earthworms revealed the presence of Bt toxins in guts and casts, but there were no significant differences in mortality, weight of these organisms, or in the total number of other soil organisms (including nematodes, protozoa, bacteria, and fungi) between the soil rhizosphere of *Bt* and non-*Bt* plants.

Studies on Food Safety The studies on food safety included the following:

(i) Compositional analysis No difference in the composition of *Bt* and non-*Bt* cotton seeds, with respect to proteins, carbohydrates, oil, calories, and ash content, was observed.

(ii) Allergenicity studies These studies were conducted on Brown Norway rats and showed no significant differences in feed consumption, weight gain, and general health of animals fed with *Bt* and non-*Bt* cotton seeds.

(iii) Toxicological studies A goat feeding study was conducted for understanding the toxicological effects of *Bt* cotton seed. The animals were assessed for gross pathology and histopathology and no significant differences were found between animals fed with *Bt* and non-*Bt* cotton seed.

(iv) Detecting the presence of Cry 1Ac protein in *Bt* cotton seed oil Investigation of refined oil obtained from *Bt* cotton seeds indicated the absence of Cry 1Ac protein.

(v) Nutritional and safety assessment by feeding studies on animals The feeding experiments using *Bt* cotton seed meal were conducted on lactating cows at the National Dairy Research Institute, Karnal, on lactating buffaloes at the Department of Animal Nutrition, College of Veterinary Sciences, G.B. Pant University of Agriculture and Technology, Pantnagar, on poultry at the Central Avian Research Institute, Izatnagar, and on fish at the Central Institute of Fisheries Education, Mumbai. These experiments indicated that *Bt* cotton seed meal was nutritionally as wholesome and safe as the non-*Bt* cotton seed meal.

Risk Management by 'Refuge Strategy'

The concern related to *Bt* crops is that the pest populations exposed to them continuously for several years may develop resistance to Bt toxin through natural mutation and selection. To prevent resistance build-up, it is recommended to deliberately plant sufficient non-*Bt* cotton of the same variety to serve as a 'refuge' for *Bt*-susceptible insects. This is called 'high-dose refuge' strategy and is the most widely used. The size of the refuge belt should be either five rows of non-*Bt* cotton or 20% of total sown area, whichever is more, and necessary control measures are taken against bollworms in the refuge crop as and when required. The 'refuge strategy' is designed to ensure that *Bt*-susceptible insects will be available to mate with *Bt*-resistant insects if these arise. As the available genetic data indicates that susceptibility is dominant over resistance, the offsprings of these matings would most likely be *Bt*-susceptible, thus mitigating the spread of resistance in the population. Another approach for the prevention of pest resistance is 'pyramiding strategy', i.e., using more than one resistance gene. Such strategy is expected to delay the evolution of resistance much more effectively than the presence of a single insecticidal toxin and may require smaller refuges.

Baseline Susceptibility Studies

In 1999 and 2001, Project Directorate of Biological Control, Bangalore, carried out baseline susceptibility studies of *Helicoverpa armigera* to Cry 1Ac protein. Geographical populations of *H. armigera* were collected from nine major cotton-growing states of India, *viz.*, Punjab, Haryana, Rajasthan, Madhya Pradesh, Gujarat, Maharashtra, Andhra Pradesh, Karnataka, and Tamil Nadu, and exposed to insecticidal protein Cry 1Ac. LC_{50} (mean lethal concentration) ranged from 1.02 to 6.94 μg of Cry 1Ac/ml of diet. The median molt inhibitor concentration MIC_{50} ranged from 0.05 to 0.27 and MIC_{90} from 0.25 to 1.58 μg of Cry 1Ac/ml of diet. The effective concentration (weight stunting related) EC_{50} ranged from 0.0003 to 0.008 and EC_{90} from 0.009 to 0.076 μg of Cry 1Ac/ml of diet.

Confirmation of the Absence of 'Terminator Technology'

In order to determine the absence of *cre* sequence, an integral component of 'terminator technology'(Exhibit 21.1),

Exhibit 21.1 Terminator technology

On March 3, 1998, a joint patent (U.S. patent no. 5723765) was granted to U.S. Department of Agriculture (USDA) and the Delta and Pine Land (D & PL) Company, Mississippi, in the name of 'control of plant gene expression'. Melvin Oliver, a scientist with USDA-ARS in Lubbock, Texas, mainly did the patent work through cooperative research with D & PL Company. According to the technology, a variety with the terminator gene would bear normal seeds with all its characteristic features but the seed would not germinate in the next generation. The patent claimed a very broad protection and was valid for plant cells, tissues, seeds, and whole plants of any species (both transgenic and conventional crop varieties) containing the combination of genes described in the patent. It was claimed that the genetic system described in the patent had worked well with tobacco and cotton. Under a research agreement with USDA, D & PL Company had exclusive right to license (or not) the technology to others. Soon after the patent was granted, on May 11, Monsanto Corporation took over D & PL Corporation, so at present Monsanto Corporation holds the patent. Monsanto described it as 'gene protection technology' with the future application for the development of technology protection system. On the contrary, Hope Shand, Research Director, Rural Advancement Foundation International [RAFI; a Canada based, nongovernmental organization (NGO)] and media persons dubbed the technology as terminator technology because they believed that it will terminate the farmers' independence and threaten the food security of over a billion resource-poor farmers in developing countries, where farmers' saved seed accounts for 80% of the total seed requirement. Thus, this technology had put a serious question to the age-old cycle of plant-to-seed.

Mechanism of Terminator Technology In several self-pollinated crops like wheat, rice, barley, beans, etc., the commercially grown cultivars are actually 'pure lines' so that the yield does not decline and harvested seeds can be used for sowing the next crop. On the other hand, in case of hybrid varieties (e.g., maize, sunflower, etc.), there is a built-in protection, which forces the farmers to purchase hybrid seeds every year. This is because the increased yield is exhibited only in the F_1 seed that is sold to the farmers and the performance declines in F_2 and subsequent generations. This discourages the farmers to harvest and re-use the harvested seed of hybrid varieties for sowing the next crop. Hence, the genetic elements used in the terminator technology differ in pure lines and hybrid seed production, which are discussed in the following:

Technology for Pure Line Seed Production in Self-pollinated Crops The patented method for terminator technology is based on a gene that produces a protein that is toxic to the plant and therefore does not allow the seed to germinate. The technology involves two gene systems brought together to stop the normal process of embryo development, leading to failure of germination of the seed. The gene system I consists of terminator gene encoding ribosome-inactivating protein (RIP). The gene is placed under the control of transiently active *Lea* promoter (embryogenesis-specific), permitting terminator gene to express only during late embryogenesis, thus affecting only the embryo development. This is because the expression of terminator gene does not allow protein synthesis to take place. This terminator gene does not express in the first generation because its expression is blocked through the use of a spacer or a blocking sequence between the promoter and the lethal terminator gene. A recombinase-specific excision sequence (*lox* sequence from P1 bacteriophage) flanks the blocking sequence on either side, whose function is to excise the spacer. Gene system II consists of a *cre* recombinase gene encoding Cre recombinase, which is specific to the *lox* sequence of the gene system I. The *cre* recombinase gene is under the control of a promoter (*P*). A third gene produces a repressor protein (R). Repressor binds to operator region (*O*) of *cre* recombinase gene and prevents its expression. The *cre* recombinase gene can be derepressed by exogenous application of tetracycline. To develop a variety of seeds with functional terminator system, two cells of the same crop are transfected with the gene system I and II separately. As a result, one transgene is obtained with unexpressed terminator gene due to the presence of the blocking sequence between terminator gene and its *Lea* promoter, while the other transgene is obtained with gene system II having repressor of *cre* recombinase gene. To recombine these two systems into one, the obtained transgenics are hybridized and normal hybrid seeds are obtained. In such seeds, repressor is constitutively expressed, which represses the transcription of *cre* recombinase gene. Thus there is no Cre recombinase to catalyze excision of spacer sequence from gene system I and the terminator gene remains unexpressed. Consequently, toxin is not synthesized and the seeds obtained remain stable and viable. Such viable seeds are sold to farmers after spraying them with tetracycline. Upon treatment of the seeds with tetracycline, recombinase is produced. This is because tetracycline is absorbed by the seedling tissue and acts as an inducer of *cre* recombinase gene. The Cre recombinase excises the intervening spacer sequence between terminator gene and its *Lea* promoter. Thus, *Lea* promoter comes in proper orientation with terminator gene, leading to its transcription. Note that *Lea* promoter specifically expresses during early embryo development

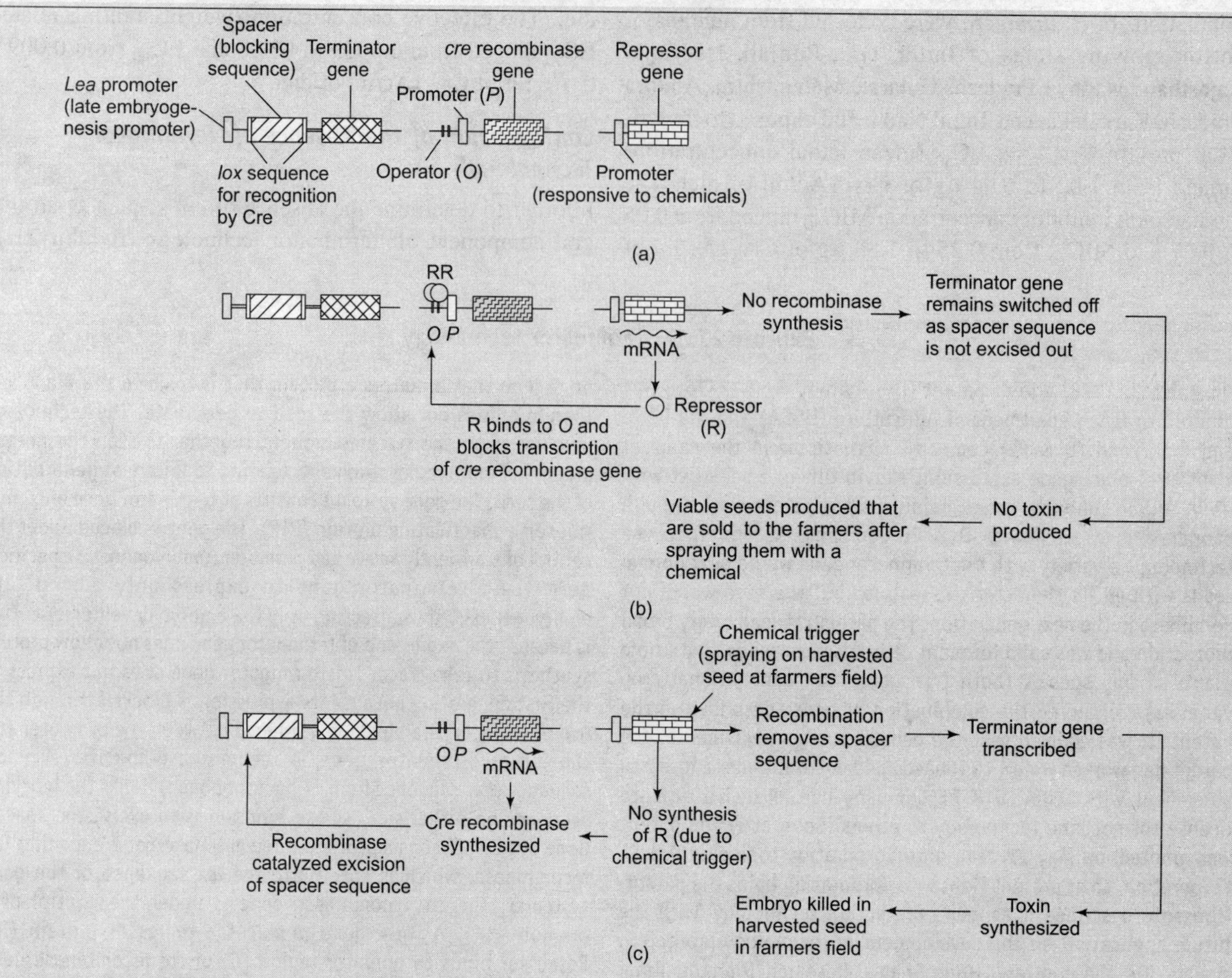

Figure A Genetic basis of terminator technology envisaged for pure line seed production in self-pollinated crops; (a) Arrangement of terminator gene, *cre* recombinase gene, and repressor gene, (b) Production of viable seeds with embryos and endosperms at farmers field, (c) Production of noninviable seeds with no embryos, but with full endosperm at farmers field

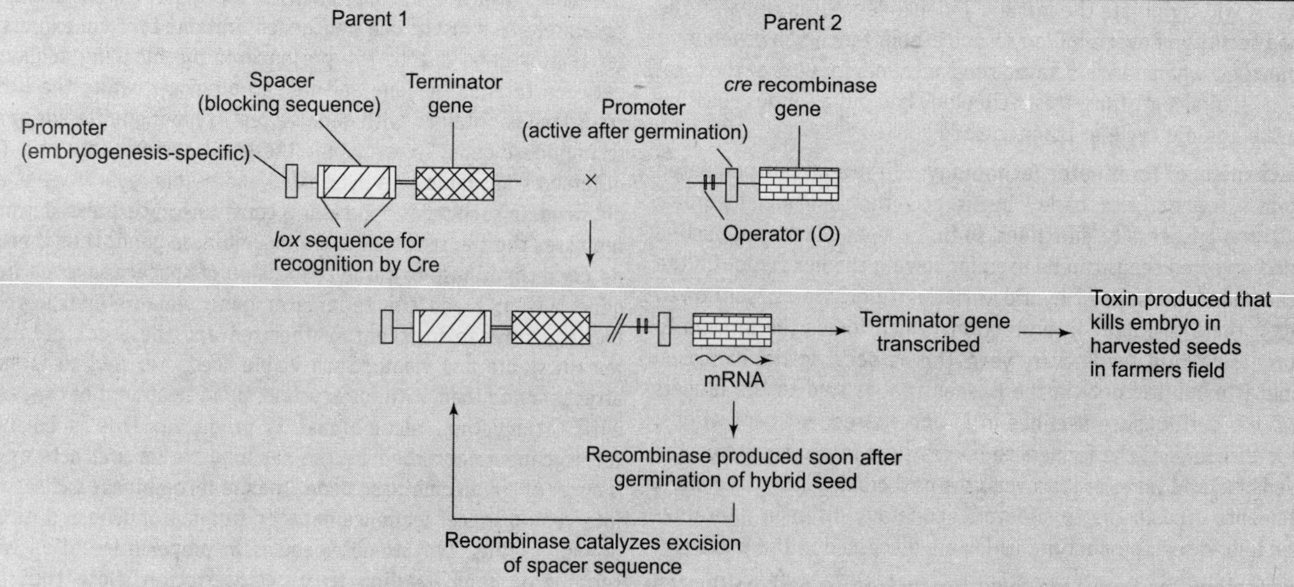

Figure B Genetic basis of terminator technology envisaged for hybrid seed production both in cross-pollinated and self-pollinated crops

and the seeds germinate normally in that generation, giving rise to normal crop and seeds. Hence upon receiving chemical trigger, the embryo gets aborted due to the expression of terminator gene and consequently the seeds obtained do not germinate. A seed thus produced will carry the endosperm, but not the embryo, so that it can be used or sold as grain, but cannot be used for sowing. So long as *cre* recombinase gene remains repressed in absence of tetracycline, terminator gene is not expressed, leading to production of viable seeds.

Technology for Hybrid Seed Production In case of hybrid seed production, a different strategy utilizing only two genes (terminator and recombinase) is employed. Thus, one gene in each of the two parents of the hybrid is used and no repressor gene is needed. One of the parental lines contains the recombinase gene, which becomes active only after germination and the other parent contains the lethal terminator gene separated from its promoter by a spacer (blocking sequence). The hybrid progeny, which is the technology-protected hybrid seed bought and planted by the farmer, thus contains both the elements of the system in every cell. The recombinase expressed right after germination excises the spacer blocking sequence, bringing the promoter and terminator gene together. Since the promoter is embryo-specific, the lethal gene does not express till seed development starts. During seed development, the lethal gene expresses during late embryogenesis and kills the embryo. Thus, the seeds harvested from the first generation hybrid crop will be normal in all essential respects except that these will not germinate if sown as a crop.

Monsanto, D & PL Company, and USDA Views About the Technology Monsanto Corporation called this technology as 'gene protection technology' or 'control of gene expression'. According to S. M. Hayes, another scientist from USDA, the above technology was developed for the study of gene expression. For instance, the technology may be used for the expression of any desired trait in one situation, but not in the other. Such possible traits include male sterility, drought or insect resistance, time of seed germination, or flower development. Harry B. Collins of D & PL argues that the farmers still have the option to use the traditional varieties having no protection system and that the new technology will encourage plant breeders to invest and develop new varieties of crop. D & PL company also claims that the technology will help the food security system as follows: First, the technology will stimulate investment and interest by breeders for which hybrids are not feasible (e.g., wheat, rice, barley, beans, etc.); second, it will provide the farmers access to continuous development of new improved cultivars; third, the incentives for development of new varieties will enhance genetic diversity in many important crops; and finally, the escape of transgenic to other wild and nontargeted plants will be impossible since unwanted pollination will give seeds that will be nonviable, thereby biosafety will be ensured. Further, the control of the expression of terminator gene lies with the growers of the seeds. They can encourage as many selfings as they wish and also terminate the viability of the seeds at any time by giving tetracycline treatment. Upon establishing a stable line with both gene systems I and II, these may be transferred easily to any variety through suitable breeding schemes. Thus it will be possible to develop any variety with the terminator gene. The other possible uses of the 'gene protection technology' advocated by Monsanto are minimization of out-crossing with related species, increased choice of seed varieties for the farmers, help in protecting farmers' rights, development of best variety seeds by yearly improvement, and maintenance of desirable characteristics in varieties that are grown for more than one year.

Apprehensions About the Technology Several NGOs including RAFI and GRAIN in Europe do not agree with the company and believe that terminator technology will do more harm than good. It was apprehended that the only motive behind this technology was to have complete biological control over the seed supply and to prevent the farmers from saving nonhybrid, open-pollinated, or genetically engineered seeds developed and sold by the TNCs. The concern was that the producer of the seeds (the TNCs) may grow sufficient quantity of seeds of the specific variety and before selling them to the farmers, they may give tetracycline treatment and ensure that the farmers will not be able to grow the seeds in the next season. This ensures full protection of Intellectual Property Rights of the TNCs without going for legal suits. Moreover, if this terminator gene gets transferred to other varieties and related species of a specific crop through cross-pollination, it will bring large-scale sterility and will disturb the ecological balance by destroying our biodiversity. It is also argued that pollen from crops carrying terminator trait will infect fields of farmers who either reject or cannot afford this technology, thereby harming such poor farmers. The neighboring crops will not be affected in that season but due to the induced nonviability of certain percentage of seeds, yield of successive crop may be affected. It is feared that the yielding pattern will become irregular if such technology is governed by recessive gene(s). This will bring about reduction in food production. In case of self-pollinated crops, the situation may not be very serious because in these crops, cases of cross-pollination are around 5%. So even if a farmer grows a variety with the terminator trait, at most 5% of the seeds of the neighboring field will be affected. Only plants in the border rows will be affected since the pollen dispersal in self-pollinated crops is not very long. To avoid this risk, the farmers may collect seeds from the middle of the field where chances of cross-pollination are less. In case of cross-pollinated crop, this technology may cause significant loss to traditional varieties, particularly synthetic and composite ones. But hybrid seed industry is not going to be benefited from this technology because in case of hybrid varieties, the farmers are forced to change the seed after every crop as the variety suffers severe inbreeding depression upon selfing. This lethal technology will also not find field in vegetatively propagated crops like sugarcane, potato, etc. The potential threat of induced nonviability of seeds may not cause any havoc to the biodiversity also because the plant species produce seeds in large quantities, the seed production may not be severely affected with such gene transfer. However, it is always better to be cautious and hence GMOs with terminator technology are not acceptable. Indian Government has banned the entry of any seed material that may carry the 'terminator gene' into the country.

Source: Rakshit S (1998) *Current Science* 75(8) 747–749; Gupta PK (1998) *Current Science* 75(12): 1319–1323.

PCR analysis of DNA samples isolated from individual seedlings derived from *Bt* cotton hybrids was performed at the Department of Genetics, University of Delhi (South Campus), Delhi. The study revealed that these lines were positive for *cry* 1Ac genes but negative for *cre* sequence, thereby conclusively indicating the absence of 'terminator gene' in *Bt* cotton hybrids.

Field Performance and Socioeconomic Impact

Several studies have been conducted on field performance of *Bt* cotton in India, initially by the seed companies as a part of the approval procedure of RCGM and GEAC and later by research scientists/organizations as independent studies or by CSOs. Mahyco conducted two sets of field experiments in 1998–99 under the monitoring of RCGM. In one set, MECH-12 *Bt*, MECH-162 *Bt*, and MECH-184 *Bt* along with their non-*Bt* counterparts were tested in replicated field trials at 15 sites in nine cotton-growing states, while in the other set, one *Bt* and one non-*Bt* hybrid along with the check were tested on large plots at 25 sites under typical farm conditions. Results of the first set of experiments indicated a 40% higher yield of *Bt* hybrids (14.64 q/ha) over their non-*Bt* counterparts (10.45 q/ha). Further, there was a significantly lower incidence of bollworm damage to fruiting bodies in *Bt* hybrids (2.5% at 61–90 days from planting) than in non-*Bt* hybrids (11.4% at 61–90 days from planting). The large-field trials at 21 sites (four trials were damaged) yielded similar results with *Bt* hybrids, showing 37% (range 14–59%) higher yield over their non-*Bt* counterparts. The overall pesticide requirement for controlling bollworm was reduced considerably. The data generated from the above-detailed multilocation tests were analyzed to assess the potential economic advantage of *Bt* cotton in India. The results showed that there was 78.8% increase in the value due to yield and 14.7% reduction in pesticide cost with the growing of *Bt* cotton as compared to non-*Bt* cotton. When compared with the prevalent farmer practices, the benefit from *Bt* cultivation increased to 100%. Taking into account the additional cost of *Bt* seeds, the farmer would still get more than 70% greater benefits. Moreover, the reduction in expenditure on pesticides would adequately compensate for the seed/technology cost increase. Hence the total cost to cultivation of *Bt* cotton would not increase and even small farmers could adopt the technology. ICAR conducted multilocation field trials in 2001 on the three Mahyco *Bt* hybrids specifically to make a cost benefit analysis of *Bt* cotton. Yield increases over local check and national check were recorded to the magnitude of 67% advantage from average Rs. 14.112/ha in local and national check to average Rs. 23.604/ha in the *Bt* hybrids. After adjusting cost of *Bt* hybrid seed, the net economic advantage of *Bt* cotton ranged between Rs. 4,633/ha and Rs. 10,205/ha. Mahyco made a survey of the performance of three *Bt* hybrids grown on over 1,000 farmers' fields in five states after their release in 2002. Yields of *Bt* hybrids (average 13.25 q/ha) were higher (by an average of about 30%) than non-*Bt* hybrids. Further, there was significant decrease in the number of insecticide sprays (from an average of 3.10 to 1.17) associated with the use of *Bt* cotton. These two factors added to the total economic benefit, providing an average additional income of more than Rs. 18,000/ha for *Bt* cotton compared to non-*Bt* cotton.

21.4.4 Regulatory Mechanism for *Bt* Cotton

MoEF had constituted a subcommittee on *Bt* cotton under the Chairmanship of Dr. C.D. Mayee, Chairman ASRB, and Co-chair GEAC, to look into the existing processes, protocols, and other related issues. The Committee analyzed the constraints in the current regulatory framework such as a case-by-case approval, which follows extensive testing under RCGM/GEAC/ICAR trials even if the hybrid contains a gene/event cleared from biosafety angle, limited parameters being conducted while selecting hybrids, nonplanting of refugia, and issues regarding affordable prices. Major recommendations of the Committee are as follows: (i) The Committee recommends 'an event based approval system', i.e., extensive biosafety and agronomic testing not necessary for approved gene/event; (ii) Due consideration to the agronomic value of the hybrid should be given; (iii) For sale/commercialization of any transgenic hybrid/variety testing by ICAR is not mandatory for an approved event that has been declared biosafe and being cultivated extensively; (iv) Involvement of the State Agricultural Universities (SAUs) and State Agriculture Department to monitor the performance of the agricultural crops in their jurisdiction; (v) Need to strengthen the enforcement mechanism, disseminate information regarding the field trails and enhance the awareness and extension work at the field level; (vi) Responsibility of both pre- and post-release monitoring should be entrusted to the SAU under the direct supervision of Director of Research and Director of Agriculture Extension respectively of each SAU; (vii) To review alternatives with respect to refugia, IRM practice, IPM strategy, appropriate packaging practice, etc.; (viii) While considering the parameters for deciding the superiority of the hybrids it may be compared with the released non-*Bt* hybrids check of respective group, *viz*., early/medium/quality of a hybrid, CIRCOT guidelines/norms should be followed; and (ix) Other recommendations are related to issues such as strengthening the enforcement mechanism to address various issues reported by the NGOs, permission for LST/commercial release based on agroclimate conditions rather than the zonal concept as recommended by ICAR and rationalization of biosafety studies with a view to promote the development of transgenic crops from public institutions. For detailed information of the subcommittee report, see http://www.envfor.nic.in.

The MoEF and Ministry of Agriculture also issued various notifications to ensure the seed quality of GM crops particularly *Bt* cotton: (i) As the 'Seeds Act, 1966' does not cover transgenic seeds, there were difficulties faced by State Department of Agriculture in the quality enforcement of the *Bt* cotton seeds. The Department of Agriculture and Cooperation, Ministry of Agriculture, therefore, had requested the MoEF to notify the seed inspectors under Section 13 of the 'Seeds Act' and Section 12 of the 'Seeds (Control) Order' to draw the seed samples of transgenic seeds as mentioned under Section 10 of the EPA 1986. MoEF in consultation with Ministry of Law and Justice has issued six Gazette Notifications [G.S.R. 584 (E) and 589 (E) dated September 1, 2006] wherein seed inspectors have been given adequate power to draw seed samples of transgenic seeds for the purpose of quality control and to get it tested in the notified seed testing laboratories and prosecute in case of spurious *Bt* cotton seeds. With the promulgation of the said notifications, the seed law enforcement agencies are empowered to take necessary punitive action against the offenders; (ii) Ministry of Agriculture has issued a Notification on November 12, 2003 nominating Central Institute of Cotton Research (CICR) to act as a referral laboratory for ascertaining the presence or absence of *cry* 1Ac gene in cotton seeds for the whole of India; and (iii) In addition to the above, seeds division issued minimum limits of purity in respect of *Bt* cotton seeds as 90% (Bt protein toxin) under Section 6 of the 'Seeds Act' in the Gazette Notification issued vide SO No. 1567 (E) dated November 5, 2005.

21.5 REGULATIONS RELATED TO STEM CELL RESEARCH AND HUMAN CLONING

Human cloning is expected to produce several benefits in basic research and gene therapy. Technically, human cloning falls in two categories, *viz*., therapeutic cloning for somatic cell gene therapy and reproductive cloning required for germ line gene therapy. It is considered that human therapeutic cloning could provide genetically identical cells for regenerative medicine and tissues and organs for transplantation. Such cells, tissues, and organs would neither trigger an immune response nor require the use of immunosuppressive drugs. As such both basic research and therapeutic development for serious diseases such as cancer, heart disease, and diabetes, as well as improvements in burn treatment and reconstructive and cosmetic surgery, are areas that might benefit from such new technology. Human reproductive cloning might also produce benefits. The cloning of human embryos may lead to development of fertility treatment and help infertile couples conceive, thereby allowing them to have children with at least some of their DNA. Some scientists suggest that human cloning might obviate the human aging process. Two terms are associated with reproductive cloning, *viz*., 'replacement cloning', which describes the generation of a clone of a previously living person and 'persistence cloning', which describes the production of a cloned body for the purpose of obviating aging. However, these potential applications are still considered as science fiction.

21.5.1 Concerns Related to Stem Cell Research and Human Cloning

Despite several above-mentioned benefits associated with human cloning, it is a topic of debate on biomedical and genetic ethics. Today stem cell research raises many ethical, legal, scientific, and policy issues that are of concern to the policy makers and public at large. These concerns were further intensified with the unfortunate death of Jesse Gelsinger in a human gene transfer experiment. Further, safeguards have to be in place to protect research participants receiving stem cell transplants and patients at large from unproven therapies/remedies. Currently human reproductive cloning is banned in a number of countries because of genetic consequences for future generations and several associated risks. However, India is supportive of stem cell (both adult and embryonic) research though the concerns are still persisting. The concerns raised are a consequence of extrapolation of results of animal cloning experiments and some are related to human ethical values, which are listed: (i) The stem cell research and therapy is objectionable as it involves destruction of human embryos to create human embryonic stem (hES) cell lines; (ii) Stem cell research has the potential for introducing commodification in human tissues and organs with inherent barriers of access to socioeconomically deprived and possible use of technology for germ line engineering and reproductive cloning; (iii) Safety and rights of those donating gametes/blastocysts/somatic cells for derivation of stem cells or fetal tissues/umbilical cord cells/adult tissue (or cells) for use as stem cells are at stake; (iv) The animal cloning experiments have proved to be inefficient (only about 1 or 2 viable offspring for every 100 experiments); consequently, more than 100 nuclear transfer procedures could be required to produce one viable clone; (v) Cloned animals tend to have more compromised immune function and higher rates of infection, tumor growth, and other disorders. About 30% of animal clones born alive are affected with 'large offspring syndrome' and other debilitating conditions, leading to premature death; (vi) Scientific data from animal experiments indicate the method is not physically safe for mother or baby; (vii) There is a lack of knowledge on reproductive cloning. The impact on mental development of cloned human is not yet clear. While factors such as intellect and mood may not be as important for a cow or a mouse, these are crucial for the development of healthy humans; (viii) Reproductive cloning is expensive; (ix) The individuality, autonomy, and kinship of the resulting children are questionable. The rights of parents to control their own embryos and other issues concerning reproductive rights and privacy are endangered; (x) There can be specific expectations regarding the cloned individual's talents and abilities. Illtreatment or discrimination against a cloned individual is also a possibility. However, some people point out that this altered perception does not occur today with identical twins, the naturally produced clones; (xi) Human reproductive cloning could affect society's perception of what it means to be a human being. Cloning would diminish respect for human life in general and for cloned individuals in particular, since the cloned person might simply be replaced with another clone; and (xii) The possibility of human cloning is offensive to the religious and other deeply held beliefs of many people. *In toto*, there is a need to generate public confidence in potential benefit of stem cell research to human health and disease.

21.5.2 Regulatory Mechanism for Stem Cell Research and Human Cloning

The area of stem cell research being new and associated with rapid scientific developments and complicated ethical, social, and legal issues require extra care and expertise in scientific and ethical evaluation of research proposals. Hence, a separate mechanism for review and monitoring is essential for research and therapy in the field of human stem cells. Stem cell research policy varies significantly throughout the world. There are overlapping jurisdictions of international organizations, nations, states, or provinces. Some government policies determine what is allowed vs. prohibited, whereas others outline what research can be publicly financed. Some organizations have issued recommended guidelines for how stem cell research is to be conducted. In India, the review and monitoring of stem cell research and therapy is governed by two committees, National Apex Committee for Stem Cell Research and Therapy (NAC-SCRT) at the national level and Institutional Committee for Stem Cell Research and Therapy (IC-SCRT) at the institutional level. The composition and functions of NAC-SCRT and IC-SCRT are as follows.

Regulatory Bodies and Their Functions

NAC-SCRT is a multidisciplinary multiagency body with a secretariat. Members include chairman, deputy chairman, member secretary, nominees from DBT, DST, CSIR, ICMR, Drug Controller General of India (DCGI), DAE, and biomedical experts drawn from various disciplines like pharmacology, immunology, cell biology, hematology, genetics, developmental biology, clinical medicine, and nursing. Other members are legal expert, social scientist, and women's representative. In addition, consultants/experts could be consulted for specific topics and advice. The Committee is independent and is respected by both general public and scientific communities. The functions of NAC-SCRT include: (i) To examine the scientific, technical, ethical, legal, and social issues in the area of stem cell research and therapy; (ii) To review and approve specific research protocols falling under restricted category; (iii) To address to new unforeseen issues of public interest from time to time; (iv) To register all institutions involved in any type of stem cell research and therapy; (v) To monitor periodic and annual reports from all centers involved in stem cell research and to ensure adherence to standards by site visits as and when required; (vi) To approve, monitor, and oversee research in the restricted areas. Every scientific proposal using embryonic stem cells (ES cells) under restrictive category has to be cleared through IC-SCRT/Institutional Ethics Committee (IEC) before referring to NAC-SCRT; (vii) To approve the use of chimeric tissue for research after clearance from IC-SCRT/IEC; (viii) To revise and update guidelines periodically, considering scientific developments at the national or international level; and (ix) To set up standards for safety, quality control, procedures for collection and its schedule, processing or preparation, expansion, differentiation, preservation for storage, removal from storage to assure quality, and/or sterility of human tissue, prevention of infectious contamination, or cross contamination during processing, carcinogenicity, and xenotransplantation, etc.

All research institutions conducting stem cell research are expected to set up a special multidisciplinary review body, namely IC-SCRT, to oversee this emerging field of research. The Committee is registered with and has to submit annual report to NAC-SCRT. It includes representatives of scientists with expertise in clinical medicine, developmental biology, stem cell research, molecular biology, and assisted reproduction technology. The scope of IC-SCRT includes: (i) To provide overview to all issues related to stem cell research and therapy; (ii) To review and approve the scientific merit of research protocols; (iii) To review compliance with all relevant regulations and guidelines; (iv) To maintain registries of hES cell research conducted at the institution and hES cell lines derived or imported by institutional investigators; and (v) To facilitate education of investigators involved in stem cell research.

Stem Cell Research and Therapy Guidelines

The stem cell research and therapy (SCRT) guidelines provide a mechanism to ensure that research with hES cells is conducted in a responsible and ethically sensitive manner and complies with all regulatory requirements pertaining to biomedical research in general and stem cell research in particular. Further, any research on human beings, including human embryos, as subjects of medical or scientific research, shall adhere to the general principles outlined in the 'Ethical Guidelines for Biomedical Research on Human Participants' issued by the ICMR in 2006 (www.icmr.nic.in/bioethics). The basic aims of these guidelines are: (i) Essentiality of research with potential health benefits; (ii) Respect for human dignity, human rights, and fundamental freedom; (iii) Individual autonomy with respect to informed consent, privacy, and confidentiality in harmony with the individual's cultural sensitivity and environment; (iv) Justice with equitable distribution of burden and benefits; (v) Beneficence with regard to improvement of health of individuals and society; (vi) Nonmaleficence with the aim of minimization of risk and maximization of benefit; and (vii) Freedom of conducting research with due respect to the above within the regulatory framework.

Categorization of Research Activities

According to the source of stem cells and nature of experiments, the research on ES cells is categorized into following three areas, *viz.*, permissible, restricted, and prohibited.

Permissible Areas of Research These include the following: (i) *In vitro* studies on established cell lines from any type of stem cell, *viz.*, hES, human embryonic germ cells (hEG), human somatic stem cells (hSS), or fetal/adult stem cells with notification to IC-SCRT, provided the cell line is registered with the IC-SCRT/NAC-SCRT and good laboratory practice (GLP) is followed; (ii) *In vivo* studies with established cell lines from any type of stem cells, *viz.*, hES, hEG, hSS, including differentiated derivatives of these cells on animals with prior approval of IC-SCRT, provided such animals are not allowed to breed. This includes preclinical evaluation of efficacy and safety of hES cell lines; (iii) *In vivo* studies on experimental animals (other than primates) using fetal/adult somatic stem cells from bone marrow, peripheral blood, umbilical cord blood, skin, limbal cells, dental cells, bone cells, cartilage cells, or any other organ (including placenta), with prior approval of the IC-SCRT/IEC provided appropriate consent is obtained from the donor; (iv) Establishment of new hES cell lines from spare, supernumerary embryos with prior approval of the IC-SCRT/IEC provided appropriate consent is obtained from the donor. Once the cell line is established, it should be registered with the IC-SCRT and NAC-SCRT; (v) Establishment of umbilical cord stem cell bank with prior approval of the IC-SCRT/IEC and DCGI following guidelines for collection, processing, and storage etc. Appropriate standard operating procedures (SOPs) have to be approved by the IC-SCRT/IEC; (vi) Cells for clinical trials must be processed as per the national good tissue practice (GTP)/good manufacturing practice (GMP) guidelines and may be carried out with prior approval of IC-SCRT/IEC or DCGI where applicable. All clinical trials on stem cells shall be registered with NAC-SCRT through IC-SCRT/IEC; and (vii) All proposals involving fetuses or fetal tissue for research or therapy are permissible. However, termination of pregnancy should not be sought with a view to donate fetal tissue in return for possible financial or therapeutic benefits. Informed consent to have a termination of pregnancy and the donation of fetal material for purpose of research or therapy should be taken separately. The medical person responsible for the care of pregnant woman planning to undergo termination of pregnancy and the person who will be using the fetal material should not be the same. The woman shall not have the option to specify the use of the donated material for a particular person or in a particular manner. The identity of the donor and the recipient should be kept confidential.

Restricted Areas of Research Under this category, experiments included are: (i) Creation of a human zygote by *in vitro* fertilization (IVF), somatic cell nuclear transfer (SCNT), or any other method with the specific aim of deriving a hES cell line for any purpose. Specific justification would be required to consider the request for approval by NAC-SCRT through IC-SCRT/IEC. It would be required to establish that creation of zygote is critical and essential for the proposed research. Informed consent procedure for donation of ova, sperm, somatic cell, or other cell types has to be followed; (ii) Clinical trials sponsored by multinationals involving stem cell products imported from abroad shall require prior approval of the NAC-SCRT through IC-SCRT/IEC, DCGI, and respective funding agency. The process of this consideration must be completed within three months of the receipt of the proposal; (iii) Research involving introduction of hES/hEG/hSS cells into animals including primates at embryonic or fetal stage of development for studies on pattern of differentiation and integration of human cells into nonhuman animal tissues. If there is a possibility that human cells could contribute in a major way to the development of brain or gonads of the recipient animal, the scientific justification for the experiments must be strong. The animals derived from these experiments shall not be allowed to breed. Such proposals would need approval of the NAC-SCRT through Institutional Animal Ethics Committee (IAEC) and IC-SCRT/IEC; (iv) Studies on chimeras where stem cells from two or more species are mixed and introduced into animals including primates at any stage of development, *viz.*, embryonic, fetal, or postnatal, for studies on pattern of development and differentiation; and (v) Research in which the identity of the donors of blastocysts, gametes, or somatic cells from which the hES cells were derived is readily ascertainable or might become known to the investigator.

Prohibited Areas of Research This category includes: (i) Any research related to human germ line genetic engineering or reproductive cloning; (ii) Any *in vitro* culture of intact human embryo, regardless of the method of its derivation, beyond 14 days or formation of primitive streak, whichever is earlier; (iii) Transfer of human blastocysts generated by SCNT or parthenogenetic or androgenetic techniques into a human or nonhuman uterus; (iv) Any research involving implantation of human embryo into uterus after *in vitro* manipulation at any stage of development in humans or primates; (v) Animals in which any of the human stem cells have been introduced at any stage of development should not be allowed to breed; and (vi) Research involving directed nonautologous donation of any stem cells to a particular individual is also prohibited.

Thus to summarize, proposals involving following research activities shall be considered for approval: (i) Increase knowledge about embryo development and causes of miscarriages and birth defects; (ii) Developing methods to detect abnormalities in embryos before implantation; (iii) Advance knowledge for infertility treatment or improving contraception techniques; (iv) Increase knowledge about serious diseases and using this knowledge to develop treatments including tissue therapies; (v) Developing methods of therapy for diseased or damaged tissues or organs; and (vi) Developing ethnically diverse hES cell lines.

Review Questions

1. Write in brief about the following:
   (i) Physical containment and biological containment
   (ii) Disabled or especially disabled vectors
   (iii) Biosafety levels
   (iv) Risk assessment factors for environmental release of GMOs
   (v) Social and ethical concerns related to human cloning
   (vi) Assessment of potential risks in handling of pathogenic organisms
2. Describe various categories of rDNA research activities. Give five examples of self-cloning strains exempted from notification or approval.
3. Discuss the roles of RCGM and GEAC in genetic engineering research activities.
4. Give an elaborative account of the national regulatory mechanism for implementation of rDNA research in India, giving details of the tasks at hand of each regulatory body.
5. Give a brief account of possible advantages and concerns related to human cloning. Discuss the review and monitoring mechanism for stem cell research and therapy in India.

Reference

Losey, J.E., L.S., Rayor, M.E. Carter (1999). Transgenic paller harms monarch larvae. *Nature* **399**:214.

Glossary

λ Terminase An enzyme that packages concatemers of phage DNA into unit length genomes with protruding 5′-termini, 12 bp in length; the site of action of terminase is called the cohesive site or *cos* site.

β-Carotene Precursor provitamin A from which vitamin A is synthesized.

α-Complementation Assembly of functional β-galactosidase from N-terminal α fragment plus rest of the protein.

β-form of DNA A right-handed double helix with 10.5 base pairs per complete turn (360°) of the helix; this is the form found under physiological conditions, the structure of which was proposed by Watson and Crick.

β-Galactosidase An enzyme encoded by the *lac* Z gene of *E. coli* that catalyzes the hydrolysis of lactose to glucose and galactose.

β-Lactam antibiotics Family of antibiotics that inhibit cross-linking of the peptidoglycan of the bacterial cell wall; includes penicillins and cephalosporins.

β-Lactamase Enzyme that inactivates β-lactam antibiotics such as ampicillin by cleaving the lactam ring.

–10 sequence The consensus sequence centered about 10 bp before the transcription start site of a bacterial gene; it is involved in melting DNA during the initiation reaction.

1-Aminocyclopropane-1-carboxylate (ACC) An immediate precursor of ethylene in plants.

2-μm plasmid (2-μ circle or 2 micron plasmid) A naturally occurring, circular ds DNA plasmid found in the nuclei of *Saccharomyces cerevisiae*, which is used as the basis for a series of cloning vectors.

3′-poly A tail A stretch of A residues present at the 3′-end of the mature eukaryotic nuclear mRNA.

3′-end (3′-OH end) The end of polynucleotide that terminates with a hydroxyl group attached to the 3′-carbon of sugar.

3′-extension (3′ protruding end or 3′ sticky end or 3′ overhang) A short single stranded nucleotide sequence on the 3′-OH end of a ds DNA molecule.

3′-UTR The untranslated region of an mRNA downstream of the termination codon.

–35 sequence The consensus sequence centered about 35 bp upstream of the start site of a bacterial gene, which is involved in initial recognition by RNA polymerase.

4-Methylumbelliferyl phosphate An artificial substrate that is cleaved by alkaline phosphatase, releasing a fluorescent molecule.

5′-Cap The chemical modification at the 5′-end of most eukaryotic mRNA molecules. The structure at the 5′-end of eukaryotic mRNA, introduced after transcription by linking the terminal phosphate of 5′ GTP to the terminal base of the mRNA. The added G (and sometimes some other bases) are methylated.

5′-end (5′-P terminus) The end of a polynucleotide that terminates with a mono-, di- or triphosphate attached to the 5′ carbon of the sugar.

5′-extension (5′ protruding end or 5′ sticky end or 5′ overhang) A short single stranded nucleotide sequence on the 5′-phosphate end of a ds DNA molecule.

5′-UTR The untranslated region of an mRNA upstream of the initiation codon.

Absorbance (optical density) The absorption of part of the visible spectrum by an object or solution.

Abundant mRNA A small number of individual mRNA species, each present in hundreds of copies per cytoplasm.

Acridine Mutagen that intercalates in the ds DNA and causes the insertion or deletion of a single base pair; this dye was used in defining the triplet nature of the genetic code.

Activation Enhancing the rate of transcription.

Active site The part of an enzyme involved in binding its substrate or the site of catalytic action.

Adaptor A synthetic double stranded oligonucleotide that is blunt at one end and at the other has a nucleotide extension that can base pair with a cohesive end created by cleavage of a DNA molecule with a specific type II restriction enzyme.

Addiction system A survival mechanism used by plasmids. The mechanism kills the bacterium upon loss of the plasmid.

Adenine (A) One of the nitrogenous bases found in either DNA or RNA.

Adenosine triphosphate (ATP) The principal compound used for the storage and transfer of energy in cellular systems.

Adenovirus Medium-sized, nonenveloped icosahedral virus composed of a nucleocapsid and a double stranded linear DNA genome, used as cloning vector.

Adhesins (colonization factor) Proteins that enable bacteria to attach themselves to the cell surface.

Aerobe An organism that grows in the presence of molecular oxygen, which it uses as a terminal electron acceptor in aerobic respiration.

Affinity capture A method based on affinity, which allows easy identification and isolation of macromolecules from cell extract; involves either polyclonal antibody or epitope tag.

Affinity tag (peptide tag, protein tag) A short sequence of amino acid that is engineered as part of a recombinant protein and binds to a specific element, compound, or macromolecule which facilitates the identification or purification of the recombinant protein.

Agar An extract of red algae (family Rhodophyceae) used as a solidifying agent for microbiological medium.

Agarase An enzyme that hydrolyze (or liquefies) agarose polymer to oligosaccharide and disaccharide subunits.

Agarose A highly purified linear polysaccharide isolated from seaweeds (red algae) such as *Gelidium* and *Gracilaria* that is ~600–800 residues long polymer comprising of repeating units of *D*-galactose and 3,6-anhydro-*L*-galactose joined by β (1 → 4) linkage, and each repeating unit further linked by α (1 → 3) glycosidic linkage.

Agarose gel electrophoresis Electrophoresis carried out in an agarose gel, which is used to separate DNA molecules.

Agrobacterium tumefaciens A rod-shaped spore-forming Gram negative pathogenic soil bacterium, which has the natural capability of genetically transforming a plant cell.

Aliquot To dispense a specific amount of liquid.

Alkaline phosphatase An enzyme that removes phosphate groups from the 5′-ends of DNA molecules; it can be conjugated to biotin to function as part of the detection system for biotinylated probes, as the reaction of enzyme with the substrate BCIP (5-bromo-4-chloro-3-indolyl phosphate) generates a colored product.

Allele One of several alternative forms of a gene occupying a given locus on a chromosome.

***Alu*-PCR** A clone fingerprinting technique that uses PCR to detect the relative positions of *Alu* sequences in cloned DNA fragments.

Amber codon The triplet UAG, which acts as a termination codon to stop protein synthesis.

Amino acids Organic acids bearing both amino and carboxyl groups that act as the building blocks (monomeric unit) of proteins.

Aminoglycosides Family of antibiotics that inhibit protein synthesis by binding to the small subunit of the ribosome; includes streptomycin, kanamycin, neomycin, tobramycin, gentamycin, etc.

Amino terminus (N-terminus) The end of a polypeptide having a free amino group.

Ampicillin A semisynthetic penicillin that inhibits cell wall synthesis in bacteria, which is commonly used in molecular biology for selection of ampicillin-resistant microorganisms.

Amplified fragment length polymorphism (AFLP or amplified fragment length polymorphism PCR or AFLP-PCR) A PCR-based tool used in genetics research, DNA fingerprinting, and in the practice of genetic engineering, which uses restriction enzymes to cut genomic DNA, followed by ligation of adaptors to the sticky ends of the restriction fragments, a subset of which is then amplified using primers complementary to the adaptor and part of the restriction site fragments; the amplified fragments are then visualized on denaturing polyacrylamide gels either through autoradiography or fluorescence methodologies.

Anaerobe An organism that grows in the absence of molecular oxygen.

Anchor sequence Sequence added to primers that may be used for binding to a support or may incorporate convenient restriction sites, primer binding sites for future manipulations, or primer binding sites for subsequent PCR reactions.

Annealing The process of heating (denaturing step) and slowly cooling (renaturing step) ds DNA to allow the formation of hybrid DNA or DNA:RNA molecules.

Attachment of an oligonucleotide primer to a DNA or RNA template.

Anode The positive electrode, usually colored red in a gel electrophoresis apparatus.

Antibiotic A microbially produced substance (or a synthetic derivative) that has antimicrobial properties.

Antibody A protein (immunoglobulin) produced by B lymphocyte cells, which recognizes a particular 'foreign antigen', and thus triggers the immune response. It is a protein of high binding specificity produced by the immune systems of higher animals in response to infection by a foreign organism.

Antifreeze protein A type of protein that binds to ice crystals and derepresses the freezing temperature of the crystal below its melting temperature. Antifreeze proteins have been found in fish, insects, plants, fungi, and bacteria and protect cells from being damaged by ice crystals.

Antigen Any foreign substance whose entry into an organism provokes an immune response by stimulating the synthesis of an antibody.

Antigenic determinant (epitope) The portion of an antigen that is recognized by the antigen receptor on lymphocytes.

Antiparallel orientation The arrangement of the two strands of a duplex DNA molecule, which are oriented in opposite directions so that the 5′-phosphate end of one strand is aligned with the 3′-OH end of the complementary strand.

Antiporter A type of carrier protein that simultaneously moves two different types of solutes in opposite directions across the plasma membrane.

Antisense gene Gene that codes for an antisense RNA having a sequence complementary to the target RNA.

Antisense RNA An RNA sequence that is complementary to all or part of a functional RNA.

Antisense strand (template strand) Strand that is complementary to the sense strand, and is the one that acts as the template for synthesis of mRNA.

Antisense therapy The *in vivo* treatment of a genetic disease by blocking translation of a protein with a DNA or an RNA sequence that is complementary to a specific mRNA.

Antitermination A mechanism of transcriptional control in which termination is prevented at a specific terminator site, allowing RNA polymerase to read into the genes beyond it. Antitermination proteins allow RNA polymerase to transcribe through certain terminator sites.

Aphid A plant-sucking insect of the family aphididae.

Apoptosis (programmed cell death) The capacity of a cell to respond to a stimulus by initiating a pathway that leads to its death by a characteristic set of reactions.

Arabidopsis thaliana A plant with a very small genome that is used as a model organism for the study of plant growth and development.

Archaea A group of prokaryotes that diverged from all others at an early stage in evolution, and that show a number of significant differences from them. One of the three domains of life.

Aseptic technique A set of practical measures designed to prevent the growth of unwanted contaminants from the environment.

Aspirate To remove a liquid layer such as a supernatant from a sample, using a pipette or equivalent attached to a vacuum source with a trap to catch effluent.

Asymmetrically tailed plasmid as primer A plasmid primer developed to prepare highly complex cDNA libraries that allow directional and full-length cDNA cloning.

***att* sites** Loci on a phage and the bacterial chromosome at which site-specific recombination integrates the phage into, or excises it from, the bacterial chromosome.

Attenuated vaccine A virulent organism that has been modified to produce a less virulent form but retains the ability to elicit antibodies against the virulent form.

Autoclave An appliance that uses steam under pressure to achieve sterilization.

Autonomous replication (self-replication) Self-replication within the host cell due to the presence of own origin of replication.

Autonomously replication sequence (ARS) An origin for replication in eukaryotes that initiates and supports extrachromosomal replication of a DNA molecule in a host cell; often used in yeast cells.

Autoradiography A simple and sensitive photochemical technique that captures the image formed in a photographic emulsion as a result of the emission of either light or radioactivity from a labeled component that is placed next to unexposed film. The technique is used to record the spatial distribution of radiolabeled compounds within a specimen or an object.

Auxotroph A mutant that lacks the ability to synthesize an important nutrient such as an amino acid or vitamin, and must therefore have it supplied with a nutrient that is not needed by the wild-type.

Avidin A glycoprotein component of egg-white that binds strongly to biotin.

***Bacillus* (pl. bacilli)** A rod-shaped, sporulating, nonpathogenic, Gram positive, obligate aerobic bacterium having low GC content; used to study sporulation and prokaryotic differentiation.

Bacmid A shuttle vector based on the *Autographa californica* multiple nuclear polyhedrosis virus (AcMNPV) genome, which can be propagated in both *E. coli* and insect cells.

Bacteria Primitive, relatively simple, single-celled organisms that lack a nucleus; one of the two main groups of prokaryotes.

Bacterial artificial chromosome (BAC) A single copy vector system based on the *E. coli* F factor that contains the sequences needed for replication and segregation in bacteria, and is used for cloning large DNA inserts; widely used in the human genome project.

Bacteriocin A toxic protein produced by one bacterium that kills closely related bacterium.

Bacteriocinogenic plasmid Plasmids harbouring genes responsible for colicin/toxin production.

Bacteriophage λ A temperate, head-and-tail coliphage consisting of linear ds DNA as genetic material packaged into its head, and follows both lytic and lysogenic alternatives to its life cycle; widely used as cloning vector.

Bacteriophage (phage) A virus that infects a bacterium.

Bacteriophage M13 A filamentous, male-specific coliphage composed of small circular ss DNA encapsulated in approximately 2,700 copies of the major coat protein pVIII, capped with ≤5 copies of two different minor coat proteins (pVII and pIX) on distal end and tailed with 5 copies of two different minor coat proteins (pIII and pVI) on proximal end.

Bacteriophage P1 Generalized trasducing phage of *E. coli*.

Baculovirus A virus that infects insects.

Bait The fusion between the DNA binding domain of a transcriptional activator protein and another protein as used in two-hybrid screening.

Basal promoter The position within a eukaryotic promoter where the initiation complex is assembled.

Basal rate of transcription The number of productive initiations of transcription occurring per unit time at a particular promoter.

Base pair A pair of two complementary bases (A with T or G with C) held together by hydrogen bonds. A thousand base pairs is often called a kilo base pair (kbp).

Basic Local Alignment Search Tool (BLAST) A very fast search algorithm, which is used by the blastn, blastp, and blastx programs to separately search protein or DNA databases; it is best used for sequence similarity searching, rather than for motif searching.

Bed volume The volume of the beads plus the void volume, i.e., the volume occupied by the packed matrix in the column.

Beer Lambert's law It describes that the amount of light absorbed by a substance at a given wavelength (A) is given by the formula $A = \log I_0/I = \varepsilon cl$, where A is absorbance, I_0 and I are, respectively, the intensities of the incident and transmitted light, c is the molar concentration of the sample, l is the length of the light path through the sample in cm, and ε is the molar extinction coefficient of the molecule.

Bidirectional replication A system in which an origin generates two replication forks that proceed away from the origin in opposite directions.

Binary vector system A two-plasmid system in *Agrobacterium* sp. used for transferring a T-DNA region that carries cloned genes into plant cells; the virulence genes are on one plasmid, and the engineered T-DNA region is on the other plasmid.

Biolistics (microprojectile bombardment or gene gun or bioblaster) A means of introducing DNA into plant and animal cells and organelles by means of DNA-coated gold or tungsten microparticles that are fired under pressure at high speed.

Biological containment Prevention of the possibility of gene escape by preventing self-catalyzed transfer of recombinant DNA molecule from one host to another.

Bioluminescence The production of light by biological organisms such as insects and bacteria, usually catalyzed by the enzyme luciferase.

Biopanning Method of screening phage display library for a desired displayed protein by binding to bait molecules attached to solid support.

Bioreactor A vessel in which cells, cell extracts, or enzymes carry out a biological reaction.

Bioremediation A process that uses living organisms to remove contaminants, pollutants, or unwanted substances from soil or water.

Biotin Vitamin that is widely used to label or tag nucleic acids; binds very tightly with avidin or streptavidin.

Biotinylation of nucleic acid (biotin labeling) A nonradioactive method of labeling nucleic acids using nick translation to incorporate biotin-derivatized nucleotides (dUTP); an alternative to radioactive labeling.

***bla* gene (*amp*ʳ gene)** Gene encoding β-lactamase, which provides resistance to ampicillin.

Blocking agents in hybridization Agents added in high concentrations during nucleic acid hybridization experiments compete with the probe for nonspecific binding sites on the solid support, and hence prevent background hybridization.

Blocking agents in chemical synthesis of oligonucleotides The protecting groups that transiently block the reactive 5′- and 3′-hydroxyl groups of sugar and exocyclic amino functions of bases.

Blotting Transfer of a macromolecule from a gel to a membrane.

Blue-white screening Screening procedure based on the insertional inactivation of the gene for β-galactosidase.

Blunt end cutters Restriction endonucleases that cut straight across the axis of symmetry of palindromic restriction site, producing blunt ended DNA.

Blunt ends (flush end or flat end) The ends of a ds DNA molecule that are fully base paired and have no unpaired single stranded extensions.

Box A short DNA sequence that plays a role in regulating, facilitating, enhancing, or silencing transcription.

Broad host range plasmid A plasmid that can replicate in a number of different bacterial species.

Bromphenol blue A chemical commonly used as a tracking dye for nucleic acids in electrophoresis gels.

Buffer A conjugate acid–base pair that functions as a system for resisting changes in pH; especially important for maintaining pH during studies of biological systems *in vitro*.

Buoyant density The density possessed by a molecule or particle when suspended in an aqueous salt or sugar solution.

CaMV 35S promoter A constitutive promoter derived from cauliflower mosaic virus; commonly used for expression of cloned genes in plants.

Capping in chemical synthesis of oligonucleotides A chemical reaction used to prevent unlinked residues from linking to the next monomer during the next cycle of chemical synthesis of oligonucleotides.

Capsid The external protein coat of a virus particle.

Carcinogen A chemical that increases the frequency with which cells are converted to a cancerous condition.

Cassette A combination of DNA elements that performs a specific function and is maintained as a clonable unit.

Cathode The negative electrode, usually colored black in a gel electrophoresis apparatus.

Cauliflower mosaic virus A small plant virus with circular ds DNA.

cDNA capture (cDNA selection) Repeated hybridization probing of a pool of cDNAs with the objective of obtaining a subpool enriched in certain sequences.

cDNA clone A ds DNA molecule that is carried in a vector and is synthesized *in vitro* from an mRNA sequence by using reverse transcriptase and DNA polymerase.

cDNA library A collection of clones containing cDNA and represents the genes (lacking introns) that are expressed within a given cell or tissue type at a particular period. It represents all the mRNA present in a particular tissue, which has been reverse transcribed to a DNA template. It also represents all the genes that are transcribed in particular tissues under particular physiological, developmental, or environmental conditions.

Cell line A cell lineage that can be maintained in culture.

Cell-free translation (*in vitro* translation or cell-free protein synthesis) Protein synthesis directed by either purified DNA with bacterial extracts or mRNA with wheat germ or rabbit reticulocyte extracts that provide ribosomes, tRNAs, and protein synthesis factors; the reaction mixture is often supplemented with ATP, GTP, and amino acids.

Centromere A central region of the chromosome that ensures correct distribution of chromosomes between daughter cells during cell division.

Cephalosporins Group of antibiotics of the β-lactam type that inhibit cross-linking of the peptidoglycan of the bacterial cell wall.

Cesium chloride (CsCl) A heavy inorganic salt used for creating density gradients in a high-speed ultracentrifuge to purify nucleic acids.

Cetyl trimethyl ammonium bromide (CTAB) Detergent that forms an insoluble complex with nucleic acids and aids in removal of polysaccharide contamination.

Chain termination sequencing Method of sequencing DNA that involves enzymatic synthesis of polynucleotide chains terminating at specific nucleotide positions by using dideoxynucleotides.

Chaotropic agent Chemical compound that disrupts water structure and so helps hydrophobic groups to dissolve.

Chemical degradation method of DNA sequencing A DNA sequencing method involving the use of chemicals, which cut DNA molecules at specific nucleotide positions.

Chemiluminescence The emission of light from a chemical reaction.

Chimera A creature in mythology with the head of a lion, body of a goat, and tail of a serpent.

A recombinant DNA molecule that contains sequences from different organisms.

Usually a plant or animal that has populations of cells with different genotypes.

Chimeric plasmid (hybrid plasmid) A plasmid that includes DNA from more than one source.

Chloramphenicol A broad-spectrum antibiotic from *Streptomyces venezuelae* that binds to 23S rRNA and interferes with peptide bond formation in prokaryotes. Used as a selection agent and for amplifying relaxed plasmids.

Chloramphenicol acetyl transferase (CAT) Enzyme that inactivates chloramphenicol by adding acetyl groups.

Chloroplast DNA (cpDNA) An independent genome (usually circular) found in a plant chloroplast.

Chromogenic substrate A compound or substance that contains color-forming group.

Chromosomal integration site A chromosomal location where foreign DNA can be integrated, often without impairing any essential function in the host organism.

Chromosome A discrete unit of the genome carrying many genes, and made up of a single molecule of DNA.

Chromosome walking A technique used to construct a clone contig by identifying overlapping fragments of cloned DNA.

Clone A genetically identical population of cells that has descended from one cell.

A population of genetically identical cells or organisms that result from asexual reproduction, breeding of purebred (isogenic) organisms, or forming genetically identical organisms by nuclear transplantation.

To insert a DNA segment into a vector or host chromosome.

Clone contig A collection of clones whose DNA fragments overlap.

Clone contig approach A genome sequencing strategy in which the molecules to be sequenced are broken into manageable segments, each a few hundred kbp or few Mbp in length, which are sequenced individually.

Clone fingerprinting Techniques used to compare cloned DNA fragments in order to identify ones that overlap.

Clone library A collection of clones, possibly representing an entire genome, from which individual clones of interest are obtained.

Cloning The production of multiple copies of a specific DNA molecule.

The production of genetically identical cells or even organisms.

Incorporating a DNA molecule into a chromosomal site or a cloning vector.

Cloning vector (cloning vehicle or vector) A DNA molecule that acts as a vehicle for carrying a foreign DNA fragment when inserted into it and transports it into a host cell; acts as an efficient agent for transfer, maintenance, and amplification of target DNA.

***cm*^{r} gene** Gene that encodes an enzyme chloramphenicol acetyl transferase, which inactivates chloramphenicol by adding acetyl groups.

Coding region A part of the gene that represents a protein sequence.

Coding strand (sense strand) The strand of a DNA equivalent in sequence to the mRNA.

Codominant alleles Two alleles are said to be codominant when both are equally evident in the phenotype of the heterozygote.

Codon (coding triplet) A triplet of nucleotides that represents an amino acid or a termination signal.

Cofactor A small inorganic component (often a metal ion) that is required for the proper structure or function of an enzyme.

Cointegrate vector system A single large vector formed in *Agrobacterium tumefaciens* by the homologous recombination of the cloning vector and a resident disarmed pTi; used for transferring the genetically engineered T-DNA region to plant cells.

Col E plasmid Small multicopy plasmid that carries genes for colicins of the E group; used as the basis of many widely used cloning vectors.

Colicin Toxic protein or bacteriocin made by *E. coli* to kill closely related bacteria.

Colony A visible growth of microorganisms or cells on a solid microbiological medium.

Colony hybridization A technique that uses a nucleic acid probe to identify a bacterial colony containing cloned gene(s).

Competent A particular condition of cells such as *E. coli* following chemical treatment to make the cell envelope permeable to exogenous DNA.

Complementarity One of a pair of segments or strands of nucleic acid that hybridizes (join by H-bonding) with each other; adenine pairs with thymine (or with uracil in RNA), and guanine pairs with cytosine.

Complementary base pairs The pairing reactions in double helical nucleic acids (A with T in DNA or with U in RNA, and C with G).

Complementary DNA (cDNA) A ds DNA complement of an RNA sequence; synthesized from it by reverse transcription.

Complementary homopolymeric tailing (dG-dC tailing, dA-dT tailing) The process of adding complementary nucleotide extensions to different DNA molecules e.g., dG (deoxyguanosine) to the 3′-hydroxyl ends of one DNA molecule and dC (deoxycytidine) to the 3′-OH ends of another DNA molecule to facilitate, after mixing, the joining of the two DNA molecules by base pairing between the complementary extensions.

Complex medium Nutritional medium for microorganisms that includes ingredients such as yeast extract and Bacto Peptone, for which the exact chemical composition is not known.

Complexity It is the total length of different sequences of DNA present in a given preparation.

Concatemer (catenanes) A tandem array of repeating unit length DNA elements.

Conformation The shape of a molecule or any other object.

Conjugation (mating) A process of genetic transfer involving intimate contact between cells and direct transfer of DNA across a sex pilus.

Conjugative functions (*tra* genes) Plasmid-based genes and their products that facilitate the transfer of a plasmid from one bacterium to another.

Conjugative plasmid A plasmid having conjugative functions.

Consensus sequence An idealized sequence in which each position represents the base most often found when many actual sequences are compared.

Conserved positions Positions defined when many examples of a particular nucleic acid or protein are compared and the same individual bases or amino acids are always found at particular locations.

Constitutive promoter An unregulated promoter segment of DNA that allows continuous transcription of its cognate gene.

Constitutive synthesis Continual production of RNA or protein by an organism.

Contig A continuous stretch of genomic DNA generated by assembling cloned fragments by means of their overlaps.

Contour clamped homogeneous electric fields (CHEF) A variant of PFGE that utilizes 24 point electrodes equally spaced around the hexagonal closed contour in combination with a horizontal gel.

Controlled pore glass (CPG) A porous borosilicate material that serves as a solid-phase of choice for oligonucleotide synthesis by phosphoramidite method.

Coomassie blue A blue dye used to stain protein.

Cooperativity in protein binding An effect in which binding of the first protein enhances binding of a second protein (or another copy of the same protein).

Copy number The number of copies of a plasmid that is maintained in a bacterium (relative to the number of copies of the origin of the bacterial chromosome).

Core sequence The segment of DNA that is common to the attachment sites on both the phage λ and bacterial genomes. It is the location of the recombination event that allows λ phage to integrate.

***cos* sites (cohesive sites)** The 12 base, single stranded, complementary extensions of bacteriophage λ DNA.

Cosmid A hybrid cloning vector that contains the *cos* end sequences of bacteriophage λ; widely used for genomic libraries construction.

Cosuppression The ability of a transgene (usually in plants) to inhibit expression of the corresponding endogenous gene.

Cotransfection The introduction of two different DNA molecules into a eukaryotic cell. In baculovirus expression systems, the procedure by which the baculovirus and the transfer vector are simultaneously introduced into insect cells in culture.

Covalent bond A bond formed by the sharing of a pair of electrons between atoms.

CP-mediated resistance Transgenic expression of viral coat protein used for generation of virus resistance plants.

Cre-*lox* system A recombination system that involves the targeting of a specific sequence of DNA and splicing it with the help of an enzyme called Cre recombinase. By using this technology, specific tissue types or cells of organisms can be genetically modified.

Crossed field gel electrophoresis (CFGE) A variant of PFGE, in which the gel is simply placed on a mobile platform that can be rotated in order to change the orientation of the electric field relative to the DNA. At each change in field direction, a DNA molecule takes off in the new direction of the field by a movement, which is led by what was formerly its back end. The effect of this ratcheting motion is to subtract from the DNA molecule's forward movement, at each step, an amount, which is proportional to its length.

Cross-over The reciprocal exchange of DNA between two chromosomes or DNA molecules by a breakage-and-reunion process. Also called recombination or recombination event.

Cross-over hotspot instigator (*chi* site or χ site) An octameric sequence of dsDNA with a sequence 5′-GCTGGTGG-3′, which serves as a recombinational hotspot for recombination promoted by Rec BCD.

Crown gall disease (crown gall tumor) A tumor that can be induced in many plants by infection with the bacterium *Agrobacterium tumefaciens*.

Cryptic plasmids A plasmid that confers no identified characteristics or phenotypic properties.

C-terminus The end of a polypeptide that has a free carboxyl group.

Culture A population of cells or microorganisms that are grown under controlled conditions.

Culture medium A solid or liquid mixture that is used to grow microorganisms, organisms, or cells.

Cuvette A small plastic or quartz tube of specific dimensions and light-absorbing qualities designed to hold a sample for spectrophotometry.

Cytokinin A plant hormone that stimulates cell division.

Cytosine (C) One of the organic bases found in either DNA or RNA.

Daughter cells The two cells that result from a cell division are referred to as daughter cells: In budding yeast only the cell derived from the bud is called the daughter cell.

Daughter strand The newly synthesized DNA.

***de novo* methylases or *de novo* methyltransferases** Methylases that add methyl groups to unmethylated DNA at specific sites, and initiate a pattern of methylation.

Defined medium (synthetic medium) A nutrient medium for microorganisms for which the exact chemical composition is known.

Degenerate primer Primer with several alternative bases at certain positions.

Degradative plasmid A type of plasmid that specifies a set of genes involved in biodegradation of an organic compound.

Delayed early genes in λ phage In λ phage, these genes are equivalent to the middle genes of other phages; these cannot be transcribed until regulator protein(s) coded by the immediate early genes have been synthesized.

Deletion Removal of a sequence of DNA, followed by joining together of the regions on either side.

Denaturation Disruption of the conformation of a macromolecule without breaking covalent bonds.

Denaturation of nucleic acids Strand separation of double stranded nucleic acid caused by heating or treating with alkali to enable hybridization of the resulting single stranded species (such as a labeled probe) to another single stranded nucleic acid target.

Denaturation of protein Conversion from the physiological conformation to some other (inactive) conformation.

Denaturing agarose gel electrophoresis Electrophoresis on denaturing (formaldehyde or glyoxal containing) agarose gel.

Denaturing agents Chemicals that denature macromolecules.

Denhardt's reagent A solution containing 1% Ficoll, 1% BSA, and 1% PVP. Used as blocking agent in Southern hybridization.

Density gradient centrifugation A centrifugation technique used to separate macromolecules on the basis of differences in their density; prepared from a heavy soluble compound such as CsCl.

Density gradient gel electrophoresis (DGGE) An electrophoresis based on principles of gel electrophoresis with DNA denaturation; allows separation of DNA molecules differing in sequence by only a single base.

Deoxynucleoside triphosphate (dNTP or deoxyribonucleotide) Precursor molecules used in the enzymatic synthesis of DNA.

Deoxyribonuclease (DNase) An enzyme that specifically digests DNA; it may cut only one or both strands.

Deoxyribonuclease I (DNase I) An enzyme that degrades DNA. It is used to remove DNA from RNA preparations and cell-free extracts.

Deoxyribonucleic acid (DNA) A high molecular weight biopolymer comprising of mononucleotides as their repeating units joined by 3′ → 5′ phosphodiester bonds that serves as the genetic information carrier.

Deoxyribonucleoside 3′-phosphoramidite The P (III) compounds with *P*-alkylamine bond, which serves as a substrate for chain elongation during chemical synthesis of oligonucleotides; it contains 5′-DMTr group, a diisopropylamine group attached to a 3′-phosphite group protected by a β-CE (or Me) residue, and a benzoyl/isobutyryl group blocking the exocyclic amino group of base, and hence is nonreactive unless activated.

Deoxyribose sugar The 5-carbon sugar component of DNA.

Deproteination The process of removing proteins clinging to the surface of the DNA molecule and those found in the core of the DNA molecule.

Depurination/depyrimidination Cleavage of the glycosidic bond linking base to sugar under acidic conditions.

Derivatization of slide surface Coating the slide surface with a chemical group from which the growth of the oligonucleotide chain can be initiated. These groups are also called cross-linkers.

Detritylation Removal of DMTr group at 5′-end of growing oligonucleotide chain by treatment with weak acid to yield a reactive 5′-hydroxyl group.

Dialysis A technique used to remove impurities such as salts from a solution of macromolecules by allowing diffusion of smaller molecules across a semipermeable membrane into water or an appropriate buffer.

Dideoxynucleotide (ddNTP) A nucleoside triphosphate that lacks hydroxyl groups on both the 2′ and 3′ carbons of the pentose sugar.

Dideoxyribose Derivative of ribose that lacks the oxygen at both the 2′ and 3′-OH groups.

Diethyl pyrocarbonate (DEPC) A highly reactive alkylating agent, which acts as a potent and efficient nonspecific RNase inhibitor.

Differential display PCR Variant of RT-PCR that specifically amplifies mRNA from eukaryotic cells using oligo (dT) primers.

Digoxigenin (DIG) A steroid hapten derived from *Digitalis* (purple foxglove). Widely used for nucleic acid labeling.

Dihydrofolate reductase (DHFR) An enzyme that catalyzes the formation of tetrahydrofolic acid.

Dimethoxytrityl (DMTr) 4,4′-dimethoxytrityl that serves as 5′-protecting group during oligonucleotide synthesis.

Diploid A cell or organism having two homologous copies of each chromosome, usually one from the mother and one from the father.

Direct repeats Identical (or closely related) sequences present in two or more copies in the same orientation in the same molecule of DNA; these are not necessarily adjacent.

Directed mutagenesis (*in vitro* mutagenesis) The process of generating nucleotide changes in cloned genes by any one of several procedures, including site-specific and random mutagenesis.

Disarm Deletion of cytotoxic or crown gall inducing genes from a plasmid or virus.

Disinfection The elimination or inhibition of pathogenic microorganisms in or on an object so that these no longer pose a threat.

Dithiothreitol (DTT) A low molecular weight thiol-containing reducing agent, which is added to buffers in low concentrations to prevent protein sulfhydryl groups from being oxidized. At higher concentrations, it is used to reduce disulfide linkages in proteins.

DNA adenine methylase (Dam) Genome-wide methylases that methylate adenine residues in the sequence 5′-GATC-3′ at the N^6 position to form N^6-methyladenosine.

DNA amplification fingerprinting (DAF) Single, arbitrary primer-based DNA amplification technique.

DNA chip (DNA microarray or biochip or DNA microchip) Miniaturized 2-D array of different DNA molecules (oligonucleotides/cDNAs/PCR products) anchored to a flat surface of nylon wafer or silicon glass in a ~1 cm wide square grid.

DNA construct (recombinant DNA molecule) A cloning vector with a DNA insert.

DNA cytosine methylase (Dcm) Genome-wide methylases that methylate internal cytosine residues in the sequence 5′-CCAGG-3′ or 5′-CCTGG-3′ at the C5 position to form 5-methylcytosine.

DNA delivery system Generic term for any procedure that facilitates the uptake of DNA by a recipient cell.

DNA fingerprint A set of DNA fragments that are characteristic for a particular source of DNA such as an insert of a clone.

DNA fingerprinting Technique used to analyze the differences between individuals of the fragments generated by using restriction enzymes to cleave regions that contain short repeated sequences.

Because these are unique to every individual, the presence of a particular subset in any two individuals can be used to define their common inheritance (e.g. a parent–child relationship).

A comparative diagnostic technique that characterizes the DNA of an organism or a sample.

DNA library A collection of cloned DNA fragments.

DNA ligase An enzyme that makes a bond between an adjacent 3′-OH and 5′-phosphate end where there is a nick in one strand of duplex DNA.

***dna* mutant of bacteria** A temperature-sensitive mutant; it cannot synthesize DNA at 42°C, but can do so at 37°C.

DNA polymerase An enzyme that links an incoming deoxyribonucleotide, which is determined by complementarity to a deoxyribonucleotide in a template DNA strand, with a phosphodiester bond to the 3′-OH group of the last incorporated nucleotide of the growing strand during replication.

DNA polymerase I The *E. coli* DNA polymerase enzyme that contains an excision repair function that is taken advantage of in nick translation reactions to incorporate labeled nucleotides into a DNA probe.

DNA size markers DNA fragments of known sizes that are used in sizing and approximate band quantification of DNA fragments on agarose gels.

DNA synthesizer (gene synthesizing machine or automated solid-phase synthesis) Machine that automates the chemical reactions for DNA synthesis. Used to produce single stranded oligonucleotides (~100 deoxyribonucleotides).

DNA topoisomerase An enzyme that changes the number of times the two strands in a closed DNA molecule cross each other. It does this by cutting the DNA, passing DNA through the break, and resealing the DNA.

DNA vaccine Third generation vaccines. Represent the expression constructs whose products stimulate the immune system to protect the organism against diseases.

DNase footprinting Method for analyzing binding of a protein to DNA on the basis of protection of DNA from DNase I degradation.

DNase I Nonspecific nuclease that cuts DNA between any two nucleotides.

Domain A segment of a protein that has a discrete function or conformation. At the protein level, a domain can be as small as a few amino acid residues and as large as half of the entire protein.

The highest level of taxonomic grouping.

Domain of a chromosome It refers either to a discrete structural entity defined as a region within which supercoiling is independent of other domains or to an extensive region including an expressed gene that has heightened sensitivity to degradation by the enzyme DNase I.

Dominant When a heterozygous and homozygous genotype determines the same phenotype.

Dominant allele An allele that produces the same phenotype whether the genotype is heterozygous or homozygous.

Dominant gene One of a pair of alleles that is sufficient to produce a phenotype in a heterozygote.

Dominant marker selection A gene encoding a product that enables only the cells that carry the gene to grow under certain conditions.

Dot/slot blotting Dot blot (or slot blot) is a technique in molecular biology used to detect biomolecules.

Double restriction digestion Cleavage by restriction enzymes with two different enzymes.

Double strand break (DSB) Cleavage of both the strands of a DNA duplex at the same site, which serves as a site for initiation of genetic recombination or DNA repair.

Doubling time (generation time) The time that it takes for a population of single celled organisms to double its cell number.

Downstream sequences The stretch of nucleotides of DNA that lie in the 3′ direction of the site of initiation of transcription, designated as +1. Downstream nucleotides are marked with plus signs, e.g., +2 and +10. Also refers to the 3′-side of a particular gene or sequence of nucleotides.

Early genes Genes transcribed before the replication of phage DNA, which encode for regulators and other proteins needed for later stages of infection.

Early infection The part of the phage lytic cycle between entry and replication of the phage DNA. During this time, the phage synthesizes the enzymes needed to replicate its DNA.

Edible vaccine Edible vaccines represent production of protective antigens in transgenic plants such that eating raw fruits or vegetables can induce neutralizing antibodies and trigger immune response to pathogen.

EDTA (disodium salt) A chelating agent that sequesters divalent metal cations such as Mg^{2+} and Ca^{2+}, and hence disrupts the overall structure of cell envelope. Removal of Mg^{2+} leads to inactivation of cellular DNases, and hence prevents degradation of DNA.

Electrophoresis A method used to separate charged molecules that migrate in response to the application of an electrical field. In gel electrophoresis, heterogeneously sized molecules in a sample are drawn through an inert matrix such as agarose and separate during migration according to size.

Electroporation Electrical treatment of cells that induces transient pores, through which DNA is taken into the cell.

Electrospray ionization Type of mass spectrometry in which gas-phase ions are generated from ions in solution.

Electrotransfer Transfer of a macromolecule by an electric field from a gel to a membrane.

Elution volume (Ve) The volume of solvent necessary to elute a solute from the time the solute enters the gel bed to the time it begins to emerge at the bottom of the column.

Embryonic stem cells (ES cells) Cells of an early embryo that can give rise to all differentiated cells, including germ line cells.

Encode To specify, after decoding by transcription and translation, the sequence of amino acids in a protein.

End-labeling The addition of a radioactively labeled group to one end (5′ or 3′) of a DNA strand.

Endonucleases Enzymes that cleave bonds within a nucleic acid chain; these may be specific for RNA or for single stranded or double stranded DNA.

Endoprotease An enzyme that cleaves the peptide bonds between amino acids within a protein. Cleavage is usually at one or more specific sites.

Endotoxin A toxin that is present on the surface of Gram negative bacteria.

Enhancer A *cis*-acting sequence that increases the utilization of (some) eukaryotic promoters, and can function in either orientation and in any location (upstream or downstream) relative to the promoter.

Enzyme A cellular catalyst (usually protein), specific to a particular reaction or group of reactions.

Episome A plasmid able to integrate into bacterial DNA.

Epitope (antigenic determinant) The portion of an antigen that is recognized by the antigen receptor on lymphocytes.

Epitope tag An affinity tag that is recognized by an antibody.

Error-prone synthesis Use of the polymerase chain reaction under conditions that promote the insertion of an incorrect nucleotide every few hundred nucleotides of the template; used as a method of random mutagenesis.

Escherichia coli A small, straight, rod-shaped, Gram negative, unicellular, facultatively anaerobic chemoorganotroph capable of both respiratory and fermentative metabolism, mesophilic, nonphotosynthetic, and nonsporulating eubacteria.

Ethidium bromide (EtBr) An orange dye (and strong mutagen) that intercalates between the stacked bases of nucleic acids (double stranded most efficiently) and causes the molecules to fluoresce under ultraviolet light. Fluorescence allows visualization of fragments in gels and cesium chloride density gradients and also provides a way to quantitate nucleic acids whereby the intensity of fluorescence is compared to known standards.

Ethylene A gaseous compound that acts as a plant hormone. It is important in fruit ripening, flower senescence, seed germination, rooting of cuttings, root elongation, and the response of the plant to environmental stress.

Euchromatin It comprises all of the genome in the interphase nucleus except for the heterochromatin. The euchromatin is less tightly coiled than heterochromatin, and contains the active or potentially active genes.

Eukaryotes Organisms, including animals, plants, fungi, and some algae, that have chromosomes enclosed within a membrane-bound nucleus and functional organelles, such as mitochondria and chloroplasts, in the cytoplasm of their cells.

European Molecular Biology Laboratory (EMBL) A nucleotide sequence database maintained by the European Bioinformatics Institute (EBI).

Excision The natural or *in vitro* enzymatic release (removal) of a DNA segment from a chromosome or cloning vector.

Excision of phage or episome or other sequence Release of phage or episome or other sequence from the host chromosome as an autonomous DNA molecule.

Exclusion limit of the gel The molecular mass of the smallest molecule unable to penetrate the pores of a given gel; it gives an indication of the molecular size of a molecule expected to elute at an elution volume equal to the bed volume, and the molecular size of a molecule expected to be totally excluded from the column, and to elute at the void volume.

Exogenous Derived externally or foreign.

Exogenous DNA DNA that has been derived from a source organism and has been cloned into a vector and introduced into a host cell. Also referred to as foreign or heterologous DNA.

Exon A coding region of a gene.

Exon trapping Experimental procedure for isolating exons by using their flanking splice recognition sites.

Exonucleases Enzymes that cleave nucleotides one at a time from the end of a polynucleotide chain. These may be specific for either the 5′ or 3′ end of DNA or RNA.

Expressed sequence tag (EST or expressed sequence tagged site or eSTS) A short sequence of DNA taken from a cDNA copy of an mRNA. The EST is complementary to the mRNA and can be used to identify genes corresponding to the mRNA.

Expression Transcription and translation of a gene.

Expression library A population of different DNA molecules cloned into an expression vector.

Expression vector A vector that allows the transcription and translation of a foreign gene inserted into it.

Extension (protruding end or sticky end or overhang or cohesive end) A single stranded DNA region consisting of one or more nucleotides at the end of a strand of duplex DNA.

Extrachromosomal DNA A replicable DNA element that is not part of a chromosome.

Extrachromosomal genome In a bacterium it represents a self-replicating set of genes that is not part of the bacterial chromosome.

Facultative anaerobe An organism that can grow in the absence of oxygen, but utilizes it when available.

False negative A test result that does not recognize a target when it is present in a sample.

False positive A test result that indicates the presence of a target when it is not present in a sample.

Fermentation A microbial process by which an organic substrate (usually a carbohydrate) is broken down without the involvement of oxygen or an electron transport chain, generating energy by substrate-level phosphorylation.

Fertility plasmid (F plasmid) A plasmid containing genes that code for the construction of the sex pilus, across which it is transferred to a recipient cell. It is an episome that can be free or integrated in *E. coli,* and which in either form can sponsor conjugation.

Field inversion gel electrophoresis (FIGE) A variant of PFGE based on homogeneous electric field in a horizontal gel generated by a single pair of parallel electrodes, and periodic inversion of a uniform electric field in one dimension, thereby switching the polarity of the field through 180° per cycle.

Filling-in Adding nucleotides complementary to protruding tails to get blunt ended ds DNA.

Filter sterilize To remove contaminating microorganisms from liquids by suction through a particle-retaining membrane into a sterile container.

Flp recombinase (flippase) Enzyme encoded by the 2-μm plasmid of yeast that catalyzes recombination between inverted repeats (*frt* sites).

Flp recombination target (*frt* site) Recognition site for Flp recombinase.

Fluorescein A fluorescent dye often used to label antibodies so that these may be visualized after these have reacted with antigens in cells.

Fluorescence A luminescence mostly found as an optical phenomenon in cold bodies, in which the molecular absorption of a photon triggers the emission of a photon with a longer (less energetic) wavelength; the energy difference between the absorbed and emitted photons ends up as molecular rotations, vibrations or heat; sometimes the absorbed photon is in the ultraviolet range, and the emitted light is in the visible range, but this depends on the absorbance curve and Stokes shift of the particular fluorophore.

Fluorescence resonance energy transfer (FRET) Transfer of energy from short-wavelength fluorophore to long-wavelength fluorophore so quenching the short wave emission.

Fluorescent *in situ* hybridization (FISH) A cytogenetic technique used to detect and localize the presence or absence of specific DNA sequences on chromosomes by using fluorescent probes that bind to only those parts of the chromosome with which these show a high degree of sequence similarity, and the location of binding of fluorescent probe to the chromosomes is then detected by fluorescence microscopy.

Fluorography A method used to visualize substances present in gels, blots, or other biochemical separations; in this method, radioactively labeled substances emit radiation that excites a molecule known as a fluor or scintillator present in the gel, and when the excited molecule relaxes to its ground state, it emits a photon of visible or ultraviolet light that is detected by photographic film.

Fluorophore Portion of a molecule that can fluoresce.

Footprinting A technique for identifying the site on DNA bound by some protein by virtue of the protection of bonds in this region against attack by nucleases.

Foreign DNA A DNA molecule that is incorporated into either a cloning vector or a chromosomal site.

Fosmid A chimeric plasmid that contains the F plasmid origin of replication and a λ *cos* site; useful for constructing stable libraries from complex genomes.

Frameshift mutation In chromosomal DNA, an insertion or deletion of base pairs that changes the reading frame of a gene.

Functional gene cloning A cloning strategy for isolating a gene that depends on information about its gene product. As long as the expressed protein is functional, it can be exploited to screen an expression library to identify the corresponding clone.

Fusion protein The product of two or more coding sequences from different genes that have been cloned together and that, after translation, form a single polypeptide sequence.

Gamete A haploid reproductive cell arising from meiosis. One gamete fuses with another gamete to form a diploid zygote.

Gap A missing internal nucleotide of one strand of ds DNA.

Gel filtration (size exclusion chromatography or SEC or molecular sieving) A technique in which the particles are separated on the basis of their size, or in more technical terms, their hydrodynamic volume. It is usually applied to large molecules or macromolecular complexes.

Gel matrix A semisolid macromolecular lattice that is used for the electrophoretic fractionation of macromolecules. It represents the stationary phase of chromatography, and is chosen for its chemical and physical stability, inertness (lack of adsorptive properties), and a controlled range of pore sizes.

Gel retardation assay (electrophoretic mobility shift assay or mobility shift assay (EMSA) or gel shift assay or gel mobility shift assay or band shift assay) A technique used to determine if a protein or mixture of proteins is capable of binding to a given DNA or RNA sequence, and can sometimes indicate if more than one protein molecule is involved in the binding complex. Used to study protein–DNA or protein–RNA interactions.

Gel Star® stain A fluorescent nucleic acid stain suitable for staining ds or ss DNA as well as ds or ss RNA, and offers an increase in sensitivity.

Gemini virus (twin virus) Quasi-isometric virus particles having a diameter of just 15–20 nm and a small circular DNA genome found usually in pairs.

Gene A segment of nucleic acid that encodes a functional protein or RNA. It is the unit of inheritance.

Gene (cistron) The segment of DNA involved in producing a polypeptide chain; it includes regions preceding and following the coding region (leader and trailer) as well as intervening sequences (introns) between individual coding segments (exons).

Gene bank A population of organisms, each of which carries DNA molecule that was inserted into a cloning vector. Ideally, all of the cloned DNA molecules represent the entire genome of another organism.

Gene cloning (recombinant DNA technology or genetic engineering or gene splicing or gene transplantation or molecular cloning or cloning) Insertion of a gene into a DNA vector (often a plasmid) to form a new DNA molecule that can be perpetuated in a host cell.

Gene cluster A group of adjacent genes, which are identical or related.

Gene family A set of genes whose exons are related; the members were derived by duplication and variation from some ancestral gene.

Gene fusion Structure in which pairs of two genes are joined together, in particular when the regulatory region of one gene is joined to the coding region of reporter gene.

Gene library (recombinant DNA library) A collection of random clones that is representative of the entire starting population of DNA (genomic DNA/cDNA) in an organism in a stable form.

Gene map The linear array of genes of a chromosome.

Gene targeting Gene targeting is a genetic technique that utilizes homologous recombination to alter an endogenous gene.

Gene therapy The insertion of genes into an individual's cells and tissues to treat a disease, such as a hereditary disease in which a deleterious mutant allele is replaced with a functional one.

Generalized transduction A process that can transfer any part of the bacterial genome during the lytic cycle of virulent temperate phages. During the assembly stage, when the phage chromosomes are packaged into capsids, random fragments of the partially degraded bacterial chromosome may also be packaged by mistake and as the capsid can accommodate only a limited quantity of DNA, the phage DNA is left behind.

Generally regarded as safe (GRAS) In the United States, this designation has been given to foods, drugs, and other materials that have been used for a considerable period of time and have a history of not causing illness to humans, even though extensive toxicity testing has not been conducted. Certain organisms, for example, yeast, have been given this status in recombinant DNA experiments.

Gene-specific primers Short nucleotide sequence complementary to the sequence of gene.

Genetic code The correspondence between triplets in DNA (or RNA) and amino acids in protein.

Genetic code dictionary The complete set of 64 triplet codons that code for all 20 amino acids and 3 termination codons.

Genetic complementation (mutant complementation) When two genomes or DNA molecules produce a function that neither genome nor DNA molecule can supply on its own.

Genetic heterogeneity The condition or state of a population in which different mutant genes produce the same phenotype.

Genetic immunization Delivery of a cloned gene that encodes an antigen to a host organism. After the cloned gene is expressed, it elicits an antibody response that protects the organism from infection by a virus, bacterium, or other disease-causing organism.

Genetic manipulation The use of *in vitro* techniques to produce DNA molecules containing novel combinations of genes or altered sequences, and the insertion of these into vectors for their incorporation into host organisms or cells in which these are capable of continued propagation of the modified genes.

Genetic map (linkage map) The linear array of genes on a chromosome based on recombinant frequencies.

Genetic polymorphism Situation in which two or more alleles of a locus in a population of individuals occur at a frequency of 1% or greater.

Genetically modified organism (GMO) An organism in which the basic genetic material (DNA) has been artificially altered or modified to improve the attributes or make it perform new functions.

Genome The complete set of sequences in the genetic material of an organism. It includes the sequence of each chromosome plus any DNA in organelles.

Genome-wide methylases Site-specific methylases that methylate specific A or C residues throughout the genome; if these residues are the part of the recognition sequence of a restriction enzyme, such methylated sites remain uncleaved by that particular restriction enzyme; includes two types, *viz.*, *de novo* methylases and maintenance methylases.

Genomic library A set of clones, packaged in the same vector that together represents all regions of the particular genome. Physically it is a collection of plasmid clones, or phage lysates, which contain recombinant DNA molecules.

Genomics The study and development of genetic and physical maps, large-scale DNA sequencing, gene discovery, and computer-based systems for managing and analyzing genomic data.

Genotype The genetic constitution of an organism.

Genotyping (DNA typing or haplotyping) The determination of the alleles of a chromosome of an individual.

Germ line cells Cells that produce gametes.

Germ line gene therapy The delivery of a gene(s) to a fertilized egg or an early embryonic cell. The transferred gene(s) is present in all the cells of the mature individual, including the reproductive cells, and alters the phenotype of the developed individual.

Glutathione-*S*-transferase (GST) Enzyme that binds to the tripeptide, glutathione; used in making fusion proteins.

Glycosidic linkage A covalent linkage formed between monosaccharides.

Gram negative bacterium Bacterium that does not retain the first stain (crystal violet) used in the Gram staining technique; it retains the second stain (safranin) and therefore has a pink color when viewed under a light microscope.

Gram positive bacterium Bacterium that retains the first stain (crystal violet) used in the Gram staining technique because of the high amount of peptidoglycan in the cell wall; it does not take up the counterstain (safranin).

Gratuitous inducer A substance that can induce transcription of a gene(s) but is not a substrate for the induced enzyme(s).

Green fluorescent protein A protein, originally from jellyfish, whose green fluorescence makes it useful as a reporter molecule.

Guanine (G) One of the four residues in DNA or RNA.

GUS (β-glucuronidase) The bacterial enzyme β-glucuronidase, which is commonly used as a reporter gene in the production of transgenic plants.

Haploid One copy of each autosome and one sex chromosome.

Heat shock genes A set of loci that are activated in response to an increase in temperature (and other abuses to the cell). Usually encodes molecular chaperones that act on denatured proteins.

Heat shock response element (*HSE*) A sequence in a promoter or enhancer used to activate a gene by an activator induced by heat shock.

Helicase An enzyme that uses energy provided by ATP hydrolysis to separate the strands of a nucleic acid duplex.

Helper plasmid A plasmid that provides a function(s) to another plasmid in the same cell. Some helper plasmids are used to mobilize nonconjugative plasmids from a donor cell into a recipient cell.

Helper virus A virus that provides essential functions for defective viruses, satellite viruses and satellite RNA.

Hemimethylated site A palindromic sequence that is methylated on only one strand of DNA.

Herpes simplex virus A family of spherical animal DNA viruses consisting of a large and linear ds DNA genome encased within an icosahedral capsid wrapped in an envelope.

Heterochromatin Regions of the genome that are highly condensed, are not transcribed, and are late-replicating. Divided into two types constitutive and facultative.

Heteroduplex DNA (hybrid DNA) It is generated by base pairing between complementary single strands derived from the different parental duplex molecules. It occurs during genetic recombination.

Heterogeneous nuclear RNA (hnRNA) It comprises transcripts of nuclear genes made by RNA polymerase II. It has a wide size distribution and low stability.

Heterohypekomer (methylation-sensitive isoschizomers) Isoschizomers with different methylation sensitivity.

Heterokaryon A cell containing two (or more) nuclei in a common cytoplasm, generated by fusing somatic cells.

Heterologous probe A DNA probe derived from one organism and used to screen a similar DNA sequence in a clone bank derived from another organism.

Heterozygote An individual that has different alleles at the same locus in its two homologous chromosomes.

Hfr strain Bacterial strain that transfers chromosomal genes at high frequency due to an integrated fertility plasmid.

High performance liquid chromatography (HPLC) A form of column chromatography used frequently in biochemistry and analytical chemistry to separate, identify, and quantify compounds. The technique utilizes a column that holds chromatographic packing material (stationary phase), a pump that moves the mobile phase(s) through the column, and a detector that shows the retention times of the molecules.

Histidine tag (His tag) Six tandem histidine residues that are fused to proteins so allowing its purification by binding to nickel ions attached to a solid support.

Homogenization A process for breaking apart cells, releasing organelles and cytoplasm. Usually perfoimed in blender, ultrasonicator, mechanical tissue disruptor, or by pestle and mortar under liquid nitrogen.

Homologous From the same source, or having the same evolutionary function or structure.

Homologous chromosome Chromosomes that normally synapse during zygotene.

Homologous recombination (generalized recombination) A reciprocal exchange of sequences of DNA, e.g. between two chromosomes that carry the same genetic loci.

Homopolymer tailing The *in vitro* addition of the same nucleotide by the enzyme terminal transferase to the 3′-OH ends of a duplex DNA molecule.

Homozygote An individual that has identical alleles at the same locus in its two homologous chromosomes.

Horizontal gene transfer (Lateral transfer) A process in which an organism incorporates genetic material from another organism without being the offspring of that organism.

Host In recombinant DNA experiments, it represents a microorganism, organism, or cell that maintains a cloning vector.

Hotspot A site in the genome at which the frequency of mutation (or recombination) is very much increased.

Housekeeping genes (constitutive genes) Genes those (theoretically) expressed in all cells because these provide basic functions needed for sustenance of all cell types.

Human artificial chromosomes (HAC or microchromosome) A chromosome that is assembled from telomere, centromere, and human genomic DNA sequences.

Human cytomegalovirus A virus belonging to the herpes group.

Human genome project (HGP) An international research effort dedicated to developing high-resolution human genetic and physical maps and the complete genomic DNA sequences of humans and other organisms.

Human minisatellite DNA Human DNA that is noncoding and generally G+C rich and contains tandem repeats of short (9 to 40 bp) stretches of DNA.

Hybrid arrest translation (HART) A method of identifying recombinant DNA clones by their ability to hybridize to, and thus prevent the translation of, a specific messenger RNA in a cell-free system.

Hybrid release translation (HRT) A method in which the cloned DNA bound to a solid support is used to isolate the complementary mRNA, which can then be eluted and translated *in vitro*.

Hybrid selection A protocol for determining genomic clone that hybridizes to a cDNA or mRNA molecule.

Hybridization The pairing of two polynucleotide strands by H-bonding between complementary nucleotides.

Hybridize To allow complementary strands of nucleic acid to anneal to yield a double stranded product.

Hydrogen bond (H-bond) A relatively weak bond that forms between covalently bonded hydrogen and any electronegative atom, most commonly oxygen or nitrogen.

Hydrophilic 'Water-loving'; having an affinity for water.

Hydrophobic 'Water-hating'; repelled by water.

Hydroxyapatite column chromatography The most stable form of crystalline calcium phosphates ($Ca_{10}(PO_4)_6(OH)_2$) precipitated from aqueous solution, inert with very large surface area in relation to particle size. Widespread use in separating single stranded nucleic acids from double stranded nucleic acids, chromatographic purification and fractionation of complex nucleic acids by thermal elution according to their G+C content, to investigate the reassociation kinetics of DNAs from many different sources, to construct transcription maps, and to measure the copy number of specific sequences in complex genomes, to prepare cDNA for subtractive cloning, as well as for the removal of contaminants from DNA preparations.

Icosahedral symmetry A symmetry typical of viruses having capsids that are polyhedrons.

Immediate early gene A viral gene expressed promptly after infection.

Immunity to superinfection The ability of a prophage to prevent another phage of the same type from infecting a cell. It results from the synthesis of phage repressor by the prophage genome.

Immunity region A segment of the phage genome that enables a prophage to inhibit additional phage of the same type from infecting the bacterium. This region has a gene that encodes for the repressor, as well as the sites to which the repressor binds.

Immunoaffinity chromatography A purification technique in which an antibody is bound to a matrix and is subsequently used to bind a specific protein and separate it from a complex mixture.

Immunoassay A protocol that uses antibody specificity to detect the presence of a particular compound in a biological sample.

Immunoglobulin (antibody) A class of protein that is produced by B cells in response to antigen.

Immunological screening Screening procedure that relies on the specific binding of antibodies to the target protein.

Importins Transport receptors that bind cargo molecules in the cytoplasm and translocate into the nucleus, where these release the cargo.

***In situ* hybridization (cytological hybridization)** It is performed by denaturing the DNA of cells squashed on a microscope slide so that hybridization is possible with an added ss RNA or DNA; the added preparation is radioactively labeled and its hybridization is followed by autoradiography.

***In situ* oligonucleotide synthesis (*de novo*/*on-chip* synthesis)** Building of oligonucleotides base-by-base on the surface of the solid substrate on the basis of light-directed photolithography, printing, microfluidics technology, or microelectrodes.

In vitro 'In glass', or 'outside of the living organism', or 'in test tubes'.

***In vitro* complementation** A functional assay used to identify components of a process; the reaction is reconstructed using extracts from a mutant cell; fractions from wild-type cells are then tested for restoration of activity.

***In vitro* packaging** Encapsidation of λ DNA using packaging extract under *in vitro* conditions to assemble an infectious phage particle capable of naturally infecting the bacterial host cell.

In vivo 'In life', or 'within the living organism'.

***In vivo* gene therapy** The delivery of a gene(s) to a tissue or an organ of an individual to alleviate a genetic disorder.

Incompatibility group A classification scheme indicating which plasmids can coexist within a single cell. Plasmids belonging to different incompatibility groups coexist within the same cell.

Indirect end-labeling A technique for examining the organization of DNA by making a cut at a specific site and isolating all fragments containing the sequence adjacent to one side of the cut; it reveals the distance from the cut to the next break(s) in DNA.

Indole 3-acetic acid (IAA) An auxin, a plant hormone synthesized from tryptophan.

Inducer A small molecule that triggers gene transcription by binding to a regulator protein.

Inducible operon An operon comprising of structural genes that are expressed only in the presence of a specific small molecule (the inducer).

Inducible promoter (regulatable promoter) The promoters that do not express a transgene unless induced by an inducer (chemical agent/environmental stimuli/stage of plant development).

Induction Turning on transcription of a specific gene or operon by the interaction of an inducer with the regulator protein.

Induction of prophage The entry of phage into the lytic cycle as a result of destruction of the lysogenic repressor, which leads to excision of free phage DNA from the bacterial chromosome.

Inoculum The cells used to 'seed' a new culture.

Insertion The incorporation of an additional nucleotide or a stretch of base pairs in DNA.

Insertion λ vector A cloning vector based on phage λ that has nonessential genes removed and like plasmid has a unique restriction enzyme site(s) for cloning of foreign DNA.

Insertional inactivation The insertion of foreign DNA within the gene sequence of a selectable marker in order to inactivate its expression.

Int recombinase (λ integrase or pInt) An enzyme that catalyzes site-specific recombination at *att* P (λ phage) and *att* B (bacteria) sites allowing integration of λ genome into host chromosome.

Intasome A protein DNA complex between the phage λ integrase (Int) and the phage λ attachment site (*att* P).

Integrant (stable transfectant) A cell line in which a gene introduced by transfection has become integrated into the genome.

Integrating vector A vector designed to integrate cloned DNA into the host cell chromosomal DNA.

Integration Insertion of a DNA molecule (usually by homologous recombination) into a chromosomal site.

Integration host factor (IHF) A 20 kDa heterodimeric protein (IHFα and IHFβ) with the capacity to bend the DNA duplex forming structures compatible with the required DNA–protein interactions, and hence facilitating proper positioning of other proteins (e.g., pInt at *att* P site in lysogenic cycle of λ bacteriophage; Par B at *par* S site during partitioning of single copy plasmid).

Integration-excision region (I/E region) The region of λ DNA that encodes proteins required for integration of λ DNA into a specific site in the *E. coli* chromosome and excision from this site at a later stage.

Intein Internal protein segment that is removed from a protein by protein splicing.

Intensifying screen A solid inorganic scintillator consisting of a flexible polyester base coated with inorganic phosphors that is placed behind the X-ray film during indirect autoradiography to convert the emitted energy to light for improved autoradiographic detection.

Intron (intervening sequence) A segment of a gene that is transcribed but is then excised from the primary transcript during processing to a functional RNA molecule.

Inverse PCR Method for using PCR to amplify unknown sequences by circularizing the template DNA molecule.

Ion channel A transmembrare protein usually oligomer with a central aqueous pore that selectively allows the passage of one type of ion across the membrane.

Ion exchange chromatography Technique that allows the separation of ions and polar molecules based on the charge of the molecules; used for almost any kind of charged molecule including large proteins, small nucleotides and amino acids.

Isocaudomers Restriction enzymes that recognize different target sequences but cleavage leads to the same termini, which are compatible.

Isoelectric focusing Technique for separating proteins according to their charge by means of electrophoresis through a pH gradient.

Isopropyl β-D-thiogalactoside (IPTG) An inducer of the *lac* (lactose) operon; in recombinant DNA technology, IPTG is often used to induce cloned genes that are under the control of the *lac* repressor-*lac* promoter system.

Isopycnic centrifugation (equilibrium sedimentation, equilibrium density gradient centrifugation) Centrifugation technique used for separation of DNA molecules of different densities but having same or different masses.

Isoschizomers Restriction endonucleases from different sources that recognize the same restriction site and cleave identically at same position.

Isotopes Atoms having the same number of protons and electrons but differing in the number of neutrons.

Ketose A sugar molecule that contains a ketone group.

Kilo base (kb) A unit of length for nucleic acid strands equal to 1,000 bases or nucleotides in RNA.

Kilobase pair (kbp) A measure of length used to refer 1,000 base pairs of DNA.

Klenow fragment A product of proteolytic digestion of the DNA polymerase I from *E. coli* that retains both polymerase and 3′ exonuclease activities but not 5′ exonuclease activity.

Knock-in (gene knock-in) A genetic engineering method that involves the insertion of a protein coding cDNA sequence at a particular locus in an organism's chromosome.

Knock-out (gene knock-out) The targeted disruption of a gene by homologous recombination.

Label (tag) A compound or atom that is either attached to or incorporated into a macromolecule and is used to detect the presence of a compound, substance, or macromolecule in a sample.

***lac* Z** Gene encoding β-galactosidase, widely used as reporter gene.

Late genes Genes that are transcribed when phage DNA is being replicated; these code for components of the phage particle.

Leader sequence (signal sequence) A short hydrophobic N-terminal sequence responsible for initiating passage into or through a membrane.

Lectin One of a group of plant proteins that can bind to specific oligosaccharides on the surface of cells; often found in seeds where these act as toxin against certain pathogenic agents.

Lentivirus A type of complex retrovirus having the ability to infect nondividing cells; lentiviral vectors permit the stable transduction of multiple cell types.

Library A set of cloned fragments together representing the entire genome (genomic library) or all the expressed genes (cDNA library).

Ligand An extracellular molecule that binds to the receptor on the plasma membrane of a cell, thereby effecting a change in the cytoplasm.

Ligation Joining of two DNA molecules by the formation of phosphodiester bonds; *in vitro*, this reaction is usually catalyzed by the enzyme DNA ligase.

Linker A synthetic double stranded oligonucleotide that carries the sequence for one or more restriction endonuclease sites.

Lipofection Delivery into eukaryotic cells of DNA, RNA, or other compounds that have been encapsulated in an artificial phospholipid vesicle.

Liposome A spherical particle of lipid molecules in which the hydrophobic portions of the molecule are facing inward; a lipid vesicle with an aqueous interior that can carry nucleic acids, drugs, or other therapeutic agents.

Locus (pl. loci) The position on a chromosome at which the gene for a particular trait resides; a locus may be occupied by any one of the alleles for the gene.

Log phase Stage of exponential microbial growth, following the lag phase, when cells are dividing at a constant rate.

Lon protease An ATP-dependent protease, which contains both ATPase and proteolytic activities, and is found in virtually all the living organisms, from Archaea to eubacteria to humans.

Long template A DNA strand that is synthesized during the PCR, has a primer sequence at one end, and is extended beyond the site that is complementary to the second primer at the other end.

Long terminal repeats (LTRs) The sequence that is repeated at each end of the integrated retroviral genome.

Loop An apparatus used in microbiology consisting of a straight handle with a wire approximately 4 inches long; the wire can be flame sterilized and the end is twisted into a loop, which is convenient for inoculating cultures, streaking plates, etc.

A single stranded region at the end of a hairpin in RNA (or ss DNA); it corresponds to the sequence between inverted repeats in ds DNA.

Loss-of-function mutation Mutation that inactivates a gene; it is recessive.

Low electroendosmosis agarose (Low EEO agarose) Low EEO agarose is prepared by addition of few positively charged groups to neutralize a few sulfates on agarose.

Low melting/gelling point agarose (LMP agarose) Agarose that melts as well as gels at temperatures lower than that of standard agaroses; this property is attributable to hydroxyethylation, and the degree of solubilization determines the exact melting and gelling temperature.

***luc* gene** Gene encoding luciferase from eukaryotes; widely used as reporter gene.

Luciferase The enzyme that catalyzes the oxidation of luciferin and subsequent emission of light in bioluminescent reactions.

Luciferin Chemical substrate used by luciferase to emit light.

Luria Bertani medium (LB) A complex rich medium for culturing bacteria. It comprises of tryptone (10 g/l), yeast extract (5 g/l), NaCl (10 g/l), and pH adjusted to 7.0.

***lux* gene** Gene encoding luciferase from eubacteria; widely used as reporter gene.

Lycopene A carotene photosynthetic pigment.

Lysis The destruction or breakage of cells by viruses or chemical or physical treatment.

Lysogeny A form of bacteriophage replication in which the viral genome is integrated into that of the host and is replicated along with it; it describes the ability of a phage to survive in a bacterium as a stable prophage component of the bacterial genome.

Lysozyme An enzyme that removes the bacterial cell wall and causes cell lysis in the presence of EDTA; used in plasmid DNA isolation and purification procedures.

Lytic cycle A process of viral replication involving the bursting of the host cell and release of new viral particles.

M13 (+) strand The ss DNA molecule present in infective M13 phage.

Maintenance methylases (maintenance methyltransferases) Methylases that add methyl groups to hemimethylated DNA and thus perpetuate patterns of methylation through successive rounds of replication.

Maltose binding protein (MBP) An abundant bacterial protein located within the periplasmic space and involved in the uptake of maltose; used in making fusion proteins.

Marker (marker site or marker locus or genetic marker) An identifiable DNA sequence on a chromosome.

Maskless photolithography Photolithography without the use of photomasks.

Mass spectrometry (MS) Technique for measuring the mass of molecular ions derived from volatilized molecules.

Master plate A plate that contains the original microbial colonies from which replica plates were made.

Matrix assisted laser desorption ionization-time of flight (MALDI-TOF) Type of mass spectrometry in which gas-phase ions are generated from a solid sample by a pulsed laser.

Maxi-prep A large-scale procedure for the isolation and purification of plasmid DNA for 100 ml to 1 liter cultures of *E. coli* yielding milligram quantities of pure plasmid.

Mcr and Mrr system Mcr A, Mcr BC and Mrr restriction enzymes are methylation-requiring systems that attack DNA only when it is methylated at specific positions.

Megabase pair (Mbp) 1 million base pairs of DNA.

Melittin A 26 amino acid peptide with antimicrobial and hemolytic activity that is the major component of bee venom.

Melting Separation of ds DNA into two strands.

Melting temperature (T_m) The temperature at which the two strands of a DNA molecule are half unpaired.

Messenger RNA (mRNA) An RNA molecule carrying the information that, during translation, specifies the amino acid sequence of a protein molecule.

Metabolite A low molecular weight biological compound that is usually synthesized by an enzyme; essential for a metabolic process.

Methotrexate A drug that inhibits DHFR (dihydrofolate reductase).

Methylation The addition of a methyl group(s) to a macromolecule; for example, the addition of a methyl group to specific cytosine, and occasionally, adenine residues in DNA.

Methyltransferase (MTase or methylase) An enzyme that adds a methyl group to a substrate, which can be a small molecule, a protein, or a nucleic acid.

Microcentrifuge A small tabletop centrifuge with a radius of 40–50 mm and rotor holes for microcentrifuge tubes that is capable of high speeds.

Microinjection The introduction of DNA or other compounds into a single eukaryotic cell with a fine, microscopic needle.

MicroRNAs (miRNA) Very short RNA molecules of eukaryotic cells that may regulate gene expression.

Microsatellite DNAs DNA sequences that consist of repetitions of extremely short (typically <10 bp) units.

Middle genes Phage genes that are regulated by the proteins coded by early genes; some proteins coded by middle genes catalyze replication of the phage DNA; others regulate the expression of a later set of genes.

Minichromosome Minichromosome of SV40 or polyoma is the nucleosomal form of the viral circular DNA.

Minimum inhibitory concentration (MIC) The lowest concentration of an antimicrobial substance that prevents growth of a given organism.

Mini-prep A small-scale (1.5 ml) procedure for the isolation and purification of plasmid DNA from an *E. coli* host, which entails growth and lysis of the bacteria, differential centrifugation in the microcentrifuge, and purification.

Minisatellite DNAs DNA sequences that consist of ~10 copies of a short repeating sequence, the length of the repeating unit is measured in 10s of base pairs; the number of repeats varies between individual genomes.

Minus strand (–) DNA The ss DNA sequence that is complementary to the viral RNA genome of a plus strand virus.

Mismatch The lack of base pairing between one or more nucleotides of two hybridized nucleic acid strands.

Mitochondrial DNA (mtDNA) An independent DNA genome, usually circular, that is located in the mitochondrion.

Mitosis (M phase) The process by which the cell nucleus divides, resulting in daughter cells that contain the same amount of genetic material as the parent cell.

Mobilizable plasmid A plasmid that is not self-transmissible, but transferred from one bacterial cell to another with the help of conjugative plasmid.

Mobilizing functions The genes on a plasmid that facilitate the transfer of either a nonconjugative or conjugative plasmid from one bacterium to another.

Modification Enzymatic methylation of a restriction site.

Specific nucleotide changes in DNA or RNA molecules.

Modified bases All the bases, except the usual four, from which DNA (T, C, A, G) or RNA (U, C, A, G) are synthesized; these result from postsynthetic changes in the nucleic acid.

Molecular beacon A fluorescent probe molecule that contains both a fluorophore and a quencher and that fluoresces only when it binds to a specific DNA target sequence.

Molecular markers Molecular markers are specific fragments of DNA that can be identified within the whole genome.

Molecular mass The sum of the masses of all the atoms in a molecule, usually expressed in daltons or kilodaltons.

Molecular pharming (molecular farming) Application of molecular biological techniques for the synthesis of commercial products in plants.

Monocistronic mRNA mRNA that codes for one protein.

Monoclonal antibody (McAb) A single type of antibody that is directed against a specific epitope (antigenic determinant) and is produced by a hybridoma cell line, which is formed by the fusion of a lymphocyte with a myeloma cell; some myeloma cells synthesize single antibodies naturally.

Monophasic reagent Lysis buffer containing acidified phenol, guanidinium or ammonium thiocyanate, and a phenol solubilizer for single-step simultaneous isolation of DNA, RNA, and proteins.

mRNA abundance Average number of mRNA molecules per cell. Abundant mRNAs consist of a small number of individual species, each present in a large number of copies per cell.

Multicopy plasmid A plasmid is said to be under multicopy control when the control system allows the plasmid to exist in more than one copy per individual bacterial cell.

Multiple cloning site (MCS or polylinker) Chemically synthesized cluster of restriction enzyme sites in vector used for cloning of target DNA.

Multiplex assay Simultaneous determination of a large number of different targets in one reaction vessel or by one analytical procedure.

Mutagen A chemical or physical agent capable of inducing mutations.

Mutagenesis Chemical or physical treatment that changes the nucleotides of the DNA of an organism.

Mutant (variant) An organism that differs from the wild-type because it carries one or more genetic changes in its DNA.

Mutation Any heritable alteration in a DNA sequence; it may or may not have an effect on the phenotype.

Narrow host range plasmid A plasmid that can replicate in one, or at most a few, different bacterial species.

National Center for Bioinformatics Information (NCBI) US government-funded national resource for molecular biology information. It provides access to many public databases and other references, including the draft human genome.

National regulatory mechanism for biosafety National regulatory mechanism developed for the implementation of biosafety guidelines to regulate the research, testing, safe use and handling of GMOs and products thereof.

Nebulization A form of hydrodynamic shearing performed by collecting the fine mist created by forcing DNA in solution through a small hole in the nebulizer unit; used to get DNA fragments from the large genomic DNA.

Negative complementation Occurs when interallelic complementation allows a mutant subunit to suppress the activity of a wild-type subunit in a multimeric protein.

Negative control A system of regulation of transcription that requires the removal of a repressor from an operator.

Neomycin phosphotransferase An enzyme that inactivates the antibiotics neomycin and kanamycin; often used as a selective marker in the production of transgenic plants.

Neoschizomers (heteroschizomers) Restriction endonucleases from different sources that recognize the same target site but cleave at different positions.

Nick A break in the backbone of one of the strands of duplex DNA due to cleavage of phosphodiester bond.

Nick translation The removal of a short stretch of a DNA starting from a nick and its replacement by a newly synthesized DNA; catalyzed by 5′ → 3′ exonuclease activity of DNA polymerase I.

Nitrocellulose A transfer medium on which nucleic acids or proteins are immobilized by blotting.

Nonsense codon Stop codon or termination codon (UAG, UGA, UAA).

Nonsense mutation A mutation that results in a 'stop' codon being inserted into the mRNA leading to premature termination of translation.

Nopaline plasmids Ti plasmids of *Agrobacterium tumefaciens* that carry genes for synthesizing the opine, nopaline; these retain the ability to differentiate into early embryonic structures.

Northern blot The piece of transfer membrane containing RNA sequences transferred on it.

Northern blotting Similar to Southern blotting, except that RNA that has been separated by gel electrophoresis is transferred from a gel onto a matrix such as nitrocellulose or nylon membrane, and the presence of a specific RNA molecule is detected by DNA:RNA hybridization.

North-western blotting Technique used to isolate sequence specific RNA-binding proteins by using a ss RNA probe.

***npt* II gene** Gene encoding neomycin phosphotransferase that inactivates neomycin or kanamycin by phosphorylation.

N-terminus (amino terminus) The first amino acid(s) of a protein.

Nuclear cloning (Nuclear transfer) Production of an organism by placing a nucleus from a somatic cell into an enucleated fertilized egg.

Nuclear envelope A layer of two concentric membranes (inner and outer nuclear membranes) that surrounds the nucleus and its underlying intermediate filament lattice, the nuclear lamina. It is penetrated by nuclear pores.

Nuclear localization signal (NLS) A domain of a protein, usually a short amino acid sequence, that interacts with an importin, allowing the protein to be transported into the nucleus.

Nuclear pore complex (NPC) A very large, proteinaceous structure that extends through the nuclear envelope, providing a channel for bidirectional transport of molecules and macromolecules between the nucleus and the cytosol.

Nucleic acid aptamer Nucleic acid species that have been engineered through repeated rounds of *in vitro selection* or equivalently, SELEX (systematic evolution of ligands by exponential enrichment) to bind to various molecular targets such as small molecules, proteins, nucleic acids, and even cells, tissues and organisms.

Nucleic acids Molecules that encode genetic information; these include DNA (deoxyribonucleic acid) and RNA (ribonucleic acid); these consist of a series of nitrogenous bases bonded by glycosidic linkage to ribose or deoxyribose sugar molecules linked with each other by phosphodiester bonds in RNA and DNA, respectively.

Nucleocapsid The genome of a virus and its surrounding protein coat.

Nucleoid (bacterial chromosome) The site of most of a prokaryotic cell's DNA.

Nucleolus A discrete region of the eukaryotic nucleus, where ribosomes are assembled.

Nucleoside A base (purine or pyrimidine) that is covalently linked to a five-carbon (pentose) sugar; when the sugar is ribose, the nucleoside is a ribonucleoside; when it is deoxyribose, the nucleoside is a deoxyribonucleoside.

Nucleosome The basic structural subunit of chromatin, consisting of 200 bp of DNA and an octamer of five types histone proteins.

Nucleotide The building block of nucleic acids, comprising a pentose sugar, a nitrogenous base and one or more phosphate groups.

Nucleus The central, membrane-bound structure in eukaryotic cells that contains the genetic material.

Nylon membrane A transfer medium for blotting nucleic acids or protein that is stronger and more versatile than nitrocellulose.

Ochre codon The triplet UAA, one of the three termination codons that end protein synthesis.

Octopine plasmids pTi plasmids of *Agrobacterium tumefaciens* that carry genes coding the synthesis of opine of the octopine type; the tumors are undifferentiated.

Oleosins Hydrophobic oil body proteins associated with plant seeds.

Oligo-dT primer A chemically synthesized stretch of 16–20 adenosine residues that are complementary to poly (A) tail of mRNA.

Oligonucleotide (oligodeoxyribonucleotides or oligomers) A short molecule of ss DNA.

Oligonucleotide array detector (DNA array, DNA chip) Chip used to simultaneously detect and identify many short DNA fragments by DNA:DNA hybridization.

Oligonucleotide primer A chemically synthesized oligonucleotide chain typically used for DNA sequencing or amplification by PCR.

Oncogenes Genes whose products have the ability to transform eukaryotic cells so that these grow in a manner analogous to tumor cells.

***o*-Nitrophenyl galactoside (ONPG)** Artificial substrate that is split by β-galactosidase, releasing yellow *o*-nitrophenol.

***o*-Nitrophenyl phosphate** Artificial substrate that is split by alkaline phosphatase, releasing yellow *o*-nitophenol.

Opal codon The triplet UGA, one of the three termination codons that end protein synthesis.

Open reading frame (ORF) A sequence of DNA consisting of triplets that can be translated into amino acids starting with an initiation codon and ending with a termination codon.

Operator The region of DNA that is upstream from a prokaryotic gene(s) and to which a repressor or activator binds.

Operon A unit of bacterial gene expression and regulation, including structural genes and control elements in DNA recognized by regulator gene product(s).

Opine A derivative of arginine that is synthesized by plant cells infected with crown gall disease.

Optical mapping The direct location of restriction sites by visualization of cut DNA molecules with a microscope.

Origin of replication (*ori* or origin) A nucleotide sequence of DNA at which DNA replication or DNA synthesis is initiated.

Orthogonal field alternation gel electrophoresis (OFAGE) An electrophoretic system that uses double inhomogeneous electric fields generated with two pairs of electrodes, placed to produce electric field diagonally across the gel, and at an angle greater than 90° to each other.

Orthologs Corresponding proteins in two species as defined by sequence homologies.

Osmolyte A compound that regulates the osmotic pressure within a cell.

Osmoprotectant Osmoprotectants are small molecules that act as osmolytes and help organisms survive extreme osmotic stress; examples include betaines, amino acids, and the sugar trehalose.

Outer membrane The outermost part of the Gram negative cell wall, comprising phospholipids and lipopolysaccharide.

Oxidation A chemical reaction in which an electron is lost.

P element A type of transposon in *Drosophila melanogaster.*

P1 cloning system A plasmid vector system based on bacteriophage P1 that uses *in vitro* bacteriophage P1 packaging for introducing a vector with a large DNA insert into *E. coli*.

P1-derived artificial chromosome (PAC) A plasmid vector cloning system based on bacteriophage P1 for introducing a vector with a large DNA insert into *E. coli*.

Packaging extract An extract containing phage head precursor proteins, packaging or assembly proteins, and phage tail proteins.

Packing of column The pouring or pipetting of slurry of a suitable gel matrix into the closed column.

Palindromic sequences (palindrome) Complementary DNA sequences that are the same when each strand is read in the same direction (e.g., 5′ to 3′); palindrome consists of adjacent inverted repeats; these types of sequences serve as recognition sites for type II restriction endonucleases.

Paralogs Highly similar proteins coded by the same genome.

Parasporal crystals Tightly packed insect protoxin molecules produced by strains of *Bacillus thuringiensis* during the formation of resting spores.

Parental strand of DNA duplex DNA strand that will be replicated.

Partial restriction digestion Cleavage by restriction enzymes at limited number of sites.

Passaging Subculturing cells that are growing *in vitro*.

Pathogen An organism with the potential to cause disease.

Pathogenesis related (PR) protein Proteins synthesized in some plants in response to stress.

pCI repressor (λ repressor) The protein having negative regulatory functions at O_L and O_R, and aids in lysogeny; it acts as a positive regulator of its own gene.

Pentose sugar A five-carbon sugar.

Peptide vaccine A short chain of amino acids that can induce antibodies against a specific infectious agent.

Peptidoglycan A polymer comprising alternate units of *N*-acetyl-muramic acid and *N*-acetylglucosamine that forms the major constituent of bacterial cell walls.

Periplasm (periplasmic space) The region between the inner and outer membranes in the bacterial envelope.

***Pfu* DNA polymerase** Thermostable DNA polymerase isolated from *Pyrococcus furiosus* that is used to quickly amplify DNA in PCR with high fidelity.

Phage display Fusion of a protein or peptide to the coat protein of a bacteriophage whose genome also carries the cloned gene encoding the protein; the proteins are displayed on the outside of the phage particle and the corresponding gene is carried on the inside.

Phage display library (surface display library) Collection of a large number of modified phages displaying different peptide or protein sequences.

Phages T4 and T7 Viruses that infect *E. coli* causing its lysis.

Phagemid A hybrid cloning vector, comprising elements of plasmid and filamentous coliphages (e.g., M13, F1, fd); the elements of filamentous coliphages include *cis* elements such as origin of replication, terminator of replication and packaging signal.

Phasmid An example of chimeric vector comprising of λ genome into which a phagemid (or plasmid) vector carrying λ attachment site (*att* P) has been inserted by site-specific recombination mechanism of phage.

Phenotype The observable characteristics of an organism.

***pho* gene** Gene encoding alkaline phosphatase; used as reporter gene.

Phosphodiester bond The linkage of a phosphate group to the 3′- carbon of one nucleotide and the 5′-carbon of another nucleotide; also defined as the linkage between nucleotides of the same nucleic acid strand.

Phosphodiesterase An enzyme that hydrolyzes 5′ mononucleotides from 3′ hydroxy terminated DNA and RNA.

Phospholipid A lipid having a positively charged head that is linked by a phosphate group to the fatty acid tails.

Phosphorimager A form of solid-state liquid scintillation where radioactive material can be both localized and quantified; it produces digital images of the gels and blots by capturing images from them and allows complete filmless, electronic analysis, and archiving.

Phosphorothioate linkage The linkage between nucleotides after a sulfur group replaces available oxygen of a phosphodiester linkage.

Photolithography (optical lithography) A process that uses light to transfer a geometric pattern from a photomask (or mirror in maskless photolithography) to a light-sensitive chemical on the substrate.

Photomask An opaque plate with holes or transparencies that allow light to shine through in a defined pattern, and hence defines chip exposure sites; it is typically transparent fused silica blank covered with a pattern defined with a chrome metal absorbing film, and used at wavelengths of 365 nm, 248 nm, and 193 nm.

Physical map A map of the positions of chromosome sites, such as restriction endonucleases recognition or sequence tagged site, on a chromosome; the distance between sites is measured in base pairs.

Phytohormones A substance that stimulates growth or other processes in plants; a plant hormone, e.g., auxin, cytokinin, gibberellin, ethylene, or abscissic acid.

Pichia pastoris A methylotrophic yeast; used for development of many expression vectors.

Piezoelectric oligonucleotide synthesizer and microarrayer (POSaM) A technique that uses piezoelectric jetting, high quality motion controllers and standard phosphoramidite oligonucleotide synthesis chemistry to synthesize arrays of any nucleic acid sequence at specific, closely spaced features on suitable solid substrates.

Pilus (pl. pili) A surface appendage on a bacterium that allows the bacterium to attach to other bacterial cells; it appears like a short, thin, flexible rod; used to transfer DNA from one bacterium to another during conjugation.

Plaque A zone of clearance produced as the phage lyses the bacterial cells, and the resulting progeny move on to infect and eventually lyse neighboring bacteria; it is represented as a clear circular area in an otherwise turbid background of bacterial growth that is visible to the naked eye after several hours of incubation.

Plaque hybridization A procedure used to identify and isolate specific recombinants from libraries of bacteriophage λ.

Plasma membrane The membrane that surrounds a cell.

Plasmid A small, self-replicating loop of extrachromosomal DNA, found in bacteria and some yeasts; specially engineered forms are used as vectors in gene cloning.

Plasmid addiction system A survival mechanism used by plasmids; the mechanism kills the bacterium upon loss of the plasmid.

Plasmid amplification The production of additional copies of extrachromosomal DNA upon inhibition of protein synthesis.

Plasmid curing Displacement of plasmids from bacteria.

Plate To spread cells onto a solid nutrient medium in a petridish.
A petridish.

Plus strand (+) DNA The strand of the duplex sequence that has the same sequence as that of the RNA.
A plus strand virus has a single stranded nucleic acid genome whose sequence directly codes for the protein products.

Plus-minus screening A screening technique used to identify differentially regulated genes with respect to topography and developmental stage; in this screening technique the cDNA library is constructed from cells expressing the gene of interest and duplicate copies of the library are screened separately with labeled cDNA probes synthesized from two preparations of mRNA.

Point mutation (single site mutation) A change in DNA sequence involving a single base pair.

Polar Having unequal charge distribution, caused by unequal sharing of atoms.

Pollen Microspores of plants that carry male gametes.

Poly (A) binding protein (PABP) The protein that binds to the 3′ stretch of poly (A) on a eukaryotic mRNA.

Poly (A) tail A stretch of multiple adenosine residues present at the 3′ end of mRNA.

Poly $(A)^+$ mRNA mRNA that has a 3′-terminal stretch of poly (A).

Poly A polymerase An enzyme which catalyzes the template independent polyadenylation at the 3′-terminus of a wide variety of RNAs.

Poly-(3-hydroxybutyric acid) The best-studied and best-characterized of the polyhydroxyalkanoates; used as biodegradable plastics.

Polyacrylamide Polymer used in separation of proteins or very small nucleic acid molecules by gel electrophoresis.

Polyacrylamide gel electrophoresis (PAGE) Technique for separating proteins by electrophoresis on a polyacrylamide gel.

Polyadenylation (poly (A) tailing) The addition of adenosine residues to the 3′-end of eukaryotic mRNAs.

Polyadenylation signal A sequence that terminates transcription and provides a recognition site at the end of an mRNA for the enzymatic addition of adenosine residues.

Polycistronic mRNA mRNA that includes coding regions representing more than one gene; an mRNA that encodes two or more proteins.

Polyethylene glycol (PEG) A flexible, inert, water-soluble polymer used to alter the plasma membrane of some types of cells to facilitate transformation; also used as crowding agent and for applying osmotic pressure, to concentrate viruses, and to induce complete fusion in liposomes reconstituted *in vitro*.

Polyhydroxyalkanoate Biodegradable polymers produced by microorganisms as a carbon and energy storage material.

Polymer A macromolecule made up of a series of covalently linked monomers.

Polymerase chain reaction (PCR) A technique used for exponential amplification of a selected region of a DNA molecule.

Polymorphic site (polymorphic locus) A chromosome location having two or more identifiable allelic DNA sequences that each occur with a frequency of 1% or greater in a large population.

Polynucleotide A linear series of 20 or more nucleotides linked by phosphodiester bonds.

Polynucleotide kinase (PNK) An enzyme that catalyzes the transfer of a γ-phosphate from ATP to the 5' hydroxyl termini of polynucleotides of either DNA or RNA.

Polypeptide A chain of many amino acids joined together by peptide bonds.

Polyribosome (polysome) An mRNA that is simultaneously being translated by several ribosomes.

Polysaccharide A carbohydrate polymer of monosaccharide units.

Positional cloning Cloning procedure that employs mapped position of a gene to obtain a clone containing it, and is done when there is no biological information about a gene (often the case for human disease genes), but its position can be mapped relative to other genes or markers; using this information, the nearest physical markers can be located, which is followed by chromosome walking.

Positive control Regulation of transcription that requires the addition of a protein activator to an activator site on the DNA.

Positive–negative selection A protocol that both selects for cells carrying a DNA insert integrated at a specific targeted chromosomal location (positive selection) and selects against cells that carry a DNA insert integrated at a nontargeted chromosomal site (negative selection).

Posttranslational modification The specific addition of phosphate groups, sugars (glycosylation), or other molecules to a protein after it has been synthesized.

Potato virus X (PVX) A member of the Potexvirus family having a monopartite RNA genome of ~6.5 kbp packaged in a filamentous particle.

Poxviruses A family of large and complex ds DNA animal viruses with 150 to 200 genes.

Precipitation The process of bringing compounds out of solution.

Prey The fusion between the activator domain of a transcriptional activator protein and another protein as used in two-hybrid screening.

Primary antibody The antibody that binds to the target molecule in an ELISA or other immunological assay.

Primary transcript Unprocessed RNA that is transcribed from a eukaryotic structural gene having exons and introns.

Primer A short oligonucleotide that hybridizes with a template strand and provides a 3′-OH end for the initiation of nucleic acid synthesis.

Primer extension Method to locate the 5′ start site of transcription by using reverse transcriptase to extend a primer bound to mRNA so locating the 5′-end of the transcript.

Primer walking A method for sequencing long (>1 kbp) cloned pieces of DNA; the initial sequencing reaction reveals the sequence of the first few hundred nucleotides of the cloned DNA; on the basis of these data, a primer that contains about 20 nucleotides and is complementary to a sequence near the end of sequenced DNA is synthesized and used for sequencing of the next few hundred nucleotides of the cloned DNA; the procedure is repeated until the complete nucleotide sequence of the cloned DNA is determined.

Printing Delivery of nanoliter sized droplets of reagents to the proper site on a chip for drop-by-drop synthesis of oligonucleotides using a device similar to an inkjet printer.

Probe A piece of labeled (radioactive or fluorescent) DNA or RNA used to locate immobilized sequences on a blot by hybridizing under optimal conditions of salt and temperature.

Processing of RNA Changes that occur after its transcription, including modification of the 5′- and 3′-ends, internal methylation, splicing, or cleavage.

Prodrug An inactive compound that is converted into a pharmacological agent by an *in vivo* metabolic process.

Productivity The amount of product that is produced in a bioreactor within a given period of time.

Progeny The offspring of a mating.

Programmable autonomously controlled electrodes (PACE) A computer-driven variant of PFGE in which all electric field parameters (number and angle of electric fields, voltage gradients, pulse time, and time ramping) are adjustable, and many types of PFGE can be chosen (OFAGE, FIGE, CHEF, PHOGE, etc.).

Prokaryotes Single celled organisms, including bacteria, which do not have a true nucleus and membrane-bound organelles.

Promoter A sequence upstream of a gene, where RNA polymerase binds to initiate transcription.

Proofreading Any mechanism for correcting errors in protein or nucleic acid synthesis that involves scrutiny of individual units after these have been added to the chain.

Prophage A repressed or inactive state of a bacteriophage genome that is maintained in a bacterial host cell as part of the chromosomal DNA.

Prosthetic group A tightly bound nonpeptide component of a protein; it may be lipids, carbohydrates, metal ions, or inorganic groups such as phosphates.

Protease (proteinase or proteolytic enzyme) An enzyme that hydrolyzes peptide bond linkages and cleaves proteins into smaller peptides.

Protease inhibitor A protein that can form a tight complex with a protease and block its activity.

Proteasome A large complex with an interior cavity that degrades cytosolic proteins previously marked by covalent addition of ubiquitin.

Protein microarray Microarray of immobilized proteins used for proteome analysis and normally screened by fluorescent or radioactive labeling.

Protein replacement therapy Treatment of an inherited disorder with a structural protein that restores normal function.

Protein translocation The movement of a protein across a membrane. This occurs across the membranes of organelles in eukaryotes, or across the plasma membrane in bacteria. Each membrane across which proteins are translocated has a channel specialized for the purpose.

Proteolysis Hydrolytic breakage of a peptide bond of a protein.

Proteome The total set of proteins encoded by a genome or the total protein complement of an organism.

Proteomics Proteomics refers to any procedure that characterizes large sets of proteins; it focuses on identifying when and where proteins are expressed in a cell and to ascertain their physiological roles in an organism.

Protocol Set of directions for a lab procedure.

Protoplast A bacterial, yeast, or plant cell with its cell wall removed either chemically or enzymatically.

Protoxin A latent, inactive precursor of a toxin.

Pseudomonas A common, widely-distributed Gram negative bacterium; many of the soil forms produce a pigment that fluoresces under ultraviolet light; hence, the descriptive term fluorescent pseudomonads.

Pulsed field gel electrophoresis (PFGE) Type of gel electrophoresis that involves switching (pulsing) of fields to separate large DNA molecules.

Pulsed homogeneous orthogonal field gel electrophoresis (PHOGE) A variant of PFGE, which uses a 90° reorientation angle, but the DNA molecules undergo four reorientations per cycle instead of two, which helps in overcoming the problem of poor resolution associated with the use of electric fields at an angle of 90°; migration of DNA molecules thus proceeds in homogeneous electric fields placed perpendicularly to one another.

Purine Fusion of a pyrimidine and an imidazole ring, e.g., adenine or guanine.

Pyrimidine A heterocyclic ring, e.g., thymine, cytosine, or uracil.

Quantitative trait locus (QTL) A quantitative trait locus is a region of DNA that is associated with a particular phenotypic trait.

Quencher Molecule that prevents fluorescence by binding to the fluorophore and absorbing its activation energy.

Random amplified polymorphic DNA (RAPD) Technique for analyzing genetic relatedness using PCR to amplify arbitrarily chosen sequences.

Random labeling Widely used method for labeling of probe in which a mixture of small oligonucleotide sequences, acting as primers, anneal to a heat-denatured, double stranded template.

Random mutagenesis A nondirected change of a nucleotide pair(s) in a DNA molecule.

Random primers (random hexamers) A collection of chemically synthesized oligonucleotides of random sequence, usually hexamers.

Rapid amplification of cDNA ends (RACE) PCR methods used to generate ds DNA (cDNA) from the 3′-end (3′ RACE) or the 5′-end (5′ RACE) of a specific mRNA.

Reading frame One of the three possible ways of reading a nucleotide sequence. Each reading frame divides the sequence into a series of successive triplets; there are three possible reading frames in any sequence, depending on the starting point; if the first frame starts at position 1, the second frame starts at position 2, and the third frame starts at position 3.

Read-through Transcription or translation that proceeds beyond the normal stopping point because the transcription or translation termination signal of a gene is absent or mutated.

Rec A A protein, found in most bacteria, that is essential for DNA repair and DNA recombination.

***rec* mutations** Mutated *E. coli* that cannot undertake general recombination.

Receptor A transmembrane protein, located in the plasma membrane that binds a ligand in a domain on the extracellular side, and as a result has a change in activity of the cytoplasmic domain.

Recessive gene An allele that does not demonstrably contribute to the phenotype in a heterozygote.

Recognition site (restriction site/cognate DNA/binding site/DNA site/host specificity site/target site) Specific sequences on ds DNA that are recognized by restriction enzymes.

Recombinant An individual with two or more linked genes that is a consequence of one or more cross-over events.

Recombinant DNA molecule DNA that comprises material from more than one source.

Recombinant genotype One that consists of a new combination of genes produced by crossing-over.

Recombinant progeny Daughter molecule having a different genotype from that of either parent.

Recombinant protein A protein whose amino acid sequence is encoded by a cloned gene.

Recombinant toxin A single multifunctional toxic protein that has been created by combining the coding regions of various genes.

Recombination Any process that results in new combinations of genes.

Recombination frequency The number of recombinant individuals (or chromosomes) among a set of offspring divided by the total number of individuals (or chromosomes) that constitute the offspring population.

Reduction A chemical reaction in which an electron is gained.

Regulator gene Gene encoding a product (typically protein) that controls the expression of other genes (usually at the level of transcription).

Regulatory protein A protein that plays a role in either turning on or turning off transcription.

Relaxase An enzyme that cuts one strand of DNA, and binds to the free 5′-end.

Relaxed plasmid High copy number plasmids.

Renaturation The reassociation of denatured complementary single strands of a DNA double helix.

Repetitive DNA DNA sequences that behave in a reassociation reaction as though many (related or identical) sequences are present in a component, allowing any pair of complementary sequences to reassociate.

Replacement therapy The administration of metabolites, cofactors, or hormones to overcome their deficiency caused as the result of a genetic disease.

Replacement λ vector (substitutional λ vector) A cloning vectors based on λ phage, in which a central nonessential DNA flanked on either side by recognition sites for a particular enzyme(s) and is replaced by the insert DNA during cloning.

Replica plating The transfer of cells from bacterial colonies on one petriplate to another petriplate; the locations of the colonies that grow on the second plate correspond to those on the original (master) petriplate.

Replication of duplex DNA The process of DNA synthesis.

Replication-defective virus Virus that cannot perpetuate an infective cycle because some of the necessary genes are absent (replaced by host DNA in a transducing virus) or mutated.

Replicative form (RF) Double stranded form of the genome of ss DNA virus; RF first replicates itself and is then used to generate the ss DNA to pack into the virus particle.

Replicon A unit of the genome in which DNA is replicated; each replicon contains an origin for initiation of replication.

Replisome The multiprotein structure that assembles at the bacterial replicating fork to undertake synthesis of DNA; it contains DNA polymerase and other enzymes.

Reporter gene A gene that is used in genetic analysis because its product is convenient to assay or easy to detect.

Repressible operon Structural genes are expressed unless the corepressor is present.

Repression The ability of bacteria to prevent synthesis of certain enzymes when their products are present; more generally, refers to inhibition of transcription (or translation) by binding of repressor protein to a specific site on DNA (or mRNA).

Repressor A protein that binds to the operator and prevents transcription by blocking the binding of RNA polymerase.

Reproductive cloning (adult cloning) Use of cloning technology to produce one or more complete and genetically identical individuals (apart from the genes in mitochondria and chloroplasts); relies on the technique of cell nuclear replacement, which leads to the production of an individual, by transferring the nucleus of differentiated adult cells into an oocyte from which the nucleus has been removed.

Resistance plasmid (R plasmid) Plasmid harboring antibiotic resistance gene(s).

Response element (initiator element or signal region) A sequence of deoxyribonucleotides, which acts as a binding site for a protein (transcription factor) that regulates transcription.

Restrict To digest with a restriction enzyme.

Restriction digestion *In vitro* cleavage catalyzed by restriction endonucleases.

Restriction endonuclease (restriction enzyme) An enzyme that cleaves ds DNA at specific recognition sequences of nucleotides; used extensively in standard molecular biology procedures such as mapping or modifying DNA for cloning.

Restriction enzyme (type II) An enzyme that recognizes a specific duplex DNA sequence and cleaves phosphodiester bonds on both strands between definite nucleotides.

Restriction fragment length polymorphism (RFLP) A difference in restriction sites between two related DNA molecules that result in the production of restriction fragments of different lengths and are detected by DNA hybridization with DNA probes after separation by gel electrophoresis.

Restriction map A physical map showing the relative positions of the restriction sites for a number of different restriction endonucleases.

Restriction mapping To determine the linear order of restriction sites on a piece of DNA by performing a series of digests and analyzing the fragment banding pattern on a gel.

Restriction site (recognition site) The sequence of nucleotide pairs in duplex DNA that is recognized by a restriction endonuclease.

Restriction-modification system (R-M system) A combination of restriction enzyme and its cognate methylase; R-M system protects the bacteria from invading phage by restricting or cleaving its DNA, and prevents autorestriction (i.e., cleavage of its own DNA by its own restriction enzyme) by methylating it.

Retention time The elapsed time between the time of injection of a solute and the time of elution of the peak maximum of that solute; it is a unique characteristic of the solute and can be used for identification purposes; it varies depending on the interactions between the stationary phase, the molecules being analyzed, and the solvent(s) used.

Retroposon (retrotransposon) A transposon that mobilizes via an RNA form; the DNA element is transcribed into RNA, and then reverse-transcribed into DNA, which is inserted at a new site in the genome; the difference from retroviruses is that the retroposon does not have an infective (viral) form.

Retrovirus A class of eukaryotic RNA viruses that can form ds DNA copies of their genomes; the double stranded forms can integrate into chromosomal sites of an infected cell.

Reverse transcriptase An RNA-dependent DNA polymerase that uses an RNA molecule as a template for the synthesis of complementary DNA strand.

Reverse transcription Synthesis of DNA on a template of RNA accomplished by reverse transcriptase enzyme.

Reverse transcription-polymerase chain reaction (RT-PCR) A two-step protocol for synthesizing cDNA molecules. First, cDNA strand is synthesized *in vitro* by reverse transcriptase with oligo (dT) as a primer and mRNA as the template; second, a specific cDNA strand is amplified by the PCR, with one primer directed to a sequence of the first cDNA strand and the other to a sequence of the complementary cDNA strand (second strand) that is synthesized during the first PCR cycle.

Revertants Organisms derived by reversion of a mutant cell or organism.

Rho factor A protein involved in assisting *E. coli* RNA polymerase to terminate transcription at certain terminators (called rho-dependent terminators); rho-dependent terminators are sequences that terminate transcription by bacterial RNA polymerase in the presence of the rho factor.

Ri plasmids Plasmids found in *Agrobacterium rhizogenes* that carry genes responsible for causing hairy root disease.

Ribonuclease H (RNase H) A ribonuclease of bacterial cells that is specific for RNA:DNA hybrids.

Ribonucleases (RNases) Small, very stable, and omnipresent enzymes that degrade RNA; these may be specific for ss RNA or ds RNA, and may be either endonucleases or exonucleases.

Ribonucleic acid (RNA) Nucleic acid that differs from DNA in having ribose in place of deoxyribose and uracil in place of thymine.

Ribonucleoprotein A complex of RNA with proteins.

Ribose sugar A pentose sugar present in RNA.

Ribosomal DNA (rDNA) Usually a tandemly repeated series of genes coding for a precursor to the two large rRNAs.

Ribosomal RNA (rRNA) A form of RNA that forms part of the structure of ribosomes.

Ribosome An organelle made up of protein and RNA, found in both prokaryotes and eukaryotes; it is the site of protein synthesis.

Ribosome binding site (RBS or Shine-Dalgarno sequence) A purine-rich sequence at the 5′ end of bacterial mRNA that is bound by a 30S subunit of ribosome due to complementary base pairing between it and pyrimidine-rich sequence at the 3′ end of 16S rRNA during the initiation phase of protein synthesis.

Ribozyme An RNA having catalytic activity.

RNA interference (RNAi) Situations in which antisense and sense RNAs apparently are equally effective in inhibiting expression of a target gene; it is caused by the ability of double stranded sequences to cause degradation of sequences that are complementary to them.

RNA ligase An enzyme that functions in tRNA splicing to make a phosphodiester bond between the two exon sequences that are generated by cleavage of the intron; T4 RNA ligase catalyzes the ATP-dependent intra- and intermolecular formation of phosphodiester bonds between 5′-phosphate and 3′-hydroxyl termini of oligonucleotides, ss RNA and DNA and the minimal substrate is a nucleoside 3′,5′-biphosphate in intermolecular reaction and oligonucleotide of 8 bases in intramolecular reaction.

RNA polymerases Enzymes that synthesize RNA using a DNA template (formally described as DNA-dependent RNA polymerases).

RNA silencing The ability of a ds RNA to suppress expression of the corresponding gene systemically.

RNA size markers RNA standard markers of known molecular weights are loaded in the gel for accurate measurement of the size of RNA of interest.

RNA splicing The process of excising the sequences in RNA that correspond to introns, so that the sequences corresponding to exons are connected into a continuous mRNA.

RNase protection assay A technique used to quantify the abundance of RNA, and to map the positions of 5′ and 3′ ends of mRNAs as well as positions of 5′ and 3′ splice sites.

Robotic deposition Deposition of presynthesized oligonucleotides on a glass slide with the help of mechanized robots.

Rolling circle amplification technology (RCAT) Technique based on rolling circle replication that uses DNA polymerase to amplify target DNA at normal temperature.

Rolling circle replication (θ replication) Mechanism of replicating circular ds DNA that starts by nicking and unrolling one strand and using the other, still circular, strand as a template for DNA synthesis.

Rotating field electrophoresis (ROFE) A variant of PFGE, in which the electrodes are carried on a rotor that rotates around the stationary gel to force the migrating DNA to a new direction.

Rotating gel electrophoresis (RGE) A variant of PFGE, in which the electrodes are positioned along opposite sides of the buffer chamber with their polarity fixed with respect to the horizontal gel, and to force the migrating DNA to a new direction, the magnetic drive simply rotates the gel to the new angle, thereby generating homogeneous electric fields.

S1 nuclease Endonuclease isolated from *Aspergillus oryzae* that specifically degrades ss DNA.

S1 nuclease mapping Method using S1 nuclease to locate the 5′-end or 3′-end of a transcript.

***Saccharomyces cerevisiae* (budding yeast)** A widely distributed, unicellular, polymorphic fungus (eukaryote) comprising of ~12 Mbp linear ds DNA genome; also called Brewer's yeast, Ale yeast, top-fermenting yeast, and Baker's yeast; used in expression cloning of eukaryotic gene.

Satellite DNA (simple sequence DNA) It consists of many tandem repeats (identical or related) of a short basic repeating unit.

Scale-up Conversion of a process such as fermentation from a small-scale to a larger scale.

Scarce mRNA (rare abundance mRNA or complex mRNA) It consists of a large number of individual mRNA species, each present in very few copies per cell; this accounts for most of the sequence complexity in RNA.

Scorpion primer DNA primer joined to a molecular beacon by an inert linker; when the probe sequence binds target DNA, the quencher and fluorophore are separated allowing fluorescence.

Second messenger A small molecule generated upon activation of a signal transduction pathway.

Secondary antibody In an ELISA or other immunological assay system, the antibody that binds to the primary antibody; it is often conjugated with an enzyme such as alkaline phosphatase.

Secretion The passage of a molecule from the inside of a cell through a membrane into the periplasmic space or the extracellular medium.

Selectable Having a gene product, that, when present, enables a researcher to identify and preferentially propagate a particular cell type.

Selectable marker A gene that allows cells containing it to be identified by the expression of a recognizable characteristic.

Selection A system for either isolating or identifying specific organisms in a mixed culture.

Selective medium A medium that favors the growth of a particular organism or group of organisms, often by suppressing the growth of others.

Self-replicating elements Extrachromosomal DNA elements that have origins of replication for the initiation of their own DNA synthesis.

Semiconservative replication The process of DNA replication by which each strand acts as a template for the synthesis of a new complementary strand; each resultant double stranded molecule thus comprises one original strand and one new one.

Semliki forest virus An alphavirus comprising of plus (+) strand RNA with icosahedral capsid which is enveloped by a lipid bilayer, derived from the host cell; the size of the virus genome is ~13,000 base pairs.

Sensitivity The ratio of all true positive test results over all positive test results, i.e., true positives plus false negatives.

Sequenase Genetically modified DNA polymerase from bacteriophage T7 used for sequencing DNA.

Sequence characterized amplified region (SCAR) A genomic DNA fragment at a single genetically defined locus, i.e., identified by oligonucleotides.

Sequence tagged site (STS) A short (200 to 500 bp) DNA sequence that occurs once in the genome and is identified by PCR amplification.

Sequence tagged site content mapping Markers based on unique PCR primers that are used to determine shared sites among clones of a library, which in turn facilitates the assembly of a contig.

Sex pilus (pl. sex pili) A narrow extension of the bacterial cell, through which genetic material is transferred during conjugation.

Shearing (mechanical shearing) The sliding of one layer across another, with deformation and fracturing in the direction parallel to the movement; this term usually refers to the forces that cells are subjected to in a bioreactor or a mechanical device used for cell breakage.

Short tandem repeat (STR) A DNA sequence with a sequential repeating set of two (di-), three (tri-), or four (tetra-) nucleotide pairs.

Short template A DNA strand that is synthesized during the PCR and has a primer sequence at one end and a sequence complementary to the second primer at the other end.

Shot gun sequencing Approach in which the genome is broken into may random short fragments for sequencing; the complete genome sequence is then assembled by computerized searching for overlaps between individual sequences.

Shotgun cloning Technique used to analyze an entire genome in the form of randomly generated fragments.

Shuttle vector (Bifunctional vector) The vector containing two different origins of replication or a broad host-range origin of replication that allows multiplication in two different hosts.

Siderophore A low molecular weight substance that binds very tightly to iron; it is synthesized by a variety of soil microorganisms and plants to ensure that these organisms obtain sufficient amounts of iron from the environment.

Sigma factor The subunit of bacterial RNA polymerase needed for recognition and binding to promoter for initiation of transcription.

Signal sequence (signal peptide or leader peptide) A short region of a protein that directs in to extracellular medium or to the plasma membrane or to one of the cell's membranous organelles.

Signal transduction The process by which a receptor interacts with a ligand at the surface of the cell and then transmits a signal to trigger a pathway within the cell.

Signal-to-noise ratio Ratio of the extent of the response to an assay when the target entity is present (signal) in a sample to the extent when it is absent (noise) from the sample.

Silencing The repression of gene expression in a localized region, usually as the result of a structural change in chromatin.

Silver staining An important technique especially to stain proteins and DNA; used in temperature gradient gel electrophoresis and in polyacrylamide gels.

Simian virus 40 (SV40) Small, spherical ds DNA virus that causes cancer in monkeys by inserting its DNA into the host chromosome.

Sindbis virus A type of alphavirus comprising of icosahedral capsid; its genome is a ss RNA approximately 11.7 kbp long having a 5' cap and 3' polyadenylated tail.

Single cell protein (SCP) Dried mass of a pure sample of a protein-rich microorganism, which may be used either as feed (for animals) or as a food (for humans).

Single copy plasmid Plasmid that replicates under the control system analogous to the bacterial chromosome that allows only one copy to exist in an individual bacterial cell.

Single nucleotide polymorphism (SNP) A polymorphism (variation in sequence between individuals) caused by a change in a single nucleotide; responsible for most of the genetic variation between individuals.

Single strand binding protein (SSB) Protein that attaches to ss DNA, and hence prevents annealing of two strands of DNA from forming duplex or secondary structures and decrease the susceptibility of ss DNA to attack by nucleases.

Single strand conformation polymorphism (SSCP) A technique for detecting polymorphisms using electrophoresis of ss DNA whose conformation differs due to point substitutions, deletions, or insertions.

Site-directed mutagenesis (site-specific mutagenesis or oligonucleotide-directed mutagenesis) A technique to change one or more specific nucleotides in a cloned gene in order to create an altered form of a protein with a specific amino acid change(s).

Site-specific recombination (specialized recombination) It occurs between two specific (not necessarily homologous) sequences, as in phage integration/excision or resolution of cointegrate structures during transposition.

Size fractionation Chromatographic method in which particles are separated based on their size, or in more technical terms, their hydrodynamic volume.

Size markers A set of macromolecules of known molecular masses that are used to calculate the molecular masses of electrophoretically fractionated macromolecules.

Small nuclear RNA (snRNA) One of many small RNA species confined to the nucleus; several of the snRNAs are involved in splicing or other RNA processing reactions.

Sodium dodecyl sulfate (SDS) An anionic detergent that denatures proteins.

Solid-phase synthesis of oligonucleotides Synthesis of oligonucleotides after anchoring or immobilizing one of the ends of the growing oligonucleotide to the solid support.

Solution-phase synthesis of oligonucleotides Synthesis of oligonucleotides in solution.

Somatic cell Any cell of a multicellular organism that does not produce gametes.

Somatic cell gene therapy The delivery of a gene(s) to a tissue other than reproductive cells of an organism, with the aim of correcting a genetic defect.

Somatic cell nuclear transfer (SCNT) A laboratory technique for creating a clonal embryo, using an ovum with a donor nucleus.

Sonication (ultrasonication) A form of hydrodynamic shearing by exposure to brief pulses of high frequency sound waves; used to disrupt cells or to get DNA fragments from the large genomic DNA.

Source DNA The DNA from an organism that contains a target gene; this DNA is used as starting material in a cloning experiment.

Source organism An organism (e.g., a bacterium, plant, or animal) from which DNA is purified and used in a cloning experiment.

Southern blot The piece of transfer medium containing the DNA sequences transferred on it.

Southern blotting A technique for transferring denatured DNA molecules that have been separated electrophoretically from a gel to a matrix (such as nitrocellulose or nylon membrane) on which a hybridization assay can be performed.

South-western blotting Detection technique in which a DNA probe binds to a protein target; used to detect DNA-binding proteins.

SP6 promoter Promoter recognized by SP6 RNA polymerase.

Spacer A sequence in a gene cluster that separates the repeated copies of the transcription unit.

Specialist vector (special purpose vector) Vectors designed for some special purposes such as secretion of exogenous protein, tagging for easy purification and protection from protease attack, or surface display for easy identification of recombinants.

Specialized transduction (restricted transduction) The transfer of a limited selection of bacterial genes due to imprecise excision of a prophage in a lysogenic infection cycle; the transducing phage genome usually is defective, lacks some part of its attachment site and can inject bacterial genes into another bacterium, even though the defective phage cannot reproduce without assistance.

Specificity The ratio of all true negative test results to all negative test results, i.e., true negatives plus false positives.

Spectrophotometry A technique based on the quantitative relationship of absorbance as a function of the length of the light path and the concentration of the absorbing species.

Splice site The sequence immediately surrounding the exon:intron boundaries.

Spore A small, protected reproductive form of a microorganism often produced when nutrient levels are low.

Sporulation Formation of spores or resting structures, usually after the near depletion of nutrients from the growth medium, by some bacteria or fungi.

Spotted DNA array Immobilized DNA clones (full-length or nearly full-length), more usually PCR products derived from them, or 20–70 residue oligonucleotides individually onto a solid support (nylon membrane or glass slide) formed by transferring or spotting.

Star activity (relaxed specificity or altered specificity) The recognition and cleavage of nucleotide sequences differing from the canonical site by restriction endonucleases under suboptimal reaction conditions, or a reduction in the specificity of restriction enzyme such that only part of the normal recognition site is recognized.

Stem The base paired segment of a hairpin structure in RNA.

Stem cell A precursor cell that undergoes division and gives rise to lineages of differentiated cells.

Sterilization The process of eradicating all life forms; culture medium, glassware, and utensils are sterilized before use to prevent contamination; accomplished by autoclaving, irradiating, heating, membrane filtration, chemical treatment, etc.

Sticky end (cohesive end or staggered end) The ends generated by two off the center cuts (cleavage of phosphodiester bonds) at palindromic sites, but between the same two bases on the opposite strands.

Sticky end cutters The restriction enzymes for which the cleavage sites are displaced from the axis of symmetry and the cuts are asymmetric that do not lie exactly opposite to each other.

Stop buffer A buffer containing a component such as the metal-chelating agent EDTA that will stop an enzymatic reaction; added to restriction digests at the end of incubation or sometimes included in the tracking dye solution.

Stop codon (termination codon or nonsense codon or translational stop signal) One of three triplets [UAG (amber), UAA (ochre), UGA (opal)] that cause protein synthesis to terminate.

Strain A microorganism or multicellular organism that is a genetic variant of a standard parental stock.

Strand A linear series of nucleotides linked to each other by phosphodiester bonds.

Streak To use a loop, toothpick, or other sterile utensil to dilute out microorganisms on solid medium in an effort to obtain isolated colonies.

Streptavidin A biotin-binding protein isolated from *Streptomyces*; useful in detection procedures for molecules labeled with biotin.

Stringency Conditions used in the hybridization of nucleic acids that reflect the degree of complementarity between the hybridizing strands. Parameters of high stringency are low salt and high temperature, wherein highly complementary strands hybridize; low stringency conditions of high salt and low temperature increase overall hybridization.

Stringent plasmid Copy numbers of these types of plasmids are strictly controlled and correlated with the number of chromosomal DNA molecules, and these are maintained at a limited number of copies per cell.

Structural gene A sequence of DNA that encodes a protein.

Subcloning Transfer of a cloned fragment of DNA from one vector to another for investigation of a short region of a large cloned fragment in more detail or to transfer a gene to a vector designed to express it in a particular species or for transferring in M13 vectors after manipulation to get ss DNA for further applications.

Subculture To reinoculate fresh culture medium with cells from an existing culture, such as from an overnight culture or streak plate.

Subspecies Population(s) of organisms sharing certain characteristics that are not present in other populations of the same species.

Substitutive therapy Treatment of an inherited disorder with a cofactor that restores enzyme function.

Substrate A compound that is altered by an enzyme.

A food source for growing cells or microorganisms.

Subtilisin A proteolytic enzyme usually found in *Bacillus subtilis*.

Subtractive hybridization Technique used to remove unwanted DNA or RNA by hybridization so leaving behind the DNA or RNA molecule of interest.

Subunit vaccine An immunogenic protein(s) either purified from the disease-causing organism or produced from a cloned gene.

Sucrose density gradient centrifugation A procedure used to fractionate mRNAs or DNA fragments according to size.

Suicide gene A plasmid-borne, inducible sequence that produces a protein that directly or indirectly kills the host cell.

Superbug Jargon for the bacterial strain *Pseudomonas* developed by Dr Anand Mohan Chakrabarty, who combined hydrocarbon-degrading genes carried on different plasmids into one organism; although this genetically engineered microorganism is neither 'super' nor a 'bug', it is a landmark example because it showed how genetically modified microbial strains could be used in a novel way and because it was the basis for the precedent-setting legal decision that, in the United States, genetically engineered organisms are patentable.

Supercoiling The coiling of a closed duplex DNA in space so that it crosses over its own axis.

Suppression The occurrence of changes that eliminate the effects of a mutation without reversing the original change in DNA.

Suppressor tRNA An abnormal tRNA that inserts an amino acid where a mutant mRNA specifies a stop codon in the middle of the coding portion of a gene; insertion of this amino acid allows a normal-sized rather than a shortened protein to be synthesized.

SYBR Gold The most sensitive fluorescent stain available for detecting ds or ss DNA or RNA in electrophoretic gels, using standard ultraviolet transilluminator.

SYBR Green I A DNA-binding fluorescent dye that binds only to ds DNA and becomes fluorescent only when bound.

Symbiosis A close biological relationship between two organisms, in which neither organism is extremely harmful to the other; in some cases, the relationship is mutually beneficial.

Syndrome A constellation of features that together make up the symptoms of a disorder of disease.

Systematic evolution of ligands by exponential enrichment (SELEX) A technique for generating single stranded oligonucleotides, which bind specifically to the 3-D structures of target molecules with high affinity and specificity.

Systemic acquired resistance Resistance, in plants, to pathogenic agents that occur following an initial exposure to the same or another pathogenic agent; this resistance extends to plant tissues that are far from the site of the initial infection, and may last for weeks to months.

T3 promoter A promoter recognized by T3 RNA polymerase.

T4 DNA ligase An enzyme from bacteriophage T4-infected cells that catalyzes the joining of duplex DNA molecules and repairs nicks in DNA molecules; the enzyme joins a 5′-phosphate group to a 3′-hydroxyl group.

T7 promoter A promoter recognized by T7 RNA polymerase.

TA cloning A cloning procedure that uses *Taq* DNA polymerase to generate single 3′-A overhangs on the ends of DNA segments that are used to clone DNA into a vector with matching 3′-T overhangs.

Tandem array Usually a DNA molecule that contains two or more identical nucleotide sequences in series.

Tandem mass spectrometry (MS/MS or tandem MS) An apparatus that determines the constituents of complex mixtures using two successive rounds of mass spectrometry in which a parent ion is first isolated and then fragmented into daughter ions for more detailed analysis.

***Taq* DNA polymerase** Thermostable DNA polymerase from *Thermus aquaticus*.

TaqMan probe Fluorescent probe consisting of two fluorophores linked by a DNA probe sequence; fluorescence increases only after the fluorophores are separated by degradation of the linking DNA.

Target gene (target DNA or foreign DNA or passenger DNA or exogenous DNA or gene of interest or insert DNA) A descriptive term for a gene that is to be either cloned or specifically mutated.

Targeting vector A cloning vector carrying a DNA segment capable of participating in a cross-over event at a specified chromosomal location in the host cell.

TATA box (Hogness box) A conserved A-T-rich septamer found about 25 bp before the start point of each eukaryotic RNA polymerase

II transcription unit; may be involved in positioning the enzyme for correct initiation.

TE buffer A buffer comprising of 10 mM Tris-HCl and 1 mM EDTA (pH 7.5-8.0), which is used in molecular biology for operations such as dialyzing or solubilizing DNA, where only a dilute buffer is required.

Telomerase The ribonucleoprotein enzyme that creates repeating units of one strand at the telomere, by adding individual bases to the DNA 3' end, as directed by an RNA sequence in the RNA component of the enzyme.

Telomere The defined end of a chromosome containing specific DNA sequence.

Temperate phage A bacteriophage with a lysogenic replication cycle.

Temperature-sensitive protein A protein that is functional at one temperature but loses function at another (usually higher) temperature.

Template strand The polynucleotide strand that a DNA polymerase uses for determining the sequence of nucleotides during the synthesis of a new nucleic acid strand.

Strand that is complementary to the sense strand, and is the one that acts as the template for synthesis of mRNA.

Terminal nucleotidyl transferase (Tdt or terminal transferase) An enzyme that adds a series of nucleotides onto 3′-OH termini of ds DNA molecule.

Terminase Enzyme that cleaves multimers of a viral genome and then uses hydrolysis of ATP to provide the energy to translocate the DNA into an empty viral capsid starting with the cleaved end.

Termination A separate reaction that ends a macromolecular synthesis reaction (replication, transcription, or translation), by stopping the addition of subunits, and (typically) causing dissolution of the synthetic apparatus.

Terminator of transcription A sequence of DNA at the 3′-end of the gene that causes RNA polymerase to terminate transcription.

Terminator technology (gene protection technology) Genetic use restriction technology (GURT), colloquially known as terminator technology, is the name given to proposed methods for restricting the use of genetically modified plants by causing second generation seeds to be sterile.

Terminus of replication A segment of DNA at which replication ends.

tet^{r} **gene** Tetracycline resistant genes.

Thaumatin A plant protein having a sweet taste; it is also synthesized in some plants in response to infection by pathogens.

Therapeutic Refers to treatment of a disease.

Therapeutic agent (pharmaceutical agent, drug, or protein drug) A compound that is used for the treatment of a disease and for improving the well-being of an organism.

Therapeutic cloning (research cloning) Cloning of embryos containing DNA from an individual's own cell to generate a source of ES cell-progenitor cells that can differentiate into different cell types of the body with the aim to produce healthy replacement tissues.

Thermocycler Machine used to rapidly shift samples between several temperatures in a pre-set order for PCR.

Thermophile A microorganism that grows optimally at high temperatures, usually above 50°C; some thermophiles can grow at temperatures of 90 to 100°C.

Thermostable DNA polymerase Heat-resistant DNA polymerase that is used for PCR.

Thermus aquaticus Thermophilic bacterium found in hot springs and used as a source of thermostable DNA polymerase.

Thioredoxin A small protein that acts as an electron carrier.

Thymine (T) One of the four bases of DNA.

Ti plasmid (tumor-inducing plasmid) A large extrachromosomal element found in strains of *Agrobacterium tumefaciens* and is responsible for crown gall formation.

Time of flight (TOF) Time for an ion to fly from the ion source to the detector.

Tissue plasminogen activator (tPA) A protein involved in dissolving blood clots.

T_m **(melting temperature)** The melting temperature of a nucleic acid hybrid; it is related to the G+C content, and also to the temperature and ionic strength of the washing buffer.

Tobacco mosaic virus (TMV) A filamentous ss RNA virus that infects a wide range of plants.

TOPO cloning A molecular biology technique in which DNA fragments amplified by either *Taq* or *Pfu* DNA polymerases are cloned into specific vectors without the requirement for DNA ligases.

Topology Level of supercoiling or catenation.

Totipotent Ability of single cell to give rise to the whole multicellular organism from which it is derived.

***tra* genes** Genes needed for plasmid transfer.

Tra^{+} Transfer positive (refers to a plasmid capable of self-transfer).

Tracking dye A low molecular weight, visible colored compound that moves with an ion front during gel electrophoresis.

Transcript An RNA molecule that has been synthesized from a specific DNA template.

Transcription The process of RNA synthesis that is catalyzed by RNA polymerase; uses a DNA strand as a template.

Transcription factor Accessory protein required for RNA polymerase to initiate transcription at specific promoter(s), but is not itself part of the enzyme.

Transcription mapping (transcript mapping or transcriptional mapping) Assigning gene transcripts, in the form of cDNA clones or ESTs, to specific chromosome regions by FISH, PCR, analysis of somatic cell hybrid panels, or other strategies.

Transcription unit The distance between sites of initiation and termination by RNA polymerase; may include more than one gene.

Transcriptome The complete set of RNAs present in a cell, tissue, or organism, under any particular set of conditions.

Transducing virus Virus that carries part of the host genome in place of part of its own sequence; the best-known examples are retroviruses in eukaryotes and DNA phages in *E. coli*.

Transduction The bacteriophage-mediated transfer of genetic material between bacteria.

Transfection Introduction of bacteriophage DNA into competent *E. coli* cells.

Introducing nucleic acids into cells by non-viral methods.

Transfer DNA (T-DNA) The segment of a Ti plasmid that is transferred and integrated into chromosomal sites in the nuclei of plant cells.

Transfer region (*tra* locus) Segment on the F plasmid that is required for bacterial conjugation.

Transfer RNA (tRNA) A form of RNA that carries specific amino acids to the site of protein synthesis and decode the sequence information contained in an mRNA molecule during the translation process.

Transformation The uptake of naked DNA from the environment and its integration into the host genome.

Conversion, by various means, of animal cells in tissue culture from controlled to uncontrolled cell growth.

The acquisition of new genetic markers by incorporation of added DNA.

Transformation efficiency The number of cells that take up foreign DNA as a function of the amount of added DNA; expressed as transformants per microgram of added DNA.

Transformation frequency The fraction of a cell population that takes up foreign DNA; expressed as the number of transformed cells divided by the total number of cells in a population.

Transgene A gene that is introduced into a cell from an external source.

Transgenesis The introduction of a gene(s) into animal or plant cells that leads to the transmission of the input gene (transgene) to successive generations.

Transgenic animal A fertile animal that carries an introduced gene(s) in its germ line.

Transgenic plant A fertile plant that carries an introduced gene(s) in its germ line.

Transient Of short duration.

Transition A mutation in which a purine replaces a purine or a pyrimidine replaces a pyrimidine.

Translation The process of protein (polypeptide) synthesis in which mRNA determines the amino acid sequence of a protein, mediated by tRNA molecules, and carried out on ribosomes.

Translocation The transfer of chromosomal material from one chromosome to another.

Transporter A type of receptor that moves small molecules across the plasma membrane; it binds the molecules on its extracellular surface, and releases them into the cytoplasm.

Transposase An enzyme that is encoded by a transposon gene and that facilitates the insertion of the transposon into a new chromosomal site and excision from a site.

Transposition The movement of a transposon to a new site in the genome.

Transposon (transposable element or mobile element) A DNA sequence able to insert itself at a new location in the genome, without having any sequence relationship with the target locus.

Transposon mutagenesis A process that allows genes to be transferred to a host organism's chromosome, interrupting or modifying the function of a gene on the chromosome and causing mutation.

Transverse alternating field electrophoresis (TAFE) A simple, high-resolution pulsed field system that uses a vertical gel, and simple electrode arrangement (straight lane geometry).

Transversion A mutation in which a purine replaces a pyrimidine, or a pyrimidine replaces a purine.

Trimming back Digesting protruding strands to get blunt ends.

Tripartite mating (three-way cross or triparental mating) A process in which conjugation is used to transfer a plasmid vector to a target cell when the plasmid vector is not self-mobilizable; when cells that have a plasmid with conjugative and mobilizing functions are mixed with cells that carry the plasmid vector and target cells, mobilizing plasmids enter the cells with the target cells; following tripartite mating, the target cells with the plasmid vector are separated from the other cell types in the mixture by various selection procedures.

Tris [Tris (hydroxymethyl) aminomethane]; an organic buffer used in molecular biology, biochemistry, and cell biology.

Triton X-100 A nonionic detergent.

Type I topoisomerase An enzyme that changes the topology of DNA by nicking and resealing one strand of DNA.

Type II topoisomerase An enzyme that changes the topology of DNA by nicking and resealing both strands of DNA.

UV transilluminator The transilluminator that is equipped with UV filter; used for gel visualization and documentation.

Ultracentrifugation A centrifuge optimized for spinning a rotor at very high speeds, capable of generating acceleration as high as 1,000,000 g (9,800 km/s²).

Ultrafiltration A type of membrane filtration in which hydrostatic pressure forces a liquid against a semipermeable membrane.

Undefined medium (complex medium) A medium whose precise chemical composition is not known.

Unidirectional replication The movement of a single replication fork from a given origin.

Unit of activity Generally the amount of an enzyme that catalyzes the conversion of a certain quantity of substrate to product in a fixed time period.

Unit of restriction enzyme activity One unit of restriction enzyme activity is the amount of enzyme required to fully digest 1 μg of bacteriophage λ DNA in 1 hour at 37°C, or under the conditions of the experiment.

Upstream The stretch of nucleotides (in coding strand) that lie in the 5′ direction from the site of initiation of transcription; usually, the first transcribed base is designated +1 and the upstream nucleotides are indicated with minus signs, e.g., –1 and –10.

***URA* 3** A gene used as a nutritional auxotrophic selectable marker for yeast.

Uracil (U) One of the four bases in RNA.

Uracil–DNA glycosylase (UDG) An enzyme that catalyzes the hydrolysis of the *N*-glycosidic bond between the uracil and sugar, leaving an apyrimidinic site in uracil-containing single or dsDNA; shows no activity on RNA.

Vaccination Inoculation with a vaccine to provide protective immunity.

Vaccine A preparation of dead or inactivated living pathogens or their products used to provide protective immunity.

***Vaccinia* virus** A large, complex, enveloped virus belonging to the poxvirus family. It has a linear, ds DNA genome of ~190 kbp in length.

Variable number tandem repeat (VNTR) A location in a genome where a short nucleotide sequence is organized as a tandem repeat; these can be found on many chromosomes, and often show variations in length between individuals; each variant acts as an inherited allele, allowing them to be used for personal or parental identification; their analysis is useful in genetics and biology research, forensics, and DNA fingerprinting.

Vector (cloning vehicle) A self-replicating DNA molecule used in gene cloning; the sequence to be cloned is inserted into the vector, and replicated along with it.

Vegetative phase The period of normal growth and division of a bacterium.
For a bacterium that can sporulate, this contrasts with the sporulation phase, when spores are being formed.

Velocity sedimentation A transport technique, akin to diffusion or viscosity; the rate at which a particle settles is measured.

***vir* genes** A set of genes on the Ti plasmid that prepares the T-DNA segment for transfer into a plant cell.

Virion An infectious virus particle.

Virulence The degree of pathogenicity of an organism.

Virulence plasmids Plasmids carrying genes that confer pathogenicity on host cells (plant or animal).

Virulent phage A bacteriophage with a lytic replication cycle.

Virulent phage mutants Phages that are unable to establish lysogeny.

Virus A submicroscopic, noncellular parasite, comprising protein and RNA or DNA.

Void volume (Vo) The total volume of buffer present in between the beads, i.e., the space surrounding and outside the particles of gel.

Western blot The piece of membrane onto which the protein is immobilized and probed with antibodies.

Western blotting Transfer of protein from a gel to a membrane and detection of specific protein by using antibody.

Wild-type A genetic term that denotes the most commonly observed phenotype, or the normal state, in contrast to a mutant condition.

Xenobiotic A chemical compound that is not produced by living organisms; a manufactured chemical compound.

Xenotransplantation The transplantation of cells, tissues or organs from an animal into a human being.

X-gal (5-bromo-4-chloro-3-indolyl-β-*D*-galactoside) Artificial substrate that is split by β-galactosidase, releasing a blue dye.

X-gluc (5-bromo-4-chloro-3-indolyl-β-*D*-glucuronic acid) The chromogenic substrate used in a variety of applications for the detection of the β-glucuronidase enzyme; upon reduction, X-gluc produces a localized color, making it useful in identifying GUS gene in most cell types and for the detection of the GUS gene fusion marker in plants.

X-phos (5-bromo-4-chloro-3-indolyl phosphate) Artificial substrate that is split by alkaline phosphatase, releasing a blue dye.

Yeast artificial chromosome (YAC) A yeast-based vector system that contains an origin for replication, a centromere to support segregation, and telomeres to seal the ends; able to accommodate inserts of several hundred kbp in size.

Yeast centromeric plasmid (YCp) Yeast cloning vector that contains a short segment of size 1.6 kbp, which functions as centromere; this centromere is required for, segregation of the YCp during the yeast division. It behaves like a small functional chromosome hence it is also called as minichromosome vector.

Yeast episomal vector (YEp) A cloning vector for the *Saccharomyces cerevisiae* that uses the 2-μm plasmid origin of replication and is maintained as an extrachromosomal nuclear DNA molecule.

Yeast integrative plasmid (YIp) Yeast vectors that rely on integration into the host chromosome for survival and replication, and are usually used when studying the functionality of a gene or when the gene is toxic.

Yeast replicative plasmid (YRp) Yeast cloning vector that incorporates a yeast genomic replication origin into a circular duplex DNA.

Yeast two-hybrid system Assay to detect interaction between two proteins by means of their ability to bring together a DNA-binding domain and a transcription-activating domain; the assay is performed in yeast using a reporter gene that responds to the interaction.

Zero-integrated-field electrophoresis (ZIFE) A variant of PFGE having the ability to resolve larger molecules (10–10,000 kbp) of DNA; in ZIFE, the product of field strength and pulse times is approximately the same for backward and forward pulses.

Zinc finger A DNA-binding motif containing domains that bind Zn^{2+}; it typifies a class of transcription factor.

Zonal centrifugation Centrifugation using a rotating chamber of large capacity to separate cell organelles by density gradient centrifugation.

Zoo blot The use of Southern blotting to test the ability of a DNA probe from one species to hybridize with the DNA from the genomes of a variety of other species.

Index